HAND BOOK OF

ENGINEERING MATHEMATICS

With GATE TUTOR

Vol. 1

CBS Engineering Series

HAND BOOK OF

ENGINEERING MATHEMATICS

With GATE TUTOR

Vol. 1

Useful for

- B.E./B.Tech. and all other Engineering courses
- Aspirants of GATE

Dr. SUDHIR KUMAR PUNDIR

Associate Professor

Department of Mathematics

S.D. (P.G.) College,

Muzaffarnagar (U.P.)

CBS Publishers & Distributors Pvt. Ltd.

New Delhi • Bengaluru • Chennai • Kochi • Kolkata • Mumbai • Pune

Hyderabad • Nagpur • Patna • Vijayawada

Disclaimer

Science and technology are constantly changing fields. New research and experience broaden the scope of information and knowledge. The author has tried his best in giving information available to him while preparing the material for this book. Although, all efforts have been made to ensure optimum accuracy of the material, yet it is quite possible some errors might have been left uncorrected. The publisher, printer and author will not be held responsible for any inadvertent errors or inaccuracies.

HAND BOOK OF
ENGINEERING MATHEMATICS
With **GATE TUTOR** **Vol. 1**

ISBN: 978-81-239-2810-4

First Edition: 2016

Published by:
Satish Kumar Jain for CBS Publishers & Distributors Pvt. Ltd.,
4819/XI Prahlad Street, 24 Ansari Road, Daryaganj, New Delhi - 110002
delhi@cbspd.com, cbspubs@airtelmail.in • www.cbspd.com
Ph.: 23289259, 23266861, 23266867 • Fax: 011-23243014

Corporate Office: 204 FIE, Industrial Area, Patparganj, Delhi - 110 092
Ph: 49344934 • Fax: 011-49344935
E-mail: publishing@cbspd.com • publicity@cbspd.com

Branches:

- ***Bengaluru:*** 2975, 17th Cross, K.R. Road, Bansankari 2nd Stage, Bengaluru - 70 • Ph: +91-80-26771678/79 • Fax: +91-80-26771680 E-mail: cbsbng@gmail.com, bangalore@cbspd.com
- ***Chennai:*** No. 7, Subbaraya Street, Shenoy Nagar, Chennai - 600030 Ph: +91-44-26681266, 26680620 • Fax: +91-44-42032115 E-mail: chennai@cbspd.com
- ***Kochi:*** Ashana House, 39/1904, A.M. Thomas Road, Valanjambalam, Ernakulum, Kochi Ph: +91-484-4059061-65 Fax: +91-484-4059065 • E-mail: cochin@cbspd.com
- ***Kolkata:*** 6-B, Ground Floor, Rameshwar Shaw Road, Kolkata - 700014 Ph: +91-33-22891126/7/8 • E-mail: kolkata@cbspd.com
- ***Mumbai:*** 83-C, Dr. E. Moses Road, Worli, Mumbai - 400018 Ph: +91-9833017933, 022-24902340/41 • E-mail: mumbai@cbspd.com
- ***Pune:*** Bhuruk Prestige, Sr. No. 52/12/2+1+3/2, Narhe, Haveli (Near Katraj-Dehu Road Bypass), Pune - 411041 Ph: +91-20-64704058/59, 32342277 • E-mail: pune@cbspd.com

Representatives:

- Hyderabad: 0-9885175004
- Nagpur: 0-9021734563
- Patna: 0-9334159340
- Vijayawada: 0-9000660880

Printed at:
Neekunj Print Process, Delhi

Preface

The book entitled 'Hand Book of ENGINEERING MATHEMATICS *with* GATE Tutor-(Vol-I)' meant for B.E, B.Tech., and students of other engineering courses. Besides, it will also be very useful for those students preparing for GATE.

In the absence of a good and comprehensive text book of Mathematics in single unit, this book serve as a complete course of B.E./B.Tech. students of all technical universities. Special and conscious efforts have been made to keep the writing style simple. The book contains more than usual solved examples together with properly graded problems. A large number of problems with solutions have been provided to assist one get a firm grip on the ideas developed. There is plenty of scope in the form of exercise for the readers to try and solve the problem on his own. To make the book self-contained and GATE oriented, a GATE tutor has been given at the end of each unit.

I wish to sincerely thank **Sh S.K. Jain**, Managing Director, CBS Publishers and Distributors, New Delhi for his encouragement and help in bringing out this publication in a present nice form.

My special thanks to Sh. Y.N. Arjuna, Sh. B.M. Singh, Sh. Sunil Dutt and entire team of CBS publishers & distributors, New Delhi whose encouragement and unstinted support enabled me to complete my book. Mr. Peeyush Goel, M/s Dreamshapers also deserve special mention for nice type setting.

I must also record my appreciation due to my wife Dr. Rimple, daughter Rijuta and son Shrish for their understanding and love during the long period that I have taken to complete this book.

Above all I am thankful to The Almighty God, without whose grace nothing is possible for any one.

Readers are welcomed to point out errors, if any and send their valuable suggestions for improving the quality of the book.

DR. SUDHIR KUMAR PUNDIR
email : skpundir05@yahoo.co.in
skpundir05@gmail.com

Also Available
by the
Same Author

(For B.E. / B.Tech.)

1. Hand Book of '**ENGINEERING MATHEMATICS**' with GATE Tutor **Vol. 2**

Contents:

- Method of Least Squares
- Statistical Methods
- Probability and Distributions
- Sampling and Inferences
- Numerical Solution of Equations
- The Calculus of Finite Differences and Interpolation
- Numerical Differentiation and Integrations
- Difference Equations
- Numerical Solution of Ordinary Differential Equations
- Numerical Solution of Partial Differential Equations
- Complex Numbers
- Calculus of Complex Functions

2. Special Topics in '**ENGINEERING MATHEMATICS**'

Contents:

- Linear Programming for Engineers
- Calculus of Variation for Engineers
- Fuzzy Mathematics for Engineers
- Integral Equations for Engineers
- Discrete Mathematics for Engineers

Contents

DIFFERENTIAL CALCULUS

UNIT OUTLINE

- Successive Differentiations
- Mean Value Theorems and Expansion of Functions
- Indeterminate Forms
- Tangents and Normals
- Curvature
- Envelope and Evolutes
- Asymptotes
- Curve Tracing

PRE-REQUISITE

List of Derivatives of Elementary functions

S. No.	Function	Derivative
1.	c, a, constant	0
2.	x^n	nx^{n-1}
3.	$\sin x$	$\cos x$
4.	$\cos x$	$-\sin x$
5.	$\tan x$	$\sec^2 x$
6.	$\cot x$	$-\text{cosec}^2 x$
7.	$\sec x$	$\sec x \tan x$
8.	$\text{cosec}\, x$	$-\text{cosec}\, x \cot x$
9.	a^x	$a^x \log_e a$
10.	e^x	e^x
11.	$\log_e x$	$1/x$
12.	$\sin^{-1} x$	$\frac{1}{\sqrt{1-x^2}}$
13.	$\cos^{-1} x$	$\frac{-1}{\sqrt{1-x^2}}$

PRE-REQUISITE

List of Derivatives of Elementary functions

S. No.	Function	Derivative
14.	$\tan^{-1} x$	$\frac{1}{1+x^2}$
15.	$\cot^{-1} x$	$\frac{-1}{1+x^2}$
16.	$\sec^{-1} x$	$\frac{1}{x\sqrt{x^2-1}}$
17.	$\text{cosec}^{-1} x$	$-\frac{1}{x\sqrt{x^2-1}}$
18.	$\sinh x$	$\cosh x$
19.	$\cosh x$	$\sinh x$
20.	$\tanh x$	$\text{sech}^2 x$
21.	$\coth x$	$-\text{cosech}^2 x$
22.	$\text{sech}\, x$	$-\text{sech} x \tanh x$
23.	$\text{cosech}\, x$	$-\text{cosech} x \coth x$
24.	$\sinh^{-1} x$	$\frac{1}{\sqrt{1+x^2}}$ $x \in R$
25.	$\cosh^{-1} x$	$\frac{1}{\sqrt{x^2-1}}$, $x \geq 1$
26.	$\tanh^{-1} x$	$\frac{1}{1-x^2}$, $-1 < x < 1$
27.	$\coth^{-1} x$	$\frac{1}{x^2-1}$
28.	$\text{sech}^{-1} x$	$-\frac{1}{x\sqrt{1-x^2}}$
29.	$\text{cosech}^{-1} x$	$-\frac{1}{x\sqrt{x^2-1}}$

CHAPTER 1 Successive Differentiations

1.1 INTRODUCTION

Let $y = f(x)$ be a function, then the differential coefficient of $f(x)$ denoted by $f'(x)$ is defined as follows

$$f'(x) = \lim_{\delta x \to 0} \frac{f(x+\delta x) - f(x)}{\delta x} = \frac{dy}{dx}$$

If the limit exists (*i.e.*, limit is finite and unique), then $f'(x)$ is called *first differential coefficient of* $f(x)$ *with respect to* x. Similarly, if $f(x)$ is differentiable twice, it is denoted by $f''(x)$, if it is differentiable thrice, it is denoted by $f'''(x)$, *i.e.*,

$$f''(x) = \frac{d}{dx}\left(\frac{dy}{dx}\right) = \frac{d^2y}{dx^2}$$

$$f'''(x) = \frac{d}{dx}\left(\frac{d^2y}{dx^2}\right) = \frac{d^3y}{dx^3}$$

If $y = f(x)$ be a function of x, then we adopt the following notations.

$$y_1 = f'(x) = \frac{dy}{dx} = Df(x) = \frac{d}{dx}(f(x))$$

$$y_2 = f''(x) = \frac{d^2y}{dx^2} = D^2f(x) = \frac{d^2}{dx^2}(f(x))$$

$$y_3 = f'''(x) = \frac{d^3y}{dx^3} = D^3f(x) = \frac{d^3}{dx^3}(f(x))$$

$$\ldots \quad \ldots \quad \ldots \quad \ldots \quad \ldots$$

$$\ldots \quad \ldots \quad \ldots \quad \ldots \quad \ldots$$

$$y_n = f^n(x) = \frac{d^ny}{dx^n} = D^nf(x) = \frac{d^n}{dx^n}(f(x))$$

This process of finding the differential coefficients of a function is called successive differentiation.

1.2 nth DIFFERENTIATION OF SOME STANDARD FUNCTIONS

(i) $y = f(x) = x^n$.

We have

$$y = f(x) = x^n$$

$$y_1 = f'(x) = nx^{n-1}$$

$$y_2 = f''(x) = n(n-1)x^{n-2}$$

$$y_3 = f'''(x) = n(n-1)(n-2)x^{n-3}$$

....

....

$$y_n = f^n(x) = n(n-1)(n-2)\ldots 3.2.1.x^0$$

$$\Rightarrow \quad \frac{d^n}{dx^n}(x^n) = y_n = n!$$

(ii) $y = f(x) = x^m$.

We have $y_1 = f'(x) = mx^{m-1}$, $y_2 = f''(x) = m(m-1)x^{m-2}$,

$$y_3 = f'''(x) = m(m-1)(m-2)x^{m-3}, \ldots, y_n = f^n(x) = m(m-1)(m-2)\ldots(m-n+1).x^{m-n}$$

$$= \left[\frac{m(m-1)(m-2)\ldots(m-n+1)(m-n)\ldots3.2.1}{(m-n)(m-n-1)\ldots3.2.1}\right]x^{m-n}$$

$$\Rightarrow \quad y_n = \frac{d^n}{dx^n}(x^m) = \frac{m!}{(m-n)!}x^{m-n}$$

(iii) $y = f(x) = \dfrac{1}{(ax+b)}$.

We have $y_1 = f'(x) = -\dfrac{a}{(ax+b)^2}$, $y_2 = f''(x) = \dfrac{a^2.2}{(ax+b)^3}$,

$$y_3 = f'''(x) = \frac{a^3.2.3}{(ax+b)^4}, \ldots\ldots, y_n = f^n(x) = \frac{(-1)^n a^n.2.3.4\ldots n}{(ax+b)^{n+1}}$$

$$\Rightarrow \quad y_n = \frac{d^n}{dx^n}\left(\frac{1}{ax+b}\right) = \frac{(-1)^n.a^n.n!}{(ax+b)^{n+1}}$$

(iv) $y = f(x) = \dfrac{1}{(ax+b)^m}$.

We have $y_1 = f'(x) = -\dfrac{a.m}{(ax+b)^{m+1}}$, $y_2 = f''(x) = \dfrac{a^2.m(m+1)}{(ax+b)^{m+2}}$,

$$y_3 = f'''(x) = -\frac{a^3.m(m+1)(m+2)}{(ax+b)^{m+3}}, \ldots, y_n = f^n(x) = (-1)^n\frac{a^n.m(m+1)(m+2)\ldots(m+n-1)}{(ax+b)^{m+n}}$$

$$\Rightarrow \quad y_n = \frac{d^n}{dx^n}\left(\frac{1}{(ax+b)^m}\right) = (-1)^n\frac{a^n.(m+n-1)!}{(m-1)!(ax+b)^{m+n}}$$

(v) $y = f(x) = \sin(ax+b)$.

We have

$$y_1 = f'(x) = a\cos(ax+b) = a\sin\left(\frac{\pi}{2}+ax+b\right), \; y_2 = f''(x) = a^2\cos\left(\frac{\pi}{2}+ax+b\right) = a^2\sin\left(2.\frac{\pi}{2}+ax+b\right)$$

$$y_3 = f'''(x) = a^3\cos\left(2.\frac{\pi}{2}+ax+b\right) = a^3\sin\left(3.\frac{\pi}{2}+ax+b\right)$$

....

$$y_n = f^n(x) = a^n\cos\left((n-1)\frac{\pi}{2}+ax+b\right) = a^n\sin\left(n.\frac{\pi}{2}+ax+b\right)$$

$$\Rightarrow \quad y_n = \frac{d^n}{dx^n}[\sin(ax+b)] = a^n\sin\left(\frac{n\pi}{2}+ax+b\right)$$

(vi) $y = f(x) = \cos(ax+b)$.

We have

$$y_1 = f'(x) = -a\sin(ax+b) = a\cos\left(\frac{\pi}{2}+ax+b\right), \; y_2 = f''(x) = -a^2\sin(\frac{\pi}{2}+ax+b) = a^2\cos\left(\frac{2\pi}{2}+ax+b\right)$$

$$y_3 = f'''(x) = -a^3\sin(2.\frac{\pi}{2}+ax+b) = a^3\cos\left(3.\frac{\pi}{2}+ax+b\right)$$

....

....

$$y_n = f^n(x) = -a^n\sin\left((n-1)\frac{\pi}{2}+ax+b\right) = a^n\cos\left(\frac{n\pi}{2}+ax+b\right)$$

$$\Rightarrow \quad y_n = \frac{d^n}{dx^n}[\cos(ax+b)] = a^n\cos\left(\frac{n\pi}{2}+ax+b\right)$$

(vii) y = f(x) = e^{ax+b}.

We have $y_1 = f'(x) = a.e^{ax+b}$

$y_2 = f''(x) = a^2.e^{ax+b}$

$y_3 = f'''(x) = a^3.e^{ax+b}$

....

....

$y_n = f^n(x) = a^n.e^{ax+b}$

$\Rightarrow \quad y_n = \frac{d^n}{dx^n}(e^{ax+b}) = a^n e^{ax+b}$

(viii) y = f(x) = log (ax + b).

We have $y_1 = f'(x) = -\frac{a}{ax+b}$

Now using result (iii), we get

$y_n = f^n(x) = (-1)^{n-1}\frac{a^n(n-1)!}{(ax+b)^n}$

$\Rightarrow \quad y_n = \frac{d^n}{dx^n}[\log(ax+b)] = (-1)^{n-1}\frac{a^n(n-1)!}{(ax+b)^n}$

(ix) y = f(x) = $e^{ax}\sin(bx+c)$.

We have $y_1 = f'(x) = ae^{ax}.\sin(bx+c) + be^{ax}\cos(bx+c)$

$= e^{ax}[a\sin(bx+c) + b\cos(bx+c)]$

Put $a = r\cos\theta, b = r\sin\theta \Rightarrow r^2 = a^2 + b^2$

and $\tan\theta = b/a$ *i.e.*, $\theta = \tan^{-1} b/a$

Therefore, $y_1 = f'(x) = r.e^{ax}\sin(bx+c+\theta)$

$= (a^2+b^2)^{1/2}.e^{ax}\sin\left(bx+c+\tan^{-1}\frac{b}{a}\right)$

Similarly, $y_2 = f''(x) = (a^2+b^2)^{1/2}(a^2+b^2)^{1/2}.e^{ax}\sin(bx+c+\tan^{-1}b/a+\tan^{-1}b/a)$

$= (a^2+b^2)^{2/2}.e^{ax}\sin(bx+c+2\tan^{-1}b/a)$

$y_3 = f'''(x) = (a^2+b^2)^{3/2}.e^{ax}\sin(bx+c+3\tan^{-1}b/a)$

....

....

$y_n = f^n(x) = (a^2+b^2)^{n/2}.e^{ax}\sin(bx+c+n\tan^{-1}b/a)$

$\Rightarrow \quad y_n = \frac{d^n}{dx^n}[e^{ax}\sin(bx+c)] = (a^2+b^2)^{n/2}.e^{ax}\sin(bx+c+n\tan^{-1}b/a)$

(x) y = f(x) = $e^{ax}\cos(bx+c)$.

We have $y_1 = f'(x) = ae^{ax}.\cos(bx+c) - be^{ax}\sin(bx+c) = e^{ax}[a\cos(bx+c) - b\sin(bx+c)]$

Put $a = r\cos\theta$, $b = r\sin\theta \Rightarrow \theta = \tan^{-1}b/a$ and $r = (a^2+b^2)^{1/2}$

$\therefore \quad y_1 = f'(x) = r.e^{ax}[\cos\theta\cos(bx+c) - \sin\theta\sin(bx+c)]$

$= re^{ax}\cos(bx+c+\theta) = (a^2+b^2)^{1/2}.e^{ax}\cos(bx+c+\tan^{-1}b/a)$

Similarly, $y_2 = f''(x) = (a^2+b^2)^{2/2}.e^{ax}\cos(bx+c+2\tan^{-1}b/a)$

$y_3 = f'''(x) = (a^2+b^2)^{3/2}.e^{ax}\cos(bx+c+3\tan^{-1}b/a)$

....

$y_n = f^n(x) = (a^2+b^2)^{n/2}.e^{ax}\cos(bx+c+n\tan^{-1}b/a)$

$\Rightarrow \quad y_n = \frac{d^n}{dx^n}[e^{ax}\cos(bx+c)] = (a^2+b^2)^{n/2}.e^{ax}\cos(bx+c+n\tan^{-1}b/a)$

Recapitulations

- $\frac{d^n}{dx^n}(x^n) = n!$
- $\frac{d^n}{dx^n}(x^m) = \frac{m!}{(m-n)!}x^{m-n}$
- $\frac{d^n}{dx^n}\left(\frac{1}{(ax+b)^m}\right) = (-1)^n\frac{a^n(m+n-1)!}{(m-1)!(ax+b)^{m+n}}$
- $\frac{d^n}{dx^n}(\sin(ax+b)) = a^n\sin\left(\frac{n\pi}{2}+ax+b\right)$
- $\frac{d^n}{dx^n}(\cos(ax+b)) = a^n\cos\left(\frac{n\pi}{2}+ax+b\right)$
- $\frac{d^n}{dx^n}(e^{ax+b}) = a^n e^{ax+b}$
- $\frac{d^n}{dx^n}[\log(ax+b)] = (-1)^{n-1}\frac{a^n(n-1)!}{(ax+b)^n}$
- $\frac{d^n}{dx^n}[e^{ax}\sin(bx+c)]$
  $= (a^2+b^2)^{n/2}.e^{ax}.\sin(bx+c+n\tan^{-1}b/a)$
- $\frac{d^n}{dx^n}[e^{ax}\cos(bx+c)]$
  $= (a^2+b^2)^{n/2}.e^{ax}.\cos(bx+c+n\tan^{-1}b/a)$

Solved Examples

Example 1. *Find the nth differential coefficient of* $\tan^{-1}\dfrac{x}{a}$.

Solution. We have $y = \tan^{-1}\dfrac{x}{a}$

$$\Rightarrow \quad y_1 = \frac{a}{x^2+a^2} = \frac{a}{(x+ia)(x-ia)}$$

Let us suppose

$$\frac{a}{(x+ia)(x-ia)} = \frac{A}{(x+ia)} + \frac{B}{(x-ia)}$$

(Using partial fractions)

$$\Rightarrow \quad a = A(x-ia) + B(x+ia)$$

To find the value of A, put $x = -ia$

We get $A = -\dfrac{1}{2i}$

and for B, put $x = ia$, which gives $B = \dfrac{1}{2i}$

therefore, we have

$$y_1 = \frac{1}{2i}\left[\frac{1}{x-ia} - \frac{1}{x+ia}\right]$$

$$= \frac{1}{2i}[(x-ia)^{-1} - (x+ia)^{-1}]$$

Differentiating $(n-1)$ times, we get

$$y_n = \frac{1}{2i}[(-1)^{n-1}(n-1)!(x-ia)^{-n} - (-1)^{n-1}(n-1)!(x+ia)^{-n}]$$

$$= \frac{(-1)^{n-1}(n-1)!}{2i}[(x-ia)^{-n} - (x+ia)^{-n}]$$

Put $x = r\cos\theta$, $a = r\sin\theta$, we have

$$y_n = \frac{(-1)^{n-1}(n-1)!}{2i}[r^{-n}(\cos\theta - i\sin\theta)^{-n} - r^{-n}(\cos\theta + i\sin\theta)^{-n}]$$

$$= \frac{(-1)^{n-1}(n-1)!}{2i}r^{-n}[(\cos n\theta + i\sin n\theta) - (\cos n\theta - i\sin n\theta)]$$

$$= \frac{(-1)^{n-1}(n-1)!}{2i}r^{-n}.2i\sin n\theta$$

$$[\because \sin(-n\theta) = -\sin n\theta]$$

$$= (-1)^{n-1}(n-1)!r^{-n}.\sin n\theta$$

$$= (-1)^{n-1}(n-1)!\left(\frac{a}{\sin\theta}\right)^{-n}\sin n\theta$$

$$\left[\text{since } r = \frac{a}{\sin\theta}\right]$$

$$= (-1)^{n-1}(n-1)!a^{-n}\sin^n\theta.\sin n\theta$$

Example 2. *Find the nth differential coefficient of* $\log(ax + x^2)$.

Solution. Let $y = \log(ax + x^2) = \log[x(a + x)]$

$$= \log x + \log(a + x)$$

Differentiating n times, we get

$$y_n = \frac{d^n}{dx^n}(\log x) + \frac{d^n}{dx^n}\log(a+x)$$

$$= \frac{(-1)^{n-1}(n-1)!.1^n}{x^n} + \frac{(-1)^{n-1}(n-1)!.1^n}{(x+a)^n}$$

$$= (-1)^{n-1}(n-1)!\left[\frac{1}{x^n} + \frac{1}{(x+a)^n}\right].$$

Example 3. *Find the nth differential coefficients of*

(i) $e^{ax}\sin bx\cos cx$ (ii) $e^{2x}\sin^3 x$

Solution. (i) Let $y = e^{ax}\sin bx\cos cx$

$$= \frac{1}{2}e^{ax}[2\sin bx\cos cx]$$

$$= \frac{1}{2}e^{ax}[\sin(bx+cx) + \sin(bx-cx)]$$

$$= \frac{1}{2}[e^{ax}\sin(b+c)x + e^{ax}\sin(b-c)x] \quad \ldots(1)$$

Differentiating (1) n times, we get

$$\frac{d^n}{dx^n}[y] = y_n$$

$$= \frac{1}{2}[\{a^2+(b+c)^2\}^{n/2}e^{ax}\sin\{(b+c)x + n\tan^{-1}(b+c)/a\} + \{a^2+(b-c)^2\}^{n/2} e^{ax}\sin\{(b-c)x + n\tan^{-1}(b-c)/a\}]$$

(ii) Let $y = e^{2x}\sin^3 x$.

Now using the result

$$\text{Sin}3x = 3\sin x - 4\sin^3 x$$

We have

$$4\sin^3 x = 3\sin x - \sin 3x$$

$$\Rightarrow \quad \sin^3 x = \frac{1}{4}(3\sin x - \sin 3x)$$

Therefore,

$$y = \frac{1}{4}e^{2x}[3\sin x - \sin 3x]$$

$$= \frac{3}{4}e^{2x}\sin x - \frac{1}{4}e^{2x}\sin 3x.$$

Now, differentiating n times, we get

$$y_n = \frac{3}{4}[(2^2+1^2)^{1/2}]^n e^{2x}\sin[x + n\tan^{-1}1/2] - \frac{1}{4}[(2^2+3^2)^{1/2}]^n e^{2x}\sin[3x + n\tan^{-1}3/2].$$

Example 4. *Find the nth differential coefficieats of* $\sin^5 x\cos^3 x$.

Solution. First we reduce $\sin^5 x\cos^3 x$ into a function consisting sine function of multble of x.

Let $z = \cos x + i \sin x.$

Then $z^{-1} = \cos x - i \sin x$

$\therefore \quad z + z^{-1} = 2\cos x \quad \text{and } z - z^{-1} = 2i\sin x$

Also, by De-Moivre's theorem, we have

$$z^m + z^{-m} = 2\cos mx$$

and $z^m - z^{-m} = 2i\sin mx$

Now $(2i\sin x)^5(2\cos x)^3$

$$= (z - z^{-1})^5 + (z + z^{-1})^3$$

$$\Rightarrow \quad 2^8 i \sin^5 x \cos^3 x = (z^8 - z^{-8}) - 2(z^6 - z^{-6}) - 2(z^4 - z^{-4}) + 6(z^2 - z^{-2})$$

$$= 2i \sin 8x - 4i \sin 6x - 4i \sin 4x + 12 i \sin 2x$$

$$\Rightarrow \quad \sin^5 x \cos^3 x = 2^{-7}[\sin 8x - 2\sin 6x - 2\sin 4x + 6 \sin 2x].$$

Dfferentiating both sides n times w.r.t. x, we get

$$D^n(\sin^5 x \cos^3 x)$$

$$= 2^{-7}\left[8^n \sin\left(8x + \frac{n\pi}{2}\right) - 2.6^n \sin\left(6x + \frac{n\pi}{2}\right) - 2.4^n \sin\left(4x + \frac{n\pi}{2}\right) + 6.2^n \sin\left(2x + \frac{n\pi}{2}\right)\right]$$

1.3 USE OF PARTIAL FRACTIONS

To determine the nth derivative of any rational function, we have to split it into partial fractions.

Partial fractions for

(i) $$\frac{f(x)}{(x-a)(x-b)(x-c)} = \frac{A}{(x-a)} + \frac{B}{(x-b)} + \frac{C}{(x-c)}$$

(ii) $$\frac{f(x)}{(x-a)^2(x-b)} = \frac{A}{(x-a)} + \frac{B}{(x-a)^2} + \frac{C}{(x-b)}$$

(iii) $$\frac{f(x)}{(x-a)^3(x-b)} = \frac{A}{(x-a)} + \frac{B}{(x-a)^2} + \frac{C}{(x-a)^3} + \frac{D}{(x-b)}$$

(iv) $$\frac{f(x)}{(x-a)(x-b)(px^2+qx+r)} = \frac{A}{(x-a)} + \frac{B}{(x-b)} + \frac{Cx+D}{(px^2+qx+r)}$$

To find A, B, C, D etc., we put each linear factor of LCM equal to zero. The remaining constants are obtained by comparing coefficients of like powers on both sides.

Remarks

- Forming partial fractions is converse process of taking LCM.
- To resolve a fraction into partial fractions, the degree of the numerator must be less than the degree of denominator.

Solved Examples

Example 1. *Find the nth differential coefficients of*

(i) $\dfrac{1}{1-5x+6x^2}$ (ii) $\dfrac{x^2}{[(x+2)(2x+3)]}$

(UKTU 2011)

Solution. (i) Let

$$y = \frac{1}{1-5x+6x^2} = \frac{1}{(3x-1)(2x-1)}$$

$$= \frac{2}{2x-1} - \frac{3}{3x-1}$$

(By resolving into partial fractions)

$$= 2(2x-1)^{-1} - 3(3x-1)^{-1}$$

Differentaittng, n times, we get

$$y_n = 2(-1)^n n! 2^n (2x-1)^{-n-1} - 3(-1)^n n! 3^n (3x-1)^{-n-1}$$

$$= (-1)^n . n![2^{n+1}(2x-1)^{-n-1} - 3^{n+1}(3x-1)^{-n-1}]$$

(ii) Let $y = \dfrac{x^2}{[(x+2)(2x+3)]}$

Since, the given fraction is not a proper one so, divide the Nr. by Dr., we observe that the quotient will be 1/2.

So let

$$\frac{x^2}{(x+2)(2x+3)} = \frac{1}{2} + \frac{A}{x+2} + \frac{B}{2x+3}$$

which gives $A = -4, B = 9/2$

Therefore,

$$y = \frac{1}{2} - \frac{4}{x+2} + \frac{9}{2(2x+3)}$$

$$= \frac{1}{2} - 4(x+2)^{-1} + \frac{9}{2}(2x+3)^{-1}$$

Differentiating n times, we get

$$y_n = -4(-1)^n n!(x+2)^{-n-1} + \frac{9}{2}(-1)^n .n!2^n(2x+3)^{-n-1}$$

$$= (-1)^n n!\left[\frac{9.2^{n-1}}{(2x+3)^{n+1}} - \frac{4}{(x+2)^{n+1}}\right]$$

REMARK

- If none of the standard formulae is applicable to find y_n in any problem, then find y_1, y_2, y_3 and generalise.

MORE SOLVED EXAMPLES

Example 1. *If $y = \sqrt{x+a}$, find y_n.*

Solution . We have

$$y = \sqrt{x+a} = (x+a)^{1/2}$$

$$y_1 = \frac{1}{2}(x+a)^{-1/2}$$

$$y_2 = \left(\frac{1}{2}\right)\left(-\frac{1}{2}\right)(x+a)^{-3/2}$$

$$y_3 = \left(\frac{1}{2}\right)\left(-\frac{1}{2}\right)\left(-\frac{3}{2}\right)(x+a)^{-5/2}$$

$$= (-1)^2\frac{1.3}{2^3}(x+a)^{-5/2}$$

....

....

$$y_n = (-1)^{n-1}\frac{1.3.5\ldots\text{upto }(n-1)\text{ times}}{2^n}(x+a)^{-\frac{(2n-1)}{2}}$$

$$y_n = (-1)^{n-1}\frac{1.3\ldots(2n-3)}{2^n}(x+a)^{-\left(\frac{2n-1}{2}\right)}$$

Example 2. *If $y = \tan^{-1}\left\{\frac{\sqrt{(1+x^2)}-1}{x}\right\}$, show that*

$$y_n = \frac{1}{2}(-1)^{n-1}(n-1)!\sin^n\theta\sin n\theta,$$

where $\theta = \cot^{-1}x$.

Solution . We have $y = \tan^{-1}\left\{\frac{\sqrt{(1+x^2)}-1}{x}\right\}$.

Put $x = \tan\phi$, then

$$y = \tan^{-1}\left\{\frac{\sqrt{(1+\tan^2\phi)}-1}{\tan\phi}\right\} = \tan^{-1}\left[\frac{\sec\phi-1}{\tan\phi}\right]$$

$$= \tan^{-1}\left(\frac{1-\cos\phi}{\sin\phi}\right) = \tan^{-1}\left(\frac{2\sin^2(\phi/2)}{2\sin(\phi/2)\cos(\phi/2)}\right)$$

$$= \tan^{-1}\tan(\phi/2) = \phi/2 = \frac{1}{2}\tan^{-1}x$$

$$\Rightarrow \quad y_1 = \frac{1}{2(1+x^2)} = \frac{1}{2(x-i)(x+i)}$$

$$= \frac{1}{4i}\left(\frac{1}{x-i} - \frac{1}{x+i}\right)$$

Differentiating $(n-1)$ times, we get

$$y_n = \frac{(-1)^{n-1}(n-1)!}{4i}[(x-i)^{-n} - (x+i)^{-n}]$$

Now putting $x = r\cos\theta$, $1 = r\sin\theta$, we have

$$y_n = \frac{(-1)^{n-1}(n-1)!}{4i}\left[\begin{array}{l} r^{-n}(\cos\theta - i\sin\theta)^{-n} \\ -r^{-n}(\cos\theta + i\sin\theta)^{-n}\end{array}\right]$$

$$= \frac{(-1)^{n-1}(n-1)!}{4i}r^{-n}\left[\begin{array}{l}(\cos n\theta + i\sin n\theta) \\ -(\cos n\theta - i\sin n\theta)\end{array}\right]$$

$$= \frac{1}{2}(-1)^{n-1}(n-1)!r^{-n}\sin n\theta$$

$$= \frac{1}{2}(-1)^{n-1}(n-1)!\left(\frac{1}{\sin\theta}\right)^{-n}\sin n\theta \qquad \left[\because r = \frac{1}{\sin\theta}\right]$$

$$= \frac{1}{2}(-1)^{n-1}(n-1)!\sin^n\theta\sin n\theta$$

where $\theta = \tan^{-1}\frac{1}{x} = \cot^{-1}x$.

Example 3. *If $y = \sin mx + \cos mx$, prove that $y_n = m^n[1 + (-1)^n \sin 2mx]^{1/2}$.* (MTU–2012)

Solution . We have

$$y_n = \frac{d^n}{dx^n}(\sin mx) + \frac{d^n}{dx^n}(\cos mx)$$

$$= m^n\sin\left(mx + n\frac{\pi}{2}\right) + m^n\cos\left(mx + n\frac{\pi}{2}\right)$$

$$= m^n\left[\left\{\sin\left(mx+n\frac{\pi}{2}\right) + \cos\left(mx+n\frac{\pi}{2}\right)\right\}^2\right]^{1/2}$$

$$= m^n\left[1 + 2\sin\left(mx+n\frac{\pi}{2}\right).\cos\left(mx+n\frac{\pi}{2}\right)\right]^{1/2}$$

$$= m^n[1+\sin(2mx+n\pi)]^{1/2}$$

$$= m^n[1\pm\sin 2mx]^{1/2}$$

$$= m^n[1+(-1)^n\sin 2mx]^{1/2}.$$

Example 4. *Find the nth difference coefficient of* $\log[(ax+b)(cx+d)]$.

Solution . Let

$$y = \log[(ax+b)(cx+d)]$$

$$= \log(ax+b) + \log(cx+d)$$

We know that

$$D^n \log(ax+b) = (-1)^{n-1}(n-1)!\,a^n(ax+b)^{-n}$$

$$\therefore\ y_n = (-1)^{n-1}(n-1)!\,a^n(ax+b)^{-n} + (-1)^{n-1}(n-1)!\,c^n(cx+d)^{-n}$$

$$= (-1)^{n-1}(n-1)!\left[\frac{a^n}{(ax+b)^n} + \frac{c^n}{(cx+d)^n}\right]$$

Example 5. *Find the nth derivative of $y = \cos^4 x$.*

Solution . Let $y = \cos^4 x = (\cos^2 x)^2$

$$= [(1/2)(1+\cos 2x)]^2$$

$$= 1/4(1 + 2\cos 2x + \cos^2 2x)$$

$$= 1/4[1 + 2\cos 2x + (1/2)(1+\cos 4x)]$$

$$= 1/4[3/2 + 2\cos 2x + 1/2\cos 4x]$$

$$= 3/8 + (1/2)\cos 2x + (1/8)\cos 4x.$$

Now,

$$y_n = 0 + \frac{1}{2}.2^n \cos\left(2x + \frac{1}{2}n\pi\right) + \frac{1}{8}.4^n\cos\left(4x+\frac{1}{2}n\pi\right)$$

$$= 2^{n-1}.\cos\left(2x+\frac{1}{2}n\pi\right) + 2^{2n-3}\cos\left(4x+\frac{1}{2}n\pi\right)$$

Example 6. *Find the nth derivative of $x^{n-1}\log x$.* (UPTU–2010, 2012)

Solution . Let $y = x^{n-1}\log x$ ...(1)

Differentiating (1) w.r.t. x we get

$$y_1 = x^{n-1}.\frac{1}{x} + (n-1)x^{n-2}\log x$$

$$= x^{n-1}.\frac{1}{x} + (n-1)\frac{x^{n-1}}{x}\log x$$

$$\Rightarrow xy_1 = x^{n-1} + (n-1)y \qquad ...(2)$$

Finally, differentiating (2) both the sides $(n-1)$ times w.r.t. x, we get

$$y_n x + (n-1)y_{n-1}.1 = (n-1)! + (n-1)y_{n-1}$$

Hence, $y_n = \dfrac{(n-1)!}{x}$.

Example 7. *If $y = x\log\dfrac{x-1}{x+1}$, show that*

$$y_n = (-1)^2(n-2)!\left[\frac{x-n}{(x-1)^n} - \frac{x+n}{(x+1)^n}\right].$$

(UPTU–2003)

Solution . Let $y = x\log\dfrac{x-1}{x+1}$

$$\Rightarrow\ y = x\log(x-1) - x\log(x+1) \qquad ...(1)$$

Differentiating (1) w.r.t. x we get

$$y_1 = \frac{x}{x-1} + \log(x-1) - \frac{x}{x+1} - \log(x+1)$$

$$= 1 + \frac{1}{x-1} + \log(x-1) - 1 + \frac{1}{x+1} - \log(x+1)$$

$$= \frac{1}{x-1} + \frac{1}{x+1} + \log(x-1) - \log(x+1) \qquad ...(2)$$

Differentiating both sides of (2) w.r.t. x, $(n-1)$ times we get

$$y_n = \frac{(-1)^{n-1}(n-1)!}{(x-1)^n} + \frac{(-1)^{n-1}(n-1)!}{(x+1)^n} + \frac{(-1)^{n-2}(n-2)!}{(x-1)^{n-1}} - \frac{(-1)^{n-2}(n-2)!}{(x+1)^{n-1}}$$

$$= (-1)^{n-2}(n-2)!\left\{-\frac{(n-1)+n-1}{(x-1)^n}\right\} + (-1)^{n-2}(n-2)!\left\{\frac{-(n-1)-(x+1)}{(x+1)^n}\right\}$$

$$\Rightarrow\quad y_n = (-1)^{n-2}(n-2)!\left\{\frac{x-n}{(x-1)^n} - \frac{x+n}{(x+1)^n}\right\}$$

Example 8. *Find the nth derivative of $\dfrac{1}{x^2+a^2}$.* (UKTU–2011)

Solution . Let $y = \dfrac{1}{x^2+a^2} = \dfrac{1}{(x+ia)(x-ia)}$

$$= \frac{1}{2ia}\left[\frac{1}{x-ia} - \frac{1}{x+ia}\right] \qquad ...(1)$$

Differentiating (1) n times w.r.t. x we get

$$y_n = \frac{1}{2ia}\left[\frac{(-1)^n n!}{(x-ia)^{n+1}} - \frac{(-1)^n n!}{(x+ia)^{n+1}}\right]$$

$$= \frac{(-1)^n.n!}{2ia}\left[\frac{1}{(x-ia)^{n+1}} - \frac{1}{(x+ia)^{n+1}}\right] \qquad ...(2)$$

Let $x = r\cos\theta$ and $a = r\sin\theta$ *i.e.*, $\theta = \tan^{-1}\dfrac{a}{x}$ in (2), we get

$$y_n = \frac{(-1)^n.n!}{2iar^{n+1}}\left[\frac{1}{(\cos\theta - i\sin\theta)^{n+1}} - \frac{1}{(\cos\theta + i\sin\theta)^{n+1}}\right]$$

$$= \frac{(-1)^n.n!}{2iar^{n+1}}\left[\frac{1}{\cos(n+1)\theta - i\sin(n+1)\theta} - \frac{1}{\cos(n+1)\theta + i\sin(n+1)\theta}\right]$$

$$= \frac{(-1)^n.n!}{2iar^{n+1}}\left[\{\cos(n+1)\theta + i\sin(n+1)\theta\} - \{\cos(n+1)\theta - i\sin(n+1)\theta\}\right]$$

$$= \frac{(-1)^n.n!}{2iar^{n+1}}[2i\sin(n+1)\theta]$$

$$= \frac{(-1)^n.n!\sin(n+1)\theta}{a\left(\dfrac{a^{n+1}}{\sin^{n+1}\theta}\right)} \qquad [\because a = r\sin\theta]$$

$$= \frac{(-1)^n.n!\sin(n+1)\theta\sin^{n+1}\theta}{a^{n+2}},$$

$\theta = \tan^{-1}\dfrac{a}{x}$.

EXERCISE 1.1

1. Find the nth derivatives of

(i) $\sin^3 x$ (ii) $\cos x \cos 2x \cos 3x$

(iii) $e^{ax}\cos^2 x \sin x$ (SVTU–2009)

(iv) $\sin^5 x \cos^3 x$ (v) $\sin ax \cos bx$

(vi) $\sin^2 x \sin 2x$

2. Find the nth derivatives of

(i) $\dfrac{x^4}{(x-1)(x-2)}$ (ii) $\dfrac{x}{1+3x+2x^2}$

(iii) $\dfrac{1}{(x-2)(x-1)^3}$ (iv) $\dfrac{1}{x^2-a^2}$

(v) $\dfrac{x^2}{(x-a)(x-b)}$ (vi) $\dfrac{17x^2+26x-42}{6x^3-25x^2-29x+20}$

3. Find the nth derivatives of

(i) $\tan^{-1}\left(\dfrac{1+x}{1-x}\right)$ (ii) $\tan^{-1}\left(\dfrac{2x}{1-x^2}\right)$ (UPTU–2002)

4. Show that the value of the nth differential coefficients of $\dfrac{x^3}{x^2-1}$ for $x = 0$, is zero if n is even and is $-n!$, if n is odd and greater than 1.

5. If $y = x(a^2+x^2)^{-1}$, show that

$y_n = (-1)^n n! a^{-n-1} \sin^{n+1}\theta \cos(n+1)\theta$ where $\theta = \tan^{-1}\left(\dfrac{a}{x}\right)$. (Mumbai–2007)

6. (i) If $x = a(t - \sin t)$ and $y = a(1 + \cos t)$, prove that $\dfrac{d^2y}{dx^2} = \dfrac{1}{4a}\text{cosec}^4\left(\dfrac{t}{2}\right)$. (Madurai–1990, 2004)

(ii) If $x = a(\cos\theta + \theta\sin\theta), y = a(\sin\theta - \theta\cos\theta)$, find $\dfrac{d^2y}{dx^2}$.

(iii) If $y = \sin(\sin x)$, show that $\left(\dfrac{d^2y}{dx^2}\right) + \left(\dfrac{dy}{dx}\right)\tan x + y\cos^2 x = 0$.

(iv) If $y = A\sin mx + B\cos mx$, show that $\dfrac{d^2y}{dx^2} + m^2 y = 0$.

(v) If $y = e^{ax}\sin bx$, show that $\dfrac{d^2y}{dx^2} - 2a\dfrac{dy}{dx} + (a^2+b^2)y = 0$.

7. If $p^2 = a^2\cos^2\theta + b^2\sin^2\theta$, prove that $p + \dfrac{d^2p}{d\theta^2} = \dfrac{a^2.b^2}{p^3}$.

8. Prove that the value when $x = 0$ of $\dfrac{d^n}{dx^n}(\tan^{-1} x)$ is 0, ([illegible] – 1)! or –(n – 1)! according as n is of the form $2p$, $4p+1$ [illegible] $4p+3$ respectively.

Hint to Selected Problems

2. (i) Resolving into partial fractions, we get

$$y = (x^2 + 3x + 7) + \frac{15x-14}{(x-1)(x-2)}.$$

Now differentiate successively.

(ii) $y = \dfrac{1}{(1+x)} - \dfrac{1}{(1+2x)}$. Now differentiate successively.

3. (i) $y = \tan^{-1}\dfrac{1+x}{1-x} \Rightarrow y_1 = \dfrac{1}{(1+x^2)}$

$$= \frac{1}{(x-i)(x+i)} = \frac{1}{2i}\left(\frac{1}{x-i} + \frac{1}{x+i}\right)$$

Then differentiate successively.

4. The given function can be written as $y = x + \dfrac{1}{2}\left(\dfrac{1}{x-1} + \dfrac{1}{x+1}\right)$.

5. The given function can be written as $y = \dfrac{1}{2}\left(\dfrac{1}{x-a} + \dfrac{1}{x+ai}\right)$.

6. (i) Find $\dfrac{dx}{dt}$ and $\dfrac{dy}{dt}$ such that $\dfrac{dx}{dt} = a(1-\cos t)$ ar[illegible] $\dfrac{dy}{dt} = -a\sin t$.

Then $\dfrac{dy}{dx} = \dfrac{dy/dt}{dx/dt} = \dfrac{-a\sin t}{a(1-\cos t)} = -\cot t/$[illegible]

Now differentiate successively.

8. Let $y = \tan^{-1}x$. Then $y_1 = \dfrac{1}{1+x^2}$

$$= \frac{1}{(x-i)(x+i)} = \frac{1}{2i}\left[\frac{1}{x-i} - \frac{1}{x+i}\right].$$

Now differentiate successively.

ANSWERS

1.(i) $y_n = \dfrac{3}{4}\sin\left(x + \dfrac{n\pi}{2}\right) - \dfrac{1}{4}.3^n\sin\left(3x + \dfrac{n\pi}{2}\right)$

(ii) $y_n = \dfrac{1}{4}\left\{6^n\cos\left(6x + \dfrac{1}{2}n\pi\right) + 4^n\cos\left(4x + \dfrac{n\pi}{2}\right) + 2^n\right.$ [illegible] $\left.\left(2x + \dfrac{n\pi}{2}\right)\right\}$

(iii) $y_n = \dfrac{1}{4}[(a^2+1)^{n/2}e^{ax}\sin\{x + n\tan^{-1}1/a\} + (a^2+9)^{n/2}e^{ax}\sin(3x + n\tan^{-1}3/a)]$

(iv) $y_n = 2^{-7}\left[8^n\sin\left(8x + \dfrac{1}{2}n\pi\right) - 2.6^n\sin\left(6x + \dfrac{1}{2}n\pi\right) - 2.4^n\sin\left(4x + \dfrac{1}{2}n\pi\right) + 6.2^n\sin\left(2x + \dfrac{1}{2}n\pi\right)\right]$

(v) $y_n = \dfrac{1}{2}\left[(a+b)^n\sin\left\{(a+b)x + \dfrac{1}{2}n\pi\right\} + (a-b)^n\sin\left\{(a-b)x + \dfrac{1}{2}n\pi\right\}\right]$

(vi) $y_n = 2^{n-1}\sin\left(2\right.$[illegible]$\left.\dfrac{1}{2}n\pi\right) - 4^{n-1}\sin\left(4x + \dfrac{1}{2}n\pi\right)$

2.(i) $y_n = (-1)^n n![16(x-2)^{-n-1} - (x-1)^{-n-1}]$

(ii) $y_n = (-1)^n n!\left[\dfrac{1}{(x+1)^{n+1}} - \dfrac{[illegible]}{(2[illegible]1)^{n+1}}\right]$

(iii) $y_n = (-1)^{n+1}.n!\left[\frac{(n+2)(n+1)}{2(x-1)^{n+3}} + \frac{(n+1)}{(x-1)^{n+2}} + \frac{1}{(x-1)^{n+1}} - \frac{1}{(x-2)^{n+1}}\right]$ (iv) $y_n = \frac{1}{2a}(-1)^n.n!\{(x-a)^{-n-1} - (x+a)^{-n-1}\}$

(v) $y_n = \frac{(-1)^n.n!}{(a-b)}\left[\frac{a^2}{(x-a)^{n+1}} - \frac{b^2}{(x-b)^{n+1}}\right]$ (vi) $y_n = (-1)^n.n!\left[\frac{2^n}{(2x-1)^{n+1}} - \frac{2.3^n}{(3x+4)^{n+1}} + \frac{3}{(x-5)^{n+1}}\right]$

3.(i) $y_n = (-1)^{n-1}.(n-1)!\sin^n\theta \sin n\theta$, where $\theta = \tan^{-1}\frac{1}{x}$

(ii) $y_n = 2(-1)^{n-1}(n-1)!\sin^n\theta \sin n\theta$, where $\theta = \tan^{-1}\frac{1}{x}$ **6.**(ii) $y_2 = \frac{1}{a}.\frac{\sec^3\theta}{\theta}$

1.4 LEIBNITZ'S THEOREM

(UPTU–2008, GBTU–2011)

This theorem help us to find the nth differential coefficient of the product of two functions in terms of the successive derivatives of the functions.

Statement. *If u, v be two functions of x, having derivative of nth order, then*

$$D^n(nv) = u_n v + {}^nC_1 u_{n-1}v_1 + {}^nC_2 u_{n-2}v_2 + \ldots + + {}^nC_r u_{n-r}v_r + \ldots + {}^nC_n uv_n$$

where suffixes of u and v denote differentiations w.r.t. x.

STEP OUTLINES	We shall prove this theorem by mathematical induction in the following steps :
Step-1	Prove the result for $n = 1, 2$.
Step-2	Assume that result is true for $n = m$.
Step-3	Prove the result for $n = m + 1$.

Step 1. Let $y = uv$

$\Rightarrow$ $y_1 = u_1 v + uv_1$

and $y_2 = u_2 v + u_1 v_1 + u_1 v_1 + uv_2 = u_2 v + 2u_1 v_1 + uv_2$

$= u_2 v + {}^2C_1 u_1 v_1 + {}^2C_2 uv_2.$

Thus the theorem is true for $n = 1, 2$.

Step 2. Let us assume that the theorem is true for a particular value of n say m, then we have

$$y_m = u_m v + {}^mC_1 u_{m-1}v_1 + {}^mC_2 u_{m-2}v_2 + \ldots + {}^mC_{r-1}u_{m-r+1}v_{r-1} + {}^mC_r u_{m-r}v_r + \ldots + {}^mC_m uv_m. \quad \ldots(1)$$

Step 3. Now, differentiating (1), we have

$$y_{m+1} = u_{m+1}v + u_m v_1 + {}^mC_1 u_m v_1 + {}^mC_1 u_{m-1}v_2 + {}^mC_2 u_{m-1}v_2 + {}^mC_2 u_{m-2}v_3 + \ldots + {}^mC_{r-1}u_{m-r+2}v_{r-1}$$
$$+ {}^mC_{r-1}u_{m-r+1}v_r + {}^mC_r u_{m-r+1}v_r + {}^mC_r u_{m-r}v_{r+1} + \ldots + {}^mC_m u_1 v_m + {}^mC_m uv_{m+1}.$$
$$= u_{m+1}.v + ({}^mC_1 + 1)u_m v_1 + ({}^mC_2 + {}^mC_1)u_{m-1}v_2 + \ldots + ({}^mC_r + {}^mC_{r-1})u_{m-r+1}v_r + \ldots + {}^mC_m uv_{m+1}.$$

Now using Pascal's law, given by

$${}^mC_{r-1} + {}^mC_r = {}^{m+1}C_r$$

For r = 1, 2, 3, ...

We have ${}^mC_0 + {}^mC_1 = {}^{m+1}C_1 \Rightarrow 1 + {}^mC_1 = {}^{m+1}C_1$

${}^mC_1 + {}^mC_2 = {}^{m+1}C_2$

....

....

and ${}^mC_m = 1 = {}^{m+1}C_{m+1}$

Therefore,

$$y_{m+1} = u_{m+1}.v + {}^{m+1}C_1 u_m v_1 + {}^{m+1}C_2 u_{m-1}v_2 + \ldots + {}^{m+1}C_r u_{m-r+1}v_r + \ldots + {}^{m+1}C_{m+1}uv_{m+1}$$

... is true for $n = m$, then it is also true for the next higher value $n = m + 1$.

... we can say that theorem is true for any positive integer n.

Solved Examples

Example 1. *Find the nth derivative of $x^2 \sin x$.* (UPTU–2009)

Solution. Let $u = \sin x$ and $v = x^2$.

Then,
$$u_n = \sin\left[x + \frac{n\pi}{2}\right]$$
$$u_{n-1} = \sin\left(x + (n-1)\frac{\pi}{2}\right)$$
$$u_{n-2} = \sin\left[x + (n-2)\frac{\pi}{2}\right]$$

Also, $v_1 = 2x,\ v_2 = 2,\ v_3 = 0$

Now, by Leibnitz's theorem, we have
$$\frac{d^n}{dx^n}(uv) = u_n.v + {}^nC_1 u_{n-1}.v_1 + {}^nC_2 u_{n-2}.v_2$$
$$\Rightarrow \frac{d^n}{dx^n}(x^2 \sin x) = \sin\left(x + \frac{n\pi}{2}\right)x^2 + {}^nC_1 \sin\left[x + (n-1)\frac{\pi}{2}\right]2x + {}^nC_2 \sin\left[x + (n-2)\frac{\pi}{2}\right]2$$
$$= x^2 \sin\left(x + \frac{n\pi}{2}\right) + 2nx \sin\left[x + (n-1)\frac{\pi}{2}\right] + n(n-1)\sin\left[x + (n-2)\frac{\pi}{2}\right]$$

Example 2. *Find the nth derivative of $x^3 \cos x$.*

Solution. Let $u = \cos x$ and $v = x^3$.

Then,
$$u_n = \cos\left[x + \frac{n\pi}{2}\right], \qquad v_1 = 3x^2$$
$$u_{n-1} = \cos\left(x + (n-1)\frac{\pi}{2}\right), \qquad v_2 = 6x$$
$$u_{n-2} = \cos\left[x + (n-2)\frac{\pi}{2}\right], \qquad v_3 = 6$$
$$u_{n-3} = \cos\left[x + (n-3)\frac{\pi}{2}\right], \qquad v_4 = 0$$

Now, by Leibnitz's theorem, we have
$$\frac{d^n}{dx^n}(uv) = u_n.v + {}^nC_1 u_{n-1}.v_1 + {}^nC_2 u_{n-2}.v_2 + {}^nC_3 u_{n-3}.v_3$$
$$\Rightarrow \frac{d^n}{dx^n}(x^3 \cos x) = \cos\left(x + \frac{n\pi}{2}\right)x^3 + {}^nC_1 \cos\left[x + (n-1)\frac{\pi}{2}\right]3x^2 + {}^nC_2 \cos\left(x + (n-2)\frac{\pi}{2}\right)6x + {}^nC_3 \cos\left(x + (n-3)\frac{\pi}{2}\right)6$$
$$= x^3 \cos\left(x + \frac{n\pi}{2}\right) - 3x^2.n \sin\left(x + \frac{n\pi}{2}\right) - 3n(n-1)x \cos\left(x + \frac{n\pi}{2}\right) - n(n-1)(n-2)\sin\left(x + \frac{n\pi}{2}\right)$$
$$= [x^3 - 3n(n-1)x]\cos\left(x + \frac{n\pi}{2}\right) + [3x^2 n - n(n-1)(n-2)]\sin\left(x + \frac{n\pi}{2}\right).$$

Example 3. *If $y = a \cos(\log x) + b \sin(\log x)$, show that*
$$x^2 y_2 + xy_1 + y = 0$$
and $x^2 y_{n+2} + (2n+1)xy_{n+1} + (n^2+1)y_n = 0.$

(UPTU–2004, 2012, Madras–2000)

Solution. We have
$$y = a \cos(\log x) + b \sin(\log x) \qquad \ldots(1)$$

Differentiating (1) with respect to x, we have
$$y_1 = -\frac{a}{x}\sin(\log x) + \frac{b}{x}\cos(\log x)$$
$$xy_1 = -a \sin(\log x) + b \cos(\log x)$$

Again, differentiating w.r.t. x, we get
$$xy_2 + y_1 = -\frac{a}{x}\cos(\log x) - \frac{b}{x}\sin(\log x)$$
$$\Rightarrow \quad x^2 y_2 + xy_1 = -a\cos(\log x) - b\sin(\log x) = -y$$
$$\Rightarrow \quad x^2 y_2 + xy_1 + y = 0 \qquad \ldots(2)$$

Now, differentiating (2) both sides n times by Leibnitz's theorem, we get
$$D^n(x^2 y_2) + D^n(xy_1) + D^n(y) = 0$$
$$\Rightarrow \quad (D^n y_2)x^2 + {}^nC_1(D^{n-1}y_2)(Dx^2) + {}^nC_2(D^{n-2}y_2)(D^2x^2) + (D^n y_1)x + {}^nC_1(D^{n-1}y_1)(Dx) + D^n y = 0$$
$$\Rightarrow x^2 y_{n+2} + 2nxy_{n+1} + \frac{n(n-1)}{2}2y_n + xy_{n+1} + ny_n + y_n = 0$$
$$\Rightarrow \quad x^2 y_{n+2} + (2n+1)xy_{n+1} + (n^2+1)y_n = 0$$

Example 4. *If $y = e^{a\sin^{-1}x}$, show that*
$$(1-x^2)y_{n+2} - (2n+1)xy_{n+1} - (n^2+a^2)y_n = 0.$$

(MDU–1998, VTU–2003, KU–1999)

Solution. We have
$$y = e^{a\sin^{-1}x} \Rightarrow y_1 = e^{a\sin^{-1}x}.\frac{a}{\sqrt{1-x^2}}$$
$$y_1\sqrt{1-x^2} = ae^{a\sin^{-1}x} = ay$$
$$\Rightarrow y_1^2(1-x^2)$$

Now differentiating (1) with respect to x, we get

$$2y_1y_2(1-x^2)+y_1^2(-2x)=2a^2yy_1$$

$$\Rightarrow 2y_1[y_2(1-x^2)-xy_1-a^2y]=0$$

$$[\because 2y_1 \neq 0]$$

$$\Rightarrow [y_2(1-x^2)-xy_1-a^2y]=0 \quad \ldots(2)$$

Using Leibnitz's theorem, differentiating (2), n times, we get

$$D^n[y_2(1-x^2)]-D^n(y_1x)-a^2D^ny=0$$

$$\Rightarrow \left[y_{n+2}(1-x^2)+ny_{n+1}(-2x)+\frac{n(n-1)}{2}y_n(-2)\right] -[y_{n+1}x+ny_n]-a^2y_n=0$$

$$\Rightarrow (1-x^2)y_{n+2}-(2n+1)xy_{n+1}-(n^2+a^2)y_n=0$$

Example 5. *If* $\cos^{-1}\left(\frac{y}{b}\right)=\log\left(\frac{x}{n}\right)^n$. *Prove that*

$$x^2y_{n+2}-(2n+1)xy_{n+1}+2n^2y_n=0.$$

Solution . We have

$$\cos^{-1}\left(\frac{y}{b}\right)=\log\left(\frac{x}{n}\right)^n$$

$$=n\log\frac{x}{n}=n(\log x-\log n)$$

Now, differentiating with respect to x, we get

$$-\frac{1}{\sqrt{\left[1-\frac{y^2}{b^2}\right]}}\frac{y_1}{b}=\frac{n}{x}$$

or $$-\frac{y_1}{\sqrt{b^2-y^2}}=\frac{n}{x}$$

or $$y_1^2x^2=n^2(b^2-y^2)$$

Again, differentiating, with respect to x, we get

$$2y_1y_2^2+2xy_1^2=-2n^2yy_1$$

or $$y_2x^2+y_1x+n^2y=0. \qquad [\because 2y_1\neq 0]$$

Using Leibnitz's theorem, differentiating n times, we get

$$y_{n+2}x^2+{}^nC_1y_{n+1}(2x)+{}^nC_2y_n(2) +y_{n+1}x+{}^nC_1y_n+n^2y_n=0$$

$$\Rightarrow x^2y_{n+2}+(2n+1)xy_{n+1}+2n^2y_n=0.$$

Example 6. *If* $y=(x^2-1)^n$, *Prove that*

$$(x^2-1)y_{n+2}+2xy_{n+1}-n(n+1)y_n=0.$$

(VTU–2010, UPTU–2010)

Hence if $P_n=\frac{d^n}{dx^n}(x^2-1)^n$ *show that*

$$\frac{d}{dx}\left\{(1-x^2)\frac{dP_n}{dx}\right\}+n(n+1)P_n=0$$

Solution . We have $y=(x^2-1)^n \quad \ldots(1)$

Therefore $y_1=n(x^2-1)^{n-1}.2x$

or $(x^2-1)y_1=n(x^2-1)^n.2x$

$$\Rightarrow (x^2-1)y_1=2nxy. \quad \ldots(2)$$

Differentiating (2), $(n+1)$ times by Leibnitz's theorem, we get

$$D^{n+1}[y_1(x^2-1)]-2nD^{n+1}(yx)=0$$

or $$y_{n+2}(x^2-1)+(n+1)y_{n+1}.2x +\frac{n(n+1)}{2}.y_n.2-2ny_{n+1}.x -2n(n+1)y_n.1=0$$

or $$(x^2-1)y_{n+2}+2xy_{n+1}-n(n+1)y_n=0$$

Hence, the first result from (2), we get

$$(x^2-1)D^2y_n+2xDy_n-n(n+1)y_n=0.\ldots(3)$$

Putting $y_n=\frac{d^n}{dx^n}(x^2-1)^n=P_n$; equation (3) becomes

$$(x^2-1)D^2P_n+2xDP_n-n(n+1)P_n=0$$

or $$-(1-x^2)D^2P_n+2xD(P_n)-n(n+1)P_n=0$$

or $$-\frac{d}{dx}\left\{(1-x^2)DP_n\right\}-n(n+1)P_n=0$$

or $$\frac{d}{dx}\left\{(1-x^2)\frac{d}{dx}P_n\right\}+n(n+1)P_n=0$$

Example 7. *If* $y=\sin(m\sin^{-1}x)$, *Prove that*

$$(1-x^2)y_2-xy_1+m^2y=0$$

and

$$(1-x^2)y_{n+2}-(2n+1)xy_{n+1}-(n^2-m^2)y_n=0$$

(UKTU–2012, UPTU–2007, 2009, GBTU–2011)

Solution . Let $y=\sin(m\sin^{-1}x) \quad \ldots(1)$

Differentiating w.r.t. x we get

$$y_1=\cos(m\sin^{-1}x).\frac{m}{\sqrt{1-x^2}}$$

$$\Rightarrow y_1\sqrt{1-x^2}=m\cos(m\sin^{-1}x)$$

$$\Rightarrow y_1^2(1-x^2)=m^2\cos^2(m\sin^{-1}x)$$

$$=m^2[1-\sin^2(m\sin^{-1}x)]$$

$$=m^2(1-y^2) \quad \ldots(2)$$

Again, differentiating both sides of (2) w.r.t. x we get

$$(1-x^2)2y_1y_2-2xy_1^2=-2m^2yy_1$$

$$\Rightarrow (1-x^2)y_2-xy_1=-m^2y$$

$$\Rightarrow (1-x^2)y_2-xy_1+m^2y=0 \quad \ldots(3)$$

Finally, differentiating (3) n times, by Leibnitz's theorem, we get

$$\left[y_{n+2}(1-x^2)+{}^nC_1y_{n+1}(-2x)+{}^nC_2y_n(-2)\right] -\left[y_{n+1}x+{}^nC_1y_n\right]+m^2y_n=0$$

$$\Rightarrow (1-x^2)y_{n+2}-2nxy_{n+1}$$
$$-n(n+1)y_n - xy_{n+1} - ny_n + m^2y_n = 0$$

or $$(1-x^2)y_{n+2}-(2n+1)xy_{n+1} -(n^2-m^2)y_n = 0$$

Example 8. *If* $\cos^{-1}\left(\frac{y}{b}\right) = \log\left(\frac{x}{m}\right)^m$, *Show that* $x^2y_{n+2} + (2n+1)xy_{n+1} + (n^2+m^2)y_n = 0.$ (UPTU–2006)

Solution . We have

$$\cos^{-1}\left(\frac{y}{b}\right) = \log\left(\frac{x}{m}\right)^m$$

$$\Rightarrow \quad y = b\cos\left(m\log\left(\frac{x}{m}\right)\right)$$

$$\therefore \quad y_1 = -b\sin\left(m\log\left(\frac{x}{m}\right)\right).m\frac{1}{(x/m)}.\frac{1}{m}$$

$$\Rightarrow \quad xy_1 = -bm\sin\left(m\log\left(\frac{x}{m}\right)\right)$$

Again differentiating, we get

$$xy_2 + y_1 = -bm\cos\left\{m\log\left(\frac{x}{m}\right)\right\}.m.\frac{1}{(x/m)}.\frac{1}{m}$$

$$\Rightarrow x^2y_2 + xy_1 = -m^2b\cos\left\{m\log\left(\frac{x}{m}\right)\right\} = -m^2y$$

$$\therefore \quad x^2y_2 + xy_1 + m^2y = 0$$

Differentiating both sides of the above equation, n times by Leibnitz's theorem, we get

$$[y_{n+2}.x^2 + {}^nC_1y_{n+1}(2x) + {}^nC_2y_n(2)] + [y_{n+1}(x) + {}^nC_1y_n(1)] + m^2y_n = 0$$

$$\Rightarrow x^2y_{n+2} + (2n+1)xy_{n+1} + (n^2+m^2)y_n = 0$$

Example 9. *If* $x = \cosh\left(\frac{1}{m}\log y\right)$, *prove that* $(x^2-1)y_{n+2} + (2n+1)xy_{n+1} + (n^2-m^2)y_n = 0.$

Solution . We have

$$x = \cosh\left(\frac{1}{m}\log y\right)$$

$$\Rightarrow \quad \cosh^{-1}x = \frac{1}{m}\log y$$

$$\Rightarrow \quad y = e^{m\cosh^{-1}x} \qquad \ldots(1)$$

$$\Rightarrow \quad y_1 = e^{m\cosh^{-1}x}.m.\frac{1}{\sqrt{x^2-1}}$$

$$\Rightarrow \quad \left(\sqrt{x^2-1}\right)y_1 = me^{m\cosh^{-1}x} = my$$

$$\Rightarrow \quad (x^2-1){y_1}^2 = m^2y^2 \qquad \ldots(2)$$

Differentiating (2) n times by Leibnitz's theorem, we get

$$(x^2-1)2y_1y_2 + {y_1}^2(2x) = 2m^2yy_1$$

$$\Rightarrow \quad (x^2-1)y_2 + xy_1 - m^2y = 0 \qquad \ldots(3)$$

Differentiating (3), n times by Leibnitz theorem, we get

$$\left[y_{n+2}(x^2-1) + {}^nC_1y_{n+1}(2x) + {}^nC_2y_n(2)\right] + \left[y_{n+1}(x) + {}^nC_1y_n(1)\right] - m^2y_n = 0$$

$$\Rightarrow (x^2-1)y_{n+2} + (2n+1)xy_{n+1} + (n^2-m^2)y_n = 0$$

Example 10. *If* $x = \tan(\log y)$, *prove that* $(1+x^2)y_{n+1} + (2nx-1)y_n + n(n-1)y_{n-1} = 0.$ (UPTU–2007)

Solution . Let $x = \tan(\log y)$

$$\Rightarrow \quad y = e^{\tan^{-1}x} \qquad \ldots(1)$$

$$\Rightarrow \quad y_1 = e^{\tan^{-1}x}.\frac{1}{(1+x^2)}$$

$$\therefore \quad (1+x^2)y_1 = y \qquad \ldots(2)$$

Differentiating (2) n times by Leibnitz's theorem, we get

$$y_{n+1}(1+x^2) + {}^nC_1y_n(2x) + {}^nC_2y_{n-1}(2) = y_n$$

$$\Rightarrow (1+x^2)y_{n+1} + (2nx-1)y_n + n(n-1)y_{n-1} = 0$$

Example 11. *If* $y = (1-x)^{\alpha}e^{-\alpha x}$, *prove that* $(1-x)y_{n+1} - (n+\alpha x)y_n - n\alpha y_{n-1} = 0.$ (UKTU–2011)

Solution . We have

$$y = (1-x)^{\alpha}e^{-\alpha x} \qquad \ldots(1)$$

$$\Rightarrow \quad y_1 = (1-x)^{-\alpha}(-\alpha e^{-\alpha x}) + e^{-\alpha x}(-\alpha)(1-x)^{-\alpha-1}(-1)$$

$$= e^{-\alpha x}(1-x)^{-\alpha}\left(-\alpha + \frac{\alpha}{1-x}\right)$$

$$\Rightarrow \quad y_1(1-x) = \alpha xy \qquad \ldots(2)$$

Differentiating (2) n times by Leibnitz's theorem, we get

$$y_{n+1}(1-x) + {}^nC_1y_n(-1) = \alpha[y_n(x) + {}^nC_1y_{n+1}(1)]$$

$$\therefore \quad (1-x)y_{n+1} + (-n-\alpha x)y_n - n\alpha y_{n-1} = 0.$$

$$\Rightarrow \quad (1-x)y_{n+1} - (n+\alpha x)y_n - n\alpha y_{n-1} = 0$$

Example 12. *If* $y = \sin\log(x^2+2x+1)$, *prove that* $(1+x^2)y_{n+2} + (2n+1)(1+x)y_{n+1} - (n^2+4)y_n = 0.$

Solution . We have

$$y = \sin\log(x^2+2x+1) \qquad \ldots(1)$$

$$\Rightarrow y_1 = \cos\log(x^2+2x+1).\frac{1}{x^2+2x+1}.(2x+2)$$

$$= \cos\log(x^2+2x+1).\frac{2}{x+1}$$

$$\Rightarrow (x+1)y_1 = 2\cos\log(x^2+2x+1)$$

$$\Rightarrow (x+1)^2 y_1^{\,2} = 4\cos^2\log(x^2+2x+1)$$

$$= 4(1-y^2) \quad \text{(Using (1))} \quad \ldots(2)$$

Now, differentiating (2) w.r.t. x, we get

$$(x+1)^2.2y_1y_2 + y_1^{\,2}.2(x+1) = 4(-2yy_1)$$

$$\Rightarrow (x+1)^2 y_2 + (x+1)y_1 + 4y = 0 \quad \ldots(3)$$

Differentiating (3), n times by Leibnitz's theorem, we get

$$\left[y_{n+2}(x+1)^2 + {}^nC_1 y_{n+1}.2(x+1) + {}^nC_2 y_n.2\right] + \left[y_{n+1}(x+1) + {}^nC_1 y_n(1)\right] + 4y_n = 0$$

$$\Rightarrow \quad (x+1)^2 y_{n+2} + (2nx+2n+x+1)y_{n+1} + (n^2-n+n+4)y_n = 0$$

$$\Rightarrow \quad (1+x^2)y_{n+2} + (2n+1)(1+x)y_{n+1} + (n^2+4)y_n = 0$$

Example 13. *If* $y = \tan^{-1}\left(\dfrac{a+x}{a-x}\right)$, *prove that*

$(a^2+x^2)y_{n+2} + 2(n+1)xy_{n+1} + n(n+1)y_n = 0.$ (GBTU–2012)

Solution. We have

$$y = \tan^{-1}\left(\frac{a+x}{a-x}\right) \quad \ldots(1)$$

Differentiating (1) w.r.t. x, we get

$$y_1 = \frac{a}{(a^2+x^2)} \Rightarrow (a^2+x^2)y_1 = a$$

Again differentiating w.r.t. x, we get

$$(a^2+x^2)y_2 + 2xy_1 = 0 \quad \ldots(2)$$

Now, differentiating (2), n times by Leibnitz's theorem, we get

$$\left[(x^2+a^2)y_{n+2} + {}^nC_1(2x)y_{n+1} + {}^nC_2(2)y_n\right] + \left[2xy_{n+1} + {}^nC_1.2.y_n\right] = 0$$

$$\Rightarrow \quad (x^2+a^2)y_{n+2} + 2nxy_{n+1} + n(n-1)y_n + 2xy_{n+1} + 2ny_n = 0$$

$$\Rightarrow \quad (x^2+a^2)y_{n+2} + 2x(n+1)y_{n+1} + [n(n-1)+2n]y_n = 0$$

$$\Rightarrow \quad (x^2+a^2)y_{n+2} + 2x(n+1)y_{n+1} + n(n+1)y_n = 0$$

EXERCISE 1.2

1. Use Leibnitz's theorem, to find y_n in the following cases :
   (i) x^3e^{ax} (ii) x^2e^x
   (iii) $x^3\sin ax$ (iv) $x^3\log x$
   (v) $x^2e^x\cos x$ (vi) $e^x\log x$
   (vii) $x^n\log x$ (UPTU 2008) (viii) $x^2\tan^{-1}x$

2. If $I_n = \dfrac{d^n}{dx^n}(x^n\log x)$, prove that $I_n = nI_{n-1} + (n-1)!$ and hence show that $I_n = n!\left(\log x + 1 + \dfrac{1}{2} + \dfrac{1}{3} + \ldots + \dfrac{1}{n}\right)$ (MTU–2011, VTU–2001, Mumbai–2008)

3. If $y = e^{\tan^{-1}x}$, prove that $(1+x^2)y_{n+2} + [2(n+1)x-1]y_{n+1} + n(n+1)y_n = 0.$

4. If $y = (\sin^{-1}x)^2$, prove that $(1-x^2)y_2 - xy_1 - 2 = 0$ and $(1-x^2)y_{n+2} - x(2n+1)y_{n+1} - n^2y_n = 0.$ (UPTU–2005, 2009, 2010)

5. If $y = \dfrac{\sin^{-1}x}{\sqrt{(1-x^2)}}$, prove that $(1-x^2)y_{n+1} - (2n+1)xy_n - n^2y_{n-1} = 0.$

6. If $y = [\log\{x+\sqrt{(1+x^2)}\}]^2$, prove that $(1+x^2)y_{n+2} + (2n+1)xy_{n+1} + n^2y_n = 0.$ (VTU–2007, Bhillai–2005, GBTU(Ag)–2010)

7. Differentiating n times the equation :
   (i) $(1+x^2)\dfrac{d^2y}{dx^2} - x\dfrac{dy}{dx} + a^2y = 0.$
   (ii) $x^2\dfrac{d^2y}{dx^2} + x\dfrac{dy}{dx} + y = 0.$

8. If $y = [x+\sqrt{(1+x^2)}]^m$, prove that $(1+x^2)y_{n+2} + (2n+1)xy_{n+1} + (n^2-m^2)y_n = 0.$ (VTU–2009, Madras–2000, 2004)

9. If $y^{1/m} + y^{-1/m} = 2x$, prove that $(x^2-1)y_{n+2} + (2n+1)xy_{n+1} + (n^2-m^2)y_n = 0.$ (UPTU–2008, UKTU–2011, VTU–2008, SVTU–2007, SRM-2006, 10, Mumbai–2007)

10. If $y = \cos(\log x)$, prove that $x^2y_{n+2} + (2n+1)xy_{n+1} + (n^2+1)y_n = 0.$

11. If $x + y = 1$, prove that $\dfrac{d^n}{dx^n}(x^n y^n) = n!\left[y^n - \left({}^nC_1\right)^2 y^{n-1}x + \left({}^nC_2\right)^2 y^{n-2}x^2 \ldots + (-1)^n x^n\right].$

12. If $y = x\cos(\log x)$, prove that $x^2y_{n+2} + (2n+1)xy_{n+1} + (n^2-2n+2)y_n = 0.$ (UPTU–2006)

13. If $y = \left(\dfrac{1+x}{1-x}\right)^{1/2}$, prove that $(1-x^2)y_n - [2(n-1)x+1]y_{n-1} - (n-1)(n-2)y_{n-2} = 0.$ (MTU –2012)

14. If $y = \dfrac{\sinh^{-1}x}{\sqrt{1+x^2}}$, prove that $(1+x^2)y_{n+2} + (2n+3)xy_{n+1} + (n+1)^2y_n = 0.$ (VTU–2010)

15. If $x = \sin t, y = \cos pt$, prove that $(1-x^2)y_2 - xy_1 + p^2y = 0$.

Hence, show that

$(1-x^2)y_{n+2} - (2n+1)xy_{n+1} - (n^2-p^2)y_n = 0$.

(VTU 2005, Raipur– 2005)

16. If $y = \sinh[m\log(x+\sqrt{x^2+1})]$, prove that

$(x^2+1)y_{n+2} + (2n+1)xy_{n+1} + (n^2-m^2)y_n = 0$. (VTU– 2010)

17. If $\sin^{-1} y = 2\log(x+1)$, prove that

$(x+1)^2 y_{n+2} + (2n+1)(x+1)y_{n+1} + (x^2+4)y_n = 0$

(VTU –2003)

18. Prove the following

$$\frac{d^n}{dx^n}\left[\frac{\log x}{x}\right] = \frac{(-1)^n.n!}{x^{n+1}}\left(\log x - 1 - \frac{1}{2} - \frac{1}{3} - \ldots - \frac{1}{n}\right)$$

(VTU –2006)

Hint to Selected Problems

2. Since $I_n = \frac{d^n}{dx^n}(x^n \log x)$

$$\Rightarrow I_n = \frac{d^{n-1}}{dx^{n-1}}\left[\frac{d}{dx}(x^n \log x)\right] = nI_{n-1} + (n-1)!$$

Now replace $(n-1)$, $(n-2)$ in place of n successively.

3. $y = e^{\tan^{-1} x} \Rightarrow y_1 = e^{\tan^{-1} x}.\frac{1}{(1+x^2)} \Rightarrow y_1(1+x^2) = y$.

Differentiating, we get

$(1+x^2)y_2 + (2x-1)y_1 = 0$.

Now apply Leibnitz theorem.

4. $y = (\sin^{-1}x)^2 \Rightarrow y_1 = 2\sin^{-1} x.\frac{1}{\sqrt{1-x^2}}$

$$\Rightarrow y_1^2 = \frac{4(\sin^{-1} x)^2}{(1-x^2)} \Rightarrow (1-x^2)y_1^2 = 4(\sin^{-1} x)^2$$

$\Rightarrow (1-x^2)y_1^2 = 4y$. Again differentiating, we get

$(1-x^2)2y_1.y_2 - 2xy_1^2 = 4y_1 \Rightarrow (1-x^2)y_2 - xy_1 - 2 = 0$.

Now apply Leibnitz's theorem.

5. $y = \frac{\sin^{-1} x}{\sqrt{1-x^2}} \Rightarrow y_1 = \frac{1}{(1-x^2)} + \frac{x}{(1-x^2)}\frac{\sin^{-1} x}{\sqrt{1-x^2}}$

$\Rightarrow \quad (1-x^2)y_1 = 1 + xy$.

Now apply Leibnitz's theorem.

6. $y = [\log\{x+\sqrt{1+x^2}\}]^2 \Rightarrow \sqrt{1+x^2}y_1 = 2[\log\{x+\sqrt{1+x^2}\}]$

On squaring, we get

$(1+x^2)y_1 = 4[\log\{x+(1+x^2)\}]^2 \Rightarrow (1+x^2)y_1^2 = 4y$.

Again differentiating, we get

$(1+x^2)y_2 + xy_1 - 2 = 0$.

Now applying Leibnitz's theorem.

7. Apply directly Leibnitz's theorem.

8. $y = [x+\sqrt{1+x^2}]^m \Rightarrow y_1 = \frac{m[x+\sqrt{1+x^2}]^m}{\sqrt{1+x^2}}$

$\Rightarrow \sqrt{1+x^2}y_1 = my$.

On squaring, we get $(1+x^2)y_1^2 = m^2y^2$

Again differentiating, we get

$(1+x^2)y_2 + xy_1 - m^2y = 0$.

Now apply Leibnitz theorem.

9. $y^{1/m} + y^{-1/m} = 2x \Rightarrow 2xy^{1/m} = y^{2/m} + 1$

Let $t = y^{1/m}$. Then $t^2 + 1 = 2xt$.

Solving for t, we get $t = x \pm \sqrt{x^2-1} \Rightarrow y = [x+\sqrt{x^2-1}]^m$.

On differentiating, we get $y_1 = \frac{m[x+\sqrt{x^2-1}]^m}{\sqrt{x^2-1}}$.

Differentiating w.r.t. x after squaring, we get

$(x^2-1)y_2 + xy_1 - m^2y = 0$. Now apply Leibnitz theorem.

10. $y = \cos(\log x) \Rightarrow y_1 = -\sin(\log x).\frac{1}{x} \Rightarrow xy_1 = -\sin(\log x)$.

Again differentiating, we get

$xy_2 + y_1 = -\cos(\log x).\frac{1}{x} \Rightarrow x^2y_2 + xy_1 + y = 0$.

Now apply Leibnitz's theorem.

ANSWERS

1. (i) $e^{ax}a^{n-3}[a^3x^3 + 3na^2x^2 + 3n(n-1)ax + n(n-1)(n-2)]$ (ii) $e^x[x^2+2nx+n(n-1)]$

(iii) $a^{n-3}\left[a^3x^3\sin\left(ax+\frac{n\pi}{2}\right) + 3na^2x^2\sin\left(ax+(n-1)\frac{\pi}{2}\right) + 3n(n-1)ax\sin\left\{ax+(n-2)\frac{\pi}{2}\right\} + n(n-1)(n-2)\sin\left(ax+(n-3)\frac{\pi}{2}\right)\right]$

(iv) $\frac{(-1)^{n-1}n!}{x^{n-3}}\left[\frac{1}{n} - \frac{3}{n-1} + \frac{3}{n-2} - \frac{1}{n-3}\right]$

(v) $e^x\left[2^{n/2}x^2\cos\left(x+\frac{n\pi}{4}\right) + 2^{(n-1)/2}2nx\cos\left(x+(n-1)\frac{\pi}{4}\right) + 2^{(n-2)/2}n(n-1)\cos\left(x+(n-2)\frac{\pi}{4}\right)\right]$

(vi) $e^x[\log x + {}^nC_1x^{-1} - {}^nC_2x^{-2} + {}^nC_3 2!x^{-3} - \ldots + {}^nC_n(-1)^{n-1}(n-1)!x^{-n}]$ (vii) $y_{n+1} = \frac{n!}{x}$

(viii) $(-1)^{n-1}(n-3)![(n-1)(n-2)x^2\sin^n\phi\sin n\phi - 2n(n-1)\sin^{n-1}\phi\sin(n-1)\phi + n(n-1)\sin^{n-2}\phi\sin(n-2)\phi]$ where $\phi = \tan^{-1}\frac{1}{x}$

7. (i) $(1-x^2)y_{n+2} - (2n+1)xy_{n+1} - (n^2-a^2)y_n = 0$ (ii) $x^2y_{n+2} + (2n+1)xy_{n+1} + (n^2+1)y_n = 0$.

1.5 DETERMINATION OF THE VALUE OF nth DERIVATIVE OF A FUNCTION AT x = 0

WORKING PROCEDURE

Step 1. *Put the given function equal to y.*

Step 2. *Find* $y_1 = \frac{dy}{dx}$. *Then*

(i) *Take L.C.M. (if required).*

(ii) *Square both sides, if square roots are there.*

(iii) *Try to get y in R.H.S. (if possible).*

Step 3. *Again differentiating both sides w.r.t. x and get an equation in* y_2, y_1 *and y.*

Step 4. *Differentiate both sides n times w.r.t. x by Leibnitz's theorem.*

Step 5. *Put x = 0 in equations of step 1, 2, 3, 4.*

Step 6. *Put n = 1, 2, 3, 4, ... in last equation of step 5.*

Step 7. *Discuss the two cases, when n is even and when n is odd.*

Solved Examples

Example 1. *If* $y = e^{a\cos^{-1}x}$, *show that*

$$(1-x^2)y_{n+2} - (2n+1)xy_{n+1} - (n^2+a^2)y_n = 0$$

and hence calculate y_n *at x = 0.*

(GBTU–2010, MDU–1997)

Solution. We have

$$y = e^{a\cos^{-1}x} \quad \ldots(1)$$

$$\therefore \quad y_1 = e^{a\cos^{-1}x} \cdot \frac{-a}{\sqrt{1-x^2}} = -\frac{ya}{\sqrt{1-x^2}} \quad \ldots(2)$$

$$\Rightarrow \quad y_1\sqrt{1-x^2} = -ya$$

Now squaring both sides we get

$$y_1^2(1-x^2) = y^2a^2$$

Differentiating w.r.t. x, we have

$$(1-x^2)2y_1y_2 - 2xy_1^2 = 2a^2yy_1$$

$$\Rightarrow \quad (1-x^2)y_2 - xy_1 = a^2y \quad \ldots(3)$$

Now, using Leibnitz's theorem, differentiating (3), n times, we get

$$(1-x^2)y_{n+2} - 2nxy_{n+1} - n(n-1)y_n - xy_{n+1} - ny_n = a^2y_n$$

$$\Rightarrow (1-x^2)y_{n+2} - (2n+1)xy_{n+1} - (n^2+a^2)y_n = 0 \quad \ldots(4)$$

By putting $x = 0$ in (1), (2), (3) and (4), we get

$$y(0) = e^{a.\pi/2}$$

$$y_1(0) = -ae^{a.\pi/2}$$

$$y_2(0) = a^2y(0) = a^2.e^{a.\pi/2}$$

$$\Rightarrow y_{n+2}(0) = (n^2+a^2)y_n(0) \ldots(5)$$

Put $n-2$ for n in (5), we get

$$y_n(0) = [(n-2)^2+a^2]y_{n-2}(0) \quad \ldots(6)$$

Again put $n-4$ for n in (5), we get

$$y_{n-2}(0) = [(n-4)^2+a^2]y_{n-4}(0) \quad \ldots(7)$$

From (6) and (7), we get

$$y_n(0) = [(n-2)^2+a^2][(n-4)^2+a^2]y_{n-4}(0) \quad \ldots(8)$$

Again put $n-6$ for n in (5), we get

$$y_{n-4}(0) = [(n-6)^2+a^2]y_{n-6}(0) \quad \ldots(9)$$

From (8) and (9), we get

$$y_n(0) = [(n-2)^2+a^2][(n-4)^2+a^2][(n-6)^2+a^2]y_{n-6}(0) \cdots(10)$$

Now there are following two cases :

Case I. When n is even.

$$y_n(0) = [(n-2)^2+a^2][(n-4)^2+a^2][(n-6)^2+a^2]\ldots[2^2+a^2]a^2e^{a\pi/2}$$

Case II. When n is odd.

$$y_n(0) = [(n-2)^2+a^2][(n-4)^2+a^2][(n-6)^2+a^2]\ldots[1^2+a^2](-ae^{a\pi/2})$$

Example 2. *If* $y = \tan^{-1}x$, *prove that*

$$(1+x^2)y_{n+1} + 2nxy_n + n(n-1)y_{n-1} = 0.$$

Hence, determine the values of all the derivatives of y with respect to x when x = 0. (Mumbai–2008)

Solution. We have

$$y = \tan^{-1}x. \quad \ldots(1)$$

$$\therefore \quad y_1 = \frac{1}{1+x^2} \quad \ldots(2)$$

$$\Rightarrow \quad y_1(1+x^2) = 1.$$

Differentiating, n times by Leibnitz's theorem, we have

$$y_{n+1}(1+x^2) + ny_n.2x + \frac{n(n-1)}{2}y_{n-1}.2 = 0$$

$$\Rightarrow (1+x^2)y_{n+1} + 2nxy_n + n(n-1)y_{n-1} = 0 \quad \ldots(3)$$

Putting $x = 0$ in (1), (2) and (3), we get

$$y(0) = 0$$
$$y_1(0) = 1$$

....

$$y_{n+1}(0) = -n(n-1)y_{n-1}(0) \quad ...(4)$$

Put $n = 1$ in (4), we get

$$y_2(0) = 0.$$

Put $n - 1$ for n in (4), we get

$$y_n(0) = -(n-1)(n-2)y_{n-2}(0) \quad ...(5)$$

Put $n - 3$ for n in (4), we get

$$y_{n-2}(0) = -(n-3)(n-4)y_{n-4}(0) \quad ...(6)$$

From (5) and (6), we get

$$y_n(0) = (n-1)(n-2)(n-3)(n-4)y_{n-4}(0) \quad ...(7)$$

There arise following two cases :

Case I. When n is even.

$$y_n(0) = (-1)^{(n-2)/2}(n-1)(n-2)(n-3)(n-4) ...4.2y_2(0)$$
$$= (-1)^{(n-2)/2}(n-1)(n-2)(n-3)(n-4) ...3.2.0$$
$$= 0 \qquad [\because y_2(0) = 0]$$

Case II. When n is odd.

$$y_n(0) = (-1)^{(n-1)/2}(n-1)(n-2)(n-3) ...3.2.1y_1(0)$$
$$= (-1)^{(n-1)/2}(n-1)!\,y_1(0)$$
$$= (-1)^{(n-1)/2}(n-1)! \qquad [\because y_1(0) = 1]$$

Example 3. *If $y = [x + \sqrt{1+x^2}]^m$, find $(y_n)_0$.*

Solution . We have

$$y = [x + \sqrt{1+x^2}]^m. \quad ...(1)$$

Differentiating both sides w.r.t. x, we get

$$y_1 = m[x + \sqrt{1+x^2}]^{m-1}\left(1 + \frac{x}{\sqrt{1+x^2}}\right)$$

or $\quad y_1 = \dfrac{m}{\sqrt{1+x^2}}[x + \sqrt{1+x^2}]^m$

or $\quad \sqrt{1+x^2} \cdot y_1 = m[x + \sqrt{1+x^2}]^m$

or $\quad \sqrt{1+x^2} \cdot y_1 = my.$

Squaring both sides, we get

$$y_1^2(1+x^2) = m^2y^2. \quad ...(2)$$

Again differentiating both sides, we get

$$2y_1(1+x^2)y_2 + 2xy_1^2 = 2m^2yy_1.$$

or $\quad (1+x^2)y_2 + xy_1 - m^2y = 0. \quad ...(3)$

Applying Leibnitz's theorem to differentiate n times, we get

$$D^n[(1+x^2)y_2] + D^n(xy_1) - m^2D^2y = 0$$
$$(1+x^2)y_{n+2} + {}^nC_1y_{n+1}D(1+x^2) + {}^nC_2y_nD^2(1+x^2) + xy_{n+1} + {}^nC_1y_nD(x) - m^2y_n = 0$$

or $\quad (1+x^2)y_{n+2} + ny_{n+1}2x + \dfrac{n(n-1)}{2}y_n.2 + xy_{n+1} + ny_n - m^2y_n = 0$

or

$$(1+x^2)y_{n+2} + x(2n+1)y_{n+1} + (n^2 - m^2)y_n = 0. \quad ...(4)$$

Putting x = 0 in (1), (2), (3) and (4), we get

$$(y)_0 = 1$$
$$(y_1)_0 = m(y_0) = m$$
$$(y_2)_0 = m^2(y)_0 = m^2$$

and $\quad (y_{n+2})_0 = (m^2 - n^2)(y_n)_0. \quad ...(5)$

Put $n - 2$ for n in (5), we get

$$(y_n)_0 = [m^2 - (n-2)^2](y_{n-2})_0 \quad ...(6)$$

Put $n - 4$ for n in (5), we get

$$(y_{n-2})_0 = [m^2 - (n-4)^2](y_{n-4})_0 \quad ...(7)$$

From (6) and (7), we get

$$(y_n)_0 = [m^2 - (n-2)^2][m^2 - (n-4)^2](y_{n-4})_0. \quad ...(8)$$

There arise two cases :

Case I. When n is even.

$$(y_n)_0 = [m^2 - (n-2)^2][m^2 - (n-4)^2] ...(m^2 - 2^2)(y_2)_0$$
$$= [m^2 - (n-2)^2][m^2 - (n-4)^2] ...[m^2 - 2^2]m^2$$
$$[\because (y_2)_0 = m^3]$$

Case II. When n is odd.

$$(y_n)_0 = [m^2 - (n-2)^2][m^2 - (n-4)^2] ...(m^2 - 1^2)(y_1)_0$$
$$= [m^2 - (n-2)^2][m^2 - (n-4)^2] ...(m^2 - 1^2)m$$
$$[\because (y_1)_0 = m]$$

Example 4. *If $y = \sin(a\sin^{-1}x)$, then, prove that $(1-x^2)y_2 - xy_1 + a^2y = 0$ and $(1-x^2)y_{n+2} - (2n+1)xy_{n+1} + (a^2 - n^2)y_n = 0$. Hence, find $y_n(0)$.* (UPTU–2009, 2011)

Solution . We have

$$y = \sin(a\sin^{-1}x) \quad ...(1)$$

Differentiating (1) w.r.t. x we get

$$y_1 = \cos(a\sin^{-1}x).\frac{a}{\sqrt{1-x^2}}$$

$$\Rightarrow y_1 = \frac{a}{\sqrt{1-x^2}}\cos(a\sin^{-1}x)$$

$$\Rightarrow \quad (\sqrt{1-x^2})y_1 = a\cos(a\sin^{-1}x)$$

$$\Rightarrow \quad (1-x^2){y_1}^2 = a^2\cos^2(a\sin^{-1}x)$$

$$= a^2(1-\sin^2(a\sin^{-1}x))$$

$$\Rightarrow \quad (1-x^2){y_1}^2 = a^2(1-y^2) \qquad \ldots(2)$$

(Using (1))

Differentiating (2) w.r.t. x, we get

$$(1-x^2)2y_1y_2 - 2x{y_1}^2 = a^2(-2yy_1)$$

$$\Rightarrow (1-x^2)y_2 - xy_1 + a^2y = 0 \qquad \ldots(3)$$

Now differentiating (3) n times by Leibnitz's theorem, we get

$$\left[(1-x^2)y_{n+2} + {}^nC_1(-2x)y_{n+1} + {}^nC_2(-2)y_n\right] - \left[xy_{n+1} + {}^nC_1(1)y_n\right] + a^2y_n = 0$$

$$\Rightarrow (1-x^2)y_{n+2} + n(-2x)y_{n+1} + \frac{n(n-1)}{2}(-2)y_n - xy_{n+1} - n.1.y_n + a^2y_n = 0$$

$$\Rightarrow \quad (1-x^2)y_{n+2} - (2n+1)xy_{n+1} + (a^2 - n^2 - n + n)y_n = 0$$

$$\Rightarrow (1-x^2)y_{n+2} - (2n+1)xy_{n+1} + (a^2 - n^2)y_n = 0 \qquad \ldots(4)$$

From (1),

$$y(0) = \sin(a\sin^{-1}0) = 0$$

From (2),

$$y_1(0) = \frac{a}{\sqrt{1-0}}\cos(a\sin^{-1}0) = a\cos 0 = a$$

From (3),

$$(1-0^2)y_2(0) - 0.y_1(0) + a^2y(0) = 0$$

$$\Rightarrow \quad y_2(0) = 0$$

Form (4),

$$(1-0^2)y_{n+2}(0) - (2n+1).0 + (a^2-n^2)y_n(0) = 0$$

$$\Rightarrow \quad y_{n+2}(0) = (n^2 - a^2)y_n(0) \qquad \ldots(5)$$

Case I. If n is even.

Put $n = 2$ in equation (5), we get

$$y_4(0) = (2^2 - a^2)y_2(0) = 0$$

Put $n = 4$ in equation (5), we get

$$y_6(0) = (4^2 - a^2)y_4(0) = 0$$

Put $n = 6$ in equation (5), we get

$$y_8(0) = (6^2 - a^2)y_6(0) = 0$$

$$\Rightarrow \quad y_n(0) = 0\text{, if } n \text{ is even}$$

Case II. If n is odd.

Put $n = 1$ in equation (5), we get

$$y_3(0) = (1^2 - a^2)y_1(0) = (1^2 - a^2).a$$

Put $n = 3$ in equation (5), we get

$$y_5(0) = (3^2 - a^2)y_3(0) = (1^2 - a^2)(3^2 - a^2).a$$

Put $n = 5$ in equation (5), we get

$$y_7(0) = (5^2 - a^2)y_5(0)$$

$$= (1^2 - a^2)(3^2 - a^2)(5^2 - a^2).a$$

$$\Rightarrow y_n(0) = (1^2 - a^2)(3^2 - a^2)(5^2 - a^2) \ldots[(n-2)^2 - a^2)]a$$

if n is odd and $n \neq 1$

Hence,

$$y_n(0) = \begin{cases} 0 & \text{if } n \text{ is even} \\ (1^2 - a^2)(3^2 - a^2)(5^2 - a^2) & \text{if } n \text{ is odd} \\ \ldots[(n-2)^2 - a^2]a & \text{and } n \neq 1 \end{cases}$$

EXERCISE 1.3

1. If $y = \sin^{-1}x$, prove that $(1-x^2)y_{n+2} - (2n+1)xy_{n+1} - n^2y_n = 0$ and also find the value of $y_n(0)$. (SVTU–2009)

2. (i) If $y = [\log\{x + \sqrt{(1+x^2)}\}]^2$, find all the derivatives of y w.r.t. x when $x = 0$.

   (ii) If $y = (\sinh^{-1}x)^2$, prove that $(1+x^2)y_{n+2} + (2n+1)xy_{n+1} + n^2y_n = 0$ Hence, find $y_n(0)$.

3. If $y = [x + \sqrt{1+x^2}]^m$, find $y_n(0)$.

4. (i) If $y = \sin(m\sin^{-1}x)$, then prove that $(y_{n+2})_0 = (n^2 - m^2)(y_n)_0$ and find $y_n(0)$. (Mumbai- 2008)

   (ii) If $y = \cos(m\sin^{-1}x)$, find $y_n(0)$. (Cochin- 2005, VTU-2009)

5. If $y = e^{a\sin^{-1}x}$, show that $(1-x^2)y_{n+2} - x(2n+1)y_{n+1} - (n^2+a^2)y_n = 0$ and hence, find the value of $y_n(0)$.

6. If $x = \sin\left(\frac{1}{a}\log y\right)$, find $(y_n)0$.

Hint to Selected Problems

1. $y=\sin^{-1}x \Rightarrow y_1=\dfrac{1}{\sqrt{1-x^2}} \Rightarrow (1-x^2)y_1^2=1$

Differentiating, we get $(1-x^2)y_2-xy_1=0$

Now apply Leibnitz's theorem and put $x = 0$.

2. (i) $y=[\log x+\sqrt{1+x^2}]^2$

$$\Rightarrow y_1=\frac{2}{\sqrt{1+x^2}}\log(x+\sqrt{1+x^2}) \Rightarrow (1+x^2){y_1}^2=4y$$

Again differentiating, we get

$$(1+x^2)y_2+xy_1-2=0.$$

Now apply Leibnitz's theorem to find y_n and then put $x = 0$.

(ii) $\sinh^{-1}x=\log[x+\sqrt{1+x^2}]$

3. $y=[x+\sqrt{1+x^2}]^m \Rightarrow y_1=\dfrac{m[x+\sqrt{1+x^2}]^m}{\sqrt{1+x^2}}$

i.e., $(1+x^2){y_1}^2=m^2y^2$.

Again differentiating, we get $(1+x^2)y_2+xy_1-m^2y=0$

Now apply Leibnitz's theorem.

4. (i) $y=\sin(m\sin^{-1}x)$

$$\Rightarrow y_1=\frac{m\cos(m\sin^{-1}x)}{\sqrt{1-x^2}} \Rightarrow y_1\sqrt{(1-x^2)}=m\cos(m\sin^{-1}x)$$

Again differentianting, we get $(1-x^2)y_2-xy_1-m^2y=0$

Now apply Leibnitz's theorem.

5. $y=e^{a\sin^{-1}x} \Rightarrow y_1=\dfrac{ay}{\sqrt{1-x^2}} \Rightarrow (1-x^2){y_1}^2=a^2y^2$

Differentiating w.r.t. x, we get

$$(1-x^2)y_2-xy_1-a^2y=0$$

Now apply Leibnitz's theorem.

ANSWERS

1. When n is even, $y_n(0)=0$; When n is odd $y_n(0)=1^2.3^2.5^2...(n-2)^2$

2. (i),(ii) when n is even, $y_n(0)=(-1)^{n/2-1}.2.2^2.4^2...(n-2)^2$, when n is odd $y_n(0)=0$

3. When n is even, $y_n(0)=[m^2-(n-2)^2][m^2-(n-4)^2]...(m^2-2^2)m^2$

When n is odd, $y_n(0)=[m^2-(n-2)^2][m^2-(n-4)^2]...(m^2-1^2)m$

4. (i) When n is even, $y_n(0)=0$, When n is odd, $y_n(0)=[(n-2)^2-m^2][(n-4)^2-m^2][(n-6)^2-m^2]...[(3^2-m^2)(1^2-m^2)]m$

(ii) When n is even, $y_n(0)=-[(n-2)^2-m^2][(n-4)^2-m^2]...[(2^2-m^2)m^2$; When n is odd, $y_n(0)=0$

5. When n is even, $y_n(0)=[(n-2)^2+a^2][(n-4)^2+a^2]...(4^2+a^2)(2^2+a^2).a^2$

When n is odd, $y_n(0)=[(n-2)^2+a^2][(n-4)^2+a^2]...(3^2+a^2)(1^2+a^2).a$

Objective Evaluations

Fill in the Blanks

1. $D^n(\log x)$ is equal to ________.
2. To find the nth derivative of the product of two functions we use ________ theorem.
3. If $y = \sin(ax + b)$, then $D^n(ax + b) =$ ________.
4. If $y = (ax + b)^{-1}$, then $D^n(ax + b)^{-1} =$ ________.
5. If $y = e^{ax} \sin bx$, then $y_2 - 2ay_1 =$ ________.
6. If $y = e^x \sin^2 x$, then $D^n(y) =$ ________.
7. $D^3(x^3) =$ ________.
8. $D^n(x^{n-1}) =$ ________.
9. $D^n(\sin^3 x) =$ ________.
10. If $y = \tan^{-1}x$, then $(y_5)_0$ is equal to ________.

True/False

Write 'T' for True and 'F' for False statement.

1. To find the nth derivative of the product of two functions we use Leibnitz's theorem. **(T/F)**
2. If we observe that one of the two functions is such that all its differential coefficients after a certain steps, become zero, then we should take this function as second function. **(T/F)**
3. If $y = a\cos(\log x) + b\sin(\log x)$, then $x^2y_2 + xy_1 = y$. **(T/F)**
4. $D^n(\log x) = \dfrac{(n-1)!}{x^{n+1}}$. **(T/F)**
5. The nth differential coefficient of y_k is the $(n + k)^{th}$ differential coefficient of y. **(T/F)**

Multiple Choice Questions

Choose the most appropriate one.

1. $D^n(e^{ax+b})$ is equal to :
   (a) $a^n e^{ax}$ (b) e^{ax+b}
   (c) $a^n b^n e^{ax+b}$ (d) $a^n e^{ax+b}$

2. $D^n \log x$ is equal to :
   (a) $\dfrac{(n-1)!}{x^n}$ (b) $\dfrac{(-1)^n(n-1)!}{x^n}$
   (c) $\dfrac{(-1)^{n-1}(n-1)!}{x^n}$ (d) $\dfrac{(-1)^n n!}{x^n}$

3. If $p^2 = a^2\cos^2\theta + b^2\sin^2\theta$ then $p + \dfrac{d^2p}{d\theta^2}$ is equal to :
   (a) $\dfrac{a^2b^2}{p^2}$ (b) $\dfrac{a^2b^2}{p^3}$
   (c) $\dfrac{a^2b^2}{p}$ (d) $\dfrac{a^2b^2}{p^4}$

4. If $y = A\sin mx + B\cos mx$ then $y_2 + m^2y$ is equal to :
   (a) 0 (b) 1
   (c) 2 (d) 3

5. If $y = e^{ax}\sin bx$ then $y_2 - 2ay_1$ is equal to :
   (a) $(a^2 + b^2)y$ (b) $-(a^2 + b^2)y$
   (c) 0 (d) 1

6. If $y = \sin^{-1}x$ then $(1-x^2)\dfrac{d^2y}{dx^2}$ is equal to :
   (a) $\dfrac{dy}{dx}$ (b) $x^2\dfrac{dy}{dx}$
   (c) $x\dfrac{dy}{dx}$ (d) $\dfrac{1}{x}\dfrac{dy}{dx}$

7. If $x = a(t - \sin t)$ and $y = a(1 + \cos t)$, then $\dfrac{d^2y}{dx^2}$ is equal to :
   (a) $4a\,\text{cosec}^4 t$ (b) $\dfrac{1}{4a}\text{cosec}^4(t/2)$
   (c) $\dfrac{1}{4a}\sin^4(t/2)$ (d) $4a\sin^4 t$

8. If $x = a(\cos\theta + \theta\sin\theta)$ and $y = a(\sin\theta - \theta\cos\theta)$ then $\dfrac{d^2y}{dx^2}$ is equal to :
   (a) $\dfrac{1}{a}\sec^3\theta$ (b) $a\sec^3\theta$
   (c) $\dfrac{1}{a\theta\cos^3\theta}$ (d) $a\theta\sec^3\theta$

9. $D^n(\sin^3 x)$ is equal to :
   (a) $\sin\left(x + \dfrac{n\pi}{2}\right)$
   (b) $\dfrac{3}{4}\sin\left(x + \dfrac{n\pi}{2}\right) - \dfrac{3^n}{4}\sin\left(3x + \dfrac{n\pi}{2}\right)$
   (c) $\dfrac{3^n}{4}\cos\left(3x + \dfrac{n\pi}{2}\right)$
   (d) none of these

10. $[\cos^2 x\sin^3 x]$ is equal to :
    (a) $\dfrac{1}{4}[2\sin x + \sin 3x - \sin 5x]$
    (b) $\dfrac{1}{16}[2\sin x + \sin 3x - \sin 5x]$
    (c) $\dfrac{1}{16}[\sin x + \sin 3x - \sin 5x]$
    (d) none of these

11. $D^n(ax+b)^m$ where m is positive integer is equal to :

(a) $m!a^n(ax+b)^{m-n}$ (b) $\frac{m!}{(m-n)!}(ax+b)^{m-n}$

(c) $\frac{m!}{(m-n)!}a^n(ax+b)^{m-n}$ (d) none of these

12. If $y=\sin(ax+b)$, then y_3 is equal to :

(a) $a^3\sin(ax+b+3\pi)$ (b) $a\sin(ax+b+3\pi/2)$

(c) $a^3\sin(ax+b+3\pi/2)$ (d) $a^3\sin(ax+b+2\pi)$

13. $D^n(ax+b)^{-1}$ is equal to :

(a) $n!a^n(ax+b)^{-n}$ (b) $(-1)^n a^n(ax+b)^{-n}$

(c) $(-1)^n a^n(ax+b)^{-n-1}$ (d) $(-1)^n a^n n!(ax+b)^{-n-1}$

14. $D^n[\sin 2x]$ is equal to :

(a) $\sin[2x+n\pi/2]$ (b) $2^n\sin[2x+n\pi]$

(c) $2^n\sin[2x+n\pi/2]$ (d) $2^n\sin[x+n\pi/2]$

15. If $y=x^n\log x$, then y_{n+1} is equal to :

(a) $(n-1)!$ (b) $n!$

(c) $\frac{n!}{x}$ (d) $(n+1)!x$

16. If $y=a\cos(\log x)+b\sin(\log x)$, then $x^2y_2+xy_1$ is equal to :

(a) y (b) 1

(c) $-y$ (d) -1

17. If $y=(\sin^{-1}x)^2$ then $(1-x^2)y_2-xy_1$ is equal to :

(a) 1 (b) 2

(c) -1 (d) -2

18. If $I_n=\frac{d^n}{dx^n}[x^n\log x]$ then I_n-nI_{n-1} is equal to :

(a) $n!$ (b) $(n+1)!$

(c) $(n-1)!$ (d) $(n+2)!$

19. If $y=(\tan^{-1}x)$, then $(y_5)_0$ is equal to :

(a) 2! (b) 3!

(c) 4! (d) 5!

20. By Leibnitz's theorem, we find n^{th} differential coefficient of the ...... of two functions :

(a) sum (b) difference

(c) product (d) quotient

ANSWERS

Fill in the Blanks

1. $\frac{(-1)^{n-1}(n-1)!}{x^n}$ **2.** Leibnitz's **3.** $a^n\sin\left(ax+b+\frac{n\pi}{2}\right)$ **4.** $(-1)^n.n!a^n(ax+b)^{-n-1}$

5. $-(a^2+b^2)y$ **6.** $\frac{1}{2}[e^x-(5)^{x/2}e^x\cos(2x+n\tan^{-1}x)]$ **7.** 3! **8.** 0 **9.** $\frac{3}{4}\sin\left(x+\frac{n\pi}{4}\right)-\frac{3^n}{4}\sin\left(3x+\frac{n\pi}{2}\right)$ **10.** 4!

True/False

1. T **2.** T **3.** F **4.** F **5.** T

Multiple Choice Questions

1. (d)	**2.** (c)	**3.** (b)	**4.** (a)	**5.** (b)	**6.** (c)	**7.** (b)	**8.** (c)	**9.** (b)	**10.** (b)
11. (c)	**12.** (c)	**13.** (d)	**14.** (c)	**15.** (c)	**16.** (c)	**17.** (b)	**18.** (c)	**19.** (c)	**20.** (c)

FFFFFF

Mean Value Theorems and Expansion of Functions

2.1 INTRODUCTION

In this chapter we shall discuss some important theorems namely, Rolle's, Lagrange's, Cauchy mean value and Taylor's theorem. We shall also discuss Maclaurin's series expansion of some standard functions like e^x, $\log(1+x)$, $\sin x$, $\cos x$ etc.

Before discussing the main topic, let us recall the following cocepts.

2.2 ROLLE'S THEOREM

If a function f defined on [a,b] is such that it is

(i) *continuous in* $[a,b]$, (ii) *differentiable in* $]a,b[$ *and* (iii) $f(a) = f(b)$,

then there exists at least one vlaue of x, say c,$(a<c<b)$ *such that* $f'(c) = 0$

Proof. Since, the function $f(x)$ is continuous on $[a, b]$

$\Rightarrow$ $f(x)$ is bounded

[$\because$ Every continuous function is bounded.]

$\Rightarrow$ $f(x)$ attains its bounds [$\because$ A function, which is continuous on a closed bounded interval $[a, b]$, then it attains its bound on $[a, b]$.]

Let M and m are the supremum and infimum of $f(x)$ respectively.

Now there are two possibilities

(i) If $M=m$, then obviously $f(x)$ is a constant function, and therefore its derivative is zero, *i.e.,* $f'(x)= 0$ $\forall\, x \in\,]a, b[$.

(ii) If $M \neq m$, then at least one of the numbers M and m must be different from the equal values $f(a)$ and $f(b)$.

Let us assume $M \neq f(a)$.

Now, since, every continuous function on a closed interval attains its supremum, therefore, there exists a real number c in $[a,b]$ such that $f(c)=M$. Also since $f(a)\neq M\neq f(b)$. Therefore $c\neq a$ and $c\neq b$, this implies that $c\in]a,b[$.

Now, $f(c)$ is the supremum of f on $[a, b]$

$$\therefore \quad f(x) \leq f(c)\ \forall x\in[a, b] \quad ...(1)$$

[By the definition of supremum]

In particular, $$f(c-h) \leq f(c) \quad h>0$$

$$\Rightarrow \quad \frac{f(c-h)-f(c)}{-h} \geq 0 \quad ...(2)$$

Since $f'(x)$ exists at each point of $]a, b[$, and hence, $f'(c)$ exists.

Therefore, from (2)

$$Lf'(c) \geq 0 \quad ...(3)$$

Similarly, from (1)

$$f(c+h) \leq f(c) \quad h>0$$

Then by the same arguments

$$Rf'(c) \leq 0. \quad \text{...(4)}$$

Since $f(x)$ is differentiable in $]a, b[$ $\Rightarrow f'(c)$ exist

$\Rightarrow$ $$Lf'(c) = f'(c) = Rf'(c) \quad \text{...(5)}$$

Now from (3), (4) and (5) $f'(c)=0$.

Similarly we can consider the case $M=f(a) \neq m$.

REMARKS

- Converse of Rolle's theorem is not true, *i.e.*, $f'(x)$ may vanish at a point $c \in]a, b[$ without $f(x)$ satisfying the three conditions of Rolle's theorem.
- There may be more than one point like c at which $f'(x)$ vanishes but Rolle's theorem ensures the existence of at least one such c.
- Rolle's theorem will not hold good if
  (a) $f(x)$ is discontinuous at some point in the interval $[a, b]$
  (b) $f'(x)$ does not exist at some point in the interval $]a, b[$
  (c) $f(a) \neq f(b)$
- The hypothesis of Rolle's theorem cannot be weakened.
  For example, if $f(x)=1-|x|, -1 \leq x \leq 1$, then $f(-1)=f(1)=0$ and f is continuous on $[-1,1]$. Also if $f'(x)$ exist $\forall x \in]-1, 1[$ except at $x=0$. Then, f satisfies all the condition of Rolle's theorem except that f is not differentiable at $x=0$. For this f, there is no c in $]-1,1[$ for which $f'(c)=0$.

2.2.1 GEOMETRICAL INTERPRETATION OF ROLLE'S THEOREM

Geometrically, Rolle's theorem means that if the curve $y=f(x)$ is continuous from $x=a$ to $x=b$, has a definite tangent at each point of $]a,b[$ and the ordinates at the extremities are equal, then there exists at least one point between a and b at which the tangent is parallel to x-axis.

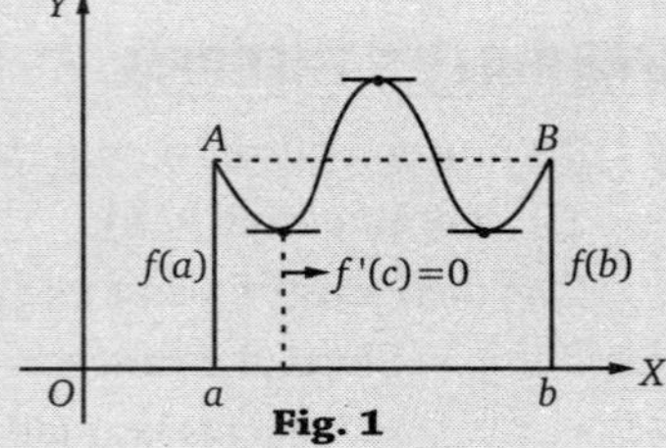

Fig. 1

2.2.2 ALGEBRAIC INTERPRETATION OF ROLLE'S THEOREM

Algebraically, Rolle's theorem means that if $f(x)$ is a polynomial function in x and $x=a$ and $x=b$ are two roots of the equation $f(x)=0$, then, there is at least one root of the equation $f'(x)=0$ which lies between a and b.

2.3 LAGRANGE'S MEAN VALUE THEOREM

Let f be a function defined on $[a, b]$ such that

(i) f is continuous on $[a, b]$, *(ii) f is differentiable on $]a, b[$.*

Then, there exists a real number $c \in]a,b[$ such that $\dfrac{f(b)-f(a)}{b-a} = f'(c)$

Proof. Let us define a function $F(x)$ such that

$$F(x)=f(x)+Ax \ \forall x \in [a,b] \quad \text{...(1)}$$

where A is a constant to be suitably chosen such that $F(a)=F(b)$.

Now

(i) Since, f is continuous on $[a,b]$ and Ax is continuous on $[a,b]$ therefore, F is continuous on $[a,b]$

[$\because$ Sum of two continuous functions is again continuous.]

(ii) Similarly F is differentiable on (a, b)

(iii) $F(a) - F(b)$ $\Rightarrow$ $$-A = \frac{f(b)-f(a)}{b-a} \quad \text{...(2)}$$

Hence, we find that F satisfy all the conditions of Rolle's Theorem on $[a,b]$ and consequently, there exists a real number $c \in]a,b[$ such that $F'(c)=0$, this gives

$$f'(c)+A=0$$

$\Rightarrow$ $$-A=f'(c). \quad \text{...(3)}$$

Now, from (2) and (3), we have

$$\frac{f(b)-f(a)}{b-a} = f'(c)$$

REMARKS

- If we take $b=a+h$ and c can be written as $a+\theta h$, where θ is some real number such that $0<\theta<1$. Lagrange's theorem then read as follows :
  " Let f be defined and continuous on $[a, a+h]$ and differentiable on $]a, a+h[$, then for some real number $\theta(0<\theta<1)$
$$\frac{f(a+h)-f(a)}{h} = f'(a+\theta h).$$
- The hypothesis of the Lagrange's mean value theorem cannot be weakened, as it is clear from the following examples :
  " Let f be the function defined on $[-1,2]$ by setting $f(x)=|x|, \quad \forall x \in [-1,2]$.
  Here, f is continuous on $[-1,2]$ and differentiable at all points of $]-1, 2[$ except at $x=0$ (so that second condition is violated.)
  Now $\quad f'(x)=\begin{cases}-1 & \text{if } x \in]-1,0[\\ 1 & \text{if } x \in]0,2[\end{cases}$; $\quad$ Also $\quad \dfrac{f(2)-f(-1)}{2-(-1)} \neq f'(x)$ for any x in $]-1, 2[$.
- Lagrange's mean value theorem is known as first mean value theorem.
- The result $f(b)-f(a)=f(b-a)f'(c)$ is also known as the formula for finite increment.
- For $f(a)=f(b)$, the Lagrange's mean value theorem yields Rolle's theorem.

2.3.1 GEOMETRICAL INTERPRETATION OF LAGRANGE'S MEAN VALUE THEOREM

If the curve $y=f(x)$ is continuous from $x=a$ and $x=b$ and has a tangent at each point on the curve between $x=a$ and $x=b$, then, geometrically, the first mean value theorem means that there is at least one point between $x=a$ and $x=b$ on the curve where the tangent to the curve parallel to the chord joining the points $(a, f(a))$ and $(b, f(b))$.

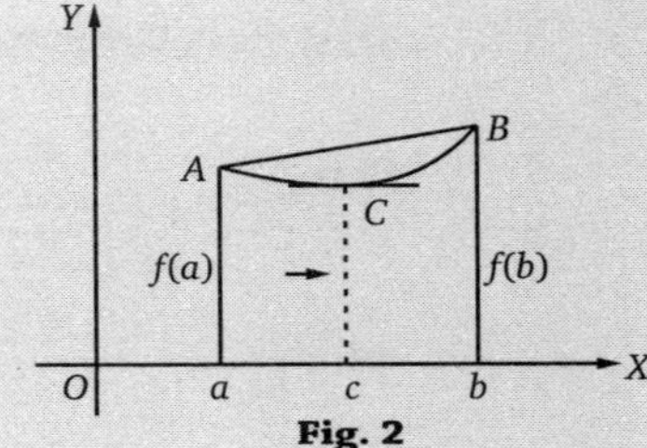

Fig. 2

Let ACB be the graph of the function $y= f(x)$ then the co-ordinate of the points A and B are given by $(a, f(a))$ and $(b, f(b))$ respectively. If the chord AB makes an angle θ with the x-axis, then
$$\tan\theta = \frac{f(b)-f(a)}{b-a} = f'(c), \text{ where } a < c < b.$$

2.3.2 DEDUCTIONS FROM THE FIRST MEAN VALUE THEOREM

THEOREM 1. *If a function $f(x)$ satisfies the conditions of mean value theorem then*

(i) $f'(x)= 0\ \forall x \in]a, b[\Rightarrow f$ is constant on $[a, b]$.

(ii) $f'(x)>0\ \forall x\in]a,b[\Rightarrow f$ is strictly increasing on $[a,b]$.

and (iii) $f'(x)<0\ \forall x\in]a,b[\Rightarrow f$ is strictly decreasing on $[a,b]$.

Proof. (i) Let x_1, x_2 (where $x_1>x_2$) be any two distinict points of $[a,b]$, then by Lagrange's mean value theorem,
$$\frac{f(x_2)-f(x_1)}{x_2-x_1} = f'(c) = 0,\ x_1 < c < x_2 \qquad \ldots(1)$$
$\Rightarrow \quad f(x_2)= f(x_1)$.

$\Rightarrow$ function keeps the same value. Therefore $f(x)$ is constant on $[a,b]$.

(ii) From (1), we have
$$\frac{f(x_2)-f(x_1)}{x_2-x_1} = f'(c) \text{ for some } c \in]x_1,\ x_2[$$
But $\quad f'(c) > 0 \qquad [\because f'(x) > 0\ \forall x \in [a, b]]$

$\Rightarrow \quad f(x_2)-f(x_1)>0$

$\Rightarrow \quad f(x_2)>f(x_1)$

Thus $\quad x_2>x_1 \Rightarrow f(x_2)>f(x_1)\ \forall x_1,x_2 \in [a,b]$

Hence, f is strictly increasing on $[a,b]$.

(iii) Same as (ii).

REMARK

- For a strictly increasing function f, the derivative $f'(x)$ need not be strictly positive. For example, consider $f(x)=x^3$, $x \in]-1, 1[$. Here, $f(x)$ is strictly increasing but $f'(x)=3x^2$, which is zero at $x=0 \in]-1, 1[$.

2.4 CAUCHY'S MEAN VALUE THEOREM

Let f and g be two functions defined on [a,b] such that

(i) f and g are continuous on $[a, b]$, (ii) f and g are differentiabfe on $]a, b[$,

and (iii) $g'(x) \neq 0$ for any point of $]a, b[$. (PTU–2007, 11, VTU–2006)

Then, there exists a real number $c \in]a, b[$ such that

$$\frac{f(b)-f(a)}{g(b)-g(a)}=\frac{f'(c)}{g'(c)}$$

Proof. Let us define a function

$$F(x)=f(x)+A.g(f) \quad ...(1)$$

where A is a constant, to be suitably chosen such that

$$F(a)=F(b) \quad ...(2)$$

Now, the function F is the sum of two continuous and differentiable functions. Therefore

(i) F is continuous on $[a,b]$,

(ii) F is differentiable on $]a,b[$,

and (iii) $F(a)=F(b)$.

Then, by Rolle's theorem, there must exists a real number c between a and b such that

$$F'(c)=0$$

Here,
$$F'(x)=f'(x)+Ag'(x)$$
$$F'(c)=0 \quad \Rightarrow f'(c)+Ag'(c)=0$$

$$\Rightarrow \quad -A=\frac{f'(c)}{g'(c)} \quad ...(3)$$

Now
$$F(a) = F(b) \quad \Rightarrow f(a)+Ag(a)=f(b) + Ag(b)$$

$$\Rightarrow \quad -A=\frac{f(b)-f(a)}{g(b)-g(a)} \quad ...(4)$$

From (3) and (4), we have

$$\frac{f(b)-f(a)}{g(b)-g(a)}=\frac{f'(c)}{g'(c)}.$$

REMARKS

- If we put $b=a+h$, then c can be written as $a+\theta h$, where $\theta \in R$ such that $0<\theta<1$, then Cauchy's mean value theorem can be restated as
"If f and g are continuous on $[a, a+h]$ and are differentiable on $]a, a+h[$ and $g'(x)\neq 0$ for any $x \in]a, a+h[$ then, $\exists$ a $\theta \in R : 0<\theta<1$ such that
$$\frac{f(a+h)-f(a)}{g(a+h)-g(a)}=\frac{f'(a+\theta h)}{g'(a+\theta h)}.$$
- If we take $g(a)=g(b)$, then the function g would satisfy all the conditions of Rolle's theorem and consequently for some x in $]a,b[$, we would have $g'(x)=0$. In view of this we take $g(a)\neq g(b)$.
- In some cases, the Lagrange's mean value theorem is a particular case of Cauchy's mean value theorem (*e.g.*, take $g(x)=k$).
- Cauchy's mean value theorem cannot be deduced by applying Lagrange's mean value theorem to two functions f and g seperately and then dividing. It can be easily seen that the desired result cannot be obtained in this manner. In this way, we get
$$\frac{f(b)-f(a)}{g(b)-g(a)}=\frac{f'(c_1)}{g'(c_2)}$$
where $a<c_1<b$, and $a<c_2<b$. But, it is not necessary that c_1 and c_2 are equal. Hence, Cauchy's means value theorem is not directly deduceable from the first one.
- The conditions in the theorem are sufficient one. The conclusion may still hold even when the function involved do not satisfy the condition on $[a,b]$.

Recapitulations

- **Rolle's Theorem :** If a function f defined on $[a, b]$ is such that it is continuous in $[a, b]$ differentiable in $]a, b[$ and $f(a) = f(b)$ then there exists $c \in]a, b[$ such that $f'(c) = 0$.
- **Lagrange's Theorem :** If a function f defined on $[a, b]$ is such that it is continuous in (a, b) and differentiable in $]a, b[$ then there exists $c \in]a, b[$ such that
$$f'(c)=\frac{f(b)-f(a)}{b-a}$$
- **Cauchy Mean Value Theorem:** Let f and g be two functions defined on $[a, b]$ such that both are continuous in $[a, b]$, differentiable in $]a, b[$ then $\exists$ $c \in]a, b[$ such that
$$\frac{f(b)-f(a)}{g(b)-g(a)}=\frac{f'(c)}{g'(c)} \quad (g'(c)\neq 0)$$

2.4.1 GEOMETRICAL INTERPRETATION OF CAUCHY'S MEAN VALUE THEOREM

(1) Under suitable conditions, Cauchy's mean value theorem geometrically means that there is an ordinate $x=c$ between $x=a$ and $x=b$, such that the tangents at the points where $x=c$ cut the graphs of the function $f(x)$ and $\frac{f(b)-f(a)}{g(b)-g(a)}g(x)$ are mutually parallel.

(2) The ratio of the mean rates of increase of two functions in an interval is equal to the ratio of the actual rates of increase of the functions at some point within the interval.

Solved Examples

Example 1. *Discuss the applicability of Rolle's theorem in the interval [–1,1] to the function $f(x)=|x|$.*

Solution . Here, we have $f(x)=|x|$

$\Rightarrow$ $f(-1)=1$

and $f(1)=1$

$\Rightarrow$ $f(1)=f(-1)$

Now, the function $f(x)$ is continuous throughout the closed interval [–1,1] but $f(x)$ is not differentiable at $x=0 \in]-1,1[$. Hence, Rolle's theorem is not satisfied (due to the second condition).

Example 2. *Verify Rolle's theorem the function $f(x)=x^3-4x$ on [–2, 2].*

Solution . The function $f(x)=x^3-4x$ is a polynomial and so it is continuous and differentiable at all $x\in R$. In particular it is continuous in the closed interval [–2,2] and differentiable in the open interval]–2,2[. Also $f(-2)=0=f(2)$.

Thus, $f(x)$ satisfies all the three conditions of Rolle's theorem in [–2,2]. Therefore, there must exist at least one real number 'x' in the open interval]–2,2[for which

$$f'(x)=0.$$

Also $f'(x)=x^3-4x$

Now $f'(x)=0$ gives $3x^2-4=0$

or $x=\pm\frac{2}{\sqrt{3}}=\pm 1.55$.

Both these values lie in the open interval]–2, 2[and thus the conclusion of Rolle's theorem is verified.

Example 3. *Discuss the applicability of Rolle's theorem to the function*

$$f(x)=\log\left[\frac{x^2+ab}{(a+b)x}\right] \text{ in the interval } [a, b].$$

(VTU–2005)

Solution . Here, we have

$$f(a)=\log\left[\frac{a^2+ab}{(a+b)a}\right]=\log 1=0$$

and $$f(b)=\log\left[\frac{b^2+ab}{(a+b)b}\right]=\log 1=0$$

Also, it can be easily seen that $f(x)$ is continuous on $[a,b]$ and differentiable on $]a,b[$.

Thus all the three conditions of Rolle's theorem are satisfied. Hence $f'(x)=0$ for at least one value of x in $]a, b[$.

Now $f'(x)=0 \Rightarrow \frac{2x}{x^2+ab}-\frac{1}{x}=0$

$\Rightarrow$ $2x^2-(x^2+ab)=0$

$\Rightarrow$ $x^2=ab$ or $x=\sqrt{ab}$.

Obviously $\sqrt{ab}\in]a,b[$

[Being the geometric mean of a and b]

Hence, the Rolle's theorem is verified.

Example 4. *Verify Rolle's theorem for the function $f(x)=2x^3+x^2-4x-2$.*

Solution . Since, $f(x)$ is a rational integral function of x, therefore it is continuous and differentiable for all real values of x.

Hence, the first two conditions of Rolle's theorem are satisfied in any interval.

Hence, $f(x)=0$ gives $2x^3+x^2-4x-2=0$

i.e., $x=\pm\sqrt{2},-\frac{1}{2}$

$\Rightarrow$ $f(\sqrt{2})=f(-\sqrt{2})=f\left(-\frac{1}{2}\right)=0$

Now take the interval $[-\sqrt{2},\sqrt{2}]$, then, all the conditions of Rolle's theorem are satisfied in this interval. Then, $\exists$ at least one value of c in $]-\sqrt{2},\sqrt{2}[$ such that $f'(c)=0$

$f'(x)=0$ $\Rightarrow$ $6x^2+4x-4=0$

$\Rightarrow$ $x=-1, 2/3$

Since, both the points –1 and 2/3 lies in the open interval $]-\sqrt{2},\sqrt{2}[$. Hence, Rolle's theorem is verified.

Example 5. *Verify Rolle's theorem for $f(x)=x(x+3)e^{-x/2}$ in [–3, 0].*

Solution . Here, we have

$$f(x)=x(x+3)e^{-x/2}$$

$$\therefore \quad f'(x) = (2x+3)e^{-x/2} + \left(x^2+3x\right)e^{-x/2}.\left(-\frac{1}{2}\right)$$

$$= e^{-x/2}\left[2x+3-\frac{1}{2}\left(x^2+3x\right)\right]$$

$$= -\frac{1}{2}\left[x^2-x-6\right]e^{-x/2}$$

$\Rightarrow$ $f'(x)$ exist for every value of x in the interval $[-3, 0]$. Hence, $f(x)$ is differentiable and continuous in the interval $[-3, 0]$. Also, we have

$$f(-3) = f(0) = 0$$

$\Rightarrow$ All the three conditions of Rolle's theorem are satisfied. So

$$f'(x) = 0 \quad \Rightarrow \quad \frac{1}{2}\left(x^2-x-6\right)e^{-x/2} = 0$$

$$\Rightarrow \quad x^2-x-6=0 \Rightarrow x=3, -2$$

Since, the values $x = -2$ lies in the open interval $]-3, 0[$, the Rolle's theorem is verified.

Example 6. *Show that there is no real number p for which the equation $x^3-3x+p=0$ has two distinct roots in $]0,1[$.*

Solution . Let, if possible, there are two distinct roots a and b of the given equation in $]0, 1[$, such that $0 < a < b < 1$.

Now, let $\quad f(x) = x^3-3x+p$

Obviously, $f(x)$ is continuous and differentiable for all values of x (Being a polynomial).

Also, we have

$$f(a) = f(b) = 0$$

$\Rightarrow$ f satisfies all the conditions of Rolle's theorem in $[a,b]$ hence, $\exists$ a point $c \in]a,b[$ such that $f'(c)=0$.

Now $\quad f'(x) = 0 \Rightarrow 3x^2-3 = 0$

$\Rightarrow \quad x = \pm 1$

which is a contradiction

$(\because a < c < b \text{ as } 0 < a < b < 1)$

$\Rightarrow$ our assumption is wrong. Hence, there cannot be two distinct roots of $f(x) = 0$ in $]0, 1[$ for any value of p.

Example 7. *If $a+b+c = 0$, then show that the quadratic equation $3ax^2+2bx+c = 0$ has at least one root in $]0, 1[$.*

Solution . Let us define a function $f(x)$ such that

$$f(x) = ax^3+bx^2+cx+d.$$

Here we have

$$f(0) = d$$

and $f(1) = a+b+c+d = d \quad (\because a+b+c = 0)$

Obviously, $f(x)$ is continuous and differentiable in $]0, 1[$ (Being a polynomial).

Thus, $f(x)$ satisfies all the three conditions of Rolle's theorem in $[0, 1]$. Hence, there is at least one value of x in the open interval $]0, 1[$ where $f'(x) = 0$

i.e., $3ax^2+2bx+c = 0$ has at least one root in $]0, 1[$.

SIMILAR PROBLEMS

(1) Verify the Rolle's theorem for the function $f(x) = x^2$ in $[-1, 1]$.
(2) Verify the Rolle's theorem for the function $f(x) = x^3 - 3x + 2$ in $[1, 2]$.
(3) Verify the Rolle's theorem for the function $f(x) = x^{2/3}$ in $[-1, 1]$.
(4) Verify the Rolle's theorem for the function $f(x) = x^3 - 6x + 11x - 6$.
(5) Verify the Rolle's theorem for the function $f(x) = 10x - x^2$.

Example 8. *Discuss the applicability of Rolle's theorem to the function*

$$f(x) = \begin{cases} x^2+1, & \text{when } 0 \le x < 1 \\ 3-x, & \text{when } 1 < x \le 2 \end{cases}$$

Solution . Here $\quad f(0)=0^2+1$ and $f(2)=3-2=1$.

We shall show that $f(x)$ is continuous for all x in the range $(0,2)$ except at $x=1$.

Also $\quad f(1)=1^2+1=2$

Again, $\quad f(1+0) = \lim_{x\to 1+0} (3-x)$

$$= \lim_{x\to 1+h} \left[3-(1+h)\right], \text{ when } h\to 0$$

$$= \lim_{h\to 0} (2-h) = 2$$

and $\quad f(1-0) = \lim_{x\to 1-0} \left(x^2+1\right)$

$$= \lim_{x\to(1-h)} \left[(1-h)^2+1\right], \text{ when } h\to 0$$

$$= \lim_{h\to 0} \left(2-2h+h^2\right) = 2$$

Hence, $f(1-0)=f(1)=f(1+0)$ and so the function $f(x)$ is continuous at $x=1$ and the continuous in the whole interval $(0,2)$.

Again, $\quad f'(x) = \begin{cases} 2x, & \text{when } 0 \le x < 1 \\ -1, & \text{when } 1 < x \le 2 \end{cases}$

$\therefore$ $f(x)$ is differentiable in the interval $(0,2)$ except at $x=1$.

Now $Rf'(1) = \lim_{h\to 0} \frac{f(1+h)-f(1)}{h}$

$= \lim_{h\to 0} \frac{\{3-(1+h)\}-2}{h}$

$= \lim_{h\to 0} \frac{2-h-2}{h} = \lim_{h\to 0}(-1) = -1$

and $Lf'(1) = \lim_{h\to 0} \frac{f(1-h)-f(1)}{-h}$

$= \lim_{h\to 0} \frac{\left[(1-h)^2+1\right]-2}{-h}$

$= \lim_{h\to 0} \frac{2h-h^2}{h} = \lim_{h\to 0}(2-h) = 2$

∴ Thus $Rf'(1) \neq Lf'(1)$ and so $f'(1)$ does not exist.

Hence, the function $f(x)$ is not differentiable in the entire range (0, 2) and therefore Rolle's theorem is not applicable to the given function $f(x)$ in (0, 2).

Example 9. *Find 'c' of the mean value theorem, if $f(x)=x(x-1)(x-2)$; $a=0, b=1/2$.* (Gorakhpur–1999)

Solution . Here, we have $f(a)=f(0)=0$

and $f(b)=f\left(\frac{1}{2}\right)=\frac{3}{8}$

∴ $\frac{f(b)-f(a)}{b-a} = \frac{\frac{3}{8}-0}{\frac{1}{2}-0} = \frac{3}{4}$

Now $f(x)= x^3-3x^2+2x$

∴ $f'(x)=3x^2-6x+2$

⇒ $f'(c)=3c^2-6c+2$

Putting all these values in the Lagrange's mean value theorem

$$\frac{f(b)-f(a)}{b-a} = f'(c), (a<c<b)$$

We get $\frac{3}{4} = 3c^2-6c+2$ or $c = 1 \pm \frac{\sqrt{21}}{6}$

Hence, $c=\frac{1-\sqrt{21}}{6}$ lies in the open interval $]0, \frac{1}{2}[$ therefore, it is the required value.

Example 10. *If $f(x)=\log x$, find all numbers strictly between e^2 and e^3 such that*

$$f'(x)=\frac{f(e^3)-f(e^2)}{e^3-e^2}$$ (Burdwan–2003)

Solution . Obviously $f(x)=\log x$ is continuous in $[e^2, e^3]$ and differentiable in $]e^2, e^3[$.

Then by Lagrange's mean value theorem. There exist $c \in]e^2, e^3]$, such that

$$f'(c)=\frac{f(e^3)-f(e^2)}{e^3-e^2}$$

⇒ $\frac{1}{c}=\frac{3-2}{e^3-e^2}$

∴ $c=(e^3-e^2)$.

There exists only one value $c=(e^3-e^2)$ in $]e^2, e^3[$.

Example 11. *Show that any chord of the parabola $y=Ax^2+Bx+C$ is parallel to the tangent at the point whose abscissa is same as that of the middle point of the chord.*

Solution . Let a and b (where $a<b$) be the abscissae of the ends of the chord and let $f(x)=Ax^2+Bx+C$. Obviously, $f(x)$ is continuous on $[a,b]$ and differentiable in $]a,b[$ (Being a polynomial).

By Lagrange's mean value theorem there exists $c \in]a,b[$ such that

$$\frac{f(b)-f(a)}{b-a} = f'(c)$$

i.e., $Ab^2+Bb+C-Aa^2-Ba-C = (b-a)(2Ac+B)$

which gives $c=\frac{1}{2}(a+b)$ *i.e.*, abscissa of the point at which the tangent is parallel to the chord is same as that of the middle point of the chord.

Example 12. *Separate the intervals in which the polynomial $2x^3-15x^2+36x+1$ is increasing or decreasing.*

Solution . Here, we have $f(x)=2x^3-15x^2+36x+1$

∴ $f'(x)=6x^2-30x+36=6(x-2)(x-3)$.

Here $f'(x)>0$ for $x<2$ or for $x>3$.

$f'(x)<0$ for $2<x<3$

and $f'(x)=0$ for $x=2,3$

$f'(x)$ is positive in the intervals $]-\infty,2]$ and $[3,\infty[$ and negative in the interval $]2,3[$

Hence, the function $f(x)$ is monotonically increasing in the interval $]-\infty,2]$, $[3,\infty[$ and monotonically decreasing in $]2, 3[$.

Example 13. *Use the function $f(x)=x^{1/x}$, $x>0$ show that $e^{\pi}>\pi^{e}$.*

Solution . Here $f(x)=x^{1/x}$, $x>0$

∴ $\log f(x) = \frac{1}{x}\log_e x$

Differentiating w.r.t. x, we get

$$\frac{1}{f(x)} f'(x) = \frac{1}{x}.\frac{1}{x} - \frac{1}{x^2}\log_e x$$

$$f'(x) = \frac{x^{1/x}}{x^2}\left[1-\log_e x\right]$$

For $x>e$, $f'(x)<0$

$[\because \log_e x>1 \text{ for } x>e]$

$\therefore$ $f(x)$ is a decreasing function of x for $x>e$.

Hence $\pi > e \Rightarrow f(\pi) > f(e) \Rightarrow \pi^{1/\pi} < e^{1/e}$

$$\Rightarrow \left(\pi^{1/\pi}\right)^{e\pi} < \left(e^{1/e}\right)^{e\pi}$$

$$\Rightarrow \pi^e < e^\pi$$

$$\Rightarrow e^\pi > \pi^e$$

Example 14. *Show that* $\frac{x}{1+x} < \log(1+x) < x$, *for* $x>0$. (Mumbai–2008)

Solution . Let, $f(x)=\log(1+x)-\frac{x}{1+x}$

Obviously, $f(0)=0$.

and $f'(x) = \frac{1}{1+x} - \frac{1.(1+x)-x.1}{(1+x)^2}$

$$= \frac{1}{1+x} - \frac{1}{(1+x)^2} = \frac{x}{(1+x)^2}$$

Here, we observe that $f'(x)>0$, for $x>0$.

$\Rightarrow$ $f(x)$ is monotonically increasing in the interval $[0,\infty[$. Therefore

$f(x)>f(0)$, for $x>0$

$\Rightarrow$ $\left[\log(1+x) - \frac{x}{1+x}\right] > 0$, for $x>0$

$\Rightarrow$ $\log(1+x) > \frac{x}{1+x}$, for $x>0$ ...(1)

Now let $F(x) = x - \log(1+x)$

Obviously $F(0)=0$

Then $F'(x) = 1 - \frac{1}{1+x} = \frac{x}{1+x}$

Here, we observe that $F'(x)>0$, for $x>0$. Hence $F(x)$ is monotonically increasing in the interval $[0,\infty[$.

$\therefore$ $F(x)>F(0)$, for $x>0$

$\Rightarrow$ $[x-\log(1+x)]>0$, for $x>0$

$\Rightarrow$ $x>\log(1+x)$, for $x>0$ ...(2)

Now from (1) and (2), we get

$$\frac{x}{1+x} < \log(1+x) < x, \text{ for } x>0$$

Example 15. *Prove that* $(1+x)<e^x<1+xe^x$, $\forall\, x>0$.

Solution . Let us consider the function $f(x)=e^x$ in $[0, x]$. Obviously $f(x)$ is continuous as well as differentiable in $]0, x[$.

Then, by Lagrange's theorem $\exists\ c \in\ ]0, x[$, such that

$$f'(c) = \frac{f(x)-f(0)}{x-0}$$

or $e^c = \frac{e^x-1}{x}$ ...(1)

$0<c<x \Rightarrow e^0<e^c<e^x$

$(\because e^x$ is an increasing function.) ...(2)

Now, from (1) and (2), we have

$$e^0 < \frac{e^x-1}{x} < e^x, \forall x>0$$

$\Rightarrow$ $1 < \frac{e^x-1}{x} < e^x \Rightarrow x < e^x-1 < xe^x$

$\Rightarrow$ $(1+x) < e^x < xe^x$

Example 16. *Let f be continuous on* $[a-h, a+h]$ *and differentiable* $]a-h, a+h[$. *Prove that there is a real number* θ *between 0 and 1 such that*

$$f(a+h)-2f(a)+f(a-h)=h[f'(a+\theta h)-f'(a-\theta h)].$$

Solution . Consider the function ϕ defined on $[0, 1]$ by $\phi = f(a+ht)+f(a-ht)\ \forall t\in[0, 1]$.

Obviously ϕ is continuous on $[0,1]$ and differentiable on $]0, 1[$.

Then, by Lagrange's mean value theorem, there is a number θ lying between 0 and 1 such that

$$\phi(1)-\phi(0)=(1-0)\phi'(\theta)$$

i.e., $f(a+h)-2f(a)+f(a-h)$

$$=h[f'(a+\theta h)-f'(a-\theta h)].$$

which is the required result.

Example 17. *Show that Lagrange's mean value theorem does not holds for the function* $f(x)=|x|$ *in the interval* $[-1,1]$.

Solution . Since $f(x)=|x|$ is a continuous function on $[-1,1]$ but it is not differentiable at $x=0\in]-1,1[$. Hence, Lagrange's mean value theorem does not hold for the function $f(x)=|x|$ in the interval $[-1,1]$.

Example 18. *Verify Lagrange's mean value theorem for the function* $f(x)=\sin x$ *in* $\left[0, \frac{\pi}{2}\right]$. (Nagpur–2008)

Solution . The function $f(x)= \sin x$ is continuous and differentiable on R. Hence it is continuous as well as differentiable in $[0, \pi/2]$. Then, by Lagrange's mean value theorem, there must exist at least one c in $]0,\pi/2[$ such that

$$\frac{f(\pi/2)-f(0)}{\pi/2-0} = f'(c) \quad ...(1)$$

Here $f(0)=0, f(\pi/2)=1$

$f'(x)= \cos x \Rightarrow f'(c)= \cos c$

Put all these values in (1), we have

$$\frac{1-0}{\pi/2} = \cos c \Rightarrow \cos c = \frac{2}{\pi}$$

$$\Rightarrow c = \cos^{-1}\left(\frac{2}{\pi}\right)$$

Since, $0<2/\pi<1$, therefore the value of $c=\cos^{-1}\left(\frac{2}{\pi}\right)$ lies in $\left]0,\frac{\pi}{2}\right[$, which is the required value of c. Hence, Lagrange's mean value theorem is satisfied.

Example 19. *If $f''(x)$ exists for all points in $[a, b]$ and $\frac{f(c)-f(a)}{c-a}=\frac{f(b)-f(c)}{b-c}$ where $a<c<b$, then, there is a number l such that*

$$a<l<b \text{ and } f''(l)=0.$$

Solution . Since $f''(x)$ exist for all points in $[a, b]$,

$\Rightarrow$ $f'(x)$ is continuous in $[a, b]$

$\Rightarrow$ $f(x)$ is continuous in $[a, b]$.

Now, applying Lagrange's mean value theorem to $f(x)$ in $[a, c]$ and $[c, b]$ respectively, we get

$$\frac{f(c)-f(a)}{c-a}=f'(l_1),\ a<l_1<c \qquad ...(1)$$

and
$$\frac{f(b)-f(c)}{b-c}=f'(l_2),\ c<l_2<b \qquad ...(2)$$

Then, from (1) and (2), we get

$$f'(l_1)=f'(l_2) \qquad \left[\because \frac{f(c)-f(a)}{c-a}=\frac{f(b)-f(c)}{b-c}\right]$$

Now $f'(x)$ satisfies all the conditions of Rolle's theorem in the interval $[l_1, l_2]$.

Hence, $f''(l)=0$ where $l\in\,]l_1, l_2[$ and $l\in\,]a, b[$.

Example 20. *If $f(x)=(x-1)(x-2)(x-3)$ and $a=0$, $b=4$, find 'c' using Langrange's mean value theorem.* (VTU–2009)

Solution . We have $f(x)=(x-1)(x-2)(x-3)$

$$=x^3-6x^2+11x-6$$

$$f(a)=f(0)=-6$$

and
$$f(b)=f(4)=6$$

$$\therefore \quad \frac{f(b)-f(a)}{b-a}=\frac{6-(-6)}{4-0}=\frac{12}{4}=3.$$

Also $f'(x)=3x^2-12x+11$ gives

$$f'(c)=3c^2-12c+11.$$

Putting these values in Lagrange's mean value theorem,

$$\frac{f(b)-f(a)}{b-a}=f'(c) \text{ where } a<c<b$$

we get $\quad 3=3c^2-12c+11$

or $\quad 3c^2-12c+8=0$

$$c=\frac{12\pm\sqrt{(144-96)}}{6}=2\pm\frac{2\sqrt{3}}{3}$$

As the value of c lies in the open interval $]0,4[$. Hence both of these are the required values of c.

Example 21. *Examine if mean value theorem applies to $f(x)=x^3+3x^2-5x$ in the interval $[1,2]$. If it does, then find the intermediate point whose existence is asserted by the theorem.*

Solution . Given $\quad f(x)=x^3+3x^2-5x \qquad ...(1)$

$\therefore \quad f'(x)=3x^2+6x-5$

and $\quad f'(c)=3c^2+6c-5 \qquad ...(2)$

Let $a=1$ and $b=2$, then from (1), we have

$$f(a)=f(1)=1^3+3(1)^2-5(1)=-1$$

$$f(b)=f(2)=2^3+3(2)^2-5(2)=10$$

From mean value theorem, we have

$$f(b)-f(a)=(b-a).f'(c)$$

$\Rightarrow \quad f(2)-f(1)=(2-1).f'(c)$

$\Rightarrow \quad 10-(-1)=(2-1).f'(c)$

$\Rightarrow \quad 3c^2+6c-5=11 \qquad$ [Using(2)]

$\Rightarrow \quad 3c^2+6c-16=0$

$\therefore \quad c=-1\pm 2.55$

i.e., $\quad c=-3.55, 1.55$

Example 22. *Verify Cauchy's mean value theorem for the functions $f(x)=x^2-2x+3$, $g(x)=x^3-7x^2+26x-5$ in the interval* $[-1, 1]$.

Solution . Since $f(x)$ and $g(x)$ are polynomial in x, so these are continuous in the closed interval $[-1, 1]$ and also differentiable and continuous in the open interval $(-1,1)$.

Also $\quad g'(x)=3x^2-14x+26$

$$g'(-1)=3(-1)^2-14(-1)+26=43=+\text{ve}$$

$$g'(1)=3(1)^2-14(1)+26=15=+\text{ve}.$$

Therefore, $g'(x)\neq 0$ for any value of x in $(-1,1)$. Hence all the conditions of Cauchy Mean Value Theorem are satisfied.

Then, by using,

$$\frac{f(b)-f(a)}{g(b)-g(a)}=\frac{f'(c)}{g'(c)}$$

Putting $a=-1$, $b=1$ (given), we have

$$\frac{f(1)-f(-1)}{g(1)-g(-1)}=\frac{f'(c)}{g'(c)}$$

$$\frac{\left[1^2-2(1)+3\right]-\left[(-1)^2-2(-1)+3\right]}{\left[1^3-7(1)^2+26(1)-5\right]-\left[(-1)^3-7(-1)^2+26(-1)-5\right]}$$

$$=\frac{2c-2}{2c^2-14c+26} \qquad [\because f'(x)=2x-2]$$

or
$$\frac{2-6}{15-(-39)}=\frac{2c-2}{3c^2-14c+26}$$

or $\quad -4(3c^2-14c+26)=54\times 2(c-1)$

or $3c^2+14c+26=-27(c-1)$

or $3c^2+13c-1=0$

$\Rightarrow \quad c=\frac{-13\pm\sqrt{(181)}}{6}=\frac{-13\pm 13.454}{6}$

i.e., $c=0.076, -4.409$

Since the value 0.076 lies in [–1,1]. Hence, Cauchy mean value theorem is verified.

Example 23. *Verify Cauchy's mean value theorem for the function x^2 and x^3 in the interval* [1,2].

Solution . Let us suppose $f(x)=x^2$ and $g(x)=x^3$.

Then, obviously $f(x)$ and $g(x)$ are continuous in [1,2] and differentiable in]1,2[.

Also $g'(x)=3x^2\neq 0$ for any point in]1,2[.

Then, by Cauchy's mean value theorem there exists at least one real number $c\in]1,2[$, such that

$$\frac{f(2)-f(1)}{g(2)-g(1)}=\frac{f'(c)}{g'(c)} \qquad \text{....(1)}$$

After solving, we get $c=\frac{14}{9}$, which lies in the open interval]1,2[. Hence, Cauchy's mean value theorem is verified.

Example 24. *Use Cauchy's mean value theorem, to evaluate*

$$\lim_{x\to 1}\left[\frac{\cos\frac{\pi x}{2}}{\log(1/x)}\right].$$

Solution . Let us suppose

$$f(1)=\cos\left(\frac{1}{2}\pi x\right),\ g(x)=\log x$$

$$a=x \text{ and } b=1$$

Putting all these values in Cauchy's mean value theorem

$$\frac{f(b)-f(a)}{g(b)-g(a)}=\frac{f'(c)}{g'(c)},\ a<c<b$$

we get

$$\frac{\cos\frac{\pi}{2}-\cos\frac{n\pi}{2}}{\log 1-\log x}=\frac{-\frac{1}{2}\pi\sin\left(\frac{\pi c}{2}\right)}{1/c};\ x<c<1$$

Now, taking the limit as $x\to 1$, which give that $c\to 1$, therefore

$$\lim_{x\to 1}\left\{\frac{0-\cos\left(\frac{1}{2}\pi x\right)}{\log(1/x)}\right\}=\lim_{c\to 1}\left\{\frac{-\frac{1}{2}\pi\sin\left(\frac{1}{2}\pi c\right)}{(1/c)}\right\}$$

$$\text{or } \lim_{x\to 1}\left\{\frac{-\cos\left(\frac{1}{2}\pi x\right)}{\log(1/x)}\right\}=-\frac{1}{2}\pi$$

$$\left(\because \sin\frac{1}{2}\pi c\to 1 \text{ as } c\to 1\right)$$

$$\text{or } \lim_{x\to 1}\left\{\frac{\cos\left(\frac{1}{2}\pi x\right)}{\log(1/x)}\right\}=\frac{\pi}{2}.$$

Example 25. *If in the Cauchy's mean value theorem, we write $f(x)=e^x$ and $g(x)=e^{-x}$, show that 'c' is the arithmetic mean between a and b.* (Mumbai-2008)

Solution . Since, we have

$$f(x)=e^x \text{ and } g(x)=e^{-x}$$

$$\therefore \quad \frac{f(b)-f(a)}{g(b)-g(a)}=\frac{e^b-e^a}{e^{-b}-e^{-a}}=-e^a e^b=-e^{a+b}$$

$$\text{and} \quad \frac{f'(x)}{g'(x)}=\frac{e^x}{-e^{-x}}$$

$$\text{so that } \frac{f'(c)}{g'(c)}=\frac{e^c}{-e^{-c}}=-e^{2c}$$

After putting all these values in Cauchy's mean value theorem, we get

$$-e^{a+b}=-e^{2c}\Rightarrow \quad a+b=2c$$

$$\Rightarrow \quad c=\frac{a+b}{2}$$

Hence, c is the arithmetic mean between a and b.

Example 26. *If $f(x)$, $g(x)$ and $h(x)$ are functions such that*

(i) $f(x)$, $g(x)$ and $h(x)$ are continuous on $[a,b]$

(ii) $f(x)$, $g(x)$ and $h(x)$ are differentiable on $]a,b[$,

then show that $\begin{vmatrix} f'(c) & g'(c) & h'(c)\\ f(b) & g(b) & h(b)\\ f(a) & g(a) & h(a)\end{vmatrix}=0$ *where* $c\in]a,b[$

Solution . Consider the function $F(x)$ such that

$$F(x)=\begin{vmatrix} f(x) & g(x) & h(x)\\ f(b) & g(b) & h(b)\\ f(a) & g(a) & h(a)\end{vmatrix}=0 \qquad \text{...(1)}$$

Obviously, $F(x)$ is of the form $A\,f(x)+B\,g(x)+C\,h(x)$, where A, B, C are some real numbers. From the condition (i) and (ii), $F(x)$ is continuous on. $[a,b]$ and differentiable on $]a,b[$.

Also $\quad F(a)=F(b)=0.$

$\Rightarrow$ $F(x)$ satisfies all the conditions of Rolle's theorem. Hence, there exists a $c\in]a,b[$ such that $F'(c)=0$

$$\text{i.e.,} \quad \begin{vmatrix} f'(c) & g'(c) & h'(c)\\ f(b) & g(b) & h(b)\\ f(a) & g(a) & h(a)\end{vmatrix}=0.$$

Example 27. *Verify Cauchy's mean value for $f(x)=\sin x$ and $g(x)=\cos x$ in* $\left[-\frac{\pi}{2},0\right]$. (JNTU–2006)

Solution. It can be easily seen that $f(x)$ and $g(x)$ both are continuous on $\left[-\frac{\pi}{2},0\right]$ and differentiable on $\left]-\frac{\pi}{2},0\right[$.

Also, $g'(x)=-\sin x \neq 0$ for any point in the interval $\left]-\frac{\pi}{2},0\right[$.

Then, by Cauchy's mean value theorem, $\exists$ at least one $c\in\left]-\frac{\pi}{2},0\right[$ such that

$$\frac{f(0)-f\left(-\frac{\pi}{2}\right)}{g(0)-g\left(-\frac{\pi}{2}\right)}=\frac{f'(c)}{g'(c)}$$

Putting all the values and after simplification, we have

$$\cot c=-1 \Rightarrow c=-\pi/4.$$

Since $c=-\pi/4$ lies in $]-\pi/2,0[$, hence, Cauchy mean value theorem is verified.

Example 28. *Show that* $\frac{\sin\alpha-\sin\beta}{\cos\beta-\cos\alpha}=\cot\theta$.

Solution . Let $f(x)=\sin x$ and $g(x)=\cos x$, for $x\in[\alpha,\beta]$, where $0<\alpha<\beta<\pi/2$.

$\therefore\ f'(x)=\cos x$ and $g'(x)=-\sin x$.

It can be easily seen that both the function $f(x)$ and $g(x)$ are continuous in the closed interval $[\alpha,\beta]$ and differentiable in the open interval $]\alpha,\beta[$.

Hence, by Cauchy's mean value theorem there exists at least one $\theta\in R$, $\theta\in]\alpha,\beta[$ such that

$$\frac{f(\beta)-f(\alpha)}{g(\beta)-g(\alpha)}=\frac{f'(\theta)}{g'(\theta)}$$

$$\Rightarrow \frac{\sin\beta-\sin\alpha}{\cos\beta-\cos\alpha}=\frac{\cos\theta}{-\sin\theta}=-\cot\theta$$

$$\Rightarrow \frac{\sin\alpha-\sin\beta}{\cos\beta-\cos\alpha}=\cot\theta,\text{ where } 0<\alpha<\theta<\beta<\pi/2.$$

Example 29. *Show that the function f and g defined on* $\left[0,\frac{1}{2}\right]$, *by* $f(x)=x(x-1)(x-2)$ *and* $g(x)=x(x-2)(x-3)$ *satisfy the condition of Cauchy's mean value theorem.*

Solution . Here, we have

$$f(x)=x(x-1)(x-2)=x^3-3x^2+2x$$

and $g(x)=x(x-2)(x-3)=x^3-5x^2+6x$

$\Rightarrow f'(x)=3x^2-6x+2$ and $g'(x)=3x^2-10x+6$

By Cauchy's mean value theorem, we have

$$\frac{f'(c)}{g'(c)}=\frac{f\left(\frac{1}{2}\right)-f(0)}{g\left(\frac{1}{2}\right)-g(0)},\ c\in\left]0,\frac{1}{2}\right[$$

or $$\frac{3c^2-6c+2}{3c^2-10c+6}=\frac{\frac{3}{8}-0}{\frac{15}{8}-0}=\frac{1}{5}$$

$$\Rightarrow 12c^2-20c+4=0$$

$$\Rightarrow c=\frac{5\pm\sqrt{13}}{6}$$

The value $\frac{5-\sqrt{13}}{6}$ of c belongs to $\left]0,\frac{1}{2}\right[$.

Hence, the Cauchy mean value theorem is satisfied.

Example 30. *Find 'c' of Cauchy's mean value theorem for the functions*

$$f(x)=\sqrt{x},\ \phi(x)=\frac{1}{x}\ \textit{in } [a,b]$$

and show that it is the G.M. of a and b.

Solution . We have

(i) $f(x)$ and $\phi(x)$ are continuous in the closed interval $[a,b]$.

(ii) $f'(x)$ and $\phi'(x)$ exist in the open interval (a,b).

(iii) $\phi'(x)=-1/2\,x^{-3/2}\neq 0$ for any x in $]a,b[$.

Therefore $f(x)$ and $\phi(x)$ satisfies all the conditions of Cauchy's mean value theorem. Hence there exist a point $c\in]a,b[$ such that

$$\frac{f(b)-f(a)}{\phi(b)-\phi(a)}=\frac{f'(c)}{\phi'(c)} \quad \ldots(1)$$

Also here

$$f'(x)=\frac{1}{2}x^{-1/2},\ \phi'(x)=-\frac{1}{2}x^{-3/2}$$

From (1), we get

$$\frac{\sqrt{b}-\sqrt{a}}{\left(1/\sqrt{b}\right)-\left(1/\sqrt{a}\right)}=\frac{1/2c^{-1/2}}{-1/2c^{-3/2}}$$

or $$\frac{\left(\sqrt{b}-\sqrt{a}\right)\sqrt{a}.\sqrt{b}}{\sqrt{a}-\sqrt{b}}=-\frac{c^{3/2}}{c^{1/2}}$$

$$\therefore \quad c=\sqrt{ab}.$$

2.5 TAYLOR'S THEOREM

Let f(x) be a single valued function defined on [a,a+h] such that

(i) all the derivative of f(x) upto $(n-1)^{th}$ *order are continuous in [a, a+h], and*

(ii) $f^n(x)$ *exists in (a,a+h), then there exists a real number* $\theta, 0<\theta<1$, *such that*

$$f(a+h)=f(a)+hf'(a)+\frac{h^2}{2!}f''(a)+\ldots+\frac{h^{n-1}}{(n-1)!}f^{n-1}(a)+\frac{h^n(1-\theta)^{n-p}}{p(n-1)!}f^n(a+\theta h)$$

where p is a given positive integer.

Proof. Since, f^n exists, all the derivative $f', f'' ... f^{n-1}$ exist and continuous on $[a, a+h]$, consider a function f defined on $[a, a+h]$ such that

$$\phi(x) = f(x) + (a+h-x)f'(x) + \frac{(a+h-x)^2}{2!}f''(x) + ... + \frac{(a+h-x)^{n-1}}{(n-1)!}f^{n-1}(x) + A(a+h-x)^p \quad ...(1)$$

where A is a constant to be determined such that $\phi(a+h) = \phi(a)$

Now $$\phi(a) = f(a) + hf'(a) + \frac{h^2}{2!}f''(a) + ... + \frac{h^{n-1}}{(n-1)!}f^{n-1}(a) + Ah^p$$

and $$\phi(a) = f(a+h)$$

$$\Rightarrow \quad f(a+h) = f(a) + hf'(a) + \frac{h^2}{2!}f''(a) + ... + \frac{h^{n-1}}{(n-1)!}f^{n-1}(a) + Ah^p \quad ...(2)$$

Now (i) $f, f', f'', ..., f^{n-1}$ being all continuous on $[a, a+h]$ the function ϕ is continuous on $[a, a+h]$,

(ii) Similarly the function ϕ is differentiable on $]a, a+h[$,

and (iii) $\phi(a+h) = \phi(a)$

Thus, the function ϕ satisfies all the conditions of Rolle's theorem and hence $\exists$ a real number $\theta (0 < \theta < 1)$ such that $\phi'(a+\theta h) = 0$.

Here $$\phi'(x) = f'(x) + (-f'(x) + (a+h-x)f''(x))$$

$$+ \frac{1}{2!}\left[-2(a+h-x)f''(x) + (a+h-x)^2 f'''(x)\right] + ...$$

$$+ \frac{1}{(n-1)!}\left[-(n-1)(a+h-x)^{n-2}f^{n-1}(x) + (a+h-x)^{n-1}f^n(x)\right] - Ap(a+h-x)^{p-1}$$

$$= \frac{(a+h-x)^{n-1}}{(n-1)!}f^n(x) - Ap(a+h-x)^{p-1}$$

[Other terms canceled in pairs]

$$\therefore \quad 0 = \phi'(a+\theta h) = \frac{h^{n-1}(1-\theta)^{n-1}}{(n-1)!}f^n(a+\theta h) - Aph^{p-1}(1-\theta)^{p-1}$$

$$\Rightarrow \quad A = \frac{h^{n-p}(1-\theta)^{n-p}}{p(n-1)!}f^n(a+\theta h), h \neq 0, \theta \neq 1$$

Now, putting the values of A in (2), we get

$$f(a+h) = f(a) + hf'(a) + \frac{h^2}{2!}f''(a) + ... + \frac{h^{n-1}}{(n-1)!}f^{n-1}(a) + \frac{h^n(1-\theta)^{n-p}}{p(n-1)!}f^n(a+\theta h)$$

2.5.1 FORMS OF REMAINDER AFTER n TERMS

(i) The term $R_n = \frac{h^n(1-\theta)^{n-1}}{(n-1)!}f^n(a+\theta h)$ which occur after n terms, is called the Taylor's remainder after n terms. The theorem with this form of remainder is called Taylor's theorem with Schlomilch and Roche form of remainder.

(ii) For $p=1$, we get

$$R_n = \frac{h^n(1-\theta)^{n-1}}{(n-1)!}f^n(a+\theta h)$$

Then, R_n is called Cauchy's form of remainder.

(iii) For $p=n$, we get

$$R_n = \frac{h^n}{n!}f^n(a+\theta h)$$

then, R_n is called Lagrange's form of remainder.

2.5.2 ANOTHER FORM OF TAYLOR'S THEOREM

Replacing h by $(x-a)$ in Taylor's theorem, we get

$$f(x) = f(a) + (x-a)f'(a) + \frac{(x-a)^2}{2!}f''(a) + ... + \frac{(x-a)^{n-1}}{(n-1)!}f^{n-1}(a) + \frac{(x-a)^n}{p(n-1)!}f^n(1-\theta)^{n-p}$$

The remainder, after n terms can be written as

$$R_n = \frac{(x-a)^n(1-\theta)^{n-p}}{p(n-1)!}f^n(c), a < c < x.$$

DEDUCTIONS

Putting $a=0$ in second form of Taylor's theorem, we get (Maclaurin's theorem)

$$f(x)=f(0)+x f'(0)+\frac{x^2}{2!}f''(0)+\ldots+\frac{x^{n-1}}{(n-1)!}f^{n-1}(0)+R_n \qquad \ldots(1)$$

(i) If $R_n=\dfrac{x^n(1-\theta)^{n-p}}{p(n-1)!}f^n(\theta x)$, then (1) is known as Maclaurin's theorem with Schlomilch and Roche's form of remainder.

(ii) For $p=1$, $R_n=\dfrac{x^n(1-\theta)^{n-p}}{p(n-1)!}f^n(\theta x)$ is called Cauchy's form of remainder.

(iii) For $p=n$, $R_n=\dfrac{x^n}{n!}f^n(\theta x)$, is called Lagrange's form of remainder.

2.5.3 TAYLOR'S SERIES

Let $f(x)$ possesess continuous derivatives of all orders in the interval $[a, a+h]$, then for every positive integral value of n, we have

$$f(a+h)=f(a)+hf'(a)+\frac{h^2}{2!}f''(a)+\ldots+\frac{h^{n-1}}{(n-1)!}f^{n-1}(a)+R_n$$

where,

$$R_n=\frac{h^n}{n!}f^n(a+\theta h),(0<\theta<1). \qquad \ldots(1)$$

Equation (1) can also be written as

$$S_n=f(a)+hf'(a)+\frac{h^2}{2!}f''(a)+\ldots+\frac{h^{n-1}}{(n-1)!}f^{n-1}(a)$$

Then

$$f(a+h)=S_n+R_n.$$

Let us suppose $R_n\to 0$ as $n\to\infty$, then $\lim_{n\to\infty} S_n=f(a+h)$

i.e., the series $f(a)+hf'(a)+\dfrac{h^2}{2!}f''(a)+\ldots+\dfrac{h^{n-1}}{(n-1)!}f^{n-1}(a)+\ldots$ converges to $f(a+h)$.

Thus,

(i) If f possess a continuous derivative of every order in $[a, a+h]$.

(ii) The remainder after n terms $R_n\to 0$ as $n\to\infty$, then

$$f(a+h)=f(a)+hf'(a)+\frac{h^2}{2!}f''(a)+\ldots+\frac{h^n}{n!}f^n(a)+\ldots$$

This series is known as Taylor's series for the expansion of $f(a+h)$ as a power series in h.

Maclaurin's series. If we put $a=0$ and replace h by x in Taylor's series, we get

$$f(x)=f(0)+x f'(0)+\frac{x^2}{2!}f''(0)+\ldots+\frac{x^n}{n!}f^n(0)+\ldots$$

This Series is known as Maclaurin's series for the expansion of $f(x)$ as a power series in x.

REMARKS

- Maclaurin's series is a particular case of Taylor's series.
- Maclaurin's expansions of $f(x)$ fails if any of the functions $f(x), f'(x), f''(x)\ldots$ becomes infinite or discontinuous at any point of the interval $[0, x]$ or if R_n does not tends to zero as n tends to infinity.

2.6 MACLAURIN'S THEOREM

Let $f(x)$ be a function of x which possesses continuous derivatives of all orders in the interval $[0, x]$ and can be expanded as an infinite series in x, then

$$f(x)=f(0)+x f'(0)+\frac{x^2}{2!}f''(0)+\ldots+\frac{x^n}{n!}f^n(0)+\ldots$$

Proof. Let us define

$$f(x)=A_0+A_1x+A_2x^2+A_3x^3+\ldots \qquad \ldots(1)$$

Let the expression (1) be differentiable term by term any number of times. Then by successive differentiation, we have

$$f'(x) = A_1 + 2A_2x + 3A_3x^2 + 4A_4x^3 + ...$$
$$f''(x) = 2.1.A_2 + 3.2.A_3x + 4.3.A_4x^2 + ...$$
$$f'''(x) = 3.2.A_3 + 4.3.2.A_4x + ...$$
....

Putting $x=0$, we get

$$f(0)=A_0, f'(0)=A_1, f''(0)=2!A_2, f'''(0)=3!A_3....$$

$$\Rightarrow \quad A_0 = f(0), A_1 = f'(0), A_2 = \frac{f''(0)}{2!}, A_3 = \frac{f'''(0)}{3!} ...$$

Substitute all these values in (1), we get

$$f(x) = f(0) + x f'(0) + \frac{x^2}{2!} f''(0) + ... + \frac{x^n}{n!} f^n(0) + ...$$

REMARKS

- The Maclaurin's theorem is a particular case of Taylor's Theorem, and can be obtained by replacing $a=0$ and $h=x$ in Taylor's theorem.
- If the function $f(x)$ is denoted by y, then the expansion may be written in the form

$$y = y(0) + x y_1(0) + \frac{x^2}{2!} y_2(0) + ... + \frac{x^n}{n!} y_n(0) + ...$$

where $y(0), y_1(0), y_2(0),..., y_n(0)$ etc. denotes values of $y, y_1, y_2, ... y_n$ respectively for $x=0$.

2.7 FAILURE OF TAYLOR'S AND MACLAURIN'S THEOREM

(a) *Taylor's theorem fails to expand $f(a+h)$ in an infinite power series in the following cases :*
- If any of the function $f(x), f'(x), f''(x)...$ become infinite or does not exist for any value of x in the given interval.
- If R_n does not tends to zero as $n \to \infty$.

(b) *Maclaurin's theorem fails to expand $f(x)$ in an infinite power series in the following cases :*
- If any of the function $f(x), f'(x), f''(x)...$ becomes infinite or does not exist in interval $[0, x]$.
- If R_n does not tends to zero as $n \to \infty$.

REMARKS

- Before expanding a given function as an infinite Taylor's or Maclaurin's series, it is essential to examine the behaviour of R_n as $n \to \infty$, which is not simple in many cases. We, therefore, generally obtain the expansion by assuming the possibility of expanding it in an infinite series by assuming that $R_n \to 0$ as $n \to \infty$.

Recapitulations

- **Taylor's series :** $f(a+h) =$
$$f(a) + hf'(a) + \frac{h^2}{2!} f''(a) + ... + \frac{h^n}{n!} f^n(a) + ...$$
- **Maclaurin's series :** $f(x) =$
$$f(0) + x f'(0) + \frac{x^2}{2!} f''(0) + ... + \frac{x^n}{n!} f^n(0) + ...$$

2.8 POWER SERIES EXPANSION OF SOME STANDARD FUNCTIONS

WORKING PROCEDURE

To find the power series expansion we shall use the following procedure :

Step 1. *Put the given function equal to $f(x)$.*

Step 2. *Differentiate $f(x)$, a number of times and obtain $f'(x), f''(x), f'''(x)$... and so on.*

Step 3. *Put $x = 0$ and find $f(0), f'(0), f''(0)$... and so on.*

Step 4. *Now substitute the values of $f(0), f'(0), f''(0), f'''(0)$,... in* $f(x) = f(0) + x f'(0) + \frac{x^2}{2!} f''(0) + ...$

We shall now consider Maclaurin's series expansions of the function e^x, $\sin x$, $\cos x$, $(1+x)^m$ and $\log x$.

(i) Expansion of e^x. (Cochin–2005)

Let $\qquad f(x) = e^x \ \forall x \in R$

Then $\qquad f^n(x) = e^x \ \forall x \in R$

Thus, for each positive n, f^n is defined in the interval $[-h, h]$.

Writing, Lagrange's form of remainder, after n terms

$$R_n(x)=\frac{x^n}{n!}f^n(\theta x),\ \theta\in R,\ 0<\theta<1=\frac{x^n}{n!}e^{\theta x}$$

Now, we shall show that $\lim_{n\to\infty} R_n(x)=0$. Here, it is enough to show that $e^{\theta x}$ is bounded in $[-h, h]$ and $\lim_{n\to\infty}\frac{x^n}{n!}=0$.

Since, $0<\theta<1$ and $x\in[-h,h]$, therefore $|\theta x|<h$ and consequently, $0<e^{\theta x}<e^h$, hence $e^{\theta x}$ is bounded.

Now, let us write $$a_n=\frac{x^n}{n!}\ \forall n\in N$$

Then $$\frac{a_{n+1}}{a_n}=\frac{x}{n+1}\Rightarrow \lim_{n\to\infty}\frac{a_{n+1}}{a_n}=0$$

$\Rightarrow$ $\lim_{n\to\infty} a_n$ exists and equal to zero.

Now, $$\lim_{n\to\infty} R_n(x)=e^{\theta x}\left[\lim \frac{x^n}{n!}\right]=0$$

Hence, we find that the function $f(x)$ has a Maclaurin's series expansions for each $x\in[-h,h]$. This implies

$$f(x)=f(0)+xf'(0)+\frac{x^2}{2!}f''(0)+\ldots+\frac{x^{n-1}}{(n-1)!}f^{n-1}(0)+\ldots\quad \forall x\in R.$$

Substituting $f(x)=e^x, f'(x)=e^x,\ldots, f^n(x)=e^x$, we have

$$e^x=1+x+\frac{x^2}{2!}+\frac{x^3}{3!}+\ldots+\frac{x^{n-1}}{(n-1)!}+\ldots\ \forall x\in R$$

(ii) Expansion of sin *x*.

Let $$f(x)=\sin x,\ \forall x\in R$$ (PTU–2005)

$$\Rightarrow\quad f^n(x)=\sin\left(x+\frac{n\pi}{2}\right),\quad \forall x\in R$$

Writing, Lagrange's form of remainder after n terms, we have

$$R_n(x)=\frac{x^n}{n!}f^n(\theta x),\text{ where } 0<\theta<1=\frac{x^n}{n!}\sin\left(\theta x+\frac{n\pi}{2}\right)$$

Now, for all $x\in R$,

$$|R_n(x)|\le\left|\frac{x^n}{n!}\right|\quad\text{and}\quad \lim_{n\to\infty}\frac{x^n}{n!}=0$$ [As in (i)]

Thus, we find that, the function $f(x)$ has a Maclaurin's series expansions for each x in $[-h,h]$. Hence, we have

$$f(x)=f(0)+xf'(0)+\frac{x^2}{2!}f''(0)+\ldots+\frac{x^{n-1}}{(n-1)!}f^{n-1}(0)+\ldots\ \forall x\in R.$$

Now, substituting $f(x)=\sin x$, $f^n(x)=\sin\frac{n\pi}{2}$, we have

$$\sin x=x-\frac{x^3}{3!}+\frac{x^5}{5!}-\ldots\ \forall x\in R.$$

(iii) Expansion of cos *x*.

Let $f(x)=\cos x$, $\forall x\in R$, then $f^n(x)=\cos\left(\theta x+\frac{n\pi}{2}\right)$

Thus, for eaeh n, f^n is defined in every interval $[-h, h]$.

Writing, Lagrange's remainder after n terms, we have

$$R_n(x)=\frac{x^n}{n!}f^n(\theta x)=\frac{x^n}{n!}\cos\left(\theta x+\frac{n\pi}{2}\right),$$ where $0<\theta<1$

Now, for all $x\in R$, $$|R_n(x)|\le\left|\frac{x^n}{n!}\right|\quad\text{and}\quad \lim_{n\to\infty}\frac{x^n}{n!}=0$$ [As in (i)]

Thus, we find that, the function f has a Maclaurin's series expansions for each $x\in[-h, h]$, which gives

$$f(x)=f(0)+x f'(0)+\frac{x^2}{2!}f''(0)+\ldots+\frac{x^n}{n!}f^n(0)+\ldots\quad \forall x\in R.$$

Now, substituting $f(x)=\cos x\ldots$, $f^n(0)=\cos\frac{n\pi}{2}$, we have $\cos x=1-\frac{x^2}{2!}+\frac{x^4}{4!}-\ldots\ \forall x\in R.$

(iv) Expansion of $(1+x)^m$.

Case (i). Let m is a positive integer, then letting

$$f(x)=(1+x)^m, \ \forall x \in R$$

We find that for each $n \in N$, $f^n(x)$ exists for all $x \in R$, and whenever $n>m$, $f^n(x)=0 \ \forall x \in R$.

$$\Rightarrow \qquad R_n(x)=0, \text{ whenever } n>m$$

Hence, $\lim_{n\to\infty} R_n(x)=0$ and for all $x \in R$, we have

$$f(x)=f(0)+x f'(0)+\ldots+\frac{x^m}{m!}f^m(0), \qquad (\because \text{ All other terms must vanish.})$$

Substituting the value of $f(x), f(0), \ldots, f^m(0)$, We have

$$(1+x)^m=1+mx+\frac{m(m-1)}{2!}x^2+\ldots+x^m.$$

Case (ii). Let m not be a positive integer (may be a fraction or negative integer).

Here, we find that, if we write

$$f(x)=(1+x)^m, \text{ whenever } x \neq -1$$

then

$$f''(x)=m(m-1)\ldots(m-n+1)(1+x)^{m-n}, \text{ whenever } x \neq -1$$

Thus, for each positive integer n, f^n is defined in $[-h,h]$ for each h between 0 and 1.

Now, writing Cauchy's form of remainder after n terms, we have

$$R_n(x)=\frac{x^n(1-\theta)^{n-1}}{(n-1)!}f^n(\theta x), \text{ where } 0<\theta<1$$

$$=\frac{x^n(1-\theta)^{n-1}}{(n-1)!}m(m-1)\ldots(m-n+1)(1+\theta x)^{m-n}$$

$$=\frac{m(m+1)\ldots(m+n+1)}{(n-1)!}x^n\left(\frac{1-\theta}{1+\theta x}\right)^{n-1}.(1+\theta x)^{m-1}$$

Now, we observe that

(a) $\lim_{n\to\infty}\frac{m(m-1)\ldots(m-n+1)}{(n-1)!}x^n=0$

If we write
$$a_n=\frac{m(m+1)\ldots(m-n+1)}{(n-1)!}x^n$$

Then, we have
$$\frac{a_{n+1}}{a_n}=\frac{(m-n)x}{n} \Rightarrow \lim_{n\to\infty}\frac{a_{n+1}}{a_n}=-x$$

If follows that if $|x|<1$, then $\lim_{n\to\infty}a_n=0$.

(b) $\lim_{n\to\infty}\left(\frac{1-\theta}{1+\theta x}\right)^{n-1}=0$

In fact, since $0<\theta<1$ and $-1<x<1$, therefore, $0<\left[\frac{1-\theta}{1+\theta x}\right]<1$ and hence $\lim_{n\to\infty}\left[\frac{1-\theta}{1+\theta x}\right]=0$

(c) If $m>1$, then
$$(1+\theta x)^{m-1}<(1-|x|)^{m-1}$$

For (a), (b) and (c), we find that for all x in $]-1,1[$ $\lim_{n\to\infty}R_n(x)=0$

Thus, we find that for each h between 0 and 1, the function f has Maclaurin's series expansion for all $x \in [-h,h]$.
Hence, we have

$$f(x)=f(0)+xf'(0)+\frac{x^2}{2!}f''(0)+\ldots+\frac{x^{n-1}}{(n-1)!}f^{n-1}(0)+\ldots \quad \forall x \in]-1,1[.$$

Substituting the values of $f(x), f'(0), \ldots, f^{n-1}(0)$, we have

$$(1+x)^m=1+mx+\frac{m(m-1)}{2!}x^2+\frac{m(m-1)(m-2)}{3!}x^3+\ldots$$
$$+\frac{m(m-1)\ldots(m-n+1)}{n!}x^n+\ldots \text{whenever } -1<x<1$$

(v) Expansion of $\log_e(1+x)$.

Let $$f(x)=\log(1+x),\ -1<x<1.$$

Then $$f^n(x)=\frac{(-1)^{n-1}(n-1)!}{(1+x)^n},\text{ whenever } x>-1.$$

Now, we shall consider the following cases :

Case (a) Let $0\le x\le 1$. Writing Lagrange's form of remainder after n terms, we have

$$R_n=\frac{x^n}{n!}f^n(\theta x)=\frac{x^n}{n!}(-1)^{n-1}\frac{(n-1)!}{(1+\theta x)^n}=\frac{(-1)^{n-1}}{n}\cdot\left(\frac{x}{1+\theta x}\right)^n$$

Since, $0\le x\le 1, 0<\theta<1$, therefore $$0<\frac{x}{1+\theta x}<1$$

$\therefore$ $$|R_n|<\frac{1}{n},\text{ and }\frac{1}{n}\to 0\text{ as } n\to\infty$$

Therefore $$\lim_{n\to\infty} R_n=0.$$

Case (b) Let $-1<x<0$. Since in this case $\left|\frac{x}{1+\theta x}\right|$ need not be less than unity, therefore, we may not be able to show easily that $R_n\to 0$ as $n\to\infty$ by considering Lagrange's remainder.

Now, writing Cauchy's form of remainder, we have

$$R_n=\frac{x^n}{(n-1)!}(1-\theta)^{n-1}f^n(\theta x)=(-1)^{n-1}x^n\left(\frac{1-\theta}{1+\theta x}\right)^{n-1}\cdot\frac{1}{1+\theta x}$$

since $|x|<1$

therefore $\left|\frac{1-\theta}{1+\theta x}\right|<1$, so that $\left|\left(\frac{1-\theta}{1+\theta x}\right)^{n-1}\right|<1$ and $\left|\frac{1}{1+\theta x}\right|<\frac{1}{1-|x|}$

Thus $$|R_n|<\frac{|x|^n}{1-|x|}$$

This implies that $\lim_{n\to\infty} R_n=0.$, since $|x|<1$. Thus we find that if $-1\le x\le 1$,

then $$\lim_{n\to\infty} R_n=0.$$

$$f(x)=f(0)+xf'(0)+\frac{x^2}{2!}f''(0)+\ldots+\frac{x^{n-1}}{(n-1)!}f^{n-1}(0)+\ldots\text{ whenever } -1<x\le 1.$$

Substituting the values of $f(x), f(0), f'(0), \ldots, f^{n-1}(0), \ldots$, we get

$$\log(1+x)=x-\frac{x^2}{2}+\frac{x^3}{3}-\ldots,\text{ whenever } -1<x\le 1.$$

Recapitulations

- $e^x=1+x+\frac{x^2}{2!}+\frac{x^3}{3!}+\ldots+\frac{x^n}{n!}+\ldots$
- $\cos x=1-\frac{x^2}{2!}+\frac{x^4}{4!}-\frac{x^6}{6!}+\ldots$
- $\sin x=x-\frac{x^3}{3!}+\frac{x^5}{5!}-\ldots$
- $(1+x)^m=1+mx+\frac{m(m-1)}{2!}x^2+\ldots$
- $\log_e(1+x)=x-\frac{x^2}{2}+\frac{x^3}{3}-\frac{x^4}{4}+\ldots$
- $a^x=1+x\log_e a+\frac{x^2}{2!}(\log_e a)^2+\ldots$

Solved Examples

Example 1. *Show that*

$$a^x=1+x\log a+\frac{x^2}{2!}(\log a)^2+\ldots+\frac{x^{n-1}}{(n-1)!}(\log a)^{n-1}+\frac{x^n}{n!}a^{\theta x}(\log a)^n, 0<\theta<1.$$

Solution . Let $$f(x)=a^x \quad\ldots(1)$$

Then $$f^n(x)=a^x(\log a)^n\ \forall n\in N \text{ and } \forall x\in R \quad\ldots(2)$$

Now, putting $x=0$, in (1) and (2), we get

$$f(0)=1, f^n(0)=(\log a)^n\ \forall\ n\in N$$

From (2), $f^n(\theta x)=a^{\theta x}(\log a)^n$

Now, by Maclaurin's series with Lagrange's form of remainder after n terms we have

$$f(x)=f(0)+xf'(0)+\frac{x^2}{2!}f''(0)+\ldots+\frac{x^{n-1}}{(n-1)!}f^{n-1}(0)+\frac{x^n}{n!}a^{\theta x}(\log a)^n\ldots \quad\ldots(3)$$

Now, substituting the above values in (3), we get

$$a^x = 1 + x\log a + \frac{x^2}{2!}(\log a)^2 + \ldots$$

$$+\frac{x^{n-1}}{(n-1)!}(\log a)^{n-1} + \frac{x^n}{n!}a^{\theta x}(\log a)^n.$$

Here, Lagrange's form of remainder after n terms

$$R_n = \frac{x^n}{n!}a^{\theta x}(\log a)^n \text{ where } 0<\theta<1.$$

Example 2. *Expand $e^{a\sin^{-1}x}$ by Maclaurin's series and find the general term. Hence, show that*

$$e^{\theta} = 1 + \sin\theta + \frac{1}{2!}\sin^2\theta + \frac{2}{3!}\sin^3\theta + \ldots$$

Solution. Here $\quad y = e^{a\sin^{-1}x}$ ...(1)

Then $y_1 = e^{a\sin^{-1}x}.\dfrac{a}{\sqrt{1-x^2}} = \dfrac{ay}{\sqrt{1-x^2}}$ ...(2)

$$\Rightarrow \left(\sqrt{1-x^2}\right)y_1 = ay$$

$$\Rightarrow \left(1-x^2\right)y_1^2 - a^2y^2 = 0 \quad \ldots(3)$$

Now, differentiating both the sides, we have

$$\Rightarrow \left(1-x^2\right)2y_1y_2 - 2xy_1^2 - 2a^2yy_1 = 0$$

$$2y_1\left[\left(1-x^2\right)y_2 - xy_1 - a^2y\right] = 0 \quad \ldots(4)$$

Since $2y_1 \neq 0$ hence

$$[(1-x^2)y_2 - xy_1 - a^2y] = 0.$$

Now, differentiating n times by Leibnitz's theorem, we get

$$\left(1-x^2\right)y_{n+2} + ny_{n+1}(-2x)$$

$$+\frac{n(n-1)}{2}y_n(-2) - y_{n+1}x - ny_n.1 - a^2y_n = 0$$

$$\Rightarrow \left(1-x^2\right)y_{n+2} - (2n+1)xy_{n+1} - \left(n^2+a^2\right)y_n = 0 \quad \ldots(5)$$

Now, we can easily find, (from (1) to (5)) the following values

$$(y)_0 = 1, \quad (y_1)_0 = a, \; (y_2)_0 = a^2$$

$$(y_{n+2})_0 = (n^2+a^2)(y_n)_0 \quad \ldots(6)$$

Replacing n by $(n-2)$ in (6), we get

$$(y_n)_0 = [(n-2)^2+a^2](y_{n-2})_0$$

$$= [(n-2)^2+a^2][(n-4)^2+a^2](y_{n-4})_0$$

If n is odd, then

$$(y_n)_0 = [(n-2)^2+a^2][(n-4)^2+a^2]$$

$$\ldots(3^2+a^2)(1^2+a^2)(y_1)_0$$

$$= [(n-2)^2+a^2][(n-4)^2+a^2]$$

$$\ldots[(3^2+a^2)(1^2+a^2)].a$$

If n is even, then

$$(y_n)_0 = [(n-2)^2+a^2][(n-4)^2+a^2]$$

$$\ldots(4^2+a^2)(2^2+a^2)(y_2)_0$$

$$= [(n-2)^2+a^2][(n-4)^2+a^2]$$

$$\ldots[(4^2+a^2)(2^2+a^2)].a^2$$

Hence,

$$y_n(0) = \begin{cases} a\left(1^2+a^2\right)\left(3^2+a^2\right)\ldots\left[(n-2)^2+a^2\right], & \text{if } n \text{ is odd} \\ a^2\left(2^2+a^2\right)\left(4^2+a^2\right)\ldots\left[(n-2)^2+a^2\right], & \text{if } n \text{ is even} \end{cases}$$

Putting $n = 1,2,3,4,\ldots$ in (6), we get

$$(y_3)_0 = (3^2+a^2)(1^2+a^2)a,$$

$$(y_6)_0 = (4^2+a^2)(2^2+a^2)a^2 \text{ etc.}$$

Now putting all these values in the Maclaurin's theorem

$$y = (y)_0 + x.(y_1)_0 + \frac{x^2}{2!}(y_2)_0 + \ldots + \frac{x^n}{n!}(y_n)_0 + \ldots$$

We have

$$e^{a\sin^{-1}x} = 1 + ax + \frac{a^2}{2!}x^2 + \frac{a(1^2+a^2)}{3!}x^3$$

$$+\frac{a(2^2+a^2)}{4!}x^4 + \ldots$$

The general term is $\dfrac{x_n}{n!}(y_n)_0$.

Now putting $x = \sin\theta$ and $a = 1$, in the above equation, we get

$$e^{\theta} = 1 + \sin\theta + \frac{1}{2!}\sin^2\theta + \frac{2}{3!}\sin^3\theta + \ldots$$

Example 3. *Expand log sin(x+h) in powers of h by Taylor's theorem.*

Solution. Let $\quad f(x) = \log\sin(x)$

$$\Rightarrow \quad f(x+h) = \log\sin(x+h)$$

Expanding $f(x+h)$ by Taylor's theorem in powers of h, we have

$$f(x+h) = f(x) + hf'(x) + \frac{h^2}{2!}f''(x) + \frac{h^3}{3!}f'''(x) + \ldots \quad \ldots(1)$$

Now $\quad f(x) = \log\sin x$

$$\Rightarrow \quad f'(x) = \cot x$$

$$f''(x) = -\text{cosec}^2 x$$

$$\Rightarrow \quad f'''(x) = 2\,\text{cosec}^2 x \cot x \text{ etc.}$$

Substituting all these values in equation (1), we get

$$\log\sin(x+h) = \log\sin x + h\cot x - \frac{h^2}{2!}\text{cosec}^2 x$$

$$+\frac{2h^3}{3!}\text{cosec}^2 x\cot x + \ldots$$

Example 4. *Expand sin x in powers of $\left(x-\frac{\pi}{2}\right)$ with the help of Taylor's theorem.*

Solution . Let $f(x)=\sin x$.

Since, we want to expand $f(x)$ in powers of $\left(x-\frac{\pi}{2}\right)$, hence, we can write

$$f(x)=f\left[\frac{\pi}{2}+\left(x-\frac{\pi}{2}\right)\right]$$

Now, expanding by Taylor's theorem, we get

$$f(x)=f\left[\left(\frac{\pi}{2}\right)+\left(x-\frac{\pi}{2}\right)\right]$$

$$=f\left(\frac{\pi}{2}\right)+\left(x-\frac{\pi}{2}\right)f'\left(\frac{\pi}{2}\right)+\frac{1}{2!}\left(x-\frac{\pi}{2}\right)^2 f''\left(\frac{\pi}{2}\right)+\frac{1}{3!}\left(x-\frac{\pi}{2}\right)^3 f'''\left(\frac{\pi}{2}\right)+\ldots \quad \ldots(1)$$

Now $f(x)=\sin x \Rightarrow f\left(\frac{\pi}{2}\right)=1$

$f'(x)=\cos x \Rightarrow f'\left(\frac{\pi}{2}\right)=0$

$f''(x)=-\sin x \Rightarrow f''\left(\frac{\pi}{2}\right)=-1$

$f'''(x)=-\cos x \Rightarrow f'''\left(\frac{\pi}{2}\right)=0$

and so on.

Substituting all these values in (1), we get

$$\sin x=1+\left(x-\frac{\pi}{2}\right).0+\frac{1}{2!}\left(x-\frac{\pi}{2}\right)^2.(-1)+\frac{1}{3!}\left(x-\frac{\pi}{2}\right)^3.0+\frac{1}{4!}\left(x-\frac{\pi}{2}\right)^4+\ldots$$

$$=1-\frac{1}{2!}\left(x-\frac{\pi}{2}\right)^2+\frac{1}{4!}\left(x-\frac{\pi}{2}\right)^4+\ldots$$

Example 5. *If $f(x)=(x-a)^{5/2}$ and*

$$f(x+h)=f(x)+hf'(x)+\frac{h^2}{2!}f''(x+\theta h),$$

find the value of θ.

Solution . Here, we have

$$f(x)=(x-a)^{5/2} \Rightarrow f(x+h)=(x+h-a)^{5/2}$$

$$\Rightarrow f'(x)=\frac{5}{2}(x-a)^{3/2} \text{ and } f''(x)=\frac{15}{4}(x-a)^{1/2}$$

$$\therefore f''(x+\theta h)=\frac{15}{4}(x+\theta h-a)^{1/2}$$

Putting all these values in the given relation, we have

$$(x+h-a)^{5/2}=(x-a)^{5/2}+\frac{5}{2}h(x-a)^{3/2}+\frac{15}{4}(x+\theta h-a)^{1/2}\frac{h^2}{2!}$$

Now, taking limit as $x\to a$, we have

$$h^{5/2}=\frac{15}{4}(\theta h)^{1/2}\frac{h^2}{2!}$$

$$\Rightarrow \theta=\frac{64}{225}.$$

Example 6. *Let f is twice differentiable function and $|f|<\alpha$, $|f''|<\beta$, for $x>a$, then show that $|f'|<2\sqrt{\alpha\beta}$ $\forall x>a$.*

Solution . Let us suppose $x>a$ and $h>0$, then

$$f(x+h)=f(x)+hf'(x)+\frac{h^2}{2!}f''(x+\theta h), \quad 0<\theta<1$$

$$\Rightarrow hf'(x)=f(x+h)-f(x)-\frac{h^2}{2!}f''(x+\theta h)$$

$$\Rightarrow |hf'(x)|=\left|f(x+h)-f(x)-\frac{h^2}{2!}f''(x+\theta h)\right|$$

$$\le |f(x+h)|+|-f(x)|+\frac{h^2}{2!}|-f''(x+\theta h)|$$

[By using triangular inequality]

$$<\alpha+\alpha+\frac{h^2}{2}\beta=2\alpha+\frac{h^2}{2}\beta$$

$$\Rightarrow |f'(x)|<\frac{2\alpha}{h}+\frac{h}{2}\beta=F(h)(\text{say})$$

Now, $|f'(x)|$ is independent of h and also less than $F(h)$ for all values of h.

Therefore $|f'(x)|$ must be less than the minimum value of $F(h)$.

For, maxima or minima of $F(h)$, we have

$$F'(h)=0$$

$$\Rightarrow -\frac{2\alpha}{h^2}+\frac{\beta}{2}=0 \Rightarrow h=\pm 2\sqrt{\frac{\alpha}{\beta}}$$

and $F''(h)=\frac{2\alpha}{h^3}>0$ for $h=2\sqrt{\frac{\alpha}{\beta}}$

Hence $f(h)$ is minimum for $h=2\sqrt{\frac{\alpha}{\beta}}$, the minimum value of $F(h)$ is

$$=2\alpha.\frac{1}{2}\sqrt{\frac{\alpha}{\beta}}+\frac{\beta}{2}.2\sqrt{\frac{\alpha}{\beta}}=2\sqrt{\alpha\beta}$$

Hence $|f'(x)|<2\sqrt{\alpha\beta}$.

EXERCISE 2.1

1. Discuss the applicability of,Rolle's theorem of the following functions :
   (a) $f(x)=2+(x-1)^{2/3}$ in the interval [0,2]
   (b) $f(x)=x^2$ in $2\le x\le 3$
   (c) $f(x)=\tan x$ in $0\le x\le \pi$
   (d) $f(x)=x^4-3x^2+4$ in the interval [–4,4]
   (e) $f(x)=1/(x^2+1)$ in the interval [–3,3]
   (f) $f(x)=e^x \sin x$ in the interval $[0,\pi]$ (JNTU–2003)
   (g) $f(x)=|x|$ in the interval [–1,1]
   (h) $f(x)=(x-2)\sqrt{x}$ in the interval [0,2]
   (i) $f(x)=(x-a)^m(x-b)^n, m,n\in Z^+$ in the interval $[a,b]$. (VTU–2010, Nagarjuna–2008)
2. Show that between any two roots of $e^x \cos x = 1$, there exists at least one root of $e^x \sin x - 1 = 0$.
3. Let $\frac{a_0}{n+1}+\frac{a_1}{n}+\frac{a_2}{n-1}+\ldots+\frac{a_{n-1}}{2}+a_n=0.$
   Show that there exists at least one real x between 0 and 1 such that $a_0x^n+a_1x^{n-1}+\ldots+a_n=0$.
4. Verify the Rolle's theorem for the following functions:
   (a) $f(x)=x^4-1$ on the interval [–1,1]
   (b) $f(x)=e^x(\sin x-\cos x)$ in $\left(\frac{\pi}{4},\frac{5\pi}{4}\right)$
5. If $f(x)=\begin{vmatrix}\sin x & \sin\alpha & \sin\beta\\ \cos x & \cos\alpha & \cos\beta\\ \tan x & \tan\alpha & \tan\beta\end{vmatrix}$ where $0<\alpha<\beta<\frac{\pi}{2}$. Show that $f'(l)=0$, where $\alpha<l<\beta$.
6. A function $f(x)$ is continuous in the closed interval [0,1] and differentiable in the open interval]0,1[prove that
   $$f'(x_1)=f(1)-f(0),\ 0<x_1<1.$$
7. Show that the set of all x for which $\log(1+x)\le x$ is equal to $[0,\infty[$.
8. Compute the value of θ in the first mean value theorem $f(x+h)=f(x)+hf'(x+\theta h)$
   if $f(x)=ax^2+bx+c$.
9. Show that $x^n-a=0$ has atmost one real positive root if n is a positive integer.
10. Show that the function f', if it exists in an interval, cannot have an ordinary or removable discontinuity in that interval.
11. Verify the Lagrange's mean value theorem for the following functions :
    (a) $f(x)=x^3$ in [–1,1]
    (b) $f(x)=\sin x$ in $[0,\pi/2]$ (Nagpur–2008)
    (c) $f(x)=x^n$ in [–1,1], $n\in Z^+$
    (d) $f(x)=2x^2-7x+10, x\in[2,5]$
12. Find the value of c, of mean value theorem, when
    (a) $f(x)=\sqrt{x^2-4}$ in the interval [2,4]
    (b) $f(x)=2x^2+3x+4$ in the interval [1,2]
    (c) $f(x)=x(x-1)$ in the interval [1,2]
13. (a) If $f(x)=\sqrt{x}$ and $g(x)=1/\sqrt{x}$, then show by Cauchy's mean value theorem that c is the geometric mean between a and b.
    (b) If $f(x)=\frac{1}{x^2}$ and $g(x)=\frac{1}{x}$, then show that c is the harmonic mean between a and b.
14. If f'' exists and continuous on $[a,b]$ and differentiable on $]a,b[$, then prove that
    $$f(b)-f(a)-\frac{1}{2}(b-a)\{f'(a)-f'(b)\}=-\frac{(b-a)^3}{12}f''(d)$$
    where $d\in R$ such that $d\in]a,b[$.
15. Prove that
    $$\sin ax= ax-\frac{a^3x^3}{3!}+\frac{a^5x^5}{5!}-\ldots+\frac{a^{n-1}x^{n-1}}{(n-1)!}\sin\left(\frac{n-1}{2}.\pi\right)+\frac{a^nx^n}{n!}\sin\left(a\theta x+\frac{n\pi}{2}\right)$$
16. If $f(x)=f(0)+xf'(0)+\frac{x^2}{2!}f''(\theta x)$
    find the value of θ as $x\to 1$, $f(x)$ being $(1-x)^{5/2}$.
17. Show that the number θ which occurs in the Taylor's Theorem with Lagrange's form of remainder after n terms approaches the limit $\frac{f^{n+1}(a)}{(n+1)}$ as $h\to 0$ provided that $f^{n+1}(x)$ is continuous and different from zero as $x\to a$.
18. Show that the function x^3-3x^2+3x+2 is monotonically increasing in every interval.
19. Obtain by Maclaurin's theorem the expansion of $e^{\sin x}$.
20. If $f(x)=\exp\left[-\frac{1}{x^2}\right]$, for $x\ne 0$ and $f(0)=0$, then show that :
    (i) $f^n(0)=0\ \forall n=0,1,2,\ldots$ and
    (ii) The Taylor's series for f about 0 agrees with $f(x)$ only at $x=0$.
21. Expand "log sec x" by Maclaurin's series expansion, upto the term containing x^6. (VTU–2009, Mumbai–2000)
22. If $x>0$, show that
    $$x-\frac{x^2}{2}+\frac{x^3}{3(1+x)}<\log(1+x)<x-\frac{x^2}{2}+\frac{x^3}{3}.$$

Hint to Selected Problems

1. (a) Since $f'(x)$ does not exist at $x=1$, the second condition of Rolle's theorem is not satisfied.

2. Let a,b be two distinct roots of $e^x\cos x - 1 = 0$.

Then $e^a\cos a = 1$ and $e^b\cos b = 1$

Define a function $f(x) = e^{-x} - \cos x$.

5. $f(x)$ can be written as

$$f(x) = (\cos\alpha\tan\beta - \cos\beta\tan\alpha)\sin x - (\sin\alpha\tan\beta - \sin\beta\tan\alpha)\cos x + (\sin\alpha\cos\beta - \sin\beta\cos\alpha)\tan x$$

Since $\sin x, \cos x, \tan x$ have finite derivatives in $]0,\pi/2[$

$\Rightarrow f'(x)$ exists.

Also, $f(\alpha) = f(\beta)$. Hence, all the conditions of Rolle's theorem are satisfied.

7. Let us suppose $f(x) = \log(1+x) - x$

$\Rightarrow \quad f(0) = 0$

$$f'(x) = \frac{1}{1+x} - 1 = \frac{-x}{1+x} \le 0$$

$\Rightarrow f(x)$ is a decreasing function.

$\Rightarrow \quad f(x) \le f(0)\ \forall x \ge 0$

$\Rightarrow \quad \log(1+x) - x \le 0$

$\Rightarrow \quad \log(1+x) \le x$

9. $f'(x) = nx^{n-1}$. Clearly $f(x)$ is an increasing function.

Let $x_1, x_2 \in]0,\infty[$ and $0 < x < r < x_2$ such that $f(r) = 0$.

Then $f(x_1) < f(r) < f(x_2) \Rightarrow f(x_1) < 0 < f(x_2)$

$\Rightarrow$ If $x \ne r$, $f(x) \ne 0$ on $(0,\infty)$.

$\Rightarrow x^n - a$ has at most one real positive root.

14. Define two functions $g(x)$ and $h(x)$ such that

$$g(x) = f(x) - f(a) - \frac{1}{2}(x-a)\{f'(a) + f'(x)\} + A(x-a)^3$$

and
$$h(x) = \frac{1}{2}[f'(x) - f'(a)]^{-1/2}(x-a)f''(x) + 3A(x-a)^2$$

Clearly, $g(x)$ and $h(x)$ satisfying all conditions of Rolle's theorem. Then use Rolle's theorem for both the above functions.

18. Since $f'(x) \ge 0$ in $]-\infty,1]$. Hence, it is monotonically increasing.

22. $$f'(x) = \frac{1}{1+x} - 1 + x - \frac{3x^2 + 2x^3}{3(1+x^2)}$$

$$= \frac{x^3}{3(1+x)^2} > 0$$

f is increasing $\Rightarrow f(x) > f(0) = 0$ for $x > 0$

$$x - \frac{x^2}{2} + \frac{x^3}{3(1+x)} < \log(1+x) \text{ if } x > 0$$

Now, $$g'(x) = 1 - x + x^2 - \frac{1}{1+x} = \frac{x^3}{1+x} > 0$$

g is increasing $\Rightarrow g(x) > g(0)$

$$\log(1+x) < x - \frac{x^2}{2} + \frac{x^3}{3}$$

ANSWERS

1. (a) Not applicable (b) Not applicable (c) Not applicable (d) Verified (e) Verified (f) Verified (g) Not applicable (h) Verified (i) Verified

4. (a) Verified (b) Verified

8. $\theta = \frac{1}{2}$

11. (a) Verified (b) Verified (c) Verified (d) Verified

12. (a) $c = \sqrt{6}$ (b) $c = 3/2$

16. $\theta = \frac{9}{25}$

19. $y = 1 + x + \frac{x^2}{2} - \frac{x^4}{8} + \ldots$

21. $y = \frac{x^2}{2} + \frac{x^4}{12} + \frac{x^6}{45} + \ldots$

2.9 SOME MORE EXPANSIONS

Example 1. *Expand* $\tan^{-1} x$.

Solution . Let $f(x) = \tan^{-1} x \Rightarrow f(0) = 0$

$$f'(x) = \frac{1}{1+x^2} \Rightarrow f'(0) = 1$$

$$= (1+x^2)^{-1} = 1 - x^2 + x^4 - x^6 + \ldots$$

(By binomial expansion)

$$f''(x) = -2x + 4x^3 - 6x^5 + \ldots \Rightarrow f''(0) = 0$$

$$f'''(x) = -2 + 12x^2 - 30x^4 + \ldots \Rightarrow f'''(0) = -2$$

$$f^{iv}(x) = 24x - 120x^3 + \ldots \quad \Rightarrow f^{iv}(0) = 0$$

$$f^{v}(x) = 24 - 360x^2 + \ldots \quad \Rightarrow f^{v}(0) = 24$$

....

Put all these vaiues in Maclaurin's series, we get

$$\tan^{-1} x = x - \frac{x^3}{3} + \frac{x^5}{5} - \frac{x^7}{7} + \ldots$$

REMARKS

- To expand an alone inverse function, find its first derivative, expand by binomial theorem and then find other derivatives.
- The expansion of $\tan^{-1} x$ is valid only if $-1<x<1$.
- This expansion for $\tan^{-1} x$ known as Gregory's series, which is very useful in finding the value of π.
- In a like manner, we may get

$$\sin^{-1} x = x + \frac{1}{2}\cdot\frac{x^3}{3} + \frac{1.3}{2.4}\cdot\frac{x^5}{5} + \frac{1.3.5}{2.4.6}\cdot\frac{x^7}{7} + \ldots$$

Example 2. *If $y = \sin(m \sin^{-1} x)$, then show that*

$$\left(1-x^2\right)\frac{d^2y}{dx^2} - x\frac{dy}{dx} + m^2 y = 0$$

Hence, or otherwise expand $\sin m\theta$ in powers of $\sin \theta$. (SVTU–2008)

Solution . Here, we have

$$y = f(x) = \sin(m \sin^{-1} x) \qquad \ldots(1)$$

$$\Rightarrow \quad y_1 = \cos\left(m\sin^{-1} x\right)\cdot\frac{m}{\sqrt{1-x^2}} \qquad \ldots(2)$$

$$\Rightarrow (1-x^2){y_1}^2 = m^2\cos^2(m \sin^{-1}x)$$

$$\Rightarrow \quad (1-x^2){y_1}^2 = m^2[1-\sin^2(m \sin^{-1}x)]$$

$$\Rightarrow \quad (1-x^2){y_1}^2 = m^2(1-y^2)[\because y=\sin(m \sin^{-1} x)]$$

$$\Rightarrow \quad (1-x^2){y_1}^2 + m^2y^2 - m^2 = 0 \qquad \ldots(3)$$

Differentiating w.r.t. x, we get

$$(1-x^2)2y_1y_2 - 2x{y_1}^2 + 2m^2yy_1 = 0$$

$$\Rightarrow \quad 2y_1[(1-x^2)y_2 - xy_1 + m^2y] = 0$$

$$\Rightarrow \quad (1-x^2)y_2 - xy_1 + m^2y = 0 \qquad \ldots(4)$$

Now, differentiating (4) n times, using Leibnitz's theorem, we get

$$\left(1-x^2\right)y_{n+2} + n.y_{n+1}(-2x) + \frac{n(n-1)}{1.2}y_n(-2)$$

$$\Rightarrow \qquad - xy_{n+1} - n.y_n + m^2y_n = 0$$

$$\left(1-x^2\right)y_{n+2} - (2n+1)xy_{n+1} - \left(n^2 - m^2\right)y_n = 0 \qquad \ldots(5)$$

Now, put $x=0$ in (1), (2), (4) and (5), we get

$$y(0)=0,\ y_1(0)=m,\ y_2(0)+m^2y(0)=0$$

$$\Rightarrow \quad y_2(0)=0$$

and $\quad y_{n+2}(0)=(n^2-m^2)y_n(0). \qquad \ldots(6)$

Putting $n=2,4,6,\ldots$ in (6), we get

$$y_4(0)=(2^2-m^2)y_2(0)=0$$

$$y_6(0)=(4^2-m^2)y_4(0)=0$$

$$y_8(0)=0$$

.... and so on.

Here, we observe that $y_n(0)=0$ if n is even.

Now, putting $n=1,3,5,\ldots$ in (6), we get

$$y_3(0)=(1^2-m^2)y_1(0)=(1^2-m^2).m$$

$$y_5(0)=(3^2-m^2)y_3(0)=(3^2-m^2)(1^2-m^2).m$$

....

Putting all these values in Maclaurin's series, we get

$$\sin\left(m\sin^{-1} x\right) = mx + \frac{m\left(1^2-m^2\right)}{3!}x^3 + \frac{m\left(1^2-m^2\right)\left(3^2-m^2\right)}{5!}x^5 + \ldots$$

Let $\quad \theta=\sin^{-1}x \Rightarrow x=\sin\theta$

Then, we get

$$\sin m\theta = m\sin\theta + \frac{m(1^2-m^2)}{3!}\sin^3\theta + \frac{m(1^2-m^2)(3^2-m^2)}{5!}\sin^5\theta + \ldots$$

Example 3. *Expand $\tan x$ by Macluarin's theorem as far as x^5 and hence find the value of $\tan 46°30'$ upto four decimal places.* (VTU–2006)

Solution . Let $\quad f(x)=\tan x$

$$\Rightarrow \quad f(0)=0$$

$$f'(x)=\sec^2x=1+\tan^2x$$

$$\Rightarrow \quad f'(0)=1$$

$$f''(x)=2\tan x\sec^2x=2\tan x(1+\tan^2x)$$

$$=2\tan x+2\tan^3x$$

$$\Rightarrow \quad f''(0)=0$$

$$f'''(x)=2\sec^2x+6\tan^2x\sec^2x$$

$$=2(1+\tan^2x)+6\tan^2x(1+\tan^2x)$$

$$=2+8\tan^2x+6\tan^4x$$

$$\Rightarrow \quad f'''(0)=2$$

$$f^{iv}(x)=16\tan x\sec^2x+24\tan^3x\sec^2x$$

$$=8\sec^2x(2\tan x+3\tan^3x)$$

$$=8(1+\tan^2x)(2\tan x+3\tan^3x)$$

$$=16\tan x+40\tan^3x+24\tan^5x$$

$\Rightarrow \quad f^{iv}(0)=0$

and $\quad f^{v}(x)=16\sec^2 x+120\tan^2 x\sec^2 x$

$+120\tan^4 x\sec^2 x$

$=8\sec^2 x(2+15\tan^2 x+15\tan^4 x)$

$\Rightarrow \quad f^{v}(0)=16$

Now, putting all these values in Maclaurin's series'

$$f(x)= f(0)+xf'(0)+\frac{x^2}{2!}f''(0)+\frac{x^3}{3!}f'''(0)+\frac{x^4}{4!}f^{iv}(0)+\frac{x^5}{5!}f^{v}(0)+...$$

we get

$$\tan x = 0+x+\frac{x^3}{3!}.2+\frac{x^5}{5!}.16+...$$

$$\Rightarrow \tan x = x+\frac{x^3}{3}+\frac{2}{5}x^5+...$$

Deduction. Here

$$x=46°30'=\left(46\frac{1}{2}\right)^{\circ}=\left(\frac{93}{2}\right)^{\circ}=\frac{93}{2}\times\frac{\pi}{180}\text{ Radians}$$

$$=\frac{31}{120}\times\frac{22}{7}=\frac{31\times 11}{60\times 7}=\frac{314}{420}=0.812$$

Now, putting $x=46°30'=0.812$ in (1) , we get

$$\tan 46°30'=0.812+\frac{(0.812)^3}{3}+\frac{2}{15}(0.812)^5$$

$$=0.812+0.1784+0.047$$

$$=1.0374$$

Example 4. *Expand* $\log\{x+\sqrt{(1+x^2)}\}$ *in ascending powers of x and find the general term.*

Solution . Let $\quad y=\log\{x+\sqrt{(1+x^2)}\}$ ...(1)

$$\Rightarrow \quad y_1=\frac{1}{x+\sqrt{1+x^2}}.\left[1+\frac{2x}{2\sqrt{(1+x^2)}}\right]=\frac{1}{\sqrt{1+x^2}} \quad ...(2)$$

$$\Rightarrow \quad y_1^2(1+x^2)-1=0.$$

Differentiating again w.r.t. x, we get

$$2y_1[(1+x^2)y_2+xy_1]=0$$

$$\Rightarrow \quad [(1+x^2)y_2+xy_1]=0 \ (\because 2y_1\neq 0) \quad ...(3)$$

Using Leibnitz's theorem, differentiating (3) n times, we get

$$(1-x^2)y_{n+2}+n.y_{n+1}.2x+\frac{n(n-1)}{1.2}y_2.2+y_{n+1}.x+n.y_n=0$$

$$\Rightarrow \quad (1+x^2)y_{n+2}+(2n+1)xy_{n+1}+n^2y_n=0 \quad ...(4)$$

Putting $x=0$ in (1),(2),(3) and (4), we have

$$y(0)=0,\ y_1(0)=1,\ y_2(0)=0$$

$$y_{n+2}(0)=n^2y_n(0) \quad ...(5)$$

From (5), we have

$$y_3(0)=-1^2y_1(0)=-1^2$$

$$y_5(0)=(-3^2)y_3(0)=(-3^2)(-1^2)=3^2.1^2$$

$$y_7(0)=(-5^2)y_5(0)=(-5^2)(-3^2)(-1^2)$$

$$=-5^2.3^2.1^2 \quad \text{.... and so on.}$$

Putting n–2 for n in (5), we get

$$y_n(0)=\{-(n-2)^2\}y_{n-2}(0) \quad ...(6)$$

$$=[-(n-2)^2][-(n-4)^2]y_{n-4}(0).$$

Here we observe that

If n is odd, then

$$y_n(0)=[-(n-2)^2][-(n-4)^2]-...(-5^2)(-3^2)(-1^2).1$$

$$=[-1]^{(n-1)/2}(n-2)^2(n-4)^2...5^2.3^2.1^2 \quad ...(7)$$

Also from (5), we get

$$y_4(0)=-2^2.y_2(0)=0$$

$$y_6(0)=-4^2.y_4(0)=0 \text{ ... and so on.}$$

If n is even,

then, $\quad y_n(0)=0.$

Putting all these values in Maclaurin's series

$$y=y(0)+\frac{x}{1!}y_1(0)+\frac{x^2}{2!}y_2(0)+\frac{x^3}{3!}y_3(0)+...$$

we get $\log\left[x+\sqrt{(1+x^2)}\right]$

$$=x-\frac{x^3}{3!}.1^2+\frac{x^5}{5!}(3^2.1^2)-\frac{x^7}{7!}(5^2.3^2.1^2)+...$$

General term. The general term $=\frac{x^n}{n!}y_n(0)$

where

$$y_n(0)=\begin{cases}(-1)^{(n-1)/2}(n-2)^2(n-4)^2...5^2.3^2.1^2, & \text{if } n \text{ is odd}\\ 0 & \text{, if } n \text{ is even}\end{cases}$$

Example 5. *Expand log sin* $(x+h)$ *in power of h by Taylor's theorem.*

Solution . Let $f(x+h)=\log\sin(x+h)$

$$\left.\begin{aligned}\Rightarrow \quad f(x)&=\log\sin x\\ f'(x)&=\frac{1}{\sin x}.\cos x=\cot x\\ f''(x)&=-\text{cosec}^2 x\\ f'''(x)&=2\,\text{cosec}\,x\,\text{cosec}\,x\cot x\\ &=2\,\text{cosec}^2 x\cot x\\ ... \quad & ... \quad ... \quad ... \quad ...\end{aligned}\right\} \quad ...(1)$$

Now by Taylor's theorem, we have

$$f(x+h)= f(x)+hf'(x)+\frac{h^2}{2!}f''(x)+\frac{h^3}{3!}f'''(x)+... \quad ...(2)$$

Putting all the values from (1) in (2), we get

$$\log\sin(x+h)=\log\sin x+h\cot x-\frac{h^2}{2}\text{cosec}^2 x+\frac{h^3}{3}\text{cosec}^2\cot x+...$$

Example 6. *Expand sin x in powers of $\left(x-\frac{\pi}{2}\right)$ by using Taylor's series.*

Solution . Let $f(x)=\sin x$

We may write $f(x)=f\left[\frac{\pi}{2}+\left(x-\frac{\pi}{2}\right)\right]$

Now, expanding $f\left[\left(\frac{\pi}{2}\right)+\left(x-\frac{\pi}{2}\right)\right]$ by Taylor's theorem in powers of $\left(x-\frac{\pi}{2}\right)$ we get

$$f(x)=f\left[\left(\frac{\pi}{2}\right)+\left(x-\frac{\pi}{2}\right)\right]$$

$$=f\left(\frac{\pi}{2}\right)+\left(x-\frac{\pi}{2}\right)f'\left(\frac{\pi}{2}\right)+\frac{1}{2!}\left(x-\frac{\pi}{2}\right)^2 f''\left(\frac{\pi}{2}\right)+\frac{1}{3!}\left(x-\frac{\pi}{2}\right)^3 f'''\left(\frac{\pi}{2}\right)+... \quad ..(1)$$

Now $f(x)=\sin x \Rightarrow f\left(\frac{\pi}{2}\right)=\sin\frac{\pi}{2}=1$

$f'(x)=\cos x \Rightarrow f'\left(\frac{\pi}{2}\right)=\cos\frac{\pi}{2}=0$

$f''(x)=-\sin x \Rightarrow f''\left(\frac{\pi}{2}\right)=-\sin\frac{\pi}{2}=-1$

$f'''(x)=-\cos x \Rightarrow f'''\left(\frac{\pi}{2}\right)=-\cos\frac{\pi}{2}=0$

$f^{iv}(x)=\sin x$

$\Rightarrow f^{iv}\left(\frac{\pi}{2}\right)=\sin\frac{\pi}{2}=1$

....

Putting all these values in (1), we get

$$\sin x=1-\frac{1}{2!}\left(x-\frac{\pi}{2}\right)^2+\frac{1}{4!}\left(x-\frac{\pi}{2}\right)^4-...$$

Example 7. *Prove by Maclaurin's theorem, that*

$$e^{\sin x}=1+x+\frac{x^2}{1.2}-\frac{3.x^4}{1.2.3.4}+...$$

(VTU–2011, Bhopal–2009)

Solution . Let $f(x)=e^{\sin x} \Rightarrow f(0)=e^0=1$

$f'(x)=e^{\sin x}.\cos x \Rightarrow f'(0)=e^0\cos 0=1$

$f''(x)=e^{\sin x}(-\sin x)+\cos x\, e^{\sin x}\cos x$

$=e^{\sin x}[\cos^2 x-\sin x]$

$\Rightarrow f''(0)=e^0[1-0]=1$

$f'''(x)=e^{\sin x}[2\cos x(-\sin x)-\cos x]+e^{\sin x}\cos x.[\cos^2 x-\sin x]$

$=e^{\sin x}\cos x[-2\sin x-1+\cos^2 x-\sin x]$

$=-e^{\sin x}\cos x[3\sin x+\sin^2 x] \Rightarrow f'''(0)=0$

$f^{iv}(x)=-e^{\sin x}\cos x[3\cos x+2\sin x\cos x]+e^{\sin x}\sin x[3\sin x+\sin^2 x]-[3\sin x+\sin^2 x]\cos x\, e^{\sin x}\cos x$

$\Rightarrow f^{iv}(0)=-3$

Putting all these values in Maclaurin's theorem, given by

$$f(x)=f(0)+xf'(0)+\frac{x^2}{2!}f''(0)+\frac{x^3}{3!}f'''(0)+\frac{x^4}{4!}f^{iv}(0)+...$$

we get,

$$e^{\sin x}=1+x+\frac{x^2}{1.2}-\frac{3.x^4}{1.2.3.4}+...$$

Example 8. *Expand $e^{a\sin^{-1}x}$ by Macluarin's theorem and find the general term. Hence, show that*

$$e^{\theta}=1+\sin\theta+\frac{1}{2!}\sin^2\theta+\frac{2}{3!}\sin^3\theta+...$$

Solution . Let $y=e^{a\sin^{-1}x} \Rightarrow y(0)=1$

$$y_1=e^{a\sin^{-1}x}.\frac{a}{\sqrt{1-x^2}} \Rightarrow y_1(0)=a$$

Similarly, $y_2(0)=a^2$

and $y_{n+2}(0)=(n^2+a^2)y_n(0)$ ...(1)

Putting $n=1,2,3,...$in (1), we get

$y_3(0)=(1^2+a^2)y_1(0)=(1^2+a^2)a$

$y_4(0)=(2^2+a^2)y_2(0)=(2^2+a^2)a^2$

$y_5(0)=(3^2+a^2)y_3(0)$

$=(3^2+a^2)(1^2+a^2)a$

$y_6(0)=(4^2+a^2)y_4(0)$

$=(4^2+a^2)(2^2+a^2)a$....so on

In general

$$y_n(0)=\begin{cases}a^2(2^2+a^2)(4^2+a^2)...[(n-2)^2+a^2], \text{if } n \text{ is even}\\ a(1^2+a^2)(3^2+a^2)...[(n-2)^2+a^2], \text{ if } n \text{ is odd}\end{cases} \quad ...(2)$$

Substituting these values in Maclaurin's series

$$y=y(0)+x\,y_1(0)+\frac{x^2}{2!}y_2(0)+...+\frac{x^n}{n!}y_n(0)+...$$

we get $e^{a\sin^{-1}x} = 1 + ax + \frac{a^2}{2!}x^2 + \frac{a(1^2+a^2)}{3!}x^3 + \frac{a^2(2^2+a^2)}{4!}x^4 + \ldots$

General term : The general term$= \frac{x^n}{n!}y_n(0)$ where $y_n(0)$ is given by (2).

Now, putting $x=\sin\theta$ and $a=1$ in (3), we get

$$e^\theta = 1 + \sin\theta + \frac{1}{2!}\sin^2\theta + \frac{2}{3!}\sin^3\theta + \ldots$$

Example 9. (i) *If $f(x)=x^3+8x^2+15x-24$, calculate the value of $\left(\frac{11}{10}\right)$ by Taylor's series.*

(ii) *If $f(x)=x^3-2x+5$, find the value of $f(2.001)$ with the help of Taylor's theorem. Find the approximate change in the value of $f(x)$ when x changes from 2 to 2.001.*

Solution . (i) By Taylor's Theorem, we have

$$f(x+h) = f(x) + hf'(x) + \frac{h^2}{2!}f''(x) + \frac{h^3}{3!}f'''(x) + \ldots \quad ..(1)$$

We want to find $f\left(\frac{11}{10}\right)$ *i.e.*, $f\left(1+\frac{1}{10}\right)$

Put $x=1$ and $h=\frac{1}{10}$, and expand by Taylor's series, we get

$$f\left(\frac{11}{10}\right) = f\left(1+\frac{1}{10}\right) = f(1) + \frac{1}{10}f'(1) + \frac{1}{10^2}\cdot\frac{1}{2!}f''(1) + \frac{1}{3!}\frac{1}{(10)^3}f'''(1) + \ldots \quad \ldots(2)$$

Now $f(x)=x^3+8x^2+15x-24 \Rightarrow f(1)=0$

$f'(x)=3x^2+16x+15 \Rightarrow f'(1)=34$

$f''(x)=6x+16 \Rightarrow f''(1)=22$

$f'''(x)=6 \Rightarrow f'''(1)=6$

$f^{iv}(x)=0 \Rightarrow f^{iv}(1)=0$

Put all these values in (2), we get

$$f\left(1+\frac{1}{10}\right) = 0 + \frac{1}{10}.34 + \frac{11}{100} + \frac{1}{1000}$$

$$= 3.4+0.11+0.001=3.511.$$

(ii) Here put $x=2$ and $h=0.001$ in Taylor's series, we get

$$f(2.001) = f(2) + (0.001)f'(2) + \frac{(0.001)^2}{2!}f''(2) + \frac{(0.001)^3}{3!}f'''(2) + \ldots \quad \ldots(3)$$

Now $f(x)=x^3-2x+5$

$\Rightarrow f(2)=9$

$f'(x)=3x^2-2 \Rightarrow f'(2)=10$

$f''(x)=6x \Rightarrow f''(2)=12$

$f'''(x)=6 \Rightarrow f'''(2)=6$

$f^{iv}(x)=0 \Rightarrow f^{iv}(2)=0$

Put all these values in (2), we get

$$f(2.0001) = 9 + (0.001)10 + \frac{1}{2!}(0.001)^2(12) + \frac{1}{3!}(0.001)^3.6 + \ldots$$

$$=9+0.01+0.000006+0.000000001$$

$$= 9.010006001 = 9.01 \text{ approximately.}$$

Approximate value of

$f(2.001)-f(2)=9.01-9=0.01$ approximately.

Example 10. *Expand log $(1+\sin x)$ by Maclaurin's theorem in ascending power of x upto first five terms.*

(SVTU–2009, JNTU–2006)

Solution . Let $y = f(x)=\log(1+\sin x)$

By Maclaurin's expansion for $f(x)$, we have

$$y=f(x)= (y)_0 + \frac{x}{1!}(y_1)_0 + \frac{x^2}{2!}(y_2)_0 + \frac{x^3}{3!}(y_3)_0 + \frac{x^4}{4!}(y_4)_0 + \ldots \quad \ldots(1)$$

Now $y = \log(1+\sin x) \quad \therefore \quad (y)_0=0$

$$y_1 = \frac{\cos x}{1+\sin x} \Rightarrow (y_1)_0 = 1$$

$$y_2 = \frac{-\sin x(1+\sin x) - \cos^2 x}{(1+\sin x)^2} = -\frac{(1+\sin x)}{(1+\sin x)^2} = -\frac{1}{1+\sin x}$$

$\Rightarrow (y_2)_0=-1$

$$y_3 = \frac{\cos x}{(1+\sin x)^2} = -\frac{\cos x}{(1+\sin x)}\cdot\frac{1}{(1+\sin x)} = -y_1y_2$$

$\Rightarrow (y_3)_0=-1(-1)=1$

$y_4=-y_1y_3-y_2^2$

$\Rightarrow (y_4)_0= -1.1-(-1)^2=-1-1=-2$

$y_5=-y_1y_4-y_2y_3-2y_2y_3=-y_1y_4-3y_2y_3$

$\Rightarrow (y_5)_0=-1.(-2)-3(-1).1=2+3=5$ and so on.

Therefore, $\log(1+\sin x)$

$$= 0 + \frac{x}{1!}.1 + \frac{x^2}{2!}.(-1) + \frac{x^3}{3!}.1 + \frac{x^4}{4!}.(-2) + \ldots$$

$$= x - \frac{x^2}{2} + \frac{x^3}{6} - \frac{x^4}{12} + \frac{x^5}{25}\ldots$$

Example 12. *Expand sin (π/4+9) in powers of θ.*

Solution. Let $f(\theta)=\sin(\pi/4+\theta)$

$\Rightarrow$ $f(0)=\sin \pi/4=1/\sqrt{2}$

$f'(\theta)=\cos(\pi/4+\theta)$

$\Rightarrow$ $f'(0)=\cos \pi/4=1/\sqrt{2}$

$f''(\theta)=-\sin(\pi/4+\theta)$

$\Rightarrow$ $f''(0)=-\sin \pi/4=-1/\sqrt{2}$

$f'''(\theta)=-\cos(\pi/4+\theta)$

$\Rightarrow$ $f'''(0)=\cos \pi/4=-1/\sqrt{2}$

$f^{iv}(\theta)=\sin(\pi/4+\theta)$

$\Rightarrow$ $f^{iv}(0)=1/\sqrt{2}$ and so on.

The nth derivative of $f(\theta)$ is given by

$$f^n(\theta)=\sin\left(\theta+\frac{\pi}{4}+\frac{n\pi}{4}\right)$$

The Maclaurin's expansion of $f(\theta)$ with Lagrange's form of remainder is

$$f(\theta)=f(0)+\frac{\theta}{1!}f'(0)+\frac{\theta^2}{2!}f''(0)+\frac{\theta^3}{3!}f'''(0)+\ldots+\frac{\theta^{n-1}}{(n-1)!}f^{n-1}(0)+R_n \quad \ldots(1)$$

where $R_n=\frac{\theta^n}{n!}f^n(t.\theta)$

$$=\frac{\theta^n}{n!}\sin\left(t.\theta+\frac{\pi}{4}+\frac{n\pi}{2}\right),\quad 0<t<1.$$

Now $|R_n|=\left|\frac{\theta^n}{n!}\sin\left(t.\theta+\frac{\pi}{4}+\frac{n\pi}{2}\right)\right|$

$$=\left|\frac{\theta^n}{n!}\right|.\left|\sin\left(t.\theta+\frac{\pi}{4}+\frac{n\pi}{2}\right)\right|\le\left|\frac{\theta^n}{n!}\right|$$

$$\therefore \quad \lim_{n\to\infty}|R_n|\le\lim_{n\to\infty}\left|\frac{\theta^n}{n!}\right|=0$$

$$\left[\because \lim_{n\to\infty}\frac{\theta^n}{n!}=0\right]$$

$$\therefore \quad \lim_{n\to\infty}R_n=0$$

Thus all the conditions of Maclaurin's series expansion are satisfied. Hence, from (1), the expansion of sin (θ+π/4) is given by

$$\sin\left(\theta+\frac{\pi}{4}\right)=\frac{1}{\sqrt{2}}+\frac{\theta}{1!}\frac{1}{\sqrt{2}}+\frac{\theta^2}{2!}\left(-\frac{1}{\sqrt{2}}\right)+\frac{\theta^3}{3!}\left(-\frac{1}{\sqrt{2}}\right)+\ldots$$

$$\sin\left(\theta+\frac{\pi}{4}\right)=\frac{1}{\sqrt{2}}\left[1+\frac{\theta}{1!}-\frac{\theta^2}{2!}-\frac{\theta^3}{3!}+\frac{\theta^4}{4!}+\frac{\theta^5}{5!}-\frac{\theta^6}{6!}-\frac{\theta^7}{7!}+\ldots\right]$$

EXERCISE 2.2

1. Expand the following functions by Maclaurin's theorem :

(i) $\sec x$ (ii) $e^{x\cos x}$

(iii) $e^x\sec x$ (iv) $\log_e(1+e^x)$ (Bhopal–2008)

(v) $\log(1+\tan x)$

2. Apply Maclaurin's theorem to prove that

$$\log\sec x=\frac{1}{2}x^2+\frac{1}{12}x^4+\frac{1}{45}x^6+\ldots$$

3. If $y=\sin^{-1}x=a_0+a_1x+a_2x^2+\ldots$ Prove that

$$(n+1)(n+2)a_{n+2}=n^2a_n.$$

4. Show that :

(i) $e^x\cos x=1+x-\frac{2x^3}{3!}-\frac{2^2x^4}{4!}-\frac{2^2x^5}{5!}+\frac{2^3x^7}{7!}+\ldots+\cos\left(\frac{n\pi}{4}\right)\frac{2^{n/2}}{n!}x^n+\ldots$

(ii) $e^x\sin x=x+x^2+\frac{2x^3}{3!}-\frac{2^2x^5}{5!}-\ldots+\sin\left(\frac{n\pi}{4}\right)\frac{2^{n/2}}{n!}x^n+\ldots$

(iii) $e^{ax}\sin bx=bx+abx^2+\frac{3a^2b-b^3}{3!}x^3+\ldots+\frac{\left(a^2+b^2\right)^{\frac{n}{2}}}{n!}x^n\sin\left(n\tan^{-1}\frac{b}{a}\right)+\ldots$

(iv) $e^{ax}\cos bx=1+ax+\frac{a^2-b^2}{2!}x^2+\frac{a\left(a^2-3b^2\right)}{3!}x^3+\ldots+\frac{\left(a^2+b^2\right)^{\frac{n}{2}}}{n!}x^n\cos\left(n\tan^{-1}\frac{b}{a}\right)+\ldots$

5. Expand the following :

(i) $\tan^{-1}x$ in powers of $\left(x-\frac{\pi}{4}\right)$.

(ii) $2x^3+7x^2+x-1$ in powers of $x-2$. (Burdwan–2003)

(iii) $\sin^{-1}(x+h)$ in power of x.

(iv) $\log\sin x$ in power of $(x-a)$.

6. Show that $\log(x+h)=\log h+\frac{x}{h}-\frac{x^2}{2h^2}+\frac{x^3}{3h^3}-\ldots$

7. Use Taylor's theorem to prove that

$$\tan^{-1}(x+h)=\tan^{-1}x+h\sin\theta\frac{\sin\theta}{1}-(h\sin\theta)^2\frac{\sin 2\theta}{2}+(h\sin\theta)^3\frac{\sin 3\theta}{3}+\ldots+(-1)^{n-1}(h\sin\theta)^n\frac{\sin n\theta}{n}+\ldots$$

where $\theta=\cot^{-1}x$.

8. If $y=e^{\tan^{-1}x}$, show that

$(1+x^2)y_{n+2}+[2(n+1)x-1]y_{n+1}+n(n+1)y_n=0.$

Hence, or otherwise, find out the coefficient of x^5 if $e^{\tan^{-1}x}$ is expanded in powers of x.

9. Expand $(\sin^{-1}x)^2$ in ascending powers of x and deduce that

$$\theta^2=2.\frac{\sin^2\theta}{2!}+2^2.\frac{2\sin^4\theta}{4!}+2^2.4^2\frac{2\sin^6\theta}{6!}+\ldots$$

10. If $y=e^{m\tan^{-1}x}=a_0+a_1x+a_2x^2+\ldots+a_nx^n+\ldots$, show that $(n+1)a_{n+1}+(n-1)a_{n-1}=ma_n$.

11. If $e^{e^x}=a_0+a_1x+a_2x^2+\ldots+a_nx^n+\ldots$ show that

$$a_{n+1}=\frac{1}{n+1}\left[a_n+\frac{a_{n-1}}{1!}+\frac{a_{n-2}}{2!}+\ldots+\frac{a_{n-r}}{r!}+\ldots\frac{a_0}{n!}\right]$$

12. Show that

$$f(mx)=f(x)+(m-1)xf'(x)+\frac{(m-1)^2}{2!}x^2f''(x)+\frac{(m-1)^3}{3!}x^3f'''(x)+\ldots$$

13. By Maclaurin's theorem find the expansion of $y=\sin(e^x-1)$ upto and including the term in x. Find also the first non-vanishing terms in the expansion of x as a series ascending powers of y.

14. Prove that

$$f\left(\frac{x^2}{1+x}\right)=f(x)-\frac{x}{1+x}f'(x)+\left(\frac{x}{1+x}\right)^2\frac{1}{2!}f''(x)-\left(\frac{x}{1+x}\right)^3f'''(x)+\ldots$$

15. Calculate the approximate value of :

(i) $\sqrt{17}$ to four decimal places.

(ii) $\sqrt{26}$ to three decimal places

by Taylor's expansion.

Hint to Selected Problems

1. (i) $y=\sec x \Rightarrow y_1=\sec x\tan x=y\tan x$

$\Rightarrow y_1^2=y^2\tan^2x=y^2(\sec^2x-1)$

$=y^2(y^2-1)=y^4-y^2.$

Again differentiating, we get

$2y_1y_2=4y^3y_1-2yy_1\Rightarrow y_2=2y^3-y$

Similarly $y_3=6y^2y_1-y_1$

$y_4=12yy_1+6y^2y_2-y_2$

$y_5=12y_1^3-36y_1y_2-y_3$

... ...

Now putting $x=0$ in the above equation and use Maclaurin's series.

2. $y=\log\sec x \Rightarrow y_1=\tan x$

$y_2=\sec^2x=1+\tan^2x=1+y_1^2$

$y_3=2y_1y_2$

$y_4=2y_2^2+2y_1y_3$

$y_5=4y_2y_3+2y_2y_3+2y_1y_4$

$y_6=8y_2y_4+6y_3^2+2y_1y_5$

... ...

Now putting $x=0$ in the above equations and use Maclaurin's series.

3. $y=\sin^{-1}x$

$\Rightarrow y_1=\frac{1}{\sqrt{1-x^2}}\Rightarrow(1-x^2)y_1^2=1$

Again differentiating, we get

$(1-x^2)y_2-xy_1=0$

Now apply Leibnitz's theorem to differentiating n times.

4. $y=e^x\cos x$

$\Rightarrow y_1=e^x\cos x-e^x\sin x=e^x(\cos x-\sin x)$

$=re^x(\cos\theta\cos x-\sin\theta\sin x),$

where $r=\sqrt{2},\theta=\frac{\pi}{4}$

$\Rightarrow y_1=re^x\cos(x+\theta)$

$\Rightarrow y_2=r^2e^x\cos(x+2\theta)$

... ...

$y_n=r^ne^x\cos(x+n\theta)=(2)^{n/2}e^x\cos\left(x+\frac{n\pi}{4}\right)$

Now putting $x=0$ and use Maclaurin's series.

(ii) Here $y_n=2^{n/2}e^x\sin\left(x+\frac{n\pi}{4}\right)$

(iii) $y_n=\left(a^2+b^2\right)^{n/2}e^{ax}\sin\left[bx+n\tan^{-1}\left(\frac{b}{a}\right)\right]$

5. (i) $y=f(x)=\tan^{-1}x$

$\Rightarrow y_1=f'(x)=\frac{1}{1+x^2}\Rightarrow y_2=\frac{-2.x}{\left(1+x^2\right)^2}$

Putting $x=\pi/4$ and find $f(\pi/4), f'(\pi/4), f''(\pi/4)$... and so on.

Now $f(x)=f\left(\frac{\pi}{4}+x-\frac{\pi}{4}\right)$

Then expand by Taylor's theorem.

6. $y=f(x)=\log x$

$\Rightarrow f(x+h)=\log(x+h)$

$\Rightarrow f'(x)=\frac{1}{x}, f''(x)=-\frac{1}{x^2}, f'''(x)=\frac{2}{x^3}...$ and so on.

$\Rightarrow$Now putting $x=h$ in above derivatives and expand $f(x+h)$ by Taylor's theorem.

7. Take $y=f(x)=\tan^{-1}x$

$$\Rightarrow f'(x)=\frac{1}{1+x^2}=\frac{1}{2i}\left(\frac{1}{x-i}-\frac{1}{x+i}\right)$$

$$\Rightarrow f''(x)=\frac{1}{2i}\left[(-1)(x-i)^{2}-(-1)(x+i)^{-2}\right]$$

... ...

$$f^n(x)=\frac{1}{2i}\left[(-1)^{n-1}(n-1)!(x-i)^{-n}-(-1)^{n-1}(n-1)!(x+i)^{-n}\right]$$

Put $\theta=\cot^{-1}x$, we get

$f^n(x)=(-1)^{n-1}(n-1)!\sin^n\theta \sin n\theta$

Now use Taylor's series.

8. $y=f(x)=e^{\tan^{-1}x}$

$$\Rightarrow y_1=\frac{e^{\tan^{-1}x}}{1+x^2}=\frac{y}{1+x^2}\Rightarrow(1+x^2)y_1=y$$

$$(1+x^2)y_2+(2x-1)y_1=0$$

Now to find y_n, use Leibnitz's theorem.

9. Let $y=f(x)=(\sin^{-1}x)^2$

$$\Rightarrow y_1=\frac{2\sin^{-1}x}{\sqrt{1-x^2}}\Rightarrow(1-x^2)y_1^{\,2}=4\left(\sin^{-1}x\right)^2=4y$$

Now differentiating n times by Leibnitz's rule.

10. $y=f(x)=e^{m\tan^{-1}x}$

$$\Rightarrow y_1=\frac{me^{\tan^{-1}x}}{1+x^2}\Rightarrow(1+x^2)y_1=my$$

Now differentiating n times by using Leibnitz's theorem.

11. Let $y=f(x)=e^{e^x}\Rightarrow y_1=e^xe^{e^x}=e^x.y$

Now to find nth derivative, using Leibnitz's theorem.

12. Write $f(mx)=f[x+(m-1)x]$. Now expand by Taylor's theorem.

13. $y=f(x)=\sin(e^x-1) \Rightarrow y_1=e^x\cos(e^x-1)$ Now differentiating successively.

14. Write

$$f\left(\frac{x^2}{1+x}\right)=f\left(\frac{x^2}{1+x}-x+x\right)=f\left(x-\frac{x}{1+x}\right)$$

Now expand by Taylor's theorem.

15. Let $y=f(17)=\sqrt{17}=\sqrt{x}$

$y=f(16+1)$.

Now expand by Taylor's theorem.

ANSWERS

1. (i) $1+\frac{x^2}{2!}+\frac{5x^4}{4!}+\frac{61x^6}{6!}+...$ (ii) $1+x+\frac{x^2}{2}-\frac{x^3}{3}-\frac{11x^4}{24}-\frac{x^5}{5}+...$ (iii) $1+x+\frac{2x^2}{2!}+\frac{4x^3}{3!}+...$ (iv) $\log 2+\frac{x}{2}+\frac{x^2}{8}-\frac{x^4}{192}+...$

(v) $x-\frac{x^2}{2}+\frac{2}{3}x^3-\frac{7x^4}{12}+...$ **5.** (i) $\tan^{-1}\left(\frac{\pi}{4}\right)+\left(x-\frac{\pi}{4}\right)\Big/\left(1+\frac{\pi^2}{16}\right)-\pi\left(x-\frac{\pi}{4}\right)^2\Big/\left[4\left(1+\frac{\pi^2}{16}\right)^2\right]+...$

(ii) $45+53(x-2)+19(x-2)^2+2(x-2)^3+...$ (iii) $\sin^{-1}h+x\left(1-h^2\right)^{-1/2}+\frac{x^2}{2!}h\left(1-h^2\right)^{-3/2}+\frac{x^3}{3!}\left[\left(1-h^2\right)^{-5/2}\left(1+2h^2\right)\right]+...$

(iv) $\log\sin a+(x-a)\cot a-\frac{(x-a)^2}{2!}\text{cosec}^2 a+\frac{(x-a)^3}{3!}2\,\text{cosec}^2 a\cot a+...$

8. $\frac{1}{24}$ **9.** $2.\frac{x^2}{2!}+\frac{2.2^2}{4!}x^4+\frac{2.2^2.4^2}{6!}x^6+...+\frac{2.2^2.4^2...(2n-2)^2}{(2n)!}x^{2n}+...$ **13.** $x+\frac{x^2}{2!}-\frac{5x^4}{24}+...,y-\frac{y^2}{2}+...$ **15.** (i) 4.123 (ii) 5.099

Objective Evaluations

Fill in the Blanks

1. The function $f(x) = \frac{1}{x}$ is __________ in $-1<x<0$.
2. In some cases, Lagrange's mean value theorem is a particular case of __________ .
3. The first mean value theorem is also known as __________ .
4. If $f'(x)>0$ then $f(x)$ is __________ function.
5. If f is a finitely differentiable in a closed interval $[a,b]$ and $f'(a), f'(b)$ are of opposite sign then $f'(c) =$ __________ for at least one value of $c \in]a,b[$.

True/False

Write 'T' for True and 'F' for False statement.

1. Rolle's theorem is a particular case of Lagrange's theorem. **(T/F)**
2. Cauchy theorem is directly deducible from the Lagrange's mean value theorem. **(T/F)**
3. Lagrange's mean value theorem is a particular case of Cauchy's mean value theorem. **(T/F)**
4. Maclaurin's theorem is a particular case of Taylor's theorem. **(T/F)**

Multiple Choice Questions

Choose the most appropriate one.

1. If $f(x)$ is an even function then the value of $f'(0)$, if exists is equal to :
   (a) 1 (b) 0
   (c) 2 (d) ∞
2. If a function f is continuous on $[a, b]$, differentiable on $]a, b[$ and if $f'(x) = 0 \ \forall x \in]a, b[$ then $f(x)$ has a :
   (a) constant value throughout $[a,b]$
   (b) constant value only on the end points
   (c) constant value throughout $]a,b[$
   (d) none of the above
3. If $f(x)$ and $g(x)$ are continuous on $[a,b]$ and differentiable on $]a,b[$ and if $f'(x)=g'(x)$ throughout the interval $]a, b[$ then :
   (a) $f(x)=g(x) \ \forall x \in]a,b[$
   (b) $f(x) \neq g(x) \ \forall x \in]a,b[$
   (c) $f(x)$ and $g(x)$ differ only by a constant
   (d) none of these
4. If f is continuous on $[a,b]$ and $f'(x) \geq 0$ on $]a,b[$ then :
   (a) f is decreasing on $]a,b[$ (b) f is decreasing on $[a,b]$
   (c) f is increasing on $[a,b]$ (d) none of the above
5. If $f(x)$ is an increasing function on x, then :
   (a) $f'(x) \leq 0$ (b) $f'(x)=0$
   (c) $f'(x)>0$ (d) none of these
6. If $f'(x)$ is positive at a point $x=a$ then in the nbd of a :
   (a) $f(x)$ is positive (b) $f(x)$ is increasing
   (c) $f(x)$ is negative (d) none of these
7. The function $f(x)$ has equal values at the point $x=a$ and $x=b$ then :
   (a) there is a maximum of $f(x)$ between a and b
   (b) there is a minimum of $f(x)$ between a and b
   (c) there is a maximum or minimum of $f(x)$ between a and b
   (d) none of these
8. If $f''(x)>0$ at points in $]a,b[$ then the function f is :
   (a) strictly increasing (b) strictly decreasing
   (c) constant (d) none of these
9. If a function $f(x)$ satisfy the condition of mean value theorem and $f'(x)=0 \ \forall x \in]a,b[$ then :
   (a) $f(x)=0$
   (b) $f(x)$ is an increasing function
   (c) $f(x)$ is constant
   (d) none of these
10. The value of c of Rolle's theorem for the function $f(x)=\sin x$ in $[0,\pi]$ is given by :
    (a) $\pi/3$ (b) $\pi/2$
    (c) π (d) none of these
11. The value of c of Lagrange's mean value theorem for $f(x)=x(x-1)$ in $[1,2]$ is given by :
    (a) $\frac{1}{4}$ (b) $\frac{3}{2}$
    (c) $\frac{5}{4}$ (d) none of these
12. Lagrange's form of remainder after n terms in Taylor's development of the function e^x in a finite form in the interval $[a,a+h]$ is :
    (a) $\frac{h^n}{n!} e^{a+\theta h}$ (b) $\frac{h^{n+1}}{(n+1)!} e^{a+\theta h}$
    (c) $\frac{h^n}{n!} e^{\theta h}$ (d) none of these

13. Let f:R→R be a differentiable function. Which of the following will follow from mean value theorem ?

(a) for all $a,b \in R\ \exists$ some $c \in]a,b[$ such that $\frac{f(b)-f(a)}{b-a}=f'(c)$

(b) for all $a,b \in R\ \exists$ some $c \in]a,b[$ such that $\frac{f(b)+f(a)}{b+a}=f'(c)$

(c) both (a) and (b) are true

(d) none of these

14. If f is continuous on $[a,b]$ and f' exists at each $x \in]a,b[$ if $f'>0$ in $]a,b[$ then :

(a) f is strictly increasing (b) f is strictly decreasing

(c) f is constant (d) none of these

15. If f is continuous on $[a,b]$ and f' exists at each $x \in]a,b[$ then if $f'=0$:

(a) f is increasing (b) f is constant

(c) f is decreasing (d) none of these

16. Let $f:[a,b]\to R$ be continuous on $[a,b]$ and if f is differentiable on $]a,b[$. If $f(a)=f(b)=0\ \exists\ c$ such that $f'(c)=0$ then :

(a) $c \in]a,b[$ (b) $c \in [a,b]$

(c) $c \in]a,b]$ (d) none of the above

17. If f is continuous on $I(=[a,b])$ then there exists a real number u such that :

(a) $f(x)=u\ \forall x \in I$ (b) $f(x) \le u\ \forall x \in I$

(c) $f(x) \ge u\ \forall x \in I$ (d) none of these

18. Let $f:[a,b]\to R$ is continuous on $[a,b]$ and f is differentiable on $]a,b[$. If $f(a)=f(b)=0\ \exists\ c \in]a,b[$ such that $f'(c)=0$. It is known as :

(a) Rolle's theorem (b) Lagrange's theorem

(c) Cauchy's theorem (d) none of these

19. If f satisfies the conditions of Lagrange's mean value theorem and if $f'(x)>0\ \forall x \in]a,b[$, then which one is true ?

(a) f is increasing in $[a,b]$

(b) f is decreasing in $[a,b]$

(c) f is constant

(d) none of these

20. Which one of the following is correct by mean value theorem ?

(a) $1+x>e^x>1+xe^x\ \forall x>0$

(b) $1+x=e^x=1+xe^x\ \forall x>0$

(c) Both (a) and (b) are true

(d) none of the above

21. Assuming Rolle's theorem for the function $f(x)=\cos x$ in $\left[\frac{-\pi}{2},\frac{\pi}{2}\right]$. If there exists a real number $c \in \left]\frac{-\pi}{2},\frac{\pi}{2}\right[$. Then value of c is :

(a) $\pi/2$ (b) $\pi/4$

(c) π (d) 0

22. If f satisfies the conditions of Lagranges's mean theorem and if $f'(x)=0\ \forall x \in]a,b[$ then which one of the following statement is true ?

(a) f is constant on $[a,b]$

(b) f is strictly increasing on $[a,b]$

(c) f is strictly decreasing on $[a,b]$

(d) none of the above

23. In Lagrange's mean value theorem, the differential function exists :

(a) in $]a,b[$ (b) in $[a,b]$

(c) in $]a,b]$ (d) none of these

24. Does the mean value theorem apply to $f(x)=x^{1/3}$; $-1 \le x \le 1$:

(a) yes (b) no

(c) can't say (d) none of these

25. The value of $c=ax^2+bx+1$, $a \ne 0$ in $[1,p]$ by Lagrange's mean value theorem is :

(a) 5 (b) 15

(c) 18 (d) none of these

ANSWERS

Fill in the Blanks

1. decreasing **2.** Cauchy's mean value theorem **3.** Lagrange's mean value theorem **4.** increasing **5.** 0

True/False

1. T **2.** F **3.** T **4.** T

Multiple Choice Questions

1. (b)	**2.** (a)	**3.** (c)	**4.** (c)	**5.** (c)	**6.** (b)	**7.** (c)	**8.** (a)	**9.** (c)	**10.** (b)
11. (b)	**12.** (a)	**13.** (a)	**14.** (a)	**15.** (b)	**16.** (a)	**17.** (a)	**18.** (a)	**19.** (b)	**20.** (b)
21. (b)	**22.** (a)	**23.** (a)	**24.** (b)	**25.** (a)					

FFFFFF

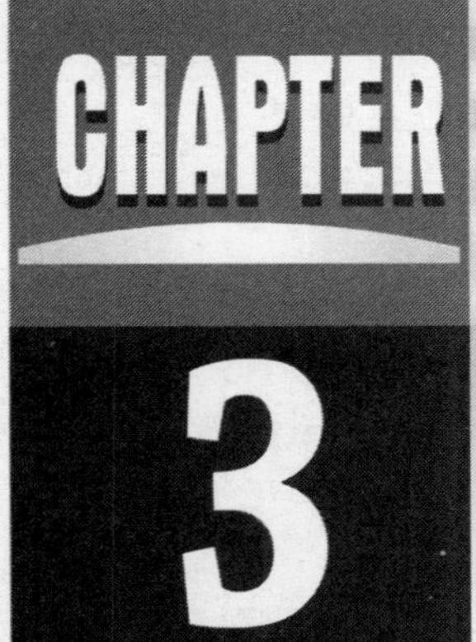

Indeterminate Forms

3.1 INTRODUCTION

When a function involves the independent variable in such a manner that for a certain assigned value of that variable, its value cannot be found by simply substituting that value of the variable, the function is said to take an indeterminate form.

The most common cases occuring is that of a fraction whose numerator and denominator both vanish for the value of the variable involved.

As $f(x) \to 0$ and $g(x) \to 0$ when $x \to a$, then the quotient $\dfrac{f(x)}{g(x)}$ is said to have attained the indeterminate form $\dfrac{0}{0}$.

Similarly if $\lim_{x\to a} f(x) = \infty$ and $\lim_{x\to a} g(x) = \infty$, then the fraction $\dfrac{f(x)}{g(x)}$ is said to have attained the indeterminate form $\dfrac{\infty}{\infty}$.

The other important indeterminate forms are $0 \times \infty$, $\infty - \infty$, 0^0, 1^∞ and ∞^0.

REMARKS

- The limiting value of the indeterminate forms is also called the true value.
- The most standard form among all indeterminate forms is $\dfrac{0}{0}$. We reduce all other cases of limits to this form.
- It will always be assumed that $f(x)$, $g(x)$, etc. and their respective derivatives are all continuous functions.
- The true value of the indeterminate form $\dfrac{0}{0}$ and $\dfrac{\infty}{\infty}$ is determined by the application of L' Hospital Rule.

3.2 L'HOSPITAL RULE FOR INDETERMINATE FORM $\dfrac{0}{0}$

To find $\lim_{x\to a} \dfrac{f(x)}{g(x)}$ *when* $\lim_{x\to a} f(x) = 0 = \lim_{x\to a} g(x)$.

Let us assume $f(x)$ and $g(x)$ be continuous at $x = a$, then, we have

$$f(a) = \lim_{x\to a} f(x) = 0,\ g(a) = \lim_{x\to a} g(x) = 0$$

By Taylor's theorem, we have

$$f(a+h) = f(a) + hf'(a+\theta_1 h) = hf'(a+\theta_1 h),\ 0 < \theta_1 < 1$$
$$g(a+h) = g(a) + hg'(a+\theta_2 h) = hg'(a+\theta_2 h),\ 0 < \theta_2 < 1$$

Therefore

$$\lim_{x\to a}\frac{f(x)}{g(x)} = \lim_{h\to 0}\frac{f(a+h)}{g(a+h)} = \lim_{h\to 0}\frac{hf'(a+\theta_1 h)}{hg'(a+\theta_2 h)} = \lim_{h\to 0}\frac{f'(a+\theta_1 h)}{g'(a+\theta_2 h)} = \frac{f'(a)}{g'(a)} \qquad \text{(Provided } g'(a) \neq 0\text{)}$$

$$= \lim_{x\to a}\frac{f'(x)}{g'(x)}$$

$$\Rightarrow \qquad \lim_{x \to a} \frac{f(x)}{g(x)} = \lim_{x \to a} \frac{f'(x)}{g'(x)}, \text{ provided } g'(a) \neq 0.$$

If $g'(a) = 0$, then this argument fails. The case when $g'(a) = 0$ but $f'(0) \neq 0$.

$$\lim_{x \to a} \frac{f'(x)}{g'(x)} \to +\infty \text{ or } -\infty$$

If $f'(a) = 0 = g'(a)$, then by Taylor's theorem, we have

$$f(a+h) = f(a) + hf'(a) + \frac{h^2}{2!} f''(a+\theta_3 h) = \frac{h^2}{2!} f''(a+\theta_3 h), \quad 0 < \theta_3 < 1$$

$$g(a+h) = g(a) + hg'(a) + \frac{h^2}{2!} g''(a+\theta_4 h) = \frac{h^2}{2!} g''(a+\theta_4 h), \quad 0 < \theta_4 < 1$$

$$\Rightarrow \qquad \lim_{x \to a} \frac{f(x)}{g(x)} = \lim_{h \to a} \frac{f(a+h)}{g(a+h)} = \lim_{h \to a} \frac{f''(a+\theta_3 h)}{g''(a+\theta_4 h)} = \frac{f''(a)}{g''(a)}, \text{ provided } g''(a) \neq 0.$$

The case of failure, when $g''(a) = 0$, the limit can be determined as before.

Now, in general if

$$f(a) = f'(a) = f''(a) = \ldots = f^{n-1}(a) = 0$$

$$g(a) = g'(a) = g''(a) = \ldots = g^{n-1}(a) = 0$$

and $\quad g^n(a) \neq 0.$

Then, by Taylor's theorem, we get

$$f(a+h) = f(a) + hf'(a) + \ldots + \frac{h^{n-1}}{(n-1)!} f^{n-1}(a) + \frac{h^n}{n!} f^n(a+\theta_n h), \quad 0 < \theta_n < 1$$

$$= \frac{h^n}{n!} f^n(a+\theta_n h).$$

and

$$g(a+h) = g(a) + hg'(a) + \ldots + \frac{h^{n-1}}{(n-1)!} g^{n-1}(a) + \frac{h^n}{n!} g^n(a+\theta_n' h),$$

$$= \frac{h^n}{n!} g^n(a+\theta_n' h). \qquad 0 < \theta_n' < 1$$

Therefore,

$$\lim_{x \to a} \frac{f(x)}{g(x)} = \lim_{h \to a} \frac{f(a+h)}{g(a+h)} = \lim_{x \to a} \frac{f^n(a+\theta_n h)}{g^n(a+\theta_n' h)} = \frac{f^n(a)}{g^n(a)}, \text{ if } g^n(a) \neq 0$$

$$\Rightarrow \qquad \lim_{x \to a} \frac{f(x)}{g(x)} = \lim_{x \to a} \frac{f^n(x)}{g^n(x)}, \text{ provided } g^n(a) \neq 0.$$

3.3 L'HOSPITAL RULE FOR INDETERMINATE FORM $\frac{\infty}{\infty}$

If $\lim_{x \to a} f(x) = \infty$ *and* $\lim_{x \to a} g(x) = \infty$, *then to prove that*

$$\lim_{x \to a} \frac{f(x)}{g(x)} = \lim_{x \to a} \frac{f'(x)}{g'(x)} \textit{ provided } \lim_{x \to a} \frac{f'(x)}{g'(x)} \textit{ exists.}$$

Proof. Consider

$$\lim_{x \to a} \frac{f(x)}{g(x)} = \lim_{x \to a} \frac{\frac{1}{g(x)}}{\frac{1}{f(x)}} = \lim_{x \to a} \left\{ \frac{-\frac{g'(x)}{[g(x)]^2}}{-\frac{f'(x)}{[f(x)]^2}} \right\} \qquad \left[\frac{0}{0} \text{form}\right]$$

[By L' Hospital rule]

$$\Rightarrow \qquad \lim_{x \to a} \frac{f(x)}{g(x)} = \lim_{x \to a} \frac{g'(x)}{f'(x)} \cdot \lim_{x \to a} \left[\frac{f(x)}{g(x)}\right]^2 \qquad \ldots(1)$$

Now, let $$\lim_{x\to a}\frac{f(x)}{g(x)}=l. \qquad \text{...(2)}$$

Then there are following three cases :

Case (i) If $l \neq 0$ and $l \neq \infty$.

In this case, (1) becomes

$$l=\lim_{x\to a}\frac{g'(x)}{f'(x)}.l^2$$

Dividing by l^2, we get $$\frac{1}{l}=\lim_{x\to a}\frac{g'(x)}{f'(x)}$$

$$\Rightarrow \qquad \lim_{x\to a}\frac{f'(x)}{g'(x)}=l=\lim_{x\to a}\frac{f(x)}{g(x)} \qquad \text{[Using (2)]}$$

Case (ii) If $l = 0$.

In this case, adding 1 to each side of (2), we get

$$l+1=\lim_{x\to a}\frac{f(x)}{g(x)}+1=\lim_{x\to a}\frac{f(x)+g(x)}{g(x)}=\lim_{x\to a}\frac{f'(x)+g'(x)}{g'(x)} \qquad \text{[By case (i)]}$$

$$=\lim_{x\to a}\frac{f'(x)}{g'(x)}+1$$

$$\Rightarrow \qquad l=\lim_{x\to a}\frac{f'(x)}{g'(x)}$$

Case (iii) Let $l=\infty$

In this case, by reciprocating, we have

$$\lim_{x\to a}\frac{g(x)}{f(x)}=0$$

By case (ii)

$$0=\lim_{x\to a}\frac{g(x)}{f(x)}=\lim_{x\to a}\frac{g'(x)}{f'(x)}$$

Therefore, $$\lim_{x\to a}\frac{g'(x)}{f'(x)}=\infty$$

Hence, the result $\lim_{x\to a}\frac{f(x)}{g(x)}=\lim_{x\to a}\frac{g'(x)}{f'(x)}$ has been established in every case.

REMARKS

- The above result can be extended to the case when $x\to\infty$, *i.e.*, we can show that

$$\lim_{x\to\infty}\frac{f(x)}{g(x)}=\lim_{x\to\infty}\frac{f'(x)}{g'(x)}$$

Let $x=\frac{1}{y}$ then

$$\lim_{x\to\infty}\frac{f(x)}{g(x)}=\lim_{y\to 0}\frac{f\left(\frac{1}{y}\right)}{g\left(\frac{1}{y}\right)}=\lim_{y\to 0}\frac{f'\left(\frac{1}{y}\right)\left(-\frac{1}{y^2}\right)}{g'\left(\frac{1}{y}\right)\left(-\frac{1}{y^2}\right)}=\lim_{y\to 0}\frac{f'\left(\frac{1}{y}\right)}{g'\left(\frac{1}{y}\right)}=\lim_{x\to\infty}\frac{f'(x)}{g'(x)}$$

- While evaluating $\lim_{x\to\infty}\frac{f(x)}{g(x)}$ when it is of the form $\frac{\infty}{\infty}$, care must be taken to change over to the form $\frac{0}{0}$ as early as possible, otherwise process of differentiating the numerator and denominator may never terminate.
- While appplying L' Hospital rule, we are not to differentiate $\frac{f(x)}{g(x)}$ by the rule for finding the differential coefficient of the quotient of two functions, but we are to differentiate the numerator and denominator separately.
- It must be remember that $\log 1=0$, $\log 0=-\infty$, and $\log\infty=\infty$.

Solved Examples

Example 1. *Find* $\lim_{x\to 0}\dfrac{e^x - e^{\sin x}}{x-\sin x}$.

Solution. We have $\lim_{x\to 0}\dfrac{e^x - e^{\sin x}}{x-\sin x}$ $\left|\dfrac{0}{0}\right.$ form

$$= \lim_{x\to 0}\frac{e^x - e^{\sin x}.\cos x}{1-\cos x} \quad \left|\frac{0}{0}\right. \text{form}$$

$$= \lim_{x\to 0}\frac{e^x - [\cos x\cdot e^{\sin x}\cdot\cos x + e^{\sin x}(-\sin x)]}{\sin x}$$

$$= \lim_{x\to 0}\frac{e^x - e^{\sin x}[\cos^2 x - \sin x]}{\sin x} \quad \left|\frac{0}{0}\right. \text{form}$$

$$= \lim_{x\to 0}\frac{e^x - e^{\sin x}[2\cos x(-\sin x) - \cos x] - [(\cos^2 x - \sin x)e^{x\sin x}\cos x]}{\sin x}$$

$$= \lim_{x\to 0}\frac{e^x - e^{\sin x}[-\sin 2x - \cos x + \cos^3 x - \sin x\cos x]}{\cos x}$$

$$= \frac{1-1(-1+1)}{1} = \frac{1}{1} = 1$$

Example 2. *Find* $\lim_{x\to 0}\dfrac{x\cos x - \log(1+x)}{x^2}$

Solution. We have $\lim_{x\to 0}\dfrac{x\cos x - \log(1+x)}{x^2}$

$$= \lim_{x\to 0}\frac{x\left(1-\frac{x^2}{2!}+\frac{x^4}{4!}-....\right)-\left(x-\frac{x^2}{2}+\frac{x^3}{3}-...\right)}{x^2} \quad \left|\frac{0}{0}\right. \text{form}$$

$$= \lim_{x\to 0}\left(\frac{\frac{x^2}{2}-\frac{5}{6}x^3+....}{x^2}\right)$$

$$= \lim_{x\to 0}\left(\frac{1}{2}-\frac{5}{6}x + \text{terms containing } x\right) = \frac{1}{2}.$$

Example 3. *Find* $\lim_{x\to 0}\dfrac{\cosh x - \cos x}{x\sin x}$.

Solution. Since we have $\lim_{x\to 0}\dfrac{\cosh x - \cos x}{x\sin x}$ $\left|\dfrac{0}{0}\right.$ form

$$= \lim_{x\to 0}\left[\left(\frac{\cosh x - \cos x}{x^2}\right)\left(\frac{x}{\sin x}\right)\right]$$

$$= \lim_{x\to 0}\frac{\cosh x - \cos x}{x^2} \quad \left|\frac{0}{0}\right. \text{form}$$

$$= \lim_{x\to 0}\frac{\sinh x + \sin x}{2x} \quad \left|\frac{0}{0}\right. \text{form}$$

$$= \lim_{x\to 0}\frac{\cosh x + \cos x}{2} = \frac{1+1}{2} = 1.$$

Example 4. *Find* $\lim_{x\to 0}\dfrac{(1+x)^{1/x} - e}{x}$.

Solution. We have $\lim_{x\to 0}\dfrac{(1+x)^{1/x} - e}{x}$ $\left|\dfrac{0}{0}\right.$ form

Evaluating the limit of expansion for $(1+x)^{1/x}$ in ascending power of x, we get

$$\lim_{x\to 0}\frac{(1+x)^{1/x} - e}{x}$$

$$= \lim_{x\to 0}\frac{e\left[1-\frac{1}{2}x+\frac{11}{24}x^2+...\right]-e}{x}$$

$$= \lim_{x\to 0}\frac{e\left[-\frac{1}{2}x+\frac{11}{24}x^2+...\right]}{x} = -\frac{1}{2}e.$$

Example 5. *Find* $\lim_{x\to 0}\dfrac{\log\sin 2x}{\log\sin x}$.

Solution. We have $\lim_{x\to 0}\dfrac{\log\sin 2x}{\log\sin x}$ $\left|\dfrac{\infty}{\infty}\right.$ form

$$= \lim_{x\to 0}\frac{\left(\frac{2}{\sin 2x}\right)\cos 2x}{\left(\frac{1}{\sin x}\right).\cos x} = \lim_{x\to 0}\frac{2\cot 2x}{\cot x} \quad \left|\frac{\infty}{\infty}\right. \text{form}$$

$$= \lim_{x\to 0}\frac{-4\,\text{cosec}^2 2x}{-\text{cosec}^2 x} \quad \left|\frac{\infty}{\infty}\right. \text{form}$$

$$= \lim_{x\to 0}\frac{4\sin^2 x}{\sin^2 2x} \quad \left|\frac{0}{0}\right. \text{form}$$

$$= \lim_{x\to 0}\frac{4\sin^2 x}{(2\sin x\cos x)^2} = \lim_{x\to 0}\frac{1}{\cos^2 x} = 1.$$

Example 6. *Find* $\lim_{x\to 0}\dfrac{\log\log(1-x^2)}{\log\log\cos x}$.

Solution. We have $\lim_{x\to 0}\dfrac{\log\log(1-x^2)}{\log\log\cos x}$ $\left|\dfrac{\infty}{\infty}\right.$ form

$$= \lim_{x\to 0}\frac{\frac{1}{\log(1-x^2)}\cdot\frac{1}{(1-x^2)}(-2x)}{\frac{1}{\log\cos x}\cdot\frac{1}{\cos x}\cdot(-\sin x)}$$

$$= 2\lim_{x\to 0}\frac{x\cos x\log\cos x}{\sin x(1-x^2)\log(1-x^2)}$$

$$= \left(2\lim_{x\to 0}\frac{x}{\sin x}\right)\left(\lim_{x\to 0}\frac{\cos x}{1-x^2}\right)\cdot\left(\lim_{x\to 0}\frac{\log\cos x}{\log(1-x^2)}\right)$$

$$= 2\times 1\times 1\times\lim_{x\to 0}\frac{\log\cos x}{\log(1-x^2)} \quad \left|\frac{0}{0}\right. \text{form}$$

$$= 2 \lim_{x \to 0} \frac{\frac{1}{\cos x}\cdot(-\sin x)}{\frac{1}{(1-x^2)}\cdot(-2x)}$$

$$= 2 \times \frac{1}{2} \cdot \lim_{x \to 0}\left(\frac{\sin x}{x} \cdot \frac{1-x^2}{\cos x}\right) = 1.$$

Example 7. *Find* $\lim_{x \to \infty} \frac{x^n}{e^x}$, *where n is a positive integer.*

Solution. We have

$$\lim_{x \to \infty} \frac{x^n}{e^x} = \lim_{x \to \infty} \frac{nx^{n-1}}{e^x} \qquad \left| \frac{\infty}{\infty} \text{ form} \right.$$

$$= \lim_{x \to \infty} \frac{n(n-1)x^{n-2}}{e^x} \qquad \left| \frac{\infty}{\infty} \text{ form} \right.$$

Repeating this process, we get

$$= \lim_{x \to \infty} \frac{[n(n-1)(n-2)\ldots n \text{ factors}]}{e^x}$$

$$= \lim_{x \to \infty} \frac{n!}{e^x} = \frac{n!}{e^\infty} = \frac{n!}{\infty} = 0.$$

Example 8. *Find* $\lim_{x \to \frac{\pi}{2}} \frac{\log\left(x - \frac{\pi}{2}\right)}{\tan x}$.

Solution. We have $\lim_{x \to \frac{\pi}{2}} \frac{\log\left(x - \frac{\pi}{2}\right)}{\tan x}$

$$= \lim_{x \to \frac{\pi}{2}} \left(\frac{\frac{1}{x - \pi/2}}{\sec^2 x}\right) \qquad \left| \frac{\infty}{\infty} \text{ form} \right.$$

$$= \lim_{x \to \frac{\pi}{2}} \left(\frac{\frac{1}{x - \pi/2}}{\frac{1}{\cos^2 x}}\right) = \lim_{x \to \frac{\pi}{2}} \left(\frac{\cos^2 x}{x - \pi/2}\right) \qquad \left| \frac{0}{0} \text{ form} \right.$$

$$= \lim_{x \to \frac{\pi}{2}} \left(\frac{-2\cos x \sin x}{1}\right) = -2\cos\frac{\pi}{2}.\sin\frac{\pi}{2} = 0.$$

Example 9. *Find the following limits :*

(*i*) $\lim_{x \to 0} \frac{\log x}{\cot x}$ (*ii*) $\lim_{x \to 0} \frac{\tan x - x}{x^2 \tan x}$

Solution. (i) We have $\lim_{x \to 0} \frac{\log x}{\cot x}$ $\left| \frac{\infty}{\infty} \text{ form} \right.$

$$= \lim_{x \to 0} \frac{1/x}{-\operatorname{cosec}^2 x} = -\lim_{x \to 0} \frac{\sin^2 x}{x}$$

$$= -\lim_{x \to 0}\left(\frac{\sin x}{x}\right) \cdot \sin x \qquad \left[\because \lim_{x \to 0}\left(\frac{\sin x}{x}\right) = 1\right]$$

$$= -1 \times 0 = 0.$$

(ii) We have $\lim_{x \to 0} \frac{\tan x - x}{x^2 \tan x}$

$$\left[\because \tan x = x + \frac{x^3}{3} + \frac{2}{15}x^5 + \ldots\right]$$

$$= \lim_{x \to 0} \frac{\left(x + \frac{x^3}{3} + \frac{2}{15}x^5 + \ldots\right) - x}{x^2\left(x + \frac{x^3}{3} + \frac{2}{15}x^5 + \ldots\right)}$$

$$= \lim_{x \to 0} \frac{\frac{x^3}{3} + \frac{2}{15}x^5 + \ldots}{x^3 + \frac{x^5}{3} + \frac{2}{15}x^7 + \ldots}$$

$$= \lim_{x \to 0} \frac{x^3\left(\frac{1}{3} + \frac{2}{15}x^2 + \ldots\right)}{x^3\left(1 + \frac{x^2}{3} + \frac{2}{15}x^4 + \ldots\right)}$$

$$= \lim_{x \to 0} \frac{\frac{1}{3} + \frac{2}{15}x^2 + \ldots}{1 + \frac{x^2}{3} + \frac{2}{15}x^4 + \ldots} = \frac{\frac{1}{3}}{1} = \frac{1}{3}$$

Example 10. *Evaluate* : $\lim_{x \to 1} \frac{x^x - x}{x - 1 - \log^x}$.

Solution.

$$\lim_{x \to 1} \frac{x^x - x}{x - 1 - \log^x} \qquad \left| \frac{0}{0} \text{ form} \right.$$

$$= \lim_{x \to 1} \frac{x^x(1 + \log x) - 1}{1 - \frac{1}{x}} \qquad \text{(By L' Hospital Rule)}$$

$$\left| \frac{0}{0} \text{ form} \right.$$

$$= \lim_{x \to 1} \frac{\frac{d}{dx}(x^x)(1 + \log x) + x^x\left(\frac{1}{x}\right) - 0}{1/x^2}$$

$$= \lim_{x \to 1} \frac{x^x(1 + \log x)^2 + x^x\left(\frac{1}{x}\right)}{x^{-2}}$$

$$= \frac{1.(1+0)^2 + 1.1}{1} = 2.$$

Working Aid

- Let $y = x^x$

$\Rightarrow \log y = x \log x$

$\therefore \frac{1}{y}\frac{dy}{dx} = x.\frac{1}{x} + 1.\log x$

$\Rightarrow \frac{d}{dx}(x^x) = x^x(1 + \log x)$

EXERCISE 3.1

1. Evaluate the following limits :

(i) $\lim_{x\to 0} \frac{x - \sin x}{x^3}$ (ii) $\lim_{x\to 0} \frac{1-\cos x}{x^2}$

(iii) $\lim_{x\to 0} \frac{a^x - b^x}{x}$ (iv) $\lim_{x\to 1} \frac{\log x}{x-1}$

(v) $\lim_{x\to 0} \frac{(1+x)^n - 1}{x}$

(vi) $\lim_{x\to 0} \frac{xe^x - \log(1+x)}{x^2}$ (VTU–2004, Osmania–2000)

(vii) $\lim_{x\to a} \frac{a^x - x^a}{x^x - a^a}$

(viii) $\lim_{x\to 0} \frac{5\sin x - 7\sin 2x + 3\sin 3x}{\tan x - x}$

(ix) $\lim_{x\to 0} \frac{\sin 2x + a\sin x}{x^2}$

(x) $\lim_{x\to 0} \frac{[\cosh x + \log(1-x) - 1 + x]}{x^2}$

(xi) $\lim_{x\to 1} \frac{x^5 - 2x^3 - 4x^2 + 9x - 4}{x^4 - 2x^3 + 2x - 1}$

2. Evaluate $\lim_{x\to 0} \frac{\sin x \sin^{-1} x - x^2}{x^6}$.

3. Evaluate $\lim_{x\to 0} \frac{(1+x)^{1/x} - e + \frac{1}{2}ex}{x^2}$.

4. Evaluate the following limits :

(i) $\lim_{x\to\infty} \frac{a^{1/x} - b^{1/x}}{\log\left(\frac{x}{x-1}\right)}$ (ii) $\lim_{x\to\pi/2} \frac{\left(\frac{\pi}{2} - x\right)^2 \sin x}{\cos^2 x}$

(iii) $\lim_{x\to 0} \frac{e^x + \log\left(\frac{1-x}{e}\right)}{\tan x - x}$ (iv) $\lim_{x\to 0+} \frac{3^x - 2^x}{\sqrt{x}}$

5. (i) If $\lim_{y\to 0} \frac{re^y - q\cos y + pe^{-y}}{y\tan y} = 3$, find the vlaues of p, q, and r. (Mumbai–2009)

(ii) Find the values of a and b in order that

$$\lim_{x\to 0} \frac{x(1 + a\cos x) - b\sin x}{x^3}$$

may be equatl to 1. (Mumbai–2007)

(iii) If $\lim_{x\to 0} \frac{\sin 2x + a\sin x}{x^3}$ be finite,

find the value of a and the limit. (Nagpur–2009)

6. Evaluate the following limits :

(i) $\lim_{x\to 0} \frac{\log x^2}{\cot x^2}$ (ii) $\lim_{x\to a} \frac{\log(x-a)}{\log(e^x - e^a)}$

(iii) $\lim_{x\to 1-0} \frac{\log(1-x)}{\cot \pi x}$ (iv) $\lim_{x\to\infty} \frac{\log x}{a^x}, a > 1$

(v) $\lim_{x\to\pi/2} \frac{\tan x}{\tan 3x}$

Hint to Selected Problems

1. (i) $\lim_{x\to 0} \frac{x - \sin x}{x^3} = \lim_{x\to 0} \frac{x - \left(x - \frac{x^3}{3!} + \frac{x^5}{5!} - \dots\right)}{x^3}$

$$= \lim_{x\to 0} \left\{ \frac{\frac{x^3}{3!} - \frac{x^5}{5!} + \dots}{x^3} \right\}$$

$$= \lim_{x\to 0} \left[\frac{1}{3!} - \frac{x^2}{5!} + \dots \right] = \frac{1}{3!} = \frac{1}{6}.$$

(vii) $\lim_{x\to a} \frac{a^x - x^a}{x^x - a^a} = \lim_{x\to a} \frac{a^x \log a - a.x^{a-1}}{x^x(1 + \log x)}$

$$= \frac{a^a \log a - a^a}{a^a(1 + \log a)}$$

$$= \frac{\log a - 1}{\log a + 1}.$$

5. (i) If $r - q + p = 0$, ...(1)

we obtained $\frac{0}{0}$ form, Then we have

$$\lim_{y\to 0} \frac{re^y + q\sin y + pe^{-y}}{y\sec^2 y + \tan y}$$

Again if $r - p = 0$, ...(2)

We obtained $\frac{0}{0}$ form. Then we have

$$\lim_{y\to 0} \frac{re^y + q\cos y + pe^{-y}}{\sec^2 y + 2y\sec^2 y\tan y + \sec^2 y}$$

$\Rightarrow$ $r + q + p = 6$ ...(3)

Now solving (1), (2) and (3).

6. (iv) $\lim_{x\to\infty} \frac{\log x}{a^x}, a > 1$ | form $\frac{\infty}{\infty}$

$$= \lim_{x\to\infty} \frac{1/x}{a^x \log a} = \frac{1}{\log a} \lim_{x\to\infty} \frac{1}{xa^x} = \frac{1}{\log a}.0 = 0.$$

ANSWERS

1. (i) $\frac{1}{6}$ (ii) $\frac{1}{2}$ (iii) $\log\frac{a}{b}$ (iv) 1 (v) n (vi) $\frac{3}{2}$ (vii) $\frac{\log a-1}{\log a+1}$ (viii) – 15 (ix) $\begin{cases}\infty \text{ if } a\neq -2\\ 0 \text{ if } a=-2\end{cases}$ (x) 0 (xi) 4

2. $\frac{1}{18}$ **3.** $\frac{11e}{24}$ **4.** (i) $\log\frac{a}{b}$ (ii) 1 (iii) $-\frac{1}{2}$ (iv) 0

5. (i) $p=\frac{3}{2}, q=3, r=\frac{3}{2}$ (ii) $a=-\frac{5}{2}, b=-\frac{3}{2}$ (iii) $\begin{cases}-1 \text{ if } a=-2\\ \infty \text{ if } a\neq -2\end{cases}$ **6.** (i) 0 (ii) 1 (iii) 0 (iv) 0 (v) 3

3.4 THE INDETERMINATE FORM $0\times\infty$

To find $\lim_{x\to a}[f(x).g(x)]$, *when* $\lim_{x\to a} f(x)=0$ *and* $\lim_{x\to a} g(x)=\infty$.

To determine this limit, the product may be transformed into the form $\frac{0}{0}$ or $\frac{\infty}{\infty}$, using any one of the following relations

$$f(x)\cdot g(x)=\frac{f(x)}{\dfrac{1}{g(x)}} \quad\text{or}\quad f(x)\cdot g(x)=\frac{g(x)}{\dfrac{1}{f(x)}}$$

and then apply previous method.

Solved Examples

Example 1. *Evaluate* $\lim_{x\to 0^+}(x\log x)$.

Solution.
$$\lim_{x\to 0^+}(x\log x)=\lim_{x\to 0^+}\frac{\log x}{1/x} \qquad \left|\frac{\infty}{\infty}\right. \text{ form}$$
$$=\lim_{x\to 0^+}\frac{1/x}{-1/x^2}$$
$$=\lim_{x\to 0^+}(-x)=0.$$

Example 2. *Evaluate* $\lim_{x\to 0} x\log\sin x$.

Solution.
$$\lim_{x\to 0} x\log\sin x \qquad |\,0\times\infty \text{ from}$$
$$=\lim_{x\to 0}\left(\frac{\log\sin x}{1/x}\right) \qquad \left|\frac{\infty}{\infty}\right. \text{ form}$$
$$=\lim_{x\to 0}\frac{(1/\sin x)\cdot\cos x}{-1/x^2} \qquad \left|\frac{\infty}{\infty}\right. \text{ form}$$
$$=\lim_{x\to 0}\frac{-x^2\cos x}{\sin x} \qquad \left|\frac{0}{0}\right. \text{ form}$$
$$=\lim_{x\to 0}\frac{x^2\sin x-2x\cos x}{\cos x}=0.$$

3.5 THE INDETERMINATE FORM $\infty - \infty$

To determine $\lim_{x\to a}[f(x)-g(x)]$, *when* $\lim_{x\to a} f(x)=\infty=\lim_{x\to\infty} g(x)$.

Here, this can be reduced to the form $\frac{0}{0}$ by the relation

$$f(x)-g(x)=\left\{\frac{\left[\dfrac{1}{g(x)}-\dfrac{1}{f(x)}\right]}{\dfrac{1}{f(x)\cdot g(x)}}\right\}$$

and then evaluate by previous method

WORKING PROCEDURE

Step 1. *Change all trigonometric-ratio into sin x and cos x (if T-ratio are present)*

Step 2. *Take L.C.M.*

Now the indeterminate form is reduced into $\frac{0}{0}$ form.

Solved Examples

Example 1. *Evaluate* $\lim_{x\to 0}\left(\frac{1}{x^2}-\frac{1}{\sin^2 x}\right)$.

Solution.
$$\lim_{x\to 0}\left(\frac{1}{x^2}-\frac{1}{\sin^2 x}\right) \quad \Big| \infty-\infty \text{ form}$$
$$= \lim_{x\to 0}\frac{\sin^2 x - x^2}{x^2\sin^2 x} \quad \Big|\frac{0}{0}\text{ form}$$
$$= \lim_{x\to 0}\frac{\left(x-\frac{x^3}{3!}+\ldots\right)^2 - x^2}{x^2\left(x-\frac{x^3}{3!}+\ldots\right)}$$
$$= \lim_{x\to 0}\frac{-\frac{2x^4}{3!}+\text{terms containing higher powers of } x}{x^4+\text{terms containing higher power of } x}$$
$$= \lim_{x\to 0}\frac{-\frac{2}{3!}+\text{terms containing } x \text{ in the numerator}}{1+\text{terms containing } x \text{ in the numerator}}$$
$$= -\frac{2}{3!} = -\frac{1}{3}$$

Example 2. *Evaluate* $\lim_{x\to\pi/2}(\sec x-\tan x)$.

Solution. We have $\lim_{x\to\pi/2}(\sec x-\tan x) \quad |\infty-\infty$ form
$$= \lim_{x\to\pi/2}\left(\frac{1}{\cos x}-\frac{\sin x}{\cos x}\right) \quad \Big|\frac{0}{0}\text{ form}$$
$$= \lim_{x\to\pi/2}\left(\frac{1-\sin x}{\cos x}\right)$$
$$= \lim_{x\to\pi/2}\frac{-\cos x}{-\sin x}$$
$$= \lim_{x\to\pi/2}\cot x = 0.$$

Example 3. *Evaluate* $\lim_{x\to\pi/2}\left(\sec x-\frac{1}{1-\sin x}\right)$.

Solution. We have $\lim_{x\to\pi/2}\left(\sec x-\frac{1}{1-\sin x}\right) \quad |\infty-\infty$ form
$$= \lim_{x\to\pi/2}\left(\frac{1}{\cos x}-\frac{1}{1-\sin x}\right) \quad |\infty-\infty \text{ form}$$
$$= \lim_{x\to\pi/2}\left(\frac{1-\sin x-\cos x}{\cos x-\cos x\sin x}\right) \quad \Big|\frac{0}{0}\text{ form}$$
$$= \lim_{x\to\pi/2}\frac{-\cos x+\sin x}{-\sin x+\sin^2 x-\cos^2 x}$$
$$= \frac{-0+1}{-1+1-0} = \infty.$$

Example 4. *Evaluate* $\lim_{x\to 0}(\tan x\log x)$. (VTU–2009)

Solution. We can write
$$\lim_{x\to 0}(\tan x\log x) = \lim_{x\to 0}\left(\frac{\log x}{\cot x}\right) \quad \Big|\text{form } \frac{\infty}{\infty}$$
$$= \lim_{x\to 0}\frac{1/x}{-\operatorname{cosec}^2 x}$$
(By L' Hospital Rule)
$$= -\lim_{x\to 0}\left(\frac{\sin^2 x}{x}\right)$$
$$= \lim_{x\to 0}\frac{2\sin x\cos x}{1} = 0.$$

3.6 THE INDETERMINATE FORM $0^0, 1^\infty, \infty^0$

To determine $\lim_{x\to a}[f(x)]^{g(x)}$ *when the limit is of the form* 0^0, 1^∞, ∞^0

Let
$$y = [f(x)]^{g(x)}$$
Taking logs;
$$\log y = g(x)\log f(x)$$
The RHS assumes the indeterminate forms $0\times\infty$ in each of these above cases. The limit can, therefore, be determined by the method used in the article 3.4.

Suppose $\lim_{x\to a}[g(x)\log f(x)] = l$ (say)

$\Rightarrow \quad \lim_{x\to a}\log y = l \Rightarrow \lim_{x\to a}[\log y] = l$

$\Rightarrow \quad \lim_{x\to a} y = e^l \Rightarrow \lim_{x\to a}[f(x)]^{g(x)} = e^l.$

WORKING PROCEDURE

Step 1. *Let the given limit* $= y$.

Step 2. *Take logs on both sides to get the forms* $0\times\infty$ *and proceed by the method of the type* $0\times\infty$.

Solved Examples

Example 1. *Evaluate* $\lim_{\theta\to\frac{\pi}{2}} (\cos\theta)^{\cos\theta}$.

Solution. Let $y = (\cos\theta)^{\cos\theta}$ | 0^0 form

Taking logs,

$$\log y = \cos\theta \log\cos\theta$$

$$\therefore \lim_{\theta\to\frac{\pi}{2}} (\log y) = \lim_{\theta\to\frac{\pi}{2}} \cos\theta\log\cos\theta \quad | \; 0\times\infty \text{ form}$$

$$= \lim_{\theta\to\frac{\pi}{2}} \frac{\log\cos\theta}{\sec\theta} \quad \left| \frac{\infty}{\infty} \text{ form}\right.$$

$$= \lim_{\theta\to\frac{\pi}{2}} \frac{\frac{1}{\cos\theta}\times -\sin\theta}{\sec\theta\tan\theta}$$

$$= \lim_{\theta\to\frac{\pi}{2}} (-\cos\theta) = 0$$

$$\Rightarrow \lim_{\theta\to\frac{\pi}{2}} (\log y) = 0$$

$$\Rightarrow \log\left(\lim_{\theta\to\frac{\pi}{2}} y\right) = 0 \Rightarrow \lim_{\theta\to\frac{\pi}{2}} y = e^0 = 1$$

$$\Rightarrow \lim_{\theta\to\frac{\pi}{2}} (\cos\theta)^{\cos\theta} = 1.$$

Example 2. *Find* $\lim_{x\to 0}\left(\frac{\tan x}{x}\right)^{1/x^2}$.

Solution. Let $y = \left(\frac{\tan x}{x}\right)^{1/x^2}$ | 1^∞ form for $x = 0$

$$\Rightarrow \log y = \frac{1}{x^2}\log\frac{\tan x}{x}$$

$$\Rightarrow \lim_{x\to 0}\log y = \lim_{x\to 0}\frac{1}{x^2}\log\frac{\tan x}{x}$$

$$= \lim_{x\to 0}\frac{\log\frac{\tan x}{x}}{x^2} \quad \left|\frac{0}{0}\text{ form}\right.$$

$$= \lim_{x\to 0}\left\{\frac{\frac{1}{\left(\frac{\tan x}{x}\right)}\left[\frac{x\sec^2 x - \tan x}{x^2}\right]}{2x}\right\}$$

$$= \lim_{x\to 0}\frac{x\sec^2 x - \tan x}{2x^3} \quad \left|\because \lim_{x\to 0}\frac{\tan x}{x} = 1\right|$$

$$= \lim_{x\to 0}\frac{x.2\sec x\sec x\tan x + \sec^2 x - \sec^2 x}{6x^2}$$

$$= \lim_{x\to 0}\frac{2x\tan x\sec^2 x}{6x^2} = \lim_{x\to 0}\frac{\tan x\sec^2 x}{3x}$$

$$= \lim_{x\to 0}\left(\frac{1}{3}\cdot\frac{\tan x}{x}\cdot\sec^2 x\right)$$

$$= \lim_{x\to 0}\frac{1}{3}\times 1\times\sec^2 x = \frac{1}{3}$$

$$\therefore \lim_{x\to 0} y = e^{1/3} \Rightarrow \lim_{x\to 0}\left(\frac{\tan x}{x}\right)^{1/x^2} = e^{1/3}.$$

Example 3. *Evaluate* $\lim_{x\to 0}\left(\frac{\sin x}{x}\right)^{1/x}$.

Solution. Let $y = \lim_{x\to 0}\left(\frac{\sin x}{x}\right)^{1/x}$

$$\therefore \log y = \lim_{x\to 0}\left(\frac{1}{x}\log\frac{\sin x}{x}\right)$$

$$= \lim_{x\to 0}\frac{1}{x}\log\left\{\frac{x - \frac{x^3}{3!} + \frac{x^5}{5!} - \dots}{x}\right\}$$

$$= \lim_{x\to 0}\frac{1}{x}\log\left(1 - \frac{x^2}{3!} + \frac{x^4}{5!} - \dots\right)$$

$$= \lim_{x\to 0}\frac{1}{x}\log\left[1 - \left(\frac{x^2}{6} - \frac{x^4}{120} + \dots\right)\right]$$

$$= \lim_{x\to 0}\frac{1}{x}\log(1 - z) \text{ where } z = \frac{x^2}{6} - \frac{x^4}{120} + \dots$$

$$= \lim_{x\to 0}\frac{1}{x}\left(-z - \frac{z^2}{2} - \dots\right)$$

$$= \lim_{x\to 0}\frac{1}{x}\left[-\left(\frac{x^2}{6} - \frac{x^4}{120} + \dots\right) - \frac{1}{2}\left(\frac{x^2}{6} - \frac{x^4}{120} + \dots\right)^2 - \dots\right]$$

$$= \lim_{x\to 0}\frac{1}{x}\left[-\frac{x^2}{6} + \left(\frac{x^4}{120} - \frac{x^4}{72}\right) + \dots\right]$$

$$= \lim_{x\to 0}\frac{1}{x}\left[-\frac{x^2}{6} + \frac{x^4}{180} + \dots\right]$$

$$= \lim_{x\to 0}\left[-\frac{x}{6} + \frac{x^3}{180} + \dots\right] = 0$$

Hence, $y = e^0 = 1.$

Example 4. *Evaluate* $\lim_{x\to 0} (\operatorname{cosec} x)^{1/\log x}$.

Solution. We have

$$y = \lim_{x\to 0} (\operatorname{cosec} x)^{1/\log x} \qquad |\infty^0 \text{ form}$$

$$\therefore \quad \log y = \lim_{x\to 0} \frac{1}{\log x}(\log \operatorname{cosec} x) \qquad \left|\frac{\infty}{\infty}\right. \text{form}$$

$$= \lim_{x\to 0} \frac{\left(\frac{1}{\operatorname{cosec} x}\right)(-\operatorname{cosec} x \cot x)}{1/x}$$

$$= \lim_{x\to 0}\left(-\frac{x}{\tan x}\right) \qquad \left|\frac{0}{0}\right. \text{form}$$

$$= \lim_{x\to 0}\left(-\frac{1}{\sec^2 x}\right) = -1$$

$$\Rightarrow \quad y = e^{-1} = \frac{1}{e}.$$

Example 5. *Evaluate* $\lim_{x\to 0}\left(\frac{\sin x}{x}\right)^{1/x^3}$. (VTU–2001)

Solution. Let

$$y = \lim_{x\to 0}\left(\frac{\sin x}{x}\right)^{1/x^3}$$

$$\Rightarrow \quad \log y = \lim_{x\to 0}\left(\frac{1}{x^3}\log\frac{\sin x}{x}\right)$$

$$= \lim_{x\to 0}\frac{1}{x^3}\log\left\{\frac{x - \frac{x^3}{3!} + \frac{x^5}{5!} - \dots}{x}\right\}$$

$$= \lim_{x\to 0}\frac{1}{x^3}\log\left(1 - \frac{x^2}{3!} + \frac{x^4}{5!} - \dots\right)$$

$$= \lim_{x\to 0}\frac{1}{x^3}\log(1-z)$$

$$\left[\text{where } z = \frac{x^2}{6} - \frac{x^4}{120} - \dots\right]$$

$$= \lim_{x\to 0}\frac{1}{x^3}\left(-z - \frac{z^2}{2} - \dots\right)$$

$$= \lim_{x\to 0}\frac{1}{x^3}\left[-\left(\frac{x^2}{6} - \frac{x^4}{120} + \dots\right) - \frac{1}{2}\left(\frac{x^2}{6} - \frac{x^4}{120} + \dots\right)^2 - \dots\right]$$

$$= \lim_{x\to 0}\frac{1}{x^3}\left[-\frac{x^2}{6} + \left(\frac{x^4}{120} - \frac{x^4}{72}\right) + \dots\right]$$

$$= \lim_{x\to 0}\frac{1}{x^3}\left[-\frac{x^2}{6} - \frac{x^4}{180} + \dots\right]$$

$$= \lim_{x\to 0}\frac{1}{x^3}\left[-\frac{1}{6x} - \frac{x}{180} + \dots\right] = \infty$$

$$\Rightarrow \quad y = e^{\infty} = \infty.$$

Example 6. (*a*) *Evaluate the following limits :*

(*i*) $\lim_{x\to 0} (\cos x)^{\cot x}$

(*ii*) $\lim_{x\to 0} \frac{e^x - e^{-x} - 2\log(1+x)}{x\sin x}$

(*b*) *If* $f(x) = (x-1)(x-3)(x-5)$, $a = 0$, $b = 4$, *find the value c such that* $f'(c)$ *has the same value as the slope of the chord joining the points for which* $x = 0$ *and* $x = 4$.

Solution. (a) (i) Let $y = \lim_{x\to 0} (\cos x)^{\cot x}$

$$\log y = \lim_{x\to 0} \log(\cos x)^{\cot x}$$

$$= \lim_{x\to 0} \cot \log \cos x$$

$$\log y = \lim_{x\to 0}\frac{\log\cos x}{\tan x}$$

$$= \lim_{x\to 0}\frac{(-\sin x)}{\cos x \cdot \sec^2 x}$$

$$= \lim_{x\to 0}\frac{-\sin x}{\sec x}$$

$$= \lim_{x\to 0} -\sin x\cos x$$

$$= \lim_{x\to 0} -\frac{\sin 2x}{2\cdot x}.x$$

$$= -\lim_{x\to 0}\left(\frac{\sin 2x}{2x}\right)\cdot \lim_{x\to 0} x = -1\times 0.$$

$$\therefore \log y = 0 \Rightarrow y = e^0 \Rightarrow y = 1.$$

(ii) We have

$$\lim_{x\to 0}\frac{e^x - e^{-x} - 2\log(1-x)}{x\sin x} \qquad \left|\frac{0}{0}\right. \text{form}$$

$$= \lim_{x\to 0}\frac{e^x + e^{-x} - \frac{2}{1+x}}{x\cos x + \sin x} \qquad \left|\frac{0}{0}\right. \text{form}$$

$$= \lim_{x\to 0}\frac{e^x - e^{-x} + \frac{2}{(1+x)^2}}{2\cos x - x\sin x}$$

$$= \frac{e^0 - e^0 + \frac{2}{(1+0)^2}}{2\cdot\cos 0 - 0}$$

$$= \frac{1 - 1 + \frac{2}{(1+0)^2}}{2\cdot\cos 0 - 0} = \frac{2}{2} = 1.$$

(b) Same as (a).

Example 7. *Find* $\lim_{x\to 0}\left(\frac{\tan x}{x}\right)^{1/x^3}$.

Solution. Let $y = \lim_{x\to 0}\left(\frac{\tan x}{x}\right)^{1/x^3}$

$$\log y = \lim_{x\to 0}\frac{1}{x^3}\log_e\left(\frac{\tan x}{x}\right)$$

$$= \lim_{x\to 0}\frac{\log_e \tan x - \log_e x}{x^3} \quad \left|\frac{0}{0}\right. \text{ form}$$

$$= \lim_{x\to 0}\frac{\frac{\sec^2 x}{\tan x}-\frac{1}{x}}{3x^2} = \lim_{x\to 0}\frac{\frac{2x}{\sin 2x}-\frac{1}{x}}{3x^2}$$

$$= \lim_{x\to 0}\frac{2x-\sin 2x}{3x^2\sin 2x} \quad \left|\frac{0}{0}\right. \text{ form}$$

$$= \lim_{x\to 0}\frac{2-2\cos 2x}{6x^2\sin 2x+6x^3\cos 2x} \quad \left|\frac{0}{0}\right. \text{ form}$$

$$= \lim_{x\to 0}\frac{2\sin 2x}{15x^2\cos 2x+6x\sin 2x-6x^3\sin 2x} \quad \left|\frac{0}{0}\right. \text{ form}$$

$$= \lim_{x\to 0}\frac{4\cos 2x}{-30x^2\sin 2x+30x\cos 2x+6\sin 2x+12x\cos 2x-18x^2\sin 2x-12x^3\cos 2x}$$

$$= \frac{4}{0} = \infty$$

Hence, $y = e^{\infty} = \infty$.

EXERCISE 3.2

1. Evaluate the following limits :

(i) $\lim_{x\to 0} x\log\tan x$ (ii) $\lim_{x\to 0}\tan\left(\frac{\pi}{2}-x\right)$

(iii) $\lim_{x\to\infty} 2^x\sin\frac{a}{2^x}$ (iv) $\lim_{x\to\infty}(a^{1/x}-1)\cdot x$

(v) $\lim_{x\to 1}\sec\frac{\pi}{2x}\log x$

(vi) $\lim_{x\to 0} x^m(\log x)^n \; m;n\in Z^+$.

2. Evaluate the following limits :

(i) $\lim_{x\to 0}\left[\frac{1}{x}-\frac{1}{x^2}\log(1+x)\right]$ (ii) $\lim_{x\to 2}\left[\frac{1}{x-2}-\frac{1}{\log(x-1)}\right]$

(iii) $\lim_{x\to 0}\left[\frac{1}{x^2}-\text{cosec}^2 x\right]$ (iv) $\lim_{x\to 0}\left(\frac{1}{x^2}-\cot^2 x\right)$

(v) $\lim_{x\to 0}\left(\frac{1}{x^2}-\frac{1}{x\tan x}\right)$ (vi) $\lim_{x\to 0}\left(\text{cosec}\, x-\frac{1}{x}\right)$.

3. Evaluate the following limits :

(i) $\lim_{x\to 0}\left(\frac{1}{x}\right)^{\tan x}$ (ii) $\lim_{x\to 0} x^x$

(iii) $\lim_{x\to\infty}\left(\frac{\pi}{2}-\tan^{-1}x\right)^{1/x}$

(iv) $\lim_{x\to 0}\left(\frac{\tan x}{x}\right)^{1/x}$ (v) $\lim_{x\to\pi/2}(\sin x)^{\tan x}$

(vi) $\lim_{x\to\pi/4}(\tan x)^{\tan 2x}$ (VTU–2004)

(vii) $\lim_{x\to 0}\left[\frac{2(\cosh x-1)}{x^2}\right]^{1/x^2}$

(viii) $\lim_{x\to 0}(\text{cosec}\, x)^{1/\log x}$

(ix) $\lim_{x\to\alpha}\left(2-\frac{x}{a}\right)^{\tan\left(\frac{\pi x}{2a}\right)}$ (VTU–2010, Nagpur–2009)

(x) $\lim_{x\to\infty}\left(1+\frac{k}{x}\right)^x$

4. Evaluate the following limits :

(i) $\lim_{x\to 0}\left[\frac{a^x+b^x}{2}\right]^{1/x}$ (VTU–2007)

(ii) $\lim_{x\to 0}\left[\frac{a_1^x+a_2^x+\ldots+a_n^x}{n}\right]^{1/x}$ (VTU–2011)

(iii) $\lim_{x\to\infty}\left(1+\frac{a}{x}\right)^x$.

Hint to Selected Problems

1. (iii) $\lim_{x\to\infty} 2^x.\sin\frac{a}{2^x} = \lim_{x\to\infty}\frac{\sin\left(\frac{a}{2^x}\right)}{\frac{1}{2^x}}$

$$= a\lim_{x\to\infty}\left[\frac{\sin\left(\frac{a}{2^x}\right)}{\frac{a}{2^x}}\right] = a\cdot 1 = a.$$

(v) $\lim_{x\to 1}\sec\left(\frac{\pi}{2x}\right).\log x = \lim_{x\to 1}\frac{\log x}{\cos\left(\frac{\pi}{2x}\right)}$

$$= \lim_{x\to 1}\frac{\frac{1}{x}}{\frac{\pi}{2x^2}\sin\left(\frac{\pi}{2}x\right)} = \lim_{x\to 1}\frac{2x}{\pi\sin\left(\frac{\pi}{2x}\right)} = \frac{2\times 1}{\pi\cdot\sin\left(\frac{\pi}{2}\right)} = \frac{2}{\pi}.$$

2. (v) $\lim\limits_{x\to 0}\left(\dfrac{1}{x^2}-\dfrac{1}{x\tan x}\right)=\lim\limits_{x\to 0}\dfrac{\tan x - x}{x^2\tan x}$

$$=\lim_{x\to 0}\frac{\left(x+\dfrac{x^3}{3}+\dfrac{2}{15}x^5+\ldots\right)-x}{x^2\left(x+\dfrac{x^3}{3}+\dfrac{2}{15}x^5+\ldots\right)}$$

$$=\lim_{x\to 0}\frac{x^3\left(\dfrac{1}{3}+\dfrac{2}{15}x^2+\ldots\right)}{x^3\left(1+\dfrac{x^2}{3}+\dfrac{2}{15}x^4+\ldots\right)}$$

$$=\lim_{x\to 0}\frac{\dfrac{1}{3}+\dfrac{2}{15}x^2+\ldots}{1+\dfrac{x^2}{3}+\dfrac{2}{15}x^4+\ldots}=\frac{1}{3}$$

3. (i) $\log y=\lim\limits_{x\to 0}\left(\dfrac{1}{x}\right)^{\tan x}$

$$\Rightarrow \log y=\lim_{x\to 0}\tan x\log\left(\frac{1}{x}\right)$$

$$=-\lim_{x\to 0}\tan x\log x=-\lim_{x\to 0}\frac{\log x}{\cot x}$$

$$=-\lim_{x\to 0}\frac{1/x}{-\text{cosec}^2 x}=\lim_{x\to 0}\frac{\sin^2 x}{x}$$

$$=\lim_{x\to 0}\left(\frac{\sin x}{x}\right).\sin x=1\times 0=0$$

$$\Rightarrow \quad y=e^0=1.$$

ANSWERS

1. (i) 0 (ii) ∞ (iii) a (iv) $\log a$ (v) $-\dfrac{2}{\pi}$ (vi) 0.

2. (i) $\dfrac{1}{2}$ (ii) $-\dfrac{1}{2}$ (iii) $-\dfrac{1}{3}$ (iv) $\dfrac{2}{3}$ (v) $\dfrac{1}{3}$ (vi) 0.

3. (i) 1 (ii) 1 (iii) 1 (iv) 1 (v) 1. (vi) $\dfrac{1}{e}$ (vii) $e^{1/12}$

(viii) $\dfrac{1}{e}$ (ix) $e^{2/\pi}$ (x) e^k. **4.** (i) $\sqrt{ab}$ (ii) $(a_1.\,a_2\ldots.a_n)^{1/n}$ (iii) e^a.

Objective Evaluations

Fill in the Blanks

1. $\lim_{x\to 1} \frac{\log x}{x-1} =$ ________
2. $\lim_{x\to 0} \frac{\sin ax}{\cos ax} =$ ________
3. $\lim_{x\to \infty} \frac{x^2+2x}{5-3x^2} =$ ________
4. $\lim_{x\to 0} \frac{\tan x}{x} =$ ________
5. $\lim_{x\to 0} x \log x =$ ________
6. $\lim_{x\to 1} \left(\sec \frac{\pi}{2x}\right) \log x =$ ________
7. $\lim_{x\to 0} x^x =$ ________
8. $\lim_{x\to \infty} \left(\frac{1}{x}\right)^{1/x} =$ ________
9. $\lim_{x\to 1} x^{\frac{1}{x-1}} =$ ________
10. $\lim_{x\to 0} (1+x)^{1/x} =$ ________

True/False

Write 'T' for True and 'F' for False statement.

1. The indeterminate form $\frac{\infty}{\infty}$ can be converted into the form $\frac{0}{0}$. **(T/F)**
2. 1° is not an indeterminate form. **(T/F)**
3. 0° is not an indeterminte form. **(T/F)**
4. $\infty+\infty$ and $\infty \times \infty$ are indeterminate form. **(T/F)**
5. By L' Hospital rule to find $\lim_{x\to a} \frac{f(x)}{g(x)}$, if the form is $\frac{0}{0}$, we are differentiate $\frac{f(x)}{g(x)}$ is a fraction. **(T/F)**
6. $\lim_{x\to 0} (1+nx)^{1/x}$ is e^n. **(T/F)**
7. $\lim_{x\to 0} \frac{x^2+2x-2}{x\sin^3 x}$ is $\frac{1}{6}$. **(T/F)**
8. If $\lim_{x\to a} f(x) = \infty = \lim_{x\to a} g(x)$ then $\lim_{x\to a} \frac{f(x)}{g(x)} = \lim_{x\to a} \frac{f'(x)}{g'(x)}$. **(T/F)**

Multiple Choice Questions

Choose the most appropriate one.

1. $\lim_{x\to 0} \frac{\tan x}{x}$ is :
   (a) 0 (b) ∞ (c) 1 (d) – 1
2. $\lim_{x\to 0} (1+nx)^{1/x}$ is :
   (a) 1 (b) e^{-n} (c) e^2 (d) e^n
3. $\lim_{x\to 1} \frac{x^5-2x^3-4x^2+9x-4}{x^4-2x^3+2x-1}$ is:
   (a) 1 (b) 2 (c) 3 (d) 4
4. $\lim_{x\to 0} \frac{\log(1-x^2)}{\log\cos x}$ is :
   (a) 1 (b) 2 (c) 3/2 (d) 2/3
5. $\lim_{x\to 0} \frac{\sin ax}{\sin bx}$ is :
   (a) 1 (b) 0 (c) a/b (d) b/a
6. $\lim_{x\to 0} \frac{x-\sin x}{x^3}$ is :
   (a) 1/2 (b) 1/3 (c) 1/5 (d) 1/6
7. $\lim_{x\to 0} \frac{x^2+2\cos x-2}{x\sin x}$ is :
   (a) 1/12 (b) 1/6 (c) 2/5 (d) 1
8. $\lim_{x\to 1} \frac{\log x}{x-1}$ is :
   (a) 0 (b) –1 (c) 1 (d) ∞
9. $\lim_{x\to 1} \frac{\sin x \sin^{-1} x}{x^2}$ is :
   (a) 0 (b) – 1 (c) 1 (d) ∞
10. $\lim_{x\to 0} \frac{x\cos x - \log(1+x)}{x^2}$ is :
    (a) 0 (b) 1 (c) 1/2 (d) 1/3

11. $\lim_{x \to a} \left(\frac{a^x - b^x}{x} \right)$ is :

(a) 1 (b) ∞ (c) $\log b/a$ (d) $\log a/b$

12. $\lim_{x \to 0} \frac{\log x}{\cot x}$ is :

(a) 1 (b) -1 (c) 0 (d) ∞

13. $\lim_{x \to \pi/2} \frac{\log(x - \pi/2)}{\tan x}$ is :

(a) 1 (b) 0 (c) -1 (d) ∞

14. $\lim_{x \to \pi/2} \frac{\tan 5x}{\tan x}$ is :

(a) 1 (b) 5 (c) 1/5 (d) -1

15. $\lim_{x \to 1} (1-x)\tan\frac{\pi x}{2}$ is:

(a) 1 (b) 2π (c) π (d) $2/\pi$

16. $\lim_{x \to \infty} (a^{1/x} - 1)x$ is :

(a) 1 (b) $\log a$ (c) a (d) 0

17. $\lim_{x \to 0} x \log x$ is :

(a) 1 (b) -1 (c) 0 (d) ∞

18. $\lim_{x \to 0} \left(\frac{1}{x} - \cot x \right)$ is :

(a) 1 (b) -1 (c) 0 (d) ∞

19. $\lim_{x \to 0} \left[\frac{1}{e^x - 1} - \frac{1}{x} \right]$ is :

(a) 1 (b) 1/2 (c) -1 (d) $-1/2$

20. $\lim_{x \to 0} \left[\frac{\cot x - 1/x}{x} \right]$ is :

(a) 1 (b) -1 (c) 3 (d) $-1/3$

21. $\lim_{x \to \pi/2} \left[\sec x - \frac{1}{(1 - \sin x)} \right]$ is :

(a) 0 (b) -1 (c) ∞ (d) $-\infty$

22. $\lim_{x \to \infty} \left[1 + \frac{a}{x} \right]^x$ is :

(a) 0 (b) 1 (c) e^a (d) e^{-a}

23. $\lim_{x \to 0} (1+x)^{1/x}$ is :

(a) 0 (b) 1 (c) $1/e$ (d) e

24. $\lim_{x \to \pi/2} (\sin x)^{\tan x}$ is :

(a) 0 (b) $e^{\pi/2}$ (c) 1 (d) -1

25. $\lim_{x \to 1} (x)^{1/x-1}$ is :

(a) e^{-2} (b) e^2 (c) e (d) e^{-1}

26. $\lim_{x \to 0} \left(\frac{\tan x}{x} \right)^{1/x}$ is :

(a) 0 (b) 1 (c) -1 (d) ∞

27. $\lim_{x \to 0} (a^x + x)^{1/x}$ is :

(a) $2ae$ (b) a/e (c) ae (d) $(ae)^2$

28. $\lim_{x \to 0} x^x$ is :

(a) 0 (b) 1 (c) -1 (d) ∞

29. $\lim_{x \to \infty} (1/x)^{1/x}$ is :

(a) 0 (b) 1 (c) -1 (d) ∞

30. $\lim_{x \to 0} (\cot x)^{\sin x}$ is :

(a) 0 (b) 1 (c) -1 (d) ∞

ANSWERS

Fill in the Blanks

1. 1	**2.** a/b	**3.** $-1/3$	**4.** 1	**5.** 0	**6.** $2/\pi$	**7.** 1	**8.** 1	**9.** e^{-1}	**10.** e.

True/False

1. T	**2.** T	**3.** F	**4.** F	**5.** F	**6.** T	**7.** F	**8.** T

Multiple Choice Questions

1. (c)	**2.** (d)	**3.** (d)	**4.** (b)	**5.** (c)	**6.** (d)	**7.** (a)	**8.** (c)	**9.** (c)	**10.** (c)
11. (d)	**12.** (c)	**13.** (b)	**14.** (c)	**15.** (d)	**16.** (b)	**17.** (c)	**18.** (d)	**19.** (d)	**20.** (d)
21. (c)	**22.** (c)	**23.** (d)	**24.** (c)	**25.** (c)	**26.** (b)	**27.** (c)	**28.** (b)	**29.** (b)	**30.** (b)

FFFFFF

4 Tangent and Normal

4.1 INTRODUCTION

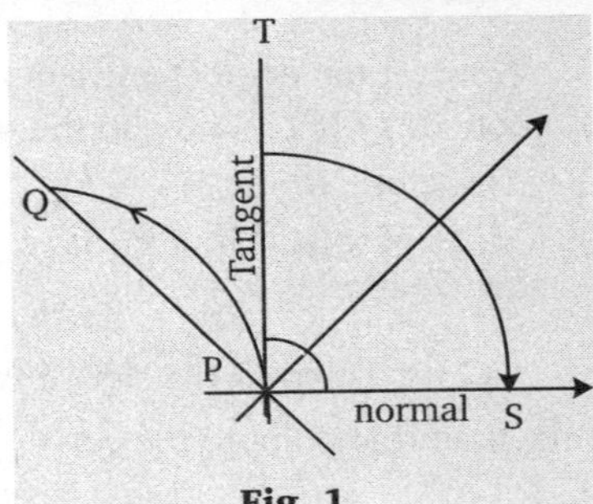

Fig. 1

Let P be a given point and Q be any other point on it. Let Q travel towards P along the curve.

Let Q travel towards P along the curve. T

hen, the limiting position PT of the secant PQ is known as the tangent to the curve.

The line PS through P which is perpendicular to the tangent PT is called the normal of the curve.

4.1.1 SOME FUNDAMENTAL CONCEPTS

(i) Slope of a line, $m = \tan\theta$, where θ is the angle which the line makes with the positive direction of x-axis.

(ii) Slope of the line $ax + by + c = 0$ is given by $m = -\frac{a}{b}$

(iii) Slope of the line joining the points (x_1, y_1) and (x_2, y_2) is $= \frac{y_2 - y_1}{x_2 - x_1}$

(iv) Slope of x-axis $= 0$, Slope of y-axis $= \infty$

(v) Two lines are parallel iff $m_1 = m_2$.

(vi) Two lines are perpendicular iff $m_1 m_2 = -1$.

(vii) Angle between two lines having slopes m_1 and m_2 is given by $\theta = \tan^{-1}\left(\frac{m_1 - m_2}{1 + m_1 m_2}\right)$

(viii) Equation of the line (one point form)

$$y - y_1 = m(x - x_1)$$

passing through the point (x_1, y_1).

(ix) Perpendicular distance formula $= \frac{|ax_1 + by_1 + c|}{\sqrt{a^2 + b^2}}$

4.1.2 EQUATION OF THE TANGENT

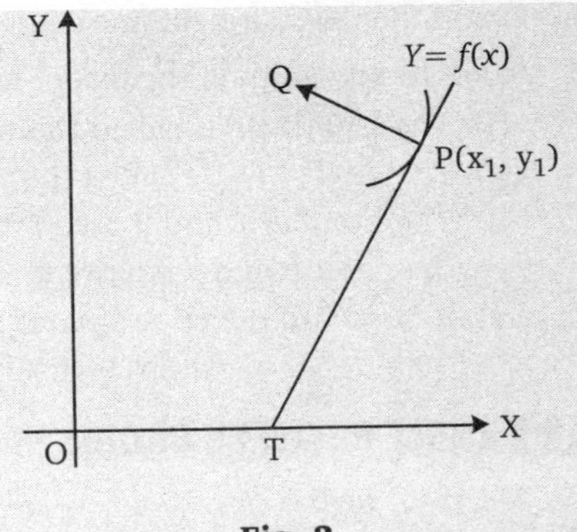

Fig. 2

Let $y = f(x)$ be the equation of the curve, and $P(x_1, y_1)$ be any given point (x_1, y_1) on this curve.

Let $Q = Q(x + \delta x, y + \delta y)$ be any neighbouring point of P. Let PT be the tangent at the point (x_1, y_1).

The slope of the tangent at $(x_1, y_1) = \frac{dy_1}{dx_1}$.

Now, tangent is a line through the point $P(x_1, y_1)$ and its slope $m = \frac{dy_1}{dx_1}$.

Hence, by Co-ordinate Geometry, the equation of the tangent is

$$y - y_1 = \frac{dy_1}{dx_1}(x - x_1).$$

REMARKS

- It should be clearly understood that by $\frac{dy_1}{dx_1}$ we mean the value of $\frac{dy}{dx}$ at (x_1, y_1) and not as derivative of y_1 with respect to x_1.
- The equation of the tangent at a point t_1 to the curve $x = f(t), y = g(t)$ is given by

$$y - g(t_1) = \frac{g'(t_1)}{f'(t_1)}[x - f(t_1)].$$

4.1.2 GEOMETRICAL MEANING OF $\frac{dy}{dx}$

Let $y = f(x)$ be the given function and let it be represented by the curve AB. Take two neighbouring points $P(x, y)$ and $Q(x+\delta x, y+\delta y)$ on the curve AB. Join PQ and let PQ be produced to meet OX at the point R.

Slope of the secant PQ

$$= \frac{y + \delta y - y}{x + \delta x - x} = \frac{\delta y}{\delta x}.$$

Now, let the point Q move along the curve and approach the point P in the limiting position. $\delta x \to 0$, $\delta y \to 0$ and the secant PQ becomes the tangent PT at P.

Therefore, from (1)

Slope of the tangent PT at $(x, y) = \lim_{\substack{\delta x \to 0 \\ \delta y \to 0}} \frac{\delta y}{\delta x} = \frac{dy}{dx}$

i.e., the value of the derivative at a point P of the curve is equal to the slope of tangent at that point to the curve.

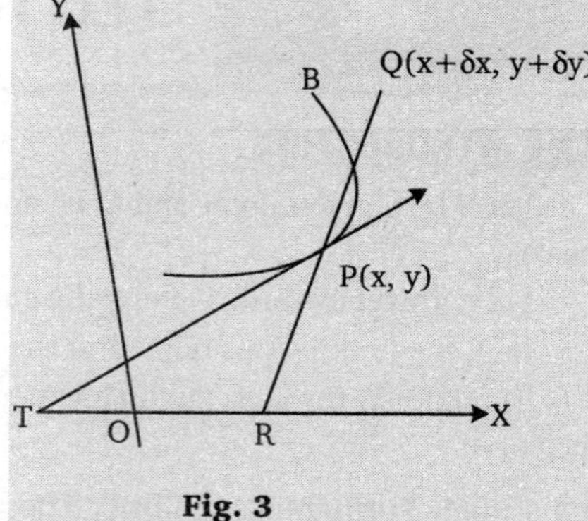

Fig. 3

REMARKS

- If the tangent at a point on the curve $y = f(x)$ is parallel to x-axis, its slope is zero *i.e.*, $\frac{dy}{dx}$ at the point $= 0$.
- If the tangent at a point on the curve is prependicular to x-axis, *i.e.*, parallel to y-axis. Its slope is ∞, *i.e.*, $\frac{dy}{dx}$ at the point $= \infty$.

4.1.3 EQUATION OF THE NORMAL

The normal to a curve at a given point is a line perpendicular to the tangent at that point and passes through the point. The slope of the normal at point $P(x_1, y_1)$ will be negative reciprocal of the slope of the tangent.

Hence, the slope of the normal at $(x_1, y_1) = -\frac{1}{dy_1 / dx_1}$

$\therefore$ The equation of the normal at $P(x_1, y_1)$ is

$$y - y_1 = -\frac{1}{dy_1 / dx_1}(x - x_1)$$

4.2 POLAR CO-ORDINATES

Let OX be a fixed straight line through fixed point O. The fixed point O is called the pole, or the origin and the fixed straight line OX is called initial line or the polar axis.

Let P be any point in the plane through the line OX. Join OP, then

(i) The length OP is called the radius vector of the point P and is denoted by r.

(ii) The angle XOP is called the vectorial angle of the point P and denoted by θ.

(iii) The number r and θ taken together in this order and called p, the polar-co-ordinates of the point P and we write it as $P(r, \theta)$.

(iv) If (x, y) are the co-ordinates of P referred to cartesian system, then it can be easily found that $x = r\cos\theta, y = r\sin\theta$.

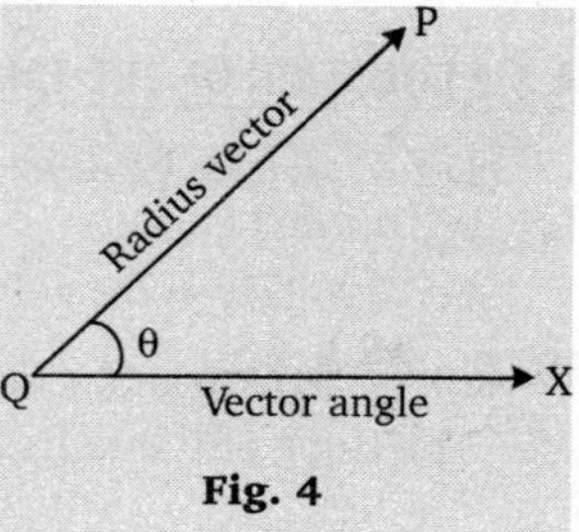

Fig. 4

4.3 ANGLE BETWEEN RADIUS VECTOR AND TANGENT

Let (r, θ) be the co-ordinate of any point P' on the curve $r = f(\theta)$. Let the tangent at P makes an angle ψ with OX.

Let ϕ be the angle between the radius vector and the tangent at P, *i.e.*, $\angle MPN = \phi$ is the angle between the radius vector OP and the tangent at P to the curve $r = f(\theta)$.

To show that for any point (r, θ) of the curve $r = f(\theta)$, the angle ϕ between the radius vector and tangent is given by

$$\tan\phi = r\frac{d\theta}{dr}.$$

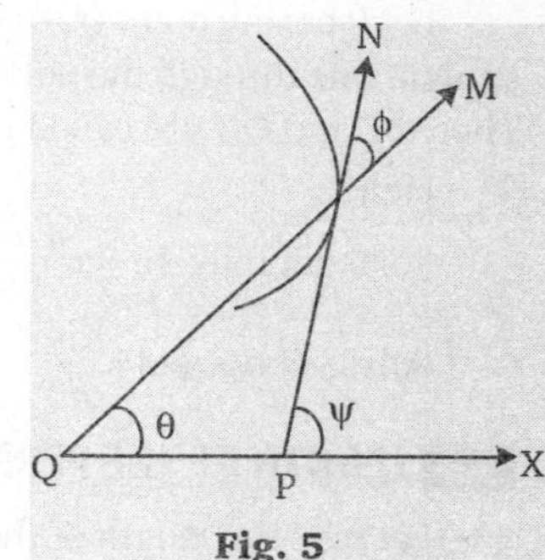

Fig. 5

Let $P(r, \theta)$ be any point on the given curve

$$r = f(\theta) \text{ or } f(r, \theta) = 0.$$

Let us suppose $Q(r + \delta r, \theta + \delta\theta)$ be the point in the neighbourhood of P on the curve.

Join OP, OQ, PQ, then

$$OP = r,\ OQ = r + \delta r$$

$$\angle XOP = \theta, \qquad \angle XOQ = \theta + \delta\theta \text{ and } \angle POQ = \delta\theta.$$

Draw $PR \perp OQ$ and $\angle PQR = \alpha$.

Now, let the angle between the radius vector OP and the tangent PT is ϕ *i.e.*,

$$\angle OPT = \phi$$

Also, we have

$$\frac{PR}{OP} = \sin\delta\theta \quad \Rightarrow PR = r\sin\delta\theta$$

$$RQ = OQ - OR = (r + \delta r) - OP\cos\delta\theta$$

$$= r + \delta r - r\cos\delta\theta$$

$$= \delta r + r(1 - \cos\delta\theta) - \delta r + 2r\sin^2\frac{\delta\theta}{2}.$$

$$\tan\alpha = \frac{PR}{QR} = \frac{r\sin\delta\theta}{\delta r + 2r\sin^2\delta\theta/2}$$

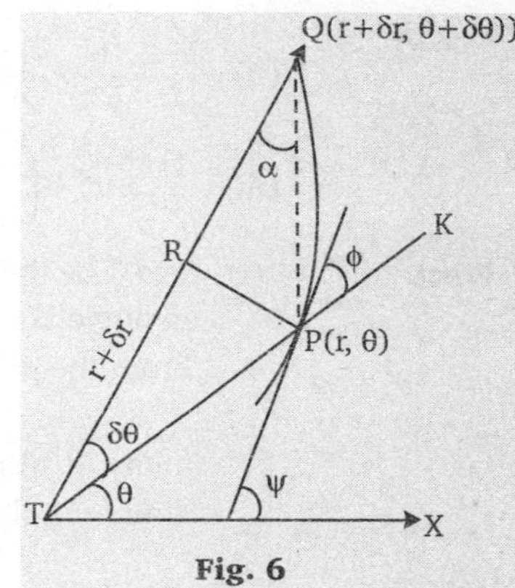

Fig. 6

Dividing the numerator and denominator by $\delta\theta$.

Then, $$\tan\alpha = \frac{r.\dfrac{\sin\delta\theta}{\delta\theta}}{\dfrac{\delta r}{\delta\theta} + r.\dfrac{\sin\delta\theta/2}{\delta\theta/2}\sin\dfrac{\delta\theta}{2}}$$

when $Q \to P$ along the curve $\alpha \to \phi$ ($\because$ PQ becomes the tangent PT and OQ coincides with OP).

$$\tan\phi = \lim_{Q\to P}\tan\alpha = \lim_{\delta\theta\to 0}\frac{r.\dfrac{\sin\delta\theta}{\delta\theta}}{\dfrac{\delta r}{\delta\theta} + r.\dfrac{\sin\delta\theta/2}{\delta\theta/2}\sin\dfrac{\delta\theta}{2}} = \frac{r.1}{dr/d\theta + r.1.0} = \frac{r}{dr/d\theta}$$

Hence, $$\tan\phi = r\frac{d\theta}{dr}.$$

REMARKS

- ϕ is the angle between the radius vector and tangent and taken to be positive when measured in the anticlockwise direction.
- Relation between θ, ϕ and ψ is $\psi = \theta + \phi$.

4.4 ANGLE OF INTERSECTION OF TWO CURVES

If the tangent to the two curves make angle ϕ_1 and ϕ_2 with the common radius vector to their point of intersection, then angle between the curves.

$$= \text{angle between tangents } = |\phi_1 - \phi_2|.$$

REMARKS

- The two curves intersect orthogonally if $\tan\phi_1 \tan\phi_2 = -1$.
- If $\dfrac{\tan\phi_1 - \tan\phi_2}{1 + \tan\phi_1.\tan\phi_2}$ is positive, we shall get acute angle of intersection at P and if $\dfrac{\tan\phi_1 - \tan\phi_2}{1 + \tan\phi_1.\tan\phi_2}$ is negative, we get the obtuse angle of intersection at P.

4.5 LENGTH OF SUBTANGENT AND SUBNORMAL

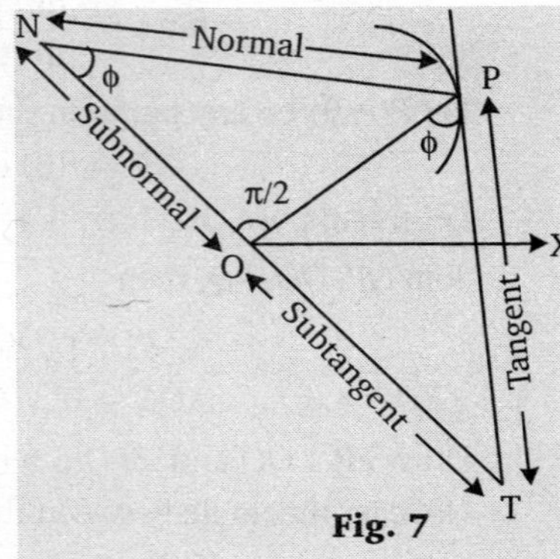

Fig. 7

Let P be any point (r, θ) on a curve $f(r, \theta) = 0$. Let the tangent and normal at P meet the straight line through the pole O perpendicular to the radius vector OP in T and N respectively. Then OT and ON are called polar subtangent and polar subnormal at P.

Hence,

Polar subtangent $= r^2 \dfrac{d\theta}{dr}$

Polar subnormal $= \dfrac{dr}{d\theta}$

4.6 LENGTH OF THE PERPENDICULAR FROM POLE TO THE TANGENT

Let p be the length of the perpendicular from the pole to the tangent at any point (r, θ) of a curve $r = f(\theta)$, then

(i) $p = r \sin\phi$

(ii) $\dfrac{1}{p^2} = \dfrac{1}{r^2} + \dfrac{1}{r^4} \cdot \left(\dfrac{dr}{d\theta}\right)^2$

(iii) $\dfrac{1}{p^2} = u^2 + \left(\dfrac{du}{d\theta}\right)^2$ where $u = \dfrac{1}{r}$

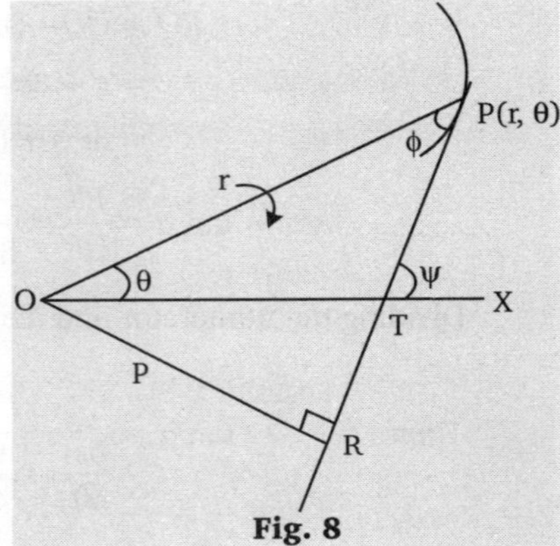

Fig. 8

Proof. (i) Let PT be the tangent at any point $P(r, \theta)$ on the curve $r = f(\theta)$ making an angle ψ with the initial line OX.

From the pole O, draw $OR \perp$ to the tangent PT.

$\therefore \quad OR = p.$

Joint OP, also, $\angle OPT = \phi$.

Now from figure, we have

$$\frac{OR}{OP} = \sin\phi \Rightarrow \frac{p}{r} = \sin\phi$$

$$\Rightarrow \quad p = r\sin\phi$$

(ii) From (i), we have

$$\frac{1}{p^2} = \frac{1}{r^2 \sin^2\phi} = \frac{1}{r^2}\operatorname{cosec}^2\phi \qquad ...(1)$$

Also, $\tan\phi = r\dfrac{d\theta}{dr}$.

$$\therefore \quad \operatorname{cosec}^2\phi = 1 + \cot^2\phi = 1 + \frac{1}{r^2}\left(\frac{dr}{d\theta}\right)^2$$

Substitute it in (1), we get

$$\frac{1}{p^2} = \frac{1}{r^2}\left[1 + \frac{1}{r^2}\left(\frac{dr}{d\theta}\right)^2\right] \Rightarrow \frac{1}{p^2} = \frac{1}{r^2} + \frac{1}{r^4}\left(\frac{dr}{d\theta}\right)^2$$

(iii) Put $r = \dfrac{1}{u}$ in (ii),

$$\frac{1}{p^2} = \frac{1}{r^2} + \frac{1}{r^4}\left(\frac{dr}{d\theta}\right)^2 \Rightarrow u^2 + u^4 \cdot \frac{1}{u^4}\left(\frac{du}{d\theta}\right)^2 \qquad \left(\because r = \frac{1}{u} \Rightarrow \frac{dr}{d\theta} = -\frac{1}{u^2}\cdot\frac{du}{d\theta}\right)$$

$$\Rightarrow \quad \frac{1}{p^2} = u^2 + \left(\frac{du}{d\theta}\right)^2$$

4.7 THE PEDAL EQUATION

Let r be the distance of any point on the curve from the origin (or pole), and p, is the length prependicular from the origin to the tangent at that point, then

The relation between p and r, where r is the distance of any point on the curve from the origin (or pole) and p is perpendicular from origin (or pole) to the tangent at that point is called the Pedal equation of the curve.

REMARK

- The Pedal equation is also called per equation of the curve.

4.7.1 PEDAL EQUATION OF A CURVE WHOSE CARTESIAN EQUATION IS GIVEN

Let the equation of the curve is

$$f(x, y) = 0 \qquad \ldots(1)$$

Then, the equation of the tangent at any point (x, y) is

$$Y - y = \frac{dy}{dx}(X - x) = y_1(X - x) \text{ where } y_1 = \frac{dy}{dx}$$

$$\Rightarrow \quad Xy_1 - Y + y - xy_1 = 0.$$

If p be the length prependicular from the origin to this tangent, then

$$p = \frac{y - xy_1}{\sqrt{1 + y_1^2}} \qquad \ldots(2)$$

Also, $$r^2 = x^2 + y^2 \qquad \ldots(3)$$

Eliminating x, y from the equation (1), (2) and (3), we get the required pedal equation of the curve (1).

4.7.2 PEDAL EQUATION OF A CURVE WHOSE POLAR EQUATION IS GIVEN

Let $r = f(\theta)$ be the polar curve. Find ϕ in terms of θ.

Eliminating θ and ϕ from (i), (ii) and $p = r \sin \phi$, we get the required pedal equation of curve (1).

REMARK

- The pedal equation is sometimes more conveniently obtained by eliminating θ between (4) and $\frac{1}{p^2} = \frac{1}{r^2} + \frac{1}{r^4}\left(\frac{dr}{d\theta}\right)^2$.

4.8 DIFFERENTIAL COEFFICIENT OF ARC LENGTH (CARTESIAN FORM)

Let $y = f(x)$ be the given curve and s denote the length of the arc, then

$$\frac{ds}{dx} = \pm\sqrt{\left[1 + \left(\frac{dy}{dx}\right)^2\right]}$$

REMARKS

- If the equation of the curve is $x = f(y)$, then $\frac{ds}{dy} = \pm\sqrt{\left[1 + \left(\frac{dx}{dy}\right)^2\right]}$
- If the given equation is in parametric form *i.e.*, $x = f_1(t), y = f_2(t)$, then $\frac{dx}{dt} = \pm\sqrt{\left[\left(\frac{dx}{dt}\right)^2 + \left(\frac{dy}{dt}\right)^2\right]}$

4.9 DIFFERENTIAL COEFFICIENT OF ARC LENGTH (POLAR FORM)

To prove that $\frac{ds}{d\theta} = \sqrt{r^2 + \left(\frac{dr}{d\theta}\right)^2}$ *where* $r = f(\theta)$ *is the polar form of curve :*

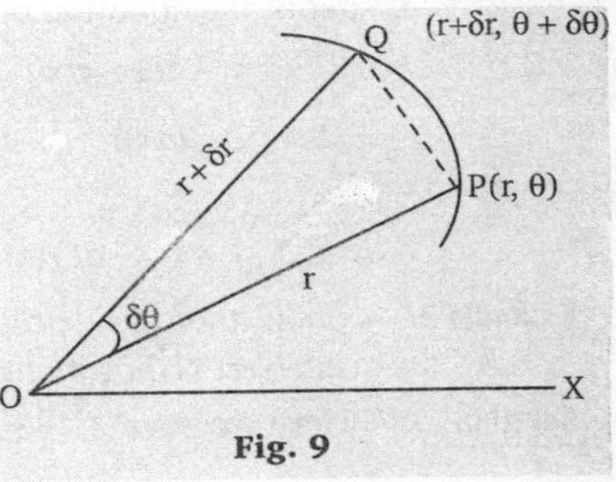

Fig. 9

Let $r = f(\theta)$ be the equation of the curve and s denote the length of arc AP. Obviously s is a function of θ. Let Q be the neighbouring point of P such that

$$AQ = s + \delta s \quad \Rightarrow \quad PQ = \delta s.$$

As $Q \to P$, $\delta\theta \to 0$ and $\delta r \to 0$

From ΔOPQ, we have

$$(\text{chord } PQ)^2 = OP^2 + OQ^2 - 2OP.OQ \cos(\angle QOP) = r^2 + (r + \delta r)^2 - 2r(r + \delta r)\cos\delta\theta$$

$$= (\delta r)^2 + 2r\delta r(1 - \cos\delta\theta) + 2r^2(1 - \cos\delta\theta)$$

Dividing by $(\delta\theta)^2$, we get

$$\left(\frac{\text{chord } PQ}{\delta\theta}\right)^2 = \left(\frac{\delta r}{\delta\theta}\right)^2 + r\left(\frac{\sin\frac{\delta\theta}{2}}{\frac{\delta\theta}{2}}\right)^2 .\delta r + r^2\left(\frac{\sin\frac{\delta\theta}{2}}{\frac{\delta\theta}{2}}\right)^2$$

$$= \left(\frac{\text{chord } PQ}{\delta s}\right)^2 = \left(\frac{\delta s}{\delta\theta}\right)^2 + r\left(\frac{\sin\frac{\delta\theta}{2}}{\frac{\delta\theta}{2}}\right)^2 .\delta r + r^2\left(\frac{\sin\frac{\delta\theta}{2}}{\frac{\delta\theta}{2}}\right)^2$$

Taking limit as $Q \to P$, we have

$$\left(\frac{ds}{d\theta}\right)^2 = \left(\frac{ds}{d\theta'}\right)^2 + r.1.0 + r^2.1 \qquad \left[\because \lim_{Q\to P}\frac{\text{chord } PQ}{PQ(=\delta s)} = 1 \text{ and } \lim_{\delta\theta\to\theta}\frac{\delta r}{\delta\theta} = \frac{dr}{d\theta}\right]$$

$$\Rightarrow \left(\frac{ds}{d\theta}\right)^2 = r^2 + \left(\frac{dr}{d\theta}\right)^2 \Rightarrow \frac{ds}{d\theta} = \pm\sqrt{\left\{r^2 + \left(\frac{dr}{d\theta}\right)^2\right\}}$$

REMARKS

- Here + or – sign is to be taken according as s increases or decreases as θ increases, we have
$$\frac{ds}{d\theta} = \pm\sqrt{\left\{r^2 + \left(\frac{dr}{d\theta}\right)^2\right\}}$$
- If $\theta = f(r)$ is the given equation of the curve, then
$$\frac{ds}{dr} = \pm\sqrt{\left\{1 + r^2\left(\frac{d\theta}{dr}\right)^2\right\}}$$
- The result $\cos\phi = \frac{dr}{ds}$ and $\sin\phi = r\frac{d\theta}{ds}$ can be remember with the help of adjoining figure(10).

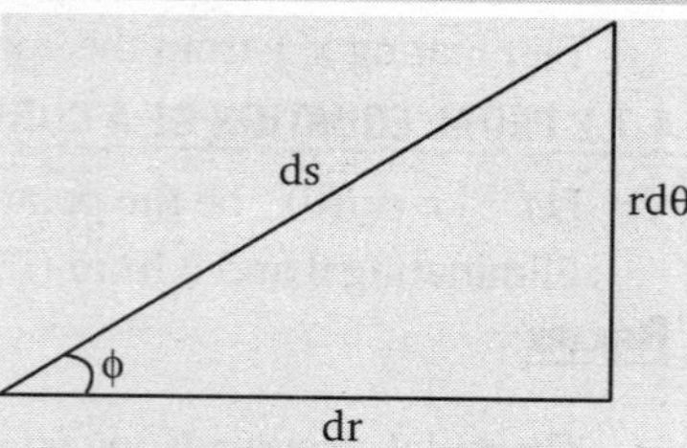

Fig. 10

Solved Examples

Example 1. *Find the equations on the tangent at the point t to the cycloid $x = a(t + \sin t)$, $y = a(1 - \cos t)$.*

Solution . We have

$$x = a(t + \sin t) \Rightarrow \frac{dx}{dt} = a(1 + \cos t)$$

and $$y = a(1 - \cos t) \Rightarrow \frac{dy}{dt} = a\sin t$$

Therefore,

$$\frac{dy}{dx} = \frac{dy/dt}{dx/dt} = \frac{a\sin t}{a(1+\cos t)} = \frac{2\left(\frac{\sin t/2}{2\cos t/2}\right)}{2\cos^2 t/2} = \tan\frac{t}{2}$$

Now, the equation of the tangent at '*t*' is

$$y - a(1 - \cos t) = \tan\frac{t}{2}[x - a(t + \sin t)]$$

$$\Rightarrow \quad y - 2a\sin^2 t/2 = (x - at)\tan\frac{t}{2} a\sin t.\tan t/2$$

$$\Rightarrow \quad y - 2a\sin^2 t/2 = (x - at)\tan t/2 - 2a\sin^2 t/2$$

$$\Rightarrow \quad y = (x - at)\tan t/2.$$

Example 2. *Show that the parabolas $r = \frac{a}{(1+\cos\theta)}$ and $r = \frac{b}{(1-\cos\theta)}$ intersect orthogonally.*

Solution . Here we have

$$r = \frac{a}{(1+\cos\theta)} \qquad \ldots(1)$$

and $$r = \frac{b}{(1-\cos\theta)} \qquad \ldots(2)$$

Recapitulations

- Equation of tangent : $y - y_1 = \frac{dy_1}{dx_1}(x - x_1)$
- Equation of normal: $y - y_1 = -\frac{1}{dy_1/dx_1}(x - x_1)$
- For any point (r, θ) of the curve $r = f(\theta)$
- The angle ϕ between the radius vector and tangent is given by $\tan\phi = r\frac{d\theta}{dr}$.
- If p is the length of the perpendicular from the pole to the tangent at any point (r, θ) of a curve $r = f(\theta)$ then
  (i) $p = r\sin\phi$
  (ii) $\frac{1}{p^2} = \frac{1}{r^2} + \frac{1}{r^4}\cdot\left(\frac{dr}{d\theta}\right)^2$
  (iii) $\frac{1}{p^2} = u^2 + \left(\frac{du}{d\theta}\right)^2,\ u = \frac{1}{r}$
- $\frac{ds}{dx} = \pm\sqrt{\left[1 + \left(\frac{dy}{dx}\right)^2\right]}$
- $\frac{ds}{d\theta} = \sqrt{r^2 + \left(\frac{dr}{d\theta}\right)^2},\ r = f(\theta)$

Taking log of both sides of (1), we get

$$\log r = \log a - \log (1 + \cos \theta)$$

Differentiating with respect to θ, we get

$$\frac{1}{r}\cdot\frac{dr}{d\theta} = \frac{(-\sin\theta)}{(1+\cos\theta)} = \frac{2\sin\theta/2\cos\theta/2}{2\cos^2\theta/2} = \tan\frac{\theta}{2}$$

$$\Rightarrow \quad \cot\phi = \tan\frac{\theta}{2} = \cot\left(\frac{\pi}{2}-\frac{\theta}{2}\right)$$

$$\Rightarrow \quad \phi_1 = \frac{\pi}{2}-\frac{\theta}{2}$$

Now, from (2), we get

$$\log r = \log b - \log (1 - \cos \theta)$$

Differentiating with respect to θ, we get

$$\frac{1}{r}\cdot\frac{dr}{d\theta} = \frac{\sin\theta}{1-\cos\theta} = -\frac{2\sin\theta/2\cdot\cos\theta/2}{2\sin^2\theta/2} = -\cot\frac{\theta}{2}$$

$$\therefore \quad \cot\phi = -\cot\frac{1}{2}\theta = \cot\left(\pi-\frac{1}{2}\theta\right)$$

$$\Rightarrow \quad \phi = \pi-\frac{1}{2}\theta \Rightarrow \phi_2 = \pi-\frac{1}{2}\theta$$

Now, the angle of intersection $= \phi_1 \sim \phi_2$

$$= \left(\pi-\frac{1}{2}\theta\right)-\left(\frac{1}{2}\pi-\frac{1}{2}\theta\right) = \frac{\pi}{2}$$

Both curves intersect orthogonally.

Example 3. *Show that the pedal equation of the ellipse* $\frac{x^2}{a^2}+\frac{y^2}{b^2}=1$ is $\frac{1}{p^2}=\frac{1}{a^2}+\frac{1}{b^2}-\frac{r^2}{a^2b^2}$.

Solution . Here, the equation of the curve is $\frac{x^2}{a^2}+\frac{y^2}{b^2}=1$.

Let $x = a \cos t, y = b \sin t$.

$$\therefore \quad \frac{dx}{dt} = -a\sin t, \frac{dy}{dt} = b\cos t$$

$$\Rightarrow \quad \frac{dy}{dx} = -\frac{b\cos t}{a\sin t}$$

Therefore, the equation of the tangent at '*t*' is

$$Y - b\sin t = -\frac{b\cos t}{a\sin t}(X - a\cos t)$$

$$\Rightarrow \quad ab - b\cos t.X - a\sin t.Y = 0 \quad \ldots(1)$$

Since p denote the length prependicular from (0, 0) to (1), therefore

$$p = \frac{ab}{\sqrt{a^2\sin^2 t + b^2\cos^2 t}}$$

$$\frac{1}{p^2} = \frac{a^2\sin^2 t + b^2\cos^2 t}{a^2b^2} \quad \ldots(2)$$

Now, $\quad r^2 = x^2+y^2 = a^2\cos^2 t + b^2\sin^2 t$

$$= a^2+b^2-a^2\sin^2 t - b^2\cos^2 t \ldots(3)$$

From (3) $a^2\sin^2 t + b^2\cos^2 t = (a^2+b^2) - r^2$.

Therefore, from (3), we get

$$\frac{1}{p^2} = \frac{(a^2+b^2)-r^2}{a^2b^2} = \frac{1}{a^2}+\frac{1}{b^2}-\frac{r^2}{a^2b^2}.$$

Example 4. *Find the pedal equation of* $r^n = a^n \sin n\theta$.

Solution . Here, the given curve is

$$r^n = a^n \sin n\theta \quad \ldots(1)$$

Taking logarithm of both the sides of (1), we get

$$n\log r = n\log a + \log\sin n\theta. \quad \ldots(2)$$

Differentiating w.r.t. θ, we get

$$\frac{n}{r}\cdot\frac{dr}{d\theta} = n\frac{\cos n\theta}{\sin n\theta} = n\cot n\theta$$

$$\Rightarrow \quad \cot\phi = \frac{1}{r}\cdot\frac{dr}{d\theta} = \cot n\theta$$

$$\therefore \quad \phi = n\theta$$

Also, $\quad p = r\sin\phi \Rightarrow p = r\sin n\theta \quad \ldots(3)$

Now from (1) and (3), we have

$$\sin n\theta = \frac{p}{r}$$

Putting the value in (1), we get

$$pa^n = r^{n+1}.$$

Example 5. *Find the angle at which the radius vector cuts the curves* $\frac{l}{r} = 1+e\cos\theta$.

Solution . Here, the given equation of the curve is

$$\frac{l}{r} = 1+e\cos\theta$$

$$\Rightarrow \quad \log l - \log r = \log (1 + e\cos\theta).$$

Diff. w.r.t. θ, we get

$$-\frac{1}{r}\cdot\frac{dr}{d\theta} = \frac{1}{(1+e\cos\theta)}(-e\sin\theta)$$

$$\therefore \quad \cot\phi = \frac{1}{r}\cdot\frac{dr}{d\theta} = \frac{e\sin\theta}{1+e\cos\theta}$$

$$\Rightarrow \quad \tan\phi = \frac{1+e\cos\theta}{e\sin\theta}$$

$$\Rightarrow \quad \phi = \tan^{-1}\left[\frac{1+e\cos\theta}{e\sin\theta}\right].$$

Example 6. *For the cardiod* $r = a(1 - \cos\theta)$, *prove that*

(*i*) $\phi = \frac{1}{2}\theta$ (VTU–2004)

(*ii*) $2ap^2 = r^3$

Solution . Here the given curve is

$$r = a(1-\cos\theta) \quad \ldots(1)$$

$$\Rightarrow \quad \frac{dr}{d\theta} = a\sin\theta$$

(a) Since, we have

$$\tan\phi = r\frac{d\theta}{dr} = \frac{a(1-\cos\theta)}{a\sin\theta}$$

$$= \frac{2a\sin^2\theta/2}{2a\sin\theta/2\cdot\cos\theta/2} = \tan\frac{\theta}{2}$$

$$\Rightarrow \quad \phi = \frac{\theta}{2}$$

(b) Since, we have $p = r\sin\phi = r\sin\theta/2$

$$\Rightarrow \quad r = 2a\sin^2\frac{\theta}{2} = 2a\frac{p^2}{r^2}$$

$$\therefore \quad 2ap^2 = r^3$$

Example 7. *Find the pedal equation of the curve* $x^{2/3} + y^{2/3} = a^{2/3}$.

Solution . Here, the given curve is

$$x^{2/3} + y^{2/3} = a^{2/3} \qquad \ldots(1)$$

Let $x = a\cos^3 t, y = a\sin^3 t$

$$\Rightarrow \quad \frac{dy}{dx} = \frac{dy/dt}{dx/dt} = \frac{3a\sin^2 t\cos t}{-3a\cos^2 t\sin t} = -\frac{\sin t}{\cos t}.$$

Hence, the equation of tangent of (1) is

$$y - a\sin^3 t = -\frac{\sin t}{\cos t}(x - a\cos^3 t)$$

$$\Rightarrow x\sin t + y\cos t = a\sin t\cos t(\cos^2 t + \sin^2 t)$$

$$= a\sin t\cos t \qquad \ldots(2)$$

p = the length of the prependicular from (0, 0) to (2)

$$= \frac{a\sin t\cos t}{\sqrt{\sin^2 t + \cos^2 t}} = a\sin t\cos t.$$

Now, $r^2 = x^2 + y^2 = a^2\cos^6 t + a^2\sin^6 t$

$$= a^2[(\cos^2 t)^3 + (\sin^2 t)^3]$$

$$= a^2[(\cos^2 t + \sin^2 t)^3 - 3\cos^2 t\sin^2 t(\cos^2 t + \sin^2 t)]$$

$$= a^2[1 - 3(p^2/a^2).1] = a^2 - 3p^2.$$

Example 8. *Show that for any curve*

$$\sin^2\phi\left(\frac{d\phi}{d\theta}\right) + r\left(\frac{d^2r}{ds^2}\right) = 0.$$

(Meerut–1998)

Solution . We have

$$\frac{dr}{ds} = \cos\phi$$

$$\Rightarrow \quad \frac{d^2r}{ds^2} = -\sin\phi\left(\frac{d\phi}{ds}\right) = -\sin\phi\left(\frac{d\phi}{d\theta}\right)\left(\frac{d\theta}{ds}\right)$$

$$\Rightarrow r\left(\frac{d^2r}{ds^2}\right) = -\sin\phi\left(\frac{d\phi}{d\theta}\right).r\left(\frac{d\theta}{ds}\right)$$

$$\Rightarrow r\left(\frac{d^2r}{ds^2}\right) = -\sin\phi\left(\frac{d\phi}{d\theta}\right).\sin\phi \left(\because r\frac{d\theta}{ds} = \sin\phi\right)$$

$$\therefore \quad r\left(\frac{d^2r}{ds^2}\right) + \sin^2\phi\left(\frac{d\phi}{d\theta}\right) = 0.$$

EXERCISE 4.1

1. Find the angle of intersection of the curve $r^2 = 16\sin 2\theta$ and $r^2\sin 2\theta = 4$.
2. Show that in the curve $r = a\theta$, the polar subnormal is constant and in the curve $r\theta = a$, the polar subtangent is constant.
3. Show that the curves $r = a(1 + \cos\theta)$ and $r = b(1 - \cos\theta)$ intersect at right angles. (VTU–2011)
4. Show that the spiral $r^n = a^n\cos n\theta$ and $r^n = b^n\sin n\theta$ intersect orthogonally. (VTU–2010)
5. Find the angle ϕ for the curve $a\theta = (r^2 - a^2)^{1/2} - a\cos^{-1} a/r$.
6. Show that the curves $r = (1 + \sin\theta)$ and $r = a(1 - \sin\theta)$ cut orthogonally.
7. Show that the curves $r = 2\sin\theta$ and $r = 2\cos\theta$ intersect at right angles.
8. Find the angle of intersection between the pair of curves $r = 6\cos\theta$ and $r = 2(1 + \cos\theta)$.
9. Show that the pedal equation of the
   (i) conic $\frac{l}{r} = 1 + e\cos\theta$ is $\frac{1}{p^2} = \frac{1}{l^2}\left(\frac{2l}{r} - 1 + e^2\right)$
   (ii) curve $r = a\theta$ is $p^2 = \frac{r^4}{r^2 + a^2}$
   (iii) cosine spiral $r^n = a^n\cos n\theta$ is $pa^n = r^{n+1}$. (VTU–2009)
   (iv) cardiod $r = a(1 + \cos\theta)$ is $r^3 = 2ap^2$.
   (v) spiral $r = a\,\text{sech}\, n\theta$ is $\frac{1}{p^2} = \frac{A}{r^2} + B$.
   (vi) hyperbola $r^2\cos 2\theta = a^2$ is $pr = a^2$.
   (vii) lemniscate $r^2 = a^2\cos 2\theta$ is $r^3 = a^2p$.
10. Show that the normal at any point (r, θ) to the curve $r^n = a^n\cos n\theta$ makes an angle $(n + 1)\theta$ with the initial line.
11. Show that in the equiangular spiral $r = ae^{\theta\cot\alpha}$, the tangent is inclined at a constant angle α to the radius vector.
12. For the curve $r = ae^{\theta\cot a}$, prove that $\frac{s}{r}$ = constant, s being measured from the pole.
13. Show that
   (i) $\frac{ds}{d\theta} = \frac{r^2}{p}$ (ii) $\frac{ds}{dr} = \frac{r}{\sqrt{r^2 - p^2}}$

14. For the ellipse $x = a \cos t, y = b \sin t$, prove that

$$\frac{ds}{dt} = a(1 - e^2 \cos^2 t)^{1/2}.$$

15. For the curve $r^n = a^n \cos n\theta$, show that

$$a^{2n}\frac{d^2r}{ds^2} + nr^{2n-1} = 0.$$

16. For the cycloid $x = a(1 - \cos t), y = a(t + \sin t)$, show that

(i) $\frac{ds}{dt} = 2a\cos\frac{t}{2}$ (ii) $\frac{ds}{dx} = \text{cosec}\frac{t}{2}$ (iii) $\frac{ds}{dy} = \sec\frac{t}{2}$

17. Show that for the curve $r^m = a^m \cos m\theta$, $\frac{ds}{d\theta} = \frac{a^m}{r^{m-1}}$.

18. Show that the pedal equation of the parabola $y^2 = 4a(x + a)$ is $p^2 = ar$.

19. Prove that for the ellipse $\frac{x^2}{a^2} + \frac{y^2}{b^2} = 1, f = \frac{a^2b^2}{p^3}$, p being the perpendicular from centre upon the tangent (x, y).

ANSWERS

1. $\frac{2\pi}{3}$ **5.** $\cos^{-1}\frac{a}{r}$ **8.** $\frac{\pi}{6}$

Objective Evaluations

Fill in the Blanks

1. The pedal equation of the curve $y^2 = 4a(x + a)$ is __________
2. If ϕ is the angle between the radius vector and the tangent of a curve, then $\tan \phi =$ __________
3. Polar subtangent for the curve $r = a\theta$ is __________
4. For the curve $r = f(\theta)$, the value of $\frac{ds}{d\theta} =$ __________
5. Polar subnormal for the curve $r = a\theta$, __________
6. For the cycloid $x = a\ (1 - \cos t),\ y = a(1 + \sin t)$, we have $\frac{ds}{dt} =$ __________
7. In the equiangular spiral $r = ae^{\theta \cot a}$, the tangent is inclined to the radius vector with angle __________
8. For the curve $r^2 = a^2 \cos 2\theta$, $\frac{ds}{d\theta}$ __________

True/False

Write 'T' for True and 'F' for False statement.

1. The relation between p and r is called pedal equation. **(T/F)**
2. The relation between p and r is called polar equation. **(T/F)**
3. The pedal equation of the curve $r = a/\theta$ is $\frac{1}{p^2} = \frac{1}{r^2} + \frac{1}{a^2}$. **(T/F)**
4. The pedal equation of the curve $r^m = a^m \cos m\theta$ is $r^{m+1} = a^m.p$. **(T/F)**
5. For the curve $r = f(\theta)$, we have $\left(\frac{dr}{ds}\right)^2 + \left(r\frac{d\theta}{ds}\right)^2 = 1.$ **(T/F)**
6. For any curve $r = f(\theta)$, the value of $\frac{ds}{d\theta}$ is $\frac{r^2}{p}$. **(T/F)**
7. If p be the length of perpendicular drawn from the pole O to the tangent at any point $P(r, \theta)$ on the curve $r = f(\theta)$ then $$\frac{1}{p^2} \neq \frac{1}{r^2} + \frac{1}{r^4}\left(\frac{dr}{d\theta}\right)^2.$$ **(T/F)**
8. The pedal equation of the cardiod $r = a(1 - \cos\theta)$ is $r^3 = 2ap$. **(T/F)**

Multiple Choice Questions

Choose the most appropriate one.

1. Two curves cut orthogonally if $\tan \phi_1.\tan\phi_2$ is equal to:
   (a) 1 (b) –1 (c) 0 (d) none of these
2. For the curve $r = f(\theta)$, the value of $\cos \phi$ is :
   (a) $r\frac{d\theta}{ds}$ (b) $r\frac{ds}{d\theta}$ (c) $\frac{ds}{dr}$ (d) $\frac{dr}{ds}$
3. The pedal equation of the curve $y^2 = 4a(x + a)$ is :
   (a) $p = a^2 r^2$ (b) $p^2 = ar$
   (c) $p^2 = r$ (d) $2ap^2 = r^3$
4. The angle at which the radius vector cuts the curve $r = a(1 - \cos\theta)$ is :
   (a) θ (b) $\theta/2$ (c) $\theta/3$ (d) $\theta/4$
5. In the equiangular spiral $r = ae^{\theta \cot a}$ the tangent is inclined to which angle to the radius vector :
   (a) $\alpha/2$ (b) $\alpha/3$ (c) α (d) 2α
6. Polar subtangent for the curve $r = a\theta$ is :
   (a) $r^2 a$ (b) r^2 (c) r^2/a (d) $(r/a)^2$
7. Polar subtangent for the curve $\frac{2a}{r} = 1 - \cos\theta$ is :
   (a) $2a \sin\theta$ (b) $-2a\cos\theta$
   (c) $2a\tan\theta$ (d) $-2a \operatorname{cosec}\theta$
8. The angle of intersection of the curve $r = a\cos\theta, 2r = a$ is :
   (a) $\pi/2$ (b) $\pi/4$ (c) $\pi/3$ (d) π
9. Polar subnormal for the curve $r = a\theta$ is :
   (a) $r^2 a$ (b) a (c) r^2/a (d) r^2
10. For the cardiod $r = a(1 - \cos\theta)$, the value of ϕ is :
   (a) θ (b) $\frac{\theta}{2}$ (c) $-\frac{\theta}{2}$ (d) none of these

ANSWERS

Fill in the Blanks

1. $p^2 = ar$ **2.** $r\frac{d\theta}{dr}$ **3.** $\frac{r^2}{a}$ **4.** $\sqrt{\left[r^2 + \left(\frac{dr}{d\theta}\right)^2\right]}$ **5.** a **6.** $2a\cos\frac{t}{2}$ **7.** a **8.** $\frac{a^2}{r}$

True/False

1. T **2.** F **3.** T **4.** T **5.** T **6.** T **7.** T **8.** F

Multiple Choice Questions

1. (b) **2.** (d) **3.** (b) **4.** (b) **5.** (c) **6.** (c) **7.** (d) **8.** (c) **9.** (b) **10.** (b)

FFFFFF

Curvature

5.1 INTRODUCTION

In figure (1), curve PQ bends more sharply than the curve AB. Then measure of the sharpness of the bending of a curve at a particular point is called curvature of the curve at the point. In this chapter, we shall find mathematical expressions for the curvature of a curve at a given point.

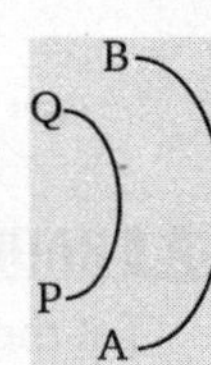

Fig. 1

5.2 CURVATURE

Let P, Q be two neighbouring points on a curve AB.

Also, let $AP = s$, arc $AQ = s + \delta s$ and arc $PQ = \delta s$.

Let the tangent to the curve at points P and Q makes angle ψ and $\psi + \delta\psi$ respectively with a fixed line say X-axis, then

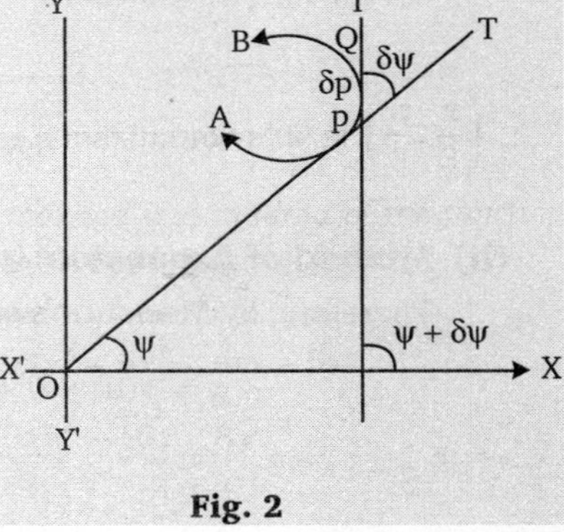

Fig. 2

(i) The angle $\delta\psi$ through which the tangent turns as its points of contact travels along the arc PQ is called the total bending or total curvature of arc PQ.

(ii) The ratio $\frac{\delta\psi}{\delta s}$ is called the mean or average curvature of arc PQ.

(iii) The limiting value of the mean curvature when Q tends to P is called the curvature of the curve at the point P. Therefore, the curvature K at point P is

$$\lim_{Q\to P}\frac{\delta\psi}{\delta s} = \lim_{\delta s\to 0}\frac{\delta\psi}{\delta s} = \frac{d\psi}{ds}$$

(iv) The reciprocal of the curvature of the given curve at P. (provided this curvature is not equal to zero), is called the radius of curvature of the curve at P. This is denoted by ρ.

$$\rho = \frac{1}{K} = \frac{ds}{d\psi}$$

5.3 FORMULA FOR RADIUS OF CURVATURE (CARTESIAN FORM)

Let $y = f(x)$ be the equation of curve. Then the slope of the tangent at any point $= \tan\psi = \frac{dy}{dx}$

Differentiating both sides, w.r.t. s, we get

$$\sec^2\psi\frac{d\psi}{ds} = \frac{d}{ds}\left(\frac{dy}{dx}\right) \quad\Rightarrow\quad \sec^2\psi.\frac{1}{\rho} = \frac{d}{dx}\left(\frac{dy}{dx}\right)\frac{dx}{ds}$$

$$\Rightarrow \qquad \sec^2\psi.\frac{1}{\rho} = \frac{d^2y}{dx^2}.\cos\psi \qquad\qquad \left(\because \frac{dx}{ds} = \cos\psi\right)$$

Therefore
$$\rho = \frac{\sec^2\psi}{\cos\psi\,\frac{d^2y}{dx^2}} = \frac{\sec^3\psi}{\frac{d^2y}{dx^2}} = \frac{(1+\tan^2\psi)^{3/2}}{\frac{d^2y}{dx^2}} \Rightarrow \qquad \rho = \frac{\left[1+\left(\frac{dy}{dx}\right)^2\right]^{3/2}}{\frac{d^2y}{dx^2}}$$

REMARKS

- The positive root is taken in numerator of above formula, therefore, radius of curvature ρ, will be positive when $\frac{d^2y}{dx^2}$ is positive (*i.e.*, when the curve is concave upward) and negative when $\frac{d^2y}{dx^2}$ is negative (*i.e.*, when the curve is concave downward).
- At a point of inflexion, the curvature of a curve is not defined. $\left(\because \text{at the point of inflexion } \frac{d^2y}{dx^2} = 0\right)$
- When the equation of the curve is given in the form $x = f(y)$ then by interchanging x and y (It is justify because curvature is a length, and its value is independent of the choice of axis), we get

$$\rho = \frac{\left[1+(dx/dy)^2\right]^{3/2}}{d^2x/dy^2}$$

- When the equation of curve is given is paraetric form *i.e.*, $x = f(t)$ and $y = g(t)$, then radius of curvature is given by $\rho = \frac{(x'^2 + y'^2)^{3/2}}{x'y'' - y'x''}$, where dash (′) denote the derivative w.r.t., ′*t*′.

$$\frac{1}{\rho^2} = \left(\frac{d^2x}{ds^2}\right) + \left(\frac{d^2y}{ds^2}\right)^2$$

5.4 RADIUS OF CURVATURE AT THE ORIGIN

Let the curve $y = f(x)$ passes through the origin. Then, we may use the following methdos, to find the radius of curvature.

(i) Method of direct substitution. Since $y = f(x)$ be given. Calculate the values of $\frac{dy}{dx}$ and $\frac{d^2y}{dx^2}$ at origin and then use the following formula

$$\rho = \frac{\left[1+\left(\frac{dy}{dx}\right)^2\right]^{3/2}}{d^2y/dx^2}$$

(ii) Method of Expansion. Let $y = f(x)$ be the equation of curve. Since, it passes through the origin, therefore $f(0) = 0$.

Therefore, by Maclaurin's series expansion, we have

$$y = f(0) + xf'(0) + \frac{x^2}{2!}f''(0) + \frac{x^3}{3!}f'''(0) + \ldots \Rightarrow y = xf'(0) + \frac{x^2}{2!}f''(0) + \frac{x^3}{3!}f'''(0) + \ldots \quad [\because f(0) = 0]$$

$$\Rightarrow \quad y = p_1x + \frac{1}{2!}p_2x^2 + \frac{1}{3!}p_3x^3 + \ldots \quad \text{where} \quad p_1 = f'(0) = y_1(0), p_2 = f''(0) = y_2(0)\text{, etc.}$$

Now, differentiating (1) with respect to x, we get

$$y_1 = p_1 + \frac{2p_2x}{2!} + \frac{3p_2x^2}{3!} + \ldots$$

Again differentiating w.r.t. x. we get

$$y_2 = \frac{2p_2}{2!} + \frac{6p_3x^2}{3!} + \ldots$$

At the origin (*i.e.*, $x = 0$), we have

$$y_1 = p_1 \text{ and } y_2 = \frac{2p_2}{2!} = p_2$$

Now putting these values of y_1 and y_2 in the formula $\rho = \frac{(1+y_1^2)^{3/2}}{y_2}$, We have $\rho = \frac{(1+p_1^2)^{3/2}}{p_2}$

REMARKS

- We can find the values of p and q in the following manner:

Put the value of $y = p_1x + \frac{p_2x^2}{2!} + \frac{p_3x^3}{3!} + \ldots$ in the given equation of the curve and equating the coefficients of the powers of x.

(iii) Newton's Method. If a curve passes through the origin, and axis of x is the tangent at the origin, then radius of curvature ρ at origin

$$= \lim_{\substack{x\to 0\\ y\to 0}} \frac{x^2}{2y}$$

Since the axis of x is the tangent at the origin, therefore, we have

$$y_1(0) = \left(\frac{dy}{dx}\right)_{(0,0)} = 0$$

Here, we observed that $\frac{x^2}{2y}$ is of the indeterminate form $\left(\frac{0}{0}\right)$ as $x \to 0, y \to 0$.

Using L' Hospital rule, we have

$$\lim_{\substack{x\to 0\\ y\to 0}} \frac{x^2}{2y} = \lim_{\substack{x\to 0\\ y\to 0}} \frac{2x}{2y_1} = \lim_{\substack{x\to 0\\ y\to 0}} \frac{x}{y_1} = \lim_{\substack{x\to 0\\ y\to 0}} \frac{1}{y_2} = \frac{1}{y_2(0)} \quad \ldots(1)$$

Now, ρ at origin $= \dfrac{[1+y_1^2(0)]^{3/2}}{y_2(0)} = \dfrac{(1+0)^{3/2}}{y_2(0)} = \dfrac{1}{y_2(0)} \quad \ldots(2)$

From (1) and (2), we have

$$\rho_{(\text{at origin})} = \lim_{\substack{x\to 0\\ y\to 0}} \frac{x^2}{2y}$$

Recapitulations

- $\rho = \dfrac{ds}{d\psi}$
- $\rho = \dfrac{\left[1+\left(\frac{dy}{dx}\right)^2\right]^{3/2}}{\frac{d^2y}{dx^2}}$
- If a curve passes through the origin and axis of x is the tangent at the origin, then radius of curvature $\rho = \lim_{\substack{x\to 0\\ y\to 0}} \dfrac{x^2}{2y}$

REMARK

- If a curve passes through the origin and axis of y is the tangent, then radius of curvature at the origin is given by $\lim_{\substack{x\to 0\\ y\to 0}} \dfrac{y^2}{2x}$.

Solved Examples

Example 1. *In the cycloid $x = a(t + \sin t)$, $y = a(1 - \cos t)$, prove that $\rho = 4a\cos\frac{t}{2}$.* (SRM-2010)

Solution . We have

$$x = a(t+\sin t) \Rightarrow \frac{dx}{dt} = a(1+\cos t)$$

and

$$y = a(1-\cos t) \Rightarrow \frac{dy}{dt} = a\sin t$$

$$\Rightarrow \quad \frac{dy}{dx} = \frac{dy/dt}{dx/dt} = \frac{a\sin t}{a(1+\cos t)}$$

$$= \frac{2\sin t/2\,\cos t/2}{2\cos^2 t/2} = \tan\frac{t}{2}$$

Also $\dfrac{d^2y}{dx^2} = \dfrac{d}{dx}\left(\dfrac{dy}{dx}\right) = \dfrac{d}{dx}\left(\tan\dfrac{t}{2}\right)$

$$= \frac{1}{2}\sec^2\frac{t}{2}.\frac{dt}{dx}$$

$$= \frac{1}{2}\sec^2\frac{t}{2}.\frac{1}{a(1+\cos t)} = \frac{1}{4a}\sec^4\frac{t}{2}$$

Now, putting the values of $\frac{dy}{dx}$ and $\frac{d^2y}{dx^2}$ in

$$\rho = \frac{\left[1+\left(\frac{dy}{dx}\right)^2\right]^{3/2}}{\frac{d^2y}{dx^2}}$$

We get $\rho = \dfrac{[1+\tan^2 t/2]^{3/2}}{\frac{1}{4a}\sec^4 t/2}$

$$= \frac{4a\sec^3 t/2}{\sec^4 t/2} = 4a\cos t/2$$

Example 2. *Find the curvature of the curve $x^3 + y^3 = 3axy$ at the point $(3a/2, 3a/2)$.*

(Anna–2009, VTU–2008, Kurukshatra–2009, Kerela–2005)

Solution . The equation of the curve is

$$x^3 + y^3 = 3axy \quad \ldots(1)$$

Differentiating w.r.t. x, we get

$$3x^2 + 3y^2\frac{dy}{dx} = 3ay + 3ax\frac{dy}{dx}$$

$$\Rightarrow \quad x^2 + y^2\frac{dy}{dx} = ay + ax\frac{dy}{dx}$$

$$\Rightarrow \quad \frac{dy}{dx} = \frac{x^2 - ay}{ax - y^2} \quad \ldots(2)$$

$$\Rightarrow \quad \left(\frac{dy}{dx}\right)_{at\left(\frac{3}{2}a, \frac{3}{2}a\right)} = -1$$

From (2), we have

$$2x + 2y\left(\frac{dy}{dx}\right)^2 + y^2\frac{d^2y}{dx^2} = a\frac{dy}{dx} + a\frac{dy}{dx} + ax\frac{d^2y}{dx^2}$$

$$\Rightarrow \quad (ax - y^2)\frac{d^2y}{dx^2} = 2x + 2y\left(\frac{dy}{dx}\right)^2 - 2a\frac{dy}{dx} \quad \ldots(3)$$

Putting $x = \frac{3a}{2}, y = \frac{3a}{2}$ and $\left(\frac{dy}{dx}\right)_{\left(\frac{3a}{2},\frac{3a}{2}\right)} = -1$, we get

$$\left[\frac{d^2y}{dx^2}\right]_{\left(\frac{3a}{2},\frac{3a}{2}\right)} = -\frac{32}{3}.\frac{1}{a}$$

Hence, the radius of curvature ρ at $\left(\frac{3a}{2},\frac{3a}{2}\right)$, we get

$$\rho = \left[\frac{\left[1+\left(\frac{dy}{dx}\right)^2\right]^{3/2}}{\frac{d^2y}{dx^2}}\right]_{\text{at}\left(\frac{3a}{2},\frac{3a}{2}\right)} = \frac{(1+1)^{3/2}}{-\frac{32}{3}.\frac{1}{a}} = -\frac{3a}{8\sqrt{2}}$$

Therefore, the curvature $\frac{1}{\rho} = +\frac{8\sqrt{2}}{3a}$.

(By ignoring the negative sign)

Example 3. *Show that the radii of curvature of the curve* $y^2 = x^2\left(\frac{a+x}{a-x}\right)$ *at the origin are* $a\sqrt{2}$.

Solution . The equation of the curve is

$$y^2 = x^2\left(\frac{a+x}{a-x}\right)$$

$$\Rightarrow y = \pm\frac{x(a+x)^{1/2}}{(a-x)^{1/2}} = \pm x\frac{a^{1/2}\left(1+\frac{x}{a}\right)^{1/2}}{a^{1/2}\left(1-\frac{x}{a}\right)^{1/2}}$$

$$\Rightarrow y = \pm x\left(1+\frac{x}{a}\right)^{1/2}\left(1-\frac{x}{a}\right)^{-1/2}$$

$$\Rightarrow y = \pm x\left(1+\frac{x}{2a}+...\right)\left(1+\frac{x}{2a}+...\right)$$

(Expanding by Binomial Expansions)

$$\text{or } y = \pm x\left(1+\frac{x}{2a}+\frac{x}{2a}+\frac{x^2}{4a^2}+...\right)$$

$$\Rightarrow y = \pm\left(x+\frac{x^2}{a}+\frac{x^3}{4a^2}+...\right)$$

Therefore,

$$\frac{dy}{dx} = y_1 = \pm\left(1+\frac{2x}{a}+\frac{3x^2}{4a^2}+...\right)$$

and $$\frac{d^2y}{dx^2} = y_2 = \pm\left(\frac{2}{a}+\frac{6x}{4a^2}+...\right)$$

At(0, 0) $y_1 = \pm1$ and $y_2 = \pm\frac{2}{a}$

$$\therefore \quad \rho = \frac{(1+y_1^2)^{3/2}}{y_2} = \frac{(1+1)^{3/2}}{\pm2/a} = \pm2\sqrt{2}.\frac{a}{2}$$

$$\Rightarrow \rho = \pm\sqrt{2}\cdot a = \sqrt{2}\cdot a. \qquad \text{(Numerically)}$$

Example 4. *Apply Netwon's formula, find the radius of curvature at the origin for the curve*

$$x^3 - 2x^2y + 3xy^2 - 4y^3 + 5x^2 - 6xy + 7y^2 - 8y = 0.$$

Solution . Since, the curve passes through the origin. Equating to zero, the lowest degree terms, we may find $y = 0$

⇒ x axis is the tangent at the origin.

Therefore, by Newton's formula, , ρ at (0, 0)

$$= \lim_{\substack{x\to0\\y\to0}}\frac{x^2}{2y}$$

Dividing the equation of the curve by $2y$, we get

$$x.\frac{x^2}{2y} - x^2 + \frac{3}{2}xy - 2y^2 + 5.\frac{x^2}{2y} - 3x + \frac{7}{2}y - 4 = 0$$

Taking lim $x \to 0$ and $y \to 0$, we get

$$\lim_{\substack{x\to0\\y\to0}}\frac{x^2}{2y} - 4 = 0 \Rightarrow 5\rho - 4 = 0 \Rightarrow \rho = \frac{4}{5}.$$

Example 5. *If CP, CD be a pair of conjugate semi-diameters of an ellipse, show that the radius of curvature at P is* $\frac{CD^3}{ab}$ *where a and b are the lengths of the semi -axes of the ellipse.*

Solution . Let CP and CD be a conjugate of semi-diameters of the ellipse $\frac{x^2}{a^2} + \frac{y^2}{b^2} = 1$

Let the coordinate of P are

$$x = a\cos t, y = b\sin t. \qquad ...(1)$$

Also, the coordinate of D are

$$\left[a\cos\left(\frac{\pi}{2}+t\right), b\sin\left(\frac{\pi}{2}+t\right)\right] = (-a\sin t, b\cos t)$$

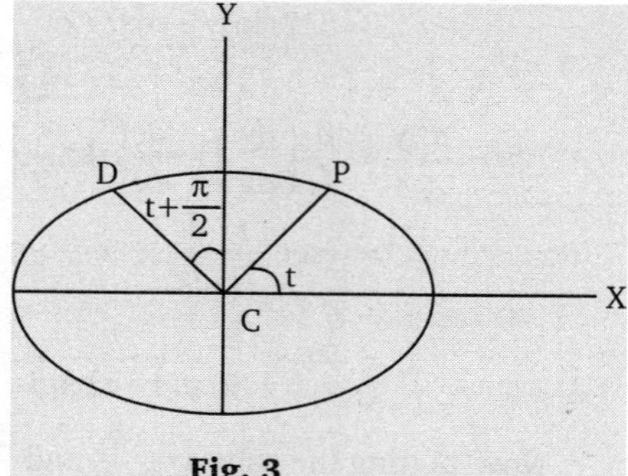

Fig. 3

From (1), we have

$$\frac{dx}{dt} = -a\sin t, \frac{dy}{dt} = b\cos t$$

$$\Rightarrow \frac{dy}{dx} = \frac{dy/dt}{dx/dt} = -\frac{b}{a}\cot t$$

$$\Rightarrow \frac{d^2y}{dx^2}=\frac{d}{dx}\left(\frac{dy}{dx}\right)=\frac{d}{dx}\left(-\frac{b}{a}\cot t\right)$$

$$=\frac{d}{dt}\left(-\frac{b}{a}\cot t\right)\frac{dt}{dx}$$

$$=\left(\frac{b}{a}\text{cosec}^2 t\right)\left(-\frac{1}{a}\text{cosec}\, t\right)$$

$$=-\frac{b}{a^2}\text{cosec}^3 t$$

Therefore, radius of curvature is given by

$$\rho=\frac{\left[1+\left(\frac{dy}{dx}\right)^2\right]^{3/2}}{-\frac{b}{a^2}\text{cosec}^3 t}=\frac{\left[1+\frac{b^2}{a^2}\cdot\frac{\cos^2 t}{\sin^2 t}\right]^{3/2}}{-\frac{b}{a^2}\text{cosec}^3 t}$$

$$=\frac{(a^2\sin^2 t+b^2\cos^2 t)^{3/2}}{ab}$$

$$\Rightarrow \rho=\frac{(a^2\sin^2 t+b^2\cos^2 t)^{3/2}}{ab} \quad ...(2)$$

(By neglecting the negative sign)

From figure

$$CD=\sqrt{(-a\sin t-0)^2+(b\cos t-0)^2}$$

$$=(a^2\sin^2 t+b^2\cos^2 t)^{1/2}$$

$$\therefore \frac{CD^3}{ab}=\frac{(a^2\sin^2 t+b^2\cos^2 t)^{3/2}}{ab} \quad ...(3)$$

Now from (2) and (3), we have $\rho=\frac{CD^3}{ab}$.

Example 6. *For the curve* $y=\frac{ax}{a+x}$, *if* ρ *is the radius of curvature at any point* (x, y), *show that* $\left(\frac{2\rho}{a}\right)^{2/3}=\left(\frac{y}{x}\right)^2+\left(\frac{x}{y}\right)^2$. (VTU–2008)

Solution. Let
$$y=\frac{ax}{a+x} \quad ...(1)$$

Therefore, $\frac{dy}{dx}=a\frac{a+x-x}{(a+x)^2}=a^2(a+x)^{-2}$

Now, again $\frac{d^2y}{dx^2}=\frac{d}{dx}\left(\frac{dy}{dx}\right)$

$$=-2a^2(a+x)^{-3}=\frac{-2a^2}{\left(\frac{ax}{y}\right)^3}$$

$$\Rightarrow \frac{d^2y}{dx^2}=\frac{-2y^3}{ax^3}$$

$$\therefore 1+\left(\frac{dy}{dx}\right)^2=1+\frac{a^4}{(a+x)^4}$$

$$=1+\frac{a^4}{\left(\frac{ax}{y}\right)^2}=1+\frac{y^4}{x^4}$$

$$\therefore \rho=\frac{\left[1+\left(\frac{dy}{dx}\right)^2\right]^{3/2}}{d^2y/dx^2}$$

$$=\frac{[(x^4+y^4)/x^4]^{3/2}}{(2y^3/ax^3)}$$

$$=\frac{a(x^4+y^4)^{3/2}}{2x^6(y^3/x^3)}=\frac{a}{2}\frac{(x^4+y^4)^{3/2}}{x^3y^3}$$

Hence,

$$\left(\frac{2\rho}{a}\right)^{2/3}=\frac{x^4+y^4}{x^2y^2}=\frac{x^2}{y^2}+\frac{y^2}{x^2}$$

$$\Rightarrow \left(\frac{2\rho}{a}\right)^{2/3}=\left(\frac{x}{y}\right)^2+\left(\frac{y}{x}\right)^2.$$

Example 7. *Find the radius of curvature at origin for the curve* $x^3+y^3-2x^2+6y=0$. (Burdwan 2003)

Solution. The curve passes through origin. Equating to zero the lowest degree terms we get $y=0$ *i.e.*, x axis as tangent to the curve at origin.

$\therefore$ By Newtons method, ρ (at origin) $=\lim_{x\to 0}\frac{x^2}{2y}$

Dividing by $2y$, the equation of the curve can be written as

$$x.\frac{x^2}{2y}+\frac{1}{2}y^2-2.\frac{x^2}{2y}+3=0$$

Taking limit as $x\to 0, y\to 0$ and $\lim_{x\to 0}\frac{y^2}{2x}=\rho$, we get

$$0.\rho+0-2\rho+3=0 \quad i.e.,\ \rho=3/2$$

Example 8. *Show that the radius of curvature of the curve* $y^2=x^2(a+x)/(a-x)$ *at the origin are* $\pm a\sqrt{2}$.

Solution. The curve passes through the origin and tangent at origin are $y^2=x^2$ *i.e.*, $y=\pm x$.

Thus neither of the co-ordinates axes is tangent at the origin. Therefore we can not apply by Newtons method. But the equation of the curve can be written as

$$y=\pm x\frac{(a+x)^{1/2}}{(a-x)^{1/2}}$$

or $$y=\pm x\left(1+\frac{x}{a}\right)^{1/2}\left(1-\frac{x}{a}\right)^{-1/2}$$

$$=\pm x\left\{1+\frac{1}{2}.\frac{x}{a}+...\right\}\left\{1+\frac{1}{2}.\frac{x}{a}+...\right\}$$

or $$y=\pm x\left\{1+\frac{x}{a}+...\right\}$$

On comparing this equation with

$$y = px + q.\frac{x^2}{2} + ...,\text{ we get}$$

$$p = 1,\ q = \frac{2}{a}\ \text{ or }\ p = -1, q = -\frac{2}{a}$$

But ρ (at origin) $= \dfrac{(1+\rho^2)^{3/2}}{4}$

Therefore ,

$$\rho\text{ (at origin)} = \frac{(1+1)^{3/2}}{2/a} = a\sqrt{2}$$

Also when $p = -1, q = -2/a$,

$$\rho\text{ (at origin)} = \frac{(1+1)^{3/2}}{-2/a} = -a\sqrt{2}.$$

Example 9. *If ρ_1 and ρ_2 be the radii of curvature of the extremities of two conjugate diameters of an ellipse prove that*

$$(\rho_1^{2/3} + \rho_2^{2/3})(ab)^{2/3} = a^2 + b^2.$$

Solution . Let the equation of an ellipse be

$$\frac{x^2}{a^2} + \frac{y^2}{b^2} = 1. \quad ...(1)$$

Let $P(a\cos\theta, b\sin\theta)$ and $Q(-a\sin\theta, b\cos\theta)$ be the extremities of two conjugate diameters of (1).

Differentiating both sides of (1) w.r.t x we get

$$\frac{2x}{a^2} + \frac{2y}{b^2}.\frac{dy}{dx} = 0$$

or

$$\frac{dy}{dx} = -\frac{b^2x}{a^2y} \quad ...(2)$$

Again differentiating, we get

$$\frac{d^2y}{dx^2} = -\frac{b^2}{a^2}\left[\frac{y - x\frac{dy}{dx}}{y^2}\right]$$

$$= -\frac{b^2}{a^2}\left[\frac{y - x\left(-\frac{b^2x}{a^2y}\right)}{y^2}\right] = -\frac{b^2}{a^2}\left[\frac{\left(\frac{y^2}{b^2} + \frac{x^2}{a^2}\right)b^2}{y^3}\right]$$

$$= -\frac{b^4}{a^2y^3} \quad \text{[Using (1)]}$$

We know that

$$\rho = \frac{\left[1 + \left(\frac{dy}{dx}\right)^2\right]^{3/2}}{\frac{d^2y}{dx^2}} = \frac{\left[1 + \left(\frac{b^2x}{a^2y}\right)^2\right]^{3/2}}{-b^4/a^2y^3}$$

$$\rho = \frac{(a^4y^2 + b^4x^2)^{3/2}}{-a^4b^4}$$

At $P(a\cos\theta, b\sin\theta)$, $\quad \rho = \rho_1$

$$\therefore\quad \rho_1 = \frac{(a^4.b^2\sin^2\theta + b^4a^2\cos^2\theta)^{3/2}}{-a^4b^4}$$

or

$$\rho_1 = \frac{(a^2\sin^2\theta + b^2\cos^2\theta)^{3/2}}{-ab}$$

or

$$\rho_1(-ab) = (a^2\sin^2\theta + b^2\cos^2\theta)^{3/2}$$

or

$$\rho_1^{2/3}(ab)^{2/3} = a^2\sin^2\theta + b^2\cos^2\theta \quad ...(3)$$

At $Q(-a\sin\theta, b\cos\theta), \rho = \rho_2$

$$\therefore\quad \rho_2^{2/3}(ab)^{2/3} = a^2\cos^2\theta + b^2\sin^2\theta \quad ...(4)$$

Adding (3) and (4), we get

$$(\rho_1^{2/3} + \rho_2^{2/3})(ab)^{2/3} = a^2 + b^2$$

Example 10. *Prove that for the ellipse $\dfrac{x^2}{a^2} + \dfrac{y^2}{b^2} = 1$, $\rho = \dfrac{a^2b^2}{p^3}$ p being the perpendicular from centre upon the tangent at (x, y).* (JNTU–2002)

Solution . We have $\dfrac{x^2}{a^2} + \dfrac{y^2}{b^2} = 1 \Rightarrow \dfrac{dy}{dx} = -\dfrac{b^2x}{a^2y}$

and

$$\frac{d^2y}{dx^2} = -\frac{b^2}{a^2}\left[\frac{y - x\frac{dy}{dx}}{y^2}\right] = -\frac{b^4}{a^2y^3}$$

Let $(a\cos\theta, b\sin\theta)$ be any point on the ellipse. The equation of the tangent at this point is

$$y - b\sin\theta = \frac{-b\cos\theta}{a\sin\theta}(x - a\cos\theta)$$

or

$$bx\cos\theta + ay\sin\theta - ab = 0 \quad ...(2)$$

We are given that

$p =$ Perpendicular from (0, 0) to the tangent (2)

or

$$p = \frac{-ab}{\sqrt{b^2\cos^2\theta + a^2\sin^2\theta}} \quad ...(3)$$

Now the radius of curvature ρ is

$$\rho = \frac{\left[1 + \left(\frac{dy}{dx}\right)^2\right]^{3/2}}{\frac{d^2y}{dx^2}} = \frac{a^2y^3\left(1 + \frac{b^4x^2}{a^4y^2}\right)}{-b^4}$$

$$= \frac{(a^4y^2 + b^4x^2)^{3/2}}{a^4b^4}$$

The ρ at $(a\cos\theta, b\sin\theta)$ is given by

$$\rho = -\frac{(a^4b^2\sin^2\theta + b^4a^2\cos^2\theta)^{3/2}}{a^4b^4}$$

$$= -\frac{(a^2\sin^2\theta + b^2\cos^2\theta)^{3/2}}{ab}$$

$$= -\frac{(-ab/p)^3}{ab} \quad \text{[Using (3)]}$$

$$\rho = \frac{a^2b^2}{p^3}.$$

Example 11. *Show that the radius of curvature at any point of the cycloid $x = a(\theta + \sin\theta)$, $y = a(1 - \cos\theta)$ is $4a\cos\frac{\theta}{2}$.* (PTU–2006, VTU–2011)

Solution . We have $x = a(\theta + \sin\theta) \Rightarrow \frac{dx}{d\theta} = a(1+\cos\theta)$

$$y = a(1-\cos\theta) \Rightarrow \frac{dy}{d\theta} = a\sin\theta$$

Now,
$$\frac{dy}{dx} = \frac{dy}{d\theta}\Big/\frac{dx}{d\theta} = \frac{a\sin\theta}{a(1+\cos\theta)}$$
$$= \frac{2a\sin\theta/2\cos\theta/2}{2\cos^2\theta/2} = \tan\theta/2$$

and
$$\frac{d^2y}{dx^2} = \frac{d}{d\theta}\left(\frac{dy}{dx}\right)\frac{d\theta}{dx} = \frac{1}{2}\sec^2\frac{\theta}{2}\cdot\frac{1}{a(1+\cos\theta)}$$
$$= \frac{1}{2}\sec^2\frac{\theta}{2}\cdot\frac{1}{2a\cos^2\frac{\theta}{2}} = \frac{1}{4a}\sec^4\frac{\theta}{2}$$

Hence,
$$\rho = \frac{\left[1+\left(\frac{dy}{dx}\right)^2\right]^{3/2}}{\frac{d^2y}{dx^2}} = \frac{4a\left(1+\tan^2\frac{\theta}{2}\right)^{3/2}}{\sec^4\frac{\theta}{2}}$$
$$= 4a\left(\sec^2\frac{\theta}{2}\right)^{3/2}.\cos^4\frac{\theta}{2} = 4a\cos\frac{\theta}{2}.$$

Example 12. *If ρ_1 and ρ_2 be the radii of curvature at the ends of a focal chord of the parabola $y^2 = 4ax$, then show that $\rho_1^{-2/3} + \rho_2^{-2/3} = (2a)^{-2/3}$.*

(Kurukshetra–2005, Rohtak–2006)

Solution . We have $y^2 = 4ax$ ...(1)

Parametric form of (1) is given by

$$x = at^2, y = 2at$$

$\therefore$ $x' = 2at, y' = 2a$

and $x'' = 2a, y'' = 0$

Therefore, radius of curvature ρ at $(at^2, 2at)$ is given by

$$\rho = \frac{(x'^2+y'^2)^{3/2}}{x'y''-x''y'}$$
$$= \frac{(4a^2t^2+4a^2)^{3/2}}{0-4a^2} = 2a(1+t^2)^{3/2}$$

If $P(t_1)$ and $Q(t_2)$ be the extremities of the focal chord of the parabola, then

$$t_1t_2 = -1 \Rightarrow t_2 = -\frac{1}{t_1}$$

So, ρ_1 at $P(t_1) = 2a(1+t_1^2)^{3/2}$

ρ_2 at $Q(t_2) = 2a(1+t_2^2)^{3/2}$

$$\therefore \rho_1^{-2/3} + \rho_2^{-2/3}$$
$$= (2a)^{-2/3}.[(1+t_1^2)^{-1} + (1+t_2^2)^{-1}]$$
$$= (2a)^{-2/3}.\left[\frac{1}{1+t_1^2} + \frac{t_1^2}{1+t_1^2}\right] = (2a)^{-2/3}.$$

Example 13. *In the ellipse $\frac{x^2}{a^2} + \frac{y^2}{b^2} = 1$, show that the radius of curvature of an end of the major axis is equal to its latusrectrum of the ellipse.* (Osmania–2000)

Solution . We have $\frac{x^2}{a^2} + \frac{y^2}{b^2} = 1$ ...(1)

$$\Rightarrow \frac{2x}{a^2} + \frac{2y}{b^2}\frac{dy}{dx} = 0 \Rightarrow \frac{dy}{dx} = -\frac{b^2}{a^2}\left(\frac{x}{y}\right)$$

Therefore,
$$\frac{d^2y}{dx^2} = -\frac{b^2}{a^2}\left[\frac{y.1 + x(dy/dx)}{y^2}\right]$$
$$= -\frac{b^2}{a^2y^2}\left[y - x\left(-\frac{b^2x}{a^2y}\right)\right]$$
$$= -\frac{b^2}{a^2y^2}\left(\frac{a^2y^2+b^2x^2}{a^2y}\right)$$
$$= -\frac{b^2}{a^2y^2}\left(\frac{a^2b^2}{a^2y}\right) \quad \text{[Using (1)]}$$
$$= -\frac{b^4}{a^2y^3}$$

Hence, ρ at (x, y)

$$= \frac{\left[1+\left(\frac{dy}{dx}\right)^2\right]^{3/2}}{\frac{d^2y}{dx^2}} = \frac{\left[1+\left(-b^2x/a^2y\right)^2\right]^{3/2}}{b^4/a^2y^3}$$
$$= \frac{\left(a^4y^2+b^4.x^2\right)^{3/2}}{a^4b^4}$$

Now, the coordinate of one end of major axis are $(a, 0)$,

$$\therefore \rho_{at(a,0)} = \frac{\left(a^4.0+b^4.x^2\right)^{3/2}}{a^4b^4} = \frac{b^2}{a},$$

semi-latusrectrum of the ellipse.

EXERCISE 5.1

1. Find the radius of curvature of the following curves:

(i) $x^{1/2} + y^{1/2} = a^{1/2}$ (ii) $a^2 y = x^3 - a^3$

(iii) $x^{2/3} + y^{2/3} = a^{2/3}$ (JNTU-2005)

(iv) $x^m + y^m = 1$

(v) $\sqrt{x} + \sqrt{y} = 1$ at $\left(\frac{1}{4}, \frac{1}{4}\right)$ (JNTU-2006)

(vi) $s = 4a \sin \psi$ at (s, ψ) (vii) $ay^2 = x^3$

(viii) $y = e^x$ at the point where it cuts the y-axis.

(ix) $x^{2/3} + y^{2/3} = a^{2/3}$ at $(a \cos^3 \theta, a \sin^3 \theta)$ (Anna–2009)

(x) $y = 4 \sin x - \sin 2x$ at $x = \frac{\pi}{2}$ (VTU–2009)

(xi) $y = x^3(x - a)$ at $(a, 0)$ (VTU–2010)

2. Find the radius of curvature at the origin of the following curves :

(i) $x^3 + y^3 = 3axy$ (ii) $y = x^3 + 5x^2 + 6x$

(iii) $5x^3 + 7y^3 + 4x^2y + xy^2 + 2x^2 + 3xy + y^2 + 4x = 0$

(iv) $a(y^2 - x^2) = x^3$ (v) $y - x = x^2 + 2xy + y^2$

(vi) $2x^4 + 4x^3 + xy^2 + 6y^3 - 3x^2 - 2xy + y^2 - 4x = 0$

(vii) $\sqrt{x} + \sqrt{y} = a$ at $\left(\frac{a}{4}, \frac{a}{4}\right)$ (JNTU–2006)

3. Show that the curvature at a point of the curve $y = f(x)$ is given by $\frac{d^2y}{dx^2} \cos^3 \psi$, where ψ is the inclination of the tangent at the point to the axis of x.

4. Show that for the curve $s = ae^{x/a}$, $a\rho = s(s^2 - a^2)^{1/2}$.

5. Show that if ρ be the radius of curvature at any point P on the parabola $y^2 = 4ax$ and S be its focus, then ρ varies as $(SP)^2$. (Kurukshetra–2006)

6. Show that for any curve $\frac{1}{\rho} = \frac{d}{dx}\left(\frac{dy}{dx}\right)$.

Hint to Selected Problems

1. (i) Differentiating two times the given equation w.r.t. x, we get
$$\frac{d^2y}{dx^2} = \frac{x^{1/2} + y^{1/2}}{2x^{3/2}} = \frac{a^{1/2}}{2x^{3/2}}.$$
Then put this value in the required formula.

2. (i) The given curve is passes through the origin. Therefore, equating the lowest degree term equal to zero, *i.e.*, $x = 0$ and $y = 0$ are the required trangent.

Then use $\rho = \lim_{x \to 0}\left(\frac{y^2}{x^2}\right)$.

3. We have $\tan \psi = \left(\frac{dy}{dx}\right)$. Put this value in the formula of curvature.

4. Find $\frac{dy}{dx}$ and $\frac{d^2y}{dx^2}$ by using $\frac{ds}{dx} = \sqrt{1 + \left(\frac{dy}{dx}\right)^2}$.

$$\therefore \quad \frac{d^2y}{dx^2} = \frac{s^2}{a^2\sqrt{s^2 - a^2}}.$$

Then use the formula of radius of curvature.

5. Let $P(at^2, 2at)$ be any point on $y^2 = 4ax$ and $S = (a, 0)$ be the coordionate of its focus. Then

$$SP = \sqrt{(at - a)^2 + (2at)^2} = \sqrt{(at^2 + a)^2} = t^2 + 1$$

Now find $\frac{d^2y}{dx^2}$ by the equation of the parabola.

Then find the relation between ρ and SP, by using formula of radius of curvature.

6. Since we have $\frac{ds}{dy} = \sqrt{1 + \left(\frac{dx}{dy}\right)^2} \Rightarrow \frac{dy}{ds} = \frac{ds/dx}{\sqrt{1 + \left(\frac{dy}{dx}\right)^2}}$

$$\therefore \quad \frac{d}{dx}\left(\frac{dy}{ds}\right) = \frac{\left(\frac{d^2y}{dx^2}\right)}{\left[1 + \left(\frac{dy}{dx}\right)^2\right]^{3/2}} = \frac{1}{\rho}.$$

ANSWERS

1. (i) $\frac{2(x + y)^{3/2}}{\sqrt{y}}$ (ii) $\frac{(a^4 + 9x^4)^{3/2}}{6a^4x}$ (iii) $3a^{1/3}x^{1/3}y^{1/3}$ (iv) $\frac{(x^{2m-2} + y^{2m-2})^{3/2}}{(1 - m)x^{m-2}y^{m-2}}$ (v) $\frac{1}{\sqrt{2}}$

(vi) $4a \cos \psi$ (vii) $\frac{1}{6a}(4a + 9x)^{3/2}x^{1/2}$ (viii) $\sqrt{8}$ (ix) $3a \sin \theta \cos \theta$ (x) $\frac{5\sqrt{5}}{4}$ (xi) $(1 + a^3)^{3b}/6a^2$

2. (i) $\frac{3a}{2}$ (ii) $\frac{37\sqrt{37}}{10}$ (iii) -2 (iv) $2a\sqrt{2}$ (v) $\frac{1}{2\sqrt{2}}$ (vi) 2 (vii) $\frac{a}{\sqrt{2}}$

5.5 RADIUS OF CURVATURE FOR PEDAL EQUATIONS

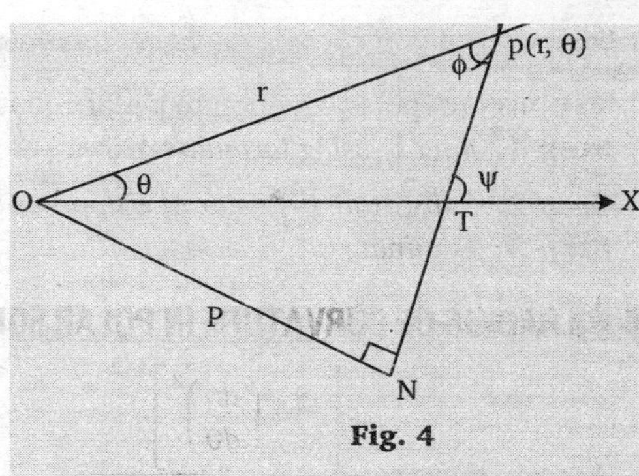

Fig. 4

To prove that $\rho = r\dfrac{dr}{dp}$

Proof. Let the pedal equation of the curve be $p = f(r)$.

Form the adjoining figure, we have

$$\psi = \theta + \phi$$

$$\Rightarrow \quad \frac{d\psi}{ds} = \frac{d\theta}{ds} + \frac{d\phi}{ds} \Rightarrow \frac{1}{\rho} = \frac{d\theta}{ds} + \frac{d\phi}{ds} \qquad \ldots(1)$$

Since, we know that

$$p = r\sin\phi$$

$$\therefore \quad \frac{dp}{dr} = \sin\phi + r\cos\phi\frac{d\phi}{dr}$$

$$= r.\frac{d\theta}{ds} + r\frac{dr}{ds}.\frac{d\phi}{dr} \qquad \left[\because \sin\phi = r.\frac{d\theta}{ds} \text{ and } \cos\phi = \frac{dr}{ds}\right]$$

$$= r\left[\frac{d\theta}{ds} + \frac{d\phi}{ds}\right] = r\frac{1}{\rho}$$

or $$\frac{dp}{dr} = r\frac{1}{\rho} \qquad \therefore \rho = \frac{r}{dp/dr} = r.\frac{dr}{dp} \quad \Rightarrow \quad \rho = r\frac{dr}{dp}.$$

5.6 RADIUS OF CURVATURE FOR TANGENTIAL POLAR EQUATIONS $p = f(\psi)$

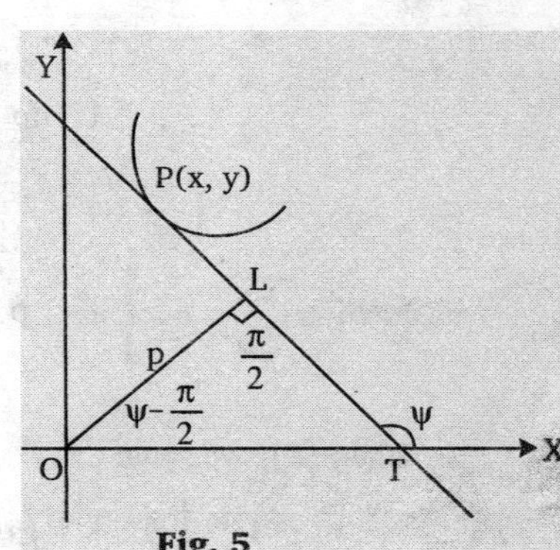

Fig. 5

To prove that $\rho = p + \dfrac{d^2p}{d\psi^2}$

Proof. Let p be the length of the perpendicular drawn from the origin on the tangent to curve at the point $P(x, y)$. Also, let ψ be the angle which the tangent makes with X-axis.

Here we observe that OL makes an angle $\psi - \dfrac{\pi}{2}$ with the positive direction of X-axis.

$\therefore$ Equation of the tangent PT is

$$p = X\cos\left(\psi - \frac{\pi}{2}\right) + \psi\sin\left(\psi - \frac{\pi}{2}\right)$$

$$\Rightarrow \quad p = X\sin\psi - Y\cos\psi \qquad [\text{Normal form}, x\cos\alpha + y\sin\alpha = p]$$

where X and Y are cartesian co-ordinates of any point on the tangent PT.

Since, $P(x, y)$ lies on PT, therefore

$$p = x\sin\psi - y\cos\psi \qquad \ldots(1)$$

$$\Rightarrow \quad \frac{dp}{d\psi} = x\cos\psi + \sin\psi\frac{dx}{d\psi} + y\sin\psi - \cos\psi.\frac{dy}{d\psi}$$

$$= x\cos\psi + y\sin\psi + \sin\psi\frac{dx}{ds}.\frac{ds}{d\psi} - \cos\psi.\frac{dy}{ds}.\frac{ds}{d\psi}$$

$$= x\cos\psi + y\sin\psi + \sin\psi.\rho.\cos\psi - \cos\psi.\rho.\sin\psi \qquad \left(\because \frac{dx}{ds} = \cos\psi \text{ and } \frac{dy}{ds} = \sin\psi\right)$$

$$= x\cos\psi + y\sin\psi$$

Differentiating again w.r.t. ψ, we get

$$\frac{d^2p}{d\psi^2} = -x\sin\psi + \cos\psi.\frac{dx}{d\psi} + y\cos\psi + \sin\psi.\frac{dy}{d\psi}$$

$$= -x\sin\psi + y\cos\psi + \cos\psi.\frac{dx}{ds}.\frac{ds}{d\psi} + \sin\psi.\frac{dy}{ds}.\frac{ds}{d\psi}$$

$$= (-x\sin\psi + y\cos\psi) + \cos\psi.\cos\psi.\rho + \sin\psi.\sin\psi.\rho$$

$$= -p + \rho[\cos^2\psi + \sin^2\psi] \qquad \text{(Using (1))}$$

$$\Rightarrow \quad \rho = p + \frac{d^2p}{d\psi^2}.$$

WORKING PROCEDURE

To transform polar equation to pedal equation, proceed as follows :

Step 1. *Find* ϕ, *using formula* $\tan\phi = \dfrac{r}{dr/d\theta}$.

Step 2. *Substitute the value of* ϕ *in* $p = r\sin\phi$.

Step 3. *Eliminate* θ.

5.7 RADIUS OF CURVATURE IN POLAR FORM

To prove that $\rho = \dfrac{\left[r^2+\left(\dfrac{dr}{d\theta}\right)^2\right]^{3/2}}{r^2+2\left(\dfrac{dr}{d\theta}\right)^2 - r\dfrac{d^2r}{d\theta^2}}$

Proof. We know that $$\frac{1}{p^2} = \frac{1}{r^2} + \frac{1}{r^4}\left(\frac{dr}{d\theta}\right)^2. \qquad \ldots(1)$$

Differentiating (1) w.r.t. r, we get

$$-\frac{2}{p^3}\frac{dp}{dr} = -\frac{2}{r^3} - \frac{4}{r^5}\left(\frac{dr}{d\theta}\right)^2 + \frac{1}{r^4}\left\{\frac{d}{dr}\left(\frac{dr}{d\theta}\right)^2\right\}$$

$$= -\frac{2}{r^3} - \frac{4}{r^5}\left(\frac{dr}{d\theta}\right)^2 + \frac{1}{r^4}\left[\frac{d}{d\theta}\left(\frac{dr}{d\theta}\right)^2\right]\cdot\frac{d\theta}{dr} = -\frac{2}{r^3} - \frac{4}{r^5}\left(\frac{dr}{d\theta}\right)^2 + \frac{2}{r^4}\frac{d^2r}{d\theta^4}$$

$$\frac{1}{p^3}\cdot\frac{dp}{dr} = \frac{1}{r^5}\left[r^2 + 2\left(\frac{dr}{d\theta}\right)^2 - r\frac{d^2r}{d\theta^2}\right]$$

Now $$\rho = r\frac{dr}{dp} = \frac{r\cdot\dfrac{1}{p^3}}{\dfrac{1}{r^5}\left[r^2+2\left(\dfrac{dr}{d\theta}\right)^2 - r\dfrac{d^2r}{d\theta^2}\right]}$$

Form (1), we have

$$\frac{1}{p^3} = \left[\frac{1}{r^2} + \frac{1}{r^4}\left(\frac{dr}{d\theta}\right)^2\right]^{3/2} = \frac{1}{r^6}\left[r^2 + \left(\frac{dr}{d\theta}\right)^2\right]^{3/2}$$

Hence, $$\rho = \frac{r^6\cdot\dfrac{1}{r^6}\left[r^2+\left(\dfrac{dr}{d\theta}\right)^2\right]^{3/2}}{r^2+2\left(\dfrac{dr}{d\theta}\right)^2 - r\dfrac{d^2r}{d\theta^2}} \Rightarrow \rho = \frac{\left[r^2+\left(\dfrac{dr}{d\theta}\right)^2\right]^{3/2}}{r^2+2\left(\dfrac{dr}{d\theta}\right)^2 - r\dfrac{d^2r}{d\theta^2}}.$$

Recapitulations

Radius of curvature

- $\rho = r\dfrac{dr}{dp}$ (For Pedal equation)
- $\rho = p + \dfrac{d^2p}{d\psi^2}$ (For tangential polar equation)
- $\rho = \dfrac{\left[r^2+\left(\dfrac{dr}{d\theta}\right)^2\right]^{3/2}}{r^2+2\left(\dfrac{dr}{d\theta}\right)^2 - r\dfrac{d^2r}{d\theta^2}}$ (For polar form)

Solved Examples

Example 1. *Find the radius of curvature for the curve* $r^n = a^n\cos n\theta$. (JNTU–2006, PTU–2010)

Solution. We have $r^n = a^n\cos n\theta$

$\Rightarrow \quad n\log r = n\log a + \log\cos n\theta.$

Now differentiating w.r.t. θ, we get

$$\frac{n}{r}\cdot\frac{dr}{d\theta} = 0 + \frac{1}{\cos n\theta}(-n\sin n\theta) = -n\tan n\theta \qquad \ldots(1)$$

$\Rightarrow \quad r_1 = -r\tan n\theta$

Again diiferentiating, we get

$$r_2 = -r.n.\sec^2 n\theta - r_1.\tan n\theta$$
$$= -rn\sec^2 n\theta + r\tan^2 n\theta. \qquad \ldots(2)$$

Putting all these values in

$$\rho = \frac{[r^2+r_1^2]^{3/2}}{r^2+2r_1^2 - rr_2}$$

$$= \frac{(r^2+r^2\tan^2 n\theta)^{3/2}}{r^2+2r^2\tan^2 n\theta + r^2.n\sec^2 n\theta - r^2\tan^2 n\theta}$$

$$= \frac{r^3 \sec^3 n\theta}{(n+1)r^2 \sec^2 n\theta} = \frac{r \sec n\theta}{(n+1)} = \frac{r}{n+1} \cdot \frac{1}{\cos n\theta}$$

$$= \frac{r}{(n+1)\frac{r^n}{a^n}} = \frac{a^n}{(n+1)r^{n-1}}$$

Example 2. *Show that in the rectangular hyperbola $r^2 \cos 2\theta = a^2$, the radius of curvature $\rho = \frac{r^3}{a^2}$.*

Solution . The given curve is

$$r^2 = \cos 2\theta = a^2 \quad \ldots(1)$$

$\Rightarrow$ $2 \log r + \log \cos 2\theta = 2 \log a$

Differentiating w.r.t. θ, we get

$$\frac{2}{r}\frac{dr}{d\theta} + \frac{1}{\cos 2\theta}(-2\sin 2\theta) = 0$$

$$\Rightarrow \quad \frac{1}{r}\frac{dr}{d\theta} = \cot\phi = \tan 2\theta = \cot\left(\frac{\pi}{2} - 2\theta\right)$$

$$\Rightarrow \quad \phi = \frac{\pi}{2} - 2\theta$$

Now $$p = r\sin\phi = r\sin\left(\frac{\pi}{2} - 2\theta\right)$$

$$= r\cos 2\theta = r.\frac{a^2}{r^2} = \frac{a^2}{r}$$

$$\Rightarrow \quad \frac{dp}{dr} = -\frac{a^2}{r^2}$$

Hence, $$\rho = r\frac{dr}{dp} = -\frac{r^3}{a^2} = \frac{r^3}{a^2}.$$

(By neglecting the negative sign)

Example 3. *Show that for the hypercycloid $P = A \sin B\psi$, ρ varies as P.*

Solution . Differentiating $P = A \sin \psi$ By with respect to ψ, we get $\frac{dP}{d\psi} = AB\cos B\psi,$

$$\therefore \quad \frac{d^2P}{d\psi^2} = -AB^2 \sin B\psi$$

Therefore,

$$\rho = P + \frac{d^2P}{d\psi^2} = A\sin B\psi - AB\sin B\psi = (1 - B^2)P$$

Hence, $\rho \propto P$.

Example 4. *Find the radius of curvature at the point (p, r) on the spiral*

$$p^2 = r^4/(r^2 + a^2).$$

Solution . The equation of curve is $p^2 = r^4/(r^2 + a^2)$.

$$\therefore \quad \frac{1}{p^2} = \frac{r^2 + a^2}{r^4} = \frac{1}{r^2} + \frac{a^2}{r^4}$$

$$\Rightarrow \quad \frac{1}{p^2} = \frac{1}{r^2} + \frac{a^2}{r^4} \quad \ldots(1)$$

Differentiating equation (1) w.r.t. r, we have

$$-\frac{2}{p^3}\frac{dp}{dr} = -\frac{2}{r^3} - \frac{4a^2}{r^5} = -2.\frac{r^2 + 2a^2}{r^5}$$

$$\therefore \quad \frac{dp}{dr} = p^3\frac{(r^2 + 2a^2)}{r^5} = (p^2)^{3/2}\left(\frac{r^2 + 2a^2}{r^5}\right)$$

$$= \left(\frac{r^4}{r^2 + a^2}\right)^{3/2} . \frac{(r^2 + 2a^2)}{r^5}$$

$$\frac{dp}{dr} = \frac{r(r^2 + 2a^2)}{(r^2 + a^2)^{3/2}}$$

Hence, $$\rho = r\frac{dr}{dp} = r.\frac{(r^2 + a^2)^{3/2}}{r(r^2 + 2a^2)}$$

$$\Rightarrow \quad \rho = \frac{(r^2 + a^2)^{3/2}}{r^2 + 2a^2}.$$

Example 5. *Prove that for any curve $\frac{r}{\rho} = \sin\phi\left(1 + \frac{d\phi}{d\theta}\right)$, where ρ is the radius of curvature and $\tan\phi = r\frac{d\theta}{dr}$.*

Solution . We know that the $\psi = \theta + \phi$. ...(1)

Differentiating (1) w.r.t. to s, we get

$$\frac{d\psi}{ds} = \frac{d\theta}{ds} + \frac{d\phi}{ds} = \frac{d\theta}{ds} + \frac{d\phi}{d\theta}.\frac{d\theta}{ds} = \frac{d\theta}{ds}\left(1 + \frac{d\phi}{d\theta}\right).$$

$$\therefore \quad \frac{1}{\rho} = \frac{\sin\phi}{r}\left(1 + \frac{d\phi}{d\theta}\right)$$

$$\left[\because \rho = \frac{ds}{d\psi} \text{ and } \sin\phi = r\frac{d\theta}{ds}\right]$$

or $$\frac{r}{\rho} = \sin\phi\left(1 + \frac{d\phi}{d\theta}\right).$$

Example 6. *Show that at any point on the equiangular spiral $r = ae^{\theta \cot \alpha}$, $\rho = r \operatorname{cosec} \alpha$ and that it subtends a right angle at the pole.*

Solution . The given equation is $r = ae^{\theta \cot \alpha}$. ...(1)

Differentiating (1) w.r.t. θ, we have

$$\frac{dr}{d\theta} = ae^{\theta\cot\alpha}.\cot\alpha = r\cot\alpha.$$

$$\therefore \quad (1/r)\frac{dr}{d\theta} = \cot\alpha$$

or $$\cot\phi = \cot\alpha \Rightarrow \phi = \alpha.$$

Now, $p = r \sin \phi$, thus the pedal equation of (1) is $p = r \sin \alpha$.

Therefore, $$\frac{dp}{dr} = \sin\alpha.$$

Now $$\rho = r\frac{dr}{dp} = \frac{r}{\sin\alpha} = r\operatorname{cosec}\alpha.$$

Second part. Let $P(r, \theta)$ be any point on the point given curve. PQ is the tangent and PR is the normal to the curve at P. Let R be center of curvatrure of the point P of the curve. Then PR = the radius of curvature of the curve at $P = r \operatorname{cosec} \alpha$.

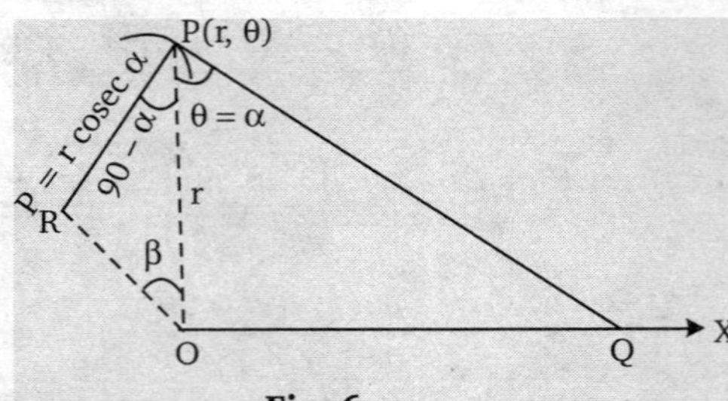

Fig. 6

Intersect OP and OR, where O is the pole.

Let $\angle POR = \beta$. Then to show that $\beta = 90°$.

We have

$$\angle OPQ = \phi = \alpha$$

$$\angle OPR = 90° - \alpha, \text{ (since } PR \text{ is normal at } P)$$

i.e., perpendicular to the tangent PQ.

Now in ΔOPR, we have

$$\angle ORP = 180° - (90° - \alpha + \beta) = 90° + \alpha - \beta.$$

Therefore, applying the sine theorem for ΔOPR, we get

$$\frac{OP}{\sin \angle ORP} = \frac{PR}{\sin \beta} \text{ or } \frac{r}{\sin(90 + \alpha - \beta)} = \frac{\rho}{\sin \beta}$$

$$\text{or } \frac{r}{\cos(\alpha - \beta)} = \frac{r \operatorname{cosec} \alpha}{\sin \beta} \qquad (\because \rho = r \operatorname{cosec} \alpha)$$

$$\sin \alpha \sin \beta = \cos(\alpha - \beta)$$

or $\sin \alpha \sin \beta = \cos \alpha \cos \beta + \sin \alpha \sin \beta$

or $\cos \alpha \cos \beta = 0$ or $\cos \beta = 0$.

Hence, $\beta = 90°$.

EXERCISE 5.2

1. FInd the radius of curvature in polar form on each of the following curves :
   (i) $r = a(1 - \cos \theta)$ (VTU–2003)
   (ii) $r(1 + \cos \theta) = 2a$ (iii) $r^2 = a^2 \cos 2\theta$
   (iv) $r^m = a^m \sin m\theta$ (v) $r = ae^{\theta \cot \alpha}$
2. Find the radius of curvature at any point (p,r) on the following curves :
   (i) $p^2 = ar$ (ii) $r^2 = a^2 - b^2 + \dfrac{a^2b^2}{p^2}$
   (iii) $2ap^2 = r^3$ (iv) $pa^2 = r^3$
3. Show that for the cardoid $r = a(1 + \cos \theta)$, $\rho = \dfrac{2}{3}\sqrt{2ar}$.
4. Show that the radius of curvature of the cardoid $r = a(1 + \cos \theta)$ at the origin is 0.
5. Show that the radius of curvature at any point on the curve $r = a(1 \pm \cos \theta)$ varies as square root of the radius vector.
6. If ρ_1, ρ_2 be the radii of curvature at the extrimities of any chord of the cardoid $r = a(1 + \cos \theta)$, which passes through the pole, then ${\rho_1}^2 + {\rho_2}^2 = 16a^2/9$.
7. Show that the radius of curvature at the point (p, r) of the ellipse $\dfrac{1}{p^2} = \dfrac{1}{a^2} + \dfrac{1}{b^2} - \dfrac{r^2}{a^2b^2}$ is $\dfrac{a^2b^2}{p^3}$. (VTU–2010)
8. Show that the radius of curvature for the hyperbola $p^2 = a^2 \cos^2 \psi + b^2 \sin^2 \psi$ is $\dfrac{a^2b^2}{p^3}$.
9. Show that the curvature of the curves $r = a\theta$ and $r\theta = a$ at their common point are in the ratio 3 : 1.
10. By Newton's method, show that the radius of curvature of the curve $r = a \sin n\theta$ at the origin is $\dfrac{na}{2}$.
11. Show that the radius of curvature at each point of the curve $x = a\left(\cos t + \log \tan \dfrac{t}{2}\right)$, $y = a \sin t$ is inversely proportional to the length of the normal intercepted between the point on the curve and the x-axis. (JNTU–2003)

Hint to Selected Problems

1. (i) Find $\dfrac{d^2r}{d\theta^2} (= -a\cos\theta)$ to the given equation and use the formula $\rho = \dfrac{\left[r^2 + \left(\dfrac{dr}{d\theta}\right)^2\right]^{3/2}}{r^2 + 2\left(\dfrac{dr}{d\theta}\right)^2 - r\dfrac{d^2r}{d\theta^2}}$.
2. For the curve of the type $r = f(p)$, the radius of curvature is $\rho = r\dfrac{dr}{dp}$.
3. $r = a(1 + \cos \theta) \Rightarrow \dfrac{dr}{d\theta} = -a\sin\theta \Rightarrow \dfrac{d^2r}{d\theta^2} = -a\cos\theta$.
   Put these values in the formula for radius of curvature.
4. Proceed as question (3). Finally, put $r = 0$.
5. Proceed same as question (3).
6. Let $P(r, \theta)$ and $Q(r, \pi + \theta)$ be the extremities of any chord of the cardoid $r = a(1 + \cos \theta)$
   Then for ρ_1, use $r = a(1 + \cos \theta)$
   $\Rightarrow \dfrac{dr}{d\theta} = -a\sin\theta$ and $\dfrac{d^2r}{d\theta^2} = -a\cos\theta$
   and for ρ_2, use $r = a[1 + \cos(\pi + \theta)] = a(1 - \cos \theta)$
   $\Rightarrow \dfrac{dr}{d\theta} = a\sin\theta$ and $\dfrac{d^2r}{d\theta^2} = a\cos\theta$
   Then find ρ_1 and ρ_2 and form a relation between ρ_1 and ρ_2.
7. Find $r\dfrac{dr}{dp}$ from the given equation
   i.e., $r\dfrac{dr}{dp} = \dfrac{a^2b^2}{p^3}$. Then use the formula $\rho = r\dfrac{dr}{dp}$.

8. If the curve is $p = f(\psi)$, then the radius of curvature is

$$\rho = p + \frac{d^2p}{d\psi^2}. \qquad \ldots(1)$$

Obtain the value of $\frac{d^2p}{d\psi^2}$ form the given equation and substitute in (1).

9. Clearly, $(a, 1)$ and $(a, -1)$ are the common points. Now find the radius of curvature for both the above points.

10. $r = a \sin n\theta$ is the equation of the given curve. At $r = 0 \Rightarrow \theta = 0$. Then, use the formula given below

$$\rho_{\text{at origin}} = \lim_{x \to 0} \frac{x^2}{2y}.$$

ANSWERS

1.(i) $\frac{2}{3}\sqrt{2ar}$ (ii) $2\sqrt{(r^3/a)}$ (iii) $\frac{a^2}{3r}$ (iv) $\frac{a^m}{(m+1)r^{m-1}}$ (v) $r \operatorname{cosec} \alpha$ **2.** (i) $\frac{2r^{3/2}}{\sqrt{a}}$ (ii) $\frac{a^2b^2}{p^3}$ (iii) $\frac{2}{3}\sqrt{2ar}$ (iv) $\frac{a^2}{3r}$

5.8 CENTER OF CURVATURE

For any point P of a curve, the center of curvature is the point on the positive direction of the normal at P, at a distance ρ from it.

Let PD be the normal curve at P and C be a point on it such that $PC = \rho$, then C is said to be the center of curvature at P.

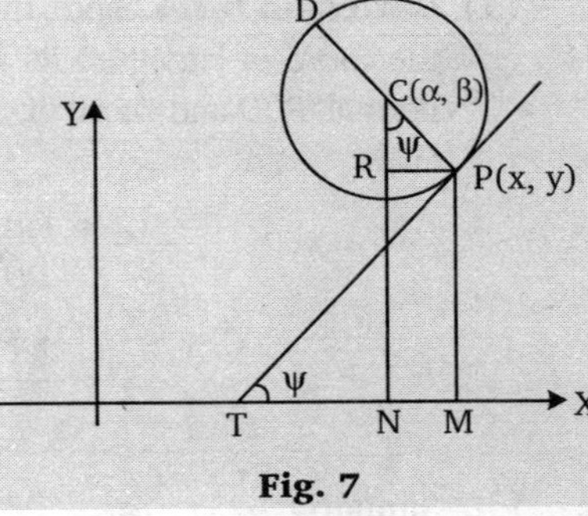

Fig. 7

5.8.1 EVOLUTE OF A CURVE

The locus of the center of curvature of the given curve is called the evolute of the curve.

5.8.2 CIRCLE OF CURVATURE

The circle with its center at the center of curvature c and radius equal to ρ is called the circle of curvature.

REMARK

- The circle of curvature touches the curve at P and both the curve and the circle of curvature have the same curvature at this point.

5.9 CO-ORDINATES OF THE CENTRE OF CURVATURE

Let $y = f(x)$ be the given curve and $P(x, y)$ be any given point.

Let $C(\alpha, \beta)$ be the center of curvature corresponding to any point $P(x, y)$ on the given curve, then from above fig. (7), we have $PC = \rho$.

Suppose, the tangent TP makes an angle ψ with positive direction of x-axis. Draw PM and CN perpendicular to x-axis and draw perpendicular to CN. Then

$$\angle PCN = 90^\circ - \angle CPR = 90^\circ - (90^\circ - \angle RPT) = \angle RPT = \angle PTX = \psi$$

$\therefore$
$$\alpha = ON = OM - NM = OM - RP = x - CP \sin\psi = x - \rho \sin\psi \qquad \ldots(1)$$

Also,
$$\beta = NC = NR + RC = MP + RC = y + CP \cos\psi = y + \rho\cos\psi \qquad \ldots(2)$$

Since, we know that

$$y_1 = \tan\psi$$

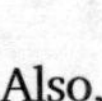

$$\sin\psi = \frac{y_1}{\sqrt{1+y_1^2}} \text{ and } \cos\psi = \frac{1}{\sqrt{1+y_1^2}}.$$

Also,
$$\rho = \frac{(1+y_1^2)^{3/2}}{y_2}$$

Putting all these values in (1) and (2), we get

$$\alpha = x - \frac{y_1(1+y_1^2)}{y_2} \text{ and } \beta = y + \frac{(1+y_1^2)}{y_2}.$$

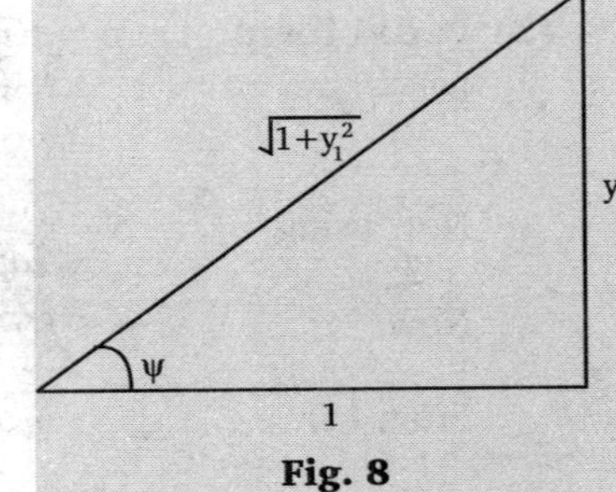

Fig. 8

REMARKS

- From (1) and (2) we have $\alpha = x - \rho \sin\psi$ and $\beta = y + \rho\cos\psi$. Since x, y, ρ, ψ depends upon s, therefore the above equations may be treated as parametric equations of the evolute.
- The equation of the circle of curvature at the given point is $(x-\alpha)^2 + (y-\beta)^2 = \rho^2$.

5.10 CHORD OF CURVATURE

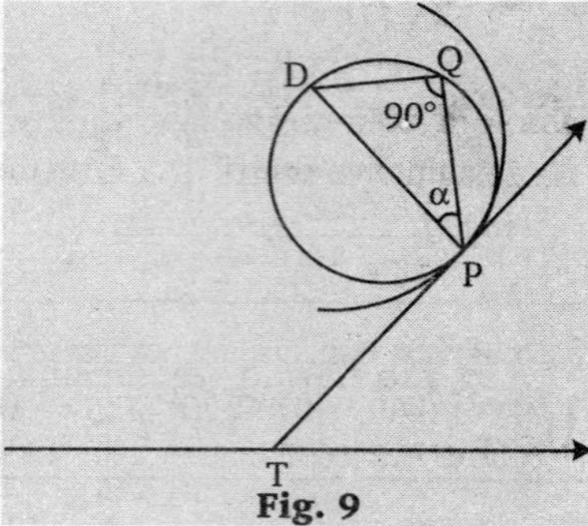

Fig. 9

The length intercepted by the circle of curvature of the curve at P, on a straight line drawn through P in any given direction is called chord of curvature through P in that direction.

Let the chord of curvature PQ makes an angle α, with the normal PD, then its length PQ is given by

$$PQ = PD \cos \alpha \qquad (\because \angle DQP, \text{ being a semicircle is a right angle.})$$

$$= 2\rho \cos \alpha.$$

5.11 LENGTH OF THE CHORD OF CURVATURE

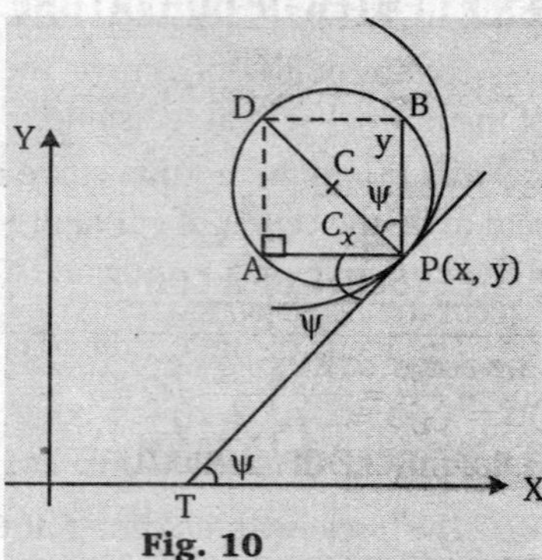

Fig. 10

(1) Cartesian form. Since, the tangernt at P makes an angle ψ with the x-axis therefore, the chord of curvature PA is parallel to x-axis, which makes an angle $90 - \psi$ with the normal PCD and chord of curvature PB parallel to y-axis makes angle ψ with the normal PCD.

$\therefore$ C_x = length of the chord of curvature PA, parallel to x-axis.

$$= PD \cos(90 - \psi) = 2\rho \sin \psi$$

$$= \frac{2(1+y_1^2)^{3/2}}{y_2} \cdot \frac{y_1}{\sqrt{1+y_1^2}} = \frac{2y_1(1+y_1^2)}{y_2}$$

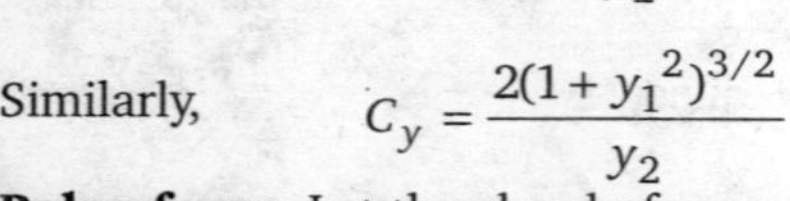

Similarly, $$C_y = \frac{2(1+y_1^2)^{3/2}}{y_2}.$$

(2) Polar form. Let the chord of curvature PL makes an angle $90 - \phi$ with PCD, the normal of the curve at P, and PM, the chord of curvature perpendicular to the radius vector OP, makes an angle ϕ with the normal PCD.

$\therefore$ C_o = Length of the chord of curvature PL through origin (or pole)

$$= PD(\cos 90 - \phi)$$

$$= 2\rho \sin\phi = \frac{2(r^2+r_1^2)^{3/2}}{r^2+2r_1^2-rr_2} \cdot \frac{r}{\sqrt{r^2+r_1^2}}$$

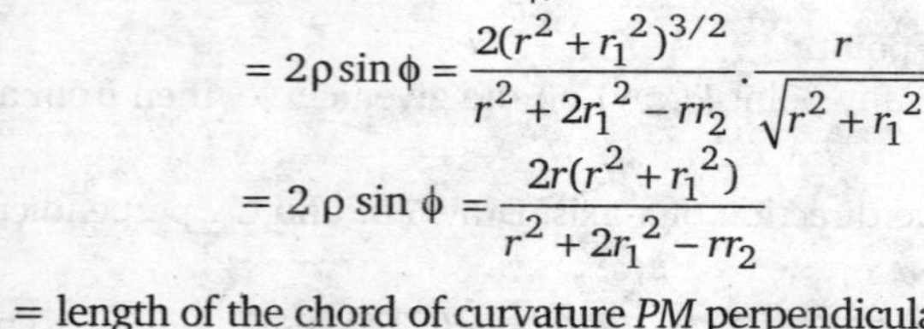

$$= 2\rho \sin \phi = \frac{2r(r^2+r_1^2)}{r^2+2r_1^2-rr_2}$$

and C_p = length of the chord of curvature PM perpendicular to radius vector.

$$= PD \cos \phi = 2\rho \cos \phi = \frac{2(r^2+r_1^2)^{3/2}}{r^2+2r_1^2-rr_2} \cdot \frac{r}{\sqrt{r^2+r_1^2}} = \frac{2r(r^2+r_1^2)}{r^2+2r_1^2-rr_2}$$

Fig. 11

(3) Pedal form. Let $p = f(r)$ be the given equation of the curve.

Let C_o = length of the chord of curvature through pole along radius vector

$$= PD \cos(90 - \phi) = 2\rho \sin \phi \qquad \ldots(1)$$

Now using $\rho = r\dfrac{dr}{dp}$ and $\sin\phi = \dfrac{p}{r}$ in (1), we get $C_o = 2r\dfrac{dr}{dp}\cdot\dfrac{p}{r} = 2p\cdot\dfrac{dr}{dp}$ ...(2)

Now $p = f(r) \Rightarrow \dfrac{dp}{dr} = f'(r)$ and $\sin\phi = \dfrac{p}{r} = \dfrac{f(r)}{r}$

$\therefore$ From (1), $$C_o = 2\rho\sin\phi = 2.r.\frac{dr}{dp}.\sin\phi = 2r.\frac{1}{f'(r)}.\frac{f(r)}{r} = \frac{2f(r)}{f'(r)}$$

Also C_p = length of the chord perpendiular to the radius vector

$$= DP \cos \phi = 2\rho \cos \phi = 2.r.\frac{dr}{dp}\frac{\sqrt{r^2-p^2}}{r}$$

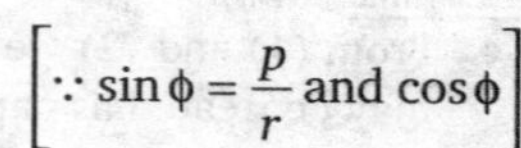

$\left[\because \sin\phi = \dfrac{p}{r} \text{ and } \cos\phi\right]$

$$= 2.\sqrt{r^2-p^2}.\frac{dr}{dp}.$$

Solved Examples

Example 1. *Find the chord of curvature through the pole of the cardioid $r = a(1 + \cos\theta)$.*

Solution. We have $r = a(1 + \cos\theta)$

$$\Rightarrow \frac{dr}{d\theta} = -a\sin\theta$$

$$\therefore \tan\phi = r\frac{d\theta}{dr} = \frac{a(1+\cos\theta)}{-a\sin\theta}$$

$$= -\cot\frac{1}{2}\theta = \tan\left(\frac{\pi}{2}+\frac{\theta}{2}\right)$$

Now $p = r\sin\phi = r\sin\left(\frac{\pi}{2}+\frac{\theta}{2}\right) = r\cos\frac{\theta}{2}$

$$\therefore \quad 2p^2 = r^2\left(2\cos^2\frac{\theta}{2}\right)$$

$$= r^2(1+\cos\theta) = r^2\frac{r}{a} = \frac{r^3}{a}$$

$\Rightarrow 2p^2a = r^3$ is the pedal equation of the curve.

On differentiating w.r.t. r we get

$$4ap\frac{dp}{dr} = 3r^2$$

$$\therefore \quad \rho = r\frac{dr}{dp} = r.\frac{4ap}{3r^2} = \frac{4ap}{3r}$$

Therefore, the chord of curvature through the pole

$$= 2\rho\sin\phi = 2.\frac{4ap}{3r}.\frac{p}{r} \quad [\because p = r\sin\phi]$$

$$= \frac{8ap^2}{3r^2} = \frac{8}{3r^2}.\frac{r^3}{2} = \frac{4r}{3} \quad [\because 2ap^2 = r^3].$$

Example 2. *Show that the chord of curvature through the pole of the curve $r^n = a^n\cos n\theta$ is $\frac{2r}{n+1}$.*

Solution. The given curve is

$$r^n = a^n\cos n\theta$$

$$\Rightarrow \quad n\log r = n\log a + \log\cos n\theta$$

Differentiating w.r.t. θ, we have

$$\frac{n}{r}\frac{dr}{d\theta} = -\frac{n}{\cos n\theta}.\sin n\theta$$

$$\Rightarrow \quad \cot\phi = -\tan n\theta = \cot\left(\frac{\pi}{2}+n\theta\right)$$

$$\therefore \quad \phi = \frac{\pi}{2}+n\theta$$

Now $p = r\sin\phi = r\sin\left(\frac{\pi}{2}+n\theta\right) = r\cos n\theta$

$\because$ Pedal equation of the curve is $p = \frac{r^{n+1}}{a^n}$.

$$\therefore \quad \frac{dp}{dr} = \frac{(n+1)r^n}{a^n}$$

Also, $$\rho = r\frac{dr}{dp} = \frac{a^n}{(n+1)r^{n-1}}$$

Therefore, the chord of curvature through pole is

$$= 2\rho\sin\phi = 2\rho\sin\left(\frac{\pi}{2}+n\theta\right) = 2\rho\cos n\theta$$

$$= 2\frac{a^n}{(n+1)r^{n-1}}.\frac{r^n}{a^n} = \frac{2r}{(n+1)}$$

Example 3. *Find the co-ordinate of the centre of curvature at any point of the parabola $y^2 = 4ax$. Hence, show that its evolute is $27ay^2 = 4(x-2a)^3$.* (VTU–2000)

Solution. We have $y^2 = 4ax$

$$\Rightarrow \quad 2yy_1 = 4a \text{ i.e., } y_1 = \frac{2a}{y}$$

and $$y_2 = -\frac{2a}{y^2}.y_1 = -\frac{4a^2}{y^3}$$

If $(\bar{x},\bar{y})$ be the centre of curvature, then

$$\bar{x} = x - \frac{y_1(1+y_1^2)}{y_2} = x - \frac{\frac{2a}{y}\left(1+\frac{4a^2}{y^2}\right)}{-4a^2/y^3}$$

$$= x + \frac{y^2+4a^2}{2a}$$

$$= x + \frac{4ax+4a^2}{2a} = 3x+2a \quad \ldots(1)$$

and $$\bar{y} = y + \frac{1+y_1^2}{y_2} = y + \frac{1+4a^2/y^2}{-4a^2/y^2}$$

$$= y - \frac{y(y^2+4a^2)}{4a^2} = \frac{-y^3}{4a^2} = -\frac{2x^{3/2}}{\sqrt{a}} \quad \ldots(2)$$

To find the required evolute, eliminate x from (1) and (2), we have

$$(\bar{y})^2 = \frac{4x^3}{a} = \frac{4}{a}\left(\frac{\bar{x}-2a}{3}\right)^3$$

$$\Rightarrow \quad 27a(\bar{y})^2 = 4(\bar{x}-2a)^3 \quad \ldots(3)$$

Now, locus of $(\bar{x},\bar{y})$ is $27ay^2 = 4(x-2a)^3$ which is the required equation of evolute.

Example 4. *Show that the evolute of the cycloid $x=a(\theta-\sin\theta)$, $y = a(1-\cos\theta)$ is another equal cycloid.* (Madras–2006)

Solution. We have $x = a(\theta-\sin\theta)$ and $y = a(1-\cos\theta)$

$$\Rightarrow \quad y_1 = \frac{dy}{d\theta}.\frac{d\theta}{dx} = \frac{a\sin\theta}{a(1-\cos\theta)} = \cot\frac{\theta}{2}$$

Now $$y_2 = \frac{d}{dx}(y_1) = \frac{d}{d\theta}\left(\cot\frac{\theta}{2}\right).\frac{d\theta}{dx}$$

$$= -\text{cosec}^2\frac{\theta}{2}.\frac{1}{2}.\frac{1}{a(1-\cos\theta)}$$

$$= -\frac{1}{4a\sin^4\theta/2}$$

If $(\bar{x},\bar{y})$ be the center of curvature, then

$$\bar{x} = x - \frac{y_1(1+y_1^2)}{y_2}$$

$$= a(\theta-\sin\theta) + \cot\frac{\theta}{2}(-4a\sin^2\frac{\theta}{2})(1+\cot^2\frac{\theta}{2})$$

$$= a(\theta-\sin\theta) + \frac{\cos\theta/2}{\sin\theta/2}.4a\sin^4\frac{\theta}{2}.\text{cosec}^2\frac{\theta}{2}$$

$= a(\theta - \sin\theta) + 4a\sin\frac{\theta}{2}.\cos\frac{\theta}{2}$

$= a(\theta - \sin\theta) + 2a\sin\theta$

$= a(\theta + \sin\theta)$

and $\bar{y} = y + \frac{1+y_1^2}{y_2}$

$= a(1-\cos\theta) + (1+\cot^2\frac{\theta}{2})(-4a\sin^4\frac{\theta}{2})$

$= a(1-\cos\theta) - 4a\sin^4\theta/2.\text{cosec}^2\theta/2$

$= a(1-\cos\theta) - 4a\sin^2\frac{\theta}{2}$

$= a(1-\cos\theta) - 2a(1-\cos\theta) = -a(1-\cos\theta)$

Hence, the required evolute is given by

$x = a(\theta + \sin\theta), y = -a(1-\cos\theta)$

which is another equal cycloid.

EXERCISE 5.3

1. Find the co-ordinates of the centre of curvature for the point (x, y) on the parabola $y^2 = 4ax$. Also find the equation of the evolute of the parabola.
2. In the curve $y = a\log\sec\left(\frac{x}{a}\right)$, show that the chord of curvature parallel to the axis of y is of constant length.
3. Prove that the centre of curvature (α, β) for the curve $x = 3t, y = t^2 - 6$ is $\alpha = -\frac{4}{3}t^3, \beta = 3t^2 - \frac{3}{2}$.
4. If C_x and C_y be the chords of curvature parallel to the axis at any point of the curve $y = ae^{x/a}$, show that $\frac{1}{C_x^2} + \frac{1}{C_y^2} = \frac{1}{2aC_x}$.
5. Show that the centre of curvature (α, β) at the point determined by t on the ellipse $x = a\cos t, y = b\sin t$, is given by $\alpha = \frac{a^2-b^2}{a}\cos^3 t, \beta = -\left(\frac{a^2-b^2}{b}\right)\sin^3 t.$
6. Show that in any curve the chord of curvature prependicular to the radius vector is $2\rho\sqrt{(r^2-p^2)}/r$.
7. Show that the chord of curvature through the pole of the equiangular spiral $r = ae^{m\theta}$ is $2r$.
8. Find the coordinates of the centre of curvatrue of ellipse $\frac{x^2}{a^2} + \frac{y^2}{b^2} = 1$ or $x = a\cos\theta, y = b\sin\theta$. Hence, show that the equation of its evolute is $(ax)^{2/3} + (by)^{2/3} = (a^2-b^2)^{2/3}$.
9. Find the chord of curvature through the pole of the curve $a\theta = \sqrt{r^2-a^2} - a\cos^{-1}(a/r)$.
10. If C_r and C_θ be the chords of curvature of the curve $r = a(1+\cos\theta)$ through the pole and perpendicular to the radius vector, then prove that $3(C_r^2 + C_\theta^2) = 8rC_r$.

Hint to Selected Problems

1. Here $\frac{d^2y}{dx^2} = -\frac{1}{2}a^{1/2}x^{-3/2}$

   If (α, β) be the centre of curvature for the point (x, y), then use

   $$\alpha = x - \left[\frac{\frac{dy}{dx}\left\{1+\left(\frac{dy}{dx}\right)^2\right\}}{\frac{d^2y}{dx^2}}\right] \text{ and } \beta = y + \left[\frac{1+\left(\frac{dy}{dx}\right)^2}{\frac{d^2y}{dx^2}}\right]$$

2. Given that $y = a\log\sec\frac{x}{a} \Rightarrow \frac{dy}{dx} = \tan\frac{x}{a}$ and $\frac{d^2y}{dx^2} = \frac{1}{a}\sec^2\frac{2}{a}$

   Put all these values in the radius of curvature, we get $\rho = a\sec\frac{x}{a}$.

   Then, chord of curvature parallel to y-axis is $= 2\rho\cos\psi$.

3. $\frac{dx}{dt} = 3, \frac{dy}{dt} = 2t \Rightarrow \frac{dy}{dx} = \frac{2t}{3}$ and $\frac{d^2y}{dx^2} = \frac{2}{3}\frac{dt}{dx} = \frac{2}{3}.\frac{1}{3} = \frac{2}{9}$.

   Putting these values in the formula of centre of curvature.

4. Since $C_x = 2\rho\sin\psi$, $C_y = 2\rho\cos\psi$.

   Then, find the value of ρ using the given curve and by putting the value of ρ in the above expression, find a relation betwen C_x and C_y.

5. $x = a\cos t, y = b\sin t$

   $\Rightarrow \frac{dx}{dt} = -a\sin t, \frac{dy}{dt} = b\cos t \Rightarrow \frac{dy}{dx} = -\frac{b}{a}\cot t$

   $\therefore \frac{d^2y}{dx^2} = -\frac{b}{a}(-\text{cosec}^2 t)\frac{dt}{dx} = -\frac{b}{a^2}\text{cosec}^3 t$

   Then use the formulae for the centre of curvature.

6. The chord of curvature perpendicular to the radius vector is $2\rho\cos\phi$.

   Since $p = r\sin\phi \Rightarrow \frac{dp}{dr} = \sin\phi \therefore \rho = r\frac{dr}{dp} = \frac{r^2}{p}$

   Now $\cos\phi = \sqrt{1-\sin^2\phi} = \sqrt{1-\frac{p^2}{n}}$. Put this value in $2\rho\cos\phi$.

7. Here, $\frac{dr}{d\theta} = mr, \frac{d^2r}{d\theta^2} = m^2r$.

   Then we may get $\rho = r(1+m^2)^{1/2}$.

   To find the chord of curvature through the pole is given by $2\rho\sin\phi$.

   where value of $\sin\phi$ can be obtained by using $r\frac{d\theta}{dr} = \tan\phi$.

8. $x = a\cos\theta, y = b\sin\theta \Rightarrow \frac{dx}{d\theta} = -a\sin\theta, \frac{dy}{d\theta} = b\cos\theta$.

   Therefore, $\frac{dy}{dx} = -\left(\frac{b}{a}\right)\cot\theta$ and $\frac{d^2y}{dx^2} = -\frac{b}{a^2}\text{cosec}^3\theta$

   Put these values in the formulae of centre of curvature (α, β). Also, to find the equation of evolute, find the locus of α and β.

ANSWERS

1. $\left((3x+2a), -2x\sqrt{\frac{x}{a}}\right), 27ay^2 = 4(x-2a)^3$ **8.** $\left(\frac{a^2-b^2}{a}\cos^3\theta, -\frac{a^2-b^2}{b}\sin^3\theta\right)$ **9.** $\frac{2(r^2-a^2)}{r}$

Objective Evaluations

Fill in the Blanks

1. $\rho = \frac{ds}{d\psi}$ is intrinsic formula for __________ of curvature.
2. The relation between __________ is called the intrinsic equation of a curve.
3. The relation between s and ψ for any curve is called __________ equation.
4. The curvature of the curve at any point P is defined as the __________ of the radius of curvature at P.
5. For a curve $y = f(x)$, the radius of curvature $\rho =$ __________.
6. If the curve is in pedal form *i.e.*, $p = f(r)$, then $\rho =$ __________.
7. Locus of centre of curvature is known as __________ of that curve.
8. Chord of curvature through origin is __________.
9. When curve is in tangential polar form $\rho =$ __________.
10. The curvature of the curve at any point P is equal to ______.

True/False

Write 'T' for True and 'F' for False statement.

1. The curvature of the curve at any point P is defined as the reciprocal of the radius of curvature of P. **(T/F)**
2. If the given curve is in parametric form $x = f(t)$ and $y = \phi(t)$ then
$$\rho = \left[\left(\frac{dx}{dt}\right)^2 + \left(\frac{dy}{dt}\right)^2\right]^{3/2} \Bigg/ \left[\frac{dx}{dt}\cdot\frac{d^2y}{dt^2} - \frac{dy}{dt}\cdot\frac{d^2x}{dt^2}\right].$$ **(T/F)**
3. If the curve is $r = f(\theta)$, then $\rho = \dfrac{\left[r^2 + \left(\frac{dr}{d\theta}\right)^2\right]^{3/2}}{r^2 + 2\left(\frac{dr}{d\theta}\right)^2 - r\frac{d^2r}{d\theta^2}}$. **(T/F)**
4. The chord of curvature parallel to x-axis is $2\rho \cos\psi$. **(T/F)**
5. The chord of curvature parallel to y-axis is $2\rho\cos\psi$. **(T/F)**
6. If y-axis is the tangent to the given curve at the origin, then radius of curvature at the origin is equal to $\lim_{x\to 0}\frac{y^2}{2x}$. **(T/F)**
7. The curvature of the circle and circle of curvature, both are the same. **(T/F)**
8. The tangential polar formula for radius of curvature is $\rho = p + \frac{d^2p}{d\psi^2}$. **(T/F)**
9. The chord of curvature through the origin is $2\rho \sin\phi$. **(T/F)**
10. The chord of curvature at the origin of the curve $3x^2 + 4x - 12y = 0$ is zero. **(T/F)**

Multiple Choice Questions

Choose the most appropriate one.

1. The radius of curvature of the curve $y = e^x$ at the point where it crosses the y-axis is :
   (a) 2 (b) $\sqrt{2}$
   (c) $2\sqrt{2}$ (d) 1
   (I.A.S.–1995, P.C.S.–1994)
2. For the curve $xy = a^2$ the radius of curvature at (2, 2) is :
   (a) 4 (b) 16
   (c) 10 (d) none of these
3. Radius of curvature at any point (s, ψ) of the curve $s = c \log \sec\psi$ is :
   (a) $c \sec\psi$ (b) $c \cot\psi$
   (c) $c \operatorname{cosec}\psi$ (d) $c \tan\psi$
4. Radius of curvature at any pomt (s, ψ) of the curve $S = a\log\cot(\pi/2 - \psi/2) + a\frac{\sin\psi}{\cos^2\psi}$ is :
   (a) $2a\cos^2\psi$ (b) $a\tan^2\psi$
   (c) $2a\sec^2\psi$ (d) $a\cot^2\psi$
5. Radius of curvature at (x, y) of the curve $y = 1/2c[e^{x/c} + e^{-x/c}] = c\cosh x/c$ is :
   (a) y^2/c (b) x^2/c
   (c) y/c (d) x/c
6. Radius of curvature at point (p, r) on curve $p^2 = ar$ is :
   (a) $2p^3/a^2$ (b) $2p^2/a^2$
   (c) p^3/a^2 (d) p^2a^2
7. Radius of curvature at point (p, r) on curve $r^3 = a^2p$ is :
   (a) $\frac{2}{5}\sqrt{ar}$ (b) $\frac{a^2}{3r}$
   (c) $\frac{2}{3}\sqrt{2ar}$ (d) $\sqrt{2ar}$
8. Radius of curvature at (r, θ) of curve $r = a(1 - \cos\theta)$ is :
   (a) $\frac{2}{3}\sqrt{2ar}$ (b) ar
   (c) $\sqrt{2ar}$ (d) $\frac{2}{3}\sqrt{ar}$
9. The radius of curvature is :
   (a) square of the curvature (b) reciprocal of curvature
   (c) equal to curvature (d) none of these

10. The radii of curvature at the origin for the curve $x^3 + y^3 = 3axy$ are each equal to :

(a) $2a/3$ (b) $a/3$
(c) $3a/2$ (d) none of these

11. A point is called point of inflexion if :

(a) $\frac{dy}{dx} = 0$ (b) $\frac{dy}{dx} = \infty$
(c) $\frac{d^2y}{dx^2} = 0$ (d) $\frac{d^2y}{dx^2} = \infty$

12. The radius of curvature at any point of the ellipse $\frac{1}{p^2} = \frac{1}{a^2} + \frac{1}{b^2} - \frac{r^2}{a^2b^2}$ is :

(a) $\frac{a^2b^2}{p^3}$ (b) $\frac{ab^2}{p}$
(c) $\frac{a^2b^2}{p^2}$ (d) $\frac{ab}{p}$

13. The radius of curvature of a curve for which the relation between p and ψ is given is :

(a) $\rho = p + \frac{dp}{d\psi}$ (b) $\rho = p + \left(\frac{dp}{d\psi}\right)^2$
(c) $\rho = p + \frac{d^2p}{d\psi^2}$ (d) $\rho = p^2 + \frac{d^2p}{d\psi^2}$

14. For any curve $\frac{d}{dx}\left(\frac{dy}{ds}\right)$ is equal to :

(a) ρ (b) $\frac{1}{\rho}$
(c) ρ^2 (d) $\frac{1}{\rho^2}$

15. The radius of curvature at the origin for the curve $3x^2 + 4x^3 - 12y = 0$ is :

(a) 1 (b) 2
(c) 3 (d) 4

16. The locus of the centre of curvature of all points of a given plane curve is called :

(a) radius of curvature (b) envelopes
(c) ervolute (d) normal

17. The centre of curvature of the curve $y = x^3 - 6x^2 + 3x + 1$ at $(1, -1)$ is :

(a) $\left(-18, \frac{-17}{c}\right)$ (b) $\left(0, \frac{-43}{3}\right)$
(c) $\left(-36, \frac{-43}{6}\right)$ (d) none of these

18. Chord of curvature through the pole is :

(a) $2\rho \sin \phi$ (b) $2\rho \cos \phi$
(c) $2\rho \tan \phi$ (d) $2\rho \sec \phi$

19. Chord of curvature perpendicular to the radius vector is :

(a) $2\rho \sin \phi$ (b) $2\rho \cos \phi$
(c) $2\rho \tan \phi$ (d) $2\rho \sec \phi$

20. Chord of curvature parallel to the axis of x is :

(a) $2\rho \sin \psi$ (b) $2\rho \cos \psi$
(c) $2\rho \tan \psi$ (d) $2\rho \sec \psi$

21. Chord of curvature to the axis of y is :

(a) $2\rho \sin \psi$ (b) $2\rho \cos \psi$
(c) $2\rho \tan \psi$ (d) $2\rho \sec \psi$

22. The chord of curvature through the pole of the curve $r = ae^{M\theta}$ is :

(a) r (b) $2r$
(c) $3r$ (d) $4r$

ANSWERS

Fill in the Blanks

1. Radius	**2.** s and ψ	**3.** intrinsic	**4.** reciprocal	**5.** $\frac{(1+y_1^2)^{3/2}}{y_2}$
6. $r\frac{dr}{dp}$	**7.** Evolute	**8.** $2\rho \sin\phi$	**9.** $p + \frac{d^2p}{d\psi^2}$	**10.** $\frac{d\psi}{ds}$

True/False

1. T	**2.** T	**3.** T	**4.** F	**5.** T
6. T	**7.** F	**8.** T	**9.** T	**10.** F

Multiple Choice Questions

1. (c)	**2.** (d)	**3.** (d)	**4.** (c)	**5.** (a)	**6.** (a)	**7.** (b)	**8.** (a)	**9.** (b)	**10.** (c)
11. (c)	**12.** (a)	**13.** (c)	**14.** (a)	**15.** (b)	**16.** (c)	**17.** (b)	**18.** (a)	**19.** (b)	**20.** (a)
21. (b)	**22.** (b)								

FFFFFF

CHAPTER 6 Envelope and Evolutes

6.1 INTRODUCTION

(i) Family of curves with one parameter. An equation in two variables x and y of the form $F(x, y, \lambda) = 0$ where λ is any constant, is known as a curve.
If λ takes all real values, then the equation $F(x, y, \lambda) = 0$ is known as a family of curves with one parameter λ.

(ii) Family of curves with two parameters. An equation in two variables x and y of the form $F(x, y, \lambda, \mu) = 0$ is known as a family of curves with two parameter λ and μ if λ and μ take all real values.

For Example :

(1) The equation $x\cos\lambda + y\sin\lambda = p$ represents a family of straight line with one parameter λ.

(2) The equation $y = mx + a/m$ represents a family of straight lines which are the tangents to parabola $y^2 = 4ax$ with one parameter m.

(3) The equation $(x - \alpha)^2 + (y - \beta)^2 = a^2$ represents a family of circles with centred at (α, β) and radius a with two parameters α and β.

6.1.1 SOME STANDARD EQUATIONS

(i) $\frac{x}{a} + \frac{y}{b} = 1$ (Equation of straight line)

(ii) System of Concentric Coaxial ellipses : $\frac{x^2}{a^2} + \frac{y^2}{b^2} = 1$

(iii) Family of parabola: $\left(\frac{x}{a}\right)^{1/2} + \left(\frac{y}{b}\right)^{1/2} = 1$

(iv) Equation of tractrix: $x = a\,(\cos t + \log\tan\frac{1}{2}t)$ $y = a\sin t$

(v) Pedal equation of equiangular spiral : $p = r\sin\alpha$

(vi) Equation of cardioid : $r = a(1 + \cos\theta)$

(vii) Other equations of straight line:

(a) $y = mx + \frac{a}{m}$ (b) $y = m^2x + \frac{1}{m^2}$ (c) $\frac{x}{a}\cos\theta + \frac{y}{b}\sin\theta = 1$

6.2 ENVELOPE OF A FAMILY OF CURVES WITH ONE PARAMETER

Let $F(x, y, \lambda) = 0$ be a family of curves with parameter λ and let $F(x, y, \lambda) = 0$ and $F(x, y, \lambda+\delta\lambda) = 0$ be two members of a family of curves $F(x, y, \lambda) = 0$ corresponding to the parameter λ and $\lambda + \delta\lambda$, suppose P is a point of intersection of two members $F(x, y, \lambda) = 0$ and $F(x, y, \lambda+\delta\lambda) = 0$. As $\delta\lambda \to 0$, the point P tends to a definite point Q which depends upon λ. Thus the locus of such points Q gives an envelop of the family.

Definition. *The locus of the limiting positions of the points of intersection of any two members of the family of curves $F(x, y, \lambda) = 0$, when one of them tends to coincide with the other fixed point is called envelope.*

REMARK

- The envelope of a family of curves is the locus of the points intersection of consecutive members of the family.

6.3 PROCEDURE FOR FINDING THE ENVELOPE

Let $F(x, y, \lambda) = 0$ be a family of the curve with one parameter λ.

Suppose $F(x, y, \lambda) = 0$ and $F(x, y, \lambda + \delta\lambda) = 0$ are two consecutive members of the family of curves corresponding to λ and $\lambda + \delta\lambda$. Thus the co-ordinates of the point of the intersection of these two members are obtained by the equations

$$F(x, y, \lambda) = 0 \quad \text{...(1)}$$

and $$F(x, y, \lambda) - F(x, y, \lambda + \delta\lambda) = 0 \quad \text{...(2)}$$

Divide the equation (2) by $\delta\lambda$, we get

$$\frac{F(x,y,\lambda) - F(x,y,\lambda+\delta\lambda)}{\delta\lambda} = 0 \quad \text{or} \quad \frac{F(x,y,\lambda+\delta\lambda) - F(x,y,\lambda)}{\delta\lambda} = 0$$

Taking limit as $\delta\lambda \to 0$, we get

$$\frac{\partial F(x,y,\lambda)}{\partial\lambda} = 0 \quad \text{...(3)}$$

Now eliminating λ between $F(x, y, \lambda) = 0$ and $\frac{\partial F(x,y,\lambda)}{\partial\lambda} = 0$, we therefore, obtain the envelop of the family of curves $F(x, y, \lambda) = 0$.

WORKING PROCEDURE

To obtain an envelope of the family of curve $F(x, y, \lambda) = 0$, we use following steps :

Step 1. *Differentiate partially $F(x, y, \lambda) = 0$ with respect to λ, to get* $\frac{\partial F}{\partial\lambda} = 0$.

Step 2. *Now eliminating λ between $F(x, y, \lambda) = 0$ and $\frac{\partial F}{\partial\lambda} = 0$, we therefore obtain envelope of the given family of curves.*

Solved Examples

Example 1. *Find the envelope of the family of straight lines $y = mx + \frac{a}{m}$, the parameter being m.*

Solution . Here, the family of straight lines is

$$y = mx + \frac{a}{m}. \quad \text{...(1)}$$

Differentiating (1) partially with respect to m, we get

$$0 = x - \frac{a}{m^2} \quad \text{...(2)}$$

Eliminating m between (1) and (2), we get

From (2), we have

$$m^2 = \frac{a}{x}.$$

From (1), we have

$$ym = m^2x + a \Rightarrow y^2m^2 = (m^2x + a)^2$$

$$\Rightarrow \quad y^2\left(\frac{a}{x}\right) = \left(\frac{a}{x}.x + a\right)^2. \quad \left(\because m^2 = \left(\frac{a}{x}\right)\right)$$

$$\Rightarrow \quad \frac{ay^2}{x} = (2a)^2 \Rightarrow y^2 = 4ax.$$

This is the required envelope.

Example 2. *Find the envelope of the family of straight lines : $x \operatorname{cosec}\theta - y\cot\theta = c$, the parameter being θ.*

Solution . Since the family of straight lines is

$$x \operatorname{cosec}\theta - y\cot\theta = c. \quad \text{...(1)}$$

Diff. (1) partially with respect to θ, we get

$$-x \operatorname{cosec}\theta\cot\theta + y\operatorname{cosec}^2\theta = 0$$

or $$x\cot\theta - y\operatorname{cosec}\theta = 0. \quad \text{...(2)}$$

Eliminating θ between (1) and (2), we get

$$(x \operatorname{cosec}\theta - y\cot\theta)^2 - (x\cot\theta - y\operatorname{cosec}\theta)^2 = c^2$$

or $$x^2(\operatorname{cosec}^2\theta - \cot^2\theta) - y^2(\operatorname{cosec}^2\theta - \cot^2\theta) - 2xy\operatorname{cosec}\theta\cot\theta + 2xy\operatorname{cosec}\theta\cot\theta = c^2$$

or $$x^2 - y^2 = c^2. (\because \operatorname{cosec}^2\theta - \cot^2\theta = 1)$$

This is the required envelope.

6.4 ENVELOPE OF THE FAMILY OF CURVES OF THE FORM $A\lambda^2 + B\lambda + C = 0$

The family of curve is

$$A\lambda^2 + B\lambda + C = 0 \quad \text{...(1)}$$

where the parameter being λ.

Differentiating (1) partially w.r.t. to λ, we get

$$2A\lambda + B = 0 \quad \text{...(2)}$$

Eliminating λ between (1) and (2), we get

$$A\left[-\frac{B}{2A}\right]^2+B\left[-\frac{B}{2A}\right]+C=0 \quad \text{or} \quad \frac{B^2}{4A}-\frac{B^2}{2A}+C=0 \quad \text{or} \quad B^2-4AC=0.$$

This is the required equation of an envelope.

REMARK

- If the equation of the family of curves is a quadratic equation in parameter, then its envelope is obtained by $D = 0$, where D is the discriminant of the quadratic.

6.5 ENVELOPE OF THE FAMILY OF CURVES WITH TWO PARAMETERS CONNECTED BY A RELATION

Let $F(x, y, \lambda, \mu) = 0$ be a family of curves with two parameters λ and μ. Let $f(\lambda, \mu) = 0$ be a relation between λ and μ. To obtain the envelope, we proceed as follows :

WORKING PROCEDURE

Differentiating the equations $F(x, y, \lambda, \mu) = 0$ and $f(\lambda, \mu) = 0$ with respect to λ regarding x and y as constants and μ as a function of λ, we get two equations. Now elimainting λ, μ between the given equations and two obtained equations. We therefore obtain the envelope.

6.6 GEOMETRICAL INTERPRETATION OF THE ENVELOPE

Let the equation of the family of curves be

$$F(x, y, \lambda) = 0\ ;\ \lambda \text{ is the parameter.} \qquad \ldots(1)$$

Thus the envelope of (1) is obtained by eliminating between (1) and

$$\frac{\partial F}{\partial \lambda}=0 \qquad \ldots(2)$$

Therefore, we can say that (2) is taken as the equation of the envelope of (1), if λ is a function of x and y but not constant. Now from (1), we have

$$\left(\frac{\partial F}{\partial x}+\frac{\partial F}{\partial \lambda}\cdot\frac{\partial \lambda}{\partial x}\right)+\left(\frac{\partial F}{\partial y}+\frac{\partial F}{\partial \lambda}\cdot\frac{\partial \lambda}{\partial y}\right)\frac{dy}{dx}=0 \quad \Rightarrow \quad \frac{dy}{dx}=-\frac{\dfrac{\partial F}{\partial x}+\dfrac{\partial F}{\partial \lambda}\cdot\dfrac{\partial \lambda}{\partial x}}{\dfrac{\partial F}{\partial y}+\dfrac{\partial F}{\partial \lambda}\cdot\dfrac{\partial \lambda}{\partial y}}. \qquad \ldots(3)$$

This gives the slope of the tangent to the envelope of (1) at any point (x, y), where (x, y) is a common point to the member 'λ' of the family of curves and the envelope.

If $\frac{\partial F}{\partial x} \neq 0$ and $\frac{\partial F}{\partial y} \neq 0$ at (x, y), then the slope of the tangent to the member $F(x, y, \lambda) = 0$ is

$$\frac{dy}{dx}=-\frac{\partial F/\partial x}{\partial F/\partial y}. \qquad \ldots(4)$$

But $F(x, y, \lambda) = 0$ is also the equation of the envelope if λ is a function of x and y which is given by $\frac{\partial F}{\partial \lambda}=0$

Since at every point of the envelope, $\frac{\partial F}{\partial \lambda} = 0$, then the slopes given by (3) and (4) are same.

Hence the curve of the family and its envelope have the same tangent lines at the common point. Consequently the envelope of a family of curves touch each member of the family.

REMARK

- If $\frac{\partial F}{\partial x} = 0$ and $\frac{\partial F}{\partial y} = 0$ at any points on the curve, then the envelope may not touch curve at that points.

Solved Examples

Example 1. *Find the envelope of the family of straight lines $y= mx + a\sqrt{1+m^2}$, the parameter being m.* (Anna–2006)

Solution . The given equation of the family can be written as :

$$(y - mx)^2 = a^2(1 + m^2)$$

or $\quad (x^2 - a^2)m^2 - 2mxy - a^2 + y^2 = 0 \quad \ldots(1)$

This equation is quadratic in m. Then the envelope of (1) is obtained by equating the discriminant of (1) to zero, we get

$$(-2xy)^2 - 4(x^2 - a^2)(-a^2) = 0$$

$$(\because B^2 - 4AC = 0)$$

or $\quad 4x^2y^2 - 4[x^2y^2 - x^2a^2 - a^2y^2 + a^4] = 0$

or $\quad x^2a^2 + a^2y^2 = a^4$

or $\quad x^2 + y^2 = a^2.$

This is the required equation of envelope.

Example 2. *Find the envelope of the family of circles*

$$(x-c)^2 + y^2 = r^2$$

where the parameter being c.

Solution . Equation of family of circles is

$$(x-c)^2 + y^2 = r^2. \qquad ...(1)$$

It can also be written as

$$c^2 - 2xc + x^2 + y^2 - r^2 = 0. \qquad ...(2)$$

This is quadratic in c, so that the envelope is

$$(-2x)^2 - 4.1.\,(x^2 + y^2 - r^2) = 0$$

$$(\because B^2 - 4AC = 0)$$

or $\quad x^2 - x^2 - y^2 + r^2 = 0$

or $\quad y^2 = r^2$

or $\quad y = r, y = -r.$

These are the required envelopes.

Example 3. *Find the envelope of the family of circles $x^2 + y^2 - 2ax\cos\alpha - 2ay\sin\alpha = c^2$ where α being the parameter. Also interpret the result.*

Solution . The equation of the family of circles is

$$x^2 + y^2 - 2ax\cos\alpha - 2ay\sin\alpha = c^2. \qquad ...(1)$$

Differentiating (1) partially w.r.t, α we get

$$2ax\sin\alpha - 2ay\cos\alpha = 0$$

or $\quad x\sin\alpha - y\cos\alpha = 0. \qquad ...(2)$

Eliminating α between (1) and (2), we get

$$4a^2(x\sin\alpha - y\cos\alpha)^2 + 4a^2(x\cos\alpha + y\sin\alpha)^2 = 0 + (x^2 + y^2 - c^2)^2$$

or $\quad 4a^2(x^2 + y^2) = (x^2 + y^2 - c^2)^2.$

This is the required envelope.

Intrepretation. The equation of envelope can be written as

$$(x^2 + y^2)^2 - (4a^2 + 2c^2)(x^2 + y^2) + c^4 = 0.$$

It is quadratic in $(x^2 + y^2)$ so solving, we get

$$x^2 + y^2 = \frac{2(2a^2 + c^2) \pm \sqrt{4(2a^2 + c^2)^2 - 4c^4}}{2}$$

$$= (2a^2 + c^2) \pm 2a\sqrt{c^2 + a^2}$$

$$= (\sqrt{a^2 + c^2} \pm a)^2$$

Thus, the equation of the envelope contains two circles with centred at (0, 0) and radius $\sqrt{a^2 + c^2} \pm a.$

Example 4. *Find the envelope of the circles drawn on the radii vectors of the parabola $y^2 = 4ax$ as diameter.* (Madras–2003)

Solution . Let $(at^2, 2at)$ be any point on the parabola $y^2 = 4ax$. Then the equation of circles drawn on the line joining (0, 0) and $(at^2, 2at)$ as diameter is

$$(x-0)(x-at)^2 + (y-0)(y-2at) = 0$$

or $\quad x^2 + y^2 - axt^2 - 2aty = 0 \qquad ...(1)$

where t being the parameter.

Differentiating (1) partially with respect to t, we get

$$-2axt - 2ay = 0$$

or $\quad xt + y = 0. \qquad ...(2)$

Eliminating t between (1) and (2), we get

$$x^2 + y^2 - ax\left(-\frac{y}{x}\right)^2 - 2a\left(-\frac{y}{x}\right)y = 0$$

or $\quad x^2 + y^2 - \dfrac{ay^2}{x} + \dfrac{2ay^2}{x} = 0$

or $\quad x^2 + y^2 + \dfrac{ay^2}{x} = 0$

or $\quad x(x^2 + y^2) + ay^2 = 0.$

This is the required envelope.

Example 5. *Find the envelope of the straight lines drawn at right angles to the radii vectors of the cardioid $r = a(1+\cos\theta)$ through their extremities.*

Solution . Let P be any point on the cardioid $r = a(1+\cos\theta)$, whose vectorial angle is α. Then $OP = a(1+\cos\alpha)$. Let Q be any point (r, θ) on the straight line drawn through P and perpendicular to OP.

In ΔOPQ, $\quad \angle P = 90°$, then

$$\frac{OP}{OQ} = \cos(\theta - \alpha)$$

or $\quad OP = OQ\cos(\theta - \alpha)$

or $\quad a(1 + \cos\alpha) = r\cos(\theta - \alpha)$

or $\quad r\cos(\theta - \alpha) = a(1 + \cos\alpha) \qquad ...(1)$

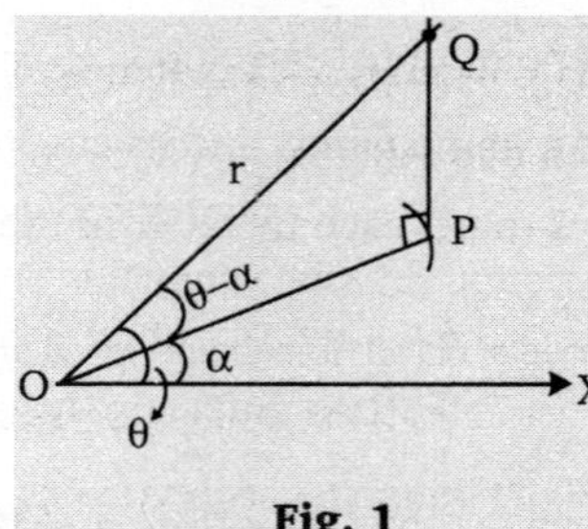

Fig. 1

This is the equation of family of straight lines drawn through P at right angle to OP.

Differentiating (1) partially with respect to α, we get

$$r\sin(\theta - \alpha) = -a\sin\alpha$$

or $\quad \dfrac{a(1+\cos\alpha)\sin(\theta - \alpha)}{\cos(\theta - \alpha)} = -a\sin\alpha$ [From (1)] or $\tan(\theta-\alpha) = -\dfrac{\sin\alpha}{1+\cos\alpha}$

$= -\dfrac{2\sin(\alpha/2)\cos(\alpha/2)}{2\cos^2(\alpha/2)}$

$= -\tan(\alpha/2) = \tan(\pi-\alpha/2)$.

$\therefore \quad \theta-\alpha = \pi-\alpha/2$ or $\dfrac{\alpha}{2} = \theta-\pi$

or $\quad \alpha = 2(\theta-\pi)$

Putting this value of α in (1), we get

$r\cos(\theta-2\theta+2\pi) = a[1+\cos(2\theta-2\pi)]$

$r\cos(2\pi-\theta) = 2a\cos^2(\theta-\pi)$

$r\cos\theta = 2a\cos^2\theta$

or $\quad r = 2a\cos\theta$.

This is the required envelope.

Example 6. *Find the envelope of the family of straight lines*

$$\frac{x}{a}+\frac{y}{b}=1$$

where a, b are connected by a relation $a^2+b^2=c^2$, *c is a constant.*

Solution . Since the equation of family of straight lines is

$$\frac{x}{a}+\frac{y}{b}=1 \qquad ...(1)$$

and $\quad a^2+b^2=c^2. \qquad ...(2)$

Differentiating (1) and (2) w.r.t. a treating x and y as constant and 'b' as a function of 'a', we get

$$-\frac{x}{a^2}-\frac{y}{b^2}\frac{db}{da}=0$$

or $\quad \dfrac{db}{da} = -\dfrac{x/a^2}{y/b^2} \qquad ...(3)$

and $\quad 2a+2b\dfrac{db}{da}=0$

or $\quad \dfrac{db}{da} = -\dfrac{a}{b}. \qquad ...(4)$

From (3) and (4), we get

$$-\frac{x/a^2}{y/b^2} = -\frac{a}{b}$$

or $\quad \dfrac{x/a}{y/b} = \dfrac{a^2}{b^2}$

or $\quad \dfrac{x/a}{a^2} = \dfrac{y/b}{b^2} \qquad ...(5)$

Eliminating a and b between (1), (2) and (5), we get

$$\frac{x/a}{a^2} = \frac{y/b}{b^2} = \frac{x/a+y/b}{a^2+b^2} = \frac{1}{c^2}.$$

[Using (1) and (2)]

$\therefore \quad \dfrac{x/a}{a^2} = \dfrac{1}{c^2} \Rightarrow xc^2 = a^3$

$\Rightarrow \quad a = (xc^2)^{1/3}$

and $\quad \dfrac{y/b}{b^2} = \dfrac{1}{c^2} \Rightarrow yc^2 = b^3$

$\Rightarrow \quad b = (yc^2)^{1/3}$.

Putting these values of a and b in (2), we get

$$(xc^2)^{2/3} + (yc^2)^{2/3} = c^2$$

or $\quad x^{2/3} + y^{2/3} = c^{2/3}$.

This is the required envelope.

EXERCISE 6.1

1. Find the envelope of the family of straight lines
   $ax\sec\theta - by\,\text{cosec}\,\theta = a^2 - b^2$
   where θ being the parameter.
2. Find the envelope of the following families of straight lines :
   (i) $y = mx + am^3$, the parameter being m.
   (ii) $y = mx + am^p$, the parameter being m.
   (iii) $x\cos^3\alpha + y\sin^3\alpha = a$, α being the parameter.
   (iv) $y = mx + \dfrac{a}{m}$ (Madras–2006)
   (v) $y = mx - 2am - m^3$
3. Find the envelope of the family of straight lines
   $x\cos\alpha + y\sin\alpha = a$
   where α being the parameter, and interpret the result.
4. Find the envelope of the family of circles
   $x^2+y^2-2ax\cos\alpha-2ay\sin\alpha+c^2=0,\ (a^2>c^2)$
   where α being the parameter and interpret the result.
5. Find the envelope of the straight lines
   $x\cos\alpha + y\sin\alpha = t\sin\alpha\cos\alpha$,
   where α being the parameter and interpret the result.
6. Find the envelope of the family of straight lines $\dfrac{x}{a}+\dfrac{y}{b}=1$, where two parameters a and b are connected by a relation $a+b=c$, c being the constant.
7. Find the envelope of the family of curves $tx^2+t^2y=a$, t being the parameter.
8. Find the envelope of following families of circles :
   (i) $(x-\alpha)^2+(y-\alpha)^2=2\alpha$, α being the parameter.
   (ii) $(x-c)^2+y^2=r^2$, where c being the parameter.
9. Find the envelope of the family of curves given by $\dfrac{x^2}{\alpha^2}+\dfrac{y^2}{k^2-\alpha^2}=1$, α being the parameter.
10. Find the envelope of the ellipse
    $x = a\sin(\theta-\alpha),\ y = b\cos\theta$
    where α being the parameter.

11. $x^{2/3}+y^{2/3}=k^{2/3}$ is the envelope of the lines $\frac{x}{a}+\frac{y}{b}=1$, then find the necessary relation between a, b and k.

12. Find the envelope of the family of the curves $x^2 \sin\alpha + y^2 \cos\alpha = a^2$, α being the parameter.

13. Show that the envelope of the family of straight lines $y = mx + \sqrt{a^2m^2+b^2}$, m being the parameter is $\frac{x^2}{a^2}+\frac{y^2}{b^2}=1$.

(Anna–2006)

14. Find the envelope of the circles which pass through the origin and whose centres lie on the ellipse $x^2/a^2 + y^2/b^2 = 1$.

15. Find the envelope of the circles drawn upon the radii vectors of the ellipse $x^2/a^2 + y^2/b^2 = 1$ as diameter.

16. Circles are described on the double ordinates of the parabola $y^2 = 4ax$ as diameter; prove that their envelope is the parabola $y^2=4a(x+a)$.

17. Show that the envelope of the straight line joining the extremities of a pair of conjugate diameters of the ellipse $x^2/a^2 + y^2/b^2 = 1$ is the ellipse $x^2/a^2 + y^2/b^2 = \frac{1}{2}$.

18. Find the envelopes of the straight lines drawn at right angles to the radii vectors of the following curves through their extremities :

(i) $r = ae^{\theta \cot \alpha}$ (ii) $r^n = a^n \cos n\theta$

(iii) $r = a + b\cos\theta$.

19. Find the envelope of the straight line $x/a + y/b = 1$ where parameter a and b are connected by the relation $a^n + b^n = c^n$, c being the constant.

20. Find the envelope of the straight lines $\frac{x}{a}+\frac{y}{b}=1$ when $a^m b^n = c^{m+n}$, c being a constant.

21. Find the envelope of the family of curves $\frac{x^m}{a^m}+\frac{y^m}{b^m}=1$

where the parameters a and b are connected by the relation $a^p + b^p = c^p$.

22. Find the envelope of the family of parabola $y = x\tan\alpha - \frac{gx^2}{2u^2\cos\alpha}$, α being the parameter.

Hint to Selected Problems

1. Differentiating the given equation w.r.t. θ, we get

$$\frac{ax\sin\theta}{\cos^2\theta}+\frac{by\cos\theta}{\sin^2\theta}=0. \quad \ldots(1)$$

On eliminating θ between (1) and the given equation, we get

$$\tan\theta = -\frac{(by)^{1/3}}{(ax)^{1/3}} \Rightarrow \sin\theta = \frac{(by)^{1/3}}{\sqrt{(ax)^{2/3}+(by)^{2/3}}},$$

and $$\cos\theta = -\frac{(ax)^{1/3}}{\sqrt{(ax)^{2/3}+(by)^{2/3}}}.$$

Put all these values in the given equation.

2. (ii) Differentiating the given equation partially w.r.t. m, we get $m^{p-1} = -\frac{x}{pa}$. Put this value in the given equation.

3. Differentiating the given equation partially w.r.t. α we get

$$-x\sin\alpha + y\cos\alpha = 0. \quad \ldots(1)$$

Eliminating α between (1) and the given equation.

5. The given equation can be written as

$$x\,\text{cosec}\,\alpha + y\sec\alpha = 0. \quad \ldots(2)$$

Differentiating (1) partially w.r.t. α, we get

$$\tan\alpha = x^{1/3}\,.\,y^{1/3}$$

$$\Rightarrow \text{cosec}\,\alpha = \frac{\sqrt{x^{2/3}+y^{2/3}}}{x^{1/3}} \text{ and } \sec\alpha = \frac{\sqrt{x^{2/3}+y^{2/3}}}{y^{1/3}}$$

Put all these values in equation (1).

9. The given equation can be written as

$$\alpha^4 + \alpha^2(y^2 - x^2 - k^2) + x^2k^2 = 0$$

which is a quadratic equation in α^2. Then envelope of the family is given by

$$b^2 - 4ac = 0$$

i.e., $(y^2 - x^2 - k^2)^2 = 4k^2x^2$.

10. The given equation can be written as

$$x = a\left\{\sqrt{\left(1-\frac{y^2}{b^2}\right)}\cos\alpha - \left(\frac{y}{b}\right)\sin\alpha\right\}.$$

15. Consider the point $(a\cos\theta, b\sin\theta)$ on the ellipse.

Then we get $x^2 + y^2 = ax\cos\theta + by\sin\theta$.

16. The co-ordinate of the extremities of a double ordinate of the parabola may be written as

$$(at^2, 2at) \text{ and } (at^2, -2at).$$

Therefore, the given equation of the circle becomes

$$a^2t^4 - 2at^2(x + 2a) + x^2 + y^2 = 0$$

which is a quadratic equation in t^2. Hence, the required envelope is given by

$$4a^2(x + 2a)^4 - 4a^2(x^2 + y^2) = 0.$$

ANSWERS

1. $(ax)^{2/3} + (by)^{2/3} = (a^2 - b^2)^{2/3}$ **2.** (i) $4x^3 + 27ay^2 = 0$ (ii) $(p-1)^{p-1}x^p + p^p a y^{p-1} = 0$ (iii) $a^2(x^2 + y^2) = x^2y^2$ (iv) $y^2 = 4ax$ (v) $\left(\frac{x}{a}\right)^2 + \left(\frac{y}{b}\right)^2 = 1$ **3.** $x^2 + y^2 = a^2$ **4.** $(x^2 + y^2 + c^2)^2 = 4a^2(x^2 + y^2)$, circles with centre at origin and radii $a \pm \sqrt{a^2 - c^2}$.

5. $x^{2/3} + y^{2/3} = t^{2/3}$ **6.** $x^{1/2} + y^{1/2} = c^{1/2}$ **7.** $x^4 + 4ay = 0$

8. (i) $(x + y + 1)^2 = 2(x^2 + y^2)$ (ii) $y = \pm r$ **9.** $x \pm y = \pm k$ **10.** $x = \pm a$ **11.** $a^2 + b^2 = k^2$

12. $x^4 + y^4 = a^4$ **14.** $(x^2 + y^2)^2 = 4(a^2x^2 + b^2y^2)$ **15.** $(x^2 + y^2)^2 = a^2x^2 + b^2y^2$

18. (i) $r \sin\alpha = ae^{(\alpha-\pi/2)\cot\alpha}e^{\theta\cot\alpha}$ (ii) $r^{n/(1-n)} = a^{n/(1-n)}\cos[n\theta/(1-n)]$ (iii) $r^2 - 2br\cos\theta + b^2 - a^2 = 0$

19. $x^{n/(n+1)} + y^{n/(n+1)} = c^{n/(n+1)}$ **20.** $\{(m+n)^{m+n}x^m y^n\}/m^m n^n = c^{m+n}$

21. $x^{mp/(p+m)} + y^{mp/(p+m)} = c^{mp/(p+m)}$ **22.** $y = \dfrac{u^2}{2g} - \dfrac{gx^2}{2u^2}$.

6.7 EVOLUTE

Definition. *The evolute of a curve is the envelope of the normals to that curve.*

In other words, The locus of the centre of curvature of a curve is called evolute for that curve.

Since the centre of curvature of a curve for a given point P on it is the limiting position of the intersection of the normal at P and the normal at other point Q as Q tends to P. Thus the envelope of the normals to a given curve is called an evolute of that curve.

6.8 EVOLUTE OF PEDAL FORM OF CURVES

Let the pedal equation of the given curve be

$$p = f(r) \quad \ldots(1)$$

and let C be the centre of curvature of (1) at the point P.

Then $PC = \rho$ (radius of curvature) and the equation joining P and C is the normal to the curve (1) at P. The point C will be on evolute corresponding to the point P on the curve. Since the evolute of the given curve $p = f(r)$ is the envelope of the normals at P of that curve, so that the normal PC of the given curve is a tangent to the evolute at C.

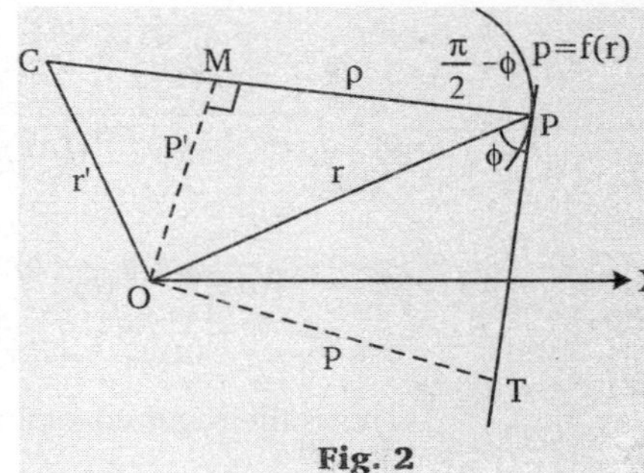

Fig. 2

Here PT is the tangent at P to the given curve $p = f(r)$ and OT is perpendicular to PT such that $OT = p$ and $OP = r$. Now draw a perpendicular OM from O to PC such that $OM = p'$ and $CO = r'$. Then in triangle OPC, we have

$$\cos\angle OPC = \frac{r^2 + \rho^2 - r'^2}{2r\rho}$$

$$\therefore \quad r'^2 = r^2 + \rho^2 - 2r\rho\cos\angle OPC = r^2 + \rho^2 - 2r\rho\cos\left(\frac{\pi}{2} - \phi\right)$$

$$= r^2 + \rho^2 - 2r\rho\sin\phi$$

$$r'^2 = r^2 + \rho^2 - 2p\rho \quad (\because p = r\sin\phi)$$

$$r'^2 = r^2 + \rho'^2 - 2\rho \quad \ldots(2)$$

Since $OTPM$ is a rectangle, so that $OM = TP = p'$, then in ΔPTO, $r^2 = p^2 + p'^2$

$$\Rightarrow \quad p'^2 = r^2 - p^2 \quad \ldots(3)$$

Also, we have

$$\rho = r\frac{dr}{dp}. \quad \ldots(4)$$

Now eliminating r, p and ρ between (1), (2), (3) and (4), we get the pedal equation of the evolute of the curve $p = f(r)$.

REMARK

- In above formulation the relation between p' and r' gives the evolute of the curve $p = f(r)$.

Solved Examples

Example 1. *Find the evolute of the hyperbola* $x^2/a^2 - y^2/b^2 = 1$.

Solution . Let $P(a \sec\theta, b\tan\theta)$ be any point on the hyperbola

$$x^2/a^2 - y^2/b^2 = 1.$$

The equation of the normal at P to the given hyperbola is

$$ax\cos\theta + by\cot\theta = a^2 + b^2. \quad \ldots(1)$$

Differentiating (1) partially w.r.t. θ, we get

$$-ax\sin\theta - by\operatorname{cosec}^2\theta = 0$$

or $\qquad \sin^3\theta = -\dfrac{by}{ax}$ or $\sin\theta = \left(-\dfrac{by}{ax}\right)^{1/3}$

$\therefore \qquad \cos\theta = \sqrt{1-\sin^2\theta} = \sqrt{1-\left(\dfrac{by}{ax}\right)^{2/3}}$

and $\qquad \cot\theta = \dfrac{\sqrt{1-\left(\dfrac{by}{ax}\right)^{2/3}}}{\left(-\dfrac{by}{ax}\right)^{1/3}}$

Putting the values of cos θ and cot θ in (1), we get

$$ax\left[\sqrt{1-\left(\frac{by}{ax}\right)^{2/3}}\right] + by\frac{\sqrt{1-\left(\frac{by}{ax}\right)^{2/3}}}{\left(-\frac{by}{ax}\right)^{1/3}} = (a^2+b^2)$$

or $\quad \dfrac{ax}{(ax)^{1/3}}\sqrt{ax^{2/3} - by^{2/3}}$

$$-\frac{by}{(by)^{1/3}}\sqrt{ax^{2/3} - by^{2/3}} = (a^2+b^2)$$

or $\quad \sqrt{ax^{2/3} - by^{2/3}}\,[(ax)^{2/3} - (by)^{2/3}]$

$$= (a^2+b^2)$$

or $\quad [(ax)^{2/3} - (by)^{2/3}]^{3/2} = (a^2+b^2)$

or $\quad (ax)^{2/3} - (by)^{2/3} = (a^2+b^2)^{2/3}.$

This is the required evolute of the given curve.

Example 2. *Show that the evolute of an equiangular spiral is an equiangular spiral.*

Solution . The pedal equation of an equiangular spiral is

$$p = r\sin\alpha. \qquad ...(1)$$

So that $\quad \dfrac{dp}{dr} = \sin\alpha.$

$\therefore \quad \rho = r\dfrac{dr}{dp} = r.\dfrac{1}{\sin\alpha} = r\,\text{cosec}\,\alpha$

or $\quad \rho = r\,\text{cosec}\,\alpha \qquad ...(2)$

Let (p', r') be any point on the evolute corresponding to the point (p, r) on the curve (1). Then we have,

$$r'^2 = r^2 + \rho^2 - 2\rho p$$

$$= r^2 + r^2\text{cosec}^2\,\alpha - 2r\,\text{cosec}\,\alpha.r\sin\alpha$$

[Using (1) and (2)]

$$= r^2\,\text{cosec}^2\,\alpha - r^2$$

$$r'^2 = r^2\cot^2\alpha.$$

Also, we have

$$p'^2 = r^2 - p^2 = r^2 - r^2\sin^2\alpha$$

$$= r^2(1-\sin^2\alpha)$$

$$p'^2 = r^2\cos^2\alpha. \qquad ...(4)$$

Dividing (4) by (3), we get

$$\frac{p'^2}{r'^2} = \frac{r^2\cos^2\alpha}{r^2\cot^2\alpha} = \sin^2\alpha.$$

$\therefore \quad p'^2 = r'^2\sin^2\alpha$

or $\quad p' = r'\sin\alpha.$

Thus the locus of the point (p', r') is $p = r\sin\alpha$, which is an equiangular spiral.

EXERCISE 6.2

1. Find the equation of the evolute of the parabola $y^2 = 2ax$.
2. Show that the equation of the evolute of the ellipse $x^2/a^2 + y^2/b^2 = 1$ is $(ax)^{2/3} + (by)^{2/3} = (a^2 - b^2)^{2/3}$.
(JNTU–2006, Anna–2005)
3. Show that the evolute of the tractrix
$x = a(\cos t + \log\tan(t/2)),\ y = a\sin t$
is the catenary $y = a\cosh(x/a)$.
4. Find the evolute of the curve
$x^{2/3} + y^{2/3} = a^{2/3}$.
5. Prove that the evolute of the ellipse $x^2/a^2 + y^2/b^2 = 1$ is the envelope of the family of the ellipses given by
$a^2x^2\sec^4\alpha + b^2y^2\,\text{cosec}^4\alpha = (a^2 - b^2)^2$,
α being the parameter.
6. Prove that the evolute of the hyperbola $2xy = a^2$ is
$(x+y)^{2/3} + (x-y)^{2/3} = 2a^{2/3}$.
7. Show that the evolute of the curve whose pedal equation is $r^2 - a^2 = mp^2$ is the curve whose pedal equation is
$r^2 - (1-m)a^2 = mp^2$.
8. Show that the evolute of the cardiod $r = a(1+\cos\theta)$, is the cardiod $r = \dfrac{1}{3}a(1-\cos\theta)$, the pole of the latter equation being at the point $\left(\dfrac{2}{3}a, 0\right)$.
9. Show that the whole length of the evolute of the ellipse $\dfrac{x^2}{a^2} + \dfrac{y^2}{b^2} = 1$ is $4\left(\dfrac{a^2}{b} - \dfrac{b^2}{a}\right)$.
10. Find the evolute of the parabola $y^2 = 4ax$.

Hint to Selected Problems

1. Equation of any normal to the parabola $y^2 = 2ax$ is
$y = mx - am - \frac{1}{2}am^3$, where m is a parameter.
Now differentiating partially w.r.t. m.

2. Let $x = a\cos\theta, y = b\sin\theta$ Then $\frac{dy}{dx} = \frac{b\cos\theta}{-a\sin\theta}$.

Therefore, slope of the normal to the given ellipse at the point
$(x, y) = -\frac{dx}{dy} = \frac{a\sin\theta}{b\cos\theta}$

$\therefore$ The equation of the normal is given by
$y - b\sin\theta = \frac{a\sin\theta}{b\cos\theta}(x - a\cos\theta)$. Now proceed as usual.

4. Take the point $x = a\cos^3\theta, y = a\sin^3\theta$, θ being the parameter.
Then proceed as in (2).

6. Let $$xy = \frac{a^2}{2} = c^2. \qquad ...(1)$$

Let $P\left(ct, \frac{c}{t}\right)$ be any point on the hyperbola (1).

Then $$\frac{dy}{dx} = -\frac{c^2}{x^2}.$$

$\therefore$ Slope of the normal $= -\frac{dx}{dy} = -\frac{x^2}{c^2} = t^2$.

Hence, the equation of the normal at P is given by
$$y - \frac{c}{t} = t^2(x - ct)$$
Then proceed as usual.

7. The pedal equation of the given curve is
$$r^2 - a^2 = mp^2 \Rightarrow \rho = r\frac{dr}{dp} = mp.$$
Now use the relation $r'^2 = r^2 + \rho^2 - 2\rho p$.

ANSWERS

1. $27ay^2 = 8(x - a)^3$ 4. $(x + y)^{2/3} + (x - y)^{2/3} = 2a^{2/3}$ 10. $27ay^2 = 4(x - 2a)^3$

Objective Evaluations

Fill in the Blanks

1. The equation of a family of curves having one parameter is ________ .
2. The equation $F(x, y, \lambda, \mu) = 0$ represents ________ with two parameters.
3. If the equation of the family of curves is $A\lambda^2 + B\lambda + C = 0$, where A, B, C are functions of x, y, then its envelope is ________ .
4. The envelope of a family of curves ________ each member of the family.
5. $xm^2 - 2ym + a = 0$ is a family of straight lines, where m being the parameter, then its envelope is ________ .
6. The envelope of the normals to the curve is ________ .

True/False

Write 'T' for True and 'F' for False statement.

1. The equation $F(x, y, \lambda) = 0$ represents a family of curve with one parameter. **(T/F)**
2. The envelope of a family of curves intersects each member of the family. **(T/F)**
3. If the equation of a family of curves is $A\lambda^2 + B\lambda + C = 0$ then its envelope is $B^2 - 4AC = 0$. **(T/F)**
4. Evolute of a curve is the locus of the centre of curvature for that curve. **(T/F)**
5. The envelope of the normals to the curve is evolute. **(T/F)**

Multiple Choice Questions

Choose the most appropriate one :

1. The envelope of a family of curves ________ each member of the family :
   (a) intersect (b) touches
   (c) is perpendicular to (d) none of these
2. The envelope of the family of curves $xm^2 - 2ym + a = 0$, m being the parameter is :
   (a) $y^2 = 4ax$ (b) $y^2 = 2ax$
   (c) $y^2 = ax$ (d) $x^2 = ay$
3. The locus of the centre of curvature for a curve is :
   (a) envelope (b) evolute
   (c) radius of curvature (d) none of these
4. $F(x, y, \lambda) = 0$ represents :
   (a) family of curves with parameter λ
   (b) curves
   (c) surface
   (d) none of these

ANSWERS

Fill in the Blanks

1. $F(x, y, \lambda) = 0$ 2. family of curves with two parameters 3. $B^2 - 4AC = 0$ 4. touches
5. $y^2 = ax$ 6. evolute

True/False

1. T 2. F 3. T 4. T 5. T

Multiple Choice Questions

1. (b) 2. (c) 3. (b) 4. (a)

FFFFFF

Asymptotes

7.1 INTRODUCTION

In calculus, there are some curves whose branches seem to go to infinity. It is not necessary that there always exists a definite straight line for all such curves which seems to touch the branch of the curves at infinite but more or less there are some certain curves for which this type of definite straight line exists, this straight line is therefore known as asymptote.

Definition. *A definite straight line whose distance from branch of the curve continuously decreases as we move away from the origin along the branch of the curve and seems to touch the branch at infnity, provided the distance of this line from origin should be finite initially, is called an asymptote of the curve.*

Suppose in the equalion of a curve, two or more than two values of y exists for every value of x, then we obtain different branches of the curve corresponding to these distinct values of y. If each branch have its own separate asymptote, then we can say that a curve may have more than one asymptote.

7.2 DETERMINATION OF ASYMPTOTES

Consider a curve

$$f(x, y) = 0 \qquad ...(1)$$

and also consider that there are no asymptotes parallel to y-axis. Thus we shall take the equation which is not parallel to y-axis. in the form of

$$y = mx + c \qquad ...(2)$$

Let us take a point $P(x, y)$ on the curve (1), therefore this point as tends to infinity along the straight line (2), x must tend to infinity. Now find the tangent to the curve $f(x, y) = 0$ at the point $P(x, y)$.

$\therefore$ The equation of tangent at $P(x, y)$ is

$$Y - y = \frac{dy}{dx}(X - x) \quad \text{or} \quad Y = \frac{dy}{dx}X + \left(y - x\frac{dy}{dx}\right). \qquad ...(3)$$

The equation (3) is of the form $y = mx + c$ so in order to exist the asymptote of the curve there must both $\frac{dy}{dx}$ and $\left(y - x\frac{dy}{dx}\right)$ tend to finite limits as x tends to infinity. Therefore, if the equation (3) tends to the straight line given in (2) as x tends to infinity, then the line (2) will be an asymptote of the curve $f(x, y) = 0$ and also we have

$$m = \lim_{x\to\infty}\frac{dy}{dx} \quad \text{and} \quad c = \lim_{x\to\infty}\left(y - x\frac{dy}{dx}\right)$$

Since c is finite, then we have

$$\lim_{x\to\infty}\left(\frac{y - x\frac{dy}{dx}}{x}\right) = \lim_{x\to\infty}\frac{c}{x} = 0 \quad \text{or} \quad \lim_{x\to\infty}\left(\frac{y}{x} - \frac{dy}{dx}\right) = 0$$

or

$$\lim_{x\to\infty}\left(\frac{y}{x}\right) = \lim_{x\to\infty}\frac{dy}{dx} \quad \text{or} \quad \lim_{x\to\infty}\frac{y}{x} = m.$$

Also $$c = \lim_{x\to\infty}\left(y - x\frac{dy}{xx}\right) \quad \text{or} \quad c = \lim_{x\to\infty}(y - mx).$$

Hence, if $y = mx + c$ is an asymptote to the curve $f(x, y) = 0$, then we obtain

$$m = \lim_{x\to\infty}\frac{dy}{dx} = \lim_{x\to\infty}\frac{y}{x} \text{ and } \quad c = \lim_{x\to\infty}(y - mx).\ .$$

7.3 ASYMPTOTES OF GENERAL EQUATION

Let the general rational algebraic equation of a curve be

$$\{a_0y^n + a_1y^{n-1}x + a_2y^{n-2}x^2 + \ldots + a_{n-1}yx^{n-1} + a_nx^n\} + \{b_1y^{n-1} + b_2y^{n-2}x + \ldots + b_{n-1}yx^{n-2} + b_nx^{n-1}\}$$
$$+ \{c_2y^{n-2} + c_3y^{n-3} + \ldots + c_{n-1}yx^{n-3} + c_nx^{n-2}\} + \ldots = 0 \quad \ldots(1)$$

or $$x^n\left\{a_0\left(\frac{y}{x}\right)^n + a_1\left(\frac{y}{x}\right)^{n-1} + a_2\left(\frac{y}{x}\right)^{n-2} + \ldots + a_{n-1}\left(\frac{y}{x}\right) + a_n\right\} + x^{n-1}\left\{b_1\left(\frac{y}{x}\right)^{n-1} + b_2\left(\frac{y}{x}\right)^{n-2} + \ldots + b_n\right\}$$
$$+ x^{n-2}\left\{c_2\left(\frac{y}{x}\right)^{n-2} + c_3\left(\frac{y}{x}\right)^{n-3} + \ldots + c_n\right\} + \ldots = 0$$

or $$x^n\phi_n\left(\frac{y}{x}\right) + x^{n-1}\phi_{n-1}\left(\frac{y}{x}\right) + x^{n-2}\phi_{n-2}\left(\frac{y}{x}\right) + \ldots + x\phi_1\left(\frac{y}{x}\right) + \phi_0\left(\frac{y}{x}\right) = 0 \quad \ldots(2)$$

where $\phi_k\left(\frac{y}{x}\right)$ is a polynomial of degree k in $\left(\frac{y}{x}\right)$.

Divide (2) by x^n, we get

$$\phi_n\left(\frac{y}{x}\right) + \frac{1}{x}\phi_{n-1}\left(\frac{y}{x}\right) + \frac{1}{x^2}\phi_{n-2}\left(\frac{y}{x}\right) + \ldots + \frac{1}{x^{n-1}}\phi_1\left(\frac{y}{x}\right) + \frac{1}{x^n}\phi_0\left(\frac{y}{x}\right) = 0$$

Now taking limit as $x \to \infty = 0$, and assuming there is no asymptote parallel to y-axis then $m = \lim_{x\to\infty}\left(\frac{y}{x}\right)$, we get

$$\phi_n(m) = 0. \quad \ldots(3)$$

This equation (3) is of degree n in m so it has at most n roots, real as well as imaginary. Out of these n roots some roots may be identical. Thus we get n values of m corresponding to the n branches of the curve (1). Since, we will have only real values of m so ignore all imaginary roots of (3) if they exists. Further if $y = mx + c$ is an asymptote of (1), then we have

$$c = \lim_{x\to\infty}(y - mx), \text{ for each specified value of } m.$$

Determination of c. For the determination of c corresponding to each distinct value of m, we put $y = mx + p$ in the equation of curve (2), where $p \to c$ as $x \to \infty$.

Now putting $y = mx + p$ *i.e.*, $\frac{y}{x} = m + \frac{p}{x}$, in the (2), we get

$$x^n\phi_n\left(m + \frac{p}{x}\right) + x^{n-1}\phi_{n-1}\left(m + \frac{p}{x}\right) + x^{n-2}\phi_{n-2}\left(m + \frac{p}{x}\right) + \ldots + x\phi_1\left(m + \frac{p}{x}\right) + \phi_0\left(m + \frac{p}{x}\right) = 0.$$

Expand each term by Taylor's expansion, we get

$$x^n\left[\phi_n(m) + \frac{p}{x}\phi_n'(m) + \frac{p^2}{2!x^2}\phi_n''(m) + \ldots\right] + x^{n-1}\left[\phi_{n-1}(m) + \frac{p}{x}\phi_{n-1}'(m) + \ldots\right] + x^{n-2}\left[\phi_{n-2}(m) + \frac{p}{x}\phi_{n-2}'(m) + \ldots\right] + \ldots = 0$$

or $$x^n\phi_n(m) + x^{n-1}[p\phi_n'(m) + \phi_{n-1}(m)] + x^{n-2}\left[\frac{p^2}{2!}\phi_n''(m) + \frac{p}{1!}\phi_{n-1}'(m) + \phi_{n-2}(m)\right] + \ldots = 0$$

Since we know that $\phi_n(m) = 0$, then

$$x^{n-1}[p\phi_n'(m) + \phi_{n-1}(m)] + x^{n-2}\left[\frac{p^2}{2!}\phi_n''(m) + \frac{p}{1!}\phi_{n-1}'(m) + \phi_{n-2}(m)\right] + \ldots = 0$$

Dividing by x^{n-1} and taking limit as $x \to \infty$, we get

$$\lim_{x\to\infty} [p\phi_n'(m) + \phi_{n-1}(m)] = 0 \quad \text{or} \quad \left(\lim_{x\to\infty} p\right)\phi_n'(m) + \phi_{n-1}(m) = 0$$

or
$$c\phi_n'(m) + \phi_{n-1}(m) = 0 \qquad \left(\because \lim_{x\to\infty} p = c\right)$$

Hence, from above relation we can determine the value of c for each distinct value of m.

REMARK

- To find the polynomial $\phi_n(m)$. We should put $y = m$ and $x = 1$ in the nth degree terms of the curve. Similarly to get $\phi_{n-1}(m)$ we put $y = m$ and $x = 1$ in the $(n-1)^{th}$ degree terms of the curve. Therefore in general, to get $\phi_k(m)$ we should put $y = m$ and $x = 1$ in the kth degree terms of the curves.

7.4 EXISTENCE OF ASYMPTOTES

From the equation $\phi_n(m) = 0$, if we obtain one or more than one values of m such that $\phi_n'(m) = 0$ and $\phi_{n-1}(m) \neq 0$, then from the equation for the determining of c. we obtain $0.c + \phi_{n-1}(m) = 0$

Thus we get c is either, $+\infty$ or $-\infty$. Hence, we can say that corresponding to such values of m no asymptotes will exists.

7.5 DETERMINATION OF *c* CORRESPONDING TO SOME IDENTICAL VALUES OF *m*

Let us suppose some of the roots of the equation $\phi_n(m) = 0$ are identical and let these identical values be r in number which will make $\phi_n'(m), \phi_n''(m), \ldots \phi_m^{r-1}(m)$ equal to zero. Now for the existence of the asymptotes $\phi_{n-1}(m)$ must be zero corresponding to the identical values of m. Also, if it will make $\phi_{n-1}'(m), \phi_{n-1}''(m), \ldots \phi_{n-1}^{r-2}(m); \phi_{n-2}(m), \phi_{n-2}'(m), \ldots \phi_{n-2}^{r-3}(m); \phi_{n-3}'(m), \phi_{n-3}'(m), \ldots \phi_{n-3}^{r-4}(m) \ldots;$ $\phi_{n-r+2}(m), \phi_{n-r+2}'(m)$ and $\phi_{n-r+1}(m)$ equal to zero, then the equation to determine c will become

$$0.c^{r-1} + 0.c^{r-2} + \ldots + 0.c + 0 = 0$$

and thus we cannot find the value of c in this way.

So to determine c let us put $\phi_n(m), \phi'(m), \ldots, \phi_n^{r-1}(m); \phi_{n-1}(m), \phi_{n-1}'(m), \ldots \phi_{n-1}^{r-2}(m); \phi_{n-2}(m), \phi_{n-2}'(m), \ldots \phi_{n-2}^{r-3}(m); \phi_{n-3}(m),$ $(\phi_{n-3}'(m), \ldots \phi_{n-3}^{r-4}(m) \ldots \phi_{n-r+2}(m), \phi_{n-r+2}'(m)$ and $\phi_{n-r+1}(m)$ equal to zero in the following equation

$$x^n\phi_n(m) + x^{n-1}[p\phi_n'(m) + \phi_{n-1}(m)] + x^{n-2}\left[\frac{p^2}{2!}\phi_n''(m) + \frac{p}{1!}\phi_{n-1}'(m) + \phi_{n-2}(m)\right] + \ldots + x^{n-r+1}\left[\frac{p^{r-1}}{r-1!}\phi_n^{r-1}(m) + \right.$$
$$\left.\frac{p^{r-2}}{r-2!}\phi_{n-1}^{r-2}(m) + \ldots + \frac{p}{1!}\phi_{n-r+2}'(m) + \phi_{n+r+1}'(m)\right] + x^{n-r}\left[\frac{p^r}{r!}\phi_n^r(m) + \frac{p^{r-1}}{r-1!}\phi_{n-1}^{r-1}(m) + \frac{p^{r-2}}{r-2!}\phi_{n-2}^{r-2}(m) + \ldots\right.$$
$$\left. + \frac{p}{1!}\phi_{n-r+1}'(m) + \phi_{n-r}(m)\right] = 0$$

We have

$$x^{n-r}\left[\frac{p^r}{r!}\phi_n^r(m) + \frac{p^{r-1}}{r-1!}\phi_{n-1}^{r-1}(m) + \ldots + \frac{p}{1!}\phi_{n-r+1}'(m) + \phi_{n-r}(m)\right] + x^{n-r-1}\left[\frac{p^{r+1}}{r+1!}\phi_n^{r+1}(m) + \frac{p^r}{1!}\phi_{n-1}^r(m) + \ldots\right].$$

Now dividing above equation by x^{n-r} and taking the limit as $x \to \infty$, we get

$$\frac{c^r}{r!}\phi_n^r(m) + \frac{c^{r-1}}{r-1!}\phi_{n-1}^{r-1}(m) + \ldots + \frac{c}{1!}\phi_{n-r+1}'(m) + \phi_{n-r}(m) = 0 \text{ where } c = \lim_{x\to\infty} p.$$

Therefore this equation gives r values of c corresponding to the identical values of m. Hence, we obtain r parallel asymptotes.

7.6 NUMBER OF ASYMPTOTES OF A CURVE

Suppose the degree of an algebraic curve is n, then we find a polynomial $\phi_n(m)$ by putting $y = m$ and $x = 1$ in the nth degree terms of the curve. Thus the equation $\phi_n(m) = 0$ is of degree n in m and which gives almost n values of m real as well as imaginary. These n values of m are nothing but the slopes of the asymptotes, which are not parallel to y axis. If there are some asymptotes, parallel to y-axis, then the degree of $\phi_n(m)$ will be smaller than n by the same number of parallel asymptotes. Suppose all the roots of $\phi_n(m) = 0$ are distinct and real, then to each value of m we obtain one value of c. Hence, we obtain n asymptotes. In case, there some roots say r (out of n) of $\phi_n(m) = 0$ are same, then we can find the values of c for these same roots the following equation

$$\frac{c}{r!}\phi_n^r(m)+\frac{c^{r-1}}{r-1!}\phi_{n-1}^{r-1}(m)+\ldots+\phi_{n-r}(m)=0$$

This equation in c is of degree r so we get r distinct values of c for the same roots, hence, again we obtain n asymptotes. Therefore we can say that the total number of asymptotes of a curve are equal to the degree of the curve, These asymptotes are real as well as imaginary but we have required only real asymptotes so we ignore all the imaginary asymptotes.

7.7 ASYMPTOTES PARALLEL TO CO-ORDINATES AXES

(a) Asymptotes parallel to x-axis. Let the general equation of an algebraic curve in decreasing powers of x be

$$x^n\phi(y)+x^{n-1}\phi_1(y)+x^{n-2}\phi_2(y)+\ldots=0 \quad \ldots(1)$$

where $\phi(y),\phi_1(y),\phi_2(y),\ldots$ are the function of y only.

Now divide (1) by x^n, we get

$$\phi(y)+\frac{1}{x}\phi_1(y)+\frac{1}{x^2}\phi_2(y)+\ldots=0. \quad \ldots(2)$$

If $y = k$ is an asymptote parallel to x-axis, then we can say that x alone tends to infinity as a point $P(x, y)$ on the curve tends to infinity along the line $y = k$ and also we have $k=\lim\limits_{x\to\infty} y$.

Now taking the limit of both sides of (2) as $x\to\infty$ and $y\to k$, we get $\phi(k)=0$.

Thus k is a root of the equation $\phi(y) = 0$. If k_1, k_2, etc. are the roots of $\phi(y) = 0$, then the asymptotes parallel to x-axis are given by $y = k_1, y = k_2$, etc. Since k is a root of the equation $\phi(y) = 0$, then $(y - k)$ is a factor of the equation $\phi(y) = 0$. Also $\phi(y)$ is the coefficient of the highest power of x *i.e.*, x^n in the equation of the curve. Hence, we obtain the asymptotes parallel to x-axis by taking the coefficient of highest power of x in the equation of the curve equal to zero.

(b) Asymptotes parallel to y-axis. Similarly, we may obtain the asymptotes parallel to y-axis by taking the coefficient of highest power of y in the equation of the curve equal to zero.

REMARK

- If the coefficient of highest power of x or y or both are constant, then no asymptotes parallel to either x or y or both axes exists respectively.

Solved Examples

Example 1. *Find the asymptotes of the curve* $x^3+y^3-3axy=0$.

Solution . Obviously, the degree of the curve is 3, so it will have 3 asymptotes real as well as imaginary. Here the coefficient of highest degree in x and y are constant so no asymptote parallel to co-ordinate axis exist. Let

$$y = mx + c \quad \ldots(1)$$

be the asymptote of the curve.

So putting $y = m$ and $x = 1$ in the highest degree terms of the curve, we get

$$\phi_3(m) = 1 + m^3.$$

Solving the equation $\phi_3(m) = 0$

i.e., $1 + m^3 = 0$

or $(1 + m)(m^2 - m + 1) = 0$

or $m = -1$

is only real root and other two roots are imaginary so ignore them.

Next, putting $y = m$ and $x = 1$ is second degree terms in the equation of the curve (1), we get

$$\phi_2(m) = -3am.$$

Now we find value of c by the following equation

$$c\phi_n'(m)+\phi_{n-1}(m)=0$$

or $c\phi_3'(m)+\phi_2(m)=0$

or $c(3m^2) + (-3am) = 0$

$[\because \phi_3(m) = 1 + m^3, \Rightarrow \phi_3'(m)=3m^2]$

If $m = -1$, then

$$c[3(-1)^2]+[-3a(-1)] = 0$$

$$3c + 3a = 0$$

or $c = -a$.

Hence, the asymptote is $y=-x-a$

or $x + y + a = 0$.

Example 2. *Find all the asymptotes of the curve* $x^3 + x^2y - xy^2 - y^3 - 3x - y - 1 = 0$.

Solution . The degree of the curve is 3 so it has 3 asymptotes which are real as well as imaginary. Since the coefficients of highest degree *i.e.*, 3rd degree of x and y are constant so there are no asymptotes parallel to co-ordinate axes. Thus there are oblique asymptotes of the form $y = mx + c$.

Now putting $y = m$ and $x = 1$ in the third degree terms of the curve, we get

$$\phi_3(m) = 1 + m - m^2 - m^3.$$

Solving the equation

$$\phi_3(m) = 0 \text{ i.e, } 1 + m - m^2 - m^3 = 0,$$

we get $(1 + m)(1 - m^2) = 0$

or $m = -1, -1, 1.$

Determination of c. For $m = 1$, we use the following equation

$$c\phi_n'(m) + \phi_{n-1}(m) = 0$$

or $$c\phi_3'(m) + \phi_2(m) = 0 \qquad ...(1)$$

Putting $y = m$ and $x = 1$ in the second degree terms of the equation we get

$$\phi_2(m) = 0.$$

From (1), we get

$$c(1-2m-3m^2)+0=0$$

at $m = 1$

$$c(1 - 2 - 3) + 0 = 0$$

or $-4c = 0$

or $c = 0$

Thus one of the asymptote is $y = x$

Determination of c for m = –1, –1. Since two out of three roots of the equation $\phi_3(m) = 0$ are same, then we use the following formula to determine c

$$\frac{c^2}{2!}\phi_3''(m) + \frac{c}{1!}\phi_2'(m) + \phi_1(m) = 0. ...(2)$$

Putting $y = m$ and $x = 1$ in the first degree terms of the equation we obtain $\phi_1(m) = -3 - m$.

From (2), we have

$$\frac{c^2}{2!}(-2-6m) + \frac{c}{1!}.0 + (-3-m) = 0$$

at $m = -1$

$$\frac{c^2}{2}(-2+6) - 3 + 1 = 0$$

or $2c^2 - 2 = 0$

or $c = \pm 1$

Thus other two asymptotes are

$$y = -x + 1,\ y = -x - 1.$$

Hence, all the asymptotes of the given curve are

$$y = x, x + y - 1 = 0, x + y + 1 = 0.$$

Example 3. *Find all the asymptotes of the curve*

$$(x - 2y)^2 (x - y) - 4y(x - 2y) - (8x + 7y) = 0.$$

Solution . Simplifying the equation of curve

$$(x^2+4y^2-4xy)(x - y) - 4xy+8y^2 - 8x - 7y = 0$$

or $$x^3+8xy^2-5x^2y-4y^3-4xy+8y^2-8x-7y = 0. \qquad ...(1)$$

The degree of the curve (1) is 3 so it has 3 asymptotes which are real as well as imaginary. Obviously there are no asymptotes parallel to co-ordinate axis. Thus there are only oblique asymptotes of the form $y = mx + c$.

Putting $y = m$ and $x = 1$ in the highest degree i.e., third degree terms of the curve (1), we obtain

$$\phi_3(m) = 1 - 5m + 8m^2 - 4m^3.$$

Solving the equation $\phi_3(m) = 0$

i.e., $1 - 5m + 8m^2 - 4m^3 = 0$

or $(1-m)(1-2m)^2 = 0$ or $m = \frac{1}{2}, \frac{1}{2}, 1.$

Determination of c for m = 1 :

Putting $y = m$ and $x = 1$ in the second degree terms of the curve (1), we obtain

$$\phi_2(m) = -4m + 8m^2.$$

Applying the formula

$$c.\phi_3'(m) + \phi_2(m) = 0$$

or $c(-5 + 16m - 12m^2) - 4m + 8m^2 = 0.$

Substitute $m = 1$, we get

$$c(-5 + 16 - 12) - 4 + 8 = 0$$

or $-c + 4 = 0$

or $c = 4.$

Thus the asymptote is $y = x + 4$

or $x - y + 4 = 0.$

Determination of c for $m = \frac{1}{2}, \frac{1}{2}$:

Putting $y = m$ and $x = 1$ in the first degree terms of the curve (1) we obtain

$$\phi_1(m) = -8 - 7m.$$

Since $m = \frac{1}{2}, \frac{1}{2}$ are two repeated roots of $\phi_3(m) = 0$, then apply the following formula to determine c,

$$\frac{c^2}{2!}[\phi_3''(m)] + \frac{c}{1!}\phi_2'(m) + \phi_1(m) = 0$$

or $$\frac{c^2}{2!}(16-24m) + c(-4+16m) - 8 - 7m = 0$$

At $m = \frac{1}{2}$

$$\frac{c^2}{2}(16-12) + c(-4+8) - 8 - \frac{7}{2} = 0$$

or $$2c^2 + 8c - \frac{23}{2} = 0$$

or $$4c^2 + 8c - 23 = 0 \Rightarrow c = \frac{-2 \pm 3\sqrt{3}}{2}.$$

Thus the other asymptotes are

$$y = \frac{1}{2}x + \frac{-2 \pm 3\sqrt{3}}{2}$$

or $$2y = x - 2 \pm 3\sqrt{3}.$$

Hence, all the three asymptotes of the curve are

$$x - y + 4 = 0,\ 2y = x - 2 \pm 3\sqrt{3}.$$

Example 4. *Find all the aysmptotes of the curve*

$$y^2(x^2 - a^2) = x^2(x^2 - 4a^2).$$

Solution . Obviously the degree of the curve is 4 so its has at most 4 asymptotes. Since highest degree term of x is x^4 whose coefficient is a constant so there is no asymptote parallel to x-axis but highest degree term of y is y^2 and whose coefficient is $x^2 - a^2$. Thus the asymptotes parallel to y-axis are

$$x^2 - a^2 = 0$$

or $\quad x = \pm a \Rightarrow x = -a, x = a.$

Putting $y = m$ and $x = 1$ in the highest *i.e.*, 4th degree terms of the curve we obtain

$$\phi_4(m) = m^2 - 1.$$

Solving the equation $\phi_4(m) = 0$ *i.e.*, $m^2 - 1 = 0$.

then $\quad m = \pm 1.$

Now putting $y = m$ and $x = 1$ in the third degree terms of the curve we obtain $\phi_3(m) = 0$

Determination of c for m=± 1. To determine the values of c at $m = \pm 1$ we use the following formula

$$c.\phi_4'(m) + \phi_3(m) = 0$$

or $\quad c(2m) + 0 = 0$

At $\quad m = \pm 1,\ c = 0.$

Thus other two asymptotes are $y = \pm x$.

Hence, all the four asymptotes are

$$x = \pm a, y = \pm x.$$

Example 5. *Find all the asymptotes of the curve*

$$y^3 - xy^2 - x^2y + x^3 + x^2 - y^2 - 1 = 0.$$

Solution . Obviously the degree of the curve is 3, therefore it has at most 3 asymptotes.

Putting $y = m$ and $x = 1$, third degree terms of the equation of the curve, we get

$$\phi_3(m) = m^3 - m^2 - m + 1 = 0$$

or $\quad (m - 1)^2(m + 1) = 0$

$\therefore \quad m = 1, 1, -1.$

Now, again putting $y = m$, and $x = 1$ in the second degree terms, we get

$$\phi_2(m) = 1 - m^2.$$

To determine c, we have $c\phi_3'(m) + \phi_2(m) = 0.$

$$c(3m^2 - 2m - 1) + (1 - m^2) = 0. \quad ...(1)$$

when $m = -1$, we have $C = 0$ and the corresponding asymptotes is

$$y = -x + 0 \Rightarrow y + x = 0$$

when $m = 1$, the equation (1) reduces to the identity $c.0 + 0 = 0$ and we cannot determine c from it. In this case c is to be determined from the equation

$$\frac{c^2}{2!}\phi_3''(m) + \frac{c}{1!}\phi_2'(m) + \phi_1(m) = 0.$$

Putting $y = m$ and $x = 1$ in the first degree terms in the equation of the curve, we get

$$\phi_1(m) = 0.$$

Hence for $m = 1$, c is to given by

$$\frac{c^2}{2}(6m - 2) + c(-2m) = 0$$

i.e., $\quad (3m - 1)c^2 - 2mc = 0$

if $m = 1$, this $\quad 2c^2 - 2c = 0$

$\Rightarrow \quad c = 0$ and $c = 1$.

Hence, $y = x + 1$ and $y = x = 0$ are three parallel asymptotes corresponding to the slope $m = 1$. Therefore the required asymptotes are $y + x = 0, y - x = 0$ and $y - x - 1 = 0$.

Example 6. *Find asymptotes of the curve*

$$x^2y^2 - x^2y - xy^2 + x + y + 1 = 0$$

Solution . Degree of the given curve is 4, so it has at most 4 asymptotes (Real and imaginary).

Asymptote parallel to *x*-axis :

Equating the coefficient of highest degree term of x (*i.e.*, x^2) to zero, we get

$$y^2 - y = 0 \Rightarrow \quad y(y - 1) = 0$$

$\Rightarrow \quad y = 0$ and $y = 1$

Thus, $y = 0$ and $y = 1$ are two asymptotes parallel to x-axis.

Asymptote parallel to *y*-axis :

Equating the coefficient of highest degree term of y (*i.e.*, y^2) to zero, we get

$$x^2 - x = 0 \quad \Rightarrow \quad x(x - 1) = 0$$

$\Rightarrow \quad x = 0 \quad$ and $\quad x = 1$

Thus, $y = 0$ and $x = 1$ are two asymptotes parallel to x-axis.

Hence, $x=0$, $y = 0$, $x = 1$ and $y = 1$ are the required asymptotes.

Example 7. *Find asymptotes parallel to axes for the curve*

$$y^2(x^2 - a^2) = x.$$

Solution . The given curve is a degree 4, so it cannot have more than four asymptotes. Now, equating to zero the coefficient of the highest power of y (*i.e.*, of y^2), the asymptotes parallel to y-axis are given by

$$x^2 - a^2 = 0 \Rightarrow \quad x = \pm a.$$

Again equating to zero the coefficient of the highest power of x (*i.e.*, of x^2), the asymptotes parallel to x-axis are given by

$$y^2 = 0 \Rightarrow y = 0, y = 0.$$

Hence, all the four asymptotes are given by $x = \pm a, y = 0, y = 0$.

EXERCISE 7.1

Find all the asymptotes of the following curves :

1. $a^2/x^2 - b^2/y^2 = 1$
2. $a^2/x^2 + b^2/y^2 = 1$
3. $y^2(a^2 - x^2) = x^4$
4. $x^2y^2 = a^2(x^2 + y^2)$
5. $x^2y^2 - x^2y - xy^2 - y + 1 = 0$
6. $3x^3 + 2x^2y - 7xy^2 + 2y^3 + 14xy + 7y^2 + 4x + 5y = 0$
7. $2x^3 - x^2y - 2xy^2 + y^3 - 4x^2 + 8xy - 4x + 1 = 0$
8. $x^3 + 2x^2y + xy^2 - x^2 - xy + 2 = 0$
9. $y^3 - 5xy^2 + 8x^2y - 4x^3 - 3y^2 + 9xy - 6x^2 + 2y - 2x + 1 = 0$
10. $y^3 - x^2y - 2xy^2 + 2x^3 - 7xy + 3y^2 + 2x^2 + 2x + 2y + 1 = 0$ (MTU–2012)
11. $y^3 - xy^2 - x^2y + x^3 + x^2 - y^2 - 1 = 0$
12. $(x^2 - y^2)(y^2 - 4x^2) - 6x^3 + 5yx^2 + 3xy^2 - 2y^3 - x^2 + 3xy - 1 = 0$
13. $y^3 = x^3 + ax^2$
14. $x^2y^3 + x^3y^2 = x^3 + y^3$.
15. $(y - x)(y - 2x)^2 + (y + 3x)(y - 2x) + 2x + 2y - 1 = 0$.
16. $x^3 + 2x^2y - xy^2 - 2y^3 + 4y^2 + 2xy + y - 1 = 0$.
17. $(x + y)^2(x + 2y + 2) = x + 9y + 2$ (MDU–2005)
18. $x^2(x - y)^2 + a^2(x^2 - y^2) - a^2xy = 0$
19. $y^3 - 2y^2x - yx^2 + 2x^3 + y^2 - 6xy + 5x^2 - 2y + 2x + 1 = 0$
20. $x^3 + 3x^2y - 4y^3 - x + y + 3 = 0$
21. $x^3 - 5x^2y + 8xy^2 - 4y^3 + x^2 - 3xy + 2y^2 - 1 = 0$
22. $xy^2 = 4a^2(2a + x)$.
23. $x^3 + 2x^2y - xy^2 - 2y^3 + xy - y^2 - 1 = 0$
24. $y^3 + x^2y + 2xy^2 - y + 1 = 0$
25. $(2x - 3y + 1)^2(x + y) - 8x + 2y - 9 = 0$
26. $(x^2 - y^2)^2 - 4y^2 + y = 0$
27. $y^2(x - 2a) = x^3 - a^3$
28. $(x^3 + a^3)y = bx^3$

ANSWERS

1. $x = \pm a$ **2.** $x = \pm a, y = \pm b$ **3.** $x = \pm a$
4. $x = \pm a, y = \pm a$ **5.** $y = 0; y = 1; x = 0; x = 1$
6. $x + 2y = 1, 2x - 2y = -7, 6x - 2y = 15$
7. $x + y - 2 = 0; x - y + 2 = 0; 2x - y - 4 = 0$
8. $x = 0; x + y = 0; x + y - 1 = 0$
9. $x - y = 0; 2x - y + 2 = 0; 2x - y + 1 = 0$
10. $x - y - 1 = 0; x + y + 2 = 0; 2x - y = 0$
11. $x + y = 0; x - y = 0; x - y + 1 = 0$
12. $x - y = 0; 2x - y = 0; x + y + 1 = 0; 2x + y + 1 = 0$
13. $3x - 3y + a = 0$
14. $y = \pm 1; x = \pm 1; x + y = 0$
15. $2x - y - 2 = 0; 2x - y - 3 = 0; x - y + 4 = 0$
16. $x - y + 1 = 0; x + y - 1 = 0; x + 2y = 0$
17. $x + 2y + 2 = 0;\ x + y \pm 2\sqrt{2} = 0$
18. $y = \pm a; x - y = \pm a$
19. $x - y = 0; 2x - y + 1 = 0; x + y + 2 = 0$
20. $x - y = 0; x + 2y - 1 = 0; x + 2y + 1 = 0$
21. $x - y = 0; x - 2y = 0; x - 2y + 1 = 0$
22. $x = 0, y = \pm 2a$
23. $x + 2y - 1 = 0; x - y = 0; x + y + 1 = 0$
24. $y = 0; x + y - 1 = 0; x + y + 1 = 0$
25. $x + y = 0; 2x - 3y + 3 = 0; 2x - 3y - 1 = 0$
26. $x + y = \pm 1; x - y = \pm 1$
27. $x = 2a, y = x + a, y = -x - a$
28. $x + a = 0, y - b = 0$

7.8 OTHER METHODS FOR FINDING THE ASYMPTOTE OF AN ALGEBRAIC CURVE

THEOREM 1. *The asymptotes of an algebraic curve are parallel to the lines which obtained by equating to zero the linear factors of the highest degree terms of the equation of curve.*

Proof. Let us suppose the equation of the curve is of degree n and let $y - mx$ be a linear factor of the nth degree term in the equation of the curve. Since $\phi_n(m)$ is a polynomial of degree n in m and obtained by putting $y = m$ and $x = 1$ in the nth degree terms of the curve, then $(m - m_1)$ is a factor of $\phi_n(m)$. Thus m_1 is a root of the equation $\phi_n(m) = 0$ which gives the slope of the asymptote. Hence, there is an asymptote parallel to the line $y = m_1x = 0$.

Conversely, let m_1 be a root of the equation $\phi_n(m) = 0$ so that there is an asymptote which is parallel to the line $y - m_1x = 0$, then $(m_1 - m)$ must be a factor of $\phi_n(m)$ and therefore, $(y/x - m_1)$ will be a linear factor of $\phi_n(y/x)$. Hence $(y - m_1x)$ is a linear factor of $x^n\phi_n(y/x)$ which is the highest degree terms in the equation of the curve. Hence the theorem is proved.

Since we know that if $y = mx + c$ is an asymptote of the curve $f(x, y) = 0$, then we have

$$\left. m = \lim_{x\to\infty} \frac{y}{x} \text{ and } c = \lim_{x\to\infty}(y - mx) = \lim_{x\to\infty,\ \frac{y}{x}\to\infty}(y - mx) \right\} \quad \ldots(1)$$

With the help of (1) and above theorem we may find the asymptotes of an algebraic curves.

WORKING PROCEDURE

Step 1. *First we collect all the highest degree terms in the equation of the curve and then resolve into linear factors.*

Step 2. *After getting linear factors there may arise some cases.*

Case I. *If the linear factor $(y - m_1x)$ of the highest degree i.e., n^{th} degree terms in the equation of the curve is simple (non-repeated). Then the given equation of the curve can be written as*

$$(y - m_1x)F_{n-1} + P_{n-1} = 0. \quad \ldots(2)$$

where F_{n-1} contains only terms of degree $n - 1$ and P_{n-1} contains the terms of various degree not exceeding $n - 1$. Therefore $y - m_1x = c$ is an asymptote of the curve where c is to be determined. Let us take a point (x, y) on the curve (1), then we have

$$y - m_1x = -\frac{P_{n-1}}{F_{n-1}}.$$

Now taking the limit as $x \to \infty$, $y/x \to m_1$, then we have

$$\lim_{x\to\infty,\frac{y}{x}\to m_1} (y - m_1x) = \lim_{x\to\infty,\frac{y}{x}\to m_1} \left(-\frac{P_{n-1}}{F_{n-1}}\right) \quad \text{or} \quad c = \lim_{x\to\infty,\frac{y}{x}\to m_1} \left(-\frac{P_{n-1}}{F_{n-1}}\right).$$

Now substitute this value of c in the equation $y = m_1x + c$
we obtained the asymptote which is parallel to the line $y - m_1x = 0$ corresponding to the linear factor $(y - m_1x)$. Similarly we may obtain other asymptotes.

Case II. *If $(y - m_1x)$ is a linear factor of the nth degree terms of order two but $(y - m_1x)$ is not a factor of the $(n - 1)^{th}$ degree terms of the curve, then we have $\phi'_n(m_1) = 0$ and $\phi_{n-1}(m_1) \neq 0$. Therefore, no asymptotes corresponding to $(y - m_1x)^2$ will exist. On the other hand if there are no terms of $(n - 1)^{th}$ degree in the equation of the curve, then make them by adding with zero coefficient and thus we can say that $(y - m_1x)$ is now a factor of $(n - 1)^{th}$ degree terms, then we have the case III.*

Case III. *If $(y - m_1x)^2$ is a linear factor of nth degree terms and $(y - m_1x)$ is a factor of $(n - 1)^{th}$ degree terms, then the equation of the curve can be written as*

$$(y - m_1x)^2 F_{n-2} + (y - m_1x)G_{n-2} + P_{n-2} = 0 \quad \ldots(3)$$

where F_{n-2} and G_{n-2} contain only the terms of degree $n - 2$, and P_{n-2} contains various degree terms not exceeding $n - 2$. Now divide (2) by F_{n-2} and taking the limit as $x \to \infty$ and $y/x \to m_1$, we get

$$\lim_{x\to\infty,y/x\to m_1} (y - m_1x)^2 + \lim_{x\to\infty,y/x\to m_1} (y - m_1x)\left(\frac{G_{n-2}}{F_{n-2}}\right) + \lim_{x\to\infty,y/x\to m_1} \left(\frac{P_{n-2}}{F_{n-2}}\right) \quad \ldots(4)$$

Since we know that $$c = \lim_{x\to\infty,y/x\to m_1} (y - m_1x)$$

and $$A = \lim_{x\to\infty,y/x\to m_1} \left(\frac{G_{n-2}}{F_{n-2}}\right) \quad \text{and} \quad B = \lim_{x\to\infty,y/x\to m_1} \left(\frac{P_{n-2}}{F_{n-2}}\right)$$

then (4) becomes

$$c^2 + Ac + B = 0.$$

This is a quadratic equation in C so it has two roots let C_1 and C_2 be these two roots. Then we obtain two asymptotes $y - m_1x = c_1$ and $y - m_1x = c_2$ corresponding to m_1.

REMARK

- As a consequence we can say that the two asymptotes corresponding to the factor $(y - m_1x)^2$ may obtain by solving the quadratic equation $(y - m_1x)^2 + A(y - m_1x) + B = 0$.
  Similarly, we can also find the asymptotes corresponding to the factor $(y - m_1x)^3$, etc. of the nth degree terms in the equation of the curve.

Case IV. *Suppose the equation of the curve is of the form*

$$(ax + by + c)P_{n-1} + Q_{n-1} = 0 \quad \ldots(5)$$

where P_{n-1} and Q_{n-1} contain various degree term not exceeding the degree $(n - 1)^{th}$, and P_{n-1} contains atleast one term of degree $(n - 1)$ such that (5) becomes of degree n. Therefore, we can say that $(ax+ by)$ is a linear factor of

n^{th} *degree terms in the equation (5). Thus (5) can also be written as*

$$(ax + by)\, P_{n-1} + cP_{n-1} + Q_{n-1} = 0.$$

Divide this equation by P_{n-1} and taking the limit as $x \to \infty$ and $y/x \to -a/b$, we obtain

$$(ax + by + c) + \lim_{x\to\infty,\, y/x\to(-a/b)} (Q_{n-1} / P_{n-1}) = 0$$

This the required equation of the asymptote.

Case V. *Let the equation of the curve of nth degree be of the form*

$$F_n + P = 0 \qquad \text{...(1)}$$

where F_n is of degree n and P is of degree n – 2 or lower and if $F_n = 0$ can be expressed as the product of n linear factors which give n straight lines such that no two of them are parallel or coincident, then all the asymptotes of the curve (1) are obtained by equating to zero the linear factors of F_n.

Solved Examples

Example 1. *Find the asymptotes of*

$(x-y)^2(x^2+y^2) - 10(x-y)x^2+12y^2+2x+y = 0.$

Solution. We have

$$(x-y)^2 - 10(x-y) \lim_{x\to\infty\; y/x\to 1} \frac{x^2}{x^2+y^2}$$

$$+12 + \lim_{x\to\infty\; y/x\to 1} \frac{y^2}{x^2+y^2} = 0$$

or $$(x-y)^2 - 5(x-y) + 6 = 0$$

which gives parallel asymptotes $x - y = 2$ and $x - y = 3$.

The other two asymptotes are imaginary. Since the remaining linear factors of the four degree terms in the equation to the curve are imaginary.

Example 2. *Find the asymptotes of* $(x - y - 1)^2(x^2 + y^2 + 2) + 6(x - y - 1)(xy + 7) - 8x^2 - 2x - 1 = 0.$

Solution. Dividing by the coefficient of $(x - y - 1)^2$ and taking limits, we see that the asymptotes parallel to $x - y - 1 = 0$ are

$$(x-y-1)^2 + 6(x-y-1) \lim_{x\to\infty\; \frac{y}{x}\to 1} \frac{xy+7}{x^2+y^2+2}$$

$$+ \lim_{x\to\infty\; \frac{y}{x}\to 1} \frac{-8x^2-2x-1}{x^2+y^2+2} = 0$$

$$\Rightarrow \quad (x-y-1)^2 + 3(x-y-1) - 4 = 0$$

$$\Rightarrow \quad x-y-1 = \frac{-3 \pm \sqrt{9+16}}{2} = 1, -4.$$

Hence, the two asymptotes are $x - y - 2 = 0$ and $x - y + 3 = 0$ the remaining two asymptotes are imaginary.

7.9 ASYMPTOTES BY EXPANSION

THEOREM. *Let the equation of the curve be of the form* $y = mx + c + \dfrac{A_1}{x} + \dfrac{A_2}{x^2} + \dfrac{A_3}{x^3} + \ldots$ ...(1)

then $y = mx + c$ is the asymptote of (1).

Proof. Since the equation of the curve is

$y = mx + c + \dfrac{A_1}{x} + \dfrac{A_2}{x^2} + \dfrac{A_3}{x^3} + \ldots$; where $\dfrac{A_1}{x} + \dfrac{A_2}{x^2} + \dfrac{A_3}{x^3} + \ldots$ is convergent for sufficiently large values of x.

Differentiating (1) w.r.t. 'x', we get $\dfrac{dy}{dx} = m - \dfrac{A_1}{x^2} - \dfrac{2A_2}{x^3} - \dfrac{3A_3}{x^4} - \ldots$

Now the equation of the tangent to (1) at the point $P(x, y)$ is

$$Y - y = \left(m - \frac{A_1}{x^2} - \frac{2A_2}{x^3} - \frac{3A_3}{x^4} - \ldots\right)(X - x)$$

or $$Y = \left(m - \frac{A_1}{x^2} - \frac{2A_2}{x^3} - \frac{3A_3}{x^4} - \ldots\right)X + c + \frac{2A_1}{x} + \frac{3A_2}{x^2} + \ldots \qquad \text{[Using(1)]}$$

Now taking the limit as $x \to \infty$, we get

$$Y = mX + c.$$

Hence $y = mx + c$ is an asymptote of the curve $y = mx + c + \dfrac{A_1}{x} + \dfrac{A_2}{x^2} + \dfrac{A_3}{x^3} + \ldots$

Solved Examples

Example 1. *Find the asymptotes of the hyperbola* $\frac{x^2}{a^2}-\frac{y^2}{b^2}=1.$

Solution . The equation of the curve can be written as

$$y^2 = b^2\left(-1+\frac{x^2}{a^2}\right)$$

or $$y = \pm b\sqrt{\left(-1+\frac{x^2}{a^2}\right)} = \pm\frac{b}{a}x\sqrt{\left(1-\frac{a^2}{x^2}\right)}$$

$$y = \pm\frac{b}{a}x\left[1-\frac{1}{2}\frac{a^2}{x^2}-\frac{1}{8}\frac{a^4}{x^4}+\ldots\right]$$

[Using binomial expansion]

Since we know that $y = mx + c$ is an asymptote of the curve

$$y = mx + c + \frac{A_1}{x}+\frac{A_2}{x^2}+\ldots$$

Hence, $y = \pm\frac{b}{a}x$ are the asymptotes of the given curve.

Example 2. *Find all the asymptotes of the curve* $(y^2-x^2)(y-2x)-7xy+3y^2+2x^2+2x+2y+1=0.$

Solution . The given equation can be written as

$$(y-x)(y+x)(y-2x) - 7xy + 3y^2 + 2x^2 + 2x + 2y + 1 = 0. \quad \ldots(1)$$

The slope of the asymptote corresponding to the factor $y - x$ is 1. Thus the asymptote corresponding to this factor is

$$y-x = \lim_{x\to\infty,\frac{y}{x}\to 1}\frac{7xy-3y^2-2x^2-2x-2y-1}{(y+x)(y-2x)}$$

$$= \lim_{x\to\infty,\frac{y}{x}\to 1}\frac{7\left(\frac{y}{x}\right)-3\left(\frac{y}{x}\right)^2-2-\frac{2}{x}-2\frac{y}{x}\left(\frac{1}{x}\right)-\frac{1}{x^2}}{\left(\frac{y}{x}-1\right)\left(\frac{y}{x}-2\right)}$$

$$= \frac{7-3-2}{2(1-2)} = \frac{2}{-2} = -1.$$

$\therefore \quad y - x + 1 = 0$

Similarly the second asymptote corresponding to the factor $(y + x)$ is

$$x+y = \lim_{x\to\infty,\frac{y}{x}\to -1}\frac{7xy-3y^2-2x^2-2x-2y-1}{(y-x)(y-2x)}$$

$$= \lim_{x\to\infty,\frac{y}{x}\to 1}\frac{7\left(\frac{y}{x}\right)-3\left(\frac{y}{x}\right)^2-2-\frac{2}{x}-2\left(\frac{y}{x}\right)\left(\frac{1}{x}\right)-\frac{1}{x^2}}{\left(\frac{y}{x}-1\right)\left(\frac{y}{x}\right)-2}$$

$$= \frac{7(-1)-3(-1)^2-2}{(-1-1)(-1-2)} = \frac{-7-3-2}{(-2)(-3)} = -2$$

$\therefore \quad x + y + 2 = 0$

and the third asymptote corresponding to the factor $y - 2x$ is

$$y-2x = \lim_{x\to\infty,\frac{y}{x}\to 2}\frac{7xy-3y^2-2x^2-2x-2y-1}{(y-x)(y+x)}$$

$$= \lim_{x\to\infty,\frac{y}{x}\to 2}\frac{7\left(\frac{y}{x}\right)-3\left(\frac{y}{x}\right)^2-2-2\left(\frac{1}{x}\right)-2\left(\frac{y}{x}\right)\left(\frac{1}{x}\right)-\frac{1}{x^2}}{\left(\frac{y}{x}+1\right)\left(\frac{y}{x}+1\right)}$$

$$= \frac{7(2)-3(2)^2-2}{(2-1)(2+1)} = \frac{14-12-2}{3} = 0.$$

$\Rightarrow y - 2x = 0$

Hence, all the asymptotes are $y - x + 1 = 0$, $x + y + 2 = 0$ and $y - 2x = 0$.

Example 3. *Find all the asymptotes of the curve* $(y-x)(y-2x)^2+(y+3x)(y-2x)+2x+2y-1=0.$

Solution . The equation of the curve is

$$(y-x)(y-2x)^2+(y+3x)(y-2x)+2x+2y-1 = 0$$

The asymptotes corresponding to the factor $(y-2x)^2$ are

$$(y-2x)^2+(y-2x)\lim_{x\to\infty,y/x\to 2}\frac{y+3x}{y-x} + \lim_{x\to\infty,y/x\to 2}\frac{2x+2y-1}{(y-x)} = 0$$

or

$$(y-2x)^2+(y-2x)\lim_{x\to\infty,y/x\to 2}\left(\frac{y/x+3}{y/x-1}\right) + \lim_{x\to\infty,y/x\to 2}\frac{2+2(y/x)-1/x}{(y/x-1)} = 0$$

or $$(y-2x)^2+5(y-2x)+6 = 0$$

or $$(y-2x) = \frac{-5\pm\sqrt{(25-24)}}{2} = \frac{-5\pm 1}{2}$$

or $y - 2x = -2$ and $y - 2x = -3$

or $y - 2x + 2 = 0$ and $y - 2x + 3 = 0$

And the asymptote corresponding to the factor $(y - x)$ is

$$(y-x)+\lim_{x\to\infty,y/x\to 1}\frac{(y+3x)(y-2x)}{(y-2x)^2} + \lim_{x\to\infty,y/x\to 1}\frac{2x+2y-1}{(y-2x)^2} = 0$$

$$\text{or } (y-x)+\lim_{x\to\infty, y/x\to 1}\frac{(y/x+3)(y/x-2)}{(y/x-2)^2}+\lim_{x\to\infty, y/x\to 1}\frac{2+2(y/x)-1/x}{x(y/x-2)^2}=0$$

$$\text{or } (y-x)+\frac{(1+3)(1-2)}{(1-2)^2}+0=0$$

$$\text{or } y-x-4=0$$

Hence, all the asymptotes of the given curve are $y-2x+2=0$, $y-2x+3=0$ and $y-x-4=0$

EXERCISE 7.2

Find all the asymptotes of the following curves :

1. $(x^2-y^2)(x+2y+1)+x+y+1=0$
2. $x^5-y^5=a^3xy$
3. $(x^2-y^2)(y^2-4x^2)-6x^3+5x^2y+3xy^2-2y^3-x^2+3xy-1=0$
4. $x^2(x^2-y^2)(x-y)+2x^3(x-y)-4y^3=0$
5. $xy(x^2-y^2)(x^2-4y^2)+xy(x^2-y^2)+x^2+y^2-7=0$
6. $(x-2y)^2(x-y)-4y(x-2y)-(8x+7y)=0$
7. $(x-y)^2(x^2+y^2)-10(x-y)x^2+12y^2+2x+y=0$
8. $(x-y-1)^2(x^2+y^2+2)+6(x-y-1)(xy+7)-8x^2-2x-1=0$
9. $(\alpha_1 x+\beta_1 y+\gamma_1)(\alpha_2 x+\beta_2 y+\gamma_2)+\gamma_3=0$
10. $(x-y+2)(2x-3y+4)(4x-5y+6)+5x-6y+7=0$
11. $(x-y+1)(x-y-2)(x+y)=8x-1$
12. $(x^2-3x+2)(x+y-2)+1=0$
13. $x(y-3)^3-4y(x-1)^3=0$
14. $x^2(x+y)(x-y)^2+ax^3(x-y)-a^2y^3=0$
15. $(y-a)^2(x^2-a^2)=x^4+a^4$

Hint to Selected Problems

1. Asymptotes corresponding to the factor $(x-y)$ is

$$(x-y)+\lim_{\substack{x\to\infty\\ y/x\to 1}}\frac{(x+y+1)}{(x+y)(x+2y+1)}=0$$

which gives $x-y=0$.

Similarly, asymptotes corresponding to the factor $(x+y)$ is given by $x+y=0$ and so on.

Note. Apply the same procedure to all other questions.

ANSWERS

1. $x-y=0, x+y=0, x+2y+1=0$
2. $y-x=0$
3. $x-y=0, 2x-y=0, x+y+1=0, 2x+y+1=0$
4. $x-y+2=0, x-y-1=0, x+y+1=0, x+2=0$
5. $x=0, y=0, x-y=0, x+y=0, x-2y=0$ and $x+2y=0$
6. $x-y+4=0, x-2y=2\pm3\sqrt{3}$
7. $x-y-2=0, x-y-3=0$
8. $x-y-2=0, x-y+3=0$
9. $\alpha_1 x+\beta_1 x+\gamma_1=0, \alpha_2 x+\beta_2 y+\gamma_2=0$
10. $x-y+2=0, 2x-3y+4=0, 4x-5y+6=0$
11. $y+x=0, x-y-2=0, x-y+1=0$
12. $x=1, x=2, x+y-2=0$
13. $x=0, y=0, 4x-2y+3=0, 4x+2y-15=0$
14. $x\pm a=0, x-y+a=0, y=\pm x-\frac{1}{2}a$
15. $x\pm a=0, x-y+a=0, x+y-a=0$

7.10 INTERSECTION OF A CURVE WITH ITS ASYMPTOTES

Let the equation

$$y=mx+c \quad \text{...(1)}$$

be an asymptote of the curve

$$x^2\phi_n\left(\frac{y}{x}\right)+x^{n-1}\phi_{n-1}\left(\frac{y}{x}\right)+x^{n-2}\phi_{n-2}\left(\frac{y}{x}\right)+\ldots=0. \quad \text{...(2)}$$

Solving (1) and (2) to find the intersection points so eliminating y between (1) and (2), we get

$$x^n\phi_n\left(m+\frac{c}{x}\right)+x^{n-1}\phi_{n-1}\left(m+\frac{c}{x}\right)+x^{n-2}\phi_{n-2}\left(m+\frac{c}{x}\right)+\ldots=0.$$

Now expand each term of above equation by Taylor's theorem, we have

$$x^n\left[\phi_n(m)+\frac{c}{x}\phi_n'(m)+\frac{c^2}{x^2}\cdot\frac{1}{2!}\phi_n''(m)+\ldots\right]+x^{n-1}\left[\phi_{n-1}(m)+\frac{c}{x}\phi_{n-1}'(m)+\ldots\right]+x^{n-2}\left[\phi_{n-2}(m)+\frac{c}{x}\phi_{n-2}'(m)+\ldots\right]=0$$

or $$x^n\phi_n(m)+[c\phi_n'(m)+\phi_{n-1}(m)x^{n-1}]+\left[\frac{c^2}{2!}\phi_n''(m)+\frac{c}{1!}\phi_{n-1}'(m)+\phi_{n-2}(m)\right]x^{n-2}+\ldots=0. \quad \ldots(3)$$

Since $y = mx + c$ is an asymptotes of the curve (2), then we have $\phi_n(m) = 0$ and $c\phi_n'(m)+\phi_{n-1}(m)=0$.
Thus (3) becomes

$$\left[\frac{c^2}{2!}\phi_n''(m)+\frac{c}{1!}\phi_{n-1}'(m)+\phi_{n-2}(m)\right]x^{n-2}+\ldots=0. \quad \ldots(4)$$

This is a equation of degree $n-2$ in x so it will have atmost $n-2$ values of x provided there is no asymptote parallel to $y = mx + c$ of the given curve.
Hence, in general we can say that any asymptote of a curve of the nth degree cuts the curve in $(n-2)$ points.

REMARKS

- Since one asymptote of the curve of nth degree cuts the curve in $(n-2)$ points so n asymptotes of that curve will cut in $n(n-2)$ points.
- If the equation of the curve of degree n can be written as $F_n + P = 0$, where F_n contains n non-repeated linear factors and P contains the terms almost of degree $n-2$, then $n(n-2)$ points of intersection of the curve will lie on the curve $P = 0$.

Solved Examples

Example 1. *Show that the four asymptotes of the curve*
$(x^2-y^2)(y^2-4x^2)+6x^3-5x^2y$
$-3xy^2+2y^3-x^2+3xy-1=0.$
cut the curve in eight points which lie on the circle $x^2+y^2=1$.

Solution. The given equation of the curve can be written as

$$(x-y)(x+y)(y-2x)(y+2x)+6x^3-5x^2y-3xy^2+2y^3-x^2+3xy-1=0 \quad \ldots(1)$$

The asymptote corresponding to the factor $x-y$ is

$$x-y+\lim_{x\to\infty,y/x\to1}\frac{6x^3-5x^2y-3xy^2+2y^3-x^2+3xy-1}{(x+y)(y-2x)(y+2x)}=0$$

or

$$x-y+\lim_{x\to\infty,\frac{y}{x}\to1}\frac{6-5\left(\frac{y}{x}\right)-3\left(\frac{y}{x}\right)^2+2\left(\frac{y}{x}\right)^3-\frac{1}{x}+3\left(\frac{y}{x}\right)\left(\frac{1}{x}\right)-\frac{1}{x^3}}{\left(1+\frac{y}{x}\right)\left(\frac{y}{x}-2\right)\left(\frac{y}{x}+2\right)}=0$$

or

$$x-y+\lim_{x\to\infty,y/x\to1}\frac{6-5-3+2}{(1+1)(1-2)(1+2)}=0$$

or $x-y=0$.

The asymptote corresponding to the factor $x+y$ is

$$x+y+\lim_{x\to\infty,y/x\to-1}\frac{6x^3-5x^2y-3xy^2+2y^3-x^2+3xy-1}{(x-y)(y-2x)(y+2x)}=0$$

or

$$x+y+\lim_{x\to\infty,y/x\to-1}\frac{6-5(y/x)-3(y/x)^2+2(y/x)^3-(1/x)+3(y/x)(1/x)-(1/x^3)}{(1-y/x)(y/x-2)(y/x+2)}=0$$

or $$x+y+\frac{6-5(-1)-3(-1)^2+2(-1)^3}{(1+1)(-1-2)(-1+2)}=0$$

or $x+y-1=0$.

Now the asymptote corresponding to the factor $y-2x$ is

$$y-2x+\lim_{x\to\infty,y/x\to2}\frac{6x^3-5x^2y-3xy^2+2y^3-x^2+3xy-1}{(x-y)(x+y)(y+2x)}=0$$

or

$$y-2x+\lim_{x\to\infty,\ y/x\to2}\frac{6-5(y/x)-3(y/x)^2+2(y/x)^3-(1/x)+3(y/x)(1/x)-(1/x^3)}{(1-y/x)(1+y/x)(y/x+2)}=0$$

or $$y-2x+\frac{6-5(2)-3(2)^2+2(2)^3}{(1-2)(1+2)(2+2)}=0$$

or $y-2x=0$.

The asymptote corresponding to the factor $y+2x$ is

$$y+2x+\lim_{x\to\infty,\ y/x\to2}\frac{6x^3-5x^2y-3xy^2+2y^3-x^2+3xy-1}{(x-y)(x+y)(y-2x)}=0$$

or

$$y+2x+\lim_{\substack{x\to\infty,\\ y/x\to 2}} \frac{6-5(y/x)-3(y/x)^2 +2(y/x)^3-(1/x) +3(y/x)(1/x)-(1/x^3)}{(1-y/x)(1+y/x)(y/x-2)}=0$$

or $$y+2x+\frac{6-5(-2)-3(-2)^2+2(-2)^3}{(1+2)(1-2)(-2-2)}=0$$

or $\quad y + 2x - 1 = 0.$

Hence, all the four asymptotes are $x - y = 0$, $x + y - 1 = 0, y - 2x = 0$ and $y + 2x - 1 = 0$.

Since one asymptote cuts the curve in $(4 - 2) = 2$ points so all the four asymptotes cut the curve in $4 \times 2 = 8$ points. Now combine all the asymptotes, we get

$$(x-y)(x+y-1)(y-2x)(y+2x-1)=0$$

or $[x^2-y^2-(x-y)][y^2-4x^2-(y-2x)]=0$

or $(x^2-y^2)(y^2-4x^2)-(x^2-y^2)(y-2x)$
$-(x-y)(y^2-4x^2)+(x-y)(y-2x)=0$

or $(x^2-y^2)(y^2-4x^2)$
$-(x^2y-2x^3-y^3+2xy^2)$
$-(xy^2-4x^3-y^3+4x^2y)$
$+xy-2x^2-y^2-2xy=0$

or $(x^2-y^2)(y^2-4x^2)+6x^3-5x^2y$
$-3xy^2+2y^3-2x^2-y^2+3xy=0.$ ...(2)

Now subtract (2) from (1), we get

$$x^2+y^2=1.$$

Hence, all the eight points of intersection lie on the circle $x^2 + y^2 = 1$.

EXERCISE 7.3

1. Show that the asymptotes of the curve
$$4(x^4+y^4)-17x^2y^2-4x(4y^2-x^2)+2(x^2-2)=0$$
cut the curve in eight points which lie on the ellipse $x^2 + 4y^2 = 4$.
2. Find the asymptotes of the curve $x^2y-xy^2+xy+y^2+x-y=0$ and show that they cut the curve again in three points which lie on the straight line $x + y = 0$.
3. Show that the eight points of intersection of the curve
$$x^4-5x^2y^2+4y^4+x^2-y^2+x+y+1=0$$
and its asymptotes lie on a rectangular hyperbola.
4. Show that the asymptotes of the cubic
$$x^3-2y^3+xy(2x-y)+y(x-y)+1=0$$
cut the curve in three points which lie on the straight line $x-y+1=0$.
5. Find the equation of the cubic which has the same asymptotes as the curve
$$x^3-6x^2y+11xy^2-6y^3+x+y+1=0$$
and which passes through the points (0, 0), (1, 0)and (0,1).
6. Show that the asymptotes of the curve $y^2(x^2-a^2)=x^2(x^2-4a^2)$ form two right angle triangles with the x-axis. $(y > 0)$.

ANSWERS

2. $y = 0, x = 1, x - y + 2 = 0$
5. $x^3-6x^2y+11xy^2-6y^3-x+6y=0.$

7.11 ASYMPTOTES OF NON-ALGEBRAIC CURVES

Definition. *A curve in which there are some terms involving cosine, sine, etc. is called non-algebraic curve.*

The method for finding the asymptotes of non-algebraic curves can be explained by following example.

Example. Let the equation of the curve be $y = \sec x$, then differentiating this w.r.t. 'x', we get

$$\frac{dy}{dx}=\sec x\tan x.$$

Therefore, the tangent at $P(x, y)$ on the curve is

$$Y-\sec x=\frac{dy}{dx}(X-x)$$

or $$Y-\sec x=\sec x\tan x(X-x)$$

or $$Y\cos^2x-\cos x=(X-x)\sin x. \qquad ...(1)$$

Now taking the distance of $P(x, y)$ from (0, 0) infinity as $x\to\pi/2$ and $y\to\infty$, we get

$$Y.0-0=(X-\pi/2).1 \text{ or } X=\pi/2.$$

This is one asymptote and the other asymptotes are $X = -\pi/2, \pm 3/2\pi,...$

7.12 ASYMPTOTES OF POLAR CURVES

(i) Equation of a line in polar form. Let O be the pole and OX the initial line and let $P(r, \theta)$ be any point the line whose equation is to be required as shown in fig. 1.

Draw a perpendicular OM from O to the line such that $OM = p$ and $\angle MOX = \alpha$ (say).

$\therefore$ In ΔOPM

$$\angle POM = \theta - \alpha$$

then,
$$\frac{OM}{OP} = \cos \angle POM$$

or
$$\frac{p}{r} = \cos(\theta - \alpha)$$

or
$$p = r\cos(\theta - \alpha).$$

This is the equation of line in polar form, where p is the perpendicular length from pole to this line and α is an angle which the perpendicular makes with initial line.

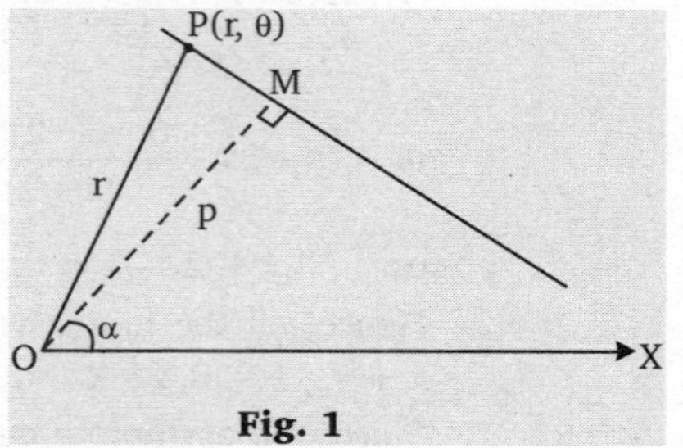

Fig. 1

(ii) Asymptotes of polar curves.

THEOREM 1. *If $\theta = \alpha$ is a root of the equation $f(\theta) = 0$, then $r\sin(\theta - \alpha) = 1/f'(\alpha)$ is an asymptote of the curve $1/r = f(\theta)$.*

Proof. Since the equation of a curve in polar form is $\frac{1}{r} = f(\theta)$. ...(1)

Let $P(r, \theta)$ be any point on this curve and draw a line through O perpendicular to OP, then radius vector which meets that tangent at P in T as show in fig. 2.

Then OT is a polar subtangent of the curve at P.

$$OT = r^2 \frac{d\theta}{dr} \qquad \text{(From calculus)}$$

Now differentiating (1) w.r.t. 'θ', we get

$$-\frac{1}{r^2}\frac{dr}{d\theta} = f'(\theta).$$

$$\therefore \qquad OT = r^2\frac{d\theta}{dr} = -\frac{1}{f'(\theta)}.$$

Since α is a root of $f(\theta) = 0$ as $\theta \to \alpha$, then $r \to \infty$ from (1) and the tangent PT tends to the asymptote and
$$OT \to \left[-\frac{1}{f'(\theta)}\right]_{\theta=\alpha}, f'(\alpha) \neq 0.$$

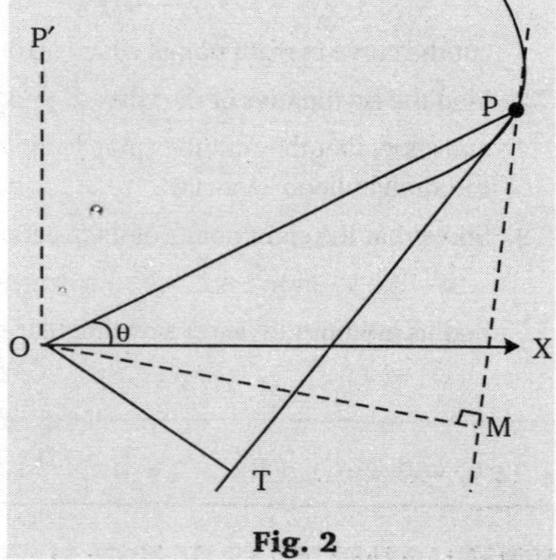

Fig. 2

And OP, PT will become parallel to lines shown dotted in the above fig. Thus $\angle OTP \to \pi/2$ and $OT \to OM$, where OM is a perpendicular distance from O to the asymptote.

$$\therefore \qquad OM = -\frac{1}{f'(\alpha)}$$

when $\theta \to \alpha$ *i.e.,* $OP \to OP'$. Then $\angle XOP' = \alpha$

$$\therefore \qquad \angle MOX = -\left(\frac{\pi}{2} - \alpha\right) \qquad \text{(In the clockwise direction)}$$

Therefore the equation of the asymptote is

$$r\cos\left[\theta - \left\{-\left(\frac{\pi}{2} - \alpha\right)\right\}\right] = -\frac{1}{f'(\alpha)} \qquad [\text{using } p = r\cos(1 - \alpha)]$$

or
$$r\cos\left(\frac{\pi}{2} + \theta - \alpha\right) = -\frac{1}{f'(\alpha)} \qquad \text{or} \qquad -r\sin(\theta - \alpha) = -\frac{1}{f'(\alpha)}$$

or
$$r\sin(\theta - \alpha) = \frac{1}{f'(\alpha)}$$

WORKING PROCEDURE

To find the asymptotes of polar curves, we use the follows steps :

Step 1. *Convert the equation of the given curve in the form* $\frac{1}{r} = f(\theta)$.

Step 2. *Find the roots of the equation* $f(\theta) = 0$ *i.e., values of* θ. *Suppose* α, β, *etc. are the roots of* $f(\theta) = 0$.

Step 3. *Now the asymptote corresponding to* $\theta = \alpha$ *is*

$$r \sin(\theta - \alpha) = \frac{1}{f'(\alpha)}$$

where $f'(\alpha)$ = *value of* $f'(\theta)$ *at* $\theta = \alpha$.

Solved Examples

Example 1. *Find the asymptotes of the curve* $r \sin n\theta = a$.

Solution . **Step I.** Convert the given curve into the form

$$\frac{1}{r} = f(\theta).$$

$$\therefore \quad \frac{1}{r} = \frac{\sin n\theta}{a} = f(\theta). \quad \ldots(1)$$

Step II. Solve the equation $f(\theta) = 0$.

i.e., $\quad \frac{\sin n\theta}{a} = 0.$

or $\quad \sin n\theta = \sin r\pi, \quad r = 0, 1, 2, \ldots,$

or $\quad n\theta = r\pi$ or $\theta = \frac{r\pi}{n}, \quad r = 1, 2, 3, \ldots$

Let $\quad \alpha = \frac{r\pi}{n}.$

Now differentiating (1) w.r.t. 'θ', we get

$$f'(\theta) = +\frac{n \cos n\theta}{a}.$$

$$\therefore \quad f'(\alpha) = \frac{n \cos n\alpha}{a} = \frac{n}{a} \cos r\pi = \frac{n}{a}(-1)^r.$$

Step III. Therefore, the asymptotes of the curve are

$$r \sin(\theta - \alpha) = \frac{1}{f'(\alpha)}$$

or
$$r \sin\left(\theta - \frac{r\pi}{n}\right) = \frac{a}{n(-1)^r},$$

where r is any integer.

Example 2. *Find the asymptotes of the curve* $r \sin \theta = a \cos 2\theta$.

Solution . First put the equation in the form of $\frac{1}{r} = f(\theta)$.

i.e., $\quad \frac{1}{r} = \frac{\sin\theta}{a \cos 2\theta}.$

$$\therefore \quad f(\theta) = \frac{\sin\theta}{a \cos 2\theta}. \quad \ldots(1)$$

Now solve the equation $f(\theta) = 0$. Then

$$\frac{\sin\theta}{a \cos 2\theta} = 0$$

or $\quad \sin\theta = \sin n\pi$ or $\theta = n\pi$.

Let $\alpha = n\pi$ be the root of the equation $f(\theta) = 0$.

Now differentiating (1) w.r.t. 'θ', we get

$$f'(\theta) = \frac{1}{a}\left[\frac{\cos 2\theta . \cos\theta + 2\sin 2\theta \sin\theta}{\cos^2 2\theta}\right]$$

$$\therefore f'(\alpha) = \frac{1}{a}\left[\frac{\cos 2\alpha . \cos\alpha + 2\sin 2\alpha \sin\alpha}{\cos^2 2\alpha}\right]$$

$$= \frac{1}{2a}\left[\frac{\cos 2n\pi . \cos n\pi + 2\sin 2n\pi \sin n\pi}{\cos^2 2n\pi}\right] \quad (\because \alpha = n\pi)$$

$$= \frac{1}{a} \cos n\pi.$$

The asymptote corresponding to $\alpha = n\pi$ is

$$r \sin(\theta - n\pi) = \frac{1}{f'(\alpha)} = \frac{a}{\cos n\pi}$$

or $\quad r(\sin\theta \cos n\pi - \cos\theta \sin n\pi) = \frac{a}{\cos n\pi}$

or $\quad r \sin\theta \cos n\pi = \frac{a}{\cos n\pi} \quad (\because \sin n\pi = 0)$

or $\quad r \sin\theta \cos^2 n\pi = a$

or $\quad r \sin\theta = a \quad (\because \cos n\pi = 1)$

Example 3. *Find the asymptotes of the curve* $r\theta = a$.

Solution . First putting the equation of curve in the form $\frac{1}{r} = f(\theta)$ so we have

$$\frac{1}{r} = \frac{\theta}{a}.$$

$$\therefore \quad f(\theta) = \frac{\theta}{a}. \quad \ldots(1)$$

Putting $f(\theta) = 0$, we get $\theta = 0$.

Then $\alpha = 0$ is the root of $f(\theta) = 0$.

Now differentiating (1) w.r.t. 'θ', we get

$$f'(\theta) = \frac{1}{a}. \Rightarrow f'(\alpha) = \frac{1}{a}.$$

Thus the asymptote corresponding to $\theta = \alpha$ is

$$r\sin(\theta-\alpha) = \frac{1}{f'(\alpha)}.$$

$$\therefore \quad r\sin(\theta-0) = \frac{1}{(1/a)}$$

or $$r\sin\theta = a.$$

Example 4. *Find the circular asymptotes of the curve* $r = a.\dfrac{\theta}{\theta-1}$.

Solution . The circular asymptote is given by

$$\lim_{r=a,\theta\to\infty} \frac{\theta}{\theta-1} = a\ .$$

Thus $r = a$ is the circular asymptote.

EXERCISE 7.4

Find the asymptotes of the following curves :

1. $y = \tan x$.
2. $r = a \operatorname{cosec}\theta + b$
3. $r \sin 2\theta = a$
4. $r \sin\theta = 2\cos 2\theta$
5. $r\sin\theta = 2\cos\theta$
6. $r\theta\cos\theta = a\cos 2\theta$
7. $r(1 - 2\cos\theta) = 2a$
8. $r = 4(\sec\theta + \tan\theta)$
9. $r\cos\theta = 4\sin^2\theta$
10. $r(e^\theta - 1) = a(e^\theta + 1)$
11. $r\cos\theta = a\sin\theta$
12. $r(1 + 2\sin\theta) = 2$
13. $r\sin\theta = 2\theta$

Hint to Selected Problems

1. (i) $y = \tan x \Rightarrow \dfrac{dy}{dx} = \sec^2 x$

Tangent at (x, y)

$Y - \tan x = \sec^2 x(Y - x) \Rightarrow Y\cos^2 x - \sin x\cos x = (X - x)$.

Now as $x \to \pi/2, y \to \infty$ and the distance of (x, y) from $(0, 0) \to \infty$.

$\therefore\ Y.0 - 0 = (X - \pi/2) \Rightarrow X = \pi/2$.

2. $\dfrac{1}{r} = f(\theta) = \dfrac{\sin\theta}{a + b\sin\theta}$. Solving, $f(0) = 0$. we get $\theta = n\pi = \alpha$ (say)

$\Rightarrow f'(\alpha) = \dfrac{1}{a}\cos n\pi$.

Now required asymptotes are given by $r\sin(\theta-\alpha) = \dfrac{1}{f'(\alpha)}$.

3. $\dfrac{1}{r} = f(\theta) = \dfrac{\sin 2\theta}{a}$.

Now on solving $f(0) = 0$ we get $\theta = \dfrac{n\pi}{2} = \alpha$ (say).

Also, $f'(\alpha) = \dfrac{2\cos n\pi}{a}$.

Therefore, the asymptotes of the given curve is

$$r\sin(\theta-\alpha) = \frac{1}{f'(\alpha)}.$$

4. $\dfrac{1}{r} = f(\theta) = \dfrac{\sin\theta}{2\cos 2\theta}$

Now, $f'(\alpha) = \dfrac{\cos n\pi}{2}$. Therefore, the asymptotes of the given curve is given by $r\sin(\theta-\alpha) = \dfrac{1}{f'(\alpha)}$.

5. Here $\theta = n\pi = \alpha, f'(\alpha) = \dfrac{1}{2}$. Then use the required formula.

6. $\theta = 0, \left(k\pi + \dfrac{\pi}{2}\right)$, $\quad f'(\alpha) = \dfrac{1}{a}$

7. $\theta = \pm\dfrac{\pi}{3}, f'(\theta_1) = \dfrac{-\sqrt{3}}{2a}$, $\quad f'(\theta_2) = \dfrac{\sqrt{3}}{2a}$

8. $\theta = \left(2n\pi + \dfrac{\pi}{2}\right) = \alpha(\text{say})$, $\quad f'(\alpha) = -\dfrac{1}{8}$

9. $\theta = \left(2n\pi + \dfrac{\pi}{2}\right) = \alpha(\text{say})$, $\quad f'(\alpha) = -\dfrac{1}{4}$

10. $\theta = 2n\pi = \alpha$, $\quad f'(\alpha) = \dfrac{1}{2a}$

11. $\theta = \left(n\pi + \dfrac{\pi}{2}\right) = \alpha$, $\quad f'(\alpha) = \dfrac{-1}{a}$

12. $\theta = \dfrac{-\pi}{6} = \alpha$, $\quad f'(\alpha) = \dfrac{\sqrt{3}}{2}$

13. $\theta = n\pi = \alpha$, $\quad f'(\alpha) = \dfrac{1}{2}\left(\dfrac{\cos n\pi}{n\pi}\right)$

ANSWERS

1. $x = \pm\pi/2, \pm 3\pi/2...$
2. $r\sin\theta = a$
3. $r\sin\theta = \pm\dfrac{1}{2}a, r\cos\theta = \pm\dfrac{1}{2}a$
4. $r\sin\theta = 2$
5. $r\sin\theta = \pm 2$
6. $r\sin\theta = a, r\cos\theta = \dfrac{a}{\left(k+\dfrac{1}{2}\right)\pi}$, k is any integer
7. $r\sin\left(\theta - \dfrac{\pi}{3}\right) = \dfrac{2a}{\sqrt{3}}, r\sin\left(\theta + \dfrac{\pi}{3}\right) = -\dfrac{2a}{\sqrt{3}}$
8. $r\cos\theta = 8$
9. $r\cos\theta = 4$
10. $r\sin\theta = 2a$
11. $r\cos\theta = \pm a$
12. $r\sin\left(\theta \pm \dfrac{\pi}{6}\right) = \dfrac{2}{\sqrt{3}}$
13. $r\sin\theta = 2n\pi, n = \pm 1, \pm 2, ...$

Objective Evaluations

Fill in the Blanks

1. If $y = mx + c$ is an asymptote of the curve $f(x, y) = 0$, then m = __________ and c = __________ .
2. The equation $\phi_n(m) = 0$ gives the __________ of the asymptotes.
3. If one or more values of m obtained from $\phi_n(m) = 0$ are such that $\phi'_n(m) = 0$ and $\phi_{n-1}(m)$, then the asymptotes __________ .
4. If the coefficients of highest degree terms of y are constant, then there are no asymptotes __________ .
5. If the coefficients of highest degree terms of x are not constant, then there will exist the asymptotes parallel to __________.
6. The number of asymptotes of nth degree curve cannot exceed __________ .
7. The asymptotes parallel to y-axis of the curve $y^2(x^2 - a^2) = x$ are __________ .
8. The curve $y^2 = 4ax$ has __________ asymptotes.
9. The n asymptotes of a curve of the nth degree cut if in __________ points.
10. If α is a root of the equation $f(\theta) = 0$, then $r \sin (\theta - \alpha)$ = __________ is an asymptote of the curve $\frac{1}{r} = f(\theta)$.

True/False

Write 'T' for True and 'F' for False statement.

1. The line $y = mx + c$ is an asymptote of the curve $y = mx + c + \frac{A}{x} + \frac{B}{x^2} + \frac{C}{x^3} + \dots$. **(T/F)**
2. The polynomial $\phi_n(m)$ is obtained by putting $y = m$ and $x = m$ in the nth degree terms of the curve. **(T/F)**
3. If $y = mx + c$ is an asymptote of the curve $f(x, y) = 0$ then $m = \lim_{x\to\infty}\left(\frac{y}{x}\right)$. **(T/F)**
4. The curve $x^5 - y^5 = a^3xy$ has at most five asymptotes real as well as imaginary. **(T/F)**
5. The numbers of asymptotes of the curve of nth degree can exceed n. **(T/F)**
6. The one asymptote of a curve of the nth degree cuts it in $(n - 1)$ points. **(T/F)**
7. The curve $x^2/a^2 + y^2/b^2 = 1$ has no real asymptotes. **(T/F)**
8. The curve $y^2 = 4ax$ has two real asymptotes. **(T/F)**
9. The asymptote parallel to x-axis of the curve $xy = c^2$ is $y = 0$. **(T/F)**
10. If α is a root of the equation $f(\theta) = 0$, then $r \sin(\theta - \alpha) = f'(\alpha)$ is an asymptote of the curve $\frac{1}{r} = f(\theta)$. **(T/F)**

Multiple Choice Questions

Choose the most appropriate one.

1. If $y = mx + c$ is an asymptote of the curve $f(x, y) = 0$, then $\lim_{x\to\infty}(y / x)$ equals :
   (a) c (b) m
   (c) $-m$ (d) $-c$
2. If $y = mx + c$ is an asymptote of the curve $f(x, y) = 0$, then $\lim_{x\to\infty, y/x\to m}(y - mx)$ equals :
   (a) m (b) $-c$
   (c) c (d) $-m$
3. The n asymptotes of a curve of the nth degree cut it in how many points :
   (a) 2 (b) n
   (c) $n - 1$ (d) $n(n - 2)$
4. For non existence of the asymptotes of the curve for some values of m obtained by $\phi_n(m) = 0$ such that $\phi_{n-1}(m) \neq 0$ and $\phi'_n(m)$ equals :
   (a) 0 (b) 1
   (c) m (d) non-zero
5. The number of asymptotes of a curve of the nth degree can not exceed :
   (a) $n - 1$ (b) n
   (c) $n - 2$ (d) $n + 1$
6. The asymptote of the curve $y = mx + c + \frac{A}{x} + \frac{B}{x^2} + \dots$ is :
   (a) $y = mx$ (b) $y = mx + c$
   (c) $y = m$ (d) $y = c$
7. The curve $y^2 = 4ax$ has how many real asymptotes ?
   (a) 1 (b) 2
   (c) Zero (d) none of these
8. The asymptotes of the curve $r(e^\theta - 1) = a(e^\theta + 1)$ are :
   (a) $r \sin \theta = 2a$ (b) $r \cos \theta = 2a$
   (c) $r \sin \theta = a$ (d) $r \cos \theta = a$

9. The number of real asymptotes of the curve $y^3 = x^3 + 3$ are :

(a) 1 (b) 0

(c) 3 (d) 2

10. For the curve $x^3 + y^3 - 3axy = 0$, $\phi_3(m)$ is :

(a) $m^2 + 1$ (b) $m + 1$

(c) $m - 1$ (d) $m^3 + 1$

ANSWERS

Fill in the Blanks

1. $\lim_{x\to\infty} y/x$, $\lim_{x\to\infty y/x\to m} (y - mx)$ **2.** slopes **3.** will not exist **4.** parallel to y-axis **5.** x-axis

6. n **7.** $x = \pm a$ **8.** No **9.** $n(n-2)$ **10.** $1/f'(\alpha)$

True/False

1. T **2.** F **3.** T **4.** T **5.** F **6.** F **7.** T **8.** F **9.** T **10.** F

Multiple Choice Questions

1. (b) **2.** (c) **3.** (d) **4.** (a) **5.** (b) **6.** (b) **7.** (c) **8.** (a) **9.** (b) **10.** (d)

FFFFFF

CHAPTER 8

Curve Tracing

8.1 INTRODUCTION

If P is any point on a curve and CD is any given line which does not passes through this point P. Then the curve is said to be concave at P with respect to the line CD if the small arc of the curve containing P lies entirely within the acute angle between the tangent at P to the curve and the line CD and the curve is said to be convex at P if the arc of the curve containing P lies wholly outside the acute angle between that tangent at P and the line CD which are shown in figures below :

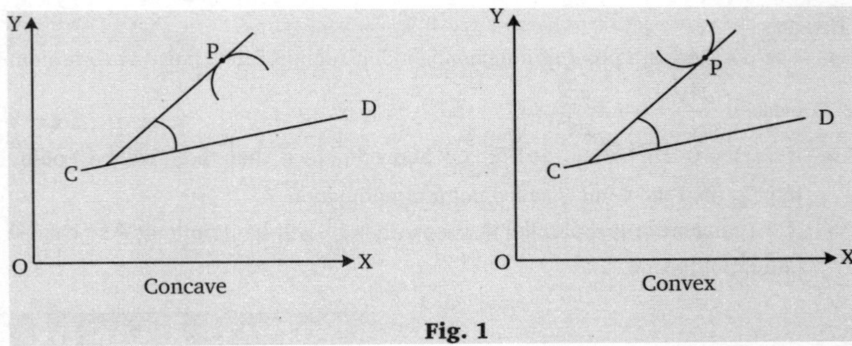

Fig. 1

8.2 POINT OF INFLEXION

A point P on the curve is said to be the point of inflexion, if the curve in one side of P is concave and other side of P is convex with respect to the line CD which does not passes through the point P as shown in fig. 2.

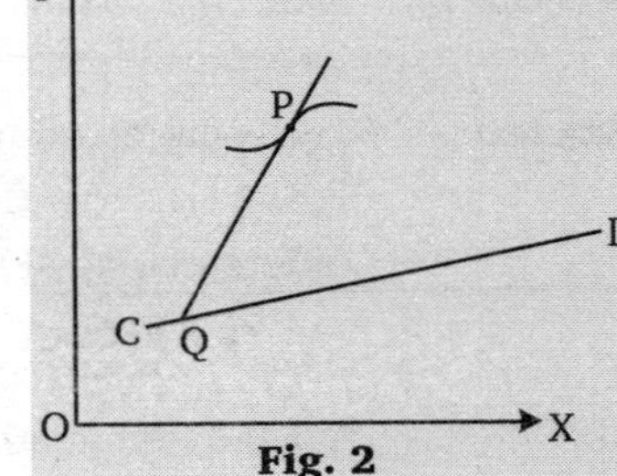

Fig. 2

Inflexion tangent. The tangent at the point of inflexion of a curve is said to be inflexion tangent. In the fig. 2 the line PQ is the inflexion tangent.

8.3 DETERMINATION OF THE POINTS OF INFLEXION

Let $y = f(x)$ be the equation of a curve and let $P(x, y)$ be any point on the curve and assuming that the tangent at P is not parallel to y-axis as shown in fig. 3.

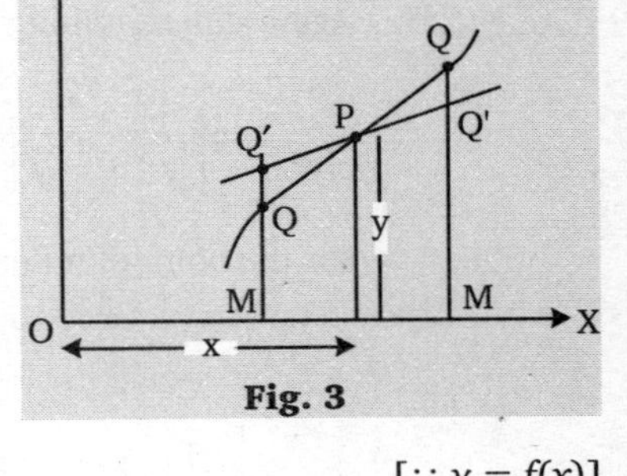

Fig. 3

Since the tangent is taken not to be parallel to y-axis, then $\frac{dy}{dx} = f'(x)$ must be finite.

Let $Q(x + h, y + k)$ be any point on the curve in the neighbourhood of P. We may take this point Q either side of P. Suppose the ordinate OM of Q intersects the tangent line at Q'.

$$Y - y = f'(x)(X - x) \qquad \ldots(1)$$

Since at point $Q(x + h, x + k)$ we have $X = x + h$ so putting $X = x + h$ in (1), we get

$$Q'M - y = f'(x)(x + h - x) \qquad [\because Y = Q'M]$$

or $$Q'M = y + hf'(x)$$

or $$Q'M = f(x) + hf'(x). \qquad [\because y = f(x)]$$

But we know that $$QM = f(x+h) = f(x) + hf'(x) + \frac{h^2}{2!}f''(x) + \frac{h^3}{3!}f'''(x) + \ldots \quad \text{(Using Taylor's theorem)}$$

$$\therefore \qquad QM - Q'M = \frac{h^2}{2!}f''(x) + \frac{h^3}{3!}f'''(x) + \ldots + \frac{h^n}{n!}f^{(n)}(x+\theta h) \text{ where } 0<\theta<1. \qquad \ldots(2)$$

Let us suppose $f''(x) \neq 0$ and taking h sufficiently small, then $(QM - Q'M)$ will have the same sign as $\frac{h^2}{2!}f''(x)$. But $\frac{h^2}{2!}f''(x)$ will have invariable sign because h^2 will always be positive. This means that on both sides of P the curve will be either concave or convex. Hence, we can say that the necessary condition for the existence of a point of inflexion at P is given by

$$f''(x) = 0 \text{ or } \frac{d^2y}{dx^2} = 0.$$

Thus (2) now becomes

$$QM - Q'M = \frac{h^2}{3!}f'''(x) + \frac{h^4}{4!}f^{iv}(x) + \ldots + \frac{h^n}{n!}f^{(n)}(x+\theta h) \qquad \ldots(3)$$

Further, if $f'''(x) \neq 0$ and taking h to be very small, then $(QM - Q'M)$ will have the same sign as $\frac{h^3}{3!}f'''(x)$ and this changes sign when h changes sign. Thus we can say that the curve with respect to the x-axis is concave on one side of P and convex on other side of P. Hence, there will exist a point of inflexion at P.

Consequently, we can have a point of inflexion at P, if $\frac{d^2y}{dx^2} = 0$ but $\frac{d^3y}{dx^3} \neq 0$.

REMARKS

- The position of a point of inflexion is independent of the choice of co-ordinate axes so we can say that a point of inflexion at P exists if $\frac{d^2y}{dx^2} = 0$ but $\frac{d^3y}{dx^3} \neq 0$.
- If $f''(x) = 0 = f'''(x) = \ldots = f^{(n-1)}(x)$ and $f^{(n)}(x) \neq 0$, then there will be a point of inflexion if n is odd and if n is even and greater than 2, then the point is called point of undulation.
- If the tangent at P is parallel to y-axis, then $\frac{dy}{dx}$ will be infinite at P so change the curve to the form $x = f(y)$ and then find the point of inflexion.

Solved Examples

Example 1. *Find the points of inflexion of the curve* $x = (\log y)^3$.

Solution. The equation of the curve is

$$x = (\log y)^3 \qquad \ldots(1)$$

Differentiating (1) with respect to 'y', we get

$$\frac{dx}{dy} = 3(\log y)^2 . \frac{1}{y}$$

Again differentiating *w.r.t.* y

$$\frac{d^2x}{dy^2} = 3\left[\frac{2\log y}{y^2} - \frac{(\log y)^2}{y^2}\right]. \qquad \ldots(2)$$

Again differentiating w.r.t. 'y', we get

$$\frac{d^3x}{dy^3} = 3\left[\frac{2}{y^3} - \frac{4\log y}{y^3} - \frac{2\log y}{y^3} - \frac{2(\log y)^2}{y^2}\right]. \qquad \ldots(3)$$

For the point of inflexion, we have

$$\frac{d^2x}{dy^2} = 0.$$

$$\therefore \quad 3\left[\frac{2\log y - (\log y)^2}{y^2}\right] = 0$$

or $\quad 3(\log y)(2 - \log y) = 0$

or $\quad \log y = 0, \log y = 2$

or $\quad y = 1, y = e^2$

From (3) it is obvious that at $y = 1, y = e^2$,

$$\frac{d^3x}{dy^3} \neq 0.$$

Hence, the points of inflexion are$(0, 1)(8, e^2)$.

Example 2. *Find the points of inflexion of the curve* $y^2 = x(x + 1)^2$.

Solution. The equation of the curve can be written as

$$y = (x+1)\sqrt{x}. \qquad \ldots(1)$$

Differentiating (1) w.r.t. 'x', we get

$$\frac{dy}{dx} = \frac{3}{2}.x^{1/2} + \frac{1}{2\sqrt{x}}.$$

Again differentiating w.r.t. 'x'

$$\frac{d^2y}{dx^2} = \frac{3}{4\sqrt{x}} - \frac{1}{4x^{3/2}}. \qquad \ldots(2)$$

and again differentiating w.r.t. 'x', we get

$$\frac{d^3y}{dx^3} = -\frac{3}{8x^{3/2}} + \frac{3}{8x^{5/2}}. \qquad \ldots(3)$$

For the point of inflexion, we have

$$\frac{d^2y}{dx^2} = 0.$$

$$\therefore \quad \frac{3}{4\sqrt{x}} - \frac{1}{4x\sqrt{x}} = 0$$

or $\quad \left(3 - \frac{1}{x}\right) = 0 \quad$ or $x = 1/3$.

From (3) it is obvious that at $x = 1/3$, $\frac{d^3y}{dx^3} \neq 0$.

Thus, the point of inflexion are given by $(1/3, \pm 4/3\sqrt{3})$.

EXERCISE 8.1

1. Find the points of inflexion of the curve $x = \log(y/x)$.
2. Find the points of inflexion of the curve $y(a^2 + x^2) = x^3$.
3. Find the points of inflexion of the curve $y = (x-1)^4(x-2)^3$.
4. Find the points of inflexion of the curve $xy = a^2 \log(y/a)$.
5. Show that the points of inflexion of the curve
$$y^2 = (x-a)^2(x-b)$$
lie on the line $3x + a = 4b$.
6. Show that the origin is a point of inflexion of the curve $a^{m-1}.y = x^m$, if m is odd and greater than 2.
7. Show that the points of inflexion of the curve $x^2y = a^2(x-y)$ are given by $x = 0$, $x = \pm a\sqrt{3}$.
8. Prove that the curve $y = (1-x)/(1+x^2)$ has three points of inflexion which lie on a straight line.
9. Show that the abscissae of the points of inflexion on the curve $y^2 = f(x)$ satisfy the equation
$$[f'(x)]^2 = 2f(x)f''(x).$$
10. Show that the points of inflexion on the curve $y = be^{-(x/a)^2}$ are given by $x = \pm a/\sqrt{2}$.
11. Find the points of inflexion on the curve $r(\theta^2 - 1) = a\theta^2$.
12. Show that the points of inflexion of the curve $r = b\theta^n$ are given by $r = b\{-n(n+1)\}^{n/2}$.
13. Find the points of inflexion of the curve
$$x = a(2\theta - \sin\theta),\ y = a(2 - \cos\theta).$$
14. Find the points of inflexion of the curve $y = 3x^4 - 4x^3 + 1$.

Hint to Selected Problems

1. Given that $x = \log\left(\frac{y}{x}\right) \Rightarrow y = xe^x$

$$\Rightarrow \frac{dy}{dx} = xe^x + e^x, \frac{d^2y}{dx^2} = xe^x + 2e^x \text{ and } \frac{d^3y}{dx^3} = xe^x + 3e^x$$

For the point of inflexion, putting $\frac{d^2y}{dx^2} = 0$ and $\frac{d^3y}{dx^3} \neq 0$.

5. $y = (x-a)\sqrt{x-b}$

$$\Rightarrow \quad \frac{dy}{dx} = \frac{3x - a - 2b}{2\sqrt{x-b}} \text{ and } \frac{d^2y}{dx^2} = \frac{3x + a - 4b}{4(x-b)^{3/2}}$$

Also, $\frac{d^3y}{dx^3} = \frac{-3x - 3a + 6b}{4(x-b)^{5/2}}$

By putting $\frac{d^2y}{dx^2} = 0$, we get $3x + a = 4b$ at which $\frac{d^3y}{dx^3} \neq 0$.

9. $y^2 = f(x) \Rightarrow \frac{dy}{dx} = \frac{f'(x)}{2y} = \frac{f'(x)}{2\sqrt{f(x)}}$

$$\frac{d^2y}{dx^2} = \frac{2f(x).f''(x) - [f'(x)]^2}{4[f(x)]^{3/2}}$$

and $\frac{d^3y}{dx^3} = \frac{4[f(x)]^2.f'''(x) - 6f(x)f'(x)f''(x) + 3[f'(x)]^3}{8[f(x)]^{5/2}}$.

For point of inflexion, put $\frac{d^2y}{dx^2} = 0$.

11. Given that $r(\theta^2 - 1) = a\theta^2$

$$\Rightarrow \frac{dr}{d\theta} = -\frac{2a\theta}{(\theta^2-1)^2} \text{ and } \frac{d^2r}{d\theta^2} = \frac{2a(1+3\theta^2)}{(\theta^2-1)^3}.$$

At the point of inflexion, use

$$r^2 + 2\left(\frac{dr}{d\theta}\right)^2 - r\frac{d^2r}{d\theta^2} = 0.$$

ANSWERS

1. $(-2, -2/e^2)$ 2. $(0,0), \left(\sqrt{3}a, \frac{3\sqrt{3}}{4}a\right), \left(-\sqrt{3}a, \frac{-3\sqrt{3}}{4}a\right)$ 3. Point of inflection at $x = 2$, $(11 \pm \sqrt{2})/7$

4. $\left(\frac{3}{2}ae^{-3/2}, ae^{3/2}\right)$ 11. $\theta = \pm\sqrt{3}$ 13. $\left[\left(4n\pi \pm \frac{2\pi}{3} \mp \frac{\sqrt{3}}{2}\right)a, \frac{3a}{2}\right]$ 14. $\left(\frac{2}{3}, \frac{11}{27}\right), (0,1)$

8.4 MULTIPLE AND SINGULAR POINTS

Definition 1. *A point on the curve is said to be multiple points if through this point more than one branches of a curve passes.*

Definition 2. *A point on the curve is called a double point if through it two branches of the curve passes.*

Definition 3. *If three branches of the curve passes through a point, then this point is called triple point.*

Definition 4. *If n branches passes through a point on the curve, then this point is called a multiple point of nth order.*

Definition 5. *The point of inflexion and multiple points are also called the singular points. Or An unusual point on the curve is basically called a singular point.*

8.5 TYPES OF DOUBLE POINT

(i) **Node.** A double point on a curve is said to be a node, if through this double point two branches of the curve passes which are real and having two different tangents at that point (Fig. 4).

(ii) **Cusp.** A double point on a curve is called a cusp if through this double point two real branches of the curve passes and have real coincident tangents at that point (Fig. 5).

(iii) **Conjugate point.** A point *P* on the curve is said to be conjugate point if there are no real points on the curve in the neighbourhood of that point and having no real tangent at that point (Fig. 6).

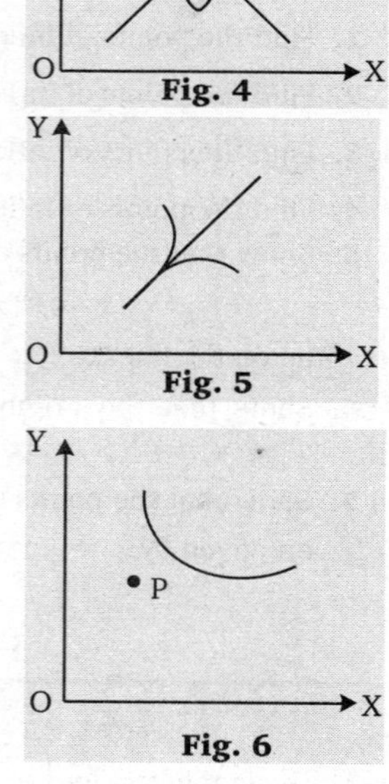

Fig. 4
Fig. 5
Fig. 6

8.6 SPECIES OF CUSP

Definition. *A cusp is said to be single if the curve lies entirely on one side of the common tangent* (Fig. 7(ii)).

Definition. *A cusp is said to be double if the curve lies on both sides of the common tangent* (Fig. 7(i)).

Definition. *A cusp is said to be of first species if the two branches of the curve lie on opposite sides of common tangent* (Fig. 7(iii)).

Definition. *A cusp is said to be of second species if the two branches of the curve lie on same side of the common tangent* (Fig. 7(ii)).

There are five different types of cusp :

(i) Single cusp of first species
(ii) Single cusp of second species
(iii) Double cusp of first species
(iv) Double cusp of second species
(v) Double cusp with change of species.

These all five types of cusp are shown below respectively :

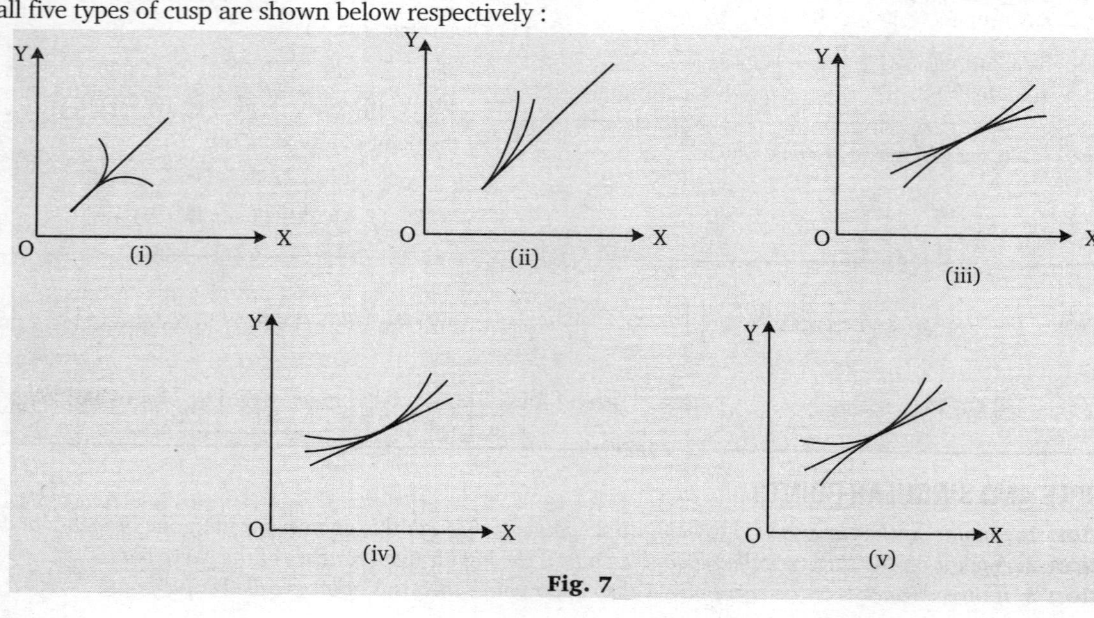

Fig. 7

8.7 TANGENTS AT THE ORIGIN

The nature of a double point depends on the tangents so we find the tangent or tangents there. If a curve passes through the origin, then the equation of the tangent or tangents at the origin are obtained by equating to zero the lowest degree terms in the equation of the curve.

8.8 CHANGE OF ORIGIN (SHIFT OF ORIGIN)

Let $P(x, y)$ be any point with respect to the co-ordinate axes OX and OY and let O' (h, k) be any other point with respect to the same co-ordinate system with origin O'.

Now draw $O'X$ and $O'Y$ parallel to the OX and OY axis respectively through $O'(h, k)$ and let co-ordinates of P with respect to the axes OX and OY be (X, Y). Then

$$M'P = PM - M'M.$$

$\therefore$ $$Y = y - k \text{ or } y = Y + k$$

and $$N'P = PN - NN'.$$

$\therefore$ $$X = x - h \text{ or } x = X + h.$$

Thus using the transformations $x = X + h$, and $y = Y + k$, the origin O is shifted to $O'(h, k)$.

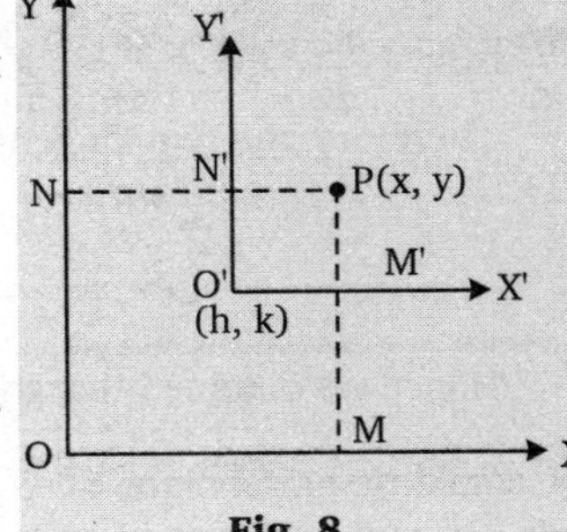

Fig. 8

8.9 TANGENT AT THE POINT (h, k) TO A CURVE

In order to find the tangent at (h, k) to the given curve, we first shift the origin at (h, k) and then find the tangent at the origin to the transformed curve by equating to zero the lowest degree terms.

8.10 POSITION AND NATURE OF DOUBLE POINTS

Let $P(x, y)$ be any point on the curve $f(x, y) = 0$, we have

$$\frac{dy}{dx} = -\frac{\partial f / \partial x}{\partial f / \partial y} \quad \text{or} \quad \frac{\partial f}{\partial x} + \frac{\partial f}{\partial y}\frac{dy}{dx} = 0. \quad \ldots(1)$$

Therefore, the slope of the tangent at $P(x, y)$ is equal to dy/dx which is given above.

Since by the definition of a multiple point we know that the curve has atleast two tangents so $\frac{dy}{dx}$ has atleast two values at a multiple point. But the equation (1) is of first degree in dy/dx. Therefore dy/dx will have two values or more than one value, if and only if

$$\frac{\partial f}{\partial x} = 0, \frac{\partial f}{\partial y} = 0$$

Thus the necessary and sufficient condition for any point of the curve $f(x, y) = 0$ to be a multiple point are that

$$\frac{\partial f}{\partial x} = 0 = \frac{\partial f}{\partial y}.$$

Hence, to find the multiple point of the curve $f(x, y) = 0$ we shall simultaneously solve the following equations

$$f(x, y) = 0, \frac{\partial f}{\partial x} = 0, \frac{\partial f}{\partial y} = 0.$$

Next, differentiating (1) w.r.t. 'x', we get

$$\frac{d}{dx}\left(\frac{\partial f}{\partial x}\right) + \frac{d}{dx}\left(\frac{\partial f}{\partial y} \cdot \frac{\partial y}{\partial x}\right) = 0$$

$$\frac{\partial}{\partial x}\left(\frac{\partial f}{\partial x}\right) + \frac{\partial}{\partial y}\left(\frac{\partial f}{\partial x}\right)\frac{dy}{dx} + \frac{d}{dx}\left(\frac{\partial f}{\partial y}\right)\frac{dy}{dx} + \frac{\partial f}{\partial y} \cdot \frac{d^2 y}{dx^2} = 0$$

or $$\frac{\partial^2 f}{\partial x^2} + \frac{\partial}{\partial y}\left(\frac{\partial f}{\partial x}\right)\frac{dy}{dx} + \left[\frac{\partial}{\partial x}\left(\frac{\partial f}{\partial y}\right) + \frac{\partial}{\partial y}\left(\frac{\partial f}{\partial y}\right) \cdot \frac{dy}{dx}\right]\frac{dy}{dx} + \frac{\partial f}{\partial y} \cdot \frac{d^2 y}{dx^2} = 0.$$

Since at the multiple point $\frac{\partial f}{\partial y} = 0$. Therefore, $$\frac{\partial^2 f}{\partial x^2} + \frac{\partial^2 f}{\partial y \partial x} \cdot \frac{dy}{dx} + \frac{\partial^2 f}{\partial x \partial y}\frac{dy}{dx} + \frac{\partial^2 f}{\partial y^2}\left(\frac{dy}{dx}\right)^2 = 0$$

or $$\frac{\partial^2 f}{\partial x^2} + 2\frac{\partial^2 f}{\partial x \partial y}\frac{dy}{dx} + \frac{\partial^2 f}{\partial y^2}\left(\frac{dy}{dx}\right)^2 = 0 \quad \ldots(2)$$

$$\left(\because \frac{\partial^2 f}{\partial x \partial y} = \frac{\partial^2 f}{\partial y \partial x}\right)$$

This is a quadratic equation in $\frac{dy}{dx}$ and the multiple point will be double point if the equation (2) will remain quadratic in $\frac{dy}{dx}$, and for the quadratic in $\frac{dy}{dx}$ it is assumed that $\frac{\partial^2 f}{\partial x^2}, \frac{\partial^2 f}{\partial x \partial y}, \frac{\partial^2 f}{\partial y^2}$ are not all zero. From the equation (2) it is obvious that the two values of dy/dx will be real and distinct, coincident, or imaginary according as

$$\left[\left(\frac{\partial^2 f}{\partial x \partial y}\right)^2 - \frac{\partial^2 f}{\partial x^2}\frac{\partial^2 f}{\partial y^2}\right] >, = \text{or} < 0.$$

Therefore, the two tangents will be real and distinct, coincident or imaginary according as

$$\left[\left(\frac{\partial^2 f}{\partial x \partial y}\right)^2 - \frac{\partial^2 f}{\partial x^2}\frac{\partial^2 f}{\partial y^2}\right] >, = \text{or} < 0.$$

Hence we obtained that the double point will be node, cusp or conjugate point according as

$$\left(\frac{\partial^2 f}{\partial x \partial y}\right)^2 > \text{or} = \text{or} < \frac{\partial^2 f}{\partial x^2} \cdot \frac{\partial^2 f}{\partial y^2}.$$

REMARK

- If $\frac{\partial^2 f}{\partial x^2}, \frac{\partial^2 f}{\partial x \partial y}, \frac{\partial^2 f}{\partial y^2}$ are all zero, then the point $P(x, y)$ will be a multiple point of order greater that two.

8.11 NATURE OF A CUSP AT THE ORIGIN

Let (0, 0) be a cusp of the curve. Then there will be two coincident tangents at (0, 0). Therefore, the curve will be of the form

$$(ax + by)^2 + \text{terms of degree greater then two} = 0 \quad \text{...(1)}$$

Thus the common tangent to the curve (1) at the origin is

$$ax + by = 0. \quad \text{...(2)}$$

Let us suppose p is perpendicular from any point $P(x, y)$ to the equation (2), then

$$p = \frac{ax + by}{\sqrt{a^2 + b^2}} \quad \text{...(3)}$$

where $P(x, y)$ is any point in the neighbourhood of (0,0).

From the equation (3) it is obvious that p is proportional to $ax + by$ so let us take

$$p = ax + by. \quad \text{...(4)}$$

Now eliminating either x or y between (1) and (4), we get the equation involving p and x. Since p is small and there are two branches of the curve passes through the origin, therefore, neglecting all those terms having the degree of p greater than two. Thus we obtain a quadratic in p of the form

$$Ap^2 + Bp + C = 0 \quad \text{...(5)}$$

where A, B, C are the functions of x only.

Now solving (5), we get

$$p = -\frac{B \pm \sqrt{(B^2 - 4AC)}}{2A} \quad \text{also } p_1 p_2 = C/A$$

where p_1 and p_2 are the roots of (5).

Now there arises following cases :

Case I. If for all numerically small values of x either negative or positive, the values of p obtained from (5) are imaginary, then the origin will be a conjugate point.

Case II. If the values of p are real for all numerically small values of x, then the origin will be a double cusp.

Case III. If the reality of p depends on the sign of x, then origin will be a single cusp.

Case IV. If p is real for numerically small values of x and if $p_1 p_2 > 0$, then p_1 and p_2 will have same sign. Therefore the origin will be a cusp of second species because the two perpendiculars p_1 and p_2 lie on the same side of the common tangent. On the other hand if $p_1 p_2 < 0$,then p_1 and p_2 are of opposite signs. Then the origin will be a cusp of the first species because the two perpendicular line on the opposite sides of the common tangent.

8.12 NATURE OF A CUSP AT ANY POINT

In order to find the nature of the cusp at any point (h, k). We first shift the origin at (h, k) and then apply above process discussed in § 8.11.

Solved Examples

Example 1. *Show that the origin is a node on the curve*

$$x^3 + y^3 - 3axy = 0.$$

Solution. The tangent at the origin are obtained by equating to zero the lowest degree terms *i.e.*, second degree term in the given equation of the curve.

$$\therefore \quad -3axy = 0 \text{ or } x = 0, y = 0.$$

Thus at the origin there are two real and distinct tangents. Hence (0, 0) is a node.

Example 2. *Find the double point of the curve* $(x-2)^2 = y(y-1)^2$.

Solution. Let $f(x, y) \equiv (x-2)^2 - y(y-1)^2 = 0 \quad \ldots(1)$

Differentiating (1) partially w.r.t. x and y, we get

$$\frac{\partial f}{\partial x} = 2(x-2) \quad \ldots(2)$$

and
$$\frac{\partial f}{\partial y} = -(y-1)^2 - 2y(y-1). \ldots(3)$$

Since the necessary and sufficient condition for a double points are

$$\frac{\partial f}{\partial x} = 0, \frac{\partial f}{\partial y} = 0, \Rightarrow 2(x-2) = 0 \quad \ldots(4)$$

$$-(y-1)^2 - 2y(y-1) = 0. \quad \ldots(5)$$

Now solving $f(x, y) = 0, \dfrac{\partial f}{\partial x} = 0$ and $\dfrac{\partial f}{\partial y} = 0$ simultaneously.

From (4), we get $x = 2$ and from (5), we get

$$-(y-1)(-y-1+2y) = 0$$

or $-(y-1)(2y-1) = 0$ or $y = 1$ and $y = 1/3$.

$\therefore$ Possible double points are (2, 1) and (2, 1/3)

But (2, 1/3) does not satisfy $f(x, y) = 0$. Hence only double point is (2, 1).

Example 3. *Examine the nature of the origin on the following curve :* $y^2 = a^2x^2 + bx^3 + cy^2$.

Solution. The given curve is

$$f(x, y) \equiv y^2 - a^2x^2 - bx^3 - cy^2 = 0. \quad \ldots(1)$$

Equating to zero the lowest degree terms in the equtaion of curve (1), we get

$$y^2 - a^2x^2 = 0 \text{ or } y = \pm ax.$$

Thus we have obtained two real and distinct tangents at (0, 0). Hence (0, 0) is a node.

Example 4. *Find the position and nature of the double points on the curve* $x^2y^2 = (a+y)^2(b^2 - y^2)$ if

(*i*) $b > a$ (*ii*) $b = a$ (*iii*) $b < a$.

Solution. Let $f(x, y) \equiv x^2y^2 - (a+y)^2(b^2 - y^2) = 0. \ldots(1)$

Differentiating (1) partially w.r.t. 'x' and 'y' respectively, we get

$$\frac{\partial f}{\partial x} = 2xy^2 \quad \ldots(2)$$

and
$$\frac{\partial f}{\partial y} = 2x^2y - 2(a+y)(b^2 - y^2) + 2y(a+y)^2. \quad \ldots(3)$$

Again differentiating, we get

$$\frac{\partial^2 f}{\partial x^2} = 2y^2 \; ; \quad \frac{\partial^2 f}{\partial x \partial y} = 4xy$$

and
$$\frac{\partial^2 f}{\partial y^2} = 2x^2 - 2(b^2 - y^2) + 4(a+y)y + 2(a+y)^2 + 4y(a+y)$$

For double point, we have

$$\frac{\partial f}{\partial x} = 0, \frac{\partial f}{\partial y} = 0, \therefore 2xy^2 = 0 \quad \ldots(4)$$

$$2x^2y - 2(a+y)(b^2 - y^2) + 2y(a+y)^2 = 0 \quad \ldots(5)$$

From (4) we get $x = 0, y = 0$

From (5) and $x = 0$, we get

$$2(a+y)[-(b^2 - y^2) + y(a+y)] = 0$$

or
$$2(a+y)(2y^2 + ay - b^2) = 0$$

or
$$y = -a \text{ and } y = \frac{-a \pm \sqrt{(a^2 + 8b^2)}}{4}$$

Thus we obtain $(0, -a)$ and $\left(0, \dfrac{-a \pm \sqrt{(a^2 + 8b^2)}}{4}\right)$

and from (5) and $y = 0$, we get two points.

Hence, $(0, -a)$ and $\left(0, \dfrac{-a \pm \sqrt{(a^2 + 8b^2)}}{4}\right)$ are possible double points. But only $(0, -a)$ satisfies the equation $f(x, y) = 0$. Hence, $(0, -a)$ is only the double point.

$$\left(\frac{\partial^2 f}{\partial x^2}\right)_{(0,-a)} = (2y^2)_{(0,-a)} = 2a^2$$

$$\left(\frac{\partial^2 f}{\partial x \partial y}\right)_{(0,-a)} = (4xy)_{(0,-a)} = 0$$

$$\left(\frac{\partial^2 f}{\partial y^2}\right)_{(0,-a)} = [2x^2 - 2(b^2 - y^2) + 4y(a+y) + 2a(a+y)^2 + 4y(a+y)]_{(0,-a)}$$

$$= 2(a^2 - b^2)$$

Then $\left(\frac{\partial^2 f}{\partial x \partial y}\right)^2 - \frac{\partial^2 f}{\partial x^2}\cdot\frac{\partial^2 f}{\partial y^2}$

$$= 0 - 2a^2[2(a^2 - b^2)]$$

$$= +4a^2(b^2 - a^2).$$

(i) If $b > a$, then $\left(\frac{\partial^2 f}{\partial x \partial y}\right)^2 > \frac{\partial^2 f}{\partial x^2}\cdot\frac{\partial^2 f}{\partial y^2}$ and thus $(0, -a)$ is a node.

(ii) If $b = a$, then $\left(\frac{\partial^2 f}{\partial x \partial y}\right)^2 = \frac{\partial^2 f}{\partial x^2}\cdot\frac{\partial^2 f}{\partial y^2}$ and thus $(0, -a)$ is a cusp.

(iii) If $b < a$, then $\left(\frac{\partial^2 f}{\partial x \partial y}\right)^2 < \frac{\partial^2 f}{\partial x^2}\cdot\frac{\partial^2 f}{\partial y^2}$ and thus $(0, -a)$ is a conjugate point.

Example 5. *Find the nature of origin on the curve* $x^4 + y^3 + 2x^2 + 3y^2 = 0$.

Solution. Let $f(x, y) = x^4 + y^3 + 2x^2 + 3y^2 = 0$

Then $\frac{\partial f}{\partial x} = 4x^3 + 4x, \frac{\partial f}{\partial y} = 3y^2 + 6y$

$$\frac{\partial^2 f}{\partial x^2} = 12x^2 + 4, \frac{\partial^2 f}{\partial y^2} = 6y + 6$$

and $\frac{\partial^2 f}{\partial x \partial y} = 0.$

At (0, 0) $\frac{\partial^2 f}{\partial x^2} = 4, \frac{\partial^2 f}{\partial y^2} = 6, \frac{\partial^2 f}{\partial x \partial y} = 0.$

$$\therefore \left(\frac{\partial^2 f}{\partial x \partial y}\right)^2 = 0 < \left(\frac{\partial^2 f}{\partial x^2}\right)\left(\frac{\partial^2 f}{\partial y^2}\right).$$

Hence, the origin is a conjugate point.

EXERCISE 8.2

1. Find the equation of the tangents at the origin to the following curves :
   (a) $(x^2 + y^2)(2a - x) = b^2x$ (b) $a^4y^2 = x^4(x^2 - a^2)$
   (c) $x^4 + 3x^3y + 2xy - y^2 = 0$ (d) $x^3 + y^3 = 3axy$
2. Examine the nature of the origin on the curve $(2x + y)^2 - 6xy(2x + y) - 7x^3 = 0$.
3. Show that the origin is a conjugate point on the curve $a^2x^2 + b^2y^2 = (x^2 + y^2)^2$.
4. Show that the origin is a conjugate point on the curve $y^2 = 2x^2y + x^4y - 2x^4$.
5. Find the position and nature of double points of the curve $y^3 = x^3 + ax^2$.
6. Examine the nature of the double points of the curve $2(x^3 + y^3) - 3(3x^2 + y^2) + 12x = 4$.
7. Find the position and nature of the double points of the curve $a^4y^2 = x^4(2x^2 - 3a^2)$.
8. Find the position and nature of the double points of the curve $x^4 - 2y^3 - 3y^2 - 2x^2 + 1 = 0$.
9. Determine the existence and nature of the double points on the curve $y^2 = (x - 2)^2(x - 1)$.
10. Prove that the curve $ay^2 = (x - a)^2(x - b)$ has at $x = a$, a conjugate point if $a < b$, a node if $a > b$ and a cusp if $a = b$.
11. Examine the curve $x^3 + 2x^2 + 2xy - y^2 + 5x - 2y = 0$ for singular points and show that it has a cusp of the first kind at the point (–1, – 2).
12. Show that the curve $y^2 = bx \tan(x/a)$ has a node or a conjugate point at the origin according as a and b have like or unlike signs.
13. Determine the position and nature of the double points of the curves :
    (a) $y(y - 1)^2 = (x - 2)^2$. (b) $x^3 - y^2 - 7x^2 + 4y + 15x - 13 = 0$.
    (c) $y^2 = x(x - a)^2, a > 0$ (d) $y^2 = x^2(a - x^2)$
    (e) $y(y - 6) = x^2(x - 2)^3 - 9$.
14. Discuss the nature of the double points of the curve $(x + y)^3 - \sqrt{2}(y - x + 2)^2 = 0$.
15. Show that the origin is a conjugate point on the curve $x^4 - ax^2y + axy^2 + a^2y^2 = 0$.
16. Show that curve $(xy + 1)^2 + (x - 1)^3(x - 2) = 0$ has a single cusp of the first species at the point (1, –1).
17. Show that the curve $y^3 = (x - a)^2(2x - a)$ has a single cusp of the first species at the point $(a, 0)$.
18. Find the nature and position of double points of the curve $a^4y^2 = x^4(a^2 - x^2)$.

ANSWERS

1. (a) $x = 0$ (b) $y = 0, y = 0$ (c) $y = 0, 2x - y = 0$ (d) $x = 0, y = 0$ **2.** Origin is a single cusp of first species
5. A cusp at (0, 0) **6.** Node at (2, 0) **7.** Cusp at (0, 0) **8.** Double points (0, –1), (1, 0) and (–1, 0) are nodes
9. Node at (2, 0) **13.** (a) Node at (2, 1) (b) Node at (3, 2) (c) Node at $(a, 0)$ (d) Node at (0, 0)
(e) Conjugate at (0, 3) and a single cusp of the first species at (2, 3) **14.** A single cusp of first species at (–1, 1)
18. Double cusp of the first species at (0, 0).

8.13 CURVE TRACING : CARTESIAN FORM

To trace any curve of cartesian form we should apply following process :

(a) Symmetricity. In order to find the symmetry of the curve we should apply following rules :

(i) If the powers of y in the equation of the curve are all even, then curve is symmetrical about x-axis.

(ii) If the powers of x in the equation of the curve are all even, then the curve is symmetrical about y-axis.

(iii) If the powers of x as well as y in the equation of the curve are all even, the curve is symmetrical about both axes.

(iv) If the equation of curve remains unchanged when x is replaced by $-x$ and y is replaced by $-y$, then the curve is symmetrical in opposite quadrants.

(v) If the equation of the curve remains unchanged when x and y are interchanged, then the curve is symmetrical about the line $y = x$.

(b) Nature of the origin on the curve. If the curve passes through the origin, then find the tangent at (0, 0) by equating to zero the lowest degree terms of the curve. If we obtain two tangent at the origin, then origin will be a double point and then find the nature of this double point.

(c) Intersection of curve with co-ordinate axes. We should check whether the curve cuts the co-ordinate axes or not, for this put $y = 0$ in the equation of the curve and find the values of x, then we get the points at which the curve cuts the x-axis. Similarly if the curve cuts the y-axis, then put $x = 0$ in the equation of the curve and obtain the points on the y-axis. Hence in this way we obtain the points of intersection of the curve with co-ordinate axis. Thereafter we should find the tangents at these points of intersection. For this first we shift the origin at these points and then obtain the tangent at these new origin by equating to zero the lowest degree terms in the new equation of the curve. On the other hand the value of dy/dx at these points of intersection can also be used to find the slope of the tangent at that point.

(d) Nature of y or x in the curve. We should now solve the equation of the curve either for y or for x whichever is convenient. Suppose we solve for y and see that nature of y as x increases from 0 to $+\infty$. Similarly see the nature of y as x decreases from 0 to $-\infty$ and finally collect those values of x for which $y = 0$ or $y \to \infty$ or $-\infty$.

REMARK

- If the curve is symmetrical about x-axis in opposite quadrants then there is no need to take the values of x of both positive and negative. We can take only positive values of x to see the variation in y.

(e) Regions in which curve does not exist. In order to find the regions where the curve does not exist we should solve the equation of curve for one variable in terms of the other. Therefore, the curve will not exist for those values of one variable which make the other variable imaginary.

(f) Asymptotes. Next, we should find all the asymptotes of the curve because the branches of the curve approach to the asymptotes if they exist.

(g) Sign of dy/dx. Next, we should find the value of dy/dx from the equation of the curve and find the points on the curve at which $dy/dx = 0$ or $dy/dx = \infty$. Therefore at these points we obtain the nature of tangents. Suppose in any region $a < x < b$, dy/dx remains positive throughout, then in this region y increase continuously as x increases. On the other hand if dy/dx remains negative, they y decreases continuously as x increases.

(h) Special points. If necessary, we should find the some special point on the curve.

(i) Points of inflexion. If necessary, we should find the point of inflexion to know the position of the curves at that point.

Now taking all above considerations in mind, draw an approximate shape of the curve.

SOLVED EXAMPLES

Example 1. *Trace the curve* $y^2(2a - x) = x^3$.

(UPTU–2007, 2011, PTU–2010, VTU–2008, Rajasthan–2006)

Solution. (i) Obviously the given curve is symmetrical about x-axis.

(ii) The curve passes through the origin and the tangents at the origin are obtained by equating to zero the lowest degree terms *i.e.*, $2ay^2$ in the equation of the curve.

$$\therefore \quad 2ay^2 = 0 \text{ or } y = 0, y = 0.$$

Thus at the origin we obtained two coincident tangents $y = 0, y = 0$ *i.e.*, x-axis. Therefore (0, 0) is a cusp.

(iii) From the equation of the curve it is obvious that the curve does not cut the co-ordinate axes.

(iv) Now solving the equation of the curve for y, we get $y^2 = x^3/(2a - x)$ when $x = 0$, $y^2 = 0$ and when $x = 2a$, thus $y^2 \to \infty$ thus $x = 2a$ is an asymptotes of the curve.
It is observed that y increases as x increases from 0 to $2a$.

(v) When x lies between 0 and $2a$, y^2 will be positive and the curve will exist in this region. When $x > 2a$, y^2 will be negative so the curve will not exist beyond the line $x = 2a$. When $x < 0$ again y^2 will be negative and thus the curve will also not exist for $x < 0$. Hence we can say that the curve only exists in the region $0 < x < 2a$.

(vi) In order to find the asymptotes, putting $y = m$ and $x = 1$ in the third degree terms in the equation of the curve, we get

$$\phi_3(m) = m^2 + 1.$$

Therefore the equation $\phi_3(m) = 0$ gives both its roots imaginary so ignore them. Consequently $x = 2a$ is only the asymptote of the curve.

(vii) Differentiating the equation of the curve

$$y = \frac{x^{3/2}}{\sqrt{(2a-x)}}$$

We get $\dfrac{dy}{dx} = \dfrac{(3a-x)x^{1/2}}{(2a-x)^{3/2}}$.

In the region $0 < x < 2a$, $\dfrac{dy}{dx}$ will be positive, so therefore in this region y increase continuously as x increases.
Now taking all above points of consideration in the mind and draw the curve whose shape is shown in fig. 9.

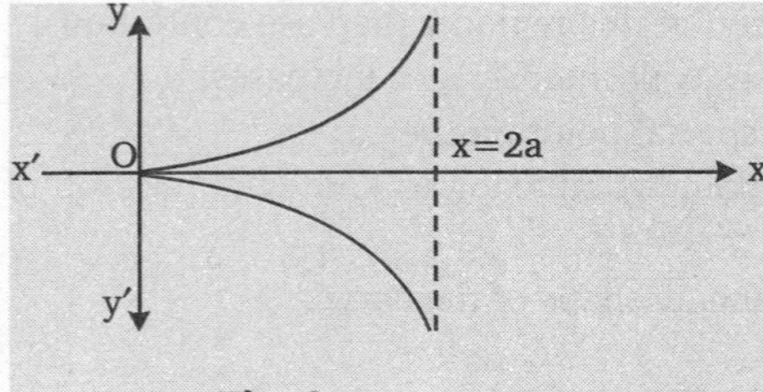

Fig. 9

Some Standard Figures and their Equations

S. No.	Name	Equation
1.	Cubical Parabola	$y = x^3$
2.	Semi cubical parabola	$ay^2 = x^3$
3.	Cissoid	$y^2(2a - x) = x^3$
4.	Folium of Descartes	$x^3 + y^3 = 3axy$
5.	Circle	$x^2 + y^2 = a^2$ $r = 2a\cos\theta$
6.	Cardioid	$r = a(1 - \cos\theta)$ $r = a(1 + \cos\theta)$
7.	Limacon	$r = a + b\cos\theta$
8.	Equiangular spiral	$r = ae^{m\theta}$
9.	Cycloid	$x = a(t + \sin t)$ $y = a(t + \cos t)$
10.	Tractrix	$x = a\cos t + \frac{1}{2}a\log\tan^2\frac{t}{2}$ $y = a\sin t$
11.	Astroid	$x^{2/3} + y^{2/3} = a^{2/3}$
12.	Inverted Cycloid	$x = a(t + \sin t)$ $y = a(1 - \cos t)$
13.	Strophoid	$y^2(a - x) = x^2(a + x)$
14.	Four leaved rose	$r = a\sin 2\theta$
15.	Spiral of Archimedes	$r\theta = a$

Example 2. *Trace the curve* $y^2(1 - x^2) = x^2(1 + x^2)$.

Solution . (i) In the equation of the curve the powers of both x and y are all even so the curve is symmetrical about both axes.

(ii) The curve passes through the origin. The tangents at the origin are obtained by equating to zero the lowest degree terms in the equation of the curve.

$$\therefore \quad y^2 - x^2 = 0 \text{ or } y = \pm x.$$

Thus there are two real and distinct tangent at the origin so (0, 0) is a node.

(iii) From the equation of the curve it is clear that curve does not cut any co-ordinate axes.

(iv) Solving the equation of the curve for y, we get

$$y^2 = \frac{x^2(1+x^2)}{(1-x^2)}.$$

When $x = 0$, $y = 0$ and when $x = \pm 1$, $y \to \infty$ so $x = \pm 1$ are two asymptotes parallel to y-axis.

(v) When $-1 < x < 1$, y^2 is positive, so the curve exists in this region. When $x > 1$, y will be negative thus curve will not exist beyond the line $x = 1$. Also when $x < -1$, y^2 will be negative so that curve will not exist for $x < -1$.

(vi) In order to find the asymptotes, putting $y = m$ and $x = 1$ in the fourth degree terms of the curve, we get $\phi_4(m) = m^2 + 1$.

Solving $\phi_4(m) = 0$, we get both values of m imaginary so ignore them. Consequently $x = \pm 1$ are only two real asymptotes.

(vii) Since, we have

$$y = x\sqrt{\left(\frac{1+x^2}{1-x^2}\right)}.$$

$$\therefore \frac{dy}{dx} = \frac{\sqrt{1-x^2}\left(\sqrt{1+x^2} + \dfrac{x^2}{\sqrt{1+x^2}}\right) - x\sqrt{(1+x^2)}\left[\dfrac{-x}{\sqrt{1-x^2}}\right]}{(1-x^2)}$$

$$= \frac{(1-x^2)(1+x^2+x^2) - x(1+x^2)(-x)}{(1-x^2)^{3/2}(1+x^2)^{1/2}}$$

$$= \frac{2x^2+1-x^4}{(1-x^2)^{3/2}(1+x^2)^{1/2}}.$$

When $-1 < x < 0$, $\frac{dy}{dx}$ is negative this means that when x decreases from -1 to 0, y decreases. When $0 < x < 1$; $\frac{dy}{dx}$ is positive this implies that when x increases from 0 to 1, y increases.

Now taking all the above facts in mind and draw the shape of the curve we get the following figure:

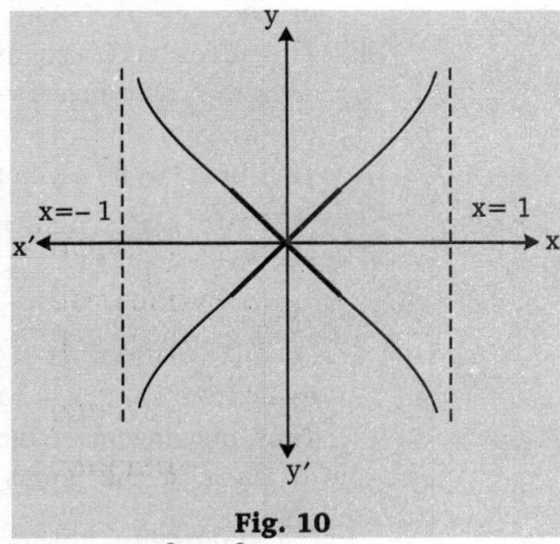

Fig. 10

Example 3. *Trace the curve $ay^2 = x^2 (a - x)$.*

(SVTU–2004, Kurukshetra–2009)

Solution. (i) The curve is symmetrical about x-axis because the powers of y are all even.

(ii) This curve passes through the origin. The tangents at the origin are obtained by equating to zero the lowest degree terms in the equation of the curve, we get

$$ay^2 - ax^2 = 0 \text{ or } y = \pm x.$$

Thus there are two real and distinct tangents at (0, 0). Therefore (0, 0) is a node.

(iii) The curve cuts the x-axis only at the point where $y = 0$.

$\therefore x^2(a - x) = 0$ or $x = 0, x = a$.

Thus the curve cuts the x-axis at $(a, 0)$.

Now $\quad y = x\sqrt{\dfrac{a-x}{a}}$

$$\therefore \quad \frac{dy}{dx} = \frac{1}{\sqrt{a}}\left[\sqrt{a-x} - \frac{x}{2\sqrt{a-x}}\right]$$

$$= \frac{2(a-x)-x}{2\sqrt{a(a-x)}} = \frac{2a-3x}{2\sqrt{a(a-x)}}.$$

At $(a, 0)$, $\frac{dy}{dx} = \infty$. Therefore the tangent at $(a, 0)$ is perpendicular to x-axis.

(iv) Since we have

$$y = x\sqrt{\frac{a-x}{a}}$$

when $x = 0, y^2 = 0$ and when $x = a, y^2$ also equals zero. Also when x increases from 0 to $a/2$ y increases and when x increases from $a/2$ to a, y decreases.

(v) When $0 < x < a$, y^2 is always positive so the curve will exist in this region. When $x < 0$, y^2 is also positive so that the curve will also exist for $x < 0$. When $x > a$, y^2 will be negative and therefore in this region the curve will not exist.

Taking all above facts into consideration and draw the shape of the curve we obtain as shown below.

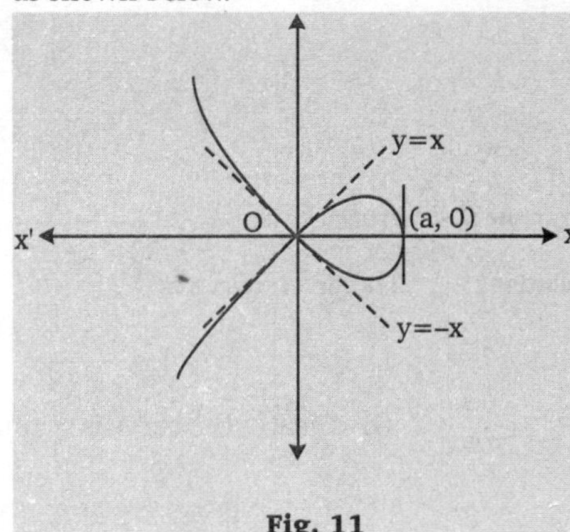

Fig. 11

Example 4. *Trace the curve* $x^3 + y^3 = 3axy$.

(UPTU–2007, Kurukshetra–2005)

Solution . (i) The given curve is symmetrical about the line $y = x$.

(ii) Curve is passing through the origin so the trangents at (0, 0) are

$xy = 0 \Rightarrow \quad x = 0$ and $y = 0$.

Thus (0, 0) is a node.

(iii) The curve does not intersect the axes.

(iv) If x is replaced by $-x$ and that of y by $-y$, the equation of the curve is changed. Thus the curve does not exist in third quadrant.

(v) The given curve has only one real asymptote which is $x + y + a = 0$.

(vi) The curve cuts the line $y = x$ at the point $\left(\frac{3a}{2}, \frac{3a}{2}\right)$.

From the curve, $\frac{dy}{dx} = \frac{ay - x^2}{y^2 - ax}$.

so at $\left(\frac{3a}{2}, \frac{3a}{2}\right)$ $\frac{dy}{dx} = -1$

Thus, the tangent at $\left(\frac{3a}{2}, \frac{3a}{2}\right)$ makes an angle 135° with the positive axis of x.

Now taking all above facts into consideration and draw the curve, we get figure (12).

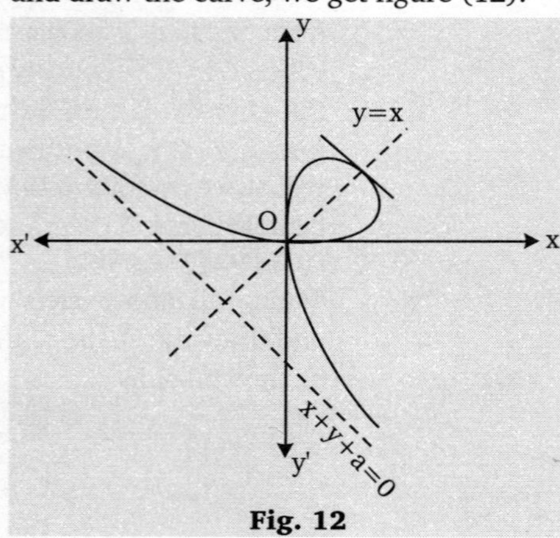

Fig. 12

Example 5. *Trace the curve* $y = 1 - \frac{1}{1+x^2}$. (UPTU–2008)

Solution . The given curve is

$$y = 1 - \frac{1}{1+x^2}. \qquad \ldots(1)$$

(i) Clearly, the curve is symmetric about y-axis.

(ii) The curve passes through the origin, because (0, 0) satisfies (1).

(iii) Tangent at origin is given by $y = 0$.

(iv) Asymptotes parallel to x-axis is $y - 1 = 0$ $\Rightarrow y = 1$.

(v) Clearly the given curve meets the axes only at the origin.

Special points

x	–2	–1	0	1	2
y	4/5	1/2	0	1/2	4/5

Now taking all the above facts in find, we draw the curve as in figure (13).

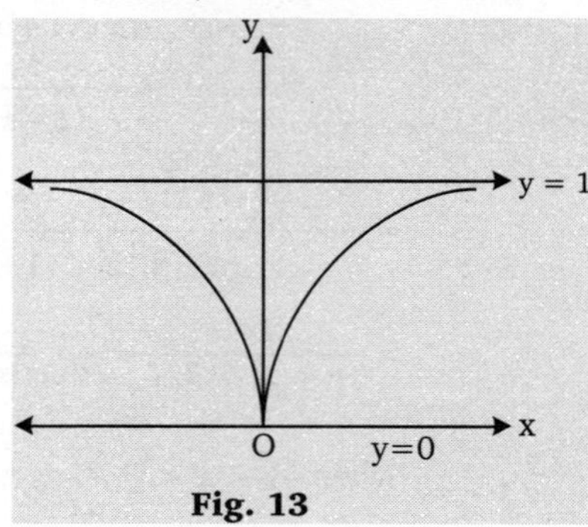

Fig. 13

Example 6. *Trace the curve* $x^{2/3} + y^{2/3} = a^{2/3}$.

(UPTU–2006, VTU–2003, PTU–2009)

Solution . The given curve is $x^{2/3} + y^{2/3} = a^{2/3}$ ...(1)

(i) The curve is symmetric about both the axes. Also, the curve is symmetric about the line $y = x$ and $y = -x$.

(ii) The curve does not passes through the origin.

(iii) The curve meets x-axis at $(a, 0)$ and $(-a, 0)$. Also, the curve meets y-axis at $(0, a)$ and $(0, -a)$.

(iv) The curve has no asymptotes.

(v) From (1) $\frac{dy}{dx} = -\frac{y^{1/3}}{x^{1/3}}$

$\frac{dy}{dx} = 0$, when $y = 0$

Again from (1) $x = \pm a$, when $y = 0$

Thus, the tangents to the curve are parallel to x-axis at the points $(\pm a, 0)$. Further, $\frac{dy}{dx} \to 0$ when $x = 0$.

From (1) when $x = 0, y = \pm a$.

Therefore, the tangents to the curve are parallel to y-axis at the points $(0, \pm a)$

(vi) From (1) we can write $y^{2/3} = a^{2/3} - x^{2/3}$.

If $|x| > a$, $y^{2/3}$ is negative $\Rightarrow y^2$ is negative.

$\Rightarrow$ y is imaginary.

$\Rightarrow$ Curve does not lie beyond the lines $x = \pm a$.

Similarly, the curve does not lie beyond the lines $y = \pm a$.

Further, when $x = 0$, $y = a$. As x-increases from 0 to a, y decreases from a to 0.

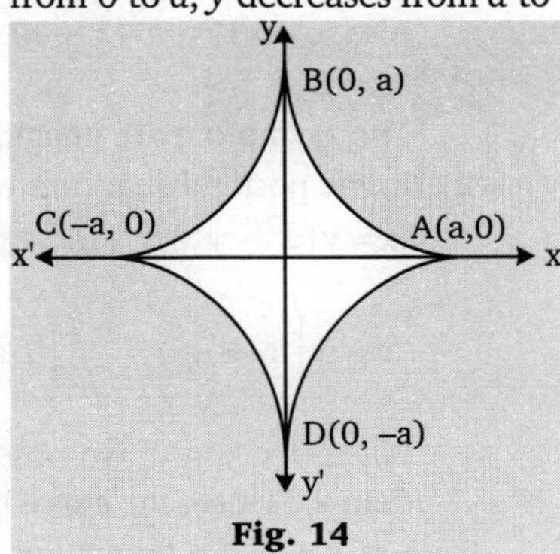

Fig. 14

Example 7. *Trace the curve* $y = x^3 - 12x - 16$. (PTU–2008)

Solution . (i) The curve has no symmetry.

(ii) The curve does not passes through the origin.

(iii) The curve has no asymptotes.

(iv) The curve cuts x-axis at (–2, 0), (4, 0) and y-axis at (0, –16).

(v) We have $\frac{dy}{dx} = 3x^2 - 12$.

At (–2, 0), $\frac{dy}{dx} = 0 \Rightarrow$ tangent is parallel to x-axis at (–2, 0)

At (4, 0), $\frac{dy}{dx} = 36 \Rightarrow$ tangent makes an acute angle $\tan^{-1}36$ with x-axis at (4, 0)

Also, $\frac{dy}{dx} = 0$ at $3x^2 - 12 = 0 \Rightarrow x = \pm 2$

$\Rightarrow$ tangent is parallel to x-axis at (2, –32).

(vi) $y \to \infty$ as $x \to \infty$ and $y \to -\infty$ as $x \to -\infty$: y is positive for $x > 4$ and y is –ve for $x < 4$.

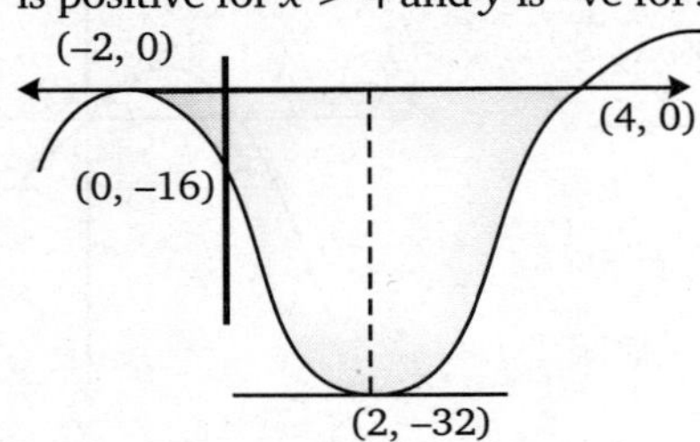

Fig. 15

Example 8. *Trace the curve* $9ay^2 = (x - 2a)(x - 5a)^2$. (JNTU–2008)

Solution . (i) The curve is symmetric about x-axis.

(ii) The curve does not passes through the origin.

(iii) The curve has no asymptotes.

(iv) The curve cuts the x-axis at $x = 2a$ and $x = 5a$ *i.e.*, at $A(2a, 0)$ and $B(5a, 0)$. It cuts the y-axis at $y^2 = -50\frac{a^2}{9}$

$\Rightarrow y$ is imaginary. Therefore curve does not cut the x-axis

(v) $y = \frac{(x-5a)\sqrt{x-2a}}{3\sqrt{a}}$

$\Rightarrow$ y is imaginary for $x < 2a$.

$\Rightarrow$ curve exists only for $x \geq 2a$.

and $\frac{dy}{dx} = \pm\frac{(x-3a)}{2\sqrt{a}\sqrt{x-2a}}$

At $A(2a, 0)$, $\frac{dy}{dx} = \infty$ *i.e.*, tangent is parallel to y-axis.

At $B(5a, 0)$, $\frac{dy}{dx} = \pm\frac{1}{\sqrt{3}}$ *i.e.*, there are two distinct tangents.

$\Rightarrow$ There is a node at $B(5a, 0)$.

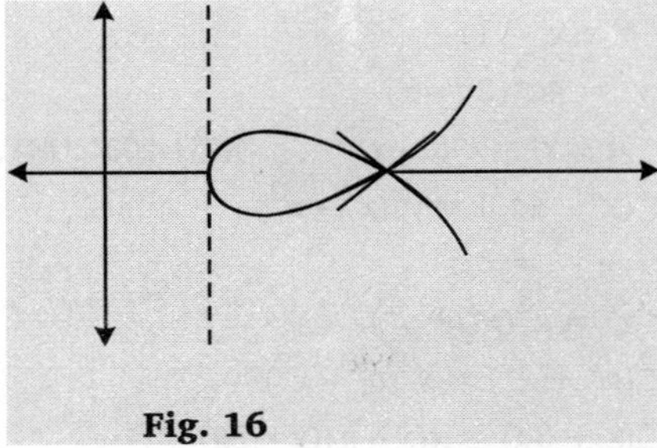

Fig. 16

Example 9. *Trace the curve* : $y^2(x^2 + y^2) + a^2(x^2 - y^2) = 0$

or $x^2(y^2 + a^2) + y^2(y^2 - a^2) = 0$ (GBTU–2010)

Solution . (i) The curve is symmetric about both the axes.

(ii) The curve passes through the origin and $a^2(x^2 - y^2) = 0$ *i.e.*, $y = \pm x$ are two tangents at origin. So, origin is a node.

(iii) The curve intersects the x-axis only at origin. It intersects the y-axis at (0, 0), (0, a) and (0, $-a$).

(iv) Shifting the origin at (0, a) the equation of the curve becomes

$(y+a)^2\{x^2 + (y+a)^2\} + a^2\{x^2 - (y+a)^2\} = 0$

$\Rightarrow$ $(y^2 + 2ay + a^2)(x^2 + y^2 + 2ay + a^2)$
$+ a^2(x^2 - y^2 - 2ay - a^2) = 0$

Equating to zero the the lowest degree terms, we get

$$2a^3y + 2a^3y - 2a^3y = 0 \Rightarrow y = 0,$$

which is the tangent at new origin. Here we need not find the tangent at (0, –a) as the curve is symmetric about x-axis.

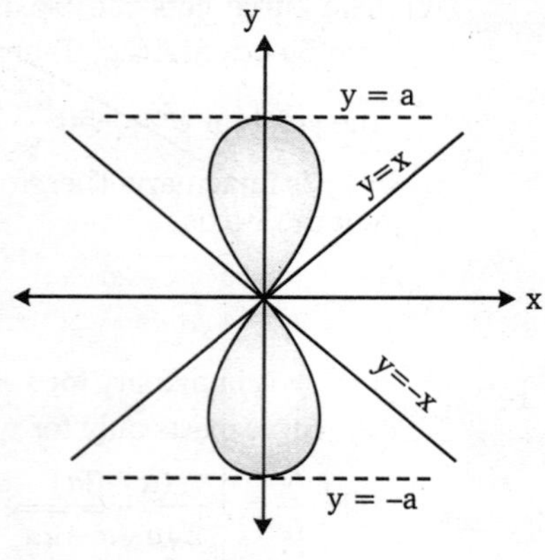

Fig. 17

(v) On solving the given equation for x, we get $x^2 = y^2(a^2 - y^2)/(a^2 + y^2)$ when $y = 0, x^2 = 0$ and when $y = a, x^2 = 0$.
When $0 < y < a$, x^2 is positive. Therefore, the curve exists in the region $0 < y < a$. When $y > a$, x^2 is negative, so the curve does not exist in the region $y > a$.

(vi) The asymptotes parallel to x-axis are given by $a^2 + y^2 = 0$ *i.e.*, $y = \pm ai$.
Also, $\phi_4(m) = m^2(1 + m^2)$. Its roots are $m = 0, 0, i, -i$.
The asymptote are imaginary.

(vii) In the positive quadrant we have

$$x = y(a^2 - y^2)^{1/2}/(a^2 + y^2)^{1/2}, y > 0$$

$$x = y\left(1 - \frac{y^2}{a^2}\right)^{1/2} \Big/ \left(1 + \frac{y^2}{a^2}\right)^{1/2}$$

When $0 < y < a$, we observe that x is less than y. Hence, the curve lies above the line $y = x$ which is tangent at origin.

EXERCISE 8.3

Trace the following curves :

1. $ay^2 = x^3$
2. $a^2y = x^3$
3. $y = x(x^2 - 1)$
4. $xy^2 = 4a^2(2a - x)$
5. $y^2(a + x) = x^2(a - x)$ (VTU–2004, UPTU–2006, SVTU–2008)
6. $x^2(x^2 - 4a^2) = y^2(x^2 - a^2)$. (UPTU–2009)
7. $x^3 + y^3 = x$
8. $x^2y^2 = a^2(x^2 + y^2)$
9. $y^2(a^2 + x^2) = x^2(a^2 - x^2)$ (VTU–2010)
10. $y^2(x + 3a) = x(x - a)(x - 2a)$
11. $y^2(x^2 + y^2) + a^2(x^2 - y^2) = 0$
12. $a^2y^2 = x^3(2a - x)$
13. $9ay^2 = x(x - 3a)^2$
14. $x^2y^2 = (1 + y)^2 (4 - y^2)$
15. $y^3 + x^3 = a^2x$
16. $x^4 + y^4 = 4a^2xy$
17. $y^2 = (x - a)(x - b)(x - c), a > b > c$
18. $y^2(x - a) = x^2(x + a)$ (UPTU–2007, VTU–2010, BPTU–2005)
19. $y^2(x^2 - 1) = x$
20. $y(x^2 - 1) = (x^2 + 1)$
21. $a^2y^2 = x^2(a^2 - x^2)$ (PTU–2009, VTU–2008)
22. $y(x^2 + 4a^2) = 8a^3$
23. $x^3y = x + 1$
24. $a^3y^2 = (x - a)^4(x - b), a > b$

ANSWERS

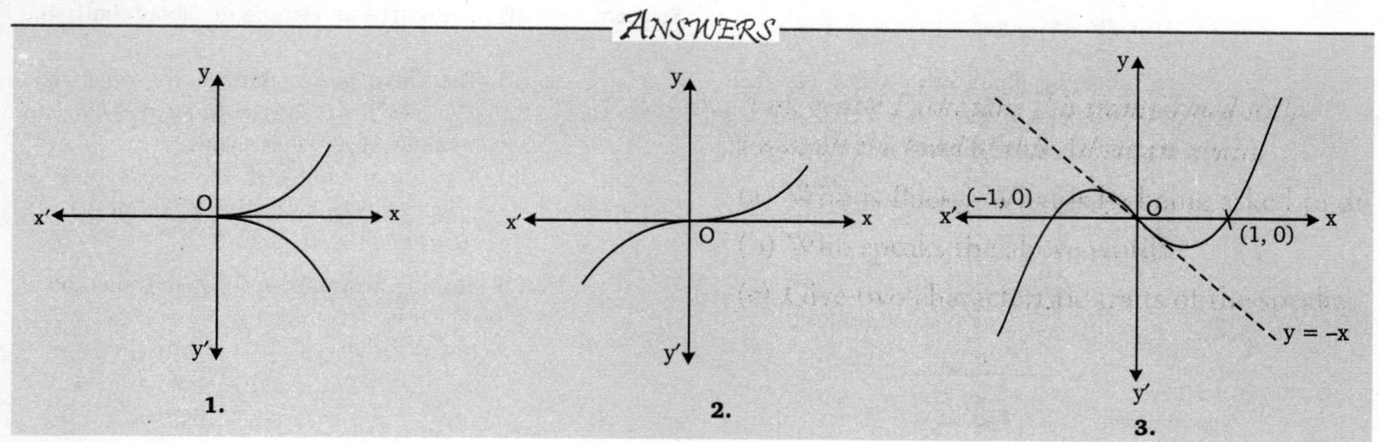

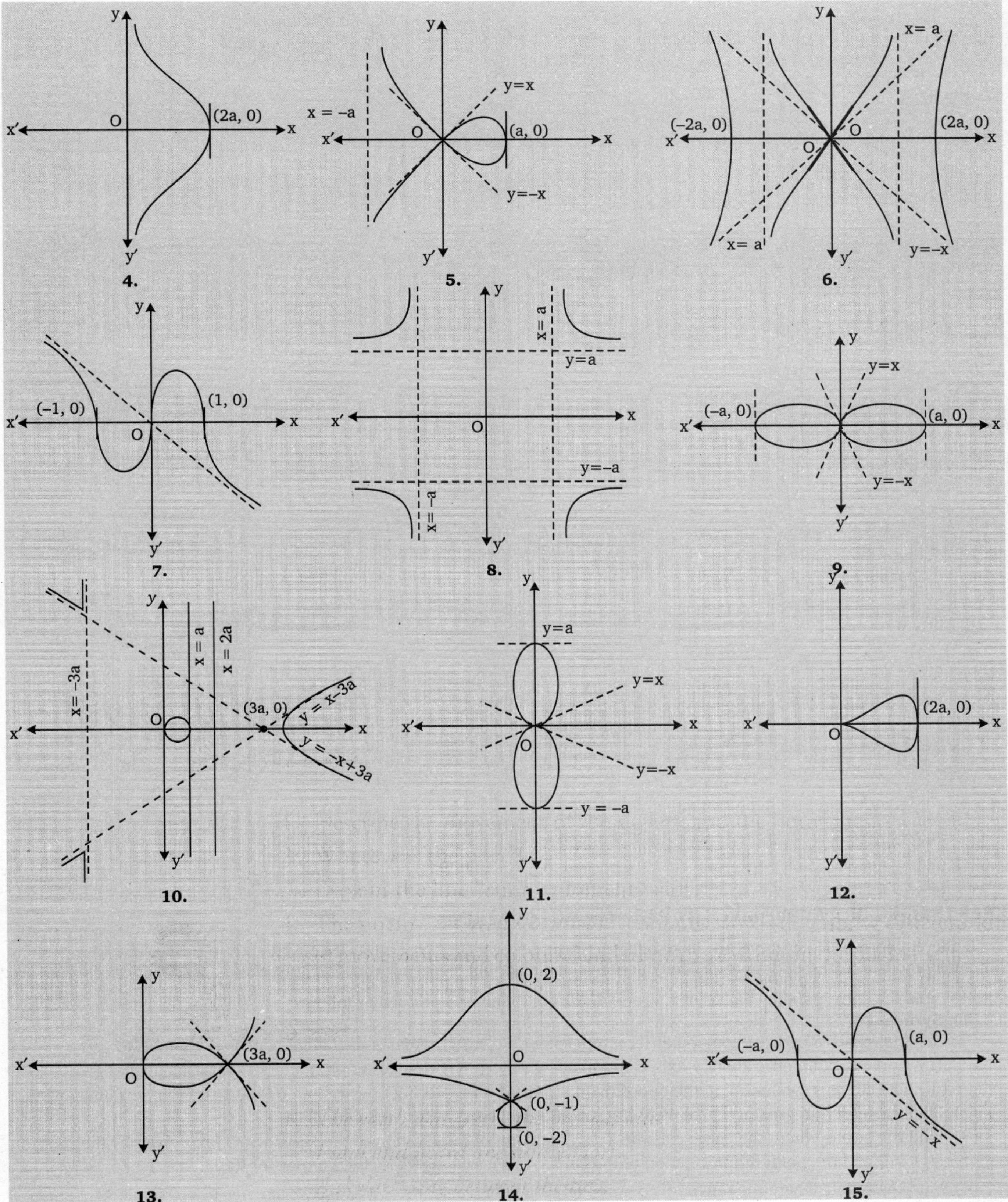

y
x′
O
(2a, 0)
x
y′
4.
y
x = –a
y=x
x′
O
(a, 0)
x
y=–x
y′
5.
y
x= a
(–2a, 0)
x′
O
(2a, 0)
x
x= a
y′
y=–x
6.
y
(–1, 0)
(1, 0)
x′
O
x
y′
7.
y
x = a
y=a
x′
O
x
y=–a
x = –a
y′
8.
y
y=x
(–a, 0)
(a, 0)
x′
x
y=–x
y′
9.
y
x = –3a
x = a
x = 2a
O
(3a, 0)
y = x–3a
x′
x
y = –x+3a
y′
10.
y
y=a
y=x
x′
x
O
y=–x
y = –a
y′
11.
y
(2a, 0)
x′
O
x
y′
12.
y
x′
O
(3a, 0)
x
y′
13.
y
(0, 2)
x′
O
x
(0, –1)
(0, –2)
y′
14.
y
(–a, 0)
(a, 0)
x′
O
x
y = –x
y′
15.

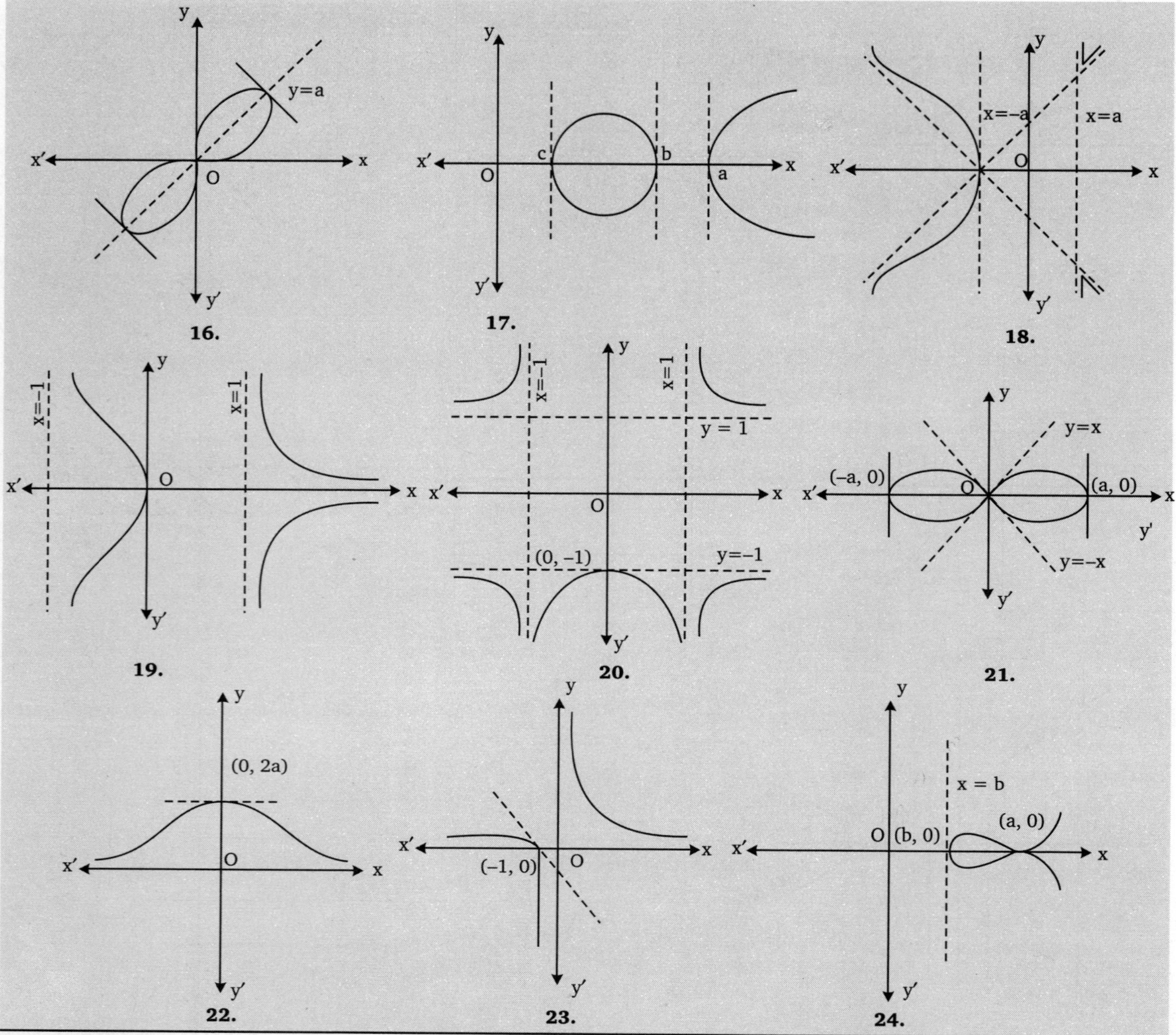

8.14 TRACING OF A CURVE GIVEN BY PARAMETRIC EQUATIONS

If the equations of the curve are in parametric form *i.e.*, $x = f(t)$, $y = g(t)$. If conveniently possible, the parameter is eliminated and the corresponding cartesian equation is obtained. But if it is not convenient to eliminate t, a series of values are given to t and the corresponding values of x, y, and dy/dx are found and proceed as follows :

(1) Symmetry

(i) Curve is symmetric about x-axis if on replacing t by $-t$, $f(t)$ remains unchanged and $g(t)$ changes to $-g(t)$.

(ii) Curve is symmetric about y-axis if on replacing t by $-t$, $f(t)$ changes to $-f(t)$ and $g(t)$ remains unchanged.

(iii) The curve is symmetric in the opposite quadrants if on replacing t by $-t$, both $f(t)$ and $g(t)$ remains unchanged.

(2) Find the least and greatest value of x and y to find the region where the curve lies.

(3) Find the points where the curve cuts the axes. The point of intersection of the curve with x-axis given by the roots of $g(t) = 0$ and the point of intersection of the curve with y-axis are given by the roots of $f(t) = 0$.

(4) Find the points where the tangent is parallel or perpendicular to the x-axis (*i.e.*, where $dy/dy = 0$ or $\to \infty$)

Solved Examples

Example 1. *Trace the curve $x = a(t+\sin t)$ and $y=a(1+\cos t)$*

Solution . (i) Given $x = a(t+\sin t) \Rightarrow \frac{dx}{dt} = a(1+\cos t)$

$$y = a(1+\cos t) \Rightarrow \frac{dy}{dt} = -a\sin t$$

$$\therefore \quad \frac{dy}{dx} = \frac{dy/dt}{dx/dt} = \frac{-a\sin t}{a(1+\cos t)}$$

$$= \frac{-2a\sin t/2\cos t/2}{2a\cos^2\frac{t}{2}} = -\tan\frac{t}{2}$$

(ii) We have $y = 0$,
when $\cos t = -1$, *i.e*, $t = -\pi, \pi$
When $t = \pi$, $x = a\pi$, $dy/dx = -\infty$
At the point $(a\pi, 0)$, the tangent to the curve is perpendicular to the x-axis. Also when $t = -\pi$, $x = -a\pi$, $dy/dx = \infty$

(iii) y is maximum when $\cos t = 1$, *i.e.*, when $t = 0$
When $t = 0$, $x = 0$, $y = 2a$ and $dy/dx = 0$.
So at the point $(0, 2a)$, the tangent to the curve is parallel to x-axis.

(iv) y can be negative. Also no part of the curve lies in the region $y > 2a$.

t	$-\pi$	$-\pi/2$	0	$\pi/2$	π
x	$-a\pi$	$-a\left(\frac{\pi}{2}+1\right)$	0	$a\left(\frac{\pi}{2}+1\right)$	$a\pi$
y	0	a	$2a$	a	0
dy/dx	∞	1	0	-1	$-\infty$

At $(-a\pi, 0)$ the tangent inclined to x-axis at the angle $\psi = \frac{\pi}{2}$

Also curve is symmetric about the y-axis.

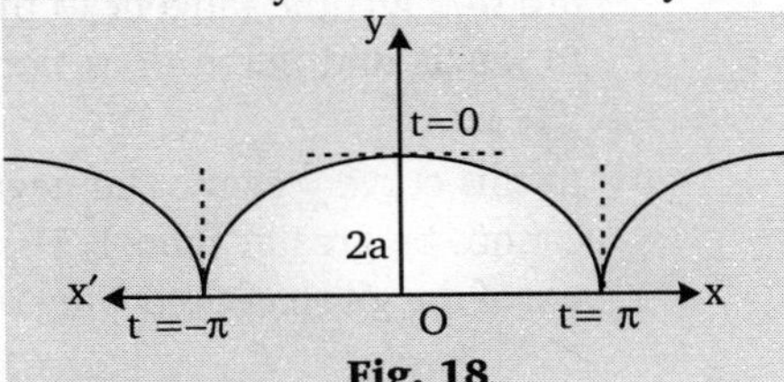

Fig. 18

Example 2. *Trace the curve*

$$x = a\cos t + \frac{1}{2}a\log\tan^2\frac{t}{2}, y = a\sin t$$

Solution . (i) Put $-t$ for t in the given equation of the curve, we get

$$x = a\cos t + \frac{1}{2}a\log\tan^2\frac{t}{2} \text{ and } y = -a\sin t$$

Therefore, for every value of x there are two equal and opposite value of $y \Rightarrow$ curve is summetric about x-axis.

Further, put $\pi - t$ for t in the given equation of the curve, we get

$$x = -a\cos t + \frac{1}{2}a\log\cot^2\frac{t}{2}$$
$$= -a\cos t - \frac{1}{2}a\tan^2\frac{t}{2}$$

and $\quad y = a\sin t$

$\Rightarrow$ For every value of y, there are two equal and opposite values of x, so curve is symmetric about y-axis.

(ii) Differentiating the given equation w.r.t. t we get

$$\frac{dx}{dt} = -a\sin t + \frac{1}{2}a\frac{1}{\tan^2\frac{t}{2}}(2\tan\frac{t}{2}\sec^2\frac{t}{2}).\frac{1}{2}$$

$$= -a\sin t + \frac{a}{2\sin\frac{t}{2}\cos\frac{t}{2}}$$

$$= -a\sin t + \frac{a}{\sin t} = \frac{a(1-\sin^2 t)}{\sin t} = \frac{a\cos^2 t}{\sin t}$$

and $\quad \frac{dy}{dt} = a\cos t$

$$\therefore \quad \frac{dy}{dx} = \frac{dy/dt}{dx/dt} = \frac{a\cos t.\sin t}{a\cos^2 t} = \tan t$$

(iii) We have $y = 0$ when $\sin t = 0$, *i.e.*, $t = 0$, when $t \to 0$, $x \to -\infty$.
Therefore, $x \to -\infty$ when $y \to 0$ showing that the line $y = 0$ is an asymptotes of the curve.

(iv) Clearly y is maximum when $\sin t = 1$ *i.e.*, $t = \pi/2$. When $t = \pi/2$, $x = 0$, $y = a$

and $\frac{dy}{dx} = \tan\frac{\pi}{2} = \infty$

$\Rightarrow$ Curve passes through the point $(0, a)$ and the tangent at this point is the x-axis.

(v) Clearly, the numerical value of y cannot be greater than a. Therefore, curve does exist in the region $y > a$ and $y < -a$.

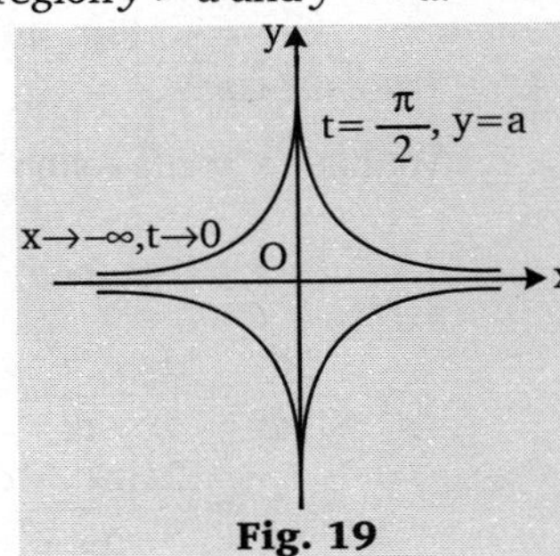

Fig. 19

Example 3. *Trace the curve $x = a(t + \sin t), y = a(1 - \cos t)$ when $-\pi \le t \le \pi$.*

Solution. We have $x = a(t + \sin t) \Rightarrow \dfrac{dx}{dt} = a(1+\cos t)$

$$y = a(1 - \cos t) \Rightarrow \frac{dy}{dt} = a\sin t$$

$$\therefore \quad \frac{dy}{dx} = \frac{dy/dt}{dx/dt} = \frac{a\sin t}{a(1+\cos t)} = \tan\frac{t}{2}$$

(i) Clearly, $y = 0$ when $\cos t = 0$ *i.e.*, $t = 0$

When $t = 0, x = 0, \dfrac{dy}{dx} = \tan 0 = 0$

$\Rightarrow$ Curve passes through the origin and the axis of x is tangent at the origin.

(ii) y is maximum when $\cos t = -1$ *i.e.*, $t = \pi$ and $-\pi$. When $t = \pi, x = a\pi, y = 2a$ and $\dfrac{dy}{dx} = \infty$.

So at the point $t = \pi$, whose coordinates are $(a\pi, 2a)$, the tangent is perpendicular to the x-axis. When $t = -\pi, x = -a\pi, y = 2a$, $\dfrac{dy}{dx} = -\infty$.

(iii) Here y can not be negative, so curve lies entirely above the axis of x and no portion of the curve lies in the region $y > 2a$.

(iv)

t	$-\pi$	$-\pi/2$	0	$\pi/2$	π
x	$-a\pi$	$-a\left(\frac{\pi}{2}+1\right)$	0	$a\left(\frac{\pi}{2}+1\right)$	$a\pi$
y	$2a$	a	0	a	$2a$
dy/dx	$-\infty$	-1	0	1	∞

Replace t by $-t$ in the given curve, we get $x = -a(t + \sin t)$ and $y = a(1 - \cos t)$. Therefore, for every value of y, there are two equal and opposite value of x. So, curve is symmetric about y-axis.

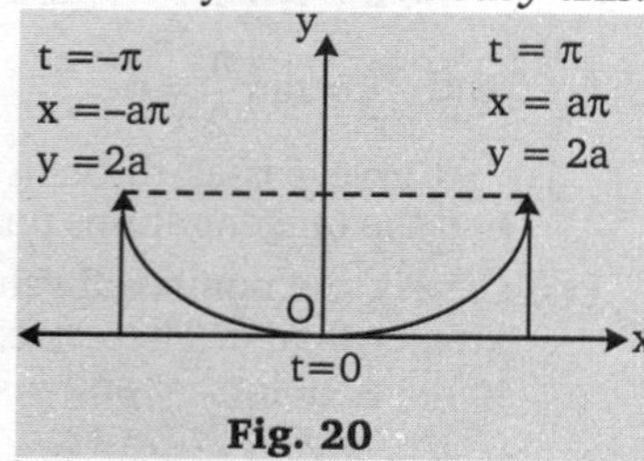

Fig. 20

Example 4. *Trace the curve $x = a(t - \sin t), y = a(t - \cos t)$.*

Solution. We have $x = a(t - \sin t) \Rightarrow \dfrac{dx}{dt} = a(1-\cos t)$

$$y = a(t - \cos t) \Rightarrow \frac{dy}{dt} = a\sin t$$

$$\therefore \quad \frac{dy}{dx} = \frac{dy/dt}{dx/dt} = \frac{a\sin t}{a(1-\cos t)}$$

$$= \frac{2\sin t/2\cos t/2}{2\sin^2 t/2} = \cot t/2$$

Here, $y = 0$ when $\cos t = 1$ *i.e.*, $t = 0, 2\pi$. When $t = 0, x = 0, y = 0$ and $\dfrac{dy}{dx} = \cot 0 = \infty$. Therefore, the curve passes through the origin and axis of y is tangent to the curve at this point.

Also, y is maximum when $\cos t = -1$ *i.e.*, $t = \pi$

When $t = \pi, x = a(\pi - \sin\pi) = a\pi, y = 2a$, $\dfrac{dy}{dx} = \cot\dfrac{\pi}{2} = 0$

Therefore, at $t = \pi$, whose cartesian coordinates are $(a\pi, 2a)$ the tangent to the curve is parallel to x-axis and curve does not lie in the region $y > 2a$.

In this curve y cannot be negative because cost cannot be greater than 1. Hence, one complete arc of the given cycloid lying between $0 \le t \le 2\pi$.

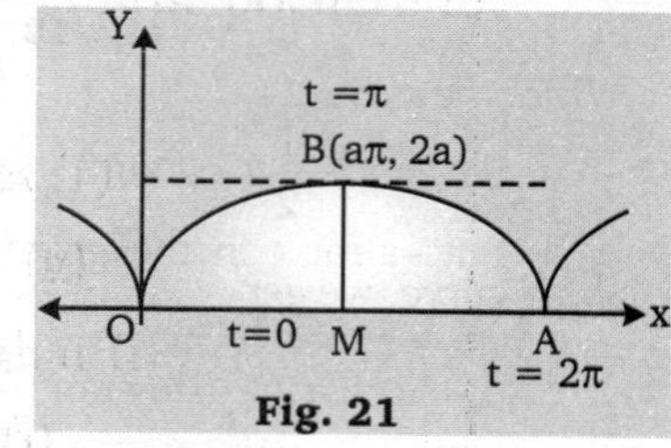

Fig. 21

Some Transformations

- $x = a\cos t, y = b\sin t$ then $\dfrac{x^2}{a^2} + \dfrac{y^2}{b^2} = 1$ (Ellipse)
- $x = a\cos t, y = b\sin t$ Then $x^2 + y^2 = a^2$ (Circle)
- $x = a\cos^3 t, y = b\sin^3 t$ Then $\left(\dfrac{x}{a}\right)^{2/3} + \left(\dfrac{y}{b}\right)^{2/3} = 1$ (Hypo–Cycloid)
- $x = a\cos^3 t, y = a\sin^3 t$ Then $x^{2/3} + y^{2/3} = a^{2/3}$ (Astroid)
- $x = t^2, y = t - \dfrac{t^3}{3}$ $y^2 = x(1 - x/3)^2$
- $x = a\sin^3 t, y = a\dfrac{\sin^3 t}{\cos t}$ Then $y^2(a - x) = x^3$ (Cissoid)
- $x = \dfrac{1-t^2}{1+t^2}, y = \dfrac{2t}{1+t^2}$ Then $x^2 + y^2 = 1$ (Circle)
- $x = \dfrac{3at}{1+t^3}, y = \dfrac{3at^2}{1+t^3}$ Then $x^3 + y^3 = 3axy$ (Follium of Descarte's)

8.15 TRACING OF EQUATION IN POLAR FROM

To trace the curve in polar form, we use the following procedure :

(a) Symmetry. In order to find the symmetry we use the following rules :

(i) If the curve $r = f(\theta)$ remains unchanged when θ is replaced by $-\theta$, then the curve is symmetrical about the initial line.

(ii) If the curve $r = f(\theta)$ remains unchanged when r is replaced by $-r$ then the curve is symmetrical about the pole. (origin).

(b) Special points on the curve. If r becomes zero for some values of θ, then the curve will pass through the pole. Therefore, if $r = 0$ when $\theta = \alpha$ (say), then $\theta = \alpha$ is the tangent to the curve at the pole.

Next we should find the maximum and minimum values of r which will exist for some values of θ.

(c) Solve these equations for r and observe the variation of r as θ varies for 0 to $+\infty$ and also we have to observe the variation as θ decreases from 0 to $-\infty$.

Therefore, we should from the table for the values of r corresponding to the values of θ.

(d) Regions where curve does not exist. If we obtain the values of r, imaginary for $\alpha < \theta < \beta$ then the curve will not exist in this region.

(e) Asymptotes. Next we should find the asymptotes if exist. For this if $r \to \infty$ for $\theta = \alpha$, then $\theta = \alpha$ is an asymptotes of the curve.

(f) Direction of tangents. Find $r\dfrac{d\theta}{dr}$ from $r = f(\theta)$. But we have that $\tan\phi = r\dfrac{d\theta}{dr}$. If for some $\theta = \alpha$, ϕ comes to be zero at any point then the line $\theta = \alpha$ will be the tangent at that point. Therefore, if ϕ comes out to be zero $\theta = \pi/2$ then $\theta = \pi/2$ is the tangent perpendicular to the radius vector $\theta = \alpha$.

Now taking all above facts into consideration and draw the shape of the curve.

REMARK

- Sometimes we face some problem to trace the curve of the form $r = f(\theta)$. Then for conveniently change the polar form of the curve into cartesian form by the following transformation : $x = r\cos\theta$, $y = r\sin\theta$ and then trace the curve.

Solved Examples

Example 1. *Trace the following curve $r^2 = a^2\cos 2\theta$.*

(VTU–2007, GBTU–2012, MTU–2011, UPTU–2009, BPTU–2005, Kurukshetra–2006)

Solution . (i) The curve is symmetrical about both the initial line and the pole.

(ii) $r=0$ when $\cos 2\theta=0$ *i.e.*, when $\theta=\pm\pi/4$. Thus $\theta = -\pi/4$ and $\theta = \pi/4$ are two real and distinct tangent at the pole so that the pole is a node.

(iii) The maximum value of r is a when $\cos 2\theta = 1$ *i.e.*, when $\theta = 0$ and π.

(iv) $$\tan\phi = r\frac{d\theta}{dr} = -\cot 2\theta$$

$$\tan\phi = \tan\left(\frac{\pi}{2} + 2\theta\right).$$

$$\therefore \quad \phi = \frac{\pi}{2} + 2\theta.$$

At the points $(a, 0)$ and (a, π), ϕ comes out to be $\pi/2$ and $2\pi+\pi/2$. Thus at these points tangents the perpendicular to the initial line.

(v) The following table gives the corresponding values of r and θ.

θ	0	$\pi/6$	$\pi/4$	$\pi/3$	$3\pi/4$	π
r	a	$a/\sqrt{2}$	0	imag.	0	$-a$

(vi) **Region.** When $-\pi/4 < \theta < \pi/4$, r^2 is negative so the curve will not exist in this region and when $\dfrac{5\pi}{4} < \theta < \dfrac{7\pi}{4}$, r^2 is negative. Also the curve will not exist in this region.

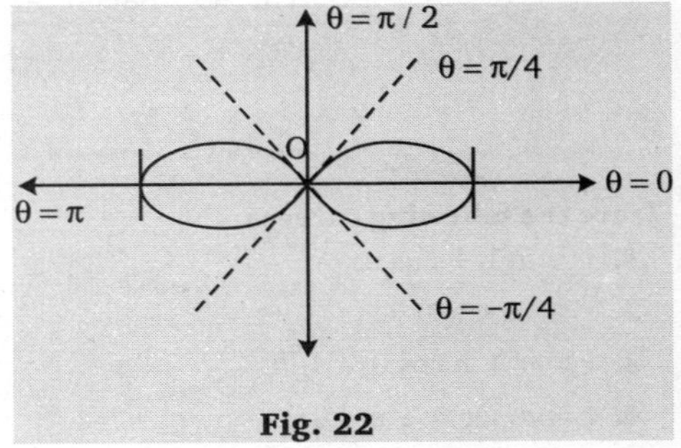

Fig. 22

Example 2. *Trace the curve $r = a\sin 4\theta$.*

Solution . (i) The curve is symmetrical about the initial line, since on putting $(\pi - \theta)$ in place of θ and $-r$ in place of r, the equation of the curve remains unchanged.

(ii) The curve is symmetrical about the pole because on replacing θ by $(\pi + \theta)$ the equation of the curves remains unchanged.

(iii) Since the curve is finite, there are no asymptotes.

(iv) Draw a table for values of r and θ.

It is evident that the greatest numerical value of r is a. Hence, the curve, between $\theta = 0$ and $\theta = \pi$, is as shown in fig. 23.

If θ increases from π to 2π, the corresponding branches of the curve are known because of symmetry about the pole. Since the curve is periodic, the values of θ outside the range $(0, 2\pi)$ need not be considered.

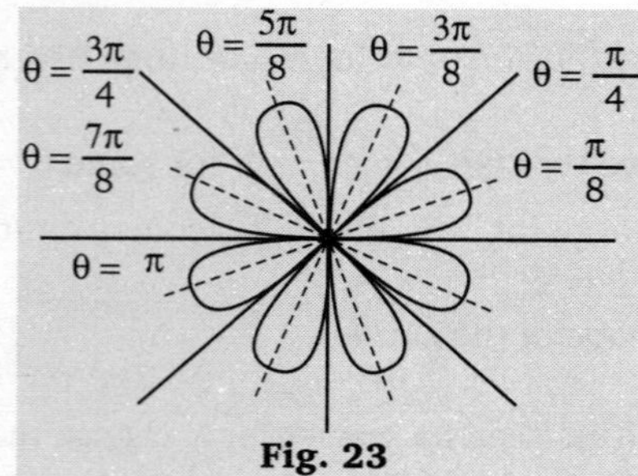

Fig. 23

Example 3. *Trace the curve $r = a + b \cos \theta$, $a < b$.*

Solution . (i) The curve is symmetrical about the initial line.

(ii) $r=0$, when $a+b\cos\theta=0$ *i.e.*,

$\cos\theta = \left(-\frac{a}{b}\right)$ or $\theta = \cos^{-1}\left(-\frac{a}{b}\right)$ but $a < b$ *i.e.*, $\frac{a}{b} < 1$, therefore $\cos^{-1}\left(-\frac{a}{b}\right)$ comes out to be real so that $\theta = \cos^{-1}\left(-\frac{a}{b}\right)$ is the trangent at the pole.

(iii) r is maximum when $\cos\theta = 1$ *i.e.*, $\theta = 0$. Then the maximum value of $r = a + b$ and the minimum value of $r = a - b$ when $\cos\theta = -1$ *i.e.*, $\theta = \pi$.

(iv) Since we have

$$r = a + b\cos\theta.$$

$$\therefore \quad \frac{dr}{d\theta} = -b\sin\theta$$

then $\tan\phi = r\frac{d\theta}{dr} = -\frac{(a+b\cos\theta)}{b\sin\theta}$.

Now if $\theta = 0$ and π, $\phi = 90°$, thus at the points $(a + b, 0)$, $(a - b, \pi)$ the tangents are perpendicular to the initial line.

(v) The following table gives the corresponding value of r and θ

θ	0	$\pi/2$	$\cos^{-1}\left(-\frac{a}{b}\right)$	$\cos\left(\frac{a}{b}\right) < \theta < \pi$	π
r	$a+b$	a	0	r is negative	$a-b$

Thus from above facts the shape of the curve is shown below :

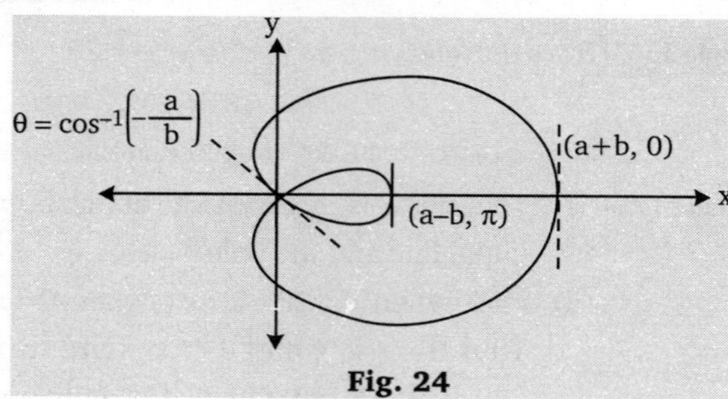

Fig. 24

EXERCISE 8.4

Trace the following curves :

1. $r = a(1 + \cos\theta)$
2. $r = 2a\cos\theta$
3. $r = a + b\cos\theta$, $a > b$
4. $r = a(\sec\theta + \cos\theta)$
5. $r\cos\theta = 2a\sin^2\theta$
6. $r^2 = a^2\sin 2\theta$
7. $r = a\sin 3\theta$ (GBTU–2011, UPTU–2002)
8. $r = ae^{m\theta}$
9. $2a/r = 1 + \cos\theta$.
10. $r = a\cos 3\theta$.
11. $r = a(1 - \cos\theta)$.
12. $r = a\sin 2\theta$. (VTU–2009)
13. $2r = 1 + 2\cos 2\theta$.
14. $x = a(\theta - \sin\theta)$, $y = a(1 - \cos\theta)$. (SRM-2008)
15. $x = a\cos\theta + \frac{1}{2}a\log\tan^2\frac{\theta}{2}$, $y = a\sin\theta$.
16. $x = a(\theta + \sin\theta)$, $y = a(1 - \cos\theta)$. (JNTU–2009)
17. $r = a\cos 2\theta$.
18. $x = a(t - \sin t)$, $y = a(1 - \cos t)$.

ANSWERS

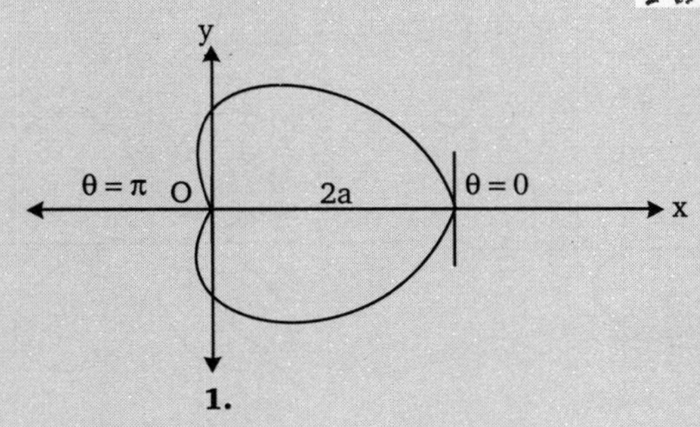

1.

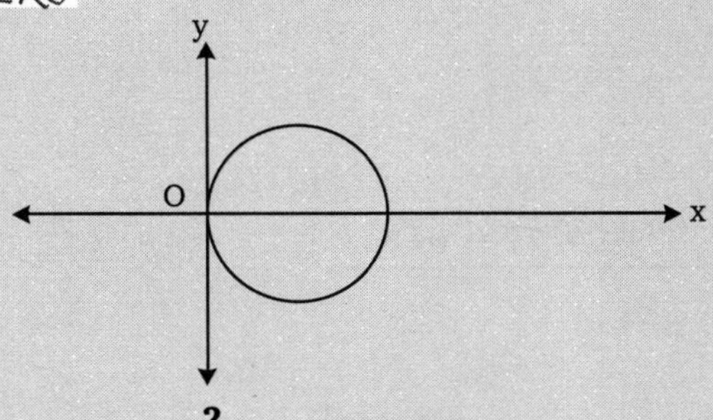

2.

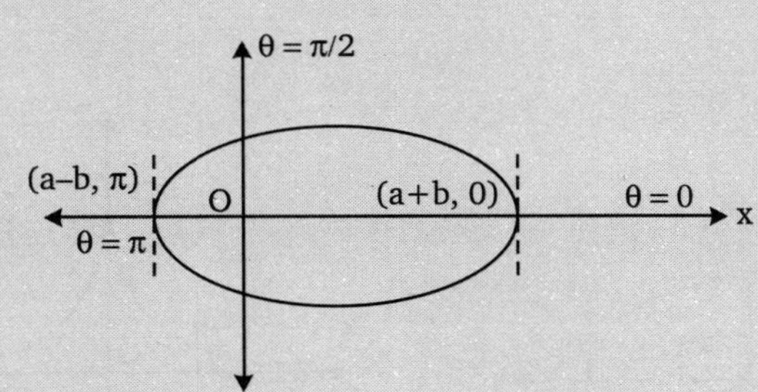

3.

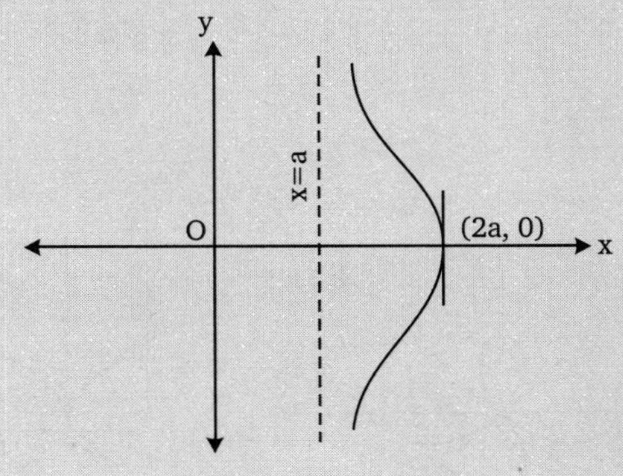

4.

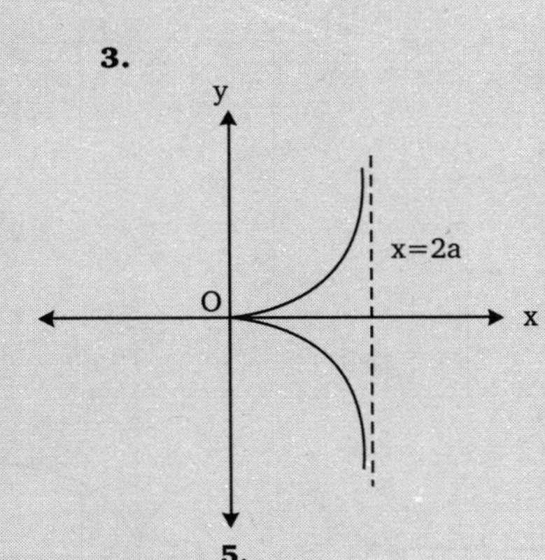

5.

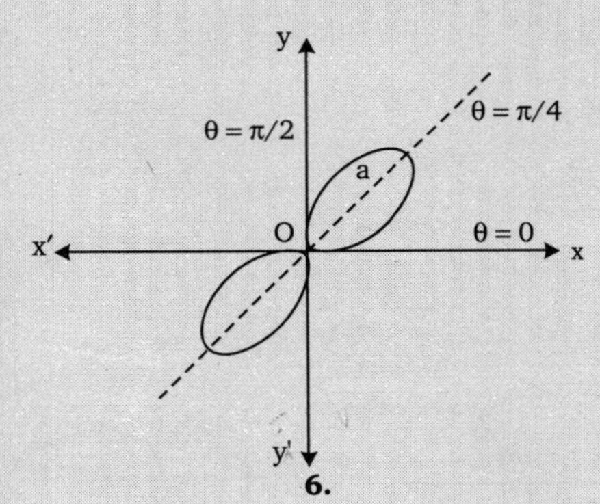

6.

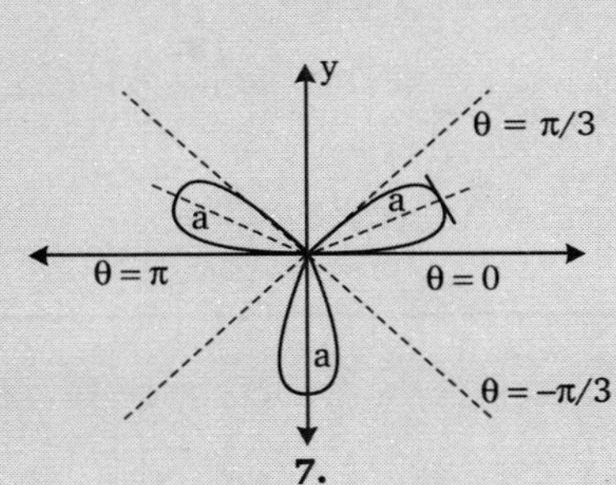

7.

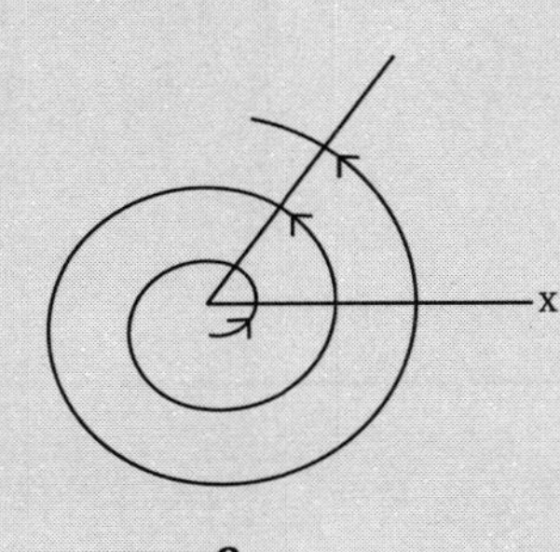

8.

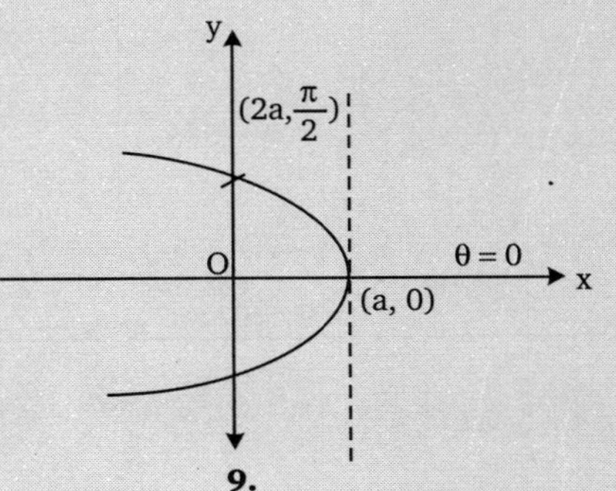

9.

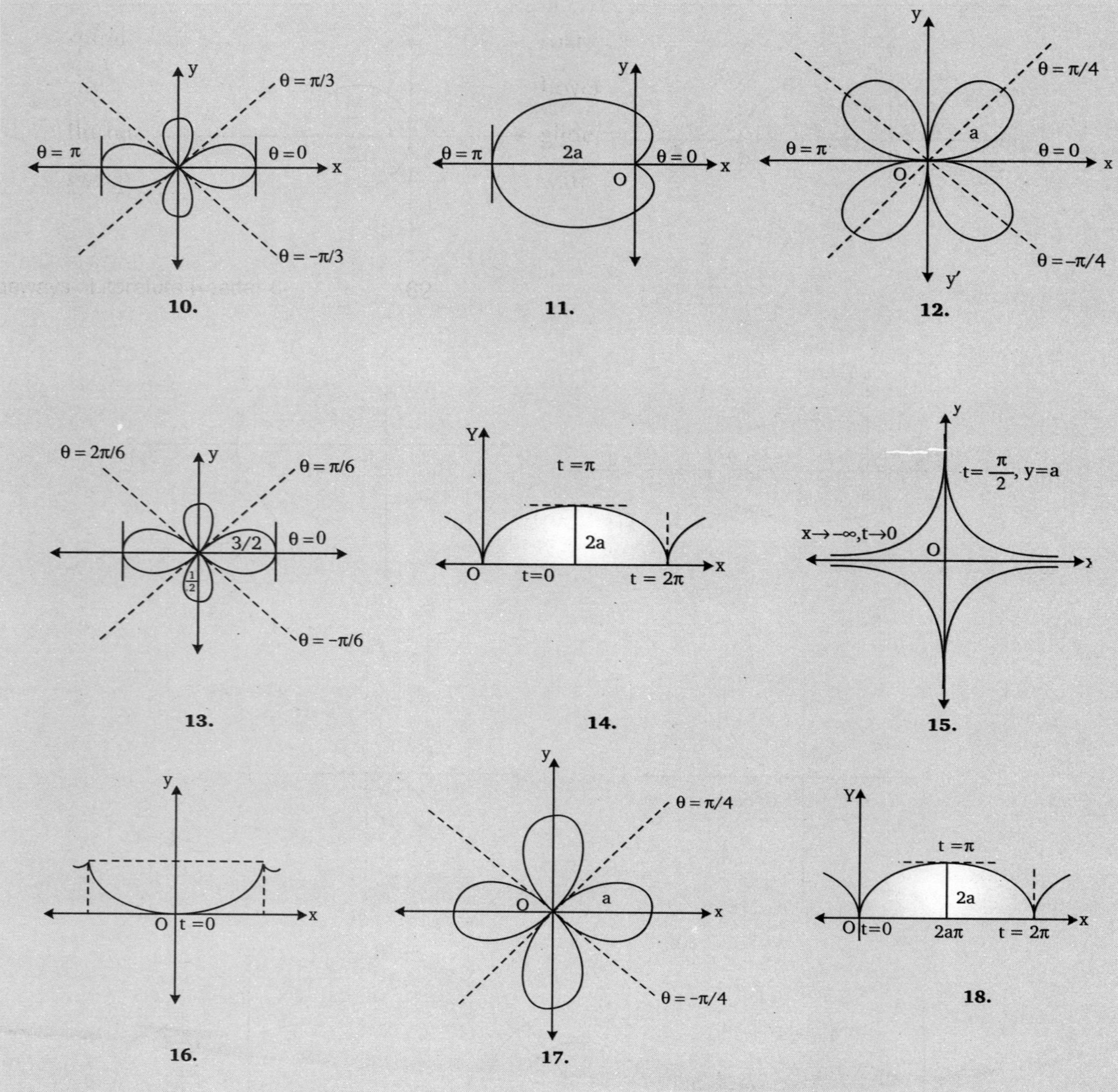

y
θ = π/3
θ = π
θ = 0
x
θ = −π/3
10.
y
θ = π
2a
θ = 0
O
x
11.
y
θ = π/4
a
θ = π
θ = 0
O
x
θ = −π/4
y′
12.
θ = 2π/6
y
θ = π/6
3/2
θ = 0
1/2
θ = −π/6
13.
Y
t = π
2a
O
t=0
t = 2π
x
14.
y
t= π/2, y=a
x→−∞, t→0
O
15.
y
O
t = 0
x
16.
y
θ = π/4
O
a
x
θ = −π/4
17.
Y
t = π
2a
O
t=0
2aπ
t = 2π
x
18.

Objective Evaluations

Fill in the Blanks

1. A point $P(x, y)$ on the curve $y = f(x)$ will be a point of inflexion if $\frac{d^2x}{dy^2} = 0$ but $\frac{d^3x}{dy^3}$ is not equal to ________ .
2. A double point is a node if at this point there are ________ distinct tangents.
3. A double point is a cusp if at this point there are two ________ tangents.
4. The curve lies entirely on one side of the common tangent, then the cusp is ________ .
5. The tangents at the origin of the curve $x^3 + y^3 - 3axy = 0$ are ________ .
6. The necessary and sufficient condition for any point of the curve $f(x, y) = 0$ to be a multiple point are that ________ .
7. A double point on the curve $f(x, y) = 0$ is conjugate point if $\left(\frac{\partial^2 f}{\partial x \partial y}\right)^2$ is less than________ .
8. In the equation of a curve $y^2(a^2 + x^2) = x^2(a^2 - x^2)$, (0, 0) is a ________ .
9. The curve is symmetrical about the line $y = x$ if the curve remains ________ when x and y are interchanged.
10. The curve is symmetrical about x-axis if the powers of y in the equation of the curve are all ________ .
11. If the curve $r = f(\theta)$ is unchanged when θ is replaced by $-\theta$, the curve is symmetrical about ________ .
12. The total loops in the curve $r = a \sin 6\theta$ is equal to ________ .

True/False

Write 'T' for True and 'F' for False statement.

1. For the point of inflexion of the curve $y = f(x)$, $\frac{d^2x}{dy^2} = 0$ but $\frac{d^3y}{dx^3} \neq 0$. **(T/F)**
2. If we obtained two real and distinct tangents at the double point, then double point is a cusp. **(T/F)**
3. If at the double point, we obtain two imaginary tangents, then double point is an isolated point. **(T/F)**
4. The tangents at origin are obtained by equating to zero the lowest degree terms of the curve. **(T/F)**
5. The curve is symmetrial about y-axis if the powers of y in the equation of the curve are all even. **(T/F)**
6. The curve is symmetrical in the opposite quadrant if the curve remains unchanged when x and y are replaced by $-x$ and $-y$ respectively. **(T/F)**
7. The tangents at the origin of the curve $y^2(a^2 + x^2) = x^2(a^2 - x^2)$ are $y = \pm x$. **(T/F)**
8. The curve $r = f(\theta)$ is symmetrical about the pole when r is replaced by $-r$ and the curve remains unchanged. **(T/F)**
9. In the curve $r = f(\theta)$, $r = 0$ when $\theta = \alpha$, then $\theta = \alpha$ is the tangent at the pole. **(T/F)**
10. In some region $0 < x < a$, if y^2 is negative, then the curve will not exist in this region. **(T/F)**
11. The curve $y^2 = 4ax$ is symmetric about x-axis. **(T/F)** (UPTU–2009)
12. The curve $x^3 + y^3 = 3axy$ is symmetric about $y = x$. **(T/F)** (UPTU–2009)
13. The curve $x^3 - y^3 = 3axy$ is symmetric about the line $y = x$. **(T/F)** (UPTU–2009)
14. The curve $x^2 + y^2 = a^2$ is symmetric about both the axis. **(T/F)** (UPTU–2009)

Multiple Choice Questions

Choose the most appropriate one.

1. At the point of inflexion of the curve $x = f(y)$, $\frac{d^2x}{dy^2} = 0$ and $\frac{d^3x}{dy^3}$ is not equal to :
   (a) 1 (b) 0
   (c) –1 (d) none of these
2. The points of inflexion of the curve $y^2 = (x - a)^2(x - b)$ lie on the line :
   (a) $3x + a = 4b$ (b) $3x + a = b$
   (c) $3x + b = 4a$ (d) $3x = a$
3. At the double point, if we obtain two real and distinct tangents, then double point is :
   (a) cusp (b) node
   (c) conjugate (d) none of these
4. For the point (x, y) on the curve $f(x, y) = 0$, if $\left(\frac{\partial^2 f}{\partial x \partial y}\right)^2 = \frac{\partial^2 f}{\partial x^2} \cdot \frac{\partial^2 f}{\partial y^2}$ then the point is :
   (a) node (b) conjugate
   (c) cusp (d) none of these

5. The tangents at the origin of the curve $y^2(a-x) = x^2(a+x)$ are :

(a) $y = \pm x$ (b) $y = \pm a$
(c) $x = \pm a$ (d) $y = 0$

6. The number of tangents at the origin of the curve $x^4 + 3x^3y + 2xy - y^2 = 0$ are :

(a) 1 (b) 3
(c) 2 (d) 0

7. The curve $x^3 + y^3 = 3axy$ is symmetrical about the line :

(a) $y = -x$ (b) $y = a$
(c) $x = a$ (d) $y = x$

8. The curve $r = a + b \cos \theta$, $b > a$ is symmetrical about :

(a) $\theta = \pi/2$ (b) $\theta = 0$
(c) $\theta = \pi/3$ (d) $\theta = \pi/4$

9. If $\theta = \alpha$ is a tangent at the pole of the curve $r = f(\theta)$ the r is equal to :

(a) 1 (b) 0
(c) –1 (d) α

10. The point (2, 0) on the curve $y^2 = (x-2)^2(x-1)$ is :

(a) cusp (b) conjugate
(c) node (d) none of these

ANSWERS

Fill in the Blanks

1. zero	**2.** two	**3.** coincident	**4.** single	**5.** $x = 0, y = 0$
6. $\frac{\partial f}{\partial x} = 0, \frac{\partial f}{\partial y} = 0$	**7.** $\frac{\partial^2 f}{\partial x^2} \frac{\partial^2 f}{\partial y^2}$	**8.** node	**9.** unchanged	**10.** even
11. initial line	**12.** 12			

True/False

1. T	**2.** F	**3.** T	**4.** T	**5.** F
6. T	**7.** T	**8.** T	**9.** T	**10.** T
11. F	**12.** F	**13.** F	**14.** T	

Multiple Choice Questions

1. (b)	**2.** (a)	**3.** (b)	**4.** (c)	**5.** (a)
6. (c)	**7.** (d)	**8.** (b)	**9.** (b)	**10.** (c)

FFFFFF

GATE tutor

KEY Terms and Results

- **First Differential Coefficients :** Let $y = f(x)$ be a given function then first differential coefficient $f'(x)$ of $f(x)$ is defined by
$$f'(x) = \lim_{h \to 0} \frac{f(x+h) - f(x)}{h}$$
- **Successive Differentiation :** The process of finding the differential coefficients $f', f'', f''', \ldots$ of a function is called successive differentiation.
- **Leibnitz's Theorem :** Let u and v be two functions having derivative of nth order, then
$$D^n(u.v) = u_n v + {}^nC_1 u_{n-1} v_1 + {}^nC_2 u_{n-2} v_2 + \ldots + {}^nC_r u_{n-r} v_r + \ldots + {}^nC_n u v_n$$
This theorem help us to find the nth differential coefficient of the product of two functions in terms of the successive derivatives of the functions.
- To find the nth derivative of a fraction, we have to split it into partial fraction.
- To resolve a quotient into partial fraction, the degree of the numerator must be less than the degree of denominator.
- **Continuity in a Closed interval:** To test the continuity of $f(x)$ in a closed interval $[a,b]$ we have the following two methods :

  Method 1: Here we check following three conditions:
  (i) $f(x)$ is not infinite in the closed interval $[a,b]$.
  (ii) $f(x)$ is not imaginary in the closed interval $[a,b]$.
  (iii) $f(x)$ has no break in the closed interval $[a,b]$.

  Method 2: Use the following results:
  (i) If $f(x)$ is a polynomial function of x, then it is continuous for all real x.
  (ii) If $f(x)$ is not a polynomial function, then we find $f'(x)$. If $f'(x)$ is finite, definite and real in $[a,b]$ then $f(x)$ is differentiable in $[a,b]$ and hence continuous in $[a,b]$.
- **Differentiability in Open Interval:**
  (i) If $f(x)$ is a polynomial function of x then $f(x)$ is differentiable in $]a,b[$ as a polynomial function is always differentiable for all $x \in R$.
  (ii) If $f(x)$ is not a polynomial function , find $f'(x)$. If $f'(x)$ is finite, definite and real in $]a,b[$ then $f(x)$ is differentiable in $]a,b[$.
- **Rolle's Theorem:** If a function f defined on $[a,b]$ is :
  (i) continuous on $[a,b]$.
  (ii) differentiable on $]a,b[$.
  (iii) $f(a) = f(b)$.
  then $\exists$ $c \in]a,b[$ such that $f'(c) = 0$.
- Geometrically Rolle's theorem states that 'Between two points with equal ordinates on the graph of f, there exists at least one point where the tangent is parallel to x-axis'.
- Algebraically, Rolle's theorem states that 'Between two zeroes of $f(x)$ there exists at least one zero of $f'(x)$'.
- Between two consecutive zeroes of $f'(x)$ there exists atmost one zero of $f(x)$.
- **Lagrange's Mean Value Theorem:** If a function f defined on $[a,b]$ is
  (i) continuous on $[a,b]$.
  (ii) differentiable on $]a,b[$.
  then there exists at least one real number $c \in]a,b[$ such that
$$\frac{f(b) - f(a)}{b - a} = f'(c)$$
- If a function $f(x)$ satisfies the condition of mean value theorem and $f'(x) = 0$ for all $x \in]a,b[$ then $f(x)$ is constant on $[a,b]$.
- If two functions have equal derivatives at all points of $]a,b[$ then they differ only by a constant.
- If a function f is continuous on $[a,b]$, differentiable on $]a,b[$ and $f'(x) > 0$ $\forall$ $x \in]a,b[$, then f is strictly increasing function.
- If f' exists and is bounded on some interval I then f is uniformly continuous on I.
- Geometrically, Lagrange's theorem state that between two points of the graph f there exists at least one point where the tangent is parallel to the chord.
- **Cauchy's Mean Value Theorem:** If two functions f and g defined on $[a,b]$ are
  (i) continuous on $[a,b]$.
  (ii) differentiable on $]a,b[$.
  (iii) $g'(x) \neq 0$ for any $x \in]a,b[$.
  then there exists at least one $c \in]a,b[$ such that
$$\frac{f(b) - f(a)}{g(b) - g(a)} = \frac{f'(c)}{g'(c)}$$
- Lagrange's mean value theorem can be deduce by Cauchy's mean value theorem as a particular case for $g(x) = x$.
- Geometrically, Cauchy's mean value theorem states that the mean rates of increase of two functions in an interval is equal to the ratio of actual rates of increase of the functions at some points within the interval.
- When a function involves the independent in such a manner that for a certain assigned value of that variable, its value cannot be found by simply substituting that value of the variable, the function is said to take an indeterminate form.
- Important indeterminate forms are $0 \times \infty$, $\infty - \infty$, 0^0, 1^∞ and ∞^0.
- If $\lim_{x \to a} f(x) = \lim_{x \to a} g(x)$ Then $\lim_{x \to a} \frac{f(x)}{g(x)} = \lim_{x \to a} \frac{f'(x)}{g'(x)}$
- If $\lim_{x \to a} f(x) = \infty$ and $\lim_{x \to a} g(x) = \infty$, then
$$\lim_{x \to a} \frac{f(x)}{g(x)} = \lim_{x \to a} \frac{f'(x)}{g'(x)}, \text{ provided } \lim_{x \to a} \frac{f'(x)}{g'(x)} \text{ exists.}$$

- If $\lim_{x\to a} f(x) = 0$ and $\lim_{x\to a} g(x) = \infty$

$$\lim_{x\to a} [f(x).g(x)] = \lim_{x\to a} \frac{g(x)}{\frac{1}{f(x)}}.$$

- If $\lim_{x\to a} f(x) = \infty = \lim_{x\to a} g(x)$

then $\lim_{x\to a} [f(x) - g(x)] = \lim_{x\to a} \frac{\left[\frac{1}{g(x)} - \frac{1}{f(x)}\right]}{\frac{1}{f(x).g(x)}}$

- If limit of the form 0°, 1^∞, ∞°, such that

$\lim_{x\to a} .[f(x)]^{g(x)} = e^l$ where $l = \lim_{x\to a} [g(x)\cdot \log f(x)]$.

- Let P be a given point on a curve and Q be any point on it. Let Q move towards P along the curve, then the limiting position PT of the secant PQ is called the tangent to the curve and the line through P which is perpendicular to the tangent PT is called the normal of the curve.
- The value of the derivative at a point P of the curve is equal to the slope of tangent at that point to the curve.
- Equation of tangent : $y - y_1 = dy/dx(x - x_1)$
- Equation of normal : $y - y_1 = -\frac{1}{dy_1/dx_1}(x - x_1)$
- The length of the perpendicular from the origin (0, 0) on the tangent at (x_1, y_1) to the curve $y = f(x)$ is

$$\frac{-x_1(dy_1/dx_1) + y_1}{\sqrt{1 + (dy_1/dx_1)^2}}$$

- The angle of intersection of two curves at a point of intersection is the angle between the tangents to the two curves at the point.
- $\tan\theta = \frac{m_1 - m_2}{1 + m_1 m_2}$
- Condition of tangency : $m_1 = m_2$
- Condition of orthogonality : $m_1 m_2 = -1$
- Length of tangent : $y = \frac{\sqrt{1 + (dy/dx)^2}}{dy/dx}$
- Length of the normal = $y\sqrt{1 + (dy/dx)^2}$
- $p = r\sin\theta;\ \frac{1}{p^2} = \frac{1}{r^2} + \frac{1}{r^4}\left(\frac{dr}{d\theta}\right)^2;\ \frac{1}{p^2} = u^2 + \left(\frac{du}{d\theta}\right)^2, u = \frac{1}{r}$
- The relation between p and r, where r is the distance of any point on the curve from the origin and p is perpendicular from origin to the tangent at that point is called the pedal equation of the curve.
- The measure of the sharpness of the bending of a curve at a particular point is called curvature.
- The reciprocal of the curvature of the curve at any point is called the radius of curvature.
- The curvature of a circle is constant and equal to the reciprocal of its radius.
- A relation between the length of the arc s of a curve measured from a given fixed point on the curve and the angle ψ between the tangents at its extremities is called the intrinsic equation of the curve.
- A relation between perpendicular p from the origin on any tangent to a curve and angle ψ which this tangent makes with x-axis is called the tangential polar equation.
- For any point P of a curve is the point on the positive direction of the normal at P, at a distance ρ from it.
- The circle with its centre at the centre of curvature C and radius equal to ρ is called the circle of curvature of the curve at the point p.
- The length intercepted by the circle of curvature of the curve at P on a straight line drawn through P, on a striaght line drawn through P in any given direction is called chord of curvature.
- The normal at any point of a curve is the tangent to its evolute at the corresponding center of curvature.
- The length of the arc of the evolute between any two points is equal to the difference between the radii of curvature at the corresponding points of original curve.
- **Envelope:** The envelope of a one parameter family of curves is the locus of the limiting position of the points of intersection of any two members of the family when one of them tends to coincide with the other which is kept fixed.
- The envelope of a family of curves touches each member of the family and at each point is touched by some member of the family.
- The envelope of the family of curve $A\alpha^2 + B\alpha + C = 0$ is $B^2 - 4AC = 0$.
- The envelope of a family of straight lines or of conics touches each member of the family at all their common points.
- **Evolute:** Evolute of a curve is the locus of the centre of curvature for that curve.
- The evolute of a curve is the envelope of the normals to that curve.
- The normal at any point of a curve is a tangent to its evolute touching at the corresponding centre of curvature.
- The difference between the radii of curvature at any two points of a curve is equal to the length of the arc of the evolute between the two corresponding sides.
- **Involutes:** If one curve is the evolute of another, then the latter is called an involute of the former.
- Every curve has an infinite number of evolutes.
- A straight line, at a finite distance from the origin is said to be asymptotes to an infinite branch of curve, if the perpendicular distance of a point P on the branch from the straight line tends to zero as $P \to \infty$.
- The asymptotes may be parallel to either x-axis or y-axis and accordingly they are called horizontal and vertical asymptotes. If an asymptote is not parallel to y-axis, it is called an oblique asymptotes.
- To find asymptotes parallel to y-axis, equate to zero the coefficient of highest power of y in the given equation of the curve.
- To find asymptotes parallel to x-axis, equate to zero the coefficient of highest power of x in the given equation of the curve.
- A curve of degree n can never have more than n asymptotes.
- If a curve of nth degree has n asymptotes, they cut the curve in $n(n-2)$ points.

- **Point of Inflexion :** A point P on the curve is said to be the point of inflexion, if the curve in one side of P is concave and other side of P is convex.
- **Multiple Points :** A point on the curve is said to be multiple point if through this point, more than one branches of a curve pass.
- **Double Point :** A point on the curve is called a double point if through it, two branches of the curve passes.
- **Node :** A double point on a curve is said to be a node, if through this double point two branches of the curve pass which are real and having two different tangents at that point.
- **Cusp :** A double point on a curve is called a cusp if through this double point two real branches of the curve pass and have real coincident tangents at that point.
- **Conjugate Point :** A point P on the curve is said to be conjugate point if there are no real points on the curve in the neighbourhood of that point and having no real tangent at that point.
- If the powers of y in the equation of the given curve are all even, then curve is symmetric about x-axis.
- If the powers of x in the equation of the given curve are all even, then the curve is symmetric about y-axis.
- If the equation of curve remains unchanged when x is replaced by $-x$ and y is replaced by $-y$, then the given curve is symmetrical in opposite quadrants.
- If the equation of the curve remains unchanged when x and y are interchanged, then the curve is symmetrical about the line $y = x$.

Review Questions and Project Work

1. If $y = \cosh 2x$, then show that $y_n = \begin{cases} 2^n \sinh(2x) & ; \quad n \text{ is odd} \\ 2^n \cosh(2x) & ; \quad n \text{ is even} \end{cases}$

2. If $y = (x-1)^n$, prove that $x_n = y + y_1 + \frac{y_2}{2!} + \frac{y_3}{3!} + \ldots + \frac{y_n}{n!}$.

3. If $y = e^x(\sin x + \cos x)$, prove that $y_n = 2^{\frac{n+1}{2}}.e^x.\sin\left(x + (n+1)\frac{\pi}{4}\right)$

4. If $Y = sX$, $Z = tX$, all the variables being functions of x, then prove that

$$\begin{vmatrix} X & Y & Z \\ X_1 & Y_1 & Z_1 \\ X_2 & Y_2 & Z_2 \end{vmatrix} = X^3 \begin{vmatrix} s_1 & t_1 \\ s_2 & t_2 \end{vmatrix}$$

5. If $y = x(x+1)\log(x+1)^3$, prove that

$$y_n = \frac{3(-1)^{n-1}(n-3)!(2x+n)}{(x+1)^{n-1}}, \ n \geq 3.$$

6. If $f(x) = \tan x$, then show that

$$y_n(0) - {}^nC_2 y_{n-2}(0) + {}^nC_4 y_{n-4}(0) \ldots = \sin\frac{n\pi}{2}.$$

7. If $f(x) = \cot x$, then show that

$${}^nC_1 y_{n-1}(0) - {}^nC_3 y_{n-3}(0) - {}^nC_5 y_{n-5}(0) - \ldots = \cos\frac{n\pi}{2}.$$

8. By forming in two different ways the nth derivative of x^{2n}, show that

$$1 + \frac{n^2}{1^2} + \frac{n^2(n-1)^2}{1^2.2^2} + \frac{n^2(n-1)^2(n-2)^2}{1^2.2^2.3^2} + \ldots = \frac{2n!}{(n!)^2}$$

9. If $y = \sec^{-1}x$, show that

$$x(x^2-1)y_{n+2} + [(2+3n)x^2 - (n+1)]y_{n+1} + n(3n+1)xy_n + n^2(n-1)y_{n-1} = 0$$

10. If $y = A\cosh(\log x^m) + B\sinh(\log x^m)$, prove that

$$x^2 y_{n+2} + x.(2n+1)y_{n+1} + (n^2 - m^2)y_n = 0$$

11. If $x = \tan(\log y)$, prove that

$$(1+x^2)y_{n+2} + [2(n+1)x - 1]y_{n+1} + n(n+1)y_n = 0$$

12. If $y = A[x + \sqrt{x^2-1}]^n + B[x - \sqrt{x^2-1}]^{-n}$, prove that

$$(x^2-1)y_{n+2} + (2n+1)xy_{n+1} = 0.$$

13. If $y = x\sin x$, prove that $y_n = x\sin\left(x + \frac{n\pi}{2}\right) - n\cos\left(x + \frac{n\pi}{2}\right)$.

14. Show that for any $c \in R$, the polynoimial $f(x) = x^3 + x + c$ has exactly one real root.

15. Show that $\frac{\sin\alpha - \sin\beta}{\cos\alpha - \cos\beta} = \cot\theta, \ 0 < \alpha < \theta < \beta < \frac{\pi}{2}$.

16. Show that $\frac{\tan\theta}{\theta} > \frac{\theta}{\sin\theta}$ for $0 < \theta < \frac{\pi}{2}$.

17. Apply Lagrange's mean value theorem to the function $\log(1+x)$ to show that

$$0 < [\log(1+x)]^{-1} - x^{-1} < 1 \quad \forall x > 0$$

18. If $f(x) = 0$ has two equal roots, show that $f'(x) = 0$ has one root equal to either.

19. Show that the number θ which occur in the Taylor's theorem with Lagrange's form of remainder after n terms approaches the limit $\frac{1}{n+1}$ as $h \to 0$, provided that $f^{n+1}(x)$ is continuous and non-zero at $x = a$.

20. Use Taylor's theorem to show that

(i) $x - \frac{x^3}{6} < \sin x < x$ for $x > 0$

(ii) $x - \frac{x^3}{6} < \sin x < x - \frac{x^3}{6} + \frac{x^5}{120} \quad \forall x > 0$

21. If $f(x)$ is real valued and differentiable on R and

$$f(x+y) = \frac{f(x) + f(y)}{1 - f(x)f(y)}$$

then show that $f(x) = \tan(xf'(0))$.

22. Show that there exists an $m \in N$ such that $\forall n \geq m$ the function

$$f(x) = \frac{x^{2n+2} - \sin x - 1}{x^{2n} + 1}$$

has no zero in [1,2] even though $f(1)f(2) < 0$.

23. If $a = 0$ and $b \geq 2$, show that f defined by $f(x) = \frac{1}{|x-1|}$ when $x \neq 1$ and $f(1) = 24$ does not satisfy the conditions of Lagrange's mean value theorem. However, show that the conclusion holds true if $b > 2 + \sqrt{2}$.

24. Show that $\lim\limits_{x \to 0} \frac{(\tan^{-1}x)^2}{\log(1+x^2)} = 1$.

25. Show that $\lim_{x\to\infty} \dfrac{a^{1/x} - b^{1/x}}{\log \dfrac{x}{x-1}} = \log\left(\dfrac{a}{b}\right)$

26. Show that $\lim_{x\to a^+} \dfrac{\log(x-a)}{\log(e^x - e^a)} = 1$

27. Show that $\lim_{x\to 1} \sec\dfrac{\pi}{2x}\log x = \dfrac{2}{\pi}$

28. Show that $\lim_{x\to 0} (\cos x)^{\cot^2 x} = e^{-1/2}$

29. Show that $\lim_{x\to 0}\left(\dfrac{\sinh x}{x}\right)^{1/x^2} = e$

30. Show that $\lim_{x\to\infty} \dfrac{a_0x^n + a_1x^{n-1} + \ldots + a_n}{b_0x^m + b_1x^{m-1} + b_2x^{m-2} + \ldots + b_m}$

$$= \begin{cases} \infty; \text{ for } n > m \\ \dfrac{a_0}{b_0}; \text{ for } n = m \\ 0 \ ; \text{ for } n < m \end{cases}$$

31. Show that $\lim_{x\to 0^+} (\cos x)^{1/x^3} = 0$

32. Show that $\lim_{x\to a}\left(2 - \dfrac{x}{a}\right)^{\tan\left(\frac{\pi x}{2a}\right)} = e^{2/\pi}$

33. Show that $\lim_{x\to 0} \dfrac{\log\log(1-x^2)}{\log\log\cos x} = 1$

34. Prove that the pedal equation of the hyperbola $\dfrac{x^2}{a^2} - \dfrac{y^2}{b^2} = 1$ is $\dfrac{a^2b^2}{p^2} = r^2 - a^2 + b^2$

35. Prove that the pedal equation of the astroid $x = a\cos^3 t$, $y = a\sin^3 t$ is $r^2 = a^2 - 3p^2$.

36. For the polar curve $r = f(\theta)$, show that

$$\frac{d\phi}{d\theta} + r\,\text{cosec}^2\,\theta\frac{d^2r}{ds^2} = 0$$ (VTU–2000)

37. Find the pedal equation of the following curves :

(i) $r^m \cos m\theta = a^m$ (VTU–2004)(**Ans :** $pa^m = r^{m+1}$)

(ii) $r^m = a^m(\cos m\theta + \sin m\theta)$ (VTU–2010)
(**Ans :** $r^{m+1} = \sqrt{2}.a^m.p$)

(iii) $r = ae^{m\theta}$ (VTU–2007)(**Ans :** $(1 + m^2)p^2 = r^2$)

38. For the curve $\theta = \cos^{-1}\left(\dfrac{r}{K}\right) - \sqrt{\dfrac{K^2 - r^2}{r}}$, prove that

$r\dfrac{ds}{dr}$ = constant (VTU–2005)

39. For the curve $y = c\cosh\left(\dfrac{x}{c}\right)$, prove that

$$\frac{ds}{dx} = \cosh\left(\frac{x}{c}\right)$$

40. If the equation of a curve is given in polar co-ordinates and if $u = \dfrac{1}{r}$, prove that curvature is given by $\left(\dfrac{d^2u}{d\theta^2} + u\right)\sin^3\phi$.

41. The curve $r = ae^{\theta\cot\alpha}$ cuts any radius vector in the consecutive points $A_1, A_2, \ldots, A_n, \ldots$ If ρ_n denotes the radius of curvature at A_n, prove that

$$\frac{1}{m-n}\log\frac{\rho_m}{\rho_n} = 2\pi\cot\alpha.$$

42. Prove that for any curve $\dfrac{d^2r}{ds^2} = \dfrac{\sin^2\phi}{r} - \dfrac{\sin\phi}{\rho}$.

43. Show that at any point on the equiangular spiral $r = ae^{\theta\cot\alpha}$, $\rho = r\,\text{cosec}\,\alpha$ and show that radius of curvature sustends a right angle at the pole.

44. Show that the radius of curvature of the curve $y^2 = x^2\left(\dfrac{a+x}{a-x}\right)$ at the origin is $a\sqrt{2}$.

45. Pvove that in the curve $r^2 = a^2\sin 2\theta$

(i) the curvature varies as the radius vector.

(ii) the tangent turns three times as fast as the radius vector.

46. If ϕ be the angle which the radius vector of the curve $r = f(\theta)$ makes with tangent, prove that

$\dfrac{r}{\rho} = \sin\phi\left(1 + \dfrac{d\phi}{d\theta}\right)$, where ρ is the radius of curvature.

47. Show that for the curve $p = ae^{br}$, the chord of curvature through the pole is of constant length.

48. Show that the envelope of the family of curves

$$(y-c)^2 - \frac{2}{3}(x-c)^3 = 0$$

are $x - y = 0$ and $x - y = 2/9$.

49. Find the envelope of the family of straight lines $x\cos^n\theta + y\sin^n\theta = a$, for different vlues of θ.

(**Ans :** $x^{2/(2-n)} + y^{2/(2-n)} = a^{2/(2-n)}$)

50. If $x^{2/3} + y^{2/3} = k^{2/3}$ is the envelope of the lines $\dfrac{x}{a} + \dfrac{y}{b} = 1$. Then show that $a^2 + b^2 = k^2$.

Multiple Choice Questions

Choose the most appropriate one.

1. $\lim_{x\to 3} \dfrac{2x^2 - 7x + 3}{5x^2 - 12x - 9} =$ (GATE(ME)–2006)

(a) –1/3 (b) 5/18

(c) 1 (d) none of these

2. $\lim_{x\to 0} \dfrac{e^x - (1 + x + \frac{x^2}{2})}{x^3} =$ (GATE(ME)–2007)

(a) 1/6 (b) 1/3

(c) 0 (d) none of these

3. The value of $\lim_{x\to 8}\frac{x^{1/3}-2}{(x-8)}=$ (GATE(ME)–2008)

(a) 1/8 (b) 1/2
(c) 1/12 (d) none of these

4. The value of $\lim_{\theta\to 0}\frac{\sin\theta}{\theta}=$ (GATE(ME)–2011)

(a) 1 (b) 2
(c) 0 (d) none of these

5. $\lim_{\theta\to 0}\frac{\sin\theta/2}{\theta}=$ (GATE(EC)–2007)

(a) 1 (b) 1/2
(c) 0 (d) none of these

6. $\lim_{x\to\infty}\frac{x-\sin x}{x+\cos x}=$ (GATE(CS)–2008)

(a) 0 (b) 1
(c) ∞ (d) none of these

7. $\lim_{n\to\infty}\left(1-\frac{1}{n}\right)^{2n}=$ (GATE(CS)–2010)

(a) $1/e$ (b) $1/e^2$
(c) $e^{-1/2}$ (d) none of these

8. $\lim_{x\to 0}\left(\frac{1-\cos x}{x^2}\right)=$ (GATE(ME)–2012)

(a) 1 (b) 1/2
(c) 1/4 (d) none of these

9. A rail engine accelerates from its stationary position for 8 seconds and travels a distance of 280 m. According to the mean value theorem, the speedometer at a certain time during acceleration must read exactly : (GATE(CE)–2005)

(a) 100 Km/h (b) 126 Km/h
(c) 125 Km/h (d) none of these

10. If $x = a(\theta + \sin\theta)$ and $y = a(1 - \cos\theta)$ then $\frac{dy}{dx}=$ (GATE(ME)–2005)

(a) $\sin\frac{\theta}{2}$ (b) $\cos\frac{\theta}{2}$
(c) $\tan\frac{\theta}{2}$ (d) none of these

11. In the Taylor's series of e^x about $x = 2$, the coefficient of $(x-2)^4$ is : (GATE(ME)–2008)

(a) e^2 (b) $\frac{e^2}{4!}$
(c) $\frac{e^2}{2!}$ (d) none of these

12. A series expansion for the function $\sin\theta$ is : (GATE(ME)–2011)

(a) $\theta-\frac{\theta^3}{3!}+\frac{\theta^5}{5!}-\ldots$
(b) $1-\theta+\frac{\theta^3}{3!}-\frac{\theta^5}{5!}+\ldots$
(c) both (a) and (b) are trrue
(d) none of these

13. For the function e^{-x}, the linear approximation around $x = 2$ is : (GATE(EC)–2007)

(a) $(3-x)e^{-2}$ (b) $(3-x)e^2$
(c) $(x-3)e^2$ (d) none of these

14. Which of the following functions would have only odd powers of x in its Taylor's series expansion about the point $x = 0$? (GATE(EC)–2008)

(a) $\sin(x^3)$ (b) $\sin(x^2)$
(c) $\cos(x^3)$ (d) none of these

15. In the Taylor's series expansion of $e^x + \sin x$ about the point $x = \pi$, the coefficient of $(x-\pi)^2$ is : (GATE(EC)–2008)

(a) $\exp(\pi)$ (b) $1/2\exp(\pi)$
(c) $\exp(\pi) + 1$ (d) none of these

16. The Taylor's series expansion of $\frac{\sin x}{x-\pi}$ at $x = \pi$ is : (GATE(EC)–2009)

(a) $1+\frac{(x-\pi)^2}{3!}+\ldots$ (b) $-1-\frac{(x-\pi)^2}{3!}+\ldots$
(c) $1-\frac{(x-\pi)^2}{3!}+\ldots$ (d) none of these

17. The infinite series $1+x+\frac{x^2}{2!}+\frac{x^3}{3!}+\ldots$ corresponds to : (GATE(CE)–2012)

(a) e^x (b) $\cos x$
(c) $\sin x$ (d) none of these

ANSWERS

1. (b)	2. (a)	3. (c)	4. (a)	5. (b)	6. (b)	7. (b)	8. (b)	9. (b)	10. (c)
11. (b)	12. (a)	13. (a)	14. (a)	15. (b)	16. (b)	17. (a)			

Hint to Selected Problems

1. $\lim_{x\to 3}f(x)=\lim_{x\to 3}\left(\frac{2x^2-7x+3}{5x^2-12x-9}\right)$ $\left(\frac{0}{0}\text{ form}\right)$

$=\lim_{x\to 3}\frac{4x-7}{10x-12}$ (By L' Hospital Rule)

$=\frac{5}{18}$

2. $\lim_{x\to 0}\frac{e^x-(1+x+\frac{x^2}{2})}{x^3}$ $\left(\frac{0}{0}\text{ form}\right)$

$=\lim_{x\to 0}\frac{e^x-(1+x)}{3x^2}$ (By L' Hospital Rule)

$=\lim_{x\to 0}\frac{e^x-1}{6x}=\lim_{x\to 0}\frac{e^x}{6}=\frac{1}{6}$

3. Let $x - 8 = h$

Then $\lim\limits_{x\to 8} \dfrac{x^{1/3}-2}{x-8}$

$$= \lim_{h\to 0} \frac{(8+h)^{1/3}-2}{h} \qquad \left(\frac{0}{0}\text{ form}\right)$$

$$= \lim_{h\to 0} \frac{\frac{1}{3}(8+h)^{\frac{1}{3}-1}}{1} \qquad \text{(By L Hospital Rule)}$$

$$= \frac{1}{3}8^{2/3} = \frac{1}{12}$$

5. $\lim\limits_{\theta\to 0} \dfrac{\sin\theta/2}{\theta}$

$$= \lim_{\theta\to 0} \frac{1}{2}\left(\frac{\sin\theta/2}{\theta/2}\right) = \frac{1}{2}\lim_{\theta\to 0}\frac{\sin\theta/2}{\theta/2} = \frac{1}{2}.1 = \frac{1}{2}$$

6. $$\lim_{x\to\infty} \frac{x-\sin x}{x+\cos x} = \lim_{x\to\infty} \frac{1-\frac{\sin x}{x}}{1+\frac{\cos x}{x}} = \frac{\lim\limits_{x\to\infty}\left(1-\frac{\sin x}{x}\right)}{\lim\limits_{x\to\infty}\left(1+\frac{\cos x}{x}\right)}$$

$$= \frac{1-\lim\limits_{x\to\infty}\frac{\sin x}{x}}{1+\lim\limits_{x\to\infty}\frac{\cos x}{x}} = \frac{1-0}{1+0} = 1$$

7. $\lim\limits_{n\to\infty}\left(1-\dfrac{1}{n}\right)^{2n}$

$$= \lim_{n\to\infty}\left[\left(1-\frac{1}{n}\right)^{n}\right]^2 = \lim_{n\to\infty}\left[\left(1-\frac{1}{n}\right)\right]^2 = (e^{-1})^2 = e^{-2}$$

8. $\lim\limits_{x\to 0}\left(\dfrac{1-\cos x}{x^2}\right)$

$$= \lim_{x\to 0}\frac{\sin x}{2x} \qquad \text{(By L Hospital Rule)}$$

$$= \lim_{x\to 0}\frac{\cos x}{2} = \frac{1}{2}$$

9. Let $s(t)$ be the function which denote the position of rail engine. Clearly, $s(t)$ is continuous and differentiable function.

Then by Lagrange's mean value theorem $\exists\, t$, $0 \le t \le 8$ such that

$s'(t) = v(t)$(= Rate of change of displacement)

$$= \frac{s(8)-s(0)}{8-0} = \frac{280}{8}\times\frac{3600}{1000}\text{Km/ h}$$
$$= 126\text{ Km/ h}$$

10. It is given that $x = a(\theta + \sin\theta)$, $y = a(1 - \cos\theta)$

$$\Rightarrow \frac{dx}{d\theta} = a(1+\cos\theta), \frac{dy}{d\theta} = a\sin\theta$$

$$\therefore \frac{dy}{dx} = \frac{dy/d\theta}{dx/d\theta} = \frac{a\sin\theta}{a(1-\cos\theta)} = \frac{2a\sin\left(\frac{\theta}{2}\right)\cos\left(\frac{\theta}{2}\right)}{a.2\cos^2\left(\frac{\theta}{2}\right)} = \tan\frac{\theta}{2}$$

11. We know that $f(x) = \sum\limits_{n=0}^{\infty} b_n(x-a)^n$ where $b_n = \dfrac{f^n(a)}{n!}$

Clearly, $f^{iv}(x) = e^x$ for $f(x) = e^x$

$\Rightarrow f^{iv}(2) = e^2$

$\therefore$ Coefficient of $(x-2)^4 = b_4 = \dfrac{f^{iv}(2)}{4!} = \dfrac{e^2}{4!}$

13. By Taylor's series expansion around $x = 2$ is given by

$$f(x) = f(2) + (x-2)f'(2) + \frac{(x-2)^2}{2!}f''(2) + ...$$

For linear approximation let us take only first two terms such that

$$f(x) = f(2) + (x-2)f'(2)$$

Here, $f(x) = e^{-x} \Rightarrow f'(x) = -e^{-x}$

$$\therefore\ f(x) = e^{-2} + (x-2)(-e^{-2}) = (3-x)e^{-2}$$

15. $f(x) = e^x + \sin x$

By Taylor's series expansion, we have

$$f(x) = f(a) + (x-a)f'(a) + \frac{(x-a)^2}{2!}f''(a) + \frac{(x-a)^3}{3!}f'''(a) + ...$$

For $a = \pi$

$$f(x) = f(\pi) + (x-\pi)f'(\pi) + \frac{(x-\pi)^2}{2!}f''(\pi) + \frac{(x-\pi)^3}{3!}f'''(\pi) + ...$$

Clearly the coefficient of $(x-\pi)^2$ is $\dfrac{f''(\pi)}{2!}$.

Now, $f(x) = e^x + \sin x$

$\Rightarrow\quad f'(x) = e^x + \cos x$

$\Rightarrow\quad f''(x) = e^x - \sin x$

$\Rightarrow\quad f''(\pi) = e^\pi - \sin\pi = e^\pi$

$\therefore$ The coefficient of $(x-\pi)^2$ is given by $\dfrac{e^\pi}{2!}$.

16. Taylor's series expansion of $f(x)$ around $x = \pi$ is given by

$$f(x) = f(\pi) + \frac{(x-\pi)}{1!}f'(\pi) + \frac{(x-\pi)^2}{2!}f''(\pi) + ...$$

Here $f(\pi) = \lim\limits_{x\to\pi}\dfrac{\sin x}{x-\pi} = \lim\limits_{x\to\pi}\dfrac{\cos x}{1}$ (By L Hospital Rule)

$\therefore$ Similarly, $f'(\pi) = 0$; $f''(\pi) = -1/6...$

Hence, the required expansion is

$$f(x) = -1 + \left(\frac{-1}{6}\right)(x-\pi)^2 + ... = -1 - \frac{(x-\pi)^2}{3!} + ...$$

Self Assessment Test

1. Find the nth derivative of the following functions :

(i) $\sinh 2x \sin 4x$ (VTU–2010)

(ii) $\log(4x^2 - 1)$ (VTU–2010)

(iii) $\dfrac{x+3}{(x-1)(x+2)}$ (VTU–2009)

(iv) $\dfrac{x^2}{2x^2+7x+6}$ (VTU–2005)

(v) $\dfrac{x^n-1}{x-1}$ (MTU–2012)

(vi) $x^{n-1}\log x$ (GBTU–2012)

(vii) $e^{\sin^{-1}x}$ at $x = 0$. (UPTU–2009)

(viii) $\sin^3 x$ (UPTU Model paper–2009)

2. Find the nth derivative of the following functions :

(i) $2^x \cos^9 x$ (Mumbai–2009)

(ii) $x^2 \sin x$ at $x = 0$ (UPTU–2009)

3. If $y^{1/m} + y^{-1/m} = 2x$, prove that

$(x^2 - 1)y_{n+2} + (2n + 1)xy_{n+1} + (n^2 - m^2)y_n = 0$

(UPTU–2008, SVTU–2007, Mumbai–2007)

4. If $y = \log\left[x+\sqrt{1+x^2}\right]^2$, prove that

$(1 + x^2)y_{n+2} + (2n + 1)xy_{n+1} + n^2 y_n = 0$ (VTU–2007)

5. Verify the Rolle's theorem for the following functions :

(i) $f(x) = (x - a)^m (x - b)^n$, m, n are positive integers in $[a, b]$.

(Nagarjuna–2008, VTU–2010)

(ii) $f(x) = \dfrac{\sin x}{e^x}$ in $[0, \pi[$ (GNTU–2010)

6. If $f(x) = f(0) + kf_1(0) + \dfrac{k^2}{2!}f_2(\theta k), 0 < \theta < 1$, then find the value of θ where $k = 1$ and $f(x) = (1-x)^{5/2}$. (GBTU–2010)

7. If $0 < a < b < 1$, prove that

$\dfrac{b-a}{1+b^2} < \tan^{-1}b - \tan^{-1}a < \dfrac{b-a}{1+a^2}$

(VTU–2006, Mumbai–2009)

8. By applying mean value theorem to

$f(x) = \log 2 . \sin\dfrac{\pi x}{2} + \log x$, prove that $\dfrac{\pi}{2}\log 2 . \cos\dfrac{\pi x}{2} + \dfrac{1}{x} = 0$

for some x between 1 and 2.

9. If x is positive, show that $x > \log(1 + x) > x - \dfrac{x^2}{2}$.

(VTU–2000)

10. If $f(x) = \sin^{-1}x$, $0 < a < b < 1$, use mean value theorem to prove that $\dfrac{b-a}{\sqrt{1-a^2}} < \sin^{-1}b - \sin^{-1}a < \dfrac{b-a}{\sqrt{1-b^2}}$.

11. Prove that $\dfrac{b-a}{b} < \log\left(\dfrac{b}{a}\right) < \dfrac{b-a}{b}$, for $0 < a < b$. Hence, show that $\dfrac{1}{4} < \log\dfrac{4}{3} < \dfrac{1}{3}$.

12. Expand $\log_e x$ in powers of $(x - 1)$ and hence evaluate $\log_e 1.1$ correct to four decimal places.

(Bhopal–2007, Kurukshetra–2006)

13. Prove that $x \operatorname{cosec} x = 1 + \dfrac{x^2}{6} + \dfrac{7x^4}{360} + \ldots$ (Mumbai -2007)

14. Show that the coordinates for the line $x \cos\alpha + y \sin\alpha = p$ to touch the curve $\left(\dfrac{x}{a}\right)^m + \left(\dfrac{y}{b}\right)^m = 1$ is

$(a\cos\alpha)^{m/m-1} + (b\sin\alpha)^{m/m-1} = p^{m/m-1}$.

15. If x, y be the parts of the axes x and y intercepted by the tangent at any point (x, y) on the curve $\left(\dfrac{x}{a}\right)^{2/3} + \left(\dfrac{y}{b}\right)^{2/3} = 1$ then show that $\left(\dfrac{x_1}{a}\right)^2 + \left(\dfrac{y_1}{b}\right)^2 = 1$. (Bhopal–2008)

16. Show that in the exponential curve $y = be^{x/a}$, the subtangent is of constant length and that the subnormal varies as the square of the ordinate. (Madras–2000)

17. Show that the tangent to the cardoid $r = a(1 + \cos\theta)$ at the point $\theta = \dfrac{\pi}{3}$ and $\theta = \dfrac{2\pi}{3}$ are respectively parallel and perpendicular to the initial line. (VTU–2006)

18. With the usual meaning for r, s, θ and ϕ for the polar curve $r = f(\theta)$, show that $\dfrac{d\phi}{d\theta} + r\operatorname{cosec}^2\theta\dfrac{d^2r}{ds^2} = 0.$. (VTU–2000)

19. Show that the radius of curvature at $(a, 0)$ on the curve $y^2 = \dfrac{a^2(a-x)}{x}$ is $\dfrac{a}{2}$. (VTU–2000)

20. Show that the radius of curvature at the point t on the curve $x = e^t \cos t, y = e^t \sin t$ is $\sqrt{2}e^t$. (Calicut–2005)

21. Show that the circle of convergence at the point $\left(\dfrac{3}{2}, \dfrac{3}{2}\right)$ of the curve $x^3 + y^3 = 3xy$ is $x^2 + y^2 - \dfrac{21}{8}(x+y) + \dfrac{432}{128} = 0$

(Anna–2009, Madras–2006, Calicut–2005)

22. Show that the circle of curvature at the origin for the curve $x + y = ax^2 + by^2 + cx^3$ is $(a + b)(x^2 + y^2) = 2(x + y)$.

(Nagpur–2009)

23. Find the asymptotes of the following curves :

(i) $(x^2 - a^2)(y^2 - b^2) = a^2b^2$ (Osmania–2002)

(ii) $x^2(x - y)^2 - a^2(x^2 + y^2) = 0$ (Kurukshetra–2006)

(iii) $r = a\tan\theta$ (Rohtak–2006)

(iv) $r\sin\theta = 2\cos 2\theta$ (Kurukshetra–2009)

24. Choose the most appropriate answer :

(i) The curve $r = a/(1 + \cos\theta)$ intersect orthogonally with the curve :

a. $r = \dfrac{b}{1-\cos\theta}$ b. $r = \dfrac{b}{1+\sin\theta}$

c. $r = \dfrac{b}{1+\sin^2\theta}$ c. $r = \dfrac{b}{1+\cos^2\theta}$ (VTU–2010)

(ii) If $f(x)$ is continuous in the closed interval $[a, b]$, differentiable in $]a, b[$ and $f(a) = f(b)$ then $\exists\, c \in\,]a, b[$ such that $f'(c) =$

a. 1 b. 0

c. –1 d. none of these (VTU–2009)

(iii) If the angle between the radius vector and the tangent is constant then the curve is

a. $r = a\cos\theta$ b. $r^2 = a^2\cos^2\theta$

c. $r = ae^{5\theta}$ d. none of these (VTU–2009)

ANSWERS

1. (i) $\frac{20^{n/2}}{2}(e^{2x}\sin(2x+n\tan^{-1}2)) - e^{-2x}(4x-n\tan^{-1}2)$ (ii) $(-1)^{n-1}(n-1)!2^n[(2x+1)^{-n}+(2x-1)^{-n}]$

(iii) $\frac{(-1)^n n!}{3}\left[\frac{4}{(x-1)^{n+1}}-\frac{1}{(x+2)^{n+1}}\right]$ (iv) $(-1)^n n!\left[\frac{9(2)^{n-1}}{(2x+3)^{n+1}}-\frac{4}{(x+2)^{n+1}}\right]$ (v) 0 (vi) $\frac{(n-1)!}{x}$

(vii) $y_n(0) = \begin{cases} 1^2.(1^2+2^2).(1^2+4^2)...[1^2+(n-2)^2], & \text{if } n \text{ is even} \\ 1^2.(1^2+1^2).(1^2+3^2)...[1^2+(n-2)^2], & \text{if } n \text{ is odd}\end{cases}$ (viii) $\frac{3}{4}\sin\left(x+\frac{n\pi}{2}\right)-\frac{1}{4}(3)^n\sin\left(3x+\frac{n\pi}{2}\right)$

2.(i) $\frac{1}{256}\Big[(\log 2)^n.2^x(\cos 9\theta+9\cos 7\theta+3\cos 5\theta+84\cos 3\theta+126\cos\theta)+{}^nC_1(\log 2)^{n-1}.2^x\left(\cos 9\theta+\frac{\pi}{2}\right)+9\cos\left(7\theta+\frac{n\pi}{2}\right)+36\cos\left(5\theta+\frac{\pi}{2}\right)$
$+84\cos\left(3\theta+\frac{\pi}{2}\right)+126\cos\left(\theta+\frac{\pi}{2}\right)+...+2^x\left(\cos 9\theta+\frac{n\pi}{2}\right)+9\cos\left(7\theta+\frac{n\pi}{2}\right)+36\cos\left(5\theta+\frac{n\pi}{2}\right)+84\cos\left(3\theta+\frac{n\pi}{2}\right)+126\cos\left(\theta+\frac{n\pi}{2}\right)\Big]$

(ii) $(n-n^2)\sin\frac{n\pi}{2}$ **12.** $\log x = (x-1)-\frac{(x-1)^2}{2}+\frac{(x-1)^3}{3}-\frac{(x-1)^4}{4}+...;0.0953$

23. (i) $x = \pm a, y = \pm b$ (ii) $x + a = 0, x - a = 0, x - y + \sqrt{2}a = 0$ (iii) $r\cos\theta = a, r\cos\theta = -a$ (iv) $r\sin\theta = 2$

24. (i) a (ii) b (iii) c

CCCCCC

DIFFERENTIAL CALCULUS-II

UNIT OUTLINES

- **Partial Differentiation**
- **Expansion of Function of Several Variables**
- **Jacobians**
- **Extrema of Function of Several Variables**
- **Error and Approximations**

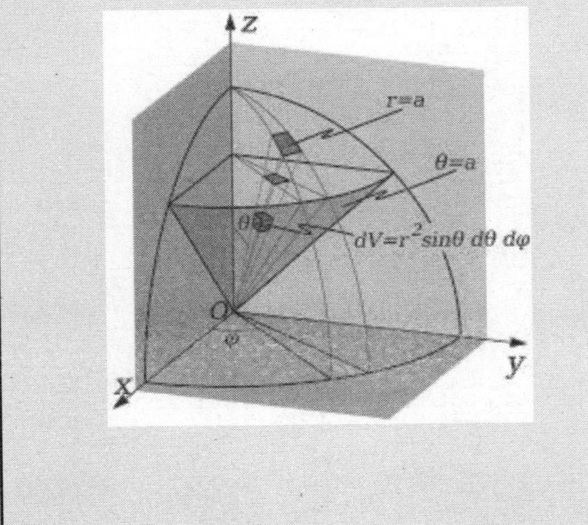

PRE-REQUISITE

- $f'(a) = \lim_{x \to a} \frac{f(x) - f(a)}{x - a}$
- $\frac{d^2y}{dx^2} = \frac{d}{dx}\left(\frac{dy}{dx}\right)$
- Continuity is a necessary condition for differentiability.
- Concepts of expansion of functions of one variable
- Concepts of maxima and minima of single variable

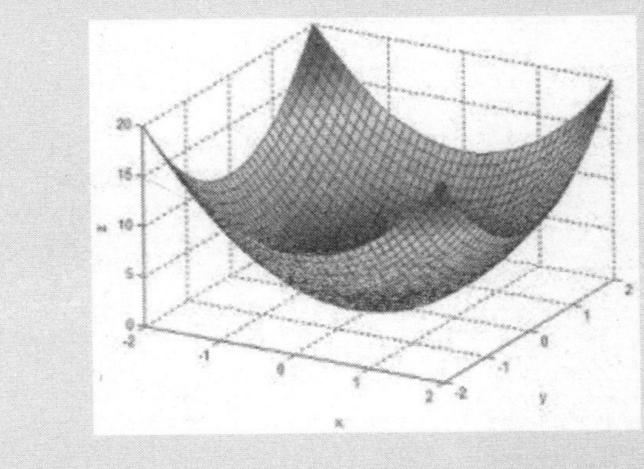

CHAPTER 9 Partial Differentiation

9.1 INTRODUCTION

We know that the differential coefficient of $f(x)$ with respect to x is $\lim_{\delta x \to 0} \frac{f(x+\delta x)-f(x)}{\delta x}$, provided this limit exists, and it is denoted by

$$f'(x) \quad \text{or} \quad \frac{d}{dx}[f(x)]$$

If $u = f(x, y)$ be a continuous function of two independent variables x and y, then the differential coefficient of u w.r.t. x (regarding y as constant) is called the partial derivative or partial differential co-efficient of u w.r.t. x and is denoted by various symbols such as

$$\frac{\partial u}{\partial x}, \frac{\partial f}{\partial x}, f_x(x, y), f_x$$

Symbolically, if $u = f(x, y)$, then $\lim_{\delta x \to 0} \frac{f(x+\delta x, y)-f(x,y)}{\delta x}$

if it exists, is called the partial derivative or partial differential co-efficient of u w.r.t. x and is denoted by

$$\frac{\partial u}{\partial x} \quad \text{or} \quad \frac{\partial f}{\partial x} \quad \text{or} \quad f_x \quad \text{or} \quad u_x.$$

Similarly, by keeping x constant and allowing y alone to vary, we can define the partial derivative or partial differential coefficient of u w.r.t. y. It is denoted by any one of the symbols $\frac{\partial u}{\partial y}, \frac{\partial f}{\partial y}, f_y(x, y), f_y$.

Symbolically,
$$\frac{\partial u}{\partial y} = \lim_{\delta y \to 0} \frac{f(x, y+\delta y)-f(x,y)}{\delta y}$$
provided this limit exists.

For Example :

If $u = ax^2 + 2hxy + by^2$ then $\frac{\partial u}{\partial x} = 2ax + 2hy$ and $\frac{\partial u}{\partial y} = 2hx + 2by$.

9.2 RULES OF PARTIAL DIFFERENTIATION

Rule (1) :

(a) If u is a function of x, y and we are to differentiate partially w.r.t. x then, y is treated as constant.

(b) Similarly, if we are to differentiate u partially w.r.t. y then x is treated as constant.

(c) If u is a function of x, y, z and we are to differentiate partially w.r.t. x, then y and z are treated as constant.

Rule (2) : If $z = u \pm v$, where u and v are functions of x and y, then

$$\frac{\partial z}{\partial x} = \frac{\partial u}{\partial x} \pm \frac{\partial v}{\partial x} \quad \text{and} \quad \frac{\partial z}{\partial y} = \frac{\partial u}{\partial y} \pm \frac{\partial v}{\partial y}.$$

Rule (3) : If $z = uv$, where u and v are functions of x and y, then

$$\frac{\partial z}{\partial x} = \frac{\partial}{\partial x}(uv) = u\frac{\partial v}{\partial x} + v\frac{\partial u}{\partial x} \quad \text{and} \quad \frac{\partial z}{\partial y} = \frac{\partial}{\partial y}(uv) = u\frac{\partial v}{\partial y} + v\frac{\partial u}{\partial y}.$$

Rule (4) : If $z = \frac{u}{v}$, where u, v are functions of x and y, then

$$\frac{\partial z}{\partial x} = \frac{\partial}{\partial x}\left(\frac{u}{v}\right) = \frac{v\frac{\partial u}{\partial x} - u\frac{\partial v}{\partial x}}{v^2} \quad \text{and} \quad \frac{\partial z}{\partial y} = \frac{\partial}{\partial y}\left(\frac{u}{v}\right) = \frac{v\frac{\partial u}{\partial y} - u\frac{\partial v}{\partial y}}{v^2}.$$

Rule (5) : If $z = f(u)$, where u is a function of x and y, then

$$\frac{\partial z}{\partial x} = \frac{\partial z}{\partial u}\cdot\frac{\partial u}{\partial x} \quad \text{and} \quad \frac{\partial z}{\partial y} = \frac{\partial z}{\partial u}\cdot\frac{\partial u}{\partial y}.$$

Remarks

- Partial means a 'part of'.
- If z is a function of one variable x, then $\frac{\partial z}{\partial x} = \frac{dz}{dx}$.
- If z is a function of two variables x_1 and x_2, we get $\frac{\partial z}{\partial x_1}$ and $\frac{\partial z}{\partial x_2}$.
- If z is a function of n variables $x_1, x_2, \ldots, x_n$ we can find $\frac{\partial z}{\partial x_1}, \frac{\partial z}{\partial x_2}, \ldots, \frac{\partial z}{\partial x_n}$.

Symmetric Function of *x* and *y*. A function $u = u(x, y)$ is said to be symmetric if, on interchanging x and y, u remains unchanged.

9.3 PARTIAL DERIVATIVES OF THE HIGHER ORDER

We can find partial derivative of $\frac{\partial u}{\partial x}$ and $\frac{\partial u}{\partial y}$ just as we found those of u for $\frac{\partial u}{\partial x}$ and $\frac{\partial u}{\partial y}$ are itself functions of x and y.

The four derivatives, thus obtained, called the second order partial derivatives of u or $f(x, y)$ are

$$\frac{\partial}{\partial x}\left(\frac{\partial u}{\partial x}\right), \frac{\partial}{\partial y}\left(\frac{\partial u}{\partial x}\right), \frac{\partial}{\partial x}\left(\frac{\partial u}{\partial y}\right), \frac{\partial}{\partial y}\left(\frac{\partial u}{\partial y}\right)$$

and are denoted as

$$\frac{\partial^2 u}{\partial x^2}, \frac{\partial^2 u}{\partial y \partial x}, \frac{\partial^2 u}{\partial x \partial y}, \frac{\partial^2 u}{\partial y^2}$$

or $\qquad f_{xx}, f_{yx}, f_{xy}, f_{yy}$.

Recapilatutions

- $\frac{\partial u}{\partial x} = \lim_{\delta x \to 0} \frac{f(x+\delta x, y) - f(x, y)}{\delta x}$
- $\frac{\partial u}{\partial y} = \lim_{\delta y \to 0} \frac{f(x, y+\delta y) - f(x, y)}{\delta y}$
- $\frac{\partial^2 u}{\partial x^2} = \frac{\partial}{\partial x}\left(\frac{\partial u}{\partial x}\right)$
- $\frac{\partial^2 u}{\partial y^2} = \frac{\partial}{\partial y}\left(\frac{\partial u}{\partial y}\right)$
- $\frac{\partial^2 u}{\partial x \partial y} = \frac{\partial^2 u}{\partial y \partial x}$

Remarks

- $\frac{\partial^2 u}{\partial x \partial y} = \frac{\partial}{\partial x}\left(\frac{\partial u}{\partial y}\right)$ and $\frac{\partial^2 u}{\partial y \partial x} = \frac{\partial}{\partial y}\left(\frac{\partial u}{\partial x}\right)$
- $\frac{\partial^2 u}{\partial x \partial y} \neq \frac{\partial u}{\partial x}\cdot\frac{\partial u}{\partial y}$
- The partial derivatives $\frac{\partial^2 u}{\partial x \partial y}$ and $\frac{\partial^2 u}{\partial y \partial x}$ are distinguished by the order in which u is successively differentiated by the order in which u is successively differntiated w.r.t. x and y, but it will be seen that , in general, that are equal.

Solved Examples

Example 1. *Verify that* $\frac{\partial^2 u}{\partial x \partial y} = \frac{\partial^2 u}{\partial y \partial x}$, *where* $u = x\sin y + y\sin x$.

Solution . We have $u = x \sin y + y \sin x$. ...(1)

Differentiating partially both sides of (1) *w.r.t.* x and y respectively, we get

$$\frac{\partial u}{\partial x} = \sin y + y\cos x \qquad \ldots(2)$$

and $\frac{\partial u}{\partial y} = x\cos y + \sin x$. ...(3)

Again differentiating (2) partially *w.r.t.* y and (3) *w.r.t.* x, we get

$$\frac{\partial^2 u}{\partial y \partial x} = \cos y + \cos x \qquad \ldots(4)$$

and $\dfrac{\partial^2 u}{\partial x \partial y} = \cos y + \cos x.$ ...(5)

Form (4) and (5), we obtain

$$\frac{\partial^2 u}{\partial y \partial x} = \frac{\partial^2 u}{\partial x \partial y}.$$

Example 2. *If* $u = x^2y + y^2z + z^2x$, *then show that* $\dfrac{\partial u}{\partial x} + \dfrac{\partial u}{\partial y} + \dfrac{\partial u}{\partial z} = (x+y+z)^2.$ (UPTU(AG)–2006)

Solution . Given that $u = x^2y + y^2z + z^2x.$ ...(1)

Differentiating partially both sides of (1) *w.r.t.* x, y and z respectively, we get

$$\frac{\partial u}{\partial x} = 2xy + z^2 \quad \text{...(2)}$$

$$\frac{\partial u}{\partial y} = x^2 + 2yz \quad \text{...(3)}$$

and $\dfrac{\partial u}{\partial z} = y^2 + 2zx.$ ...(4)

Adding (2), (3) and (4), we get

$$\frac{\partial u}{\partial x} + \frac{\partial u}{\partial y} + \frac{\partial u}{\partial z} = 2xy + z^2 + x^2 + 2yz + y^2 + 2zx$$
$$= x^2 + y^2 + z^2 + 2xy + 2yz + 2zx$$
$$= (x+y+z)^2.$$

Example 3. *If* $z = f(x + ay) + \phi(x - ay)$, *prove that*

$$\frac{\partial^2 z}{\partial y^2} = a^2 \frac{\partial^2 z}{\partial x^2}.$$ (SRM-2006)

Solution . Given that $z = f(x + ay) + \phi(x - ay).$ ...(1)

Differentiating partially both sides of (1) w.r.t. x and y respectively, we get

$$\frac{\partial z}{\partial x} = f'(x+ay) + \phi'(x-ay) \quad \text{...(2)}$$

and $\dfrac{\partial z}{\partial y} = af'(x+ay) - a\phi'(x-ay).$ ...(3)

Again differentiating partially both sides of (2) w.r.t. x and (3) *w.r.t.* y, we get

$$\frac{\partial^2 z}{\partial x^2} = f''(x+ay) + \phi''(x-ay) \quad \text{...(4)}$$

and $\dfrac{\partial^2 z}{\partial y^2} = a^2 f''(x+ay) - a^2\phi''(x-ay).$ ...(5)

Form (4) and (5), we get

$$\frac{\partial^2 z}{\partial y^2} = a^2 \frac{\partial^2 z}{\partial x^2}.$$

Example 4. *If* $u = \log(x^3 + y^3 + z^3 - 3xyz)$, *show that*

$$\left(\frac{\partial}{\partial x} + \frac{\partial}{\partial y} + \frac{\partial}{\partial z}\right)^2 u = -\frac{9}{(x+y+z)^2}.$$

(UKTU–2011, PTU–2010, Anna–2007, UPTU–2006, Bhopal–2008, VTU–2003)

Solution . We have $u = \log(x^3 + y^3 + z^3 - 3xyz)$

Differentiating partially with respect to x, we have

$$\frac{\partial u}{\partial x} = \frac{1}{x^3 + y^3 + z^3 - 3xyz}(3x^2 - 3yz)$$

$$\Rightarrow \frac{\partial u}{\partial x} = \frac{3(x^2 - yz)}{x^3 + y^3 + z^3 - 3xyz}. \quad \text{...(1)}$$

Similarly,

$$\frac{\partial u}{\partial y} = \frac{3(y^2 - zx)}{x^3 + y^3 + z^3 - 3xyz} \quad \text{...(2)}$$

and $\dfrac{\partial u}{\partial z} = \dfrac{3(z^2 - xy)}{x^3 + y^3 + z^3 - 3xyz}$ ...(3)

Adding (1), (2) and (3), we get

$$\frac{\partial u}{\partial x} + \frac{\partial u}{\partial y} + \frac{\partial u}{\partial z} = \frac{3(x^2 + y^2 + z^2 - yz - zx - xy)}{x^3 + y^3 + z^3 - 3xyz}$$
$$= \frac{3(x^2 + y^2 + z^2 - yz - zx - xy)}{(x+y+z)(x^2 + y^2 + z^2 - yz - zx - xy)}$$
$$= \frac{3}{(x+y+z)}.$$

Also,

$$\left(\frac{\partial}{\partial x} + \frac{\partial}{\partial y} + \frac{\partial}{\partial z}\right)^2 u$$
$$= \left(\frac{\partial}{\partial x} + \frac{\partial}{\partial y} + \frac{\partial}{\partial z}\right)\left(\frac{\partial}{\partial x} + \frac{\partial}{\partial y} + \frac{\partial}{\partial z}\right)u$$
$$= \left(\frac{\partial}{\partial x} + \frac{\partial}{\partial y} + \frac{\partial}{\partial z}\right)\left(\frac{\partial u}{\partial x} + \frac{\partial u}{\partial y} + \frac{\partial u}{\partial z}\right)$$
$$= \left(\frac{\partial}{\partial x} + \frac{\partial}{\partial y} + \frac{\partial}{\partial z}\right)\left(\frac{3}{x+y+z}\right)$$
$$= 3\left[\frac{\partial}{\partial x}\left(\frac{1}{x+y+z}\right) + \frac{\partial}{\partial y}\left(\frac{1}{x+y+z}\right) + \frac{\partial}{\partial z}\left(\frac{1}{x+y+z}\right)\right]$$
$$= 3\left[-\frac{1}{(x+y+z)^2} - \frac{1}{(x+y+z)^2} - \frac{1}{(x+y+z)^2}\right]$$
$$= -\frac{9}{(x+y+z)^2}$$

Example 5. *If* $u = \sin^{-1}\dfrac{x}{y} + \tan^{-1}\dfrac{y}{x}$, *show that*

$$x\frac{\partial u}{\partial x} + y\frac{\partial u}{\partial y} = 0.$$ (UPTU–2007)

Solution . We have

$$u = \sin^{-1}\frac{x}{y} + \tan^{-1}\frac{y}{x}$$

$$\Rightarrow \frac{\partial u}{\partial x} = \frac{1}{\sqrt{\left(1-\left(\frac{x}{y}\right)^2\right)}}\cdot\frac{1}{y} + \frac{1}{1+\left(\frac{y}{x}\right)^2}\cdot\left(-\frac{y}{x^2}\right)$$

$$= \frac{1}{\sqrt{y^2-x^2}} - \frac{y}{(x^2+y^2)}$$

$$\Rightarrow x\frac{\partial u}{\partial x} = \frac{x}{\sqrt{(y^2-x^2)}} - \frac{xy}{x^2+y^2} \quad \ldots(1)$$

Also,

$$\frac{\partial u}{\partial y} = \frac{1}{\sqrt{\left(1-\left(\frac{x}{y}\right)^2\right)}}\cdot\left(-\frac{x}{y^2}\right) + \frac{1}{1+\left(\frac{y}{x}\right)^2}\cdot\left(\frac{1}{x}\right)$$

$$= -\frac{x}{y\sqrt{y^2-x^2}} + \frac{x}{x^2+y^2}$$

$$\Rightarrow y\frac{\partial u}{\partial y} = -\frac{x}{\sqrt{(y^2-x^2)}} + \frac{xy}{x^2+y^2} \quad ..(2)$$

On adding (1) and (2), we get $x\frac{\partial u}{\partial x} + y\frac{\partial u}{\partial y} = 0.$

Example 6. *If $u = f(r)$, where $r^2 = x^2 + y^2$, show that*

$$\frac{\partial^2 u}{\partial x^2} + \frac{\partial^2 u}{\partial y^2} = f''(r) + \frac{1}{r}f'(r).$$

(UPTU–2005, SVTU–2008, Rajasthan–2006)

Solution . We have $r^2 = x^2 + y^2$

$$\left.\begin{aligned} \Rightarrow \quad & 2r\frac{\partial r}{\partial x} = 2x \text{ or } \frac{\partial r}{\partial x} = \frac{x}{r} \\ \text{and} \quad & 2r\frac{\partial r}{\partial y} = 2y \text{ or } \frac{\partial r}{\partial y} = \frac{y}{r} \end{aligned}\right] \quad \ldots(1)$$

Since, $u = f(r)$

$$\Rightarrow \frac{\partial u}{\partial x} = [f'(r)]\cdot\frac{\partial r}{\partial x} = \frac{x}{r}f'(r)$$

and

$$\frac{\partial^2 u}{\partial x^2} = \frac{\partial}{\partial x}\left(\frac{\partial u}{\partial x}\right) = \frac{\partial}{\partial x}\left[x\cdot\frac{1}{r}f'(r)\right]$$

$$= 1\cdot\frac{1}{r}\cdot f'(r) + [xf'(r)]\left[-\frac{1}{r^2}\frac{\partial r}{\partial x}\right] + \frac{x}{r}[f''(r)]\frac{\partial r}{\partial x}$$

$$= \frac{1}{r}\cdot f'(r) - \frac{x}{r^2}\cdot\frac{x}{r}f'(r) + \frac{x^2}{r^2}f''(r)$$

$$= \frac{1}{r}\cdot f'(r) - \frac{x^2}{r^3}f'(r) + \frac{x^2}{r^2}f''(r). \quad \ldots(2)$$

Similarly, we may get

$$\frac{\partial^2 u}{\partial y^2} = \frac{1}{r}\cdot f'(r) - \frac{y^2}{r^3}f'(r) + \frac{y^2}{r^2}f''(r) \quad \ldots(3)$$

Adding (2) and (3), we get

$$\frac{\partial^2 u}{\partial x^2} + \frac{\partial^2 u}{\partial y^2} = \frac{2}{r}\cdot f'(r) - \frac{x^2+y^2}{r^3}f'(r) + \frac{x^2+y^2}{r^2}f''(r)$$

$$= \frac{2}{r}\cdot f'(r) - \frac{r^2}{r^3}f'(r) + \frac{r^2}{r^2}f''(r)$$

$$= \frac{2}{r}\cdot f'(r) - \frac{1}{r}f'(r) + f''(r)$$

$$= f''(r) + \frac{1}{r}\cdot f'(r).$$

Example 7. *If $x^x y^y z^z = c$. Show that at $x = y = z$,*

$$\frac{\partial^2 z}{\partial x \partial y} = -[x\log ex]^{-1}$$

(Bhopal–2008)

Solution . We have $x^x y^y z^z = c.$ ...(1)

We observe that z can be regarding as a function of two independent variables x and y.

Taking log of both sides of (1), we have

$$x\log x + y\log y + z\log z = \log c. \quad \ldots(2)$$

Differentiating (2) partially w.r.t. x, we get

$$x\cdot\frac{1}{x} + 1\cdot\log x + \left[z\cdot\frac{1}{z} + 1\cdot\log z\right]\frac{\partial z}{\partial x} = 0$$

$$\Rightarrow \frac{\partial z}{\partial x} = -\frac{(1+\log x)}{(1+\log z)}. \quad \ldots(3)$$

Similarly differentiating (2), w.r.t. y, we get

$$\frac{\partial z}{\partial y} = -\frac{(1+\log y)}{(1+\log z)}. \quad \ldots(4)$$

Also,

$$\frac{\partial^2 z}{\partial x\partial y} = \frac{\partial}{\partial x}\left(\frac{\partial z}{\partial y}\right) = \frac{\partial}{\partial x}\left[-\left(\frac{1+\log y}{1+\log z}\right)\right]$$

$$= -(1+\log y)\frac{\partial}{\partial x}[(1+\log z)^{-1}]$$

$$= -(1+\log y)\cdot\left[-(1+\log z)^{-2}\frac{1}{z}\cdot\frac{\partial z}{\partial x}\right]$$

$$= \frac{(1+\log y)}{z(1+\log z)^2}\cdot\left[-\left(\frac{1+\log x}{1+\log z}\right)\right].$$

For $x = y = z$, we have

$$\frac{\partial^2 z}{\partial x\partial y} = -\frac{(1+\log x)^2}{x(1+\log x)^3} = -\frac{1}{x(1+\log x)}$$

$$= \frac{-1}{x[\log e + \log x]} \quad [\because \log e = 1]$$

$$= \frac{-1}{x\log(ex)} = -[x\log(ex)]^{-1}.$$

Example 8. *If $u = f\left(\frac{y}{x}\right)$, show that $x\frac{\partial u}{\partial x} + y\frac{\partial u}{\partial y} = 0.$*

Solution . We have $u = f\left(\frac{y}{x}\right)$ ...(1)

Differentiating (1) partially w.r.t. x and y respectively, we get

$$\frac{\partial u}{\partial x} = f'\left(\frac{y}{x}\right)\cdot\left(-\frac{y}{x^2}\right)$$

$$\Rightarrow x\frac{\partial u}{\partial x} = -\frac{y}{x}f'\left(\frac{y}{x}\right) \quad \ldots(2)$$

and

$$\frac{\partial u}{\partial y} = f'\left(\frac{y}{x}\right)\cdot\frac{1}{x}$$

$$\Rightarrow \quad y\frac{\partial u}{\partial y} = \frac{y}{x} f'\left(\frac{y}{x}\right) \quad \ldots(3)$$

Adding (2) and (3), we get $x\dfrac{\partial u}{\partial x} + y\dfrac{\partial u}{\partial y} = 0.$

Example 9. *If* $u = (1 - 2xy + y^2)^{-1/2}$, *prove that*

$$\frac{\partial}{\partial x}\left\{(1 - x^2)\frac{\partial u}{\partial x}\right\} + \frac{\partial}{\partial y}\left\{y^2\frac{\partial u}{\partial y}\right\} = 0.$$

(Rohtak–2006)

Solution. We have $u = (1 - 2xy + y^2)^{-1/2}$ ...(1)

Differentiating (1) partially with respect to x, we get

$$\frac{\partial u}{\partial x} = -\frac{1}{2}(1 - 2xy + y^2)^{-3/2}(-2y)$$

or
$$\frac{\partial u}{\partial x} = y(1 - 2xy + y^2)^{-3/2}$$

$$\Rightarrow \quad (1 - x^2)\frac{\partial u}{\partial x} = y(1 - x^2)(1 - 2xy + y^2)^{-3/2}$$

Again differentiating partially *w.r.t.* x, we get

$$\frac{\partial}{\partial x}\left\{(1 - x^2)\frac{\partial u}{\partial x}\right\} = y\begin{bmatrix} -2x(1 - 2xy + y^2)^{-3/2} \\ +(1 - x^2)\left(-\dfrac{3}{2}\right)(-2y)(1 - 2xy + y^2)^{-5/2} \end{bmatrix}$$

$$= -2xy(1 - 2xy + y^2)^{-3/2} + 3y^2(1 - x^2)(1 - 2xy + y^2)^{-5/2}$$

$$\therefore \quad \frac{\partial}{\partial x}\left\{(1 - x^2)\frac{\partial u}{\partial x}\right\} = -2xyu^3 + 3y^2(1 - x^2)u^5$$

[Using (1)] ...(2)

Differentiating (1) partially *w.r.t.* y, we get

$$\frac{\partial u}{\partial y} = -\frac{1}{2}(1 - 2xy + y^2)^{-3/2}(-2x + 2y)$$

or
$$\frac{\partial u}{\partial y} = (x - y)(1 - 2xy + y^2)^{-3/2}$$

$$\Rightarrow \quad y^2\frac{\partial u}{\partial y} = (x - y)y^2(1 - 2xy + y^2)^{-3/2}$$

Again differentiating partially *w.r.t.* y, we get

$$\frac{\partial}{\partial y}\left(y^2\frac{\partial u}{\partial y}\right) = (2xy - 3y^2)(1 - 2xy + y^2)^{-3/2} + (xy^2 - y^3)\left(-\frac{3}{2}\right)(-2x + 2y)(1 - 2xy + y^2)^{-5/2}$$

$$= 2xy(1 - 2xy + y^2)^{-3/2} - 3y^2(1 - 2xy + y^2)^{-3/2} + 3y^2(x - y)^2(1 - 2xy + y^2)^{-5/2}$$

$$= 2xy(1 - 2xy + y^2)^{-3/2} - 3y^2(1 - 2xy + y^2)^{-5/2}\{(1 - 2xy + y^2) - (x - y)^2\}$$

$$= 2xy(1 - 2xy + y^2)^{-3/2} - 3y^2(1 - x^2)(1 - 2xy + y^2)^{-5/2}$$

$$\therefore \quad \frac{\partial}{\partial y}\left\{y^2\frac{\partial u}{\partial y}\right\} = 2xyu^3 - 3y^2(1 - x^2)u^5$$

[Using (1)] ...(3)

Adding (2) and (3), we get

$$\frac{\partial}{\partial x}\left\{(1 - x^2)\frac{\partial u}{\partial x}\right\} + \frac{\partial}{\partial y}\left\{y^2\frac{\partial u}{\partial y}\right\} = 0.$$

Example 10. *If* $u = (x^2 + y^2 + z^2)^{-1/2}$, *show that*

(i) $x\dfrac{\partial u}{\partial x} + y\dfrac{\partial u}{\partial y} + z\dfrac{\partial u}{\partial z} = -u$

(ii) $\dfrac{\partial^2 u}{\partial x^2} + \dfrac{\partial^2 u}{\partial y^2} + \dfrac{\partial^2 u}{\partial z^2} = 0$

(GBTU–2011, VTU–2006, Osmania–2003)

Solution. (i) We have $u = (x^2 + y^2 + z^2)^{-1/2}$ ...(1)

Differentiating (1) partially *w.r.t.* x, y and z respectively, we get

$$\frac{\partial u}{\partial x} = \left(-\frac{1}{2}\right)(x^2 + y^2 + z^2)^{-3/2}(2x)$$

or
$$\frac{\partial u}{\partial x} = \frac{-x}{(x^2 + y^2 + z^2)^{3/2}}$$

$$\Rightarrow \quad x\frac{\partial u}{\partial x} = \frac{-x^2}{(x^2 + y^2 + z^2)^{3/2}} \quad \ldots(2)$$

Similarly,

$$y\frac{\partial u}{\partial y} = \frac{-y^2}{(x^2 + y^2 + z^2)^{3/2}} \quad \ldots(3)$$

and

$$z\frac{\partial u}{\partial z} = \frac{-z^2}{(x^2 + y^2 + z^2)^{3/2}} \quad \ldots(4)$$

Adding (2), (3) and (4), we get

$$x\frac{\partial u}{\partial x} + y\frac{\partial u}{\partial y} + z\frac{\partial u}{\partial z} = \frac{-(x^2 + y^2 + z^2)}{(x^2 + y^2 + z^2)^{3/2}}$$

$$= -(x^2 + y^2 + z^2)^{-1/2}$$

$$\therefore \quad x\frac{\partial u}{\partial x} + y\frac{\partial u}{\partial y} + z\frac{\partial u}{\partial z} = -u$$

(ii) We have

$$\frac{\partial^2 u}{\partial x^2} = \frac{\partial}{\partial x}\left(\frac{\partial u}{\partial x}\right)$$

$$= \frac{\partial}{\partial x}\left\{\frac{-x}{(x^2 + y^2 + z^2)^{3/2}}\right\}$$

$$= -\left[\frac{1}{(x^2 + y^2 + z^2)^{3/2}} + x\left\{\left(-\frac{3}{2}\right)(2x)(x^2 + y^2 + z^2)^{-5/2}\right\}\right]$$

$$= -\left[\frac{1}{(x^2 + y^2 + z^2)^{3/2}} - \frac{3x^2}{(x^2 + y^2 + z^2)^{5/2}}\right]$$

$$= -\frac{(y^2 + z^2 - 2x^2)}{(x^2 + y^2 + z^2)^{5/2}}$$

$$\frac{\partial^2 u}{\partial x^2} = \frac{2x^2 - y^2 - z^2}{(x^2+y^2+z^2)^{5/2}} \quad \ldots(5)$$

Similarly,

$$\frac{\partial^2 u}{\partial y^2} = \frac{2y^2 - x^2 - z^2}{(x^2+y^2+z^2)^{5/2}} \quad \ldots(6)$$

and $$\frac{\partial^2 u}{\partial z^2} = \frac{2z^2 - y^2 - x^2}{(x^2+y^2+z^2)^{5/2}} \quad \ldots(7)$$

Adding (5), (6) and (7), we get

$$\frac{\partial^2 u}{\partial x^2} + \frac{\partial^2 u}{\partial y^2} + \frac{\partial^2 u}{\partial z^2} = 0$$

Example 11. *If* $\theta = t^n e^{-r^2/4t}$, *find the value of n for which* $\frac{1}{r^2} \cdot \frac{\partial}{\partial r}\left(r^2 \frac{\partial \theta}{\partial r}\right) = \frac{\partial \theta}{\partial t}$.

(UPTU–2006, Kurukshetra–2006, Nagpur–2009)

Solution . We have $\theta = t^n e^{-r^2/4t}$ ...(1)

Then $$\frac{\partial \theta}{\partial r} = t^n \left[e^{-r^2/4t}\left(-\frac{2r}{4t}\right)\right] = -\frac{r}{2} t^{n-1} e^{-r^2/4t}$$

$$\Rightarrow r^2 \frac{\partial \theta}{\partial r} = -\frac{r^3}{2} t^{n-1} e^{-r^2/4t}$$

Now $$\frac{\partial}{\partial r}\left(r^2 \frac{\partial \theta}{\partial r}\right)$$

$$= -\frac{1}{2} t^{n-1}\left[3r^2 e^{-r^2/4t} + r^3 e^{-r^2/4t}\left(\frac{-2r}{4t}\right)\right]$$

$$= -\frac{3}{2} r^2 t^{n-1} e^{-r^2/4t} + \frac{1}{4} r^4 t^{n-2} e^{-r^2/4t}$$

$$\therefore \frac{1}{r^2} \cdot \frac{\partial}{\partial r}\left(r^2 \frac{\partial \theta}{\partial r}\right)$$

$$= -\frac{3}{2} t^{n-1} e^{-r^2/4t} + \frac{1}{4} r^2 t^{n-2} e^{-r^2/4t} \quad \ldots(2)$$

Again from (1), we get

$$\frac{\partial \theta}{\partial t} = n t^{n-1} e^{-r^2/4t} + t^n . e^{-r^2/4t} . \left(\frac{r^2}{4t^2}\right)$$

or $$\frac{\partial \theta}{\partial t} = n t^{n-1} e^{-r^2/4t} + \frac{1}{4} r^2 t^{n-2} . e^{-r^2/4t} \quad \ldots(3)$$

Since, $$\frac{1}{r^2} \cdot \frac{\partial}{\partial r}\left(r^2 \frac{\partial \theta}{\partial r}\right) = \frac{\partial \theta}{\partial t}$$

Then form (2) and (3), we have

$$-\frac{3}{2} t^{n-1} e^{-r^2/4t} + \frac{1}{4} r^2 t^{n-2} e^{-r^2/4t}$$

$$= n t^{n-1} e^{-r^2/4t} + \frac{1}{4} r^2 t^{n-2} . e^{-r^2/4t}$$

$$\Rightarrow \quad n = -\frac{3}{2}.$$

Example 12. *If* $v = (x^2 + y^2 + z^2)^{m/2}$, *show that*

$$\frac{\partial^2 v}{\partial x^2} + \frac{\partial^2 v}{\partial y^2} + \frac{\partial^2 v}{\partial z^2} = m(m+1)(x^2+y^2+z^2)^{(m-2)/2}$$

Solution . Let $r^2 = x^2 + y^2 + z^2$, then

$$\frac{\partial r}{\partial x} = \frac{x}{r}, \frac{\partial r}{\partial y} = \frac{y}{r}, \frac{\partial r}{\partial z} = \frac{z}{r}$$

Now $v = (x^2 + y^2 + z^2)^{m/2} = r^m$

Then, we have

$$\frac{\partial v}{\partial x} = m r^{m-1} \frac{\partial r}{\partial x} = m r^{m-1} . \frac{x}{r} = m x r^{m-2}$$

and $$\frac{\partial^2 v}{\partial x^2} = m(m-2) x^2 r^{m-4} + m r^{m-2}$$

Similarly,

$$\frac{\partial^2 v}{\partial y^2} = m(m-2) y^2 r^{m-4} + m r^{m-2}$$

and $$\frac{\partial^2 v}{\partial z^2} = m(m-2) z^2 r^{m-4} + m r^{m-2}$$

$$\therefore \frac{\partial^2 v}{\partial x^2} + \frac{\partial^2 v}{\partial y^2} + \frac{\partial^2 v}{\partial z^2}$$

$$= m(m-2)(x^2+y^2+z^2) r^{m-4} + 3 m r^{m-2}$$

$$= m(m-2) r^2 . r^{m-4} + 3 m r^{m-2}$$

$$= m(m+1) r^{m-2}$$

$$= m(m+1)(x^2+y^2+z^2)^{(m-2)/2}$$

Example 13. *If* $z = \sin^{-1}\frac{y}{x} + x \sin\frac{y}{x}$, *show that*

$$x^2 \frac{\partial^2 z}{\partial x^2} + 2xy \frac{\partial^2 z}{\partial x \partial y} + y^2 \frac{\partial^2 z}{\partial y^2} = 0.$$

Solution . We have $z = \sin^{-1}\frac{y}{x} + x \sin\frac{y}{x}$

Then

$$\frac{\partial z}{\partial x} = \frac{1}{\sqrt{1 - \frac{y^2}{x^2}}}\left(-\frac{y}{x^2}\right) + \sin\left(\frac{y}{x}\right) + x\cos\left(\frac{y}{x}\right).\left(-\frac{y}{x^2}\right)$$

$$= -\frac{y}{x\sqrt{x^2 - y^2}} + \sin\frac{y}{x} - \frac{y}{x}\cos\frac{y}{x} \quad \ldots(2)$$

and

$$\frac{\partial^2 z}{\partial x^2} = -y\left[-\frac{1}{x^2\sqrt{x^2-y^2}} + \frac{1}{x}\left(-\frac{x}{(x^2-y^2)^{3/2}}\right)\right]$$

$$+ \cos\frac{y}{x} . \left(-\frac{y}{x^2}\right)$$

$$-y\left[-\frac{1}{x^2}\cos\frac{y}{x} + \frac{1}{x}\left(-\sin\frac{y}{x}\right)\left(-\frac{y}{x^2}\right)\right]$$

$$= -y\left[-\frac{1}{x^2\sqrt{x^2-y^2}} - \frac{1}{(x^2-y^2)^{3/2}}\right]$$

$$-\frac{y}{x^2}\cos\frac{y}{x} - y\left[-\frac{1}{x^2}\cos\frac{y}{x} + \frac{y}{x^3}\sin\frac{y}{x}\right]$$

$$= \frac{y}{x^2\sqrt{x^2-y^2}} + \frac{y}{(x^2-y^2)^{3/2}}$$

$$-\frac{y}{x^2}\cos\frac{y}{x} + \frac{y}{x^2}\cos\frac{y}{x} - \frac{y^2}{x^3}\sin\frac{y}{x}$$

$$\therefore x^2\frac{\partial^2 z}{\partial x^2} = \frac{y}{\sqrt{x^2-y^2}}$$

$$+\frac{x^2y}{(x^2-y^2)^{3/2}} - \frac{y^2}{x}\sin\frac{y}{x} \quad \ldots(3)$$

Also $\dfrac{\partial z}{\partial y} = \dfrac{1}{\sqrt{1-\dfrac{y^2}{x^2}}}\left(\dfrac{1}{x}\right) + x\cos\dfrac{y}{x}\cdot\left(\dfrac{1}{x}\right)$

or $\dfrac{\partial z}{\partial y} = \dfrac{y}{\sqrt{x^2-y^2}} + \cos\dfrac{y}{x}$

and

$$\frac{\partial^2 z}{\partial y^2} = \left(-\frac{1}{2}\right)(x^2-y^2)^{-3/2}(-2y) - \sin\frac{y}{x}\cdot\left(\frac{1}{x}\right)$$

or $\dfrac{\partial^2 z}{\partial y^2} = \dfrac{y}{(x^2-y^2)^{3/2}} - \dfrac{1}{x}\sin\dfrac{y}{x}$

$$\therefore y^2\frac{\partial^2 z}{\partial y^2} = \frac{y^3}{(x^2-y^2)^{3/2}} - \frac{y^2}{x}\sin\frac{y}{x} \quad \ldots(4)$$

Also $\dfrac{\partial^2 z}{\partial x\partial y} = \dfrac{\partial}{\partial x}\left(\dfrac{\partial z}{\partial y}\right)$

$$= \frac{\partial}{\partial x}\left[\frac{y}{\sqrt{x^2-y^2}} + \cos\frac{y}{x}\right]$$

$$= -\frac{x}{(x^2-y^2)^{3/2}} - \sin\frac{y}{x}\cdot\left(-\frac{y}{x^2}\right)$$

$$= -\frac{x}{(x^2-y^2)^{3/2}} + \frac{y}{x^2}\sin\frac{y}{x}$$

$$\therefore 2xy\frac{\partial^2 z}{\partial x\partial y} = -\frac{2x^2y}{(x^2-y^2)^{3/2}} + \frac{2y^2}{x}\sin\frac{y}{x} \ldots(5)$$

Adding (3), (4) and (5), we get

$$x^2\frac{\partial^2 z}{\partial x^2} + 2xy\frac{\partial^2 z}{\partial x\partial y} + y^2\frac{\partial^2 z}{\partial y^2}$$

$$= \frac{y}{\sqrt{x^2-y^2}} + \frac{x^2y}{(x^2-y^2)^{3/2}} - \frac{y^2}{x}\sin\frac{y}{x}$$

$$-\frac{2x^2y}{(x^2-y^2)^{3/2}} + \frac{2y^2}{x}\sin\frac{y}{x}$$

$$+\frac{y^3}{(x^2-y^2)^{3/2}} - \frac{y^2}{x}\sin\frac{y}{x}$$

$$= \frac{y}{\sqrt{x^2-y^2}} + \frac{x^2y - 2x^2y + y^3}{(x^2-y^2)^{3/2}}$$

$$= \frac{y}{\sqrt{x^2-y^2}} + \frac{y^3 - x^2y}{(x^2-y^2)^{3/2}}$$

$$= \frac{y}{\sqrt{x^2-y^2}} - \frac{y(x^2-y^2)}{(x^2-y^2)\sqrt{x^2-y^2}}$$

$$= \frac{y}{\sqrt{x^2-y^2}} - \frac{y}{\sqrt{x^2-y^2}} = 0$$

Hence $\quad x^2\dfrac{\partial^2 z}{\partial x^2} + 2xy\dfrac{\partial^2 z}{\partial x\partial y} + y^2\dfrac{\partial^2 z}{\partial y^2} = 0.$

Example 14. *If* $z = f(x + ct) + \phi(x - ct)$, *prove that*

$$\frac{\partial^2 z}{\partial t^2} = c^2\frac{\partial^2 z}{\partial x^2}.$$

(GBTU–2011, VTU–2003, JNTU–2006)

Solution . We have $z = f(x + ct) + \phi(x - ct)$

$$\Rightarrow \quad \frac{\partial z}{\partial x} = f'(x+ct) + \phi'(x-ct)$$

and $\quad \dfrac{\partial^2 z}{\partial x^2} = f''(x+ct) + \phi''(x-ct) \quad \ldots(1)$

Also, $\quad \dfrac{\partial z}{\partial t} = cf'(x+ct) - c\phi'(x-ct)$

and $\quad \dfrac{\partial^2 z}{\partial t^2} = c^2f''(x+ct) + c^2\phi''(x-ct) \ldots(2)$

Form (1) and (2), we get

$$\frac{\partial^2 z}{\partial t^2} = c^2\frac{\partial^2 z}{\partial x^2}.$$

Example 15. *If* $u(x, y, z) = \log(\tan x + \tan y + \tan z)$, *show that* $\sin 2x\dfrac{\partial u}{\partial x} + \sin 2y\dfrac{\partial u}{\partial y} + \sin 2z\dfrac{\partial u}{\partial z} = 2.$

(UPTU–2007, MTU–2011, GBTU–2012)

Solution . Differentiating the given function partially w.r.t. x, we get

$$\frac{\partial u}{\partial x} = \frac{1}{\tan x + \tan y + \tan z}(\sec^2 x) \quad \ldots(1)$$

Similarly,

$$\frac{\partial u}{\partial y} = \frac{1}{\tan x + \tan y + \tan z}(\sec^2 y) \quad \ldots(2)$$

and

$$\frac{\partial u}{\partial z} = \frac{1}{\tan x + \tan y + \tan z}(\sec^2 z) \quad \ldots(3)$$

Now multiply (1) by sin2x, (2) by sin2y, (3) by sin2z and then adding we get

$$\sin 2x\frac{\partial u}{\partial x} + \sin 2y\frac{\partial u}{\partial y} + \sin 2z\frac{\partial u}{\partial z}$$

$$= \frac{\sin 2x\sec^2 x + \sin 2y\sec^2 y + \sin 2z\sec^2 z}{\tan x + \tan y + \tan z}$$

$$= \frac{2(\tan x + \tan y + \tan z)}{\tan x + \tan y + \tan z} = 2.$$

Example 16. *If* $z = \tan(y + ax) + (y - ax)^{3/2}$, *show that*

$$\frac{\partial^2 z}{\partial x^2} = a^2\frac{\partial^2 z}{\partial y^2}.$$

(Mumbai–2009)

Solution . We have $z = \tan(y+ax)+(y-ax)^{3/2}$

$$\Rightarrow \quad \frac{\partial z}{\partial x} = (\sec^2(y+ax)).a + \frac{3}{2}(y-ax)^{1/2}(-a)$$

and $$\frac{\partial^2 z}{\partial x^2} = 2a^2\tan(y+ax)\sec^2(y+ax) + \frac{3}{4}a^2(y-ax)^{-1/2} \quad \ldots(1)$$

Also, $$\frac{\partial z}{\partial y} = \sec^2(y+ax)+\frac{3}{2}(y-ax)^{1/2}$$

and $$\frac{\partial^2 z}{\partial y^2} = 2\sec^2(y+ax)\tan(y+ax) + \frac{3}{4}(y-ax)^{-1/2} \quad \ldots(2)$$

From (1) and (2) we conclude that

$$\frac{\partial^2 z}{\partial x^2} = a^2\left(\frac{\partial^2 z}{\partial y^2}\right).$$

EXERCISE 9.1

1. Find $\frac{\partial u}{\partial x}$ and $\frac{\partial u}{\partial y}$ when:

(i) $u = \log(x^2+y^2)$ (ii) $u = \cos^{-1}\left(\frac{x}{y}\right)$

(iii) $u = \frac{x^2}{a^2}+\frac{y^2}{b^2}-1$ (iv) $u = \tan^{-1}\left(\frac{x^2+y^2}{x+y}\right)$

2. Find the second order partial derivatives of $\log(e^x + e^y)$.

3. Verify that $\frac{\partial^2 u}{\partial x\partial y} = \frac{\partial^2 u}{\partial y\partial x}$, where

(i) $u = \log(y\sin x + x\sin y)$

(ii) $u = \log\left(\frac{x^2+y^2}{xy}\right)$ (UPTU–2008)

(iii) $u = \log\left(\frac{x^2+y^2}{x+y}\right)$ (iv) $u = \sin^{-1}\frac{x}{y}$

(v) $u = x^y$ (Anna–2009)

(vi) $u = \log\tan\left(\frac{y}{x}\right)$ (vii) $u = x^4 + x^2y^2 + y^4$

(viii) $u = \log\left(\frac{xy}{x^2+y^2}\right)$ (ix) $u = x\log y$

4. If $x = r\cos\theta$, $y = r\sin\theta$, show that $\frac{\partial r}{\partial x} = \frac{\partial x}{\partial r}, \frac{\partial x}{r\partial\theta} = r\frac{\partial\theta}{\partial x}$.

5. If $u = \log(\tan x + \tan y)$, prove that $\sin 2x\frac{\partial u}{\partial x}+\sin 2y\frac{\partial u}{\partial y} = 2$.

6. If $u = x^2\tan^{-1}\frac{y}{x} - y^2\tan^{-1}\frac{x}{y}$, prove that $\frac{\partial^2 u}{\partial x\partial y} = \frac{x^2-y^2}{x^2+y^2}$.

(UKTU–2012, Mumbai–2008, Madras–2000)

7. If $u = 2(ax + by)^2 - (x^2 + y^2)$ and $a^2 + b^2 = 1$, prove that $\frac{\partial^2 u}{\partial x^2}+\frac{\partial^2 u}{\partial y^2} = 0$.

8. If $u = \log(x^3 + y^3 - x^2y - xy^2)$, prove that

(i) $\frac{\partial u}{\partial x}+\frac{\partial u}{\partial y} = 2(x+y)^{-1}$

(ii) $\frac{\partial^2 u}{\partial x^2}+2\frac{\partial^2 u}{\partial x\partial y}+\frac{\partial^2 u}{\partial y^2} = -4(x+y)^{-2}$

9. If $u = f(x + 2y) + g(x - 2y)$, show that $4\frac{\partial^2 u}{\partial x^2} = \frac{\partial^2 u}{\partial y^2}$.

10. If $u = e^{xyz}$, show that $\frac{\partial^3 u}{\partial x\partial y\partial z} = (1+3xyz+x^2y^2z^2)e^{xyz}$.

(UPTU–2007, 2009)

11. If $u(x+y) = x^2+y^2$, show that $\left(\frac{\partial u}{\partial x}-\frac{\partial u}{\partial y}\right)^2 = 4\left(1-\frac{\partial u}{\partial x}-\frac{\partial u}{\partial y}\right)$.

(VTU–2003)

12. If $\tan u = \frac{\cos x}{\sinh y}$ and $\tanh v = \frac{\sinh x}{\cosh y}$ show that $\frac{\partial u}{\partial x} = \frac{\partial v}{\partial y}$ and $\frac{\partial u}{\partial y} = -\frac{\partial v}{\partial x}$.

13. Show that $\frac{\partial^2 u}{\partial x^2}+\frac{\partial^2 u}{\partial y^2} = 0$, if

(i) $u = e^{my}\cos mx$ (ii) $u = \tan^{-1}\frac{y}{x}$.

14. If $\frac{x^2}{a^2+u}+\frac{y^2}{b^2+u}+\frac{z^2}{c^2+u} = 1$, show that

$$\left(\frac{\partial u}{\partial x}\right)^2+\left(\frac{\partial u}{\partial y}\right)^2+\left(\frac{\partial u}{\partial z}\right)^2 = 2\left(x\frac{\partial u}{\partial x}+y\frac{\partial u}{\partial y}+z\frac{\partial u}{\partial z}\right)$$

(UPTU–2003)

15. Find the value of $\frac{1}{a^2}\frac{\partial^2 z}{\partial x^2}+\frac{1}{b^2}\frac{\partial^2 z}{\partial y^2}$, when $a^2x^2+b^2y^2-c^2z^2 = 0$.

16. If $z = e^{ax+by}f(ax - by)$, show that $b\frac{\partial z}{\partial x}+a\frac{\partial z}{\partial y} = 2abz$.

(UPTU–2006, VTU–2010)

17. If $u = \sqrt{x^2+y^2+z^2}$, show that $\left(\frac{\partial u}{\partial x}\right)^2+\left(\frac{\partial u}{\partial y}\right)^2+\left(\frac{\partial u}{\partial z}\right)^2 = 1$.

18. If $x = r\cos\theta$, $y = r\sin\theta$, prove that

(i) $\frac{\partial^2\theta}{\partial x^2}+\frac{\partial^2\theta}{\partial y^2} = 0$ except when $x = 0, y = 0$

(ii) $\left(\frac{\partial r}{\partial x}\right)^2+\left(\frac{\partial r}{\partial y}\right)^2 = 1$ (Burdwan–2003)

(iii) $\frac{\partial^2 r}{\partial x^2}+\frac{\partial^2 r}{\partial y^2} = \frac{2}{r}\left\{\left(\frac{\partial r}{\partial x}\right)^2+\left(\frac{\partial r}{\partial y}\right)^2\right\}$.

19. If $u = \log(x^2 + y^2 + z^2)$, then prove that

$$x\frac{\partial^2 u}{\partial y\partial z} = y\frac{\partial^2 u}{\partial z\partial x} = z\frac{\partial^2 u}{\partial x\partial y}.$$

20. If $x^2(y-z)+y^2(z-x)+z^2(x-y)$, prove that $\frac{\partial u}{\partial x}+\frac{\partial u}{\partial y}+\frac{\partial u}{\partial z} = 0$.

21. (i) If $u = \sqrt{x^2+y^2+z^2}$, then prove that $\frac{\partial^2 u}{\partial x^2}+\frac{\partial^2 u}{\partial y^2}+\frac{\partial^2 u}{\partial z^2} = \frac{2}{u}$.

(ii) If $u = \log\sqrt{x^2+y^2+z^2}$, show that

$$(x^2+y^2+z^2)\left(\frac{\partial^2 u}{\partial x^2}+\frac{\partial^2 u}{\partial y^2}+\frac{\partial^2 u}{\partial z^2}\right) = 1.$$

(UKTU–2011)

22. (i) If $u = \sin^{-1}\left(\dfrac{x^{1/3}+y^{1/3}}{x^{1/2}-y^{1/2}}\right)^{1/2}$, show that

$$x\frac{\partial u}{\partial x}+y\frac{\partial u}{\partial y} = -\frac{1}{12}\tan u.$$ (MTU–2012)

(ii) If $u = x\sin^{-1}\left(\dfrac{x}{y}\right)+y\sin^{-1}\left(\dfrac{y}{x}\right)$, show that

$$x^2\frac{\partial^2 u}{\partial x^2}+2xy\frac{\partial^2 u}{\partial x\partial y}+y^2\frac{\partial^2 u}{\partial y^2} = 0.$$ (UPTU–2008)

(iii) If $u = x^2\tan^{-1}\left(\dfrac{y}{x}\right)-y^2\tan^{-1}\left(\dfrac{x}{y}\right)$, show that

$$x^2\frac{\partial^2 u}{\partial x^2}+2xy\frac{\partial^2 u}{\partial x\partial y}+y^2\frac{\partial^2 u}{\partial y^2} = 2u.$$

(UPTU–2009, UKTU–2010, Hisar–2003)

ANSWERS

1. (i) $\dfrac{2x}{x^2+y^2}, \dfrac{2y}{x^2+y^2}$ (ii) $-\dfrac{1}{\sqrt{y^2-x^2}}, \dfrac{x}{y\sqrt{y^2-x^2}}$ (iii) $\dfrac{2x}{a^2}, \dfrac{2y}{b^2}$ (iv) $\dfrac{(x^2+2xy-y^2)}{(x+y)^2+(x^2+y^2)^2}, \dfrac{(y^2+2xy-x^2)}{(x+y)^2+(x^2+y^2)^2}$

2. (i) $\dfrac{e^{x+y}}{(e^x+e^y)^2}, -\dfrac{e^{x+y}}{(e^x+e^y)^2}, \dfrac{e^{x+y}}{(e^x+e^y)^2}$ **15.** $\dfrac{1}{c^2 z}$

9.4 HOMOGENEOUS FUNCTIONS

A function $f(x, y)$ is said to be homogeneous function of degree n, if the degree of each of its terms in x and y is equal to n. Thus

$$a_0x^n + a_1x^{n-1}y + a_2x^{n-2}y^2 + \ldots + a_{n-1}xy^{n-1} + a_ny^n \qquad \ldots(1)$$

is homogeneous function in x and y of order n.

REMARKS

- This definition of homogeneity applies to polynomial functions only. To widen the concept of homogeneity so as to bring even transcendental functions within its scope, we define u as a homogeneous function in x and y of order or degree n, if it can be expressed in the form of $x^n f\left(\dfrac{y}{x}\right)$.
- This definition also covers the polynomial function (1), which can be written as

$$x^n\left[a_0 + a_1\frac{y}{x} + a_2\left(\frac{y}{x}\right)^2 + \ldots + a_n\left(\frac{y}{x}\right)^n\right] = x^n f\left(\frac{y}{x}\right).$$

$\therefore$ It is a homogeneous function of order n.

- To test whether a given function $f(x, y)$, is homogeneous or not we put $x = hx$ and $y = hy$ in it. If we get $f(hx, hy) = h^n f(x, y)$, the function $f(x, y)$ is homogeneous of degree n, otherwise $f(x, y)$ is not a homogeneous function.
- A homogeneous function in x and y of degree n can also be written as $y^n f\left(\dfrac{x}{y}\right)$.
- A function u of three variables x, y, z is said to be homogeneous function of degree n, if it can be expressed in the form

$$u = x^n f_1\left(\frac{y}{x}, \frac{z}{x}\right) \quad \text{or} \quad y^n f_2\left(\frac{x}{y}, \frac{z}{y}\right) \quad \text{or} \quad z^n f_3\left(\frac{x}{z}, \frac{y}{z}\right).$$

In general, f function u of several variables $x_1, x_2, \ldots, x_n$ is said to be homogeneous function of degree m if it can be expressed in the form

$$u = x_1^m f_1\left(\frac{x_2}{x_1}, \frac{x_3}{x_1}, \ldots, \frac{x_n}{x_1}\right) \quad \text{or} \quad x_2^m f_2\left(\frac{x_1}{x_2}, \frac{x_3}{x_2}, \ldots, \frac{x_n}{x_2}\right) \quad \text{or etc.}$$

THEOREM 1. *If u is a homogeneous function of x and y of degree n, then $\dfrac{\partial u}{\partial x}$ and $\dfrac{\partial u}{\partial y}$ are homogeneous function of degree $(n-1)$ each.*

Proof. Since, u is a homogeneous function of x and y of degree n therefore, u can be expressed as

$$u = x^n f\left(\frac{y}{x}\right). \qquad \ldots(1)$$

Now from (1)

$$\frac{\partial u}{\partial x} = nx^{n-1}f\left(\frac{y}{x}\right) + x^n f'\left(\frac{y}{x}\right)\left(-\frac{y}{x^2}\right) = x^{n-1}\left[nf\left(\frac{y}{x}\right) + f'\left(\frac{y}{x}\right)\left(-\frac{y}{x}\right)\right]$$

$$= x^{n-1} \times \text{a function of } \frac{y}{x} = x^{n-1}g\left(\frac{y}{x}\right) \text{ (say).}$$

which is a homogeneous function of degree $(n-1)$.

Also, $$\frac{\partial u}{\partial y} = x^n f'\left(\frac{y}{x}\right).\left(\frac{1}{x}\right) = x^{n-1} f'\left(\frac{y}{x}\right) = x^{n-1} \times \text{a function of} \frac{y}{x}$$
$$= x^{n-1} g\left(\frac{y}{x}\right) \text{(say).}$$

which is a homogeneous function of x and y of degree $(n-1)$.

THEOREM 2. [Euler's Theorem on Homogeneous Functions].

If u be a homogeneous function of x and y of degree n, then $x\frac{\partial u}{\partial x} + y\frac{\partial u}{\partial y} = nu.$

(UPTU–2006, 07, GBTU–2010, UKTU–2011)

Proof. Since, u is a homogeneous function of x and y of degree n therefore, u can be expressed as

$$u = x^n f\left(\frac{y}{x}\right).$$

$\therefore$ $$\frac{\partial u}{\partial x} = nx^{n-1} f\left(\frac{y}{x}\right) + x^n f'\left(\frac{y}{x}\right)\left(-\frac{y}{x^2}\right) = nx^{n-1} f\left(\frac{y}{x}\right) - yx^{n-2} f'\left(\frac{y}{x}\right).$$

Also, $$\frac{\partial u}{\partial y} = x^n f'\left(\frac{y}{x}\right).\left(\frac{1}{x}\right) = x^{n-1} f'\left(\frac{y}{x}\right).$$

Now, $$\text{L.H.S.} = x\frac{\partial u}{\partial x} + y\frac{\partial u}{\partial y} = x\left[nx^{n-1} f\left(\frac{y}{x}\right) - yx^{n-2} f'\left(\frac{y}{x}\right)\right] + yx^{n-1} f'\left(\frac{y}{x}\right)$$
$$= nx^n f\left(\frac{y}{x}\right) - yx^{n-1} f'\left(\frac{y}{x}\right) + yx^{n-1} f'\left(\frac{y}{x}\right) = nx^n f\left(\frac{y}{x}\right) = nu = \text{R.H.S.}$$

REMARK

- Euler's theorem can be extended to a homogeneous functions of several variables. Thus, if u be the function of m independent variables $x_1, x_2, \ldots, x_m$ of degree n then, Euler's theorem states that
$$x_1 \frac{\partial u}{\partial x_1} + x_2 \frac{\partial u}{\partial x_2} + \ldots + x_m \frac{\partial u}{\partial x_m} = nu \cdot$$

THEOREM 3. *If u is a homogeneous function in x and y of degree n, then prove that* $x^2 \frac{\partial^2 u}{\partial x^2} + 2xy \frac{\partial^2 u}{\partial x \partial y} + y^2 \frac{\partial^2 u}{\partial y^2} = n(n-1)u.$

(UPTU–2006, 07, VTU–2007, Anna–2009)

Proof. Since, u is a homogeneous function in x and y of degree n therefore, by Euler's theorem

$$x\frac{\partial u}{\partial x} + y\frac{\partial u}{\partial y} = nu \quad \ldots(1)$$

Differentiating (1) partially *w.r.t.* x, we get

$$\frac{\partial}{\partial x}\left(x\frac{\partial u}{\partial x}\right) + \frac{\partial}{\partial x}\left(y\frac{\partial u}{\partial y}\right) = \frac{\partial}{\partial x}(nu) \qquad \left(\because \text{Each of } \frac{\partial u}{\partial x} \text{ and } \frac{\partial u}{\partial y} \text{ is a function of both } x \text{ and } y\right)$$

$\Rightarrow$ $$x\frac{\partial^2 u}{\partial x^2} + \frac{\partial u}{\partial x}.1 + y\frac{\partial^2 u}{\partial x \partial y} = n\frac{\partial u}{\partial x}$$

$\Rightarrow$ $$x\frac{\partial^2 u}{\partial x^2} + y\frac{\partial^2 u}{\partial x \partial y} = (n-1)\frac{\partial u}{\partial x} \quad \ldots(2)$$

ifferentiating (2) partially w.r.t. y, we get

$$y\frac{\partial^2 u}{\partial y^2} + x\frac{\partial^2 u}{\partial x \partial y} = (n-1)\frac{\partial u}{\partial y} \quad \ldots(3)$$

ply (2) by x, (3) by y and then adding, we get

$$^2 \frac{\partial^2 u}{\partial x^2} + 2xy\frac{\partial^2 u}{\partial x \partial y} + y^2 \frac{\partial^2 u}{\partial y^2} = (n-1)\left[x\frac{\partial u}{\partial y} + y\frac{\partial u}{\partial y}\right] = (n-1)nu = n(n-1)u.$$

REMARK

- If z is a homogeneous function of x and y of degree n and if $z = f(u)$, then we have the following results :

(i) $x\frac{\partial u}{\partial x} + y\frac{\partial u}{\partial y} = n\frac{f(u)}{f'(u)} = G(u)$

(ii) $x^2\frac{\partial^2 u}{\partial x^2} + 2xy\frac{\partial^2 u}{\partial x \partial y} + y^2\frac{\partial^2 u}{\partial y^2} = G(u)[G'(u) - 1]$

Recapitulations

- **Euler's Theorem.** If u is a homogeneous function of x and y of degree n then
$$x\frac{\partial u}{\partial x} + y\frac{\partial u}{\partial y} = nu.$$
- $$x^2\frac{\partial^2 u}{\partial x^2} + 2xy\frac{\partial^2 u}{\partial x \partial y} + y^2\frac{\partial^2 u}{\partial y^2} = n(n-1)u.$$
- If u is a homogeneous function of x and y of degree n, then $\frac{\partial u}{\partial x}$ and $\frac{\partial u}{\partial y}$ are homogeneous function of degree $(n-1)$ each.

SOLVED EXAMPLES

Example 1. *Verify the Euler's theorem for the function*
$$u = axy + byz + czx.$$

Solution. We have $u = axy + byz + czx.$ ...(1)

which is a homogeneous function of x, y and z of degree 2.

To verify the Euler's theorem, we must show
$$x\frac{\partial u}{\partial x} + y\frac{\partial u}{\partial y} + z\frac{\partial u}{\partial z} = 2u$$

Now, $\frac{\partial u}{\partial x} = ay + cz, \frac{\partial u}{\partial y} = ax + bz, \frac{\partial u}{\partial z} = by + cx.$

$$\therefore \quad x\frac{\partial u}{\partial x} + y\frac{\partial u}{\partial y} + z\frac{\partial u}{\partial z}$$
$$= x(ay + cz) + y(ax + bz) + z(by + cx).$$
$$= 2(axy + byz + czx) = 2u.$$

Hence, Euler's theorem is verified.

Example 2. *If* $u = \sin^{-1}\left[\frac{x^2 + y^2}{x + y}\right]$,

show that $x\frac{\partial u}{\partial x} + y\frac{\partial u}{\partial y} = \tan u.$

(UPTU–2007, VTU–2003, Bhopal–2009)

Solution. We have $\sin u = \left[\frac{x^2 + y^2}{x + y}\right]$

Let $v = \frac{x^2 + y^2}{x + y}$

$\Rightarrow$ v is a homogeneous of x and y of degree 1.

Then, by Euler's theorem, we have
$$x\frac{\partial v}{\partial x} + y\frac{\partial v}{\partial y} = v \quad \text{...(1)}$$
$$v = \sin u \Rightarrow \frac{\partial v}{\partial x} = \cos u\frac{\partial u}{\partial x}$$

and $\frac{\partial v}{\partial y} = \cos u\frac{\partial u}{\partial y}.$

Put these values in (1), we get
$$x\cos u\frac{\partial u}{\partial x} + y\cos u\frac{\partial u}{\partial y} = v$$
$$\Rightarrow \quad x\frac{\partial u}{\partial x} + y\frac{\partial u}{\partial y} = \frac{v}{\cos u} = \frac{\sin u}{\cos u} = \tan u.$$

Example 3. *If* $u = \tan^{-1}\frac{x^3 + y^3}{x - y}$, *prove that*
$$x^2\frac{\partial^2 u}{\partial x^2} + 2xy\frac{\partial^2 u}{\partial x \partial y} + y^2\frac{\partial^2 u}{\partial y^2} = (1 - 4\sin^2 u)\sin 2u.$$

Solution. We have $u = \tan^{-1}\frac{x^3 + y^3}{x - y}$

$$\therefore \tan u = \frac{x^3 + y^3}{x - y} = \frac{x^3\left[1 + \left(\frac{y}{x}\right)^3\right]}{x\left[1 - \frac{y}{x}\right]} = x^2 f\left(\frac{y}{x}\right)$$

$\tan u$ is of the form $x^n f\left(\frac{y}{x}\right)$ with $n = 2$.

$\therefore$ $\tan u$ is a homogeneous function in x, y of degree 2. Then, by Euler's theorem
$$x\frac{\partial}{\partial x}(\tan u) + y\frac{\partial}{\partial y}(\tan u) = 2\tan u$$
$$\Rightarrow \quad x\sec^2 u\frac{\partial u}{\partial x} + y\sec^2 u\frac{\partial u}{\partial y} = 2\tan u$$
$$\Rightarrow x\frac{\partial u}{\partial x} + y\frac{\partial u}{\partial y} = \frac{2\tan u}{\sec^2 u} = 2\sin u\cos u = \sin 2u \quad \text{...(1)}$$

Differentiate (1) partially *w.r.t.* x, we get
$$\left(x\frac{\partial^2 u}{\partial x^2} + \frac{\partial u}{\partial x}\right) + y\frac{\partial^2 u}{\partial x \partial y} = 2\cos 2u\frac{\partial u}{\partial x}$$
$$\therefore \quad x\frac{\partial^2 u}{\partial x^2} + y\frac{\partial^2 u}{\partial x \partial y} = (2\cos 2u - 1)\frac{\partial u}{\partial x} \quad \text{...(2)}$$

Interchanging x and y in (2), we get
$$y\frac{\partial^2 u}{\partial y^2} + x\frac{\partial^2 u}{\partial x \partial y} = (2\cos 2u - 1)\frac{\partial u}{\partial y} \quad \text{...(3)}$$

Now multiplying (2) by x, (3) by y and then adding, we get

$$x^2\frac{\partial^2 u}{\partial x^2}+2xy\frac{\partial^2 u}{\partial x\partial y}+y^2\frac{\partial^2 u}{\partial y^2}$$

$$=(2\cos 2u-1)\left[x\frac{\partial u}{\partial x}+y\frac{\partial u}{\partial y}\right]$$

$$=(2\cos 2u-1).\sin 2u$$

$$=[2(1-2\sin^2 u)-1]\sin 2u$$

$$=(1-4\sin^2 u)\sin 2u.$$

Example 4. *If* $u=\sin^{-1}\left(\frac{x}{y}\right)+\tan^{-1}\left(\frac{y}{x}\right)$, *show that*

$$x\frac{\partial u}{\partial x}+y\frac{\partial u}{\partial y}=0\times u=0$$

(UPTU–2007, Hazaribagh–2009, Osmania–2003)

Solution. We have

$$u=\sin^{-1}\left(\frac{x}{y}\right)+\tan^{-1}\left(\frac{y}{x}\right)$$

$$=x^0\left[\sin^{-1}\left(\frac{1}{y/x}\right)+\tan^{-1}\left(\frac{y}{x}\right)\right]=x^0$$

$\Rightarrow$ u is a homogeneous function of order 0.

Then, by Euler's theorem,we have

$$x\frac{\partial u}{\partial x}+y\frac{\partial u}{\partial y}=0\times u=0$$

Example 5. *If* $u=\left(x^{1/4}+y^{1/4}\right)\left(x^{1/5}+y^{1/5}\right)$. *Apply Euler's theorem to find the value of* $x\frac{\partial u}{\partial x}+y\frac{\partial u}{\partial y}$.

(MTU–2011)

Solution. Here, we have

$$u(x,y)=\left(x^{1/4}+y^{1/4}\right)\left(x^{1/5}+y^{1/5}\right)$$

$$\Rightarrow u(tx,ty)=t^{1/4}\left(x^{1/4}+y^{1/4}\right)t^{1/5}\left(x^{1/5}+y^{1/5}\right)$$

$$=t^{9/20}\left(x^{1/4}+y^{1/4}\right)\left(x^{1/5}+y^{1/5}\right)$$

$$=t^{9/20}u(x,y)$$

Clearly, u is a homogeneous function of degree $\frac{9}{20}$.

Hence, by Euler's theorem we have

$$x\frac{\partial u}{\partial x}+y\frac{\partial u}{\partial y}=\frac{9}{20}u.$$

Example 6. *Verify Euler's theoerm for* $f(x,y,z)=3x^2yz+5xy^2z+4z^4$. (JNTU–1999)

Solution. Let $f(x,y,z)=3x^2yz+5xy^2z+4z^4$.

$$\therefore\quad \frac{\partial f}{\partial x}=6xyz+5y^2z;\ \frac{\partial f}{\partial y}=3x^2z+10xyz$$

and $$\frac{\partial f}{\partial z}=3x^2y+5xy^2+16z^3$$

$$\therefore\ x\frac{\partial f}{\partial x}+y\frac{\partial f}{\partial y}+z\frac{\partial f}{\partial z}=x(6xyz+5y^2z)$$

$$+y(3x^2z+10xyz)$$

$$+z(3x^2y+5xy^2+16z^3)$$

$$=4(3x^2yz+5xy^2z+4z^4)=4f\quad\ldots(1)$$

Also,

$$f(x,y,z)=x^4\left[3.\frac{y}{x}.\frac{z}{x}+5\left(\frac{y}{x}\right)^2\left(\frac{z}{x}\right)+4\left(\frac{z}{x}\right)^4\right]$$

is a homogeneous function of x, y, z of degree 4.

Hence, by Euler's theorem

$$x\frac{\partial f}{\partial x}+y\frac{\partial f}{\partial y}+z\frac{\partial f}{\partial z}=4f.\quad\ldots(2)$$

From (1) and (2) we conclude that Euler's theorem is verified.

Example 7. *If* $u=f\left(\frac{y}{x}\right)+\sqrt{x^2+y^2}$, *show that*

$$x\frac{\partial u}{\partial x}+y\frac{\partial u}{\partial y}=\sqrt{x^2+y^2}.$$

(Mumbai–2008)

Solution. Let us write $u=v+w$

where $$v=f\left(\frac{y}{x}\right)=x^0f\left(\frac{y}{x}\right)$$

and $$w=\sqrt{x^2+y^2}=x\sqrt{1+\left(\frac{y}{x}\right)^2}$$

Therefore, v and w are homogeneous function of degree 0 and 1 in x and y respectively. Hence, by Euler's theorem

$$x\frac{\partial v}{\partial x}+y\frac{\partial v}{\partial y}=0.v=0\quad\ldots(1)$$

and $$x\frac{\partial w}{\partial x}+y\frac{\partial w}{\partial y}=1.w=\sqrt{x^2+y^2}\quad\ldots(2)$$

On adding (1) and (2), we get

$$x\left(\frac{\partial v}{\partial x}+\frac{\partial w}{\partial x}\right)+y\left(\frac{\partial v}{\partial y}+\frac{\partial w}{\partial y}\right)=\sqrt{x^2+y^2}\quad\ldots(3)$$

Now, since $u=v+w$, then using (3) we get

$$x\frac{\partial u}{\partial x}+y\frac{\partial u}{\partial y}=\sqrt{x^2+y^2}.$$

Example 8. *If* $z=x^nf_1\left(\frac{y}{x}\right)+y^{-n}f_2\left(\frac{x}{y}\right)$, *then show that*

$$x^2\frac{\partial^2 z}{\partial x^2}+2xy\frac{\partial^2 z}{\partial x\partial y}+y^2\frac{\partial^2 z}{\partial y^2}+x\frac{\partial z}{\partial x}+y\frac{\partial z}{\partial y}=n^2z$$

(Rohtak(MDU)–2003, Kurukshetra–2009)

Solution. Let $$u=x^nf_1\left(\frac{y}{x}\right),v=y^{-n}f_2\left(\frac{x}{y}\right)\quad\ldots(1)$$

$$\therefore \qquad z = u + v \qquad \ldots(2)$$

Clearly, u and v are homogeneous functions of degree n and $-n$ respectively. Then by Euler's theorem, we get

$$x\frac{\partial u}{\partial x} + y\frac{\partial u}{\partial y} = nu \qquad \ldots(3)$$

$$x\frac{\partial v}{\partial x} + y\frac{\partial v}{\partial y} = (-n).v \qquad \ldots(4)$$

$$x^2\frac{\partial^2 u}{\partial x^2} + 2xy\frac{\partial^2 u}{\partial x \partial y} + y^2\frac{\partial^2 u}{\partial y^2} = n(n-1)u \qquad \ldots(5)$$

and
$$x^2\frac{\partial^2 v}{\partial x^2} + 2xy\frac{\partial^2 v}{\partial x \partial y} + y^2\frac{\partial^2 v}{\partial y^2} = (-n)(-n-1)v = n(n+1)v \qquad \ldots(6)$$

Since $z = u + v \Rightarrow \dfrac{\partial z}{\partial x} = \dfrac{\partial u}{\partial x} + \dfrac{\partial v}{\partial x}$

and $\dfrac{\partial z}{\partial y} = \dfrac{\partial u}{\partial y} + \dfrac{\partial v}{\partial y} \qquad \ldots(7)$

Adding (3) and (4) and using (7) we get

$$x\frac{\partial z}{\partial x} + y\frac{\partial z}{\partial y} = n(u-v) \qquad \ldots(8)$$

Similarly, adding (5) and (6) and using (7) we get

$$x^2\left[\frac{\partial^2 u}{\partial x^2} + \frac{\partial^2 v}{\partial x^2}\right] + 2xy\left[\frac{\partial^2 u}{\partial x \partial y} + \frac{\partial^2 v}{\partial x \partial y}\right] + y^2\left[\frac{\partial^2 u}{\partial y^2} + \frac{\partial^2 v}{\partial y^2}\right] = n(n-1)u + n(n+1)v$$

$$\Rightarrow \quad x^2\frac{\partial^2 z}{\partial x^2} + 2xy\frac{\partial^2 z}{\partial x \partial y} + y^2\frac{\partial^2 z}{\partial y^2} = n^2(u+v) - n(u-v)$$

$$= n^2 z - \left(x\frac{\partial z}{\partial x} + y\frac{\partial z}{\partial y}\right) \qquad \text{(Using 8)}$$

$$\Rightarrow \quad x^2\frac{\partial^2 z}{\partial x^2} + 2xy\frac{\partial^2 z}{\partial x \partial y} + y^2\frac{\partial^2 z}{\partial y^2} + x\frac{\partial z}{\partial x} + y\frac{\partial z}{\partial y} = n^2 z$$

Example 9. *If* $u = \operatorname{cosec}^{-1}\sqrt{\dfrac{x^{1/2} + y^{1/2}}{x^{1/3} + y^{1/3}}}$, *then prove that*

$$x^2\frac{\partial^2 u}{\partial x^2} + 2xy\frac{\partial^2 u}{\partial x \partial y} + y^2\frac{\partial^2 u}{\partial y^2} = \frac{\tan u}{12}\left(\frac{13}{12} + \frac{\tan^2 u}{12}\right)$$

(Rohtak–2006, Mumbai–2008)

Solution. We have $z = \operatorname{cosec} u = x^{1/2}\sqrt{\dfrac{1 + \left(\dfrac{y}{x}\right)^{1/2}}{1 + x^{1/3}}}$

$\Rightarrow$ $z = f(u) = \operatorname{cosec} u$ is a homogeneous function of degree $\dfrac{1}{2}$.

Then by result (ii) of remark of theorem-3, we have

$$x^2\frac{\partial^2 u}{\partial x^2} + 2xy\frac{\partial^2 u}{\partial x \partial y} + y^2\frac{\partial^2 u}{\partial y^2} = G(u)[G'(u) - 1] \qquad \ldots(1)$$

where $G(u) = \dfrac{nf(u)}{f'(u)} \Rightarrow G(u) = \dfrac{1}{12}\dfrac{f(u)}{f'(u)}$

$\because$ $f(u) = \operatorname{cosec} u$ therefore

$$G(u) = \frac{1}{12}\frac{\operatorname{cosec} u}{(-\operatorname{cosec} u.\cot u)} = -\frac{1}{12}\tan u$$

$$\Rightarrow \quad G'(u) = -\frac{1}{12}\sec^2 u$$

Putting all these values in equation (1), we get

$$x^2\frac{\partial^2 u}{\partial x^2} + 2xy\frac{\partial^2 u}{\partial x \partial y} + y^2\frac{\partial^2 u}{\partial y^2} = -\frac{1}{12}\tan u\left[-\frac{1}{12}\sec^2 u - 1\right]$$

$$= \frac{1}{12}\tan u\left[\frac{1}{12}(1 + \tan^2 u) + 1\right]$$

$$= \frac{1}{12}\tan u\left[\frac{1}{12}\tan^2 u + \frac{13}{12}\right]$$

Example 10. *If* $u = \sin^{-1}\dfrac{x + 2y + 3z}{x^8 + y^8 + z^8}$, *find the value of* $x\dfrac{\partial u}{\partial x} + y\dfrac{\partial u}{\partial y} + z\dfrac{\partial u}{\partial z}$. (UKTU–2010, UPTU–2004)

Solution. We can write

$$w = \sin u = \frac{x + 2y + 3z}{x^8 + y^8 + z^8} = x^{-7}\frac{1 + 2\left(\dfrac{y}{x}\right) + 3\left(\dfrac{z}{x}\right)}{1 + \left(\dfrac{y}{x}\right)^8 + \left(\dfrac{z}{x}\right)^8}$$

$\Rightarrow$ w is a homogeneous function of degree -7.

Then by Euler's theorem, we get

$$x\frac{\partial w}{\partial x} + y\frac{\partial w}{\partial y} + z\frac{\partial w}{\partial z} = (-7)w \qquad \ldots(1)$$

Here,

$$\frac{\partial w}{\partial x} = \cos u\frac{\partial u}{\partial x}, \frac{\partial w}{\partial y} = \cos u\frac{\partial u}{\partial y}, \frac{\partial w}{\partial z} = \cos u\frac{\partial u}{\partial z}$$

Then from (1)

$$x\cos u\frac{\partial u}{\partial x} + y\cos u\frac{\partial u}{\partial y} + z\cos u\frac{\partial u}{\partial z} = -7\sin u$$

$$\Rightarrow \quad x\frac{\partial u}{\partial x} + y\frac{\partial u}{\partial y} + z\frac{\partial u}{\partial z} = -7\tan u.$$

Example 11. If $u = \dfrac{x^3y^3z^3}{x^3+y^3+z^3} + \log\left(\dfrac{xy+yz+zx}{x^2+y^2+z^2}\right)$, find the value of $x\dfrac{\partial u}{\partial x} + y\dfrac{\partial u}{\partial y} + z\dfrac{\partial u}{\partial z}$. (Mumbai–2009)

Solution. Let $v = \dfrac{x^3y^3z^3}{x^3+y^3+z^3}$

and $w = \log\left(\dfrac{xy+yz+zx}{x^2+y^2+z^2}\right)$

Clearly, $v = x^6\left[\dfrac{\left(\frac{y}{x}\right)^3\left(\frac{z}{x}\right)^3}{1+\left(\frac{y}{x}\right)^3+\left(\frac{z}{x}\right)^3}\right]$

is a homogeneous function of degree 6.

$\therefore$ By Euler's theorem

$$x\frac{\partial v}{\partial x} + y\frac{\partial v}{\partial y} + z\frac{\partial v}{\partial z} = 6v \quad \ldots(1)$$

Further, $w = \log\left[\dfrac{\frac{y}{x}+\frac{y}{x}\cdot\frac{z}{x}+\frac{z}{x}}{1+\left(\frac{y}{x}\right)^2+\left(\frac{z}{x}\right)^2}\right]$

is a homogeneous function of degree zero.

Then, by Euler's theorem

$$x\frac{\partial w}{\partial x} + y\frac{\partial w}{\partial y} + z\frac{\partial w}{\partial z} = 0 \quad \ldots(2)$$

Adding (1) and (2), we get

$$x\left(\frac{\partial v}{\partial x}+\frac{\partial w}{\partial x}\right) + y\left(\frac{\partial v}{\partial y}+\frac{\partial w}{\partial y}\right) + z\left(\frac{\partial v}{\partial z}+\frac{\partial w}{\partial z}\right) = 6v$$

$$\Rightarrow x\left(\frac{\partial u}{\partial x}\right) + y\left(\frac{\partial u}{\partial y}\right) + z\left(\frac{\partial u}{\partial z}\right) = 6.\frac{x^3y^3z^3}{x^3+y^3+z^3}$$

EXERCISE 9.2

1. Verify the Euler's theorem for the following functions :

(i) $u = \dfrac{x(x^3-y^3)}{x^3+y^3}$ (ii) $u = x^n \sin\left(\dfrac{y}{x}\right)$

(iii) $u = x^n \log\left(\dfrac{y}{x}\right)$ (iv) $u = \dfrac{1}{\sqrt{x^2+y^2}}$

(v) $u = x^n \sin\dfrac{y}{x}$ (vi) $x^4 \log\dfrac{y}{x}$

(vii) $u = \log\left(\dfrac{x^2+y^2}{xy}\right)$ (UPTU–2006, UKTU–2011)

(viii) $u = \dfrac{x^{1/3}+y^{1/3}}{x^{1/2}+y^{1/2}}$ (GBTU–2010)

2. (i) If $u = xf\left(\dfrac{y}{x}\right)$, prove that $x\dfrac{\partial u}{\partial x} + y\dfrac{\partial u}{\partial y} = u$.

(ii) If $u = f\left(\dfrac{y}{x}\right)$, prove that $x\dfrac{\partial u}{\partial x} + y\dfrac{\partial u}{\partial y} = 0$. (GBTU–2011)

(iii) If $u = xyf\left(\dfrac{y}{x}\right)$, prove that $x\dfrac{\partial u}{\partial x} + y\dfrac{\partial u}{\partial y} = 2u$.

(iv) If $u = \log\left(\dfrac{x^2+y^2}{x+y}\right)$, show by Euler's theorem :

$x\dfrac{\partial u}{\partial x} + y\dfrac{\partial u}{\partial y} = 1$. (UPTU–2009)

3. If $u = \tan^{-1}\left(\dfrac{x^3+y^3}{x+y}\right)$, show that $x\dfrac{\partial u}{\partial x} + y\dfrac{\partial u}{\partial y} = \sin 2u$

and $x^2\dfrac{\partial^2 u}{\partial x^2} + 2xy\dfrac{\partial^2 u}{\partial x\partial y} + y^2\dfrac{\partial^2 u}{\partial y^2} = 2\cos 3u \sin u$.

(PTU–2009, SVTU–2009, Bhopal–2008, Mumbai–2009)

4. If $u = \tan^{-1}\dfrac{y}{x}$, show that(using Euler's theorem)

$x\dfrac{\partial u}{\partial x} + y\dfrac{\partial u}{\partial y} = 0$.

5. If $u = \sin^{-1}\dfrac{x+y}{\sqrt{x}+\sqrt{y}}$, show that

(i) $x\dfrac{\partial u}{\partial x} + y\dfrac{\partial u}{\partial y} = \dfrac{1}{2}\tan u$ (Rajasthan–2006, Calicut–2005)

(ii) $x^2\dfrac{\partial^2 u}{\partial x^2} + 2xy\dfrac{\partial^2 u}{\partial x\partial y} + y^2\dfrac{\partial^2 u}{\partial y^2} = -\dfrac{\sin u\cos 2u}{4\cos^3 u}$ (PTU–2006)

6. If $u = \sin^{-1}\dfrac{\sqrt{x}-\sqrt{y}}{\sqrt{x}+\sqrt{y}}$, show that $x\dfrac{\partial u}{\partial x} + y\dfrac{\partial u}{\partial y} = 0$.

7. (i) If $u = \log\dfrac{x^4+y^4}{x+y}$, show that $x\dfrac{\partial u}{\partial x} + y\dfrac{\partial u}{\partial y} = 3$.

(UPTU–2009, UKTU–2012)

(ii) If $u = \log\dfrac{x^3+y^3}{x+y}$, show that $x\dfrac{\partial u}{\partial x} + y\dfrac{\partial u}{\partial y} = 2$.

8. If $\sin u = \dfrac{x^2y^2}{x+y}$, show that $x\dfrac{\partial u}{\partial x} + y\dfrac{\partial u}{\partial y} = 3\tan u$.

(VTU–2003, Kottayam–2005)

9. (i) If $u = \dfrac{x^2y^2}{x+y}$, show that $y\dfrac{\partial^2 u}{\partial y^2} + x\dfrac{\partial^2 u}{\partial x\partial y} = 2\dfrac{\partial u}{\partial y}$.

(ii) If $u = \dfrac{xy}{x+y}$, show that $x\dfrac{\partial^2 u}{\partial x^2} + 2xy\dfrac{\partial^2 u}{\partial x\partial y} + y^2\dfrac{\partial^2 u}{\partial y^2} = 0$.

(iii) If $u = \dfrac{x^2y^2}{x+y}$, show that $x\dfrac{\partial^2 u}{\partial x^2} + y\dfrac{\partial^2 u}{\partial y\partial x} = 2\dfrac{\partial u}{\partial x}$.

10. If $u = xf_1\left(\dfrac{y}{x}\right) + f_2\left(\dfrac{y}{x}\right)$, show that

$x^2\dfrac{\partial^2 u}{\partial x^2} + 2xy\dfrac{\partial^2 u}{\partial x\partial y} + y^2\dfrac{\partial^2 u}{\partial y^2} = 0$. (UPTU–2006, SVTU–2009)

11. (i) If $u = \log\left(\sqrt{x}+\sqrt{y}\right)$, show that $x\dfrac{\partial u}{\partial x} + y\dfrac{\partial u}{\partial y} = \dfrac{1}{2}$.

(ii) If $u = \log \dfrac{x^4 + y^4 + x^2y^2}{x + y + \sqrt{xy}}$, show that $x\dfrac{\partial u}{\partial x} + y\dfrac{\partial u}{\partial y} = 3.$

12. If z be a homogeneous function of degree n, show that $x\dfrac{\partial^2 z}{\partial x^2} + y\dfrac{\partial^2 z}{\partial x \partial y} = (n-1)\dfrac{\partial z}{\partial x}.$

13. If $u = \cos^{-1}\dfrac{x+y}{\sqrt{x}+\sqrt{y}}$, prove that $x\dfrac{\partial u}{\partial x} + y\dfrac{\partial u}{\partial y} = -\dfrac{1}{2}\cot u.$ (VTU–2004, GBTU–2010)

14. If $\sin u = \dfrac{x+2y+3z}{\sqrt{x^8+y^8+z^8}}$, show that $x\dfrac{\partial u}{\partial x} + y\dfrac{\partial u}{\partial y} + z\dfrac{\partial u}{\partial z} = -3\tan u.$ (VTU–2009, SVTU–2009)

15. If $u = \dfrac{x}{y+z} + \dfrac{y}{z+x} + \dfrac{z}{x+y}$, show that $x\dfrac{\partial u}{\partial x} + y\dfrac{\partial u}{\partial y} + z\dfrac{\partial u}{\partial z} = 0.$ (VTU–2000)

16. If $u = \tan^{-1}\left(\dfrac{y^2}{x}\right)$, show that

$$x^2\frac{\partial^2 u}{\partial x^2} + 2xy\frac{\partial^2 u}{\partial x \partial y} + y^2\frac{\partial^2 u}{\partial y^2} = -\sin^2 u.\sin 2u.$$

(PTU–2005, Bhillai–2005)

17. Show that $x\dfrac{\partial u}{\partial x} + y\dfrac{\partial u}{\partial y} = 2u\log u$ where $u = e^{x^2+y^2}$. (PTU–2010)

18. If $\log u = \dfrac{x^3+y^3}{3x+4y}$, show that $x\dfrac{\partial u}{\partial x} + y\dfrac{\partial u}{\partial y} = 2u\log u$ (UKTU–2011)

19. If $u = x^3 + y^3 + z^3 + 3xyz$, show that $x\dfrac{\partial u}{\partial x} + y\dfrac{\partial u}{\partial y} + z\dfrac{\partial u}{\partial z} = 3u.$ (GBTU–2010)

20. If $u = \sec^{-1}\left(\dfrac{x^3-y^3}{x+y}\right)$, show that $x\dfrac{\partial u}{\partial x} + y\dfrac{\partial u}{\partial y} = 2\cot u.$ (UPTU–2008)

9.5 TOTAL DIFFERENTIAL

Let $\qquad u = f(x, y) \qquad$ ...(1)

be the given function of x and y, which have continuous partial derivatives of first order *w.r.t.* x and y.

Let δx and δy be the increments in x and y respectively and let δu be the consequent change in u, then we have

$$u + \delta u = f(x+\delta x, y+\delta y)$$

$$\therefore \quad \delta u = f(x+\delta x, y+\delta y) - f(x,y) \qquad \text{...(2)}$$

$$= [f(x+\delta x, y+\delta y) - f(x, y+\delta y)] + [f(x, y+\delta y) - f(x,y)]$$

$$\Rightarrow \quad \frac{\delta u}{\delta t} = \frac{[f(x+\delta x, y+\delta y) - f(x, y+\delta y)]}{\delta t} + \frac{[f(x, y+\delta y) - f(x,y)]}{\delta t}$$

Now,

$$\frac{du}{dt} = \lim_{\delta t \to 0}\frac{\delta u}{\delta t}$$

$$= \lim_{\delta t \to 0}\left[\frac{f(x+\delta x, y+\delta y) - f(x, y+\delta y)}{\delta x}\frac{\delta x}{\delta t} + \frac{f(x, y+\delta y) - f(x,y)}{\delta y}\frac{\delta y}{\delta t}\right] \qquad \text{...(3)}$$

Since δx and δy tends to zero, when $\delta t \to 0$ so we have

$$\lim_{\delta x \to 0}\frac{f(x+\delta x, y+\delta y) - f(x, y+\delta y)}{\delta x} = \frac{\partial f}{\partial x} = \frac{\partial u}{\partial x}.$$

Similarly,

$$\lim_{\delta y \to 0}\frac{f(x, y+\delta y) - f(x,y)}{\delta y} = \frac{\partial f}{\partial y} = \frac{\partial u}{\partial y} \text{ and } \lim_{\delta t \to 0}\frac{\delta x}{\delta t} = \frac{dx}{dt}, \lim_{\delta t \to 0}\frac{\delta y}{\delta t} = \frac{dy}{dt}.$$

Therefore, from (3), we get

$$\frac{du}{dt} = \frac{\partial u}{\partial x}.\frac{dx}{dt} + \frac{\partial u}{\partial y}.\frac{dy}{dt}.$$

REMARKS

- This result can be extended as follows :
  If $u = f(x_1, x_2, ..., x_m)$ and $x_1, x_2, ..., x_m$ all are functions of t, then $\dfrac{du}{dt} = \dfrac{\partial u}{\partial x_1}.\dfrac{dx_1}{dt} + \dfrac{\partial u}{\partial x_2}.\dfrac{dx_2}{dt} + ... + \dfrac{\partial u}{\partial x_m}.\dfrac{dx_m}{dt}.$
- The differentials dx and dy of the independent variables x and y are the actual changes δx and δy but the differential du of the dependent variable u is not the same as the change δu, it being the principal part of the increment δu.

9.6 IMPLICIT RELATION OF x AND y

In most of the cases, we are mainly concerned with the case in which y is expressed explicity *i.e.*, directly in terms of x. There are so many cases in which y is not expreesed directly in terms of x, but functionally it is implied by an algebraic relation $f(x, y) = 0$ connecting x and y.

The relation of the type $f(x, y) = c$, where y is not explicity in terms of x are called implicit function.

9.7 DIFFERENTIATION OF IMPLICIT FUNCTIONS

To find $\frac{dy}{dx}$ for an implicit function $f(x, y) = 0$ or $f(x, y) = c$:

Let $f(x, y)$ be a function of two variables x and y and y itself is a function of x *i.e.*, $f(x, y)$ may be consider as a composite function of x. Then, we have

$$\frac{df}{dx} = \frac{\partial f}{\partial x}\cdot\frac{dx}{dx} + \frac{\partial f}{\partial y}\cdot\frac{dy}{dx} \quad \Rightarrow \quad \frac{df}{dx} = \frac{\partial f}{\partial x} + \frac{\partial f}{\partial y}\cdot\frac{dy}{dx} \quad ...(1)$$

Since $f(x, y) = 0$, therefore $\frac{df}{dx} = 0$.

Now from (1), we have
$$\frac{\partial f}{\partial x} + \frac{\partial f}{\partial y}\cdot\frac{dy}{dx} = 0$$

$$\Rightarrow \quad \frac{dy}{dx} = -\frac{\partial f}{\partial x},\frac{\partial f}{\partial y} = -\frac{f_x}{f_y}, \text{provided } f_y \neq 0.$$

Solved Examples

Example 1. *If $x^y + y^x = a^b$. Find $\frac{dy}{dx}$.*

Solution . Let $f(x, y) = x^y + y^x - a^b$

$\Rightarrow \quad f(x, y) = 0$

Therefore
$$\frac{dy}{dx} = -\frac{\partial f/\partial x}{\partial f/\partial y} = -\frac{yx^{y-1} + y^x \log y}{x^y \log x + xy^{x-1}}.$$

Example 2. *If $u = \log [(x^2 + y^2)/xy]$, find du.*

Solution . Let $u = \log (x^2 + y^2) - \log x - \log y$.

$$\therefore \quad \frac{\partial u}{\partial x} = \frac{2x}{x^2+y^2} - \frac{1}{x} = \frac{2x^2 - x^2 - y^2}{x(x^2+y^2)} = \frac{x^2-y^2}{x(x^2+y^2)}$$

$$\text{and} \quad \frac{\partial u}{\partial y} = \frac{2y}{x^2+y^2} - \frac{1}{y} = \frac{2y^2 - x^2 - y^2}{y(x^2+y^2)} = \frac{y^2-x^2}{y(x^2+y^2)}$$

$$\text{Now,} \quad du = \frac{\partial u}{\partial x}dx + \frac{\partial u}{\partial y}dy = \frac{(x^2-y^2)}{x(x^2+y^2)}dx + \frac{(y^2-x^2)}{y(x^2+y^2)}dy = \frac{(x^2-y^2)}{xy(x^2+y^2)}(ydx - xdy).$$

Example 3. *If $f(x, y) = 0$ and $g(y, z) = 0$, show that*
$$\frac{\partial f}{\partial y}\cdot\frac{\partial g}{\partial z}\cdot\frac{dz}{dx} = \frac{\partial f}{\partial x}\cdot\frac{\partial g}{\partial y}.$$

Solution . Let $f(x, y) = 0$, then we have
$$\frac{dy}{dx} = -\frac{\partial f/\partial x}{\partial f/\partial y}. \quad ...(1)$$

Also, let $g(y, z) = 0$
$$\Rightarrow \quad \frac{dz}{dy} = -\frac{\partial g/\partial y}{\partial g/\partial z}. \quad ...(2)$$

Now, from (1) and (2), we have
$$\frac{dy}{dx}\cdot\frac{dz}{dy} = \left(\frac{\partial f}{\partial x}\cdot\frac{\partial g}{\partial y}\right)\Big/\left(\frac{\partial f}{\partial y}\cdot\frac{\partial g}{\partial z}\right)$$

$$\Rightarrow \quad \frac{dz}{dx}\cdot\frac{\partial f}{\partial y}\cdot\frac{\partial g}{\partial z} = \frac{\partial f}{\partial x}\cdot\frac{\partial g}{\partial y}$$

Example 4. *If $u = x^2y$, where $x^2 + xy + y^2 = 1$. Find $\frac{du}{dx}$.*

Solution . We know that
$$\frac{du}{dx} = \frac{\partial u}{\partial x} + \frac{\partial u}{\partial y}\cdot\frac{\partial y}{\partial x}. \quad ...(1)$$

Given that $u = x^2y$
$$\frac{\partial u}{\partial x} = 2xy \text{ and } \frac{\partial u}{\partial y} = x^2$$

$$\therefore \quad f(x, y) = x^2 + xy + y^2 - 1$$

$$\text{Then} \quad \frac{dy}{dx} = -\frac{\partial f/\partial x}{\partial f/\partial y} = -\frac{2x+y}{x+2y}$$

Putting all these values in (1), we get
$$\frac{du}{dx} = 2xy + x^2.\left(-\frac{2x+y}{x+2y}\right) = 2xy - \frac{x^2(2x+y)}{x+2y}$$

Example 5. *If the curves $f(x, y) = 0$ and $\phi(x, y) = 0$ touch each other, show that at the point of contact*
$$\frac{\partial f}{\partial x}\cdot\frac{\partial \phi}{\partial y} - \frac{\partial f}{\partial y}\cdot\frac{\partial \phi}{\partial x} = 0.$$

Solution . We have $f(x, y) = 0$
$$\Rightarrow \quad \frac{\partial f}{\partial x} + \frac{\partial f}{\partial y}\cdot\frac{dy}{dx} = 0 \quad \Rightarrow \quad \frac{dy}{dx} = -\frac{\partial f/\partial x}{\partial f/\partial y} \quad ...(1)$$

$$\text{Also,} \quad \phi(x, y) = 0 \quad \Rightarrow \quad \frac{\partial \phi}{\partial x} + \frac{\partial \phi}{\partial y}\cdot\frac{dy}{dx} = 0$$

$$\Rightarrow \quad \frac{dy}{dx} = -\frac{\partial \phi/\partial x}{\partial \phi/\partial y} \quad ...(2)$$

At the point of contact, the slope of the tangents to both curve must coincide, therefore, from (2) and (3) we get

$$-\left(\frac{\partial f}{\partial x}\Big/\frac{\partial f}{\partial y}\right) = -\left(\frac{\partial \phi}{\partial x}\Big/\frac{\partial \phi}{\partial y}\right)$$

$$\Rightarrow \quad \frac{\partial f}{\partial x}.\frac{\partial \phi}{\partial y} = \frac{\partial f}{\partial y}.\frac{\partial \phi}{\partial x}$$

$$\Rightarrow \quad \frac{\partial f}{\partial x}.\frac{\partial \phi}{\partial y} - \frac{\partial f}{\partial y}.\frac{\partial \phi}{\partial x} = 0$$

Example 6. *If* $u = x \log (xy)$, *where* $x^3 + y^3 + 3xy = 1$. *Find* $\frac{du}{dx}$. (VTU–2009, UPTU–2006)

Solution . We have $u = x \log (xy)$. ...(1)

$$\Rightarrow \quad \frac{\partial u}{\partial x} = x\left(\frac{1}{xy}.y\right) + \log xy = 1 + \log xy$$

and $\frac{\partial u}{\partial y} = x\left(\frac{1}{xy}.x\right) = \frac{x}{y}$

Also it is given that

$$x^3 + y^3 + 3xy = 1 \quad \text{...(2)}$$

Differentiating (2) we get

$$3x^2 + 3y^2\frac{dy}{dx} + 3\left(x\frac{dy}{dx} + y\right) = 0$$

$$\Rightarrow \quad \frac{dy}{dx} = -\left(\frac{x^2 + y}{x + y^2}\right)$$

Now,

$$\frac{du}{dx} = \frac{\partial u}{\partial x} + \frac{\partial u}{\partial y}.\frac{dy}{dx}$$

$$= 1 + \log(xy) + \frac{x}{y}\left\{-\frac{(x^2 + y)}{(y^2 + x)}\right\}$$

$$= 1 + \log(xy) - \frac{x(x^2 + y)}{y(y^2 + x)}$$

Example 7. *If* $u = \sin^{-1}(x - y)$, $x = 3t$, $y = 4t^3$, *show that* $\frac{du}{dt} = \frac{3}{\sqrt{1 - t^2}}$.

Solution . Clearly, u is a composite function of t.

$$\therefore \quad \frac{du}{dt} = \frac{\partial u}{\partial x}.\frac{dx}{dt} + \frac{\partial u}{\partial y}.\frac{dy}{dt}$$

$$= \frac{1}{\sqrt{1 - (x - y)^2}}.3 + \frac{1}{\sqrt{1 - (x - y)^2}}.(-1).12t^2$$

$$= \frac{3(1 - 4t^2)}{\sqrt{1 - (3t - 4t^3)^2}} = \frac{3(1 - 4t^2)}{\sqrt{1 - 9t^2 + 24t^4 - 16t^6}}$$

$$= \frac{3(1 - 4t^2)}{\sqrt{(1 - t^2)(1 - 8t^2 + 16t^4)}}$$

$$= \frac{3(1 - 4t^2)}{\sqrt{(1 - t^2)(1 - 4t^2)^2}} = \frac{3}{\sqrt{1 - t^2}}.$$

Example 8. *Show that* $\frac{\partial^2 z}{\partial u^2} + \frac{\partial^2 z}{\partial v^2} = \frac{\partial^2 z}{\partial u^2} + \frac{\partial^2 z}{\partial v^2}$ *where* $x = u \cos \alpha - v \sin \alpha$, $y = u \sin \alpha + v \cos \alpha$. (UPTU–2008)

Solution . We have z is a composite function of u and v. Therefore,

$$\frac{\partial z}{\partial u} = \frac{\partial z}{\partial x}.\frac{\partial x}{\partial u} + \frac{\partial z}{\partial y}.\frac{\partial y}{\partial u} = \cos\alpha\frac{\partial z}{\partial x} + \sin\alpha\frac{\partial z}{\partial y}$$

$$\Rightarrow \quad \frac{\partial}{\partial u} = \cos\alpha\frac{\partial}{\partial x} + \sin\alpha\frac{\partial}{\partial y} \quad \text{...(1)}$$

Also, $\frac{\partial z}{\partial v} = \frac{\partial z}{\partial x}.\frac{\partial x}{\partial v} + \frac{\partial z}{\partial y}.\frac{\partial y}{\partial v}$

$$\Rightarrow \quad = -\sin\alpha\frac{\partial z}{\partial x} + \cos\alpha\frac{\partial z}{\partial y}$$

$$\frac{\partial}{\partial v} = -\sin\alpha\frac{\partial}{\partial x} + \cos\alpha\frac{\partial}{\partial y} \quad \text{...(2)}$$

Now,

$$\frac{\partial^2 z}{\partial u^2} = \frac{\partial}{\partial u}\left(\frac{\partial z}{\partial u}\right)$$

$$= \left(\cos\alpha\frac{\partial}{\partial x} + \sin\alpha\frac{\partial}{\partial y}\right)\left(\cos\alpha\frac{\partial z}{\partial x} + \sin\alpha\frac{\partial z}{\partial y}\right)$$

$$= \cos^2\alpha\frac{\partial^2 z}{\partial x^2} + \cos\alpha\sin\alpha\frac{\partial^2 z}{\partial x\partial y} + \sin\alpha\cos\alpha\frac{\partial^2 z}{\partial y\partial x} + \sin^2\alpha\frac{\partial^2 z}{\partial y^2}$$

$$= \cos^2\alpha\frac{\partial^2 z}{\partial x^2} + 2\cos\alpha\sin\alpha\frac{\partial^2 z}{\partial x\partial y} + \sin^2\alpha\frac{\partial^2 z}{\partial y^2} \quad \text{...(3)}$$

Also, $\frac{\partial^2 z}{\partial v^2} = \frac{\partial}{\partial v}\left(\frac{\partial z}{\partial v}\right)$

$$= \left(-\sin\alpha\frac{\partial}{\partial x} + \cos\alpha\frac{\partial}{\partial y}\right)\left(-\sin\alpha\frac{\partial z}{\partial x} + \cos\alpha\frac{\partial z}{\partial y}\right)$$

$$= \sin^2\alpha\frac{\partial^2 z}{\partial x^2} - \sin\alpha\cos\alpha\frac{\partial^2 z}{\partial x\partial y} - \cos\alpha\sin\alpha\frac{\partial^2 z}{\partial x\partial y} + \cos^2\alpha\frac{\partial^2 z}{\partial y^2}$$

$$= \sin^2\alpha\frac{\partial^2 z}{\partial x^2} - 2\cos\alpha\sin\alpha\frac{\partial^2 z}{\partial x\partial y} + \cos^2\alpha\frac{\partial^2 z}{\partial y^2} \quad \text{...(4)}$$

Adding (3) and (4), we get

$$\frac{\partial^2 z}{\partial u^2} + \frac{\partial^2 z}{\partial v^2} = \frac{\partial^2 z}{\partial u^2} + \frac{\partial^2 z}{\partial v^2}$$

Example 9. If $u = u\left(\frac{y-x}{xy}, \frac{z-x}{xz}\right)$, show that

$$x^2\frac{\partial u}{\partial x} + y^2\frac{\partial u}{\partial y} + z^2\frac{\partial u}{\partial z} = 0. \quad \text{(UPTU–2005)}$$

Solution. Suppose $v = \frac{y-x}{xy} = \frac{1}{x} - \frac{1}{y}$

and $w = \frac{z-x}{xz} = \frac{1}{x} - \frac{1}{z}$ ...(1)

Then clearly, $u = u(v, w)$

$$\therefore \quad \frac{\partial u}{\partial x} = \frac{\partial u}{\partial v}\cdot\frac{\partial v}{\partial x} + \frac{\partial u}{\partial w}\cdot\frac{\partial w}{\partial x}$$

$$= \frac{\partial u}{\partial v}\left(-\frac{1}{x^2}\right) + \frac{\partial u}{\partial w}\left(-\frac{1}{x^2}\right)$$

$$\Rightarrow \quad x^2\frac{\partial u}{\partial x} = -\frac{\partial u}{\partial v} - \frac{\partial u}{\partial w} \quad \text{...(2)}$$

Further, $\frac{\partial u}{\partial y} = \frac{\partial u}{\partial v}\cdot\frac{\partial v}{\partial y} + \frac{\partial u}{\partial w}\cdot\frac{\partial w}{\partial y}$

$$= \frac{\partial u}{\partial v}\left(\frac{1}{y^2}\right) + \frac{\partial u}{\partial w}(0)$$

$$\Rightarrow \quad y^2\frac{\partial u}{\partial y} = \frac{\partial u}{\partial v} \quad \text{...(3)}$$

Similarly,

$$\frac{\partial u}{\partial z} = \frac{\partial u}{\partial v}\cdot\frac{\partial v}{\partial z} + \frac{\partial u}{\partial w}\cdot\frac{\partial w}{\partial z}$$

$$= \frac{\partial u}{\partial v}(0) + \frac{\partial u}{\partial w}\left(\frac{1}{z^2}\right)$$

$$\Rightarrow \quad z^2\frac{\partial u}{\partial z} = \frac{\partial u}{\partial w} \quad \text{...(4)}$$

Finally, adding (2), (3) and (4), we get

$$x^2\frac{\partial u}{\partial x} + y^2\frac{\partial u}{\partial y} + z^2\frac{\partial u}{\partial z} = 0.$$

Example 10. If $f(x, y) = 0$, show that $\frac{\partial^2 y}{\partial x^2} = -\frac{q^2 r - 2pqs + p^2 t}{q^3}$

(Kurukshetra–2006)

Solution. We have $\frac{dy}{dx} = -\frac{\partial f/\partial x}{\partial f/\partial y} = \frac{-p}{q}$

$$\Rightarrow \quad \frac{d^2y}{dx^2} = -\frac{d}{dx}\left(\frac{dy}{dx}\right) \quad \text{...(1)}$$

$$= -\frac{d}{dx}\left(\frac{p}{q}\right) = \frac{q\frac{dp}{dx} - p\frac{dq}{dx}}{q^2}$$

Now, $\frac{dp}{dx} = \frac{\partial p}{\partial x} + \frac{\partial p}{\partial y}\cdot\frac{dy}{dx} = r + s\left(-\frac{p}{q}\right) = -\frac{qr - ps}{q}$

and $\frac{dq}{dx} = \frac{\partial q}{\partial x} + \frac{\partial q}{\partial y}\cdot\frac{dy}{dx} = s + t\left(-\frac{p}{q}\right) = \frac{qs - pt}{q}$

Putting all these value in (1), we get

$$\frac{d^2y}{dx^2} = -\frac{1}{q^2}\left[q\left(\frac{qr - ps}{q}\right) - p\left(\frac{qs - pt}{q}\right)\right]$$

$$= -\frac{q^2 r - 2pqs + p^2 t}{q^3}$$

Here, $p = \frac{\partial f}{\partial x}, q = \frac{\partial f}{\partial y}, r = \frac{\partial^2 f}{\partial x^2} = \frac{\partial p}{\partial x}$

$$s = \frac{\partial^2 f}{\partial x \partial y} = \frac{\partial q}{\partial x}, \; t = \frac{\partial^2 f}{\partial y^2} = \frac{\partial q}{\partial y}$$

Example 11. If $u = f(r, s, t)$ and $r = \frac{x}{y}, s = \frac{y}{z}, t = \frac{z}{x}$, prove that $x\frac{\partial u}{\partial x} + y\frac{\partial u}{\partial y} + z\frac{\partial u}{\partial z} = 0$.

(JNTU–1990, 2007, UKTU–2011, 12)

Solution. We have $u = f(r, s, t)$ ...(1)

then $\frac{\partial u}{\partial x} = \frac{\partial u}{\partial r}\cdot\frac{\partial r}{\partial x} + \frac{\partial u}{\partial s}\cdot\frac{\partial s}{\partial x} + \frac{\partial u}{\partial t}\cdot\frac{\partial t}{\partial x}$

$$= \frac{1}{y}\cdot\frac{\partial u}{\partial r} + 0.\frac{\partial u}{\partial s} - \frac{z}{x^2}\cdot\frac{\partial u}{\partial t} \quad \text{...(2)}$$

Also, $\frac{\partial u}{\partial y} = \frac{\partial u}{\partial r}\cdot\frac{\partial r}{\partial y} + \frac{\partial u}{\partial s}\cdot\frac{\partial s}{\partial y} + \frac{\partial u}{\partial t}\cdot\frac{\partial t}{\partial y}$

$$= -\frac{x}{y^2}\cdot\frac{\partial u}{\partial r} + \frac{1}{z}\cdot\frac{\partial u}{\partial s} + 0.\frac{\partial u}{\partial t} \quad \text{...(3)}$$

and $\frac{\partial u}{\partial z} = \frac{\partial u}{\partial r}\cdot\frac{\partial r}{\partial z} + \frac{\partial u}{\partial s}\cdot\frac{\partial s}{\partial z} + \frac{\partial u}{\partial t}\cdot\frac{\partial t}{\partial z}$

$$= 0.\frac{\partial u}{\partial r} + \left(\frac{-y}{z^2}\right)\cdot\frac{\partial u}{\partial s} + \frac{1}{x}\cdot\frac{\partial u}{\partial t}$$

...(4)

Now multiplying (2) by x, (3) by y and (4) by z and then adding we get

$$x\frac{\partial u}{\partial x} + y\frac{\partial u}{\partial y} + z\frac{\partial u}{\partial z} = 0$$

EXERCISE 9.3

1. If $(\tan x)^y + (y)^{\cot x} = a$. Find the value of $\frac{dy}{dx}$.
2. If $u = \sin(x^2 + y^2)$, where $a^2x^2 + b^2y^2 = c^2$. Find the value of $\frac{du}{dx}$.
3. If $u = f(y - z, z - x, x - y)$, prove that $\frac{\partial u}{\partial x} + \frac{\partial u}{\partial y} + \frac{\partial u}{\partial z} = 0$.
(UKTU–2010, GBTU–2010)
4. If z is a function of x and y; where $x = e^u + e^{-v}$ and $y = e^{-u} - e^v$, show that $\frac{\partial z}{\partial u} - \frac{\partial z}{\partial v} = x\frac{\partial z}{\partial x} - y\frac{\partial z}{\partial y}$. (VTU–2003, 06)

5. Find the total derivative of u with respect to t, when

(i) $u = \cosh\left(\frac{y}{x}\right)$, where $x = t^2, y = e^t$

(ii) $u = e^x \sin y$, where $x = \log t, y = t^2$

6. If $u = \sqrt{(x^2+y^2)}$ and $x^3 + y^3 + 3axy = 5a^2$. Find the value of $\frac{du}{dx}$ at $x = a, y = a$.

7. Find $\frac{dy}{dx}$ and $\frac{d^2y}{dx^2}$ from the following implicit relations.

(i) $x^2 + y^2 = a^2$ (ii) $x^{2/3} + y^{2/3} = a^{2/3}$

8. If $f(x, y, z) = 0$, show that

$$\left(\frac{\partial y}{\partial z}\right)_{x\text{ const.}} \cdot \left(\frac{\partial z}{\partial x}\right)_{y\text{ const.}} \cdot \left(\frac{\partial x}{\partial y}\right)_{z\text{ const.}} = -1.$$

9. If $x + y = 2e^\theta \cos\phi, x - y = 2ie^\theta \sin\phi$, where $i = \sqrt{(-1)}$,

show that $\frac{\partial^2 u}{\partial\theta^2} + \frac{\partial^2 u}{\partial\phi^2} = 4xy\frac{\partial^2 u}{\partial x \partial y}$. (Lucknow–2005)

ANSWERS

1. $-\frac{y(\tan x)^{y-1}\sec^2 x - y^{\cot x}.\log y.\operatorname{cosec}^2 x}{(\tan x)^y \log\tan x + \cot x\, y^{\cot x - 1}}$ **2.** $2x[\cos(x^2+y^2)]\left(1 - \frac{a^2}{b^2}\right)$ **5.** (i) $\frac{du}{dt} = \frac{1}{x^2}(xe^t - 2yt)\sinh\frac{y}{x}$

5. (ii) $\frac{du}{dt} = \frac{e^x}{t}(\sin y + 2t^2\cos y)$, where $x = \log t, y = e^t$ **6.** 0 **7.** (i) $-\frac{x}{y}, \frac{-a^2}{y^3}$ (ii) $\frac{dy}{dx} = -\frac{y^{1/3}}{x^{1/3}}, \frac{d^2y}{dx^2} = \frac{a^{1/3}}{3x^{4/3}.y^{1/3}}$

9.8 CHANGE OF VARIABLES

9.8.1 CHANGE OF INDEPENDENT VARIABLE INTO DEPENDENT VARIABLES

Let $y = f(x)$ be a function with x dependent and y is independent variable. Then

$$\frac{dy}{dx} = 1\Big/\left(\frac{dx}{dy}\right) = \left(\frac{dx}{dy}\right)^{-1} \quad \ldots(1)$$

$\therefore$

$$\frac{d^2y}{dx^2} = \frac{d}{dx}\left(\frac{dy}{dx}\right) = \frac{d}{dx}\left[\left(\frac{dx}{dy}\right)^{-1}\right] = \frac{d}{dy}\left[\left(\frac{dx}{dy}\right)^{-1}\right]\frac{dy}{dx}$$

$$= -\left(\frac{dx}{dy}\right)^{-2}.\frac{d^2x}{dy^2}.\frac{dy}{dx} = -\left(\frac{dx}{dy}\right)^{-2}.\frac{d^2x}{dy^2}\left(\frac{dx}{dy}\right)^{-1} \quad \text{[From (1)]}$$

or

$$\frac{d^2y}{dx^2} = -\left(\frac{dx}{dy}\right)^{-3}.\frac{d^2x}{dy^2} \quad \ldots(2)$$

and

$$\frac{d^3y}{dx^3} = -\frac{d}{dx}\left(\frac{d^2y}{dx^2}\right) = \frac{d}{dx}\left[\left(-\frac{dx}{dy}\right)^{-3}\frac{d^2x}{dy^2}\right] \quad \text{[From (2)]}$$

$$= \frac{d}{dy}\left[-\left(\frac{dx}{dy}\right)^{-3}\frac{d^2x}{dy^2}\right]\frac{dy}{dx} = \left\{3\left(\frac{dx}{dy}\right)^{-4}\frac{d^2x}{dy^2}\right\}\frac{d^2x}{dy^2}.\frac{dy}{dx} - \left(\frac{dx}{dy}\right)^{-3}\frac{d^3x}{dy^3}\frac{dy}{dx}$$

$$= 3\left(\frac{dx}{dy}\right)^{-4}\left(\frac{d^2x}{dy^2}\right)^2\left(\frac{dx}{dy}\right)^{-1} - \left(\frac{dx}{dy}\right)^{-3}\frac{d^3x}{dy^3}\left(\frac{dx}{dy}\right)^{-1} \quad \text{[From (1)]}$$

or

$$\frac{d^3y}{dx^3} = 3\left(\frac{dx}{dy}\right)^{-5}\left(\frac{d^2x}{dy^2}\right)^2 - \left(\frac{dx}{dy}\right)^{-4}\frac{d^3x}{dy^3} \quad \ldots(3)$$

Similarly, we can find $\frac{d^4y}{dx^4}, \frac{d^5y}{dx^5}$, etc.

Solved Examples

Example . *Show that the equation* $\frac{dy}{dx}.\frac{d^3y}{dx^3} - 3\left(\frac{d^2y}{dx^2}\right)^2 = 0$ *can be written in the form* $\frac{d^3x}{dy^3} = 0$.

Solution . Here, we have

$$\frac{dy}{dx} = \left(\frac{dx}{dy}\right)^{-1}, \frac{d^2y}{dx^2} = -\left(\frac{dx}{dy}\right)^{-3}\frac{d^2x}{dy^2}$$

and

$$\frac{d^3y}{dx^3} = 3\left(\frac{dx}{dy}\right)^{-5}\left(\frac{d^2x}{dy^2}\right)^2 - \left(\frac{d^3x}{dy^3}\right)\left(\frac{dx}{dy}\right)^{-4}$$

Making these substitutions in the given equation, we have

$$\left(\frac{dx}{dy}\right)^{-1}\left[3\left(\frac{dx}{dy}\right)^{-5}\left(\frac{d^2x}{dy^2}\right)^2-\left(\frac{d^3x}{dy^3}\right)\left(\frac{dx}{dy}\right)^{-4}\right]$$

$$-3\left[-\left(\frac{dx}{dy}\right)^{-3}\frac{d^2x}{dy^2}\right]^2=0$$

$$\Rightarrow \quad -\frac{d^3x}{dy^3}\left(\frac{dx}{dy}\right)^{-3}=0$$

$$\Rightarrow \quad \frac{d^3x}{dy^3}=0 \text{ since } \frac{dx}{dy}\neq 0.$$

9.8.2 CHANGE OF INDEPENDENT VARIABLE INTO ANOTHER VARIABLE z, GIVEN x = f(z)

We have $$\frac{dy}{dx}=\frac{dy}{dz}\cdot\frac{dz}{dx}=\frac{dy}{dz}\left(\frac{dx}{dz}\right)^{-1} \quad \text{...(1)}$$

or $$\frac{d}{dx}(y)=\left(\frac{dx}{dz}\right)^{-1}\frac{d}{dz}(y) \quad \text{...(2)}$$

i.e., the operator $\frac{d}{dx}$ is equivalent to the operator $\left(\frac{dx}{dz}\right)^{-1}\frac{d}{dz}$ or $\frac{d}{dx}\equiv\left(\frac{dx}{dz}\right)^{-1}\frac{d}{dz}$.

Therefore, $$\frac{d^2y}{dx^2}=\frac{d}{dx}\left(\frac{dy}{dx}\right)=\left(\frac{dx}{dz}\right)^{-1}\frac{d}{dz}\left(\frac{dy}{dx}\right), \text{ with the help of (2)}$$

$$=\left(\frac{dx}{dz}\right)^{-1}\frac{d}{dz}\left[\frac{dy}{dz}\left(\frac{dx}{dz}\right)^{-1}\right] \quad \text{[From (1)]}$$

$$=\left(\frac{dx}{dz}\right)^{-1}\left[\frac{dy}{dz}\left\{-\left(\frac{dx}{dz}\right)^{-2}\cdot\frac{d^2x}{dz^2}\right\}+\left(\frac{dx}{dz}\right)^{-1}\frac{d^2y}{dz^2}\right]$$

or $$\frac{d^2y}{dx^2}=\left(\frac{dx}{dz}\right)^{-3}\left[\frac{dx}{dz}\cdot\frac{d^2y}{dz^2}-\frac{dy}{dz}\cdot\frac{d^2x}{dz^2}\right]=\frac{\frac{dx}{dz}\cdot\frac{d^2y}{dz^2}-\frac{dy}{dz}\cdot\frac{d^2x}{dz^2}}{\left(\frac{dx}{dz}\right)^3} \quad \text{...(3)}$$

and $$\frac{d^3y}{dx^3}=\frac{d}{dx}\left[\frac{d^2y}{dx^2}\right]=\left(\frac{dx}{dz}\right)^{-1}\frac{d}{dz}\left(\frac{d^2y}{dx^2}\right), \quad \text{[From (2)]}$$

$$=\left(\frac{dx}{dz}\right)^{-1}\frac{d}{dz}\left[\frac{\frac{dx}{dz}\cdot\frac{d^2y}{dz^2}-\frac{dy}{dz}\cdot\frac{d^2x}{dz^2}}{\left(\frac{dx}{dz}\right)^3}\right], \quad \text{[From (3)]}$$

$$=\frac{\left(\frac{dx}{dz}\right)^{-1}\left[\left(\frac{dx}{dz}\right)^3\left\{\frac{dx}{dz}\cdot\frac{d^3y}{dz^3}+\frac{d^2x}{dz^2}\cdot\frac{d^2y}{dz^2}-\frac{d^2y}{dz^2}\cdot\frac{d^2x}{dz^2}-\frac{dy}{dz}\cdot\frac{d^3x}{dz^3}\right\}-3\left(\frac{dx}{dz}\right)^2\cdot\frac{d^2x}{dz^2}.N_r z\right]}{\left(\frac{dx}{dz}\right)^6}$$

where $$N_r=\frac{dx}{dz}\cdot\frac{d^2y}{dz^2}-\frac{dy}{dz}\cdot\frac{d^2x}{dz^2}$$

or $$\frac{d^3y}{dx^3}=\left(\frac{dx}{dz}\right)^{-5}\left[\left(\frac{dx}{dz}\frac{d^3y}{dz^3}-\frac{dy}{dz}\frac{d^3x}{dz^3}\right)\frac{dx}{dz}-3\frac{d^2x}{dz^2}\left(\frac{dx}{dz}\frac{d^2y}{dz^2}-\frac{dy}{dz}\frac{d^2x}{dz^2}\right)\right]$$

Solved Examples

Example 1. *Transform the equation*

$$\frac{d}{dx}\left\{(1-x^2)\frac{dy}{dx}\right\}+n(n+1)y=0,$$

by the substitution $x=\frac{1}{2}[z+(1/z)]$.

Solution. Given $\quad x=\frac{1}{2}[z+(1/z)]$

$$\therefore \quad \frac{dx}{dz}=\frac{1}{2}\left(1-\frac{1}{z^2}\right)=\frac{z^2-1}{2z^2} \qquad \ldots(1)$$

$$\therefore \quad \frac{dy}{dx}=\frac{dy}{dz}\cdot\frac{dz}{dx}=\frac{2z^2}{z^2-1}\cdot\frac{dy}{dz}$$

[From (1)]

or $$(1-x^2)\frac{dy}{dx}=\left[1-\frac{1}{4}\left(z+\frac{1}{z}\right)^2\right]\frac{2z^2}{z^2-1}\frac{dy}{dz},$$

(substituting value of x)

$$=\left[1-\frac{1}{4}z^2-\frac{1}{4z^2}-\frac{1}{2}\right]\frac{2z^2}{z^2-1}\frac{dy}{dz}$$

$$=\frac{2z^2-z^4-1}{4z^2}\cdot\frac{2z^2}{z^2-1}\frac{dy}{dz}=-\frac{1}{2}(z^2-1)\frac{dy}{dz}$$

$\therefore$ Putting this value of $(1-x^2)\frac{dy}{dx}$ in given equation, we get

$$\frac{d}{dx}\left\{-\frac{1}{2}(z^2-1)\frac{dy}{dz}\right\}+n(n+1)y=0$$

or $$-\frac{1}{2}\frac{d}{dz}\left\{(z^2-1)\frac{dy}{dz}\right\}\cdot\frac{dz}{dx}+n(n+1)y=0$$

or $$-\frac{1}{2}\left[(z^2-1)\frac{d^2y}{dz^2}+2z.\frac{dy}{dz}\right]\cdot\frac{2z^2}{z^2-1}+n(n+1)y=0$$

or $$z^2(z^2-1)\frac{d^2y}{dz^2}+2z^3\frac{dy}{dz}-n(n+1)(z^2-1)y=0.$$

Example 2. *Transform the equation* $(1+x^2)^2y_2+2x(1+x^2)y_1+y=0$ *by the substitution* $x=\tan z$.

Solution. Given $x=\tan z$. $\qquad \ldots(1)$

$$\therefore \quad \frac{dx}{dz}=\sec^2 z$$

or $$\frac{dz}{dx}=\cos^2 z \qquad \ldots(2)$$

Now $$y_1=\frac{dy}{dx}=\frac{dy}{dz}\cdot\frac{dz}{dx}=\cos^2 z.\frac{dy}{dz}$$

[From (1)]

$$y_2=\frac{d^2y}{dx^2}=\frac{d}{dx}\left(\frac{dy}{dx}\right)=\frac{d}{dx}\left[\cos^2 z\frac{dy}{dz}\right]$$

[From (2)]

$$=\frac{d}{dz}\left[\cos^2 z\frac{dy}{dz}\right]\frac{dz}{dx}$$

$$=\left[\cos^2 z\frac{d^2y}{dz^2}+2\cos z(-\sin z)\frac{dy}{dz}\right]\cos^2 z$$

[From (1)]

or $$\frac{d^2y}{dx^2}=\cos^4 z\frac{d^2y}{dz^2}-2\sin z\cos^3 z\frac{dy}{dz} \qquad \ldots(3)$$

Substituting the value of x, y_1 and y_2 in the given equation, we get

$$(1+\tan^2 z)^2\left[\cos^4 z\frac{d^2y}{dz^2}-2\sin z\cos^3 z\frac{dy}{dz}\right]+2\tan z(1+\tan^2 z)\cos^2 z\frac{dy}{dz}+y=0$$

or $$\frac{d^2y}{dz^2}-2\tan z\frac{dy}{dz}+2\tan z\frac{dy}{dz}+y=0$$

or $$\frac{d^2y}{dz^2}+y=0.$$

Example 3. *Transform the equation*

$$y_2+[1-(1/x)]y_1+4x^2ye^{-2x}=4(x^2+x^3)e^{-3x}$$

by the substitution $z=(1+x)e^{-x}$.

Solution. Given $\quad z=(1+x)e^{-x}$

$$\therefore \quad \frac{dz}{dx}=(1+x)(-e^{-x})+e^{-x}=-xe^{-x} \qquad \ldots(1)$$

Now, $$\frac{dy}{dx}=\frac{dy}{dz}\cdot\frac{dz}{dx}=-xe^{-x}\frac{dy}{dz} \quad \text{[From (1)]} \qquad \ldots(2)$$

and $$\frac{d^2y}{dx^2}=\frac{d}{dx}\left(\frac{dy}{dx}\right)=\frac{d}{dx}\left[-xe^{-x}\frac{dy}{dz}\right].$$

[From (2)]

$$=-\left[e^{-x}\frac{dy}{dz}+x(-e^{-x})\frac{dy}{dz}+xe^{-x}\frac{d}{dx}\left(\frac{dy}{dz}\right)\right]$$

$$=-e^{-x}\left[\frac{dy}{dz}-x\frac{dy}{dz}+x\frac{d^2y}{dz^2}\frac{dz}{dx}\right]$$

$$\left[\because \frac{d}{dx}\left(\frac{dy}{dz}\right)=\frac{d^2y}{dz^2}\frac{dz}{dx}\right]$$

$$=-e^{-x}\left[\frac{dy}{dz}-x\frac{dy}{dz}-x^2e^{-x}\frac{d^2y}{dz^2}\right]$$

[From (1)]

Substituting these values of dy/dx, d^2y/dx^2 in the given equation, we get

$$-e^{-x}\left[\frac{dy}{dz}-x\frac{dy}{dz}-x^2e^{-x}\frac{d^2y}{dz^2}\right]+\left(1-\frac{1}{x}\right)\left(-xe^{-x}\frac{dy}{dz}\right)+4x^2ye^{-2x}=4x^2(1+x)e^{-3x}$$

or $$x^2e^{-2x}\frac{d^2y}{dz^2}+(x-1-x+1)e^{-x}\frac{dy}{dx}+4x^2ye^{-2x}=4x^2(1+x)e^{-3x}$$

or $$\left(d^2y/dz^2\right)+4y=4(1+x)e^{-x}=4z \quad [\because z=(1+x)e^{-x}]$$

or $$\left(d^2y/dz^2\right)+4y=4z$$

which is the required equation.

9.8.3 TRANSFORMATION INVOLVING CHANGE OF DEPENDENT AS WELL AS INDEPENDENT VARIABLES

Such transformations will be clear from the examples given below.

Example 1. *Transform into cartesian the polar formula* $\tan\phi=\dfrac{r d\theta}{dr}$.

Solution . We know $x=r\cos\theta, y=r\sin\theta$.

Hence we get $r^2=x^2+y^2$ and $\theta=\tan^{-1}(y/x)$.

$$\therefore \quad 2r\frac{dr}{dx}=2x+2y\frac{dy}{dx}$$

or $$r\frac{dr}{dx}=x+y\frac{dy}{dx} \quad \ldots(1)$$

and $$\frac{d\theta}{dx}=\frac{1}{1+(y/x)^2}\cdot\frac{x(dy/dx)-y.1}{x^2}=\frac{1}{x^2+y^2}\left(x\frac{dy}{dx}-y\right) \quad \ldots(2)$$

Now, $$\tan\phi=r\frac{d\theta}{dr}=\frac{r\frac{d\theta}{dx}}{\frac{dr}{dx}}=\frac{r\left[1/(x^2+y^2)\right]\left(x\frac{dy}{dx}-y\right)}{(1/r)\left(x+y\frac{dy}{dx}\right)}$$

[From (1) and (2)]

or $$\tan\phi=\frac{x(dy/dx)-y}{x+y(dy/dx)},\text{ using } r^2=x^2+y^2.$$

This is the required formula in cartesian form.

Example 2. *Transform cartesian formula*

$$\rho=\frac{\left[1+(dy/dx)^2\right]^{3/2}}{d^2y/dx^2} \text{ into polar form.}$$

Solution . We know $x=r\cos\theta, y=r\sin\theta$.

$$\therefore \quad \frac{dx}{d\theta}=r(-\sin\theta)+\cos\theta.\frac{dr}{d\theta} \quad \ldots(1)$$

and $$\frac{dy}{d\theta}=r\cos\theta+\sin\theta\frac{dr}{d\theta} \quad \ldots(2)$$

Now $$\frac{dy}{dx}=\frac{dy/d\theta}{dx/d\theta}=\frac{r\cos\theta+\sin\theta(dr/d\theta)}{\cos\theta(dr/d\theta)-r\sin\theta} \quad \ldots(3)$$

Again $$\frac{d^2y}{dx^2}=\frac{d}{dx}\left(\frac{dy}{dx}\right)=\frac{d}{d\theta}\left(\frac{dy}{dx}\right)\frac{d\theta}{dx}$$

$$=\frac{d}{d\theta}\left[\frac{\left(r\cos\theta+\sin\theta.\frac{dr}{d\theta}\right)}{\cos\theta(dr/d\theta)-r\sin\theta}\right]\frac{1}{d\theta/dx}.$$

[From (3)]

$$=\frac{\left\{\frac{dr}{d\theta}\cos\theta-r\sin\theta\right\}\left\{\frac{dr}{d\theta}\cos\theta-r\sin\theta+\frac{dr}{d\theta}\cos\theta+\sin\theta\frac{d^2r}{d\theta^2}\right\}-\left\{r\cos\theta+\sin\theta.\frac{dr}{d\theta}\right\}\times\left\{\cos\theta\frac{d^2r}{d\theta^2}-\sin\theta\frac{dr}{d\theta}-\frac{dr}{d\theta}\sin\theta-r\cos\theta\right\}}{[(dr/d\theta)\cos\theta-r\sin\theta]^3}$$

$$=\frac{\left\{\frac{dr}{d\theta}\cos\theta-r\sin\theta\right\}\left\{\sin\theta\frac{d^2r}{d\theta^2}+2\frac{dr}{d\theta}\cos\theta-r\sin\theta\right\}-\left\{r\cos\theta+\sin\theta.\frac{dr}{d\theta}\right\}\times\left\{\cos\theta\frac{d^2r}{d\theta^2}-2\sin\theta\frac{dr}{d\theta}-r\cos\theta\right\}}{[(dr/d\theta)\cos\theta-r\sin\theta]^3}$$

$$=\left[2\left(\frac{dr}{d\theta}\right)^2-r\frac{d^2r}{d\theta^2}+r^2\right]\Bigg/\left[\frac{dr}{d\theta}\cos\theta-r\sin\theta\right]^3.$$

After simplification putting these values in cartesian formula, we get

$$\rho=\frac{\left[1+(dy/dx)^2\right]^{3/2}}{d^2y/dx^2}$$

$$= \frac{\left[1+\left\{\frac{r\cos\theta+\sin\theta(dr/d\theta)}{\cos\theta(dr/d\theta)-r\sin\theta}\right\}^2\right]^{3/2}}{\left[2\left(\frac{dr}{d\theta}\right)^2-r\frac{d^2r}{d\theta^2}+r^2\right]\Big/\left[\frac{dr}{d\theta}\cos\theta-r\sin\theta\right]^3}$$

$$= \left[\left(\frac{dr}{d\theta}\right)^2+r^2\right]^{3/2}\Big/\left[2\left(\frac{dr}{d\theta}\right)^2-r\frac{d^2r}{d\theta^2}+r^2\right]$$

which is required polar form.

9.8.4 TRANSFORMATION IN THE CASE OF TWO INDEPENDENT VARIABLES

Here we use the following important results :

1. If $z = f(x, y)$, then $dz = \frac{\partial z}{\partial x}dx + \frac{\partial z}{\partial y}dy$. ...(1)

2. If $z = f(x, y)$, where $r = \phi(t)$ and $y = \psi(t)$, then $\frac{dz}{dt} = \frac{\partial z}{\partial x}\frac{dx}{dt} + \frac{\partial z}{\partial y}\frac{dy}{dt}$. ...(2)

3. If $z = f(x, y)$, where $x = \phi(t_1, t_2)$ and $y = \phi(t_1, t_2)$, then $\frac{\partial z}{\partial t_1} = \frac{\partial z}{\partial x}\frac{\partial x}{\partial t_1} + \frac{\partial z}{\partial y}\frac{\partial y}{\partial t_1}$ and $\frac{\partial z}{\partial t_2} = \frac{\partial z}{\partial x}\frac{\partial x}{\partial t_2} + \frac{\partial z}{\partial y}\frac{\partial y}{\partial t_2}$ ...(3)

4. In case $x = \phi(t_1, t_2)$ and $y = \psi(t_1, t_2)$ can easily be solved for t_1 and t_2 in terms of x and y, say $t_1 = F_1(x, y)$ and $t_2 = F_2(x, y)$, then the following formulae are used $\frac{\partial z}{\partial x} = \frac{\partial x}{\partial t_1}\frac{\partial t_1}{\partial x} + \frac{\partial x}{\partial t_2}\frac{\partial t_2}{\partial x}$ and $\frac{\partial z}{\partial y} = \frac{\partial z}{\partial t_1}\frac{\partial t_1}{\partial y} + \frac{\partial z}{\partial t_2}\frac{\partial t_2}{\partial y}$. ...(4)

Solved Examples

Example 1. *If $z = f(x,y), x^2 = uv$ and $y^2 = u/v$, change the independent variables to u, v in the equation*

$$x^2\frac{\partial^2 z}{\partial x^2} - 2xy\frac{\partial^2 z}{\partial x\partial y} + y^2\frac{\partial^2 z}{\partial y^2} + 2y\frac{\partial z}{\partial y} = 0.$$

Solution . Solving $x^2 = uv$ and $y^2 = u/v$, we get

$$u = xy \text{ and } v = x/y. \quad ...(1)$$

$$\therefore \quad \frac{\partial u}{\partial x} = y, \frac{\partial u}{\partial y} = x, \frac{\partial v}{\partial x} = \frac{1}{y}, \frac{\partial v}{\partial y} = -\frac{x}{y^2}. \quad ...(2)$$

Now, $\frac{\partial z}{\partial x} = \frac{\partial z}{\partial u}.\frac{\partial u}{\partial x} + \frac{\partial z}{\partial v}.\frac{\partial v}{\partial x} = \frac{\partial z}{\partial u}(y) + \frac{\partial z}{\partial v}\left(\frac{1}{y}\right)$

[From (2)]

or $x\frac{\partial z}{\partial x} = (xy)\frac{\partial z}{\partial u} + \left(\frac{x}{y}\right)\frac{\partial z}{\partial v} = u\frac{\partial z}{\partial u} + v\frac{\partial z}{\partial v}$

[From (1)]

$$\therefore \quad x\frac{\partial}{\partial x} \equiv u\frac{\partial}{\partial u} + v\frac{\partial}{\partial v} \quad ...(3)$$

and similarly

$$y\frac{\partial}{\partial y} \equiv u\frac{\partial}{\partial u} - v\frac{\partial}{\partial v} \quad ...(4)$$

Now $\left(x\frac{\partial}{\partial x} - y\frac{\partial}{\partial y}\right)^2 z$

$$= \left(x\frac{\partial}{\partial x} - y\frac{\partial}{\partial y}\right)\left(x\frac{\partial z}{\partial x} - y\frac{\partial z}{\partial y}\right)z$$

$$= x\frac{\partial}{\partial x}\left(x\frac{\partial z}{\partial x} - y\frac{\partial z}{\partial y}\right) - y\frac{\partial}{\partial y}\left(x\frac{\partial z}{\partial x} - y\frac{\partial z}{\partial y}\right)z$$

$$= x\left[x\frac{\partial^2 z}{\partial x^2} + \frac{\partial z}{\partial x} - y\frac{\partial^2 z}{\partial x\partial y}\right] - y\left[x\frac{\partial^2 z}{\partial x\partial y} - y\frac{\partial^2 z}{\partial y^2} - \frac{\partial z}{\partial y}\right]z$$

or $\left(x\frac{\partial}{\partial x} - y\frac{\partial}{\partial y}\right)^2 z = x^2\frac{\partial^2 z}{\partial x^2} - 2xy\frac{\partial^2 z}{\partial x\partial y} + y^2\frac{\partial^2 z}{\partial y^2} + x\frac{\partial z}{\partial x} + y\frac{\partial z}{\partial y}$.

$\therefore$ The given equation

$$\left(x\frac{\partial}{\partial x} - y\frac{\partial}{\partial y}\right)^2 z + \left(y\frac{\partial z}{\partial y} - x\frac{\partial z}{\partial x}\right) = 0$$

which with the help of (3) and (4) reduces to

$$\left[\left(u\frac{\partial}{\partial u} + v\frac{\partial}{\partial v}\right) - \left(u\frac{\partial}{\partial u} - v\frac{\partial}{\partial v}\right)\right]^2 z + \left[\left(u\frac{\partial z}{\partial u} - v\frac{\partial z}{\partial v}\right) - \left(u\frac{\partial z}{\partial u} + v\frac{\partial z}{\partial v}\right)\right] = 0$$

or $$4\left(v\frac{\partial}{\partial v}\right)^2 z - 2v\frac{\partial z}{\partial v} = 0$$

or $$2v\frac{\partial}{\partial v}\left(v\frac{\partial z}{\partial v}\right) - v\frac{\partial z}{\partial v} = 0$$

or $$2\left[v\frac{\partial^2 z}{\partial v^2} + \frac{\partial z}{\partial v}\right] - \frac{\partial z}{\partial v} = 0$$

or $$2v\frac{\partial^2 z}{\partial v^2} + \frac{\partial z}{\partial v} = 0$$

9.8.5 TRANSFORMATION FROM CARTESIAN TO POLAR CO-ORDINATES AND VICE-VERSA

Transform the Laplace equation $\dfrac{\partial^2 u}{\partial x^2}+\dfrac{\partial^2 u}{\partial y^2}=0$ *to polars.*

We know $$x = r\cos\theta,\ y = r\sin\theta. \qquad \text{...(1)}$$

$\therefore$ $$r^2 = x^2 + y^2. \qquad \text{...(2)}$$

and $$\theta = \tan^{-1}(y/x) \qquad \text{...(3)}$$

From (2), we get $2r\dfrac{\partial r}{\partial x} = 2x$ or $$\frac{\partial r}{\partial x} = \frac{x}{r} = \frac{r\cos\theta}{r} \qquad \text{[From (1)]}$$

or $$\frac{\partial r}{\partial x} = \cos\theta \qquad \text{...(4)}$$

Similarly, $$\frac{\partial r}{\partial y} = \frac{y}{r} = \frac{r\sin\theta}{r} = \sin\theta \qquad \text{...(5)}$$

Also, from (3), $$\frac{\partial\theta}{\partial x} = \frac{1}{1+(y/x)^2}\left(-\frac{y}{x^2}\right) = -\frac{y}{x^2+y^2} = -\frac{r\sin\theta}{r^2} \text{ or } \frac{\partial\theta}{\partial x} = -\frac{\sin\theta}{r} \qquad \text{...(6)}$$

and $$\frac{\partial\theta}{\partial y} = \frac{1}{1+(y/x)^2}\left(\frac{1}{x}\right) = \frac{x}{x^2+y^2} = \frac{r\cos\theta}{r^2} = \frac{\cos\theta}{r} \qquad \text{...(7)}$$

Now $$\frac{\partial u}{\partial x} = \frac{\partial u}{\partial r}\cdot\frac{\partial r}{\partial x} + \frac{\partial u}{\partial\theta}\cdot\frac{\partial\theta}{\partial x} \qquad \text{...(8)}$$

$$= \frac{\partial u}{\partial r}.(\cos\theta) + \frac{\partial u}{\partial\theta}\left(-\frac{\sin\theta}{r}\right) \qquad \text{[From (4) and (6)]}$$

or $$\frac{\partial}{\partial x}(u) = \cos\theta\frac{\partial}{\partial r}(u) - \frac{\sin\theta}{r}\frac{\partial}{\partial\theta}(u) \qquad \text{...(9)}$$

Again $$\frac{\partial u}{\partial y} = \frac{\partial u}{\partial r}\cdot\frac{\partial r}{\partial y} + \frac{\partial u}{\partial\theta}\cdot\frac{\partial\theta}{\partial y} \qquad \text{[From (5) and (7)]}$$

or $$\frac{\partial}{\partial y}(u) = \sin\theta\frac{\partial}{\partial r}(u) + \frac{\cos\theta}{r}\frac{\partial}{\partial\theta}(u) \qquad \text{...(10)}$$

$\therefore$ $$\frac{\partial^2 u}{\partial x^2} = \frac{\partial}{\partial x}\left(\frac{\partial u}{\partial x}\right) = \cos\theta\frac{\partial}{\partial r}\left(\frac{\partial u}{\partial x}\right) - \frac{\sin\theta}{r}\frac{\partial}{\partial\theta}\left(\frac{\partial u}{\partial x}\right) \qquad \text{replacing } u \text{ by } \frac{\partial u}{\partial x} \text{ in (9)}$$

$$= \cos\theta\frac{\partial}{\partial r}\left[\cos\theta\frac{\partial u}{\partial r} - \frac{\sin\theta}{r}\frac{\partial u}{\partial\theta}\right] - \frac{\sin\theta}{r}\frac{\partial}{\partial\theta}\left[\cos\theta\frac{\partial u}{\partial r} - \frac{\sin\theta}{r}\frac{\partial u}{\partial\theta}\right]$$

substituting from (9) the polar equivalent of $\dfrac{\partial u}{\partial x}$

$$= \cos\theta\left[\cos\theta\frac{\partial^2 u}{\partial r^2} - \sin\theta\frac{\partial}{\partial r}\left(\frac{1}{r}\frac{\partial u}{\partial\theta}\right)\right]$$

$$-\frac{\sin\theta}{r}\left[\left(\cos\theta\frac{\partial^2 u}{\partial\theta\partial r} - \sin\theta\frac{\partial u}{\partial r}\right) - \frac{1}{r}\frac{\partial}{\partial\theta}\left(\sin\frac{\partial u}{\partial\theta}\right)\right]$$

$$= \cos\theta\left[\cos\theta\frac{\partial^2 u}{\partial r^2} - \sin\theta\left\{\frac{1}{r}\frac{\partial^2 u}{\partial r\partial\theta} - \frac{1}{r^2}\frac{\partial u}{\partial\theta}\right\}\right]$$

$$-\frac{\sin\theta}{r}\left[\left(\cos\theta\frac{\partial^2 u}{\partial r\partial\theta} - \sin\theta\frac{\partial u}{\partial r}\right) - \frac{1}{r}\left(\sin\theta\frac{\partial^2 u}{\partial\theta^2} + \cos\theta\frac{\partial u}{\partial\theta}\right)\right]$$

or $$\frac{\partial^2 u}{\partial x^2} = \cos^2\theta\frac{\partial^2 u}{\partial r^2} - \frac{2\sin\theta\cos\theta}{r}\frac{\partial^2 u}{\partial r\partial\theta} + \frac{\sin^2\theta}{r^2}\frac{\partial^2 u}{\partial\theta^2}$$

$$+\frac{\sin^2\theta}{r}\frac{\partial u}{\partial r} + \frac{2\cos\theta\sin\theta}{r^2}\frac{\partial u}{\partial\theta} \qquad \text{...(11)}$$

Similarly, from $\frac{\partial^2 u}{\partial y^2} = \frac{\partial}{\partial y}\left(\frac{\partial u}{\partial y}\right) = \sin\theta \frac{\partial}{\partial r}\left(\frac{\partial u}{\partial y}\right) + \frac{\cos\theta}{r}\frac{\partial}{\partial \theta}\left(\frac{\partial u}{\partial y}\right)$, from (10), we get

$$\frac{\partial^2 u}{\partial y^2} = \sin^2\theta \frac{\partial^2 u}{\partial r^2} + \frac{2\sin\theta\cos\theta}{r}\frac{\partial^2 u}{\partial r\partial\theta} + \frac{\cos^2\theta}{r^2}\frac{\partial^2 u}{\partial\theta^2}$$

$$+ \frac{\cos^2\theta}{r}\frac{\partial u}{\partial r} - \frac{2\cos\theta\sin\theta}{r^2}\frac{\partial u}{\partial\theta}. \quad \text{...(12)}$$

Adding (11) and (12), we get

$$\frac{\partial^2 u}{\partial x^2} + \frac{\partial^2 u}{\partial y^2} = (\cos^2\theta + \sin^2\theta)\frac{\partial^2 u}{\partial r^2} + \frac{1}{r^2}(\sin^2\theta + \cos^2\theta)\frac{\partial^2 u}{\partial\theta^2}$$

$$+ \frac{1}{r}(\sin^2\theta + \cos^2\theta)\frac{\partial u}{\partial r} = \frac{\partial^2 u}{\partial r^2} + \frac{1}{r^2}\frac{\partial^2 u}{\partial\theta^2} + \frac{1}{r}\frac{\partial u}{\partial r}.$$

Hence, the given differential equation

$$\frac{\partial^2 u}{\partial x^2} + \frac{\partial^2 u}{\partial y^2} = 0 \text{ transforms into } \frac{\partial^2 u}{\partial r^2} + \frac{1}{r^2}\frac{\partial^2 u}{\partial\theta^2} + \frac{1}{r}\frac{\partial u}{\partial r} = 0.$$

Case II. Transformation from Polar to Cartesian.

Now let us consider the converse of Case I, *i.e.*, let us transform the polar differential equation $\frac{\partial^2 u}{\partial r^2} + \frac{1}{r^2}\frac{\partial^2 u}{\partial\theta^2} + \frac{1}{r}\frac{\partial u}{\partial r} = 0$ to cartesian.

As before,
$$x = r\cos\theta,\ y = r\sin\theta. \quad \text{...(1)}$$

Then,
$$\frac{\partial u}{\partial r} = \frac{\partial u}{\partial x}.\frac{\partial x}{\partial r} + \frac{\partial u}{\partial y}.\frac{\partial y}{\partial r} \quad \text{...(2)}$$

Now, from (1)
$$\left.\begin{aligned} \frac{\partial x}{\partial r} &= \cos\theta = \frac{x}{r};\ \frac{\partial x}{\partial\theta} = -r\sin\theta = -y \\ \frac{\partial y}{\partial r} &= \sin\theta = \frac{y}{r};\ \frac{\partial y}{\partial\theta} = r\cos\theta = x \end{aligned}\right] \quad \text{...(3)}$$

Hence, from (2), we get
$$\frac{\partial u}{\partial r} = \frac{\partial u}{\partial x}\left(\frac{x}{r}\right) + \frac{\partial u}{\partial y}\left(\frac{y}{r}\right) \quad \text{[Using (3)]}$$

or
$$r\frac{\partial}{\partial r}(u) = x\frac{\partial}{\partial x}(u) + y\frac{\partial}{\partial y}(u) \quad \text{...(4)}$$

Also we have
$$\frac{\partial u}{\partial\theta} = \frac{\partial u}{\partial x}\frac{\partial x}{\partial\theta} + \frac{\partial u}{\partial y}\frac{\partial y}{\partial\theta} = \frac{\partial u}{\partial x}(-y) + \frac{\partial u}{\partial y}(x) \quad \text{[From (3)]}$$

or
$$\frac{\partial}{\partial\theta}(u) = x\frac{\partial}{\partial y}(u) - y\frac{\partial}{\partial x}(u). \quad \text{...(5)}$$

Now
$$r\frac{\partial}{\partial r}\left(r\frac{\partial u}{\partial r}\right) = x\frac{\partial}{\partial x}\left(r\frac{\partial u}{\partial r}\right) + y\frac{\partial}{\partial y}\left(r\frac{\partial u}{\partial r}\right) \quad \left(\text{replacing } u \text{ by } r\frac{\partial u}{\partial r} \text{ in (4)}\right)$$

$$= x\frac{\partial}{\partial x}\left[x\frac{\partial u}{\partial x} + y\frac{\partial u}{\partial y}\right] + y\frac{\partial}{\partial y}\left(x\frac{\partial u}{\partial x} + y\frac{\partial u}{\partial y}\right)$$

Substituting the value of $r\frac{\partial u}{\partial r}$ from (4)

or
$$r\left[r\frac{\partial^2 u}{\partial r^2} + \frac{\partial u}{\partial r}\right] = x\left[x\frac{\partial^2 u}{\partial x^2} + \frac{\partial u}{\partial x} + y\frac{\partial^2 u}{\partial x\partial y}\right] + y\left[x\frac{\partial^2 u}{\partial y\partial x} + y\frac{\partial^2 u}{\partial y^2} + \frac{\partial u}{\partial y}\right]$$

or
$$r^2\frac{\partial^2 u}{\partial r^2} + r\frac{\partial u}{\partial r} = x^2\frac{\partial^2 u}{\partial x^2} + 2xy\frac{\partial^2 u}{\partial x\partial y} + y^2\frac{\partial^2 u}{\partial y^2} + x\frac{\partial u}{\partial x} + y\frac{\partial u}{\partial y}$$

or
$$r^2\frac{\partial^2 u}{\partial r^2} = x^2\frac{\partial^2 u}{\partial x^2} + 2xy\frac{\partial^2 u}{\partial x\partial y} + y^2\frac{\partial^2 u}{\partial y^2} \quad \text{...(6)}$$

Since $r\frac{\partial u}{\partial r} = x\frac{\partial u}{\partial x} + y\frac{\partial u}{\partial y}$, (from (4)) and $$\frac{\partial^2 u}{\partial\theta^2} = \frac{\partial}{\partial\theta}\left(\frac{\partial u}{\partial\theta}\right) = x\frac{\partial}{\partial y}\left(\frac{\partial u}{\partial\theta}\right) - y\frac{\partial}{\partial x}\left(\frac{\partial u}{\partial\theta}\right) \text{ replacing } u \text{ by } \frac{\partial u}{\partial\theta} \text{ in (5)}$$

$$= x\frac{\partial}{\partial y}\left(x\frac{\partial u}{\partial y} - y\frac{\partial u}{\partial x}\right) - y\frac{\partial}{\partial x}\left(x\frac{\partial u}{\partial y} - y\frac{\partial u}{\partial x}\right)$$

(Substituting the value of $\frac{\partial u}{\partial\theta}$ from (5))

$$= x\left(x\frac{\partial^2 u}{\partial y^2} - y\frac{\partial^2 u}{\partial x\partial y} - \frac{\partial u}{\partial x}\right) - y\left(x\frac{\partial^2 u}{\partial x\partial y} + \frac{\partial u}{\partial y} - y\frac{\partial^2 u}{\partial x^2}\right)$$

$$= x\frac{\partial^2 u}{\partial y^2} - 2xy\frac{\partial^2 u}{\partial x\partial y} + y^2\frac{\partial^2 u}{\partial x^2} - \left(x\frac{\partial u}{\partial x} + y\frac{\partial u}{\partial y}\right)$$

or $$\frac{\partial^2 u}{\partial\theta^2} + r\frac{\partial u}{\partial r} = x^2\frac{\partial^2 u}{\partial y^2} - 2xy\frac{\partial^2 u}{\partial x\partial y} + y^2\frac{\partial^2 u}{\partial x^2} \quad \ldots(7)$$

Adding (6) and (7), we get

$$r^2\frac{\partial^2 u}{\partial r^2} + \frac{\partial^2 u}{\partial\theta^2} + r\frac{\partial u}{\partial r} = (x^2+y^2)\frac{\partial^2 u}{\partial x^2} + (x^2+y^2)\frac{\partial^2 u}{\partial y^2} = r^2\left(\frac{\partial^2 u}{\partial x^2} + \frac{\partial^2 u}{\partial y^2}\right) \quad [\because r^2 = x^2+y^2]$$

or $$\frac{\partial^2 u}{\partial r^2} + \frac{1}{r^2}\frac{\partial^2 u}{\partial\theta^2} + \frac{1}{r}\frac{\partial u}{\partial r} = \frac{\partial^2 u}{\partial x^2} + \frac{\partial^2 u}{\partial y^2}$$

Hence, the given differential equation

$$\frac{\partial^2 u}{\partial r^2} + \frac{1}{r^2}\frac{\partial^2 u}{\partial\theta^2} + \frac{1}{r^2}\frac{\partial u}{\partial r} = 0, \text{ transforms into } \frac{\partial^2 u}{\partial x^2} + \frac{\partial^2 u}{\partial y^2} = 0.$$

Solved Examples

Example 1. *If $x = r\cos\theta$, $y = r\sin\theta$, prove that*

$$\frac{\partial^2 r}{\partial x^2}\frac{\partial^2 r}{\partial y^2} = \left(\frac{\partial^2 r}{\partial x\partial y}\right)^2.$$

Solution . We know that $\frac{\partial r}{\partial x} = \frac{x}{r}, \frac{\partial r}{\partial y} = \frac{y}{r}$.

Now $$\frac{\partial^2 r}{\partial x^2} = \frac{\partial}{\partial x}\left(\frac{\partial r}{\partial x}\right) = \frac{\partial}{\partial x}\left(\frac{x}{r}\right) = \frac{r.1 - x(\partial r/\partial x)}{r^2}$$

$$= \frac{r - x(x/r)}{r^2} = \frac{r^2 - x^2}{r^3} = \frac{y^2}{r^3}.$$

$$[\because x^2 + y^2 = r^2] \quad \ldots(1)$$

Similarly, we can get

$$\frac{\partial^2 r}{\partial y^2} = \frac{x^2}{r^3} \quad \ldots(2)$$

Also, $$\frac{\partial^2 r}{\partial x\partial y} = \frac{\partial}{\partial x}\left(\frac{\partial r}{\partial y}\right) = \frac{\partial}{\partial x}\left(\frac{y}{r}\right)$$

$$= \frac{r.0 - y(\partial r/\partial x)}{r^2}$$

$$= -\frac{y(x/r)}{r^2} = -\frac{xy}{r^3}.$$

$$\therefore \left(\frac{\partial^2 r}{\partial x\partial y}\right)^2 = \left(-\frac{xy}{r^3}\right)^2 = \frac{x^2}{r^3}.\frac{y^2}{r^3} = \frac{\partial^2 r}{\partial y^2}.\frac{\partial^2 r}{\partial x^2}.$$

[From (1) an d (2)]

Example 2. *If $x = r\cos\theta$, $y = r\sin\theta$, prove that*

$$\frac{\partial^2 r}{\partial x^2} + \frac{\partial^2 r}{\partial y^2} = \frac{1}{r}\left\{\left(\frac{\partial r}{\partial x}\right)^2 + \left(\frac{\partial r}{\partial y}\right)^2\right\}.$$

Solution . As in example 1 above, we can get

$$\frac{\partial^2 r}{\partial x^2} + \frac{\partial^2 r}{\partial y^2} = \frac{y^2}{r^3} + \frac{x^2}{r^3} = \frac{x^2+y^2}{r^3} = \frac{r^2}{r^3} = \frac{1}{r}$$

$$[\because x^2 + y^2 = r^2] \quad \ldots(1)$$

Also, we can get $\frac{\partial r}{\partial x} = \frac{x}{r}, \frac{\partial r}{\partial y} = \frac{y}{r}$.

$$\therefore \frac{1}{r}\left\{\left(\frac{\partial r}{\partial x}\right)^2 + \left(\frac{\partial r}{\partial y}\right)^2\right\} = \frac{1}{r}\left\{\frac{x^2}{r^2} + \frac{y^2}{r^2}\right\}$$

$$= \frac{1}{r}\left\{\frac{x^2+y^2}{r^2}\right\} = \frac{1}{r}\left\{\frac{r^2}{r^2}\right\}. \quad \ldots(2)$$

Hence, from (1) and (2), we get

$$\frac{\partial^2 r}{\partial x^2} + \frac{\partial^2 r}{\partial y^2} = \frac{1}{r}\left\{\left(\frac{\partial r}{\partial x}\right)^2 + \left(\frac{\partial r}{\partial y}\right)^2\right\}.$$

Example 3. *Transform* $\frac{\partial^2 u}{\partial x^2}+\frac{\partial^2 u}{\partial y^2}=0$ *into polars and show that* $u=\left(Ar^n+Br^{-n}\right)\sin n\theta$ *satisfies the above equation.*

Solution. We can transform the given equation

$$\frac{\partial^2 u}{\partial x^2}+\frac{\partial^2 u}{\partial y^2}=0 \text{ into } \frac{\partial^2 u}{\partial r^2}+\frac{1}{r^2}\frac{\partial^2 u}{\partial \theta^2}+\frac{1}{r^2}\frac{\partial u}{\partial r}=0,.$$

Now $u=\left(Ar^n+Br^{-n}\right)\sin n\theta$

$$\therefore \quad \frac{\partial u}{\partial r}=\left(nAr^{n-1}-Bnr^{-n-1}\right)\sin n\theta$$

$$\frac{\partial^2 u}{\partial r^2}=n[(n-1)Ar^{n-2}+Bn(n+1)r^{-n-2}]\sin n\theta$$

$$\frac{\partial u}{\partial \theta}=n[Ar^n+Br^{-n}]\cos n\theta;\ \frac{\partial^2 u}{\partial \theta^2}=-n^2[Ar^n+Br^{-n}]\sin n\theta.$$

$$\therefore \quad \frac{\partial^2 u}{\partial r^2}+\frac{1}{r^2}\frac{\partial^2 u}{\partial \theta^2}+\frac{1}{r}\frac{\partial u}{\partial r}$$
$$=n[(n-1)Ar^{n-2}+B(n+1)r^{-n-2}]\sin n\theta+\frac{1}{r^2}[-n^2(Ar^n+Br^{-n})\sin n\theta]$$
$$+\frac{1}{r}n(Ar^{n-1}-Br^{-n-1})\sin n\theta$$
$$=[A\{n(n-1)-n^2+n\}r^{n-2}+\{Bn(n+1)-n^2-n\}r^{-n-2}]\sin n\theta=0$$

Hence, the equation $\frac{\partial^2 u}{\partial r^2}+\frac{1}{r^2}\frac{\partial^2 u}{\partial \theta^2}+\frac{1}{r^2}\frac{\partial u}{\partial r}=0$, *i.e.*, $\frac{\partial^2 u}{\partial x^2}+\frac{\partial^2 u}{\partial y^2}=0$ is satisfied by $(Ar^n+Br^{-n})\sin n\theta$.

9.8.6 TO TRANSFORM $\nabla^2 V$ INTO POLAR CO-ORDINATES, WHERE THE OPERATOR ∇^2 STANDS FOR $\frac{\partial^2}{\partial x^2}+\frac{\partial^2}{\partial y^2}+\frac{\partial^2}{\partial z^2}$

For polar transformation (in three dimensions), we have $x = r\sin\theta\cos\phi$, $y = r\sin\theta\sin\phi$, $z = r\cos\theta$.
Let $r\sin\theta = u$, then $x = u\cos\phi$, $y = u\sin\phi$.
Then, as in § 9.8.5 Case I, we can have

$$\frac{\partial^2 V}{\partial x^2}+\frac{\partial^2 V}{\partial y^2}=\frac{\partial^2 V}{\partial z^2}+\frac{1}{u}\cdot\frac{\partial V}{\partial u}+\frac{1}{u^2}\cdot\frac{\partial^2 V}{\partial \phi^2} \quad \ldots(1)$$

Again, we have $z = r\cos\theta$, $u = r\sin\theta$.

$$\therefore \quad \frac{\partial^2 V}{\partial z^2}+\frac{\partial^2 V}{\partial u^2}=\frac{\partial^2 V}{\partial r^2}+\frac{1}{r}\cdot\frac{\partial V}{\partial r}+\frac{1}{r^2}\cdot\frac{\partial^2 V}{\partial \theta^2} \quad \ldots(2)$$

Also,
$$\frac{\partial V}{\partial u}=\frac{\partial V}{\partial r}\cdot\frac{\partial r}{\partial u}+\frac{\partial V}{\partial \theta}\cdot\frac{\partial \theta}{\partial u} \quad \ldots(3)$$

Now $u = r\sin\theta$, $z = r\cos\theta$, whence we get

$$r^2=u^2+z^2 \quad \text{and} \quad \theta=\tan^{-1}(x/z).$$

$$\therefore \quad 2r\frac{\partial r}{\partial u}=2u \quad \text{or} \quad \frac{\partial r}{\partial u}=\frac{u}{r}=\frac{r\sin\theta}{r}=\sin\theta$$

and
$$\frac{\partial \theta}{\partial u}=\frac{1}{1+(x/z)^2}\cdot\frac{1}{z}=\frac{z}{u^2+z^2}=\frac{r\cos\theta}{r^2}=\frac{\cos\theta}{r}$$

Substituting these values of $\partial r/\partial u$ and $\partial\theta/\partial u$ in (3), we get

$$\frac{\partial V}{\partial u}=\frac{\partial V}{\partial r}(\sin\theta)+\frac{\partial V}{\partial \theta}\left(\frac{\cos\theta}{r}\right)$$

$$\text{or } \frac{1}{u}\left(\frac{\partial V}{\partial u}\right)=\frac{1}{r\sin\theta}\left(\frac{\partial V}{\partial u}\right)=\frac{1}{r\sin\theta}\left[\sin\theta\frac{\partial V}{\partial r}+\frac{\cos\theta}{r}\frac{\partial V}{\partial \theta}\right] \text{ or } \frac{1}{u}\frac{\partial V}{\partial u}=\frac{1}{r}\frac{\partial V}{\partial r}+\frac{\cot\theta}{r^2}\frac{\partial V}{\partial \theta}. \quad \ldots(4)$$

Now adding (1) and (2), we get

$$\nabla^2 V=\frac{\partial^2 V}{\partial x^2}+\frac{\partial^2 V}{\partial y^2}+\frac{\partial^2 V}{\partial z^2}=\frac{1}{u}\frac{\partial V}{\partial u}+\frac{1}{u^2}\frac{\partial^2 V}{\partial \phi^2}+\frac{\partial^2 V}{\partial r^2}+\frac{1}{r}\frac{\partial V}{\partial r}+\frac{1}{r^2}\frac{\partial^2 V}{\partial \theta^2}$$
$$=\frac{1}{r}\frac{\partial V}{\partial r}+\frac{\cot\theta}{r^2}\frac{\partial V}{\partial \theta}+\frac{1}{r^2\sin^2\theta}\cdot\frac{\partial^2 V}{\partial \phi^2}+\frac{\partial^2 V}{\partial r^2}+\frac{1}{r}\frac{\partial V}{\partial r}+\frac{1}{r^2}\frac{\partial^2 V}{\partial \theta^2}$$

i.e.,
$$\nabla^2 V=\frac{\partial^2 V}{\partial r^2}+\frac{2}{r}\frac{\partial V}{\partial r}+\frac{1}{r^2}\frac{\partial^2 V}{\partial \theta^2}+\frac{\cot\theta}{r^2}\frac{\partial V}{\partial \theta}+\frac{1}{r^2\sin^2\theta}\cdot\frac{\partial^2 V}{\partial \phi^2}$$

Solved Examples

Example 1. *If $x = r\cos\theta$, $y = r\sin\theta$ and $r = e^t$, prove that*

$$x^2\frac{\partial^2 u}{\partial x^2} + 2xy\frac{\partial^2 u}{\partial y\partial x} + y^2\frac{\partial^2 u}{\partial y^2} = r\frac{\partial}{\partial r}\left(r\frac{\partial}{\partial r}-1\right)u = \frac{\partial}{\partial z}\left(\frac{\partial}{\partial z}-1\right)u$$

and $$x^2\frac{\partial^2 u}{\partial y^2} - 2xy\frac{\partial^2 u}{\partial x\partial y} + y^2\frac{\partial^2 u}{\partial x^2} = \frac{\partial^2 u}{\partial \theta^2} + r\frac{\partial u}{\partial r} = \frac{\partial^2 u}{\partial \theta^2} + \frac{\partial u}{\partial z}$$

Solution. As in § 9.8.5 Case II, we can show that

$$r\frac{\partial u}{\partial r} = x\frac{\partial u}{\partial x} + y\frac{\partial u}{\partial y}$$

Therefore, $r\frac{\partial}{\partial r}\left(r\frac{\partial}{\partial r}-1\right)u$

$$= \left(x\frac{\partial}{\partial x} + y\frac{\partial}{\partial y}\right)\left[x\frac{\partial u}{\partial x} + y\frac{\partial u}{\partial y} - u\right]$$

$$= x\left[x\frac{\partial^2 u}{\partial x^2} + \frac{\partial u}{\partial x} + y\frac{\partial^2 u}{\partial x\partial y} - \frac{\partial u}{\partial x}\right] + y\left[x\frac{\partial^2 u}{\partial y\partial x} + y\frac{\partial^2 u}{\partial y^2} + \frac{\partial u}{\partial y} - \frac{\partial u}{\partial y}\right]$$

$$= x^2\frac{\partial^2 u}{\partial x^2} + 2xy\frac{\partial^2 u}{\partial x\partial y} + y^2\frac{\partial^2 u}{\partial y^2}.$$

Also as $r = e^z$ or $z = \log r$.

$\therefore$ Using $\frac{\partial u}{\partial r} = \frac{\partial u}{\partial z}\cdot\frac{\partial z}{\partial r} = \frac{\partial u}{\partial z}\cdot\frac{1}{r}$

$$\therefore \quad r\frac{\partial u}{\partial r} = \frac{\partial u}{\partial z}$$

or $$r\frac{\partial}{\partial r} = \frac{\partial}{\partial z} \qquad \ldots(1)$$

$$\therefore \quad r\frac{\partial}{\partial r}\left(r\frac{\partial}{\partial r}-1\right)u = \frac{\partial}{\partial z}\left(\frac{\partial}{\partial z}-1\right)u.$$

Again, as in § 9.8.6 Case I, we can prove that

$$\frac{\partial u}{\partial \theta} = x\frac{\partial u}{\partial y} - y\frac{\partial u}{\partial x}.$$

$$\therefore \frac{\partial^2 u}{\partial \theta^2} = \frac{\partial}{\partial \theta}\left(\frac{\partial u}{\partial \theta}\right)$$

$$= \left(x\frac{\partial}{\partial y} - y\frac{\partial}{\partial x}\right)\left(x\frac{\partial u}{\partial y} - y\frac{\partial u}{\partial x}\right)$$

$$= x\frac{\partial}{\partial y}\left(x\frac{\partial u}{\partial y} - y\frac{\partial u}{\partial x}\right) - y\frac{\partial}{\partial x}\left(x\frac{\partial u}{\partial y} - y\frac{\partial u}{\partial x}\right)$$

$$= x\left[x\frac{\partial^2 u}{\partial y^2} - \frac{\partial u}{\partial x} - y\frac{\partial^2 u}{\partial y\partial x}\right] - y\left[x\frac{\partial^2 u}{\partial x\partial y} + \frac{\partial u}{\partial y} - y\frac{\partial^2 u}{\partial x^2}\right]$$

$$= x^2\frac{\partial^2 u}{\partial y^2} - 2xy\frac{\partial^2 u}{\partial x\partial y} + y^2\frac{\partial^2 u}{\partial x^2} - \left(x\frac{\partial u}{\partial x} + y\frac{\partial u}{\partial y}\right)$$

or $$\frac{\partial^2 u}{\partial \theta^2} + r\frac{\partial u}{\partial r} = x^2\frac{\partial^2 u}{\partial y^2} - 2xy\frac{\partial^2 u}{\partial x\partial y} + y^2\frac{\partial^2 u}{\partial x^2}. \qquad \text{[From (1)]}$$

Also, from (1), we get $\frac{\partial^2 u}{\partial \theta^2} + r\frac{\partial u}{\partial r} = \frac{\partial^2 u}{\partial \theta^2} + \frac{\partial u}{\partial z}$.

Example 2. *If V be a function of r along where $r^2 = x^2 + y^2 + z^2$, show that* $\frac{\partial^2 V}{\partial x^2} + \frac{\partial^2 V}{\partial y^2} + \frac{\partial^2 V}{\partial z^2} = \frac{\partial^2 V}{\partial r^2} + \frac{2}{r}\frac{dV}{dr}$.

Solution. As V is given to be a function of r alone, so we have

$$\frac{\partial V}{\partial x} = \frac{dV}{dr}\frac{\partial r}{\partial x} \qquad \ldots(1)$$

Also, from $r^2 = x^2 + y^2 + z^2$,

we get $2r\frac{\partial r}{\partial x} = 2x$.

$\therefore$ From (1), $\frac{\partial V}{\partial x} = \frac{dV}{dr}\frac{x}{r}$

$$\therefore \quad \frac{\partial^2 V}{\partial x^2} = \frac{\partial}{\partial x}\left(\frac{\partial V}{\partial x}\right) = \frac{\partial}{\partial x}\left[\frac{dV}{dr}\frac{x}{r}\right]$$

$$= \frac{dV}{dr}\frac{\partial}{\partial x}\left(\frac{x}{r}\right) + \frac{x}{r}\frac{\partial}{\partial x}\left(\frac{dV}{dr}\right)$$

$$= \frac{dV}{dr}\left[\frac{1}{r} + x\left(-\frac{1}{r^2}\frac{\partial r}{\partial x}\right)\right] + \frac{x}{r}\left(\frac{d^2V}{dr^2}\frac{\partial r}{\partial x}\right)$$

$$= \frac{1}{r}\frac{dV}{dr} - \frac{x}{r^2}\frac{dV}{dr}\left(\frac{x}{r}\right) + \frac{x}{r}\frac{d^2V}{dr^2}\left(\frac{x}{r}\right) \qquad \left(\because \frac{\partial r}{\partial x} = \frac{x}{r}\right)$$

or $$\frac{\partial^2 V}{\partial x^2} = \frac{1}{r}\frac{dV}{dr} - \frac{x^2}{r^3}\frac{dV}{dr} + \frac{x^2}{r^2}\frac{d^2V}{dr^2}$$

Similarly, $$\frac{\partial^2 V}{\partial y^2} = \frac{1}{r}\frac{dV}{dr} - \frac{y^2}{r^3}\frac{dV}{dr} + \frac{y^2}{r^2}\frac{d^2V}{dr^2}$$

and $$\frac{\partial^2 V}{\partial z^2} = \frac{1}{r}\frac{dV}{dr} - \frac{z^2}{r^3}\frac{dV}{dr} + \frac{z^2}{r^2}\frac{d^2V}{dr^2}$$

On adding, we get

$$\frac{\partial^2 V}{\partial x^2} + \frac{\partial^2 V}{\partial y^2} + \frac{\partial^2 V}{\partial z^2} = \frac{3}{r}\frac{dV}{dr} - \frac{1}{r}\frac{dV}{dr} + \frac{d^2V}{dr^2} = \frac{2}{r}\frac{dV}{dr} + \frac{d^2V}{dr^2}$$

EXERCISE 9.4

1. Reduce the equation $\frac{d^2x}{dy^2} = a$ to the form $\frac{d^2y}{dx^2} + a\left(\frac{dy}{dx}\right)^3 = 0$.

2. Transform the equation $x^4\left(\frac{d^2y}{dx^2}\right) + a^2y = 0$ by the substitution $x = 1/z$.

3. Transform the equation $\sin^2 2z\frac{d^2y}{dz^2} + \sin 4z\frac{dy}{dz} + 4y = 0$ the substitution $\tan z = e^z$.

4. If $x^2 + z^2 = 1$, show that the equation
$$\frac{d}{dx}\left\{(1-x^2)\frac{dy}{dx}\right\} + n(n+1)y = 0 \text{ becomes}$$
$$z(z^2-1)\frac{d^2y}{dz^2} + (2z^2-1)\frac{dy}{dz} - n(n+1)zy = 0.$$

5. Show that the equation $x^2\frac{d^2y}{dx^2} + x\frac{dy}{dx} + y = 0$ becomes $\frac{d^2y}{dz^2} + y = 0$ by substituting e^z for x.

6. Transform $\frac{d^2y}{dx^2}$ to new variables u and v by taking u as independent variable such that $y = uv$, $xy = 1$.

7. Show that $\frac{\partial^2 u}{\partial x^2} + \frac{\partial^2 u}{\partial y^2} = \frac{\partial^2 u}{\partial \xi^2} + \frac{\partial^2 u}{\partial \eta^2}$ where $x = \xi\cos\alpha - \eta\sin\alpha$, $y = \xi\sin\alpha + \eta\cos\alpha$.

8. If $x = e^\theta, y = e^\phi$, show that
$$e^{2\theta}\frac{\partial^2 v}{\partial x^2} + e^{2\phi}\frac{\partial^2 v}{\partial y^2} + e^\theta\frac{\partial v}{\partial x} + e^\phi\frac{\partial v}{\partial y} = \frac{\partial^2 v}{\partial \theta^2} + \frac{\partial^2 v}{\partial \phi^2}.$$

9. If $f(x, y)$ has continuous partial derivatives of first two orders and $x + y = (u+v)^3$, $(x-y) = (u-v)^3$, then show that
$$9(x^2-y^2)\left(\frac{\partial^2 f}{\partial x^2} - \frac{\partial^2 f}{\partial y^2}\right) = (u^2-v^2)\left\{\frac{\partial^2 f}{\partial u^2} - \frac{\partial^2 f}{\partial v^2}\right\}.$$

10. If z is a function of u and v, where $u = x^2 - y^2 - 2xy$ and $v = y$, show that the equation $(x+y)\frac{\partial z}{\partial x} + (x-y)\left(\frac{\partial z}{\partial y}\right) = 0$ is transformed into $\frac{\partial v}{\partial z} = 0$.

11. Show that the equation $xy\left(\frac{\partial^2 u}{\partial x^2} - \frac{\partial^2 u}{\partial y^2}\right) - (x^2-y^2)\frac{\partial^2 u}{\partial x \partial y} = 0$ becomes $r.\frac{\partial^2 u}{\partial r \partial \theta} - \frac{\partial u}{\partial \theta} = 0$, when transformed to polar.

12. If $x = r\cos\theta$, $y = r\sin\theta$ and $z = f(x, y)$, show that
$$\frac{\partial z}{\partial x} = \frac{\partial z}{\partial r}\cos\theta - \frac{1}{r}\frac{\partial z}{\partial \theta}\sin\theta.$$
Also show that $\frac{\partial^2(r^n \cos n\theta)}{\partial x \partial y} = -n(n-1)r^{n-2}\sin(n-2)\theta$.

13. If $x + y = 2e^\theta\cos\phi$ and $x - y = 2e^\theta\sin\phi$, show that
$$\frac{\partial^2 V}{\partial \theta^2} + \frac{\partial^2 V}{\partial \phi^2} = 4xy\frac{\partial^2 V}{\partial x \partial y}.$$ (UPTU–2002)

ANSWERS

2. $\frac{d^2y}{dz^2} + \frac{2}{z}\frac{dy}{dz} + a^2y = 0$

3. $\frac{d^2y}{dx^2} + y = 0$

6. $4v^4\left(\frac{dv}{du}\right)^{-1} + 2uv^3 - v^5\left(\frac{d^2v}{du^2}\right)\left(\frac{dv}{du}\right)^{-3}$

Objective Evaluations

Fill in the Blanks

1. $\cos^{-1}\dfrac{y}{x}$ is a homogeneous functionof degree ________.

2. If $\phi = \sin^{-1}\left(\dfrac{x^2+y^2}{x+y}\right)$, then $x\dfrac{\partial\phi}{\partial x}+y\dfrac{\partial\phi}{\partial y}$ is ________.

3. If $u = e^{my}\cos mx$, then $\dfrac{\partial^2 u}{\partial x^2}+\dfrac{\partial^2 u}{\partial y^2}=$ ________.

4. A function is said to be homogeneous if every term is of ________.

5. An expression in which every term is of the same degree is called ________ function.

6. If $z = f(y/x)$ then $x\left(\dfrac{\partial z}{\partial x}\right)+y\left(\dfrac{\partial z}{\partial y}\right)$ is ________.

7. If $z = xy\,f(y/x)$ then $x\left(\dfrac{\partial z}{\partial x}\right)+y\left(\dfrac{\partial z}{\partial y}\right)$ is ________.

8. If $u = e^{xyz}$, then $\dfrac{\partial^2 z}{\partial y\partial z}$ is ________.

9. If $u(x, y)$ is a homogeneous function of x and y of degree n, then
$x\dfrac{\partial}{\partial x}(u_x)+y\dfrac{\partial}{\partial y}(u_x)=$ ________ where $u_x=\dfrac{\partial u}{\partial x}$.

10. If $u = f(x, y)$, and its partial derivatives are continuous, then order of differentiation is ________.

True/False

Write 'T' for True and 'F' for False statement.

1. An expression in which every term is of same degree is called homogeneous function. **(T/F)**

2. In homogeneous function every term is not necessarily of same degree. **(T/F)**

3. If u is a homogeneous function of x and y of degree n, then $\dfrac{\partial u}{\partial x}$ and $\dfrac{\partial u}{\partial y}$ are also homogeneous function of degree n. **(T/F)**

4. If x and y are connected by an equation of the form $f(x, y) = 0$, then $\dfrac{dy}{dx}$ is $-\dfrac{\partial f/\partial x}{\partial f/\partial y}$. **(T/F)**

5. If u is a homogeneous function of degree n, then $x\dfrac{\partial u}{\partial x}+y\dfrac{\partial u}{\partial y}$ is equal to n. **(T/F)**

6. If $f(x, y)$ be an implicit function of x and y and $p=\dfrac{\partial f}{\partial x}$, $q=\dfrac{\partial f}{\partial y}$, $r=\dfrac{\partial^2 f}{\partial x^2}$, $s=\dfrac{\partial^2 f}{\partial x\partial y}$ and $t=\dfrac{\partial^2 f}{\partial y^2}$, then $\dfrac{d^2y}{dx^2}=-\dfrac{(q^2r-2pqs+p^2t)}{q^3}$. **(T/F)**

7. If $u=\sqrt{(x^2+y^2+z^2)}$, then $x\dfrac{\partial u}{\partial x}+y\dfrac{\partial u}{\partial y}+z\dfrac{\partial u}{\partial z}$ is equal to u. **(T/F)**

8. The Euler's theorem for homogeneous function is not true for a function of more than two variables. **(T/F)**

9. If $u = f(x, y)$, where $x = g(t)$ and $y = \phi(t)$, then $\dfrac{\partial u}{\partial t}=\dfrac{\partial u}{\partial x}\cdot\dfrac{\partial x}{\partial t}+\dfrac{\partial u}{\partial y}\cdot\dfrac{\partial y}{\partial t}$. **(T/F)**

10. If $u=\sin^{-1}\left(\dfrac{x}{y}\right)+\tan^{-1}\left(\dfrac{y}{x}\right)$, then $x\dfrac{\partial u}{\partial x}+y\dfrac{\partial u}{\partial y}=0$. **(T/F)**

Multiple Choice Questions

Choose the most appropriate one.

1. $\sin^{-1}(y/x)$ is a homogeneous function of degree :
   (a) 1 (b) 2
   (c) 3 (d) 0

2. If $z = xy\,f\left(\dfrac{y}{x}\right)$ then $x\dfrac{\partial z}{\partial x}+y\dfrac{\partial z}{\partial y}$ is equal to :
   (a) z (b) $2z$
   (c) xy (d) yz

3. If $f=\sin^{-1}\left(\dfrac{x^2+y^2}{x+y}\right)$ then $x\dfrac{\partial f}{\partial x}+y\dfrac{\partial f}{\partial y}$ is :
   (a) f (b) $2f$
   (c) $\tan f$ (d) $\sin f$

4. A function $f(x, y)$ is said to be homogeneous of degree n if :
   (a) $f(x, ty) = t^{2n}f(xy)$ (b) it is of the form $x^n f(x/y)$
   (c) it is of the form $x^n f(y/x)$ (d) $\dfrac{\partial f}{\partial y}=\dfrac{\partial f}{\partial x}$

5. If $z = e^{ax}\sin by$, then $\dfrac{\partial^2 z}{\partial y\partial x}$ is :
   (a) $ae^{ax}\cos by$ (b) $be^{ax}\sin by$
   (c) $abe^{ax}\cos by$ (d) $abe^{ax}\sin by$

6. If $z = f(y/x)$ then $x\left(\dfrac{\partial z}{\partial x}\right)+y\left(\dfrac{\partial z}{\partial y}\right)$ is :
   (a) 1 (b) 2
   (c) –2 (d) 0

7. If $z = f(x + ay) + \phi(x - ay)$, then $\frac{\partial^2 z}{\partial y^2}$ is :

(a) $\frac{\partial^2 z}{\partial x^2}$ (b) $a^2 \frac{\partial^2 z}{\partial y^2}$

(c) $a^2 \frac{\partial^2 z}{\partial x^2}$ (d) $a^2 \frac{\partial^2 z}{\partial x \partial y}$

8. If $u = \log(x^3 + y^3 + z^3 - 3xyz)$ then $\frac{\partial u}{\partial x} + \frac{\partial u}{\partial y} + \frac{\partial u}{\partial z}$ is :

(a) $\frac{-9}{(x+y+z)^2}$ (b) $\frac{3}{x+y+z}$

(c) $\frac{9}{(x+y+z)^2}$ (d) $\frac{-3}{x+y+z}$

9. If $x = r\cos\phi$, $y = r\sin\theta$ then $\left(\frac{\partial r}{\partial x}\right)^2 + \left(\frac{\partial r}{\partial y}\right)^2$ is :

(a) r (b) $-r$

(c) 1 (d) -1

10. If $u = \tan^{-1} y/x$ then $\frac{\partial^2 u}{\partial x^2} + \frac{\partial^2 u}{\partial y^2}$ is :

(a) 0 (b) 1

(c) $\sin 2u$ (d) $\cos 2u$

11. If u is a homogeneous function of x and y of degree n then $x\frac{\partial u}{\partial x} + y\frac{\partial u}{\partial y}$ is :

(a) n (b) $n(n-1)$

(c) nu (d) $n(n-1)u$

12. If $u = \frac{x^{1/4} + y^{1/4}}{x^{1/5} + y^{1/5}}$ then $x\frac{\partial u}{\partial x} + y\frac{\partial u}{\partial y}$ is :

(a) $\frac{1}{10}u$ (b) $-\frac{1}{10}u$

(c) $\frac{1}{20}u$ (d) $-\frac{1}{20}u$

13. If $u = \frac{1}{\sqrt{x^2 + y^2}}$ then $x\frac{\partial u}{\partial x} + y\frac{\partial u}{\partial y}$ is :

(a) u (b) $-u$

(c) $\frac{1}{2u}$ (d) $-\frac{1}{2u}$

14. If $u = \sin^{-1}\left(\frac{x^2 + y^2}{x + y}\right)$ then $x\frac{\partial u}{\partial x} + y\frac{\partial u}{\partial y}$ is :

(a) $\tan u$ (b) $\sin u$

(c) $\cos u$ (d) $\sec u$

15. If $u = \tan^{-1}\left(\frac{x^3 + y^3}{x + y}\right)$ then $x\frac{\partial u}{\partial x} + y\frac{\partial u}{\partial y}$ is :

(a) $\tan 2u$ (b) $\cos 2u$

(c) $\sin 2u$ (d) $\sec 2u$

16. If $u = (y - z, z - x, x - y)$, then $\frac{\partial u}{\partial x} + \frac{\partial u}{\partial y} + \frac{\partial u}{\partial z}$ is :

(a) 0 (b) 1

(c) 2 (d) 3

17. If $u = y^x$ then $\frac{\partial u}{\partial x}$ is :

(a) xy^{x-1} (b) $y^x \log x - 1$

(c) $y^x \log y$ (d) yx^{y-1}

18. If $u = \sqrt{x^2 + y^2 + z^2}$ then $x\frac{\partial u}{\partial x} + y\frac{\partial u}{\partial y} + z\frac{\partial u}{\partial z}$ is :

(a) 1 (b) u

(c) $2u$ (d) $3u$

19. If $z = \log(x^2 + y^2)$ then $x\frac{\partial z}{\partial x} + y\frac{\partial z}{\partial y}$ is :

(a) 1 (b) 2

(c) 3 (d) 4

20. If $u = e^{xyz}$, then $\frac{\partial^2 z}{\partial y \partial z}$ is :

(a) $uy + ux^2yz$ (b) $ux + uxy^2z$

(c) $ux + uxyz$ (d) $ux + ux^2yz$

21. If $u = x^4y^2$ where $x = t^2$ and $y = t^3$ then $\frac{du}{dt}$ is :

(a) $5t^{13}$ (b) t^{22}

(c) $14t^{13}$ (d) $12t^6$

22. If $u = 3x^2yz + 2yz^3 + 6x^4$ then $x\frac{\partial u}{\partial x} + y\frac{\partial u}{\partial y} + z\frac{\partial u}{\partial z}$ is :

(a) $6x^2yz + 4xy^3 + 12x^4$ (b) $3x^2yz + 2yz^3 + 6x^4$

(c) $12x^2yz + 6yz^3 + 12x^4$ (d) $12x^2yz + 8yz^3 + 24x^4$

23. If $f(x, y) = x^4 + x^2y^2 + y^4$ then $\frac{\partial^2 f}{\partial x \partial y}$ is :

(a) $4xy$ (b) $12x^2 + 2y^2$

(c) $2x^2 + 12y^2$ (d) $4x^3 + 2xy^2$

24. The second order partial derivatives $\frac{\partial^2 z}{\partial x \partial y} = \frac{\partial^2 z}{\partial y \partial x}$ are :

(a) always equal

(b) never equal

(c) may or may not be equal

(d) equal only when these are continuous

25. If x and y are connected by an equation of the form $f(x, y) = 0$, then the value of $\frac{dy}{dx}$ is :

(a) $\frac{\partial f/\partial x}{\partial f/\partial y}$ (b) $(-1)^n \frac{\partial f/\partial x}{\partial f/\partial y}$

(c) $\frac{\partial f/\partial y}{\partial f/\partial x}$ (d) $\frac{\partial f/\partial x}{\partial f/\partial y}$

26. If $x = r\cos y$, $y = r\sin\theta$, then the value of $\frac{\partial\theta}{\partial x}$ is :

(a) $-\frac{\sin\theta}{r}$ (b) $\frac{r}{\sin\theta}$

(c) $\frac{\sin\theta}{r}$ (d) $-\frac{1}{r\sin\theta}$

27. If $f(x, y)$ is a homogeneous function of x and y of degree n, then :

(a) $x\frac{\partial f}{\partial x} + y\frac{\partial f}{\partial y} = f$ (b) $y\frac{\partial f}{\partial x} + x\frac{\partial f}{\partial y} = f$

(c) $y\frac{\partial f}{\partial x} + x\frac{\partial f}{\partial y} = nf$ (d) $x\frac{\partial f}{\partial x} + y\frac{\partial f}{\partial y} = nf$

ANSWERS

Fill in the Blanks

1. 0	**2.** tan ϕ	**3.** 0	**4.** same degree	**5.** homogeneous
6. 0	**7.** $2z$	**8.** $4x + 4x^2yz$	**9.** $(n-1)4x$	**10.** immaterial

True/False

1. T	**2.** F	**3.** F	**4.** T	**5.** F	**6.** T	**7.** F	**8.** F	**9.** T	**10.** T

Multiple Choice Questions

1. (d)	**2.** (b)	**3.** (c)	**4.** (b)	**5.** (c)	**6.** (d)	**7.** (c)	**8.** (a)	**9.** (c)	**10.** (a)
11. (c)	**12.** (c)	**13.** (b)	**14.** (a)	**15.** (c)	**16.** (a)	**17.** (c)	**18.** (b)	**19.** (b)	**20.** (d)
21. (c)	**22.** (d)	**23.** (a)	**24.** (d)	**25.** (d)	**26.** (a)	**27.** (d)			

FFFFFF

CHAPTER 10 Expansion of Function of Several Variables

10.1 INTRODUCTION

We know that the polynomial approximation of the function $z = f(x, y)$ is of great importance to engineering problems. We can find the approximate value of a function of several variables at a point by using mathematical tools. In this chapter we shall discuss the expansion of functions of several variables by using Taylor's and Maclaurin's series.

10.2 TAYLOR'S THEOREM FOR TWO VARIABLES

Let $f(x, y)$ be a function of two independent variables and h, k are small increment in x and y respectively, then

$$f(x+h, y+k) = f(x, y) + \left(h\frac{\partial}{\partial x} + k\frac{\partial}{\partial y}\right) f(x, y) + \frac{1}{2!}\left[h\frac{\partial}{\partial x} + k\frac{\partial}{\partial y}\right]^2 f(x, y) + \ldots$$

Proof. Consider the function of two variables $f(x + h, y + k)$.

Expand $f(x + h, y + k)$ using Taylor's series by considering $f(x + h, y + k)$ as a function of single variable x, we get

$$f(x+h, y+k) = f(x, y+k) + h\frac{\partial f(x, y+k)}{\partial x} + \frac{h^2}{2!}\frac{\partial^2 f(x, y+k)}{\partial x^2} + \ldots \qquad \ldots(1)$$

Now expand $f(x, y + k)$ using Taylor's series by considering $f(x, y + k)$ as a function of single variable y. Then

$$f(x, y+k) = f(x, y) + k\frac{\partial f(x, y)}{\partial y} + \frac{k^2}{2!}\frac{\partial^2 f(x, y)}{\partial y^2} + \ldots \qquad \ldots(2)$$

Using (2) in (1), we get

$$f(x+h, y+k) = f(x, y) + k\frac{\partial f(x, y)}{\partial y} + \frac{k^2}{2!}\frac{\partial^2 f(x, y)}{\partial y^2} + \ldots + h\frac{\partial}{\partial x}\left[f(x, y) + k\frac{\partial f(x, y)}{\partial y} + \frac{k^2}{2!}\frac{\partial^2 f(x, y)}{\partial y^2} + \ldots\right]$$
$$+ \frac{h^2}{2!}\frac{\partial^2}{\partial x^2}\left[f(x, y) + k\frac{\partial f(x, y)}{\partial y} + \frac{k^2}{2!}\frac{\partial^2 f(x, y)}{\partial y^2} + \ldots\right] + \ldots$$

$$= f(x, y) + \left(h\frac{\partial}{\partial x} + k\frac{\partial}{\partial y}\right) f(x, y) + \frac{1}{2!}\left(h^2\frac{\partial^2}{\partial x^2} + 2hk\frac{\partial^2}{\partial x \partial y} + k^2\frac{\partial^2}{\partial y^2}\right) f(x, y) + \ldots$$

$$\Rightarrow f(x+h, y+k) = f(x, y) + \left(h\frac{\partial}{\partial x} + k\frac{\partial}{\partial y}\right) f(x, y) + \frac{1}{2!}\left(h\frac{\partial}{\partial x} + k\frac{\partial}{\partial y}\right)^2 f(x, y) + \ldots \qquad \ldots(3)$$

which is the required Taylor's series for functions of two variables.

DEDUCTIONS

(1) On putting $x = a$ and $y = b$ in (3), we get

$$f(a+h, b+k) = f(a, b) + hf_x(a, b) + kf_y(a, b) + \frac{1}{2!}\left[h^2 f_{xx}(a, b) + 2hkf_{xy}(a, b) + k^2 f_{yy}(a, b)\right]$$
$$+ \frac{1}{3!}\left[h^3 f_{xxx}(a, b) + 3h^2 k f_{xxy}(a, b) + 3hk^2 f_{xyy}(a, b) + k^3 f_{yyy}(a, b)\right] + \ldots \qquad \ldots(4)$$

(2) On putting $h = x - a$ and $k = y - b$ in (4), we get

$$f(x,y) = f(a,b) + \left[(x-a)f_x(a,b) + (y-b)f_y(a,b)\right] + \frac{1}{2!}\left[(x-a)^2 f_{xx}(a,b) + 2(x-a)(y-b)f_{xy}(a,b) + (y-b)^2 f_{yy}(a,b)\right] + \ldots \quad \ldots(5)$$

(3) On putting $a = 0$ and $b = 0$ in (5), we get

$$f(x,y) = f(0,0) + \left[xf_x(0,0) + yf_y(0,0)\right] + \frac{1}{2!}\left[x^2 f_{xx}(0,0) + 2xyf_{xy}(0,0) + y^2 f_{yy}(0,0)\right] + \ldots \quad \ldots(6)$$

This is called Maclaurin's series for two variables.

Solved Examples

Example 1. *Expand $e^x \cos y$ about the point $(1, \pi/4)$.* (UPTU–2008)

Solution. Let $f(x, y) = e^x \cos y$

By Taylor's theorem, we have

$$f(x+h, y+k) = f(x,y) + \left(h\frac{\partial}{\partial x} + k\frac{\partial}{\partial y}\right)f(x,y) + \frac{1}{2!}\left(h\frac{\partial}{\partial x} + k\frac{\partial}{\partial y}\right)^2 f(x,y) + \ldots \quad \ldots(1)$$

Now

$$e^x \cos y = f(x,y) = f\left[1 + (x-1).\frac{\pi}{4} + \left(y - \frac{\pi}{4}\right)\right] = f\left(1+h, \frac{\pi}{4} + k\right)$$

where $h = x - 1$, $k = y - \pi/4$

Therefore,

$$f(x,y) = e^x \cos y \Rightarrow f\left(1, \frac{\pi}{4}\right) = e.\frac{1}{\sqrt{2}}$$

$$\frac{\partial f}{\partial x} = e^x \cos y \Rightarrow \left(\frac{\partial f}{\partial x}\right)_{(1,\pi/4)} = \frac{e}{\sqrt{2}}$$

$$\frac{\partial f}{\partial y} = -e^x \sin y \Rightarrow \left(\frac{\partial f}{\partial y}\right)_{(1,\pi/4)} = -\frac{e}{\sqrt{2}}$$

$$\frac{\partial^2 f}{\partial x^2} = e^x \cos y \Rightarrow \left(\frac{\partial^2 f}{\partial x^2}\right)_{\left(1,\frac{\pi}{4}\right)} = \frac{e}{\sqrt{2}}$$

$$\frac{\partial^2 f}{\partial y^2} = -e^x \cos y \Rightarrow \left(\frac{\partial^2 f}{\partial y^2}\right)_{\left(1,\frac{\pi}{4}\right)} = -\frac{e}{\sqrt{2}}$$

$$\frac{\partial^2 f}{\partial x \partial y} = -e^x \sin y \Rightarrow \left(\frac{\partial^2 f}{\partial x \partial y}\right)_{\left(1,\frac{\pi}{4}\right)} = -\frac{e}{\sqrt{2}}$$

Putting all these values in (1), we get

$$e^x \cos y = \frac{e}{\sqrt{2}} + \left[(x-1)\frac{e}{\sqrt{2}} + \left(y - \frac{\pi}{4}\right)\left(-\frac{e}{\sqrt{2}}\right)\right] + \frac{1}{2!}\left[(x-1)^2 \frac{e}{\sqrt{2}} + 2(x-1)\left(y-\frac{\pi}{4}\right)\left(-\frac{e}{\sqrt{2}}\right) + \left(y - \frac{\pi}{4}\right)^2\left(-\frac{e}{\sqrt{2}}\right)\right] + \ldots$$

$$= \frac{e}{\sqrt{2}} + \left[1 + (x-1) + \left(y - \frac{\pi}{4}\right) + \frac{(x-1)^2}{2} + (x-1)\left(y - \frac{\pi}{4}\right) + \ldots\right]$$

Example 2. *Expand $\tan^{-1}\frac{y}{x}$ in the neighbourhood of (1, 1) upto and inclusive of second degree term. Hence, compute $f(1.1, 0.9)$ approximately.*

(UPTU–2005, 06, 07, GBTU–2010, UKTU–2011, 12, VTU–2010, JNTU–2006)

Solution. We have

$$f(x,y) = \tan^{-1}\frac{y}{x} \Rightarrow f(1,1) = \tan^{-1} 1 = \frac{\pi}{4}$$

$$\frac{\partial f}{\partial x} = \frac{1}{1 + \frac{y^2}{x^2}}\left(-\frac{y}{x^2}\right) = -\frac{y}{x^2 + y^2}$$

$$\Rightarrow \left(\frac{\partial f}{\partial x}\right)_{(1,1)} = -\frac{1}{2}$$

$$\frac{\partial f}{\partial y} = \frac{1}{1 + \frac{y^2}{x^2}}.\frac{1}{x} = \frac{x}{x^2 + y^2}$$

$$\Rightarrow \left(\frac{\partial f}{\partial y}\right)_{(1,1)} = \frac{1}{2}$$

$$\frac{\partial^2 f}{\partial x^2} = -y(-1)(x^2 + y^2)^{-2}.2x = \frac{2xy}{(x^2 + y^2)^2}$$

$$\Rightarrow \left(\frac{\partial^2 f}{\partial x^2}\right)_{(1,1)} = \frac{1}{2}$$

$$\frac{\partial^2 f}{\partial x \partial y} = \frac{(x^2 + y^2) - x.2x}{(x^2 + y^2)^2} = \frac{y^2 - x^2}{(x^2 + y^2)^2}$$

$$\Rightarrow \left(\frac{\partial^2 f}{\partial x \partial y}\right)_{(1,1)} = 0$$

$$\frac{\partial^2 f}{\partial y^2} = x(-1)(x^2 + y^2)^{-2}.2y = -\frac{2xy}{(x^2 + y^2)^2}$$

$$\Rightarrow \left(\frac{\partial^2 f}{\partial y^2}\right)_{(1,1)} = -\frac{1}{2}$$

Similarly, we may get

$$\left(\frac{\partial^3 f}{\partial x^3}\right)_{(1,1)} = -\frac{1}{2}, \left(\frac{\partial^3 f}{\partial x^2 \partial y}\right)_{(1,1)} = -\frac{1}{2}$$

$$\left(\frac{\partial^3 f}{\partial x \partial y^2}\right)_{(1,1)} = \frac{1}{2}$$

We know that

$$f(x,y) = f(1,1) + [(x-1)f_x(1,1) + (y-1)f_y(1,1)] + \frac{1}{2!}\Big[(x-1)^2 f_{xx}(1,1) + 2(x-1)(y-1)f_{xy}(1,1) + (y-1)^2 f_y(1,1)\Big] + \frac{1}{3!}\Big[(x-1)^3 f_{xxx}(1,1) + 3(x-1)^2(y-1)f_{xxy}(1,1) + 3(x-1)(y-1)^2 f_{xyy}(1,1) + (y-1)^3 f_{yyy}(1,1)\Big] + \ldots$$

Hence

$$\tan^{-1}\frac{y}{x} = \frac{\pi}{4} + \left[(x-1)\left(-\frac{1}{2}\right) + (y-1)\frac{1}{2}\right] + \frac{1}{2!}\left[(x-1)^2\frac{1}{2} + 2(x-1)(y-1).0 + (y-1)^2\left(-\frac{1}{2}\right)\right] - \frac{1}{12}[(x-1)^3 + 3(x-1)^2(y-1) - 3(x-1)(y-1)^2 - (y-1)^3] + \ldots$$

$$= \frac{\pi}{4} - \frac{1}{2}(x-1) + \frac{1}{2}(y-1) + \frac{1}{4}(x-1)^2 - \frac{1}{4}(y-1)^2 - \frac{1}{12}[(x-1)^3 + 3(x-1)^2(y-1) - 3(x-1)(y-1)^2 - (y-1)^3] + \ldots.$$

Further,

$$f(1.1, 0.9) = \frac{\pi}{4} - \frac{1}{2}(.1) + \frac{1}{2}(-.1) + \frac{1}{4}(.1)^2 - \frac{1}{4}(-.1)^2 - \frac{1}{12}[(.1)^3 + 3(.1)^2(-.1) - 3(.1)(-.1)^2 - (-.1)^3] + \ldots$$

$$= \frac{\pi}{4} - \frac{(.2)}{2} - \frac{1}{12}[.001 - .003 - .003 + .001] + \ldots$$

$$\approx 0.6857$$

Example 3. *Find Taylor's series expansion of function on $f(x,y) = e^{-x^2-y^2}\cos xy$ about the point $x_0 = 0$, $y_0 = 0$ upto three terms.* (UPTU–2007)

Solution . We have

$$f(x,y) = e^{-x^2-y^2}\cos xy \quad \Rightarrow f(0,0) = 1$$

$$\frac{\partial f}{\partial x} = e^{-x^2-y^2} y \sin xy + e^{-x^2-y^2}(-2x)\cos xy$$

$$\Rightarrow \left(\frac{\partial f}{\partial x}\right)_{(0,0)} = 0$$

and $\dfrac{\partial f}{\partial y} = -xe^{-x^2-y^2}\sin xy - 2ye^{-x^2-y^2}\cos xy$

$$\Rightarrow \left(\frac{\partial f}{\partial y}\right)_{(0,0)} = 0$$

$$\frac{\partial^2 f}{\partial x^2} = -y\left[e^{-x^2-y^2} y\cos xy - 2xe^{-x^2-y^2}\sin xy\right] - 2\left[e^{-x^2-y^2}\cos xy\right] - 2x\left[-ye^{-x^2-y^2}\sin xy - 2xe^{-x^2-y^2}\cos xy\right]$$

$$\Rightarrow \left(\frac{\partial^2 f}{\partial x^2}\right)_{(0,0)} = -2$$

$$\frac{\partial^2 f}{\partial y^2} = -x\left[-2ye^{-x^2-y^2}\sin xy + e^{-x^2-y^2} x\cos xy\right] - 2e^{-x^2-y^2}\cos xy - 2y\left[-2ye^{-x^2-y^2}\cos xy - e^{-x^2-y^2} x\sin xy\right]$$

$$\Rightarrow \left(\frac{\partial^2 f}{\partial y^2}\right)_{(0,0)} = -2$$

$$\frac{\partial^2 f}{\partial x \partial y} = e^{-x^2-y^2}\sin xy - x\left[-2xe^{-x^2-y^2}\sin xy + e^{-x^2-y^2} y\cos xy\right] - 2y\left[-2xe^{-x^2-y^2}\cos xy - e^{-x^2-y^2} y\sin xy\right]$$

$$\Rightarrow \left(\frac{\partial^2 f}{\partial x \partial y}\right)_{(0,0)} = 0$$

Putting all these values in Taylor's theorem

$$f(x,y) = f(0,0) + \left(x\frac{\partial}{\partial x} + y\frac{\partial}{\partial y}\right) f(0,0) + \frac{1}{2!}\left(x\frac{\partial}{\partial x} + y\frac{\partial}{\partial y}\right)^2 f(0,0) + \ldots$$

we get

$$f(x,y) = 1 + 0 + \frac{1}{2!}[x(-2) + 2xy(0) + y(-2)] + \ldots$$

$$= 1 - (x+y) + \ldots$$

Example 4. *Expand $x^2y + 3y - 2$ in powers of $(x-1)$ and $(y+2)$ using Taylor's theorem.*
(PTU–2010, VTU–2008, Anna–2005, UPTU(AG)–2006, AMIETE–2003, AMIE–1999)

Solution . We have

$$f(x, y) = x^2y + 3y - 2$$

$a + h = x$ and $h = x - 1 \Rightarrow a = 1$

$b + k = y$ and $k = y + 2 \Rightarrow b = -2$

Now

$$f(x, y) = x^2y + 3y - 2 \Rightarrow f(1, -2) = -10$$
$$f_x(x, y) = 2xy \Rightarrow f_x(1, -2) = -4$$
$$f_y(x, y) = x^2 + 3 \Rightarrow f_y(1, -2) = 4$$
$$f_{xx}(x, y) = 2y \Rightarrow f_{xx}(1, -2) = -4$$
$$f_{xy}(x, y) = 2x \Rightarrow f_{xy}(1, -2) = 2$$
$$f_{yy}(x, y) = 0 \Rightarrow f_{yy}(1, -2) = 0$$
$$f_{xxx}(x, y) = 0 \Rightarrow f_{xxx}(1, -2) = 0$$
$$f_{xxy}(x, y) = 2 \Rightarrow f_{xxy}(1, -2) = 2$$
$$f_{xyy}(x, y) = 0 \Rightarrow f_{xyy}(1, -2) = 0$$
$$f_{yyy}(x, y) = 0 \Rightarrow f_{yyy}(1, -2) = 0$$

Putting all these values in Taylor's theorem given by

$$f(a + h, b + k) = f(a,b) + \left(h\frac{\partial f}{\partial x} + k\frac{\partial f}{\partial y}\right)_{(a,b)} + \frac{1}{2!}\left(h^2\frac{\partial^2 f}{\partial x^2} + 2hk\frac{\partial^2 f}{\partial x \partial y} + k^2\frac{\partial^2 f}{\partial y^2}\right)_{(a,b)}$$
$$+\frac{1}{3!}\left(h^3\frac{\partial^3 f}{\partial x^3} + 3h^2k\frac{\partial f}{\partial x^2 \partial y} + 3hk^2\frac{\partial^3 f}{\partial x \partial y^2} + k^3\frac{\partial^3 f}{\partial y^3}\right)_{(a,b)} + \ldots$$

We get $x^2y + 3y - 2$

$$= -10 + ((x-1)(-4) + (y+2)4) + \frac{1}{2!}\left((x-1)^2(-4) + 2(x-1)y(y+2)2 + (y+2)^2(0)\right)$$
$$+\frac{1}{3!}\left((x-1)^3.0 + 3(x-1)^2(y+2).2 + 3(x-1)(y+2)^2.0 + (y+2)^3.0\right) + \ldots$$
$$= -10 - 4(x-1) + 4(y+2) - 2(x-1)^2 + 2(x-1)(y+2) + (x-1)^2(y+2)$$

Example 5. *Expand $e^x \sin y$ in the powers of x and y in the neighbourhood of $(0, \pi/4)$ upto the third degree terms.* (Anna–2009)

Solution . We have

$$f(x, y) = e^x \sin y \Rightarrow f\left(0, \frac{\pi}{4}\right) = \frac{1}{\sqrt{2}}$$
$$f_x(x, y) = e^x \sin y \Rightarrow f_x\left(0, \frac{\pi}{4}\right) = \frac{1}{\sqrt{2}}$$
$$f_y(x, y) = e^x \cos y \Rightarrow f_y\left(0, \frac{\pi}{4}\right) = \frac{1}{\sqrt{2}}$$
$$f_{xx}(x, y) = e^x \sin y \Rightarrow f_{xx}\left(0, \frac{\pi}{4}\right) = \frac{1}{\sqrt{2}}$$
$$f_{xy}(x, y) = e^x \cos y \Rightarrow f_{xy}\left(0, \frac{\pi}{4}\right) = \frac{1}{\sqrt{2}}$$
$$f_{yy}(x, y) = -e^x \sin y \Rightarrow f_{yy}\left(0, \frac{\pi}{4}\right) = -\frac{1}{\sqrt{2}}$$
$$f_{xxx}(x, y) = e^x \sin y \Rightarrow f_{xxx}\left(0, \frac{\pi}{4}\right) = \frac{1}{\sqrt{2}}$$
$$f_{xxy}(x, y) = e^x \cos y \Rightarrow f_{xxy}\left(0, \frac{\pi}{4}\right) = \frac{1}{\sqrt{2}}$$
$$f_{xyy}(x, y) = -e^x \sin y \Rightarrow f_{xyy}\left(0, \frac{\pi}{4}\right) = -\frac{1}{\sqrt{2}}$$
$$f_{yyy}(x, y) = -e^x \cos y \Rightarrow f_{yyy}\left(0, \frac{\pi}{4}\right) = -\frac{1}{\sqrt{2}}$$

By Taylor's theorem, we get

$$f(x,y) = f\left(0, \frac{\pi}{4}\right) + \left(x\frac{\partial}{\partial x} + \left(y - \frac{\pi}{4}\right)\frac{\partial}{\partial y}\right)f\left(0, \frac{\pi}{4}\right) + \frac{1}{2!}\left(x\frac{\partial}{\partial x} + \left(y - \frac{\pi}{4}\right)\frac{\partial}{\partial y}\right)^2 f\left(0, \frac{\pi}{4}\right)$$
$$+\frac{1}{3!}\left(x\frac{\partial}{\partial x} + \left(y - \frac{\pi}{4}\right)\frac{\partial}{\partial y}\right)^3 f\left(0, \frac{\pi}{4}\right) + \ldots \quad \ldots(1)$$

Putting all the above values in (1), we get

$$f(x,y) = \frac{1}{\sqrt{2}} + x.\frac{1}{\sqrt{2}} + \left(y - \frac{\pi}{4}\right)\frac{1}{\sqrt{2}} + \frac{1}{2!}\left[x^2\frac{1}{\sqrt{2}} + 2x\left(y - \frac{\pi}{4}\right)\frac{1}{\sqrt{2}} - \left(y - \frac{\pi}{4}\right)^2\frac{1}{\sqrt{2}}\right]$$
$$+\frac{1}{3!}\left[x^3\frac{1}{\sqrt{2}} + 3x^2\left(y - \frac{\pi}{4}\right)\frac{1}{\sqrt{2}} + 3x\left(y - \frac{\pi}{4}\right)^2\left(-\frac{1}{\sqrt{2}}\right) + \left(y - \frac{\pi}{4}\right)^3\left(-\frac{1}{\sqrt{2}}\right)\right] + \ldots$$
$$= \frac{1}{\sqrt{2}}\left[1 + x + \left(y - \frac{\pi}{4}\right) + \frac{1}{2!}\left\{x^2 + 2x\left(y - \frac{\pi}{4}\right) - \left(y - \frac{\pi}{4}\right)^2\right\} + \frac{1}{3!}\left\{x^3 + 3x^2\left(y - \frac{\pi}{4}\right) - 3x\left(y - \frac{\pi}{4}\right)^2 - \left(y - \frac{\pi}{4}\right)^3 -\right\} + \ldots\right]$$

Example 6. *Expand $e^x \log(1 + y)$ in the neighbourhood of the point (0, 0).* (VTU–2010, PTU–2009, JNTU–2006)

Solution . We have

$$f(x, y) = e^x \log(1+y) \Rightarrow f(0, 0) = 0$$
$$f_x(x, y) = e^x \log(1+y) \Rightarrow f_x(0, 0) = 0$$
$$f_y(x, y) = \frac{e^x}{1+y} \Rightarrow f_y(0, 0) = 1$$

$f_{xx}(x,y) = e^x \log(1+y) \quad \Rightarrow f_{xx}(0,0) = 0$

$f_{xy}(x,y) = \dfrac{e^x}{1+y} \quad \Rightarrow f_{xy}(0,0) = 1$

$f_{yy}(x,y) = -\dfrac{e^x}{(1+y)^2} \quad \Rightarrow f_{yy}(0,0) = -1$

$f_{xxx}(x,y) = e^x \log(1+y) \quad \Rightarrow f_{xxx}(0,0) = 0$

$f_{xxy}(x,y) = \dfrac{e^x}{1+y} \quad \Rightarrow f_{xxy}(0,0) = 1$

$f_{xyy}(x,y) = -\dfrac{e^x}{(1+y)^2} \quad \Rightarrow f_{xyy}(0,0) = -1$

$f_{yyy}(x,y) = \dfrac{2e^x}{(1+y)^3} \quad \Rightarrow f_{yyy}(0,0) = 2$

Putting all these values in Taylor's series given by $f(x, y)$

$$= f(0,0) + [xf_x(0,0) + yf_y(0,0)] + \frac{1}{2!}\left[x^2 f_{xx}(0,0) + 2xyf_{xy}(0,0) + y^2 f_{yy}(0,0)\right]$$
$$+ \frac{1}{3!}\left[x^3 f_{xxx}(0,0) + 3x^2 y f_{xxy}(0,0) + 3xy^2 f_{xyy}(0,0) + y^3 f_{yyy}(0,0)\right] + \ldots.$$

we get

$e^x \log(1+y)$

$$= 0 + [x.0 + y.1] + \frac{1}{2}\left[x^2.0 + 2xy.1 + y^2(-1)\right] + \frac{1}{6}\left[x^3.0 + 3x^2y.1 + 3xy^2(-1) + y^3.2\right] + \ldots$$
$$= y + xy - \frac{1}{2}y^2 + \frac{1}{2}x^2y - \frac{1}{2}xy^2 + \frac{1}{3}y^3 + \ldots$$

Example 7. *Expand x^y in powers of $(x - 1)$ and $(y - 1)$ upto the third degree terms.*

(UPTU–2004, UKTU–2010, VTU–2009)

Solution . We have

$f(x, y) = x^y \quad \Rightarrow \quad f(1, 1) = 1$

$f_x(x, y) = yx^{y-1} \quad \Rightarrow \quad f_x(1, 1) = 1$

$f_y(x, y) = x^y \log x \quad \Rightarrow \quad f_y(1, 1) = 0$

$f_{xx}(x, y) = y(y-1)x^{y-2} \quad \Rightarrow f_{xx}(1, 1) = 0$

$f_{xy}(x, y) = x^{y-1} + yx^{y-1}\log x \Rightarrow f_{xy}(1, 1) = 1$

$f_{yy}(x, y) = x^y(\log x)^2 \quad \Rightarrow f_{yy}(1, 1) = 0$

$f_{xxx}(x, y) = y(y-1)(y-2)x^{y-3} \Rightarrow f_{yyy}(1, 1) = 0$

$f_{xxy}(x, y) = (y-1)x^{y-2} + y(y-1)x^{y-2}\log_e x + yx^{y-2}$

$\Rightarrow \quad f_{xxy}(1, 1) = 1$

$f_{xyy}(x, y) = yx^{y-1}(\log x)^2 + x^4 . \dfrac{2\log x}{x}$

$= yx^{y-1}(\log x)^2 + 2x^{y-1}\log x$

$\Rightarrow \quad f_{xyy}(1, 1) = 0$

$f_{yyy}(x, y) = x^4(\log x)^3$

$\Rightarrow \quad f_{yyy}(1, 1) = 0$

Putting all these values in Taylor's series given by $f(x, y)$

$$f(x,y) = f_x(a,b) + (x-a)f_x(a,b) + (y-b)f_y(a,b) + \frac{1}{2!}\left[(x-a)^2 f_{xx}(a,b) + 2(x-a)(y-b)f_{xy}(a,b) + (y-b)^2 f_{yy}(a,b)\right]$$
$$+ \frac{1}{3!}\left[(x-a)^3 f_{xxx}(a,b) + 3(x-a)^2(y-b)f_{xxy}(a,b) + 3(x-a)(y-b)^2 f_{xyy}(a,b) + (y-b)^3 f_{yyy}(a,b)\right] + \ldots$$

we get $f(x, y) = x^y$

$$= 1 + (x-1).1 + (y-1).0 + \frac{1}{2!}\left[(x-1)^2.0 + 2(x-1)(y-1)^2.1 + (y-1)^2.0\right]$$
$$+ \frac{1}{3!}\left[(x-1)^3.0 + 3(x-1)^2(y-1).1 + 3(x-1)(y-1)^2.0 + (y-1)^3.0\right] + \ldots$$
$$= 1 + (x-1) + (x-1)(y-1) + \frac{1}{2}(x-1)^2(y-1) + \ldots$$

Example 8. *Expand $\dfrac{(x+h)(y+k)}{x+h+y+k}$ in powers of h and k upto and inclusive of the second degree terms.*

(AMIE–1999, 2001)

Solution . We have $f(x+h, y+k) = \dfrac{(x+h)(y+k)}{x+h+y+k}$

Putting $h = k = 0 \quad \Rightarrow \quad f(x,y) = \dfrac{xy}{x+y}$

Now $f_x(x,y) = \dfrac{(x+y).y - xy.1}{(x+y)^2} = \dfrac{y^2}{(x+y)^2}$,

$f_y(x,y) = \dfrac{x^2}{(x+y)^2}$, $f_{xx}(x,y) = -\dfrac{2y^2}{(x+y)^3}$,

$f_{yy}(x,y) = -\dfrac{2x^2}{(x+y)^3}$,

$f_{xy}(x,y) = \dfrac{(x+y)^2.2x - x^2.2(x+y)}{(x+y)^4}$

$= \dfrac{2xy}{(x+y)^3}$

Therefore,

$$\frac{(x+h)(y+k)}{x+h+y+k} = f(x+h, y+k)$$
$$= f(x,y) + (hf_x + kf_y) + \frac{1}{2!}\left[h^2 f_{xx} + 2hkf_{xy} + k^2 f_{yy}\right] + \ldots$$

$$= \frac{xy}{x+y} + \left[h.\frac{y^2}{(x+y)^2} + k\frac{x^2}{(x+y)^2}\right] + \frac{1}{2}\left[h^2\frac{(-2y^2)}{(x+y)^3} + 2hk\frac{2xy}{(x+y)^3} + k^2\frac{(-2x^2)}{(x+y)^3}\right] + \ldots$$

$$= \frac{xy}{x+y} + \frac{y^2}{(x+y)^2}.h + \frac{x^2}{(x+y)^2}.k - \frac{y^2}{(x+y)^3}h^2 + 2xy.hk - \frac{x^2}{(x+y)^3}k^2 + \ldots$$

Example 9. *Expand e^{ax} sin by in powers of x and y upto third degree term.* (GBTU–2011)

Solution . Since, the point is not given, so we expand it about the point (0, 0).

Let

$f(x, y) = e^{ax} \sin by \Rightarrow f(0, 0) = 0$

$f_x(x, y) = ae^{ax} \sin by \Rightarrow f_x(0, 0) = 0$

$f_{xx}(x, y) = a^2e^{ax} \sin by \Rightarrow f_{xx}(0, 0) = 0$

$f_{xxx}(x, y) = a^3e^{ax} \sin by \Rightarrow f_{xxx}(0, 0) = 0$

$f_y(x, y) = be^{ax} \cos by \Rightarrow f_y(0, 0) = b$

$f_{yy}(x, y) = -b^2e^{ax} \sin by \Rightarrow f_{yy}(0, 0) = 0$

$f_{yyy}(x, y) = -b^3e^{ax} \cos by \Rightarrow f_{yyy}(0, 0) = -b^3$

$f_{xy}(x, y) = abe^{ax} \cos by \Rightarrow f_{xy}(0, 0) = ab$

$f_{xxy}(x, y) = a^2be^{ax} \cos by \Rightarrow f_{xxy}(0, 0) = a^2b$

$f_{xyy}(x, y) = -ab^2e^{ax} \sin by \Rightarrow f_{xyy}(0, 0) = 0$

Putting all these values in Maclaurin's theorem given by

$$f(x,y) = f(0,0) + [xf_x(0,0) + yf_y(0,0)] + \frac{1}{2!}\left[x^2 f_{xx}(0,0) + 2xyf_{xy}(0,0) + y^2 f_{yy}(0,0)\right] + \frac{1}{3!}\left[x^3 f_{xxx}(0,0) + 3x^2yf_{xxy}(0,0) + 3xy^2 f_{xyy}(0,0) + y^3 f_{yyy}(0,0)\right] + \ldots.$$

We get

$$e^{ax} \sin by = by + \frac{1}{2!}ab(2xy) + \frac{1}{3!}\left[3x^2y(a^2b - b^3y^3)\right] + \ldots$$

$$= by + abxy + \frac{1}{2}a^2bx^2y - \frac{b^3}{6}y^3 + \ldots$$

Example 10. *Expand $f(x, y) = e^{xy}$ in Taylor's series at (1, 1) upto second degree term.*

Solution . Let $f(x, y) = e^{xy}$

$f(x, y) = e^{xy} \Rightarrow f(1, 1) = e$

$f_x(x, y) = ye^{xy} \Rightarrow f_x(1, 1) = e$

$f_y(x, y) = xe^{xy} \Rightarrow f_y(1, 1) = e$

$f_{xx}(x, y) = y^2e^{xy} \Rightarrow f_{xx}(1, 1) = e$

$f_{xy}(x, y) = xye^{xy} \Rightarrow f_{xy}(1, 1) = e$

$f_{yy}(x, y) = x^2e^{xy} \Rightarrow f_{yy}(1, 1) = e$

Putting all these values in Taylor's theorem given by

$$f(x,y) = f(a,b) + [(x-a)f_x(a,b) + (y-b)f_y(a,b)] + \frac{1}{2!}\left[(x-a)^2 f_{xx}(a,b) + 2(x-a)(y-b)f_{xy}(a,b) + (y-b)^2 f_{yy}(a,b)\right] + \ldots$$

we get $f(x, y) = e^{xy}$

$$= e + [(x-1)e + (y-1)e] + \frac{1}{2!}[(x-1)^2e + 2(x-1)(y-1)e + (y-1)^2e] + \ldots$$

$$\Rightarrow e^{xy} = e\left[1 + (x-1) + (y-1)^2 + \frac{(x-1)^2 + 2(x-1)(y-1) + (y-1)^2}{2!} + \ldots\right]$$

Example 11. *Expand e^{x+y} in powers of (x – 1) and (y + 1) upto first degree.*

Solution . Let $f(x, y) = e^{x+y}$

Here $a = 1, b = -1$

$f(x, y) = e^{x+y} \Rightarrow f(1, -1) = e^0 = 1$

$f_x(x, y) = e^{x+y} \Rightarrow f_x(1, -1) = 1$

$f_y(x, y) = e^{x+y} \Rightarrow f_y(1, -1) = 1$

... and so on.

Putting all these values in Taylor's series expansion given by

$$f(x, y) = f(a,b) + [(x-a)f_x(a,b) + (y-b)f_y(a,b)] + \ldots$$

we get

$$f(x, y) = 1 + (x-1).1 + (y+1).1 + \ldots$$

$$\Rightarrow e^{xy} = 1 + (x-1) + (y+1) + \ldots$$

Example 12. *Expand $x^2y + \sin y + e^x$ in Taylor's series about (1, π) upto second degree.*

Solution . Let $f(x, y) = x^2y + \sin y + e^x$

Here, $a = 1, b = \pi$

Now, $f(x, y) = x^2y + \sin y + e^x$

$\Rightarrow f(1, \pi) = \pi + \sin \pi + e = \pi + e$

$f_x(x, y) = 2xy + e^x \Rightarrow f_x(1, \pi) = 2\pi + e$

$f_y(x, y) = x^2 + \cos y \Rightarrow f_y(1, \pi) = 0$

$f_{xx}(x, y) = 2y + e^x \Rightarrow f_{xx}(1, \pi) = 2\pi + e$

$f_{xy}(x, y) = 2x \Rightarrow f_{xy}(1, \pi) = 2$

$f_{yy}(x, y) = -\sin y \Rightarrow f_{yy}(1, \pi) = 0$

Putting all these values in Taylor's series expansion given by

$$f(x,y) = f(a,b)+[(x-a)f_x(a,b) + (y-b)f_y(a,b)]+...$$

we get

$$f(x,y) = (\pi+e)+[(x-1)(2\pi+e)+(y-\pi).0] + \frac{1}{2!}[(x-1)^2(2\pi+e) + 2(x-1)(y-\pi)(2)+(y-\pi)^2.0]+...$$

Hence, $x^2y + \sin y + e^x$

$$= (\pi+e)+(2\pi+e)(x-1)+\frac{1}{2}(x-1)^2(2\pi+e) + 2(x-1)(y-\pi)+...$$

Example 13. *Find the Taylor's series expansion of $f(x, y) = x^3 + xy^2$ about the point (2, 1).* (UPTU–2012)

Solution. We have $f(x, y) = x^3 + xy^2$

$a = 2, b = 1$

Let

$f(x, y) = x^3 + xy^2 \quad \Rightarrow \quad f(2, 1) = 10$

$f_x(x, y) = 3x^2 + y^2 \quad \Rightarrow \quad f_x(2, 1) = 13$

$f_{xx}(x, y) = 6x \quad \Rightarrow \quad f_{xx}(2, 1) = 12$

$f_{xxx}(x, y) = 0 \quad \Rightarrow \quad f_{xxx}(2, 1) = 0$

$f_y(x, y) = 2xy \quad \Rightarrow \quad f_y(2, 1) = 4$

$f_{yy}(x, y) = 2x \quad \Rightarrow \quad f_{yy}(2, 1) = 4$

$f_{yyy}(x, y) = 0 \quad \Rightarrow \quad f_{yyy}(2, 1) = 0$

$f_{xy}(x, y) = 2y \quad \Rightarrow \quad f_{xy}(2, 1) = 2$

Putting all these values in Taylor's series expansion given by

$$f(x,y) = f(a,b)+\left[(x-a)f_x(a,b)+(y-b)f_y(a,b)\right] + \frac{1}{2!}\left[(x-a)^2 f_{xx}(a,b)+(y-b)^2 f_{yy}(a,b) + 2(x-a)(y-b)f_{xy}(a,b)\right]+...$$

we get

$$f(x,y) = 10+[(x-2).13+(y-1).4] + \frac{1}{2}\left[(x-2)^2.12+(y-1)^2.4 + 2(x-2)(y-1).2\right]+...$$

Hence,

$$x^3 + xy^2 = 10+13(x-2)+4(y-1) + 6(x-2)^2+2(y-1)^2 + 2(x-2)(y-1)+...$$

Example 14. *Using Maclaurin's series expansion, show that*

$$\log(1+x+y) = (x+y)-\frac{1}{2}(x+y)^2 + \frac{1}{3}(x+y)^3+...$$

Solution. We have $f(x, y) = \log(1+x+y)$

$a = 0, b = 0$

Now $f(x, y) = \log(1 + x + y)$

$\Rightarrow \quad f(0, 0) = \log 1 = 0$

$$f_x(x,y) = \frac{1}{1+x+y} \quad \Rightarrow f_x(0, 0) = 1$$

$$f_y(x,y) = \frac{1}{1+x+y} \quad \Rightarrow f_y(0, 0) = 1$$

$$f_{xx}(x,y) = -\frac{1}{(1+x+y)^2} \Rightarrow f_{xx}(0, 0) = -1$$

$$f_{xy}(x,y) = -\frac{1}{(1+x+y)^2} \Rightarrow f_{xy}(0, 0) = -1$$

$$f_{yy}(x,y) = -\frac{1}{(1+x+y)^2} \Rightarrow f_{yy}(0, 0) = -1$$

Putting all these values in Taylor's series expansion given by

$$f(x,y) = f(a,b)+[(x-a)f_x(a,b) + (y-b)f_y(a,b)]+...$$

We get $f(x, y)$

$$= 0+[(x-0).1+(y-0).1]+\frac{1}{2!}[(x-0)^2(-1) + 2(x-0)(y-0)(-1)+(y-0)^2(-1)]+...$$

$$= (x+y)-\frac{1}{2}(x^2+2xy+y^2)+...$$

$$\Rightarrow \log(1+x+y) = (x+y)-\frac{1}{2}(x+y)^2 + \frac{1}{3}(x+y)^3+...$$

Example 15. *If $f(x, y) = \tan^{-1}xy$, compute an approximate value of $f(0.9, -1.2)$* (AMIE–1996, 2007)

Solution. Let $f(x, y) = \tan^{-1}xy$

Let us expand it near the point (1, –1)

$f(0.9, -1.2) = f(1 - 0.1, -1 - 0.2)$

$$= f(1,-1)+\left[(-0.1)\frac{\partial f}{\partial x}+(-0.2)\frac{\partial f}{\partial y}\right] + \frac{1}{2!}\left[(-0.1)^2\frac{\partial^2 f}{\partial x^2}+2(-0.1)(-0.2)\frac{\partial^2 f}{\partial x \partial y} + (0.2)^2\frac{\partial^2 f}{\partial y^2}\right]+... \quad ...(1)$$

Now $f(x, y) = \tan^{-1}xy$

$$\Rightarrow \quad f(1,-1) = \tan^{-1}(-1) = -\frac{\pi}{4}$$

$$f_x(x,y) = \frac{y}{1+x^2y^2}$$

$$\Rightarrow \quad f_x(1,-1) = \frac{-1}{1+1} = -\frac{1}{2}$$

$$f_y(x,y) = \frac{x}{1+x^2y^2}$$

$$\Rightarrow f_y(1,-1) = \frac{1}{1+1} = \frac{1}{2}$$

$$f_{xx}(x,y) = -\frac{2x.y}{(1+x^2y^2)^2}$$

$$\Rightarrow f_{xx}(1,-1) = \frac{(-2)(-1)}{(1+1)^2} = \frac{1}{2}$$

$$f_{xy}(x,y) = \frac{1-x^2y^2}{(1+x^2y^2)^2} \Rightarrow f_{xy}(1,-1) = 0$$

$$\text{and } f_{yy}(x,y) = \frac{-x(2x^2y)}{(1+x^2y^2)^2}$$

$$\Rightarrow f_{yy}(1,-1) = \frac{2}{(1+1)^2} = \frac{1}{2}$$

Putting all these values in (1), we get

$f(0.9, -1.2)$

$$= -\frac{\pi}{4} + (-0.1)\left(-\frac{1}{2}\right) + (-0.2)\left(\frac{1}{2}\right) + \frac{1}{2}\left[(-0.1)^2\left(\frac{1}{2}\right) + 2(-0.1)(-0.2).0 + (-0.2)^2\left(\frac{1}{2}\right)\right] + \ldots$$

$$= -\frac{\pi}{4} + 0.05 - 0.1 + \frac{1}{2}(0.005 + 0.02) + \ldots$$

$$= -\frac{\pi}{4} + 0.05 - 0.1 + 0.0125 = -0.823$$

EXERCISE 10.1

1. Expand $(1 + x + y^2)^{1/2}$ at $(1, 0)$.
2. Show that $e^y \log(1 + x) = x + xy - \frac{x^2}{2} + \ldots$
3. Expand $\sin xy$ in powers of $(x - 1)$ and $\left(y - \frac{\pi}{2}\right)$ upto the second degree terms. (UPTU–2008, UKTU–2011)
4. Expand $x^2y + \sin y + e^x$ in powers of $(x - 1)$ and $(y - \pi)$.
5. Expand $\sin y$ about the origin upto third order terms.
6. Expand $x^2 + 3y^2 - 9x - 9y + 26$ in powers of $(x - 2)$ and $(y - 2)$ using Taylor's series expansion.
7. Evaluate $\log[(1.03)^{1/3} + (0.98)^{1/4} - 1]$ approximately using Taylor's series expansion.

ANSWERS

1. $\sqrt{2}\left[1 + \frac{x-1}{4} - \frac{(x-1)^2}{32} + \frac{y^2}{4} + \ldots\right]$
3. $1 - \frac{\pi^2}{8}(x-1)^2 - \frac{\pi}{2}(x-1)\left(y - \frac{\pi}{2}\right) - \frac{1}{2}\left(y - \frac{\pi}{2}\right)^2$
4. $\pi + e + (x-1)(2\pi + e) + \frac{1}{2}(x-1)^2(2\pi + e) + 2(x-1)(y-\pi) + \ldots$
5. $xy + \ldots$
6. $6 - 5(x-2) + 3(y-2) + (x-2)^2 + 3(y-2)^2 + \ldots$
7. 0.005

FFFFFF

Jacobian

11.1 INTRODUCTION

Sometimes a function is not exclusively defined in terms of the independent variable and we assumed that a functional equation $f(x,y)=0$ gives y as a function of x. Some times y cannot be expressed in terms of x then we say that function is implicit.

Definition. *Let $f(x,y)$ be a function of two variables and $y=g(x)$ be a function of x such that for every value of x for which $g(x)$ is defined, $f(x,g(x))$ vanish identically, i.e, $y=g(x)$ is a root of the equation $f(x,y)=0$. Then $y=g(x)$ is called the implicit function defined by the functional equation $f(x,y)=0$.*

REMARK

- A functional equation in general may or may not define an implicit function.

11.2 EXISTENCE AND DERIVABILITY OF IMPLICIT FUNCTIONS

If $f_x(x,y)$, $f_y(x,y)$ are continuous in a nbd of (a,b) and $f(a,b)\neq 0$ then $\exists$ a rectangle R : $[a-h,a+h,b-k,b+k]$ about (a,b) such that

(i) for each $x\in[a-h,a+h]$, the equation $f(x,y)=0$ determine unique solution $y=g(x)$ in $(b-k, b+k)$

and (ii) $g'(x)$ is continuous in $[a-h,a+h]$ and $f_y(x,g(x))\neq 0$ and $g'(x)=\dfrac{-f_x(x,g(x))}{f_y(x,g(x))}$.

11.2.1 GENERAL CASE

If $f(x_1,x_2,\ldots,x_n,y)$ be a function of $(n+1)$ variables $x_1, x_2,\ldots,x_n,y$ and $(a_1,a_2,\ldots,a_n,b)$ be a point of its domain such that

(i) $f(a_1,a_2,\ldots,a_n,b)=0$

(ii) the partial derivatives w.r.t. all $(n+1)$ variables exists and are continuous in a nbd of $(a_1,a_2,\ldots,a_n,b)$

(iii) $f_y(a_1,a_2,\ldots,a_n,b)\neq 0$
then $\exists$ a nbd $(a_1-h_1, a_1+h_1; a_2-h_2, a_2+h_2; \ldots; a_n-h_n, a_n+h_n; b-k, b+k)$ of $(a_1, a_2, \ldots, a_n, b)$ such that for every point $(x_1, x_2, \ldots, x_n)$ of the nbd $R:(a_1-h_1,a_1+h_1;a_2-h_2,a_2+h_2;\ldots;a_n-h_n,a_n+h_n)$.

The equation $f(x_1,x_2,\ldots,x_n,y)=0$ gives only one value $y=g(x_1,x_2,\ldots,x_n)$ in $[b-k,b+k]$ satisfying the following conditions :

(i) $b=g(a_1,a_2,\ldots,a_n)$

(ii) $f(x_1,x_2,\ldots,x_n,g)=0$ for every point $(x_1,x_2,\ldots,x_n)$ in R.

(iii) g is continuous and having continuous first order partial derivatives w.r.t. $x_1,x_2,\ldots x_n$ in R.

11.2.2 DERIVATIVE OF IMPLICIT FUNCTION

If the equation $f(x,y)=0$ defines y as a function of x. Then derivative $\dfrac{dy}{dx}$ can be obtained simply by differentiating the equation w.r.t. x assuming y as a function of $g(x)$.

Therefore,
$$f_x+f_y\frac{dy}{dx}=0$$
$$f_{xx}+f_{xy}\frac{dy}{dx}+\left(f_{xy}+f_{yy}\frac{dy}{dx}\right)\frac{dy}{dx}+f_y\frac{d^2y}{dx^2}=0$$

or $$f_{xx}+2f_{xy}\frac{dy}{dx}+f_{yy}\left(\frac{dy}{dx}\right)^2+f_y\frac{d^2y}{dx^2}=0 \qquad \text{provided } f_y\neq 0$$

In a similar manner, we may find the other higher order derivatives.

11.3 JACOBIAN

Here we shall discuss some important definitions related to Jacobian and its properties.

(i) If u and v are the functions of two independent variables x and y, then the determinant

$$\begin{vmatrix} \frac{\partial u}{\partial x} & \frac{\partial u}{\partial y} \\ \frac{\partial v}{\partial x} & \frac{\partial v}{\partial y} \end{vmatrix}$$

is called the Jacobian of u and v with respect to x and y. It is denoted by $\frac{\partial(u,v)}{\partial(x,y)}$ or $J(u,v)$.

i.e., $$\frac{\partial(u,v)}{\partial(x,y)}=\begin{vmatrix} \frac{\partial u}{\partial x} & \frac{\partial u}{\partial y} \\ \frac{\partial v}{\partial x} & \frac{\partial v}{\partial y} \end{vmatrix}$$

(ii) If u,v and w are the functions of three independent variables x, y and z, then the determinant

$$\begin{vmatrix} \frac{\partial u}{\partial x} & \frac{\partial u}{\partial y} & \frac{\partial u}{\partial z} \\ \frac{\partial v}{\partial x} & \frac{\partial v}{\partial y} & \frac{\partial v}{\partial z} \\ \frac{\partial w}{\partial x} & \frac{\partial w}{\partial y} & \frac{\partial w}{\partial z} \end{vmatrix}$$

is called the Jacobian of u, v and w with respect to x, y and z. It is denoted by $\frac{\partial(u,v,w)}{\partial(x,y,z)}$ or $J(u,v,w)$.

(iii) If $u_1,u_2,\ldots,u_n$ are n functions of independent variables $x_1,x_2,\ldots,x_n$, then the determinant

$$\begin{vmatrix} \frac{\partial u_1}{\partial x_1} & \frac{\partial u_1}{\partial x_2} & \frac{\partial u_1}{\partial x_3} & \cdots & \frac{\partial u_1}{\partial x_n} \\ \frac{\partial u_2}{\partial x_1} & \frac{\partial u_2}{\partial x_2} & \frac{\partial u_2}{\partial x_3} & \cdots & \frac{\partial u_2}{\partial x_n} \\ \vdots & \vdots & \vdots & & \vdots \\ \frac{\partial u_n}{\partial x_1} & \frac{\partial u_n}{\partial x_2} & \frac{\partial u_n}{\partial x_3} & \cdots & \frac{\partial u_n}{\partial x_n} \end{vmatrix}$$

is called the Jacobian of $u_1,u_2,\ldots,u_n$ with respect to $x_1,x_2,\ldots,x_n$. It is denoted by $\frac{\partial(u_1,u_2,\ldots,u_n)}{\partial(x_1,x_2,\ldots,x_n)}$ or $J(u_1,u_2,\ldots,u_n)$.

(iv) If the functions $u_1,u_2,\ldots,u_n$ of n independent variables $x_1,x_2,\ldots,x_n$ are of the following form

$$u_1=f_1(x_1), u_2=f_2(x_1,x_2), \ldots, u_n=f_n(x_1,x_2,\ldots,x_n)$$

Then $$\frac{\partial(u_1,u_2,\ldots,u_n)}{\partial(x_1,x_2,\ldots,x_n)}=\begin{vmatrix} \frac{\partial u_1}{\partial x_1} & \frac{\partial u_1}{\partial x_2} & \frac{\partial u_1}{\partial x_3} & \cdots & \frac{\partial u_1}{\partial x_n} \\ \frac{\partial u_2}{\partial x_1} & \frac{\partial u_2}{\partial x_2} & \frac{\partial u_2}{\partial x_3} & \cdots & \frac{\partial u_2}{\partial x_n} \\ \vdots & \vdots & \vdots & & \vdots \\ \frac{\partial u_n}{\partial x_1} & \frac{\partial u_n}{\partial x_2} & \frac{\partial u_n}{\partial x_3} & \cdots & \frac{\partial u_n}{\partial x_n} \end{vmatrix}$$

11.4 IMPORTANT THEOREMS ON JACOBIAN

THEOREM 1. $$\frac{\partial(u,v)}{\partial(x,y)}\cdot\frac{\partial(x,y)}{\partial(u,v)}=1$$ [UPTU -2006; UPTU (Sum) 2008, 2009]

Proof. Let $u = u(x, y)$ and $v = v(x, y)$ be the given functions.
Differentiating partially each of these functions, we get

$$1 = \frac{\partial u}{\partial x}\cdot\frac{\partial x}{\partial u} + \frac{\partial u}{\partial y}\cdot\frac{\partial y}{\partial u}$$

$$0 = \frac{\partial u}{\partial x}\cdot\frac{\partial x}{\partial v} + \frac{\partial u}{\partial y}\cdot\frac{\partial y}{\partial v}$$

and

$$0 = \frac{\partial v}{\partial x}\cdot\frac{\partial x}{\partial u} + \frac{\partial v}{\partial y}\cdot\frac{\partial y}{\partial u}$$

$$1 = \frac{\partial v}{\partial x}\cdot\frac{\partial x}{\partial v} + \frac{\partial v}{\partial y}\cdot\frac{\partial y}{\partial v}$$

Now

$$\frac{\partial(u,v)}{\partial(x,y)}\cdot\frac{\partial(x,y)}{\partial(u,v)} = \begin{vmatrix} \frac{\partial u}{\partial x} & \frac{\partial u}{\partial y} \\ \frac{\partial v}{\partial x} & \frac{\partial v}{\partial y} \end{vmatrix} \begin{vmatrix} \frac{\partial x}{\partial u} & \frac{\partial x}{\partial v} \\ \frac{\partial y}{\partial u} & \frac{\partial y}{\partial v} \end{vmatrix} = \begin{vmatrix} \frac{\partial u}{\partial x}\cdot\frac{\partial x}{\partial u} + \frac{\partial u}{\partial y}\cdot\frac{\partial y}{\partial u} & \frac{\partial u}{\partial x}\cdot\frac{\partial x}{\partial v} + \frac{\partial u}{\partial y}\cdot\frac{\partial y}{\partial v} \\ \frac{\partial v}{\partial x}\cdot\frac{\partial x}{\partial u} + \frac{\partial v}{\partial y}\cdot\frac{\partial y}{\partial u} & \frac{\partial v}{\partial x}\cdot\frac{\partial x}{\partial v} + \frac{\partial v}{\partial y}\cdot\frac{\partial y}{\partial v} \end{vmatrix}$$

$$= \begin{vmatrix} \frac{\partial u}{\partial x} & \frac{\partial u}{\partial y} \\ \frac{\partial v}{\partial x} & \frac{\partial v}{\partial y} \end{vmatrix} \cdot \begin{vmatrix} \frac{\partial x}{\partial u} & \frac{\partial y}{\partial u} \\ \frac{\partial x}{\partial v} & \frac{\partial y}{\partial v} \end{vmatrix} = \begin{vmatrix} 1 & 0 \\ 0 & 1 \end{vmatrix} = 1$$

THEOREM 2. *If the function $u_1, u_2, \ldots, u_n$ of n independent variables $x_1, x_2, \ldots, x_n$ are of the following form*

$$u_1 = f_1(x_1)$$
$$u_2 = f_2(x_1, x_2)$$
$$\vdots \quad \vdots \quad \vdots$$
$$u_n = f_n(x_1, x_2, \ldots, x_n)$$

Then $$\frac{\partial(u_1, u_2, \ldots, u_n)}{\partial(x_1, x_2, \ldots, x_n)} = \frac{\partial u_1}{\partial x_1}\cdot\frac{\partial u_2}{\partial x_2}\cdot\frac{\partial u_3}{\partial x_3}\cdots\frac{\partial u_n}{\partial x_n}$$

Proof. We know that

$$\frac{\partial(u_1, u_2, \ldots, u_n)}{\partial(x_1, x_2, \ldots, x_n)} = \begin{vmatrix} \frac{\partial u_1}{\partial x_1} & \frac{\partial u_1}{\partial x_2} & \frac{\partial u_1}{\partial x_3} & \cdots & \frac{\partial u_1}{\partial x_n} \\ \frac{\partial u_2}{\partial x_1} & \frac{\partial u_2}{\partial x_2} & \frac{\partial u_2}{\partial x_3} & \cdots & \frac{\partial u_2}{\partial x_n} \\ \vdots & \vdots & \vdots & & \vdots \\ \frac{\partial u_n}{\partial x_1} & \frac{\partial u_n}{\partial x_2} & \frac{\partial u_n}{\partial x_3} & \cdots & \frac{\partial u_n}{\partial x_n} \end{vmatrix} \qquad \ldots(1)$$

(i) u_1 is a function of x_1 only, therefore $\frac{\partial u_1}{\partial x_1}$ exists and $\frac{\partial u_1}{\partial x_2} = 0, \frac{\partial u_1}{\partial x_3} = 0, \ldots, \frac{\partial u_1}{\partial x_n} = 0$

(ii) u_2 is a function of x_1 and x_2 only, therefore $\frac{\partial u_2}{\partial x_1}$ and $\frac{\partial u_2}{\partial x_2}$ exist and $\frac{\partial u_2}{\partial x_3} = 0, \frac{\partial u_2}{\partial x_4} = 0, \ldots, \frac{\partial u_2}{\partial x_n} = 0$

(iii) u_3 is a function of x_1, x_2 and x_3 only,

therefore, $\frac{\partial u_3}{\partial x_1}, \frac{\partial u_3}{\partial x_2}$ and $\frac{\partial u_3}{\partial x_3}$ exist and $\frac{\partial u_3}{\partial x_4} = 0, \frac{\partial u_3}{\partial x_5} = 0, \ldots, \frac{\partial u_3}{\partial x_n} = 0$

Preceding in the same manner, we have u_n is a function of $x_1, x_2, \ldots, x_n$,

therefore $\frac{\partial u_n}{\partial x_1}, \frac{\partial u_n}{\partial x_2}, \ldots, \frac{\partial u_n}{\partial x_n}$ all exist.

Putting all these values in (1), we get

$$\frac{\partial(u_1,u_2,\ldots,u_n)}{\partial(x_1,x_2,\ldots,x_n)} = \begin{vmatrix} \frac{\partial u_1}{\partial x_1} & 0 & 0 & \cdots & 0 \\ \frac{\partial u_2}{\partial x_1} & \frac{\partial u_2}{\partial x_2} & 0 & \cdots & 0 \\ \vdots & \vdots & \vdots & \vdots & \vdots \\ \frac{\partial u_n}{\partial x_1} & \frac{\partial u_n}{\partial x_2} & \frac{\partial u_n}{\partial x_3} & \cdots & \frac{\partial u_n}{\partial x_n} \end{vmatrix}$$

Now expanding the determinant along the first row, we get

$$\frac{\partial(u_1,u_2,\ldots,u_n)}{\partial(x_1,x_2,\ldots,x_n)} = \frac{\partial u_1}{\partial x_1}\cdot\frac{\partial u_2}{\partial x_2}\cdot\frac{\partial u_3}{\partial x_3}\cdots\frac{\partial u_n}{\partial x_n}$$

THEOREM 3. *If u_1,u_2 are functions of y_1, y_2 and y_1,y_2 are functions of x_1,x_2 then* $\dfrac{\partial(u_1,u_2)}{\partial(x_1,x_2)} = \dfrac{\partial(u_1,u_2)}{\partial(y_1,y_2)}\cdot\dfrac{\partial(y_1,y_2)}{\partial(x_1,x_2)}$

Proof. Since u_1,u_2 are functions of y_1,y_2. Also y_1,y_2 are functions of x_1,x_2; therefore, we get

$$\left.\begin{aligned} \frac{\partial u_1}{\partial x_1} &= \frac{\partial u_1}{\partial y_1}\cdot\frac{\partial y_1}{\partial x_1} + \frac{\partial u_1}{\partial y_2}\cdot\frac{\partial y_2}{\partial x_1} \\ \frac{\partial u_1}{\partial x_2} &= \frac{\partial u_1}{\partial y_1}\cdot\frac{\partial y_1}{\partial x_2} + \frac{\partial u_1}{\partial y_2}\cdot\frac{\partial y_2}{\partial x_2} \\ \frac{\partial u_2}{\partial x_1} &= \frac{\partial u_2}{\partial y_1}\cdot\frac{\partial y_1}{\partial x_1} + \frac{\partial u_2}{\partial y_2}\cdot\frac{\partial y_2}{\partial x_1} \\ \frac{\partial u_2}{\partial x_2} &= \frac{\partial u_2}{\partial y_1}\cdot\frac{\partial y_1}{\partial x_2} + \frac{\partial u_2}{\partial y_2}\cdot\frac{\partial y_2}{\partial x_2} \end{aligned}\right] \qquad \ldots(1)$$

We have $\dfrac{\partial(u_1,u_2)}{\partial(y_1,y_2)}\cdot\dfrac{\partial(y_1,y_2)}{\partial(x_1,x_2)} = \begin{vmatrix} \frac{\partial u_1}{\partial y_1} & \frac{\partial u_1}{\partial y_2} \\ \frac{\partial u_2}{\partial y_1} & \frac{\partial u_2}{\partial y_2} \end{vmatrix} \times \begin{vmatrix} \frac{\partial y_1}{\partial x_1} & \frac{\partial y_1}{\partial x_2} \\ \frac{\partial y_2}{\partial x_1} & \frac{\partial y_2}{\partial x_2} \end{vmatrix} = \begin{vmatrix} \frac{\partial u_1}{\partial y_1}\frac{\partial y_1}{\partial x_1} + \frac{\partial u_1}{\partial y_2}\frac{\partial y_2}{\partial x_1} & \frac{\partial u_1}{\partial y_1}\frac{\partial y_1}{\partial x_2} + \frac{\partial u_1}{\partial y_2}\frac{\partial y_2}{\partial x_2} \\ \frac{\partial u_2}{\partial y_1}\frac{\partial y_1}{\partial x_1} + \frac{\partial u_2}{\partial y_2}\frac{\partial y_2}{\partial x_1} & \frac{\partial u_2}{\partial y_1}\frac{\partial y_1}{\partial x_2} + \frac{\partial u_2}{\partial y_2}\frac{\partial y_2}{\partial x_2} \end{vmatrix}$

Now, using relation (1), we get

$$\frac{\partial(u_1,u_2)}{\partial(y_1,y_2)}\cdot\frac{\partial(y_1,y_2)}{\partial(x_1,x_2)} = \begin{vmatrix} \frac{\partial u_1}{\partial x_1} & \frac{\partial u_1}{\partial x_2} \\ \frac{\partial u_2}{\partial x_1} & \frac{\partial u_2}{\partial x_2} \end{vmatrix} = \frac{\partial(u_1,u_2)}{\partial(x_1,x_2)}$$

THEOREM 4. *If u_1,u_2,u_3 are functions of y_1, y_2, y_3 and y_1,y_2, y_3 are functions of x_1,x_2,x_3, then*

$$\frac{\partial(u_1,u_2,u_3)}{\partial(x_1,x_2,x_3)} = \frac{\partial(u_1,u_2,u_3)}{\partial(y_1,y_2,y_3)}\cdot\frac{\partial(y_1,y_2,y_3)}{\partial(x_1,x_2,x_3)}$$

Proof. Since u_1,u_2 and u_3 are functions of y_1,y_2 and y_3. Also y_1,y_2 and y_3 are functions of x_1,x_2 and x_3 therefore, we get

$$\frac{\partial u_1}{\partial x_1} = \frac{\partial u_1}{\partial y_1}\cdot\frac{\partial y_1}{\partial x_1} + \frac{\partial u_1}{\partial y_2}\cdot\frac{\partial y_2}{\partial x_1} + \frac{\partial u_1}{\partial y_3}\cdot\frac{\partial y_3}{\partial x_1} = \sum_{i=1}^{3}\frac{\partial u_1}{\partial y_i}\cdot\frac{\partial y_i}{\partial x_1}$$

$$\frac{\partial u_1}{\partial x_2} = \frac{\partial u_1}{\partial y_1}\cdot\frac{\partial y_1}{\partial x_2} + \frac{\partial u_1}{\partial y_2}\cdot\frac{\partial y_2}{\partial x_2} + \frac{\partial u_1}{\partial y_3}\cdot\frac{\partial y_3}{\partial x_2} = \sum_{i=1}^{3}\frac{\partial u_1}{\partial y_i}\cdot\frac{\partial y_i}{\partial x_2}$$

Similarly, $\dfrac{\partial u_1}{\partial x_3} = \sum_{i=1}^{3}\dfrac{\partial u_1}{\partial y_i}\cdot\dfrac{\partial y_i}{\partial x_3}$, $\dfrac{\partial u_2}{\partial x_1} = \sum_{i=1}^{3}\dfrac{\partial u_2}{\partial y_i}\cdot\dfrac{\partial y_i}{\partial x_1}$,

$$\frac{\partial u_2}{\partial x_2} = \sum_{i=1}^{3}\frac{\partial u_2}{\partial y_i}\cdot\frac{\partial y_i}{\partial x_2}, \quad \frac{\partial u_2}{\partial x_3} = \sum_{i=1}^{3}\frac{\partial u_2}{\partial y_i}\cdot\frac{\partial y_i}{\partial x_3},$$

$$\frac{\partial u_3}{\partial x_1} = \sum_{i=1}^{3}\frac{\partial u_3}{\partial y_i}\cdot\frac{\partial y_i}{\partial x_1}, \quad \frac{\partial u_3}{\partial x_2} = \sum_{i=1}^{3}\frac{\partial u_3}{\partial y_i}\cdot\frac{\partial y_i}{\partial x_2}, \text{ and } \frac{\partial u_3}{\partial x_3} = \sum_{i=1}^{3}\frac{\partial u_3}{\partial y_i}\cdot\frac{\partial y_i}{\partial x_3}$$

Now,

$$\frac{\partial(u_1,u_2,u_3)}{\partial(y_1,y_2,y_3)}\cdot\frac{\partial(y_1,y_2,y_3)}{\partial(x_1,x_2,x_3)}=\begin{vmatrix}\frac{\partial u_1}{\partial y_1} & \frac{\partial u_1}{\partial y_2} & \frac{\partial u_1}{\partial y_3}\\ \frac{\partial u_2}{\partial y_1} & \frac{\partial u_2}{\partial y_2} & \frac{\partial u_2}{\partial y_3}\\ \frac{\partial u_3}{\partial y_1} & \frac{\partial u_3}{\partial y_2} & \frac{\partial u_3}{\partial y_3}\end{vmatrix}\begin{vmatrix}\frac{\partial y_1}{\partial x_1} & \frac{\partial y_1}{\partial x_2} & \frac{\partial y_1}{\partial x_3}\\ \frac{\partial y_2}{\partial x_1} & \frac{\partial y_2}{\partial x_2} & \frac{\partial y_2}{\partial x_3}\\ \frac{\partial y_3}{\partial x_1} & \frac{\partial y_3}{\partial x_2} & \frac{\partial y_3}{\partial x_3}\end{vmatrix}=\begin{vmatrix}\sum\frac{\partial u_1}{\partial y_i}\cdot\frac{\partial y_i}{\partial x_1} & \sum\frac{\partial u_1}{\partial y_i}\cdot\frac{\partial y_i}{\partial x_2} & \sum\frac{\partial u_1}{\partial y_i}\cdot\frac{\partial y_i}{\partial x_3}\\ \sum\frac{\partial u_2}{\partial y_i}\cdot\frac{\partial y_i}{\partial x_1} & \sum\frac{\partial u_2}{\partial y_i}\cdot\frac{\partial y_i}{\partial x_2} & \sum\frac{\partial u_2}{\partial y_i}\cdot\frac{\partial y_i}{\partial x_3}\\ \sum\frac{\partial u_3}{\partial y_i}\cdot\frac{\partial y_i}{\partial x_1} & \sum\frac{\partial u_3}{\partial y_i}\cdot\frac{\partial y_i}{\partial x_2} & \sum\frac{\partial u_3}{\partial y_i}\cdot\frac{\partial y_i}{\partial x_3}\end{vmatrix}$$

Putting the values of each element of the determinant from the above relations, we get

$$\frac{\partial(u_1,u_2,u_3)}{\partial(y_1,y_2,y_3)}\cdot\frac{\partial(y_1,y_2,y_3)}{\partial(x_1,x_2,x_3)}=\begin{vmatrix}\frac{\partial u_1}{\partial x_1} & \frac{\partial u_1}{\partial x_2} & \frac{\partial u_1}{\partial x_3}\\ \frac{\partial u_2}{\partial x_1} & \frac{\partial u_2}{\partial x_2} & \frac{\partial u_2}{\partial x_3}\\ \frac{\partial u_3}{\partial x_1} & \frac{\partial u_3}{\partial x_2} & \frac{\partial u_3}{\partial x_3}\end{vmatrix}=\frac{\partial(u_1,u_2,u_3)}{\partial(x_1,x_2,x_3)}$$

Generalization. If $u_1,u_2,\ldots,u_n$ are functions of $y_1,y_2,\ldots,y_n$ and $y_1,y_2,\ldots,y_n$ are functions of $x_1,x_2,\ldots,x_n$, then

$$\frac{\partial(u_1,u_2,u_3,\ldots,u_n)}{\partial(x_1,x_2,x_3,\ldots,x_n)}=\frac{\partial(u_1,u_2,u_3,\ldots,u_n)}{\partial(y_1,y_2,y_3,\ldots,y_n)}\cdot\frac{\partial(y_1,y_2,y_3,\ldots,y_n)}{\partial(x_1,x_2,x_3,\ldots,x_n)}$$

The proof may be easily extended as in the case of two and three variables.

THEOREM 5. *If the functions u, v, w of three independent variables x, y and z are not independent, then the Jacobian of u, v, w with respect to x, y, z vanishes.*

Proof. Since, the functions u, v and w (of three independent variables x, y and z) are not independent. Then there will be a relation $\quad F(u,v,w) = 0 \quad$ ...(A)

which will connect these independent variables.

Differentiating (A), with respect to x, y and z, we get

$$\frac{\partial F}{\partial u}\cdot\frac{\partial u}{\partial x}+\frac{\partial F}{\partial v}\cdot\frac{\partial v}{\partial x}+\frac{\partial F}{\partial w}\cdot\frac{\partial w}{\partial x}=0 \qquad \text{...(1)}$$

$$\frac{\partial F}{\partial u}\cdot\frac{\partial u}{\partial y}+\frac{\partial F}{\partial v}\cdot\frac{\partial v}{\partial y}+\frac{\partial F}{\partial w}\cdot\frac{\partial w}{\partial y}=0 \qquad \text{...(2)}$$

$$\frac{\partial F}{\partial u}\cdot\frac{\partial u}{\partial z}+\frac{\partial F}{\partial v}\cdot\frac{\partial v}{\partial z}+\frac{\partial F}{\partial w}\cdot\frac{\partial w}{\partial z}=0 \qquad \text{...(3)}$$

Eliminating $\frac{\partial F}{\partial u}$, $\frac{\partial F}{\partial v}$ and $\frac{\partial F}{\partial w}$ from (1), (2) and (3), we get

$$\begin{vmatrix}\frac{\partial u}{\partial x} & \frac{\partial v}{\partial x} & \frac{\partial w}{\partial x}\\ \frac{\partial u}{\partial y} & \frac{\partial v}{\partial y} & \frac{\partial w}{\partial y}\\ \frac{\partial u}{\partial z} & \frac{\partial v}{\partial z} & \frac{\partial w}{\partial z}\end{vmatrix}=0\Rightarrow\begin{vmatrix}\frac{\partial u}{\partial x} & \frac{\partial u}{\partial y} & \frac{\partial u}{\partial z}\\ \frac{\partial v}{\partial x} & \frac{\partial v}{\partial y} & \frac{\partial v}{\partial z}\\ \frac{\partial w}{\partial x} & \frac{\partial w}{\partial y} & \frac{\partial w}{\partial z}\end{vmatrix}=0 \text{ (Interchanging rows and columns)}$$

Hence,

$$\frac{\partial(u,v,w)}{\partial(x,y,z)}=0$$

THEOREM 6. $\frac{\partial(u,v,w)}{\partial(x,y,z)}\times\frac{\partial(x,y,z)}{\partial(u,v,w)}=1$

Proof. Let us suppose $\quad u=f_1(x,y,z);\ v=f_2(x,y,z)$ and $w=f_3(x,y,z)$.

Then we may write these equations as

$$x=\phi_1(u,v,w);\ y=\phi_2(u,v,w) \text{ and } z=\phi_3(u,v,w)$$

Now, differentiating $u=f_1(x, y, z)$ partially w.r.t. u, v and w respectively, we get

$$\frac{\partial u}{\partial u}=\frac{\partial u}{\partial x}\cdot\frac{\partial x}{\partial u}+\frac{\partial u}{\partial y}\cdot\frac{\partial y}{\partial u}+\frac{\partial u}{\partial z}\cdot\frac{\partial z}{\partial u} \Rightarrow 1=\frac{\partial u}{\partial x}\cdot\frac{\partial x}{\partial u}+\frac{\partial u}{\partial y}\cdot\frac{\partial y}{\partial u}+\frac{\partial u}{\partial z}\cdot\frac{\partial z}{\partial u} \quad ...(1)$$

and

$$\frac{\partial u}{\partial v}=\frac{\partial u}{\partial x}\cdot\frac{\partial x}{\partial v}+\frac{\partial u}{\partial y}\cdot\frac{\partial y}{\partial v}+\frac{\partial u}{\partial z}\cdot\frac{\partial z}{\partial v} \Rightarrow 0=\frac{\partial u}{\partial x}\cdot\frac{\partial x}{\partial v}+\frac{\partial u}{\partial y}\cdot\frac{\partial y}{\partial v}+\frac{\partial u}{\partial z}\cdot\frac{\partial z}{\partial v} \quad ...(2)$$

Similarly

$$0=\frac{\partial u}{\partial x}\cdot\frac{\partial x}{\partial w}+\frac{\partial u}{\partial y}\cdot\frac{\partial y}{\partial w}+\frac{\partial u}{\partial z}\cdot\frac{\partial z}{\partial w} \quad ...(3)$$

Now differentiating $v=f_2(x,y,z)$ and $w=f_3(x,y,z)$ partially with respect to u, v and w respectively, we get

$$\left.\begin{aligned} 0&=\frac{\partial v}{\partial x}\cdot\frac{\partial x}{\partial u}+\frac{\partial v}{\partial y}\cdot\frac{\partial y}{\partial u}+\frac{\partial v}{\partial z}\cdot\frac{\partial z}{\partial u}\\ 1&=\frac{\partial v}{\partial x}\cdot\frac{\partial x}{\partial v}+\frac{\partial v}{\partial y}\cdot\frac{\partial y}{\partial v}+\frac{\partial v}{\partial z}\cdot\frac{\partial z}{\partial v}\\ 0&=\frac{\partial v}{\partial x}\cdot\frac{\partial x}{\partial w}+\frac{\partial v}{\partial y}\cdot\frac{\partial y}{\partial w}+\frac{\partial v}{\partial z}\cdot\frac{\partial z}{\partial w}\end{aligned}\right] \quad ...(4)$$

and

$$\left.\begin{aligned} 0&=\frac{\partial w}{\partial x}\cdot\frac{\partial x}{\partial u}+\frac{\partial w}{\partial y}\cdot\frac{\partial y}{\partial u}+\frac{\partial w}{\partial z}\cdot\frac{\partial z}{\partial u}\\ 0&=\frac{\partial w}{\partial x}\cdot\frac{\partial x}{\partial v}+\frac{\partial w}{\partial y}\cdot\frac{\partial y}{\partial v}+\frac{\partial w}{\partial z}\cdot\frac{\partial z}{\partial v}\\ 1&=\frac{\partial w}{\partial x}\cdot\frac{\partial x}{\partial w}+\frac{\partial w}{\partial y}\cdot\frac{\partial y}{\partial w}+\frac{\partial w}{\partial z}\cdot\frac{\partial z}{\partial w}\end{aligned}\right] \quad ...(5)$$

We have

$$\frac{\partial(u,v,w)}{\partial(x,y,z)}\times\frac{\partial(x,y,z)}{\partial(u,v,w)}=\begin{vmatrix}\frac{\partial u}{\partial x} & \frac{\partial u}{\partial y} & \frac{\partial u}{\partial z}\\ \frac{\partial v}{\partial x} & \frac{\partial v}{\partial y} & \frac{\partial v}{\partial z}\\ \frac{\partial w}{\partial x} & \frac{\partial w}{\partial y} & \frac{\partial w}{\partial z}\end{vmatrix}\begin{vmatrix}\frac{\partial x}{\partial u} & \frac{\partial x}{\partial v} & \frac{\partial x}{\partial w}\\ \frac{\partial y}{\partial u} & \frac{\partial y}{\partial v} & \frac{\partial y}{\partial w}\\ \frac{\partial z}{\partial u} & \frac{\partial z}{\partial v} & \frac{\partial z}{\partial w}\end{vmatrix}=\begin{vmatrix}\Sigma\frac{\partial u}{\partial x}\cdot\frac{\partial x}{\partial u} & \Sigma\frac{\partial u}{\partial x}\cdot\frac{\partial x}{\partial v} & \Sigma\frac{\partial u}{\partial x}\cdot\frac{\partial x}{\partial w}\\ \Sigma\frac{\partial v}{\partial x}\cdot\frac{\partial x}{\partial u} & \Sigma\frac{\partial v}{\partial x}\cdot\frac{\partial x}{\partial v} & \Sigma\frac{\partial v}{\partial x}\cdot\frac{\partial x}{\partial w}\\ \Sigma\frac{\partial w}{\partial x}\cdot\frac{\partial x}{\partial u} & \Sigma\frac{\partial w}{\partial x}\cdot\frac{\partial x}{\partial v} & \Sigma\frac{\partial w}{\partial x}\cdot\frac{\partial x}{\partial w}\end{vmatrix}$$

$$=\begin{vmatrix}1 & 0 & 0\\ 0 & 1 & 0\\ 0 & 0 & 1\end{vmatrix}=1$$

(Using the relations (1) to (5))

Hence

$$\frac{\partial(u,v,w)}{\partial(x,y,z)}\times\frac{\partial(x,y,z)}{\partial(u,v,w)}=1$$

11.5 JACOBIAN OF IMPLICIT FUNCTIONS

THEOREM 1. *If u_1, u_2, are implicit functions of x_1, x_2 that is $F_1(u_1,u_2,x_1,x_2)=0$ and $F_2(u_1,u_2,x_1,x_2)=0$, then*

$$\frac{\partial(u_1,u_2)}{\partial(x_1,x_2)}=(-1)^2\left[\frac{\partial(F_1,F_2)}{\partial(x_1,x_2)}\Big/\frac{\partial(F_1,F_2)}{\partial(u_1,u_2)}\right]$$

Proof. We have

$$\left.\begin{aligned}F_1(u_1,u_2,x_1,x_2)&=0\\ F_2(u_1,u_2,x_1,x_2)&=0\end{aligned}\right] \quad ...(1)$$

Differentiating relation (1), partially w.r.t. x_1 and x_2 respectively, we get

$$\left.\begin{aligned}\frac{\partial F_1}{\partial x_1}+\frac{\partial F_1}{\partial u_1}\cdot\frac{\partial u_1}{\partial x_1}+\frac{\partial F_1}{\partial u_2}\cdot\frac{\partial u_2}{\partial x_1}&=0\\ \frac{\partial F_1}{\partial x_2}+\frac{\partial F_1}{\partial u_1}\cdot\frac{\partial u_1}{\partial x_2}+\frac{\partial F_1}{\partial u_2}\cdot\frac{\partial u_2}{\partial x_2}&=0\\ \frac{\partial F_2}{\partial x_1}+\frac{\partial F_2}{\partial u_1}\cdot\frac{\partial u_1}{\partial x_1}+\frac{\partial F_2}{\partial u_2}\cdot\frac{\partial u_2}{\partial x_1}&=0\\ \frac{\partial F_2}{\partial x_2}+\frac{\partial F_2}{\partial u_1}\cdot\frac{\partial u_1}{\partial x_2}+\frac{\partial F_2}{\partial u_2}\cdot\frac{\partial u_2}{\partial x_2}&=0\end{aligned}\right] \quad ...(2)$$

We have $\dfrac{\partial(F_1,F_2)}{\partial(u_1,u_2)}\times\dfrac{\partial(u_1,u_2)}{\partial(x_1,x_2)}=\begin{vmatrix}\dfrac{\partial F_1}{\partial u_1} & \dfrac{\partial F_1}{\partial u_2}\\ \dfrac{\partial F_2}{\partial u_1} & \dfrac{\partial F_2}{\partial u_2}\end{vmatrix}\times\begin{vmatrix}\dfrac{\partial u_1}{\partial x_1} & \dfrac{\partial u_1}{\partial x_2}\\ \dfrac{\partial u_2}{\partial x_1} & \dfrac{\partial u_2}{\partial x_2}\end{vmatrix}=\begin{vmatrix}\dfrac{\partial F_1}{\partial u_1}.\dfrac{\partial u_1}{\partial x_1}+\dfrac{\partial F_1}{\partial u_2}.\dfrac{\partial u_2}{\partial x_1} & \dfrac{\partial F_1}{\partial u_1}.\dfrac{\partial u_1}{\partial x_2}+\dfrac{\partial F_1}{\partial u_2}.\dfrac{\partial u_2}{\partial x_2}\\ \dfrac{\partial F_2}{\partial u_1}.\dfrac{\partial u_1}{\partial x_1}+\dfrac{\partial F_2}{\partial u_2}.\dfrac{\partial u_2}{\partial x_1} & \dfrac{\partial F_2}{\partial u_1}.\dfrac{\partial u_1}{\partial x_2}+\dfrac{\partial F_2}{\partial u_2}.\dfrac{\partial u_2}{\partial x_2}\end{vmatrix}$

$$=\begin{vmatrix}-\dfrac{\partial F_1}{\partial x_1} & -\dfrac{\partial F_1}{\partial x_2}\\ -\dfrac{\partial F_2}{\partial x_1} & -\dfrac{\partial F_2}{\partial x_2}\end{vmatrix}=(-1)^2\dfrac{\partial(F_1,F_2)}{\partial(x_1,x_2)}\qquad \text{Using relation (2),}$$

$$\Rightarrow\qquad \frac{\partial(u_1,u_2)}{\partial(x_1,x_2)}=(-1)^2\left[\frac{\partial(F_1,F_2)}{\partial(x_1,x_2)}\Big/\frac{\partial(F_1,F_2)}{\partial(u_1,u_2)}\right]$$

THEOREM 2. *If u_1,u_2 and u_3 be the implicit functions of x_1,x_2,x_3 that is*

$$F_1(u_1,u_2,u_3,x_1,x_2,x_3)=0,\ F_2(u_1,u_2,u_3,x_1,x_2,x_3)=0,\ F_3(u_1,u_2,u_3,x_1,x_2,x_3)=0$$

then $\dfrac{\partial(u_1,u_2,u_3)}{\partial(x_1,x_2,x_3)}=(-1)^3\left[\dfrac{\partial(F_1,F_2,F_3)}{\partial(x_1,x_2,x_3)}\Big/\dfrac{\partial(F_1,F_2,F_3)}{\partial(u_1,u_2,u_3)}\right]$

Proof. We have

$$\left.\begin{aligned}F_1(u_1,u_2,u_3,x_1,x_2,x_3)=0\\ F_2(u_1,u_2,u_3,x_1,x_2,x_3)=0\\ F_3(u_1,u_2,u_3,x_1,x_2,x_3)=0\end{aligned}\right]\qquad ...(1)$$

Differentiating (1), partially w.r.t. x_1, x_2 and x_3 respectively, we get

$$\frac{\partial F_1}{\partial x_1}+\frac{\partial F_1}{\partial u_1}.\frac{\partial u_1}{\partial x_1}+\frac{\partial F_1}{\partial u_2}.\frac{\partial u_2}{\partial x_1}+\frac{\partial F_1}{\partial u_3}.\frac{\partial u_3}{\partial x_1}=0$$

$\Rightarrow$ $\displaystyle\sum_{r=1}^{3}\frac{\partial F_1}{\partial u_r}.\frac{\partial u_r}{\partial x_1}=-\frac{\partial F_1}{\partial x_1},\ \sum_{r=1}^{3}\frac{\partial F_1}{\partial u_r}.\frac{\partial u_r}{\partial x_2}=-\frac{\partial F_1}{\partial x_2},\ \sum_{r=1}^{3}\frac{\partial F_1}{\partial u_r}.\frac{\partial u_r}{\partial x_3}=-\frac{\partial F_1}{\partial x_3},$

Similarly $\displaystyle\sum_{r=1}^{3}\frac{\partial F_2}{\partial u_r}.\frac{\partial u_r}{\partial x_1}=-\frac{\partial F_2}{\partial x_1},\ \sum_{r=1}^{3}\frac{\partial F_2}{\partial u_r}.\frac{\partial u_r}{\partial x_2}=-\frac{\partial F_2}{\partial x_2},\ \sum_{r=1}^{3}\frac{\partial F_2}{\partial u_r}.\frac{\partial u_r}{\partial x_3}=-\frac{\partial F_2}{\partial x_3},$

$\displaystyle\sum_{r=1}^{3}\frac{\partial F_3}{\partial u_r}.\frac{\partial u_r}{\partial x_1}=-\frac{\partial F_3}{\partial x_1},\ \sum_{r=1}^{3}\frac{\partial F_3}{\partial u_r}.\frac{\partial u_r}{\partial x_2}=-\frac{\partial F_3}{\partial x_2},\ \sum_{r=1}^{3}\frac{\partial F_3}{\partial u_r}.\frac{\partial u_r}{\partial x_3}=-\frac{\partial F_3}{\partial x_3},$

and

Now consider

$$\frac{\partial(F_1,F_2,F_3)}{\partial(u_1,u_2,u_3)}\times\frac{\partial(u_1,u_2,u_3)}{\partial(x_1,x_2,x_3)}=\begin{vmatrix}\frac{\partial F_1}{\partial u_1} & \frac{\partial F_1}{\partial u_2} & \frac{\partial F_1}{\partial u_3}\\ \frac{\partial F_2}{\partial u_1} & \frac{\partial F_2}{\partial u_2} & \frac{\partial F_2}{\partial u_3}\\ \frac{\partial F_3}{\partial u_1} & \frac{\partial F_3}{\partial u_2} & \frac{\partial F_3}{\partial u_3}\end{vmatrix}\times\begin{vmatrix}\frac{\partial u_1}{\partial x_1} & \frac{\partial u_1}{\partial x_2} & \frac{\partial u_1}{\partial x_3}\\ \frac{\partial u_2}{\partial x_1} & \frac{\partial u_2}{\partial x_2} & \frac{\partial u_2}{\partial x_3}\\ \frac{\partial u_3}{\partial x_1} & \frac{\partial u_3}{\partial x_2} & \frac{\partial u_3}{\partial x_3}\end{vmatrix}=\begin{vmatrix}\Sigma\frac{\partial F_1}{\partial u_r}.\frac{\partial u_r}{\partial x_1} & \Sigma\frac{\partial F_1}{\partial u_r}.\frac{\partial u_r}{\partial x_2} & \Sigma\frac{\partial F_1}{\partial u_r}.\frac{\partial u_r}{\partial x_3}\\ \Sigma\frac{\partial F_2}{\partial u_r}.\frac{\partial u_r}{\partial x_1} & \Sigma\frac{\partial F_2}{\partial u_r}.\frac{\partial u_r}{\partial x_2} & \Sigma\frac{\partial F_2}{\partial u_r}.\frac{\partial u_r}{\partial x_3}\\ \Sigma\frac{\partial F_3}{\partial u_r}.\frac{\partial u_r}{\partial x_1} & \Sigma\frac{\partial F_3}{\partial u_r}.\frac{\partial u_r}{\partial x_2} & \Sigma\frac{\partial F_3}{\partial u_r}.\frac{\partial u_r}{\partial x_3}\end{vmatrix}$$

Now, using (2), we get

$$=\begin{vmatrix}-\frac{\partial F_1}{\partial x_1} & -\frac{\partial F_1}{\partial x_2} & -\frac{\partial F_1}{\partial x_3}\\ -\frac{\partial F_2}{\partial x_1} & -\frac{\partial F_2}{\partial x_2} & -\frac{\partial F_2}{\partial x_3}\\ -\frac{\partial F_3}{\partial x_1} & -\frac{\partial F_3}{\partial x_2} & -\frac{\partial F_3}{\partial x_3}\end{vmatrix}=(-1)^3\begin{vmatrix}\frac{\partial F_1}{\partial x_1} & \frac{\partial F_1}{\partial x_2} & \frac{\partial F_1}{\partial x_3}\\ \frac{\partial F_2}{\partial x_1} & \frac{\partial F_2}{\partial x_2} & \frac{\partial F_2}{\partial x_3}\\ \frac{\partial F_3}{\partial x_1} & \frac{\partial F_3}{\partial x_2} & \frac{\partial F_3}{\partial x_3}\end{vmatrix}=(-1)^3\frac{\partial(F_1,F_2,F_3)}{\partial(x_1,x_2,x_3)}$$

Hence, $\dfrac{\partial(u_1,u_2,u_3)}{\partial(x_1,x_2,x_3)}=(-1)^3\left[\dfrac{\partial(F_1,F_2,F_3)}{\partial(x_1,x_2,x_3)}\Big/\dfrac{\partial(F_1,F_2,F_3)}{\partial(u_1,u_2,u_3)}\right]$

Generalization. Let $u_1, u_2, \ldots, u_n$ be the implicit functions of $x_1, x_2, \ldots, x_n$, that is

$$F_1(u_1,u_2,\ldots,u_n,x_1,x_2,\ldots,x_n)=0$$
$$F_2(u_1,u_2,\ldots,u_n,x_1,x_2,\ldots,x_n)=0$$
$$\vdots \quad \vdots \quad \vdots \quad \vdots \quad \vdots$$
$$F_n(u_1,u_2,\ldots,u_n,x_1,x_2,\ldots,x_n)=0$$

Then, $$\frac{\partial(u_1,u_2,\ldots,u_n)}{\partial(x_1,x_2,\ldots,x_n)}=(-1)^n\left[\frac{\partial(F_1,F_2,\ldots,F_n)}{\partial(x_1,x_2,\ldots,x_n)}\Big/\frac{\partial(F_1,F_2,\ldots,F_n)}{\partial(u_1,u_2,\ldots,u_n)}\right]$$

The proof may be easily extended as in case of two and three implicit functions. (Theorem 7 and Theorem 8).

11.6 NECESSARY AND SUFFICIENT CONDITIONS FOR A JACOBIAN TO BE VANISH

THEOREM 1. *If $v_1, v_2, \ldots, v_n$ be the functions of n independent variables $x_1, x_2, \ldots, x_n$. For $F(v_1, v_2, \ldots, v_n)=0$, it is necessary and sufficient that the Jacobian $\dfrac{\partial(v_1,v_2,\ldots,v_n)}{\partial(x_1,x_2,\ldots,x_n)}$ should vanish identically.*

Proof. **Necessary Condition.** Suppose that there exists a relation of $v_1, v_2, \ldots, v_n$ such that

$$F(v_1,v_2,\ldots,v_n)=0 \qquad \ldots(1)$$

To show that Jacobian is necessarily zero.

Differentiating (1), partially w.r.t. $x_1, x_2, \ldots, x_n$ respectively, we get

$$\frac{\partial F}{\partial v_1}\cdot\frac{\partial v_1}{\partial x_1}+\frac{\partial F}{\partial v_2}\cdot\frac{\partial v_2}{\partial x_1}+\ldots+\frac{\partial F}{\partial v_n}\cdot\frac{\partial v_n}{\partial x_1}=0$$
$$\frac{\partial F}{\partial v_1}\cdot\frac{\partial v_1}{\partial x_2}+\frac{\partial F}{\partial v_2}\cdot\frac{\partial v_2}{\partial x_2}+\ldots+\frac{\partial F}{\partial v_n}\cdot\frac{\partial v_n}{\partial x_2}=0$$
$$\ldots \quad \ldots \quad \ldots \quad \ldots \quad \ldots \quad \ldots$$
$$\frac{\partial F}{\partial v_1}\cdot\frac{\partial v_1}{\partial x_n}+\frac{\partial F}{\partial v_2}\cdot\frac{\partial v_2}{\partial x_n}+\ldots+\frac{\partial F}{\partial v_n}\cdot\frac{\partial v_n}{\partial x_n}=0$$

Now eliminating $\dfrac{\partial F}{\partial v_1}, \dfrac{\partial F}{\partial v_2}, \ldots, \dfrac{\partial F}{\partial v_n}$ from these equations, we get

$$\begin{vmatrix} \frac{\partial v_1}{\partial x_1} & \frac{\partial v_2}{\partial x_1} & \cdots & \frac{\partial v_n}{\partial x_1} \\ \frac{\partial v_1}{\partial x_2} & \frac{\partial v_2}{\partial x_2} & \cdots & \frac{\partial v_n}{\partial x_2} \\ \cdots & \cdots & \cdots & \cdots \\ \frac{\partial v_1}{\partial x_n} & \frac{\partial v_2}{\partial x_n} & \cdots & \frac{\partial v_n}{\partial x_n} \end{vmatrix}=0$$

$$\Rightarrow \qquad \frac{\partial(v_1,v_2,\ldots,v_n)}{\partial(x_1,x_2,\ldots,x_n)}=0$$

Sufficient Condition. If the Jacobian $J(v_1, v_2, \ldots, v_n)$ is zero, then to show that there must exist a relation between $v_1, v_2, \ldots, v_n$.

The equation connecting the functions $v_1, v_2, \ldots, v_n$ and the variables $x_1, x_2, \ldots, x_n$ can be written as

$$g_1(x_1,x_2,\ldots,x_n,v_1)=0$$
$$g_2(x_2,x_3,\ldots,x_n,v_1,v_2)=0$$
$$\ldots \quad \ldots \quad \ldots \quad \ldots \quad \ldots$$
$$g_k(x_k,x_{k+1},\ldots,x_n,v_1,v_2,\ldots,v_k)=0$$
$$\ldots \quad \ldots \quad \ldots \quad \ldots \quad \ldots$$
$$g_n(x_n,v_1,v_2,\ldots,v_n)=0$$

Then, we have

$$J=\frac{\partial(v_1,v_2,\ldots,v_n)}{\partial(x_1,x_2,\ldots,x_n)}=(-1)^n\frac{\left[\frac{\partial(g_1,g_2,\ldots,g_n)}{\partial(x_1,x_2,\ldots,x_n)}\right]}{\left[\frac{\partial(g_1,g_2,\ldots,g_n)}{\partial(v_1,v_2,\ldots,v_n)}\right]}=(-1)^n\frac{\left(\frac{\partial g_1}{\partial x_1}\cdot\frac{\partial g_2}{\partial x_2}\cdots\frac{\partial g_n}{\partial x_n}\right)}{\left(\frac{\partial g_1}{\partial v_1}\cdot\frac{\partial g_2}{\partial v_2}\cdots\frac{\partial g_n}{\partial v_n}\right)}$$

If $J=0$, then $\quad \frac{\partial g_1}{\partial x_1}\cdot\frac{\partial g_2}{\partial x_2}\cdots\frac{\partial g_r}{\partial x_r}\cdots\frac{\partial g_n}{\partial x_n}=0$

$\Rightarrow$ At least one of $\frac{\partial v_1}{\partial x_1},\frac{\partial v_2}{\partial x_2},\ldots,\frac{\partial v_n}{\partial x_n}$ is zero.

$\Rightarrow$ $\frac{\partial g_k}{\partial x_k}=0$ for some value of k between 1 and n.

$\Rightarrow$ For that particular value of k, the function g_k must not contain x_k and hence

$$g_k(x_{k+1},\ldots,x_n,v_1,v_2,\ldots,v_k)=0 \qquad \ldots(2)$$

Now we may easily eliminate the variables $x_{k+1},x_{k+2},\ldots,x_n$ between (2) and $g_{r+1}=0$, $g_{r+2}=0,\ldots,g_n=0$ and an equation between $v_1, v_2,\ldots,v_n$ alone, can be obtained.

Solved Examples

Example 1. *If $x = r\cos\theta$, $y = r\sin\theta$, show that*

(i) $\frac{\partial(x,y)}{\partial(r,\theta)}=r$ (ii) $\frac{\partial(r,\theta)}{\partial(x,y)}=\frac{1}{r}$ (GBTU (SUM) 2010)

Solution. (a) We have

$$\frac{\partial(x,y)}{\partial(r,\theta)}=\begin{vmatrix}\partial x/\partial r & \partial x/\partial\theta\\ \partial y/\partial r & \partial y/\partial\theta\end{vmatrix}=\begin{vmatrix}\cos\theta & -r\sin\theta\\ \sin\theta & r\cos\theta\end{vmatrix}$$
$$=r\cos^2\theta+r\sin^2\theta=r$$

(b) From the given relations, we get

$$r^2=x^2+y^2 \text{ and } \tan\theta=y/x$$

Now differentiating partially w.r.t. x and y, we obtain

$$2r\frac{\partial r}{\partial x}=2x \quad\text{or}\quad \frac{\partial r}{\partial x}=\frac{x}{r}$$
$$2r\frac{\partial r}{\partial y}=2y \quad\text{or}\quad \frac{\partial r}{\partial y}=\frac{y}{r}$$

and $\tan\theta=y/x \Rightarrow \sec^2\theta\frac{\partial\theta}{\partial x}=-\frac{y}{x^2}$

$$\text{or}\quad \frac{\partial\theta}{\partial x}=-\frac{y}{x^2\sec^2\theta}=-\frac{y}{r^2\cos^2\theta\sec^2\theta}=-\frac{y}{r^2}$$

$$\text{and}\quad \sec^2\theta\frac{\partial\theta}{\partial y}=-\frac{1}{x}$$

$$\text{or}\quad \frac{\partial\theta}{\partial y}=\frac{1}{x\sec^2\theta}=\frac{\cos^2\theta}{x}=\frac{x^2}{r^2}\cdot\frac{1}{x}=\frac{x}{r^2}$$

$$\frac{\partial(r,\theta)}{\partial(x,y)}=\begin{vmatrix}\frac{\partial r}{\partial x} & \frac{\partial r}{\partial y}\\ \frac{\partial\theta}{\partial x} & \frac{\partial\theta}{\partial y}\end{vmatrix}=\begin{vmatrix}x/r & y/r\\ -y/r^2 & x/r^2\end{vmatrix}$$
$$=\frac{x^2}{r^3}+\frac{y^2}{r^3}=\frac{x^2+y^2}{r^3}=\frac{r^2}{r^3}=\frac{1}{r}$$

Example 2. *If $x=r\sin\theta\cos\phi$, $y=r\sin\theta\sin\phi$, $z=r\cos\theta$, show that*

$$\frac{\partial(x,y,z)}{\partial(r,\theta,\phi)}=r^2\sin\theta \quad \text{(GBTU (C.O) 2010)}$$

Solution. We know that

$$\frac{\partial(x,y,z)}{\partial(r,\theta,\phi)}=\begin{vmatrix}\frac{\partial x}{\partial r} & \frac{\partial x}{\partial\theta} & \frac{\partial x}{\partial\phi}\\ \frac{\partial y}{\partial r} & \frac{\partial y}{\partial\theta} & \frac{\partial y}{\partial\phi}\\ \frac{\partial z}{\partial r} & \frac{\partial z}{\partial\theta} & \frac{\partial z}{\partial\phi}\end{vmatrix}$$
$$=\begin{vmatrix}\sin\theta\cos\phi & r\cos\theta\cos\phi & -r\sin\theta\sin\phi\\ \sin\theta\sin\phi & r\cos\theta\sin\phi & r\sin\theta\cos\phi\\ \cos\theta & -r\sin\theta & 0\end{vmatrix}$$

[expanding the determinant along the third row]

$$=\cos\theta\,(r^2\sin\theta\cos\theta\cos^2\phi+r^2\sin\theta\cos\theta\sin^2\phi)$$
$$+r\sin\theta(r\sin^2\theta\cos^2\phi+r\sin^2\theta\sin^2\phi)$$
$$=r^2\sin\theta\cos^2\theta+r^2\sin^3\theta=r^2\sin\theta(\cos^2\theta+\sin^2\theta)$$
$$=r^2\sin\theta.$$

Example 3. *If $x=c\cos u\cosh v$ and $y=c\sin u\sinh v$ prove that*

$$\frac{\partial(x,y)}{\partial(u,v)}=\frac{1}{2}c^2(\cos 2u-\cosh 2v)$$

Solution. We have

$$x=c\cos u\cosh v \quad\text{and}\quad y=c\sin u\sinh v$$

$$\Rightarrow\quad \frac{\partial x}{\partial u}=-c\sin u\cosh v,\qquad \frac{\partial x}{\partial v}=c\cos u\sinh v$$

$$\text{and}\quad \frac{\partial y}{\partial u}=c\cos u\sinh v,\qquad \frac{\partial y}{\partial v}=c\sin u\cosh v$$

$$\text{Now}\quad \frac{\partial(x,y)}{\partial(u,v)}=\begin{vmatrix}\frac{\partial x}{\partial u} & \frac{\partial x}{\partial v}\\ \frac{\partial y}{\partial u} & \frac{\partial y}{\partial v}\end{vmatrix}$$

$$= \begin{vmatrix} -c\sin u\cosh v & c\cos u\sinh v \\ c\cos u\sinh v & c\sin u\cosh v \end{vmatrix}$$

$$= -c^2\sin^2 u\cosh^2 v - c^2\cos^2 u\sinh^2 v$$

$$= -\frac{c^2}{2}\left[2\sin^2 u\cosh^2 v + 2\cos^2 u\sinh^2 v\right]$$

$$= -\frac{c^2}{2}\left[(1-\cos 2u)\cosh^2 v + (1+\cos 2u)\sinh^2 v\right]$$

$$= -\frac{c^2}{2}\left[\cos 2u\left(\sinh^2 v - \cosh^2 v\right) + \cosh^2 v + \sinh^2 v\right]$$

$$= \frac{c^2}{2}[\cos 2u - \cos 2v]$$

Example 4. *If $y_1 = r\sin\theta_1\sin\theta_2$, $y_2 = r\sin\theta_1\cos\theta_2$, $y_3 = r\cos\theta_1\sin\theta_3$, $y_4 = r\cos\theta_1\cos\theta_3$, find the value of Jacobian.*

Solution . We have

$$y_1 = r\sin\theta_1\sin\theta_2 \qquad ...(1)$$
$$y_2 = r\sin\theta_1\cos\theta_2 \qquad ...(2)$$
$$y_3 = r\cos\theta_1\sin\theta_3 \qquad ...(3)$$
$$y_4 = r\cos\theta_1\cos\theta_3 \qquad ...(4)$$

Squaring and adding the given four relations, we get

$$y_1^2 + y_2^2 + y_3^2 + y_4^2 = r^2$$

$$\left.\begin{aligned} \therefore \quad & y_1\frac{\partial y_1}{\partial r} + y_2\frac{\partial y_2}{\partial r} + y_3\frac{\partial y_3}{\partial r} + y_4\frac{\partial y_4}{\partial r} = r \\ \text{and} \quad & y_1\frac{\partial y_1}{\partial \theta_i} + y_2\frac{\partial y_2}{\partial \theta_i} + y_3\frac{\partial y_3}{\partial \theta_i} + y_4\frac{\partial y_4}{\partial \theta_i} = 0, \\ & i = 1,2,3 \end{aligned}\right\} \quad ...(5)$$

Also $y_3^2 + y_4^2 = r^2\cos^2\theta_1$, so that

$$\left.\begin{aligned} & y_1\frac{\partial y_3}{\partial \theta_1} + y_4\frac{\partial y_4}{\partial \theta_1} = -r^2\cos\theta_1\sin\theta_1 \\ & y_3\frac{\partial y_3}{\partial \theta_j} + y_4\frac{\partial y_4}{\partial \theta_j} = 0, j = 2,3 \end{aligned}\right\} \quad ...(6)$$

Now the required Jacobian

$$J = \begin{vmatrix} \partial y_1/\partial r & \partial y_1/\partial\theta_1 & \partial y_1/\partial\theta_2 & \partial y_1/\partial\theta_3 \\ \partial y_2/\partial r & \partial y_2/\partial\theta_1 & \partial y_2/\partial\theta_2 & \partial y_2/\partial\theta_3 \\ \partial y_3/\partial r & \partial y_3/\partial\theta_1 & \partial y_3/\partial\theta_2 & \partial y_3/\partial\theta_3 \\ \partial y_4/\partial r & \partial y_4/\partial\theta_1 & \partial y_4/\partial\theta_2 & \partial y_4/\partial\theta_3 \end{vmatrix}$$

Operating, $y_1R_1 + (y_2R_2 + y_3R_3 + y_4R_4)$ and using result (5), we obtain

$$J = \frac{1}{y_1}\begin{vmatrix} r & 0 & 0 & 0 \\ \partial y_2/\partial r & \partial y_2/\partial\theta_1 & \partial y_2/\partial\theta_2 & \partial y_2/\partial\theta_3 \\ \partial y_3/\partial r & \partial y_3/\partial\theta_1 & \partial y_3/\partial\theta_2 & \partial y_3/\partial\theta_3 \\ \partial y_4/\partial r & \partial y_4/\partial\theta_1 & \partial y_4/\partial\theta_2 & \partial y_4/\partial\theta_3 \end{vmatrix}$$

$$= \frac{r}{y_1}\begin{vmatrix} \frac{\partial y_2}{\partial\theta_1} & \frac{\partial y_2}{\partial\theta_2} & \frac{\partial y_2}{\partial\theta_3} \\ \frac{\partial y_3}{\partial\theta_1} & \frac{\partial y_3}{\partial\theta_2} & \frac{\partial y_3}{\partial\theta_3} \\ \frac{\partial y_4}{\partial\theta_1} & \frac{\partial y_4}{\partial\theta_2} & \frac{\partial y_4}{\partial\theta_3} \end{vmatrix}$$

$$= \frac{r}{y_1 y_3}\begin{vmatrix} \frac{\partial y_2}{\partial\theta_1} & \frac{\partial y_2}{\partial\theta_2} & \frac{\partial y_2}{\partial\theta_3} \\ -r^2\cos\theta_1\sin\theta_1 & 0 & 0 \\ \frac{\partial y_4}{\partial\theta_1} & \frac{\partial y_4}{\partial\theta_2} & \frac{\partial y_4}{\partial\theta_3} \end{vmatrix}$$

[Adding y_4R_3 to y_3R_2 and using the result (6)]

$$= \frac{r}{y_1 y_3}.r^2\cos\theta_1\sin\theta_1\left[\frac{\partial y_2}{\partial\theta_2}.\frac{\partial y_4}{\partial\theta_3} - \frac{\partial y_4}{\partial\theta_2}.\frac{\partial y_2}{\partial\theta_3}\right]$$

$$= \frac{r^3\cos\theta_1\sin\theta_1}{y_1 y_3}$$

$$\left[(-r\sin\theta_1\sin\theta_2)(-r\cos\theta_1\sin\theta_3) - 0\right]$$

$$= \frac{r^5\sin^2\theta_1\cos^2\theta_1\sin\theta_2\sin\theta_3}{r^2\sin\theta_1\cos\theta_1\sin\theta_2\sin\theta_3}$$

$$= r^3\sin\theta_1\cos\theta_1.$$

Example 5. *If $y_1 = 1 - x_1, y_2 = x_1(1-x_2), y_3 = x_1x_2(1-x_3)$... $y_n = x_1x_2...x_{n-1}(1-x_n)$. Prove that $J(y_1,y_2,...,y_n) = (-1)^n x_1^{n-1}x_2^{n-2}...x_{n-1}$.*

Solution . In the above relations

y_1 is a function of x_1

y_2 is a function of x_1, x_2

y_3 is a function of $x_1x_2x_3$

... ...

and y_n is a function of $x_1x_2,...,x_n$.

$$\therefore \frac{\partial(y_1,y_2,...,y_n)}{\partial(x_1,x_2,...,x_n)}$$

$$= \begin{vmatrix} \frac{\partial y_1}{\partial x_1} & 0 & 0 & \cdots & 0 \\ \frac{\partial y_2}{\partial x_1} & \frac{\partial y_2}{\partial x_2} & 0 & \cdots & 0 \\ \vdots & \vdots & \vdots & \vdots & \vdots \\ \frac{\partial y_n}{\partial x_1} & \frac{\partial y_n}{\partial x_2} & \frac{\partial y_n}{\partial x_3} & \cdots & \frac{\partial y_n}{\partial x_n} \end{vmatrix}$$

$$= \frac{\partial y_1}{\partial x_1}.\frac{\partial y_2}{\partial x_2}.\frac{\partial y_3}{\partial x_3}...\frac{\partial y_n}{\partial x_n}$$

$$= (-1)(-x_1)(-x_1x_2)...(-x_1x_2x_3...x_{n-1})$$

$$= (-1)^n x_1^{n-1}x_2^{n-2}...x_{n-1}.$$

Example 6. *If $u^3 + v^3 = x + y$ and $u^2 + v^2 = x^3 + y^3$, then prove that*

$$\frac{\partial(u,v)}{\partial(x,y)} = \frac{1}{2}\frac{y^2 - x^2}{uv(u-v)}$$

Solution . Here we can write above relations, as

$$F_1 = u^3 + v^3 - x - y = 0$$
$$F_2 = u^2 + v^2 - x^3 - y^3 = 0$$

Now $\dfrac{\partial(u,v)}{\partial(x,y)} = (-1)^2 \dfrac{\partial(F_1,F_2)}{\partial(x,y)} \Big/ \dfrac{\partial(F_1,F_2)}{\partial(u,v)}$ ...(1)

We have

$$\frac{\partial(F_1,F_2)}{\partial(x,y)} = \begin{vmatrix} \frac{\partial F_1}{\partial x} & \frac{\partial F_1}{\partial y} \\ \frac{\partial F_2}{\partial x} & \frac{\partial F_2}{\partial y} \end{vmatrix} = \begin{vmatrix} -1 & -1 \\ -3x^2 & -3y^2 \end{vmatrix} \quad ...(2)$$

$$=3y^2-3x^2=3(y^2-x^2)$$

and $$\frac{\partial(F_1,F_2)}{\partial(x,y)} = \begin{vmatrix} 3u^2 & 3v^2 \\ 2u & 2v \end{vmatrix} \quad ...(3)$$

$$= 6u^2v - 6uv^2 = 6uv(u-v)$$

From equations (1), (2) and (3), we get

$$\frac{\partial(u,v)}{\partial(x,y)} = \frac{3(y^2-x^2)}{6uv(u-v)} = \frac{1}{2}\frac{y^2-x^2}{uv(u-v)}$$

Example 7. *If* $u^3+v^3+w^3=x+y+z;\ u^2+v^2+w^2=x^3+y^3+z^3$ *and* $u+v+w=x^2+y^2+z^2$

Prove that $\dfrac{\partial(u,v,w)}{\partial(x,y,z)} = \dfrac{(y-z)(z-x)(x-y)}{(u-v)(v-w)(w-u)}$

Solution. Here, given relation can be written as

$$F_1 = u^3+v^3+w^3-x-y-z=0$$
$$F_2 = u^2+v^2+w^2-x^3-y^3-z^3=0$$
$$F_3 = u+v+w-x^2-y^2-z^2=0.$$

Now

$$\frac{\partial(u,v,w)}{\partial(x,y,z)} = (-1)^3 \frac{\partial(F_1,F_2,F_3)}{\partial(x,y,z)} \Big/ \frac{\partial(F_1,F_2,F_3)}{\partial(u,v,w)}$$

We have

$$\frac{\partial(F_1,F_2,F_3)}{\partial(x,y,z)} = \begin{vmatrix} -1 & -1 & -1 \\ -3x^2 & -3y^2 & -3z^2 \\ -2x & -2y & -2z \end{vmatrix}$$

$$= -6\begin{vmatrix} 1 & 1 & 1 \\ x^2 & y^2 & z^2 \\ x & y & z \end{vmatrix} = 6\begin{vmatrix} 1 & 1 & 1 \\ x & y & z \\ x^2 & y^2 & z^2 \end{vmatrix}$$

$$= 6\begin{vmatrix} 1 & 0 & 0 \\ x & y-x & z-x \\ x^2 & y^2-x^2 & z^2-x^2 \end{vmatrix}$$

$$= 6(y-x)(z-x)\begin{vmatrix} 1 & 1 \\ y+x & z+x \end{vmatrix}$$

$$=6(y-x)(z-x)(z-y)$$
$$=6(x-y)(y-z)(z-x)$$

Also

$$\frac{\partial(F_1,F_2,F_3)}{\partial(u,v,w)} = \begin{vmatrix} 3u^2 & 3v^2 & 3w^2 \\ 2u & 2v & 2w \\ 1 & 1 & 1 \end{vmatrix} = \begin{vmatrix} 1 & 1 & 1 \\ u & v & w \\ u^2 & v^2 & w^2 \end{vmatrix}$$

$$=-6(u-v)(v-w)(w-u)$$

From equations (1), (2) and (3), we get

$$\frac{\partial(u,v,w)}{\partial(x,y,z)} = -\frac{6(x-y)(y-z)(z-x)}{-6(u-v)(v-w)(w-u)}$$
$$= \frac{(y-z)(z-x)(x-y)}{(u-v)(v-w)(w-u)}$$

Example 8. *Prove that*

$$\frac{\partial(y_1,y_2,...,y_n)}{\partial(x_1,x_2,...,x_n)} \cdot \frac{\partial(x_1,x_2,...,x_n)}{\partial(y_1,y_2,...,y_n)} = 1$$

Solution. Let

$$\left.\begin{array}{l} y_1 = f_1(x_1,x_2,...,x_n), y_2 = f_2(x_1,x_2,...,x_n),... \\ y_n = f_n(x_1,x_2,...,x_n) \end{array}\right] \quad ...(1)$$

Above relations can be written as

$$x_1=F_1(y_1,y_2,...,y_n),\ x_2=F_2(y_1,y_2,...,y_n)...$$
$$x_n=F_n(y_1,y_2,...,y_n)$$

Differentiating (1) partially w.r.t. $y_1, y_2,..., y_n$, we have

$$\left[\begin{array}{l} 1 = \frac{\partial y_1}{\partial x_1}\cdot\frac{\partial x_1}{\partial y_1} + \frac{\partial y_1}{\partial x_2}\cdot\frac{\partial x_2}{\partial y_1} + ... + \frac{\partial y_1}{\partial x_n}\cdot\frac{\partial x_n}{\partial y_1} = \sum \frac{\partial y_1}{\partial x_r}\cdot\frac{\partial x_r}{\partial y_1} \\ 0 = \frac{\partial y_1}{\partial x_1}\cdot\frac{\partial x_1}{\partial y_2} + \frac{\partial y_1}{\partial x_2}\cdot\frac{\partial x_2}{\partial y_2} + ... + \frac{\partial y_1}{\partial x_n}\cdot\frac{\partial x_n}{\partial y_2} = \sum \frac{\partial y_1}{\partial x_r}\cdot\frac{\partial x_r}{\partial y_2} \\ ... \\ ... \\ 0 = \frac{\partial y_1}{\partial x_1}\cdot\frac{\partial x_1}{\partial y_n} + \frac{\partial y_1}{\partial x_2}\cdot\frac{\partial x_2}{\partial y_n} + ... + \frac{\partial y_1}{\partial x_n}\cdot\frac{\partial x_n}{\partial y_n} = \sum \frac{\partial y_1}{\partial x_r}\cdot\frac{\partial x_r}{\partial y_n} \end{array}\right.$$

Similarly other relations from $y_2, y_3,...y_n$ can be obtained.

Now, $\dfrac{\partial(y_1,y_2,...,y_n)}{\partial(x_1,x_2,...,x_n)} \cdot \dfrac{\partial(x_1,x_2,...,x_n)}{\partial(y_1,y_2,...,y_n)}$

$$= \begin{vmatrix} \frac{\partial y_1}{\partial x_1} & \frac{\partial y_1}{\partial x_2} & \cdots & \frac{\partial y_1}{\partial x_n} \\ \frac{\partial y_2}{\partial x_1} & \frac{\partial y_2}{\partial x_2} & \cdots & \frac{\partial y_2}{\partial x_n} \\ \cdots & \cdots & & \cdots \\ \frac{\partial y_n}{\partial x_1} & \frac{\partial y_n}{\partial x_2} & \cdots & \frac{\partial y_n}{\partial x_n} \end{vmatrix} \times \begin{vmatrix} \frac{\partial x_1}{\partial y_1} & \frac{\partial x_1}{\partial y_2} & \cdots & \frac{\partial x_1}{\partial y_n} \\ \frac{\partial x_2}{\partial y_1} & \frac{\partial x_2}{\partial y_2} & \cdots & \frac{\partial x_2}{\partial y_n} \\ \cdots & \cdots & \cdots & \cdots \\ \frac{\partial x_n}{\partial y_1} & \frac{\partial x_n}{\partial y_2} & \cdots & \frac{\partial x_n}{\partial y_n} \end{vmatrix}$$

$$= \begin{vmatrix} \sum \frac{\partial y_1}{\partial x_r}\cdot\frac{\partial x_r}{\partial y_1} & \sum \frac{\partial y_1}{\partial x_r}\cdot\frac{\partial x_r}{\partial y_2} & \cdots & \sum \frac{\partial y_1}{\partial x_r}\cdot\frac{\partial x_r}{\partial y_n} \\ \sum \frac{\partial y_2}{\partial x_r}\cdot\frac{\partial x_r}{\partial y_1} & \sum \frac{\partial y_2}{\partial x_r}\cdot\frac{\partial x_r}{\partial y_2} & \cdots & \sum \frac{\partial y_2}{\partial x_r}\cdot\frac{\partial x_r}{\partial y_n} \\ \cdots & & \cdots & \cdots \\ \sum \frac{\partial y_n}{\partial x_r}\cdot\frac{\partial x_r}{\partial y_1} & \sum \frac{\partial y_n}{\partial x_r}\cdot\frac{\partial x_r}{\partial y_2} & \cdots & \sum \frac{\partial y_n}{\partial x_r}\cdot\frac{\partial x_r}{\partial y_n} \end{vmatrix}$$

(Operating row by column multiplication)

$$= \begin{vmatrix} 1 & 0 & \cdots & 0 \\ 0 & 1 & \cdots & 0 \\ \cdots & \cdots & \cdots & \cdots \\ 0 & 0 & \cdots & 1 \end{vmatrix}$$

Example 9. *If x,y,z are connected by a functional relation f(x,y,z)=0 then prove that*

$$\frac{\partial(y,z)}{\partial(x,z)}=\left(\frac{\partial y}{\partial x}\right)_{z=\text{constant}}$$

Solution. We have $f(x,y,z)=0 \Rightarrow y$ is a function of x and z. Also from this relation z may be regarded as a function of x and z.

$$\therefore \quad \frac{\partial(y,z)}{\partial(x,z)}=\begin{vmatrix}\left(\frac{\partial y}{\partial x}\right)_{z=\text{const.}} & \left(\frac{\partial y}{\partial z}\right)_{x=\text{const.}} \\ \frac{\partial z}{\partial x} & \frac{\partial z}{\partial z}\end{vmatrix}$$

$$=\begin{vmatrix}\left(\frac{\partial y}{\partial x}\right)_{z=\text{const.}} & \left(\frac{\partial y}{\partial z}\right)_{x=\text{const.}} \\ 0 & 1\end{vmatrix}$$

$$=\left(\frac{\partial y}{\partial x}\right)_{z=\text{const.}} \qquad \left[\because \frac{\partial z}{\partial x}=0, \frac{\partial z}{\partial z}=1\right]$$

Example 10. *Show thal $ax^2+2hxy+by^2$ and $Ax^2+2Hxy+By^2$ are independent unless*

$$\frac{a}{A}=\frac{h}{H}=\frac{b}{B}.$$

Solution. Let $u= ax^2+2hxy+by^2, v= Ax^2+2Hxy+By^2$. If the functions u, v are not independent, then

$$\frac{\partial(u,v)}{\partial(x,y)}=0$$

or

$$\begin{vmatrix}\frac{\partial u}{\partial x} & \frac{\partial u}{\partial y} \\ \frac{\partial v}{\partial x} & \frac{\partial v}{\partial y}\end{vmatrix}=0$$

or

$$\begin{vmatrix}2(ax+hy) & 2(hx+by) \\ 2(Ax+Hy) & 2(Hx+By)\end{vmatrix}=0$$

or $(ax+hy)(Hx+By)-(hx+by)(Ax+Hy)=0$

or $(aH-Ah)x^2+(aB-Ab)xy+(Bh-bH)y^2=0$

Since the variables x, y are independent, the coefficients of x and y in above equation must be separately zero. Hence we have $aH-Ah=0$ and $Bh-bH=0$

Hence, $\quad \dfrac{a}{A}=\dfrac{h}{H}=\dfrac{b}{B}.$

Example 11. *If $u=x+2y+z$, $v=x-2y+3z$ and $w=2xy-xz+4yz-2z^2$, then prove that they are not independent. Find the relation between u, v and w.*

[UPTU (SUM) 2007; G.B.T.U(C.O.) 2011]

Solution. We have

$$\frac{\partial(u,v,w)}{\partial(x,y,z)}=\begin{vmatrix}1 & 2 & 1 \\ 1 & -2 & 3 \\ 2y-z & 2x+4z & -x+4y-4z\end{vmatrix}$$

$$=\begin{vmatrix}1 & 0 & 0 \\ 1 & -4 & 2 \\ 2y-z & -x+2y-6z & -x+2y-3z\end{vmatrix}$$

by c_2-2c_1 and c_3-c_1

$$=-2\begin{vmatrix}1 & 0 & 0 \\ 1 & 2 & 2 \\ 2y-z & -x+2y-3z & -x+2y-3z\end{vmatrix}=0$$

Here last two columns are identical. So the Jacobian of the functions u, v, w is zero, therefore these functions are not independent so there must exist a relation between them.

We have $\quad u^2-v^2=(x+2y+z)^2-(x-2y+3z)^2$

$$=(2x+4z)(4y-2z)$$

$$=4(x+2z)(2y-z)$$

By simplification $= 4(2xy-xz+4yz-2z^2)=4w$

Therefore $u^2-v^2=4w$, which is the required relation between u, v and w.

Example 12. *Show that the functions*

$u=x+y+z$, $v=xy+yz+zx$, $w=x^3+y^3+z^3-3xyz$ are not independent. Also find the relation between u, v and w.

Solution. We have

$$\frac{\partial(u,v,w)}{\partial(x,y,z)}=\begin{vmatrix}1 & 1 & 1 \\ y+z & z+x & x+y \\ 3(x^2-yz) & 3(y^2-zx) & 3(z^2-xy)\end{vmatrix}$$

$$=3\begin{vmatrix}1 & 1 & 1 \\ y+z & z+x & x+y \\ x^2-yz & y^2-zx & z^2-xy\end{vmatrix}$$

$$=3\begin{vmatrix}1 & 0 & 0 \\ y+z & x-y & x-z \\ x^2-yz & -(y-x)(x+y+z) & (z-x)(x+y+z)\end{vmatrix}$$

operating c_2-c_1 and c_3-c_1

$$=3(x-y)(x-z)\begin{vmatrix}1 & 0 & 0 \\ y+z & 1 & 1 \\ x^2-yz & -(x+y+z) & -(x+y+z)\end{vmatrix}$$

[Last two columns being identical]

since Jacobian is zero, therefore functions are not independent, therefore, a relation set up between them.

We have

$$w=x^3+y^3+z^3-3xyz$$

$$=(x+y+z)(x^2+y^2+z^2-yz-zx-xy)$$

$$=(x+y+z)[(x+y+z)^2-3(yz+zx+xy)]$$

$$=u(u^2-3v)=u^3-3uv.$$

$\therefore \quad u^3=3uv+w$ is the required relation.

Example 13. *If $u=(x+y)/(1-xy)$ and $v=\tan^{-1}x+\tan^{-1}y$, find $\frac{\partial(u,v)}{\partial(x,y)}$. Are u and v functionally related? If so, find their relationship.*

Solution . We have

$$\frac{\partial u}{\partial x}=\frac{1.(1-xy)-(-y)(x+y)}{(1-xy)^2}=\frac{1+y^2}{(1-xy)^2}$$

$$\frac{\partial u}{\partial y}=\frac{1.(1-xy)-(-x)(x+y)}{(1-xy)^2}=\frac{1+x^2}{(1-xy)^2}$$

$$\frac{\partial v}{\partial x}=\frac{1}{(1+x^2)},\frac{\partial v}{\partial y}=\frac{1}{(1+y^2)}$$

Now

$$\frac{\partial(u,v)}{\partial(x,y)}=\begin{vmatrix}\partial u/\partial x & \partial u/\partial y\\ \partial v/\partial x & \partial v/\partial y\end{vmatrix}$$

$$=\frac{\partial u}{\partial x}\cdot\frac{\partial v}{\partial y}-\frac{\partial u}{\partial y}\cdot\frac{\partial v}{\partial x}$$

$$=\frac{1+y^2}{(1-xy)^2}\cdot\frac{1}{1+y^2}-\frac{1+x^2}{(1-xy)^2}\cdot\frac{1}{1+x^2}$$

$$=\frac{1}{(1-xy)^2}-\frac{1}{(1-xy)^2}=0$$

Since the Jacobian of the function u, v is zero, therefore these functions are not independent and so they must be functionally related.

Now, we have

$$y=\tan^{-1}x+\tan^{-1}y=\tan^{-1}\frac{x+y}{1-xy}=\tan^{-1}u.$$

Thus $v=\tan^{-1}u$ or $\tan v=u$ which is the required relation, between u and v.

EXERCISE 11.1

1. If $u=\frac{y^2}{2x}$ and $v=\frac{x}{2}+\frac{y^2}{2x}$, find $\frac{\partial(u,v)}{\partial(x,y)}$.

2. (a) If $u_1=\frac{x_2x_3}{x_1}$, $u_2=\frac{x_3x_1}{x_2}$, $u_3=\frac{x_1x_2}{x_3}$,

then show that $J(u_1,u_2,u_3)=4$.

(GBTU 2011, 2012; GBTU(AG) SUM 2010; UKTU 2011)

(b) If $x=u(1+v)$, $y=v(1+u)$, then show that

$$\frac{\partial(x,y)}{\partial(u,v)}=1+u+v.$$ (UPTU (SUM) 2009)

3. If $x=\sin\theta\sqrt{1-a^2\sin^2\phi}$, $y=\cos\theta\cos\phi$, then show that

$$\frac{\partial(x,y)}{\partial(\theta,\phi)}=-\sin\phi\frac{\left[(1-a^2)\cos^2\theta+a^2\cos^2\phi\right\}}{\sqrt{1-a^2\sin^2\phi}}$$

4. If $y_1=x_1(1-x_1)$, $y_2=x_1x_2(1-x_3)$,...,

$y_{n-1}=x_1x_2...x_{n-1}(1-x_n)$, $y_n=x_1x_2...x_n$.

Then prove that

$$\frac{\partial(y_1,y_2,...,y_n)}{\partial(x_1,x_2,...,x_n)}=x_1^{\ n-1}.x_2^{\ n-2}.....x_{n-1}.$$

5. If $y_1=\cos x_1$, $y_2=\sin x_1\cos x_2$, $y_3=\sin x_1\sin x_2\cos x_3$,..., $y_n=\sin x_1\sin x_2\sin x_3...\sin x_{n-1}\cos x_n$. Then find the Jacobian of $y_1, y_2, ..., y_n$ with respect to $x_1,x_2,...,x_n$.

6. Show that $\frac{\partial(u,v)}{\partial(x,y)}\cdot\frac{\partial(x,y)}{\partial(u,v)}=1$.

7. If $u^3=xyz$, $\frac{1}{v}=\frac{1}{x}+\frac{1}{y}+\frac{1}{z}$, $w^2=x^2+y^2+z^2$, show that

$$\frac{\partial(u,v,w)}{\partial(x,y,z)}=\frac{-v(y-z)(z-x)(x-y)(x+y+z)}{3u^2w(yz+zx+xy)}$$

8. If $u=2xy$, $v=x^2-y^2$, and $x=r\cos\theta$, $y=r\sin\theta$, show that

$$\frac{\partial(u,v)}{\partial(r,\theta)}=-4r^3.$$

9. (i) If $u_1=x_1+x_2+x_3+x_4$, $u_1u_2=x_2+x_3+x_4$, $u_1u_2u_3=x_3+x_4$, $u_1u_2u_3u_4=x_4$. Show that

$$\frac{\partial(x_1,x_2,x_3,x_4)}{\partial(u_1,u_2,u_3,u_4)}=u_1^3u_2^2u_3.$$

(ii) If $x+y+z=u$, $y+z=uv$, $z=uvw$, show that

$$\frac{\partial(x,y,z)}{\partial(u,v,w)}=u^2v.$$

10. If $y_1.(x_1-x_2)=0$, $y_2.(x_1^{\ 2}+x_1x_2+x_2^{\ 2})=0$, show that

$$\frac{\partial(y_1,y_2)}{\partial(x_1,x_2)}=3y_1y_2\frac{x_1+x_2}{x_1^{\ 3}-x_2^{\ 3}}.$$

11. If l, m, n are the roots of the equation in k.

$$\frac{x}{a+k}+\frac{y}{b+k}+\frac{z}{c+k}=1$$

Prove that $\frac{\partial(x,y,z)}{\partial(l,m,n)}=-\left[\frac{(m-n)(n-l)(l-m)}{(b-c)(c-a)(a-b)}\right].$

12. If the roots of the equation $(\lambda-x)^3+(\lambda-y)^3+(\lambda-z)^3=0$ are u, v, w. Prove that

$$\frac{\partial(u,v,w)}{\partial(x,y,z)}=-2\frac{(y-z)(z-x)(x-y)}{(v-w)(w-u)(u-v)}.$$ [UKTU 2011]

13. Show that the functions

$u=x+y-z$, $v=x-y+z$, $w=x^2+y^2+z^2-2yz$

are not independent of one another. Also find the relation between them.

14. If $u=x^2+y^2+z^2$, $v=x+y+z$, $w=xy+yz+zx$. Show that the Jacobian $\frac{\partial(u,v,w)}{\partial(x,y,z)}$ vanish identically. Also find the relation between u, v and w. [UKTU 2010; UPTU 2009]

15. If $u=x+y+z+t$, $v=x+y-z-t$, $w=xy-zt$, $r=x^2+y^2-z^2-t^2$. Show that $\frac{\partial(u,v,w,r)}{\partial(x,y,z,t)}=0$.

16. If $f(0)=0$ and $f'(x)=\frac{1}{1+x^2}$. Then prove that

$$f(x)+f(y)=f\left(\frac{(x+y)}{(1-xy)}\right)$$

(without using the method of integration).

17. $u=x\left(1-r^2\right)^{-\frac{1}{2}}$, $v=y\left(1-r^2\right)^{-\frac{1}{2}}$,

$w=z\left(1-r^2\right)^{-\frac{1}{2}}$ where $r^2=x^2+y^2+z^2$, then

show that $\frac{\partial(u,v,w)}{\partial(x,y,z)}=\left(1-r^2\right)^{-5/2}$ [UPTU(SUM) 2009]

18. If $u=\frac{x+y}{z}, v=\frac{y+z}{x}, w=y\frac{(x+y+z)}{xz}$

then show that u, v, w are not independent. (GBTU 2010)

19. If

$$x_1+x_2+...+x_n=y_1$$
$$x_2+x_3+...+x_n=y_1y_2$$
$$x_3+x_4+...+x_n=y_1y_2y_3$$
$$\dots\quad\dots\quad\dots$$
$$x_n=y_1y_2y_3...y_n$$

then show that

$$\frac{\partial(x_1,x_2,...,x_n)}{\partial(y_1,y_2,...,y_n)}=y_1^{n-1}y_2^{n-2}...y_{n-2}^2y_{n-1}.$$

20. If $u_1=\frac{x_1}{x_n}, u_2=\frac{x_2}{x_n}, u_3=\frac{x_3}{x_n},..., u_{n-1}=\frac{x_{n-1}}{x_n}$ and

$x_1^2+x_2^2+...x_{n-1}^2+x_n^2=1$, find the value of $\frac{\partial(u_1,u_2,...,u_n)}{\partial(x_1,x_2,...,x_n)}$.

21. (i) Calculate the Jacobian $\frac{\partial(x,y,z)}{\partial(u,v,w)}$ of the following :

$u=x+2y+z$, $v=x+2y+3z$, $w=2x+3y+5z$. (UPTU 2008)

(ii) If $u=xyz$, $v=xy+yz+zx$, $w=x+y+z$, then compute the jacobian $\frac{\partial(x,y,z)}{\partial(u,v,w)}$. (UPTU (SUM) 2008)

22. (i) Verify the chain rule for Jacobians if $x=u$, $y=u\tan v$, $z=w$. (UPTU 2009)

(ii) If $x=\sqrt{vw}$, $y=\sqrt{wu}$, $z=\sqrt{uv}$ and $u=r\sin\theta\cos\phi$, $v=r\sin\theta\sin\phi$, $w=r\cos\theta$ then calculate the Jacobian $\frac{\partial(x,y,z)}{\partial(u,v,w)}$ (UPTU, Model paper 2008)

23. If $u=xyz$, $v=x^2+y^2+z^2$, $w=x+y+z$, find the Jacobian $\frac{\partial(x,y,z)}{\partial(u,v,w)}$ (UKTU 2012)

24. If $x=e^v\sec u$, $y=e^v\tan u$, then evaluate $\frac{\partial(x,y)}{\partial(u,v)}$. (GBTU 2010)

25. If $u^3+v^3=x+y$, $u^2+v^2=x^3+y^3$, then show that

$$\frac{\partial(u,v)}{\partial(x,y)}=\frac{y^2-x^2}{2uv(u-v)}$$ (UPTU 2007)

Hint to Selected Problems

1. $\frac{\partial u}{\partial x}=\frac{-y^2}{2x^2}$, $\frac{\partial u}{\partial y}=\frac{y}{x}$

$\frac{\partial v}{\partial x}=\frac{1}{2}-\frac{y^2}{2x^2}$, $\frac{\partial v}{\partial y}=\frac{y}{x}$

2. $\frac{\partial u_1}{\partial x_1}=-\frac{x_2x_3}{x_1^2}$, $\frac{\partial u_1}{\partial x_2}=\frac{x_3}{x_1}$, $\frac{\partial u_1}{\partial x_3}=\frac{x_2}{x_1}$

$\frac{\partial u_2}{\partial x_1}=\frac{x_3}{x_2}$, $\frac{\partial u_2}{\partial x_2}=-\frac{x_3x_1}{x_2}$, $\frac{\partial u_2}{\partial x_3}=\frac{x_1}{x_2}$, $\frac{\partial u_3}{\partial x_1}=\frac{x_2}{x_3}$, $\frac{\partial u_3}{\partial x_2}=\frac{x_1}{x_3}$.

3. $\frac{\partial x}{\partial\theta}=\cos\theta\sqrt{\left(1-a^2\sin^2\phi\right)}$

$\frac{\partial x}{\partial\phi}=\frac{-a^2\sin\theta\sin\phi\cos\phi}{\sqrt{1-a^2\sin^2\phi}}$

$\frac{\partial y}{\partial\theta}=-\sin\theta\cos\phi$

$\frac{\partial y}{\partial\phi}=-\cos\theta\sin\phi$

9. $u_1=x_1+x_3+x_4$, $u_2=x_2+x_3+x_4$, $u_1u_2u_3=x_3+x_4$, $u_1u_2u_3u_4=x_4$. Therefore $x_3=u_1u_2u_3(1-u_4)$, $x_2=u_1u_2(1-u_3)$, $x_1=u_1(1-u_2)$.

Now find required partial derivative of x_1, x_2 and x_3 w.r.t. u_1, u_2 and u_3.

13. If $\frac{\partial(u,v,w)}{\partial(x,y,z)}=0$ Then u, v, w are not independent.

Answers

1. $-\frac{y}{2x}$ **5.** $(-1)^n\sin^n x_1\sin^{n-1}x_2\sin^{n-2}x_3...\sin x_n$ **13.** $u^2+v^2=2w$ **14.** $v^2=u+2w$ **15.** $uv=r+2w$ **20.** $\frac{1}{x_n^n}$

21. (i) 2 (ii) $(x-y)(y-z)(z-x)$ **22.** (ii) $\frac{1}{4}r^2\sin\theta$ **23.** $\frac{-1}{2(x-y)(y-z)(z-x)}$ **24.** $-e^{2v}\sec u$

Objective Evaluations

Fill in the Blanks

1. If u and v are the functions of two independent variables x and y, then Jacobian of u and v with respect to x and y is denoted by ________.

2. The function u,v and w of three independent variables x,y and z will not ________ if $\frac{\partial(u,v,w)}{\partial(x,y,z)} = 0$.

3. It is ________ that $\frac{\partial(u,v)}{\partial(x,y)} = \frac{\partial(x,y)}{\partial(u,v)}$.

4. If $x = r\cos\theta$, $y = r\sin\theta$, then the value of $\frac{\partial(x,y)}{\partial(r,\theta)} =$ ________.

5. The value of $\frac{\partial(u,v)}{\partial(x,y)} \cdot \frac{\partial(x,y)}{\partial(u,v)} =$ ________.

True/False

Write 'T' for True and 'F' for False statement.

1. If u_1, u_2 are functions of y_1, y_2 and y_1, y_2 are functions of x_1, x_2, then,
$\frac{\partial(u_1,u_2)}{\partial(x_1,x_2)} = \frac{\partial(u_1,u_2)}{\partial(y_1,y_2)} \cdot \frac{\partial(y_1,y_2)}{\partial(x_1,x_2)}$ **(T/F)**

2. $\frac{\partial(u_1,u_2)}{\partial(x_1,x_2)} \cdot \frac{\partial(x_1,x_2)}{\partial(u_1,u_2)} = 0$. **(T/F)**

3. If the functions u, v, w of three independent variables x, y, z are not independent then the Jacobian of u, v, w with respect to x, y, z vanishes. **(T/F)**

4. If u_1, u_2 be the implicit function of x_1, x_2, *i.e.*, $F_1(u_1, u_2, x_1, x_2) = 0$ and $F_2(u_1, u_2, x_1, x_2) = 0$ then
$\frac{\partial(F_1,F_2)}{\partial(u_1,u_2)} \times \frac{\partial(u_1,u_2)}{\partial(x_1,x_2)} = (-1)^2 \frac{\partial(F_1,F_2)}{\partial(x_1,x_2)}$. **(T/F)**

5. The necessary and sufficient condition for the existence of a relation of the form $F(u_1, u_2, ..., u_n) = 0$ is that the Jacobian should vanish identically. **(T/F)**

Multiple Choice Questions

Choose the most appropriate one.

1. If the functions u,v,w of three independent variables x,y,z and $\frac{\partial(u,v,w)}{\partial(x,y,z)} = 0$ then the functions are :

   (a) independent (b) not independent
   (c) may be independent (d) none of these

2. The value of $\frac{\partial(u,v)}{\partial(x,y)} \cdot \frac{\partial(x,y)}{\partial(u,v)}$ is :

   (a) 1 (b) 0
   (c) ∞ (d) none of these

3. The necessary and sufficient condition for the existence of a relation $F(u_1,u_2,...,u_n) = 0$ is that the Jacobian must be :

   (a) equal to 1 (b) equal to 2
   (c) vanish identically (d) none of the above

4. If x,y,z are connected by a functional relation $f(x, y, z) = 0$ then the value of $\frac{\partial(y,z)}{\partial(x,z)}$ is equal to :

   (a) $\frac{\partial x}{\partial y}$ (b) $\frac{\partial y}{\partial x}$
   (c) $\left(\frac{\partial y}{\partial x}\right)_{z=\text{constant}}$ (d) $\left(\frac{\partial x}{\partial y}\right)_{z=\text{constant}}$

5. If $x = r\cos\theta$, $y = r\sin\theta$, then the value of $\frac{\partial(r,\theta)}{\partial(x,y)}$:

   (a) 1 (b) r
   (c) $1/r$ (d) none of the above

ANSWERS

Fill in the Blanks

1. $\dfrac{\partial(u,v)}{\partial(x,y)}$ or $J(u,v)$ **2.** independent **3.** necessary **4.** r **5.** 1

True/False

1. T **2.** F **3.** T **4.** T **5.** T

Multiple Choice Questions

1. (b) **2.** (a) **3.** (c) **4.** (c) **5.** (c)

FFFFF

CHAPTER

12 Extrema of Function of Several Variables

12.1 INTRODUCTION

If $y = f(x)$ be a continuous function. At a point $x = x_1$, if the function $f(x)$ does not increase and begins to decrease then $f(x)$ has its maximum value at $x = x_1$ and if at a point $x = x_2$, $f(x)$ does not decrease and begins to increase, then $f(x)$ has its minimum value at $x = x_2$.

If $f(x)$ is maximum at a point $x = x_1$ then $f(x)$ is an increasing function for the preceding values of x_1 and is a decreasing function for those value of x just below x_1 or we can say derivative of the function $\left(i.e., \frac{dy}{dx}\right)$ will be positive before $x = x_1$ and will be negative after $x = x_1$. But $\frac{dy}{dx}$ is a continuous function and $\frac{dy}{dx}$ changes the sign from positive to negative. So, $\frac{dy}{dx}$ will be zero at any point.

Therefore, for a maximum value of $y = f(x)$ at a point, we have $\frac{dy}{dx} = 0$ and $\frac{dy}{dx}$ changes the sign from positive to negative. On the other hand, for a minimum value of $y = f(x)$ at point we have $\frac{dy}{dx} = 0$ and $\frac{dy}{dx}$ changes the sign negative to positive.

REMARKS

- If $\frac{dy}{dx}$ changes the sign positive to negative; it means that $f(x)$ is a decreasing function of x *i.e.*, $\frac{d^2y}{dx^2} < 0$.
- If $\frac{dy}{dx}$ changes the sign from negative to positive, it means that the $f(x)$ is an increasing function of x, *i.e.*, $\frac{d^2y}{dx^2} > 0$.
- A function may have more than one maximum and minimum value.
- Any minimum value of the function $f(x)$ can be greater than any maximum value.
- Maximum and minimum values of the function occur alternately.
- Maximum and minimum values of the function are sometimes known as extreme value.
- From the definition of maxima and minima, it is clear that $\frac{dy}{dx} = 0$ is the necessary condition for maximum or minimum.
- $\frac{d^2y}{dx^2} < 0$ is sufficient condition for maximum and $\frac{d^2y}{dx^2} > 0$ is sufficient condition for minimum.

WORKING PROCEDURE

Step 1. *Find the derivative of the given function i.e.,* $\frac{dy}{dx}$.

Step 2. *Put* $\frac{dy}{dx} = 0$ *and find all the real values of x. (say* x_1, x_2, x_3 ...).

Step 3. *Find* $\frac{d^2y}{dx^2}$.

Step 4. *Put* $x = x_i$ in $\frac{d^2y}{dx^2}$ *and find the result. If result is negative then the function f(x) is maximum at* $x = x_i$ *and max.* $f(x) = f(x_i)$. *On the other hand, if result is positive then the function f(x) is minimum at* $x = x_i$ *and minimum* $f(x) = f(x_i)$.

REMARKS

- In a continuous function, maxima and minima values occur alternately, *i.e.*, between two successive maxima there is one minimum and between two successive minima, there is one maximum.
- If $\frac{d^2y}{dx^2}$ is equal to 0 at any point $x = x_i$ then find $\frac{d^3y}{dx^3}, \frac{d^4y}{dx^4}$, and find the values of these derivatives at $x = x_i$ successively and cheek the sign.

Solved Examples

Example 1. *Find the value of x for which* $f(x) = y = x^4 + 2x^3 - 3x^2 - 4x + 4$ *is maximum or minimum and also find those value of f(x).*

Solution. Here, the given function is

$$y = f(x) = x^4 + 2x^3 - 3x^2 - 4x + 4 \quad ...(1)$$

So $\frac{dy}{dx} = 4x^3 + 6x^2 - 6x - 4$

$= 2(x+2)(2x+1)(x-1)$

Now, put $\frac{dy}{dx} = 0$, we have

$2(x+2)(2x+1)(x-1) = 0$

So, $x = -2, -\frac{1}{2}, 1$

Again differentiating (2) *w.r.t.* to x, we get

$$\frac{d^2y}{dx^2} = 12x^2 + 12x - 6$$

At $x = -2$, we have

$$\frac{d^2y}{dx^2} = 12(-2)^2 + 12(-2) - 6$$

$= 48 - 24 - 6 = 18 > 0$

Since, $\frac{d^2y}{dx^2} > 0$ (*i.e.*, positive). So $f(x)$ is minimum at $x = -2$. The minimum value of $f(x)$ at $x = -2$ is given by

$$f(-2) = (-2)^4 + 2(-2)^3 - 3(-2)^2 - 4(-2) + 4 = 0$$

Now, at $x = -\frac{1}{2}$, we have

$$\frac{d^2y}{dx^2} = 12\left(-\frac{1}{2}\right)^2 + 12\left(-\frac{1}{2}\right) - 6 = 3 - 6 - 6 = -9 < 0$$

Since, $\frac{d^2y}{dx^2} < 0$ (*i.e.*, negative). So, $f(x)$ is maximum at $x = -\frac{1}{2}$ and maximum value of $f(x)$ at $x = -\frac{1}{2}$ is

$$f\left(-\frac{1}{2}\right) = \left(-\frac{1}{2}\right)^4 + 2\left(-\frac{1}{2}\right)^3 - 3\left(-\frac{1}{2}\right)^2 - 4\left(-\frac{1}{2}\right) + 4$$

$$= \frac{1}{16} - \frac{1}{4} - \frac{3}{4} + 2 + 4 = \frac{81}{16}$$

Similarly, at $x = 1$, we have

$$\frac{d^2y}{dx^2} = 12(1)^2 + 12(1) - 6 = 12 + 12 - 6 = 18 > 0$$

Since, $\frac{d^2y}{dx^2} > 0$ (*i.e.*, positive). So $f(x)$ is minimum at $x = 1$ and minimum value of $f(x)$ at $x = 1$ is

$$f(1) = (1)^4 + 2(1)^3 - 3(1)^2 - 4(1) + 4$$

$= 1 + 2 - 3 - 4 + 4 = 0.$

Example 2. *Find the maximum and minimum value of the function*

$$y = f(x) = x^3 - 12x^2 + 36x + 21$$

Solution. Here, the given function is

$$y = x^3 - 12x^2 + 36x + 21$$

Now, differentiating *w.r.t.* x, we get

$$\frac{dy}{dx} = 3x^2 - 24x + 36$$

Puting $\frac{dy}{dx} = 0$, we get

$3x^2 - 24x + 36 = 0$ or $x^2 - 8x + 12 = 0$

or $(x-2)(x-6) = 0$ or $x = 2, 6$

Again, differentiating *w.r.t.* x, we get

$$\frac{d^2y}{dx^2} = 6x - 24$$

At $x = 2$, we have $\frac{d^2y}{dx^2} = 6(2) - 24 = -12 < 0$

Since, $\frac{d^2y}{dx^2} < 0$ so $f(x)$ is maximum at $x = 2$.

The maximum value of $f(x)$ at $x = 2$ is given by

$$f(2) = (2)^3 - 12(2)^2 + 36(2) + 21$$

$= 8 - 48 + 72 + 21 = 53.$

Similarly, at $x = 6$, we have

$$\frac{d^2y}{dx^2} = 6 \times 6 - 24 = 36 - 24 = 12 > 0$$

Since, $\frac{d^2y}{dx^2} > 0$ so, $f(x)$ is minimum at $x = 6$ and minimum value of $f(x)$ at $x = 6$ is

$$f(6) = (6)^3 - 12(6)^2 + 36(6) + 21$$

$= 216 - 432 + 216 + 21 = 453 - 432$

$= 21.$

12.2 MAXIMA AND MINIMA OF A FUNCTION OF SEVERAL INDEPENDENT VARIABLES

Let $f(x, y, z, ...)$ be a function of several independent variables x, y, z.... If f is continuous and finite for all values of x, y, z, ... in the neighbourhood of $x = a, y = b, z = c$, ... respectively, then the value of $(a, b, c, ...)$ is said to be a maximum or minimum if $f(a+h, b+k, c+l, ...)$ is less than or greater than $f(a, b, c, ...)$ for all values of h, k, l, ... (where h, k, l, ...) are sufficiently small, may be positive or negative provided they are not all zero.

In other words we can say, the value of $f(a, b, c, ...)$ is said to be a maximum or minimum if $f(a+h, b + k, c + l, ...) - f(a, b, c, ...)$ maintain an invariant sign (may be positive or negative) for all values of h, k, l, ... positive or negative provided they are taken sufficiently small and finite.

12.2.1 STATIONARY AND EXTREME POINTS

A point $(a_1, a_2, ..., a_n)$ is called a stationary point, if all the first order partial derivative of the function $f(x_1, x_2, ..., x_n)$ vanish at the point. A stationary point, if it is maximum or minimum is known as extreme point and the value of the function at an extreme point is known as an extreme value.

REMARK

- A stationary point may be a maximum or minimum or neither of these two.

12.3 NECESSARY CONDITION FOR THE EXISTENCE OF MAXIMA OR MINIMA

Let $f(x, y, z, ...)$ be a function of several independent variables x, y, z,... It is clear from the definition of maxima and minima that maximum or minimum of $f(x, y, z, ..)$ will occur for those values of x, y, z, ..., for which the expression $f(x+h, y+k, z+l, ...) - f(x, y, z, ...)$ maintain an invariant sign for all sufficiently small and finite values of h, k, l, ... positive or negative.

Now, expanding $f(x+h, y+k, z+l, ...)$ by Taylor's theorem, we have

$$f(x+h, y+k, z+l...) = f(x,y,z) + \left(h\frac{\partial f}{\partial x} + k\frac{\partial f}{\partial y} + l\frac{\partial f}{\partial z} + ...\right) + \text{ terms of second and higher order.}$$

$$\Rightarrow \quad f(x+h, y+k, z+l...) - f(x,y,z,...) = \left(h\frac{\partial f}{\partial x} + k\frac{\partial f}{\partial y} + l\frac{\partial f}{\partial z} + ...\right) + \text{ terms of second and higher orders.} \qquad ...(1)$$

Now, since h, k, l, ... are sufficiently small, the first degree expression

$$\left(h\frac{\partial f}{\partial x} + k\frac{\partial f}{\partial y} + l\frac{\partial f}{\partial z} + ...\right)$$

of the equation (1) can be made to govern the sign of right hand side and hence, of the left hand side as well as. Thus, by changing the sign of the left hand side of the equation (1) will also change.

Since, left hand side is to preserve an invariable sign for maxima or minima, therefore, as a necessary condition for maximum and minimum values, we must have

$$h\frac{\partial f}{\partial x} + k\frac{\partial f}{\partial y} + l\frac{\partial f}{\partial z} + ... = 0 \qquad ...(2)$$

Now, since h, k, l, ... are arbitrary and independent of each other, we must have

$$\frac{\partial f}{\partial x} = 0, \ \frac{\partial f}{\partial y} = 0, \ \frac{\partial f}{\partial z} = 0, \text{ etc.} \qquad ...(3)$$

If the number of independent variables be n, we shall get n simultaneous equations in these n variables, which will give the values a, b, c, ... of the n variables x, y, z, ... respectively for which $f(x, y, z, ...)$ will have a maximum or a minimum values.

REMARKS

- The necessary condition for a function $f(x, y, z, ...)$ of the independent variables x, y, z, ... to be maximum or minimum is given by

$$\frac{\partial f}{\partial x} = 0, \ \frac{\partial f}{\partial y} = 0, \ \frac{\partial f}{\partial z} = 0,$$

- The conditions given above is only a necessary condition for the maxima and minima of the function $f(x, y, z, ...)$. These conditions are not sufficient.

12.3.1 MAXIMA AND MINIMA FOR A FUNCTION OF TWO INDEPENDENT VARIABLES

(1) *To find the condition which governs the sign of a quadratic expression.*

Consider, a binary expression

$$I = ax^2 + 2hxy + by^2$$

of two variables x and y. Then I can be written as

$$I = ax^2 + 2hxy + by^2 = \frac{1}{a}[(ax+hy)^2 + (ab-h^2)y^2].$$

If $(ab - h^2)$ is positive, the sign of I will be the same as that of a.

But if $(ab - h^2)$ is negative, then, the expression within the brackets may be positive or negative and therefore we cannot say anything about the sign of expression I.

(2) *Stationary and extreme points (For the function of two independent variables)*:

Let $f(x, y)$ be a function of two independent variables x and y. A point (a, b) is called a stationary point, if both the first order partial derivatives $\left(\frac{\partial f}{\partial a} \text{ and } \frac{\partial f}{\partial b}\right)$ of the function $f(x, y)$ at (a, b) vanish.

A stationary point which is either a maximum or minimum is called an extreme point.

REMARKS

- A stationary point is not necessarily an extreme point, hence a stationary point may be a maximum or a minimum or neither of these two.
- The value of the function at extreme point is called extreme value.
- A point at which function is neither maximum nor minimum, is known as saddle point.

12.4 NECESSARY CONDITION FOR MAXIMA AND MINIMA

Let $f(x, y)$ be a function of two independent variables x and y.Then, we have the maximum or minimum of $f(x, y)$ at $x = a$ and $x = b$ if the expression $f(a+ h, b + k) - f(a, b)$ is of invariable sign for all sufficiently small independent variables h and k provided both of them are not equal to zero.

We observe that,

(i) If the sign of $f(a+h, b+k) - f(a, b)$ is negative, then we have a maximum of $f(x, y)$ at $x = a, y = b$.

(ii) If the sign of $f(a+h, b+k) - f(a, b)$ is positive, we have a minimum of $f(x, y)$ at $x = a, y = b$.

Expand $f(a+h, b+k)$ by Taylor's theorem, we have

$$f(a+h,b+k) = f(a,b) + \left(h\frac{\partial f}{\partial x} + k\frac{\partial f}{\partial y}\right)_{\substack{x=a\\y=b}} + \frac{1}{2!}\left(h^2\frac{\partial^2 f}{\partial x^2} + 2hk\frac{\partial^2 f}{\partial x\,\partial y} + k^2\frac{\partial^2 f}{\partial y^2}\right)_{\substack{x=a\\y=b}} + \ldots$$

$$\Rightarrow \quad f(a+h,b+k) - f(a,b) = h\left(\frac{\partial f}{\partial x}\right)_{\substack{x=a\\y=b}} + k\left(\frac{\partial f}{\partial y}\right)_{\substack{x=a\\y=b}} + \text{term of the second and higher orders in } h \text{ and } k.$$

Now, since h and k are sufficiently small, the expression $h\left(\frac{\partial f}{\partial x}\right)_{\substack{x=a\\y=b}} + k\left(\frac{\partial f}{\partial y}\right)_{\substack{x=a\\y=b}}$ of the equation (1) can be made to govern the sign of right hand side and hence of the left hand side as well. Thus by changing the sign of h and k, the sign of the left hand side of the equation (1) will also change.

Since L.H.S. is to preserve an invariable sign for maximum or minimum, therefore as a necessary condition for maximum and minimum values, we must have

$$h\left(\frac{\partial f}{\partial x}\right)_{\substack{x=a\\y=b}} + k\left(\frac{\partial f}{\partial y}\right)_{\substack{x=a\\y=b}} = 0. \qquad \ldots(2)$$

If $k = 0$, we find that if $\left(\frac{\partial f}{\partial x}\right)_{\substack{x=a\\y=b}} \neq 0$, the R.H.S. of (2) changes sign when h changes sign. Therefore $f(x, y)$ cannot have a maximum or minimum at $x = a$, $y = b$ if $\left(\frac{\partial f}{\partial x}\right)_{\substack{x=a\\y=b}} \neq 0$.

Similarly, taking $h = 0$, we see that $f(x, y)$ cannot have a maximum or a minimum at $x = a$, $y = b$ if $\left(\frac{\partial f}{\partial y}\right)_{\substack{x=a\\y=b}} \neq 0$.

Thus, a set of necessary conditions that $f(x, y)$ should have a maximum or minimum at $x=a, y=b$ is that

$$\left(\frac{\partial f}{\partial x}\right)_{\substack{x=a\\y=b}} = 0 \text{ and } \left(\frac{\partial f}{\partial y}\right)_{\substack{x=a\\y=b}} = 0.$$

12.5 SUFFICIENT CONDITION FOR MAXIMA AND MINIMA: THE LAGRANGE'S CONDITION

Let $f(x, y)$ be a function of two variables x and y.

Let $$r = \frac{\partial^2 f}{\partial x^2},\ s = \frac{\partial^2 f}{\partial x\,\partial y},\ t = \frac{\partial^2 f}{\partial y^2} \text{ at } x = a \text{ and } y = b.$$

As a set of necessary conditions for a maximum or minimum at (a, b) we have

$$\frac{\partial f}{\partial x} = 0 \text{ and } \frac{\partial f}{\partial y} = 0 \text{ at } (a, b)$$

then $$f(a + h, b + k) - f(a, b) = \frac{1}{2!}[rh^2 + 2shk + tk^2] + R \quad \text{...(1)}$$

Where R consists of terms of third and higher order of small quantities h and k.

Now, by taking h and k sufficiently small, the second degree terms in R.H.S. of (1) may be made to govern the sign of R.H.S. and therefore of the L.H.S. also *i.e.*, for sufficiently small values of h and k, the sign of $\frac{1}{2}(rh^2 + 2shk + tk^2) + R$ is same as that of $rh^2 + 2shk + tk^2$.

If the sign is negative, then the function is maximum at (a, b) and if the sign is positive, then the function is minimum at (a, b).

Case (i) *If* $(rt - s^2) > 0$.

Then, neither r nor t can be zero. Hence, we can write

$$rh^2 + 2shk + tk^2 = \frac{1}{2}[r^2h^2 + 2rshk + rtk^2] = \frac{1}{2}[(rh + sk)^2 + (rt - s^2)k^2]$$

since $rt - s^2 > 0$, therefore $(rh + sk)^2 + (rt - s^2)k^2 > 0$ for all values of h and k except when $rh + sk = 0$, $k = 0$ *i.e.*, at $h = 0$, $k = 0$, which is not possible.

Hence, in this case the expression $rh^2 + 2shk + tk^2$ will have the same sign for all values of h and k, and the sign is determined by the sign of r.

Thus, the function $f(x, y)$ will have a maximum or minimum at $x = a$ and $y = b$. If $rt - s^2 > 0$. The function $f(x, y)$ is maximum or minimum according as r is negative or positive.

Case (ii) *If* $(rt - s^2) < 0$.

If $rt - s^2$ is negative, we are not sure about the sign of second degree term of R.H.S. of (1) and hence there is neither a maximum nor a minimum value.

Case (iii) *If* $rt - s^2 = 0$.

If $rt = s^2$, then quadratic expression $rh^2 + 2shk + tk^2$ becomes $\frac{1}{r}(hr + ks)^2$.

So that, the quadratic expression will be of the same sign as that of r or t unless

$$\frac{h}{k} = -\frac{s}{r} = \alpha \text{ (say) } i.e., \ rh + sk = 0.$$

If this condition is safisfied, then the second degree expression in R.H.S. of (1) vanishes and hence, the sign of the R.H.S. of (1) depends upon third degree expression in h and k, which change sign with the change of sign of h and k and hence, the sign of L.H.S. of (1) will also change and hence, there will be neither maximum nor minimum.

Thus, the necessary condition for the existence of maxima and minima now is that the cubic terms must vanish collectively in R.H.S. of (1) when $\frac{h}{k} = -\frac{s}{r} = \alpha$; and then the biquadratic terms of R.H.S. of (1) must collecctively be of the same sign as r and t, when

$$\frac{h}{k} = -\frac{s}{r} = \alpha \quad i.e., \quad hr + ks = 0$$

Hence, the case is doubtful.

Thus, if $rt - s^2 = 0$, the case is doubtful and further, investigation is needed to determine the maxima and minima of $f(x, y)$ at (a, b).

WORKING PROCEDURE

To discuss the maxima and minima at $x = a, y = b$, we must find

$$r = \left(\frac{\partial^2 u}{\partial x^2}\right)_{\substack{x=a\\y=b}}, s = \left(\frac{\partial^2 u}{\partial x \partial y}\right)_{\substack{x=a\\y=b}}, t = \left(\frac{\partial^2 u}{\partial y^2}\right)_{\substack{x=a\\y=b}}$$

Then, calculate $rt - s^2$.

Now following cases arise :

(i) *If $rt - s^2 > 0$, then*

(A) *If r is negative then, $f(x, y)$ is maximum at $x = a, y = b$.*

(B) *If r is positive then, $f(x, y)$ is minimum at $x = a, y = b$.*

(ii) *If $rt - s^2 < 0, f(x, y)$ is neither maximum nor minimum at $x = a, y = b$.*

(iii) *If $rt - s^2 = 0$, the case is doubtful, and further investigation will be required.*

REMARK

- While solving problems, we frequently used the identity, given by Lagrange.

$$\{(a^2+b^2+c^2)(p^2+q^2+r^2)-(ap+bq+cr)^2\} = \{(br-cq)^2+(cp+ar)^2+(aq-bp)^2\}.$$

Solved Examples

Example 1. *Find all maximum or minimum values of the function :*

$$f(x,y) = y^2 + x^2y + x^4.$$

Solution. Since, we have

$$f(x,y) = y^2 + x^2y + x^4.$$

$$\therefore \quad \frac{\partial f}{\partial x} = 2xy + 4x^3 \quad \text{and} \quad \frac{\partial f}{\partial y} = 2y + x^2.$$

For a maximum or minimum of $f(x, y)$, we must have

$$\frac{\partial f}{\partial x} = 0 \text{ and } \frac{\partial f}{\partial y} = 0$$

$$\therefore \quad \frac{\partial f}{\partial x} = 0 \Rightarrow 2xy + 4x^3 = 0$$

$$\Rightarrow 2x(y + 2x^2) = 0 \quad \text{...(1)}$$

$$\frac{\partial f}{\partial y} = 0 \Rightarrow 2y + x^2 = 0$$

Solving (1) and (2), we get $x = 0, y = 0$.

Thus (0, 0) is the only point of maximum or minimum.

$$\text{Now} \quad r = \left(\frac{\partial^2 f}{\partial x^2}\right)_{(0,0)} = [2y + 12x^2]_{(0,0)} = 0$$

$$s = \left(\frac{\partial^2 f}{\partial x \partial y}\right)_{(0,0)} = [2x]_{(0,0)} = 0$$

$$\text{and} \quad t = \left(\frac{\partial^2 f}{\partial y^2}\right)_{(0,0)} = [2]_{(0,0)} = 2$$

$$\therefore rt - s^2 = 0\,(2) - 0^2 = 0.$$

Thus, the case is doubtful and further investigation will be required.

Example 2. *Find the maximum or minimum values of the function $x^3y^2(1-x-y)$.*

(Anna-2009, JNTU-2006, 08, Bhopal-2012)

Solution. Let $\quad u = x^3y^2(1-x-y)$

$$\Rightarrow \quad \frac{\partial u}{\partial x} = 3x^2y^2(1-x-y) - x^3y^2$$

$$\text{and} \quad \frac{\partial u}{\partial y} = 2x^3y(1-x-y) - x^3y^2.$$

For a maximum or minimum of u, we must have $\frac{\partial u}{\partial x} = 0$ and $\frac{\partial u}{\partial y} = 0$

$$\Rightarrow \quad 3x^2y^2(1-x-y) - x^3y^2 = 0 \quad \text{...(1)}$$

$$\text{and} \quad 2x^3y(1-x-y) - x^3y^2 = 0. \quad \text{...(2)}$$

Now, subtracting (2) from (1), we have

$$x^2y(1-x-y)(3y-2x) = 0$$

which gives $\quad y = \frac{2}{3}x.$

Putting the value of y in (1), we get $x = \frac{1}{2}$

So $\left(\frac{1}{2}, \frac{1}{3}\right)$ be the point of maxima or minima.

$$\text{Now} \quad r = \frac{\partial^2 u}{\partial x^2} = 6xy^2 - 12x^2y^2 - 6xy^3$$

$$= -\frac{1}{9}, \text{ at } \left(\frac{1}{2}, \frac{1}{3}\right)$$

$$t = \frac{\partial^2 u}{\partial y^2} = 2x^3 - 2x^4 - 6x^3y$$

$$= -\frac{1}{8}, \text{ at } \left(\frac{1}{2}, \frac{1}{3}\right)$$

$$s = \frac{\partial^2 u}{\partial x \partial y} = 6x^2y - 8x^3y - 9x^2y^2$$

$$= -\frac{1}{12} \text{ at } \left(\frac{1}{2}, \frac{1}{3}\right).$$

Now, $rt - s^2 =$ positive.

Also, r is negative, hence the function u has a maximum at $x = \frac{1}{2}, y = \frac{1}{3}$.

The maximum value is

$$= \left(\frac{1}{2}\right)^3 \left(\frac{1}{3}\right)^2 \left(1 - \frac{1}{2} - \frac{1}{3}\right) = \frac{1}{432}.$$

Example 3. *Discuss the maximum or minimum values of u, where* $u = 2a^2xy - 3ax^2y - ay^3 + x^3y + xy^3$.

Solution. We have

$$u = 2a^2xy - 3ax^2y - ay^3 + x^3y + xy^3$$

which gives

$$\frac{\partial u}{\partial x} = 2a^2y - 6axy + 3x^2y + y^3$$

and $$\frac{\partial u}{\partial y} = 2a^2x - 3ax^2 - 3ay^2 + x^3 + 3xy^2$$

For a maximum and minima of u, we have

$$\frac{\partial u}{\partial x} = 0, \frac{\partial u}{\partial y} = 0$$

which gives,

$$y(2a^2 - 6ax + 3x^2 + y^2) = 0 \quad ...(1)$$

and $2a^2x - 3ax^2 - 3ay^2 + x^3 + 3xy^2 = 0$ ...(2)

Equation (1) and (2) gives the following values of x and y :

$$x = 0,\ y = 0;\ x = a, y = 0;$$

$$x = 2a,\ y = 0;\ x = \frac{3}{2}a,\ y = \pm\frac{1}{2}a;$$

$$x = a, y = a,\ x = \frac{1}{2},\ y = \frac{1}{2}a;$$

$$x = a, y = -a;\ x = \frac{1}{2}\ a,\ y = -\frac{1}{2}a.$$

Then, we get the following pairs of values of x and y which make the function u stationary.

$$(0,0), (a,0), (2a,0), \left(\frac{3}{2}a, \frac{1}{2}a\right), \left(\frac{3}{2}a, -\frac{1}{2}a\right)$$

$$(a,a), \left(\frac{1}{2}a, \frac{1}{2}a\right), (a,-a), \left(\frac{1}{2}a, -\frac{1}{2}a\right).$$

Also $$r = \frac{\partial^2 u}{\partial x^2} = -6ay + 6xy,$$

$$s = \frac{\partial^2 u}{\partial x \partial y} = 2a^2 - 6ax + 3x^2 + 3y^2,$$

and $$t = \frac{\partial^2 u}{\partial y^2} = -6ay + 6xy.$$

For (0, 0).

$$r = 0, s = 2a^2, t = 0$$

$\Rightarrow$ $rt - s^2$, is negative.

Therefore, we have neither maximum nor a minimum of u at (0, 0).

Similarly, we can easily shown that u has neither a maximum nor a minimum at $(a, 0)$, $(2a, 0)$, (a, a), $(a, -a)$.

For $\left(\frac{3a}{2}, \frac{a}{2}\right)$.

$$r = \frac{3}{2}a^2,\ s = \frac{1}{2}a^2,\ t = \frac{3}{2}a^2,$$

$\Rightarrow$ $rt - s^2$ is positive.

Here, since r is positive, therefore u has minimum at $\left(\frac{3a}{2}, \frac{a}{2}\right)$.

Similarly, we can check the maxima and minima at all other points.

Example 4. *Find the maximum and minimum values of* $xy(a - x - y)$.

Solution. Let $u = xy(a - x - y)$

Then $$\frac{\partial u}{\partial x} = ay - 2xy - y^2$$

and $$\frac{\partial u}{\partial y} = ax - x^2 - 2xy.$$

For a maximum or minimum of u, we have

$$\frac{\partial u}{\partial x} = 0 \text{ and } \frac{\partial u}{\partial y} = 0.$$

Thus, we have

$$ay - 2xy - y^2 = 0 \Rightarrow y(a - 2x - y) = 0 \quad ...(1)$$

$$ax - x^2 - 2xy = 0 \Rightarrow x(a - x - 2y) = 0 \quad ...(2)$$

Solving (1) and (2), we get the following pairs of values x and y which makes the function stationary

$$(0,0), (0,a), (a,0), \left(\frac{1}{3}a, \frac{1}{3}a\right).$$

Here $$r = \frac{\partial^2 u}{\partial x^2} = -2y,\ s = \frac{\partial^2 u}{\partial x \partial y} = a - 2x - 2y,$$

and $$t = \frac{\partial^2 u}{\partial y^2} = -2x.$$

For (0, 0). $r = 0, s = a, t = 0$

$\Rightarrow$ $rt - s^2$ is negative.

$\therefore$ We have neither a maximum nor a minimum of u at (0, 0).

For (0, *a*). $r = -2a, s = -a, t = 0$

$\Rightarrow$ $rt - s^2$ is negative.

$\therefore$ We have neither a maximum nor a minimum of u at $(a, 0)$.

Similarly, we have neither a maximum nor a minimum of u at $(a, 0)$.

For $\left(\frac{1}{3}a, \frac{1}{3}a\right)$.

$$r = -\frac{2}{3}a,\ s = -\frac{1}{3}a,\ t = -\frac{2}{3}a$$

$\Rightarrow$ $rt - s^2$ is positive.

Since $\quad rt - s^2 > 0.$

$\therefore \quad u$ has an extreme value at $\left(\frac{1}{3}a, \frac{1}{3}a\right)$

$\Rightarrow$ u has a maximum if r is negative, *i.e.*, if a is positive and u has a minimum if r is positive, *i.e.*, if a is negative.

Example 5. *Find a point within a triangle such that the sum of the squares of its distances from the vertices is a minimum.*

Solution. Let us suppose $[(x_r, y_r) : r = 1, 2, 3]$ be the vertices of the triangle and (x, y) be any point inside the triangle.

Now, let us define a function

$$u = \sum_{r=1}^{3} [(x - x_r)^2 + (y - y_r)^2].$$

Then, we have

$$\frac{\partial u}{\partial x} = \Sigma 2(x - x_r)$$

$$= 2[(x - x_1) + (x - x_2) + (x - x_3)]$$

and $\frac{\partial u}{\partial y} = \Sigma 2(y - y_r)$

$$= 2[(y - y_1) + (y - y_2) + (y - y_3)].$$

For a maximum or minimum of u, we must have

$$\frac{\partial u}{\partial x} = 0 \Rightarrow (x - x_1) + (x - x_2) + (x - x_3) = 0$$

$$\Rightarrow x = \frac{x_1 + x_2 + x_3}{3}$$

and $\frac{\partial u}{\partial y} = 0 \Rightarrow (y - y_1) + (y - y_2) + (y - y_3) = 0$

$$\Rightarrow y = \frac{y_1 + y_2 + y_3}{3}.$$

Thus, we have $\left(\frac{x_1 + x_2 + x_3}{3}, \frac{y_1 + y_2 + y_3}{3}\right)$ is the only point at which u have a maximum or minimum.

Now $r = \frac{\partial^2 u}{\partial x^2} = 6, s = \frac{\partial^2 u}{\partial x \partial y} = 0, t = \frac{\partial^2 u}{\partial y^2} = 6.$

Now, at $\left[\frac{x_1 + x_2 + x_3}{3}, \frac{y_1 + y_2 + y_3}{3}\right]$

$$r = 6, s = 0, t = 6$$

$\Rightarrow \quad rt - s^2 = 36 > 0.$

Also, since $\quad r > 0.$

Therefore u have a minimum value at $\left[\frac{x_1 + x_2 + x_3}{3}, \frac{y_1 + y_2 + y_3}{3}\right].$

Hence, the point $\left(\frac{x_1 + x_2 + x_3}{3}, \frac{y_1 + y_2 + y_3}{3}\right)$ is the required point at which u is minimum.

REMARK

- The point $\left(\frac{x_1 + x_2 + x_3}{3}, \frac{y_1 + y_2 + y_3}{3}\right)$ is the centroid of the given triangle.

Example 6. *Show that the minimum value of* $u = xy + \left(\frac{a^3}{x}\right) + \left(\frac{a^3}{y}\right)$ *is* $3a^2$.

Solution. We have

$$u = xy + \left(\frac{a^3}{x}\right) + \left(\frac{a^3}{y}\right)$$

$$\Rightarrow \quad \frac{\partial u}{\partial x} = y - \frac{a^3}{x^2} \text{ and } \frac{\partial u}{\partial y} = x - \frac{a^3}{y^2}.$$

For a maximum or minimum of u, we have

$$\frac{\partial u}{\partial x} = 0 \text{ and } \frac{\partial u}{\partial y} = 0$$

Now, $\frac{\partial u}{\partial x} = 0 \Rightarrow y - \frac{a^3}{x^2} = 0 \quad \text{... (1)}$

and $\frac{\partial u}{\partial y} = 0 \Rightarrow x - \frac{a^3}{y^2} = 0. \quad \text{... (2)}$

Solving (1) and (2), we get, $x = a, y = a$

Now $r = \frac{\partial^2 u}{\partial x^2} = \frac{2a^3}{x^3}, \; s = \frac{\partial^2 u}{\partial x \partial y} = 1$

and $t = \frac{\partial^2 u}{\partial y^2} = \frac{2a^3}{y^3}.$

At $\quad x = y = a$, we have

$$r = 2, s = 1, t = 2$$

$\Rightarrow \quad rt - s^2 = 3 > 0.$

Thus, at (a, a), $rt - s^2 > 0$ and $r > 0$. Therefore u is minimum at $x = a$, $y = a$.

The minimum value of

$$u = a.a + \left(\frac{a^3}{a}\right) + \left(\frac{a^3}{a}\right) = 3a^2.$$

Example 7. *Determine the points where a function* $x^3 + y^3 - 3axy$ *has maximum or minimum.*

(UPTU -2009, GBTU-2010, 12, MTU -2011)

Solution. Here, we have

$$u = x^3 + y^3 - 3axy$$

$$\Rightarrow \quad \frac{\partial u}{\partial x} = 3x^2 - 3ay \text{ and } \frac{\partial u}{\partial y} = 3y^2 - 3ax.$$

For a maximum or minimum of u, we must have

$$\frac{\partial u}{\partial x} = 0 \text{ and } \frac{\partial u}{\partial y} = 0$$

which gives, $\quad x^2 - ay = 0 \quad \text{... (1)}$

and $\quad y^2 - ax = 0 \quad \text{... (2)}$

Solving (1) and (2), we get

$$x = 0, y = 0; x = a, y = a.$$

Thus (0, 0) and (a, a) are the stationary points of u.

Now

$$r = \frac{\partial^2 u}{\partial x^2} = 6x,\ s = \frac{\partial^2 u}{\partial x\,\partial y} = -3a,\ t = \frac{\partial^2 u}{\partial y^2} = 6y.$$

For $x = 0, y = 0$

$$r = 0, s = -3a \text{ and } t = 0$$

$\therefore \quad rt - s^2 = -9a^2 < 0$, for all values of a.

$\Rightarrow \quad u$ is neither maximum nor minimum at $x = 0, y = 0$.

For $x = a, y = a$

$$r = 6\,a, s = -3a \text{ and } t = 6a$$

$\Rightarrow \quad rt - s^2 = 27a^2 > 0$, for all values of a.

Also $r = 6a$, which is positive if $a > 0$.

Thus (i) u is maximum at $x = a, y = a$ if $a < 0$

and (ii) u is minimum at $x = a, y = a$ if $a > 0$.

Example 8. *Discuss the maxima and minima of the function* $u = \sin x \sin y \sin (x+y)$. (UPTU-2009)

Solution. Here, we have

$$u = \sin x \sin y \sin (x + y)$$

$$\Rightarrow \quad \frac{\partial u}{\partial x} = \sin y[\sin x\cos(x+y) + \cos x\sin(x+y)]$$

and
$$\frac{\partial u}{\partial y} = \sin x[\sin y\cos(x+y) + \cos y\sin(x+y)].$$

For a maxima and minima of u, we must have

$$\frac{\partial u}{\partial x} = 0 \text{ and } \frac{\partial u}{\partial y} = 0.$$

$\Rightarrow \sin y\,[\sin x \cos (x + y) + \cos x \sin (x+y)] = 0$

and $\sin x\,[\sin y \cos (x + y) + \cos y \sin (x+y)] = 0$.

Equation (1) and (2) gives

$$\tan (x + y) = -\tan x \qquad \text{...(1)}$$

$$\Rightarrow \quad \tan x = \tan y$$

and
$$\tan (x + y) = -\tan y \qquad \text{...(2)}$$

$$\Rightarrow \quad x = y$$

From (1) and (2), we have

$$\tan 2x = -\tan x = \tan (\pi - x)$$

$$\Rightarrow \quad 2x = \pi - x$$

$$\Rightarrow \quad 3x = \pi \Rightarrow x = \frac{\pi}{3} = y.$$

Moreover, $\frac{\partial u}{\partial x} = 0$, gives $\sin y = 0 \Rightarrow y = 0$

and $\frac{\partial u}{\partial y} = 0$, gives $\sin x = 0 \Rightarrow x = 0$.

Thus, we get the following pair of values, which makes the function u stationary

$$(0,0), \left(\frac{\pi}{3}, \frac{\pi}{3}\right).$$

Now
$$r = \frac{\partial^2 u}{\partial x^2} = 2\sin y\cos(2x+y),$$

$$s = \frac{\partial^2 u}{\partial x\,\partial y} = \sin 2(x+y),$$

and
$$t = \frac{\partial^2 u}{\partial y^2} = 2\sin x\cos(2y+x).$$

For (0, 0).

$$r = 0, s = 0, t = 0$$

$$\Rightarrow \quad rt - s^2 = 0.$$

$\therefore$ this case is doubtful and need further investigation.

For $\left(\frac{\pi}{3}, \frac{\pi}{3}\right)$.

$$r = 2\sin\frac{1}{3}\pi.\cos\pi = -\sqrt{3},$$

$$s = \sin\left(\frac{4\pi}{3}\right) = -\sin\frac{\pi}{3} = -\frac{\sqrt{3}}{2},$$

and
$$t = 2\sin\frac{1}{3}\pi\cos\pi = -\sqrt{3}.$$

$$\therefore \quad rt - s^2 = \frac{9}{4} = \text{positive}.$$

Also $\quad r = -\sqrt{3}.$

Hence, u has a maximum value at $\left(\frac{\pi}{3}, \frac{\pi}{3}\right)$.

Example 9. *Discuss the maxima and minima of the function* $u = x^2y^2 - 5x^2 - 8xy - 5y^2$.

Solution. Here, we have

$$u = x^2y^2 - 5x^2 - 8xy - 5y^2$$

$$\Rightarrow \quad \frac{\partial u}{\partial x} = 2xy^2 - 10x - 8y$$

and
$$\frac{\partial u}{\partial y} = 2x^2y - 8x - 10y.$$

For a maximum or minimum of u, we must have

$$\frac{\partial u}{\partial x} = 0 \text{ and } \frac{\partial u}{\partial y} = 0.$$

which implies

$$2xy^2 - 10x - 8y = 0, \qquad \text{... (1)}$$

and $\quad 2x^2y - 8x - 10y = 0. \qquad$ ... (2)

From equation (2) we have $y = \dfrac{4x}{x^2 - 5}$

Put this value of y in equation (1), we get

$$x.\frac{16x^2}{(x^2-5)^2} - 5x - \frac{16}{x^2-5} = 0$$

$$\Rightarrow \quad x[-5x^4 + 50x^2 - 45] = 0$$

$$\Rightarrow \quad x[x^4 - 10x^2 + 9] = 0$$

$$\Rightarrow \quad x = 0, \pm 1, \pm 3.$$

Also from (2), $\quad$ for $x = 0$, $\quad y = 0$,

for $x = 1$, $\quad y = -1$,

for $x = -1, \quad y = 1$,
for $x = 3, \quad y = 3$,
and for $x = -3, \quad y = -3$.

Hence, the function u is stationary at the points (0, 0), (1, –1), (–1, 1), (3, 3) and (–3, –3).

Now $r = \dfrac{\partial^2 u}{\partial x^2} = 2y^2 - 10, \quad s = \dfrac{\partial^2 u}{\partial y\,\partial x} = 4xy - 8,$

and $t = \dfrac{\partial^2 u}{\partial y^2} = 2x^2 - 10.$

For (0, 0). $r = -10, s = -8, t = -10$

$\Rightarrow \quad rt - s^2 = 36 = +$ ve.

Since $r = -10 < 0$. Hence, u is maximum at (0, 0).

For (1, –1). $r = -8, s = -12, t = -8$

$\Rightarrow \quad rt - s^2 = -80 < 0.$

Hence, the stationary value of u at (1, –1) is neither maximum nor minimum.

Similarly at (–1, 1), (3, 3) and (–3, 3) the function u is neither maximum nor minimum.

Example 10. *Find the minimum value of $x^2 + y^2 + z^2$ when $ax + by + cz = p$.*

Solution. Here, $u = x^2 + y^2 + z^2$ ...(1)

Also $ax + by + cz = p$

$\Rightarrow \quad z = \dfrac{p - ax - by}{c}.$

Put this value of z in equation (1), we get

$$u = x^2 + y^2 + \frac{(p - ax - by)^2}{c^2}$$

$$\Rightarrow \quad \frac{\partial u}{\partial x} = 2x - \frac{2a}{c^2}(p - ax - by)$$

$$\Rightarrow \quad \frac{\partial u}{\partial y} = 2y - \frac{2b}{c^2}(p - ax - by).$$

For a maxima and minima of u, we must have

$$\frac{\partial u}{\partial x} = 0 \text{ and } \frac{\partial u}{\partial y} = 0$$

$$\Rightarrow \quad x = \frac{ap}{a^2 + b^2 + c^2} \text{ and } y = \frac{bp}{a^2 + b^2 + c^2}.$$

Now, $r = \dfrac{\partial^2 u}{\partial x^2} = 2 + \dfrac{2a^2}{c^2}, \quad s = \dfrac{\partial^2 u}{\partial x\,\partial y} = \dfrac{2ab}{c^2}$

and $t = \dfrac{\partial^2 u}{\partial y^2} = 2 + \dfrac{2b^2}{c^2}$

$$\Rightarrow \quad rt - s^2 = 4\left(1 + \frac{a^2}{c^2}\right)\left(1 + \frac{b^2}{c^2}\right) - \frac{4a^2b^2}{c^4}$$

$$= 4\left(1 + \frac{a^2}{c^2} + \frac{b^2}{c^2}\right) = \text{positive}.$$

Since r is positive and $rt - s^2 > 0$, therefore u is minimum for the above values of x and y.

The minimum value is $\dfrac{p^2}{a^2 + b^2 + c^2}$.

Example 11. *Find the stationary point of $x^4 + y^4 - 2x^2 + 4xy - 2y^2$ and determine their nature.* (JNTU-2009)

Solution. Here, we have

$$u = x^4 + y^4 - 2x^2 + 4xy - 2y^2$$

$$\Rightarrow \quad \frac{\partial u}{\partial x} = 4x^3 - 4x + 4y$$

and $\dfrac{\partial u}{\partial y} = 4y^3 + 4x - 4y.$

For, a maxima and minima of u, we must have

$$\frac{\partial u}{\partial x} = 0 \Rightarrow 4x^3 - 4x + 4y = 0 \quad ...(1)$$

and $\dfrac{\partial u}{\partial y} = 0 \Rightarrow 4y^3 + 4x - 4y = 0.$ ...(2)

Solving (1)and (2), we get

$$4x^3 + 4y^3 = 0 \Rightarrow x^3 + y^3 = 0$$

$\Rightarrow \quad (x + y)(x^2 - xy + y^2) = 0$

$\Rightarrow$ either $x + y = 0$ or $x^2 - xy + y^2 = 0$ ...(3)

$$x + y = 0 \Rightarrow y = -x.$$

Put $y = -x$ in equation (1), we get

$4x^3 - 8x = 0$

$\Rightarrow \quad x(x^2 - 2) = 0$

$\Rightarrow \quad x = 0, \sqrt{2}, -\sqrt{2}.$

Then $y = 0, -\sqrt{2}, \sqrt{2}$

Also from (3) $x = 0, y = 0$(only real solution)

Hence, the stationary points u are given by $(0,0), (\sqrt{2}, -\sqrt{2}), (-\sqrt{2}, \sqrt{2})$.

Now, we have $r = \dfrac{\partial^2 u}{\partial x^2} = 12x^2 - 4,$

$s = \dfrac{\partial^2 u}{\partial x \partial y} = 4,$

and $t = \dfrac{\partial^2 u}{\partial y^2} = 12y^2 - 4.$

For (0, 0). $r = -4, s = 4, t = -4$

$\Rightarrow \quad rt - s^2 = 0.$

$\Rightarrow$ At the point (0, 0) the case is doubtful, and there is a need of further investigation.

For $(\sqrt{2}, -\sqrt{2})$.

$$r = 20, s = 4, t = 20$$

$\Rightarrow \quad rt - s^2 = 400 - 16 = 384 > 0.$

Also $r > 0.$

$\Rightarrow$ u has a minimum value at $(\sqrt{2}, -\sqrt{2})$.

Similarly u has a minimum value at $(-\sqrt{2}, \sqrt{2})$.

Example 12. *Prove that the maxima or minima of the function*

$$u=\left[\frac{ax^2+by^2+2hxy+2gx+2fy+c}{a'x^2+b'y^2+2h'xy+2g'x+2f'y+c'}\right]$$

are given by the roots of the equation

$$\begin{vmatrix} a-a'u & h-h'u & g-g'u \\ h-h'u & b-b'u & f-f'u \\ g-g'u & f-f'u & c-c'u \end{vmatrix}=0.$$

Solution. Here, we have

$$u=\left[\frac{ax^2+by^2+2hxy+2gx+2fy+c}{a'x^2+b'y^2+2h'xy+2g'x+2f'y+c'}\right]$$

$$\Rightarrow u[a'x^2+b'y^2+2h'xy+2g'x+2f'y+c']$$
$$=[ax^2+by^2+2hxy+2gx+2fy+c]. \quad ...(1)$$

Differentiating (1) partially w.r.t. x and y, we have

$$\frac{\partial u}{\partial x}[a'x^2+b'y^2+2h'xy+2g'x+2f'y+c']+u[2a'x+2h'y+2g']$$
$$=2ax+2hy+2g \quad ...(2)$$

and

$$\frac{\partial u}{\partial y}[a'x^2+b'y^2+2h'xy+2g'x+2f'y+c']+u[2b'y+2h'x+2f']$$
$$=2by+2hx+2f. \quad ...(3)$$

For the maxima and minima of u, we must have

$$\frac{\partial u}{\partial x}=0 \Rightarrow u[a'x+h'y+g']=ax+hy+g \quad ...(4)$$

and

$$\frac{\partial u}{\partial y}=0 \Rightarrow u[h'x+b'y+f']=hx+by+f. \quad ...(5)$$

Now, multiplying (4) by x, (5) by y and adding, we have

$$u[a'x^2+b'y^2+2h'xy+g'x+f'y]$$
$$=ax^2+by^2+2hxy+gx+fy \quad ...(6)$$

Subtracting (6) from (1), we get

$$u(g'x+f'y+c')=gx+fy+c \quad ...(7)$$

Now, from (4), (5) and (7), we have

$$(a-a'u)x+(h-h'u)y+(g-g'u)=0 \quad ...(8)$$
$$(h-h'u)x+(b-b'u)y+(f-f'u)=0 \quad ...(9)$$
$$(g-g'u)x+(f-f'u)y+(c-c'u)=0. \quad ...(10)$$

By eliminating x and y from (8), (9) and (10), we get

$$\begin{vmatrix} a-a'u & h-h'u & g-g'u \\ h-h'u & b-b'u & f-f'u \\ g-g'u & f-f'u & c-c'u \end{vmatrix}=0.$$

which is a cubic equation in u. The roots of this equation gives the required maxima and minima.

REMARK

- If there is a function of two variables x and y connected by a relation $g(x, y) = 0$. Then we find the maxima and minima of the function in the following manner.

Let $$u=f(x,y) \quad ...(1)$$
and $$g(x,y)=0. \quad ...(2)$$

Generally, it is possible to eliminate one of the variables x and y from (1) and (2), then u is expressed in terms of a single variable and we can proceed in the usual way. But if it is not convenient to take the value of one variable in terms of the other from (2), then we should proceed as follows :

From (2), we get $$\frac{dg}{dx}=-\frac{\partial g/\partial x}{\partial g/\partial y} \quad ...(3)$$

Now, differentiating (1) with respect to x, we get $$\frac{du}{dx}=\frac{\partial f}{\partial x}+\frac{\partial f}{\partial y}\frac{dg}{dx} \quad ...(4)$$

Now, from (3) and (4), we get $$\frac{du}{dx}=0$$

Solve (5) with the help of (2), and get the required values of x and y for which u will have maximum or minimum values.

Example 13. *Test the function* $u=x^2y-y^2x-x+y$ *for maximum and minimum.*

Solution. For maximum of u, we must have

$$\frac{\partial u}{\partial x}=2xy-y^2-1=0$$

and $$\frac{\partial u}{\partial y}=x^2-2xy+1=0.$$

Solving these equations, we get

$$x=1, y=1, x=-1, y=-1.$$

Now $$r=\frac{\partial^2 u}{\partial x^2}=2y,\ s=\frac{\partial^2 u}{\partial x\,\partial y}=2x-2y,$$

$$t=\frac{\partial^2 u}{\partial y^2}=-2x.$$

For $x = 1, y = 1$, we have $r = 2, s = 0, t = -2$ so that $rt-s^2=-4$, which is negative.

Hence, u has neither a maximum nor a minimum at (1, 1). Thus (1, 1) is a saddle point.

For $x=-1, y=-1$. We have $r=-2$, $s=0$, $t=2$. so that $rt-s^2=-4$, which is negative.

Hence, u has neither a maximum nor a minimum at (–1, –1).

Thus (–1, –1) is a saddle point.

Example 14. *Show that distance l of any point (x, y, z) on the plane* $2x+3y-z=12$ *from the origin is given by*

$$l=\sqrt{[x^2+y^2+(2x+3y-12)^2]}.$$

Hence, find the point on the plane that is nearest to the origin.

Solution. If l is the distance from (0, 0, 0) of any point (x, y, z) then $l=\sqrt{(x^2+y^2+z^2)}$. If the point (x, y, z) lies on the plane $2x+3y-z=12$, then

$$l=\sqrt{[x^2+y^2+(2x+3y-12)^2]}$$

[$\because z = 2x + 3y - 12$, from the equation of the plane]

$$\therefore\ l^2=x^2+y^2+(2x+3y-12)^2$$

$$=5x^2+10y^2+12xy-48x+72y+144=u(\text{say}).$$

Now l is maximum or minimum according as l^2 *i.e.*, u is maximum or minimum.

For a maximum or minimum of u, we get

$$\frac{\partial u}{\partial x}=10x+12y-48=0$$

and $$\frac{\partial u}{\partial y}=20y+12x-72=0$$

Solving these equations, we get

$$x=\frac{12}{7}\text{ and }y=\frac{18}{7}.$$

Also

$$r=\frac{\partial^2u}{\partial x^2}=10,\ s=\frac{\partial^2u}{\partial x\,\partial y}=12\text{ and }t=\frac{\partial^2u}{\partial y^2}=20.$$

Therefore $rt-s^2 = 10\times 20-(12)^2 = +\text{ ve}$, since $rt-s^2>0$

and $r>0$, then u is minimum and hence l is minimum.

When $x=\frac{12}{7}$ and $y=\frac{18}{7}$. Putting these values of x and y in the equation of the plane, we get

$$z=2\cdot\left(\frac{12}{7}\right)+3\cdot\left(\frac{18}{7}\right)-12=-\frac{6}{7}.$$

Hence, the required point is $\left(\frac{12}{7},\frac{18}{7},-\frac{6}{7}\right)$.

Example 15. *Find the points on* $z^2 = xy+1$ *nearest to the origin.*

Solution. Let l be the distance from the origin (0, 0, 0) of any point (x, y, z) on the surface

$$z^2=xy+1 \quad ...(1)$$

Then $l=\sqrt{x^2+y^2+z^2}=\sqrt{(x^2+y^2+xy+1)}$ [Using equation (1)]

Since l is always greater than zero, therefore l is maximum or minimum according as l^2, *i.e.*, u is maximum or minimum, where $u=l^2$.

For a maximum or minimum of u, we must have

$$\frac{\partial u}{\partial x}=2x+y=0 \quad ...(2)$$

and $$\frac{\partial u}{\partial y}=2y+x=0. \quad ...(3)$$

Solving the equation (2) and (3), we get

$$x=0,y=0$$

Also $$r=\frac{\partial^2u}{\partial x^2}=2,\ s=\frac{\partial^2u}{\partial x\,\partial y}=1,\ t=\frac{\partial^2u}{\partial y^2}=2.$$

$$\therefore\quad rt-s^2=2.\,2-1=3>0.$$

Since at $x=0, y=0$, then $rt-s^2>0$ and $r>0$.

Therefore u is minimum at $x=0, y=0$. Hence l is minimum, when $x=0, y=0$.

Putting $x=0, y=0$ in the equation (1), we get $z^2=1$ *i.e.*, $z=\pm 1$.

Hence, the required points are (0, 0, 1) and (0, 0, –1).

EXERCISE 12.1

1. Find the points (x, y) where the function $f(x, y) = xy(1-x-y)$ is maximum or minimum. Also find the maximum value of $f(x, y)$.
2. Discuss the maxima and minima of the function $f(x,y)=x^2+y^2+\frac{2}{x}+\frac{2}{y}$.
3. Find the values of x and y for which the expression
$$(a_1x+b_1y+c_1)^2+(a_2x+b_2y+c_2)^2+\ldots+(a_nx+b_ny+c_n)^2$$
is minimum.
4. Discuss the maxima and minima of the function $f(x, y) = x^4+2x^2y-x^2+3y^2$.
5. Examine for maximum and minimum values of the function $f(x, y) = x^2-3xy+y^2+2x$.
6. Examine the function $f(x, y) = x^2y-y^2x-x+y$ for maxima and minima.
7. Discuss the maxima and minima of the function
$$f(x,y)=2\sin\frac{1}{2}(x+y)\cos\frac{1}{2}(x-y)+\cos(x+y).$$
8. Find points on $z^2 = xy+1$ nearest to the origin.
9. Show that the distance l of any point (x, y, z) on the plane $2x + 3y - z = 12$ from the origin is given by
$$l=\sqrt{(x^2+y^2)+(2x+3y-12)^2}.$$
10. Find the maximum and minimum values of $u = 6xy + (47-x-y)(4x+3y)$.
11. Examine for extreme values
(i) $x^2+y^2+6x+12$ (GBTU 2012) (ii) $x^3+y^3-63(x+y)+12xy$ (UKTU 2011)

Hint to Selected Problems

1. $\frac{\partial f}{\partial x} = 0 \Rightarrow y - 2xy - y^2 = 0$

$\frac{\partial f}{\partial y} = 0 \Rightarrow x - 2xy - x^2 = 0.$

On solving above two equations, we get (0, 0), (1, 0), (0, 1) and (1/3, 1/3) are the extreme points.

At (0, 0), $rt - s^2$ is negative $\Rightarrow f(x, y)$ is neither maximum nor minimum.

At (1, 0), $rt - s^2$ is negative $\Rightarrow f(x, y)$ is neither maximum nor minimum.

At (0, 1), $rt - s^2$ is negative $\Rightarrow f(x, y)$ is neither maximum nor minimum.

At (1/3, 1/3), $rt - s^2$ is positive and $r\left(=-\frac{2}{3}\right)$ is negative.

Hence, at $\left(\frac{1}{3}, \frac{1}{3}\right)$, $f(x, y)$ is maximum.

7. $\frac{\partial f}{\partial x} = \cos x - \sin(x+y)$, $\frac{\partial f}{\partial y} = \cos y - \sin(x+y)$

$\frac{\partial f}{\partial x} = 0$, $\frac{\partial f}{\partial y} = 0$, we get $\cos x = \sin(x+y)$, and $\cos y = \sin(x+y)$.

The extreme points are given by

$$\left(-\frac{\pi}{2}, \frac{\pi}{2}\right), \left(\frac{3\pi}{2}, \frac{\pi}{2}\right) \text{ and } \left(\frac{\pi}{2}, \frac{\pi}{2}\right).$$

ANSWERS

1. $f(x, y)$ is maximum at the point $\left(\frac{1}{3}, \frac{1}{3}\right)$; maximum value $= \frac{1}{27}$.

2. $f(x, y)$ is minimum at (1, 1).

3. $f(x, y)$ is minimum for the value of x and y which are obtained by $\Sigma(a_1^2)x + (a_1b_1)y + a_1c_1 = 0$ and $\Sigma(a_1b_1)x + (b_1^2)y + b_1c_1 = 0$.

4. $f(x, y)$ is minimum for $\left(\frac{\sqrt{3}}{2}, \frac{-1}{4}\right)$ and $\left(-\frac{\sqrt{3}}{2}, -\frac{1}{4}\right)$.

5. Stationary point is $x = \frac{4}{5}$, $y = \frac{6}{5}$. The function $f(x, y)$ is neither maximum nor minimum at $\left(\frac{4}{5}, \frac{6}{5}\right)$.

6. At (1, 1) and (–1, –1) function is neither maximum nor minimum.

7. $x = y = 2n\pi \pm \pi/2$; neither maximum nor minimum ; $x = y = n\pi + (-1)^n \pi/6$; f is maximum.

8. (0, 0, 1) and (0, 0, –1). **10.** Maximum value is 3384.

11. (i) At x = –3, y = 0, minimum (ii) max at (–7, –7) min. at (3, 3) neither max nor min. at (5, –1) and (–1, 5).

12.6 MAXIMA AND MINIMA OF THE FUNCTION OF THREE INDEPENDENT VARIABLES

(1) *To find the condition, which governs the sign of the quadratic equation of three independent variables.*

Let I be the expression of three independent variables x,y and z given by

$$I = ax^2 + by^2 + cz^2 + 2fyz + 2gzx + 2hxy$$

I can be written as

$$I = \frac{1}{a}\left[a^2x^2 + aby^2 + acz^2 + 2afyz + 2agzx + 2ahxy\right] (a \neq 0)$$

$$= \frac{1}{a}\left[a^2x^2 + 2ax(gz + hy) + aby^2 + acz^2 + 2afyz\right]$$

$$= \frac{1}{a}\left[(ax + hy + gz)^2 + aby^2 + acz^2 + 2afyz - (gz + hy)^2\right]$$

$$= \frac{1}{a}\left[(ax + hy + gz)^2 + \left(ab - h^2\right)y^2 + 2yz(af - gh) + \left(ac - g^2\right)z^2\right]$$

Here, we observe that I be of the same sign as provided the expression within the square brackets is positive which will of course be so if $ab-h^2$ and $\{(ah-h^2)(ac-g^2)-(af-gh)^2\}$ are positive *i.e.*, if

$$ab-h^2 \text{ and } a[abc+2fgh-af^2-bg^2-ch^2] \text{ are both positive.}$$

Hence, I will be positive if

$$a, \begin{vmatrix} a & h \\ h & b \end{vmatrix}, \begin{vmatrix} a & h & g \\ h & b & f \\ g & f & c \end{vmatrix}$$

be all positive and will be negative if these three expression are alternately negative and positive.

12.7 MAXIMA AND MINIMA FOR A FUNCTION OF THREE INDEPENDENT VARIABLES: THE LAGRANGE'S CONDITION

Let $f(x,y,z)$ be a given function of three independent variables x, y and z.

Let A, B, C, F, G, H stand for $\frac{\partial^2 f}{\partial x^2}, \frac{\partial^2 f}{\partial y^2}, \frac{\partial^2 f}{\partial z^2}, \frac{\partial^2 f}{\partial y \partial z}, \frac{\partial^2 f}{\partial z \partial x}, \frac{\partial^2 f}{\partial x \partial y}$ respectively.

Let a set of the values of x, y, z obtained by solving the equations

$$\frac{\partial f}{\partial x} = \frac{\partial f}{\partial y} = \frac{\partial f}{\partial z} = 0 \text{ be } a, b, c.$$

By Taylor's theorem, we have

$$f(a+h,b+k,c+l),-f(a,b,c) = \frac{1}{2!}\left[Ah^2 + Bk^2 + Cl + 2Fkl + 2Glh + 2Hhk\right] + R \qquad ...(1)$$

where, remainder term R consist of third and higher order of same quantity (*i.e.*, h, k, l).

Now, by taking h, k, l sufficiently small the second term of R.H.S. of (1) can be made to govern the sign of R.H.S. and therefore of L.H.S. also.

If for all such values of h, k and l, these terms be of permanent sign, then we shall have a maximum or minimum of $f(x,y,z)$ according as that sign is negative or positive.

Hence, the function will be minimum if the expression

$$A, \begin{vmatrix} A & H \\ H & B \end{vmatrix}, \begin{vmatrix} A & H & G \\ H & B & F \\ G & F & C \end{vmatrix} \text{ be all positive.}$$

The function will have a maximum value, if the above three quantities are alternately negative and positive. If these conditions are not satisfied, we have neither a maximum nor a minimum.

WORKING PROCEDURE

Let $f(x, y, z)$ be a function of three independent variables x, y and z. Find the values of triads (a,b,c) of the value x, y and z by putting $\frac{\partial f}{\partial x} = 0, \frac{\partial f}{\partial y} = 0, \frac{\partial f}{\partial z} = 0$. The values of triads (a,b,c) will give the stationary values of $f(x, y, z)$.

Now, to discuss maximum and minimum values, at (a, b, c) we find the following six partial derivatives of second order

$$A = \frac{\partial^2 f}{\partial x^2}, B = \frac{\partial^2 f}{\partial y^2}, C = \frac{\partial^2 f}{\partial z^2}, F = \frac{\partial^2 f}{\partial y \partial z}, G = \frac{\partial^2 f}{\partial z \partial x}, \text{and } H = \frac{\partial^2 f}{\partial x \partial y}$$

Now, we have the following cases :

Case (i) *The function $f(x,y,z)$ will be minimum at (a,b,c) if the expressions*

$$A, \begin{vmatrix} A & H \\ H & B \end{vmatrix}, \begin{vmatrix} A & H & G \\ H & B & F \\ G & F & C \end{vmatrix} \text{ be all positive at } (a, b, c).$$

Case (ii) *The function $f(x, y, z)$ will be maximum at (a, b, c) if the expressions*

$$A, \begin{vmatrix} A & H \\ H & B \end{vmatrix}, \begin{vmatrix} A & H & G \\ H & B & F \\ G & F & C \end{vmatrix}$$

be alternately negative and positive.

Case (iii) *If the expression, using in case (i) and (ii) neither be all positive nor having alternately negative and positive sign at (a,b,c). Then $f(x, y, z)$ is neither maximum nor minimum at (a,b,c).*

REMARK

- To find the maximum and minimum of the function at stationary point, it is sufficient to find the value of a second order partial derivative of function with respect to any of the independent variables. Then, the value of the function is maximum or minimum according as the value of this second order partial derivative at the stationary point under consideration is negative or positive.

Solved Examples

Example 1. *Find the maximum value of u, where*

$$u = \frac{xyz}{(a+x)(x+y)(y+z)(z+b)}.$$

Solution. We have

$$u = \frac{xyz}{(a+x)(x+y)(y+z)(z+b)}$$

Taking, log of both the sides, we have

$$\log u = \log x + \log y + \log z - \log(a+x) - \log(x+y) - \log(y+z) - \log(z+b).$$

Differentiating w.r.t. x, we have

$$\frac{1}{u}\frac{\partial u}{\partial x}=\frac{1}{x}-\frac{1}{a+x}-\frac{1}{x+y}=\frac{ay-x^2}{x(a+x)(x+y)}$$

$$\Rightarrow \quad \frac{\partial u}{\partial x}=\frac{\left(ay-x^2\right)u}{x(a+x)(x+y)}$$

Similarly

$$\frac{\partial u}{\partial y}=\frac{\left(xz-y^2\right)u}{y(x+y)(y+z)} \text{ and } \frac{\partial u}{\partial z}=\frac{\left(by-z^2\right)u}{z(y+z)(z+b)}$$

For, a maxima and minima of u, we must have

$$\frac{\partial u}{\partial x}=0 \Rightarrow ay-x^2=0\,;\ \frac{\partial u}{\partial y}=0$$

$$\Rightarrow \quad xz-y^2=0$$

$$\text{and} \quad \frac{\partial u}{\partial z}=0 \Rightarrow by-z^2=0$$

Here, we observe that $x^2=ay$, $y^2=xz$, $z^2=by$ which implies that a, x, y, z and b are in G.P. Let r be the common ratio of this G.P.

Then $\quad ar^4=b \quad$ or $\quad r=\left(\frac{b}{a}\right)^{1/4}$

Also $\quad x=ar, y=ar^2, z=ar^3.$

Hence, we have

$$u=\frac{ar.ar^2.ar^3}{a(1+r)ar(1+r)ar^2(1+r)ar^3(1+r)}$$

$$=\frac{1}{a(1+r)^4}=\frac{1}{a\left[1+\left(\frac{b}{a}\right)^{1/4}\right]^4}=\frac{1}{\left(a^{1/4}+b^{1/4}\right)^4}$$

which gives a stationary value of u. Now, to decide whether this value of u is a maximum or a minimum, we proceed to find the second order partial derivative of u.

$$\text{Here} \quad \frac{\partial^2 u}{\partial x^2}=\frac{-2ux}{x(a+x)(x+y)}$$

$$+\left(ay-x^2\right)\frac{\partial}{\partial x}\left[\frac{u}{x(a+x)(x+y)}\right]$$

When $x=ar$, $y=ar^2$, $z=ar^3$, we have

$$A=\frac{\partial^2 u}{\partial x^2}=-\frac{2u}{a^2r(1+r)^2}<0$$

Hence, the above stationary value of u is maximum.

Example 2. *Find the maxima and minima value of the function*

$$u = \sin x \sin y \sin z$$

where x,y and z are the vertex angles of a triangle.

Solution. Here, we have

$$u=\sin x \sin y \sin z\,;\ \text{where } x+y+z=\pi \quad \ldots(1)$$

$$\therefore \quad u=\sin x \sin y \sin\,[\pi-(x+y)]$$

$$=\sin x \sin y \sin(x+y)$$

$$\therefore \quad \frac{\partial u}{\partial x}=\cos x \sin y \sin(x+y)$$

$$+\sin x \sin y \cos(x+y)$$

$$=\sin y \sin(2x+y). \quad \ldots(2)$$

$$\text{Similarly} \quad \frac{\partial u}{\partial y}=\sin x \sin(2y+x) \quad \ldots(3)$$

For a maxima and minima, we must have

$$\frac{\partial u}{\partial x}=0,\ \frac{\partial u}{\partial y}=0$$

$$\text{So,} \quad \frac{\partial u}{\partial x}=0 \Rightarrow \sin y \sin(2x+y)=0$$

$$\Rightarrow \sin y=0 \text{ or } \sin(2x+y)=0$$

$$\Rightarrow \quad y=0 \quad \text{or } \sin(x+x+y)=0$$

$$\Rightarrow \quad y=0$$

$$\text{or } \sin x \cos(x+y)+\cos x \sin\,(x+y)=0$$

$$\Rightarrow \quad \tan\,(x+y)=-\tan x$$

$$\Rightarrow \quad \tan(x+y)=\tan(-x)=\tan(\pi-x) \quad \ldots(4)$$

$$\Rightarrow \quad x+y=\pi-x$$

$$\Rightarrow \quad 2x+y=\pi \quad \ldots(5)$$

Similarly, from (3)

$$x=0$$

$$\text{or} \quad \tan(x+y)=-\tan y \quad \ldots(6)$$

Now, by (4) and (6), we have

$$\tan x=\tan y \Rightarrow x=y.$$

Hence, by (5), we have

$$3y=\pi \quad \Rightarrow y=\frac{\pi}{3} \text{ and } x=\frac{\pi}{3}$$

Therefore, the stationary points are $\left(\frac{\pi}{3},\frac{\pi}{3}\right)$ and (0, 0).

For (0,0): $u=0$.

For $\left(\frac{\pi}{3},\frac{\pi}{3}\right)$

$$r=\frac{\partial^2 u}{\partial x^2}=2\sin y\cos(2x+y)$$

$$=2\sin\frac{\pi}{3}\cos\left(\frac{2\pi}{3}+\frac{\pi}{3}\right)=-\sqrt{3}<0$$

$$\text{and } s=\frac{\partial^2 u}{\partial x\partial y}=\sin(2x+2y)$$

$$=\sin\left(\frac{2\pi}{3}+\frac{2\pi}{3}\right)$$

$$=\sin\left(\frac{4\pi}{3}\right)=-\frac{\sqrt{3}}{2}<0$$

$$t=\frac{\partial^2 u}{\partial y^2}=2\sin x\cos(x+2y)$$

$$=2\sin\frac{\pi}{3}\cos\pi=-\sqrt{3}<0$$

Now $rt - s^2 = \left(-\sqrt{3}\right)\left(-\sqrt{3}\right) - \left(\frac{\sqrt{3}}{2}\right)^2 = \frac{9}{4} > 0$

Thus $rt - s^2 > 0$ and $r < 0$.

Hence, the function u will be maximum at $\left(\frac{\pi}{3}, \frac{\pi}{3}\right)$.

Example 3. *Show that the points such that the sum of the squares of its distances from n given points shall be minimum, is the centre of the mean position of the given points.*

Solution . Let n given points be (a_1, b_1, c_1), (a_2, b_2, c_2),...., (a_n, b_n, c_n) and let (x, y, z) be the coordinates of the required point.

If u denotes the sum of the squares of the distances of (x, y, z) from the n given points, then

$$u = \Sigma[(x-a_1)^2+(y-b_1)^2+(z-c_1)^2]$$
$$= \Sigma(x-a_1)^2+\Sigma(y-b_1)^2+\Sigma(z-c_1)^2$$

$$\Rightarrow \quad \frac{\partial u}{\partial x} = 2\Sigma\left(x - a_1\right) = 2nx - 2\Sigma a_1$$
$$\frac{\partial u}{\partial y} = 2\Sigma\left(y - b_1\right) = 2ny - 2\Sigma b_1 \qquad ...(1)$$
$$\text{and} \quad \frac{\partial u}{\partial z} = 2\Sigma\left(z - c_1\right) = 2nz - 2\Sigma c_1$$

For the maxima and minima of u, we must have

$$\frac{\partial u}{\partial x} = 0, \frac{\partial u}{\partial y} = 0 \text{ and } \frac{\partial u}{\partial z} = 0 \qquad ...(2)$$

Now from (1) and (2), we have

$$x = \frac{\Sigma a_1}{n}, y = \frac{\Sigma b_1}{n}, z = \frac{\Sigma c_1}{n}$$

Now

$$A = \frac{\partial^2 u}{\partial x^2} = 2n, B = \frac{\partial^2 f}{\partial y^2} = 2n, C = \frac{\partial^2 f}{\partial z^2} = 2n,$$

$$F = \frac{\partial^2 f}{\partial y \partial z} = 0, G = \frac{\partial^2 f}{\partial z \partial x} = 0, H = \frac{\partial^2 f}{\partial x \partial y} = 0.$$

Here, we have

$$A = 2n, \begin{vmatrix} A & H \\ H & B \end{vmatrix} = \begin{vmatrix} 2n & 0 \\ 0 & 2n \end{vmatrix} = 4n^2$$

$$\text{and} \begin{vmatrix} A & H & G \\ H & B & F \\ G & F & C \end{vmatrix} = \begin{vmatrix} 2n & 0 & 0 \\ 0 & 2n & 0 \\ 0 & 0 & 2n \end{vmatrix} = 8n^3$$

Since, these expressions are all positive, therefore u is minimum when

$$x = \frac{\Sigma a_1}{n}, y = \frac{\Sigma b_1}{n}, z = \frac{\Sigma c_1}{n}.$$

Hence, the function u is minimum when the point (x, y, z) is the centre of the mean position of n given points.

Example 4. *Show that the function $u=(x+y+z)^3-3(x+y+z)-24xyz+a^3$ has minimum at (1,1,1) and maximum at (–1,–1,–1).*

Solution . Here we have

$$u=(x+y+z)^3-3(x+y+z)-24xyz+a^3$$

$$\Rightarrow \quad \frac{\partial u}{\partial x} = 3(x+y+z)^2-3-24yz \qquad ...(1)$$
$$\frac{\partial u}{\partial y} = 3(x+y+z)^2-3-24xz \qquad ...(2)$$
$$\text{and} \quad \frac{\partial u}{\partial z} = 3(x+y+z)^2-3-24xy \qquad ...(3)$$

For the maxima and minima of u, we must have

$$\frac{\partial u}{\partial x} = 0, \frac{\partial u}{\partial y} = 0 \text{ and } \frac{\partial u}{\partial z} = 0$$

The equations (1), (2) and (3) are satisfied when $x=y=z$.

Putting $y=x$ and $z=x$ in (1), we get

$$27.x^2 - 3 - 24x^2 = 0$$
$$\Rightarrow \quad x = \pm 1$$

$\Rightarrow$ $x = y = z = 1$ and $x = y = z = -1$ are the solutions of (1), (2) and (3).

Hence, the stationary points are (1,1,1) and (–1,–1,–1).

Now,

$$A = \frac{\partial^2 u}{\partial x^2} = 6(x+y+z), B = \frac{\partial^2 u}{\partial y^2} = 6(x+y+z),$$

$$C = \frac{\partial^2 u}{\partial z^2} = 6(x+y+z),$$

$$F = \frac{\partial^2 u}{\partial y \partial z} = 6(x+y+z) - 24x,$$

$$G = \frac{\partial^2 u}{\partial z \partial x} = 6(x+y+z) - 24y,$$

$$H = \frac{\partial^2 u}{\partial x \partial y} = 6(x+y+z) - 24z.$$

For (1,1,1). $A=18$, $B=18$, $C=18$, $F=-6$, $G=-6$, $H=-6$.

$\therefore$ At the point (1,1,1), we have $A=18. >0$

$$\begin{vmatrix} A & H \\ H & B \end{vmatrix} = \begin{vmatrix} 18 & -6 \\ -6 & 18 \end{vmatrix} = 288 > 0$$

$$\text{and} \quad \begin{vmatrix} A & H & G \\ H & B & F \\ G & F & C \end{vmatrix} = \begin{vmatrix} 18 & -6 & -6 \\ -6 & 18 & -6 \\ -6 & -6 & 18 \end{vmatrix} = 3426 > 0$$

Since, all these three expressions are positive, therefore u is minimum at the point (1,1,1).

For (–1,–1,–1).

$A=-18$, $B=-18$, $C = -18$, $F=6$, $G=6$, $H=6$.

$\therefore$ At the point (–1,–1,–1), we have

$$A = -18 < 0$$

$$\begin{vmatrix} A & H \\ H & B \end{vmatrix} = \begin{vmatrix} -18 & 6 \\ 6 & -18 \end{vmatrix} = 288 > 0$$

and $\begin{vmatrix} A & H & G \\ H & B & F \\ G & F & C \end{vmatrix} = \begin{vmatrix} -18 & 6 & 6 \\ 6 & -18 & 6 \\ 6 & 6 & -18 \end{vmatrix} = -3426 < 0$

Here, the above three expressions are alternately negative and positive. Hence, u is maximum at the point (–1,–1,–1).

EXERCISE 12.2

1. Prove that the function $u = x^2 + y^2 + x - 2z - xy$ is minimum at $\left(-\frac{2}{3}, -\frac{1}{3}, 1\right)$.

2. Find the maximum and minimum values of $u = y^2 + 2z^2 - 5x^4 + 4x^5$.

3. Find the maximum or minimum values of the function u, where $u = axy^2z^3 - x^2y^2z^3 - xy^3z^3 - xy^2z^4$

4. Find the maximum value of

$$(ax+by+cz)\, e^{-(\alpha^2 x^2+\beta^2 y^2+\gamma^2 z^2)}.$$

5. A rectangle box is placed on x-y plane. The one end of the box is at the origin. If the vertex opposite to the origin be on the plane $6x+4y+3z=24$, then find the maximum value of this box.

6. In a plane triangle xyz, find the maximum value of $\sin x \sin y \sin z$.

7. A rectangular box, open at the top is to have a given capacity. Show that the domain of the box requiring least material for its construction $x = y = (2v)^{1/3}$, where $v = xyz$.

(UPTU 2006, GBTU 2010)

ANSWERS

2. Minimum at (1,0,0), neither maximum nor minimum at (0,0,0). **3.** Maximum at $\left(\frac{a}{7}, \frac{2a}{7}, \frac{3a}{7}\right)$, max. value$=\frac{108a^7}{7^7}$

4. Maximum at $\left(\frac{a}{2\alpha^2 k}, \frac{b}{2\beta^2 k}, \frac{c}{2\gamma^2 k}\right)$ where $k=\sqrt{\left\{\frac{1}{2}\left(\frac{a^2}{\alpha^2}+\frac{b^2}{\beta^2}+\frac{c^2}{\gamma^2}\right)\right\}}$, Maximum value $=\sqrt{\left\{\frac{1}{2e}\left(\frac{a^2}{\alpha^2}+\frac{b^2}{\beta^2}+\frac{c^2}{\gamma^2}\right)\right\}}$

5. Maximum at $\left(\frac{4}{3}, 2\right)$. maximum value $= \frac{64}{9}$ cube units. Neither maximum nor minimum at (0,0).

6. Maximum at $\left(\frac{\pi}{3}, \frac{\pi}{3}, \frac{\pi}{3}\right)$, value$=\frac{3\sqrt{3}}{8}$

12.8 LAGRANGE'S METHOD OF UNDETERMINED MULTIPLIERS

Let $u=f(x_1, x_2, ..., x_n)$ be a function of n variables $x_1, x_2, ..., x_n$.
Let us suppose these variables $x_1, x_2, ..., x_n$ are connected by k equations

$$g_1(x_1, x_2, ..., x_n) = 0$$
$$g_2(x_1, x_2, ..., x_n) = 0$$
$$\cdots \cdots \cdots$$
$$g_k(x_1, x_2, ..., x_n) = 0$$

so, that there are $n-k$ independent variables out of these n variables. For the maxima and minima of u, we find

$$du = \frac{\partial u}{\partial x_1}dx_1 + \frac{\partial u}{\partial x_2}dx_2 + \ldots + \frac{\partial u}{\partial x_n}dx_n = 0 \qquad \ldots(1)$$

Also

$$dg_1 = \frac{\partial g_1}{\partial x_1}dx_1 + \frac{\partial g_1}{\partial x_2}dx_2 + \ldots + \frac{\partial g_1}{\partial x_n}dx_n = 0 \qquad \ldots(2)$$

$$dg_2 = \frac{\partial g_2}{\partial x_1}dx_1 + \frac{\partial g_2}{\partial x_2}dx_2 + \ldots + \frac{\partial g_2}{\partial x_n}dx_n = 0 \qquad \ldots(3)$$

$$\vdots$$

$$dg_k = \frac{\partial g_k}{\partial x_1}dx_1 + \frac{\partial g_k}{\partial x_2}dx_2 + \ldots + \frac{\partial g_k}{\partial x_n}dx_n = 0 \qquad \ldots(k+1)$$

Multiplying equation (1),(2),(3)...(k+1) by $1, l_1, l_2, ..., k$ respectively and adding, we get the result, which can be written as

$$P_1dx_1+P_2dx_2+P_3dx_3+\ldots+P_ndx_n=0 \qquad \ldots(4)$$

where

$$P_k = \frac{\partial u}{\partial x_k} + l_1\frac{\partial g_1}{\partial x_k} + l_2\frac{\partial g_2}{\partial x_k} + \ldots + l_k\frac{\partial g_k}{\partial x_k}$$

Now we have at our choice k multiple *viz* $l_1, l_2, ..., l_k$ and can be chosen such that

$$P_1=0, P_2=0, ..., P_k=0$$

Then, the equation (4) reduces to

$$P_{k+1}dx_{k+1}+P_{k+2}dx_{k+2}+P_{k+3}dx_{k+3}+...+P_ndx_n=0 \qquad ...(5)$$

Now, let us suppose that out of n variables, the $(n-k)$ variables $x_{k+1}, x_{k+2}, ..., x_n$ are independent.

Then, since $n-k$ quantities $dx_{k+1}, dx_{k+2}, ..., dx_n$ are independent so their coefficients must be separately zero. Hence, we have

$$P_{k+1}=0, P_{k+2}=0, ..., P_n=0$$

Thus, we have $k+n$ equations

$$P_1=0, P_2=0, ..., P_n=0$$

and

$$g_1=0, g_2=0, ..., g_k=0.$$

Hence, we get $(n+k)$ equations which determine the k multipliers $l_1, l_2, ..., l_k$ and get the possible value of u.

REMARKS

- The Lagrange's method of undetermined multipliers is very convenient to apply. It gives the maximum and minimum values of the function without actually determining the values of the multipliers $l_1, l_2, ..., l_k$.
- It does not determine the nature of stationary point, which is the only drawback of this method.

10.24.1 APPLICATIONS OF THE METHOD OF UNDETERMINED MULTIPLIERS

The Lagrange's method of undetermined multipliers can be applied to determine the extreme values of the given functions, it does not detemine the nature of stationary point. Now, it is more convenient to find out the extreme values of a function F with the help of new function, given by

$$V=g+l_1f_1+l_2f_2+...+l_mf_m$$

and use the following method. Here, we give the method for four variables x,y,u,v connected by the following two relations.

Let $F=g(x, y, u, v)$ be subjected to the conditions

$$f_1(x,y,u,v)=0 \qquad ...(1)$$

and

$$f_2(x,y,u,v)=0. \qquad ...(2)$$

For the maxima and minima of F, we have

$$dF = \frac{\partial g}{\partial x}dx + \frac{\partial g}{\partial y}dy + \frac{\partial g}{\partial u}du + \frac{\partial g}{\partial v}dv = 0 \qquad ...(3)$$

Now, from (1) and (2), we have

$$df_1 = \frac{\partial f_1}{\partial x}dx + \frac{\partial f_1}{\partial y}dy + \frac{\partial f_1}{\partial u}du + \frac{\partial f_1}{\partial v}dv = 0 \qquad ...(4)$$

and

$$df_2 = \frac{\partial f_2}{\partial x}dx + \frac{\partial f_2}{\partial y}dy + \frac{\partial f_2}{\partial u}du + \frac{\partial f_2}{\partial v}dv = 0 \qquad ...(5)$$

Multiplying (4) by l_1, (5) by l_2 and adding their sum to (3), we get

$$\left(\frac{\partial g}{\partial x} + l_1\frac{\partial f_1}{\partial x} + l_2\frac{\partial f_2}{\partial x}\right)dx + \left(\frac{\partial g}{\partial y} + l_1\frac{\partial f_1}{\partial y} + l_2\frac{\partial f_2}{\partial y}\right)dy$$

$$+\left(\frac{\partial g}{\partial u} + l_1\frac{\partial f_1}{\partial u} + l_2\frac{\partial f_2}{\partial u}\right)du + \left(\frac{\partial g}{\partial v} + l_1\frac{\partial f_1}{\partial v} + l_2\frac{\partial f_2}{\partial v}\right)dv = 0 \qquad ...(6)$$

Here, we have l_1 and l_2 are arbitrary, therefore we can choose them to satisfy the two linear equations

$$\frac{\partial g}{\partial x} + l_1\frac{\partial f_1}{\partial x} + l_2\frac{\partial f_2}{\partial x} = 0 \qquad ...(7)$$

and

$$\frac{\partial g}{\partial y} + l_1\frac{\partial f_1}{\partial y} + l_2\frac{\partial f_2}{\partial y} = 0 \qquad ...(8)$$

Using (7) and (8), equation (6) reduces to

$$\left(\frac{\partial g}{\partial u} + l_1\frac{\partial f_1}{\partial u} + l_2\frac{\partial f_2}{\partial u}\right)du + \left(\frac{\partial g}{\partial v} + l_1\frac{\partial f_1}{\partial v} + l_2\frac{\partial f_2}{\partial v}\right)dv = 0$$

Since, the given function contains four variables (namely x, y, u and v) and we are given two equations of conditions, therefore, only two of the variables are independent and it is immaterial which two of the four variables are regarded as independent. Let them be u and v then du and dv are also independent, therefore, their coefficients must be zero separately. Thus

$$\frac{\partial g}{\partial u} + l_1\frac{\partial f_1}{\partial u} + l_2\frac{\partial f_2}{\partial u} = 0 \qquad ...(9)$$

$$\frac{\partial g}{\partial v}+l_1\frac{\partial f_1}{\partial v}+l_2\frac{\partial f_2}{\partial v}=0 \qquad ...(10)$$

Now, we have six equations namely (1),(2),(7),(8),(9) and (10) to determine the two multipliers l_1,l_2 and values of the four variables x,y,u and v for which maximum and minimum values of F are possible.

Now, defined a new function $V(x, y, u, v)$ such that

$$V(x,y,u,v)=g(x, y, u, v)+l_1f_1(x, y, u, v)+l_2f_2(x, y, u, v).$$

Assuming that x, y, u, v are now all independent variables. Hence, for the maxima and minima of V, we must have

$$\frac{\partial V}{\partial x}=\frac{\partial g}{\partial x}+l_1\frac{\partial f_1}{\partial x}+l_2\frac{\partial f_2}{\partial x}=0 \qquad ...(11)$$

$$\frac{\partial V}{\partial y}=\frac{\partial g}{\partial y}+l_1\frac{\partial f_1}{\partial y}+l_2\frac{\partial f_2}{\partial y}=0 \qquad ...(12)$$

$$\frac{\partial V}{\partial u}=\frac{\partial g}{\partial u}+l_1\frac{\partial f_1}{\partial u}+l_2\frac{\partial f_2}{\partial u}=0 \qquad ...(13)$$

and

$$\frac{\partial V}{\partial v}=\frac{\partial g}{\partial v}+l_1\frac{\partial f_1}{\partial v}+l_2\frac{\partial f_2}{\partial v}=0 \qquad ...(14)$$

Equations (11), (12), (13) and (14) are exactly the same as the equations (7). (8), (9) and (10). Hence, the maxima and minima of $V(x, y, u, v)$ are same as those of $F(x, y, u, v)$ assuming that $V(x, y, u, v)$ the variables x, y, u, v are now all independent.

Now, we proceed to find whether the values of F obtained with the help of above equations are maximum or minimum. For this, adopt the procedure, which is discussed ahead.

From (3), we get
$$d^2F=\left(\frac{\partial}{\partial x}dx+\frac{\partial}{\partial y}dy+\frac{\partial}{\partial u}du+\frac{\partial}{\partial y}dy\right)^2 g+\left(\frac{\partial g}{\partial x}d^2x+\frac{\partial g}{\partial y}d^2y+\frac{\partial g}{\partial u}d^2u+\frac{\partial g}{\partial y}d^2v\right)... \qquad ...(15)$$

Also
$$d^2f_1=\left(\frac{\partial}{\partial x}dx+\frac{\partial}{\partial y}dy+\frac{\partial}{\partial u}du+\frac{\partial}{\partial v}dv\right)^2 f_1+\frac{\partial f_1}{\partial x}d^2x+\frac{\partial f_1}{\partial y}d^2y+\frac{\partial f_1}{\partial u}d^2u+\frac{\partial f_1}{\partial v}d^2v=0 \qquad ...(16)$$

and
$$d^2f_2=\left(\frac{\partial}{\partial x}dx+\frac{\partial}{\partial y}dy+\frac{\partial}{\partial u}du+\frac{\partial}{\partial v}dv\right)^2 f_2+\frac{\partial f_2}{\partial x}d^2x+\frac{\partial f_2}{\partial y}d^2y+\frac{\partial f_2}{\partial u}d^2u+\frac{\partial f_2}{\partial v}d^2v=0 \qquad ...(17)$$

Multiplying (16) by l_1 and (17) by l_2 and adding their sum to (15) and using the result (11), (12),(13) and (14), we have

$$d^2F=\left(\frac{\partial}{\partial x}dx+\frac{\partial}{\partial y}dy+\frac{\partial}{\partial u}du+\frac{\partial}{\partial v}dv\right)^2(g+l_1f_1+l_2f_2)=\left(\frac{\partial}{\partial x}dx+\frac{\partial}{\partial y}dy+\frac{\partial}{\partial u}du+\frac{\partial}{\partial v}dv\right)^2 V=d^2V.$$

Hence d^2F is equal to d^2V, where d^2V is obtained by assuming all the variables x,y,u and v as independent. Therefore, it is clear that d^2F and d^2V have the same sign. Hence, F will be minimum or maximum according as V is minimum or maximum.

Remark

- This method has the advantage over the Lagrange's methods that it enables us to decide whether the values are maximum or minimum.

Solved Examples

Example 1. *Find the maxima and minima of $x^2+y^2+z^2$ subject to the conditions :*
$ax^2+by^2+cz^2=1$ and $lx+my+nz=0$

(UKTU- 2011)

Solution. Here, we have $u=x^2+y^2+z^2$ ...(1)

where, the relations between the variables x, y and z are given by

$$ax^2+by^2+cz^2=1 \qquad ...(2)$$

and

$$lx+my+nz=0 \qquad ...(3)$$

For the maxima and minima of u, we must have

$$du=0$$

$$\Rightarrow \quad 2xdx+2ydy+2zdz=0$$

$$\Rightarrow \quad xdx+ydy+zdz=0 \qquad ...(4)$$

From (2) and (3), we get

$$ax\,dx+by\,dy+cz\,dz=0 \qquad ...(5)$$

$$l\,dx+m\,dy+n\,dz=0 \qquad ...(6)$$

Now, multiplying (4) by 1, (5) by $l_1$and (6) by l_2 and adding, we get

$$(x\,dx+y\,dy+z\,dz)+l_1(ax\,dx+by\,dy+cz\,dz)$$
$$+l_2(l\,dx+m\,dy+n\,dz)=0$$

$$\Rightarrow \quad (x+al_1x+ll_2)dx+(y+bl_1y+ml_2)$$

$$dy+(z+cl_1z+nl_2)dz = 0$$

Now equating the coefficient of dx, dy, dz to zero, we get

$$x+l_1ax+l_2l=0 \quad ...(7)$$

$$y+bl_1y+ml_2=0 \quad ...(8)$$

and $$z+cl_1z+nl_2=0 \quad ...(9)$$

Multiplying the equations (7), (8) and (9) by x, y and z respectively, and adding we get

$$x^2+y^2+z^2+l_1(ax^2+by^2+cz^2) +l_2(lx+my+nz)=0$$

or $$u+l_1.1+l_2.0 = 0$$

[By using (1), (2) and (3)]

$$\Rightarrow \quad l_1 = -u$$

Substituting for l_1 in the equations (7), (8) and (9), we get

$$x = \frac{l_2 l}{au-1}, y = \frac{l_2 m}{bu-1}, z = \frac{l_2 n}{cu-1} \quad ...(10)$$

Now from (10) and (3), we get

$$\frac{l_2 l^2}{au-1}+\frac{l_2 m^2}{bu-1}+\frac{l_2 n^2}{cu-1}=0$$

or $$\frac{l^2}{au-1}+\frac{m^2}{bu-1}+\frac{n^2}{cu-1}=0 \quad ...(11)$$

which gives the maximum and minimum of $u=x^2+y^2+z^2$.

REMARKS

- Equation (11) is a quadratic in u. So it gives two stationary values of u.
- Geometrically, the surface $ax^2+by^2+cz^2=1$ represents an ellipsoid whose centre is origin, and $lx+my+nz=0$ represents a plane passing through the origin. The points (x, y, z) satisfying both the conditions (2) and (3) lies on the conic in which (2) and (3) intersect. $x^2+y^2+z^2$ gives the square of the distance (x, y, z) from the origin, which is also the centre of the conic of intersection. The maximum value of this distance is the major axis of this conic, and the minimum value of this distance is the minor axis of this conic. Hence, equation (11) gives the squares of the lengths of the semi-axis of the conic of intersection.

Example 2. *Find the maxima and minima of $x^2+y^2+z^2$, where $ax^2+by^2+cz^2+2fyz+2gzx+2hxy=1$.*

Solution. Let $$u= x^2+y^2+z^2 \quad ...(1)$$

where the relation between the variables x,y and z is

$$ax^2+by^2+cz^2+2fyz+2gzx+2hxy=1. \quad ...(2)$$

For a maximum or minima of u, we must have

$$du=0$$

$$\Rightarrow \quad x\,dx+y\,dy+z\,dz=0. \quad ...(3)$$

From (2), we have

$$2ax\,dx+2by\,dy+2cz\,dz+2fy\,dz+2fz\,dy+2gz\,dx+2gx\,dz+2hx\,dy+2hy\,dx=0$$

$$\Rightarrow \quad (ax+hy+gz)dx+(hx+by+fz)dy+(gx+fy+cz)dz=0. \quad ...(4)$$

Now, multiplying (3) by 1 and (4) by l_1, adding, and then equating the coefficient of dx, dy, dz to zero, we have

$$x+l_1(ax+hy+gz)=0. \quad ...(5)$$

$$y+l_1(hx+by+fz)=0. \quad ...(6)$$

$$z+l_1(gx+fy+cz)=0. \quad ...(7)$$

Multiplying (5) by x, (6) by y, (7) by z and adding, we get

$$x^2+y^2+z^2+l_1(ax^2+by^2+cz^2 +2fyz+2gzx+2hxy)=0$$

$$\Rightarrow \quad u+l_1.1=0$$

[From (1) and (2)]

$$\therefore \quad l_1 =-u.$$

Hence, from (5), we have

$$x-u(ax+hy+gz)=0$$

$$\Rightarrow \quad \left(a-\frac{1}{u}\right)x+hy+gz=0 \quad ...(8)$$

Similarly from (6) and (7), we get

$$hx+\left(b-\frac{1}{u}\right)y+fz=0 \quad ...(9)$$

and $$gx+fy+\left(c-\frac{1}{u}\right)z=0 \quad ...(10)$$

Eliminating x,y,z from (8), (9) and (10), we get

$$\begin{vmatrix} \left(a-\frac{1}{u}\right) & h & g \\ h & \left(b-\frac{1}{u}\right) & f \\ g & f & \left(c-\frac{1}{u}\right) \end{vmatrix} = 0 \quad ...(11)$$

Hence, the maximum or minimum values of u are the roots of the equation (11).

Example 3. *Find the maximum and minima of $u=x^2+y^2$ subject to the condition*

$$ax^2+2hxy+by^2=1.$$

Solution . Here, we have

$$u = x^2+y^2 \quad ...(1)$$

where the relation between the variables x and y is

$$ax^2+2hxy+by^2=1. \quad ...(2)$$

For the maxima and minima of u, we must have

$$du = 0$$

$$\Rightarrow \quad 2x\,dx+2y\,dy = 0$$

$$\Rightarrow \quad x\,dx+y\,dy = 0. \quad ...(3)$$

Now, from (2), we get

$$2ax\,dx+2hx\,dy+2hy\,dx+2by\,dy=0$$

$$\Rightarrow \quad (ax+hy)dx+(hx+by)dy=0 \qquad ...(4)$$

Now, multiplying (3) by 1, (4) by l_1, adding and then equating the coefficients of dx, dy to zero, we have

$$x+l_1(ax+hy)=0 \qquad ...(5)$$

and

$$y+l_1(hx+by)=0 \qquad ..(6)$$

Multiplying (5) by x, (6) by y and adding, we get

$$x^2+y^2+l_1(ax^2+2hxy+by^2)=0$$

$$\Rightarrow \quad u+l_1.1=0$$

[using (1) and (2)]

$$\Rightarrow \quad u=-l_1$$

Therefore, from (5), we have

$$x-u(ax+hy)=0$$

$$\Rightarrow \left(a-\frac{1}{u}\right)x+hy=0 \qquad ...(7)$$

Similarly from (6), we have

$$hx+\left(b-\frac{1}{u}\right)y=0 \qquad ...(8)$$

Eliminating x and y from (7) and (8), we get

$$\begin{vmatrix} a-\frac{1}{u} & h \\ h & b-\frac{1}{u} \end{vmatrix}=0$$

Hence, the maximum or minimum values of u are the roots of the equation (9).

Example 4. *Find the maximum value of $u=x^my^nz^p$ subject to the condition $x+y+z=a$.* (Anna-2009)

Solution . Here, we have $u=x^my^nz^p$ ...(1)

and x, y, z connected by the relation given by

$$x+y+z=a \qquad ...(2)$$

Taking log of both the sides of (1), we get

$$\log u = m\log x+n\log y+p\log z.$$

On differentiating, we get

$$\frac{1}{u}du=\frac{m}{x}dx+\frac{n}{y}dy+\frac{p}{z}dz$$

For the maxima and minima of u, we must have

$$du=0$$

$$\Rightarrow \quad \frac{m}{x}dx+\frac{n}{y}dy+\frac{p}{z}dz=0 \qquad ...(3)$$

Now, differentiating (2), we get

$$dx+dy+dz=0. \qquad ...(4)$$

Now, multiplying (3) by 1 and (4) by l, and equating the coefficient of dx, dy, dz to zero (after adding), we get

$$\frac{m}{x}+l=0,\quad \frac{n}{y}+l=0 \text{ and } \frac{p}{z}+l=0$$

which implies $x=-\frac{m}{l},\ y=-\frac{n}{l},\ z=-\frac{p}{l}$

Putting the values of x, y and z in (2), we get

$$l=-\left(\frac{m+n+p}{a}\right)$$

therefore, we can say that, u is stationary when

$$x=\frac{am}{m+n+p},\ y=\frac{an}{m+n+p},\ z=\frac{ap}{m+n+p}$$

Now, we find the nature of this stationary value of u.

Let us regard x and y as independent variable and z is a function of x and y given by (2) [It is justify, because the variables x, y and z are connected by the relation (2), any two of them may be regarded as independent].

Now from (1), we get

$$\log u = m\log x+n\log y+p\log z$$

$$\therefore \quad \frac{1}{u}\frac{\partial u}{\partial x}=\frac{m}{x}+\frac{p}{z}\frac{\partial z}{\partial x}$$

Now, differentiating (2) partially *w.r.t* x (treating y as constant), we get

$$1+\frac{\partial z}{\partial x}=0 \Rightarrow \frac{\partial z}{\partial x}=-1$$

Put this value in (5), we get

$$\frac{1}{u}\frac{\partial u}{\partial x}=\frac{m}{x}-\frac{p}{z}$$

$$\Rightarrow \quad \frac{1}{u}\frac{\partial^2 u}{\partial x^2}-\frac{1}{u^2}\left(\frac{\partial u}{\partial x}\right)^2$$

$$=-\frac{m}{x^2}+\frac{p}{z^2}\frac{\partial z}{\partial x}=-\frac{m}{x^2}-\frac{p}{z^2}$$

At stationary point $\frac{\partial u}{\partial x}=0$

Therefore, $\frac{1}{u}\frac{\partial^2 u}{\partial x^2}=\frac{-m}{x^2}-\frac{p}{z^2}$

$$\Rightarrow \quad \frac{\partial^2 u}{\partial x^2}=u\left[-\frac{m}{x^2}-\frac{p}{z^2}\right]$$

$$=-x^my^nz^p\left[-\frac{m}{x^2}-\frac{p}{z^2}\right]$$

which is negative for the obtained values of x, y and z.

Hence, at the stationary point, u is maximum and maximum value is

$$=\left(\frac{am}{m+n+p}\right)^m\left(\frac{an}{m+n+p}\right)^n\left(\frac{ap}{m+n+p}\right)^p$$

Example 5. *Find the maximum and minimum value of $u=\frac{5xyz}{(x+2y+4z)}$ subject to the condition $xyz=8$.*

Solution. Here, we have $u=\frac{5xyz}{(x+2y+4z)}$ ...(1)

The variables x,y,z are connected by the relation $xyz=8$. ...(2)

From (1) and (2), we get

$$u = \frac{40}{x + 2y + 4z}$$

$$\Rightarrow \quad du = \frac{-40}{(x+2y+4z)^2}(dx + 2dy + 4dz)$$

For the maxima or minima of u, we must have $du=0$

$$\Rightarrow \quad dx+2dy+4dz=0 \qquad ...(3)$$

From (2), we get

$$\log x+\log y+\log z=\log 8.$$

On differentiating, we get

$$\frac{1}{x}dx + \frac{1}{y}dy + \frac{1}{z}dz = 0 \qquad ...(4)$$

Now, multiplying (3) by 1, (4) by l, adding and then equating to zero the coefficients of dx, dy and dz, we get

$$1+\frac{l}{x} = 0, 2+\frac{l}{y} = 0, 4+\frac{l}{z} = 0$$

Now using (2), we get $l = -4$

$\therefore$ u is stationary at the point given by $x=4$, $y=2$, $z=1$.

Regard x and y as independent variables and z is a function of x and y given by (2). From (1)

$$\frac{\partial u}{\partial x} = -\frac{40}{(x+2y+4z)^2}\left[1+4\frac{\partial z}{\partial x}\right]$$

From (2), we get

$$\log x + \log y + \log z = \log 8$$

$$\therefore \quad \frac{1}{x} + \frac{1}{z}\frac{\partial z}{\partial x} = 0$$

$$\Rightarrow \quad \frac{\partial z}{\partial x} = -\frac{z}{x}$$

$$\therefore \quad \frac{\partial u}{\partial x} = -\frac{40}{(x+2y+4z)^2}\left[1-4\frac{z}{x}\right]$$

$$\Rightarrow \quad \frac{\partial^2 u}{\partial x^2} = \frac{80}{(x+2y+4z)^3}\left[1+4\frac{\partial z}{\partial x}\right]\left[1-4\frac{z}{x}\right] - \frac{40}{(x+2y+4z)^2}\left[\frac{4z}{x^2} - \frac{4}{x}\frac{\partial z}{\partial x}\right]$$

Now using $x = 4, y = 2, z = 1$.

We get $\dfrac{\partial^2 u}{\partial x^2} = -\text{ve}$

$\therefore$ u is maximum at the point given by

$$x=4, y=2, z=1.$$

The maximum value is given by

$$u = \frac{5\times4\times2\times1}{(4+2\times2+4\times1)} = \frac{40}{12} = \frac{10}{3}.$$

Example 6. *In a plane triangle ABC, find the maximum value of $u=\cos A \cos B \cos C$.* (VTU-2010, Anna-2006)

Solution . Here, we have

$$u = \cos A \cos B \cos C \qquad ...(1)$$

Since, we know that the sum of the angles of a triangle is always 180°.

$\therefore$ The variables A, B and C are connected by the relation $A + B + C = \pi$ ...(2)

From (1), we get

$$\log u = \log \cos A + \log \cos B + \log \cos C$$

$$\Rightarrow \frac{1}{u}du = -\tan A\, dA - \tan B\, dB - \tan C\, dC.$$

For the maxima and minima of u, we must have $du=0$

$$\Rightarrow \tan A\, dA + \tan B\, dB + \tan C\, dC = 0 \quad ...(3)$$

Also from (2),

$$dA + dB + dC = 0 \qquad ...(4)$$

Now, multiply (3) by 1, (4) by l, adding, and equating the coefficients of dA, dB and dC to zero, we get

$$\tan A + l = 0$$
$$\tan B + l = 0$$
$$\tan C + l = 0$$

$$\Rightarrow \quad l = -\tan A = -\tan B = -\tan C$$

$$\Rightarrow \quad A = B = C.$$

Now from (2), $A = B = C = \dfrac{\pi}{3}$ *i.e.*, the triangle is equilateral.

Now to show that the stationary value of u given by $A = B = C = \dfrac{\pi}{3}$ is maximum.

Let C be a function of A and B, regarding A and B as independent variables. From (1),

$$\log u = \log \cos A+\log \cos B+\log \cos C$$

$$\Rightarrow \quad \frac{1}{u}\frac{\partial u}{\partial A} = -\tan A - \tan C\frac{\partial C}{\partial A}$$

Now, differentiating (2), partially w.r.t. A, we get

$$1+\frac{dC}{dA} = 0 \Rightarrow \frac{\partial C}{\partial A} = -1$$

$$\therefore \quad \frac{1}{u}\frac{\partial u}{\partial A} = -\tan A + \tan C$$

$$\Rightarrow \quad \frac{1}{u}\frac{\partial^2 u}{\partial^2 A} - \frac{1}{u^2}\left(\frac{\partial u}{\partial A}\right)^2 = -\sec^2 A + \sec^2 C.\frac{\partial C}{\partial A}$$

$$= -\left(\sec^2 A + \sec^2 C\right)$$

At stationary point $\dfrac{\partial u}{\partial A} = 0$

$$\because \quad \frac{\partial^2 u}{\partial^2 A} = -u\left(\sec^2 A + \sec^2 C\right) = -\text{ve}$$

for $A=B=C=\dfrac{\pi}{3}$.

Hence, u is maximum at $A=B=C=\dfrac{\pi}{3}$ and the maximum value is given by

$$u=\left(\cos\frac{\pi}{3}\right)^3=\left(\frac{1}{2}\right)^3=\frac{1}{8}.$$

EXERCISE 12.3

Using Lagrange's method of undetermined multiplirers:

1. Find the maximum and minimum values of

$$\frac{x^2}{a^4}+\frac{y^2}{b^4}+\frac{z^2}{c^4}$$

where $lx+my+nz=0$ and $\frac{x^2}{a^2}+\frac{y^2}{b^2}+\frac{z^2}{c^2}=1$.

2. Find the maximum and minimum values of

$$f=a^2x^2+b^2y^2+c^2z^2$$

where $x^2+y^2+z^2=1$ and $lx+my+nz=0$.

3. Show that the maximum and minimum values of $u=x^2+y^2+z^2$ subject to the conditions

$$px+qy+rz = 0 \text{ and } \frac{x^2}{a^2}+\frac{y^2}{b^2}+\frac{z^2}{c^2}=1$$

are given by $\frac{a^2p^2}{u-a^2}+\frac{b^2q^2}{u-b^2}$.

4. Find the minimum value of $u=x+y+z$ subject to the condition $\frac{a}{x}+\frac{b}{y}+\frac{c}{z}=1$.

5. Find the minimum value of $u=x^2+y^2+z^2$, subject to the condition $ax+by+cz=p$. (UKTU-2012, UPTU-2009)

6. Find the minimum value of $x+y+z$ where $xyz=c^3$.

7. Find the extreme values of $x^py^qz^r$ subject to the condition $\frac{a}{x}+\frac{b}{y}+\frac{c}{z}=1$.

8. Show that the maximum and minimum values of the radii vectors of the sections of the surface

$$(x^2+y^2+z^2)^2=\frac{x^2}{a^2}+\frac{y^2}{b^2}+\frac{z^2}{c^2}$$

by the plane $\lambda x+\mu y+\nu z=0$

are given by $\frac{a^2\lambda^2}{1-a^2r^2}+\frac{b^2\mu^2}{1-b^2r^2}+\frac{c^2\nu^2}{1-c^2r^2}=0$

9. Find the stationary points of the function $u=ax^p+by^q+cz^r$ subject to the condition

$$x^l+y^m+z^n=k.$$

10. If two variables x and y are connected by the relation $ax^2+by^2=ab$, show that the maximum and minimum values of the function $u=x^2+y^2+xy$ will be the roots of the equation

$$4(u-a)(u-b)=ab.$$

11. Prove that of all rectangular parallelopipeds of the same volume, the cube has the least surface. (Kurukshetra-2006, UPTU-2004)

12. Prove that if $x+y+z=1$, $ayz+bzx+cxy$ has an extreme value equal to

$$\frac{abc}{2bc+2ca+2ab-a^2-b^2-c^2}$$

Also, prove if a, b, c are all positive and c lies between $a+b-2\sqrt{ab}$ and $a+b+2\sqrt{ab}$ this value is true maximum and that if a, b, c are all negative and c lies between $a+b\pm2\sqrt{ab}$. It is true minimum.

13. Find the maximum value of u, when

$$u=\sin x \sin y \sin z$$

and x,y,z are the angles of a triangle.

14. Find the triangle of maximum area inscribed in a circle.

15. Prove that the rectangular solid of maximum volume which can be inscribed in a sphere is a cube.

16. Find a plane triangle ABC such that

$$u=\sin^aA \sin^bB \sin^cC$$

has maximum value.

17. Find the rectangular parallelopiped of maximum volume that can be inscribed in the ellipsoid

$$\frac{x^2}{a^2}+\frac{y^2}{b^2}+\frac{z^2}{c^2}=1$$

(UKTU-2010, Anna-2009, Madras-2006)

18. Divide a number n into three parts x,y,z such that $ayz+bzx+cxy$ shall have maximum or minimum and determine which it is.

19. Prove that a rectangular solid of maximum volume which can be inscribed in a sphere is a cube.

20. Find the maximum or minimum value of $x^py^qz^r$ subject to the condition $ax+by+cz=p+q+r$.

21. Show that the maximum and minimum value of

$$u=ax^2+by^2+cz^2+2fyz+2gzx+2hxy$$

subject to the conditions $lx+my+nz=0$

and $x^2+y^2+z^2=1$

are given by the equation

$$\begin{vmatrix} a-u & h & g & l \\ h & b-u & f & m \\ g & f & c-u & n \\ l & m & n & o \end{vmatrix}=0$$

22. Show that of the perimeter of a triangle is constant, its area is maximum when it is equilateral.

23. Show that the volume of the largest rectangular parallelopiped that can be inscribed in the ellipsoid $\frac{x^2}{a^2}+\frac{y^2}{b^2}+\frac{z^2}{c^2}=1$ is $\frac{8abc}{3\sqrt{3}}$. (UKTU-2010)

24. Show that the maximum and minimum distances from the origin to the curve $x^2+4xy+6y^2 = 140$ are respectively given by 21.6589 and 4.5706. (MTU-2012)

25. A rectangular box which is open at the top, has a capacity of 32 c.c. Find the dimension of the box such that the least material is required for the construction of the box.

(UPTU-2008, GBTU-2011, PTU-2006, Kurukshetra-2006)

ANSWERS

1. The maximum and minimum values of the given function is given by the equation $\dfrac{l^2a^4}{a^2u-1}+\dfrac{m^2b^4}{b^2u-1}+\dfrac{n^2c^4}{c^2u-1}=0$

2. The maximum and minimum values of the given function is given by the equation $\dfrac{l^2}{u-a^2}+\dfrac{m^2}{u-b^2}+\dfrac{m^2}{u-c^2}=0$

4. Stationary points are $x=\sqrt{a}\left(\sqrt{a}+\sqrt{b}+\sqrt{c}\right), y=\sqrt{b}\left(\sqrt{a}+\sqrt{b}+\sqrt{c}\right), z=\sqrt{c}\left(\sqrt{a}+\sqrt{b}+\sqrt{c}\right)$, minimum value is $\left(\sqrt{a}+\sqrt{b}+\sqrt{c}\right)^2$.

5. Minimum value is $\dfrac{p^2}{\left(a^2+b^2+c^2\right)}$ **6.** u is minimum at the point $x=y=z=c$. Value is $=3c^4$.

7. u is stationary when $\dfrac{px}{a}=\dfrac{qy}{b}=\dfrac{rc}{c}=p+q+r$, Minimum value is $\dfrac{a^pb^qc^r}{p^pq^qr^r}(p+q+r)^{p+q+r}$.

9. Stationary points are given by $\dfrac{x^{p-1}}{l/pa}=\dfrac{y^{q-m}}{m/qb}=\dfrac{z^{r-n}}{n/rc}$ **13.** u is maximum, when $x=y=z=\dfrac{\pi}{3}$. Maximum value is $\dfrac{3\sqrt{3}}{8}$.

14. Equilateral. **16.** u is maximum when, A, B, C are given by $\dfrac{\tan A}{a}=\dfrac{\tan B}{b}=\dfrac{\tan C}{c}$.

17. Stationary points are $x=\dfrac{a}{\sqrt{3}}, y=\dfrac{b}{\sqrt{3}}, z=\dfrac{c}{\sqrt{3}}$, Maximum value $=\dfrac{8abc}{3\sqrt{3}}$. **25.** $x=4, y=4, z=2$

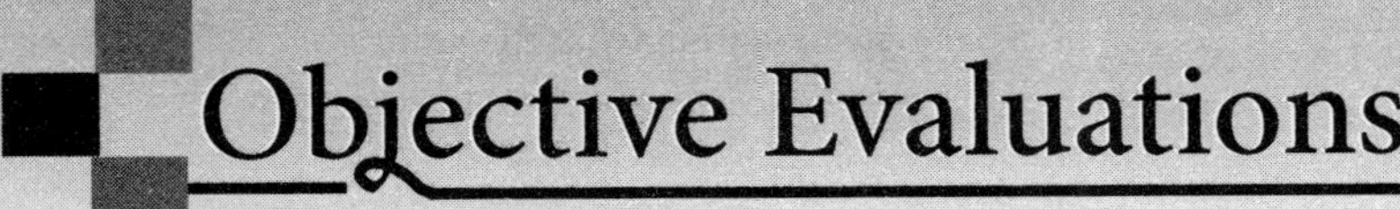

Objective Evaluations

Fill in the Blanks

1. A function of two variables x and y may be writen as ________ .
2. $\lim_{(x,y)\to(a,b)} f(x,y)$, if exists, is ________ .
3. If a function $f(x, y)$ is totally differentiable, then the partial derivatives f_x and f_y ________ .
4. If $u = f\left(\frac{y}{x}\right)$, then $x\frac{\partial u}{\partial x} + y\frac{\partial u}{\partial y} =$ ________ .
5. If u is a homogeneous function of x and y of degree n, then $x\frac{\partial u}{\partial x} + y\frac{\partial u}{\partial y} =$ ________ .
6. The statement given in (5) is known as ________ .
7. If u is a function of variables x and y, and x and y both are the functions of the variable t, then u is said to be ________.
8. If $f_{xy}(a, b)$ and $f_{yx}(a, b)$, both are continuous, then f_{xy} ________ .

True/False

Write 'T' for True and 'F' for False statement.

1. The limit of a function is unique. **(T/F)**
2. $\lim_{(x,y)\to(1,1)} f(x,y)$ does not exist, where $f(x, y) = (x^2 + 2y)$. **(T/F)**
3. The two iterated limits obtained by reversing the order of limits, are always equal. **(T/F)**
4. The function $f(x, y) = xy$ is not continuous at (2, 3). **(T/F)**
5. In case of a function of two variables, the continuity is a necessary condition for differentiability. **(T/F)**

Multiple Choice Questions

Choose the most appropriate one.

1. If both the first order partial derivatives of the function $f(x,y)$ vanish at that point, then this point is called :
   (a) stationary point (b) saddle point
   (c) maxima point (d) minima point
2. The function $f(x,y)$ will have a maximum or minimum at $x=a, y=b$ if :
   (a) $rt>s^2$ (b) $rt<s^2$
   (c) $rt=s^2$ (d) none of these
3. If $rt-s^2<0$, then $f(x,y)$ is :
   (a) maximum
   (b) minimum
   (c) neither maximum nor minimum
   (d) none of the above
4. If at a point (a,b), $\frac{\partial f}{\partial x}=0, \frac{\partial f}{\partial y}=0$, then $f(x,y)$ is maximum at (a,b) if at (a,b) :
   (a) $rt-s^2>0$
   (b) $rt-s^2>0$ and $r<0$
   (c) $rt-s^2>0$ and $r>0$
   (d) $rt-s^2<0$ and $r<0$
5. The condition $\frac{\partial f}{\partial x}=0, \frac{\partial f}{\partial y}=0, \frac{\partial f}{\partial z}=0$ for the maxima and minima of a function is :
   (a) necessary condition
   (b) sufficient condition
   (c) necessary and sufficient both
   (d) none of the above
6. The volume of the greatest rectangular parallelopiped that can be inscribed in the ellipsoid is :
   (a) $8abc$ (b) $\frac{8abc}{3\sqrt{3}}$
   (c) $\frac{8abc}{3}$ (d) $\frac{8abc}{\sqrt{3}}$
7. The maximum and minimum values of $u=a^2x^2+b^2y^2+c^2z^2$, where $x^2+y^2+z^2=1$ and $lx+my+nz=0$ are the roots of the equation :
   (a) $\frac{l^2}{u-a^2}+\frac{m^2}{u-b^2}+\frac{n^2}{u-c^2}=0$
   (b) $\frac{l}{u-a}+\frac{m}{u-b}+\frac{n}{u-c}=0$
   (c) $\frac{l}{u-a^2}+\frac{m}{u-b^2}+\frac{n}{u-c^2}=0$
   (d) none of the above
8. If we divide a number a into three parts such that their product will be maximum then parts of this number are :
   (a) $\frac{a}{3},\frac{a}{3},\frac{a}{3}$ (b) $\frac{a}{2},\frac{a}{2}$
   (c) $\frac{a}{\sqrt{3}},\frac{a}{\sqrt{3}},\frac{a}{\sqrt{3}}$ (d) none of these

9. If $u = x^2y^3z^4$ and $2x + 3y + 4z = a$ then maximum value of u is given by :

(a) $\frac{a}{9}$ (b) $\left(\frac{a}{9}\right)^2$

(c) $\left(\frac{a}{9}\right)^9$ (d) none of these

10. The function $u = \sin x \sin y \sin z$, where x, y, z are the angles of a triangle is stationary at the point :

(a) $x = y = z = \frac{\pi}{3}$ (b) $x = \frac{\pi}{2}, y = \frac{\pi}{4}, z = \frac{\pi}{4}$

(c) $x = 0, y = \frac{\pi}{2} = 2$ (d) none of the above

ANSWERS

Fill in the Blanks

1. $f(x, y)$ **2.** unique **3.** both exist and equal **4.** 0 **5.** nu
6. Euler's theorem **7.** composite function **8.** f_{yx}

True/False

1. T **2.** F **3.** F **4.** F **5.** T

Multiple choice questions

1. (a) **2.** (a) **3.** (c) **4.** (b) **5.** (a) **6.** (b) **7.** (a) **8.** (a) **9.** (c) **10.** (a)

FFFFF

CHAPTER 13 Error and Approximations

13.1 INTRODUCTION

It is a well known fact that error play an important role in daily life as well as in computational mathematics. In this chapter, we shall discuss the methods to estimate errors in any measurement process by applying differentiation method.

13.2 ERROR AND APPROXIMATIONS

Let $y = f(x)$ be a function and Δx be the small error in x. Then Δy be the corresponding error in y. Therefore, we have

$$y + \Delta y = f(x + \Delta x)$$

$$\Rightarrow \quad \Delta y = f(x + \Delta x) - y = f(x + \Delta x) - f(x)$$

$$= \left[f(x) + \Delta x . f'(x) + \frac{(\Delta x)^2}{2!} f''(x) + \ldots \right] - f(x) \quad \text{(on expanding by Taylor's series)}$$

On neglecting the second and higher powers of Δx, we get

$$\Delta y = \Delta x . f'(x) \qquad \ldots(1)$$

$$= \Delta x . \frac{df(x)}{dx} \qquad \ldots(2)$$

Similarly, when y is a function of more than one variable, *i.e.*, $y = f(x_1, x_2)$

Then $\Delta y = f(x_1 + \Delta x_1, x_2 + \Delta x_2)$

where Δx_1 and Δx_2 are the errors in x_1 and x_2 respectively.

Now using Taylor's theorem, we get

$$\Delta y = \left(\left(f(x_1, x_2) + \frac{\partial f}{\partial x_1} \Delta x_1 + \frac{\partial f}{\partial x_2} \Delta x_2 + \ldots \right) - f(x_1, x_2) \right)$$

On neglecting the second and higher degree terms we get

$$\Delta y = \frac{\partial f}{\partial x_1} \Delta x_1 + \frac{\partial f}{\partial x_2} \Delta x_2 \qquad \ldots(3)$$

and so on.

In general, if $y = f(x_1, x_2, \ldots, x_n)$ and $\Delta x_1, \Delta x_2, \ldots, \Delta x_n$ be the errors in $x_1, x_2, \ldots, x_n$ respectively and Δy is the corresponding error in y.

Then we have

$$\Delta y = \frac{\partial f}{\partial x_1} \Delta x_1 + \frac{\partial f}{\partial x_2} \Delta x_2 + \ldots + \frac{\partial f}{\partial x_n} \Delta x_n . \qquad \ldots(4)$$

which is the required error relation of the function $y = f(x_1, x_2, \ldots, x_n)$.

Here, we observe that

(1) Δy is the absolute error in y.

(2) $\frac{\Delta y}{y}$ is the relative error in y.

(3) $\frac{\Delta y}{y} \times 100$ is percentage error in y.

Solved Examples

Example 1. *Find the percentage error in the area of an ellipse when an error of one percent is made in measuring the major and minor axes.* (GBTU–2011)

Solution . Let x and y are semi-major and semi-minor axes of the ellipse. Then area of the ellipse is

$$A = \pi xy$$

Taking log of both the sides, we get

$$\log A = \log \pi + \log x + \log y$$

On differentiating, we get

$$\frac{\Delta A}{A} = 0 + \frac{\Delta x}{x} + \frac{\Delta y}{y}$$

$$\frac{\Delta A}{A} \times 100 = \frac{\Delta x}{x} \times 100 + \frac{\Delta y}{y} \times 100$$

$$= 1 + 1 = 2$$

which is the required error in area.

Example 2. *If the base radius and height of a cone are measured as 4 and 8 inches with a possible error of 0.04 and 0.08 inches respectively. Calculate the percentage error in calculating volume of the cone.* (GBTU–2012)

Solution . We know that

$$\text{Volume, } V = \frac{1}{3}\pi r^2 h$$

$$\Rightarrow \quad \log V = \log\frac{1}{3} + \log \pi + 2\log r + \log h$$

Differentiating, we get

$$\frac{\Delta V}{V} = 2\frac{\Delta r}{r} + \frac{\Delta h}{h}$$

$$\Rightarrow \quad \frac{\Delta V}{V} = 2\left(\frac{0.4}{4}\right) + \left(\frac{0.8}{8}\right) = 0.03$$

Hence, percentage error in volume

$$= 0.03 \times 100 = 3\ \%$$

Example 3. *Let A be the area of a triangle, prove that the error A resulting from a small error in c is given by*

$$\Delta V = \frac{A}{4}\left[\frac{1}{s} + \frac{1}{s-a} + \frac{1}{s-b} + \frac{1}{s-c}\right]\Delta c$$

(UPTU–2007, 08)

Solution . Let a, b, c be the sides of a traiangle, then

$$A = \sqrt{s(s-a)(s-b)(s-c)}$$

where $s = \dfrac{a+b+c}{2}$

Taking log of both sides, we get

$$\log = \frac{1}{2}\log(s(s-a)(s-b)(s-c))$$

$$= \frac{1}{2}(\log s + \log(s-a) + \log(s-b) + \log(s-c))$$

Differentiating, we get

$$\frac{1}{A}\cdot\frac{dA}{dc} = \frac{1}{2}\left[\frac{1}{s}\frac{ds}{dc} + \frac{1}{s-a}\frac{d(s-a)}{dc} + \frac{1}{s-b}\frac{d(s-b)}{dc} + \frac{1}{s-c}\frac{d(s-c)}{dc}\right] \quad \ldots(1)$$

We have $\dfrac{ds}{dc} = \dfrac{1}{2}$ $\quad\left[\because s = \dfrac{a+b+c}{2}\right]$

$$\frac{d(s-c)}{dc} = \frac{d}{dc}\left(\frac{a+b-c}{2}\right) = -\frac{1}{2}$$

$$\frac{d(s-a)}{dc} = \frac{1}{2}, \frac{d(s-b)}{dc} = \frac{1}{2}$$

Putting all these values in (1), we get

$$\frac{1}{A}\cdot\frac{dA}{dc} = \frac{1}{2}\left[\frac{1}{s}\times\frac{1}{2} + \frac{1}{s-a}\times\frac{1}{2} + \frac{1}{s-b}\times\frac{1}{2} + \frac{1}{s-c}\times\left(\frac{-1}{2}\right)\right]$$

$$\Rightarrow \quad \frac{dA}{dc} = \frac{A}{4}\left[\frac{1}{s} + \frac{1}{s-a} + \frac{1}{s-b} - \frac{1}{s-c}\right]$$

$$\Rightarrow \quad \frac{\Delta A}{\Delta c} = \frac{A}{4}\left[\frac{1}{s} + \frac{1}{s-a} + \frac{1}{s-b} - \frac{1}{s-c}\right]$$

$$\Rightarrow \quad \Delta A = \frac{A}{4}\left[\frac{1}{s} + \frac{1}{s-a} + \frac{1}{s-b} - \frac{1}{s-c}\right]\Delta c.$$

Example 4. *The angles of a triangle are calculated from the sides a, b, c if small changes δa, δb and δc are made in the sides. Find δa, δb and δc approximately, where Δ is the area of the triangle and A, B, C are angles opposite to sides a, b, c respectively. Also, show that δa + δb + δc = 0.* (GBTU–2012)

Solution . From trigonometry, we have

$$a^2 = b^2 + c^2 - 2bc \cos A \quad \ldots(1)$$

Differentiating, we get

$$2a\delta a = 2b\delta b + 2c\delta c - 2b\delta c \cos A - 2c\delta b \cos A + 2bc \sin A\delta A$$

$$\Rightarrow bc \sin A.\delta A = a\delta a - (b - c\cos A)\delta b - (c - b\cos A)\delta c$$

$$\Rightarrow 2\Delta\,\delta A = a\delta a - (a\cos C + c\cos A - c\cos A)\,\delta b - (a\cos B + b\cos A - b\cos A)\,\delta c$$

$$\Rightarrow \quad \delta A = \frac{a}{2\Delta}(\delta a - \delta b\cos C - \delta c\cos B) \quad \ldots(2)$$

In a similar manner, we may get

$$\delta B = \frac{b}{2\Delta}(\delta b - \delta c\cos A - \delta a\cos C) \quad \ldots(3)$$

and $\delta C = \dfrac{c}{2\Delta}(\delta c - \delta a\cos B - \delta b\cos A) \quad \ldots(4)$

On adding (2), (3) and (4), we get

$$\delta A + \delta B + \delta C = \frac{1}{2\Delta}[(a - b\cos C - c\cos B)\delta a + (b - c\cos A - a\cos C)\delta b + (c - a\cos B - b\cos A)\delta c]$$

$$= \frac{1}{2\Delta}[(a-a)\delta a + (b-b)\delta b + (c-c)\delta c]$$

$$= \frac{1}{2\Delta}.0 = 0.$$

Example 5. *Find the possible percentage error in computing the parallel resistance r of three resistances r_1, r_2, r_3 from the formula* $\frac{1}{r}=\frac{1}{r_1}+\frac{1}{r_2}+\frac{1}{r_3}$. *if r_1, r_2, r_3 are each in error by +1.2%.* (UKTU–2011, 13)

Solution. Given that

$$\frac{1}{r}=\frac{1}{r_1}+\frac{1}{r_2}+\frac{1}{r_3}$$

$$\Rightarrow -\frac{1}{r^2}\Delta r=-\frac{1}{r_1^2}\Delta r_1-\frac{1}{r_2^2}\Delta r_2-\frac{1}{r_3^2}\Delta r_3$$

$$\Rightarrow \quad \frac{1}{r}\left(\frac{\Delta r}{r}\times 100\right)=\frac{1}{r_1}\left(\frac{\Delta r_1}{r_1}\times 100\right)+\frac{1}{r_2}\left(\frac{\Delta r_2}{r_2}\times 100\right)+\frac{1}{r_3}\left(\frac{\Delta r_3}{r_3}\times 100\right)$$

$$=\frac{1}{r_1}(1.2)+\frac{1}{r_2}(1.2)+\frac{1}{r_3}(1.2)$$

$$=1.2\left(\frac{1}{r_1}+\frac{1}{r_2}+\frac{1}{r_3}\right)=1.2\left(\frac{1}{r}\right)$$

$\Rightarrow \frac{\Delta r}{r}\times 100=1.2$, which is the required error in r.

Example 6. (i) *In estimating the number of bricks in a pile which is measured to be (5m × 10m × 5m), the count of bricks is taken as 100 bricks per m^3. Find the error in the cost when the tape is stretched 2% beyond its standard length. The cost of bricks is Rs 2000 per thousand bricks.* (UPTU–2001, 07)

(ii) *In estimating the cost of a pile of bricks as 6m × 50m × 4m, the tape is stretched 1% beyond the standard length. If the count is 12 bricks is 1 m^3 and bricks cost Rs 1000, find the approximate error in the cost.* (UKTU–2010)

Solution. (i) Let length, breadth and height of the pile be x, y and z respectively.

The volume of the pile, $V = xyz$

$\Rightarrow \log V = \log x + \log y + \log z$

Differentiating, we get

$$\frac{\Delta V}{V}\times 100=\left(\frac{\Delta x}{x}\times 100\right)+\left(\frac{\Delta y}{y}\times 100\right)+\left(\frac{\Delta z}{z}\times 100\right)$$

$$= 2 + 2 + 2 = 6$$

Therefore,

$$\Delta V=\frac{6}{100}V=\frac{6}{100}(250)=15\text{ m}^3$$

$$[\because V = 5\times 10\times 5 = 250\text{ m}^3]$$

Hence, number of bricks in ΔV

$$=15\times 100=1500$$

and error in the cost

$$=1500\times\frac{2000}{1000}=\text{Rs } 3000$$

(ii) Proceed same as in part (i), we get

$$\frac{\Delta V}{V}\times 100=\left(\frac{\Delta x}{x}\times 100\right)+\left(\frac{\Delta y}{y}\times 100\right)+\left(\frac{\Delta z}{z}\times 100\right)$$

$$= 1 + 1 + 1 = 3$$

$$\therefore \quad \Delta V=\frac{3}{100}V=\frac{3}{100}\times 1200=36\text{ m}^3$$

$$[\because V = 6\times 50\times 4 = 1200\text{ m}^3]$$

$$\therefore \quad \Delta V = 36\times 12 = 432$$

and error in the cost

$$=432\times\frac{100}{1000}=\text{Rs } 43.20$$

Example 7. *The period of a simple pendulum is* $T=2\pi\sqrt{\frac{l}{g}}$. *Find the maximum percentage error in T due to the possible errors upto 1% in l and 2.5% in g.* (UPTU–2004)

Solution. We have $\quad T=2\pi\sqrt{\frac{l}{g}}$

$$\Rightarrow \quad \log T=\log 2\pi+\frac{1}{2}\log l-\frac{1}{2}\log g$$

$$\Rightarrow \quad \frac{1}{T}\Delta T=\frac{1}{2}\frac{\Delta l}{l}-\frac{1}{2}\frac{\Delta g}{g}$$

$$\Rightarrow \quad \frac{\Delta T}{T}\times 100=\frac{1}{2}\left[\frac{\Delta l}{l}\times 100-\frac{\Delta g}{g}\times 100\right]$$

$$=\frac{1}{2}[1\pm 2.5]$$

Hence, maximum error in T

$$=\frac{1}{2}(1+2.5)=1.75\%.$$

Example 8. *The height h and the semi-vertical angle α of a cone are measured and from them, the total surface area of the cone including the base is calculated. If h and α are in error by small quantities δh and δα respectively. Find the corresponding error in the area. Show further that if* $\alpha=\frac{\pi}{6}$ *an error of +1% in h will be approximately compensated by an error of – 0.33° in α.* (UPTU–2009)

Solution. We have

Base radius, $r = h\tan\alpha$

Slant height, $l = h\sec\alpha$

Total area, $A = \pi r^2 + \pi rl$

$= \pi r(r + l)$

$= \pi h\tan\alpha(h\tan\alpha + h\sec\alpha)$

$= \pi h^2(\tan^2\alpha + \sec\alpha\tan\alpha)$

So, $$\delta A=\frac{\partial A}{\partial h}\delta h+\frac{\partial A}{\partial \alpha}\delta\alpha$$

$$= 2\pi h(\tan^2\alpha + \sec\alpha\tan\alpha)\,\delta h + \pi h^2(2\tan\alpha\sec^2\alpha + \sec^3\alpha + \sec\alpha\tan^2\alpha)\delta\alpha \quad ...(1)$$

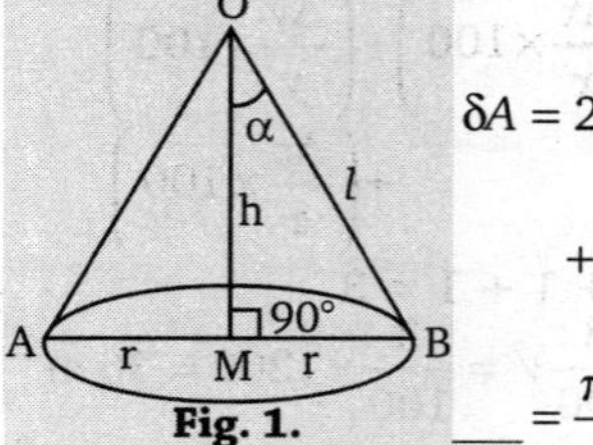

Fig. 1.

Let us set $\alpha = \frac{\pi}{6}, \delta h = \frac{h}{100}$ in (1), we get

$$\delta A = 2\pi h\left(\frac{1}{3}+\frac{2}{3}\right)\frac{h}{100} + \pi h^2\left[\frac{2}{\sqrt{3}}\left(\frac{4}{3}\right) + \frac{8}{3\sqrt{3}} + \frac{2}{3\sqrt{3}}\right]\delta\alpha$$

$$= \frac{\pi h^2}{50} + 2\sqrt{3}\pi h^2\delta\alpha$$

Now, since the error in h is to be compensated by the error in α.

Therefore, $\delta A = 0$

$$\therefore \quad \delta\alpha = \frac{-1}{100\sqrt{3}} \text{ radians} = -\frac{57.3^\circ}{173.2} = -0.33^\circ.$$

$$(\because 1 \text{ radian} \approx 57.3^\circ)$$

Example 9. *Evaluate* $[(4.85)^2 + 2(2.5)^3]^{1/5}$.

Solution . Let $f(x, y) = [x^2 + 2y^3]^{1/5}$

Taking $x = 5$, $\delta x = 4.85 - 5 = -0.15$, $y = 2$, $\delta y = 2.5 - 2 = 0.5$

Now $\frac{\partial f}{\partial x} = \frac{1}{5}(x^2 + 2y^3)^{-4/5}(2x)$

$$= \frac{1}{5}(5^2 + 2\times 2^3)^{-4/5}(2\times 5)$$

$$= \frac{10}{5}(41)^{-4/5}$$

Fig. (1)

and $\frac{\partial f}{\partial y} = \frac{1}{5}(x^2 + 2y^3)^{-4/5}.6y^2$

$$= \frac{1}{5}(5^2 + 2\times 2^3)^{-4/5}(6\times 2^2)$$

$$= \frac{24}{5}(41)^{-4/5}$$

We know that

$$df = \frac{\partial f}{\partial x}\delta x + \frac{\partial f}{\partial y}\delta y$$

$$= \frac{10}{5}(41)^{-4/5}(-0.15) + \frac{24}{5}(41)^{-4/5}(0.5)$$

$$= -\frac{1.5}{5}(41)^{-4/5} + \frac{12}{5}(41)^{-4/5}$$

$$= \frac{10.5}{5}(41)^{-4/5} = 2.1(41)^{-4/5}$$

Therefore, we have

$f(x + \delta x, y + \delta y)$

$$= f(x, y) + \left(\frac{\partial f}{\partial x}\delta x + \frac{\partial f}{\partial y}\delta y\right) = f(x, y) + df$$

Hence

$[(4.85)^2 + 2(2.5)^3]^{1/5}$

$$= f(5, 2) + \delta f$$

$$= [5^2 + 2\times 2^3]^{1/5} + 2.1(41)^{-4/5}$$

$$= (41)^{1/5} + 2.1(41)^{-4/5}$$

$$= 2.10163 + 2.1(0.05126) = 2.15289.$$

Example 10. *Compute an approximate value of* $(1.05)^{3.02}$.

Solution . Let $f(x, y) = x^y$

$$\Rightarrow \frac{\partial f}{\partial x} = yx^{y-1}, \frac{\partial f}{\partial y} = x^y \log x$$

Let us take $x = 1$, $\delta x = 0.05$, $y = 3$, $\delta y = 0.02$

We know that

$$df = \frac{\partial f}{\partial x}\delta x + \frac{\partial f}{\partial y}\delta y$$

$$= yx^{y-1}\delta x + x^y \log x\,\delta y$$

$$\Rightarrow \quad df = 3(1)(0.05) + 1^3.\log 1.(0.02) = 0.15$$

Hence, $(1.05)^{3.01} = f(1.3) + df$

$$= 1 + 0.15 = 1.15.$$

Example 11. *Find the approximate value of* $[(0.98)^2 + (2.01)^2 + (1.94)^2]^{1/2}$.

Solution . Let $f(x, y, z) = (x^2 + y^2 + z^2)^{1/2} \quad ...(1)$

Taking $x = 1$, $y = 2$ and $z = 2$ so that $dx = -0.02$, $dy = 0.01$, $dz = -0.06$

From (1), we get

$$\frac{\partial f}{\partial x} = x(x^2 + y^2 + z^2)^{-1/2}$$

$$\frac{\partial f}{\partial y} = y(x^2 + y^2 + z^2)^{-1/2},$$

$$\frac{\partial f}{\partial z} = z(x^2 + y^2 + z^2)^{-1/2}$$

We know that

$$df = \frac{\partial f}{\partial x}dx + \frac{\partial f}{\partial y}dy + \frac{\partial f}{\partial z}dz$$

$$= (x^2 + y^2 + z^2)^{-1/2}(xdx + ydy + zdz)$$

$$= \frac{1}{3}(-0.02 + 0.02 - .12) = -0.04$$

Hence, $[(0.98)^2 + (2.01)^2 + (1.94)^2]^{1/2}$

$$= f(1, 2, 2) + df$$

$$= 3 + (-0.04) = 2.96.$$

Example 12. *The angles of a triangle are calculated from its sides a, b, c. If small changes δa, δb, δc are made in the sides. Show that approximately*

$$\delta A = \frac{a}{2\Delta}[da - db\cos C - dc\cos B]$$

where Δ is the area of the triangle and A, B and C are angles opposite to sides a, b, c respectively. Verify δA + δB + δC = 0. (UPTU–2002)

Solution . By trigonometry, we have

$$\cos A = \frac{b^2 + c^2 - a^2}{2bc} \Rightarrow 2\cos A = \frac{b^2 + c^2 - a^2}{bc} \quad ...(1)$$

Differentiating (1), we get

$$-2\sin A\,\delta A = \frac{bc[2b\delta b + 2c\delta c - 2a\delta a] - (b^2+c^2-a^2)(b\delta c + c\delta b)}{(bc)^2}$$

$$= \frac{(2b^2c - b^2c - c^3 + a^2c)\delta b + (2bc^2 - b^3 - c^2b + a^2b)\delta c - 2abc\delta a}{(bc)^2}$$

$$= \frac{(b^2c - c^3 + a^2c)\delta b + (bc^2 - b^3 + a^2b)\delta c - 2abc\delta a}{(bc)^2}$$

$$= \frac{2a}{bc}\left[\frac{b^2+a^2-c^2}{2ab}\delta b + \frac{a^2+c^2-b^2}{2ac}\delta c - \delta a\right]$$

$$= \frac{2a}{bc}[\cos C\delta b + \cos B\delta c - \delta a] \qquad \left[\because \Delta = \frac{1}{2}bc\sin A\right] \quad \ldots(2)$$

$$\Rightarrow \quad \delta A = \frac{a}{bc\sin A}[\delta a - \delta b\cos C - \delta a\cos B]$$

$$= \frac{a}{2\Delta}[\delta a - \delta b\cos C - \delta a\cos B] \quad \ldots(3)$$

Similarly, $\delta B = \dfrac{b}{2\Delta}[\delta b - \delta c\cos A - \delta a\cos B]$

and $\delta C = \dfrac{c}{2\Delta}[\delta c - \delta a\cos B - \delta b\cos A] \ldots(4)$

On adding (2), (3) and (4), we get

$$2\Delta[\delta A + \delta B + \delta C] = [a - b\cos C - c\cos B]\delta a + [b - a\cos C - c\cos A]\delta b + [c - a\cos B - b\cos A]\delta c$$

$$= (a-a)\delta a + (b-b)\delta b + (c-c)\delta c = 0.$$

Example 13. *Find the percentage error in calculating the area of an ellipse $\frac{x^2}{a^2} + \frac{y^2}{b^2} = 1$ when an error of +1% is made in measuring the major and minor axes.* (UPTU–2011)

Solution. We know that if $2a$ and $2b$ are the length of major and minor axes respectively of the ellipse $\frac{x^2}{a^2} + \frac{y^2}{b^2} = 1$, then the area of the ellipse is

$$A = \pi ab$$

$$\Rightarrow \quad \log A = \log \pi + \log a + \log b$$

$$\Rightarrow \quad \frac{\delta A}{A} = 0 + \frac{\delta a}{a} + \frac{\delta b}{b}$$

$$\Rightarrow \frac{\delta A}{A}\times 100 = \frac{\delta a}{a}\times 100 + \frac{\delta b}{b}\times 100 \quad \ldots(1)$$

Now, percentage error in measuring the major and minor axes respectively is

$$\frac{\delta(2a)}{2a}\times 100 = \frac{\delta a}{a}\times 100 = 1$$

and $$\frac{\delta(2b)}{2a}\times 100 = \frac{\delta b}{b}\times 100 = 1$$

Putting the values in (1), we get

$$\frac{\delta A}{A} = 1 + 1 = 2.$$

Hence, percentage error in the area of ellipse is 2%.

Example 14. *If $f(x, y) = x^2y^{1/10}$. Compute the value of f when $x = 1.99$ and $y = 3.01$.* (UPTU–2008)

Solution. We have $f(x, y) = x^2y^{1/10}$.

We have to calculate $f(1.99, 3.01)$

Let $x = 2, y = 3$.

Then $x + \delta x = 1.99 \Rightarrow \delta x = -0.01$

$y + \delta y = 3.01 \Rightarrow \delta y = 0.01$

Now $f = x^2y^{1/10}$

$$\Rightarrow \quad \delta f = \frac{\partial f}{\partial x}\delta x + \frac{\partial f}{\partial y}\delta y$$

$$= 2xy^{1/10}\delta x + \frac{1}{10}x^2y^{-1/10}\delta y$$

$$= 2(2)3^{1/10}(-0.01) + \frac{1}{10}(2)^2(3)^{-1/10}(0.01)$$

$$= 3^{1/10}\left[-0.04 + \frac{1}{10}(4)(3)^{-1}(0.01)\right]$$

$$= 3^{3/10}\left(-0.04 + \frac{0.004}{3}\right)$$

$$= 3^{-9/10}(-0.12 + 0.004)$$

$$= 3^{-9/10}(-0.116)$$

Hence, $f(1.99, 3.01) = f(2, 3) + \delta f$

$$= 2^2(3)^{3/10} + (-0.116)3^{-9/10}$$

$$= 3^{1/10}\left[4 - \frac{0.116}{3}\right]$$

$$= 3^{-9/10}(12 - 0.116)$$

$$= 3^{-9/10}\times 11.884$$

$$= 0.3720\times 11.884 = 4.4213$$

Example 15. *In the facturing of closed rectangular boxes with specified sides a, b, c ($a \neq b \neq c$), small changes of A%, B%, C% occurred in a, b and c respectively from box to box from the specified dimension, show that* $\frac{A}{a(b-c)} = \frac{B}{b(c-a)} = \frac{C}{c(a-b)}$

Solution. We have $V = abc$

$$\Rightarrow \quad \delta V = bc\delta a + ca\delta b + ab\delta c$$

$$0 = bc\delta a + ca\delta b + ab\delta c$$

$$0 = \frac{\delta a}{a} + \frac{\delta b}{b} + \frac{\delta c}{c}$$

$$= \left(\frac{\delta a}{a}\times 100\right) + \left(\frac{\delta b}{b}\times 100\right) + \left(\frac{\delta c}{c}\times 100\right)$$

$$\Rightarrow \quad A + B + C = 0 \quad \ldots(1)$$

Now, surface area

$$s = 2(ab + bc + ca)$$

$$\Rightarrow \quad \frac{1}{2}\delta s = (b+c)\delta a + (c+a)\delta b + (a+b)\delta c$$

$$0=\left(\frac{b+c}{bc}\right)\frac{\delta a}{a}+\left(\frac{c+a}{ca}\right)\frac{\delta b}{b}+\left(\frac{a+b}{ab}\right)\frac{\delta c}{c}$$

$$0=\left(\frac{b+c}{bc}\right)\left(\frac{\delta a}{a}\times 100\right)+\left(\frac{c+a}{ca}\right)\left(\frac{\delta b}{b}\times 100\right)+\left(\frac{a+b}{ab}\right)\left(\frac{\delta c}{c}\times 100\right)$$

$$\Rightarrow \left(\frac{b+c}{bc}\right)A+\left(\frac{c+a}{ca}\right)B+\left(\frac{a+b}{ab}\right)C=0 \quad \ldots(2)$$

Solving (1) and (2) by cross multiplication method, we have

$$\frac{A}{\left(\frac{a+b}{ab}\right)-\left(\frac{a+c}{ac}\right)}=\frac{B}{\left(\frac{b+c}{bc}\right)-\left(\frac{a+b}{ab}\right)}=\frac{C}{\left(\frac{c+a}{ca}\right)-\left(\frac{b+c}{bc}\right)}$$

$$\Rightarrow \quad \frac{A}{a(b-c)}=\frac{B}{b(c-a)}=\frac{C}{c(a-b)}$$

EXERCISE 13.1

1. The time T of a complete oscillation of a simple pendulum of length l is governed by the equation $T=2\pi\sqrt{\frac{l}{g}}$, g is constant. Find the approximate error in the calculated value of T corresponding to an error of 2% in the value of l. (UPTU–2009)
2. The diameter and height of a right circular cylinder are found by measurement to be 8.0 cm and 12.5 cm respectively with possible errors of 0.05 in each measurement. Find the maximum possible approximate error in the computed volume. (GBTU–2011)
3. The work that must be done to propel a ship of displacement D for a distance S in time t is proportional to $\frac{S^2D^{2/3}}{t^2}$. Find approximately the increase of work necessary when the displacement is increased by 1%, the time diminished by 1% and the distance diminished by 2%.
4. The power P required to propel a steamer of length l at a speed u is given by $P=\lambda u^3 l^3$ where λ is constant. If u is increased by 3% and l is decreased by 1%, find the corresponding increase in P. (GBTU–2010, UKTU–2012)
5. The diameter and altitude of a can in the shape of a right circualr cylinder are measured as 4 cm and 6 cm respectively. The possible error in each measurement is 0.1 cm. Find approximately the maximum possible error in the value computed for the volume and lateral surface.
6. What error in the common logarithm of a number will be produced by an error of 1% in the number.
7. Find the possible percentage error in computing the parallel resistance r of two resistances r_1 and r_2 from the formula $\frac{1}{r}=\frac{1}{r_1}+\frac{1}{r_2}$ where the error in both r_1 and r_2 is +2% each. (UKTU–2011)
8. Compute an approximate value of $[(3.82)^2+2(2.1)^3]^{1/5}$. (MTU–2012)
9. Evaluate $\log[(1.01)^{1/3}+(0.99)^{1/4}-1]$. (UPTU–2009, MTU–2011)
10. The resistance R of a circuit was found by formula $I=\frac{E}{R}$. If there is an error of 0.1 amp in reading I, 0.5 volts in E. Find the corresponding possible percentage error in R when reading are I = 15 amp and E = 100 volts. (MTU–2011)
11. Prove that the relative error of a quotient does not exceed the sum of the relative errors of dividend and the divisor.
12. The side a and the opposite angle A of a ΔABC remain constant. Show that when the other sides and angles are slightly varied
$$\frac{\delta b}{\cos B}+\frac{\delta c}{\cos C}=0.$$
13. In determining the specific gravity by the formula $S=\frac{A}{A-W}$ where A is the weight in air and W is the weight in water; A can be read within 0.01 gm and W within 0.02 gm. Find approximately the maximum error in S if the readings are A = 1.1 gm W = 0.6 gm. Find also the maximum relative error.
14. If the kinetic energy T is given by $T=\frac{1}{2}mv^2$, find approximate the change in T as the mass m changes from 49 to 49.5 and the velocity v changes from 1600 to 1590.
15. The deflection at the centre of a rod of length l and diameter d supported at its ends and loaded at the centre with a weight w varies as wl^2d^{-4}. What is the increase in the deflection corresponding to p% increase in w, q% decrease in l and r% increase in d.

ANSWERS

1. 1% **2.** 3.3 π cu cm **3.** $-\frac{4}{3}$% **4.** 6% **5.** 1.6 π cu cm, π sq cm
6. 0.0043429 **7.** 2% **8.** 2.012 **9.** 0.00083
10. –16.66% **11.** $\left|\frac{\delta z}{z}\right|\le\left|\frac{\delta x}{x}\right|+\left|\frac{\delta y}{y}\right|$ **13.** 0.112, 0.05091 **14.** 144000 units **15.** $(p-3q-4r)$%

FFFFF

GATEtutor

KEY Terms and Results

- **Partial Derivatives :** If $u = f(x, y)$ be a continuous function of two independent variables x and y, then the differential coefficient of u w.r.t. x (regarding y as constant) is called the partial derivative of u w.r.t. x. Similarly, by keeping x constant and allowing y alone to vary, we can define the partial derivative of u w.r.t. y.
- **Symmetric function :** A function $u = u(x, y)$ is said to be symmetric if $u(x, y) = u(y, x)$.
- **Homogeneous function :** A function $f(x, y)$ is said to be homogeneous function of degree n, if the degree of each term of $f(x, y)$ is equal to n.
- If u is a homogeneous function of degree n, then $x\frac{\partial u}{\partial x}+y\frac{\partial u}{\partial y}=nu$ **(Euler's Theorem)**

  and $x^2\frac{\partial^2 u}{\partial x^2}+2xy\frac{\partial^2 u}{\partial x\partial y}+y^2\frac{\partial^2 u}{\partial y^2}=n(n-1)u.$
- The relation of the type $f(x, y) = c$, where y is not explicity in terms of x are called implicit function.
- **Implicit function:** If $f(x, y)$ be a function of two variables and $y = g(x)$ be a function of x such that $f(x,g(x))=0$ then $y=g(x)$ is called an implicit function defined by the functional equation $f(x,y)=0$.
- **Jacobian:** Let $u_1,u_2,\ldots,u_n$ be n functions of n variables $x_1,x_2,\ldots,x_n$, possessing partial derivatives of the first order, then the determinant

$$\begin{vmatrix} \frac{\partial u_1}{\partial x_1} & \frac{\partial u_1}{\partial x_2} & \cdots & \frac{\partial u_1}{\partial x_n} \\ \frac{\partial u_2}{\partial x_1} & \frac{\partial u_2}{\partial x_2} & \cdots & \frac{\partial u_2}{\partial x_n} \\ \vdots & \vdots & \vdots & \vdots \\ \frac{\partial u_n}{\partial x_1} & \frac{\partial u_n}{\partial x_2} & \cdots & \frac{\partial u_n}{\partial x_2} \end{vmatrix}$$

  is called the Jacobian.
- A functional equation may or may not define an implicit function.
- If the functions $u_1,u_2,\ldots,u_n$ of n independent variables $x_1,x_2,\ldots,x_n$ are of the following form $u_1=f_1(x_1)$, $u_2=f_2(x_1, x_2),\ldots,u_n=f_n(x_1, x_2,\ldots,x_n)$ then

$$\frac{\partial(u_1,u_2,\ldots,u_n)}{\partial(x_1,x_2,\ldots,x_n)}=\frac{\partial u_1}{\partial x_1}\cdot\frac{\partial u_2}{\partial x_2}\cdot\frac{\partial u_3}{\partial x_3}\cdots\frac{\partial u_n}{\partial x_n}$$

- If u_1,u_2 are functions of y_1, y_2 and y_1,y_2 are functions of x_1,x_2 then

$$\frac{\partial(u_1,u_2)}{\partial(x_1,x_2)}=\frac{\partial(u_1,u_2)}{\partial(y_1,y_2)}\cdot\frac{\partial(y_1,y_2)}{\partial(x_1,x_2)}$$

- If u_1, u_2, ..., u_n are functions of y_1, y_2, ..., y_n and y_1, y_2, ..., y_n are functions of x_1, x_2, ..., x_n, then

$$\frac{\partial(u_1,u_2,u_3,\ldots,u_n)}{\partial(x_1,x_2,x_3,\ldots,x_n)}=\frac{\partial(u_1,u_2,u_3,\ldots,u_n)}{\partial(y_1,y_2,y_3,\ldots y_n)}\cdot\frac{\partial(y_1,y_2,y_3,\ldots,y_n)}{\partial(x_1,x_2,x_3,\ldots,x_n)}$$

- If the functions u, v, w of three variables x,y,z are not independent, then the Jacobian of u,v,w w.r.t. x, y, z vanishes.
- $\frac{\partial(u,v,w)}{\partial(x,y,z)}\times\frac{\partial(x,y,z)}{\partial(u,v,w)}=1$
- If u_1,u_2, are implicit functions of x_1,x_2, *i.e.*, $F_1(u_1,u_2,x_1,x_2)=0$ and $F_2(u_1,u_2,x_1,x_2)=0$ then

$$\frac{\partial(u_1,u_2)}{\partial(x_1,x_2)}=(-1)^2\left[\frac{\partial(F_1,F_2)}{\partial(x_1,x_2)}\bigg/\frac{\partial(F_1,F_2)}{\partial(u_1,u_2)}\right]$$

- If $v_1,v_2,\ldots,v_n$ be the functions of n independent variables $x_1,x_2,\ldots,x_n$ such that $F(v_1,v_2,\ldots,v_n)=0$ it is necessary and sufficient that the Jacobian $\frac{\partial(v_1,v_2,\ldots,v_n)}{\partial(x_1,x_2,\ldots,x_n)}$ should vanish identically.

Review Questions and Project Work

1. Find the value of the parameter n so that $v = r^n(3\cos^2\theta - 1)$ satisfies $\frac{\partial}{\partial r}\left(r^2\frac{\partial V}{\partial r}\right)+\frac{1}{\sin\theta}\frac{\partial}{\partial\theta}\left(\sin\theta\frac{\partial V}{\partial\theta}\right)=0.$

2. If $x = r\cos\theta$, $y = r\sin\theta$, prove that $\frac{\partial^2 r}{\partial x^2}\cdot\frac{\partial^2 r}{\partial y^2}=\left(\frac{\partial^2 r}{\partial x\partial y}\right)^2.$

3. Verify Euler's theorem for the function

$$u=(x^{1/2}+y^{1/2})(x^n+y^n).$$

4. If $u=\sin^{-1}(x^2+y^2)^{1/5}$, prove that

$$x\frac{\partial^2 u}{\partial x^2}+2xy\frac{\partial^2 u}{\partial y\partial x}+y^2\frac{\partial^2 u}{\partial y^2}=\frac{2}{25}\tan u(2\tan^2 u-3).$$

5. If $z=\sqrt{x^2+y^2}$ and $x^3+y^3+3axy=5a^2$. Show that at $x=a, y=a$, $\frac{dz}{dx}=0$.

6. If $x = u + v$, $y = uv$ and z is function of x and y, show that $u\frac{\partial z}{\partial u}+v\frac{\partial z}{\partial v}=x\frac{\partial z}{\partial x}+2y\frac{\partial z}{\partial y}$.

7. If $u=\frac{y}{z}+\frac{z}{x}+\frac{x}{y}$, prove that $x\frac{\partial u}{\partial z}+y\frac{\partial u}{\partial y}+z\frac{\partial u}{\partial z}=0.$

8. The conduction of heat along a bar satisfies the differential equation $\frac{\partial u}{\partial t} = \mu \frac{\partial^2 u}{\partial x^2}$. Show that if $u = Ae^{-gx}\sin(nt - gx)$ where A, g, n are positive constants then $g = \sqrt{\frac{n}{2\mu}}$.

9. If $u = \sin^{-1}(x - y), x = 3t, y = 4t^3$, show that $\frac{du}{dt} = \frac{3}{\sqrt{1-t^2}}$.

10. If u and v are functions of x and y defined by $x = u + e^{-v}\sin u, y = v + e^{-v}\cos u$, prove that $\frac{\partial u}{\partial y} = \frac{\partial v}{\partial x}$.

11. Show that the least positive root of $xy = \sec y$ is a continuous function of $x \in [0, \infty[$ and it increases monotonically from 0 to $\frac{\pi}{2}$ as x increases from 0 to ∞.

12. Show that the equation $y^3 \sin x + y^2 \cos^2 x = 7$ determine unique implicit functions in a *nbd* of that point $\left(\frac{\pi}{6}, 0\right)$.

13. Show that none of x, y, z can be expressed as a function of the other two in a nbd of any point when

$$\sin x + \sin y + 2\sec z = 0.$$

14. If $f(1)=0$ and $f'(x) = \frac{1}{x}$ without using integration, prove that

$$f(x) + f(y) = f(x, y)$$

15. If $xyz = \lambda^3$ and a, b, c are positive, show that $(x+a)(y+b)(z+c)$ is minimum when $\frac{x}{a} = \frac{y}{b} = \frac{z}{c} = \frac{\lambda}{(abc)^{1/3}}$.

16. Show that $u=x+y+z$, $v=x-y+z$ and $w=x^2+y^2+z^2-2yz$ are not functionally independent.

17. If roots of the equation in t

$(t-x)^3+(t-y)^3+(t-z)^3=0$ are u, v, w, show that

$$\frac{\partial(u,v,w)}{\partial(x,y,z)} = -2\frac{(x-y)(y-z)(z-x)}{(u-v)(v-w)(w-u)}$$

Multiple Choice Questions

Choose the most appropriate one.

1. The maximum value of $u = \sin x \sin y \sin(x+y)$ is :

(a) $\frac{3\sqrt{3}}{8}$ (b) $\frac{\sqrt{3}}{8}$

(c) $\frac{3}{8}$ (d) none of these

2. When $x = y = \frac{\pi}{6}$ then function

$u = 2\sin\frac{x+y}{2}\cos\frac{x-y}{2} + \cos(x+y)$ is :

(a) minimum

(b) maximum

(c) neither maximum nor minimum

(d) none of the above

3. The minimum value of $u = xy + a^3\left(\frac{1}{x} + \frac{1}{y}\right)$ is :

(a) a^2 (b) $2a^2$

(c) $3a^2$ (d) $4a^2$

4. The function $u = x^3y^2(1-x-y)$ attains its maximum value at :

(a) $x = \frac{1}{2}, y = \frac{1}{2}$ (b) $x = \frac{1}{2}, y = \frac{1}{3}$

(c) $x = \frac{1}{3}, y = \frac{1}{2}$ (d) $x = 1, y = 1$

5. The function $u = 2a^2xy - 3ax^2y - ay^3 + x^3y + xy^3$ is:

(a) maximum at $\left(\frac{3a}{2}, \frac{-a}{2}\right)$ and $\left(\frac{a}{2}, \frac{a}{2}\right)$

(b) minimum at $\left(\frac{a}{2}, -\frac{a}{2}\right)$

(c) both (a) and (b) are true

(d) none of the above

6. The function $u = x^3 + y^3 - 3axy$ is

(a) minimum at $x=0$

(b) maximum at $x = 2^{1/3}$

(c) both (a) and (b) are true

(d) none of the above

7. The function $u = x^2y^2 - 5x^2 - 8xy - 5y^2$ is maximum at:

(a) $x=y=1$ (b) $x=y=0$

(c) $x=y=5$ (d) none of these

8. The function $u = x^4 + 2x^2y - x^2 + 3y^2$ is minimum at:

(a) $x = \pm\frac{\sqrt{3}}{2}, y = \frac{1}{4}$ (b) $x=0, y=0$

(c) $x = \frac{1}{2}, y = \frac{1}{4}$ (d) none of the above

9. The function $xy^2(3x+6y-2)$ is:

(a) maximum at (0, 0)

(b) minimum at (0, 0)

(c) neither minimum nor maximum at (0, 0)

(d) none of the above

10. The function $y^2 + 4xy + 3x^2 + x^3$ is:

(a) minimum at $\left(\frac{2}{3}, -\frac{4}{3}\right)$

(b) neither minimum nor maximum at (0,0)

(c) both (a) and (b) are true

(d) none of the above

11. The function $\frac{(x+y-1)}{x^2+y^2}$ is:

(a) maximum at (1, 1)

(b) minimum at (1, 1)

(c) neither minimum nor maximum at (1, 1)

(d) none of the above

12. The function $\frac{(x+y)}{x^2+2y^2+6}$ is:

(a) maximum at (2, 1)

(b) minimum at (–2, –1)

(c) both (a) and (b) are true

(d) none of the above

13. The maximum value of
$u = axy^2z^3 - x^2y^2z^3 - xy^3z^3 - xy^2z^4$ is:

(a) $108a^7$ (b) $\frac{107a^7}{7^7}$

(c) $\frac{a^7}{7^7}$ (d) none of these

14. The function $u=y^2+2z^2-5x^4+4x^5$ is:
(a) minimum at (1, 0, 0)
(b) neither minimum nor maximum at (0, 0, 0)
(c) both (a) and (b) are true
(d) none of the above

15. The function
$(x+y+z)^3-3(x+y+z)-24xyz+a^3$ has a:
(a) minimum at (1, 1, 1)
(b) maximum at (–1, –1, –1)
(c) both (a) and (b) are true
(d) none of the above

16. The function
$u=x^2+y^2+z^2+x-2z-xy$ is:

(a) minimum at $\left(-\frac{2}{3},-\frac{1}{3},1\right)$

(b) maximum at $\left(-\frac{2}{3},-\frac{1}{3},1\right)$

(c) neither maximum nor minimum at $\left(-\frac{2}{3},-\frac{1}{3},1\right)$
(d) none of the above

17. The maximum value of $xy(z-h)\left[\frac{x^2}{a^2}+\frac{y^2}{b^2}-\frac{z^2}{c^2}\right]$ is :

(a) $\left(\frac{2h}{5}\right)^5$ (b) $\frac{ab}{c^4}$

(c) $\left(\frac{2h}{5}\right)^5\left(\frac{ab}{c^4}\right)$ (d) none of these

18. The point such that the sum of the squares of its distance from n given points is:
(a) centroid
(b) centre of the mean position of the given points
(c) both (a) and (b) are true
(d) none of the above

19. The maximum value of
$-\alpha^2x^2-\beta^2y^2-\gamma^2z^2+(ax+by+cz).e$

(a) $\frac{1}{2e}\left(\frac{a^2}{\alpha^2}+\frac{b^2}{\beta^2}+\frac{c^2}{\gamma^2}\right)^{1/2}$

(b) $\left[\frac{1}{2e}\left(\frac{a^2}{\alpha^2}+\frac{b^2}{\beta^2}+\frac{c^2}{\gamma^2}\right)\right]^{1/2}$

(c) $\frac{1}{2e}$

(d) none of the above

20. The function $u=ax^3y^2-x^4y^2-x^3y^3$ is :

(a) maximum at $x=\frac{a}{2}, y=\frac{a}{3}$

(b) minimum at $x=\frac{a}{2}, y=\frac{a}{3}$

(c) neither maximum nor minimum at $x=\frac{a}{2}, y=\frac{a}{3}$

(d) none of the above

21. If x, y, z are the angles of a triangle, then the function $u=\sin x \sin y \sin z$ is :

(a) maximum at $x=y=z=\frac{\pi}{3}$

(b) minimum at $x=y=z=\frac{\pi}{3}$

(c) neither maximum nor minimum at $x=y=z=\frac{\pi}{3}$

(d) none of the above

22. The triangle of maximum area which can be inserted in a circle is:
(a) equilateral (b) isosceles
(c) right angled (d) none of the above

23. If the parameter of a triangle is constant, then its area is maximum when triangle is
(a) equilateral (b) isosceles
(c) right angled (d) none of the above

24. The function $f(x, y) = 2x^4 - 3x^2y + y^2$ has
(a) maximum at (0, 0)
(b) minimum at (0, 0)
(c) neither maximum nor minimum at (0, 0)
(d) none of the above

25. The function $f(x, y) = x^2 - 2xy + y^2 + x^3 - y^3 + x^5$ has:
(a) maximum value at origin
(b) minimum value at origin
(c) neither maximum nor minimum at origin
(d) none of the above

36. The function $f(x,y)=x^3+y^2-63(x+y)+12xy$ has:
(a) four stationary points
(b) maximum at (–7,–7)
(c) minimum at (3,3)
(d) all are true

27. Let $f = y^x$ then the value of $\frac{\partial^2 f}{\partial x \partial y}$ at $x = 2, y = 1$:
[GATE (ME)-2008]
(a) four stationary points
(b) maximum at (–7,–7)
(c) minimum at (3,3)
(d) all are true

ANSWERS

1. (a)	**2.** (b)	**3.** (c)	**4.** (b)	**5.** (c)	**6.** (c)	**7.** (b)	**8.** (a)	**9.** (c)	**10.** (c)
11. (a)	**12.** (c)	**13.** (b)	**14.** (c)	**15.** (c)	**16.** (a)	**17.** (c)	**18.** (c)	**19.** (b)	**20.** (a)
21. (a)	**22.** (a)	**23.** (a)	**24.** (c)	**25.** (c)	**26.** (d)	**27.** (a)			

Self Assessment Test

Verify each of the following :

1. If $u = \log(x^2+y^2) + \tan^{-1}\left(\frac{y}{x}\right)$, prove that $\frac{\partial^2 u}{\partial x^2} + \frac{\partial^2 u}{\partial y^2} = 0$ (Anna-2009)

2. If $r^2 = x^2+y^2+z^2$ and $u = r^m$, prove that
$u_{xx} + u_{yy} + u_{zz} = m(m+1)r^{m-2}$ (SRM, 2009; Raipur -2005, 11)

3. If $u = e^{xyz}$ show that $\frac{\partial^3 u}{\partial x \partial y \partial z} = e^{xyz}(x^2y^2z^2 + 3xyz + 1)$.
(Rajsthan-2005, 07; Osmania-2003)

4. If $x = r\cos\theta, y = r\sin\theta$, prove that $\left(\frac{\delta r}{\delta y}\right)^2 + \left(\frac{\delta r}{\delta y}\right)^2 = 1$
(Burdwan-2003)

5. If $z = x\log(x+r) - r$ where $r = x^2+y^2$, prove that
$\frac{\partial^2 z}{\partial x^2} + \frac{\partial^2 z}{\partial yz} = \frac{1}{x+y}, \frac{\partial^3 z}{\partial x^3} = \frac{-x}{r^3}$. (Mumbai-2008)

6. If $u = y^2 - 4ax, x = at^2, y = 2at$, show that $\frac{du}{dt} = 0$.
(Anna-2009)

7. If $x + y = 2e^{\theta}\cos\phi$ and $x - y = 2ie^{\theta}\sin\phi$, show that
$\frac{\partial^2 u}{\partial \theta^2} + \frac{\partial^2 u}{\partial \theta^2} = 4xy\frac{\partial^2 u}{\partial x \partial y}$ (UPTU-2002, 09; Nagpur-2009)

8. If $u = f(x,y)$ and $x = r\cos\theta, y = r\sin\theta$, prove that
$$\left(\frac{\partial u}{\partial x}\right)^2 + \left(\frac{\partial u}{\partial y}\right)^2 = \left(\frac{\partial u}{\partial r}\right)^2 + \frac{1}{r^2}\left(\frac{\partial u}{\partial \theta}\right)^2$$
(VTU–2010, Rohtak–2005, Mumbai-2006)

9. If $u = f(2x-3y), 3y-4z, 4z-2x$, prove that
$\frac{1}{2}\frac{\partial u}{\partial x} + \frac{1}{3}\frac{\partial u}{\partial y} + \frac{1}{4}\frac{\partial u}{\partial z} = 0$ (UPTU-2006; Raipur-2005)

10. If $u = x^2 - y^2, v = 2xy$ and $x = r\cos\theta, y = r\sin\theta$
show that $\frac{\partial(u,v)}{\partial(r,\theta)} = 4r^3$ (VTU-2009, Madras-2006)

11. If $u = x + 3y^2 - z^3, V = 4x^2yz, w = 2z^2 - xy$, show that at $(1,-1,0)$; $\frac{\partial(u,v,w)}{\partial(x,y,z)} = 20$ (VTU-2006)

12. If $u = x+y+z$, $uv = y+z$, $uvw = z$, show that $\frac{\partial(x,y,z)}{\partial(u,v,w)} = u^2v$
(PTU–2009,VTU–2003, Kurukshetra-2009)

13. In estimating the cost of a pile of bricks measured as 2m×15m×1.2m, the tape is streched 1% beyond the standard length. If the count is 450 bricks to 1 m^3 and bricks cost Rs. 530 per 1000. Show that the approximate error in the cost is Rs. 257.58.

14. In a place triangle, show that the maximum value of $\cos A \cos B \cos C$ is $\frac{1}{8}$. (VTU-2010, Nagpur-2009)

15. If $u = a^3x^2 + b^3y^2 + c^3z^2$, where $\frac{1}{x} + \frac{1}{y} + \frac{1}{z} = 1$, show that the stationary value of u is given by
$x = \frac{\Sigma a}{a}, y = \frac{\Sigma a}{b}, z = \frac{\Sigma c}{c}$ (SRM-2011, Kerala–2006)

16. Show that, if the perimeter of a triangle is constant, the triangle has maximum area when it is equivalent.

17. The temperature T at any point (x, y, z) in space is $T = 400xyz^2$. Show that the highest temperature on the surface of the unit sphere $x^2+y^2+z^2 = 1$ is 50.
(VTU-2009, Hissar-2005)

cccccc

INTEGRAL CALCULUS-I

UNIT OUTLINES

- **Reduction Formulae**
- **Definite Integrals**
- **Rectification of Curves**
- **Quadrature (Area of Curves)**
- **Surface and Volume of Solids of Revolutions**

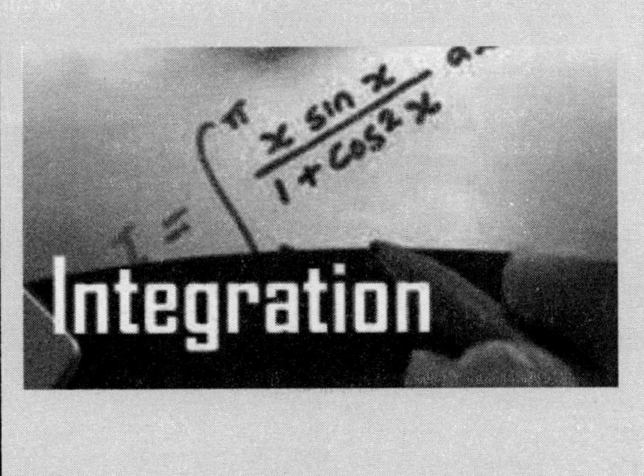

PRE-REQUISITE

S. No.	
1.	$\int \frac{1}{a^2+x^2}dx = \frac{1}{a}\tan^{-1}\frac{x}{a}$
2.	$\int \frac{1}{a^2-x^2}dx = \frac{1}{2a}\log\left\lvert\frac{a+x}{a-x}\right\rvert = \frac{1}{a}\tanh^{-1}\frac{x}{a}, \lvert x\rvert < a$
3.	$\int \frac{1}{x^2-a^2}dx = \frac{1}{2a}\log\left\lvert\frac{x-a}{x+a}\right\rvert = -\frac{1}{a}\cosh^{-1}\frac{x}{a}, \quad \lvert x\rvert < a$
4.	$\int \frac{1}{\sqrt{x^2-a^2}}dx = \cosh^{-1}\frac{x}{a} = \log\left\lvert\frac{x+\sqrt{x^2-a^2}}{a}\right\rvert$
5.	$\int \frac{1}{x^2+a^2}dx = \sinh^{-1}\frac{x}{a} = \log\left\lvert\frac{x+\sqrt{x^2+a^2}}{a}\right\rvert$
6.	$\int \sqrt{x^2+a^2}dx = \frac{x\sqrt{a^2+x^2}}{2} + \frac{a^2}{2}\sinh^{-1}\frac{x}{a}$
7.	$\int \sqrt{a^2-x^2}dx = \frac{x\sqrt{a^2-x^2}}{2} - \frac{a^2}{2}\cosh^{-1}\frac{x}{a}$
8.	$\int \sqrt{a^2-x^2}dx = \frac{x\sqrt{a^2-x^2}}{2} + \frac{a^2}{2}\sin^{-1}\frac{x}{a}$

PRE-REQUISITE

9.	$\int e^{ax}\cos bx\,dx = \frac{e^{ax}}{a^2+b^2}(a\cos bx + b\sin bx) = \frac{e^{ax}}{\sqrt{a^2+b^2}}\cos\left[bx - \tan^{-1}\frac{b}{a}\right]$
10.	$\int e^{ax}\sin bx\,dx = \frac{e^{ax}}{a^2+b^2}(a\sin bx - b\cos bx) = \frac{e^{ax}}{\sqrt{a^2+b^2}}\sin\left[bx - \tan^{-1}\frac{b}{a}\right]$
11.	$\int \sin^{-1} ax\,dx = x\sin^{-1} ax + \frac{1}{a}\sqrt{1-a^2x^2}$
12.	$\int \cos^{-1} ax\,dx = x\cos^{-1} ax - \frac{1}{a}\sqrt{1-a^2x^2}$
13.	$\int \tan^{-1} ax\,dx = x\tan^{-1} ax - \frac{1}{2a}\log(1+a^2x^2)$

Reduction Formulae

14.1 INTRODUCTION

The process relating one intregral to one or more integrals of the same type but simpler is called a reduction and this relation, between integrals is called a reduction formula.

The reduction formula is derived by two different methods one is integration by parts and other is differentiation of a suitable functions, says $P(x)$. Here the function $P(x)$ is chosen in such a way that $\frac{dP}{dx}$ has atleast one function, which is the integral of the given integral, whose reduction formula is required. It consists of the following steps :

(i) Selection of $P(x)$ (ii) Differentiation of $P(x)$ (iii) Integral of (ii).

14.2 REDUCTION FORMULAE FOR $\int \sin^m x \cos^n x dx$

(i) $I_{m,n} = \int \sin^m x \cos^n xdx = \frac{\cos^{n-1} x \sin^{m+1} x}{m+n} + \frac{n-1}{m+n} I_{m,n-2}$ $(m+n \neq 0)$

(ii) $I_{m,n} = \int \sin^m x \cos^n xdx = -\frac{\sin^{m-1} x \cos^{n+1} x}{m+n} + \frac{m-1}{m+n} I_{m-2,n}$ $(m+n \neq 0)$

Proof . (i) Here

$$I_{m,n} = \int \sin^m x \cos^n xdx = \cos^{n-1} x \int \sin^m x \cos xdx - \int \left[\frac{d}{dx}(\cos^{n-1} x) \int \sin^m x \cos xdx \right] .dx$$

$$= \frac{\cos^{n-1} x \sin^{m+1} x}{m+1} + \frac{n-1}{m+1} \int \cos^{n-2} x \sin^{m+2} xdx$$

$$= \frac{\cos^{n-1} x \sin^{m+1} x}{m+1} + \frac{n-1}{m+1} \int \cos^{n-2} x \sin^m x(1-\cos^2 x)dx$$

$$= \frac{\cos^{n-1} x \sin^{m+1} x}{m+1} + \frac{n-1}{m+1} .I_{m,n-2} - \frac{n-1}{m+1} .I_{m,n}$$

or $$\left(1+\frac{n-1}{m+1}\right) I_{m,n} = \frac{\cos^{n-1} x \sin^{m+1} x}{m+1} + \frac{n-1}{m+1} .I_{m,n-2}$$

$\Rightarrow$ $$I_{m,n} = \frac{\cos^{n-1} x \sin^{m+1} x}{m+n} + \frac{n-1}{m+n} I_{m,n-2}, m \neq 0.$$

REMARKS

- The above formula reduces the power or exponent m of cosine in each successive steps by 2. So, to establish the relation, one cosine is separated from the product so that $\sin^m x \cos x$ can be integrated.
- Here, one sine is separated from the product, so that $\cos^n x \sin x$ can be integrated and can be treated as a second function of the integration by parts. Therefore, following the same procedure as above, we can easily find

$$I_{m,n} = -\frac{\sin^{m-1} x \cos^{n+1} x}{m+n} + \frac{m-1}{m+n} I_{m-2,n}, m+n \neq 0.$$

- Put $m = 0$ in reduction formula (i), we may find

$$I_n = \int \cos^n x dx = \frac{\sin x \cos^{n-1} x}{n} + \frac{n-1}{n} I_{n-2}, n \neq 0.$$

- Put $n = 0$ in (ii), we may get

$$I_m = \int \sin^m x dx = -\frac{\sin^{m-1} x \cos x}{m} + \frac{m-1}{m} I_{m-2}, m \neq 0.$$

14.3 REDUCTION FORMULA FOR $\int \frac{dx}{\sin^m x \cos^n x} = \int \operatorname{cosec}^m x \sec^n x dx$

Replace n by $n + 2$ in § 14.2 (i), we get

$$I_{m,n+2} = \int \sin^m x \cos^{n+2} x dx = \frac{\sin^{m+1} x \cos^{n+1} x}{m+n+2} + \frac{n+1}{m+n+2} I_{m,n}$$

Let us replace m by $-m$ and n by $-n$, we get

$$I_{-m,-n+2} = \int \frac{dx}{\sin^m x \cos^{n-2} x} = -\frac{1}{m+n-2} \cdot \frac{1}{\sin^{m-1} x \cos^{n-1} x} + \frac{n-1}{m+n-2} I_{-m,-n}$$

or

$$I_{-m,-n} = \int \frac{dx}{\sin^m x \cos^n x} = \frac{1}{(n-1)\sin^{m-1} x \cos^{n-1} x} + \frac{m+n-2}{n-1} I_{-m,-n+2}, n \neq 1 \qquad \ldots(1)$$

Similarly from § 14.2 (ii), we get

$$I_{-m,-n} = \int \frac{dx}{\sin^m x \cos^n x} = \frac{1}{(m-1)\sin^{m-1} x \cos^{n-1} x} + \frac{m+n-2}{m-1} I_{-m+2,-n}, m \neq 1 \qquad \ldots(2)$$

Here, from (1) and (2), we observed that powers of cosines and sines are respectively reduced by 2 in each step of reduction.
Now, putting $m = 0$ in (1), and $n = 0$ in (2), we can obtain

$$I_{-n} = \int \frac{1}{\cos^n x} dx = \int \sec^n x dx = \frac{\sin x}{(n-1)\cos^{n-1} x} + \frac{n-2}{n-1} I_{-n+2}, n \neq 1 \qquad \ldots(3)$$

and

$$I_{-m} = \int \frac{1}{\sin^m x} dx = \int \operatorname{cosec}^m x dx = -\frac{\cos x}{(m-1)\sin^{m-1} x} + \frac{m-2}{m-1} I_{-m+2}, m \neq 1 \qquad \ldots(4)$$

From equations (1), (2), (3) and (4), we may get

$$I_{m,n} = \int \operatorname{cosec}^m x \sec^n x dx = \frac{\operatorname{cosec}^{m-1} x \sec^{n-1} x}{n-1} + \frac{m+n-2}{n-1} I_{m,n-2}, n \neq 1$$

$$I_{m,n} = \int \operatorname{cosec}^m x \sec^n x dx = \frac{\operatorname{cosec}^{m-1} x \sec^{n-1} x}{m-1} + \frac{m+n-2}{m-1} I_{m-2n}, m \neq 1$$

$$I_n = \int \sec^n x dx = \frac{\tan x \sec^{n-2} x}{n-1} + \frac{n-2}{n-1} I_{n-2}, n \neq 1$$

and

$$I_n = \int \operatorname{cosec}^m x dx = -\frac{\cot x \operatorname{cosec}^{m-2} x}{m-1} + \frac{m-2}{m-1} I_{m-2}, m \neq 1$$

14.4 REDUCTION FORMULA FOR $\int \sin^m x \sec^n x dx$ AND $\int \cos^n x \operatorname{cosec}^m x dx$

Let

$$I_{m,-n} = \int \frac{\sin^m x}{\cos^n x} dx = \int \frac{\sin^{m-1} x \sin x}{\cos^n x} dx = \frac{\sin^{m-1} x}{(n-1)\cos^{n-1} x} - \frac{m-1}{n-1} \int \frac{\sin^{m-2} x}{\cos^{n-2} x} dx$$

$$= \frac{\sin^{m-1} x}{(n-1)\cos^{n-1} x} - \frac{m-1}{n-1} I_{m-2,-n+2}, n \neq 1$$

$\Rightarrow$

$$\int \sin^m x \sec^n x dx = \frac{\sin^{m-1} x}{(n-1)\cos^{n-1} x} - \frac{m-1}{n-1} I_{m-2,-n+2}, n \neq 1 \qquad \ldots(1)$$

Similarly we can find

$$I_{-m,n} = \int \frac{\cos^n x}{\sin^m x} dx = -\frac{\cos^{m-1} x}{(m-1)\sin^{m-1} x} - \frac{n-1}{m-1} I_{-m+2,n-2}, m \neq 1 \quad \ldots(2)$$

Here, we observed that, in the above two cases powers of the numerator have been reduced. In the former case, a sine and in the latter case a cosine of the numerator has been separated and integrated separately. Finally, the powers of both numerator and denominator have been reduced by two in both cases.

The formuale (1) and (2) can also be written as

$$I_{m,n} = \int \sin^m x \sec^n x dx = \frac{\sin^{m-1} x \sec^{n-1} x}{n-1} - \frac{m-1}{n-1} I_{m-2,n-2}, n \neq 1 \quad \ldots(3)$$

and

$$I_{m,n} = \int \operatorname{cosec}^m x \cos^n x dx = -\frac{\operatorname{cosec}^{m-1} x \cos^{n-1} x}{m-1} - \frac{n-1}{m-1} I_{m-2,n-2}, n \neq 1 \quad \ldots(4)$$

14.5 REDUCTION FORMULAE FOR $\int \tan^n x dx$ AND $\int \cot^n x dx$

(i)

$$I_n = \int \tan^n x dx = \int \tan^{n-2} x \tan^2 x dx = \int \tan^{n-2} x (\sec^2 x - 1) dx$$

$$= \int \tan^{n-2} x \sec^2 x dx - \int \tan^{n-2} x dx \quad \ldots(1)$$

Now $\int \tan^{n-2} x \sec^2 x dx = \int t^{n-2} dt$, putting $\tan x = t$ and $\sec^2 t \, dt = dx$

$$\int \tan^{n-2} x \sec^2 x dx = \frac{t^{n-1}}{n-1} = \frac{\tan^{n-1} x}{n-1} \quad \ldots(2)$$

From equations (1) and (2), we get the reduction formula :

$$\therefore \quad \int \tan^n x dx = \frac{\tan^{n-1} x}{n-1} - \int \tan^{n-2} x dx, n \neq 1$$

(ii) Similarly,

$$I_n = \int \cot^n x dx = \int \cot^{n-2} x \cot^2 x dx = \int \cot^{n-2} x (\operatorname{cosec}^2 x - 1) dx$$

$$= \int \cot^{n-2} x \operatorname{cosec}^2 x dx - \int \cot^{n-2} x dx$$

$$\therefore \quad \int \cot^n x dx = -\frac{\cot^{n-1} x}{n-1} - \int \cot^{n-2} x dx \cdot$$

14.6 REDUCTION FORMULAE FOR $\int \sec^n x dx$ AND $\int \operatorname{cosec}^n x dx$

(i) Let

$$I_n = \int \sec^n x dx = \int \sec^{n-2} x \sec^2 x dx$$

$$= \sec^{n-2} x \int \sec^2 x dx - \int \left\{ \frac{d}{dx} (\sec^{n-2} x) \int \sec^{n-2} x dx \right\} dx$$

$$= \tan x \sec^{n-2} x - (n-2) \int \sec^{n-2} x \tan^2 x dx$$

$$= \tan x \sec^{n-2} x - (n-2) \int \sec^{n-2} x (\sec^2 x - 1) dx$$

or

$$I_n = \tan x \sec^{n-2} x - (n-2) \int \sec^n x dx + (n-2) \int \sec^{n-2} x dx$$

or

$$I_n = \tan x \sec^{n-2} x - (n-2) I_n + (n-2) I_{n-2}$$

or

$$I_n (1 + n - 2) = \tan x \sec^{n-2} x + (n-2) I_{n-2}$$

$$\therefore \quad I_n = \frac{1}{n-1} \tan x \sec^{n-2} x + \frac{n-2}{n-1} I_{n-2}, n \neq 1$$

Hence

$$\int \sec^n x dx = \frac{\sec^{n-2} x \tan x}{n-1} + \frac{n-2}{n-1} \int \sec^{n-2} x dx$$

Similarly, we obtain $\int \operatorname{cosec}^n x dx = -\frac{\operatorname{cosec}^{n-2} x \cot x}{n-1} + \frac{n-2}{n-1} \int \operatorname{cosec}^{n-2} x dx$

14.7 REDUCTION FORMULAE FOR $\int \tan^m x\sec^n x dx$ AND $\int \cot^m x \operatorname{cosec}^n x dx$

(i) Let

$$I_{m,n} = \int \tan^m x \sec^n x dx = \int \tan^{m-1} x(\sec^{n-1} x \sec x \tan x)dx$$

$$= \tan^{m-1} x \frac{\sec^n x}{n} - \frac{m-1}{n}\int \tan^{m-2} x \sec^2 x \sec^n x dx$$

$$= \frac{\tan^{m-1} x\sec^n x}{n} - \frac{m-1}{n} I_{m-2,n} - \frac{m-1}{m+n-1} I_{m,n}$$

$$\Rightarrow \quad I_{m,n} = \int \tan^m x\sec^n x dx = \frac{\tan^{m-1} x \sec^n x}{m+n-1} - \frac{m-1}{m+n-1} I_{m-2,n},\ m+n \neq 1$$

(ii) Similarly, the power of $\sec x$ is reduced by two during integrating $\int \tan^m x \sec^2 xdx$ by parts and then, we get

$$I_{m,n} = \int \tan^m x \sec^n x dx = \frac{\tan^{m+1} x \sec^{n-2} x}{m+n-1} - \frac{n-2}{m+n-1} I_{m,n-2}, (m+n \neq 1)$$

REMARKS

- $I_{m,n} = \int \cot^m x \operatorname{cosec}^n x dx = \frac{\cot^{m-1} x \operatorname{cosec}^n x}{m-n+1} - \left(\frac{m-1}{m-n+1}\right) I_{m-2,n}, m+1 \neq n$
- $I_{m,n} = \int \cot^m x \operatorname{cosec}^n x dx = \frac{\cot^{m+1} x \operatorname{cosec}^{n-2} x}{n-m+1} - \frac{n-2}{n-m+1} I_{m,n-2}, n+1 \neq m$

14.8 REDUCTION FORMULAE FOR $\int \cos^m x \cos nx dx$, $\int \cos^m x \sin nx dx$, $\int \sin^m x \cos nx dx$ ETC.

(i) Let

$$I_{m,n} = \int \cos^m x \cos nx dx = \cos^m x \frac{\sin nx}{n} + \frac{m}{n}\int \cos^{m-1} x \sin x \sin nx dx \quad \ldots(1)$$

$$= \frac{\cos^m x \sin nx}{n} + \frac{m}{n}\int \cos^{m-1} x\{\cos(n-1)x - \cos nx \cos x\}dx$$

$$[\because \cos(n-1)x = \cos nx \cos x + \sin nx \sin x]$$

$$= \frac{\cos^m x \sin nx}{n} + \frac{m}{n} I_{m-1,n-1} - \frac{m}{n} I_{m,n}$$

$$\Rightarrow \quad \left(1+\frac{m}{n}\right) I_{m,n} = \frac{\cos^m x \sin nx}{n} + \frac{m}{n} I_{m-1,n-1}$$

$$\Rightarrow \quad I_{m,n} = \frac{\cos^m x \sin nx}{n} + \frac{m}{n} I_{m-1,n-1}, m+n \neq 0 \quad \ldots(2)$$

Intetgrate (1) by parts and using (2), we get

$$I_{m,n} = \frac{\cos^m x \sin nx}{n} + \frac{m}{n}\left\{\cos^{m-1} x \sin x \left(-\frac{\cos nx}{n}\right) + \int [\cos^m x - (m-1)\cos^{m-2} x \sin^2 x]\frac{\cos nx}{n} dx\right\}$$

$$= \frac{\cos^m x \sin nx}{n} - \frac{m}{n^2}\cos^{m-1} x \sin x \cos nx + \frac{m^2}{n^2} I_{m,n} - \frac{m(m-1)}{n^2} I_{m-2,n}$$

$$\Rightarrow \quad \left(1-\frac{m^2}{n^2}\right) I_{m,n} = \frac{\cos^m x \sin nx}{n} - \frac{m}{n^2}\cos^{m-1} x \sin x \cos nx - \frac{m(m-1)}{n^2} I_{m-2,n}$$

$$\Rightarrow \quad I_{m,n} = -\frac{n\cos^m x \sin nx}{m^2-n^2} + \frac{m}{m^2-n^2}\cos^{m-1} x \sin x \cos nx + \frac{m(m-1)}{m^2-n^2} I_{m-2,n}$$

$$= -\frac{\cos^{m-1} x}{m^2-n^2}[n\cos x \sin nx - m \sin x \cos nx] + \frac{m(m-1)}{m^2-n^2} I_{m-2,n}$$

$$\Rightarrow \quad I_{m,n} = -\frac{\cos^2 nx}{m^2-n^2}\frac{d}{dx}\left(\frac{\cos^m x}{\cos nx}\right) + \frac{m(m-1)}{m^2-n^2} I_{m-2,n}$$

REMARK

- Here the power of cosine is reduced by two only.

(ii) Let

$$I_{m,n} = \int \cos^m x \sin nx dx = \cos^m x\left(-\frac{\cos nx}{n}\right) - \int m\cos^{m-1} x(-\sin x)\left(-\frac{\cos nx}{n}\right)dx \quad \ldots(1)$$

$$= -\frac{\cos^m x \cos nx}{n} - \frac{m}{n}\int \cos^{m-1} x \sin x \cos nx\, dx$$

$$= -\frac{\cos^m x \cos nx}{n} - \frac{m}{n}\int \cos^{m-1} x[\sin nx \cos x - \sin(n-1)x]dx$$

$$[\because \sin(n-1)x = \sin nx \cos x - \cos nx \sin x]$$

$$= -\frac{\cos^m x \cos nx}{n} - \frac{m}{n} I_{m,n} + \frac{m}{n} I_{m-1,n-1}$$

$$\Rightarrow \quad \left(1+\frac{m}{n}\right) I_{m,n} = -\frac{\cos^m x \cos nx}{m+n} + \frac{m}{m+n} I_{m-1,n-1}$$

$$\Rightarrow \quad I_{m,n} = -\frac{\cos^m x \cos nx}{m+n} + \frac{m}{m+n} I_{m-1,n-1}, m+n \neq 0$$

On intetgrating by parts the right hand integral of (1), we get

$$I_{m,n} = -\frac{\cos^m x \cos nx}{n} - \frac{m}{n}\left[(\cos^{m-1} x \sin x)\frac{\sin nx}{n} - \int\{(m-1)\cos^{m-2} x(-\sin^2 x) + \cos^m x\}\frac{\sin nx}{n} dx\right]$$

$$= -\frac{\cos^m x \cos nx}{n} - \frac{m}{n^2}\cos^{m-1} x \sin x \sin nx + \frac{m}{n^2}\int\{-(m-1)\cos^{m-2} x(1-\cos^2 x) + \cos^m x\}\sin nx dx$$

$$= -\frac{\cos^m x \cos nx}{n} - \frac{m}{n^2}\cos^{m-1} x \sin x \sin nx - \frac{m(m-1)}{n^2} I_{m-2,n} + \frac{m^2}{n^2} I_{m,n}$$

$$\Rightarrow \quad \left(1-\frac{m^2}{n^2}\right) I_{m,n} = -\frac{\cos^m x \cos nx}{n} - \frac{m}{n^2}\cos^{m-1} x \sin x \sin nx - \frac{m(m-1)}{n^2} I_{m-2,n}$$

$$\Rightarrow \quad I_{m,n} = \frac{n\cos^m x \cos nx}{m^2-n^2} + \frac{m\cos^{m-1} x \sin x \sin nx}{m^2-n^2} - \frac{m(m-1)}{m^2-n^2} I_{m-2,n}$$

(iii) Here, let

$$I_{m,n} = \int \sin^m x \cos nx dx = \sin^m x \frac{\sin nx}{n} - \int m\sin^{m-1} x \cos x \frac{\sin nx}{n} dx$$

$$= \frac{\sin^m x \sin nx}{n} - \frac{m}{n}\int \sin^{m-1} x \cos x \sin nx dx$$

$$= \frac{\sin^m x \sin nx}{n} - \frac{m}{n}\left[(\sin^{m-1} x \cos x)\left(-\frac{\cos nx}{n}\right) - \int\{(m-1)\sin^{m-2} x \cos^2 x - \sin^m x\}\left(-\frac{\cos nx}{n}\right)dx\right]$$

[Integrating by parts]

$$= \frac{\sin^m x \sin nx}{n} + \frac{m}{n^2}\sin^{m-1} x \cos x \cos nx - \frac{m}{n^2}\int\{(m-1)\sin^{m-2} x(1-\sin^2 x) - \sin^m x\}\cos nx dx$$

$$= \frac{\sin^m x \sin nx}{n} + \frac{m}{n^2}\sin^{m-1} x \cos x \cos nx - \frac{m(m-1)}{n^2} I_{m-2,n} + \frac{m^2}{n^2} I_{m,n}$$

$$\Rightarrow \quad I_{m,n} = -\frac{n\sin^m x\sin nx}{m^2-n^2} - \frac{m\sin^{m-1} x\cos x\cos nx}{m^2-n^2} + \frac{m(m-1)}{m^2-n^2} I_{m-2,n}$$

(iv) Let $I_{m,n} = \int \sin^m x \sin nx dx$

Proceed as above, we may easily get

$$\Rightarrow \quad I_{m,n} = -\frac{n\sin^m x\cos nx}{m^2-n^2} - \frac{m\sin^{m-1} x\cos x\sin nx}{m^2-n^2} + \frac{m(m+1)}{m^2-n^2} I_{m-2,n}.$$

14.9 REDUCTION FORMULAE FOR $\int x^n \sin mx dx$ AND $\int x^n \cos mx dx$

(i) Let $I(m,n) = \int x^n \sin mx dx = x^n \int \sin mx\, dx - \int \left\{ \frac{d}{dx}(x^n) \int \sin mx\, dx \right\} dx$ (Kanpur-2002)

$$= -\frac{x^n \cos mx}{m} + \frac{n}{m} \int x^{n-1} \cos mx\, dx$$

$$= -\frac{x^n \cos mx}{m} + \frac{n}{m}\left[\frac{x^{n-1}\sin mx}{m} - \frac{n-1}{m}\int x^{n-2} \sin mx\, dx \right]$$

or $$I(m,n) = -\frac{x^n \cos mx}{m} + \frac{nx^{n-1}\sin mx}{m^2} - \frac{n(n-1)}{m^2}\int x^{n-2}\sin mx\, dx$$

Hence, $$\int x^n \sin mx\, dx = -\frac{x^n \cos mx}{m} + \frac{nx^{n-1}\sin mx}{m^2} - \frac{n(n-1)}{m^2}\int x^{n-2}\sin mx\, dx$$

Similarly, $$\int x^n \cos mx\, dx = \frac{x^n \sin mx}{m} + \frac{nx^{n-1}\cos mx}{m^2} - \frac{n(n-1)}{m^2}\int x^{n-2}\cos mx\, dx.$$

14.10 REDUCTION FORMULAE FOR $\int x\sin^n x dx$ AND $\int x\cos^n x dx$

(i) Let $I_n = \int x \sin^n x dx$

$I_n = \int (x\sin^{n-1} x)\sin x dx$ (Integrating by part)

$$= x\sin^{n-1} x \int \sin x dx - \int \{\sin^{n-1} x + x(n-1)\sin^{n-2} x\cos x\}(-\cos x)dx$$

$$= -x\sin^{n-1} x\cos x + \int \sin^{n-1} x\cos x dx + (n-1)\int x\sin^{n-2} x\cos^2 x dx$$

$$= -x\sin^{n-1} x\cos x + \int \sin^{n-1} x\cos x dx + (n-1)\int x\sin^{n-2} x(1-\sin^2 x)dx$$

$\therefore$ $$I_n = -x\sin^{n-1} x\cos x + \frac{\sin^n x}{n} + (n-1)\int x\sin^{n-2} x dx - (n-1)\int x\sin^n x dx$$

or $$I_n = -x\sin^{n-1} x\cos x + \frac{\sin^n x}{n} + (n-1)\int x\sin^{n-2} x dx - (n-1)I_n$$

or $$I_n(1+n-1) = -x\sin^{n-1} x\cos x + \frac{\sin^n x}{n} + (n-1)\int x\sin^{n-2} x dx$$

Hence, $$\int x\sin^n x\, dx = \frac{-x\sin^{n-1} x\cos x}{n} + \frac{\sin^n x}{n^2} + \frac{n-1}{n}\int x\sin^{n-2} x\, dx$$

Similarly, $$\int x\cos^n x\, dx = \frac{x\cos^{n-1} x\sin x}{n} + \frac{\cos^n x}{n^2} + \frac{n-1}{n}\int x\cos^{n-2} x\, dx.$$

14.11 REDUCTION FORMULAE FOR $\int e^{ax}\cos^n x dx$ AND $\int e^{ax}\sin^n x dx$

(i) Let $$I_n = \int e^{ax}\cos^n x dx = \frac{e^{ax}}{a}\cos^n x + \frac{n}{a}\int e^{ax}\cos^{n-1} x\sin x dx$$

$$= \frac{e^{ax}}{a}\cos^n x + \frac{n}{a}\left\{ \frac{e^{ax}}{a}\cos^{n-1} x\sin x - \int \frac{e^{ax}}{a}[\cos^n x - (n-1)\cos^{n-2} x\sin^2 x]dx \right\}$$

$$= \frac{e^{ax}}{a}\cos^n x + \frac{n}{a^2}e^{ax}\cos^{n-1} x\sin x - \frac{n}{a^2}I_n + \frac{n(n-1)}{a^2}\int e^{ax}\cos^{n-2} x(1-\cos^2 x)dx$$

$$= \frac{e^{ax}}{a}\cos^n x + \frac{n}{a^2}e^{ax}\cos^{n-1} x\sin x - \frac{n}{a^2}I_n + \frac{n(n-1)}{a^2}I_{n-2} - \frac{n(n-1)}{a^2}I_n$$

$$\Rightarrow \quad \left(1+\frac{n}{a^2}+\frac{n(n-1)}{a^2}\right)I_n = e^{ax}\cos^{n-1} x\frac{(a\cos x+n\sin x)}{a^2} + \frac{n(n-1)}{a^2}I_{n-2}$$

$$\Rightarrow \quad I_n = \frac{e^{ax}\cos^{n-1} x(a\cos x+n\sin x)}{(n^2+a^2)} + \frac{n(n-1)}{n^2+a^2}I_{n-2}.$$

(ii) Similarly, we can find that

$$I_n = \int e^{ax}\sin^n xdx = \frac{e^{ax}\sin^{n-1} x}{n^2+a^2}(a\sin x - n\cos x) + \frac{n(n-1)}{n^2+a^2}I_{n-2}.$$

14.12 REDUCTION FORMULAE FOR $\int \cos nx \operatorname{cosec} x dx$ AND $\int \sin nx \sec x dx$

(i) Let

$$I_n = \int \cos nx \operatorname{cosec} xdx = \frac{\sin nx}{n}\operatorname{cosec} x + \int\frac{\sin nx}{n}\operatorname{cosec} x\cot xdx$$

$$= \frac{\sin nx}{n}\operatorname{cosec} x + \int\frac{1}{n}\frac{\sin nx\cos x}{\sin^2 x}dx$$

$$= \frac{\sin nx}{n}\operatorname{cosec} x + \int\frac{1}{n}\frac{\sin(n-1)x+\cos nx\sin x}{\sin^2 x}dx$$

$$= \frac{\sin nx\operatorname{cosec} x}{n} + \frac{1}{n}\int\sin(n-1)x\operatorname{cosec}^2 xdx + \frac{1}{n}I_n$$

$$\Rightarrow \quad \left(1-\frac{1}{n}\right)I_n = \frac{\sin nx\operatorname{cosec} x}{n} + \frac{1}{n}[\sin(n-1)x(-\cot x) - (n-1)\int\cos(n-1)x(-\cot x)dx]$$

$$\Rightarrow \quad \frac{n-1}{n}I_n = \frac{\sin nx\operatorname{cosec} x}{n} - \frac{\sin(n-1)x\cot x}{n} + \frac{(n-1)}{n}\int\frac{\cos(n-1)x\cos x}{\sin x}dx$$

$$\Rightarrow \quad I_n = \frac{\sin nx\operatorname{cosec} x}{n-1} - \frac{\sin(n-1)x\cot x}{n-1} + \int\frac{\cos(n-2)x-\sin(n-1)x\sin x}{\sin x}dx$$

$$= \frac{\sin nx\operatorname{cosec} x}{n-1} - \frac{\sin(n-1)x\cot x}{n-1} + I_{n-2} - \int\sin(n-1)xdx$$

$$= \frac{\sin nx\operatorname{cosec} x}{n-1} - \frac{\sin(n-1)x\cot x}{n-1} + I_{n-2} + \frac{1}{n-1}\cos(n-1)x$$

$$I_n - I_{n-2} = \frac{1}{n-1}\left[\frac{\sin nx}{\sin x} - \frac{\sin(n-1)x\cos x}{\sin x} + \cos(n-1)x\right]$$

$$= \frac{1}{n-1}\left[\frac{\sin nx - \sin(n-1)x\cos x}{\sin x} + \cos(n-1)x\right]$$

$$= \frac{1}{n-1}\left[\frac{\sin[(n-1)+1]x - \sin(n-1)x\cos x}{\sin x} + \cos(n-1)x\right] = \frac{2\cos(n-1)x}{n-1}$$

$$I_n = \frac{2\cos(n-1)x}{n-1} + I_{n-2}$$

(ii) Similarly,

$$I_n = \int\sin nx\sec xdx = -\frac{2\cos(n-1)x}{n-1} - I_{n-2}.$$

14.13 REDUCTION FORMULAE FOR $\int(\sin^{-1}x)^n dx$ AND $\int(\cos^{-1}x)^n dx$

(i) Let

$$I_n = \int(\cos^{-1} x)^n dx = x(\cos^{-1} x)^n - \int n(\cos^{-1} x)^{n-1}\left(\frac{-1}{\sqrt{1-x^2}}\right)xdx$$

$$= x(\cos^{-1} x)^n - n\int(\cos^{-1} x)^{n-1}\frac{x}{\sqrt{1-x^2}}dx$$

$$= x(\cos^{-1}x)^n - n(\cos^{-1}x)^{n-1}\sqrt{1-x^2}$$

$$- n\int (n-1)(\cos^{-1}x)^{n-2}\left(-\frac{1}{\sqrt{1-x^2}}\right)\sqrt{1-x^2}\,dx$$

$\Rightarrow$
$$I_n = x(\cos^{-1}x)^n - n(\cos^{-1}x)^{n-1}\sqrt{1-x^2} + n(n-1)I_{n-2}$$

(ii) Proceeding as same manner we get

$$I_n = \int (\sin^{-1}x)^n dx = x(\sin^{-1}x)^n + n(\sin^{-1}x)^{n-1}\sqrt{1-x^2} - n(n-1)I_{n-2}.$$

14.14 REDUCTION FORMULAE FOR $\int \frac{dx}{(a+b\cos x)^n}$ AND $\int \frac{dx}{(a+b\sin x)^n}$

(i) Let
$$I_n = \int \frac{dx}{(a+b\cos x)^n}$$

Also, let
$$P(x) = \frac{\sin x}{(a+b\cos x)^{n-1}} = \frac{\sin x}{t^{n-1}}, \text{ where } t = a + b\cos x$$

$\Rightarrow$
$$\cos x = \frac{t-a}{b}$$

Differentiating with respect to x, we get

$$\frac{dP(x)}{dx} = \frac{\cos x}{t^{n-1}} - (n-1)\frac{\sin x}{t^n}\frac{dt}{dx} = \frac{\cos x}{t^{n-1}} - (n-1)\frac{\sin x}{t^n}(-b\sin x)$$

$$= \frac{t-a}{bt^{n-1}} - \frac{b(n-1)}{t^n}(1-\cos^2 x) = \frac{t-a}{bt^{n-1}} + \frac{b(n-1)}{t^n}\left[1-\left(\frac{t-a}{b}\right)^2\right]$$

$$= \frac{t-a}{bt^{n-1}} + \frac{b(n-1)}{t^n}\left[1-\frac{t^2-2at+a^2}{b^2}\right] = \frac{(n-1)(b^2-a^2)}{bt^n} + \frac{a(2n-3)}{bt^{n-1}} - \frac{n-2}{bt^{n-2}}$$

$$= \frac{(n-1)(b^2-a^2)}{b}\cdot\frac{1}{(a+b\cos x)^n} + \frac{a(2n-3)}{b}\cdot\frac{1}{(a+b\cos x)^{n-1}} - \frac{n-2}{b}\cdot\frac{1}{(a+b\cos x)^{n-2}}$$

On integrating *w.r.t.* x, we get

$$P(x) = \frac{(n-1)(b^2-a^2)}{b}I_n + \frac{a(2n-3)}{b}I_{n-1} - \frac{n-2}{b}I_{n-2}$$

$\Rightarrow$
$$I_n = \frac{b\sin x}{(n-1)(b^2-a^2)(a+b\cos x)^{n-1}} - \frac{a(2n-3)}{(n-1)(b^2-a^2)}I_{n-1} + \frac{n-2}{(n-1)(b^2-a^2)}I_{n-2}$$

(ii) Similarly, by setting $P(x) = \frac{\cos x}{(a+b\sin x)^{n-1}}$, we may easily get

$$I_n = \int \frac{dx}{(a+b\sin x)^n} = \frac{b}{(n-1)(a^2-b^2)}\cdot\frac{\cos x}{(a+b\sin x)^{n-1}} + \frac{(2n-3)a}{(n-1)(a^2-b^2)}I_{n-1} - \frac{n-2}{(n-1)(a^2-b^2)}I_{n-2}.$$

14.15 REDUCTION FORMULA FOR $\int x^m(\log x)^n dx$

Let
$$I(m,n) = \int x^m(\log x)^n dx \qquad \text{(Integrating by part)}$$

$$= (\log x)^n \int x^m dx - \int\left\{\frac{n(\log x)^{n-1}}{x}\cdot\frac{x^{m+1}dx}{m+1}\right\}$$

$$= \frac{x^{m+1}(\log x)^n}{m+1} - \frac{n}{m+1}\int x^m(\log x)^{n-1}dx$$

Hence,
$$\int x^m(\log x)^n dx = \frac{x^{m+1}(\log x)^n}{m+1} - \frac{n}{m+1}\int x^m(\log x)^{n-1}dx\,.$$

14.16 REDUCTION FORMULA FOR $\int \frac{x^n}{\sqrt{ax^2+bx+c}}dx$

Let
$$I_n = \int \frac{x^n}{\sqrt{ax^2+bx+c}}dx = \frac{1}{2a}\int \frac{[(2ax+b)-b]x^{n-1}}{\sqrt{ax^2+bx+c}}dx$$
$$= \frac{1}{2a}\int \frac{(2ax+b)x^{n-1}}{\sqrt{ax^2+bx+c}}dx - \frac{b}{2a}\frac{x^{n-1}}{\sqrt{ax^2+bx+c}}dx$$
$$= \frac{1}{2a}\left[x^{n-1}.2\sqrt{ax^2+bx+c} - \int (n-1)x^{n-2}2\sqrt{ax^2+bx+c}dx\right] - \frac{b}{2a}I_{n-1}$$
$$= \frac{1}{a}x^{n-1}\sqrt{ax^2+bx+c} - \frac{(n-1)}{a}\int \frac{x^{n-2}(ax^2+bx+c)}{\sqrt{ax^2+bx+c}}dx - \frac{b}{2a}I_{n-1}$$
$$= \frac{1}{a}x^{n-1}\sqrt{ax^2+bx+c} - (n-1)I_n - \frac{b(n-1)}{a}I_{n-1} - \frac{(n-1)}{a}I_{n-2} - \frac{b}{2a}I_{n-1}$$
$$\Rightarrow \quad I_n = \frac{1}{na}x^{n-1}\sqrt{ax^2+bx+c} - \frac{b(2n-1)}{2an}I_{n-1} - \frac{c(n-1)}{an}I_{n-2}.$$

14.17 REDUCTION FORMULA FOR $\int \frac{px+q}{(ax^2+bx+c)^n}dx$

Let
$$I_n = \int \frac{px+q}{(ax^2+bx+c)}dx = \frac{p}{2a}\int \frac{2ax}{(ax^2+bx+c)^n}dx + \left(q - \frac{bp}{2a}\right)\int \frac{dx}{(ax^2+bx+c)^n}$$
$$= -\frac{p}{2a(n-1)}\frac{1}{(ax^2+bx+c)^{n-1}} + \left(q-\frac{bp}{2a}\right).\frac{1}{a^n}\int \frac{dx}{\left[\left(x+\frac{b}{2a}\right)^2 + \frac{4ac-b^2}{4a^2}\right]^n}$$
$$= -\frac{p}{2a(n-1)}.\frac{1}{(ax^2+bx+c)^{n-1}} + \frac{2qa-pb}{2a^{n+1}}\int \frac{dt}{(r^2+r)^n} \quad \ldots(1)$$

where $t = x + \frac{b}{2a}$ and $r = \frac{4ac-b^2}{4a^2}$

Integrating right hand integral of (1), by parts, we have
$$P_n = \int -\frac{dt}{(r^2+r)^n} = \frac{t}{(r^2+r)^n} + \int \frac{2t^2}{(r^2+r)^{n+1}}dt = \frac{t}{(t^2+r)^n} + 2n\int \frac{(t^2+r)-r}{(r^2+r)^{n+1}}dt$$
$$= \frac{t}{(t^2+r)^n} + 2n\int \frac{dt}{(r^2+r)^n} - 2nr\int \frac{dt}{(r^2+r)^{n+1}}$$

Now, replacing n by $n-1$ in this relation and solving for $\int \frac{dt}{(r^2+r)^n}$, we get
$$\int \frac{dt}{(r^2+r)^n} = \frac{1}{2(n-r)^r}.\frac{t}{(t^2+r)^{n-1}} + \frac{2n-3}{2(n-1)r}\int \frac{dt}{(t^2+r)^{n-1}} \quad \ldots(2)$$

From (1) and (2), we conclude that
$$I_n = -\frac{p}{2a(n-1)}.\frac{1}{(ax^2+bx+c)^{n-1}} + \frac{2aq-pb}{2a^{n+1}}\left[\frac{1}{2(n-r)r}.\frac{1}{(r^2+r)^{n-1}} + \frac{2n-3}{2(n-1)r}P_{n-1}\right]$$
$$\Rightarrow \quad I_n = \frac{-p}{2a(n-1)}.\frac{1}{(ax^2+bx+c)^{n-1}} + \frac{2aq-qb}{4(n-1)r}\left[\frac{1}{(r^2+r)^{n-1}} + (2n-3)P_{n-1}\right]$$

where $P_n = \int \frac{dt}{(r^2+r)^n}, t = x + \frac{b}{2a}$ and $r = \frac{4ac-b^2}{4a^2}$.

14.18 SOME MORE IMPORTANT REDUCTION FORMULAE

The reduction formulae for $I_{m,p} = \int x^m (a+bx^n)^p dx$ are :

(i) $I_{m,p} = \dfrac{x^{m-n+1}.(a+bx^n)^{p+1}}{(np+m+1)^b} - \dfrac{(m-n+1)}{(np+m+1)^b} I_{m-n,p}, (np+m+1 \neq 0)$

(ii) $I_{m,p} = \dfrac{x^{m+1}(a+bx^n)^p}{np+m+1} + \dfrac{anp}{np+m+1} I_{m,p-1}, (np+m+1 \neq 0)$

(iii) $I_{m,p} = \dfrac{x^{m+1}(a+bx^n)^{p+1}}{(m+1)a} - \dfrac{(np+n+m+1)b}{(m+1)a} I_{m+n,p}, (m+1 \neq 0)$

(iv) $I_{m,p} = \dfrac{-x^{m+1}(a+bx^n)^{p+1}}{n(p+1)a} + \dfrac{(np+n+m+1)}{n(p+1)a} I_{m,p+1}, (n, p+1 \neq 0)$

Solved Examples

Example 1. *Use reduction formula to integrate* $\sin^{1/2} x \cos^{7/2} x$.

Solution . Since, we have

$$I_{m,n} = \int \sin^m x \cos^n x dx$$

$$= \frac{\cos^{n-1} x \sin^{m+1} x}{m+n} + \frac{n-1}{m+n} I_{m,n-2},$$

$$m+n \neq 0$$

Put $m = 1/2$, $n = 7/2$, we get

$$I_{\frac{1}{2},\frac{7}{2}} = \int \sin^{1/2} x \cos^{7/2} x$$

$$= \frac{\cos^{5/2} x \sin^{3/2} x}{4} + \frac{5}{8} I_{1/2,3/2}.$$

Also, $$I_{\frac{1}{2},\frac{3}{2}} = \frac{\cos^{1/2} x \sin^{3/2} x}{2} + \frac{1}{4} I_{1/2,-1/2}$$

$$\Rightarrow \quad I_{\frac{1}{2},\frac{7}{2}} = \frac{\cos^{5/2} x \sin^{3/2} x}{4}$$

$$+ \frac{5}{16} \cos^{1/2} x \sin^{3/2} x + \frac{5}{32} I_{1/2,-1/2}.$$

Here, further reduction is not possible because

$$m+n = \frac{1}{2} - \frac{1}{2} = 0.$$

But $$I_{\frac{1}{2},-\frac{1}{2}} = \int \frac{\sqrt{\sin x}}{\sqrt{\cos x}} dx = \int \sqrt{\tan x} dx$$

$$= \frac{1}{2} \tan^{-1}\left(\frac{\tan x - 1}{\sqrt{2 \tan x}}\right)$$

$$+ \frac{1}{2\sqrt{2}} \log \frac{\tan x - \sqrt{2 \tan x}}{\tan x + \sqrt{2 \tan x}}$$

Therefore,

$$I_{\frac{1}{2},\frac{7}{2}} = \frac{1}{4} \sqrt{\cos x \sin^{3/2} x} \left(\cos^2 x + \frac{5}{4}\right)$$

$$+ \frac{5}{32}\left[\frac{1}{2} \tan^{-1}\left(\frac{\tan x - 1}{\sqrt{2 \tan x}}\right)\right.$$

$$\left. + \frac{1}{2\sqrt{2}} \log \frac{\tan x - \sqrt{2 \tan x}}{\tan x + \sqrt{2 \tan x}}\right] + c.$$

Example 2. *Compute* $\int \sin^4 x \cos^5 x dx$. (Cochin–2005)

Solution . We know that

$$I_{m,n} = \int \sin^m x \cos^n x dx$$

$$= -\frac{\sin^{m-1} x \cos^{n+1} x}{m+n}$$

$$+ \frac{m-1}{m+n} I_{m-2,n}, m+n \neq 0.$$

Put $m = 4$, $n = 5$, we get

$$I_{4,5} = \int \sin^4 x \cos^5 x dx$$

$$= -\frac{\sin^3 x \cos^6 x}{9} + \frac{1}{3} I_{2,5}$$

Now $$I_{2,5} = \int \sin^2 x \cos^5 x dx$$

$$= -\frac{\sin x \cos^6 x}{5} + \frac{1}{7} I_{0,5}$$

Here, $$I_{0,5} = I_5 = \int \cos^5 x dx$$

$$= \frac{\sin x \cos^4 x}{5} + \frac{4}{5} I_3$$

and $$I_3 = \int \cos^3 x dx = \frac{\sin x \cos^2 x}{3} + \frac{2}{3} I_1$$

$$I_1 = \int \cos x dx = \sin x$$

Therefore,

$$I_{4,5} = -\frac{\sin^3 x \cos^6 x}{9} - \frac{1}{21} \sin x \cos^6 x$$

$$+ \frac{1}{105} \sin x \cos^4 x + \frac{4}{315} \sin x \cos^2 x$$

$$+ \frac{8}{315} \sin x + c$$

$$= \frac{\sin x}{315}[35 \cos^8 x - 50 \cos^6 x$$

$$+ 3 \cos^4 x + 4 \cos^2 x + 8] + c.$$

Example 3. *Compute* $\int \cos^4 x \cos 3x dx$

Solution . Since we know that

$$I_{m,n} = \int \cos^m x \cos nx dx$$

$$= \frac{\cos^2 nx}{m^2 - n^2} \frac{d}{dx}\left(\frac{\cos^m x}{\cos nx}\right)$$
$$+ \frac{m(m-1)}{m^2 - n^2} I_{m-2,n}$$

Let $m = 4, n = 3$.

Then we have

$$I_{4,3} = \int \cos^4 x \cos 3x dx$$
$$= -\frac{\cos^2 3x}{7} \frac{d}{dx}\left(\frac{\cos^4 x}{\cos 3x}\right)$$
$$+ \frac{4(4-1)}{7} I_{2,3}$$
$$= \frac{4}{7} \cos 3x \cos^3 x \sin x$$
$$- \frac{3}{7} \cos^4 x \sin 3x + \frac{12}{7} I_{2,3}$$
$$I_{2,3} = \int \cos^2 x \cos 3x dx$$
$$= -\frac{\cos^2 3x}{5} \frac{d}{dx}\left(\frac{\cos^2 x}{\cos 3x}\right)$$
$$+ \frac{2(2-1)}{-5} I_{0,3}$$
$$= -\frac{2}{5} \cos 3x \sin x \cos x$$
$$+ \frac{3}{5} \cos^3 x \sin 3x - \frac{2}{5} I_{0,3}$$
$$I_{0,3} = \int \cos^0 x \cos 3x dx = \int \cos 3x dx$$
$$= \frac{1}{3} \sin 3x + c.$$

Therefore,

$$I_{4,3} = \int \cos^4 x \cos 3x \, dx$$
$$= \frac{4}{7} \cos 3x \cos^3 x \sin x - \frac{3}{7} \cos^4 x \sin 3x$$
$$- \frac{24}{35} \cos 3x \sin x \cos x + \frac{36}{35} \cos^2 x \sin 3x$$
$$- \frac{8}{35} \sin 3x + c$$
$$= \frac{4}{35} \cos 3x \cos x \sin x (5\cos^2 x - 6)$$
$$- \frac{1}{35} \sin 3x (15 \cos^4 x$$
$$- 36 \cos^2 x + 8) + c.$$

Example 4. *Compute* $\int \frac{dx}{(2 + \cos x)^5}$.

Solution . Since, we know that

$$I_n = \int \frac{dx}{(a + b\cos x)^n}$$
$$= \frac{b \sin x}{(n-1)(b^2 - a^2)(a + b\cos x)^{n-1}}$$
$$- \frac{a(2n-3)}{(n-1)(b^2 - a^2)} I_{n-1}$$
$$+ \frac{(n-2)}{(n-1)(b^2 - a^2)} I_{n-2}.$$

Here $a = 2, b = 1, n = 5$

Therefore,

$$I_5 = \int \frac{dx}{(2 + \cos x)^5}$$
$$= -\frac{\sin x}{4.3(2 + \cos x)^4} + \frac{2(10-3)}{4.3} . I_4$$
$$- \frac{3}{4.3} I_3.$$

Also,

$$I_4 = \int \frac{dx}{(2 + \cos x)^4}$$
$$= -\frac{\sin x}{3.3(2 + \cos x)^3} + \frac{2(8-3)}{3.3} . I_3$$
$$- \frac{2}{3.3} I_2$$
$$I_3 = \int \frac{dx}{(2 + \cos x)^3}$$
$$= -\frac{\sin x}{2.3(2 + \cos x)^2} + \frac{2(6-3)}{2.3} . I_2$$
$$- \frac{1}{2.3} I_1$$
$$I_2 = \int \frac{dx}{(2 + \cos x)^2}$$
$$= -\frac{\sin x}{1.3(2 + \cos x)} + \frac{2(4-3)}{1.3} . I_1$$

and

$$I_1 = \int \frac{dx}{(2 + \cos x)}$$
$$= 2 \tan^{-1}\left(\frac{1}{\sqrt{3}} \tan \frac{x}{2}\right) + c_1.$$

Therefore,

$$I_5 = -\frac{1}{12} \frac{\sin x}{(2 + \cos x)^4}$$
$$+ \frac{7}{6}\left[-\frac{1}{9} \frac{\sin x}{(2 + \cos x)^3}\right.$$
$$\left. + \frac{10}{9} I_3 - \frac{2}{9} I_2\right] - \frac{1}{3} I_3$$
$$= -\frac{1}{12} \frac{\sin x}{(2 + \cos x)^4} - \frac{7}{54} . \frac{\sin x}{(2 + \cos x)^3}$$
$$+ \frac{113}{108} I_3 - \frac{7}{27} I_2$$
$$= -\frac{1}{12} \frac{\sin x}{(2 + \cos x)^4} - \frac{7}{54} . \frac{\sin x}{(2 + \cos x)^3}$$
$$+ \frac{113}{108}\left[-\frac{1}{6} . \frac{\sin x}{(2 + 3\cos x)^2} + I_2 - \frac{I_1}{6}\right]$$
$$- \frac{7}{27} I_2$$

$$= -\frac{1}{12}\frac{\sin x}{(2+\cos x)^4} - \frac{7}{54}.\frac{\sin x}{(2+\cos x)^3} - \frac{113}{660}.\frac{\sin x}{(2+3\cos x)^2} + \frac{20}{27}\left[-\frac{1}{3}.\frac{\sin x}{2+\cos x} + \frac{2}{3}I_1 - \frac{1}{3}I_0\right] - \frac{1}{6}I_1$$

$$= -\frac{1}{12}\frac{\sin x}{(2+\cos x)^4} - \frac{7}{54}.\frac{\sin x}{(2+\cos x)^3} - \frac{113}{648}.\frac{\sin x}{(2+\cos x)^2} + \frac{85}{324}.\frac{\sin x}{(2+\cos x)} + \frac{170}{324}I_1 - \frac{113}{648}I_0$$

$$= -\frac{1}{12}\frac{\sin x}{(2+\cos x)^4} - \frac{7}{54}.\frac{\sin x}{(2+\cos x)^3} - \frac{113}{648}.\frac{\sin x}{(2+\cos x)^2} - \frac{85}{324}.\frac{\sin x}{(2+\cos x)} + \frac{227}{324}\tan^{-1}\left(\frac{1}{\sqrt{3}}\tan\frac{x}{2}\right).$$

Example 5. *Compute* $\int \frac{1-7x^2+14x^4-8x^6}{\sqrt{2x^2-1}}dx$.

Solution . Since we know that

$$I_n = \int \frac{x^n}{\sqrt{ax^2+bx+c}}dx = \frac{1}{na}x^{n-1}\sqrt{ax^2+bx+c} - \frac{b(2n-1)}{2an}I_{n-1} - \frac{c(n-1)}{an}I_{n-2}.$$

Let $I_n = \int \frac{x^n}{\sqrt{2x^2-1}}dx$, therefore

$$I = \int \frac{1-7x^2+14x^4-8x^6}{\sqrt{2x^2-1}}dx = I_0 - 7I_2 + 14I_4 - 8I_6$$

$$I_n = \frac{1}{2n}x^{n-1}\sqrt{2x^2-1} - \frac{0(2n-1)}{4n}I_{n-1} + \frac{(n-1)}{2n}I_{n-2}.$$

Therefore,

$$I_6 = \frac{1}{12}x^5\sqrt{2x^2-1} + \frac{5}{12}I_4$$

$$I_4 = \frac{1}{8}x^3\sqrt{2x^2-1} + \frac{3}{8}I_2$$

$$I_2 = \frac{1}{4}x\sqrt{2x^2-1} + \frac{1}{4}I_0$$

and

$$I_0 = \int \frac{1}{\sqrt{2x^2-1}}dx = \frac{1}{\sqrt{2}}\int \frac{1}{\sqrt{x^2-1/2}}dx = \frac{1}{\sqrt{2}}\cosh^{-1}(\sqrt{2}x)$$

$$\Rightarrow \quad I = I_0 - 7I_2 + 14I_4 - \frac{2}{3}x^5\sqrt{2x^2-1} - \frac{10}{3}I_4$$

$$= I_0 - 7I_2 + \frac{3}{3}\left(\frac{1}{8}x^3\sqrt{2x^2-1} + \frac{3}{8}I_2\right) - \frac{2}{3}x^5\sqrt{2x^2-1}$$

$$= I_0 - 3\left(\frac{1}{4}x\sqrt{2x^2-1} + \frac{1}{4}I_0\right) + \frac{4}{3}x^3\sqrt{2x^2-1} - \frac{2}{3}x^5\sqrt{2x^2-1}$$

$$= \frac{1}{4}I_0 - \frac{3}{4}x\sqrt{2x^2-1} + \frac{4}{3}x^3\sqrt{2x^2-1} - \frac{2}{3}x^5\sqrt{2x^2-1}$$

$$= \frac{1}{4\sqrt{2}}\cosh^{-1}(\sqrt{2}x) - \left(\frac{3}{4}x - \frac{4}{3}x^3 - \frac{2}{3}x^5\right)\sqrt{2x^2-1} + c.$$

Example 6. *Compute* $\int \frac{x^3}{\sqrt{1-x^2}}dx$.

Solution . Here we use the following formula

$$I_{m,p} = \frac{x^{m-n+1}.(abx^n)^{p+1}}{(np+m+1)^b} - \frac{(m-n+1)}{(np+m+1)^b}I_{m-n,p}$$

Here $m = 3, n = 2, p = -1/2, a = 1, b = 1$

Therefore,

$$I_{3,-\frac{1}{2}} = \int x^3(1-x^2)^{-1/2}dx$$

$$= \frac{x^{3-2+1}.(1-x^2)^{-1/2+1}}{[2.(1-1/2)+b+1](-1)} - \frac{(3-2+1).1}{-1[2(-1/2)+3+1]}I_{1,-1/2}$$

$$= -\frac{1}{3}x^2(1-x^2)^{1/2} + \frac{2}{3}I_1 - \frac{1}{2}$$

$$= -\frac{1}{3}x^2(1-x^2)^{1/2} + \frac{2}{3}\int \frac{x}{\sqrt{1-x^2}}dx$$

$$= -\frac{x^2}{3}\sqrt{1-x^2} - \frac{2}{3}\sqrt{1-x^2} + c$$

$$= -\frac{1}{3}(x^2+2)\sqrt{1-x^2} + c$$

which is the required integral.

Example 7. *If n is a positive integer, prove that*

$$\int_0^{\pi/2}\cos^n x\cos nx dx = \frac{\pi}{2^{n+1}}.$$

Solution . We know that

$$I(m,n) = \int \cos^m x \cos nx dx$$

$$= \frac{\cos^m x \sin nx}{m+n} + \frac{m}{m+n}\int \cos^{m-1} x \cos(n-1)x\,dx$$

Then

$\int_0^{\pi/2} \cos^n x \cos nxdx$

$$= \left[\frac{\cos^m x \sin nx}{m+n}\right]_0^{\pi/2} + \frac{m}{m+n}\int_0^{\pi/2} \cos^{m-1} x \cos(n-1)xdx$$

$$= \frac{m}{m+n}\int_0^{\pi/2} \cos^{m-1} x \cos(n-1)xdx$$

i.e., $I(m,n) = \frac{m}{m+n} I(m-1,n-1)$

Now putting $m = n$, we get

$$I(n,n) = \frac{n}{n+n} I(n-1,n-1)$$

or $I(n,n) = \frac{1}{2} I(n-1,n-1)$ ...(1)

Now replacing n by $n-1$ on both sides of (1), we get

$I(n-1,n-1) = \frac{1}{2} I(n-2,n-2)$ ...(2)

Putting the value of $I(n-1, n-1)$ in (1), we get

$$I(n,n) = \frac{1}{2}.\frac{1}{2} I(n-2,n-2) = \frac{1}{2^2} I(n-2,n-2)$$

Continuing in the same way, we get

$$I(n,n) = \frac{1}{2^n} I(0,0)$$

Now $I(0,0) = \int_0^{\pi/2} \cos^0 x \cos 0xdx$

$$= \int_0^{\pi/2} dx = [x]_0^{\pi/2} = \frac{\pi}{2}$$

$$\therefore \quad I(n,n) = \frac{\pi}{2^{n+1}}$$

Hence $\int_0^{\pi/2} \cos^n x \cos nxdx = \frac{\pi}{2^{n+1}}$.

Example 8. *If* $I_n = \int_0^{\pi/4} \tan^n xdx$, *show that*

$$I_n + I_{n-2} = \frac{1}{n-1}.$$ *Hence, deduce the value of* I_5.

Solution. We know that

$$\int \tan^n xdx = \frac{\tan^{n-1} x}{n-1} - \int \tan^{n-2} xdx$$

Then $I_n = \int_0^{\pi/4} \tan^n xdx$

$$= \left[\frac{\tan^{n-1} x}{n-1}\right]_0^{\pi/4} - \int_0^{\pi/4} \tan^{n-2} xdx$$

or $I_n = \frac{1}{n-1} - I_{n-2}$

$\therefore I_n + I_{n-2} = \frac{1}{n-1}$ ...(1)

Next, putting $n = 3$ and 5 successively, we get

$$I_3 + I_1 = \frac{1}{2},\ I_5 + I_3 = \frac{1}{4}$$

Form these equations, we get

$$I_5 = \frac{1}{4} - I_3 = \frac{1}{4} - \left(\frac{1}{2} - I_1\right)$$

or $I_5 = -\frac{1}{4} + I_1$

or $I_5 = -\frac{1}{4} + \int_0^{\pi/4} \tan xdx$

$$= -\frac{1}{4} + [\log \sec x]_0^{\pi/4}$$

$$= -\frac{1}{4} + \left[\log \sec \frac{\pi}{4} - \log \sec 0\right]$$

$$= -\frac{1}{4} + \left[\log \sqrt{2} - \log 1\right]$$

$$= -\frac{1}{4} + \log \sqrt{2} - 0$$

Hence, $I_5 = \frac{1}{2}\left[\log 2 - \frac{1}{2}\right]$.

Example 9. *If* $u_n = \int_0^{\pi/2} x^n \sin xdx$ *and* $n > 1$, *prove that*

$$u_n + n(n-1)u_{n-2} = n\left(\frac{\pi}{2}\right)^{n-1}$$

Hence evaluate $\int_0^{\pi/2} x^5 \sin xdx$. (Madras–2000)

Solution. By reduction formula, we have

$\int x^n \sin mxdx$

$$= -\frac{x^n \cos mx}{m} + \frac{nx^{n-1} \sin mx}{m^2} - \frac{n(n-1)}{m^2}\int x^{n-2} \sin mxdx$$

Putting $m = 1$ in both sides, we get

$$\int x^n \sin xdx = -x^n \cos x + nx^{n-1} \sin x - n(n-1)\int x^{n-2} \sin xdx$$

Then $u_n = \int_0^{\pi/2} x^n \sin nxdx$

$$= \left[-x^n \cos x + nx^{n-1} \sin x\right]_0^{\pi/2} - n(n-1)\int_0^{\pi/2} x^{n-2} \sin xdx$$

or $u_n = \left[u\left(\frac{\pi}{2}\right)^{n-1}\right] - n(n-1)u_{n-2}$

$\therefore u_n + n(n-1)u_{n-2} = n\left(\frac{\pi}{2}\right)^{n-1}$ ...(1)

Next, putting $n = 3$ and 5 successively in (1), we get

$$u_{5+5(5-1)}u_3 = 5\left(\frac{\pi}{2}\right)^4$$

and $u_{3+3(3-1)}u_1 = 3\left(\frac{\pi}{2}\right)^2$

From these equations, we get

$$u_5 + 20\left[3\left(\frac{\pi}{2}\right)^2 - 6u_1\right] = 5\left(\frac{\pi}{2}\right)^4$$

Now, $u_1 = \int_0^{\pi/2} x \sin x dx$

$$= [-x\cos x]_0^{\pi/2} - \int_0^{\pi/2}(-\cos x)dx$$
$$= 1$$
$$= [0] + [\sin x]_0^{\pi/2}$$

$$\therefore u_5 + 20\left[3\left(\frac{\pi}{2}\right)^2 - 6(1)\right] = 5\left(\frac{\pi}{2}\right)^4$$

or $\quad u_5 = \frac{5}{16}\pi^4 - 15\pi^2 + 120$

Example 10. *If $u_n = \int_0^{\pi/2} x \sin^n x dx$ and $n > 1$, prove that*

$$u_n = \frac{n-1}{n}u_{n-2} + \frac{1}{n^2}$$

Hence deduce that $u_5 = \frac{149}{225}$.

Solution. By reduction formula, we have

$$\int x \sin^n x dx = -\frac{x\sin^{n-1} x \cos x}{n} + \frac{\sin^n x}{n^2} + \frac{n-1}{n}\int x \sin^{n-2} x dx$$

Then $\quad u_n = \int_0^{\pi/2} x \sin^n x dx$

$$= \left[\frac{-x\sin^{n-1} x \cos x}{n} + \frac{\sin^n x}{n^2}\right]_0^{\pi/2} + \frac{n-1}{n}\int_0^{\pi/2} x \sin^{n-2} x dx$$

or $\quad u_n = \left[0 + \frac{1}{n^2} - 0 - 0\right] + \frac{n-1}{n}u_{n-2}$

$$\therefore \quad u_n = \frac{n-1}{n}u_{n-2} + \frac{1}{n^2} \qquad \ldots(1)$$

Next, putting $n = 3$ and 5 successively in (1), we get

$$u_3 = \frac{2}{3}u_1 + \frac{1}{9}, \; u_5 = \frac{4}{5}u_3 + \frac{1}{25}$$

From these equations, we get

$$u_5 = \frac{4}{5}\left[\frac{2}{3}u_1 + \frac{1}{9}\right] + \frac{1}{25}$$
$$= \frac{8}{15}u_1 + \frac{4}{45} + \frac{1}{25} = \frac{8}{15}u_1 + \frac{29}{225}$$

Now, $\quad u_1 = \int_0^{\pi/2} x \sin x dx = 1$

(As done in example 9)

$$\therefore \quad u_5 = \frac{8}{15}(1) + \frac{29}{225} = \frac{120+29}{225} = \frac{149}{225}.$$

Example 11. *Evaluate $\int_0^1 x^m (\log x)^n dx$, when $m \geq 0$ and n is positive integer.* (SVTU–2009, Bhillai–2005)

Solution. We know by reduction formula

$$\int x^m (\log x)^n dx = \frac{x^{m+1}(\log x)^n}{m+1} - \frac{n}{m+1}\int x^m (\log x)^{n-1} dx$$

Let $I(m,n) = \int_0^1 x^m (\log x)^n dx$, then we have

$$I(m,n) = \left[\frac{x^{m+1}(\log x)^n}{m+1}\right]_0^1 - \frac{n}{m+1}I(m,n-1)$$

or $I(m,n) = [0-0] - \frac{n}{m+1}I(m,n-1)$

or $I(m,n) = -\frac{n}{m+1}I(m,n-1) \qquad \ldots(1)$

Replace n by $n-1$ in (1), we get

$$I(m,n-1) = -\frac{n-1}{m+1}I(m,n-2) \qquad \ldots(2)$$

From (2) and (1), we get

$$I(m,n) = (-1)^2\left(\frac{n}{m+1}\right)\left(\frac{n-1}{m+1}\right)I(m,n-2) \qquad \ldots(3)$$

Again by repeated application of (1), we get

$$I(m,n) = (-1)^n\left(\frac{n}{m+1}\right)\left(\frac{n-1}{m+1}\right)\left(\frac{n-2}{m+1}\right)\ldots\left(\frac{1}{m+1}\right)I(m,0)$$

or $\quad I(m,n) = (-1)^n \frac{n!}{(m+1)^n}I(m,0)$

Now $I(m,0) = \int_0^1 x^m dx = \left[\frac{x^{m+1}}{m+1}\right]_0^1 = \frac{1}{m+1}$

$$\therefore \quad I(m,n) = (-1)^n \frac{n!}{(m+1)^n}\cdot\frac{1}{m+1}$$
$$= \frac{(-1)^n n!}{(m+1)^{n+1}}$$

Hence, $\int_0^1 x^m (\log x)^n dx = \frac{(-1)^n n!}{(m+1)^{n+1}}.$

Example 12. *If $u_n = \int \frac{\sin nx}{\sin x}dx$, prove that*

$$u_n = \frac{2\sin(n-1)x}{n-1} + u_{n-2}$$

Hence, evalutae $\int_0^{\pi/2} \frac{\sin 7x}{\sin x}dx$.

Solution . We know that

$$\sin nx - \sin (n-2)x = 2\cos (n-1)x \sin x$$

$$\Rightarrow \frac{\sin nx}{x} - \frac{\sin(n-2)x}{\sin x} = 2\cos(n-1)x$$

Then $u_n = \int \frac{\sin nx}{\sin x} dx$

$$= \int \left[\frac{\sin(n-2)x}{\sin x} + 2\cos(n-1)x\right] dx$$

$$= \int \frac{\sin(n-2)x}{\sin x} dx + 2\int \cos(n-1)x dx$$

or $u_n = u_{n-2} + 2\int \cos(n-1)x dx$

or $u_n = u_{n-2} + \frac{2\sin(n-1)x}{n-1}$

Hence, $u_n = \frac{2\sin(n-1)x}{\sin x} + u_{n-2}$ ...(1)

Next putting $n = 3, 5$ and 7 in (1) respectively, we get

$$u_3 = \frac{2\sin 2x}{\sin x} + u_1,\ u_5 = \frac{2\sin 4x}{\sin x} + u_3$$

$$u_7 = \frac{2\sin 6x}{\sin x} + u_5$$

Adding all three equations

$$u_7 = \frac{2}{\sin x}[\sin 2x + \sin 4x + \sin 6x] + u_1$$

Now $u_1 = \int \frac{\sin x}{\sin x} dx = \int dx = x$

$$u_7 = \frac{2}{\sin x}[\sin 2x + \sin 4x + \sin 6x] + x$$

Then $\int_0^{\pi/2} \frac{\sin 7x}{\sin x} dx$

$$= \left[\frac{2}{\sin x}\{\sin 2x + \sin 4x + \sin 6x\}\right]_0^{\pi/2} + [x]_0^{\pi/2}$$

$$= 0 + \frac{\pi}{2}$$

$$\therefore \int_0^{\pi/2} \frac{\sin 7x}{\sin x} dx = \frac{\pi}{2}.$$

Example 13. *Evaluate* : $\int_0^{\pi/2} \sin^n x dx$

Solution . Let $I_n = \int_0^{\pi/2} \sin^n x dx$

$$= \int_0^{\pi/2} \sin^{n-1} x \sin x dx$$

$$= \left[-\sin^{n-1} x \cos x\right]_0^{\pi/2} + (n-1)\int_0^{\pi/2} \sin^{n-2} x dx - (n-1)I_n$$

$$\Rightarrow I_n(1 + n - 1) = [0] + (n-1)I_{n-2}$$

$$\Rightarrow I_n = \left(\frac{n-1}{n}\right) I_{n-2} \quad ...(1)$$

Replacing n by $n-2$, $n-4$, respectively, we get

$$\Rightarrow I_n = \left(\frac{n-1}{n}\right)\left(\frac{n-3}{n-2}\right)\left(\frac{n-5}{n-4}\right) I_{n-6} \quad ...(2)$$

Now we have two cases :

Case I : When n is odd

$$I_n = \left(\frac{n-1}{n}\right)\left(\frac{n-3}{n-2}\right)\left(\frac{n-5}{n-4}\right)\ldots\frac{4}{5}.\frac{2}{3}.I_1$$

$$= \frac{(n-1)(n-3)(n-5)\ldots4.2}{n(n-2)(n-4)\ldots5.3}$$

$$\left[\because I_1 = \int_0^{\pi/2} \sin x dx = 1\right]$$

Case II : When n is even

$$I_n = \frac{(n-1)(n-3)(n-5)\ldots3.1}{n(n-2)(n-4)\ldots4.2}.I_2$$

$$= \frac{(n-1)(n-3)(n-5)\ldots3.1}{n(n-2)(n-4)\ldots4.2}\left(\frac{\pi}{2}\right)$$

$$\left[\because I_2 = \int_0^{\pi/2} \sin^2 x dx = \frac{\pi}{2}\right]$$

EXERCISE 14.1

Use reduction formulae, compute the following :

1. $\int \cos^3 x \operatorname{cosec}^2 x dx$
2. $\int \sqrt{\cos\theta}.\sin^3\theta d\theta$
3. $\int \cos^{-3}\theta \sin^{-1}\theta d\theta$
4. $\int \frac{x^4}{\sqrt{1-x^2}} dx$
5. $\int (1+x^2)^{3/2} dx$
6. $\int \frac{x^4}{(a^2+x^2)^2} dx$
7. $\int \frac{x^2}{\sqrt{2ax-x^2}} dx$ (Madras–2000)
8. $\int \frac{x^5}{(a+bx^2)^4} dx$
9. $\int \frac{dx}{x^{1/2}(1+x^2)^{5/4}}$
10. $\int \frac{x^3}{\sqrt{4x-x^2}} dx$
11. $\int \frac{x^3+5x^2-3x+4}{\sqrt{x^2+x+1}} dx$
12. $\int x^3 \cos 3x dx$
13. $\int x^3 e^{ax} dx$
14. Prove that : $I_n = \int x^n e^{-x} dx = -x^n e^{-x} + nI_{n-1}$.
15. Find the following integrals using reduction formula
    (i) $\int \frac{x+1}{(x^2+1)^3} dx$ (ii) $\int \frac{x^2-a^2}{(x^2+a^2)^3} dx$

16. (i) If $I_n = \int \frac{dx}{(x^2+a^2)^n}$, then show that

$$I_{n+1} = \frac{1}{2na^2}\cdot\frac{x}{(x^2+a^2)^n} + \frac{2n-1}{2n}\cdot\frac{1}{a^2}I_n.$$

(ii) If $I_n = \int (\log x)^n dx$, then show that $I_n = x(\log x)^n - nI_{n-1}$.

(iii) If $I_n = \int x^n e^x dx$, then show that $I_n = x^n e^x - nI_{n-1}$. (Madras–2000)

(iv) If $I_n = \int e^{ax}\sin^n x dx$, then show that

$$I_n = \frac{e^{ax}}{a^2+n^2}\sin^{n-1}x(a\sin x - n\cos x) + \frac{n(n-1)}{a^2+n^2}I_{n-2}.$$

(Gorakhpur–1999)

17. Evaluate the following integrals :

(i) $\int_0^{\pi/4}\tan^5\theta d\theta$ (ii) $\int_0^{\pi/4}\tan^7 xdx$

18. If $I_n = \int_0^{\pi/4}\tan^n xdx$, prove that $n(I_{n-1}+I_{n+1}) = 1$. (VTU–2009)

19. If $I_n = \int_0^{\pi/3}\tan^n xdx$, prove that $(n-1)(I_n + I_{n-2}) = \left(\sqrt{3}\right)^{n-1}$.

20. Evaluate $\int \frac{d\theta}{\sin^4\frac{\theta}{2}}$

21. Prove that $\int_0^{\pi/4}\sec^3 xdx = \frac{1}{3}\left\{\sqrt{2}+\log(\sqrt{2}+1)\right\}$.

22. Prove that $\int_0^{\pi}\sin^m x\sin nxdx = \frac{m(m-1)}{m^2-n^2}\int_0^{\pi}\sin^{m-2}x\sin nxdx$.

23. If n is a positive integer greater than 1, prove that

$$\int_0^{\pi/2}\cos^{n-2}x\sin nxdx = \frac{1}{n-1}.$$

24. If $u_n = \int_0^{\pi/2}x^n\sin mx\, dx$, prove that

$$u_n = \frac{n}{m^2}\left(\frac{\pi}{2}\right)^{n-1} - \frac{n(n-1)}{m^2}u_{n-2}$$

where m is of the form $4r+1$. (Marathwada–2008)

25. Evaluate the following integrals :

(i) $\int_0^{\pi/2}x^3\sin 3xdx$ (ii) $\int_0^{\pi}x\sin^3 xdx$

(iii) $\int_0^{\pi/2}x^5\sin xdx$ (iv) $\int_0^{\pi}\theta\sin^2\theta\cos\theta d\theta$

26. If $I_m = \int_0^{\infty}e^{-x}\sin^m xdx$, where $m \ge 2$, prove that

$$(1+m^2)I_m = m(m-1)I_{m-2}.$$

Hence deduce I_4.

27. If $I_n = \int_{-\pi/2}^{\pi/2}e^{ax}\cos^n xdx$, prove that $I_n = \frac{n(n-1)}{a^2+n^2}I_{n-2}$.

28. Evaluate $\int_0^1 x^{n-1}(\log x)^m dx$, when m and n are positive integers. (SVTU–2009)

29. Evaluate the following integrals :

(i) $\int_0^1 x^m(\log x)^5 dx$ (ii) $\int_0^1 x^m(\log x)^4 dx$

30. If $I_n = \int x^n(a-x)^{1/2}dx$, prove that

$$(2n+3)I_n = 2anI_{n-1} - 2x^n(a-x)^{3/2}.$$

31. If $I_n = \int_0^a (a^2-x^2)^n dx$, prove that $I_n = \frac{2na^2}{(2n+1)}I_{n-1}$. Hence deduce I_3.

32. If $I_n = \int_0^1 x^p(1-x^q)^n dx$, where p, q and n are positive, prove that $(qn+p+1)I_n = qnI_{n-1}$.

33. Prove that $\int_0^1 x^{-1/4}(1-x^{1/2})^{5/2}dx = \frac{5}{16}\int_0^1 x^{-1/4}(1-x^{1/2})^{1/2}dx$.

34. Prove that $\int_0^{\pi}\frac{\sin nx}{\sin x}dx = \begin{cases}0, & \text{when } n \text{ is even positive integer}\\ \pi, & \text{when } n \text{ is odd positive integer}\end{cases}$

35. If $I_n = \int_0^1 e^{-1}x^n dx$, find the reduction formula and hence show that $I_n = n!$.

ANSWERS

1. $-\frac{1}{3}\sin x[\cot^2 x + 2\cos^2 x + 4] + c$ **2.** $\frac{2}{7}\cos^{7/2}\theta - \frac{2}{3}\cos^{3/2}\theta + c$ **3.** $\frac{1}{2}\tan^2\theta + \log\tan\theta + c$ **4.** $\frac{1.3}{2.4}\sin^{-1}x - \frac{x\sqrt{1-x^2}}{8}(3+2x^2) + c$

5. $\frac{\sin\theta}{4\cos^4\theta} + \frac{3\sin\theta}{8\cos^2\theta} + \frac{3}{8}\log(\sec\theta+\tan\theta) + c\ (x = \tan\theta)$ **6.** $c - \frac{x^3}{2(a^2+x^4)} + \frac{3}{2}\left(x - a\tan^{-1}\frac{x}{a}\right)$ **7.** $c - \sqrt{2ax-x^2}\left(\frac{x}{2}+\frac{3a}{2}\right) + 3a^2\sin^{-1}\frac{x-a}{a}$

8. $\frac{1}{6a}\cdot\frac{x^4}{(a+bx^2)^3} + \frac{1}{12a^2}\cdot\frac{x^4}{(a+bx^2)^2} + c$ **9.** $2\sqrt{x}(1+x^2)^{-1/4} + c$ **10.** $c - \frac{1}{3}(x^2+5x+30)\sqrt{4x-x^2} + 10\cos^{-1}\left(1-\frac{x-2}{2}\right)$

11. $\left(\frac{1}{3}x^3 + \frac{25}{12}x - \frac{163}{24}\right)\sqrt{x^2+x+1} + \frac{85}{16}\sin^{-1}\left(\frac{2x+1}{\sqrt{3}}\right) + c$ **12.** $\frac{1}{27}(9x^2-2)\cos 3x + \frac{1}{9}(3x^2-2x)\sin 3x + c$

13. $\frac{e^{ax}}{a^4}(a^3x^3 - 3a^2x^2 + 6ax - 6) + c$ **15.** (i) $\frac{x-1}{4(x^2+1)^2} + \frac{3x}{8(x^2+1)} + \frac{3}{8}\tan^{-1}x + c$ (ii) $-\frac{x}{4a^2(x^2+a^2)} - \frac{x}{2(x^2+a^2)^2} - \frac{1}{4a^3}\tan^{-1}\frac{x}{a} + c$

17. (i) $\frac{1}{2}\left(\log 2 - \frac{1}{2}\right)$ (ii) $\frac{5}{12} - \frac{1}{2}\log 2$ **20.** $-\frac{2}{3}\left[\text{cosec}^2\frac{\theta}{2}\cot\frac{\theta}{2} + 2\cot\frac{\theta}{2}\right]$

25. (i) $\frac{2}{27} - \frac{\pi^2}{12}$ (ii) $\frac{2\pi}{3}$ (iii) $\frac{5}{16}\pi^4 - 15\pi^2 + 120$ (iv) $-\frac{4}{9}$ **26.** $I_4 = \frac{24}{85}$ **28.** $\frac{(-1)^m m!}{n^{m+1}}$

29. (i) $-\frac{120}{(m+1)^6}$ (ii) $\frac{24}{(m+1)^5}$ **34.** $\frac{16a^7}{35}$ **35.** $I_n = nI_{n-1}$

FFFFF

15 Definite Integrals

15.1 INTRODUCTION

If $f(x)$ is continuous and non-negative function over a closed interval $[a, b]$ then $\int_a^b f(x)dx$ is called the definite integral of $f(x)$ between the limits a and b $(b > a)$.

If $\int f(x)dx = F(x)+c$, then $\int_a^b f(x)dx = [F(x)+c]_a^b = F(b) - F(a)$ is a definite value.

Here, a is called lower limit and b is called the upper limit and the interval $[a, b]$ is called the range of integration.

REMARKS

- $\int_a^b f(x)dx$ represents the area bounded by the lines $x = a$ and $x = b$.
- If $F(b) - F(a)$ is not a definite value, then the integral $\int_a^b f(x)dx$ is indefinite.

15.2 PROPERTIES OF DEFINITE INTEGRALS

PROPERTY 1. $\int_a^a f(x)dx = 0.$

Proof. Let $\int f(x)dx = F(x)$ then $\int_a^a f(x)dx = [F(x)]_a^a = F(a) - F(a) = 0.$

PROPERTY 2. *The value of definite integral is independent of the variable of integration. i.e.,* $\int_a^b f(x)dx = \int_a^b f(u)du.$

Proof. Let $\int f(x)dx = F(x)$ then $\int_a^b f(x)dx = [F(x)]_a^b = F(b) - F(a) = \int_a^b f(u)du.$

PROPERTY 3. $\int_a^b f(x)dx = -\int_b^a f(x)dx.$

Proof. Let $\int f(x)dx = F(x)$

then $\int_a^b f(x)dx = [F(x)]_a^b = F(b) - F(a) = -[F(x)]_b^a = \int_b^a f(x)dx.$

PROPERTY 4. $\int_a^c f(x)dx + \int_c^b f(x)dx = \int_a^b f(x)dx$, *where* $a < c < b$.

Proof. Let $\int f(x)dx = F(x)$

then $\int_a^c f(x)dx + \int_c^b f(x)dx = [F(x)]_a^c + [F(x)]_c^b = [F(c) - F(a)] + [F(b) - F(c)] = F(b) - F(a) = [F(x)]_a^b = \int_a^b f(x)dx.$

PROPERTY 5. $\int_0^a f(a-x)dx = \int_0^a f(x)dx\cdot$

Proof. We have $\int_0^a f(a-x)dx\cdot$

Let $a - x = t$, then $-dx = dt$. If $x = 0$ $\Rightarrow t = a$ and $x = a$, $t = 0$.

So, $\int_0^a f(a-x)dx = -\int_a^0 f(t)dt = \int_0^a f(t)dt = \int_0^a f(x)dx$ (By property 2)

PROPERTY 6. *If $f(x)$ is an even function of x, then* $\int_{-a}^a f(x)dx = 2\int_0^a f(x)dx$ *and if $f(x)$ is an odd function then* $\int_{-a}^a f(x)dx = 0.$

(Bhopal–2008)

Proof. Consider $\int_{-a}^a f(x)dx$. Then, $\int_{-a}^a f(x)dx = \int_{-a}^0 f(x)dx + \int_0^a f(x)dx.$

Now, let $I = \int_{-a}^0 f(x)dx$

put $x = -t$, so $x = -a \Rightarrow t = a$ and $x = 0 \Rightarrow t = 0$.

So, $\int_{-a}^{0} f(x)dx = -\int_{a}^{0} f(-t)dt = \int_{0}^{a} f(-t)dt = \int_{0}^{a} f(-x)dx$ (by 2)

Thus, we have $\int_{-a}^{a} f(x)dx = \int_{0}^{a} f(-x)dx + \int_{0}^{a} f(x)dx = \int_{0}^{a} [f(-x)] + f(x)dx$.

Now, if $f(x)$ is an even function *i.e.*, $f(-x) = f(x)$, then we get $\int_{-a}^{a} f(x)dx = \int_{0}^{a} [f(x) + f(x)]dx = 2\int_{0}^{a} f(x)dx$

and, if $f(x)$ is an odd function *i.e.*, $f(-x) = -f(x)$, then we get $\int_{-a}^{a} f(x)dx = \int_{0}^{a} [-f(x) + f(x)]dx = 0$.

PROPERTY 7. $\int_{0}^{2a} f(x)dx = 2\int_{0}^{a} f(x)dx$ *if* $f(2a - x) = f(x)$ *and* $\int_{0}^{2a} f(x)dx = 0$ *if* $f(2a - x) = -f(x)$.

Proof. This integral can be written as $\int_{0}^{2a} f(x)dx = \int_{0}^{a} f(x)dx + \int_{a}^{2a} f(x)dx$ ...(1)

Now, consider the integral $\int_{0}^{2a} f(x)dx$.

Put $x = 2a - t$, then $dx = -dt$ and if $x = a$ then $t = a$ and if $x = 2a$ then $t = 0$.

So, $\int_{0}^{2a} f(x)dx = -\int_{a}^{0} f(2a - t)dt = \int_{0}^{a} f(2a - t)dt = \int_{0}^{a} f(2a - x)dx$ (by 2)

Therefore, from (1), we have

$$\int_{0}^{2a} f(x)dx = \int_{0}^{a} f(x)dx + \int_{0}^{a} f(2a - x)dx = \int_{0}^{a} [f(x) + f(2a - x)]dx \quad ...(2)$$

Now, if $f(2a - x) = f(x)$ then from (2), we get $\int_{0}^{2a} f(x)dx = 2\int_{0}^{a} f(x)dx$

and if $f(2a - x) = -f(x)$ then from (2), we get $\int_{0}^{2a} f(x)dx = \int_{0}^{a} [f(x) - f(x)]dx = 0$

PROPERTY 8. $\int_{0}^{na} f(x)dx = n\int_{0}^{a} f(x)dx$, *if* $f(x + ma) = f(x)$ *for all integral values of m.*

Proof. The given integral can be written as

$$\int_{0}^{na} f(x)dx = \int_{0}^{a} f(x)\,dx + \int_{0}^{2a} f(x)\,dx + ... + \int_{(m-1)a}^{ma} f(x)dx + ... + \int_{(n-1)a}^{na} f(x)\,dx \quad ...(1)$$

Now consider the integral $\int_{(m-1)a}^{ma} f(x)dx$.

put $x = y + ma \Rightarrow dx = dy$.

but $f(y + ma) = f(y)$ is given, so we have

$$\int_{(m-1)a}^{ma} f(x)dx = \int_{0}^{a} f(y)dy = \int_{0}^{a} f(x)dx.$$

Now from (1), we have

$$\int_{0}^{na} f(x)dx = \int_{0}^{a} f(x)dx + \int_{0}^{a} f(x)dx + ... + \int_{0}^{a} f(x)dx = n\int_{0}^{a} f(x)dx.$$

Hence, $\int_{0}^{na} f(x)dx = n\int_{0}^{a} f(x)dx$, if $f(x + ma) = f(x)$ for all integral values of m.

Recapitulations

- $\int_{a}^{a} f(x)dx = 0$
- $\int_{a}^{b} f(x)dx = \int_{a}^{b} f(u)du$
- $\int_{a}^{b} f(x)dx = \int_{a}^{c} f(x)dx + \int_{c}^{b} f(x)dx, a < c < b$
- $\int_{0}^{a} f(a - x)dx = \int_{0}^{a} f(x)dx$
- $\int_{-a}^{a} f(x)dx = \begin{cases} 2\int_{0}^{a} f(x)dx & \text{, if } f(x) \text{ is even function} \\ 0 & \text{, if } f(x) \text{ is odd function} \end{cases}$
- $\int_{0}^{2a} f(x)dx = 2\int_{0}^{a} f(x)dx$ if $f(2a-x) = f(x)$
- $\int_{0}^{na} f(x)dx = n\int_{0}^{a} f(x)dx$, if $f(x + ma) = f(x)$

Solved Examples

Example 1. *Evaluate the following integrals.*

(i) $\int_{0}^{\pi/2} \log \tan x dx$

(ii) $\int_{0}^{\pi/4} \log(1 + \tan\theta)d\theta$ (Madras–2000)

(iii) $\int_{0}^{\pi} \frac{x \sin x}{1 + \sin x} dx$

(iv) $\int_{0}^{\pi/2} \log \sin x dx$ (Anna–2005)

Solution. (i) Consider, $I = \int_{0}^{\pi/2} \log \tan x dx$.

Now, $I = \int_{0}^{\pi/2} \log \tan\left(\frac{\pi}{2} - x\right)dx$ $\left(\because \int_{0}^{a} f(x)dx = \int_{0}^{a} f(a - x)dx\right)$

$I = \int_{0}^{\pi/2} \log \cot x dx$

On adding, we get

$$2I = \int_{0}^{\pi/2} \log \tan x dx + \int_{0}^{\pi/2} \log \cot x dx$$
$$= \int_{0}^{\pi/2} \log(\tan x \cdot \cot x)dx$$
$$= \int_{0}^{\pi/2} \log 1 = 0$$

Hence $2I = 0 \Rightarrow I = 0$.

(ii) Consider, $I = \int_0^{\pi/4} \log(1+\tan\theta)d\theta$

Now, $I = \int_0^{\pi/4} \log\left[1+\tan\left(\frac{\pi}{4}-\theta\right)\right]d\theta$

$$\left[\because \int_0^a f(x)dx = \int_0^a f(a-x)dx\right]$$

$$= \int_0^{\pi/4} \log\left[1+\frac{1-\tan\theta}{1+\tan\theta}\right]d\theta$$

$$= \int_0^{\pi/4} \log\left(\frac{2}{1+\tan\theta}\right)d\theta$$

$$= \int_0^{\pi/4} [\log 2 - \log(1+\tan\theta)]d\theta$$

$$= \int_0^{\pi/4} \log 2 d\theta - \int_0^{\pi/4} \log(1+\tan\theta)d\theta$$

$$= \frac{\pi}{4}\log 2 - I$$

So, $2I = \frac{\pi}{4}\log 2 \Rightarrow I = \frac{\pi}{8}\log 2.$

(iii) Here, $I = \int_0^{\pi} \frac{x\sin x}{1+\sin x}dx$

$$\Rightarrow \quad I = \int_0^{\pi} \frac{(\pi-x)\sin(\pi-x)}{1+\sin(\pi-x)}dx$$

$$\left(\because \int_0^a f(x)dx = \int_0^a f(a-x)dx\right)$$

$$= \int_0^{\pi} \frac{(\pi-x)\sin x}{1+\sin x}dx$$

$$= \int_0^{\pi} \frac{\pi\sin x}{1+\sin x}dx - \int_0^{\pi} \frac{x\sin x}{1+\sin x}dx$$

$$= \int_0^{\pi} \frac{\pi\sin x}{1+\sin x}dx - I$$

So, $2I = \int_0^{\pi} \frac{\pi\sin x}{1+\sin x}dx = \pi\int_0^{\pi} \frac{\sin x}{1+\sin x}dx$

$$= \pi\int_0^{\pi}\left(1-\frac{1}{1+\sin x}\right)dx$$

$$= \pi\int_0^{\pi}\left(1-\frac{1-\sin x}{\cos^2 x}\right)dx$$

$$= \pi\int_0^{\pi}[1-\sec^2 x + \sec x\tan x]dx$$

$$= \pi[x - \tan x + \sec x]_0^{\pi} = \pi(\pi-2)$$

$\therefore \quad 2I = \pi(\pi-2)$

Hence, $I = \pi\left(\frac{\pi}{2}-1\right)$.

(iv) Here, $I = \int_0^{\pi/2} \log\sin x dx$.

Also, $I = \int_0^{\pi/2} \log\sin\left(\frac{\pi}{2}-x\right)dx$

$$\left(\because \int_0^a f(x)dx = \int_0^a f(a-x)dx\right)$$

$\therefore \quad I = \int_0^{\pi/2} \log\cos x dx$.

On adding, we get

$$2I = \int_0^{\pi/2} (\log\sin x)dx + \int_0^{\pi/2} (\log\cos x)dx$$

$$= \int_0^{\pi/2} \log\left(\frac{\sin 2x}{2}\right)dx$$

$$= \int_0^{\pi/2} \log\sin 2x dx - \int_0^{\pi/2} \log 2 dx$$

Let $2x = t$ for first integral, then on differentiating, we get $2dx = dt$

Now, $2I = \frac{1}{2}\int_0^{\pi} \log\sin t dt - [(x\log 2)]_0^{\pi/2}$

$$= \frac{1}{2}\int_0^{\pi} \log\sin t dt - \frac{\pi}{2}\log 2$$

$$= \int_0^{\pi/2} \log\sin t dt - \frac{\pi}{2}\log 2$$

$\therefore \quad 2I = I - \frac{\pi}{2}\log 2$

Hence, $I = -\frac{\pi}{2}\log 2$.

Example 2. *Evaluate the following integrals :*

(i) $\int_0^3 |3x-1|\,dx$ *(ii)* $\int_0^{\pi} |\cos x|\,dx$

(iii) $\int_0^6 |x+2|\,dx$

Solution. (i) Given integral is $\int_0^3 |3x-1|\,dx$.

Now, $|3x-1| = \begin{cases} 3x-1, & \text{when } x \geq 1/3 \\ -(3x-1), & \text{when } x < 1/3 \end{cases}$

So, $\int_0^3 |3x-1|\,dx$

$$= \int_0^{1/3} -(3x-1)dx + \int_{1/3}^3 (3x-1)dx$$

$$= \left[-\frac{3x^2}{2}+x\right]_0^{1/3} + \left[\frac{3x^2}{2}-x\right]_{1/3}^3$$

$$= \frac{65}{6}.$$

(ii) Here, the given intregral is $\int_0^{\pi} |\cos x|\,dx$.

Now, $|\cos x| = \begin{cases} \cos x, & 0 \leq x \leq \pi/2 \\ -\cos x, & \pi/2 \leq x \leq \pi \end{cases}$

So, $\int_0^{\pi} |\cos x|\,dx$

$$= \int_0^{\pi/2} \cos x dx + \int_{\pi/2}^{\pi} (-\cos x)dx$$

$$= \int_0^{\pi/2} \cos x dx - \int_{\pi/2}^{\pi} \cos x dx$$

$$= [\sin x]_0^{\pi/2} - [\sin x]_{\pi/2}^{\pi} = 2.$$

(iii) Let $\int_0^6 |x+2|dx$

$$= \int_0^6 (x+2)dx = \int_0^6 x dx + \int_0^6 2dx$$

$$= \left(\frac{x^2}{2}\right)_0^6 + (2x)_0^6 = \frac{36}{7} - 0 + 2\times 6$$

$$= 18 + 12 = 30.$$

Example 3. *Evaluate the integral* $\int_0^{\pi} \log(1+\cos x)dx$.

Solution. We have, $I = \int_0^{\pi} \log(1+\cos x)dx$

Now, $I = \int_0^{\pi} \log\{1+\cos(\pi - x)\}dx$

$= \int_0^{\pi} \log(1-\cos x)dx$

On adding,

$$2I = \int_0^{\pi} \log(1+\cos x)dx + \int_0^{\pi/2} \log(1-\cos x)dx$$

$$= \int_0^{\pi} \log(1+\cos x)(1-\cos x)dx$$

$$= \int_0^{\pi} \log \sin^2 x dx = 2\int_0^{\pi} \log \sin x dx$$

$$= 4\int_0^{\pi/2} \log \sin x dx$$

(Using property 7)

$\therefore \quad 2I = 4\left(-\frac{\pi}{2}\log 2\right)$

So, $I = -\pi \log 2$

Hence, $I = \pi \log\left(\frac{1}{2}\right)$.

Example 4. *Evaluate the integral* $\int_0^1 \frac{\sin^{-1} x}{x} dx$.

Solution. We have, $I = \int_0^1 \frac{\sin^{-1} x}{x} dx.$

Putting $x = \sin\theta \Rightarrow dx = \cos\theta\, d\theta$.

So, $I = \int_0^{\pi/2} \theta \cot\theta d\theta$.

Now integrating by parts *w.r.t.* θ, we get

$$I = [\theta.\log\sin\theta]_0^{\pi/2} - \int_0^{\pi/2} \log\sin\theta d\theta$$

$$= 0 - \left[-\frac{\pi}{2}\log 2\right] = \frac{\pi}{2}\log 2.$$

Example 5. *Evaluate the integral* $\int_0^{\infty} \log\left(\frac{1+x^2}{x}\right)\frac{dx}{1+x^2}$.

Solution. We have $I = \int_0^{\infty} \log\left(\frac{1+x^2}{x}\right)\frac{dx}{1+x^2}$.

Now, putting $x=\tan\theta \Rightarrow \theta=\tan^{-1}x$, $d\theta = \frac{dx}{1+x^2}$.

So, $I = \int_0^{\pi/2} \log\left(\frac{\sec^2\theta}{\tan\theta}\right)d\theta$

$$= \int_0^{\pi/2} \log\left(\frac{1}{\sin\theta.\cos\theta}\right)d\theta$$

$$= \int_0^{\pi/2} \log\left(\frac{2}{2\sin\theta.\cos\theta}\right)d\theta$$

$$= \int_0^{\pi/2} \log 2 d\theta - \int_0^{\pi/2} \log\sin 2\theta d\theta$$

$$= \frac{\pi}{2}\log 2 - \int_0^{\pi/2} \log\sin 2\theta d\theta$$

Let $2\theta = t \Rightarrow 2d\theta = dt$,

then $I = \frac{\pi}{2}\log 2 - \frac{1}{2}\int_0^{\pi} \log\sin t dt$

$$= \frac{\pi}{2}.\log 2 - \int_0^{\pi/2} \log\sin x dx$$

(Using property 7)

$$= \frac{\pi}{2}.\log 2 - \left(-\frac{\pi}{2}\log 2\right) = \pi\log 2$$

$$\left(\because \int_0^{\pi/2} \log\sin x dx = -\frac{\pi}{2}\log 2\right)$$

Example 6. *Evaluate the integral* $\int_{-1}^{1} f(x)dx$, *where*

$$f(x) = \begin{cases} e^x & , \quad -1 \le x \le 0 \\ 1 & , \quad 0 \le x \le \frac{1}{2} \\ 3^x & , \quad \frac{1}{2} \le x \le 1 \end{cases}$$

Solution. We can write

$$\int_{-1}^{1} f(x)dx = \int_{-1}^{0} f(x)dx + \int_0^{1/2} f(x)dx + \int_{1/2}^{1} f(x)dx$$

$$= \int_{-1}^{0} e^x dx + \int_0^{1/2} 1 dx + \int_{1/2}^{1} 3^x dx$$

$$= \left[e^x\right]_{-1}^{0} + [x]_0^{\frac{1}{2}} + \left[\frac{3^x}{\log 3}\right]_{\frac{1}{2}}^{1}$$

$$= \left[e^0 - e^{-1}\right] + \frac{1}{2} + \frac{1}{\log 3}(3-\sqrt{3})$$

$$= 1 - \frac{1}{e} + \frac{1}{2} + \frac{1}{\log 3}(3-\sqrt{3}).$$

Example 7. *Evaluate the integral* $\int_0^{\pi} \frac{x\tan x}{\sec x + \tan x} dx$.

Solution. Here, $I = \int_0^{\pi} \frac{x\tan x}{\sec x+\tan x}dx$ ...(1)

Now, $I = \int_0^{\pi} \frac{(\pi - x)\tan(\pi - x)}{\sec(\pi - x) + \tan(\pi - x)}dx$

$$\left(\because \int_0^a f(x)dx = \int_0^a f(a-x)dx\right)$$

$$= \int_0^{\pi} \frac{(\pi - x)\tan x}{\sec x+\tan x}dx \quad ...(2)$$

On adding (1) and (2), we get

$$2I = \int_0^{\pi} \frac{x\tan x}{\sec x+\tan x}dx + \int_0^{\pi} \frac{(\pi - x)\tan x}{\sec x + \tan x}dx$$

$$= \int_0^{\pi} \frac{\pi\tan x}{\sec x+\tan x}dx$$

$$= \int_0^{\pi} \frac{\pi\tan x(\sec x - \tan x)}{\sec^2 x - \tan^2 x}dx$$

$$\left(\because \sec^2 x - \tan^2 x = 1\right)$$

$$= \pi\int_0^{\pi}[\sec x\tan x - \sec^2 x + 1]dx$$
$$= \pi\left[\sec x - \tan x + x\right]_0^{\pi}$$
$$= \pi[\{\sec\pi - \tan\pi + \pi\} - \{\sec 0 - \tan 0 + 0\}]$$
$$= \pi(-1+\pi-1) = \pi(\pi-2)$$

Hence, $I = \pi\left(\frac{\pi}{2}-1\right)$.

Example 8. *Evaluate* $\int_0^{\pi/2}\frac{dx}{5+4\sin x}$.

Solution. Let $I = \int_0^{\pi/2}\frac{dx}{5+8\sin x/2\cos x/2}$

Divide Nr and Dr by $\cos^2 x/2$ we have

$$I = \int_0^{\pi/2}\frac{\sec^2(x/2)dx}{5\sec^2\frac{x}{2}+8\tan\frac{x}{2}}$$
$$= \int_0^{\pi/2}\frac{\sec^2(x/2)dx}{5\left(1+\tan^2\frac{x}{2}\right)+8\tan\frac{x}{2}}$$

Put $\tan\frac{x}{2} = t \Rightarrow \frac{1}{2}\sec^2\frac{x}{2}dx = dt$

$$\Rightarrow \sec^2\frac{x}{2}dx = 2dt$$

When $x = 0 \Rightarrow t = 0$

$x = \frac{\pi}{2} \Rightarrow t = 1$

Then $I = \int_0^1\frac{2dt}{5t^2+8t+5} = \frac{2}{5}\int_0^1\frac{dt}{t^2+\frac{8}{5}t+1}$

$$= \frac{2}{5}\int_0^1\frac{dt}{\left(t+\frac{4}{5}\right)^2+\left(\frac{3}{4}\right)^2}$$
$$= \frac{2}{5}\frac{4}{3}\left[\tan^{-1}\left(\frac{t+\frac{4}{5}}{\frac{3}{4}}\right)\right]_0^1$$
$$= \frac{8}{15}\left[\tan^{-1}\frac{36}{15} - \tan^{-1}\frac{16}{15}\right]$$

Example 9. *Evaluate* $\int_0^1\sin^{-1}x\,dx$.

Solution. Let $I = \int_0^1\sin^{-1}x\,dx$

Putting $x = \sin\theta \Rightarrow dx = \cos\theta\,d\theta$

When $x = 0 \Rightarrow \theta = 0$

$x = 1 \Rightarrow \theta = \frac{\pi}{2}$

Then $I = \int_0^{\pi/2}\theta\cos\theta d\theta$;

Integrating by parts, we get

$$I = (\theta\sin\theta)_0^{\pi/2} - \int_0^{\pi/2}\sin\theta d\theta$$
$$= \frac{\pi}{2} + [\cos\theta]_0^{\pi/2} = \frac{\pi}{2} - 1 = \frac{\pi-2}{2}$$

Example 10. *Evaluate* $\int_0^a\frac{x^4dx}{\sqrt{a^2-x^2}}$.

Solution. Let $I = \int_0^a\frac{x^4dx}{\sqrt{a^2-x^2}}$.

Put $x = a\sin\theta \Rightarrow dx = a\cos\theta\,d\theta$

When $x = 0 \Rightarrow \theta = 0$

$x = a \Rightarrow \theta = \frac{\pi}{2}$

Then $I = \int_0^{\pi/4}\frac{a^4\sin^4\theta.a\cos\theta d\theta}{\sqrt{a^2-a^2\sin^2\theta}}$

$$= a^4\int_0^{\pi/2}\sin^4 d\theta$$
$$= a^4.\frac{3.1}{4.2}\frac{\pi}{2}$$ By Walli's formula
$$= \frac{3\pi}{16}.a^4.$$

Example 11. *Evaluate* $\int_2^3\frac{dx}{\sqrt{5x-6-x^2}}$.

Solution. Let $I = \int_2^3\frac{dx}{\sqrt{5x-6-x^2}}$

$$= \int_2^3\frac{dx}{\sqrt{-(x^2-5x+6)}}$$
$$= \int_2^3\frac{dx}{\sqrt{-\left\{\left(x-\frac{5}{2}\right)^2-\frac{1}{4}\right\}}}$$
$$= \int_2^3\frac{dx}{\sqrt{\left(\frac{1}{2}\right)^2-\left(x-\frac{5}{2}\right)^2}}$$
$$= \left[\sin^{-1}\left(\frac{x-5/2}{1/2}\right)\right]_2^3$$
$$= \left[\sin^{-1}(2x-5)\right]_2^3$$
$$= [\sin^{-1}1 - \sin^{-1}(-1)]$$
$$= [\sin^{-1}1 - \sin^{-1}1]$$
$$= 2\sin^{-1}1 = 2.\frac{\pi}{2} = \pi$$

Example 12. *Evaluate* $\int_0^{\infty}\frac{dx}{(x^2+a^2)(x^2+b^2)}$.

Solution. Let $\frac{1}{(x^2+a^2)(x^2+b^2)} = \frac{Ax+B}{(x^2+a^2)} + \frac{Cx+D}{(x^2+b^2)}$

$$\Rightarrow 1 = (Ax+B)(x^2+b^2) + (Cx+D)(x^2+a^2) \quad ...(1)$$

Equating like powers of x on both sides of (1) we get

$$0 = A + C, 0 = B + D$$

$$0 = b^2 A + a^2 C$$

$$1 = b^2 B + a^2 D$$

Solving these equations, we get

$$A = C = 0 \text{ and } B = -D = \frac{1}{b^2 - a^2}$$

Thus, we can write

$$\int_0^\infty \frac{dx}{(x^2+a^2)(x^2+b^2)} = \frac{1}{b^2-a^2}\int_0^\infty \frac{dx}{x^2+a^2} + \frac{1}{a^2-b^2}\int_0^\infty \frac{dx}{x^2+b^2}$$

$$= \frac{1}{a(b^2-a^2)}\left[\tan^{-1}\frac{x}{a}\right]_0^\infty + \frac{1}{b(a^2-b^2)}\left[\tan^{-1}\frac{x}{b}\right]_0^\infty$$

$$= \frac{1}{a(b^2-a^2)}\left(\tan^{-1}\frac{\infty}{a} - \tan^{-1}\frac{0}{a}\right) + \frac{1}{b(a^2-b^2)}\left(\tan^{-1}\frac{\infty}{b} - \tan^{-1}\frac{0}{b}\right)$$

$$= \frac{1}{a(b^2-a^2)}\left[\frac{\pi}{2} - 0\right] + \frac{1}{b(a^2-b^2)}\left(\frac{\pi}{2} - 0\right)$$

$$= \frac{\pi}{2ab(a+b)}.$$

Example 13. *Evaluate* $\int_{-1}^{1} \log\left(\frac{2-x}{2+x}\right)dx$.

Solution. Let $I = \int_{-1}^{1} \log\left(\frac{2-x}{2+x}\right)dx$

Clearly $f(x) = \log\left(\frac{2-x}{2+x}\right)$ is an odd function because $f(-x) = -f(x)$

$$\int_{-1}^{1} \log\left(\frac{2-x}{2+x}\right)dx = 0.$$

EXERCISE 15.1

1. Evaluate the integral $\int_0^{\pi/2} \frac{\sin^4 x dx}{\sin^4 x + \cos^4 x}$.
2. Evaluate the integral $\int_0^{\pi/2} \frac{\sin x - \cos x}{1 + \sin x \cos x} dx$.
3. Evaluate the integral $\int_0^4 f(x)dx$ where $f(x) = \begin{cases} 2x+3 & , \ 0 \le x \le 3 \\ 3x & , \ 3 \le x \le 4 \end{cases}$
4. Evaluate $\int_0^{\pi/2} \frac{x \sin x \cos x}{\cos^4 x + \sin^4 x} dx$.
5. Evaluate the integral $\int_0^{\pi} \frac{x \sin x}{1 + \cos^2 x} dx$.
6. Show that $\int_0^{\pi} \frac{x dx}{a^2 \cos^2 x + b^2 \sin^2 x} = \frac{\pi^2}{2ab}$.
7. Evaluate the integral $\int_0^{\pi/2} \log(\tan x + \cot x)dx$.
8. Evaluate $\int_0^1 e^{\sin^{-1} x} dx$.

Hint to Selected Problems

1.

$$I = \int_0^{\pi/2} \frac{\sin^4 x dx}{\sin^4 x + \cos^4 x} \quad \ldots(1)$$

$$= \int_0^{\pi/2} \frac{\sin^4\left(\frac{\pi}{2} - x\right)}{\sin^4\left(\frac{\pi}{2} - x\right) + \cos^4\left(\frac{\pi}{2} - x\right)} dx$$

$$= \int_0^{\pi/2} \frac{\cos^4 x}{\sin^4 x + \cos^4 x} dx \quad \ldots(2)$$

Adding (1) and (2) we get

$$2I = \int_0^{\pi/2} dx = \frac{\pi}{2} \Rightarrow I = \frac{\pi}{4}$$

2.

$$I = \int_0^{\pi/2} \frac{\sin x - \cos x}{1 + \cos x \sin x} dx \quad \ldots(1)$$

$$= \int_0^{\pi/2} \frac{\sin\left(\frac{\pi}{2} - x\right) - \cos\left(\frac{\pi}{2} - x\right)}{1 + \cos\left(\frac{\pi}{2} - x\right)\sin\left(\frac{\pi}{2} - x\right)} dx$$

$$= \int_0^{\pi/2} \frac{\cos x - \sin x}{1 + \cos x \sin x} dx \quad \ldots(2)$$

Adding (1) and (2) $2I = 0 \Rightarrow I = 0.$

3.

$$I = \int_0^4 f(x)dx = \int_0^3 (2x+3)dx + \int_3^4 3x dx$$

$$= [x^2 + 3x]_0^3 + \left[\frac{3x^2}{2}\right]_3^4$$

$$= 18 + 24 - \frac{27}{2} = 42 - \frac{27}{2} = \frac{57}{2}.$$

4. $$I=\int_0^{\pi/2}\frac{x\sin x\cos x dx}{\cos^4 x+\sin^4 x}$$

$$=\int_0^{\pi/2}\frac{\left(\frac{\pi}{2}-x\right)\sin\left(\frac{\pi}{2}-x\right)\cos\left(\frac{\pi}{2}-x\right)}{\cos^4\left(\frac{\pi}{2}-x\right)+\sin^4\left(\frac{\pi}{2}-x\right)}dx$$

$$=\int_0^{\pi/2}\frac{\pi}{2}\frac{\sin x\cos x}{\sin^4 x+\cos^4 x}dx-\int_0^{\pi/2}\frac{x\sin x\cos x}{\sin^4 x+\cos^4 x}dx$$

$\Rightarrow$ $$2I=\frac{\pi}{2}\int_0^{\pi/2}\frac{\sin x\cos x}{\sin^4 x+\cos^4 x}dx$$

divide Nr and Dr by $\cos^4 x$

$$2I=\frac{\pi}{2}\int_0^{\pi/2}\frac{\tan x\sec^2 x}{1+\tan^4 x}dx$$

Put $\tan^2 x=t \Rightarrow 2\tan x\sec^2 x dx = dt$

$\Rightarrow \tan x\sec^2 x dx=\frac{dt}{2}$

$\therefore$ $$2I=\frac{\pi}{2}\int_0^{\infty}\frac{1}{2}\frac{dt}{1+t^2}$$

When $x=0 \Rightarrow t=0$

When $x=\pi/2 \Rightarrow t=\infty$

$$=\frac{\pi}{4}\left[\tan^{-1}t\right]_0^{\infty}=\frac{\pi}{4}\left(\frac{\pi}{2}-0\right)=\frac{\pi^2}{8}.$$

So, $I=\frac{\pi^2}{16}$.

5. $$I=\int_0^{\pi}\frac{x\sin x}{1+\cos^2 x}dx \quad \ldots(1)$$

$$=\int_0^{\pi}\frac{(\pi-x)\sin(\pi-x)}{1+\cos^2(\pi-x)}dx \quad \ldots(2)$$

Adding (1) and (2)

$$2I=\int_0^{\pi}\frac{(x+\pi-x)\sin x}{1+\cos^2 x}dx=\pi\int_0^{\pi}\frac{\sin x}{1+\cos^2 x}dx$$

Let $\cos x=t \Rightarrow \sin x\,dx=-dt$

$x=0 \Rightarrow t=1$

$x=\pi \Rightarrow t=-1$

Now $$2I=\pi\int_1^{-1}\frac{dt}{1+t^2}=-\pi\left[\tan^{-1}t\right]_1^{-1}=\pi\left[-\frac{\pi}{4}-\frac{\pi}{4}\right]$$

$$2I=\frac{\pi^2}{2} \Rightarrow I=\frac{\pi^2}{4}.$$

6. $$I=\int_0^{\pi}\frac{x dx}{a^2\cos^2 x+b^2\sin^2 x} \quad \ldots(1)$$

$$=\int_0^{\pi}\frac{(\pi-x)dx}{a^2\cos^2(\pi-x)+b^2\sin^2(\pi-x)}$$

$$=\int_0^{\pi}\frac{(\pi-x)dx}{a^2\cos^2 x+b^2\sin^2 x} \quad \ldots(2)$$

Adding (1) and (2) we get

$$2I=\int_0^{\pi}\frac{(x+\pi-x)dx}{a^2\cos^2 x+b^2\sin^2 x}$$

$$=\pi\int_0^{\pi}\frac{dx}{a^2\cos^2 x+b^2\sin^2 x}$$

$$=2\pi\int_0^{\pi/2}\frac{dx}{a^2\cos^2 x+b^2\sin^2 x}$$

$$\left[\int_0^{2a}f(x)dx=2\int_0^{a}f(x)dx;\ f(2a-x)=f(x)\right]$$

Divide Nr and Dr by $\cos^2 x$

$$I=\pi\int_0^{\pi/2}\frac{\sec^2 x dx}{a^2+b^2\tan^2 x}$$

Put $\tan x=t$

$\sec^2 x\,dx=dt$

When $x=0 \Rightarrow t=0$

$x=\pi/2 \Rightarrow t=\infty$

Then, $$I=\pi\int_0^{\infty}\frac{dt}{a^2+b^2t^2}=\frac{\pi}{b^2}\int_0^{\infty}\frac{dt}{\left(\frac{a}{b}\right)^2+t^2}$$

$$=\frac{\pi}{b^2}\frac{1}{a/b}\left[\tan^{-1}\frac{t}{a/b}\right]_0^{\infty}$$

$$I=\frac{\pi}{ab}\left[\tan^{-1}\frac{bt}{a}\right]_0^{\infty}=\frac{\pi}{ab}\left(\frac{\pi}{2}-0\right)=\frac{\pi^2}{2ab}$$

7. $$I=\int_0^{\pi/2}\log(\tan x+\cot x)dx$$

$$=\int_0^{\pi/2}\log\left(\frac{\sin^2 x+\cos^2 x}{\sin x\cos x}\right)dx$$

$$=\int_0^{\pi/2}\log\left(\frac{1}{\sin x\cos x}\right)dx$$

$$=\int_0^{\pi/2}-(\log\cos x+\log\sin x)dx$$

$$=-\left[\int_0^{\pi/2}\log\sin x dx+\int_0^{\pi/2}\log\cos x dx\right]$$

$$=-2\int_0^{\pi/2}\log\sin x dx$$

$$\left[\because \int_0^{\pi/2}\log\sin x dx=\int_0^{\pi/2}\log\cos x dx=-\frac{\pi}{2}\log 2\right]$$

$$=-2\left(-\frac{\pi}{2}\log 2\right)=\pi\log 2.$$

8. $$I=\int_0^1 e^{\sin^{-1}x}dx$$

Put $\sin^{-1}x=t \Rightarrow \sin t=x \Rightarrow \cos t dt=dx$

When $x=0 \Rightarrow t=0$

$x=1 \Rightarrow t=\pi/2$

$$I=\int_0^{\pi/2}e^t\cos t dt$$

$$=\left[\frac{1}{1^2+t^2}\int[e^t\cos t+e^t\sin t]\right]_0^{\pi/2}$$

$$=\frac{1}{2}[e^{\pi/2}(0+1)-1]$$

$$\left[\text{Since}\int_0^{\pi/2}e^{ax}\cos bx dx=\frac{1}{a^2+b^2}[ae^{ax}\cos bx+be^{ax}\sin bx=\frac{1}{2}(e^{\pi/2}-1)]\right]$$

Answers

1. $\frac{\pi}{4}$ **2.** 0 **3.** $\frac{57}{2}$ **4.** $\frac{\pi^2}{16}$ **5.** $\frac{\pi^2}{4}$ **7.** $\pi\log 2$ **8.** $\frac{e^{\pi/2}-1}{2}$

MORE SOLVED EXAMPLES

Example 1. *Evaluate* $\int_0^\pi \frac{x\sin x}{(1+\cos^2 x)}dx$

Solution. Let $I = \int_0^\pi \frac{x\sin x}{(1+\cos^2 x)}dx$ ...(1)

$$= \int_0^\pi \frac{(\pi - x)\sin(\pi - x)}{1+\cos^2(\pi - x)}dx$$

$$\left[\because \int_0^a f(x)dx = \int_0^a f(a-x)dx\right]$$

$$= \int_0^\pi \frac{(\pi - x)\sin x}{1+\cos^2 x}dx \quad ...(2)$$

Adding (1) and (2), we get

$$2I = \int_0^\pi \frac{\pi \sin x}{1+\cos^2 x}dx$$

$$= 2\pi\int_0^{\pi/2} \frac{\sin x}{1+\cos^2 x}dx$$

Now put $\cos x = t \Rightarrow -\sin x\, dx = dt$

Also, $t = 1$ at $x = 0$ and $t = 0$ at $x = \pi/2$.

Therefore, we have

$$I = \pi\int_1^0 -\frac{dt}{1+t^2} = \pi\int_0^1 \frac{dt}{1+t^2}$$

$$= \pi\left[\tan^{-1} t\right]_0^1 = \frac{\pi^2}{4}$$

Example 2. *Evaluate* $\int_0^\pi x\sin^6 x\cos^4 x dx$ (VTU–2001)

Solution. Here $I = \int_0^\pi x\sin^6 x\cos^4 x dx$

$$= \int_0^\pi (\pi - x)\sin^6(\pi - x)\cos^4(\pi - x)dx$$

$$= \int_0^\pi (\pi - x)\sin^6 x\cos^4 x dx$$

$$= \int_0^\pi \pi\sin^6 x\cos^4 x dx - \int_0^\pi x\sin^6 x\cos^4 x dx$$

$$= \pi\int_0^\pi \sin^6 x\cos^4 x dx - I$$

Hence, $2I = \pi\int_0^\pi \sin^6 x\cos^4 x dx$

$$= 2\pi\int_0^{\pi/2} \sin^6 x\cos^4 x dx$$

$$I = \pi\int_0^{\pi/2} \sin^6 x\cos^4 x dx$$

$$= \pi\frac{5.3.1.3.1}{10.8.6.4.2}.\frac{\pi}{2} = \frac{3\pi^2}{512}$$

(By Walli's formula)

Example 3. *Evaluate* $\int_0^\pi \sin^3\theta(1+2\cos\theta)(1+\cos\theta)^2 d\theta$

Solution. Let $I = \int_0^\pi \sin^3\theta(1+2\cos\theta)(1+\cos\theta)^2 d\theta$

$$= \int_0^\pi \sin^3\theta(1+2\cos\theta)(1+2\cos\theta+\cos^2\theta)d\theta$$

$$= \int_0^\pi (\sin^3\theta + 4\sin^3\theta\cos\theta + 5\sin^3\theta\cos^2\theta + 2\sin^3\theta\cos^3\theta)d\theta$$

Now $\int_0^\pi \sin^m\theta\cos^n\theta d\theta$

$= 2\int_0^{\pi/2} \sin^m\theta\cos^n\theta d\theta$, if n is even

$= 0$, if n is odd

Hence, $I = 2.\frac{2}{3.1} + 10.\frac{2.1}{5.3.1} = \frac{4}{3} + \frac{4}{3} = \frac{8}{3}$.

Example 4. *Evaluate* $\int_0^{\pi/2} \log\sin 2x dx$.

Solution. Let $I = \int_0^{\pi/2} \log\sin 2x dx$. ...(1)

Put $2x = t \Rightarrow 2dx = dt$, we get

$$I = \frac{1}{2}\int_0^\pi \log\sin t dt$$

$$= \frac{1}{2}.2\int_0^{\pi/2} \log\sin t dt$$

$$= \int_0^{\pi/2} \log\sin t dt$$

$$= \int_0^{\pi/2} \log\sin\left(\frac{\pi}{2} - t\right)dt$$

$$= \int_0^{\pi/2} \log\cos t dt \quad ...(2)$$

Adding (1) and (2), we get

$$\Rightarrow \quad 2I = \int_0^{\pi/2} \log\sin t dt + \int_0^{\pi/2} \log\cos t dt$$

$$= \int_0^{\pi/2} \log\left(\frac{\sin 2t}{2}\right)dt$$

$$= \int_0^{\pi/2} \log\sin 2t dt - \int_0^{\pi/2} \log 2 dt$$

$$= I - \frac{\pi}{2}\log 2$$

Hence, $I = -\frac{\pi}{2}\log 2$.

Example 5. *Show that* $\int_0^{\pi/2} \frac{\sqrt{\sin x}}{\sqrt{\sin x} + \sqrt{\cos x}}dx = \frac{\pi}{4}$.

Solution. Let $I = \int_0^{\pi/2} \frac{\sqrt{\sin x}}{\sqrt{\sin x} + \sqrt{\cos x}}dx$ ...(1)

$$= \int_0^{\pi/2} \frac{\sqrt{\sin\left(\frac{\pi}{2} - x\right)}}{\sqrt{\sin\left(\frac{\pi}{2} - x\right)} + \sqrt{\cos\left(\frac{\pi}{2} - x\right)}}dx$$

$$= \int_0^{\pi/2} \frac{\sqrt{\cos x}}{\sqrt{\cos x} + \sqrt{\sin x}}dx \quad ...(2)$$

Adding (1) and (2), we get

$$2I = \int_0^{\pi/2} \left[\frac{\sqrt{\sin x} + \sqrt{\cos x}}{\sqrt{\cos x} + \sqrt{\sin x}}\right]dx$$

$$= \int_0^{\pi/2} 1 dx = \frac{\pi}{2}$$

Hence, $I = \frac{\pi}{4}$.

Example 6. *Show that*

$$\int_0^{\pi/2} \frac{\sin^2 x}{(\sin x + \cos x)} dx = \frac{1}{\sqrt{2}} \log(\sqrt{2}+1).$$

Solution. Let

$$I = \int_0^{\pi/2} \frac{\sin^2 x}{(\sin x + \cos x)} dx \quad \ldots(1)$$

$$= \int_0^{\pi/2} \frac{\sin^2\left(\frac{\pi}{2} - x\right)}{\sin\left(\frac{\pi}{2} - x\right) + \cos\left(\frac{\pi}{2} - x\right)} dx$$

$$= \int_0^{\pi/2} \frac{\cos^2 x}{\cos x + \sin x} dx \quad \ldots(2)$$

Adding (1) and (2), we get

$$2I = \int_0^{\pi/2} \frac{\sin^2 x}{\sin x + \cos x} dx + \int_0^{\pi/2} \frac{\cos^2 x}{\cos x + \sin x} dx$$

$$= \int_0^{\pi/2} \frac{dx}{\sin x + \cos x}$$

$$= \int_0^{\pi/2} \frac{(1/\sqrt{2})dx}{\left(\frac{1}{\sqrt{2}}\sin x + \frac{1}{\sqrt{2}}\cos x\right)}$$

$$= \frac{1}{\sqrt{2}} \int_0^{\pi/2} \frac{dx}{\cos(x - \pi/4)}$$

$$= \frac{1}{\sqrt{2}} \int_0^{\pi/2} \sec\left(x - \frac{\pi}{4}\right) dx$$

$$= \frac{1}{\sqrt{2}} \log\left[\sec\left(x - \frac{\pi}{4}\right) + \tan\left(x - \frac{\pi}{4}\right)\right]_0^{\frac{\pi}{2}}$$

$$= \frac{1}{\sqrt{2}}\left[\log\left(\sec\frac{\pi}{4} + \tan\frac{\pi}{4}\right) - \log\left\{\sec\left(-\frac{\pi}{4}\right) + \tan\left(-\frac{\pi}{4}\right)\right\}\right]$$

$$= \frac{1}{\sqrt{2}} \log\left[\frac{(\sqrt{2}+1)(\sqrt{2}+1)}{(\sqrt{2}-1)(\sqrt{2}+1)}\right]$$

$$= \frac{1}{\sqrt{2}} \log(\sqrt{2}+1)^2$$

$$= \frac{1}{\sqrt{2}} .2\log(\sqrt{2}+1)$$

Hence, $I = \frac{1}{\sqrt{2}} \log(\sqrt{2}+1)$.

EXERCISE 15.2

Prove the following :

1. $\int_0^{\pi} x \log \sin x dx = \frac{1}{2}\pi^2 \log\frac{1}{2}$
2. $\int_0^{\pi/2} x \cot x dx = \frac{\pi}{2}\log 2$
3. $\int_0^{\pi/2} \left[\frac{\theta}{\sin\theta}\right]^2 d\theta = \pi \log 2$
4. $\int_0^{1} \frac{\sin^{-1} x}{x} dx = \frac{\pi}{2}\log 2$
5. $\int_0^{\pi/4} \log(1+\tan\theta)d\theta = \frac{\pi}{8}\log 2$
6. $\int_0^{\infty} \frac{xdx}{(1+x)(1+x^2)} = \frac{\pi}{4}$
7. $\int_0^{\pi/2} \frac{\sqrt{\tan x}}{\sqrt{\tan x}+\sqrt{\cot x}} dx = \frac{\pi}{4}$
8. $\int_0^{\pi/2} \frac{\cos^2 x dx}{(\sin x + \cos x)} = \frac{1}{\sqrt{2}}\log(\sqrt{2}+1)$
9. $\int_0^{\pi} \sin^m x \cos^{2m+1} dx = 0$
10. $\int_0^{\pi} \frac{x^2 \sin 2x \sin\left(\frac{\pi}{2}\cos x\right)}{2x - \pi} dx = \frac{8}{\pi}$

Hint to Selected Problems

1.

$$I = \int_0^{\pi} x \log \sin x dx = \int_0^{\pi} (\pi - x)\log\sin(\pi - x)dx$$

$$= \int_0^{\pi} \pi \log \sin x dx - \int_0^{\pi} x \log\sin x dx$$

$$\Rightarrow \quad 2I = \int_0^{\pi} \pi\log\sin x dx = 2\pi\int_0^{\pi/2} \log\sin x dx$$

$$= 2\pi\left(\frac{\pi}{2}\log 2\right) \text{ if } f(2a-x) = f(x)$$

$$= \pi^2 \log\frac{1}{2} \Rightarrow I = \frac{\pi^2}{2}\log\frac{1}{2}$$

3.

$$I = \int_0^{\pi/2} \left(\frac{\theta}{\sin\theta}\right)^2 d\theta$$

$$= \int_0^{\pi/2} \theta^2 \operatorname{cosec}^2\theta d\theta - \left(-\theta^2\cot\theta\right)_0^{\pi/2} - \int_0^{\pi/2} 2\theta\cot\theta d\theta$$

$$= (-0+0) - 2\int_0^{\pi/2} -\theta\cot\theta d\theta$$

$$= 2\left[(\theta \log\sin\theta)_0^{\theta/2} - \int_0^{\pi/2} \log\sin\theta\, d\theta\right]$$

$$= 2\left[(0-0) - \int_0^{\pi/2}\log\sin\theta d\theta\right]$$

$$= 2\left(\frac{\pi}{2}\log 2\right) \quad \left[\because \int_0^{\pi/2} \log\sin\theta d\theta = \pi\log 2\right]$$

$$= \pi\log 2$$

5.

$$I = \int_0^{\pi/4} \log(1+\tan\theta)d\theta$$

$$= \int_0^{\pi} \log\left\{1+\log\left(\frac{\pi}{4}-\theta\right)\right\}d\theta$$

$$= \int_0^{\pi/4} \log\left\{1+\frac{\tan\pi/4-\tan\theta}{1+\tan\pi/4\tan\theta}\right\}d\theta$$

$$= \int_0^{\pi/4} \log\left\{1+\frac{1-\tan\theta}{1+\tan\theta}\right\}d\theta$$

$$= \int_0^{\pi/4} \log\left\{\frac{2}{1+\tan\theta}\right\}d\theta$$

$$= \int_0^{\pi/4} \log 2d\theta - \int_0^{\pi/4} \log(1+\tan\theta)d\theta$$

$$2I = \log 2[\theta]_0^{\pi/4} = \frac{\pi}{4}\log 2 \Rightarrow I = \frac{\pi}{8}\log 2$$

7.

$$I = \int_0^{\pi/2} \frac{\sqrt{\tan x}}{\sqrt{\tan x}+\sqrt{\cot x}}dx \qquad \ldots(1)$$

$$= \int_0^{\pi/2} \frac{\sqrt{\tan\left(\frac{\pi}{2}-x\right)}}{\sqrt{\tan\left(\frac{\pi}{2}-x\right)}+\sqrt{\cot\left(\frac{\pi}{2}-x\right)}}dx$$

$$= \int_0^{\pi/2} \frac{\sqrt{\cot x}}{\sqrt{\cot x}+\sqrt{\tan x}}dx \qquad \ldots(2)$$

Adding (1) and (2)

$$2I = \int_0^{\pi/2} dx = \frac{\pi}{2} \Rightarrow I = \pi/4$$

8.

$$I = \int_0^{\pi/2} \frac{\cos^2 x dx}{\sin x+\cos x} \qquad \ldots(1)$$

$$= \int_0^{\pi/2} \frac{\cos^2\left(\frac{\pi}{2}-x\right)}{\sin\left(\frac{\pi}{2}-x\right)+\cos\left(\frac{\pi}{2}-x\right)}dx$$

$$= \int_0^{\pi/2} \frac{\sin^2 x}{\cos x+\sin x}dx \qquad \ldots(2)$$

Adding (1) and (2)

$$2I = \int_0^{\pi/2} \frac{\cos^2 x+\sin^2 x}{\cos x+\sin x}dx = \int_0^{\pi/2} \frac{dx}{\cos x+\sin x}$$

$$= \int_0^{\pi/2} \frac{dx}{1-2\sin^2\frac{x}{2}+2\sin\frac{x}{2}\cos\frac{x}{2}}$$

Divide Nr and Dr by $\cos^2\frac{x}{2}$

$$2I = \int_0^{\pi/2} \frac{\sec^2 x/2dx}{1+2\tan\frac{x}{2}-\tan^2\frac{x}{2}}$$

Let $\tan\frac{x}{2} = t \Rightarrow \sec^2\frac{x}{2}dx = 2dt$

Also when $x = 0 \Rightarrow t = 0 \qquad x = \frac{\pi}{2} \Rightarrow t = 1$

$$2I = \int_0^1 \frac{2dt}{2t+1-t^2} = 2\int_0^1 \frac{dt}{(\sqrt{2})^2-(t-1)^2}$$

$$= 2\times\frac{1}{2\sqrt{2}}\left[\log\left\{\frac{\sqrt{2}+t-1}{\sqrt{2}-t+1}\right\}\right]_0^1$$

$$= \frac{1}{\sqrt{2}}\left[\log\left(\frac{\sqrt{2}}{\sqrt{2}}\right)-\log\left(\frac{\sqrt{2}-1}{\sqrt{2}+1}\right)\right]$$

$$= -\frac{1}{\sqrt{2}}\log\left\{\frac{\sqrt{2}-1}{\sqrt{2}+1}\right\} = \frac{1}{\sqrt{2}}\log\frac{\sqrt{2}+1}{\sqrt{2}-1}$$

$$= \frac{1}{\sqrt{2}}\log\left\{\frac{(\sqrt{2}+1)(\sqrt{2}+1)}{(\sqrt{2}-1)(\sqrt{2}+1)}\right\}$$

$$= \frac{1}{\sqrt{2}}\log(\sqrt{2}+1)^2 = \frac{2}{\sqrt{2}}\log(\sqrt{2}+1)$$

So, $\quad I = \frac{1}{\sqrt{2}}\log(\sqrt{2}+1)$

9.

$$I = \int_0^{\pi} \sin^m x\cos^{2m+1} x dx = f(x), \text{say}$$

Here $\quad f(x) = \sin^m x\cos^{2m+1} x$

$$f(\pi-x) = \sin^m(\pi-x)\cos^{2m+1}(\pi-x)$$

$$= -\sin^m x\cos^{2m+1} x = f(-x)$$

So, $\quad I = 0$

Since $\int_0^{2a} f(x)dx = 0$ if $f(2a-x) = -f(x)$

15.3 DEFINITE INTEGRAL AS THE LIMIT OF SUM

It is always possible to regard a definite integral as the limit of the sum of certain number of terms, when the number of terms tends to infinity and each term tends to zero.

Here, we define the definite integral as follows :

$$\int_0^a f(x)dx = \lim h[f(a)+f(a+h)+f(a+2h)+\ldots+f\{a+(n-1)h\}]$$

when $n \to \infty, h \to 0$ and $nh \to b-a$.

Solved Examples

Example 1. *Evaluate $\int_a^b x^2 dx$, directly from the definition of integral as the limit of a sum.*

Solution. We know that

$$\int_0^a f(x)dx = \lim_{n\to\infty} h[f(a)+f(a+h)+f(a+2h)+\ldots+f\{a+(n-1)h\}] \qquad \ldots(1)$$

Here $f(x) = x^2 \qquad f(a) = a^2$

$$f(a+h) = (a+h)^2$$

$\ldots \quad \ldots$ and so on.

Put all these values in (1), we get

$$\int_a^b x^2 dx = \lim_{n\to\infty} h[a^2+(a+h)^2+(a+2h)^2+\ldots+\{a+(n-1)h\}^2]$$

when $h \to 0, n \to \infty$ and $nh \to b - a$.

$$= \lim_{n\to\infty} h[na^2 + 2ah\{1+2+3+...+(n-1)\}] + h^2[1^2+2^2+...+(n-1)^2]$$

Using $\Sigma n = \frac{n(n+1)}{2}$ and $\Sigma n^2 = \frac{n(n+1)(2n+1)}{6}$

$$\therefore \int_a^b x^2dx = \lim_{n\to\infty} h\left[na^2 + 2ah\frac{(n-1)n}{2} + \frac{h^2}{6}(n-1)n(2n-1)\right]$$

$$= \lim_{n\to\infty}\left[(nh)a^2 + a(nh)(n-1)h + \frac{1}{6}(nh)(n-1)h(2n-1)h\right]$$

$$= \lim_{n\to\infty}\left[(nh)a^2 + a(nh)^2\left(1-\frac{1}{n}\right) + \frac{1}{6}2(nh)^3\left(1-\frac{1}{n}\right)\left(1-\frac{1}{2n}\right)\right]$$

$$= (b-a)a^2 + a(b-a)^2 + \frac{1}{3}(b-a)^3$$

$(\therefore$ as $n \to \infty, h \to 0, nh \to b - a)$

$$= \frac{1}{3}(b-a)[3a^2 + 3(b-a)a + b^2 - 2ab + a^2]$$

$$= \frac{1}{3}(b-a)(a^2+ab+b^2)$$

$$= \frac{1}{3}(b^3-a^3).$$

Example 2. *From the definition of a definite integral as the limit of a sum, evaluate* $\int_a^b e^x dx$.

Solution. Here we have $f(x) = e^x$

Therefore, $f(a) = e^a$

$f(a+h) = e^{a+h}$

... ... etc.

Now $\int_a^b e^x dx$

$$= \lim_{h\to 0} h[e^a + e^{a+h} + e^{a+2h} + ... + e^{a+(n-1)h}]$$

where, $nh \to b - a$ and $n \to \infty$ as $h \to 0$.

$$= \lim_{h\to\infty} he^a[1 + e^h + e^{2h} + ... + e^{(n-1).h}]$$

$$= \lim_{h\to 0} he^a\left\{\frac{(e^h)^n - 1}{e^h - 1}\right\} = \lim_{h\to 0} he^a\left\{\frac{e^{nh}-1}{e^h-1}\right\}$$

$$= \lim_{h\to 0} he^a\left[\frac{e^{b-a}-1}{e^h-1}\right] = \lim_{h\to 0} e^a\left[\frac{e^{b-a}-1}{\frac{e^h-1}{h}}\right]$$

$$= e^b - e^a \qquad \left(\because \lim_{h\to 0}\frac{e^h-1}{h} = 1\right)$$

15.4 SUMMATION OF SERIES WITH THE HELP OF DEFINITE INTEGRAL

We know that

$$\int_a^b f(x)dx = \lim_{n\to\infty} h[f(a) + f(a+h) + f(a+2h) + ... + f\{a+(n-1)h\}]$$

$$= \lim_{n\to\infty}\sum_{r=0}^{n-1} f(a+rh). \qquad \text{where, } na = b - a$$

Now putting $a = 0$ and $b = 1$ so that $h = 1/n$, we get

$$\int_0^1 f(x)dx = \lim_{n\to\infty}\frac{1}{n}\sum_{r=0}^{n-1} f\left(\frac{r}{n}\right).$$

WORKING PROCEDURE

Step 1. *Write the* r^{th} *terms of the series.*

Step 2. *Write the* r^{th} *term in the form of* $\frac{1}{n}f\left(\frac{r}{n}\right)$.

Step 3. *Replace* $\frac{r}{n}$ *by x,* $\frac{1}{n}$ *by dx and* $\lim_{x\to\infty}$ *by f.*

Then lower limit of the definite integral will be value of $\frac{r}{n}$ *for the first term as* $n \to \infty$ *and the upper limit will be the value of* $\frac{r}{n}$ *for the last term as* $n \to \infty$.

Solved Examples

Example 1. *Evaluate the following*

$$\lim_{n\to\infty}\left[\frac{1}{n+1}+\frac{1}{n+2}+\ldots+\frac{1}{2n}\right].$$

Solution . The general term r^{th} term $=\dfrac{1}{n+r}$.

We have to find

$$\lim_{n\to\infty}\sum_{r=1}^{n}\frac{1}{n+r}=\lim_{n\to\infty}\frac{1}{n[1+r/n]}.$$

$$=\lim_{n\to\infty}\frac{1}{n}\sum_{r=1}^{n}\frac{1}{1+(r/n)}.$$

Since the limit of r in the summation are 1 to n, therefore, the lower limit of integration

$$=\lim_{n\to\infty}\frac{1}{n}=0.$$

Also, the upper limit of integration

$$=\lim_{n\to\infty}\frac{n}{n}=1.$$

Hence, the required limit

$$\int_0^1\frac{1}{1+x}dx=[\log(1+x)]_0^1=\log 2.$$

Example 2. *Evaluate*

$$\lim_{n\to\infty} n\left[\frac{1}{(n+1)(n+2)}+\frac{1}{(n+2)(n+4)}+\ldots+\frac{1}{6n^2}\right].$$

Solution . Here, the given limit

$$=\lim_{n\to\infty} n\sum_{r=1}^{n}\frac{1}{(n+r)(n+2r)}.$$

$$=\lim_{n\to\infty}\frac{n}{n^2}\sum_{r=1}^{n}\frac{1}{(1+(r/n))(1+(2r/n))}$$

$$=\lim_{n\to\infty}\frac{1}{n}\sum_{r=1}^{n}\frac{1}{(1+(r/n))(1+(2r/n))}$$

$$=\int_0^1\frac{1}{(1+x)(1+2x)}dx=\int_0^1\left[\frac{-1}{1+x}+\frac{2}{1+2x}\right]dx$$

(Resolving into partial fraction)

$$=[-\log(1+x)+\log(1+2x)]_0^1$$

$$=\left[\log\frac{(1+2x)}{(1+x)}\right]_0^1=\log\frac{3}{2}-\log 1=\log\frac{3}{2}.$$

Example 3. *Evaluate*

$$\lim_{n\to\infty}\left[\left(1+\frac{1}{n^2}\right)\left(1+\frac{2^2}{n^2}\right)\left(1+\frac{3^2}{n^2}\right)\ldots\left(1+\frac{n^2}{n^2}\right)\right]^{\frac{1}{n}}$$

(Bhopal–2008)

Solution . Let

$$A=\lim_{n\to\infty}\left[\left(1+\frac{1}{n^2}\right)\left(1+\frac{2^2}{n^2}\right)\left(1+\frac{3^2}{n^2}\right)\ldots\left(1+\frac{n^2}{n^2}\right)\right]^{\frac{1}{n}}$$

$$\log A=\lim_{n\to\infty}\frac{1}{n}\left[\log\left(1+\frac{1}{n^2}\right)+\log\left(1+\frac{2^2}{n^2}\right)+\log\left(1+\frac{3^2}{n^2}\right)+\ldots+\log\left(1+\frac{n^2}{n^2}\right)\right]$$

$$=\lim_{n\to\infty}\frac{1}{n}\sum_{r=1}^{\infty}\log\left(1+\frac{r^2}{n^2}\right)$$

$$=\int_0^1\log(1+x^2)dx=\int_0^1\log(1+x^2)1.dx$$

$$=\left[x\log(1+x^2)\right]_0^1-\int_0^1\frac{2x.xdx}{1+x^2}$$

$$=\log 2-2\int_0^1\frac{(1+x^2)-1}{1+x^2}dx$$

$$=\log 2-2\int_0^1\left[1-\frac{1}{(1+x^2)}\right]dx$$

$$=\log 2-2\left[x-\tan^{-1}x\right]_0^1$$

$$=\log 2-2\left(1-\frac{\pi}{4}\right)$$

Therefore, $\log A=\log 2+\dfrac{1}{2}(\pi-4)$

$\Rightarrow$ $\log\dfrac{A}{2}=\dfrac{1}{2}(\pi-4)$

$\Rightarrow$ $A=2e^{(\pi-4)/2}.$

Example 4. *Find the limit of* $\left[\dfrac{n!}{n^n}\right]^{1/n}$ *when* $n\to\infty$.

Solution . Let

$$A=\lim_{n\to\infty}\left[\frac{n!}{n^n}\right]^{1/n}$$

$$=\lim_{n\to\infty}\left[\frac{1.2.3.4\ldots n}{n.n.n\ldots n}\right]^{1/n}$$

$$\Rightarrow\quad \log A=\lim_{n\to\infty}\frac{1}{n}\left[\log\left(\frac{1}{n}\right)+\log\left(\frac{2}{n}\right)+\log\left(\frac{3}{n}\right)+\ldots+\log\left(\frac{n}{n}\right)\right]$$

$$=\lim_{n\to\infty}\sum_{r=1}^{n}\frac{1}{n}\log\left(\frac{r}{n}\right)$$

$$=\int_0^1\log x dx=\int_0^1\log x.1dx$$

$$=[(\log x).x]_0^1-\int_0^1\frac{1}{x}.xdx$$

$$=0-[x]_0^1=-1$$

Hence, $A=e^{-1}=\dfrac{1}{e}.$

EXERCISE 15.3

1. Show that the limit of the sum $\frac{1}{n}+\frac{1}{n+1}+\frac{1}{n+2}+\ldots+\frac{1}{6n}$ when n is indefinitely increased is log 6.
2. Evaluate $\int_a^b x^2dx$ directly from the definition of the integral as the limit of the sum.
3. Evaluate by summation $\int_1^2 xdx$.
4. Evaluate by summation $\int_a^b \sin xdx$.
5. Evaluate by summation $\int_0^{\pi/2} \sin xdx$.
6. Show that the limit (when $n \to \infty$) of the series $\frac{n}{(n+1)^2}+\frac{n}{(n+2)^2}+\ldots+\frac{n}{(n+n)^2}$ is $\frac{1}{2}$.
7. Show that
$$\lim_{n\to\infty}\left[\frac{n}{n^2}+\frac{n}{n^2+1^2}+\frac{n}{n^2+2^2}+\ldots+\frac{n}{n^2+(n+1)^2}\right]=\frac{\pi}{4}.$$
8. Show that
$$\lim_{n\to\infty}\left[\frac{n}{n^2+1^2}+\frac{n}{n^2+2^2}+\ldots+\frac{1}{2n}\right]=\frac{\pi}{4}.$$
9. Show that $\lim_{n\to\infty}\left[\frac{1}{n^3}(1+4+9+\ldots+n^2)\right]=\frac{1}{3}$.
10. Show that
$$\lim_{n\to\infty}\left[\frac{1}{n}+\frac{n^2}{(n+1)^3}+\frac{n^2}{(n+2)^2}+\ldots+\frac{1}{8n}\right]=\frac{3}{8}.$$
11. Show that
$$\lim_{n\to\infty}\left[\frac{1}{n}+\frac{1}{\sqrt{n^2-1^2}}+\frac{1}{\sqrt{n^2-2^2}}+\ldots+\frac{1}{\sqrt{n^2-(n-1)^2}}\right]=\frac{\pi}{2}.$$
12. Show that $\lim_{n\to\infty}\left[\frac{1}{n^2}\sec^2\frac{1}{n^2}+\frac{2}{n^2}\sec^2\frac{4}{n^2}+\frac{3}{n^2}\sec^2\frac{9}{n^2}+\ldots+\frac{1}{n}\sec^2 1\right]=\frac{1}{2}\tan 1.$
13. Show that
$$\lim_{n\to\infty}\left[\frac{1}{\sqrt{n^2-1^2}}+\frac{1}{\sqrt{n^2-2^2}}+\ldots+\frac{1}{\sqrt{n^2-(n-1)^2}}\right]=\frac{\pi}{2}.$$
14. Show that $\lim_{n\to\infty}\left[\frac{n^{1/2}}{n^{3/2}}+\frac{n^{1/2}}{(n+3)^{3/2}}+\frac{n^{1/2}}{(n+6)^{3/2}}+\ldots+\frac{n^{1/2}}{[n+3(n+1)]^{3/2}}\right]=\frac{1}{3}.$

Hint to Selected Problems

1. $$\lim_{n\to\infty}\left[\frac{1}{n}+\frac{1}{n+1}+\frac{1}{n+2}+\ldots+\frac{1}{6n}\right]=\lim_{n\to\infty}\sum_{r=0}^{5n}\left[\frac{1}{n+r}\right]$$
$$=\lim_{n\to\infty}\frac{1}{n}\sum_{r=0}^{5n}\left[\frac{1}{1+\frac{r}{n}}\right]=\int_0^5\frac{1}{1+x}dx$$
$$=[\log(1+x)]_0^5=\log 6-\log 1=\log 6$$

4. Let $I=\int_a^b \sin xdx$. Here $f(x) = \sin x$

Let $h=\frac{b-a}{n}, n\in N$

$$I=\lim_{n\to\infty} h[f(a)+f(a+h)+\ldots+f(a+(n-1)h)]$$
$$=\lim_{n\to\infty} h[\sin a+\sin(a+h)+\ldots+\sin(a+(n-1)h)]$$

Now $\sin a + \sin (a + h)+\ldots+ \sin (a + (n - 1)h)$
$$=\frac{1}{2\sin h/2}\left[2\sin a\sin\frac{h}{2}+2\sin(a+h).\sin\frac{h}{2}+\ldots+2\sin(a+(n-1)\sin\frac{h}{2}\right]$$
$$=\frac{1}{2\sin h/2}\left[\left\{\cos\left(a-\frac{h}{2}\right)-\cos\left(a+\frac{h}{2}\right)\right\}+\left\{\cos\left(a+\frac{h}{2}\right)-\cos\left(a+\frac{3h}{2}\right)\right\}+\ldots+\left\{\cos\left(a+\left(n-\frac{3}{2}\right)h\right)-\cos\left(a+\left(n-\frac{1}{2}\right)h\right)\right\}\right]$$
$$=\frac{1}{2\sin h/2}\left[\cos\left(a-\frac{h}{2}\right)-\cos\left(a+\left(n-\frac{1}{2}\right)h\right)\right]$$
$$=\frac{1}{2\sin h/2}\left[\cos\left(a-\frac{h}{2}\right)-\cos\left(b-\frac{h}{2}\right)\right]$$

$b = a + nh$

$$I=\lim_{h\to\infty} h.\frac{1}{2\sin h/2}\left[\cos\left(a-\frac{h}{2}\right)-\cos\left(b-\frac{h}{2}\right)\right]$$
$$=\lim_{h\to\infty}\frac{h/2}{\sin h/2}\lim_{h\to\infty}\left[\cos\left(a-\frac{h}{2}\right)-\cos\left(b-\frac{h}{2}\right)\right]$$
$$=1.[\cos(a-0)-\cos(b-0))]$$
$$=\cos a-\cos b$$

6. $$\lim_{n\to\infty}\left[\frac{n}{(n+1)^2}+\frac{n}{(n+2)^2}+\frac{n}{(n+3)^2}+\ldots+\frac{n}{(n+n)^2}\right]$$
$$=\lim_{n\to\infty}\sum_{r=0}^{n+1}\frac{n}{(n+r)^2}=\lim_{n\to\infty}\frac{1}{n}\sum_{r=0}^{n+1}\frac{1}{\left(1+\frac{r}{n}\right)^2}$$
$$=\int_0^1\frac{1}{(1+x)^2}dx=\left[-\frac{1}{1+x}\right]_0^1=\frac{1}{2}$$

7. $$\lim_{n\to\infty}\left[\frac{n}{n^2}+\frac{n}{n^2+1^2}+\frac{n}{n^2+2^2}+\ldots+\frac{n}{n^2+(n+1)^2}\right]$$
$$=\lim_{n\to\infty}\sum_{r=0}^{n}\frac{n}{n^2+r^2}=\lim_{n\to\infty}\frac{1}{n}\sum_{r=0}^{n+1}\frac{1}{1+\left(\frac{r}{n}\right)^2}$$
$$=\int_0^1\frac{1}{1+x^2}dx=\left(\tan^{-1}x\right)_0^1=\tan^{-1}1-\tan^{-1}0=\pi/4.$$

8. $\lim_{n\to\infty}\left[\frac{1}{n^3}(1+4+9+\ldots+n^2)\right]$

$$= \lim_{n\to\infty}\left[\frac{1}{n^3}(1^2+2^2+3^2+\ldots+n^2)\right] = \lim_{n\to\infty}\sum_{r=0}^{n}\frac{1}{n^3}r^2$$

$$= \lim_{n\to\infty}\frac{1}{n}\sum_{r=0}^{n}\left(\frac{r}{n}\right)^2 = \int_0^1 x^2dx = \left(\frac{x^3}{3}\right)_0^1 = \frac{1}{3}.$$

11. $\lim_{n\to\infty}\left[\frac{1}{n}+\frac{1}{\sqrt{n^2-1^2}}+\frac{1}{\sqrt{n^2-2^2}}+\ldots+\frac{1}{\sqrt{n^2-(n-1)^2}}\right]$

$$= \lim_{n\to\infty}\sum_{r=0}^{n-1}\frac{1}{\sqrt{n^2-r^2}} = \lim_{n\to\infty}\frac{1}{n}\sum_{r=0}^{n-1}\frac{1}{\sqrt{1-\left(\frac{r}{n}\right)^2}}$$

$$= \int_0^1\frac{1}{\sqrt{1-x^2}}dx = \left(\sin^{-1}x\right)_0^1$$

$$= \sin^{-1}(1)-\sin^{-1}(0) = \pi/2.$$

12. $\lim_{n\to\infty}\left[\frac{1}{n^2}\sec^2\frac{1}{n^2}+\frac{2}{n^2}\sec^2\frac{4}{n^2}+\frac{3}{n^2}\sec^2\frac{9}{n^2}+\ldots+\frac{1}{n}\sec^2 1\right]$

$$= \lim_{n\to\infty}\sum_{r=0}^{n}\frac{r}{n^2}\sec^2\frac{r^2}{n^2} = \lim_{n\to\infty}\frac{1}{n}\sum_{r=0}^{n}\left(\frac{r}{n}\right)\sec^2\left(\frac{r}{n}\right)^2$$

$$= \int_0^1 x\sec^2 xdx.$$

$$= \frac{1}{2}\int_0^1\sec^2 tdt = \frac{1}{2}(\tan t)_0^1 = \frac{1}{2}\tan 1$$

(Let $x^2 = t \Rightarrow 2xdx = dt$)

14. $\lim_{n\to\infty}\left[\frac{n^{1/2}}{n^{3/2}}+\frac{n^{1/2}}{(n+3)^{3/2}}+\frac{n^{1/2}}{(n+6)^{3/2}}+\ldots+\frac{n^{1/2}}{\{n+3(n+1)\}^{3/2}}\right]$

$$= \lim_{n\to\infty}\sum_{r=0}^{n}\frac{n^{1/2}}{(n+3r)^{3/2}} = \lim_{n\to\infty}\frac{1}{n}\sum_{r=0}^{n}\frac{1}{\left(1+\frac{3r}{n}\right)^{3/2}}$$

$$= \int_0^1\frac{1}{(1+3x)^{3/2}}dx = \int_0^1(1+3x)^{-3/2}dx$$

$$= \left[\frac{(1+3x)^{-1/2}}{(-1/2)\times 3}\times 3\right]_0^1 = -\frac{2}{3}\left[(1+3x)^{-1/2}\right]_0^1$$

$$= -\frac{2}{3}[4^{-1/2}-1^{-1/2}] = -\frac{2}{3}\left[\frac{1}{2-1}\right] = -\frac{2}{3}\left[-\frac{1}{2}\right] = \frac{1}{3}.$$

ANSWERS

1. $\frac{1}{3}(b^3-a^3)$ **3.** $\frac{3}{2}$ **4.** $\cos b-\cos a$ **5.** 1

FFFFF

Rectification of Curves

16.1 INTRODUCTION

Rectification is a process for finding the length of an arc of a plane curve between two given points on a curve.

16.2 FORMULAE FOR FINDING THE LENGTH OF THE CURVES

(a) Let the equation of a curve be $y = f(x)$ and let A and B be two points on this curve between A and B, the length of curve is to be required. Let s be the length of an arc from a fixed point on the curve to any point on it. Therefore, we have

$$\frac{ds}{dx} = \pm\sqrt{\left[1+\left(\frac{dy}{dx}\right)^2\right]} \quad \text{or} \quad ds = \pm\sqrt{\left[1+\left(\frac{dy}{dx}\right)^2\right]}dx$$

where positive and negative sign will have to take according as x increases and decreases as s increases. Thus, the length of an arc between the points A and B where at A, $x = a$ and at B, $x = b$ is given by

$$s = \int_a^b \sqrt{\left[1+\left(\frac{dy}{dx}\right)^2\right]}dx \qquad (a < b)$$

(b) If the equation of the curve is $x = f(y)$, then the length of an arc between c and d is given by

$$s = \int_c^d \sqrt{\left[1+\left(\frac{dx}{dy}\right)^2\right]}dy \qquad (c < d)$$

(c) If the equation of the curve is in parametric form, *i.e.*, $x = f(t)$, $y = g(t)$, then we have

$$\frac{ds}{dt} = \sqrt{\left[\left(\frac{dx}{dt}\right)^2+\left(\frac{dy}{dt}\right)^2\right]} \quad \text{or} \quad ds = \sqrt{\left(\frac{dx}{dt}\right)^2+\left(\frac{dy}{dt}\right)^2}\,dt$$

Thus the length of an arc between A and B where A, $t = t_1$ and B, $t = t_2$ is given by

$$s = \int_{t_1}^{t_2} \sqrt{\left[\left(\frac{dx}{dt}\right)^2+\left(\frac{dy}{dt}\right)^2\right]}dt$$

(d) If the equation of the curve is $r = f(\theta)$ (in polar form), then

$$\frac{ds}{d\theta} = \sqrt{\left[r^2+\left(\frac{dr}{d\theta}\right)^2\right]} \quad \text{or} \quad ds = \sqrt{r^2+\left(\frac{dr}{d\theta}\right)^2}\,d\theta$$

and s is measured in the direction of θ increasing. Let at point A, $\theta = \theta_1$ and at B, $\theta = \theta_2$. Therefore, the length of an arc between A and B is given by

$$s = \int_{\theta_1}^{\theta_2} \sqrt{\left[r^2+\left(\frac{dr}{d\theta}\right)^2\right]}d\theta$$

(e) If the equation of the curve is $\theta = f(r)$, then the length of an arc between A and B is given by

$$s = \int_{r_1}^{r_2} \sqrt{\left[1+\left(r\frac{d\theta}{dr}\right)^2\right]}dr$$

(f) If the equaton of the curve is in pedal form *i.e.*, $p = f(r)$. Since we know that

$$\frac{ds}{dr} = \frac{r}{\sqrt{(r^2 - p^2)}}$$

Thus the length of an arc between $A(r = r_1)$ and $B(r = r_2)$ is given by

$$s = \int_{r_1}^{r_2} \frac{rdr}{\sqrt{(r^2 - p^2)}}$$

REMARK

- If the curve is symmetrical about some lines, then in order to find the length of an arc, we first find the length of one of the symmetrical part and multiply this length by the number of symmetrical parts.

Recapitulations

- The length of the curve $y = f(x)$ between $x = a$ and $x = b$ is

$$s = \int_a^b \sqrt{\left[1+\left(\frac{dy}{dx}\right)^2\right]}dx$$

- The length of the curve $x = f(y)$ between $y = a$ and $y = b$ is

$$s = \int_a^b \sqrt{\left[1+\left(\frac{dx}{dy}\right)^2\right]}dy$$

- If $x = f(t), y = g(t)$, then

$$s = \int_a^b \sqrt{\left[\left(\frac{dx}{dt}\right)^2+\left(\frac{dy}{dt}\right)^2\right]}dt$$

- The length of the arc of the curve $r = f(\theta)$ between the points $\theta = \alpha$ and $\theta = \beta$ is

$$s = \int_\alpha^\beta \sqrt{\left[r^2+\left(\frac{dr}{d\theta}\right)^2\right]}d\theta$$

Solved Examples

Example 1. *Find the length of the arc of the parabola $y^2 = 4ax$ cut off by its latus rectum.* (VTU–2008, Mumbai–2006)

Solution . **Latus rectum.** A line which passes through the focus of the given parabola and perpendicular to the axis of that parabola.

Here the equation of the parabola is $y^2=4ax$ whose trace is shown above in the figure, LL' is the latus rectum, the co-ordinates of L and L' are respectively $(a, 2a)$ and $(a, -2a)$. Since $y^2=4ax$ is symmetrical about the line OX. Therefore the required arc length

$$= 2\times \text{ arc length } OL.$$

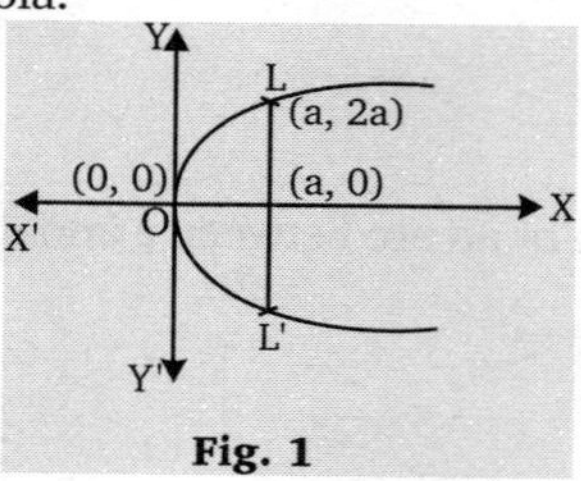

Fig. 1

Since $\quad y^2 = 4ax.$

$\therefore \quad y = 2\sqrt{a}\sqrt{x}$

$\therefore \quad \frac{dy}{dx} = \frac{\sqrt{a}}{\sqrt{x}}.$

Now arc length

$$OL = \int_0^a \sqrt{\left[1+\left(\frac{dy}{dx}\right)^2\right]}dx$$

($\because$ At point $O, x = 0$ and at point $L, x = a$)

$$= \int_0^a \sqrt{\left(1+\frac{a}{x}\right)}dx = \int_0^a \frac{\sqrt{x+a}}{\sqrt{x}}dx$$

$$= \int_0^a \frac{x+a}{\sqrt{x^2+ax}}dx = \frac{1}{2}\int_0^a \frac{2x+2a}{\sqrt{(x^2+ax)}}dx$$

$$= \frac{1}{2}\int_0^a \frac{(2x+a)dx}{\sqrt{x^2+ax}} + \frac{a}{2}\int_0^a \frac{dx}{\sqrt{(x^2+ax)}}$$

$$= \frac{1}{2}\int_0^a \frac{(2x+a)dx}{\sqrt{x^2+ax}} + \frac{a}{2}\int_0^a \frac{dx}{\sqrt{\left[\left(x+\frac{a}{2}\right)^2-\left(\frac{a}{2}\right)^2\right]}}$$

$$= \frac{1}{2}\left(2\sqrt{x^2+ax}\right)_0^a + \frac{a}{2}\left[\log\left\{\left(x+\frac{a}{2}\right)+\sqrt{x^2+ax}\right\}\right]_0^a$$

$$= a\sqrt{2}+\frac{a}{2}\log(3+2\sqrt{2})$$

$$= a\sqrt{2}+\frac{a}{2}\log(1+\sqrt{2})^2$$

Arc length $\quad OL = a\sqrt{2}+a\log(1+\sqrt{2}).$

Hence the required arc length

$$= 2\times \text{ arc length } OL$$

$$= 2\sqrt{2}a+2a\log(1+\sqrt{2}).$$

Example 2. *Find the length of the curve $y = \log\sec x$ between the points $x = 0$ and $x = \pi/3$.* (VTU-2010, PTU-2007)

Solution . Since the equation of the curve is

$$y = \log\sec x$$

$$\therefore \quad \frac{dy}{dx} = \frac{1}{\sec x}.\sec x\tan x = \tan x.$$

Now

$$\sqrt{\left[1+\left(\frac{dy}{dx}\right)^2\right]} = \sqrt{(1+\tan^2 x)}$$

$$= \sqrt{\sec^2 x} = \sec x.$$

Therefore the length of the given curve between $x = 0$ and $x = \pi/3$ is

$$s = \int_0^{\pi/3}\sqrt{\left[1+\left(\frac{dy}{dx}\right)^2\right]}dx$$

$$= \int_0^{\pi/3} \sec x dx$$

$$= \left[\log\left\{\tan\left(\frac{\pi}{4}+\frac{x}{2}\right)\right\}\right]_0^{\pi/3}$$

$$s = \log\left[\tan\left(\frac{\pi}{4}+\frac{\pi}{6}\right)\right] - \log\left[\tan\left(\frac{\pi}{4}\right)\right]$$

$$= \log\left[\frac{\tan\frac{\pi}{4}+\tan\frac{\pi}{6}}{1-\tan\frac{\pi}{4}\tan\frac{\pi}{6}}\right] - \log 1$$

$$\left[\because \tan\left(\frac{\pi}{4}+\theta\right) = \frac{1+\tan\theta}{1-\tan\theta}\right]$$

$$= \log\left[\frac{1+1/\sqrt{3}}{1-1/\sqrt{3}}\right] - 0 \quad (\because \log 1 = 0)$$

$$= \log\left(\frac{\sqrt{3}+1}{\sqrt{3}-1}\right)$$

$$s = \log(2+\sqrt{3})$$

Example 3. *Find the whole length of the astroid*

$$x^{2/3} + y^{2/3} = a^{2/3}.$$

or $x = a\cos^3\theta, y = a\sin^3\theta$

(VTU–2010, Rajasthan–2006, Marathwada–2008)

Solution. The equation of the curve is

$$x^{2/3} + y^{2/3} = a^{2/3} \quad \ldots(1)$$

Since the curve is symmetrical about both the axis, *i.e.*, curve lies in all the four quadrants as shown in fig. 2.

Therefore the whole length of the given curve $= 4 \times$ arc length of the curve in first quadrant.

Since the co-ordinates of A and B are $(a, 0)$ and $(0, a)$ respectively. Thus in the first quadrant x varies from 0 to a.

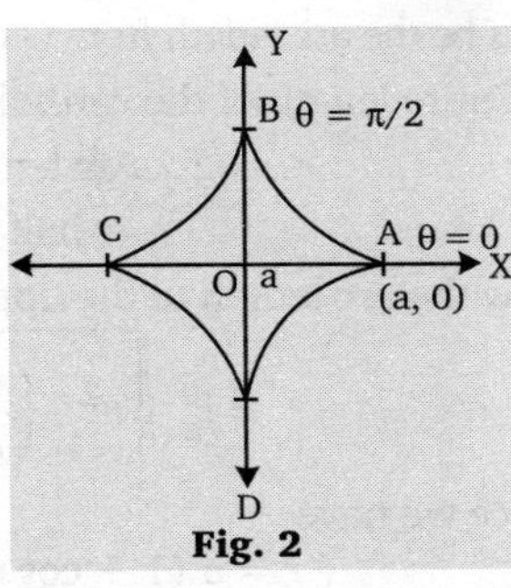

Fig. 2

Now differentiating (1) *w.r.t.* x, we get

$$\frac{2}{3}x^{-1/3} + \frac{2}{3}y^{-1/3}\frac{dy}{dx} = 0$$

or $$\frac{dy}{dx} = -\left(\frac{y}{x}\right)^{1/3}$$

Therefore, the length of the curve in the first quadrant

$$= \int_0^a \sqrt{\left[1+\left(\frac{dy}{dx}\right)^2\right]}dx = \int_0^a \sqrt{\left(1+\frac{y^{2/3}}{x^{2/3}}\right)}dx$$

$$= \int_0^a \sqrt{\left(\frac{x^{2/3}+y^{2/3}}{x^{2/3}}\right)}dx = \int_0^a \sqrt{\left(\frac{a^{2/3}}{x^{2/3}}\right)}dx$$

[Using (1)]

$$= a^{1/3}\int_0^a x^{-1/3}dx = a^{1/3}\left[\frac{3}{2}x^{2/3}\right]_0^a$$

$$= a^{1/3}\left[\frac{3}{2}a^{2/3}\right] = \frac{3}{2}a$$

Hence, the whole length of the astroid

$$= 4 \times \frac{3a}{2} = 6a$$

Example 4. *Find the entire length of the cardioid* $r = a(1+\cos\theta)$.

(PTU–2010, Kurukshetra–2005, Bhopal–2008, Bhillai–2005)

Solution. From the equation of the cardioid $r = a(1+\cos\theta)$ it is obvious, that the curve is symmetrical about the initial line (x-axis) and r will become zero, when $\theta = \pi$ and maximum, *i.e.*, $r = 2a$, when $\theta = 0$. Thus r varies from 0 to $2a$. As θ varies from π to 0. Therefore curve is shown in the fig.3.

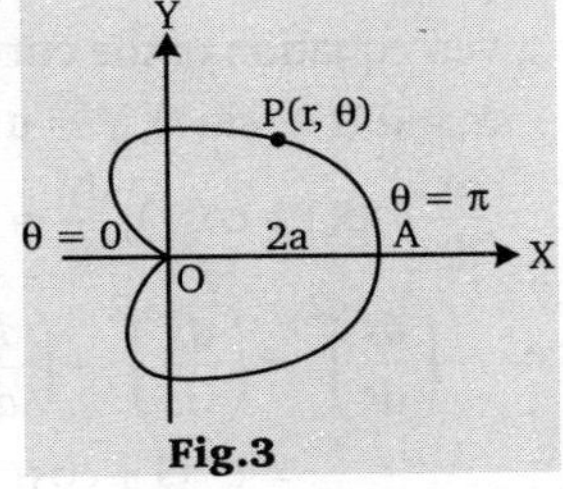

Fig.3

Let s be the arc length from O to any point $P(r, \theta)$.

$\therefore$ Entire length of the cardioid

$= 2 \times$ Arc length of the upper half of the cardioid.

Now the arc length of the upper half

$$= \int_{\pi}^{0} \sqrt{\left[r^2 + \left(\frac{dr}{d\theta}\right)^2\right]} d\theta \qquad \ldots(1)$$

Since we have

$$r = a(1 + \cos\theta).$$

$$\therefore \quad \frac{dr}{d\theta} = -a\sin\theta$$

and
$$\frac{ds}{d\theta} = \sqrt{\left[r^2 + \left(\frac{dr}{d\theta}\right)^2\right]}$$

$$= \sqrt{a^2(1+\cos\theta)^2 + a^2\sin^2\theta}$$

$$= \sqrt{[2a^2(1+\cos\theta)]}$$

$$= \sqrt{\left[2a^2\left(2\cos^2\frac{\theta}{2}\right)\right]}$$

$$\frac{ds}{d\theta} = 2a\cos\theta/2.$$

Here we have to measure the arc length 's' from the cusp O where $\theta = \pi$ to any point $P(r, \theta)$ in the direction of θ decreasing, then the arc length 's' increases as θ decreases. Therefore, we will take $\frac{ds}{d\theta}$ to be negative. Thus from (1), we obtain

$$s = \int_{\pi}^{0}\left(-2a\cos\frac{\theta}{2}\right)d\theta$$

$$= -2a\int_{\pi}^{0}\cos\theta/2\,d\theta$$

$$= -2a\left[2\sin\frac{\theta}{2}\right]_{\pi}^{0}$$

$$= -4a\left[0 - \sin\frac{\pi}{2}\right] = 4a.$$

Hence, the entire length of the given cardioid

$$= 2s = 8a.$$

Example 5. *Find the length of an arc of the curve*

$x = a(t + \sin t)$, $y = a(1 - \cos t)$

(PTU–2009, VTU–2004)

Solution . The given equation of the curve is

$$x = a(t + \sin t),\ y = a(1 - \cos t).$$

$$\therefore \quad \frac{dx}{dt} = a(1+\cos t), \frac{dy}{dt} = a\sin t.$$

Since,
$$\left(\frac{ds}{dt}\right)^2 = \left(\frac{dx}{dt}\right)^2 + \left(\frac{dy}{dt}\right)^2$$

$$= a^2(1+\cos t)^2 + a^2\sin^2 t$$

$$= 2a^2(1+\cos t)$$

$$\therefore \quad \left(\frac{ds}{dt}\right)^2 = 4a^2\cos^2 t/2$$

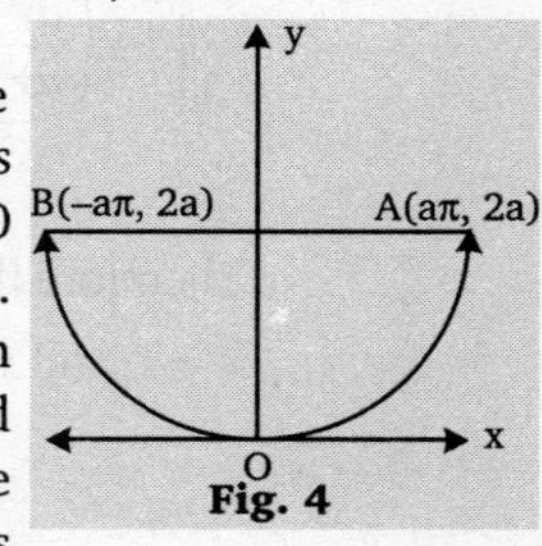

Fig. 4

From the equation of the curve it is obvious when $\cos t = 1$, $y = 0$ *i.e.*, when $t=0$, $y=0$. This implies when $t = 0$, $x = 0$ and $y = 0$. Therefore the curve will passes through the point (0, 0) and at the point (0, 0) tangent is the x-axis. Also y will always positive. Thus the tracing of the given curve is shown in fig. 4.

From the fig. 4 it is obvious that the curve is symmetrical about the y-axis. Therefore, the entire length of the arc of the given curve will be twice of the arc OA. At 0, $t = 0$ and at A, $t = \pi$.

Let s be the arc length measured from O to any point P on the curve towards the point A, then s increases as t increases so that $\frac{ds}{dt}$ will be taken positive.

$\therefore$ Arc length OA

$$\int_0^{\pi}\frac{ds}{dt}.dt = \int_0^{\pi} 2a\cos t/2\,dt \qquad \text{[Using(1)]}$$

$$= 2a\left[2\sin t/2\right]_0^{\pi} = 4a.$$

Hence, the entire length of the given curve

$$= 2 \times \text{arc } OA = 2 \times 4a = 8a.$$

Example 6. *Find the length of the curve* $x\sin\theta + y\cos\theta = f'(\theta)$ *and* $x\cos\theta - y\sin\theta = f''(\theta)$.

Solution . We have

$$x\sin\theta + y\cos\theta = f'(\theta) \qquad \ldots(1)$$

$$x\cos\theta - y\sin\theta = f''(\theta) \qquad \ldots(2)$$

Solving (1) and (2), we have

$$x = \sin\theta f'(\theta) + \cos\theta f''(\theta)$$

and
$$y = \cos\theta f'(\theta) - \sin\theta f''(\theta)$$

$$\therefore \quad \frac{dx}{d\theta} = \cos\theta f'(\theta) + \sin\theta f''(\theta) - \sin\theta f''(\theta) + \cos\theta f'''(\theta)$$

$$= \cos\theta f'(\theta) + \cos\theta f'''(\theta)$$

and
$$\frac{dy}{d\theta} = \cos\theta f''(\theta) - \sin\theta f'(\theta) - \cos\theta f''(\theta) - \sin\theta f'''(\theta)$$

$$= -\sin\theta f'(\theta) - \sin\theta f'''(\theta)$$

If s is the arc length of the curve in the direction of θ increasing, then

$$\left(\frac{ds}{d\theta}\right)^2=\left(\frac{dx}{d\theta}\right)^2+\left(\frac{dy}{d\theta}\right)^2$$

$$=\{\cos\theta[f'(\theta)+f''(\theta)]\}^2$$

$$+\{-\sin\theta[f'(\theta)+f'''(\theta)]\}^2$$

$$=[f'(\theta)+f'''(\theta)]^2(\cos^2\theta+\sin^2\theta)$$

$$=[f'(\theta)+f'''(\theta)]^2$$

$$\therefore \quad \frac{ds}{d\theta}=f'(\theta)+f'''(\theta)$$

Now integrating both sides *w.r.t.* θ, *we get*

$$s=\int[f'(\theta)+f'''(\theta)]d\theta+c$$

$$=f(\theta)+f''(\theta)+c,$$

where c is a constant of integration.

Example 7. *Find the length of the curve* $y=\log\dfrac{e^x-1}{e^x+1}$ *from* $x=1$ *to* $x=2$.

Solution . We have $\quad y=\log\dfrac{e^x-1}{e^x+1}$

$$=\log(e^x-1)-\log(e^x+1).$$

Differentiating w.r.t. x, we get

$$\frac{dy}{dx}=\frac{e^x}{e^x-1}-\frac{e^x}{e^x+1}=\frac{2e^x}{e^{2x}-1}.$$

If s is the arc length of the curve in the direction of x increasing, then

$$\left(\frac{ds}{dx}\right)^2=1+\left(\frac{dy}{dx}\right)^2=1+\left(\frac{2e^x}{e^{2x}-1}\right)^2$$

$$=1+\frac{4e^{2x}}{(e^{2x}-1)^2}$$

$$=\frac{(e^{2x}-1)^2+4e^{2x}}{(e^{2x}-1)^2}=\left(\frac{e^{2x}+1}{e^{2x}-1}\right)^2.$$

$$\therefore \quad \frac{ds}{dx}=\frac{e^{2x}+1}{e^{2x}-1}.$$

Integrating *w.r.t.* x from $x=1$ to $x=2$, *we get*

$$s=\int_1^2\frac{e^{2x}+1}{e^{2x}-1}dx=\int_1^2\frac{e^x+e^{-x}}{e^x-e^{-x}}dx$$

$$=\left[\log(e^x-e^{-x})\right]_1^2$$

$$=\log(e^2-e^{-2})-\log(e^1-e^{-1})$$

$$=\frac{\log(e^2-e^{-2})}{\log(e^1-e^{-1})}$$

$$=\log\frac{(e^1-e^{-1})(e^1+e^{-1})}{(e^1-e^{-1})}$$

$$[\because a^2-b^2=(a-b)(a+b)]$$

$$\therefore \quad s=\log(e^1+e^{-1}).$$

Example 8. *Find the length of the arc of the equiangular spiral* $r=ae^{\theta\cot\alpha}$ *between the points for which radii vectors are* r_1 *and* r_2.

Solution . Since we have

$$r=ae^{\theta\cot\alpha} \qquad ...(1)$$

Now differentiating (1) *w.r.t.* θ, *we get*

$$\frac{dr}{d\theta}=a\cot\alpha e^{\theta\cot\alpha}=r\cot\alpha\,.$$

If s is the arc length of the spiral in the direction of r increasing, then

$$\frac{ds}{dr}=\sqrt{\left[1+\left(r\frac{d\theta}{dr}\right)^2\right]}$$

$$=\sqrt{(1+\tan^2\alpha)}$$

$$\left(\because r\frac{d\theta}{dr}=\tan\alpha\right)$$

$$=\sqrt{(\sec^2\alpha)}$$

$$\frac{ds}{dr}=\sec\alpha\Rightarrow ds=\sec\alpha\,dr$$

Now integrating *w.r.t.* r from $r=r_1$ to $r=r_2$, *we get*

$$s=\int_{r_1}^{r_2}\sec\alpha\,dr=\sec\alpha\int_{r_1}^{r_2}dr$$

$$=\sec\alpha\left[r\right]_{r_1}^{r_2}=(r_2-r_1)\sec\alpha$$

Example 9. *Find the length of the loop of the curve* $3ay^2=x(x-a)^2$.

Solution . The equation of the curve is

$$3ay^2=x(x-a)^2. \qquad ...(1)$$

This curve is symmetrical about the x-axis and passes through the origin. The tangent at (0, 0) is the y-axis. The curve cuts the x-axis only at the point $(a, 0)$.

Now differentiating (1) *w.r.t.* x, *we get*

$$\frac{dy}{dx}=\frac{3x-a}{2\sqrt{3ax}}. \qquad ...(2)$$

$\therefore$ At $(a, 0)$ $\dfrac{dy}{dx}=\pm\dfrac{1}{\sqrt{3}}$. Thus at $(a, 0)$, tangents make the angles $\pi/6$ and $-\pi/6$ with positive x-axis.

Therefore, we have

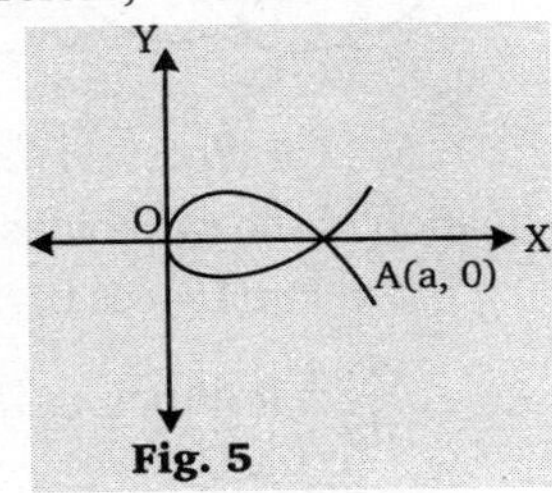

Fig. 5

If s is the arc length measured from 0 to any point on the curve in the direction of x-increasing. Then we will take $\frac{ds}{dx}$ positive.

$$\therefore \quad \frac{ds}{dx} = \sqrt{\left[1+\left(\frac{dy}{dx}\right)^2\right]} = \sqrt{\left[1+\frac{(3x-a)^2}{12ax}\right]}$$

$$= \sqrt{\left(\frac{12ax+9x^2+a^2-6xa}{12ax}\right)}$$

$$\Rightarrow \quad \frac{ds}{dx} = \frac{3x+a}{2\sqrt{3ax}}.$$

If s_1 denotes the length of the loop of the curve between the points $x = 0$ to $x = a$. Therefore, we have

$$s_1 = 2\int_0^a \left(\frac{ds}{dx}\right).dx$$

or
$$s_1 = 2\int_0^a \frac{3x+a}{2\sqrt{3ax}}dx$$

$$= \frac{1}{\sqrt{3a}}\int_0^a \frac{3x+a}{\sqrt{x}}dx$$

$$= \frac{1}{\sqrt{3a}}\int_0^a (3x^{1/2}+ax^{-1/2})dx$$

$$= \frac{1}{\sqrt{3a}}\left[2x^{3/2}+2ax^{1/2}\right]_0^a$$

$$= \frac{1}{\sqrt{3a}}[2a\sqrt{a}+2a\sqrt{a}] = \frac{4a\sqrt{a}}{\sqrt{3}\sqrt{a}}$$

Hence, $s_1 = \frac{4a}{\sqrt{3}}$

Example 10. *Find the length of the cardioid $r = a\ (1 - \cos\theta)$ lying outside the circle $r = a\cos\theta$.*

Solution . Both curves intersect. Therefore, we have

$$a(1 - \cos\theta) = a\cos\theta$$

or $2\cos\theta = 1$ or $\cos\theta = \frac{1}{2}$ or $\theta = \pm\frac{\pi}{3}$

$\therefore$ The intersection points are $\left(\frac{a}{2}, \frac{\pi}{3}\right)$ and $\left(\frac{a}{2}, -\frac{\pi}{3}\right)$:

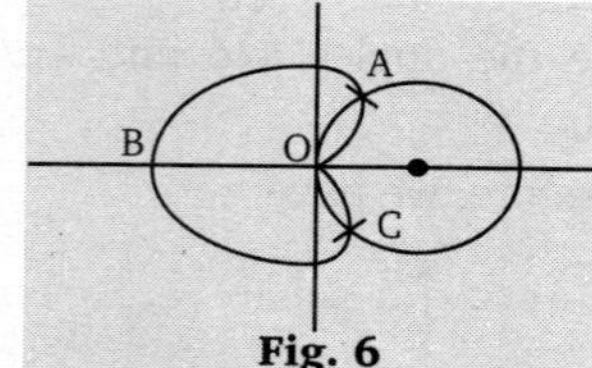

Fig. 6

The equation of the cardioid is

$$r = a(1 - \cos\theta)$$

$$\therefore \quad \frac{ds}{d\theta} = a\sin\theta$$

$$\Rightarrow \quad \frac{ds}{d\theta} = \sqrt{\left[r^2+\left(\frac{dr}{d\theta}\right)^2\right]}$$

$$= \sqrt{[a^2(1-\cos\theta)^2 + a^2\sin^2\theta]}$$

or
$$\frac{ds}{d\theta} = 2a\sin\theta/2.$$

Since s is the arc length of the cardioid to any point $P(r, \theta)$ in the direction of θ increasing so we will take $\frac{ds}{d\theta}$ positive. But we have to find the length of the cardioid lying outside the circle. Therefore if s_1 is the required length, then

$s_1 = 2 \times$ upper portion of the cardioid from A to B.

At the point A, $\theta = \pi/3$ and at B, $\theta = \pi$.

$$\therefore \quad s_1 = 2\int_{\pi/3}^{\pi}\left(\frac{ds}{d\theta}\right).d\theta$$

$$= 2\int_{\pi/3}^{\pi} 2a\sin\theta/2\,d\theta$$

$$= 4a[-2\cos\theta/2]_{\pi/3}^{\pi}$$

$$= 4a\left[-2\cos\frac{\pi}{2}+2\cos\frac{\pi}{6}\right]$$

$$= 4a\left[2.\frac{\sqrt{3}}{2}\right]$$

Hence, $s_1 = 4a\sqrt{3}$.

Example 11. *Prove that the line $4r\cos\theta = 3a$ divides the cardioid $r = a\ (1 + \cos\theta)$ into two equal lengths of the arc.*

Solution . The line $4r\cos\theta = 3a$ and the cardioid $r = a(1 + \cos\theta)$ intersect, then we have

$$4\,a(1 + \cos\theta)\cos\theta = 3a$$

or
$$4\cos^2\theta + 4\cos\theta - 3 = 0$$

or
$$(2\cos\theta + 3)(2\cos\theta - 1) = 0$$

or $\cos\theta = -\frac{3}{2}$ which is not possible and $\cos\theta = \frac{1}{2}$ which is possible and gives $\theta = \pm\pi/3$. Therefore the given line intersect the given cardioid in two points $(3a/2, \pi/3)$ and $(3a/2, -\pi/3)$ as shown in fig. 7.

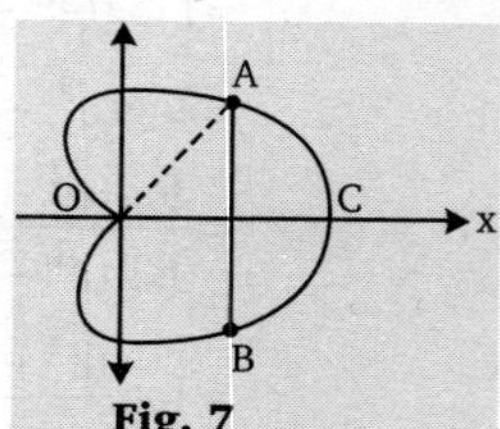

Fig. 7

Since the equation of the cardioid is

$$r = a\ (1 + \cos\theta)$$

$$\therefore \quad \frac{dr}{d\theta} = -a\sin\theta.$$

Now, $\dfrac{dr}{d\theta} = \pm\sqrt{\left[r^2 + \left(\dfrac{dr}{d\theta}\right)^2\right]}$

$= \pm\sqrt{[a^2(1+\cos\theta)^2 + a^2\sin^2\theta]}$

$= \pm 2a\cos\theta/2.$

Since s is the length of arc measured from 0 to any point on the cardioid in the decreasing direction of θ so we will take $\dfrac{ds}{d\theta}$ negative. Let s_1 be the arc length of the cardioid in one side and s_2 be on the other sides.

$\therefore \quad s_1 = 2 \text{ arc } OA = 2\int_{\pi}^{\pi/3}\left(\dfrac{ds}{d\theta}\right).d\theta$

$= 2\int_{\pi}^{\pi/3} -2a\cos(\theta/2)\,d\theta$

$= 4a\int_{\pi/3}^{\pi}\cos(\theta/2)\,d\theta$

$= 4a\left[2\sin(\theta/2)\right]_{\pi/3}^{\pi}$

$= 4a\left[2\sin(\pi/2) - 2\sin(\pi/6)\right]$

$= 4a\left[2 - 2\dfrac{1}{2}\right] = 4a(2-1) = 4a$

and $\quad s_2 = 2 \text{ arc } AC = 2\int_{\pi/3}^{0}\left(\dfrac{ds}{d\theta}\right).d\theta$

($\because$ at A, $\theta = \pi/3$ and at C, $\theta = 0$)

$= 2\int_{\pi/3}^{0} -2a\cos(\theta/2)\,d\theta$

$= -4a\int_{\pi/3}^{0}\cos(\theta/2)\,d\theta$

$= 4a\int_{0}^{\pi/3}\cos(\theta/2)\,d\theta$

$= 4a\left[2\sin(\theta/2)\right]_{0}^{\pi/3}$

$= 4a\left[2\sin\pi/6 - 0\right]$

$= 4a\left(2.\dfrac{1}{2}\right) = 4a$

Hence, $\quad s_1 = s_2$

Example 12. *If s be the length of the curve $r = a\tanh\dfrac{\theta}{2}$ between the origin and $\theta = 2\pi$, and Δ be the area under the curve between the same two points, prove that*
$\Delta = a(s - a\pi).$

Solution . At the origin $r = 0$, we have

$$\tanh\frac{\theta}{2} = 0 \Rightarrow \theta = 0.$$

Therefore, s is the length of the curve $r = a\tanh\dfrac{\theta}{2}$ between the points $\theta = 0$ and $\theta = 2\pi$ and Δ is the area under the curve between $\theta = 0$ and $\theta = 2\pi$. Since the equation of the curve is

$$r = a\tanh\frac{\theta}{2}.$$

$\Rightarrow \quad \dfrac{dr}{d\theta} = \dfrac{1}{2}a\,\text{sech}^2\dfrac{\theta}{2}$

Now $\quad \dfrac{ds}{d\theta} = \sqrt{\left[r^2 + \left(\dfrac{dr}{d\theta}\right)^2\right]}$

$= \sqrt{\left[a^2\tanh^2\dfrac{\theta}{2} + \dfrac{1}{4}a^2\,\text{sech}^4\dfrac{\theta}{2}\right]}$

$= \dfrac{1}{2}a\sqrt{\left[4\tanh^2\dfrac{\theta}{2} + \text{sech}^4\dfrac{\theta}{2}\right]}$

$\dfrac{ds}{d\theta} = \dfrac{1}{2}a\left(2 - \text{sech}^2\dfrac{\theta}{2}\right)\cdot \quad \ldots(1)$

The length of arc is measured in the direction of θ increasing. Therefore, we will take $\dfrac{ds}{d\theta}$ positive.

Since s is the length of an arc of $r = a\tanh\dfrac{\theta}{2}$ between $\theta = 0$ to $\theta = 2\pi$, we have

$s = \int_0^{2\pi}\left(\dfrac{ds}{d\theta}\right)d\theta$

$= \int_0^{2\pi}\dfrac{1}{2}a\left(2 - \text{sech}^2\dfrac{\theta}{2}\right)d\theta$

$= \dfrac{a}{2}\int_0^{2\pi}\left(2 - \text{sech}^2\dfrac{\theta}{2}\right)d\theta$

$= \dfrac{a}{2}\left[2\theta - 2\tanh\dfrac{\theta}{2}\right]_0^{2\pi}$

$s = 2a\pi - a\tanh\pi\cdot \quad \ldots(2)$

Next, we have that Δ is the area under the curve $r = a\tanh\dfrac{\theta}{2}$ between $\theta = 0$ and $\theta = 2\pi$.

Then $\quad \Delta = \int_0^{2\pi}\dfrac{1}{2}r^2 d\theta$

$\left(\because \text{Area} = \int_{\theta_1}^{\theta_2}\dfrac{1}{2}r^2 d\theta\right)$

$= \dfrac{1}{2}\int_0^{2\pi}a^2\tanh^2\dfrac{\theta}{2}d\theta$

$\left(\because r = a\tanh\dfrac{\theta}{2}\right)$

$= \dfrac{a^2}{2}\int_0^{2\pi}\left(1 - \text{sech}^2\dfrac{\theta}{2}\right)d\theta$

$\left(\because \tanh^2\dfrac{\theta}{2} = 1 - \text{sech}^2\dfrac{\theta}{2}\right)$

$= \dfrac{a^2}{2}\left[\theta - 2\tanh\dfrac{\theta}{2}\right]_0^{2\pi}$

$\Delta = a^2\pi - a^2\tanh\pi.$

Now subtract (2) from (3) after multiplying (2) by a, we get

$\Delta - as = -a^2\pi$ or $\Delta = as - a^2\pi$ or $\Delta = a(s - a\pi)$.

EXERCISE 16.1

1. Find the whole length of the curve $x^2 + y^2 = b^2$.
2. Find the length of the arc of the curve $x^2 = 8y$ from the vertex to an extremity of the latusrectum.
3. Find the arc length of the curve $y = \frac{1}{2}x^2 - \frac{1}{4}\log x$ from $x = 1$ to $x = 2$.
4. Find the length of the arc from $\theta = 0$ to $\theta = 2\pi$ of the curve
$$x = a(\cos\theta + \theta\sin\theta), y = a(\sin\theta - \theta\cos\theta).$$
5. Find the length of the arc of the curve $ay^2 = x^3$ between the points (0, 0) and (a, a).
6. Show that the length of the curve
$$x = e^\theta\left(\sin\frac{\theta}{2} + 2\cos\frac{\theta}{2}\right), y = e^\theta\left(\cos\frac{\theta}{2} - 2\sin\frac{\theta}{2}\right)$$
between $\theta = 0$ to $\theta = \pi$ is $\frac{5}{2}(e^\pi - 1)$.
7. Show that the length of the arc of the curve $y^2 = 4ax$ which is intercepted between the points of intersection of the curve and the line $3y = 8x$ is $a(\log 2 + 15/16)$.
8. Find the length of the loop of the curve
$$9ay^2 = (x - 2a)(x - 5a)^2.$$
9. Find the length of an arc of the curve
$$x = a\left(\sin t + \frac{1}{3}\sin 3t\right), y = a\left(\cos t - \frac{1}{3}\cos 3t\right)$$
between $t = 0$ and $t = \pi/4$.
10. Find the whole length of the hypo-cycloid $x = a\cos^3 t, y = b\sin^3 t$.
11. Find the length of the loop of the curve $x = t^2, y = t - \frac{1}{3}t^3$.
12. Show that the whole length of the curve $x^2(a^2 - x^2) = 8a^2y^2$ is $\pi a^2\sqrt{2}$.
13. Find the length of an arc of the cycloid
$$x = a(\theta - \sin\theta), y = a(1 - \cos\theta).$$ (VTU–2004, PTU–2009)
14. Find the length of the arc of the curve $x = e^\theta\sin\theta, y = e^\theta\cos\theta$ between $\theta = 0$ to $\theta = \pi/2$.
15. Find the entire length of the cardioid $r = a(1 - \cos\theta)$.
16. Show that the arc of the upper half of the curve $r = a(1 - \cos\theta)$ is bisected by $\theta = 2\pi/3$.
17. In the ellipse $x = a\cos\theta$, $y = b\sin\theta$, show that $ds = a\sqrt{(1 - e^2\cos^2\theta)}d\theta$, and hence show that the whole length of the ellipse is
$$2\pi a\left[1 - \left(\frac{1}{2}\right)^2 . \frac{e^2}{1} - \left(\frac{1.3}{2.4}\right)^2 . \frac{e^4}{3} - \left(\frac{1.3.5}{2.4.6}\right)^2 . \frac{e^6}{5} - \ldots\right].$$
18. Find the perimeter of curve $r = a(1 + \cos\theta)$ and show that the arc of the upper half is bisected by $\theta = \pi/3$. (JNTU–2003)
19. Prove that the perimeter of the limacon $r = a + b\cos\theta, a > b$ is approximately $2\pi a(1 + b^2/4a^2)$.
20. Find the length of the arc of the curve $r = ae^{\theta\cot\alpha}$ taking $s = 0$ when $\theta = 0$.
21. Show that $\theta = \pi/3$ divides the length of the cycloid
$$x = a(\theta - \sin\theta), y = a(1 - \cos\theta)$$
in the ratio 1:3.
22. Find the length of the arc of the curve
$$x = a(3\sin\theta - \sin^3\theta), y = a\cos^3\theta$$
between $\theta = 0$ and $\theta = \pi/2$.
23. Find the length of the arc of the curve
$$x = a\sin 2\theta(1 + \cos 2\theta), y = a\cos 2\theta(1 - \cos 2\theta)$$
between $\theta = 0$ and $\theta = \pi/2$.
24. Show that $\theta = \pi/6$ divides the arc in the first quadrant of the curve $x = a\cos^3\theta, y = a\sin^3\theta$ in the ratio 1:3.
25. Find the length of any arc of the cissoid
$$r = \frac{a\sin^2\theta}{\cos\theta}.$$
26. Show that the ratio of the lengths of the cardioid $r=a(1-\cos\theta)$ lying inside and outside the circle $r=a\cos\theta$ is $(2-\sqrt{3}):\sqrt{3}$.
27. Find the perimeter of the loop of the curve $3ay^2 = x^2(a - x)$.
28. Show that the length of the arc of the curve $x\cos\theta - y\sin\theta = f'(\theta)$, $x\sin\theta + y\cos\theta = f(\theta)$ is given by $S = f(\theta) = $ constant.

Hint to Selected Problems

1. The given equation of the curve $x^2 + y^2 = b^2$, which is a circle.

$\therefore$ Therefore, $s = 4\int_0^b \frac{ds}{dx}dx$.

Here $\frac{dy}{dx} = -\left(\frac{x}{y}\right)$ which implies $\frac{ds}{dx} = \sqrt{1 + \left(\frac{dy}{dx}\right)^2} = \frac{b}{\sqrt{b^2 - x^2}}$.

2. Let s be the length from the vertex (0, 0) to $L(4, 2)$. Then, we have
$$s = \int_0^4 \frac{ds}{dx}dx \text{ and using } \int\sqrt{16 + x^2}dx = \frac{x}{2}\sqrt{16 + x^2}$$
$$= 8\log\left(x + \sqrt{16x^2}\right)$$
Then proceed as in (1).

4. $\frac{dx}{d\theta} = a\theta, \cos\theta, \frac{dy}{d\theta} = a\theta\sin\theta$
$$\Rightarrow \frac{dy}{dx} = \tan\theta \text{ . Therefore, } \frac{ds}{dx} = \sqrt{1 + \left(\frac{dy}{dx}\right)^2} = \sqrt{1 + \tan^2\theta}$$
$$= \sec\theta.$$
Then obtained the arc using the following formula
$$s = \int_0^{2\pi} \frac{ds}{dx}.dx .$$

5. $\frac{dy}{dx} = \frac{3x^2}{2ay}$
$$\therefore \frac{ds}{dx} = \sqrt{1 + \left(\frac{dy}{dx}\right)^2} = \sqrt{1 + \frac{9x}{4a}}$$
Now using the following formula $s = \int_0^a \frac{ds}{dx}.dx$.

6. $\frac{dx}{d\theta} = \frac{5}{2}e^{\theta}\cos\frac{\theta}{2}, \quad \frac{dy}{d\theta} = -\frac{5}{2}e^{\theta}\sin\frac{\theta}{2}$

$\Rightarrow \quad \frac{dy}{dx} = \frac{dy/d\theta}{dx/d\theta} = -\tan\frac{\theta}{2}$

$\therefore \quad \frac{ds}{dx} = \sqrt{1+\tan^2\theta/2} = \sec\frac{\theta}{2}.$

Now $\frac{ds}{d\theta} = \frac{ds}{dx}\cdot\frac{dx}{d\theta} = \frac{5}{2}e^{\theta}$. Now using the formula $s = \int_0^{\pi}\frac{ds}{d\theta}d\theta$.

7. The points of intersection of the given curves are (0, 0) and $\left(\frac{9a}{16}, \frac{3a}{2}\right)$.

Also $\frac{dy}{dx} = \frac{2a}{y}$. Now find $\frac{ds}{dx}$ and then use the formula, given below

$$s = \int_0^{9a/16}\frac{ds}{dx}.dx\,.$$

10. $\frac{dx}{dt} = -3a\cos^2 t\sin^2 t, \frac{dy}{dt} = 3b\sin^2 t\cos t$

$\therefore \quad \frac{ds}{dt} = \sqrt{\left(\frac{dx}{dt}\right)^2 + \left(\frac{dy}{dt}\right)^2} \quad \therefore s = 4\int_0^{\pi/2}\frac{ds}{dt}.dt$

11. Eliminate t between the given curve, we get $9y^2 = x(3-x)^2$.

Now $\quad \frac{dx}{dt} = 2t, \frac{dy}{dt} = (1-t^2).$

Then use $\quad \frac{ds}{dt} = \sqrt{\left(\frac{dx}{dt}\right)^2 + \left(\frac{dy}{dt}\right)^2}$ and $s = 2\int_0^{\sqrt{3}}\frac{ds}{dt}.dt$

15. $\frac{dr}{d\theta} = a\sin\theta.$

Since the given curve is symmetrical about the initial line, therefore, the entire length is twice the arc measure from 0 to π.

Here, $\frac{ds}{d\theta} = \sqrt{r^2 + \left(\frac{dr}{d\theta}\right)^2} = 2a\sin\frac{\theta}{2}$. Then using the formula, we get $s = 2\int_0^{\pi}\frac{ds}{d\theta}.d\theta$.

26. The points of the intersection of the given curve are $\theta = \pm\frac{\pi}{3}$.

Let s_1 be the arc length of the cardioid inside the circle and s_2 be the arc length of the cardioid outside of the circle.

Therefore, $s_1 = 2\int_0^{\pi/3}\frac{ds}{d\theta}.d\theta$ and $s_2 = 2\int_{\pi/3}^{\pi}\frac{ds}{d\theta}.d\theta$

27. Do same as ex. 9 by taking $a = 1$.

28. Do same as ex. 6.

ANSWERS

1. $2\pi b$ **2.** $2[\sqrt{2}+\log(1+\sqrt{2})]$ **3.** $\frac{3}{2}+\frac{1}{4}\log 2$ **4.** $2a\pi^2$ **5.** $\frac{1}{27}a[13\sqrt{13}-8]$

8. $4a\sqrt{3}$ **9.** a **10.** $4(a^2+ab+b^2)/(a+b)$ **11.** $4\sqrt{3}$ **13.** $8a$

14. $\sqrt{2}[e^{\pi/2}-1]$ **15.** $8a$ **18.** $8a$ **20.** $a\sec\alpha\,(e^{\theta\cot\alpha}-1)$ **22.** $3a\pi/4$ **23.** $4a/3$

25. $s_1(\theta_2) - s_1(\theta_1)$ where $s_1(\theta) = a\sqrt{(\sec^2\theta+3)} - a\sqrt{3}[\log\{\cos\theta + \sqrt{(\cos^2\theta + \frac{1}{3})}\}]$ **27.** $\frac{4a}{\sqrt{3}}$

16.3 INTRINSIC EQUATION OF CURVES

The relation between s and ψ is called an intrinsic equation of any curve, where s is the length of an arc of the curve measured from a fixed point on the curve to any point P on it and ψ is the angle which the tangent to the curve at P makes with the positive x-axis. Thus the co-ordinates (s, ψ) is known as Intrinsic co-ordinates.

16.4 DERIVATION OF INTRINSIC EQUATION OF THE CURVES

(a) If the equation of a curve is in cartesian form. Let the equation of the curve be $y = f(x)$ as shown in fig. 8. Let us consider a point A as fixed on this curve and let $P(x, y)$ be any point on this curve such that $AP = s$. Suppose the tangent at P makes the angle ψ with the fixed straight line (*i.e.*, x-axis). Since, we have

$$\frac{dy}{dx} = \tan\psi\,.$$

$$\therefore \quad \tan\psi = \frac{d}{dx}f(x) = f'(x)\,. \quad \ldots(1)$$

Further since we know that $\frac{ds}{dx} = \sqrt{\left[1+\left(\frac{dy}{dx}\right)^2\right]}. \quad \ldots(2)$

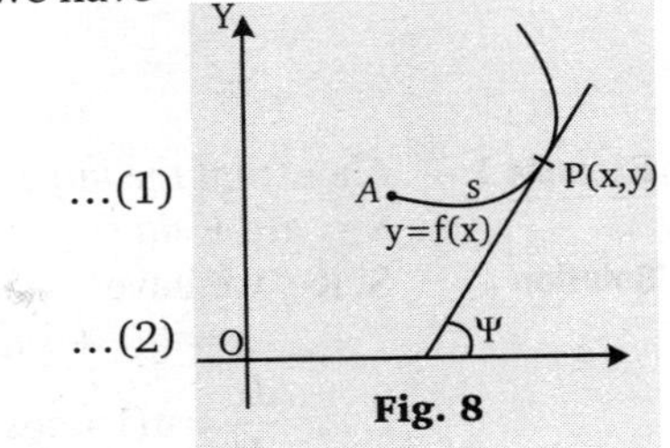

Fig. 8

Let x_1 be the x-co-ordinate of the fixed point A. Therefore, we have

$$s = \int_a^{x_1}\left(\frac{ds}{dx}\right)dx \quad \text{or} \quad s = \int_a^{x_1}\sqrt{\left[1+\left(\frac{dy}{dx}\right)^2\right]}dx \quad \text{[Using (2)]}$$

or $\quad s = \int_a^{x_1}\sqrt{1+[f'(x)]^2}\,dx \quad$ [Using (1)] $\quad \ldots(3)$

Now eliminate x and $f'(x)$ between (1) and (3), we obtain the relation between s and ψ and thus obtain the intrinsic equations.

REMARK

- If $x = f(t), y = g(t)$, then $\frac{dy}{dx} = \frac{dy}{dt} \Big/ \frac{dx}{dt}$.

(b) Equations of a curve in polar form. Let the equations of a curve be $r = f(\theta)$ and consider a point A as fixed on this curve. Let $P(r, \theta)$ be any point on this curve such that arc $AP = s$ and the tangent at P makes an angle ψ with the positive x-axis as shown in fig. 9. Let ϕ be the angle between the radius vector and the tangent at P. Thus, we have

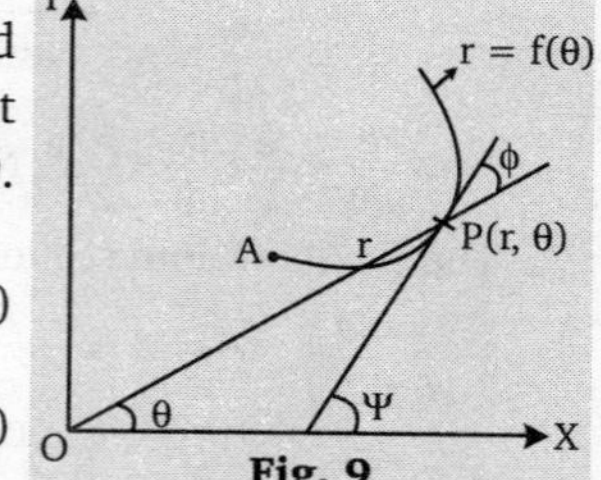

Fig. 9

$$\psi = \theta + \phi \quad \ldots(1)$$

Since we know that

$$\tan\phi = r\frac{d\theta}{dr}. \quad \ldots(2)$$

But, the equation of a curve is

$$\left.\begin{aligned} r &= f(\theta) \\ \therefore \quad \frac{dr}{d\theta} &= f'(\theta) \end{aligned}\right\} \quad \ldots(3)$$

Now using (2) and (3), we get

$$\tan\phi = \frac{f(\theta)}{f'(\theta)}. \quad \ldots(4)$$

Now $$\frac{ds}{d\theta} = \sqrt{\left[r^2 + \left(\frac{dr}{d\theta}\right)^2\right]}. \Rightarrow \quad \frac{ds}{d\theta} = \sqrt{[[f(\theta)]^2 + [f'(\theta)]^2]}. \quad \ldots(5)$$

Let the vectorial angle of the point A be α. Then, we have

$$s = \int_\alpha^\theta \left(\frac{ds}{d\theta}\right)d\theta \quad \Rightarrow \quad s = \int_\alpha^\theta \sqrt{[\{f(\theta)\}^2 + \{f'(\theta)\}^2]}d\theta \quad \ldots(6)$$

[Using (5)]

Now eliminate θ and ϕ between (1), (4) and (6), we obtain the required intrinsic equation of the curve.

(c) Equation of curve in Pedal form. Let the equation of a curve in pedal form be $p=f(r)$ and let A be the fixed point on the curve such that at A, $r = a$ (say), then, we have

$$s = \int_a^r \frac{rdr}{\sqrt{(r^2 - p^2)}}. \Rightarrow \quad s = \int_a^r \frac{rdr}{\sqrt{[r^2 - \{f(r)\}^2]}} \quad \ldots(1)$$

where s is an arc measured from A to any point $P(p, r)$. Let f be the radius of curvature of the given curve at P, then we have

$$\rho = \frac{ds}{d\psi} = r\frac{dr}{dp} = r / f'(r). \quad \ldots(2)$$

Now eliminating r between (1) and (2), we obtain the required intrinsic equation.

Solved Examples

Example 1. *Show that the intrinsic equation of the curve $x = a(t + \sin t), y = a(1 - \cos t)$ is $s = 4a \sin \psi$.*

Solution . Since, we have

$$x = a(t + \sin t), y = a(1 - \cos t)$$

$$\therefore \quad \frac{dx}{dt} = a(1+\cos t), \frac{dy}{dt} = a\sin t$$

$$\therefore \quad \frac{dy}{dx} = \frac{dy}{dt} \Big/ \frac{dx}{dt} = \frac{\sin t}{1+\cos t}$$

$$= \frac{2\sin t/2\cos t/2}{2\cos^2 t/2} = \tan t/2$$

Since, we know that $\frac{dy}{dx} = \tan\psi$.

$\tan\psi = \tan t/2$ or $\psi = t/2$ or $t = 2\psi$.

If s is the length of the arc of the curve measured from the vertex $A(0, 0)$ to any point $P(x, y)$ in the direction of t increasing, then

$$s = \int_0^t \left(\frac{ds}{dt}\right).dt \quad \text{or} \quad s = \int_0^t \sqrt{\left[\left(\frac{dx}{dt}\right)^2 + \left(\frac{dy}{dt}\right)^2\right]}.dt$$

where $\left(\frac{ds}{dt}\right)^2 = \left(\frac{dx}{dt}\right)^2 + \left(\frac{dy}{dt}\right)^2$

$$\therefore \quad s = \int_0^t \sqrt{[a^2(1+\cos t)^2 + a^2\sin^2 t]}.dt$$

$$= \int_0^t 2a\cos t/2dt$$

$$s = 2a\left[2\sin t/2\right]_0^t = 4a\sin t/2$$

But $t = 2\psi$.

$\therefore$ $s = 4a\sin\psi$.

Example 2. *Show that the intrinsic equation of $3ay^2 = 2x^3$ taking its cusp as the fixed point is $9s=4a(\sec^3\psi-1)$.*

Solution . The equation of the curve is

$$3ay^2 = 2x^3 \text{ or } y = \frac{\sqrt{2}x^{3/2}}{\sqrt{3a}}$$

$$\therefore \quad \frac{dy}{dx} = \sqrt{\frac{2}{3a}}\cdot\frac{3}{2}x^{1/2} = \sqrt{\frac{3x}{2a}}.$$

Further since, we have

$$\frac{dy}{dx} = \tan\psi$$

$$\therefore \quad \tan\psi = \sqrt{\frac{3x}{2a}} \text{ or } x = \frac{2a}{3}\tan^2\psi \qquad \ldots(1)$$

$$\text{But} \quad \frac{ds}{dx} = \sqrt{\left[1+\left(\frac{dy}{dx}\right)^2\right]} = \sqrt{\left[1+\frac{3x}{2a}\right]}.$$

If s is the arc length of the given curve measure from the cusp at which $x = 0$ to any point $P(x, y)$ in the direction of x decreasing. Then

$$\frac{ds}{dx} = \sqrt{\left(1+\frac{3x}{2a}\right)} \qquad \ldots(2)$$

$$\therefore \quad s = \int_0^x\left(\frac{ds}{dx}\right)dx \text{ or } s = \int_0^x\sqrt{\left(1+\frac{3x}{2a}\right)}dx$$

[Using (2)]

$$= \left[\frac{\frac{2}{3}\left(1+\frac{3x}{2a}\right)^{3/2}}{3/2a}\right]_0^x$$

$$\Rightarrow \quad s = \frac{4a}{9}\left[\left(1+\frac{3x}{2a}\right)^{3/2}-1\right] \qquad \ldots(3)$$

Now eliminating x between (1) and (3), we get

$$s = \frac{4a}{9}\left[(1+\tan^2\psi)^{3/2}-1\right]$$

$$s = \frac{4a}{9}[\sec^3\psi-1] \text{ or } 9s = 4a[\sec^3\psi-1].$$

Example 3. *Find the intrinsic equation of the cardioid $r = a(1-\cos\theta)$.*

Solution . The equation of the curve is

$$r = a(1-\cos\theta). \qquad \ldots(1)$$

$$\therefore \quad \frac{dr}{d\theta} = a\sin\theta.$$

$$\text{Now,} \quad \frac{ds}{d\theta} = \sqrt{\left[r^2+\left(\frac{dr}{d\theta}\right)^2\right]}$$

$$= \sqrt{[a^2(1-\cos\theta)^2 + a^2\sin^2\theta]}$$

$$= \sqrt{2a^2(1-\cos\theta)} = \pm 2a\sin\theta/2.$$

If s is the length of the arc of the curve measured from pole (0, 0) to any point $P(r, \theta)$ in the direction of θ increasing, then we will take the sign of $\frac{ds}{d\theta}$ positive.

$$s = \int_0^\theta\left(\frac{ds}{d\theta}\right)d\theta = \int_0^\theta 2a\sin(\theta/2)d\theta$$

$$= 2a\left[-2\cos(\theta/2)\right]_0^\theta$$

$$s = 4a(1-\cos(\theta/2)) \qquad \ldots(2)$$

Further since, we know that

$$\tan\phi = r\frac{d\theta}{dr} = a(1-\cos\theta).\frac{1}{a\sin\theta}$$

$$= \frac{1-\cos\theta}{\sin\theta} = \frac{2\sin^2(\theta/2)}{2\sin(\theta/2)\cos(\theta/2)}$$

$$\therefore \quad \tan\phi = \tan\theta/2$$

$$\therefore \quad \phi = \theta/2 \text{ or } \theta = 2\phi.$$

But we have

$$\psi = \theta+\phi = \theta+\theta/2 \qquad (\because \phi = \theta/2)$$

$$\psi = \frac{3}{2}\theta \text{ or } \theta = \frac{2}{3}\psi.$$

Substitute the value of θ in (2) , we get

$$s = 4a\left(1-\cos\frac{2}{6}\psi\right) = 4a\left(2\sin^2\frac{1}{6}\psi\right)$$

$$\text{Hence,} \quad s = 8a\sin^2\left(\frac{\psi}{6}\right).$$

Example 4. *Find the intrinsic equation of the curve $p = r\sin\alpha$.*

Solution . Since the equation of the curve is in pedal form

i.e., $p = r\sin\alpha$

$$\therefore \quad \frac{dp}{dr} = \sin\alpha$$

$$\text{Now,} \quad \rho = \frac{ds}{d\psi} = r\frac{dr}{dp}$$

$$\therefore \quad \frac{ds}{d\psi} = r.\frac{1}{\sin\alpha} = r\,\text{cosec}\,\alpha \qquad \ldots(1)$$

If s is the length of arc measured from the point $r = 0$ to any point P in the direction of r increasing. Then, we have

$$s = \int_0^r\frac{rdr}{\sqrt{r^2-p^2}}$$

$$= \int_0^r\frac{rdr}{\sqrt{r^2-r^2\sin^2\alpha}} \qquad (\because p = r\sin\alpha)$$

$$= \int_0^r\sec\alpha dr = \sec\alpha[r]_0^r$$

$$s = r\sec\alpha \qquad \ldots(2)$$

Now eliminate r between (1) and (2), we get

$$\frac{ds}{d\psi} = s\cot\alpha \quad \text{or} \quad \frac{ds}{s} = \cot\alpha\, d\psi.$$

Integrating, we get

$$\log s = \psi\cot\alpha + \log c \text{ or } s = ce^{\psi\cot\alpha}$$

where c is the constant of integration.

This is the required intrinsic equation of the curve.

EXERCISE 16.2

1. Show that the intrinsic equation of the curve $y^2 = 4ax$ is $s = a \cot \psi \operatorname{cosec} \psi + a \log (\cot \psi + \operatorname{cosec} \psi)$.
2. Find the intrinsic equation of the curve $y = c \cosh (x/c)$.
3. Find the intrinsic equation of the curve $x^2 = 4ay$.
4. Find the intrinsic equation of the curve $x = a(1 + \sin t)$, $y = a(1 + \cos t)$.
5. Find the intrinsic equation of the curve $x^{2/3} + y^{2/3} = a^{2/3}$.
   (i) If s is measured from the vertex.
   (ii) If s is measured from the cusp on x-axis.
6. Find the intrinsic equation of the cardioid $r = a(1 + \cos \theta)$, $\theta = 0$ being the fixed point.
7. Find the intrinsic equation of the curve $r = ae^{\theta \cot \alpha}$, where s is measured from the point $(a, 0)$.
8. Find the intrinsic equation of the curve $r = a\theta$, s being measured from $(0, 0)$.
9. Find the intrinsic equation of the curve $p = \sqrt{(r^2 - a^2)}$.
10. Prove that the intrinsic equation of the curve $x^2 = 4ay$ is $s = a \tan \psi \sec \psi + a \log(\tan \psi + \sec \psi)$.
11. Find the intrinsic equation of the curve $x = e^t \sin t$, $y = e^t \cos t$, where $t = \pi/4$ being the fixed point.
12. Find the intrinsic equation of the curve $y = a \log \sec \dfrac{x}{a}$.

Hint to Selected Problems

1. Since $y^2 = 4\,ax$. Therefore, $\dfrac{dy}{dx} = \dfrac{2a}{y}$. $\left[\because \dfrac{dy}{dx} = \tan \psi\right]$

$\therefore \quad y = 2a \cot \psi$

Now using $\dfrac{ds}{dy} = \sqrt{1 + \left(\dfrac{dx}{dy}\right)^2} = \dfrac{1}{2a}\sqrt{4a^2 + y^2}$

Then integrating after separating the variables.

2. $\dfrac{dy}{dx} = \sinh\left(\dfrac{x}{c}\right) \therefore \dfrac{ds}{dx} = \cosh \dfrac{x}{c}$.

Now using the formula given below

$$\int_0^s ds = \int_0^x \cosh\left(\frac{x}{c}\right) dx.$$

3. Do same as (1)

4. $\dfrac{dx}{dt} = a \cos t, \dfrac{dy}{dt} = -a \sin t$

$\therefore \dfrac{dy}{dx} = -\tan t = \tan \psi \Rightarrow t = -\psi.$

Now $\dfrac{ds}{dt} = \sqrt{\left(\dfrac{dx}{dt}\right)^2 + \left(\dfrac{dy}{dt}\right)^2} = a$. Then use $\int_0^s ds = a\int_0^t dt$.

6. $\dfrac{dr}{d\theta} = -a \sin \theta$. Also, $\tan \phi = r\dfrac{d\theta}{dr} = -\cot \dfrac{\theta}{2} = \tan\left(\dfrac{\pi}{2} + \dfrac{\theta}{2}\right)$.

Now using $\psi = \theta + \phi = \dfrac{3\theta}{2} + \dfrac{\pi}{2} \Rightarrow \theta = \dfrac{2}{3}\left(\psi - \dfrac{\pi}{2}\right)$

$$\frac{ds}{d\theta} = \sqrt{r^2 + \left(\frac{dr}{d\theta}\right)^2} = 2a \cos \frac{\theta}{2}.$$

$\therefore \quad \int_0^s ds = \int_0^\theta 2a \cos \dfrac{\theta}{2} d\theta.$

7. $\dfrac{dr}{d\theta} = r \cot \alpha$. Also, we have $\tan \phi = r\dfrac{d\theta}{dr} = r.\dfrac{1}{r \cot \alpha} = \tan \alpha$

$\Rightarrow \tan \phi = \tan \alpha \Rightarrow \phi = \alpha$.

Now $\psi = \theta + \phi \Rightarrow \phi = \psi - \alpha$

$\therefore \quad \dfrac{ds}{d\theta} = \sqrt{r^2 + \left(\dfrac{dr}{d\theta}\right)^2}$

$\Rightarrow \quad ds = a \operatorname{cosec} \alpha e^{\theta \cot \alpha} d\theta$

then integrating w.r.t. θ.

8. Do same as (7).

11. $\dfrac{dx}{dt} = e^t \sin t + e^t \cos t, \dfrac{dy}{dt} = e^t \cos t - e^t \sin t$.

$\therefore \dfrac{dy}{dx} = \dfrac{1 - \tan t}{1 + \tan t} = \tan\left(\dfrac{\pi}{4} - t\right) \Rightarrow \psi = \dfrac{\pi}{4} - t$ i.e., $t = \dfrac{\pi}{4} - \psi$.

Now proceed as above.

12. $y = a \log \sec (x/a) \quad \Rightarrow \quad \dfrac{dy}{dx} = \tan(x/a)$

$\Rightarrow \quad \tan \psi = \tan (x/a) \Rightarrow \quad x = a\psi$

$\Rightarrow \quad \dfrac{ds}{dx} = \sqrt{1 + \left(\dfrac{dy}{dx}\right)^2} = \sqrt{1 + \tan^2 \dfrac{x}{a}} \Rightarrow \dfrac{ds}{dx} = \sec \dfrac{x}{a}$

$\Rightarrow \quad s = \int_0^x \dfrac{ds}{dx}.dx.$

ANSWERS

2. $s = c \tan \psi$
3. $s = a \tan \psi \sec \psi + a \log (\tan \psi + \sec \psi)$
4. $s + a\psi = 0$
5. (i) $4s = 3a \cos 2\psi$ (ii) $2s = 3a \sin^2 \psi$
6. $s = 4a \sin\left(\dfrac{\psi}{3} - \dfrac{\pi}{6}\right)$
7. $s = a \sec \alpha\, [e^{(\psi - \alpha)\cot \alpha} - 1]$
8. $s = \dfrac{1}{2}a[\theta\sqrt{(1+\theta)^2} + \log\{\theta + \sqrt{(1+\theta)^2}\}]$
9. $s = \dfrac{1}{2}a(\psi^2 + 1)$
11. $s = \sqrt{2}e^{\pi/4}[e^{-\psi} - 1]$
12. $s = a \log [\tan \psi + \sec \psi]$
13. $s = a \tan \psi$

FFFFF

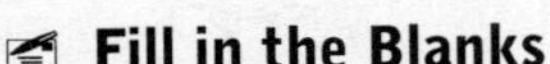

Objective Evaluations

Fill in the Blanks

1. The curve $y = f(x)$ is rectified between $x = a$ and $x = b$ if $f(x)$ is ________.
2. The whole length of the arc of the curve $x^2 + y^2 = a^2$ is ________.
3. If $x = f(y)$, then the formula for finding the length of the curve between $y = a$ and $y = b$ is ________.
4. If $\theta = f(r)$, then the required length is ________.
5. If $p = f(r)$, the required length is ________.
6. The whole length of the astroid $x^{2/3} + y^{2/3} = a^{2/3}$ is ________.
7. The length of an arc of the cycloid $x = a(t - \sin t)$, $y = a(1 - \cos t)$ is________.
8. The whole length of the cardioid $r = a(1 + \cos\theta)$ ________.
9. The perimeter of $r = a + b\cos\theta$ $(a > b)$ is ________.
10. The length of an arc of the curves, $x \sin t + y \cos t = f'(t)$, $x\cos t - y\sin t = f''(t)$ is ________.
11. The equation $x = f(\psi)$ is called ________.

True/False

Write 'T' for True and 'F' for False statement.

1. The length of the loop of the curve $x = t^2, y = t - \frac{1}{3}t^3$ is $4\sqrt{3}$. **(T/F)**
2. If $r = f(\theta)$, then the required length between $\theta = \alpha$ and $\theta = \beta$ is $s = \int_\alpha^\beta \sqrt{\{r^2 + [f'(\theta)]^2\}}d\theta$. **(T/F)**
3. The whole length of the astroid $x^{2/3} + y^{2/3} = a^{2/3}$ is $7a$. **(T/F)**
4. The perimeter of the cardioid $r = a(1 - \cos\theta)$ is $8a$. **(T/F)**
5. The length of the arc of the cycloid $x = a(t + \sin t)$, $y = a(1 - \cos t)$ is equal to the perimeter of the cardioid $r = a(1 + \cos\theta)$. **(T/F)**
6. The length of the arc of the curve $r = ae^{\theta \cot \alpha}$ between r_1 and r_2 is $(r_1 - r_2)$ cosec α. **(T/F)**
7. The length of the curve $y = \frac{e^x - 1}{e^x + 1}$ from $x = 1$ to $x = 2$ is $\log[e + (1/e)]$. **(T/F)**
8. The length of an arc of the curve $x^2 + y^2 = [f'(t)]^2 + [f''(t)]^2$ is given by $s = f(t) + f''(t) + c$ where c is some constant. **(T/F)**
9. The intrinsic equation of the cardioid $r = a(1 - \cos\theta)$ is $s = 8a \sin^2 \frac{\psi}{6}$. **(T/F)**
10. $s = 4a \tan \psi$ is the intrinsic equation of the curve $y = 4a \cosh(x/4a)$. **(T/F)**
11. The intrinsic equation of the curve $p^2 = r^2 - a^2$ is $s = \frac{1}{2}a\psi$. **(T/F)**
12. The intrinsic equation of $3ay^2 = 2x^3$ taking its cusp as the fixed point is $s = \frac{4a}{9}(\sec^3 \psi - 1)$. **(T/F)**

Multiple Choice Questions

Choose the most appropriate one.

1. The whole length of the curve $x^2 + y^2 - 2ax = 0$ is :
   (a) πa (b) $2\pi a$ (c) πa^2 (d) $3\pi a$
2. The perimeter of the cardioid $r = a(1 + \cos\theta)$ is :
   (a) $8a$ (b) $4a$ (c) $6a$ (d) $7a$
3. The whole length of the curve $x = a\cos^3 t$, $y = a\sin^3 t$ is :
   (a) $6a$ (b) $8a$ (c) $3a$ (d) $2a$
4. The length of an arc of the cycloid $x = a(\theta - \sin\theta)$, $y = a(1 - \cos\theta)$ is :
   (a) $6a$ (b) $5a$ (c) $3a$ (d) $8a$
5. The length of an arc of the curve $r = ae^{\theta/\sqrt{2}}$ between $r = 1$ and $r = 2$ is :
   (a) 1 (b) $\sqrt{2}$ (c) $2\sqrt{2}$ (d) none of these
6. The length of an arc of the curve $x^2 + y^2 = [f'(t)]^2 + [f''(t)]^2$ where t is a parameter is :
   (a) $f(t) + f''(t) + c$ (b) $f(t) + c$
   (c) $f''(t) + c$ (d) $f'(t) + f''(t) + c$
7. The length of an arc of the curve $x = t^2, y = t - \frac{1}{3}t^3$ is :
   (a) $\sqrt{3}$ (b) $2\sqrt{3}$ (c) $3\sqrt{3}$ (d) $4\sqrt{3}$
8. The intrinsic equation of the curve $y = c\cosh(x/c)$ is :
   (a) $s = c\cot\psi$ (b) $s = c\tan\psi$
   (c) $s = c\tan^2\psi$ (d) none of these
9. The intrinsic equation of the curve $x = a(1 + \sin t)$, $y = a(1 + \cos t)$ is :
   (a) $s + a\psi = 0$ (b) $s + a^2\psi = 0$
   (c) $s - a\psi = 0$ (d) $s + a\psi^2 = 0$
10. The intrinsic equation of the cardioid $r = a(1 - \cos\theta)$ is :
    (a) $s = a\psi$ (b) $s = 8a\sin^2(\psi/6)$
    (c) $s = 8a\sin(\psi/6)$ (d) $s = 8a\sin^2(\psi/3)$

ANSWERS

Fill in the Blanks

1. Continuous **2.** $2\pi a$ **3.** $\int_a^b \sqrt{\left[1+\left(\frac{dx}{dy}\right)^2\right]}dy$ **4.** $\int_{r_1}^{r_2} \sqrt{\left[1+\left(\frac{rd\theta}{dr}\right)^2\right]}dr$ **5.** $\int_{r_1}^{r_2} \frac{rdr}{\sqrt{(r^2-p^2)}}$

6. $6a$ **7.** $8a$ **8.** $8a$ **9.** $2\pi a(1 + b^2/4a^2)$ **10.** $f(t) + f''(t) + c$

11. Intrinsic equation

True/False

1. T	**2.** T	**3.** F	**4.** T	**5.** T	**6.** F	**7.** T	**8.** T	**9.** T	**10.** T
11. F	**12.** T								

Multiple Choice Questions

1. (b)	**2.** (a)	**3.** (a)	**4.** (d)	**5.** (b)	**6.** (a)	**7.** (d)	**8.** (b)	**9.** (a)	**10.** (b)

FFFFF

CHAPTER

17 Quadrature (Area of Bounded Curves)

17.1 AREA OF CURVE IN CARTESIAN FORM

Let $y = f(x)$ be a continuous curve in cartesian form and let A be the area of the region bounded by the curve $y = f(x)$, the axis of x and the two ordinates $x = a$ and $y = b$. Then

$$A = \int_a^b y\,dx = \int_a^b f(x)\,dx$$

Proof. Let BC be the arc of the curve $y = f(x)$ cut by the lines $x = a$ and $y = b$ as shown in fig. 1.

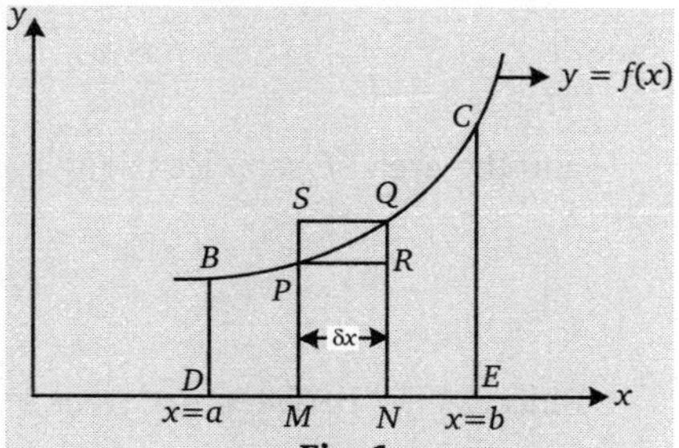

Let $P(x, y)$ and $Q(x + \delta x, y + \delta y)$ be two neighbouring points on the cuve between B and C. Now draw the perpendicular PM and QN on the axis of x such that $PM = y$ and $QN = y+dy$ and $MN = dx$. Since we observe that the area of $DMPB$ increases as P moves along arc BC from B to C. Draw the perpendicular PR on QN and QS to MP produced to S. Let δA be the the area of $MNQP$, this area lies between the area $MNRP$ and the area $MNQS$.

Fig. 1

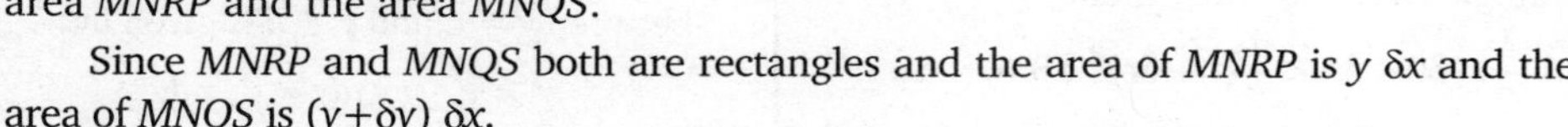

Since $MNRP$ and $MNQS$ both are rectangles and the area of $MNRP$ is $y\ \delta x$ and the area of $MNQS$ is $(y+\delta y)\ \delta x$.

$\therefore$ Area of $MNRP < \delta A <$ area of $MNQS$

or $$y\,\delta x < \delta A < (y + \delta y)\,\delta x \text{ or } y < \frac{\delta A}{\partial x} < y + \delta y.$$

Taking the limit as $\delta x \to 0$ and $\delta y \to 0$, where $Q \to P$.

$$\therefore \quad y < \lim_{\delta x \to 0} \frac{\delta A}{\partial x} < y \text{ or } y < \frac{dA}{dx} < y$$

$$\Rightarrow \quad \frac{dA}{dx} = y \text{ or } \frac{dA}{dx} = f(x) \qquad [\because y = f(x)]$$

or $$dA = f(x)\,dx.$$

Now integrating *w.r.t.* 'x' from $x = a$ to $x = b$, *we get*

$$\int_{x=a}^{x=b} dA = \int_a^b f(x)\,dx$$

$$\Rightarrow \quad [A]_{x=a}^{x=b} = \int_a^b f(x)\,dx$$

or $$\text{Area } DECB = \int_a^b f(x)\,dx$$

or $$A = \int_a^b f(x)\,dx = \int_a^b y\,dx.$$

Similarly, the area bounded by the curve $x = f(y)$, the axis of y and the ordinates $y = a$ and $y = b$ is given by

$$A = \int_a^b f(y)\,dy = \int_a^b x\,dy.$$

REMARKS

- If the given curve is symmetrical either about x-axis or about y-axis or both, then find the area of one of the symmetrical part and multiply this area by the number of symmetrical parts, we get the whole area of the bounded region.
- Area bounded by two curves = | Area bounded by one curve – Area bounded by other curve |.

17.2 AREA OF CURVE IN POLAR FORM

Let $r = f(\theta)$ be the equation of a curve in polar form where $f(\theta)$ is a continuous function of θ, then the area of the sector bounded by the curve $r = f(\theta)$ and two radii vectors $\theta = \theta_1$ and $\theta = \theta_2$ such that $\theta_2 > \theta_1$ is given by

$$A = \frac{1}{2}\int_{\theta_1}^{\theta_2} r^2\, d\theta \cdot$$

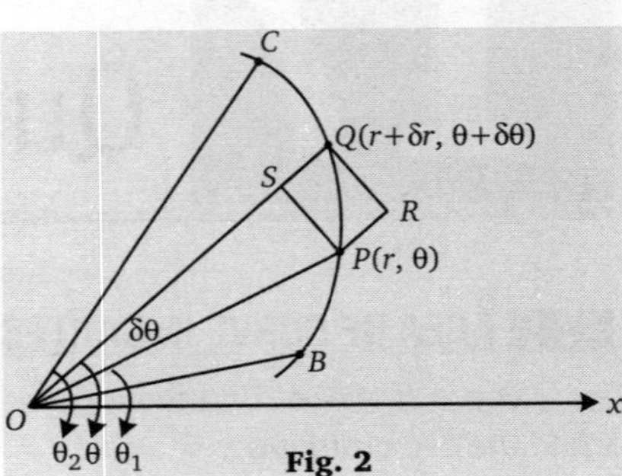

Fig. 2

Proof. Since the equation of curve in polar form is $r = f(\theta)$.

Let A be the area of the sector OBC bounded by the curve and two radii vector $\theta = \theta_1$ and $\theta = \theta_2$ as shown in fig. 2.

Let $P(r, \theta)$ and $Q\ (r + \delta r, \theta + \delta\theta)$ be two neighbouring points on the curve $r = f(\theta)$ such that $OP = r$, $OQ = r + \delta r$ and $\angle POQ = \delta\theta$. Draw the perpendicular PS to OQ and QR to OR where OP producted to R. Let δA be the area of the sector OPQ. Obviously, this area δA lies between the area of isosceles triangle OPS and the area of isosceles triangle ORQ.

Now, the area of the isosceles triangle $OPS = \frac{1}{2} OP \times OS \sin\delta\theta = \frac{1}{2} r \times r \cdot \delta\theta$

$(\therefore \sin\delta\theta \approx \delta\theta)$

$$= \frac{1}{2} r^2 \delta\theta \qquad (\delta\theta \text{ is very small})$$

and the area of isosceles triangle, $ORQ = \frac{1}{2} OR \times OQ \sin\delta\theta$

$$= \frac{1}{2}(r + \delta r)^2 \delta\theta$$

$(\therefore \delta\theta$ is very small so $\sin\delta\theta \approx \delta\theta)$

Since δA lies between the areas of triangle OPS and ORQ, then

$$\frac{1}{2} r^2 \delta\theta < \delta A < \frac{1}{2}(r + \delta r)^2\, \delta\theta .$$

Divide by $\delta\theta$, we have

$$\frac{1}{2} r^2 < \frac{\delta A}{\delta\theta} < \frac{1}{2}(r + \delta r)^2 .$$

As $Q \to P$, $\delta\theta > 0$ and $\delta r \to 0$ so taking the limit as $\delta\theta \to 0$, we get

$$\frac{1}{2} r^2 < \lim_{\delta\theta \to 0} \frac{\delta A}{\delta\theta} < \frac{1}{2} r^2$$

or
$$\frac{1}{2} r^2 < \frac{dA}{d\theta} < \frac{1}{2} r^2$$

or
$$\frac{dA}{d\theta} = \frac{1}{2} r^2 \text{ or } dA = \frac{1}{2} r^2 d\theta.$$

Integrating $w.r.t.$ θ from $\theta = \theta_1$ to $\theta = \theta_2$, we get

$$\int_{\theta=\theta_1}^{\theta=\theta_2} dA = \int_{\theta_1}^{\theta_2} \frac{1}{2} r^2\, d\theta$$

Hence,
$$A = \frac{1}{2}\int_{\theta_1}^{\theta_2} r^2 d\theta \qquad \left(\because \int_{\theta_1}^{\theta_2} dA = \int_B^C dA = A\right)$$

Recapitulations

- Area bounded by the curve $y = f(x)$ between $x = a$ and $x = b$ is given by
  $A = \int_a^b f(x)\,dx = \int_a^b y\,dx$
- Area bounded by the curve $x = f(y)$ between $y = a$ and $y = b$ is given by
  $A = \int_a^b f(y)\,dy = \int_a^b x\,dy$
- If the given curve is $r = f(\theta)$. Then area $= \int_{\theta_1}^{\theta_2} r^2\, d\theta$

REMARKS

- In case of $r = a\cos n\theta$ or $r = a \sin n\theta$, the number of loops are n and $2n$ according as n is odd and n is even.
- Area by double integration bounded by $r = f(\theta)$ and $\theta = \theta_1$ and $\theta = \theta_2$ is given by $\int_{\theta_1}^{\theta_2}\int_{r=0}^{r=f(\theta)} r\, d\theta\, dr$.

Solved Examples

Example 1. *Find the area of the region bounded by the line $x = 2$ and the parabola $y^2 = 8x$.*

Solution. The equation of the parabola is

$$y^2 = 8x$$

which is symmetrical about x-axis and the line $x=2$ intersects the parabola $y^2 = 8x$ in two points (2,4) and (2, –4) as shown in fig. 3.

$\therefore$ The required area

$$= 2\int_0^2 y\,dx$$

$$= 2\int_0^2 \sqrt{8x}\,dx$$

$$= 4\sqrt{2}\left[\frac{2}{3}x^{3/2}\right]_0^2$$

$$= \frac{32}{3} \text{ sq. units.}$$

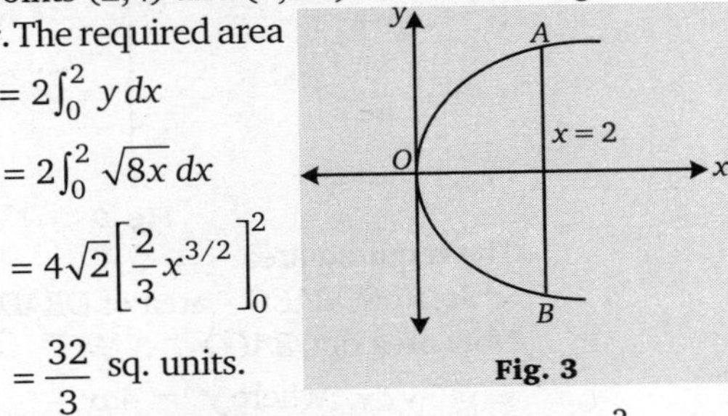

Fig. 3

Example 2. *Find the area bounded by the parabola $y^2 = 4ax$ and its latusrectum.*

Solution. We know that a line through the focus of the parabola and perpendicular to its axis is called latus-rectum. Since the equation of the parabola is

$$y^2 = 4\,ax.$$

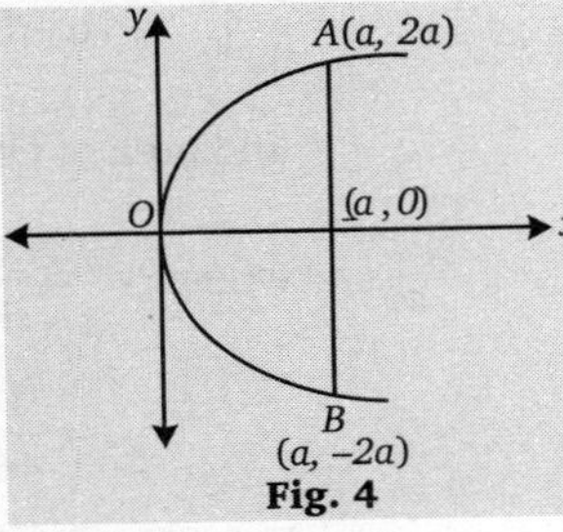

Fig. 4

$\therefore$ Extremities of latusrectum are $(a, 2a)$ and $(a, -2a)$ which is shown in fig. 4.

$\therefore$ The required area

$$= 2\int_0^a y\,dx = 2\int_0^a \sqrt{4ax}\,dx$$

$$= 4\sqrt{a}\left[\frac{2}{3}x^{3/2}\right]_0^a = 4\sqrt{a}\left[\frac{2}{3}a\sqrt{a}\right]$$

$$= \frac{8}{3}a^2 \text{ sq. units.}$$

Example 3. *Find the whole area of the ellipse*

$$\frac{x^2}{a^2}+\frac{y^2}{b^2}=1.$$

Solution. The ellipse $\frac{x^2}{a^2}+\frac{y^2}{b^2}=1$ is symmetrical about both axes and whose trace is shown in fig. 5.

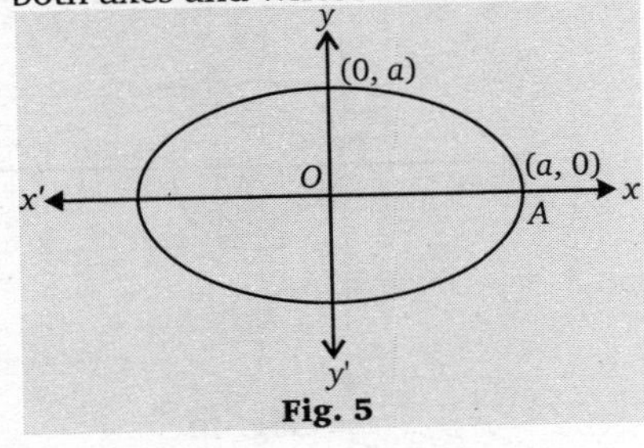

Fig. 5

$\therefore$ Required area $= 4\int_0^a y\,dx$

$$= 4\int_0^a b\sqrt{\left(1-\frac{x^2}{a^2}\right)}dx = 4b\int_0^a \sqrt{\left(1-\frac{x^2}{a^2}\right)}dx.$$

Let us put $x = a\sin\theta$.

$\therefore$ $dx = a\cos\theta\,d\theta$ and θ varies from 0 to $\pi/2$.

$$\therefore \quad A. = 4b\int_0^{\pi/2}\cos\theta . a\cos\theta\,d\theta$$

$$= 4ab\int_0^{\pi/2}\cos^2\theta\,d\theta$$

$$= 4ab\left[\frac{(2-1)}{2}.\frac{\pi}{2}\right] \quad \text{(By Walli's formula)}$$

$$= \pi\,ab \text{ sq. units.}$$

Example 4. *Find the area of the loop of the curve*

$ay^2 = x^2(a-x)$. (SVTU-2009, Osmania-2000)

Solution. Since the curve is symmetrical about x-axis and $y = 0$ when $x = 0$ and $x = a$ so the loop exists between $x = 0$ and $x = a$ as shown in fig. 6.

$\therefore$ Required area

$$= 2\int_0^a y\,dx$$

$$= 2\int_0^a x\sqrt{\frac{a-x}{a}}\,dx.$$

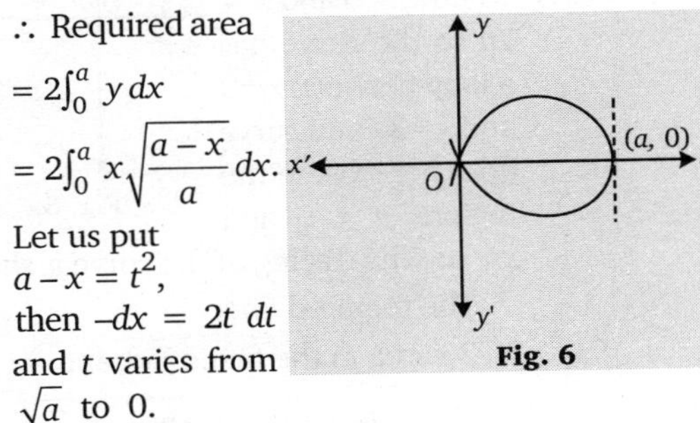

Fig. 6

Let us put $a - x = t^2$, then $-dx = 2t\,dt$ and t varies from $\sqrt{a}$ to 0.

$$\therefore \quad = \frac{2}{\sqrt{a}}\int_{\sqrt{a}}^0 (a-t^2)t(-2t\,dt)$$

$$= \frac{2}{\sqrt{a}}.2\int_0^{\sqrt{a}}(at^2-t^4)\,dt$$

$$= \frac{4}{\sqrt{a}}\left[\frac{at^3}{3}-\frac{t^5}{5}\right]_0^{\sqrt{a}} = \frac{4}{\sqrt{a}}\left[\frac{a^2\sqrt{a}}{3}-\frac{a^2\sqrt{a}}{5}\right]$$

$$= \frac{8}{15}a^2 \text{ units.}$$

Example 5. *Find the area of the one loop of the curve*

$$y^2 = x^2(a^2 - x^2).$$

Solution. Obviously the given curve is symmetrical about both axes and $y = 0$ when $x = 0, x = \pm a$. Thus the curve has two loops one of them lies between $x = 0$ and $x = a$ and other lies between $x = -a$ and $x=0$ as shown in fig. 7.

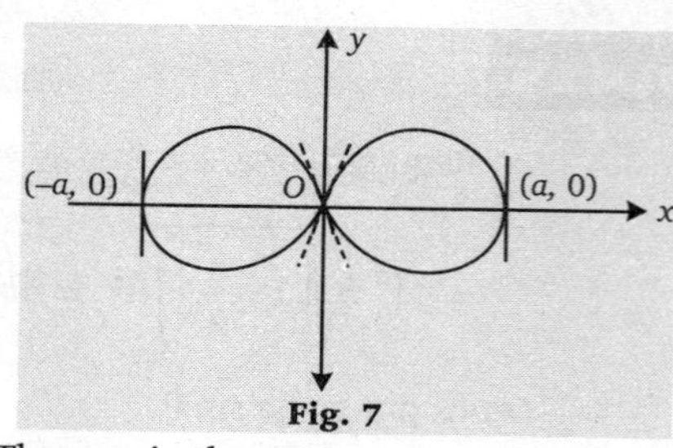

Fig. 7

∴ The required area

$$= 2\int_0^a y\,dx = 2\int_0^a x\sqrt{(a^2 - x^2)}\,dx$$

Let us put $a^2 - x^2 = t^2$, then $x\,dx = -t\,dt$ and t lies between a and 0. Thus the required area is

$$= 2\int_0^a t(-t\,dt) = -2\int_0^a t^2 dt$$

$$= 2\int_0^a t^2 dt = 2\left[\frac{t^3}{3}\right]_0^a = \frac{2}{3}a^3 \text{ sq. units.}$$

Example 6. *Find the whole area of the curve $a^2y^2 = x^3(2a-x)$.*

Solution . Clearly, the curve is symmetrical about x-axis and $y=0$ when $x = 0$ and $x = 2a$ so the curve has a loop between $x=0$ and $x=2a$ and curve does not exist in the regions $x < 0$ and $x>2a$. The tracing of the curve is shown in fig. 8.

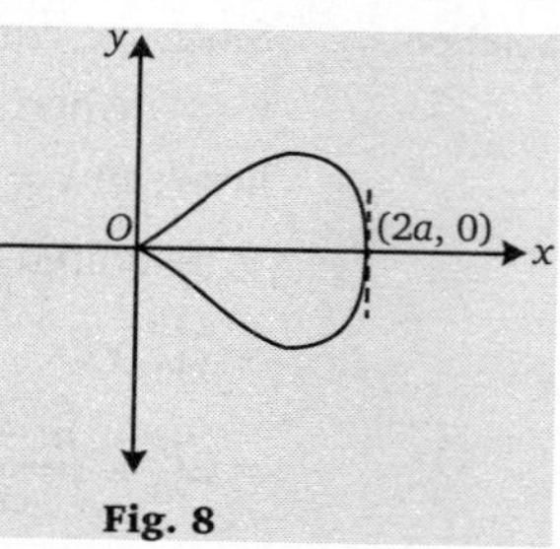

Fig. 8

∴ The required area

= 2×area in the first quadrant

$$= 2\int_0^{2a} y\,dx = 2\int_0^{2a} \frac{x}{a}\sqrt{(2ax - x^2)}\,dx$$

$$= \frac{2}{a}\int_0^{2a} x\sqrt{(2ax - x^2)}\,dx.$$

Let us put $x=2a\sin^2\theta$, then $dx=4a\sin\theta\cos\theta d\theta$ and θ varies from $\theta=0$ to $\theta=\pi/2$.

Thus, we have

$$= \frac{2}{a}\int_0^{\pi/2} 2a\sin^2\theta . 2a\sin\theta\cos\theta . 4a\sin\theta\cos\theta\,d\theta$$

$$= 32a^2\int_0^{\pi/2} \sin^4\theta\cos^2\theta\,d\theta$$

$$= 32a^2\left[\frac{(4-1)(4-3)}{6.4.(6-4)}.(2-1).\frac{\pi}{2}\right]$$

[By Walli's formula]

$$= \pi a^2 \text{ sq. units.}$$

Example 7. *Find the common area between the curves $y^2=4ax$ and $x^2 = 4ay$.* (SVTU-2008, Kurukshetra-2005)

Solution . The curve $y^2 = 4ax$ is symmetrical about x-axis and the curve $x^2 = 4ay$ is symmetrical about y-axis. Both curves intersects at two points (0,0) and (4a, 4a) which are obtained by solving both equations of curves. The tracing of the curves is shown in fig. 9.

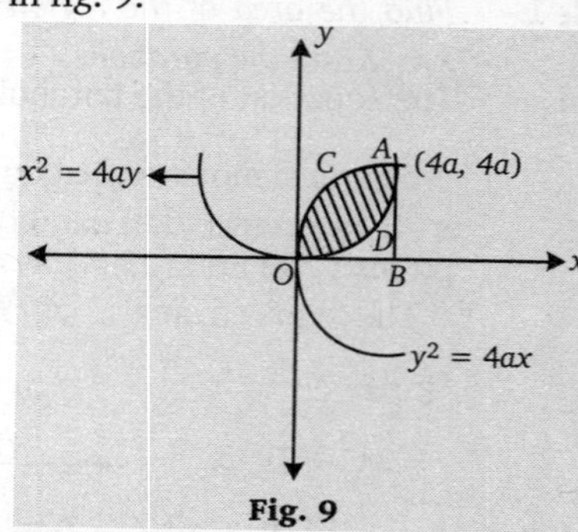

Fig. 9

The required area

= area of $OBACO$ – area of $OBADO$ ...(1)

Now area of $OBACO$

$$= \int_0^{4a} y\,dx, \text{ where } y^2 = 4ax$$

$$= \int_0^{4a} \sqrt{4ax}\,dx = 2\sqrt{a}\left[\frac{2}{3}x^{3/2}\right]_0^{4a} = \frac{32}{3}a^2$$

and area of $OBADO = \int_0^{4a} y\,dx$,

$$\text{where } y^2 = \frac{x^2}{4a} = \int_0^{4a}\frac{x^2}{4a}dx$$

$$= \frac{1}{4a}\left[\frac{x^3}{3}\right]_0^{4a} = \frac{16a^2}{3}.$$

From (1), we get

$$\text{Required area} = \frac{32}{3}a^2 - \frac{16}{3}a^2 = \frac{16}{3}a^2 \text{ sq. units.}$$

Example 8. *Find the area common of the curves $y^2 =ax$, $x^2 + y^2 = 4ax$.*

Solution . Both the curves are symmetrical about x-axis and intersect, then we have

$$x^2+y^2 = 4ax, y^2 = ax.$$

∴ $x^2+ax = 4ax$

or $x^2 - 3ax = 0$

or $x = 0, x = 3a$

∴ If $x = 0$, $y = 0$ and $x = 3a, y = \pm a\sqrt{3}$

Thus both curves intersect at three points (0,0) and $(3a \pm a\sqrt{3})$. The tracing of the curves are shown in fig. 10.

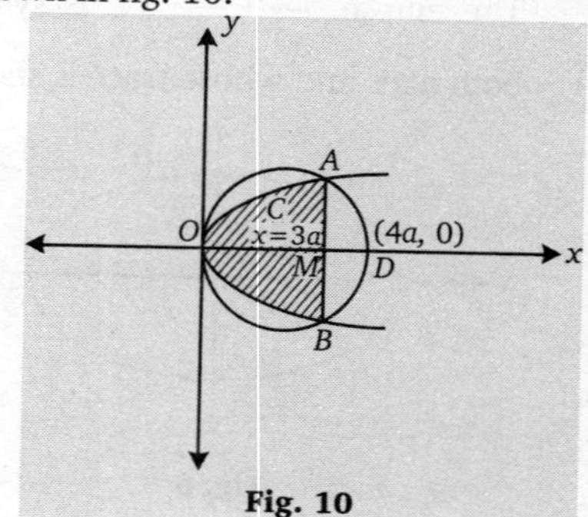

Fig. 10

The required area = 2[area $OMACO$+area $MDAM$] ...(1)

Now area $OMACO = \int_0^{3a} y\,dx$, where $y^2 = ax$

$$= \int_0^{3a} \sqrt{ax}\,dx$$

$$= \sqrt{a}\left[\frac{2}{3}x^{3/2}\right]_0^{3a} = 2\sqrt{3}\,a^2$$

and area $MDAM = \int_{3a}^{4a} y\,dx$

(where $y = \sqrt{4ax - x^2}$ and at M, $x=3a$, at D, $x=4a$.)

$$= \int_{3a}^{4a} \sqrt{4ax - x^2}\,dx$$

$$= \int_{3a}^{4a} \sqrt{[(2a)^2 - (x-2a)^2]}\,dx$$

$$= \left[\frac{x-2a}{2}\sqrt{[(2a)^2-(x-2a)^2]} + \frac{(2a)^2}{2}\sin^{-1}\frac{x-2a}{2a}\right]_{3a}^{4a}$$

$$= 2a^2\sin^{-1}1 - \frac{a}{2}.a\sqrt{3} - 2a^2\sin^{-1}\frac{1}{2}$$

$$= \left(a^2\pi - \frac{\sqrt{3}}{2}a^2 - \frac{a^2\pi}{3}\right) = \frac{2\pi a^2}{3} - \frac{\sqrt{3}}{2}a^2.$$

From (1), we get

$$\text{Required area} = 2\left[2\sqrt{3}a^2 + \frac{2\pi}{3}a^2 - \frac{\sqrt{3}}{2}a^2\right]$$

$$= 2\left(\frac{2\pi}{3}a^2 + \frac{3\sqrt{3}}{2}a^2\right)$$

$$= a^2\left(\frac{4\pi}{3} + 3\sqrt{3}\right) \text{ sq. units.}$$

Example 9. *Find the area included between the parabola $x^2 = 4ay$ and the curve $y = 8a^3/(x^2+4a^2)$.*

Solution . Both the curves are symmetrical about y-axis and intersect, then we have

$$(x^2 + 4a^2).\frac{x^2}{4a} = 8a^3$$

or $\quad x^4 + 4a^2x^2 - 32a^4 = 0$

$$(x^2 + 8a^2)(x^2 - 4a^2) = 0$$

or $\quad x^2 + 8a^2 = 0, x^2 - 4a^2 = 0$

$$x = \pm 2\sqrt{2}ai \text{ or } x = \pm 2a.$$

$x = \pm 2\sqrt{2}\,ai$ are not possible point, therefore both curves intersect at two points $(2a, a)$ and $(-2a, a)$. The tracing of the curves are shown Fig. 11.

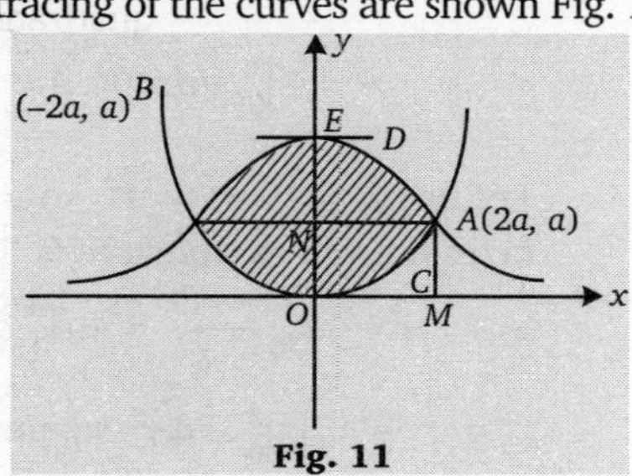

Fig. 11

The required area

= 2[area $OCANO$ + area $NADEN$] ...(1)

Now area $OCANO$ + area $NADEN$

= – area $OMACO$ + aera $OMADEO$

∴ From (1), we get

Required area = 2[–area $OMACO$+area $OMADEO$] ...(2)

Now, area $OMACO = \int_0^{2a} y\,dx$, where $y = \dfrac{x^2}{4a}$

$$= \int_0^{2a}\frac{x^2}{4a}dx = \frac{1}{4a}\left[\frac{x^3}{3}\right]_0^{2a}$$

$$= \frac{1}{4a}\left[\frac{8a^3}{3}\right] = \frac{2}{3}a^2$$

and area $OMADEO = \int_0^{2a} y\,dx$, $y = \dfrac{8a^3}{(x^2+4a^2)}$

$$= \int_0^{2a}\frac{8a^3}{x^2+4a^2}dx$$

$$= 8a^3\int_0^{2a}\frac{dx}{x^2+4a^2} = 8a^3\left[\frac{1}{2a}\tan^{-1}\frac{x}{2a}\right]_0^{2a}$$

$$= 8a^3\left[\frac{1}{2a}\tan^{-1}1 - 0\right] = 4a^2.\frac{\pi}{4} = a^2\pi.$$

From (2), we get

$$\text{Required area} = 2\left[-\frac{2}{3}a^2 + a^2\pi\right] = a^2\left(-\frac{4}{3} + 2\pi\right)$$

$$= a^2\left(2\pi - \frac{4}{3}\right) \text{ sq. units.}$$

Example 10. *Find the area included between the curve $x = a(t+\sin t)$, $y = a(1-\cos t)$ and its base.* (VTU- 2000)

Solution . Since the equation of the curve is

$$x = a(t + \sin t)$$
$$y = a(1 - \cos t)\ .$$

Obviously, $y = 0$ when $\cos t = 1$ *i.e.*, $t = 0$ and $x = 0$. Thus the curve passes through the point (0,0). Also

$$\frac{dy}{dx} = \tan t/2 \text{ at } t = 0,\ \frac{dy}{dx} = 0.$$

Therefore, the tangent at (0,0) is the x-axis. The tracing of the curve is shown in fig 12.

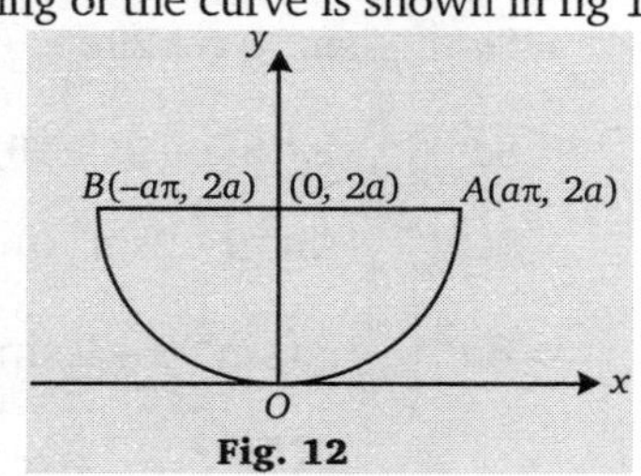

Fig. 12

Obviously, curve is symmetrical about y-axis so the required area is given by

$A = 2\times$area bounded by the axis of y and $y=2a$

$A = 2\int_{y=0}^{y=2a} x\,dy.$

Since $x = a(t+\sin t), y=a(1-\cos t)$

$\therefore\ dy = a \sin t\, dt.$

$$\therefore\ A = 2\int_0^{\pi} a(t+\sin t).a\sin t\,dt$$

$$= 2a^2\int_0^{\pi}(t\sin t + \sin^2 t)dt$$

$$= 2a^2\left[\int_0^{\pi} t\sin t\,dt + \int_0^{\pi}\sin^2 t\,dt\right]$$

$$= 2a^2\left[(-t\cos t + \sin t)_0^{\pi} + \frac{1}{2}\left(t - \frac{\sin 2t}{2}\right)_0^{\pi}\right]$$

$$= 2a^2\left[\pi + \frac{\pi}{2}\right] = 3\pi a^2 \text{ sq. units.}$$

Example 11. *Find the area bounded by the curve $y^2(a+x)=x^2(a-x)$ and its asymptotes.*

Solution . Obviously, the given curve is symmetrical about x-axis and cuts the x-axis only at two points (0,0) and $(a, 0)$ and $x = -a$ is the asymptote of this curve. This curve has a loop lying between $x = 0$ and $x = a$. Therefore, the tracing of this curve is shown in fig. 13.

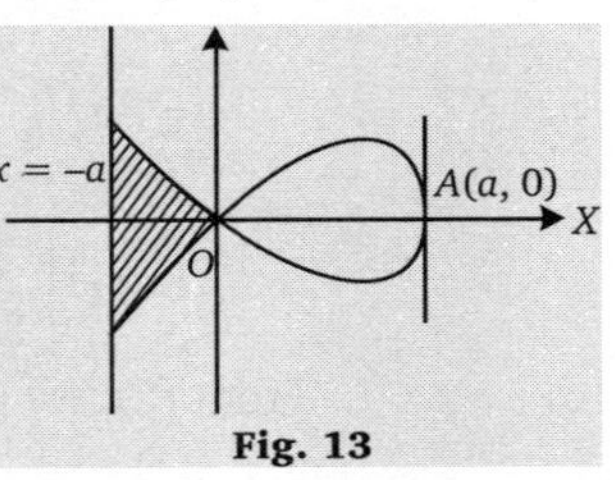

Fig. 13

The required area $= 2\int_{-a}^{0} y\,dx$

$$A = 2\int_{-a}^{0} x\sqrt{\frac{a-x}{a+x}}dx.$$

Let us put $x = a\cos 2\theta$ so $dx = -2a\sin 2\theta\, d\theta$ and θ will take the value from $\pi/2$ to $\pi/4$. Therefore, we have

$$A = 2\int_{\pi/2}^{\pi/4} a\cos 2\theta.\sqrt{\frac{1-\cos 2\theta}{1+\cos 2\theta}}.(-2a\sin 2\theta)d\theta$$

$$= -4a^2\int_{\pi/4}^{\pi/2}\frac{\sin\theta}{\cos\theta}.\cos 2\theta.\sin 2\theta\, d\theta$$

$$= 4a^2\int_{\pi/4}^{\pi/2}\frac{\sin\theta}{\cos\theta}.\cos 2\theta.2\sin\theta\cos\theta\, d\theta$$

$$= 8a^2\int_{\pi/4}^{\pi/2}\sin^2\theta\cos 2\theta\, d\theta$$

$$= 8a^2\int_{\pi/4}^{\pi/2}\sin^2\theta(1-2\sin^2\theta)d\theta$$

$$= 8a^2\int_{\pi/4}^{\pi/2}(\sin^2\theta - 2\sin^4\theta)d\theta$$

$$= 8a^2\left[\int_{\pi/4}^{0}(\sin^2\theta - 2\sin^4\theta)d\theta + \int_0^{\pi/2}(\sin^2\theta - 2\sin^4\theta)d\theta\right]$$

$$= 8a^2\left[\int_{\pi/4}^{0}\sin^2\theta(1-2\sin^2\theta)d\theta + \int_0^{\pi/2}\sin^2\theta\, d\theta - 2\int_0^{\pi/2}\sin^4\theta\, d\theta\right]$$

$$= 8a^2\left[\int_{\pi/4}^{0}\frac{(1-\cos 2\theta)}{2}.\cos 2\theta\, d\theta + \frac{(2-1)}{2}.\frac{\pi}{2} - 2.\frac{(4-1)(4-3)}{4.2}\frac{\pi}{2}\right]$$

(Walli's formula)

$$= 8a^2\left[\frac{1}{2}\int_{\pi/4}^{0}(\cos 2\theta - \cos^2 2\theta)d\theta + \frac{\pi}{4} - \frac{3\pi}{8}\right].$$

Now let us put $2\theta = \phi$ so $d\theta = \frac{1}{2}d\phi$ and ϕ takes the value from $\pi/2$ to 0. Then

$$= 8a^2\left[\frac{1}{2}\int_{\pi/2}^{0}(\cos\phi - \cos^2\phi)\frac{1}{2}d\phi - \frac{\pi}{8}\right]$$

$$= 8a^2\left[\frac{1}{4}\int_{\pi/2}^{0}\cos\phi\, d\phi - \frac{1}{4}\int_{\pi/2}^{0}\cos^2\phi\, d\phi - \frac{\pi}{8}\right]$$

$$= 8a^2\left[\frac{1}{4}(\sin\phi)_{\pi/2}^{0} + \frac{1}{4}\left(\frac{2-1}{2}.\frac{\pi}{2}\right) - \frac{\pi}{8}\right]$$

$$= 8a^2\left[-\frac{1}{4} + \frac{\pi}{16} - \frac{\pi}{8}\right] = 8a^2\left[-\frac{1}{4} - \frac{\pi}{16}\right]$$

$$= -a^2\left(2 + \frac{\pi}{2}\right).$$

Hence, $A = a^2\left(2 + \frac{\pi}{2}\right)$. (Taking magnitude value)

Example 12. *Find the whole area between the curve $x^2y^2=a^2(y^2-x^2)$ and its asymptotes.* (VTU-2007)

Solution . The given curve is symmetrical about both axes and curve passes through only (0,0). The tangents at (0, 0) are given by

$$y^2 - x^2 = 0$$

or $\quad y = \pm x.$

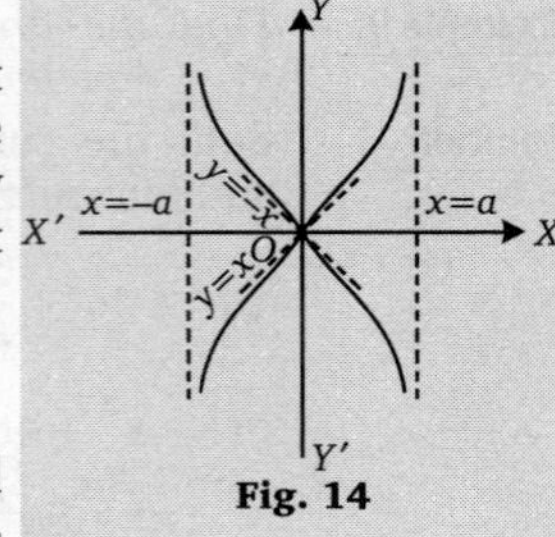

Fig. 14

Thus we obtained two distinct tangents at (0, 0) so (0, 0) is a node and its two real asymptotes are $x = \pm a$. The tracing of the curve is shown in fig. 14.

The required area = 4 × area between curve and asymptote in the first quadrant.

$$\therefore\quad A = 4\int_0^a y\,dx \text{ or } A = 4\int_0^a \frac{ax\,dx}{\sqrt{(a^2-x^2)}}$$

Let us put $a^2 - x^2 = t^2$, so $x\,dx = -t\,dt$ and t takes the values from a to 0. Then

$$A = 4a\int_0^a \frac{-t\,dt}{t} = 4a\int_0^a dt$$

$$= 4a[t]_0^a = 4a^2 \text{ sq. units.}$$

Example 13. *Show that the area bounded by the cissoid* $x = a\sin^2 t,\ y = \dfrac{a\sin^3 t}{\cos t}$ *and its asymptote is* $3\pi a^2/4$.

Solution . Equation of the curve is

$$x = a\sin^2 t,\quad y = a\frac{\sin^3 t}{\cos t}.$$

Eliminating t between above equations, we get

$$y^2(a-x) = x^3 \qquad ...(1)$$

Obviously, the curve is symmetrical about x-axis and passes through only (0, 0). The tangents at (0, 0) are $y^2 = 0$

or $y = 0$

and $y = 0$.

That is, there are two same tangents at (0, 0). Thus (0, 0) is a cusp. Also $x=a$ is the asymptotes of the curve.

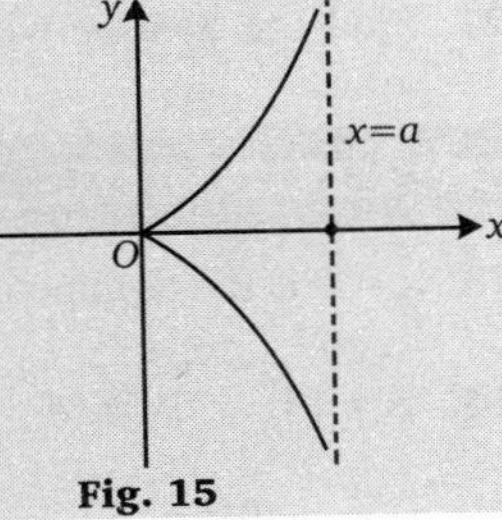

Fig. 15

The tracing of the curve is shown in fig. 15.

The required area = 2 × area bounded by the curve and its asymptote in the first quadrant.

$\therefore \quad A = 2\int_0^a y\,dx$

or $\quad A = 2\int_0^a x\sqrt{\dfrac{x}{a-x}}\,dx \qquad$ [Using (1)]

Let us put $x = a\sin^2\theta$ so $dx = 2a\sin\theta\cos\theta\,d\theta$ and θ takes the value from $\theta = 0$ to $\theta = \pi/2$. Then, we have

$$A = 2\int_0^{\pi/2} a\sin^2\theta.\frac{\sin\theta}{\cos\theta}.2a\sin\theta\cos\theta\,d\theta$$

$$= 4a^2\int_0^{\pi/2}\sin^4\theta\,d\theta$$

$$= 4a^2\left[\frac{(4-1)(4-3)}{4.2}.\frac{\pi}{2}\right] \text{ [By Walli's formula]}$$

$\therefore\ A = 3\pi a^2/4.$

Example 14. *Find by double integration the area of the region closed by the curves* $x^2 + y^2 = a^2,\ x + y = a$ *(in the first quadrant).*

Solution . The given equations of the circle

$$x^2 + y^2 = a^2.$$

The straight line $x + y = a$ shows in fig. 16.

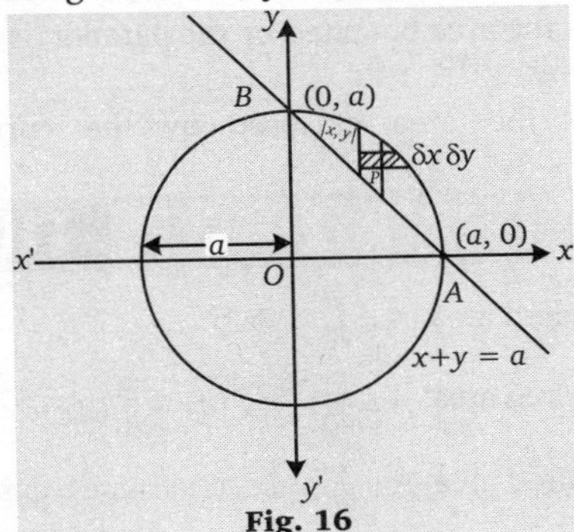

Fig. 16

To find the required area of the arc AB and the line AB by double integration take any point $P(x, y)$ and consider an area $\delta x\,\delta y$ at P. The arc of the circle $x^2 + y^2 = a^2$ and then moving x from 0 to a.

$\therefore$ The required area $= \int_{x=0}^{a}\int_{y=(a-x)}^{\sqrt{a^2-x^2}} dx\,dy.$

$$= \int_0^a [y]_{a-x}^{\sqrt{a^2-x^2}} dx = \int_0^a \left[\sqrt{a^2-x^2} - (a-x)\right]dx$$

$$= \left[\left\{\frac{1}{2}x\sqrt{(a^2-x^2)} + \frac{1}{2}a^2\sin^{-1}\left(\frac{x}{a}\right)\right\} - ax + \frac{1}{2}x^2\right]_0^a$$

$$= \frac{1}{2}a^2.\left(\frac{\pi}{2}\right) - a^2 + \frac{1}{2}a^2$$

$$= \frac{1}{a}a^2\left(\frac{\pi}{2} - 1\right) = \frac{1}{4}a^2(\pi - 2).$$

EXERCISE 17.1

1. Find the area of the region bounded by the following curves, and the axis of x and the given ordinates :
   (i) $y = \log x;\quad x = a,\quad x = b$
   (b) $y = c\cosh(x/c);\ x = 0,\quad x = a$
   (c) $y = \sin^2 x;\quad x = 0,\quad x = \pi/2$
2. Find the area of the region bounded by the parabola $y^2 = 4x$ and the line $y = 2x$.
3. Find the area of a quadrant of the ellipse $x^2/a^2 + y^2/b^2 = 1$. (VTU-2003, Kerala-2005)
4. Find the area of a loop of the curve $xy^2 + (x+a)^2(x+2a) = 0$.
5. Find the area of the loop of the curve $3ay^2 = x(x-a)^2$. (Rajasthan-2005)
6. Find the whole area of the curve $a^2x^2 = y^3(2a-y)$. (Nagpur-2009)
7. Prove that the area of a loop of the curve $a^4y^2 = x^4(a^2-x^2)$ is $\pi a^2/8$.
8. Find the area bounded by the curve $xy^2 = a^2(a-x)$ and y-axis.
9. Find the area of the loop of the curve $y^2 = x(x-1)^2$.
10. Find the whole area of the curve $y^2 = x^2(a^2 - x^2)$.
11. Find the area between the curve $y^2(a-x) = x^2$ and its asymptote.
12. Find the area between the curve $y^2(2a-x) = x^3$ and its asymptote. (VTU 2003)
13. Find the area between the curve $xy^2 = 4a^2(2a-x)$ and its asymptote.

14. Find the area between the $y^2(a-x)=x^3$ and its asymptote. Also find the ratio in which the ordinate $x=a/2$ divides this area.

15. Find the area of the region bounded by the parabola $y^2= 4ax$ and $x^2=4by$.

16. Find the area bounded by the curves $x^2+y^2 \le 2ax$ and $y^2 \ge ax$, $a>0, x>0, y>0$.

17. Find the area bounded by the curves $y \ge x^2$ and $y \le |x|$.

18. Find the area of the segment cut off from the parabola $y^2= 4x$ by the line $y = 8x-1$.

19. Find the area bounded by the parabola $4y = 3x^2$ and the line $3x - 2y + 12 = 0$.

20. Find the area enclosed by the curves $y^2 \le 3x$ and $3x^2 + 3y^2 \ge 16$.

21. Find the area included between the cycloid $x = a(\theta - \sin\theta)$, $y = a(1-\cos\theta)$ and its base. (Gorakhpur-1999)

22. Find the area enclosed between the curve $y = x^3$ and the line $y = x$.

23. Find whole area of the curve $x^{2/3}+ y^{2/3} = a^{2/3}$. (VTU- 2005)

24. Find the area of the loop of the curve $x^3+ y^3 = 3axy$.

25. Find the area enclosed between the parabola $y^2 =4a(x+a)$ and $y^2 = -4a(x-a)$.

26. Find the area of the smaller portion enclosed by the curves $x^2+y^2=9$ and $y^2=8x$.

27. Find the area between the parabola $y= 4x-x^2$ and the line $y=x$. (VTU-2010, SVTU-2008, UPTU-2008)

Hint to Selected Problems

1. Required area $A = \int_a^b y\,dx$.

2. Required area $A = \int_0^1 (\sqrt{4x})dx - \int_0^1 (2x)dx$.

3. Required area $A = \int_0^a y\,dx$. (Do same example 3.)

4. $A = 2\int_{-2a}^{a} y\,dx = \pm 2\int_{-2a}^{a} \frac{\sqrt{(x+a)^2(x+2a)}}{-x}dx$.

Put $-x = t$. Then integrate.

5. $3ay^2 = x(x-a)^2 \Rightarrow y = (x-a)\sqrt{\frac{x}{3a}}$. Then $A = 2\int_0^a y\,dx$.

6. Here $x = 0$, 0 be two real tangents. Therefore $A = 2\int_0^{2a} x\,dy$.

8. $A = 2\int_0^{\infty} x\,dy$.

9. $A = 2\int_0^1 y\,dx$.

12. $x = 2a$ is the asymptotes of the given curve, therefore, the required area $A = 2\int_0^{2a} y\,dx$.

13. $x = 0$ is the asympotoes. Therefore, $A = 2\int_0^{2a} y\,dx$.

14. $A = 2\int_0^a y\,dx \Rightarrow A = 2\int_0^a x\sqrt{\frac{x}{a-x}}dx$

$A_1 = 2\int_0^{a/2} y\,dx \Rightarrow A_1 = 2\int_0^{a/2} x\sqrt{\frac{x}{a-x}}dx$

Put $x = a\sin^2\theta$ and integrate.

16. $A = \int_0^a (\sqrt{2ax - x^2})dx - \int_0^a (\sqrt{ax})dx$.

17. $A = 2\int_0^1 x\,dx - \int_0^1 x^2\,dx$.

21. Do same as example 10.

24. Put $x = r\cos\theta$, $y = r\sin\theta$, in the equation of the given curve, we get

$$r = \frac{3a\sin\theta\cos\theta}{(\cos^3\theta + \sin^3\theta)}.$$

Required area $= \int_0^{\pi/2} \frac{r^2}{2} d\theta$.

ANSWERS

1.(i) $b\log(b/e) - a\log(a/e)$ (ii) $c^2\sinh(a/c)$ (iii) $\pi/4$ **2.** 1/3 **3.** $\frac{1}{4}\pi ab$ **4.** $2a^2(1-\pi/4)$ **5.** $8a^2/(15\sqrt{3})$ **6.** πa^2 **8.** πa^2 **9.** 8/15 **10.** $4a^3/3$ **11.** $(8/3)a^3$ **12.** $3\pi a^2$ **13.** $4\pi a^2$ **14.** $3\pi a^2/4$; $(3\pi-8) : (3\pi+8)$ **15.** (16/3) ab **16.** $(a^2/12)(3\pi-8)$ **17.** 1/3 **18.** 9/64 **19.** 27 **20.** $\frac{4}{3}a^{3/2} + \frac{8\pi}{3} - \frac{a}{2}\sqrt{\left(\frac{16}{3}\right) - a^2} - \frac{8}{3}\sin^{-1}\left(\frac{a}{4\sqrt{3}}\right)$, where $a = (-a + \sqrt{273})/6$ **21.** $3\pi a^2$ **22.** 1/2 **23.** $\frac{3}{8}\pi a^2$ **24.** $3a^2/2$ **25.** $(16/3)a^2$ **26.** $2\left[\frac{\sqrt{2}}{3} + \frac{9\pi}{4} - \frac{9}{2}\sin^{-1}\left(\frac{1}{3}\right)\right]$ **27.** 9/2

17.3 PROBLEM BASED ON POLAR FORM

Example 1. *Find the area of the cardioid $r = a(1+\cos\theta)$.* (VTU-2008)

Solution . The equation of the cardioid $r = a(1+\cos\theta)$ is symmetrical about the initial line and $r = 0$ when $\cos\theta = -1$ *i.e.*, $\theta = \pi$ and r is maximum when $\cos\theta = 1$ *i.e.*, $\theta = 0$. Thus the tracing of the curve is shown in fig. 17.

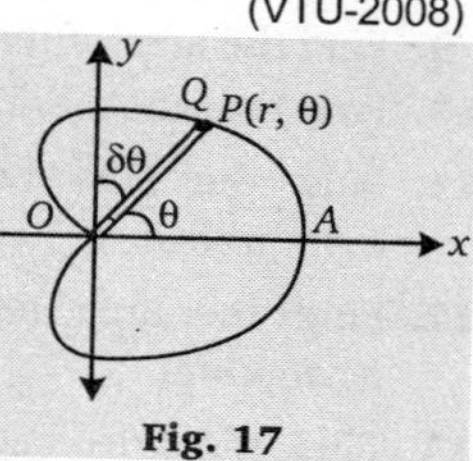

Fig. 17

Let A be the area of the cardiod $r = a(1+\cos\theta)$. This area A is the twice the area of the upper half of the curve between $\theta = 0$ and $\theta = \pi$.

Now the required area

$$A = 2\int_0^{\pi} \frac{1}{2}r^2\,d\theta = \int_0^{\pi} a^2(1+\cos\theta)^2 d\theta$$

$$= a^2\int_0^{\pi} (2)^2\cos^4\frac{\theta}{2}d\theta. .$$

$$= 4a^2\int_0^{\pi}\cos^4\theta/2d\theta.$$

Let us put $\theta/2 = \phi$ so $d\theta=2\,d\phi$ and ϕ runs from 0 to $\pi/2$.

$$\therefore\ A = 8a^2\int_0^{\pi/2}\cos^4\phi\, d\phi$$

$$= 8a^2\left[\frac{(4-1)(4-3)}{4.2}\cdot\frac{\pi}{2}\right] \text{ [By Walli's formula]}$$

$$= \frac{3}{2}\pi a^2.$$

Example 2. *Find the area of a loop of the curve $r^2 = a^2\cos 2\theta$.* (VTU- 2006)

Solution . The curve is symmetrical about pole and initial line both and $r=0$ when $\cos 2\theta = 0$ *i.e.*, $\theta = \pm\pi/4$. Thus a loop of the curve lies betwen $\theta = -\pi/4$ and $\theta = \pi/4$.

Let A be the area of this loop. Then

$$A = \int_{-\pi/4}^{\pi/4}\frac{1}{2}r^2 d\theta = \frac{1}{2}\int_{-\pi/4}^{\pi/4} a^2\cos 2\theta\, d\theta$$

$$= \frac{a^2}{2}\int_{-\pi/4}^{\pi/4}\cos 2\theta\, d\theta$$

$$= \frac{a^2}{2}\left[\frac{1}{2}\sin 2\theta\right]_{-\pi/4}^{\pi/4} = \frac{a^2}{2}\left[\frac{1}{2}+\frac{1}{2}\right] = \frac{a^2}{2}.$$

Example 3. *Find the whole area of the curve $r = a\sin 3\theta$.*

Solution . The given curve is not symmetrical and $r = 0$ when $\sin 3\theta = 0$ *i.e.*, $\theta = 0$, $\theta = \pm\pi/3$ and r is maximum when $\sin 3\theta = 1$ *i.e.*, $\theta = \pi/6$. Also this curve has three loops so the whole area of the curve is thrice the area of one loop. Let whole area be A. Then

$$A = 3\int_{\theta=0}^{\theta=\pi/3}\frac{1}{2}r^2 d\theta$$

$$= \frac{3}{2}\int_0^{\pi/3} a^2\sin^2 3\theta\, d\theta \quad (\because r = a\sin 3\theta)$$

$$= \frac{3}{2}a^2\int_0^{\pi/3}\left(\frac{1-\cos 6\theta}{2}\right)d\theta$$

$$= \frac{3}{4}a^2\left[\theta - \frac{1}{6}\sin 6\theta\right]_0^{\pi/3} = \frac{3}{4}a^2\left[\frac{\pi}{3}\right]$$

$$A = \frac{1}{4}\pi a^2.$$

Example 4. *Find the area common to the circle $r = a$ and the cardioid $r = a(1+\cos\theta)$.*

Solution . The points of intersection of the given curves $r = a$ and $r = a(1+\cos\theta)$ are given by $\theta = \pm\pi/2$.

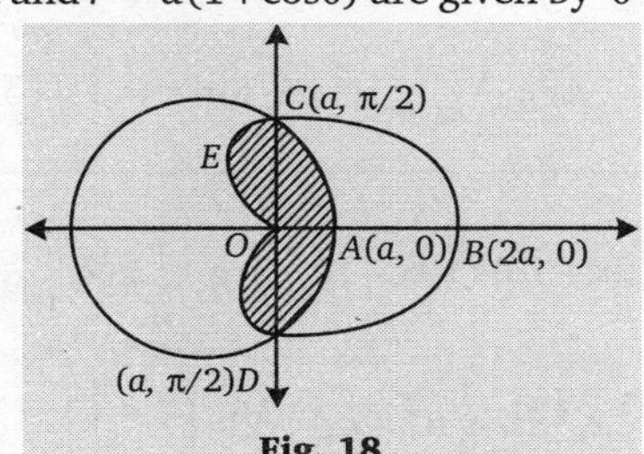

Fig. 18

Required area $= 2$ [area ACO + area $CEOC$]

Now area $ACO = \int_0^{\pi/2}\frac{1}{2}r^2 d\theta$ for $r = a$

$$= \frac{1}{2}a^2\int_0^{\pi/2}d\theta = \frac{\pi a^2}{4}$$

Area $CEOC = \int_{\pi/2}^{\pi}\frac{1}{2}r^2 d\theta$ for $r = a(1+\cos\theta)$

$$= \frac{a^2}{2}\int_{\pi/2}^{\pi}(1+\cos\theta)^2 d\theta$$

$$= \frac{a^2}{2}\int_{\pi/2}^{\pi}(1+\cos^2\theta+2\cos\theta)d\theta$$

$$= \frac{a^2}{2}\left[\theta+\frac{1}{2}\left\{\theta+\frac{\sin 2\theta}{2}\right\}+2\sin\theta\right]_{\pi/2}^{\pi}$$

$$= \frac{a^2}{2}\left[\frac{\pi}{2}+\frac{1}{2}\left(\frac{\pi}{2}\right)-2\right] = \frac{3\pi a^2}{8}-a^2.$$

Hence, required area $= 2\left(\frac{\pi a^2}{4}+\frac{3\pi a^2}{8}-a^2\right)$

$$= a^2\left(\frac{5\pi}{4}-2\right).$$

Example 5. *Find the area of common to the cardioids $r = a(1+\cos\theta)$ and $r = a(1-\cos\theta)$.* (VTU-2006, Kurukshetra-2006)

Solution . Clearly both the cardioids are symmetrical about the initial line OX and intersect at B and B' as shown in the adjoining figure.

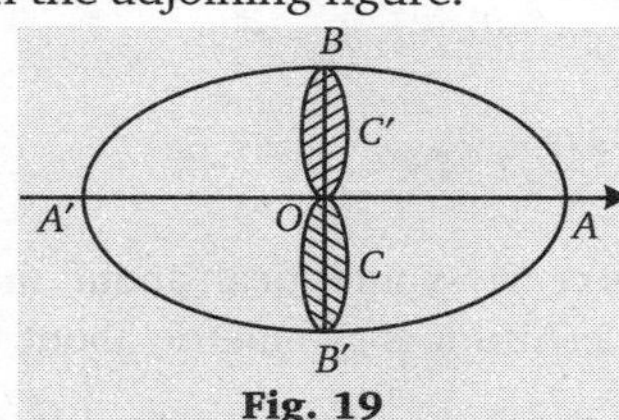

Fig. 19

Required area $= 2$. Area $OC'BCO$

$= 2$[area $OC'BO$ + area $OBCO$]

$$= 2\left\{\left[\int_0^{\pi/2}\frac{1}{2}r^2 d\theta\right]_{r=a(1-\cos\theta)} + \left[\int_{\pi/2}^{\pi}\frac{1}{2}r^2 d\theta\right]_{r=a(1+\cos\theta)}\right\}$$

$$= a^2\int_0^{\pi/2}(1-\cos\theta)^2 d\theta + a^2\int_{\pi/2}^{\pi}(1+\cos\theta)^2 d\theta$$

$$= a^2\left\{\int_0^{\pi/2}(1-2\cos\theta+\cos^2\theta)d\theta + \int_{\pi/2}^{\pi}(1+2\cos\theta+\cos^2\theta)d\theta\right\}$$

$$= a^2\left\{\int_0^{\pi}(1+\cos^2\theta)d\theta - 2\int_0^{\pi/2}\cos\theta\, d\theta + 2\int_{\pi/2}^{\pi}\cos\theta\, d\theta\right\}$$

$$= a^2\left\{\int_0^{\pi}\left(1+\frac{1+\cos 2\theta}{2}\right)d\theta - 2|\sin\theta|_0^{\pi/2} + 2|\sin\theta|_{\pi/2}^{\pi}\right\}$$

$$= a^2\left\{\left|\frac{3}{2}\theta+\frac{\sin 2\theta}{4}\right|_0^{\pi} - 2(1-0)+2(0-1)\right\}$$

$$= \left(\frac{3\pi}{2}-4\right)a^2.$$

EXERCISE 17.2

1. Find the area of the parabola $r(1 + \cos\theta) = l$ belwecn $\theta = 0$, $\theta = \alpha$.
2. Find the area of one loop of the curve $r = a\cos 4\theta$.
3. Find the whole area of the curve $r^2 = a^2\cos^2\theta + b^2\sin^2\theta$.
4. Find ihe area of a loop of the curve $x^3 + y^3 = 3axy$.
5. Show that the area of the limacon $r=a+b\cos\theta$, $(a > b)$ is $\pi\left(a^2+\frac{1}{2}b^2\right)$.
6. Prove that the sum of the areas of Ihe two loops of the limacon $r = a + b\cos\theta$ $(a<b)$ is $\pi(2a^2 + b^2)/2$.
7. Find the ratio of the two parts into which the parabola $2a=r(1+\cos\theta)$ divides the area of the cardioid $r=2a(1+\cos\theta)$.
8. Find the area outside the circle $r=2a\cos\theta$ and inside the cardioid $r = a(1 + \cos\theta)$. (Kurukshetra-2006)
9. Find the area between the curve $r = a(\sec\theta + \cos\theta)$ and its asymptote.
10. Find the area of a loop of the curve $x^4 + y^4 = 4a^2xy$.
11. Prove that the area of a loop of the curve $x^6 + y^6 = a^2x^2y^2$ is $\pi a^2/12$.
12. Find the area lying between the cardioid $r = (1 - \cos\theta)$ and its double tangent. (VTU -2004)
13. Find the area common to the circles $r = a\sqrt{2}$ and $r = 2a\cos\theta$.
14. If O is the pole of the lemniscate $r^2 = a^2\cos 2\theta$ and PQ is a common tanget to its two loops. Find the area bounded by the line PQ and the arcs OP and OQ of the curve.
15. Find the area of a loop of the curve $x^4+3x^2y^2+2y^4=a^2xy$.
16. Find the total area inside $r = \sin\theta$ and outside $r = 1 - \cos\theta$. (Anna-2009)
17. Find the area of a loop of the curve $r = a\cos 3\theta + b\sin 3\theta$.

Hint to Selected Problems

1. $A = \frac{1}{2}\int_0^{\alpha} r^2\,d\theta.$

2. $A = \frac{1}{2}\int_{-\pi/8}^{\pi/8} r^2\,d\theta.$

3. The curve is symmetrical about the initial line and line $\theta = \pi/2$. Also it is symmetric about the pole $A = 4 \times$ Area lying in the first quadrant $= 4\times\frac{1}{2}\int_0^{\pi/2} r^2\,d\theta.$

4. In polar form the given equation becomes $r = \dfrac{3a\cos\theta\sin\theta}{\cos^3\theta+\sin^3\theta}$

   Then $A = \int_0^{\pi/2}\frac{1}{2}r^2\,d\theta.$

5. $A = 2\int_0^{\pi}\frac{1}{2}r^2\,d\theta.$

6. $A = 2\left[\int_0^{\cos^{-1}\left(-\frac{a}{b}\right)}\frac{1}{2}r^2 d\theta + \int_{\cos^{-1}\left(-\frac{a}{b}\right)}^{\pi}\frac{1}{2}r^2 d\theta\right].$

7. Smaller area

   $$A_1 = \frac{1}{2}\int_0^{\pi/2}\frac{4a^2}{(1+\cos\theta)^2}d\theta + \frac{1}{2}\int_{\pi/2}^{\pi}4a^2(1+\cos\theta)^2\,d\theta.$$

   Larger area A_2 = whole area $- A_1$. Then find $\dfrac{A_1}{A_2}$.

13. The points of intersection are given by $\theta = \pm\frac{\pi}{4}$.

   Required area $= 2(A_1 + A_2)$

   where $A_1 = \int_{\pi/4}^{\pi/2}\frac{1}{2}r^2\,d\theta$ for $r = 2a\cos\theta$

   and $A_2 = \int_0^{\pi/4}\frac{1}{2}r^2\,d\theta$ for $r = a\sqrt{2}$.

16. The both curve intersect at $(0, 0)$ and $(1, \pi/2)$.

   Therefore, the required area $= A_1 - A_2$

   where $A_1 = \frac{1}{2}\int_0^{\pi/2} r^2\,d\theta$ or $r = \sin\theta$

   and $A_2 = \frac{1}{2}\int_0^{\pi/2} r^2\,d\theta$ for $r = 1 - \cos\theta$.

ANSWERS

1. $\frac{1}{4}l^2\left[\tan\frac{\alpha}{2}+\frac{1}{3}\tan^3\frac{\alpha}{2}\right]$ 2. $\frac{1}{16}\pi a^2$ 3. $\frac{1}{2}\pi(a^2+b^2)$ 4. $\frac{3a^2}{2}$ 7. $(9\pi+16):(9\pi-16)$ 8. $\frac{1}{2}\pi a^2$ 9. $\frac{5}{4}\pi a^2$

10. $\frac{1}{2}\pi a^2$ 12. $\frac{1}{16}(15\sqrt{3}-8\pi)a^2$ 13. $a^2(\pi-1)$ 14. $\frac{1}{8}a^2(3\sqrt{3}-4)$ 15. $\frac{1}{4}a^2\log 2$ 16. $\left(1-\frac{\pi}{4}\right)$ 17. $(a^2+b^2)\frac{\pi}{12}$.

Objective Evaluations

Fill in the Blanks

1. The integral $\int_a^b f(x)dx$ represents ....
2. The integral $\int_{\theta_1}^{\theta_2} \frac{1}{2} r^2 \, d\theta$ represents the area of..........
3. The area of the ellipse $x^2 / a^2 + y^2 / b^2 = 1$ is ....
4. The area of the curve $x^2 + y^2 - 2ax - 2ay = 0$ is ....
5. The area common to the curves $y^2 = 4ax, x^2 = 4ay$ is ....
6. The area bounded by the curve $y = x^3$, the y-axis and the lines $y = 1$ and $y = 8$ is ....
7. The area of the curve $x^2 + 4y^2 = a$ is..........
8. The area of the cardioid $r = a\,(1 + \cos\theta)$ is ....
9. The whole area of the curve $r = a \sin 50\theta$, if A is the area of one loop of this curve, is ....
10. The area of the curve $r\sqrt{\theta} = a$ between $\theta = \alpha$ and $\theta = \beta$ is ....
11. The area of the curve $r = 2a \cos\theta$ is ......

True/False

Write 'T' for true and 'F' for false statement.

1. The integral $\int_a^b f(y)dy$ represents the area of a plane bounded by the curve $x = f(y)$ the y-axis and the lines $y = a$ and $y = b$. **(T/F)**
2. If the curve is symmetrical about x-axis, then the area of bounded portion is equal to twice the area of the upper portion of the curve above x-axis. **(T/F)**
3. The whole area of the curve $a^2y^2 = x^3\,(2a - x)$ is $\frac{\pi}{2} a^2$. **(T/F)**
4. The area included between the cycloid $x = a\,(\theta - \sin\theta)$, $y = a(1 - \cos\theta)$ and its base is $3\pi a^2$. **(T/F)**
5. The area of the ellipse $x^2/a^2 + y^2/b^2 = 1$ is πab. **(T/F)**
6. The area of the curve $x^2 + y^2 = \sqrt{2}$ is 2π. **(T/F)**
7. The area of the loop of the curve $y^2 = x\,(x - 1)^2$ is 8/15. **(T/F)**
8. The area of the segment cut off from the parabola $y^2 = 4x$ by the line $y = 8x - 1$ is 9/32. **(T/F)**
9. The curve $r = a \sin 5\theta$ has five leaves. **(T/F)**
10. The area of the cardioid $r = a(1 - \cos\theta) = \frac{3}{2}\pi a^2$. **(T/F)**
11. The area of the curve $r = a\cos\theta$ is $\frac{\pi}{2} a^2$. **(T/F)**

Multiple Choice Questions

Choose the most appropriate one.

1. The value of the integral $\int_1^2 f(x)dx$ if $f(x) = \frac{1}{x}$, is :
   (a) log 2 (b) log (2 –1)
   (c) 0 (d) 1
2. The integral $\int_a^b f(x)dy$ represents the area of bounded region between the lines :
   (a) $x = a, x = b$ (b) $x = 1, x = b$
   (c) $x = -a, y = b$ (d) $y = a, y = b$
3. The area bounded by the curve $y = \sin^2 x$, x-axis and the lines $x = 0, x = \pi/2$ is :
   (a) $\pi/2$ (b) $\pi/4$
   (c) π (d) $\pi/3$
4. The area of the curve $x^2 + 4y^2 = 16$ is :
   (a) π (b) 6π
   (c) 8π (d) $8\pi - 1$
5. The area of the loop of the curve $y^2 = x\,(x - 1)^2$ is :
   (a) 7/15 (b) 8/15
   (c) 2/15 (d) 1/15
6. The area of the loop of the curve $ay^2 = x^2\,(a - x)$ is :
   (a) $8a^2/15$ (b) $7a^2/15$
   (c) $a^2/15$ (d) none of these
7. The area of one loop of the curve $r^2 = a^2 \cos 2\theta$ is :
   (a) a^2 (b) $\frac{3}{2}a^2$
   (c) $\frac{a^2}{2}$ (d) $\frac{1}{3}a^2$
8. The area of the cardioid $r = a(1 + \cos\theta)$ is :
   (a) a^2 (b) $\frac{3}{2}\pi a^2$
   (c) $\frac{1}{2}\pi a^2$ (d) πa^2
9. The area of the curve $r = 2a \cos\theta$ is :
   (a) πa^2 (b) $\frac{1}{2}\pi a^2$
   (c) a^2 (d) $\frac{3}{2}\pi a^2$

10. The area of one loop of the curve $r = a \sin 6\theta$ is A, then whole area of the curve is :

(a) $6A$ (b) $5A$

(c) $12A$ (d) $10A$

ANSWERS

Fill in the blanks

1. area of bounded region **2.** a piane bounded by $r = f(\theta)$ and two radii **3.** πab **4.** $2\pi a^2$ **5.** $\frac{16}{3}a^2$

6. 45/4 **7.** 9/2 **8.** $\frac{3}{2}\pi a^2$ **9.** 100 A **10.** $\frac{1}{2}a^2 \log\left(\frac{\beta}{\alpha}\right)$ **11.** πa^2

True/False

1. T **2.** T **3.** F **4.** T **5.** T **6.** F **7.** T **8.** F **9.** T **10.** T **11.** F

Multiple Choice Questions

1. (a) **2.** (d) **3.** (b) **4.** (c) **5.** (b) **6.** (a) **7.** (c) **8.** (b) **9.** (a) **10.** (c)

FFFFF

CHAPTER 18 Surface and Volume of Solid of Revolutions

18.1 INTRODUCTION

When a plane curve is revolved about a certain fixed line lying in its own plane, a surface is generated. This surface is called a surface of revolution. Also the fixed line is called the axis of revolution.

18.2 REVOLUTION ABOUT X-AXIS

Let S be the surface area (curved surface) of a solid which is generated by the revolution of the curve $y = f(x)$ about x-axis between the ordinates $x = a$ and $x = b$ and let s be the arc length measured from the point a, f(a) to any point P(x, y). Then

$$S = \int_a^b 2\pi y\ ds = \int_a^b 2\pi y \frac{ds}{dx}.dx.$$

Proof. Let $A\ (a, f(a))$ and $B\ (b, f(b))$ be the points on the curve $y = f(x)$and assuming that the curve $y = f(x)$ is continuous in (a, b) and does not intersect the axis of x. Let $P(x, y)$ be any point on the curve and s be the arc length of the curve measured from A as shown in fig. 1.

Let $Q(x + \delta x, y + \delta y)$ be any other point very near to $P(x, y)$. Then $PQ = \delta s$, because $AP = s$ and $AQ = s + \delta s$. Draw the perpendiculars PM and QN to the axis of x from P and Q respectively. As the curve revolves about x-axis, the arc length $PQ = \delta s$ also revolves and from a right circular cydinder of thickness δs of radii y and $y + \delta y$. Let δS be the surface area of this cylinderical element which lies between the surface areas $2\pi y \delta s$ and $2\pi\ (y + \delta y)\delta s$. That is,

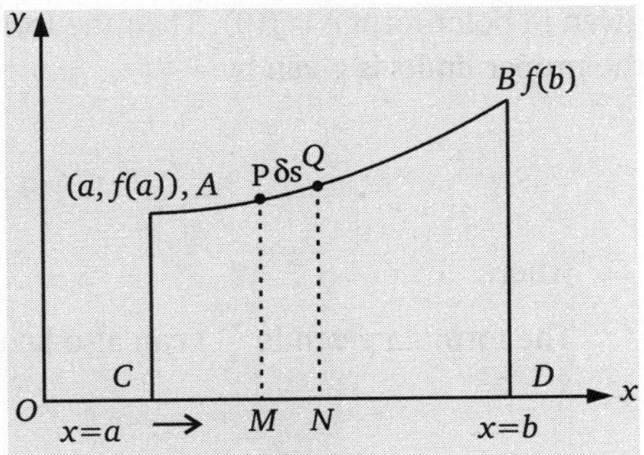

Fig. 1

$$2\pi y\ \delta s < \delta S < 2\pi(y + \delta y)\ \delta s.$$

Divide by δs, we get

$$2\pi y < \frac{\delta S}{\delta s} < 2\pi(y + \delta y).$$

As $Q \to P$, $\delta s \to 0$ and $\delta y \to 0$, then taking the limit as $\delta s \to 0$, we obtain

$$2\pi y < \lim_{\delta s \to 0} \frac{\delta S}{\delta s} < 2\pi y.$$

$$\therefore \quad \lim_{\delta s \to 0} \frac{\delta S}{\delta s} = 2\pi y \text{ or } \frac{dS}{ds} = 2\pi y \text{ or } dS = 2\pi y\, ds.$$

Now integrating, we get

$$\int_{x=a}^{x=b} dS = \int_{x=a}^{x=b} 2\pi y\, ds$$

or

$$S = \int_a^b 2\pi y\, ds = \int_a^b 2\pi y \frac{ds}{dx}.dx$$

where $\frac{ds}{dx} = \sqrt{\left[1 + \left(\frac{dy}{dx}\right)^2\right]}$ and S is the surface area of the solid of revolution of the curve $y = f(x)$ about x-axis between $x=a$ and $x=b$.

18.3 REVOLUTION ABOUT Y-AXIS

Let S be the surface area of a solid generated by the revolution of the curve $x = f(y)$ about y-axis between $y = a$ and $y=b$. Then

$$S = \int_a^b 2\pi x\,ds = \int_a^b 2\pi x \frac{ds}{dy}.dy$$

where $\frac{ds}{dy} = \sqrt{\left[1+\left(\frac{dx}{dy}\right)^2\right]}$ and s the arc length being measured from the point $(f(a), a)$.

Proof. Similar as before given in §18.2.

18.4 REVOLUTION ABOUT ANY LINE

Let S be the surface area of a solid generated by the curve about any line between certain points. Let s be the arc length of the curve measured from one of the two given points to any point P on the curve and let Q be any point very near to P such that $PQ = \delta s$. Now draw a perpendicular PM from the point P to the line of axis of the revolution. Then

$$S = \int 2\pi (PM)\, ds$$

Here the limits of integration are taken the given points between them a solid is formed by the revolution.

18.5 SURFACE FORMULAE FOR DIFFERENT FORM OF EQUATIONS

(a) Equation of a curve in parametric form. Suppose the equation of a curve is given in parametric form $x = f(t)$, $y=g(t)$ where t is the parameter, then the surface area of a solid generated by the revolution of the given curve about x-axis between the suitable limits is

$$S = \int 2\pi y\left(\frac{ds}{dt}\right) dt$$

where

$$\frac{ds}{dt} = \sqrt{\left[\left(\frac{dx}{dt}\right)^2+\left(\frac{dy}{dt}\right)^2\right]}.$$

Similarly for y-axis as the axis of revolution we may find the surface area.

(b) Equation of a curve in polar form. Suppose the equation of a curve is given in polar form $r = f(\theta)$. Then the formula for finding the surface area between the proper limits is given by

$$S = \int 2\pi (r\sin\theta)\frac{ds}{d\theta}.d\theta \quad ...(1)$$

where

$$\frac{ds}{d\theta} = \sqrt{\left[r^2+\left(\frac{dr}{d\theta}\right)^2\right]}.$$

The formula given in (1) can also be taken as

$$S = \int 2\pi (r\sin\theta)\frac{ds}{dr}.dr \quad ...(2)$$

where

$$\frac{ds}{dr} = \sqrt{\left[1+\left(r\frac{d\theta}{dr}\right)^2\right]}.$$

Recapitulations

Surface formulae

- $S = \int_a^b 2\pi y\,ds = \int_a^b 2\pi y\frac{ds}{dx}.dx$ (Revolution about x-axis)
- $S = \int_a^b 2\pi x\,ds = \int_a^b 2\pi x\frac{ds}{dy}.dy$ (Revolution about y-axis)
- $S = \int 2\pi (PM)\,ds$ (Revolution about any line PM)
- $S = \int 2\pi y\left(\frac{ds}{dt}\right)dt$ (Parametric form)
- $S = \int 2\pi (r\sin\theta)\frac{ds}{dr}.dr$ where $\frac{ds}{dr} = \sqrt{1+\left(r\frac{d\theta}{dr}\right)^2}$ (Polar form)

WORKING PROCEDURE

To find the surface area of a solid generated by the revolution of the curve about any line, use the following steps :

Step 1. *Take any point P on the curve between the given points.*

Step 2. *Draw the perpendicular from P to the line of axis which meets the axis at the point M (say).*

Step 3. *Find the perpendicular PM.*

Step 4. *And use the formula given in* 18.4.

Solved Examples

Example 1. *Find the surface of a sphere of radius a.*

Solution. The sphere is generated, if a semi-circle is revolved about its diameter. Let the equation of a circle of radius a is

$$x^2 + y^2 = a^2. \quad ...(1)$$

$$\therefore \quad 2x + 2y\frac{dy}{dx} = 0 \text{ or } \frac{dx}{dy} = -\frac{x}{y}.$$

$$\Rightarrow \quad \frac{ds}{dx} = \sqrt{\left[1+\left(\frac{dy}{dx}\right)^2\right]} = \sqrt{\left(1+\frac{x^2}{y^2}\right)}$$

$$= \sqrt{\left(\frac{x^2+y^2}{y^2}\right)} = \frac{a}{y} \quad \text{[Using (1)]}$$

Let $A(-a, 0)$ and $B(a, 0)$ be the bounding points of the semi-circle as shown in fig. 2.

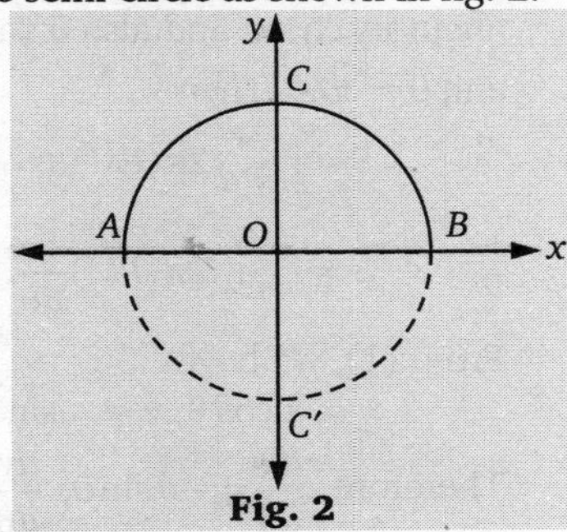

Fig. 2

Here the diameter is taken as x-axis. Let S be the surface area of the sphere, then

$$S = \int_{-a}^{a} 2\pi y.\frac{ds}{dx}.dx$$

$$= \int_{-a}^{a} 2\pi y.\frac{a}{y}.dx \quad \left(\because \frac{ds}{dx} = \frac{a}{y}\right)$$

$$= 2\pi a\int_{-a}^{a} dx = 2\pi a\left[x\right]_{-a}^{a}$$

$$\therefore \quad S = 4\pi a^2.$$

Example 2. *Find the surface of the solid generated by the revolution of the ellipse $x^2 + 4y^2 = 16$ about its major axis.*

Solution. The equation of the curve is

$$x^2 + 4y^2 = 16 \quad ...(1)$$

or

$$\frac{x^2}{16} + \frac{y^2}{4} = 1.$$

The end points of major axis are $A(-4, 0)$ and $B(4, 0)$ which are on the x-axis so the major axis is the axis of x. Thus the curve is revolved about the x-axis.

Now differentiating (1) *w.r.t.* 'x', we get

$$2x + 8y\frac{dy}{dx} = 0 \text{ or } \frac{dy}{dx} = -\frac{x}{4y}.$$

$$\therefore \quad \frac{ds}{dx} = \sqrt{\left[1+\left(\frac{dy}{dx}\right)^2\right]}$$

$$= \sqrt{\left(1+\frac{x^2}{16y^2}\right)} = \frac{\sqrt{(16y^2+x^2)}}{4y}$$

$$= \frac{\sqrt{14(16-x^2)+x^2]}}{4y} \quad \text{[Using (1)]}$$

$$\therefore \quad \frac{ds}{dx} = \frac{\sqrt{(64-3x^2)}}{4y}.$$

Let S be the surface area of the solid so formed by the revolution of the ellipse given in (1) about its major axis (x-axis is)

$$S = \int_{-4}^{4} 2\pi y\frac{ds}{dx}.dx = \int_{-4}^{4} 2\pi y.\frac{\sqrt{(64-3x^2)}}{4y}dx$$

$$= \frac{\pi}{2}\int_{-4}^{4}\sqrt{(64-3x^2)}\,dx = \pi\int_0^4\sqrt{(64-3x^2)}\,dx$$

$$= \sqrt{3}\pi\int_0^4\sqrt{\left[\left(\frac{8}{\sqrt{3}}\right)^2 - x^2\right]}\,dx$$

$$= \sqrt{3}\pi\left[\frac{x}{2}\sqrt{\left[\left(\frac{8}{\sqrt{3}}\right)^2 - x^2\right]} + \frac{32}{3}\sin^{-1}\frac{\sqrt{3}x}{8}\right]_0^4$$

$$= \sqrt{3}\pi\left[2\sqrt{\left(\frac{64}{3}-16\right)} + \frac{32}{3}\sin^{-1}\left(\frac{\sqrt{3}}{2}\right)\right]$$

$$= \sqrt{3}\pi\left[2.\frac{4}{\sqrt{3}} + \frac{32}{(\sqrt{3})^2}.\frac{\pi}{3}\right] = \pi\left[8+\frac{32}{3\sqrt{3}}\pi\right]$$

$$\therefore \quad S = 8\pi\left[1+\frac{4\pi}{3\sqrt{3}}\right].$$

Example 3. *Find the surface of the solid generated by the revolution of the lemniscate $r^2 = a^2\cos 2\theta$ about the initial line.*

Solution. The equation of the curve is

$$r^2 = a^2\cos 2\theta. \quad ...(1)$$

From (1) $r = 0$, when $\cos 2\theta = 0$ *i.e.*, $2\theta = \pm\pi/2$ or $\theta = \pm\pi/4$ and maximum value of r is a, when $\cos 2\theta = 1$ is $\theta = 0$. Thus the tracing of this curve is as fig. 3.

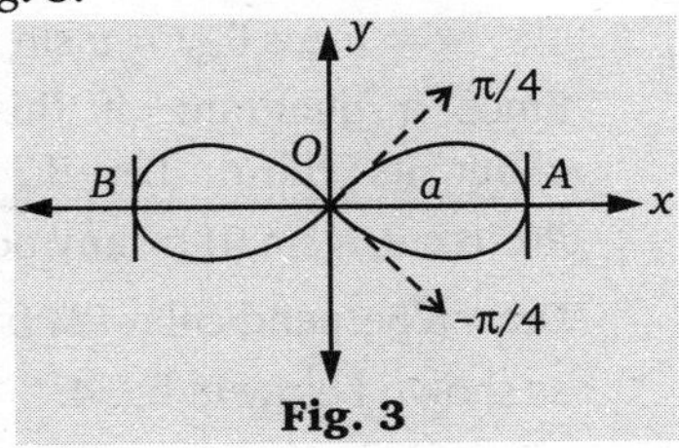

Fig. 3

Now differentiating (1) w.r.t. θ, we get

$$2r\frac{dr}{d\theta} = -2a^2 \sin 2\theta$$

or $$\frac{dr}{d\theta} = -\frac{a^2 \sin 2\theta}{r}. \quad ...(2)$$

$$\therefore \quad \frac{ds}{d\theta} = \sqrt{\left[r^2 + \left(\frac{dr}{d\theta}\right)^2\right]}$$

$$= \sqrt{\left[a^2 \cos 2\theta + \frac{a^4 \sin^2 2\theta}{r^2}\right]}$$

[Using (1) and (2)]

$$= \frac{\sqrt{(r^2a^2 \cos 2\theta + a^4 \sin^2 2\theta)}}{r}$$

$$= \frac{\sqrt{(a^4 \cos^2 2\theta + a^4 \sin^2 2\theta)}}{r} \quad \text{[Using (1)]}$$

$$= \frac{a^2}{r}.$$

Since there are two loops in the curve and one loop of the curve lies between $\theta = -\pi/4$ and $\theta = \pi/4$. Also the curve is symmetrical about the pole as well as about the initial line. Let S be the surface of the solid generated by the revolution of the given curve. Then

$$S = 2\int_0^{\pi/4} 2\pi y \frac{ds}{d\theta}.d\theta, \text{ where } y = r \sin\theta$$

$$= 2\int_0^{\pi/4} 2\pi(r \sin\theta).\frac{a^2}{r}.d\theta \quad \text{[Using (2)]}$$

$$= 4\pi a^2 \int_0^{\pi/4} \sin\theta \, d\theta = 4\pi a^2 [-\cos\theta]_0^{\pi/4}$$

$$= 4\pi a^2 \left[-\cos\frac{\pi}{4} + \cos 0\right]$$

$$= 4\pi a^2 \left(-\frac{1}{\sqrt{2}} + 1\right)$$

$$\therefore \quad S = 4\pi a^2 \left(1 - \frac{1}{\sqrt{2}}\right).$$

Example 4. *A quadrant of a circle of radius a revolves about its chord. Show that the surface of the spindle generated is* $2\pi a^2\left(1-\frac{\pi}{4}\right)\sqrt{2}$.

Solution . Let the parametric equation of a circle of radius a be

$$x = a \cos\theta, y = a \sin\theta. \quad ...(1)$$

Since a quadrant of this circle is revolved about its chord. Let BC be the chord and $P(a \cos\theta, a \sin\theta)$ be any point on this curve.

Draw a perpendicular PM from P to the chord BC as shown below in fig. 4.

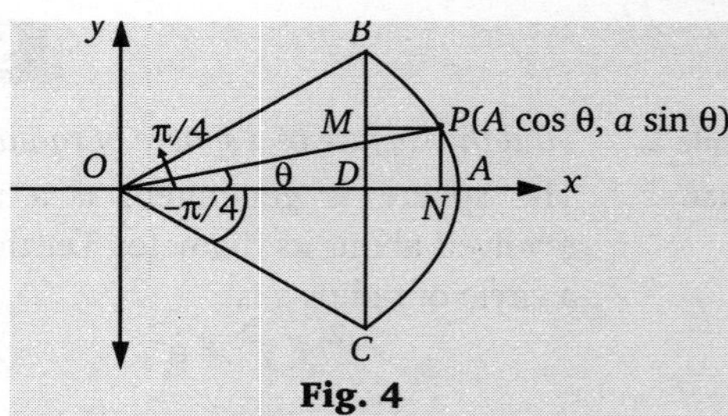

Fig. 4

Since the angle $\angle BOC = \pi/2$ and let $\angle AOB = \angle AOC = \pi/4$. In ΔPON, $PN = a \sin\theta$, $ON = a \cos\theta$. Also in $\angle OBD$, $OB = a$ (radius) and $\angle BOD = \pi/4$.

$$\therefore \quad OD = OB\cos\frac{\pi}{4} = \frac{a}{\sqrt{2}}$$

$$\therefore \quad PM = DN = ON - OD = a\cos\theta - \frac{a}{\sqrt{2}}.$$

Let S be the surface generated by the quadrant about its chord and also θ varies from $\theta = -\pi/4$ and $\theta = \pi/4$. Then

$$S = \int_{-\pi/4}^{\pi/4} 2\pi(PM)\,ds$$

or $$S = \int_{-\pi/4}^{\pi/4} 2\pi(PM)\frac{ds}{d\theta}.d\theta. \quad ...(2)$$

From (1), we have

$$x = a \cos\theta, y = a \sin\theta.$$

Therefore $$\frac{dx}{d\theta} = -a\sin\theta, \quad \frac{dy}{d\theta} = a\cos\theta.$$

Now, $$\frac{ds}{d\theta} = \sqrt{\left[\left(\frac{dx}{d\theta}\right)^2 + \left(\frac{dy}{d\theta}\right)^2\right]}$$

$$= \sqrt{[(-a\sin\theta)^2 + (a\cos\theta)^2]}$$

$$= \sqrt{(a^2 \sin^2\theta + a^2 \cos^2\theta)} = a.$$

From (2), we have

$$S = \int_{-\pi/4}^{\pi/4} 2\pi(PM)\,a.d\theta$$

$$= 2\pi a \int_{-\pi/4}^{\pi/4} (a\cos\theta - a/\sqrt{2})\,d\theta$$

$$= 2\pi a^2 \left[\int_{-\pi/4}^{\pi/4} \left(\cos\theta - \frac{1}{\sqrt{2}}\right)d\theta\right]$$

$$= 2\pi a^2 \left[\sin\theta - \frac{\theta}{\sqrt{2}}\right]_{-\pi/4}^{\pi/4}$$

$$= 2\pi a^2 \left[2\left(\sin\frac{\pi}{4} - \frac{\pi}{4\sqrt{2}}\right)\right]$$

$$= 4\pi a^2 \left[\frac{1}{\sqrt{2}} - \frac{\pi}{4\sqrt{2}}\right]$$

$$\therefore \quad S = 2\sqrt{2}\pi a^2 \left(1 - \frac{\pi}{4}\right).$$

Example 5. *The lemniscate $r^2 = a^2\cos 2\theta$ revolves about a tangent at the pole. Find the surface of the solid thus generated.*

Solution . We have $\quad r^2 = a^2 \cos 2\theta.$

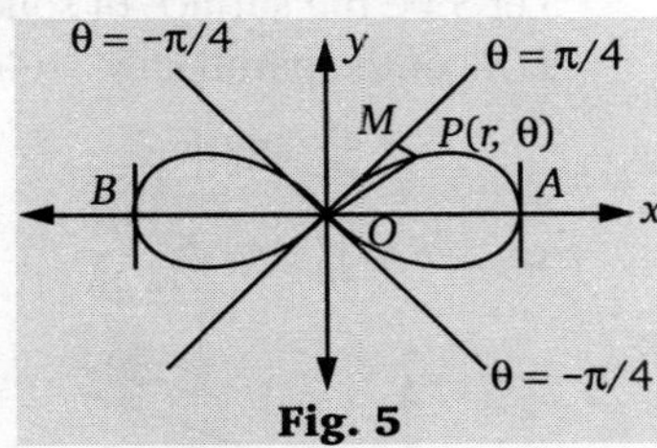

Fig. 5

Clearly, $\theta = \pi/4$ is one of the tangent at the pole. Let $P(r, \theta)$ be any point on the curve. Draw PM as perpendicular from P to the line $\theta = \pi/4$. Then

$$\angle POM = \frac{\pi}{4} - \theta.$$

So in $\Delta\, PMO$, $\dfrac{PM}{PO} = \sin\left(\dfrac{\pi}{4} - \theta\right)$

$$\therefore \quad PM = r\sin\left(\frac{\pi}{4} - \theta\right).$$

Now differentiating (1) w.r.t. θ, we get

$$2r\frac{dr}{d\theta} = -2a^2 \sin 2\theta$$

or
$$\frac{dr}{d\theta} = -\frac{a^2 \sin 2\theta}{r}$$

$$\therefore \quad \frac{ds}{d\theta} = \sqrt{r^2 + \left(\frac{dr}{d\theta}\right)^2}$$

$$= \sqrt{a^2 \cos 2\theta + \frac{a^4 \sin^2 2\theta}{r^2}}$$

$$= \frac{1}{r}\sqrt{a^4 \cos^2 2\theta + a^4 \sin^2 2\theta}$$

$$= \frac{a^2}{r}$$

There are two loops in the curve and loop lies between $\theta = -\pi/4$ and $\theta = \pi/4$. Also the curve is symmetrical about the pole as well as about the initial line.

Let S be the surface of the solid generated by the revolution of the given curve about the line $\theta = \pi/4$. Then

$$S = 2\int_{-\pi/4}^{\pi/4} 2\pi(PM)\frac{ds}{d\theta}d\theta$$

$$= 4\pi\int_{-\pi/4}^{\pi/4} r\sin\left(\frac{\pi}{4} - \theta\right)\frac{a^2}{r}d\theta$$

$$= 4\pi a^2 \int_{-\pi/4}^{\pi/4} \sin\left(\frac{\pi}{4} - \theta\right)d\theta$$

$$= 4\pi a^2\left[\cos\left(\frac{\pi}{4} - \theta\right)\right]_{-\pi/4}^{\pi/4}$$

$$= 4\pi a^2\left[1 - \cos\frac{\pi}{2}\right] = 4\pi a^2(1-0)$$

$$= 4\pi a^2.$$

Example 6. *Find the surface of the solid generated by the revolution of the curve $x = a\cos^3 t$ and $y = a\sin^3 t$ about the x-axis.*

Solution . We have $\quad x = a\cos^3 t$ and $y = a\sin^3 t.$

This curve is symmetrical about both the axes.

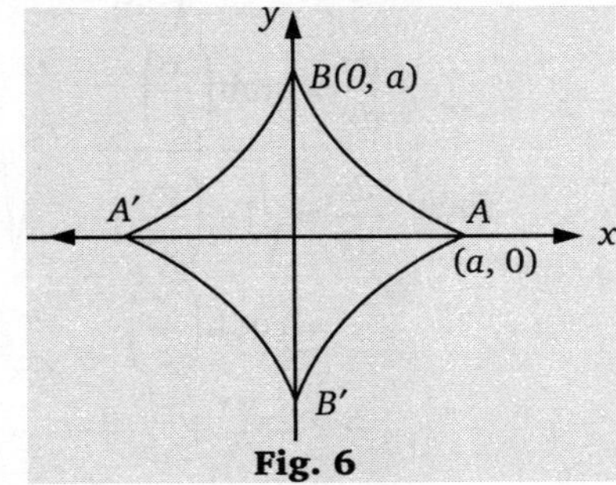

Fig. 6

At $A(a, 0)$ $t = 0$ and $B(0, a)$ $t = \pi/2$.

Now $\quad \dfrac{dx}{dt} = -3a\cos^2 t\sin t$

$$\frac{dy}{dt} = 3a\sin^2 t\cos t$$

$$\therefore \quad \frac{ds}{dt} = \sqrt{\left(\frac{dx}{dt}\right)^2 + \left(\frac{dy}{dt}\right)^2}$$

$$= 3a\sqrt{\cos^4 t\sin^2 t + \sin^4 t\cos^2 t}$$

$$= 3a\sin t\cos t = \frac{3a}{2}\sin 2t.$$

Let S be the surface of a solid of revolution of the given curve about x-axis. Then

$$S = 2\int_0^{\pi/2} 2\pi y\frac{ds}{dt}dt$$

$$= 4\pi\int_0^{\pi/2} a\sin^3 t.\frac{3a}{2}\sin 2t\,dt$$

$$= 12\pi a^2\int_0^{\pi/2}\sin^4 t\cos t\,dt$$

$$= 12\pi a^2\left[\frac{\sin^5 t}{5}\right]_0^{\pi/2} = \frac{12\pi a^2}{5}.$$

Example 7. *Find the surface generated by the revolution of an arc of the catenary $y = c\cosh\left(\dfrac{x}{c}\right)$ about the x-axis.*

Solution . Let S be the surface generated by the revolution of an arc DP of $y = c\cosh\left(\dfrac{x}{c}\right)$ about x-axis.

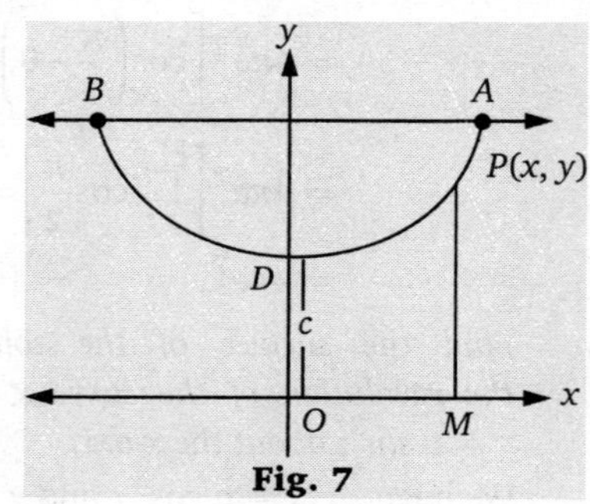

Fig. 7

From $y = c\cosh\left(\frac{x}{c}\right)$, we have

$$\frac{dy}{dx} = \sinh\left(\frac{x}{c}\right)$$

Then $$\frac{ds}{dx} = \sqrt{1+\left(\frac{dy}{dx}\right)^2} = \sqrt{1+\sinh^2\left(\frac{x}{c}\right)}$$

$$= \cosh\left(\frac{x}{c}\right)$$

$$S = 2\pi\int_0^x y\frac{ds}{dx}dx$$

$$= 2\pi\int_0^x c\cosh\left(\frac{x}{c}\right)\cosh\left(\frac{x}{c}\right)dx$$

$$= 2\pi c\int_0^x \cosh^2\left(\frac{x}{c}\right)dx$$

$$= \pi c\int_0^x\left[1+\cosh\left(\frac{2x}{c}\right)\right]dx$$

$$= \pi c\left[x+\frac{c}{2}\sinh\frac{2x}{c}\right]_0^x$$

$$= \pi c\left[x+\frac{c}{2}\sinh\frac{2x}{c}\right].$$

Example 8. *Find the surface of the solid generated by revolving the arc of the parabola $y^2 = 4ax$ bounded by its latusrectum about x-axis.* (Rohtak 2003)

Solution . Let S be the surface of solid of revolution of arc LOL' of the parabola $y^2 = 4ax$ about x-axis. Then

$$S = \int_0^a 2\pi y\, ds$$

or $$S = \int_0^a 2\pi y\, dx.\frac{ds}{dx} \qquad ...(1)$$

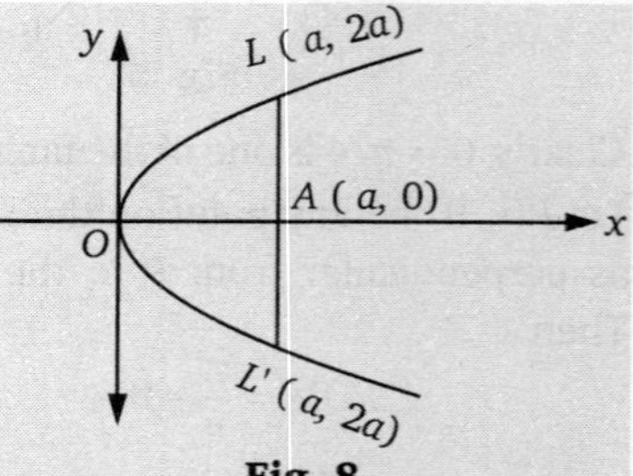

Fig. 8

We have

$$y^2 = 4ax \Rightarrow y = 2\sqrt{ax} \Rightarrow \frac{dy}{dx} = \frac{\sqrt{a}}{\sqrt{x}}$$

$$\therefore \frac{ds}{dx} = \sqrt{1+\left(\frac{dy}{dx}\right)^2} = \sqrt{1+\frac{a}{x}} = \sqrt{\frac{x+a}{x}}$$

Therefore,

$$S = 2\pi\int_0^a 2\sqrt{ax}\left[\sqrt{\frac{x+a}{x}}\right]dx$$

$$= 4\pi\sqrt{a}\int_0^a\sqrt{x+a}\,dx = 4\pi\sqrt{a}\left[\frac{2}{3}(x+a)^{3/2}\right]_0^a$$

$$= \frac{8\pi\sqrt{a}}{3}[(2a)^{3/2}-a^{3/2}]$$

$$= \frac{8\pi}{3}(2^{3/2}-1)a^2 = \frac{8\pi}{3}(2\sqrt{2}-1)a^2$$

EXERCISE 18.1

1. Find the curved surface of a hemi-sphere of radius a.
2. Find the area of the surface formed by the revolution of the parabola $y^2 = 4ax$ about the x-axis by arc from the vertex to one of the latusrectum.
3. For a catenary $y = a\cosh(x/a)$, prove that
$$aS = \pi a\,(ax + sy).$$
where s is the length of the arc measured from the vertex, S is the area of curved surface of the solid generated by the revolution of the arc about x-axis.
4. Find the surface of the solid generated by the revolution of the astroid $x^{2/3} + y^{2/3} = a^{2/3}$ aboul x-axis.
5. Find the surface of the solid formed by the revolution, about the axis of y, of the part of the curve $ay^2 = x^3$ from $x = 0$ to $x = 4a$ which is above the x-axis.
6. Prove that the surface of the prolate spheroid formed by the revolution of the ellipse of essentricity e about its major axis is equal to $2\pi ab\left[\sqrt{(1-e^2)}+\frac{1}{e}\sin^{-1}e\right]$.
7. Prove that the surface of the oblate spheroid formed by the revolution of the ellipse of essentricity e about its minor axis is $2\pi a^2\left[1+\frac{1-e^2}{2e}\log\left(\frac{1+e}{1-e}\right)\right]$.
8. Find the surface area of the solid generated by revolution of the cycloid $x = a\,(\theta - \sin\theta)$, $y = a(1-\cos\theta)$about x-axis. (VTU-2003)
9. The portion between two consecutive cusps of the cycloid, $x = a\,(\theta + \sin\theta)$, $y = a\,(1+\cos\theta)$ is revolved about x-axis. Prove that the area of the surface so formed is to the area of the cycloid as 64 : 9.
10. Prove that the surface area of the solid generated by the revolution of the loop of the curve $x = t^2$, $y = t - \frac{1}{3}t^3$ about x-axis is 3π.

11. Find the surface of the solid formed by the revolution of the cardioid $r = a(1 + \cos\theta)$ about the initial line. (VTU-2009, JNTU-2003, Rajasthan-2006)

12. The arc of the cardioid $r = a(1 + \cos\theta)$ included between $-\frac{\pi}{2} \le \theta \le \frac{\pi}{2}$ is rotated about the line $\theta = \pi/2$. Find the area of the surface thus generated.

13. Find the area of the surface of revolution formed by revolving the curve $r = 2a\cos\theta$ about the initial line. (VTU 2009)

14. The lemniscate $r^2 = a^2\cos 2\theta$ revolves about the initial line. Find ihe surface of the solid thus generated.

15. A circular arc revolves aboul its chord. Find the area of the surface generated, when 2α is the angle subtended by the arc at the centre.

Hint to Selected Problems

1. $S = 2\pi\int_0^a 2\pi x\,ds = 2\pi\int_0^a \sqrt{a^2 - y^2}.\frac{ds}{dy}.dy.$

2. $y^2 = 4ax \Rightarrow \frac{dy}{dx} = \frac{2a}{y}.$

$\therefore \frac{ds}{dx} = \sqrt{1 + \frac{4a^2}{y^2}} = \frac{\sqrt{4ax + 4a^2}}{y}.$

Now, $S = 2\pi\int_0^a y.\frac{ds}{dx}.dx.$

3. $y = a\cosh\frac{x}{a} \Rightarrow \frac{dy}{dx} = \sinh\left(\frac{x}{a}\right)$

$\Rightarrow \frac{ds}{dx} = \cosh\frac{x}{a} \Rightarrow s = a\sinh\frac{x}{a}.$

Now $S = 2\pi\int_0^x y\frac{ds}{dx}.dx = 2\pi\int_0^x a\cosh\frac{x}{a}.\cosh\frac{x}{a}.dx.$

4. From the given curve, we obtained $\frac{dy}{dx} = -\left(\frac{y}{x}\right)^{1/3}$

$\therefore \frac{ds}{dx} = \sqrt{1 + \left(\frac{dy}{dx}\right)^2} = \frac{a^{1/3}}{x^{1/3}}.$

Now use $S = 4\pi\int_0^a y.\frac{ds}{dx}.dx$

$= 4\pi\int_0^a (a^{2/3} - x^{2/3})^{3/2}\frac{a^{1/3}}{x^{1/3}}dx.$

Then put $x^{2/3} = a^{2/3}\sin^2\theta.$

5. $S = \int_0^{8a} 2\pi x.\frac{ds}{dy}.dy.$ ...(1)

Also $\frac{ds}{dy} = \sqrt{1 + \left(\frac{dx}{dy}\right)^2} = \frac{\sqrt{9x^4 + 4a^2y^2}}{3x^2}.$

Put in (1) and then solve.

6. Given that $\frac{x^2}{a^2} + \frac{y^2}{b^2} = 1$ ...(1)

and $b^2 = a^2(1 - e^2)$

Then $S = 2\int_0^a 2\pi y\,ds.$

From (1), $\frac{ds}{dx} = -\left(\frac{b^2x}{a^2y}\right).$

Then find $\frac{ds}{dx}$ and put in $S = 2\int_0^a 2\pi y\,ds.$

10. $x = t^2,\ y = t - \frac{t^3}{3} \Rightarrow \frac{dx}{dt} = 2t, \frac{dy}{dt} = 1 - t^2.$

$\therefore \frac{dy}{dx} = \frac{1 - t^2}{2t}.$

Now by using the formula

$\left(\frac{ds}{dt}\right)^2 = \left(\frac{dx}{dt}\right)^2 + \left(\frac{dy}{dt}\right)^2.$

We obtained $\frac{ds}{dt} = 1 + t^2.$

Then $S = \int_0^{\sqrt{3}} 2\pi y.\frac{ds}{dt}.dt.$

11. $r = a(1 + \cos\theta) \Rightarrow \frac{dr}{d\theta} = -a\sin\theta$

$\Rightarrow \frac{ds}{d\theta} = \sqrt{r^2 + \left(\frac{dr}{d\theta}\right)^2} = 2a\cos\frac{\theta}{2}$

Then use $S = \int_0^\pi 2\pi(r\sin\theta).\frac{ds}{d\theta}d\theta.$

12. Here, we have $\frac{dr}{d\theta} = -a\sin\theta$

$\Rightarrow \frac{ds}{d\theta} = \sqrt{r^2 + \left(\frac{dr}{d\theta}\right)^2} = 2a\cos\frac{\theta}{2}$

Finally use $\frac{ds}{d\theta} = 2a\cos\frac{\theta}{2}.$

13. Here, we have $\frac{dr}{d\theta} = -2a\sin\theta.$

$\therefore \frac{ds}{d\theta} = \sqrt{r^2 + \left(\frac{dr}{d\theta}\right)^2} = 2a.$

Then we use $S = \int_0^{\pi/2} 2\pi y\frac{ds}{d\theta}d\theta.$

ANSWERS

1. $2\pi a^2$ **2.** $\frac{8}{3}\pi a^2(2\sqrt{2} - 1)$ **4.** $\frac{12\pi a^2}{5}$ **5.** $\frac{128}{1215}\pi a^2[125\sqrt{10} + 1]$ **8.** $\frac{64}{3}\pi a^2$ **11.** $\frac{32}{5}\pi a^2$

12. $\frac{48}{5}\sqrt{2}\pi a^2$ **13.** $4\pi a^2$ **14.** $4\pi a^2\left(1 - \frac{1}{\sqrt{2}}\right)$ **15.** $4\pi a^2(\sin\alpha - \alpha\cos\alpha).$

18.6 VOLUME OF SOLID OF REVOLUTIONS

When a plane area is revolved about any fixed line lying in the same plane a solid (body) is generated. This solid (body) is called a solid of revolution. Also the x-axis, two right circular cylinders are formed of volumes $\pi y^2 \delta x$ and $\pi(y+\delta y)^2\,\delta x$ respectively. Since the plane area $PMNP$ lies between $PMNP'$ and $Q'MNQ$. Therefore δV_1 lies between $\pi y^2\delta x$ and $\pi(y+\delta y)^2\delta x$. Then

$$\pi y^2\,\delta x < \delta V_1 < \pi\,(y+\delta y)^2\,\delta x.$$

Divide by δx, we get

$$\pi y^2 < \frac{\delta V_1}{\delta x} < \pi(y+\delta y)^2.$$

As $Q \to P$, $\delta x \to 0$ and $\delta y \to 0$ so taking the limit as $\delta x \to 0$, we get

$$\pi y^2 < \lim_{\delta x\to 0}\frac{\delta V_1}{\delta x} < \pi y^2 \quad\text{or}\quad \lim_{\delta x\to 0}\frac{\delta V_1}{\delta x} = \pi y^2$$

or
$$\frac{dV_1}{dx} = \pi y^2 \qquad\text{or}\qquad dV_1 = \pi y^2\,dx.$$

Now integrating w.r.t. x between the limits $x = a$ to $x = b$, we get $\int_a^b dV_1 = \int_a^b \pi y^2\,dx$.

$\therefore$ Volume generated by the plane area $ABDC = \int_a^b \pi y^2 dx$

or
$$V = \int_a^b \pi y^2\,dx$$

Where V is the required volume of a solid formed by the plane area bounded by the curve $y = f(x)$, the ordinates $x = a$ and $x = b$ and x-axis about x-axis fixed line is called axis of revolution.

18.6.1 REVOLUTION ABOUT X-AXIS

Let V be the volume of a solid which is generated by the revolution of a plane area bounded by the curve $y = f(x)$, the ordinates $x = a$, $x = b$ and x-axis about the x-axis. Then

$$V = \int_a^b \pi y^2\,dx$$

where $y = f(x)$ is a continuous and single valued function defined on $[a, b]$.

Proof. Let us assume that the curve $y = f(x)$does not cut the x-axis and let AB be the arc of $y = f(x)$ between the ordinates $x = a$ and $x = b$ as shown in fig. 9.

Let $P(x, y)$ and $Q\,(x+\delta x, y+\delta y)$ be two neighbouring points on the curve $y = f(x)$. Draw the perpendiculars PM and QN to the x-axis about which the plane area $ACDB$ is revolved. Also PP' is the perpendicular to QN and QQ' is the perpendicular to PM, where (Q' is a point on MP when P produced to Q'. Let V_1 be the volume of the solid formed by the revolution of plane area $ACMP$ about x-axis and $(V_1+\delta V_1)$ be the volume of the solid formed by the revolution of the plane area $ACNQ$. Then δV_1 is the volume of a solid formed by the revolution of the plane area $PMNQ$ about x-axis. The rectangular plane areas $PMNP'$and $Q'MNQ$ when revolve about.

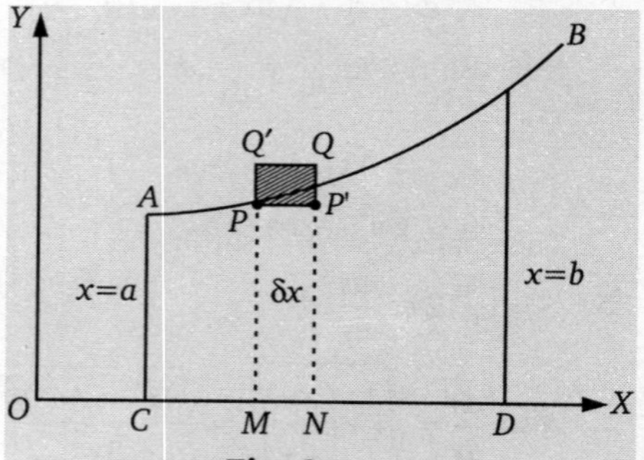

Fig. 9

18.6.2 REVOLUTION ABOUT Y-AXIS

The volume of a solid formed by the revolution of a plane area bounded by the curve $x = f(y)$and the lines $y = a$ and $y = b$ and y-axis about y-axis is

$$V = \int_a^b \pi x^2\,dy.$$

18.6.3 REVOLUTION ABOUT ANY LINE

The volume of a solid formed by the revolution of a plane area bounded by the arc AB and the lines AC and AD and the axis CD about any line CD (different from x-axis and y-axis) is

$$V = \int_{OC}^{OD} \pi (PM)^2\,d(OM)$$

where PM is the length of perpendicular from any point P on the arc AB to the axis CD and O be any fixed point on the axis CD.

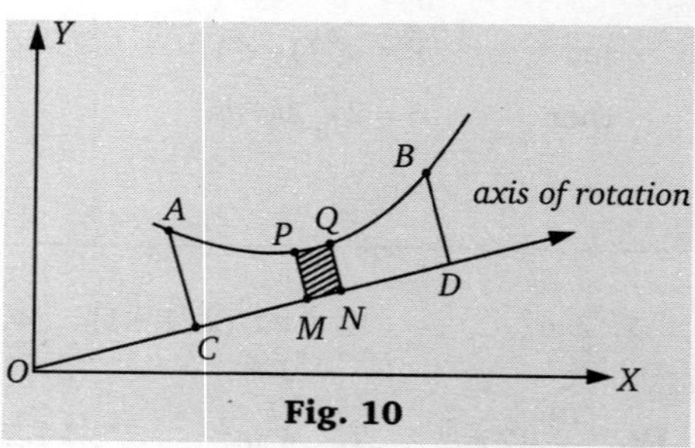

Fig. 10

18.7 VOLUME OF A SOLID OF REVOLUTION WHEN THE EQUATION OF THE CURVE ARE IN DIFFERENT FORMS

18.7. 1 EQUATION OF CURVE IN PARAMETRIC FORM.

Suppose the equation of a generating curve are in parametric form $x = f(t)$ and $y = g(t)$, then the volume of the solid generated by ihe revolution of the plane area bounded by the given curve, axis of x and the ordinates at the points where $t = a$ and $t = b$, about x-axis is

$$V = \int_a^b \pi y^2\left(\frac{dx}{dt}\right).dt = \int_a^b \pi[g(t)]^2 \frac{dx}{dt}.dt \text{ where } \frac{dx}{dt} = \frac{d}{dt}[f(t)].$$

Similarly, the volume of a solid formed by the revolution of a plane area bounded by $x = f(t)$, $y = g(t)$ axis of y and the two absciassae where $t = a$ and $t = b$ about y-axis is

$$V = \int_a^b \pi x^2\left(\frac{dy}{dt}\right).dt = \int_a^b \pi[f(t)]^2 \frac{dy}{dt}.dt \text{ where } \frac{dy}{dt} = \frac{d}{dt}[g(t)].$$

18.7.2 EQUATION OF CURVE IN POLAR FORM.

Suppose the equation of a curve in polar form is $r = f(\theta)$ where $x = r\cos\theta, y = r\sin\theta$, then the volume of a solid generated by the revolution of the plane area of the curve about the initial line (x-axis) between the lines $\theta = \alpha$ and $\theta = \beta$ is given by

$$V = \int_\alpha^\beta \pi y^2 \frac{dx}{d\theta}.d\theta = \int_\alpha^\beta \pi(r\sin\theta)^2 \frac{d}{d\theta}(r\cos\theta)\,d\theta \qquad \text{where} \qquad r = f(\theta).$$

Similarly, the volume of a solid formed by the revolution of the plane area bounded by the curve $r = f(\theta)$ and the lines $\theta = \alpha$ and $\theta = \beta$ about the line $\theta = \pi/2$ (y-axis) is given by

$$V = \int_\alpha^\beta \pi x^2 \frac{dy}{d\theta}.d\theta = \int_\alpha^\beta \pi(r\cos\theta)^2 \frac{d}{dr}(r\sin\theta)\,d\theta \qquad \text{where} \qquad r = f(\theta).$$

18.7.3 FORMULAE FOR FINDING THE VOLUME IN CASE OF POLAR FORM.

(i) The volume of a solid formed by the revolution of the plane area bounded by the curve $r = f(\theta)$ and the radii vectors $\theta = \alpha$ and $\theta = \beta$ about the initial line *i.e.*, $\theta = 0$ (x-axis) is

$$V = \int_\alpha^\beta \frac{2}{3}\pi r^3 \sin\theta\,d\theta.$$

(ii) The volume of a solid formed by the revolution of a plane area bounded by $r = f(\theta)$ and $\theta = \alpha$, $\theta = \beta$ about the line $\theta = \pi/2$ (y-axis) is given by

$$V = \int_\alpha^\beta \frac{2}{3}\pi r^3 \cos\theta\,d\theta.$$

(iii) The volume of a solid formed by the revolution of a plane area bounded by $r = f(\theta)$ and $\theta = \alpha$, $\theta = \beta$ about any line $\theta = \gamma$ is given by

$$V = \int_\alpha^\beta \frac{2}{3}\pi r^3 \sin(\theta - \gamma)\,d\theta.$$

Recapitulations

Volume formulae

- $V = \int_a^b \pi y^2 dx$ (Revolution about x-axis)
- $V = \int_a^b \pi x^2 dy$ (Revolution about y-axis)
- $V = \int \pi(PM)^2 d(OM)$ (Revolution about any line PM)
- $V = \int_a^b \pi y^2\left(\frac{dx}{dt}\right)dt$ (Parametric form)
- $V = \int_\alpha^\beta \pi y^2 \frac{dx}{d\theta}.d\theta$ $= \int_\alpha^\beta \pi(r\sin\theta)^2.\frac{d}{d\theta}(r\cos\theta)d\theta$ (Polar form)

REMARKS

- If the curve is symmetrical about x-axis. then the portion of the curve above x-axis overlaps the other portion of the curve below x-axis during the revolution. So that the volume shall not double between the bounding points.
- If the curve is symmeirieai about x-axis and the volume of a solid generated by the revolution of the plane area about y-axis is required, then the required volume will be double the volume which is obtained by the revolution of half of the symmetrical curve.

Solved Examples

Example 1. *Find the volume of a spherical cap of height h cut off from a sphere of radius a .*

Solution . The spherical cap is generated by the revolution of a plane area bounded by the curve $x^2+y^2=a^2$, the axis of y and the line $y = a-h$ and $y = a$ in the first quadrant about y-axis as shown in fig. 11.

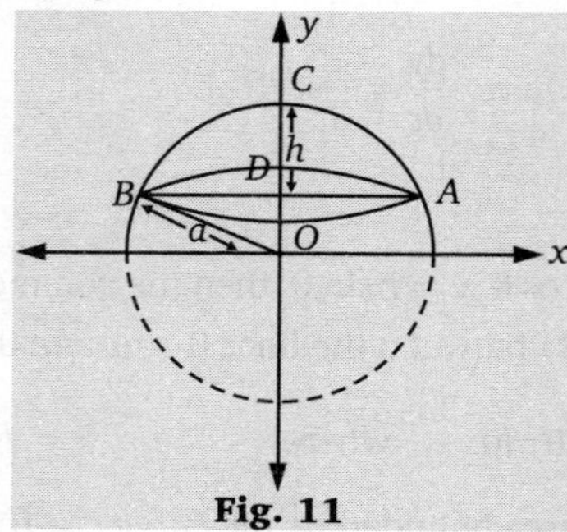

Fig. 11

Let V be the volume of this spherical cap, then

$$V=\int_{a-h}^{a}\pi x^2\,dy$$

$$=\pi\int_{a-h}^{a}(a^2-y^2)dy \quad (\therefore\ x^2+y^2=a^2)$$

$$=\pi\left[a^2y-\frac{y^3}{3}\right]_{a-h}^{a}$$

$$=\pi\left[\left(a^3-\frac{a^3}{3}\right)-(a-h)\left\{a^2-\frac{(a-h)^2}{3}\right\}\right]$$

$$=\pi\left[\frac{2a^3}{3}-\frac{a-h}{3}\{3a^2-(a-h)^2\}\right]$$

$$=\frac{\pi}{3}[2a^3-(a-h))\{2a^2-h^2+2ah\}]$$

$$=\frac{\pi}{3}[2a^3-2a^2(a-h)+(a-h)h^2-2ah(a-h)]$$

$$=\frac{\pi}{3}(3ah^2-h^3)$$

Hence, $V=\pi h^2\left(a-\frac{h}{3}\right)$.

Example 2. *Find the volume of the solid generated by the revolution of the curve $y = a^3/(a^2+x^2)$ about its asymptote.*

Solution . Clearly the curve cuts only y-axis at the point $(0, a)$ and $y = 0$ *i.e.*, x-axis is *its asymptote.* Therefore the solid is generated by the revolution of the curve about x-axis. The tracing of the curve is shown in fig. 12 .

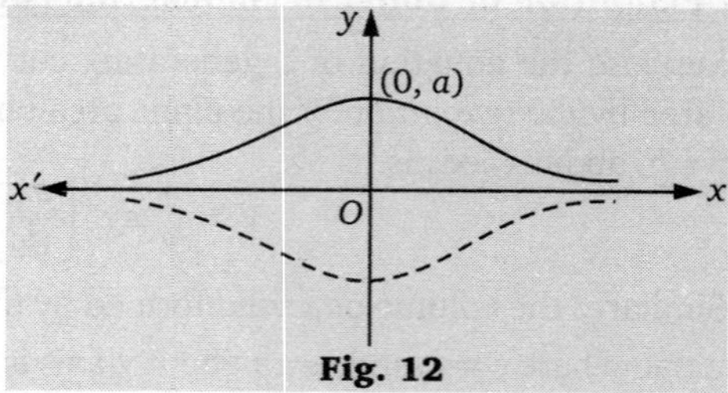

Fig. 12

Let V be the volume of the solid generated by the revolution of the curve about x-axis from $x=-\infty$ to $x=\infty$, then

$$V=\int_{-\infty}^{\infty}\pi y^2\,dx$$

$$=\pi\int_{-\infty}^{\infty}\frac{a^6}{(a^2+x^2)^2}dx \quad \left[\because y=\frac{a^3}{(a^2+x^2)}\right]$$

$$=2\pi\int_0^{\infty}\frac{a^6}{(a^2+x^2)^2}dx.$$

Let us put $x = a\tan\theta$, then $dx = a\sec^2\theta\, d\theta$ and θ varies from $\theta = 0$ to $\theta = \pi/2$.

$$\therefore\ V=2\pi a^6\int_0^{\pi/2}\frac{a\sec^2\theta\,d\theta}{a^4\sec^4\theta}$$

$$=2\pi a^3\int_0^{\pi/2}\cos^2\theta\,d\theta$$

$$=2\pi a^3\left[\frac{(2-1)}{2}\cdot\frac{\pi}{2}\right] \quad \text{(By walli's formula)}$$

Hence, $V=\frac{1}{2}\pi^2a^3$.

Example 3. *Find the volume of the solid formed by revolving the cycloid*

$$x = a(\theta - \sin\theta),\ y = (1-\cos\theta)$$

(i) about its base *(ii) about y-axis.*

(VTU- 2003, 05; Kurukshetra- 2006)

Solution . Since the equation of the cyloid is

$$x = a(\theta-\sin\theta),\ y = a(1-\cos\theta) \quad ...(1)$$

The tracing of this curve is given below :

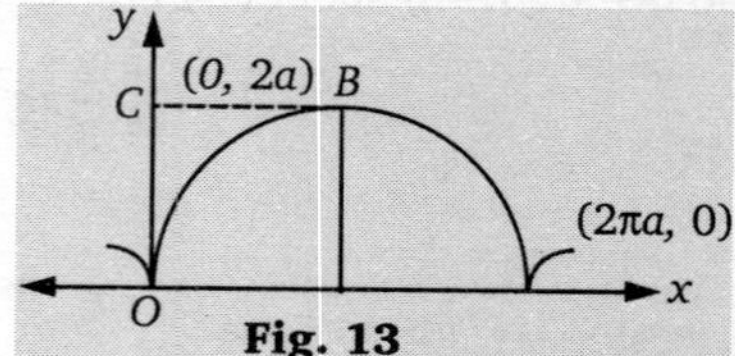

Fig. 13

(i) In the above fig. base is the axis of x so the volume of the solid formed by revolving the cylcoid given in (1) about its base (x-axis) between $x = 0$ to $x = 2\pi a$ where $\theta = 0$ to $\theta = 2\pi$, is given by

$$\Rightarrow V=\int_0^{2\pi a}\pi y^2dx=\pi\int_0^{2\pi}y^2\left(\frac{dx}{d\theta}\right).d\theta$$

$$= \pi \int_0^{2\pi} a^2(1-\cos\theta)^2 . a(1-\cos\theta)\, d\theta$$

[Using (1)]

$$= \pi a^3 \int_0^{2\pi} (1-\cos\theta)^3\, d\theta$$

$$\Rightarrow \quad V = \pi a^3 \int_0^{2\pi} 8\sin^6(\theta/2)\, d\theta. \qquad ...(2)$$

Let us put $\theta/2 = \phi$ so $d\theta = 2d\phi$ and ϕ varies from 0 to π, then (2) becomes

$$V = 8\pi a^3 \int_0^{\pi} \sin^6\phi (2d\phi)$$

$$= 16\pi a^3 \int_0^{\pi} \sin^6\phi\, d\phi$$

$$= 16\pi a^3 . 2\int_0^{\pi/2} \sin^6\phi\, d\phi$$

$$= 32\pi a^3 \int_0^{\pi/2} \sin^6\phi\, d\phi$$

$$= 32\pi a^3 \left[\frac{(6-1)(6-3)(6-5)}{6.4.2}.\frac{\pi}{2}\right]$$

(By Walli's formula)

$$V = 5\pi^2 a^3.$$

(ii) Let V be the volume of the solid formed by revolving of the cycloid about y-axis. This volume is the difference of the volume generated by the revolution of the area $OABCO$ and the volume generated by the revolution of the area $OBCO$. Since we have that at A, $\theta = 2\pi$, at O, $\theta = 0$ and at B, $\theta = \pi$. Therefore, the volume generated by the revolution of the area $OABCO$ about y-axis is given by V_1 (say)

$$V_1 = \int_{2\pi}^{\pi} \pi x^2 \frac{dy}{d\theta}.d\theta$$

$$= \pi\int_{2\pi}^{\pi} a^2(\theta-\sin\theta)^2 . a\sin\theta\, d\theta$$

$$= \pi a^3 \int_{2\pi}^{\pi} (\theta^2+\sin^2\theta-2\theta\sin\theta)\sin\theta\, d\theta$$

$$= \pi a^3 \int_{2\pi}^{\pi} (\theta^2\sin\theta+\sin^3\theta-2\theta\sin^2\theta)\, d\theta$$

$$= \pi a^3 \int_{2\pi}^{\pi} \left(\theta^2\sin\theta+\frac{3}{4}\sin\theta-\frac{1}{4}\sin 3\theta-\theta+\theta\cos 2\theta\right) d\theta$$

$$= \pi a^3 \left[-\theta^2\cos\theta+2\theta\sin\theta+2\cos\theta-\frac{3}{4}\cos\theta + \frac{1}{12}\cos 3\theta-\frac{\theta^2}{2}+\frac{1}{2}\theta\sin 2\theta+\frac{1}{4}\cos 2\theta\right]_{2\pi}^{\pi}$$

(where $\int\theta^2\sin\theta\, d\theta$ and $\int\theta\cos 2\theta\, d\theta$ are solved by integration by parts.)

$$= \pi a^3\left[\left(\pi^2-2+\frac{3}{4}-\frac{1}{12}-\frac{\pi^2}{2}+\frac{1}{4}\right) - \left(-4\pi^2+2-\frac{3}{4}+\frac{1}{12}-2\pi^2+\frac{1}{4}\right)\right]$$

$$= \pi a^3\left(\frac{13}{2}\pi^2-\frac{8}{3}\right)$$

and let V_2 be the volume generated by the revolution of the area $OBCO$ about y-axis, then

$$V_2 = \int_0^{\pi} \pi x^2 \frac{dy}{d\theta} d\theta$$

$$= \pi\int_0^{\pi} a^2(\theta-\sin\theta)^2 a\sin\theta\, d\theta$$

$$= \pi a^3 \int_0^{\pi} (\theta^2\sin\theta+\sin^3\theta-2\theta\sin^2\theta)\, d\theta$$

$$= \pi a^3 \left[-\theta^2\cos\theta+2\theta\sin\theta+2\cos\theta-\frac{3}{4}\cos\theta + \frac{1}{12}\cos 3\theta-\frac{\theta^2}{2}+\frac{1}{2}\theta\sin 2\theta+\frac{1}{4}\cos 2\theta\right]_0^{\pi}$$

$$= \pi a^3\left[\left(\pi^2-2+\frac{3}{4}-\frac{1}{12}-\frac{\pi^2}{2}+\frac{1}{4}\right) - \left(2-\frac{3}{4}+\frac{1}{12}+\frac{1}{4}\right)\right]$$

$$V_2 = \pi a^3\left(\frac{1}{2}\pi^2-\frac{8}{3}\right).$$

$$\therefore \quad V = V_1 - V_2$$

$$= \pi a^3\left(\frac{13}{2}\pi^2-\frac{8}{3}\right) - \pi a^3\left(\frac{1}{2}\pi^2-\frac{8}{3}\right)$$

$$= \pi a^3(6\pi^2)$$

$$V = 6\pi^3 a^3.$$

Example 4. *Find the volume of a solid formed by the revolution of the loop of the curve $y^2(a+x) = x^2(a-x)$ about x-axis.*

Solution . Clearly, the given curve is symmetrical about x-axis and the curve cuts the x-axis only at the points (0, 0) and (a, 0) so the loop exists between these points. The tracing of this curve is shown in fig. 14.

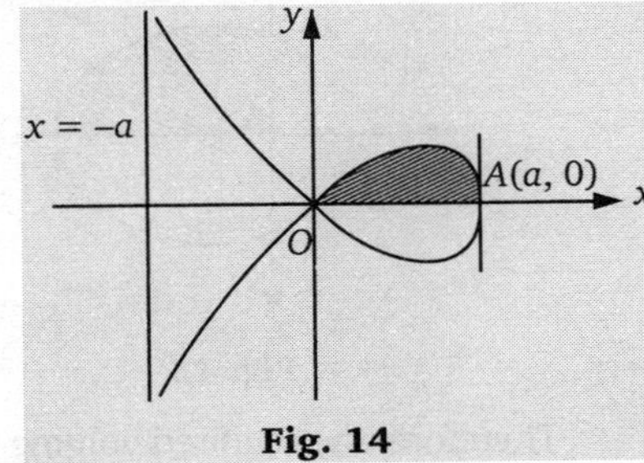

Fig. 14

Therefore, the required volume is the volume of a solid formed by the revolution of upper half of the loop of the curve about x-axis where x varies from 0 to a. Then

$$V = \int_0^a \pi y^2\, dx$$

$$= \pi\int_0^a \frac{x^2(a-x)}{a+x} dx \quad [\because y^2(a+x) = x^2(a-x)]$$

$$= \pi\int_0^a \frac{x^2(2a-a-x)}{a+x}dx$$

$$= \pi\int_0^a \frac{2ax^2}{a+x}dx - \pi\int_0^a x^2\,dx$$

$$= 2a\pi\int_0^a \frac{(x^2-a^2+a^2)}{a+x}dx - \pi\int_0^a x^2dx$$

$$= 2a\pi\left[\int_0^a (x-a)\,dx + a^2\int_0^a \frac{dx}{a+x}\right] - \pi\int_0^a x^2dx$$

$$= 2a\pi\left[\frac{x^2}{2} - ax + a^2\log(a+x)\right]_0^a - \pi\left[\frac{x^3}{3}\right]_0^a$$

$$= 2a\pi\left[\frac{a^2}{2} - a^2 + a^2\log 2a - a^2\log a\right] - \frac{\pi a^3}{3}$$

$$= 2a\pi\left[-\frac{a^2}{2} + a^2\log\frac{2a}{a}\right] - \frac{\pi a^3}{3}$$

$$= -\pi a^3 + 2a^3\pi\log 2 - \frac{\pi a^3}{3}$$

$$= 2a^3\pi\log 2 - \frac{4\pi}{3}a^3$$

$$\Rightarrow \quad V = 2\pi a^3\left[\log 2 - \frac{2}{3}\right].$$

Example 5. *Find the volume of the solid generated by the revolution of the cardioid $r = a(1+\cos\theta)$ about the initial line.* (VTU-2010; Kurukshatra-2009)

Solution . Obviously, the curve is symmetrical about the initial line and $r = 0$ when $\cos\theta = -1$ *i.e.*, $\theta = \pi$ and the maximum value of $r=2a$ when $\cos\theta=1$ *i.e.*, $\theta = 0$. Thus the tracing of the curve is as under :

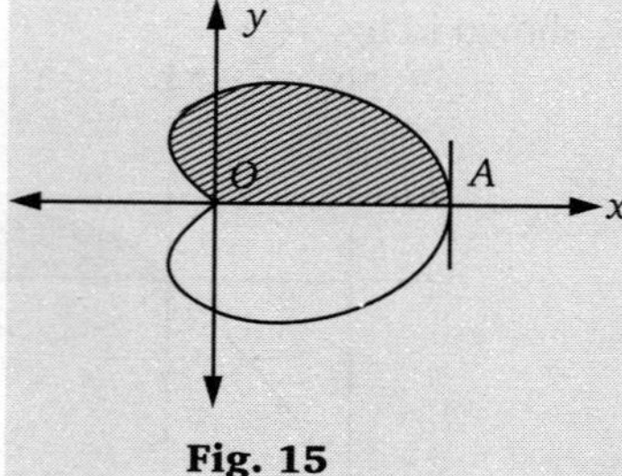

Fig. 15

Therefore, the required volume is the volume of solid generated by the revolution of the upper half of the curve between $\theta = 0$ and $\theta = \pi$ about initial line (x-axis). Let this volume be V. Then

$$V = \int_{\theta=0}^{\theta=\pi} \pi y^2 \frac{dy}{d\theta}.d\theta$$ (as θ increases x decreases so $\frac{dx}{d\theta}$ will have to take negative.)

$$= \pi\int_0^\pi (r\sin\theta)^2.\frac{d}{d\theta}(r\cos\theta)\,d\theta$$

($\therefore x = r\cos\theta,\ y = r\sin\theta$)

$$= -\pi\int_0^\pi a^2(1+\cos\theta)^2\sin^2\theta\frac{d}{d\theta}[a(1+\cos\theta)\cos\theta]d\theta$$

[$\therefore r = a\ (1+\cos\theta)$]

$$= -\pi a^3\int_0^\pi (1+\cos\theta)^2\sin^2\theta.(-\sin\theta - 2\cos\theta\sin\theta)d\theta$$

$$= +\pi a^3\int_0^\pi (1+\cos^2\theta + 2\cos\theta)(1+2\cos\theta).\sin^3\theta d\theta$$

$$= +\pi a^3\int_0^\pi (\sin^3\theta + 4\cos\theta\sin^3\theta + 5\cos^2\theta\sin^3\theta + 2\cos^3\theta\sin^3\theta)\,d\theta$$

$$= +\pi a^3\left[\int_0^\pi \sin^3\theta d\theta + 4\int_0^\pi \cos\theta\sin^3\theta d\theta + 5\int_0^\pi \cos^2\theta\sin^3\theta d\theta + 2\int_0^\pi \cos^3\theta\sin^3\theta d\theta\right]$$

$$= \pi a^3\left[\int_0^\pi \sin^3\theta d\theta + 5\int_0^\pi \cos^2\theta\sin^3\theta d\theta\right]$$

(The second and fourth integral vanish by the property of definite integral.)

$$= \pi a^3\left[2\int_0^{\pi/2} \sin^3\theta d\theta + 10\int_0^{\pi/2} \cos^2\theta\sin^3\theta d\theta\right]$$

(By the property of definite integral)

$$= \pi a^3\left[2.\frac{(3-1)}{3.1}.1 + 10.\frac{(2-1)(3-1)}{5.3.1}.1\right]$$

(By Walli's formula)

$$= \pi a^3\left[\frac{4}{3} + \frac{4}{3}\right] = \frac{8}{3}\pi a^3.$$

$$\therefore \quad V = \frac{8}{3}\pi a^3.$$

Aliter. The required volume is also taken as

$$V = \int_0^\pi \frac{2}{3}\pi r^3\sin\theta d\theta$$

$$= \frac{2}{3}\pi\int_0^\pi a^3(1+\cos\theta)^3\sin\theta d\theta$$

$$= \frac{2}{3}\pi a^3\int_0^\pi (1+\cos\theta)^3\sin\theta d\theta$$

$$= \frac{2}{3}\pi a^3\left[-\frac{(1+\cos\theta)^4}{4}\right]_0^\pi = \frac{2}{3}\pi a^3\left[\frac{16}{4}\right].$$

$$\therefore \quad V = \frac{8}{3}\pi a^3.$$

Example 6. *Find the volume of the solid generated by the revolution of the tractrix*

$$x = a\cos t + \frac{a}{2}\log\tan^2\left(\frac{t}{2}\right),\ y = a\sin t$$

about its asymptote.

Solution . We have

$$x = a\cos t + \frac{a}{2}\log\tan^2\left(\frac{t}{2}\right)$$

and $y = a \sin t$.

Now $\frac{dx}{dt} = -a\sin t + \frac{a}{\sin t} = \frac{a\cos^2 t}{\sin t}$.

Here, x-axis is the asymptote of the given curve so that for x-axis *i.e.*, $y = 0$, $t = \pi/2$ and at $A(0, a)$, $t = 0$.

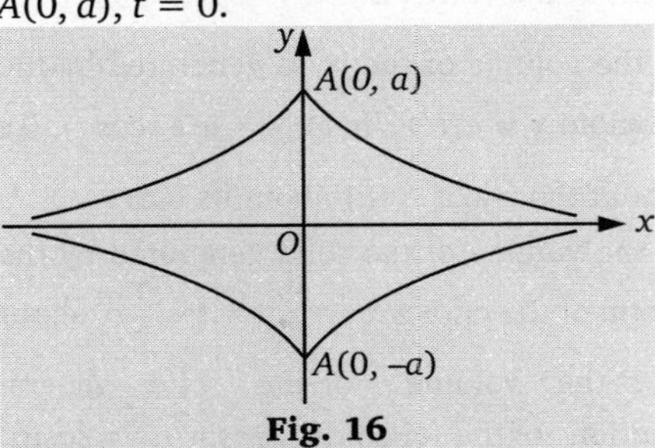

Fig. 16

Let V be the volume of the solid generated by the revolution of the tractrix. Then

$$V = 2\int_{\pi/2}^{0} \pi y^2 \frac{dx}{dt} dt$$

$$= 2\pi\int_{\pi/2}^{0} a^2 \sin^2 t \frac{a\cos^2 t}{\sin t} dt$$

$$= 2\pi a^3 \int_{\pi/2}^{0} \cos^2 t \sin t \, dt$$

$$= 2\pi a^3\left[\frac{\cos^3 t}{3}\right]_{\pi/2}^{0} = 2\pi a^3\left[\frac{1}{3} - 0\right] = \frac{2}{3}\pi a^3.$$

Example 7. *Find the volume of the solid generated by revolution of one loop o the lemniscate $r^2 = a^2 \cos 2\theta$ about the line $\theta = \pi/2$.* (UKTU 2006)

Solution . We have $r^2 = a^2 \cos 2\theta$. ...(1)

Clearly one loop of the curve lies between $\theta = -\pi/4$ and $\theta = \pi/4$.

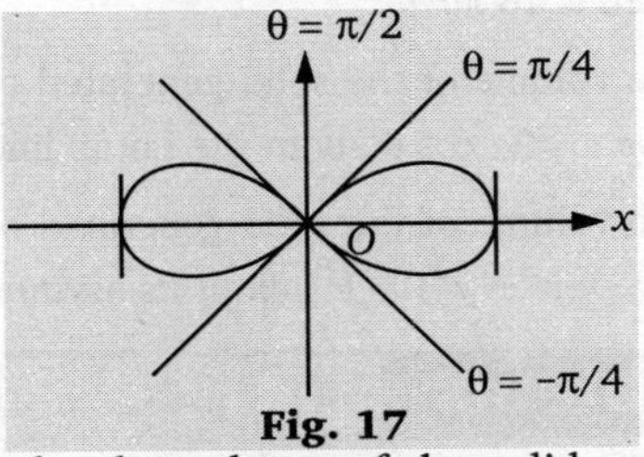

Fig. 17

Let V be the volume of the solid generated by revolving one loop of the curve (1) about the line $\theta = \pi/2$. Then

$$V = \int_{-\pi/4}^{\pi/4} \frac{2}{3}\pi r^3 \cos\theta \, d\theta$$

$$= \frac{2}{3}\pi\int_{-\pi/4}^{\pi/4} a^3(\cos 2\theta)^{3/2} \cos\theta \, d\theta$$

$$= \frac{2\pi a^3}{3}\int_{-\pi/4}^{\pi/4}(1 - 2\sin^2\theta)^{3/2}\cos\theta \, d\theta$$

$$= \frac{4\pi a^3}{3}\int_0^{\pi/4}(1 - 2\sin^2\theta)^{3/2}\cos\theta \, d\theta$$

$$= \frac{4\pi a^3}{3}\int_0^{\pi/4}(1 - \sin^2\phi)^{3/2}\frac{1}{\sqrt{2}}\cos\phi \, d\phi$$

put $\sqrt{2}\sin\theta = \sin\phi$

$$= \frac{4\pi a^3}{3\sqrt{2}}\int_0^{\pi/2}\cos^4\phi \, d\phi$$

$$= \frac{4\pi a^3}{3\sqrt{2}}\left[\frac{(4-1)(4-3)}{4.(4-2)}\cdot\frac{\pi}{2}\right] = \frac{\pi^2 a^3}{4\sqrt{2}}.$$

Example 8. *Find the volume of the solid generated by the revolution of the cissoid $y^2(2a–x) = x^3$ about its asymptotes.* (VTU -2000)

Solution . Let $P(x, y)$ be any point on the curve and let V be the volume of the solid generated by revolution of $y^2(2a–x) = x^3$ about its asymptote $x = 2a$ then

$$V = \int_{-\infty}^{\infty} \pi(PM)^2 dy$$

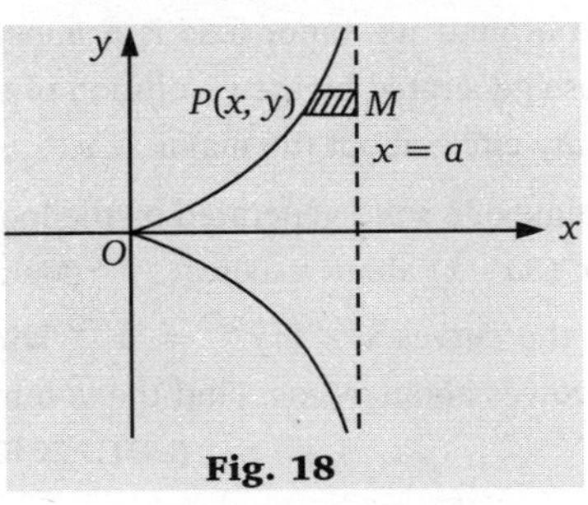

Fig. 18

$$= 2\int_0^{\infty} \pi(PM)^2 dy$$

$$V = 2\pi\int_0^{\infty}(2a - x)^2 dy \qquad ...(1)$$

Now, $y^2(2a - x) = x^3$

$$\Rightarrow \quad y = \frac{x^{3/2}}{\sqrt{2a - x}} \Rightarrow \frac{dy}{dx} = \frac{\sqrt{x}.(3a - x)}{(2a - x)^{3/2}}$$

Then $V = 2\pi\int_0^{2a}\frac{(2a - x)^2(3a - x)\sqrt{x}}{(2a - x)^{3/2}}dx$

[as $y \to 0$ to ∞, $x \to 0$ to $2a$]

$$= 2\pi\int_0^{2a}\sqrt{2ax - x^2}(3a - x)dx$$

$$= \pi\left[\int_0^{2a} 2(a - x)\sqrt{2ax - x^2}\,dx + 4a\int_0^{2a}\sqrt{a^2 - (x - a)^2}\,dx\right]$$

$$= \pi\left[\frac{2}{3}(2ax - x^2) + 4a\left\{\frac{(x - a)}{2}\sqrt{a^2(x - a)^2} + \frac{a^2}{2}\sin^{-1}\frac{(x - a)}{a}\right\}\right]_0^{2a}$$

$$= \pi\left\{0 + 4a\left\{\frac{a^2\pi}{2}\right\}\right\} = 2\pi^2 a^3.$$

EXERCISE 18.2

1. Show that the volume of sphere of radius a is $\frac{4}{3}\pi a^3$.

2. Find the volume of a hemi-sphere, (SVTV- 2007)

3. (i) The part of a parabola $y^2 = 4ax$ cut off by the latus rectum revolves about the tangent at the vertex. Find the volume of a solid thus generated. (VTU- 2009)

   (ii) Find the volume of the paraboloid generated by the revolution about the axis of x, of the parabola $y^2 = 4ax$ from $x = 0$ to x to $x = h$.

4. Find the volume of the solid generated by the revolution of an arc of the catenary $y = c \cosh (x/c)$ about x-axis.

5. Find the volume of solid generated by the revolution of the loop of the curve $y^2 = x^2 (a - x)$ about the axis of x.

6. Prove that the volume of this solid generated by the revolution of an ellipse around its minor axis is a mean proportional between those generated by the revolution of the ellipse and of the auxiliary circle about the major axis.

7. Find the volume of a solid generated by the loop of the curve $y^2(a + x) = x^2(3a - x)$ about x-axis. (Marathwada-2008)

8. The area of the curves $x^{2/3} + y^{2/3} = a^{2/3}$ lying in the first quadrant revolves about x-axis. Find the volume of the solid generated. (UPTU-2010, SVTU-2008)

9. Show that the volume of the solid generated by the revolution of the upper half of the loop of the curve $y^2 = x^2(2-x)$ about x-axis is $\frac{4}{3}\pi$.

10. Find the volume of the solid generated by revolving the loop of the curve $a^2y^2 = x^2(2a - x)(x - a)$ about x-axis.

11. Show that the volume of the solid generated by the revolution of the curve $(a-x)y^2 = a^2 x$, about its asymptote is $\frac{1}{2}\pi^2 a^3$.

12. The area cut off from the parabola $y^2 = 4ax$ by the chord joining the vertex to an end of the latusrectum is rotated through four right angles about the chord. Find the volume of the solid thus generated.

13. Prove that the volume of the reel formed by the revolution of the cycloid $x = a(t + \sin t)$, $y = a(1 - \cos t)$ about the tangent at the vertex is $\pi^2 a^3$.

14. Find the volume of the solid generated by the revolution of the cycloid $x = a(t + \sin t), y = a(t - \cos t), 0 \le t \le n$.

    (i) about the x-axis (ii) about its base.

15. Find the voiume of the solid generated by the revolution of the loop of the curves $x = t^2$, $y = t - \frac{1}{2}t^3$ about x-axis.

16. Find the volume of ihe solid generated by the revolution of the cissoid $x = 2a\sin^2 t, y = 2a\sin^3 t / \cos t$ about its asymptote. (PTU-2001)

17. Find the volume of the solid generated by the revolution of the cardioid $r = a(1 - \cos\theta)$ about the initial line. (PTU-2006)

18. (i) Find the volume of the solid formed by revolving one loop of the curve $r^2 = a^2 \cos 2\theta$ about the initial line.

    (ii) The lemniscate $r^2 = a^2 \cos 2\theta$ revolves about a tangent at the pole. Show that the volume generated is $\frac{\pi^2 a^3}{4}$.

19. Show that the volume of the solid formed by the revolution of the curve $r = a + b\cos\theta (a > b)$ about the initial line is $\frac{4}{3}\pi a(a^2 + b^2)$.

20. Show that if the area lying within the cardioid $r = 2a(1 + \cos\theta)$ and without the parabola $r(1 + \cos\theta) = 2a$ revolves about the initial line, the volume of a solid thus generated is $18\pi a^3$.

21. Find the volume of the solid generated by the revolution of the curve $r = 2a\cos\theta$ about the initial line.

22. Find the volume of the solid generated by the revolution of the curve $x(b^2 + y^2) = b^3$ about its asymptote.

Hint to Selected Problems

1. We know that, when a circle $x^2 + y^2 = a^2$ is revolved about its diameter a sphere is formed, therefore

$$V = \int_{-a}^{a} \pi y^2 dx = \pi\int_{-a}^{a} (a^2 - x^2)dx.$$

2. When a quadrant of a circle $x^2 + y^2 = a^2$ is revolved about its one of bounding radius, a hemi-sphere is generated.

$$\therefore\ V = \int_0^a \pi x^2 dy = \pi\int_0^a (a^2 - y^2)dy.$$

4. Use the formula $V = \int_0^x \pi y^2\, dx$.

6. The equation of the ellipse is given by $\frac{x^2}{a^2} + \frac{y^2}{b^2} = 1$.

Therefore, $V = \int_{-b}^{b} \pi x^2 dy = \frac{4\pi a^2 b}{3}$.

Let V_1 be the volume generated by an ellipse about major axis between $x = -a$ to $x = a$. Therefore

$$V_1 = \int_{-a}^{a} \pi y^2\, dx = \pi\int_{-a}^{a} b^2\left(1 - \frac{x^2}{a^2}\right)dx = \sqrt{\frac{4\pi}{3}b^2 a}.$$

Now let V_2 be the volume, when the circle revolves about major axis.

$\therefore$ $V_2 = \int_{-a}^{a} \pi y^2\,dx = \dfrac{4\pi}{3a^2}.$

7. Let V be the volume of a solid generated by the revolution of the loop about x-axis between $x = 0$ and $x = 3a$. Then, we have

$$V = \int_0^{3a} \pi y^2\,dx = \pi\int_0^{3a} \frac{x^2(3a-x)}{a+x}dx.$$

8. Here $V = \int_0^a \pi y^2\,dx$. Change the given equation into polar form by assuming

$$x = a\cos^3\theta \text{ and } y = a\sin^3\theta.$$

13. $\dfrac{dx}{dt} = a(1+\cos t), \dfrac{dy}{dt} = a(1-\cos t) \Rightarrow \dfrac{dy}{dx} = \tan\dfrac{t}{2}.$

Then, the required volume $V = \int_{-a\pi}^{a\pi} \pi y^2\,dx.$

17. The required volume is given by $V = \int_0^{\pi} \dfrac{2}{3}\pi r^3 \sin\theta\,d\theta.$

18. The required volume is given by $V = 2\int_0^{\pi/4} \dfrac{2}{3}\pi r^3 \sin\theta\,d\theta.$

20. $V = \int_0^{\pi/2} \dfrac{2}{3}\pi r^3 \sin\theta\,d\theta - \int_0^{\pi/2} \dfrac{2\pi}{3} r^3 \sin\theta\,d\theta.$

21. The required volume is given by

$$V = \int_0^{\pi/2} \frac{2}{3}\pi r^3 \sin\theta\,d\theta.$$

ANSWERS

2. $\dfrac{2}{3}\pi a^3$ **3. (i)** $\dfrac{4}{5}\pi a^3$ **(ii)** $2ah^2$ **4.** $\dfrac{\pi c^2}{2}\left[x + \dfrac{c}{2}\sinh\left(\dfrac{2x}{c}\right)\right]$ **5.** $\dfrac{1}{12}\pi a^4$ **7.** $\pi a^3[8\log 2 - 3]$ **8.** $\dfrac{16}{105}\pi a^3$ **10.** $\dfrac{23}{60}\pi a^3$

12. $\dfrac{2}{75}\sqrt{5}\,\pi a^3$ **14.(i)** $\dfrac{1}{2}\pi^2 a^3$ **(ii)** $\dfrac{5}{2}\pi^2 a^3$ **15.** $\dfrac{4}{3}\pi$ **16.** $2\pi^2 a^3$ **17.** $\dfrac{8}{3}\pi a^3$ **18.** $\dfrac{\pi a^3}{24}\sqrt{2}[3\log(\sqrt{2}+1) - \sqrt{2}]$ **21.** $\dfrac{4}{3}\pi a^3$ **22.** $\dfrac{1}{2}\pi^2 b^3$

18.8 PAPPUS AND GULDIN'S THEOREMS

18.8.1 FOR THE VOLUME OF A SOLID OF REVOLUTION.

Statement. *When a closed plane curve revolves about any straight line lying in the same plan e but not intersecting the closed curve, then the volume of a ring shape solid thus generated, is equal to the product of the area of a closed curve and the perimeter of a cirlce described by the centroid of the closed curve.*

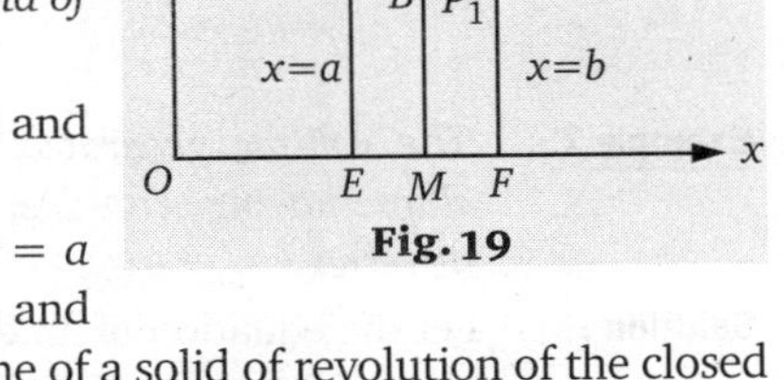

Fig. 19

Proof. Let $ABCDA$ be a closed curve which revolves about x-axis. Let the lines $x = a$ and $x = b$ touch the closed plane curve at the points A and C as shown in fig. 19.

Now draw a line through the centroid of the closed curve and parallel to the line $x = a$ and $x = b$. This line intersects the given curve at two points P_1 and P_2 such that $MP_1 = y_1$ and $MP_2 = y_2$ provided y_1 and y_2 both are the functions of x only. Therefore the required volume of a solid of revolution of the closed plane curve about x-axis is equal to the difference of the volume generated by the plane area $AEFCDA$ and the volume generated by the plane area $AEFCBA$ about x-axis.

Now the volume generated by $AEFCDA = \int_a^b \pi y_2^2\,dx$

and the volume generated by $AEFCBA = \int_a^b \pi y_1^2\,dx.$

$$\therefore \quad \text{Required volume, } V = \int_a^b \pi y_2^2\,dx - \int_a^b \pi y_1^2\,dx = \pi\int_a^b (y_2^2 - y_1^2)dx. \qquad \text{...(1)}$$

Let $(\bar{x}, \bar{y})$ be the co-ordinates of the centroid of the closed plane curve $ABCDA$. Then by the method of finding the centre of gravity, we have

$$\bar{y} = \frac{\int_a^b \frac{1}{2}(y_2+y_1)(y_2-y_1)dx}{A}$$ where A is the area of the closed plane curve.

$$\therefore \quad \bar{y}A = \frac{1}{2}\int_a^b (y_2^2 - y_1^2)dx \qquad \text{...(2)}$$

Now from (1) and (2), we get

Required volume $= 2\pi\bar{y}A = A \times 2\pi\bar{y}$

$=$ (area of the closed curve) $\times$ (perimeter of the circle whose radius is $\bar{y}$)

$V =$ (area of the closed curve)

$\times$ (perimeter of the circle described by the centroid of the closed curve)

18.8.2 FOR THE SURFACE OF A SOLID OF REVOLUTION.

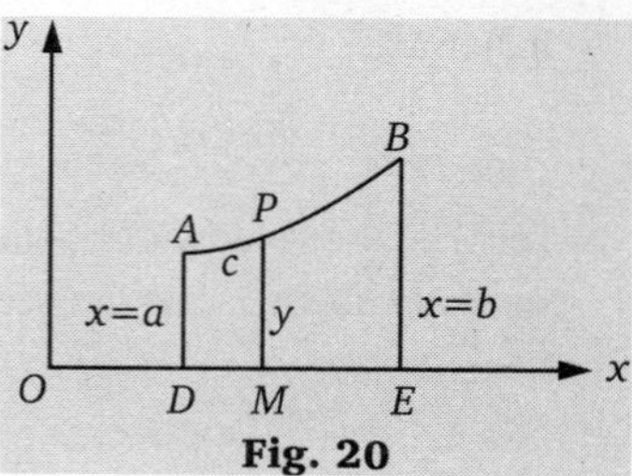

Fig. 20

Statement. *When an arc of a plane curve revolves about any straight line lying in the plane of curve but not intersecting it, then the surface area of a solid of revolution thus generated is equal to the product of the length of the arc and the perimeter of a cirlce described by the centroid of that arc.*

Proof. Let ACB be an arc of a plane curve cut off by the lines $x = a$ and $x = b (a<b)$. Let $P(x, y)$ be any point on the arc. Now draw a line parallel to the lines $x = a$ and $x = b$ through P which meets the x-axis at M such that $PM = y$. Suppose the arc revolves about the x-axis as shown in fig. 20.

Let l be the length of an arc ACB. Therefore the surface area is given by

$$S = \int_a^b 2\pi y \, ds \qquad ...(1)$$

where s is the iength of the arc measured from A to any point P.

Let $(\bar{x}, \bar{y})$ be the co-ordinates of the centroid of the arc ACB, then we know that

$$\bar{y} = \frac{\int_a^b y \, ds}{\int_a^b ds} = \frac{\int_a^b y \, ds}{l}$$

or $$l\bar{y} = \int_a^b y \, ds. \qquad ...(2)$$

From (1) and (2), we get

$$S = 2\pi \bar{y} \, l = l \times 2\pi \bar{y} = \text{length of the arc } ACB \times \text{perimeter of the circle of radius } y$$

$\therefore$ $S=$ (length of an arc ACB) $\times$ (perimeter of the circle described by the centroid of that arc)

REMARKS

- These theorems are only applicable when the closed curve or arc donot intersect the line of revolution.
- These theorems are also used to find the centroid of the closed curve or an arc only when volume and surface area of revolution are known.

Solved Examples

Example 1. *The volume generated by the revolution of an ellipse having semi-axes a and b about a tangent at vertex.*

Solution . Let the equation of an ellipse be

$$\frac{x^2}{a^2} + \frac{y^2}{b^2} = 1$$

The centroid of this ellipse is (0, 0)and the area is πab. There are four vertices $(\pm a, 0)$and $(0, \pm b)$. Now first we revolve the ellipse about tangent at $(a, 0)$, then the distance of the centroid (0, 0) from this tangent is a.

Thus the generated volume

= (area of ellipse)

$\times$ (perimeter of the circle of radius a)

$= \pi ab \times 2\pi a = 2\pi^2 a^2 b.$

Similarly, if we revolve the ellipse about the tangent at $(0, b)$, then the distance of the centroid (0, 0) from this tangent is b. Thus the required volume is

= (area of an ellipse)

$\times$ (perimeter of the circle of radius b)

$= \pi ab \times 2\pi b = 2\pi^2 ab^2.$

Example 2. *Find the position of the centroid of a semi-circular area.*

Solution . Let the equation of a circle whose radius is a and centre is (0,0) be

$$x^2 + y^2 = a^2$$

Let the semi-circular area be obtained by the circle $x^2 + y^2 = a^2$ and the x-axis as shown in fig. 21.

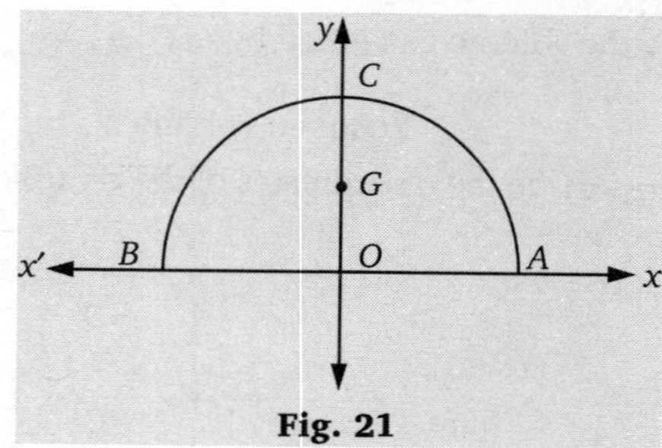

Fig. 21

Let V be the volume of a solid generated by the semi-circular area about the x-axis, between $x = -a$ to $x = a$. Then

$$V = \int_{-a}^{a} \pi y^2 \, dx$$

$$= \pi \int_{-a}^{a} (a^2 - x^2)\,dx$$

$$= \pi \left[a^2 x - \frac{x^3}{3} \right]_{-a}^{a} = \frac{4}{3}\pi a^3$$

and let A be the area of this semi-circle. Then

$$A = \frac{1}{2}(\pi a^2) = \frac{\pi a^2}{2}.$$

Let G be the centroid of this semi-circular area whose position be $\bar{y}$ from x-axis (axis of revolution).

Then by Pappus theorem, we have

$$V = (\text{area of the semi-circle}) \times 2\pi \bar{y}$$

$$\Rightarrow \frac{4}{3}\pi a^3 = \frac{\pi a^2}{2} \times 2\pi\bar{y}$$

$$\therefore \quad \bar{y} = \frac{4a}{3\pi}.$$

EXERCISE 18.3

1. Find the volume and surface area of the anchor-ring generated by the revolution of a circle of radius a about an axis in its own plane distance b from its centre ($b < a$).
2. The volume of the ring generated by the revolution of an ellipse of eccentricity $1/\sqrt{2}$ about a straight line parallel to the minor axis and situated at a distance from the centre equal to three times the major axis.
3. The loop of the curve $2ay^2 = x(x - a)^2$ revolves about the straight line $y = a$. Find the volume of the solid generated.
4. The volume of a ring generated by the revolution of the cardioid $r = a(1 + \cos\theta)$ about the line $r\cos\theta + a = 0$, given that the centroid of the cardioid is at a distance $5a/6$ from the pole.
5. Using Pappus theorem, determine the position of the centre of gravity of the quadrant of a uniform circular lamina of radius a, where the volume of the solid generated by the revolution of the quadrant of circular lamina about the tangent at either of its extremities is $\frac{\pi(3\pi - 4)}{6}a^3$.

Hint to Selected Problems

1. The area of the circle of radius A is given by $A = \pi a^2$.

   Circular circumference $= 2\pi b$.

   Then by Pappus and Guldin's theorem, we have

   Volume, $V = A \times 2\pi b = 2\pi^2 a^2 b$

   and surface $= 2\pi a \times 2\pi b = 4\pi^2 ab$.

2. Here $A = \pi a \left(\frac{a}{\sqrt{2}}\right) = \frac{1}{\sqrt{2}}\pi a^2$.

   Let l be the distance of straight line parallel to the minor axis from C.G. of the ellipse.

   Then $l = 3$ (Major axis) $= 6a$

   length of the circular path $= 12\,\pi a$

   $\therefore$ Required volume $V = \frac{1}{\sqrt{2}}\pi a^2 \times 12\pi a.$

4. $A = 2\int_0^{\pi} \frac{1}{2} r^2 d\theta = \frac{11\pi a}{3}.$

   The volume $V = A \times \frac{11\pi a}{3}.$

ANSWERS

1. Volume $= 2\pi^2 a^2 b$, surface $= 4\pi^2 ab$ **2.** $6\sqrt{2}\,\pi^2 a^3$, a beging the semi-major axis. **3.** $\frac{8}{15}\sqrt{2}\,\pi a^3$ **4.** $\frac{11}{2}\pi^2 a^3$

5. $d = \frac{(3\pi - 4)a}{3\pi}$, d being the distance of centroid from the tangent.

Objective Evaluations

Fill in the Blanks

1. The integral $\int_a^b 2\pi y\, ds$ represents .... of a solid generated by the revolution about x-axis of the area bounded by the curve $y = f(x)$, the ordinates $x = a$, $x = b$ and x-axis.
2. The integral $\int_\alpha^\beta 2\pi(r\sin\theta)\dfrac{ds}{d\theta}d\theta$ represents the surface area about.......... of a solid generated by the revolution of the curve $r = f(\theta)$ between $\theta = \alpha$ and $\theta = \beta$.
3. The surface area of a hemi-sphere of radius a is ....
4. The surface area of revolution of the curve $r = 2a\cos\theta$ about the initial line is ....
5. The surface area of a solid generated by the revolution of the cardioid $r = a(1 + \cos\theta)$ about initial line is..........
6. The integral $\int_a^b \pi[f(x)]^2 dx$ represents .... of a solid generated by the plane area bounded by $y = f(x)$, the lines $x = a$, $x = b$ and x-axis about x-axis.
7. The volume of a sphere of radius a is ....
8. The volume of a solid generated by the revolution of the cardioid $r = a(1 + \cos\theta)$ about initial line is ....
9. The volume of the reel formed by the revolution of the cycloid $x = a(t + \sin t)$, $y = a(1 - \cos t)$ about the tangent at the vertex is ....
10. The volume of the solid generated by the revolution of the curve $r = f'(\theta)$ and radii vectors $\theta = \theta_1$, $\theta = \theta_2$ about the initial line is.........

True/False

Write 'T' for True and 'F' for False statement.

1. The volume of a hemi-sphere of radius a is $\dfrac{2}{3}\pi a^3$.. **(T/F)**
2. The surface area of a sphere of radius a is $2\pi a^2$. **(T/F)**
3. The surface area of a solid generated by the revolution of the curve $x^{2/3} + y^{2/3} = a^{2/3}$ is $\dfrac{12}{5}\pi a^2$. **(T/F)**
4. The surface of the solid formed by the revolution of the cardioid $r = a(1 + \cos\theta)$ about the initial line is $\dfrac{31}{5}\pi a^2$. **(T/F)**
5. The surface area of the solid formed by the revolution of the curve $r = 2a\cos 2\theta$ about the line $\theta = 0$ is $4\pi a^2$. **(T/F)**
6. The volume of the solid generated by the revolution of the cardioid $r = a(1 - \cos\theta)$ about $\theta = 0$ is $\dfrac{8}{3}\pi a^3$. **(T/F)**
7. The volume of the solid generated by the revolution of the loop of the curve $y^2 = x^2(a - x)$ about x-axis is $\dfrac{1}{12}\pi a^3$. **(T/F)**
8. The volume of the solid generated by the revolution of the curve $y^2(2a - x) = x^3$ about its asymptote is $\pi^2 a^3$. **(T/F)**

Multiple Choice Questions

Choose the most appropriate one.

1. The surface area of a sphere of radius r is :
   (a) $2\pi r^2$ (b) $4\pi r^2$
   (c) πr^2 (d) $3\pi r^2$
2. The surface area of a solid generated by the revolution of the curve $r = 2a\cos\theta$ about $\theta = 0$ is :
   (a) $4\pi a^2$ (b) $3\pi a^2$
   (c) πa^2 (d) $2\pi a^2$
3. The surface area of a solid generated by the revolution of the curve $r = a(1 + \cos\theta)$ about $\theta = 0$ is:
   (a) $\dfrac{31}{5}\pi a^2$ (b) $\dfrac{32}{5}\pi a^3$
   (c) $\dfrac{32}{5}\pi a^2$ (d) πa^2
4. The surface area of a solid generated by fhe revolution of the curve $x = a\cos t + \dfrac{1}{2}a\log\tan^2 1/2$, $y = a\sin t$ about its asymptote is :
   (a) $3\pi a^2$ (b) $2\pi a^2$
   (c) πa^2 (d) $4\pi a^2$
5. The volume of a hemi-sphere of radius a is :
   (a) $\dfrac{2}{3}\pi a^2$ (b) $\dfrac{2}{3}\pi a^3$
   (c) $\dfrac{1}{3}\pi a^3$ (d) πa^3
6. The volume of a solid generated by the revolution of the cardioid $r = a(1 + \cos\theta)$ is :
   (a) $\dfrac{7}{3}\pi a^3$ (b) $\dfrac{8}{3}\pi a^2$
   (c) $\dfrac{8}{3}\pi a^3$ (d) $\dfrac{1}{3}\pi a^3$

7. The volume of the solid generated by the revolution of the curve $r = 2a \cos \theta$ about $\theta = 0$ is :

(a) $\frac{2}{3}\pi a^3$ (b) $\frac{4}{3}\pi a^3$

(c) $\frac{1}{3}\pi a^3$ (d) πa^3

8. The Volume of the solid formed by the revolution of the cycloid $x = a(\theta - \sin \theta), y = a(1 - \cos \theta)$ about its base is :

(a) $5\pi^2 a^3$ (b) $\pi^2 a^3$

(c) $6\pi^2 a^3$ (d) $5\pi^2 a^2$

9. The volume of the solid generated by the revolution of one loop of the curve $r^2 = a^2 \cos 2\theta$ about $\theta = \pi/2$ is:

(a) $\frac{1}{4}\pi^2 a^3$ (b) $\frac{1}{4\sqrt{2}}\pi^2 a^3$

(c) $\pi^2 a^3$ (d) a^3

ANSWERS

Fill in the Blanks

1. surface **2.** initial line **3.** $2\pi a^2$ **4.** $4\pi a^2$ **5.** $\frac{32}{5}\pi a^2$ **6.** volume **7.** $\frac{4}{3}\pi a^3$

8. $\frac{8}{3}\pi a^3$ **9.** $\pi^2 a^3$ **10.** $\int_{\theta_1}^{\theta_2} \frac{2}{3}\pi r^3 \sin\theta \, d\theta.$

True/False

1. T **2.** F **3.** T **4.** F **5.** T **6.** T **7.** T **8.** F

Multiple Choice Questions

1. (b) **2.** (a) **3.** (c) **4.** (d) **5.** (b) **6.** (c) **7.** (b) **8.** (a) **9.** (b)

FFFFF

GATE tutor

Key Terms and Results

- **Reduction formula :** The process relating one integral to one or more integrals of the same type is called a reduction and this relation between integrals is called a reduction formula.
- **Definite integral :** If $f(x)$ is continuous and non-negative function over a closed interval $[a, b]$ then $\int_a^b f(x)dx$ is called the definite integral of $f(x)$ between the limit a and b.
- **Rectification :** The process for finding the length of an arc of a plane curve between two given points on a curve is called rectification.
- **Quadrative :** The process for finding the area of the curve between the given range is called quadrative.

Results

- $\int \sin^m x \cos^n x dx = \dfrac{\cos^{n-1} x \sin^{m+1} x}{m+n} + \dfrac{n-1}{m+n} I_{m,n-2}$
- $\int \sin^m x dx = -\dfrac{\sin^{m-1} x \cos x}{m} + \dfrac{m-1}{m} I_{m-2}, m \neq 0$
- $\int \cos^n x dx = \dfrac{\sin x \cos^{n-1} x}{n} + \dfrac{n-1}{n} I_{n-2}, n \neq 0$
- $\int \sec^n x dx = \dfrac{\tan x \sec^{n-2} x}{n-1} + \dfrac{n-2}{n-1} I_{n-2}, n \neq 1$
- $\int \text{cosec}^m x dx = -\dfrac{\cot x \,\text{cosec}^{m-2} x}{m-1} + \dfrac{m-2}{m-1} I_{m-2}, m \neq 1$
- $\int \text{cosec}^m x \cos^n x dx = -\dfrac{\text{cosec}^{m-1} x \cot^{n-1} x}{m-1} - \dfrac{n-1}{m-1} I_{m-2,n-2}, m \neq 1$
- $\int \tan^m x \sec^n x dx = \dfrac{\tan^{m+1} x \sec^{n-2} x}{m+n-1} + \dfrac{n-2}{m+n-1} I_{m,n-2}, (m+n \neq 1)$
- $\int x^n \sin mx dx = -\dfrac{x^n \cos mx}{m} + \dfrac{nx^{n-1} \sin mx}{m^2} - \dfrac{n(n-1)}{m^2} I_{n-2}, m \neq 0$
- $\int x^n \cos nx dx = \dfrac{x^n \sin mx}{m} + \dfrac{nx^{n-1} \cos mx}{m^2} - \dfrac{n(n-1)}{m^2} I_{n-2}, m \neq 0$
- **Properties of Definite Integrals :**
  - $\int_a^b f(x)dx = 0$
  - $\int_a^b f(x)dx = \int_a^b f(u)du$ *i.e.*, The value of the definite integral is independent of the variable of integration.
  - $\int_a^b f(x)dx = -\int_b^a f(x)dx$
  - $\int_a^b f(x)dx = \int_a^c f(x)dx + \int_c^b f(x)dx \quad a < c < b$
  - $\int_0^a f(x)dx = \int_0^a f(a-x)dx$
  - If $f(x)$ is an even function of x then $\int_{-a}^a f(x)dx = 2\int_0^a f(x)dx$ and if $f(x)$ is an odd function of x then $\int_{-a}^a f(x)dx = 0$
  - $\int_0^{2a} f(x)dx = 2\int_0^a f(x)dx$ if $f(2a-x) = f(x)$.
  - $\int_0^{na} f(x)dx = n\int_0^a f(x)dx$ if $f(x + ma) = f(x)$ for all integral value of m and n.
- Length of curve $y = f(x)$ between $x = a$ and $x = b$ is

$$s = \int_a^b \left[\sqrt{1+\left(\frac{dy}{dx}\right)^2}\right] dx$$

- If curve is of the form $x = f(y)$ then

$$s = \int_c^d \left[\sqrt{1+\left(\frac{dx}{dy}\right)^2}\right] dy$$

- If the equation of the curve is $r = f(\theta)$ then

$$s = \int_{\theta_1}^{\theta_2} \left[\sqrt{r^2+\left(\frac{dr}{d\theta}\right)^2}\right] d\theta$$

- If the given curve is of the form $\theta = f(r)$ then

$$s = \int_{r_1}^{r_2} \left[\sqrt{1+\left(r\frac{d\theta}{dr}\right)^2}\right] dr$$

- If the given curve is of the form $p = f(r)$ *i.e.*, in pedal form then

$$s = \int_{r_1}^{r_2} \frac{rdr}{\sqrt{r^2-p^2}}$$

- Let $y = f(x)$ be a continuous curve in cartesian form and A be the area of the region bounded by the curve $y = f(x)$, then

$$A = \int_a^b ydx = \int_a^b f(x)dx$$

- If the given curve is in polar form *i.e.*, $r = f(\theta)$ then area

$$A = \frac{1}{2}\int_{\theta_1}^{\theta_2} r^2 d\theta$$

- Let S be the surface area of a solid which is generated by the reduction of the curve $y = f(x)$ about x-axis between the ordinate $x = a$ and $x = b$ and if s be the arc length then surface area

$$S = \int_a^b 2\pi y ds = \int_a^b 2\pi y \frac{ds}{dx}.dx \quad \text{where} \quad \frac{ds}{dx} = \int_a^b \sqrt{1+\left(\frac{dy}{dx}\right)^2}$$

- If the given equation in polar form then

$$S = \int 2\pi(r\sin\theta)\frac{ds}{d\theta}.d\theta \quad \text{where} \quad \frac{ds}{d\theta} = \sqrt{r^2+\left(\frac{dr}{d\theta}\right)^2}$$

◂ Some formulae for Volume

➢ $V = \int_a^b \pi y^2 dx$ if $y = f(x)$

➢ $V = \int_a^b \pi x^2 dy$ if $x = f(y)$

➢ $V = \int_a^b \pi y^2 \left(\frac{dx}{dt}\right).dt = \int_a^b \pi [g(t)]^2 \frac{dx}{dt}.dt$

if $x = f(t)$ and $y = g(t)$

Review Questions and Project Work

Descriptive Type Questions

1. Prove the following :

(i) $\int_0^a \frac{x^7}{\sqrt{a^2 - x^2}} dx = \frac{16}{35} a^7.$ (VTU–2006)

(ii) $\int_0^a \sin^4 x \cos^2 x dx = -\frac{\sin^3 x \cos^3 x}{6} + \frac{1}{2}\left\{-\frac{\sin x \cos^3 x}{4} + \frac{1}{16}(2x + \sin 2x)\right\}.$ (Raipur–2005)

(iii) $\int_0^\infty \frac{x^2}{(1+x^2)^{7/2}} dx = \frac{5\pi}{2048}.$ (VTU–2010)

(iv) $\int_0^{2a} x^2 \sqrt{2ax - x^2} dx = \frac{5\pi a^4}{8}.$ (VTU–2010)

(v) $\int_{\pi/3}^{\pi/2} \text{cosec}^3 \theta d\theta = \frac{1}{3} + \frac{1}{4}\log 3.$ (VTU–2008)

2. $I_n = \int_{\pi/4}^{\pi/2} \cot^n \theta d\theta$, $n > 2$, prove that $I_n = \frac{1}{n-1} - I_{n-1}.$ (Marathwada–2008)

3. Show that

$$\int_0^{\pi/2} \cos^m x \cos nx dx = \frac{m}{m+n}\int_0^{\pi/2} \cos^{m-1} x \cos(n-1)x dx$$

and hence deduce that $\int_0^{\pi/2} \cos^n x \cos nx dx = \frac{\pi}{2^{n+1}}.$ (SVTU–2008)

4. Prove the following :

(i) $\int_0^1 \frac{\log(1+x)}{1+x^2} dx = \frac{\pi}{8}\log 2$ (Cochin–2005)

(ii) $\int_0^\pi \frac{x \sin^3 x}{1+\cos^2 x} dx = \frac{\pi^2}{2} - \pi$ (Madras–2006)

(iii) $\int_0^\pi \log(1+\cos\theta) d\theta = -\pi \log_e 2$ (Madras–2003)

5. Show that the length of the arc of the parabola $x^2 = 4ay$ measured from the vertex to one extremities of the latus rectum is $a[\sqrt{2} + \log(1+\sqrt{2})]$. (Delhi-2002)

6. Show that the volume of solid formed by revolving about x-axis, the area enclosed by the parabola $y^2 = 4ax$ its evolute $27ay^2 = 4(x - 2a)^3$ and the x-axis is $80\pi a^3$.

7. Show that the volume of the solid obtained by revolving the lemicon $r = a + b \cos\theta$ $(a > b)$ about the initial line is $\frac{4}{3}\pi a(a^2 + b^2)$. (Gorakhpur–1999)

8. Show that the volume of the solid formed by revolving a loop of the leminscate $r^2 = a^2 \cos 2\theta$ about the initial line is $\frac{\pi a^3}{4}\left[\frac{1}{\sqrt{2}}\log(\sqrt{2}+1) - \frac{1}{3}\right]$ (Delhi–2002, JNTU–2003)

9. Show that the surface area of the solid generated by the revolution of the ellipse $\frac{x^2}{a^2} + \frac{y^2}{b^2} = 1$ about the x-axis is

$$2\pi ab\left[\frac{b}{a} + \frac{a}{\sqrt{a^2 - b^2}}\sin^{-1}\frac{\sqrt{a^2 - b^2}}{a}\right].$$

(Bhopal–2002, Raipur–2005)

10. Show that the surface area of the solid generated by revolving the cycloid $x = a(t - \sin t)$, $y = a(1 - \cos t)$ about the base is $\frac{64}{3}\pi a^2$.

Multiple Choice Questions

Choose the most appropriate one.

1. The value of $\int_0^{\pi/2} \sin^{99} x \cos x dx =$ (VTU–2009)

(a) 1/99 (b) $\pi/100$
(c) 9/100 (d) none of these

2. The value of $\int_0^1 x^{3/2}(1-x)^{3/2} dx =$ (VTU–2010)

(a) $3\pi/128$ (b) $\pi/32$
(c) $\pi/64$ (d) none of these

3. The entire length of the cardiod $r = 5(1 - \cos\theta) =$ (VTU–2009)

(a) 30 (b) 40
(c) 50 (d) none of these

4. The volume generated by the revolution of the curve $y = a^3(a^2 + x^2)^{-1}$ about its asymptotes is : (VTU–2010)

(a) $\pi a^3/2$ (b) $\pi^2 a^3/2$
(c) πa^2 (d) none of these

5. The length of the arc of the curve $y = \log \sec x$ from $x = 0$ to $x = \pi/4$ is : (Bhopal–2008)

(a) $\log_e 2$ (b) $\log_e 3$
(c) $\log_e(1+\sqrt{2})$ (d) none of these

6. The value of the integral $\int_0^a \frac{\sqrt{x}}{\sqrt{x} + \sqrt{a-x}} dx =$ (GATE(CE)–2011)

(a) 0 (b) $a/2$
(c) a (d) none of these

7. If $f(x)$ is an even function and a is a positive real number then $\int_{-a}^a f(x)dx =$ (GATE(ME)–2011)

(a) 0 (b) $2\int_0^a f(x)dx$
(c) $\int_0^a f(x)dx$ (d) None of these

8. The value of $\int_0^{\pi/6} \cos^4 3\theta \sin^3 6\theta d\theta =$ (GATE(CE)–2013)

(a) 1/15 (b) 15
(c) 3/8 (d) none of these

9. The value of the definite integral $\int_1^e \sqrt{x}\log(x)dx =$ **(GATE(ME)–2013)**

(a) $\sqrt{e^3}+\frac{2}{9}$ (b) $\sqrt{e^3}-\frac{2}{9}$

(c) $\frac{2}{9}\sqrt{e^3}+\frac{4}{9}$ (d) none of these

10. The area common to the circles $r = a$ and $r = 2a\cos\theta$ is equal to : **(GATE(ME)–2013)**

(a) $12.28\,a^2$ (b) $1.228\,a^2$

(c) $1.567\,a^2$ (d) none of these

11. A path AB in the form of one quarter of a circle of unit radius is shown in the figure. Integration of $(x + y)^2$ on path AB in a counter clockwise sense is : **(GATE(ME)–2009)**

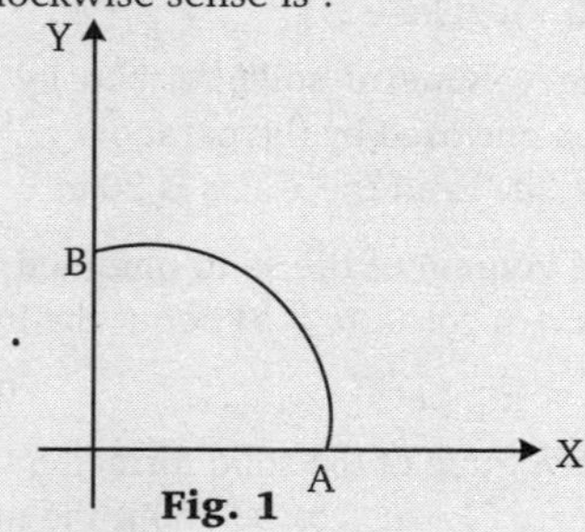

Fig. 1

(a) $\frac{\pi}{2}$ (b) $\frac{\pi}{2}+1$

(c) $\frac{\pi}{2}-1$ (d) none of these

12. A parabolic cable is held between two supports at the same level. The horizontal span between the supports is L. The sag at the mid span is h. The equation of the parabola is $y = 4h(x^2/L^2)$ where x is the horizontal coordinate and y is the vertical coordinate with the origin at the centre of the cable. The expression for the total length of the cable is : **(GATE(CE)–2012)**

(a) $\int_0^L \sqrt{1+64\frac{h^2x^2}{L^4}}\,dx$

(b) $2\int_0^{L/2} \sqrt{1+64\frac{h^2x^2}{L^4}}\,dx$

(c) both (a) and (b) are true

(d) none of these

13. The length of the curve $y = \frac{2}{3}x^{3/2}$ between $x = 0$ and $x = 1$ is : (GATE(ME)–2008)

(a) 1.21 (b) 1.20

(c) 1.22 (d) none of these

14. The parabolic arc $y = \sqrt{x}$, $1 \le x \le 2$ is revolved around the x-axis. The volume of the solid of revolution is : **(GATE(ME)–2010)**

(a) $3\pi/2$ (b) $\pi/2$

(c) $3\pi/4$ (d) none of these

15. The area enclosed between $y^2 = 4x$ and $x^2 = 4y$ is : **(GATE(ME)–2009)**

(a) 32/3 (b) 11/3

(c) 16/3 (d) none of these

16. The expression $V = \int_0^H \pi R^2\left(1-\frac{h}{H}\right)^2 dh$ for the volume of a cone is equal to : **(GATE(EE)–2006)**

(a) $\int_0^R \pi rH\left(1-\frac{r}{R}\right)^2 dr$ (b) $\int_0^H 2\pi rH\left(1-\frac{r}{R}\right)dh$

(c) $\int_0^R \pi R^2\left(1-\frac{h}{H}\right)^2 dr$ (d) none of these

17. The following plot shows a function y which varies linearity with x. The value of the integral $I = \int_1^2 y\,dx$: **(GATE(EC)–2007)**

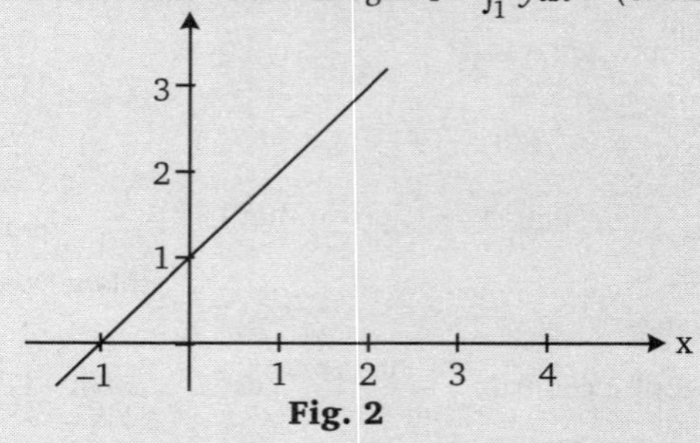

Fig. 2

(a) 5.0 (b) 4.0

(c) 2.5 (d) none of these

18. The area enclosed between the straight line $y = x$ and the parabola $y = x^2$ in the xy-plane : **(GATE(ME)–2012)**

(a) 1/6 (b) 1/3

(c) 5/6 (d) None of these

19. The value of the integral of the function $f(x, y) = 4x^3 + 10y^4$ along the straight line segment from the point (0, 0) to the point (1, 2) in the xy-plane is : **(GATE(EC)–2008)**

(a) 33 (b) 23

(c) 43 (d) none of these

20. If $f(r, \theta) = f(-r, \theta)$ then the curve is symmetrical about : **(VTU–2010)**

(a) pole (b) origin

(c) either (a) or (b) (d) none of these

ANSWERS

1. (d) **2.** (a) **3.** (b) **4.** (b) **5.** (c) **6.** (b) **7.** (b) **8.** (a) **9.** (c) **10.** (b)
11. (b) **12.** (b) **13.** (c) **14.** (a) **15.** (c) **16.** (a) **17.** (a) **18.** (a) **19.** (a) **20.** (c)

Hint to Selected Problems

1. $I = \int_0^a \frac{\sqrt{x}}{\sqrt{x}+\sqrt{a-x}}dx$ ...(1)

We know that $\int_0^a f(x)dx = \int_0^a f(a-x)dx$

$\therefore\ I = \int_0^a \frac{\sqrt{a-x}}{\sqrt{a-x}+\sqrt{x}}dx$ ...(2)

On adding (1) and (2), we get

$$2I = \int_0^a \frac{\sqrt{x}+\sqrt{a-x}}{\sqrt{x}+\sqrt{a-x}}dx = \int_0^a 1dx = a$$

$\Rightarrow$ $I = a/2.$

7. By the property of definite integral, if $f(x)$ is an even function then

$$\int_{-a}^{a} f(x)dx = 2\int_0^a f(x)dx$$

8. $\int_0^{\pi/6} \cos^4 3\theta \sin^3 6\theta d\theta = 8\int_0^{\pi/6} \cos^6 3\theta \sin^3 3\theta \cos 3\theta d\theta$

Let $\sin 3\theta = t \Rightarrow 3\cos 3\theta\, d\theta = dt$

$$\therefore\ I = \frac{8}{3}\int_0^1 (1-t^2)^3 t^3 dt = \frac{8}{3}\int_0^1 (t^6 - 2t^2 + 3t^4)t^3 dt$$

$$= \frac{8}{3}\int_0^1 (t^3 - t^9 - 3t^5 + 3t^7)dt = \frac{8}{3}\left[\frac{t^4}{4} - \frac{t^{10}}{10} - \frac{3t^6}{6} + \frac{3t^8}{8}\right]_0^1$$

$$= \frac{8}{3}\left[\frac{1}{4} - \frac{1}{10} - \frac{3}{6} + \frac{3}{8}\right] = \frac{1}{15}.$$

11. $x^2 + y^2 = 1$

$x = \cos\theta, y = \sin\theta$

Along path AB, θ varies from 0° to 90°

$$\int_{\text{path } AB} (x+y)^2 (rd\theta) = \int_0^{\pi/2} (\cos\theta + \sin\theta)^2 d\theta$$

$$= \int_0^{\pi/2} (\sin^2\theta + \cos^2\theta + 2\sin\theta\cos\theta)d\theta$$

$$= \int_0^{\pi/2} (1 + \sin 2\theta)d\theta$$

$$= \left[\theta + \frac{(-\cos 2\theta)}{2}\right]_0^{\pi/2}$$

$$= \frac{\pi}{2} - \frac{1}{2}\left(\cos 2\frac{\pi}{2} - \cos 0\right)$$

$$= \frac{\pi}{2} - \frac{1}{2}(-1-1) = \frac{\pi}{2} + 1$$

12. We know that

$$s = \int_a^b \left(\sqrt{1 + \left(\frac{dy}{dx}\right)^2}\right)dx$$

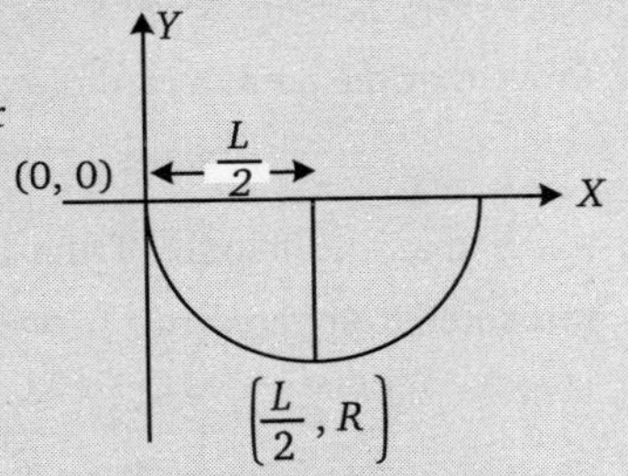

Here $y = 4h\dfrac{x^2}{L^2}$

$\Rightarrow \dfrac{dy}{dx} = 8h\dfrac{x}{L^2}$

Also $y = 0$ at $x = 0$

$\Rightarrow \quad y = h$ at $x = \dfrac{L}{2}$

(0, 0) $\dfrac{L}{2}$ X Y $\left(\dfrac{L}{2}, R\right)$

Fig. 3

$$\frac{1}{2}(\text{length of Cable}) = \int_0^{L/2}\left(\sqrt{1 + \left(\frac{dy}{dx}\right)^2}\right)dx$$

$$= \int_0^{L/2}\sqrt{1 + \left(\frac{8hx}{L^2}\right)^2}dx$$

$$\Rightarrow \text{length of cable} = 2\int_0^{L/2}\sqrt{1 + 64\frac{h^2x^2}{L^4}}dx$$

13. Since $y = \dfrac{2}{3}x^{3/2} \Rightarrow \dfrac{dy}{dx} = x^{1/2}$

$$\therefore\ s = \int_0^1\left(\sqrt{1 + \left(\frac{dy}{dx}\right)^2}\right)dx = \int_0^1 \sqrt{1+x}dx$$

$$= \left[\frac{2}{3}(1+x)^{3/2}\right]_0^1 = 1.22.$$

14. Volume $= \int_a^b \pi y^2 dx$

Here $a = 1$, $b = 2$ and $y = \sqrt{x} \Rightarrow y^2 = x$

$$\therefore\ V = \int_1^2 \pi x dx = \pi\left[\frac{x^2}{2}\right]_1^2 = \frac{\pi}{2}[x^2]_1^2 = \frac{3}{2}\pi.$$

15. Clearly point of intersection of the given parabolas are (4, 4), (0, 0).

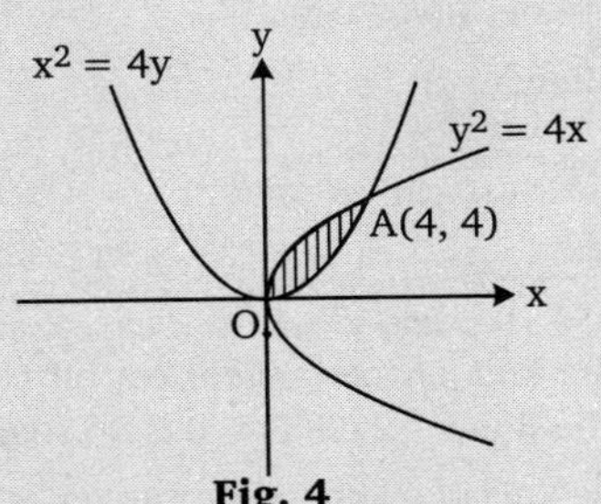

Fig. 4

Required area

$$= \int_{x_1}^{x_2} y_1 dx - \int_{x_1}^{x_2} y_2 dx = -\int_0^4 \frac{x^2}{4}dx = 2\left.\frac{x^{3/2}}{3/2}\right|_0^4 - \left.\frac{x^3}{3\times 4}\right|_0^4$$

$$= \frac{4}{3}(4)^{3/2} - \frac{(4)^3}{3\times 4} = \frac{16}{3}.$$

17. Equation of the line with slope1 and y-intercept of 1 is given by

$y = x + 1$

Therefore, $I = \int_1^2 ydx = \int_1^2 (x+1)dx$

$$= \left.\frac{(x+1)^2}{2}\right|_1^2 = \frac{1}{2}(9-4) = 2.5$$

18. Point of intersection of the given curves are (0, 0), (1, 1).

$\therefore$ Required area

$$= \int_0^1 xdx - \int_0^1 x^2 dx$$

$$= \left[\frac{x^2}{2}\right]_0^1 - \left[\frac{x^3}{3}\right]_0^1 = \frac{1}{2} - \frac{1}{3} = \frac{1}{6}$$

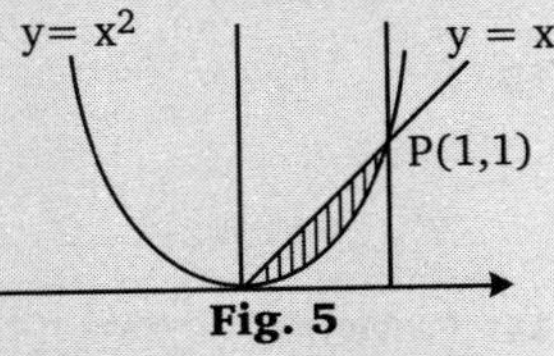

Fig. 5

19. Equation of the straight line from the point (0, 0) to (1, 2) is

$$y - 0 = \frac{2-0}{1-0}(x-0) \quad \Rightarrow \quad y = 2x$$

$\therefore\ f(x, y) = 4x^3 + 10y^4 = 4x^3 + 10(2x)^4 = 4x^3 + 160x^4$

$$\therefore\ \int_0^1 (4x^3 + 160x^4)dx = \left.\left(\frac{4x^4}{4} + \frac{160x^5}{5}\right)\right|_0^1 = 1 + 32 = 33$$

Self Assessment Test

1. Find the reduction formulae for

   (i) $\int \tan^n x dx$ (ii) $\int \cot^n x dx$

2. Prove that $\int_0^{\pi/2} \frac{\sqrt{\sin x}}{\sqrt{\sin x}+\sqrt{\cos x}} dx = \frac{\pi}{2}$.

3. If $I_n = \int_0^{\pi/2} x \cos^n x dx, n > 1$ prove that $I_n = \frac{n-1}{n} I_{n-2} - \frac{1}{n^2}$.

4. Prove that $\int_0^{\pi} \theta \sin^2\theta \cos^4\theta d\theta = \frac{\pi^2}{32}$.

5. Prove the following :

   (i) $\int_0^{\pi} \frac{xdx}{a^2\cos^2 x + b^2 \sin^2 x} = \frac{\pi^2}{2ab}$.

   (ii) $\int_0^{\pi} \frac{xdx}{a^2 - \cos^2 x} = \frac{\pi^2}{2a\sqrt{a^2-1}}, a > 1$.

6. Show that the area of the tangent cut off from the parabola $x^2 = 8y$ by the line $x - 2y + 8 = 0$ is 36 square units.

7. Show that the area of a loop of the curve $x^3+y^3=3axy$ is $\frac{3a^2}{2}$.

8. Show that the area included between the folium $x^3+y^3=3axy$ and its asymptotes is equal to the area of the loop.

9. Show that the whole length of the curve $x^2(a^2-x^2) = 8a^2y^2$ is $\pi a\sqrt{2}$.

10. Show that the volume of the solid formed by the revolution of the curve $(a^2 + x^2) = a^2$ about its asymptotes is $\frac{1}{2}\pi^2 a^3$.

11. Show that the surface area of the solid generated by the revolution of the curve $x = a\cos^3 t, y = a \sin^3 t$ about x-axis is $12\frac{\pi^2}{5}$.

12. Show that the surface of the solid generated by the revolution of the tractrix $x = a\cos t + \frac{a}{2}\log\tan^2 t/2, y = a\sin t$ about x-axis is $4\pi a^2$.

13. If $\int_0^{\pi} \frac{dx}{a+b\cos x} = \frac{\pi}{\sqrt{a^2-b^2}}$ $(a > b)$ then show that $\int_0^{\pi} \frac{\cos x}{(a+b\cos x)^2} dx = \frac{\pi b}{(a^2-b^2)^{3/2}}$. (Madras-2006)

14. Prove that $\int_0^1 \frac{\log(1+x)}{1+x^2} dx = \frac{\pi}{8}\log_e 2$. (Hissar-2005)

15. Prove that $\int_0^{\infty} \frac{\tan^{-1} ax}{x(1+x^2)} dx = \frac{\pi}{2}\log(1+a), a \geq 0$

    (VTU-2010, SVTU-2009, Rohtak-2006, Anna-2005, 09)

16. Prove that $\int_0^{\pi/2} \frac{\log(1+y\sin^2 x)}{\sin^2 x} dx = \pi[\sqrt{1+y} - 1]$. (SVTU-2008)

17. Using reduction formula, prove that $\int_0^a \frac{x^7}{\sqrt{a^2-x^2}} dx = \frac{1}{a^{2n-1}} \frac{(2n-3)(2n-5)...3.1}{(2n-2)(2n-4)...4.2}\left(\frac{\pi}{2}\right)$. (VTU-2006)

18. Prove that $\int_0^{\infty} \frac{x^2}{(1+x^2)^{7/2}} dx = \frac{2}{15}$. (VTU-2010)

19. If $I_n = \int x^n \sqrt{a-x}\, dx$, prove that $(2n+3) I_n = 2anI_{n-1} - 2x^n(a-x)^{3/2}$. (Marathwada-2008)

20. Prove that $\int_0^{\pi/2} \cos^n x \cos nx\, dx = \frac{\pi}{2^{n+1}}$. (SVTU-2008)

21. Prove that $\int_0^{\pi} \theta \sin^2\theta \cos^4\theta\, d\theta = \frac{\pi^2}{32}$. (VTU-2009)

22. Prove that the area of cardioid $r = a(1-\cos\theta) = \frac{3\pi a^2}{2}$. (VTU-2004)

23. Show that the length of the arc of the parabola $x^2=4ay$ measured from the vertex to one extremity of the latusrectum is given by $a(\sqrt{2} + \log(1+\sqrt{2}))$. (Delhi-2002)

24. Show that the surface of the solid generated by the revolution of the ellipse $2\pi ab\left\{\frac{b}{a} + \frac{a}{\sqrt{a^2-b^2}}\sin^{-1}\sqrt{\frac{(a^2-b^2)}{a}}\right\}$. (SRM-2008)

25. Show that the surface area of the solid generated by revolving the cyloid $x = a(t - \sin t)$, $y = a(1-\cos t)$ about the base is given by $\frac{64}{3}\pi a^2$.

    (Marathwada-2008, cohin-2005, Kurukshetra-2005)

CCCCCC

INTEGRAL CALCULUS-II

UNIT OUTLINES

- **Multiple Integrals**
- **Change of order of Integrations**
- **Beta and Gamma Functions**

PRE-REQUISITE

- **Definite Integral :** The concept of definite integral $\int_a^b f(x)\,dx$ is physically the area under a curve $y = f(x)$, the x-axis and the two ordinates $x = a$ and $x = b$.
- **Double Integration:** Let $f(x, y)$ be a single valued and bounded function of two idependent variables x and y defined in a closed region A in xy-plane. Let A be devided into n elementry area $\delta A_1, \delta A_2, \ldots, \delta A_n$.

  Then limit of the sum $\sum_{r=1}^{n} f(x, y)\,\delta Ar$ if exists as $n \to \infty$ and each sub elementry area approaches to zero is called double integral of $f(x, y)$.

CHAPTER 19 Multiple Integrals

19.1 INTRODUCTION

Double integral is an extension of a definite integral in two-dimensional space.

Let (x, y) be a single valued function of x and y, bounded and defined in the region R of XY-plane, and A be the area of region R and let R be divided in any manner into n-sub regions $\alpha_1, \alpha_2, ..., \alpha_n$, whose areas are $\delta s_1, \delta s_2, ... \delta s_n$ respectively. If $p_r(\xi_r, \eta_r)$ is any point inside the region α_n. $\beta_n = f(\xi_1)$.

Let $B_n = \sum_{r=1}^{n} f(\xi_r, \mu_r)\delta s_r$ then the limits of B_n which is assumed to exists as $n \to \infty$ such that every $\alpha_r \to 0$ in all its dimensions is known as double integral of $f(x, y)$ over the region R and is denoted by

$$\int_R f(x,y)ds \quad \text{or} \quad \iint_R f(x,y)dx\,dy .$$

Y

R

$B_n(\xi_r, \eta_r)$

δs_r

O

X

Fig. 1

Hence, the area R is called the region or field of integration for the double integral and ds is called element of area.

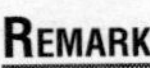

REMARK

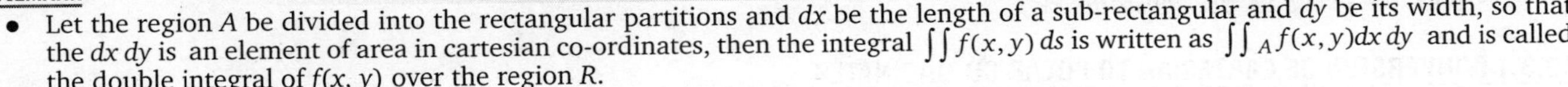

- Let the region A be divided into the rectangular partitions and dx be the length of a sub-rectangular and dy be its width, so that the $dx\,dy$ is an element of area in cartesian co-ordinates, then the integral $\iint f(x,y)\,ds$ is written as $\iint_A f(x,y)dx\,dy$ and is called the double integral of $f(x, y)$ over the region R.

19.2 PROPERTIES OF DOUBLE INTEGRALS

(1) When the region R is partitioned into two parts say R_1 and R_2 then

$$\iint_R f(x,y)dx\,dy = \iint_{R_1} f(x,y)dx\,dy + \iint_{R_2} f(x,y)dx\,dy$$

Similarly, we divide the region into three or more parts.

(2) The double integral of a algebraic sum of a fixed number of functions is equal to the algebraic sum of double integrals taken for each term separately. Thus

$$\iint_R [f_1,(x,y) + f_2(x,y) + f_3(x,y) + ...]dx\,dy$$
$$= \iint_R f_1(x,y)\,dx\,dy + \iint_R f_2(x,y)\,dx\,dy + \iint_R f_3(x,y)\,dx\,dy + ...$$

(3) A constant factor may be taken outside the integral sign. Thus

$$\iint_R mf(x,y)dx\,dy = m\iint_R f(x,y)dx\,dy \text{ where } m \text{ is a constant.}$$

19.3 EVALUATION OF DOUBLE INTEGRALS

(i) *Over a rectangular region R.* If the region R be given by the inequalities $a \le x \le b$, $c \le y \le d$, then the double integral

$$\iint_R f(x,y)dx\,dy = \int_a^b \int_c^d f(x,y)dx\,dy$$
$$= \int_a^b \left[\int_c^d f(x,y)dy\right]dx . \quad ...(1)$$

We first evaluate $\int_c^d f(x,y)dy$ *i.e.*, integrate $f(x, y)$ with respect to y regarding x as constant and then resulting function of x is to be integrated with respect to x between the limits a and b

$$\iint_R f(x,y)dx\,dy = \int_c^d \int_a^b f(x,y)\,dx\,dy$$

or

$$= \int_c^d \left[\int_a^b f(x,y)dx\right]dy \cdot \quad \text{... (2)}$$

Now, we integrate $\int_a^b f(x,y)dx$ and then integrate with respect to y.

(ii) *Over the regions which are not rectangular.* Let the region R be described by $a \le x \le b$ and $\phi_1(x) \le y \le \phi_2(x)$ so that $y = \phi_1(x)$ and $y = \phi_2(x)$ respectively, the boundary of R then

$$\iint_R f(x,y)dx\,dy = \int_a^b \left[\int_{\phi_1(x)}^{\phi_2(x)} f(x,y)dy\right]dx$$

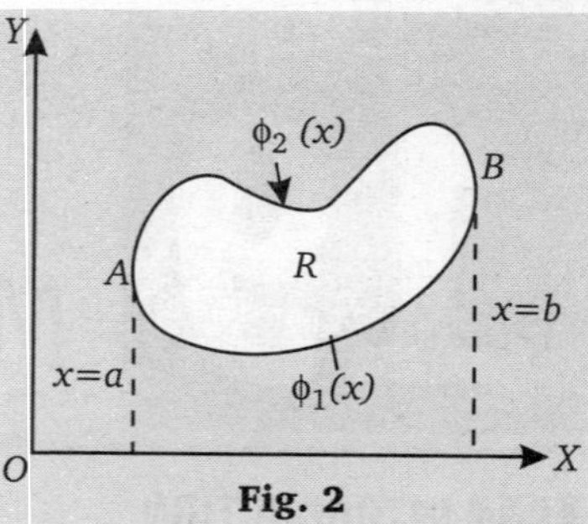

Fig. 2

Here, the inner integral $\int_{\phi_1(x)}^{\phi_2(x)} f(x,y)dy$ is integrated first and in this integral the result of integration is a function of x, say $\phi_1(x)$ then $\phi_1(x)$ is integrated with respect to x between the limits a and b to obtain the value of double integral.

In a similar way, if R can be described by

$$c \le y \le d, \;\; \phi_3(y) \le x \le \phi_4(y)$$

then we get

$$\iint_R f(x,y)dx\,dy = \int_c^d \left[\int_{\phi_3(y)}^{\phi_4(y)} f(x,y)dx\right]dy\,.$$

Here, the result of integration

$$\int_{\phi_3(y)}^{\phi_4(y)} f(x,y)dx\,.$$

which is evaluated first, is a function of y say $\phi_2(y)$, then $\phi_2(y)$ is integrated with respect to y between the limits c to d.

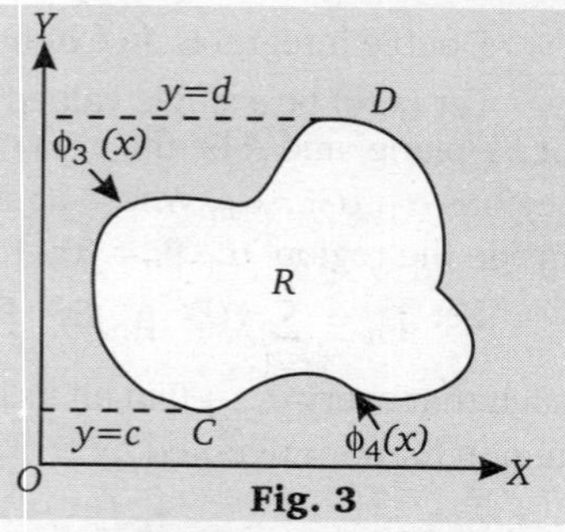

Fig. 3

WORKING PROCEDURE

- *While evaluating double integrals, first integrate with respect to variable having variable limits and treating the other variable as constant and then integrate with respect to variable with constant limits. In case the limits of integration of both the variables are constants.*

19.3.1 CONVERSION OF CARTESIAN TO POLAR CO-ORDINATES

The transformation formula required is $x = r\cos\theta, y = r\sin\theta$ and elementary area $\delta A = r\delta\theta.\,\delta r$ so that

$$\iint f(x,y)dx\,dy = \iint f(x,y)dA = \iint f(r,\theta)r\,d\theta\,dr.$$

Solved Examples

Example 1. *Evaluate* $\int_1^2\int_0^{y/2} y\,dy\,dx$.

Solution. We have

$$\int_1^2\int_0^{y/2} y\,dy\,dx = \int_1^2 y(x)_0^{y/2}\,dy = \int_1^2 y\left(\frac{1}{2}y\right)dy$$

$$= \frac{1}{2}\int_1^2 y^2 dy = \frac{1}{2}\left[\frac{1}{3}y^3\right]_1^2 = \frac{1}{6}(2^3 - 1^3) = 7/6\,.$$

Example 2. *Evaluate* $\int_1^2\int_0^x \frac{1}{x^2+y^2}dx\,dy$.

Solution. We have

$$\int_1^2\int_0^x \frac{dx\,dy}{x^2+y^2} = \int_1^2\left[\int_0^x \frac{dy}{x^2+y^2}\right]dx$$

$$= \int_1^2\left[\frac{1}{x}\tan^{-1}\frac{y}{x}\right]_{y=0}^{x} dx$$

$$= \int_1^2\left[\frac{1}{x}(\tan^{-1}1 - \tan^{-1}0)\right]dx$$

$$= \int_4^\pi \int_1^2 \frac{dx}{x} = \frac{\pi}{4}[\log x]_1^2$$

$$= \frac{\pi}{4}.[\log 2 - \log 1] = \frac{1}{4}\pi\log 2.$$

Example 3. *Show that* $\int_1^2\int_0^{y/2} y\,dy\,dx = \int_1^2\int_0^{x/2} x\,dx\,dy$.

Solution. We have

$$\int_1^2\int_0^{y/2} y\,dy\,dx = \int_1^2 [y]\left[\int_0^{y/2} dx\right]dy$$

$$= \int_1^2 y[x]_0^{y/2}dy = \int_1^2 y[y/2 - 0]dy$$

$$= \frac{1}{2}\int_1^2 y^2 dy = \frac{1}{2}\left[\frac{y^3}{3}\right]_1^2 = \frac{7}{6}.$$

Again

$$\int_1^2\int_0^{x/2} x\,dx\,dy = \int_1^2 x\left[\int_0^{x/2} dy\right]dx$$

$$= \int_1^2 x[y]_0^{x/2}dx$$

$$= \int_1^2 x\left[\frac{x}{2}-0\right]dx = \frac{1}{2}\int_1^2 x^2 dx$$

$$= \frac{1}{2}\left[\frac{x^3}{3}\right]_1^2 = \frac{1}{6}(8-1) = \frac{7}{6}.$$

Hence, $\int_1^2\int_0^{y/2} y\,dy\,dx = \int_1^2\int_0^{x/2} x\,dx\,dy$.

Example 4. *Evaluate the double integral of x^2y^3 over the rectangle bounded by x=2, x=3, y=2, y=4.*

Solution. The required integral

$$= \int_2^3\int_0^4 x^2y^3 dx\,dy = \int_2^3\left[\frac{1}{4}y^4\right]_2^4 x^2 dx$$

$$= \frac{1}{4}(4^4-2^4)\int_2^3 x^2 dx$$

$$= 60\left[\frac{1}{3}x^3\right]_2^3 = 20[27-8] = 380.$$

Example 5. *Evaluate $\int_0^3\int_1^2 xy(1+x+y)\,dx\,dy$*

Solution. We have $\int_0^3\int_1^2 xy(1+x+y)\,dx\,dy$

$$= \int_0^3\left[x\frac{y^2}{2}+x^2\frac{y^2}{2}+x\frac{y^3}{3}\right]_{y=1}^2 dx$$

$$= \int_0^3\left[\frac{x}{2}(4-1)+\frac{x^2}{2}(4-1)+\frac{x}{3}(8-1)\right]dx$$

$$= \int_0^3\left[\left(\frac{3}{2}+\frac{7}{3}\right)x+\frac{3}{2}x^2\right]dx$$

$$= \left[\frac{23}{6}.\frac{x^2}{2}+\frac{3}{2}.\frac{x^2}{3}\right]_0^3$$

$$= \frac{23}{6}.\frac{9}{2}+\frac{27}{2} = \frac{123}{4}.$$

Example 6. *Evaluate $\iint_A (x^2+y^2)\,dx\,dy$, where A is the region bounded by x=0, y=0, x+y=1.*

Solution. Let R be the region of integration x+y=1 and the limit of itegration can be expressed as $0 \le x \le 1, 0 < y < 1-x$.

From the equation x+y =1, we have $x = 1$ for $y = 0$ and for the positive quadrant x varies from 0 to 1 and for y which varies from $y = 0$ to $y = 1-x$. First integrate with respect to y, treated x as constant and then integrate with respect to 'x'.

Hence, the integral

$$= \int_0^1\int_0^{1-x}(x^2+y^2)\,dx\,dy = \int_0^1\left(x^2y+\frac{1}{3}y^3\right)_0^{1-x}dx$$

$$= \int_0^1\left[x^2(1-x)+\frac{1}{3}(1-x)^3\right]dx$$

$$= \int_0^1(1-x)\left\{x^2+\frac{1}{3}(1-x)^2\right\}dx$$

$$= \int_0^1\frac{1}{3}[1-3x+6x^2-4x^3]dx$$

$$= \frac{1}{3}\left[x-\frac{3}{2}x^2+2x^3-x^4\right]_0^1$$

$$= \frac{1}{3}\left[1-\frac{3}{2}+2-1\right] = \frac{1}{6}$$

Example 7. *Find the area by double integration the region bounded by circle $x^2+y^2=a^2$.*

Solution. The area of a small element at any point (x, y) is $dx\,dy$. Now to find the area bounded by the circle $x^2+y^2=a^2$, the region of integration R can be expressed as

$-a \le y \le a, -\sqrt{a^2-y^2} \le x \le \sqrt{(a^2-y^2)}$.

Now, first integration is to be performed w.r. to x regarding y as constant.

∴ The required area

$$= \iint_R dx\,dy = \int_{y=-a}^{a}\int_{x=-\sqrt{(a^2-y^2)}}^{\sqrt{(a^2-y^2)}} 1.dy\,dx$$

$$= \int_{-a}^{a}\left[2\int_0^{\sqrt{(a^2-y^2)}} 1.dx\right]dy,$$

by property of definite integral

$$= 2\int_{-a}^{a}[x]_0^{\sqrt{(a^2-y^2)}}dy = 2\int_{-a}^{a}\sqrt{(a^2-y^2)}dy$$

$$= 2.2\int_0^a\sqrt{(a^2-y^2)}dy$$

$$= 4\left[\frac{y\sqrt{(a^2-y^2)}}{2}+\frac{a^2}{2}\sin^{-1}\frac{y}{a}\right]_0^a$$

$$= 4\left[0+\frac{a^2}{2}\sin^{-1}1\right] = 4.\frac{1}{2}a^2.\frac{1}{2}\pi = \pi a^2.$$

Example 8. *Evaluate $\iint(x+y)^2 dx\,dy$ over the region bounded by ellipse $\frac{x^2}{a^2}+\frac{y^2}{b^2}=1$. Hence find the mass of an elliptic plate whose density per unit area is given by $\rho = k(x+y)^2$.* (UKTU-2011)

Solution. Since the region is bounded by ellipse $\frac{x^2}{a^2}+\frac{y^2}{b^2}=1$, we expressed it as:

$x = -a$ and $x = a$

$y = -b\sqrt{(1-x^2/a^2)}, y = b\sqrt{(1-x^2/a^2)}$.

$\therefore \iint(x+y)^2 dx\,dy$

$$= \int_{-a}^{a}\int_{-b\sqrt{(1-x^2/a^2)}}^{b\sqrt{(1-x^2/a^2)}}(x^2+y^2+2xy)\,dx\,dy$$

$$= \int_{-a}^{a} 2\int_0^{b\sqrt{(1-x^2/a^2)}} (x^2+y^2)\,dx\,dy$$

[∵ $2xy$ being an odd function of f, its integration under the given limits of y is 0]

$$= 2\int_{-a}^{a}\left[x^2y+\frac{y^3}{3}\right]_0^{b\sqrt{1-x^2/a^2}} dx$$

$$= 2\int_{-a}^{a}\left\{x^2 b\sqrt{\left(1-\frac{x^2}{a^2}\right)}+\frac{b^3}{3}\left(1-\frac{x^2}{a^2}\right)^{3/2}\right\}dx$$

$$= 2\times 2\int_{0}^{a}\left\{x^2 b\sqrt{\left(1-\frac{x^2}{a^2}\right)}+\frac{b^3}{3}\left(1-\frac{x^2}{a^2}\right)^{3/2}\right\}dx$$

$$= 4b\int_0^{\pi/2}\left\{a^2\sin^2\theta\cos\theta+\frac{b^2}{3}\cos^3\theta dx\right\}a\cos\theta\, d\theta$$

(By putting $x = a\sin\theta$ so that $dx = a\cos\theta\, d\theta$)

$$= 4ab\int_0^{\pi/2}\left[a^2\sin^2\theta\cos^2\theta+\frac{b^2}{3}\cos^4\theta\right]d\theta$$

$$= 4ab\left[a^2\int_0^{\pi/2}\sin^2\theta\cos^2\theta\, d\theta+\frac{b^2}{3}\int_0^{\pi/2}\cos^4\theta\, d\theta\right]$$

$$= 4ab\left[\frac{1}{16}\pi a^2+\frac{1}{16}\pi b^2\right]=\frac{1}{4}\pi ab(a^2+b^2)$$

= the mass of elliptic plate whose density is given by $\rho = k(x+y)^2$

$$= \iint_R k(x+y)^2\,dx\,dy$$

(where integration is to be performed over the area A of ellipse.)

$$= k.\frac{1}{4}\pi ab(a^2+b^2)$$

Example 9. *Evaluate* $\iint xy(x+y)\,dx\,dy$ *over the region between* $y = x^2$ *and* $y = x$. (UKTU-2011, 12)

Solution. When we draw the given curve, the parabola $y = x^2$ and line $y = x$ intersect at the point (0, 0) and (1, 1) ,

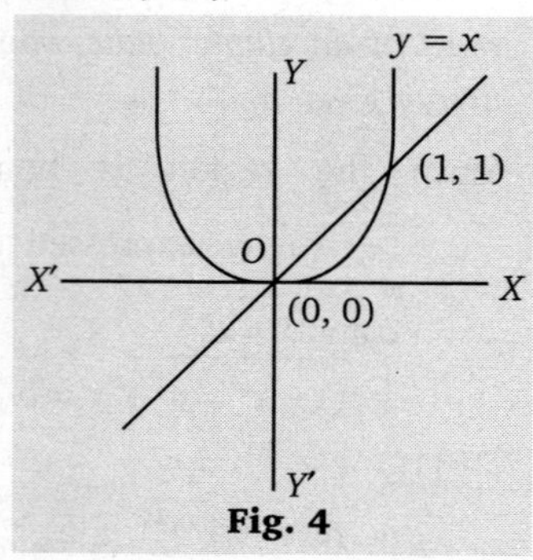

Fig. 4

Here $x^2 = x$ or $x(x-1) = 0$ *i.e.*, $x=0$ or 1, when $x=0, y=0$, and $x=1, y=1$].

So the area of integration for x is from $x=0$ to $x=1$ and for y from x^2 to x.

∴ Given integral

$$= \int_0^1\int_{x^2}^{x} xy(x+y)\,dx\,dy$$

$$= \int_0^1\int_{x^2}^{x} x(yx+y^2)\,dx\,dy$$

$$= \int_0^1 x\left[x.\frac{y^2}{2}+\frac{1}{3}y^3\right]_{x^2}^{x} dx$$

$$= \int_0^1 x.\left[\left(x.\frac{x^2}{2}+\frac{1}{3}x^3\right)-\left(x.\frac{x^4}{2}+\frac{1}{3}x^6\right)\right]dx$$

$$= \int_0^1 x\left[\frac{5x^3}{6}-\frac{1}{2}x^5-\frac{1}{3}x^6\right]dx$$

$$= \int_0^1\left[\frac{5}{6}x^4-\frac{1}{2}x^6-\frac{1}{3}x^7\right]dx$$

$$= \left[\frac{1}{6}x^5-\frac{1}{14}x^7-\frac{1}{24}x^8\right]_0^1$$

$$= \frac{1}{6}-\frac{1}{14}-\frac{1}{24}=\frac{3}{56}$$

Example 10. *When the region of integration A is the triangle given by y=0, y=x and x=1, show that* $\iint_A\sqrt{4x^2-y^2}\,dx\,dy=\frac{1}{3}\left(\frac{\pi}{3}+\frac{\sqrt{3}}{2}\right)$.

Solution. Here we draw straight lines $y = 0$, $y = x$ and $x = 1$. We can express the region of integration as $0 \le y \le x,\ 0 \le x \le 1$.

$$\therefore \iint_A\sqrt{4x^2-y^2}\,dx\,dy$$

$$= \int_{x=0}^{1}\int_{y=0}^{x}\sqrt{(4x^2-y^2)}\,dx\,dy$$

$$= \int_0^1\left[\frac{y}{2}\sqrt{4x^2-y^2}+2x^2\sin^{-1}\frac{y}{2x}\right]_{y=0}^{x} dx$$

Integrating w.r. to y treating x as constant.

$$= \int_0^1\left[\frac{x}{2}\sqrt{4x^2-x^2}+2x^2\sin^{-1}\frac{1}{2}-0\right]dx$$

$$= \int_0^1\left[\frac{\sqrt{3}}{2}x^2+\frac{\pi}{3}x^2\right]dx=\left[\frac{\sqrt{3}}{2}.\frac{x^3}{3}+\frac{\pi}{3}.\frac{x^3}{3}\right]_0^1$$

$$= \frac{1}{3}\left(\frac{\sqrt{3}}{2}+\frac{\pi}{3}\right).$$

EXERCISE 19.1

1. Evaluate $\int_2^3 dx \int_0^1 (x^2+3y^2)dy$.

2. Evaluate $\int_0^2 \int_0^{\sqrt{4+x^2}} \frac{dx\,dy}{(4+x^2+y^2)}$.

3. Evaluate $\int_0^{\pi/2} \int_{\pi/2}^{\pi} \cos(x+y)dx\,dy$.

4. Evaluate $\int_0^2 \int_0^{\sqrt{2x-x^2}} x\,dx\,dy$. (SRM-2010)

5. Evaluate $\int_0^1 \int_0^{x^2} e^{y/x}\,dx\,dy$.

6. Evaluate $\int_0^1 \int_0^1 \frac{dx\,dy}{\sqrt{(1-x^2)(1-y^2)}}$ (UPTU (AG)-2006)

7. Evaluate $\iint e^{2x+3y}dx\,dy$ over the triangle bounded by $x=0$, $y=0$ and $x+y=1$.

8. Evaluate $\iint_p x\sin(x+y)dx\,dy$, where p is a rectangle $[0 \le x \le \pi, 0 \le y \le \pi/2]$.

9. Show that $\int_1^2 \int_3^4 (xy+e^y)dx\,dy = \int_3^4 \int_1^2 (xy+e^y)dy\,dx$.

10. Evaluate $\iint x^2y^2dx\,dy$ over the region bounded by $x=0$, $y=0$, where A is the region bounded by $x^2+y^2=1$.

11. Find the area of the ellipse $\frac{x^2}{a^2}+\frac{y^2}{b^2}=1$ by double integration.

12. Show that by double integration that the area between the parabolas $y^2 = 4ax$ and $x^2 = 4by$ is $(16/3)\,ab$.

13. Find by double integration the region included between the parabola $x^2 = 4ay$ and the curve $y = 8a^3/(x^2+4a^2)$.

14. Evaluate $\iint y\,dx\,dy$ over the region between the parabolas $y^2 = 4x$ and $x^2 = 4y$.

15. Find the double integration the region lying between the parabola $y = 4x - x^2$, and the line $y = x$.

16. Find by double integration the area of the region lying between the parabola $y^2=4ax$, and line $y=mx$.

17. Find by double integration the area of the region lying between the semi-cubical parbola $y^2 = x^3$, and line $y=mx$.

18. Find by double integration the area of the region lying between the circle $x^2+y^2=a^2$, and line $x+y=a$ (in first quadrant)

19. Find by double integration the area of the region lying between the curves $(x^2+4a^2)\,y = 8a^3$, $2y = x$ and $x=0$.

20. Evaluate

(i) $\int_0^1 \int_0^{\sqrt{1+x^2}} \frac{dx\,dy}{1+x^2+y^2}$ (UPTU-2006)

(ii) $\int_0^a \int_0^{\sqrt{a^2+y^2}} (a^2-x^2-y^2)dx\,dy$

21. (i) Evaluate $\iint_R \left(1-\frac{x^2}{a^2}+\frac{y^2}{b^2}\right)dx\,dy$ over the first quadrant of the ellipse $\frac{x^2}{a^2}+\frac{y^2}{b^2}=1$. (GBTU-2010)

(ii) Evaluate $\iint xy\,dx\,dy$ where A is the domain bounded by x-axis, ordinate $x = 2a$ and the curve $x^2=4ay$.

(MTU-2012, GBTU-2010)

Hint to Selected Problems

5. $I = \int_0^1 \int_0^{x^2} e^{y/x}dx\,dy = \int_0^1 x[e^{y/x}]_{y=0}^{x^2}dx$

$= \int_0^1 x(e^x-1)dx = \int_0^1 xe^x dx - \int_0^1 x\,dx$

$= \left(xe^x\right)_0^1 - \left[e^x\right]_0^1 - \left(\frac{x^2}{2}\right)_0^1$

$= (e-0)-(e-1)-\frac{1}{2} = \frac{1}{2}$

10. $I = \iint_A x^2y^2dx\,dy = \int_0^1 \int_0^{\sqrt{1-x^2}} x^2y^2dx\,dy$

$= \int_0^1 x^2\left[\frac{y^2}{3}\right]_0^{\sqrt{1-x^2}} dx = \frac{1}{2}\int_0^1 x^2(1-x^2)^{3/2}dx$

Now put $x^2 = t$.

11. $I = \iint_A dx\,dy = \int_{-a}^{a} \int_{-b/a\sqrt{a^2-x^2}}^{b/a\sqrt{a^2-x^2}} dx\,dy$

12. $A = \int_0^{4a^{1/3}b^{2/3}} \int_{x^2/4b}^{\sqrt{4ax}} dx\,dy$.

13. $A = \int_{-2a}^{2a} \int_{x^2/4a}^{8a^3/x^2+4a^2} dx\,dy$.

14. The curves $y^2 = 4a$ and $x^2 = 4y$ intersect at the points where $x = 0$ and $x = 4$. Also, when

$$0 < x < 4, \sqrt{4x} > \frac{x^2}{4}.$$

$\therefore \quad I = \int_0^4 \int_{x^2/4}^{\sqrt{4x}} y\,dx\,dy.$

16. Since, the two corners cut at the point where $x = 0$ and $x = \frac{4a}{m^2}$

$\therefore \quad I = \int_0^{4a/m^2} \int_{mx}^{\sqrt{4ax}} dx\,dy$

ANSWERS

1. $\frac{22}{3}$ 2. $\frac{\pi}{4}\log(1+\sqrt{2})$ 3. -2 4. $\frac{\pi}{2}$ 5. $\frac{1}{2}$ 6. $\frac{\pi^2}{4}$ 7. $\frac{1}{6}(e-1)^2(2e+1)$ 8. $\pi+2$ 9. $\frac{21}{4}+e^4-e^3$

10. $\pi/96$ 11. πab 13. $\left(2\pi-\frac{4}{3}\right)a^2$ 14. $48/5$ 15. $9/2$ 16. $8a^2/3m^2$ 17. $1/10\,m^5$ 18. $\frac{1}{4}(\pi-2)a^2$

19. $(\pi-1)a^2$ 20. (i) $\frac{\pi}{4}\log(1+\sqrt{2})$ (ii) $\frac{\pi a^4}{8}$ 21. (i) $\frac{\pi ab}{4}$ (ii) $\frac{a^4}{3}$.

19.4 DOUBLE INTEGRAL IN POLAR CO-ORDINATES

Let us consider a function $f(r, \theta)$ of polar co-ordinates r, θ over a certain area A with whose boundary is also given in terms of polar co-ordinates. We divide the area into n parts of elementary areas $\delta A_1, \delta A_2, \delta A_3, \ldots \delta A_n$ and let

$$s_n = \sum_{r=1}^{n} f(r,\theta)\delta A$$

where (r_1, θ_1) is a point inside the elementary area δA_1, the dobule integral of $f(r, \theta)$ is then defined as

$$\iint_A f(r,\theta)dA = \lim_{\substack{n\to\infty \\ \delta A_i \to 0}} \sum_{i=1}^{n} f(r_i,\theta_i)\delta A_i ,$$

provided limit toward right hand side exists.

REMARK

- In case of cartesian co-ordinates when the double integral $\iint_A f(x,y)dA$ is expressed in the form of repeated integral, dA representes the area of the rectangle with sides dx and dy and hence $dA = dx\, dy$.

If the radius vector of OS and OP are r and $r + \delta r$ respectively and $\angle POQ = d\theta \Rightarrow RS = r\,d\theta$ as RS and PQ are arcs of circles. Then

$$dA = RS \times SP$$
$$= r\,d\theta .\, dr = r\,dr\,d\theta$$

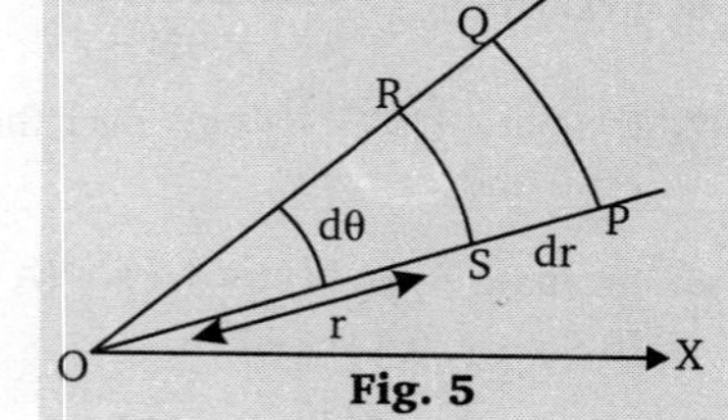

Fig. 5

Solved Examples

Example 1. *Evaluate the double integral* $\int_0^{\pi/2}\int_0^{2a\cos\theta} r^2 \sin\theta.\cos\theta.d\theta\, dr$.

Solution. We have $\int_0^{\pi/2}\int_0^{2a\cos\theta} r^2 \sin\theta.\cos\theta.d\theta\, dr$

$$= \int_0^{\pi/2}\int_0^{2a\cos\theta} (r^2 \sin\theta.\cos\theta)\, dr\, d\theta$$

$$= \int_0^{\pi/2}\left[\frac{r^3}{3}.\sin\theta.\cos\theta\right]_0^{2a\cos\theta} d\theta$$

$$= \frac{1}{3}\int_0^{\pi/2}(2a\cos\theta)^3.\sin\theta.\cos\theta\, d\theta$$

$$= \frac{8a^3}{3}\int_0^{\pi/2}\sin\theta.\cos^4\theta\, d\theta$$

$$= -\frac{8a^3}{3}\int_0^{\pi/2}\cos^4\theta\, d(\cos\theta)$$

$$= -\frac{8a^3}{3}\left[\frac{\cos^5\theta}{5}\right]_0^{\pi/2}$$

$$= -\frac{8a^3}{3}\left[0-\frac{1}{5}\right] = \frac{8a^3}{15}.$$

Example 2. *Evaluate* $\iint \frac{r\,d\theta\,dr}{\sqrt{a^2+r^2}}$ *over one loop of the lemniscate* $r^2 = a^2 \cos 2\theta$

Solution. In lemniscate, there are two loops.

We see that when $-\pi/4 < \theta < \pi/4$ or $3\pi/4 < \theta < 5\pi/4$, where r is real.

We want to evaluate the given integral over the right loop of the leminscate. [Fig. (6)]

Therefore, $\iint \frac{r\,d\theta\,dr}{\sqrt{a^2+r^2}}$

$$= \int_{-\pi/4}^{\pi/4}\int_{-a\sqrt{\cos 2\theta}}^{a\sqrt{\cos 2\theta}} \frac{r}{\sqrt{r^2+a^2}}dr\, d\theta$$

as $r^2 = a^2 \cos 2\theta$.

$\therefore$ $r = \pm a\sqrt{\cos 2\theta}$

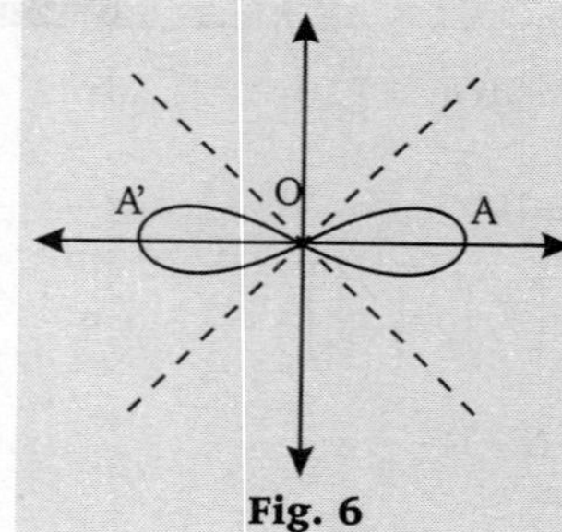

Fig. 6

Thus, r varies from $-a\sqrt{\cos 2\theta}$ to $a\sqrt{\cos 2\theta}$.

Since, there is a symmetry about X-axis, we should evaluate the double integral over half of the right loop as follows:

$$\iint \frac{r\,d\theta\,dr}{\sqrt{a^2+r^2}} = \int_0^{\pi/4}\int_0^{a\sqrt{\cos 2\theta}} \frac{r}{\sqrt{a^2+r^2}}dr\, d\theta$$

$$= \frac{1}{2}\int_0^{\pi/4}\int_0^{a\sqrt{\cos 2\theta}} \frac{2r}{\sqrt{a^2+r^2}}dr\, d\theta$$

$$= \frac{1}{2}\int_0^{\pi/4}\int_0^{a\sqrt{\cos 2\theta}} \frac{d(a^2+r^2)}{\sqrt{a^2+r^2}}d\theta$$

$$= \frac{1}{2}\int_0^{\pi/4}\left[2\sqrt{a^2+r^2}\right]_0^{a\sqrt{\cos 2\theta}} d\theta$$

$$= \int_0^{\pi/4}\left[\sqrt{a^2+a^2\cos 2\theta} - \sqrt{a^2+0}\right]d\theta$$

$$= \int_0^{\pi/4}[\sqrt{2}.a\cos\theta - a]d\theta$$

$$= \sqrt{2}.a\int_0^{\pi/4}\cos\theta\, d\theta - a\int_0^{\pi/4} d\theta$$

$$= \sqrt{2}.a[\sin\theta]_0^{\pi/4} - a[\theta]_0^{\pi/4}$$

$$= \sqrt{2}.a.\frac{1}{\sqrt{2}} - a.\frac{\pi}{4} = a - \frac{a}{4}\pi$$

$\therefore$ Value of double integral over the complete right loop

$$= 2a - \frac{2a\pi}{4} = 2a(1-\pi/4).$$

REMARK

- If we evaluate the double integral as $\int_{-\pi/4}^{\pi/4}\int_{-a\sqrt{\cos 2\theta}}^{a\sqrt{\cos 2\theta}} \frac{r}{\sqrt{a^2+r^2}} d\theta\, dr$, then it will become zero due to oddness of function $\frac{r}{\sqrt{a^2+r^2}}$ therefore we must not calculate the double integral over the complete loop.

Example 3. *Evaluate* $\iint r^2 d\theta\, dr$ *over the area of circle* $r = a\cos\theta$.

Solution. In the given region of circle, θ varies from $-\pi/2$ to $\pi/2$ and r varies from 0 to a.

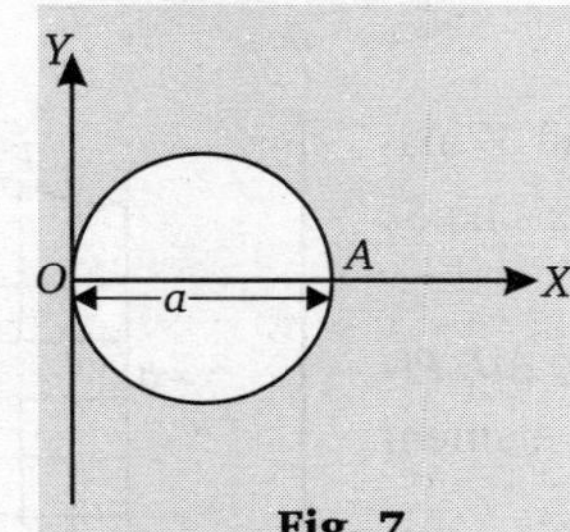

Fig. 7

$$\therefore \iint r^2 d\theta\, dr = \int_{-\pi/2}^{\pi/2}\int_0^{a\cos\theta} r^2 d\theta\, dr$$

$$= \int_{-\pi/2}^{\pi/2}\left[\frac{r^3}{3}\right]_0^{a\cos\theta} d\theta$$

$$= \frac{a^3}{3}\int_{-\pi/2}^{\pi/2}\cos^3\theta\, d\theta$$

$$= \frac{a^3}{3}\int_{-\pi/2}^{\pi/2}\left(\frac{3}{4}\cos\theta + \frac{1}{4}\cos 3\theta\right)d\theta$$

$$= \frac{a^3}{3}\times\frac{3}{4}\int_{-\pi/2}^{\pi/2}[\cos\theta d\theta] + \frac{a^3}{12}\int_{-\pi/2}^{\pi/2}\cos 3\theta\, d\theta$$

$$= \frac{a^3}{4}[\sin\theta]_{-\pi/2}^{\pi/2} + \frac{a^3}{12}\left[\frac{\sin 3\theta}{3}\right]_{-\pi/2}^{\pi/2}$$

$$= \frac{a^3}{4}.2 + \frac{a^3}{12}\times\frac{1}{3}\left[\sin\frac{3\pi}{2} + \sin\frac{3\pi}{2}\right]$$

$$= \frac{a^3}{4} - \frac{a^3}{18} = \frac{4}{9}a^3.$$

Example 4. *Evaluate* $\int_0^{\pi}\int_0^{a(1+\cos\theta)} r^2\cos\theta\, d\theta\, dr$.

Solution. We have $\int_0^{\pi}\int_0^{a(1+\cos\theta)} r^2\cos\theta\, d\theta\, dr$

$$= \int_0^{\pi}\cos\theta\left[\frac{r^3}{3}\right]_0^{a(1+\cos\theta)} d\theta$$

$$= \frac{1}{3}\int_0^{\pi}\cos\theta.a^3(1+\cos\theta)^3 d\theta$$

$$= \frac{a^3}{3}\int_0^{\pi}\cos\theta(1+3\cos\theta+3\cos^2\theta+\cos^3\theta)d\theta$$

$$= \frac{a^3}{3}\int_0^{\pi}[\cos\theta+3\cos^2\theta+3\cos^3\theta+\cos^4\theta)d\theta$$

$$= 2.\frac{a^3}{3}\int_0^{\pi}[3\cos^2\theta+\cos^4\theta]d\theta$$

$$\left[\because \int_0^{\pi}\cos^n\theta\, d\theta = 0, \text{since } n \text{ is odd}\right]$$

$$= \frac{2a^3}{3}\left[3.\frac{1}{2}.\frac{\pi}{2} + \frac{3}{4}.\frac{1}{2}.\frac{\pi}{2}\right] = \frac{2a^3}{3}.\frac{3\pi}{4}\left[1+\frac{1}{4}\right]$$

$$= \frac{2a^3}{3}.\frac{3\pi}{4}.\frac{5}{4} = \frac{5\pi a^3}{8}$$

Example 5. *Evaluate* $\int_0^{\pi/2}\int_0^{\sin\theta} r\, d\theta\, dr$.

Solution. We have

$$\int_0^{\pi/2}\int_0^{\sin\theta} r\, d\theta\, dr = \int_0^{\pi/2} d\theta\int_0^{\sin\theta} r\, dr$$

$$= \int_0^{\pi/2}\left(\frac{1}{2}r^2\right)_0^{\sin\theta} d\theta$$

$$= \frac{1}{2}\int_0^{\pi/2}\sin^2\theta\, d\theta = \frac{1}{4}\int_0^{\pi/2}(1-\cos 2\theta)\, d\theta$$

$$= \frac{1}{4}\left[\theta - \frac{1}{2}\sin 2\theta\right]_0^{\pi/2} = \frac{1}{4}\left[\frac{1}{2}\pi - \frac{1}{2}(0)\right] = \frac{\pi}{8}$$

Example 6. *Evaluate* $\int_0^{\pi/2}\int_{a(1-\cos\theta)}^{a\sin\theta} r\, d\theta\, dr$.

Solution. We have

$$\int_0^{\pi/2}\int_{a(1-\cos\theta)}^{a\sin\theta} r d\theta\, dr = \int_0^{\pi/2}\left[\frac{1}{2}r^2\right]_{a(1-\cos\theta)}^{a\sin\theta} d\theta$$

$$= \frac{1}{2}a^2\int_0^{\pi/2}[\sin^2\theta - (1-\cos\theta)^2]d\theta$$

$$= \frac{1}{2}a^2\int_0^{\pi/2}[2\cos\theta - \cos^2\theta - (1-\sin^2\theta)]d\theta$$

$$= a^2\int_0^{\pi/2}(\cos\theta - \cos^2\theta)\, d\theta$$

$$= a^2 \int_0^{\pi/2} \left[\cos\theta - \frac{1}{2}(1 + \cos 2\theta) \right] d\theta$$

$$= a^2 \left[\sin\theta - \frac{1}{2}\theta - \frac{1}{4}\sin 2\theta \right]_0^{\pi/2}$$

$$= a^2 \left[1 - \frac{\pi}{4} \right] = \frac{1}{4} a^2 (4 - \pi).$$

EXERCISE 19.2

1. Integrate $r \sin\theta$ over the area of cardiod $r = a(1+\cos\theta)$ lying above the initial line.
2. Find by double integration that the area lying inside the cardiod $r = a(1+\cos\theta)$ and outside the circle $r = a$.
3. Find by double integration the area lying inside the cardiod $r = 1 + \cos\theta$ and outside the parabola $r(1+\cos\theta) = 1$.
4. Find by double integration the area lying inside the circle $r = a \sin\theta$ and outside the cardioid $r = a(1 - \cos\theta)$.

ANSWERS

1. $\frac{4a^3}{3}$ **2.** $\frac{1}{4}a^2(\pi + 8)$ **3.** $\frac{9\pi + 16}{12}$ **4.** $\frac{a^2}{4}(4 - \pi)$.

19.5 APPLICATIONS OF DOUBLE INTEGRATION

Double integration is generally used to find the area of curves, volume and surface of solids of revolution.

(1) *Area of curves.* Let AD be an arc of the curve $y = f(x)$.

Let area $ABCD$ be divided into sub-area by drawing lines parallel to X and Y axis respectively such that distance between two adjoining lines drawn parallel to Y-axis be δx and those drawn parallel to X-axis be δy.

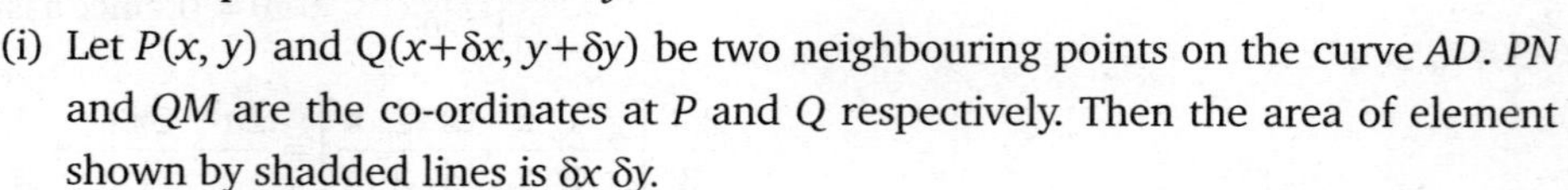

(i) Let $P(x, y)$ and $Q(x+\delta x, y+\delta y)$ be two neighbouring points on the curve AD. PN and QM are the co-ordinates at P and Q respectively. Then the area of element shown by shadded lines is $\delta x\ \delta y$.

Therefore, the area of strip PN

$$= \int_{y=0}^{f(x)} dx\, dy \text{ where } y = f(x).$$

Fig. 8

The required area

$$ABCD = \int_{x=a}^{b} \int_{y=0}^{f(x)} dx\, dy.$$

(ii) We can find the area bounded by the two curves $y = f_1(x)$ and $y = f_2(x)$ and the ordinates $x = a$ and $x = b$

$$\int_{x=a}^{\beta} \int_{y=f_2(x)}^{f_1(x)} dx\, dy.$$

(iii) *In polar co-ordinates.* The area bounded by curve $r = f(\theta)$ where $f(\theta)$ is a single valued function of θ in the domain (α, β) and the radii vector $\theta = \alpha$ and $\theta = \beta$ is

$$\int_{\theta=\alpha}^{\beta} \int_{r=\theta}^{f(\theta)} r\, d\theta\, dr.$$

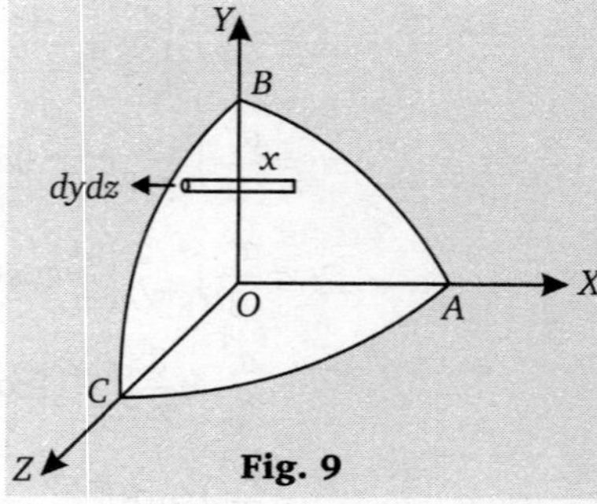

Fig. 9

(2) *Volume of a solid.* Consider the area $dy\, dz$ on the plane $x=0$ through each point on the boundary of this small area. Draw the lines parallel to X-axis and thus construct a small cylinder whose base is area to X-axis. This cylinder cuts the given surface, and volume of this cylinder

$$= x\, dy\, dz.$$

$$\therefore \quad \text{Volume of solid} = \iint x\, dy\, dz.$$

REMARKS

- By considering area $dx\, dy$ on plane $z=0$ the volume of solid $= \iint z\, dx\, dy$.
- By considering area $dx\, dz$ on plane $y=0$ the volume of solid $= \iint y\, dx\, dz$.

(3) *Area of surface of a solid.* Let the equatioin of surface be $z = f(x, y)$. Consider a point $P(x, y, z)$ on this surface surrounding this point P. Consider an element of area δs of the surface. Let $\delta x\, \delta y$ be the projection of this area δs on the plane $z = 0$, then we have

$$\delta x\, \delta y = \delta s \cos\alpha \quad \text{... (1)}$$

where α is the angle between the tangent plane to the given surface at $P(x, y, z)$ and the plane $z=0$ then by co-ordinate geometry, we have

$$\sec\alpha = \sqrt{\left[1+\left\{\frac{\partial z}{\partial x}\right\}^2+\left\{\frac{\partial z}{\partial y}\right\}^2\right]} \quad \text{...(2)}$$

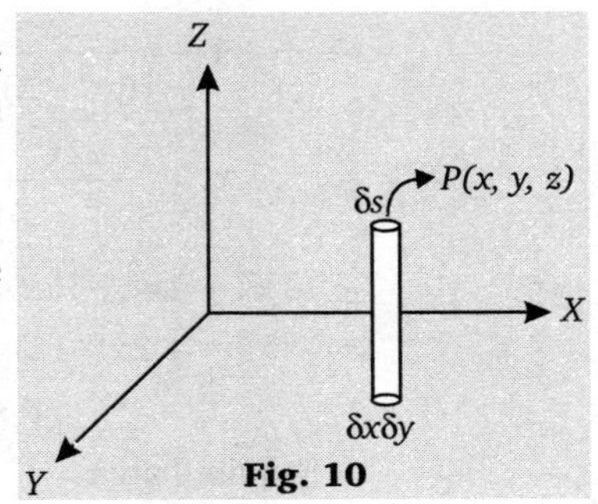

Fig. 10

From (1) we have $\delta s = \delta x\, \delta y \sec\alpha$

$$= \delta x \delta y \sqrt{\left[1+\left\{\frac{\partial z}{\partial x}\right\}^2+\left\{\frac{\partial z}{\partial y}\right\}^2\right]} \quad \text{[From (2)]}$$

$\therefore$ The required area of surface $= \iint \sqrt{\left[1+\left\{\frac{\partial z}{\partial x}\right\}^2+\left\{\frac{\partial z}{\partial y}\right\}^2\right]}\, dx\, dy$.

Solved Examples

Example 1. *Find the area of ellipse* $\frac{x^2}{a^2}+\frac{y^2}{b^2}=1$.

Solution. Required area of ellipse

$= 4$ (area of quadrants $OABO$ of ellipse)

$= 4\int_{x=0}^{a}\int_{y=0}^{f(x)} dx\, dy,$

where $y = f(x) = \frac{b}{a}\sqrt{(a^2-x^2)}$

$= 4\int_0^a [y]_0^{f(x)} dx = 4\int_0^a f(x) dx$

$= 4\int_0^a \frac{b}{a}\sqrt{(a^2-x^2)}\, dx$

$= \frac{4b}{a}\left[\frac{1}{2}x\sqrt{(a^2-x^2)}+\frac{1}{2}a^2\sin^{-1}\left(\frac{x}{a}\right)\right]_0^a$

$= \frac{2b}{a}[0+a^2\sin^{-1}(1)] = \frac{2b}{a}a^2.\frac{\pi}{2} = ab\pi$

Example 2. *Find the whole area of curve* $a^2x^2 = y^3(2a-y)$.

Solution. The shape of curve is shown in fig. 11.

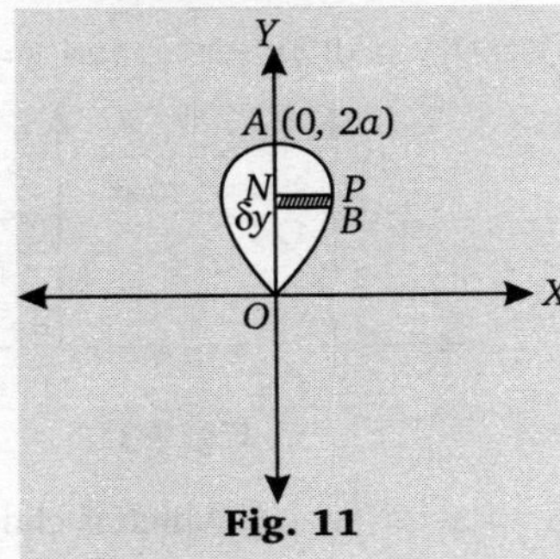

Fig. 11

The required area $= 2\times$ area OAB

$= 2\int_{y=0}^{2a}\int_{x=0}^{f(y)} dy\, dx$

where $x = f(y)$ *i.e.*, $x = y^{3/2}\frac{\sqrt{2a-y}}{a}$ is the equation of curve.

$\therefore$ The required area

$= 2\int_{y=0}^{2a}[x]_0^{f(y)} dy$

$= 2\int_0^{2a} f(y)\, dy$

$= 2\int_0^{2a} \frac{y^{3/2}\sqrt{2a-y}}{a} dy$

$\left[\because f(y) = x = y^{3/2}\frac{\sqrt{2a-y}}{a}\right]$

Put $y = 2a\sin^2\theta$

$\Rightarrow$ $dy = 4a\sin\theta\cos\theta\, d\theta$

at $y = 0, \theta = 0$ and $y = 2a, \theta = \pi/2$

$\therefore$ Required area

$= \frac{2}{a}\int_0^{\pi/2}(2a\sin^2\theta)^{3/2}\sqrt{(2a-2a\sin^2\theta)}\; 4a\sin\theta\cos\theta\, d\theta$

$= 32a^2\int_0^{\pi/2}\sin^4\theta\cos^2\theta\, d\theta$

$= \frac{32a^2\Gamma(5/2)\Gamma(3/2)}{2\Gamma 4}$

$= \frac{32a^2.(3/2).(1/2).\sqrt{\pi}.(1/2).\sqrt{\pi}}{2.3.2.1} = \pi a^2$

Example 3. *Find by double integration the area between* $y = \frac{3x}{(x^2+2)}$ *and* $4y = x^2$

Solution. We have

$$4y = x^2, \quad \text{and} \quad y = \frac{3x}{(x^2+2)}$$

$$\Rightarrow \quad 4y = \frac{12x}{(x^2+2)}, \text{ and } 4y = x^2$$

$$\Rightarrow \quad x^2 = \frac{12x}{(x^2+2)}$$

$$\Rightarrow \quad x^4 + 2x^2 - 12x = 0$$

$$\Rightarrow \quad x(x^3 + 2x - 12) = 0$$

$$x = 0, 2$$

$$\therefore \text{ Required area } = \int_{x=0}^{2}\int_{y=x^2/4}^{3x/(x^2+2)} dx\,dy$$

$$= \int_0^2 [y]_{x^2/4}^{3x/(x^2+2)} dx = \int_0^2 \left[\frac{3x}{x^2+2} - \frac{x^2}{4}\right] dx$$

$$= \frac{3}{2}\int_0^2 \frac{2x\,dx}{x^2+2} - \frac{1}{4}\int_0^2 x^2\,dx$$

$$= \frac{3}{2}\left[\log(x^2+2)\right]_0^2 - \frac{1}{4}\left(\frac{1}{3}x^3\right)_0^2$$

$$= \frac{3}{2}[\log(6) - \log(2)] - \frac{1}{12}(8-0)$$

$$= \frac{3}{2}\log 3 - \frac{2}{3}.$$

Example 4. *Find the area of curve r=a (1+cos θ).*

Solution. The required area

$$= 2 \times \text{area } OABO = 2\int_{\theta=0}^{\pi}\int_{r=0}^{f(\theta)} r\,d\theta\,dr,$$

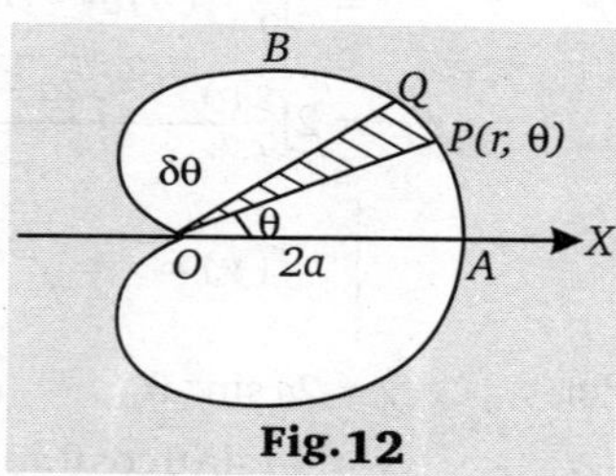

Fig.12

(where $r = f(\theta)$ and $r = a\,(1+\cos\theta)$ is the equation of the curve.)

$$= 2\int_{\theta=0}^{\pi}\left[\frac{1}{2}r^2\right]_{r=0}^{f(\theta)} d\theta = \int_0^{\pi}[f(\theta)]^2\,d\theta$$

$$= \int_0^{\pi} a^2(1+\cos\theta)^2 d\theta = a^2\int_0^{\pi}(2\cos^2\theta/2)^2 d\theta$$

$$= 4a^2\int_0^{\pi}\cos^4\frac{\theta}{2}d\theta$$

$$= 8a^2\int_0^{\pi/2}\cos^4\phi\,d\phi, \qquad (\text{Putting } \theta = 2\phi)$$

$$= 8a^2.\frac{3}{4}.\frac{1}{2}.\frac{1}{2}\pi = (3/2)a^2\pi.$$

Example 5. *Find the volume bounded by co-ordinates planes and the plane* $\frac{x}{a}+\frac{y}{b}+\frac{z}{c} = 1$.

Solution. The plane cuts X, Y and Z- axis at point (a, 0, 0), (0, b, 0) and (0, 0, c) respectively. The surface ABCD of co-ordinates planes will be equal to

$$\int_0^a\int_0^{b(1-x/a)}\int_0^{c(1-x/a-y/b)} dx\,dy\,dz$$

$$= \int_0^a\int_0^{b(1-x/a)} c\left(1-\frac{x}{a}-\frac{y}{b}\right)dy\,dx$$

$$= c\int_0^a\int_0^{b(1-x/a)}\left(1-\frac{x}{a}-\frac{y}{b}\right)dy\,dx$$

$$= c\int_0^a\left[y - \frac{x}{a}.y - \frac{y^2}{2b}\right]_0^{b(1-x/a)} dx$$

$$= c\int_0^a\left[b\left(1-\frac{x}{a}\right) - \frac{x}{a}.b\left(1-\frac{x}{a}\right) - \frac{1}{2b}b^2\left(1-\frac{x}{a}\right)^2\right]dx$$

Example 6. *Find the volume bounded by the cylinder* $x^2+y^2 = 4$ *and the hyperboloid* $-x^2-y^2+z^2 = 1$.

Solution. Here, surfaces $x^2+y^2 = 4$ and $-x^2-y^2+z^2 = 1$ are symmetrical about all the three axes. Therefore, volume

$$V = \iint z\,dx\,dy = 8\int_0^2\int_0^{\sqrt{4-x^2}}\sqrt{(x^2+y^2+1)}\,dx\,dy.$$

Change to polar co-ordinates and change the limits of integrations for the region of quadrant of circle $r = 2$ and $\theta = 0$ to $\pi/2$

$$V = 8\int_0^{\pi/2}\int_0^2\sqrt{(r^2+1)}r\,d\theta\,dr$$

$$= 8\int_0^{\pi/2}\left[\frac{1}{3}(r^2+1)^{3/2}\right]_0^2 d\theta$$

$$= 8\int_0^{\pi/2}\frac{1}{3}(5\sqrt{5}-1)d\theta = \frac{4\pi}{3}(5\sqrt{5}-1).$$

Example 7. *Transform the integral*

$$\int_0^2\int_0^{\sqrt{2x-x^2}}\frac{x\,dx\,dy}{\sqrt{(x^2+y^2)}}$$

by changing into polar co-ordinates and hence evaluate it.

Solution. We have the limit of integration be

$$y = 0, y = \sqrt{(2x-x^2)} \text{ and } x = 0, x = 2.$$

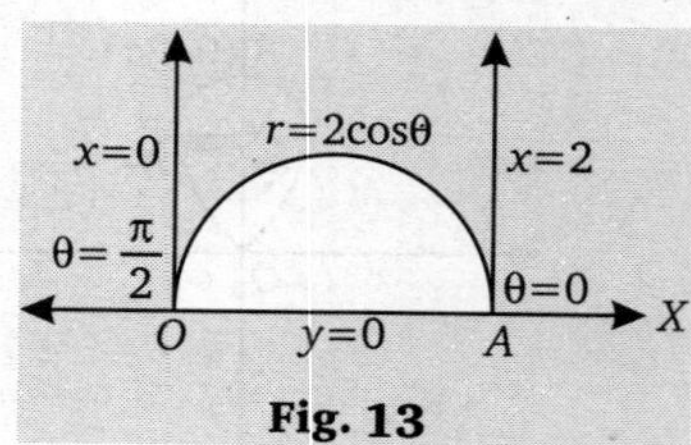

Fig. 13

$x^2+y^2-2x = 0$ which is change into

$$r^2(\cos^2\theta+\sin^2\theta) - 2r\cos\theta = 0$$

or $\qquad r = 2\cos\theta.$

Now r varies form 0 to 2 cosθ and θ varies from 0 to $\pi/2$.

Note that at the point A of the circle, $\theta = 0$ and at point O, $r = 0$ and so from $r = 2\cos\theta$, we get

$$\theta = \frac{\pi}{2} \text{ at } O$$

the polar equivalent of the elementary area $dx\,dy$ is $r\,d\theta\,dr$.

$$\therefore \iint_A f(x,y)dx\,dy$$
$$= \iint_A f(r\cos\theta, r\sin\theta) r\,d\theta\,dr$$

where A is the region of integration.

Therefore, transforming to polar co-ordinates, the given double integral

$$= \int_{\theta=0}^{\pi/2}\int_{r=0}^{2\cos\theta} \frac{r\cos\theta}{r} r\,d\theta\,dr$$

$$= \int_0^{\pi/2} \cos\theta \left[\frac{r^2}{2}\right]_0^{2\cos\theta} d\theta$$

$$= \int_0^{\pi/2} \frac{1}{2}\cos\theta . 4\cos^2\theta\,d\theta = 2\int_0^{\pi/2}\cos^3\theta d\theta$$

$$= 2.\frac{2}{3} = \frac{4}{3}.$$

Example 8. *Find the area of the surface $z^2 = 2xy$ included between planes $x=0$, $x=a$, $y=0$, $y=b$.*

Solution. The given surface is $z^2 = 2xy$.

$$\therefore \quad 2z\frac{\partial z}{\partial x} = 2y \text{ or } \frac{\partial z}{\partial x} = \frac{y}{z}$$

Similarly $$\frac{\partial z}{\partial y} = \frac{x}{z}$$

Then required area of the surface

$$= \iint \sqrt{\left[1 + \left(\frac{\partial z}{\partial x}\right)^2 + \left(\frac{\partial z}{\partial y}\right)^2\right]}\,dx\,dy$$

$$= \int_{x=0}^{a}\int_{y=0}^{b} \sqrt{\left\{1 + \left(\frac{y}{z}\right)^2 + \left(\frac{x}{z}\right)^2\right\}}\,dx\,dy$$

$$= \int_{x=0}^{a}\int_{y=0}^{b} \sqrt{\left(\frac{z^2+y^2+x^2}{2xy}\right)}\,dx\,dy$$

$$= \int_{x=0}^{a}\int_{y=0}^{b} \sqrt{\left(\frac{x^2+y^2+z^2}{2xy}\right)}\,dx\,dy$$

$$= \int_{x=0}^{a}\int_{y=0}^{b} \sqrt{\frac{x^2+y^2+z^2}{2xy}}\,dx\,dy$$

$$= \int_{x=0}^{a}\int_{y=0}^{b} \frac{(x+y)}{\sqrt{2}\sqrt{(xy)}}\,dx\,dy$$

$$= \frac{1}{\sqrt{2}}\int_{x=0}^{a}\int_{y=0}^{b}\left(\sqrt{x}.\frac{1}{\sqrt{y}} + \sqrt{y}.\frac{1}{\sqrt{x}}\right)dx\,dy$$

$$= \frac{1}{\sqrt{2}}\int_{x=0}^{a}\sqrt{x}(2\sqrt{y})_0^b dx$$

$$+ \frac{1}{\sqrt{2}}\int_{x=0}^{a}\frac{1}{\sqrt{x}}\left(\frac{2}{3}y^{3/2}\right)_0^b dx$$

$$= \sqrt{(2b)}\int_0^a \sqrt{x}dx + \frac{\sqrt{2}}{3}b^{3/2}\int_0^a \frac{1}{\sqrt{x}}dx$$

$$= \sqrt{2b}\left[\frac{2}{3}x^{3/2}\right]_0^a + \frac{1}{3}\sqrt{2b^3}\,(2x^{1/2})_0^a$$

$$= \frac{2}{3}\sqrt{2}\sqrt{(ab)}(a+b).$$

Example 9. *Show that the area of the surface of paraboloid $x^2 + y^2 = a^2$ which lies between the planes $z = 0$ and $z = a$ is $(\pi/6)(5.\sqrt{5} - 1)a^2$.*

Solution. The projection of given suface between the planes $z=0$ and $z=a$ on the x-y planes is circle $x^2 + y^2 = a^2, z = 0$.

Also, $$\frac{\partial z}{\partial x} = \frac{2x}{a}, \frac{\partial z}{\partial y} = \frac{2y}{a}$$

$$\therefore \quad S = \iint_A \sqrt{1 + \left(\frac{\delta z}{\partial x}\right)^2 + \left(\frac{\partial z}{\partial y}\right)^2}\,dx\,dy$$

$$= \iint_A \sqrt{1 + \frac{4x^2}{a^2} + \frac{4y^2}{a^2}}\,dx\,dy$$

$$= \frac{1}{a}\iint_A \sqrt{a^2 + 4x^2 + 4y^2}\,dx\,dy$$

where A is the circle $x^2 + y^2 = a^2$ in the xy-plane.

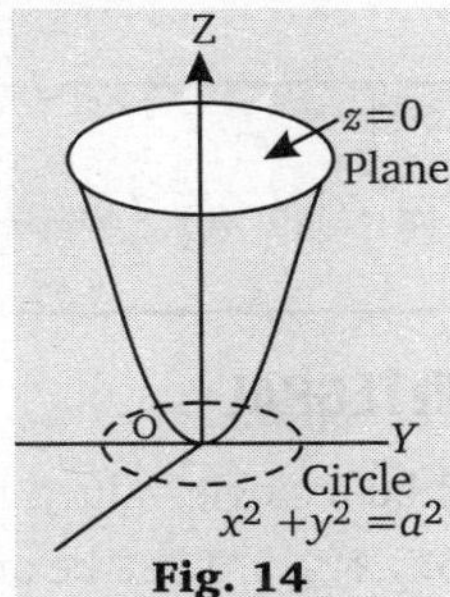

Fig. 14

The equation of the circle $x^2 + y^2 = a^2$ in polar co-ordinates is $r = a$. Hence ,transforming the above double integral into polar co-ordinates, we have

$$S = \frac{1}{a}\int_0^{2\pi}\int_0^a \sqrt{a^2 + 4r^2}\,(r\,d\theta\,dr)$$

$$= \frac{1}{a}\int_0^{2\pi}\left[\int_0^a r\sqrt{a^2+4r^2}\,dr\right]d\theta$$

$$= \frac{1}{8a}\int_0^{2\pi}\left[\int_0^a \sqrt{a^2+4r^2}\,d(a^2+4r^2)\right]d\theta$$

$$= \frac{1}{8a}\int_0^{2\pi}\left[\frac{2}{3}.(a^2+4r^2)^{3/2}\right]_0^a d\theta$$

$$= \frac{1}{8}\times\frac{2}{3}\times\frac{1}{a}\int_0^{2\pi}[(5a^2)^{3/2}-(a^2)^{3/2}]d\theta$$

$$= \frac{1}{12a}\int_0^{2\pi}(5\sqrt{5}a^3-a^3)d\theta$$

$$= \frac{1}{12a}(5\sqrt{5}-1)a^3\int_0^{2\pi}d\theta$$

$$= \frac{1}{12a}(5\sqrt{5}-1)a^3.2\pi = \frac{\pi}{6}(5\sqrt{5}-1)a^2$$

EXERCISE 19.3

1. Find by double integration, the area of the region enclosed by curves

(a) $y = 4x - x^2, y = x$

(b) $(x^2+4a^2)y = 8a^3, 2y = x$ and $x = 0$

(c) $y = \dfrac{3x}{(x^2+2)}, 4y = x^2$.

2. Show that by double integration that the area between the parabolas $y^2=4ax$ and $x^2 = 4by$ is $(16/3)ab$. (GBTU-2010)

3. Find by double integration the area included between the parabola $x^2 = 4ay$ and the curve

$$y = \frac{8x^3}{(x^2+4a^2)}.$$

4. Find the volume of the region bounded by $z= x^2+y^2$ and $z= 2x$.

5. Find the volume cut off the sphere $x^2+y^2+z^2=a^2$ by the cone $x^2+y^2 = z^2$

6. Transforms the following double integrals to polar co-ordinates and hence, evaluate them

(a) $\int_{y=0}^{a}\int_{x=0}^{\sqrt{a^2-y^2}}(a^2-x^2-y^2)dx\,dy$

(b) $\int_0^1\int_x^{\sqrt{2x-x^2}}(x^2+y^2)dx\,dy$

(c) $\int_0^a\int_0^{\sqrt{a^2-x^2}}y^2\sqrt{(x^2+y^2)}dx\,dy$

7. Evaluate $\iint r^2 d\theta\,dr$ over the area of circle $r = a\cos\theta$.

Hint to Selected Problems

2. Required area $= \int_{x=0}^{4a^{1/3}b^{2/3}}\int_{y=x^2/4b}^{2\sqrt{ax}}dx\,dy$

3. Given that $x^2 = 4ay$ and $y = \dfrac{8a^3}{x^2+4a^2}$

After simplification, we get

$$x = \pm i2\sqrt{2}a, \pm 2a$$

$\therefore$ Required area $= \int_{x=-2a}^{2a}\int_{y=x^2/4a}^{8a^3/x^2+4a^2}dx\,dy$

4. $I = \int_0^{\pi}\int_0^{a(1+\sin\theta)} r\sin\theta.rd\theta\,dr$

5. Since the two curves intersect at the points where $\cos\theta = 0$ i.e., $\theta = \dfrac{\pi}{2}$

$\therefore\ I = \int_{\theta=-\pi/2}^{\pi/2}\int_{r=a}^{a(1+\cos\theta)} rd\theta\,dr$.

7. The required volume is given by

$$V = \int_0^{2\pi}\int_0^{a/\sqrt{2}}\int_r^{\sqrt{a^2-r^2}}dz(r\,d\theta\,dr).$$

ANSWERS

1. (a) $\frac{9}{2}$ (b) $(\pi-1)a^2$ (c) $\frac{3}{2}\log 3-\frac{2}{3}$ **3.** $\left(2\pi-\frac{4}{3}\right)a^2$ **4.** $\frac{\pi^3}{-2}$ **5.** $\frac{(2-\sqrt{2})\pi a^3}{3}$ **6.** (a) $\frac{\pi a^4}{8}$ (b) $\frac{3\pi}{8}-1$ (c) $\frac{\pi a^5}{20}$ **7.** $\frac{4a^3}{9}$.

19.6 TRIPLE INTEGRAL

Let $f(x, y, z)$ be a single-valued function of the independent variables x, y, z in finite region V. Divide the region V into n subregions $\delta V_1, \delta V_2, \delta V_3, \ldots$ Let P be any point on the boundary or inside.

Take a point in each part and form the sum

$$S_n = f(x_1,y_1,z_1)\delta V_1 + f(x_2,y_2,z_2)\delta V_2 + \ldots + f(x_n,y_n,z_n)\delta V_n$$

$$= \sum_{r=1}^{n} f(x_r,y_r,z_r)\delta V_r \qquad \ldots(1)$$

when n tends to infinity. The limit of sum (1) tends to zero is called the triple integral of function $f(x, y, z)$ over the region V and is denoted by

$$\iiint_V f(x,y,z)dv.$$

The triple integral can be utilised in evaluating a number of physical quantities like, $f(x, y, z) = 1$

We find volume, $V = \iiint_V dV$ and putting $f(x, y, z)=\rho$

We get, $mass = \iiint_V \rho dV$.

19.6.1 EVALUATION OF TRIPLE INTEGRALS

The region V divide into elementary cuboids by drawing parallel co-ordinate planes. The volume V can then be considered as the sum of number of columns parallel to z-axis extending from the lower surface of V say $z=z_1(x, y)$ to the upper surface of V say $z = z_2(x, y)$ the bases of these as column (only one column has been shown in fig. 15) are the elementary area δs_r, which cover a certain area S in x- y plane i.e. plane $z=0$.

$\therefore$ Summing up over the elementary cuboids in the same column first and then taking the sum of all such columns we can write

$$\sum_{r=1}^{n} f(x_r, y_r, z_r) \text{ as } \sum_{r}^{n}\sum_{m} f(x_r, y_r, z_m)\delta_z]\delta s_r$$

where (x_r, y_r, z_r) is a point in the mth cuboid.

When δS_r and δz tend to zero this becomes equal to

$$\iint_S \left\{\int_{z=z_1(x,y)}^{z_2(x,y)} f(x,y,z)dz\right\}ds$$

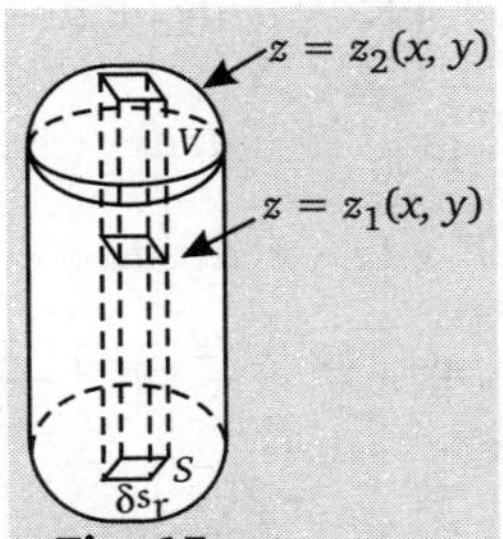

$z = z_2(x, y)$

$z = z_1(x, y)$

Fig. 15

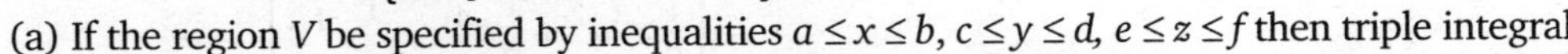

(a) If the region V be specified by inequalities $a \le x \le b,\ c \le y \le d,\ e \le z \le f$ then triple integral

$$\iiint_V f(x,y,z)\,dx\,dy\,dz = \int_a^b\int_c^d\int_e^f f(x,y,z)\,dx\,dy\,dz$$
$$= \int_a^b dx \int_c^d dy \int_e^f f(x,y,z)dz.$$

Here, we integrate first with respect to z keeping x and y constant and then the remaining integration is done as in the case of doube integrals.

The integration with respect to z is performed first regarding x and y as constant then integration w.r to y regarding x as a constant and then integrate w.r to x.

(b) If the limits of z are function of x and y and y as function of x and x takes the constant values as from $x = a$ to $x = b$.

$$\iiint_V f(x,y,z)\,dx\,dy\,dz = \int_a^b dx \int_{y_1(x)}^{y_2(x)} f(x,y,z)dz$$

The integration with respect to z perform first regarding x and y as constant then integral w.r.t. y regarding x as a constant and then integrate w.r.t. x.

Solved Examples

Example 1. *Evaluate* $\int_0^1\int_{y^2}^1\int_0^{1-x} x\,dy\,dx\,dz$.

Solution. We have

$$I = \int_0^1\int_{y^2}^1 (z)_0^{1-x}\, x\,dy\,dx$$
$$= \int_0^1\int_{y^2}^1 x(1-x)\,dy\,dx$$
$$= \int_0^1\int_{y^2}^1 (x - x^2)dy\,dx = \int_0^1\left[\frac{1}{2}x^2 - \frac{1}{3}x^3\right]_{y^2}^1 dy$$
$$= \int_0^1\left[\left\{\frac{1}{2}(1)^2 - \frac{1}{3}(1)^3\right\} - \left\{\frac{1}{2}(y^2)^2 - \frac{1}{3}(y^2)^3\right\}\right]dy$$
$$= \int_0^1\left[\left(\frac{1}{2} - \frac{1}{3}\right) - \left(\frac{1}{2}y^4 - \frac{1}{3}y^6\right)\right]dy$$
$$= \int_0^1\left(\frac{1}{6} - \frac{1}{2}y^4 + \frac{1}{3}y^6\right)dy$$
$$= \left(\frac{1}{6}y - \frac{1}{10}y^5 + \frac{1}{21}y^7\right)_0^1$$
$$= \frac{1}{6} - \frac{1}{10} + \frac{1}{21} = \frac{4}{35}.$$

Example 2. *Evaluate* $\int_{x=0}^1\int_{y=0}^{\sqrt{1-x^2}}\int_{z=0}^{\sqrt{1-x^2-y^2}} xyz\,dx\,dy\,dz$

Solution. The given integral

$$I = \int_{x=0}^1\int_0^{\sqrt{1-x^2}} xy\left(\frac{1}{2}z^2\right)_0^{\sqrt{1-x^2-y^2}} dx\,dy$$
$$= \frac{1}{2}\int_{x=0}^1\int_{y=0}^{\sqrt{1-x^2}} xy(1-x^2-y^2)dx\,dy$$
$$= \frac{1}{2}\int_{x=0}^1\int_{y=0}^{\sqrt{1-x^2}} x[y(1-x^2) - y^3)]dx\,dy$$
$$= \frac{1}{2}\int_{x=0}^1 x\left[\frac{1}{2}(1-x^2)y^2 - \frac{1}{4}y^4\right]_0^{\sqrt{1-x^2}} dx$$
$$= \frac{1}{2}\int_0^1 x\left[\frac{1}{2}(1-x^2)(1-x^2) - \frac{1}{4}(1-x^2)^2\right]dx$$
$$= \frac{1}{2}\int_0^1 x\left(\frac{1}{2} - \frac{1}{4}\right)(1-x^2)^2 dx$$
$$= \frac{1}{8}\int_0^1 (x - 2x^3 + x^5)dx$$
$$= \frac{1}{8}\left[\frac{1}{2}x^2 - \frac{1}{2}x^4 + \frac{1}{6}x^6\right]_0^1$$
$$= \frac{1}{8}\left(\frac{1}{2} - \frac{1}{2} + \frac{1}{6}\right) = \frac{1}{48}.$$

Example 3. *Evaluate* $\int_0^4 \int_0^{2\sqrt{z}} \int_0^{\sqrt{4z-x^2}} dz\,dx\,dy$.

Solution. The given triple integral

$$I = \int_0^4 \int_0^{2\sqrt{z}} \left[\int_0^{\sqrt{4z-x^2}} dy\right] dz\,dx$$

$$= \int_0^4 \int_0^{2\sqrt{z}} [y]_0^{\sqrt{4z-x^2}} dz\,dx$$

$$= \int_0^4 \left[\int_0^{2\sqrt{z}} \sqrt{4z-x^2}\,dx\right] dz$$

$$= \int_0^4 \left[\frac{x}{2}\sqrt{4z-x^2} + \frac{4z}{2}\sin^{-1}\frac{x}{2\sqrt{z}}\right]_0^{2\sqrt{z}} dz$$

$$= \int_0^4 \left[0 + \frac{4z}{2}\sin^{-1}\frac{2\sqrt{z}}{2\sqrt{z}}\right] dz$$

$$= \int_0^4 2z.\frac{\pi}{2} dz = \int_0^4 \pi z\,dz$$

$$= \pi\left[\frac{z^2}{2}\right]_0^4 = \frac{\pi}{2}[16] = 8\pi$$

Example 4. *Evaluate* $\int_0^{\log a} \int_0^x \int_0^{x+y} e^{x+y+z} dx\,dy\,dz$.

Solution. Let

$$I = \int_0^{\log a} \int_0^x \int_0^{x+y} e^{x+y} e^z dx\,dy\,dz$$

$$= \int_0^{\log a} \int_0^x e^{x+y} (e^z)_0^{x+y} dx\,dy$$

$$= \int_0^{\log a} \int_0^x e^{x+y} [e^{x+y} - 1] dx\,dy$$

$$= \int_0^{\log a} \int_0^x e^{2(x+y)} dx\,dy - \int_0^{\log a} \int_0^x e^{x+y} dx\,dy$$

$$= \int_0^{\log a} \int_0^x e^{2x}.e^{2y} dx\,dy - \int_0^{\log a} \int_0^x e^{x+y} dx\,dy$$

$$= \int_0^{\log a} e^{2x}\left(\frac{1}{2}e^{2y}\right)_0^x dx - \int_0^{\log a} e^x (e^y)_0^x dx$$

$$= \frac{1}{2}\int_0^{\log a} e^{2x}(e^{2x} - e^0)dx - \int_0^{\log a} e^x(e^x - e^0)dx$$

$$= \frac{1}{2}\int_0^{\log a} (e^{4x} - e^{2x})dx - \int_0^{\log a} (e^{2x} - e^x)dx$$

$$= \frac{1}{2}\int_0^{\log a} (e^{4x} - 3e^{2x} + 2e^x)dx$$

$$= \frac{1}{2}\left[\frac{1}{4}e^{4x} - \frac{3}{2}e^{2x} + 2e^x\right]_0^{\log a}$$

$$= \frac{1}{8}(e^{4\log a} - e^0) - \frac{3}{4}(e^{2\log a} - e^0) + (e^{\log a} - e^0)$$

$$= \frac{1}{8}(a^4 - 1) - \frac{3}{4}(a^2 - 1) + (a - 1)$$

$$= \frac{1}{8}a^4 - \frac{3}{4}a^2 + a - \frac{3}{8} = \frac{1}{8}[a^4 - 6a^2 + 8a - 3].$$

Example 5. *Evaluate* $\iiint_V (x^2 + y^2 + z^2) dx\,dy\,dz$, *where V is the volume of cube bounded by the co-ordinates planes and the planes x=y=z=a.*

Solution. Here, the limits of x, y and z are varies from 0 to a.

Therefore, the given integral

$$I = \int_0^a \int_0^a \int_0^a (x^2 + y^2 + z^2) dx\,dy\,dz$$

$$= \int_0^a \int_0^a \left[x^2 z + y^2 z + \frac{1}{3}z^3\right]_0^a dx\,dy$$

$$= \int_0^a \int_0^a \left(x^2 a + y^2 a + \frac{1}{3}a^3\right) dx\,dy$$

$$= \int_0^a \left[x^2 ay + \frac{1}{3}y^3 a + \frac{1}{3}ya^3\right]_0^a dx$$

$$= \int_0^a \left(x^2 a^2 + \frac{1}{3}a^4 + \frac{1}{3}a^4\right) dx$$

$$= \left[\frac{1}{3}x^3 a^2 + \frac{1}{2}a^4 x + \frac{1}{3}a^4 x\right]_0^a = a^5.$$

Example 6. *Evaluate the volume of tetrahedron bounded by the co-ordinate planes and the planes x+y+z=1.*

Solution. The volume of tetrahedron can be expressed as

$$0 \le x \le 1, 0 \le y \le 1 - x, 0 \le z \le 1 - x - y.$$

$\therefore$ The integral

$$I = \iiint dx\,dy\,dz = \int_0^1 \int_0^{1-x} \int_0^{1-x-y} dx\,dy\,dz$$

$$= \int_0^1 \int_0^{1-x} [z]_0^{1-x-y} dx\,dy$$

$$= \int_0^1 \int_0^{1-x} (1 - x - y) dx\,dy$$

$$= \int_0^1 \left[(1-x)y - \frac{y^2}{2}\right]_0^{1-x} dx$$

$$= \int_0^1 \left[(1-x)^2 - \frac{(1-x)^2}{2}\right] dx = \int_0^1 \frac{1}{2}(1 - x^2) dx$$

$$= \frac{1}{2}\left[\frac{(1-x)^3}{3.(-1)}\right]_0^1 = -\frac{1}{6}(0 - 1) = \frac{1}{6}.$$

Example 7. *Evaluate* $\iiint_V zy^2 dx\,dy\,dz$, *where V is the region bounded between the xy plane and the sphere,* $x^2 + y^2 + z^2 = 1$.

Solution. Here the column parallel to z-axis is bounded by the plane z=0 and the surface of sphere $x^2 + y^2 + z^2 = 1$

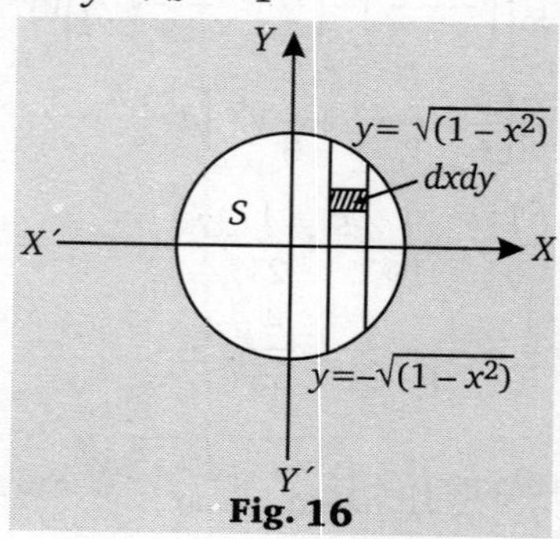

Fig. 16

i.e., $z = \sqrt{1-x^2-y^2}$.

The region S above which the volume V stands, is the area of circle of intersection of sphere $x^2+y^2+z^2=1$ by the *xy* plane.

Hence, the region S is the circle $x^2+y^2=1$.

It is clear from the figure that limits of integration for *y* are $-\sqrt{1-x^2}$ to $\sqrt{1-x^2}$ and for *x* are -1 to 1.

Hence, the given integral.

$$I = \int_{x=-1}^{1}\int_{y=-\sqrt{1-x^2}}^{\sqrt{1-x^2}}\int_{z=0}^{\sqrt{1-x^2-y^2}} zy^2\,dx\,dy\,dz$$

$$= \int_{x=-1}^{1}\int_{y=-\sqrt{1-x^2}}^{\sqrt{1-x^2}} y^2\left(\frac{1}{2}z^2\right)_0^{\sqrt{1-x^2-y^2}} dx\,dy$$

$$= \frac{1}{2}\int_{x=-1}^{1}\int_{y=-\sqrt{1-x^2}}^{\sqrt{1-x^2}} y^2(1-x^2-y^2)\,dx\,dy$$

$$= \frac{1}{2}\int_{x=-1}^{1}\int_{y=-\sqrt{1-x^2}}^{\sqrt{1-x^2}} (y^2-x^2y^2-y^4)\,dx\,dy$$

$$= \frac{1}{2}\int_{x=-1}^{1}\left(\frac{1}{3}y^3-\frac{1}{3}x^2y^3-\frac{1}{5}y^5\right)_{-\sqrt{1-x^2}}^{\sqrt{1-x^2}} dx$$

$$= \frac{1}{2}\int_{x=-1}^{1}\left[\begin{array}{l}\left(\frac{2}{3}\right)(1-x^2)^{3/2}\\ -\frac{2}{3}x^2(1-x^2)^{3/2}-\frac{2}{5}(1-x^2)^{5/2}\end{array}\right]dx$$

$$= \int_{x=-1}^{1}\left[\frac{1}{3}(1-x^2)^{5/2}-\frac{1}{5}(1-x^2)^{5/2}\right]dx$$

$[\because (1-x^2)^{3/2}-x^2(1-x^2)^{3/2}=(1-x^2)^{5/2}]$

$$= \frac{2}{15}\int_{x=-1}^{1}(1-x^2)^{5/2}\,dx = \frac{4}{15}\int_0^1(1-x^2)^{5/2}dx$$

$$= \frac{4}{15}\int_0^{\pi/2}(1-\sin^2\phi)^{5/2}\cos\phi\,d\phi$$

(putting $x=\sin\phi$)

$$= \frac{4}{15}\int_0^{\pi/2}\cos^6\phi\,d\phi = \frac{4}{15}\cdot\frac{5}{6}\cdot\frac{3}{4}\cdot\frac{1}{2}\cdot\frac{\pi}{2} = \frac{\pi}{24}.$$

Example 8. *Find the volume of region bounded by the cylinder* $x^2+y^2=16$ *and point* $z=0$ *to* $z=3$.

Solution. Here the limits are given from $z=0$ to $z=3$.

Also, from the equation of cylinder $x^2+y^2=16$ we find that the limits of *y* are from $-\sqrt{16-x^2}$ to $+\sqrt{16-x^2}$ and limit of *x* are from $-\sqrt{16}$ to $\sqrt{16}$ *i.e.*, from -4 to $+4$.

$\therefore$ Required volume

$$V = \int_{x=-4}^{4}\int_{y=-\sqrt{16-x^2}}^{\sqrt{16-x^2}}\int_{z=0}^{3} n\,dx\,dy\,dz$$

$$= 4\int_0^4\int_0^{\sqrt{16-x^2}}\int_0^3 dx\,dy\,dz$$

$$= 4\int_0^4\int_0^{\sqrt{16-x^2}}(z)_0^3\,dx\,dy$$

$$= 12\int_0^4\int_0^{\sqrt{16-x^2}} dx\,dy$$

$$= 12\int_0^4(y)_0^{\sqrt{16-x^2}}dx = 12\int_0^4\sqrt{4^2-x^2}\,dx$$

$$= 12\left[\frac{1}{2}x\sqrt{4^2-x^2}+\frac{1}{2}4^2\sin^{-1}\left(\frac{x}{4}\right)\right]_0^4$$

$$= 12.\left[\frac{1}{2}.16\sin^{-1}(1)\right] = 96(\pi/2) = 48\pi.$$

EXERCISE 19.4

1. Evaluate $\int_{x=0}^{1}\int_{y=0}^{2}\int_{z=1}^{2} x^2yz\,dx\,dy\,dz$.
2. Evaluate $\int_{-a}^{a}\int_{-b}^{b}\int_{-c}^{c}(x^2+y^2+z^2)\,dx\,dy\,dz$.
3. Evaluate $\int_{-1}^{1}\int_0^{z}\int_{x-z}^{x+z}(x+y+z)\,dy\,dx\,dz$. (GBTU-2010)
4. Evaluate $\int_0^1\int_0^{1-x}\int_0^{1-x-y}\frac{dy\,dx\,dz}{(1+x+y+z)^3}$.
5. Evaluate $\int_0^{\pi/2}d\theta\int_0^{a\sin\theta}dr\int_0^{(a^2-r^2)/a} r\,dz$.
6. Evaluate $\int_0^a\int_0^{a-x}\int_0^{a-x-y} x^2\,dx\,dy\,dz$.
7. Evaluate $\int_0^2\int_0^x\int_0^{x+y} e^x(y+2z)\,dx\,dy\,dz$.
8. Evaluate $\int_0^{\log 2}\int_0^x\int_0^{x+\log y} e^{x+y+z}\,dx\,dy\,dz$.
9. Evaluate the integral $\iiint xyz\,dx\,dy\,dz$ over the the volume enclosed by three co-ordinates plane and the plane $x+y+z=1$
10. Evaluate $\iiint\frac{dx\,dy\,dz}{(x+y+z+1)^2}$ over the region $x\ge 0, y\ge 0, z\ge 0, x+y+z\le 1$.
11. Evaluate $\iiint(z^5+z)\,dx\,dy\,dz$ over the sphere $x^2+y^2+z^2=1$.
12. Evaluate $\iiint_R u^2v^2w\,du\,dv\,dw$, where R is the region $u^2+v^2\le 1, 0\le w\le 1$. (VTU-2011; SRM-2009)
13. Find the volume of the tetrahedron bounded by the plane $\frac{x}{a}+\frac{y}{b}+\frac{z}{c}=1$ and $(x+z=a)$ and coordinate plane.
14. Evaluate $\int_0^2\int_0^x\int_0^{x+y} e^x(y+2z)dx\,dy\,dz$.
15. Evaluate $\iiint_R(x-2y+z)$ where R is the region determined by $0\le x\le 1$, $0<y\le x^2$, $0\le z\le x+y$ (UPTU-2009)

ANSWERS

1. 1 **2.** $\frac{2}{3}abc(a^2+b^2+c^2)$ **3.** 0 **4.** $\frac{1}{2}\left(\log 2-\frac{5}{8}\right)$ **5.** $\frac{5a^3\pi}{64}$ **6.** $\frac{a^5}{60}$ **7.** $19[(1/3)e^2+1]$

8. $\frac{8}{3}\log 2-\frac{19}{9}$ **9.** 0 **10.** $\frac{1}{2}\left(\log 2-\frac{5}{8}\right)$ **11.** 0 **12.** $\pi/48$ **13.** $abc/5$ **14.** $\frac{19}{3}(e^2+3)$ **15.** 8/35

19.7 DIRICHLET'S THEOREM FOR THREE VARIABLES

Statements. *Let V be the region given by* $x \geq 0, y \geq 0, z \geq 0, x+y+z \leq 1, l, m, n$ *are positive. Then*

$$\int_V x^{l-1}y^{m-1}z^{n-1}dx\,dy\,dz = \frac{\Gamma(l)\Gamma(m)\Gamma(n)}{\Gamma(l+m+n+1)}.$$

Proof. We evaluate the given integral over the volume enclosed by the three co-ordinates planes and the plane $x+y+z=1$, $x=0, y=0, z=0$. The limits of integration for this region can be expressed as $0 \leq x \leq 1, 0 \leq y \leq 1-x, 0 < z \leq 1-x-y$.

Hence we may write the given triple integral as

$$\int_0^1\int_0^{1-x}\int_0^{1-x-y} x^{l-1}y^{m-1}z^{n-1}dx\,dy\,dz = \int_0^1\int_0^{1-x} x^{l-1}y^{m-1}[z^n/n]_0^{1-x-y}\,dx\,dy$$

$$= \frac{1}{n}\int_0^1\int_0^{1-x} x^{l-1}y^{m-1}(1-x-y)^n\,dx\,dy$$

$$= \frac{1}{n}\int_0^1\int_0^1 x^{l-1}\{(1-x)t\}^{m-1}[1-x-(1-x)t]^n(1-x)dx\,dt$$

(Putting $y=(1-x)t, \Rightarrow dy=(1-x)dt$))

$$= \frac{1}{n}\int_0^1\int_0^1 x^{l-1}(1-x)^{m-1}t^{m-1}(1-x)^n(1-t)^n(1-x)dx\,dt$$

$$= \frac{1}{n}\int_0^1\int_0^1 x^{l-1}(1-x)^{m+n}t^{m-1}(1-t)^n\,dx\,dt$$

$$= \frac{1}{n}\int_0^1 x^{l-1}(1-x)^{m+n}dx \times \int_0^1 (t)^{m-1}(1-t)^n\,dt$$

$$= \frac{1}{n}B(l, m+n+1)B(m, n+1) \quad \text{(By definition of Beta function)}$$

$$= \frac{1}{n}\cdot\frac{\Gamma(l)\Gamma(m+n+1)}{\Gamma(l+m+n+1)}\cdot\frac{\Gamma(m)\Gamma(n+1)}{\Gamma(m+n+1)} \quad \left[\because B(m,n)=\frac{\Gamma(m)\Gamma(n)}{\Gamma(m+n)}\right]$$

$$= \frac{\Gamma(l)\Gamma(m)}{\Gamma(l+m+n+1)}\cdot\frac{n\Gamma(n)}{n} \quad [\because \Gamma(n+1)=n\,\Gamma(n)]$$

$$= \frac{\Gamma(l)\Gamma(m)\Gamma(n)}{\Gamma(l+m+n+1)}.$$

Aliter. Here we solve it by first consider the double integral

$$I_2 = \iint x^{l-1}y^{m-1}dx\,dy$$

where, integral extended to all positive values of variables. The condition is $x+y \leq 1$.

Here, 2-dimensional Euclidean space, (*i.e.*, the region of integration of I_2), is bounded by the straight lines $x=0, y=0$ and $x+y=1$, and the region expressed as $0 \leq x \leq 1, 0 \leq y \leq 1-x$.

$$\therefore \quad I_2 = \int_{x=0}^1\int_{y=0}^{1-x} x^{l-1}y^{m-1}dx\,dy = \int_0^1 x^{l-1}\left[\frac{y^m}{m}\right]_0^{1-x}dx = \int_0^1 \frac{1}{m}x^{l-1}(1-x)^m dx$$

$$= \frac{1}{m}\int_0^1 x^{l-1}(1-x)^{m+1-1}dx = \frac{1}{m}B(l, m+1) \quad \text{[By Beta function]}$$

$$= \frac{1}{m}\frac{\Gamma(l)\Gamma(m+1)}{\Gamma(l+m+1)} = \frac{1}{m}\frac{\Gamma(l)\Gamma(m)\Gamma(m)}{\Gamma(l+m+1)}$$

$$= \frac{\Gamma(l)\Gamma(m)}{\Gamma(l+m+1)}. \quad \ldots (1)$$

It is for two variables.

Now, we consider the double integral as follows

$$U_2 = \iint x^{l-1} y^{m-1} dx\, dy, x + y \le h.$$

We have $x + y \le h \Rightarrow \frac{x}{h} + \frac{y}{h} \le 1$

So putting $\frac{x}{h} = u$ and $\frac{y}{h} = v$ so that $dx = h\,du$ and $dy = h\,dv$ the integrals U_2 becomes

$$U_2 = \iint (hu)^{l-1} (hv)^{m-1} h^2 du\, dv$$

$$= h^{l+m} \iint u^{l-1} v^{m-1} du\, dv, \text{where } v + u \le 1$$

$$= h^{l+m} \frac{\Gamma(l)\Gamma(m)}{\Gamma(l+m+1)} \qquad \text{[By (1)]} \qquad \dots (2)$$

Now, we consider the triple integral

$$I_3 = \iiint x^{l-1} y^{m-1} z^{n-1} dx\, dy\, dz.$$

Condition $x + y + z \le 1$ *i.e.*, $y + z \le 1 - x$. We have

$$I_3 = \int_{x=0}^{1} \left[\iint y^{m-1} z^{n-1} dy\, dz\right] x^{l-1} dx, \qquad \text{where } y+z \le 1-x$$

$$= \int_0^1 (1-x)^{m+n} \frac{\Gamma(m)\Gamma(n)}{\Gamma(m+n+1)} x^{l-1} dx \qquad \text{[By (2)]}$$

$$= \frac{\Gamma(m)\Gamma(n)}{\Gamma(m+n+1)} \int_0^1 x^{l-1} (1-x)^{m+n+1-1} dx$$

$$= \frac{\Gamma(m)\Gamma(n)}{\Gamma(m+n+1)} B(l, m+n+1) = \frac{\Gamma(m)\Gamma(n)}{\Gamma(m+n+1)} \cdot \frac{\Gamma(l)\Gamma(m+n+1)}{\Gamma(l+m+n+1)} = \frac{\Gamma(l)\Gamma(m)\Gamma(n)}{\Gamma(l+m+n+1)}.$$

REMARKS

- Dirichlet's theorem holds good even if the conditons is taken as $x+y+z<1$ in place of $x + y + z \le 1$.
- The triple integral $\iiint x^{l-1} y^{m-1} z^{n-1} dx\, dy\, dz = h^{l+m+n} \frac{\Gamma(l)\Gamma(m)\Gamma(n)}{\Gamma(l+m+n+1)}$ where the integral is extended to all positive values of the variables x, y and z, when $x+y+z \le h$.

19.8 DIRICHLET'S THEOREM FOR *n* VARIABLES

Statement. *If the integral is extended to all positive values of the variables* x_1, x_2,.., x_n *subject to the condition* $x_1 + x_2 + \dots + x_n \le 1$. *Then*

$$\iint \dots \int x_1^{l_1-1} x_2^{l_2-1} \dots x_n^{l_n-1} dx_1 dx_2 \dots dx_n = \frac{\Gamma(l_1)\Gamma(l_2)\dots\Gamma(l_n)}{\Gamma(1 + l_1 + l_2 + \dots + l_n)}$$

Proof. We shall prove this theorem by mathematical induction.

First we prove the theorem for 2-variables *i.e.*, $n=2$

Let us consider the integral

$$I_2 = \iint x_1^{l_1-1} x_2^{l_2-1} dx_1 dx_2 \text{ such that } x_1 + x_2 \le 1.$$

Now, using previous theorem, we have

$$I_2 = \frac{\Gamma(l_1)\Gamma(l_2)}{\Gamma(1 + l_1 + l_2)} \qquad \dots (2)$$

Equation (1) is true for two variables. Now assume that theorem is true for n variables. Therefore

$$I_n = \iint \dots \int x_1^{l_1-l} x_2^{l_2-l} \dots x_n^{l_n-l} dx_1 . dx_2 \dots . dx_n$$

$$= \frac{\Gamma(l_1)\Gamma(l_2)\dots\Gamma(l_n)}{\Gamma(1 + l_1 + l_2 + \dots + l_n)} \qquad \dots (2)$$

with condition $x_1 + x_2 + \dots + x_n \le 1$.

If the condition $x_1 + x_2 + \dots + x_n \le h$, then putting $\frac{x_1}{h} = u_1, \frac{x_2}{h} = u_2 \dots \frac{x_n}{h} = u_n$ so that

$$dx_1 = h\,du_1, dx_2 = h\,du_2, \dots dx_n = h\,du_n$$

We have $\iint\ldots\int x_1^{l_1-1}\ x_2^{l_2-1}\ldots x_n^{l_n-1}dx_1dx_2\ldots dx_n\ = h^{l_1+l_2+\ldots+l_n}\iint\ldots\int u_1^{l_1-1}u_2^{l_2-1}\ldots u_n^{l_n-1}du_1du_2\ldots du_n$

subject to the condition $u_1+u_2+\ldots+u_n\le 1$

$$= h^{l_1+l_2+\ldots+l_n}\frac{\Gamma(l_1)\Gamma(l_2)\ldots\Gamma(l_n)}{\Gamma(1+l_1+l_2+\ldots+l_n)} \quad \ldots (3)$$

(Using the assumed result (2))

Now for $n+1$ variables the conditions are $x_1+x_2+\ldots+x_n+x_{n+1}\le 1$ *i.e.*, $x_2+x_3+\ldots+x_n+x_{n+1}\le 1-x_1$ and $0\le x_1\le 1$. We have

$$\iint\ldots\int x_1^{l_1-1}x_2^{l_2-1}\ldots x_n^{l_n-1}x_{n+1}^{l_{(n+1)}-1}dx_1dx_2\ldots dx_ndx_{n+1} \quad \text{where } x_1+x_2+\ldots x_{n+1}\le 1$$

$$=\int_{x_1=0}^{1}x_1^{l_1-1}\left[\iint\ldots\int x_2^{l_2-1}\ldots x_{n+1}^{l_{n+1}-1}dx_2\ldots dx_{n+1}\right]dx_1$$

$$=\frac{\Gamma(l_2)\Gamma(l_3)\ldots\Gamma(l_{n+1})}{\Gamma(l_1+1+l_2+\ldots l_n+l_{n+1})}\cdot\int_0^1 x_1^{l_1-1}(1-x_1)^{(1+l_2+l_3+\ldots+l_{n+1})-1}dx_1$$

Using (3)

$$=\frac{\Gamma(l_2)\Gamma(l_3)\ldots\Gamma(l_{n+1})}{\Gamma(1+l_2+\ldots+l_{n+1})}\cdot\frac{\Gamma(1+l_2+\ldots+l_{n+1})}{\Gamma(1+l_1+l_2+\ldots+l_n+l_{n+1})}$$

$$=\frac{\Gamma(l_1)\Gamma(l_2)\ldots\Gamma(l_{n+1})}{\Gamma(1+l_1+l_2+\ldots+l_{n+1})} \quad \ldots (4)$$

The result (4) shows that the theorem hold for $(n+1)$ variables. Hence, by principle of mathematical induction, theorem is true for all values of n.

Solved Examples

Example 1. *Evaluate* $\iiint x^{l-1}y^{m-1}z^{n-1}dx\,dy\,dz$ *in which* $x\ge 0, y\ge 0, z=0$ *and* $(x/a)^{1/2}+(y/b)^{1/2}+(z/c)^{1/2}\le 1.$

Solution. Let $(x/a)^{1/2}=u$, $(y/b)^{1/2}=v$

and $(z/b)^{1/2}=w$

Then $x=au^2, y=bv^2, z=cw^2$

$dx=2au\,du;\ dy=2bv\,dv;\ dz=2cw\,dw,$

$u\ge 0, v\ge 0, w\ge 0$ and $u+v+w\le 1$

Hence, $\iiint(au^2)^{l-1}(bv^2)^{m-1}(cw^2)^{n-1}$

$.2au.2bv.2cw\,du\,dv\,dw$

$$=8a^lb^mc^n\iiint u^{2l-1}v^{2m-1}w^{2n-1}\,du\,dv\,dw$$

$$=8a^lb^mc^n\frac{\Gamma(2l)\Gamma(2m)\Gamma(2n)}{\Gamma(2l+2m+2n-1)}$$

Example 2. *Evaluate* $\iint dx\,dy$ *over the region in the positive quadrant for which* $x+y\le 1$.

Solution. We have $x+y\le 1$ and $x\ge 0, y\ge 0$ and so by Dirichlet's theorem, we get

$$I=\iint x^{2-1}y^{2-1}dx\,dy=\frac{\Gamma(2)\Gamma(2)}{\Gamma(2+2+1)}=\frac{1}{4}\cdot\frac{1}{\Gamma(3)}$$

$$=\frac{1}{4.3.2\Gamma(1)}=\frac{1}{24\times 1}=\frac{1}{24}.$$

Example 3. *Show that the integral* $\iiint x^{l-1}y^{m-1}z^{n-1}\,dx\,dy\,dz$ *integrand over the region in the first octant below the surface* $(x/a)^p+(y/b)^q+(z/c)^r=1$ *is* $\dfrac{a^lb^mc^n}{pqr}\cdot\dfrac{\Gamma(l/p)\Gamma(m/q)\Gamma(n/r)}{\Gamma[(l/p)+(m/q)+(n/r)+1]}$.

(UPTU-2006, 2009, GBTU-2011)

Solution. Putting

$$\left(\frac{x}{a}\right)^p=u \text{ or } x=au^{1/p}$$

$$\Rightarrow \quad dx=a(1/p)u^{(1/p)-1}.du.$$

Similarly putting $(y/b)^q=v$ and $(z/c)^r=w$, we get

$dy=b(1/q)v^{(1/q)-1}dv$

and $dz=c(1/r)w^{(1/r)-1}dw$.

$$\therefore\ x^{l-1}dx=a^{l-1}u^{(l-1)/p}a(1/p)u^{(1-p)/p}du$$
$$=a^l(1/p)u^{(l/p)-1}du$$

Similarly, $y^{m-1}dy=b^m(1/q)v^{(m/q)-1}dv$;

$z^{n-1}dz=c^n(1/r)w^{(n/r)-1}dw$

Hence, subject to the condition $u+v+w\le 1$, the given integral

$$=\frac{a^lb^mc^n}{pqr}\iiint u^{(l/p)-1}v^{(m/q)-1}w^{(n/r)-1}du\,dv\,dw$$

$$=\frac{a^lb^mc^n}{pqr}\cdot\frac{\Gamma(l/p)\Gamma(m/q)\Gamma(n/r)}{\Gamma[(l/p)+(m/q)+(n/r)+1]}$$

Example 4. *Find the value of* $\iint\ldots\int dx_1\,dx_2\ldots dx_n$ *extended to all positive values of variables subject to the condition* $x_1^2+x_2^2+\ldots+x_n^2<R^2$.

Solution. To find the value of integral I extended to all positive values of $x_1,x_2,\ldots x_n$ subject to the condition

$$\frac{x_1^2}{R^2}+\frac{x_2^2}{R^2}+\ldots+\frac{x_n^2}{R^2}=1$$

We put $\left(\frac{x_1}{R}\right)^2=u_1 \Rightarrow x_1=Ru_1^{1/2}$

$\Rightarrow$ $dx_1=(1/2)Ru_1^{-1/2}du_1$

$$\left(\frac{x_2}{R}\right)^2=u_2 \Rightarrow x_2=Ru_2^{1/2}$$

so that $dx_2=(1/2)Ru_2^{-1/2}du_2$ and so on.

Then the required integral

$$I=\iint\ldots\int(1/2)^n R^n u_1^{-1/2}\ldots u_n^{-1/2}du_1.du_2\ldots du_n$$

$$=\left(\frac{R}{2}\right)^2\iint\ldots\int u_1^{(1/2)-1}u_2^{(1/2)-1}\ldots u_n^{(1/2)-1}du_1\,du_2\ldots du_n$$

Condition is $u_1+u_2+\ldots+u_n<1$

$$=\left(\frac{R}{2}\right)^2\frac{\{\Gamma(1/2)\}^n}{\Gamma(1+n/2)} \quad \text{(By Dirichlet's theorem)}$$

$$=\left(\frac{R}{2}\right)^2\cdot\frac{\pi^{n/2}}{\Gamma\left(1+\frac{n}{2}\right)} \qquad [\because \Gamma(1/2)=\sqrt{\pi}]$$

Example 5. *Evaluate* $\iiint dx\,dy\,dz$ *where* $\frac{x^2}{a^2}+\frac{y^2}{b^2}+\frac{z^2}{c^2}\le 1$ *or find the volume of*

$$(x^2/a^2)+(y^2/b^2)+(z^2/c^2)=1.$$

Solution. Let $\frac{x^2}{a^2}=u, x=au^{1/2}$ so that $dx=\frac{1}{2}au^{-1/2}du$.

Similarly, putting $\frac{y^2}{b^2}=v$ and $\frac{z^2}{c^2}=w$, we get

$$dy=\frac{1}{2}bv^{-1/2}dv \quad \text{and} \quad dz=\frac{1}{2}cw^{-1/2}dw.$$

$\therefore \iiint dx\,dy\,dz$

$$=\iiint\frac{1}{2}au^{-1/2}du.\frac{1}{2}bv^{-1/2}.dv.\frac{1}{2}cw^{-1/2}dw$$

$$=\frac{1}{8}abc\iiint u^{1/2-1}v^{1/2-1}w^{1/2-1}du\,dv\,dw$$

$$=\frac{1}{8}abc\frac{\Gamma(1/2)\Gamma(1/2)\Gamma(1/2)}{\Gamma(1/2+1/2+1/2+1)}$$

$$=\frac{1}{8}abc\frac{\sqrt{\pi}\sqrt{\pi}\sqrt{\pi}}{\Gamma(5/2)}=\frac{\pi\sqrt{\pi}\,bca}{8\cdot 3/2\cdot 1/2\sqrt{\pi}}$$

$$=\frac{1}{6}\pi abc$$

Example 6. *Find the volume enclosed by the surface* $(x/a)^{2n}+(y/b)^{2n}+(z/c)^{2n}=1$.

Solution. The given surface is symmetrical in all the eight octants. Now we want to find the volume V in the positive octant. Clearly

$$V=\iiint dx\,dy\,dz$$

where the integral is extended to all positive values of the variables x, y, z subject to the condition $(x/a)^{2n}+(y/b)^{2n}+(z/c)^{2n}\le 1$.

Now put

$$(x/a)^{2n}=u,(y/b)^{2n}=v,(z/c)^{2n}=w$$

$\Rightarrow$ $x=au^{1/2n}, y=bv^{1/2n}, z=cw^{1/2n}$

So that $dx=\frac{a}{2n}u^{(1/2n)-1}du$

$$\therefore \; V=\frac{abc}{8n^3}\iiint u^{(1/2n)-1}v^{(1/2n)-1}w^{(1/2n)-1}du\,dv\,dw$$

$$=\frac{abc}{8n^3}\frac{[\Gamma(1/2n)]^3}{\Gamma\{(3/2n)+1\}}=\frac{abc}{8n^3}\frac{[\Gamma(1/2n)]^3}{(3/2n).\Gamma(3/2n)}$$

$$=\frac{abc}{12n^2}\frac{[\Gamma(1/2n)]^3}{\Gamma(3/2n)}.$$

Hence, the total volume enclosed by given surface

$$=8V=\frac{2}{3}.\frac{abc}{n^2}\frac{[\Gamma(1/2n)]^3}{\Gamma(3/2n)}.$$

Example 7. *Find the volume of the tetrahedron bounded by the plane* $\frac{x}{a}+\frac{y}{b}+\frac{z}{c}=1$ *and the co-ordinate planes.*

(GBTU-2012, MTU-2012)

Solution. The volume of a small element at a point $(x,y,z)=dx\,dy\,dz$

$\therefore$ the volume of the given tetrahedron $=\iiint dx\,dy\,dz$ where the integral is extended to all positive values of variables x, y, z.

Put $x/a=u, y/b=v, z/c=w$ subject to the condition so that $\frac{x}{a}+\frac{y}{b}+\frac{z}{c}\le 1$

$$dx=a.du, dy=b.dv \text{ and } dz=c.dw$$

then the required volume

$$=\iiint abc\,du\,dv\,dw \quad \text{where } u+v+w\le 1$$

$$=abc\iiint u^{1-1}v^{1-1}w^{1-1}du\,dv\,dw$$

$$=abc\frac{[\Gamma(1)]^3}{\Gamma(1+1+1+1)} \quad \text{[By Dirichlet's theorem]}$$

$$=abc\frac{1}{\Gamma(4)}=\frac{abc}{3.2.1}=\frac{abc}{6}$$

19.9 LIOUVILLE'S EXTENSION OF DIRICHLET'S THEOREM

Statement. *If* x, y, z *are all positive and such that* $h_1<x+y+z\le h_2$ *then*

$$\iiint f(x+y+z)x^{l-1}y^{m-1}z^{n-1}dx\,dy\,dz=\frac{\Gamma(l)\Gamma(m)\Gamma(n)}{\Gamma(l+m+n)}\int_{h_1}^{h_2}f(u)u^{l+m+n-1}du.$$

Proof. From Dirichlet's theorem, we have

$$I = \iiint x^{l-1}y^{m-1}z^{n-1}dx\,dy\,dz = \frac{\Gamma(l)\Gamma(m)\Gamma(n)}{\Gamma(l+m+n)}u^{(l+m+n)} \quad \text{... (1)}$$

subject to the condition that $x, y, z \geq 0$ and $x+y+z \leq u$.

Now if $x, y, z \geq 0$ and $x+y+z \leq u+\delta u$, then we have

$$I = \iiint x^{l-1}y^{m-1}z^{n-1}dx\,dy\,dz = \frac{\Gamma(l)\Gamma(m)\Gamma(n)}{\Gamma(l+m+n+1)}(u+\delta u)^{(l+m+n)} \quad \text{... (2)}$$

So the value of integral given above extended to all such positive value of x, y, z such that $x+y+z$ lies between u and $u+\delta u$, is given by

$$I = \iiint x^{l-1}y^{m-1}z^{n-1}dx\,dy\,dz$$

$$= \frac{\Gamma(l)\Gamma(m)\Gamma(n)}{\Gamma(l+m+n+1)}[(u+\delta u)^{l+m+n} - u^{l+m+n}] = \frac{\Gamma(l)\Gamma(m)\Gamma(n)}{\Gamma(l+m+n+1)}u^{l+m+n}\left[\left(1+\frac{\delta u}{u}\right)^{l+m+n} - 1\right]$$

$$= \frac{\Gamma(l)\Gamma(m)\Gamma(n)}{\Gamma(l+m+n+1)}u^{l+m+n}\left[1+(l+m+n)\frac{\delta u}{u}+\ldots-1\right] \quad \text{[On expanding by Taylor's series]}$$

$$= \frac{\Gamma(l)\Gamma(m)\Gamma(n)}{\Gamma(l+m+n+1)}(l+m+n)u^{(l+m+n-1)}\delta u$$

[Neglecting the second and higher degree terms of δu]

$$= \frac{\Gamma(l)\Gamma(m)\Gamma(n)}{\Gamma(l+m+n)}u^{(l+m+n-1)}\delta u \quad [\because \Gamma(l+m+n+1) = (l+m+n)\Gamma(l+m+n)]$$

Now, consider the intergral $\iiint f(x+y+z)x^{l-1}y^{m-1}z^{n-1}dx\,dy\,dz$.

Since $u \leq x+y+z \leq \delta u$, so the function $f(x+y+z)$ will differ by a small quantity of same order of solution. Hence, the integral

$$\iiint f(x+y+z)x^{l-1}y^{m-1}z^{n-1}dx\,dy\,dz = \frac{\Gamma(l)\Gamma(m)\Gamma(n)}{\Gamma(l+m+n)}F(u)u^{(l+m+n-l)}\delta u$$

subject to the condition that $x, y, z \geq 0$ and $u \leq x+y+z \leq u+\delta u$, to the first approximation.

So finally for the given condition that for positive x, y, z such that $h_1 < x+y+z \leq h_2$,

we get $\iiint f(x+y+z)x^{l-1}y^{m-1}z^{n-1}dx\,dy\,dz = \dfrac{\Gamma(l)\Gamma(m)\Gamma(n)}{\Gamma(l+m+n)}\displaystyle\int_{h_1}^{h_2} f(u)^{(l+m+n-1)}du$

Solved Examples

Example 1. *Evaluate $\iiint e^{x+y+z}dx\,dy\,dz$ taken over the positive octant such that $x+y+z \leq 1$.* (UPTU-2008)

Solution. In the positive octant x, y, z are all positive and therefore $0 < (x+y+z) \leq 1$.

Therefore, we have

$$\iiint e^{x+y+z}dx\,dy\,dz$$

$$= \frac{\Gamma(1)\Gamma(1)\Gamma(1)}{\Gamma(1+1+1)}\int_0^1 e^h h^{1+1+1-1}dh$$

[By Liouville's theorem]

$$= \frac{1}{\Gamma(3)}\int_0^1 h^2e^h dh = \frac{1}{2!}\left[(h^2e^h)_0^1 - \int_0^1 2he^h dh\right]$$

$$= \frac{1}{2}\left[e - 2\left\{(he^h)_0^1 - \int_0^1 e^h dh\right\}\right]$$

$$= \frac{1}{2}\left[e - 2\left\{e-(e^h)_0^1\right\}\right] = \frac{1}{2}[e-2\{e-e+1\}]$$

$$= \frac{1}{2}(e-2).$$

Example 2. *Evaluate $\iiint \log(x+y+z)\,dx\,dy\,dz$ taken over all positive values of x, y, z subject to the condition $x+y+z \leq 1$.*

Solution. Since x, y, z are to be taken positive value only, we have $0 < (x+y+z) \leq 1$.

Therefore, we have

$$\iiint \log(x+y+z)\,dx\,dy\,dz$$

$$= \iiint \log(x+y+z)x^{1-1}y^{1-1}z^{1-1}\,dx\,dy\,dz$$

$$= \frac{\Gamma(1)\Gamma(1)\Gamma(1)}{\Gamma(1+1+1)}\int_0^1(\log h)h^{1+1+1-1}dh,$$

[By Liouville's theorem]

$$= \frac{1}{\Gamma(3)}\int_0^1 h^2(\log h)dh$$

$$= \frac{1}{2!}\left[\left\{(\log h)\frac{1}{3}h^3\right\}_0^1 - \int_0^1\frac{1}{h}\cdot\frac{1}{3}h^3 dh\right]$$

$$= \frac{1}{6}\left[h^3\log h - \frac{h^3}{3}\right]_0^1 = -\frac{1}{18}$$

Example 3. *Evaluate* $\iiint (x+y+z)dx\,dy\,dz$ *over the tetrahedron* $x = 0, y = 0, z = 0$ *and* $x+y+z \le 1$

Solution. We are given that : $0 \le x+y+z \le 1$ therefoe, by Liouville's extension of Dirichlets theorem

$$\iiint (x+y+z)dx\,dy\,dz$$

$$= (x+y+z)x^{1-1}y^{1-1}z^{1-1}dx\,dy\,dz$$

$$= \frac{\Gamma(1)\Gamma(1)\Gamma(1)}{\Gamma(1+1+1)}\int_0^1 u.u^{1+1+1-1}du = \frac{1}{\Gamma(3)}\int_0^1 u^3 du$$

$$= \frac{1}{2!}\left[\frac{u^4}{4}\right]_0^1 = \frac{1}{2}\left[\frac{1}{4}\right] = \frac{1}{8}.$$

Example 4. *Prove that* $\iiint \frac{dx\,dy\,dz}{\sqrt{(1-x^2-y^2-z^2)}} = \frac{\pi^2}{8}$; *the integral being extended to a positive values of variables for which the expression is real.*

Solution. Since x, y, z are to be taken positive values only, we have $0 < x^2 + y^2 + z^2 < 1$.

Put $x^2 = u$ or $x = u^{1/2}$ so that $dx = 1/2u^{-1/2}\,du$.

Similarly putting $y^2 = v$ and $z^2 = w$, we get $dy = 1/2\, v^{-1/2}\,dv$, $dz = \frac{1}{2}w^{-1/2}dw$.

Now, given integral

$$= \iiint \frac{(1/2)u^{-1/2}(1/2)v^{-1/2}(1/2)w^{-1/2}}{\sqrt{1-(u+v+w)}}\,du\,dv\,dw$$

where $0<u+v+w<1$

$$= \frac{1}{8}\iiint \frac{u^{1/2-1}v^{1/2-1}w^{1/2-1}}{\sqrt{1-(u+v+w)}}du\,dv\,dw$$

$$= \frac{1}{8}\frac{\Gamma(1/2)\Gamma(1/2)\Gamma(1/2)}{\Gamma(1/2+1/2+1/2)}$$

$$\int_0^1 \frac{1}{\sqrt{1-h}}h^{1/2+1/2+1/2-1}dh$$

$$= \frac{1}{8}\frac{\sqrt{\pi}\sqrt{\pi}\sqrt{\pi}}{\Gamma(3/2)}\int_0^1 \sqrt{\left(\frac{h}{1-h}\right)}dh$$

$$= \frac{1}{4}\pi\int_0^{\pi/2}\sqrt{\left(\frac{\sin^2\theta}{\cos^2\theta}\right)}2\sin\theta\cos\theta d\theta$$

Putting $h = \sin^2\theta$

$$= \frac{1}{4}\pi\int_0^{\pi/2}2\sin^2\theta d\theta = \frac{1}{4}\pi\int_0^{\pi/2}(1-\cos 2\theta)d\theta$$

$$= \frac{1}{4}\pi\left[\theta - \frac{1}{2}\sin 2\theta\right]_0^{\pi/2} = \frac{1}{4}\pi(1/2\pi) = \frac{1}{8}\pi^2.$$

Example 5. *Evaluate* $\iiint x^\alpha y^\beta z^\gamma (1-x-y-z)^\lambda\,dx\,dy\,dz$ *over the interior of tetrahedron formed by the co-ordinate plane and the plane* $x+y+z=1$.

Solution. The region of integration is bounded by the plane $x = 0, y = 0, z = 0$ and $x+y+z=1$. So, the variable x, y, z take all positive values subject to the condition

$$0<x+y+z<1.$$

Therefore the given integral

$$= \iiint x^{(\alpha+1)-1}y^{(\beta+1)-1}z^{(\gamma+1)-1}[1-(x+y+z)]^\lambda dx\,dy\,dz$$

$$= \frac{\Gamma(\alpha+1)\Gamma(\beta+1)\Gamma(\gamma+1)}{\Gamma(\alpha+\beta+\gamma+3)}\int_0^1 u^{\alpha+1+\beta+1+\gamma+1-1}(1-u)^\lambda du$$

[By Liouville's extension of Dirichlet's theorem]

$$= \frac{\Gamma(\alpha+1)\Gamma(\beta+1)\Gamma(\gamma+1)}{\Gamma(\alpha+\beta+\gamma+3)}\int_0^1 u^{(\alpha+\beta+\gamma+3)-1}(1-u)^{(\lambda+1)-1}du$$

$$= \frac{\Gamma(\alpha+1)\Gamma(\beta+1)\Gamma(\gamma+1)}{\Gamma(\alpha+\beta+\gamma+3)}B(\alpha+\beta+\gamma+3, \lambda+1)$$

$$= \frac{\Gamma(\alpha+1)\Gamma(\beta+1)\Gamma(\gamma+1)}{\Gamma(\alpha+\beta+\gamma+3)} \cdot \frac{\Gamma(\alpha+\beta+\gamma+3)\Gamma(\lambda+1)}{\Gamma(\alpha+\beta+\gamma+4)}$$

$$= \frac{\Gamma(\alpha+1)\Gamma(\beta+1)\Gamma(\gamma+1)\Gamma(\lambda+1)}{\Gamma(\alpha+\beta+\gamma+\lambda+4)}$$

Example 6. *Evaluate*

$$\iiint \sqrt{(a^2b^2c^2 - b^2c^2x^2 - c^2a^2y^2 - a^2b^2z^2)}dx\,dy\,dz$$

taken throughout the ellipsoid

$$\frac{x^2}{a^2}+\frac{y^2}{b^2}+\frac{z^2}{c^2} = 1.$$

Solution. Let us first evaluate the given integral over the region of ellipsoid which lie in the positive octants the given ellipsoid $\frac{x^2}{a^2}+\frac{y^2}{b^2}+\frac{z^2}{c^2} = 1$ is symmetrical in all the eight octants.

Put $\frac{x^2}{a^2} = u, \frac{y^2}{b^2} = v, \frac{z^2}{c^2} = w$ then $x = au^{1/2}$

$\Rightarrow$ $dx = 1/2au^{-1/2}du$

$y = bv^{1/2}$ $\Rightarrow$ $dx = \frac{1}{2}bv^{-1/2}dv$

and $z = cw^{1/2}$ $\Rightarrow$ $dz = \frac{1}{2}cw^{-1/2}dw$

The given integral

$$I = abc\iiint \sqrt{\left(1-\frac{x^2}{a^2}-\frac{y^2}{b^2}-\frac{z^2}{c^2}\right)}dx\,dy\,dz$$

where $0 < x^2/a^2 + y^2/b^2 + z^2/c^2 \le 1$

$$= abc\iiint \sqrt{1-u-v-w}(1/8)\, abcu^{-1/2}v^{-1/2}w^{-1/2}du\,dv\,dw$$

where $0 < u+v+w \le 1$

$$= \frac{a^2b^2c^2}{8}\iiint u^{(1/2)-1}v^{(1/2)-1}w^{(1/2)-1}\sqrt{1-(u+v+w)}\,du\,dv\,dw$$

$$= \frac{a^2b^2c^2}{8}\frac{[\Gamma(1/2)]^3}{\Gamma(3/2)}\int_0^1 \sqrt{1-t}.t^{1/2+1/2+1/2-1}dt$$

[By Liouville's theorem]

$$= \frac{a^2b^2c^2}{8}\frac{(\sqrt{\pi})^3}{1/2\sqrt{\pi}}\int_0^1 (1-t)^{(3/2)-1}t^{(3/2)-1}dt$$

$$= \frac{a^2b^2c^2}{8}.2\pi\frac{\Gamma(3/2)\Gamma(3/2)}{\Gamma(3)} = \frac{\pi^2a^2b^2c^2}{32}$$

Hence, if the integration is extended throughout the ellipsoid then the given integral

$$= 8I = 8.\frac{\pi^2a^2b^2c^2}{32} = \frac{\pi^2a^2b^2c^2}{4}.$$

Example 7. *Evaluate* $\iiint_R (x+y+z+1)^2 dx\,dy\,dz$ *where R defined by* $x \ge 0, y \ge 0, z \ge 0, x+y+z \le 1$.

Solution. As given x, y, z are all positive such that $0 \le x+y+z \le 1$.

$$\therefore \iiint (x+y+z+1)^2 dx\,dy\,dz$$

$$= \iiint x^{1-1}y^{1-1}z^{1-1}\{(x+y+z)+1\}^2 dx\,dy\,dz$$

$$= \frac{\Gamma(1)\Gamma(1)\Gamma(1)}{\Gamma(1+1+1)}\int_0^1 (u+1)^2.u^{1+1+1-1}du$$

[By Liouville's extension of Dirichlet's theorem]

$$= \frac{1}{2}\int_0^1 (u^2+2u+1)u^2du = \frac{1}{2}\left[\frac{u^2}{5}+\frac{3u^4}{4}+\frac{u^3}{3}\right]_0^1$$

$$= \frac{1}{2}\left[\frac{1}{5}+\frac{1}{2}+\frac{1}{3}\right] = \frac{1}{2}\frac{(6+15+10)}{5\times2\times3} = \frac{1}{2}.\frac{31}{30} = \frac{31}{60}$$

Example 8. *Evaluate*

$$\iiint x^{-1/2}y^{-1/2}z^{-1/2}(1-x-y-z)^{1/2}dx\,dy\,dz$$

extended to all positive values of variable subject to the condition $x+y+z<1$

Solution. The given condition is $0 < x+y+z < 1$

∴ the given integral

$$= \iiint x^{1/2-1}y^{1/2-1}z^{1/2-1}[1-(x+y+z)^{1/2}dx\,dy\,dz]$$

$$= \frac{\Gamma(1/2)\Gamma(1/2)\Gamma(1/2)}{\Gamma(1/2+1/2+1/2)}\int (1-h)^{1/2}h^{1/2+1/2+1/2}dh$$

$$= \frac{\sqrt{\pi}\sqrt{\pi}\sqrt{\pi}}{\Gamma(3/2)}\int_0^1 h^{1/2}(1-h)^{1/2}dh$$

$$= \frac{\pi\sqrt{\pi}}{1/2\sqrt{\pi}}\int_0^1 h^{3/2-1}(1-h)^{3/2-1}dh$$

$$= 2\pi B(3/2, 3/2)$$

$$= \frac{2\pi.\Gamma(3/2)\Gamma(3/2)}{\Gamma(3/2+3/2)} = \frac{2.\pi.(1/2\sqrt{\pi}).(1/2\sqrt{\pi})}{\Gamma(3)}$$

$$= \frac{\pi^2}{2\Gamma(3)} = \frac{\pi^2}{2.2.1} = \frac{\pi^2}{4}.$$

Example 9. *Prove that when x and y are positive and* $x+y<h$

$$\iint f'(x+y)x^{l-1}y^{-1}dx\,dy = \frac{\pi}{\sin \pi l}[f(h)-f(0)]$$

Solution. The given integral

$$I = \iint f'(x+y)x^{l-1}y^{(1-l)-1}dx\,dy \text{ where } 0 < x+y < h$$

$$= \frac{\Gamma(l)\Gamma(l-1)}{\Gamma(l+1-l)}\int_0^h f'(u)u^{l+(1-l)-1}du$$

[By Liouville's extension]

$$= \frac{\Gamma(l)\Gamma(l-1)}{\Gamma(1)}\int_0^h f'(u)du = \frac{\pi}{\sin \pi.l}[f(u)]_0^h$$

$$= \frac{\pi}{\sin \pi l}[f(h)-f(0)]$$

Example 10. *Show that*

$$\iint\left(\frac{1-x^2-y^2}{1+x^2+y^2}\right)^{1/2}dx\,dy = \frac{\pi}{8}(\pi-2)$$

over the positive quadrant of circle $x^2+y^2=1$. (UPTU-2008)

Solution. The given integral is to be extended to all positive values of x and y such that

$$0 \le x^2+y^2 \le 1 \quad \text{... (1)}$$

Put $x^2 = u, y^2 = v \Rightarrow x = u^{1/2}, y = v^{1/2}$

so that

$$dx = \frac{1}{2}u^{-1/2}du,\ dy = \frac{1}{2}v^{-1/2}dv$$

With these substitution, the condition (1) become $0 \le u \le v \le 1$

Therefore the integral

$$= \iint\left[\frac{1-(u+v)}{1+(u+v)}\right]^{1/2}\frac{1}{4}u^{-1/2}v^{-1/2}du\,dv$$

$$= \frac{1}{4}\iint\left[\frac{1-(u+v)}{1+(u+v)}\right]^{1/2}u^{(1/2)-1}v^{(1/2)-1}du\,dv$$

where $0 \le u+v \le 1$

$$= \frac{1}{4}\frac{\Gamma(1/2)\Gamma(1/2)}{\Gamma(1/2+1/2)}\int_0^1\left[\frac{1-h}{1+h}\right]^{1/2}.h^{(1/2)+(1/2)-1}dh$$

[By Liouville's extension of Dirichlet's theorem]

$$= \frac{1}{4}\frac{\sqrt{\pi}\sqrt{\pi}}{\Gamma(1)}\int_0^1\frac{1-h}{\sqrt{(1-h^2)}}dh$$

$$= \frac{\pi}{4}\int_0^1\frac{(1-\sin\theta)}{\cos\theta}\cos\theta\,d\theta$$

Putting $h = \sin\theta$, so that $dh = \cos\theta\,d\theta$

$$= \frac{\pi}{4}[\theta+\cos\theta]_0^{\pi/2}$$

$$= \frac{\pi}{4}\left[\frac{\pi}{2}-1\right] = \frac{\pi}{8}(-2).$$

Example 11. *Prove that* $I = \iiiint dx\,dy\,dz\,dw$, *for all positive values of the variables for which* $x^2+y^2+z^2+w^2$ *is not less than* a^2 *and not greater than* b^2, *is* $\pi^2(b^4-a^4)/32$.

Solution. As per given, we have

$$a^2 < x^2 + y^2 + z^2 + w^2 < b^2.$$

Put $x^2 = u_1$ or $x = u_1^{1/2}, \Rightarrow dx = 1/2u_1^{-1/2}du_1$.

Similarly, putting $y^2 = u_2$, $z^2 = u_3$, $w^2 = u_4$, we get

$$dy = \frac{1}{2}u_2^{-1/2}du_2, dz = \frac{1}{2}u_3^{-1/2}du_3, dw = \frac{1}{2}u_4^{-1/2}du_4$$

$\therefore$ Then

$$I = \iiiint \frac{1}{2}u_1^{-1/2}\frac{1}{2}u_2^{-1/2} \cdot \frac{1}{2}u_3^{-1/2}\frac{1}{2}u_4^{-1/2}du_1du_2du_3du_4$$

$$= \frac{1}{16}\iiiint u_1^{1/2-1}u_2^{1/2-1}u_3^{1/2-1}u_4^{1/2-1}du_1\,du_2\,du_3\,du_4$$

$$= \frac{1}{16}\frac{\Gamma(1/2)\Gamma(1/2)\Gamma(1/2)\Gamma(1/2)}{\Gamma(1/2+1/2+1/2+1/2)} \cdot \int_{a^2}^{b^2} h^{\frac{1}{2}+\frac{1}{2}+\frac{1}{2}+\frac{1}{2}-1}dh$$

$$= \frac{1}{16}\frac{(\sqrt{\pi})^4}{\Gamma(2)}\int_{a^2}^{b^2} h\,dh \qquad [\because \Gamma(1/2) = \sqrt{\pi}]$$

$$= \frac{\pi^2}{16}\left(\frac{1}{2}h^2\right)_{a^2}^{b^2} = \frac{\pi^2}{32}(b^4-a^4).$$

EXERCISE 19.5

1. Show that if l, m, n are all positive, then
$$\iiint x^{l-1}y^{m-1}z^{n-1}dx\,dy\,dz = \frac{a^lb^mc^n}{8}\cdot\frac{\Gamma(l/2)\Gamma(m/2)\Gamma(n/2)}{\Gamma(l/2+m/2+n/2+1)}$$
where the triple integral is taken throughout the part of the ellipsoid $\frac{x^2}{a^2}+\frac{y^2}{b^2}+\frac{z^2}{c^2}=1$ which lies in the positive octant.
2. Evaluate $\iiint x^py^qz^r(1-x-y-z)^s dx\,dy\,dz$ over the interior of the tetrahedron formed by four planes $x=0, y=0, z=0, x+y+z=1$.
3. Find the volume in the positive octant of the ellipsoid $\frac{x^2}{a^2}+\frac{y^2}{b^2}+\frac{z^2}{c^2}=1$.
4. Find the volume of ellipsoid $\frac{x^2}{a^2}+\frac{y^2}{b^2}+\frac{z^2}{c^2}=1$.
5. Evaluate $\iint x^{2l-1}y^{2m-1}dx\,dy$ such that $x^2+y^2 \le c^2$ for all positive values of x and y.
6. Find the volume of solid surrounded by the surface $\left(\frac{x}{a}\right)^{2/3}+\left(\frac{y}{b}\right)^{2/3}+\left(\frac{z}{c}\right)^{2/3}=1$.
7. Evaluate the double integral $\iint_p x^{1/2}y^{1/2}(1-x-y)^{2/3}dx\,dy$ over the domain D bounded by lines $x=0$, $y=0$, $x+y=1$.
8. Evaluate $\iint_T x^{1/2}y^{1/2}(1-x-y)^{3/2}dx\,dy$, where T is the region bounded by $x \ge 0, y \ge 0, x+y \le 1$.
9. Find the volume of the tetrahedron bounded by $\frac{x}{a}+\frac{y}{b}+\frac{z}{c}=1$ and the co-odinates axes.
10. Evaluate $\iiint \sqrt{\left(\frac{1-x^2-y^2-z^2}{1+x^2+y^2+z^2}\right)}dx\,dy\,dz$, integral being taken over all positive values of x, y, z such that $x^2+y^2+z^2 \le 1$.
11. Evaluate $\iint \sqrt{\left\{\frac{1-(x^2/a^2-y^2/b^2)}{1+(x^2/a^2+y^2/b^2)}\right\}}dx\,dy$ where $\frac{x^2}{a^2}+\frac{y^2}{b^2} \le 1$.
12. Evaluate $\iint_R \sqrt{(x^2+y^2)}dx\,dy$, where R is the region in the xy plane bounded by $x^2+y^2=4$ and $x^2+y^2=9$.
13. Prove that $\iiint \frac{dx\,dy\,dz}{(x+y+z+1)^2} = \frac{1}{2}\left[\log 2 - \frac{5}{8}\right]$ throughout the volume bounded by the co-ordinates planes and plane $x+y+z=1$.
14. Evaluate the integral $\iiint_R (ax^2+by^2+cz^2)dx\,dy\,dz$ where R is the region given by $x^2+y^2+z^2 \le d^2$
15. Find the value of $\iiint xyz\sin(x+y+z)\,dx\,dy\,dz$, the integral being extended to all positive values of variables subject to the condition $x+y+z \le \pi/2$. (UKTU-2011)
16. Evaluate $\iiint_R x^2y^2z^2\,dx.dy.dz$ where R is the region given by $x^2+y^2<1, 0 \le z \le 1$.
17. Find the value of $\iint x^{l-1}y^{-1}e^{x+y}\,dx,dy, 0<l<1$ to all positive values subject to $x+y<h$.
18. A triangular prism is formed by planes whose equations are $ay=bx$, $y=0$ and $x=a$. Show that the volume of the prism between the planes $z=0$ and surface $z=c+xy$ is $\frac{ab}{8}(4c+ab)$ (GBTU-2015)
19. Show that the volume of the paraboloid of revolution $x^2+y^2=4z$ cut off by the plane $z=4$ is 32π. (UPTU-2006)
20. Show that the volume common to the cylinder $x^2+y^2=a^2$ and $x^2+z^2=a^2$ is $\frac{16a^3}{3}$. (UPTU-2009)

21. Show that the volume of the solid which is bounded by the surface $2z = x^2+y^2$ and $z = x$ is $\pi/4$. (MTU-2012)

22. Show that the volume enclosed between the two surfaces $z = 8-x^2-y^2$ and $z = x^2 + 3y^2$ is $8\pi\sqrt{2}$. (GBTU-2011)

23. Show that the volume bounded by the elliptic paraboloids $z = x^2 + 9y^2$ and $z = 18-x^2-9y^2$ is 27π. (GBTU-2012)

24. Apply Dirichlet's integral to find the mass of an octant of the ellipsoid $\frac{x^2}{a^2}+\frac{y^2}{b^2}+\frac{z^2}{c^2}=1$, the density at any point being $\rho = kxyz$. (UPTU-2006, 07, GBTU-2011)

Hint to Selected Problems

1. Put $\frac{x^2}{a^2} = u$ *i.e.*, $x = au^{1/2}, \frac{y^2}{b^2} = v$ *i.e.*, $y = bv^{1/2}$ and $z = cw^{1/2}$

Then apply Dirichlet's theorem.

3. Do same as (1).

5. Put $x^2 = c^2u, y^2 = c^2v$

6. Put $\left(\frac{x}{a}\right)^{2/3} = u, \left(\frac{y}{b}\right)^{2/3} = v, \left(\frac{z}{c}\right)^{2/3} = w$.

7. The given integral can be written as

$$I = \iint x^{(3/2)-1} y^{(3/2)-1} [1-(x+y)]^{2/3} dx\, dy$$.

Then apply Liouville's extension of Dirichlet's theorem.

9. Put $\frac{x}{a} = u, \frac{y}{b} = v, \frac{z}{c} = w$ and apply Liouville's extension of Dirichlet's theorem.

10. Put $x^2 = u, y^2 = v, z^2 < w$ and apply Liouville's extension of Dirichlet's theorem.

11. Put $\frac{x^2}{a^2} = u$ and $\frac{y^2}{b^2} = v$ and apply Liouville's extension of Dirichlet's theorem.

12. Put $x^2 = u, y^2 = v$ and Liouville's extension of Dirichlet's theorem.

14. Put $x^2 = d^2u, y^2 = d^2v, z^2 = d^2w$.

ANSWERS

2. $\frac{\Gamma(p+1)\Gamma(q+1)\Gamma(r+1)\Gamma(s+1)}{\Gamma(p+q+r+s+4)}$ **3.** $\frac{\pi abc}{6}$ **4.** $\frac{\pi abc}{6}$ **5.** $\frac{c^{2l+2m}}{4}\frac{\Gamma(l)\Gamma(m)}{\Gamma(l+m+1)}$ **6.** $\frac{4}{35}\pi abc$ **7.** $\frac{27\pi}{1760}$ **8.** $\frac{2\pi}{315}$ **9.** $\frac{abc}{6}$

10. $\frac{\pi}{8}\left[B\left(\frac{3}{4},\frac{1}{2}\right) - B\left(\frac{5}{4},\frac{1}{2}\right)\right]$ **11.** $\pi ab\left[\frac{\pi}{2}-1\right]$ **12.** $\frac{38\pi}{3}$ **14.** $\frac{4}{15}\pi(a+b+c)d^5$ **15.** $\frac{1}{384}[\pi^4 - 48\pi^2 + 384]$ **16.** $\frac{\pi}{48}$

17. $\frac{\pi}{\sin l\pi}(e^n - 1)$ **18.** $\frac{ka^2b^2c^2}{48}$.

19.10 CHANGE OF VARIABLES

Some time we change the variables from one system to another system for more convenient way to find the double integrals. The variables x, y in $\iint_R f(x,y)dx\, dy$ are changed to u, v by means of the relations $x = f_1(u, v), y = f_2(u, v)$ then the double integral is transformed into

$$\iint f\{f_1(u,v), f_2(u,v)\}\,|J|\,du\,dv$$

where $J = \begin{vmatrix} \frac{\partial x}{\partial u} & \frac{\partial x}{\partial v} \\ \frac{\partial y}{\partial u} & \frac{\partial y}{\partial v} \end{vmatrix}$ and R' is the region in the u-v plane corresponding to region R in the x-y plane.

WORKING PROCEDURE

- Replace x, y by their equivalent in terms of u and v, the element of area $dx\, dy$ by $(J)\, du\, dv$ and the region R of integration in xy plane by the region R', in the uv plane.

19.10.1 CHANGE TO POLAR CO-ORDINATES

To change the variable from cartesian to polar form we put $x = r\cos\theta, y = r\sin\theta$.

Then $J = \begin{vmatrix} \frac{\partial x}{\partial r} & \frac{\partial x}{\partial \theta} \\ \frac{\partial y}{\partial r} & \frac{\partial y}{\partial \theta} \end{vmatrix} = \begin{vmatrix} \cos\theta & -r\sin\theta \\ \sin\theta & r\cos\theta \end{vmatrix} = r$

$$\int_R f(x,y)dx\,dy = \iint_{R'} f(r\cos\theta, r\sin\theta)\,|J|\,dr\,d\theta = \iint_{R'} f(r\cos\theta, r\sin\theta)\,|r\,dr\,d\theta$$

Solved Examples

Example 1. *Transform $\iint f(x,y)dx\,dy$, by the substitution $x+y=u$, $y=vu$.*

Solution. We have $x+y=u$ and $y=uv$ therefore,

$$x = u - y = u - uv \text{ and } y = uv.$$

$$\therefore \quad \frac{\partial x}{\partial u} = 1 - v, \frac{\partial x}{\partial v} = -u, \frac{\partial y}{\partial u} = v \text{ and } \frac{\partial y}{\partial v} = u \quad ...(1)$$

$$\therefore \quad J = \frac{\partial(x,y)}{\partial(u,v)} = \begin{vmatrix} \frac{\partial x}{\partial u} & \frac{\partial x}{\partial v} \\ \frac{\partial y}{\partial u} & \frac{\partial y}{\partial v} \end{vmatrix} = \begin{vmatrix} 1-v & -u \\ v & u \end{vmatrix} = u$$

$$\therefore \quad dx\,dy = J\,du\,dv = u\,du\,dv.$$

Hence, the given integral transforms to $\iint F(u, vu)du\,dv$

Example 2. *Transform to polar co-ordinate and integrate*

$$\iint \sqrt{\left(\frac{1-x^2-y^2}{1+x^2+y^2}\right)} dx\,dy$$

the integral being extended over all positive values of x and y subject to $x^2+y^2 \le 1$.

Solution. Here, x varies from 0 to 1 and y varies from 0 to $\sqrt{1-x^2}$ in the first quadrant where x and y are both positive.

$\therefore$ Given integral

$$I = \int_0^1 \int_0^{\sqrt{1-y^2}} \sqrt{\frac{(1-x^2-y^2)}{1+x^2+y^2}} dx\,dy$$

Now change it into polar form by putting $x = r\cos\theta, y = r\sin\theta$ then the circle $x^2+y^2=1$ transform into $r^2 = 1$ to $r=1$ and its first quadrant θ varies from 0 to $\pi/2$ and r varies from 0 to 1, then integral

$$I = \int_{\theta=0}^{\pi/2} \int_{r=0}^{1} \sqrt{\left(\frac{1-r^2}{1+r^2}\right)} r\,d\theta\,dr$$

$$= \int_0^{\pi/2} d\theta \int_0^1 \sqrt{\left(\frac{1-r^2}{1+r^2}\right)} r\,dr$$

$$= [\theta]_0^{\pi/2} \left[\frac{1}{2}\left(\frac{\pi}{2} - 1\right)\right]$$

$$= \frac{1}{2}\pi . \frac{1}{2}\left(\frac{1}{2}\pi - 1\right) = \frac{1}{4}\pi\left(\frac{1}{2}\pi - 1\right).$$

Example 3. *Evaluate $\iint \sqrt{(a^2-x^2-y^2)}\,dx\,dy$, over the semi-circle $x^2+y^2=ax$ in the positive quadrant.*

Solution. The region of integration is a semi-circle, we change it into the polar co-ordinate by putting $x=r\cos\theta$, and $y = r\sin\theta$ in $x^2+y^2=ax$, then we have

$$r^2\cos^2\theta + r^2\sin^2\theta = ar\cos\theta$$

$$r^2(\sin^2\theta + \cos^2\theta) = aR\cos\theta$$

$$r = a\cos\theta.$$

The equation $r = a\cos\theta$ represent a circle which passing through the pole for the given region where r varies from 0 to $a\cos\theta$ and θ varies from 0 to $\pi/2$.

$$\therefore \iint \sqrt{(a^2-x^2-y^2)}dx\,dy$$

$$= \int_0^{\pi/2} \int_0^{a\cos\theta} \sqrt{a^2-r^2}.r\,d\theta\,dr$$

$$[\because x^2+y^2=r^2 \text{ and } dx\,dy = r\,d\theta\,dr]$$

$$= \int_0^{\pi/2} \left[\int_0^{a\cos\theta} \left\{-\frac{1}{2}(a^2-r^2)^{1/2}(-2r)\right\} dr\right] d\theta$$

$$= \int_0^{\pi/2} \left[-\frac{1}{2}.\frac{2}{3}(a^2-r^2)^{3/2}\right]_0^{a\cos\theta} d\theta$$

$$= \frac{-1}{3}\int_0^{\pi/2} (a^3\sin^3\theta - a^3)d\theta$$

$$= \frac{-a^3}{3}\left[\frac{2}{3.1} - \frac{\pi}{2}\right] = \frac{1}{3}a^3\left(\frac{\pi}{2} - \frac{2}{3}\right).$$

Example 4. *By using the transformation $x+y = u$, $y = vu$, show that*

$$\int_0^1 \int_0^{1-x} e^{y/(x+y)} dx\,dy = \frac{1}{2}(e-1)$$

Solution. We have $\quad dx\,dy = u\,du\,dv$

The region of integration is bounded by the lines $y = 0, y = 1-x, x=0$ and $x=1$

Changing these equations into new variables u and v by using the relation

$$x = u - y = u - uv = u(1-v)$$

and $y = uv$, we have $uv = 0$, $uv = 1 - u(1-v)$, $u(1-v) = 0$ and $u(1-v)=1$

giving $\quad v=0$ to $v=1$, $u = 0$ to $u=1$.

Therefore for the given region v varies from 0 to 1 and u varies from 0 to 1.

and $\quad e^y/(x+y) = e^{uv/u} = e^v.$

Changing the variables to u, v the given integral becomes

$$I = \int_0^1 \int_0^1 e^v u\,du\,dv = \int_0^1 [e^v]_0^1 u\,du$$

$$= \int_0^1 (e^1 - e^0)u\,du$$

$$= (e-1)\int_0^1 du = (e-1)\left[\frac{u^2}{2}\right]_0^1 = \frac{1}{2}(e-1).$$

Example 5. *Evaluate the integral $\int_0^a \int_0^{\sqrt{a^2-y^2}} (x^2+y^2)dy\,dx$ by changing into polar co-ordinates.* (UPTU-2009)

Solution. Putting $x = r\cos\theta$, $y = r\sin\theta$, we have

$$J = \begin{vmatrix} \frac{\partial x}{\partial r} & \frac{\partial x}{\partial \theta} \\ \frac{\partial y}{\partial r} & \frac{\partial y}{\partial \theta} \end{vmatrix} = \begin{vmatrix} \cos\theta & -r\sin\theta \\ \sin\theta & r\cos\theta \end{vmatrix} = r$$

$\therefore$ $dx\,dy$ is to be replaced by $J\,dr\,d\theta$.

$$x^2+y^2=r^2\cos^2\theta+r^2\sin^2\theta=r^2.$$

Again we find that in the upper limit $x=\sqrt{a^2-y^2}$, y varies from 0 to a and x varies from 0 to any point on the circle $x^2+y^2=a^2$.

In the polar form of the circle $r^2=a^2$ i.e., $r=a$ we find that r varies from 0 to a and θ varies from 0 to $\pi/2$.

$$\therefore \int_0^a\int_0^{\sqrt{a^2-y^2}}(x^2+y^2)dy\,dx=\int_{\theta=0}^{\pi/2}\int_{r=0}^{a}r^2.r\,d\theta\,dr$$

$$=\int_{\theta=0}^{\pi/2}\left[\frac{r^4}{4}\right]_0^a d\theta=\frac{1}{4}a^4(\theta)_0^{\pi/2}=\left(\frac{1}{8}\right)\pi a^4.$$

Example 6. *Prove that* $\iint_D e^{-x^2-y^2}\,dx\,dy=\frac{\pi}{4}(1-e^{-R^2})$

where D is the region defined by $x\ge 0$, $y\ge 0$ *and* $x^2+y^2\le R^2$.

Solution. Let $x=r\cos\theta$, $y=r\sin\theta$, then

$$dx\,dy=\begin{vmatrix}\frac{\partial x}{\partial r} & \frac{\partial x}{\partial\theta}\\ \frac{\partial y}{\partial r} & \frac{\partial y}{\partial\theta}\end{vmatrix}dr\,d\theta=\begin{vmatrix}\cos\theta & -r\sin\theta\\ \sin\theta & r\cos\theta\end{vmatrix}dr\,d\theta=r\,dr\,d\theta$$

$$\iint_D e^{-x^2-y^2}dx\,dy=\int_{\theta=0}^{\pi/2}\int_{r=0}^{R}r\,dr\,d\theta$$

$$=\int_0^{\pi/2}\left[-\frac{e^{-r^2}}{2}\right]_0^R d\theta$$

$$=\frac{1}{2}(1-e^{-R^2})\int_0^{\pi/2}d\theta$$

$$=\frac{1}{2}(1-e^{-R^2})[\theta]_0^{\pi/2}$$

$$=\frac{1}{2}(1-e^{-R^2})\left[\frac{\pi}{2}\right]$$

$$=\frac{\pi}{4}(1-e^{-R^2}).$$

EXERCISE 19.6

1. Transform $\int_0^a\int_0^{a-x}f(x,y)dx\,dy$, by the substitution $x+y=u$, $y=uv$.
2. By using the transformation $x+y=u$, $y=uv$ show that $\iint\{xy(1-x-y)\}^{1/2}dx\,dy$ taken over the area of the triangle bounded by lines $x=0$, $y=0$, $x+y=1$ is $\frac{2\pi}{105}$. (GBTU-2011)
3. Transform the integral $\int_0^a\int_0^{\sqrt{a^2-x^2}}y\sqrt{x^2+y^2}dx\,dy$ by changing to polar co-ordinates and hence solve it.
4. Evaluate $\int_0^2\int_0^{\sqrt{2x-x^2}}\frac{x\,dx\,dy}{\sqrt{x^2+y^2}}$, by changing to polar co-ordinates.
5. Evaluate $\iint(x^2+y^2)^{7/2}dx\,dy$, over the circle $x^2+y^2=1$.
6. Evaluate $\iint xy(x^2+y^2)^{3/2}dx\,dy$, over the positive axes of circle $x^2+y^2=1$.
7. Transform the integral $\int_0^{\pi/2}\int_0^{\pi/2}\sqrt{\frac{\sin\phi}{\sin\theta}}d\phi\,d\theta$ by the substitutions $x=\sin\phi\cos\theta$, $y=\sin\phi\sin\theta$ and show that its value is π. (UPTU-2007)
8. Evaluate $\int_0^a\int_0^{\sqrt{a^2-y^2}}y^2\sqrt{x^2+y^2}\,dxdy$. (UPTU-2008)

Hint to Selected Problems

1. $x+y=u$ and $y=uv \Rightarrow x=u-uv$ and $y=uv$

Now,

$$J=\frac{\partial(x,y)}{\partial(x,v)}=\begin{vmatrix}\frac{\partial x}{\partial u} & \frac{\partial x}{\partial v}\\ \frac{\partial y}{\partial u} & \frac{\partial y}{\partial v}\end{vmatrix}=\begin{vmatrix}1-v & -u\\ v & u\end{vmatrix}=u$$

$\Rightarrow$ $dx\,dy=u\,du\,dv$.

2. Proceed same as (1), we have $dx\,dy=u\,du\,dv$

$\therefore$ $\{xy(1-x-y)\}^{1/2}=[u(1-v)uv(1-u)]^{1/2}$
$=u(1-u)^{1/2}v^{1/2}(1-v)^{1/2}$.

3. $x=r\cos\theta$, $y=r\sin\theta$, $J=r$

$\Rightarrow$ $dx\,dy=r\,d\theta\,dr$

$$\therefore \int_0^a\int_0^{\sqrt{(a^2-x^2)}}y\sqrt{(x^2+y^2)}dx\,dy$$

$$=\int_{\theta=0}^{\pi/2}\int_{r=0}^{a}r\sin\theta.r\,d\theta\,dr$$

ANSWERS

1. $\int_0^a\int_0^1 F(u,v)u\,du\,dv$ 3. $\frac{a^4}{4}$ 4. $\frac{4}{3}$ 5. $\frac{2\pi}{9}$ 6. $\frac{1}{14}$ 7. $\int_0^1\int_0^{\sqrt{1-y^2}}\frac{dx\,dy}{\sqrt{y-y(x^2+y^2)}}$ 8. $\frac{\pi a^5}{20}$

19.11 CHANGE OF ORDER OF INTEGRATIONS

If the limits of integration are constants in the double integration then the value of integration can be obtained by integrating with respect to any independent variable.

When the limits are not constant but are the function of x and y then firstly we integrate with respect to first independent variable and then with respect to second.

In this case the limits of integration are determined in the given region by drawing the strips parallel to Y-axis or X-axis.

In this case limits of y are function of x then we find the new limits of x as function of y and new constant.

WORKING PROCEDURE

Step 1. *If we perform the integration first with respect to y, we take the elementary strip parallel to y-axis and determine the limits of y and add up the vertical strip from extreme left to the extreme right of the region.*

Step 2. *If the order of integration is performed first with respect to x, we take the elementary strip parallel to x-axis and proceed.*

Solved Examples

Example 1. *Evaluate the following integral by changing the order of integration*

$$\int_0^{2a}\int_0^{\sqrt{2ax-x^2}} a-\sqrt{(a^2-y^2)}\,dx\,dy.$$

Solution. Here, we have the figure (17).

The limits of y are from $y = 0$ to $y=\sqrt{2ax-x^2}$ or between x-axis and semicircle

$$x^2+y^2-2ax=0$$

or

$$(x-a)^2+y^2=a^2$$

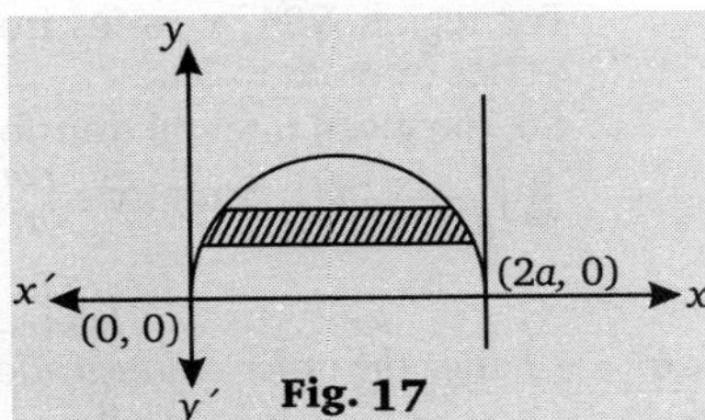

Fig. 17

In the figure, the region x varies from one end of the circle to other end

i.e., $(x-a)^2=a^2-y^2 \Rightarrow x=a\pm\sqrt{a^2-y^2}$.

The strips are taken parallel to x-axis and y varies from 0 to a. So the order of integration is changed as

$$I=\int_0^{2a}\int_0^{\sqrt{2ax-x^2}}(a-\sqrt{a^2-y^2})\,dx\,dy$$

$$=\int_0^{a}\int_{a-\sqrt{a^2-y^2}}^{a+\sqrt{a^2-y^2}}(a-\sqrt{a^2-y^2})\,dx\,dy.$$

Example 2. *Change the order of integration in the integral* $\int_0^a\int_0^x f(x,y)\,dx\,dy$.

Solution. The given limits shows that the region of integration is bounded by the curve $y=0$, $y=x$, $x=0$, $x=a$.

Hence $y = 0$ represent X-axis and $y=x$ represent a straight line through the origin. Also $x = 0$ and $x = a$, represent straight lines parallel to y-axis therefore the region of the integration is the triangle OAB in the figure 18 and B is (a, a).

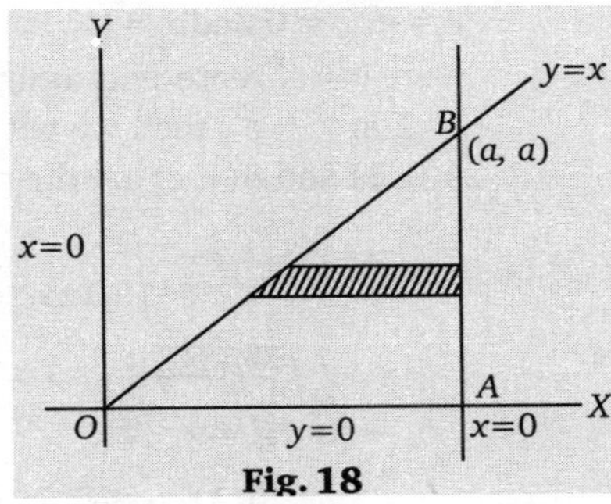

Fig. 18

In the given integral, the limits of integration of y being variable, we are required to integrate first w.r. to y regarding x as constant and then w.r. to x. To change the order of integration drawn parallel strip along x-axis, straight from the line OB and terminating on the line AB.

Thus in region OBA, x varies from y to a and y varies from 0 to a.

Hence, by changing the order of integration we have

$$\int_0^a\int_0^x f(x,y)\,dx\,dy=\int_0^a\int_y^a f(x,y)\,dy\,dx.$$

Example 3. *Evaluate* $\iint xy(x+y)\,dx\,dy$ *over the area between* $y=x^2$ *and* $y=x$.

Solution. Here $x^2 = y$ represents a parabola whose vertex is the origin and axis is the axis of y. The equation $y=x$ is a line through origin making an angle of 45° with x-axis. Solving $y=x^2$ and $y = x$ we find that the parabola $y = x^2$ and the line $y = x$ intersect at the point (0, 0) and (1, 1). When we integrate with respect to x-along a strip parallel to x-axis the strip starts from the line $y = x$ and ends on the parabola $y=x^2$ and A is (1, 1).

$\therefore$ Required value $= \int_{x=0}^{1}\int_{y=x}^{x^2} xy(x+y)\,dx\,dy$

$= \int_{x=0}^{1}\int_{y=x}^{x^2} (x^2y+xy^2)\,dx\,dy$

$= \int_{x=0}^{1}\left(\frac{1}{2}x^2y^2+\frac{1}{3}xy^3\right)_0^{x^2} dx$

$= \int_0^1\left(\frac{1}{2}x^6+\frac{1}{3}x^7\right)-\left(\frac{1}{2}x^4+\frac{1}{3}x^4\right)dx$

$= \int_0^1\left(\frac{1}{2}x^6+\frac{1}{3}x^7-\frac{5}{6}x^4\right)dx$

$= \left[\left(\frac{1}{14}\right)x^7+\left(\frac{1}{24}\right)x^8-\left(\frac{1}{6}\right)x^5\right]_0^1$

$= \left(\frac{1}{14}\right)+\left(\frac{1}{24}\right)-\left(\frac{1}{6}\right)=\frac{3}{56}.$

Example 4. *Change the order of integration and evaluate* $\int_0^1\int_{e^x}^{e}\frac{dx\,dy}{\log y}$.

Solution. The region of integration is bounded by $e^x = y$, $y = e$, $x = 0$ and $x = 1$.

Here $y = e^x$ represents a curve. Putting $x = 0$ and $x=1$ in $y = e^x$, then we get $y = 1$ and $y = e$. So $A(0, 1)$ and $B(1, e)$ are the points on this curve.

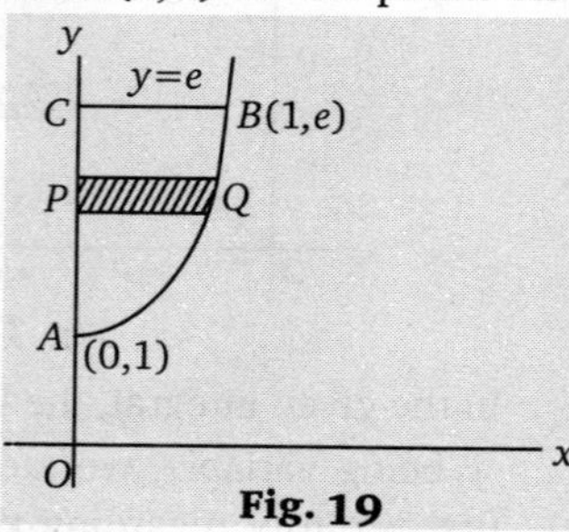

Fig. 19

When we integrate with respect to x first drawn a strip parallel to x-axis.

The strip starts from $x = 0$ and exends upto the curve $y = e^x$ *i.e.*, $x = \log y$. Also for the given region y varies from $y = 1$ to $y = e$. On changing the order of integration, the given integral

$I = \int_1^e\int_0^{\log y}\frac{dy\,dx}{(\log y)} = \int_1^e\frac{1}{\log y}(x)_0^{\log y}\,dy$

$= \int_1^e\frac{1}{\log y}(\log y-0)dy = \int_1^e dy = (y)_1^e = e-1.$

Example 5. *Change the order of integration in* $\int_0^a\int_{\sqrt{a^2+x^2}}^{x+2a} f(x,y)\,dxdy$.

Solution. The area of integration is bounded by the curves $y=\sqrt{a^2-x^2}$ *i.e.*, $x^2+y^2=a^2$.

This is the equation of the circle with centre (0, 0) and radius a. Also $y = x + 2a$ represents a straight line which passing through (0, 2a) *i.e.*, the Y-axis and the line $x=a$ which is parallel to Y-axis.

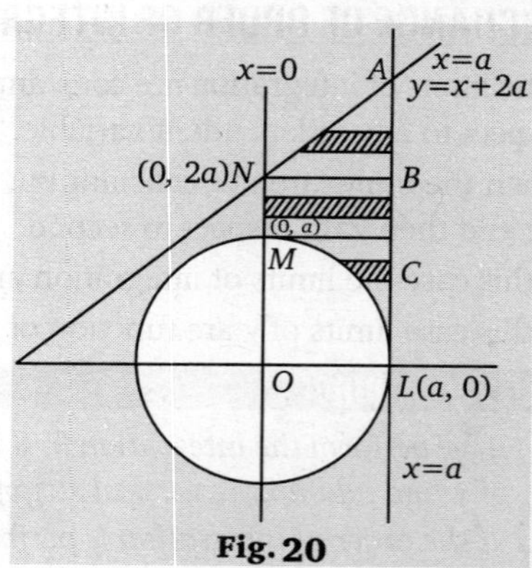

Fig. 20

We draw the curves $x^2+y^2=a^2$, $y = x + 2a$, $x = 0$ and $x = a$. We observe that the region of integration is the area *MLANM*.

To change the order of integration we draw a strip parallel to x-axis. Draw the lines *MC* and *MB* parallel to X-axis. So the area of integration is divided into three portions *MLEC*, *NNCB* and *NAB*.

For region *MLC*, x varies from $x^2+y^2=a^2$ circle's arc to line $x = a$ or $x=\sqrt{a^2-y^2}$ to a and y varies from 0 to a.

For region *NMCB*, x varies from 0 to a and y varies from a to 2a.

For region *NBA*, x varies from y–2a to a and y varies from 2a to 3a.

So, the given integral transform to

$\int_0^a\int_{\sqrt{a^2-y^2}}^{a} f(x,y)\,dy\,dx + \int_a^{2a}\int_0^a f(x,y)\,dy\,dx$

$+ \int_{2a}^{3a}\int_{y-2a}^{a} f(x,y)\,dy\,dx.$

Example 6. *Change the order of integration in the integral* $\int_0^{\pi/2}\int_0^{2a\cos\theta} f(r,\theta)\,d\theta\,dr$.

Solution. The limits are given by $\theta = 0$ to $\theta=\pi/2$ and $r = 0$ to $r = 2a\cos\theta$. Also the curve $r=2a\cos\theta$ is the circle.

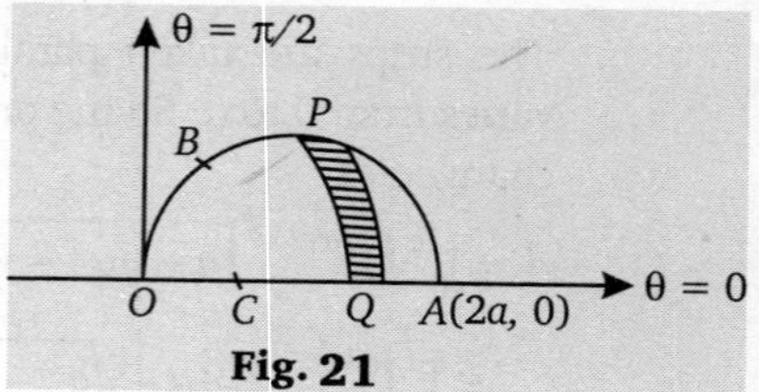

Fig. 21

The region of integration is the area *OABO* of circle. In the given integral the limits of integration of r is variable while limit of θ are constant. Now draw a strip parallel to θ (pole) such strip extends from the points O to the point A *i.e.*, $r = 0$ to $r = 2a$ and for a particular circular strip of this type we observe that θ varies from $\theta = 0$ to θ of curve *i.e.*, $\theta=\cos^{-1}(r/2a)$.

Hence, the given integral

$= \int_{r=0}^{2a}\int_{\theta=0}^{\cos^{-1}(r/2a)} f(r,\theta)\,dr\,d\theta.$

Example 7. *Change the order of integration in*

$$\int_0^a \int_x^{a^2/x} \phi(x, y)\, dx\, dy.$$

Solution. We observe that the region is bounded by $y = x$ and $y = a^2/x$ and $x = 0$ to $x = a$. Clearly the region is *OAC*.

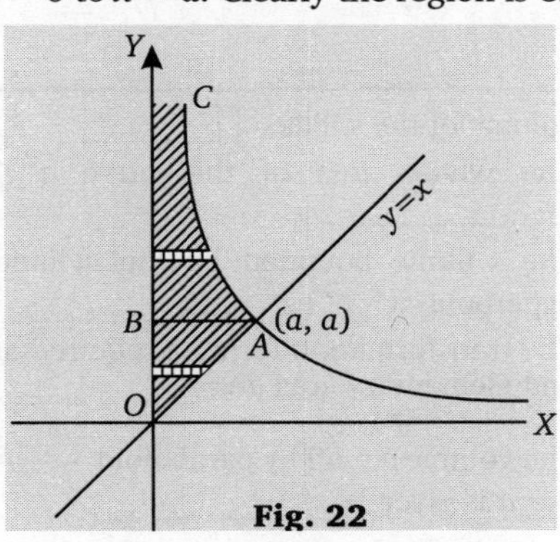

Fig. 22

Now draw a line *AB* parallel to *x*-axis, which divides the given region into two parts, *OAB* and *BAC*, Therefore, we draw a strip parallel to *x*-axis in *OAB*, where the left end of the strip is at *y*-axis and the right end is at $y = x$ and y takes the values from 0 to a. Also in region *BAC*, the left end of this strip is at *y*-axis whereas the right end is at the curve $y = a^2/x$, and y takes the values from $y = a$ to $y = \infty$. Hence ,the given integral becomes

$$\int_0^a \int_x^{a^2/x} \phi(x, y)\, dx\, dy = \int_0^a \int_0^y \phi(x, y)\, dx\, dy + \int_a^\infty \int_0^{a^2/y} \phi(x, y)\, dx\, dy.$$

EXERCISE 19.7

Change the order of integration in the following integral (Ques. 1-9)

1. $\int_0^{a/2} \int_{x^2/2}^{x - x^2/a} dy\, dx.$
2. $\int_0^a \int_x^{a^3/x} f(x, y)\, dx\, dy.$
3. $\int_0^1 \int_x^{x(2-x)} f(x, y)\, dx\, dy.$
4. $\int_0^{a\cos\alpha} \int_{x\tan\alpha}^{\sqrt{a^2 - x^2}} f(x, y)\, dx\, dy.$
5. $\int_0^a \int_{-mx}^{lx} f(x, y)\, dx\, dy.$
6. $\int_0^a \int_{(b/a)\sqrt{a^2 - x^2}}^b f(x, y)\, dx\, dy$, where $c < a$.
7. $\int_0^{2a} \int_{\sqrt{2a - x^2}}^{\sqrt{2ax}} f(x, y)\, dx\, dy.$
8. $\int_0^{\pi/2} \int_0^{2a\cos\theta} f(r, \theta)\, r d\theta\, dr.$
9. $\int_0^a \int_0^{b/(b+x)} f(x, y)\, dx\, dy.$
10. Change the order of integration in $\int_0^\infty \int_0^\infty \frac{e^{-y}}{y} dx\, dy$ and hence find its value.
11. Change the order of integration in $\int_0^\infty \int_x^\infty f(x, y)\, dx\, dy.$
12. Change the order of integration in $\int_0^{2a} \int_{x(2a-x)/2a}^{\sqrt{2a - x^2}} f(x, y)\, dx dy.$
13. Change the order of integration and evaluate $\int_0^\infty \int_0^x xe^{-x^2/y} dx\, dy.$
14. Change the order of integration in $\int_0^a \int_{\sqrt{ax - x^2}}^{\sqrt{ax}} f(x, y)\, dx\, dy.$

11. Change the order of integration to evaluate $\int_0^1 \int_{2y}^2 e^{x^2} dx\, dy$. (GBTU-2010)

ANSWERS

1. $\int_0^{a/4} \int_{1/2\left[a - \sqrt{a^2 - 4ay}\right]}^{\sqrt{ay}} f(x, y)\, dy\, dx.$ 2. $\int_0^a \int_0^y f(x, y)\, dy\, dx + \int_a^\infty \int_0^{a^2/y} f(x, y)\, dy\, dx$ 3. $\int_0^1 \int_{1 - \sqrt{1-y}}^y f(x, y)\, dy\, dx.$

4. $\int_0^{a\sin\alpha} \int_0^{y\cot\alpha} f(x, y)\, dy\, dx + \int_{a\sin\alpha}^a \int_0^{\sqrt{a^2 - y^2}} f(x, y)\, dy\, dx$ 5. $\int_0^{am} \int_{y/l}^{y/m} f(x, y)\, dy\, dx + \int_{am}^{al} \int_{y/l}^a f(x, y)\, dy\, dx$

6. $\int_0^{b\sqrt{1 - (c^2/a^2)}} \int_{a\sqrt{1 - y^2/b^2}}^a f(x, y)\, dy\, dx + \int_{b\sqrt{1 - c^2/a^2}}^b \int_c^a f(x, y)\, dy\, dx$

7. $\int_0^a \int_{y^2/2a}^{a - \sqrt{a^2 - y^2}} f(x, y)\, dy\, dx + \int_0^a \int_{a + \sqrt{a^2 - y^2}}^{2a} f(x, y)\, dy\, dx + \int_a^{2a} \int_{y^2/2a}^{2a} f(x, y)\, dy\, dx$ 8. $\int_0^{2a} \int_0^{\cos^{-1}(r/2a)} f(r, \theta)\, dr\, d\theta$

9. $\int_0^{b/(a+b)} \int_0^a f(x, y)\, dy\, dx + \int_{b/(a+b)}^1 \int_0^{b(1-y)/y} f(x, y)\, dy\, dx$ 10. 1 11. $\int_0^\infty \int_0^y f(x, y)\, dy\, dx$

12. $\int_0^{a/2} \int_{a - \sqrt{a^2 - y^2}}^{a - \sqrt{a^2 - 2ay}} f(x, y)\, dx\, dy + \int_{a/2}^a \int_{a - \sqrt{a^2 - y^2}}^{a + \sqrt{a^2 - y^2}} f(x, y)\, dx\, dy + \int_a^{a/2} \int_{a + \sqrt{a^2 - 2ay}}^{a + \sqrt{a^2 - y^2}} f(x, y)\, dx\, dy$

13. $\int_0^\infty \int_0^y xe^{-x^2/y} dx\, dy$ 14. $\int_0^{a/2} \int_{y^2/a}^{a/2 - \sqrt{(a^2/4) - y^2}} f(x, y)\, dx\, dy + \int_{a/2}^a \int_{y^2/a}^a f(x, y)\, dx\, dy + \int_0^{a/2} \int_{a/2 + \sqrt{a^2/4 - y^2}}^a f(x, y)\, dx\, dy$ 15. $e^2 - 3$

Objective Evaluations

Fill in the Blanks

1. Double integral is an extension of a definite integral in ____________ dimensional space.
2. The element of the area is generally denoted by____________.
3. To change the integral from cartesian to polar, we put $x=r\cos\theta$ and $y=$____________.
4. The value of $\int_0^{\pi/2}\int_0^{\sin\theta} r\,d\theta\,dr=$ ____________.
5. The value of the double integration of the region bounded by the circle $x^2+y^2=a^2$ is ____________.
6. If we consider the area $dx\,dy$ on the plane $z=0$ then the volume of the solids$=\iint$____________.
7. The whole area of the curve $a^2x^2=y^3(2a-y)$ is given by____________.
8. The volume bounded by the cylinder $x^2+y^2=4$ and the hyperbola $-x^2-y^2+z^2=1$ is ____________.
9. The transformation formula required as $x=r\cos\theta$, $y=r\sin\theta$ and elementary area $dA=$____________.
10. The volume cut off by paraboloid $\frac{y^2}{b}+\frac{y^2}{c}=2x$ and the plane $x=a$ is given by____________.

True/False

Write 'T' for True and 'F' for False statement.

1. The transformation formula required as $x=r\cos\theta$, $y=r\sin\theta$ and elementary area $\delta A=\ r\,\delta\theta\,\delta r$. **(T/F)**
2. The volume inside the paraboloid $x^2+4z^2+8y=16$ and on the positive side of xz-plane is π. **(T/F)**
3. In evaluating the double integral, if we perform the integration first with respect to y, we take the elementary strip parallel to y-axis and determine the limits for y and add up the vertical strips from extreme left to the extreme right of the region A. **(T/F)**
4. If the order of integration is performed first with respect to x, we take the elementary strip parallel to x-axis. **(T/F)**
5. The value of $\frac{\partial(x,y)}{\partial(u,v)}$ is known as Jacobian of x and y with respect to u and v. **(T/F)**
6. The value of $\frac{\partial(x,y)}{\partial(u,v)}$ is known as Jacobian of u and v with respect to x and y. **(T/F)**

Multiple Choice Questions

Choose the most appropriate one.

1. The value of integration $\int_0^{\pi/2}\int_0^{2a\cos\theta} r\sin\theta\,d\theta\,dr$ is :
   (a) a^2 (b) $2a^2$
   (c) $\frac{2a^2}{3}$ (d) $4a^2$
2. If we consider the integration $\int_R f(x,y)\,ds$ then R is called :
   (a) region (b) field of integration
   (c) both (a) and (b) (d) none of these
3. In question (2), ds is called :
   (a) elementary strip (b) element of area
   (c) volume (d) none of these
4. If A be a region bounded by the areas $y=f_1(x)$ and $y=f_2(x)$, $x=a$ and $x=b$, then $\iint_A f(x,y)\,dA$ is equal to :
   (a) $\int_a^b\left\{\int_{f_1}^{f_2} f(x,y)\,dy\right\}dx$ (b) $\int_a^b\left\{\int_{f_1}^{f_2} f(x,y)\,dx\right\}dy$
   (c) both are true (d) none of these
5. To transform the cartesian equation into polar, we put $x=r\cos\theta$, $y=r\sin\theta$, then the value of $\iint f(x,y)\,dx\,dy$ is :
   (a) $\iint f(r,\theta)\,r\,d\theta\,dr$ (b) $\iint f(r,\theta)\,d\theta\,dr$
   (c) $\iint f(r,\theta)\,r\theta\,d\theta\,dr$ (d) none of these

ANSWERS

Fill in The Blanks

1. two 2. dS 3. $r\sin\theta$ 4. $\frac{\pi}{8}$ 5. πa^2
6. $z\,dx\,dy$ 7. πa^2 8. $\frac{4\pi}{3}(5\sqrt{5}-1)$ 9. $r\,d\theta\,dr$ 10. $a^2\left(1-\frac{\pi}{4}\right)$.

True/False

1. T 2. F 3. T 4. T 5. T 6. F

Multiple Choice Questions

1. (c) 2. (c) 3. (b) 4. (a) 5. (a)

FFFFF

Beta and Gamma Functions

20.1 INTRODUCTION

The definite integral $\int_0^\infty e^{-x} x^{n-1}\,dx$, for $n > 0$ is known as the gamma function and is denoted by $\Gamma(n)$ ['read as Gamma n']. Gamma function is also called the Eulerian integral of second kind. Weierstrass defined it as infinite product as

$$\frac{1}{\Gamma(z)} = z\,e^{2/z} \prod_{n=1}^{\infty}\left[\left(1+\frac{z}{n}\right)e^{-z/n}\right]$$

for non-zero and non-negative number z and n is an Euler's constant. (UPTU-2006, 2008, GBTU-2010)

REMARK

- The integral is valid only for $n > 0$ because it is for just those values of m and n that the above integral are convergent.

20.2 PROPERTIES OF GAMMA FUNCTION

(1) $\Gamma(1) = 1$.

Proof. We have $\Gamma(n) = \int_0^\infty e^{-x} x^{n-1} dx,\ n > 0.$ ...(1)

Put $n = 1$ in equation (1), we get

$$= \int_0^\infty e^{-x}\,dx = \left[-e^{-x}\right]_0^\infty = 1.$$

$\therefore$ $\Gamma(1) = 1$

(2) $\Gamma(n+1) = n\,\Gamma(n),\ n > 0.$

Proof. We have $\Gamma(n) = \int_0^\infty e^{-x} x^{n-1}\,dx$, for $n > 0$

Replacing n by $(n+1)$, we have

$$\Gamma(n+1) = \int_0^\infty e^{-x} x^{n+1-1} dx = \int_0^\infty e^{-x} x^n\,dx$$

$$= \left[x^n.(-e^{-x})\right]_0^\infty - \int_0^\infty (nx^{n-1})(-e^{-x})\,dx$$

$$\therefore \quad \Gamma(n+1) = -\lim_{x\to\infty} \frac{x^n}{e^x} + 0 + n\int_0^\infty e^{-x} x^{n-1} dx \qquad ...(1)$$

$\left(\because \lim_{x\to\infty} x^n e^{-x} = 0 \text{ as } n > 0\right)$

But

$$\lim_{x\to\infty} \frac{x^n}{e^x} = \lim_{x\to\infty} \frac{x^n}{1+\frac{x}{1!}+\frac{x^2}{2!}+\ldots+\frac{x^n}{n!}+\frac{x^{n+1}}{(n+1)!}+\ldots}$$

$$= \lim_{x\to\infty} \frac{1}{\frac{1}{x^n}+\frac{1}{1!x^{n-1}}+\ldots+\frac{1}{n!}+\frac{x}{(n+1)!}+\ldots}$$

$$= 0 \qquad ...(2)$$

Also, by definition, we have $\Gamma(n) = \int_0^\infty e^{-x} x^{n-1}\,dx$. ...(3)

Using (2) and (3), (1) redcuces to $\Gamma(n+1) = n\,\Gamma(n)$.

REMARKS

- The formula $\Gamma(n+1) = n\,\Gamma(n)$ is known as a recurrence formula for gamma function.
- The gamma function can be generalized to $n < 0$ by using recurrence formula in the form of $\Gamma(n) = \dfrac{\Gamma(n+1)}{n}$ This process is known as analytic continuation.

(3) *If n is a non-negative integer, then* $\Gamma(n+1) = n\,!$.3

Proof. We know that for $n > 0$.

$$\begin{aligned}\Gamma(n+1) &= n\,\Gamma(n) = n\,\Gamma(n-1+1)\\ &= n(n-1)\Gamma(n-1) && \text{[By property 2]}\\ &= n(n-1)(n-2)\,\Gamma(n-2)\\ &= n(n-1)(n-2)\ldots 3.\,2.\,1.\,\Gamma(1)\\ &= n\,! && [\because \Gamma(1)=1]\end{aligned}$$

REMARK

- Gauss's Pi-function is denoted by $\pi(n)$ and is defined by $\pi(n) = \Gamma(n+1)$, when n is +ve integer.

(4) $\Gamma(1/2) = \sqrt{\pi}$. (UPTU-2006, GBTU-2010)

Proof. By definition, we have

$$\Gamma(n) = \int_0^\infty e^{-t} t^{n-1}\,dt,\ n > 0 \qquad \ldots(1)$$

Replacing n by 1/2 in equation (1), we get

$$\Gamma(1/2) = \int_0^\infty e^{-t} t^{-1/2} dt = 2\int_0^\infty e^{-u^2} du \qquad \ldots(2)$$

[Putting $t = u^2$, *i.e.*, $dt = 2u\,du$]

$$\therefore \quad \Gamma(1/2) = 2\int_0^\infty e^{-x^2} dx \text{ and } \Gamma(1/2) = 2\int_0^\infty e^{-y^2} dy \qquad \ldots(3)$$

(Limits remaining same)

Multiplying the corresponding sides of two equations of (3), we get

$$[\Gamma(1/2)]^2 = \left(2\int_0^\infty e^{-x^2} dx\right)\left(2\int_0^\infty e^{-y^2} dy\right) = 4\int_0^\infty\int_0^\infty e^{-(x^2+y^2)} dx\,dy.$$

Now, changing the variables to polar co-ordinates (r, θ) where $x = r\cos\theta$, $y = r\sin\theta$

$$\Rightarrow \quad x^2+y^2 = r^2 \text{ and } dx\,dy = r\,d\theta\,dr$$

we have $\quad [\Gamma(1/2)]^2 = 4\int_{\theta=0}^{\pi/2}\int_{r=0}^\infty e^{-r^2}\, r\,d\theta\,dr$.

The area of integration in the positive quadrant of plane is

$$= 2\int_0^{\pi/2}\left\{\int_0^\infty 2e^{-r^2} r.dr\right\}d\theta. \qquad \text{(Putting } r^2 = v, \text{ so that } 2r\,dr = dv)$$

$$= 2\int_0^{\pi/2}\left[-e^{-v}\right]_0^\infty d\theta = 2\int_0^{\pi/2} d\theta = 2[\theta]_0^{\pi/2} = \pi$$

Therefore, $\quad [\Gamma(1/2)]^2 = \pi$ so that $\Gamma(1/2) = \sqrt{\pi}$.

(5) $\Gamma(n) = \int_0^1 (\log 1/y)^{n-1} dy$.

Proof. By definition of gamma function, we have

$$\Gamma(n) = \int_0^\infty e^{-x} x^{n-1}\,dx, n > 0$$

Putting $x = \log(1/y)$ in gamma function, we get

$$\Gamma(n) = -\int_1^0 (\log 1/y)^{n-1} dy = \int_0^1 (\log 1/y)^{n-1} dy.$$

20.3 SOME TRANSFORMATIONS OF GAMMA FUNCTION

Gamma function is given by

$$\Gamma(n) = \int_0^\infty x^{n-1} e^{-x}\,dx. \qquad \ldots(1)$$

(1) $\dfrac{\Gamma(n)}{a^n} = \int_0^\infty e^{-ay}\, y^{n-1}\,dy,\ n > 0,\ a > 0$

Proof. We have $\Gamma(n) = \int_0^\infty x^{n-1} e^{-x}\, dx$, $n > 0$.

Put $x = ay$, so that $dx = a\, dy$.

When $x = 0, y = 0$ and when $x \to \infty, y \to \infty$.

$\therefore$ $\Gamma(n) = \int_0^\infty e^{-ay} (ay)^{n-1}.a\, dy$.

Hence, $\int_0^\infty e^{-ay} y^{n-1} dy = \dfrac{\Gamma(n)}{a^n}$.

(2) $\Gamma(n) = \dfrac{1}{n}\int_0^\infty e^{-x^{1/n}}\, dx, n > 0$

Proof. We have $\Gamma(n) = \int_0^\infty e^{-x} x^{n-1} dx, n > 0$ ...(1)

Put $x^n = t$. *i.e.,* $nx^{n-1}\, dx = dt$, then (1) gives

$$\Gamma(n) = \frac{1}{n}\int_0^\infty e^{-t^{1/n}}\, dt$$

$\Rightarrow$ $\Gamma(n) = \dfrac{1}{n}\int_0^\infty e^{-x^{1/n}}\, dx$ [By the property of definite integral]

(3) $\Gamma(n) = 2\int_0^\infty e^{-x^2} x^{2n-1} dx, n > 0$

Proof. We have $\Gamma(n) = \int_0^\infty e^{-x} x^{n-1} dx$, ...(1)

Put $x = t^2$, so that $dx = 2t\, dt$,

Therefore, $\Gamma(n) = \int_0^\infty e^{-t^2} (t^2)^{n-1} 2t\, dt$

or $\Gamma(n) = 2\int_0^\infty e^{-t^2} t^{2n-1}\, dt$

$\Rightarrow$ $\Gamma(n) = 2\int_0^\infty e^{-x^2} x^{2n-1}\, dx$.

Solved Examples

Example 1. *Evaluate* :

(i) $\int_0^\infty e^{-x} x^4\, dx$ (ii) $\int_0^\infty x^6 e^{-2x}\, dx$

Solution. (i) We have

$$\int_0^\infty e^{-x} x^4\, dx = \int_0^\infty e^{-x} x^{5-1}\, dx$$

[By definition of gamma function]

$= \Gamma(5) = (4)! = 24$.

(ii) Let $I = \int_0^\infty x^6 e^{-2x} dx$ ...(1)

Put $2x = t$, so that $dx = 1/2\, dt$ then

$$I = \int_0^\infty \left(\frac{t}{2}\right)^6 e^{-t} . \frac{1}{2} dt = \frac{1}{2^7}\int_0^\infty e^{-t}\, t^{7-1} dt$$

$$= \frac{1}{2^7}\Gamma(7)$$

[By definition of gamma function]

$$= \frac{1}{2^7} \times (6!) = \frac{45}{8}.$$

Example 2. *Show that* $\int_0^1 \dfrac{dx}{\sqrt{(-\log x)}} = \sqrt{\pi}$.

Solution. We know that

$$\Gamma(n) = \int_0^1 (-\log x)^{n-1} dx$$

Putting $n = 1/2$, we have

$$\Gamma(1/2) = \int_0^1 (-\log x)^{(1/2)-1} dx$$

or $\sqrt{\pi} = \int_0^1 (-\log x)^{-1/2}\, dx$

or $\sqrt{\pi} = \int_0^1 \dfrac{dx}{\sqrt{(-\log x)}}$.

Example 3. *If n is a positive integer, prove that*

$$2n\, \Gamma(n+1/2) = 1.\,3.\,5.\, ...(2n-1)\sqrt{\pi}$$

Solution. We know that

$\Gamma(n+1) = n\Gamma(n)$ ...(1)

Now $\Gamma(n + 1/2)$

$= \Gamma\,(n - 1/2 + 1)$

$= (n - 1/2)\,\Gamma(n - 1/2)$ [Using (1)]

$= (n - 1/2)\,\Gamma\,(n - 3/2 + 1)$

$= (n - 1/2)(n - 3/2)\,\Gamma\,(n - 3/2)$

$= \dfrac{2n-1}{2}.\dfrac{2n-3}{2}.\Gamma\left(\dfrac{2n-3}{2}\right)$

$= \dfrac{2n-1}{2}.\dfrac{2n-3}{2}.\dfrac{5}{2}.\dfrac{3}{2}.\dfrac{1}{2}.\Gamma\left(\dfrac{1}{2}\right)$

$= \dfrac{(2n-1)(2n-3)...5.3.1}{2^n}\sqrt{\pi}$ [$\because \Gamma(1/2) = \sqrt{\pi}$]

Hence,

$2^n\Gamma(n+1/2) = (2n-1)(2n-3)...5.3.1.\sqrt{\pi}$.

Example 4. *Show that* $\int_0^\infty \exp(2ax - x^2)\,dx = \frac{1}{2}\sqrt{\pi}\exp a^2.$

Solution. Consider $\int_0^\infty \exp(2ax - x^2)\,dx$

$$= \int_0^\infty e^{2ax-x^2}dx = \int_0^\infty e^{a^2-(x-a)^2}dx$$

$$= e^{a^2}\int_0^\infty e^{-(x-a)^2}dx = e^{a^2}\int_0^\infty e^{-t^2}dt.$$

Put $x - a = t$, $\therefore\ dx = dt$

$$\Rightarrow \int_0^\infty \exp(2ax - x^2)\,dx = \exp a^2\int_0^\infty e^{-t^2}dt \quad ...(1)$$

Now $\Gamma(n) = \int_0^\infty e^{-u}u^{n-1}du.$ ...(2)

Putting $n=1/2$ in (2), we have

$$\Gamma(1/2) = \int_0^\infty e^{-u}u^{-1/2}du. \quad ...(3)$$

Putting $u=t^2$ so that $du= 2t\,dt$ in (3), we get

$$\Gamma(1/2) = \int_0^\infty e^{-t^2}t^{-1}.2t\,dt$$

or $\sqrt{\pi} = 2\int_0^\infty e^{-t^2}dt$ or $\int_0^\infty e^{-t^2}dt = \sqrt{\pi}/2$ ...(4)

Using (4), (1) reduces to

$$\int_0^\infty \exp(2ax - x^2)\,dx = \frac{1}{2}\sqrt{\pi}\exp a^2.$$

Example 5. *Evaluate* $\int_0^\infty t^{-3/2}(1-e^{-t})\,dt$.

Solution. We have $\int_0^\infty t^{-3/2}(1-e^{-t})\,dt$

$$= (1-e^{-t})\left[\frac{t^{-1/2}}{-1/2}\right]_0^\infty - \int_0^\infty (e^{-t})\left(\frac{t^{-1/2}}{-1/2}\right)dt$$

$$= 0 + 2\int_0^\infty e^{-t}t^{(1/2)-1}dt$$

$= 2\Gamma(1/2)$ [By definition of gamma function]

$= 2\sqrt{\pi}$.

Example 6. *Prove that*

(i) $\int_0^\infty xe^{-\alpha x}\cos\beta x\,dx = \frac{\alpha^2-\beta^2}{(\alpha^2+\beta^2)^2},\ \alpha > 0$

(ii) $\int_0^\infty xe^{-\alpha x}\sin\beta x\,dx = \frac{2\alpha\beta}{(\alpha^2+\beta^2)^2},\ \alpha > 0$

(Remember)

Solution. We know that

$$\int_0^\infty e^{-kx}x^{n-1}dx = \frac{\Gamma(n)}{k^n}, n > 0, k > 0. \quad ...(1)$$

Putting $k= \alpha - i\beta$ and $n =2$ in (1), we get

$$\int_0^\infty e^{-(\alpha-i\beta)x}x\,dx = \frac{\Gamma(2)}{(\alpha-i\beta)^2}$$

or $\int_0^\infty xe^{-\alpha x}e^{i\beta x}dx = \frac{(\alpha+i\beta)^2}{(\alpha-i\beta)^2(\alpha+i\beta)^2}$

[as $\Gamma(2)=1$]

$$\int_0^\infty xe^{-\alpha x}e^{i\beta x}dx = \frac{\alpha^2-\beta^2+2i\alpha\beta}{[(\alpha+i\beta)(\alpha-i\beta)]^2}$$

$$\Rightarrow \int_0^\infty xe^{-\alpha x}(\cos\beta x + i\sin\beta x)\,dx = \frac{\alpha^2-\beta^2+2i\alpha\beta}{(\alpha^2+\beta^2)^2}$$

or $\int_0^\infty xe^{-\alpha x}\cos\beta x\,dx + i\int_0^\infty xe^{-\alpha x}\sin\beta x$

$$= \frac{\alpha^2-\beta^2}{(\alpha^2+\beta^2)^2} + i\frac{2\alpha\beta}{(\alpha^2+\beta^2)^2}.$$

Equating real and imaginary parts of both sides, we get

$$\int_0^\infty xe^{-\alpha x}\cos\beta x\,dx = \frac{\alpha^2-\beta^2}{(\alpha^2+\beta^2)^2}$$

and $\int_0^\infty xe^{-\alpha x}\sin\beta x\,dx = \frac{2\alpha\beta}{(\alpha^2+\beta^2)^2}.$

Example 7. *Show that*

$$\int_0^\infty \frac{x^c}{c^x}dx = \frac{\Gamma(c+1)}{(\log c)^{c+1}}, c > 0.$$

Solution. We have

$$\int_0^\infty \frac{x^c}{c^x}dx = \int_0^\infty x^c c^{-x}\,dx$$

$$= \int_0^\infty x^c[e^{\log_e c}]^{-x}dx$$

$[\because c = e^{\log_e c}$ if $c \geq 0]$

$$= \int_0^\infty x^{(c+1)-1}e^{-x\log_e c}dx$$

$$= \frac{\Gamma(c+1)}{(\log_e c)^{c+1}}$$

$$\left[\because \int_0^\infty x^{n-1}e^{-kx}dx = \frac{\Gamma(n)}{k^n};\ n>0,\ k>0\right]$$

Example 8. *With certain limitations on the values of a, b, m and n, prove that*

$$\int_0^\infty\int_0^\infty e^{-(ax^2+by^2)}x^{2m-1}y^{2n-1}dx\,dy = \frac{\Gamma(m)\Gamma(n)}{4a^m b^n}.$$

Solution. Let $I = \int_0^\infty\int_0^\infty e^{-(ax^2+by^2)}x^{2m-1}y^{2n-1}dx\,dy$ ...(1)

$$\Rightarrow I = \int_0^\infty e^{-ax^2}x^{2m-1}dx$$

$$\times\int_0^\infty e^{-by^2}y^{2n-1}dy = I_1 \times I_2 \quad ...(2)$$

where $I_1 = \int_0^\infty e^{-ax^2}x^{2m-1}dx$ ...(3)

$I_2 = \int_0^\infty e^{-by^2}y^{2n-1}dy$ ...(4)

Put $ax^2 = t,\ x = \left(\frac{t}{a}\right)^{1/2}$ so that $dx = \frac{dt}{2\sqrt{at}}$

then equation (3) becomes

$$I_1 = \int_0^\infty e^{-t}\left[\frac{t}{a}\right]^{\frac{(2m-1)}{2}}\frac{dt}{2\sqrt{at}}$$

$$= \frac{1}{2a^m}\int_0^\infty e^{-t} t^{m-1}\, dt$$

$$= \frac{\Gamma(m)}{2a^m}$$

[By definition of gamma function taking $n>0$, $a>0$]

Then $I_2 = \frac{\Gamma(n)}{2b^n}$ if $n > 0, b > 0$.

$\therefore$ from (1) and (2), we get

$$I = I_1 \times I_2 = \frac{\Gamma(m)\Gamma(n)}{4a^m b^n}.$$

20.4 BETA FUNCTION

Definition. *The definite integral* $\int_0^1 x^{m-1}(1-x)^{n-1}\,dx$, *for* $m > 0, n > 0$ *is known as the Beta function and denoted by* $B(m, n)$ *which is read as "Beta m, n", where m, n are positive number or integers. Thus* $B(m,n) = \int_0^1 x^{m-1}(1-x)^{n-1}\,dx$.

(UPTU-2006, 2008, GBTU(AG)-2010)

REMARK

- Beta function is also called the Eulerian integral of first kind.

20.5 PROPERTIES OF BETA FUNCTION

(1) *Symmetry of Beta function i.e.,* $B(m, n) = B(n, m)$ (UPTU-2006)

Proof. By definition of beta function, we have

$$B(m,n) = \int_0^1 x^{m-1}(1-x)^{n-1}\,dx$$

$$= \int_0^1 (1-x)^{m-1}[1-(1-x)]^{n-1} dx \qquad \left[\because \int_0^a f(x)\,dx = \int_0^a f(a-x)\,dx\right]$$

$$= \int_0^1 (1-x)^{m-1} x^{n-1} dx = \int_0^1 x^{n-1}(1-x)^{m-1}\,dx$$

$$= B(n,m) \qquad \text{[By definition of Beta function]}$$

$$B(m,n) = B(n,m)$$

i.e., the interchange of position of m and n does not change the value of beta function.

REMARK

- This is the fundamental property of Beta function and also called symmetric property of Beta function.

(2) *Beta function B(m, n) can be evaluated in an explicit form if m or n is a positive integer.*

Proof. Case I. *When 'n' is a positive integer.*

If $n = 1$, then by definition of Beta function, we have

$$B(m,n) = \int_0^1 x^{m-1}(1-x)^{n-1}\,dx \qquad \text{...(1)}$$

$$\Rightarrow \quad B(m,1) = \int_0^1 x^{m-1}(1-x)^{1-1}\,dx$$

$$= \int_0^1 x^{m-1} dx = \left[\frac{x^m}{m}\right]_0^1 = \frac{1}{m}. \qquad \text{...(2)}$$

Now, let $n > 1$, then from (1), we have

$$B(m,n) = \int_0^1 (1-x)^{n-1} x^{m-1}\,dx = \left[(1-x)^{n-1}.\frac{x^m}{m}\right]_0^1 - \int_0^1 (n-1)(1-x)^{n-2}.(-1)\frac{x^m}{m}dx.$$

Integrating by parts taking x^{m-1} as second function, we have

$$B(m, n) = 0 + \frac{n-1}{m}\int_0^1 x^m(1-x)^{n-2}dx \qquad \left[\because n > 1 \text{ and } \lim_{x\to 0}(1-x)^{n-1}\frac{x^m}{m} = 0\right]$$

$$= \frac{n-1}{m}\int_0^1 x^{(m+1)-1}(1-x)^{(n-1)-1}dx = \frac{n-1}{m}B(m+1,n-1)$$

Thus
$$B(m,n) = \frac{n-1}{m}B(m+1,n-1) \qquad \text{...(3)}$$

Now replacing m by $m+1$ and n by $n-1$ in (3), we get

$$B(m+1,n-1)=\frac{n-1-1}{m+1}B(m+2,n-2) \quad ...(4)$$

Using equation (4), the equation (3) becomes

$$B(m,n)=\frac{n-1}{m}.\frac{n-2}{m+1}B(m+2,n-2) \quad ...(5)$$

After applying the above process successively, we get

$$B(m,n)=\frac{n-1}{m}.\frac{n-2}{m+1}.\frac{n-3}{m+2}...\frac{1}{m+n-2}B(m+n-1,1) \quad ...(6)$$

$$=\frac{n-1}{m}.\frac{n-2}{m+1}.\frac{n-3}{m+2}...\frac{1}{m+n-2}\int_0^1 x^{m+n-2}(1-x)^0\,dx$$

$$=\frac{n-1}{m}.\frac{n-2}{m+1}.\frac{n-3}{m+2}...\frac{1}{m+n-2}\left[\frac{x^{m+n-1}}{m+n-1}\right]_0^1$$

$$=\frac{n-1}{m}.\frac{n-2}{m+1}.\frac{n-3}{m+2}...\frac{1}{m+n-2}.\frac{1}{m+n-1}$$

$$\Rightarrow \quad B(m,n)=\frac{n-1}{m}.\frac{n-2}{m+1}.\frac{n-3}{m+2}...\frac{1}{m+n-2}.\frac{1}{m+n-1}$$

$$\therefore \quad B(m,n)=\frac{(n-1)!}{m(m+1)(m+2)...(m+n-2)(m+n-1)} \quad ...(7)$$

Case II. *When m is a positive integer.*

Since the beta function is symmetrical in m and n *i.e.*, $B(m, n) = B(n, m)$ therefore by interchanging m and n in Case I we get

$$B(m,n)=\frac{(m-1)!}{n(n+1)(n+2)...(n+m-2)(n+m-1)}. \quad ...(8)$$

Case III. *When both m and n are positive integers.*

We have, by Case I

$$B(m,n)=\frac{(n-1)!}{m(m+1)(m+2)...(m+n-2)(m+n-1)}.$$

$$=\frac{[1.2.3...(m-1)](n-1)!}{1.2.3....m(m+1)(m+2)...(m+n-2)(m+n-1)}$$

Multiplying both numerator and denominator by 1. 2. 3. .. .(m–1) !, we get

$$B(m,n)=\frac{(m-1)!(n-1)!}{(m+n-1)!}$$

20.6 TRANSFORMATION OF BETA FUNCTION

The Beta function

$$B(m,n)=\int_0^1 x^{m-1}(1-x)^{n-1}\,dx \quad ...(A)$$

can be transformed into many forms given below :

(I) $B(m,n)=\int_0^\infty \frac{x^{n-1}}{(1+x)^{m+n}}dx=\int_0^\infty \frac{x^{m-1}}{(1+x)^{m+n}}dx$.

Proof. Put $x=\frac{1}{(1+y)}$ and $dx=-\frac{dy}{(1+y)^2}$ and $y\to 0$ when $x=1$, $y\to\infty$, when $x=0$.

$$\therefore \quad B(m,n)=\int_\infty^0\left(\frac{1}{1+y}\right)^{m-1}\left[1-\frac{1}{1+y}\right]^{n-1}\left[\frac{-dy}{(1+y)^2}\right]$$

$$=\int_0^\infty \frac{(y)^{n-1}}{(1+y)^{m+1}}\left(\frac{1}{1+y}\right)^{n-1}.dy=\int_0^\infty \frac{y^{n-1}}{(1+y)^{m+n}}dy$$

$$\Rightarrow \quad B(m,n)=\int_0^\infty \frac{x^{n-1}\,dx}{(1+x)^{m+n}}. \quad ...(1)$$

Since m and n are interchangeable in beta function by symmetric property therefore (1) gives

$$B(m,n)=\int_0^{\infty}\frac{x^{m-1}}{(1+x)^{m+n}}dx$$

thus
$$B(m,n)=\int_0^{\infty}\frac{x^{n-1}dx}{(1+x)^{m+n}}=\int_0^{\infty}\frac{x^{m-1}dx}{(1+x)^{m+n}}.$$

(II) $B(m,n)=2\int_0^{\pi/2}\cos^{2m-1}\theta\sin^{2n-1}\theta\,d\theta$.

Proof. Put $x=\sin 2\theta$ and $dx=2\sin\theta\cos\theta\,d\theta$ and when $x=0,\ \theta=\pi/2$ when $x=1$ in (A) we get

$$B(m,n)=2\int_0^{\pi/2}\sin^{2m-1}\theta\cos^{2n-1}\theta\,d\theta$$

$$=2\int_0^{\pi/2}\cos^{2m-1}\theta\sin^{2n-1}\theta\,d\theta$$ [By symmetric property of beta function]

(III) $B(m,n)=\dfrac{1}{a^{m+n-1}}\int_0^a x^{m-1}(a-x)^{n-1}dx$.

Proof. Put $x=\dfrac{y}{a}$, *i.e.*, $dx=\dfrac{1}{a}dy$ and when $x\to 0$, then $y\to 0$, when $x=1$ then $y\to a$.

So
$$B(m,n)=\frac{1}{a^{m+n-1}}\int_0^a y^{m-1}(a-y)^{n-1}dy=\frac{1}{a^{m+n-1}}\int_0^a x^{m-1}(a-x)^{n-1}dx.$$

(IV) $\dfrac{B(m,n)}{a^n(1+a)^m}=\int_0^1\dfrac{x^{m-1}(1-x)^{n-1}dx}{(x+a)^{m+n}}$.

Proof. Let $\dfrac{x}{1+a}=\dfrac{t}{t+a}\Rightarrow dx=a(1+a)\dfrac{dt}{(t+a)^2}$

then we have

$$B(m,n)=\int_0^1(1+a)^{m-1}\left(\frac{t}{t+a}\right)^{m-1}a^{n-1}\left(\frac{1-t}{a+t}\right)^{n-1}\frac{a(a+1)}{(t+a)^2}dt$$

$$=a^n(1+a)^m\int_0^1\frac{t^{m-1}(1-t)^{n-1}}{(t+a)^{m+n}}dt=a^n(1+a)^m\int_0^1\frac{x^{m-1}(1-x)^{n-1}}{(x+a)^{m+n}}dx$$

Hence,
$$\frac{B(m,n)}{a^n(1+a)^m}=\int_0^1\frac{x^{m-1}(1-x)^{n-1}}{(x+a)^{m+n}}dx$$

(V) $B(m,n)(a-b)^{m+n-1}=\int_b^a(x-b)^{m-1}(a-x)^{n-1}dx$.

Proof. Put $x=\dfrac{t-b}{a-b}$ so that $dx=\dfrac{dt}{a-b}$. in (A), we get

$$B(m,n)=\int_b^a\left(\frac{t-b}{a-b}\right)^{m-1}\left(\frac{a-t}{a-b}\right)^{n-1}\frac{dt}{a-b}=\frac{1}{(a-b)^{m+n-1}}\int_b^a(t-b)^{m-1}(a-t)^{n-1}dt$$

$$=\frac{1}{(a-b)^{m+n-1}}\int_b^a(x-b)^{m-1}(a-x)^{n-1}dx.$$

$\therefore$
$$B(m,n)(a-b)^{m+n-1}=\int_b^a(x-b)^{m-1}(a-x)^{n-1}dx.$$

(VI) $\dfrac{1}{a^nb^m}B(m,n)=\int_0^1\dfrac{x^{m-1}(1-x)^{n-1}dx}{\{a+(b-a)x\}^{m+n}}$

Proof. We put

$$\frac{a}{y}-\frac{b}{x}=a-b.\ \text{(Remember)} \qquad ...(1)$$

$\therefore$
$$\frac{b}{x}=\frac{a}{y}+(b-a)=\frac{a+(b-a)y}{y}$$

$\Rightarrow$
$$x=\frac{by}{a+(b-a)y} \qquad ...(2)$$

$$\Rightarrow \qquad dx = \frac{b[a+(b-a)y] - by(b-a)}{\{a+(b-a)y\}^2}dy$$

i.e.,
$$dx = \frac{ab\,dy}{[a+(b-a)y]^2}. \qquad ...(3)$$

Again from (1), we see that when $x = 1, y = 1$ and $x = 0, y = 0$.
and from (2), we have

$$1 - x = 1 - \frac{by}{a+by-ay} = \frac{a(1-y)}{a+(b-a)y}. \qquad ...(4)$$

Using (2), (3) and (4), (1) gives

$$B(m,n) = \int_0^1 \left\{\frac{by}{a+(b-a)y}\right\}^{m-1} \left\{\frac{a(1-y)}{a+(b-a)y}\right\}^{n-1} \frac{ab\,dy}{\{a+(b-a)y\}^2}$$

$$= a^n b^m \int_0^1 \frac{y^{m-1}(1-y)^{n-1}}{\{a+(b-a)y\}^{m+n}}dy = a^n b^m \int_0^1 \frac{x^{m-1}(1-x)^{n-1}\,dx}{\{a+(b-a)x\}^{m+n}}$$

$$\Rightarrow \qquad \frac{1}{a^n b^m}B(m,n) = \int_0^1 \frac{x^{m-1}(1-x)^{n-1}\,dx}{\{a+(b-a)x\}^{m+n}}.$$

Solved Examples

Example 1. *Express* $\int_0^1 x^m(1-x^p)^n dx$ *in terms of beta function and hence evaluate* $\int_0^1 x^5(1-x^3)^{10}dx$.

Solution. Put $x^p = t$ so that $dx = \left(\frac{1}{p}\right)t^{1/p-1}\,dt$.

$\therefore \quad \int_0^1 x^m(1-x^p)^n dx$

$$= \int_0^1 t^{m/p}(1-t)^n(1/p)t^{1/p-1}\,dt$$

$$= \frac{1}{p}\int_0^1 t^{(m+1)/p-1}(1-t)^{n+1-1}dt$$

$$= \frac{1}{p}B\left(\frac{m+1}{p}, n+1\right).$$

Putting $m = 5$, $n = 10$ and $p = 3$, we have

$$\int_0^1 x^5(1-x^3)^{10}dx = \frac{1}{3}B\left(\frac{5+1}{3}, 11\right) = \frac{1}{3}B(2,11)$$

$$= \frac{1}{3}\frac{\Gamma(2)\Gamma(11)}{\Gamma(13)} = \frac{1}{3}\frac{1.\Gamma(11)}{12.11\Gamma(11)}$$

$$= \frac{1}{3.12.11} = \frac{1}{396}.$$

Example 2. *Evaluate the following integrals by expressing them in terms of Beta function*

(i) $\int_0^1 x^m(1-x^2)^n\,dx,\ m > 1, n > -1$

(ii) $\int_0^1 \frac{x^2 dx}{\sqrt{(1-x^5)}}$.

Solution. (i) We have

$$\int_0^1 x^m(1-x^2)^n dx = \int_0^1 x^{m-1}(1-x^2)^n x.dx$$

$$= \int_0^1 y^{\frac{(m-1)}{2}}(1-y)^n.\frac{dy}{2}$$

(Putting $x^2 = y$ so that $2x\,dx = dy$)

$$= \frac{1}{2}\int_0^1 y^{\frac{(m-1)}{2}}(1-y)^n dy$$

$$= \frac{1}{2}\int_0^1 y^{\frac{m+1}{2}-1}(1-y)^{(n+1)-1}dy$$

$$= \frac{1}{2}B\left[\frac{1}{2}(m+1), n+1\right].$$

(ii) We have

$$\int_0^1 \frac{x^2}{\sqrt{(1-x^5)}}dx = \int_0^1 x^2(1-x^5)^{-1/2}dx$$

$$= \int_0^1 x^2.\frac{1}{x^4}(1-x^5)^{-1/2}x^4 dx$$

$$= \int_0^1 x^{-2}(1-x^5)^{-1/2}x^4 dx$$

$$= \int_0^1 y^{-2/5}(1-y)^{-1/2}\frac{1}{5}dy$$

(Putting $x^5 = y$, i.e., $5x^4\,dx = dy$)

$$= \frac{1}{5}\int_0^1 y^{-2/5}(1-y)^{-1/2}dy$$

$$= \frac{1}{5}\int_0^1 y^{(3/5)-1}(1-y)^{(1/2)-1}dy$$

$$= \frac{1}{5}B\left(\frac{3}{5}, \frac{1}{2}\right).$$

Example 3. *Show that*

$$\int_0^1 \frac{x^{m-1}(1-x)^{n-1}}{(a+bx)^{m+n}}dx = \frac{1}{(a+b)^m a^n}B(m,n).$$

Solution. Let $I = \int_0^1 \frac{x^{m-1}(1-x)^{n-1}}{(a+bx)^{m+n}}dx$

$$= \int_0^1 \left(\frac{x}{a+bx}\right)^{m-1} \cdot \left(\frac{1-x}{a+bx}\right)^{n-1} \frac{1}{(a+bx)^2}dx.$$

Put $\frac{x}{a+bx} = \frac{y}{a+b}$

i.e. $\frac{(a+bx)-x.b}{(a+bx)^2}dx = \frac{dy}{a+b}$

$$\Rightarrow \quad \frac{1}{(a+bx)^2}dx = \frac{dy}{a(a+b)}$$

$$\Rightarrow \quad \frac{1-x}{a+bx} = \frac{1}{a}\left(\frac{a-ax}{a+bx}\right)$$

$$= \frac{1}{a}\left[\frac{a+bx-ax-bx}{a+bx}\right]$$

$$= \frac{1}{a}\left[1-\frac{x(a+b)}{a+bx}\right] = \frac{1-y}{a}.$$

Also when $x = 0$, $y = 0$, and when $x = 1$, $y = 1$. Therefore,

$$I = \int_0^1 \left(\frac{y}{a+b}\right)^{m-1} \left(\frac{1-y}{a}\right)^{n-1} \cdot \frac{dy}{a(a+b)}$$

$$= \frac{1}{(a+b)^m . a^n}\int_0^1 y^{m-1}(1-y)^{n-1}dy$$

$$= \frac{B(m,n)}{(a+b)^m . a^n}.$$

Example 4. *Evaluate* $\int_0^\infty x^m e^{-ax^n}dx$, *when m, n and a are all positive constant.*

Solution. Let $ax^n = y \Rightarrow nax^{n-1} = dy$

and at $x = 0, y = 0$ and at $x = \infty, y = \infty$.

$$\therefore \int_0^\infty x^m e^{-ax^n}dx$$

$$= \frac{1}{na}\int_0^\infty \left(\frac{y}{a}\right)^{m/n}\left(\frac{y}{a}\right)^{1/n-1} e^{-y}dy$$

$$= \frac{1}{na^{(m+1)/n}}\int_0^\infty y^{\left(\frac{m+1}{n}\right)-1} . e^{-y}dy$$

$$= \frac{1}{na^{(m+1)/n}}\Gamma\left(\frac{m+1}{n}\right).$$

Example 5. *Prove that*

$$\int_0^\infty \frac{x^{m-1}-x^{n-1}}{(1+x)^{m+n}}dx = 0, m > 0, n > 0.$$

Solution. We have

$$\int_0^\infty \frac{x^{m-1}}{(1+x)^{m+n}}dx - \int_0^\infty \frac{x^{n-1}}{(1+x)^{m+n}}dx$$

$$= B(m,n) - B(n,m)$$

$$= B(m,n) - B(m,n) = 0.$$

20.7 RELATION BETWEEN BETA AND GAMMA FUNCTION

We have $$B(m,n) = \frac{\Gamma(m)\Gamma(n)}{\Gamma(m+n)}, m > 0, n > 0 = \int_0^\infty \frac{y^{n-1}dy}{(1+y)^{m+n}}.$$ (UPTU-2009, 10 ; VKTU-2010)

Proof. We have $$\int_0^\infty y^{n-1}e^{-xy}dy = \frac{\Gamma(n)}{x^n}$$

or $$\Gamma(n) = \int_0^\infty x^n y^{n-1}e^{-xy}dy \quad ...(1)$$

Also $$\Gamma(m) = \int_0^\infty x^{m-1}e^{-x}dx \quad ...(2)$$

Multiplying both sides of (1) by $x^{m-1}e^{-x}$, we have

$$\Gamma(n).x^{m-1}e^{-x} = \int_0^\infty x^{n+m-1}y^{n-1}e^{-(y+1)x}dy.$$

Integrating both sides with respect to x within limits $x = 0$ to $x = \infty$, we have

$$\Gamma(n)\int_0^\infty x^{m-1}e^{-x}dx = \int_0^\infty \left[\int_0^\infty x^{n+m-1}e^{-(y+1)x}dx\right]y^{n-1}dy \quad ...(3)$$

But $$\int_0^\infty x^{(n+m)-1}e^{-(y+1)x}dx = \frac{\Gamma(n+m)}{(1+y)^{m+n}}.$$

Hence with the help of this result and (2), we get from (3)

$$\Gamma(n)\Gamma(m) = \int_0^\infty \Gamma(n+m)\frac{y^{n-1}}{(1+y)^{n+m}}dy$$

$$= \Gamma(n+m)\int_0^\infty \frac{y^{n-1}}{(1+y)^{n+m}}dy = \Gamma(n+m)B(m,n)$$

or $$B(m,n) = \frac{\Gamma(m)\Gamma(n)}{\Gamma(n+m)}.$$

Deduction 1. $\Gamma(n)\Gamma(1-n) = \dfrac{\pi}{\sin n\pi}$, *where* $0 < n < 1$.

Proof. We have

$$B(m,n) = \int_0^\infty \frac{x^{n-1}dx}{(1+x)^{m+n}},\ m > 0,\ n > 0.$$

Therefore the relation between beta and gamma functions becomes

$$\int_0^\infty \frac{x^{n-1}dx}{(1+x)^{m+n}} = \frac{\Gamma(m)\Gamma(n)}{\Gamma(m+n)}.$$

Taking $m + n = 1$, so that $m = 1 - n$, we get

$$\int_0^\infty \frac{x^{n-1}}{1+x}dx = \frac{\Gamma(1-n)\Gamma(n)}{\Gamma(1)},\ 0 < n < 1. \qquad [\because m > 0 \Rightarrow 1-n > 0 \Rightarrow n < 1. \text{ Also } n > 0]$$

But also we know that $\int_0^\infty \dfrac{x^{n-1}}{1+x}dx = \dfrac{\pi}{\sin n\pi}$ and $\Gamma(1) = 1$

$$\therefore \qquad \frac{\pi}{\sin n\pi} = \Gamma(1-n)\Gamma(n),\ 0 < n < 1.$$

Deduction 2. $\Gamma(1/2) = \sqrt{\pi}$.

Proof. We have just proved that

$$\Gamma(n)\Gamma(1-n) = \frac{\pi}{\sin n\pi} \qquad ...(1)$$

Putting $n = 1/2$ in (1), we obtain

$$\Gamma\left(\frac{1}{2}\right)\Gamma\left(1-\frac{1}{2}\right) = \frac{\pi}{\sin \pi/2} \quad \text{or} \quad \left[\Gamma\left(\frac{1}{2}\right)\right]^2 = \pi$$

$$\Rightarrow \qquad \Gamma\left(\frac{1}{2}\right) = \sqrt{\pi}.$$

Aliter. We know $B(m,n) = \dfrac{\Gamma(m)\Gamma(n)}{\Gamma(m+n)}$.

Putting $m = n = 1/2$, we get

$$B(1/2, 1/2) = \frac{\Gamma(1/2)\Gamma(1/2)}{\Gamma(1/2+1/2)} = \frac{\{\Gamma(1/2)\}^2}{\Gamma(1)} \qquad [\because \Gamma(1)=1]$$

or

$$\{\Gamma(1/2)\}^2 = B(1/2, 1/2) = \int_0^1 x^{(1/2)-1}(1-x)^{(1/2)-1}dx$$

$$= \int_0^1 x^{-1/2}(1-x)^{-1/2}dx = \int_0^1 \frac{dx}{\sqrt{x}\sqrt{1-x}}$$

$$= \int_0^{\pi/2} \frac{2\sin\theta\cos\theta\, d\theta}{\sin\theta\sqrt{(1-\sin^2\theta)}} \qquad (\text{By putting } x = \sin^2\theta)$$

$$= 2\int_0^{\pi/2} d\theta = 2[\theta]_0^{\pi/2} = 2\left(\frac{\pi}{2}\right) = \pi$$

$$\Rightarrow \qquad \left\{\Gamma\left(\frac{1}{2}\right)\right\}^2 = \pi \Rightarrow \Gamma\left(\frac{1}{2}\right) = \sqrt{\pi}.$$

Deduction 3. $\int_0^1 e^{-x^2}dx = \dfrac{1}{2}\sqrt{\pi}$.

Proof. We have

$$\int_0^1 e^{-x^2}dx = \int_0^1 e^{-y}.\frac{1}{2\sqrt{y}}dy, \text{ putting } x^2 = y,\ 2x\,dx = dy$$

$$= \frac{1}{2}\int_0^\infty e^{-y}\,y^{-1/2}dy = \frac{1}{2}\int_0^1 e^{-y}\,y^{(1/2)-1}dy$$

$$= \frac{1}{2}\Gamma\left(\frac{1}{2}\right) \qquad \left[\because \int_0^\infty e^{-x}.x^{n-1}dx = \Gamma n\right]$$

$$= \frac{1}{2}\sqrt{\pi} \qquad \left[\because \Gamma\left(\frac{1}{2}\right) = \sqrt{\pi}\right]$$

$$\Rightarrow \qquad \int_0^{\infty} e^{-x^2} dx = \frac{1}{2}\sqrt{\pi}.$$

Deduction 4. $\int_0^{\pi/2} \cos^m \theta \sin^n \theta d\theta = \dfrac{\Gamma\left(\frac{m+1}{2}\right)\Gamma\left(\frac{n+1}{2}\right)}{2\Gamma\left(\frac{m+n+2}{2}\right)}$ *for all values of m and n such that m > –1, n > –1.*

Proof. We put $\sin^2\theta = x, \quad \Rightarrow \quad 2\sin\theta \cos\theta \, d\theta = dx$

$$\Rightarrow \qquad 2\sin\theta.\sqrt{(1-\sin^2\theta)}\, d\theta = dx \qquad \Rightarrow \quad 2x^{1/2}\sqrt{1-x}\, d\theta = dx$$

$$\Rightarrow \qquad d\theta = \frac{dx}{2x^{1/2}(1-x)^{1/2}}.$$

Also, when $\theta = \pi/2, x = 1$ and $\theta = 0, x = 0$.
Putting these values in L.H.S. of the given equation, we get

$$\int_0^{\pi/2} \cos^m \theta \sin^n \theta d\theta = \int_0^{\pi/2} (1-\sin^2\theta)^{m/2}.\sin^n \theta d\theta$$

$$= \int_0^1 (1-x)^{m/2}.x^{n/2}.\frac{dx}{2x^{1/2}(1-x)^{1/2}} = \frac{1}{2}\int_0^1 x^{\frac{(n-1)}{2}}(1-x)^{\frac{(m-1)}{2}} dx$$

$$= \frac{1}{2}\int_0^1 x^{\left\{\frac{(n+1)}{2}\right\}-1}(1-x)^{\left\{\frac{(m+1)}{2}\right\}-1} dx = \frac{1}{2}B\left(\frac{m+1}{2},\frac{n+1}{2}\right)$$

$$= \frac{\frac{1}{2}\Gamma\frac{1}{2}(m+1)\Gamma\frac{1}{2}(n+1)}{\Gamma\frac{1}{2}(m+n+1+1)} \qquad \left(\text{Because } B(m,n) = \frac{\Gamma(m)\Gamma(n)}{\Gamma(m+n)}\right)$$

$$= \frac{\Gamma\frac{1}{2}(m+1)\Gamma\frac{1}{2}(n+1)}{2\Gamma\frac{1}{2}(m+n+2)}.$$

Deduction 5. $\int_0^{\pi/2} \sin^{p-1} \theta \cos^{q-1} d\theta = \dfrac{\Gamma(p/2)\Gamma(q/2)}{2\Gamma\left(\frac{p+q}{2}\right)}$.

Proof. By definition of Beta function, we have

$$B(m,n) = \int_0^1 x^{m-1}(1-x)^{n-1}dx = 2\int_0^{\pi/2} \cos^{2m-1}\theta \sin^{2n-1}\theta d\theta = \frac{\Gamma(m).\Gamma(n)}{\Gamma(m+n)} \qquad ...(1)$$

Let $2m = p$ and $2n = q$. So that $m = p/2$ and $n = q/2$
Put in equation (1), we get

$$\int_0^{\pi/2} \sin^{p-1}\theta \cos^{q-1}\theta d\theta = \frac{\Gamma(p/2)\Gamma(q/2)}{2\Gamma\left(\frac{p+q}{2}\right)}.$$

Deduction 6. $\int_0^{\pi/2} \sin^{p-1}\theta d\theta = \int_0^{\pi/2} \cos^{p-1}\theta d\theta = \dfrac{\Gamma\left(\frac{p}{2}\right)\Gamma\left(\frac{1}{2}\right)}{2\Gamma\left(\frac{p+1}{2}\right)} = \dfrac{\sqrt{\pi}}{2}\dfrac{\Gamma(p/2)}{\Gamma\left(\frac{p+1}{2}\right)}$. $\quad \left(\because \Gamma\left(\frac{1}{2}\right) = \sqrt{\pi}\right)$

Proof. Replacing q by 1 in deduction 5, we get

$$\int_0^{\pi/2} \sin^{p-1}\theta d\theta = \frac{\Gamma\left(\frac{p}{2}\right)\Gamma\left(\frac{1}{2}\right)}{2\Gamma\left(\frac{p+1}{2}\right)}.$$

Next, replacing p by 1 and q by p in equation of deduction 5, we have

$$\int_0^{\pi/2} \cos^{p-1}\theta d\theta = \frac{\Gamma\left(\frac{1}{2}\right)\Gamma\left(\frac{p}{2}\right)}{2\Gamma\left(\frac{1+p}{2}\right)} = \frac{\sqrt{\pi}}{2}.\frac{\Gamma\left(\frac{p}{2}\right)}{\Gamma\left(\frac{p+1}{2}\right)}$$

Solved Examples

Example 1. *Evaluate the following integrals*:

(i) $\int_0^1 x^4(1-x^2)dx$ (ii) $\int_0^a y^4\sqrt{a^2-y^2}\,dy$

(iii) $\int_0^2 x(8-x^3)^{1/3}dx$ (iv) $\int_0^\infty \frac{x\,dx}{1+x^6}$

Solution. (i) We have

$$\int_0^1 x^4(1-x)^2dx$$

$$=\int_0^1 x^{5-1}(1-x)^{3-1}dx$$

$$=\frac{\Gamma(5)\Gamma(3)}{\Gamma(5+3)}=\frac{4!2!}{7!}$$

$$=\frac{4!\times 2}{7\times 5\times 4!\times 6}=\frac{1}{105}.$$

(ii) $\int_0^a y^4\sqrt{a^2-y^2}\,dy$.

Let $y^2 = a^2 t$ so that $dy=\frac{a^2\,dt}{2y}=\frac{a\,dt}{2\sqrt{t}}$, then

$$I=\int_0^1 (a^2t)^2\sqrt{(a^2-ta^2)}\left(\frac{a\,dt}{2\sqrt{t}}\right)$$

$$=\frac{a^6}{2}\int_0^1 t^{3/2}(1-t)^{1/2}dt$$

$$=\frac{a^6}{2}\int_0^1 t^{(5/2)-1}(1-t)^{(3/2)-1}dt$$

$$=\frac{a^6}{2}\frac{\Gamma\left(\frac{5}{2}\right)\Gamma\left(\frac{3}{2}\right)}{\Gamma\left(\frac{5}{2}+\frac{3}{2}\right)}$$

$$=\frac{a^6}{2}\frac{\frac{3}{2}.\frac{1}{2}\sqrt{\pi}.\frac{1}{2}\sqrt{\pi}}{3!}=\frac{\pi a^6}{32}.$$

(iii) Let $\int_0^2 x(8-x^3)^{1/3}\,dx = I$

Put $x^3 = 8t$ or $x = 2t^{1/3}$ so that $dx=\frac{2}{3}t^{-2/3}dt.$, we get

$$I=\int_0^1 (2t^{1/3})(8-8t)^{1/3}\left(\frac{2}{3}t^{-2/3}\,dt\right)$$

$$=\frac{8}{3}\frac{\Gamma\left(\frac{2}{3}\right)\Gamma\left(\frac{4}{3}\right)}{\Gamma\left(\frac{2}{3}+\frac{4}{3}\right)}=\frac{8}{3}\frac{\Gamma\left(1-\frac{1}{3}\right)\Gamma\left(1+\frac{1}{3}\right)}{\Gamma(2)}$$

$$=\frac{8}{3}\Gamma\left(1-\frac{1}{3}\right).\frac{1}{3}\Gamma\left(\frac{1}{3}\right)$$

$$=\frac{8}{9}\frac{\pi}{\sin\pi/3}=\frac{16\pi}{2\sqrt{3}}.$$

$(\because \Gamma 2 = 1! = 1, \Gamma(1+p) = p\Gamma(p)$ and $\Gamma(1-n)\Gamma n=\frac{\pi}{\sin n\pi})$

(iv) Let $I=\int_0^\infty \frac{x\,dx}{1+x^6}$.

Put $x^6 = y$ or $x = y^{1/6}$.

$$\Rightarrow dx=\frac{1}{6}.y^{-5/6}dy$$

$$\therefore \quad I=\frac{1}{6}\int_0^\infty \frac{y^{1/6}.y^{-5/6}}{1+y}dy$$

$$=\frac{1}{6}\int_0^\infty \frac{y^{-2/3}}{1+y}dy$$

$$=\frac{1}{6}\int_0^\infty \frac{y^{(1/3)}-1}{(1+y)^{2/3+1/3}}dy$$

$$=\frac{1}{6}B\left(\frac{1}{3},\frac{2}{3}\right)=\frac{1}{6}\frac{\Gamma\left(\frac{1}{3}\right)\Gamma\left(\frac{2}{3}\right)}{\Gamma\left(\frac{1}{3}+\frac{2}{3}\right)}$$

$$=\frac{1}{6}\frac{\Gamma\left(\frac{1}{3}\right)\Gamma\left(1-\frac{1}{3}\right)}{\Gamma 1}=\frac{1}{6}\frac{\pi}{\sin\frac{\pi}{3}}$$

$$\left[\because \Gamma(n)\Gamma(1-n)=\frac{\pi}{\sin n\pi}\right]$$

$$=\frac{1}{6}.\frac{\pi}{(\sqrt{3}/2)}=\frac{1}{6}\frac{2\pi}{\sqrt{3}}=\frac{\pi}{3\sqrt{3}}.$$

Example 2. *Prove that* $\int_0^2 (8-x^3)^{-1/3}dx=\frac{2\pi}{3\sqrt{3}}$

Solution. Let $x^3 = 8t$, then $x = 2t^{1/3}$

$$\Rightarrow \quad dx=\frac{2}{3}t^{-2/3}dt$$

and when $x = 0$ to $x = 2$, $t = 0$ to $t = 1$

So

$$\int_0^2 (8-x^3)^{-1/3}dx=\int_0^1 (8-8t)^{-1/3}.\frac{2}{3}t^{-2/3}dt$$

$$=(8)^{-1/3}.\frac{2}{3}\int_0^1 t^{-2/3}(1-t)^{-1/3}dt$$

$$=\frac{1}{3}\int_0^1 t^{(1/3)-1}(1-t)^{(2/3)-1}dt$$

$$=\frac{1}{3}B\left(\frac{1}{3},\frac{2}{3}\right)=\frac{1}{3}\frac{\Gamma\left(\frac{1}{3}\right)\Gamma\left(\frac{2}{3}\right)}{\Gamma\left(\frac{1}{3}+\frac{2}{3}\right)}$$

$$=\frac{1}{3}\frac{\Gamma\left(\frac{1}{3}\right)\Gamma\left(1-\frac{1}{3}\right)}{\Gamma(1)}=\frac{1}{3}\frac{\pi}{\sin\frac{\pi}{3}}=\frac{2\pi}{3\sqrt{3}}.$$

Example 3. *Show that* $\int_0^1 \frac{dx}{(1-x^n)^{1/2}}=\frac{\sqrt{\pi}\,\Gamma(1/n)}{n\Gamma\left(\frac{1}{n}+\frac{1}{2}\right)}.$

Solution. Let $x^n = \sin^2\theta \Rightarrow x = \sin^{2/n}\theta$

so that $dx = 2.\frac{1}{n}.\sin^{\left(\frac{2}{n}-1\right)}\theta\cos\theta\,d\theta$

then $\int_0^1 \frac{dx}{\sqrt{(1-x^n)}}$

$$= \frac{2}{n}\int_0^{\pi/2} \frac{\sin^{(2/n)-1}\theta\cos\theta\,d\theta}{\cos\theta}$$

$$= \frac{2}{n}\int_0^{\pi/2} \sin^{(2/n)-1}\theta\cos^0\theta\,d\theta$$

$$= \frac{2}{n}.\frac{\Gamma\left(\frac{1}{n}\right)\Gamma\left(\frac{1}{2}\right)}{2\Gamma\left(\frac{1}{n}+\frac{1}{2}\right)} = \frac{\sqrt{\pi}}{n}.\frac{\Gamma\left(\frac{1}{n}\right)}{\Gamma\left(\frac{1}{n}+\frac{1}{2}\right)}.$$

Example 4. *Evaluate* $\int_0^\infty \frac{x^8(1-x^6)}{(1+x)^{24}}dx$.

Solution. We have $\int_0^\infty \frac{x^8(1-x^6)}{(1+x)^{24}}dx$

$$= \int_0^\infty \frac{x^8dx}{(1+x)^{24}} - \int_0^\infty \frac{x^{14}}{(1+x)^{24}}dx$$

$$= \int_0^\infty \frac{x^{9-1}}{(1+x)^{9+15}}dx - \int_0^\infty \frac{x^{15-1}}{(1+x)^{15+9}}dx$$

$$= B(9, 15) - B(15, 9) = 0$$

$$[\because B(9, 15) = B(15, 9)]$$

Example 5. *Prove that* $\int_0^\infty \frac{y^{n-1}}{1+x}dx = \frac{\pi}{\sin n\pi}$.

Solution. We know that

$$B(m,n) = \int_0^\infty \frac{x^{n-1}}{(1+x)^{m+n}}dx$$

Put $m = 1-n$, we get

$$B(1-n,n) = \int_0^\infty \frac{x^{n-1}}{1+x}dx$$

$$\Rightarrow \quad \int_0^\infty \frac{x^{n-1}}{1+x}dx = B(n,1-n)$$

$$[\because B(m, n) = B(n, m)]$$

$$= \frac{\Gamma(n)\Gamma(1-n)}{\Gamma(n+1-n)} = \Gamma(n)\Gamma(1-n)$$

$$= \frac{\pi}{\sin n\pi}.$$

Example 6. *Prove that*

(a) $\int_0^{\pi/2} \sqrt{\tan\theta}\,d\theta = \frac{1}{2}\Gamma\left(\frac{1}{4}\right)\Gamma\left(\frac{3}{4}\right) = \frac{\pi\sqrt{2}}{2}$.

(UKTU-2011)

(b) $\int_0^{\pi/2} \tan^n x\,dx = \frac{\pi}{2}\sec\frac{n\pi}{2}, -1 < n < 1.$

Solution. (a) We have

$$\int_0^{\pi/2} \sqrt{\tan\theta}\,d\theta = \int_0^{\pi/2} \left(\frac{\sin\theta}{\cos\theta}\right)^{1/2} d\theta$$

$$= \int_0^{\pi/2} \sin^{1/2}\theta\cos^{-1/2}\theta\,d\theta$$

$$= \frac{\Gamma\left(\frac{1+\frac{1}{2}}{2}\right)\Gamma\left(\frac{1-\frac{1}{2}}{2}\right)}{2\Gamma\left(\frac{\frac{1}{2}-\frac{1}{2}+2}{2}\right)}$$

$$\because \int_0^{\pi/2} \sin^n\theta\cos^m\theta\,d\theta$$

$$\left[= \frac{\Gamma\left(\frac{n+1}{2}\right)\Gamma\left(\frac{m+1}{2}\right)}{2\Gamma\left(\frac{n+m+2}{2}\right)}, n > -1, m > -1\right]$$

$$= \frac{\Gamma\left(\frac{3}{4}\right)\Gamma\left(\frac{1}{4}\right)}{2\Gamma(1)} = \frac{1}{2}\Gamma\left(\frac{3}{4}\right)\Gamma\left(\frac{1}{4}\right)$$

$$= \frac{1}{2}\Gamma\left(\frac{1}{4}\right)\Gamma\left(1-\frac{1}{4}\right) = \frac{1}{2}\frac{\pi}{\sin\left(\frac{\pi}{4}\right)} = \frac{\pi\sqrt{2}}{2}.$$

(b) Consider L.H.S.

$$\int_0^{\pi/2} \tan^n x\,dx = \int_0^{\pi/2} \sin^n x\cos^{-n} x\,dx$$

$$= \frac{\Gamma\left(\frac{1+n}{2}\right)\Gamma\left(\frac{1-n}{2}\right)}{2\Gamma\left(\frac{n-n+2}{2}\right)}$$

$$\left[\begin{array}{l}\text{Here } \frac{1+n}{2} > 0, \ \frac{1-n}{2} > 0 \\ \Rightarrow \quad n > -1 \text{ and } n < 1\end{array}\right]$$

$$= \frac{1}{2}\Gamma\left(\frac{1+n}{2}\right)\Gamma\left(1-\frac{1+n}{2}\right)$$

$$= \frac{1}{2}\frac{\pi}{\sin\left(\frac{1+n}{2}\right)\pi} = \frac{\pi}{2\sin\left(\frac{\pi}{2}+\frac{n\pi}{2}\right)} = \frac{\pi}{2\cos\frac{n\pi}{2}}$$

$$= \frac{\pi}{2}\sec\frac{n\pi}{2}, \quad \text{where } -1 < n < 1.$$

Example 7. *Evaluate* $\int_{-1}^1 \left(\frac{1+x}{1-x}\right)^{1/2} dx$.

Solution. Le $I = \int_{-1}^1 \left(\frac{1+x}{1-x}\right)^{1/2} dx \quad \text{...(1)}$

Putting $t = \frac{1}{2}(1+x)$ so that $x = 2t-1$, $dx = 2\,dt$ in (1), we get

$$I = \int_0^1 \left(\frac{1+2t-1}{1-2t+1}\right)^{1/2} 2dt = 2\int_0^1 \left(\frac{t}{1-t}\right)^{1/2} dt$$

$$= 2\int_0^1 t^{(3/2)-1}(1-t)^{(1/2)-1} dt = 2B\left(\frac{3}{2},\frac{1}{2}\right)$$

$$= \frac{2\Gamma\left(\frac{3}{2}\right)\Gamma\left(\frac{1}{2}\right)}{\Gamma\left(\frac{3}{2}+\frac{1}{2}\right)} = 2\frac{\frac{1}{2}\sqrt{\pi}\sqrt{\pi}}{\Gamma(2)} = \pi.$$

Example 8. *Simplify* $\int_0^1 \frac{35x^3\,dx}{32\sqrt{1-x}}$.

Solution. Let $I = \int_0^1 \frac{35x^3\,dx}{32\sqrt{1-x}}$ ...(1)

Putting $x = \sin^2\theta$ so that $dx = 2\sin\theta\cos\theta\,d\theta$ then (1) gives

$$I = \int_0^{\pi/2} \frac{35\sin^6\theta\, 2\sin\theta\cos\theta}{32\cos\theta}$$

$$= \frac{35}{16}\int_0^{\pi/2} \sin^7\theta\,d\theta$$

$$= \frac{35}{16}\cdot\frac{6}{7}\cdot\frac{4}{5}\cdot\frac{2}{3} = 1.$$

Example 9. *If $p > 0$, $q > 0$, $m+1 > 0$, $n+1 > 0$, then prove that*

$$\int_0^p x^m(p^q - x^q)^n\,dx = \frac{p^{nq+m+1}}{q}.B\left(n+1, \frac{m+1}{q}\right)$$

Solution. Let $I = \int_0^p x^m(p^q - x^q)^n\,dx$ ...(1)

Putting $x^q = p^q t \Rightarrow dx = \left(\frac{p}{q}\right)(t)^{(1/q)-1}dt$ in (1) then we have

$$I = \int_0^1 (pq^{1/q})^m(p^q - p^q t)^n \left(\frac{p}{q}\right) t^{1/q-1} dt$$

$$= \frac{p^m.p^{nq}.p}{q}\int_0^1 t^{(m/q)+(1/q)-1}(1-t)^{(n+1)-1}dt$$

$$= \frac{p^{nq+m+1}}{q} B\left(\frac{m+1}{q}, n+1\right)$$

$$= \frac{p^{nq+m+1}}{q} B\left(n+1, \frac{m+1}{q}\right).$$

Example 10. *Show that* $B(n,n+1) = \frac{1}{2}\frac{\Gamma(n)^2}{\Gamma(2n)}$ *and hence deduce that*

$$\int_0^{\pi/2}\left(\frac{1}{\sin^3\theta} - \frac{1}{\sin^2\theta}\right)^{1/4}\cos\theta\,d\theta = \frac{\left\{\Gamma\left(\frac{1}{4}\right)\right\}^2}{2\sqrt{\pi}}.$$

Solution. We have

$$B(n,n+1) = \frac{\Gamma(n)\Gamma(n+1)}{\Gamma(n+n+1)} = \frac{\Gamma(n).n\Gamma n}{(2n)\Gamma(2n)}$$

$$\therefore B(n,n+1) = \frac{1}{2}\frac{\{\Gamma(n)\}^2}{\Gamma(2n)} \quad [\because \Gamma(p+1) = p\Gamma(p)]$$

Let $I = \int_0^{\pi/2}\left(\frac{1}{\sin^3\theta} - \frac{1}{\sin^2\theta}\right)^{1/4}\cos\theta\,d\theta$ ...(2)

Putting $x = \sin\theta$, so that $dx = \cos\theta\,d\theta$ in (2), we get

$$I = \int_0^1\left(\frac{1}{x^3} - \frac{1}{x^2}\right)^{1/4} dx = \int_0^1\left(\frac{1-x}{x^3}\right)^{1/4} dx$$

$$= \int_0^1 x^{-3/4}(1-x)^{1/4}dx$$

$$= \int_0^1 x^{(1/4)-1}(1-x)^{(5/4)-1}dx$$

$$= B\left(\frac{1}{4},\frac{5}{4}\right)$$

$$= B\left(\frac{1}{4},\frac{1}{4}+1\right) = \frac{1}{2}\frac{\left\{\Gamma\left(\frac{1}{4}\right)\right\}^2}{\Gamma\left(\frac{1}{2}\right)} = \frac{\left\{\Gamma\left(\frac{1}{4}\right)\right\}^2}{2\sqrt{\pi}}.$$

Example 11. *Show that* $I = \int_0^{\pi/2}\sqrt{\sin n\theta}\,d\theta.\int_0^{\pi/2}\frac{d\theta}{\sqrt{\sin\theta}} = \pi.$.

Solution. We know that

$$\int_0^{\pi/2}\sin^p d\theta = \frac{\Gamma\left(\frac{p+1}{2}\right)\sqrt{\pi}}{2\Gamma\left(\frac{p+2}{2}\right)} \quad ...(1)$$

Now, $I = \int_0^{\pi/2}\sin^{1/2}\theta.d\theta.\int_0^{\pi/2}\sin^{-1/2}\theta\,d\theta$

$$= \frac{\Gamma\left(\frac{1/2+1}{2}\right)\sqrt{\pi}.\Gamma\left(\frac{-1/2+1}{2}\right)\sqrt{\pi}}{2\Gamma\left(\frac{1/2+2}{2}\right).2\Gamma\left(\frac{-1/2+2}{2}\right)}$$

$$= \frac{\Gamma\left(\frac{3}{4}\right)\sqrt{\pi}}{2\Gamma\left(\frac{5}{4}\right)}.\frac{\Gamma\left(\frac{1}{4}\right)\sqrt{\pi}}{2\Gamma\left(\frac{3}{4}\right)} = \frac{\pi\Gamma\left(\frac{1}{4}\right)}{4\Gamma\left(1+\frac{1}{4}\right)}$$

$$= \frac{\pi\Gamma\left(\frac{1}{4}\right)}{4.\frac{1}{4}.\Gamma\left(\frac{1}{4}\right)} = \pi.$$

Example 12. Prove that

(a) $\int_0^{\pi}\frac{\sin^{n-1}x\,dx}{(a+b\cos x)^n} = \frac{2^{n-1}}{(a^2-b^2)^{n/2}}B\left(\frac{n}{2},\frac{n}{2}\right).$

(b) $\int_0^{\pi}\frac{\sqrt{\sin x}}{[5+3\cos x]^{3/2}} = \frac{\left[\Gamma\left(\frac{3}{4}\right)\right]^2}{2\sqrt{2}\pi}.$

Solution. (a) Let

$$I = \int_0^{\pi}\frac{\sin^{n-1}x\,dx}{(a+b\cos x)^n} = \int_0^{\pi}\frac{(\sin x)^{n-1}dx}{(a+b\cos x)^n} \quad ...(1)$$

$$= \int_0^{\pi} \frac{\left(2\sin\frac{x}{2}\cos\frac{x}{2}\right)^{n-1} dx}{\left[\begin{array}{l} a\left\{\cos^2\left(\frac{x}{2}\right)+\sin^2\left(\frac{x}{2}\right)\right\} \\ +b\left\{\cos^2\left(\frac{x}{2}\right)-\sin^2\left(\frac{x}{2}\right)\right\}\end{array}\right]^n}$$

[by (1)]

$$= 2^{n-1}\int_0^{\pi} \frac{\sin^{n-1}\left(\frac{x}{2}\right)\cos^{n-1}\left(\frac{x}{2}\right)dx}{\left[(a+b)\cos^2\left(\frac{x}{2}\right)+(a-b)\sin^2\left(\frac{x}{2}\right)\right]^n}$$

$$= \frac{2^{n-1}}{(a+b)^n}\int_0^{\pi} \frac{\sin^{n-1}\left(\frac{x}{2}\right)\cos^{n-1}\left(\frac{x}{2}\right)dx}{\cos^{2n}\left(\frac{x}{2}\right)\left[1+\frac{a-b}{a+b}\tan^2\frac{x}{2}\right]^n}$$

$$= \frac{2^{n-1}}{(a+b)^n}\int_0^{\infty} \frac{\tan^{n-1}\left(\frac{x}{2}\right)\sec^2\left(\frac{x}{2}\right)dx}{\left[1+\frac{a-b}{a+b}\tan^2\frac{x}{2}\right]^n}$$

$$= \frac{2^{n-1}}{(a+b)^n}\int_0^{\infty} \frac{\left[\frac{a+b}{a-b}t\right]^{\frac{(n-2)}{2}} \cdot \frac{a+b}{a-b}dt}{(1+t)^n}$$

$$\left[\begin{array}{l}\text{Put } \frac{a-b}{a+b}\tan^2\frac{x}{2}=t, \\ i.e.,\ 2\frac{a-b}{a+b}\tan\frac{x}{2}\sec^2\frac{x}{2}\cdot\frac{dx}{2}=dt\end{array}\right]$$

$$I = \frac{2^{n-1}}{[(a+b)(a-b)]^{n/2}}\int_0^{\infty}\frac{t^{(n/2)-1}}{(1+t)^{n/2+n/2}}dt$$

$$= \frac{2^{n-1}}{(a^2-b^2)^{n/2}}B\left(\frac{n}{2},\frac{n}{2}\right).$$

(b) Taking $n = 3/2$, $a = 5$ and $b = 3$, in part (a), we get

$$\int_0^{\pi}\frac{\sin^{(3/2)-1}x\,dx}{(5+3\cos x)^{3/2}} = \frac{(2)^{(3/2)-1}}{(25-9)^{3/4}}B\left(\frac{3}{4},\frac{3}{4}\right)$$

$$= \frac{\sqrt{2}}{2^3}\cdot\frac{\Gamma\left(\frac{3}{4}\right)\Gamma\left(\frac{3}{4}\right)}{\Gamma\left(\frac{3}{4}+\frac{3}{4}\right)} = \frac{\sqrt{2}\left\{\Gamma\left(\frac{3}{4}\right)\right\}^2}{8\Gamma\left(\frac{3}{2}\right)}$$

$$= \frac{\sqrt{2}\left\{\Gamma\left(\frac{3}{4}\right)\right\}^2}{8.\frac{1}{2}\sqrt{\pi}} = \frac{\left\{\Gamma\left(\frac{3}{4}\right)\right\}^2}{2\sqrt{2\pi}}.$$

EXERCISE 20.1

1. Show that $\int_0^{\infty} e^{-4x}x^{3/2}\,dx = \frac{3\sqrt{\pi}}{128}$.

2. Show that $\int_0^{\infty} e^{-x^2}.\, x^2\,dx = \frac{\sqrt{\pi}}{4}$.

3. Show that $\int_0^1 \frac{dx}{\sqrt{(-\log x)}} = \sqrt{\pi}$.

4. Show that $\Gamma\left(-\frac{15}{2}\right) = \frac{2^8\sqrt{\pi}}{1.3.5.7.9.11.13.15}$.

5. Show that $\int_0^1 x^{n-1}\left(\log\frac{1}{x}\right)^{m-1}dx = \frac{\Gamma(m)}{n^m}$ $m > 0, n > 0$.

6. Show that $\int_0^{\pi/2}\sin^m\theta\cos^n\theta\,d\theta = \frac{\Gamma\left(\frac{m+1}{2}\right)\Gamma\left(\frac{n+1}{2}\right)}{2\Gamma\left(\frac{m+n+2}{2}\right)}$
$= \frac{1}{2}B\left(\frac{m+1}{2},\frac{n+1}{2}\right)$.

7. Show that $\int_0^{\infty} 3^{-4x^2}dx = \frac{\sqrt{\pi}}{4\sqrt{(\log 3)}}$.

8. Show that
$\int_0^a (a-x)^{m-1}x^{n-1}dx = \frac{a^{m+n-1}\Gamma(m)\Gamma(n)}{\Gamma(m+n)}$.

9. Show that
$\int_0^{\pi/2}\sin^p\theta\cos^q\theta\,d\theta = \frac{1}{2}B\left(\frac{p+1}{2},\frac{q+1}{2}\right)$, $p > -1, q > -1$.
Deduce that $\int_0^2 x^4(8-x^3)^{-1/3}dx = \frac{16}{3}B\left(\frac{5}{3},\frac{2}{3}\right)$.

10. Show that
$\int_0^1\left(\frac{1}{x}-1\right)^{1/4}dx = B\left(\frac{5}{4},\frac{3}{4}\right) = \frac{\pi}{2\sqrt{2}}$.

11. Prove that
(i) $B(l, m)\, B(l+m, n) = B(m, n)B(m+n, l)$
$= B(n, l)\, B(n+l, m)$
(ii) $B(l, m)B(l+m, n)\, B(l+m+n, p) = \frac{\Gamma(l)\Gamma(m)\Gamma(n)\Gamma(p)}{\Gamma(l+m+n+p)}$.

12. Show that
$\int_0^1 x^m(1-x^n)^p dx = \frac{1}{n}B\left(\frac{m+1}{n},p+1\right)$.

13. Show that $\int_0^1\left(1-x^n\right)^{1/n}dx = \frac{1[\Gamma(1/n)]^2}{n\,2\Gamma(2/n)}$.

14. Show that $\int_0^1\frac{x^2dx}{(1-x^4)^{1/2}}\times\int_0^1\frac{dx}{(1+x^4)^{1/2}} = \frac{\pi}{4\sqrt{2}}$.

15. Show that $B(m,n) = B(m+1, n) + B(m, n+1)$ for $m > 0, n > 0$.

16. Prove the following

(i) $\int_0^\infty x^{1/4}e^{-\sqrt{x}}dx = \frac{3}{2}\sqrt{\pi}$ (UPTU-2008)

(ii) $\int_0^1 \left(\frac{x^3}{1-x^3}\right)^{1/2} dx = \sqrt{3\pi}.\frac{\Gamma(2/3)}{\Gamma(1/6)}$ (UPTU-2007)

(iii) $\int_0^1 x^5(1-x^3)^{10}dx = \frac{1}{396}$ (UPTU-2008)

Hint to Selected Problems

1. $I = \int_0^\infty e^{-4x}x^{3/2}dx = \int_0^\infty e^{-4x}x^{(5/2)-1}dx$

$$= \frac{\Gamma\left(\frac{5}{2}\right)}{4^{5/2}} = \frac{\frac{3}{2}\cdot\frac{1}{2}\sqrt{\pi}}{2^5} = \frac{3\sqrt{\pi}}{128}.$$

3. $I = \int_0^1 \frac{dx}{\sqrt{\log\frac{1}{x}}} = \int_0^1 \left[\log\left(\frac{1}{x}\right)\right]^{-1/2} dx \cdot$

Now put $\log(1/x) = t$.

5. $I = \int_0^1 x^{n-1}\left(\log\frac{1}{x}\right)^{m-1} dx.$

Put $\frac{1}{x} = t, i.e., x = e^{-t} \Rightarrow dx = -e^{-t}\, dt.$

6. The given integral can be written as

$$I = \int_0^{\pi/2} (\sin^2\theta)^{(m-1)/2}(1-\sin^2)^{(n-1)/2}\sin\theta\cos\theta\, d\theta$$

Now put $\sin^2\theta = x$ i.e., $dx = 2\sin\theta\cos\theta\, d\theta$.

7. Put $3^{-4x^2} = e^{-t}$ i.e., $x = \frac{\sqrt{t}}{2\sqrt{\log 3}}$.

8. Put $x = at$ in the LHS.

9. Do same as (6).

12. Put $x^n = t$ i.e., $x = (t)^{1/n} \Rightarrow dx = \frac{1}{n}(t)^{1/n-1}dt.$

20.8 DUPLICATION FORMULA

We have

$$\Gamma(n)\Gamma\left(n+\frac{1}{2}\right) = \frac{\sqrt{\pi}}{2^{2n-1}}\Gamma(2n),\ n > 0.$$

Proof. We know that $$B(m,n) = \frac{\Gamma(m)\Gamma(n)}{\Gamma(m+n)} \text{ where } m > 0, n > 0. \quad ...(1)$$

Now putting $m = n$ in equation (1), we get

$$B(n,n) = \frac{[\Gamma(n)]^2}{\Gamma(2n)}. \quad ...(2)$$

By definition of Beta function, we get

$$B(n,n) = \int_0^1 x^{n-1}(1-x)^{n-1}dx.$$

Putting $x = \sin^2\theta$ so that $dx = 2\sin\theta\cos\theta\, d\theta$ in (1), we get

$$B(n,n) = \int_0^{\pi/2} (\sin^2\theta)^{n-1}(1-\sin^2\theta)^{n-1}.2\sin\theta\cos\theta\, d\theta$$

$$= 2\int_0^{\pi/2} (\sin\theta\cos\theta)^{2n-1}d\theta = 2\int_0^{\pi/2}\left(\frac{\sin 2\theta}{2}\right)^{2n-1} d\theta$$

$$= \frac{1}{2^{2n-2}}\int_0^{\pi/2} \sin^{2n-1} 2\theta\, d\theta$$

$$= \frac{1}{2^{2n-2}}\int_0^{\pi} \sin^{2n-1}\phi\frac{d\phi}{2} \text{ (By putting } 2\theta=\phi \Rightarrow d\theta = \frac{1}{2}d\phi)$$

$$= \frac{1}{2^{2n-1}}\int_0^{\pi} \sin^{2n-1}\phi\, d\phi = \frac{1}{2^{2n-2}}\int_0^{\pi/2} \sin^{2n-1}\phi\, d\phi$$

$$\left[\because \int_0^{2a} f(x)dx = 2\int_0^a f(x)dx \text{ when } f(2a-x) = f(x)\right]$$

$$= \frac{1}{2^{2n-2}}\int_0^{\pi/2} \sin^{2n-1}\phi(\cos\phi)^0 d\phi$$

$$= \frac{1}{2^{2n-2}} \frac{\Gamma\left(\frac{2n-1+1}{2}\right)\Gamma\left(\frac{0+1}{2}\right)}{2\Gamma\left(\frac{2n-1+0+2}{2}\right)}$$

$$\therefore \qquad B(n,n) = \frac{1}{2^{2n-1}} \cdot \frac{\Gamma(n)\sqrt{\pi}}{\Gamma\left(n+\frac{1}{2}\right)} \text{ as } \Gamma\left(\frac{1}{2}\right) = \sqrt{\pi}. \qquad \ldots(3)$$

Equating two values of $B(n, n)$ given by (2) and (3), we obtain

$$\frac{[\Gamma(n)]^2}{\Gamma(2n)} = \frac{1}{2^{2n-1}} \frac{\Gamma(n)\sqrt{\pi}}{\Gamma\left(n+\frac{1}{2}\right)}$$

or

$$\Gamma(n)\Gamma\left(n+\frac{1}{2}\right) = \frac{\sqrt{\pi}}{2^{2n-1}}\Gamma(2n). \qquad \ldots(4)$$

Deduction 1. *For all positive real value of p, we have* $2^p\Gamma\left(\frac{p+1}{2}\right)\Gamma\left(\frac{p+2}{2}\right) = \sqrt{\pi}\,\Gamma(p+1)$

Proof. We know that

$$\Gamma(n)\Gamma\left(n+\frac{1}{2}\right) = \frac{\sqrt{\pi}}{2^{2n-1}}\Gamma(2n). \qquad \ldots(1)$$

Putting $2n - 1 = p$, so that $n = \frac{1}{2}(p+1)$ in equation (1), we get

$$\Gamma\left(\frac{p+1}{2}\right)\Gamma\left(\frac{p+1}{2}+\frac{1}{2}\right) = \frac{\sqrt{\pi}}{2^p}\Gamma(p+1) \Rightarrow \quad 2^p\Gamma\left(\frac{p+1}{2}\right)\Gamma\left(\frac{p+2}{2}\right) = \sqrt{\pi}\,\Gamma(p+1).$$

Deduction 2. *For any positive integer n, we have* $\Gamma\left(n+\frac{1}{2}\right) = \frac{(2n)!}{2^{2n}.n!}\sqrt{\pi}.$

Proof. Let n be positive integer, then we have

$$\frac{\Gamma(2n)}{\Gamma(n)} = \frac{(2n-1)!}{(n-1)!} = \frac{(2n)(2n-1)!}{2.n.(n-1)!} = \frac{(2n)!}{2.(n)!} \qquad \ldots(1)$$

Now, from the duplication formula (4), we have

$$\Gamma\left(n+\frac{1}{2}\right) = \frac{\sqrt{\pi}}{2^{2n-1}} \cdot \frac{\Gamma(2n)}{\Gamma(n)} = \frac{\sqrt{\pi}}{2^{2n-1}} \cdot \frac{(2n)!}{2.n!} \qquad \text{[By (1)]}$$

$$= \frac{(2n)!}{2^{2n}.\,n!}\sqrt{\pi}.$$

Deduction 3. *For any integer n, we have* $\Gamma\left(\frac{1}{n}\right)\Gamma\left(\frac{2}{n}\right)\Gamma\left(\frac{3}{n}\right)\ldots\Gamma\left(\frac{n-1}{n}\right) = \frac{(2\pi)^{(n-1)/2}}{n^{1/2}}.$

Proof. Let

$$X = \Gamma\left(\frac{1}{n}\right)\Gamma\left(\frac{2}{n}\right)\Gamma\left(\frac{3}{n}\right)\ldots\Gamma\left(\frac{n-2}{n}\right)\Gamma\left(\frac{n-1}{n}\right). \qquad \ldots(1)$$

Writing the above expression in the reversed order, we get

$$X = \Gamma\left(\frac{n-1}{n}\right)\Gamma\left(\frac{n-2}{n}\right)\ldots\Gamma\left(\frac{2}{n}\right)\Gamma\left(\frac{1}{n}\right)$$

$$X = \Gamma\left(1-\frac{1}{n}\right)\Gamma\left(1-\frac{2}{n}\right)\ldots\Gamma\left(1-\frac{n-2}{2}\right)\Gamma\left(1-\frac{n-1}{n}\right). \qquad \ldots(2)$$

Multiplying (1) and (2) and arranging in products of terms in the $\Gamma(n)\,\Gamma(1-n)$, we have

$$X^2 = \left[\Gamma\left(\frac{1}{n}\right)\Gamma\left(1-\frac{1}{n}\right)\right]\cdot\left[\Gamma\left(\frac{2}{n}\right)\Gamma\left(1-\frac{2}{n}\right)\right]\ldots.$$

$$\ldots\left[\Gamma\left(\frac{n-2}{n}\right)\Gamma\left(1-\frac{n-2}{n}\right)\right]\left[\Gamma\left(\frac{n-1}{n}\right)\Gamma\left(1-\frac{n-1}{n}\right)\right]$$

$$= \frac{\pi.\pi}{\sin\frac{\pi}{n}.\sin\frac{2\pi}{n}} \cdots \frac{\pi}{\sin\frac{n-2}{n}\pi\sin\frac{n-1}{n}\pi} \quad \left[\because \Gamma(m)\Gamma(1-m) = \frac{\pi}{\sin m\pi}\right]$$

$$\therefore \qquad X^2 = \frac{\pi^{n-1}}{\sin\left(\frac{\pi}{n}\right)\sin\left(\frac{2\pi}{n}\right)\ldots\sin\left(\frac{(n-1)\pi}{n}\right)}$$

Now, using the following trigonometrical identity :

$$2^{n-1}\sin\left(\theta+\frac{\pi}{n}\right)\sin\left(\theta+\frac{2\pi}{n}\right)\sin\left(\theta+\frac{n-1}{n}\pi\right) = \frac{\sin n\theta}{\sin\theta} \qquad \ldots(4)$$

and, taking limit as $\theta \to 0$, equation (4) gives

$$2^{n-1}\sin\frac{\pi}{n}\sin\frac{2\pi}{n}\ldots\sin\left(\frac{n-1}{n}\pi\right) = \lim_{\theta\to 0}\frac{\sin n\theta}{\sin\theta} = n\lim_{\theta\to 0}\left[\frac{\sin n\theta}{n\theta}.\frac{\theta}{\sin\theta}\right] = n.$$

$$\therefore \qquad \sin\left(\frac{\pi}{n}\right)\sin\left(\frac{2\pi}{n}\right)\ldots\sin\left\{\frac{(n-1)\pi}{n}\right\} = \frac{n}{2^{n-1}}. \qquad \ldots(5)$$

Using (5), (3) reduces to

$$X^2 = \frac{\pi^{n-1}}{(n/2)^{n-1}} = \frac{(2\pi)^{n-1}}{n} \text{ or } X = \frac{(2\pi)^{(n-1)/2}}{n^{1/2}} \qquad \ldots(6)$$

From (1) and (6), we get

$$\Gamma\left(\frac{1}{n}\right)\Gamma\left(\frac{2}{n}\right)\ldots\Gamma\left(\frac{n-1}{n}\right) = \frac{(2\pi)^{\frac{(n-1)}{2}}}{n^{1/2}}$$

Deduction 4. *To prove that*

(i) $\int_0^\infty e^{-ax}\cos bx\, x^{m-1}dx = \frac{\Gamma(m)}{r^m}\cos m\theta,\ r^2 = a^2 + b^2.$

(ii) $\int_0^\infty e^{-ax}\sin bx\, x^{m-1}dx = \frac{\Gamma(m)}{r^m}\sin m\theta$ *where* $r = (a^2+b^2)^{1/2}$ *and* $\theta = \tan^{-1}\left(\frac{b}{a}\right).$

Proof. We know that

$$\int_0^\infty e^{-kx}x^{m-1}dx = \frac{\Gamma(m)}{k^m},\ m > 0,\ k > 0 \qquad \ldots(1)$$

Putting $k = a - ib$ in both sides of (1), we get

$$\int_0^\infty e^{-(a-ib)x}x^{m-1}dx = \frac{\Gamma(m)}{(a-ib)^m}$$

or

$$\int_0^\infty e^{-ax}e^{ibx}x^{m-1}dx = \frac{\Gamma(m)(a+ib)^m}{[(a+ib)(a-ib)]^m}$$

$$\Rightarrow \qquad \int_0^\infty e^{-ax}x^{m-1}(\cos bx + i\sin bx)dx = \frac{\Gamma(m)(a+ib)^m}{(a^2+b^2)^m} \qquad \ldots(2)$$

Let $\qquad a + ib = r(\cos\theta + i\sin\theta) \qquad \ldots(3)$

Equating real and imaginary parts of both sides, we get

$$r^2 = a^2 + b^2, \qquad \tan\theta = \frac{b}{a} \qquad \ldots(4)$$

Now $\qquad (a+ib)^m = [r(\cos\theta + i\sin\theta)]^m \qquad$ [By (3)]

$$(a+ib)^m = r^m(\cos m\theta + i\sin m\theta) \qquad \ldots(5)$$

[By De'Moivre's theorem]

Using (4), (5) and (2) reduces to

$$\int_0^\infty e^{-ax}x^{m-1}(\cos bx + i\sin bx)dx = \frac{\Gamma(m)r^m(\cos m\theta + i\sin m\theta)}{r^{2m}}.$$

Equating real and imaginary parts of both sides, we get

$$\int_0^\infty e^{-ax} x^{m-1} \cos bx\, dx = \frac{\Gamma(m)}{r^m} \cos m\theta \quad \text{...(6)}$$

and
$$\int_0^\infty e^{-ax} x^{m-1} \sin bx\, dx = \frac{\Gamma(m)}{r^m} \sin m\theta. \quad \text{...(7)}$$

Deduction 5. Let $m = 1$, then $\Gamma(m) = \Gamma(1) = 1$, so (6), (7) reduces to

$$\int_0^\infty e^{-ax} \cos bx\, dx = \frac{\cos\theta}{r} \quad \text{...(8)}$$

$$\int_0^\infty e^{-ax} \sin bx\, dx = \frac{\sin\theta}{r}. \quad \text{...(9)}$$

But $\tan\theta = \frac{b}{a}$, so that $\sin\theta = \frac{b}{\sqrt{a^2+b^2}}$ and $\cos\theta = \frac{a}{\sqrt{(a^2+b^2)}}$

Also, $r^2 = (a^2 + b^2)$. Hence (8) and (9) becomes

$$\int_0^\infty e^{-ax} \cos bx\, dx = \frac{a}{a^2+b^2} \quad \text{...(10)}$$

and
$$\int_0^\infty e^{-ax} \sin bx\, dx = \frac{b}{a^2+b^2} \quad \text{...(11)}$$

Solved Examples

Example 1. *Express $\Gamma(1/6)$ in terms of $\Gamma(1/3)$.*

Solution. By duplication formula, we have

$$\Gamma(n)\Gamma\left(n+\frac{1}{2}\right) = \frac{\sqrt{\pi}}{2^{n-1}}\Gamma(2n) \quad \text{...(1)}$$

Put $n = 1/6$ in (1), we get

$$\Gamma\left(\frac{1}{6}\right)\Gamma\left(\frac{2}{3}\right) = \frac{\sqrt{\pi}\,\Gamma\left(\frac{1}{3}\right)}{2^{-2/3}}$$

$$\Rightarrow \quad \Gamma\left(\frac{1}{6}\right) = \frac{\sqrt{\pi}\,\Gamma\left(\frac{1}{3}\right)}{2^{-2/3}\Gamma\left(\frac{2}{3}\right)} \quad \text{...(2)}$$

Also, we know that

$$\Gamma(n)\Gamma(1-n) = \frac{\pi}{\sin n\pi}. \quad \text{...(3)}$$

Putting $n = 1/3$ in (3), we get

$$\Gamma\left(\frac{1}{3}\right)\Gamma\left(\frac{2}{3}\right) = \frac{\pi}{\sin(\pi/3)} = \frac{2\pi}{\sqrt{3}}$$

$$\Gamma\left(\frac{2}{3}\right) = \frac{2\pi}{\sqrt{3}\,\Gamma(1/3)} \quad \text{...(4)}$$

Substituting the value of $\Gamma(2/3)$ given by (4) in (2), we get

$$\Gamma\left(\frac{1}{6}\right) = \frac{\sqrt{\pi}\,\Gamma\left(\frac{1}{3}\right)}{2^{-2/3}} \cdot \frac{\sqrt{3}\,\Gamma\left(\frac{1}{3}\right)}{2\pi} = \frac{\sqrt{3}}{2^{1/3}\sqrt{\pi}}\left[\Gamma\left(\frac{1}{3}\right)\right]^2$$

Example 2. *Find the value of* $\Gamma\left(\frac{1}{9}\right)\Gamma\left(\frac{2}{9}\right)\Gamma\left(\frac{3}{9}\right)\ldots\Gamma\left(\frac{8}{9}\right)$.

Solution. We know that

$$\Gamma\left(\frac{1}{n}\right)\Gamma\left(\frac{2}{n}\right)\Gamma\left(\frac{3}{n}\right)\ldots\Gamma\left(\frac{n-1}{n}\right) = \frac{(2\pi)^{(n-1)/2}}{n^{1/2}}.$$

Putting $n = 9$, in the above relation, we get

$$\Gamma\left(\frac{1}{9}\right)\Gamma\left(\frac{2}{9}\right)\Gamma\left(\frac{3}{9}\right)\ldots = \frac{(2\pi)^{(9-1)/2}}{9^{1/2}}$$

$$= \frac{(2\pi)^4}{3} = \frac{16}{3}\pi^4.$$

Example 3. *Prove that* $B(m,m)\,B\left(m+\frac{1}{2}, m+\frac{1}{2}\right) = \frac{\pi m^{-1}}{2^{4m-1}}$.

Solution.

$$\text{L.H.S.} = \frac{\Gamma(m)\Gamma(m)}{\Gamma(m+m)} \cdot \frac{\Gamma\left(m+\frac{1}{2}\right)\Gamma\left(m+\frac{1}{2}\right)}{\Gamma\left(m+\frac{1}{2}+m+\frac{1}{2}\right)}$$

$$= \frac{[\Gamma(m)\Gamma(m+1/2)]^2}{\Gamma(2m)\Gamma(2m+1)}$$

$$= \frac{[\Gamma(m)\Gamma(m+1/2)]^2}{\Gamma(2m).2m\,\Gamma(2m)}$$

$[\because \Gamma(p+1) = p\Gamma(p)]$

$$= \frac{1}{2m}\left[\frac{\Gamma(m)\Gamma(m+1/2)}{\Gamma(2m)}\right]^2$$

$$= \frac{1}{2m}\cdot\left(\frac{\sqrt{\pi}}{2^{2m-1}}\right)^2$$

(By duplication formula)

$$= \frac{\pi}{2m.2^{4m-2}} = \frac{\pi m^{-1}}{2^{4m-1}}.$$

Example 4. *Prove that* $\int_{-\infty}^{\infty} \cos\frac{\pi}{2}x^2\,dx = 1.$

Solution. Let $I = \int_{-\infty}^{\infty} \cos\frac{1}{2}\pi x^2\,dx$. ...(1)

Since $\cos\frac{1}{2}\pi x^2$ is an even function therefore (1) gives

$$I = 2\int_0^{\infty} \cos\frac{1}{2}\pi x^2\,dx \quad ...(2)$$

Putting $x^2 = t$ so that $x = t^{1/2}$ and $dx = \left(\frac{1}{2}\right)t^{-1/2}dt$ then equation (2) reduces to

$$I = 2\int_0^{\infty} \cos\frac{1}{2}\pi t.\frac{1}{2}t^{-1/2}dt$$

$$= \int_0^{\infty} (t)^{1/2-1}\cos\frac{1}{2}\pi t\,dt$$

$$= \frac{\Gamma\left(\frac{1}{2}\right)}{\left(\frac{\pi}{2}\right)^{1/2}}\cos\left(\frac{1}{2}.\frac{\pi}{2}\right)$$

$$\left[\because \int_0^{\infty} x^{m-1}\cos bx\,dx = \frac{\Gamma(m)}{b^m}\cos\frac{m\pi}{2}\right]$$

$$= \frac{\Gamma(1/2)}{(\pi/2)^{1/2}}\cos\left(\frac{1}{2}.\frac{\pi}{2}\right) = \frac{\sqrt{\pi}}{\sqrt{\pi/2}}.\frac{1}{\sqrt{2}} = 1.$$

Example 5. *Show that*

(i) $2^n\Gamma\left(n+\frac{1}{2}\right) = 1.3.5....(2n-1)\sqrt{\pi}$, *where n is a positive integer.*

(ii) $\Gamma\left(\frac{3}{2}-x\right)\Gamma\left(\frac{3}{2}+x\right) = \left(\frac{1}{4}-x^2\right)\pi\sec\pi x$, *provided* $-1 < 2x < 1$.

Solution. (i) We have

$$\Gamma\left(n+\frac{1}{2}\right) = \left(n-\frac{1}{2}\right)\Gamma\left(n-\frac{1}{2}\right)$$

$$= \left(n-\frac{1}{2}\right)\left(n-\frac{3}{2}\right)\Gamma\left(n-\frac{3}{2}\right)$$

$$= \left(n-\frac{1}{2}\right)\left(n-\frac{3}{2}\right)\left(n-\frac{5}{2}\right)...\frac{3}{2}.\frac{1}{2}\Gamma\left(\frac{1}{2}\right)$$

$$= \frac{2n-1}{2}.\frac{2n-3}{2}.\frac{2n-5}{2}...\frac{3}{2}.\frac{1}{2}.\sqrt{\pi}$$

$$= \frac{1}{2^n}(2n-1)(2n-3)(2n-5)...3.1.\sqrt{\pi}.$$

Hence, $2^n\Gamma\left(n+\frac{1}{2}\right) = 1.3.5....(2n-1)\sqrt{\pi}.$

(ii) We have

$$\Gamma\left(\frac{3}{2}-x\right)\Gamma\left(\frac{3}{2}+x\right)$$

$$= \left(\frac{1}{2}-x\right)\Gamma\left(\frac{1}{2}-x\right).\left(\frac{1}{2}+x\right)\Gamma\left(\frac{1}{2}+x\right)$$

$$= \left(\frac{1}{4}-x^2\right)\Gamma\left(\frac{1-2x}{2}\right)\Gamma\left(\frac{1+2x}{2}\right)$$

$$= \left(\frac{1}{4}-x^2\right)\Gamma\left(\frac{1-2x}{2}\right)\Gamma\left(1-\frac{1-2x}{2}\right)$$

$$= \left(\frac{1}{4}-x^2\right)\frac{\pi}{\sin\left(\frac{1-2x}{2}\pi\right)}$$

$$= \left(\frac{1}{4}-x^2\right)\frac{\pi}{\sin\left(\frac{\pi}{2}-\pi x\right)}$$

$$= \left(\frac{1}{4}-x^2\right).\frac{\pi}{\cos\pi x} = \left(\frac{1}{4}-x^2\right)\sec\pi x.\pi.$$

Example 6. *Prove that* $\int_0^{\infty} \cos(bx^{1/n})\,dx = \frac{\Gamma(n+1)}{b^n}\cos\frac{n\pi}{2}.$

Solution. Let $I = \int_0^{\infty} \cos(bx^{1/n})\,dx.$ ...(1)

Putting $x = t^n$ so that $dx = nt^{n-1}dt$ then (1) gives

$$I = n\int_0^{\infty} \cos(bt).t^{n-1}\,dt = \frac{n\Gamma(n)}{b^n}\cos\frac{n\pi}{2}$$

$$= \frac{\Gamma(n+1)}{b^n}\cos\frac{n\pi}{2}.$$

Aliter.

$$\int_0^{\infty} \cos(bz^{1/n})\,dz = \frac{1}{b^n}\Gamma(n+1)\cos\frac{n\pi}{2}.$$

Put $z^{1/n} = x$

$\Rightarrow$ $z = x^n$.

so that $dz = nx^{n-1}\,dx.$

$$\therefore \int_0^{\infty} \cos(bz^{1/n})\,dz = \int_0^{\infty} \cos(bx).nx^{n-1}dx$$

$$= n\int_0^{\infty} x^{n-1}\cos(bx)\,dx$$

$$= \text{real part of } n\int_0^{\infty} e^{-bxi}x^{n-1}dx$$

$$= \text{real part of } n\frac{\Gamma(n)}{(bi)^n}$$

$$= \text{real part of} \frac{n\Gamma(n)}{b^n}\left(\cos\frac{\pi}{2}+i\sin\frac{\pi}{2}\right)^{-n}$$

$$= \text{real part of} \frac{\Gamma(n+1)}{b^n}\left(\cos\frac{n\pi}{2}-i\sin\frac{n\pi}{2}\right)$$

$$= \frac{1}{b^n}\Gamma(n+1)\cos\left(\frac{n\pi}{2}\right).$$

Example 7. *Evaluate*

(i) $\int_0^{\infty} \cos(c^2x^2)dx$ (ii) $\int_0^{\infty} \sin x^2\,dx.$

Solution. (i) Let $I = \int_0^{\infty} \cos(c^2x^2)\,dx$ ...(1)

Putting $x^2 = t$ so that $x = t^{1/2}$

$$\Rightarrow \quad dx = \left(\frac{1}{2}\right)t^{-1/2}dt$$

Then, (1) reduces to

$$I = \int_0^\infty \cos(c^2 t)\frac{1}{2}t^{-1/2}dt$$

$$= \frac{1}{2}\int_0^\infty (t)^{(1/2)-1}\cos(c^2 t)\,dt$$

$$= \frac{1}{2}\frac{\Gamma(1/2)}{(c^2)^{1/2}}\cos\left(\frac{1}{2}\cdot\frac{\pi}{2}\right)$$

$$= \frac{\sqrt{\pi}}{2c}\cdot\frac{1}{\sqrt{2}} = \frac{1}{2c}\sqrt{\frac{\pi}{2}}.$$

(ii) Let $\quad I = \int_0^\infty \sin x^2\,dx$ ...(2)

Putting $x^2 = t$, so that $x = t^{1/2}$

and $dx = \left(\frac{1}{2}\right)t^{-1/2}dt$, then (2) reduces to

$$I = \frac{1}{2}\int_0^\infty t^{(1/2)-1}\cos t\,dt$$

$$= \frac{1}{2}\frac{\Gamma(1/2)}{1}\sin\left(\frac{1}{2}\cdot\frac{\pi}{2}\right)$$

$$\left[\because \int_0^\infty x^{m-1}\sin bx\,dx = \frac{\Gamma(m)}{b^m}\sin\frac{m\pi}{2}\right]$$

$$= \frac{\sqrt{\pi}}{2}\cdot\frac{1}{\sqrt{2}} = \frac{1}{2}\sqrt{\frac{\pi}{2}}.$$

EXERCISE 20.2

1. Show that

$$\Gamma(n)\Gamma\left(\frac{1-n}{2}\right) = \frac{\sqrt{\pi}\,\Gamma(n/2)}{2^{1-n}\cos\left(\frac{n\pi}{2}\right)}, 0 < n < 1.$$

2. Show that $\int_0^1 \frac{dx}{\sqrt{(1-x^6)}} = \frac{\sqrt{3}}{2}\int_0^1 \frac{dx}{\sqrt{1-x^3}}$

$$= \frac{1}{2^{7/3}\pi}\left[\Gamma\left(\frac{1}{3}\right)\right]^3.$$

3. Show that

$$\Gamma(0.1).\Gamma(0.2).\Gamma(0.3)...\Gamma(0.9) = \frac{(2\pi)^{9/2}}{\sqrt{10}}.$$

4. Show that $\int_0^1 \frac{dx}{\sqrt{(1-x^4)}} = \frac{\sqrt{2}}{8\sqrt{\pi}}\left[\Gamma\left(\frac{1}{4}\right)\right]^2.$

5. Show that if $m > -1$ then

$$\int_0^\infty x^m e^{-n^2x^2}dx = \frac{1}{2n^{m+1}}\Gamma\left(\frac{m+1}{2}\right).$$

6. Proe the following

(i) $B(l,m).B(l+m,n).B(l+m+n,p) = \frac{\Gamma(l)\Gamma(m)\Gamma(n)}{\Gamma(l+m+n+p)}$ (UPTU-2008)

(ii) $\int_0^\infty c^{-x}x^l dx = \frac{\Gamma(c+1)}{(\log c)^{c+1}}, c > 1$ (UPTU-2007)

(iii) $B(m,n) = 2^{1-2m}\left(m,\frac{1}{2}\right)$ (UPTU-2006, 2010)

(iv) $\int_0^{\pi/2}\tan^n x\,dx = \frac{\pi}{2}\sec\frac{n\pi}{2}$ (UPTU-2006)

7. Prove that $B(m,n) = \int_0^1 \frac{x^{m-1}+x^{n-1}}{(1+x)^{m+n}}dx$ (MTU-2012)

Hint to Selected Problems

2. Let $A = \int_0^1 \frac{dx}{\sqrt{1-x^6}}$. Now putting $x^6 = \sin^2\theta$ *i.e.*, $x = \sin^{1/3}\theta \Rightarrow\ dx = \frac{1}{2}\sin^{-2/3}\theta\cos\theta\,d\theta$.

$$\therefore \quad A = \frac{1}{3}\int_0^{\pi/2}\sin^{-2/3}\theta\cos^0\theta\,d\theta = \frac{1}{3}\frac{\Gamma\left(\frac{1}{6}\right)\Gamma\left(\frac{1}{2}\right)}{2\Gamma\left(\frac{2}{3}\right)} = \frac{1}{6}\frac{\Gamma\left(\frac{1}{6}\right).\sqrt{\pi}}{\Gamma\left(\frac{2}{3}\right)}.$$ Now find the values of $\Gamma\left(\frac{1}{6}\right)$ and $\Gamma\left(\frac{2}{3}\right)$ separately.

4. Let $I = \int_0^1 \frac{dx}{\sqrt{1-x^4}}$. Then solve after putting $x^4 = t$.

5. Let $I = \int_0^\infty x^m e^{-n^2x^2}dx$. Then solve after putting $n^2x^2 = t$.

Objective Evaluations

Fill in the Blanks

1. $\Gamma(1) =$ ______________ .
2. $\Gamma(n+1) =$ ______________ $\Gamma(n), n>0$.
3. $\Gamma\left(\frac{1}{2}\right) =$ ______________ .
4. $\int_0^\infty x^{n-1}e^{-x}dx =$ ______________ .
5. $\int_0^1 x^{m-1}(1-x)^{n-1}dx =$ ______________ $m>0, n>0$.
6. Beta function is also called ______________ integral of first kind.
7. $\dfrac{\Gamma(m)\Gamma(n)}{\Gamma(m+n)} =$ ______________ .

True/False

Write 'T' for True and 'F' for False statement.

1. Beta function is also called Eulerian integral of first kind. **(T/F)**
2. Beta function is also called Eulerian integral of second kind. **(T/F)**
3. $\int_0^\infty e^{-ax}\sin bx\,dx = \dfrac{b}{a^2+b^2}$. **(T/F)**
4. $\int_0^{\pi/2} e^{-ax}\sin bx\,dx = \dfrac{b}{a^2+b^2}$. **(T/F)**
5. $\int_{-\infty}^\infty \cos\frac{\pi}{2}x^2.dx = 1$. **(T/F)**

Multiple choice Questions

Choose the most appropriate one :

1. The value of $\Gamma\left(\frac{1}{2}\right)$ is equal to :
   (a) π (b) $\sqrt{\pi}$
   (c) 1 (d) 2
2. $\Gamma(x) = \int_0^\infty e^{-x}x^{n-1}dx$ if :
   (a) $n > 0$ (b) $n < 0$
   (c) $n = 0$ (d) none of these
3. The value of $\Gamma(n+1)$, if $n > 0$ is:
   (a) n (b) $\Gamma(n)$
   (c) $n\,\Gamma(n)$ (d) none of these
4. If n is a non-negative integer then value of $\Gamma(n+1)$ is equal to :
   (a) n (b) $n\,!$
   (c) $(n+1)\,!$ (d) none of these
5. $\Gamma(m)\,\Gamma(1-m) =$ ________ , $(0<m<1)$.
   (a) $\dfrac{1}{\sin m\pi}$ (b) $\dfrac{\pi}{\sin m\pi}$
   (c) $\dfrac{m}{\sin m\pi}$ (d) none of these

ANSWERS

Fill in The Blanks

1. 1 **2.** n **3.** $\sqrt{\pi}$ **4.** $\Gamma(n)$ **5.** $B(m, n)$ **6.** Eulerian **7.** $B(m,n)$

True/False

1. T **2.** F **3.** T **4.** F **5.** T

Multiple Choice Questions

1. (b) **2.** (a) **3.** (c) **4.** (b) **5.** (b)

FFFFF

GATE tutor

Key Terms and Results

- **Double Integral:** Double integral is an extension of a definite integral in two-dimensional space.
- **Triple Integral:** Let $f(x, y, z)$ be a single valued function of the independent variables x, y, z in a finite region V. Then if limit of the sum $\sum_{r=1}^{n} f(x_r, y_r, z_r)\ \delta V_r$ tends to zero then it is called a triple integral of $f(x, y, z)$.
- **Gamma Function:** The definite integral $\int_0^\infty e^{-x} x^{n-1} dx, n > 0$ is called the gamma function.
- **Beta Function :** The definite integral $\int_0^1 x^{m-1}(1-x)^{n-1} dx$, $m > 0, n > 0$ is called beta function.

Results

- The double integral of a algebraic sum of a fixed number of functions is equal to the algebraic sum of double integrals taken for each term separately.
- A constant factor may be taken outside the integral sign.
- $\int_V x^{l-1} y^{m-1} z^{n-1} dx\, dy\, dz = \dfrac{\Gamma(l)\Gamma(m)\Gamma(n)}{\Gamma(l+m+n+1)}$
- $\underbrace{\int\int\ldots\int}_{n\text{ times}} x_1^{l_1-1}.x_2^{l_2-1}\ldots x_n^{l_n-1} dx_1 dx_2 \ldots dx_n$

$$= \frac{\Gamma(l_1)\Gamma(l_2)\ldots\Gamma(l_n)}{\Gamma(1+l_1+l_2+\ldots+l_n)}$$

- If x, y, z are all positive and such that $h_1 < x+y+z \le h_2$ then

$$\iiint f(x+y+z) x^{l-1} y^{m-1} z^{n-1} dx\, dy\, dz$$
$$= \frac{\Gamma(l)\Gamma(m)\Gamma(n)}{\Gamma(l+m++n)} \int_{h_1}^{h_2} f(u) u^{l+m+n-1} du$$

- If the limits of integration are constants in the double integration then the value of integration can be obtained by integrating with respect to any independent variable.
- If $\phi(x) = \int_{u_1}^{u_2} f(x, \alpha) d\alpha, a \le \alpha \le b$ and where u_1 and u_2 depend on the parameter α then

$$\frac{d\phi}{d\alpha} = \int_{u_1}^{u_2} \frac{\partial f}{\partial \alpha} + f(u_2, \alpha)\ \frac{du_2}{d\alpha} - f(u_1, \alpha)\frac{du_1}{d\alpha}$$

- If $\phi(\alpha) = \int_{u_1}^{u_2} f(x, \alpha) dx,\ a \le \alpha \le b$ and $f(x, \alpha)$ is continuous in x and α is a region including $u_1 \le x \le u_2$, $a \le \alpha \le b$ then if u_1 and u_2 are constants then

$$\int_a^b \phi(\alpha) d\alpha = \int_a^b \left[\int_{u_1}^{u_2} f(x, \alpha) dx\right] d\alpha$$
$$= \int_{u_1}^{u_2} \left\{\int_a^b f(x, \alpha) d\alpha\right\} dx$$

- $\Gamma(n+1) = n\Gamma(n), \quad \Gamma(1) = 1$
- $\Gamma(n+1) = n!$
- $\Gamma\left(\frac{1}{2}\right) = \sqrt{\pi}$
- $B(m, n) = B(n, m)$
- $B(m,n) = 2\int_0^{\pi/2} \cos^{2m-1}\theta \sin^{2n-1}\theta\, d\theta$
- $B(m,n) = \dfrac{\Gamma(m)\Gamma(n)}{\Gamma(m+n)}, m > 0, n > 0$
- $\Gamma(n)\Gamma(1-n) = \dfrac{\pi}{\sin n\pi}, 0 < n < 1$
- $\int_0^{\pi/2} \cos^m\theta \sin^n\theta\, d\theta = \dfrac{\Gamma\left(\frac{m+1}{2}\right)\Gamma\left(\frac{n+1}{2}\right)}{2\Gamma\left(\frac{m+n+2}{2}\right)}$
- **Duplication formula:**

$$\Gamma(n)\Gamma\left(n+\frac{1}{2}\right) = \frac{\sqrt{\pi}}{2^{2n-1}}\Gamma(2n),\ n > 0$$

- $\Gamma\left(\frac{1}{n}\right)\Gamma\left(\frac{2}{n}\right)\Gamma\left(\frac{3}{n}\right)\ldots\Gamma\left(\frac{n-1}{n}\right) = \dfrac{(2\pi)^{(n-1)/2}}{n^{1/2}}, n \in Z.$

Review Questions and Project Work

1. Prove that

$$\int_1^2\int_3^4 \left(xy + e^y\right) dy dx = \int_3^4\int_1^2 \left(xy + e^y\right) dx dy.$$

2. Show that the value of $\iint \sqrt{a^2 - x^2 - y^2}\, dx\, dy$ over the semi-circle $x^2+y^2=ax$ in the positive quadrant is $\dfrac{a^3}{3}\left[\dfrac{\pi}{2} - \dfrac{2}{3}\right]$.

3. Show that $\displaystyle\iiint_{x^2+y^2+z^2 \le 1} (x^2 + y^2 + z^2) dxdydz = \frac{4\pi}{5}$.

4. If n is a positive integer, show that

$$\iiint_{x^2+y^2+z^2 \le 1} (x^2 + y^2 + z^2)^n dxdydz = \frac{4\pi}{2n+3}.$$

5. Show that $\iint_D x^{m-1} y^{n-1} dx\, dy = \dfrac{\Gamma(m)\Gamma(n)}{\Gamma(1+m+n)} .h^{m+n}$ where D is the region $x \ge 0, y \ge 0$ and $x+y \le h$.

6. Show that the volume of the solid whose surface is represented by the equation $\frac{x^4}{a^4}+\frac{y^4}{b^4}+\frac{z^4}{c^4}=1$ is $\frac{abc}{6\sqrt{2}.\pi}\left[\Gamma\left(\frac{1}{4}\right)\right]^2$

7. Show that the volume enclosed by the surface

$$\left(\frac{x}{a}\right)^{2n}+\left(\frac{y}{b}\right)^{2n}+\left(\frac{z}{c}\right)^{2n}=1,$$

is $\frac{2abc\left(\Gamma\left(\frac{1}{2n}\right)\right)^3}{3n^2\Gamma\left(\frac{3}{2n}\right)}$. n being an integer.

8. Show that mass of an octant of the ellipsoid $\frac{x^2}{a^2}+\frac{y^2}{b^2}+\frac{z^2}{c^2}=1$, the density at any point being $r=kxyz$ is $\frac{ka^2b^2c^2}{48}$.

9. If $p>0$, $q>0$ then show that

$$\int_{-1}^{1}\frac{(1+x)^{2p-1}(1-x)^{2q-1}}{(1+x^2)^{p+q}}dx=2^{p+q-2}B(p,q).$$

10. Show that $\Gamma(p)=\lim_{n\to\infty} n^pB(p,n), p>0$

11. Show that $\sqrt{\pi}.\Gamma(2p)=2^{2p-1}\Gamma(p)\Gamma\left(\pi+\frac{1}{2}\right), p>0$.

12. Show that $\int_0^1\frac{(1-x^4)^{3/4}}{(1+x^4)^2}dx=\frac{B\left(\frac{7}{4},\frac{1}{4}\right)}{2^{9/4}}$.

13. Show that for $x>0$, $\lim_{n\to\infty}\frac{\Gamma(n+x)}{n^x\Gamma(n)}=1$.

14. Show that the gamma function is continuous on $]0,\infty[$.

15. Show that $\int_0^{\pi/2}\sin^p x\,dx\int_0^{\pi/2}\sin^{p+1}x\,dx=\frac{\pi}{2(p+1)}$ if $p>-1$.

16. Show that the beta function $B(x,y)$ is a convex function of x on $]0,\infty[$ for each fixed $y>0$.

Multiple Choice Questions

Choose the most appropriate one.

1. The volume of an object expressed in spherical coordinates is given by $V=\int_0^{2\pi}\int_0^{\pi/3}\int_0^1 r^2\sin\phi\,dr\,d\phi\,d\theta$.
The volume of the integral is : [GATE (ME)-2004]
(a) π (b) $\pi/3$
(c) $2\pi/3$ (d) none of these

2. The value of $\int_0^3\int_0^x(6-x-y)\,dx\,dy$ is : [GATE (CE)-2008]
(a) 12.5 (b) 13.5
(c) 20.5 (d) none of these

3. Changing the order of integration in the double integral $I=\int_0^8\int_{x/4}^2 f(x,y)\,dx\,dy$ tends to $I=\int_r^s\int_p^q f(x,y)\,dx\,dy$. What is q? [GATE(ME)–2005]
(a) $4y$ (b) $4x$
(c) x (d) none of these

4. Consider the shaded triangular region P shown in the figure.

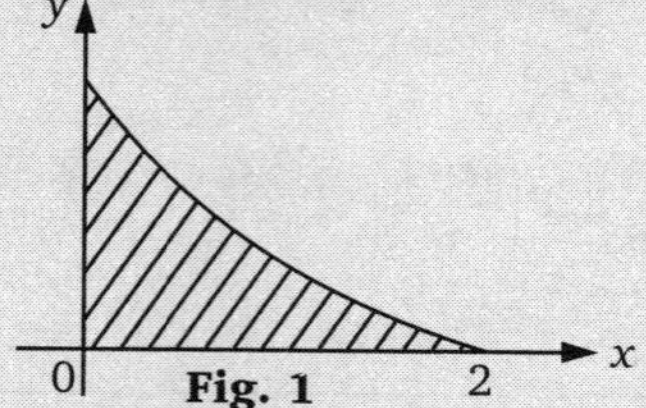

Fig. 1

What is $\iint_P xy\,dx\,dy$? [GATE(ME)-2008]
(a) 2/7 (b) 7/16
(c) 1/6 (d) none of these

5. By a change of variable $x(u,v)=uv$, $y(u,v)=\frac{v}{u}$ is double integral, the integral $f(x,y)$ changes to $f\left(uv,\frac{v}{u}\right)\phi(u,v)$. Then $\phi(u,v)$ is : [GATE(ME)–2005]
(a) $\frac{u}{v}$ (b) $\frac{3u}{v}$
(c) $\frac{2u}{v}$ (d) $2uv$

6. $f(x,y)$ is a continuous function defined over $(x,y)\in[0,1]\times[0,1]$. Given the two constraints $x>y^2$ and $y>x^2$ the volume under $f(x,y)$ is : [GATE(EE)–2009]
(a) $\int_{y=0}^{1}\int_{x=y^2}^{\sqrt{y}} f(x,y)\,dx\,dy$ (b) $\int_{y=x^2}^{1}\int_{x=y^2}^{1} f(x,y)\,dx\,dy$
(c) $\int_{y=0}^{1}\int_{x=0}^{1} f(x,y)\,dx\,dy$ (d) none of these

7. A surface $S(x,y)=2x+5y-3$ is integrated once over a path consisting of the points that satisfy $(x+1)^2+(y-1)^2=\sqrt{2}$. The integral evaluates to : (GATE(EE)–2006)
(a) $17\sqrt{2}$ (b) $17/\sqrt{2}$
(c) 0 (d) none of these

8. If $\vec{A}=xy\,\hat{a}_x+x^2\hat{a}_y$, then $\oint_c \vec{A}.d\vec{l}$ over the path shown in he figure is : [GATE(EC)–2010]

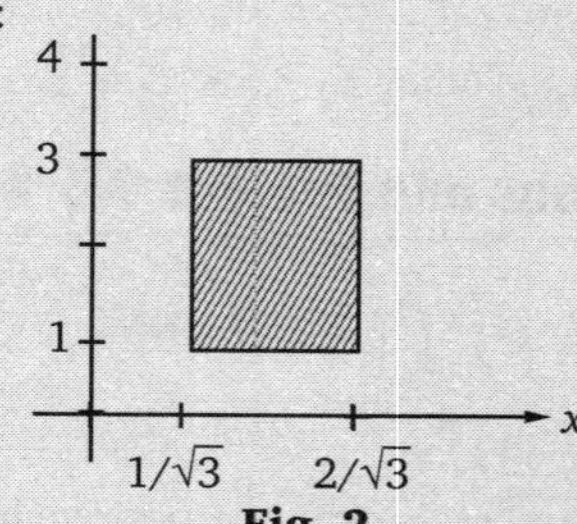

Fig. 2

(a) 0 (b) 1
(c) $\sqrt{3}$ (d) none of these

ANSWERS

1. (b)	**2.** (b)	**3.** (a)	**4.** (c)	**5.** (c)	**6.** (a)	**7.** (c)	**8.** (b)

Hint to Selected Problems

1. $V = \int_0^{2\pi}\int_0^{x/3}\int_0^1 r^2 \sin\phi \, dr\, d\phi\, d\theta$

$$= \int_0^{2\pi}\int_0^{\pi/3}\left[\frac{r^3}{3}\right]_0^1 \sin\phi \, d\phi\, d\theta$$

$$= \frac{1}{3}\int_0^{2\pi}[-\cos\phi]_0^{\pi/3} \, d\theta$$

$$= \frac{1}{3}\times\frac{1}{2}\int_0^{2\pi} d\theta \quad = \frac{1}{3}\times\frac{1}{2}\times 2\pi = \frac{\pi}{3}$$

2. We have

$$\int_0^3\int_0^x (6-x-y)\,dx\,dy = \int_0^3\left[(6-x)y - \frac{y^2}{2}\right]_0^x dx$$

$$= \int_0^3\left[(6-x).x - \frac{x^2}{2}\right]dx = 13.5$$

3. When $I = \int_0^8\int_{x/4}^2 f(x,y)\,dy\,dx$

Then, we have this figure

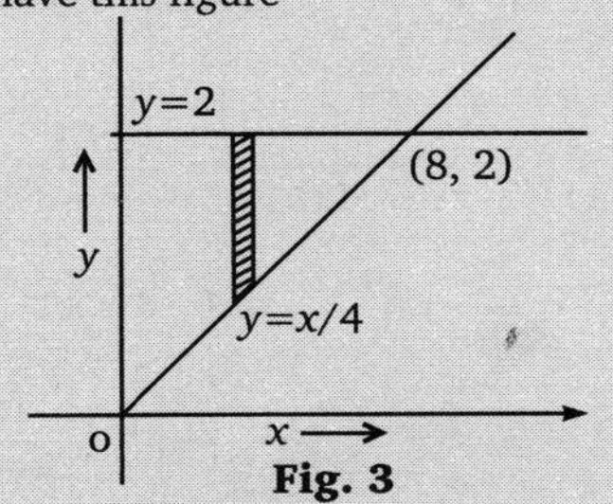

Fig. 3

Now

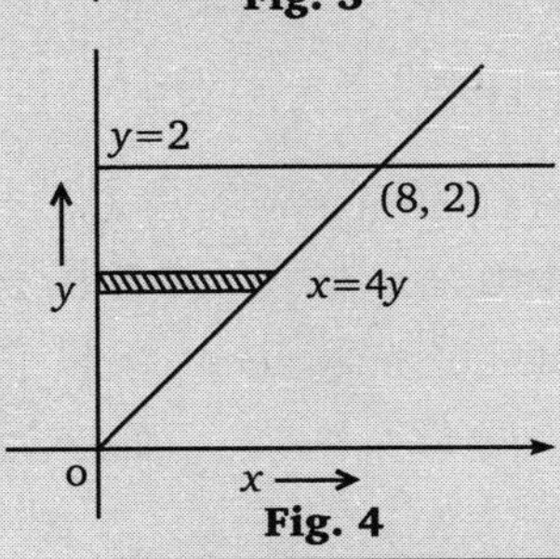

Fig. 4

$$I = \int_0^2\int_0^{4y} f(x,y)\,dy\,dx$$

$$\Rightarrow \qquad q = 4y$$

4. The equation of the straight line with x-intercept = 2 and y-intercept = 1 is

$$\frac{x}{2}+\frac{y}{1} = 1$$

$$\Rightarrow \qquad y = 1 - \frac{x}{2}$$

i.e., $\qquad x = 2 - 2y$

Now, $\int_0^1\int_0^{2-2y}(xy\,dx)\,dy = \int_0^1\left[\left(\frac{yx^2}{2}\right)_0^{2-2y}\right]dy$

$$= \int_0^1 \frac{y}{2}(2-2y)^2 dy$$

$$= \int_0^1 2y(1-y)^2\,dy = \frac{1}{6}$$

5. $\frac{\partial x}{\partial u} = v, \; \frac{\partial x}{\partial v} = u$

$$\frac{\partial y}{\partial u} = -\frac{v}{u^2}, \; \frac{\partial y}{\partial v} = \frac{1}{u}$$

and $\phi(u,v) = \begin{vmatrix} \frac{\partial x}{\partial u} & \frac{\partial x}{\partial v} \\ \frac{\partial y}{\partial x} & \frac{\partial y}{\partial v} \end{vmatrix} = \begin{vmatrix} v & u \\ -\frac{v}{u^2} & \frac{1}{u} \end{vmatrix}$

$$= \frac{v}{u}+\frac{v}{u} = \frac{2v}{u}$$

Self Assessment Test

1. Prove that $\int_0^{a/\sqrt{2}}\int_y^{\sqrt{a^2-y^2}} \log(x^2+y^2)dxdy\ (a>0)$
$$=\frac{\pi a^2}{4}\left(\log a-\frac{1}{2}\right).$$

2. Prove that $\int_0^a\int_{x^2/a}^{2a-x} xydydx = \frac{3}{8}a^4.$

3. Prove that $\iiint_V \sqrt{1-(x^2+y^2+z^2)}dx\,dy\,dz = \frac{\pi^2}{3^2},$ where V is the region interior to the sphere $x^2+y^2+z^2=1$.

4. Prove that $\iiiint dx\,dy\,dz\,dw = \frac{\pi^2}{32}(b^4-a^4)$, where $a^2 < x^2+y^2+z^2+w^2 < b^2, a < b$.

5. Prove that the volume of the solid bounded by the surface .
$$\left(\frac{x}{a}\right)^{2/3}+\left(\frac{y}{b}\right)^{2/3}+\left(\frac{z}{c}\right)^{2/3}=1 \text{ is } \frac{4\pi abc}{35}$$

6. Prove that the volume determined by the surface $x^n+y^n+z^n=a^n$, $n>0$ in the positive octant is $\dfrac{a^3\left[\Gamma\left(1+\frac{1}{n}\right)\right]^3}{\sqrt{\Gamma\left(1+\frac{3}{n}\right)}}$.

7. Prove that the volume bounded by the surface
$$\frac{x^2}{a^2}+\frac{y^2}{b^2}+\frac{z^2}{c^2}=1 \text{ is } \frac{8\pi abc}{5}.$$

8. Prove that the volume of solid whose surface is represent by the equation
$$\frac{x^4}{a^4}+\frac{y^4}{b^4}+\frac{z^4}{c^4}=1 \text{ is } \frac{abc}{6\sqrt{2}.\pi}\left[\Gamma\left(\frac{1}{4}\right)\right]^4.$$

9. Prove that the volume of the ellipsoid
$$\frac{x^2}{a^2}+\frac{y^2}{b^2}+\frac{z^2}{c^2}=1 \text{ is } \frac{4}{3}\pi abc.$$

10. Prove that the area of the ellipse $\frac{x^2}{9}+\frac{y^2}{4}=1$ is 6π.

11. Prove that the volume of the tetrahedron bounded by the co-ordinate planes and the plane $x+y+z=1$ is $\frac{1}{6}$.

12. Prove that the area of ellipse $\frac{x^2}{a^2}+\frac{y^2}{b^2}=1$ is πab.

13. Prove that the $\iiint(x+y+z)dx\,dy\,dz = \frac{1}{8}$ over the tetrahedron bounded by the planes $x=0$, $y=0$, $z=0$ and $x+y+z=1$.

14. Prove that the value of $\iint\sqrt{a^2-x^2-y^2}dx\,dy$ over the semi-circle $x^2+y^2=ax$ in the positive quadrant is $\frac{a^3}{3}\left(\frac{\pi}{2}-\frac{2}{3}\right)$.

15. Prove that if R is a region bounded by the curves $x=f_1(y)$, $x=f_2(y)$, $y=c$ $y=d$ then
$$\iint_R f(x,y)dA = \int_c^d\left[\int_{f_1(y)}^{f_2(y)} f(x,y)dx\right]dy.$$

16. Prove that if $p>0$, $B(p,p)=B\left(p+\frac{1}{2},p+\frac{1}{2}\right)=\frac{\pi}{2^{4p-1}.p}$.

17. Prove that if $\int_0^1 x^{-1/3}(1-x)^{-2/3}(1-2x)^{-1}dx = \frac{1}{9^{1/3}}B\left(\frac{2}{3},\frac{1}{3}\right)$

18. Prove that if $\int_{-\infty}^{\infty}\frac{e^{pt}}{1+e^t}dt = \Gamma(p)\Gamma(1-p), 1>p>0$

19. Prove that if $p>0$, $q>1$ then $\sum_{n=0}^{\infty} B(p+r,q)$ converges to $B(p, q-1)$.

20. Prove that if $\sin\frac{\pi}{2n}.\sin\frac{2\pi}{2n}.\sin\frac{3\pi}{2n}...\sin\frac{(n-1)\pi}{2\pi} = \sqrt{n}.2^{-n+1}$.

21. Prove that if $p>0$, $q>0$, $a>0$, $b>0$, then
$$\int_0^{\pi/2}\frac{\cos^{2p-1}\theta\sin^{2q-1}\theta}{(a\cos^2\theta+b\sin^2\theta)^{p+q}}d\theta = \frac{B(p,q)}{2a^p.b^q}.$$

22. Prove that if $n>0$, then $\dfrac{\Gamma\left(n+\frac{1}{2}\right)}{\Gamma(n)} = \dfrac{(2n)!}{2^{2n}.n!}$.

23. Prove that $\int_0^{\pi/2}\frac{d\theta}{\sqrt{\sin\theta}}\times\int_0^{\pi/2}\sqrt{\sin\theta}\,d\theta = \pi$.

24. Prove that $\int_0^1\sqrt{(1-x^4)}\,dx = \frac{1}{12}\sqrt{\frac{2}{\pi}}\left[\Gamma\left(\frac{1}{4}\right)\right]^2$.

25. Prove that $\int_0^{\infty} x^{m-1}\cos bx = \frac{\Gamma(m)}{b^m}\cos\left(\frac{m\pi}{2}\right)$.

26. Change the order of integration in $I=\int_0^1\int_{x^2}^{2-x} xy\,dx\,dy$.
(Bhopal-2008, VTU-2008, SVTU-2007, PTU-200, UPTU-2005)

27. Change the order of integration $I=\int_0^{4a}\int_{x^2/4a}^{2\sqrt{ax}} dy\,dx$ and hence show that $I=\frac{16a^2}{3}$. (Nagpur-2009, PTU-2009)

28. Show that $\int_0^1\int_x^{\sqrt{2-x^2}}\frac{x}{\sqrt{x^2+y^2}}dydx = 1-\frac{1}{\sqrt{2}}$
(PTU-2010, Marathwada-2008)

29. Show that the value of $\int_0^{\infty}\int_0^x x e^{-x^2/y}dy\,dx = 1$
(SVTU-2006; VTU-2004, 10)

30. Show that the value of $\iint r\sin dr\,d\theta$ over $r=a(1-\cos\theta)$ above the initial line is $\frac{4a^2}{3}$. (Kerala-2005)

VECTOR CALCULUS

UNIT OUTLINE

- **Differentiation of Vectors**
- **Integration of Vectors**
- **Gradient of a Scalar Field**
- **Divergence of a Vector Field**
- **Curl of a Vector Field**
- **Gauss's Theorem**
- **Stoke's Theorem**
- **Green's Theorem**
- **Applications of Gauss, Stoke and Green's Theorems**

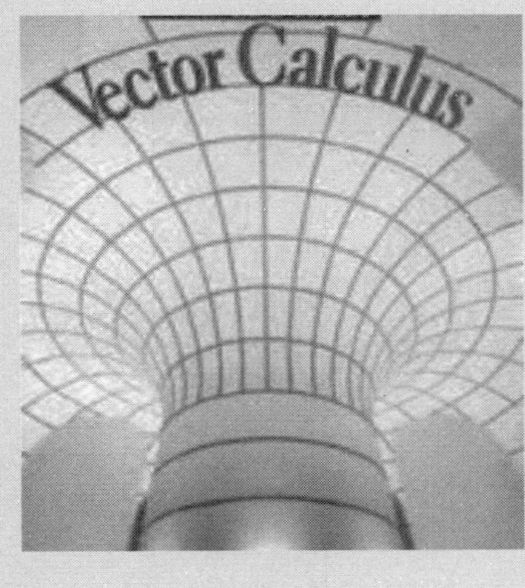

PRE-REQUISITE

- The quantities which have only magnitude and are not related to any direction in space are called scalars. On the other hand, those quantities which have both magnitude and direction are called vector quantities.
- **Scalar or dot product of two Vectors**
  If $\mathbf{a} = a_1\hat{i} + a_2\hat{j} + a_3\hat{k}$ and $\mathbf{b} = b_1\hat{i} + b_2\hat{j} + b_3\hat{k}$
  then $\mathbf{a} \cdot \mathbf{b} = ab\cos\theta$
  $= a_1a_2 + b_1b_2 + c_1c_2$
- **Vector or Cross product**
  $\mathbf{a} \times \mathbf{b} = ab\sin\theta$

Differentiation and Integration of Vectors

21.1 SCALAR FUNCTION

Since we know that the quantity which is associated with the magnitude but not associated with direction is known as scalar quantity. Therefore every real number is a scalar quantity.

Let D be a subset of a set of real numbers. Then a function f defined over the subset D such that for all $t \in D$, $f(t)$ is obtained as a scalar quantity, is called a scalar function.

21.2 VECTOR FUNCTION

If the scalar fucntion $f(t)$ for all $t \in D$ is associated with some direction then this function is called a vector function and is therefore, denoted by $\mathbf{f}(t)$ or $\mathbf{f}$.

Let $f_1(t), f_2(t), f_3(t)$ be three components of a vector function $\mathbf{f}(t)$, then this function can be uniquely expressed as a linear combination of these three fixed non-coplanar vectors $f_1(t)\hat{i}, f_2(t)\hat{j}, f_3(t)\hat{k}$.

$$\therefore \qquad \mathbf{f}(t) = f_1(t)\hat{i} + f_2(t)\hat{j} + f_3(t)\hat{k}$$

where $\hat{i}, \hat{j}, \hat{k}$ are three mutually perpendicular non-coplanar unit vectors.

21.3 SCALAR AND VECTOR FIELDS

Scalar fields. A scalar point function f defined over some region R such that to each point $P(x, y, z)$ in space, there corresponds a unique scalar $f(P)$, is called a scalar field. For example

$$f(x, y, z) = x^2 + y^2 + z^2 - 3xyz.$$

Vector fields. A vector point function f defined over a region R such that to each point $P(x, y, z)$ there exists a unique vector $\mathbf{f}(P)$, is called vector field. For example

$$\mathbf{f}(x, y, z) = x^2y\hat{i} + x^3z\hat{j} - y^3z\hat{k}.$$

21.4 LIMIT AND CONTINUITY OF A VECTOR FUNCTION

Limit. A vector function $f(t)$ is said to have the limit l as t tends to t_0, for given $\varepsilon > 0$ there exists a positive number δ such that

$$|\mathbf{f}(t) - l| < \varepsilon \text{ whenever } 0 < |t - t_0| < \delta \text{ i.e., } \lim_{t \to t_0} \mathbf{f}(t) = l.$$

Continuity. A vector function $\mathbf{f}(t)$ is said to be continuous at t_0, if for given $\varepsilon > 0$ there must exists a positive number δ such that

$$|\mathbf{f}(t) - \mathbf{f}(t_0)| < \varepsilon \text{ whenever } |t - t_0| < \delta, \text{provided } \mathbf{f}(t_0) \text{ is defined.}$$

REMARK

- A vector function $\mathbf{f}(t)$ is said to be continuous for every value of t in the domain over which $\mathbf{f}(t)$ is defined.

21.5 SOME RESULTS RELATED TO THE LIMIT AND CONTINUITY OF A VECTOR FUNCTION

1. The necessary and sufficient condition for a vcector function $\mathbf{f}(t)$ to be continuous at t_0 is that $\lim_{t \to t_0} \mathbf{f}(t) = \mathbf{f}(t_0)$.

2. If $\mathbf{f}(t) = f_1(t)\hat{i} + f_2(t)\hat{j} + f_3(t)\hat{k}$, then $\mathbf{f}(t)$ is continuous iff $f_1(t), f_2(t), f_3(t)$ are continuous.

3. If $\mathbf{f}(t) = f_1(t)\hat{i} + f_2(t)\hat{j} + f_3(t)\hat{k}$ and $\mathbf{l} = l_1\hat{i} + l_2\hat{j} + l_3\hat{k}$, then $\lim_{t \to t_0} \mathbf{f}(t) = \mathbf{l}$ iff $\lim_{t \to t_0} f_1(t) = l_1$, $\lim_{t \to t_0} f_2(t) = l_2$ and $\lim_{t \to t_0} f_3(t) = l_3$.

4. If $\mathbf{f}(t)$ and $\mathbf{g}(t)$ are vector functions of scalar variable t and $\phi(t)$ is a scalar function, then

(i) $\lim_{t\to t_0}[\mathbf{f}(t)\pm\mathbf{g}(t)] = \lim_{t\to t_0}\mathbf{f}(t) \pm \lim_{t\to t_0}\mathbf{g}(t)$

(ii) $\lim_{t\to t_0}[\mathbf{f}(t).\mathbf{g}(t)] = \left[\lim_{t\to t_0}\mathbf{f}(t)\right].\left[\lim_{t\to t_0}\mathbf{g}(t)\right]$

(iii) $\lim_{t\to t_0}[\mathbf{f}(t)\times\mathbf{g}(t)] = \left[\lim_{t\to t_0}\mathbf{f}(t)\right]\times\left[\lim_{t\to t_0}\mathbf{g}(t)\right]$

(iv) $\lim_{t\to t_0}|\mathbf{f}(t)| = \left|\lim_{t\to t_0}\mathbf{f}(t)\right|$

(v) $\lim_{t\to t_0}[\phi(t)\mathbf{f}(t)] = \left[\lim_{t\to t_0}\phi(t)\right]\left[\lim_{t\to t_0}\mathbf{f}(t)\right].$

21.6 DIFFERENTIATION OF A VECTOR FUNCTION WITH RESPECT TO A SCALAR

Definition (1) *Let* $\mathbf{f}(t)$ *be a vector function of scalar variable t. The function* $\mathbf{f}(t)$ *is differentiable with respect to t if*

$$\lim_{\delta t\to 0}\frac{\mathbf{f}(t+\delta t)-\mathbf{f}(t)}{\delta t} \text{ exists.}$$

and it is denoted by $\frac{d\mathbf{f}(t)}{dt}$.

Definition (2). *If* $\frac{d\mathbf{f}(t)}{dt}$ *exists, then* $\mathbf{f}(t)$ *is differentiable and* $\frac{d\mathbf{f}(t)}{dt}$ *is also a vector function of variable t. If* $\frac{d\mathbf{f}(t)}{dt}$ *is differentiable, then* $\frac{d^2\mathbf{f}(t)}{dt^2}$ *is called second derivative of* $\mathbf{f}(t)$*. Similarly we can find third, fourth, etc. derivaties of* $\mathbf{f}(t)$*.*

REMARK

- If $\mathbf{r} = \mathbf{f}(t)$, then $\frac{d\mathbf{r}}{dt}, \frac{d^2\mathbf{r}}{dt^2}$, etc. are the first, second etc. derivaties of $r = \mathbf{f}(t)$ and also denoted by $\dot{\mathbf{r}}, \ddot{\mathbf{r}}$ respectively etc.

21.7 DIFFERENTIATION FORMULAE FOR THE VECTOR FUNCTION

Let **a**, **b**, **c** be differentiable vector function of a scalar variable f and ϕ be a differentiable scalar function of t, then

(i) $\frac{d}{dt}(\mathbf{a}\pm\mathbf{b}) = \frac{d\mathbf{a}}{dt}\pm\frac{d\mathbf{b}}{dt}$

(ii) $\frac{d}{dt}(\mathbf{a}.\mathbf{b}) = \mathbf{a}.\frac{d\mathbf{b}}{dt}+\frac{d\mathbf{a}}{dt}.\mathbf{b}$

(iii) $\frac{d}{dt}(\mathbf{a}\times\mathbf{b}) = \mathbf{a}\times\frac{d\mathbf{b}}{dt}+\frac{d\mathbf{a}}{dt}\times\mathbf{b}$

(iv) $\frac{d}{dt}(\phi\mathbf{a}) = \phi\frac{d\mathbf{a}}{dt}+\frac{d\phi}{dt}\mathbf{a}$

(v) $\frac{d}{dt}[\mathbf{a}\,\mathbf{b}\,\mathbf{c}] = \left[\frac{d\mathbf{a}}{dt}\mathbf{b}\,\mathbf{c}\right]+\left[\mathbf{a}\frac{d\mathbf{b}}{dt}\mathbf{c}\right]+\left[\mathbf{a}\,\mathbf{b}\frac{d\mathbf{c}}{dt}\right]$

(vi) $\frac{d}{dt}\{\mathbf{a}\times(\mathbf{b}\times\mathbf{c})\} = \frac{d\mathbf{a}}{dt}\times(\mathbf{b}\times\mathbf{c})+\mathbf{a}\times\left[\frac{d\mathbf{b}}{dt}\times\mathbf{c}\right]+\mathbf{a}\times\left[\mathbf{b}\times\frac{d\mathbf{c}}{dt}\right].$

Proof. (i) Let $\mathbf{p} = \mathbf{a} + \mathbf{b}.$

$\Rightarrow$ $\mathbf{p} + \delta\mathbf{p} = \mathbf{a} + \mathbf{b} + \delta\mathbf{a} + \delta\mathbf{b}$

$\therefore$ $\delta\mathbf{p} = (\mathbf{p} + \delta\mathbf{p}) - \mathbf{p}$

$\delta\mathbf{p} = \delta\mathbf{a} + \delta\mathbf{b}.$

Divide by δt and taking the limit of both sides as $\delta t \to 0$, we get

$$\lim_{\delta t\to 0}\frac{\delta\mathbf{p}}{\delta t} = \lim_{\delta t\to 0}\left(\frac{\delta\mathbf{a}+\delta\mathbf{b}}{\delta t}\right) = \lim_{\delta t\to 0}\frac{\delta\mathbf{a}}{\delta t}+\lim_{\delta t\to 0}\frac{\delta\mathbf{b}}{\delta t}$$

$\therefore$ $\frac{d\mathbf{p}}{dt} = \frac{d\mathbf{a}}{dt}+\frac{d\mathbf{b}}{dt}$

or $\frac{d}{dt}(\mathbf{a}+\mathbf{b}) = \frac{d\mathbf{a}}{dt}+\frac{d\mathbf{b}}{dt}$

Similarly, $\frac{d}{dt}(\mathbf{a}-\mathbf{b}) = \frac{d\mathbf{a}}{dt}-\frac{d\mathbf{b}}{dt}$

Hence, $\frac{d}{dt}(\mathbf{a}\pm\mathbf{b}) = \frac{d\mathbf{a}}{dt}\pm\frac{d\mathbf{b}}{dt}$

(ii) Let $\mathbf{p} = \mathbf{a}\,.\,\mathbf{b}$ ...(1)

$\therefore$ $\mathbf{p} + \delta\mathbf{p} = (\mathbf{a} + \delta\mathbf{a})\cdot(\mathbf{b} + \delta\mathbf{b})$ ...(2)

From (1) and (2), we get

$$\delta\mathbf{p} = (\mathbf{a} + \delta\mathbf{a})\cdot(\mathbf{b} + \delta\mathbf{b}) - \mathbf{a}\cdot\mathbf{b} = \mathbf{a}\cdot\mathbf{b} + \mathbf{a}\cdot\delta\mathbf{b} + \delta\mathbf{a}\cdot\mathbf{b} + \delta\mathbf{a}\cdot\delta\mathbf{b} - \mathbf{a}\cdot\mathbf{b}$$

$$\delta\mathbf{p} = \mathbf{a}\cdot\delta\mathbf{b} + \delta\mathbf{a}\cdot\mathbf{b} + \delta\mathbf{a}\cdot\delta\mathbf{b}$$

$$\therefore \quad \frac{\delta\mathbf{p}}{\delta t} = \mathbf{a}\cdot\frac{\delta\mathbf{b}}{\delta t} + \frac{\delta\mathbf{a}}{\delta t}\cdot\mathbf{b} + \left(\frac{\delta\mathbf{a}}{\delta t}\cdot\frac{\delta\mathbf{b}}{\delta t}\cdot\delta t\right).$$

Now taking the limit of both sides as $\delta t \to 0$, we get

$$\lim_{\delta t\to 0}\frac{\delta\mathbf{p}}{\delta t} = \mathbf{a}\cdot\lim_{\delta t\to 0}\frac{\delta\mathbf{b}}{\delta t} + \lim_{\delta t\to 0}\frac{\delta\mathbf{a}}{\delta t}\cdot\mathbf{b} + \left(\lim_{\delta t\to 0}\frac{\delta\mathbf{a}}{\delta t}\cdot\lim_{\delta t\to 0}\frac{\delta\mathbf{b}}{\delta t}\right)\lim_{\delta t\to 0}\delta t$$

or $$\frac{d\mathbf{p}}{dt} = \mathbf{a}\cdot\frac{d\mathbf{b}}{dt} + \frac{d\mathbf{a}}{dt}\cdot\mathbf{b} \qquad \left[\because \lim_{\delta t\to 0}\delta t = 0\right]$$

or $$\frac{d}{dt}(\mathbf{a}\cdot\mathbf{b}) = \mathbf{a}\cdot\frac{d\mathbf{b}}{dt} + \frac{d\mathbf{a}}{dt}\cdot\mathbf{b} \qquad (\because \mathbf{p} = \mathbf{a}.\mathbf{b})$$

(iii) Let $$\mathbf{p} = \mathbf{a}\times\mathbf{b} \qquad \ldots(1)$$

and $$\mathbf{p} + \delta\mathbf{p} = (\mathbf{a} + \delta\mathbf{a})\times(\mathbf{b} + \delta\mathbf{b})$$

$$= \mathbf{a}\times\mathbf{b} + \mathbf{a}\times\delta\mathbf{b} + \delta\mathbf{a}\times\mathbf{b} + \delta\mathbf{a}\times\delta\mathbf{b} \qquad \ldots(2)$$

Subtract (1) from (2), we get

$$\delta\mathbf{p} = \mathbf{a}\times\delta\mathbf{b} + \delta\mathbf{a}\times\mathbf{b} + \delta\mathbf{a}\times\delta\mathbf{b} \qquad \ldots(3)$$

$$\Rightarrow \quad \frac{\delta\mathbf{p}}{\delta t} = \mathbf{a}\times\frac{\delta\mathbf{b}}{\delta t} + \frac{\delta\mathbf{a}}{\delta t}\times\mathbf{b} + \frac{\delta\mathbf{a}}{\delta t}\times\frac{\delta\mathbf{b}}{\delta t}\delta b$$

$$\therefore \quad \frac{\delta\mathbf{p}}{\delta t} = \mathbf{a}\times\frac{\delta\mathbf{b}}{\delta t} + \frac{\delta\mathbf{a}}{\delta t}\times\mathbf{b} + \frac{\delta\mathbf{a}}{\delta t}\times\frac{\delta\mathbf{b}}{\delta t}\delta t.$$

Now taking the limit of both sides as $\delta t \to 0$, we get

$$\lim_{\delta t\to 0}\frac{\delta\mathbf{p}}{\delta t} = \lim_{\delta t\to 0}\left[\mathbf{a}\times\frac{\delta\mathbf{b}}{\delta t} + \frac{\delta\mathbf{a}}{\delta t}\times\mathbf{b} + \frac{\delta\mathbf{a}}{\delta t}\times\frac{\delta\mathbf{b}}{\delta t}\delta t\right]$$

$$= \lim_{\delta t\to 0}\left(\mathbf{a}\times\frac{\delta\mathbf{b}}{\delta t}\right) + \lim_{\delta t\to 0}\left(\frac{\delta\mathbf{a}}{\delta t}\times\mathbf{b}\right) + \lim_{\delta t\to 0}\left(\frac{\delta\mathbf{a}}{\delta t}\times\frac{\delta\mathbf{b}}{\delta t}\delta t\right)$$

$$= \mathbf{a}\times\lim_{\delta t\to 0}\frac{\delta\mathbf{b}}{\delta t} + \lim_{\delta t\to 0}\frac{\delta\mathbf{a}}{\delta t}\times\mathbf{b} + \left(\lim_{\delta t\to 0}\frac{\delta\mathbf{a}}{\delta t}\times\lim_{\delta t\to 0}\frac{\delta\mathbf{b}}{\delta t}\right)\lim_{\delta t\to 0}\delta t$$

$$\therefore \quad \frac{d\mathbf{p}}{dt} = \mathbf{a}\times\frac{d\mathbf{b}}{dt} + \frac{d\mathbf{a}}{dt}\times\mathbf{b} + \left(\frac{d\mathbf{a}}{dt}\times\frac{d\mathbf{b}}{dt}\right)(0) = \mathbf{a}\times\frac{d\mathbf{b}}{dt} + \frac{d\mathbf{a}}{dt}\times\mathbf{b}$$

or $$\frac{d}{dt}(\mathbf{a}\times\mathbf{b}) = \mathbf{a}\times\frac{d\mathbf{b}}{dt} + \frac{d\mathbf{a}}{dt}\times\mathbf{b} \qquad (\because \mathbf{p} = \mathbf{a}\times\mathbf{b})$$

(iv) $$\frac{d}{dt}(\phi\mathbf{a}) = \phi\frac{d\mathbf{a}}{dt} + \frac{d\phi}{dt}\mathbf{a}$$

Let $$\mathbf{p} = \phi\mathbf{a} \qquad \ldots(1)$$

and $$\mathbf{p} + \delta\mathbf{p} = (\phi + \delta\phi)(\mathbf{a} + \delta\mathbf{a}) = \phi\mathbf{a} + \phi\delta\mathbf{a} + \delta\phi\,\mathbf{a} + \delta\phi\,\delta\mathbf{a} \qquad \ldots(2)$$

Subtract (1) from (2), we get

$$\delta\mathbf{p} = \phi\,\delta\mathbf{a} + \delta\phi\,\mathbf{a} + \delta\phi\,\delta\mathbf{a} \qquad \ldots(3)$$

$$\Rightarrow \quad \frac{\delta\mathbf{p}}{\delta t} = \phi\frac{\delta\mathbf{a}}{\delta t} + \frac{\delta\phi}{\delta t}\mathbf{a} + \frac{\delta\phi}{\delta t}\times\frac{\delta\mathbf{a}}{\delta t}\delta t\,.$$

Now taking the limit of both sides as $\delta t \to 0$, we get

$$\lim_{\delta t\to 0}\frac{\delta\mathbf{p}}{\delta t} = \phi\lim_{\delta t\to 0}\frac{\delta\mathbf{a}}{\delta t} + \lim_{\delta t\to 0}\frac{\delta\phi}{\delta t}\mathbf{a} + \lim_{\delta t\to 0}\frac{\delta\phi}{\delta t}\lim_{\delta t\to 0}\frac{\delta\mathbf{a}}{\delta t}\lim_{\delta t\to 0}\delta t$$

$$\therefore \quad \frac{d\mathbf{p}}{dt} = \phi\frac{d\mathbf{a}}{dt} + \frac{d\phi}{dt}\mathbf{a} + 0$$

or $$\frac{d}{dt}(\phi\mathbf{a}) = \phi\frac{d\mathbf{a}}{dt} + \frac{d\phi}{dt}\mathbf{a} \qquad (\because p = \phi\mathbf{a})$$

(v) Let $\mathbf{p} = [\mathbf{a}\,\mathbf{b}\,\mathbf{c}] = \mathbf{a}\,.\,(\mathbf{b}\times\mathbf{c})$ ...(1)

and let $\mathbf{b}\times\mathbf{c} = \mathbf{q}$, then $\mathbf{p} = \mathbf{a}\,.\,\mathbf{q}$

$\therefore$ $$\frac{d\mathbf{p}}{dt} = \frac{d}{dt}(\mathbf{a}.\mathbf{q}) = \mathbf{a}.\frac{d\mathbf{q}}{dt} + \frac{d\mathbf{a}}{dt}.\mathbf{q} \quad \text{[Using (ii)]} \quad ...(2)$$

Now $$\frac{d\mathbf{q}}{dt} = \frac{d}{dt}(\mathbf{b}\times\mathbf{c}) = \mathbf{b}\times\frac{d\mathbf{c}}{dt} + \frac{d\mathbf{b}}{dt}\times\mathbf{c} \quad \text{[Using (iii)]} \quad ...(3)$$

From (2) and (3), we get

$$\frac{d\mathbf{p}}{dt} = \mathbf{a}.\left(\mathbf{b}\times\frac{d\mathbf{c}}{dt} + \frac{d\mathbf{b}}{dt}\times\mathbf{c}\right) + \frac{d\mathbf{a}}{dt}.\mathbf{b}\times\mathbf{c}$$

$$= \mathbf{a}.\mathbf{b}\times\frac{d\mathbf{c}}{dt} + \mathbf{a}.\frac{d\mathbf{b}}{dt}\times\mathbf{c} + \frac{d\mathbf{a}}{dt}.\mathbf{b}\times\mathbf{c}$$

$$\frac{d\mathbf{p}}{dt} = \left[\mathbf{a}\,\mathbf{b}\frac{d\mathbf{c}}{dt}\right] + \left[\mathbf{a}\frac{d\mathbf{b}}{dt}\mathbf{c}\right] + \left[\frac{d\mathbf{a}}{dt}\mathbf{b}\,\mathbf{c}\right]$$

or $$\frac{d}{dt}[\mathbf{a}\,\mathbf{b}\,\mathbf{c}] = \left[\mathbf{a}\,\mathbf{b}\frac{d\mathbf{c}}{dt}\right] + \left[\mathbf{a}\frac{d\mathbf{b}}{dt}\mathbf{c}\right] + \left[\frac{d\mathbf{a}}{dt}\mathbf{b}\,\mathbf{c}\right] \quad (\because \mathbf{p} = [\mathbf{a}\,\mathbf{b}\,\mathbf{c}])$$

(vi) $$\frac{d}{dt}\{\mathbf{a}\times(\mathbf{b}\times\mathbf{c})\} = \frac{d\mathbf{a}}{dt}\times(\mathbf{b}\times\mathbf{c}) + \mathbf{a}\times\left(\frac{d\mathbf{b}}{dt}\times\mathbf{c}\right) + \mathbf{a}\times\left(\mathbf{b}\times\frac{d\mathbf{c}}{dt}\right)$$

Let $\mathbf{p} = \mathbf{a}\times(\mathbf{b}\times\mathbf{c})$ ...(1)

and let $\mathbf{b}\times\mathbf{c} = \mathbf{q}$

$\therefore$ $\mathbf{p} = \mathbf{a}\times\mathbf{q}$

$\therefore$ $$\frac{d\mathbf{p}}{dt} = \frac{d}{dt}(\mathbf{a}\times\mathbf{q}) = \mathbf{a}\times\frac{d\mathbf{q}}{dt} + \frac{d\mathbf{a}}{dt}\times\mathbf{q} \quad \text{[Using (iii)]} \quad ...(2)$$

Now $$\frac{d\mathbf{q}}{dt} = \frac{d}{dt}(\mathbf{b}\times\mathbf{c})$$

$$\frac{d\mathbf{q}}{dt} = \mathbf{b}\times\frac{d\mathbf{c}}{dt} + \frac{d\mathbf{b}}{dt}\times\mathbf{c} \quad \text{[Using (iii)]} \quad ...(3)$$

From (2) and (3), we get

$$\frac{d\mathbf{p}}{dt} = \mathbf{a}\times\left(\mathbf{b}\times\frac{d\mathbf{c}}{dt} + \frac{d\mathbf{b}}{dt}\times\mathbf{c}\right) + \frac{d\mathbf{a}}{dt}(\mathbf{b}\times\mathbf{c})$$

$$\frac{d\mathbf{p}}{dt} = \mathbf{a}\times\left(\mathbf{b}\times\frac{d\mathbf{c}}{dt}\right) + \mathbf{a}\times\left(\frac{d\mathbf{b}}{dt}\times\mathbf{c}\right) + \frac{d\mathbf{a}}{dt}\times(\mathbf{b}\times\mathbf{c})$$

or $$\frac{d}{dt}\{\mathbf{a}\times(\mathbf{b}\times\mathbf{c})\} = \mathbf{a}\times\left(\mathbf{b}\times\frac{d\mathbf{c}}{dt}\right) + \mathbf{a}\times\left(\frac{d\mathbf{b}}{dt}\times\mathbf{c}\right) + \frac{d\mathbf{a}}{dt}\times(\mathbf{b}\times\mathbf{c}) \quad [\because \mathbf{p} = \mathbf{a}\times(\mathbf{b}\times\mathbf{c})]$$

21.8 DERIVATIVE OF A CONSTANT VECTOR

Definition. *A vector is said to be constant vector if its magnitude as well as direction are fixed.*

Let $\mathbf{r}$ be a constant vector, then $\mathbf{r} = \mathbf{c}$ (a constant vector) ...(1)

$\therefore$ $\mathbf{r} + \delta\mathbf{r} = \mathbf{c}.$ ...(2)

Subtract (1) from (2), we get $\delta\mathbf{r} = 0.$

Divide by δt and taking the limit as $\delta t \to 0$, we get $\lim\limits_{\delta t\to 0}\frac{\delta\mathbf{r}}{\delta t} = 0$ or $\frac{d\mathbf{r}}{dt} = 0.$

Hence the derivative of a constant vector is a zero vector.

21.9 DERIVATIVE OF A VECTOR FUNCTION IN TERMS OF ITS COMPONENTS

Let $P(x, y, z)$ be any point in space and its position vector with respect to the origin O be $\mathbf{r}$ and let x, y, z be the function of scalar variable t, then we have

$$\mathbf{r} = x\hat{i} + y\hat{j} + z\hat{k} \quad ...(1)$$

where $\hat{i}, \hat{j}, \hat{k}$ are constant vectors.

$\therefore$ $$\mathbf{r} + \delta\mathbf{r} = (x+\delta x)\hat{i} + (y+\delta y)\hat{j} + (z+\delta z)\hat{k} \quad ...(2)$$

Subtract (1) from (2), we get

$$\delta\mathbf{r} = \delta x\hat{i} + \delta y\hat{j} + \delta z\hat{k}$$

Now divide this equation by δt and taking the limit as $\delta t \to 0$, we have

$$\lim_{\delta t\to 0}\frac{\delta \mathbf{r}}{\delta t} = \lim_{\delta t\to 0}\left(\frac{\delta x}{\delta t}\hat{i}+\frac{\delta y}{\delta t}\hat{j}+\frac{\delta z}{\delta t}\hat{k}\right)$$

$$\frac{d\mathbf{r}}{dt} = \left(\lim_{\delta t\to 0}\frac{\delta x}{\delta t}\right)\hat{i}+\left(\lim_{\delta t\to 0}\frac{\delta y}{\delta t}\right)\hat{j}+\left(\lim_{\delta t\to 0}\frac{\delta z}{\delta t}\right)\hat{k}$$

$$\therefore \quad \frac{d\mathbf{r}}{dt} = \frac{dx}{dt}\hat{i}+\frac{dy}{dt}\hat{j}+\frac{dz}{dt}\hat{k}.$$

Similarly, we can find $\dfrac{d^2\mathbf{r}}{dt^2}, \dfrac{d^3\mathbf{r}}{dt^3}$, etc.

21.10 DERIVATIVE OF A VECTOR FUNCTION OF FUNCTION

Let $\mathbf{r}$ be a function of a scalar variable u, and u is also a scalar function of scalar variable t.

$\therefore \quad \mathbf{r} = \mathbf{f}(u)$ ...(1)

and $\quad u = g(t)$ ...(2)

$\therefore \quad \mathbf{r} + \delta\mathbf{r} = \mathbf{f}(u + \delta u)$ ...(3)

and $\quad u + \delta u = g(t + \delta t)$ ...(4)

Subtract (2) from (3), we get

$$\delta\mathbf{r} = \mathbf{f}(u + \delta u) - \mathbf{f}(u) \quad \text{...(5)}$$

and subtract (2) from (4), we get

$$\delta u = g(t + \delta t) - g(t) \quad \text{...(6)}$$

Now divide (5) by δt, we have

$$\frac{\delta\mathbf{r}}{\delta t} = \frac{\mathbf{f}(u+\delta u)-\mathbf{f}(u)}{\delta t} = \frac{\mathbf{f}(u+\delta u)-\mathbf{f}(u)}{\delta u}\cdot\frac{\delta u}{\delta t}$$

$$\frac{\delta\mathbf{r}}{\delta t} = \frac{\mathbf{f}(u+\delta u)-\mathbf{f}(u)}{\delta u}\cdot\frac{g(t+\delta t)-g(t)}{\delta t} \quad \text{[Using (6)]}$$

Taking the limit $\delta t \to 0$, when $\delta t \to 0$, $\delta\mathbf{r} \to 0$ and $\delta u \to 0$, we get

$$\lim_{\delta t\to 0}\frac{\delta\mathbf{r}}{\delta t} = \lim_{\delta u\to 0}\frac{\mathbf{f}(u+\delta u)-\mathbf{f}(u)}{\delta u}\cdot\lim_{\delta t\to 0}\frac{g(t+\delta t)-g(t)}{\delta t}$$

$$\frac{d\mathbf{r}}{dt} = \frac{d\mathbf{f}}{du}\frac{dg}{dt} \quad \text{or} \quad \frac{d\mathbf{r}}{dt} = \frac{d\mathbf{r}}{du}\frac{du}{dt} \qquad [\because \mathbf{r} = \mathbf{f}(u), u = g(t)]$$

THEOREM 1. *The vector* $\mathbf{a}(t)$ *has a constant magnitude if and only if* $\mathbf{a}.\dfrac{d\mathbf{a}}{dt} = 0$.

Proof. Let us suppose $\mathbf{a}(t)$ has a constant magnitude. Therefore, $|\mathbf{a}(t)| = a(\text{constant})$

or $\quad \mathbf{a}\cdot\mathbf{a} = a^2(\text{constant})$

$$\therefore \quad \frac{d}{dt}(\mathbf{a}\cdot\mathbf{a}) = \mathbf{a}\cdot\frac{d\mathbf{a}}{dt}+\frac{d\mathbf{a}}{dt}\cdot\mathbf{a} = 2\mathbf{a}\cdot\frac{d\mathbf{a}}{dt} \qquad (\because \mathbf{a}\cdot\mathbf{b} = \mathbf{b}\cdot\mathbf{a})$$

Since $\quad \mathbf{a}\cdot\mathbf{a} = a^2$

$$\frac{d}{dt}(\mathbf{a}\cdot\mathbf{a}) = \frac{d}{dt}(a^2) = 0.$$

$$\therefore \quad 2\mathbf{a}\cdot\frac{d\mathbf{a}}{dt} = 0 \quad \text{or} \quad \mathbf{a}\cdot\frac{d\mathbf{a}}{dt} = 0$$

Conversely, suppose $\mathbf{a}\cdot\dfrac{d\mathbf{a}}{dt} = 0$, then we get

$$\frac{d}{dt}(\mathbf{a}\cdot\mathbf{a}) = \mathbf{a}\cdot\frac{d\mathbf{a}}{dt}+\frac{d\mathbf{a}}{dt}\cdot\mathbf{a} = 2\mathbf{a}\cdot\frac{d\mathbf{a}}{dt} \qquad |\mathbf{a}.\mathbf{b} = \mathbf{b}.\mathbf{a}|$$

$$\frac{d}{dt}(\mathbf{a}.\mathbf{a}) = 0 \quad \text{or} \quad \mathbf{a}.\mathbf{a} = \text{constant}$$

or $\quad |\mathbf{a}|^2 = \text{constant} \quad$ or $\quad |\mathbf{a}| = \text{constant}$

THEOREM 2. *The vector function* $\mathbf{a}(t)$ *is constant vector if and only if* $\frac{d\mathbf{a}}{dt} = 0$.

Proof. Let us suppose first $\mathbf{a}(t)$ is a constant vector such that $\mathbf{a}(t) = \mathbf{c}$ where c is a constant vector, then

$$\mathbf{a}(t + \delta t) = \mathbf{c}$$

$$\therefore \quad \mathbf{a}(t + \delta t) - \mathbf{a}(t) = \mathbf{c} - \mathbf{c} = \mathbf{0}.$$

Divide by δt and taking the limit as $\delta t \to 0$, we get

$$\lim_{\delta t \to 0} \frac{\mathbf{a}(t+\delta t) - \mathbf{a}(t)}{\delta t} = \mathbf{0}. \Rightarrow \quad \frac{d\mathbf{a}}{dt} = \mathbf{0}.$$

Conversely, suppose $\frac{d\mathbf{a}}{dt} = \mathbf{0}$.

Let
$$\mathbf{a}(t) = a_1(t)\hat{i} + a_2(t)\hat{j} + a_3(t)\hat{k}$$

$$\therefore \quad \frac{d\mathbf{a}}{dt} = \frac{da_1(t)}{dt}\hat{i} + \frac{da_2(t)}{dt}\hat{j} + \frac{da_3(t)}{dt}\hat{k}$$

$$\therefore \quad \frac{da_1(t)}{dt}\hat{i} + \frac{da_2(t)}{dt}\hat{j} + \frac{da_3(t)}{dt}\hat{k} = \mathbf{0} \qquad \left(\because \frac{d\mathbf{a}}{dt} = 0\right)$$

This implies
$$\frac{da_1}{dt} = 0, \frac{da_2}{dt} = 0, \frac{da_3}{dt} = 0.$$

Therefore, a_1, a_2, a_3 are all constants.

Hence $\mathbf{a}(t) = a_1\hat{i} + a_2\hat{j} + a_3\hat{k}$ is a constant vector.

THEOREM 3. *If the vector* $\mathbf{a}$ *has a constant magnitude* $\mathbf{a}$, *then* $\mathbf{a}$ *and* $\frac{d\mathbf{a}}{dt}$ *are perpendicular, provided* $\left|\frac{d\mathbf{a}}{dt}\right| \neq 0$.

Proof. Since, we have that $|\mathbf{a}| = a$ (constant), then

$$\mathbf{a} \cdot \mathbf{a} = |\mathbf{a}|^2 = a^2 \text{ (constant)}$$

$$\therefore \quad \frac{d}{dt}(\mathbf{a} \cdot \mathbf{a}) = \mathbf{a} \cdot \frac{d\mathbf{a}}{dt} + \frac{d\mathbf{a}}{dt} \cdot \mathbf{a} = 2\mathbf{a} \cdot \frac{d\mathbf{a}}{dt}$$

and
$$\frac{d}{dt}(\mathbf{a} \cdot \mathbf{a}) = \frac{d}{dt}(a^2) = 0$$

$$\therefore \quad 2\mathbf{a}\frac{d\mathbf{a}}{dt} = 0 \quad \text{or} \quad \mathbf{a}\frac{d\mathbf{a}}{dt} = 0.$$

This implies that vector $\mathbf{a}$ is perpendicular to $\frac{d\mathbf{a}}{dt}$, provided $\left|\frac{d\mathbf{a}}{dt}\right| \neq 0$.

THEOREM 4. *If a vector* $\mathbf{a}$ *is a differentiable vector function of t, then* $\frac{d}{dt}\left(\mathbf{a} \times \frac{d\mathbf{a}}{dt}\right) = \mathbf{a} \times \frac{d^2\mathbf{a}}{dt^2}$.

Proof. Since, we have
$$\frac{d}{dt}(\mathbf{a} \times \mathbf{b}) = \mathbf{a} \times \frac{d\mathbf{b}}{dt} + \frac{d\mathbf{b}}{dt} \times \mathbf{b}.$$

$$\therefore \quad \frac{d}{dt}\left(\mathbf{a} \times \frac{d\mathbf{a}}{dt}\right) = \mathbf{a} \times \frac{d}{dt}\left(\frac{d\mathbf{a}}{dt}\right) + \frac{d\mathbf{a}}{dt} \times \frac{d\mathbf{a}}{dt} = \mathbf{a} \times \frac{d^2\mathbf{a}}{dt^2} + \frac{d\mathbf{a}}{dt} \times \frac{d\mathbf{a}}{dt} = \mathbf{a} \times \frac{d^2\mathbf{a}}{dt^2}$$

$$\left(\because \frac{d\mathbf{a}}{dt} \times \frac{d\mathbf{a}}{dt} = 0 \text{ i.e., cross product of two same vector is zero.}\right)$$

THEOREM 5. *The vector* $\mathbf{a}(t)$ *has a constant direction if and only if* $\mathbf{a} \times \frac{d\mathbf{a}}{dt} = \mathbf{0}$.

Proof. Suppose $\mathbf{a}(t)$ has a constant direction. Let $\hat{a}$ be the unit vector along $\mathbf{a}(t)$ and $|\mathbf{a}(t)| = a$, then

$$\mathbf{a}(t) = a\hat{a}.$$

$$\therefore \quad \frac{d\mathbf{a}}{dt} = \frac{d}{dt}(a\hat{a})$$

$$\frac{d\mathbf{a}}{dt} = \frac{da}{dt}\hat{a} + a\frac{d\hat{a}}{dt}$$

$$\therefore \quad \mathbf{a} \times \frac{d\mathbf{a}}{dt} = \mathbf{a} \times \left(\frac{da}{dt}\hat{a} + a\frac{d\hat{a}}{dt}\right) = \frac{da}{dt}\mathbf{a} \times \hat{a} + a\mathbf{a} \times \frac{d\hat{a}}{dt}$$

or
$$\mathbf{a} \times \frac{d\mathbf{a}}{dt} = a\left(\mathbf{a} \times \frac{d\hat{a}}{dt}\right). \qquad \left(\because \mathbf{a} \times \hat{a} = \mathbf{a} \times \frac{\mathbf{a}}{a} = 0\right) \qquad \ldots(1)$$

Since, **a** has a constant direction, then $\hat{a}$ is a constant vector, and thus we have $\dfrac{d\hat{a}}{dt} = \mathbf{0}$.

$$\therefore \quad \mathbf{a} \times \frac{d\mathbf{a}}{dt} = a(\mathbf{a} \times \mathbf{0}) = \mathbf{0}.$$

Conversely, suppose $\mathbf{a} \times \dfrac{d\mathbf{a}}{dt} = 0$, then from (1)

$$a\left(\mathbf{a} \times \frac{d\hat{a}}{dt}\right) = 0 \quad \text{or} \quad \mathbf{a} \times \frac{d\hat{a}}{dt} = 0 \quad \text{or} \quad \hat{a} \times \frac{d\hat{a}}{dt} = 0 \qquad \ldots(2)$$

$[\because \mathbf{a} = a\hat{a}]$

Since $\hat{a}$ has a constant magnitude, then by theorem (1)

$$\hat{a} \times \frac{d\hat{a}}{dt} = 0 \cdot \qquad \ldots(3)$$

From (2) and (3), we get $\dfrac{d\hat{a}}{dt} = \mathbf{0}$.

This implies $\hat{a}$ is a constant vector hence **a** has a constant directions.

21.11 CURVES IN THREE DIMENSIONAL SPACE

Let $f(x, y, z) = 0$ and $\phi(x, y, z) = 0$ be two surfaces, then a curve in three dimensional space is obtained by the intersection of the surfaces $f(x, y, z) = 0$ and $\phi(x, y, z) = 0$. Therefore, the equation is of the form

$$x = f_1(t), y = f_2(t), z = f_3(t) \qquad \ldots(1)$$

and it also represents a curve in three dimensional space. Where t takes the value between a and b, *i.e.* $a \le t \le b$. Let (x, y, z) be any point on the curve (1) and let **r** be its position vector, then we have

$$\mathbf{r} = x\hat{i} + y\hat{j} + z\hat{k} \quad \text{and} \quad \mathbf{f}(t) = f_1(t)\hat{i} + f_2(t)\hat{j} + f_3(t)\hat{k}.$$

$\therefore$ From (1), we have

$$x\hat{i} + y\hat{j} + z\hat{k} = f_1(t)\hat{i} + f_2(t)\hat{j} + f_3(t)\hat{k} \quad \text{or} \quad \mathbf{r} = \mathbf{f}(t). \qquad \ldots(2)$$

Thus equation (2) represents a curve in three dimensional space.

21.11.1 GEOMETRICAL INTERPRETATION OF $\dfrac{d\mathbf{r}}{dt}$

Let $\mathbf{r} = \mathbf{f}(t)$ be a curve in three dimensional space and let P and Q be two neighbouring points on this curve and **r** and $\mathbf{r} + \delta\mathbf{r}$ be the position vectors of P adn Q respectively as shown in fig. 1.

$$\therefore \quad \mathbf{r} = \overrightarrow{OP} = \mathbf{f}(t) \quad \text{and} \quad \overrightarrow{OQ} = \mathbf{r}\delta\mathbf{r} = \mathbf{f}(t + \delta t)$$

$$\therefore \quad \overrightarrow{PQ} = \overrightarrow{OQ} - \overrightarrow{OP} = \mathbf{r} + \delta\mathbf{r} - \mathbf{r} = \delta\mathbf{r}.$$

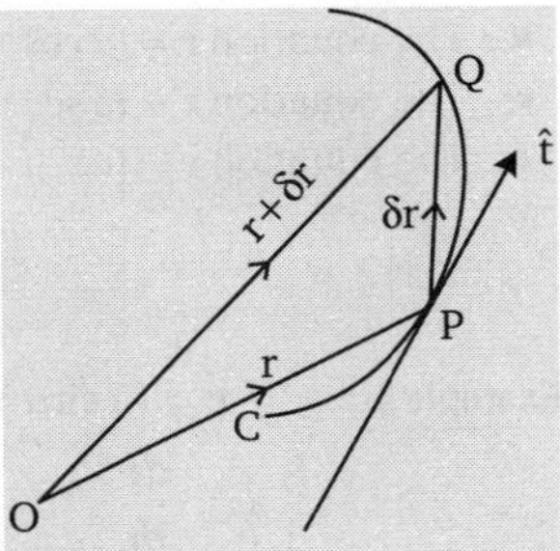

Thus $\dfrac{\delta\mathbf{r}}{\delta t}$ is a vector parallel to the vector $\overrightarrow{PQ}$ (which is a chord).

As $\delta t \to 0$ *i.e.*, $Q \to P$, then the chord PQ tending to a line at P to the curve. This line is known as tangent.

$$\therefore \quad \lim_{\delta t \to 0} \frac{\delta\mathbf{r}}{\delta t} = \frac{d\mathbf{r}}{dt}.$$

Hence, $\dfrac{d\mathbf{r}}{dt}$ is obtained a vector which is parallel to the line (tangent) at P to the curve $\mathbf{r} = \mathbf{f}(t)$.

Fig. 1

21.11.2 SIGNIFICANCE OF $\dfrac{d\mathbf{r}}{ds}$

Let C be any fixed point on the curve as shown in fig. 1 at $r = f(s)$, where s is the arc length measured from the point C. Then

$$CP = s, CQ = s + \delta s.$$

Thus $\dfrac{d\mathbf{r}}{ds}$ is a vector along the tangent at P to the curve $\mathbf{r} = \mathbf{f}(s)$ in the direction of s increasing.

$$\therefore \quad \frac{d\mathbf{r}}{ds} = \lim_{\delta x \to 0} \frac{\delta\mathbf{r}}{\delta s} = \lim_{Q \to P} \frac{\delta r}{\text{arc } PQ} \qquad (\because \text{ as } \delta s \to 0, Q \to P \text{ and } \delta s = \text{arc length } PQ)$$

or

$$\left|\frac{d\mathbf{r}}{ds}\right| = \lim_{Q \to P} \frac{|\delta r|}{\text{arc } PQ} = \lim_{Q \to P} \frac{\text{chord } PQ}{\text{arc } PQ} \qquad (\because \delta\mathbf{r} = \overrightarrow{PQ})$$

Thus $\dfrac{d\mathbf{r}}{ds}$ is a unit vector along the tangent at P to the curve $\mathbf{r} = \mathbf{f}(s)$. Thus unit vector is known as unit tangent vector and is denoted by $\hat{\mathbf{t}}$.

Hence,

$$\hat{\mathbf{t}} = \frac{d\mathbf{r}}{ds}.$$

21.12 VELOCITY AND ACCELERATION VECTORS

(GBTU–2010)

Let a particle be moving along the curve $\mathbf{r} = \mathbf{f}(t)$. At any instant t the moving particle is at P whose position vector is $\mathbf{r}$. In time interval δt the moving particle reached to the point Q whose position vector is $\mathbf{r} + \delta\mathbf{r}$.

$\therefore$ $\delta\mathbf{r}$ is the displacement of the moving particle in the time interval δt.

Thus $\frac{\delta\mathbf{r}}{\delta t}$ gives an average velocity of the particle during the interval δt.

If the vector $\mathbf{v}$ represents the velocity vector at P, then

$$\mathbf{v} = \lim_{\delta t \to 0} \frac{\delta\mathbf{r}}{\delta t} = \frac{d\mathbf{r}}{dt}.$$

Since $\frac{d\mathbf{r}}{dt}$ is a vector along the tangent at P to the curve. Hence, the vector $\mathbf{v}$ of the particle always along the tangent.

Further, if $\delta\mathbf{v}$ is the change in velocity during the time interval δt, then $\frac{\delta\mathbf{v}}{\delta t}$ represents a vector which gives an average acceleration of the particle.

Let $\mathbf{a}$ be the acceleration vector of the particle, then we have

$$\mathbf{a} = \lim_{\delta t \to 0} \frac{\delta\mathbf{v}}{\delta t} = \frac{d\mathbf{v}}{dt}.$$

Since $\mathbf{v} = \frac{d\mathbf{r}}{dt}$ therefore $\mathbf{a} = \frac{d}{dt}\left(\frac{d\mathbf{r}}{dt}\right) = \frac{d^2\mathbf{r}}{dt^2}.$

Hence $\mathbf{a} = \frac{d\mathbf{v}}{dt} = \frac{d^2\mathbf{r}}{dt^2}.$

Recapitulations

- A vector is said to be constant if its magnitude as well as direction are fixed.
- $\frac{d}{dt}[\mathbf{a}\times(\mathbf{b}\times\mathbf{c})] = \mathbf{a}\times\left(\mathbf{b}\times\frac{d\mathbf{c}}{dt}\right) + \mathbf{a}\times\left(\frac{d\mathbf{b}}{dt}\times\mathbf{c}\right) + \frac{d\mathbf{a}}{dt}\times(\mathbf{b}\times\mathbf{c})$
- $\frac{d\mathbf{r}}{dt} = \frac{dx}{dt}\hat{i} + \frac{dy}{dt}\hat{j} + \frac{dz}{dt}\hat{k}$
- The vector $\mathbf{a}(t)$ has a constant magnitude if and only if $\mathbf{a}.\frac{d\mathbf{a}}{dt} = 0$.
- A vector function $\mathbf{a}(t)$ is constant if and only if $\frac{d\mathbf{a}}{dt} = 0$.
- If the vector $\mathbf{a}$ has a constant magnitude a then $\mathbf{a}$ and $\frac{d\mathbf{a}}{dt}$ are perpendicular provided $\left|\frac{d\mathbf{a}}{dt}\right| \neq 0$.
- $\frac{d}{dt}\left(\mathbf{a}\times\frac{d\mathbf{a}}{dt}\right) = \mathbf{a}\times\frac{d^2\mathbf{a}}{dt^2}$.
- The vector $\mathbf{a}(t)$ has a constant direction if and only if $\mathbf{a}\times\frac{d\mathbf{a}}{dt} = 0$.
- Velocity $\mathbf{v} = \frac{d\mathbf{r}}{dt}$; Acceleration $\mathbf{a} = \frac{d\mathbf{v}}{dt} = \frac{d^2\mathbf{r}}{dt^2}$

REMARKS

- The equation $\mathbf{r} = (a\cos t)\hat{i} + (b\sin t)\hat{j} + 0.\hat{k}$ represents the equation of an ellipse.
- The equation $\mathbf{r} = (a\sec t)\hat{i} + (b\tan t)\hat{j} + 0.\hat{k}$ represents the equation of hyperbola.
- The equation $\mathbf{r} = (at^2)\hat{i} + (2at)\hat{j} + 0.\hat{k}$ represents the equation of a parabola.

SOLVED EXAMPLES

Example 1. *If* $\mathbf{r} = (2\sin t)\hat{i} + (3\cos t)\hat{j} + t\hat{k}$, *find*

(i) $\frac{d\mathbf{r}}{dt}$ (ii) $\left|\frac{d\mathbf{r}}{dt}\right|$

(iii) $\frac{d^2\mathbf{r}}{dt^2}$ (iv) $\left|\frac{d^2\mathbf{r}}{dt^2}\right|$

Solution . Since we know that $\hat{i}, \hat{j}, \hat{k}$ are constant vectors, so

$$\frac{d\hat{i}}{dt} = \mathbf{0}, \frac{d\hat{j}}{dt} = \mathbf{0} \text{ and } \frac{d\hat{k}}{dt} = \mathbf{0}.$$

(i) $\mathbf{r} = (2\sin t)\hat{i} + (3\cos t)\hat{j} + t\hat{k}$,

$$\therefore \frac{d\mathbf{r}}{dt} = (2\cos t)\hat{i} - (3\sin t)\hat{j} + \hat{k}$$

(ii) $\left|\frac{d\mathbf{r}}{dt}\right| = \sqrt{4\cos^2 t + 9\sin^2 t + 1}$

$$= \sqrt{5(1 + \sin^2 t)}.$$

(iii) $\frac{d^2\mathbf{r}}{dt^2} = \frac{d}{dt}\left(\frac{d\mathbf{r}}{dt}\right)$

$$= \frac{d}{dt}(2\cos t\hat{i} - 3\sin t\hat{j} + \hat{k})$$

$$= -2\sin t\hat{i} - 3\cos t\hat{j}.$$

(iv) $\left|\frac{d^2\mathbf{r}}{dt^2}\right| = \sqrt{(4\sin^2 t + 9\cos^2 t)}$

$$= \sqrt{(4 + 5\cos^2 t)}.$$

Example 2. *If $\hat{r}$ be a unit vector in the direction of* $\mathbf{r}$, *prove that* $\hat{r}\times\frac{d\hat{r}}{dt} = \frac{1}{r^2}\mathbf{r}\times\frac{d\mathbf{r}}{dt}$, *where* $|\mathbf{r}| = r$.

Solution . Since $\hat{r}$ is a unit vector along the vector $\mathbf{r}$, so we have

$$\mathbf{r} = r\hat{r} \qquad \ldots(1)$$

$$\because \quad |\mathbf{r}| = r.$$

Differentiating w.r.t. t of both sides, we get

$$\frac{d\mathbf{r}}{dt} = \frac{d}{dt}(r\hat{r})$$

$$\therefore \quad \frac{d\mathbf{r}}{dt} = r\frac{d\hat{r}}{dt} + \frac{dr}{dt}\hat{r} \qquad \ldots(2)$$

Now $\mathbf{r} \times \frac{d\mathbf{r}}{dt} = \mathbf{r} \times \left(r\frac{d\hat{r}}{dt} + \frac{dr}{dt}\hat{r}\right)$

$$= r\mathbf{r} \times \frac{d\hat{r}}{dt} + \frac{dr}{dt}\mathbf{r} \times \hat{r}$$

$$= r(r\hat{r}) \times \frac{d\hat{r}}{dt} + \frac{dr}{dt} r\hat{r} \times \hat{r} \quad (\because \mathbf{r} = r\hat{r})$$

$$= r^2\hat{r} \times \frac{d\hat{r}}{dt} + 0$$

($\because$ Cross product of same vector is zero, *i.e.*, $\hat{r} \times \hat{r} = \mathbf{0}$)

$$= r^2\hat{r} \times \frac{d\hat{r}}{dt}$$

$$\therefore \quad \hat{r} \times \frac{d\hat{r}}{dt} = \frac{1}{r^2}\mathbf{r} \times \frac{d\mathbf{r}}{dt}$$

Example 3. *If* $\mathbf{r} = \mathbf{a} \sin \omega t + \mathbf{b} \cos \omega t + \frac{\mathbf{c}t}{\omega^2} \sin \omega t$, *prove that*

$$\frac{d^2\mathbf{r}}{dt^2} + \omega^2\mathbf{r} = \frac{2\mathbf{c}}{\omega} \cos \omega t,$$

where **a**, **b**, **c**, *are constant vectors and* ω *is a constant scalar.*

Solution. Since **a**, **b**, **c**, are constant vectors so

$$\frac{d\mathbf{a}}{dt} = 0, \frac{d\mathbf{b}}{dt} = 0 \text{ and } \frac{d\mathbf{c}}{dt} = 0.$$

and $\mathbf{r} = \mathbf{a} \sin \omega t + \mathbf{b} \cos \omega t + \frac{\mathbf{c}t}{\omega^2} \sin \omega t \quad \ldots(1)$

$$\therefore \quad \frac{d\mathbf{r}}{dt} = \omega\mathbf{a} \cos \omega t - \omega\mathbf{b} \sin \omega t + \frac{\mathbf{c}}{\omega^2} \sin \omega t + \frac{\mathbf{c}t}{\omega} \cos \omega t$$

and $\frac{d^2\mathbf{r}}{dt^2} = \frac{d}{dt}\left(\frac{d\mathbf{r}}{dt}\right)$

$$= -\omega^2\mathbf{a} \sin \omega t - \omega^2\mathbf{b} \cos \omega t + \frac{\mathbf{c}}{\omega} \cos \omega t + \frac{\mathbf{c}}{\omega} \cos \omega t - \mathbf{c}t \sin \omega t$$

$$= -\omega^2\left(\mathbf{a} \sin \omega t + \mathbf{b} \cos \omega t + \frac{\mathbf{c}t}{\omega^2} \cos \omega t\right) + \frac{2\mathbf{c}}{\omega} \cos \omega t$$

$$= -\omega^2\mathbf{r} + \frac{2\mathbf{c}}{\omega} \cos \omega t$$

$$\therefore \quad \frac{d^2\mathbf{r}}{dt^2} + \omega^2\mathbf{r} = \frac{2\mathbf{c}}{\omega} \cos \omega t.$$

Example 4. *If* $\mathbf{r} = (\cos nt)\hat{i} + (\sin nt)\hat{j}$, *where n is a constant and t varies, show that*

$$\mathbf{r} \times \frac{d\mathbf{r}}{dt} = n\hat{k}.$$

Solution. Since $\hat{i}$ and $\hat{j}$ are constant vectors so

$$\frac{d\hat{i}}{dt} = \mathbf{0}, \frac{d\hat{j}}{dt} = \mathbf{0} \text{ and}$$

$$\mathbf{r} = (\cos nt)\hat{i} + (\sin nt)\hat{j} \qquad \ldots(1)$$

Differentiating (1) w.r.t. 't', we get

$$\frac{d\mathbf{r}}{dt} = -n(\sin nt)\hat{i} + n(\cos nt)\hat{j} \qquad \ldots(2)$$

Now

$$\mathbf{r} \times \frac{d\mathbf{r}}{dt} = \mathbf{r} \times [-n(\sin nt)\hat{i} + n(\cos nt)\hat{j}]$$

$$= [(\cos nt)\hat{i} + (\sin nt)\hat{j}] \times [-n(\sin nt)\hat{i} + n(\cos nt)\hat{j}]$$

[From (1)]

$$= (n\cos^2 nt)\hat{i} \times \hat{j} - n(\sin^2 nt)\hat{j} \times \hat{i}$$

$$= n(\cos^2 nt)\hat{k} + n(\sin^2 nt)\hat{k}$$

$[\because \hat{j} \times \hat{i} = -\hat{k} \text{ and } \hat{i} \times \hat{j} = \hat{k}]$

$$= (\cos^2 nt + \sin^2 nt)n\hat{k} = n\hat{k}$$

$(\because \cos^2 nt + \sin^2 nt = 1)$

Hence, $\mathbf{r} \times \frac{d\mathbf{r}}{dt} = n\hat{k}$.

Example 5. *If* $\frac{d\mathbf{a}}{dt} = \mathbf{c} \times \mathbf{a}, \frac{d\mathbf{b}}{dt} = \mathbf{c} \times \mathbf{b}$ *show that*

$$\frac{d}{dt}(\mathbf{a} \times \mathbf{b}) = \mathbf{c} \times (\mathbf{a} \times \mathbf{b}).$$

Solution. Since we know that

$$\frac{d}{dt}(\mathbf{a} \times \mathbf{b}) = \mathbf{a} \times \frac{d\mathbf{b}}{dt} + \frac{d\mathbf{a}}{dt} \times \mathbf{b}$$

$$= \mathbf{a} \times (\mathbf{c} \times \mathbf{b}) + (\mathbf{c} \times \mathbf{a}) \times \mathbf{b}$$

$$\left(\because \frac{d\mathbf{a}}{dt} = \mathbf{c} \times \mathbf{a}, \frac{d\mathbf{b}}{dt} = \mathbf{c} \times \mathbf{b}\right)$$

$$= [(\mathbf{a} \cdot \mathbf{b})\mathbf{c} - (\mathbf{a} \cdot \mathbf{c})\mathbf{b}] - [(\mathbf{b} \cdot \mathbf{a})\mathbf{c} - (\mathbf{b} \cdot \mathbf{c})\mathbf{a}]$$

$$= (\mathbf{a} \cdot \mathbf{b})\mathbf{c} - (\mathbf{a} \cdot \mathbf{c})\mathbf{b} - (\mathbf{a} \cdot \mathbf{b})\mathbf{c} + (\mathbf{b} \cdot \mathbf{c})\mathbf{a}$$

$(\because \mathbf{a} \cdot \mathbf{b} = \mathbf{b} \cdot \mathbf{a})$

$$= (\mathbf{b} \cdot \mathbf{c})\mathbf{a} - (\mathbf{a} \cdot \mathbf{c})\mathbf{b}$$

$$= (\mathbf{c} \cdot \mathbf{b})\mathbf{a} - (\mathbf{c} \cdot \mathbf{a})\mathbf{b}$$

($\because$ dot products are commutative.)

$$= \mathbf{c} \times (\mathbf{a} \times \mathbf{b}).$$

$$\therefore \frac{d}{dt}(\mathbf{a} \times \mathbf{b}) = \mathbf{c} \times (\mathbf{a} \times \mathbf{b}).$$

Example 6. *If* **r** *is a vector function of a scalar variable t,* $|\mathbf{r}| = r$ *and* **a**, **b** *are constant vectors, then differentiate the following with respect to t :*

(i) $\mathbf{r}^2 + \frac{1}{\mathbf{r}^2}$ (ii) $r^n\mathbf{r}$

(iii) $(a\mathbf{r} + r\mathbf{b})^2$.

Solution. Since **a**, **b** are constant vectors, so

$$\frac{d\mathbf{a}}{dt} = 0, \frac{d\mathbf{b}}{dt} = 0 \text{ and } |\mathbf{a}| = a \text{ and } |\mathbf{r}| = r.$$

(i) Let $\mathbf{R} = \mathbf{r}^2 + \frac{1}{\mathbf{r}^2}$

$$\therefore \quad \frac{d\mathbf{R}}{dt} = \frac{d}{dt}\left(\mathbf{r}^2 + \frac{1}{\mathbf{r}^2}\right)$$

$$= \frac{d}{dt}(\mathbf{r}^2) + \frac{d}{dt}\left(\frac{1}{\mathbf{r}^2}\right)$$

$$= \frac{d}{dt}(\mathbf{r}.\mathbf{r}) + \frac{d}{dt}\left(\frac{1}{\mathbf{r}.\mathbf{r}}\right)$$

$$= \frac{d}{dt}(r^2) + \frac{d}{dt}\left(\frac{1}{r^2}\right)$$

$$= 2r\frac{dr}{dt} - \frac{2}{r^3}\frac{dr}{dt}$$

(ii) Let $\mathbf{R} = r^n\mathbf{r}$.

$$\frac{d\mathbf{R}}{dt} = \frac{d}{dt}(r^n\mathbf{r}) = r^n\frac{d\mathbf{r}}{dt} + \frac{d(r^n)}{dt}\mathbf{r}$$

$$= r^n\frac{d\mathbf{r}}{dt} + nr^{n-1}\frac{d\mathbf{r}}{dt}\mathbf{r}.$$

(iii) Let $\mathbf{R} = (a\mathbf{r} + r\mathbf{b})^2$.

$$\therefore \frac{d\mathbf{R}}{dt} = \frac{d}{dt}(a\mathbf{r} + r\mathbf{b})^2$$

$$= 2(a\mathbf{r} + r\mathbf{b}).\frac{d}{dt}(a\mathbf{r} + r\mathbf{b})$$

$$= 2(a\mathbf{r} + r\mathbf{b}).\left(a\frac{d\mathbf{r}}{dt} + \frac{da}{dt}\mathbf{r} + r\frac{d\mathbf{b}}{dt} + \frac{dr}{dt}\mathbf{b}\right)$$

$$= 2(a\mathbf{r} + r\mathbf{b}).\left(a\frac{d\mathbf{r}}{dt} + \frac{dr}{dt}\mathbf{b}\right)$$

$$\left(\because \frac{d\mathbf{a}}{dt} = 0, \frac{d\mathbf{b}}{dt} = 0\right)$$

Example 7. *Show that*

$$\frac{d}{dt}\left[\mathbf{a}.\left(\frac{d\mathbf{a}}{dt} \times \frac{d^2\mathbf{a}}{dt^2}\right)\right] = \mathbf{a}.\left(\frac{d\mathbf{a}}{dt} \times \frac{d^3\mathbf{a}}{dt^3}\right).$$

Solution . Let $A = \left[\mathbf{a}.\left(\frac{d\mathbf{a}}{dt} \times \frac{d^2\mathbf{a}}{dt^2}\right)\right]$

Then A is the scalar triple product of three vectors $\mathbf{a}$, $\frac{d\mathbf{a}}{dt}$ and $\frac{d^2\mathbf{a}}{dt^2}$. Therefore using the rule for finding the derivative of a scalar triple product, we have

$$\frac{dA}{dt} = \frac{d\mathbf{a}}{dt}.\left(\frac{d\mathbf{a}}{dt} \times \frac{d^2\mathbf{a}}{dt^2}\right) + \mathbf{a}.\left(\frac{d^2\mathbf{a}}{dt^2} \times \frac{d^2\mathbf{a}}{dt^2}\right) + \mathbf{a}.\left(\frac{d\mathbf{a}}{dt} \times \frac{d^3\mathbf{a}}{dt^3}\right)$$

$$= \mathbf{a}.\left(\frac{d\mathbf{a}}{dt} \times \frac{d^3\mathbf{a}}{dt^3}\right),$$

since scalar triple products having two equal vectors vanish.

Example 8. *A particle moves along the curve*

$$x = 2t^2, y = t^2 - 4t, z = 3t - 5,$$

where t is the time. Find the magnitude of the velocity and acceleration at $t = 0$. (VTU–2008)

Solution . Since $r = x\hat{i} + y\hat{j} + z\hat{k}$

$$= 2t^2\hat{i} + (t^2 - 4t)\hat{j} + (3t - 5)\hat{k}$$

Velocity vector

$$\mathbf{v} = \frac{d\mathbf{r}}{dt} = 4t\hat{i} + (2t - 4)\hat{j} + 3\hat{k}$$

Acceleration vector

$$\mathbf{a} = \frac{d\mathbf{v}}{dt} = 4\hat{i} + 2\hat{j} + 0\hat{k}$$

$$\therefore \text{Velocity} = |\mathbf{v}| = \sqrt{(4t)^2 + (2t-4)^2 + 3^2}$$

and

$$\text{Acceleration} = |\mathbf{a}| = \sqrt{(4)^2 + (2)^2 + 0} = \sqrt{20}$$

$\therefore$ at $t = 0$, velocity $= \sqrt{0 + (0-4)^2 + 9} = 5$

and acceleration $= \sqrt{20} = 2\sqrt{5}$.

Example 9. *If* $\mathbf{a} = \sin\theta\hat{i} + \cos\theta\hat{j} + \theta\hat{k}$, $\mathbf{b} = \cos\theta\hat{i} - \sin\theta\hat{j} - 3\hat{k}$ *and* $\mathbf{c} = 2\hat{i} + 3\hat{j} - 3\hat{k}$, *find* $\frac{d}{d\theta}\{\mathbf{a} \times (\mathbf{b} \times \mathbf{c})\}$ *at* $\theta = \pi/2$.

Solution . Since we have

$$\mathbf{a} \times (\mathbf{b} \times \mathbf{c}) = (\mathbf{a}.\mathbf{c})\mathbf{b} - (\mathbf{a}.\mathbf{b})\mathbf{c} \qquad \ldots(1)$$

$$\therefore \quad \mathbf{a}.\mathbf{c} = (\sin\theta\hat{i} + \cos\theta\hat{j} + \theta\hat{k}).(2\hat{i} + 3\hat{j} - 3\hat{k})$$

$$= (2\sin\theta + 3\cos\theta - 3\theta)$$

$$(\because \hat{i}\cdot\hat{i} = \hat{j}\cdot\hat{j} = \hat{k}\cdot\hat{k} = 1, \hat{i}\cdot\hat{j} = 0, \text{etc.})$$

and $\mathbf{a}.\mathbf{b} =$

$$(\sin\theta\hat{i} + \cos\theta\hat{j} + \theta\hat{k}).(\cos\theta\hat{i} - \sin\theta\hat{j} - 3\hat{k})$$

$$= \sin\theta\cos\theta - \cos\theta\sin\theta - 3\theta = -3\theta$$

From (1), we get

$$\mathbf{a} \times (\mathbf{b} \times \mathbf{c}) = (2\sin\theta + 3\cos\theta - 3\theta)(\cos\theta\hat{i} - \sin\theta\hat{j} - 3\hat{k}) + 3\theta(2\hat{i} + 3\hat{j} - 3\hat{k})$$

$$= (2\sin\theta\cos\theta + 3\cos^2\theta - 3\theta\cos\theta + 6\theta)\hat{i} + (-2\sin^2\theta - 3\cos\theta\sin\theta + 3\theta\sin\theta + 9\theta)\hat{j} + (-6\sin\theta - 9\cos\theta)\hat{k}$$

$$\therefore \frac{d}{d\theta}\{\mathbf{a} \times (\mathbf{b} \times \mathbf{c})\}$$

$$= (2\cos 2\theta - 6\cos\theta\sin\theta - 3\cos\theta + 3\theta\sin\theta + 6)\hat{i} + (-4\sin\theta\cos\theta - 3\cos 2\theta + 3\theta\cos\theta + 9)\hat{j} + (-6\cos\theta + 9\sin\theta)\hat{k}$$

At $\theta = \pi/2$

$$\frac{d}{d\theta}\{\mathbf{a} \times (\mathbf{b} \times \mathbf{c})\}$$

$$= \left(-2 + \frac{3\pi}{2} + 6\right)\hat{i} + (3 + 3 + 9)\hat{j} + 9\hat{k}$$

$$= \left(4 + \frac{3\pi}{2}\right)\hat{i} + 15\hat{j} + 9\hat{k}.$$

Example 10. *A particle moves along the curve* $x = t^3 + 1$, $y = t^2$, $z = 2t + 5$, *where t is the time. Find the components of its velocity and acceleration at* $t = 1$ *in the direction* $\hat{i} + \hat{j} + 3\hat{k}$.

Solution. Let $\mathbf{r}$ be the position vector of a moving particle at any time t at the point (x, y, z) on the curve, then

$$\mathbf{r} = x\hat{i} + y\hat{j} + z\hat{k}.$$

$$\mathbf{r} = (t^3+1)\hat{i} + t^2\hat{j} + (2t+5)\hat{k} \quad \ldots(1)$$

where $\hat{i}, \hat{j}, \hat{k}$ are constant vector.

$$\therefore \quad \frac{d\mathbf{r}}{dt} = 3t^2\hat{i} + 2t\hat{j} + 2\hat{k} \quad \ldots(2)$$

At t = 1

$$\frac{d\mathbf{r}}{dt} = 3\hat{i} + 2\hat{j} + 2\hat{k}$$

It is a velocity vector whose components are 3, 2, 2.

Again from (2)

$$\frac{d^2\mathbf{r}}{dt^2} = 6t\hat{i} + 2\hat{j}.$$

At t = 1

$$\frac{d^2\mathbf{r}}{dt^2} = 6\hat{i} + 2\hat{j}.$$

This is the acceleration vector whose components are 6, 2, 0.

Now we have to find the components in the direction of $\hat{i} + \hat{j} + 3\hat{k}$.

$\therefore$ Unit vector along in the direction of $\hat{i} + \hat{j} + 3\hat{k}$

$$= \frac{\hat{i} + \hat{j} + 3\hat{k}}{\sqrt{11}}.$$

Thus the component of velocity in the direction of $\hat{i} + \hat{j} + 3\hat{k}$ is

$$= (3\hat{i} + 2\hat{j} + 2\hat{k}) \cdot \frac{\hat{i} + \hat{j} + 3\hat{k}}{\sqrt{11}}$$

$$= \frac{3+2+6}{\sqrt{11}} = \frac{11}{\sqrt{11}} = \sqrt{11} \text{ units}$$

and the component of acceleration in the direction of $\hat{i} + \hat{j} + 3\hat{k}$ is

$$= (6\hat{i} + 2\hat{j}) \cdot \frac{\hat{i} + \hat{j} + 3\hat{k}}{\sqrt{11}} = \frac{6+2}{\sqrt{11}} = \frac{8}{\sqrt{11}} \text{ units.}$$

Example 11. *If* $\mathbf{A} = x^2yz\hat{i} - 2xz^3\hat{j} + xz^2\hat{k}$ *and* $\mathbf{B} = 2z\hat{i} + y\hat{j} - x^2\hat{k}$ *find the value of the* $\frac{\partial^2}{\partial x \partial y}(\mathbf{A} \times \mathbf{B})$ *at the point* (1, 0, 2).

Solution. We have

$$\mathbf{A} \times \mathbf{B} = \begin{vmatrix} \hat{i} & \hat{j} & \hat{k} \\ x^2yz & -2xz^3 & xz^2 \\ 2z & y & -x^2 \end{vmatrix}$$

$$= \hat{i}(2x^3z^3 - xyz^2) - \hat{j}(-x^4yz - 2xz^3) + \hat{k}(x^2y^2z + 2xz^4)$$

Then

$$\frac{\partial}{\partial y}(\mathbf{A} \times \mathbf{B}) = \hat{i}(-xz^2) - \hat{j}(-x^4z) + \hat{k}(2x^2yz)$$

and

$$\frac{\partial^2}{\partial x \partial y}(\mathbf{A} \times \mathbf{B}) = \hat{i}(-z^2) - \hat{j}(-4x^3z) + \hat{k}(4xyz)$$

$\therefore$ At the point (1, 0, 2)

$$\frac{\partial^2}{\partial x \partial y}(\mathbf{A} \times \mathbf{B}) = -4\hat{i} + 8\hat{j}$$

Example 12. *Show that* $\frac{d^2}{dt^2}\left(\mathbf{r} \times \frac{d\mathbf{r}}{dt}\right) = \frac{d\mathbf{r}}{dt} \times \frac{d^2\mathbf{r}}{dt^2} + \mathbf{r} \times \frac{d^3\mathbf{r}}{dt^3}$.

Solution.

$$\frac{d}{dt}\left(\mathbf{r} \times \frac{d\mathbf{r}}{dt}\right) = \frac{d\mathbf{r}}{dt} \times \frac{d\mathbf{r}}{dt} + \mathbf{r} \times \frac{d^2\mathbf{r}}{dt^2}$$

$$= \mathbf{r} \times \frac{d^2\mathbf{r}}{dt^2} \quad \left[\because \frac{d\mathbf{r}}{dt} \times \frac{d\mathbf{r}}{dt} = \mathbf{0}\right]$$

$$\therefore \frac{d^2}{dt^2}\left(\mathbf{r} \times \frac{d\mathbf{r}}{dt}\right) = \frac{d}{dt}\left(\mathbf{r} \times \frac{d^2\mathbf{r}}{dt^2}\right)$$

$$= \frac{d\mathbf{r}}{dt} \times \frac{d^2\mathbf{r}}{dt^2} + \mathbf{r} \times \frac{d^3\mathbf{r}}{dt^3}$$

Example 13. *Show that the radial and transverse acceleration of a particle moving in a plane are* $\frac{d^2\mathbf{r}}{dt^2} - \mathbf{r}\left(\frac{d\theta}{dt}\right)^2$ *and* $2\frac{d\mathbf{r}}{dt} + \mathbf{r}\frac{d^2\theta}{dt^2}$ *respectively.*

(UPTU–2008, Kurukshetra–2006, Rajasthan–2005, 2006)

Solution. Let $\hat{e}_r$ and $\hat{e}_\theta$ be two unit vectors along and perpendicular to radius vector OP where $\hat{i}$ and $\hat{j}$ be two mutually perpendicular unit vectors in the plane.

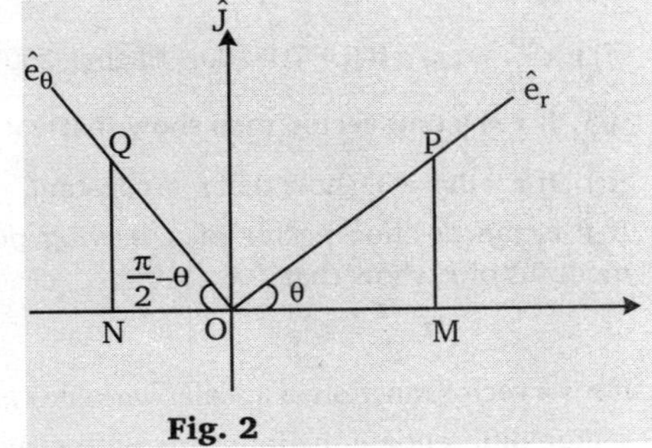

Fig. 2

Then

$$\hat{e}_r = \overrightarrow{OM} = \overrightarrow{OM} + \overrightarrow{MP}$$

$$= \cos\theta\hat{i} + \sin\theta\hat{j} \quad \ldots(1)$$

and

$$\hat{e}_\theta = \cos\left(\frac{\pi}{2} + \theta\right)\hat{i} + \sin\left(\frac{\pi}{2} + \theta\right)\hat{j}$$

$$= -\sin\theta\hat{i} + \cos\theta\hat{j} \quad \ldots(2)$$

Also

$$\mathbf{r} = r\hat{e}_r = r\cos\theta\hat{i} + r\sin\theta\hat{j} \quad \ldots(3)$$

$$\frac{d\hat{e}_r}{dt} = (-\sin\theta\hat{i} + \cos\theta\hat{j})\frac{d\theta}{dt}$$

$$= \hat{e}_\theta \frac{d\theta}{dt} = \dot{\theta}\hat{e}_\theta \quad \ldots(4)$$

$$\frac{d\hat{e}_\theta}{dt} = (-\cos\theta\hat{i} - \sin\theta\hat{j})\frac{d\theta}{dt}$$

$$= -\hat{e}_r \frac{d\theta}{dt} = -\dot{\theta}\hat{e}_r \quad \ldots(5)$$

Now, $\frac{d\mathbf{r}}{dt} = \frac{d}{dt}(\mathbf{r}\cos\theta\hat{i} + \mathbf{r}\sin\theta\hat{j})$

$$= \left(\frac{d\mathbf{r}}{dt}\cos\theta - \mathbf{r}\sin\theta\frac{d\theta}{dt}\right)\hat{i} + \left(\frac{d\mathbf{r}}{dt}\sin\theta + \mathbf{r}\cos\theta\frac{d\theta}{dt}\right)\hat{j}$$

$$= \frac{d\mathbf{r}}{dt}(\cos\theta\hat{i} + \sin\theta\hat{j}) + \mathbf{r}\frac{d\theta}{dt}(-\sin\theta\hat{i} + \cos\theta\hat{j})$$

$$= \dot{\mathbf{r}}\hat{e}_r + \mathbf{r}\dot{\theta}\hat{e}_\theta$$

$\therefore$ Acceleration

$$\frac{d^2\mathbf{r}}{dt^2} = \ddot{\mathbf{r}}\hat{e}_r + \dot{\mathbf{r}}\frac{d\hat{e}_r}{dt} + \dot{\mathbf{r}}\dot{\theta}\hat{e}_\theta + \mathbf{r}\ddot{\theta}\hat{e}_\theta + \mathbf{r}\dot{\theta}\frac{d\hat{e}_\theta}{dt}$$

$$= \ddot{\mathbf{r}}\hat{e}_r + 2\dot{\mathbf{r}}\dot{\theta}\hat{e}_\theta + \mathbf{r}\ddot{\theta}\hat{e}_\theta - \mathbf{r}\dot{\theta}^2\hat{e}_r$$

$$= (\ddot{\mathbf{r}} - \mathbf{r}\dot{\theta}^2)\hat{e}_r + (2\dot{\mathbf{r}}\dot{\theta} + \mathbf{r}\ddot{\theta})\hat{e}_\theta$$

$$= \left[\frac{d^2\mathbf{r}}{dt^2} - \mathbf{r}\left(\frac{d\theta}{dt}\right)^2\right]\hat{e}_r + \frac{1}{\mathbf{r}}\frac{d}{dt}\left(\mathbf{r}^2\frac{d\theta}{dt}\right)\hat{e}_\theta$$

Hence, Radial acceleration

$$= \text{Acceleration along } \hat{e}_r = \frac{d^2\mathbf{r}}{dt^2} - \mathbf{r}\left(\frac{d\theta}{dt}\right)^2.$$

Transverse acceleration

$$= \text{Acceleration along } \hat{e}_\theta = \frac{1}{\mathbf{r}}\frac{d}{dt}\left(\mathbf{r}^2\frac{d\theta}{dt}\right).$$

EXERCISE 21.1

1. If $\mathbf{r} = (t+1)\hat{i} + (t^2+t+1)\hat{j} + (t^3+t^2+t+1)\hat{k}$, find $\frac{d\mathbf{r}}{dt}, \frac{d^2\mathbf{r}}{dt^2}$.
2. If **a**, **b** are constant vectors, ω is a constant, and **r** is a vector function of the scalar variable t given by
   $$\mathbf{r} = \cos\omega t\,\mathbf{a} + \sin\omega t\,\mathbf{b}$$
   Show that
   (i) $\frac{d^2\mathbf{r}}{dt^2} + \omega^2\mathbf{r} = \mathbf{0}$
   (ii) $\mathbf{r} \times \frac{d\mathbf{r}}{dt} = \omega\mathbf{a} \times \mathbf{b}$ (UPTU–2007, Bhopal–2007)
3. (i) If **r** is a unit vector, then show that $\left|\mathbf{r} \times \frac{d\mathbf{r}}{dt}\right| = \left|\frac{d\mathbf{r}}{dt}\right|$.
   (ii) If $\mathbf{r} \times d\mathbf{r} = \mathbf{0}$, show that $\hat{r}$ = constant.
4. If **r** is the position vector of a moving point and r is the modulus of **r**, show that
   $$\mathbf{r}.\frac{d\mathbf{r}}{dt} = r\frac{dr}{dt}.$$
5. If **r** is a vector function of a scalar variable t and **a** is a constant vector, differentiate the following with respect to t :
   (i) $\mathbf{r} \times \mathbf{a}$ (ii) $\mathbf{r} \times \frac{d\mathbf{r}}{dt}$
   (iii) $\frac{\mathbf{r} \times \mathbf{a}}{\mathbf{r}.\mathbf{a}}$ (iv) $r^3\mathbf{r} + \mathbf{a} \times \frac{d\mathbf{r}}{dt}$
6. (i) If $\mathbf{r} = \sin t\hat{i} + \cos t\hat{j} + t\hat{k}$, find $\left|\frac{d\mathbf{r}}{dt} \times \frac{d^2\mathbf{r}}{dt^2}\right|, \left|\frac{d^2\mathbf{r}}{dt^2}\right|$.
   (ii) If $\mathbf{r} = a\cos t\hat{i} + a\sin t\hat{j} + at\tan\alpha\hat{k}$, find $\left|\frac{d\mathbf{r}}{dt} \times \frac{d^2\mathbf{r}}{dt^2}\right|$ and $\frac{d\mathbf{r}}{dt}\cdot\left(\frac{d^2\mathbf{r}}{dt^2} \times \frac{d^3\mathbf{r}}{dt^3}\right)$. (Rohtak–2005)
7. If $\mathbf{r} = r^3\hat{i} + \left(2t^3 - \frac{1}{5t^2}\right)\hat{j}$, show that $\mathbf{r} \times \frac{d\mathbf{r}}{dt} = \hat{k}$.
8. Show that if **a**, **b**, **c** are constant vectors, then $\mathbf{r} = \mathbf{a}t^2 + \mathbf{b}t + \mathbf{c}$ is the path of a particle moving with constant acceleration.
9. If $\mathbf{r} = e^{nt}\mathbf{a} + e^{-nt}\mathbf{b}$, where **a**, **b** are constant vectors, show that
   $$\frac{d^2\mathbf{r}}{dt^2} - n^2\mathbf{r} = \mathbf{0}.$$
10. Show that $\mathbf{r} = \mathbf{a}e^{nt} + \mathbf{b}e^{nt}$ is the solution of the differential equation
    $$\frac{d^2\mathbf{r}}{dt^2} - (m+n)\frac{d\mathbf{r}}{dt} + mn\mathbf{r} = 0.$$
    Hence solve the equation $\frac{d^2\mathbf{r}}{dt^2} - \frac{d\mathbf{r}}{dt} - 2\mathbf{r} = 0$.
    where $\mathbf{r} = \hat{i}$ and $\frac{d\mathbf{r}}{dt} = \hat{j}$ at $t = 0$.
11. If $\mathbf{a} = 5t^2\hat{i} + t\hat{j} - t^3\hat{k}$ and $\mathbf{b} = \sin t\hat{i} - \cos t\hat{j}$, then find
    (i) $\frac{d}{dt}(\mathbf{a}.\mathbf{b})$ (ii) $\frac{d}{dt}(\mathbf{a} \times \mathbf{b})$
    (iii) $\frac{d}{dt}(\mathbf{a}.\mathbf{a})$
12. A particle moves along the curve $x = 4\cos t$, $y = 4\sin t$, $z = 6t$. Find the velocity and acceleration at time $t = 0$ and $t = \frac{\pi}{2}$.
13. A particle moves along the curve $x = e^{-t}$, $y = 2\cos 3t$, $z = 2\sin 3t$. Find the velocity and acceleration at any time t and their magnitudes at $t = 0$. (PTU–2003, VTU–2003)
14. Find the unit tangent vector to any point on the curve $x = a\cos t$, $y = a\sin t$, $z = bt$.
15. A particle P is moving on a circle of radius r with constant angular velocity $\omega = \frac{d\theta}{dt}$ show that the acceleration is $-\omega^2\mathbf{r}$.
16. The position vector of a moving particle at a time t is $\mathbf{r} = 3\cos t\hat{i} + 3\sin t\hat{j} + 4t\hat{k}$. Find the tangent and normal components of its acceleration at $t = 1$. (Marathwada–2008)

Hint to Selected Problems

1. $\mathbf{r} = (t+1)\hat{i} + (t^2+t+1)\hat{j} + (t^3+t^2+t+1)\hat{k}$

$\therefore \quad \frac{d\mathbf{r}}{dt} = \hat{i} + (2t+1)\hat{j} + (3t^2+2t+1)\hat{k}$

$\frac{d^2\mathbf{r}}{dt^2} = 2\hat{j} + (6t+2)\hat{k}$.

2. $\mathbf{r} = (\cos\omega t)\mathbf{a} + (\sin\omega t)\mathbf{b}$

$\frac{d\mathbf{r}}{dt} = -\omega\sin\omega t\mathbf{a} + \omega\cos\omega t\mathbf{b}$

$\frac{d^2\mathbf{r}}{dt^2} = -\omega^2[\cos\omega t\mathbf{a} + \sin\omega t\mathbf{b}] = -\omega^2\mathbf{r}$

$\therefore \quad \frac{d^2\mathbf{r}}{dt^2} + \omega^2\mathbf{r} = \mathbf{0}$.

Similarly

$\mathbf{r}\times\frac{d\mathbf{r}}{dt} = (\cos\omega t\,\mathbf{a} + \sin\omega t\mathbf{b})\times(-\omega\sin\omega t\,\mathbf{a} + \omega\cos\omega t\mathbf{b})$

$= \omega\mathbf{a}\times\mathbf{b}$ ·

3. (i) Since $\mathbf{r}$ is a unit vector, then

$|\mathbf{r}| = 1 \quad \Rightarrow \quad \mathbf{r}.\mathbf{r} = 1$

$\Rightarrow \quad \frac{d}{dt}(\mathbf{r}.\mathbf{r}) = 0 \quad \Rightarrow \quad \mathbf{r}.\frac{d\mathbf{r}}{dt} = 0$

$\Rightarrow$ $\mathbf{r}$ is perpendicular to $\frac{d\mathbf{r}}{dt}$

$\therefore \quad \left|\mathbf{r}\times\frac{d\mathbf{r}}{dt}\right| = |\mathbf{r}|\left|\frac{d\mathbf{r}}{dt}\right|\sin\frac{\pi}{2} = \left|\frac{d\mathbf{r}}{dt}\right|$.

(ii) $\mathbf{r} = r\hat{r}$

$\Rightarrow \quad d\mathbf{r} = d(r\hat{r}) = rd\hat{r} + \hat{r}dr$

$\therefore \quad \mathbf{r}\times d\mathbf{r} = r\hat{r}\times(rd\hat{r} + \hat{r}dr) = r^2\hat{r} + d\hat{r}$

$\Rightarrow \quad \mathbf{0} = r^2\hat{r} + d\hat{r} \quad \Rightarrow \quad \hat{r}\times d\hat{r} = \mathbf{0}$

$\therefore \quad \mathbf{r}\times d\mathbf{r} = 0 \quad$ and $\quad \hat{r}\times d\hat{r} = 0$

$\Rightarrow \quad \hat{r}$ = constant.

4. Since $\mathbf{r} = r\hat{r}$

$\Rightarrow \quad \frac{d\mathbf{r}}{dt} = \frac{d}{dr}(r\hat{r}) = \frac{rd\hat{r}}{dt} + \hat{r}\frac{dr}{dt}$

$\therefore \quad \mathbf{r}.\frac{d\mathbf{r}}{dt} = \mathbf{r}.\left(r\frac{d\hat{r}}{dt} + \hat{r}\frac{dr}{dt}\right) = r\mathbf{r}.\frac{d\hat{r}}{dt} + \mathbf{r}.\hat{r}\frac{dr}{dt}$

$= r^2\hat{r}.\frac{d\hat{r}}{dt} + r(r\hat{r})\frac{dr}{dt} = 0 + r\frac{dr}{dt} \qquad [\because \hat{r}.\hat{r} = 1]$

$= r\frac{dr}{dt}$.

6. (i) $\mathbf{r} = \sin t\hat{i} + \cos t\hat{j} + t\hat{k}$

$\therefore \quad \frac{d\mathbf{r}}{dt} = \cos t\hat{i} - \sin t\hat{j} + \hat{k}$

$\frac{d^2\mathbf{r}}{dt^2} = -\sin t\hat{i} - \cos t\hat{j}$

$\therefore \quad \frac{d\mathbf{r}}{dt}\times\frac{d^2\mathbf{r}}{dt^2} = |\cos t\hat{i} - \sin t\hat{j} - \hat{k}| = \sqrt{2}.$

11. (i) $\mathbf{a} = 5t^2\hat{i} + t\hat{j} - t^3\hat{k}$ and $\mathbf{b} = \sin t\hat{i} - \cos t\hat{j}$

$\mathbf{a}.\mathbf{b} = 5t^2\sin t - t\cos t$

$\therefore \quad \frac{d}{dt}(\mathbf{a}.\mathbf{b}) = \frac{d}{dt}(5t^2\sin t - t\cos t)$

$= 10t\sin t + 5t^2\cos t - \cos t + t\sin t$

$= 5t^2\cos t + 11t\sin t - \cos t$

Similarly, we can find $\frac{d}{dt}(\mathbf{a}\times\mathbf{b})$ and $\frac{d}{dt}(\mathbf{a}\cdot\mathbf{b})$

13. $x = e^{-t}, y = 2\cos 3t, z = 2\sin 3t$

$\therefore \quad \mathbf{r} = e^{-t}\hat{i} + 2\cos 3t\hat{j} + 2\sin 3t\hat{k}$

so $\quad \frac{d\mathbf{r}}{dt} = -e^{-t}\hat{i} - 6\sin 3t\hat{j} + 6\cos 3t\hat{k}$

and $\quad \frac{d^2\mathbf{r}}{dt^2} = e^{-t}\hat{i} - 18\cos 3t\hat{j} - 18\sin 3t\hat{k}$

Velocity = $\mathrm{v} = \left(\frac{d\mathbf{r}}{dt}\right)_{\text{at } t=0} = -\hat{i} + 6\hat{k}$

$\therefore \quad |\mathrm{v}| = \sqrt{1+36} = \sqrt{37}$

and $\quad$ acceleration $= \left(\frac{d^2\mathbf{r}}{dt^2}\right)_{\text{at } t=0} = \hat{i} - 18\hat{j}$

$\therefore \quad \left|\frac{d^2\mathbf{r}}{dt^2}\right| = \sqrt{1+324} = \sqrt{325}.$

15. Let i and j be the unit vector two perpendicular radii of the circle.

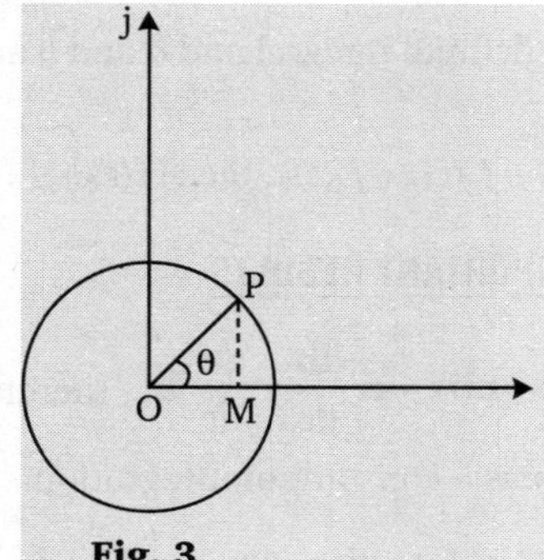

Fig. 3

If P be any point on the circle such that OP makes an angle θ with the direction of i, then the position vector of P is given by

$$\mathbf{r} = \overrightarrow{OP} = \overrightarrow{OM} + \overrightarrow{MP}$$

$$= (r\cos\theta)i + (r\sin\theta)j$$

where r is the radius of circle and hence constant and further

$$\frac{d\mathbf{r}}{dt}, \frac{d^2\mathbf{r}}{dt^2} = -\omega^2\mathbf{r}.$$

ANSWERS

1. $\frac{d\mathbf{r}}{dt} = \hat{i} + (2t+1)\hat{j} + (3t^2+2t+1)\hat{k}$, $\frac{d^2\mathbf{r}}{dt^2} = 2\hat{j} + (6t+2)\hat{k}$ 5. (i) $\frac{d\mathbf{r}}{dt}\times\mathbf{a}$ (ii) $\mathbf{r}\times\frac{d^2\mathbf{r}}{dt^2}$ (iii) $\frac{\frac{d\mathbf{r}}{dt}\times\mathbf{a}}{\mathbf{r.a}} - \frac{\frac{d\mathbf{r}}{dt}.\mathbf{a}}{(\mathbf{r.a})^2}(\mathbf{r}\times\mathbf{a})$

(iv) $3r^2\frac{dr}{dt}\mathbf{r} + r^3\frac{d\mathbf{r}}{dt} + \mathbf{a}\times\frac{d^2\mathbf{r}}{dt^2}$ 6.(i) $\sqrt{2}, 1$ (ii) $a^2 \sec\alpha$, $a^3\tan\alpha$ 10. $\mathbf{r} = \frac{1}{3}(e^{2t}+2e^{-t})\hat{i} + \frac{1}{3}(e^{2t}-e^{-t})\hat{j}$

1. (i) $(5t^2-1)\cos t + 11t\sin t$ (ii) $(t^3\sin t - 3t^2\cos t)\hat{i} - (t^3\cos t + 3t^2\sin t)\hat{j} + (5t^2\sin t - 11t\cos t - \sin t)\hat{k}$ (iii) $100t^3 + 2t + 6t^5$

12. $\mathbf{v} = 4\hat{j}+6\hat{k}, \mathbf{a} = -4\hat{i}$ and $\mathbf{v} = -4\hat{i}+6\hat{k}, \mathbf{a} = -4\hat{j}$ 13. $|\mathbf{v}| = \sqrt{37}, \mathbf{a} = \sqrt{(325)}$ 14. $\frac{1}{\sqrt{(a^2+b^2)}}[-a\sin t\hat{i} + a\cos\hat{j} + b\hat{k}]$ 16. 0, 3

21.13 INTEGRATION OF A VECTOR FUNCTION

Let $\mathbf{F}(t)$ be a differentiable vector function and let $\mathbf{f}(t)$ be its differential coefficient, then

$$\frac{d}{dt}(\mathbf{F}(t)) = \mathbf{f}(t). \qquad \text{...(1)}$$

Therefore the integral of $\mathbf{f}(t)$ is $\mathbf{F}(t)$. Consequently we can say that integration is the reverse process of differentiation.

$$\therefore \qquad \int \mathbf{f}(t)dt = \mathbf{F}(t). \qquad \text{...(2)}$$

This is an indefinite integral and the function $\mathbf{f}(t)$ which is being integrated is known as integrand.

Moreover, let $\mathbf{c}$ be a constant vector which is independent of t, then (1) can also be written as

$$\frac{d}{dt}[\mathbf{F}(t)+\mathbf{c}] = \mathbf{f}(t) \qquad \text{...(3)}$$

$$\therefore \qquad \int \mathbf{f}(t)dt = \mathbf{F}(t)+\mathbf{c}. \qquad \text{...(4)}$$

This constant vector $\mathbf{c}$ is called constant of integration, since this vector $\mathbf{c}$ is taken to be arbitrary so the integral given by (4) is therefore known as indefinite integral.

If $\mathbf{f}(t)$ is defined over the closed interval $[a, b]$, then the integral given in (5)

$$\int_a^b \mathbf{f}(t)dt = [\mathbf{F}(t)+\mathbf{c}]_a^b = \mathbf{F}(b) - \mathbf{F}(a) \qquad \text{...(5)}$$

is called the definite integral and a and b are called limits of integration.

REMARK

- If $\mathbf{f}(t) = f_1(t)\hat{i} + f_2(t)\hat{j} + f_3(t)\hat{k}$, then $\int \mathbf{f}(t)dt = \hat{i}\int f_1(t)dt + \hat{j}\int f_2(t)dt + \hat{k}\int f_3(t)dt$.

21.13.1 SOME IMPORTANT RESULTS

1. Since $\frac{d}{dt}(\mathbf{a.b}) = \mathbf{a}.\frac{d\mathbf{b}}{dt} + \frac{d\mathbf{a}}{dt}.\mathbf{b}$, therefore $\int\left(\mathbf{a}.\frac{d\mathbf{b}}{dt} + \frac{d\mathbf{a}}{dt}.\mathbf{b}\right)dt = \mathbf{a.b} + \mathbf{c}$

   where, $\mathbf{c}$ is a constant of integration.

2. $\frac{d}{dt}(\mathbf{a}\times\mathbf{b}) = \mathbf{a}\times\frac{d\mathbf{b}}{dt} + \frac{d\mathbf{a}}{dt}\times\mathbf{b}$, so $\int\left(\mathbf{a}\times\frac{d\mathbf{b}}{dt} + \frac{d\mathbf{a}}{dt}\times\mathbf{b}\right)dt = (\mathbf{a}\times\mathbf{b}) + \mathbf{c}$, where, $\mathbf{c}$ is a constant vector.

3. Since we have that $\mathbf{a.a} = a^2 \Rightarrow \frac{d}{dt}(\mathbf{a.a}) = \mathbf{a}.\frac{d\mathbf{a}}{dt} + \frac{d\mathbf{a}}{dt}.\mathbf{a} = 2\mathbf{a}.\frac{d\mathbf{a}}{dt}$

   $\therefore \qquad \int\left(2\mathbf{a}.\frac{d\mathbf{a}}{dt}\right)dt = (\mathbf{a.a}) + \mathbf{c}$ or $\int\left(2\mathbf{a}.\frac{d\mathbf{a}}{dt}\right)dt = a^2 + \mathbf{c}$. Here, $\mathbf{c}$ is a scalar quantity.

4. Since we have that

   $$\frac{d}{dt}\left(\mathbf{a}\times\frac{d\mathbf{a}}{dt}\right) = \mathbf{a}\times\frac{d^2\mathbf{a}}{dt^2} + \frac{d\mathbf{a}}{dt}\times\frac{d\mathbf{a}}{dt} = \mathbf{a}\times\frac{d^2\mathbf{a}}{dt^2} \qquad \left(\because \frac{d\mathbf{a}}{dt}\times\frac{d\mathbf{a}}{dt} = 0\right)$$

   $$\therefore \qquad \int\left(\mathbf{a}\times\frac{d^2\mathbf{a}}{dt^2}\right)dt = \left(\mathbf{a}\times\frac{d\mathbf{a}}{dt}\right) + \mathbf{c}$$

   Here, $\mathbf{c}$ is a constant vectror of integration.

5. Since we have that

$$\frac{d}{dt}\left(\mathbf{a}\times\frac{d\mathbf{b}}{dt}\right)=\mathbf{a}\times\frac{d^2\mathbf{b}}{dt^2}+\frac{d\mathbf{a}}{dt}\times\frac{d\mathbf{b}}{dt}$$

If **a** is a constant vector, then $\frac{d\mathbf{a}}{dt}=0$ and $\frac{d}{dt}\left(\mathbf{a}\times\frac{d\mathbf{b}}{dt}\right)=\mathbf{a}\times\frac{d^2\mathbf{b}}{dt^2}$

$$\therefore \quad \int\left(\mathbf{a}\times\frac{d^2\mathbf{b}}{dt^2}\right)dt=\left(\mathbf{a}\times\frac{d\mathbf{b}}{dt}\right)+\mathbf{c}.$$

Here, **c** is a constant vector of integration.

6. If **a** is a constant vector, we have that

$$\frac{d}{dt}(\mathbf{a}\times\mathbf{b})=\mathbf{a}\times\frac{d\mathbf{b}}{dt} \qquad \left(\because \frac{d\mathbf{a}}{dt}=0\right)$$

$$\therefore \quad \int\left(\mathbf{a}\times\frac{d\mathbf{b}}{dt}\right)dt=(\mathbf{a}\times\mathbf{b})+\mathbf{c}.$$

Here, **c** is a constant vector which is constant of integration.

7. If c is constant scalar and **a** is a vector function of t, then $\int c\mathbf{a}dt = c\int \mathbf{a}dt$.

Solved Examples

Example 1. *Interpret the relations*

$$\mathbf{r}.\frac{d\mathbf{r}}{ds}=0 \quad \textit{and} \quad \mathbf{r}\times\frac{d\mathbf{r}}{ds}=0$$

Solution . For $\mathbf{r}.\frac{d\mathbf{r}}{ds}=0 \Rightarrow 2\mathbf{r}.\frac{d\mathbf{r}}{ds}=0$

Integrating w.r.t. s we get $\int\left(2\mathbf{r}.\frac{d\mathbf{r}}{ds}\right)ds=\int 0ds$

or $\mathbf{r}^2 = a$ (constant) $\Rightarrow$ **r** has constant magnitude.

Thus **r** describe a circle.

Again, for $\mathbf{r}\times\frac{d\mathbf{r}}{ds}=0 \Rightarrow$ **r** and $\frac{d\mathbf{r}}{ds}$ are parallel.

Also $\frac{d\mathbf{r}}{ds}$ is a unit vector along tangent.

$\therefore$ **r** has constant direction that the tangent at every point is along **r**. Thus **r** describes a straight line.

Example 2. *If* $\mathbf{f}(t)=(t+1)\hat{i}+(t^2+t+1)\hat{j}+(t^3+t^2+t+1)\hat{k}$ *find* $\int_0^1\mathbf{f}(t)dt$.

Solution . Since

$$\mathbf{f}(t)=(t+1)\hat{i}+(t^2+t+1)\hat{j}+(t^3+t^2+t+1)\hat{k},$$

then

$$\int_0^1\mathbf{f}(t)dt$$

$$=\int_0^1[(t+1)\hat{i}+(t^2+t+1)\hat{j}+(t^3+t^2+t+1)\hat{k}]dt$$

$$=\hat{i}\int_0^1(t+1)dt+\hat{j}\int_0^1(t^2+t+1)dt+\hat{k}\int_0^1(t^3+t^2+t+1)dt$$

$$=\hat{i}\left(\frac{t^2}{2}+t\right)_0^1+\hat{j}\left(\frac{t^3}{3}+\frac{t^2}{2}+t\right)_0^1+\hat{k}\left(\frac{t^4}{4}+\frac{t^3}{3}+\frac{t^2}{2}+t\right)_0^1$$

$$=\frac{3}{2}\hat{i}+\frac{11}{6}\hat{j}+\frac{25}{12}\hat{k}.$$

Example 3. *If* $\mathbf{r}=5t^2\hat{i}+t\hat{j}-t^3\hat{k}$*, then prove that*

$$\int_1^2\left(\mathbf{r}\times\frac{d^2\mathbf{r}}{dt^2}\right)dt=-14\hat{i}+75\hat{j}-15\hat{k}.$$

Solution . Since $\mathbf{r}=5t^2\hat{i}+t\hat{j}-t^3\hat{k}$, then

$$\frac{d\mathbf{r}}{dt}=10t\hat{i}+\hat{j}-3t^2\hat{k}$$

again $$\frac{d^2\mathbf{r}}{dt^2}=10\hat{i}-6t\hat{k}.$$

$$\therefore\ \mathbf{r}\times\frac{d^2\mathbf{r}}{dt^2}=(5t^2\hat{i}+t\hat{j}-t^3\hat{k})\times(10\hat{i}-6t\hat{k})$$

$$=-30t^3\hat{i}\times\hat{k}+10t\hat{j}\times\hat{i}-6t^2\hat{j}\times\hat{k}-10t^3\hat{k}\times\hat{i}$$

$$=30t^3\hat{j}-10t\hat{k}-6t^2\hat{i}-10t^3\hat{j}$$

$$=-6t^2\hat{i}+20t^3\hat{j}-10t\hat{k}.$$

Now

$$\int_1^2\left(\mathbf{r}\times\frac{d^2\mathbf{r}}{dt^2}\right)=\int_1^2(-6t^2\hat{i}+20t^3\hat{j}-10t\hat{k})dt$$

$$=\left[-2t^3\hat{i}+5t^4\hat{j}-5t^2\hat{k}\right]_1^2$$

$$=-14\hat{i}+75\hat{j}-15\hat{k}.$$

Example 4. *Find the value of* **r** *satisfying the equation*

$$\frac{d^2\mathbf{r}}{dt^2}=6t\hat{i}-24t^2\hat{j}+4\sin t\hat{k}$$

given that $\mathbf{r}=2\hat{i}+\hat{j}$ *and* $\frac{d\mathbf{r}}{dt}=-\hat{i}-3\hat{k}$ *at* $t=0$.

Solution . We know that

$$\frac{d^2\mathbf{r}}{dt^2}=\frac{d}{dt}\left(\frac{d\mathbf{r}}{dt}\right). \text{ so } \frac{d\mathbf{r}}{dt}=\int\left(\frac{d^2\mathbf{r}}{dt^2}\right)dt+\mathbf{c}$$

(where **c** is constant vector taken to be as constant of integration.)

$$\frac{dr}{dt}=\int(6t\hat{i}-24t^2\hat{j}+4\sin t\hat{k})dt+\mathbf{c}$$

$$\frac{d\mathbf{r}}{dt} = 3t^2\hat{i} - 8t^3\hat{j} - 4\cos t\hat{k} + \mathbf{c} \qquad \ldots(1)$$

Initially at $t = 0, \frac{d\mathbf{r}}{dt} = 0 \Rightarrow 0 = -4\hat{k} + \mathbf{c}$

$$\therefore \quad \mathbf{c} = 4\hat{k} \qquad \ldots(2)$$

Form (1) and (2), we get

$$\frac{d\mathbf{r}}{dt} = 3t^2\hat{i} - 8t^3\hat{j} - 4\cos t\hat{k} + 4\hat{k}$$

Again integrating, we get

$$\mathbf{r} = \int(3t^2\hat{i} - 8t^3\hat{j} - 4\cos t\hat{k} + 4\hat{k}) + \mathbf{d}$$

Hence, $\mathbf{d}$ is constant vector of integration.

$$\therefore \quad \mathbf{r} = (t^3\hat{i} - 2t^4\hat{j} - 4\sin t\hat{k} + 4t\hat{k}) + \mathbf{d} \qquad \ldots(3)$$

Again initially, at $t = 0, \mathbf{r} = 2\hat{i} + \hat{j}$.

$$\therefore \quad 2\hat{i} + \hat{j} = \mathbf{d} \qquad \ldots(4)$$

Thus, form (3) and (4), we get the required result

$$\mathbf{r} = t^3\hat{i} - 2t^4\hat{j} + 4(1 - \sin t)\hat{k} + 2\hat{i} + \hat{j}$$

or $\quad \mathbf{r} = (t^3 + 2)\hat{i} + (1 - 2t^4)\hat{j} + 4(1 - \sin t)\hat{k}$

Example 5. *If* $\mathbf{F}(t) = 3t^2\hat{i} + t\hat{j} + 2\hat{k}$ *and* $\mathbf{G}(t) = 6t^2\hat{i} + (t-1)\hat{j} + 3t\hat{k}$ *then find*

$$\int_0^1\left(\frac{d\mathbf{F}}{dt}.\mathbf{G} + \mathbf{F}.\frac{d\mathbf{G}}{dt}\right)dt \text{ and}$$

$$\int_0^1\left(\mathbf{F}\times\frac{d\mathbf{G}}{dt} + \frac{d\mathbf{F}}{dt}\times\mathbf{G}\right)dt$$

Solution . We have

$$\mathbf{F.G} = 18t^4 + t(t-1) + 6t$$

$$\mathbf{F}\times\mathbf{G} = \begin{vmatrix} \hat{i} & \hat{j} & \hat{k} \\ 3t^2 & t & 2 \\ 6t^2 & (t-1) & 3t \end{vmatrix}$$

$$= \hat{i}(3t^2 - 2t + 2) - \hat{j}(9t^3 - 12t^2) + \hat{k}(3t^3 - 3t - 6t^3)$$

$$= (3t^2 - 2t + 2)\hat{i} - (9t^3 - 12t^2)\hat{j} - (3t^3 + 3t)\hat{k}$$

Now, $\int_0^1\left(\frac{d\mathbf{F}}{dt}.\mathbf{G} + \mathbf{F}.\frac{d\mathbf{G}}{dt}\right)dt$

$$= \left[\mathbf{F.G}\right]_0^1 = \left[18t^4 + t(t-1) + 6t\right]_0^1$$

$$= 18 + 6 = 24$$

$$\int_0^1\left(\mathbf{F}\times\frac{d\mathbf{G}}{dt} + \frac{d\mathbf{F}}{dt}\times\mathbf{G}\right)dt$$

$$= \left[\mathbf{F}\times\mathbf{G}\right]_0^1$$

$$= \left[\begin{matrix}(3t^2 - 2t + 2)\hat{i} - (9t^3 - 12t^2)\hat{j} \\ - (3t^3 + 3t)\hat{k}\end{matrix}\right]_0^1$$

$$= (3\hat{i} + 3\hat{j} - 6\hat{k}) - (2\hat{i})$$

$$= \hat{i} + 3\hat{j} - 6\hat{k}.$$

EXERCISE 21.2

1. If $\mathbf{f}(t) = (t - t^2)\hat{i} + 2t^3\hat{j} - 3\hat{k}$, find
   (i) $\int \mathbf{f}(t)dt$ (ii) $\int_1^2 \mathbf{f}(t)dt$
2. Integrate $\mathbf{a}\times\frac{d^2\mathbf{r}}{dt^2} = \mathbf{b}$, where $\mathbf{a}$ and $\mathbf{b}$ are constant vectors.
3. Find the value of $\mathbf{r}$ satisfying the equation $\frac{d^2\mathbf{r}}{dt^2} = t\mathbf{a} + \mathbf{b}$ where $\mathbf{a}$ and $\mathbf{b}$ are constant vectors.
4. Given that $\mathbf{r}(t) = \begin{cases} 2\hat{i} - \hat{j} + 2\hat{k}, & t = 2 \\ 4\hat{i} - 2\hat{j} + 3\hat{k}, & t = 3 \end{cases}$
   Show that $\int_2^3\left(\mathbf{r}.\frac{d\mathbf{r}}{dt}\right)dt = 10$.
5. Find $\int_0^1\left(e^t\hat{i} + e^{-2t}\hat{j} + t\hat{k}\right)dt$
6. If $\mathbf{r} = t\hat{i} - t^2\hat{j} + (t-1)\hat{k}$ and $\mathbf{s} = 2t^2\hat{i} + 6t\hat{k}$, evaluate
   (i) $\int_0^2 \mathbf{r}.\mathbf{s}dt$ (ii) $\int_0^2 \mathbf{r}\times\mathbf{s}dt$.
7. Solve the equation $\frac{d^2\mathbf{r}}{dt^2} = \mathbf{a}$ where $\mathbf{a}$ is a constant vector given that $\mathbf{r} = 0$ and $\frac{d\mathbf{r}}{dt} = 0$ when $t = 0$.
8. If $\mathbf{f}(t) = t\hat{i} + (t^2 - 2t)\hat{j} + (3t^2 + 3t^3)\hat{k}$, find $\int_0^1 \mathbf{f}(t)dt$.
9. If $\mathbf{a} = t\hat{i} - 3\hat{j} + 2t\hat{k}, \mathbf{b} = \hat{i} - 2\hat{j} + 2\hat{k}$ and $\mathbf{c} = 3\hat{i} + t\hat{j} - \hat{k}$, then evaluate $\int_1^2 \mathbf{a}\cdot(\mathbf{b}\times\mathbf{c})dt$.
10. The acceleration of a particle at any time $t \geq 0$ is given by
    $$\mathbf{a} = \frac{d\mathbf{v}}{dt} = 12\cos 2t\hat{i} - 8\sin 2t\hat{j} + 16t\hat{k}$$
    if the velocity $\mathbf{v}$ and displacement $\mathbf{r}$, are zero at $t = 0$, find $\mathbf{v}$ and $\mathbf{r}$ at any time t.
11. The acceleration of a particle at any time t is given by $\mathbf{a} = \frac{d\mathbf{v}}{dt} = e^t\hat{i} + e^{2t}\hat{j} + \hat{k}$, find v if $\mathbf{v} = \hat{i} + \hat{j}$ at $t = 0$.
12. Find the value of $\mathbf{r}$ satisfying the equation $\frac{d^2\mathbf{r}}{dt^2} = \mathbf{a}$, where $\mathbf{a}$ is a constant vector. Also it is given that when $t = 0$, $\mathbf{r} = 0$ and $\frac{d\mathbf{r}}{dt} = \mathbf{u}$.
13. If $\mathbf{A} = \hat{i} + u^2\hat{j} - 2u\hat{k}$ and $\mathbf{B} = e^u\hat{i} - u\hat{j} - \hat{k}$, find $\int(\mathbf{A}\times\mathbf{B})du$.

Hint to Selected Problems

1. $\mathbf{f}(t) = (t - t^2)\hat{i} + 2t^3\hat{j} - 3\hat{k}$

$\therefore \int \mathbf{f}(t)dt = \int[(t-t^2)\hat{i} + 2t^3\hat{j} - 3\hat{k}]dt$

$= \left(\frac{t^2}{2} - \frac{t^3}{3}\right)\hat{i} + \frac{t^4}{2}\hat{j} - 3t\hat{k}$

and $\int_1^2 \mathbf{f}(t)dt = \left[\left(\frac{t^2}{2} - \frac{t^3}{3}\right)\hat{i} + \frac{t^4}{2}\hat{j} - 3t\hat{k}\right]_1^2 = \frac{-5}{6}\hat{i} + \frac{15}{2}\hat{j} - 3\hat{k}$

4. $\mathbf{r}(t) = \begin{cases} 2\hat{i} - \hat{j} + 2\hat{k}, & t = 2 \\ 4\hat{i} - 2\hat{j} + 3\hat{k}, & t = 3 \end{cases}$

Since $\int\left(\mathbf{r}.\frac{d\mathbf{r}}{dt}\right)dt = \frac{1}{2}\int\frac{d}{dt}(\mathbf{r}.\mathbf{r})dt = \frac{1}{2}(\mathbf{r}.\mathbf{r})$

$\therefore \int_2^3\left(\mathbf{r}.\frac{d\mathbf{r}}{dt}\right)dt = \frac{1}{2}[\mathbf{r}(t).\mathbf{r}(t)]_2^3 = \frac{1}{2}[\mathbf{r}(3).\mathbf{r}(3) - \mathbf{r}(2).\mathbf{r}(2)]$

$= \frac{1}{2}[(4\hat{i} - 2\hat{j} + 3\hat{k}).(4\hat{i} - 2\hat{j} + 3\hat{k}) - (2\hat{i} - \hat{j} + 2\hat{k}).(2\hat{i} - \hat{j} + 2\hat{k})]$

$= \frac{1}{2}[(16+4+9) - (4+1+4)]$

$= \frac{1}{2}[29 - 9] = \frac{1}{2}(20) = 10.$

9. $\mathbf{a} = t\hat{i} - 3\hat{j} + 2t\hat{k}, \mathbf{b} = \hat{i} - 2\hat{j} + 2\hat{k}, \mathbf{c} = 3\hat{i} + t\hat{j} - \hat{k},$

then $\mathbf{a}.(\mathbf{b}\times\mathbf{c}) = \begin{vmatrix} t & -3 & 2t \\ 1 & -2 & 2 \\ 3 & t & -1 \end{vmatrix}$

$= t(2-2t) + 3(-1-6) + 2t(t+6)$

$= 2t - 2t^2 - 21 = 2t^2 + 12t = 14t - 21$

$\therefore \int_1^2 \mathbf{a}.(\mathbf{b}\times\mathbf{c})dt = \int_1^2(14t - 21)dt = \left[7t^2 - 21t\right]_1^2$

$= (28 - 42) - (7 - 21) = 0.$

ANSWERS

1. (i) $\left(\frac{t^2}{2} - \frac{t^3}{3}\right)\hat{i} + \frac{t^4}{2}\hat{j} - 3t\hat{k} + \mathbf{c}$ (ii) $-\frac{5}{6}\hat{i} + \frac{15}{2}\hat{j} - 3\hat{k}$ **2.** $\mathbf{a}\times\mathbf{r} = \frac{1}{2}t^2\mathbf{b} + t\mathbf{c} + \mathbf{d}$ **3.** $\mathbf{r} = \frac{1}{6}t^3\mathbf{a} + \frac{1}{2}t^2\mathbf{b} + t\mathbf{c} + \mathbf{d}$

5. $(e-1)\hat{i} - \frac{1}{2}(e^{-2} - 1)\hat{j} + \frac{1}{2}\hat{k}$ **6.** (i) 12 (ii) $-24\hat{i} - \frac{40}{3}\hat{j} + \frac{64}{5}\hat{k}$ **7.** $\mathbf{r} = \frac{1}{2}t^2\mathbf{a}$ **8.** $\frac{1}{2}\hat{i} - \frac{2}{3}\hat{j} + \frac{7}{4}\hat{k}$ **9.** 0

10. $\mathbf{v} = 6\sin 2t\hat{i} + (4\cos 2t - 4)\hat{j} + 8t^2\hat{k}$ **11.** $\mathbf{v} = e^t\hat{i} + \frac{1}{2}(e^{2t} + 1)\hat{j} + t\hat{k}$ **12.** $\mathbf{r} = \frac{1}{2}t^2\mathbf{a} + t\mathbf{u}$

$\mathbf{r} = (3 - 3\cos 2t)\hat{i} + (2\sin 2t - 4t)\hat{j} + \frac{8}{3}t^3\hat{k}$

13. $-\frac{u^3}{3}\hat{i} - \hat{j}\left(-u + 2ue^u - 2e^u\right) + \hat{k}\left(-\frac{u^2}{2} - u^2e^u + 2ue^u + 2e^u\right)$

Objective Evaluations

Fill in the Blanks

1. $\mathbf{F}(x, y, z)$ is a __________.
2. $f(x, y, z) = x^2 + y^2 + z^2 - 3xyz$ is a __________.
3. The necessary and sufficient condition for a vector function $\mathbf{F}(t)$ to be continuous at $t = t_0$ is that __________.
4. $\lim_{t\to t_0} |\mathbf{F}(t)| =$ __________.
5. $\lim_{\delta t\to 0} \frac{\delta \mathbf{r}}{\delta t} =$ __________.
6. $\frac{d}{dt}(\mathbf{a}\times\mathbf{b}) = \mathbf{a}\times\frac{d\mathbf{b}}{dt} +$ __________.
7. The derivative of a constant vector is equal to the __________.
8. If $\mathbf{a}$ has constant length and $\left|\frac{d\mathbf{a}}{dt}\right| \neq 0$, then $\mathbf{a}$ and $\frac{d\mathbf{a}}{dt}$ are __________.
9. The vector $\mathbf{a}(t)$ has constant magnitude iff $\mathbf{a}.\frac{d\mathbf{a}}{dt}$ is equal to __________.
10. The vector equation $\mathbf{r}(t) = a\cos t\hat{i} + b\sin t\hat{j} + 0\hat{k}$ represents __________.
11. $\frac{d\mathbf{r}}{dt}$ represents __________.
12. If $\mathbf{r} = \sin t\hat{i} + \cos t\hat{j} + \hat{k}$, then $\left|\frac{d^2\mathbf{r}}{dt^2}\right|$ equals __________.
13. $\frac{d}{dt}(\mathbf{r}\times\mathbf{a}) =$ __________ if $\mathbf{a}$ is a constant vector.
14. $\frac{d}{dt}\left(\mathbf{r}.\frac{d\mathbf{r}}{dt}\right) = \frac{d\mathbf{r}}{dt}.\frac{d\mathbf{r}}{dt} +$ __________.
15. If $\mathbf{a}$ is a constant vcector, then $\int\left(\mathbf{a}\times\frac{d\mathbf{r}}{dt}\right)dt =$ __________.
16. If $\mathbf{r}(t) = 5t^2\hat{i} + t\hat{j} - t^3\hat{k}$, then $\int_1^2 \mathbf{r}\times\frac{d^2\mathbf{r}}{dt^2}dt =$ __________.
17. If $\mathbf{a}, \mathbf{b}$ are constant vectors, then the solution of $\frac{d^2\mathbf{r}}{dt^2} = t\mathbf{a} + \mathbf{b}$ is__________.
18. If $\mathbf{F} = (x^2 + y^2)\hat{i} - 2xy\hat{j}$, then $\mathbf{F}.d\mathbf{r} =$ __________.
19. If $\mathbf{r}\times d\mathbf{r} = 0$, then $\hat{r} =$ __________.
20. $\frac{d}{dt}\left(\mathbf{a}\times\frac{d\mathbf{a}}{dt}\right) =$ __________.

True/False

Write 'T' for True and 'F' for False statement.

1. $\mathbf{F}(x, y, z)$ is a scalar point function. **(T/F)**
2. $\left|\lim_{t\to 0}\mathbf{F}(t)\right| = \lim_{t\to 0}|\mathbf{F}(t)|$. **(T/F)**
3. If $\mathbf{f}(t) = f_1(t)\hat{i} + f_2(t)\hat{j} + f_3(t)\hat{k}$, then $f(t)$ is continuous iff $f_1(t)$, $f_2(t), f_3(t)$ are continuous. **(T/F)**
4. $\frac{d}{dt}(\phi\mathbf{a}) = \phi.\frac{d\mathbf{a}}{dt}$, if $\phi = \phi(t)$. **(T/F)**
5. If $\mathbf{r}$ is a constant vector function of scalar variable t, then $(d\mathbf{r}/dt)$ is a zero vector. **(T/F)**
6. If $\mathbf{a}.\frac{d\mathbf{a}}{dt} = 0$ then $|\mathbf{a}|$ is constant. **(T/F)**
7. If $|\mathbf{a}|$ = constant and $\left|\frac{d\mathbf{a}}{dt}\right| \neq 0$, then $\mathbf{a}.\frac{d\mathbf{a}}{dt} \neq 0$. **(T/F)**
8. If $|\mathbf{a}|$ = constant, then $\mathbf{a}\times\frac{d\mathbf{a}}{dt} = \mathbf{0}$. **(T/F)**
9. If $\mathbf{r}(t) = at^2\hat{i} + 2at\hat{j}$, then $\frac{d\mathbf{r}}{dt} = 2at\hat{i} + 2a\hat{j}$. **(T/F)**
10. $\frac{d\mathbf{r}}{dt}$ represents a velocity vector. **(T/F)**
11. $\frac{d^2\mathbf{r}}{dt^2}$ represents rate of change of velocity. **(T/F)**
12. If $\mathbf{r} = \cos 3t\hat{i} + \sin 3t\hat{j}$, then $\left|\mathbf{r}\times\frac{d\mathbf{r}}{dt}\right| = 3$. **(T/F)**
13. If $\mathbf{a}, \mathbf{b}, \mathbf{c}$ are constant vectors and $\mathbf{r} = \mathbf{a}t^2 + \mathbf{b}t + \mathbf{c}$, then acceleration of the particle whose path is given by $\mathbf{r}$ is not constant. **(T/F)**
14. If $\mathbf{r}\times d\mathbf{r} = \mathbf{0}$, then $\hat{r}$ will be constant vector. **(T/F)**
15. $\int\left(2\mathbf{r}.\frac{d\mathbf{r}}{dt}\right)dt = \mathbf{r}^2 + c$. **(T/F)**
16. $\int\left(\mathbf{r}\times\frac{d^2\mathbf{r}}{dt^2}\right)dt = \mathbf{r}\times\frac{d\mathbf{r}}{dt} + c$. **(T/F)**
17. Integration of a vector function is the reverse process of the differentiation of that vector function. **(T/F)**
18. The value of the integration of the vector function $\mathbf{F}(t) = 2t\hat{i} + 3t^2\hat{j} + 4t^3\hat{k}$ from $t = 0$ to $t = 1$ is $\hat{i} + \hat{j} + \hat{k}$. **(T/F)**
19. The necessary and sufficient condition for the vector $\mathbf{a}(t)$ to have a constant magnitude is that $\mathbf{a}\times\frac{d\mathbf{a}}{dt} = \mathbf{0}$. **(T/F)**

Multiple Choice Questions

Choose the most appropriate one.

1. If $\lim_{t\to t_0} \mathbf{f}(t) = l$, then $\lim_{t\to t_0} |\mathbf{f}(t)|$:

(a) l (b) $|l|$

(c) $-l$ (d) none of these

2. $\lim_{t\to 2}(t\hat{i}+t^2\hat{j}-t^3\hat{k}).(t^2\hat{i}+t\hat{j}+\hat{k})$ equals :

(a) 6 (b) 4

(c) 8 (d) 1

3. If **a**, **b**, **c** are constant vectors, then $\frac{d}{dt}[\mathbf{a}\ \mathbf{b}\ \mathbf{c}]$ equals:

(a) $\left[\frac{d\mathbf{a}}{dt}\mathbf{b}\ \mathbf{c}\right]$ (b) $\left[\mathbf{a}\frac{d\mathbf{b}}{dt}\mathbf{c}\right]$

(c) 0 (d) [a b c]

4. If ϕ is not a function of t, then $\frac{d}{dt}(\phi\mathbf{a})$ equals :

(a) $\frac{d\phi}{dt}\mathbf{a}$ (b) $\frac{d\phi}{dt}$

(c) $\phi\frac{d\mathbf{a}}{dt}$ (d) $\frac{d\mathbf{a}}{dt}$

5. If $|\mathbf{a}|$ = constant, then $\mathbf{a}.\frac{d\mathbf{a}}{dt}$ is equal to :

(a) 1 (b) 0

(c) 2 (d) –1

6. If $|\mathbf{a}| = a$, then $\mathbf{a}.\frac{d\mathbf{a}}{dt}$ equals :

(a) $a\frac{da}{dt}$ (b) a

(c) $\frac{da}{dt}$ (d) 1

7. If $|\mathbf{a}|$ = constant, then $\mathbf{a}\times\frac{d\mathbf{a}}{dt}$ is equal to :

(a) 0 (b) –1

(c) 1 (d) a

8. The velocity of a particle whose path is $\mathbf{r} = at^2\hat{i}+2at\hat{j}+0\hat{k}$ at $t = 1/2a$ is :

(a) $\hat{i}-\hat{j}$ (b) $\hat{j}$

(c) $\hat{i}$ (d) $\hat{i}+\hat{j}$

9. The acceleration of a particle whose path is $\mathbf{r} = \mathbf{a}t^2 + \mathbf{b}t + \mathbf{c}$ where **a**, **b**, **c** are constant at any time is :

(a) **a** (b) 2**a**

(c) **b** (d) **c**

10. If **a**, **b** are constant vectors, ω is a constant, then for $\mathbf{r} = \cos\omega t\ \mathbf{a} + \sin\omega t\ \mathbf{b}$, $\frac{d^2\mathbf{r}}{dt^2}$ equals :

(a) $-\omega\mathbf{r}$ (b) $\omega\mathbf{r}$

(c) $-\omega^2\mathbf{r}$ (d) $\omega^2\mathbf{r}$

11. If $\mathbf{r} = a\cos t\hat{i} + a\sin t\hat{j} + at\tan\alpha\hat{k}$, then $\left|\frac{d\mathbf{r}}{dt}\times\frac{d^2\mathbf{r}}{dt^2}\right|$ equals :

(a) $a\sec\alpha$ (b) $a\sec^2\alpha$

(c) $a\tan\alpha$ (d) $a\tan^2\alpha$

12. $\frac{d}{dt}\left[\mathbf{r}, \frac{d\mathbf{r}}{dt}, \frac{d^2\mathbf{r}}{dt^2}\right]$ is equal to :

(a) **r** (b) $\left[\mathbf{r}, \frac{d\mathbf{r}}{dt}, \frac{d^3\mathbf{r}}{dt^3}\right]$

(c) $\frac{d\mathbf{r}}{dt}$ (d) $\frac{d^3\mathbf{r}}{dt^3}$

13. If **r** is a unit vector, then $\left|\mathbf{r}\times\frac{d\mathbf{r}}{dt}\right|$ equals :

(a) $\frac{d\mathbf{r}}{dt}$ (b) **r**

(c) $\left|\frac{d\mathbf{r}}{dt}\right|$ (d) 1

14. $\int\left(2\frac{d\mathbf{r}}{dt}.\frac{d^2\mathbf{r}}{dt^2}\right)dt$ is equal to :

(a) 0 (b) $\frac{d\mathbf{r}}{dt}+\mathbf{c}$

(c) $\frac{d^2\mathbf{r}}{dt^2}+\mathbf{c}$ (d) $\left(\frac{d\mathbf{r}}{dt}\right)^2+\mathbf{c}$

15. If $\mathbf{r} = 5t^2\hat{i}+t\hat{j}-t^3\hat{k}$, then the value of $\int_1^2\left(\mathbf{r}\times\frac{d^2\mathbf{r}}{dt^2}\right)dt$ is :

(a) $-14\hat{i}+75\hat{j}-15\hat{k}$ (b) $14\hat{i}+75\hat{j}-15\hat{k}$

(c) $\hat{i}$ (d) 0

16. At $t = 2$, $\mathbf{r}(t) = 2\hat{i}-\hat{j}+2\hat{k}$ and at $t = 3$, $\mathbf{r}(t) = 4\hat{i}-2\hat{j}+3\hat{k}$, then the value of $\int_2^3\left(\mathbf{r}.\frac{d\mathbf{r}}{dt}\right)dt$ is equal to :

(a) 0 (b) 1 (c) 10 (d) –10

17. If $\mathbf{p} = \mathbf{A}\cos kt + \mathbf{B}\sin kt$ where **A** and **B** are constant vectors and k is a constant scalar, then which is correct ?

(a) $\frac{d^2\mathbf{p}}{dt^2}+k^2\mathbf{p}=0$ (b) $\frac{d^2\mathbf{p}}{dt^2}-k^2\mathbf{p}=0$

(c) $\frac{d^2\mathbf{p}}{dt^2}+k\mathbf{p}=0$ (d) $\frac{d^2\mathbf{p}}{dt^2}-k\mathbf{p}=0$

18. If $\mathbf{f}(t) = (t-t^2)\hat{i}+2t^3\hat{j}-3\hat{k}$, then value of $\int_1^2\mathbf{f}(t)dt$ is :

(a) $-\frac{5}{6}\hat{i}+\frac{15}{2}\hat{j}-3\hat{k}$ (b) $\frac{5}{6}\hat{i}-\frac{15}{2}\hat{j}-3\hat{k}$

(c) $\frac{5}{6}\hat{i}+\frac{15}{2}\hat{j}+3\hat{k}$ (d) $2i$

ANSWERS

Fill in the Blanks

1. Vector function
2. scalar point function
3. $\lim_{t \to t_0} \mathbf{F}(t) = t$
4. $\left|\lim_{t \to 0} \mathbf{F}(t)\right|$
5. $\frac{d\mathbf{r}}{dt}$
6. $\frac{d\mathbf{a}}{dt} \times \mathbf{b}$
7. null vector
8. perpendicular
9. 0
10. ellipse
11. velocity vector
12. 1
13. $\frac{d\mathbf{r}}{dt} \times \mathbf{b}$
14. $\mathbf{r} + \frac{d^2\mathbf{r}}{dt^2}$
15. $\mathbf{a} \times \mathbf{r} + \mathbf{c}$
16. $-14\hat{i} + 75\hat{j} - 15\hat{k}$
17. $\frac{1}{6}t^3\mathbf{a} + \frac{1}{2}t^2\mathbf{b} + t\mathbf{c} + \mathbf{d}$
18. $(x^2 + y^2)dx - 2xy\,dy$
19. constant
20. $\mathbf{a} \times \frac{d^2\mathbf{a}}{dt^2}$

True/False

1. F	2. T	3. T	4. F	5. T	6. T	7. F	8. T	9. T	10. T
11. T	12. T	13. F	14. T	15. T	16. T	17. T	18. T	19. T	

Multiple Choice Questions

1. (b)	2. (c)	3. (a)	4. (c)	5. (b)	6. (a)	7. (a)	8. (d)	9. (b)	10. (c)
11. (a)	12. (b)	13. (c)	14. (d)	15. (a)	16. (c)	17. (a)	18. (b)		

FFFFF

Gradient, Divergence and Curl

22.1 PARTIAL DERIVATIVE OF VECTORS

Let $\mathbf{r} = \mathbf{f}(x, y, z)$ be a vector function of three scalar variables x, y, z. The first order partial derivative of $\mathbf{r}$ with respect to x is given by

$$\frac{\partial \mathbf{r}}{\partial x} = \lim_{\delta x \to 0} \frac{\mathbf{f}(x+\delta x, y, z) - \mathbf{f}(x, y, z)}{\delta x}, \text{ if this limit exists.}$$

Similarly we can find first order partial derivatives of $\mathbf{r}$ with respect to y and z respectively and are denoted by $\frac{\partial \mathbf{r}}{\partial y}, \frac{\partial \mathbf{r}}{\partial z}$.

During the differentiation if y and z are treating as constant, then $\frac{\partial \mathbf{r}}{\partial x}$ is regarded as ordinary derivative. Likewise we can find higher order partial derivatives.

22.2 VECTOR DIFFERENTIAL OPERATOR

The vector differential operator is defined by the formula $\nabla = \frac{\partial}{\partial x}\hat{i} + \frac{\partial}{\partial y}\hat{j} + \frac{\partial}{\partial z}\hat{k}$.

Obviously, ∇ is a vector quantity. This vector ∇ is read as **nabla** or **del**.

22.3 GRADIENT OF A SCALAR FIELD

(UPTU–2006, 07, 08)

Let $f(x, y, z)$ be a scalar point function which is defined over some region R in space and also differentiable at each point (x, y, z) in R, then the gradient of $f(x, y, z)$ is defined as

$$\text{grad } f = \frac{\partial f}{\partial x}\hat{i} + \frac{\partial f}{\partial y}\hat{j} + \frac{\partial f}{\partial z}\hat{k} \quad \text{or} \quad \text{grad } f = \left(\frac{\partial}{\partial x}\hat{i} + \frac{\partial}{\partial y}\hat{j} + \frac{\partial}{\partial z}\hat{k}\right) f = \nabla f$$

Thus gradient of f can also be written in terms of vector differential operator(∇). Since ∇ is a vector quantity, thus ∇f is a vector whose components are $\frac{\partial f}{\partial x}, \frac{\partial f}{\partial y}, \frac{\partial f}{\partial z}$. Hence, gradient of a scalar field is a vector field.

22.4 SOME FORMULAE RELATED TO GRADIENT

1. If f and g are two scalar point functions, then grad $(f + g)$ = grad f + grad g

or $\nabla(f+g) = \nabla f + \nabla g$.

Proof. Since we know that

$$\nabla f = \frac{\partial f}{\partial x}\hat{i} + \frac{\partial f}{\partial y}\hat{j} + \frac{\partial f}{\partial z}\hat{k}$$

$$\therefore \quad \nabla(f+g) = \frac{\partial}{\partial x}(f+g)\hat{i} + \frac{\partial}{\partial y}(f+g)\hat{j} + \frac{\partial}{\partial z}(f+g)\hat{k} = \frac{\partial f}{\partial x}\hat{i} + \frac{\partial g}{\partial x}\hat{i} + \frac{\partial f}{\partial y}\hat{j} + \frac{\partial g}{\partial y}\hat{j} + \frac{\partial f}{\partial z}\hat{k} + \frac{\partial g}{\partial z}\hat{k}$$

$$= \left(\frac{\partial f}{\partial x}\hat{i} + \frac{\partial f}{\partial y}\hat{j} + \frac{\partial f}{\partial z}\hat{k}\right) + \left(\frac{\partial g}{\partial x}\hat{i} + \frac{\partial g}{\partial y}\hat{j} + \frac{\partial g}{\partial z}\hat{k}\right) = \nabla f + \nabla g.$$

Hence, $\nabla(f+g) = \nabla f + \nabla g$.

2. *If f and g are two scalar point functions, then* $\nabla(fg) = f\nabla g + g\nabla f$.

or grad $(fg) = f(\text{grad } g) + g(\text{grad } f)$.

Proof. Since we know that

$$\nabla f = \frac{\partial f}{\partial x}\hat{i} + \frac{\partial f}{\partial y}\hat{j} + \frac{\partial f}{\partial z}\hat{k}$$

$$\therefore \quad \nabla(fg) = \frac{\partial}{\partial x}(fg)\hat{i} + \frac{\partial}{\partial y}(fg)\hat{j} + \frac{\partial}{\partial z}(fg)\hat{k}$$

$$= \left(f\frac{\partial g}{\partial x} + g\frac{\partial f}{\partial x}\right)\hat{i} + \left(f\frac{\partial g}{\partial y} + g\frac{\partial f}{\partial y}\right)\hat{j} + \left(f\frac{\partial g}{\partial z} + g\frac{\partial f}{\partial z}\right)\hat{k}$$

$$= f\left(\frac{\partial g}{\partial x}\hat{i} + \frac{\partial g}{\partial y}\hat{j} + \frac{\partial g}{\partial z}\hat{k}\right) + g\left(\frac{\partial f}{\partial x}\hat{i} + \frac{\partial f}{\partial y}\hat{j} + \frac{\partial f}{\partial z}\hat{k}\right) = f\nabla g + g\nabla f.$$

Hence, $\nabla(fg) = f\nabla g + g\nabla f$.

3. *If f and g are scalar point functions and $g \neq 0$ for all point in the region R, then* $\nabla\left(\frac{f}{g}\right) = \frac{g\nabla f - f\nabla g}{g^2}$.

Proof. Since

$$\nabla f = \frac{\partial f}{\partial x}\hat{i} + \frac{\partial f}{\partial y}\hat{j} + \frac{\partial f}{\partial z}\hat{k}$$

$$\therefore \quad \nabla\left(\frac{f}{g}\right) = \frac{\partial}{\partial x}\left(\frac{f}{g}\right)\hat{i} + \frac{\partial}{\partial y}\left(\frac{f}{g}\right)\hat{j} + \frac{\partial}{\partial z}\left(\frac{f}{g}\right)\hat{k}$$

$$= \frac{1}{g^2}\left(g\frac{\partial f}{\partial x} - f\frac{\partial g}{\partial x}\right)\hat{i} + \frac{1}{g^2}\left(g\frac{\partial f}{\partial y} - f\frac{\partial g}{\partial y}\right)\hat{j} + \frac{1}{g^2}\left(g\frac{\partial f}{\partial z} - f\frac{\partial g}{\partial z}\right)\hat{k}$$

$$= \frac{1}{g^2}\left[g\left(\frac{\partial f}{\partial x}\hat{i} + \frac{\partial f}{\partial y}\hat{j} + \frac{\partial f}{\partial z}\hat{k}\right) - f\left(\frac{\partial g}{\partial x}\hat{i} + \frac{\partial g}{\partial y}\hat{j} + \frac{\partial g}{\partial z}\hat{k}\right)\right] = \frac{1}{g^2}[g\nabla f - f\nabla g]$$

$$= \frac{g(\nabla f) - f(\nabla g)}{g^2}.$$

Hence, $\nabla\left(\frac{f}{g}\right) = \frac{g\nabla f - f\nabla g}{g^2}$.

4. *If f is a scalar point function, then f is constant if and only if* $\nabla f = \mathbf{0}$.

Proof. Suppose f is constant, then

$$\frac{\partial f}{\partial x} = 0, \frac{\partial f}{\partial y} = 0, \frac{\partial f}{\partial z} = 0 \qquad (\because f(x, y, z) = c)$$

$$\therefore \quad \nabla f = \frac{\partial f}{\partial x}\hat{i} + \frac{\partial f}{\partial y}\hat{j} + \frac{\partial f}{\partial z}\hat{k} = 0\hat{i} + 0\hat{j} + 0\hat{k} = \mathbf{0}.$$

Conversely, suppose $\nabla f = \mathbf{0}$. Then we have $\nabla f = \frac{\partial f}{\partial x}\hat{i} + \frac{\partial f}{\partial y}\hat{j} + \frac{\partial f}{\partial z}\hat{k} = \mathbf{0}$. So, $\frac{\partial f}{\partial x} = 0, \frac{\partial f}{\partial y} = 0, \frac{\partial f}{\partial z} = 0$.

Hence, $f(x, y, z) = c$ (constant)

REMARKS

- $\nabla(f - g) = \nabla f - \nabla g$
- $\nabla(cf) = c\nabla f$, where c is a constant.
- $\nabla\left(\frac{1}{f}\right) = -\frac{\nabla f}{f^2}$, where $f \neq 0\ \forall\ (x, y, z) \in \mathbf{R}$.

22.5 DIRECTIONAL DERIVATIVES

Let us consider a scalar field given by a scalar point function $f(P) = f(x, y, z)$ where P is any point in space whose co-ordinates are (x, y, z). Since we know that the first order partial derivatives of f are the rates of change of f in the direction of co-ordinate axes. Now we shall have to discuss the rate of change of f in any direction this leads the notion of a directional derivative.

Let us choose a point P in space and a direction at P, given by a unit vector $\hat{a}$. Let C be the ray from P in the direction of $\hat{a}$

and let Q be any point on this ray C such that PQ is s shown in fig. 1.

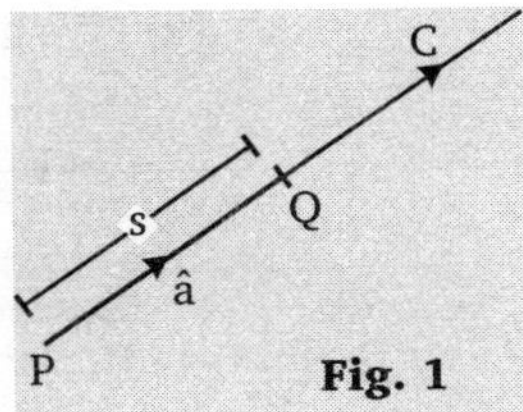

Fig. 1

Then the limit

$$\frac{\partial f}{\partial s} = \lim_{s \to 0} \frac{f(Q) - f(P)}{s}, \text{ where } s = PQ$$

if exists is called the directional derivative of f at P in the direction of $\hat{a}$. In fact there are infinitely many directional derivatives of f at P, each corresponding to a certain direction. But if a cartesian co-ordinates system is given, then we may represent any such derivative in terms of the first order partial derivatives of f at P. If the position vector P is $\mathbf{p}$, then the ray C can be written as

$$\mathbf{r}(s) = x(s)\hat{i} + y(s)\hat{j} + z(s)\hat{k} \qquad \ldots(1)$$
$$= \mathbf{p} + s\hat{a} \quad (s \geq 0)$$

and $\frac{\partial f}{\partial s}$ is the derivative of the function $f[x(s), y(s), z(s)]$ with respect to the arc length s of C. Hence, assuming that f has continuous partial derivative of first order, we have

$$\frac{\partial f}{\partial s} = \frac{\partial f}{\partial x}\frac{dx}{ds} + \frac{\partial f}{\partial y}\frac{dy}{ds} + \frac{\partial f}{\partial z}\frac{dz}{ds} \qquad \ldots(2)$$

Form (1)

$$\frac{d\mathbf{r}}{ds} = \frac{dx}{ds}\hat{i} + \frac{dy}{ds}\hat{j} + \frac{dz}{ds}\hat{k} = \hat{a} \qquad \ldots(3)$$

Since we have

$$\text{grad } f = \frac{\partial f}{\partial x}\hat{i} + \frac{\partial f}{\partial y}\hat{j} + \frac{\partial f}{\partial z}\hat{k}. \qquad \ldots(4)$$

Thus, equation (2) becomes

$$\frac{\partial f}{\partial s} = \left(\frac{\partial f}{\partial x}\hat{i} + \frac{\partial f}{\partial y}\hat{j} + \frac{\partial f}{\partial z}\hat{k}\right) \cdot \left(\frac{\partial x}{\partial s}\hat{i} + \frac{\partial y}{\partial s}\hat{j} + \frac{\partial z}{\partial s}\hat{k}\right)$$
$$= (\text{grad } f).\hat{a}$$

[From (3) and (4)]

or

$$\frac{\partial f}{\partial s} = \hat{a} \cdot \text{grad } f = \hat{a} \cdot \nabla f.$$

Hence, the directional derivative $\frac{\partial f}{\partial s}$ is given as $\hat{a} \cdot \nabla f$.

Recapitulations

- $\text{grad } f = \left(\frac{\partial}{\partial x}\hat{i} + \frac{\partial}{\partial y}\hat{j} + \frac{\partial}{\partial z}\hat{k}\right)$
- $\nabla(f \pm g) = \nabla f \pm \nabla g$
- $\nabla(fg) = f\nabla g + g\nabla f$
- $\nabla\left(\frac{f}{g}\right) = \frac{g\nabla f - f\nabla g}{g^2}$
- If f is a scalar point function, then f is constant if and only if $\nabla f = \mathbf{0}$.
- The limit
$$\frac{\partial f}{\partial s} = \lim_{s \to 0} \frac{f(Q) - f(P)}{s}, \text{ where } s = PQ$$
if exists is called the directional derivative of f at P

Remarks

- If $\hat{\mathbf{a}} = \hat{i}$, then $\frac{\partial f}{\partial s} = \hat{i}.\nabla f = \hat{i}.\left(\frac{\partial f}{\partial x}\hat{i} + \frac{\partial f}{\partial y}\hat{j} + \frac{\partial f}{\partial z}\hat{k}\right) = \frac{\partial f}{\partial x}$. Similarly, if $\hat{\mathbf{a}} = \hat{i}, \hat{k}$, then $\frac{\partial f}{\partial s} = \frac{\partial f}{\partial y}, \frac{\partial f}{\partial s} = \frac{\partial f}{\partial z}$.
- Maximum value of the directional derivative is $|\text{grad } f|$.

Solved Examples

Example 1. *If $r = |\mathbf{r}|$ where $\mathbf{r} = x\hat{i} + y\hat{j} + z\hat{k}$, prove that*

(i) $\nabla f(r) = f'(r)\nabla r$ (GBTU–2011)

(ii) $\nabla r = \frac{\mathbf{r}}{r}$ (GBTU–2011)

(iii) $\nabla f(r) \times \mathbf{r} = 0$

(iv) $\nabla r^n = nr^{n-2}\mathbf{r}$. (UPTU–2008, GBTU–2011, Anna–2003, Bhopal–2007, VTU–2000)

(v) $\nabla r^{-3} = -3r^{-5}\mathbf{r}$.

Solution. (i) Since we know that

$$\nabla f = \frac{\partial f}{\partial x}\hat{i} + \frac{\partial f}{\partial y}\hat{j} + \frac{\partial f}{\partial z}\hat{k} \qquad \ldots(1)$$

$$\therefore \nabla f(r) = \frac{\partial}{\partial x}(f(r))\hat{i} + \frac{\partial}{\partial y}(f(r))\hat{j} + \frac{\partial}{\partial z}(f(r))\hat{k}$$

$$\text{or } \nabla f(r) = f'(r)\frac{\partial r}{\partial x}\hat{i} + f'(r)\frac{\partial r}{\partial y}\hat{j} + f'(r)\frac{\partial r}{\partial z}\hat{k}$$

$$\nabla f(r) = f'(r)\left[\frac{\partial r}{\partial x}\hat{i} + \frac{\partial r}{\partial y}\hat{j} + \frac{\partial r}{\partial z}\hat{k}\right]$$

$$\nabla f(r) = f'(r)\nabla r. \text{ [Using (i)]}$$

(ii) We have $\nabla r = \frac{\partial r}{\partial x}\hat{i} + \frac{\partial r}{\partial y}\hat{j} + \frac{\partial r}{\partial z}\hat{k}$

Since $\mathbf{r} = x\hat{i} + y\hat{j} + z\hat{k}$

$\therefore \quad |\mathbf{r}|^2 = x^2 + y^2 + z^2$

or $\quad r^2 = x^2 + y^2 + z^2 \qquad (\because |\mathbf{r}| = r)$

$\therefore \quad \frac{\partial r}{\partial x} = \frac{x}{r}, \frac{\partial r}{\partial y} = \frac{y}{r}, \frac{\partial r}{\partial z} = \frac{z}{r}$

$\therefore \quad \nabla r = \frac{x}{r}\hat{i} + \frac{y}{r}\hat{j} + \frac{z}{r}\hat{k}$

$= \frac{1}{r}(x\hat{i} + y\hat{j} + z\hat{k})$

or $\quad \nabla r = \frac{\mathbf{r}}{r}.$

(iii) $\quad \nabla f(r) = f'(r)\nabla r \qquad$ [From (i)]

$= f'(r)\frac{\mathbf{r}}{r} \qquad$ [From (ii)]

Now

$\nabla f(r) \times \mathbf{r} = \frac{f'(r)}{r}\mathbf{r} \times \mathbf{r} = \mathbf{0} \qquad (\because \mathbf{r} \times \mathbf{r} = \mathbf{0})$

(iv) Since

$\nabla f(r) = f'(r)\nabla r \qquad$ (From (i))

Let $f(r) = r^n$.

$\therefore \quad \nabla r^n = nr^{n-1}\nabla r$

$= nr^{n-1}\left(\frac{\mathbf{r}}{r}\right) \qquad \left(\because \nabla r = \frac{\mathbf{r}}{r}\right)$

$= nr^{n-2}\mathbf{r}$

or $\quad \nabla r^n = nr^{n-2}\mathbf{r}.$

(v) From part (iv), $\nabla r^n = nr^{n-2}\mathbf{r}$

$\Rightarrow \nabla r^{-3} = -3r^{-3-2}\mathbf{r} = -3r^{-5}\mathbf{r}.$

Example 2. *If $f(x, y, z) = 3x^2y - y^3z^2$, find grad f and |grad f| at (1, –2, 1).* (UPTU–2007)

Solution. Since we know that

$$\text{grad } f = \nabla f = \frac{\partial f}{\partial x}\hat{i} + \frac{\partial f}{\partial y}\hat{j} + \frac{\partial f}{\partial z}\hat{k}$$

$$= \frac{\partial}{\partial x}(3x^2y - y^3z^2)\hat{i} + \frac{\partial}{\partial y}(3x^2y - y^3z^2)\hat{j} + \frac{\partial}{\partial z}(3x^2y - y^3z^2)\hat{k}$$

$$= 6xy\hat{i} + (3x^2 - 3y^2z^2)\hat{j} + (-2y^3z)\hat{k}$$

At (1, –2, –1)

$$\text{grad } f = -12\hat{i} - 9\hat{j} - 16\hat{k}$$

and $\quad |\text{grad } f| = \sqrt{144 + 81 + 256} = \sqrt{481}.$

Example 3. *If $\phi(x, y, z) = xy^2z$ and $\mathbf{f}(x, y, z) = xz\hat{i} - xy\hat{j} + yz\hat{k}$, show that $\frac{\partial^3}{\partial x^2 \partial z}(\phi\mathbf{f})$ at (2, –1, 1) is $4\hat{i} + 2\hat{j}$.* (Bhopal–2008)

Solution. We have $\quad \phi\mathbf{f} = x^2y^2z^2\hat{i} - x^2y^3z\hat{j} + xy^3z^2\hat{k}.$

$\therefore \quad \frac{\partial}{\partial z}(\phi\mathbf{f}) = 2x^2y^2z\hat{i} - x^2y^3\hat{j} + 2xy^3z\hat{k}$

$\frac{\partial^2}{\partial x \partial z}(\phi\mathbf{f}) = 4xy^2z\hat{i} - 2xy^3\hat{j} + 2y^3z\hat{k}$

$\frac{\partial^3}{\partial x^2 \partial z}(\phi\mathbf{f}) = 4y^2z\hat{i} - 2y^3\hat{j}$

At (2, –1, 1)

$$\frac{\partial^3(\phi\mathbf{f})}{\partial x^2 \partial z} = 4(-1)^2(1)\hat{i} - 2(-1)^3\hat{j} = 4\hat{i} + 2\hat{j}.$$

Example 4. *Find $\nabla\phi$ and $|\nabla\phi|$ when*

$$\phi = (x^2 + y^2 + z^2)e^{-(x^2+y^2+z^2)^{1/2}}.$$

Solution. Let $r^2 = x^2 + y^2 + z^2$, then ϕ can be written as

$$\phi = r^2e^{-r} \qquad \ldots(1)$$

Now $\nabla\phi = \frac{\partial \phi}{\partial x}\hat{i} + \frac{\partial \phi}{\partial y}\hat{j} + \frac{\partial \phi}{\partial z}\hat{k} \qquad \ldots(2)$

By (1), $\quad \frac{\partial \phi}{\partial x} = \frac{\partial \phi}{\partial r}\frac{\partial r}{\partial x} = (2re^{-r} - r^2e^{-r})\frac{\partial r}{\partial x}.$

Again by $r^2 = x^2 + y^2 + z^2$,

$2r\frac{\partial r}{\partial x} = 2x \Rightarrow \frac{\partial r}{\partial x} = \frac{x}{r}.$

$\therefore \quad \frac{\partial \phi}{\partial x} = r(2-r)e^{-r}\frac{x}{r} = (2-r)e^{-r}x$

Similarly, $\frac{\partial \phi}{\partial y} = (2-r)e^{-r}y$ and $\frac{\partial \phi}{\partial z} = (2-r)e^{-r}z$

$\therefore$ By (2), we have

$$\nabla\phi = (2-r)e^{-r}(x\hat{i} + y\hat{j} + z\hat{k})$$
$$= (2-r)e^{-r}\mathbf{r}.$$

Also, $\quad |\nabla\phi| = \left|(2-r)e^{-r}\mathbf{r}\right| = (2-r)e^{-r}|\mathbf{r}|$

$= (2-r)e^{-r}r = (2-r)re^{-r}$

Example 5. *Prove that*

(i) $\nabla(\mathbf{r.a}) = \mathbf{a}$ (UPTU–2008)

(ii) $\nabla[\mathbf{r\ a\ b}] = \mathbf{a} \times \mathbf{b}.$

*where **a** and **b** are constant vectors.*

Solution. Suppose $\quad \mathbf{a} = a_1\hat{i} + a_2\hat{j} + a_3\hat{k}$ and

$\mathbf{r} = x\hat{i} + y\hat{j} + z\hat{k}, \mathbf{b} = b_1\hat{i} + b_2\hat{j} + b_3\hat{k},$

then $\quad \mathbf{r.a} = xa_1 + a_2y + a_3z$

and $\mathbf{r}.(\mathbf{a} \times \mathbf{b}) = \begin{vmatrix} x & y & z \\ a_1 & a_2 & a_3 \\ b_1 & b_2 & b_3 \end{vmatrix}$

$= x(a_2b_3 - a_3b_2) + y(a_3b_1 - a_1b_3) + z(a_1b_2 - a_2b_1)$

(i) $\nabla(\mathbf{r}.\mathbf{a}) = \nabla(xa_1 + a_2y + a_3z)$

$$= a_1\nabla(x) + a_2\nabla(y) + a_3\nabla(z)$$
$$= a_1\hat{i} + a_2\hat{j} + a_3\hat{k}$$
$$(\because \nabla(x) = \hat{i}, \nabla(y) = \hat{j}, \nabla(z) = \hat{k})$$
$$= \mathbf{a}$$

(ii) $\nabla[\mathbf{r}\ \mathbf{a}\ \mathbf{b}] = \nabla(\mathbf{r}.(\mathbf{a}\times\mathbf{b}))$

$$= \nabla[x(a_2b_3 - a_3b_2) + y(a_3b_1 - a_1b_3) + z(a_1b_2 - a_2b_1)]$$
$$= \nabla[x(a_2b_3 - a_3b_2)] + \nabla[y(a_3b_1 - a_1b_3)] + \nabla[z(a_1b_2 - a_2b_1)]$$
$$= (a_2b_3 - a_3b_2)\nabla(x) + (a_3b_1 - a_1b_3)\nabla(y) + (a_1b_2 - a_2b_1)\nabla(z)$$
$$= (a_2b_3 - a_3b_2)\hat{i} + (a_3b_1 - a_1b_3)\hat{j} + (a_1b_2 - a_2b_1)\hat{k}$$
$$= \begin{vmatrix} \hat{i} & \hat{j} & \hat{k} \\ a_1 & a_2 & a_3 \\ b_1 & b_2 & b_3 \end{vmatrix} = \mathbf{a}\times\mathbf{b}$$

$\therefore \quad \nabla[\mathbf{r}\ \mathbf{a}\ \mathbf{b}] = \mathbf{a}\times\mathbf{b}$.

Example 6. *If* $\phi = (3r^2 - 4r^{1/2} + 6r^{-1/3})$, *show that* $\nabla\phi = 2(3 - r^{-3/2} - r^{-7/3})\mathbf{r}$.

Solution . Let $\mathbf{r} = x\hat{i} + y\hat{j} + z\hat{k}$

then $|\mathbf{r}| = r = \sqrt{(x^2 + y^2 + z^2)}$.

Now
$$\nabla\phi = \frac{\partial\phi}{\partial x}\hat{i} + \frac{\partial\phi}{\partial y}\hat{j} + \frac{\partial\phi}{\partial z}\hat{k}$$
$$= \frac{\partial\phi}{\partial r}\frac{\partial r}{\partial x}\hat{i} + \frac{\partial\phi}{\partial r}\frac{\partial r}{\partial y}\hat{j} + \frac{\partial\phi}{\partial r}\frac{\partial r}{\partial z}\hat{k}$$
$$\nabla\phi = \frac{\partial\phi}{\partial r}\left(\frac{\partial r}{\partial x}\hat{i} + \frac{\partial r}{\partial y}\hat{j} + \frac{\partial r}{\partial z}\hat{k}\right) \quad \ldots(1)$$

Since $\phi = 3r^2 - 4r^{1/2} + 6r^{-1/3}$

$\therefore$ $\frac{\partial\phi}{\partial r} = (6r - 2r^{-1/2} - 2r^{-4/3})$

and $r^2 = x^2 + y^2 + z^2$ $\quad \ldots(2)$

$\therefore$ $\frac{\partial r}{\partial x} = \frac{x}{r}, \frac{\partial r}{\partial y} = \frac{y}{r}, \frac{\partial r}{\partial z} = \frac{z}{r}$

Using (2), equation (1) becomes

$$\therefore \quad \nabla\phi = (6r - 2r^{-1/2} - 2r^{-4/3})\frac{1}{r}(x\hat{i} + y\hat{j} + z\hat{k})$$
$$= 2(3 - r^{-3/2} - r^{-7/3})\,\mathbf{r}.$$

Example 7. *Find the directional derivative of* $f(x, y, z) = x^2yz + 4xz^2$ *at the point* $(1, -2, -1)$ *in the direction of the vector* $2\hat{i} - \hat{j} - 2\hat{k}$.

(UPTU–2006, JNTU–2006, VTU–2007, Rohtak–2006)

Solution . Let $\mathbf{a} = 2\hat{i} - \hat{j} - 2\hat{k}$, then

$$\hat{a} = \frac{\mathbf{a}}{|\mathbf{a}|} = \frac{2\hat{i} - \hat{j} - 2\hat{k}}{\sqrt{4+1+4}} = \frac{1}{3}(2\hat{i} - \hat{j} - 2\hat{k}).$$

Since $f(x, y, z) = x^2yz + 4xz^2$

$$\therefore \quad \frac{\partial f}{\partial x} = 2xyz + 4z^2$$
$$\frac{\partial f}{\partial y} = x^2z, \frac{\partial f}{\partial z} = x^2y + 8xz$$
$$\therefore \quad \nabla f = \frac{\partial f}{\partial x}\hat{i} + \frac{\partial f}{\partial y}\hat{j} + \frac{\partial f}{\partial z}\hat{k}$$
$$= (2xyz + 4z^2)\hat{i} + x^2z\hat{j} + (x^2y + 8xz)\hat{k}$$

At (1, –2, –1)
$$\nabla f = 8\hat{i} - \hat{j} - 10\hat{k}.$$

Now directional derivative of f at (1, –2, –1) in the direction of $2\hat{i} - \hat{j} - 2\hat{k}$ is

$$\nabla f\cdot\hat{a} = (8\hat{i} - \hat{j} - 10\hat{k}).\left(\frac{1}{3}(2\hat{i} - \hat{j} - 2\hat{k})\right)$$
$$= \frac{1}{3}(16 + 1 + 20) = \frac{37}{3}.$$

Example 8. *For the function* $f = y/(x^2 + y^2)$, *find the value of the directional derivative making an angle* 30° *with the positive x-axis at the point* (0, 1).

Solution . Let $\hat{a}$ be the unit vector which makes an angle 30° with the positive *x*-axis, then

$$\hat{a} = \cos 30^\circ\,\hat{i} + \sin 30^\circ\,\hat{j}$$
$$= \frac{\sqrt{3}}{2}\hat{i} + \frac{1}{2}\hat{j} = \frac{1}{2}(\sqrt{3}\hat{i} + \hat{j}).$$

Since
$$f = \frac{y}{(x^2 + y^2)}$$
$$\therefore \quad \frac{\partial f}{\partial x} = -\frac{2xy}{(x^2 + y^2)^2}$$
$$\frac{\partial f}{\partial y} = \frac{x^2 - y^2}{(x^2 + y^2)^2}$$
$$\therefore \quad \nabla f = \frac{\partial f}{\partial x}\hat{i} + \frac{\partial f}{\partial y}\hat{j}$$
$$= -\frac{2xy}{(x^2 + y^2)^2}\hat{i} + \frac{x^2 - y^2}{(x^2 + y^2)^2}\hat{j}$$

At (0, 1), $\nabla f = 0\cdot\hat{i} - \hat{j}$.

Now directional derivative of *f* is

$$\nabla f\cdot\hat{a} = -\hat{j}.\frac{1}{2}(\sqrt{3}\hat{i} + \hat{j}) = -\frac{1}{2}$$

Example 9. *Find the directional derivative of* $\phi = xy + yz + zx$ *in the direction of* $\hat{i} + 2\hat{j} + 2\hat{k}$ *at* (1, 2, 0).

Solution . Let $\mathbf{a} = \hat{i} + 2\hat{j} + 2\hat{k}$, then

$$\hat{a} = \frac{\mathbf{a}}{|\mathbf{a}|} = \frac{\hat{i} + 2\hat{j} + 2\hat{k}}{3} \quad (\because |\mathbf{a}| = 3)$$

Since $\phi = xy + yz + zx$

$$\frac{\partial\phi}{\partial x} = y + z, \frac{\partial\phi}{\partial y} = x + z, \frac{\partial\phi}{\partial z} = y + x.$$

$$\therefore \quad \nabla\phi = \frac{\partial\phi}{\partial x}\hat{i} + \frac{\partial\phi}{\partial y}\hat{j} + \frac{\partial\phi}{\partial z}\hat{k}$$
$$= (y+z)\hat{i} + (x+z)\hat{j} + (x+y)\hat{k}$$

At (1, 2, 0), $\nabla\phi = 2\hat{i} + \hat{j} + 3\hat{k}$

Now the directional derivative of ϕ is

$$\nabla\phi.\hat{a} = (2\hat{i} + \hat{j} + 3\hat{k}).\frac{1}{3}(\hat{i} + 2\hat{j} + 2\hat{k})$$
$$= \frac{(2+2+6)}{3} = \frac{10}{3}$$

Example 10. *Find the maximum value of the directional derivative of $\phi = x^2yz$ at the point (1, 4, 1).*

Solution . Since $\phi = x^2yz$

$$\therefore \quad \nabla\phi = \frac{\partial\phi}{\partial x}\hat{i} + \frac{\partial\phi}{\partial y}\hat{j} + \frac{\partial\phi}{\partial z}\hat{k}$$
$$= 2xyz\hat{i} + x^2z\hat{j} + x^2y\hat{k}$$

At (1, 4, 1),

$$\nabla\phi = 8\hat{i} + \hat{j} + 4\hat{k}.$$

Maximum value of directional derivative of ϕ at

(1, 4, 1) $= |\nabla\phi| = \sqrt{64+1+16}$
$= \sqrt{81} = 9.$

Example 11. *Prove that $\nabla\phi.\overrightarrow{dr} = d\phi$.*

Solution .
$$\nabla\phi = \frac{\partial\phi}{\partial x}\hat{i} + \frac{\partial\phi}{\partial y}\hat{j} + \frac{\partial\phi}{\partial z}\hat{k}$$
$$\overrightarrow{dr} = dx\hat{i} + dy\hat{j} + dz\hat{k}$$
$$[\because \mathbf{r} = x\hat{i} + y\hat{j} + z\hat{k}]$$
$$\therefore \quad \nabla\phi.\overrightarrow{dr} = \left(\left(\frac{\partial\phi}{\partial x}\hat{i} + \frac{\partial\phi}{\partial y}\hat{j} + \frac{\partial\phi}{\partial z}\hat{k}\right).(dx\hat{i} + dy\hat{j} + dz\hat{k})\right)$$
$$= \frac{\partial\phi}{\partial x}dx + \frac{\partial\phi}{\partial y}dy + \frac{\partial\phi}{\partial z}dz = d\phi.$$

Example 12. *Find grad ϕ, if $\phi = r^n = (x^2+y^2+z^2)^{n/2}$.*

Solution . We have
$$\frac{\partial\phi}{\partial x} = \frac{n}{2}(x^2+y^2+z^2)^{(n/2)-1}.2x$$
$$= n.(r^2)^{(n-2)/2}.x = nr^{n-2}.x$$

Similarly, $\frac{\partial\phi}{\partial y} = nr^{n-2}.y$, $\frac{\partial\phi}{\partial z} = nr^{n-2}.z$

$$\text{grad }\phi = \frac{\partial\phi}{\partial x}\hat{i} + \frac{\partial\phi}{\partial y}\hat{j} + \frac{\partial\phi}{\partial z}\hat{k}$$
$$= nr^{n-2}.x\hat{i} + nr^{n-2}.y\hat{j} + nr^{n-2}.z\hat{k}$$
$$= nr^{n-2}(x\hat{i} + y\hat{j} + z\hat{k})$$
$$= nr^{n-2}.\vec{r}.$$

Example 13. *If $\phi(x, y) = \log\sqrt{x^2+y^2}$ show that*

$$\text{grad }\phi = \frac{r - (\hat{k}\cdot\vec{r})\hat{k}}{\{\vec{r} - (\hat{k}\cdot\vec{r}\cdot\hat{k})\}.\{\vec{r} - (\hat{k}\cdot\vec{r})\hat{k}\}}.$$

Solution . We have $\mathbf{r} = x\hat{i} + y\hat{j} + z\hat{k}$

$$\phi = \frac{1}{2}\log(x^2+y^2)$$
$$\frac{\partial\phi}{\partial x} = \frac{1}{2(x^2+y^2)}.2x = \frac{x}{x^2+y^2}$$

Similarly, $\frac{\partial\phi}{\partial y} = \frac{y}{x^2+y^2}, \frac{\partial\phi}{\partial z} = 0$

$$\text{grad }\phi = \hat{i}\frac{\partial\phi}{\partial x} + \hat{j}\frac{\partial\phi}{\partial y} + \hat{k}\frac{\partial\phi}{\partial z}$$
$$= \frac{x}{x^2+y^2}\hat{i} + \frac{y}{x^2+y^2}\hat{j} + 0\hat{k}$$
$$= \frac{x\hat{i} + y\hat{j}}{x^2+y^2} = \frac{\mathbf{r} - z\hat{k}}{(x\hat{i} + y\hat{j}).(x\hat{i} + y\hat{j})}$$
$$= \frac{\mathbf{r} - z\hat{k}}{(\mathbf{r} - z\hat{k}).(\mathbf{r} - z\hat{k})} \quad \text{[By (1)]}$$

Now by replacing z by $\hat{k}\cdot\mathbf{r}$ we get

$$\text{grad }\phi = \frac{\mathbf{r} - (\hat{k}.\mathbf{r})\hat{k}}{\{\mathbf{r} - (\hat{k}.\mathbf{r})\hat{k}\}.\{\mathbf{r} - (\hat{k}.\mathbf{r})\hat{k}\}}.$$

Example 14. *Find the directional derivative of $f(x, y, z) = xy^2 + yz^3$ at the point (2, –1, 1) in the direction of the vector $\hat{i} + 2\hat{j} + 2\hat{k}$.* (Rohtak–2003, Bhopal–2008, Kurukshetra–2006, UPTU–2011, SVTU–2009, GBTU–2011)

Solution . Given $f(x, y, z) = xy^2 + yz^3$.

Therefore,
$$\nabla f = \hat{i}\frac{\partial}{\partial x}(xy^2 + yz^3) + \hat{j}\frac{\partial}{\partial y}(xy^2 + yz^3) + \hat{k}\frac{\partial}{\partial x}(xy^2 + yz^3)$$
$$= \hat{i}(y^2) + \hat{j}(2xy + z^3) + \hat{k}(3yz^2)$$
$$\nabla f_{\text{at }(2,-1,1)} = \hat{i} - 3\hat{j} - 3\hat{k}$$

$\therefore$ Directional derivative of f in the direction

$$\hat{i} + 2\hat{j} + 2\hat{k} = (\hat{i} - 3\hat{j} - 3\hat{k})\frac{\hat{i} + 2\hat{j} + 2\hat{k}}{\sqrt{1^2+2^2+2^2}}$$
$$= (1.1 - 3.2 - 3.2)/3 = -3\frac{2}{3}.$$

Example 15. *Find the directional derivative of $\phi = 5x^2y - 5y^2z + \frac{5}{2}z^2x$ at the point P(1, 1, 1) in the direction of the line $\frac{x-1}{2} = \frac{y-3}{2} = \frac{z}{1}$.* (GBTU–2010, UPTU–2004, Bhopal–2008)

Solution . We have $\phi = 5x^2y - 5y^2z + \frac{5}{2}z^2x$

$$\therefore \quad \text{grad }\phi = \hat{i}\frac{\partial\phi}{\partial x} + \hat{j}\frac{\partial\phi}{\partial y} + \hat{k}\frac{\partial\phi}{\partial z}$$
$$= (10xy + \frac{5}{2}z^2)\hat{i} + (5x^2 - 10yz)\hat{j} + (-5y^2 + 5zx)\hat{k}$$

$$= \frac{25}{2}\hat{i} - 5\hat{j} \text{ at the point } (1, 1, 1)$$

$\therefore$ Required direction derivative

$$= \left(\frac{25}{2}\hat{i} - 5\hat{j}\right) \cdot \left(\frac{2}{3}\hat{i} - \frac{2}{3}\hat{j} + \frac{1}{3}\hat{k}\right) z$$

$$= \frac{25}{3} + \frac{10}{3} = \frac{35}{3}.$$

Example 16. *Find the directional derivative of $\phi = (x^2 + y^2 + z^2)^{-\frac{1}{2}}$ at the point (3, 1, 2) in the direction of the vector $y = \hat{i} + zx\hat{j} + xy\hat{k}$.*

(UPTU–2007)

Solution . We have $\phi = (x^2 + y^2 + z^2)^{-\frac{1}{2}}$

Therefore,

$$\text{grad } \phi = \hat{i}\frac{\partial \phi}{\partial x} + \hat{j}\frac{\partial \phi}{\partial y} + \hat{k}\frac{\partial \phi}{\partial z}$$

$$= \hat{i}\left(-\frac{1}{2}(x^2 + y^2 + z^2)^{-3/2}.2x\right) + \hat{j}\left(-\frac{1}{2}(x^2 + y^2 + z^2)^{-3/2}.2y\right) + \hat{k}\left(-\frac{1}{2}(x^2 + y^2 + z^2)^{-3/2}.2z\right)$$

$$= -\frac{(x\hat{i} + y\hat{j} + z\hat{k})}{(x^2 + y^2 + z^2)^{3/2}}$$

$$= -\frac{3\hat{i} + \hat{j} + 2\hat{k}}{14\sqrt{14}} \text{ at } (3, 1, 2).$$

Let $\hat{a}$ be the unit vector in the given direction

then $$\hat{a} = \frac{yz\hat{i} + zx\hat{j} + xy\hat{k}}{\sqrt{y^2z^2 + z^2x^2 + x^2y^2}}$$

$$= \frac{2\hat{i} + 6\hat{j} + 3\hat{k}}{7} \text{ at } (3, 1, 2).$$

Hence $$\frac{d\phi}{ds} = \hat{a}.\text{grad } \phi$$

$$= \frac{2\hat{i} + 6\hat{j} + 3\hat{k}}{7}\left(-\frac{3\hat{i} + \hat{j} + 2\hat{k}}{14\sqrt{14}}\right)$$

$$= -\frac{2(3) + 6 \cdot 1 + 3 \cdot 2}{7 \cdot 14 \cdot \sqrt{14}} = -\frac{9}{49\sqrt{14}}$$

Example 17. *If the directional derivative of $\phi = ax^2y + by^2z + cz^2x$ at the point (1, 1, 1) has maximum magnitude 15 in the direction parallel to the line $\frac{x-1}{2} = \frac{y-3}{-2} = \frac{z}{1}$. Find the values of a, b and c.*

Solution . We have

$$\phi = ax^2y + by^2z + cz^2x$$

$$\Rightarrow \quad \text{grad } \phi = \hat{i}\frac{\partial \phi}{\partial x} + \hat{j}\frac{\partial \phi}{\partial y} + \hat{k}\frac{\partial \phi}{\partial z}$$

$$= (2axy + cz^2)\hat{i} + (ax^2 + 2byz)\hat{j} + (by^2 + 2czx)\hat{k}$$

$$= (2a + c)\hat{i} + (a + 2b)\hat{j} + (b + 2c)\hat{k} \quad \text{at } (1, 1, 1)$$

Now the directional derivative is maximum along the normal to the surface, *i.e.* along grad ϕ

$$|\text{grad } \phi| = \sqrt{(2a + c)^2 + (a + 2b)^2 + (b + 2c)^2}$$

$$\Rightarrow \quad 15 = \sqrt{(2a + c)^2 + (a + 2b)^2 + (b + 2c)^2}$$

$$\Rightarrow (2a + c)^2 + (a + 2b)^2 + (b + 2c)^2 = 225 \quad \ldots(1)$$

Since the directional derivative is maximum in the direction parallel to the line $\frac{x-1}{2} = \frac{y-3}{-2} = \frac{z}{1}$, *i.e.*, parallel to the vector $2\hat{i} - 2\hat{j} + \hat{k}$, therefore,

$$\frac{2a + c}{2} = \frac{a + 2b}{-2} = \frac{2c + b}{1}$$

$$\Rightarrow \quad 2a + c = -a - 2b \quad \Rightarrow 3a + 2b + c = 0 \quad \ldots(2)$$

and $2b + a = -4c - 2b \Rightarrow a + 4b + 4c = 0 \quad \ldots(3)$

On solving (1), (2) and (3) by cross multiplication we get

$$\frac{a}{4} = \frac{b}{-11} = \frac{c}{10} = k \text{ (say)}$$

$$\Rightarrow \quad a = 4k, b = -11k, c = 10k$$

Then from (1)

$$(8k + 10k)^2 + (4k - 22k)^2 + (-11k + 20k)^2 = 225$$

$$\Rightarrow \quad k = \pm\frac{5}{9}$$

Hence $a = \pm\frac{20}{9}, b = \mp\frac{55}{9}$ and $c = \pm\frac{50}{9}$.

EXERCISE 22.1

1. If $\phi(x, y, z) = x^2y + y^2x + z^2$, find $\nabla\phi$ at the point (1, 1, 1).
2. If $f(x, y, z) = x^2yz\hat{i} - 2xz^3\hat{j} + xz^2\hat{k}, \phi(x, y, z) = 2z\hat{i} + y\hat{j} - x^2\hat{k}$, find the value of $\frac{\partial^2}{\partial x \partial y}(\mathbf{f} \times \phi)$ at (1, 0, –2).
3. If $|\mathbf{r}| = r$ where $\mathbf{r} = x\hat{i} + y\hat{j} + z\hat{k}$, prove that

   (i) $\nabla\left(\frac{1}{r}\right) = -\frac{\mathbf{r}}{r^3}$ (GBTU–2011)

   (ii) $\nabla \log r = \frac{\mathbf{r}}{r^2}$ (GBTU–2011)
4. Prove that $f(r)\nabla r = \nabla \int f(r)dr$.
5. (i) Interpret the symbol $\mathbf{a}.\nabla$. (ii) Prove that $(\mathbf{a}.\nabla)\phi = \mathbf{a}.\nabla\phi$.

   (iii) Prove that $(\mathbf{a}.\nabla)\mathbf{r} = \mathbf{a}$.
6. Find the grad f, where f is given by $f(x, y, z) = x^3 - y^3 + xz^2$, at the point (1, –1, 2).

7. If $u = x + y + z$, $v = x^2 + y^2 + z^2$, $w = yz + zx + xy$, prove that (grad u) . [(grad v) × (grad w)] = 0. (UKTU–2010, UPTU–2002)

8. f and p are two scalar point functions such that f is a function of p, show that

$$\nabla f = \frac{df}{dp}\nabla p.$$

9. If $\mathbf{F} = \left(y\frac{\partial f}{\partial z} - z\frac{\partial f}{\partial y}\right)\hat{i} + \left(z\frac{\partial f}{\partial x} - x\frac{\partial f}{\partial z}\right)\hat{j} + \left(x\frac{\partial f}{\partial y} - y\frac{\partial f}{\partial x}\right)\hat{k}$. Prove that

(i) $\mathbf{F} = \mathbf{r} \times \nabla f$ (ii) $\mathbf{F} . \mathbf{r} = 0$

(iii) $\mathbf{F} . \nabla f = 0$

10. Prove that the directional derivative of a scalar field f at a point $P(x, y, z)$ in the direction of a unit vector $\hat{a}$ is given by

$$\frac{\partial f}{\partial s} = \nabla f.\hat{a}$$

11. Find the directional derivative of the function

$$f(x, y, z) = x^2 - y^2 + 2z^2.$$

at the point $P(1, 2, 3)$ in the direction of the line PQ where Q is the point $(5, 0, 4)$. (UKTU–2011, GBTU–2010)

12. In what direction from the point (1, 1, –1) is the directional derivative of $f = x^2 - 2y^2 + 4z^2$ a maximum? Also find the value of this maximum directional derivative.

13. Find the directional derivative of the function $f = xy + yz + zx$ in the direction of the vector $2\hat{i} + 3\hat{j} + 6\hat{k}$ at the point (3, 1, 2).

14. Find the greatest value of the derivative of the function $f = 2x^2 - y - z^4$ at the point (2, –1, 1).

15. Find the directional derivative $\partial f/\partial s$ of $f(x, y, z) = 2x^2 + 3y^2 + z^2$ at the point $P(2, 1, 3)$ in the direction of the vector $\mathbf{a} = \hat{i} - 2\hat{k}$. (UPTU–2009)

16. Find the directional derivative of $f = x^2 + y^2 + z^2$ at (1, 2, 3) in the direction of the line $\frac{x}{3} = \frac{y}{4} = \frac{z}{5}$.

17. Find the directional derivative of the function

$$f(x, y, z) = 4e^{x + 5y - 13z} \text{ at the point } (1, 2, 3)$$

in the direction towards the point (–3, 5, 7). (UPTU–2009)

18. In what direction from (3, 1, –2) is the directional derivative of $\phi = x^2y^2z^4$ maximum and what is its magnitude. (Rohtak–2003)

19. Show that grad $(e^{r^2}) = 2e^{r^2} \cdot \mathbf{r}$.

20. Show that grad $f(r) \times \mathbf{r} = 0$. (GBTU–2011)

21. Show that the directional derivative of $\frac{1}{r}$ in the direction of $\mathbf{r}$ where $\bar{r} = x\hat{i} + y\hat{j} + z\hat{k}$ is $-\frac{1}{r^2}$. (UKTU–2011)

22. Show that the directional derivative of $\frac{1}{r^2}$ in the direction of $\mathbf{r}$ where $\bar{r} = x\hat{i} + y\hat{j} + z\hat{k}$ is $-\frac{2}{r^3}$. (UPTU–2006)

Hint to Selected Problems

1. $\phi(x,y,z) = x^2y + y^2x + z^2$

$\therefore$ $\nabla\phi = \frac{\partial\phi}{\partial x}\hat{i} + \frac{\partial\phi}{\partial y}\hat{j} + \frac{\partial\phi}{\partial z}\hat{k}$

$= (2xy + y^2)\hat{i} + (x^2 + 2xy)\hat{j} + 2z\hat{k}$

$\therefore$ $[\nabla\phi]_{(1,1,1)} = 3\hat{i} + 3\hat{j} + 2\hat{k}$.

3. (ii) Since $\nabla f(r) = f(r)\nabla r = f'(r)\frac{\mathbf{r}}{r}$.

$\therefore$ $\nabla \log r = \frac{d}{dt}(\log r)\frac{\mathbf{r}}{r} = \frac{1}{r}\frac{\mathbf{r}}{r} = \frac{\mathbf{r}}{r^2}$.

(iv) $\mathbf{r} = x\hat{i} + y\hat{j} + z\hat{k}$ $\therefore$ $d\mathbf{r} = dx\hat{i} + dy\hat{j} + dz\hat{k}$.

$\therefore$ $\nabla\phi.d\mathbf{r} = \left(\frac{\partial\phi}{\partial x}\hat{i} + \frac{\partial\phi}{\partial y}\hat{j} + \frac{\partial\phi}{\partial z}\hat{k}\right).(dx\hat{i} + dy\hat{j} + dz\hat{k})$

$= \frac{\partial\phi}{\partial x}dx + \frac{\partial\phi}{\partial y}dy + \frac{\partial\phi}{\partial z}dz = d\phi.$

5. Let $\mathbf{a} = a_1\hat{i} + a_2\hat{j} + a_3\hat{k}, \mathbf{r} = x\hat{i} + y\hat{j} + z\hat{k}.$

$\therefore$ $\mathbf{a}\cdot\nabla = a_1\frac{\partial}{\partial x} + a_2\frac{\partial}{\partial y} + a_3\frac{\partial}{\partial z}$

Now $(\mathbf{a}\cdot\nabla)\mathbf{r} = a_1\frac{\partial\mathbf{r}}{\partial x} + a_2\frac{\partial\mathbf{r}}{\partial y} + a_3\frac{\partial\mathbf{r}}{\partial z}$

$= a_1\hat{i} + a_2\hat{j} + a_3\hat{k} = \mathbf{a}$

7. $u = x + y + z$, $v = x^2 + y^2 + z^2$, $w = yz + zx + xy$

$\text{grad } v = \frac{\partial v}{\partial x}\hat{i} + \frac{\partial v}{\partial y}\hat{j} + \frac{\partial v}{\partial z}\hat{k} = 2x\hat{i} + 2y\hat{j} + 2z\hat{k}$

$\text{grad } w = \frac{\partial w}{\partial x}\hat{i} + \frac{\partial w}{\partial y}\hat{j} + \frac{\partial w}{\partial z}\hat{k}$

$= (z + y)\hat{i} + (z + x)\hat{j} + (y + x)\hat{k}$

$$(\text{grad } v)\times(\text{grad } w) = \begin{vmatrix} \hat{i} & \hat{j} & \hat{k} \\ 2x & 2y & 2z \\ z+y & z+x & y+x \end{vmatrix}$$

$= \hat{i}(2y^2 + 2xy - 2z^2 - 2zx)$

$+ \hat{j}(2z^2 + 2yz - 2xy - 2x^2)$

$+ \hat{k}(2xz + 2x^2 - 2yz - 2y^2)$

$\text{grad } u = \frac{\partial u}{\partial x}\hat{i} + \frac{\partial u}{\partial y}\hat{j} + \frac{\partial u}{\partial z}\hat{k} = \hat{i} + \hat{j} + \hat{k}$

$\therefore$ (grad u)·[(grad v) × (grad w)]

$= (2y^2 + 2xy - 2z^2 - 2zx)$

$+ (2z^2 + 2yz - 2xy - 2x^2)$

$+ (2xz + 2x^2 - 2yz - 2y^2)$

$= 0.$

8. Since $f = f(p)$

$\therefore$ $\nabla f = \nabla(f(p))$

$= \frac{\partial}{\partial x}(f(p))\hat{i} + \frac{\partial}{\partial y}(f(p))\hat{j} + \frac{\partial}{\partial z}(f(p))\hat{k}$

$= \frac{df}{dp}\frac{\partial p}{\partial x}\hat{i} + \frac{df}{dp}\frac{\partial p}{\partial y}\hat{j} + \frac{df}{dp}\frac{\partial p}{\partial z}\hat{k}$

$= \frac{df}{dp}\left(\frac{\partial p}{\partial x}\hat{i} + \frac{\partial p}{\partial y}\hat{j} + \frac{\partial p}{\partial z}\hat{k}\right) = \frac{df}{dp}\nabla p.$

11. $f(x,y,z) = x^2 - y^2 + 2z^2$

$\therefore \quad \dfrac{\partial f}{\partial x} = 2x, \dfrac{\partial f}{\partial y} = -2y, \dfrac{\partial f}{\partial z} = 4z.$

$\therefore \quad \nabla f = \dfrac{\partial f}{\partial x}\hat{i} + \dfrac{\partial f}{\partial y}\hat{j} + \dfrac{\partial f}{\partial z}\hat{k} = 2x\hat{i} - 2y\hat{j} + 4z\hat{k}$

at $P(1, 2, 3)$ $\quad \nabla f = 2\hat{i} - 4\hat{j} + 12\hat{k}$

Now $\quad \overline{PQ} = 4\hat{i} - 2\hat{j} + \hat{k}$

$\therefore$ directional derivative along $\overline{PQ}$ is

$$(\nabla f).\frac{\overline{PQ}}{|\overline{PQ}|} = \frac{(8+8+12)}{\sqrt{16+4+1}}$$

$$= \frac{28}{\sqrt{21}} = \frac{28\sqrt{21}}{21} = \frac{4}{3}\sqrt{21}.$$

12. Same as **11.**

13. Same as **11.**

14. Same as **11.**

15. Same as **11.**

Answers

1. $3\hat{i} + 3\hat{j} + 2\hat{k}$ **2.** $-4\hat{i} - 8\hat{j}$ **6.** $7\hat{i} - 3\hat{j} + 4\hat{k}$ **11.** $\frac{4}{3}\sqrt{21}$ **12.** $2\hat{i} - 4\hat{j} - 8\hat{k}, 2\sqrt{(21)}$

13. 45/7 **14.** 9 **15.** $-4/\sqrt{5}$ **16.** $\frac{52}{\sqrt{50}}$ **17.** $-4\sqrt{41}e^{-28}$ **18.** $96(\hat{i} + 3\hat{j} - 3\hat{k}), 96\sqrt{19}$

22.6 LEVEL SURFACES

Let us consider a scalar function $f(x, y, z)$ and suppose that for each constant c the equation

$$f(x, y, z) = c = \text{constant}$$

represents a surface in space (in three dimensional space). Then assuming c takes all values, we obtain a family of surfaces, which are known as level surfaces.

THEOREM 1. *Let $f(x, y, z)$ be a scalar point function over some region R, then show that through any point on R there passes one and only one, level surface of f.*

Proof. Let (x_1, y_1, z_1) be any point in space (on R). Then the level surface $f(x, y, z)$ will pass through this point

$$\therefore \quad f(x, y, z) = f(x_1, y_1, z_1). \quad \ldots(1)$$

Let us suppose that the level surfaces $f(x, y, z) = c_1$ and $f(x, y, z) = c_2$ passes through the point (x_1, y_1, z_1), then

$$f(x_1, y_1, z_1) = c_1 \quad \text{and} \quad f(x_2, y_2, z_2) = c_2.$$

Using (1), we get

$$c_1 = c_2.$$

Hence, through (x_1, y_1, z_1) there passes one and only one level surface.

THEOREM 2. *grad $(f) = \nabla f$ is a normal vector to the surface $f(x, y, z) = c$, where c is a constant.*

Proof. The equation of a curve in space can be represented in the form

$$\mathbf{r}(t) = x(t)\hat{i} + y(t)\hat{j} + z(t)\hat{k}. \quad \ldots(1)$$

If the curve lies on the surface $f(x, y, z) = c$, then we have $f[x(t), y(t), z(t)] = c$.
Differentiating with respect to t, we get

$$\frac{\partial f}{\partial x}.\frac{dx}{dt} + \frac{\partial f}{\partial y}.\frac{dy}{dt} + \frac{\partial f}{\partial z}.\frac{dz}{dt} = 0$$

or
$$\left(\frac{\partial f}{\partial x}\hat{i} + \frac{\partial f}{\partial y}\hat{j} + \frac{\partial f}{\partial z}\hat{k}\right).\left(\frac{dx}{dt}\hat{i} + \frac{dy}{dt}\hat{j} + \frac{dz}{dt}\hat{k}\right) = 0$$

or
$$(\text{grad } f).\frac{d\mathbf{r}}{dt} = 0 \quad \text{[Using (1)]} \quad \ldots(2)$$

where the vector $\dfrac{d\mathbf{r}}{dt} = \dfrac{dx}{dt}\hat{i} + \dfrac{dy}{dt}\hat{j} + \dfrac{dz}{dt}\hat{k}$ is a vector parallel to the tangent at the point $P(x, y, z)$ to the surface.
Equation (2) implies that (grad f) is perpendicular to the tangent plane at $P(x, y, z)$ to the surface $f(x, y, z) = c$.
Hence, (grad f) or ∇f is a vector normal to the surface $f(x, y, z) = c$.

22.7 TANGENT AND NORMAL TO THE LEVEL SURFACE

(i) Tangent Plane. Let $f(x, y, z) = c$ be the equation of a level surfaces and let $P(x, y, z)$ be any point on this surface whose position vector be $\mathbf{r}$.

$$\therefore \qquad \mathbf{r} = x\hat{i} + y\hat{j} + z\hat{k}. \qquad \ldots(1)$$

Since ∇f is perpendicular to the tangent plane at $P(x, y, z)$. Let Q be any variable point on the surface tangent plane to the whose co-ordinates are (X, Y, Z) and whose position vector is $\mathbf{R}$.

$$\therefore \qquad \overrightarrow{PQ} = \mathbf{R} - \mathbf{r} = (X-x)\hat{i} + (Y-y)\hat{j} + (Z-z)\hat{k}.$$

Since $\overrightarrow{PQ}$ is along the tangent plane at P, then ∇f is perpendicular to $\overrightarrow{PQ}$.

$$\therefore \qquad \nabla f . \overrightarrow{PQ} = 0$$

$$\left(\frac{\partial f}{\partial x}\hat{i} + \frac{\partial f}{\partial y}\hat{j} + \frac{\partial f}{\partial z}\hat{k}\right)((X-x)\hat{i} + (Y-y)\hat{j} + (Z-z)\hat{k}) = 0$$

or
$$(X-x)\frac{\partial f}{\partial x} + (Y-y)\frac{\partial f}{\partial y} + (Z-z)\frac{\partial f}{\partial z} = 0$$

This is the equation of a tangent plane at $P(x, y, z)$ to the level surface $f(x, y, z) = c$.

(ii) Normal. In this case the point Q is taken on the normal to the surface $f(x, y, z) = c$. So that the direction ratios of the line PQ are $X-x, Y-y, Z-z$

or
$$\overrightarrow{PQ} = (X-x)\hat{i} + (Y-y)\hat{j} + (Z-z)\hat{k}$$

Thus the vector $\overrightarrow{PQ}$ is now parallel to the ∇f.

$$\therefore \qquad \nabla f \times \overrightarrow{PQ} = \mathbf{0}$$

or
$$\left(\frac{\partial f}{\partial x}\hat{i} + \frac{\partial f}{\partial y}\hat{j} + \frac{\partial f}{\partial z}\hat{k}\right) \times ((X-x)\hat{i} + (Y-y)\hat{j} + (Z-z)\hat{k}) = \mathbf{0}$$

$$\left[\frac{\partial f}{\partial y}(Z-z) - \frac{\partial f}{\partial z}(Y-y)\right]\hat{i} + \left[\frac{\partial f}{\partial z}(X-x) - \frac{\partial f}{\partial x}(Z-z)\right]\hat{j} + \left[\frac{\partial f}{\partial x}(Y-y) - \frac{\partial f}{\partial y}(X-x)\right]\hat{k} = \mathbf{0}$$

or
$$\frac{\partial f}{\partial y}(Z-z) - \frac{\partial f}{\partial z}(Y-y) = 0$$

$$\frac{\partial f}{\partial z}(X-x) - \frac{\partial f}{\partial x}(Z-z) = 0 \qquad \text{and} \qquad \frac{\partial f}{\partial x}(Y-y) - \frac{\partial f}{\partial y}(X-x) = 0.$$

From these three equations we obtain the equation to the normal which is as follows :

$$\frac{X-x}{\frac{\partial f}{\partial x}} = \frac{Y-y}{\frac{\partial f}{\partial y}} = \frac{Z-z}{\frac{\partial f}{\partial z}} = 0.$$

REMARK

- If $f(x, y, z) = c$, then $\frac{\partial f}{\partial x}, \frac{\partial f}{\partial y}, \frac{\partial f}{\partial z}$ are the direction ratios of the normal to the surface.

Solved Examples

Example 1. *Find a unit normal vector to the level surface $x^2y + 2xz = 4$ at the point (2, –2, 3).*

Solution . Let $f(x, y, z) \equiv x^2y + 2xz - 4 = 0$.

$$\therefore \quad \frac{\partial f}{\partial x} = 2xy + 2z, \frac{\partial f}{\partial y} = x^2, \frac{\partial f}{\partial z} = 2x.$$

$$\therefore \quad \nabla f = \frac{\partial f}{\partial x}\hat{i} + \frac{\partial f}{\partial y}\hat{j} + \frac{\partial f}{\partial z}\hat{k}$$

$$= (2xy + 2z)\hat{i} + x^2\hat{j} + 2\,x\hat{k}$$

At (2, –2, 3), $\nabla f = -2\hat{i} + 4\hat{j} + 4\hat{k}$.

The unit normal vector is given by

$$\frac{\nabla f}{|\nabla f|} = \frac{-2\hat{i} + 4\hat{j} + 4\hat{k}}{\sqrt{(4+16+16)}} = \frac{-\hat{i} + 2\hat{j} + 2\hat{k}}{3}.$$

Example 2. *Find the unit normal to the surface $x^4 - 3xyz + z^2 + 1 = 0$ at the point (1, 1, 1).*

Solution . Suppose $f(x, y, z) = x^4 - 3xyz + z^2 + 1 = 0$, then

$$\frac{\partial f}{\partial x} = 4x^3 - 3yz, \frac{\partial f}{\partial y} = -3xz, \frac{\partial f}{\partial z} = -3xy + 2z.$$

$$\therefore \quad \nabla f = \frac{\partial f}{\partial x}\hat{i} + \frac{\partial f}{\partial y}\hat{j} + \frac{\partial f}{\partial z}\hat{k}$$

or $\quad \nabla f = (4x^3 - 3yz)\hat{i} - 3xz\hat{j} + (2z - 3xy)\hat{k}$

At (1, 1, 1), $\nabla f = \hat{i} - 3\hat{j} - \hat{k}$.

The unit normal vector is given by

$$\frac{\nabla f}{|\nabla f|} = \frac{\hat{i} - 3\hat{j} - \hat{k}}{\sqrt{(1+9+1)}} = \frac{1}{\sqrt{11}}(\hat{i} - 3\hat{j} - \hat{k}).$$

Example 3. *Find a unit vector normal to the surface $xy^3z^2 = 4$ at the point (–1, –1, 2).* (UPTU–2008, Mumbai–2008)

Solution. We have $\phi = xy^3z^2$

$$\therefore \quad \text{grad}\,\phi = \hat{i}\frac{\partial \phi}{\partial x} + \hat{j}\frac{\partial \phi}{\partial y} + \hat{k}\frac{\partial \phi}{\partial z}$$

$$= \hat{i}\frac{\partial}{\partial x}(xy^3z^2) + \hat{j}\frac{\partial}{\partial y}(xy^3z^2) + \hat{k}\frac{\partial}{\partial z}(xy^3z^2)$$

$$= \hat{i}(y^3z^2) + \hat{j}(3xy^2z^2) + \hat{k}(2xy^3z)$$

$$= -4\hat{i} - 12\hat{j} + 4\hat{k} \text{ at the point } (-1, -1, 2)$$

Hence, required unit normal vector to the surface is

$$= \frac{-4\hat{i} - 12\hat{j} + 4\hat{k}}{\sqrt{(-4)^2 + (-12)^2 + 4^2}}$$

$$= -\frac{1}{\sqrt{11}}(\hat{i} + 3\hat{j} - \hat{k}).$$

Example 4. *Find a unit normal vector $\hat{n}$ of the cone of revolution $z^2 = 4(x^2 + y^2)$ at the point (1, 0, 2).* (UPTU-2010)

Solution. We have $\phi = z^2 - 4x^2 - 4y^2$.

Then $\quad \nabla\phi = \hat{i}\frac{\partial \phi}{\partial x} + \hat{j}\frac{\partial \phi}{\partial y} + \hat{k}\frac{\partial \phi}{\partial z}$

$$= -8x\hat{i} - 8y\hat{j} + 2z\hat{k}$$

$$= -8\hat{i} + 4\hat{k} \text{ at the point } (1, 0, 2)$$

$$\Rightarrow \quad |\nabla\phi| = \sqrt{64 + 16} = \sqrt{80}$$

Hence, unit normal vetor $\hat{n}$ to the given cone at (1, 0, 2) is

$$\hat{n} = \frac{\nabla\phi}{|\nabla\phi|} = \frac{-8\hat{i} + 4\hat{k}}{\sqrt{80}} = \frac{-2\hat{i} + \hat{k}}{\sqrt{5}}.$$

Example 5. *Find the equation of the tangent plane to the surface $yz - zx + xy + 5 = 0$ at the point (1, –1, 2).*

Solution. Let $f = yz - zx + xy + 5 = 0$, then

$$\frac{\partial f}{\partial x} = -z + y, \frac{\partial f}{\partial y} = z + x, \frac{\partial f}{\partial z} = y - x.$$

$$\therefore \quad \nabla f = \frac{\partial f}{\partial x}\hat{i} + \frac{\partial f}{\partial y}\hat{j} + \frac{\partial f}{\partial z}\hat{k}.$$

$$= (y - z)\hat{i} + (z + x)\hat{j} + (y - x)\hat{k}.$$

At (1, –1, 2),

$$\nabla f = -3\hat{i} + 3\hat{j} - 2\hat{k}.$$

Let $Q(X, Y, Z)$ be any point on the tangent plane to the surface and P is given as (1, –1, 2)

$$\therefore \quad \overrightarrow{PQ} = (X - 1)\hat{i} + (Y + 1)\hat{j} + (Z - 2)\hat{k}.$$

For the equation of the tangent at (1, –1, 2), we have

$$\nabla f . \overrightarrow{PQ} = 0$$

$$\Rightarrow (X - 1)(-3) + (Y + 1)(3) + (Z - 2)(-2) = 0$$

$$-3X + 3Y - 2Z + 3 + 3 + 4 = 0$$

or $\; 3X - 3Y + 2Z = 10 \;$ or $\; 3x - 3y + 2z = 10$

Example 6. *Find the equation of the normal to the surface $2xz^2 - 3xy - 4x = 7$ at the point (1, –1, 2).*

Solution. Let $f(x, y, z) \equiv 2xz^2 - 3xy - 4x - 7 = 0$

$$\therefore \quad \nabla f = \frac{\partial f}{\partial x}\hat{i} + \frac{\partial f}{\partial y}\hat{j} + \frac{\partial f}{\partial z}\hat{k}$$

$$= (2z^2 - 3y - 4)\hat{i} - 3x\hat{j} + 4xz\hat{k}$$

At (1, –1, 2)

$$\nabla f = 7\hat{i} - 3\hat{j} + 8\hat{k}$$

Now the position vector of the point (1, –1, 2) is

$$\therefore \quad \mathbf{r} = \hat{i} - \hat{j} + 2\hat{k}$$

Let $\mathbf{R} = X\hat{i} + Y\hat{j} + Z\hat{k}$ be the position vector of any variable point (X, Y, Z) on the normal, then the vector $\mathbf{R} - \mathbf{r}$ is parallel to ∇f.

$$(\mathbf{R} - \mathbf{r}) \times \nabla f = 0$$

or the equation of the normal is

$$\frac{X - 1}{7} = \frac{Y + 1}{-3} = \frac{Z - 2}{8}.$$

Example 7. *Find the angle between the surface $x^2 + y^2 + z^2 = 9$ and $z = x^2 + y^2 - 3$ at the point (2, –1, 2).*

(VTU–2010, UPTU–2003, Kottayam–2005)

Solution. Let the given surfaces be

$$f_1(x, y, z) \equiv x^2 + y^2 + z^2 = 9 \text{ as } f_1(x, y, z) = c_1 \quad \ldots(1)$$

$$f_2(x, y, z) \equiv x^2 + y^2 - z = 3 \text{ as } f_2(x, y, z) = c_2 \quad \ldots(2)$$

Normal vector to surface (1) is

$$\mathbf{n_1} = \text{grad}\, f_1$$

$$= \left(\hat{i}\frac{\partial}{\partial x} + \hat{j}\frac{\partial}{\partial y} + \hat{k}\frac{\partial}{\partial z}\right)(x^2 + y^2 + z^2)$$

$$= 2x\hat{i} + 2y\hat{j} + 2z\hat{k}$$

At point (2, –1, 2),

$$\mathbf{n_1} = 2.2\hat{i} + 2(-1)\hat{j} + 2.2\hat{k} = 4\hat{i} - 2\hat{j} + 4\hat{k}.$$

Normal vector to surface (2) is

$$\mathbf{n_2} = \text{grad}\, f_2$$

$$= \left(\hat{i}\frac{\partial}{\partial x} + \hat{j}\frac{\partial}{\partial y} + \hat{k}\frac{\partial}{\partial z}\right)(x^2 + y^2 - z)$$

$= 2x\hat{i} + 2y\hat{j} - \hat{k}$.

At point (2, –1, 2),

$\mathbf{n_2} = 2.2\hat{i} + 2(-1)\hat{j} - \hat{k} = 4\hat{i} - 2\hat{j} - \hat{k}$.

Now let θ be the angle between surfaces (1) and (2), then the angle between their normals $\mathbf{n_1}$ and $\mathbf{n_2}$ is also θ.

$$\therefore \quad \cos\theta = \frac{\mathbf{n_1}.\mathbf{n_2}}{|\mathbf{n_1}||\mathbf{n_2}|}$$

$$= \frac{4.4 + (-2)(-2) + 4(-1)}{\sqrt{(4)^2 + (-2)^2 + (4)^2}\sqrt{(4)^2 + (-2)^2 + (-1)^2}}$$

$$= \frac{16 + 4 - 4}{\sqrt{36}\sqrt{21}} = \frac{16}{6\sqrt{21}} = \frac{8}{3\sqrt{21}}$$

$$\therefore \quad \theta = \cos^{-1}\left(\frac{8}{3\sqrt{21}}\right).$$

EXERCISE 22.2

1. Find the unit vector normal to the surfce $x^2 - y^2 + z = 2$ at the point (1, –1, 2).
2. Find the vector normal to the surface $z = x^2 + y^2$ at the point (–1, –2, 5).
3. Find the unit normal to the surface $x^2 + y - z = 4$ at the point (2, 0, 0).
4. Find the equation of the tangent plane and normal to the surface $xyz = 4$ at the point (1, 2, 2).
5. Find the equation of the tangent plane and normal to the surface $x^2 + y^2 + z^2 = 25$ at the point (4, 0, 3).
6. Find the equation of the tangent plane and normal to the surface $z = x^2 + y^2$ at the point (2, –1, 5).
7. If $\hat{n}$ be a unit vector normal to the level surface $f(x, y, z) = c$ at a point $P(x, y, z)$ and n be the distance of P from some fixed point A in the direction of $\hat{n}$ so that δn represents element of normal at P in the direction of $\hat{n}$, then
$$\text{grad } f = \frac{df}{dn}\hat{n}.$$
8. Prove that grad f is a vector in the direction of which the maximum value of the directional derivative of f.
9. Find the angle between the normals to the surface $xy = z^2$ at the points (4, 1, 2) and (3, 3, –3). (UKTU–2012)

Hint to Selected Problems

1. $f \equiv x^2 - y^2 + z - 2 = 0.$

$$\therefore \quad \nabla f = \frac{\partial f}{\partial x}\hat{i} + \frac{\partial f}{\partial y}\hat{j} + \frac{\partial f}{\partial z}\hat{k} = 2x\hat{i} - 2y\hat{j} + \hat{k}$$

At (1, –1, 2)

$$\nabla f = 2\hat{i} + 2\hat{j} + \hat{k}$$

$$\therefore \text{ Unit normal to the surface } f = \frac{\nabla f}{|\nabla f|} = \frac{2\hat{i} + 2\hat{j} + \hat{k}}{3}.$$

4. $f(x, y, z) \equiv xyz - 4$

$$\therefore \quad \frac{\partial f}{\partial x} = yz, \frac{\partial f}{\partial y} = xz, \frac{\partial f}{\partial z} = xy$$

$\therefore$ at (1, 2, 2) $\quad \nabla f = 4\hat{i} + 2\hat{j} + 2\hat{k}$.

The equation of the tangent plane at (1, 2, 2) is

$$[(x-1)\hat{i} + (y-2)\hat{j} + (z-2)\hat{k}].(4\hat{i} + 2\hat{j} + 2\hat{k}) = 0$$

or $\quad 4(x-1) + 2(y-2) + 2(z-2) = 0$

or $\quad 4x + 2y + 2z - 12 = 0 \quad$ or $\quad 2x + y + z - 6 = 0$

and the equation of the normal at (1, 2, 2) is

$$\frac{x-1}{4} = \frac{y-2}{2} = \frac{z-2}{2} \quad \text{or} \quad \frac{x-1}{2} = \frac{y-2}{1} = \frac{z-2}{1}$$

6. Same as (4).

ANSWERS

1. $\frac{1}{3}(2\hat{i} + 2\hat{j} + \hat{k})$
2. $2\hat{i} + 4\hat{j} + \hat{k}$
3. $\frac{1}{3\sqrt{2}}(4\hat{i} + \hat{j} - \hat{k})$
4. $2x + y + z = 6; \frac{x-1}{2} = \frac{y-2}{1} = \frac{z-2}{1}$
5. $4x + 3z = 25; \frac{x-4}{4} = \frac{y}{0} = \frac{z-3}{3}$
6. $4x - 2y - z = 5; \frac{x-2}{4} = \frac{y+1}{-2} = \frac{z-5}{-1}$
9. $\cos^{-1}\left(\frac{1}{\sqrt{22}}\right)$

22.8 DIVERGENCE OF A VECTOR FIELD

(UPTU–2006, 07, 08)

Let $\mathbf{V}(x, y, z)$ be a differentiable vector function, where x, y, z are cartesian co-ordinates in space and let V_1, V_2, V_3 be the components of $\mathbf{V}$, then the function

$$\text{div } \mathbf{V} = \frac{\partial V_1}{\partial x} + \frac{\partial V_2}{\partial y} + \frac{\partial V_3}{\partial z} \qquad \text{...(1)}$$

is called the divergence of V.

Since we have that the differential operator $\nabla \equiv \frac{\partial}{\partial x}\hat{i} + \frac{\partial}{\partial y}\hat{j} + \frac{\partial}{\partial z}\hat{k}$ and the vector $\mathbf{V} = V_1\hat{i} + V_2\hat{j} + V_3\hat{k}$.

Then

$$\nabla.\mathbf{V} = \frac{\partial V_1}{\partial x} + \frac{\partial V_2}{\partial y} + \frac{\partial V_3}{\partial z} \qquad \text{...(2)}$$

From equation (1) and (2), we get

$$\text{div}\,\mathbf{V} = \nabla.\mathbf{V}$$

Hence, divergence of a vector function **V** can also be written as $\nabla.\mathbf{V}$. Consequently divergence of a vector function is scalar because dot product of ∇ and **V** gives a scalar quantity.

REMARKS

- Though dot product is cummulative but ∇ being operator which operates right side function only, we have $\nabla.\mathbf{f} \neq \mathbf{f}.\nabla$.
- If div $\mathbf{V} = 0$, then the vector **V** is called solenoidal vector.
- If the vector **V** is a velocity vector of a fluid and if div $\mathbf{V} = 0$, then the fluid is incompressible.
- $\text{div}\,\mathbf{V} = \nabla.\mathbf{V} = \Sigma\hat{i}.\frac{\partial \mathbf{V}}{\partial x}$.

22.9 CURL OF A VECTOR FIELD

(UPTU–2006, 07,08; gbtu–2012)

Let $\mathbf{V}(x, y, z)$ be a vector function of x, y, z where (x, y, z) are right handed cartesian co-ordinates in space and let

$$\mathbf{V}(x,y,z) = V_1(x,y,z)\hat{i} + V_2(x,y,z)\hat{j} + V_3(x,y,z)\hat{k}$$

be a differentiable vector function. Then the function

$$\text{curl}\,\mathbf{V} = \nabla\times\mathbf{V} = \begin{vmatrix} \hat{i} & \hat{j} & \hat{k} \\ \frac{\partial}{\partial x} & \frac{\partial}{\partial y} & \frac{\partial}{\partial z} \\ V_1 & V_2 & V_3 \end{vmatrix} = \left(\frac{\partial V_3}{\partial y} - \frac{\partial V_2}{\partial z}\right)\hat{i} + \left(\frac{\partial V_1}{\partial z} - \frac{\partial V_3}{\partial x}\right)\hat{j} + \left(\frac{\partial V_2}{\partial x} - \frac{\partial V_1}{\partial y}\right)\hat{k}$$

is called the curl of the vector function V or the curl of the vector field defined by Curl **V** is a vector quantity.

REMARKS

- If curl $\mathbf{V} = 0$, then the vector **V** is called irrotational.
- $\text{Curl}\,\mathbf{V} = \nabla\times\mathbf{V} = \Sigma\hat{i}\times\frac{\partial \mathbf{V}}{\partial x}$.
- Curl **V** is perpendicular to **V**.
- In the case of a rigid body rotation, the curl of the velocity field has the direction of the axis of rotation and its magnitude equals twice the angular speed of the rotation.

22.10 LAPLACIAN OPERATOR

If the function $f(x, y, z)$ is a twice differentiable scalar function, then we have

$$\text{grad}\,f = \frac{\partial f}{\partial x}\hat{i} + \frac{\partial f}{\partial y}\hat{j} + \frac{\partial f}{\partial z}\hat{k}$$

Since grad f is a vector function, then

$$\text{div (grad } f) = \left(\frac{\partial}{\partial x}\hat{i} + \frac{\partial}{\partial y}\hat{j} + \frac{\partial}{\partial z}\hat{k}\right).\left(\frac{\partial f}{\partial x}\hat{i} + \frac{\partial f}{\partial y}\hat{j} + \frac{\partial f}{\partial z}\hat{k}\right) = \frac{\partial^2 f}{\partial x^2} + \frac{\partial^2 f}{\partial y^2} + \frac{\partial^2 f}{\partial z^2} = \left(\frac{\partial^2}{\partial x^2} + \frac{\partial^2}{\partial y^2} + \frac{\partial^2}{\partial z^2}\right)f$$

$$\therefore \quad \text{div (grad } f) = \nabla^2 f. \qquad \ldots(1)$$

Thus, R.H.S. of (1) is the Laplacian of f. Consequently the Laplacian is defined as

$$\nabla^2 = \left(\frac{\partial^2}{\partial x^2} + \frac{\partial^2}{\partial y^2} + \frac{\partial^2}{\partial z^2}\right).$$

Hence, ∇^2 is a Laplacian operator.

REMARKS

- The equation $\nabla^2 f = 0$ is called Laplace's equation.
- If **f** is a scalar point function, then $\nabla^2\mathbf{f}$ is a scalar quantity.
- If **f** is a vector point function, then $\nabla^2\mathbf{f}$ is a vector quantity.
- If a function f satisfies the Laplace's equation then f is called harmonic function.

22.11 PHYSICAL INTERPRETATION OF DIVERGENCE AND CURL

(i) Physical interpretation of divergence. Let us consider the motion of a fluid in a region R having no sources or sinks in R. (GBTU–2010)

Let $\mathbf{V}(x, y, z)$ be the velocity of the fluid at any point. Now consider the flow through a small rectangular parallelopiped of edges δx, δy, δz parallel to the co-ordinate axes as shown in fig. 2

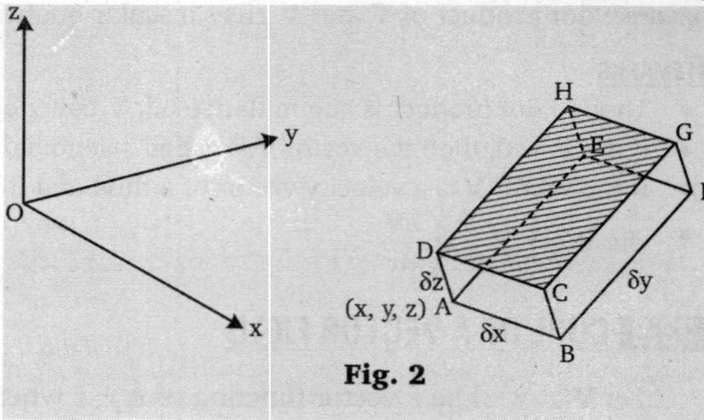

Fig. 2

The volume of this parallelopiped is $\delta x\, \delta y\, \delta z$. Let

$$\mathbf{V} = V_1\hat{i} + V_2\hat{j} + V_3\hat{k}$$

and assuming that $\mathbf{V}$ is continuous differentiable vector function of x, y, z. Let us calculate the change in the mass in the parallelopiped by considering the flux across the boundary, that is the total loss of mass leaving the parallelopiped per unit time.

Let the co-ordinates of A be (x, y, z). Now consider a flow through the face $ABCD$ whose area is $\delta x\delta z$. In this case the components V_1 and V_2 of $\mathbf{V}$ are parallel to that face so no contribution to that flow. Hence, the mass of the fluid entering through the face $ABCD$ per unit time is given approximately by $V_2(y)\delta x\delta z$ and the mass of fluid leaving the face $EFGH$ per unit time is approximately $V_2(y + \delta y)\delta x\delta z$.

The loss of mass along y-axis is given by

$$V_2(y+\delta y)\delta x\delta z - V_2(y)\delta x\delta z = [V_2(y+\delta y) - V_2(y)]\delta x\delta z$$

$$= \frac{\delta V_2}{\delta y}\delta x\delta y\delta z.$$

where $V_2(y + \delta y)$ and $V_2(y)$ are the components of $\mathbf{V}$ along y-axis at point A and E respectively.

Similarly, loss in mass per unit time in x-direction is $\frac{\partial V_1}{\partial x}\delta x\delta y\delta z$ and loss in mass per unit time in z-direction is $\frac{\partial V_3}{\partial z}\delta x\delta y\delta z$.

$\therefore$ The total loss in mass per unit time in the parallelopiped is given by

$$\left(\frac{\partial V_1}{\partial x} + \frac{\partial V_2}{\partial y} + \frac{\partial V_3}{\partial z}\right)\delta x\delta y\delta z.$$

Hence, the total loss in the mass per unit time and per unit volume is

$$\left(\frac{\partial V_1}{\partial x} + \frac{\partial V_2}{\partial y} + \frac{\partial V_3}{\partial z}\right) = \text{div } V = \nabla.V.$$

Hence, the divergence of $\mathbf{V}$ is nothing but the loss in mass of fuid per unit volume per unit time in a small parallelopiped whose edges are parallel to the co-ordinate axes.

(ii) Physical interpretation of the curl. (GBTU–2010, MTU–2011)

Circulation. The integeral $\oint_C \mathbf{V}.d\mathbf{r}$ is called circulation where C is a closed curve and $\mathbf{V}$ a velocity vector.

Let S be a circular disc of radius r and enclosed by a closed curve C (a circle). Let $\mathbf{V}(x, y, z)$ be the velocity vector and assuming that $\mathbf{V}$ is continuously differentiable in S. Then by Stoke's theorem

$$\oint_C \mathbf{V}.d\mathbf{r} = \iint_S (\text{curl } \mathbf{V}).\mathbf{n}\; dS$$

Let λ be the intermediate value between the maximum and minimum of (curl $\mathbf{V}$).$\mathbf{n}$ over S. Then by mean value theorem of integral calculus, we have

$$\oint_C \mathbf{V}.d\mathbf{r} = \lambda\iint_S dS = \lambda S$$

Recapitulations

- div $\mathbf{V} = \frac{\partial V_1}{\partial x} + \frac{\partial V_2}{\partial y} + \frac{\partial V_3}{\partial z} = \nabla.\mathbf{V}$
- curl $\mathbf{V} = \nabla \times \mathbf{V} = \begin{vmatrix} \hat{i} & \hat{j} & \hat{k} \\ \frac{\partial}{\partial x} & \frac{\partial}{\partial y} & \frac{\partial}{\partial z} \\ V_1 & V_2 & V_3 \end{vmatrix}$
- If div $\mathbf{V} = 0$, then the vector $\mathbf{V}$ is called solenoidal vector.
- If curl $\mathbf{V} = 0$, then the vector $\mathbf{V}$ is called irrotational.
- Curl $\mathbf{V}$ is perpendicular to $\mathbf{V}$.
- The equation $\nabla^2 f = 0$ is called Laplace equation.
- The divergence of $\mathbf{V}$ is the loss in mass of fluid per unit volume per unit time in a small parallelopiped whose edges are parallel to the co-ordinate axes.

or
$$\lambda = \frac{\oint_C \mathbf{V}.d\mathbf{r}}{S}$$

Now taking limit as $r \to 0$, we get
$$(\text{curl } \mathbf{V}).\mathbf{n} = \lim_{r\to 0} \frac{\oint_C \mathbf{V}.d\mathbf{r}}{S}$$

Now (curl **V**).**n** is a normal component of curl **V** at the centre of circular disc S and $\oint_C \mathbf{V}.d\mathbf{r}$ is a circulation of **V** about C. Hence the normal component of the curl is nothing but the limit of the circulation per unit area.

REMARK

- Stoke's states that $\oint_C \mathbf{A}.d\mathbf{r} = \iint_S (\text{curl } \mathbf{A}).\mathbf{n}\, dS$. This theorem will discuss later on.

SOLVED EXAMPLES

Example 1. *Prove that the followings :*
(i) *div* $\mathbf{r} = 3$ (ii) *curl* $\mathbf{r} = \mathbf{0}$
where $\mathbf{r} = x\hat{i} + y\hat{j} + z\hat{k}$.

Solution . (i) Since div $\mathbf{r} = \nabla.\mathbf{r}$

$$= \left(\frac{\partial}{\partial x}\hat{i} + \frac{\partial}{\partial y}\hat{j} + \frac{\partial}{\partial z}\hat{k}\right).(x\hat{i} + y\hat{j} + z\hat{k})$$

$$= \frac{\partial x}{\partial x} + \frac{\partial y}{\partial y} + \frac{\partial z}{\partial z} = 1 + 1 + 1 = 3.$$

(ii) Curl $\mathbf{r} = \nabla \times \mathbf{r}$

$$= \begin{vmatrix} \hat{i} & \hat{j} & \hat{k} \\ \frac{\partial}{\partial x} & \frac{\partial}{\partial y} & \frac{\partial}{\partial z} \\ x & y & z \end{vmatrix}$$

$$= \hat{i}\left(\frac{\partial z}{\partial y} - \frac{\partial y}{\partial z}\right) + \hat{j}\left(\frac{\partial x}{\partial z} - \frac{\partial z}{\partial x}\right) + \hat{k}\left(\frac{\partial y}{\partial x} - \frac{\partial x}{\partial y}\right)$$

$$= \hat{i}(0-0) + \hat{j}(0-0) + \hat{k}(0-0) = \mathbf{0}.$$

Example 2. *If* **a** *is a constant vector, find*
(i) *div* $(\mathbf{r} \times \mathbf{a})$ (ii) *curl* $(\mathbf{r} \times \mathbf{a})$
where $\mathbf{r} = x\hat{i} + y\hat{j} + z\hat{k}$.

Solution . Let $\mathbf{a} = a_1\hat{i} + a_2\hat{j} + a_3\hat{k}$ be a constant vector, then

$$\mathbf{r} \times \mathbf{a} = \begin{vmatrix} \hat{i} & \hat{j} & \hat{k} \\ x & y & z \\ a_1 & a_2 & a_3 \end{vmatrix}$$

$$= \hat{i}(ya_3 - za_2) + \hat{j}(za_1 - xa_3) + \hat{k}(xa_2 - ya_1)$$

(i) $\text{div}(\mathbf{r} \times \mathbf{a}) = \nabla.(\mathbf{r} \times \mathbf{a})$

$$= \frac{\partial}{\partial x}(ya_3 - za_2) + \frac{\partial}{\partial y}(za_1 - xa_3) + \frac{\partial}{\partial z}(xa_2 - ya_1)$$

$$= 0 + 0 + 0 = 0$$

(ii) curl $(\mathbf{r} \times a) = \nabla \times (\mathbf{r} \times \mathbf{a})$

$$= \begin{vmatrix} \hat{i} & \hat{j} & \hat{k} \\ \frac{\partial}{\partial x} & \frac{\partial}{\partial y} & \frac{\partial}{\partial z} \\ (ya_3 - za_2) & (za_1 - xa_3) & (xa_2 - ya_1) \end{vmatrix}$$

$$= \hat{i}\left[\frac{\partial}{\partial y}(xa_3 - ya_1) - \frac{\partial}{\partial z}(za_1 - xa_3)\right]$$
$$+ \hat{j}\left[\frac{\partial}{\partial z}(ya_3 - za_2) - \frac{\partial}{\partial x}(xa_2 - ya_1)\right]$$
$$+ \hat{k}\left[\frac{\partial}{\partial x}(za_1 - xa_3) - \frac{\partial}{\partial y}(ya_3 - za_2)\right]$$

$$= \hat{i}[-a_1 - a_1] + \hat{j}[-a_2 - a_2] + \hat{k}[-a_3 - a_3]$$

$$= -2(a_1\hat{i} + a_2\hat{j} + a_3\hat{k}) = -2\mathbf{a}$$

Similarly, we can show that

$$\frac{1}{2}\text{curl } (\mathbf{a} \times \mathbf{r}) = \mathbf{a}.$$

Example 3. *If* **V** *is differentiable vector function and f is a scalar point function, then show that*
(i) *div* $(f\mathbf{V}) = f$ *div* $\mathbf{V} + \mathbf{V}.(\text{grad} f)$ (UPTU–2006)
(ii) *curl* $(f\mathbf{V}) = (\nabla f) \times \mathbf{V} + f(\nabla \times \mathbf{V})$

Solution . (i) Since we know that

$$\text{div } (f\mathbf{V}) = \nabla.(f\mathbf{V})$$

$$= \left(\hat{i}\frac{\partial}{\partial x} + \hat{j}\frac{\partial}{\partial y} + \hat{k}\frac{\partial}{\partial z}\right).(f\mathbf{V})$$

$$= \hat{i}\cdot\frac{\partial}{\partial x}(f\mathbf{V}) + \hat{j}\cdot\frac{\partial}{\partial y}(f\mathbf{V}) + \hat{k}\cdot\frac{\partial}{\partial z}(f\mathbf{V})$$

$$= \hat{i}\cdot\left(\frac{\partial f}{\partial x}\mathbf{V} + f\frac{\partial \mathbf{V}}{\partial x}\right) + \hat{j}\cdot\left(\frac{\partial f}{\partial y}\mathbf{V} + f\frac{\partial \mathbf{V}}{\partial y}\right) + \hat{k}\cdot\left(\frac{\partial f}{\partial z}\mathbf{V} + f\frac{\partial \mathbf{V}}{\partial z}\right)$$

$$= \left[\hat{i}\cdot\left(\frac{\partial f}{\partial x}\mathbf{V}\right) + \hat{j}\cdot\left(\frac{\partial f}{\partial y}\mathbf{V}\right) + \hat{k}\cdot\left(\frac{\partial f}{\partial z}\mathbf{V}\right)\right] + \left[\hat{i}\cdot\left(f\frac{\partial \mathbf{V}}{\partial x}\right) + \hat{j}\cdot\left(f\frac{\partial \mathbf{V}}{\partial y}\right) + \hat{k}\cdot\left(\frac{\partial \mathbf{V}}{\partial z}f\right)\right]$$

$$= \left[\left(\frac{\partial f}{\partial x}\hat{i}\right)\cdot\mathbf{V} + \left(\frac{\partial f}{\partial y}\hat{j}\right)\cdot\mathbf{V} + \left(\frac{\partial f}{\partial z}\hat{k}\right)\cdot\mathbf{V}\right] + \left[f\left(\hat{i}\cdot\frac{\partial \mathbf{V}}{\partial x}\right) + f\left(\hat{j}\cdot\frac{\partial \mathbf{V}}{\partial y}\right) + f\left(\hat{k}\cdot\frac{\partial \mathbf{V}}{\partial z}\right)\right]$$

$$= \left[\left(\frac{\partial f}{\partial x}\cdot\hat{i}+\frac{\partial f}{\partial z}\cdot\hat{k}+\frac{\partial f}{\partial y}\cdot\hat{j}\right)\mathbf{V}\right]$$
$$+\left[f\left(\hat{i}\cdot\frac{\partial \mathbf{V}}{\partial x}+\hat{j}\cdot\frac{\partial \mathbf{V}}{\partial y}+\hat{k}\cdot\frac{\partial \mathbf{V}}{\partial z}\right)\right]$$
$$= (\nabla f).\mathbf{V} + f(\nabla.\mathbf{V})$$
$$\therefore \text{div}(f\mathbf{V}) = f(\text{div}\,\mathbf{V}) + \mathbf{V}\cdot(\text{grad}\,f)$$

(ii) Curl $(f\mathbf{V}) = \nabla \times (f\mathbf{V})$

$$= \left(\hat{i}\frac{\partial}{\partial x}+\hat{j}\frac{\partial}{\partial y}+\hat{k}\frac{\partial}{\partial z}\right)\times(f\mathbf{V})$$
$$= \left(\hat{i}\frac{\partial}{\partial x}\right)\times(f\mathbf{V})+\left(\hat{j}\frac{\partial}{\partial y}\right)\times(f\mathbf{V})+\left(\hat{k}\frac{\partial}{\partial z}\right)\times(f\mathbf{V})$$
$$= \hat{i}\times\frac{\partial}{\partial x}\times(f\mathbf{V})+\hat{j}\times\frac{\partial}{\partial y}(f\mathbf{V})+\hat{k}\times\frac{\partial}{\partial z}(f\mathbf{V})$$
$$= \hat{i}\times\left(\frac{\partial f}{\partial x}\mathbf{V}+f\frac{\partial \mathbf{V}}{\partial x}\right)+\hat{j}\times\left(\frac{\partial f}{\partial y}\mathbf{V}+f\frac{\partial \mathbf{V}}{\partial y}\right)+\hat{k}\times\left(\frac{\partial f}{\partial z}\mathbf{V}+f\frac{\partial \mathbf{V}}{\partial z}\right)$$
$$= \left[\left(\frac{\partial f}{\partial x}\hat{i}\right)\times\mathbf{V}+\left(\frac{\partial f}{\partial y}\hat{j}\right)\times\mathbf{V}+\left(\frac{\partial f}{\partial z}\hat{k}\right)\times\mathbf{V}\right]$$
$$+\left[f\left(\hat{i}\times\frac{\partial \mathbf{V}}{\partial x}\right)+f\left(\hat{j}\times\frac{\partial \mathbf{V}}{\partial y}\right)+f\left(\hat{k}\times\frac{\partial \mathbf{V}}{\partial z}\right)\right]$$
$$= \left(\frac{\partial f}{\partial x}\hat{i}+\frac{\partial f}{\partial y}\hat{j}+\frac{\partial f}{\partial z}\hat{k}\right)\times\mathbf{V}+f\left(\hat{i}\times\frac{\partial \mathbf{V}}{\partial x}+\hat{j}\times\frac{\partial \mathbf{V}}{\partial y}+\hat{k}\times\frac{\partial \mathbf{V}}{\partial z}\right)$$
$$= (\nabla f)\times\mathbf{V} + f(\nabla\times\mathbf{V}).$$
$$\therefore \text{curl}(f\mathbf{V}) = (\nabla f)\times\mathbf{V} + f(\nabla\times\mathbf{V}).$$

Example 4. *Find the divergence of the following vector funtions :*

(i) $\mathbf{f} = x^2\hat{i}+y^2\hat{j}-z\hat{k}$ (ii) $\mathbf{f} = xyz(\hat{i}+\hat{j}+\hat{k})$

(iii) $\mathbf{f} = yz^2\hat{i}-zx^2\hat{k}$

Solution . (i) We have

$$\text{div}\,\mathbf{f} = \nabla.\mathbf{f} = \left(\hat{i}\frac{\partial}{\partial x}+\hat{j}\frac{\partial}{\partial y}+\hat{k}\frac{\partial}{\partial z}\right).(x^2\hat{i}+y^2\hat{j}-z\hat{k})$$
$$= \frac{\partial}{\partial x}(x^2)+\frac{\partial}{\partial y}(y^2)+\frac{\partial}{\partial z}(-z)$$
$$= 2x + 2y - 1$$

(ii) $\mathbf{f} = xyz(\hat{i}+\hat{j}+\hat{k})$

$$\therefore \text{div}\,\mathbf{f} = \nabla.\mathbf{f} = \left(\hat{i}\frac{\partial}{\partial x}+\hat{j}\frac{\partial}{\partial y}+\hat{k}\frac{\partial}{\partial z}\right)\cdot xyz.(\hat{i}+\hat{j}+\hat{k})$$
$$= \frac{\partial}{\partial x}(xyz)+\frac{\partial}{\partial y}(xyz)+\frac{\partial}{\partial z}(xyz)$$
$$= yz + zx + xy$$

(iii) $\mathbf{f} = yz^2\hat{i}-zx^2\hat{k}$

$$\therefore \text{div}\,\mathbf{f} = \nabla.\mathbf{f} = \left(\hat{i}\frac{\partial}{\partial x}+\hat{j}\frac{\partial}{\partial y}+\hat{k}\frac{\partial}{\partial z}\right).(yz^2\hat{i}-zx^2\hat{k})$$
$$= \frac{\partial}{\partial x}(yz^2)-\frac{\partial}{\partial z}(zx^2)$$
$$= 0 - x^2 = -x^2.$$

Example 5. *Find curl* $\mathbf{f}$ *if*

(i) $\mathbf{f} = z^2\hat{i}+x^2\hat{j}+y^2\hat{k}$ (ii) $\mathbf{f} = xz\hat{i}-yz\hat{j}$

(iii) $\mathbf{f} = e^{xyz}(\hat{i}+\hat{j}+\hat{k})$

(iv) $\mathbf{f} = e^x\sin y\hat{i}+e^x\cos y\hat{j}$

Solution . (i) $\mathbf{f} = z^2\hat{i}+x^2\hat{j}+y^2\hat{k}$

We have curl $\mathbf{f} = \nabla\times\mathbf{f}$

$$= \begin{vmatrix}\hat{i} & \hat{j} & \hat{k}\\ \frac{\partial}{\partial x} & \frac{\partial}{\partial y} & \frac{\partial}{\partial z}\\ z^2 & x^2 & y^2\end{vmatrix}$$
$$= \hat{i}\left(\frac{\partial}{\partial y}(y)^2-\frac{\partial}{\partial z}(x)^2\right)+\hat{j}\left(\frac{\partial}{\partial z}(z)^2-\frac{\partial}{\partial x}(y)^2\right)+\hat{k}\left(\frac{\partial}{\partial x}(x)^2-\frac{\partial}{\partial y}(z)^2\right)$$
$$= 2y\hat{i}+2z\hat{j}+2x\hat{k}$$
$$= 2(y\hat{i}+z\hat{j}+x\hat{k}).$$

(ii) $\mathbf{f} = xz\hat{i}-yz\hat{j}$

$$\therefore \text{curl}\,\mathbf{f} = \nabla\times\mathbf{f} = \begin{vmatrix}\hat{i} & \hat{j} & \hat{k}\\ \frac{\partial}{\partial x} & \frac{\partial}{\partial y} & \frac{\partial}{\partial z}\\ xz & -yz & 0\end{vmatrix}$$
$$= \hat{i}\left(\frac{\partial}{\partial z}(yz)\right)+\hat{j}\left(\frac{\partial}{\partial z}(xz)\right)+\hat{k}\left(-\frac{\partial}{\partial x}(yz)-\frac{\partial}{\partial y}(xz)\right)$$
$$= y\hat{i}+x\hat{j}+\hat{k}(0-0)$$
$$= y\hat{i}+x\hat{j}.$$

(iii) $\mathbf{f} = e^{xyz}(\hat{i}+\hat{j}+\hat{k})$

$\therefore$ curl $\mathbf{f} = \nabla \times \mathbf{f}$

$$= \begin{vmatrix} \hat{i} & \hat{j} & \hat{k} \\ \frac{\partial}{\partial x} & \frac{\partial}{\partial y} & \frac{\partial}{\partial z} \\ e^{xyz} & e^{xyz} & e^{xyz} \end{vmatrix}$$

$$= \hat{i}\left(\frac{\partial}{\partial y}(e^{xyz}) - \frac{\partial}{\partial z}(e^{xyz})\right) + \hat{j}\left(\frac{\partial}{\partial z}(e^{xyz}) - \frac{\partial}{\partial x}(e^{xyz})\right) + \hat{k}\left(\frac{\partial}{\partial x}(e^{xyz}) - \frac{\partial}{\partial y}(e^{xyz})\right)$$

$$= \hat{i}[e^{xyz}(zx - xy)] + \hat{j}[e^{xyz}(xy - yz)] + \hat{k}[e^{xyz}(yz - zx)]$$

$$= e^{xyz}[(zx - xy)\hat{i} + (xy - yz)\hat{j} + (yz - zx)\hat{k}]$$

(iv) $\mathbf{f} = e^x \sin y\hat{i} + e^x \cos y\hat{j}$

$\therefore$ curl $\mathbf{f} = \nabla \times \mathbf{f}$

$$= \begin{vmatrix} \hat{i} & \hat{j} & \hat{k} \\ \frac{\partial}{\partial x} & \frac{\partial}{\partial y} & \frac{\partial}{\partial z} \\ e^x \sin y & e^x \cos y & 0 \end{vmatrix}$$

$$= \hat{i}\left(-\frac{\partial}{\partial z}(e^x \cos y)\right) + \hat{j}\left(\frac{\partial}{\partial z}(e^x \sin y)\right) + \hat{k}\left(\frac{\partial}{\partial x}(e^x \cos y) - \frac{\partial}{\partial y}(e^x \sin y)\right)$$

$$= 0\hat{i} + 0\hat{j} + (e^x \cos y - e^x \cos y)\hat{k}$$

$$= 0\hat{i} + 0\hat{j} + 0\hat{k} = \mathbf{0}$$

Example 6. *Prove that*
div $(\mathbf{a} \times \mathbf{b}) = \mathbf{b} \cdot$ *curl* $\mathbf{a} - \mathbf{a}.$ *curl* $\mathbf{b}$ (GBTU–2012)
or $\nabla\cdot(\mathbf{a} \times \mathbf{b}) = \mathbf{b}\cdot(\nabla \times \mathbf{a}) - \mathbf{a}\cdot(\nabla \times \mathbf{b})$

Solution . We have

div $(\mathbf{a} \times \mathbf{b}) = \nabla \cdot (\mathbf{a} \times \mathbf{b})$

$$= \left(\hat{i}\frac{\partial}{\partial x} + \hat{j}\frac{\partial}{\partial y} + \hat{k}\frac{\partial}{\partial z}\right).(\mathbf{a} \times \mathbf{b})$$

$$= \left(\hat{i}\frac{\partial}{\partial x}\right).(\mathbf{a} \times \mathbf{b}) + \left(\hat{j}\frac{\partial}{\partial y}\right).(\mathbf{a} \times \mathbf{b}) + \left(\hat{k}\frac{\partial}{\partial z}\right).(\mathbf{a} \times \mathbf{b})$$

$$= \hat{i}.\frac{\partial}{\partial x}(\mathbf{a} \times \mathbf{b}) + \hat{j}.\frac{\partial}{\partial y}(\mathbf{a} \times \mathbf{b}) + \hat{k}.\frac{\partial}{\partial z}(\mathbf{a} \times \mathbf{b})$$

$$= \hat{i}.\left(\frac{\partial \mathbf{a}}{\partial x} \times \mathbf{b} + \mathbf{a} \times \frac{\partial \mathbf{b}}{\partial x}\right) + \hat{j}.\left(\frac{\partial \mathbf{a}}{\partial y} \times \mathbf{b} + \mathbf{a} \times \frac{\partial \mathbf{b}}{\partial y}\right) + \hat{k}.\left(\frac{\partial \mathbf{a}}{\partial z} \times \mathbf{b} + \mathbf{a} \times \frac{\partial \mathbf{b}}{\partial z}\right)$$

$$= \hat{i}.\left(\frac{\partial \mathbf{a}}{\partial x} \times \mathbf{b}\right) + \hat{i}.\left(\mathbf{a} \times \frac{\partial \mathbf{b}}{\partial x}\right) + \hat{j}.\left(\frac{\partial \mathbf{a}}{\partial y} \times \mathbf{b}\right) + \hat{j}.\left(\mathbf{a} \times \frac{\partial \mathbf{b}}{\partial y}\right) + \hat{k}.\left(\frac{\partial \mathbf{a}}{\partial z} \times \mathbf{b}\right) + \hat{k}.\left(\mathbf{a} \times \frac{\partial \mathbf{b}}{\partial z}\right)$$

$$= \left[\hat{i}.\left(\frac{\partial \mathbf{a}}{\partial x} \times \mathbf{b}\right) + \hat{j}.\left(\frac{\partial \mathbf{a}}{\partial y} \times \mathbf{b}\right) + \hat{k}.\left(\frac{\partial \mathbf{a}}{\partial z} \times \mathbf{b}\right)\right] + \left[\hat{i}.\left(\mathbf{a} \times \frac{\partial \mathbf{b}}{\partial x}\right) + \hat{j}.\left(\mathbf{a} \times \frac{\partial \mathbf{b}}{\partial y}\right) + \hat{k}.\left(\mathbf{a} \times \frac{\partial \mathbf{b}}{\partial z}\right)\right]$$

Using $\mathbf{a}.(\mathbf{b} \times \mathbf{c}) = \mathbf{a} \times (\mathbf{b}.\mathbf{c})$ and $\mathbf{a}.(\mathbf{b} \times \mathbf{c}) = -\mathbf{a}.(\mathbf{c} \times \mathbf{b})$

$$= \left[\left(\hat{i} \times \frac{\partial \mathbf{a}}{\partial x}\right).\mathbf{b} + \left(\hat{j} \times \frac{\partial \mathbf{a}}{\partial y}\right).\mathbf{b} + \left(\hat{k} \times \frac{\partial \mathbf{a}}{\partial z}\right).\mathbf{b}\right] - \left[\hat{i}.\left(\frac{\partial \mathbf{b}}{\partial x} \times \mathbf{a}\right) + \hat{j}.\left(\frac{\partial \mathbf{b}}{\partial y} \times \mathbf{a}\right) + \hat{k}.\left(\frac{\partial \mathbf{b}}{\partial z} \times \mathbf{a}\right)\right]$$

$$= \left(\hat{i} \times \frac{\partial \mathbf{a}}{\partial x} + \hat{j} \times \frac{\partial \mathbf{a}}{\partial y} + \hat{k} \times \frac{\partial \mathbf{a}}{\partial z}\right).\mathbf{b} - \left(\hat{i} \times \frac{\partial \mathbf{b}}{\partial x} + \hat{j} \times \frac{\partial \mathbf{b}}{\partial y} + \hat{k} \times \frac{\partial \mathbf{b}}{\partial z}\right).\mathbf{a}$$

$$= (\nabla \times \mathbf{a}).\mathbf{b} - \nabla \times \mathbf{b}).\mathbf{a}$$

$$\therefore \quad \nabla\cdot(\mathbf{a} \times \mathbf{b}) = \mathbf{b}\cdot(\nabla \times \mathbf{a}) - \mathbf{a}\cdot(\nabla \times \mathbf{b}) \qquad (\because \mathbf{a}.\mathbf{b} = \mathbf{b}.\mathbf{a})$$

Example 7. *Prove that*
curl $(\mathbf{a} \times \mathbf{b}) = (\mathbf{b}\cdot\nabla)\mathbf{a} - \mathbf{b}$ *div* $\mathbf{a} - (\mathbf{a}\cdot\nabla)\mathbf{b} + \mathbf{a}$ *div* $\mathbf{b}$

Solution . We have curl $(\mathbf{a} \times \mathbf{b}) = \nabla \times (\mathbf{a} \times \mathbf{b})$

$$= \left(\hat{i}\frac{\partial}{\partial x} + \hat{j}\frac{\partial}{\partial y} + \hat{k}\frac{\partial}{\partial z}\right) \times (\mathbf{a} \times \mathbf{b})$$

$$= \hat{i} \times \frac{\partial}{\partial x}(\mathbf{a} \times \mathbf{b}) + \hat{j} \times \frac{\partial}{\partial y}(\mathbf{a} \times \mathbf{b}) + \hat{k} \times \frac{\partial}{\partial z}(\mathbf{a} \times \mathbf{b})$$

$$= \hat{i} \times \left(\frac{\partial \mathbf{a}}{\partial x} \times \mathbf{b} + \mathbf{a} \times \frac{\partial \mathbf{b}}{\partial x}\right) + \hat{j} \times \left(\frac{\partial \mathbf{a}}{\partial y} \times \mathbf{b} + \mathbf{a} \times \frac{\partial \mathbf{b}}{\partial y}\right) + \hat{k} \times \left(\frac{\partial \mathbf{a}}{\partial z} \times \mathbf{b} + \mathbf{a} \times \frac{\partial \mathbf{b}}{\partial z}\right)$$

$$= \hat{i} \times \left(\frac{\partial \mathbf{a}}{\partial x} \times \mathbf{b}\right) + \hat{i} \times \left(\mathbf{a} \times \frac{\partial \mathbf{b}}{\partial x}\right)$$
$$+ \hat{j} \times \left(\frac{\partial \mathbf{a}}{\partial y} \times \mathbf{b}\right) + \hat{j} \times \left(\mathbf{a} \times \frac{\partial \mathbf{b}}{\partial y}\right)$$
$$+ \hat{k} \times \left(\frac{\partial \mathbf{a}}{\partial z} \times \mathbf{b}\right) + \hat{k} \times \left(\mathbf{a} \times \frac{\partial \mathbf{b}}{\partial z}\right)$$
$$= \left[\hat{i} \times \left(\frac{\partial \mathbf{a}}{\partial x} \times \mathbf{b}\right) + \hat{j} \times \left(\frac{\partial \mathbf{a}}{\partial y} \times \mathbf{b}\right) + \right.$$
$$\left. + \hat{k} \times \left(\frac{\partial \mathbf{a}}{\partial z} \times \mathbf{b}\right)\right] + \left[\hat{i} \times \left(\mathbf{a} \times \frac{\partial \mathbf{b}}{\partial x}\right)\right.$$
$$\left. + \hat{j} \times \left(\mathbf{a} \times \frac{\partial \mathbf{b}}{\partial y}\right) + \hat{k} \times \left(\mathbf{a} \times \frac{\partial \mathbf{b}}{\partial z}\right)\right]$$
$$= \left[(\hat{i} \cdot \mathbf{b})\frac{\partial \mathbf{a}}{\partial x} - \left(\hat{i} \cdot \frac{\partial \mathbf{a}}{\partial x}\right)\mathbf{b} + (\hat{j} \cdot \mathbf{b})\frac{\partial \mathbf{a}}{\partial y}\right.$$
$$\left. - \left(\hat{j} \cdot \frac{\partial \mathbf{a}}{\partial y}\right)\mathbf{b} + (\hat{k} \cdot \mathbf{b})\frac{\partial \mathbf{a}}{\partial z} - \left(\hat{k} \cdot \frac{\partial \mathbf{a}}{\partial z}\right)\mathbf{b}\right]$$
$$- \left[(\hat{i} \cdot \mathbf{a})\frac{\partial \mathbf{b}}{\partial x} - \left(\hat{i} \cdot \frac{\partial \mathbf{b}}{\partial x}\right)\mathbf{a} + (\hat{j} \cdot \mathbf{a})\frac{\partial \mathbf{b}}{\partial y}\right.$$
$$\left. - \left(\hat{j} \cdot \frac{\partial \mathbf{b}}{\partial y}\right)\mathbf{a} + (\hat{k} \cdot \mathbf{a})\frac{\partial \mathbf{b}}{\partial z} - \left(\hat{k} \cdot \frac{\partial \mathbf{b}}{\partial z}\right)\mathbf{a}\right]$$

Using $\mathbf{a}\cdot\mathbf{b} = \mathbf{b}\cdot\mathbf{a}$, we get

$$= \left[\left(\mathbf{b} \cdot \hat{i}\frac{\partial \mathbf{a}}{\partial x} + \mathbf{b} \cdot \hat{j}\frac{\partial \mathbf{a}}{\partial y} + \mathbf{b} \cdot \hat{k}\frac{\partial \mathbf{a}}{\partial z}\right)\right.$$
$$\left. - \left(\hat{i} \cdot \frac{\partial \mathbf{a}}{\partial x} + \hat{j} \cdot \frac{\partial \mathbf{a}}{\partial y} + \hat{k} \cdot \frac{\partial \mathbf{a}}{\partial z}\right)\mathbf{b}\right]$$
$$- \left[\left(\mathbf{a} \cdot \hat{i}\frac{\partial \mathbf{b}}{\partial x} + \mathbf{a} \cdot \hat{j}\frac{\partial \mathbf{b}}{\partial y} + \mathbf{a} \cdot \hat{k}\frac{\partial \mathbf{b}}{\partial z}\right)\right.$$
$$\left. - \left(\hat{i} \cdot \frac{\partial \mathbf{b}}{\partial x} + \hat{j} \cdot \frac{\partial \mathbf{b}}{\partial y} + \hat{k} \cdot \frac{\partial \mathbf{b}}{\partial z}\right)\mathbf{a}\right]$$
$$= \left[\left(\mathbf{b} \cdot \hat{i}\frac{\partial}{\partial x} + \mathbf{b} \cdot \hat{j}\frac{\partial}{\partial y} + \mathbf{b} \cdot \hat{k}\frac{\partial}{\partial z}\right)\mathbf{a} - (\nabla \cdot a)\mathbf{b}\right]$$
$$- \left[\left(\mathbf{a} \cdot \hat{i}\frac{\partial}{\partial x} + \mathbf{a} \cdot \hat{j}\frac{\partial}{\partial y} + \mathbf{a} \cdot \hat{k}\frac{\partial}{\partial z}\right)\mathbf{b} - (\nabla \cdot b)\mathbf{a}\right]$$
$$= (\mathbf{b} \cdot \nabla)\mathbf{a} - (\nabla \cdot \mathbf{a})\mathbf{b} - (\mathbf{a} \cdot \nabla)\mathbf{b} + (\nabla \cdot \mathbf{b})\mathbf{a}$$

Hence, curl $(\mathbf{a} \times \mathbf{b}) = (\mathbf{b} \cdot \nabla)\mathbf{a} - \mathbf{b}$ div $\mathbf{a} - (\mathbf{a} \cdot \nabla)\mathbf{b} + \mathbf{a}$ div $\mathbf{b}$.

Example 8. *Prove that grad* $(\mathbf{a} \cdot \mathbf{b}) = (\mathbf{b} \cdot \nabla)\mathbf{a} + (\mathbf{a} \cdot \nabla)\mathbf{b} + \mathbf{b} \times$ *curl* $\mathbf{a}$ + $\mathbf{a}$ *curl* $\mathbf{b}$.

Solution. Since we have

$$\text{grad}\,(\mathbf{a} \cdot \mathbf{b}) = \left(\hat{i}\frac{\partial}{\partial x} + \hat{j}\frac{\partial}{\partial y} + \hat{k}\frac{\partial}{\partial z}\right)(\mathbf{a} \cdot \mathbf{b})$$
$$= \hat{i}\frac{\partial}{\partial x}(\mathbf{a} \cdot \mathbf{b}) + \hat{j}\frac{\partial}{\partial y}(\mathbf{a} \cdot \mathbf{b}) + \hat{k}\frac{\partial}{\partial z}(\mathbf{a} \cdot \mathbf{b})$$
$$= \hat{i}\left(\mathbf{a} \cdot \frac{\partial \mathbf{b}}{\partial x} + \frac{\partial \mathbf{a}}{\partial x} \cdot \mathbf{b}\right) + \hat{j}\left(\mathbf{a} \cdot \frac{\partial \mathbf{b}}{\partial y} + \frac{\partial \mathbf{a}}{\partial y} \cdot \mathbf{b}\right)$$
$$+ \hat{k}\left(\mathbf{a} \cdot \frac{\partial \mathbf{b}}{\partial z} + \frac{\partial \mathbf{a}}{\partial z} \cdot \mathbf{b}\right)$$
$$\text{grad}\,(\mathbf{a} \cdot \mathbf{b}) = \left[\hat{i}\left(\mathbf{a} \cdot \frac{\partial \mathbf{b}}{\partial x}\right) + \hat{j}\left(\mathbf{a} \cdot \frac{\partial \mathbf{b}}{\partial y}\right) + \hat{k}\left(\mathbf{a} \cdot \frac{\partial \mathbf{b}}{\partial z}\right)\right]$$
$$+ \left[\hat{i}\left(\frac{\partial \mathbf{a}}{\partial x} \cdot \mathbf{b}\right) + \hat{j}\left(\frac{\partial \mathbf{a}}{\partial y} \cdot \mathbf{b}\right) + \hat{k}\left(\frac{\partial \mathbf{a}}{\partial z} \cdot \mathbf{b}\right)\right] \quad \ldots(1)$$

Further, since we know that

$$\mathbf{a} \times (\mathbf{b} \times \mathbf{c}) = (\mathbf{a} \cdot \mathbf{c})\mathbf{b} - (\mathbf{a} \cdot \mathbf{b})\mathbf{c}$$
$$\therefore \quad (\mathbf{a} \cdot \mathbf{b})\mathbf{c} \equiv (\mathbf{a} \cdot \mathbf{c})\mathbf{b} - \mathbf{a} \times (\mathbf{b} \times \mathbf{c})$$
$$\therefore \left(\mathbf{a} \cdot \frac{\partial \mathbf{b}}{\partial x}\right)\hat{i} = (\mathbf{a} \cdot \hat{i})\frac{\partial \mathbf{b}}{\partial x} - \mathbf{a} \times \left(\frac{\partial \mathbf{b}}{\partial x} \times \hat{i}\right)$$
$$= (\mathbf{a} \cdot \hat{i})\frac{\partial \mathbf{b}}{\partial x} + \mathbf{a} \times \left(\hat{i} \times \frac{\partial \mathbf{b}}{\partial x}\right)$$

Similarly, $\left(\mathbf{a} \cdot \frac{\partial \mathbf{b}}{\partial y}\right)\hat{j} = (\mathbf{a} \cdot \hat{j})\frac{\partial \mathbf{b}}{\partial y} + \mathbf{a} \times \left(\hat{j} \times \frac{\partial \mathbf{b}}{\partial y}\right)$

and $\left(\mathbf{a} \cdot \frac{\partial \mathbf{b}}{\partial z}\right)\hat{k} = (\mathbf{a} \cdot \hat{k})\frac{\partial \mathbf{b}}{\partial z} + \mathbf{a} \times \left(\hat{k} \times \frac{\partial \mathbf{b}}{\partial z}\right)$

$$\therefore \hat{i}\left(\mathbf{a} \cdot \frac{\partial \mathbf{b}}{\partial x}\right) + \hat{j}\left(\mathbf{a} \cdot \frac{\partial \mathbf{b}}{\partial y}\right) + \hat{k}\left(\mathbf{a} \cdot \frac{\partial \mathbf{b}}{\partial z}\right)$$
$$= \left(\mathbf{a} \cdot \hat{i}\frac{\partial}{\partial x} + \mathbf{a} \cdot \hat{j}\frac{\partial}{\partial y} + \mathbf{a} \cdot \hat{k}\frac{\partial}{\partial z}\right)\mathbf{b} + \mathbf{a} \times (\nabla \times \mathbf{b})$$
$$= (\mathbf{a} \cdot \nabla)\mathbf{b} + \mathbf{a} \times (\nabla \times \mathbf{b}) \quad \ldots(2)$$

Similarly, $\hat{i}\left(\frac{\partial \mathbf{a}}{\partial x} \cdot \mathbf{b}\right) + \hat{j}\left(\frac{\partial \mathbf{a}}{\partial y} \cdot \mathbf{b}\right) + \hat{k}\left(\frac{\partial \mathbf{a}}{\partial z} \cdot \mathbf{b}\right)$

$$= (\mathbf{b} \cdot \nabla)\mathbf{a} + \mathbf{b} \times (\nabla \times \mathbf{a}) \quad \ldots(3)$$

From (1), (2) and (3), we get

$$\text{grad}\,(\mathbf{a}\cdot\mathbf{b}) = (\mathbf{a} \cdot \nabla)\mathbf{b} + \mathbf{a} \times (\nabla \times \mathbf{b}) \times (\mathbf{b} \cdot \nabla)\mathbf{a} + \mathbf{b} \times (\nabla \times \mathbf{a})$$

$(\mathbf{a}\cdot\nabla)\mathbf{b} + (\mathbf{b}\cdot\nabla)\mathbf{a} + \mathbf{b}\times(\nabla\times\mathbf{a}) + \mathbf{a}\times(\nabla\times\mathbf{b})$

Hence, grad $(\mathbf{a} \cdot \mathbf{b}) = (\mathbf{b} \cdot \nabla)\mathbf{a} + (\mathbf{a} \cdot \nabla)\mathbf{b} + \mathbf{b} \times$ curl $\mathbf{a}$ + $\mathbf{a}$ curl $\mathbf{b}$.

Example 9. *Prove that the curl of the gradient of f (scalar function) is zero, i.e.* $\nabla \times (\nabla f) = \mathbf{0}$ (GBTU–2011)

or

If $f = r^n$, *then* $\nabla \times (\nabla r^n) = 0$

Solution. Since we have

$$\nabla f = \frac{\partial f}{\partial x}\hat{i} + \frac{\partial f}{\partial y}\hat{j} + \frac{\partial f}{\partial z}\hat{k}$$
$$\therefore \nabla \times (\nabla f) = \begin{vmatrix} \hat{i} & \hat{j} & \hat{k} \\ \frac{\partial}{\partial x} & \frac{\partial}{\partial y} & \frac{\partial}{\partial z} \\ \frac{\partial f}{\partial x} & \frac{\partial f}{\partial y} & \frac{\partial f}{\partial z} \end{vmatrix}$$

$$= \hat{i}\left[\frac{\partial^2 f}{\partial y \partial z} - \frac{\partial^2 f}{\partial z \partial y}\right] + \hat{j}\left[\frac{\partial^2 f}{\partial z \partial x} - \frac{\partial^2 f}{\partial x \partial z}\right] + \hat{k}\left[\frac{\partial^2 f}{\partial x \partial y} - \frac{\partial^2 f}{\partial y \partial x}\right]$$

Since $\frac{\partial^2 f}{\partial y \partial z} = \frac{\partial^2 f}{\partial z \partial y}$ etc.

$\therefore \nabla \times (\nabla f) = 0\hat{i} + 0\hat{j} + 0\hat{k} = \mathbf{0}$

Example 10: *Prove that the div(curl* $\mathbf{V}$*) = 0.* (GBTU–2010, 11)

i.e. $\nabla \cdot (\nabla \times \mathbf{V}) = 0$.

Solution . Since we have

$$\nabla \times \mathrm{V} = \begin{vmatrix} \hat{i} & \hat{j} & \hat{k} \\ \frac{\partial}{\partial x} & \frac{\partial}{\partial y} & \frac{\partial}{\partial z} \\ V_1 & V_2 & V_3 \end{vmatrix}$$

where $\mathbf{V} = V_1\hat{i} + V_2\hat{j} + V_3\hat{k}$ (say).

$$\therefore \nabla \times \mathbf{V} = \hat{i}\left(\frac{\partial V_3}{\partial y} - \frac{\partial V_2}{\partial z}\right) + \hat{j}\left(\frac{\partial V_1}{\partial z} - \frac{\partial V_3}{\partial x}\right) + \hat{k}\left(\frac{\partial V_2}{\partial x} - \frac{\partial V_1}{\partial y}\right)$$

Now $\nabla \cdot (\nabla \times \mathbf{V})$

$$= \frac{\partial}{\partial x}\left(\frac{\partial V_3}{\partial y} - \frac{\partial V_2}{\partial z}\right) + \frac{\partial}{\partial y}\left(\frac{\partial V_1}{\partial z} - \frac{\partial V_3}{\partial x}\right) + \frac{\partial}{\partial z}\left(\frac{\partial V_2}{\partial x} - \frac{\partial V_1}{\partial y}\right)$$

$$= \frac{\partial^2 V_3}{\partial x \partial y} - \frac{\partial^2 V_2}{\partial x \partial z} + \frac{\partial^2 V_1}{\partial y \partial z} - \frac{\partial^2 V_3}{\partial y \partial x} + \frac{\partial^2 V_2}{\partial z \partial x} - \frac{\partial^2 V_1}{\partial z \partial y}$$

Since $\frac{\partial^2 V_1}{\partial y \partial z} = \frac{\partial^2 V_1}{\partial z \partial y}$ etc.

$\therefore \nabla \cdot (\nabla \times \mathbf{V}) = 0.$

Example 11. $\nabla^2 f(r) = f''(r) + \frac{2}{r} f'(r)$

(UKTU–2010, GBTU–2012, Bhopal–2008, SVTU–2008, VTU–2006)

Solution . Since we have

$\nabla^2 f(r) = \nabla \cdot (\nabla f(r)) = \nabla \cdot (f'(r) \nabla r)$ $(\because \nabla f(r) = f'(r) \nabla r)$

$$= \nabla \cdot \left(f'(r) \frac{\mathbf{r}}{r}\right) \qquad \left(\because \nabla r = \frac{\mathbf{r}}{r}\right)$$

$$= \nabla \cdot \left(\frac{f'(r)}{r} \mathbf{r}\right) = \frac{f'(r)}{r} \nabla . \mathbf{r} + \mathbf{r} . \nabla \left\{\frac{f'(r)}{r}\right\}$$

$[\because \nabla f(\mathbf{V}) = f \nabla \cdot \mathbf{V} + \mathbf{V} \cdot (\nabla f)]$

$$= \frac{3}{r} f'(r) + \mathbf{r} . \left\{\frac{f'(r)}{r}\right\}' \nabla r$$

$$= \frac{3}{r} f'(r) + \mathbf{r} . \left\{\frac{r f''(r) - f'(r)}{r^2}\right\} \frac{\mathbf{r}}{r}$$

$$= \frac{3}{r} f'(r) + \frac{r f''(r) - f'(r)}{r^2} \frac{\mathbf{r} \cdot \mathbf{r}}{r}$$

$$= \frac{3}{r} f'(r) + \frac{r f''(r) - f'(r)}{r^2} \frac{r^2}{r}$$

$$= \frac{3}{r} f'(r) + \frac{r f''(r) - f'(r)}{r}$$

$$= f''(r) + \frac{2}{r} f'(r)$$

Hence, $\nabla^2 f(r) = f''(r) + \frac{2}{r} f'(r)$.

Example 12. *Solve* $\nabla^2 f(r) = 0$.

Solution . We have

$\nabla^2 f(r) = f''(r) + \frac{2}{r} f'(r)$

Since $\nabla^2 f(r) = 0$ given

$\therefore f''(r) + \frac{2}{r} f'(r) = 0$ or $\frac{f''(r)}{f'(r)} = -\frac{2}{r}$

Integrating w.r.t. 'r', we get

$\log f'(r) = -2 \log r + \log c_1$

where c_1 is a constant of integration.

$\therefore f'(r) = \frac{c_1}{r^2}$.

Again integrating w.r.t. 'r', we get

$f(r) = -\frac{c_1}{r} + c_2$

where c_1 and c_2 are constant of integration.

Example 13. *Prove that* $\nabla^2 \left(\frac{1}{r}\right) = 0$. (UPTU–2003, PTU–2003)

Solution . Since we have

$$\nabla^2 \left(\frac{1}{r}\right) = \nabla \cdot \left(\nabla \left(\frac{1}{r}\right)\right) = \nabla \cdot \left(-\frac{1}{r^2} \nabla r\right)$$

$[\because \text{grad} f(r) = f'(r) \text{ grad } r]$

$$= \nabla \cdot \left(-\frac{1}{r^3} \mathbf{r}\right) \qquad \left[\because \nabla r = \frac{\mathbf{r}}{r}\right]$$

$$= \left(-\frac{1}{r^3}\right) \nabla \cdot \mathbf{r} + \mathbf{r} \cdot \text{grad}\left(-\frac{1}{r^3}\right)$$

$$= -\frac{3}{r^3} + \mathbf{r} \cdot \left(\frac{3}{r^4} \text{grad}\, r\right) = -\frac{3}{r^3} + \mathbf{r} \cdot \left(\frac{3}{r^4} \frac{\mathbf{r}}{r}\right)$$

$$= -\frac{3}{r^3} + \frac{3}{r^5} \mathbf{r} \cdot \mathbf{r} = -\frac{3}{r^3} + \frac{3}{r^5} r^2$$

$(\because \mathbf{r} . \mathbf{r} = r^2)$

$$= -\frac{3}{r^3} + \frac{3}{r^3} = 0.$$

$$\therefore \nabla^2 \left(\frac{1}{r}\right) = 0.$$

Example 14. *Prove that*

(i) $\nabla\times(\nabla\times a) = \nabla(\nabla.\mathbf{a}) - \nabla^2\mathbf{a}$.

(ii) $\mathbf{a}\times(\nabla\times\mathbf{r}) = \nabla(\mathbf{a}.\mathbf{r}) - (\mathbf{a}.\nabla)\mathbf{r}$.

where $\mathbf{a} = x\hat{i} + y\hat{j} + z\hat{k}$ (UPTU–2008)

Solution. (i) Let $\mathbf{a} = a_1\hat{i} + a_2\hat{j} + a_3\hat{k}$. Then

$$\nabla\times\mathbf{a} = \begin{vmatrix} \hat{i} & \hat{j} & \hat{k} \\ \frac{\partial}{\partial x} & \frac{\partial}{\partial y} & \frac{\partial}{\partial z} \\ a_1 & a_2 & a_3 \end{vmatrix}$$

$$= \left(\frac{\partial a_3}{\partial y} - \frac{\partial a_2}{\partial z}\right)\hat{i} + \left(\frac{\partial a_1}{\partial z} - \frac{\partial a_3}{\partial x}\right)\hat{j} + \left(\frac{\partial a_2}{\partial x} - \frac{\partial a_1}{\partial y}\right)\hat{k}$$

$\therefore$ $\nabla\times(\nabla\times\mathbf{a})$

$$= \begin{vmatrix} \hat{i} & \hat{j} & \hat{k} \\ \frac{\partial}{\partial x} & \frac{\partial}{\partial y} & \frac{\partial}{\partial z} \\ \frac{\partial a_3}{\partial y} - \frac{\partial a_2}{\partial z} & \frac{\partial a_1}{\partial z} - \frac{\partial a_3}{\partial x} & \frac{\partial a_2}{\partial x} - \frac{\partial a_1}{\partial y} \end{vmatrix}$$

$$= \Sigma\left[\left\{\frac{\partial}{\partial y}\left(\frac{\partial a_2}{\partial x} - \frac{\partial a_1}{\partial y}\right) - \frac{\partial}{\partial z}\left(\frac{\partial a_1}{\partial z} - \frac{\partial a_3}{\partial x}\right)\right\}\hat{i}\right]$$

$$= \Sigma\left[\left\{\left(\frac{\partial^2 a_2}{\partial x\partial y} + \frac{\partial^2 a_3}{\partial x\partial z}\right) - \left(\frac{\partial^2 a_1}{\partial y^2} - \frac{\partial^2 a_1}{\partial z^2}\right)\right\}\hat{i}\right]$$

$$= \Sigma\left[\left\{\frac{\partial}{\partial x}\left(\frac{\partial a_1}{\partial x} + \frac{\partial a_2}{\partial y} + \frac{\partial a_3}{\partial z}\right) - \left(\frac{\partial^2 a_1}{\partial x^2} + \frac{\partial^2 a_1}{\partial y^2} + \frac{\partial^2 a_1}{\partial z^2}\right)\right\}\hat{i}\right]$$

$$= \Sigma\left[\left\{\frac{\partial}{\partial x}(\nabla\cdot\mathbf{a}) - (\nabla^2\cdot a_1)\right\}\hat{i}\right]$$

$$= \left(\Sigma\hat{i}\frac{\partial}{\partial x}\right)(\nabla\cdot\mathbf{a}) - \nabla^2(\Sigma a_1\hat{i})$$

$$= \nabla(\nabla\cdot\mathbf{a}) - \nabla^2\mathbf{a}.$$

(ii) LHS $= \mathbf{a}\times(\nabla\times\mathbf{r}) = \mathbf{a}\times\mathbf{0} = 0$ [$\because$ curl $\mathbf{r} = 0$]

RHS $= \nabla(\mathbf{a}\cdot\mathbf{r}) - (\mathbf{a}\cdot\Delta)\mathbf{r}$

$$= \nabla(a_1x + a_2y + a_3z) - \left(a_1\frac{\partial}{\partial x} + a_2\frac{\partial}{\partial y} + a_3\frac{\partial}{\partial z}\right)\mathbf{r}$$

$$= \hat{i}(a_1) + \hat{j}(a_2) + \hat{k}(a_3) - (a_1\hat{i} + a_2\hat{j} + a_3\hat{k})$$

$$= \mathbf{0} = \text{LHS}$$

Example 15. *If f and g are two scalar point functions, show that*

(i) *div* $(f\nabla g) = f\nabla^2 g + \nabla f\cdot\nabla g$

(ii) *div* $(f\nabla g) -$ *div* $(g\nabla f) = f\nabla^2 g - g\nabla^2 f$

Solution. (i) Since $\nabla g = \frac{\partial g}{\partial x}\hat{i} + \frac{\partial g}{\partial y}\hat{j} + \frac{\partial g}{\partial z}\hat{k}$

$$\therefore f\nabla g = f\frac{\partial g}{\partial x}\hat{i} + f\frac{\partial g}{\partial y}\hat{j} + f\frac{\partial g}{\partial z}\hat{k}$$

and

$$\text{div}(f\nabla g) = \frac{\partial}{\partial x}\left(f\frac{\partial g}{\partial x}\right) + \frac{\partial}{\partial y}\left(f\frac{\partial g}{\partial y}\right) + \frac{\partial}{\partial z}\left(f\frac{\partial g}{\partial z}\right)$$

$$= f\frac{\partial^2 g}{\partial x^2} + \frac{\partial f}{\partial x}\frac{\partial g}{\partial x} + f\frac{\partial^2 g}{\partial y^2} + \frac{\partial f}{\partial y}\frac{\partial g}{\partial y} + f\frac{\partial^2 g}{\partial z^2} + \frac{\partial f}{\partial z}\frac{\partial g}{\partial z}$$

$$= f\left(\frac{\partial^2 g}{\partial x^2} + \frac{\partial^2 g}{\partial y^2} + \frac{\partial^2 g}{\partial z^2}\right) + \frac{\partial f}{\partial x}\frac{\partial g}{\partial x} + \frac{\partial f}{\partial y}\frac{\partial g}{\partial y} + \frac{\partial f}{\partial z}\frac{\partial g}{\partial z}$$

$$= f\nabla^2 g + \left(\frac{\partial f}{\partial x}\hat{i} + \frac{\partial f}{\partial y}\hat{j} + \frac{\partial f}{\partial z}\hat{k}\right)\cdot\left(\frac{\partial g}{\partial x}\hat{i} + \frac{\partial g}{\partial y}\hat{j} + \frac{\partial g}{\partial z}\hat{k}\right)$$

$$= f\nabla^2 g + (\nabla f)\cdot(\nabla g)$$

Hence, div $(f\nabla g) = f\nabla^2 g + (\nabla f)\cdot(\nabla g)$ ...(1)

(ii) Similarly, we may get

div $(g\nabla f) = g\nabla^2 f + (\nabla g).(\nabla f)$ ...(2)

Form (1) and (2), we get

div $(f\nabla g) -$ div $(g\nabla f) = f\nabla^2 g - g\nabla^2 f$

Example 16. *Prove that div grad* $(r^n) = n(n+1)r^{n-2}$.

(UKTU–2012, UPTU–2005, Bhopal–2008, JNTU–2006, SVTU–2006)

Solution. Since we have

grad $(r^n) = (nr^{n-1}$ grad $r)$ [$\because$ grad $f(r) = f'(r)(\nabla r)$]

$$= nr^{n-1}\frac{\mathbf{r}}{r} \quad \left[\because \text{grad } r = \frac{\mathbf{r}}{r}\right]$$

$$= nr^{n-2}\mathbf{r}$$

Now

div grad $(r^n) = \text{div}(nr^{n-2}\mathbf{r})$

$$= n\,\text{div}(r^{n-2}\mathbf{r})$$

$$= n[r^{n-2}\nabla\cdot\mathbf{r} + \mathbf{r}\cdot\text{grad}(r^{n-2})]$$

$$= n[3r^{n-2} + \mathbf{r}\cdot((n-2)r^{n-3}\text{grad } r)]$$

($\because$ $\nabla\cdot\mathbf{r} = 3$)

$$= n\left[3r^{n-2} + \mathbf{r}\cdot\left((n-2)r^{n-3}\frac{\mathbf{r}}{r}\right)\right]$$

$$= n[3r^{n-2} + (n-2)r^{n-4}\mathbf{r}\cdot\mathbf{r}]$$

$$= n[3r^{n-2} + (n-2)r^{n-4}r^2]$$
$$[\because \mathbf{r}\cdot\mathbf{r} = r^2]$$
$$= n(n+1)r^{n-2}$$

$\therefore$ div grad $r^n = n(n+1)r^{n-2}$.

Example 17. *If* $\mathbf{u} = y\hat{i} + z\hat{j} + x\hat{k}, \mathbf{v} = xy\hat{i} + yz\hat{j} + zx\hat{k}$, *find*

(i) *curl* $(\mathbf{u}\times\mathbf{v})$ (ii) $\mathbf{u}\times$ *curl* $\mathbf{v}$
(iii) $\mathbf{v}\times$ *curl* $\mathbf{u}$ (iv) *div* $(\mathbf{u}\times\mathbf{v})$

Solution. (i) Since $\mathbf{u} = y\hat{i} + z\hat{j} + x\hat{k}, \mathbf{v} = xy\hat{i} + yz\hat{j} + zx\hat{k}$.

$$\therefore \quad \mathbf{u}\times\mathbf{v} = \begin{vmatrix} \hat{i} & \hat{j} & \hat{k} \\ y & z & x \\ xy & yz & zx \end{vmatrix}$$
$$= \hat{i}(z^2x - xyz) + \hat{j}(x^2y - xyz) + \hat{k}(y^2z - xyz).$$

Then curl $(\mathbf{u}\times\mathbf{v})$

$$= \begin{vmatrix} \hat{i} & \hat{j} & \hat{k} \\ \dfrac{\partial}{\partial x} & \dfrac{\partial}{\partial y} & \dfrac{\partial}{\partial z} \\ (z^2x - xyz) & (x^2y - xyz) & (y^2z - xyz) \end{vmatrix}$$

$$= \hat{i}\left[\frac{\partial}{\partial y}(y^2z - xyz) - \frac{\partial}{\partial z}(x^2y - xyz)\right] + \hat{j}\left[\frac{\partial}{\partial z}(z^2x - xyz) - \frac{\partial}{\partial x}(y^2z - xyz)\right] + \hat{k}\left[\frac{\partial}{\partial x}(x^2y - xyz) - \frac{\partial}{\partial y}(z^2x - xyz)\right]$$

$$= \hat{i}[(2yz - xz) - (-xy)] + \hat{j}[2zx - xy + yz] + \hat{k}[2xy - yz + xz]$$

$$= (2yz - xz + xy)\hat{i} + \hat{j}(2zx - xy + yz) + \hat{k}(2xy - yz + xz).$$

(ii)
$$\text{curl } \mathbf{v} = \begin{vmatrix} \hat{i} & \hat{j} & \hat{k} \\ \dfrac{\partial}{\partial x} & \dfrac{\partial}{\partial y} & \dfrac{\partial}{\partial z} \\ xy & yz & zx \end{vmatrix}$$

$$= \hat{i}\left[\frac{\partial}{\partial y}(zx) - \frac{\partial}{\partial z}(yz)\right] + \hat{j}\left[\frac{\partial}{\partial z}(xy) - \frac{\partial}{\partial x}(zx)\right] + \hat{k}\left[\frac{\partial}{\partial x}(yz) - \frac{\partial}{\partial y}(xy)\right]$$

$$= -y\hat{i} - z\hat{j} - x\hat{k}$$
$$= -(y\hat{i} + z\hat{j} + x\hat{k}) = -\mathbf{u}$$

$$\therefore \quad \mathbf{u}\times \text{curl } \mathbf{v} = \mathbf{u}\times -\mathbf{u} = -(\mathbf{u}\times\mathbf{u}) = 0$$

(iii)
$$\text{curl } \mathbf{u} = \begin{vmatrix} \hat{i} & \hat{j} & \hat{k} \\ \dfrac{\partial}{\partial x} & \dfrac{\partial}{\partial y} & \dfrac{\partial}{\partial z} \\ y & z & x \end{vmatrix}$$
$$= \hat{i}(-1) + \hat{j}(-1) + \hat{k}(-1)$$
$$= -(\hat{i} + \hat{j} + \hat{k})$$

$\therefore$ $\mathbf{v}\times$ curl $\mathbf{u}$

$$= (xy\hat{i} + yz\hat{j} + zx\hat{k})\times\{-(\hat{i} + \hat{j} + \hat{k})\}$$
$$= -[xy\hat{k} - xy\hat{j} - yz\hat{k} + yz\hat{i} + zx\hat{j} - zx\hat{i}]$$
$$= \hat{i}(zx - yz) + \hat{j}(xy - zx) + (yz - xy)\hat{k}.$$

(iv)
$$\mathbf{u}\times\mathbf{v} = (z^2x - xyz)\hat{i} + (x^2y - xyz)\hat{j} + (y^2z - xyz)\hat{k}.$$

$\therefore$ div $(\mathbf{u}\times\mathbf{v})$

$$= \frac{\partial}{\partial x}(z^2x - xyz) + \frac{\partial}{\partial y}(x^2y - xyz) + \frac{\partial}{\partial z}(y^2z - xyz).$$
$$= z^2 - yz + x^2 - xz + y^2 - xy$$
$$= x^2 + y^2 + z^2 - xy - yz - zx.$$

Example 18. *If f and g are two scalar point functions, show that*

$$div(g\,\nabla f\times f\,\nabla g) = 0.$$

Solution. Since we have

div $(\mathbf{a}\times\mathbf{b})$ = $\mathbf{b}\cdot$ curl $\mathbf{a}$ – $\mathbf{a}$ curl $\mathbf{b}$

Let $\mathbf{a} = g\,\nabla f$ and $\mathbf{b} = f\,\nabla g$

$\therefore$ div$(g\,\nabla f\times f\,\nabla g)$

$$= f\,\nabla g\cdot\text{curl }(g\,\nabla f) - g\,\nabla f\cdot\text{curl}(f\,\nabla g)$$
$$= f\,\nabla g.g \text{ curl grad} f - g\,\nabla f\cdot f \text{ curl grad } g$$
$$= 0 - 0.$$

($\because$ curl grad $f = \mathbf{0}$, curl grad $g = \mathbf{0}$)

Hence, div$(g\,\nabla f\times f\,\nabla g) = 0$.

Example 19. *If* $\mathbf{a}$ *and* $\mathbf{b}$ *are constant vectors, show that*

(i) *div* $[(\mathbf{r}\times\mathbf{a})\times\mathbf{b}] = -2\mathbf{b}.\mathbf{a}$
(ii) *curl* $[(\mathbf{r}\times\mathbf{a})\times\mathbf{b}] = \mathbf{b}\times\mathbf{a}$

Solution. (i) Since

div $(\mathbf{a}\times\mathbf{b})$ = $\mathbf{b}\cdot$curl $\mathbf{a}$ – $\mathbf{a}\cdot$curl $\mathbf{b}$

$\therefore$ div $[(\mathbf{r}\times\mathbf{a})\times\mathbf{b}]$

$$= \mathbf{b}.\text{curl }(\mathbf{r}\times\mathbf{a}) - (\mathbf{r}\times\mathbf{a})\cdot\text{curl }\mathbf{b}$$

Since $\mathbf{b}$ is constant vector, then curl $\mathbf{b} = \mathbf{0}$

$\therefore$ div $[(\mathbf{r}\times\mathbf{a})\times\mathbf{b}]$ = $\mathbf{b}\cdot$curl $(\mathbf{r}\times\mathbf{a})$ ...(1)

Further since

curl $(\mathbf{a}\times\mathbf{b})$

$$= (\mathbf{b}\cdot\nabla)\mathbf{a} - \mathbf{b}\text{ div }\mathbf{a} - (\mathbf{a}\cdot\nabla)\mathbf{b} + \mathbf{a}\text{ div }\mathbf{b}$$

$\therefore$ curl $(\mathbf{r}\times\mathbf{a})$

$$= (\mathbf{a}\cdot\nabla)\mathbf{r} - \mathbf{a}\text{ div }\mathbf{r} - (\mathbf{r}\cdot\nabla)\mathbf{a} + \mathbf{r}\text{ div }\mathbf{a}$$

Since div $\mathbf{r}$ = 3, div $\mathbf{a}$ = 0.

$\therefore$ curl $(\mathbf{r}\times\mathbf{a})$ = $(\mathbf{a}\cdot\nabla)\mathbf{r} - 3\mathbf{a} - (\mathbf{r}\cdot\nabla)\mathbf{a}$ ...(2)

From (1) and (2), we get

div $[(\mathbf{r}\times\mathbf{a})\times\mathbf{b}]$

$$= \mathbf{b}.[(\mathbf{a}.\nabla)\mathbf{r} - 3\mathbf{a} - (\mathbf{r}.\nabla)\mathbf{a}]$$
$$= (\mathbf{a}.\nabla)\mathbf{b}.\mathbf{r} - 3\mathbf{b}.\mathbf{a} - (\mathbf{r}.\nabla)\mathbf{b}.\mathbf{a}$$
$$= \mathbf{a}.\text{grad}(\mathbf{b}.\mathbf{r}) - 3\mathbf{b}.\mathbf{a} - \mathbf{r}.\text{grad}(\mathbf{b}.\mathbf{a})$$
$$= \mathbf{a}.\mathbf{b} - 3\mathbf{b}.\mathbf{a} - \mathbf{r}.\mathbf{0}$$
$$[\because \text{grad}\,(\mathbf{b}.\mathbf{r}) = \mathbf{b},\ \text{grad}\,\mathbf{b}.\mathbf{a} = \mathbf{0}]$$
$$= -2\mathbf{b}.\mathbf{a}. \qquad (\because \mathbf{a}.\mathbf{b} = \mathbf{b}.\mathbf{a})$$

Hence, div $[(\mathbf{r}\times\mathbf{a})\times\mathbf{b}] = -2\mathbf{b}.\mathbf{a}$

(ii) Since $(\mathbf{r}\times\mathbf{a})\times\mathbf{b} = -\mathbf{b}\times(\mathbf{r}\times\mathbf{a})$

$$= -[(\mathbf{b}.\mathbf{a})\mathbf{r} - (\mathbf{b}.\mathbf{r})\mathbf{a}]$$
$$= (\mathbf{b}.\mathbf{r})\mathbf{a} - (\mathbf{b}.\mathbf{a})\mathbf{r}$$

$\therefore$ curl $[(\mathbf{r}\times\mathbf{a})\times\mathbf{b}]$

$$= \text{curl}[(\mathbf{b}.\mathbf{r})\mathbf{a} - (\mathbf{b}.\mathbf{a})\mathbf{r}]$$
$$= \text{curl}[(\mathbf{b}.\mathbf{r})\mathbf{a}] - \text{curl}[(\mathbf{b}.\mathbf{a})\mathbf{r}] \qquad \ldots(1)$$

Since we have

$$\text{curl}\,(f\mathbf{V}) = (\text{grad}\,f)\times\mathbf{V} + f\,\text{curl}\,\mathbf{V}$$

$\therefore$ curl$[(\mathbf{b}\cdot\mathbf{a})\mathbf{r}]$

$$= [\text{grad}(\mathbf{b}.\mathbf{r})]\times\mathbf{a} + (\mathbf{b}.\mathbf{r})\,\text{curl}\,\mathbf{a}$$
$$= \mathbf{b}\times\mathbf{a} + 0$$
$$[\because \text{curl}\,\mathbf{a} = 0,\ \text{grad}\,(\mathbf{b}.\mathbf{r}) = \mathbf{b}]$$
$$= \mathbf{b}\times\mathbf{a}$$

and curl$[(\mathbf{b}.\mathbf{r})\mathbf{a}]$

$$= [\text{grad}(\mathbf{b}.\mathbf{a})]\times\mathbf{r} + (\mathbf{b}.\mathbf{a})\,\text{curl}\,\mathbf{r}$$
$$= \mathbf{0} + \mathbf{0} = \mathbf{0}$$
$$[\because \text{curl}\,\mathbf{r} = 0,\ \text{grad}\,(\mathbf{b}.\mathbf{a}) = 0]$$

From (1), we get

$$\therefore \text{curl}\,[(\mathbf{r}\times\mathbf{a})\times\mathbf{b}] = \mathbf{b}\times\mathbf{a} - \mathbf{0} = \mathbf{b}\times\mathbf{a}.$$

Example 20. *Prove that $r^n\mathbf{r}$ is an irrotational vector for any value of n but is solenodial if $n + 3 = 0$.*

(GBTU–2010, VTU–2006, UPTU–2006, PTU–2005, 06, Kottayam–2005)

Solution . Since we know that if $\mathbf{a}$ is irrotational, then

$$\nabla\times\mathbf{a} = 0$$
$$\therefore\ \nabla\times(r^n\mathbf{r}) = (\text{grad}\,r^n)\times\mathbf{r} + r^n\,\text{curl}\,\mathbf{r}$$
$$= (nr^{n-1}\,\text{grad}\,r)\times\mathbf{r} + \mathbf{0}$$
$$[\because \text{curl}\,\mathbf{r} = 0]$$
$$= \left(nr^{n-1}\frac{\mathbf{r}}{r}\right)\times\mathbf{r} \qquad \left[\because \text{grad}\,r = \frac{\mathbf{r}}{r}\right]$$
$$= nr^{n-2}\,\mathbf{r}\times\mathbf{r} = \mathbf{0}$$

[$\because$ Vector product of two same vectors is zero.]

Hence $r^n\mathbf{r}$ is irrotational for any value of n.

Further since if $\mathbf{a}$ is solenoidal, then $\nabla.\mathbf{a} = 0$

$$\therefore\ \nabla\cdot(r^n\mathbf{r}) = (\text{grad}\,r^n).\mathbf{r} + r^n\,\nabla.\mathbf{r}$$
$$= (nr^{n-1}\,\text{grad}\,\mathbf{r}).\mathbf{r} + 3r^n$$
$$[\because \nabla.\mathbf{r} = 3]$$
$$= \left(nr^{n-1}\frac{\mathbf{r}}{r}\right).\mathbf{r} + 3r^n \left[\because \text{grad}\,r = \frac{\mathbf{r}}{r}\right]$$
$$= nr^{n-2}\,\mathbf{r}.\mathbf{r} + 3\,r^n$$
$$= nr^{n-2}r^2 + 3\,r^n \qquad [\because \mathbf{r}.\mathbf{r} = r^2]$$
$$= nr^n + 3\,r^n = r^n(n+3).$$

If $n + 3 = 0$, then $\nabla.(r^n\mathbf{r}) = 0$, and hence $r^n\mathbf{r}$ is solenoidal.

Example 21. *If $\mathbf{a}$ and $\mathbf{b}$ are irrotational, prove that $\mathbf{a}\times\mathbf{b}$ is solenoidal.* (GBTU–2010, Madras–2003, VTU–2001)

Solution . Since $\mathbf{a}$ and $\mathbf{b}$ are irrotational, then

$$\nabla\times\mathbf{a} = 0,\ \nabla\times\mathbf{b} = 0.$$

Now we have to prove that $\mathbf{a}\times\mathbf{b}$ is solenoidal.

$$\therefore\ \nabla\cdot(\mathbf{a}\times\mathbf{b}) = \mathbf{b}\cdot\text{curl}\,\mathbf{a} - \mathbf{a}\cdot\text{curl}\,\mathbf{b}$$
$$= \mathbf{b}\cdot(\nabla\times\mathbf{a}) - \mathbf{a}\cdot(\nabla\times\mathbf{b})$$
$$= 0 - 0 = 0.$$
$$[\because \nabla\times\mathbf{a} = 0,\ \nabla\times\mathbf{b} = 0]$$

Thus, $\nabla.(\mathbf{a}\times\mathbf{b}) = 0$.

Hence $(\mathbf{a}\times\mathbf{b})$ is solenoidal.

Example 22. *Show that the vector field $\mathbf{F} = \dfrac{\mathbf{r}}{r^3}$ is irrotaitonal as well as solenoidal. Find the scalar potential.*

(UPTU–2006)

Solution . We know that

$$\text{curl}\,(u.\mathbf{a}) = u\,\text{curl}\,\mathbf{a} + (\text{grad}\,u)\times\mathbf{a}$$

Therefore,

$$\text{curl}\left(\frac{1}{r^3}\cdot\mathbf{r}\right) = \frac{1}{r^3}\,\text{curl}\,\mathbf{r} + \left(\text{grad}\frac{1}{r^3}\right)\times\mathbf{r}$$
$$= \frac{1}{r^3}\,\mathbf{0} + \left(-\frac{3}{r^4}\hat{r}\right)\times\mathbf{r}$$
$$= \mathbf{0} - \frac{3}{r^5}(\mathbf{r}\times\mathbf{r}) = \mathbf{0} - \mathbf{0} = \mathbf{0}$$

$\Rightarrow$ **F** is irrotational.

Also, we know that for the vector field **F** to be solenoidal, div **F** = 0

We know that

$$\text{div}\,(u\mathbf{a}) = u\,\text{div}\,\mathbf{a} + \mathbf{a}.\text{grad}\,u$$
$$\therefore\ \text{div}\left(\frac{\mathbf{r}}{r^3}\right) = \frac{1}{r^3}\text{div}\,\mathbf{r} + \mathbf{r}.\text{grad}\left(\frac{1}{r^3}\right)$$
$$= \frac{3}{r^3} + \mathbf{r}\left(-\frac{3}{r^4}\frac{\mathbf{r}}{r}\right) \qquad (\because \text{div}\,\mathbf{r} = 3)$$
$$= \frac{3}{r^3} - \frac{3}{r^5}r^2 = \frac{3}{r^3} - \frac{3}{r^3} = 0$$

$\Rightarrow$ $\vec{F}$ is solenoidal.

Now, let $\vec{F} = \nabla\phi$, where ϕ is scalar potential.

$$\therefore\ \vec{F}.d\mathbf{r} = \nabla\phi.d\mathbf{r}$$
$$\vec{F}.d\mathbf{r} = d\phi$$

$$\therefore \quad d\phi = \frac{x\hat{i} + y\hat{j} + z\hat{k}}{(x^2+y^2+z^2)^{3/2}}(dx\hat{i} + dy\hat{j} + dz\hat{k})$$

$$= \frac{xdx + ydy + zdz}{(x^2+y^2+z^2)^{3/2}}$$

$$= d[-(x^2+y^2+z^2)^{-1/2}]$$

$$\Rightarrow \quad \phi = -\frac{1}{\sqrt{x^2+y^2+z^2}} + c.$$

$$\therefore \quad \phi = -\frac{1}{r} + c.$$

Recapitulations

- div $\mathbf{r} = 3$
- curl $\mathbf{r} = 0$
- $\text{div}(f\cdot\mathbf{V}) = f \,\text{div}\, \mathbf{V} + \mathbf{V} \,.\, (\text{grad}\, f)$
- $\text{curl}(f\,\mathbf{V}) = (\nabla f) \times \mathbf{V} + f(\nabla \times \mathbf{V})$
- $\text{div}(\mathbf{a} \times \mathbf{b}) = \mathbf{b}.\text{curl}\, \mathbf{a} - \mathbf{a} \,.\, \text{curl}\, \mathbf{b}$
- $\nabla\cdot(\mathbf{a} \times \mathbf{b}) = \mathbf{b}.(\nabla \times \mathbf{a}) - \mathbf{a}.(\nabla \times \mathbf{b})$
- $\text{curl}\,(\mathbf{a} \times \mathbf{b}) = (\mathbf{b}.\nabla)\mathbf{a} - \mathbf{b}.\text{div}\, \mathbf{a} - (\mathbf{a}.\nabla)\mathbf{b} + \mathbf{a}\, \text{div}\, \mathbf{b}$
- $\nabla \times \nabla f = 0$
- $\nabla\cdot(\nabla \times \mathbf{V}) = 0$
- $\nabla^2(1/r) = 0$
- $\nabla \times (\nabla \times \mathbf{a}) = \nabla(\nabla\cdot \mathbf{a}) - \nabla^2\mathbf{a}$
- $\text{div}\,(\text{grad}\, r^n) = n(n+1)\, r^{n-2}$
- $\text{div}\,(g\, \nabla f \times f\, \nabla g) = 0$

EXERCISE 22.3

1. If $\mathbf{f} = x^2y\hat{i} - 2xz\hat{j} + 2yz\hat{k}$, find
(i) div $\mathbf{f}$ (ii) curl $\mathbf{f}$
(iii) curl curl $\mathbf{f}$

2. If $\mathbf{f} = xy^2\hat{i} + 2x^2yz\hat{j} - 3yz^2\hat{k}$, then at the point (1, –1, 1) find
(i) div $\mathbf{f}$ (ii) curl $\mathbf{f}$

3. If $\mathbf{f} = \text{grad}(x^3 + y^3 + z^3 - 3xyz)$, find
(i) div $\mathbf{f}$ (ii) curl $\mathbf{f}$

4. (i) Determine the constant λ so that the vector
$$\mathbf{f} = (x+3y)\hat{i} + (y-2z)\hat{j} + (x+\lambda z)\hat{k}$$
is solenoidal.
(ii) Find the constants a, b, c so that the vector
$$\mathbf{f} = (x+2y+az)\hat{i} + (bx-3y-z)\hat{j} + (4x+cy+2z)\hat{k}$$
is irrotational, *i.e.* curl $\mathbf{f} = \mathbf{0}$.

5. Show that the vector $\mathbf{f} = (\sin y + z)\hat{i} + (x\cos y - z)\hat{j} + (x-y)\hat{k}$ is irrotational.

6. Show that $\nabla^2\left(\dfrac{x}{r^2}\right) = -\dfrac{2x}{r^4}$.

7. If $\mathbf{v} = (x\hat{i} + y\hat{j} + z\hat{k}) / (x^2+y^2+z^2)^{3/2}$, find
(i) $\nabla\cdot \mathbf{v}$ (ii) $\nabla \times \mathbf{v}$

8. If $\mathbf{v} = (x+y+1)\hat{i} + \hat{j} + (-x-y)\hat{k}$, prove that
$$\mathbf{v}\cdot(\nabla \times \mathbf{v}) = 0$$

9. If $\mathbf{v} = (y^2+z^2-x^2)\hat{i} + (z^2+x^2-y^2)\hat{j} + (x^2+y^2-z^2)\hat{k}$, find
(i) div $\mathbf{v}$ (ii) curl $\mathbf{v}$

10. Prove that
(i) div $(\mathbf{a} + \mathbf{b})$ = div $\mathbf{a}$ + div $\mathbf{b}$
(ii) curl $(\mathbf{a} + \mathbf{b})$ = curl $\mathbf{a}$ + curl $\mathbf{b}$

11. Prove that div $\nabla\phi = \nabla^2\phi$, *i.e.* $\nabla\cdot\nabla\phi = \nabla^2\phi$,
where ϕ is a scalar point function.

12. If $\mathbf{f} = x^2y\hat{i} + xz\hat{j} + 2yz\hat{k}$, prove that div curl $\mathbf{f} = 0$.

13. Prove that div $\hat{r} = 2/r$, where $\hat{r}$ is a unit vector.

14. Prove that the vector $f(r)\mathbf{r}$ is irrotational.

15. Prove that :
(i) $\nabla^2(fg) = f\,\nabla^2 g + 2\nabla f\cdot\nabla g + g\nabla^2 f$
(ii) div $(\nabla f \times \nabla g) = 0$
where f and g are two scalar point function.

16. Prove that : $\nabla\cdot\left\{r\nabla\left(\dfrac{1}{r^3}\right)\right\} = \dfrac{3}{r^4}$.

17. If $\mathbf{a}$ is a constant vector, prove that
(i) div $\{r^n(\mathbf{a} \times \mathbf{r})\} = 0$
(ii) curl $\left(\dfrac{\mathbf{a}\times\mathbf{r}}{r^3}\right) = -\dfrac{\mathbf{a}}{r^3} + \dfrac{3\mathbf{r}}{r^5}(\mathbf{a}\cdot\mathbf{r})$

18. If $\mathbf{f} = f_1\hat{i} + f_2\hat{j} + f_3\hat{k}$, show that
(i) $\nabla\cdot\mathbf{f} = \nabla f_1\cdot\hat{i} + \nabla f_2\cdot\hat{j} + \nabla f_3\cdot\hat{k}$
(ii) $\nabla\times\mathbf{f} = \nabla f_1\times\hat{i} + \nabla f_2\times\hat{j} + \nabla f_3\times\hat{k}$

19. Prove that :
(i) $\mathbf{a}.\nabla\left(\dfrac{1}{r}\right) = -\dfrac{\mathbf{a}.\mathbf{r}}{r^3}$
(ii) $\mathbf{b}\cdot\nabla\left[\mathbf{a}\cdot\nabla\left(\dfrac{1}{r}\right)\right] = -\dfrac{\mathbf{a}\cdot\mathbf{b}}{r^3} + \dfrac{3(\mathbf{a}\cdot\mathbf{r})(\mathbf{b}\cdot\mathbf{r})}{r^5}$

20. Prove that div $\left\{\dfrac{f(r)}{r}\mathbf{r}\right\} = \dfrac{1}{r^2}\dfrac{d}{dr}r^2 f(r)$.

21. Prove that curl $(g\nabla f) = \nabla g \times \nabla f = -$ curl $(f\,\nabla g)$.

22. Prove that curl $(\mathbf{a} \times r)r^n = (n+2)r^n\mathbf{a} - nr^{n-2}(\mathbf{r}\cdot\mathbf{a})\mathbf{r}$.

23. If $\hat{a}$ is a constant unit vector, prove that
$$\hat{a}\cdot\{\nabla(\mathbf{v}\cdot\hat{a}) - \nabla\times(\mathbf{v}\times\hat{a})\} = \text{div}\,\mathbf{v}$$

24. Prove that $\nabla^2\left[\nabla\cdot\left(\dfrac{\mathbf{r}}{r^2}\right)\right] = 2r^{-4}$.

25. Prove that $\dfrac{1}{2}\nabla\mathbf{a}^2 = (\mathbf{a}\cdot\nabla)\mathbf{a} + \mathbf{a}\times\text{curl}\,\mathbf{a}$

26. Prove that curl grad $r^n = \mathbf{0}$.

27. If $\mathbf{a}$ and $\mathbf{b}$ are constant vectors, prove that $\nabla[(\mathbf{a}\cdot\mathbf{b})\mathbf{r}] = \mathbf{a}\cdot\mathbf{b}$.

28. If $\mathbf{a}$ is a constant vector, prove that
(i) $\nabla\,(\mathbf{a}\cdot\mathbf{u}) = (\mathbf{a}\cdot\nabla)\mathbf{u} + \mathbf{a} \times \text{curl}\,\mathbf{u}$
(ii) $\nabla\cdot(\mathbf{a}\times\mathbf{u}) = -\,\mathbf{a}\cdot\text{curl}\,\mathbf{u}$
(iii) $\nabla\times(\mathbf{a}\times\mathbf{u}) = \mathbf{a}\,\text{div}\,\mathbf{u} - (\mathbf{a}\cdot\nabla)\mathbf{u}$

29. If $\mathbf{u} = \left(\dfrac{1}{r}\right)\mathbf{r}$, then prove that
(i) $\nabla\times\mathbf{u} = 0$ (ii) grad (div $\mathbf{u}$) $= -\left(\dfrac{2}{r^3}\right)\mathbf{r}$.

30. Prove that

(i) div $(\mathbf{a}\times\mathbf{r}) = \mathbf{r}\cdot$curl $\mathbf{a}$ (ii) div $(\mathbf{r}\times\mathbf{a}) = 0$

(iii) curl $(\mathbf{r}\times a) = -2a$ (VTU–2010, Burdwan–2009)

31. If $\mathbf{V} = \dfrac{x\hat{i}+y\hat{j}+z\hat{k}}{\sqrt{x^2+y^2+z^2}}$, show that $\nabla\mathbf{V} = \dfrac{2}{\sqrt{x^2+y^2+z^2}}$ and $\nabla\times\mathbf{V} = 0.$ (MTU–2011)

32. Show that $\mathbf{A} = (6xy+z^3)\hat{i}+(3x^2-z)\hat{j}+(3xz^2-y)\hat{k}$ is irrotational. (UKTU–2011)

33. Show that the fluid motion given by

$$\mathbf{V} = (y\sin z+\sin x)\hat{i}+(x\sin z+2yz)\hat{j}+(xy\cos z+y^2)\hat{k}$$

is irrotational. (UKTU–2011)

34. Show that the directional derivative of $\nabla(\nabla\phi)$ at the point (1, –2, 1) in the direction of the normal to the surface $xy^2z = 3x+z^2$, where $\phi = 2x^3y^2z^4$ is given by $\dfrac{1724}{\sqrt{21}}$. (UPTU–2009)

35. Show that

$$(y^2-z^2+3yz-2x)\hat{i}+(3xz+2xy)\hat{j}+(3xy-2xz+2z)\hat{k}$$

is both solenoidal and irrotational. (UPTU–2009)

Hint to Selected Problems

1. $\mathbf{f} = x^2y\hat{i}-2xz\hat{j}+2yz\hat{k}$

(i) $\nabla\cdot\mathbf{f} = \dfrac{\partial}{\partial x}(x^2y)\hat{i}+\dfrac{\partial}{\partial y}(-2xz)\hat{j}+\dfrac{\partial}{\partial z}(2yz)\hat{k}$

$= 2xy+0+2y = 2y(x+1)\cdot$

3. $\mathbf{f} = \text{grad}(x^3+y^3+z^3-3xyz)$

Let $\phi = x^3+y^3+z^3-3xyz$.

$\therefore$ $\mathbf{f} = \nabla\phi$.

(i) $\nabla\cdot\mathbf{f} = \nabla.(\nabla\phi) = \nabla^2\phi = \dfrac{\partial^2\phi}{\partial x^2}+\dfrac{\partial^2\phi}{\partial y^2}+\dfrac{\partial^2\phi}{\partial z^2}$

$= 6x+6y+6z = 6(x+y+z).$

(ii) $\nabla\times(\nabla\phi) = 0.$

4. (ii) $\mathbf{f} = (x+2y+az)\hat{i}+(bx-3y-z)\hat{j}+(4x+cy+2z)\hat{k}$

$$\nabla\times\mathbf{f} = \begin{vmatrix}\hat{i} & \hat{j} & \hat{k}\\ \dfrac{\partial}{\partial x} & \dfrac{\partial}{\partial y} & \dfrac{\partial}{\partial z}\\ x+2y+az & bx-3y-z & 4x+cy+2z\end{vmatrix}$$

$= (c+1)\hat{i}+(a-4)\hat{j}+(b-2)\hat{k}$

For irrotaional, we have $\nabla\times\mathbf{f} = 0.$

$\therefore$ $c = -1, a = 4, b = 2.$

8. $\mathbf{v} = (x+y+1)\hat{i}+\hat{j}+(-x-y)\hat{k}$, then

$$\nabla\times\mathbf{v} = \begin{vmatrix}\hat{i} & \hat{j} & \hat{k}\\ \dfrac{\partial}{\partial x} & \dfrac{\partial}{\partial y} & \dfrac{\partial}{\partial z}\\ (x+y+1) & 1 & (-x-y)\end{vmatrix}$$

$= \hat{i}[-1-0]-\hat{j}[-1-0]+\hat{k}[0-1] = -\hat{i}+\hat{j}-\hat{k}$

$\therefore$ $\mathbf{v}.\nabla\times\mathbf{v} = -(x+y+1)+1+x+y = 0\,.$

11. $\text{div}(\nabla\phi) = \nabla.(\nabla\phi) = \nabla.\left[\dfrac{\partial\phi}{\partial x}\hat{i}+\dfrac{\partial\phi}{\partial y}\hat{j}+\dfrac{\partial\phi}{\partial z}\hat{k}\right]$

$= \dfrac{\partial}{\partial x}\left(\dfrac{\partial\phi}{\partial x}\right)+\dfrac{\partial}{\partial y}\left(\dfrac{\partial\phi}{\partial y}\right)+\dfrac{\partial}{\partial z}\left(\dfrac{\partial\phi}{\partial z}\right)$

$= \dfrac{\partial^2\phi}{\partial x^2}+\dfrac{\partial^2\phi}{\partial y^2}+\dfrac{\partial^2\phi}{\partial z^2} = \nabla^2\phi.$

16. Since $\nabla f(r) = f'(r).\nabla r = f'(r)\left(\dfrac{\mathbf{r}}{r}\right).$

$\therefore$ $\nabla\left(\dfrac{1}{r^3}\right) = -\dfrac{3}{r^5}\mathbf{r}.$

$\therefore$ $\nabla\left\{r\nabla\left(\dfrac{1}{r^3}\right)\right\} = \nabla\cdot\left(-\dfrac{3}{r^4}\mathbf{r}\right) = \left(-\dfrac{3}{r^4}\right)\nabla\cdot\mathbf{r}+\mathbf{r}\nabla\left(-\dfrac{3}{r^4}\right)$

$= -\dfrac{9}{r^4}+\mathbf{r}\cdot\left(\dfrac{12}{r^6}\mathbf{r}\right) = -\dfrac{9}{r^4}+\dfrac{12}{r^6}(\mathbf{r}\cdot\mathbf{r}).$

$= -\dfrac{9}{r^4}+\dfrac{12}{r^6}(r^2) = -\dfrac{9}{r^4}+\dfrac{12}{r^4} = \dfrac{3}{r^4}.$

17. (ii) $\text{curl}\left\{\dfrac{\mathbf{a}\times\mathbf{r}}{r^3}\right\} = \nabla\left(\dfrac{1}{r^3}\right)\times(\mathbf{a}\times\mathbf{r})+\dfrac{1}{r^3}\nabla\times(\mathbf{a}\times\mathbf{r})$

$= -\dfrac{3}{r^3}\left(\dfrac{\mathbf{r}}{r}\right)\times(\mathbf{a}\times\mathbf{r})+\dfrac{1}{r^3}(0-0-\mathbf{a}+3\mathbf{a})$

$= -\dfrac{3}{r^5}[(\mathbf{r}\cdot\mathbf{r})\mathbf{a}-(\mathbf{r}\cdot\mathbf{a})\mathbf{r})+\dfrac{2\mathbf{a}}{r^3}$

$= -\dfrac{3\mathbf{a}}{r^3}+\dfrac{3\mathbf{r}}{r^5}(\mathbf{r}\cdot\mathbf{a})+\dfrac{2\mathbf{a}}{r^3} = -\dfrac{\mathbf{a}}{r^3}+\dfrac{3\mathbf{r}}{r^5}(\mathbf{r}\cdot\mathbf{a})$

19. (i) $\mathbf{a}\cdot\nabla\left(\dfrac{1}{r}\right) = \mathbf{a}\cdot\left[-\dfrac{1}{r^2}\dfrac{\mathbf{r}}{r}\right] = \dfrac{\mathbf{a}\cdot\mathbf{r}}{r^3}$

(ii) $\mathbf{b}\cdot\nabla\left[\mathbf{a}\cdot\nabla\left(\dfrac{1}{r}\right)\right] = \mathbf{b}\cdot\nabla\left[-\dfrac{\mathbf{a}.\mathbf{r}}{r^3}\right]$

$= \mathbf{b}\cdot\left[-\dfrac{\mathbf{a}}{r^3}+\dfrac{3(\mathbf{a}\cdot\mathbf{r})}{r^5}\mathbf{r}\right] = -\dfrac{\mathbf{a}\cdot\mathbf{b}}{r^3}+\dfrac{3(\mathbf{a}\cdot\mathbf{r})(\mathbf{b}\cdot\mathbf{r})}{r^5}$

24. $\nabla\cdot\left(\dfrac{\mathbf{r}}{r^2}\right) = \nabla\left(\dfrac{1}{r^2}\right)\cdot\mathbf{r}+\dfrac{1}{r^2}(\nabla\cdot\mathbf{r})$

$= -\dfrac{2}{r^3}\left(\dfrac{\mathbf{r}}{r}\right)\cdot\mathbf{r}+\dfrac{3}{r^2} = -\dfrac{2}{r^2}+\dfrac{3}{r^2} = \dfrac{1}{r^2}$

$\therefore$ $\nabla^2\left[\nabla\cdot\left(\dfrac{\mathbf{r}}{r^2}\right)\right] = \nabla^2\cdot\left(\dfrac{1}{r^2}\right) = \nabla\left(\nabla\left(\dfrac{1}{r^2}\right)\right) = \nabla\cdot\left[-\dfrac{2\mathbf{r}}{r^4}\right]$

$= \nabla\left(-\dfrac{2}{r^4}\right)\cdot\mathbf{r}+\left(-\dfrac{2}{r^4}\right)\nabla\cdot\mathbf{r} = \dfrac{8}{r^4}-\dfrac{6}{r^4} = \dfrac{2}{r^4} = 2r^{-4}.$

ANSWERS

1. (i) $2y(x+1)$ (ii) $(2x+2z)\hat{i}-(x^2+2z)\hat{k}$ (iii) $(2x+2)\hat{j}$ **2.** (i) div $\mathbf{f} = 9$ (ii) curl $\mathbf{f} = -\hat{i}-2\hat{k}$

3. (i) div $\mathbf{f} = 6(x+y+z)$ (ii) curl $\mathbf{f} = 0$ **4.** (i) $\lambda = -2$ (ii) $a = 4, b = 2\ c = -1$

7. (i) $\dfrac{3}{2}\cdot\dfrac{1}{(x^2+y^2+z^2)^{3/2}}$ (ii) **0** **9.** (i) $-2x-2y-2z$ (ii) $2(y-z)\hat{i}+2(z-x)\hat{j}+2(x-y)\hat{k}$

Objective Evaluations

Fill in the Blanks

1. $\nabla \equiv \frac{\partial}{\partial x}\hat{i} + \frac{\partial}{\partial y}\hat{j} + \frac{\partial}{\partial z}\hat{k}$ is a vector ________.
2. The gradient of a scalar field is a ________.
3. Gradient of a constant scalar field is ________.
4. If f and g are two scalar point function, then $f\,\nabla g + g\,\nabla f =$
5. If $|\mathbf{r}| = r$, then $\nabla f(r) \times \mathbf{r}$ is equal to ________.
6. If $|\mathbf{r}| = r$, then ∇r is equal to ________.
7. If $|\mathbf{r}| = r$, then $\frac{\nabla f(r)}{\nabla r}$ is equal to ________.
8. If **a**, **b** are constant vectors, then grad [**r a b**] is equal to ________.
9. If **a** is a constant vector, then grad (**r**·**a**) = ________.
10. The family of surfaces $f(x, y, z) = c$ (constant) are called ________.
11. If $f(x, y, z) = c$ (constant), the vector normal to f is ________.
12. The directional derivative of a scalar field f at a point P in the direction of a unit vector $\hat{a}$ is ________.
13. If $\hat{n}$ be a unit vector normal to the level surface $f(x, y, z) = c$ at a point P and n be the distance of P from any fixed point in the direction of $\hat{n}$, then grad f = ________.
14. The directional derivatives of a scalar point function at any point along any tangnet line to the level surface at the point is ________.
15. The equation of a tangent plane to the surface $f(x, y, z) = c$ is ________.
16. The unit normal vector to the surface $f = 0$ is ________.
17. Divergence of a vector is a ________.
18. If $\mathbf{F} = F_1\hat{i} + F_2\hat{j} + F_3\hat{k}$, then $\frac{\partial F_1}{\partial x} + \frac{\partial F_2}{\partial y} + \frac{\partial F_3}{\partial z} =$ ________.
19. Curl of a vector is a ________.
20. If the divergence of a vector is zero, then vector is ________.
21. If **F** is a irrotational vector then $\nabla \times \mathbf{F} =$ ________.
22. $\nabla.\mathbf{r} =$ ________, where $\mathbf{r} = x\hat{i} + y\hat{j} + z\hat{k}$.
23. $\nabla \times \mathbf{r} =$ ________, where $\mathbf{r} = x\hat{i} + y\hat{j} + z\hat{k}$.
24. The curl of a gradient of ϕ is ________.
25. The divergence of a curl of a vector is ________.
26. The divergence of gradient of ϕ is equal to ________.
27. $\mathbf{b}\cdot(\nabla \times \mathbf{a}) - \mathbf{a}\cdot(\nabla \times \mathbf{b}) =$ ________.
28. The vector $f(r)\mathbf{r}$ is a ________.
29. The vector $\frac{\mathbf{r}}{r^3}$ is a ________.
30. $\nabla^2\left(\frac{1}{r}\right) =$ ________.
31. div $(\nabla\phi \times \nabla\psi) =$ ________.
32. If **a** and **b** are irrotational, then **a** × **b** is ________.
33. $\nabla(fg) = f\,\nabla g$ ________.
34. If $\mathbf{r} = x\mathbf{i} + y\mathbf{j} + z\mathbf{k}$ then div $\mathbf{r} =$ ________.

True/False

Write 'T' for True and 'F' for False statement.

1. The differential operator ∇ is called nabla. **(T/F)**
2. The gradient of a scalar is a scalar. **(T/F)**
3. Curl of a vector is a scalar. **(T/F)**
4. Divergence of a vector is a vector. **(T/F)**
5. ∇(cf) = c∇f, where c is a constant. **(T/F)**
6. $\nabla f(u) = f'(u)\nabla u$. **(T/F)**
7. $\nabla\phi.d\mathbf{r} = d\phi$. **(T/F)**
8. ∇f is a vector along the tangent to the surface $f = 0$. **(T/F)**
9. If **V** is solenoidal, then curl **V** = 0. **(T/F)**
10. If **V** is irrotational, then div **V** = 0. **(T/F)**
11. The equation $\nabla^2 f = 0$ is called Laplace's equation. **(T/F)**
12. If $\mathbf{r} = x\hat{i} + y\hat{j} + z\hat{k}$, then
    (i) div **r** = 3. **(T/F)**
    (ii) curl **r** = **0**. **(T/F)**
13. The divergence of a constant vector is zero vector. **(T/F)**
14. The curl of a constant vector is zero vector. **(T/F)**
15. The directional derivative of f along $\hat{a}$ is grad $f.\hat{a}$. **(T/F)**
16. $\nabla(\mathbf{a} \times \mathbf{r}) = \mathbf{a}$, where **a** is a constant vector. **(T/F)**
17. $\nabla(\mathbf{r} \times \mathbf{a}) = -2\mathbf{a}$, where **a** is a constant vector. **(T/F)**
18. curl (ϕ**a**) = grad ϕ × **a** + ϕ curl **a**. **(T/F)**
19. curl grad $\phi \neq 0$. **(T/F)**
20. $\nabla\cdot(\nabla \times \mathbf{a}) = 0$ **(T/F)**
21. $\nabla \times (\nabla \times \mathbf{a}) + \nabla^2\mathbf{a} = \nabla(\nabla\cdot\mathbf{a})$. **(T/F)**
22. curl (ϕ grad ϕ) = 0. **(T/F)**
23. div (**a** × **r**) = **r** curl **a**. **(T/F)**
24. The vector $f(r)\mathbf{r}$ is a solenoidal vector. **(T/F)**
25. The vector $r^n\mathbf{r}$ is solenoidal vector if $n = -3$. **(T/F)**
26. If **V** is a constant vector, then div **V** = 0. **(T/F)**

Multiple Choice Questions

Choose the most appropriate one.

1. Gradient of a constant scalar field is equal to :
   (a) 0 (b) 1 (c) –1 (d) 3
2. ∇r^2 is equal to :
   (a) **r** (b) 2**r** (c) 3**r** (d) –2**r**
3. If $|\mathbf{r}| = r$, then $r\nabla r$ is equal to :
   (a) –**r** (b) 2**r** (c) **r** (d) 0
4. If **a** is a constant vector, then grad (**r**·**a**) is equal to :
   (a) **r** (b) –**a** (c) 0 (d) **a**

5. If **a**, **b** are constant vectors, then grad [**r a b**] equals :
(a) **a** (b) **b** (c) $\mathbf{a} \times \mathbf{b}$ (d) $\mathbf{b} \times \mathbf{a}$

6. The numbers of level surfaces that pass through any point of *R* over which $f(x, y, z)$ is defined :
(a) 2 (b) 1 (c) 3 (d) 0

7. The vector normal to the surface $f = 0$ is :
(a) ∇f (b) $\nabla^2 f$ (c) 0 (d) 1

8. The grad of f along $\hat{n}$ is :
(a) $\frac{\partial f}{\partial n}\hat{n}$ (b) $\frac{\partial f}{\partial n}$ (c) 0 (d) $\hat{n}$

9. The unit vector normal to the surface $x^2y + 2xz = 4$ at (2, –2, 3) is :
(a) $\frac{1}{3}(\hat{i} - 2\hat{j} + 2\hat{k})$ (b) $\frac{1}{3}(\hat{i} - 2\hat{j} - 2\hat{k})$
(c) $\frac{1}{3}(\hat{i} + 2\hat{j} - 2\hat{k})$ (d) 2

10. The directional derivative of $\phi = xy + yz + zx$ in the direction of the vector $\hat{i} + 2\hat{j} + \hat{k}$ at (1, 2, 0) is :
(a) 11/3 (b) 11 (c) 10/3 (d) 7/3

11. If a vector **V** is solenoidal, then div **V** is equal to :
(a) 1 (b) –1 (c) 2 (d) 0

12. Curl of a constant vector is :
(a) **0** (b) 1 (c) –1 (d) 2

13. The div **r** is equal to :
(a) 2 (b) 3 (c) 1 (d) 0

14. Curl **r** is equal to :
(a) 1 (b) 3 (c) **0** (d) 0

15. If **a** is a constant vector, then div ($\mathbf{r} \times \mathbf{a}$) is equal to :
(a) 1 (b) 0 (c) –1 (d) 3

16. If $\mathbf{f}$ = grad ϕ, then curl **f** is equal to :
(a) **0** (b) 0 (c) 1 (d) –1

17. $(\mathbf{a}.\nabla)\mathbf{a} + \mathbf{a} \times \text{curl } \mathbf{a}$ is equal to :
(a) grad $\mathbf{a}^2$ (b) $\frac{1}{2}$grad $\mathbf{a}^2$
(c) $\frac{1}{2}$grad **a** (d) **a**

18. The div curl **a** is equal to :
(a) 0 (b) 1 (c) –1 (d) **a**

19. $\frac{\text{grad } f(r)}{f'(r)}$ is equal to :
(a) r (b) grad **r** (c) curl **r** (d) $f(r)$

20. The vector $r^n\mathbf{r}$ is solenoidal if n is equals :
(a) 3 (b) –3 (c) 2 (d) 0

21. Curl ($f(r)\mathbf{r}$) is equal to :
(a) 1 (b) 2 (c) –1 (d) **0**

22. $\nabla^2 f(r) - \frac{2}{r}f'(r)$ is equal to :
(a) $f'(r)$ (b) $f''(r)$ (c) $f(r)$ (d) $[f(r)]^2$

23. If **a** and **b** are irrotational, then div ($\mathbf{a} \times \mathbf{b}$) is equal to :
(a) **a** (b) **b** (c) 0 (d) **a** + **b**

24. $\nabla^2 r^2$ is equal to :
(a) 3 (b) 2 (c) 4 (d) 6

25. If **a** and **b** are constant vectors, then div [($\mathbf{r} \times \mathbf{a}) \times \mathbf{b}$] equals :
(a) 2**b.a** (b) –2**b.a** (c) **a.b** (d) 0

26. If **a** and **b** are constant vectors, then $\nabla.((\mathbf{a}.\mathbf{b})\mathbf{r})$ is equal to :
(a) **a.b** (b) **a** (c) **b** (d) 0

ANSWERS

Fill in the Blanks

1. Differential operator **2.** vector **3.** 0 **4.** $\nabla(fg)$ **5.** **0** **6.** $\frac{\mathbf{r}}{r}$ **7.** $f'(r)$ **8.** $\mathbf{a} \times \mathbf{b}$ **9.** **a**
10. Level surfaces **11.** ∇f **12.** $\nabla f.\hat{a}$ **13.** $\frac{\partial f}{\partial n}\hat{n}$ **14.** 0 **15.** $(R - r).\nabla f = 0$ **16.** $\frac{\nabla f}{|\nabla f|}$ **17.** scalar
18. div **F** **19.** vector **20.** solenoidal **21.** 0 **22.** 3 **23.** **0** **24.** 0 **25.** 0
26. $\nabla^2\phi$ **27.** $\nabla.(\mathbf{a} \times \mathbf{b})$ **28.** irrotational **29.** solenoidal **30.** 0 **31.** 0
32. solenoidal **33.** $+ g \nabla f$ **34.** 3

True/False

1. T **2.** F **3.** F **4.** F **5.** T **6.** T **7.** T **8.** F **9.** F **10.** F
11. T **12.** (i) T (ii) T **13.** F **14.** T **15.** T **16.** F **17.** T **18.** T **19.** F
20. T **21.** T **22.** T **23.** T **24.** F **25.** T **26.** T

Multiple Choice Questions

1. (a) **2.** (b) **3.** (c) **4.** (d) **5.** (c) **6.** (b) **7.** (a) **8.** (a) **9.** (b) **10.** (c)
11. (d) **12.** (a) **13.** (b) **14.** (c) **15.** (b) **16.** (a) **17.** (b) **18.** (a) **19.** (b) **20.** (b)
21. (d) **22.** (b) **23.** (c) **24.** (d) **25.** (b) **26.** (a)

FFFFF

23 Gauss', Stoke's and Green's Theorems

23.1 INTRODUCTION

In this chapter we shall discuss line integral, surface integrals and volume intetgrals and shall consider some important applications of such integrals which deal with physical and engineering problems. We shall see that line integral is a natural generalization of a definite integral, a surface integral is a generalization of double integral and volume intgral is generalization of triple integral.

Line integrals can be transformed into dounble integral and conversely. Triple integrals can be transformed into double integral and these transformations has great importance. Therefore, we shall discuss some formulas of Gauss, Green and Strokes which are powerful in many applications as well as in theoretical problems. These formulas give the physical meaning of the divergence and the curl of a vector function.

23.2 ORIENTED CURVES

Let us consider a curve C in space and oriented the curve C by choosing one of the two directions along C as the positive direction and the opposite direction along C is then called the negative direction.

Let A be the initial point and B be the terminal point of C under the chosen orientation. Therefore, we may now represent the curve C by the parametric equation

$$\mathbf{r}(s) = x(s)\hat{i} + y(s)\hat{j} + z(s)\hat{k}$$

where s is the arc length of C and for the point A, $s = a$ and for the point B, $s = b$, hence $a \leq s \leq b$.

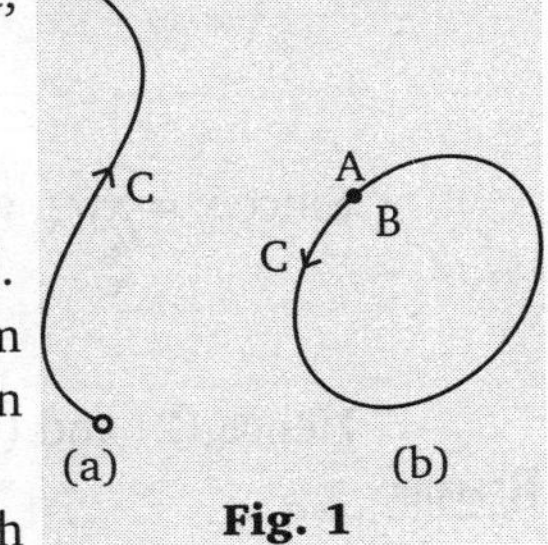

Fig. 1

(i) Closed curve. If the point A and B coincide as shown in fig. 1(b), then the curve is closed.

(ii) Smooth curve. If $\mathbf{r}(s)$ is continuously differentiable and its first derivative is different from zero vector for all values of s and the curve C has a unique tangent at each of its points, then the curve C is called smooth curve.

(iii) Piecewise smooth curve. A curve C which is the composition of a finite number of smooth curve, is called piecewise smooth curve.
In the adjoining fig. 2 the curve is composed of four smooth curves C_1, C_2, C_3 and C_4 hence the curve is piecewise smooth.

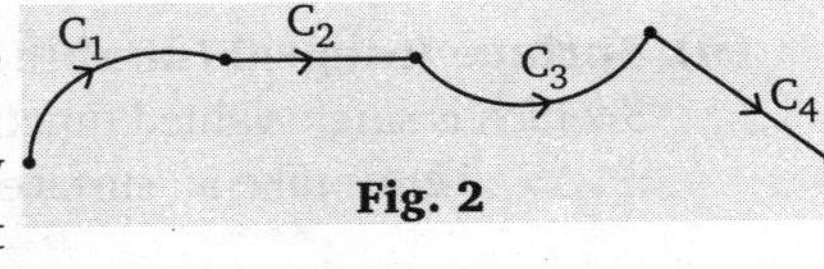

Fig. 2

(iv) Smooth surface. A surface S over each of its points a unique normal may drawn and the direction of each normal depends only on the point at which it is drawn, is called smooth surface.

(v) Piecewise smooth surface. A surface which is composed of a finite number of smooth surface, is called piecewise smooth surface.

(vi) Simply connected domain. A region (or domain) in which every closed curve can be shrink to a point without crossover the boundary of the region, is called simply connected domain. Otherwise the region is called multiply connected domain.

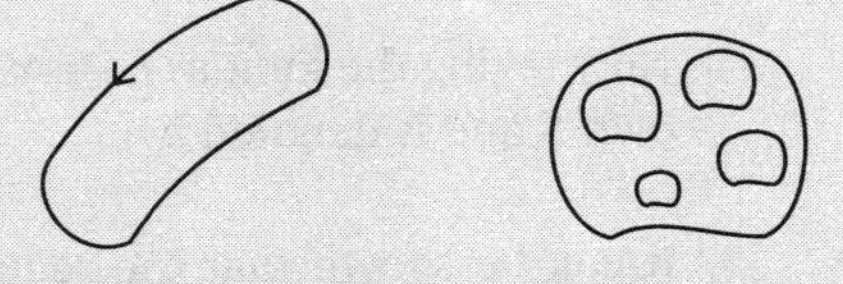
(a) Simple connected (b) Multipe connected
Fig. 3

23.3 LINE, SURFACE AND VOLUME INTEGRALS

(i) Line integrals. Let $f(x, y, z)$ be a given function which is defined at each point of the curve C and $f(x, y, z)$ is continuous function of s and let P be a point on C with co-ordintaes $(x(s), y(s), z(s))$. Thus $f(x, y, z)$ is written as $f(P)$. Now divide the curve C into n parts in an arbitrary way and letting $P_0 = A, P_1, P_2, ..., P_{n-1}, P_n = B$ where A and B are the end points of the curve C.

Let us divide in the interval $a \le s \le b$ such that

$$a = s_0 < s_1 < s_2 < ... < s_n = b.$$

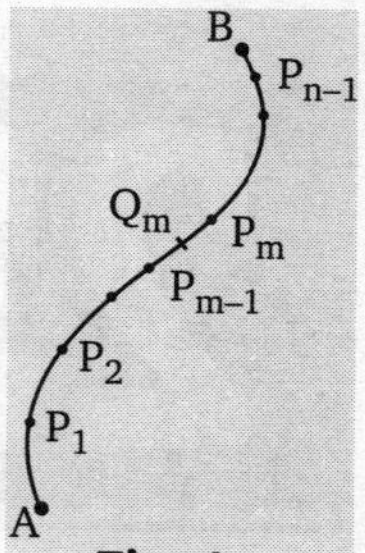

Now choose an arbitrary point between each portion, *i.e.* between A and P_1, P_1 and P_2 and so on. Let Q_1 be that point between A and P_1, Q_2 between P_1 and P_2 etc. and form the sum

$$S_n = \sum_{m=1}^{n} f(Q_m)\Delta s_m, \text{ where } \Delta s_m = s_m - s_{m-1}.$$

Now for $n = 2, 3, 4, ...,$ and the greatest $\Delta s_m \to 0$ as $n \to \infty$, we get a sequence of real numbers S_2, S_3, S_4, ... The limit of this squence $\langle s_n \rangle$ is called the line integral of f along the curve C from A to B is denoted by $\int_C f(x, y, z)ds$.

Fig. 4

In most cases the representation of C will be of the form

$$\mathbf{r}(t) = x(t)\hat{i} + y(t)\hat{j} + z(t)\hat{k}, t_0 \le t \le t_1$$

then we have

$$\int_C f(x, y, z)dS = \int_a^b f[x(s), y(s), z(s)]ds \quad ...(1)$$

and

$$\int_a^b f[x(s), y(s), z(s)] = \int_{t_0}^{t_1} f[x(t), y(t), z(t)]\frac{ds}{dt}dt \quad ...(2)$$

In particular, suppose $\mathbf{r}(t)$ is the position vector of (x, y, z), then $\mathbf{r}(t) = x\hat{i} + y\hat{j} + z\hat{k}$ and let $t = t_0$ at A and $t = t_1$ at B and suppose $\mathbf{F}(x, y, z) = f_1\hat{i} + f_2\hat{j} + f_3\hat{k}$ is a vector function and continuous along C. Let s be the arc length of the curve C *i.e.*, $s =$ arc AP, then

$$\frac{d\mathbf{r}}{ds} = \mathbf{t}$$

is a unit trangent vector at the point $P(x, y, z)$. Thus the component of $\mathbf{f}$ along this tangent is $\mathbf{f}\cdot\frac{d\mathbf{r}}{ds}$. Therefore, we have

$$\int_A^B \left(\mathbf{f}\cdot\frac{d\mathbf{r}}{ds}\right)ds = \int_A^B \mathbf{f}\cdot d\mathbf{r} = \int_C \mathbf{f}\cdot d\mathbf{r}.$$

$$\therefore \quad \int_C \mathbf{f}\cdot d\mathbf{r} = \int_C (f_1dx + f_2dy + f_3dz). \quad ...(3)$$

Since $x = x(t), y = y(t), z = z(t)$, then

$$\int_C \mathbf{f}\cdot d\mathbf{r} = \int_{t_0}^{t_1}\left[f_1\frac{dx}{dt} + f_2\frac{dy}{dt} + f_3\frac{dz}{dt}\right]dt. \quad ...(4)$$

Hence (2) and (4) are equivalent.

REMARK

- If the curve C is simple closed curve, then the integral $\int_C \mathbf{f}\cdot d\mathbf{r}$ is known as the circulation.

(ii) Surface Integral (double integral). Let S be a surface of finite area and let $f(x, y, z)$ be defined over this surface S which is single valued function. Now divide the whole surface S into n surface elements of areas $\Delta S_1, \Delta S_2, ..., \Delta S_m, ..., \Delta S_n$. Let us take an surface element of area ΔS_m and choose anarbitrary point P_m inside ΔS_m and form the sum

$$J_n = \sum_{m=1}^{n} f(P_m)\Delta S_m.$$

Now taking the limit as $n \to \infty$ in such a way that $\Delta S_m \to 0$, then this limit if exists is called the surface intregral of f over S and is denoted by

$$\iint_S f(x, y, z)dS.$$

It can be shown that the sequence $\langle J_n \rangle$ converges and its limit is independent of the choice of subdivisions and corresponding point P_m.

In particular, let S be a piecewise smooth surface and $\mathbf{f}(x, y, z)$ is a vector function which is continuous and defined over S. Let us consider a surface element of area dS enclosing a point P and let $\mathbf{n}$ be the unit vector drawn at P

outward to the element dS and normal to it which is shown in fig. 5.

Thus $\mathbf{f} \cdot \mathbf{n}$ is the normal component of $\mathbf{f}$ at P. Therefore, the integral of $\mathbf{f} \cdot \mathbf{n}$ over S can be written as

$$\iint_S \mathbf{f} \cdot \mathbf{n} dS = \iint_S \mathbf{f} \cdot d\mathbf{S}, \text{ where } d\mathbf{S} = \mathbf{n} dS \qquad \ldots(1)$$

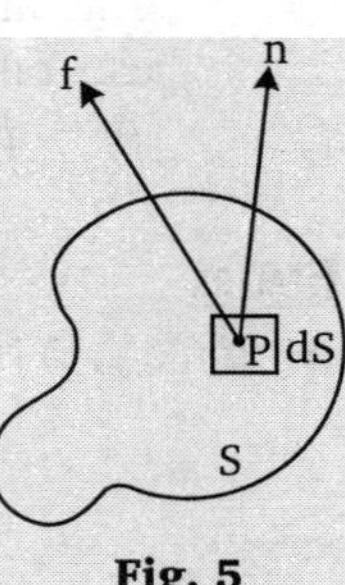

Fig. 5

If $\mathbf{n} = l\hat{i} + m\hat{j} + n\hat{k}$, where l, m, n are the direction cosines of normal which makes the angles α, β and γ with the positive axis *i.e.*, $l = \cos\alpha$, $m = \cos\beta$, $n = \cos\gamma$.

Let $\mathbf{f}(x, y, z) = f_1\hat{i} + f_2\hat{j} + f_3\hat{k}$, then

$$\iint_S \mathbf{f} \cdot \mathbf{n} dS = \iint_S (f_1 \cos\alpha + f_2 \cos\beta + f_3 \cos\gamma) dS.$$

Since we have $\cos\alpha \, dS = dydz$, $\cos\beta \, dS = dzdx$, $\cos\gamma \, dS = dxdy$

$$\therefore \qquad \iint_S \mathbf{f} \cdot \mathbf{n} dS = \iint_S (f_1 dzdy + f_2 dzdx + f_3 dxdy).$$

Evaluation of double integral. To find the value of double integral, it is taken over the orthogonal projection of S on one of the co-ordinate planes. This projection is obtained only if a line perpendicular to the closed co-ordinate planes meet the given surface S in only one point. If it is not so, then subdivide the surface S into surface which do satisfy this condition. This method can be understood by the following process : For this let us consider the surface S in such a way that a line perpendicular to xy-plane meets S in only one point. The equation of the surface S is then taken

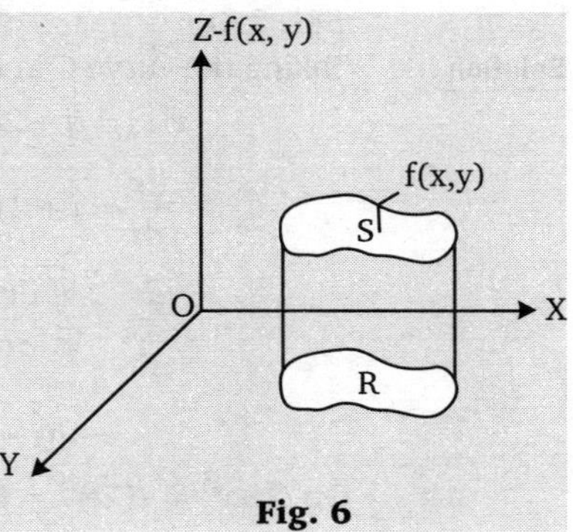

Fig. 6

$$z = f(x, y).$$

Let R be the orthogonal projection of S on xy-plane and let $\hat{\mathbf{n}}$ be the unit vector drawn at some point P on S and normal to S and let this normal $\hat{\mathbf{n}}$ makes the acute angle γ with the z-axis then

$$\cos\gamma \, dS = dxdy$$

where dS is the area of a small patch at P on S.

$$\therefore \qquad dS = \frac{dxdy}{\cos\gamma}.$$

Since $\mathbf{n} = \cos\alpha\hat{i} + \cos\beta\hat{j} + \cos\gamma\hat{k}$, therefore $\cos\gamma = \mathbf{n} \cdot \hat{k}$.

Since γ is acute, then $\cos\gamma = \left|\mathbf{n} \cdot \hat{k}\right|$, so $dS = \dfrac{dxdy}{\left|\mathbf{n} \cdot \hat{k}\right|}$

Thus
$$\iint_S \mathbf{f} \cdot \mathbf{n} dS = \iint_R \mathbf{f} \cdot \mathbf{n} \frac{dxdy}{\left|\mathbf{n} \cdot \hat{k}\right|}. \qquad \ldots(1)$$

Hence the surface integral over S can be evaluated using (1)

- If S can be described by inequalities of the form $a \le x \le b$ and $g(x) \le y \le h(x)$ where $S : z = f(x, y)$, then
$$\iint_S f dxdy = \int_a^b \left[\int_{g(x)}^{h(x)} f dy\right] dx.$$

(iii) **Volume integral.** Let V be a volume enclosed by a surface S and let $f(x, y, z)$ be a point function defined over V. Now divide the volume V into n subvolume element of volumes $\Delta V_1, \Delta V_2, \ldots, \Delta V_{n-1}, \Delta V_n$ and choose an arbitrary point $P_m(x_m, y_m, z_m)$ in each of elements ΔV_m such that $f(P_m) = f(x_m, y_m, z_m)$ and form the sum

$$J_n = \sum_{m=1}^{n} f(P_m)\Delta V_m.$$

for $n = 2, 3, 4, \ldots$, and taking greatest $\Delta V_m \to 0$ as $n \to \infty$. Then we get the sequence $J_2, J_3, J_4, \ldots$. If the limit of this sequence $\langle J_n \rangle$ exists, then this limit is called the volume integral of f over the volume V which is denoted by

$$\iiint_V f(x, y, z) dV.$$

This limit is independent of the choice of the subdivision of V, if V is piecewise smooth volume. Therefore, we can take the volume elememts in the form of urboids whose edges are parallel to the co-ordinates axis. Then $dV = dxdydz$, hence

$$\iiint_V f(x,y,z)dV = \iiint_V f(x,y,z)dxdydz.$$

REMARK

- If $f(x, y, z)$ is a vector function, then the volume integral of $f(x, y, z)$ over V is $\iiint_V \mathbf{f}(x,y,z)dV$.

SOLVED EXAMPLES

Example 1. *Evaluate* $\int_C xy^3 ds$, *where C is the segment of the line* $y = 2x$ *in the xy-plane from* $A(-1, -2, 0)$ *to* $B(1, 2, 0)$.

Solution. Taking the curve C in the following form

$$\mathbf{r}(t) = t\hat{i} + 2t\hat{j}, (-1 \le t \le 1).$$

$$\therefore \quad \frac{d\mathbf{r}}{dt} = \hat{i} + 2\hat{j}.$$

$$\frac{ds}{dt} = \sqrt{\left(\frac{d\mathbf{r}}{dt}\cdot\frac{d\mathbf{r}}{dt}\right)}$$

$$= \sqrt{(\hat{i}+2\hat{j})\cdot(\hat{i}+2\hat{j})} = \sqrt{5}$$

On C, $xy^3 = t(2t)^3 = 8t^4$ and therefore

$$\int_C xy^3 ds = 8\sqrt{5}\int_{-1}^{1} t^4 dt$$

$$= 8\sqrt{5}\left[\frac{t^5}{5}\right]_{-1}^{1} = \frac{16\sqrt{5}}{5}.$$

Example 2. *Evaluate* $\int \mathbf{F}\cdot d\mathbf{r}$ *along the curve* $C : x^2 + y^2 = 1$, $z = 1$ *in the positive direction from* $(0, 1, 1)$ *to* $(1, 0, 1)$, *where*

$$\mathbf{F}(x, y, z) = (2x + yz)\hat{i} + xz\hat{j} + (xy + 2z)\hat{k}.$$

Solution. Let $A(0, 1, 1)$ and $B(1, 0, 1)$ be the points on the curve C:

$$x^2 + y^2 = 1, z = 1 \text{ and } \mathbf{r} = x\hat{i} + y\hat{j} + z\hat{k}.$$

$$\mathbf{F}\cdot d\mathbf{r} = [(2x + yz)\hat{i} + xz\hat{j} + (xy + 2z)\hat{k}]$$

$$.(dx\hat{i} + dy\hat{j} + dz\hat{k})$$

$$= (2x + yz)dx + xz\,dy + (xy + 2z)dz.$$

$$\therefore \quad \int_C \mathbf{F}\cdot d\mathbf{r}$$

$$= \int_A^B [(2x + yz)dx + xzdy + (xy + 2z)dz]$$

$$= \int_0^1 (2x + y)dx + \int_1^0 xdy$$

$$(\because z = 1 \Rightarrow dz = 0)$$

$$= \int_0^1 (2x + \sqrt{1-x^2})dx + \int_1^0 \sqrt{1-y^2}dy$$

$$(\because x^2 + y^2 = 1)$$

$$= \int_0^1 2xdx + \int_0^1 \sqrt{1-x^2})dx - \int_0^1 \sqrt{1-y^2}dy$$

$$\left(\because \int_a^b f(x)dx = -\int_b^a f(x)dx\right)$$

$$= \left[x^2\right]_0^1 = 1. \left(\because \int_a^b f(x)dx = \int_a^b f(t)dt\right)$$

Example 3. *Evaluate* $\int_C \mathbf{F}\cdot d\mathbf{r}$, *where* $\mathbf{F} = 3xy\hat{i} - y^2\hat{j}$ *and C is the curve* $y = 2x^2$ *with xy-plane from* $(0, 0)$ *to* $(1, 2)$.

Solution. The parametric equations of the given curve *i.e.*, the parabola $y = 2x^2$ can be taken as $x = t, y = 2t^2$. At point $(0, 0)$, $x = 0$ and so $t = 0$ and at the point $(1, 2)$, $x = 1$ and so $t = 1$.

Again $\frac{dx}{dt} = 1$ and $\frac{dy}{dt} = 4t$

and $d\mathbf{r} = dx\hat{i} + dy\hat{j}$ $\quad [\because \mathbf{r} = x\hat{i} + y\hat{j}]$

$$\therefore \quad \int_C \mathbf{F}\cdot d\mathbf{r} = \int_C (3xy\hat{i} - y^2\hat{j}).(dx\hat{i} + dy\hat{j})$$

$$= \int_C (3xydx - y^2dy,$$

$$= \int_{t=0}^1 \left(3xy\frac{dx}{dt} - y^2\frac{dy}{dt}\right)dt$$

$$= \int_{t=0}^1 (3.t.2t^2.1 - 4t^4.4t)dt$$

$$= \int_{t=0}^1 (6t^3 - 16t^5)dt$$

$$= \left[6.\frac{t^4}{4} - 16.\frac{t^6}{6}\right]_{t=0}^1$$

$$= \frac{6}{4} - \frac{16}{6} = \frac{3}{2} - \frac{8}{3} = -\frac{7}{6}.$$

Example 4. *Evaluate* $\int_C \mathbf{F}\cdot d\mathbf{r}$ *where* $\mathbf{F} = xy\hat{i} + yz\hat{j} + zx\hat{k}$ *and curve C is* $\mathbf{r} = t\hat{i} + t^2\hat{j} + t^3\hat{k}$ *and* $-1 \le t \le 1$.

Solution. Since $\mathbf{r} = t\hat{i} + t^2\hat{j} + t^3\hat{k}$ is given but

$\mathbf{r} = x\hat{i} + y\hat{j} + z\hat{k}$, then $x = t, y = t^2, z = t^3$

and $\frac{d\mathbf{r}}{dt} = \hat{i} + 2t\hat{j} + 3t^2\hat{k}$

$$\therefore \quad \int_C \mathbf{F}\cdot d\mathbf{r} = \int_{-1}^1 \left(\mathbf{F}\cdot\frac{d\mathbf{r}}{dt}\right)dt \quad \text{...(1)}$$

and $\mathbf{F} = xy\hat{i} + yz\hat{j} + zx\hat{k}$

$$= t^3\hat{i} + t^5\hat{j} + t^4\hat{k}$$

$$\therefore \quad \mathbf{F}.\frac{d\mathbf{r}}{dt} = (t^3\hat{i} + t^5\hat{j} + t^4\hat{k}).(\hat{i} + 2t\hat{j} + 3t^2\hat{k})$$

$$= t^3 + 2t^6 + 3t^6 = t^3 + 5t^6.$$

$\therefore$ From (1)

$$\int_C \mathbf{F}\cdot d\mathbf{r} = \int_{-1}^{1}(t^3 + 5t^6)dt$$

$$= \left[\frac{t^4}{4} + \frac{5t^7}{7}\right]_{-1}^{1}$$

$$= \frac{1}{4} - \frac{1}{4} + \frac{10}{7} = \frac{10}{7}.$$

Example 5. *Evaluate* $\int_C \mathbf{F}\cdot d\mathbf{r}$ *where* $\mathbf{F} = (x^2 + y^2)\hat{i} - 2xy\hat{j}$, *curve C is the rectangle in the xy-plane bounded by* $y = 0, x = a, y = b, x = 0$. (UPTU–2006)

Solution. The curve C is shown in fig. 7.

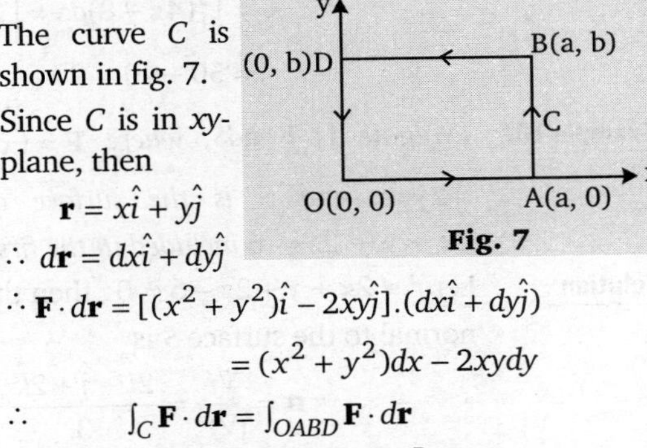

Fig. 7

Since C is in xy-plane, then

$$\mathbf{r} = x\hat{i} + y\hat{j}$$

$$\therefore d\mathbf{r} = dx\hat{i} + dy\hat{j}$$

$$\therefore \mathbf{F}\cdot d\mathbf{r} = [(x^2 + y^2)\hat{i} - 2xy\hat{j}].(dx\hat{i} + dy\hat{j})$$

$$= (x^2 + y^2)dx - 2xydy$$

$$\therefore \quad \int_C \mathbf{F}\cdot d\mathbf{r} = \int_{OABD} \mathbf{F}\cdot d\mathbf{r}$$

$$= \int_O^A \mathbf{F}\cdot d\mathbf{r} + \int_A^B \mathbf{F}\cdot d\mathbf{r} + \int_B^D \mathbf{F}\cdot d\mathbf{r} + \int_D^O \mathbf{F}\cdot d\mathbf{r} \quad ...(1)$$

Along the line OA, $y = 0$, x varies from 0 to a.

$$\therefore \quad dy = 0$$

and $$\int_O^A \mathbf{F}\cdot d\mathbf{r} = \int_O^A [(x^2 + y^2)dx - 2xydy]$$

$$= \int_0^a x^2 dx = \left[\frac{x^3}{3}\right]_0^a = \frac{a^3}{3}.$$

Along the line AB, $x = a$, y varies from 0 to b.

$$\therefore \quad dx = 0$$

and $$\int_A^B \mathbf{F}\cdot d\mathbf{r} = \int_A^B [(x^2 + y^2)dx - 2xydy]$$

$$= \int_0^b (-2ay)dy$$

$$= -2a\left[\frac{y^2}{2}\right]_0^b = -ab^2$$

Along the line BD, $y = b$, x varies from a to 0, then $dy = 0$

and $$\int_B^D \mathbf{F}\cdot d\mathbf{r} = \int_B^D [(x^2 + y^2)dx - 2xydy]$$

$$= \int_a^0 (x^2 + b^2)dx = \left[\frac{x^3}{3} + b^2 x\right]_a^0$$

$$= -\frac{a^3}{3} - ab^2.$$

And along the line DO, $x = 0$ and y varies from b to 0, then $dx = 0$

and $$\int_D^O \mathbf{F}\cdot d\mathbf{r} = \int_D^O [(x^2 + y^2)dx - 2xydy]$$

$$= \int_b^0 y^2.0 - 0.dy = 0$$

Substitute the values of these integral in (1), we get

$$\int_C \mathbf{F}\cdot d\mathbf{r} = \frac{a^3}{3} - ab^2 - \frac{a^3}{3} - ab^2 + 0$$

$$= -2ab^2$$

Example 6. *If* $\mathbf{F} = (3x^2 + 6y)\hat{i} - 14yz\hat{j} + 20xz^2\hat{k}$, *evaluate* $\int_C \mathbf{F}\cdot d\mathbf{r}$ *where C is a straight line joining (0, 0, 0) and (1, 1, 1).*

Solution. Since curve C is a straight line joining (0, 0, 0) and (1, 1, 1). Then

$$C: \frac{x}{1} = \frac{y}{1} = \frac{z}{1} = t \text{ (say)}$$

$$\therefore \quad x = t, y = t, z = t \text{ such that } 0 \le t \le 1$$

and $$\mathbf{r} = x\hat{i} + y\hat{j} + z\hat{k} = t\hat{i} + t\hat{j} + t\hat{k}$$

$$\therefore \quad d\mathbf{r} = (\hat{i} + \hat{j} + \hat{k})dt$$

and $$\mathbf{F} = (3t^2 + 6t)\hat{i} - 14t^2\hat{j} + 20t^3\hat{k}$$

$$\therefore \quad \mathbf{F}\cdot d\mathbf{r} = [(3t^2 + 6t)\hat{i} - 14t^2\hat{j} + 20t^3\hat{k}].(\hat{i} + \hat{j} + \hat{k})dt$$

$$= (3t^2 + 6t - 14t^2 + 20t^3)dt$$

$$= (20t^3 - 11t^2 + 6t)dt.$$

Now $$\int_C \mathbf{F}\cdot d\mathbf{r} = \int_0^1 \mathbf{F}\cdot d\mathbf{r} = \int_0^1 (20t^3 - 11t^2 + 6t)dt$$

$$= \left[20\frac{t^4}{4} - 11\frac{t^3}{3} + \frac{6t^2}{2}\right]_0^1 = \frac{13}{3}.$$

Example 7. *Evaluate* $\int(xdy - ydx)$ *around the circle* $x^2 + y^2 = 1$.

Solution. Let C denote the circle $x^2 + y^2 = 1$.

$\therefore$ $x = \cos t, y = \sin t$ be the paramatric equation of this circle.

To integrate around the circle C, we should vary t from $t = 0$ to $t = 2\pi$.

$$\oint_C (xdy - ydx) = \int_0^{2\pi}\left(x\frac{dy}{dt} - y\frac{dx}{dt}\right)dt$$

$$= \int_0^{2\pi}(\cos^2 t + \sin^2 t)dt$$

$$= \int_0^{2\pi} dt = 2\pi.$$

Example 8. *Evaluate* $\int_C \mathbf{F}\cdot d\mathbf{r}$ *where* $\mathbf{F} = xy\hat{i} + (x^2 + y^2)\hat{j}$ *and C is the x-axis from* $x = 2$ *to* $x = 4$ *and the line* $x = 4$ *from* $y = 0$ *to* $y = 12$.

Solution. Here the curve C consists of two lines, one of them is x-axis from $x = 2$ to $x = 4$ and other is the line $x=4$ from $y=0$ to $y = 12$. has shown in fig. 8m

Y
D(4, 12)
X
O
A(2, 0)
B(4, 0)
Fig. 8

Since $\mathbf{r} = x\hat{i} + y\hat{j}, z = 0$

$$d\mathbf{r} = dx\hat{i} + dy\hat{j}$$

and $\mathbf{F} = xy\hat{i} + (x^2 + y^2)\hat{j}$

$\therefore$ $\mathbf{F}\cdot d\mathbf{r} = [xy\hat{i} + (x^2 + y^2)\hat{j}].(dx\hat{i} + dy\hat{j})$

$= xydx + (x^2 + y^2)dy$.

Now $\int_C \mathbf{F}\cdot d\mathbf{r} = \int_{ABD}\mathbf{F}\cdot d\mathbf{r} = \int_A^B \mathbf{F}\cdot d\mathbf{r} + \int_B^D \mathbf{F}\cdot d\mathbf{r}$ ...(1)

Along the line AB, $y = 0$ and x varies from 2 to 4 and $dy = 0$

$$\therefore \int_A^B \mathbf{F}\cdot d\mathbf{r} = \int_A^B [xydx + (x^2 + y^2)dy]$$

$$= \int_2^4 0.dx = 0$$

Along the line BD, $x = 4$ and y varies from 0 to 12 and $dx = 0$

$$\therefore \int_B^D \mathbf{F}\cdot d\mathbf{r} = \int_B^D [xydx + (x^2 + y^2)dy]$$

$$= \int_0^{12}(16 + y^2)dy = \left[16y + \frac{y^3}{3}\right]_0^{12}$$

$$= 192 + 576 = 768.$$

Example 9. *Evaluate $\iint_S \mathbf{F}\cdot\mathbf{n}dS$, where $\mathbf{F} = z\hat{i} + x\hat{j} - 3y^2z\hat{k}$ and S is the surface of the cylinder $x^2 + y^2 = 16$ included in the first octant between $z = 0$ and $z = 5$.*

Solution . Since $S : x^2 + y^2 = 16$

Let $f \equiv x^2 + y^2 - 16$

then the vector normal to the surface S is the gradient of f and let $\mathbf{n}$ be the unit normal to S.

Then $\mathbf{n} = (\nabla f)/|\nabla f|$

$$= \frac{2x\hat{i} + 2y\hat{j}}{\sqrt{(4x^2 + 4y^2)}} = \frac{2x\hat{i} + 2y\hat{j}}{\sqrt{4(x^2 + y^2)}}$$

$$= \frac{2x\hat{i} + 2y\hat{j}}{\sqrt{4\times 16}} \quad (\because x^2 + y^2 = 16)$$

$$= \frac{2x\hat{i} + 2y\hat{j}}{8}$$

$$\therefore \quad \mathbf{n} = \frac{x\hat{i} + y\hat{j}}{4}.$$

Here the surface S is perpendicular to xy-plane so we will take the projection of S on zx-plane. Let R be that projection

$$\iint_S \mathbf{F}\cdot\mathbf{n}dS = \iint_R \mathbf{F}\cdot\mathbf{n}\frac{dxdz}{|\mathbf{n}\cdot\hat{j}|} \quad ...(1)$$

Now $\mathbf{F}\cdot\mathbf{n} = (z\hat{i} + x\hat{j} - 3y^2z\hat{k}).\left(\dfrac{x\hat{i} + y\hat{j}}{4}\right)$

$$= \frac{zx + xy}{4}$$

and $|\mathbf{n}\cdot\hat{j}| = \left|\left(\dfrac{x\hat{i} + y\hat{j}}{4}\right).\hat{j}\right| = \dfrac{y}{4}$

From (1), we get

$$\iint_S \mathbf{F}.\mathbf{n}dS = \iint_R \frac{zx + xy}{4}.\frac{4}{y}dxdz$$

$$= \iint_R \frac{(xz + xy)}{y}dxdz.$$

Since z varies from 0 to 5 and $y = \sqrt{16 - x^2}$ on S.

$$\therefore \iint_R \frac{(xz + xy)}{y}dxdz$$

$$= \int_{z=0}^5 \int_{x=0}^4 \left[\frac{xz}{\sqrt{16 - x^2}} + x\right]dxdz$$

$$= \int_0^5 \left[\frac{x^2}{2} - z\sqrt{16 - x^2}\right]_0^4 dz$$

$$= \int_0^5 (4z + 8)dz = \left[2z^2 + 8z\right]_0^5$$

$$= 50 + 40 = 90.$$

Example 10. *Evaluate $\iint_S \mathbf{F}\cdot\mathbf{n}dS$, where $\mathbf{F} = (x + y^2)\hat{i} - 2x\hat{j} + 2yz\hat{k}$ and S is the surface of the plane $2x + y + 2z = 6$ included in the first octant.*

Solution . Let $f \equiv 2x + y + 2z - 6 = 0$, then the unit vector normal to the surface S is

$$\mathbf{n} = \frac{\nabla f}{|\nabla f|} = \frac{2\hat{i} + \hat{j} + 2\hat{k}}{3}$$

Let R be the projection of S on the xy-plane bounded by the x-axis, y-axis and the line $2x + y = 6$, $z = 0$.

$$\therefore \iint_S \mathbf{F}\cdot\mathbf{n}dS = \iint_R \mathbf{F}\cdot\mathbf{n}\frac{dxdy}{|\mathbf{n}\cdot\hat{k}|} \quad ...(1)$$

Now $\mathbf{F}\cdot\mathbf{n} = [(x + y^2)\hat{i} - 2x\hat{j} + 2yz\hat{k}] . \dfrac{2\hat{i} + \hat{j} + 2\hat{k}}{3}$

$$= \frac{1}{3}(2x + 2y^2 - 2x + 4yz)$$

$$= \frac{1}{3}(2y^2 + 4yz) = \frac{2}{3}(y^2 + 2yz)$$

and $|\mathbf{n}\cdot\hat{k}| = \left|\dfrac{2\hat{i} + \hat{j} + 2\hat{k}}{3}.\hat{k}\right| = \dfrac{2}{3}$

Then from (1), we have

$$\iint_S \mathbf{F}\cdot\mathbf{n}dS = \iint_R \frac{2}{3}(y^2 + 2yz).\frac{dxdy}{2/3}$$

$$= \iint_R (y^2 + 2yz)dxdy$$

where R : x-axis, y-axis and $2x + y = 6$, $z = 0$

$$= \iint_R \left[y^2 + 2y\left(\frac{6 - 2x - y}{2}\right)\right]dxdy$$

$$(\because 2x + y + 2z = 6 \text{ on } S)$$

$$= \iint_R (y^2 + 6y - 2xy - y^2)dxdy$$

$$= 2\iint_R y(3 - x)dxdy$$

$$= 2\int_{y=0}^6 \int_{x=0}^{\frac{6-y}{2}} y(3 - x)dxdy$$

$$= 2\int_0^6 y\left[3x - \frac{x^2}{2}\right]_0^{\frac{6-y}{2}} dy$$

$$= 2\int_0^6 y\left[\frac{3}{2}(6-y) - \frac{1}{8}(6-y)^2\right]dy$$

$$= \int_0^6\left(9y - \frac{y^3}{4}\right)dy = \left[\frac{9y^2}{2} - \frac{y^4}{16}\right]_0^6$$

$$= (162 - 81) = 81$$

$\therefore \quad \iint_S \mathbf{F}.\mathbf{n}dS = 81.$

Example 11. *Evaluate $\iiint_V \phi dV$, where $\phi = 45x^2y$ and V is the closed region bounded by the planes $4x + 2y + z = 8$, $x = 0$, $y = 0$, $z = 0$.*

Solution. Since $\phi = 45x^2y$ and $dV = dxdydz$

$$\therefore \iiint_V \phi dV = \iiint_V 45x^2 ydxdydz$$

$$= \int_{x=0}^{2}\int_{y=0}^{4-2x}\int_{z=0}^{8-4x-2y} 45x^2 ydxdydz$$

$$= 45\int_{x=0}^{2}\int_{y=0}^{4-2x} x^2y[z]_0^{8-4x-2y} dxdy$$

$$= 45\int_{x=0}^{2} x^2\int_{y=0}^{4-2x} y(8-4x-2y)dxdy$$

$$= 45\int_{x=0}^{2} x^2\left[4y^2 - 2xy^2 - \frac{2}{3}y^3\right]_0^{4-2x} dx$$

$$= 45\int_{x=0}^{2} x^2[4(4-2x)^2 - 2x(4-2x)^2 - \frac{2}{3}(4-2x)^3]dx$$

$$= \frac{45}{3}\int_{x=0}^{2} x^2(4-2x)^3 dx$$

$$= 15\int_{x=0}^{2} x^2(64 - 8x^3 - 96x + 48x^2)dx$$

$$= 15\left[\frac{64x^3}{3} - \frac{8x^6}{6} - \frac{96x^4}{4} + \frac{48x^5}{5}\right]_0^2$$

$$= 15\left[\frac{512}{3} - \frac{256}{3} - 384 + \frac{1536}{5}\right] = 128$$

Example 12. *Evaluate $\int_C \mathbf{F}\cdot d\mathbf{r}$ where $\mathbf{F} = (x^2 - y^2)\hat{i} + xy\hat{j}$ and curve C is the arc of the curve $y = x^3$ from (0, 0) to (2, 8).*

Solution. Taking the given arc of the following form $x = t$ and $y = t^3$ and takes the value from $t = 0$ to $t = 2$ (*i.e.*, from (0, 0) to (2, 8)).

$\therefore \quad \mathbf{r}(t) = x\hat{i} + y\hat{j} = t\hat{i} + t^3\hat{j}$ so $\frac{d\mathbf{r}}{dt} = \hat{i} + 3t^2\hat{j}$.

Also, $\mathbf{F} = (x^2 - y^2)\hat{i} + xy\hat{j} = (t^2 - t^6)\hat{i} + t^4\hat{j}$

$$\therefore \int_C \mathbf{F}\cdot d\mathbf{r} = \int_C\left(\mathbf{F}\cdot\frac{d\mathbf{r}}{dt}\right)dt$$

$$= \int_0^2[(t^2 - t^6)\hat{i} + t^4\hat{j}].[\hat{i} + 3t^2\hat{j}]dt$$

$$= \int_0^2(t^2 - t^6 + 3t^6)dt = \int_0^2(t^2 + 2t^6)dt$$

$$= \left[\frac{t^3}{3} + \frac{2}{7}t^7\right]_0^2 = \frac{8}{3} + \frac{256}{7} = \frac{824}{21}.$$

Example 13. *Find the workdone when a force*

$$\mathbf{F} = (x^2 - y^2 + x)\hat{i} - (2xy + y)\hat{j}$$

moves a particle in xy-plane from (0, 0) to (1, 1) along the parabola $y^2 = x$.

Solution. Let C be the given arc of the parabola $y^2 = x$ measured form (0, 0) to (1, 1).

Taking the given arc in the following form :

$$x = t^2, y = t$$

and t takes from $t = 0$ to $t = 1$

so $\quad \mathbf{r}(t) = x\hat{i} + y\hat{j} = t^2\hat{i} + t\hat{j}$

$\therefore \quad \frac{d\mathbf{r}}{dt} = 2t\hat{i} + \hat{j}$

Also $\quad \mathbf{F} = (x^2 - y^2 + x)\hat{i} - (2xy + y)\hat{j}$

$$= (t^4 - t^2 + t^2)\hat{i} - (2t^3 + t)\hat{j}$$

$$= t^4\hat{i} - (2t^3 + t)\hat{j}$$

Now the required workdone

$$= \int_C \mathbf{F}.d\mathbf{r} = \int_C\left(\mathbf{F}.\frac{d\mathbf{r}}{dt}\right)dt$$

$$= \int_0^1[t^4\hat{i} - (2t^3 + t)\hat{j}].[2t\hat{i} + \hat{j}]dt$$

$$= \int_0^1(2t^5 - 2t^3 - t)dt$$

$$= \left[\frac{2t^6}{6} - \frac{2t^4}{4} - \frac{t^2}{2}\right]_0^1$$

$$= \frac{2}{6} - \frac{2}{4} - \frac{1}{2} = -\frac{2}{3}\text{ unit.}$$

Example 14. *Evaluate $\int_C \mathbf{F}\cdot d\mathbf{r}$ where $\mathbf{F} = xy\hat{i} + (x^2 + y^2)\hat{j}$ and curve C is the arc of $y = x^2 - 4$ from (2, 0) to (4, 12).*

Solution. Here $\quad \mathbf{F} = xy\hat{i} + (x^2 + y^2)\hat{j}$

Since $\mathbf{r} = x\hat{i} + y\hat{j}$ so $d\mathbf{r} = dx\hat{i} + dy\hat{j}$

$\therefore \quad \mathbf{F}\cdot d\mathbf{r} = xydx + (x^2 + y^2)dy$

Now x takes the values on $y = x^2 - 4$ from $x = 2$ at (2, 0) to $x = 4$ at (4, 12) and y takes $y = 0$ at (2, 0) and $y = 12$ at (4, 12).

Then we have

$$\int_C \mathbf{F}\cdot d\mathbf{r} = \int_C xydx + (x^2 + y^2)dy$$

$$= \int_{x=2}^{4} x(x^2 - 4)dx + \int_{y=0}^{12}(y + 4 + y^2)dy$$

$$(\because y = x^2 - 4)$$

$$= \left[\frac{x^4}{4} - \frac{4x^2}{2}\right]_2^4 + \left[\frac{y^2}{2} + 4y + \frac{y^3}{3}\right]_0^{12}$$

$$= \left(\frac{256}{4} - \frac{64}{2}\right) - \left(\frac{16}{4} - \frac{16}{2}\right) + \left(\frac{144}{2} + 48 + \frac{1728}{3}\right)$$

$$= 732.$$

Example 15. *Evaluate $\int_C \mathbf{F}\cdot d\mathbf{r}$ where $\mathbf{F} = z\hat{i} + x\hat{j} + y\hat{k}$ and C is the arc of the curve*

$$\mathbf{r} = \cos t\hat{i} + \sin t\hat{j} + t\hat{k} \text{ from } t = 0 \text{ to } t = 2\pi.$$

Solution . Here

$\therefore \quad \mathbf{r}(t) = \cos t\hat{i} + \sin t\hat{j} + t\hat{k}$

$\therefore \quad x = \cos t, y = \sin t, z = t$

$\therefore \quad \mathbf{F} = t\hat{i} + \cos t\hat{j} + \sin t\hat{k}$

$\therefore \quad d\mathbf{r} = (-\sin t\hat{i} + \cos t\hat{j} + \hat{k})dt$

$\therefore \quad \mathbf{F}\cdot d\mathbf{r} = (-t\sin t + \cos^2 t + \sin t)dt$

Now $\int_C \mathbf{F}\cdot d\mathbf{r} = \int_0^{2\pi}(-t\sin t + \cos^2 t + \sin t)dt$

$$= -\int_0^{2\pi} t\sin t dt + \int_0^{2\pi}\cos^2 t dt + \int_0^{2\pi}\sin t dt$$

$$= -\left[-t\cos t + \sin t\right]_0^{2\pi} + \int_0^{2\pi}\frac{1}{2}(1+\cos 2t)dt + \left[-\cos t\right]_0^{2\pi}$$

$$= -\left[-2\pi + \sin 2\pi\right] + \frac{1}{2}\left[t + \frac{1}{2}\sin 2t\right]_0^{2\pi} + \left[-\cos 2\pi + \cos 0\right]$$

$$= 2\pi + \frac{1}{2}[2\pi + 0] + [-1+1] = 3\pi.$$

Example 16. *Evaluate $\int_C \mathbf{F}\cdot d\mathbf{r}$, where $\mathbf{F} = x^2y^2\hat{i} + y\hat{j}$ and the curve C is $y^2 = 4x$ in xy-plane from (0, 0) to (4, 4).*

Solution . Here $\quad \mathbf{F} = x^2y^2\hat{i} + y\hat{j}$

Then $\quad \mathbf{F}\cdot d\mathbf{r} = x^2y^2dx + ydy \quad [\because d\mathbf{r} = dx\hat{i} + dy\hat{j}]$

$\therefore \quad \int_C \mathbf{F}\cdot d\mathbf{r} = \int_C (x^2y^2dx + ydy)$

$$= \int_0^4 x^2y^2dx + \int_0^4 ydy$$

$$= \int_0^4 x^2(4x)dx + \int_0^4 ydy \quad [\because y^2 = 4x]$$

$$= 4\left[\frac{x^4}{4}\right]_0^4 + \left[\frac{y^2}{2}\right]_0^4 = 256 + \frac{16}{2} = 264$$

Example 17. *Show that the vector field* $\mathbf{F}$ *defined by*

$$\mathbf{F} = (\sin y + z)\hat{i} + (x\cos y - z)\hat{j} + (x-y)\hat{k}$$

is conservative and find a function ϕ *such that* $\mathbf{F} = \nabla\phi$.

Solution . We have

$$\nabla\times\mathbf{F} = \begin{vmatrix} \hat{i} & \hat{j} & \hat{k} \\ \frac{\partial}{\partial x} & \frac{\partial}{\partial y} & \frac{\partial}{\partial z} \\ \sin y + z & x\cos y - z & x - y \end{vmatrix}$$

$$= \hat{i}\left[\frac{\partial}{\partial y}(x-y) - \frac{\partial}{\partial z}(x\cos y - z)\right] + \hat{j}\left[\frac{\partial}{\partial z}(\sin y + z) - \frac{\partial}{\partial x}(x-y)\right] + \hat{k}\left[-\frac{\partial}{\partial y}(\sin y + z) + \frac{\partial}{\partial x}(x\cos y - z)\right]$$

$$= \hat{i}[-1+1] + \hat{j}[1-1] + \hat{k}[-\cos y + \cos y]$$

$$= \mathbf{0}$$

Hence, the field $\mathbf{F}$ is conservative.

Since $\mathbf{F} = \nabla\phi = \frac{\partial\phi}{\partial x}\hat{i} + \frac{\partial\phi}{\partial y}\hat{j} + \frac{\partial\phi}{\partial z}\hat{k}$

$\Rightarrow F_x = \frac{\partial\phi}{\partial x}, F_y = \frac{\partial\phi}{\partial y}, F_z = \frac{\partial\phi}{\partial z}$ if $\mathbf{F} = (F_x, F_y, F_z)$

$\therefore F_x dx + F_y dy + F_z dz = \frac{\partial\phi}{\partial x}dx + \frac{\partial\phi}{\partial y}dy + \frac{\partial\phi}{\partial z}dz$

$$= d\phi$$

$\Rightarrow \quad d\phi = (\sin y + z)dx + (x\cos y - z)dy + (x-y)dz$

$\Rightarrow \quad d\phi = \sin y dx + x\cos y dy + zdx + xdz - (zdy - ydz)$

$\Rightarrow \quad d\phi = d(x\sin y) + d(zx) - d(yz)$

Integrating, we get

$$\phi = x\sin y + zx - yz + c, c \text{ is a constant}$$

Example 18. *Let $F(x, y, z) = x^3\hat{i} + y\hat{j} + z\hat{k}$ is the force field. Find the workdone* $\mathbf{F}$ *along the line from (1, 2, 3) to (3, 5, 7).* (UPTU–2006)

Solution . Equation of the line from (1, 2, 3) to (3, 5, 7) is given by

$$\frac{x-1}{3-1} = \frac{y-2}{5-2} = \frac{z-3}{7-3} = k \text{ (say)}$$

$\Rightarrow \quad x = 2k+1, y = 3k+2, z = 4k+3$

Now $\quad \mathbf{r} = x\hat{i} + y\hat{j} + z\hat{k}$

$$= (2k+1)\hat{i} + (3k+2)\hat{j} + (4k+3)\hat{k}$$

$\Rightarrow \quad d\mathbf{r} = (2\hat{i} + 3\hat{j} + 4\hat{k})dk$

Also at (1, 2, 3), $k = 0$ and at (3, 5, 7), $k = 1$

$\therefore$ Required workdone

$$= \int_C \mathbf{F}\cdot d\mathbf{r}$$

$$= \int_0^1 [(2k+1)^3\hat{i} + (3k+2)\hat{j} + (4k+3)\hat{k}].(2\hat{i} + 3\hat{j} + 4\hat{k})dk$$

$$= \int_0^1 [2(2k+1)^3 + 3(3k+2) + 4(4k+3)]dk$$

$$= \int_0^1 (16k^3 + 24k^2 + 37k + 20)dk$$

$$= \left(4k^4 + 8k^3 + \frac{37}{2}k^2 + 20k\right)_0^1$$

$$= \frac{101}{2}.$$

Example 19. *Find the workdone by the force*

$$\mathbf{F} = (2y+3)\hat{i} + xz\hat{j} + (yz - x)\hat{k}$$

when it moves a particle from the point (0, 0, 0) to the point (2, 1, 1) along the curve $x = 2t^2, y = t$ and $z = t^3$. (UPTU–2011, Madras–2010)

Solution . We have $\mathbf{F} = (2y+3)\hat{i} + xz\hat{j} + (yz - x)\hat{k}$

and $\mathbf{r} = x\hat{i} + y\hat{j} + z\hat{k} \quad \Rightarrow \quad d\mathbf{r} = dx\hat{i} + dy\hat{j} + dz\hat{k}$

$\Rightarrow \quad \mathbf{F}\cdot d\mathbf{r} = (2y+3)dx + xzdy + (yz-x)dz$

$\therefore$ Workdone $= \int_C \mathbf{F}\cdot d\mathbf{r}$

$= \int_C (2y+3)dx + xzdy + (yz-x)dz \quad \ldots(1)$

Along C, $x = 2t^2, y = t, z = t^3$

$\Rightarrow \quad dx = 4tdt, dy = dt$ and $dz = 3t^2dt$

Also t varies from 0 to 1

$$\therefore \quad W = \int_0^1 (2t+3)4tdt + \int_0^1 2t^2.t^3dt + \int_0^1 (tt^3 - 2t^2)3t^2dt$$

$$= \int_0^1 (8t^2 + 12t + 2t^5 + 3t^6 - 6t^4)dt$$

$$= \left[\frac{8t^3}{3} + \frac{12t^2}{2} + \frac{2t^6}{6} + \frac{3t^7}{7} - \frac{6t^5}{5}\right]_0^1$$

$$= \frac{8}{3} + \frac{12}{2} + \frac{2}{6} + \frac{3}{7} - \frac{6}{5} = \frac{288}{35}.$$

Example 20. *A vector field is given by* $\mathbf{F} = \sin y\hat{i} + x(1+\cos y)\hat{j}$ *Evaluate the line integral over a circular path given by* $x^2 + y^2 = a^2$. (PTU–2003, Rohtak–2006)

Solution. Let $\mathbf{r} = x\hat{i} + y\hat{j} + z\hat{k}$

Since $z = 0$ (given)

$\therefore \quad \mathbf{r} = x\hat{i} + y\hat{j} \Rightarrow d\mathbf{r} = dx\hat{i} + dy\hat{j}$

Also, the circular path is $x = a\cos t, y = a\sin t$, $z = 0$ where t varies from 0 to 2π.

$$\therefore \int_C \mathbf{F}.d\mathbf{r} = \int_C (\sin y\hat{i} + x(1+\cos y)\hat{j}).(dx\hat{i} + dy\hat{j})$$

$$= \int_C [\sin y dx + x(1+\cos y)dy]$$

$$= \int_C [\sin y dx + x\cos y dy + xdy]$$

$$= \int_C [d(x\sin y) + xdy]$$

$$= \int_0^{2\pi} [d(a\cos t\sin(a\sin t))] + a^2\cos^2 t dt$$

$$= |a\cos t\sin(a\sin t)|_0^{2\pi} + \frac{a^2}{2}\int_0^{2\pi}(1+\cos 2t)dt$$

$$= \frac{a^2}{2}\left|t + \frac{\sin 2t}{2}\right|_0^{2\pi} = \pi a^2.$$

Example 21. *Find the workdone in moving a particle in a force field* $\mathbf{F} = 3x^2\hat{i} + (2xy - y)\hat{j} + 3\hat{k}$ *along the curve* $x^2 = 4y$, $3x^2 = 8z$ *from* $x = 0$ *to* $x = 2$. (MTU–2012)

Solution. Required work done, $W = \int_C \mathbf{F}.d\mathbf{r}$

$$= \int_C [3x^2\hat{i} + (2xy - y)\hat{j} + 3\hat{k}].(dx\hat{i} + dy\hat{j} + dz\hat{k})$$

$$= \int_C (3x^2dx + (2xy - y))dy + 3dz \quad \ldots(1)$$

Put $x = t$, in $x^2 = 4y$ and $3x^2 = 8z$. The parametric equation of C are $x = t, y = t^2/4, z = (3/8)\,t^2$ (t varies from 0 to 2).

Then from (1)

$$W = \int_0^2 3t^2dt + \left[2t\left(\frac{t^2}{4}\right) - \frac{t^2}{4}\right]\frac{2t}{4}dt + 3\left(\frac{6t}{4}\right)dt$$

$$= \int_0^2 \left(3t^2 + \frac{t^4}{4} - \frac{t^3}{8} + \frac{9t}{4}\right)dt$$

$$= \left[t^3 + \frac{t^5}{20} - \frac{t^4}{32} + \frac{9t^2}{8}\right]_0^2$$

$$= \left[8 + \frac{32}{20} - \frac{16}{32} + \frac{9}{2}\right] = \frac{136}{10} = 13.6.$$

Example 22. *Find the total workdone in moving a particle in a force field given by* $\mathbf{F} = 3xy\hat{i} - 5z\hat{j} + 10x\hat{k}$ *along the curve* $x = t^2 + 1, y = 2t^2, z = t^3$ *from* $t = 1$ *to* 2. (MTU-2013)

Solution. Required workdone $W = \int_C \mathbf{F}.d\mathbf{r}$

$$= \int_C [3xy\hat{i} - 5z\hat{j} + 10x\hat{k}].(dx\hat{i} + dy\hat{j} + dz\hat{k})$$

$$= \int_C (3xydx - 5zdy + 10xdz)$$

$$= \int_1^2 (3xy\frac{dx}{dt} - 5z\frac{dy}{dt} + 10x\frac{dz}{dt})dt$$

$$= \int_1^2 [3(t^2+1)(2t^2)2t - (5t^3)(4t) + 10(t^2+1)(3t^2)]dt$$

$$= \int_1^2 (12t^5 + 12t^3 - 20t^4 + 30t^4 + 30t^2)dt$$

$$= \int_1^2 (12t^5 + 10t^4 + 12t^3 + 30t^2)dt$$

$$= \left[\frac{12t^6}{6} + \frac{10t^5}{5} + \frac{12t^4}{4} + \frac{30t^3}{3}\right]_1^2$$

$= 303.$

Example 23. *If* $\mathbf{F} = (2y+3)\hat{i} + xz\hat{j} + (yz - x)\hat{k}$, *evaluate* $\int_C \mathbf{F}\cdot d\mathbf{r}$ *where C is the path consisting of the straight lines from* (0, 0, 0) *to* (0, 0, 1) *then to* (0, 1, 1) *and then to* (2, 1, 1). (GBTU–2011)

Solution. We have

$$\mathbf{F}\cdot d\mathbf{r} = [(2y+3)\hat{i} + xz\hat{j} + (yz-x)\hat{k}].(dx\hat{i} + dy\hat{j} + dz\hat{k})$$

$$= (2y+3)dx + xzdy + (yz-x)dz$$

Let C_1 denote the straight line joining (0, 0, 0) to (0, 0, 1), C_2 denote the straight line joining (0, 0, 1) to (0, 1, 1) and C_3 denote the straight line joining (0, 1, 1) to (2, 1, 1).

Along C_1, $x = 0, y = 0 \Rightarrow dx = 0, dy = 0$ and z varies from 0 to 1.

Along C_2, $x = 0, z = 1 \Rightarrow dx = 0, dz = 0$ and y varies from 0 to 1.

Along C_3, $y = 1$, $z = 1 \Rightarrow dy = 0$, $dz = 0$ and x varies from 0 to 2.

Therefore,

$$\int_C \mathbf{F}\cdot d\mathbf{r} = \int_{C_1} \mathbf{F}\cdot d\mathbf{r} + \int_{C_2} \mathbf{F}\cdot d\mathbf{r} + \int_{C_3} \mathbf{F}\cdot d\mathbf{r}$$

$$= \int_{z=0}^{1}(0.z-0)dz + \int_{y=0}^{1}(0.1)dy + \int_{x=0}^{2}(2.1+3)dx$$

$$= 0+0+5[x]_0^2 = 10.$$

EXERCISE 23.1

1. Evaluate $\int_C \mathbf{F}\cdot d\mathbf{r}$, where $\mathbf{F} = x^2\hat{i} + y^3\hat{j}$ and curve C is the arc of the parabola $y = x^2$ in the xy-plane from (0, 0) to (1, 1).
2. If $\mathbf{F} = (3x^2+6y)\hat{i} - 14yz\hat{j} + 20xz^2\hat{k}$, then evaluate $\int_C \mathbf{F}\cdot d\mathbf{r}$ from (0, 0, 0) to (1, 1, 1) along the curve C
$x = t, y = t^2, z = t^3$. (GBTU–2010, VTU–2001)
3. If $\mathbf{F} = y\hat{i} - x\hat{j}$, evaluate $\int_C \mathbf{F}\cdot d\mathbf{r}$ from (0, 0) to (1, 1) along the following paths C :
   (i) The parabola $y = x^2$
   (ii) The straight lines form (0, 0) to (1, 0) and then to (1, 1)
   (iii) The straight line joining (0, 0) and (1, 1).
4. Find the workdone in moving a particle in a force field
$$\mathbf{F} = 3x^2\hat{i} + (2xz-y)\hat{j} + z\hat{k}$$
along the line joining (0, 0, 0) to (2, 1, 3).
(MTU–2012, Delhi–2002, JNTU–2002, SVTU–2007)
[**Hint :** Workdone= $\int_C \mathbf{F}\cdot d\mathbf{r}$]
5. Evaluate $\int_C \mathbf{F}\cdot d\mathbf{r}$,where $\mathbf{F} = yz\hat{i} + zx\hat{j} + xy\hat{k}$ and the curve C is the position of the curve $\mathbf{r} = a\cos t\hat{i} + b\sin t\hat{j} + ct\hat{k}$ from $t = 0$ to $t = \pi/2$.
6. Evaluate the integral
$$\int_C [(2xy^3 - y^2\cos x)dx + (1-2y\sin x + 3x^2y^2)dy],$$
where C is the arc of the parabola $2x = \pi y^2$ from (0, 0) to $\left(\frac{\pi}{2}, 1\right)$.
7. Evaluate $\int_C x^{-1}(y+z)ds$ where C is the arc of the circle $x^2 + y^2 = 4$, $z = 0$ from (2, 0, 0) to $(\sqrt{2}, \sqrt{2}, 0)$.
8. If $\mathbf{F} = (2x+y)\hat{i} + (3y-x)\hat{j}$, evaluate $\int_C \mathbf{F}\cdot d\mathbf{r}$ where C is the curve in the xy-plane consisting of the straight lines from (0, 0) to (2, 0) and then to (3, 2).
9. Evaluate $\iint_S \mathbf{F}\cdot\mathbf{n}dS$, where $\mathbf{F} = yz\hat{i} + zx\hat{j} + xy\hat{k}$ and S is that part of the surface of the sphere $x^2 + y^2 + z^2 = 1$ which lies in the first octant.
10. Evaluate $\iint_S \mathbf{F}\cdot\mathbf{n}dS$, where $\mathbf{F} = xy\hat{i} - x^2\hat{j} + (x+z)\hat{k}$, S is the portion of the plane $2x + 2y + z = 6$ included in the first octant.
11. Evaluate $\iint_S \mathbf{F}\cdot\mathbf{n}dS$, where $\mathbf{F} = y\hat{i} + 2x\hat{j} - z\hat{k}$ and S is the surface of the plane $2x + y = 6$ in the first octant cut off by the plane $z = 4$.
12. If $\mathbf{F} = 2y\hat{i} - 3\hat{j} + x^2\hat{k}$ and S is the surface of the parabolic cylinder $y^2 = 8x$ in the first octant bounded by the planes $y = 4$ and $z = 6$, then evaluate $\iint_S \mathbf{F}\cdot\mathbf{n}dS$.
13. If $\mathbf{F} = (2x^2-3z)\hat{i} - 2xy\hat{j} - 4x\hat{k}$, then evaluate $\iiint_V \nabla\cdot\mathbf{F}dV$ where V is the closed region bounded by the planes $x = 0$, $y = 0$, $z = 0$ and $2x + 2y + z = 4$.
14. Evaluate $\int_C \mathbf{F}\cdot d\mathbf{r}$, where $\mathbf{F} = xy\hat{i} + yz\hat{j} + zx\hat{k}$ and C is the arc of the curve $\mathbf{r} = (a\cos\theta)\hat{i} + (a\sin\theta)\hat{j} + a\theta\hat{k}$ from $\theta = 0$ to $\theta = \frac{\pi}{2}$.
15. If $\mathbf{F} = yz\hat{i} + zx\hat{j} - xy\hat{k}$, find $\int_C \mathbf{F}\cdot d\mathbf{r}$ where C is given by $x = t$, $y = t^2$, $z = t^3$ from $P(0, 0, 0)$ to $Q(2, 4, 8)$.
16. Evaluate $\int_C \mathbf{F}\cdot d\mathbf{r}$, where $\mathbf{F} = (2x+y)\hat{i} + (3y-x)\hat{j} + yz\hat{k}$ and C is the curve $x = 2t^2$, $y = t$, $z = t^3$ from $t = 0$ to $t = 1$.
17. Find the circulation of $\mathbf{F}$ round the curve C where $\mathbf{F} = y\hat{i} + z\hat{j} + x\hat{k}$ and C is the circle $x^2 + y^2 = 1$, $z = 0$.
18. Show that $\int_C \left[-\frac{y}{x^2+y^2}\hat{i} + \frac{x}{x^2+y^2}\hat{j}\right].d\mathbf{r} = 2\pi$ where C is the circle $x^2 + y^2 = 1$ in the xy-plane described in counter-clockwise sense.
19. Evaluate $\int_C \mathbf{F}\cdot d\mathbf{r}$, where
$$\mathbf{F} = c[-3a\sin^2 t\cos t\hat{i} + a(2\sin t - 3\sin^3 t)\hat{j} + b\sin 2t\hat{k}]$$
and C is given by $\mathbf{r} = a\cos t\hat{i} + a\sin t\hat{j} + bt\hat{k}$ from $t = \pi/4$ to $\pi/2$.
20. If $\mathbf{F} = (x^2+y^3)\hat{i} + (x^3-y^2)\hat{j}$, evaluate line integral $\int \mathbf{F}\cdot d\mathbf{r}$ along the path $y^2 = x$, joining (0, 0) to (1, 0).
21. If $\mathbf{A} = (x-y)\hat{i} + (x+y)\hat{j}$ show that around the curve C consistintg of $y = x^2$ and $y^2 = x$, $\int_C \mathbf{A}\cdot d\mathbf{r} = \frac{2}{3}$. (GBTU–2012)
22. If $\mathbf{F} = e^{xyz}(yz\hat{i} + zx\hat{j} + xy\hat{k})$ and $\mathbf{r} = x\hat{i} + y\hat{j} + z\hat{k}$ and C is the boundary of $0 \le x \le 1$, $0 \le y \le 1$ and $z = 1$ clockwise then show that $\int_C \mathbf{F}\cdot d\mathbf{r} = 0$. (UPTU–2008)
23. Show that the surface area of the plane $x + 2y + 2z = 12$ cut off by $x = 0$, $y = 0$ and $x^2 + y^2 = 16$ is given by 6π square units. (GBTU–2012)
24. Show that the integral $\int_C (x^2+xy)dx + (x^2+y^2)dy$ where C is the square formed by the lines $y = \pm 1$, and $x = \pm 1$ is equal to zero. (Delhi–2002)
25. If $\mathbf{F} = (5xy-6x^2)\hat{i} + (2y-4x)\hat{j}$, show that $\int_C \mathbf{F}\cdot d\mathbf{r} = 35$ along the curve C in the xy plane, $y=x^3$ from the point (1, 1) to (2, 8). (JNTU–2006)
26. Show that the total workdone by the force $\mathbf{F} = 3xy\hat{i} - y\hat{j} + 2zx\hat{k}$ in moving a particle around the circle $x^2 + y^2 = 4$ is 0. (VTU–2010)

Hint to Selected Problems

1. Let $x = t, y = t^2$ for $0 \le t \le 1$

$\mathbf{F} = x^2\hat{i} + y^3\hat{j} = t^2\hat{i} + t^6\hat{j}$ and $d\mathbf{r} = dx\hat{i} + dy\hat{j} = (\hat{i} + 2t\hat{j})dt$

$\therefore \quad \mathbf{F}\cdot d\mathbf{r} = (t^2 + 2t^7)dt$

$\therefore \quad \int_C \mathbf{F}\cdot d\mathbf{r} = \int_0^1 (t^2 + 2t^7)dt = \left[\frac{t^3}{3} + \frac{2t^8}{8}\right]_0^1 = \frac{7}{12}.$

3. (i) $C : y = x^2$ so $x = t, y = t^2$ and $0 \le t \le 1$

$\mathbf{F} = y\hat{i} - x\hat{j} = t^2\hat{i} - t\hat{j}$ and $d\mathbf{r} = dx\hat{i} + dy\hat{j} = (\hat{i} + 2t\hat{j})dt$

$\therefore \quad \mathbf{F}\cdot d\mathbf{r} = -t^2 dt$

$\therefore \quad \int_C \mathbf{F}\cdot d\mathbf{r} = -\int_0^1 t^2 dt = -\left[\frac{t^3}{3}\right]_0^1 = -\frac{1}{3}.$

(ii) C : The straight line from (0, 0) to (1, 0) and to (1, 1).
Let OA and OB be the lines where $O(0, 0), A(1, 0)$ and $B(1, 1)$

$\int_C \mathbf{F}\cdot d\mathbf{r} = \int_{OA} \mathbf{F}\cdot d\mathbf{r} + \int_{AB} \mathbf{F}\cdot d\mathbf{r}$

Along OA, $y = 0, x = 1$, then $dy = 0$

$\mathbf{F}\cdot d\mathbf{r} = ydx - xdy$

$\therefore \quad \int_{OA} \mathbf{F}\cdot d\mathbf{r} = \int_0^1 0dx = 0$

Along AB, $x = 1, 0 \le y \le 1$, then $dx = 0$

$\therefore \quad \int_{AB} \mathbf{F}\cdot d\mathbf{r} = \int_0^1 (-1)dy = (-y)_0^1 = -1.$

Hence $\int_C \mathbf{F}\cdot d\mathbf{r} = 0 - 1 = -1.$

(iii) Same as part (ii).

6. Since $2x = \pi y^2$ so $dx = \pi y dy$ and $0 \le x \le \frac{\pi}{2}, 0 \le y \le 1$

$$\int_C (2xy^3 - y^2\cos x)dx + \int_C (1 - 2y\sin x + 3x^2y^2)dy$$

$$= \int_0^{\pi/2}\left[2x\left(\frac{2x}{\pi}\right)^{3/2} - \frac{2x}{\pi}\cos x\right]dx$$

$$+ \int_0^{\pi/2}\left[1 - 2\left(\frac{2x}{\pi}\right)^{1/2}\sin x + 3x^2\left(\frac{2x}{\pi}\right)\right]\frac{1}{\pi}\left(\frac{\pi}{2x}\right)^{1/2} dx$$

$$= \frac{7}{2}\left(\frac{2}{\pi}\right)^{3/2}\int_0^{\pi/2} x^{5/2}dx - \frac{2}{\pi}\int_0^{\pi/2} x\cos x dx$$

$$+ \frac{1}{\sqrt{2\pi}}\int_0^{\pi/2} x^{-1/2}dx - \frac{2}{\pi}\int_0^{\pi/2}\sin x dx = \frac{\pi^2}{4}.$$

9. The parametric equation of S are

$S = \sin\theta\cos\phi, y = \sin\theta\sin\phi, z = \cos\theta$

where $0 \le \theta \le \pi/2$, $0 \le \phi \le \pi/2$

$\mathbf{F} = yz\hat{i} + zx\hat{j} + xy\hat{k}$

$\iint_S \mathbf{F}\cdot\mathbf{n}dS = \iint_S (yzdydz + zxdzdx + xydxdy)$

Also, $\quad dydz = \sin^2\theta\cos\phi\, d\theta\, d\phi$, $dzdx = \sin^2\theta\sin\phi\, d\theta\, d\phi$

$dxdy = \sin\theta\cos\theta\, d\theta\, d\phi$

$\therefore \quad \iint_S \mathbf{F}\cdot\mathbf{n}dS = 3\int_0^{\pi/2}\int_0^{\pi/2}\sin^3\theta\cos\theta\cos\phi\sin\phi d\theta d\phi$

$= 3\int_0^{\pi/2}\sin^3\theta\cos\theta d\theta.\int_0^{\pi/2}\cos\phi\sin\phi d\phi$

$$= 3\left[\frac{\sin^4\theta}{4}\right]_0^{\pi/2}.\left[\frac{\sin^2\phi}{2}\right]_0^{\pi/2} = 3\left(\frac{1}{4}\right)\left(\frac{1}{2}\right) = \frac{3}{8}$$

12. $f \equiv -y^2 + 8x = 0$, then $\mathbf{n} = \frac{\nabla f}{|\nabla f|} = \frac{8\hat{i} - 2y\hat{j}}{\sqrt{64 + 4y^2}} = \frac{8\hat{i} - 2y\hat{j}}{2\sqrt{16 + 8x}}$

$\therefore \quad \mathbf{F}\cdot\mathbf{n} = \frac{11y}{\sqrt{16 + 8x}}$ and $|\mathbf{n}\cdot\hat{j}| = \frac{y}{\sqrt{16 + 8x}}$

$$\iint_S \mathbf{F}\cdot\mathbf{n}dS = \iint_R \frac{11y}{\sqrt{16 + 8x}}\frac{dzdx}{|\mathbf{n}\cdot\hat{j}|}$$

$$= 11\int_R dzdx = 11\int_0^2\int_0^6 dzdx = 132.$$

14. Since $\quad \mathbf{r} = a(\cos\theta)\hat{i} + a(\sin\theta)\hat{j} + a\theta\hat{k}$

$\therefore \quad x = a\cos\theta, y = a\sin\theta, z = a\theta$

so $\quad \mathbf{F} = (a^2\cos\theta\sin\theta)\hat{i} + (a^2\theta\sin\theta)\hat{j} + (a^2\theta\cos\theta)\hat{k}$

and $\quad d\mathbf{r} = [(-a\sin\theta)\hat{i} + (a\cos\theta)\hat{j} + a\hat{k}]d\theta.$

$\therefore \quad \mathbf{F}\cdot d\mathbf{r} = (-a^3\cos\theta\sin^2\theta + a^3\theta\sin\theta\cos\theta + a^3\theta\cos\theta)d\theta.$

$$\int_C \mathbf{F}\cdot d\mathbf{r} = -a^3\int_0^{\pi/2}\sin^2\theta\cos\theta d\theta + \frac{a^3}{2}\int_0^{\pi/2}\theta\sin 2\theta d\theta + a^3\int_0^{\pi/2}\theta\cos\theta d\theta$$

$$= -\frac{1}{3}a^3 + \frac{1}{2}a^3\left(\frac{\pi}{4}\right) + a^3\left(\frac{1}{2}\pi - 1\right) = a^3\left(\frac{5}{8}\pi - \frac{4}{3}\right).$$

15. $x = t, y = t^2, z = t^3$ and $0 \le t \le 2$. $\therefore \mathbf{r} = t\hat{i} + t^2\hat{j} + t^3\hat{k}$

$\frac{d\mathbf{r}}{dt} = \hat{i} + 2t\hat{j} + 3t^2\hat{k}$ and $\mathbf{F} = t^5\hat{i} + t^4\hat{j} - t^3\hat{k}$

$$\int_C \mathbf{F}.d\mathbf{r} = \int_0^2\left(\mathbf{F}.\frac{d\mathbf{r}}{dt}\right)dt = \int_0^2 (t^5 + 2t^5 - 3t^5)dt = \int_0^2 0dt = 0.$$

16. Same as 15

18. The parametric equation of the circle are $x = \cos t, y = \sin t, z = 0$, and $0 \le t \le 2\pi$

So, $\quad \mathbf{r} = x\hat{i} + y\hat{j} = (\cos t)\hat{i} + (\sin t)\hat{j}$

$\therefore \quad \frac{d\mathbf{r}}{dt} = (-\sin t)\hat{i} + (\cos t)\hat{j}$

$$\therefore \int_C\left[-\frac{y}{x^2 + y^2}\hat{i} + \frac{x}{x^2 + y^2}\hat{j}\right].d\mathbf{r}$$

$$= \int_0^{2\pi}(-\sin t\hat{i} + \cos t\hat{j}).(-\sin t\hat{i} + \cos t\hat{j})dt$$

$$= \int_0^{2\pi}(\sin^2 t + \cos^2 t)dt = \int_0^{2\pi} dt = [t]_0^{2\pi} = 2\pi.$$

ANSWERS

1. 7/12 **2.** 5 **3.** (i) $-\frac{1}{3}$ (ii) –1 (iii) 0 **4.** 16 **5.** 0 **6.** $\pi^2/4$ **7.** $\log_e 2$ **8.** 11 **9.** 3/8

10. 27/4 **11.** 108 **12.** 132 **13.** 8/3 **14.** $a^3\left(\frac{5\pi}{8} - \frac{4}{3}\right)$ **15.** 0 **16.** 277/42 **17.** $-\pi$ **18.** 2π **19.** $\frac{1}{2}c[a^2 + b^2]$ **20.** $\frac{97}{105}$

23.4 GREEN'S THEOREM IN THE PLANE

George Green (1793 – 1841), the English mathematician, discovered a method to transform a double integral over a plane region into line integral over the boundary of the region and conversely. This method (transformation) is of practical as well as theoretical interest. This transformation is as follows :

THEOREM. *Let* **R** *be a closed and bounded region in xy-plane whose boundary C consists of finitely many smooth curves. Let P(x, y) and Q(x, y) be the continuous functions and have continuous partial derivatives* $\frac{\partial P}{\partial y}$ *and* $\frac{\partial Q}{\partial x}$ *everywhere in* **R**. *Then*

$$\iint_R \left(\frac{\partial Q}{\partial x} - \frac{\partial P}{\partial y}\right) dxdy = \oint_C (Pdx + Qdy)$$

the integration being taken along the entire boundary C of **R** *such that* **R** *is on the left as one advances in the direction of integration.*

(UPTU–2008, GBTU–2012)

Proof. We shall first prove the theorem for a special region R which is given as follows :

$$a \le x \le b,\ u(x) \le y \le v(x) \qquad \text{and} \qquad c \le y \le d,\ p(y) \le x \le q(y).$$

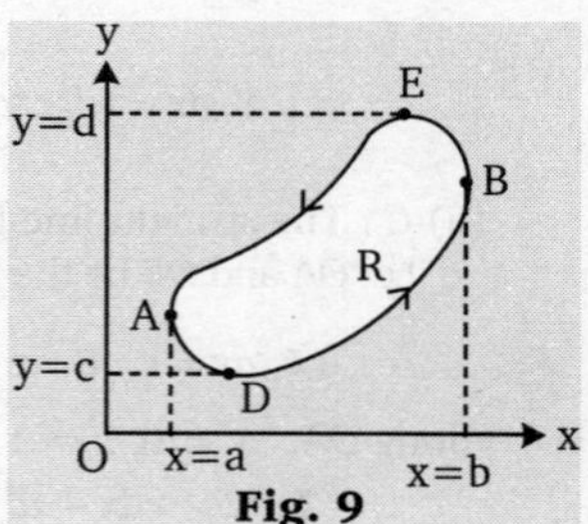

Fig. 9

This region **R** has been shown in the adjoining fig. 9.
In the above fig. the equation of the curves ADB and BEA are respectively $y = u(x)$ and $y = v(x)$ and the equation of the curves EAD and DBE are represented by $x = p(y)$ and $x = q(y)$ respectively. Therefore, we have

$$\iint_R \frac{\partial P}{\partial y} dxdy = \int_{x=a}^{x=b} \left[\int_{y=u(x)}^{y=v(x)} \frac{\partial P}{\partial y} dy\right] dx \qquad \ldots(1)$$

Now evaluate the integral

$$\int_{u(x)}^{v(x)} \frac{\partial P}{\partial y} dy = \left[P(x,y)\right]_{u(x)}^{v(x)} = P[x,v(x)] - P[x,u(x)]. \qquad \ldots(2)$$

From (1) and (2), we have

$$\iint_R \frac{\partial P}{\partial y} dxdy = \int_a^b (P[x,v(x)] - P[x,u(x)])dx = \int_a^b P(x,v(x))dx - \int_b^a P(x,u(x))dx.$$

$$\therefore \qquad \iint_R \frac{\partial P}{\partial y} dxdy = -\int_a^b P(x,u(x))dx - \int_b^a P(x,v(x))dx. \qquad \ldots(3)$$

Since $y = u(x)$ represents the oriented curve ADB and $y = v(x)$ represents oriented curve BEA. Thus the integrals on R.H.S. of (3) may be written as the line integral over ADB and BEA. Therefore, we obtain

$$\iint_R \frac{\partial P}{\partial y} dxdy = -\oint_C P(x,y)dx \quad \text{or} \quad -\iint_R \frac{\partial P}{\partial y} dxdy = \oint_C P(x,y)dx \qquad \ldots(4)$$

If the portions of the curve C are the segments parallel to y-axis as shown in fig. 10.
Then the result in (4) does not change therefore the value of the integral $\int P(x,y)dx$ along the segments ST and UV are zero. The reason for the value of the above integral to be zero are that along ST and UV, x are constant and thus $dx = 0$ and so the value of integral becomes zero.

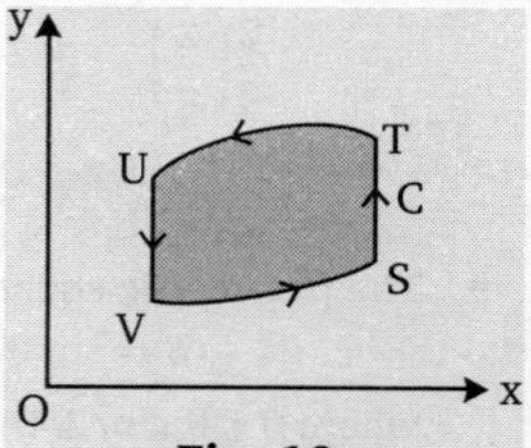

Similarly, we obtain

$$\iint_R \frac{\partial Q}{\partial x} dxdy = \int_{x=c}^{x=d} \left[\int_{x=p(y)}^{x=q(y)} \frac{\partial Q}{\partial x} dx\right] dy = \oint_C Q(x,y)dy. \qquad \ldots(5)$$

Fig. 10

From (4) and (5), we have

$$\iint_R \left(\frac{\partial Q}{\partial x} - \frac{\partial P}{\partial y}\right) dxdy = \oint_C (Pdx + Qdy).$$

Hence this is the required formula for a special region.
The proof of the above theorem can now be extended to a region R which is itself is not a special region but can be subdivided into finitely many special regions as shown below in fig 11. In particular.
The special regions are obtained by drawing the lines parallel to the co-ordinate axes. In fig. 11 ST and UV are the two lines drawn parallel to y and x-axis respectively which divide the region R_1, R_2 and R_3 and apply above theoren to each subregion R_1, R_2 and R_3 and then add the results. The sum of the left hand members will give the integral over the region R and the sum of the right hand members will give the line integral

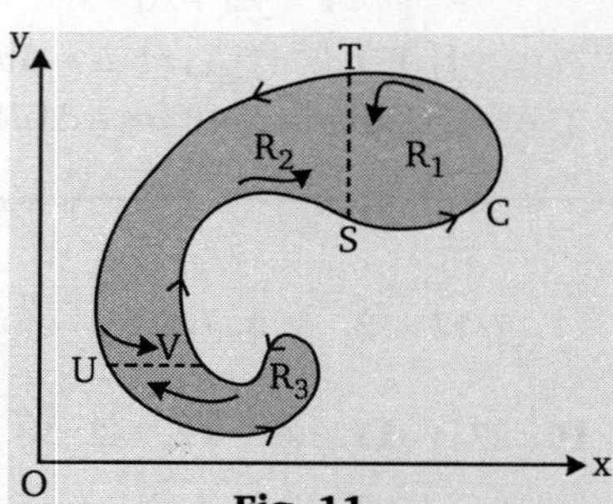

Fig. 11

over C plus integrals over the curves introduced for subdividing R. Each of the latter intetgrals occurs twice, taken in each direction, hence these two integrals will cancel each other and finally we obtain the value of the integral over C.

23.5 APPLICATIONS OF GREEN'S THEOREM

Green's theorem has various applications and important consequences, some of which may be illustrated by the subsequent examples.

(i) Area of plane region as a line integral over the boundary. Let A be the area of a plane region R and $P(x, y) = 0$ and $Q(x, y) = x$. Then from Green's theorem

$$\iint_R \left(\frac{\partial Q}{\partial x} - \frac{\partial P}{\partial y}\right) dxdy = \oint_C (Pdx + Qdy). \qquad \ldots(1)$$

Putting $P(x, y) = 0$ and $Q(x, y) = x$ into above formula, we get

$$\iint_R dxdy = \oint_C xdy.$$

Since $\quad A = \iint_R dxdy \quad \therefore \quad A = \oint_C xdy. \qquad \ldots(2)$

Similarly, let $P(x, y) = -y$ and $Q(x, y) = 0$, then from (1), we have

$$\iint_R dxdy = -\oint_C ydx. \quad \therefore \quad A = -\oint_C ydx. \qquad \ldots(3)$$

Adding (2) and (3), we get

$$A = \frac{1}{2}\oint_C (xdy - ydx). \qquad \ldots(4)$$

The integration in (4) being taken as indicated in Green's theorem. This formula gives the area of R in terms of a line integral over the boundary of C.

REMARK

- The theory of planimeters is based upon this formula.

(ii) Transformation of a double integral of the Laplacian of a function into a line integral of its normal derivative.

Let $\phi(x, y)$ be a function which is continuous and has continuous first and second derivatives in a region **R** of the xy-plane where **R** is as same as taken in Green's theorem. Let $P(x, y) = -\frac{\partial \phi}{\partial y}$ and $Q(x, y) = \frac{\partial \phi}{\partial x}$. Then $\frac{\partial P}{\partial y}$ and $\frac{\partial Q}{\partial x}$ are continuous in R and

$$\frac{\partial Q}{\partial x} - \frac{\partial P}{\partial y} = \frac{\partial^2 \phi}{\partial x^2} + \frac{\partial^2 \phi}{\partial y^2} = \left(\frac{\partial^2}{\partial x^2} + \frac{\partial^2}{\partial y^2}\right)\phi = \nabla^2 \phi. \qquad \ldots(1)$$

where ∇^2 is a Laplacian operator. The region **R** is shown in fig. 12.

Now we have

$$\int_C (Pdx + Qdy) = \int_C \left(P\frac{dx}{ds} + Q\frac{dy}{ds}\right) ds = \int_C \left(-\frac{\partial \phi}{\partial y}\frac{dx}{ds} + \frac{\partial \phi}{\partial x}\frac{dy}{ds}\right) ds \qquad \ldots(2)$$

where s is the arc length of C and

$$-\frac{\partial \phi}{\partial y}\frac{dx}{ds} + \frac{\partial \phi}{\partial x}\frac{dy}{ds} = (\text{grad } \phi) \cdot \mathbf{n} \qquad \ldots(3)$$

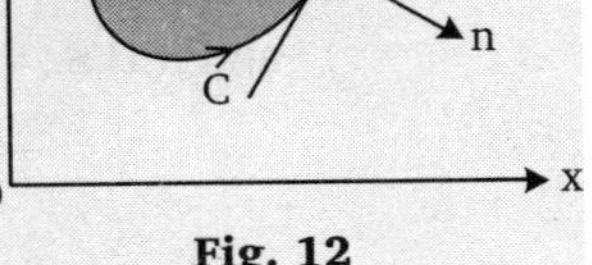

Fig. 12

Let t be the unit tangent vector to C which is given by

$$\mathbf{t} = \frac{d\mathbf{r}}{ds} = \frac{dx}{ds}\hat{i} + \frac{dy}{ds}\hat{j} \qquad (\because \mathbf{r} = x\hat{i} + y\hat{j})$$

From (3) $\quad \mathbf{n} = \frac{dy}{ds}\hat{i} - \frac{dx}{ds}\hat{j}$, then $\mathbf{t} \cdot \mathbf{n} = 0$

this implies that the vector **n** is a unit outward drawn normal to C. Thus $(\text{grad } \phi) \cdot \mathbf{n} = \frac{\partial \phi}{\partial n}$

That is the expression on R.H.S. of (3) is the derivative of ϕ in the direction of the outward normal to C. Therefore (3) becomes

$$-\frac{\partial \phi}{\partial y}\frac{dx}{ds} + \frac{\partial \phi}{\partial x}\frac{dy}{ds} = \frac{\partial \phi}{\partial n}$$

so, equation (2) becomes

$$\int_C (Pdx + Qdy) = \int_C \frac{\partial \phi}{\partial n} ds \qquad \ldots(4)$$

From Green's theorem we have

$$\iint_R \left(\frac{\partial Q}{\partial x} - \frac{\partial P}{\partial y}\right) dxdy = \int_C (Pdx + Qdy). \qquad \ldots(5)$$

From (1) and (4) and using (5), we get

$$\iint_C \nabla^2\phi dxdy = \int_C \frac{\partial\phi}{\partial n} ds.$$

This is the required transformation and is important application of Green's theorem.

Solved Examples

Example 1. *Verify Green's therorem in the plane for*

$$\int_C[(2xy - x^2)dx + (x^2 + y^2)dy]$$

where C is the boundary of the region enclosed by $y = x^2$ and $y^2 = x$ described in the positive sense.

Solution . Let **R** be the region enclosed by $y = x^2$ and $y^2 = x$ whose boundary C is traversed in the positive direction as shown in the fig. 13.

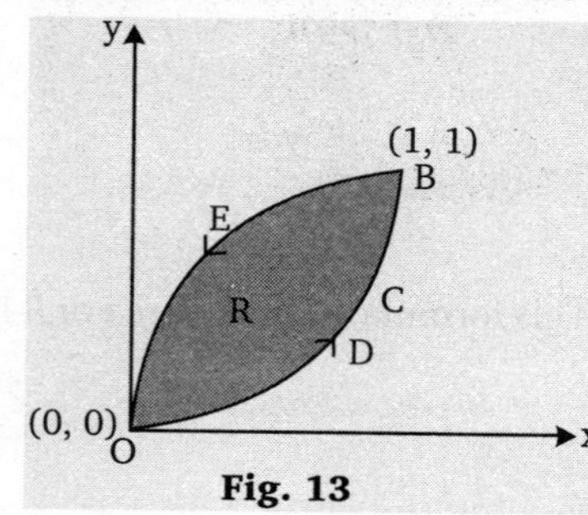

Fig. 13

The curves $y = x^2$ and $y^2 = x$ intersect at (0, 0) and (1, 1) and have

$$P(x, y) = 2xy - x^2$$

and $Q(x, y) = x^2 + y^2$.

$$\therefore \quad \frac{\partial P}{\partial y} = 2x \text{ and } \frac{\partial Q}{\partial x} = 2x$$

By Green's theorem, we have

$$\iint_R\left(\frac{\partial Q}{\partial x} - \frac{\partial P}{\partial y}\right)dxdy = \int_C(Pdx + Qdy). \quad ...(1)$$

$$\therefore \quad \text{L.H.S.} = \iint_R\left(\frac{\partial Q}{\partial x} - \frac{\partial P}{\partial y}\right)dxdy$$

$$= \iint_R(2x - 2x)dxdy = \iint_R 0.dxdy = 0$$

and R.H.S. $= \int_C(Pdx + Qdy)$

$$= \int_C[(2xy - x^2)dx + (x^2 + y^2)dy]$$

$$= \int_{ODB}[(2xy - x^2)dx + (x^2 + y^2)dy] + \int_{BEO}[(2xy - x^2)dx + (x^2 + y^2)dy] \quad ...(2)$$

($\because$ C consists of two curves ODB amd BEO.)

Along the curve BEO, we have

$y^2 = x$ and x varies from 0 to 1.

$\therefore \quad 2ydy = dx$

$$\therefore \quad \int_{BEO}[(2xy - x^2)dx + (x^2 + y^2)dy]$$

$$= \int_0^1(2x^{3/2} - x^2)dx + \int_0^1(x^2 + x)\frac{dx}{2\sqrt{x}}$$

$$= \left[2\frac{x^{5/2}}{5/2} - \frac{x^3}{3}\right]_0^1 + \frac{1}{2}\left[\frac{x^{5/2}}{5/2} + \frac{x^{3/2}}{3/2}\right]_0^1$$

$$= \left[\frac{4}{5} - \frac{1}{3}\right] + \frac{1}{2}\left[\frac{2}{5} + \frac{2}{3}\right] = 1$$

and along the curve ODB, we have

$y = x^2$ and x varies from 1 to 0.

$\therefore \quad dy = 2xdx$

$$\therefore \quad \int_{OBD}[(2xy - x^2)dx + (x^2 + y^2)dy]$$

$$= \int_1^0[2x^3 - x^2 + 2x^3 + 2x^5]dx$$

$$= \int_1^0(4x^3 - x^2 + 2x^5)dx$$

$$= -\int_0^1(4x^3 - x^2 + 2x^5)dx$$

(By the property of definite integral)

$$= -\left[x^4 - \frac{x^3}{3} + \frac{x^6}{3}\right]_0^1 = \left[1 - \frac{1}{3} + \frac{1}{3}\right] = -1$$

$\therefore$ R.H.S. $= -1 + 1 = 0$.

Thus, L.H.S. = R.H.S.

Hence, Green's theorem is verified.

Example 2. *Apply Green's theorem in the plane to evaluate*

$$\int_C[(y - \sin x)dx + \cos xdy]$$

where C is the triangle enclosed by the lines $y = 0$, $x = \pi$, $\pi y = 2x$. (Anna–2003, JNTU–2005)

Solution . By Green's theorem, we have

$$\iint_R\left(\frac{\partial Q}{\partial x} - \frac{\partial P}{\partial y}\right)dxdy = \int_C(Pdx + Qdy). \quad ...(1)$$

Here $\quad P = y - \sin x,\ Q = \cos x.$

$$\therefore \quad \frac{\partial P}{\partial y} = 1, \frac{\partial Q}{\partial x} = -\sin x.$$

Now from (1), we have

$$\int_C[(y - \sin x)dx + \cos xdy]$$

$$= \iint_R[(-\sin x - 1)dxdy]$$

$$= \int_{x=0}^{\pi}\int_{y=0}^{(2x)/\pi}(-\sin x - 1)dxdy$$

$$= -\int_0^{\pi}(1 + \sin x)[y]_0^{2x/\pi} dx$$

$$= -\frac{2}{\pi}\int_0^{\pi} x(1 + \sin x)dx$$

$$= -\frac{2}{\pi}\int_0^{\pi} xdx - \frac{2}{\pi}\int_0^{\pi} x\sin xdx$$

$$= -\frac{2}{\pi}\left[\frac{x^2}{2}\right]_0^{\pi} - \frac{2}{\pi}[-x\cos x + \sin x]_0^{\pi}$$

$$= -\frac{2}{\pi}\left[\frac{\pi^2}{2}\right] - \frac{2}{\pi}[\pi] = -\pi - 2.$$

Example 3. *Using Green's theorem evaluate*

$$\int_C [(e^x - 3y)dx + (e^y + 6x)dy]$$

where $C : x^2 + 4y^2 = 4$.

Solution . From Green's theorem, we have

$$\oint_C (Pdx + Qdy) = \iint_R \left(\frac{\partial Q}{\partial x} - \frac{\partial P}{\partial y}\right) dxdy. \qquad \ldots(1)$$

Here, $P = e^x - 3y$, $Q = e^y + 6x$.

$$\therefore \quad \frac{\partial P}{\partial y} = -3, \frac{\partial Q}{\partial x} = 6$$

Thus from (1), we get

$$\int_C [(e^x - 3y)dx + (e^y + 6x)dy] = \iint_R (6+3)dxdy$$

$$= 9\iint_R dxdy$$

Since R is an elliptic region given by $x^2 + 4y^2 = 4$

$$= 9\int_{x=-2}^{x=2} \int_{y=-\frac{1}{2}\sqrt{4-x^2}}^{y=\frac{1}{2}\sqrt{4-x^2}} dxdy$$

$$= 9\int_{x=-2}^{x=2} \left[y\right]_{y=-\frac{1}{2}\sqrt{4-x^2}}^{y=\frac{1}{2}\sqrt{4-x^2}} .dx$$

$$= 9\int_{-2}^{2} \sqrt{4-x^2}\, dx$$

$$= 18\int_0^2 \sqrt{4-x^2}\, dx$$

(By the property of definite integral)

$$= 18\left[\frac{x}{2}\sqrt{4-x^2} + \frac{4}{2}\sin^{-1}\frac{x}{2}\right]_0^2$$

$$= 18[2\sin^{-1}(1)] = 18\left[2.\frac{\pi}{2}\right] = 18\pi.$$

Example 4. *Evaluate by Green's theorem :*

$$\oint_C [(x^2 - \cosh y)dx + (y + \sin x)dy)],$$

where C is the rectangle with vertices (0, 0), (π, 0), (π, 1), (0, 1). (Nagpur–2009, PTU–2006)

Solution . By Green's theorem in the plane, we have

$$\oint_C (Pdx + Qdy) = \iint_R \left(\frac{\partial Q}{\partial x} - \frac{\partial P}{\partial y}\right) dxdy. \qquad \ldots(1)$$

Here, $P = x^2 - \cosh y$, $Q = y + \sin x$.

$$\therefore \quad \frac{\partial P}{\partial y} = -\sinh y, \frac{\partial Q}{\partial x} = \cos x.$$

Thus from (1), we get

$$\oint_C [(x^2 - \cosh y)dx + (y + \sin x)dy)]$$

$$= \iint_R (\cos x + \sinh y)dxdy$$

$$= \int_{x=0}^{\pi} \int_{y=0}^{1} (\cos x + \sinh y)dxdy$$

$$= \int_{x=0}^{\pi} \left[y\cos x + \cosh y\right]_{y=0}^{1} dx$$

$$= \int_{x=0}^{\pi} (\cos x + \cosh 1 - 1)dx$$

$$= \left[\sin x + x\cosh 1 - x\right]_{x=0}^{1} = \pi(\cosh 1 - 1)$$

Example 5. *Using Green's theorem evaluate*

$$\oint_C [(x^2 - y^2)dx + 2xydy)]$$

where C is the boundary of a rectangle formed by the lines $x = 0$, $x = a$, $y = 0$, $y = b$.

Solution . From Green's theorem, we have

$$\oint_C (Pdx + Qdy) = \iint_R \left(\frac{\partial Q}{\partial x} - \frac{\partial P}{\partial y}\right) dxdy. \qquad \ldots(1)$$

Here, $P = x^2 - y^2$, $Q = 2xy$.

$$\therefore \quad \frac{\partial P}{\partial y} = -2y, \frac{\partial Q}{\partial x} = 2y.$$

Thus (1) becomes

$$\int_C (x^2 - y^2)dx + 2xydy = \iint_R 4ydxdy \qquad \ldots(2)$$

where R is the rectangle which is shown in fig. 14.

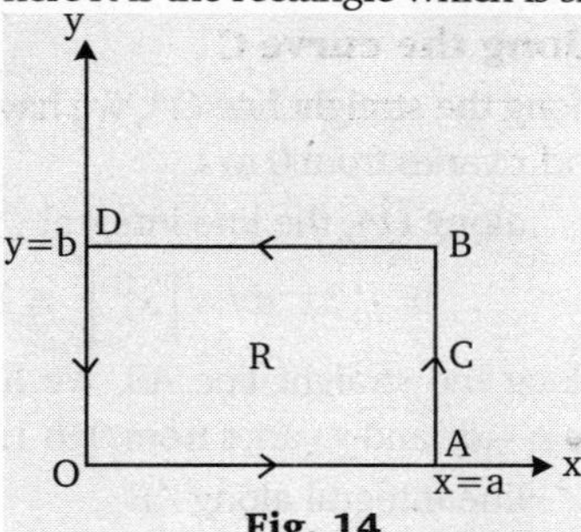

Fig. 14

$$\therefore \iint_R 4ydxdy = \int_{x=0}^{x=a} \left\{\int_{y=0}^{y=b} 4ydy\right\} dx$$

$$= \int_{x=0}^{x=a} \left[2y^2\right]_0^b dx$$

$$= 2b^2 \int_{x=0}^{x=a} dx = 2b^2 \left[x\right]_0^a = 2ab^2.$$

Hence form (2)

$$\int_C (x^2 - y^2)dx + 2xydy = 2ab^2.$$

Example 6. *Verify Green's theorem in a plane for*

$$\int_C (3x^2 - 8y^2)dx + (4y - 6xy)dy$$

where C is the boundary of the region defined by $x = 0$, $y = 0$ *and* $x + y = 1$.

Solution . We know that

$$\iint_R \left(\frac{\partial N}{\partial x} - \frac{\partial M}{\partial y}\right) dxdy = \int_C Mdx + Ndy \qquad \ldots(1)$$

(By Green's theorem)

Here, we have $M = 3x^2 - 8y^2$; $N = 4y - 6xy$

The closed curve C consist of the straight line OA, the straight line AB and the straight line BO.

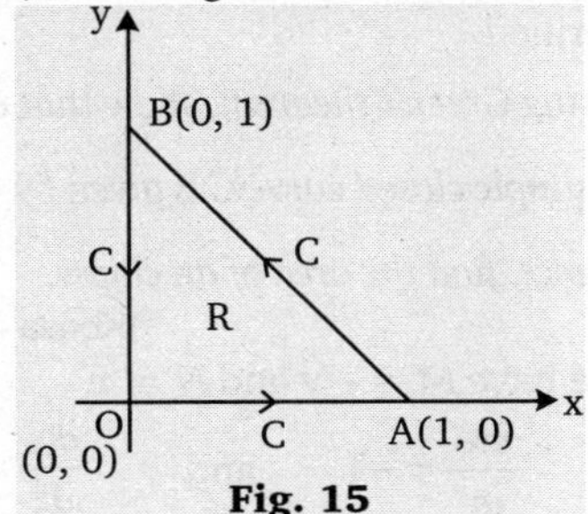

Fig. 15

$\therefore \quad \iint\left(\frac{\partial N}{\partial x}-\frac{\partial M}{\partial y}\right)dxdy$

$= \iint_R\left[\frac{\partial}{\partial x}(4y-6xy)-\frac{\partial}{\partial y}(3x^2-8y^2)\right]dxdy$

$= \iint_R[-6y+16y]dxdy = 10\iint_R ydxdy$

$= 10\int_{x=0}^{1}\int_{y=0}^{1-x} ydxdy$ (For the region R)

$= 10\int_0^1\left[\frac{y^2}{2}\right]_{y=0}^{1-x} dx = 5\int_0^1(1-x)^2dx$

$= 5\int_0^1(x-1)^2dx = \frac{5}{3}\left[(x-1)^3\right]_0^1 = \frac{5}{3}$ ...(2)

Along the curve C

Along the straight line OA, we have $y = 0$, $dy = 0$ and x varies from 0 to 1.

$\therefore$ along OA, the line integral

$= \int_0^1 3x^2dx = \left[x^3\right]_0^1 = 1$

Along the straight line AB, we have $x = 1 - y$, $dx = -dy$ and y varies from 0 to 1.

$\therefore$ line integral along AB

$= \int_0^1[\{3(1-y)^2 - 8y^2\}(-dy) + \{4y - 6y(1-y)\}]dy$

$= \int_0^1[-3(1-2y+y^2)+8y^2 + 4y - 6y + 6y^2]dy$

$= \int_0^1(11y^2+4y-3)dy$

$= \left[\frac{11}{3}y^3 + 2y^2 - 3y\right]_0^1$

$= \frac{11}{3}+2-3 = \frac{8}{3}$

Now, since along the straight line BO, we have $x = 0$, $dx = 0$ and y varies from 1 to 0

$\therefore$ line integral along BO

$= \int_1^0 4ydy = 2\left[y^2\right]_1^0 = -2$

Hence, total line integral along the closed curve C

$= 1+\frac{8}{3}-2 = \frac{5}{3}$ ...(3)

Finally, from (2) and (3), Green's theorem is verified.

Example 7. *Using Green's theorem, show that area bounded by a simple closed curve C is given by* $\frac{1}{2}\int(xdy - ydx)$. *Hence, find the area of an ellipse.*

(Kerala–2005, VTU–2000)

Solution . We have $M = -y$ and $N = x$.

$\Rightarrow \quad \frac{\partial M}{\partial y} = -1$ and $\frac{\partial N}{\partial x} = 1$ ...(1)

From Green's theorem, we have

$\int_C(Mdx+Ndy) = \iint_S\left(\frac{\partial N}{\partial x}-\frac{\partial M}{\partial y}\right)dxdy.$ ...(2)

Using (1) in (2), we get

$\int_C(-ydx+xdy) = \iint_S(1+1)dxdy$

$= 2\iint_S dxdy = 2A$

where A is the required area given by

$A = \frac{1}{2}\int_C(xdy - ydx)$

Any point (x, y) on the ellipse is given by

$x = a\cos\phi, y = b\sin\phi$, ϕ is a parameter

Hence, area of the ellipse

$= \frac{1}{2}\int_0^{2\pi}(a\cos\phi)(b\cos\phi)d\phi - (b\sin\phi)(-a\sin\phi)d\phi$

$= \frac{1}{2}ab\int_0^{2\pi}(\cos^2\phi+\sin^2\phi)d\phi$

$= \frac{1}{2}ab(2\pi) = \pi ab.$

Example 8. *Use Green's theorem to evaluate*

$\int_C(x^2+xy)dx+(x^2+y^2)dy$

where C is the square formed by the lines $y = \pm 1$, $x = \pm 1$.

(GBTU–2010, MTU–2011, SVTU–2008, SRM-2006, Marathwada–2008)

Solution . We have

$\int_C(x^2+xy)dx+(x^2+y^2)dy$

$= \iint_S\left[\frac{\partial}{\partial x}(x^2+y^2)-\frac{\partial}{\partial y}(x^2+xy)\right]dxdy$

$= \iint_S(2x-x)dxdy$

$= \int_{x=-1}^{1}\int_{y=-1}^{1}xdxdy = \int_{-1}^{1}x\left[y\right]_{-1}^{1}dx$

$= \int_{-1}^{1}2xdx = 0.$

Example 9. *Apply Green's theorem to evaluate*

$\int_C[(2x^2-y^2)dx+(x^2+y^2)dy]$

where C is the boundary of the area enclosed by the x-axis and upper half of the circle $x^2+y^2=a^2$.

(UPTU–2005, GBTU(Ag)–2010)

Solution . Let $\mathbf{F} = M\hat{i}+N\hat{j}$

Then, $\mathbf{F}.d\mathbf{r} = (M\hat{i}+N\hat{j}).(dx\hat{i}+dy\hat{j})$

$= Mdx+Ndy$

Here we have

$M = 2x^2-y^2 \Rightarrow \frac{\partial M}{\partial y} = -2y$

$N = x^2+y^2 \Rightarrow \frac{\partial N}{\partial x} = 2x$

Therefore,

$\int_C\mathbf{F}.d\mathbf{r} = \int_C[(2x^2-y^2)dx+(x^2+y^2)dy]$

$= \iint_S(2x+2y)dxdy$ (By Green's theorem)

$= 2\int_{x=-a}^{a}\int_{y=0}^{\sqrt{a^2-x^2}}(x+y)dxdy$

$$= 2\int_{-a}^{a}\left(xy + \frac{y^2}{2}\right)_0^{\sqrt{a^2-x^2}} dx$$

$$= 2\int_{-a}^{a}\left(x\sqrt{a^2 - x^2} + \frac{a^2 - x^2}{2}\right)dx$$

$$= 0 + 2\int_0^a (a^2 - x^2)dx$$

$$= 2\left(a^2 x - \frac{x^3}{3}\right)_0^a = 2\left(a^3 - \frac{a^3}{3}\right) = \frac{4}{3}a^3$$

Example 10. *Using Green's theorem, evaluate $\int_C (x^2 y + x^2 dy)$ where C is the boundary described counter clockwise of the triangle with vertices (0, 0), (1, 0), (1, 1).* (UKTU–2010)

Solution . We have $\int_C (x^2 y\, dx + x^2 dy)$

$$= \iint_S \left[\frac{\partial}{\partial x}(x^2) - \frac{\partial}{\partial y}(x^2 y)\right] dxdy$$

(By Green's theorem)

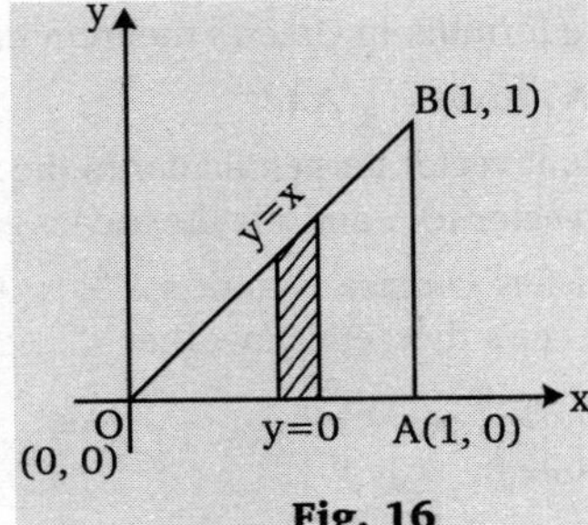

Fig. 16

$$= \int_{x=0}^{1}\int_{y=0}^{x}(2x - x^2)dxdy$$

$$= \int_0^1 (2x - x^2)(y)_0^x\, dx$$

$$= \int_0^1 (2x^2 - x^3)dx = \left(\frac{2}{3}x^3 - \frac{x^4}{4}\right)_0^1$$

$$= \frac{2}{3} - \frac{1}{4} = \frac{5}{12}$$

Example 11. *Using Green's theorem, find the area of the region in the first quadrant bounded by the curves $y = x$, $y = \frac{1}{x}, y = \frac{x}{4}$.* (UPTU–2009)

Solution . Using Green's theorem, we have

$$A = \frac{1}{2}\int_C (xdy - ydx)$$

$$A = \frac{1}{2}\left[\int_{C_1}(xdy - ydx) + \int_{C_2}(xdy - ydx) + \int_{C_3}(xdy - ydx)\right] = \frac{1}{2}(I_1 + I_2 + I_3) \quad \ldots(1)$$

Along C_1 :

We have $y = \frac{x}{4} \Rightarrow dy = \frac{1}{4}dx$, $x = 0$ to 2

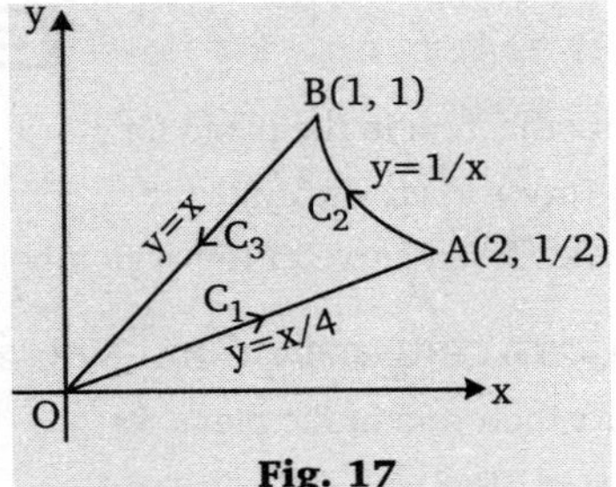

Fig. 17

$$\therefore \quad I_1 = \int_{C_1}(xdy - ydx)$$

$$= \int_{C_1}\left(x\cdot\frac{dx}{4} - \frac{x}{4}dx\right) = 0 \quad \ldots(2)$$

Along C_2 :

We have $y = \frac{1}{x} \Rightarrow dy = -\frac{1}{x^2}dx$, $x = 2$ to 1

$$\therefore \quad I_2 = \int_{C_2}(xdy - ydx)$$

$$= \int_2^1\left(x\left(\frac{-1}{x^2}\right)dx - \frac{1}{x}dx\right)$$

$$= -2\int_2^1 \frac{1}{x}dx = -2\log x\Big|_2^1 = 2\log 2 \quad \ldots(3)$$

Along C_3 :

We have $y = x \Rightarrow dy = dx$, $x = 1$ to 0

$$\therefore \quad I_3 = \int_{C_3}(xdy - ydx)$$

$$= \int xdx - xdx = 0 \quad \ldots(4)$$

Using (2), (3) and (4) in (1), we get

$$A = \frac{1}{2}(I_1 + I_2 + I_3) = \log 2$$

Example 12. *If C is a simple closed curve in the xy-plane not enclosing the origin, show that $\int_C \mathbf{F}.d\mathbf{r} = 0$ where $\mathbf{F} = \frac{-y\hat{i} - x\hat{j}}{x^2 + y^2}$.* (PTU–2005)

Solution . We have

$$\int_C \mathbf{F}.d\mathbf{r} = \int_C \frac{-y\hat{i} - x\hat{j}}{x^2 + y^2}(dx\hat{i} + dy\hat{j})$$

$$= \int_C \frac{ydx - xdy}{x^2 + y^2} = \int_C Mdx + Ndy$$

where $M = \frac{y}{x^2 + y^2}, N = \frac{-x}{x^2 + y^2}$

$$= \iint_S \left(\frac{\partial N}{\partial x} - \frac{\partial M}{\partial y}\right)dxdy$$

(By Green's theorem)

$$= \iint_S \left[\frac{-(x^2 + y^2) + x(2x)}{(x^2 + y^2)^2} - \frac{(x^2 + y^2) - y(2y)}{(x^2 + y^2)^2}\right]dxdy$$

$$= \iint \left[\frac{x^2 - y^2}{(x^2 + y^2)^2} - \frac{x^2 - y^2}{(x^2 + y^2)^2}\right]dxdy$$

$$= 0$$

EXERCISE 23.2

1. Verify Green's theorem in the plane for

$$\int_C [(xy + y^2)dx + x^2 dy]$$

where C is the closed curve of the region bounded by $y = x$ and $y = x^2$.

(UPTU–2008, GBTU–2010, VTU–2011, SVTU–2009, Rohtak–2003)

2. Verify Green's therorem in the plane for

$$\int_C [(x^2 - xy^3)dx + (y^2 - 2xy)dy]$$

where C is the square with vertices (0, 0), (2, 0),(2, 2), (0, 2).

(UPTU–2008)

3. Apply Green's theorem to evaluate :

$$\int_C [(e^{-x} \sin y dx + e^{-x} \cos y dy)$$

where C is the rectangle with vertices (0, 0), (π, 0), (π, $\pi/2$), (0, $\pi/2$). (UPTU–2007)

4. Using Green's theorem evaluate

$$\int_C [(3x^2 + y)dx + 4y^2 dy]$$

where C is the triangle with vertices (0, 0), (1, 0), (0, 2).

5. Using Green's theorem evaluate

$$\int_C [(x^2 - \cosh y)dx + (y + \sin x)dy]$$

where C is the boundary of the rectangle $0 \le x \le \pi$, $0 \le y \le 1$.

6. Using Green's theorem evaluate

$$\int_C [(\cos x \sin y - xy)dx + \sin x \cos y dy]$$

where C is the circle $x^2 + y^2 = 1$.

7. Using Green's theorem evaluate

$$\int_C [x^{-1}e^y dx + (e^y \log x + 2x)dy]$$

where C is the boundary of the region bounded by $y = x^4 + 1$ and $y = 2$.

8. Verify the Green's theorem for

$$\int_C (y^2 dx + x^2 dy)$$

where C is the boundary of the square $-1 \le x \le 1$ and $-1 \le y \le 1$.

9. Using Green's theorem in the plane, evaluate

$$\int_C [2\tan^{-1}(y/x)dx + \log(x^2 + y^2)dy]$$

where C is the boundary of the circle $(x - 1)^2 + (y + 1)^2 = 4$.

10. Using the following formula

$$\iint_R \nabla^2 \phi dxdy = \int_C \frac{\partial \phi}{\partial n} ds$$

evaluate $\int_C \frac{\partial \phi}{\partial n} ds$, where $\phi = x^2 + 3y^2$ and C is the boundary of the circle $x^2 + y^2 = 4$.

11. If $\mathbf{A} = Q\hat{i} - P\hat{j}$, show that the formula in Green's theorem may be written as

$$\iint_R \text{div } \mathbf{A} \, dxdy = \int_C \mathbf{A}.\mathbf{n} ds$$

where $\mathbf{n}$ is the outward unit normal vector to C and s is the arc length of C.

12. Show that the formula in Green's theorem may be written as

$$\iint_R (\text{curl } \mathbf{A}).\hat{k} dxdy = \int_C \mathbf{A}.\mathbf{t} ds$$

where $\hat{k}$ is a unit vector perpendicular to the xy-plane, $\mathbf{t}$ is the unit tangent vector to C and s is the arc length of C.

13. If $\phi(x, y)$ satisfies Laplace equation $\nabla^2 \phi = 0$ in a region $\mathbf{R}$, then using Green's theorem show that

$$\iint_R \left[\left(\frac{\partial \phi}{\partial x}\right)^2 + \left(\frac{\partial \phi}{\partial y}\right)^2 \right] dxdy = \int_C \phi . \frac{\partial \phi}{\partial n} ds.$$

Hint to Selected Problems

2. Here $P = x^2 - xy^3, Q = y^2 - 2xy$

$$\therefore \quad \frac{\partial P}{\partial y} = -3xy^2, \frac{\partial Q}{\partial x} = -2y$$

Then by Green's theorem

$$\iint_R \left(\frac{\partial Q}{\partial x} - \frac{\partial P}{\partial y} \right) dxdy = \int_C Pdx + Qdy.$$

$$\text{L.H.S.} = \iint_R \left(\frac{\partial Q}{\partial x} - \frac{\partial P}{\partial y} \right) dxdy = \iint_R (-2y + 3xy^2) dxdy$$

$$= \int_0^2 \int_{x=0}^{x=2} (-2y + 3xy^2) dxdy = \int_0^2 \left[-2xy + \frac{3x^2y^2}{2} \right]_0^2 dy$$

$$= \int_0^2 (-4y + 6y^2) dy = 8.$$

Since C is the boundary of a square with vertices $O(0, 0)$, $A(2, 0)$, $B(2, 2)$ and $C(0, 2)$. Therefore

$$\text{R.H.S.} = \int_C Pdx + Qdy$$

$$= \int_{OA} Pdx + Qdy + \int_{AB} Pdx + Qdy + \int_{BC} Pdx + Qdy + \int_{CO} Pdx + Qdy.$$

Along OA : $y = 0$, $0 \le x \le 2$, then

$$\int_{OA} Pdx + Qdy = \int_0^2 x^2 dx = \left[\frac{x^3}{3} \right]_0^2 = \frac{8}{3}.$$

Along AB : $x = 2$, $0 \le y \le 2$, then

$$\int_{AB} Pdx + Qdy = \int_0^2 (y^2 - 4y)dy = \left[\frac{y^3}{3} - 2y^2 \right]_0^2 = -\frac{16}{3}.$$

Along BC : $y = 2$ and x varies from 2 to 0.

$$\int_{BC} Pdx + Qdy = \int_2^0 (x^2 - 8x)dx = \frac{40}{3}.$$

Along CO : $x = 0$ and y varies from 2 to 0, then

$$\int_{CO} Pdx + Qdy = \int_2^0 y^2 dy = -\frac{8}{3}.$$

$$\text{L.H.S.} \quad = \int_C Pdx + Qdy = \frac{8}{3} - \frac{16}{3} + \frac{40}{3} - \frac{8}{3} = 8.$$

$\therefore$ L.H.S. = R.H.S.

4. Let $\mathbf{R}$ be the region enclosed by the triangle with vertices $O(0, 0)$, $A(1, 0)$ and $B(0, 2)$.

By Green's theorem

$$\int_C Pdx + Qdy = \iint_R \left(\frac{\partial Q}{\partial x} - \frac{\partial P}{\partial y} \right) dxdy$$

Here $P = 3x^2 + y$, $Q = 4y^2$

$$\text{so,} \quad \frac{\partial P}{\partial y} = 1, \frac{\partial Q}{\partial x} = 0.$$

$\therefore \quad \int_C (3x^2 + y)dx + 4y^2 dy = \iint_R (0-1)dxdy = -\iint_R dxdy$

$= -$(Area of the triangle OAB)

$= -\left(\frac{1}{2} \times 1 \times 2\right) = -1$

7. Let **R** be the region enclosed by $y = x^4 + 1$ and $y = 2$. The intersection points of $y = 2$ and $y = x^4 + 1$ are (1, 2) and (–1, 2) and $y = x^4 + 1$ cuts only y-axis at (0, 1), therefore y varies from 1 to 2 and x varies from –1 to 1.

By Green's theorem

$$\iint_R \left(\frac{\partial Q}{\partial x} - \frac{\partial P}{\partial y}\right) dxdy = \int_C P dx + Q dy$$

Here $P = x^{-1}e^y$, $Q = e^y \log x + 2x$.

$$\therefore \quad \frac{\partial P}{\partial y} = x^{-1}e^y, \frac{\partial Q}{\partial x} = \frac{e^y}{x} + 2$$

$$\int_C x^{-1}e^y dx + (e^y \log x + 2x)dy$$
$$= \iint_R \left(e^y x^{-1} + 2 - x^{-1}e^y\right) dxdy$$
$$= 2\iint_R dxdy = 2\int_{-1}^{1}\int_{y=x^4+1}^{2} dxdy$$
$$= 2\int_{-1}^{1} [y]_{x^4+1}^{2} dx = 2\int_{-1}^{1}(2 - x^4 - 1)dx$$
$$= 2\int_{-1}^{1}(1 - x^4)dx = 4\int_0^1 (1 - x^4)dx$$
$$= 4\left(x - \frac{x^5}{5}\right)_0^1 = \frac{16}{5}.$$

10. Since **R** is region enclosed by the circle $x^2 + y^2 = 4$ whose boundary is C and $\phi = x^2 + 3y^2$.

$$\therefore \quad \nabla^2\phi = \frac{\partial^2\phi}{\partial x^2} + \frac{\partial^2\phi}{\partial y^2} = \frac{\partial}{\partial x}(2x) + \frac{\partial}{\partial y}(6y) = 2 + 6 = 8$$

$$\therefore \quad \int_C \frac{\partial\phi}{\partial n} ds = \iint_R \nabla^2\phi dxdy = \iint_R 8 dxdy = 8\iint_R dxdy$$

$= 8$ (Area of the circle **R**) $= 8[\pi(2)^2] = 32\pi$

13. $\quad \mathbf{A} = Q\hat{i} - P\hat{j}$

$$\nabla \cdot \mathbf{A} = \frac{\partial Q}{\partial x} - \frac{\partial P}{\partial y}$$

By Green's theorem,

$$\int_C P dx + Q dy = \iint_R \left(\frac{\partial Q}{\partial x} - \frac{\partial P}{\partial y}\right) dxdy$$

$$\Rightarrow \quad \iint_R \nabla \cdot \mathbf{A} dxdy = \int_C \left(P\frac{dx}{ds} + Q\frac{dy}{ds}\right) ds$$

$$\Rightarrow \quad \iint_R \nabla \cdot \mathbf{A} dxdy = \int_C (Q\hat{i} - P\hat{j})\left(\frac{dy}{ds}\hat{i} - \frac{dx}{ds}\hat{j}\right) ds = \int_C \mathbf{A} \cdot \mathbf{n} ds$$

Now putting $\mathbf{A} = \phi(\nabla\phi)$

$$\therefore \quad \nabla \cdot \mathbf{A} = \nabla\phi \cdot \nabla\phi$$

$$\therefore \iint_R \nabla\phi \cdot \nabla\phi dxdy = \int_C \phi\nabla\phi \cdot \mathbf{n} ds$$

$$\Rightarrow \iint_R |\nabla\phi|^2 dxdy = \int_C \phi \frac{\partial\phi}{\partial n} ds$$

$$\Rightarrow \iint_R \left[\left(\frac{\partial\phi}{\partial x}\right)^2 + \left(\frac{\partial\phi}{\partial y}\right)^2\right] dxdy = \int_C \phi \frac{\partial\phi}{\partial n} ds.$$

ANSWERS

3. $2(e^{-\pi} - 1)$ **4.** –1 **5.** $\pi(\cosh 1 - 1)$ **6.** 0 **7.** $\frac{16}{5}$ **9.** 0 **10.** 32π

23.6 GAUSS'S DIVERGENCE THEOREM

THEOREM. *Let V be the volume enclosed by a closed and bounded piecewise smooth surface S and let* $\mathbf{F}(x, y, z)$ *be a vector function which is continuous and has continuous first partial derivatives on V. Then*

$$\iiint_V \text{div } \mathbf{F} dV = \iint_S \mathbf{F} \cdot \mathbf{n} dS \qquad \ldots(1)$$

where **n** *is the outward unit normal vector the surface S.* (UPTU–2006, 07, GBTU–2011, 12)

Cartesian form of (1). *Let* $\mathbf{F} = F_1\hat{i} + F_2\hat{j} + F_3\hat{k}$ *and suppose the outward unit normal vector* $\hat{\mathbf{n}}$ *makes the angle* α, β *and g with the positive axes of x, y, z respectively. Then* $\cos\alpha$, $\cos\beta$ *and* $\cos\gamma$ *are the direction-cosines of* $\hat{\mathbf{n}}$, *we have*

$$\hat{\mathbf{n}} = \cos\alpha\hat{i} + \cos\beta\hat{j} + \cos\gamma\hat{k}.$$

$$\therefore \quad \mathbf{F} \cdot \hat{\mathbf{n}} = F_1\cos\alpha + F_2\cos\beta + F_3\cos\gamma.$$

and $\quad \text{div } \mathbf{F} = \frac{\partial F_1}{\partial x} + \frac{\partial F_2}{\partial y} + \frac{\partial F_3}{\partial z} \quad$ and $\quad dV = dxdydz.$

Thus (1) *becomes*

$$\iiint_V \left(\frac{\partial F_1}{\partial x} + \frac{\partial F_2}{\partial y} + \frac{\partial F_3}{\partial z}\right) dxdydz = \iint_S (F_1\cos\alpha + F_2\cos\beta + F_3\cos\gamma) dS. \qquad \ldots(2)$$

Proof. We shall first prove the theorem for a special volume V which is bounded by a piecewise smooth oriented surface S and has the property that any straight line drawn parallel to any one of the co-ordinate axes and intersecting V has only one point (or one segment) in common with V.

Then V can be represented by $\quad f(x, y) \le z \le g(x, y) \qquad \ldots(3)$

where $(x, y) \in \mathbf{R}$. This R is the orthogonal projection of V in the xy-plane. Obviously $z = f(x, y)$ represents the lower part S_2 of S and $z = g(x, y)$ represents the upper part S_1 of S and there may be a remaining vertical S_3 of S has shown in fig. 18.

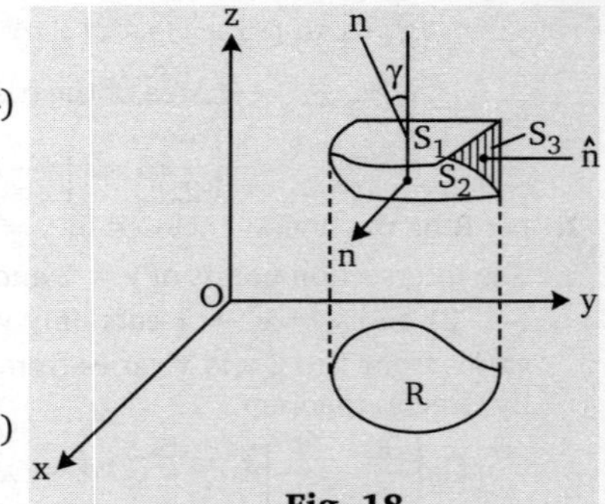

Fig. 18

First we prove that

$$\iiint_V \frac{\partial F_3}{\partial z} dxdydz = \iint_S F_3 \cos\gamma dS. \qquad \ldots(4)$$

Since $\mathbf{F}(x, y, z)$ is continuously differentiable in V and using (3), we have

$$\iiint_V \frac{\partial F_3}{\partial z} dxdydz = \iint_R \left[\int_{z=f(x,y)}^{z=g(x,y)} \frac{\partial F_3}{\partial z} dz \right] dxdy$$

$$= \iint_R \left[F_3(x, y, z) \right]_{z=f(x,y)}^{z=g(x,y)} dxdy.$$

$$\therefore \quad \iiint_V \frac{\partial F_3}{\partial z} dxdydz = \iint_R F_3[x, y, g(x, y)]dxdy - \iint_R F_3[x, y, f(x, y)]dxdy \qquad \ldots(5)$$

Now we have

$$\iint_S F_3 \cos\gamma dS = \iint_{S_1} F_3 \cos\gamma dS + \iint_{S_2} F_3 \cos\gamma dS + \iint_{S_3} F_3 \cos\gamma dS. \qquad \ldots(6)$$

Since on the portion S_3 of S the outward drawn unit normal vector makes an angle $\pi/2$ with z-axis, then $\cos\gamma = 0$ on S_3. Thus

$$\iint_{S_3} F_3 \cos\gamma dS = \iint_{S_3} 0.dS = 0. \qquad \ldots(7)$$

On the portion S_1 of S the outward drawn unit normal makes an acute angle γ with positive z-axis and the equation of S_1 is $z = g(x, y)$. Then

$$\cos\gamma \, dS = dxdy$$

$$\therefore \qquad \iint_{S_1} F_3 \cos\gamma dS = \iint_R F_3[x, y, g(x, y)]dxdy \qquad \ldots(8)$$

and on the portion S_2 of S the outward drawn unit normal vector makes obtuse angle γ with positive z-axis and the equation of S_2 is $z = f(x, y)$. Then

$$\cos\gamma \, dS = -\, dxdy$$

$$\therefore \qquad \iint_{S_2} F_3 \cos\gamma dS = -\iint_R F_3[x, y, f(x, y)]dxdy. \qquad \ldots(9)$$

Using (7), (8) and (9) the equation (6) becomes

$$\iint_S F_3 \cos\gamma dS = \iint_R F_3[x, y, g(x, y)]dxdy - \iint_R F_3[x, y, f(x, y)]dxdy. \qquad \ldots(10)$$

From (5) and (10), we obtain

$$\iiint_V \frac{\partial F_3}{\partial z} dxdydz = \iint_S F_3 \cos\gamma dS. \qquad \ldots(11)$$

Similarly taking the projection of S on the other co-ordinate planes, we have

$$\iiint_V \frac{\partial F_1}{\partial x} dxdydz = \iint_S F_1 \cos\alpha dS. \qquad \ldots(12)$$

and

$$\iiint_V \frac{\partial F_2}{\partial y} dxdydz = \iint_S F_2 \cos\beta dS. \qquad \ldots(13)$$

Now adding (11), (12) and (13), we get

$$\iiint_V \left(\frac{\partial F_1}{\partial x} + \frac{\partial F_2}{\partial y} + \frac{\partial F_3}{\partial z} \right) dxdydz = \iint_S (F_1 \cos\alpha + F_2 \cos\beta + F_3 \cos\gamma)dS. \qquad \ldots(14)$$

or

$$\iiint_V \text{div } \mathbf{F} dV = \iint_S \mathbf{F} \cdot \mathbf{n} dS.$$

Hence proved the theorem for special region V.

23.6.1 GAUSS' DIVERGENCE THEOREM FOR ANY REGION

Let V be any volume which is not a special volume but can be subdivided into finitely many special volumes by drawing auxillary surfaces. Now apply above theorem to each special volume and adding the result for each part. On the left hand side of this result we obtain the sum of volume integral over parts of V and which gives the volume integral over V. On the right hand side we obtain the sum of surface itegral over auxillary surfaces plus the sum of the remaining surface integral. In this side the surface integral over auxillary surfaces cancel in pairs and the remaining surface integrals give the surface integral over the whole boundary S of V.

REMARK

- The divergence theorem of Gauss can also be stated as the surface integral of the normal component of a vector **F** taken over a closed surface is equal to the volume integral of the divergence of **F** taken over the volume V enclosed by the surface.

23.7 APPLICATIONS OF GAUSS' DIVERGENCE THEOREM

The divergence theorem has various applications, some of which may be illustrated by the examples.

1. Representation of the divergence independent of the coordinates. By the divergence theorem, we have

$$\iiint_V \text{div } \mathbf{F} dV = \iint_S \mathbf{F}\cdot\mathbf{n} dS. \quad ...(1)$$

Dividing by the volume V of both sides of (1), we get

$$\frac{1}{V}\iiint_V \text{div } \mathbf{F} dV = \frac{1}{V}\iint_S \mathbf{F}\cdot\mathbf{n} dS. \quad ...(2)$$

Since for any continuous function $f(x, y, z)$, then by the mean value theorem for triple integral, we have

$$\iiint_V f(x,y,z)dV = f(x_0,y_0,z_0)V \quad ...(3)$$

where (x_0, y_0, z_0) is any point in V. Thus from (3), we have

$$\frac{1}{V}\iiint_V \text{div } \mathbf{F} dV = \text{div } \mathbf{F}(x_0,y_0,z_0). \quad ...(4)$$

Now let $P(x_1, y_1, z_1)$ be any fixed point in V and suppose V shrinks to the point P, so that the maximum distance $d(V)$ of the points of V from $P \to 0$, then $Q \to P$ and from (1) and (4), we have

$$\text{div } \mathbf{F}(x_1,y_1,z_1) = \lim_{d(V)\to 0}\frac{1}{V}\iint_S \mathbf{F}\cdot\mathbf{n} dS.$$

This formula is independent of the co-ordinate system while the definition of divergence involves co-ordintaes.

2. Heat flow. Since we know that in a body heat will flow from high temperature to lower temperature region and the ratio of flow is proportional to the gradient to the temperature. Let $\mathbf{F}$ be the velocity of the heat flow in a body. Then we have

$$\mathbf{F} = -k \text{ grad } U \quad ...(1)$$

where $U(x, y, z, t)$ is the temperature at the time t and k is the thermal conductivity of the body which is a constant. Let V be a volume in the body and S be its boundary surface. Then the amount of heat leaving V per unit time is

$$\iint_S \mathbf{F}\cdot\mathbf{n} dS.$$

Now using Gauss's divergence theorem

$$\iint_S \mathbf{F}\cdot\mathbf{n} dS = \iiint_V \text{div } \mathbf{F} dV.$$

Using (1), we have $\quad \text{div } \mathbf{F} = -k \text{ div (grad } U) = -k\nabla^2 U.$

$$\therefore \quad \iint_S \mathbf{F}\cdot\mathbf{n} dS = -k\iiint_V \nabla^2 U dxdydz. \qquad (\because dV = dxdydz) \quad ...(2)$$

Let H be the total amount of heat in V which is given by

$$H = \iiint_V \sigma\rho U dxdydz.$$

where the constant σ is the specific heat of the body and ρ is the density. Therefore the rate of decrease of H is

$$-\frac{\partial H}{\partial t} = -\iiint_V \sigma\rho\frac{\partial U}{\partial t} dxdydz.$$

Since $\quad -\dfrac{\partial H}{\partial t} = -\iint_S \mathbf{F}\cdot\mathbf{n} dS.$

$$\therefore \quad -\iiint_V \sigma\rho\frac{\partial U}{\partial t} dxdydz = -k\iiint_V \nabla^2 U dxdydz. \quad \text{or} \quad \iiint_V\left(\sigma\rho\frac{\partial U}{\partial t} - k\nabla^2 U\right)dxdydz = 0.$$

This equation holds for any volume V in the body, hence

$$\sigma\rho\frac{\partial U}{\partial t} - k\nabla^2 U = 0 \quad \text{or} \quad \frac{\partial U}{\partial t} = \frac{k}{\sigma\rho}\nabla^2 U \quad \text{or} \quad \frac{\partial U}{\partial t} = c^2\nabla^2 U, \quad c^2 = \frac{k}{\sigma\rho}$$

This is the heat equation.

Solved Examples

Example 1. *Let ϕ and ψ be scalar functions such that $\mathbf{F} = \phi$ grad ψ and $\mathbf{F} = \psi$ grad ϕ respectively, ϕ and ψ both are continuously differentiable in V enclosed by a surface S, then*

$$\iiint_V(\phi\nabla^2\psi - \psi\nabla^2\phi)dV = \iint_S(\phi\nabla\psi - \psi\nabla\phi)\cdot\mathbf{n} ds.$$

Solution . Since $\quad \mathbf{F} = \phi \text{ grad } \psi$, then

$$\text{div } \mathbf{F} = \text{div}(\phi \text{ grad } \psi) = \nabla\cdot(\phi\nabla\psi)$$
$$= \phi\nabla^2\psi + \nabla\phi\cdot\nabla\psi$$

and $\quad \mathbf{F.n} = \phi\nabla\psi\cdot\mathbf{n}.$

Using Gauss's divergence theorem

$$\iiint_V \text{div } \mathbf{F} dV = \iint_S \mathbf{F}\cdot\mathbf{n} dS.$$

$\therefore \iiint_V (\phi\nabla^2\psi + \nabla\psi \cdot \nabla\phi)dV = \iint_S \phi\nabla\psi \cdot \mathbf{n}ds.$ ...(1)

Now taking $\mathbf{F} = \psi$ grad ϕ,

$\therefore$ div $\mathbf{F} = \psi\nabla^2\phi + \nabla\phi \cdot \nabla\psi$

and $\mathbf{F.n} = \psi\nabla\phi\cdot\mathbf{n}.$

Again using Gauss's divergence theorem, we have

$$\iiint_V (\psi\nabla^2\phi + \nabla\phi \cdot \nabla\psi)dV = \iint_S \psi\nabla\phi \cdot \mathbf{n}ds. \quad ...(2)$$

Subtract (2) from (1), we obtain

$$\iiint_V (\phi\nabla^2\psi - \psi\nabla^2\phi)dV = \iint_S (\phi\nabla\psi - \psi\nabla\phi) \cdot \mathbf{n}ds. \quad ...(3)$$

REMARKS

- The formula given by (1) is called the first Green's formula or first form of Green's theorem.
- The formula given by (2) is called the second Green's formula or second form of Green's theorem.
- The formula obtained in (3) is called Green's theorem in symmetrical form.

Example 2. *If ϕ and ψ both are two harmonic scalar point functions and are continuously differentiable in V enclosed by S. Then*

$$\iint_S \left(\phi\frac{\partial\psi}{\partial n} - \psi\frac{\partial\phi}{\partial n}\right)dS = 0.$$

Solution. From example-1, we have

$$\iiint_V (\phi\nabla^2\psi - \psi\nabla^2\phi)dV = \iint_S (\phi\nabla\psi - \psi\nabla\phi) \cdot \mathbf{n}ds.$$

Since $\nabla\phi = \frac{\partial\phi}{\partial n}\mathbf{n}, \nabla\psi \frac{\partial\psi}{\partial n}\mathbf{n}$. Then we have

$$\iint_S \left(\phi\frac{\partial\psi}{\partial n} - \psi\frac{\partial\phi}{\partial n}\right)dS = \iiint_V (\phi\nabla^2\psi - \psi\nabla^2\phi)dV.$$

$(\because \mathbf{n.n} = 1)$

Further since ϕ and ψ both are harmonic so that $\nabla^2\phi = 0 = \nabla^2\psi$ hence we obtain

$$\iint_S \left(\phi\frac{\partial\psi}{\partial n} - \psi\frac{\partial\phi}{\partial n}\right)dS = 0.$$

Example 3. *Prove that $\iiint_V \nabla\phi dV = \iint_S \phi\mathbf{n}dS$.*

Solution. From Gauss's divergence theorem, we have

$$\iiint_V \text{div } \mathbf{F}dV = \iint_S \mathbf{F} \cdot \mathbf{n}dS. \quad ...(1)$$

Now assuming $\mathbf{F} = \phi\mathbf{a}$, where $\mathbf{a}$ is a constant vector, then

$$\text{div } \mathbf{F} = \nabla\cdot(\phi\mathbf{a}) = \nabla\phi \cdot \mathbf{a} + \phi\nabla\cdot\mathbf{a} = \nabla\phi \cdot \mathbf{a}$$

$(\because \nabla\cdot \mathbf{a} = 0)$

and $\mathbf{F}\cdot\mathbf{n} = (\phi\mathbf{a})\cdot\mathbf{n}$.

From (1), we have

$$\iiint_V \nabla\phi \cdot \mathbf{a}dV = \iint_S \phi\mathbf{a} \cdot \mathbf{n}dS.$$

or $\mathbf{a} \cdot \iiint_V \nabla\phi dV = \mathbf{a} \cdot \iint_S \phi\mathbf{n}dS.$

or $\mathbf{a} \cdot \left[\iiint_V \nabla\phi dV - \iint_S \phi\mathbf{n}dS\right] = 0.$

Since a is arbitrary so we get

$$\iiint_V \nabla\phi dV = \iint_S \phi\mathbf{n}dS.$$

Example 4. *Prove that $\int_S \nabla\phi \times \nabla\psi \cdot d\mathbf{S} = 0$.*

Solution. Since we have

$$\int_S \nabla\phi \times \nabla\psi \cdot d\mathbf{S} = \int_S (\nabla\phi \times \nabla\psi) \cdot \mathbf{n}dS.$$

Now by Gauss's divergence theorem

$$\int_S (\nabla\phi \times \nabla\psi) \cdot \mathbf{n}dS = \int_S \text{div}(\nabla\phi \times \nabla\psi)dV.$$

We know that

$$\text{div } (\nabla\phi \times \nabla\psi)dV = \nabla \cdot (\nabla\phi \times \nabla\psi).$$

$$= \nabla\psi \cdot \text{curl grad } \phi - \nabla\phi \cdot \text{curl grad } \psi.$$

$$= 0$$

$(\because$ curl grad $\phi = 0 =$ curl grad $\psi)$

$$\int_S \nabla\phi \times \nabla\psi \cdot d\mathbf{S} = \int_V 0 dV = 0.$$

Example 5. *Prove that $\iiint_V \nabla \times \mathbf{A}dV = \iint_S \mathbf{n} \times \mathbf{A}dS$.*

Solution. By Gauss's divergence theorem

$$\iiint_V \text{div } \mathbf{F}dV = \iint_S \mathbf{F} \cdot \mathbf{n}dS. \quad ...(1)$$

Let $\mathbf{F} = \mathbf{A} \times \mathbf{C}$ where $\mathbf{C}$ is an arbitrary constant vector. Then

$$\text{div } (\mathbf{A} \times \mathbf{C}) = \mathbf{C}.\text{curl } \mathbf{A} - \mathbf{A}.\text{curl } \mathbf{C} = \mathbf{C}.\text{ curl } \mathbf{A}$$

$(\because$ curl $\mathbf{C} = 0)$

and $\mathbf{F}\cdot\mathbf{n} = \mathbf{A} \times \mathbf{C}\cdot\mathbf{n} = [\mathbf{A\ C\ n}] = [\mathbf{C\ n\ A}]$

(By cyclic property of scalar triple product)

$$= \mathbf{C}\cdot\mathbf{n} \times \mathbf{A}$$

Equation (1) now becomes

$$\iiint_V \mathbf{C} \cdot \text{curl } \mathbf{A}dV = \iint_S \mathbf{C} \cdot \mathbf{n} \times \mathbf{A}dS.$$

or $\mathbf{C} \cdot \iiint_V \text{curl } \mathbf{A}dV = \mathbf{C} \cdot \iint_S \mathbf{n} \times \mathbf{A}dS.$

or $\mathbf{C} \cdot \left[\iiint_V \text{curl } \mathbf{A}dV - \iint_S \mathbf{n} \times \mathbf{A}dS\right] = 0.$

Since $\mathbf{C}$ is an arbitrary vector so that we have

$$\iiint_V \text{curl } \mathbf{A}dV - \iint_S \mathbf{n} \times \mathbf{A}dS = 0.$$

or $\iiint_V \nabla \times \mathbf{A}dV = \iint_S \mathbf{n} \times \mathbf{A}dS.$

Example 6. *Evaluate $\iint_S \mathbf{r} \cdot \mathbf{n}dS$, where S is a closed surface.*

Solution. By Gauss's divergence theorem

$$\iint_S \mathbf{r} \cdot \mathbf{n}dS = \iiint_V \text{div}\mathbf{r}dV. \quad ...(1)$$

where V is the volume enclosed by S.

Since div $\mathbf{r} = 3$. Then

$$\iint_S \mathbf{r} \cdot \mathbf{n}dS = \iiint_V 3dV = 3\iiint_V dV = 3V.$$

Example 7. *If ϕ is harmonic in V, then*

$$\iint_S \phi\frac{\partial\phi}{\partial n}dS = \iiint_V |\nabla\phi|^2 dV.$$

Solution. Since ϕ is harmonic, then $\nabla^2\phi = 0$.

Since we have

$$\iint_S \phi\frac{\partial\phi}{\partial n}dS = \iint_S \phi\text{grad}\phi.\mathbf{n}dS. \quad ...(1)$$

Using Gauss's divergence theorem

$$\iint_S \phi\text{grad}\phi.\mathbf{n}dS = \iiint_V \text{div}(\phi\text{grad } \phi)dV. \quad ...(2)$$

From (1) and (2), we get

$$\iint_S \phi \frac{\partial \phi}{\partial n} dS = \iiint_V \text{div}(\phi \text{grad } \phi) dV. \quad \ldots(3)$$

Now $\text{div}(\phi \nabla \phi) = \nabla\phi \cdot \nabla\phi + \phi \nabla^2 \phi = \nabla\phi \cdot \nabla\phi$

$$(\because \nabla^2 \phi = 0)$$

From (3), we get

$$\iint_S \phi \frac{\partial \phi}{\partial n} dS = \iiint_V (\nabla\phi \cdot \nabla\phi) dV = \iiint_V |\nabla\phi|^2 \, dV.$$

$$(\because \nabla\phi \cdot \nabla\phi = (\nabla\phi)^2 = |\nabla\phi|^2)$$

Example 8. *Verify Gauss's divergence theorem for*

$$\mathbf{F} = (2x - z)\hat{i} + x^2 y\hat{j} - xz^2 \hat{k}$$

taken over the region bounded by $x = 0$, $x = 1$, $y = 0$, $y = 1$, $z = 0$, $z = 1$.

(UPTU–2009, GBTU–2011)

Solution . By Gauss's divergence theorem, we have

$$\iiint_V \text{div } \mathbf{F} dV = \iint_S \mathbf{F} \cdot \mathbf{n} dS. \quad \ldots(1)$$

Here $\mathbf{F} = (2x - z)\hat{i} + x^2 y\hat{j} - xz^2 \hat{k}$

$$\therefore \quad \text{div } \mathbf{F} = \frac{\partial}{\partial x}(2x - z) + \frac{\partial}{\partial y} x^2 y + \frac{\partial}{\partial z}(-xz^2).$$

$$= 2 + x^2 - 2xz.$$

$$\therefore \quad \iiint_V \text{div } \mathbf{F} dV = \iiint_V (2 + x^2 - 2xz) dV$$

Here V is a cube bounded by $x = 0$, $x = 1$, $y = 0$, $y = 1$, $z = 0$, $z = 1$ is as shown in fig. 19.

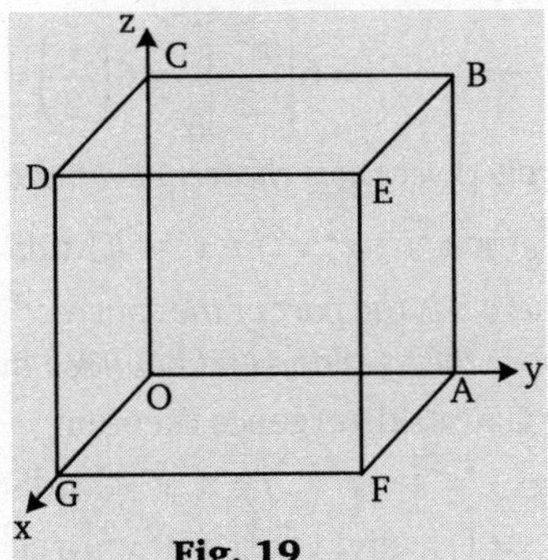

Fig. 19

$$\therefore \iiint_V (2 + x^2 - 2xz) dV$$

$$= \iiint_V (2 + x^2 - 2xz) dx dy dz$$

$$= \int_{x=0}^{1} \int_{y=0}^{1} \int_{z=0}^{1} (2 + x^2 - 2xz) dx dy dz$$

$$= \int_{x=0}^{1} \int_{y=0}^{1} \left[2z + x^2 z - xz^2 \right]_0^1 dy dx$$

$$= \int_{x=0}^{1} \int_{y=0}^{1} (2 + x^2 - x) dy dx$$

$$= \int_{x=0}^{1} \left[(2 + x^2 - x) y \right]_0^1 dx$$

$$= \int_{x=0}^{1} (2 + x^2 - x) dx$$

$$= \left(2x + \frac{x^3}{3} - \frac{x^2}{2} \right)_0^1 = \left(2 + \frac{1}{3} - \frac{1}{2} \right) = \frac{11}{6}.$$

Now we find $\iint_S \mathbf{F} \cdot \mathbf{n} dS$ over six faces of the cube.

Over the face $DEFG$: $\mathbf{n} = \hat{i}$, $x = 1$ and $dS = dydz$

$$\therefore \iint_{DEFG} \mathbf{F} \cdot \mathbf{n} dS$$

$$= \int_{z=0}^{z=1} \int_{y=0}^{y=1} [(2x - z)\hat{i} + x^2 y\hat{j} - xz^2 \hat{k}].\hat{i} dydz.$$

$$= \int_{z=0}^{z=1} \int_{y=0}^{y=1} (2x - z) dydz$$

$$= \int_{z=0}^{z=1} [(2x - z)y]_0^1 dz = \int_{z=0}^{z=1} (2x - z) dz$$

$$= \int_0^1 (2 - z) dz \quad (\because x = 1)$$

$$= \left(2z - \frac{z^2}{2} \right)_0^1 = 2 - \frac{1}{2} = \frac{3}{2}$$

Over the face $ABCO$: $\mathbf{n} = -\hat{i}$, $x = 0$, then

$$\therefore \iint_{ABCO} \mathbf{F} \cdot \mathbf{n} dS = \iint_{ABCO} (-z)\hat{i}.(-\hat{i}) dydz.$$

$$= \int_{z=0}^{z=1} \int_{y=0}^{y=1} z dydz$$

$$= \int_{z=0}^{z=1} z [y]_0^1 dz = \int_{z=0}^{z=1} z dz$$

$$= \left(\frac{z^2}{2} \right)_0^1 = \frac{1}{2}$$

Over the face $ABEF$: $\mathbf{n} = \hat{j}$, $y = 1$ and $dS = dzdx$

$$\therefore \iint_{ABEF} \mathbf{F} \cdot \mathbf{n} dS$$

$$= \iint_{ABEF} [(2x - z)\hat{i} + x^2 \hat{j} - xz^2 \hat{k}].\hat{j} dzdx.$$

$$= \int_{z=0}^{z=1} \int_{x=0}^{x=1} x^2 dzdx$$

$$= \int_{z=0}^{z=1} \left[\frac{x^3}{3} \right]_0^1 dz = \frac{1}{3} \int_0^1 dz$$

$$= \frac{1}{3} (z)_0^1 = \frac{1}{3}$$

Over the face $OCDG$: $\mathbf{n} = -\hat{j}$, $y = 0$ then

$$\therefore \iint_{OCDG} \mathbf{F} \cdot \mathbf{n} dS$$

$$= \iint_{OCDG} [(2x - z)\hat{i} - xz^2 \hat{k}].(-\hat{j}) dzdx$$

$$= \iint_{OCDG} 0 dzdx = 0$$

Over the face $BCDE$: $\mathbf{n} = \hat{k}$, $z = 1$ and $dS = dxdy$

$$\therefore \iint_{BCDE} \mathbf{F}.\mathbf{n} dS$$

$$= \iint_{BCDE} [(2x - 1)\hat{i} + x^2 \hat{j} - x\hat{k}].\hat{k} dxdy.$$

$$= \int_{y=0}^{1} \int_{x=0}^{1} (-x) dxdy$$

$$= -\int_{y=0}^{1} \left[\frac{x^2}{2} \right]_0^1 dy = -\frac{1}{2} \int_{y=0}^{1} dy$$

$$= -\frac{1}{2} (y)_0^1 = -\frac{1}{2}$$

Over the face $AOGF$: $\mathbf{n} = -\hat{k}, z = 0$ then

$$\therefore \iint_{AOGF} \mathbf{F}\cdot\mathbf{n}dS$$

$$= \iint_{AOGF}(2x\hat{i} + x^2y\hat{j}).(-\hat{k})dxdy$$

$$= \iint_{AOGF} 0.dxdy = 0$$

$$\text{Thus} \iint_S \mathbf{F}\cdot\mathbf{n}dS = \iint_{DEFG}\mathbf{F}\cdot\mathbf{n}dS + \iint_{ABCO}\mathbf{F}\cdot\mathbf{n}dS + \iint_{ABEF}\mathbf{F}\cdot\mathbf{n}dS + \iint_{OCDG}\mathbf{F}\cdot\mathbf{n}dS + \iint_{BCDE}\mathbf{F}\cdot\mathbf{n}dS + \iint_{AOGF}\mathbf{F}\cdot\mathbf{n}dS$$

$$= \frac{3}{2} + \frac{1}{2} + \frac{1}{3} + 0 - \frac{1}{2} + 0 = \frac{11}{6}$$

$$\iiint_V \text{div}\,\mathbf{F}dV = \iint_S \mathbf{F}\cdot\mathbf{n}dS = \frac{11}{6}.$$

Hence Gauss's divergence theorem is verified.

Example 9. *Prove that* $\iiint_V \frac{dV}{r^2} = \iint_S \frac{\mathbf{r}\cdot\mathbf{n}}{r^2}dS.$

Solution . Since we have $\iint_S \frac{\mathbf{r}\cdot\mathbf{n}}{r^2}dS = \iint_S \frac{\mathbf{r}}{r^2}\cdot\mathbf{n}dS$ ...(1)

Using Gauss's divergence theorem

$$\iint_S \frac{\mathbf{r}}{r^2}\cdot\mathbf{n}dS = \iiint_V \text{div}\left(\frac{\mathbf{r}}{r^2}\right)dV. \quad ...(2)$$

Since

$$\text{div}\left(\frac{\mathbf{r}}{r^2}\right) = \text{grad}\left(\frac{1}{r^2}\right).\mathbf{r} + \frac{1}{r^2}\text{div}\,\mathbf{r}$$

$$= -\frac{2}{r^3}\text{grad}r.\mathbf{r} + \frac{3}{r^2} \quad (\because \text{div}\,\mathbf{r} = 3)$$

$$= -\frac{2}{r^3}\cdot\frac{\mathbf{r}}{r}\cdot\mathbf{r} + \frac{3}{r^2} \quad \left(\because \text{grad}\,r = \frac{\mathbf{r}}{r}\right)$$

$$= \frac{-2}{r^2} + \frac{3}{r^2}$$

$$\therefore \quad \text{div}\left(\frac{\mathbf{r}}{r}\right) = \frac{1}{r^2}$$

$$\therefore \iiint_V \text{div}\left(\frac{\mathbf{r}}{r^2}\right)dV = \iiint_V \frac{dV}{r^2}. \quad ...(3)$$

From (1), (2) and (3), we get

$$\iint_S \frac{\mathbf{r.n}}{r^2}dS = \iiint_V \frac{dV}{r^2}.$$

Example 10. *Using Gauss's divergence theorem evaluate*

$$\iint_S[(x+z)dydz + (y+z)dzdx + (x+y)dxdy]$$

where S is the surface of the sphere $x^2+y^2+z^2=4$.

Solution . By Gauss's divergence theorem, we have

$$\iint_S[(x+z)dydz + (y+z)dzdx + (x+y)dxdy]$$

$$= \iiint_V\left[\frac{\partial}{\partial x}(x+z) + \frac{\partial}{\partial y}(y+z) + \frac{\partial}{\partial z}(x+y)\right]dxdydz$$

$$= \iiint_V 2dxdydz = 2\iiint_V dV,$$

where V is the volume of the sphere $x^2+y^2+z^2=4$.

$$= 2\left[\frac{4}{3}\pi(2)^3\right] = \frac{64}{3}\pi$$

Example 11. *Using divergence theorem evaluate*

$$\iint_S xyzdydz$$

where S is surface of the parallelopiped $0 \le x \le 3$, $0 \le y \le 2$, $0 \le z \le 1$.

Solution . By Gauss's divergence theorem, we have

$$\iint_S (F_1dydz + F_2dzdx + F_3dxdy)$$

$$= \iiint_V\left(\frac{\partial F_1}{\partial x} + \frac{\partial F_2}{\partial y} + \frac{\partial F_3}{\partial z}\right)dxdydz.$$

Here $F_1 = xyz, F_2 = 0, F_3 = 0$

$$\therefore \quad \frac{\partial F_1}{\partial x} = \frac{\partial}{\partial x}(xyz) = yz$$

$$\therefore \iint_S xyzdydz = \iiint_V yzdxdydz$$

$$= \int_{z=0}^1\int_{y=0}^2\int_{x=0}^3 yzdxdydz$$

$$= \int_{z=0}^1\int_{y=0}^2 yz[x]_0^3\,dydz$$

$$= 3\int_{z=0}^1\int_{y=0}^2 yzdydz$$

$$= 3\int_{z=0}^1 z\left[\frac{y^2}{2}\right]_0^2 dz = 6\int_{z=0}^1 zdz$$

$$= 6\left[\frac{z^2}{2}\right]_0^1 = 6\left(\frac{1}{2}\right) = 3.$$

Example 12. *Apply divergence theorem evaluate*

$$\iint_S(y^2z^2\hat{i} + z^2x^2\hat{j} + z^2y^2\hat{k}).\mathbf{n}dS$$

where S is the part of the sphere $x^2+y^2+z^2=1$ *above the xy-plane and bounded by this plane.*

Solution . By Gauss's divergence theorem

$$\iint_S(y^2z^2\hat{i} + z^2x^2\hat{j} + z^2y^2\hat{k}).\mathbf{n}dS$$

$$= \iiint_V \text{div}[y^2z^2\hat{i} + z^2x^2\hat{j} + z^2y^2\hat{k}]dV$$

$$= \iiint_V\left[\frac{\partial}{\partial x}(y^2z^2) + \frac{\partial}{\partial y}(z^2x^2) + \frac{\partial}{\partial z}(z^2y^2)\right]dV$$

$$= \iiint_V 2zy^2dV$$

(where V is the volume of the sphere $x^2+y^2+z^2=1$ above the xy-plane and bounded by this plane)

Now using spherical polar co-ordinate system, in this system

$$dV = dr\,r\,d\theta\,r\sin\theta\,d\phi = r^2\sin\theta\,dr\,d\theta\,d\phi$$

where $z = r\cos\theta, y = r\sin\theta\sin\phi, x = r\sin\theta\cos\phi$

The limit of integration which cover the volume V are r varies from 0 to 1, θ varies from 0 to $\pi/2$ and ϕ takes the value from 0 to 2π.

$$\iiint_V 2zy^2 dV$$

$$= 2\int_{r=0}^{1}\int_{\theta=0}^{\pi/2}\int_{\phi=0}^{2\pi}(r\cos\theta)r^2\sin^2\theta \sin^2\phi r^2 \sin\theta dr d\theta d\phi$$

$$= 2\int_{r=0}^{1}\int_{\theta=0}^{\pi/2}\int_{\phi=0}^{2\pi} r^5 \sin^3\theta\cos\theta\sin^2\phi dr d\theta d\phi$$

$$= 2\int_{r=0}^{1} r^5 dr . \int_{\theta=0}^{\pi/2}\sin^3\theta\cos\theta d\theta . \int_{\phi=0}^{2\pi}\sin^2\phi d\phi$$

(This is done because the limits of r, θ and ϕ are constant.)

$$= 2\left[\frac{r^6}{6}\right]_0^1 . \left[\frac{\sin^4\theta}{4}\right]_0^{\pi/2} . \frac{1}{2}\left(\phi - \frac{\sin 2\phi}{2}\right)_0^{2\pi}$$

$$= 2.\frac{1}{6}.\frac{1}{4}.\frac{1}{2}(2\pi - 0) = \frac{\pi}{12}.$$

Example 13. *Using divergence theorem evaluate*

$$\iint_S (e^x dydz - ye^x dzdx + 3zdxdy)$$

where S is the surface of the cylinder $x^2 + y^2 \le c^2$, $0 \le z \le h$.

Solution . By Gauss's divergence theorem , we have

$$\iint_S (F_1 dydz + F_2 dzdx + F_3 dxdy) = \iiint_V \left(\frac{\partial F_1}{\partial x} + \frac{\partial F_2}{\partial y} + \frac{\partial F_3}{\partial z}\right) dxdydz.$$

Here $F_1 = e^x, F_2 = -ye^x, F_3 = 3z$

$$\therefore \iint_S (e^x dydz - ye^x dzdx + 3zdxdy) = \iiint_V (e^x - e^x + 3) dxdydz$$

$$= 3\iiint_V dxdydz$$

where V is the volume of the cylinder $x^2 + y^2 \le c^2$, $0 \le z \le h$

$$= 3[\pi(c^2).h] = 3\pi c^2 h.$$

Example 14. *Evaluate* $\iint_S (\nabla \times \mathbf{F}) \cdot \mathbf{n} dS$ *where*

$$\mathbf{F} = (x - z)\hat{i} + (x^3 + yz)\hat{j} - 3xy^2\hat{k}$$

and S is the surface of the cone $x^2 + y^2 = (2 - z)^2$ *above the xy-plane*

Solution . The cone $x^2 + y^2 = (2 - z)^2$ meets the xy-plane in a circle whose equation is $x^2 + y^2 = 4, z = 0$. Let S_1 be the plane region bounded by the circle. If S_2 is the surface consisting of the surfaces S and S_1, then S_2 is closed surface. Then we have

$$\iint_{S_2} (\nabla \times \mathbf{F}) \cdot \mathbf{n} dS = 0$$

$$\therefore \iint_S (\nabla \times \mathbf{F}) \cdot \mathbf{n} dS + \iint_{S_1} (\nabla \times \mathbf{F}) \cdot \mathbf{n} dS = 0 \quad \ldots(1)$$

Since outward drawn unit normal to S_1 is $-\hat{k}$

Thus $\mathbf{n} = -k$ on S_1.

$\therefore$ From (1), we get

$$\iint_S (\nabla \times \mathbf{F}) \cdot \mathbf{n} dS = \iint_{S_1} (\nabla \times \mathbf{F}) \cdot \hat{k} dS. \quad \ldots(2)$$

Now find $\nabla \times \mathbf{F}$

$$\nabla \times \mathbf{F} = \begin{vmatrix} \hat{i} & \hat{j} & \hat{k} \\ \frac{\partial}{\partial x} & \frac{\partial}{\partial y} & \frac{\partial}{\partial z} \\ x - z & x^3 + yz & -3xy^2 \end{vmatrix}$$

$$= \hat{i}\left[\frac{\partial}{\partial y}(-3xy^2) - \frac{\partial}{\partial z}(x^3 + yz)\right] + \hat{j}\left[\frac{\partial}{\partial z}(x - z) - \frac{\partial}{\partial x}(-3xy^2)\right] + \hat{k}\left[\frac{\partial}{\partial x}(x^3 + yz) - \frac{\partial}{\partial y}(x - z)\right]$$

$$= \hat{i}(-6xy - y) + \hat{j}(-1 + 3y^2) + \hat{k}(3x^2)$$

and $(\nabla \times \mathbf{F}).\hat{k} = 3x^2$.

Now from (2), we get

$$\iint_S (\nabla \times \mathbf{F}).\mathbf{n} dS = \iint_{S_1} 3x^2 dS.$$

$$= \int_{\theta=0}^{2\pi}\int_{r=0}^{2} 3r^2 \cos^2\theta r d\theta dr$$

$(\because x = r\cos\theta, y = r\sin\theta, dxdy = rd\theta dr)$

$$= 3\int_{\theta=0}^{2\pi}\left[\frac{r^4}{4}\right]_0^2 \cos^2\theta d\theta$$

$$= 12\int_0^{2\pi}\cos^2\theta d\theta$$

$$= \frac{12}{2}\left[\theta + \frac{\sin 2\theta}{2}\right]_0^{2\pi}$$

$$= \frac{12}{2}(2\pi) = 12\pi$$

Example 15. *Prove that* $\int_V \nabla\phi \cdot \text{curl } \mathbf{F} dV = \int_S (\mathbf{F} \times \nabla\phi) \cdot dS$.

Solution . Since we know that

$$\int_S (\mathbf{F} \times \nabla\phi).dS = \int_S (\mathbf{F} \times \nabla\phi).\mathbf{n} dS = \int_V \text{div}(\mathbf{F} \times \nabla\phi) dV \quad \ldots(1)$$

(By Gauss's divergence theorem)

Further since,

$$\text{div } (\mathbf{F} \times \nabla\phi) = \nabla\cdot(\mathbf{F} \times \nabla\phi) = \nabla\phi \cdot \text{curl } \mathbf{F} - \mathbf{F}\cdot \text{curl } (\nabla\phi) = \nabla\phi \cdot \text{curl } \mathbf{F} \quad (\because \text{curl } (\nabla\phi) = \mathbf{0})$$

Now putting div $(\mathbf{F} \times \nabla\phi) = \nabla\phi$. curl $\mathbf{F}$ in (1), we get

$$\int_S (\mathbf{F} \times \nabla\phi) \cdot dS = \int_V (\nabla\phi \cdot \text{curl } \mathbf{F}) dV.$$

Example 16. *If* ϕ *is harmonic in V and* $\frac{\partial\phi}{\partial n} = 0$ *on S, then* ϕ *is constant in V.*

Solution . Since ϕ is harmonic in V, then $\nabla^2\phi = 0$ and we have

$$\iint_S \phi\frac{\partial\phi}{\partial n} dS = \iiint_V |\nabla\phi|^2 dV. \quad \ldots(1)$$

Since $\frac{\partial\phi}{\partial n} = 0$ on S.

$\therefore$ From (1), we get

$$\iiint_V |\nabla\phi|^2\, dV = 0.$$

$\therefore\ |\nabla\phi|^2 = 0$ in V or $\nabla\phi = \mathbf{0}$ in V

$\therefore\ \phi$ is a constant in V.

Example 17. *By Gauss divergence theorem, show that*

$$\iint_S (x^2\hat{i} + y^2\hat{j} + z^2\hat{k})\cdot \mathbf{n}dS = 0,$$

where S is the surface of the ellipsoid $\dfrac{x^2}{a^2}+\dfrac{y^2}{b^2}+\dfrac{z^2}{c^2}=1.$

Solution. By Gauss divergence theorem, we have

$$\iint_S (x^2\hat{i} + y^2\hat{j} + z^2\hat{k})\cdot \mathbf{n}dS$$

$$= \iiint_V \operatorname{div}(x^2\hat{i} + y^2\hat{j} + z^2\hat{k})dV$$

where V is the volume enclosed by S.

$$= \iiint_V (2x+2y+2z)dxdydz$$

$$= 2\iiint_V (x+y+z)dxdydz \qquad (\because dV = dxdydz)$$

$$= 2\int_{z=-c}^{c}\int_{y=-b\sqrt{1-z^2/c^2}}^{y=b\sqrt{1-z^2/c^2}}\int_{x=-a\sqrt{1-y^2/b^2-z^2/c^2}}^{x=a\sqrt{1-y^2/b^2-z^2/c^2}} (x+y+z)dxdydz$$

$$= 2\int_{z=-c}^{c}\int_{y=-b\sqrt{1-z^2/c^2}}^{y=b\sqrt{1-z^2/c^2}} \left[\frac{x^2}{2}+x(y+z)\right]_{-a\sqrt{1-y^2/b^2-z^2/c^2}}^{a\sqrt{1-y^2/b^2-z^2/c^2}} dydz$$

$$= 4a\int_{z=-c}^{c}\int_{y=-b\sqrt{1-z^2/c^2}}^{y=b\sqrt{1-z^2/c^2}} (y+z)\sqrt{1-\frac{y^2}{b^2}-\frac{z^2}{c^2}}\, dydz$$

$$= 8a\int_{z=-c}^{c}\int_{0}^{y=b\sqrt{1-z^2/c^2}} z\left(\sqrt{\left(1-\frac{z^2}{c^2}\right)-\frac{y^2}{b^2}}\right)dydz$$

$$= \frac{8a}{b}\int_{z=-c}^{c} z\left[\frac{y}{b}\sqrt{b^2\left(1-\frac{z^2}{c^2}\right)-y^2} + \frac{b^2}{2}\left(1-\frac{z^2}{c^2}\right)\sin^{-1}\left\{\frac{y}{b\sqrt{1-\frac{z^2}{c^2}}}\right\}\right]_0^{b\sqrt{1-z^2/c^2}} dz$$

$$= \frac{8a}{b}\int_{z=-c}^{c} z\left[\frac{b^2}{2}\left(1-\frac{z^2}{c^2}\right)\sin^{-1}1\right]dz$$

$$= \frac{8ab^2\pi}{4b}\int_{z=-c}^{c} z\left(1-\frac{z^2}{c^2}\right)dz = 0$$

EXERCISE 23.3

1. Verify divergence theorem for
$$\mathbf{F} = (x^2-yz)\hat{i} + (y^2-zx)\hat{j} + (z^2-xy)\hat{k}$$
taken over the region bounded by $0 \le x \le a,\ 0 \le y \le b,\ 0 \le z \le c.$ (Rohtak–2006, Madras–2000)
2. Verify Gauss's divergence theorem for $\mathbf{F} = 4xz\hat{i} - y^2\hat{j} + yz\hat{k}$ taken over the cube bounded by $x = 0, x = 1, y = 0, y = 1, z = 0, z = 1$. (Madras–2006)
3. Verify divergence theorem for $\mathbf{F} = 4x\hat{i} - 2y^2\hat{j} + z^2\hat{k}$ taken over the region bounded by the surfaces $x^2+y^2 = 4$, $z = 0$, $z = 3$. (JNTU–2006, SVTU–2007, Mumbai–2006)
4. Using Gauss's divergence theorem, evaluate
$$\iint_S (x\hat{i}+y\hat{j}+z\hat{k}).\mathbf{n}dS$$
where S is the closed surface bounded by the cone $x^2+y^2 = z^2$ and the plane $z = 1$.
5. For any closed surface S, prove that
$$\iint_S \operatorname{curl}\mathbf{F}\cdot\mathbf{n}dS = 0.$$
6. If $\mathbf{F} = \nabla\phi$ and $\nabla^2\phi = 0$, show that for a closed surface S
$$\iiint_V \mathbf{F}^2 dV = \iint_S \phi\mathbf{F}\cdot\mathbf{n}dS.$$
7. If ϕ and ψ are harmonic in V and $\dfrac{\partial\phi}{\partial n} = \dfrac{\partial\psi}{\partial n}$ on S, then $\phi = \psi + c$ in V where C is a constant.
8. For any closed surface S, show that
(i) $\iint_S \mathbf{n}dS = \mathbf{0}$. (ii) $\iint_S \mathbf{r}\times\mathbf{n}dS = \mathbf{0}$.
9. Using the divergence theorem, show that the volume V of a region bounded by a surface S is
$$V = \iint_S xdydz = \iint_S ydzdx = \iint_S zdxdy$$
$$= \frac{1}{3}\iint_S (xdydz + ydzdx + zdxdy).$$ (Kurukshetra–2008)
10. By Gauss's divergence theorem evaluate
$$\iint_S (x^2dydz + y^2dzdx + z^2dxdy)$$
where S is a cube bounded by $0 \le x \le 1,\ 0 \le y \le 1,\ 0 \le z \le 1$.
11. Using divergence theorem evaluate
$$\iint_S (yzdydz + zxdzdx + xydxdy)$$
where S is the boundary of the sphere $x^2+y^2+z^2 = 4$. (UPTU–2004, 09)
12. Apply Gauss's divergence theorem, evaluate
$$\iint_S [\sin x dydz + (2-\cos x)ydzdx]$$
where S is the parallelopiped bounded by $0 \le x \le 3,\ 0 \le y \le 2$ and $0 \le z \le 1$.
13. Evaluate $\iint_S [x^2dydz + y^2dzdx + 2z(xy-x-y)dxdy]$ where S is the surface of the cube $0 \le x \le 1,\ 0 \le y \le 1$ and $0 \le z \le 1$.

14. If $\mathbf{F} = x\hat{i} - y\hat{j} + (z^2 - 1)\hat{k}$, find the value of $\iint_S \mathbf{F}\cdot\mathbf{n}dS$ where S is the closed surface bounded by the planes $z = 0$, $z = 1$ and the cylinder $x^2 + y^2 = 4$.

15. Evaluate $\iint_S (ax^2 + by^2 + cz^2)dS$ over the sphere $x^2 + y^2 + z^2 = 1$ using divergence theorem.

16. If $\mathbf{F} = y\hat{i} + (x - 2xz)\hat{j} - xy\hat{k}$, evaluate $\iint_S (\nabla\times\mathbf{F})\cdot\mathbf{n}dS$ where S is the surface of the sphere $x^2 + y^2 + z^2 = a^2$ above the xy-plane.

17. Evaluate $\iint_S (x^2 + y^2)dS$ where S is surface of the cone $z^2 = 3(x^2 + y^2)$ bounded by $z = 0$ and $z = 3$.

18. If $\mathbf{F} = (x^2 + y - 4)\hat{i} + 3xy\hat{j} + (2xz + z^2)\hat{k}$, evaluate
$$\iint_S (\nabla\times\mathbf{F})\cdot\mathbf{n}dS$$
where S is the surface of the sphere $x^2 + y^2 + z^2 = 16$ above the xy-plane.

19. Apply Gauss's divergence theorem, evaluate
$$\iint_S [(x^3 - yz)dydz - 2x^2ydzdx + zdxdy]$$
over the surface of a cube bounded by the co-ordinate planes and the planes $x = y = z = a$.

20. Prove that $\iint_S \mathbf{n}\times(\mathbf{a}\times\mathbf{r})dS = 2V\mathbf{a}$ where $\mathbf{a}$ is a constant vector and V is the volume enclosed by the closed surface S.

21. If $\mathbf{F} = \nabla\phi, \nabla^2\phi = -4\pi\rho$, show that $\iint_S \mathbf{F}\cdot\mathbf{n}dS = -4\pi\iiint_V \rho dV$.

22. Prove that $\iiint_V \nabla\phi\cdot\mathbf{F}dV = \iint_S \phi\mathbf{F}\cdot\mathbf{n}dS - \iiint_V \nabla\phi\cdot\mathbf{F}dV$.

23. If $\mathbf{C} = \frac{1}{2}\nabla\times\mathbf{B}, \mathbf{B} = \nabla\times\mathbf{A}$, show that
$$\iiint_V \mathbf{B}^2 dV = \iint_S \mathbf{A}\times\mathbf{B}.\mathbf{n}dS + 2\iiint_V \mathbf{A}.\mathbf{C}dV.$$

24. A vector $\mathbf{B}$ is always normal to a given closed surface S. Show that $\iiint_V \nabla\times\mathbf{B}dV = \mathbf{0}$ wherer V is the region bounded by S.

25. Let $\mathbf{r}$ denoted the position vector of the point $P(x, y, z)$ with respect to the origin 0 and $|\mathbf{r}| = r$, evaluate $\iint_S \frac{\mathbf{r}}{r^3}.\mathbf{n}dS$, where S is the sphere $x^2 + y^2 + z^2 = a^2$.

26. If V is the volume enclosed by a closed surface S and $\vec{F} = x\hat{i} + 2y\hat{j} + 3z\hat{k}$, show that $\int_S \vec{F}.\hat{n}dS = 6V$.

27. Verify Gauss's Divergence theorem and show that
$$\iint_S [(x^3 - yz)\hat{i} - 2x^2y\hat{j} + 2\hat{k}]\cdot\hat{n}dS = \frac{a^5}{2}$$
where S denotes the surface of the cube bounded by the planes $x = 0, x = a, y = 0, y = a, z = 0, z = a$. (UPTU–2006)

Hint to Selected Problems

5. Since S is closed and let V be the volume enclosed by S. Then by Gauss's divergence theorem,
$$\iint_S \mathbf{F}\cdot\mathbf{n}dS = \iiint_V \nabla\cdot\mathbf{F}dV$$
$$\Rightarrow \iint_S (\text{curl}\,\mathbf{F})\cdot\mathbf{n}dS = \iiint_V \text{div}(\text{curl}\,\mathbf{F})dV = 0 \ (\because \text{div}(\text{curl}\,\mathbf{F}) = 0)$$

7. Since ϕ and ψ are harmonic so $\nabla^2\phi = 0, \nabla^2\psi = 0$
$$\Rightarrow \quad \nabla^2\phi - \nabla^2\psi = 0 \quad \Rightarrow \quad \nabla^2(\phi - \psi) = 0$$
$\Rightarrow \quad (\phi - \psi)$ is also harmonic on V.

Also $\frac{\partial\phi}{\partial n} = \frac{\partial\psi}{\partial n}$ on S $\quad \Rightarrow \quad \frac{\partial}{\partial n}(\phi - \psi) = 0$

$\Rightarrow \quad \phi - \psi$ = constant $\Rightarrow \quad \phi = \psi + c$, c being constant

10. $\iint_S (x^2dydz + y^2dzdx + z^2dxdy) = \iint_S (x^2\hat{i} + y^2\hat{j} + z^2\hat{k}).\mathbf{n}dS$
$$= \iiint_V (2x + 2y + 2z)dxdydz \quad \text{(By Gauss's divergence theorem)}$$
$$= 2\int_{x=0}^1\int_{y=0}^1\int_{z=0}^1 (x + y + z)dxdydz$$
$$= 2\int_{x=0}^1\int_{y=0}^1 \left[xz + yz + \frac{z^2}{2}\right]_{z=0}^1 dydx$$
$$= 2\int_{x=0}^1\int_{y=0}^1 \left(x + y + \frac{1}{2}\right)dydx = 2\int_{x=0}^1 \left[xy + \frac{y^2}{2} + \frac{1}{2}y\right]_{y=0}^1 dx$$
$$= 2\int_0^1 \left(x + \frac{1}{2} + \frac{1}{2}\right)dx = 2\left[\frac{x^2}{2} + x\right]_{x=0}^1 = 3.$$

12. $\iint_S [\sin x dydz + (2 - \cos x)ydzdx]$
$$= \iint_S (\sin x\hat{i} + (2 - \cos x)\hat{j}]\cdot\mathbf{n}dS = \iiint_V (\cos x + 2 - \cos x)dx$$
(By Gauss's divergence theorem)
$$= 2\iiint_V dV = 2(\text{Volume of cuboid}) = 2[(3)(2)(1)] = 12$$

15. Let $f \equiv x^2 + y^2 + z^2 - 1 = 0$
$$\mathbf{n} = \frac{\nabla f}{|\nabla f|} = (x\hat{i} + y\hat{j} + z\hat{k})$$
$$\therefore \ \iint_S (ax^2 + by^2 + cz^2)dS = \iint_S \mathbf{F}\cdot(x\hat{i} + y\hat{j} + z\hat{k})dS.$$
$\therefore \ \mathbf{F} = ax\hat{i} + by\hat{j} + cz\hat{k}$ so div $\mathbf{F} = a + b + c$.

$\therefore \iint_S \mathbf{F}\cdot\mathbf{n}dS = \iiint_V \text{div}\,\mathbf{F}dV$ [By Gauss's divergence theorem]
$$= \iiint_V (a + b + c)dV = (a + b + c)\iiint_V dV$$
$$= (a + b + c) \text{ (Volume of sphere)}$$
$$= \frac{4\pi}{3}(a + b + c).$$

19. $\iint_S [(x^3 - yz)dydz - 2x^2ydzdx + zdxdy]$
$$= \iint_S [(x^3 - yz)\hat{i} - 2x^2y\hat{j} + z\hat{k}]\cdot\mathbf{n}dS$$
$$= \iiint_V (3x^2 - 2x^2 + 1).dxdydz$$
(By Gauss's divergence theorem)
$$= \int_{z=0}^a\int_{y=0}^a\int_{x=0}^a (x^2 + 1)dxdydz.$$

21. Since $\mathbf{F} = \nabla\phi, \nabla^2\phi = -4\pi\rho$
$$\iint_S \mathbf{F}\cdot\mathbf{n}dS = \iiint_V \text{div}\,\mathbf{F}dV = \iiint_V \nabla\cdot(\nabla\phi)dV$$
$$= \iiint_V \nabla^2\phi dV = \iiint_V (-4\pi\rho)dV = -4\pi\rho\iiint_V dV.$$

24. Let $\mathbf{n}$ be the unit vector normal to the surface S and also $\mathbf{B}$ is always normal to $\mathbf{S}$, so $\mathbf{n}$ and $\mathbf{B}$ are parallel
$\therefore \mathbf{n}\times\mathbf{B} = 0$
$$\iiint_V \nabla\times\mathbf{B}dV = \iint_S \mathbf{n}\times\mathbf{B})dS = \iint_S \mathbf{0}dS = \mathbf{0}.$$

26. Using Gauss's divergence theorem
$$\iint_S \mathbf{F}\cdot\hat{n}dS = \iiint_V \text{div}\cdot\mathbf{F}dV = \iiint_V (1 + 2 + 3)dV = 6\iiint_V dV = 6V.$$

ANSWERS

4. $\frac{7\pi}{6}$ **10.** 3 **11.** 0 **12.** 12 **13.** $\frac{1}{2}$ **14.** 4π **15.** $\frac{4\pi}{3}(a + b + c)$ **16.** 0 **17.** 9π **18.** -16π **19.** $a^2\left(\frac{a^3}{3} + a\right)$ **25.** 4π

23.8 STOKE'S THEOREM

Let S be a piecewise smooth oriented surface in space bounded by a piecewise smooth simple closed curve C. Let $\mathbf{F}(x, y, z)$ be a continuous vector function having continuous first order partrial derivatives in a region of space in which S lies interior. Then

$$\iint_S (\text{curl}\,\mathbf{F})\cdot \mathbf{n}\, dS = \int_C \mathbf{F}\cdot d\mathbf{r} \quad \ldots(1)$$

where C is taken as counterclockwise direction and $\mathbf{n}$ is a outward drawn unit normal vector to S.

(UPTU–2007, 08, 09, UKTU–2012)

Proof. We shall first prove Stoke's theorem for a surface S which represented simultaneously in the forms of

$$z = f(x, y),\, y = g(x, z),\, x = h(y, z)$$

where f, g, h are continuous functions and having continuous first order partial derivatives. Let $\mathbf{n} = \cos\alpha\hat{i} + \cos\beta\hat{j} + \cos\gamma\hat{k}$ be outward drawn unit normal to the surface S which makes the angles α, β, γ with positive co-ordinate axes respectively, and let

$$\mathbf{F} = F_1\hat{i} + F_2\hat{j} + F_3\hat{k}.$$

$$\therefore \quad \nabla\times\mathbf{F} = \begin{vmatrix} \hat{i} & \hat{j} & \hat{k} \\ \dfrac{\partial}{\partial x} & \dfrac{\partial}{\partial y} & \dfrac{\partial}{\partial z} \\ F_1 & F_2 & F_3 \end{vmatrix} = \hat{i}\left(\frac{\partial F_3}{\partial y} - \frac{\partial F_2}{\partial z}\right) + \hat{j}\left(\frac{\partial F_1}{\partial z} - \frac{\partial F_3}{\partial x}\right) + \hat{k}\left(\frac{\partial F_2}{\partial x} - \frac{\partial F_1}{\partial y}\right).$$

and
$$(\nabla\times\mathbf{F})\cdot\mathbf{n} = \left(\frac{\partial F_3}{\partial y} - \frac{\partial F_2}{\partial z}\right)\cos\alpha + \left(\frac{\partial F_1}{\partial z} - \frac{\partial F_3}{\partial x}\right)\cos\beta + \left(\frac{\partial F_2}{\partial x} - \frac{\partial F_1}{\partial y}\right)\cos\gamma.$$

Let $P(x, y, z)$ be any point on C whose position vector is

$$\mathbf{r} = x\hat{i} + y\hat{j} + z\hat{k}.$$

$$\therefore \quad d\mathbf{r} = dx\hat{i} + dy\hat{j} + dz\hat{k}.$$

Thus
$$\mathbf{F}\cdot d\mathbf{r} = F_1 dx + F_2 dy + F_3 dz.$$

Now the equation (1) becomes

$$\iint_S \left[\left(\frac{\partial F_3}{\partial y} - \frac{\partial F_2}{\partial z}\right)\cos\alpha + \left(\frac{\partial F_1}{\partial z} - \frac{\partial F_3}{\partial x}\right)\cos\beta + \left(\frac{\partial F_2}{\partial x} - \frac{\partial F_1}{\partial y}\right)\cos\gamma\right] = \int_C (F_1 dx + F_2 dy + F_3 dz). \quad \ldots(2)$$

First, we shall prove that

$$\iint_S \left(\frac{\partial F_1}{\partial z}\cos\beta - \frac{\partial F_1}{\partial y}\cos\gamma\right) = \int_C F_1 dx. \quad \ldots(3)$$

Let R be the orthogonal projection of S in the xy-plane and C^* be its boundary which is oriented in positive direction as shown in fig. 20.

Using the representation $z = f(x, y)$ of S, we may write the line integral over C as a line integral over C^* as follows :

$$\int_C F_1(x,y,z)dx = \int_{C^*} F_1[x, y, f(x,y)]dx = \int_{C^*} [F_1[x,y,f(x,y)]dx + 0dy].$$

We now apply Green's theorem in the plane to the functions $F_1[x, y, f(x, y)]$ and 0. Then we have

$$\int_C F_1(x,y,z)dx = -\iint_R \frac{\partial F_1}{\partial y} dxdy.$$

But
$$\frac{\partial F_1[x,y,f(x,y)]}{\partial y} = \frac{\partial F_1(x,y,z)}{\partial y} + \frac{\partial F_1(x,y,z)}{\partial z}\cdot\frac{\partial f}{\partial y} \quad [\because z = f(x,y)]$$

$$\therefore \quad \int_C F_1(x,y,z)dx = -\iint_R \left(\frac{\partial F_1}{\partial y} + \frac{\partial F_1}{\partial z}\cdot\frac{\partial f}{\partial y}\right)dxdy. \quad \ldots(4)$$

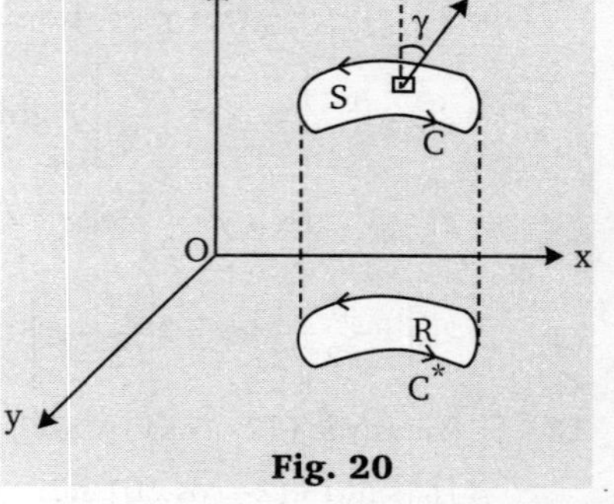

Fig. 20

Now we shall prove that the integral on R.H.S. of (4) is equal to integral on L.H.S. of (3). For this let us consider

$$\phi(x, y, z) = z - f(x, y) = 0.$$

$$\therefore \quad \text{grad}\,\phi = \frac{\partial\phi}{\partial x}\hat{i} + \frac{\partial\phi}{\partial y}\hat{j} + \frac{\partial\phi}{\partial z}\hat{k} = -\frac{\partial f}{\partial x}\hat{i} - \frac{\partial f}{\partial y}\hat{j} + \hat{k}. \quad \left(\because \frac{\partial\phi}{\partial x} = -\frac{\partial f}{\partial x}, \frac{\partial\phi}{\partial y} = -\frac{\partial f}{\partial y}, \frac{\partial\phi}{\partial z} = 1\right)$$

Let the length of grad ϕ be a

$$\therefore \quad a = |\text{grad}\,\phi|.$$

Since we know that grad ϕ is perpendicular to the surface S. Therefore, we have

$$\mathbf{n} = \pm\frac{\text{grad }\phi}{|\text{grad }\phi|} = \pm\frac{\text{grad }\phi}{a}$$

But the components of both $\mathbf{n}$ and grad ϕ in the positive direction of z-axis are positive. Thus

$$\mathbf{n} = +\frac{\text{grad }\phi}{a} = \frac{1}{a}\left(-\frac{\partial f}{\partial x}\hat{i} - \frac{\partial f}{\partial y}\hat{j} + \hat{k}\right).$$

Since $\mathbf{n} = \cos\alpha\hat{i} + \cos\beta\hat{j} + \cos\gamma\hat{k}$, therefore on comparing these twos, we get

$$\cos\alpha = -\frac{1}{a}\frac{\partial f}{\partial x}, \cos\beta = -\frac{1}{a}\frac{\partial f}{\partial y}, \cos\gamma = \frac{1}{a}$$

Since $\cos\gamma\, dS = dxdy \quad \therefore \quad dS = \dfrac{dxdy}{\cos\gamma} = adxdy \qquad \left(\because \cos\gamma = \dfrac{1}{a}\right)$

$$\therefore \iint_S\left(\frac{\partial F_1}{\partial z}\cos\beta - \frac{\partial F_1}{\partial y}\cos\gamma\right)dS = \iint_R\left[\frac{\partial F_1}{\partial z}\left(-\frac{1}{a}\frac{\partial f}{\partial y}\right) - \frac{\partial F_1}{\partial y}\frac{1}{a}\right]adxdy = -\iint_R\left(\frac{\partial F_1}{\partial y} + \frac{\partial F_1}{\partial z}\frac{\partial f}{\partial y}\right)dxdy. \qquad \ldots(5)$$

Thus from (4) and (5), we get

$$\int_C F_1 dx = \iint_S\left(\frac{\partial F_1}{\partial z}\cos\beta - \frac{\partial F_1}{\partial y}\cos\gamma\right)dS. \qquad \ldots(6)$$

Similarly using the representation $y = g(x, z)$ and $x = h(y, z)$ and having the projection on the other co-ordinates planes, we obtain

$$\int_C F_2 dy = \iint_S\left(\frac{\partial F_2}{\partial x}\cos\gamma - \frac{\partial F_2}{\partial z}\cos\alpha\right)dS. \qquad \ldots(7)$$

and

$$\int_C F_3 dz = \iint_S\left(\frac{\partial F_3}{\partial y}\cos\alpha - \frac{\partial F_3}{\partial x}\cos\beta\right)dS. \qquad \ldots(8)$$

Adding (6), (7) and (8), we get

$$\int_C(F_1dx + F_2dy + F_3dz) = \iint_S\left[\left(\frac{\partial F_3}{\partial y} - \frac{\partial F_2}{\partial z}\right)\cos\alpha + \left(\frac{\partial F_1}{\partial z} - \frac{\partial F_3}{\partial x}\right)\cos\beta + \left(\frac{\partial F_2}{\partial x} - \frac{\partial F_1}{\partial y}\right)\cos\gamma\right]dS.$$

or $\quad \int_C \mathbf{F}\cdot d\mathbf{r} = \iint_S(\nabla\times\mathbf{F})\cdot\mathbf{n}dS.$

This proves Stoke's theorem for the surface S which can be represented simultaneously by $z = f(x, y)$, $y = g(x, y)$ and $x = h(y, z)$.

The proof of this theorem can be extended to a surface S which does not satisfy above conditions but can be decomposed into finitely many surfaces $S_1, S_2, \ldots, S_n$ whose boundary are $C_1, C_2, \ldots, C_n$. To each surface this theorem is applied as follows :

$$\int_{C_1}\mathbf{F}\cdot d\mathbf{r} = \iint_{S_1}(\nabla\times\mathbf{F})\cdot\mathbf{n}dS.$$
$$\int_{C_2}\mathbf{F}\cdot d\mathbf{r} = \iint_{S_2}(\nabla\times\mathbf{F})\cdot\mathbf{n}dS.$$
$$\ldots \quad \ldots \quad \ldots \quad \ldots \quad \ldots$$
$$\ldots \quad \ldots \quad \ldots \quad \ldots \quad \ldots$$
$$\int_{C_n}\mathbf{F}\cdot d\mathbf{r} = \iint_{S_n}(\nabla\times\mathbf{F})\cdot\mathbf{n}dS.$$

On adding, we get

$$\int_{C_1}\mathbf{F}\cdot d\mathbf{r} + \int_{C_2}\mathbf{F}\cdot d\mathbf{r} + \ldots + \int_{C_n}\mathbf{F}\cdot d\mathbf{r} = \iint_{S_1}(\nabla\times\mathbf{F})\cdot\mathbf{n}dS + \iint_{S_2}(\nabla\times\mathbf{F})\cdot\mathbf{n}dS + \ldots + \iint_{S_n}(\nabla\times\mathbf{F})\cdot\mathbf{n}dS.$$

or $\quad \int_C \mathbf{F}\cdot d\mathbf{r} = \iint_S(\nabla\times\mathbf{F})\cdot\mathbf{n}dS.$

Hence Stoke's theorem is proved for any surface S enclosed by a closed curve C.

REMARK

- Stoke's theorem can also be stated as the line integral of the tangential component of the vector function $\mathbf{F}$ taken around a closed curve C is equal to the surface integral of the normal component of the curl $\mathbf{F}$ taken over any surface S enclosed by C. That is

$$\int_C \mathbf{F}\cdot\mathbf{t}\, ds = \iint_S(\nabla\times\mathbf{F})\cdot\mathbf{n}dS$$

where $\mathbf{t} = \dfrac{d\mathbf{r}}{ds}$ and s is the arc length of C.

23.9 APPLICATIONS OF STOKE'S THEOREM

The applications of Stoke's theorem may be illustrated by some examples.

1. Green's theorem in the plane as a special case of Stoke's theorem.

If $\mathbf{F} = F_1\hat{i} + F_2\hat{j}$ is a vector function which is continuously differentiable in a simply connected bounded closed surfaces in the xy-plane whose boundary is the piecewise smooth simple closed curve, then

$$(\nabla \times \mathbf{F}) \cdot \mathbf{n} = \left(\frac{\partial F_2}{\partial x} - \frac{\partial F_1}{\partial y}\right)$$

and $$\mathbf{F} \cdot d\mathbf{r} = F_1 dx + F_2 dy.$$

By Stoke's theorem, we have

$$\iint_S (\nabla \times \mathbf{F}) \cdot \mathbf{n} dS = \int_C \mathbf{F} \cdot d\mathbf{r}.$$

$$\therefore \quad \iint_S \left(\frac{\partial F_2}{\partial x} - \frac{\partial F_1}{\partial y}\right) dS = \int_C (F_1 dx + F_2 dy).$$

This is a Green's theorem in the plane which is a special case of Stoke's theorem.

2. Stoke's theorem gives the physical interpretation of curl.

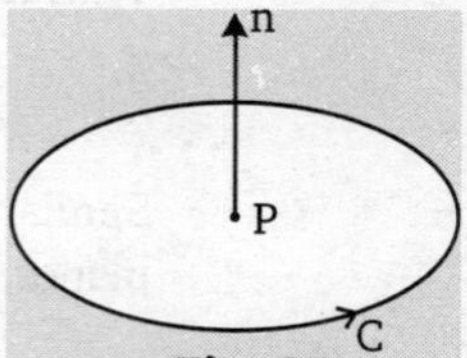

Let S be circular disc of radius r centred at P bounded by the circle C as shown in fig. 21.

Let $\mathbf{F}(Q) = \mathbf{F}(x, y, z)$ be a continuously differentiable vector function in S. Then by Stoke's theorem, we have

$$\int_C \mathbf{F} \cdot \mathbf{t}\, ds = \iint_S (\nabla \times \mathbf{F}) \cdot \mathbf{n} dS. \quad \ldots(1)$$

By mean value theorem for integrals, we have

$$\iint_S (\nabla \times \mathbf{F}) \cdot \mathbf{n} dS = [\nabla \times \mathbf{F}(P)] \cdot \mathbf{n}. \quad \ldots(2)$$

Fig. 21

where A is the area of S and P is a suitable point of S. From (1) and (2), we have

$$[\nabla \times \mathbf{F}(P)] \cdot \mathbf{n} = \frac{1}{A}\int_C \mathbf{F} \cdot d\mathbf{r} = \frac{1}{A}\int_C \mathbf{F} \cdot \mathbf{t} ds.$$

If $\mathbf{F}$ is the velocity vector of the fluid motion, then the integral $\int_C \mathbf{F} \cdot d\mathbf{r}$ is called the circulation for the flow around C. If we now let $r \to 0$, then

$$[\nabla \times \mathbf{F}(P)] \cdot \mathbf{n} = \lim_{r \to 0} \frac{1}{A}\int_C \mathbf{F} \cdot d\mathbf{r}.$$

Thus the component of the curl in the positive normal direction can be regarded as the specific circulation (circulation per unit area) of the flow in the surface at the corresponding point.

Solved Examples

Example 1. *Prove that* $\int_C \phi d\mathbf{r} = \iint_S dS \times \nabla\phi$.

Solution . Let $\mathbf{F} = \phi\mathbf{a}$, where $\mathbf{a}$ be any arbitrary constant vector. Then by Stoke's theorem

$$\int_C \mathbf{F} \cdot d\mathbf{r} = \iint_S (\nabla \times \mathbf{F}) \cdot \mathbf{n} dS.$$

$$\therefore \int_C (\phi\mathbf{a}) \cdot d\mathbf{r} = \iint_S (\nabla \times (\phi\mathbf{a})) \cdot \mathbf{n} dS.$$

$$= \iint_S (\nabla\phi \times \mathbf{a} + \phi\nabla \times \mathbf{a}) \cdot \mathbf{n} dS.$$

$$= \iint_S (\nabla\phi \times \mathbf{a}) \cdot \mathbf{n} dS. \quad (\because \nabla \times \mathbf{a} = \mathbf{0})$$

$$= \iint_S (\nabla\phi \times \mathbf{a}) \cdot d\mathbf{S}.$$

$$\therefore \int_C \mathbf{a} \cdot (\phi d\mathbf{r}) = \iint_S \mathbf{a} \cdot (d\mathbf{S} \times \nabla\phi).$$

$$\text{or } \mathbf{a} \cdot \left[\int_C \phi d\mathbf{r} - \iint_S d\mathbf{S} \times \nabla\phi\right] = 0.$$

Since $\mathbf{a}$ is an arbitrary constant vector, then

$$\int_C \phi d\mathbf{r} = \iint_S dS \times \nabla\phi.$$

Example 2. *Using Stoke's theorem prove that :*

(*i*) *div curl* $\mathbf{F} = 0$

(*ii*) *curl grad* $\phi = \mathbf{0}$. (Kerala–2005)

Solution . (i) Let V be a any volume enclosed by a closed surface S. Then by Gauss's divergence theorem,

$$\iiint_V \text{div}(\text{curl } \mathbf{F}) dV = \iint_S (\text{curl } \mathbf{F}) \cdot \mathbf{n} dS. \ldots(1)$$

Now divide the surface S into S_1 and S_2 in a closed curve C. Then

$$\iint_S (\text{curl } \mathbf{F}) \cdot \mathbf{n} dS = \iint_{S_1} (\text{curl } \mathbf{F}) \cdot \mathbf{n} dS_1 + \iint_{S_2} (\text{curl } \mathbf{F}) . \mathbf{n} dS_2.$$

Using Stoke's theorem, we get

$$\iint_S (\text{curl } \mathbf{F}) \cdot \mathbf{n} dS = \int_C \mathbf{F} \cdot d\mathbf{r} + \int_C \mathbf{F} \cdot d\mathbf{r}.$$
$$= 0.$$

(Negative sign is taken because positive direction along the boundaries of two surfaces are opposite)

Thus the equation (1) becomes

$$\iiint_V \text{div}(\text{curl } \mathbf{F}) dV = 0. \quad \ldots(2)$$

Since the equation (2) is true for all volume V, hence

$$\text{div curl } \mathbf{F} = 0.$$

(ii) Let S be any surface enclosed by a simple closed curve C. Then by Stoke's theorem

$$\iint_S(\text{curl }\mathbf{F}).\mathbf{n}dS = \int_C \mathbf{F}.d\mathbf{r}.$$

Let $\mathbf{F} = \text{grad }\phi$

$$\therefore \iint_S(\text{curl grad }\phi).\mathbf{n}dS = \int_C \text{grad }\phi.d\mathbf{r}. \quad \ldots(1)$$

Since $\mathbf{r} = x\hat{i} + y\hat{j} + z\hat{k}$.

$\therefore$ $d\mathbf{r} = dx\hat{i} + dy\hat{j} + dz\hat{k}$.

$$\text{Now grad }\phi.d\mathbf{r} = \left(\frac{\partial\phi}{\partial x}\hat{i} + \frac{\partial\phi}{\partial y}\hat{j} + \frac{\partial\phi}{\partial z}\hat{k}\right).(dx\hat{i} + dy\hat{j} + dz\hat{k}).$$

$$= \frac{\partial\phi}{\partial x}dx + \frac{\partial\phi}{\partial y}dy + \frac{\partial\phi}{\partial z}dz = d\phi.$$

$$\therefore \int_C \text{grad }\phi.d\mathbf{r} = \int_C d\phi = 0.$$

($\because$ C is closed curve)

Thus the equation (1) becomes

$$\iint_S(\text{curl grad }\phi).\mathbf{n}dS = 0. \quad \ldots(2)$$

Since S is an arbitrary surface, and the equation (2) holds for any S. Then we have

$$\text{curl grad }\phi = \mathbf{0}.$$

Example 3. *Prove that* $\oint_C \mathbf{r}\cdot d\mathbf{r} = 0$.

Solution . Using Stoke's theorem, we have

$$\oint_C \mathbf{r}\cdot d\mathbf{r} = 0 = \iint_S(\nabla\times\mathbf{r})\cdot\mathbf{n}dS = 0.$$

$$[\because \nabla\times\mathbf{r} = 0]$$

Example 4. *Prove that* $\oint_C \phi\nabla\phi\cdot d\mathbf{r} = 0$, *C being a closed curve.*

Solution . We know that

$$\nabla\times(\phi\nabla\phi) = \phi\nabla\times(\nabla\phi) + \nabla\phi\times\nabla\phi = \mathbf{0}$$

$$[\because \nabla\times(\nabla\phi) = \mathbf{0} \text{ and } \nabla\phi\times\nabla\phi = \mathbf{0}]$$

Using Stoke's theorem, we have

$$\oint_C \phi\nabla\phi\cdot d\mathbf{r} = \iint_S \nabla(\phi\nabla\phi)\cdot\mathbf{n}dS$$

$$= \iint_S \mathbf{0}.\mathbf{n}dS = 0$$

Example 5. *Prove that* $\oint_C \phi\nabla\psi.d\mathbf{r} = \iint_S(\nabla\phi\times\nabla\psi).\mathbf{n}dS$.

Solution . We know that

$$\nabla\times(\phi\nabla\psi) = \phi\nabla\times(\nabla\psi) + \nabla\phi\times\nabla\psi$$

$$= \nabla\phi\times\nabla\psi$$

$$[\because \nabla\times(\nabla\psi) = \mathbf{0}]$$

Using Stoke's theorem, we have

$$\oint_C \phi\nabla\phi\cdot d\mathbf{r} = \iint_S \nabla(\phi\nabla\psi)\cdot\mathbf{n}dS$$

$$= \iint_S(\nabla\phi\times\nabla\psi).\mathbf{n}dS$$

Example 6. *Prove that* $\oint_C \phi\nabla\psi\cdot d\mathbf{r} = -\oint_C \psi\nabla\phi\cdot d\mathbf{r}$.

Solution . We know that

$$\nabla(\phi\psi) = \phi\nabla\psi + \psi\nabla\phi \quad \ldots(1)$$

Now by Stoke's theorem, we have

$$\oint_C \nabla(\phi\psi)\cdot d\mathbf{r} = \iint_S[\nabla\times\nabla(\phi\psi)]\cdot\mathbf{n}dS = 0.$$

$$[\because \nabla\times\nabla(\phi\psi) = 0]$$

$$\Rightarrow \oint_C[\phi\nabla\psi + \psi\nabla\phi].d\mathbf{r} = 0. \text{ [using equation (1)]}$$

$$\therefore \oint_C \phi\nabla\psi\cdot d\mathbf{r} = -\oint_C \psi\nabla\phi\cdot d\mathbf{r}.$$

Example 7. *Verify Stoke's theorem for*

$$\mathbf{F} = (2x - y)\hat{i} - yz^2\hat{j} - y^2z\hat{k}$$

where S is the upper half surface of the supere $x^2 + y^2 + z^2 = 1$ *and C is its boundary.*

(SVTU–2006, Bhopal–2008, Madras–2006)

Solution . The boundary C of the surface S in the xy-plane is a circle whose equation is $x^2 + y^2 + z^2 = 1, z = 0$, therefore its parametric equations are

$$x = \cos\theta, y = \sin\theta, z = 0, 0 \le \theta \le 2\pi$$

Now

$$\mathbf{F}\cdot d\mathbf{r} = [(2x - y)\hat{i} - yz^2\hat{j} - y^2z\hat{k}].[dx\hat{i} + dy\hat{j} + dz\hat{k}]$$

$$= (2x - y)dx - yz^2dy - y^2zdz$$

$$= (2x - y)dx \qquad [\because z = 0]$$

$$\therefore \oint_C \mathbf{F}.d\mathbf{r} = \oint_C(2x - y).dx.$$

$$= \int_0^{2\pi}[2\cos\theta - \sin\theta](-\sin\theta d\theta)$$

$$= -\int_0^{2\pi}\sin 2\theta d\theta + \int_0^{2\pi}\sin^2\theta d\theta)$$

$$= -\left[-\frac{\cos 2\theta}{2}\right]_0^{2\pi} + \frac{1}{2}\left[\theta - \frac{\sin 2\theta}{2}\right]_0^{2\pi}$$

$$= -\left[-\frac{\cos 4\pi}{2} + \frac{\cos 0}{2}\right] + \frac{1}{2}\left[2\pi - \frac{\sin 4\pi}{2} - 0 + \frac{\sin 0}{2}\right]$$

$$= -\left[-\frac{1}{2} + \frac{1}{2}\right] + \frac{1}{2}[2\pi - 0] = \pi$$

and

$$\nabla\times\mathbf{F} = \begin{vmatrix} \hat{i} & \hat{j} & \hat{k} \\ \dfrac{\partial}{\partial x} & \dfrac{\partial}{\partial y} & \dfrac{\partial}{\partial z} \\ 2x - y & -yz^2 & -y^2z \end{vmatrix}$$

$$= (-2yz + 2yz)\hat{i} - (0 - 0)\hat{j} + (0 + 1)\hat{k} = \hat{k}.$$

Let S' be the surface consisting of surface S and S_1 and S_1 is the plane region by the circle C, then S' is closed. Then we have

$$\iint_{S'}(\nabla\times\mathbf{F})\cdot\mathbf{n}dS = 0$$

$$\iint_S(\nabla\times\mathbf{F})\cdot\mathbf{n}dS + \iint_{S_1}(\nabla\times\mathbf{F})\cdot\mathbf{n}dS = 0$$

Since the outward drawn unit normal vector $\mathbf{n}$ to S_1 is $-\hat{k}$, then

$$\iint_S(\nabla\times\mathbf{F})\cdot\mathbf{n}dS - \iint_{S_1}(\nabla\times\mathbf{F})\cdot\hat{k}dS = 0$$

or

$$\iint_S(\nabla\times\mathbf{F})\cdot\mathbf{n}dS = \iint_{S_1}(\hat{k}\cdot\hat{k})dS$$

$$[\because \nabla\times\mathbf{F} = \hat{k}]$$

$$= \iint_{S_1} dS$$

$$= S_1(\text{the area of the circle } C \text{ of radius } 1)$$

$$= \pi\,[1]^2 = \pi$$

$$\therefore \oint_C \mathbf{F}\cdot d\mathbf{r} = \iint_S(\nabla\times\mathbf{F})\cdot\mathbf{n}dS.$$

Hence Stoke's theorem is verified.

Example 8. *Verify Stoke's theorem for* $\mathbf{F} = 2y\hat{i} + 3x\hat{j} - z^2\hat{k}$, *where S is the upper half surface of the sphere* $x^2 + y^2 + z^2 = 9$ *and C is its boundary.*

Solution . The boundary C of S in the xy-plane is a circle whose equation is $x^2 + y^2 + z^2 = 9$, $z = 0$. Its parametric equations are $x = 3\cos\theta, y = 3\sin\theta$ and $z = 0$ where θ lies between 0 and 2π.

$$\mathbf{F}.d\mathbf{r} = (2y\hat{i} + 3x\hat{j} - z^2\hat{k}).(dx\hat{i} + dy\hat{j} + dz\hat{k})$$
$$= 2ydx + 3xdy - z^2dz.$$

$$\text{Now} \int_C \mathbf{F}.d\mathbf{r} = \int_C (2ydx + 3xdy - z^2dz).$$
$$= \int_C (2ydx + 3xdy). \qquad (\because z = 0)$$
$$= \int_0^{2\pi} [2(3\sin\theta)(-3\sin\theta)d\theta + 3(3\cos\theta)(3\cos\theta)d\theta]$$
$$(\because x = 3\cos\theta, y = 3\sin\theta)$$
$$= -18\int_0^{2\pi} \sin^2\theta d\theta + 27\int_0^{2\pi} \cos^2\theta d\theta.$$
$$= -\frac{18}{2}\left[\theta - \frac{\sin 2\theta}{2}\right]_0^{2\pi} + \frac{27}{2}\left[\theta + \frac{\sin 2\theta}{2}\right]_0^{2\pi}.$$
$$= -9[2\pi] + \frac{27}{2}[2\pi] = -18\pi + 27\pi = 9\pi.$$

$$\text{And } \nabla \times \mathbf{F} = \begin{vmatrix} \hat{i} & \hat{j} & \hat{k} \\ \dfrac{\partial}{\partial x} & \dfrac{\partial}{\partial y} & \dfrac{\partial}{\partial z} \\ 2y & 3x & -z^2 \end{vmatrix}$$
$$= \hat{i}[0 - 0] + \hat{j}[0 - 0] + \hat{k}[3 - 2] = \hat{k}.$$

Let S' be the surface consisting of surfaces S and S_1 where S_1 is the plane region bounded by the circle C, then S' is closed. Therefore we have

$$\iint_{S'} (\nabla \times \mathbf{F}) \cdot \mathbf{n}dS = 0$$

and $\iint_S (\nabla \times \mathbf{F}) \cdot \mathbf{n}dS + \iint_{S_1} (\nabla \times \mathbf{F}) \cdot \mathbf{n}dS = 0$

Since the outward drawn unit normal to S_1 is $-\hat{k}$, then

$$\therefore \quad \iint_S (\nabla \times \mathbf{F}) \cdot \mathbf{n}dS = \iint_{S_1} (\nabla \times \mathbf{F}) \cdot \hat{k}dS$$
$$= \iint_{S_1} (\hat{k}.\hat{k})dS \qquad [\because \nabla \times \mathbf{F} = \hat{k}]$$
$$= \iint_{S_1} dS \qquad (\because \hat{k}.\hat{k} = 1)$$
$$= S_1$$

(Area of the circle bounded by C)

$$= \pi\,[3]^2 = 9\pi.$$

Hence, we obtain

$$\iint_S (\nabla \times \mathbf{F}) \cdot \mathbf{n}dS = \int_C \mathbf{F} \cdot d\mathbf{r}.$$

Hence Stoke's theorem is verified.

Example 9. *Verify Stroke's theorem for the vector function*

$$\mathbf{F} = 3y\hat{i} - xz\hat{j} + yz^2\hat{k}$$

where S is the surface of the paraboloid $2z = x^2 + y^2$ *bounded by* $z = 2$ *and C its boundary.* (UPTU–2007)

Solution . C is the boundary of the surface given by $2z = x^2 + y^2$ and $z = 2$. Thus C is the boundary of the circle $x^2 + y^2 = 4$, $z = 2$ whose parametric equations are $x = 2\cos\theta$, $y = 2\sin\theta$, $z = 2$, where $0 \le \theta \le 2\pi$.

$$\therefore \quad \mathbf{F} \cdot d\mathbf{r} = (3y\hat{i} - xz\hat{j} + yz^2\hat{k}).(dx\hat{i} + dy\hat{j} + dz\hat{k})$$
$$= 3ydx - xzdy + yz^2dz$$

$$\text{Now} \int_C \mathbf{F} \cdot d\mathbf{r} = \int_C (3ydx - xzdy + yz^2dz).$$
$$= \int_C 3ydx - 2xdy. \quad (\because z = 2, \therefore dz = 0)$$
$$= \int_0^{2\pi} [3(2\sin\theta)(-2\sin\theta)d\theta - 2(2\cos\theta)(2\cos\theta)d\theta]$$
$$(\because x = 2\cos\theta, y = 2\sin\theta)$$
$$= -12\int_0^{2\pi} \sin^2\theta d\theta - 8\int_0^{2\pi} \cos^2\theta d\theta.$$
$$= -\frac{12}{2}\left[\theta - \frac{\sin 2\theta}{2}\right]_0^{2\pi} - \frac{8}{2}\left[\theta + \frac{\sin 2\theta}{2}\right]_0^{2\pi}.$$
$$= -6[2\pi] - 4[2\pi] = -12\pi - 8\pi = -20\pi.$$

$$\text{Now } \nabla \times \mathbf{F} = \begin{vmatrix} \hat{i} & \hat{j} & \hat{k} \\ \dfrac{\partial}{\partial x} & \dfrac{\partial}{\partial y} & \dfrac{\partial}{\partial z} \\ 3y & -xz & yz^2 \end{vmatrix}$$
$$= \hat{i}(z^2 + x) + \hat{j}(0 - 0) + \hat{k}(-z - 3)$$
$$= (z^2 + x)\hat{i} - (z + 3)\hat{k}.$$

Let S' be the surface consisting of surface S and S_1 where S_1 is the surface whose boundary is the curve C, then S' is a closed surface. Therefore we have

$$\iint_{S'} (\nabla \times \mathbf{F}) \cdot \mathbf{n}dS = 0$$

or $\iint_S (\nabla \times \mathbf{F}) \cdot \mathbf{n}dS + \iint_{S_1} (\nabla \times \mathbf{F}) \cdot \mathbf{n}dS = 0$

Since the unit normal vector drawn outward to S_1 is $-\hat{k}$, then

$$\iint_S (\nabla \times \mathbf{F}) \cdot \mathbf{n}dS = \iint_{S_1} (\nabla \times \mathbf{F}) \cdot \hat{k}dS$$
$$= \iint_{S_1} [(z^2 + x)\hat{i} - (z + 3)\hat{k}].\hat{k}dS$$
$$= -\iint_{S_1} (z + 3)dS$$
$$(\because \hat{k}.\hat{i} = 0, \hat{k}.\hat{k} = 1)$$
$$= -\iint_{S_1} (2 + 3)dS \qquad (\because z = 2)$$
$$= -5\iint_{S_1} dS$$
$$= -5\,S_1$$

(where S_1 is the area of the circle $x^2 + y^2 = 4$)

$$= -5[\pi(2)^2] = -20\pi.$$

Thus

$$\iint_S (\nabla \times \mathbf{F}) \cdot \mathbf{n}dS = \int_C \mathbf{F} \cdot d\mathbf{r}.$$

Hence Stoke's theorem is verified.

Example 10. *Using Stoke's theorem evaluate*

$$\int_C (\sin zdx - \cos xdy + \sin ydz)$$

where C is the boundary of the rectangle $0 \le x \le \pi$, $0 \le y \le 1$ *and* $z = 3$. (UKTU-2013; Rohtak–2005)

Solution . Since $\sin zdx + (-\cos x)dy + \sin ydz$

$$= (\sin z\hat{i} - \cos x\hat{j} + \sin y\hat{k}).(dx\hat{i} + dy\hat{j} + dz\hat{k})$$

$\therefore \quad \mathbf{F} = \sin z\hat{i} - \cos x\hat{j} + \sin y\hat{k}$

$$\text{Now } \nabla\times\mathbf{F} = \begin{vmatrix} \hat{i} & \hat{j} & \hat{k} \\ \frac{\partial}{\partial x} & \frac{\partial}{\partial y} & \frac{\partial}{\partial z} \\ \sin z & -\cos x & \sin y \end{vmatrix}$$

$$= \hat{i}(\cos y - 0) + \hat{j}(\cos z - 0) + \hat{k}(\sin x - 0)$$
$$= \cos y\hat{i} + \cos z\hat{j} + \sin x\hat{k}.$$

Since the rectangle is parallel to xy-plane at the distance $z = 3$ from the origin so the outward drawn unit normal to S bounded by C is $\hat{k}$. By Stoke's theorem

$$\iint_S (\nabla\times\mathbf{F})\cdot\mathbf{n}dS = \int_C \mathbf{F}\cdot d\mathbf{r}$$

$$\int_C \mathbf{F}\cdot d\mathbf{r} = \iint_S(\cos y\hat{i} + \cos z\hat{j} + \sin x\hat{k}).\hat{k}dS$$
$$= \iint_S \sin x dS = \iint_S \sin x dx dy$$
$$= \int_{x=0}^{\pi}\int_{y=0}^{1} \sin x dx dy$$
$$(\because\ 0 \le x \le \pi,\ 0 \le y \le 1 \text{ on } S)$$
$$= \int_{x=0}^{\pi} \sin x\left[y\right]_0^1 dx = \int_{x=0}^{\pi}\sin x dx$$
$$= \left[-\cos x\right]_0^{\pi} = -\cos\pi + \cos 0 = 2$$

$\therefore \quad \int_C (\sin z dx - \cos x dy + \sin y dz) = 2.$

Example 11. *Using Stoke's theorem, evaluate*

$$\int_C (e^x dx + 2y dy - dz)$$

where C is boundary of the circle in plane $z = 2$.

Solution . Since $e^x dx + 2y dy - dz$

$$= (e^x\hat{i} + 2y\hat{j} - \hat{k}).(dx\hat{i} + dy\hat{j} + dz\hat{k})$$

$\therefore \quad \mathbf{F} = e^x\hat{i} + 2y\hat{j} - \hat{k}, d\mathbf{r} = dx\hat{i} + dy\hat{j} + dz\hat{k}$

By Stoke's theorem, we have

$$\iint_S (\nabla\times\mathbf{F})\cdot\mathbf{n}dS = \int_C \mathbf{F}\cdot d\mathbf{r} \quad \ldots(1)$$

$$\text{Now } \nabla\times\mathbf{F} = \begin{vmatrix} \hat{i} & \hat{j} & \hat{k} \\ \frac{\partial}{\partial x} & \frac{\partial}{\partial y} & \frac{\partial}{\partial z} \\ e^x & 2y & -1 \end{vmatrix}$$
$$= \hat{i}(0-0) + \hat{j}(0-0) + \hat{k}(0-0) = \mathbf{0}$$

Since the circle is in the plane $z = 2$, so the outward drawn unit normal to the circle is $\hat{k}$.

$\therefore \quad (\nabla\times\mathbf{F})\cdot\mathbf{n} = \mathbf{0}\cdot\hat{k} = 0.$

From (1), we get

$$\int_C \mathbf{F}\cdot d\mathbf{r} = \iint_S (\nabla\times\mathbf{F})\cdot\mathbf{n}dS = \iint_S 0 dS = 0.$$

$\therefore \quad \int_C (e^x dx + 2y dy - dz) = 0.$

Example 12. *Using Stoke's theorem to prove that*

$$\int_C (y dx + z dy + x dz) = -2\sqrt{2}\pi a^2.$$

where C is the curve by $x^2+y^2+z^2-2ax-2ay = 0$, $x + y = 2a$ and begins at the point $(2a, 0, 0)$ and goes at first below the z-plane. (Bhopal–2008)

Solution . Since the intersection of the sphere $x^2 + y^2 + z^2 - 2ax - 2ay = 0$ and the plane $x + y = 2a$ is a circle. The centre of the sphere is $(a, a, 0)$ and this centre also lies on the plane $x + y = 2a$ so the obtained circle is a great circle whose radius is the radius of the sphere.

$\therefore$ Radius of the circle $= \sqrt{a^2 + a^2} = a\sqrt{2}.$

Since $y dx + z dy + x dz$

$$= (y\hat{i} + z\hat{j} + x\hat{k}).(dx\hat{i} + dy\hat{j} + dz\hat{k}).$$

$\therefore \quad \mathbf{F} = y\hat{i} + z\hat{j} + x\hat{k}.$

$$d\mathbf{r} = dx\hat{i} + dy\hat{j} + dz\hat{k}.$$

By Stoke's theorem, we have

$$\int_C \mathbf{F}\cdot d\mathbf{r} = \iint_S (\nabla\times\mathbf{F})\cdot\mathbf{n}dS. \quad \ldots(1)$$

where S is the surface whose boundary C is the circle.

$$\text{Now } \nabla\times\mathbf{F} = \begin{vmatrix} \hat{i} & \hat{j} & \hat{k} \\ \frac{\partial}{\partial x} & \frac{\partial}{\partial y} & \frac{\partial}{\partial z} \\ y & z & x \end{vmatrix}$$
$$= \hat{i}(0-1) + \hat{j}(0-1) + \hat{k}(0-1)$$
$$= -\hat{i} - \hat{j} - \hat{k}$$

Since S is the surface of the plane $x + y = 2a$ bounded by the circle C. Then

$$\mathbf{n} = \frac{\nabla(x+y-2a)}{|\nabla(x+y-2a)|} = \frac{\hat{i}+\hat{j}}{\sqrt{2}}.$$

From (1)

$$\int_C \mathbf{F}\cdot d\mathbf{r} = \iint_S -(\hat{i}+\hat{j}+\hat{k})\cdot\frac{\hat{i}+\hat{j}}{\sqrt{2}}dS.$$

$$= -\frac{2}{\sqrt{2}}\iint_S dS = -\frac{2}{\sqrt{2}}S$$

where S is the area of the circle of radius $a\sqrt{2}$

$$= -\frac{2}{\sqrt{2}}(\pi(a\sqrt{2})^2) = -\frac{4\pi a^2}{\sqrt{2}} = -2\sqrt{2}a^2.$$

Hence, $\int_C (y dx + z dy + x dz) = -2\sqrt{2}\pi a^2.$

Example 13. *Use Stoke's theorem evaluate*

$$\int_C [(x+y)dx + (2x-z)dy + (y+z)dz]$$

where C is the boundary of the triangle with vertices (2, 0, 0), (0, 3, 0) and (0, 0, 6).

(Nagpur–2009, Kurukshetra–2009, Kerala–2005)

Solution . We have $\mathbf{F} = (x+y)\hat{i} + (2x-z)\hat{j} + (y+z)\hat{k}$

Therefore,

$$\text{curl } \mathbf{F} = \begin{vmatrix} \hat{i} & \hat{j} & \hat{k} \\ \frac{\partial}{\partial x} & \frac{\partial}{\partial y} & \frac{\partial}{\partial z} \\ x+y & 2x-y & y+z \end{vmatrix} = 2\hat{i} + \hat{k}$$

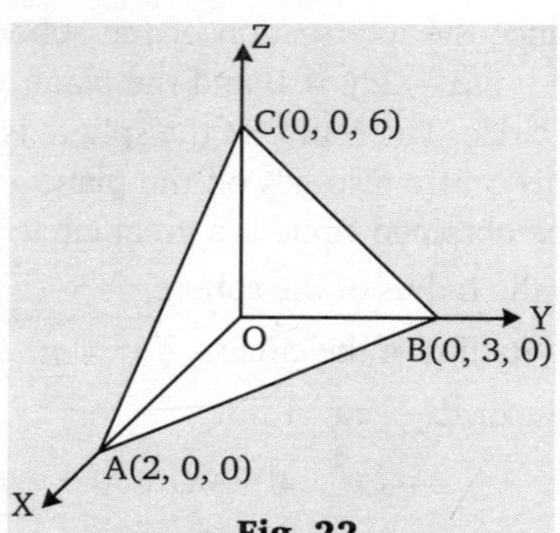

Fig. 22

Equation of the plane through A, B and C

$$\frac{x}{2}+\frac{y}{3}+\frac{z}{6}=1$$

$\Rightarrow \quad 3x + 2y + z = 6$

Now vector $\hat{n}$ normal to this plane is

$$\hat{n}=\frac{3\hat{i}+2\hat{j}+\hat{k}}{\sqrt{9+4+1}}=\frac{1}{\sqrt{14}}(3\hat{i}+2\hat{j}+\hat{k}).$$

Therefore,

$$\int_C[(x+y)dx+(2x-z)dy+(y+z)dz]=\int_C \mathbf{F}\cdot d\mathbf{r}$$

$$=\int_S \text{curl } \mathbf{F}\cdot\hat{n}dS \text{, where } S \text{ is the triangle ABC}$$

$$=\int_s(2\hat{i}+\hat{k})\cdot\left(\frac{3\hat{i}+2\hat{j}+\hat{k}}{\sqrt{14}}\right)dS$$

$$=\frac{1}{\sqrt{14}}(6+1)\int_s dS=\frac{7}{\sqrt{14}}(\text{Area of } \Delta\, ABC)$$

$$=\frac{7}{\sqrt{14}}.3\sqrt{14}=21$$

EXERCISE 23.4

1. Verify Stoke's theorem for $\mathbf{F}=z\hat{i}+x\hat{j}+y\hat{k}$ taken over the half of the sphere $x^2 + y^2 + z^2 = a^2$ lying above the xy-plane.
2. Verify Stoke's theorem for $\mathbf{F}=(x^2+y^2)\hat{i}-2xy\hat{j}$ taken round the rectangle bounded by $x = \pm a, y = 0, y = b$.

   (VTU–2007, Bhopal–2008, JNTU–2003, UPTU–2003, Mumbai–2007, BPTU–2006,)
3. Verify Stoke's theorem fot the vector function $\mathbf{F}=xy\hat{i}+xy^2\hat{j}$ taken over round the square with vertices (1, 0, 0), (1, 1, 0), (0, 1, 0) and (0, 0, 0).
4. Verify Stoke's theorem for $\mathbf{F}=x^2\hat{i}+xy\hat{j}$ taken over round the square in the plane $z = 0$, whose sides are along the lines $x = 0, y = 0, x = a, y = a$.
5. Verify Stoke's theorem for $\mathbf{F}=-y^3\hat{i}+x^3\hat{j}$, where S is the circular disc $x^2 + y^2 \le 1, z = 0$.
6. Using Stoke's theorem, evaluate $\int_C \mathbf{F}\cdot d\mathbf{r}$ where $\mathbf{F}=y^2\hat{i}+x^2\hat{j}-(x+z)\hat{k}$ and C is the boundary of the triangle with vertices at (0, 0, 0), (1, 0, 0), (1, 1, 0). (UKTU–2012)
7. If $\mathbf{F}=2y\hat{i}+2\hat{j}+3y\hat{k}$, evaluate by Stoke's theorem $\int_C \mathbf{F}\cdot d\mathbf{r}$, where C is the boundary of the intersection $x^2 + y^2 + z^2 = 6z$ and $z = x + 3$.
8. Using Stoke's theorem evaluate

   $$\int_C(yzdx+xzdy+xydz)$$

   where C is the boundary of the intersection of $x^2 + y^2 = 1$ and $z = y^2$. (JNTU–2005)
9. Using Stoke's theorem evaluate

   $$\int\int_S(\nabla\times\mathbf{F})\cdot\mathbf{n}dS$$

   where $\mathbf{F}=(x^2+y-4)\hat{i}+3xy\hat{j}+(2xz+z^2)\hat{k}$ and S is the surface of

   (i) The hemisphere $x^2 + y^2 + z^2 = 16$ above the xy-plane.

   (ii) The paraboloid $z = 4 - (x^2 + y^2)$ above the xy-plane.
10. Using Stoke's theorem evaluate

    $$\int\int_S(\nabla\times\mathbf{F})\cdot\mathbf{n}dS$$

    where $\mathbf{F}=(x-z)\hat{i}+(x^3+yz)\hat{j}-3xy^2\hat{k}$ and S is the surface of the cone $z=2-\sqrt{x^2+y^2}$ above the xy-plane.
11. Prove that the necessary and sufficient condition that $\int_C \mathbf{F}\cdot d\mathbf{r}=\mathbf{0}$ for every closed curve C lying in a simply connected region R is that $\nabla\times\mathbf{F}=\mathbf{0}$ identically.
12. Show that $\int\int_S \phi \text{ curl } \mathbf{F}.d\mathbf{S}=\int_C \phi\, \mathbf{F}.d\mathbf{r}-\int\int_S(\text{grad } \phi\times\mathbf{F}).d\mathbf{S}$.
13. If $\mathbf{F} = \nabla\phi$ and $\mathbf{G} = \nabla\psi$ such that $\nabla^2\phi = 0, \nabla^2\psi = 0$, then show that

    $$\int\int_S(\mathbf{G}.\nabla)\mathbf{F}.d\mathbf{S}=\int_C(\mathbf{F}\times\mathbf{G}).d\mathbf{r}+\int\int_S(\mathbf{F}.\nabla)\mathbf{G}.d\mathbf{S}.$$
14. Verify Stoke's theorem for the function

    $$\mathbf{F}=(x-y-z)\hat{i}+(y-z-x)\hat{j}+(z-x-y)\hat{k}$$

    over the unclosed surface of cylinder $\frac{x^2}{a^2}+\frac{y^2}{b^2}=1$ bounded by the plane $z = h$ and open at the end $z = 0$. (MTU–2011)
15. Use Stoke's theorem to evaluate

    $$\int_C[(x+2y)dx+(x-z)dy+(y-z)dz]$$

    where C is the boundary of the triangle with vertices (2, 0, 0), (0, 3, 0) and (0, 0, 6) oriented in the anticlockwise direction (UPTU–2009)
16. Verify Stoke's theorem by evaluating the line integral

    $$\int_C xe^z dx+ye^x dy+ze^y dz$$

    along the boundary of the triangle with vertices (0, 0, 1), (1, 0, 1), (1, 1, 1). (UPTU–2008)
17. Verify Stoke's theorem for $\mathbf{F}=(y-z+2)\hat{i}+(yz+4)\hat{j}-xz\hat{k}$ where S is the surface of the cube $x = 0, y = 0, z = 0, x = 2, y = 2, z = 2$ above the xy-plane. (Andhra–2000)
18. If S be the surface of the sphere $x^2 + y^2 + z^2 = 1$, prove that $\int_C \text{curl}\cdot\mathbf{F}dS=0$. (JNTU–2009)
19. Use Stoke's theorem to evaluate $(\nabla\times\mathbf{F})\cdot\hat{n}dS$ where

    $$\mathbf{F}=y\hat{i}+(x-2xz)\hat{j}-xy\hat{k}$$

    and S is the surface of the sphere $x^2 + y^2 + z^2 = a^2$ above the xy-plane. (Kottayam–2005)
20. Verify Stoke's theorem first by evaluating

    $$\int_C(3x^2-6yz)dx+(2y+3zx)dy+(1-4yz^2)dz$$

from (0, 0, 0) to (1, 1, 1) along the path C given by the straight line from (1, 0, 0) to (1, 1, 0) then from (1, 1, 0) to (1, 1, 1) then from (1, 1, 1) to (1, 0, 0) in counter clockwise and secondly evaluating the integral as surface integral, the surface being enclosed by the path C and the co-ordinate plane. (UPTU–2009)

Hint to Selected Problems

7. Let S be the surface of the intersection of the sphere $x^2 + y^2 + z^2 = 6z$ and the plane $z = x + 3$ with boundary C, then S is obtained as $2x^2 + y^2 = 9$.

$$\therefore \quad \mathbf{n} = \hat{k}$$

$$\Rightarrow \quad \nabla \times \mathbf{F} = \begin{vmatrix} \hat{i} & \hat{j} & \hat{k} \\ \frac{\partial}{\partial x} & \frac{\partial}{\partial y} & \frac{\partial}{\partial z} \\ 4y & 2 & 3y \end{vmatrix} = 3\hat{i} - 4\hat{k}.$$

By Stoke's theorem

$$\int_C \mathbf{F} \cdot d\mathbf{r} = \iint_S (\nabla \times \mathbf{F}) \cdot \mathbf{n} dS$$

$$= \iint_S (3\hat{i} - 4\hat{k}) \cdot \hat{k} dS = -4 \iint_S dS$$

$$= -4 \text{ (area of an ellipse : } 2x^2 + y^2 = 9)$$

$$= -4\left[\pi\left(\frac{\sqrt{3}}{2}\right)\right] = -18\pi\sqrt{2}.$$

9. (i) Let S be the surface of the intersection of hemisphere $x^2 + y^2 + z^2 = 16$ and $z = 0$ with boundary C, therefore C is a circle $x^2 + y^2 = 16$, $z = 0$.

So its parametric equations are

$$x = 4 \cos \theta, y = 4 \sin \theta, z = 0.$$

Also $\mathbf{F} \cdot d\mathbf{r} = (x^2 + y - 4)dx + xydy + (2xz + z^2)dz$.

By Stoke's theorem

$$\iint_S (\nabla \times \mathbf{F}) \cdot \mathbf{n} dS = \int_C \mathbf{F} \cdot d\mathbf{r}$$

$$= \int_0^{2\pi} (-16 \sin^2 \theta + 16 \sin \theta) d\theta = -16\pi$$

(ii) Same as part (i)

13. By Stoke's theorem

$$\int_C (\mathbf{F} \times \mathbf{G}) \cdot d\mathbf{r} = \iint_S [\nabla \times (\mathbf{F} \times \mathbf{G})] \cdot d\mathbf{S}. \quad \ldots(1)$$

Since

$$\nabla \times (\mathbf{F} \times \mathbf{G}) = (\mathbf{G} \cdot \nabla)\mathbf{F} - \mathbf{G} \text{ div } \mathbf{F} - (\mathbf{F} \cdot \nabla)\mathbf{G} + \mathbf{F} \text{ div } \mathbf{G} \quad \ldots(2)$$

But $\quad \mathbf{F} = \nabla\phi, \mathbf{G} = \nabla\psi, \text{div } \mathbf{F} = 0, \text{div } \mathbf{G} = 0. \quad \ldots(3)$

Using (1), (2), (3), we obtain the required result.

ANSWERS

6. $\frac{1}{3}$	**7.** $-18\pi\sqrt{2}$	**8.** 0	**9.** (i) -16π	(ii) -4π
10. 12π	**15.** 15	**19.** 0	**20.** $-\frac{\pi a^2}{\sqrt{2}}$	

Objective Evaluations

Fill in the Blanks

1. If the initial and terminal points of a curve coincide, then curve is a __________ .
2. $\frac{d\mathbf{r}}{dt}$ is a __________ to the curve at the point $\mathbf{r}$.
3. If a surface S has a unique normal at each of its points and the direction of this normal depends continuously on the points of S, then the surface is called __________ .
4. The integral $\int_C \mathbf{F}\cdot d\mathbf{r}$ is called __________ .
5. The value of $\int_C (xdy - ydx)$ around the circle $x^2 + y^2 = 1$ is __________.
6. $\iint_R \left(\frac{\partial Q}{\partial x} - \frac{\partial P}{\partial y}\right)dxdy =$ __________ .
7. $\iiint_V \nabla\cdot\mathbf{F}dV = \iint_S \mathbf{A}dS$, then $\mathbf{A}$ is equal to __________ .
8. If $\nabla^2\phi = 0$, $\nabla^2\psi = 0$, then $\iint_S \left(\phi\frac{\partial\psi}{\partial n} - \psi\frac{\partial\phi}{\partial n}\right)dS =$ __________ .
9. $\iint_S \mathbf{F}\cdot\mathbf{n}dS$ is called the __________ $d\mathbf{F}$ over S.
10. If S is a closed surface, then $\iint_S \text{curl}\,\mathbf{F}\cdot\mathbf{n}dS =$ __________ .
11. If S is a closed surface, then $\iint_S \mathbf{r}\cdot\mathbf{n}dS =$ __________ .
12. If ϕ is harmonic in V, then $\iint_S \frac{\partial\phi}{\partial n}dS =$ __________ where S is the surface enclosing V.
13. The formula $\int_C \mathbf{F}\cdot dr = \iint_S (\nabla\times\mathbf{F})\cdot\mathbf{n}dS$ is governed by __________ .
14. $\int_C \mathbf{r}\cdot d\mathbf{r} =$ __________ .

True/False

Write 'T' for True and 'F' for False statement.

1. The integral $\iint_S f(x,y,z)dS$ is a volume integral. **(T/F)**
2. The integral $\int_S f(x,y,z)dS$ is a line integral. **(T/F)**
3. The integral $\int_V f(x,y,z)dS$ is a volume integral. **(T/F)**
4. The integral $\int_C \mathbf{F}\cdot d\mathbf{r}$ is called circulation. **(T/F)**
5. The formula $\iint_S (\text{curl}\,\mathbf{F})\cdot\mathbf{n}dS = \int_C \mathbf{F}\cdot dr$ is governed by Stoke's theorem. **(T/F)**
6. $\int_S (\nabla\phi\times\nabla\psi)\cdot\mathbf{n}dS \neq 0$. **(T/F)**
7. If ϕ is harmonic in V and $\frac{\partial\phi}{\partial n} = 0$ on S, then ϕ is constant in V. **(T/F)**
8. $\int_C \phi\nabla\phi\cdot d\mathbf{r} = 0$, C being a closed curve. **(T/F)**
9. If ϕ and ψ are harmonic in V and $\frac{\partial\phi}{\partial n} = \frac{\partial\psi}{\partial n}$ on S, then $\phi = \psi + C$, where C is a constant. **(T/F)**
10. If $\mathbf{n}$ is the unit outward drawn normal to any closed surface S, then $\iiint_V \nabla\cdot\mathbf{n}dV \neq S$. **(T/F)**
11. The value of the integral $\iint_S \mathbf{r}\cdot\mathbf{n}dS$, where S is a closed surface is $3V$, where V is enclosed by S. **(T/F)**
12. If ϕ is harmonic in V, then $\iint_S \phi\frac{\partial\phi}{\partial n}dS = \iiint_V |\nabla\phi|^2\,dV$. **(T/F)**
13. Green's theorem in a plane is a special case of Stoke's theorem. **(T/F)**
14. $\iint_S (\mathbf{F}\cdot\mathbf{n})dS = \iiint \text{div}\,\mathbf{F}dV$. **(T/F)**
15. Any integral which is evaluated along a curve is called surface integral. **(T/F)**

Multiple Choice Questions

Choose the most appropriate one.

1. The formula $\iiint_V \nabla\cdot\mathbf{F}dV = \iint_S \mathbf{F}\cdot\mathbf{n}dS$ is governed by :
   (a) Stoke's theorem (b) Gauss's theorem
   (c) Green's theorem (d) none of these
2. The integral $\int_S f(x,y,z)dS$ is a :
   (a) surface integral (b) line integral
   (c) volume integral (d) none of these
3. If S is any closed surface enclosing a volume V and $\mathbf{F} = x\hat{i} + 2y\hat{j} + 3z\hat{k}$, then the value of the integral $\iint_S (\mathbf{F}\cdot\mathbf{n})dS$ is equal to :
   (a) $3V$ (b) $6V$
   (c) $2V$ (d) $6S$
4. A vector $\mathbf{F}$ is always normal to a given closed surface S enclosing V the value of the intetgral $\iiint_V \text{curl}\,\mathbf{F}dV$ is :
   (a) $\mathbf{0}$ (b) 0
   (c) V (d) S
5. If $\mathbf{a}$ is a constant vector and V is the volume enclosed by the closed surface S, then the value of $\iint \mathbf{n}\times(\mathbf{a}\times\mathbf{r})dS$ is :
   (a) 0 (b) $\mathbf{0}$
   (c) $2\mathbf{a}$ (d) $2V\mathbf{a}$
6. For any closed surface S, the value of $\iint_S \mathbf{r}\times\mathbf{n}dS$ is :
   (a) $\mathbf{r}$ (b) $\mathbf{n}$
   (c) $\mathbf{0}$ (d) $2\mathbf{n}$
7. If $\mathbf{n}$ is the unit outward drawn normal to any closed surface S, the value of $\iiint_V \text{div}\,\mathbf{n}dV$ is :
   (a) V (b) S
   (c) 0 (d) $2S$

8. The value of $\int_C \mathbf{r} \cdot d\mathbf{r}$ is :

(a) 1 (b) –1
(c) **r** (d) 0

9. The value of $\int_C \phi d\mathbf{r} - \iint_S d\mathbf{S} \times \nabla\phi$ is :

(a) **0** (b) 0
(c) ϕ (d) $\nabla\phi$

10. The value of $\int_C \phi\nabla\psi \cdot d\mathbf{r} + \int_C \psi\nabla\phi \cdot d\mathbf{r}$ is :

(a) 1 (b) 0
(c) $\nabla\phi$ (d) $\nabla\psi$

11. The value of $\iint_S \mathbf{n} dS$ is :

(a) **n** (b) S
(c) V (d) 0.7

12. If $\mathbf{F} = \nabla\phi$, $\nabla^2\phi = -4\pi\rho$, where ρ is a constant, then the value of $\iint_S \mathbf{F} \cdot \mathbf{n} dS$ is :

(a) 4π (b) $-4\pi\rho$
(c) $-4\pi\rho V$ (d) V

13. The value of $\iint_S \mathbf{F} \cdot \mathbf{n} dS$ where $\mathbf{F} = x^2\hat{i} + y^2\hat{j} + z^2\hat{k}$ and S is the surface of the cube $0 \le y \le 1$, $0 \le z \le 1$, is :

(a) 6 (b) 3
(c) 8 (d) 3π

ANSWERS

Fill in the Blanks

1. closed curve 2. tangent Vector 3. smooth surface 4. circualtion 5. 2π
6. $\int_C (Pdx + Qdy)$ 7. **F·n** 8. 0 9. flux or surface integral of normal component 10. 0
11. $3V$, V is the volume enclosed by S 12. 0 13. stoke's theorem 14. 0

True/False

1. F	2. F	3. T	4. T	5. T
6. F	7. T	8. T	9. T	10. F
11. T	12. T	13. F	14. T	15. F

Multiple Choice Questions

1. (b)	2. (a)	3. (b)	4. (a)	5. (d)
6. (c)	7. (b)	8. (d)	9. (a)	10. (b)
11. (d)	12. (c)	13. (b)		

FFFFFF

GATE tutor

KEY Terms and Results

- **Scalar Point Function :** Let R be the region of space at each point of which a scalar $\phi = \phi(x, y, z)$ is given then ϕ is called scalar function.
- **Vector Point Function :** Let R be the region of space at each point of which a vector $\mathbf{V} = \mathbf{V}(x, y, z)$ is given then $\mathbf{V}$ is called vector function.
- **Gradient :** grad $\phi = \hat{i}\frac{\partial \phi}{\partial x} + \hat{j}\frac{\partial \phi}{\partial y} + \hat{k}\frac{\partial \phi}{\partial z}$.
- **Divergence :** div.$\mathbf{V} = \nabla \cdot \mathbf{V} = \hat{i}\frac{\partial \mathbf{V}}{\partial x} + \hat{j}\frac{\partial \mathbf{V}}{\partial y} + \hat{k}\frac{\partial \mathbf{V}}{\partial z}$.
- **Curl :** curl $\mathbf{V} = \nabla \times \mathbf{V}$
- **Irrotational Vector :** If curl.$\mathbf{V} = 0$
- **Solenoidal Vector :** If div.$\mathbf{V} = 0$
- **Line Integral :** Any integral which is to be evaluated along a curve is called a line integral.
- **Surface Integral :** Any integral which is to be evaluated over a surface is called a surface integral.
- **Volume Integral :** Any integral which is to be evaluated over a volume is called a volume integral.
- **Gauss' Divergence Theorem :** If $\mathbf{F}$ is a vector function having contuinuous first order partial derivatives in the region V bounded by a closed surface S, then

$$\iint_S \mathbf{F}\cdot\hat{n}ds = \iiint_V \text{div}\,\mathbf{F}\cdot dV$$

where $\hat{n}$ is the outward drawn unit normal vector to the surface S.

- **Stroke's Theorem :** If S is an open surface bounded by a closed curve C and $\mathbf{F} = F_1\hat{i} + F_2\hat{j} + F_3\hat{k}$ is any vector function having continuous first order partial derivatives, then

$$\int_C \mathbf{F}\cdot d\mathbf{r} = \iint_S \text{curl}\,\mathbf{F}\cdot\hat{n}ds$$

where $\hat{n}$ is a unit normal vector at any point of S drawn in the sense in which a right handed screw would advance when rotated in the sense of description of C.

- **Green's Theorem :** If M and N are two scalar point functions such that they and their derivatives in any direction are uniform and continuous in a certain region V bounded by a closed serface S then

$$\int_V (M\nabla^2 N - N\nabla^2 M)dv = \int_S (M\nabla N - N\nabla M).\hat{n}dS$$

- **Green's theorem in the Plane :** If C is a regular closed curve in the xy-plane and S be the region bounded by C then

$$\int_C (Mdx + Ndy) = \iint_S \left(\frac{\partial N}{\partial x} - \frac{\partial M}{\partial y}\right)dxdy$$

where $M(x, y)$ and $N(x, y)$ are continuously differentiable functions inside and on C.

Results

- The directional derivative of a scalar point function is a scalar.
- The directional derivative of a scalar point function and a vector point function along the axes any line with direction cosines l, m, n are of the same form.
- $\nabla\phi$ is a vector normal to the surface.
- grad ϕ is a vector in the direction in which maximum value of $\frac{d\phi}{ds}$ occurs.
- Maximum directional derivative is along the normal to the surface.
- grad is associated with scalar point function ϕ and grad ϕ is a vector.
- Divergence is associated with vector point function $\mathbf{f}$ and div $\mathbf{f}$ is a scalar.
- Curl is associated with vector function $\mathbf{f}$ and curl $\mathbf{f}$ is a vector.
- If $\mathbf{f}$ and $\mathbf{g}$ are irrotational then $\mathbf{f} \times \mathbf{g}$ is solenoidal.
- The necessary and sufficient condition that a field $\mathbf{F}$ be conservative is that curl $\mathbf{F} = 0$.

Review Questions and Project Work

1. If $\mathbf{u}$ and $\mathbf{v}$ are irrotational then prove that $\mathbf{u} \times \mathbf{v}$ will be solenoidal. (MTU–2011)
2. If $\mathbf{r} = x\hat{i} + y\hat{j} + z\hat{k}$ then prove that $\nabla(\log r) = \frac{\mathbf{r}}{r^2}$.
3. Show that the vector $\mathbf{V} = 3y^4z\hat{i} + 4x^3z^2\hat{j} - 3x^2y^2\hat{k}$ is solenoidal. (MTU–2012)
4. Show that the magnitude of the gradient of the function $f = xyz^3$ at (1, 0, 2) is 8. (GBTU–2012)
5. Find the constants a and b such that the curl of the vector $\mathbf{A} = (2xy + 3yz)\hat{i} + (x^2 + axz - 4z^2)\hat{j} + (3xy + 2byz)\hat{k}$ is zero. (UPTU–2009)
6. If $\mathbf{F} = xz\hat{i} - y^2\hat{j} + 2x^2y\hat{k}$, then show that curl $\mathbf{F} = 2x^2\hat{i} + (x - 4xy)\hat{j}$. (UPTU–2009)
7. If $\mathbf{r} = a \sin \omega t + b \cos \omega t$, then show that $\frac{d^2\mathbf{r}}{dt^2} = -\omega^2\mathbf{r}$. (Bhopal–2007)
8. Show that the radial and transverse components of the acceleration are $\frac{d^2\mathbf{r}}{dt^2} - \mathbf{r}\left(\frac{d\theta}{dt}\right)^2$ and $2\frac{d\mathbf{r}}{dt}.\frac{d\theta}{dt} + \mathbf{r}\frac{d^2\theta}{dt^2}$.
(Kurukshetra–2006, Rajasthan–2006, UPTU–2010)

9. Find the value of a and b such that the surface $ax^2 - byz = (a + 2)x$ and $4x^2y + z^3 = 4$ cut orthognally at $(1, -1, 2)$. (Madras–2004)

$\left(\textbf{Ans : } a = \frac{5}{2}, b = 1\right)$

10. Show that the angle between the tangent planes to the surface $ax^2 + by^3 = 4$ orthogonally at $(1, -1, 2)$ is $\cos^{-1}\left(\frac{-1}{\sqrt{30}}\right)$.

11. Show that folllowing vectors are solenoidal :

$(-x^2 + yz)\hat{i} + (4y - z^2x)\hat{j} + (2xz - 4z)\hat{k}$. (Delhi–2002)

12. Use Stroke's theorem evaluate

$\int_C [(x+y)dx + (2x-z)dy + (y+z)dz]$

where C is the boundary of the triangle with values $(2, 0, 0)$, $(0, 3, 0)$ and $(0, 0, 6)$.

(Nagpur–2009, Kurukshetra–2009, Kerala–2005)

13. Show that $\int_S (a^2x^2 + b^2y^2 + c^2z^2)^{-1/2} dS = \frac{4\pi}{\sqrt{abc}}$ where S is the surface of ellipsoid $ax^2 + by^2 + cz^2 = 1$.

14. Prove that $\int\int_S (x^3dydz + x^2ydzdx + x^2zdxdy) = \frac{5}{4}\pi a^4 b$ where S is the closed surface consisting of the cylinder $x^2 + y^2 = a^2$ and the circular disc $z = 0$ and $z = b$. (Burdwan–2003)

Objective Type Questions

Choose the most appropriate one.

1. The directional derivative of $f(x, y, z) = 2x^2 + 3y^2 + z^2$ at the point $(2, 1, 3)$ in the direction of the vector $\mathbf{a} = \hat{i} - 2\hat{k}$ is : [GATE(CE)–2006]

(a) –1.789 (b) 1.000
(c) 2.000 (d) none of these

2. Potential function ϕ is given as $\phi = x^2 - y^2$. Then the stream function ψ with the condition $\psi = 0$ at $x = y = 0$ is : [GATE(CE)–2007]

(a) $2x^2y^2$ (b) $2xy$
(c) $x^2 + y^2$ (d) none of these

3. Function f is known at the following points.

x	0	0.3	0.6	0.9	1.2	1.5	1.8	2.1	2.4	2.7	3.0
$f(x)$	0	0.09	0.36	0.81	1.44	2.25	3.24	4.41	5.76	7.29	9.00

The value of $\int_0^3 f(x)dx =$ [GATE(CS)–2013]

(a) 9.006 (b) 9.045
(c) 9.876 (d) none of these

4. A velocity vector is given by $\mathbf{V} = 5xy\hat{i} + 2y\hat{j} + 3yz^2\hat{k}$. The divergence of this velocity vector at $(1, 1, 1)$ is : [GATE(CS)–2007]

(a) 10 (b) 14
(c) 15 (d) none of these

5. The gradient of the function $f(x, y, z) = x^2 + 3y^2 + 2z^2$ at the point $(1, 2, -1)$ is : [GATE(CE)–2009]

(a) $2\hat{i} + 12\hat{j} - 4\hat{k}$ (b) $2\hat{i} + 6\hat{j} + 4\hat{k}$
(c) $2\hat{i} + 3\hat{j} - 5\hat{k}$ (d) none of these

6. Equation of the line normal to the function $f(x) = (x-8)^{2/3} + 1$ at $(0, 5)$ is given by : [GATE(ME)–2006]

(a) $y = 3x + 5$ (b) $3y = x + 5$
(c) $y = 3x - 5$ (d) none of these

7. The curl of the gradient of the scalar field given by

$V = 2x^2y + 3y^2z + 4z^2x$ is : [GATE(EE)–2013]

(a) 1 (b) 2
(c) 0 (d) none of these

8. The directional derivative of the scalar function $f(x, y, z) = x^2 + 2y^2 + z$ at the point $(1, 1, 2)$ in the direction of the vector $\mathbf{a} = 3\hat{i} - 4\hat{j}$ is : [GATE(ME)–2008]

(a) 2 (b) –2
(c) 1 (d) none of these

9. For a scalar function $f(x, y, z) = x^2 + 3y^2 + 2z^2$ the directional derivative at the point $(1, 2, -1)$ in the direction of $\hat{i} - \hat{j} + 2\hat{k}$ is : [GATE(ME)–2009]

(a) $3\sqrt{6}$ (b) $-3\sqrt{6}$
(c) $2\sqrt{6}$ (d) none of these

10. The divergence of the vector field $\mathbf{A} = x\hat{a}_x + y\hat{a}_y + z\hat{a}_z$ is : [GATE(EC)–2013]

(a) 1 (b) 2
(c) 3 (d) 10

11. The divergence of the vector field

$(x-y)\hat{i} + (y-x)\hat{j} + (x+y+z)\hat{k}$ is : [GATE(ME)–2008]

(a) 1 (b) 2
(c) 0 (d) 3

12. The divergence of the vector field $3xz\hat{i} + 2xy\hat{j} - yz^2\hat{k}$ at a point $(1, 1, 1)$ is equal to : [GATE(ME)–2009]

(a) 3 (b) 4
(c) 7 (d) none of these

13. For the vector field $u = \frac{x^2}{2} + \frac{y^2}{3}$, magnitude of the gradient at the point $(1, 3)$ is : [GATE(EE)–2005]

(a) 5 (b) $\sqrt{5}$
(c) $5\sqrt{5}$ (d) none of these

14. Velocity vector of a flow field is given by $\mathbf{V} = 2xy\hat{i} - x^2z\hat{j}$. The vorticity vector at $(1, 1, 1)$ is : [GATE(ME)–2010]

(a) $1 - 4\hat{k}$ (b) $\hat{i} - 4\hat{k}$
(c) $\hat{i} + 4\hat{k}$ (d) none of these

15. For a vector $\mathbf{V}$ which one of the following is not true ? [GATE(IN)–2013]

(a) If $\nabla \cdot \mathbf{V} = 0 \Rightarrow E$ is solenoidal.
(b) If $\nabla \times \mathbf{V} = 0 \Rightarrow E$ is irrotational.
(c) If $\nabla \cdot \mathbf{V} = 0 \Rightarrow E$ is irrotational.
(d) none of the above

16. Divergence of the three dimensional radical vector **r** is : [GATE(EE)–2010]

(a) 3 (b) 4
(c) 1 (d) 10

17. For spherical surface $x^2 + y^2 + z^2 = 1$, the unit outward normal vector at the point $\left(\frac{1}{\sqrt{2}}, \frac{1}{\sqrt{2}}, 0\right)$ is given by : [GATE(ME)–2012]

(a) $\frac{1}{\sqrt{2}}\hat{i} + \frac{1}{\sqrt{2}}\hat{j}$ (b) $\hat{i} + \sqrt{2}\hat{j}$
(c) $\hat{k}$ (d) none of these

18. The direction of vector **A** is radically outward from the origin with $|\mathbf{A}| = Kr^n$ where $r^2 = x^2 + y^2 + z^2$ and K is a constant. The value of n for which $\nabla \cdot \mathbf{A} = 0$ is : [GATE(IN)–2012]

(a) 2 (b) 1
(c) –2 (d) 10

19. The value of $\nabla \times \mathbf{P} \times \mathbf{P}$: [GATE(EC)–2006]

(a) $\nabla(\nabla \cdot \mathbf{P}) - \nabla^2\mathbf{P}$ (b) $\nabla \cdot (\nabla \mathbf{P})$
(c) $\nabla^2\mathbf{P}$ (d) none of these

20. The line integral $\int \mathbf{V}.d\mathbf{r}$ of the vector $\mathbf{V}(\mathbf{r}) = 2xyz\hat{i} + x^2z\hat{j} + x^2y\hat{k}$ from the origin to the point (1, 1, 1) is : [GATE(ME)–2005]

(a) 1 (b) 2
(c) 3 (d) 0

21. Let $P = (1, 0)$ and $Q = (0, 1)$ be two points in xy-plane. The line integral $2\int_P^Q (xdx + ydy)$ along the semi-circle with the line segment PQ as its diameter is : [GATE(EC)–2008]

(a) 1 (b) 2
(c) 0 (d) none of these

22. Value of the integral $\int_C (xydy - y^2dx)$ where C is the square cut from the first quadrant by the lines $x = 1$ and $y = 1$ will be : [GATE(CE)-2005]

(a) $\frac{2}{3}$ (b) 1
(c) $\frac{3}{2}$ (d) 0

23. Given a vector field $\mathbf{F} = y^2x\hat{a}_x - yz\hat{a}_y - x^2\hat{a}_z$, the line integral $\int \mathbf{F} \cdot d\mathbf{r}$ evaluated along a segment on the x-axis from $x = 1$ to $x = 2$ is : [GATE(EE)–2013]

(a) 0 (b) 1
(c) 2 (d) 3

24. The following surface integral is to be evaluated over a sphere for the given steady velocity vector $\mathbf{F} = x\hat{i} + y\hat{j} + z\hat{k}$ defined w.r.t. a cartesian coordinate system having $\hat{i}, \hat{j}$ and $\hat{k}$ as unit base vectors $\iint_S \frac{1}{4}(\mathbf{F} \cdot \hat{n})dS$ where S is the surface $x^2+y^2+z^2= 1$ and $\hat{n}$ is the outward unit normal vector to the sphere. The value of the surface integral is : [GATE(ME)–2013]

(a) π (b) 2π
(c) 3π (d) None of these

25. Consider a vector field $\mathbf{A}(r)$. The closed loop line integral $\oint \mathbf{A} \cdot dl$ can be expressed as : [GATE(EC)–2013]

(a) $\iint (\nabla \times \mathbf{A})d\mathbf{s}$ over the open surface bounded by the loop
(b) $\iint (\nabla.\mathbf{A})ds$ over the open surface bounded by the loop
(c) both (a) and (b) are true
(d) none of these

26. $\iint (\nabla \times \mathbf{p})ds$ where **p** is a vector is equal to: [GATE(EC)–2006]

(a) $\oint \mathbf{p} \cdot dl$
(b) $\oint \nabla \times \mathbf{p}\, dl$
(c) both (a) and (b) are true
(d) none of the above

27. Stoke's theorem connects : (GATE(ME)–2005)

(a) a line integral and a surface integral
(b) a surface integral and a volume integral
(c) a line integral and a volume integral
(d) none of these

28. The vector $\mathbf{V} = e^x \sin y\hat{i} + e^x \cos y\hat{j}$ is :

(a) solenoidal (b) rotational
(c) irrotational (d) none of these

29. The value of the line integral $\int_C (y^2dx + x^2dy)$ where C is the boundary of the square $-1 \le x \le 1, -1 \le y \le 1$ is : (VTU–2010)

(a) 0 (b) $x + y$
(c) $2(x + y)$ (d) none of these

30. The spherical coordinate system is : (VTU–2010)

(a) coplanar (b) orthogonal
(c) non-coplanar (d) none of these

ANSWERS

1. (a)	**2.** (b)	**3.** (b)	**4.** (c)	**5.** (a)	**6.** (a)	**7.** (c)	**8.** (b)	**9.** (b)	**10.** (c)
11. (d)	**12.** (a)	**13.** (b)	**14.** (a)	**15.** (c)	**16.** (a)	**17.** (a)	**18.** (c)	**19.** (a)	**20.** (a)
21. (c)	**22.** (c)	**23.** (a)	**24.** (a)	**25.** (a)	**26.** (a)	**27.** (a)	**28.** (c)	**29.** (a)	**30.** (b)

Hint to Selected Problems

1. $\mathbf{f} = 2x^2 + 3y^2 + z^2 \quad P(1, 2, 3)$

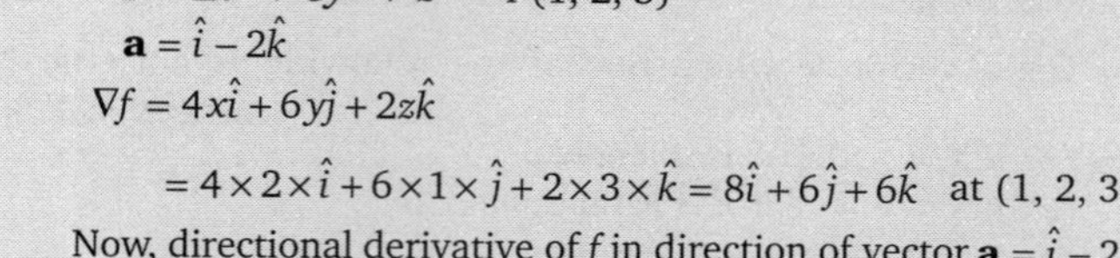

$\mathbf{a} = \hat{i} - 2\hat{k}$

$\nabla f = 4x\hat{i} + 6y\hat{j} + 2z\hat{k}$

$= 4\times2\times\hat{i} + 6\times1\times\hat{j} + 2\times3\times\hat{k} = 8\hat{i} + 6\hat{j} + 6\hat{k}$ at (1, 2, 3)

Now, directional derivative of f in direction of vector $\mathbf{a} = \hat{i} - 2\hat{k}$ is the component of grad f in the direction of vector **a** and is given by $\frac{\mathbf{a}}{|\mathbf{a}|}$ grad f

$$= \left(\frac{\hat{i} - 2\hat{k}}{\sqrt{1^2 + (-2)^2}}\right)(8\hat{i} + 6\hat{j} + 6\hat{k})$$

$$= \frac{1}{\sqrt{5}}(1\times8 + 0\times6 + (-2)\times6) = -1.789$$

4. $\mathbf{V} = 5xy\hat{i} + 2y^2\hat{j} + 3yz^2\hat{k} = V_1\hat{i} + V_2\hat{j} + V_3\hat{k}$

$div(\mathbf{V}) = \frac{dV_1}{dx} + \frac{dV_2}{dy} + \frac{dV_3}{dx} = 5y + 4y + 6yz$

At (1, 1, 1), div ($\mathbf{V}$) = $5 \times 1 + 4 \times 1 + 6 \times 1 \times 1 = 15$

5. $f = x^2 + 3y^2 + 2z^2$

$\nabla f = \text{grad } f = \hat{i}\frac{\partial f}{\partial x} + \hat{j}\frac{\partial f}{\partial y} + \hat{k}\frac{\partial f}{\partial z} = \hat{i}(2x) + \hat{j}(6y) + \hat{k}(4z)$

The gradient at $P(1, 2, -1)$ is

$= \hat{i}(2\times 1) + \hat{j}(6\times 2) + \hat{k}(4\times -1) = 2\hat{i} + 12\hat{j} - 4\hat{k}$

6. $f(x) = (x-8)^{2/3} + 1 \quad \Rightarrow \quad f'(x) = \frac{2}{3}(x-8)^{-1/3}$

Slope of tangent at point (0, 5) is

$m = \frac{2}{3}(0-8)^{-1/3} = -\frac{1}{3}$

Slope of normal at point (0, 5) is

$m_1 = -\frac{1}{m} = 3$

Equation of normal at point (0, 5) is

$y - 5 = 3(x - 0) \quad \Rightarrow \quad y = 3x + 5$

7. Curl of gradient of scalar field is always zero.

8. grad $f = 2x\hat{i} + 4y\hat{j} + \hat{k}$

At (1, 1, 2), grad $f = 2\hat{i} + 4\hat{j} + \hat{k}$

Directional derivative of f at (1, 1, 2) in the direction of vector $\mathbf{a} = 3\hat{i} - 4\hat{j}$ given by

$\frac{\mathbf{a}}{|\mathbf{a}|}\text{grad } f = \left(\frac{3\hat{i} - 4\hat{j}}{\sqrt{25}}\right).(2\hat{i} + 4\hat{j} + \hat{k})$

$= \frac{1}{5}(3\times 2 - 4\times 4 + 0) = -2$

9. $\nabla f = \hat{i}(2x) + \hat{j}(6y) + \hat{k}(4z)$

At $P(1,2,-1)$, $\nabla f = \hat{i}(2\times 1) + \hat{j}(6\times 2) + \hat{k}(4\times -1) = 2\hat{i} + 12\hat{j} - 4\hat{k}$

The directional derivative in direction of vector $\mathbf{a} = i - \hat{j} + 2\hat{k}$ is given by

$\frac{\mathbf{a}}{|\mathbf{a}|}\text{grad } f = \left(\frac{\hat{i} - \hat{j} + 2\hat{k}}{\sqrt{1^2 + (-1)^2 + 2^2}}\right).(2\hat{i} + 12\hat{j} - 4\hat{k})$

$= \frac{1}{\sqrt{6}}(1\times 2 + (-1)\times 12 + 2(-4)) = \frac{-18}{\sqrt{6}} = -3\sqrt{6}$

10. $\nabla \mathbf{A} = \frac{\partial A_x}{\partial x} + \frac{\partial A_y}{\partial y} + \frac{\partial A_z}{\partial z} = \frac{\partial}{\partial x}(x) + \frac{\partial}{\partial y}(y) + \frac{\partial}{\partial z}(z) = 1 + 1 + 1 = 3$

11. $\text{div}\left\{(x-y)\hat{i} + (y-x)\hat{j} + (x+y+z)\hat{k}\right\}$

$= \frac{\partial}{\partial x}(x-y) + \frac{\partial}{\partial y}(y-x) + \frac{\partial}{\partial z}(x+y+z) = 3$

12. $\text{Div}\cdot(\mathbf{f}) = \nabla\cdot\mathbf{f} = \frac{\partial}{\partial x}(3xz) + \frac{\partial}{\partial y}(2xy) + \frac{\partial}{\partial z}(-2yz^2) = 3z + 2x - 2zy$

$= 3(1) + 2(1) - 2(1)(1) = 3$ at (1,1,1)

13. $\text{grad } u = \hat{i}\frac{\partial}{\partial x}\left(\frac{x^2}{2} + \frac{y^2}{3}\right) + \hat{j}\frac{\partial}{\partial y}\left(\frac{x^2}{2} + \frac{y^2}{3}\right) = x\hat{i} + \frac{2}{3}y\hat{j}$

$= \hat{i} + 2\hat{j}$ at (1, 3)

$\Rightarrow |\text{grad } u| = \sqrt{1^2 + 2^2} = \sqrt{5}$

14. Vorticity vector = $\nabla \times \mathbf{V} = \begin{vmatrix} \hat{i} & \hat{j} & \hat{k} \\ \frac{\partial}{\partial x} & \frac{\partial}{\partial y} & \frac{\partial}{\partial z} \\ 2xy & -x^2z & 0 \end{vmatrix}$

$= \left[\frac{\partial}{\partial y}(0) - \frac{\partial}{\partial z}(-x^2z)\right]\hat{i} - \left[\frac{\partial}{\partial x}(0) - \frac{\partial}{\partial z}(2xy)\right]\hat{j}$

$+ \left[\frac{\partial}{\partial x}(-x^2z) - \frac{\partial}{\partial y}(2xy)\right]\hat{k}$

$= x^2\hat{i} - 2xz\hat{k} = \hat{i} - 2\hat{k}$ at (1, 1, 1)

16. $\mathbf{r} = x\hat{i} + y\hat{j} + z\hat{k}$

$\text{div}\cdot\mathbf{r} = \nabla\cdot\mathbf{r} = \left(\hat{i}\frac{\partial}{\partial x} + \hat{j}\frac{\partial}{\partial y} + \hat{k}\frac{\partial}{\partial z}\right).(x\hat{i} + y\hat{j} + z\hat{k})$

$= \frac{\partial x}{\partial x} + \frac{\partial y}{\partial y} + \frac{\partial z}{\partial z} = 1 + 1 + 1 = 3$

17. $\text{grad } f = \left(\frac{\partial f}{\partial x}\hat{i} + \frac{\partial f}{\partial y}\hat{j} + \frac{\partial f}{\partial z}\hat{k}\right) = 2x\hat{i} + 2y\hat{j} + 2z\hat{k}$

$= \frac{2}{\sqrt{2}}\hat{i} + \frac{2}{\sqrt{2}}\hat{j} + z\times 0\hat{k} \quad \text{at}\left(\frac{1}{\sqrt{2}}, \frac{1}{\sqrt{2}}, 0\right)$

$= \sqrt{2}\hat{i} + \sqrt{2}\hat{j} + 0\hat{k}$

$\Rightarrow |\text{grad } f| = \sqrt{2+2} = 2$

The unit outward normal vector at point P is given by

$\hat{n} = \frac{1}{|\text{grad } f|}(\text{grad } f)_{\text{at } P} = \frac{1}{2}(\sqrt{2}\hat{i} + \sqrt{2}\hat{j}) = \frac{1}{\sqrt{2}}\hat{i} + \frac{1}{\sqrt{2}}\hat{j}$

18. $\nabla\cdot\mathbf{A} = \frac{1}{r^2}\frac{\partial}{\partial r}(r^2\cdot Kr^n) = \frac{1}{r^2}\frac{\partial}{\partial r}(Kr^{n+2})$

$= (n+2)K\frac{r^{n+1}}{r^2} = (n+2)Kr^{n-1}$

19. Use the property for $\nabla\cdot A = 0 \Rightarrow n + 2 = 0$

$\Rightarrow \quad n = -2$

$A \times (B \times C) = (A\cdot C)B - (A.B)C$

20. $f_x = 2xyz, f_y = x^2z, f_z = x^2y$

On integrating f = potential function of $\mathbf{V} = x^2yz$.

So, line integral of the vector function from point $A(0, 0, 0)$ to the point $B(1, 1, 1)$ is

$= f(B) - f(A)$

$= (x^2yz)_{(1,1,1)} - (x^2yz)_{(0,0,0)} = 1 - 0 = 1$

21. Taking $f(x, y) = xy$, we can show that $xdx + ydy$ is exact. Therefeore, the value of the integral is independent of path

$= 2\int_1^0(xdx + ydy) = 2\int_1^0 xdx + 2\int_0^1 ydy = 2\left[\frac{x^2}{2}\right]_1^0 + \left[\frac{y^2}{2}\right]_0^1 = 0$

22. Apply Green's theorem.

23. $\int\mathbf{F}\cdot d\mathbf{r} = \int(y^2x\hat{a}_x - yz\hat{a}_y - x^2\hat{a}_z).(\hat{a}_xdx + \hat{a}_ydy + \hat{a}_zdz)$

$= \int y^2x\,dx - yzdy - x^2dz$

Putting $y = 0, z = 0 \Rightarrow dy = 0 = dz$

$\therefore \quad \int\mathbf{F}\cdot d\mathbf{r} = 0$

Self Assessment Test

1. Find the tangent vector to the curve $\mathbf{H}(t) = t^2\hat{i} + \sin t\hat{j} - t^2\hat{k}$ at $t = 0$ and $t = 1$.

2. If $\mathbf{r} = \cos nt\hat{i} + \sin nt\hat{j}$, where n is a constant and t varies, show that $\mathbf{r} \times \frac{d\mathbf{r}}{dt} = n\hat{k}$.

3. If $u = x + y + z$, $v = x^2 + y^2 + z^2$ and $w = xy + yz + zx$, show that ∇u, ∇v and ∇w are coplanar.

4. If $\mathbf{r} = x\hat{i} + y\hat{j} + z\hat{k}$, show that

(i) $\nabla(\mathbf{a}\cdot\mathbf{r}) = a$, where a is a constant vector.

(ii) $\text{grad}\,|\mathbf{r}| = \frac{\mathbf{r}}{|\mathbf{r}|}$

(iii) $\text{grad}\,\frac{1}{|r|} = -\frac{\mathbf{r}}{|r|^3}$

(iv) $\text{grad}\,|\mathbf{r}|^n = n|\mathbf{r}|^{n-2}\cdot\mathbf{r}$

5. Prove that $f(r) \times \mathbf{r} = 0$.

6. If **a** is a constant vector and **r** is a point function, prove that
$$(\mathbf{a}\cdot\nabla)\mathbf{r} = \mathbf{a}$$

7. Show that $\nabla^2\left(\frac{x}{r^3}\right) = 0$ where r is a magnitude of the position vector $\mathbf{r} = x\hat{i} + y\hat{j} + z\hat{k}$.

8. Find the divergence and curl of the vectot field
$$\mathbf{V} = x^2y^2\hat{i} + 2xy\hat{j} + (y^2 - xy)\hat{k}.$$

9. Show that the vector field $\mathbf{F} = \frac{r}{|r|^3}$ is rotational as well as solenoidal.

10. Prove : curl·curl $\mathbf{f} = \nabla\cdot$ div $\mathbf{f} - \nabla^2\mathbf{f}$.

11. Prove : grad div. $\mathbf{V}$ = curl curl $\mathbf{F} + \Sigma\frac{\partial^2\mathbf{F}}{\partial x^2}$.

12. Show that vector field **F** defined by
$$\mathbf{F} = (\sin y + z)\hat{i} + (x\cos y - z)\hat{j} + (x - y)\hat{k}$$
is conservative.

13. Show that $\iint_S (yz\hat{i} + zx\hat{j} + xy\hat{k})\,dS$, where S is the surface of the sphere $x^2 + y^2 + z^2 = 1$ in the first octant is equal to $\frac{3}{8}$.

14. Verify the Green's theorem for the functions $e^{-x}\sin y$ and $e^{-x}\cos y$.

15. Verify Green's theorem in the xy-plane for $\int_C [(3x^2 - 8y^2)dz + (4y - 6xy)dy]$ where C is the region bounded by the paraboles $y = \sqrt{x}$ and $y = x^2$.

16. Verify Gauss's divergence theorem for the function
$$\mathbf{F} = (x^2 - yz)\hat{i} + (y^2 - zx)\hat{j} + (z^2 - xy)\hat{k}$$
taken over the rectangular parallelopiped $0 \le x \le a$, $0 \le y \le b$, $0 \le z \le c$.

17. If $\mathbf{F} = 3y\hat{i} - xz\hat{j} + yz^2\hat{k}$ and S is the surface of the paraboloid $2z = x^2 + y^2$ bounded by $z = 2$, show using Stoke's theorem that $\iint_S (\nabla\cdot\mathbf{F})\cdot dS = -20\pi$.

18. Verify Stoke's theorem for the function
$$\mathbf{F} = (x^2 + y - 4)\hat{i} + 3xy\hat{j} + (2xz + z^2)\hat{k}$$
where S is the upper half of the sphere $x^2 + y^2 + z^2 = 16$ and C is its boundary.

cccccc

Unit-VI

ALGEBRA

UNIT OUTLINE

- Solution of Equations
- Horner's Synthetic Division
- Transformation of Equations
- Descarte's Rule of Signs
- Cardan's Method
- Ferrari's Method
- Matrices
- Determinants
- Eigen Values and Eigen Vectors
- Diagonalisation of a Matrix
- Quadratic Forms

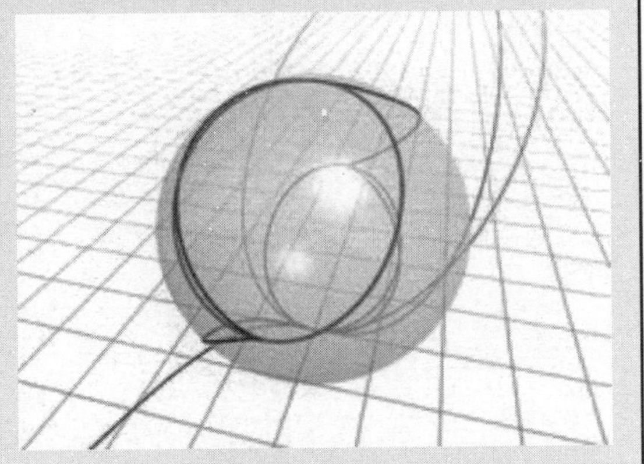

PRE-REQUISITE

- Relation between Roots and Coefficients of a Quadratic Equation
- Concepts of Theory of Equations
- Basic Concepts of Matrices
- Basic Concepts of Determinants

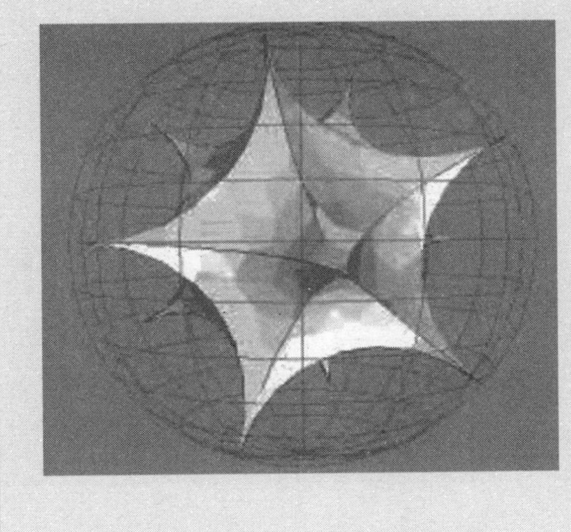

CHAPTER 24 Solution of Equations

24.1 INTRODUCTION

A function *f*(*x*) in the form $f(x) = a_0x^n + a_1x^{n-1} + a_2x^{n-2} + ... + a_{n-1}x + a_n, a_0 \neq 0$ of degree *n* is said to be a rational integral function of *x* if all the coefficients $a_0, a_1, a_2 ..., a_{n-1}, a_n$ are supposed to be rational.

Definition. *The equation f(x) = 0 is called the general form of rational integral equation of nth degree.*

Definition. *Any value of x for which the value of f(x) comes out to be zero, is called a root of the equation f(x) = 0.*

Since when *f*(*x*) is divided by the factor (*x* – *a*) then *f*(*a*) is obtained as a remainder. If this remainder *f*(*a*) becomes zero, then *a* is a root of the function *f*(*x*) = 0. Therefore, we can say that if '*a*' is a root of the equation *f*(*x*) = 0, then *f*(*a*)=0.

24.2 NUMBER OF ROOTS OF ANY EQUATION

THEOREM 1. *Every equation of degree n has n roots and no more.*

Proof. Let the equation of degree *n* be

$$f(x) = a_0x^n + a_1x^{n-1} + a_2x^{n-2} + ... + a_{n-1}x + a_n = 0; \text{ provided } a_0 \neq 0. \quad ...(1)$$

The equation *f*(*x*) = 0 has the roots, real as well as imaginary. Therefore if α_1 is any root of the equation (1), then *f*(*x*) can be written as

$$f(x) = (x - \alpha_1)(a_0x^{n-1} + ...) \quad \text{or} \quad f(x) = (x - \alpha_1)\phi_1(x) \quad ...(2)$$

where $\phi_1(x)$ is a function of *x* of degree *n* – 1, such that $\phi_1(\alpha_1) \neq 0$. Further let α_2 be a root of $\phi_1(x) = 0$, then $\phi_1(x)$ can be written

$$\phi_1(x) = (x - \alpha_2)\phi_2(x)$$

$$\therefore \quad f(x) = (x - \alpha_1)(x - \alpha_2)\phi_2(x) \quad ...(3)$$

Continuing this process upto *n* times, we obtain

$$f(x) = a_0(x - \alpha_1)(x - \alpha_2)...(x - \alpha_n) \quad ...(4)$$

From equation (4) it is clear that when *x* take the values form α_1 to α_n, $f(x)$ comes out be zero.

Hence, the equation *f*(*x*) has *n* roots. Moreover if *x* takes any value different from $\alpha_1, \alpha_2, ...\alpha_n$, $f(x)$ cannot be zero so that *f*(*x*) = 0 has exactly *n* roots.

24.3 RELATION BETWEEN THE ROOT AND COEFFICIENTS

Let the general equation of degree *n* be given by

$$a_0x^n + a_1x^{n-1} + a_2x^{n-2} + ... + a_{n-1}x + a_n = 0 \quad ...(1)$$

where $a_0, a_1, a_2 ... a_n$ are the coefficients and $a_0 \neq 0$ let $\alpha_1, \alpha_2, \alpha_3 ... \alpha_n$ be the roots of the equation (1). The equation (1) can be identically written as

$$a_0x^n + a_1x^{n-1} + a_2x^{n-2} + ... + a_{n-1}x + a_n = a_0(x - \alpha_1)(x - \alpha_2)...(x - \alpha_n)$$

or $$a_0x^n + a_1x^{n-1} + a_2x^{n-2} + ... + a_{n-1}x + a_n = a_0[x^n - (\sum\alpha_1)x^{n-1} + (\sum\alpha_1\alpha_2)x^{n-2} + ... + (-1)^n\alpha_1\alpha_2...\alpha_n]$$

where $$\sum\alpha_1 = \alpha_1 + \alpha_2 + ... + \alpha_n$$

$$\sum\alpha_1\alpha_2 = \alpha_1\alpha_2 + \alpha_1\alpha_3 + ... \text{ etc.}$$

Now equating the coefficients of like power of x of both sides we get

$$\left.\sum\alpha_1 = -\frac{a_1}{a_0};\ \sum\alpha_1\alpha_2 = \frac{a_2}{a_0};\ \sum\alpha_1\alpha_2\alpha_3 = -\frac{a_3}{a_0};...;\ \alpha_1\alpha_2\alpha_3...\alpha_n = (-1)^n\frac{a_n}{a_0}\right\} \quad ...(2)$$

Hence, the equation (2) gives the required relation between the roots and the coefficients of the equation.

REMARK

- If the equation is not complete *i.e.*, some of the terms are missing, then we should first make this equation complete by adding the missing terms with zero coefficients.

Solved Examples

Example 1. *Find the condition that the cubic $x^3 - lx^2 + mx - n = 0$ should have its roots in*
(1) arithmetic Progression
(2) geometric Progression (Madras–2000, 09)

Solution. (1) Let the roots in A.P be

$$a-d, a \text{ and } a+d$$

The sum of the roots

$$= a-d+a+a+d = 3a = l$$

$$\Rightarrow \quad a = \frac{l}{3} \qquad ...(1)$$

Since a is the root of the given equation, therefore

$$a^3 - la^2 + ma - n = 0$$

$$\Rightarrow \quad \left(\frac{l}{3}\right)^3 - l\left(\frac{l}{3}\right)^2 + m\left(\frac{l}{a}\right) - n = 0 \text{ (Using (1))}$$

$$\Rightarrow \quad 2l^3 - 9lm + 27n = 0,$$

which is the required condition.

(2) Let the roots in G.P be $\frac{a}{r}, a$ and ar

Then product of the roots $= \frac{a}{r}.a.ar = n$

$$\Rightarrow \quad a^3 = n$$

Then form (1)

$$n - ln^{2/3} + mn^{1/3} - n = 0$$

$$\Rightarrow \quad m = ln^{1/3} \quad \Rightarrow \quad m^3 = l^3 n,$$

which is the required condition.

Example 2. *Solve the equation $x^4 - 2x^3 + 4x^2 + 6x - 21 = 0$ given that the sum of two of its roots is zero.* (Madras–2003, Cochin–2005)

Solution. Let α, β, γ and δ be the roots of the given equation such that

$$\alpha + \beta = 0$$

Then $\alpha+\beta+\gamma+\delta = 2 \Rightarrow \quad \gamma+\delta = 2$

Therefore, the quadratic factor corresponding to α, β of the form $x^2 - 0x + p$ and that corresponding to γ, δ is of the form $x^2 - 2x + q$.

Therefore, we can write

$$x^4 - 2x^3 + 4x^2 + 6x - 21$$
$$= (x^2+p)(x^2-2x+q)$$

Equating the coefficient of x^2 and x from both sides of (1) we get

$$p + q = 4, -2p = 6$$

$$\Rightarrow \quad p = -3, q = 7$$

So, the given equation is equivalent to

$$(x^3-3)(x^2-2x+7) = 0$$

Hence, the required roots are $x = \pm\sqrt{3}, 1 \pm i\sqrt{6}$.

EXERCISE 24.1

1. Solve the equation $x^4 - 2x^3 - 21x^2 + 22x + 40 = 0$ whose roots are in A.P.
2. Solve the equation $2x^4 - 15x^3 + 35x^2 - 30x + 8 = 0$ whose roots are in G.P.
3. Form the equation of fourth degree whose roots are $3+i$ and $\sqrt{7}$. (Madras–2000)
4. Show that $x^7 - 3x^4 + 2x^3 - 1 = 0$ has at least four imaginary roots. (Cochin–2006, 14)
5. Solve the equation $x^3 - 4x^2 - 20x + 48 = 0$, given that the roots α and β connected by the relation $\alpha + 2\beta = 0$. (SVTU–2007)
6. Solve the equation $x^4 - 6x^3 + 13x^2 - 12x + 4 = 0$ given that it has two parts of equal roots. (Madras–2003)
7. If O, A, B, C are the four points on a straight line such that the distances A, B, C from O are the roots of the equation $ax^3 + 3bx^2 + 3cx + d = 0$. If B is the middle point of AC, show that $a^2d - 3abc + 2b^3 = 0$. (SVTU–2006)
8. Solve the equation $x^4 - 8x^3 + 21x^2 - 20x + 5 = 0$, given that the sum of two of the roots is equal to the sum of the other two.
9. Solve the equation $8x^3 - 14x^2 + 7x - 1 = 0$, roots being in G.P. (Osmania–1999, 2007)
10. Solve the equation $x^3 - 12x^2 + 39x - 28 = 0$, roots being in A.P. (Madras–2001, 2011)

ANSWERS

1. $-4, -1, 2, 5$ **2.** $\frac{1}{2}, 1, 2, 4$ **3.** $x^4 - 6x^3 + 3x^2 + 42x - 70 = 0$ **5.** $-4, 2, 6$

6. $1, 1, 2, 2$ **8.** $\frac{1}{2}(3 \pm \sqrt{5}), \frac{1}{2}(5 + \sqrt{5})$ **9.** $1, \frac{1}{2}, \frac{1}{4}$ **10.** $1, 4, 7$

24.4 IMPORTANT RESULTS

1. In an equation with real coefficients, imaginary roots occur in pair, that is if $\alpha + i\beta$ is one of the root of the equation $f(x) = 0$, then $\alpha - i\beta$ will also be a root of that equation.
2. If the equation $f(x) = 0$ has a pair of complex (imaginary) roots $\alpha \pm i\beta$ then $(x-\alpha)^2 + \beta^2$ will be a factor of $f(x)$.
3. If $\alpha + \sqrt{\beta}$ is a root of the equation $f(x) = 0$, then $\alpha - \sqrt{\beta}$ will also be a root of $f(x) = 0$.
4. Every equation of odd degree with real coefficients has at least one real root with the sign opposite to that of its last term.
5. Every equation of even degree with negative last term has at least two real roots with contrary sign.
6. If the equation $f(x) = 0$ and $g(x) = 0$ have common roots and there common roots are the roots of $h(x) = 0$, then $h(x)$ will be H.C.F. (G.C.D.) of $f(x)$ and $g(x)$.
7. If the equation $f(x) = 0$ has two roots equal, then the equation $f(x) = 0$, and $f'(x) = 0$ must have a common root.

24.5 HORNER'S SYNTHETIC DIVISION

In order to find the quotient and the remainder when a polynomial

$$f(x) = a_0x^n + a_1x^{n-1} + a_2x^{n-2} + \ldots + a_{n-1}x + a_n, (a_0 \neq 0) \quad \ldots(1)$$

of degree n is divided by a linear factor $(x-\alpha)$, we use a method given by Horner, called *synthetic division*. This method is being discussed as follows:

$$\begin{array}{c|ccccc} \alpha & a_0 & a_1 & a_2\ldots & a_{n-1} & a_n \\ & & \alpha a_0 & \alpha b_1\ldots & \alpha b_{n-2} & \alpha b_{n-1} \\ \hline & a_0 & b_1 & b_2 & b_{n-1} & R \end{array}$$

Step (1) If the equation (1) is not complete, then first make it complete by adding missing terms with zero coefficient.

Step (2) In the first horizontal line (row) we should write the coefficients $a_0, a_1, a_2, \ldots a_{n-1}, a_n$ of the polynomial $f(x)$.

Step (3) Since we have to divide to polynomial $f(x)$ by $x - \alpha$, so we should write α to the left to the vertical line as shown above.

Step (4) In the third horizontal line (row) we should write a_0 and the first term of the second horizontal line (row) is obtained by multiplying a_0 to α and then add this term with a_1 we obtain b_1 which is the second term of the third row. Next, we multiply b_1 and α and obtained the second term of the second row now adding this αb_1 with a_2 we obtain third terms of the third row. Continue the process in the same way we obtain the last term in the third row which is in fact the remainder R while the second last term in the same is b_{n-1}.

REMARK

- If the remainder R comes out be zero, then α will be root of the equation $f(x) = 0$.

24.6 TRANSFORMATION OF EQUATION

Sometimes there arises some difficulties to find the roots of a given equation. In that case a process of transformation of a given equation into another equation plays an important role for finding the roots of given equation.

In this section we shall discuss some important transformation.

(i) **To transform an equation into another equation whose roots are the roots of the given equation with different sign.**

Let the given equation be

$$f(x) = a_0x^n + a_1x^{n-1} + a_2x^{n-2} + \ldots + a_{n-1}x + a_n = 0 \quad \ldots(1)$$

and let $\alpha_1, \alpha_2, \alpha_3, \ldots, \alpha_n$ be the roots of the equation (1).

Now put $x = -y$ in (1), we get

$$f(-y) = a_0(-y)^n + a_1(-y)^{n-1} + a_2(-y)^{n-2} + \ldots + a_{n-1}(-y) + a_n = 0$$

or
$$f(-y) = (-1)^n[a_0y^n - a_1y^{n-1} + a_2y^{n-2} - \ldots + (-1)^{n-1}a_{n-1}y + (-1)^n a_n] = 0 \quad \ldots(2)$$

This is the transformed equation.

Now we shall have to show that the equations (2) has the roots $-\alpha_1, -\alpha_2, -\alpha_3, \ldots, -\alpha_n$.

Since $\alpha_1, \alpha_2, \ldots, \alpha_n$ are the roots of equation (1), then (1) can also be written as

$$a_0x^n + a_1x^{n-1} + a_2x^{n-2} + \ldots + a_{n-1}x + a_n = a_0(x-\alpha_1)(x-\alpha_2)\ldots(x-\alpha_n)$$

Now putting $x = -y$ in both sides, we get

$$a_0(-y)^n + a_1(-y)^{n-1} + a_2(-y)^{n-2} + ... + a_{n-1}(-y) + a_n = a_0(-y-\alpha_1)(-y-\alpha_2)...(-y-\alpha_n)$$

or $$(-1)^n[a_0y^n - a_1y^{n-1} + a_2y^{n-2} - ... + (-1)^{n-1}a_{n-1}y + (-1)^n a_0] = a_0(-1)^n(y+\alpha_1)(y+\alpha_2)...(y+\alpha_n)$$

Using (2) $$f(-y) = a_0(-1)^n(y+\alpha_1)(y+\alpha_2)...(y+\alpha_n)$$

Thus the roots of the equation $f(-y) = 0$ are given by

$$(y+\alpha_1)(y+\alpha_2)...(y+\alpha_n) = 0 \quad \text{or} \quad y = -\alpha_1, -\alpha_2, ... -\alpha_n$$

Hence, the roots of the transformed equation (2) are the roots of the given equation with different sign.

(ii) To transform an equation into another equation whose roots are equal to the roots of the given equation multiplied by a given constant number m.

Let the given equation be

$$f(x) = a_0x^n + a_1x^{n-1} + a_2x^{n-2} + ...a_{n-1}x + a_n = 0 \quad ...(1)$$

and let $\alpha_1, \alpha_2, ..., \alpha_n$ be its roots, then (1) can be written as

$$a_0x^n + a_1x^{n-1} + a_2x^{n-2} + ... + a_{n-1}x + a_n = a_0(x-\alpha_1)(x-\alpha_2)...(x-\alpha_n) \quad ...(2)$$

Putting $y = mx$ or $x = \dfrac{y}{m}$ in (1), we get

$$f\left(\frac{y}{m}\right) = a_0\left(\frac{y}{m}\right)^n + a_1\left(\frac{y}{m}\right)^{n-1} + a_2\left(\frac{y}{m}\right)^{n-2} + ... + a_{n-1}\left(\frac{y}{m}\right) + a_n = 0$$

or $$f\left(\frac{y}{m}\right) = \frac{1}{m^n}[a_0y^n + ma_1y^{n-1} + m^2a_2y^{n-2} + ... + m^{n-1}ya_{n-1} + m^na_n] = 0$$

or $$a_0y^n + ma_1y^{n-1} + m^2a_2y^{n-2} + ... + m^{n-1}ya_{n-1} + m^na_n = 0 \quad ...(3)$$

This is the transformed equation. Now we shall show that the transformed equation has the roots $m\alpha_1, m\alpha_2, ... m\alpha_n$.

For this let us put $x = \dfrac{y}{m}$ in (2), we get

$$a_0\left(\frac{y}{m}\right)^n + a_1\left(\frac{y}{m}\right)^{n-1} + a_2\left(\frac{y}{m}\right)^{n-2} + .. + a_{n-1}\left(\frac{y}{m}\right) + a_n = a_0\left(\frac{y}{m}-\alpha_1\right)\left(\frac{y}{m}-\alpha_2\right)...\left(\frac{y}{m}-\alpha_n\right)$$

or $$\frac{1}{m^n}[a_0y^n + ma_1y^{n-1} + m^2a_2y^{n-2} + ... + m^{n-1}a_{n-1}y + m^na_n] = a_0\frac{1}{m^n}(y-m\alpha_1)(y-m\alpha_2)...(y-m\alpha_n)$$

or $$a_0y^n + ma_1y^{n-1} + m^2a_2y^{n-2} + ... + m^{n-1}a_{n-1}y + m^na_n = a_0(y-m\alpha_1)(y-m\alpha_2)...(y-m\alpha_n)$$

This shows that the transformed equation (3) has the roots $m\alpha_1, m\alpha_2, ... m\alpha_n$.

(iii) To transform an equation into another equation whose roots are the reciprocals of the roots of the given equation.

Let the given equation be

$$f(x) = a_0x^n + a_1x^{n-1} + a_2x^{n-2} + ... + a_{n-1}x + a_n = 0 \quad ...(1)$$

And let $\alpha_1, \alpha_2 ... \alpha_n$ be its roots, then, we have

$$a_0x^n + a_1x^{n-1} + a_2x^{n-2} + ... + a_{n-1}x + a_n = a_0(x-\alpha_1)(x-\alpha_2)...(x-\alpha_n) \quad ...(2)$$

Putting $x = \dfrac{1}{y}$ in (1), we get

$$f\left(\frac{1}{y}\right) = a_0\left(\frac{1}{y}\right)^n + a_1\left(\frac{1}{y}\right)^{n-1} + a_2\left(\frac{1}{y}\right)^{n-2} + ... + a_{n-1}\left(\frac{1}{y}\right) + a_n = 0$$

or $$f\left(\frac{1}{y}\right) = \frac{1}{y^n}[a_0 + a_1y + a_2y^2 + ... + a_{n-1}y^{n-1} + a_ny^n] = 0$$

or $$a_ny^n + a_{n-1}y^{n-1} + ... + a_1y + a_0 = 0$$

This is the transformed equation. Now we shall show that this equation (3) has the roots $\dfrac{1}{\alpha_1}, \dfrac{1}{\alpha_2}, ..., \dfrac{1}{\alpha_n}$.

Let us put $x = \frac{1}{y}$ in (2), we get

$$a_0\left(\frac{1}{y}\right)^n + a_1\left(\frac{1}{y}\right)^{n-1} + a_2\left(\frac{1}{y}\right)^{n-2} + \ldots + a_{n-1}\left(\frac{1}{y}\right) + a_n = a_0\left(\frac{1}{y} - \alpha_1\right)\left(\frac{1}{y} - \alpha_2\right)\ldots\left(\frac{1}{y} - \alpha_n\right)$$

or $$\frac{1}{y^n}[a_0 + a_1 y + a_2 y^2 + \ldots + a_{n-1}y^{n-1} + a_n y^n] = \frac{a_0}{y^n}(1-\alpha_1 y)(1-\alpha_2 y)\ldots(1-\alpha_n y)$$

or $$a_0 + a_1 y + a_2 y^2 + \ldots + a_{n-1}y^{n-1} + a_n y^n = a_0(1-\alpha_1 y)(1-\alpha_2 y)\ldots(1-\alpha_n y)$$

This shows that the equation (3) has the roots $\frac{1}{\alpha_1}, \frac{1}{\alpha_2}, \ldots, \frac{1}{\alpha_n}$.

(iv) We know that an equation which remains unchanged when x is replaced by $\frac{1}{x}$ is called a reciprocal equation. Let the given equation be

$$f(x) \equiv a_0 x^n + a_1 x^{n-1} + a_2 x^{n-2} + \ldots + a_{n-1}x + a_n = 0 \qquad \ldots(1)$$

Replace x by $\frac{1}{x}$, we obtain

$$f\left(\frac{1}{x}\right) \equiv a_0 + a_1 x + a_2 x^2 + \ldots + a_{n-1}x^{n-1} + a_n x^n = 0 \qquad \ldots(2)$$

The equation (2) is an equation whose roots are the reciprocal of the roots of the equation (1). If both equations are same, then by comparing the coefficients of like powers of x we obtain

$$\frac{a_0}{a_n} = \frac{a_1}{a_{n-1}} = \frac{a_2}{a_{n-2}} \ldots \frac{a_{n-1}}{a_1} = \frac{a_n}{a_0}$$

Form first and last fraction, we get

$$\frac{a_0}{a_n} = \frac{a_n}{a_0} \quad \text{or} \quad a_n^2 = a_0^2 \quad \text{or} \quad a_n = \pm a_0$$

Therefore from this result we have $a_n = a_0, a_n = -a_0$ and thus there are two classes of the reciprocal equations.

(a) If $a_n = a_0$, then

$$\frac{a_1}{a_{n-1}} = \frac{a_2}{a_{n-2}} = \ldots 1 \quad \text{or} \quad a_1 = a_{n-1}, a_2 = a_{n-2} \ldots$$

This is, the coefficients of the terms in the equation equidistant from the beginning and the end are equal and the equation is therefore called the first class.

(b) If $a_n = -a_0$, then

$$\frac{a_1}{a_{n-1}} = \frac{a_2}{a_{n-2}} = \ldots = -1 \quad \text{or} \quad a_1 = -a_{n-1}, a_2 = -a_{n-2} \ldots$$

That is, the coefficients of the terms in the equation equidistant from the beginning and the end are equal in magnitude and opposite in sign. Therefore the reciprocal is called second class. In this case if the degree of the equation is ***2m*** (even) then, $a_m = -a_m$ or $a_m = 0$. Thus we can say that if the equation of second class and of even degree, then the middle term of the equation will be absent.

(v) Standard form of the reciprocal equation. Let $f(x) = 0$ be a reciprocal equation and if $f(x) = 0$ is of first class and of an odd degree, then one of the roots of this equation $f(x) = 0$ must be its own reciprocal so it has a root -1 and thus $f(x)$ is divisible by the factor $x + 1$. If $\phi(x)$ is the quotient, then $\phi(x) = 0$ will be a reciprocal equation of first class and of an even degree.

On the other hand if the equation $f(x) = 0$ is of second class and of an odd degree, then it will have the root $+1$, and therefore $f(x)$ is divisible by the factor $(x - 1)$. If $\phi(x)$ is the quotient, then $f(x) = (x-1)\phi(x)$

Thus $\phi(x) = 0$ is a reciprocal equation of first class and of even degree. And if the equation $f(x) = 0$ is of the second class and of an even degree, then it will have two roots -1 and $+1$. Therefore $f(x)$ is divisible by $(x + 1)$ and $(x - 1)$ or divisible by $(x^2 - 1)$. If $\phi(x)$ is the quotient, then $f(x) = (x^2 - 1)\phi(x)$.

From this equation it is obvious that $\phi(x) = 0$ will be a reciprocal equation of first class and of even degree. Hence from above discussion we can say that every reciprocal equation can be reduced to a reciprocal of first class and of even degree which is known as the standard form.

(vi) Every reciprocal equation of the standard form can be reduced to an equation of degree half of the degree of the original equation.

Let the reciprocal equation of the standard form be given by

$$a_0x^{2m} + a_1x^{2m-1} + a_2x^{2m-2} + \ldots + a_mx^m + \ldots + a_2x^2 + a_1x + a_0 = 0 \qquad \ldots(1)$$

Divide this equation by x^m, we get

$$a_0x^m + a_1x^{m-1} + a_2x^{m-2} + \ldots + a_m + \ldots + a_2\frac{1}{x^{m-2}} + \frac{a_1}{x^{m-1}} + \frac{a_0}{x^m} = 0$$

$$a_0\left(x^m + \frac{1}{x^m}\right) + a_1\left(x^{m-1} + \frac{1}{x^{m-1}}\right) + a_2\left(x^{m-2} + \frac{1}{x^{m-2}}\right) + \ldots + a_m = 0 \qquad \ldots(2)$$

Since, we know that

$$x^{k+1} + \frac{1}{x^{k+1}} = \left(x^k + \frac{1}{x^k}\right)\left(x + \frac{1}{x}\right) - \left(x^{k-1} + \frac{1}{x^{k-1}}\right)$$

Putting $x + \frac{1}{x} = y$ for k = 1, 2, 3... successively, such that

for $k = 1, x^2 + \frac{1}{x^2} = \left(x + \frac{1}{x}\right)\left(x + \frac{1}{x}\right) - (1+1) = y^2 - 2$

for $k = 2, x^3 + \frac{1}{x^3} = \left(x^2 + \frac{1}{x^2}\right)\left(x + \frac{1}{x}\right) - \left(x + \frac{1}{x}\right) = (y^2 - 2)y - y = y^3 - 3y$

for $k = 3, x^4 + \frac{1}{x^4} = \left(x^3 + \frac{1}{x^3}\right)\left(x + \frac{1}{x}\right) - \left(x^2 + \frac{1}{x^2}\right) = (y^3 - 3y)y - (y^2 - 2) = y^4 - 4y^2 + 2$

and so on, we obtain $\left(x^m + \frac{1}{x^m}\right)$ is a polynomial of degree'm'. Hence, the equation (2) is obtained an equation of degree m which is half of the degree of the equation (1).

(vii) To transform an equation into another equation whose roots are any powers of the roots of the given equation.

Let the given equation be

$$f(x) \equiv a_0x^n + a_1x^{n-1} + a_2x^{n-2} + \ldots + a_{n-1}x + a_n = 0 \qquad \ldots(1)$$

And let $\alpha_1, \alpha_2, \ldots \alpha_n$ be its roots, then we have

$$f(x) \equiv a_0x^n + a_1x^{n-1} + a_2x^{n-2} + \ldots + a_{n-1}x + a_n = a_0(x - \alpha_1)(x - \alpha_2)\ldots(x - \alpha_n) \qquad \ldots(2)$$

The equation (2) can be modified as follows

$$f(x) \equiv a_0(x - \alpha_1)\left(\frac{x^m - \alpha_1^m}{x^m - \alpha_1^m}\right)(x - \alpha_2)\left(\frac{x^m - \alpha_2^m}{x^m - \alpha_2^m}\right)\ldots(x - \alpha_n)\left(\frac{x^m - \alpha_n^m}{x^m - \alpha_n}\right)$$

or $$a_0(x^m - \alpha_1{}^m)(x^m - \alpha_2{}^m)\ldots(x^m - \alpha_n{}^m) = f(x)\left(\frac{x^m - \alpha_1^m}{x - \alpha_1}\right)\left(\frac{x^m - \alpha_2^m}{x - \alpha_2}\right)\ldots\left(\frac{x^m - \alpha_n^m}{x - \alpha_n}\right)$$

or $$a_0(x^m - \alpha_1{}^m)(x^m - \alpha_2{}^m)\ldots(x^m - \alpha_n{}^m) = f(x)[x^{m+1} + \alpha_1x^{m-2} + \ldots + \alpha_1{}^{m-1}]$$
$$[x^{m-1} + \alpha_2x^{m-2} + \ldots + \alpha_2{}^{m-1}]\ldots[x^{m-1} + \alpha_nx^{m-2} + \ldots + \alpha_n{}^{m-1}] \qquad \ldots(3)$$

Let us assume

$$\phi(x^m) = a_0(x^m - \alpha_1^m)(x^m - \alpha_2^m)\ldots(x^m - \alpha_n^m) \qquad \ldots(4)$$

It is obvious that for $x = \alpha_1, \alpha_2, \ldots \alpha_n$ equation (3) gives the identity so that the equation (4) is the transformed equation whose roots are $\alpha_1{}^m, \alpha_2{}^m, \ldots \alpha_n{}^m$. Hence, if we put $x^m = y$ in (4), we obtain (4) as follows :

$$\phi(y) = a_0(y - \alpha_1^m)(y - \alpha_2^m)\ldots(y - \alpha_n^m)$$

(viii) To transform an equation into another equation whose roots exceed the roots of the given equation by a constant h.

Let the given equation be

$$f(x) \equiv a_0x^n + a_1x^{n-1} + a_2x^{n-2} + ... + a_{n-1}x + a_n = 0 \quad ...(1)$$

and let $\alpha_1, \alpha_2, ... \alpha_n$ be its roots. We have

$$f(x) \equiv a_0x^n + a_1x^{n-1} + a_2x^{n-2} + ... + a_{n-1}x + a_n = a_0(x - \alpha_1)(x - \alpha_2)...(x - \alpha_n) \quad ...(2)$$

Putting $y = x + h$, *i.e.*, $x = y - h$ in (1), we get

$$f(y-h) = a_0(y-h)^n + a_1(y-h)^{n-1} + a_2(y-h)^{n-2} + ... + a_{n-1}(y-h) + a_n = 0$$

The equation can be written in descending powers of y as follows:

$$A_0y^n + A_1y^{n-1} + A_2y^{n-2} + ... + A_{n-1}y + A_n = 0 \quad ...(3)$$

where $A_0, A_1, A_2, ..., A_n$ are coefficients and constants and whose values depend upon $a_0, a_1, a_2, ..., a_n$

Now put $y = x + h$ in (3), we get

$$f(x) = A_0(x+h)^n + A_1(x+h)^{n-1} + ... + A_{n-1}(x+h) + A_n = 0$$

$$f(x) = (x+h)[A_0(x+h)^{n-1} + A_1(x+h)^{n-2} + ... + A_{n-1}] + A_n$$

The equation gives that if $f(x)$ is divided by $x + h$, then A_n is obtained as remainder and the quotient is

$$A_0(x+h)^{n-1} + A_1(x+h)^{n-2} + ... + A_{n-2}(x+h) + A_{n-1}$$

Similarly if this quotient is divided by $(x + h)$, then we obtain A_{n-1} as remainder.

Continuing this process until we get all the constant $A_n, A_{n-1}, ..., A_2, A_1$ and we also obtain $A_0 = a_0$

Hence, the transformed equation is

$$f(y-h) \equiv A_0y^n + A_1y^{n-1} + A_2y^{n-2} + ... + A_{n-1}y + A_n = 0$$

Now we have to show that $\alpha_1 + h, \alpha_2 + h, ..., \alpha_n + h$ are the roots of this transformed equation.

For this put $x = y - h$ in (2), we get

$$f(y-h) \equiv a_0(y - h - \alpha_1)(y - h - \alpha_2)...(y - h - \alpha_n) \equiv a_0(y - (\alpha_1 + h)(y - (\alpha_2 + h))...(y - (\alpha_n + h)).$$

Hence, $\alpha_1 + h$, $\alpha_2 + h, ...$ $\alpha_n + h$ are the roots of the transformed equation.

24.7 REMOVAL OF TERMS OF AN EQUATION

Let the given equation be

$$f(x) \equiv a_0x^n + a_1x^{n-1} + a_2x^{n-2} + ... + a_{n-1}x + a_n = 0 \quad ...(1)$$

If we put $x = y + h$, we get

$$a_0(y+h)^n + a_1(y+h)^{n-1} + a_2(y+h)^{n-2} + ... + a_{n-1}(y+h) + a_n = 0$$

This equation can be written in the descending powers of y as follows:

$$a_0y^n + (na_0h + a_1)y^{n-1} + \left\{\frac{n(n-1)}{2!}a_0h^2 + (n-1)a_1hy^{n-1} + a_2\right\} + ... = 0$$

Now, we want to remove second term, then we shall equal to zero the coefficient of y^{n-1}, we get $na_0h + a_1 = 0$ or $h = -\frac{a_1}{na_0}$

Hence, we decreased all the roots of the given equation by a constant $-\frac{a_1}{na_0}$, the second term of the given equation can be removed.

Similarly, if we want to remove third term, we put $\frac{n(n-1)}{2!}a_0h^2 + (n-1)a_1h + a_2 = 0$

Solve this equation we get two values of h and similarly we can remove any term of the given equation.

(i) To remove the second term of the equation $\mathbf{a_0x^3 + 3a_1x^2 + 3a_2x + a_3 = 0}$ and form the obtained equation with integral coefficients having leading coefficient unity.

Since the equation is

$$f(x) \equiv a_0x^3 + 3a_1x^2 + 3a_2x + a_3 = 0 \quad ...(1)$$

Let $\alpha_1, \alpha_2, \alpha_3$ be its roots.

Put $x = y + h$ in (1), we get

$$a_0(y+h)^3 + 3a_1(y+h)^2 + 3a_2(y+h) + a_3 = 0$$

or

$$a_0(y^3 + h^3 + 3y^2h + 3yh^2) + 3a_1(y^2 + h^2 + 2yh) + 3a_2(y+h) + a_3 = 0$$

or $a_0y^3+(3ha_0+3a_1)y^2+(3h^2a_0+6a_1h+3a_2)y+(a_0h^3+3a_1h^2+3a_2h+a_3)=0$ ...(2)

Now, we want to remove second term, then put $3ha_0+3a_1=0$ or $h=-\frac{a_1}{a_0}$

Substitute the value of h in (2), we get

$$a_0y^3+\left(\frac{3a_1^2}{a_0}-\frac{6a_1^2}{a_0}+3a_2\right)y+\left(-\frac{a_1^3}{a_0^2}+\frac{3a_1^2}{a_0^2}-\frac{3a_1a_2}{a_0}+a_3\right)=0$$

or $$a_0y^3+\frac{3(a_0a_2-a_1^2)}{a_0}y+\frac{(a_0^2a_3-3a_0a_1a_2+2a_1^2)}{a_0^2}=0 \quad\text{or}\quad a_0y^3+\frac{3H}{a_0}y+\frac{G}{a_0^2}=0 \qquad ...(3)$$

where $H=a_0a_2-a_1^2$, $G=a_0^2a_3-3a_0a_1a_2+2a_1^2$

Thus the equation (3) is a transformed equation. Further, make all the coefficients of (3) integers, so that (3) can be written as

$$a_0^3y^3+3Ha_0y+G=0$$

Let us put $z=a_0y$

$$\therefore \quad z^3+3Hz+G=0 \qquad ...(4)$$

This is the transformed equation with integral coefficient and having leading coefficient unity. Now the roots of (4) are obtained by the transformation

$$z=a_0y=a_0(x-h)=a_0\left(x+\frac{a_1}{a_0}\right)=a_0x+a_1 \qquad \left[\because h=-\frac{a_1}{a_0}\right]$$

Since $\alpha_1,\alpha_2,\alpha_3$ are the roots of equation (1), then the roots of (4) are $a_0\alpha_1+a_1, a_0\alpha_2+a_1, a_2\alpha_3+a_1$

Further since we know that

$$\alpha_1+\alpha_2+\alpha_3=-\frac{3a_1}{a_0} \quad\text{or}\quad \frac{a_1}{a_0}=-\frac{\alpha_1+\alpha_2+\alpha_3}{3}$$

then $$a_0\alpha_1+a_1=a_0\left(\alpha_1+\frac{a_1}{a_0}\right)=a_0\left(\alpha_1-\frac{\alpha_1+\alpha_2+\alpha_3}{3}\right)=\frac{a_0}{3}(2\alpha_1-\alpha_2-\alpha_3)$$

Similarly $$a_0\alpha_2+a_1=\frac{a_0}{3}(2\alpha_2-\alpha_1-\alpha_3) \text{ and } a_0\alpha_3+a_1=\frac{a_0}{3}(2\alpha_3-\alpha_1-\alpha_2)$$

Hence, the roots of (4) can also be taken as $\frac{a_0}{3}(2\alpha_1-\alpha_2-\alpha_3), \frac{a_0}{3}(2\alpha_2-\alpha_1-\alpha_3), \frac{a_0}{3}(2\alpha_3-\alpha_1-\alpha_2)$

Now if we put $z=a_0x+a_1$ in (4), we get

$$(a_0x+a_1)^3+3H(a_0x+a_1)+G\equiv a_0^2[a_0x^3+3a_1x^2+3a_2x+a_3]$$

(ii) To remove the second term in the equation $a_0x^4+4a_1x^3+6a_2x^2+4a_3x+a_4=0$ with the binomial coefficients and to form the equation with integral coefficients having leading coefficients unity.

Since the equation is

$$f(x)\equiv a_0x^4+4a_1x^3+6a_2x^2+4a_3x+a_4=0 \qquad ...(1)$$

And let $\alpha_1,\alpha_2,\alpha_3,\alpha_4$ be its roots

Put $x=y-h$ in (1), we obtain

$$f(y-h)\equiv a_0(y-h)^4+4a_1(y-h)^3+6a_2(y-h)^2+4a_3(y-h)+a_4=0$$

or $$f(y-h)\equiv a_0y^4+4(a_0h+a_1)y^3+6(a_0h^2+2a_1h+a_2)y^2+4(a_0h^3+3a_1h^2+3a_2h+a_3)$$
$$+(a_0h^4+4a_1h^3+6a_2h^2+4a_3h+a_4)=0 \qquad ...(2)$$

Now we want to remove second term by putting $4(a_0h+a_1)=0$ or $h=-\frac{a_1}{a_0}$

Substitute the value of h in(2), we obtain

$$a_0y^4+\frac{6H}{a_0}y^2+\frac{4G}{a_0^2}y+\frac{(a_0^2I-3H^2)}{a_0^3}=0 \qquad ...(3)$$

where $H=a_0a_1-a_1^2$, $G=a_0^2a_3-3a_0a_1a_2+2a_1^3$ and $I=a_0a_4-4a_1a_3+3a_2^2$

Equation (3) can also be written as $a_0^4y^4 + 6Ha_0^2y^2 + 4Ga_0y + (a_0^2I - 3H^2) = 0$

Let us put $z = a_0y$ we get

$$\therefore \quad z^4 + 6Hz^2 + 4Gz + (a_0^2I - 3H^2) = 0 \qquad ...(4)$$

This is transformed equation whose leading coefficients being unity and all other coefficients are integers. Since we have

$$z = a_0y = a_0(x-h) = a_0\left(x + \frac{a_1}{a_0}\right) \qquad \left[\because h = -\frac{a_1}{a_0}\right]$$

$$\therefore \quad z = a_0x + a_1$$

Thus the roots of the equation (4) are obtained by the transformation $z = a_0x + a_1$

Since $\alpha_1, \alpha_2, \alpha_3, \alpha_4$ are the roots of (1), then $a_0\alpha_1 + a_1, a_0\alpha_2 + a_2, a_0\alpha_3 + a_3$ and $a_0\alpha_4 + a_4$ are the roots of (4). Further since we know that

$$\alpha_1 + \alpha_2 + \alpha_3 + \alpha_4 = -\frac{3a_1}{a_0} \quad \text{or} \quad \frac{a_1}{a_0} = -\frac{\alpha_1 + \alpha_2 + \alpha_3 + \alpha_4}{4}$$

$$\therefore \quad a_0\alpha_1 + a_1 = a_0\left(\alpha_1 + \frac{a_1}{a_0}\right) = a_0\left(\alpha_1 - \frac{\alpha_1 + \alpha_2 + \alpha_3 + \alpha_4}{4}\right) = \frac{a_0}{4}(3\alpha_1 - \alpha_2 - \alpha_3 - \alpha_4)$$

Similarly $\quad a_0\alpha_2 + a_1 = \dfrac{a_0}{4}(3\alpha_2 - \alpha_1 - \alpha_3 - \alpha_4)$

$$a_0\alpha_3 + a_1 = \frac{a_0}{4}(3\alpha_3 - \alpha_1 - \alpha_2 - \alpha_4)$$

and $\quad a_0\alpha_4 + a_1 = \dfrac{a_0}{4}(3\alpha_4 - \alpha_1 - \alpha_2 - \alpha_3)$

Hence, the roots of equation (4) can also be taken as $\frac{a_0}{4}(3\alpha_1 - \alpha_2 - \alpha_3 - \alpha_4), \frac{a_0}{4}(3\alpha_2 - \alpha_1 - \alpha_3 - \alpha_4)$, $\frac{a_0}{4}(3\alpha_3 - \alpha_1 - \alpha_2 - \alpha_4)$ and $\frac{a_0}{4}(3\alpha_4 - \alpha_1 - \alpha_2 - \alpha_3)$.

24.8 AN IMPORTANT RELATION

In order to discuss the biquadratic equation, a function of its coefficients plays a key role. This function is taken as

$$J = a_0a_2a_4 + 2a_1a_2a_3 - a_0a_3^2 - a_1^2a_4 - a_2^3$$

which can also be written in the form of a determinant as follows :

$$J = \begin{vmatrix} a_0 & a_1 & a_2 \\ a_1 & a_2 & a_3 \\ a_2 & a_3 & a_4 \end{vmatrix}$$

Further, we have an important relation between H, G, I and J as follows :

$$G^2 + 4H^3 = a_0^2(HI - a_0J)$$

Verification :

L.H.S. $= G^2 + 4H^3 = (a_0^2a_3 - 3a_0a_1a_2 + 2a_1^3)^2 + 4(a_0a_2 - a_1^2)^3$

$= (a_0^4a_3^2 + 9a_0^2a_1^2a_2^2 + 4a_1^6 - 6a_0^3a_1a_2a_3 + 4a_0^2a_1^3a_3 - 12a_0a_1^4a_2) + 4(a_0^3a_2^3 - a_1^6 - 3a_0^2a_2^2a_1^2 + 3a_0a_2a_1^4)$

$= a_0^4a_3^2 - 3a_0^2a_1^2a_2^2 - 6a_0^3a_1a_2a_3 + 4a_0^2a_1^3a_3 + 4a_0^3a_2^3 = a_0^2(a_0^2a_3^2 - 3a_1^2a_2^2 - 6a_0a_1a_2a_3 + 4a_1^3a_3 + 4a_0a_2^3)$

R.H.S. $= a_0^2(HI - a_0J) = a_0^2[(a_0a_2 - a_1^2)(a_0a_4 - 4a_1a_3 + 3a_2^2) - a_0(a_0a_2a_4 + 2a_1a_2a_3 - a_0a_3^2 - a_1^2a_4 - a_2^3)]$

$= a_0^2[a_0^2a_2a_4 - 4a_0a_1a_2a_3 + 3a_0a_2^3 - a_0a_1^2a_4 + 4a_1^3a_3 - 3a_1^2a_2^2 - a_0^2a_2a_4 - 2a_0a_1a_2a_3 + a_0^2a_3^2 + a_0a_1^2a_4 + a_0a_2^3]$

$= a_0^2(a_0^2a_3^2 - 3a_1^2a_2^2 - 6a_0a_1a_2a_3 + 4a_1^3a_3 + 4a_0a_2^3)$

Hence, L.H.S. = R.H.S.

24.9 GENERAL METHOD OF TRANSFORMATION

Let the given equation be

$$f(x) = 0 \qquad ...(1)$$

and suppose y is a root of transformed equation such that x and y are related by some relation

$$\phi(x, y) = 0 \qquad ...(2)$$

Eliminating x between (1) and (2), we get the transformed equation.

(i) **To form the equation whose roots are $(\alpha-\beta)^2, (\beta-\gamma)^2, (\gamma-\alpha)^2$, where α, β, γ, are the roots of the given equation $a_0x^3 + 3a_1x^2 + 3a_2x + a_3 = 0$ and to discuss the nature of the roots of the given equation.**

Since the given equation is

$$a_0x^3 + 3a_1x^2 + 3a_2x + a_3 = 0 \qquad ...(1)$$

First remove the second term of (1) by diminishing it's root by h, we obtain

$$y^3 + \frac{3H}{a_0^2}y + \frac{G}{a_0^3} = 0 \quad \text{or} \quad y^3 + 2y + r = 0 \text{ where } q = \frac{3H}{a_0^2}, r = \frac{G}{a_0^3} \qquad ...(2)$$

and also $\quad h = -\frac{a_1}{a_0}, H = a_0a_2 - a_1^2, G = a_0^2a_3 - 3a_0a_1a_2 + 2a_1^3$

The roots of the transformed equation (2) are $\alpha - h, \beta - h, \gamma - h$ respectively.

For simplicity let us take $\alpha_1 = \alpha - h, \beta_1 = \beta - h$ and $\gamma_1 = \gamma - h$.

Now $\quad (\gamma-\beta)^2 = (\alpha - h - \beta + h)^2 = [(\alpha - h) - (\beta - h)]^2 = (\alpha_1 - \beta_1)^2$.

Similarly $\quad (\beta-\gamma)^2 = (\beta_1 - \gamma_1)^2$ and $(\gamma-\alpha)^2 = (\gamma_1 - \alpha_1)^2$.

Hence, the equation of the squared differences of (1) is same as that of the equation (2). Therefore if $\alpha_1, \beta_1, \gamma_1$ are the roots of (2), then we have to find the equation whose roots are $(\alpha_1-\beta_1)^2, (\beta_1-\gamma_1)^2, (\gamma_1-\alpha_1)^2$. Let z be one of the roots of required equation.

$$\therefore \quad z = (\alpha_1-\beta_1)^2 = \alpha_1^2 + \beta_1^2 - 2\alpha_1\beta_1 = (\alpha_1+\beta_1)^2 - 4\alpha_1\beta_1 = (-\gamma_1)^2 - \frac{4\alpha_1\beta_1\gamma_1}{\gamma_1} \quad [\because \alpha_1 + \beta_1 + \gamma_1 = 0]$$

$$z = \gamma_1^2 + \frac{4r}{\gamma_1} \quad \text{or} \quad \gamma_1^3 + z\gamma_1 - 4r = 0 \quad [\because \alpha_1\beta_1\gamma_1 = -r]$$

Since γ_1 is the root of equation (2) so put $\gamma_1 = y$.

$$\therefore \quad y^3 + zy - 4r = 0 \qquad ...(3)$$

Eliminating y between (2) and (3), we get

$$z^3 + 6qz^2 + 9q^2z + 27r^2 + 4q^3 = 0 \qquad ...(4)$$

Putting $q = \frac{3H}{a_0^2}, r = \frac{G}{a_0^3}$, we get $\quad z^3 + \frac{18H}{a_0^2}z^2 + \frac{81H^2}{a_0^4}z + \frac{27}{a_0^6}(G^2 + 4H^3) = 0 \qquad ...(5)$

Hence, $(\alpha-\beta)^2, (\beta-\gamma)^2, (\gamma-\alpha)^2$ are the roots of (5)

$$\therefore \quad (\alpha-\beta)^2(\beta-\gamma)^2(\gamma-\alpha)^2 = -\frac{27}{a_0^6}(G^2 + 4H^3) \qquad ...(6)$$

24.9.1 NATURE OF ROOTS OF THE GIVEN EQUATION

Since the degree of the given equation is odd so it has at least one of the roots α, β, γ say α real, and we know that complex roots lie in pair. If β, γ are also real, then $(\alpha-\beta)^2, (\beta-\gamma)^2, (\gamma-\alpha)^2$ must be positive. If β, γ are supposed to be imaginary and if $\beta = a + ib$, then $\gamma = a - ib$ where a, b are real.

$$\therefore \quad (\alpha-\beta)^2(\beta-\alpha)^2(\gamma-\alpha)^2 = (\alpha - a - ib)^2(2ib)^2(a - ib - \alpha) = -4b^2[(\alpha-a)^2 + b^2] < 0.$$

Thus $(\alpha-\beta)^2(\beta-\gamma)^2(\gamma-\alpha)^2$ is negative.

Therefore we can say that if $(\alpha-\beta)^2(\beta-\gamma)^2(\gamma-\alpha)^2$ is positive then the roots of the given equation will be real and if $(\alpha-\beta)^2(\beta-\gamma)^2(\gamma-\alpha)^2$ is negative, then two roots of the given equation will be imaginary. But we have

$$(\alpha-\beta)^2(\beta-\gamma)^2(\gamma-\alpha)^2 = -\frac{27}{a_0^6}(G^2 + 4H^3)$$

So that we can discuss the nature of the roots of the given equation as follows :

1. If $G^2 + 4H^3 > 0$ then two roots of the given equation will be imaginary.

2. If $G^2 + 4H^3 < 0$ then all the roots of the given equation will be real.

3. If $G^2 + 4H^3 = 0$ then two roots of the given equation will be equal.

4. If $G = 0, H = 0$ then all the three roots of the given equation will be equal.

On the other hand we can discuss this case as follows :

Since we know that $G^2 + 4H^3 = a_0^2(HI - a_0J)$

$\therefore$ If $G = 0, H = 0$, then $a_0^2(HI - a_0J) = 0$ or $a_0^2\Delta = 0$ or $\Delta = 0$

where Δ is the discriminant of the given equation. Hence, we can say that if the discriminant of the cubic is zero, then all the root of the equation will be equal.

Solved Examples

Example 1. *Change the signs of the roots of the equation*

$$x^7 + 5x^5 - x^3 + x^2 + 7x + 3 = 0.$$

Solution. First making the equation complete by adding missing terms with zero coefficients, we get

$$f(x) \equiv x^7 + 0.x^6 + 5x^5 + 0.x^4 - x^3 + x^2 + 7x + 3 = 0 \quad ...(1)$$

Put $x = -y$ in (1), we get

$$(-y)^7 + 0.(-y)^6 + 5(-y)^5 + 0.(-y)^4 - (-y)^3 + (-y)^2 + 7(-y) + 3 = 0$$

or

$$-y^7 + 0.y^6 - 5y^5 + 0.y^4 + y^3 + y^2 - 7y + 3 = 0$$

$$\Rightarrow \quad y^7 + 5y^5 - y^3 - y^2 + 7y - 3 = 0$$

This is the required equation whose roots are same to the roots of the given equation with contrary signs.

Example 2. *Transform the equation*

$$72x^3 - 54x^2 + 45x - 7 = 0$$

into another equation with integral coefficients and having the leading coefficient unity.

Solution. The given equation can be written as

$$x^3 - \frac{54}{72}x^2 + \frac{45}{72}x - \frac{7}{72} = 0$$

or $$x^3 - \frac{3}{4}x^2 + \frac{5}{8}x - \frac{7}{72} = 0 \quad ...(1)$$

Put $y = xm$ or $x = \frac{y}{m}$ in (1), we get

$$\left(\frac{y}{m}\right)^3 - \frac{3}{4}\left(\frac{y}{m}\right)^2 + \frac{5}{8}\left(\frac{y}{m}\right) - \frac{7}{72} = 0$$

or $$y^3 - \frac{3}{4}my^2 + \frac{5}{8}m^2y - \frac{7}{72}m^3 = 0 \quad ...(2)$$

Now to remove fractional coefficients let us put $m = 12$ in (2), we get

$$y^3 - \frac{3}{4}(12)y^2 + \frac{5}{8}(12)^2y - \frac{7}{72}(12)^3 = 0$$

or $$y^3 - 9y^2 + 90y - 168 = 0$$

This is the required equation.

Example 3. *Form the equation whose roots are the reciprocals of the roots of the equation*

$$x^4 - 3x^3 + 7x^2 + 5x - 2 = 0$$

Solution. The given equation is

$$x^4 - 3x^3 + 7x^2 + 5x - 2 = 0 \quad ...(1)$$

Putting $x = \frac{1}{y}$ in (1), we get

$$\left(\frac{1}{y}\right)^4 - 3\left(\frac{1}{y}\right)^3 + 7\left(\frac{1}{y}\right)^2 + 5\left(\frac{1}{y}\right) - 2 = 0$$

or $$1 - 3y + 7y^2 + 5y^3 - 2y^4 = 0$$

or $$2y^4 - 5y^3 - 7y^2 + 3y - 1 = 0$$

This is the required equation whose roots are the reciprocals of the roots of (1).

Example 4. *Remove the fractional coefficients from the equation* $2x^3 - \frac{3}{2}x^2 - \frac{1}{8}x + \frac{3}{16} = 0$.

Solution. The given equation is

$$2x^3 - \frac{3}{2}x^2 - \frac{1}{8}x + \frac{3}{16} = 0 \quad ...(1)$$

Putting $x = \frac{y}{m}$ in (1) we get

$$2\left(\frac{y}{m}\right)^3 - \frac{3}{2}\left(\frac{y}{m}\right)^2 - \frac{1}{8}\left(\frac{y}{m}\right) + \frac{3}{16} = 0$$

or $$2y^3 - \frac{3}{2}my^2 - \frac{1}{8}m^2y + \frac{3}{16}m^3 = 0$$

Let us put $m = 4$, we get

$$2y^3 - \frac{3}{2}(4)y^2 - \frac{1}{8}(4)^2y + \frac{3}{16}(4)^3 = 0$$

or $$2y^3 - 6y^2 - 2y + 12 = 0$$

or $$y^3 - 3y^2 - y + 6 = 0$$

This is the required equation.

Example 5. *Solve the following reciprocal equation*

$$x^4 - 10x^3 + 26x^2 - 10x + 1 = 0.$$

Solution. The given equation can be written as

$$x^4 + 1 - 10(x^3 + x) + 26x^2 = 0$$

Divide by x^2, we get

$$\left(x^2 + \frac{1}{x^2}\right) - 10\left(x + \frac{1}{x}\right) + 26 = 0 \quad \ldots(1)$$

Let us put $x + \frac{1}{x} = y$ and $x^2 + \frac{1}{x^2} = y^2 - 2$ in (1), we get

$$y^2 - 2 - 10y + 26 = 0$$

or $$y^2 - 10y + 24 = 0$$

or $$(y-6)(y-4) = 0$$

or $$y = 4, 6$$

Since $x + \frac{1}{x} = y$ if $y = 4$, then $x + \frac{1}{x} = 4$

or $x^2 - 4x + 1 = 0$

or $x = \frac{4 \pm \sqrt{16-4}}{2} = \frac{4 \pm 2\sqrt{3}}{2} = 2 \pm \sqrt{3}$

If $y = 6$, then $x + \frac{1}{x} = 6$

or $x^2 - 6x + 1 = 0$

or $x = \frac{6 \pm \sqrt{36-4}}{2} = \frac{6 \pm 4\sqrt{2}}{2} = 3 \pm 2\sqrt{2}$

Hence, the roots of the given equation are $2 \pm \sqrt{3}, 3 \pm 2\sqrt{2}$.

Example 6. *Find the equation whose roots are the cubes of the roots of the equation*

$$x^4 - x^3 + 2x^2 + 3x + 1 = 0$$

Solution. Since the equation is

$$x^4 - x^3 + 2x^2 + 3x + 1 = 0 \quad \ldots(1)$$

This equation can be written as

$$(1 - x^3) + x(x^3 + 3) + 2x^2 = 0$$

Let $P = (1 - x^3), Q = x(x^3 + 3), R = 2x^2$

Then we have

$$P + Q + R = 0 \text{ or } P + Q = -R$$

Cubing of both sides, we get

$$(P + Q)^3 = -R^3$$

or $P^3 + Q^3 + 3PQ(P + Q) = -R^3$

or $P^3 + Q^3 + 3PQ(-R) = -R^3$ $[\because P + Q = -R]$

or $P^3 + Q^3 + R^3 - 3PQR = 0$

Now substitute the value of P, Q and R in (3), we get

$$(1 - x^3)^3 + x^3(x^3 + 3)^3 + (2x^2)^3 - 3(1 - x^3)x(x^3 + 3)(2x^2) = 0$$

$$(1 - x^3)^3 + x^3(x^3 + 3)^3 + 8(x^3)^2 - 6(1 - x^3)(x^3 + 3)x^3 = 0$$

Let us put $x^3 = y$, we get

$$(1 - y)^3 + y(y + 3)^3 + 8y^2 - 6(1 - y)(y + 3)y = 0$$

$$1 - y^3 - 3y + 3y^2 + y(y^3 + 27 + 9y^2 + 27y) + 8y^2 - 6y(y + 3 - y^2 - 3y) = 0$$

$$1 - y^3 - 3y + 3y^2 + y^4 + 27y + 9y^3 + 27y^2 + 8y^2 - 6y^2 - 18y + 6y^3 + 18y^2 = 0$$

or $$y^4 + 14y^3 + 50y^2 + 6y + 1 = 0$$

This is the required equation whose roots are the cube of the roots of the given equation.

Example 7. *Find the equation whose roots are the roots of the equation* $x^5 - 4x^4 + 3x^2 - 4x + 6 = 0$ *diminished by* 3.

Solution. First complete the given equation as follows :

$$f(x) \equiv x^5 - 4x^4 + 0x^3 + 3x^2 - 4x + 6 = 0 \quad \ldots(1)$$

Suppose the required equation is

$$A_0 y^5 + A_1 y^4 + A_2 y^3 + A_3 y^2 + A_4 y + A_5 = 0 \quad \ldots(2)$$

where $A_0, A_1, A_2, \ldots A_5$ are the constants which can be determined as follows :

Use synthetic division method as follows :

3	1	− 4	0	3	− 4	6
		3	− 3	− 9	− 18	− 66
3	1	− 1	− 3	− 6	− 22	− 60=A_5
		3	6	9	9	
3	1	2	3	3	−13=A_4	
		3	15	54		
3	1	5	18	57=A_3		
		3	24			
3	1	8	42=A_2			
		3				
	1	11=A_1				
	1=A_0					

$$\therefore A_0 = 1, A_1 = 11, A_2 = 42, A_3 = 57, A_4 = -13, A_5 = -60$$

Thus the required equation is

$$y^5 + 11y^4 + 42y^3 + 57y^2 - 13y - 60 = 0.$$

Example 8. *If* α, β, γ *are the roots of the cubic* $x^3 - px^2 + qx - r = 0$, *form the equation whose roots are.*

$$\beta\gamma + \frac{1}{\alpha}, \gamma\alpha + \frac{1}{\beta}, \alpha\beta + \frac{1}{\gamma}.$$ (SVTU–2008)

Solution. Since the given equation is

$$x^3 - px^2 + qx - r = 0 \quad \ldots(1)$$

and α, β, γ are its roots, then

$$\alpha+\beta+\gamma = p,\ \alpha\beta+\beta\gamma+\alpha\gamma = q, \alpha\beta\gamma = r$$

Let y be a root of the required equation. Then

$$y = \beta\gamma + \frac{1}{\alpha} = \frac{\alpha\beta\gamma+1}{\alpha}$$

$$y = \frac{r+1}{\alpha} \quad \Rightarrow \quad y = \frac{r+1}{x} \qquad [\because x = \alpha]$$

$$\therefore \quad x = \frac{r+1}{y}$$

Substitute this value of x in (1), we get

$$\left(\frac{r+1}{y}\right)^3 - p\left(\frac{r+1}{y}\right)^2 + q\left(\frac{r+1}{y}\right) - r = 0$$

or $$\frac{(r+1)^3}{y^3} - \frac{p(r+1)^2}{y^2} + \frac{q(r+1)}{y} - r = 0$$

or $(r+1)^3 - p(r+1)^2 y + q(r+1)y^2 - ry^3 = 0$

or $ry^3 - q(r+1)y^2 + p(r+1)^2 y - (r+1)^3 = 0$

This is the required equation.

Example 9. *Remove the second term of the equation*

$$x^4 + 4x^3 + 2x^2 - 4x - 2 = 0.$$

Solution. Suppose the roots of the given equation are diminished by h so put $y = x - h$ or $x = y + h$ in the given equation, we get

$$(y+h)^4 + 4(y+h)^3 + 2(y+h)^2 - 4(y+h) - 2 = 0$$

$$(y^4 + 4hy^3 + 6h^2y^2 + 4h^3y + h^4) + 4(y^3 + 3hy^2 + 3h^2y + h^3) + 2(y^2 + 2yh + h^2) - 4y - 4h - 2 = 0$$

or $$y^4 + (4h+4)y^3 + (6h^2 + 12h + 2)y^2 + (4h^3 + 12h^2 + 4h - 4)y + (h^4 + 4h^3 + 2h^2 - 4h - 2) = 0 \qquad \ldots(1)$$

In order to remove the second term let us put

$$4h + 4 = 0 \text{ or } h = -1$$

Substitute this value of h in (1), we get

$$y^4 - 4y^2 + 1 = 0.$$

Example 10. *If α, β, γ are the roots of the equation $x^3 + qx + r = 0$. Find the equation whose roots are $(\alpha-\beta)^2, (\beta-\gamma)^2, (\gamma-\alpha)^2$.*

Solution. Since the given equation is

$$x^3 + qx + r = 0 \qquad \ldots(1)$$

And α, β, γ are its roots, then we have

$$\alpha+\beta+\gamma = 0$$

$$\alpha\beta+\beta\gamma+\alpha\gamma = q$$

$$\alpha\beta\gamma = -r$$

Let y be a root of the required equation, then

$$y = (\alpha-\beta)^2 = \alpha^2 + \beta^2 - 2\alpha\beta$$

$$= \alpha^2 + \beta^2 + \gamma^2 - \gamma^2 - 2\alpha\beta$$

$$= (\alpha+\beta+\gamma)^2 - 2(\alpha\beta+\beta\gamma+\alpha\gamma) - \gamma^2 - \frac{2\alpha\beta\gamma}{\gamma}$$

$$y = 0 - 2q - \gamma^2 + \frac{2r}{\gamma}$$

or $y + 2q = -\gamma^2 + \dfrac{2r}{\gamma}$

or $\gamma^3 + (y+2q)\gamma - 2r = 0$

Since γ is a root of (1) so we have $x = \gamma$, then

$$x^3 + (y+2q)x - 2r = 0 \qquad \ldots(2)$$

Substract (2) from (1), we get

$$-(y+q)x + 3r = 0$$

or $$x = \frac{3r}{y+q}$$

Now substitute this value of x in (1), we get

$$\left(\frac{3r}{y+q}\right)^3 + q\left(\frac{3r}{y+q}\right) + r = 0$$

or $(3r)^3 + 3qr(y+q)^2 + r(y+q)^3 = 0$

or $(y+q)^3 + 3q(y+q)^2 + 27r^2 = 0$

or $y^3 + q^3 + 3y^2q + 3yq^2 + 3qy^2 + 3q^3 + 6q^2y + 27r^2 = 0$

or $y^3 + 6qy^2 + 9q^2y + (4q^3 + 27r^2) = 0$

This is the required solution.

EXERCISE 24.2

1. Change the sign of the roots of the equation $x^5 - 4x^3 + 3x^2 + 8x - 9 = 0$.
2. Transform the equation $x^3 - 4x^2 + \frac{1}{4}x - \frac{1}{9} = 0$ into another equation with integral coefficients and having leading coefficient unity.
3. Transform the equation $3x^4 - 5x^3 + x^2 - x + 1 = 0$ into another equation with integral coefficients having leading coefficient unity.
4. Find the equation whose roots are twice the reciprocals of the roots of $x^4 + 3x^3 - 6x^2 + 2x - 4 = 0$
5. Remove the fractional coefficients from the equation $x^3 - \frac{5}{2}x^2 - \frac{7}{18}x + \frac{1}{108} = 0$
6. Remove the fractional coefficients from the equation $x^4 - \frac{5}{6}x^3 - \frac{13}{12}x^2 + \frac{1}{300} = 0$
7. Solve the following reciprocal equations:
   (i) $6x^6 - 25x^5 + 31x^4 - 31x^2 + 25x - 6 = 0$ (Madras–2003)
   (ii) $x^5 - 5x^4 + 9x^3 - 9x^2 + 5x - 1 = 0$
   (iii) $6x^5 - 41x^4 + 97x^3 - 97x^2 + 41x - 6 = 0$
   (Coimbatore–2003, 09)

8. Reduce the equation $4x^4 - 85x^3 + 357x^2 - 340x + 64 = 0$ into a reciprocal equation.

9. Find the equation whose roots are the squares of the roots of the equation $x^4 + x^3 + 2x^2 + x + 1 = 0$

10. Find the equation whose roots are the cubes of the roots of the following equations :

(i) $x^3 + ax^2 + bx + ab = 0$ (ii) $x^3 + 3x^2 + 2 = 0$

11. Remove the second term form the following equations :

(i) $x^3 - 6x^2 + 10x - 3 = 0$ (ii) $x^4 + 8x^3 + x - 5 = 0$

(iii) $x^5 + 5x^4 + 3x^3 + x^2 + x - 1 = 0$

(iv) $x^4 + 20x^3 + 143x^2 + 430x + 462 = 0$

(v) $x^6 - 12x^5 + 3x^2 - 17x + 300 = 0$

12. Find the equation each of whose roots is greater than unity then a root of the equation $x^3 - 5x^2 + 2x - 3 = 0$.

13. Transform the equation $x^3 - \frac{x}{4} - \frac{3}{4} = 0$ into an equation whose roots increased by $\frac{3}{2}$ the corresponding roots of the given equation.

14. Find the equation whose roots are the roots of $3x^3 - 2x^2 + x - 9 = 0$ each diminished by 5.

15. If α, β, γ are the roots of the equation $x^3 + xq + r = 0$ form the equation whose roots are :

(i) $\alpha(\beta+\gamma), \beta(\gamma+\alpha), \gamma(\alpha+\beta)$ (ii) $\left(\frac{\beta}{\gamma}+\frac{\gamma}{\beta}\right), \left(\frac{\gamma}{\alpha}+\frac{\alpha}{\gamma}\right), \left(\frac{\alpha}{\beta}+\frac{\beta}{\alpha}\right)$

(iii) $\left(\alpha-\frac{1}{2}\right), \left(\beta-\frac{1}{2}\right), \left(\gamma-\frac{1}{2}\right)$

16. If α, β, γ are the roots of the equation $x^3 + px^2 + qx + r = 0$ form the equation whose roots are $\alpha - \frac{1}{\beta\gamma}, \beta - \frac{1}{\gamma\alpha}, \gamma - \frac{1}{\alpha\beta}$.

17. Show that the same transformation removes both second and fourth terms of the equation $x^4 + 16x^3 + 83x^2 + 152x + 84 = 0$.

18. Find the condition that the second and third terms of the equation $a_0x^3 + 3a_1x^2 + 3a_2x + a_3 = 0$ are removed by the same transformation.

19. If α, β, γ are the roots of the equation $x^3 - 6x^2 + 11x - 6 = 0$ form the equation whose roots are $\beta^2 + \gamma^2, \gamma^2 + \alpha^2, \alpha^2 + \beta^2$.

20. If α, β, γ are the roots of the equation $2x^3 + x^2 + x + 1 = 0$ form the equation whose roots are

$$\frac{1}{\beta^2}+\frac{1}{\gamma^2}-\frac{1}{\alpha^2};\frac{1}{\gamma^2}+\frac{1}{\alpha^2}-\frac{1}{\beta^2};\frac{1}{\alpha^2}+\frac{1}{\beta^2}-\frac{1}{\gamma^2}.$$

21. If α, β, γ are the roots of the equation $x^3 + px^2 + qx + r = 0$ form the equation whose roots are :

(i) $\alpha + \frac{1}{\beta\gamma}, \beta + \frac{1}{\gamma\alpha}, \gamma + \frac{1}{\alpha\beta}$ (ii) $\frac{\alpha}{\beta+\gamma}, \frac{\beta}{\gamma+\alpha}, \frac{\gamma}{\alpha+\beta}$

22. If α, β, γ are the roots of the equation $x^3 + px^2 + qx + r = 0$ form the equation whose roots are $\alpha^2 + 2\beta\gamma, \beta^2 + 2\alpha\gamma, \gamma^2 + 2\alpha\beta$.

23. If α, β, γ are the roots of the equation $x^3 - px^2 + qx - r = 0$ form the equation whose roots are $\beta+\gamma-\alpha, \gamma+\alpha-\beta, \alpha+\beta-\gamma$ Also find the value of $(\beta+\gamma-\alpha)(\gamma+\alpha-\beta)(\alpha+\beta-\gamma)$.

24. If the roots of $x^3 + 3px^2 + 2qx + r = 0$ are in harmonic progression, show that $2q^3 = r(3pq - r)$.

25. Transform the equation $x^3 - 6x^2 + 5x + 8 = 0$ into another in which the second term is missing, find the equation of squared difference. (Cochin–2005)

26. Form the equation, whose roots are the reciprocal of the roots of $2x^5 + 4x^3 - 13x^2 + 7x - 6 = 0$. (SVTU–2009)

27. Solve the equation

(i) $4x^4 - 20x^3 + 33x^2 - 20x + 4 = 0$ (Madras–2003, 13)

(ii) $6x^5 + x^4 - 43x^3 - 43x^2 + x + 6 = 0$ (SVTU–2006)

Hint to Selected Problems

1. Put $x = -y$ in the given equation.

2. Put $y = mx$, *i.e.* $x = \frac{y}{m}$ in the given equation

7. (i) The given equation can be written as

$$6(x^6 - 1) - 25x(x^4 - 1) + 31x^2(x^2 - 1) = 0$$

9. The given equation can be written as

$$(x^4 + 2x^2 + 1) = -x(x^2 + 1)$$

On squaring both sides, we get

$$x^8 + 3x^6 + 4x^4 + 3x^2 + 1 = 0$$

Now put $x^2 = y$.

17. Suppose the roots of the given equation, are diminished by h, put $y = x - h$ or $x = y + h$ in the given equation.

21. If the roots of the given equation are α, β, γ. Then

$$\alpha + \beta + \gamma = -p$$

$$\alpha\beta + \beta\gamma + \gamma\alpha = q$$

$$\alpha\beta\gamma = -r$$

If y be the root of the required equation, the $y = \alpha + \frac{1}{\beta\gamma}$.

ANSWERS

1. $y^5 - 4y^3 - 3y^2 + 8y + 9 = 0$ **2.** $y^3 - 24y^2 + 9y - 24 = 0$ **3.** $y^4 - 5y^3 + 3y^2 - 9y + 27 = 0$ **4.** $y^4 - y^3 + 6y^2 - 6y - 4 = 0$

5. $y^3 - 15y^2 - 14y^2 + 2 = 0$ **6.** $y^4 - 25y^3 - 975y^2 + 2700 = 0$ **7.**(i) $\pm1, 2, \frac{1}{2}, \frac{5 \pm i\sqrt{11}}{6}$ (ii) $1, \frac{1}{2}(1 \pm i\sqrt{3}), \frac{1}{2}(3 \pm \sqrt{5})$ (iii) $1, \frac{1}{3}, 3, \frac{1}{2}, 2$

8. $16y^4 - 170y^3 + 357y^2 - 170y + 16 = 0$ **9.** $y^4 + 3y^3 + 4y^2 + 3y + 1 = 0$

10. (i) $y^3 + a^3y^2 + b^3y + a^3b^3 = 0$ (ii) $y^3 + 33y^2 + 12y + 8 = 0$ **11.**(i) $y^3 - 2y + 1 = 0$ (ii) $y^4 - 24y^2 + 65y - 55 = 0$

(iii) $y^5 - 7y^3 + 12y^2 - 7y + 2 = 0$ (iv) $y^4 - 7y^2 + 12 = 0$ (v) $y^6 - 60y^4 - 320y^3 - 717y^2 - 773y - 42 = 0$

12. $y^3-8y^2+19y-15=0$ **13.** $y^3-\frac{9}{2}y^2+\frac{13}{2}y-\frac{15}{4}=0$ **14.** $3y^3+43y^2+206y+321=0$

15. (i) $y^3-2qy^2+q^2y+r^2=0$ (ii) $r^3y^3+3r^2y^2+(3r^2+q^3)y+(r^2+2q^3)=0$ (iii) $8y^3+12y^2+(6+8q)y+(8r+4q+1)=0$

16. $r^3y^3+pr(1+r)y^2+q(1+r)^2y+(1+r)^3=0$ **18.** $a_0a_2-a_1^2=0$ **19.** $y^3-28y^2+245y-650=0$ **20.** $z^3+z^2-13z+19=0$

21. (i) $r^2y^3+pr(1-r)y^2+q(1-r)^2y+(1-r)^3=0$ (ii) $(pq-r)y^3+(2pq-p^3-3r)y^2+(pq-3r)y-r=0$

22. $y^3-p^2y^2+q(2p^2-3q)y-(4p^3r-18pqr+2q^3+27r^2)=0$ **23.** $y^3-py^2+(4q-p^2)y+(8r-4pq+p^3)=0;\ 4pq-p^3-8r$

25. $y^3-28y^2+245y-682=0$ **26.** $6x^5-7x^4-13x^3+4x^2-2=0$ **27.** (i) $2,2,\frac{1}{2},\frac{1}{2}$ (ii) $-1,-2,3,-\frac{1}{2},\frac{1}{3}$

24.10 DESCARTE'S RULE OF SIGNS

We know that an equation $f(x) = 0$ cannot have more positive roots than the number of changes of signs form positive to negative or from negative to positive terms of its first number and an equation $f(x) = 0$ cannot have more negative roots than the number of changes of sign $f(-x) = 0$.

We shall simply verify the above statement.

Let the signs of a polynomial be + + − + − + − −.

The given polynomial has five changes of sings. Now we shall multiply the given polynomial by binomial $x-h$ corresponding to the positive root h. The sings of this binomial are + −. We are concerned only with the signs and hence we multiply as below :

```
+ + − + − −        5 changes of sign
+ −
-----------------
+ + − + − + − −
  − − + − − + +
-----------------
+ ± − + − + − ± +
```

The resulting polynomial has two ambiguous sings and we can write in four different ways as follows :

+ + − + − + − + + 6 changes of sign
+ + − + − + − − + 6 changes of sign
+ − − + − + − − + 6 changes of sign
+ − − + − + − + + 6 changes of sign

Thus we see that in all the four possible ways the resulting polynomial has six changes of signs, *i.e.*, one more than the number of changes of signs in the original polynomial.

Hence, we conclude that corresponding to the introduction of a positive root the resulting polynomial has one more change of sign. Now if $\phi(x)$ be the product of factors corresponding to negative and complex roots and α, β, γ... be the positive roots, then if $\phi(x)$ be multiplied by $(x-\alpha),(x-\beta),(x-\gamma)\ldots$ in succession, then each multiplication will introduce one more change of sign. Hence, the number of positive roots cannot exceed the number of changes of signs in $f(x) = 0$.

24.10.1 NEGATIVE ROOTS

We know that negative roots of $f(x) = 0$ are positive roots of $f(-x) = 0$ and as such the number of negative roots of $f(x) = 0$ cannot exceed the number of changes of signs of $f(-x) = 0$.

24.10.2 COMPLEX ROOTS

If $f(x) = 0$ be an equation of nth degree and if it be complete, then the number of changes of signs in $f(x)$, *i.e.*, positive roots and numbers of sign in $f(-x)$, *i.e.*, –ve roots is equal to n the degree of the equation and as such we cannot draw any definite conclusion regarding the existence of imaginary roots. In case the equation be incomplete, then the number of changes of signs in $f(x)$, *i.e.*, positive roots and the number of changes of signs in $f(-x)$, *i.e.*, negative roots is less than the degree n of the equation. If a and b be the number of changes of signs in $f(x)$ and $f(-x)$ respectively, *i.e.*, greatest number of positive roots is a and that of negative roots is b, then $n-(a+b)$ is the least number of imaginary roots. For example, consider the equation

$$f(x)=x^7-3x^5+4x^4+2x^3-11=0.$$

The above equation has three changes of sign and as such if cannot have more than three positive roots. Again $f(-x) = 0$ *i.e.*, $-x^7+3x^5+4x^4-2x^3-11=0$ has only two changes of sign and as such $f(x) = 0$ cannot gave more than two negative roots. Thus the max. number of real roots is 3 + 2, *i.e.*, 5 and the degree of the equation being 7 we conclude that the equation must have at least two imaginary roots.

24.11 CHANGE OF SIGN

Let two real number a and b be substituted for x in the polynomial $f(x)$ and $f(a)$ and $f(b)$ are found to be of opposite signs, then at least one or an odd number of real roots of the equation $f(x) = 0$ lie between a and b. In case $f(a)$ and $f(b)$ of the same sign, then either no real root or an even number of roots of $f(x) = 0$ lie between a and b.

Case I. $f(a)$ and $f(b)$ of opposite signs.

Let $y = f(x)$ be a continuous function of x, it should assume all values between $f(a)$ and $f(b)$

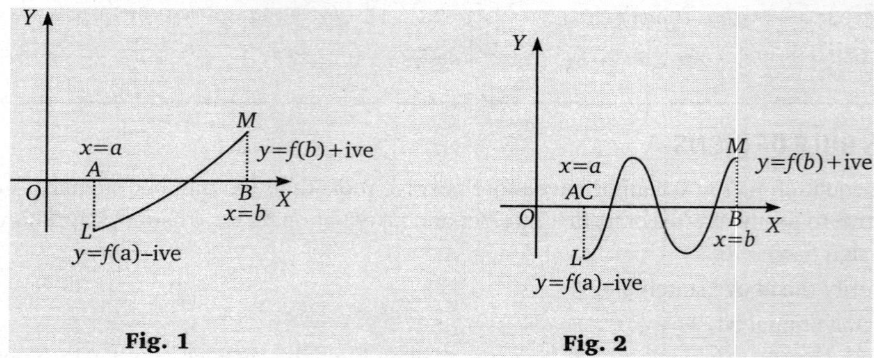

Fig. 1 **Fig. 2**

Now $f(a)$, $f(b)$ are of opposite signs, *i.e.*, values of y corresponding to the values of x, *i.e.*, a and b are of opposite signs. From one side of x-axis to the other side of x-axis the curve $y = f(x)$ must cross the axis of x at least once as in fig.1 at C or an odd number of times as in fig. 2 at C, D and E at all such points where the curve crosses the axis of x, $y = 0$, *i.e.*, $f(x) = 0$ which means that $f(x)$ vanishes either at one or an odd number of times for values of x between a and b. Hence, at least one or an odd number of roots of $f(x) = 0$ lie between a and b.

Case II. If $f(a)$ and $f(b)$ of same sign.

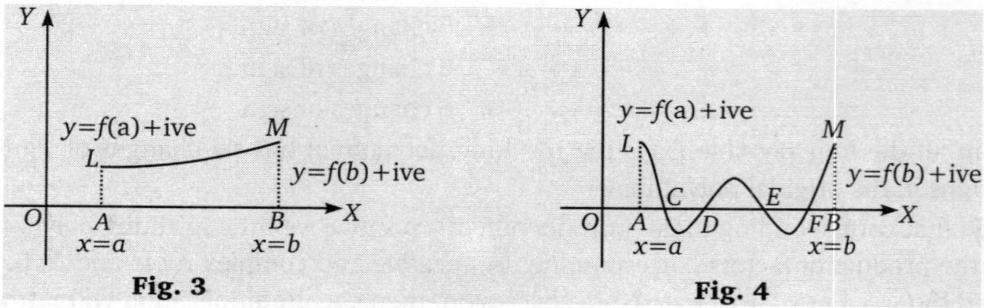

Fig. 3 **Fig. 4**

Here, $f(a)$ and $f(b)$ are of the same sign, the values of y corresponding to the values of x, *i.e.*, a and b are of the same sign which means that in passing from a point on one side of x-axis to the other point on the same side either the curve $y = f(x)$ will not cross the x-axis as in Fig.3 where $y = 0$, *i.e.*, $f(x) = 0$ or, it must cross an even number of times of Fig. 4. Hence we conclude that either $f(x)$ does not vanish for values of x between a and b. or, if it vanishes, it must vanish for even number of times.

THEOREM 1. *Every equation of an odd degree has at least one real root whose sign is opposite to that of its last term, the coefficients of the first term being positive.*

Proof. Let the equation be $x^n + a_1 x^{n-1} + a_2 x^{n-1} + \ldots + a_n = 0$ (n odd)

$$f(-\infty) = \text{negative } (n \text{ odd}),\ f(0) = a_n,\ f(\infty) = \text{positive}$$

Case I. a_n is positive. In this case $f(\infty)$ and $f(0)$ are of opposite signs and hence, at least open real root must lie between $-\infty$ and 0, *i.e.*, it must be negative opposite to the sign of a_n which is positive.

Case II. a_n is negative. Here $f(0) = a_n$ = negative and $f(\infty)$ = positive, *i.e.*, they are of opposite signs and hence at least one real root must lie between 0 and ∞. This root is clearly, positive, *i.e.*, of sign opposite to that of a_n which is negative.

THEOREM 2. *Every equation of an even degree whose last terms is negative and the coefficient of the first term positive, has at least two real roots, one positive and other negative.*

Proof. Let the equation be

$$x^n + a_1 x^{n-1} + a_2 x^{n-2} + ... + a_n = 0\ (n \text{ even})$$

$$f(-\infty) = \text{positive}\ ;\ f(0) = a_n, i.e., \text{negative}\ f(\infty) = \text{positive}$$

Hence, $f(-\infty)$ and $f(0)$ are of opposite signs, therefore at least one real root must lie between $-\infty$ and 0 and it is negative. Again $f(0)$ and $f(\infty)$ are of opposite signs and hence at least one real root must lie between 0 and ∞ and it is positive. Thus the equation must have two real roots, one positive and the other negative.

THEOREM 3. *If an equation has only one change of sign, it must have one positive root and no more.*

Proof. Taking the leading coefficient to be positive, the equation must have a set of positive terms followed by a set of negative terms (*i.e.* last term negative) since there is only one change of sign.

$\therefore$ $f(\infty)$ = positive; and $f(0)$ = last term negative, *i.e.* $f(0)$ and $f(\infty)$ are of opposite signs and as such there must lie at least one or an odd number of roots of the equation $f(x) = 0$ between 0 and ∞. But as there is only one changes of sign in $f(x)$, the number of positive roots cannot be greater than one. Hence the equation must have only one positive root.

THEOREM 4. *If all terms of an equation are positive and the equation, involves no odd power of x, then all its root are complex.*

Proof. Clearly both $f(x)$ and $f(-x)$ will have no change of signs and hence by Descarte's rule there will not be any positive or negative roots. Therefore all the roots must be complex.

REMARK

- If all the terms of an equation are positive and all involve odd powers of x, then 0 is the only real root.

Recapitulations

• An equation $f(x) = 0$ cannot have more positive roots then the number of changes of signs form positive to negative or from negative to positive in terms of it first member.	• An equation $f(x) = 0$ cannot have more negative roots than the number of changes of sign in $f(-x) = 0$
• Every equation of an odd degree has at least one real root whose sign is opposite to that if its, last term the coefficient of the first term being positive.	• Every equation of an even degree whose last term is negative and the coefficient of the first term positive has at least two real roots, one positive and one negative.
• If all the terms of an equation are positive and the equation involves no odd powers of x then all its roots are complex.	• If all the terms of an equation are positive and all involve odd powers of x then 0 is the only root.

SOLVED EXAMPLES

Example 1. *Apply Descarte's rule of signs to discuss the nature of the roots of the equation*

$$x^4 + 15x^2 + 7x - 11 = 0.$$

Solution. Let $f(x) = x^4 + 15x^2 + 7x - 11 = 0$. It has only one change of sign and hence it must have one positive root.

Now, $f(x) = x^4 + 15x^2 + 7x - 11 = 0$.

As above it must have one positive root, *i.e*, $f(x) = 0$ must have one negative root. Thus the equation has two real roots, one positive and one negative and hence, the other two roots must be imaginary.

Example 2. *Show that the equation*

$$f(x) = x^5 + x^3 - 8x - 5 = 0$$

cannot have more than three real roots and prove that it must have three real roots.

Solution. Let $f(x) = 0$ has only one changes of sign and hence, the equation must have one positive root.

Now, $f(-x) = -x^5 - x^3 + 8x - 5 = 0$

has two changes of sign and as such it can have at the most two positive roots or the maximum number of negative roots of $f(x) = 0$ is two.

Again

$$f(0) = \text{negative}, f(-1) = \text{positive}, f(-\infty) = \text{negative}$$

Since $f(0)$ and $f(-\infty)$ are of the same sign, so there lies either none or an even number of roots between 0 and $-\infty$ and now $f(0)$ and $f(-1)$ are of opposite signs and also $f(-1)$ and $f(-\infty)$ too are of opposite signs and hence one negative root lies between 0 and -1 and the other between -1 and $-\infty$. Thus we conclude that the equation must have three real roots and hence two complex.

Example 3. *Show that the equation*

$$f(x) = x^{12} - x^4 - x^3 - x^2 + 1 = 0$$

has at least six complex roots.

Solution. Here $f(x) = 0$ has four changes and hence it can have at the most four positive roots.

Now, $f(-x) = x^{12} - x^4 + x^3 - x^2 + 1 = 0$

has two changes and it can at the most have two positiive roots or $f(x) = 0$ can at the most have two negative roots. Hence the maximum number of real roots of $f(x) = 0$ can be six, but there being twelve roots, the minimum number of complex roots must be six.

Example 4. *Find the minimum number of imaginary roots which equation $f(x) = 2x^7 - x^4 + 4x^3 - 5 = 0$ must possess.*

Solution. Here $f(x) = 0$ has three changes of signs and as such it can at the most have three positive roots.

Now, $f(-x) = -2x^7 - x^4 - 4x^3 - 5 = 0$

or $\quad 2x^7 + x^4 + 4x^3 + 5 = 0$

Clearly, $f(-x) = 0$ has no changes of signs and as such it will have no positive root which means that $f(x) = 0$ has no negative roots. Hence, the given equation can at the most have three real, *i.e.*, positive roots and it being of 7th degree and hence, the minimum number of imaginary roots is 7 – 3, *i.e*, 4.

Example 5. *Find the least possible number of imaginary roots of the equation $x^9 - x^5 + x^4 + x^2 + 1 = 0$*

Solution. Clearly, $f(x) = 0$ has two changes of signs and hence 2 is the maximum number of positive roots.

$$f(-x) = -x^9 - x^5 + x^4 + x^2 + 1 = 0$$

or
$$x^9 + x^5 - x^4 - x^2 - 1 = 0$$

$f(-x) = 0$ has only one change of sign and hence it has only one positive root or $f(x) = 0$ has only one negative root. Thus the maximum number of real roots is 2 + 1 = 3 and the equation being of 9th degree will have at least 9 – 3 = 6 imaginary roots.

Example 6. *Prove that the equation $x^5 - x + 16 = 0$ has two pairs of complex roots.*

Solution. Let $f(x) = x^5 - x + 16 = 0$

$f(x)$ has got two changes of signs and as such it cannot have more than two positive roots. Again $f(0)$ = positive and $f(\infty)$ = positive, since both $f(0)$ and $f(\infty)$ are of the same sign. Also we observe that for all values of x between 0 and ∞, $f(x)$ remain positive always, *i.e.*, the graph of the curve $y = f(x)$ never crosses the x-axis which in other words means that y or $f(x)$ never becomes negative or y or $f(x)$ is always positive for all values of x. Hence, the equation $f(x) = 0$ has no positive roots.

If $f(a)$ and $f(b)$ are of the same sign, then either no root or in general even number of roots of $f(x) = 0$ lie between a and b. In the later case the curve crosses the x-axis even times; but here we have shown $f(x)$ always remains positive and hence, we have established that the given equation has no positive root even though $f(x) = 0$ has two changes of signs.

Again $\quad f(-x) = -x^5 + x + 16 = 0$

or $\quad x^5 - x - 16 = 0$

$f(-x) = 0$ has only one change of sign and as such it must have one positive root or $f(x) = 0$ must have one negative root.

Thus in all the given equation has only one real root which is negative and it being of fifth degree we conclude that the remaining four roots must be imaginary. Again since imaginary roots occur in conjugate pairs we say that the given equation has two pairs of complex roots.

Example 7. *Show that the equation $f(x) = x^3 - qx + r = 0$ where q and r are essentially positive has one negative root and that the other two roots are either imaginary or both positive.*

Solution. Clearly $f(x) = 0$ has got only two changes of signs and hence the number of positive roots cannot exceed two

Now, $f(-x) = -x^3 + qx + r = 0$

or $\quad x^3 - qx - r = 0$

and it has only one change of sign; hence $f(x)=0$ must have one negative root. Otherwise the equation must have one real root whose sign is opposite to that of the last terms and hence it should be negative.

Hence, we conclude that one root of $f(x) = 0$ is essentially negative and therefore the remaining two are either both positive or imaginary.

Example 8. *Correct the mistakes if any in the following :*

(i) *If n be the degree of f(x) and μ and μ′ the number of changes of signs in f(x) and f(–x) respectively then if $\mu + \mu' < n$ the equation f(x) = 0 has exactly $n - (\mu + \mu')$ imaginary roots.*

(ii) *If two numbers a and b be substituted for x in polynomial f(x) gives results with the same sign no real root lies between them.*

(iii) *f(x) cannot have a greater number of negative roots then there are changes of sign in the terms of polynomial of f(x).*

Solution. (i) Replace the word 'exactly' by 'at least'.

(ii) Replace 'no real root' by 'either no real root' or 'an even number of real roots lie between a and b

(iii) Replace 'polynomial $f(x)$' by 'polynomial $f(-x)$'.

Example 9. *Prove that if n be even, the equation $x^{n-1} = 0$ has two and only two real roots, one negative and one positive and the rest complex and when n is odd the real roots is unity and rest are complex.*

Solution. **If *n* even:** Then $f(x) = 0$ and $f(-x) = 0$ both have only one change of sign each and hence the equation must have one positive and one negative root and no more. Thus only two real and rest complex roots.

If *n* odd: $f(x)$ has one change of sign and hence $f(x) = 0$ has one and only one positive root. Again $f(-x) = x^{n-1} = 0$ has no change of sign and hence $f(x) = 0$ has no negative root. Thus an odd degree equation has only one positive root which is clearly seen to be unity and the rest are therefore complex which will be even in number.

Example 10. *Find the locations of the roots of the equation*

$$f(x) = x^3 + x^2 - 2x - 1 = 0$$

Solution. Clearly $f(x)$ has only one change of sign and hence it must have one positive root. Also $f(0) = -1$, $f(1) = -1$, $f(2) = +7$. Since $f(1)$ and $f(2)$ are of opposite signs therefore the positive root lies between 1 and 2.

$$F(x) = f(-x) = -x^3 + x^2 + 2x - 1 = 0$$

or $\quad F(x) = 0 \Rightarrow x^3 - x^2 - 2x + 1 = 0$

$F(x)$ has got two changes of sign; so at the most it can have two positive roots.

$$F(0) = 1, F(1) = -1, F(2) = +1.$$

$\Rightarrow F(0)$ and $F(1)$ are of opposite signs and again $F(1)$ and $F(2)$ are of opposite signs.

$\therefore \quad F(x) = 0$ has two positive roots between (0, 1) and (1, 2). But $F(x) = f(-x)$.

$f(x) = 0$ has two negative root one between (0, −1) the other between (−1, −2).

EXERCISE 24.3

1. Show that the equation $x^7 - 3x^4 + 12x^2 + 5x - 4 = 0$ has at least two imaginary roots.
2. Show that the equation $2x^7 + 3x^4 + 3x + k = 0$ has at least four imaginary roots for all values of k (constant).
3. Prove if $q > r > 0$ the cubic $x^3 + 9x + r = 0$ has one negative and two imaginary roots.
4. Prove that the equation $x^9 - x^5 + x^4 + x + 1$ has one real root which is negative and eight imaginary roots.
5. Find an equation whose roots are the squares of the roots $x^3 - x^2 + 8x - 6 = 0$ and hence deduce that the equation must have a pair of imaginary roots.
6. Find the equation whose roots are the squares of the roots of the equation $x^3 + 4x^2 + 9x + 10 = 0$ and hence, find the nature of the roots of the given equation.

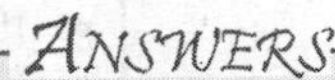

5. $y^3 + 15y^2 + 52y - 36 = 0$ 6. One real, two complex

24.12 MULTIPLE ROOTS

If an equation $f(x) = 0$ has exactly m roots equal to α, then $f(x)$ and its first $(m - 1)$ derivatives all vanish for $x = \alpha$ but the mth and all the following derivatives do not vanish and if $f(x)$ and its first $(m - 1)$ derivatives vanish for $x = \alpha$, then $f(x) = 0$ has m roots equal to α.

Proof. Let $f(x) = (x - \alpha)^m \phi(x)$ ...(1)

where $\phi(x)$ does not vanish for $x = \alpha$, for if it vanishes, then it will contain the factor $(x - \alpha)$ and then $f(x)$ will have more than m equal roots.

Differentiating both sides of (1), we get

$$f'(x) = m(x-\alpha)^{m-1}\phi(x) + (x-\alpha)^m \phi'(x) = (x-\alpha)^{m-1}[m\phi(x) + (x-\alpha)\phi'(x)]$$

$$= (x-\alpha)^{m-1}\psi_1(x) \quad ...(2)$$

where $\psi_1(x)$ does not vanish for $x = \alpha$, because $\psi_1(\alpha) = m\phi(\alpha)$ and it can be zero only if $\phi(\alpha) = 0$ which is contrary to the supposition. From (2), we observe that a root which occurs exactly m times in $f(x) = 0$ occurs exactly $(m - 1)$ times in $f'(x) = 0$. Hence we have the following :

A multiple root of order m of the equation $f(x) = 0$ is a multiple root of order $(m - 1)$ of the first derived equation $f'(x) = 0$ and hence $f(x)$ and $f'(x)$ have the common factor $(x - \alpha)^{m-1}$.

Again if we differentiate (2), we get

$$f''(x) = (x-\alpha)^{m-2}[(m-1)\psi_1(x) + (x-\alpha)\psi_1'(x)] = (x-\alpha)^{m-2}\psi_2(x)$$

where $\psi_2(x)$ does not vanish for $x = \alpha$, *i.e.*, $f''(x) = 0$ has exactly $(m - 2)$ equal roots α. Similarly we can show $f'''(x)$ will have the factor $(x-\alpha)^{m-3}$ and so on till $f^{m-1}(x)$ will have the following :

Any root which occurs m times in f(x) = 0 occurs in degree of multiplicity diminishing by it in the first derived function.

Conversely Let $f(x) = f\{\alpha + (x-\alpha)\}$

Expanding the R.H.S by Taylor's Theorem, we get

$$f(x) = f(\alpha) + (x-\alpha)f'(\alpha) + \frac{(x-\alpha)^2}{2!}f''(a) + \ldots + \frac{(x-\alpha)^{m-1}}{(m-1)!}f^{m-1}(\alpha)$$

$$+\frac{(x-\alpha)^m}{m!}f^m(\alpha) + \ldots + \frac{(x-\alpha)^n}{n!}f^n(\alpha)$$

Now $\quad f(\alpha) = f'(\alpha) = f''(\alpha) = \ldots = f^{m-1}(\alpha) = 0$ (given)

$$\therefore \quad f(x) = \frac{(x-\alpha)^m}{m!}f^m(\alpha) + \ldots + \frac{(x-\alpha)^n}{n!}f^m(\alpha)$$

Above shows that $(x-\alpha)^m$ is a factor of $f(x) = 0$ which therefore has m roots equal to α.

24.12.1 DETERMINATION OF MULTIPLE ROOTS

We have discussed above that multiple root of the order m is multiple root of order $(m - 1)$ of the equation $f'(x) = 0$. In order to find such roots we should find the H.C.F. of $f(x)$ and $f'(x)$. This H.C.F. will give the multiple root of $f(x)$ each repeated $(m - 1)$ times. Thus if $(x - 2)$ is the H.C.F. of $f(x)$ and $f'(x)$, then $f(x)$ contains $(x - 2)^2$, as a factor or 2 is a double root of $f(x) = 0$. If $(x-2)^2(x-1)^5$ is the H.C.F. of $f(x)$ and $f'(x)$, then $f(x) = 0$ has three roots equal to 2 and six roots equal to 1.

REMARK

- In case $f(x)$ and $f'(x)$ have no common factor, then clearly $f(x)$ has no equal roots.

24.12.2 CONDITION FOR TWO OR THREE EQUAL ROOTS

If α be a double root of $f(x) = 0$, then $f(\alpha) = 0$ and $f'(\alpha) = 0$ and the required condition is obtained by eliminating α between $f(\alpha) = 0$ and $f'(\alpha) = 0$. Similarly if α is a triple root the required condition can be obtained by eliminating α between $f(\alpha) = 0$, $f'(\alpha) = 0$ and $f''(\alpha) = 0$.

Recapitulations

- If an equation $f(x) = 0$ has exactly m roots equal to α then $f(x)$ and its first $(m - 1)$ derivatives all vanish for $x = \alpha$ but the mth and all the following derivatives do not vanish.
- If $f(x)$ and its $(m - 1)$ derivatives vanish for $x = \alpha$ then $f(x) = 0$ has m roots equal to α.

Solved Examples

Example 1. *Solve the equation*

$$x^5 - 15x^3 + 10x^2 + 60x - 72 = 0$$

be testing for equal roots.

Solution. Let $f(x) = x^5 - 15x^3 + 10x^2 + 60x - 72 = 0$

$$f'(x) = x^4 - 9x^2 + 4x + 12 = 0$$

Let us find the H.C.F. of $f(x) = 0$ and $f'(x) = 0$

x	$x^5 - 15x^3 + 10x^2 + 60x - 72$	$x^4 - 9x^2 + 4x + 12$	x
	$x^5 - 9x^3 + 4x^2 + 12x$	$x^4 - x^3 - 8x^2 + 12x$	
	$-6x^3 + 6x^2 + 48x - 72$	$x^3 - x^2 - 8x + 12$	1
or	$x^3 - x^2 - 8x + 12$	$x^3 - x^2 - 8x + 12$	

Thus we find that H.C.F. of $f(x) = 0$ and $f'(x) = 0$ is

$$x^3 - x^2 - 8x + 12 = 0$$

or $\quad (x-2)(x^2 + x - 6) = 0$

or $\quad (x-2)^2(x+3) = 0$

Giving $n = 2, 2, -3$. Now we know that H.C.F. gives a multiple root of $f(x) = 0$, $(m - 1)$ times.

Hence, 2 is triple root and -3 is a double root of $f(x) = 0$.

Example 2. *Solve the equation*

$$f(x) = 4x^3 + 20x^2 - 3x + 6 = 0$$

given that two of its roots are equal.

Solution. The H.C.F. of $f(x)$ and $f'(x)$ is found to be $2x - 1 = 0$ giving $x = 1/2$.

$\therefore \frac{1}{2}$ is double root of $f(x) = 0$

Let the third root be γ; then product of all the roots is

$$\frac{1}{2} \cdot \frac{1}{2}\gamma = -\frac{6}{4} \qquad \therefore \gamma = -6$$

$\therefore \quad$ Roots are $\frac{1}{2}, \frac{1}{2}, -6$

Example 3. *Solve the equation*

$$x^4 - 6x^3 + 12x^2 - 10x + 3 = 0$$

which has equal roots.

Solution. The H.C.F. is $(x-1)^2 = 0$ giving 1 as a triple root of $f(x) = 0$

Also $\alpha\beta\gamma\delta = 3$ or $1.1.1.\delta = 3$; $\therefore \delta = 3$

$\therefore$ Roots are 1, 1, 1, 3.

Example 4. *Factorize the following :*

$$2x^5 - x^4 + 5x^2 - 4x + 3.$$

Solution. The H.C.F. of $f(x)$ and $f'(x)$ is found to be $x^2 - x + 1$. Hence $(x^2 - x + 1)^2$ is a factor of $f(x)$. Let the other factor which will be linear is say $x + k$.

$\therefore f(x) = 2(x^2 - x + 1)^2(x + k)$ as the coefficient of x^5 is 2.

Comparing constant term on either side, we get

$$3 = 2k; \quad \therefore k = \frac{3}{2}.$$

$$\therefore f(x) = (x^2 - x + 1)^2(2x + 3)$$

Example 5. *Show that the equation $x^m - qx^{n-m} + r = 0$ has two equal roots if*

$$\left\{\frac{q}{n}(n-m)\right\}^n = \left\{\frac{r}{m}(n-m)\right\}^m.$$

Solution. Let $f(x) = x^m - qx^{n-m} + r = 0$ ...(1)

and $f'(x) = nx^{n-1} - q(n-m)x^{n-m-1} = 0$ ...(2)

The required condition is obtained by eliminating x between (1) and (2).

Form (2), $n = \dfrac{q(n-m)}{x^m}$ or $x^m = \dfrac{q}{n}(n-m)$ ...(3)

and from (1) with the help of (3), we get

$$x^n = \left[1 - \frac{n}{n-m}\right] + r = 0$$

or $x^n = \dfrac{r}{m}(n-m)$ ...(4)

or $(x^m)^n = (x^n)^m$ etc. (3) and (4).

Example 6. *If the equation $x^4 - 4p^3x + 1$ has a pair of equal roots, find the value of p and solve the equation completely.*

Solution. Let $f(x) = x^4 - 4p^3x + 1$

$\Rightarrow f'(x) = 4(x^3 - p^3) = 0$ giving $x = p$.

Eliminating x between $f(x)$ and $f'(x) = 0$, we get

$$p^4 - 4p^4 + 1 = 0$$

or $3p^4 = 1$ or $p = 3^{-1/4}$ ...(1)

Now let us find the H.C.F. of $f(x)$ and $f'(x)$.

$$x^3 - p^3 \overline{)\, x^4 - 4p^3x + 1 \,}\, x$$
$$x^4 - p^3x$$
$$\overline{-3p^3x + 1}$$

or $-3p^4x + p$ or $-x + p$

$\therefore$ $3p^4 = 1$, by (1)

or $-x + p \overline{)\, x^3 - p^3 \,}\, -x^2 - px - p^2$
$$x^3 - p^3$$
$$\overline{\times}$$

Thus the H.C.F. is $(x - p)$ and hence two equal roots are p and q, their sum being $2p$ and product p^2. Let the other roots be α and β.

Now sum of all roots = 0

$\therefore$ $\alpha + \beta + p + p = 0$ or $\alpha + \beta = -2p$

and $\alpha\beta pp$ = product of all the roots = 1

$\therefore$ $\alpha\beta = \dfrac{1}{p^2} = 3p^2$ from (1), $\because 3p^4 = 1$

$\therefore$ α and β are the roots of the quadratic

$$t^2 + 2pt + 3p^2 = 0$$

or $t = \dfrac{-2p \pm \sqrt{(4p^2 - 12p^2)}}{2} = p \pm ip\sqrt{2}$

Hence, the four roots are $p, p, -p \pm ip\sqrt{2}$ where $p = 3^{-(1/4)}$.

Example 7. *If the equation*

$$x^n + p_1x^{n-1} + p_2x^{n-2} + \dots + p_n = 0$$

has two roots and equal to α, prove that α is also a root of the equation.

$$p_1x^{n-1} + 2p_2x^{n-2} + 3p_3x^{n-3} \dots + np_n = 0$$

Solution. Let

$$f(x) = x^n + p_1x^{n-1} + p_2x^{n-2} + \dots + p_{n-2}x^2 + p_{n-1}x + p_n = 0$$

$$\Rightarrow f'(x) = -nx^{n-1} + (n-1)p_1x^{n-2} + (n-2)p_2x^{n-3} + \dots + 2xp_{n-2} + p_{n-1} = 0$$

and $f''(x) = n(n-1)x^{n-2} + (n-1)(n-2)p_1x^{n-3} + (n-2)(n-3)p_2x^{n-4} + \dots + 2p_{n-2} = 0$

If α is a double root of $f(x) = 0$ then it is also a root of $f'(x) = 0$. Hence, α will also be a root of $nf(x) - xf'(x) = 0$ as both $f(x)$ and $f'(x)$ vanish for $x = \alpha$. Putting the values of $f(x)$ and $f'(x)$ and simplifying, we get the required equation as given.

Example 8. *In case the given equation has three root equal to α then α is a root of the equation*

$$n^2x^{n-1} + (n-1)^2p_1x^{n-2} + (n-p)^2p_2x^{n-3} + \dots + p_{n-1} = 0$$

Solution. In case α be a triple root of $f(x) = 0$, then α is a double root of $f'(x) = 0$ and is a single root of $f''(x) = 0$. Hence, α is a root of $f'(x) + xf''(x) = 0$ as both $f'(x)$ and $f''(x)$ vanish for $x = \alpha$.

$$\text{or } \left\{nx^{n-1} + (n-1)p_1x^{n-2} + \ldots + 2xp_{n-2} + p_{n-1}\right\}$$
$$+ x\{n(n-1)x^{n-2}$$
$$+ (n-1)(n-2)p_1x^{n-3} + 2p_{n-2}\} = 0$$
$$\text{or } \{n + n^2 - n\}x^{n-1} + (n-1)(1+n-2)p_1x^{n-2}$$
$$+ (n-2)\{1+n-3)p_2x^{n-3}$$
$$+ (2+2)xp_{n-2} + p_{n-1} = 0$$
$$\text{or } n^2x^{n-1} + (n-1)^2 p_1x^{n-2} + (n-2)^2 p_2x^{n-2}$$
$$+ \ldots + 2^2 xp_{n-2} + p_{n-1} = 0$$

Example 9. *Prove that the equation* $x^5 + 5px^3 + 5p^2x + q = 0$ *has a pair of equal roots when* $q^2 + 4p^5 = 0$ *and that if it has one pair of equal roots, it must have a second pair.*

Solution. Let us find the H.C.F. of $f(x) = 0$ and $f'(x) = 0$ in order to find the repeated roots

	$x^4 + 3px^2 + p^2$	$x^5 + 5px^3 + 5p^2x + q$	
	$2p$	$x^5 + 3px^3 + p^2x$	x
	$2px^4 + 6p^2x^2 + 2p^3$	$2px^3 + 4p^2 + q$	
	$2px^4 + 4p^2x^2 + qx$	p	
x	$2p^2x^2 - qx + 2p^3$	$2p^2x^3 + 4p^3x + pq$	
		$2p^2x^3 - qx^2 + 2p^3x$	x
		$qx^2 + 2p^3x + pq$	
		$2p^2$	
		$2p^2qx^2 + 4p^5x + 2p^3q$	q
		$2p^2qx^2 - q^2x + 2p^3q$	
		$x(4p^5 + q^2)$	

In case the given equation has a pair of equal roots then the H.C.F. should be linear. Therefore $x(4p^5 + q^2) = 0$ is the H.C.F. giving that $x = 0$ is the H.C.F. If $x = 0$ is the H.C.F. then x^2 should be a factor of $f(x)$ which is not. Hence, when $x(4p^5 + q^2) = 0$, $x \neq 0$ but $4p^5 + q^2 = 0$.

Also when $4p^5 + q^2 = 0$ then the H.C.F. will be $2p^2x^2 - qx + 2p^3$ so that $(2p^2x^2 - qx + 2p^3)^2$ is a factor giving that $f(x) = 0$ has two pairs of equal roots.

Example 10. *Prove that if the coefficients of a given equation are all integers, an integral root is an exact divisor of the absolute term.*

Solution. Let the equation be

$$a_0x^n + a_1x^{n-1} + a_2x^{n-2} + \ldots + a_{n-1}x + a_n = 0$$

where all the a's are integers. If α be an integral root of above equation, then

$$a_0\alpha^n + a_1\alpha^{n-1} + a_2\alpha^{n-2} + \ldots + a_{n-1}\alpha + a_n = 0$$
$$\text{or } \alpha(a_0\alpha^{n-1} + a_1\alpha^{n-2} + \ldots + a_{n-1}) = -a_n$$
$$\text{or } \frac{a_n}{\alpha} = -(a_0\alpha^{n-1} + a_1\alpha^{n-2} + \ldots + a_{n-1})$$
$$= \text{an integer}$$

Hence $\frac{a_n}{\alpha}$ = an integer. Therefore the integral root α divides a_n exactly.

Example 11. *Find the values of a, for which the equation* $ax^3 - 9x^2 + 12x - 5 = 0$ *has two equal roots and solve the equation completely in one case.*

Solution. Let $f(x) = ax^3 - 9x^2 + 12x - 5 = 0$ ...(1)

$\Rightarrow f'(x) = 3(ax^2 - 6x + 4) = 0$

or $ax^2 - 6x + 4 = 0$ ...(2)

Multiplying (2) by x and subtracting form (1), we get

$$3x^2 - 8x + 5 = 0$$

Solving (2) and (3) by method of cross-multiplication, we get

$$\frac{x^2}{-30+32} = \frac{x}{12-5a} = \frac{1}{18-8a}$$

Eliminating x, we get

$$(12-5a)^2 = 2(18-8a)$$

or $25a^2 - 104a + 108 = 0$

or $(a-2)(25a-54) = 0 \therefore a = 2 \text{ or } \frac{54}{25}$

Putting $a = 2$, we get

$$f(x) = 2x^3 - 9x^2 + 12x + 5 = 0$$
$$\Rightarrow f'(x) = x^2 - 3x + 2 = 0$$

or $(x-1)(x-2) = 0$

whose roots are 1 and 2. Either can be a double root of the given equation $f(x) = 0$. Since (2) does not divide the absolute term of the given equation, as such it cannot be its root. Hence, (1) is a double root of the given equation corresponding to the value 2 of a.

Example 12. *If the equation* $x^4 + ax^3 + bx^2 + cx + d = 0$ *had three equal roots show that each of them is* $\frac{(6c-ab)}{(3a^2-8b)}$.

Solution. In the process of finding H.C.F. of $f(x)$ and $f'(x)$, the 1st quotient is $x+a$ and quadratic remainder is

$F(x) = (8b - 3a^2)x^2 + 2(6c - ab)x + (16d - ac)$.

If the equation is to have three equal roots then the H.C.F. should be perfect square of a linear expression in x. Hence, $F(x)$ should be the H.C.F. and must have equal roots the condition for which is

$$4(6c-ab)^2 = 4(8b-3a^2)(16d-ac)$$

Multiplying $F(x)$ by $8b - 3a^2$, it becomes with the help of the above condition,

$$[(8b-3a^2)x + (6c-ab)]^2$$

and hence the equal root is given by

$$F(x) = 0 \text{ or } x = \frac{6c - ab}{3a^2 - 8b}$$

Example 13. *Find the condition that the cubic $a_0x^3 + 3a_1x^2 + 3a_2x + a_3 = 0$ may have two roots equal and find its value. Also, find the condition that the cubic may have all the three roots equal.*

Solution. The above cubic can be reduced to the form

$$f'(z) = 3(z^2 + H) = 0 \text{ or } z^2 + H = 0$$

$$f(z) = z^3 + 3Hz + G = 0 \text{ where } z = a_0x + a_1$$

The H.C.F. of $f(z)$ and $f'(z)$ is found below :

	$z^2 + H$	$z^3 + 3Hz + G$	z
	$2H$	$z^3 + Hz$	
z	$2Hz^2 + 2H^2$	$2Hz + G$	
	$2Hz^2 + Gz$	1st quotient	
	$-Gz + 2H^2$		
	$-2H$		
	$2HGz - 4H^3$		Remainder
	$2HGz + G^2$		
	$-(G^2 + 4H^3)$		

From above the remainder is $-(G^2 + 4H^3)$ and the first quotient is $2Hz + G$. If the equation has two equal roots then the H.C.F. should be of first degree and hence the remainder should be zero. Therefore G2 + 4H3 is the required condition and the equal root is given by equating the H.C.F.

Now, $\quad 2Hz + G = 0$

$$\therefore \quad z = -\frac{G}{2H} = a_0x + a_1;$$

$$\therefore \quad x = -\frac{1}{a_0}\left[a_0 + \frac{G}{2H}\right]$$

In case all the three roots are equal then the H.C.F. should be of 2nd degree and that too a perfect square and hence the first degree remainder, *i.e.*, $2Hz + G = 0$ should be zero. Now the H.C.F. in this case will be $z^2 + H$ and it will be a perfect square only if $H = 0$ and putting $H = 0$ in $2Hz + G = 0$ we get $G = 0$. Hence, the required conditions in this case are $G = 0, H = 0$.

EXERCISE 24.4

1. Factorize : $x^4 + 12x^3 + 32x^2 - 24x + 4$
2. Solve the equation :
   (i) $x^4 - 8x^3 + 24x^2 - 32x + 16 = 0$
   (ii) $x^6 - 6x^4 - 4x^3 + 9x^2 + 12x + 4 = 0$
3. Find the multiple root of :
   (i) $x^4 + 3x^3 - 7x^2 - 15x + 18 = 0$
   (ii) $x^4 + 7x^3 + 17x^2 + 17x + 6 = 0$
   (iii) $x^5 - x^4 + 2x^3 - 2x^2 + x - 1 = 0$
4. Show that the equation $x^n - nax + (n-1)b = 0$ will have a pair of equal roots of $a^n = b^{n-1}$.
5. Show that the equation
   (i) $x^n - 1 = 0$ (ii) $1 + \frac{x}{1!} + \frac{x^2}{2!} + \ldots + \frac{x^n}{n!} = 0$
   cannot have equal roots.
6. If $x^5 + 5px^3 + 5qx^2 + r = 0$ has two equal roots prove that either of them is a root of the quadratic
   $$3qx^2 - 6p^2x - 4pq + r = 0$$

ANSWERS

1. $(x^2 + 6x - 2)^2$
2. (i) All roots are equal to 2 (ii) $(2, 2, -1, -1, -1, -1)$
3. (i) $-3, -3, 1, 2$ (ii) $-1, -1, -2, -3$ (iii) $\pm i, \pm i, 1$

24.13 MAXIMUM AND MINIMUM VALUES OF f(x)

We know from the definition of Maximum and Minumum that as x varies in the interval $(x - h, x + h)$ where h is small, then $f(x)$ is greater than both $f(x - h)$ and $f(x + h)$ if $f(x)$ be Maximum, *i.e.*, $f(x - h) - f(x)$ and $f(x + h) - f(x)$ are of the same sign, *i.e.*, negative. Similarly if $f(x)$ is Minimum, then $f(x)$ is less than both $f(x - h)$ and $f(x + h)$, *i.e.*, $f(x - h) - f(x)$ and $f(x + h) - f(x)$ are of the same sign, *i.e.* positive.

Now by Taylor's Theorem

$$f(x+h) - f(x) = hf'(x) + \frac{h^2}{2!}f''(x) + \frac{h^3}{3!}f'''(x) + \frac{h^4}{4!}f^{iv}(x) + \ldots$$

$$f(x-h) - f(x) = -hf'(x) + \frac{h^2}{2!}f''(x) - \frac{h^3}{3!}f'''(x) + \frac{h^4}{4!}f^{iv}(x) - \ldots$$

Now when h is made sufficiently small, the sign of the right hand side of each equation in (1) and therefore of the left hand side is ultimately dependent upon that of $hf'(x)$, that being the term of lowest degree in h. But both $f(x+h)-f(x)$ and $f(x-h)-f(x)$ are of opposite signs whereas they should have the same sign in $f(x)$ is either Max. of Min. It is therefore necessary that $f'(x)$ should vanish so that the lowest term of the right hand side of equation in (1) should depend upon an even power of h.

Hence $f'(x) = 0$ is the essential condition for the occurrence of Max. and Min. values of $f(x)$.

Let the roots of $f'(x) = 0$ be a, b, c etc.

Consider one of the roots of the above equation say b; then putting $x = b$ and $f'(x) = 0$ at $x = b$ in (1), we get

$$f(b+h)-f(b) = \frac{h^2}{2!}f''(b) + \frac{h^3}{3!}f'''(b) + \frac{h^4}{4!}f^{iv}(b) + \ldots$$

$$f(b-h)-f(x) = \frac{h^2}{2!}f''(b) - \frac{h^3}{3!}f'''(b) + \frac{h^4}{4!}f^{iv}(b) + \ldots$$

Now if $f''(b)$ is negative, then clearly $f(b + h)$ and $f(b - h)$ are both less than $f(b)$ so that $f(x)$ is maximum at $x = b$. In case $f''(b)$ is positive then both $f(b + h)$ and $f(b - h)$ are greater than $f(b)$, showing that $f(x)$ is minimum at $x = b$.

Solved Example

Example. *Show that the maximum and minimum values of the biquadratic $ax^4 + 4bx^3 + 6cx^2 + 4dx + e$ are the roots of the equation*

$$a^2k^3 - 3(a^2I - 9H^2)k^2 + 3(aI^2 - 18H)k - \Delta = 0$$

where Δ is discriminant of the quadratic.

Solution. We know that the discriminant Δ of the quadratic is $I^3 - 27J^2$,

where $I = ae - 4bd + 3c^2$,

and $J = ace + 2bcd - ad^2 - b^2e - c^3$.

Let $f(x) = ax^4 + 4bx^3 + 6cx^2 + 4dx + e$.

Now if the curve $y = f(x)$ be moved parallel to the axis of x through a distance k, where k is the maximum or minimum value of $f(x)$, then x-axis will become a tangent, *i.e.*, the two values of x will coincide. Now by moving the curve parallel to x-axis through a distance k, y becomes $y - k$. Then the equation $f(x) - k = 0$ will have two roots equal, the condition for which is that its discriminant should be zero.

Now,, $f(x) - k$

$$\equiv ax^4 + 4bx^3 + 6cx^2 + 4dx + (e-k) = 0$$

Its discriminate is $I'^3 - 27J'^2$ where in the usual values of I and J, we have to put e equal to $e - k$ in order to get I' and J'.

$$\therefore\ I' = a(e-k) - 4bd + 3c^2 = I - ak$$

$$J' = ac(e-k) + 2bcd - ad^2 - b^2(e-k) - c^3$$

$$= J - (ack - b^2k) = J - kH \quad \because H = ac - b^2$$

$$\therefore\ I'^3 - 27J'^2 = 0 \text{ gives}$$

$$I^3 - 3I^2ak + 3Ia^2k^3 - a^3k^3$$

$$-27J^2 + 54JkH - 27k^2H^2 = 0$$

Cancelling minus sign and putting $I^3 - 27J^2 = \Delta$, we get

$$a^3k^3 - 3(a^2I - 9H^2)k^2$$

$$+ 3(aI^2 - 18HJ)k - \Delta = 0$$

The above equation gives the values of k.

24.14 ROLLE'S THEOREM

Between two consecutive real roots of the equation $f(x) = 0$ there lies at least one real root of the equation $f'(x) = 0$.

Proof. If a and b are consecutive roots of $f(x) = 0$ and as x varies from a to b, $f(x)$ varies continuously from $f(a)$ to $f(b)$. It will therefore vary either by increasing first and then decreasing or by decreasing first and then increasing. It must be therefore pass through at least one (in general odd) maximum or minimum values during its variation from $f(a)$ to $f(b)$ and let α be the value of x between a and b where $f(x)$ is either maximum or minimum and from the definition of maximum or minimum $f'(x) = 0$ at $x = \alpha$. Thus a number α lying between a and b is a root of equation $f'(x) = 0$.

Deduction. *To prove that between any two consecutive roots of the equation $f'(x) = 0$ there lies at the most one real root of the equation $f(x) = 0$ or they may not lie any.*

Proof. Let x_1 and x_2 be the two consecutive roots of the equations $f'(x) = 0$ lying between the intervals (α, β) and (β, γ) where α, β, γ are three consecutive roots of the equation $f(x) = 0$ [Rolle's Theorem]

Hence $\alpha < x_1 < \beta < x_2 < \gamma$. This relation shows that β, a root of $f(x) = 0$ lies between x_1 and x_2 the consecutive roots of $f'(x) = 0$.

Uniqueness. If possible, it is suppose that there are two roots β and β′ of the equation $f(x) = 0$ which lie between x_1 and x_2 the consecutive roots of $f'(x) = 0$ so that $x_1 < \beta < \beta' < x_2$ then β and β′ being consecutive roots of $f(x) = 0$ must have between them a root of the equation $f'(x) = 0$ [Rolle's theorem] which is contrary to the supposition that x_1 and x_2 are consecutive roots of $f(x) = 0$. Hence there cannot lie more than one root of $f(x) = 0$ between two consecutive root of the equation $f'(x) = 0$. In this case $f(x_1)$ and $f(x_2)$ should be of opposite signs and if they are of the same sign, then no root of $f(x) = 0$ will lie between x_1 and x_2.

Solved Example

Example. *The equation $x^4 - 15x^3 + 75x^2 - 145x + 84 = 0$ has its roots 1, 2, 3, 4, 7. Locate the roots of the equation $4x^3 - 45x^2 + 150x - 145 = 0$.*

Solution. We have

$f(x) = x^4 - 15x^3 + 75x^2 - 145x + 84 = 0$ then

$\Rightarrow f'(x) = 4x^3 - 45x^2 + 150x - 145 = 0$

Now the roots of $f(x) = 0$ are 1, 3, 4, 7 arranged in order and by Rolle's theorem we know that between the consecutive roots of the equation $f(x) = 0$ there lies at least one real root of the equation $f'(x) = 0$. Hence, there will be a root each of $f'(x) = 0$ between 1, 3; 3, 4 and 4, 7.

24.15 LIMIT OF THE ROOTS OF AN EQUATION

1. **Definition 1.** *A number which is greater than all the positive roots of a given equation is called upper or superior limit of the positive roots of that equation.*
2. **Definition 2.** *A number which is less than all the positive roots of a given equation is called the inferior or lower limit of the positive roots of that equation.*
3. **Definition 3.** *A number which is greater than all the negative roots of the given equation is called an upper or superior limit of the negative roots of that equation. In other words, superior limit of negative roots of a given equation is that negative number below which (numerically) lie all the negative roots.*
4. **Definition 4.** *A number which is less than all the negative roots of the given equation is called the inferior or lower limit of negative roots of that equation. In other words, inferior limit of negative roots of a given equation is that negative number above which (numerically) lie all the negative roots.*

REMARK

- From the above definitions we can say that the superior limit of the real roots of a given equation is that number which is greater than all real roots of that equation and the inferior limit of the real roots in that number which is less than all the real roots of that equation. Thus superior limit of positive roots is the superior limit of the real roots and the inferior limit of the negative roots is the inferior limit of the real roots of a given equation.

THEOREM 1. *If the polynomial $a_0x^n + a_1x^{n-1} + a_2x^{n-2} + ... + a_{n-1}x + a_n$ the value of $\frac{a_k}{a_0} + 1$ or any greater value be substituted for x where a_k is that one of the coefficient $a_1, a_2, ..., a_n$ whose numerical value is greatest, irrespective of sign, the term containing the highest power of x will exceed the sum of all the terms which follows.*

Proof. We have a_0x^n is the term continuing the highest power of x and it will exceed the sum of all the terms that follow it if

$$a_0x^n > a_1x^{n-1} + ... + a_2x^{n-2} + a_{n-1}x + a_n$$

The above inequality is satisfied for any value of x which makes

$$a_0x^n > a_k(x^{n-1} + x^{n-2} + ... + x + 1)$$

where a_k is the greatest among the coefficients $a_1, a_2, ... a_n$ without any regard to sign

or $$a_0x^n > a_k\left(\frac{x^n - 1}{x - 1}\right) \quad \text{(Sum of G.P.)}$$

or $$x^n > \frac{a_k}{a_0}\left(\frac{x^n - 1}{x - 1}\right)$$

Now $x^n > x^n - 1$ and hence above will be satisfied if

$$\frac{a_k}{a_0(x-1)} \le 1 \quad \text{or} \quad a_k \le a_0(x-1) \quad \text{or} \quad x - 1 \ge \frac{a_k}{a_0} \quad \text{or} \quad x \ge \frac{a_k}{a_0} + 1$$

THEOREM 2. *If in the polynomials* $a_0x^n + a_1x^{n-1} + ... + a_{n-1}x + a_n$ *the value* $\frac{a_n}{a_n + a_k}$ *or any smaller value be substituted for w where* a_k *is the greatest coefficient exclusive of* a_n, *the term* a_n *will numerically be greater than sum of all the others.*

Proof. Let us put $x = \frac{1}{y}$ in the given polynomial so that it becomes $\frac{1}{y^n}[a_ny^n + a_{n-1}.y^{n-1} + a_{n-2}y^{n-2} + ... + a_1y + a_0]$

Hence, by theorem 1, we have

$$a_ny^n < a_{n-1}y^{n-1} + a_{n-2}y^{n-2} + ... + a_1y + a_0$$

for all values of $y \geq \left(\frac{a_k}{a_n} + 1\right)$ where a_k is numerically the greatest of all the coefficients $a_0, a_1, a_2, ... a_{n-1}$.

Dividing by y^n, we get from (1),

$$a_n > \frac{a_{n-1}}{y} + \frac{a_{n-2}}{y^2} + ... + \frac{a_0}{y^n} \quad ...(2)$$

or $\quad > a_k\left(\frac{1}{y} + \frac{1}{y^2} + ... + \frac{1}{y^n}\right)$ or $\quad > a_k\frac{1}{y}\left\{\frac{1-\left(\frac{1}{y^n}\right)}{1-\left(\frac{1}{y}\right)}\right\}$ or $\quad > \frac{a_k}{y^n}\left(\frac{y^n - 1}{y - 1}\right)$ or $\quad y^n > \frac{a_k}{a_n(y-1)}(y-1)$

Now y^n is greater than y^{n-1} and hence above will be satisfied if

$\frac{a_k}{a_n(y-1)} \leq 1$ or $a_k \leq a_n(y-1)$ or $y - 1 \geq \frac{a_k}{a_n}$ or $y \geq \frac{a_k}{a_n} + 1$ or $y \geq \frac{a_k + a_n}{a_n}$ or $\frac{1}{x} \geq \frac{a_k + a_n}{a_n}$

or $\quad x \leq \frac{a_n}{a_k + a_n}$

Hence, the value of x equal to or less than $\frac{a_n}{a_k + a_n}$ will make

$$a_n > a_{n-1}x + a_{n-2}x^2 + ... + a_0x^n \quad \text{[From (2)]}$$

24.16 STURM'S METHOD OF FINDING THE EXACT NUMBER OF REAL ROOTS OF AN EQUATION

By Descarte's rule of signs we cannot get the exact number of real roots of a given equation $f(x) = 0$. Sturm's theorem gives us the exact number of real roots of a given equation.

24.17 STURM'S FUNCTIONS

Let $f(x)$ be any function of x of degree n and $f'(x)$ be its first derivative. Divide $f(x)$ by $f'(x)$ and let the remainder with sign changed be denoted by $f_2(x)$. Again divide $f'(x)$ by $f_2(x)$ and let the remainder with sign changed be denoted by $f_3(x)$.

Continue the process till you get the last remainder whose sign is also to be changed.

The above process is the same as that of finding the H.C.F. of $f(x)$ and $f'(x)$ with the modification that the sign of each remainder is to be changed before it becomes the divisor. Also we know that in process of finding the H.C.F. we can multiply and divide any remainder by any constant before using it as a divisor; but here in the above process we can only multiply or divide the remainder only by a positive constant before using it as a divisor or by a polynomial of x which is always positive real values of x say of the type $x^2 + 1$ or $x^4 + x^2 + 1$, etc.

The series of functions $f(x), f'(x), f_2(x), f_3(x), ... f_r(x)$ consisting of the given function, its derivative and remainder with their sign changed in process of finding the H.C.F. of $f(x)$ and $f'(x)$ are called Sturm's functions.

The functions $f_2(x), f_3(x), ... f_r(x)$ are called auxiliary functions.

(i) $f(x) = 0$ having equal roots.

In case $f(x) = 0$ has equal roots, then we know that $f(x)$ and $f'(x)$ will have some H.C.F. and hence the last of the Sturm's remainder will be a function of x.

(ii) $f(x) = 0$ having no equal roots.

In this case evidently the last of the Sturm's remainder will be numerical for if it is some function of x, then it would mean that $f(x)$ has got equal roots. In this case there will be $(n + 1)$ Sturm's function.

i.e., $\quad f(x), f'(x), f_2(x), f_3(x), ..., f_n(x)$

24.18 STURM'S THEOREM

Case I. All roots unequal.

If f(x) is a polynomial and a and b be any two real numbers, then the number of distinct real roots of the equation f(x)=0 lying between a and b is exactly equal to the difference between the number of changes of sign when x is put equal to a and the number when x is put equal to b in the (n +1) Sturm's functions $f(x), f'(x), f_2(x), f_3(x), \ldots f_n(x)$ consisting of the given function, its derivative and the (n – 1) remainders with their sign changed in the process of finding the H.C.F. of f(x) and f´(x).

From the definition of Sturm's functions we can establish the following relations between them :

Dividend = Q × Divisor + Remainder

$$\left.\begin{aligned} f(x) &= q_1 f'(x) - f_2(x), \\ f'(x) &= q_2 f_2(x) - f_3(x), \\ f_2(x) &= q_3 f_3(x) - f_4(x), \\ &\ldots\ldots\ldots\ldots \\ &\ldots\ldots\ldots\ldots \\ f_{r-1}(x) &= q_r f_r(x) - f_{r+1}(x) \\ &\ldots\ldots\ldots\ldots \\ f_{n-2}(x) &= q_{n-1} f_{n-1}(x) - f_n(x) \end{aligned}\right] \quad \ldots(A)$$

From the relation (A) we have the following observations.

1. As $f(x) = 0$ has no equal roots, $f(x)$ and $f'(x)$ have no common factors and consequently $f_n(x)$ the last Strum's function is numerical having a definite sign + or –.
2. No two consecutive auxiliary functions vanish for the same value of x. If possible, suppose that when $x = \alpha$, both $f_2(x)$ and $f_3(x)$ vanish which shows that $f'(x)$ contains the factor $x - \alpha$ and from first of the relations (A), we find that $(x - \alpha)$ is also a factor of $f(x)$. Thus $(x - \alpha)$ is the H.C.F. of $f(x)$ and $f'(x)$ showing the existence of equal roots which is contrary to the hypothesis.
3. If any of the auxiliary functions vanishes, then the two adjacent functions, *i.e.*, one which precedes it and the one which follows it must have opposite signs. For example suppose that $f_3(x) = 0$, when $x = \alpha$, then from relation (A), $f_2(x) = -f_4(x)$ for $x = \alpha$ showing that $f_2(x)$ and $f_4(x)$ have opposite signs for $x = \alpha$.
4. The same reasoning applies if any of the q's vanishes. Since all the functions are polynomials in x, no one can changes sign as x increases continuously form a to b, when x assumes a value which causes that functions to vanish.
5. Now we shall show that when x in passing from a to b takes a values α such that $f_r(\alpha) = 0$, then no change of sign is gained or lost. Now since $f_r(\alpha) = 0$, we have from (3), $f_{r-1}(\alpha) = 0$ and $f_{r+1}(\alpha)$ must be of opposite signs. Now as $f_r(x)$ passes through zero, it changes sign either from + to – or from – to +. But $f_{r-1}(x), f_{r+1}(x)$ are continuous at $x = \alpha$; so that each of them has an invariable sign near $x = \alpha$. Thus the three functions $f_{r-1}(x), f_r(x)$ and $f_{r+1}(x)$ will have only one changes of sign just before $x = \alpha$ and just after $x = \alpha$, *i.e.*, their signs can be either + +, –, – – +, + – –, – + +.

   Thus we find that whatever sign we may place between two unlike signs, we have only one change of sign. Hence, no change of sign in either lost or gained among Sturm's functions.

   In case the value of x, *i.e.*, α be such that it causes more than one of the functions to vanish, then they cannot be consecutive, for in that case, by (2) the equation $f(x) = 0$ will have equal roots.
6. Now we shall show that when x in passing from a to b takes a values α which causes $f(x)$ to vanish, *i.e.*, α be a root of $f(x) = 0$, then a change of sign is lost. Now by Taylor's theorem,

$$f(a-h) = 0 - hf'(a) + \frac{h^2}{2!} f''(\alpha) \ldots (\because f(\alpha) = 0)$$

$$f(a+h) = 0 + hf'(a) + \frac{h^2}{2!} f''(\alpha) \ldots \because (f(\alpha) = 0)$$

   Now let h be sufficiently small so that the sign of the L.H.S. is made to depend on the first term of R.H.S. Therefore if $f'(\alpha)$ be positive then $f(\alpha - h)$ is negative and $f(\alpha + h)$ is positive, *i.e.*, in this case the signs of $f(x)$ and $f'(x)$ will – + just before $x = \alpha$ and + + just after $x = \alpha$. Thus one change of sign is lost. Again if $f'(x)$ be negative then $f(\alpha - h)$ is positive and $f(\alpha + h)$ is negative, *i.e.*, in this case the signs of $f(x)$ and $f'(x)$ will be + – just before $x = \alpha$ and – – just after $x = \alpha$. Here also one change of sign is lost. Thus we conclude that when x passes through a root α of the equation $f(x) = 0$, one change of sign is lost whether $f'(\alpha)$ be positive or negative. Hence, the number of variations lost as x goes from a real value a to a real value b is exactly equal to the number of real root; of the equation $f(x) = 0$ between a and b.

Case II. Equal roots.

If $f(x) = 0$ be an equation having equal roots and the Strum's functions be found as $f, f', f_2, ... f_r$, the last of these being the H.C.F. of $f(x)$ and $f'(x)$ then the difference between the number of changes of signs when a and b are substituted in the Strum's functions is equal to the number of real roots of the equation $f(x) = 0$ which lie between a and b each multiple root being counted once only.

Let $f(x) = (x-\alpha)^p (x-\beta)^q (x-\gamma)(x-\delta)...$, then clearly $f(x)$ and $f'(x)$ will have an H.C.F. $(x-\alpha)^{p-1}(x-\beta)^{q-1}$ which may be denoted by H.

Thus H.C.F. will be a factor of all Sturm's functions $f, f', f_2, f_3, ... f_r$. Again let $\psi(x)$ stand for $(x-\alpha)(x-\beta)(x-\gamma)(x-\delta)$ then clearly $f(x) = H.\psi(x), f'(x) = H.\psi'(x), f_2(x) = H.\psi_2(x)...$ For any value of x, the number of changes of sign in the sequence of f's is the same as that in sequence of ψ's.

Now $\psi(x) = 0$ has all its roots unequal and all its roots are the same as those of $f(x) = 0$ only with the change that the multiple roots of $f(x) = 0$ occur in $\psi(x) = 0$ only once.

Now applying the reasoning of 1st case on $\psi(x) = 0$ we can say that difference between the number of changes of signs when x is put equal to a and b in the sequence of ψ's (hence of f's) represents exactly the number of real roots of the equation $\psi(x) = 0$ lying between a and b or the number of real roots of $f(x) = 0$ that lie between a and b but each multiple roots being counted only once.

TIME SAVING TRICKS

Trick 1. When $f(x) = 0$ has no repeated root, *i.e.*, the last Sturmain function be numerical, and we are concerne only with its sign. In order to get its sign we put $f_{n-1}(x) = 0$ and find the value of x; then we know that for this value of x, $f_{n-2}(x)$ and $f_n(x)$ must have the opposite sign. Thus if the value of x obtained from $f_{n-1}(x) = 0$ makes $f_{n-2}(x)$ positive, then $f_n(x)$ is negative and if it makes $f_{n-2}(x)$ negative, then $f_n(x)$ is positive. This device saves us he labour of actually calculating the value of $f_n(x)$.

Trick 2. If any of the Sturmain function say $f_2(x)$ has all roots imaginary, we may stop further calculation and we should see $f, f', f_2, f_3, ... f_r$ functions only for Strum's theorem, because in this theorem the last of the function should be of invariable sign for all real values of x and we know from the properties of equation that if $f(x) = 0$ has all its roots imaginary, then $f(x)$ is always positive for all real values of x. The quadratic $ax^2 + bx + c = 0$ has its roots imaginary if $b^2 - 4ac < 0$.

Similarly the calculation of Sturmain function will stop at the stage when any of them become a perfect squares for it too will have an in variable sign for all real values of x.

Trick3. In case any of the Sturm's functions $f_r(x)$ vanishes for $x = \alpha$; then for counting the number of changes of sign in the sequence $f(\alpha), f'(\alpha), f_2(\alpha), f_3(\alpha), ...,$

We may regard the sign of $f_r(\alpha)$ either positive of negative because the function that precedes it and the one which follows it have opposite signs.

WORKING PROCEDURE

Step 1. *Find the Sturm's functions as explained.*

Step 2. *The last Sturm's function will be numerical in case $f(x) = 0$ has no equal roots. Otherwise it will be some function of x.*

Step 3. *The calculation of Sturm's functions should stop at the stage when any of them is either a perfect square or has all its roots imaginary.*

Step 4. *Make a table as explained below :*

(i) *In the first column write down $x, f, f', f_2, f_3, f_4, ...$*
In the first row write down x and the various values that you may give to x.

(ii) *In the various other columns write down the signs of the Sturm's functions corresponding to the value of x written at the top that column.*

(iii) *At the bottom of each column write down the number of changes of signs in that particular column.*

(iv) *If we are to find only number of real roots, we put $x = \infty$ and $-\infty$ and find out the difference of the changes of signs corresponding to the two values.*

(v) *If we want to find out the positive and negative roots, we put $x = \infty$, 0 and $-\infty$ and proceeding as above, we get the number of positive roots which will lie between 0 and ∞. Negative roots which lie between 0 and $-\infty$.*

(vi) *If we want to find interval in which the roots lie, then for positive roots, we put $x = 1, 2, 3, ...$ and for negative roots, we put $x = -1, -2, -3...$*

(vii) *If corresponding to the two values of x say 2 and 3 the changes of sign in Sturm's function be same, i.e., their differences be zero, then no root will lie between 2 and 3.*

(viii) *We know that if $f(a)$ and $f(b)$ are of opposite signs, then at least one or in general odd number of roots of the equation $f(x) = 0$ lie between a and b. In case they be of the same sign, then either no root or an even number of roots of $f(x) = 0$ lie between a and b.*

In case there be only one positive roots, then in order to find its location we need not put $x = 1, 2, 3 \ldots$ in all the Sturm's functions. We shall put only in $f(x)$ and in case of any two consecutive values of x the values of $f(x)$ are of opposite signs; then by the above theorem the root will lie between those two consecutive numbers. This will save us the labour of putting the consecutive numbers in all the series of Strum's functions.

Solved Examples

Example 1. *Find the number and position of the real roots of the equation* $x^6 - 2x^2 + 3x - 4 = 0$.

Solution. We have

$$f(x) = x^6 - 2x^2 + 3x - 4 = 0$$
$$f'(x) = 6x^5 - 4x + 3$$

Multiplying $f(x)$ be 6 and then dividing by $f'(x)$

$$\begin{array}{r|l} & x \\ 6x^5 - 4x + 3 & 6x^6 - 12x^2 + 18x - 24 \\ & 6x^6 - 4x^2 - 3x \\ \hline & -8x^2 + 15x - 24 \end{array}$$

Changing the sign of this remainder, we get

$$f_2(x) = 8x^2 - 15x + 24$$

Now since $b^2 - 4ac$, *i.e.*, $(15)^2 - 4.8.24$ is negative, we conclude that the roots of the Strum's functions $f_2(x) = 0$ are imaginary and hence we stop further calculations.

x	$-\infty$	0	∞	1	2	–1	–2
$f(x)$	+	–	+	–	+	–	+
$f'(x)$	–	+	+	+	+	+	–
$f_2(x)$	+	+	+	+	+	+	+
No. of changes of sign	2	1	0	1	0	1	2

Now, we shall discuss the nature of the roots

1. There are only two real roots (from column 2 and 4) as the difference of changes of signs when x is put $-\infty$ and ∞ in Sturm's functions $2 - 0 = 2$.
2. One of them is positive (from column 3 and 4) which lies between 1 and 2 (from column 5 and 6).
3. One of them is negative (from columns 2 and 3) which lies between –1 and –2 (from columns 7 and 8).
4. The equation being of sixth degree has only two real roots; therefore remaining four roots are imaginary.

Remark

- We have found above that the above equations has only one positive root and only one negative root. Also we know that if $f(a)$ and $f(b)$ be of opposite signs then at least one root and in general odd number of roots must lie between a and b.

Now $f(0) = -$ and $f(1) = -$

From above we could easily conclude that positive root does not lie between 0 and 1 as $f(0)$ and $f(1)$ are of the same sign. This would save us from calculating the signs of all the Sturm's functions corresponding to $x = 1$.

Again $f(1)$ = negative, $f(2)$ = positive. Since $f(1)$ and $f(2)$ are of opposite signs, hence the only positive roots of $f(x) = 0$ must lie between 1 and 2. This has saved us the labour of calculating the signs of all Sturm's functions corresponding to $x = 2$.

Again $f(0) = -$ and $f(-1) = -$.

Therefore arguing as above negative root does not lie between 0 and –1.

Again $f(-1)$ = negative and $f(-2)$ = positive

$\therefore$ Negative root lies between –1, –2.

Hence, the above chart could be easily put as :

x	$-\infty$	0	∞	1	2	–1	–2
$f(x)$	+	–	+	–	+	–	+
$f'(x)$	–	+	+				
$f_2(x)$	+	+	+				
No. of changes of sign	2	1	0				

Example 2. *Find the number and position of real roots of the equation* $x^3 - 7x + 7 = 0$.

Solution. We have $f(x) = x^3 - 7x + 7$, $f'(x) = 3x^2 - 7$

Let us find the H.C.F. of $f(x)$ and $f'(x)$ and change the sign of remainders to get Sturm's functions.

	$f(x)$	$f'(x)$	
	$x^3 - 7x + 7$	$3x^2 - 7$	
	3	2	
	$3x^3 - 21x + 21$	$6x^2 - 14$	$3x$
x	$3x^3 - 7x$	$6x^2 - 9x$	
	$-14x + 21$	$9x - 14$	
	or $-2x + 3$	2	
	$\therefore f_2(x) = 2x - 3$	$18x - 28$	9
	after changing sign	$18x - 27$	
		-1	
		$\therefore f_2(x)$	
		after changing sign	

$\therefore$ Sturm's functions are

$$x^3 - 7x + 7,\quad 3x^2 - 7, 2x - 3, 1$$

x	$-\infty$	0	∞	1	2	-1	-2	-3	-4
$f(x) = x^3 - 7x + 7$	−	+	+	+	+	+	+	+	−
$f'(x) = 3x^2 - 7$	+	−	+	−	+				
$f_2(x) = 2x - 3$	−	−	+	−	+				
$f_3(x) = 1$	+	+	+	+	+				
No. of changes of sign	3	2	0	2	0				

Above table shows that all the three roots are real out of which one is negative and two are positive which clearly lie between 1 and 2. Also the negative root lies between –3 and –4. We have not tabulated the signs of all Sturm's functions for the only negative root because $f(0)$, $f(-1)$, $f(2)$, $f(-3)$ are all of same signs and hence there arises no questions of a negative root lying between these numbers. Again $f(-3)$ = positive and $f(-4)$ = negative and hence a root lies between –3 and – 4.

Example 3. *Find the number of distinct real roots of the equation* $x^3 - 3x + 1 = 0$ *and locate them.*

Solution. We have $f(x) = x^3 - 3x + 1$,

$f'(x) = 3(x^2 - 1)$ or $(x^2 - 1)$ on dividing by 3.

$f_2(x) = 2x - 1$ [After changing sign of the remainder]

$f_3(x) = 3$ [After changing the sign.]

x	$-\infty$	0	∞	1	2	-1	-2
$f(x)$	−	+	+	+	+	+	−
$f'(x)$	+	−	+	+ or −	+	+ or −	+
$f_2(x)$	−	−	+	+	+	−	−
$f_3(x)$	+	+	+	+	+	+	+
No. of changes of sign	3	2	0	1	0	2	3

As in example 1 the above table shows that the given equation has all its roots real and distinct out of which one is negative and the other two positive. The negative root lies in the interval (–2, –1) and one of the positive roots lies in the interval (1, 2) and the other in the interval (0, 1).

Since $f(0)$ = positive and $f(-1)$ = positive, *i.e.*, they are of the same sign, hence the root does not lie between 0 and –1 and as such we need not have completed wholly the column corresponding to $x = -1$.

Again $f(-1)$ = positive and $f(-2)$ = negative hence the only negative root lies between (–1, –2) and we need not complete the column corresponding to $x = -2$.

Example 4. *Apply Sturm's theorem to the analysis of the equation* $x^4 - 2x^3 - 3x^2 + 10x - 4 = 0$.

Solution. We have

$f(x) = x^4 - 2x^3 - 3x^2 + 10x - 4$

$f'(x) = 2x^3 - 3x^2 - 3x + 5$, after cancelling 2.

$f_2(x) = 9x^2 - 27x + 11$, after changing sign.

$f_3(x) = -8x - 3$, after changing sign.

Now $f_4(x)$ will be numerical and we are concerned only with its sign.

Putting $f_3(x) = 0$, we get $x = -\frac{3}{8}$. This value of x makes $f_2(x) = 9.\frac{3}{64} + \frac{81}{8} + 11$, *i.e.*, positive and we know that when $f_3(x) = 0$, then $f_2(x)$ and $f_4(x)$ are of opposite signs. Since $f_2(x)$ is positive therefore $f_4(x)$ must be negative.

REMARK

- If we actually proceed to calculate the value of $f_4(x)$, it will be –1433 (after changing the sign).

x	$-\infty$	0	∞	1	-1	-2	3
$f(x)$	+	−	+	+	−	−	+
$f'(x)$	−	+	+				
$f_2(x)$	+	−	+				
$f_3(x)$	+	−	−				
$f_4(x)$	−	−	−				
No. of changes of sign	3	2	1				

From above table we observe that there are only $3-1=2$ real roots. One of them is positive and the other negative. Again $f(0)$ = negative and $f(1)$ = positive, *i.e.*, they are of opposite signs; so we conclude that the positive root lie between 0 and 1. Hence we need not complete the column corresponding to $x=1$.

Again $f(0)$ = negative and $f(-1)$ = negative, *i.e.* they are of the same sign. Hence, either no root or an even no. of roots of $f(x)=0$, lie between 0 and –1; but as there is only one negative root, as such it cannot lie between 0 and –1; therefore we have not completed the column corresponding to $x=-1$. Similarly for column corresponding to $x=-2$.

Again $f(-2)$ = negative and $f(-3)$ = positive.

As they are of opposite signs, therefore they only negative root lies between –2 and –3, and we need not complete the column corresponding to $x=-3$.

Example 5. *Discuss the nature and position of the roots of the equation $x^4-12x^2+12x-3=0$ by means of Sturmain functions.*

Solution. We have $f(x)=x^4-12x^2+12x-3=0$

$$f'(x)=4(x^3-6x+3) \text{ or } x^3-6x+3$$

on dividing by 4.

$$x^3-6x+3\,\big)\,x^4-12x^2+12x-3\,\big(\,x$$
$$x^4-6x^2+3x$$
$$\overline{\quad -6x^2+9x-3\quad}$$

changing the sign and dividing by 3.

$$f_2(x)=2x^2-3x+1\,\big)\,x^3-6x+3\,\big(\,x+3$$
$$2$$
$$2x^3-12x+6$$
$$2x^3-3x^2+6$$
$$\overline{3x^2-13x+6}$$
$$2$$
$$6x^2-26x+12$$
$$6x^2-9x+3$$
$$\overline{-17x+9}$$

changing the sign

$$f_3(x)=17x-9$$

Now $f_4(x)$ will be numerical and we are concerned only with its sign, $f_3(0)=0$ gives $x=\frac{9}{17}$, *i.e.*, slightly $>\frac{1}{2}$ say $\frac{1}{2}+k$ where k is small. This value of x when substituted in $f_2(x)$ makes it.

$$2\left(\frac{1}{2}+k\right)^2-3\left(\frac{1}{2}+k\right)+1=2k^2-k$$

i.e. – negative for small values of k and hence $f_4(x)$ is positive and numerical, is may be verified by actual calculation that $f_4(x)=8$.

x	$-\infty$	0	∞	1	2	3	-1	-2	-3	-4
$f(x)$	+	−	+	−	−	+	−	−	−	+
$f'(x)$	−	+	+	−	−	+	+	+	−	−
$f_2(x)$	+	+	+	+ or −	+	+	+	+	+	+
$f_3(x)$	−	−	+	+	+	+	−	−	−	−
$f_4(x)$	+	+	+	+	+	+	+	+	+	+
No. of changes of sign	4	3	0	1	1	0	3	3	3	4

The above table shows that the given equation has all its real roots; one of them is negative lying in the interval (–4, –3). The other there are positive lying in the interval (2, 3) and two in the interval (0, 1). Here we need not commute fully the columns corresponding to $x=-1,-2,-3,-4$, for in all these cases $f(0), f(-1), f(-2), f(-3)$ are all negative, *i.e.*, of the same sign and there being only one negative root, it cannot lie between any of the two consecutive numbers. Again $f(-3)$ = negative and $f(-4)$ = positive, *i.e.*, opposite signs, therefore the only negative lies between –3 and –4.

Example 6. *By Sturm's method prove that the equation $x^4 - 6x^3 + 13x^2 - 12x + 4$, has two pairs of equal roots.*

Solution. We have
$$f(x) = x^4 - 6x^3 + 13x^2 - 12x + 4$$
$$f'(x) = 2x^3 - 9x^2 + 13x - 6 \text{ (cancelling 2)},$$
$$f_2(x) = x^2 - 3x + 2, \text{ as shown below.}$$

	$f(x)$	$f'(x)$	
	$x^4 - 6x^3 + 13x^2 - 12x + 4$	$2x^3 - 9x^2 + 13x - 6$	$2x - 3$
	4	$2x^3 - 9x^2 + 13x$	
x	$2x^4 - 12x^3 + 26x^2 - 24x + 8$	$-3x^2 + 9x - 6$	
	$2x^4 - 9x^3 + 13x^2 - 6x$	$-3x^2 + 9x - 6$	
	$-3x^2 + 13x^2 - 18x + 8$	$\times$	
	2		
-3	$-3x^3 + 26x^2 - 36x + 16$		
	$-6x^3 + 27x^2 - 39x + 18$		
$f_2(x)$	$-x^2 + 3x - 2$		
	$x^2 - 3x + 2$		
	sign changed		

From above we conclude that $x^2 - 3x + 2$ or $(x-2).(x-1)$ is the H.C.F. of $f(x)$ and $f'(x)$. Hence the given equation has all the four roots real each root accruing twice, *i.e.*, 2, 2, 1, 1. But Strum's function will show only distinct type of roots as shown below :

x	$-\infty$	0	∞	1	2	
$f(x)$	+	+	+	0	0	1 and 2 are the roots of $f(x)$
$f'(x)$	–	–	+			
$f_2(x)$	+	+	+			
No. of changes of sign	2	2	0			

24.19 CONDITION FOR ALL THE ROOTS REAL AND DISTINCT

Case I. In case the equation has all its roots distinct, then the sequence of Sturm's function must in general consist of $(n + 1)$ functions, *i.e.*, the given function, its derivatives and $(n - 1)$ Sturm's remainders.

Case II. In case the equation has all its roots real, then the difference of the number of changes of signs when x is put ∞ in the sequence over the number when x is put $-\infty$ should be n. In other words it means that when x is put $+\infty$ no change of signs occurs, *i.e.*, the number of changes of signs be zero and when $x = -\infty$ in the $(n + 1)$ functions, it should be alternately positive and negative so that there be n changes of signs and the difference of these change of signs be $n - 0 = n$, the number of real and distinct roots.

The above conditions are satisfied if we say that the leading coefficients of all the Sturm's functions be positive (we always take the leading coefficients of a given equation to be positive).

Solved Example

Example. *Use Sturm's theorem to show that the equation $z^3 + Hz + G = 0$ has all the three roots real and distinct if and only if $G^2 + 4H^3 < 0$.*

Solution. The leading coefficients of all the Sturm's functions is positive we get the condition as

$-H$ positive, $-(G^2+4H^3)$ positive

i.e., H negative and $G^2 + 4H^3$ negative

But $G^2 + 4H^3$ can be negative only if H is negative and hence the former condition is implied in the latter. Hence $G^2 + 4H^3$ is negative *i.e.* < 0 if all the roots of the given cubic are real and distinct.

24.20 NATURE OF THE ROOTS OF BIQUADRATIC

By the help of Sturm's theorem, discuss the nature of the roots of the equation $ax^4 + 4bx^3 + 6cx^2 + 4dx + e = 0$

The above equation can be reduced to the form

$$f(z) = z^4 + 6Hz^2 + 4Gz + a^2I - 3H^2 = 0$$

where $z = ax + b$ and the nature of the roots of the given equation and of the transformation equation is same the Sturm's functions are

$$f(z) = z^4 + 6Hz^2 + 4Gz + a^2I - 3H^2$$

$$f'(z) = z^3 + 3Hz + G$$

$$f_2(z) = -3Hz^2 - 3Gz - (a^2I - 3H^2)$$

$$f_3(z) = -z(2HI - 3aJ) - IG$$

$$f_4(z) = I^3 - 27J^2 = \Delta \text{, the discriminate.}$$

1. **All roots real and distinct.** The leading coefficients of Sturm's functions should be positive.

   $\therefore$ H negative, $2HI - 3aJ$ negative, $I^3 - 27J^2$ positive.

2. **All roots imaginary.**

   $I^3 - 27J^2$ positive, and either H positive or $2HI - 3aJ$ positive.

   In this case corresponding to the $z = \infty$, number of changes of sign in the above sequence will be

   $z = \infty$ + + – (+ or –) + 2 changes

   $z = -\infty$ + – – (+ or –) +2 changes

   The difference corresponding to the values ∞ and $-\infty$ of z is zero and hence the equation has no real roots. Thus all its roots are imaginary.

3. **Two roots real and two imaginary.**

   $I^3 - 27J^2$ negative : In this case we shall find that the difference of the changes of signs corresponding to $z = \infty$ and $-\infty$ in the above sequence is always 2. We may give H and $2HI - 3aJ$ any sign we like. Thus the equation has two real and two imaginary roots.

4. **Two equal roots.**

   In this case clearly $I^3 - 27J^2 = 0$ and then $f_3(z) = 0$ will give the H.C.F. of $f(z)$ and $f'(z)$ and thus proving the existence of two equal roots.

5. **Three equal roots or two pairs of equal roots.**

   In this case the H.C.F. should be of 2nd degree and also a perfect square or composed of two unequal factors according as three roots are equal to two pair of equal roots exist. Hence, $f_2(z)$ should vanish identically which will happen when either.

   (1) $I = 0$ and $J = 0$ or

   (2) $G = 0$ and $2HI - 3aJ = 0$

   when $I = 0$ and $J = 0$, we get $G^2 + 4H^3 = a^2(HI - aJ) = 0$ and $f_2(z) = 0$ becomes

   $$3Hz^2 + 3Gz - 3H^2 = 0$$

   and its discriminant is $9(G^2 + 4H^3)$ which is zero and hence its roots are equal, *i.e.*, it is a perfect square. Since the H.C.F. is a perfect square, $f(z) = 0$, has three equal roots.

   When $G = 0$ and $2HI - 3aJ = 0$, then $f_2(z)$ is the H.C.F. but is not a perfect square and hence the equation will have a pair of equal roots.

REMARK

- We have $G^2 + 4h^3 = a^2HI - a^3J$. Putting $G = 0$ and $J = \frac{2HI}{3a}$, we get $4H^3 = a^2HI - a^3\frac{2HI}{3a}$ or $a^2I = 12H^2$ and this is same as the condition we had found.

6. **All roots equal.**

   If $I = 0$, $H = 0$ and $G = 0$ then $f_3(z)$ vanishes identically and the H.C.F. comes out to be $f'(z)$, *i.e.*, z^5 showing that all the roots are equal.

EXERCISE 24.5

1. Find the number and position of real roots of the following equations :

   (i) $x^4+15x^2+7x-11=0$ (ii) $x^4-6x^3+5x^2+14x-1=0$

   (iii) $x^4-4x^3+7x^2-6x-4=0$

   (iv) $x^4-8x^3+25x^3-36x+8=0$

   (v) $x^3+x^2-2x-1=0$ (vi) $x^4-7x^2+18x-8=0$

2. Apply Sturm's theorem to prove that the equation $x^3-2x-5=0$ has only one positive real root lying between 2 and 3.

3. Prove that the equation $x^5-x+16=0$ has two pair of complex roots.

4. Prove by Sturm's method that the equation $x^4-6x^3+13x^2+4=0$ has two pair of equal roots.

5. Varify by means of Sturm's remainders, the condition which must be fulfilled when the biquadratic $9x^4+4bx^3+6cx^2+4dx+e=0$ is a perfect square and prove in that case $a^3 f(x)=\{(ax+b)^2+3H\}^2$.

6. Prove that when the biquadratic of the last example has a triple factor it may be expressed in the form
$$a^3 f(x)=\{ax+b+\sqrt{(-H)}\}^3\{ax+b-3\sqrt{(-H)}\}$$

7. Find Sturm's function for the cubic $z^3+3Hz+G=0$.

ANSWERS

1. (i) One negative in (–1, –2) and one positive in (0, 1)
   (ii) One negative in (–1, –2), three positive out of which one is in (0, 1) and two in (2, 4)
   (iii) One negative in (0, –1) and one positive in (2, 3) and two imaginary
   (iv) Two imaginary, two real positive, one in (0, 1) and other two in (3, 4)
   (v) All real, one positive in (1, 2) two negative in (0, 1) and other in (–1, –2)
   (vi) Two imaginary, two real one positive in (0, 1) one negative in (3, 4)

4. 1, 1, 2, 2

7. $f(z)=z^3+3Hz+G$, $f_1(z)=z^2+H$, $f_2(z)=-2Hz-G$, $f_3(z)=-(G^2+4H^3)$

24.21 CARDAN'S METHOD TO FIND THE ROOTS OF A CUBIC EQUATION

Let the general cubic equation be $a_0x^3+3a_1x^2+3a_2x+a_3=0$ ...(1)

First reduce this equation (1) into an equation having no second degree term, *i.e.*, $3a_1x^2$. The equation (1) is reduced to the following equation

$$Z^3+3HZ+G=0 \quad ...(2)$$

where $H=a_0a_2-a_1^2, G=a_0^2a_3-3a_0a_1a_2+2a_1^3$ and $Z=a_0x+a_1$

Let as assume $z=u+v$. ...(3)

Cubing both sides of (3), we get

$$z^3=(u+v)^3=u^3+v^3+3uv(u+v)=u^3+v^3+3uv(z)$$

$$z^3=u^3+v^3+3uvz \quad \text{or} \quad z^3-3uvz-(u^3+v^3)=0 \quad ...(4)$$

Comparing (2) and (4), we get

$$uv=-H, u^3+v^3=-G \quad \text{or} \quad u^3v^3=(-H)^3, u^3+v^3=-G$$

Hence u^3, v^3 are the roots of the quadratic equation given by

$$t^2+Gt-H^3=0 \quad ...(5)$$

Solving (5), we get

$$t=\frac{-G\pm\sqrt{G^2+4H^3}}{2}$$

$$\therefore \quad u^3=\frac{-G+\sqrt{G^2+4H^3}}{2} \quad ...(6)$$

and

$$v^3=\frac{-G-\sqrt{G^2+4H^3}}{2} \quad ...(7)$$

From (3), we get

$$z=\left\{-\frac{G}{2}+\frac{1}{2}\sqrt{G^2+4H^3}\right\}^{1/3}+\left\{-\frac{G}{2}-\frac{1}{2}\sqrt{G^2+4H^3}\right\} \quad ...(8)$$

From (6) and (7) it is obvious that each u and v will have three cube roots and hence from (8) z will have nine values. But the degree of the equation (2) in z is three so it must have three roots, *i.e.*, three values of z. Since we have that $uv=-H$, therefore the cube roots are taken in pairs so that $uv=-H$. Hence we shall take the pair of cube roots as

$$u, v, u\omega, v\omega^2; u\omega^2, v\omega$$

where ω and ω^2 are the imaginary cube roots of unity. Therefore the roots of the equation (2) are

$$u+v, u\omega+v\omega^2, u\omega^2+v\omega$$

and hence we can find the roots of the equation (1) by the relation $z=a_0x+a_1$ corresponding to $u+v, u\omega+v\omega^2$ and $u\omega^2+v\omega$.

24.22 APPLICATION OF CARDAN'S METHOD

From equation (6) and (7), we have

$$u^3=\frac{-G+\sqrt{G^2+4H^3}}{2} \quad \text{and} \quad v^3=\frac{-G-\sqrt{G^2+4H^3}}{2}$$

Case I. If $G^2+4H^3>0$, *i.e.*, the cubic equation (1) has a pair of imaginary roots, then R.H.S. of above two equations are real and hence by the some method we can extract the cube root of real quantities and consequently we get the values of z from equation (8).

Case II. If $G^2+4H^3<0$, *i.e.*, if the roots of the cubic equation (1) are all real, then from above equations u^3 and v^3 are imaginary and there is no method in general for extracting the cube root of imaginary number so that Cardan's method fails to give the roots of the given cubic (1). However in this case we use De Moivre's theorem to find the cubic root of imaginary number.

Let us assume

$$-G=P \text{ and } G^2+4H^3=-Q \text{ then from equation (8), we get}$$

$$z=\left(\frac{P}{2}+\frac{i}{2}Q\right)^{1/3}+\left(\frac{P}{2}-\frac{i}{2}Q\right)^{1/3} \quad \text{where } i=\sqrt{-1}$$

Now put $\frac{P}{2}=r\cos\theta, \frac{Q}{2}=r\sin\theta$

$$\therefore \quad r^2=\frac{P^2+Q^2}{4}=\frac{G^2-G^2-4H^3}{4}=-H^3$$

and $\tan\theta=\frac{Q/2}{P/2}=\frac{Q}{P} \Rightarrow \tan\theta=\frac{-\sqrt{-(G^2+4H^3)}}{G}$

$$\therefore \quad z=(r\cos\theta+ir\sin\theta)^{1/3}+(r\cos\theta-ir\sin\theta)^{1/3}$$

$$=r^{1/3}\left[\cos\left(\frac{2n\pi+\theta}{3}\right)+i\sin\left(\frac{2n\pi+\theta}{3}\right)+\cos\left(\frac{2n\pi+\theta}{3}\right)-i\sin\left(\frac{2n\pi+\theta}{3}\right)\right]$$

$$=2r^{1/3}\cos\left(\frac{2n\pi+\theta}{3}\right) \quad \text{where n = 0, 1, 2.}$$

Hence, the roots of the equation (2) are

$$z_1=2r^{1/3}\cos\left(\frac{\theta}{3}\right)=2(-H)^{1/2}\cos\left(\frac{\theta}{3}\right); \quad z_2=2r^{1/3}\cos\left(\frac{2n\pi+\theta}{3}\right)=2(-H)^{1/2}\cos\left(\frac{2n\pi+\theta}{3}\right)$$

$$z_3=2r^{1/3}\cos\left(\frac{2\pi-\theta}{3}\right)=2(-H)^{1/2}\cos\left(\frac{2\pi-\theta}{3}\right) \quad \text{where } \theta=\tan^{-1}\left[\frac{-\sqrt{-(G^2+4H^3)}}{G}\right]$$

Consequently by the relation $z=a_0x+a$, we can find all the roots of the given cubic.

24.23 METHOD BY EXPRESSING THE EQUATION AS SUM OR DIFFERENCE OF TWO CUBES

Let the given cubic equation be

$$a_0x^3+3a_1x^2+3a_2x+a_3=0 \quad \ldots(1)$$

Let us suppose the equation (1) can be expressed as follows :

$$a_0x^3+3a_1x^2+3a_2x+a_3 \equiv A(x-a)^3+B(x-b)^3 \quad \ldots(2)$$

or

$$a_0x^3+3a_1x^2+3a_2x+a_3 \equiv (A+B)x^3-3(Aa+Bb)x^2+3(Aa^2+Bb^2)x-(Aa^3+Bb^3).$$

Now equating the coefficients of like powers of x, we get

$$A+B=a_0 \quad \ldots(3)$$

$$Aa + Bb = -a_1 \quad \ldots(4)$$

$$Aa^2 + Bb^2 = a_2 \quad \ldots(5)$$

$$Aa^3 + Bb^3 = -a_3 \quad \ldots(6)$$

Multiplying (3), (4), (5) by a and subtracting respectively (4), (5) and (6), we get

$$B(a-b) = a_0 a + a_1 \quad \ldots(7)$$

$$-Bb(a-b) = a_1 a + a_2 \quad \ldots(8)$$

$$Bb^2(a-b) = a_2 a + a_3 \quad \ldots(9)$$

From (7), (8) and (9), we get

$$(a_1 a + a_2)^2 = (a_0 a + a_1)(a_2 a + a_3) \quad \ldots(10)$$

Similarly multiply (3), (4), (5) by b subtracting respectively (4), (5) and (6), we get

$$(a_1 b + a_2)^2 = (a_0 b + a_1)(a_2 b + a_3) \quad \ldots(11)$$

From equation (10) and (11) it is concluded that a and b are the roots of the equation

$$(a_1 x + a_2)^2 = (a_0 x + a_1)(a_2 x + a_3) \quad \ldots(12)$$

or

$$(a_0 a_2 - a_1^2)x^2 + (a_0 a_3 - a_1 a_2)x + (a_1 a_3 - a_2^2) = 0 \quad \ldots(13)$$

This equation (13) is called the Hessian of the cubic.

From (13) we obtain the values of a and b and then we can find A and B from (3) and (4). Substitute these values of a, b, A and B in (2) we get the given equation as the sum or difference of two cubes as

$$A(x-a)^3 + B(x-b)^3 = 0 \quad \text{or} \quad \left(\frac{x-a}{x-b}\right)^3 = \left(-\frac{B}{A}\right) \quad \text{or} \quad \left(\frac{x-a}{x-b}\right) = \left(-\frac{B}{A}\right)^{1/3} = -\left(\frac{B}{A}\right)^{1/3} k \quad \ldots(14)$$

where $k = 1, \omega, \omega^2$.

Therefore if a and b (provided $a \neq b$) are the roots of the equation,

$$(a_1 x + a_2)^2 = (a_0 x + a_1)(a_2 x + a_3)$$

Then the given equation (1) can be reduced to the form given by

$$A(x-a)^3 + B(x-b)^3 = 0$$

Now from (14) one values of x is obtained and then divide the given equation by so obtained factor we get a quadratic as quotient and the remaining roots of the given equation can be obtained from this quadratic quotient.

Further, if the obtained values of a and b are imaginary and so A and B, then we cannot find the cube roots of complex number. Hence, by above method we can find the roots of the given cubic if the roots of the quadratic equation (13) are real and distinct.

That is, if the condition $(a_0 a_3 - a_1 a_2)^2 - 4(a_1 a_3 - a_2^2)(a_0 a_2 - a_1^2) > 0$ holds then the equation (13) will have both real and distinct roots.

24.24 HESSIAN OF THE CUBIC EQUATION

The quadratic equation (13) of §24.23 is given by $(a_0 a_2 - a_1^2)x^2 + (a_0 a_3 - a_1 a_2)x + (a_1 a_3 - a_2^2) = 0$

is called the Hessian of the cubic equation $a_0 x^3 + 3a_1 x^2 + 3a_2 x + a_3 = 0$.

This Hessian of cubic can also be obtained as follows.

First making the given cubic homogenous by introducing a new variable t as

$$f(x,t) \equiv a_0 x^3 + 3a_1 x^2 t + 3a_2 x t^2 + a_3 t^3 = 0 \quad \ldots(1)$$

Now differentiating (1) partially w.r.t. x and t and respectively, we get

$$\frac{\partial f}{\partial x} = 3a_0 x^2 + 6a_1 xt + 3a_2 t^2 \quad \ldots(2)$$

$$\frac{\partial f}{\partial t} = 3a_1 x^2 + 6a_2 xt + 3a_3 t^2 \quad \ldots(3)$$

Again differentiating (2) w.r.t. x and (3) w.r.t. t, we get

$$\frac{\partial^2 f}{\partial x^2} = 6a_0 x + 6a_1 t \quad \ldots(4)$$

$$\frac{\partial^2 f}{\partial t^2} = 6a_2 x + 6a_3 t \quad \ldots(5)$$

And differentiating (3) w.r.t. x, we get

$$\frac{\partial^2 f}{\partial x \partial t} = 6a_1 x + 6a_2 t \qquad \ldots(6)$$

Now putting $t = 1$ and observed that the expression

$$\left(\frac{\partial^2 f}{\partial x \partial t}\right)^2 = \left(\frac{\partial^2 f}{\partial x^2}\right)\cdot\left(\frac{\partial^2 f}{\partial t^2}\right) \qquad \ldots(7)$$

gives the Hessian of the cubic. Hence the Hessian of the cubic can also be obtained by (7).

REMARK

- Hessian of the cubic can also be expressed in the form of a determinant as follows :

$$\begin{vmatrix} 1 & -x & x^2 \\ a_0 & a_1 & a_2 \\ a_1 & a_2 & a_3 \end{vmatrix} = 0$$

SOLVED EXAMPLES

Example 1. *Solve the equation $x^3 - 15x - 126 = 0$ by Cardan's method.* (SVTU–2009)

Solution. The given equation is

$$x^3 - 15x - 126 = 0 \qquad \ldots(1)$$

Let the solution of (1) be

$$x = u + v$$

Cubing (2), we get

$$x^3 = (u+v)^3 = u^3 + v^3 + 3uv(u+v)$$

or $x^3 = u^3 + v^3 + 3uv(x)$ $\quad [\because x = u + v]$

or $x^3 - 3uvx - (u^3 + v^3) = 0$

The equations (1) and (3) are same so comparing the coefficients of like terms, we get

$$3uv = 15$$

or $uv = \frac{15}{3}$ or $u^3v^3 = 125$

and $u^3 + v^3 = 126$

Hence u^3, v^3 are the roots of the quadratic

$$t^2 - 126t + 125 = 0$$

$\therefore$ $(t-125)(t-1) = 0, t = 125, t = 1$

$\therefore$ $u^3 = 125, v^3 = 1$ or $u = 5, v = 1$

Thus the roots of (1) are given by

$$u + v,\ u\omega + v\omega^2,\ u\omega^2 + v\omega$$

where $\omega = -\frac{1}{2} + \frac{i\sqrt{3}}{2}$

$\therefore$ $u + v = 5 + 1 = 6$

$$u\omega + v\omega^2 = 5\omega + \omega^2 = 4\omega + \omega + \omega^2$$

$$= 4\omega - 1 \qquad (\because 1 + \omega + \omega^2 = 0)$$

$$= 4\left(-\frac{1}{2} + \frac{i\sqrt{3}}{2}\right) - 1 = -3 + i2\sqrt{3}$$

and

$$u\omega^2 + v\omega = 5\omega^2 + \omega = 4\omega^2 + \omega^2 + \omega = 4\omega^2 - 1$$

$$= 4\left(-\frac{1}{2} + \frac{i\sqrt{3}}{2}\right)^2 - 1 = -3 - i2\sqrt{3}$$

Hence, roots are $6, -3 + 2i\sqrt{3}, -3 - 2i\sqrt{3}$.

Example 2. *Solve the equation $x^3 - 15x^2 - 33x + 847 = 0$ by Cardan's method.*

Solution. The given equation is

$$x^3 - 15x^2 - 33x + 847 = 0 \qquad \ldots(1)$$

First we remove the second term, *i.e.*, $-15x^2$ by diminishing each of its roots by the constant.

$$h = -\frac{a_1}{na_0} = -\frac{-15}{3 \times 1} = 5$$

Now using synthetic division method as follows:

5	1	– 15	– 33	847
		5	– 50	– 418
	1	– 10	– 83	432
		+5	– 25	
	1	– 5	– 108	
		5		
	1	0		
	1			

Thus the transformed equation (without second degree term) is

$$z^3 - 108z + 432 = 0 \qquad \ldots(2)$$

where $z = x - 5$

Let the solution of (2) be

$$z = u + v \qquad \ldots(3)$$

Cubing (3) of both sides, we get

$$z^3 - 3uvz - (u^3 + v^3) = 0 \qquad \ldots(4)$$

The equation (2) and (4) are same so we have

$$uv = 36,\ u^3 + v^3 = -432 \text{ or } u^3v^3 = (36)^3$$

$\therefore$ u^3, v^3 are the roots of the equation

$$t^2 + 432t + (36)^3 = 0$$

$\therefore \quad t = \dfrac{-432 \pm \sqrt{(432)^2 - 4(36)^3}}{2}$

$= -\dfrac{432}{2} = -216$

$\therefore\ u^3 = -216, v^3 = -216$

$\therefore \quad u = (-216)^{1/3} = -6,\ v = (-216)^{1/3} = -6$

The roots of (2) are given by

$u + v,\ u\omega + v\omega^2,\ u\omega^2 + v\omega$,

i.e., $z_1 = u + v = -6 - 6 = -12$

$z_2 = -6\omega - 6\omega^2 = -6(\omega + \omega^2) = -6(-1) = 6$

$z_3 = -6\omega^2 - 6\omega = -6(\omega^2 + \omega) = -6(-1) = 6$

Therefore the roots of given equation (1) are

$x_1 = z_1 + 5 = -12 + 5 = -7$

$x_2 = z_2 + 5 = 6 + 5 = 11$

$x_3 = z_3 + 5 = 6 + 5 = 11$

Hence, the roots of the given cubic equation are −7, 11, 11.

Example 3. *Show that the roots of the equation* $x^3 - 3x + 1 = 0$ *are* $2\cos\dfrac{2\pi}{9}, 2\cos\dfrac{8\pi}{9}, \cos\dfrac{14\pi}{9}$.

Solution. The given equation is

$$x^3 - 3x + 1 = 0 \quad \ldots(1)$$

Let $\quad x = u + v \quad \ldots(2)$

Cubing (2) of both sides, we get

$$x^3 - 3uvx - (u^3 + v^3) = 0$$

Since (1) and (3) are same so we have

$uv = 1,\ u^3 + v^3 = -1$

or $\quad u^3v^3 = 1,\ u^3 + v^3 = -1$

$\therefore u^3, v^3$ are the roots of the equation $t^2 + t + 1 = 0$

$\therefore \quad t = \dfrac{-1 \pm \sqrt{1-4}}{2} \Rightarrow \quad t = \dfrac{-1 \pm i\sqrt{3}}{2}$

$\therefore \quad u^3 = -\dfrac{1}{2} + \dfrac{i}{2}\sqrt{3},\ v^3 = -\dfrac{1}{2} - \dfrac{i}{2}\sqrt{3}$

From (2) we get

$$x = \left(-\frac{1}{2} + \frac{i}{2}\sqrt{3}\right)^{1/3} + \left(-\frac{1}{2} - \frac{i}{2}\sqrt{3}\right)^{1/3} \quad \ldots(4)$$

Change the complex number of R.H.S of (4) into polar form by putting

$-\dfrac{1}{2} = r\cos\theta,\ \dfrac{\sqrt{3}}{2} = r\sin\theta$

$\therefore \quad r^2 = 1 \quad \Rightarrow r = 1$

and $\tan\theta = -\sqrt{3} \Rightarrow \quad \theta = \dfrac{2\pi}{3}$

$\therefore \quad x = (r\cos\theta + ir\sin\theta)^{1/3} + (r\cos\theta - ir\sin\theta)^{1/3}$

$= r^{1/3}\left[\cos\dfrac{2n\pi + \theta}{3} + i\sin\dfrac{2n\pi + \theta}{3} + \cos\dfrac{2n\pi + \theta}{3} - i\sin\dfrac{2n\pi + \theta}{3}\right]$

$= 2r^{1/3}\cos\dfrac{2n\pi + \theta}{3},\ n = 0, 1, 2$

Therefore, $x_1 = 2(1)^{1/3}\cos\dfrac{\theta}{3} = 2\cos\dfrac{2\pi}{9}$

$x_2 = 2\cos\dfrac{2\pi + \theta}{3} = 2\cos\left(\dfrac{2\pi}{3} + \dfrac{2\pi}{9}\right)$

$= 2\cos\dfrac{8\pi}{9}$

and $\quad x_3 = 2\cos\dfrac{4\pi + \theta}{3} = 2\cos\left(\dfrac{4\pi}{3} + \dfrac{2\pi}{9}\right)$

$= 2\cos\dfrac{14\pi}{9}$.

Example 4. *Solve the equation* $28x^3 - 9x^2 + 1 = 0$ *by Cardan's method.*

Solution. The given equation is

$$28x^3 - 9x^2 + 1 = 0 \quad \ldots(1)$$

First remove the second terms, i.e., $-9x^2$ by putting $x = \dfrac{1}{z}$, we get

$\dfrac{28}{z^3} - \dfrac{9}{z^2} + 1 = 0$ or $28 - 9z + z^3 = 0$

or $\quad z^3 - 9z + 28 = 0$ where $z = \dfrac{1}{x} \quad \ldots(2)$

Let the solution of (2) be

$$z = u + v \quad \ldots(3)$$

Cubing (3), we get

$$z^3 - 3uvz - (u^3 + v^3) = 0 \quad \ldots(4)$$

The equation (2) and (4) are same so, we have

$uv = 3,\ u^3 + v^3 = -28$

or $\quad u^3v^3 = 27,\ u^3 + v^3 = -28$

$\therefore \quad u^3, v^3$ are the root of the equation

$t^2 + 28t + 27 = 0$

or $\quad (t + 27)(t + 1) = 0$

$t = -27, t = -1$

$\therefore \quad u^3 = (-27),\ v^3 = -1$

or $\quad u = (-27)^{1/3} = -3,\ v = (-1)^{1/3} = -1$

Thus the roots of (2) are

$z_1 = u + v = -3 - 1 = -4$

$z_2 = u\omega + v\omega^2 = -3\omega - \omega^2$

$= -2\omega - (\omega + \omega^2) = -2\omega + 1$

$[\because 1 + \omega + \omega^2 = 0]$

$= -2\left(-\dfrac{1}{2} + \dfrac{i}{2}\sqrt{3}\right) + 1 = 2 - i\sqrt{3}$

$$z_3 = u\omega^2 + v\omega = -3\omega^2 - \omega = 2 + i\sqrt{3}$$

Hence, the roots the given cubic equation are

$$x_1 = \frac{1}{z_1} = \frac{1}{-4} = -\frac{1}{4}$$

$$x_2 = \frac{1}{z_2} = \frac{1}{2-i\sqrt{3}} = \frac{2+i\sqrt{3}}{4+3} = \frac{2}{7} + \frac{i}{7}\sqrt{3}$$

$$x_3 = \frac{1}{z_3} = \frac{1}{2+i\sqrt{3}} = \frac{2-i\sqrt{3}}{4+3} = \frac{2}{7} - \frac{i}{7}\sqrt{3}$$

Example 5. *Solve the cubic $x^3 - 3a^2x - 2a^3\cos 3A = 0$ by Cardan's method.*

Solution. Since the given cubic is

$$x^3 - 3a^2x - 2a^3\cos 3A = 0 \qquad \ldots(1)$$

Let the solution of (1) be

$$x = u + v \qquad \ldots(2)$$

Cubing of both sides of (2), we get

$$x^3 - 3uvx - (u^3 + v^3) = 0 \qquad \ldots(3)$$

The equation (1) and (3) are same so we have

$$uv = a^2,\ u^3 + v^3 = 2a^3\cos 3A$$

or $u^3v^3 = a^6,\ u^3 + v^3 = 2a^3\cos 3A$

$\therefore u^3, v^3$ are the roots of the following equation

$$t^3 - 2a^3\cos 3A + a^6 = 0 \qquad \ldots(4)$$

$$\therefore \quad t = \frac{2a^3\cos 3A \pm \sqrt{(4a^6\cos^2 3A - 4a^6)}}{2}$$

$$= \frac{2a^3\cos 3A \pm 2a^3 i\sin 3A}{2}$$

$$t = a^3(\cos 3A \pm i\sin 3A)$$

$$\therefore u^3 = a^3(\cos 3A + i\sin 3A),$$

$$v^3 = a^3(\cos 3A - i\sin 3A)$$

or $u = a(\cos 3A + i\sin 3A)^{1/3}$,

$$v = a(\cos 3A - i\sin 3A)^{1/3}$$

Substitute these values of u and v in (2), we get

$$x = a(\cos 3A + i\sin 3A)^{1/3} + a(\cos 3A - i\sin 3A)^{1/3}$$

$$= a\left[\cos\left(\frac{2n\pi + 3A}{3}\right) + i\sin\left(\frac{2n\pi + 3A}{3}\right) + \cos\left(\frac{2n\pi + 3A}{3}\right) - i\sin\left(\frac{2n\pi + 3A}{3}\right)\right]$$

$$= 2a\cos\left(\frac{2n\pi + 3A}{3}\right), \text{ where } n = 0, 1, 2$$

$$\therefore x_1 = 2a\cos A \quad \text{when } n = 0$$

$$x_2 = 2a\cos\left(\frac{2\pi + 3A}{3}\right), \quad \text{when } n = 1$$

$$= 2a\cos\left(\frac{2\pi}{3} + 1\right)$$

and $x_3 = 2a\cos\left(\frac{4\pi + 3A}{3}\right)$, when $n = 2$

$$= 2a\cos\left(\frac{4\pi}{3} + A\right) = 2a\cos\left\{2\pi - \left(\frac{2\pi}{3} - A\right)\right\}$$

$$= 2a\cos\left(\frac{2\pi}{3} - A\right)$$

Hence, the solution of the given cubic is given by

$$2a\cos A,\ 2a\cos\left(\frac{2\pi}{3} \pm A\right).$$

Example 6. *Solve the equation $9x^3 - 30x^2 + 36x - 16 = 0$ by expressing it as the sum or difference of two cubes.*

Solution. The given cubic is

$$9x^3 - 30x^2 + 36x - 16 = 0 \qquad \ldots(1)$$

Compare this equation with the equation

$$a_0x^3 + 3a_1x^2 + 3a_2x + a_3 = 0$$

we get $a_0 = 9,\ a_1 = -10,\ a_2 = 12,\ a_3 = -16$

Let the given cubic (1) can be expressed as

$$9x^3 - 30x^2 + 36x - 16$$

$$\equiv A(x-a)^3 + B(x-b)^3 = 0 \qquad \ldots(2)$$

Then a, b are the roots of the following equation

$$(a_1x + a_2)^2 = (a_0x + a_1)(a_2x + a_3)$$

or $(a_0a_2 - a_1^2)x^2 + (a_0a_3 - a_1a_2)x + (a_1a_3 - a_2^2) = 0 \qquad \ldots(3)$

Substitute the values of a_0, a_1, a_2, a_3 in (3), we get

$$(108 - 100)x^2 + (-144 + 120)x + (160 - 144) = 0$$

or $8x^2 - 24x + 16 = 0$

or $x^2 - 3x + 2 = 0$

or $(x-1)(x-2) = 0$

or $x = 1, 2$

$\therefore$ $a = 1, b = 2$

Now substitute these values of a and b in (2) and equation the coefficients of like powers of x of both sides, we get

$$9x^3 - 30x^2 + 36x - 16$$

$$\equiv A(x-1)^3 + B(x-2)^3 = 0 \qquad \ldots(4)$$

Taking the coefficients of x^3 and x^2, we get

$$A + B = 9$$

$$A + 2B = 10$$

Solving these two equation, we get $A = 8, B = 1$. Substitute the values of A and B in (4), we get the given cubic as

$$8(x-1) + (x-2)^3 = 0 \text{ or } \left(\frac{x-1}{x-2}\right)^3 = -\frac{1}{8}$$

or $\frac{x-1}{x-2} = \left(-\frac{1}{8}\right)^{1/3}$

or $\frac{x-1}{x-2} = \left(-\frac{1}{2}\right)^{1/3} (1)^{1/3} = \left(-\frac{1}{2}\right)k$

where $k = 1, \omega, \omega^2$ and $\omega = \left(\frac{-1}{2} + \frac{i}{2}\sqrt{3}\right)$

or $2(x-1) = -k(x-2)$

or $2x - 2 = -kx + 2k$

or $x(2+k) = 2k+2$

or $x = \frac{2k+2}{2+k} = \frac{2(k+1)}{(2+k)}$

when $k = 1$, $x_1 = \frac{2(1+1)}{2+1} = \frac{4}{3}$ when $k = \omega$

$$x_2 = \frac{2(\omega+1)}{(2+\omega)} = \frac{2(-\omega^2)}{(1-\omega^2)}$$

$(1+\omega+\omega^2 = 0$ and $\omega^3 = 1)$

$$= \frac{2(-\omega^3)}{(\omega-\omega^3)} = -\frac{2}{\omega-1} = \frac{2}{1-\omega}$$

$$= \frac{2}{1-\left(-\frac{1}{2}+\frac{i}{2}\sqrt{3}\right)} = \frac{2}{\frac{3}{2}-\frac{i}{2}\sqrt{3}}$$

$$= \frac{4}{3-i\sqrt{3}} = \frac{4(3+i\sqrt{3})}{9+3} = \frac{3+i\sqrt{3}}{3}$$

when $k = \omega^2$

$$x_3 = \frac{2(\omega^2+1)}{(2+\omega^2)} = \frac{2(\omega^3+\omega)}{(2\omega+\omega^3)} = \frac{2(1+\omega)}{2\omega+1}$$

$(\because \omega^3 = 1)$

$$= \frac{2\left(1-\frac{1}{2}+\frac{i}{2}\sqrt{3}\right)}{2\left(-\frac{1}{2}+\frac{i}{2}\sqrt{3}\right)+1}$$

$$\left[\because \omega = \left(-\frac{1}{2}+\frac{i}{2}\sqrt{3}\right)\right]$$

$$= \frac{1+i\sqrt{3}}{i\sqrt{3}} = \frac{i\sqrt{3}-3}{-3} = \frac{3-i\sqrt{3}}{3}$$

Hence, the roots of the given cubic are $\frac{4}{3}, \frac{3 \pm i\sqrt{3}}{3}$.

EXERCISE 24.6

Solve the following cubic equations by Cardan's method.

1. $x^3 + 6x^2 + 9x + 4 = 0$
2. $x^3 + 6x^2 - 12x + 32 = 0$
3. $x^3 - 21x - 344 = 0$
4. $x^3 - 12x^2 - 6x - 10 = 0$
5. $27x^3 + 54x^2 + 198x - 73 = 0$
6. $x^3 - 18x - 35 = 0$ (Osmania–2003)
7. $9x^3 + 6x^2 - 1 = 0$ (SVTU–2008)
8. $x^3 - 15x^2 - 357x + 5491 = 0$
9. $x^3 + 3x^2 - 27x + 104 = 0$
10. $x^3 - 6x - 9 = 0$
11. $2x^3 + 3x^2 + 3x + 1 = 0$
12. $8a^3x^3 - 6ax + 2\sin 3A = 0$
13. $64x^3 - 144x^2 + 108x - 27 = 0$
14. $x^3 + 3ax^2 + 3(a^2 - bc)x + a^3 + b^3 + c^3 - 3abc = 0$

Solve the following equations by expressing them as the sum or difference of two cubes.

15. $x^3 - 3x^2 + 33x - 1 = 0$
16. $2x^3 + 3x^2 - 21x + 19 = 0$
17. $x^3 + 3x^2 - 27x + 104 = 0$
18. $152x^3 - 60x^2 - 606x - 485 = 0$

Hint to Selected Problems

1. $h = -\frac{a_0}{na_0} = -\frac{6}{3\times 1} = -2$

The transformed equation is $z^3 - 3z + 2 = 0$

where $z = x + 2$

Assume that the solution of transformed equation is $z = u + v$

11. $h = -\frac{1}{2}$ Transformed equation is $2z^3 + \frac{3}{2}z = 0$

13. $h = \frac{3}{4}$ Transformed equation is $z^3 = 0$

14. $h = -a$ Transformed equation is $z^3 - 3bcz + b^3 + c^3 = 0$

18. The given equation can be expressed as

$$152x^3 - 60x^2 - 606x - 485 = A(x-a)^3 + B(x-b)^3$$

Then a, b are the roots of the following equation

$$(a_0a_1 - a_1^2)x^2 + (a_0a_3 - a_1a_2)x + (a_1x_3 - a_2^2) = 0$$

ANSWERS

1. $-4, -1, -1$ **2.** $-8, (1 \pm i\sqrt{3})$ **3.** $8, (-4 \pm i3\sqrt{3})$

4. $4 + 3(2)^{1/3} + 3(4)^{1/3}, 4 + 3\omega(2)^{1/3} + 3\omega^2(4)^{1/3}, 4 + 3\omega^2(2)^{1/3} + 3\omega(4)^{1/3}$ where $\omega = \left(-\frac{1}{2} \pm \frac{i}{2}\sqrt{3}\right)$ **5.** $\frac{1}{3}, \left(-\frac{7}{6} \pm \frac{3\sqrt{3}}{2}i\right)$

6. $5, \left(-\frac{5}{2} \pm \frac{i\sqrt{3}}{2}\right)$ **7.** $\frac{1}{3}, \frac{-3 \pm i\sqrt{3}}{6}$ **8.** $-19, 17, 17$ **9.** $-8, \frac{1}{2}, (5 \pm i3\sqrt{3})$ **10.** $3, \left(-\frac{3}{2} \pm \frac{i\sqrt{3}}{2}\right)$

11. $-\frac{1}{2}, \left(-\frac{1}{2} \pm \frac{i\sqrt{3}}{2}\right)$ **12.** $\frac{1}{a}\sin A, \frac{1}{a}\sin\left(\frac{\pi}{3} - A\right), -\frac{1}{a}\sin\left(\frac{\pi}{3} + A\right)$ **13.** $\frac{3}{4}, \frac{3}{4}, \frac{3}{4}$

14. $-(a+b+c), -(a+b\omega+c\omega^2), -(a+b\omega^2+c\omega)$ **15.** $\frac{7-7k(2)^{1/3}(5)^{2/3}+7k^2(2)^{2/3}(5)^{1/3}}{7}, k = 1, \omega, \omega^2$

16. $\frac{-1-3^{1/3}(5)^{2/3}\alpha-(3)^{2/3}(5)^{1/3}\alpha}{2}, \alpha = 1, \omega, \omega^2$ **17.** $-8, \frac{1}{2}(5 \pm i3\sqrt{3})$ **18.** $\frac{5}{2}, \frac{-40 \pm i7\sqrt{7}}{38}$

24.25 REDUCTION OF BIQUADRATIC EQUATION INTO EULER'S CUBIC AND REDUCING CUBIC

(i) **Reduction into Euler's Cubic.** Let the biquadratic equation be

$$a_0x^4 + 4a_1x^3 + 6a_2x^2 + 4a_3x + a_4 = 0 \quad \ldots(1)$$

First we remove the second term, *i.e.*, $4a_1x^3$ by diminishing each of its roots by a constant $h = -\frac{a_1}{4a_0}$, we get

$$z^4 + 6Hz^2 - 4Gz + (a_0^2I - 3H^2) = 0 \quad \ldots(2)$$

where $z = a_0x + a_1$, $H = a_0a_2 - a_1^2$, $G = a_0^2a_3 - 3a_0a_1a_2 + 2a_1^3$ and $I = a_0a_4 - 4a_1a_3 + 3a_2^2$

Let the solution of (2) be $z = \sqrt{a} + \sqrt{b} + \sqrt{c}$. ...(3)

Squaring of both sides of (3), we get

$$z^2 = a+b+c+2(\sqrt{a}\sqrt{b}+\sqrt{b}\sqrt{c}+\sqrt{c}\sqrt{a}) \quad \text{or} \quad z^2-(a+b+c) = 2(\sqrt{a}\sqrt{b}+\sqrt{b}\sqrt{c}+\sqrt{c}\sqrt{a})$$

Again squaring of both sides, we get

$$[z^2-(a+b+c)^2] = 4[ab+bc+ca+2\sqrt{a}\sqrt{b}\sqrt{c}+(\sqrt{a}+\sqrt{b}+\sqrt{c})]$$

or $$z^4+(a+b+c)^2-2(a+b+c)z^2 = 4(ab+bc+ca)+8\sqrt{a}\sqrt{b}\sqrt{c}z)$$

or $$z^4-2(a+b+c)z^2-8z\sqrt{a}\sqrt{b}\sqrt{c}+(a+b+c)^2-4(ab+bc+ca)^2 = 0 \quad \ldots(4)$$

The equation (2) and (4) are same, so comparing the coefficients of like powers of z, we get

$$(a+b+c) = -3H, \sqrt{a}\sqrt{b}\sqrt{c} = -\frac{G}{2} \quad \text{and} \quad (a+b+c)^2-4(ab+bc+ca) = a_0^2I-3H^2$$

or $$\sum a = -3H, abc = \frac{G^2}{4} \quad \text{and} \quad \left(\sum a\right)^2 - 4\left(\sum ab\right) = a_0^2I - 3H^2$$

or $$\Sigma ab = 3H^2 - \frac{a_0^2I}{4}$$

Therefore a, b, c are the roots of the equation

$$t^2 - \left(\sum a\right)t^2 + \left(\sum ab\right)t - abc = 0 \quad \text{or} \quad t^3+3Ht^2+\left(3H^2-\frac{a_0^2I}{4}\right)t-\frac{G^2}{4} = 0 \quad \ldots(5)$$

This equation is called Euler's cubic of biquadratic equation (1).

(ii) **Reducing cubic.** We have a relation $G^2+4H^3 = a_0^2(HI-a_0J)$ where $J = a_0a_2a_3+2a_1a_2a_3-a_0a_3^2-a_1^2a_4-a_2^3$

$$\therefore \quad \frac{G^2}{4}+H^3 = \frac{a_0^2HI}{4}-\frac{a_0^3J}{4} \quad \text{or} \quad -\frac{G^2}{4} = H^3-\frac{a_0^2HI}{4}+\frac{a_0^3J}{4}$$

Substitute the value $-\frac{G^2}{4}$ in (5), we get

$$t^3+3Ht^2+\left(3H^2-\frac{a_0^2t}{4}\right)t+H^3-\frac{a_0^2HI}{4}+\frac{a_0^3J}{4} = 0$$

or $$t^3+H^3+3H^2t+3Ht^2-\frac{a_0^2I}{4}t-\frac{a_0^2HI}{4}+\frac{a_0^3J}{4} = 0$$

or $$(t+H)^3-\frac{a_0^2I}{4}(t+H)+\frac{a_0^3J}{4} = 0$$

Let us put $t + H = a_0^2\theta$, we get

$$a_0^6\theta^3 - \frac{a_0^2 I}{4}(a_0^2\theta) + \frac{a_0^3 J}{4} = 0 \qquad \text{or} \qquad 4a_0^3\theta^3 - Ia_0\theta + J \qquad \text{...(6)}$$

This cubic equation of θ is called the reducing cubic of the biquadratic equation (1).

24.26 RELATION BETWEEN THE ROOTS OF BIQUADRATIC AND EULER'S CUBIC

The equation of the biquadratic and Euler's cubic are respectively given by

$$a_0x^4 + 4a_1x^3 + 6a_2x^2 + 4a_3x + a_4 = 0 \qquad \text{...(1)}$$

and
$$t^3 + 3Ht^2 + \left(3H^2 - \frac{a_0^2 I}{4}\right)t - \frac{G^2}{4} = 0 \qquad \text{...(2)}$$

Let α, β, γ, δ be the roots of the biquadratic and a, b, c the roots of the Euler's cubic.

Since we have taken
$$z = \sqrt{a} + \sqrt{b} + \sqrt{c} \qquad \text{...(3)}$$

and
$$z = a_0x + a_1 \qquad \text{...(4)}$$

It has been observed form equation (2) that z will have eight values because $\sqrt{a}, \sqrt{b}$ and $\sqrt{c}$ will have double signs. But we have

$$\sqrt{a}\sqrt{b}\sqrt{c} = -\frac{G}{2} \quad \text{therefore } \sqrt{c} = \frac{G}{2\sqrt{a}\sqrt{b}}$$

$\therefore$ from (2), we get
$$z = \sqrt{a} + \sqrt{b} - \frac{G}{2\sqrt{a}\sqrt{b}} \qquad \text{...(5)}$$

Now from this equation it is observed that z will have four values. The signs of $\sqrt{a}, \sqrt{b}$ and $\sqrt{c}$ are taken in such a way that equation (4) should be satisfied. Therefore it G is negative, then $-\frac{G}{2}$ will be positive. Thus we should take the signs of $\sqrt{a}, \sqrt{b}$ and $\sqrt{c}$ so as to make the quantity $\sqrt{a}, \sqrt{b}, \sqrt{c}$ positive. Let z_1, z_2, z_3 and z_4 be the four values of z as the equation (5) indicates.

$$\therefore \quad \left.\begin{aligned} z_1 &= a_0\alpha + a_1 = \sqrt{a} - \sqrt{b} - \sqrt{c} && \text{(Using (2) and (3) and } \sqrt{a} \text{ is taken positive.)} \\ z_2 &= a_0\beta + a_1 = -\sqrt{a} + \sqrt{b} - \sqrt{c} && (\sqrt{b} \text{ taken positive.}) \\ z_3 &= a_0\gamma + a_1 = -\sqrt{a} - \sqrt{b} + \sqrt{c} && (\sqrt{c} \text{ is taken positive.}) \\ z_4 &= a_0\delta + a_1 = \sqrt{a} + \sqrt{b} + \sqrt{c} && \text{(all are taken positive.)} \end{aligned}\right\} \qquad \text{...(A)}$$

Adding first two equations and adding last two equations of above system of equations respectively, we get

$$a_0(\alpha + \beta) + 2a_1 = -2\sqrt{c} \qquad \text{and} \qquad a_0(\gamma + \delta) + 2a_1 = 2\sqrt{c}$$

Now substracting these equations, we get

$$a_0(\alpha + \beta - \gamma - \delta) = -4\sqrt{c}$$

Squaring of both sides, we get

$$c = \frac{a_0^2}{16}(\alpha + \beta - \gamma - \delta)^2$$

Similarly we get
$$a = \frac{a_0^2}{16}(\beta + \gamma - \alpha - \delta)^2 \qquad \text{and} \qquad b = \frac{a_0^2}{16}(\gamma + \alpha - \beta - \delta)^2$$

Hence, we obtained the required relations as follows :

$$a = \frac{a_0^2}{16}(\beta + \gamma - \alpha - \delta)^2, b = \frac{a_0^2}{16}(\gamma + \alpha - \beta - \delta)^2, c = \frac{a_0^2}{16}(\alpha + \beta - \gamma - \delta)^2$$

It G is a positive, then $-\frac{G}{2}$ will be negative. Therefore we will take the signs of $\sqrt{a}, \sqrt{b}, \sqrt{c}$ either all negative or two positive and one negative. Thus we shall obtain

$$\left.\begin{aligned} z_1 &= a_0\alpha + a_1 = -\sqrt{a} + \sqrt{b} + \sqrt{c} \\ z_2 &= a_0\beta + a_1 = \sqrt{a} - \sqrt{b} + \sqrt{c} \\ z_3 &= a_0\gamma + a_1 = \sqrt{a} + \sqrt{b} - \sqrt{c} \\ \text{and} \quad z_4 &= a_0\delta + a_1 = -\sqrt{a} - \sqrt{b} - \sqrt{c} \end{aligned}\right\} \qquad \text{...(B)}$$

Solving these equation's we obtain the relations as the same as obtained above.

24.27 RELATION BETWEEN THE ROOTS OF BIQUADRATIC AND THE REDUCING CUBIC

The equations of biquadratic and reducing cubic are respectively given by

$$a_0x^4 + 4a_1x^3 + 6a_2x^2 + 4a_3x + a_4 = 0 \qquad ...(1)$$

and

$$4a_0^3\theta^3 - Ia_0\theta + J = 0 \qquad ...(2)$$

Let α, β, γ, δ be the roots of (1) and θ_1, θ_2, θ_3 be the roots of (2) in the system of equations (A) of §24.26, we have

$$a_0(\alpha-\beta) = 2(\sqrt{a}-\sqrt{b}) \quad \text{and} \quad a_0(\gamma-\beta) = -2(\sqrt{a}+\sqrt{b})$$

$$\therefore \quad a_0^2(\alpha-\beta)(\gamma-\delta) = 4(a-b)$$

But we have $\quad t + H = a_0^2\theta$

$$\therefore \quad a + H = a_0^2\theta_1, b + H = a_0^2\theta_2, c + H = a_0^2\theta_3$$

$$\therefore \quad (a-b) = a_0^2(\theta_1-\theta_2)$$

Similarly

$$\left.\begin{aligned} 4(a-b) &= 4a_0^2(\theta_1-\theta_2) = -a_0^2(\alpha-\beta)(\gamma-\delta) \\ 4(b-c) &= 4a_0^2(\theta_2-\theta_3) = -a_0^2(\beta-\gamma)(\alpha-\delta) \\ 4(c-a) &= 4a_0^2(\theta_3-\theta_1) = -a_0^2(\gamma-\alpha)(\beta-\delta) \end{aligned}\right\} \qquad ...(A)$$

Substracting first equation of (A) from third equation of (A), we get

$$4a_0^2(\theta_3-\theta_1-\theta_1+\theta_2) = a_0^2[(\alpha-\beta)(\gamma-\delta)-(\gamma-\alpha)(\beta-\delta)]$$

$$4a_0^2(\theta_2+\theta_3-2\theta_1) = a_0^2[(\alpha-\beta)(\gamma-\delta)-(\gamma-\alpha)(\beta-\delta)]$$

But we have $\quad \theta_1+\theta_2+\theta_3 = 0, \theta_2+\theta_3 = -\theta_1$

$$\therefore \quad 4a_0^2(-3\theta_1) = a_0^2[(\alpha-\beta)(\gamma-\delta)-(\gamma-\alpha)(\beta-\delta)]$$

or $\quad 12\theta_1 = (\gamma-\alpha)(\beta-\delta)-(\alpha-\beta)(\gamma-\delta)$

Similarly $\quad 12\theta_2 = (\alpha-\beta)(\gamma-\delta)-(\beta-\gamma)(\alpha-\delta), 12\theta_3 = (\beta-\gamma)(\alpha-\delta)-(\gamma-\alpha)(\beta-\delta)$

Hence, the required relations are

$$\theta_1 = \frac{1}{12}[(\gamma-\alpha)(\beta-\delta)-(\alpha-\beta)(\gamma-\delta)], \theta_2 = \frac{1}{12}[(\alpha-\beta)(\gamma-\delta)-(\beta-\gamma)(\alpha-\delta)]$$

$$\theta_3 = \frac{1}{12}[(\beta-\gamma)(\alpha-\delta)-(\gamma-\alpha)(\beta-\delta)]$$

24.28 DESCARTE'S METHOD FOR FINDING THE ROOTS OF A BIQUADRATIC

Let the equation of a biquadratic be

$$a_0x^4 + 4a_1x^3 + 6a_2x^2 + 4a_3x + a_4 = 0 \qquad ...(1)$$

First we remove the second term, *i.e.*, $4a_1x^3$ from (1) be diminishing each of root of (1) by a constant $h = -\frac{a_1}{na_0}$, we get

$$z^4 + 6Hz^2 + 4Gz + a_0^2I - 3H^2 = 0 \qquad ...(2)$$

where $H = a_0a_2 - a_1^2, G = a_0^2a_3 - 3a_0a_1a_2 + 2a_1^3, I = a_0a_4 - 4a_1a_3 + 3a_2^2$ and $z = a_0x + a_1$.

Let us assume

$$z^4 + 6Hz^2 + 4Gz + a_0^2I - 3H^2 \equiv (z^2 + kz + l)(z^2 - kz + m)$$

Now equating the coefficients of like powers of z, we get

$$l + m - k^2 = 6H, k(m-1) = 4G, lm = a_0^2I - 3H^2$$

Solving first two of these equations for l and m, we get

$$\left. 2l = k^2 + 6H - \frac{4G}{k} \text{ and } 2m = k^2 + 6H + \frac{4G}{k}\right\} \qquad ...(A)$$

Substitute these values of l, m in the following equation $lm = a_0^2I - 3H^2$, we get

$$\left(k^2 + 6H - \frac{4G}{k}\right)\left(k^2 + 6H + \frac{4G}{k}\right) = 4(a_0^2I - 3H^2)$$

$$\Rightarrow \quad (k^3 + 6Hk - 4G)(k^3 + 6Hk + 4G) = 4(a_0^2I - 3H^2)k^2$$

$$k^6 + 12Hk^4 + 4k^2(12H^2 - a_0^2I) - 16G^2 = 0 \qquad ...(3)$$

This is cubic equation in k^2 so it will always have one positive real value of k^2 when k^2 is know, then the values of l and m are obtained from the equation (A). Thus the biquadratic (2) is obtained as the product of quadratics $(z^2 + kz + 1)$ and $(z^2 - kz + m)$.

Now solving these two quadratics

$$(z^2 + kz + l) \text{ and } (z^2 - kz + m) = 0$$

and finally from the transformation $z = a_0x + a_1$ we obtain the solution of the given biquadratic (1) corresponding to the roots of the equations.

$$z^2 + kz + l = 0 \quad \text{and} \quad z^2 - kz + m = 0$$

24.29 FERRARI'S METHOD FOR FINDING THE ROOTS OF A BIQUADRATIC EQUATION

Let the equation of a biquadratic be

$$x^4 + 2a_1x^3 + a_2x^2 + 2a_3x + a_4 = 0 \qquad ...(1)$$

Now adding $(ax + b)^2$ to each side of (1), we get

$$x^4 + 2a_1x^3 + a_2x^2 + 2a_3x + a_4 + (ax + b)^2 = (ax + b)^2$$

or $$x^4 + 2a_1x^3 + (a_2 + a^2)x^2 + 2(a_3 + ab)x + (a_4 + b)^2 = (ax + b)^2 \qquad ...(2)$$

In order to determine a and b make the left side of above equation of perfect square.

Suppose the perfect square of left side of (2) is $(x^2 + a_1x + k)^2$, then

$$x^4 + 2a_1x^3 + (a_2 + a^2)x^2 + 2(a_3 + ab)x + (a_4 + b)^2 \equiv (x^2 + a_1x + k)^2 \qquad ...(3)$$

Comparing the coefficients of like powers of x of (3), we get

$$a_1^2 + 2k = a_2 + a^2,\ a_1k = a_3 + ab,\ k^2 = a_4 + b^2.$$

Eliminating a and b between these equations, we get

$$(2k + a_1^2 - a_2)(k^2 - a_4) = (a_1k - a_3)^2 \quad \text{or} \quad 2k^3 - a_2k^2 + 2(a_1a_3 - a_4)k - a_1^2a_4 + a_2a_4 - a_3^2 = 0 \qquad ...(4)$$

This is a cubic equation in k so it must have one real values of k. This real value is obtained by trial method. Once we obtained the value of k we thus obtain a and b and then put these values in (3) and using (2), we get

$$(x^2 + a_1x + k^2) = (ax + b)^2 \quad \text{or} \quad x^2 + a_1x + k = \pm(ax + b)$$

Thus the given biquadratic is obtained as the product of two quadratics.

$$\left.\begin{aligned} & x^2 + (a_1 - a)x + (k - b) = 0 \\ \text{and} \quad & x^2 + (a_1 + a)x + (k + b) = 0 \end{aligned}\right\} \qquad ...(5)$$

On solving these quadratics we finally obtained the solution of the given quadratic.

Solved Examples

Example 1. *Solve the equation $x^4 - 3x^2 - 42x - 40 = 0$ by Descarte's method.*

Solution. The given equation is

$$x^4 - 3x^2 - 42x - 40 = 0 \qquad ...(1)$$

Let us assume

$$x^4 - 3x^2 - 42x - 40$$
$$\equiv (x^2 + kx + l)(x^2 - kx + m) = 0 \qquad ...(2)$$

Equating the coefficients of like powers of x, we get

$$l + m - k^2 = -3 \quad \text{or} \quad l + m = -3 + k^2 \qquad ...(3)$$

and $$k(m - l) = -42 \quad \text{or} \quad m - l = -\frac{42}{k} \qquad ...(4)$$

and $$lm = -40 \qquad ...(5)$$

Solving (3) and (4) we get

$$2m = -3 + k^2 - \frac{42}{k}$$

and $$2l = -3 + k^2 + \frac{42}{k}$$

Substitute the values of l and m in (5), we get

$$\left(-3 + k^2 - \frac{42}{k}\right)\left(-3 + k^2 + \frac{42}{k}\right) = 4(-40)$$

or $$(k^3 - 3k - 42)(k^3 - 3k + 42) = -160k^2$$

or $$(k^3 - 3k)^2 - (42)^2 = -160k^2$$

or $$k^6 - 6k^4 + 169k^2 - 1764 = 0$$

Let $k^2 = t$, then we get

$$t^3 - 6t^2 + 169t - 1764 = 0$$

By trial method it is obvious that $t = 9$ satisfies above equation.

Hence $k^2 = 9$ or $k = \pm 3$

Taking $k = 3$, in (3) and (4), we get

$$l + m = 6 \text{ and } m - l = -14$$

Solving these equation for l and m, we get $l = 10$, $m = -4$ therefore from (2) we obtain the given biquadratic as the product of two quadratics

$$(x^2+3x+10)(x^2-3x-4)=0$$

Solving these quadratics respectively, we get the required solutions

$$x = 4,-1,\frac{-2\pm i\sqrt{31}}{2}$$

Example 2. *Solve the equation* $x^4+8x^3+9x^2-8x-10=0$ *by Descrate's method.*

Solution. The given equation is

$$x^4+8x^3+9x^2-8x-10=0 \qquad \ldots(1)$$

First we remove the second term, *i.e.*, by diminishing each of its roots by a constant

$$h=-\frac{a_1}{na_0}=-\frac{8}{4.1}=-2$$

Using synthetic division method as follows :

−2	1	8	9	−8	−10
		−2	−12	6	4
	1	6	−3	−2	−6
		−2	−8	22	
	1	4	−11	20	
		−2	−4		
	1	2	−15		
		−2			
	1	0			
	1				

Thus the transformed equation is

$$z^4-15z^2+20z-6=0 \qquad \ldots(2)$$

where $z = x+2$

let as assume

$$z^4-15z^2+20z-6$$
$$\equiv (z^2+kz+l)(z^2-kz+m)=0 \qquad \ldots(3)$$

Comparing the coefficients of like powers of z, we get

$$l+m-k^2=-15$$

or $$l+m=-15+k^2 \qquad \ldots(4)$$

$$k(m-l)=20$$

or $$m-l=\frac{20}{k} \qquad \ldots(5)$$

and $$lm=-6$$

Solving (4) and (5), we get

$$2l = k^2-15-\frac{20}{k}$$
$$2m = k^2-15+\frac{20}{k}$$

Substitute these values of l and m in (6), we get

$$\left(k^2-15-\frac{20}{k}\right)\left(k^2-15+\frac{20}{k}\right)=-24$$

or $(k^3-15k-20)(k^3-15k+20)=-24k^2$

or $$(k^3-15k)^2-400=-24k^2$$

or $$k^6-30k^4+249k^2-400=0$$

Let $k^2 = t$, then

$$t^3-30t^2+249t-400=0 \qquad \ldots(7)$$

From (7) it is obvious that $t = 16$ satisfies the equation (7)

$\therefore \quad k^2=16$ or $k=\pm 4$

Taking $k = 4$, in (4) and (5), we get

$$l+m=1$$
$$m-l=5$$

On solving these equations, we get

$$l=-2,\ m=3$$

Substitute the values of l, m and k in (3), we get

$$z^4-15z^2+20z-6$$
$$\equiv(z^2+4z-2)(z^2-4z+3)=0$$

$$\therefore \quad (z^2+4z-2)(z^2-4z+3)=0$$

and $$z=1,3,-2\pm\sqrt{6}$$

But $$z=x+2 \quad \Rightarrow \quad x=z-2$$

Hence, the solution of the given biquadratic are

$$x=-1,1,-4\pm\sqrt{6}$$

Example 3. *Solve the equation* $x^4-2x^3-5x^2+10x-3=0$ *by Ferrari's method.*

Solution. Since the equation is

$$x^4-2x^3-5x^2+10x-3=0 \qquad \ldots(1)$$

Adding $(ax+b)^2$ of both sides, we get

$$x^4-2x^3-5x^2+10x$$
$$-3+(ax+b)^2=(ax+b)^2$$

or $$x^4-2x^3+(a^2-5)x^2$$
$$+2(ab+5)x+b^2-3=(ax+b)^2 \quad \ldots(2)$$

Let as assume that L.H.S. of (2) must be a perfect square, let us suppose $(x^2-a_1x+k^2)$ is a perfect square of L.H.S. of (2).

$$\therefore\ x^4-2x^3+(a^2-5)x^2$$
$$+2(ab+5)x+b^2-3\equiv(x^2-x+k)^2 \quad \ldots(3)$$
$$(\because a_1=-1)$$

Equating the coefficients of like powers of x, we get

$$a^2=2k+6, ab=-k-5, b^2=k^2+3$$

Now eliminating a and b between these three equations, we get

$$(2k+6)(k^2+3)=(k+5)^2$$

or $$2k^3+5k^2-4k-7=0$$

It is a cubic in k so it must have one real root, then by trial method, we get

$$k=-1$$

and hence $\quad a^2=4,\ b^2=4,\ ab=-4$

or $$a=2,\ b=-2$$

Substitute the values of k, a and b in (3) and (4), we get

$$(x^2 - x - 1)^2 = (2x - 2)^2$$

or $\quad x^2 - x - 1 = \pm(2x - 2)$

or $\quad x^2 - 3x + 1 = 0$

and $\quad x^2 + x - 3 = 0$

Solving these quadratics, we get

$$x = \frac{3 \pm \sqrt{5}}{2}, \frac{-1 \pm \sqrt{13}}{2}$$

These are the solutions of the given biquadratic equation.

Example 4. *Solve the equation $x^4 + 2x^3 - 7x^2 - 8x + 12 = 0$ by Ferrari's method.*

Solution. Since the given biquadratic is

$$x^4 + 2x^3 - 7x^2 - 8x + 12 = 0 \qquad \ldots(1)$$

Adding $(ax + b)^2$ of both sides of (1), we get

$$x^4 + 2x^3 - 7x^2 - 8x + 12 + (ax + b)^2 = (ax + b)^2$$

or $\quad x^4 + 2x^3 + (a^2 - 7)x^2 + (2ab - 8)x + b^2 + 12 = (ax + b)^2 \qquad \ldots(2)$

In order to determine a and b make the L.H.S. of (2) a perfect square. Let the perfect square be $(x^2 + a_1x + k)^2$

$\therefore \quad x^4 + 2x^3 + (a^2 - 7)x^2 + (2ab - 8)x + b^2 + 12 \equiv (x^2 + a_1x + k)^2$

or $\quad x^4 + 2x^3 + (a^2 - 7)x^2 + (2ab - 8)x + b^2 + 12 \equiv (x^2 + x + k)^2 \qquad \ldots(3)$

$(\because a_1 = 1 \text{ (from(1))})$

Equating the coefficients of like powers of x, we get

$$a^2 - 7 = 2k + 1, 2ab - 8 = 2k, b^2 + 12 = k^2$$

Eliminating a and b between above three equations, we get

$$(k + 4)^2 = (2k + 8)(k^2 - 12)$$

or $\quad k^3 + 16 + 8k = 2k^2 + 8k^2 - 24k - 96$

or $\quad 2k^3 + 7k^2 - 32k - 112 = 0$

This is a cubic in k so it must have one real root. By trial method, $k = -7/2$ satisfies above cubic.

Then $\quad a^2 = 1, b^2 = \frac{1}{4}$

$\therefore \quad a = 1, b = \frac{1}{2}$

Now substitute the values of k, a and b in (3) and using (2), we get

$$\left(x^2 + x - \frac{7}{2}\right)^2 = \left(x + \frac{1}{2}\right)^2$$

or $\quad x^2 + x - \frac{7}{2} = \pm\left(x + \frac{1}{2}\right)$

or $\quad x^2 - 4 = 0$ and $x^2 + 2x - 3 = 0$

Solving these quadratics, we get

$$x = -2, 2, \text{ and } x = 1, -3$$

Hence, the solution of given biquadratic are

$$x = -3, -2, 1, 2.$$

Example 5. *Show that the equation*

$$x^4 + px^3 + qx^2 + rx + s = 0$$

may be solved a quadratic if $r^2 = p^2s$.

Solution. Since the given equation is

$$x^4 + px^3 + qx^2 + rx + s = 0 \qquad \ldots(1)$$

Let us assume

$$x^4 + px^3 + qx^2 + rx + s \equiv \left(x^2 + \frac{p}{2}x + l\right)^2 = 0$$

Comparing the coefficients of like power of x, we get

$$2l\left(q^2 - \frac{p^2}{4}\right), pl = r, l^2 = s$$

Eliminating l between last two equations, we get

$$pl = r \quad \Rightarrow \quad p^2l^2 = r^2$$

or $\quad p^2s = r^2 \quad \because \quad l^2 = s$

EXERCISE 24.7

Solve the following biquadratic equation by Descarte's method.

1. $x^4 - 6x^3 - 9x^2 + 66x - 22 = 0$
2. $x^4 - 8x^2 - 24x + 7 = 0$ (UPTU–2001, 10)
3. $x^4 - 10x^2 - 20x - 16 = 0$
4. $x^4 + 2x^3 - 7x^2 - 8x + 12 = 0$
5. $x^4 - 8x^3 - 12x^2 + 60x + 63 = 0$
6. $x^4 - 3x^2 - 6x - 2 = 0$
7. $x^4 - 5x^2 - 6x - 5 = 0$
8. $x^4 - 12x - 5 = 0$
9. $4x^4 - 20x^3 + 33x^2 - 20x + 4 = 0$
10. $x^4 + 8x^3 + 9x^2 - 8x - 10 = 0$
11. $x^4 - 2x^2 + 8x - 3 = 0$

Solve the following biquadratic equation by Ferrari's method.

12. $x^4 - 8x^3 - 12x^2 + 60x + 63 = 0$ (UPTU–2005)
13. $x^4 + 12x - 5 = 0$
14. $x^4 - 2x^3 - 5x^2 + 10x - 3 = 0$
15. $x^4 - 3x^2 - 42x - 40 = 0$
16. $x^4 + 9x^3 + 12x^2 - 80x - 192 = 0$
17. $x^4 - 2x^3 - 12x^2 + 10x + 3 = 0$

18. $x^4 - 10x^3 + 44x^2 - 104x + 96 = 0$

19. $x^4 + 4x^3 + 12x^2 - 8x + 95 = 0$

20. $x^4 + 3x^3 + x^2 - 2 = 0$

21. $2x^4 + 6x^3 - 3x^2 + 2 = 0$

22. $x^4 - 12x^3 + 41x^2 - 18x - 72 = 0$ (SVTU–2007)

23. $x^4 - 10x^3 + 35x^2 - 50x + 24 = 0$ (UPTU–2003, 08)

24. $x^4 + 2x^3 - 7x^2 - 8x + 12 = 0$ (UPTU–2002)

Hint to Selected Problems

1. $h = \frac{3}{2}$ The transformed equation is

$$y^4 - \frac{90}{4}y^2 + \frac{96}{8}y + \frac{665}{16} = 0$$

$\Rightarrow \quad (2y)^4 - 90(2y)^2 + 96(2y) + 665 = 0 \quad$ Put $2y = z$

5. $h = 2$

The transformed equation is

$$z^4 - 36z^2 - 52z + 87 = 0 \text{ where } z = x - 2$$

Let us write

$$z^4 - 36z^2 - 52z + 89 = (z^2 + kx + l)(z^2 - kz + m) = 0$$

11. Assume that $x^4 - 2x^2 + 8x - 3 = (x^2 + kx + l)(x^2 - kx + m) = 0$

13. Adding $(ax + b)^2$ of both the sides of the given equation.

Answers

1. $\pm\sqrt{11}, 3 \pm \sqrt{7}$ **2.** $-2 \pm i\sqrt{3}, 2 \pm \sqrt{3}$ **3.** $4, -2, -1 \pm i$ **4.** $\pm 2, -3, 1$ **5.** $-1, 3, 3 \pm \sqrt{30}$

6. $1 \pm \sqrt{2}, -1 \pm i$ **7.** $\frac{-1 \pm i\sqrt{3}}{2}, \frac{1 \pm \sqrt{21}}{2}$ **8.** $-1 \pm 2i, 1 \pm \sqrt{2}$ **9.** $2, 2, \frac{1}{2}, \frac{1}{2}$ **10.** $\pm 1, -4 \pm \sqrt{6}$

11. $1 \pm i\sqrt{2}, -1 \pm \sqrt{2}$ **12.** $-1, 3, 3 \pm \sqrt{30}$ **13.** $-1 \pm \sqrt{2}, -1 \pm 2i$ **14.** $\frac{3 \pm \sqrt{5}}{2}, \frac{-1 \pm \sqrt{13}}{2}$ **15.** $4, -1, -\frac{1}{2}(3 \pm i\sqrt{31})$

16. $-4, -4, -4, 3$ **17.** $1, -3, 2 \pm \sqrt{5}$ **18.** $2, 4, 2 \pm i2\sqrt{2}$ **19.** $-3 \pm i\sqrt{10}, 1 \pm 2i$ **20.** $-1 \pm \sqrt{3}, \frac{-1 \pm i\sqrt{3}}{2}$

21. $-2 \pm \sqrt{2}, \frac{1}{2}(1 \pm i)$ **22.** $-1, 3, 4, 6$ **23.** $1, 2, 3, 4$ **24.** $-3, 1, 2, -2$

Objective Evaluations

Fill in the Blanks

1. If $\alpha_1, \alpha_2, ... \alpha_n$ are the roots of equation $f(x) = 0$ where $f(x) \equiv a_0 x^n + a_1 x^{n-1} + ... a_n$ then the product of the roots is ________.
2. If $1, \alpha_1, \alpha_2, ... \alpha_{n-1}$ are the roots of $x^n - 1 = 0$ then the value of $(1-\alpha_1)(1-\alpha_2)...(1-\alpha_{n-1})$ is________.
3. If α, β are the roots of $ax^2 + bx + c = 0$, then the equation whose roots are $\frac{1}{\alpha}, \frac{1}{\beta}$ is ________.
4. If α, β, γ are the roots of equation $x^3 - 5x - 3 = 0$, then the equation whose roots are $-\alpha, -\beta, -\gamma$ is ________.
5. If α, β, γ are the roots of equation $x^3 - 5x^2 + 6x - 7 = 0$ the equation whose roots are $3\alpha, 3\beta, 3\gamma$ is ________.
6. If α, β, γ are the roots of equation $x^3 + qx + r = 0$, then the equation whose roots are $\frac{1}{\alpha}, \frac{1}{\beta}, \frac{1}{\gamma}$ is ________.
7. To remove the second term of the equation $x^4 - 5x^3 + 7x^2 - 17x + 11 = 0$ we diminish its all roots by diminishing $h =$ ________.
8. If the three roots of the equation $x^5 - 5x^4 + 9x^3 - 9x^2 + 5x - 1 = 0$ are $1, \frac{1}{2}(1 - i\sqrt{3}), \frac{1}{2}(3 + \sqrt{5})$ then its other roots are ________.
9. The equation $x^3 + qx + r = 0$ may have all its roots real if $27r^3 + 4q^3 <$ ________.
10. If α, β, γ be the rots of $x^3 + 2x^2 - 3x - 1 = 0$ then all the value of $\alpha^{-3} + \beta^{-3} + \gamma^{-3}$ is ________.
11. Every equation of an odd degree has at least ________ real root.
12. Every equation of even degree having its least term negative has at least ________ real roots.
13. If α, β, γ be the roots of the equation $x^3 + px + r = 0$ then the equation whose roots are $\alpha^2, \beta^2, \gamma^2$ is ________.
14. If the two roots of the equation $x^3 + px + r = 0$ are 2 and -1, then the value of p and r are ________.
15. The value of H and G for the cubic $x^3 + 3x^2 + 4x - 10 = 0$ are ________.
16. The equation whose roots are the cube of the roots of the equation $x^3 + 3x^2 + 2 = 0$ is ________.
17. Let α, β, γ, δ be the roots of $ax^4 + bx^3 + cx^2 + dx + c = 0$. Then $\Sigma\alpha =$ ________.
18. If $f(x) = x^4 - 3x^2 - 6x - 2 = (x^2 + 2x + 2)(x^2 - 2x - 1)$ then roots of $f(x) = 0$ are ________.
19. To solve the biquadratic equation
$$a_0 x^4 + 4a_1 x^3 + a_2 x^2 + a_3 x + a_4 = 0$$
by Descarte's method, we first remove its second term by diminishing its roots by $h =$ ________.
20. The biquadratic equation $a_0 x^4 + 4a_1 x^3 + 6a_2 x^2 + 4a_3 x + a_4 = 0$ reduces to the cubic $t^3 + 3Ht^2 + \left(3H^2 - \frac{a_0^2 I}{4}\right)t - \frac{G^2}{4} = 0$. Then this cubic is known as ________.
21. To solve $x^4 - 8x^3 - 12x^2 + 60x + 63 = 0$ by Descarte's method we first remove the second term by diminishing its roots by $h =$ ________.
22. If the two roots of $x^4 + 12x - 5 = 0$ are $-1 + \sqrt{2}$ and $1 - 2i$ then its other roots are ________.
23. If $x^4 - 2x^2 + 8x - 3 \equiv (x^2 + kx + l)(x^2 - kx + m)$ then $l + m - k^2 =$ ________ and $k(m - l) =$ ________ and $lm =$ ________.
24. If $x^4 - 2x^3 - 5x^2 + 10x - 3 \equiv (x^2 - x - 1)^4 - (2x - 2)^2 = 0$ then the roots of the $x^4 - 2x^3 - 5x^2 + 10x - 3 = 0$ are ________.
25. The biquadratic equation $a_0 x^4 + 4a_1 x^3 + 6a_2 x^2 + 4a_3 x + a_4 = 0$ can be reduced to $z^4 + 6Hz^2 + 4Gz + (a_0^2 I - 3H^2) = 0$ by $z =$ ________.
26. If $f(x) \equiv x^4 - 3x^2 - 6x - 2 = (x^2 - 2x + 2)(x^2 - 2x - 1)$ then the roots of $f(x) = 0$ are ________.
27. To solve the cubic equation $a_0 x^3 + 3a_1 x^2 + 3a_2 x + a_3 = 0$ by Cardan's method, we first remove the second term by diminishing its roots by $h =$ ________.
28. The cubic equation $a_0 x^3 + 3a_1 x^2 + 3a_2 x + a_3 = 0$ reduces to $z^3 + 3Hz + G = 0$ by $z = a_0 x_1 + a_1$, then G equals ________.
29. If $z = u + v$ is a solution of the cubic equation $z^3 + 3Hz + G$, then $u^3 + v^3 =$ ________ and $u^3 v^3 =$ ________.
30. If $x = u + v$ is a solution of $x^3 - 15x - 126 = 0$ then u^3 and v^3 are the roots of the qudratic equation________.
31. The roots of the cubic $z^3 + 3Hz + G = 0$ are all real if $G^2 +$ ________ ≤ 0.
32. If $G^2 + 4H^3 > 0$ then $z^3 + 3Hz + G = 0$ has two________roots.
33. If $G^2 + 4H^3 = 0$ then $z^3 + 3Hz + G = 0$ has two________roots.
34. If the two roots of $28x^3 - 9x^2 + 1 = 0$ are $\frac{1}{7}(2 \pm i\sqrt{3})$ then its third root is ________.
35. If $x^3 - 3x^2 + 33x - 1 \equiv A(x-u)^3 - B(x-v)^3 = 0$ then u and v are the roots of the equation $x^2 + x +$ ________ $\equiv 0$.
36. If the cubic equation $a_0 x^3 + 3a_1 x^2 + 3a_2 x + a_3 = 0$ has two roots equal to α, then $\alpha =$ ________.

True / False

Write 'T' for True and 'F' for False statement.

1. Every equation of odd degree has at least two real roots. **(T/F)**
2. Every equation of even degree with last term negative has at least two real roots. **(T/F)**
3. To remove the second term of the equation $a_0x^n + a_1x^{n-1} + ... + a_n = 0$, we diminish its all roots by $h = \frac{-a_1}{na_0}$. **(T/F)**
4. It α and β are the roots of $x^2 + bx + c = 0$, then the equation has $x^2 + (b-2)x + c - b + 1 = 0$ has the roots of α +1, β + 1. **(T/F)**
5. If α, β, γ are the roots of $x^3 + 3x + 3 = 0$, then the equation whose roots are $\frac{1}{\alpha}, \frac{1}{\beta}, \frac{1}{\gamma}$ is $3x^3 + 3x + 1 = 0$. **(T/F)**
6. If α, β, γ are the roots of the equation $x^3 + qr + r = 0$, then the equation whose roots are α + β, β + γ, γ + α is $x^3 + qr - r = 0$. **(T/F)**
7. If equation $z^3 + 3Hz + g = 0$ has two roots equal if $G^2 + 4H^3 \neq 0$. **(T/F)**
8. If the two roots of equation $x^4 - 3x^2 - 6x - 2 = 0$ are $-1 + i$ and $1 + \sqrt{2}$, then its other roots are $1 + i$ and $1 - \sqrt{2}$. **(T/F)**
9. The equation $a_0x^4 + 4a_1x^3 + 6a_2x^2 + 4a_3x + a_4 = 0$ reduces to $z^4 + 6Hz^2 + 4Gz + (a_0^2I - 3H^2) = 0$ by $z = a_0x + a_1$. **(T/F)**
10. Solve the equation $x^4 - 6x^3 - 9x^2 + 66x - 22 = 0$ by Descarte's method we first remove the second term by diminishing its root by $h = \frac{3}{2}$. **(T/F)**
11. The equation $z^3 + 3Hz + G = 0$ has two equal roots if $G^2 + 4H^3 = 0$. **(T/F)**
12. A cubic equation with real coefficient has at least one real root. **(T/F)**
13. To solve the cubic equation $a_0x^3 + 3a_1x^2 + 3a_2x + a_3 = 0$ by Cardan's method we first reduce the given cubic into $z^3 + 3Hz^2 + Gz = 0$. **(T/F)**
14. The equation $x^3 + 3Hx + G = 0$ all its roots real if $G^2 + 4H^3 > 0$. **(T/F)**
15. If $a_0x^3 + 3a_1x^2 + 3a_2x + a_3 \equiv A(x-u)^3 + B(x-v)^3$ then
$$(a_0a_2 - a_1^2)x^2 + (a_0a_3 - a_1a_2)x + (a_1a_3 - a_2^2) = 0$$
is known as Hessian of the given cubic. **(T/F)**
16. If $z = u + v$ is a solution of $z^3 + 3Hz + G = 0$, then $z = u\omega + \omega^2$ is also the solution of given equation where ω is cube root of unity. **(T/F)**
17. The equation $x^3 + 3bx + c = 0$ has two imaginary roots of $c^2 + 4b^2 > 0$. **(T/F)**

Multiple Choice Questions

Choose the most appropriate one

1. If 1, $\alpha_1, \alpha_2, ..., \alpha_{n-1}$ are the roots of $x^n - 1 = 0$ then the value of $(1-\alpha_1)(1-\alpha_2)...(1-\alpha_{n-1})$ is :
   (a) $n - 1$ (b) n
   (c) $n + 1$ (d) n^2
2. If α, β, γ are the roots of the equation $x^3 + qx + r = 0$, then the equation whose roots are $\frac{1}{\alpha}, \frac{1}{\beta}, \frac{1}{\gamma}$:
   (a) $rx^3 + qx^2 + 1 = 0$ (b) $rx^3 - qx^2 + 1 = 0$
   (c) $rx^3 + qx^2 - 1 = 0$ (d) $rx^3 - qx^2 + r = 0$
3. If α, β are the roots of $x^2 - x + 1 = 0$, then the equation whose roots are α^2, β^2 is :
   (a) $x^4 + x^2 + 1 = 0$ (b) $x^2 + x + 1 = 0$
   (c) $x^4 - x^2 + 1 = 0$ (d) $x^2 + x - 1 = 0$
4. To remove second term of the equation $x^4 + 8x^3 + x - 5 = 0$ we diminish its all roots by :
   (a) 2 (b) 3
   (c) – 2 (d) – 3
5. If α, β, γ are the roots of the equation $x^3 + px + r = 0$ then α + β + γ is :
   (a) p (b) $-p$
   (c) 0 (d) 1
6. If α, β, γ be the roots of the equation $x^3 + 2x^2 - 3x - 1 = 0$ then $\alpha^{-3} + \beta^{-3} + \gamma^{-3}$ is :
   (a) 42 (b) 41
   (c) – 42 (d) – 41
7. If $x^4 - 2x^2 + 8x - 3 \equiv (x^2 + 2x + l)(x^2 - 2x + m)$ then the values of l and m are :
   (a) –1, – 3 (b) –1, 3
   (c) 1, 3 (d) 1, –3
8. If two roots of $x^4 - 3x^2 - 6x - 2 = 0$ are $-1 + i$ and $1 + \sqrt{2}$ then its other two roots are :
   (a) $-1-i, -1+\sqrt{2}$ (b) $-1-i, -1-\sqrt{2}$
   (c) $-1-i, 1-\sqrt{2}$ (d) $1+i, 1-\sqrt{2}$
9. If $a_0x^4 + 4a_1x^3 + 6a_2x^2 + 4a_3x + a_4 = 0$ reduces to $z^4 + 6Hz^2 + 4Gz + (a_0^2I - 3H^2) = 0$ by $z = a_0x + a_1$ then H equals :
   (a) $a_0a_2 - a_1^2$ (b) $a_0a_1 - a_2^2$
   (c) $a_1a_2 - a_0^2$ (d) $a_1a_0 - a_2$
10. The sum of all the four roots of $a_0x^4 + a_1x^3 + a_2x^2 + a_3x + a_4 = 0$ is :
   (a) a_1/a_0 (b) $-a_1/a_0$
   (c) a_2/a_0 (d) $-a_1/a_2$
11. If $z = u + v$ is a solution of $z^3 + 3Hz + G = 0$, then $u^3 + v^3$ equals :
   (a) G (b) $-G$
   (c) H (d) $-H$

12. If $z = u + v$, $z = u\omega + v\omega^2$ are the roots of $z^3 + 3Hz + G = 0$, then its third root is :

(a) $u\omega^2 + v\omega$ (b) $u\omega^2 - v\omega$

(c) $u\omega - v\omega^2$ (d) $u - v$

13. If $z = u + v$ is a solution $z^3 - 12z - 65 = 0$, then u and v are the roots of the quadratic :

(a) $t^2 + 65t - 64 = 0$ (b) $t^2 - 65t + 64 = 0$

(c) $t^2 - 64t + 65 = 0$ (d) $t^2 + 64t + 65 = 0$

14. To remove the second term of the cubic $27x^3 + 54x^2 + 198x - 73 = 0$ we diminish its roots by diminishing h =

(a) 2/3 (b) –3/2

(c) – 2/3 (d) 3/2

15. If $x^3 + 3x^2 - 27x + 104 \equiv A(x-u)^3 + B(x-v)^3$ then the Hessian of the given cubic is:

(a) $10x^2 - 113x - 23 = 0$ (b) $10x^2 + 113x - 23 = 0$

(c) $10x^2 + 113x + 23 = 0$ (d) $10x^2 - 113x + 23 = 0$

ANSWERS

Fill in the Blanks

1. $(-1)^n \dfrac{a_n}{a_0}$ **2.** n **3.** $cx^2 + bx + a = 0$ **4.** $x^3 - 5x + 3 = 0$

5. $x^3 - 15x^2 + 54x - 189 = 0$ **6.** $rx^3 + qx^2 + 1 = 0$ **7.** $\dfrac{5}{4}$ **8.** $\dfrac{1}{2}(1+i\sqrt{3}), \dfrac{1}{2}(3-\sqrt{5})$

9. 0 **10.** – 42 **11.** One **12.** Two **13.** $y^3 + 2py^2 + p^2y - r^2 = 0$

14. $p = -3, r = -2$ **15.** $H = \dfrac{1}{3}, G = -12$ **16.** $y^3 + 33y^2 + 12y + 8 = 0$ **17.** $-b/a$ **18.** $1 \pm \sqrt{2}i$

19. $h = -\dfrac{a_1}{4a_0}$ **20.** Euler cubic **21.** 2 **22.** $-1-\sqrt{2}, 1 + 2i$ **23.** –2, 8, –3

24. $\dfrac{3 \pm \sqrt{5}}{2}, \dfrac{-1 \pm \sqrt{13}}{2}$ **25.** $a_0x + a_1$ **26.** $-1 \pm i, 1 \pm \sqrt{2}$ **27.** $\dfrac{-a_1}{na_0}$ **28.** $a_0^2a_3 - 3a_0a_1a_2 + 2a_1^3$

29. $-G, -H^3$ **30.** $t^2 - 126t + 125 = 0$ **31.** $4H^3$ **32.** imaginary **33.** equal

34. –1/4 **35.** –12 **36.** $\alpha = \dfrac{a_1a_2 - a_0a_3}{2(a_0a_2 - a_1^2)}$

True/False

1. F	**2.** T	**3.** T	**4.** T	**5.** T	**6.** T	**7.** F	**8.** F	**9.** T	**10.** T
11. T	**12.** T	**13.** F	**14.** F	**15.** T	**16.** T	**17.** T			

Multiple Choice Questions

1. (b)	**2.** (a)	**3.** (b)	**4.** (c)	**5.** (c)	**6.** (c)	**7.** (b)	**8.** (c)	**9.** (a)	**10.** (b)
11. (b)	**12.** (a)	**13.** (b)	**14.** (c)	**15.** (a)					

FFFFFF

25 Linear Algebra: Matrices and Determinants

25.1 INTRODUCTION

A set of mn numbers either real or complex arranged in the form of a reactangular array in which there are m rows and n columns, rectangular arrrangement is called a matrix of order $m \times n$ which is denoted by $[a_{ij}]_{m\times n}$ where $i = 1, 2, 3, ..., m$ represents the number of rows and $j = 1, 2, 3, ..., n$ represents the number of columns and thus a matrix of order $m \times n$ is usually written as

$$[a_{ij}]_{m\times n} = \begin{bmatrix} a_{11} & a_{12} & \cdots & a_{1n} \\ a_{21} & a_{22} & \cdots & a_{2n} \\ \vdots & \vdots & \vdots & \vdots \\ a_{m1} & a_{m2} & \cdots & a_{mn} \end{bmatrix}_{m\times n}$$

REMARK

- Sometimes, a matrix is a rectangular array of numbers enclosed in double straight lines shown as '|| ||' or enclosed in parenthesis '( )'.

25.2 TYPE OF MATRICES

(i) Null matrix (or zero matrix): A matrix of order $m \times n$ is called a *null matrix* if it contains all mn elements zero. It is denoted by O and is usually written as

$$O = \begin{bmatrix} 0 & 0 & \cdots & 0 \\ 0 & 0 & \cdots & 0 \\ \vdots & \vdots & \vdots & \vdots \\ 0 & 0 & \cdots & 0 \end{bmatrix}_{m\times n}$$

(ii) Row matrix: A matrix having only one row and n columns is called a *row matrix* of order $1 \times n$.

For example : $A = [a_{11} \quad a_{12} \quad a_{13} \quad \cdots \quad a_{1n}]_{1\times n}$

(iii) Column matrix : A matrix having m rows and only one column is called a *column matrix* of order $m \times 1$.

For example : $$A = \begin{bmatrix} a_{11} \\ a_{21} \\ a_{31} \\ \vdots \\ a_{m1} \end{bmatrix}_{m\times 1}$$

(iv) Horizontal matrix : A matrix having more columns than the number of its rows, is called *Horizontal matrix*.

For example : $$A = \begin{bmatrix} a_{11} & a_{12} & a_{13} \\ a_{21} & a_{22} & a_{23} \end{bmatrix}_{2\times 3}$$

(v) Vertical matrix : A matrix having more number of rows than its columns, *is called vertical matrix.* **For exmaple :**

$$A = \begin{bmatrix} a_{11} & a_{12} \\ a_{21} & a_{22} \\ a_{31} & a_{32} \end{bmatrix}_{3\times 2}$$

REMARK

- Row matrix is also a horizontal matrix and column matrix is also a vertical matrix.

(vi) Square matrix : A matrix having a number of rows equal to number of columns, is called *square matrix*.

For example :
$$A = \begin{bmatrix} a_{11} & a_{12} & a_{13} \\ a_{21} & a_{22} & a_{23} \\ a_{31} & a_{32} & a_{33} \end{bmatrix}_{3\times3}$$

Here, the matrix A has 3 rows and 3 columns, so it is a square matrix. Also the elements a_{11}, a_{22}, a_{33} are placed in the diagonal, so these elements are known as *diagonal elements*.

(vii) Diagonal matrix : A matrix of order $n \times n$ is called a *diagonal matrix* if it contains all its off diagonal elements equal to zero.

Suppose $A = [a_{ij}]_{n\times n}$ and if $a_{ij} = 0$ for all $i \neq j$, then A is a diagonal matrix. Diagonal matrix of order $n\times n$ is usually written as

$$\text{Diag } [a_{11} \quad a_{22} \quad a_{33} \quad \dots \quad a_{nn}]$$

For example :
$$A = \begin{bmatrix} 1 & 0 & 0 \\ 0 & 2 & 0 \\ 0 & 0 & 3 \end{bmatrix}_{3\times3} = \text{Diag } [1 \ 2 \ 3]$$

(viii) Scalar matrix : A diagonal matrix whose diagonal elements are all equal but not equal to 1 is called a *scalar matrix*.

For example :
$$A = \begin{bmatrix} k & 0 & 0 \\ 0 & k & 0 \\ 0 & 0 & k \end{bmatrix}, k \neq 1$$

(ix) Unit matrix : A square matrix of order $n \times n$ having all off diagonal elements equal to zero and each of the diagonal elements equal to 1, is called a *unit matix*. It is usually denoted by I_n and is written as

$$I_n = \begin{bmatrix} 1 & 0 & \dots & 0 \\ 0 & 1 & \dots & 0 \\ 0 & 0 & \dots & 0 \\ \vdots & \vdots & \vdots & \vdots \\ 0 & 0 & \dots & 1 \end{bmatrix}_{n\times n}$$

REMARK

- Unit matrix can also be denoted by I.

(x) Triangular matrix : A matrix in which the elements lying above or below principal diagonal are all zero, is called a *triangular matrix*. There are two kinds of triangular matrix.

(a) Upper triangular matrix : A matrix of order $n \times n$ is called an *upper triangular matrix* if it contains all its elements below the diagonal elements equal to zero.

Suppose $A = [a_{ij}]_{n\times n}$ and if $a_{ij} = 0$ for all $i > j$, then A is an upper triangular matrix.

For example : $A = \begin{bmatrix} 2 & 3 & 4 \\ 0 & 1 & 5 \\ 0 & 0 & 3 \end{bmatrix}_{3\times3}$ is an upper triangular matrix of order 3×3.

(b) Lower triangular matrix : A matrix of order $n \times n$ is called a *lower triangular matrix* if it contains all its elements above the diagonal elements equal to zero.

Suppose $A = [a_{ij}]_{n\times n}$ and if $a_{ij} = 0$ for all $i < j$, then A is called lower triangular matrix.

For example : $A = \begin{bmatrix} 1 & 0 & 0 \\ 3 & 4 & 0 \\ 5 & 6 & 7 \end{bmatrix}_{3\times3}$ is a lower triangular matrix of order 3×3.

25.3 OPERATION ON MATRICES

25.3.1 ADDITION OF MATRICES

Suppose A and B are two matrices of same order, then the addition of these two matrices is obtained by adding corresponding elements of A and B. It is denoted by $A + B$. If the order of A and B is $m\times n$, then the order of $A+B$ will be $m \times n$.

Suppose $A = [a_{ij}]_{m\times n}$ and $B = [b_{ij}]_{m\times n}$

then $A + B = [a_{ij} + b_{ij}]_{m\times n}$

For example: If $A = \begin{bmatrix} 1 & 2 & 3 \\ 5 & 1 & 4 \\ 7 & 8 & 9 \end{bmatrix}$ and $B = \begin{bmatrix} 1 & 3 & 5 \\ 5 & 0 & 1 \\ 3 & 2 & 12 \end{bmatrix}$

then $A + B = \begin{bmatrix} 1 & 2 & 3 \\ 5 & 1 & 4 \\ 7 & 8 & 9 \end{bmatrix} + \begin{bmatrix} 1 & 3 & 5 \\ 5 & 0 & 1 \\ 3 & 2 & 12 \end{bmatrix} = \begin{bmatrix} 1+1 & 2+3 & 3+5 \\ 5+5 & 1+0 & 4+1 \\ 7+3 & 8+2 & 9+12 \end{bmatrix} = \begin{bmatrix} 2 & 5 & 8 \\ 10 & 1 & 5 \\ 10 & 10 & 21 \end{bmatrix}$

REMARK

- If the orders of the matrices are different, then they are not conformable for addition.

25.3.2 SUBSTRACTION OF MATRICES

Suppose A and B are two matrices of same order, then the substraction of A and B, *i.e.*, $A–B$ is obtained by substracting each element of B from the corresponding element of A. If A and B are of order $m \times n$, then the order of $A - B$ will be of order $m \times n$.

Let $A = [a_{ij}]_{m\times n}$ and $B = [b_{ij}]_{m\times n}$

then $A - B = [a_{ij} - b_{ij}]_{m\times n}$

For example: If $A = \begin{bmatrix} 1 & 2 & 3 \\ 3 & 4 & 5 \\ 5 & 6 & 7 \end{bmatrix}$ and $B = \begin{bmatrix} 0 & 5 & 2 \\ 3 & -2 & 2 \\ 5 & 7 & 8 \end{bmatrix}$

then $A - B = \begin{bmatrix} 1 & 2 & 3 \\ 3 & 4 & 5 \\ 5 & 6 & 7 \end{bmatrix} - \begin{bmatrix} 0 & 5 & 2 \\ 3 & -2 & 2 \\ 5 & 7 & 8 \end{bmatrix} = \begin{bmatrix} 1-0 & 2-5 & 3-2 \\ 3-3 & 4-(-2) & 5-2 \\ 5-5 & 6-7 & 7-8 \end{bmatrix} = \begin{bmatrix} 1 & -3 & 1 \\ 0 & 6 & 3 \\ 0 & -1 & -1 \end{bmatrix}$

REMARK

- If the order of matrices are different, then they are not conformable for subtraction.

25.3.3 MULTIPLICATION OF A MATRIX BY A SCALAR

Suppose A is a matrix of order $m \times n$ and k is a scalar, then the multiplication of A by k, *i.e.* kA is obtained by multiplying each element of A by k.

Let $A = [a_{ij}]_{m\times n}\ \forall\ 1 \le i \le m$ and $i \le j \le m$

For example : If $A = \begin{bmatrix} 1 & 2 & 3 \\ 4 & 5 & 6 \\ 7 & 8 & 9 \end{bmatrix}$ and $k = 3$, then $3A = 3\begin{bmatrix} 1 & 2 & 3 \\ 4 & 5 & 6 \\ 7 & 8 & 9 \end{bmatrix} = \begin{bmatrix} 3\times1 & 3\times2 & 3\times3 \\ 3\times4 & 3\times5 & 3\times6 \\ 3\times7 & 3\times8 & 3\times9 \end{bmatrix} = \begin{bmatrix} 3 & 6 & 9 \\ 12 & 15 & 18 \\ 21 & 24 & 27 \end{bmatrix}$

25.3.4 EQUALITY OF MATRICES

Two matrices are said to be equal if both have same order and having same corresponding elements.

For example : The matrices $A = \begin{bmatrix} 1 & 2 \\ -3 & 4 \end{bmatrix}$ and $B = \begin{bmatrix} x & y \\ z & 4 \end{bmatrix}$ are said to be equal if $x = 1, y = 2$ and $z = -3$.

25.4 PROPERTIES OF MATRIX ADDITION

25.4.1 COMMUTATIVE LAW

If A and B are two matrices of same order $m \times n$, then $A + B = B + A$

Proof. Let $A = [a_{ij}]_{m\times n}$ and $B = [b_{ij}]_{m\times n}$ where $1 \le i \le m$ and $1 \le j \le n$. Then

$$A+B = [a_{ij}]_{m\times n} + [b_{ij}]_{m\times n}$$
$$= [a_{ij} + b_{ij}]_{m\times n} \quad \text{(By definition of addition)}$$
$$= [b_{ij} + a_{ij}]_{m\times n} \quad \text{(Real numbers are always commutative.)}$$
$$= [b_{ij}]_{m\times n} + [a_{ij}]_{m\times n} = B + A$$

Hence, $A + B = B + A$

25.4.2 ASSOCIATIVE LAW

If A ,B and C are three matrices of same order $m \times n$, then $(A + B) + C = A + (B + C)$

Proof. Let $A = [a_{ij}]_{m\times n}$ and $B = [b_{ij}]_{m\times n}$ where $1 \le i \le m$ and $1 \le j \le n$. Then

$$(A+B)+C = ([a_{ij}]_{m\times n} + [b_{ij}]_{m\times n}) + [c_{ij}]_{m\times n}$$
$$= [a_{ij} + b_{ij}]_{m\times n} + [c_{ij}]_{m\times n}$$
$$= [(a_{ij} + b_{ij}) + (c_{ij})]_{m\times n} \quad \text{(Numbers are always associative.)}$$
$$= [a_{ij}]_{m\times n} + ([b_{ij} + c_{ij}]_{m\times n}) = [a_{ij}]_{m\times n} + ([b_{ij}]_{m\times n} + [c_{ij}]_{m\times n})$$
$$= A + (B + C)$$

Hence, $(A+B) + C = A + (B+C)$

25.4.3 ADDITIVE IDENTITY

If A is a matrix of order $m \times n$ and O is a null matrix of the same order $m \times n$, then $A + O = A = O + A$

Proof. Let $A = [a_{ij}]_{m\times n}$ and $O = [0]_{m\times n}$, then

$$A + O = [a_{ij}]_{m\times n} + [0]_{m\times n} = [a_{ij} + 0]_{m\times n} = [a_{ij}]_{m\times n} = A$$

Also $O + A = [0]_{m\times n} + [a_{ij}]_{m\times n} = [0 + a_{ij}]_{m\times n} = [a_{ij}]_{m\times n} = A$

Hence $A + O = A = O + A$

Therefore, the null matrix O is treated as an additive identity.

25.4.4 ADDITIVE INVERSE

If A is a matrix of order $m\times n$ and $-A$ is the negative of A, so its order is also $m\times n$, then $-A + A = O$ (Null matrix)

Hence, $-A$ is the additive inverse of A.

25.4.5 CANCELLATION LAW

If A, B and C are three matrices of order $m \times n$ then

(i) $A+B = A+ C \Rightarrow B = C$ (Left cancellation law) (ii) $B + A = C + A \Rightarrow B = C$ (Right cancellation law)

Proof.

(i) It is given that

$$A + B = A + C \quad \text{...(1)}$$

Adding $-A$ to the left of both sides, we get

$$-A + (A + B) = -A + (A + C)$$
$$\Rightarrow (-A + A) + B = (-A + A) + C \quad \text{(From associative law)}$$
$$\Rightarrow O + B = O + C \quad \text{(By additive inverse)}$$
$$\Rightarrow B = C \quad \text{(By additive inverse)}$$

Similarly, we can prove that if $B + A = C + A$, then $B = C$.

25.5 PROPERTIES OF MULTIPLICATION OF MATRIX BY A SCALAR

Prop. (i) **Distribution law of scalar multiplication over matrix addition :** If A and B are two matrices of order $m \times n$ and k is any scalar, then $k(A + B) = kA + kB$

Prop. (ii) If A is a matrix of order $m \times n$ and a, b are two scalars, then $(a + b)A = aA + bA$

Prop. (iii) If A is a matrix of order $m \times n$ and a, b are two scalars, then $a(bA) = (ab)A$

Prop. (iv) If A is a matrix of order $m \times n$ and k is any scalar, then $(-k)A = -(kA) = k(-A)$

Solved Examples

Example 1. *Find the number of rows and columns in the following matrices :*

(i) $[1\ 2\ 3\ 4]$ (ii) $\begin{bmatrix} -1 \\ 3 \\ 0 \\ 4 \end{bmatrix}$

(iii) $\begin{bmatrix} -1 & 0 & 3 & 4 \\ 5 & 6 & 7 & 8 \end{bmatrix}$

Solution. (i) In the matrix $[1\ 2\ 3\ 4]$, there is one row and four columns.

(ii) In the matrix $\begin{bmatrix} -1 \\ 3 \\ 0 \\ 4 \end{bmatrix}$ there are four rows and one column.

(iii) In the given matrix $\begin{bmatrix} -1 & 0 & 3 & 4 \\ 5 & 6 & 7 & 8 \end{bmatrix}$.

There are 2 rows and 4 columns.

Example 2. *Find the order of the following matrix :*

$$\begin{bmatrix} -1 & 0 & 3 & 4 \\ 5 & 6 & 7 & 8 \\ 0 & 3 & -2 & 4 \end{bmatrix}$$

Solution. Here the given matrix has 3 rows and 4 columns, so its order is 3 × 4.

Example 3. *For what values of x, y and z, the matrices A and B are equal :*

$$A = \begin{bmatrix} x & 1 & 2 \\ 3 & 4 & z \\ -1 & 0 & 2 \end{bmatrix} \text{ and } B = \begin{bmatrix} 3 & 1 & 2 \\ 3 & 4 & 5 \\ -1 & y & 2 \end{bmatrix}$$

Solution. Since the matrices A and B are given equal, then comparing the corresponding elements, we get

$$x = 3, y = 0 \text{ and } z = 5.$$

Example 4. *If* $A = \begin{bmatrix} 1 & 0 \\ 2 & -1 \end{bmatrix}, B = \begin{bmatrix} 3 & 7 \\ 4 & 8 \end{bmatrix}, C = \begin{bmatrix} -1 & 1 \\ 0 & 0 \end{bmatrix}$ *then find :*

(i) 7 A; *(ii) –3B* *(iii) 2C*

(iv) A–5B *(v) 4A+3C.*

Solution. (i) $7A = 7\begin{bmatrix} 1 & 0 \\ 2 & -1 \end{bmatrix} = \begin{bmatrix} 7\times1 & 7\times0 \\ 7\times2 & 7\times-1 \end{bmatrix}$

$$= \begin{bmatrix} 7 & 0 \\ 14 & -7 \end{bmatrix}$$

(ii) $-3B = -3\begin{bmatrix} 3 & 7 \\ 4 & 8 \end{bmatrix} = \begin{bmatrix} -3\times3 & -3\times7 \\ -3\times4 & -3\times8 \end{bmatrix}$

$$= \begin{bmatrix} -9 & -21 \\ -12 & -24 \end{bmatrix}$$

(iii) $2C = 2\begin{bmatrix} -1 & 1 \\ 0 & 0 \end{bmatrix} = \begin{bmatrix} 2\times-1 & 2\times1 \\ 2\times0 & 2\times0 \end{bmatrix}$

$$= \begin{bmatrix} -2 & 2 \\ 0 & 0 \end{bmatrix}$$

(iv) $A - 5B = \begin{bmatrix} 1 & 0 \\ 2 & -1 \end{bmatrix} - 5\begin{bmatrix} 3 & 7 \\ 4 & 8 \end{bmatrix}$

$$= \begin{bmatrix} 1 & 0 \\ 2 & -1 \end{bmatrix} - \begin{bmatrix} 15 & 35 \\ 20 & 40 \end{bmatrix}$$

$$= \begin{bmatrix} 1-15 & 0-35 \\ 2-20 & -1-40 \end{bmatrix}$$

$$= \begin{bmatrix} -14 & -35 \\ -18 & -41 \end{bmatrix}$$

(v) $4A + 3C = 4\begin{bmatrix} 1 & 0 \\ 2 & -1 \end{bmatrix} + 3\begin{bmatrix} -1 & 1 \\ 0 & 0 \end{bmatrix}$

$$= \begin{bmatrix} 4 & 0 \\ 8 & -4 \end{bmatrix} + \begin{bmatrix} -3 & 3 \\ 0 & 0 \end{bmatrix}$$

$$= \begin{bmatrix} 4-3 & 0+3 \\ 8+0 & -4+0 \end{bmatrix}$$

$$= \begin{bmatrix} 1 & 3 \\ 8 & -4 \end{bmatrix}$$

Example 5. *Find the additive inverse of the matrix :*

$$A = \begin{bmatrix} 2 & -3 & -1 & -1 \\ -3 & 1 & -2 & 2 \\ 1 & -2 & -8 & 7 \end{bmatrix}$$

Solution. The additive inverse of $A = -A$

$$\therefore \quad -A = -\begin{bmatrix} 2 & -3 & -1 & -1 \\ -3 & 1 & -2 & 2 \\ 1 & -2 & -8 & 7 \end{bmatrix}$$

$$= \begin{bmatrix} -2 & 3 & 1 & 1 \\ 3 & -1 & 2 & -2 \\ -1 & 2 & 8 & -7 \end{bmatrix}$$

Example 6. *If* $A = \begin{bmatrix} 1 & 2 \\ 3 & 4 \end{bmatrix}$, $B = \begin{bmatrix} -1 & 5 \\ 5 & 7 \end{bmatrix}$

then prove that $5(A+B) = 5A + 5B$.

Solution. $A + B = \begin{bmatrix} 1 & 2 \\ 3 & 4 \end{bmatrix} + \begin{bmatrix} -1 & 5 \\ 5 & 7 \end{bmatrix}$

$$= \begin{bmatrix} 1-1 & 2+5 \\ 3+5 & 4+7 \end{bmatrix} = \begin{bmatrix} 0 & 7 \\ 8 & 11 \end{bmatrix}$$

Now $5(A+B) = 5\begin{bmatrix} 0 & 7 \\ 8 & 11 \end{bmatrix}$

$$= \begin{bmatrix} 5\times0 & 5\times7 \\ 5\times8 & 5\times11 \end{bmatrix} = \begin{bmatrix} 0 & 35 \\ 40 & 55 \end{bmatrix}$$

and $5A = 5\begin{bmatrix} 1 & 2 \\ 3 & 4 \end{bmatrix}$

$$= \begin{bmatrix} 5\times1 & 5\times2 \\ 5\times3 & 5\times4 \end{bmatrix} = \begin{bmatrix} 5 & 10 \\ 15 & 20 \end{bmatrix}$$

$$5B = 5\begin{bmatrix} -1 & 5 \\ 5 & 7 \end{bmatrix}$$

$$= \begin{bmatrix} 5\times-1 & 5\times5 \\ 5\times5 & 5\times7 \end{bmatrix} = \begin{bmatrix} -5 & 25 \\ 25 & 35 \end{bmatrix}$$

Now

$$5A + 5B = \begin{bmatrix} 5 & 10 \\ 15 & 20 \end{bmatrix} + \begin{bmatrix} -5 & 25 \\ 25 & 35 \end{bmatrix}$$

$$= \begin{bmatrix} 5-5 & 10+25 \\ 15+25 & 20+35 \end{bmatrix}$$

$$= \begin{bmatrix} 0 & 35 \\ 40 & 55 \end{bmatrix}$$

Hence $5(A+B) = 5A + 5B$

25.6 MULTIPLICATION OF MATRICES

Let A and B be two matrices of order $m \times n$ and $n \times p$ respectively. Then a matrix C of order $m \times p$ is obtained by multiplying each row of A to each column of B.

Suppose $A = [a_{ij}]_{m\times n}$, $B = [b_{jk}]_{n\times p}$, then $C = [c_{ik}]_{m\times p}$ is known as the multiplication of A and B where

$$C_{ik} = \sum_{j=1}^{n} a_{ij} b_{jk}$$

and hence, we can write $$C = AB$$

WORKING PROCEDURE

First we check whether the matrices are conformable for multiplication or not. For this we check that if the number of columns of first matrix is equal to the number of rows of the second matrix, then the matrices can be multiplied. Multiplication is operated by the rule (row × Column). In this rule, we first put the first row of the first matrix next to the first column of the second matrix and the corresponding elements are now multiplied and then summed up which gives the first element of the first row of the product matrix. This process runs till the first row of the first matrix is operated to all columns of the second matrix. After that the first process is applied to the second, third, etc. rows of the first matrix.

For example : If $A = \begin{bmatrix} 2 & 1 & 5 \\ 6 & 2 & 3 \end{bmatrix}_{2\times 3}$ and $B = \begin{bmatrix} 3 & 4 \\ 5 & 6 \\ 7 & 8 \end{bmatrix}_{3\times 2}$, then

$$AB = \begin{bmatrix} 2 & 1 & 5 \\ 6 & 2 & 3 \end{bmatrix}\begin{bmatrix} 3 & 4 \\ 5 & 6 \\ 7 & 8 \end{bmatrix} = \begin{bmatrix} 2\times3+1\times5+5\times7 & 2\times4+1\times6+5\times8 \\ 6\times3+2\times5+3\times7 & 6\times4+2\times6+3\times8 \end{bmatrix}$$

$$= \begin{bmatrix} 6+5+35 & 8+6+40 \\ 18+10+21 & 24+12+24 \end{bmatrix} = \begin{bmatrix} 49 & 54 \\ 49 & 60 \end{bmatrix}$$

REMARKS

- If the number of columns of the matrix A is equal to the number of rows of matrix B, then A and B are conformable for the multiplication AB but not for BA.
- Square matrices are always conformable for multiplication both ways.

25.7 PROPERTIES OF MATRIX MULTIPLICATION

(i) Associative law : *If A, B and C are the matrices of order $m \times n$, $n \times p$ and $p \times q$, then $(AB)C = A(BC)$.*

(JNTU 2002)

Proof. Let $A = [a_{ij}]_{m\times n}$, $B = [b_{jk}]_{n\times p}$ and $C = [c_{kl}]_{p\times q}$, then

$$AB = [a_{ij}]_{m\times n} \cdot B = [b_{jk}]_{n\times p} = [x_{ik}]_{m\times p}$$

where $$x_{ik} = \sum_{j=1}^{n} a_{ij} b_{jk} \qquad ...(1)$$

$\therefore$ $$(AB)C = [x_{ik}]_{m\times p} \cdot [c_{kl}]_{p\times q} = [u_{il}]_{m\times q}$$

where $$u_{il} = \sum_{k=1}^{p} x_{ik} c_{kl} \qquad ...(2)$$

Now from (1) and (2), we get

$$u_{il} = \sum_{k=1}^{p}\left(\sum_{j=1}^{n} a_{ij} b_{jk}\right) c_{kl} = \sum_{j=1}^{n} a_{ij}\left(\sum_{k=1}^{p} b_{jk} c_{kl}\right) \qquad ...(3)$$

Now $$BC = [b_{jk}]_{m\times p} \cdot [c_{kl}]_{p\times q} = [v_{jl}]_{m\times q}$$

where $$v_{jl} = \sum_{k=1}^{p} b_{jk}\, c_{kl} \qquad ...(4)$$

From equations (3) and (4), we get $u_{il} = \sum_{j=1}^{n} a_{ij} v_{jl}$ ...(5)

Equation (2) implies that u_{il} is the (i, l)th element in $(AB)C$ and the equation (5) implies that u_{il} is (i, l)th element in $A(BC)$. Therefore by the equality of two matrices, we obtain $(AB)C = A(BC)$.

(ii) **Distributive law :** *Matrix multiplication satisfies the distributive law over the matrix addition.*

If A, B and C are the matrices of order $m\times n$, $n\times p$ and $n\times p$ respectively, then

$$A(B+C) = AB + AC$$

Proof. Let $A = [a_{ij}]_{m\times n}$, $B = [b_{jk}]_{n\times p}$ and $C = [c_{jk}]_{n\times p}$, then

$$B+C = [b_{jk}]_{n\times p} + [c_{jk}]_{n\times p} = [b_{jk}+c_{jk}]_{n\times p}$$

$$\therefore \quad A(B+C) = [a_{ij}]_{m\times n} + [b_{jk}+c_{jk}]_{n\times p} = [x_{ik}]_{m\times p} \qquad ...(1)$$

where
$$x_{ik} = \sum_{j=1}^{n} a_{ij}(b_{jk}+c_{jk})$$
$$= \sum_{j=1}^{n} (a_{ij}b_{jk} + a_{ij}c_{jk}) = \sum_{j=1}^{n} a_{ij}b_{jk} + \sum_{j=1}^{n} a_{ij}c_{jk}$$
$= (i, k)$ the element of AB + (i, k) the element of AC
$= (i, k)$ the element of $(AB + BC)$

But equation (1) implies that x_{ik} is the (i, k) the element of $A(B+C)$, therefore, by the definition of equality of two matrices, we must have

$$A(B + C) = AB + AC.$$

REMARK

- Matrix multiplication is not commutative in general.

Solved Examples

Example 1. *If* $A = \begin{bmatrix} 2 & 3 & 4 \\ 3 & 2 & 3 \\ -1 & 1 & 2 \end{bmatrix}$, $B = \begin{bmatrix} 1 & 3 & 0 \\ -1 & 2 & 1 \\ 1 & 0 & 2 \end{bmatrix}$ *then find AB and BA and show that* $AB \neq BA$.

Solution. Since A and B are square matrices of order 3×3 so that multiplication of AB and BA is possible.

Now,

$$AB = \begin{bmatrix} 2 & 3 & 4 \\ 3 & 2 & 3 \\ -1 & 1 & 2 \end{bmatrix}\begin{bmatrix} 1 & 3 & 0 \\ -1 & 2 & 1 \\ 1 & 0 & 2 \end{bmatrix}$$

$$= \begin{bmatrix} 2\times1+3\times(-1)+4\times0 & 2\times3+3\times2+4\times0 & 2\times0+3\times1+4\times2 \\ 1\times1+2\times(-1)+3\times0 & 1\times3+2\times2+3\times0 & 1\times0+2\times1+3\times2 \\ -1\times1+1\times(-1)+2\times0 & -1\times3+1\times2+2\times0 & -1\times0+1\times1+2\times2 \end{bmatrix}$$

$$= \begin{bmatrix} 2-3+0 & 6+6+0 & 0+3+8 \\ 1-2+0 & 3+4+0 & 0+2+6 \\ -1+1+0 & -3+2+0 & 0+1+4 \end{bmatrix}$$

$$= \begin{bmatrix} -1 & 12 & 11 \\ -1 & 7 & 8 \\ -2 & -1 & 5 \end{bmatrix}_{3\times3}$$

and
$$BA = \begin{bmatrix} 1 & 3 & 0 \\ -1 & 2 & 1 \\ 0 & 0 & 2 \end{bmatrix}\begin{bmatrix} 2 & 3 & 4 \\ 1 & 2 & 3 \\ -1 & 1 & 2 \end{bmatrix}$$

$$= \begin{bmatrix} 1\times2+3\times1+0\times(-1) & 1\times3+3\times2+0\times1 & 1\times4+3\times3+0\times2 \\ -1\times2+2\times1+1\times(-1) & -1\times3+2\times2+1\times1 & -1\times4+2\times3+1\times2 \\ 0\times2+0\times1+2\times(-1) & 0\times3+0\times2+2\times1 & 0\times4+0\times3+2\times2 \end{bmatrix}$$

$$= \begin{bmatrix} 2+3-0 & 3+6+0 & 4+9+0 \\ -2+2-1 & -3+4+1 & -4+6+2 \\ 0+0-2 & 0+0+2 & 0+0+4 \end{bmatrix}$$

$$= \begin{bmatrix} 5 & 9 & 13 \\ -1 & 2 & 4 \\ -2 & 2 & 4 \end{bmatrix}_{3\times3}$$

Here AB and BA have same order but different corresponding elements. Hence $AB \neq BA$.

Example 2. *If* $A = \begin{bmatrix} 7 & 0 & 0 \\ 0 & 7 & 0 \\ 0 & 0 & 7 \end{bmatrix}$ *and* $B = \begin{bmatrix} a & b & c \\ d & e & f \\ g & h & i \end{bmatrix}$ *then prove that* $AB = 7B$.

Solution. $AB = \begin{bmatrix} 7 & 0 & 0 \\ 0 & 7 & 0 \\ 0 & 0 & 7 \end{bmatrix}\begin{bmatrix} a & b & c \\ d & e & f \\ g & h & i \end{bmatrix} = \begin{bmatrix} 7a+0+0 & 7b+0+0 & 7c+0+0 \\ 0+7d+0 & 0+7e+0 & 0+7f+0 \\ 0+0+7g & 0+0+7h & 0+0+7i \end{bmatrix}$

$= \begin{bmatrix} 7\times a+0\times d+0\times g & 7\times b+0\times e+0\times h & 7\times c+0\times f+0\times i \\ 0\times a+7\times d+0\times g & 0\times b+7\times e+0\times h & 0\times c+7\times f+0\times i \\ 0\times a+0\times d+7\times g & 0\times b+0\times e+7\times h & 0\times c+0\times f+7\times i \end{bmatrix} = \begin{bmatrix} 7a & 7b & 7c \\ 7d & 7e & 7f \\ 7g & 7h & 7i \end{bmatrix} = 7\begin{bmatrix} a & b & c \\ d & e & f \\ g & h & i \end{bmatrix} = 7B.$

25.8 DETERMINANT OF A SQUARE MATRIX

Let A be a square matrix. Then the determinant which is formed by the elements of matrix A is usually denoted by $|A|$.

For example : If $A = \begin{bmatrix} a_{11} & a_{12} & a_{13} \\ a_{21} & a_{22} & a_{23} \\ a_{31} & a_{32} & a_{33} \end{bmatrix}$, then its determinant is $A = \begin{vmatrix} a_{11} & a_{12} & a_{13} \\ a_{21} & a_{22} & a_{23} \\ a_{31} & a_{32} & a_{33} \end{vmatrix}$

REMARK

- The determinant of a matrix is reduced to a number.

25.9 PROPERTIES OF DETERMINANTS

(1) The value of a determinant is zero if all the elements of a row or column are zero.
(2) The value of a determinant remain unchanged when rows are changed into corresponding columns.
(3) If any two rows or columns of a determinant are interchanged the sign of the determinant is changed.
(4) If any two rows or columns of a determinant are identical, then the value of the determinant is zero.
(5) If every element of same columns or row is the sum of two terms then determinant is equal to the sum of two determinant are containing only the first term and other the second term only in place of each sum.
(6) If each element of a row (or column) is multiplied by a constant k, then then value of the new determinant will be k times the value of original determinant.
(7) If each element of a row (or column) of a determinant by a constant k and then added to the corresponding elements of some other row (or column) then the value of the determinant remain same.
(8) If the elements of the determinant are the polynomial in a variable x and if by putting $x = a$, the determinant vanishes then $(x - a)$ will be a factor of determinant.

25.10 EVOLUTION OF A DETERMINANT BY SARRUS DIAGRAM

$$\begin{vmatrix} a_{11} & a_{12} & a_{13} \\ a_{21} & a_{22} & a_{23} \\ a_{31} & a_{32} & a_{33} \end{vmatrix} = a_{11}(a_{22}a_{33} - a_{32}a_{23}) - a_{12}(a_{21}a_{33} - a_{31}a_{23}) + a_{13}(a_{21}a_{32} - a_{31}a_{22})$$

$$= a_{11}a_{22}a_{33} + a_{12}a_{31}a_{23} + a_{13}a_{21}a_{32} - (a_{11}a_{32}a_{23} + a_{11}a_{21}a_{33} + a_{13}a_{31}a_{22})$$

WORKING PROCEDURE

- *Write the columns of the determinant and again write the first and second columns on the right side and draw the lines as shown in the following figure :*

$$\begin{matrix} a_{11} & a_{12} & a_{13} & a_{11} & a_{12} \\ a_{21} & a_{22} & a_{23} & a_{21} & a_{22} \\ a_{31} & a_{32} & a_{33} & a_{31} & a_{32} \end{matrix}$$

$\searrow$ + , $\nearrow$ −

For example :

Let $A = \begin{vmatrix} 1 & 2 & 3 \\ 2 & 3 & 4 \\ 2 & 0 & 5 \end{vmatrix}$, *then we have*

$$\begin{matrix} 1 & 2 & 3 & 1 & 2 \\ 2 & 3 & 4 & 2 & 3 \\ 2 & 0 & 5 & 2 & 0 \end{matrix}$$

$|A| = 1\cdot3\cdot5 + 2\cdot4\cdot2 + 3\cdot2\cdot0 - (2\cdot3\cdot3 + 0\cdot4\cdot1 + 5\cdot2\cdot2)$

$= 15 + 16 + 0 - (18 + 0 + 20) = 31 - 38 = -7$

Solved Examples

Example 1. *Evaluate the following determinant*

$$\begin{vmatrix} 3 & -2 \\ 4 & 5 \end{vmatrix}$$

Solution. We have $|A| = \begin{vmatrix} 3 & -2 \\ 4 & 5 \end{vmatrix}$

$$= 3\times 5 - 4\times(-2) = 15 + 8 = 23$$

Example 2. *Find the value of the determinant of the matrix*

$$A = \begin{bmatrix} 1 & 2 & 3 \\ 2 & 3 & 1 \\ 3 & 1 & 2 \end{bmatrix}$$

Solution. We have $A = \begin{bmatrix} 1 & 2 & 3 \\ 2 & 3 & 1 \\ 3 & 1 & 2 \end{bmatrix}$

On expanding the determinant along the first row, we get

$$= 1\begin{vmatrix} 3 & 1 \\ 1 & 2 \end{vmatrix} - 2\begin{vmatrix} 2 & 1 \\ 3 & 2 \end{vmatrix} + 3\begin{vmatrix} 2 & 3 \\ 3 & 1 \end{vmatrix}$$

$$= 1.(6-1) - 2.(4-3) + 3.(2-9) = -18$$

Example 3. *Evaluate the determinant* $\begin{vmatrix} 4 & 1 & 4 \\ 0 & 1 & 0 \\ 1 & 2 & 1 \end{vmatrix}$.

Solution. We have $|A| = \begin{vmatrix} 4 & 1 & 4 \\ 0 & 1 & 0 \\ 1 & 2 & 1 \end{vmatrix}$

On expanding the determinant along the first column, we get

$$= 4\begin{vmatrix} 1 & 0 \\ 2 & 1 \end{vmatrix} - 0\begin{vmatrix} 1 & 4 \\ 2 & 1 \end{vmatrix} + 1\begin{vmatrix} 1 & 4 \\ 1 & 0 \end{vmatrix}$$

$$= 4(1-0) - 0 + 1(0-4)$$

$$= 4 - 4 = 0$$

Example 4. *Show that*

$$\begin{vmatrix} 1 & x & y \\ 0 & \cos x & \sin y \\ 0 & \sin x & \cos y \end{vmatrix} = \cos(x+y)$$

Solution. We have $\begin{vmatrix} 1 & x & y \\ 0 & \cos x & \sin y \\ 0 & \sin x & \cos y \end{vmatrix}$

On expanding the determinant along the first column, we get

$$= 1\begin{vmatrix} \cos x & \sin y \\ \sin x & \cos y \end{vmatrix} - 0\begin{vmatrix} x & y \\ \sin x & \sin y \end{vmatrix} + 0\begin{vmatrix} x & y \\ \cos x & \cos y \end{vmatrix}$$

$$= \cos x \cos y - \sin x \sin y = \cos(x+y)$$

Example 5. *Show that* $\begin{vmatrix} 1 & 1 & 1 \\ 1 & 1+x & 1 \\ 1 & 1 & 1+y \end{vmatrix} = xy$

Solution. We have LH.S. $= \begin{vmatrix} 1 & 1 & 1 \\ 1 & 1+x & 1 \\ 1 & 1 & 1+y \end{vmatrix}$

Applying $C_2 - C_1$ and $C_3 - C_1$ in the given determinant, we get

$$= \begin{vmatrix} 1 & 0 & 0 \\ 1 & x & 0 \\ 1 & 0 & y \end{vmatrix}$$

On expanding the determinant along the first row, we get

$$= 1\begin{vmatrix} x & 0 \\ 0 & y \end{vmatrix} - 0\begin{vmatrix} 1 & 0 \\ 1 & y \end{vmatrix} - 0\begin{vmatrix} 1 & x \\ 1 & 0 \end{vmatrix}$$

$$= xy = \text{R.H.S.}$$

Example 6. *Without expanding, show that*

$$\begin{vmatrix} b-c & c-a & a-b \\ c-a & a-b & b-c \\ a-b & b-c & c-a \end{vmatrix} = 0$$

Solution. We have

$$\begin{vmatrix} b-c & c-a & a-b \\ c-a & a-b & b-c \\ a-b & b-c & c-a \end{vmatrix} = \begin{vmatrix} 0 & c-a & a-b \\ 0 & a-b & b-c \\ 0 & b-c & c-a \end{vmatrix}$$

(Operating $C_1 \to C_1 + C_2 + C_3$, we get)

$$= 0$$

Example 7. *Without expanding, show that*

$$\begin{vmatrix} b^2c^2 & bc & b+c \\ c^2a^2 & ca & c+a \\ a^2b^2 & ab & a+b \end{vmatrix} = 0$$

Solution. Consider

$$\begin{vmatrix} b^2c^2 & bc & b+c \\ c^2a^2 & ca & c+a \\ a^2b^2 & ab & a+b \end{vmatrix} = \frac{abc}{abc}\begin{vmatrix} b^2c^2 & bc & b+c \\ c^2a^2 & ca & c+a \\ a^2b^2 & ab & a+b \end{vmatrix}$$

(Multiply R_1 by a, R_2 by b and R_3 by c)

$$= \frac{1}{abc}\begin{vmatrix} ab^2c^2 & abc & ab+ca \\ bc^2a^2 & abc & bc+ab \\ ca^2b^2 & abc & ca+bc \end{vmatrix}$$

(Take abc out from C_1 and C_2)

$$= \frac{abc.abc}{abc}\begin{vmatrix} bc & 1 & ab+ca \\ ca & 1 & bc+ab \\ ab & 1 & ca+bc \end{vmatrix}$$

$$= abc\begin{vmatrix} bc & 1 & ab+bc+ca \\ ca & 1 & bc+ab+ca \\ ab & 1 & ab+ca+bc \end{vmatrix}$$

(Operate $C_3 \to C_3 + C_1$)

$$= abc(ab+bc+ca)\begin{vmatrix} bc & 1 & 1 \\ ca & 1 & 1 \\ ab & 1 & 1 \end{vmatrix}$$

$$= abc(ab+bc+ca)\times 0$$

$$= 0$$

Example 8. *If a, b, c are in A.P., prove that*

$$\begin{vmatrix} x+1 & x+2 & x+a \\ x+2 & x+3 & x+b \\ x+3 & x+4 & x+c \end{vmatrix} = 0$$

Solution. Given a, b, c are in A.P. therefore, $a+c=2b$

$$\Rightarrow \quad a+c-2b=0$$

Operating $R_1 \to R_1 + R_3 - 2R_2$, we get

$$\begin{vmatrix} x+1 & x+2 & x+a \\ x+2 & x+3 & x+b \\ x+3 & x+4 & x+c \end{vmatrix} = \begin{vmatrix} 0 & 0 & a+c-2b \\ x+2 & x+3 & x+b \\ x+3 & x+4 & x+c \end{vmatrix}$$

$$= \begin{vmatrix} 0 & 0 & 0 \\ x+2 & x+3 & x+b \\ x+3 & x+4 & x+c \end{vmatrix} = 0$$

Example 9. *Prove that*

$$\begin{vmatrix} a & b & c \\ a^2 & b^2 & c^2 \\ a^3 & b^3 & c^3 \end{vmatrix} = abc\begin{vmatrix} 1 & 1 & 1 \\ a & b & c \\ a^2 & b^2 & c^2 \end{vmatrix}$$

$$= abc(a-b)(b-c)(c-a)$$

Solution. We have $|A| = \begin{vmatrix} a & b & c \\ a^2 & b^2 & c^2 \\ a^3 & b^3 & c^3 \end{vmatrix} = = abc\begin{vmatrix} 1 & 1 & 1 \\ a & b & c \\ a^2 & b^2 & c^2 \end{vmatrix}$

Now again, $|A| = abc\begin{vmatrix} 1 & 1 & 1 \\ a & b & c \\ a^2 & b^2 & c^2 \end{vmatrix}$

Applying $C_2 - C_1$ and $C_3 - C_1$, we get

$$= abc\begin{vmatrix} 1 & 0 & 0 \\ a & b-a & c-a \\ a^2 & b^2-a^2 & c^2-a^2 \end{vmatrix}$$

On expanding along the first row, we get

$$= abc\begin{vmatrix} b-a & c-a \\ b^2-a^2 & c^2-a^2 \end{vmatrix}$$

$$= abc[(b-a)(c^2-a^2)-(b^2-a^2)(c-a)]$$

$$= abc[(b-a)(c-a)\{(c+a)-(b+a)\}]$$

$$= abc(b-a)(c-a)(c+a-b-a)$$

$$= abc\ (a-b)\ (b-c)\ (c-a)$$

Example 10. *Prove that*

$$\begin{vmatrix} a+b+2c & a & b \\ c & b+c+2a & b \\ c & a & c+a+2b \end{vmatrix} = 2(a+b+c)^3$$

Solution. Let $|A| = \begin{vmatrix} a+b+2c & a & b \\ c & b+c+2a & b \\ c & a & c+a+2b \end{vmatrix}$

Applying C_2 and C_3 in C_1, we get

$$= \begin{vmatrix} 2(a+b+c) & a & b \\ 2(a+b+c) & b+c+2a & b \\ 2(a+b+c) & a & c+a+2b \end{vmatrix}$$

$$= 2(a+b+c)\begin{vmatrix} 1 & a & b \\ 1 & b+c+2a & b \\ 1 & a & c+a+2b \end{vmatrix}$$

Applying $(R_2 - R_1)$ and $(R_3 - R_1)$, we get

$$= 2(a+b+c)\begin{vmatrix} 1 & a & b \\ 0 & b+c+a & 0 \\ 0 & 0 & c+a+b \end{vmatrix}$$

On expanding determinant along the first column, we get

$$= 2(a+b+c)\begin{vmatrix} b+c+a & 0 \\ 0 & a+b+c \end{vmatrix}$$

$$= 2(a+b+c)(a+b+c)^2$$

$$= 2(a+b+c)^3$$

Example 11. *Prove that*

$$\begin{vmatrix} 1+a & 1 & 1 \\ 1 & 1+b & 1 \\ 1 & 1 & 1+c \end{vmatrix} = abc\left(1+\frac{1}{a}+\frac{1}{b}+\frac{1}{c}\right)$$

Solution. Operating $C_1 \to C_1 - C_3$ and $C_2 \to C_2 - C_3$, we get

$$\begin{vmatrix} 1+a & 1 & 1 \\ 1 & 1+b & 1 \\ 1 & 1 & 1+c \end{vmatrix} = \begin{vmatrix} a & 0 & 1 \\ 0 & b & 1 \\ -c & -c & 1+c \end{vmatrix}$$

$$= a[b.(1+c)-(-c).1]+1[0.(-c)-(-c)b]$$

$$= a(b+bc+c)+bc$$

$$= abc+bc+ca+ab$$

$$= abc\left(1+\frac{1}{a}+\frac{1}{b}+\frac{1}{c}\right)$$

Example 12. *Prove that*

$$\begin{vmatrix} a-b-c & 2a & 2a \\ 2b & b-c-a & 2b \\ 2c & 2c & c-a-b \end{vmatrix} = (a+b+c)^3$$

Solution. Operating $R_1 \to R_1 + R_2 + R_3$, we get

$$\begin{vmatrix} a-b-c & 2a & 2a \\ 2b & b-c-a & 2b \\ 2c & 2c & c-a-b \end{vmatrix}$$

$$= \begin{vmatrix} a+b+c & a+b+c & a+b+c \\ 2b & b-c-a & 2b \\ 2c & 2c & c-a-b \end{vmatrix}$$

(Taking $(a+b+c)$ out from R_1)

$$= (a+b+c)\begin{vmatrix} 1 & 1 & 1 \\ 2b & b-c-a & 2b \\ 2c & 2c & c-a-b \end{vmatrix}$$

(Operate $C_2 \to C_2 - C_1$ and $C_3 \to C_3 - C_1$)

$$= (a+b+c)\begin{vmatrix} 1 & 0 & 0 \\ 2b & -b-c-a & 0 \\ 2c & 0 & -a-b-c \end{vmatrix}$$

(expand by R_1)

$$= (a+b+c).1(-a-b-c)(-a-b-c)$$

$$= (a+b+c)^3$$

Example 13. *Show that*

$$\begin{vmatrix} a & b & c \\ a-b & b-c & c-a \\ b+c & c+a & a+b \end{vmatrix} = a^3+b^3+c^3-3abc$$

Solution. Operating $R_2 \to R_2 - R_1$ and $R_3 \to R_3 + R_1$, we get

$$\begin{vmatrix} a & b & c \\ a-b & b-c & c-a \\ b+c & c+a & a+b \end{vmatrix}$$

$$= \begin{vmatrix} a & b & c \\ -b & -c & -a \\ a+b+c & a+b+c & a+b+c \end{vmatrix}$$

[Take (a+b+c) out from R_3 and (–1) from R_2)

$$= -(a+b+c).\begin{vmatrix} a & b & c \\ b & c & a \\ 1 & 1 & 1 \end{vmatrix} \quad \text{(Expand by } R_3\text{)}$$

$$= -(a+b+c).[1.(ab-c^2) - 1(a^2-bc) + 1.(ca-b^2)]$$

$$= -(a+b+c).(ab+bc+ca-a^2-b^2-c^2)$$

$$= (a+b+c).(a^2+b^2+c^2-ab-bc-ca)$$

$$= a^3+b^3+c^3-3abc$$

Example 14. *Find the value of x if*

$$\begin{vmatrix} 3+x & 5 & 2 \\ 1 & 7+x & 6 \\ 2 & 5 & 3+x \end{vmatrix} = 0$$

Solution. We have $\begin{vmatrix} 3+x & 5 & 2 \\ 1 & 7+x & 6 \\ 2 & 5 & 3+x \end{vmatrix} = 0$

Applying $(R_1 - R_3)$, we get

$$\begin{vmatrix} 1+x & 0 & -1-x \\ 1 & 7+x & 6 \\ 2 & 5 & 3+x \end{vmatrix} = 0$$

Applying $C_3 \to C_3 + C_1$, we get

$$\begin{vmatrix} 1+x & 0 & 0 \\ 1 & 7+x & 7 \\ 2 & 5 & 5+x \end{vmatrix} = 0$$

On expanding the determinant along the first row, we get

$$(1+x)\begin{vmatrix} 7+x & 7 \\ 5 & 5+x \end{vmatrix} = 0$$

$$(1+x)[(7+x)(5+x)-35] = 0$$

or $(1+x)(x^2+12x) = 0$

or $x(1+x)(x+12) = 0$

$\therefore$ $x = 0, -1, -12$

Example 15. *Evaluate :*

$$|A| = \begin{vmatrix} 3 & 2 & 1 & 4 \\ 15 & 29 & 2 & 14 \\ 16 & 19 & 3 & 17 \\ 23 & 39 & 8 & 38 \end{vmatrix}$$

Solution. Applying $C_1 \to C_1 - 3C_2, C_2 \to C_2 - 3C_3$, $C_4 \to C_4 - 4C_3$, we get

$$|A| = \begin{vmatrix} 0 & 0 & 1 & 0 \\ 9 & 25 & 2 & 6 \\ 7 & 13 & 3 & 5 \\ 9 & 23 & 8 & 6 \end{vmatrix}$$

On expanding the determinant along first row, we get

$$= 1\begin{vmatrix} 9 & 25 & 6 \\ 7 & 13 & 5 \\ 9 & 23 & 6 \end{vmatrix}$$

Applying $R_1 \to R_1 - R_3$, we get

$$= 1\begin{vmatrix} 0 & 2 & 0 \\ 7 & 13 & 5 \\ 9 & 23 & 6 \end{vmatrix}$$

On exapnding the determinant along the first row, we get

$$= -2\begin{vmatrix} 7 & 5 \\ 9 & 6 \end{vmatrix} = -2(42-45) = 6$$

Example 16. *Using properties of determinants, solve the following determinant for x.*

$$\begin{vmatrix} a+x & a-x & a-x \\ a-x & a+x & a-x \\ a-x & a-x & a+x \end{vmatrix} = 0$$

Solution. Given $\begin{vmatrix} a+x & a-x & a-x \\ a-x & a+x & a-x \\ a-x & a-x & a+x \end{vmatrix} = 0$

(Operate $C_1 \to C_1 + C_2 + C_3$)

$$\Rightarrow \begin{vmatrix} 3a-x & a-x & a-x \\ 3a-x & a+x & a-x \\ 3a-x & a-x & a+x \end{vmatrix} = 0$$

$$\Rightarrow (3a-x)\begin{vmatrix} 1 & a-x & a-x \\ 1 & a+x & a-x \\ 1 & a-x & a+x \end{vmatrix} = 0$$

(Operate $R_2 \to R_2 - R_1, R_3 \to R_3 - R_1$)

$$\Rightarrow (3a-x)\begin{vmatrix} 1 & a-x & a-x \\ 0 & 2x & 0 \\ 0 & 0 & 2x \end{vmatrix} = 0 \text{ (Expand by } C_1\text{)}$$

$$\Rightarrow (3a-x).1.\begin{vmatrix} 2x & 0 \\ 0 & 2x \end{vmatrix} = 0$$

$$\Rightarrow (3a-x).(4x^2-0) = 0$$

$$\Rightarrow 4x^2(3a-x) = 0$$

$$\Rightarrow x^2 = 0 \text{ or } 3a-x = 0$$

$$\Rightarrow x = 0, 0, 3a$$

Hence, the values of x are 0, 0, $3a$.

Example 17. *Using properties of determinants. Prove that*

$$\begin{vmatrix} 1 & 1 & 1 \\ \alpha & \beta & \gamma \\ \beta\gamma & \gamma\alpha & \alpha\beta \end{vmatrix} = (\alpha-\beta)(\beta-\gamma)(\gamma-\alpha)$$

Solution. Operate $C_2 \to C_2 - C_1$ and $C_3 \to C_3 - C_1$, we get

$$\begin{vmatrix} 1 & 1 & 1 \\ \alpha & \beta & \gamma \\ \beta\gamma & \gamma\alpha & \alpha\beta \end{vmatrix} = \begin{vmatrix} 1 & 0 & 0 \\ \alpha & \beta-\alpha & \gamma-\alpha \\ \beta\gamma & \gamma(\alpha-\beta) & \beta(\alpha-\gamma) \end{vmatrix}$$

[Take $(\alpha-\beta)$ out from C_2 and $(\gamma-a)$ out from C_3)

$$= (\alpha-\beta)(\gamma-\alpha)\begin{vmatrix} 1 & 0 & 0 \\ \alpha & -1 & 1 \\ \beta\gamma & \gamma & -\beta \end{vmatrix}$$

(Expand by C_1)

$$= (\alpha-\beta)(\gamma-\alpha).1.\begin{vmatrix} -1 & 1 \\ \gamma & -\beta \end{vmatrix}$$

$$= (\alpha-\beta)(\gamma-\alpha)(\beta-\gamma)$$

$$= (\alpha-\beta)(\beta-\gamma)(\gamma-\alpha)$$

Example 18. *Prove that*

$$\begin{vmatrix} a^2+1 & ab & ac \\ ab & b^2+1 & bc \\ ac & bc & c^2+1 \end{vmatrix} = 1+a^2+b^2+c^2$$

Solution. We have $|A| = \begin{vmatrix} a^2+1 & ab & ac \\ ab & b^2+1 & bc \\ ac & bc & c^2+1 \end{vmatrix}$

Now multiply the column 1st, 2nd and 3rd by a, b and c respectively, we get

$$|A| = \frac{1}{abc}\begin{vmatrix} a(a^2+1) & ab^2 & ac^2 \\ a^2b & b(b^2+1) & bc^2 \\ a^2c & b^2c & c(c^2+1) \end{vmatrix}$$

To take a, b, c common from 1st, 2nd and 3rd rows respectively, we get

$$= \frac{abc}{abc}\begin{vmatrix} a^2+1 & b^2 & c^2 \\ a^2 & (b^2+1) & c^2 \\ a^2 & b^2 & c^2+1 \end{vmatrix}$$

Now applying $C_1 \to C_1 + C_2 + C_3$, we get

$$= \begin{vmatrix} a^2+b^2+c^2+1 & b^2 & c^2 \\ a^2+b^2+c^2+1 & b^2+1 & c^2 \\ a^2+b^2+c^2+1 & b^2 & c^2+1 \end{vmatrix}$$

$$= (a^2+b^2+c^2+1)\begin{vmatrix} 1 & b^2 & c^2 \\ 1 & b^2+1 & c^2 \\ 1 & b^2 & c^2+1 \end{vmatrix}$$

Now apply $R_2 \to R_2 - R_1$ and $R_3 \to R_3 - R_1$ we get

$$= (a^2+b^2+c^2)\begin{vmatrix} 1 & 0 \\ 0 & 1 \end{vmatrix}$$

$$= a^2+b^2+c^2+1$$

Example 19. *Prove that* $\begin{vmatrix} b+c & c+a & a+b \\ q+r & r+p & p+q \\ y+z & z+x & x+y \end{vmatrix} = 2\begin{vmatrix} a & b & c \\ p & q & r \\ x & y & z \end{vmatrix}$

Solution. We have

$$\text{L.H.S.} = \begin{vmatrix} b+c & c+a & a+b \\ q+r & r+p & p+q \\ y+z & z+x & x+y \end{vmatrix}$$

Applying $C_1 \to C_1 + C_2 - 2C_3$, we get

$$= \begin{vmatrix} 2c & c+a & a+b \\ 2r & r+p & p+q \\ 2z & z+x & x+y \end{vmatrix} = 2\begin{vmatrix} c & c+a & a+b \\ r & r+p & p+q \\ z & z+x & x+y \end{vmatrix}$$

Now applying $C_2 \to C_2 - C_1$, we get

$$= 2\begin{vmatrix} c & a & a+b \\ r & p & p+q \\ z & x & x+y \end{vmatrix}$$

Applying $C_3 \to C_3 - C_2$, we get

$$= 2\begin{vmatrix} c & a & b \\ r & p & q \\ z & x & y \end{vmatrix}$$

$$= 2\begin{vmatrix} a & b & c \\ p & q & r \\ x & y & z \end{vmatrix}$$

(By interchanging the columns)

= R.H.S.

Example 20. *If x, y, z are all different and* $\begin{vmatrix} x & x^2 & 1+x^3 \\ y & y^2 & 1+y^3 \\ z & z^2 & 1+z^3 \end{vmatrix} = 0$

Show that $xyz = -1$ (Andhra 1999, Asam 1999)

Solution. Given $\begin{vmatrix} x & x^2 & 1+x^3 \\ y & y^2 & 1+y^3 \\ z & z^2 & 1+z^3 \end{vmatrix} = 0$

$$\Rightarrow \begin{vmatrix} x & x^2 & 1 \\ y & y^2 & 1 \\ z & z^2 & 1 \end{vmatrix} + \begin{vmatrix} x & x^2 & x^3 \\ y & y^2 & y^3 \\ z & z^2 & z^3 \end{vmatrix} = 0$$

[Take x, y, z out from R_1, R_2 and R_3 respectively from the second determinant]

$$\Rightarrow \begin{vmatrix} 1 & x & x^2 \\ 1 & y & y^2 \\ 1 & z & z^2 \end{vmatrix} + xyz\begin{vmatrix} 1 & x & x^2 \\ 1 & y & y^2 \\ 1 & z & z^2 \end{vmatrix} = 0$$

$$\Rightarrow \begin{vmatrix} 1 & x & x^2 \\ 1 & y & y^2 \\ 1 & z & z^2 \end{vmatrix}(1+xyz) = 0$$

$$\Rightarrow (x-y)(y-z)(z-x).(1+xyz) = 0$$

$$\Rightarrow \qquad (1+xyz) = 0$$

(Because x, y, z are all distinct, so

$$x-y \neq 0, y-z \neq 0, z-x \neq 0)$$

$$\Rightarrow \qquad xyz = -1$$

Example 21. *Evaluate the value of x for which*

$$\begin{vmatrix} 4x & 6x+2 & 8x+1 \\ 6x+2 & 9x+3 & 12x \\ 8x+1 & 12x & 16x+2 \end{vmatrix} = 0$$

Solution. We have $\begin{vmatrix} 4x & 6x+2 & 8x+1 \\ 6x+2 & 9x+3 & 12x \\ 8x+1 & 12x & 16x+2 \end{vmatrix} = 0$

Applying $\left(C_2 \to C_2 - \frac{3}{2}C_1\right)$ and $C_3 \to C_3 - 2C_1$ we get

$$\begin{vmatrix} 4x & 2 & 1 \\ 6x+2 & 0 & -4 \\ 8x+1 & -3/2 & 0 \end{vmatrix} = 0$$

Now applying $R_2 \to R_2 + 4R_1$

$$\begin{vmatrix} 4x & 2 & 1 \\ 22x+2 & 8 & 0 \\ 8x+1 & -3/2 & 0 \end{vmatrix} = 0$$

On expanding the determinant along 3rd column, we get

$$1\begin{vmatrix} 22x+2 & 8 \\ 8x+1 & -3/2 \end{vmatrix} = 0$$

$$\Rightarrow \qquad -33x - 3 - 64x - 8 = 0$$

or $\qquad -97x = 11$ or $x = -\dfrac{11}{97}$

Example 22. *Without expanding show that the value of the determinant given below, in zero*

$$\begin{vmatrix} \sin\alpha & \cos\alpha & \sin(\alpha+\delta) \\ \sin\beta & \cos\beta & \sin(\beta+\delta) \\ \sin\gamma & \cos\gamma & \sin(\gamma+\delta) \end{vmatrix}$$

Solution. Let $\Delta = \begin{vmatrix} \sin\alpha & \cos\alpha & \sin(\alpha+\delta) \\ \sin\beta & \cos\beta & \sin(\beta+\delta) \\ \sin\gamma & \cos\gamma & \sin(\gamma+\delta) \end{vmatrix}$

Using $\sin(A+B) = \sin A\cos B + \cos A \sin B$

$$\Delta = \begin{vmatrix} \sin\alpha & \cos\alpha & \sin\alpha\cos\delta + \cos\alpha\sin\delta \\ \sin\beta & \cos\beta & \sin\beta\cos\delta + \cos\beta\sin\delta \\ \sin\gamma & \cos\gamma & \sin\gamma\cos\delta + \cos\gamma\sin\delta \end{vmatrix}$$

$$= \begin{vmatrix} \sin\alpha & \cos\alpha & 0 \\ \sin\beta & \cos\beta & 0 \\ \sin\gamma & \cos\gamma & 0 \end{vmatrix}$$

(Using $C_3 \to C_3(\cos\delta)C_1 - (\sin\delta)C_2$)

$= 0$

Example 23. *Show that*

$$\begin{vmatrix} (b+c)^2 & a^2 & bc \\ (c+a)^2 & b^2 & ca \\ (a+b)^2 & c^2 & ab \end{vmatrix}$$

$$= (a^2+b^2+c^2)(a+b+c)(b-c)(c-a)(a-b)$$

Solution. Let $\Delta = \begin{vmatrix} (b+c)^2 & a^2 & bc \\ (c+a)^2 & b^2 & ca \\ (a+b)^2 & c^2 & ab \end{vmatrix}$

Applying $C_1 \to C_1 - 2C_3$, we get

$$= \begin{vmatrix} b^2+c^2+a^2 & a^2 & bc \\ c^2+a^2+b^2 & b^2 & ca \\ a^2+b^2+c^2 & c^2 & ab \end{vmatrix}$$

Operating $C_1 \to C_1 + C_2$, we get

$$= (a^2+b^2+c^2)\begin{vmatrix} 1 & a^2 & bc \\ 1 & b^2 & ca \\ 1 & c^2 & ab \end{vmatrix}$$

Operating $R_2 \to R_2 - R_1$ and $R_3 \to R_3 - R_2$

$$= (a^2+b^2+c^2)\begin{vmatrix} 1 & a^2 & bc \\ 0 & b^2-a^2 & (ca-bc) \\ 0 & c^2-a^2 & (ab-bc) \end{vmatrix}$$

$$= (a^2+b^2+c^2)(b-a)(c-a)\begin{vmatrix} 1 & a^2 & bc \\ 0 & b+a & -c \\ 0 & c+a & -b \end{vmatrix}$$

$R_3 \to R_3 - R_2$, we get

$$= (a^2+b^2+c^2)(b-a)(c-a)\begin{vmatrix} 1 & a^2 & bc \\ 0 & b+a & -c \\ 0 & c-b & c-b \end{vmatrix}$$

$$= (a^2+b^2+c^2)(b-a)(c-a)(c-b)\begin{vmatrix} 1 & a^2 & bc \\ 0 & b+a & -c \\ 0 & 1 & 1 \end{vmatrix}$$

Expanding along first column, we get

$$\Delta = (a^2+b^2+c^2)(b-a)(c-a)(c-b)(a+b+c)$$

Example 24. *Show that*

$$\begin{vmatrix} a+b & b+c & c+a \\ b+c & c+a & a+b \\ c+a & a+b & b+c \end{vmatrix} = 2\begin{vmatrix} a & b & c \\ b & c & a \\ c & a & b \end{vmatrix}$$

Solution. Let $\Delta = \begin{vmatrix} a+b & b+c & c+a \\ b+c & c+a & a+b \\ c+a & a+b & b+c \end{vmatrix}$

Applying $C_1 \to C_1 + C_2 + C_3$, we get

$$= \begin{vmatrix} 2(a+b+c) & b+c & c+a \\ 2(a+b+c) & c+a & a+b \\ 2(a+b+c) & a+b & b+c \end{vmatrix}$$

$$= 2\begin{vmatrix} (a+b+c) & -a & -b \\ (a+b+c) & -b & -c \\ (a+b+c) & -c & -a \end{vmatrix}$$

Applying $C_2 \to C_2 - C_1, C_3 \to C_3 - C_1$, we get

$$= 2(-1)(-1)\begin{vmatrix} a+b+c & a & b \\ a+b+c & b & c \\ a+b+c & c & a \end{vmatrix}$$

Applying $C_1 \to C_1 - C_2 - C_3$, we get

$$= 2\begin{vmatrix} c & a & b \\ a & b & c \\ b & c & a \end{vmatrix}$$

$$= -2\begin{vmatrix} a & c & b \\ b & a & c \\ c & b & a \end{vmatrix} (C_1 \to C_2) = 2\begin{vmatrix} a & b & c \\ b & c & a \\ c & b & a \end{vmatrix}$$

Example 25. *If a, b, c (all positive) are the pth, qth and rth terms respectively of a geometric progression, show that*

$$\begin{vmatrix} \log a & p & 1 \\ \log b & q & 1 \\ \log c & r & 1 \end{vmatrix} = 0$$

Solution. Consider the terms of G.P. which are

$$A, AR, AR^2, \ldots$$

$$a = T_p = AR^{p-1}$$

$$b = T_q = AR^{q-1}$$

$$c = T_r = AR^{r-1}$$

$$\text{Consider } \begin{vmatrix} \log a & p & 1 \\ \log b & q & 1 \\ \log c & r & 1 \end{vmatrix} = \begin{vmatrix} \log AR^{p-1} & p & 1 \\ \log AR^{q-1} & q & 1 \\ \log AR^{r-1} & r & 1 \end{vmatrix}$$

$$= \begin{vmatrix} \log A + (p-1)\log R & p & 1 \\ \log A + (q-1)\log R & q & 1 \\ \log A + (r-1)\log R & r & 1 \end{vmatrix}$$

$$= \begin{vmatrix} \log A & p & 1 \\ \log A & q & 1 \\ \log A & r & 1 \end{vmatrix} + \begin{vmatrix} (p-1)\log R & p & 1 \\ (q-1)\log R & q & 1 \\ (r-1)\log R & r & 1 \end{vmatrix}$$

$$= \log A \times 0 + \log R\begin{vmatrix} p & p & 1 \\ q & q & 1 \\ r & r & 1 \end{vmatrix}$$

$$= 0 + \log R \times 0 = 0$$

Example 26. *Show that*

$$\begin{vmatrix} b^2+c^2 & ab & ac \\ ba & c^2+a^2 & bc \\ ca & cb & a^2+b^2 \end{vmatrix} = 4a^2b^2c^2$$

Solution. Let $\Delta = \begin{vmatrix} b^2+c^2 & ab & ac \\ ba & c^2+a^2 & bc \\ ca & cb & a^2+b^2 \end{vmatrix}$

Multiplying R_1, R_2, R_3 by a, b, c respectively and dividing Δ by abc, we get

$$\Delta = \frac{1}{abc}\begin{vmatrix} a(b^2+c^2) & a^2b & a^2c \\ b^2a & b(c^2+a^2) & b^2c \\ c^2a & c^2b & c(a^2+b^2) \end{vmatrix}$$

Taking a, b, c common from C_1, C_2, C_3 respectively, we get

$$\Delta = \frac{abc}{abc}\begin{vmatrix} b^2 & a^2 & a^2 \\ b^2 & c^2+a^2 & b^2 \\ c^2 & c^2 & a^2+b^2 \end{vmatrix}$$

Applying $R_1 \to R_1 + R_2 + R_3$, we get

$$= \begin{vmatrix} 2(a^2+b^2) & 2(c^2+a^2) & 2(a^2+b^2) \\ b^2 & c^2+a^2 & b^2 \\ c^2 & c^2 & a^2+b^2 \end{vmatrix}$$

Taking 2 common from R_1, we get

$$= 2\begin{vmatrix} a^2+b^2 & c^2+a^2 & a^2+b^2 \\ b^2 & c^2+a^2 & b^2 \\ c^2 & c^2 & a^2+b^2 \end{vmatrix}$$

Operating $R_2 \to R_2 - R_1, R_3 \to R_3 - R_1$, we get

$$= 2\begin{vmatrix} b^2+c^2 & c^2+a^2 & a^2+b^2 \\ -c^2 & 0 & -a^2 \\ -b^2 & -a^2 & 0 \end{vmatrix}$$

Operating $R_1 \to R_1 + R_2 + R_3$, we get

$$\Delta = 2\begin{vmatrix} 0 & c^2 & b^2 \\ -c^2 & 0 & -a^2 \\ -b^2 & -a^2 & 0 \end{vmatrix}$$

$$= 2[0 - c^2(0 - a^2b^2) + b^2(a^2c^2 - 0)]$$

(Expanding along R_1)

$$= 2[a^2b^2c^2 + a^2b^2c^2] = 4a^2b^2c^2$$

Example 27. *Show that*

$$\begin{vmatrix} (y+z)^2 & xy & zx \\ xy & (x+z)^2 & yz \\ xz & yz & (x+y)^2 \end{vmatrix} = 2xyz(x+y+z)^2$$

Solution. Let $\Delta = \begin{vmatrix} (y+z)^2 & xy & zx \\ xy & (x+z)^2 & yz \\ xz & yz & (x+y)^2 \end{vmatrix}$

Operating $R_1 \to xR_1, R_2 \to yR_2, R_3 \to zR_3$, we get

$$\Delta = \frac{1}{xyz}\begin{vmatrix} x(y+z)^2 & x^2y & x^2z \\ xy^2 & y(x+z)^2 & y^2z \\ xz^2 & yz^2 & z(x+y)^2 \end{vmatrix}$$

Taking x, y, z common from C_1, C_2, C_3 respectively, we get

$$\Delta = \frac{xyz}{xyz}\begin{vmatrix}(y+z)^2 & x^2 & x^2\\ y^2 & (x+z)^2 & y^2\\ z^2 & z^2 & (x+y)^2\end{vmatrix}$$

$$= \begin{vmatrix}(y+z)^2 & x^2 & x^2\\ y^2 & (x+z)^2 & y^2\\ z^2 & z^2 & (x+y)^2\end{vmatrix}$$

$$= \begin{vmatrix}(y+z)^2-x^2 & 0 & x^2\\ 0 & (z+x)^2-y^2 & y^2\\ z^2-(x+y)^2 & z^2-(x+y)^2 & (x+y)^2\end{vmatrix}$$

Operating $C_1 \to C_1 - C_3, C_2 \to C_2 - C_3$, we get

$$= \begin{vmatrix}(y+z+x)(y+z-x) & 0 & x^2\\ 0 & (z+x+y)(z+x-y) & y^2\\ (z+x+y)(z-x-y) & (z+x+y)(z-x-y) & (x+y)^2\end{vmatrix}$$

Taking $(x+y+z)$ common from C_1 and C_2 each, we get

$$\Delta = (x+y+z)^2\begin{vmatrix}y+z-x & 0 & x^2\\ 0 & z+x-y & y^2\\ z-x-y & z-x-y & (x+y)^2\end{vmatrix}$$

Operating $R_3 \to R_3 - R_1 - R_2$, we get

$$\Delta = (x+y+z)^2\begin{vmatrix}y+z-x & 0 & x^2\\ 0 & z+x-y & y^2\\ -2y & -2x & 2xy\end{vmatrix}$$

$$= (x+y+z)^2\begin{vmatrix}y+z & x^2/y & x^2\\ y^2/x & z+x & y^2\\ 0 & 0 & 2xy\end{vmatrix}$$

Expanding along R_1, we get

$$\Delta = (x+y+z)^2.2xy\begin{vmatrix}y+z & x^2/y\\ y^2/x & z+x^2\end{vmatrix}$$

$$= (x+y+z)^2.2xy[(y+z)(z+x)-xy)]$$

$$= (x+y+z)^2.2xy(yz+z^2+zx)$$

$$= 2xyz(x+y+z)^2$$

Example 28. *Solve the equation*

$$\begin{vmatrix}3x-8 & 3 & 3\\ 3 & 3x-8 & 3\\ 3 & 3 & 3x-8\end{vmatrix} = 0$$

Solution. The given equation is

$$\begin{vmatrix}3x-8 & 3 & 3\\ 3 & 3x-8 & 3\\ 3 & 3 & 3x-8\end{vmatrix} = 0$$

$$\Rightarrow \begin{vmatrix}3x-2 & 3x-2 & 3x-2\\ 3 & 3x-8 & 3\\ 3 & 3 & 3x-8\end{vmatrix} = 0$$

(By applying $R \to R_1 + R_2 + R_3$)

$$\Rightarrow (3x-2)\begin{vmatrix}1 & 1 & 1\\ 3 & 3x-8 & 3\\ 3 & 3 & 3x-8\end{vmatrix} = 0$$

$$\Rightarrow (3x-2)\begin{vmatrix}0 & 0 & 1\\ 0 & 3x-11 & 3\\ 11-3x & 11-3x & 3x-8\end{vmatrix} = 0$$

(Applying $C_1 \to C_1 - C_3, C_2 \to C_2 - C_3$)

$$\Rightarrow (3x-2)\times 1\begin{vmatrix}0 & 3x-11\\ 11-3x & 11-3x\end{vmatrix} = 0$$

$$\Rightarrow (3x-2)(3x-11)^2 = 0$$

$$\Rightarrow x = \frac{2}{3}, \frac{11}{3}, \frac{11}{3}$$

EXERCISE 25.1

Evaluate the following determinants. (1 to 7)

1. $\begin{vmatrix}\frac{1}{2} & 8\\ 4 & 2\end{vmatrix}$

2. $\begin{vmatrix}-2 & 3\\ 4 & -9\end{vmatrix}$

3. $\begin{vmatrix}\cos\theta & -\sin\theta\\ \sin\theta & \cos\theta\end{vmatrix}$

4. $\begin{vmatrix}x^2-x+1 & x-1\\ x+1 & x+1\end{vmatrix}$

5. $\begin{vmatrix}1 & 0 & 6\\ 3 & 4 & 15\\ 5 & 6 & 21\end{vmatrix}$

6. $\begin{vmatrix}23 & 12 & 11\\ 36 & 10 & 26\\ 63 & 26 & 37\end{vmatrix}$

7. $\begin{vmatrix}3 & 1 & -4\\ 3 & 2 & 5\\ 1 & 1 & 3\end{vmatrix}$

Write the minor and co-factor of each element of the following determinants and also evaluate the determinants in each case. (8 to 11):

8. $\begin{vmatrix}5 & -10\\ 0 & 3\end{vmatrix}$

9. $\begin{vmatrix}1 & 3 & -2\\ 4 & -5 & 6\\ 3 & 5 & 2\end{vmatrix}$

10. $\begin{vmatrix}1 & 0 & 0\\ 0 & 1 & 0\\ 0 & 0 & 1\end{vmatrix}$

11. $\begin{vmatrix}1 & 0 & 4\\ 3 & 5 & -1\\ 0 & 1 & 2\end{vmatrix}$

12. Evaluate $\begin{vmatrix}x+1 & x+2 & x+4\\ x+5 & x+6 & x+8\\ x+7 & x+10 & x+14\end{vmatrix}$

13. Evaluate $\begin{vmatrix} 1 & a & bc \\ 1 & b & ca \\ 1 & c & ab \end{vmatrix}$

14. Evaluate $\begin{vmatrix} x+\lambda & x & x \\ x & x+\lambda & x \\ x & x & x+\lambda \end{vmatrix}$

15. Evaluate $\begin{vmatrix} b+c & a & a \\ b & c+a & b \\ c & c & a+b \end{vmatrix}$

16. Prove that $\begin{vmatrix} 1 & x & x^2 \\ 1 & y & y^2 \\ 1 & z & z^2 \end{vmatrix} = (x-y)(y-z)(z-x)$

17. Prove that $\begin{vmatrix} -a^2 & ab & ac \\ ba & -b^2 & bc \\ ac & bc & -c^2 \end{vmatrix} = 4a^2b^2c^2$

18. Prove that $\begin{vmatrix} x & x^2 & yz \\ y & y^2 & zx \\ z & z^2 & xy \end{vmatrix} = (x-y)(y-z)(z-x)(xy+yz+zx)$

19. Using properties of determinants, prove that

$$\begin{vmatrix} y+z & x & y \\ z+x & z & x \\ x+y & y & z \end{vmatrix} = (x+y+z)(x-z)^2$$

20. Using properties of determinants, prove that

$$\begin{vmatrix} a-b-c & 2a & 2a \\ 2b & b-c-a & 2b \\ 2c & 2c & c-a-b \end{vmatrix} = (a+b+c)^3$$

21. Solve the following determinant

$$\begin{vmatrix} x-2 & 2x-3 & 3x-4 \\ x-4 & 2x-9 & 3x-16 \\ x-8 & 2x-27 & 3x-64 \end{vmatrix} = 0$$

22. Prove that using properties of determinants

$$\begin{vmatrix} 1+a^2-b^2 & 2ab & -2b \\ 2ab & 1-a^2+b^2 & 2a \\ 2b & -2a & 1-a^2-b^2 \end{vmatrix} = (1+a^2+b^2)^3$$

23. Prove that

$$\begin{vmatrix} x & x^2 & 1+px^3 \\ y & y^2 & 1+py^3 \\ z & z^2 & 1+pz^3 \end{vmatrix} = (1+pxyz)(x-y)(y-z)(z-x)$$

(Andhra 1999, Assam 1999)

24. Prove that using properties of determinants

$$\begin{vmatrix} 3a & -a+b & -a+c \\ -b+a & 3b & -b+c \\ -c+a & -c+b & 3c \end{vmatrix} = 3(a+b+c)(ab+bc+ca)$$

25. Prove that

$$\begin{vmatrix} \sin\alpha & \cos\alpha & \cos(\alpha+\delta) \\ \sin\beta & \cos\beta & \cos(\beta+\delta) \\ \sin\gamma & \cos\gamma & \cos(\gamma+\delta) \end{vmatrix} = 0$$

Hint to Selected Problems

12. Applying $R_3 \to R_3 - R_1$ and $R_2 \to R_2 - R_1$
$C_3 \to C_3 - C_1$ and $C_2 \to C_2 - C_1$
We get value of let = – 24

13. Applying $R_2 \to R_2 - R_1$ and $R_3 \to R_3 - R_1$
and after expansion, we get the required result.

14. Applying $R_1 \to R_1 + R_2 + R_3$
$C_2 \to C_2 - C_1 . C_3 \to C_3 - C_1$

15. Applying $R_1 \to R_1 + R_2 - R_3$ and expanding.

16. Applying $R_1 \to R_2 - R_1, R_3 \to R_3 - R_1$.

17. Taking *a, b, c* common from first, second and third column and after than applying $R_2 \to R_2 + R_1$ and $R_3 \to R_3 + R_1$.

18. Multiplying first, second and third rows of the determinant by *x, y, z* respectively, and thus Applying $C_2 \to C_2 - C_1, C_3 \to C_3 - C_1$

19. Applying $R_1 \to R_1 + R_2 + R_3$
Then $C_1 \to C_1 - C_2 - C_3$.

20. Applying $R_1 \to R_1 + R_2 + R_3$,
$C_2 \to C_2 - C_1$
$C_3 \to C_3 - C_1$

22. Applying $C_1 \to C_1 - C_3$,
$R_3 \to R_3 - R_1$

24. $C_1 \to C_1 + C_2 + C_3$
$R_2 \to R_2 - R_1, R_3 \to R_3 - R_1$

ANSWERS

1. –31 **2.** 6 **3.** 1 **4.** $x^3 - x^2 + 2$ **5.** – 18 **6.** 0 **7.** 49

8. $M_{11} = 3, M_{12} = 0, M_{21} = -10, M_{22} = 5, A_{11} = 3, A_{12} = 0, A_{21} = 10, A_{22} = 5$, 15

9. $M_{11} = -40, M_{12} = -10, M_{13} = 35, M_{21} = 16, M_{22} = 8, M_{23} = -4, M_{31} = 8, M_{32} = 14, M_{33} = -17$
$A_{11} = -40, A_{12} = 10, A_{13} = 35, A_{21} = -16, A_{22} = 8, A_{23} = 4, A_{31} = 8, A_{32} = -14, A_{33} = -17$; –80

10. $M_{11} = 1, M_{12} = 0, M_{13} = 0, M_{21} = 0, M_{22} = 1, M_{23} = 0, M_{31} = 0, M_{32} = 0, M_{33} = 1$
$A_{11} = 1, A_{12} = 0, A_{13} = 0, A_{21} = 0, A_{22} = 1, A_{23} = 0, A_{31} = 0, A_{32} = 0, A_{33} = 1$; 1

11. $M_{11} = 11, M_{12} = 6, M_{13} = 3, M_{21} = -4, M_{22} = 2, M_{23} = 1, M_{31} = -20, M_{32} = -13, M_{33} = 5$
$A_{11} = 11, A_{12} = -6, A_{13} = 3, A_{21} = 4, A_{22} = 2, A_{23} = -1, A_{31} = 20, A_{32} = 13, A_{33} = 5$; 23

12. –24 **13.** (*a–b*) (*b–c*) (*c–a*) **14.** $\lambda^2(3x+\lambda)$ **15.** 4*abc* **21.** $x = 4$

25.11 MINOR AND COFACTORS

In determinant, $$\Delta = \begin{vmatrix} a_{11} & a_{12} & a_{13} \\ a_{21} & a_{22} & a_{23} \\ a_{31} & a_{32} & a_{33} \end{vmatrix} \qquad ...(1)$$

If we leave the row and column passing through the element a_{ij} then we obtained the second order determinant, which is called the minor of the element a_{ij}. It is denoted by M_{ij} . Therefore, in a determinant of order 3, we may get 9 minors corresponding to the 9 elements of the determinant.

For example, in determinant (1)

$$\text{Minor of } a_{21} = \begin{vmatrix} a_{12} & a_{13} \\ a_{32} & a_{33} \end{vmatrix} = M_{21}$$

and $$\text{Minor of } a_{32} = \begin{vmatrix} a_{11} & a_{13} \\ a_{21} & a_{23} \end{vmatrix} = M_{32}$$

If we expand the determinant along the first row, then

$$\Delta = (-1)^{1+1} a_{11} M_{11} + (-1)^{1+2} a_{12} M_{12} + (-1)^{1+3} a_{13} M_{13}$$
$$= a_{11} M_{11} - a_{12} M_{12} + a_{13} M_{13}$$

Similarly, along second column, we can write

$$\Delta = -a_{12} M_{12} + a_{22} M_{22} - a_{32} M_{32}$$

Cofactor: If we multiply the minor M_{ij} by $(-1)^{i+j}$. Then resulting value is called cofactor of the element a_{ij}. If A_{ij} is the cofactor of then we write

$$\text{Cofactor of } a_{ij} = A_{ij} = (-1)^{i+j} M_{ij}$$

$$\text{Cofactor of } a_{21} = A_{21} = (-1)^{2+1} M_{21} = -\begin{vmatrix} a_{12} & a_{13} \\ a_{32} & a_{33} \end{vmatrix}$$

$$\text{Cofactor of } a_{32} = A_{32} = (-1)^{3+2} M_{32} = -\begin{vmatrix} a_{11} & a_{13} \\ a_{21} & a_{23} \end{vmatrix}$$

Hence, cofactor of $a_{ij} = (-1)^{i+j}$ determinant obtained by leaving row and column passing through that element. Therefore, we can write

$$\Delta = a_{11}A_{11} + a_{12}A_{12} + a_{13}A_{13}$$
$$\Delta = a_{21}A_{21} + a_{22}A_{22} + a_{23}A_{23}$$
$$\Delta = a_{31}A_{31} + a_{32}A_{32} + a_{33}A_{33}$$

and

$$a_{11}A_{21} + a_{12}A_{22} + a_{13}A_{23} = 0$$
$$a_{11}A_{31} + a_{12}A_{32} + a_{13}A_{33} = 0$$

25.12 SINGULAR AND NON-SINGULAR MATRIX

Definition. *A matrix whose determinant value is zero, is said to be singular matrix.*
If the matrix is not singular then it is said to be non-singular.

For example : If $A = \begin{bmatrix} 2 & 3 \\ 6 & 9 \end{bmatrix}$, then its determinant value $|A| = \begin{bmatrix} 2 & 3 \\ 6 & 9 \end{bmatrix} = 2\times 9 - 3\times 6 = 18 - 18 = 0$

Thus the matrix A is singular.

25.13 TRANSPOSE OF A MATRIX

Definition 1. *Consider a matrix $A = [a_{ij}]_{m\times n}$. Then a matrix which is obtained by interchanging the rows and columns of A is called the transpose of A. It is denoted by A' or A^T.*

That is , if $A = [a_{ij}]_{m\times n}$, then $A' = [a_{ji}]_{n\times m}$.

For example : If $A = \begin{bmatrix} 2 & 3 & 5 \\ 1 & 6 & 7 \end{bmatrix}_{2\times 3}$, then its transpose is $A' = \begin{bmatrix} 2 & 3 & 5 \\ 1 & 6 & 7 \end{bmatrix}' = \begin{bmatrix} 2 & 1 \\ 3 & 6 \\ 5 & 7 \end{bmatrix}_{3\times 2}$

Definition 2. *Transpose of the matrix of cofactors is called adjoint of the given matrix.*

REMARKS

- Transpose of row matrix is a column matrix and transpose of a column matrix is a row matrix.
- If a matrix is square then its transpose will be a square matrix of same order.

25.14 PROPERTIES OF TRANSPOSE OF A MATRIX

THEOREM 1. *If A′ and B′ are the transpose of the matrix A and B respectively, then :*

(i) $(A')' = A$ (ii) $(A+B)' = A' + B'$, *(here A and B must be of same order).*

(iii) $(kA)' = kA'$, *here k is any scalar.* (iv) $(AB)' = B'A'$, *here AB and B′A′ are conformable for multiplication.*

Proof.

(i) Let $A = [a_{ij}]_{m\times n}$, then $A' = [a_{ji}]_{n\times m}$ since

(i, j)th element in $(A')' = (j, i)$th element in $A' = (i, j)$th element in A

Thus by the definition of equality of matrices, we must have $(A')' = A$.

(ii) Let $A = [a_{ij}]_{m\times n}$, $B = [b_{ij}]_{m\times n}$. So, $A' = [a_{ji}]_{n\times m}$ *and* $B' = [b_{ji}]_{n\times m}$, then

(i, j)th element in $(A+B)' = (j, i)$th element in $(A+B)$
$= (j, i)$th element in A + (j, i)th element in B
$= (i, j)$th element in A' + (i, j)th element in B'
$= (i, j)$th element in $(A'+B')$

Thus by the definition of equality of matrices, we get

$$(A+B)' = A' + B'$$

(iii) Let $A = [a_{ij}]_{m\times n}$ so that $A' = [a_{ji}]_{n\times m}$ and k be a scalar, then

(i, j)th element in $(kA)' = (j, i)$th element in $(kA) = (i, j)$th element in kA'

Thus by the defintion of equality of matrices, we get

$$(kA)' = kA'$$

(iv) Let $A = [a_{ij}]_{m\times n}$ and $B = [b_{ij}]_{n\times p}$ then AB is conformable for multiplication and having the order $m\times p$. Therefore, the order of $(AB)'$ is $p \times m$. Since the orders of A' and B' are respectively $n \times m$ and $p \times n$ so $B'A'$ is conformable for multiplication and having the order $p \times m$.

Now (k, i)th element in $(AB)' = (i, k)$th element in $AB = \sum_{j=1}^{n} a_{ij}b_{jk}$

(By the definition of multiplication of matrices)

But (k, i)th element in $B'A' = \sum_{j=1}^{n} b_{kj}b_{ji} = \sum_{j=1}^{n} a_{ji}b_{kj} = (i, k)$th elemetn in AB

$\therefore$ (k, i)th element in $A'B' = (k, i)$th element in $B'A'$

Thus by the definition of equality of matrices, we must have $(AB)' = B'A'$

25.15 SYMMETRIC MATRIX

Definition: *A matrix 'A'is said to be a symmetric matrix if $A' = A$, that is, the transpose of a matrix is equal to the matrix itself.*

For exmaple : If $A = \begin{bmatrix} 1 & 2 & 3 \\ 2 & 4 & 5 \\ 3 & 5 & 6 \end{bmatrix}$, then $A' = \begin{bmatrix} 1 & 2 & 3 \\ 2 & 4 & 5 \\ 3 & 5 & 6 \end{bmatrix}$ so that $A' = A$

Hence, A is symmetric.

25.16 SKEW-SYMMETRIC MATRIX

Definition: *A matrix 'A' is said to be a skew-symmetric matrix if $A' = -A$.*

For exmaple : If $A = \begin{bmatrix} 0 & 2 & 3 \\ -2 & 0 & 4 \\ -3 & -4 & 0 \end{bmatrix}$, then $A' = \begin{bmatrix} 0 & -2 & -3 \\ 2 & 0 & -4 \\ 3 & 4 & 0 \end{bmatrix} = \begin{bmatrix} 0 & 2 & 3 \\ -2 & 0 & 4 \\ 3 & -4 & 0 \end{bmatrix} = -A$

Hence A is skew-symmetric matrix.

REMARK

- The diagonal elements of a skew-symmetric matrix are all zero.

25.17 PROPERTIES OF SYMMETRIC AND SKEW-SYMMETRIC MATRIX

(i) *If A is a symmetric (skew-symmetric) matrix, then kA is symmetric (skew-symmetric) matrix, where k is any scalar.*

Proof. Let A be a symmetric matrix, then $A' = A$, since we have

$$(kA)' = kA' = kA \qquad (\because A' = A)$$

$\Rightarrow$ $(kA)' = (kA)$

$\Rightarrow$ kA is symmetric matrix.

Also, if A is a skew-symmetric matrix, then $A' = -A$.

Since we have $(kA)' = kA' = k(-A) \qquad (\because A' = -A)$

$$= -(kA)$$

$\therefore$ $(kA') = -(kA)$

$\Rightarrow$ kA is skew-symmetric matrix.

(ii) If *A and B are symmetric (skew-symmetric) matrices then A+ B is symmetric (skew- symmetric) matrix.*

Proof. Let A and B be symmetric matrices, then $A' = A$ and $B' = B$.

Since we have $(A+B)' = A' + B'$

$\Rightarrow$ $(A+B)' = (A+B) \qquad (\because A' = A, B' = A)$

$\Rightarrow$ $A+B$ is symmetric.

Similarly, if A and B are skew-symmetric matrices, then $A' = -A$ and $B' = -B$.

But $(A+B)' = A' + B'$

$\Rightarrow$ $(A+B)' = -A - B \qquad (\because A' = -A, B' = -B)$

$\Rightarrow$ $(A+B)' = -(A+B)$

$\Rightarrow$ $A+B$ is skew-symmetric matrix.

(iii) *If A is any matrix, then AA' and A'A both are symmetric matrices.*

Proof. If $(AA')' = (A')'A' \qquad [\because (AB)' = B'A']$

$$= AA'$$

$\Rightarrow$ AA' is symmetric.

Also $(A'A)' = A'(A')' \qquad [\because (AB)' = B'A']$

$$= A'A \qquad [\because (A')' = A]$$

$\Rightarrow$ AA' is symmetric.

(iv) *If A is any square matrix, then A + A' is symmetric and A – A' is skew-symmetric.*

Proof. We have $(A+A')' = A' + (A')' \qquad [\because (A+B)' = A' + B']$

$$= A' + A \qquad [\because (A')' = A]$$

$$= A + A' \qquad \text{[Matrix addition is commutative.]}$$

$\Rightarrow$ $A + A'$ is symmetric.

Similarly, $(A - A')' = A' - (A')'$

$$= A' - A = -(A - A')$$

$\Rightarrow$ $A - A'$ is skew-symmetric.

(v) *All positive integral powers of a symmetric matrix are symmetric.*

Proof. Let A be a symmetric matrix and m be any positive integer, then

$$A_m = A \,.\, A \,.\, A \,.\, \ldots .\, A \text{ (m times)}$$

$\Rightarrow$ $(A^m)' = [A \,.\, A \,.\, A \,.\, \ldots A \text{ (m times)}]'$

$$= A' \,.\, A' \,.\, A' \ldots A' \text{ (m times)} (\because A' = A)$$

$$= A \,.\, A \,.\, A \,.\, \ldots A \text{ (m times)} = A^m$$

$\Rightarrow$ A^m is symmetric matrix.

25.18 COMPLEX MATRIX

A matrix 'A' is said to be *complex matrix* if it contains some of its elements equal to a complex number.

25.19 CONJUGATE OF A COMPLEX MATRIX

Definition. *Let A be a complex matrix, then a matrix which is obtained by replacing all the complex elements of A by their complex conjugate number, is called conjugate of a matrix. It is denoted by A.*

For example: If $A = \begin{bmatrix} 1+2i & 3i & 6 \\ 7 & 2+4i & 1+i \end{bmatrix}$, then $\bar{A} = \begin{bmatrix} 1-2i & -3i & 6 \\ 7 & 2-4i & 1-i \end{bmatrix}$

25.20 TRANSPOSE CONJUGATE OF A MATRIX

Definition. *The transpose of the conjugate matrix is called the transposed conjugate of that matrix. It is denoted by* A^θ.

Therefore, we must have $(A^\theta) = (\bar{A})'$.

25.21 PROPERTIES OF TRANSPOSE CONJUGATE OF MATRIX

(1) $(A^\theta) = A$

(2) $(A + B)^\theta = A^\theta + B^\theta$, A, B being of same order

(3) $(kA)^\theta = \bar{k}A^\theta$, k being a complex number

(4) $(AB)^\theta = B^\theta A^\theta$, A, B being conformable for multiplication.

25.22 HERMITIAN AND SKEW-HERMITIAN MATRICES

(1) Hermitian matrix : A matrix A is said to be Hermitian if $A^\theta = A$.

(2) Skew-Hermitian matrix : A matrix A is said to be skew-Hermitian if $A^\theta = -A$.

25.23 ORTHOGONAL AND UNITARY MATRICES

(i) Orthogonal matrix : A matrix A is said to be orthogonal if $A'A = I$, where I is a unit matrix of order same as of order A.

(ii) Unitary Matrix : A square matrix A is said to be unitary if $A^\theta A = I$.

For example : If $A = \begin{bmatrix} \cos\theta & \sin\theta \\ -\sin\theta & \cos\theta \end{bmatrix}$, then $A' = \begin{bmatrix} \cos\theta & -\sin\theta \\ \sin\theta & \cos\theta \end{bmatrix}$

$$\Rightarrow \quad A'A = \begin{bmatrix} \cos\theta & -\sin\theta \\ \sin\theta & \cos\theta \end{bmatrix}\begin{bmatrix} \cos\theta & \sin\theta \\ -\sin\theta & \cos\theta \end{bmatrix}$$

$$= \begin{bmatrix} \cos^2\theta + \sin^2\theta & 0 \\ 0 & \sin^2\theta + \cos^2\theta \end{bmatrix} = \begin{bmatrix} 1 & 0 \\ 0 & 1 \end{bmatrix} = I$$

Hence, A is orthogonal matrix.

For example : If $A = \begin{bmatrix} 0 & -i \\ i & 0 \end{bmatrix}$, then $\bar{A} = \begin{bmatrix} 0 & i \\ -i & 0 \end{bmatrix}$

$$A^\theta = (\bar{A})' = \begin{bmatrix} 0 & i \\ i & 0 \end{bmatrix}$$

$$A^\theta A = \begin{bmatrix} 0 & -i \\ i & 0 \end{bmatrix}\begin{bmatrix} 0 & -i \\ i & 0 \end{bmatrix} = \begin{bmatrix} 1 & 0 \\ 0 & 1 \end{bmatrix} \qquad (\because i^2 = -1)$$

$$= I$$

$\Rightarrow$ A is unitary.

25.24 PROPERTIES OF HERMITIAN AND SKEW-HERMITIAN MATRICES

(1) If A is Hermitian (skew-Hermitian) matrix, then $i\,A$ is Hermitian (skew-Hermitian) matrix.

(2) If A, B are Hermitian or skew-Hermitian, then $A + B$ is Hermitian or skew-Hermitian.

(3) If A is Hermitian or skew-Hermitian, then $\bar{A}$ is Hermitian or skew-Hermitian.

Recapitulations

- $A^\theta = (\bar{A})'$
- For Hermitian matrix $A^\theta = A$
- For skew-Hermitian $A^\theta = -A$
- For orthogonal matrix $AA' = I$
- For unitary matrix $A^\theta A = I$

Solved Examples

Example 1. *If A be any square matrix, prove that $A + A^\theta$, AA^θ, $A^\theta A$ are all Hermitian and $A - A^\theta$ is skew-Hermitian.*

Solution. (i) $(A+A^\theta)^\theta = A^\theta + (A^\theta)^\theta$ $[\because (A+B)^\theta = A^\theta + B^\theta]$

$= A^\theta + A$ $[\because (A^\theta)^\theta = A]$

$= A + A^\theta$

$\Rightarrow A + A^\theta$ is Hermitian.

(ii) $(AA^\theta)^\theta = (A^\theta)^\theta A^\theta$ $[\because (AB)^\theta = B^\theta A^\theta]$

$= AA^\theta$ $[\because (A^\theta)^\theta = A]$

$\Rightarrow$ AA^θ is Hermitian.

(iii) $(A^\theta A)^\theta = A^\theta (A^\theta)^\theta = A^\theta A$

$\Rightarrow$ $A^\theta A$ is Hermitian.

(iv) $(A - A^\theta) = A^\theta - (A^\theta)^\theta$

$= A^\theta - A = -(A - A^\theta)$

$\Rightarrow$ $A - A^\theta$ is skew-Hermitian.

Example 2. *Prove that the matrix*

$$\begin{bmatrix} \frac{1+i}{2} & \frac{-1+i}{2} \\ \frac{1+i}{2} & \frac{1-i}{2} \end{bmatrix}$$

is unitary.

Solution. Let us suppose $A = \begin{bmatrix} \frac{1+i}{2} & \frac{-1+i}{2} \\ \frac{1+i}{2} & \frac{1-i}{2} \end{bmatrix}$

$$\therefore \quad A' = \begin{bmatrix} \frac{1+i}{2} & \frac{1+i}{2} \\ \frac{-1+i}{2} & \frac{1-i}{2} \end{bmatrix}$$

$$A^\theta = \overline{A'} = \begin{bmatrix} \frac{1-i}{2} & \frac{1-i}{2} \\ \frac{-1-i}{2} & \frac{1+i}{2} \end{bmatrix}$$

Now

$$A^\theta A = \begin{bmatrix} \frac{1-i}{2} & \frac{1-i}{2} \\ \frac{-1-i}{2} & \frac{1-i}{2} \end{bmatrix}\begin{bmatrix} \frac{1+i}{2} & \frac{-1+i}{2} \\ \frac{1+i}{2} & \frac{1-i}{2} \end{bmatrix}$$

$$= \begin{bmatrix} \frac{1}{4}(1-i^2)+\frac{1}{4}(1-i)^2 & -\frac{1}{4}(1-i^2)+\frac{1}{4}(1-i)^2 \\ -\frac{1}{4}(1+i^2)+\frac{1}{4}(1+i)^2 & \frac{1}{4}(1-i^2)+\frac{1}{4}(1-i)^2 \end{bmatrix}$$

$$= \begin{bmatrix} \frac{1}{4}(1+1)+\frac{1}{4}(1+1) & 0 \\ 0 & \frac{1}{4}(1+1)+\frac{1}{4}(1+1) \end{bmatrix}$$

$$= \begin{bmatrix} 1 & 0 \\ 0 & 1 \end{bmatrix} = I$$

Hence A is unitary.

EXERCISE 25.2

1. If $A = \begin{bmatrix} 1 & 0 \\ 2 & -1 \end{bmatrix}$, $B = \begin{bmatrix} 3 & 7 \\ 4 & 8 \end{bmatrix}$, $C = \begin{bmatrix} -1 & 1 \\ 0 & 0 \end{bmatrix}$,

then prove that

$A + (B+C) = (A+B) + C$.

2. If $A = \begin{bmatrix} 2 & 3 \\ 4 & -5 \end{bmatrix}$, $B = \begin{bmatrix} 8 & 9 \\ 6 & 7 \end{bmatrix}$, then find

(i) $2A + 3B$ (ii) $5A - 3B$

3. Find $A + B$ and show that $A+B = B+A$, when

(i) $A = \begin{bmatrix} 7 & 8 \\ 9 & 2 \\ 3 & 4 \end{bmatrix}$, $B = \begin{bmatrix} 2 & 3 \\ 4 & 5 \\ 6 & 7 \end{bmatrix}$

(ii) $A = \begin{bmatrix} 1 & 2 & -3 \\ 4 & 1 & 5 \\ -3 & -2 & 2 \end{bmatrix}$, $B = \begin{bmatrix} 3 & -1 & 2 \\ 4 & 2 & 5 \\ 2 & 0 & 3 \end{bmatrix}$

4. If $A = \begin{bmatrix} \cos\alpha & -\sin\alpha \\ \sin\alpha & \cos\alpha \end{bmatrix}$,

$B = \begin{bmatrix} \cos\beta & -\sin\beta \\ \sin\beta & \cos\beta \end{bmatrix}$,

then show that $AB = BA$.

5. If $A = \begin{bmatrix} 1 & 1 & -1 \\ 2 & 0 & 3 \\ 3 & -1 & 2 \end{bmatrix}$, $B = \begin{bmatrix} 1 & 3 \\ 0 & 2 \\ -1 & 4 \end{bmatrix}$ and

$C = \begin{bmatrix} 1 & 2 & 3 & -4 \\ 2 & 0 & -2 & 1 \end{bmatrix}$, then find $A(BC)$,

$(AB)C$ and show that $A(BC) = (AB)C$.

6. (i) If $A = \begin{bmatrix} 1 & 2 \\ 3 & 4 \end{bmatrix}$, $B = \begin{bmatrix} 2 & 1 \\ 4 & 2 \end{bmatrix}$,

$C = \begin{bmatrix} 5 & 1 \\ 7 & 4 \end{bmatrix}$, then verify that :

$A(B+C) = AB + BC$.

(ii) For the matrices :

$A = \begin{bmatrix} 3 & 2 \\ 1 & 5 \end{bmatrix}$, $B = \begin{bmatrix} 2 & -4 \\ -3 & 1 \end{bmatrix}$, $C = \begin{bmatrix} 1 & 2 \\ 3 & -1 \end{bmatrix}$

Verify $(A+B)C = AC + BC$.

7. If $\begin{bmatrix} 4 & 1 & 2 \\ 0 & 5 & 3 \end{bmatrix}\begin{bmatrix} 3 & 4 & 5 \\ -1 & 0 & -2 \\ 3 & 4 & 7 \end{bmatrix} = \begin{bmatrix} 8x+3y & 6z & 32 \\ 4 & 12 & 26x-5y \end{bmatrix}$

find the values of x, y, z.

8. For what values of a, b, c, d, e the following matrices are same ?

$$A = \begin{bmatrix} a & 1 & 2 \\ c & b & 3 \\ 1 & -1 & 0 \end{bmatrix}, B = \begin{bmatrix} 0 & e & 2 \\ 7 & 9 & d \\ e & -1 & a \end{bmatrix}$$

9. If $A = \begin{bmatrix} 2 & 3 \\ 1 & -4 \end{bmatrix}, B = \begin{bmatrix} 1 & 0 \\ 0 & 1 \end{bmatrix}$, then find

(i) $10B$ (ii) $4A$

(iii) $A+B$ (iv) $A-B$

(v) $2A-3B$ (vi) $5A+3B$.

10. Find the product of

(i) $A = \begin{bmatrix} 1 & 2 \\ 3 & 4 \end{bmatrix}, B = \begin{bmatrix} 1 & 7 \\ 2 & 3 \end{bmatrix}$

(ii) $A = \begin{bmatrix} 1 & 2 & 1 \\ 4 & 0 & 2 \end{bmatrix}, B = \begin{bmatrix} 3 & -4 \\ 1 & 5 \\ -2 & 2 \end{bmatrix}$

(iii) $A = \begin{bmatrix} x \\ y \\ z \end{bmatrix}, B = \begin{bmatrix} x & y & z \end{bmatrix}$

11. If $A = \begin{bmatrix} 1 & 0 & 0 \\ 0 & 1 & 0 \\ 0 & 0 & 1 \end{bmatrix}$, then prove that $A^3 = A$.

12. Find the number of rows and columns in the following matrices :

(i) $\begin{bmatrix} 2 & 6 & 7 & 8 \\ 2 & 5 & 11 & 6 \end{bmatrix}$ (ii) $\begin{bmatrix} 1 & 1 & 1 & 1 \end{bmatrix}$

(iii) $\begin{bmatrix} 2 & -1 & 3 \\ 0 & 3 & 4 \\ 2 & 3 & 7 \\ 2 & 5 & 11 \end{bmatrix}$ (iv) $\begin{bmatrix} x \\ y \\ z \end{bmatrix}$

13. Find the order of the following matrices :

(i) $\begin{bmatrix} 1 & 1 & 1 & 1 \end{bmatrix}$ (ii) $\begin{bmatrix} 1 & a & b & 0 \\ 0 & c & d & 1 \\ 1 & a & b & 0 \end{bmatrix}$

(iii) $\begin{bmatrix} 2 \\ 3 \\ 4 \\ 5 \end{bmatrix}$ (iv) $\begin{bmatrix} 2 & 3 \\ 3 & 4 \\ 5 & 6 \\ 7 & 8 \end{bmatrix}$

14. Are the following matrices conformable for addition ?

(i) $\begin{bmatrix} 1 & 2 & 3 \\ 4 & 5 & 6 \\ 7 & 8 & 9 \end{bmatrix}, \begin{bmatrix} 6 & 4 \\ 7 & 4 \\ 7 & 3 \end{bmatrix}$ (ii) $\begin{bmatrix} 5 & 6 \\ 7 & 8 \end{bmatrix}, \begin{bmatrix} 9 & 10 \\ 11 & 12 \end{bmatrix}$

(iii) $\begin{bmatrix} 3 & 2 & -1 \\ 0 & 4 & 0 \end{bmatrix}, \begin{bmatrix} 2 & 1 \\ 5 & 2 \\ 7 & 8 \end{bmatrix}$ (iv) $\begin{bmatrix} x \\ y \\ z \end{bmatrix}, \begin{bmatrix} a \\ b \\ c \end{bmatrix}$

15. Find the additive inverse of matrix :

$$A = \begin{bmatrix} 1 & 2 & -7 & 5 \\ 0 & 5 & 0 & 8 \\ 0 & 0 & 0 & -8 \end{bmatrix}$$

16. Are the following matrices conformable for the product AB ?

(i) $A = \begin{bmatrix} 5 & 7 \\ 8 & 9 \end{bmatrix}, B = \begin{bmatrix} 3 & 5 & 7 \\ 1 & 0 & 1 \end{bmatrix}$

(ii) $A = \begin{bmatrix} -1 & 2 \\ 3 & 4 \end{bmatrix}, B = \begin{bmatrix} 3 \\ 4 \end{bmatrix}$

(iii) $A = \begin{bmatrix} 2 & 1 & 0 \\ 3 & 2 & 1 \\ 1 & 0 & 1 \end{bmatrix}, B = \begin{bmatrix} 1 & 2 & 3 & 4 \\ 2 & 0 & 1 & 2 \\ 3 & 1 & 0 & 5 \end{bmatrix}$

(iv) $A = \begin{bmatrix} 1 & 1 & 1 & 1 \end{bmatrix}, B = \begin{bmatrix} 1 \\ 1 \\ 1 \\ 1 \end{bmatrix}$

17. Find the values of x and y if :

$$\begin{bmatrix} x & 3 \\ 5 & x-y \end{bmatrix} = \begin{bmatrix} 2 & 3 \\ 5 & 1 \end{bmatrix}$$

18. Transpose the matrix :

$$A = \begin{bmatrix} 1 & 3 & 5 \\ -2 & 5 & 6 \\ 7 & 0 & 3 \end{bmatrix}.$$

19. If $A = \begin{bmatrix} 2 & 1 & 4 \\ 5 & -3 & 7 \end{bmatrix}, B = \begin{bmatrix} 1 & 2 & 3 \\ 2 & -4 & 6 \end{bmatrix}$

Find $2A - 3B$ and $B - \dfrac{A}{2}$. Also write down the unit matrix of order 3.

20. Prove that the matrix $\dfrac{1}{3}\begin{bmatrix} 1 & 2 & 2 \\ 2 & 1 & -2 \\ -2 & 2 & 1 \end{bmatrix}$ is orthogonal.

21. If $A = \begin{bmatrix} 1 & 2 & 1 \\ a & 0 & 4 \\ 1 & 1 & 1 \end{bmatrix}$ and $adj.\, adj(A) = A$, find a. (MTU–2012)

22. If $A = \begin{bmatrix} 0 & 1+2i \\ -1+2i & 0 \end{bmatrix}$ obtain the matrix $(I-N)\ (I+N)^{-1}$ and show that it is unitary. (GBTU–2011)

23. Show that the matrix $A = \begin{bmatrix} 2 & 3-4i \\ 3+4i & 2 \end{bmatrix}$ is Hermitian and iA is skew Hermitian. (UKTU–2012)

24. Show that the matrix $A = \begin{bmatrix} i & 0 & 0 \\ 0 & 0 & i \\ 0 & i & 0 \end{bmatrix}$ is skew Hermition. (MTU–2012)

25. Express the Hermitian matrix $A = \begin{bmatrix} 1 & -i & 1+i \\ i & 0 & 2-3i \\ 1-i & 2+3i & 2 \end{bmatrix}$ and $P+iQ$ where P is real symmetric and Q is real skew-symmetric matrix. (GBTU–2012)

ANSWERS

2. (i) $\begin{bmatrix} 28 & 33 \\ 26 & 11 \end{bmatrix}$ (ii) $\begin{bmatrix} -14 & -12 \\ 2 & -46 \end{bmatrix}$ **3.** (i) $\begin{bmatrix} 9 & 11 \\ 13 & 7 \\ 9 & 11 \end{bmatrix}$ (ii) $\begin{bmatrix} 4 & 1 & -1 \\ 8 & 3 & 10 \\ -1 & -2 & 5 \end{bmatrix}$ **7.** $x = 1, y = 3, z = 4$

8. $a = 0, b = 9, c = 7, d = 3, e = 1$ **9.** (i) $\begin{bmatrix} 10 & 0 \\ 0 & 10 \end{bmatrix}$ (ii) $\begin{bmatrix} 8 & 12 \\ 4 & -16 \end{bmatrix}$ (iii) $\begin{bmatrix} 3 & 3 \\ 1 & -3 \end{bmatrix}$ (iv) $\begin{bmatrix} 1 & 3 \\ 1 & -5 \end{bmatrix}$ (v) $\begin{bmatrix} 1 & 6 \\ 2 & -11 \end{bmatrix}$ (vi) $\begin{bmatrix} 13 & 15 \\ 5 & -17 \end{bmatrix}$

10. (i) $\begin{bmatrix} 5 & 13 \\ 11 & 33 \end{bmatrix}$ (ii) $\begin{bmatrix} 3 & 8 \\ 8 & -12 \end{bmatrix}$ (iii) $\begin{bmatrix} x^2 & xy & xz \\ yx & y^2 & yz \\ \delta x & zy & z^2 \end{bmatrix}$

12. (i) Rows = 2, Columns = 4, (ii)Row = 1, Columns = 4 (iii) Rows =4, Columns = 3 (ii)Rows = 3, Column = 1.

13. (i) 1 × 4 (ii) 3 × 4 (iii) 4 × 1 (iv) 4 × 2

14. (i) No (ii) Yes (iii) No (iv) Yes

15. $\begin{bmatrix} -1 & -2 & 7 & -5 \\ 0 & -5 & 0 & -8 \\ 0 & 0 & 0 & 8 \end{bmatrix}$ **16.** (i) Yes (ii) Yes (iii) Yes (iv) Yes

17. $x = 2, y = 1$ **18.** $\begin{bmatrix} 1 & -2 & 7 \\ 3 & 5 & 0 \\ 5 & 6 & 3 \end{bmatrix}$ **19.** $\begin{bmatrix} 1 & -4 & -1 \\ 4 & 6 & -4 \end{bmatrix}$, $\begin{bmatrix} 0 & \frac{3}{2} & 1 \\ -\frac{1}{2} & -\frac{5}{3} & \frac{-5}{2} \end{bmatrix}$, $\begin{bmatrix} 1 & 0 & 0 \\ 0 & 1 & 0 \\ 0 & 0 & 1 \end{bmatrix}$ **21.** 3 **22.** $\frac{1}{6}\begin{bmatrix} -4 & -2-4i \\ 2-4i & -4 \end{bmatrix}$

25.25 SUBMATRIX OF A MATRIX

Let A be a matrix of order $m \times n$, then a matrix obtained from A by removing some rows and columns, is called a submatrix of the matrix A.

For Example :

(i) Consider a matrix $A = \begin{bmatrix} 1 & 2 & 3 & 6 \\ 5 & 7 & 9 & 9 \\ 4 & 5 & 6 & 12 \end{bmatrix}$ of order 3 × 4, then a matrix $\begin{bmatrix} 1 & 2 & 6 \\ 5 & 7 & 9 \end{bmatrix}$ is a submatrix of A, which is obtained from A by removing third column and fourth row.

(ii) Consider a matrix $A = \begin{bmatrix} 1 & 5 & 2 \\ 0 & 1 & 3 \\ 0 & 0 & 1 \end{bmatrix}$ of order 3 × 3, then a matrix $\begin{bmatrix} 1 & 5 \\ 0 & 1 \\ 0 & 0 \end{bmatrix}$ is submatrix of A, which is obtained from A by removing third column.

REMARK

- If the given matrix A is a square matrix, then a square submatrix of A is known as principal submatrix.

25.26 MINORS OF A MATRIX

Let A be a matrix of order $m \times n$, then the determinant of every square submatrix of A is called a minor of A. If the order of the determinant of square submatrix of A is $r \times r$, then it is denoted as r-minor of A.

For example : (i) Consider a matrix $A = \begin{bmatrix} 1 & 2 & 3 \\ 2 & 4 & 6 \end{bmatrix}$

Then all the 2-minors of A are $\begin{vmatrix} 1 & 2 \\ 2 & 4 \end{vmatrix}, \begin{vmatrix} 2 & 3 \\ 4 & 6 \end{vmatrix}, \begin{vmatrix} 1 & 3 \\ 2 & 6 \end{vmatrix}$

(ii) Consider a matrix $A = \begin{bmatrix} 1 & 2 & 3 \\ 5 & 7 & 9 \\ 4 & 5 & 6 \\ 6 & 9 & 12 \end{bmatrix}$

Then all the 3-minors of A are

$$\begin{vmatrix} 1 & 2 & 3 \\ 5 & 7 & 9 \\ 4 & 5 & 6 \end{vmatrix}, \begin{vmatrix} 1 & 2 & 3 \\ 5 & 7 & 9 \\ 6 & 9 & 12 \end{vmatrix}, \begin{vmatrix} 1 & 2 & 3 \\ 4 & 5 & 6 \\ 6 & 9 & 12 \end{vmatrix}, \begin{vmatrix} 5 & 7 & 9 \\ 4 & 5 & 6 \\ 6 & 9 & 12 \end{vmatrix}$$

25.27 RANK OF A MATRIX

Let A be a matrix of order $m \times n$, then a non-negative integer r is said to the rank of matrix A if it possesses the following two properties :

(i) There exists at least one r-minor of A which is not equal to zero.

(ii) Every s-minor of A for all $s > r$ is zero.

We denote the rank of A by $\rho(A)$.

In other words, the rank of a matrix is the order of any highest order of a non-zero minor of the matrix.

REMARKS

- If the order of a matrix A is $m \times n$, then $\rho(A) \le \min. \{m, n\}$
- A is a null matrix iff $\rho(A) = 0$.
- If A is any non-zero matrix, then $\rho(A) \ge 1$.
- $\rho(A) \ge r$, if there exists a non-zero r-minor of A.
- For any square matrix A of order n, $\rho(A) = n$ iff A is non-singular.
- For any square matrix A of order n, $\rho(A) < n$ iff A is singular.
- $r(A) \le r$ if every s-minor of A is zero, where $s > r$.
- Every $(r+1)$- rowed minor of A can be expressed as a linear combination of its r-rowed minors, therefore if every r-minor of A is zero, then its every $(r+1)$-minor is also zero.

25.28 ECHELON FROM OF A MATRIX

A matrix A is said to be in Echelon form if :

(i) every row of A has all its entries 0 (zero) which occurs below every row having a non-zero entry and

(ii) the number of zeros before the first non-zero entry in a row is less than the number of such zeros in the next row.

REMARK

- The rank of a matrix is equal to the number of non-zero rows in Echelon form of that matrix.

For example: Consider a matrix

$$A = \begin{bmatrix} 0 & 2 & 3 & 5 \\ 0 & 0 & 3 & 2 \\ 0 & 0 & 0 & 0 \end{bmatrix}$$

Clearly, A is in Echelon form which has 2 non-zero rows, hence the rank of A is 2.

THEOREM 1. *The rank of the transpose of a matrix is equal to the rank of that matrix.*

Proof. Let A be a marix, then A' is its transpose and let $\rho(A) = r$, then there exists an r-rowed minor of A which is not equal to zero and all s-rowed minors of A are zero, where $s > r$. Let $|B|$ be a r-rowed minor of A such that $|B| \ne 0$. Since A' is the transpose of A, then $|B'|$ is the r-rowed minor of A' but $|B'| = |B| \ne 0$, therefore $\rho(A') \ge r$. Suppose there is an s-minor $|C|$ of A' such that $|C| \ne 0$, where $s > r$, then $|C'|$ will be an s-minor of A such that $|C'| = |C| \ne 0$, therefore $\rho(A) > r$ which is a contradiction, hence $\rho(A') = r$.

Recapitulations

- The rank of a matrix is the order of any highest order of a non-zero minor of the matrix.
- A matrix A is said to be Echelon form if
  (i) every row of A has all its entries 0 (zero) which occurs below every row having a non-zero entry. and
  (ii) The number of zeros before the first non-zero entry in a row in less than the number of such zeros in the next row.
- The rank of a matrix is equal to the number of non-zero rows in Echelon form.

Solved Examples

Example 1. *Find the rank of the following matrices :*

(i) $\begin{bmatrix} 3 & 0 & 0 \end{bmatrix}$ (ii) $\begin{bmatrix} 1 & 2 & 3 \\ 2 & 4 & 5 \end{bmatrix}$

(iii) $\begin{bmatrix} 1 & 2 & 3 \\ 3 & 4 & 5 \\ 4 & 5 & 6 \end{bmatrix}$ (iv) $\begin{bmatrix} 1 & 5 & 2 & 4 \\ 0 & 1 & 3 & 1 \\ 0 & 0 & 1 & 3 \end{bmatrix}$

Solution. (i) Let $A = \begin{bmatrix} 3 & 0 & 0 \end{bmatrix}$, then A is the non-zero rowed matrix, thereofore $\rho(A) \ge 1$. Also A is a matrix of order of 1×3, then $\rho(A) \le 1$, hence $\rho(A) = 1$.

(ii) Let $A = \begin{bmatrix} 1 & 2 & 3 \\ 2 & 4 & 5 \end{bmatrix}$

The order of A is 2×3, then $\rho(A) \le 2$.

Also there is a 2-minor $\begin{vmatrix} 2 & 3 \\ 4 & 5 \end{vmatrix}$ of A which is not equal to zero, then $\rho(A) \ge 2$, hence $\rho(A) = 2$.

(iii) Let $A = \begin{bmatrix} 1 & 2 & 3 \\ 3 & 4 & 5 \\ 4 & 5 & 6 \end{bmatrix}$. The order of A is 3×3, then $\rho(A) \le 3$.

Now,

$$|A| = \begin{vmatrix} 1 & 2 & 3 \\ 3 & 4 & 5 \\ 4 & 5 & 6 \end{vmatrix}$$

$$= 1(24-25) - 2(18-20) + 3(15-16) = 0$$

$\therefore$ The only 3-minor $|A|$ of A is zero, thus $\rho(A) < 3$. Further, there is a 2-minor $\begin{vmatrix} 1 & 2 \\ 3 & 4 \end{vmatrix}$ of A which is not equal to zero, hence $\rho(A) = 2$.

(iv) Let $A = \begin{bmatrix} 1 & 5 & 2 & 4 \\ 0 & 1 & 3 & 1 \\ 0 & 0 & 1 & 3 \end{bmatrix}$

The order of A is 3×4, then $\rho(A) \le 3$.

Now there is a 3-minor $\begin{vmatrix} 1 & 5 & 2 \\ 0 & 1 & 3 \\ 0 & 0 & 1 \end{vmatrix}$ of A which is not equal to zero, then $\rho(A) \ge 3$.

Hence $\rho(A) = 3$.

Example 2. *Let A and B be two square matrices of order n. If $\rho(A) = \rho(B) = n$, then prove that $\rho(AB) = n$ and conversely.*

Solution. Suppose $\rho(A) = \rho(B) = n$, then both A and B are non-singular.

$\therefore$ $|A| \ne 0$ and $|B| \ne 0$

$\Rightarrow$ $|AB| = |A|\,|B| \ne 0$

Since the order of AB is n and $|AB| \ne 0$, therefore $\rho(AB) = n$.

Conversely, suppose that

$$\rho(AB) = n$$

$\therefore$ $|AB| \ne 0$

$\Rightarrow$ $|A|\,|B| \ne 0$

$\Rightarrow$ $|A| \ne 0$ and $|B| \ne 0$

$\Rightarrow$ $\rho(A) = n,\ \rho(B) = n.$

Example 3. *Prove that every skew-symmetric matrix of odd order has rank less than its order.*

Solution. Let A be a skew-symmetric matrix of order n, where n is an odd natural number, then

$$A' = -A.$$

Now $A' = -A$

$\Rightarrow$ $|A'| = |-A|$

$\Rightarrow$ $|A| = |(-1)A|$ $[\because |A'| = |A|]$

$\Rightarrow$ $|A| = (-1)^n |A|$

$\Rightarrow$ $|A| = -|A|$ $[\because n$ is odd$]$

$\Rightarrow$ $2|A| = 0$

$\Rightarrow$ $|A| = 0$

$\Rightarrow$ $\rho(A) \ne n$

But $\rho(A) \le n$, hence

$$\rho(A) < n.$$

Example 4. *If A be a non-zero column and B is a non-zero row matrix, then show that $\rho(AB) = 1$.*

Solution. Let $A = \begin{bmatrix} a_{11} \\ a_{21} \\ \vdots \\ a_{m1} \end{bmatrix}$ and $B = \begin{bmatrix} b_{11} & b_{12} & \cdots & b_{1n} \end{bmatrix}$

be two non-zero column and row matrices respectively.

Then we have $AB = \begin{bmatrix} a_{11} \\ a_{21} \\ \vdots \\ a_{m1} \end{bmatrix} \begin{bmatrix} b_{11} & b_{12} & \cdots & b_{1n} \end{bmatrix}$

$$= \begin{bmatrix} a_{11}b_{11} & a_{11}b_{12} & \cdots & a_{11}b_{1n} \\ a_{21}b_{11} & a_{21}b_{12} & \cdots & a_{21}b_{1n} \\ \vdots & \vdots & & \vdots \\ a_{m1}b_{11} & a_{m1}b_{12} & \cdots & a_{m1}b_{1n} \end{bmatrix}$$

Clearly, AB is a matrix of order $m \times n$ and AB is a non-zero matrix since A and B are non-zero matrices,

then $\rho(AB) \ge 1$ ..(1)

also every 2-minor of AB vanishes, then

$\rho(AB) \le 1$ ...(2)

From (1) and (2) we have

$$\rho(AB) = 1.$$

Example 5. *If A is n-rowed square matrix of rank n −1, then show that adj A is non-zero matrix.*

Solution. Since the rank of A is $n-1$, *i.e.*, $\rho(A) = n-1$, then there exists a non-zero $(n-1)$ minor of A, therefore there exists at least one element of *adj A* which is non-zero, hence *adj. A* is a non-zero matrix.

Example 6. *Let A be a square matrix of order n. Show that $\rho(adj.\ A)$ is n or 0 in accordance with $\rho(A)$ is n or less than n–1.*

Solution. Suppose $\rho(A) = n$, then $|A| \ne 0$

But we know that

$$|adj.\ A| = |A|^{n-1}$$

$\Rightarrow$ $|adj.\ A| \ne 0$ $[\because |A| \ne 0]$

$\therefore$ $\rho(adj.\ A) = n,$

since the order of *adj.* $A = n$.

Next, when $\rho(A) < n-1$, then $|A| = 0$ and every r-minor of A is zero, where $r \ge n-1$, therefore every element of *adj.* A is zero so that *adj.* A is a null matrix, hence

$$\rho(adj.\ A) = 0.$$

Example 7. *Find the value of x so that $\rho(A) \le 2$, where A is the matrix given by*

$$\begin{bmatrix} 3x-8 & 3 & 3 \\ 3 & 3x-8 & 3 \\ 3 & 3 & 3x-8 \end{bmatrix}$$

Solution. Since $\rho(A) \le 2$, then $|A| = 0$ because A is a square matrix of order 3×3.

Now $|A| = 0$

$$\Rightarrow \begin{bmatrix} 3x-8 & 3 & 3 \\ 3 & 3x-8 & 3 \\ 3 & 3 & 3x-8 \end{bmatrix}$$

$\Rightarrow (3x-8)\{(3x-8)(3x-8)-9\}$

$\quad -3\{3(3x-8)-9\}+3\{9-3(3x-8)\} = 0$

$\Rightarrow \quad (3x-8)^3 - 9(3x-8) - 9(3x-8) + 27$

$\quad + 27 - 9(3x-8) = 0$

$\Rightarrow \quad (3x-8)^3 - 27(3x-8) + 54 = 0$

$\Rightarrow \quad (3x-5)^2(3x-2) = 0$

$\Rightarrow \quad 3x-2 = 0$ or $3x-5 = 0$

$\Rightarrow \quad x = \frac{2}{3}$ or $x = \frac{5}{3}$

When $x = \frac{2}{3}$, then $A = \begin{bmatrix} -6 & 3 & 3 \\ 3 & -6 & 3 \\ 3 & 3 & -6 \end{bmatrix}$, clearly there is a 2-minor $\begin{vmatrix} -6 & 3 \\ 3 & -6 \end{vmatrix}$ of A, which is non-zero, hence $\rho(A) = 2$.

Again, when $x = \frac{5}{3}$, then $A = \begin{bmatrix} -3 & 3 & 3 \\ 3 & -3 & 3 \\ 3 & 3 & -3 \end{bmatrix}$

Clearly, there is a 2-minor $\begin{vmatrix} -3 & 3 \\ 3 & -3 \end{vmatrix}$ of A, which is non-zero, hence $\rho(A) = 2$.

EXERCISE 25.3

1. Find the rank of the following matrices :

(i) $\begin{bmatrix} 0 & 0 \\ 0 & 0 \end{bmatrix}$ (ii) $\begin{bmatrix} 5 & 10 \\ 3 & 6 \end{bmatrix}$

(iii) $\begin{bmatrix} 1 & -3 & 4 & 7 \\ 9 & 1 & 2 & 0 \end{bmatrix}$ (iv) $\begin{bmatrix} 1 & 2 & 3 \\ 2 & 1 & 0 \\ 0 & 1 & 2 \end{bmatrix}$

(v) $\begin{bmatrix} 1 & 2 & -7 & 5 \\ 0 & 5 & 0 & 8 \\ 0 & 0 & 0 & -3 \end{bmatrix}$ (vi) $\begin{bmatrix} 0 & 1 & 2 & 1 \\ 1 & 2 & 3 & 2 \\ 3 & 1 & 1 & 3 \end{bmatrix}$

(vii) $\begin{bmatrix} 1 & 2 & 3 & 4 \\ 2 & 4 & 6 & 8 \\ 3 & 6 & 9 & 12 \end{bmatrix}$ (viii) $\begin{bmatrix} 1 & 5 & 4 & 6 \\ 2 & 7 & 5 & 9 \\ 3 & 9 & 6 & 12 \end{bmatrix}$

(ix) $\begin{bmatrix} 1 & x & x^2 \\ 1 & y & y^2 \\ 1 & z & z^2 \end{bmatrix}$ (x) $\begin{bmatrix} 1 & 1 & 1 & 1 \\ 1 & 1 & 1 & 1 \\ 1 & 1 & 1 & 1 \\ 1 & 1 & 1 & 1 \end{bmatrix}$

(xi) $\begin{bmatrix} 1 & 0 & 0 & 0 \\ 0 & 1 & 0 & 0 \\ 0 & 0 & 1 & 0 \\ 0 & 0 & 0 & 1 \end{bmatrix}$

2. If $A = \begin{bmatrix} 0 & 1 & 0 & 0 \\ 0 & 0 & 1 & 0 \\ 0 & 0 & 0 & 1 \\ 0 & 0 & 0 & 0 \end{bmatrix}$, find $\rho(A)$, $\rho(A^2)$, $\rho(A^3)$ and $\rho(A^4)$.

3. Show that the rank of a matrix does not alter on affixing any number of additional rows or columns of zeros.

4. Show that the rank of a matrix is greater than or equal to the rank of its every submatrix.

5. Find the rank of A, B, $A+B$ and AB, where

$$A = \begin{bmatrix} 1 & 1 & -1 \\ 2 & -3 & 4 \\ 3 & -2 & 3 \end{bmatrix} \text{ and } B = \begin{bmatrix} -1 & -2 & -1 \\ 6 & 12 & 6 \\ 5 & 10 & 5 \end{bmatrix}$$

ANSWERS

1. (i) 0 (ii) 1 (iii) 2 (iv) 2 (v) 3 (vi) 3 (vii) 1 (viii) 2 (ix) Rank =3 if $x \neq y \neq z$; Rank =2 if only two of x, y, z are different; Rank = 1 if $x = y = z$. (x) 1 (xi) 4

2. $\rho(A) = 3$, $\rho(A^2) = 2$, $\rho(A^3) = 1$, $\rho(A^4) = 0$

5. $r(A) = 2$, $\rho(B) = 1$, $\rho(A+B) = 2$, $\rho(AB) = 0$

25.29 ELEMENTARY TRANSFORMATIONS (OR E-TRANSFORMATIONS) OF A MATRIX

(UPTU 2007)

Consider the matrices

$$A = \begin{bmatrix} 1 & 2 & -3 \\ 3 & 0 & 1 \end{bmatrix}, B = \begin{bmatrix} 3 & 0 & 1 \\ 1 & 2 & -3 \end{bmatrix}, C = \begin{bmatrix} -3 & 2 & 1 \\ 1 & 0 & 3 \end{bmatrix}, D = \begin{bmatrix} 4 & 8 & -12 \\ 3 & 0 & 1 \end{bmatrix}$$

$$E = \begin{bmatrix} -3 & 2 & 7 \\ 1 & 0 & 21 \end{bmatrix}, F = \begin{bmatrix} 1 & 2 & -3 \\ 6 & 6 & -8 \end{bmatrix} \text{ and } G = \begin{bmatrix} 7 & 0 & 1 \\ -11 & 0 & -3 \end{bmatrix}$$

From above matrices, we observe that :

(1) B can be obtained from A by interchanging the first and second row.

(2) C can be obtained from A by interchanging the first and third column.

(3) D can be obtained from A by multiplying each element of the first row by 4.

(4) E can be obtained from C by multiplying each element of third column by 7.

(5) F can be obtained from A by adding to the elements of second row, 3 times the corresponding elements of the first row.

(6) G can be obtained from B by adding to the elements of first column, 4 times the corresponding elements of the third column.

Such transformations as performed above are known as elementary transformations (or E-operations or E-transformations).

Elementary transformations on rows are known as elementary row transformations whereas the transformations on columns are known as elementary column transformations.

Thus we may define E-transformations as follows:

Definition: *An elementary transformation (or E-transformation) is an operation of any one of the following types:*

(i) *The interchange of any two rows (or columns).*

(ii) *The multiplication of any row (or column) by any non-zero number.*

(iii) *The addition of non-zero scalar multiple of any row (or column) to another row (or column).*

25.30 NOTATIONS FOR E-TRANSFORMATIONS

E-transformations can be denoted by the following notations:

(i) The transformation of interchanging i^{th} and j^{th} row of a matrix is denoted by $R_i \leftrightarrow R_j$.

(ii) The transformation of interchanging i^{th} and j^{th} column is denoted by $C_i \leftrightarrow C_j$.

(iii) The transformation of multiplication of ith row of a matrix by non-zero scalar k is denoted by $R_i \to kR_i$.

(iv) The transformation of multiplication of jth column by a non-zero scalar k is denoted by $C_{ij} \to kC_j$.

(v) The transformation of addition of a non-zero scalar k multiple of jth row to another i^{th} row of a matrix is denoted by $R_i \to R_i + kR_j$ $(i \neq j)$.

(vi) The transformation of addition of a non-zero scalar k multiple of jth column to another i^{th} column of a matrix is denoted by $C_i \to C_i + kC_j$ $(i \neq j)$.

25.31 ELEMENTARY MATRICES

A matrix which is obtained from a unit (identity) matrix by a single E-transformation is known as an elementary matrix.

For example : Consider the matrices

$$\begin{bmatrix} 0 & 0 & 1 \\ 0 & 1 & 0 \\ 1 & 0 & 0 \end{bmatrix}, \begin{bmatrix} 1 & 0 & 0 \\ 0 & 1 & 0 \\ 3 & 0 & 1 \end{bmatrix}, \begin{bmatrix} 1 & 0 & 0 \\ 0 & 3 & 0 \\ 0 & 0 & 1 \end{bmatrix}$$

Clearly, these matrices are elementary matrices because these are obtained from a unit matrix I_3 (the identity matrix of order 3×3) by performing the E-transformations $C_1 \to C_3$, $R_3 \to R_3 + 3R_1$ and $R_2 \to 3R_2$.

The elementary matrices of different types can be denoted by the following notations:

(i) The elementary matrix obtained by interchanging i^{th} and j^{th} rows (or columns) of a unit matrix is denoted by E_{ij}.

(ii) The elementary matrix obtained by multiplying i^{th} row (or column) of a unit matrix by a non-zero scalar k is denoted by $E_i(k)$.

(iii) The elementary matrix obtained by adding a non-zero scalar k multiple of jth row (or column) to i^{th} row (or column) of a unit matrix is denoted by $E_{ij}(k)$.

Obviously, $\quad |E_{ij}| = -1, \; |E_i(k)| = k \neq 0$ and $|E_{ij}(k)| = 1$

Hence we can say that all the elementary matrices are non-singular and hence they possess their inverse.

THEOREM 1. *Every E-row (column) transformation of a matrix can be obtained by pre-multiplication (post-multiplication) with the corresponding elementary matrix.*

Proof. Let A be a matrix of order $m \times n$, then we can write $A = I_m A I_n$

So that any elementary row transformation can be obtained by subjecting the pre-factor I_m and the same elementary column transformation can be obtained by subjecting the post-factor I_n.

In order to prove this result we shall first prove that any E-row transformation a product AB can be obtained by subjecting the pre-factor A to the same E-row transformation and any E-column transformation of a product AB can be obtained by subjecting the post-factor B to the same E-column transformation, where B is a matrix of order $n \times p$.

Let $A = [a_{ij}]_{m \times n}$ and $B = [b_{ij}]_{n \times p}$, then the product AB is conformable.

We can write A and B as follows : $A=\begin{bmatrix} R_1 \\ R_2 \\ \vdots \\ R_m \end{bmatrix}$ and $B=\begin{bmatrix} C_1 & C_2 & \cdots & C_p \end{bmatrix}$

where $R_1, R_2, ..., R_m$ are row vectors of A and $C_1, C_2, ..., C_p$ are column vectors of B.

Now
$$AB=\begin{bmatrix} R_1C_1 & R_1C_2 & \cdots & R_1C_p \\ R_2C_1 & R_2C_2 & \cdots & R_2C_p \\ \vdots & \vdots & & \vdots \\ R_mC_1 & R_mC_2 & \cdots & R_mC_p \end{bmatrix}_{m\times p}$$

If σ be any E-row transformation, then
$$(\sigma A)B = \sigma(AB)$$

[**Note:** If σ denotes the operation $R_1 \leftrightarrow R_2$, then
$$\sigma A=\begin{bmatrix} R_2 \\ R_1 \\ \vdots \\ R_m \end{bmatrix}$$

$$\therefore \quad (\sigma A)B=\begin{bmatrix} R_2 \\ R_1 \\ \vdots \\ R_m \end{bmatrix}\begin{bmatrix} C_1 & C_2 & \cdots & C_p \end{bmatrix}=\begin{bmatrix} R_2C_1 & R_2C_2 & \cdots & R_2C_p \\ R_1C_1 & R_1C_2 & \cdots & R_1C_p \\ \vdots & & & \\ R_mC_1 & R_mC_2 & \cdots & R_mC_p \end{bmatrix}$$

Clearly
$$(\sigma A)B = \sigma(AB)$$

Similarly, if the columns $C_1, C_2, ..., C_p$ of B be subjected to any E-column transformation, the columns of AB are also subjected to the same E-column transformation.

i.e., $\sigma(\sigma B) = \sigma(AB)$ where σ denotes $C_1 \leftrightarrow C_2$.

Now we move to main theorem, if A is a matrix of order $m\times n$, then we can write
$$A = I_m A \text{ where } I_m \text{ is a unit matrix of order } n \times m.$$

If σ be any E-row transformation, then $\sigma(I_mA) = (\sigma I_m)A = EA$

where E is the elementary matrix corresponding to the same row transformation σ.

Similarly, we can also write $A = AI_m$

and if σ dentoes the E-column transformation, then
$$\sigma(AI_m) = A(\sigma I_m) = AE_1$$

where E_1 is the elementary matrix corresponding to the same column transformation.

Hence the theorem.

25.32 INVARIANCE OF RANK UNDER E-TRANSFORMATIONS

Elementary transformation (E-transformation) do not change the rank of matrix.

(i) Interchanging the rows (or columns) does not change the rank.

(ii) Multiplication of the elements of a row by a non-zero number does not change the rank.

(iii) Addition of any row to the product of any number k and other row does not change the rank.

REMARKS

- The rank of a matrix does not change by a series of E-transformation.
- The rank of a matrix does not change by a column-transformation.

25.33 NORMAL FORM

Definition: *If a matrix is reduced to the form* $\begin{pmatrix} I_r & O \\ O & O \end{pmatrix}$. *Then this form is called normal form of the given matrix.*

THEOREM 1. *Every matrix of order $m \times n$ of rank r can be reduced to the form $\begin{pmatrix} I_r & O \\ O & O \end{pmatrix}$ by a finite number of E-transformations, where I_r is the unit matrix of order $r \times r$.*

Proof. Let $A = [a_{ij}]_{m\times n}$ be a matrix of order $m \times n$ and of rank r. If A is a zero matrix, then its rank is zero and thus A can be written as $\begin{pmatrix} I_r & O \\ O & O \end{pmatrix}$.

Let us suppose A is a non-zero matrix. It means that it has at least one of its elements non-zero. Let this non-zero element be $a_{ij} = k \neq 0$.

Let B be a matrix which is obtained from A by E-transformations $R_1 \leftrightarrow R_i$ and and $C_1 \leftrightarrow C_j$ and whose leading element is k. Again using the E-transformation $R_1 \to \frac{1}{K} R_1$ on B we get a matrix C whose leading element becomes 1. Let this matrix C be

$$C = \begin{bmatrix} 1 & c_{12} & c_{13} & \cdots & c_{1n} \\ c_{21} & c_{22} & c_{23} & \cdots & c_{2n} \\ c_{31} & c_{32} & c_{33} & \cdots & c_{3n} \\ \cdots & \cdots & \cdots & \cdots & \cdots \\ c_{m1} & c_{m2} & c_{m3} & \cdots & c_{mn} \end{bmatrix}_{m\times n}$$

Now subtracting first column after multiplying by suitable number from remaining columns of C and subtracting first row after multiplying by suitable number from remaining rows of C, we obtain a matrix D whose elements of the first row and first column are zero except the leading element. Let D be given as

$$D = \begin{bmatrix} 1 & 0 & 0 & \ldots & 0 \\ 0 & & & & \\ 0 & & A_1 & & \\ \vdots & & & & \\ 0 & & & & \end{bmatrix}_{m\times n}$$

where A_1 is a matrix of order $(m-1)\times(n-1)$.

If this matrix A_1 is non-zero matrix, then we shall apply above process on A_1. Since we know that E-transformation will not effect the first row and first column of D, so that we shall apply E-transformations on D and there is no need to take A_1 separately. Continuing this process finitely we obtain a matrix M such that

$$M = \begin{pmatrix} I_k & O \\ O & O \end{pmatrix}$$

This implies that matrix M has a rank k. But M is obtained from A by a finite number of E-transformations and we know that E-tansformations do not change the rank, therefore k must be equal to r.

Hence the matrix A of order $m\times n$ of rank r can be reduced to the form $\begin{pmatrix} I_r & O \\ O & O \end{pmatrix}$ by a finite number of E-transformations.

REMARKS

- The form $\begin{pmatrix} I_r & O \\ O & O \end{pmatrix}$ of A is also called first canonical form.
- The rank of a matrix of order $m \times n$ is r if and only if it can be reduced to the form $\begin{bmatrix} I_r & O \\ O & O \end{bmatrix}$ by a finite chain of E-transformations.
- If A is a matrix of order $m \times n$ of rank r, then there exists two non-singular matrices P and Q such that $PAQ = \begin{bmatrix} I_r & O \\ O & O \end{bmatrix}$

25.34 EQUIVALENCE OF MATRICES

Definition: *Let A be a matrix of order m×n. If a matrix B of order m×n is obtained from A by a finite chain of elementary transformation on A, then A is said to be equivalent to B. We write symbolically as A ~ B which is read as 'A is equivalent to B'.*

THEOREM 1. *The relation '~' in the set of all m × n matrices is an equivalent relation.*

Proof. We shall prove that the relation '~' is

(i) reflexive (ii) symmetric (iii) transitive.

(i) Reflexivity: Let A be an $m \times n$ matrix, then A can be obtained from A itself by the elementary transformation $R_i \to R_i$, for all $i = 1, 2, \ldots, m$

$$\therefore \quad A \sim A$$

(ii) Symmetry: Let A and B be any two $m \times n$ matrices such that $A \sim B$.

Now $A \sim B \Rightarrow B$ can be obtained from A by a finite chain of elementary transformations on A

$\Rightarrow A$ can also be obtained from B by a finite chain of elementary transformations on B.

$\Rightarrow \qquad B \sim A$

i.e., If $A \sim B$, then $\qquad B \sim A$.

(iii) Transitivity: Let A, B and C be any three $m \times n$ matrices such that $A \sim B$ and $B \sim C$.

$A \sim B \Rightarrow B$ can be obtained from A by a finite chain of elementary transformations on A.

$B \sim C \Rightarrow C$ can be obtained from B by a finite chain of elementary transformations on B.

On combining these two statements we can say that C can also be obtained from A by a finite chain of elementary transformations on A.

$\therefore \qquad A \sim C$

i.e., $\qquad A \sim B, B \sim C \Rightarrow A \sim C$

Hence the relation '$\sim$' is an equivalence relation.

25.35 ROW AND COLUMN EQUIVALENCE OF MATRICES

(i) Row equivalence of matrix : A matrix A is said to be row equivalent to a matrix B, if B can be obtained from A by a finite chain of elementary row transformations on A and we write $A \overset{R}{\sim} B$.

(ii) Column equivalence of matrix : A matrix A is said to be column equivalent to a matrix B, if B can be obtained from A by a finite chain of elementary column transformations on A, we write $A \overset{C}{\sim} B$.

Theorem 1. (Employment of only row transformations) :

Let A be a matrix of order $m \times n$ of rank r, then there exists a non-singular matrix P such that $PA = \begin{bmatrix} G \\ O \end{bmatrix}$ *where G is a matrix of order $r \times n$ of rank r and O is a null matrix of order $(m-r) \times n$.*

Proof. Since A is a matrix of order $m \times n$ of rank r, then there exists two non-singular matrices P and Q such that

$$PAQ = \begin{bmatrix} I_r & O \\ O & O \end{bmatrix} \qquad ...(1)$$

Further, since every non-singular matrix can be expressed as the product of elementary matrices.

So let $\qquad Q = Q_1 Q_2 \ldots Q_t$

where $Q_1, Q_2, \ldots, Q_t$ are elementary matrices. Now (1) can be written as

$$PAQ_1 Q_2 \ldots Q_t = \begin{bmatrix} I_r & O \\ O & O \end{bmatrix} \qquad ...(2)$$

Again, every elementary column transformation of a matrix is equivalent to post-multiplication with the corresponding elementary matrix. Since no column transformation can affect the last $(m-r)$ rows of RHS of (2), therefore post-multiplying the LHS of (2) by elementary matrices $Q_t^{-1}, Q_{t-1}^{-1}, \ldots, Q_2^{-1}, Q_1^{-1}$ successively and effecting the corresponding column transformations in RHS of (2), we get

$$PA = \begin{bmatrix} G \\ O \end{bmatrix}$$

Since the elementary transformations do not change the rank, therefore the rank of PA is the same as the rank of A which is r, thus the rank of $\begin{bmatrix} G \\ O \end{bmatrix}$ is r. Hence the rank of G is r as G has r rows and the last $(m-r)$ rows of the matrix $\begin{bmatrix} G \\ O \end{bmatrix}$ consist only zero entries, *i.e.*, O is a null matrix of order $(m-r) \times n$.

THEOREM 2. (Employment of only column transformations)

Let A be a matrix of order $m \times n$ of rank r, then there exists a non-singular matrix Q such that $AQ = [H \;\; O]$ where H is a matrix of order $m \times n$ and O is a null matrix of order $m \times (n-r)$.

Proof. Since A is a matrix of order $m \times n$, then there exists two non-singular matrices P and Q such that

$$PAQ = \begin{bmatrix} I_r & O \\ O & O \end{bmatrix} \qquad ...(1)$$

Further, since every non-singular matrix can be expressed as the product of elementary matrices.

So let $\qquad P = P_1 P_2 \ldots P_s$

where $P_1 P_2 \ldots P_s$ are elementary matrices.

Now (1) can be written as

$$P_1 P_2 \dots P_s AQ = \begin{bmatrix} I_r & O \\ O & O \end{bmatrix} \qquad \dots(2)$$

Again, every elementary row transformation of a matrix is equivalent to pre-multiplication with the corresponding elementary matrix. Since no row transformation can affect the last $(n - r)$ columns of RHS of (2), therefore premultiplying the LHS of (2) by elementary matrices $P_1^{-1}, P_2^{-1}, \dots, P_s^{-1}$ successively and effecting the corresponding row transformations in RHS of (2), we get

$$AQ = [H \quad O]$$

Since the elementary transformations do not change the rank, therefore the rank of AQ is the same as the rank of A, which is r, thus the rank of $[H \quad O]$ is r, hence the rank of H is r as H has r columns and the last $(n - r)$ column of the matrix $[H \quad O]$ consists only zero entries, *i.e.*, O is a null matrix of order $m \times n - r$.

THEOREM 3. (Rank of product of matrices)

The rank of a product of two matrices cannot exceed the rank of either matrix, i.e., if A and B be two matrices conformable for the product AB, then $\rho(AB) \leq \rho(A)$, $\rho(AB) \leq \rho(B)$, *i.e.,* $\rho(AB) \leq \min\{\rho(A), \rho(B)\}$.

Proof. Let A be a matrix of order $m \times n$ of rank r_1 and B be a matrix of order $n \times p$ of order r_2, then AB is conformable and let r be the rank of AB.

We shall prove that $\qquad r \leq r_1, r \leq r_2$

Since A is a matrix of order $m \times n$ of rank r_1, then there exists a non-singular matrix P such that

$$PA = \begin{bmatrix} G \\ O \end{bmatrix} \qquad \dots(1)$$

where G is a matrix of order $r_1 \times n$ of rank r_1 and O is a null matrix of order $(m - r_1) \times n$.

Now post-multiplying both sides of (1) by B, we get

$$PAB = \begin{bmatrix} G \\ O \end{bmatrix} B \qquad \dots(2)$$

Since the rank of matrix does not change by pre-multiplying it by a non-singular, therefore

$$\rho(PAB) = \rho(AB) = r$$

$\therefore$ Rank of the matrix $\begin{bmatrix} G \\ O \end{bmatrix} B = r$

Since the rank of G is r_1 so it contains r_1 non-zero rows, therefore the matrix $\begin{bmatrix} G \\ O \end{bmatrix} B$ cannot have more than r_1 rows, hence

$$\text{Rank of } \begin{bmatrix} G \\ O \end{bmatrix} B \leq r_1$$

$\therefore \qquad r \leq r_1 \qquad \dots(3)$

i.e., $\qquad$ Rank of $AB \leq$ rank of the prefactor A

Again, since $\qquad \rho(AB) = \rho((AB)')$

$\Rightarrow \qquad \rho(AB) = \rho(B'A')\rho(B') = \rho(B)$

$\therefore \qquad r \leq r_2 \qquad \dots(4)$

i.e., $\qquad$ rank of $AB \leq$ rank of the post factor B

From (3) and (4) we conclude that

$$r \leq \min\{r_1, r_2\}$$

i.e., $\qquad \rho(AB) \leq \min\{\rho(A), \rho(B)\}$.

THEOREM 4. *Every non-singular matrix is row equivalent to a unit matrix.*

Proof. Let $A = [a_{ij}]$ be a non-singular matrix of order $n \times n$. We shall prove the theorem by mathematical induction on n. Suppose $n = 1$, then $A = [a_{11}]$, therefore in this case the theorem is trivially proved. Thus we assume that the theorem holds for all non-singular matrices of order n–1. For a non-singular matrix $A = [a_{ij}]$ of order $n \times n$ there must be at least one non-zero element in the first column of A otherwise $|A| = 0$. Suppose that $a_{i1} = s \neq 0$.

Now (if necessary), interchanging the ith and first row, we get a matrix B whose leading element is s and which is not equal to zero.

$$B = \begin{bmatrix} s & a_{i2} & a_{i3} & \cdots & a_{in} \\ a_{21} & a_{22} & a_{23} & \cdots & a_{2n} \\ \vdots & \vdots & \vdots & & \vdots \\ a_{(i-1)1} & a_{(i-1)2} & a_{(i-3)3} & \cdots & a_{(i-1)n} \\ a_{11} & a_{12} & a_{13} & \cdots & a_{1n} \\ a_{(i+1)1} & a_{(i+1)2} & a_{(i+1)3} & \cdots & a_{(i+1)n} \\ \vdots & \vdots & \vdots & & \vdots \\ a_{n1} & a_{n2} & a_{n3} & \cdots & a_{nn} \end{bmatrix}$$

Multiplying each element of the first row of B by $1/s$, we get a matrix C whose leading element is equal to unity.

Let
$$C = \begin{bmatrix} 1 & c_{12} & c_{13} & \cdots & c_{1n} \\ c_{21} & c_{22} & c_{23} & \cdots & c_{2n} \\ c_{31} & c_{32} & c_{33} & \cdots & c_{3n} \\ \vdots & \vdots & \vdots & & \vdots \\ c_{n1} & c_{n2} & c_{n3} & \cdots & c_{nn} \end{bmatrix}$$

Now applying $R_2 \to R_2 - c_{21}R_1, R_2 \to R_2 - c_{31}R_1$, etc we get a matrix D in which all elements of the first column except the leading element are equal to zero.

Let
$$D = \begin{bmatrix} 1 & d_{12} & d_{13} & \cdots & d_{1n} \\ 0 & & & & \\ 0 & & A_1 & & \\ \cdots & & & & \\ 0 & & & & \end{bmatrix}$$

where A_1 is a non-singular matrix, otherwise $|A_1| = 0 \Rightarrow |D| = 0 \Rightarrow |A| = 0$.

By hypothesis A_1, a matrix of order $(n-1) \times (n-1)$, can be transformed to a unit matrix I_{n-1} of order $(n-1) \times (n-1)$ by elementary row transformations. If these two row transformations are applied to the matrix D, they will not effect the first row and the first column of D, and we, therefore, get a matrix M such that

$$M = \begin{bmatrix} 1 & d_{12} & d_{13} & \cdots & d_{1n} \\ 0 & 1 & 0 & \cdots & 0 \\ 0 & 0 & 1 & \cdots & 0 \\ \vdots & \vdots & \vdots & \vdots & \vdots \\ 0 & 0 & 0 & 0 & 0 \end{bmatrix}$$

Again applying $R_2 \to R_2 - d_{12}R_1, R_3 \to R_3 - d_{13}R$, etc. we get a matrix I_n, the unit matrix of order $n \times n$. Hence A is reduced to I_n by elementary row transformations only and hence the theorem.

REMARKS

- If A is a non-singular matrix of order $n \times n$, there exist 8 elementary matrices $E_1, E_2, \ldots, E_s$ *such that*
$$E_s E_{s-1} \ldots E_2 E_1 A = I_n$$
- Every non-singular matrix A *is expressible as the product of elementary matrices.*
- The rank of a matrix does not change by pre-multiplication or postmultiplication with a non-singular matrix.

Solved Examples

Example 1. *Show that the matrices* $\begin{bmatrix} 1 & 2 & 3 \\ 2 & 4 & 6 \end{bmatrix}$ and $\begin{bmatrix} 0 & 3 & 2 \\ 0 & 6 & 4 \end{bmatrix}$ *are equivalent.*

Solution. Let $A = \begin{bmatrix} 1 & 2 & 3 \\ 2 & 4 & 6 \end{bmatrix}$ and $B = \begin{bmatrix} 0 & 3 & 2 \\ 0 & 6 & 4 \end{bmatrix}$

Applying $R_2 \to R_2 - 2R_1$ on A

$$A \sim \begin{bmatrix} 1 & 2 & 3 \\ 0 & 0 & 0 \end{bmatrix}$$

Again applying $C_1 \to C_1 - \frac{1}{2}C_2$

$$A \sim \begin{bmatrix} 0 & 2 & 3 \\ 0 & 0 & 0 \end{bmatrix}$$

Again applying $R_2 \to R_2 + 2R_1$

$$A \sim \begin{bmatrix} 0 & 2 & 3 \\ 0 & 4 & 6 \end{bmatrix}$$

Again applying $C_2 \leftrightarrow C_3$

$$A \sim \begin{bmatrix} 0 & 3 & 2 \\ 0 & 6 & 4 \end{bmatrix} = B$$

$$A \sim B$$

Thus B can be obtained from A by a finite number of elementary transformations on A. Hence A and B are equivalent.

Example 2. *If A and B be two equivalent matrices, then show that $\rho(A) = \rho(B)$.*

Solution. Since A and B are equivalent, therefore B can be obtained from A by a finite chain of elementary transformations on A and elementary transformations do not change the rank of the matrices, hence $\rho(A) = \rho(B)$.

Example 3. *Show that if two matrices A and B have the same size and the same rank they are equivalent.*

Solution. Let A and B be two matrices of order $m \times n$ and $\rho(A) = \rho(B) = r$. Then we have

$$A \sim \begin{bmatrix} I_r & O \\ O & O \end{bmatrix} \text{ and } B \sim \begin{bmatrix} I_r & O \\ O & O \end{bmatrix}$$

Since '$\sim$' is symmetric, then

$$B \sim \begin{bmatrix} I_r & O \\ O & O \end{bmatrix} \Leftrightarrow \begin{bmatrix} I_r & O \\ O & O \end{bmatrix} \sim B$$

Again '$\sim$' is transitive, then

$$A \sim \begin{bmatrix} I_r & O \\ O & O \end{bmatrix} \text{ and } \begin{bmatrix} I_r & O \\ O & O \end{bmatrix} \sim B$$

$\Rightarrow$ $A \sim B$

Hence A and B are equivalent.

Example 4. *Use E-transformations to reduce the following matrices to triangular form and hence find their rank.*

$$(i) \begin{bmatrix} 5 & 3 & 14 & 4 \\ 0 & 1 & 2 & 1 \\ 1 & -1 & 2 & 0 \end{bmatrix} \quad (ii) \begin{bmatrix} 8 & 1 & 3 & 6 \\ 0 & 3 & 2 & 2 \\ -8 & -1 & -3 & 4 \end{bmatrix}$$

(Kurukshetra-2005)

Solution. (i) Let
$$A = \begin{bmatrix} 5 & 3 & 14 & 4 \\ 0 & 1 & 2 & 1 \\ 1 & -1 & 2 & 0 \end{bmatrix}$$

Applying $R_1 \leftrightarrow R_3$

$$A \sim \begin{bmatrix} 1 & -1 & 2 & 0 \\ 0 & 1 & 2 & 1 \\ 5 & 3 & 14 & 4 \end{bmatrix}$$

Again applying $R_3 \to R_3 - 5R_1$

$$A \sim \begin{bmatrix} 1 & -1 & 2 & 0 \\ 0 & 1 & 2 & 1 \\ 0 & 8 & 4 & 4 \end{bmatrix}$$

Again applying $R_3 \to R_3 - 8R_2$

$$A \sim \begin{bmatrix} 1 & -1 & 2 & 0 \\ 0 & 1 & 2 & 1 \\ 0 & 0 & -12 & -4 \end{bmatrix}$$

The last equivalent matrix is in Echelon form (or triangular form) which has three non-zero rows. Hence $\rho(A) = 3$.

(i.i) Let
$$A = \begin{bmatrix} 8 & 1 & 3 & 6 \\ 0 & 3 & 2 & 2 \\ -8 & -1 & -3 & 4 \end{bmatrix}$$

Applying $C_1 \to \frac{1}{8} C_1$

$$A \sim \begin{bmatrix} 1 & 1 & 3 & 6 \\ 0 & 3 & 2 & 2 \\ -1 & -1 & -3 & 4 \end{bmatrix}$$

Again applying $R_3 \to R_3 + R_1$

$$A \sim \begin{bmatrix} 1 & 1 & 3 & 6 \\ 0 & 3 & 2 & 2 \\ 0 & 0 & 0 & 10 \end{bmatrix}$$

The last equivalent matrix is in Echelon form (or triangular form) which has three non-zero rows, hence $\rho(A) = 3$.

Example 5. *Is the matrix $\begin{bmatrix} 1 & 2 & 1 \\ -1 & 0 & 2 \\ 2 & 1 & -3 \end{bmatrix}$ equivalent to I_3 ?*

Solution. Let
$$A = \begin{bmatrix} 1 & 2 & 1 \\ -1 & 0 & 2 \\ 2 & 1 & -3 \end{bmatrix}$$

Then $|A| = 1(0-2) - 2(-3-4) + 1(-1-0)$
$= -2 + 2 + 1 = 1 \neq 0$

Therefore, A is a non-singular matrix of order 3×3, so it is row equivalent to a unit matrix. Hence $A \sim I_3$.

Example 6. *If A and B are two matrices of the same type, then* $\rho(A+B) \leq \rho(A) + \rho(B)$

Solution. Let A and B be two matrices of order $m \times n$ and let $\rho(A) = r_1$ and $\rho(B) = r_2$.

Now $\rho(A) = r_1$

$\Rightarrow$ A contains r_1 linearly independent rows.

and $\rho(B) = r_2$

$\Rightarrow$ B contains r_2 linearly independent rows.

Therefore $A + B$ will contain at most $r_1 + r_2$ linearly independent rows

$\therefore$ $\rho(A+B) \leq r_1 + r_2$

Hence $\rho(A+B) \leq \rho(A) + \rho(B)$.

Example 7. *If A and B are two n-rowed square matrices, then $\rho(AB) \geq \rho(A) + \rho(B) - n$.*

Solution. Let $\rho(A) = r$, then there exists two non-singular matrices P and Q such that

$$PAQ = \begin{bmatrix} I_r & O \\ O & O \end{bmatrix} \quad ...(1)$$

Pre-multiplying by P^{-1} and post-multiplying by Q^{-1} we get

$$A = P^{-1} \begin{bmatrix} I_r & O \\ O & O \end{bmatrix} Q^{-1} \quad ...(2)$$

Consider a matrix

$$C = P^{-1}\begin{bmatrix} O_r & O \\ O & I_{n-r} \end{bmatrix} Q^{-1}$$

Now $A + C = P^{-1}\left\{\begin{bmatrix} I_r & O \\ O & O \end{bmatrix} + \begin{bmatrix} O_r & O \\ O & I_{n-r} \end{bmatrix}\right\} Q^{-1}$

$\Rightarrow \quad A + C = P^{-1}\begin{bmatrix} I_r & O \\ O & I_{n-r} \end{bmatrix} Q^{-1}$

$\Rightarrow \quad A + C = P^{-1} I_n Q^{-1}$

$\Rightarrow \quad A + C = P^{-1} Q^{-1}$

$\therefore$ $A+C$ is non-singular matrix of order $m \times n$, since $P^{-1}Q^{-1}$ is non-singular of order $n \times n$.

Therefore,

$$\rho(A+C) = n$$

But by the definition of C, we have

$$\rho(C) = n - r$$

$\Rightarrow \quad \rho(C) = n - \rho(A) \qquad \because \rho(A) = r$

Since the rank of a matrix does not change on pre-multiplying it with a non-singular matrix, then

$$\rho((A+C))\, B = \rho(B)$$

$[\because A+C$ is non-singular matrix.$]$

$\Rightarrow \quad \rho(B) = \rho(AB + CB)$

$\Rightarrow \quad \rho(B) \leq \rho(AB) + \rho(CB)$

$[\because \rho(A+B) \leq \rho(A) + \rho(B)]$

$\Rightarrow \quad \rho(B) \leq \rho(AB) + \rho(C)$

$[\because \rho(CB) \leq \min\{\rho(C), \rho(B)\}]$

$\Rightarrow \quad \rho(B) \geq \rho(AB) + n - \rho(A)$

$[\because \rho(C) = n - \rho(A)]$

$\therefore \quad \rho(AB) \geq \rho(A) + \rho(B) - n.$

Example 8. *If A be any non-singular matrix and B a matrix such that AB exists, then show that*

$$\rho(AB) = \rho(B).$$

Solution. Since A is a non-singular matrix, then

$$B = A^{-1}(AB)$$

We know that

$\rho(AB) \leq \rho(B)$ ...(1)

Now $\rho(B) = \rho(A^{-1}(AB)) \leq \rho(AB)$

or $\rho(B) \leq \rho(AB)$ ...(2)

From (1) and (2), we have

$$\rho(AB) = \rho(B).$$

Example 9. *If A is a square matrix of order $n \times n$ such that $A^2 = A$, then show that*

$$\rho(A) + \rho(I_n - A) = n.$$

Solution. We have $A^2 = A$

$\Rightarrow \quad A - A^2 = 0$

$\Rightarrow \quad AI_n - A^2 = 0 \qquad [\because AI_n = A]$

$\Rightarrow \quad A(I_n - A) = 0$

$\therefore \quad P(A(I_n - A)) = 0$

Also we know that

$$\rho(A(I_n - A)) \geq \rho(A) + \rho(I_n - A) - n$$

$\Rightarrow \quad \rho(A) + \rho(I_n - A) - n \leq 0$

$\therefore \quad \rho(A) + \rho(I_n - A) \leq n$ ...(1)

Again we know that

$$\rho(A + I_n - A) = \rho(A) + \rho(I_n - A)$$

$\Rightarrow \quad \rho(I_n) \leq \rho(A) + \rho(I_n - A)$

$\therefore \quad \rho(A) + \rho(I_n - A) \geq n$ ...(2)

$[\because \rho(I_n) = n]$

From (1) and (2), we get

$$\rho(A) + \rho(I_n - A) = n$$

Example 10. *If A is a square matrix of order $n \times n$ and $\rho(A) = n - 1$, show that $\rho(adj.\ A) = 1$.*

Solution. Since A is an $n \times n$ matrix and $\rho(A) = n - 1$, then we have $|A| = 0$

But we know that $A(adj.\ A) = |A| I_n$

$\Rightarrow \quad A(adj.\ A) = O$

$\therefore \quad \rho(A\ adj.\ A) = 0$

Also $\quad \rho(A\ adj.\ A) \geq \rho(A) + \rho(adj.\ A) - n$

$\Rightarrow \quad \rho(A) + \rho(adj.\ A) - n \leq O$

$\Rightarrow \quad \rho(A) + \rho(adj.\ A) \leq n$

$\Rightarrow \quad \rho(adj.\ A) \leq n - \rho(A)$

$\Rightarrow \quad \rho(adj.\ A) \leq n - (n - 1)$

$\therefore \quad \rho(adj.\ A) \leq 1$ ...(1)

Since $\rho(A) = n - 1$, then there exists at least one minor of order $n - 1$ of A not equal to zero, therefore there exists at least one element of *adj. A* which is non-zero, it follows that

$\rho(adj.\ A) > 0$ ...(2)

From (1) and (2), we get

$$\rho(adj.\ A) = 1.$$

Example 11. *Find the rank of the matrix*

$$A = \begin{bmatrix} 1 & 3 & 4 & 3 \\ 3 & 9 & 12 & 9 \\ -1 & -3 & -4 & -3 \end{bmatrix}$$

Solution. We have

$$A = \begin{bmatrix} 1 & 3 & 4 & 3 \\ 3 & 9 & 12 & 9 \\ -1 & -3 & -4 & -3 \end{bmatrix}$$

Applying $R_2 \to R_2 - 3R_1$ and $R_3 \to R_3 + R_1$

$$A \sim \begin{bmatrix} 1 & 3 & 4 & 3 \\ 0 & 0 & 0 & 0 \\ 0 & 0 & 0 & 0 \end{bmatrix}$$

The last equivalent matrix is in Echelon form which has one non-zero row.

Hence $\quad \rho(A) = 1.$

Example 12. *Determine the rank of the following matrices :*

$$\text{(i)} \begin{bmatrix} 2 & -1 & 3 & 4 \\ 0 & 3 & 4 & 1 \\ 2 & 3 & 7 & 5 \\ 2 & 5 & 11 & 6 \end{bmatrix} \quad \text{(ii)} \begin{bmatrix} -2 & -1 & -3 & -1 \\ 1 & 2 & 3 & -1 \\ 1 & 0 & 1 & 1 \\ 0 & 1 & 1 & -1 \end{bmatrix}$$

Solution. (i) Let $A = \begin{bmatrix} 2 & -1 & 3 & 4 \\ 0 & 3 & 4 & 1 \\ 2 & 3 & 7 & 5 \\ 2 & 5 & 11 & 6 \end{bmatrix}$

Applying $R_3 \to R_3 - R_1, R_4 \to R_4 - R_1$

$$A \sim \begin{bmatrix} 2 & -1 & 3 & 4 \\ 0 & 3 & 4 & 1 \\ 0 & 4 & 4 & 1 \\ 0 & 6 & 8 & 2 \end{bmatrix}$$

Again applying $R_3 \to R_3 - \frac{4}{3}R_2, R_4 \to R_4 - 2R_2$

$$A \sim \begin{bmatrix} 2 & -1 & 3 & 4 \\ 0 & 3 & 4 & 1 \\ 0 & 0 & -4/3 & -1/3 \\ 0 & 0 & 0 & 0 \end{bmatrix}$$

Again applying $R_3 \to 3R_3$

$$A \sim \begin{bmatrix} 2 & -1 & 3 & 4 \\ 0 & 3 & 4 & 1 \\ 0 & 0 & -4 & -1 \\ 0 & 0 & 0 & 0 \end{bmatrix}$$

The last equaivalent matrix is in Echelon form which has 3 non-zero rows, hence $\rho(A) = 3$.

(ii) Let $A = \begin{bmatrix} -2 & -1 & -3 & -1 \\ 1 & 2 & 3 & -1 \\ 1 & 0 & 1 & 1 \\ 0 & 1 & 1 & -1 \end{bmatrix}$

Applying $R_1 \to R_2$

$$A \sim \begin{bmatrix} 1 & 2 & 3 & -1 \\ -2 & -1 & -3 & -1 \\ 1 & 0 & 1 & 1 \\ 0 & 1 & 1 & -1 \end{bmatrix}$$

Again applying $R_2 \to R_2 + 2R_1, R_3 \to R_3 - R_1$

$$A \sim \begin{bmatrix} 1 & 2 & 3 & -1 \\ 0 & 3 & 3 & -3 \\ 0 & -2 & -2 & 2 \\ 0 & 1 & 1 & -1 \end{bmatrix}$$

Again applying $R_2 \to R_2 + R_3$

$$A \sim \begin{bmatrix} 1 & 2 & 3 & -1 \\ 0 & 1 & 1 & -1 \\ 0 & -2 & -2 & 2 \\ 0 & 1 & 1 & -1 \end{bmatrix}$$

Again applying $R_3 \to R_3 + 2R_2, R_4 \to R_4 - R_2$

$$A \sim \begin{bmatrix} 1 & 2 & 3 & -1 \\ 0 & 1 & 1 & -1 \\ 0 & 0 & 0 & 0 \\ 0 & 0 & 0 & 0 \end{bmatrix}$$

The last equivalent matrix is in Echelon form which has 2 non-zero rows, hence $\rho(A) = 2$.

Example 13. *Find the rank of the matrix*

$$A = \begin{bmatrix} 6 & 1 & 3 & 8 \\ 4 & 2 & 6 & -1 \\ 10 & 3 & 9 & 7 \\ 16 & 4 & 12 & 15 \end{bmatrix}$$

Solution. We have $A = \begin{bmatrix} 6 & 1 & 3 & 8 \\ 4 & 2 & 6 & -1 \\ 10 & 3 & 9 & 7 \\ 16 & 4 & 12 & 15 \end{bmatrix}$

Applying $R_1 \to R_1 - R_2$

$$A \sim \begin{bmatrix} 2 & -1 & -3 & 9 \\ 4 & 2 & 6 & -1 \\ 10 & 3 & 9 & 7 \\ 16 & 4 & 12 & 15 \end{bmatrix}$$

Again applying $R_2 \to R_2 - 2R_1$, $R_3 \to R_3 - 5R_1$, $R_4 \to R_4 - 8R_1$

$$A \sim \begin{bmatrix} 2 & -1 & -3 & 9 \\ 0 & 4 & 12 & -19 \\ 0 & 8 & 24 & -38 \\ 0 & 12 & 0 & -57 \end{bmatrix}$$

Again applying $R_2 \to R_2 - 2R_1$, $R_3 \to R_3 - 5R_1$, $R_4 \to R_4 - 8R_2$

$$A \sim \begin{bmatrix} 2 & -1 & -3 & 9 \\ 0 & 4 & 12 & -19 \\ 0 & 8 & 0 & 0 \\ 0 & 12 & 0 & 0 \end{bmatrix}$$

The last equivalent matrix is in Echelon form which has 2 non-zero rows, hence

$\rho(A) = 2$.

Example 14. *Find the rank of the matrix*

$$A = \begin{bmatrix} 1 & a & b & 0 \\ 0 & c & d & 1 \\ 1 & a & b & 0 \\ 0 & c & d & 1 \end{bmatrix}$$

Solution. We have $A = \begin{bmatrix} 1 & a & b & 0 \\ 0 & c & d & 1 \\ 1 & a & b & 0 \\ 0 & c & d & 1 \end{bmatrix}$

Applying $R_3 \to R_3 - R_1$, $R_4 \to R_4 - R_2$

$$A \sim \begin{bmatrix} 1 & a & b & 0 \\ 0 & c & d & 1 \\ 0 & 0 & 0 & 0 \\ 0 & 0 & 0 & 0 \end{bmatrix}$$

The last equivalent matrix is in Echelon form which has 2 non-zero rows, hence

$\rho(A) = 2$.

Example 15. *Reduce the matrix* $A = \begin{bmatrix} 1 & -1 & 2 & -3 \\ 4 & 1 & 0 & 2 \\ 0 & 3 & 0 & 4 \\ 0 & 1 & 0 & 2 \end{bmatrix}$ *to the normal form* $\begin{bmatrix} I_r & O \\ O & O \end{bmatrix}$ *and hence determine its rank.*

Solution. We have $A = \begin{bmatrix} 1 & -1 & 2 & -3 \\ 4 & 1 & 0 & 2 \\ 0 & 3 & 0 & 4 \\ 0 & 1 & 0 & 2 \end{bmatrix}$

Applying $R_2 \to R_2 - 4R_1$

$$A \sim \begin{bmatrix} 1 & -1 & 2 & -3 \\ 0 & 5 & -8 & 14 \\ 0 & 3 & 0 & 4 \\ 0 & 1 & 0 & 2 \end{bmatrix}$$

Applying $C_2 \to C_2 + C_1$, $C_3 \to C_3 - 2C_1$, $C_4 \to C_4 + 3C_1$

$$A \sim \begin{bmatrix} 1 & 0 & 0 & 0 \\ 0 & 5 & -8 & 14 \\ 0 & 3 & 0 & 4 \\ 0 & 1 & 0 & 2 \end{bmatrix}$$

Applying $R_2 \leftrightarrow R_4$

$$A \sim \begin{bmatrix} 1 & 0 & 0 & 0 \\ 0 & 1 & 0 & 2 \\ 0 & 3 & 0 & 4 \\ 0 & 5 & -8 & 14 \end{bmatrix}$$

Applying $R_3 \to R_3 - 3R_2$, $R_4 \to R_4 - 5R_2$

$$A \sim \begin{bmatrix} 1 & 0 & 0 & 0 \\ 0 & 1 & 0 & 2 \\ 0 & 0 & 0 & -2 \\ 0 & 0 & -8 & 4 \end{bmatrix}$$

Applying $C_4 \to C_4 - 2C_2$

$$A \sim \begin{bmatrix} 1 & 0 & 0 & 0 \\ 0 & 1 & 0 & 0 \\ 0 & 0 & 0 & -2 \\ 0 & 0 & -8 & 4 \end{bmatrix}$$

Applying $C_3 \leftrightarrow C_4$

$$A \sim \begin{bmatrix} 1 & 0 & 0 & 0 \\ 0 & 1 & 0 & 0 \\ 0 & 0 & -2 & 0 \\ 0 & 0 & 4 & -8 \end{bmatrix}$$

Applying $R_4 \to R_4 + 2R_3$

$$A \sim \begin{bmatrix} 1 & 0 & 0 & 0 \\ 0 & 1 & 0 & 0 \\ 0 & 0 & -2 & 0 \\ 0 & 0 & 0 & -8 \end{bmatrix}$$

Applying $R_3 \to \frac{1}{2} R_3, R_4 \to -\frac{1}{8} R_4$

$$A \sim \begin{bmatrix} 1 & 0 & 0 & 0 \\ 0 & 1 & 0 & 0 \\ 0 & 0 & 1 & 0 \\ 0 & 0 & 0 & 1 \end{bmatrix}$$

$\therefore$ $A \sim I_4$

Hence $\rho(A) = 4$

Example 16. *Find the rank of the matrix*

$$A = \begin{bmatrix} 2 & -2 & 0 & 6 \\ 4 & 2 & 0 & 2 \\ 1 & -1 & 0 & 3 \\ 1 & -2 & 1 & 2 \end{bmatrix}$$

by reducing it to normal form.

Solution. We have $A = \begin{bmatrix} 2 & -2 & 0 & 6 \\ 4 & 2 & 0 & 2 \\ 1 & -1 & 0 & 3 \\ 1 & -2 & 1 & 2 \end{bmatrix}$

Applying $R_1 \to \frac{1}{2} R_1$

$$A \sim \begin{bmatrix} 1 & -1 & 0 & 3 \\ 4 & 2 & 0 & 2 \\ 1 & -1 & 0 & 3 \\ 1 & -2 & 1 & 2 \end{bmatrix}$$

Applying $R_2 \to R_2 - 4R_1$, $R_3 \to R_3 - R_1$, $R_4 \to R_4 - R_1$

$$A \sim \begin{bmatrix} 1 & -1 & 0 & 3 \\ 0 & 6 & 0 & -10 \\ 0 & 0 & 0 & 0 \\ 0 & -1 & 1 & -1 \end{bmatrix}$$

Applying $C_2 \to C_2 + C_1$, $C_4 \to C_4 - 3C_1$

$$A \sim \begin{bmatrix} 1 & -1 & 0 & 0 \\ 0 & 6 & 0 & -10 \\ 0 & 0 & 0 & 0 \\ 0 & -1 & 1 & 1 \end{bmatrix}$$

Applying $R_2 \leftrightarrow R_4$

$$A \sim \begin{bmatrix} 1 & 0 & 0 & 0 \\ 0 & -1 & 1 & 1 \\ 0 & 0 & 0 & 0 \\ 0 & 6 & 0 & -10 \end{bmatrix}$$

Applying $R_4 \to R_4 + 6R_2$

$$A \sim \begin{bmatrix} 1 & 0 & 0 & 0 \\ 0 & -1 & 1 & 1 \\ 0 & 0 & 0 & 0 \\ 0 & 0 & 6 & -4 \end{bmatrix}$$

Applying $C_3 \to C_3 + C_2$, $C_4 \to C_4 + C_2$

$$A \sim \begin{bmatrix} 1 & 0 & 0 & 0 \\ 0 & -1 & 0 & 0 \\ 0 & 0 & 0 & 0 \\ 0 & 0 & 6 & -4 \end{bmatrix}$$

Applying $C_3 \to \frac{1}{6} C_3$, $C_4 \to \frac{-1}{4} C_4$, $C_2 \to (-1) C_2$

$$A \sim \begin{bmatrix} 1 & 0 & 0 & 0 \\ 0 & 1 & 0 & 0 \\ 0 & 0 & 0 & 0 \\ 0 & 0 & 1 & 1 \end{bmatrix}$$

Applying $R_3 \leftrightarrow R_4$

$$A \sim \begin{bmatrix} 1 & 0 & 0 & 0 \\ 0 & 1 & 0 & 0 \\ 0 & 0 & 1 & 1 \\ 0 & 0 & 0 & 0 \end{bmatrix}$$

Applying $C_4 \to C_4 - C_3$

$$A \sim \begin{bmatrix} 1 & 0 & 0 & 0 \\ 0 & 1 & 0 & 0 \\ 0 & 0 & 1 & 0 \\ 0 & 0 & 0 & 0 \end{bmatrix}$$

$$A \sim \begin{bmatrix} I_3 & O \\ O & O \end{bmatrix}$$

Hence $\rho(A) = 3$.

Example 17. *Find two non-singular matrices P and Q such that PAQ is in the normal form where*

$$A = \begin{bmatrix} 1 & 1 & 1 \\ 1 & -1 & -1 \\ 3 & 1 & 1 \end{bmatrix}$$

Also find the rank of the matrix A.

(Kurukshetra 2005)

Solution. We write $A = I_3 A I_3$

$$\text{or} \begin{bmatrix} 1 & 1 & 1 \\ 1 & -1 & -1 \\ 3 & 1 & 1 \end{bmatrix} = \begin{bmatrix} 1 & 0 & 0 \\ 0 & 1 & 0 \\ 0 & 0 & 1 \end{bmatrix} A \begin{bmatrix} 1 & 0 & 0 \\ 0 & 1 & 0 \\ 0 & 0 & 1 \end{bmatrix} \quad ...(1)$$

In order to find P and Q such that $PAQ = \begin{bmatrix} I_r & O \\ O & O \end{bmatrix}$ we shall reduce the matrix on LHS of (1) by using elementary transformations, while in doing so we shall apply elementary row transformation to pre-factor of A and elementary-column transformation to post-factor of A on RHS of (1).

Now applying $R_2 \to R_2 - R_1$, $R_3 \to R_3 - 3R_1$

$$\begin{bmatrix} 1 & 1 & 1 \\ 1 & -2 & -2 \\ 0 & -2 & -2 \end{bmatrix} = \begin{bmatrix} 1 & 0 & 0 \\ -1 & 1 & 0 \\ -3 & 0 & 1 \end{bmatrix} A \begin{bmatrix} 1 & 0 & 0 \\ 0 & 1 & 0 \\ 0 & 0 & 1 \end{bmatrix}$$

Applying $C_2 \to C_2 - C_1$, $C_3 \to C_3 - C_1$

$$\begin{bmatrix} 1 & 0 & 0 \\ 0 & -2 & -2 \\ 0 & -2 & -2 \end{bmatrix} = \begin{bmatrix} 1 & 0 & 0 \\ -1 & 1 & 0 \\ -3 & 0 & 1 \end{bmatrix} A \begin{bmatrix} 1 & -1 & -1 \\ 0 & 1 & 0 \\ 0 & 0 & 1 \end{bmatrix}$$

Applying $R_2 \to \left(-\frac{1}{2}\right) R_2$

$$\begin{bmatrix} 1 & 0 & 0 \\ 0 & 1 & 1 \\ 0 & -2 & -2 \end{bmatrix} = \begin{bmatrix} 1 & 0 & 0 \\ 1/2 & -1/2 & 0 \\ -3 & 0 & 1 \end{bmatrix} A \begin{bmatrix} 1 & -1 & -1 \\ 0 & 1 & 0 \\ 0 & 0 & 1 \end{bmatrix}$$

Applying $R_3 \to R_3 + 2R_2$

$$\begin{bmatrix} 1 & 0 & 0 \\ 0 & 1 & 1 \\ 0 & 0 & 0 \end{bmatrix} = \begin{bmatrix} 1 & 0 & 0 \\ 1/2 & -1/2 & 0 \\ -2 & -1 & 1 \end{bmatrix} A \begin{bmatrix} 1 & -1 & -1 \\ 0 & 1 & 0 \\ 0 & 0 & 1 \end{bmatrix}$$

Applying $C_2 \to C_3 - C_2$

$$\begin{bmatrix} 1 & 0 & 0 \\ 0 & 1 & 0 \\ 0 & 0 & 0 \end{bmatrix} = \begin{bmatrix} 1 & 0 & 0 \\ 1/2 & -1/2 & 0 \\ -2 & -1 & 1 \end{bmatrix} A \begin{bmatrix} 1 & -1 & 0 \\ 0 & 1 & -1 \\ 0 & 0 & 1 \end{bmatrix}$$

$$\text{or} \quad \begin{bmatrix} I_2 & O \\ O & O \end{bmatrix} = PAQ$$

where

$$P = \begin{bmatrix} 1 & 0 & 0 \\ 1/2 & -1/2 & 0 \\ -2 & -1 & 1 \end{bmatrix} \text{ and } Q = \begin{bmatrix} 1 & -1 & 0 \\ 0 & 1 & -1 \\ 0 & 0 & 1 \end{bmatrix}$$

$$A \sim \begin{bmatrix} I_2 & O \\ O & O \end{bmatrix}$$

Hence $\rho(A) = 2$.

Example 18. *Determine non-singular matrices P and Q such PAQ is in the normal form* $\begin{bmatrix} I_r & O \\ O & O \end{bmatrix}$, *where*

$$A = \begin{bmatrix} 3 & 2 & -1 & 5 \\ 5 & 1 & 4 & -2 \\ 1 & -4 & 11 & -19 \end{bmatrix}$$

Solution. Since A is a matrix of order 3×4, therefore we write $A = I_3 A I_4$

$$\text{or} \begin{bmatrix} 3 & 2 & -1 & 5 \\ 5 & 1 & 4 & -2 \\ 1 & -4 & 11 & -19 \end{bmatrix} = \begin{bmatrix} 1 & 0 & 0 \\ 0 & 1 & 0 \\ 0 & 0 & 1 \end{bmatrix} A \begin{bmatrix} 1 & 0 & 0 & 0 \\ 0 & 1 & 0 & 0 \\ 0 & 0 & 1 & 0 \\ 0 & 0 & 0 & 1 \end{bmatrix}$$

Applying $R_1 \leftrightarrow R_4$

$$\begin{bmatrix} 1 & -4 & 11 & -19 \\ 5 & 1 & 4 & -2 \\ 3 & 2 & -1 & 5 \end{bmatrix} = \begin{bmatrix} 0 & 0 & 1 \\ 0 & 1 & 0 \\ 1 & 0 & 0 \end{bmatrix} A \begin{bmatrix} 1 & 0 & 0 & 0 \\ 0 & 1 & 0 & 0 \\ 0 & 0 & 1 & 0 \\ 0 & 0 & 0 & 1 \end{bmatrix}$$

Applying $R_2 \to R_2 - 5R_1$, $R_3 \to R_3 - 3R_1$

$$\begin{bmatrix} 1 & -4 & 11 & -19 \\ 0 & 21 & -51 & -93 \\ 0 & 14 & -34 & 62 \end{bmatrix} = \begin{bmatrix} 0 & 0 & 1 \\ 0 & 1 & -5 \\ 1 & 0 & -3 \end{bmatrix} A \begin{bmatrix} 1 & 0 & 0 & 0 \\ 0 & 1 & 0 & 0 \\ 0 & 0 & 1 & 0 \\ 0 & 0 & 0 & 1 \end{bmatrix}$$

Applying $C_2 \to C_2 + 4C_1$, $C_3 \to C_3 - 11C_1$, $C_4 \to C_4 + 19C_1$

$$\begin{bmatrix} 1 & 0 & 0 & 0 \\ 0 & 21 & -51 & 93 \\ 0 & 14 & -34 & 62 \end{bmatrix}$$

$$= \begin{bmatrix} 0 & 0 & 1 \\ 0 & 1 & -5 \\ 1 & 0 & -3 \end{bmatrix} A \begin{bmatrix} 1 & 4 & -11 & 19 \\ 0 & 1 & 0 & 0 \\ 0 & 0 & 1 & 0 \\ 0 & 0 & 0 & 1 \end{bmatrix}$$

Applying

$$C_2 \to \frac{1}{7} C_2,\ C_3 \to -\frac{1}{17} C_3,\ C_4 \to \frac{1}{31} C_4$$

$$\begin{bmatrix} 1 & 0 & 0 & 0 \\ 0 & 3 & 3 & 3 \\ 0 & 2 & 2 & 2 \end{bmatrix}$$

$$= \begin{bmatrix} 0 & 0 & 1 \\ 0 & 1 & -5 \\ 1 & 0 & -3 \end{bmatrix} A \begin{bmatrix} 1 & 4/7 & 11/7 & 19/31 \\ 0 & 1/7 & 0 & 0 \\ 0 & 0 & -1/7 & 0 \\ 0 & 0 & 0 & 1/31 \end{bmatrix}$$

Applying $R_2 \to \frac{1}{3} R_2$, $R_3 \to \frac{1}{2} R_3$

$$\begin{bmatrix} 1 & 0 & 0 & 0 \\ 0 & 1 & 1 & 1 \\ 0 & 1 & 1 & 1 \end{bmatrix} = \begin{bmatrix} 0 & [illegible] & [illegible] \\ 0 & [illegible] & [illegible] \\ 1/2 & [illegible] & -3/2 \end{bmatrix}$$

$$A \begin{bmatrix} 1 & 4/7 & 11/7 & 19/31 \\ 0 & 1/7 & 0 & 0 \\ 0 & 0 & -1/7 & 0 \\ 0 & 0 & 0 & 1/31 \end{bmatrix}$$

Applying $R_3 \to R_3 - R_2$

$$\begin{bmatrix} 1 & 0 & 0 & 0 \\ 0 & 1 & 1 & 1 \\ 0 & 0 & 0 & 0 \end{bmatrix} = \begin{bmatrix} 0 & 0 & 1 \\ 0 & 1/3 & -5/3 \\ 1/2 & -1/3 & 1/6 \end{bmatrix} A \begin{bmatrix} 1 & 4/7 & 11/7 & 19/31 \\ 0 & 1/7 & 0 & 0 \\ 0 & 0 & -1/7 & 0 \\ 0 & 0 & 0 & 1/31 \end{bmatrix}$$

Applying $C_3 \to C_3 - C_2$, $C_4 \to C_4 - C_2$

$$\begin{bmatrix} 1 & 0 & 0 & 0 \\ 0 & 1 & 0 & 0 \\ 0 & 0 & 0 & 0 \end{bmatrix} = \begin{bmatrix} 0 & 0 & 1 \\ 0 & 1/3 & -5/3 \\ 1/2 & -1/3 & 1/6 \end{bmatrix} A \begin{bmatrix} 1 & 4/7 & 9/119 & 19/217 \\ 0 & 1/7 & -1/7 & -1/7 \\ 0 & 0 & -1/17 & 0 \\ 0 & 0 & 0 & 1/31 \end{bmatrix}$$

or $\begin{bmatrix} I_2 & O \\ O & O \end{bmatrix} = PAQ$

where $P = \begin{bmatrix} 0 & 0 & 1 \\ 0 & 1/3 & -5/3 \\ 1/2 & -1/3 & 1/6 \end{bmatrix}$

and $Q = \begin{bmatrix} 1 & 4/7 & 9/119 & 19/217 \\ 0 & 1/7 & -1/7 & -1/7 \\ 0 & 0 & -1/17 & 0 \\ 0 & 0 & 0 & 1/31 \end{bmatrix}$

$\therefore \quad A \sim \begin{bmatrix} I_2 & O \\ O & O \end{bmatrix}$

Hence $\rho(A) = 2$.

EXERCISE 25.4

1. Are the followings pairs of matrices equivalent ?

(i) $\begin{bmatrix} 4 & 0 & 2 \\ 3 & 1 & 0 \\ 5 & 2 & 0 \end{bmatrix}, \begin{bmatrix} 3 & 9 & 0 & 2 \\ 7 & -2 & 0 & 1 \\ 8 & 1 & 1 & 5 \end{bmatrix}$

(ii) $\begin{bmatrix} 2 & -1 & 3 & 4 \\ 0 & 3 & 4 & 1 \\ 2 & 3 & 7 & 5 \\ 2 & 5 & 11 & 5 \end{bmatrix}, \begin{bmatrix} 1 & 0 & -5 & 6 \\ 3 & -2 & 1 & 2 \\ 5 & -2 & -9 & 14 \\ 4 & -2 & -4 & 8 \end{bmatrix}$

Determine the rank of the following matrices:

2. $\begin{bmatrix} 1 & 1 & 1 \\ 2 & 2 & 2 \\ 3 & 3 & 3 \end{bmatrix}$

3. $\begin{bmatrix} 2 & 1 & 3 \\ 4 & 7 & 13 \\ 4 & -3 & -1 \end{bmatrix}$

4. $\begin{bmatrix} 4 & 5 & 6 \\ 5 & 6 & 7 \\ 7 & 8 & 9 \end{bmatrix}$

5. $\begin{bmatrix} 1 & 2 & 3 \\ 2 & 3 & 4 \\ 3 & 5 & 7 \end{bmatrix}$

6. $\begin{bmatrix} 2 & 3 & 7 \\ 3 & -2 & 4 \\ 1 & -3 & -1 \end{bmatrix}$

7. $\begin{bmatrix} 3 & -1 & 2 \\ -6 & 2 & -4 \\ -3 & 1 & -2 \end{bmatrix}$

8. $\begin{bmatrix} 1 & 2 & 3 & 1 \\ 2 & 4 & 6 & 2 \\ 1 & 2 & 3 & 2 \end{bmatrix}$

9. $\begin{bmatrix} 1 & 3 & 4 & 3 \\ 3 & 9 & 12 & 9 \\ 1 & 3 & 4 & 1 \end{bmatrix}$

10. $\begin{bmatrix} 1 & 2 & -1 & 4 \\ 2 & 4 & 3 & 5 \\ -1 & -2 & 6 & -7 \end{bmatrix}$

11. $\begin{bmatrix} 1 & 2 & -4 & 5 \\ 2 & -1 & 3 & 6 \\ 8 & 1 & 9 & 7 \end{bmatrix}$

12. $\begin{bmatrix} 1 & -1 & 3 & 6 \\ 1 & 3 & -3 & -4 \\ 5 & 3 & 3 & 11 \end{bmatrix}$

13. $\begin{bmatrix} 1 & 2 & 3 & 0 \\ 2 & 4 & 3 & 2 \\ 3 & 2 & 1 & 3 \\ 6 & 8 & 7 & 5 \end{bmatrix}$

14. $\begin{bmatrix} 2 & 3 & -1 & -1 \\ 1 & -1 & -2 & -4 \\ 3 & 1 & 3 & -2 \\ 6 & 3 & 0 & -7 \end{bmatrix}$ (UPTU-2005)

15. $\begin{bmatrix} 1 & 2 & 1 & 2 \\ 1 & 3 & 2 & 2 \\ 2 & 4 & 3 & 4 \\ 3 & 7 & 4 & 6 \end{bmatrix}$

16. $\begin{bmatrix} 3 & -2 & 0 & -1 \\ 0 & 2 & 2 & 1 \\ 1 & -2 & -3 & 2 \\ 0 & 1 & 2 & 1 \end{bmatrix}$

17. $\begin{bmatrix} 0 & 1 & -3 & -1 \\ 1 & 0 & 1 & 1 \\ 3 & 1 & 0 & 2 \\ 1 & 1 & -2 & 0 \end{bmatrix}$

18. $\begin{bmatrix} 1 & 2 & -1 & 3 \\ 4 & 1 & 2 & 1 \\ 3 & -1 & 1 & 2 \\ 1 & 2 & 0 & 1 \end{bmatrix}$

19. $\begin{bmatrix} 1 & 0 & 2 & 1 \\ 0 & 1 & -2 & 1 \\ 1 & -1 & 4 & 0 \\ -2 & 2 & 8 & 0 \end{bmatrix}$

20. $\begin{bmatrix} 8 & 0 & 0 & 1 \\ 1 & 0 & 8 & 1 \\ 0 & 0 & 1 & 8 \\ 0 & 1 & 1 & 8 \end{bmatrix}$

21. $\begin{bmatrix} 6 & 1 & 3 & 8 \\ 4 & 2 & 6 & -1 \\ 10 & 3 & 9 & 7 \\ 16 & 4 & 12 & 15 \end{bmatrix}$

22. Reduce the matrix

$$\begin{bmatrix} 0 & 1 & -3 & -1 \\ 1 & 0 & 1 & 1 \\ 3 & 1 & 0 & 2 \\ 1 & 1 & -2 & 0 \end{bmatrix}$$

to normal form and find its rank. (UPTU-2007)

23. Find the rank of the matrix :

$$A = \begin{bmatrix} 1 & 2 & 3 \\ 2 & 3 & 4 \\ 3 & 5 & 7 \end{bmatrix}$$

after reducing it to normal form.

24. Reduce the matrix

$$A = \begin{bmatrix} 9 & 7 & 3 & 6 \\ 5 & -1 & 4 & 1 \\ 6 & 8 & 2 & 4 \end{bmatrix}$$

to normal form and find its rank.

25. Use elementary row or column transformations to find the rank of the matrix

$$\begin{bmatrix} 1 & 1 & 2 & 3 \\ 1 & 3 & 0 & 3 \\ 1 & -2 & -3 & -3 \\ 1 & 1 & 2 & 3 \end{bmatrix}$$

26. Find the rank of A, B, $A+B$, AB and BA where

$$A = \begin{bmatrix} 1 & 1 & -1 \\ 2 & -3 & 4 \\ 3 & -2 & 3 \end{bmatrix}, B = \begin{bmatrix} -1 & -2 & -1 \\ 6 & 12 & 6 \\ 5 & 10 & 5 \end{bmatrix}$$

27. Find two non-singular matrices P and Q such that PAQ is in the normal form where $A = \begin{bmatrix} 1 & -1 & 2 & -1 \\ 4 & 2 & -1 & 2 \\ 2 & 2 & -2 & 0 \end{bmatrix}$

Also find the rank of the matrix A.

28. Show that if A and B are equivalent matrices, then there exists non-singular matrices P and Q such that $B = PAQ$.

29. Show that the rank of a matrix is not altered if a column of a matrix is multiplied by a non-zero scalar.

30. Find matrices P and Q such that $P\begin{bmatrix} 2 & 2 & -6 \\ -1 & 2 & 2 \end{bmatrix}Q$ is in the normal form.

31. Transform $\begin{bmatrix} 1 & 3 & 3 \\ 2 & 4 & 10 \\ 3 & 8 & 4 \end{bmatrix}$ into a unit matrix by using elementary transformation. (UPTU-2011)

32. Reduce the following matrix into normal form and find its rank

$$\begin{bmatrix} 1 & 2 & -1 & 4 \\ 2 & 4 & 3 & 4 \\ 1 & 2 & 3 & 4 \\ -1 & -2 & 6 & 7 \end{bmatrix}$$ (GBTU-2010)

33. Find the rank of the following matrices

(i) $\begin{bmatrix} 1 & -3 & 1 & 2 \\ 0 & 1 & 2 & 3 \\ 3 & 4 & 1 & -2 \end{bmatrix}$ (GBTU-2010)

(ii) $\begin{bmatrix} 0 & 1 & 2 & -2 \\ 4 & 0 & 2 & 6 \\ 2 & 1 & 3 & 1 \end{bmatrix}$ (UPTU-2007)

(iii) $\begin{bmatrix} 5 & 3 & 14 & 4 \\ 0 & 1 & 2 & 1 \\ 1 & -1 & 2 & 0 \end{bmatrix}$ (GBTU-2011)

(iv) $\begin{bmatrix} 1 & 2 & 1 & 0 \\ -2 & 4 & 3 & 0 \\ 1 & 0 & 2 & 8 \end{bmatrix}$ (MTU-2011)

(v) $\begin{bmatrix} 0 & 1 & 2 & -1 \\ 1 & 0 & 1 & 1 \\ 3 & 1 & 0 & 2 \\ 1 & 1 & -2 & 0 \end{bmatrix}$ (UKTU-2011)

34. (i) Find all values of μ for which rank of the matrix

$$\begin{bmatrix} \mu & -1 & 0 & 0 \\ 0 & \mu & -1 & 0 \\ 0 & 0 & \mu & -1 \\ -6 & 11 & -6 & 1 \end{bmatrix}$$ is equal to 3. (UPTU-2009)

(ii) Find the value of k for which the matrix

$$\begin{bmatrix} 3 & k & k \\ k & 3 & k \\ k & k & 3 \end{bmatrix}$$ is of rank 1. (MTU-2012)

ANSWERS

1. (i) Not equivalent (ii) Not equivalent

2. 1 **3.** 2 **4.** 2 **5.** 2 **6.** 1 **7.** 2 **8.** 2 **9.** 2 **10.** 3 **11.** 3

12. 3 **13.** 3 **14.** 3 **15.** 4 **16.** 2 **17.** 3 **18.** 3 **19.** 4 **20.** 2 **21.** 3

22. 2 **23.** 3 **24.** 3

25. $\rho(A) = 2, \rho(B) = 1, \rho(A+B) = 2, \rho(AB) = 0, \rho(BA) = 1$ **26.** $\rho(A) = 2.29$, $R = \begin{bmatrix} 1 & 1 \\ 1/2 & 0 \end{bmatrix}, S = \begin{bmatrix} 1 & 4 & 8 \\ 0 & 0 & 1 \\ 0 & 1 & 3 \end{bmatrix}$

31. $\begin{bmatrix} 1 & 0 & 0 \\ 0 & 1 & 0 \\ 0 & 0 & 1 \end{bmatrix}$ **32.** 3 **33.** (i) 3 (ii) 2 (iii) 3 (iv) 3 (v) 3 **34.** (i) $\mu = 1, 2, 3$ (ii) $k = 3$

25.36 INVERSE OF A MATRIX

Let A be a non-singular matrix of order $n \times n$. Then it is said to be invertible if there exists a non-singular square matrix of order $n \times n$ such that $AB = I_n = BA$ where I_n is the unit matrix of order $n \times n$.

The matrix B is the inverse of A, we write $B = A^{-1}$.

THEOREM 1. *The inverse of a matrix, if it exists, is unique.*

Proof. Let A be a non-singular matrix of order $n \times n$ and if possible, let B and C be its inverses, then we have

$$AB = I_n = BA \quad ...(1)$$

and
$$AC = I_n = CA \quad ...(2)$$

From (1) and (2) we get,
$$AB = AC$$

$$\Rightarrow \quad B(AB) = B(AC)$$

$$\Rightarrow \quad (BA)B = (BA)C \qquad \text{(By associative law)}$$

$$\Rightarrow \quad I_nB = I_nC \qquad \text{[Using (1)]}$$

$$\Rightarrow \quad B = C.$$

THEOREM 2. *A square matrix is invertible if and only if it is non-singular.*

Proof. Let A be a square matrix of order $n \times n$ and suppose that A is invertible, then there exists a matrix B of order n such that

$$AB = I_n = BA$$

$$\Rightarrow \quad |AB| = |I_n|$$

$$\Rightarrow \quad |A|\,|B| = 1$$

$$\Rightarrow \quad |A| \neq 0$$

Thus, A is non-singular.

Conversely, suppose that A is non-singular matrix, then we have

$$A(adj.\ A) = |A|\ I_n = (adj.\ A)A$$

$$\Rightarrow \quad A\left(\frac{adj.A}{|A|}\right) = I_n = \left(\frac{adj.A}{|A|}\right)A \qquad \left[\because |A| \neq 0 \Rightarrow \frac{1}{|A|} \text{exists}\right]$$

$$\Rightarrow \quad AB = I_n = BA, \text{ if } B = \frac{adj.A}{|A|}$$

Thus, A is invertible.

THEOREM 3. *If A is an invertible matrix, then* $(A^{-1})^{-1} = A.$

Proof. Since A is invertible, then we have $\quad AA^{-1} = I = AA^{-1}$

$\Rightarrow \quad A$ is the inverse of A^{-1}

$$\therefore \quad (A^{-1})^{-1} = A$$

THEOREM 4. **(Reversal law):** *If A and B are invertible matrices of the same order, then AB is invertible and* $(AB)^{-1} = B^{-1}A^{-1}$.

Proof. Since A and B are invertible, therefore we have

$$|A| \neq 0,\ |B| \neq 0$$

$$\Rightarrow \quad |AB| = |A|\ |B|$$

$$\Rightarrow \quad |AB| \neq 0$$

$\therefore \quad AB$ is invertible.

Now

$$(AB)(B^{-1}A^{-1}) = A(BB^{-1})A^{-1} \qquad \text{[By associative law]}$$

$$\Rightarrow \quad (AB)(B^{-1}A^{-1}) = A(I)A^{-1} \qquad [\because BB^{-1} = I]$$

$$\Rightarrow \quad (AB)(B^{-1}A^{-1}) = (AI)A^{-1} \qquad \text{[By associative law]}$$

$$\Rightarrow \quad (AB)(B^{-1}A^{-1}) = AA^{-1} \qquad [\because AI = A]$$

$$\Rightarrow \quad (AB)(B^{-1}A^{-1}) = I \qquad [\because AA^{-1} = I]$$

Also,

$$(B^{-1}A^{-1})(AB) = B^{-1}(A^{-1}A)B \qquad \text{[By associative law]}$$

$$\Rightarrow \quad (B^{-1}A^{-1})(AB) = B^{-1}(I)B \qquad [\because A^{-1}A = I]$$

$$\Rightarrow \quad (B^{-1}A^{-1})(AB) = B^{-1}(IB) \qquad \text{[By associative law]}$$

$$\Rightarrow \quad (B^{-1}A^{-1})(AB) = B^{-1}B \qquad [\because IB = B]$$

$$\Rightarrow \quad (B^{-1}A^{-1})(AB) = I \qquad [\because B^{-1}B = I]$$

$$\therefore \quad (AB)(B^{-1}A^{-1}) = I = (B^{-1}A^{-1})(AB)$$

$$\Rightarrow \quad (AB)^{-1} = B^{-1}A^{-1}$$

REMARK

- If A, B and C are three invertible matrices of the same order, then $(ABC)^{-1} = C^{-1}B^{-1}A^{-1}$.

THEOREM 5. *If A is an invertible square matrix, then A' is also invertible and $(A')^{-1} = (A^{-1})'$, where A' is the transpose of A.*

Proof. Since A is an invertible matrix, then we have $|A| \neq 0$

Now $$|A| = |A'| \Rightarrow |A'| \neq 0 \Rightarrow A' \text{ is invertible.}$$

Also $$AA^{-1} = I = A^{-1}A$$

$$\Rightarrow \quad (AA^{-1})' = (I)' = (A^{-1}A)'$$

$$\Rightarrow \quad (A^{-1})'A' = I = A'(A^{-1})' \qquad \text{(By reversal rule of transpose)}$$

$\Rightarrow (A^{-1})'$ is the inverse of A' $\Rightarrow$ $(A')^{-1} = (A^{-1})'$

THEOREM 6. *The inverse of an invertible matrix is a symmetric matrix.*

Proof. Let A be an invertible symmetric matrix, then

$$|A| \neq 0 \text{ and } A' = A$$

Now by above theorem.

$$(A')^{-1} = (A^{-1})'$$

$$\Rightarrow \quad (A^{-1})' = A^{-1} \qquad [\because A' = A]$$

$\Rightarrow$ A^{-1} is a symmetric matrix.

THEOREM 7. *If A is an invertible matrix, then (adj. A)' = adj (A')*

Proof. Since A is an invertible matrix, then $|A| \neq 0$

Now $$|A'| = |A| \Rightarrow |A'| \neq 0$$

$\Rightarrow$ A' is invertible $\Rightarrow$ $(A')^{-1}$ exists.

We have $$A(adj.\ A) = |A|I$$

$$\Rightarrow \quad (A\ adj.\ A) = (|A|I)' = |A|I' = |A|I$$

$$\Rightarrow \quad (adj.\ A')\,A' = |A|I \qquad ...(1)$$

Also $$(adj.\ A)'\,A' = |A'|\,I$$

$$\Rightarrow \quad (adj.\ A)'A' = |A|I \qquad [\because |A'| = |A|] \qquad ...(2)$$

From (1) and (2), we get, $$(adj.\ A)'\,A' = (adj.\ A')A'$$

$$\Rightarrow \quad (adj.\ A)'\,A'\,(A')^{-1} = (adj.\ A')\,A'\,(A')^{-1}$$

$$\Rightarrow \quad (adj.\ A)'\,I = (adj.\ A')I$$

$$\Rightarrow \quad (adj.\ A)' = adj.\ (A)$$

THEOREM 8. *The adjoint of a symmetric matrix is also a symmetric matrix,*

Proof. Let A be a symmetric matrix of order $n \times n$, then $A' = A$

Now by above theorem, $$(adj.\ A)' = adj.(A')$$

$$\Rightarrow \quad (adj.\ A)' = adj.\ A \qquad [\because A' = A]$$

$\Rightarrow$ $adj.\ A$ is a symmetric matrix.

THEOREM 9. *If A is a non-singular matrix, then $|A^{-1}| = |A|^{-1}$.*

Proof. Since A is a non-singular matrix, then

$$|A| \neq 0 \qquad \Rightarrow A^{-1} \text{ exists.}$$

Also $$AA^{-1} = |I| = A^{-1}A \Rightarrow |AA^{-1}| = |I| = 1$$

$$\Rightarrow \quad |A|\,|A^{-1}| = 1 \qquad \Rightarrow \quad |A^{-1}| = \frac{1}{|A|} = |A|^{-1}$$

THEOREM 10. *If A and B are non-singular matrices of the same order, then*

$$adj.\ (AB) = (adj.\ B)\ (adj.\ A)$$

Proof. Since A and B are non-singular matrices of the same order, then AB exists.

Also $$|A| \neq 0,\ |B| \neq 0 \Rightarrow |AB| = |A|\,|B| \neq 0$$

$\Rightarrow$ $(AB)^{-1}$ exists.

Now we have $A(adj.\ A) = |A|\ I$ ...(1)

and $B(adj.\ B) = |B|\ I$ ...(2)

Also $AB\ (adj.\ B) = |AB|\ I$ ...(3)

We have $(AB)\ (adj.\ B\ adj.\ A) = A(\ adj.\ B)\ adj.\ A$ (By associative law)

$\Rightarrow$ $(AB)\ (adj.\ B\ adj.\ A) = A(\ |B|\ I\ )\ adj.\ A$ [Using (1)]

$\Rightarrow$ $(AB)\ (adj.\ B\ adj.\ A) = |B|(AI\ )\ adj.\ A$

$\Rightarrow$ $(AB)\ (adj.\ B\ adj.\ A) = |B|(A\ adj.\ A)$

$\Rightarrow$ $(AB)\ (adj.\ B\ adj.\ A) = |B|\,|A\ |\ I$

$\therefore$ $(AB)\ (adj.\ B\ adj.\ A) = |AB|\ I$ ...(4)

From (3) and (4), we get

$$(AB)\ (adj.\ AB) = (AB)\ (adj.\ B\ adj.\ A)$$

$\Rightarrow$ $(AB)^{-1}\ (AB)(adj.\ AB) = (AB)^{-1}\ (AB)\ (adj.\ B\ adj.\ A)$ [$\because (AB)^{-1}$ exists.]

$\Rightarrow$ $I(adj.\ AB) = I(adj.\ B\ adj.\ A)$

$\Rightarrow$ $adj.\ (AB) = (adj.\ B)\ (adj.\ A)$

THEOREM 11. (Cancellation laws) : *Let A, B and C be three square matrices of the same order. If A is a non-singular matrix, then*

(i) $AB = AC \Rightarrow B = C$ [Left cancellation law] *(ii) $BA = CA \Rightarrow B = C$ [Right cancellation law]*

Proof. Since A is a non-singular matrix, then $|A| \neq 0 \Rightarrow A^{-1}$ exists.

(i) We have $AB = AC$

$\Rightarrow$ $A^{-1}(AB) = A^{-1}(AC)$ [$\because A^{-1}$ exists.]

$\Rightarrow$ $(A^{-1}A)B = (A^{-1}A)C$ [By associative law]

$\Rightarrow$ $IB = IC$ [$\because A^{-1}A = I$]

$\Rightarrow$ $B = C$ [$\because IB = B, IC = C$]

(ii) We have $BA = CA$

$\Rightarrow$ $(BA)\ A^{-1} = (CA)\ A^{-1}$ [$\because A^{-1}$ exists.]

$\Rightarrow$ $B\ (AA^{-1}) = C(AA^{-1})$ [By associative law]

$\Rightarrow$ $BI = CI$ [$\because AA^{-1} = I$]

$\Rightarrow$ $B = C$ [$\because BI = B, CI = C$]

THEOREM 12. *If the product of two non-null square matrices is a null matrix, then both of them must be singular.*

Proof. Let A and B be two non-null matrices of the same order $n \times n$ such that

$$AB = O \quad ...(1)$$

where O is a null matrix of order $n \times n$.

Let, if possible B be a non-singular matrix, then B^{-1} exists.

From (1) we have $AB = O$

$\Rightarrow$ $(AB)B^{-1} = OB^{-1}$

$\Rightarrow$ $A(BB^{-1}) = O$ [By associative law and $OB^{-1} = O$]

$\Rightarrow$ $AI_n = O$ [$\because BB^{-1} = I_n$]

$\Rightarrow$ $A = O$ [$\because AI_n = A$]

which is a contradiction because A is a non-null matrix.

Therefore, B is a singular matrix.

Similarly, we can prove that A is a singular matrix.

25.37 INVERSE OF A MATRIX BY ELEMENTARY TRANSFORMATIONS

Let A be a non-singular matrix of order $n \times n$, then there exists a finite number of elementary matrices $E_1, E_2, ..., E_3, ..., E_s$ such that

$$E_sE_{s-1}\ ...E_2E_1A = I_n$$

$\Rightarrow$ $E_sE_{s-1} = E_2E_1\,AA^{-1} = I_n\,A^{-1}$ [$\because |A| \neq 0 \Rightarrow A^{-1}$ exists.]

$\Rightarrow$ $(E_sE_{s-1}\ ...\ E_2E_1\ )\ (AA^{-1}) = I_nA^{-1}$ [By associative law]

$\Rightarrow$ $(E_sE_{s-1}\ ...\ E_2E_1\ )\ I_n = A^{-1}$ [$\because AA^{-1} = I_n, A^{-1} = A^{-1}$]

Hence, $A^{-1} = (E_sE_{s-1}\ ...\ E_2E_1\ )\ I_n$ ...(1)

We know that every non-singular matrix of order $n \times n$ can be reduced to the unit matrix I_n by a finite chain of elementary row-transformations only and each elementary row-transformation of a matrix is equivalent to pre-multiplication by the corresponding elementary matrix.

From (1) it follows that if a non-singular matrix A of order $n \times n$ is reduced to the unit matrix I_n by a finite chain of elementary row-transformations only, then the same chain of elementary row-transformations applied to the unit matrix I_n gives the inverse of A.

WORKING PROCEDURE

Let A be a non-singular matrix of order $n \times n$, then we follow the following steps :

Step 1. *Write* $A = I_n A$ ...(1)

Step 2. *Apply elementary row-transformations on A on L.H.S. of (1) and reduce it to I_n and apply corresponding elementary row-transformations on the pre-factor I_n on R.H.S. of (1) till we obtain $I_n = BA$.*

Step 3. *Finally, we write $A^{-1} = B$.*

Solved Examples

Example 1. *By using elementary row-transformations find the inverse of the following matrices:*

(i) $\begin{bmatrix} 1 & 2 \\ 3 & 7 \end{bmatrix}$ (ii) $\begin{bmatrix} 1 & 2 \\ 2 & -1 \end{bmatrix}$

Solution. (i) We write $A = I_2 A$

or
$$\begin{bmatrix} 1 & 2 \\ 3 & 7 \end{bmatrix} = \begin{bmatrix} 1 & 0 \\ 0 & 1 \end{bmatrix} A$$

Applying $R_2 \to R_2 - 3R_1$, we get
$$\begin{bmatrix} 1 & 2 \\ 0 & 1 \end{bmatrix} = \begin{bmatrix} 1 & 0 \\ -3 & 1 \end{bmatrix} A$$

Again applying $R_1 \to R_1 - 2R_2$, we get
$$\begin{bmatrix} 1 & 0 \\ 0 & 1 \end{bmatrix} = \begin{bmatrix} 7 & -2 \\ -3 & 1 \end{bmatrix} A$$

$\Rightarrow$ $I_2 = BA$

$$\Rightarrow \quad A^{-1} = B = \begin{bmatrix} 7 & -2 \\ -3 & 1 \end{bmatrix}.$$

(ii) We write $A = I_2 A$

or
$$\begin{bmatrix} 1 & 2 \\ 2 & -1 \end{bmatrix} = \begin{bmatrix} 1 & 0 \\ 0 & 1 \end{bmatrix} A$$

Applying $R_2 \to R_2 - 2R_1$, we get
$$\begin{bmatrix} 1 & 2 \\ 0 & -5 \end{bmatrix} = \begin{bmatrix} 1 & 0 \\ -2 & 1 \end{bmatrix} A$$

Applying $R_2 \to -\frac{1}{5} R_2$, we get
$$\begin{bmatrix} 1 & 2 \\ 0 & 1 \end{bmatrix} = \begin{bmatrix} 1 & 0 \\ 2/5 & -1/5 \end{bmatrix} A$$

Applying $R_1 \to R_1 - 2R_2$, we get
$$\begin{bmatrix} 1 & 0 \\ 0 & 1 \end{bmatrix} = \begin{bmatrix} 1/5 & 2/5 \\ 2/5 & -1/5 \end{bmatrix} A$$

$\Rightarrow$ $I_2 = BA$

$$\Rightarrow \quad A^{-1} = B = \begin{bmatrix} 1/5 & 2/5 \\ 2/5 & -1/5 \end{bmatrix}$$

Example 2. *Find the inverse of the matrix* $A = \begin{bmatrix} 1 & 2 & 1 \\ 3 & 2 & 3 \\ 1 & 1 & 2 \end{bmatrix}$ *by using elementary row-transformation.*

Solution. We write $A = I_3 A$

or
$$\begin{bmatrix} 1 & 2 & 1 \\ 3 & 2 & 3 \\ 1 & 1 & 2 \end{bmatrix} = \begin{bmatrix} 1 & 0 & 0 \\ 0 & 1 & 0 \\ 0 & 0 & 1 \end{bmatrix} A$$

Applying $R_2 \to R_2 - 3R_1$, $R_3 \to R_3 - R_1$, we get
$$\begin{bmatrix} 1 & 2 & 1 \\ 0 & -4 & 0 \\ 0 & -1 & 1 \end{bmatrix} = \begin{bmatrix} 1 & 0 & 0 \\ -3 & 1 & 0 \\ -1 & 0 & 1 \end{bmatrix} A$$

Applying $R_2 \to \frac{-1}{4} R_2$, we get
$$\begin{bmatrix} 1 & 2 & 1 \\ 0 & 1 & 0 \\ 0 & -1 & 1 \end{bmatrix} = \begin{bmatrix} 1 & 0 & 0 \\ 3/4 & -1/4 & 0 \\ -1 & 0 & 1 \end{bmatrix} A$$

Applying $R_3 \to R_3 + R_2$, we get
$$\begin{bmatrix} 1 & 2 & 1 \\ 0 & 1 & 0 \\ 0 & 0 & 1 \end{bmatrix} = \begin{bmatrix} 1 & 0 & 0 \\ 3/4 & -1/4 & 0 \\ -1/4 & -1/4 & 1 \end{bmatrix} A$$

Applying $R_1 \to R_1 - 2R_2$, we get
$$\begin{bmatrix} 1 & 0 & 1 \\ 0 & 1 & 0 \\ 0 & 0 & 1 \end{bmatrix} = \begin{bmatrix} -1/2 & 1/2 & 0 \\ 3/4 & -1/4 & 0 \\ -1/4 & -1/4 & 1 \end{bmatrix} A$$

Applying $R_1 \to R_1 - R_3$, we get
$$\begin{bmatrix} 1 & 0 & 0 \\ 0 & 1 & 0 \\ 0 & 0 & 1 \end{bmatrix} = \begin{bmatrix} -1/4 & 3/4 & -1 \\ 3/4 & -1/4 & 0 \\ -1/4 & -1/4 & 1 \end{bmatrix} A$$

$\Rightarrow$ $I_3 = BA$

$$\Rightarrow \quad A^{-1} = B = \begin{bmatrix} -1/4 & 3/4 & -1 \\ 3/4 & -1/4 & 0 \\ -1/4 & -1/4 & 1 \end{bmatrix}$$

Example 3. *Using elementary transformations, find the inverse of the following matrix :*

$$A = \begin{bmatrix} 1 & 2 & 3 \\ 2 & 5 & 7 \\ -2 & -4 & -5 \end{bmatrix}$$

Solution. We write $A_3 = I_3 A$

or $$\begin{bmatrix} 1 & 2 & 3 \\ 2 & 5 & 7 \\ -2 & -4 & -5 \end{bmatrix} = \begin{bmatrix} 1 & 0 & 0 \\ 0 & 1 & 0 \\ 0 & 0 & 1 \end{bmatrix} A$$

Applying $R_2 \to R_2 - 2R_1$, $R_3 \to R_3 + 2R_1$, we get

$$\begin{bmatrix} 1 & 2 & 3 \\ 0 & 1 & 1 \\ 0 & 0 & 1 \end{bmatrix} = \begin{bmatrix} 1 & 0 & 0 \\ -2 & 1 & 0 \\ 2 & 0 & 1 \end{bmatrix} A$$

Applying $R_1 \to R_1 - 2R_2$, we get

$$\begin{bmatrix} 1 & 0 & 1 \\ 0 & 1 & 1 \\ 0 & 0 & 1 \end{bmatrix} = \begin{bmatrix} 5 & -2 & 0 \\ -2 & 1 & 0 \\ 2 & 0 & 1 \end{bmatrix} A$$

Applying $R_1 \to R_1 - R_3$, $R_2 \to R_2 - R_3$, we get

$$\begin{bmatrix} 1 & 0 & 0 \\ 0 & 1 & 0 \\ 0 & 0 & 1 \end{bmatrix} = \begin{bmatrix} 3 & -2 & -1 \\ -4 & 1 & -1 \\ 2 & 0 & 1 \end{bmatrix} A$$

$$\Rightarrow \quad I_3 = BA$$

$$\Rightarrow \quad A^{-1} = B = \begin{bmatrix} 3 & -2 & -1 \\ -4 & 1 & -1 \\ 2 & 0 & 1 \end{bmatrix}$$

Example 4. *Find the inverse of the matrix* $A = \begin{bmatrix} 0 & 1 & 2 & 2 \\ 1 & 1 & 2 & 3 \\ 2 & 2 & 2 & 3 \\ 2 & 3 & 3 & 3 \end{bmatrix}$ *by using elementary transformations.*

Solution. We write $A = I_4 A$

or $$\begin{bmatrix} 0 & 1 & 2 & 2 \\ 1 & 1 & 2 & 3 \\ 2 & 2 & 2 & 3 \\ 2 & 3 & 3 & 3 \end{bmatrix} = \begin{bmatrix} 1 & 0 & 0 & 0 \\ 0 & 1 & 0 & 0 \\ 0 & 0 & 1 & 0 \\ 0 & 0 & 0 & 1 \end{bmatrix} A$$

Applying $R_2 \leftrightarrow R_1$, we get

$$\begin{bmatrix} 1 & 1 & 2 & 3 \\ 0 & 1 & 2 & 2 \\ 2 & 2 & 2 & 3 \\ 2 & 3 & 3 & 3 \end{bmatrix} = \begin{bmatrix} 0 & 1 & 0 & 0 \\ 1 & 0 & 0 & 0 \\ 0 & 0 & 1 & 0 \\ 0 & 0 & 0 & 1 \end{bmatrix} A$$

Applying $R_3 \to R_3 - 2R_1$, $R_4 \to R_4 - 2R_1$, we get

$$\begin{bmatrix} 1 & 1 & 2 & 3 \\ 0 & 1 & 2 & 2 \\ 0 & 0 & -2 & -3 \\ 0 & 1 & -1 & -3 \end{bmatrix} = \begin{bmatrix} 0 & 1 & 0 & 0 \\ 1 & 0 & 0 & 0 \\ 0 & -2 & 1 & 0 \\ 0 & -2 & 0 & 1 \end{bmatrix} A$$

Applying $R_1 \to R_1 - R_2$, $R_4 \to R_4 - R_2$, we get

$$\begin{bmatrix} 1 & 0 & 0 & 1 \\ 0 & 1 & 2 & 2 \\ 0 & 0 & -2 & -3 \\ 0 & 0 & -3 & -5 \end{bmatrix} = \begin{bmatrix} -1 & 1 & 0 & 0 \\ 1 & 0 & 0 & 0 \\ 0 & -2 & 1 & 0 \\ -1 & -2 & 0 & 1 \end{bmatrix} A$$

Applying $R_3 \to -\frac{1}{2}R_3$, we get

$$\begin{bmatrix} 1 & 0 & 0 & 1 \\ 0 & 1 & 2 & 2 \\ 0 & 0 & 1 & 3/2 \\ 0 & 0 & -3 & -5 \end{bmatrix} = \begin{bmatrix} -1 & 1 & 0 & 0 \\ 1 & 0 & 0 & 0 \\ 0 & 1 & -1/2 & 0 \\ -1 & -2 & 0 & 1 \end{bmatrix} A$$

Applying $R_2 \to R_2 - 2R_3$, $R_4 \to R_4 + 3R_3$, we get

$$\begin{bmatrix} 1 & 0 & 0 & 1 \\ 0 & 1 & 0 & -1 \\ 0 & 0 & 1 & 3/2 \\ 0 & 0 & 0 & -1/2 \end{bmatrix} = \begin{bmatrix} -1 & 1 & 0 & 0 \\ 1 & -2 & 1 & 0 \\ 0 & 1 & -1/2 & 0 \\ -1 & 1 & -3/2 & 1 \end{bmatrix} A$$

Applying $R_4 \to -2R_4$, we get

$$\begin{bmatrix} 1 & 0 & 0 & 1 \\ 0 & 1 & 0 & -1 \\ 0 & 0 & 1 & 3/2 \\ 0 & 0 & 0 & 1 \end{bmatrix} = \begin{bmatrix} -1 & 1 & 0 & 0 \\ 1 & -2 & 1 & 0 \\ 0 & 1 & -1/2 & 0 \\ 2 & -2 & 3 & -2 \end{bmatrix} A$$

Applying

$R_1 \to R_1 - R_4, R_2 \to R_2 + R_4, R_3 \to R_3 - \frac{3}{2}R_4$, we get

$$\begin{bmatrix} 1 & 0 & 0 & 0 \\ 0 & 1 & 0 & 0 \\ 0 & 0 & 1 & 0 \\ 0 & 0 & 0 & 1 \end{bmatrix} = \begin{bmatrix} -3 & 3 & -3 & 2 \\ 3 & -4 & 4 & -2 \\ -3 & 4 & -5 & 3 \\ 2 & -2 & 3 & -2 \end{bmatrix} A$$

$$\Rightarrow \quad I_4 = BA$$

$$\Rightarrow \quad A^{-1} = B = \begin{bmatrix} -3 & 3 & -3 & 2 \\ 3 & -4 & 4 & -2 \\ -3 & 4 & -5 & 3 \\ 2 & -2 & 3 & -2 \end{bmatrix}$$

EXERCISE 25.5

Using elementary row-transformations, find the inverse of each of the following matrices, if it exists :

1. $\begin{bmatrix} 5 & 2 \\ 2 & 1 \end{bmatrix}$

2. $\begin{bmatrix} 2 & 3 \\ 0 & 1 \end{bmatrix}$

3. $\begin{bmatrix} 1 & 6 \\ -3 & 5 \end{bmatrix}$

4. $\begin{bmatrix} 1 & 2 & 3 \\ 2 & 4 & 5 \\ 3 & 5 & 6 \end{bmatrix}$

5. $\begin{bmatrix} 1 & 2 & -1 \\ -1 & 1 & 2 \\ 2 & -1 & 1 \end{bmatrix}$

6. $\begin{bmatrix} 1 & -1 & 0 \\ 1 & -3 & 9 \\ 8 & 9 & 2 \end{bmatrix}$

7. $\begin{bmatrix} 0 & 1 & 2 \\ 1 & 2 & 3 \\ 3 & 1 & 1 \end{bmatrix}$ **8.** $\begin{bmatrix} 2 & 3 & 1 \\ 2 & 4 & 1 \\ 3 & 7 & 2 \end{bmatrix}$ **17.** $\begin{bmatrix} -1 & -3 & 3 & -1 \\ 1 & 1 & -1 & 0 \\ 2 & -5 & 2 & -3 \\ -1 & 1 & 0 & 1 \end{bmatrix}$ **18.** $\begin{bmatrix} 1 & 1 & 2 & 0 \\ 0 & 1 & 1 & -1 \\ 2 & 1 & 2 & 1 \\ 3 & -2 & 1 & 6 \end{bmatrix}$

9. $\begin{bmatrix} 1 & -3 & 2 \\ 2 & 0 & 0 \\ 1 & 4 & 1 \end{bmatrix}$ **10.** $\begin{bmatrix} 2 & -1 & 3 \\ 1 & 2 & 4 \\ 3 & 1 & 1 \end{bmatrix}$ **19.** $\begin{bmatrix} 3 & -3 & 4 \\ 2 & -3 & 4 \\ 0 & -1 & 1 \end{bmatrix}$ (GBTU-2010)

11. $\begin{bmatrix} 1 & 1 & 1 \\ 2 & 2 & 3 \\ 2 & 4 & 9 \end{bmatrix}$ **12.** $\begin{bmatrix} 2 & 0 & -1 \\ 5 & 1 & 0 \\ 0 & 1 & 3 \end{bmatrix}$ **20.** $\begin{bmatrix} 1 & 3 & 3 \\ 1 & 4 & 3 \\ 1 & 3 & 4 \end{bmatrix}$ (UPTU-2008)

13. $\begin{bmatrix} 1 & 2 & 0 \\ 2 & 3 & -1 \\ 1 & -1 & 3 \end{bmatrix}$ **14.** $\begin{bmatrix} 1 & 1 & 2 \\ 3 & 1 & 1 \\ 2 & 3 & 1 \end{bmatrix}$ **21.** $\begin{bmatrix} \frac{1}{3} & \frac{1}{5} & \frac{1}{7} \\ \frac{1}{5} & \frac{1}{7} & \frac{1}{11} \\ -\frac{1}{7} & \frac{1}{11} & \frac{1}{13} \end{bmatrix}$ (UPTU-2008)

15. $\begin{bmatrix} 1 & 2 & 1 \\ 3 & 2 & 3 \\ 1 & 1 & 2 \end{bmatrix}$ **16.** $\begin{bmatrix} 3 & 0 & -1 \\ 2 & 3 & 0 \\ 0 & 4 & 1 \end{bmatrix}$

ANSWERS

1. $\begin{bmatrix} 1 & -2 \\ -2 & 5 \end{bmatrix}$ **2.** $\begin{bmatrix} 1 & -3 \\ 0 & 2 \end{bmatrix}$ **3.** $\frac{1}{23}\begin{bmatrix} 5 & -6 \\ 3 & 1 \end{bmatrix}$ **4.** $\begin{bmatrix} 1 & -3 & 2 \\ -3 & 3 & -1 \\ 2 & -1 & 0 \end{bmatrix}$

5. $\frac{1}{14}\begin{bmatrix} 3 & -1 & 5 \\ 5 & 3 & -1 \\ -1 & 5 & 3 \end{bmatrix}$ **6.** $\frac{1}{157}\begin{bmatrix} 87 & -2 & 9 \\ -70 & -2 & 9 \\ -33 & 17 & 2 \end{bmatrix}$ **7.** $\frac{1}{2}\begin{bmatrix} 1 & -1 & 1 \\ -8 & 6 & -2 \\ 5 & -3 & 1 \end{bmatrix}$ **8.** $\begin{bmatrix} 1 & 1 & -1 \\ -1 & 1 & 0 \\ 2 & -5 & 2 \end{bmatrix}$

9. $\frac{1}{22}\begin{bmatrix} 0 & 11 & 0 \\ -2 & -1 & 4 \\ 8 & -7 & 6 \end{bmatrix}$ **10.** $\frac{1}{30}\begin{bmatrix} 2 & -4 & 10 \\ -11 & 7 & 5 \\ 5 & 5 & -5 \end{bmatrix}$ **11.** $\frac{1}{3}\begin{bmatrix} -6 & 5 & -1 \\ 15 & -8 & 1 \\ -6 & 3 & 0 \end{bmatrix}$ **12.** $\begin{bmatrix} 3 & -1 & 1 \\ -15 & 6 & -5 \\ 5 & -2 & 2 \end{bmatrix}$ **13.** $\frac{1}{6}\begin{bmatrix} -8 & 6 & 2 \\ 7 & -3 & -1 \\ 5 & -3 & 1 \end{bmatrix}$

14. $\frac{1}{11}\begin{bmatrix} -2 & 5 & -1 \\ -1 & -3 & 5 \\ 7 & -1 & -2 \end{bmatrix}$ **15.** $\frac{1}{4}\begin{bmatrix} -1 & 3 & -4 \\ 3 & -1 & 0 \\ -1 & -1 & 4 \end{bmatrix}$ **16.** $\begin{bmatrix} 3 & -4 & -3 \\ -2 & 3 & 2 \\ 8 & -12 & 9 \end{bmatrix}$ **17.** $\begin{bmatrix} 0 & 2 & 1 & 3 \\ 1 & 1 & -1 & -2 \\ 1 & 2 & 0 & 1 \\ -1 & 1 & 2 & 6 \end{bmatrix}$ **18.** $\begin{bmatrix} 2 & -1 & 1 & -1 \\ -5 & -3 & 1 & 1 \\ 2 & 3 & -1 & 0 \\ -3 & -1 & 0 & 1 \end{bmatrix}$

19. $\begin{bmatrix} 1 & -1 & 0 \\ -2 & 3 & -4 \\ -2 & 3 & -3 \end{bmatrix}$ **20.** $\begin{bmatrix} 7 & -3 & -3 \\ -1 & 1 & 0 \\ -1 & 0 & 1 \end{bmatrix}$ **22.** $\frac{1}{2238}\begin{bmatrix} 55125 & -48510 & -45045 \\ -48510 & 105875 & -35035 \\ -45045 & -35035 & 154154 \end{bmatrix}$

25.38 SYSTEM OF LINEAR EQUATIONS

In this section we shall study the nature of solutions of a system of linear equations with the help of the theory of matrices discussed in previous chapters. Before going into details of solutions of linear equations we shall try to understand the concepts of linearly dependent and independent set of vectors.

25.39 VECTORS AND THEIR DEPENDENCE AND INDEPENDENCE

An ordered set of n numbers $(x_1, x_2, x_3, ..., x_n)$ *is known as a vector of order n.*

The n numbers $x_1, x_2, x_3, ..., x_n$ are called the components of the vector. We denote this vector by a single letter X. Conveniently, we may write the components of a vector X in the form of a row or in the form of a column.

Therefore, we may write
$$X = [x_1, x_2, x_3, ..., x_n]$$

which is known as a n-dimensional row vector or it may be written as $X = \begin{bmatrix} x_1 \\ x_2 \\ x_3 \\ \vdots \\ x_n \end{bmatrix}$ which is known as an n-dimensional column vector.

If we consider an $m \times n$ matrix, then it contains m-row vectors and n-column vectors, each row vector consists of the components of an n-vector and each column vector consists of the components of an m-vector.

For example: Consider a matrix of order 3×4 given by

$$A=\begin{bmatrix}1 & -1 & 4 & 5\\ 2 & 3 & 0 & -7\\ 3 & 2 & 2 & 6\end{bmatrix}$$

Then, $R_1=[1\ \ -1\ \ 4\ \ 5],\ R_2=[2\ \ 3\ \ 0\ \ -7],\ R_3=[3\ \ 2\ \ 2\ \ 6]$

$$C_1=\begin{bmatrix}1\\2\\3\end{bmatrix}, C_2=\begin{bmatrix}-1\\3\\2\end{bmatrix},\ C_3=\begin{bmatrix}4\\0\\2\end{bmatrix}, C_4=\begin{bmatrix}5\\-7\\6\end{bmatrix}$$

Thus A can be written as $A=\begin{bmatrix}R_1\\R_2\\R_3\end{bmatrix}$ or $A=\begin{bmatrix}C_1 & C_2 & C_3 & C_4\end{bmatrix}$

Definiton : *If all the components of a vector are zero, then it is called a null vector or a zero vector. It is usually denoted by capital lette O.*

For example : The vectors $[0\ \ 0\ \ 0\ \ 0]$ and $\begin{bmatrix}0\\0\\0\\0\end{bmatrix}$ are both null vectors.

25.39.1 SUM OF TWO VECTORS

Let $X=(x_1, x_2, x_3, ,..., x_n)$ and $Y=(y_1, y_2, y_3, ,..., y_n)$ be two vectors, then $X+Y$ is obtained by adding their corresponding components. Thus

$$X+Y=(x_1+y_1, x_2+y_2, \dots, x_n+y_n)$$

REMARK

- If two vectors are of different dimensions then they cannot be added up.

25.39.2 MULTIPLICATION OF A VECTOR BY A SCALAR

Let $X=(x_1, x_2, ,..., x_n)$ be an n-vector and λ be a scalar, then λX can be obtained on multiplication of each component of X by λ. Thus

$$\lambda X=(\lambda x_1, \lambda x_2, \dots, \lambda x_n)$$

25.39.3 LINEAR DEPENDENCE AND INDEPENDENCE OF VECTORS

Let $x_1, x_2, x_3, ..., x_m$ be m vectors. Then they are said to be linearly independent if

$$\lambda_1 x_1+\lambda_2 x_2+\dots+\lambda_m x_m=0 \quad\Rightarrow\quad \lambda_1=\lambda_2=\lambda_3=\dots=\lambda_m=0$$

If none of $\lambda_1, \lambda_2, ..., \lambda_m$ is zero, then the vectors $x_1, x_2, ..., x_m$ are called linearly dependent.

25.39.4 LINEAR COMBINATION OF VECTORS

A vector X is said to be a linear combination of the vectors $x_1, x_2, ..., x_m$ if there exists scalars $\lambda_1, \lambda_2, ..., \lambda_m$ such that

$$X=\lambda_1 x_1+\lambda_2 x_2+\dots+\lambda_m x_m$$

Suppose that the vectors $x_1, x_2, \dots, x_m$ are linearly dependent, then in the equation.

$$\lambda_1 x_1+\lambda_2 x_2+\dots+\lambda_m x_m=O \qquad \dots (1)$$

there is at least one of $\lambda_1, \lambda_2, \dots \lambda_m$ is non-zero, let it be λ_r, then equation (1) can be written as

$$\lambda_r X_r=-\lambda_1 x_1-\lambda_2 x_2-\dots\lambda_{r-1}x_{r-1}-\lambda_{r+1}x_{r+1}-\dots-\lambda_m x_m$$

$$\Rightarrow \quad X_r=\left(-\frac{\lambda_1}{\lambda_r}\right)x_1+\left(-\frac{\lambda_2}{\lambda_r}\right)x_2+\dots+\left(-\frac{\lambda_{r-1}}{\lambda_r}\right)x_{r-1}+\left(-\frac{\lambda_{r+1}}{\lambda_r}\right)x_{r+1}+\dots+\left(-\frac{\lambda_m}{\lambda_r}\right)x_m$$

$$\Rightarrow \quad X_r=k_1x_1+k_2x_2+\dots+k_{r-1}x_{r-1}+k_{r+1}x_{r+1}+\dots+k_m x_m$$

It follows that X_r is a linear combinations of vectors $x_1, x_2, .., x_{r-1}, x_{r+1}, ..., x_m$.

Hence if a set of vectors is linearly dependent, then at least one member of the set can be expressed as a linear combination of the remaining vectors.

25.39.5 LINEAR DEPENDENCE OF THE ROWS AND COLUMNS OF A SQUARE MATRIX

Consider a square matrix of order 3×3, namely $A = \begin{bmatrix} a_{11} & a_{12} & a_{13} \\ a_{21} & a_{22} & a_{23} \\ a_{31} & a_{32} & a_{33} \end{bmatrix}$ or $A = [C_1 \quad C_2 \quad C_3]$

where $C_1 = \begin{bmatrix} a_{11} \\ a_{21} \\ a_{31} \end{bmatrix}, C_2 = \begin{bmatrix} a_{12} \\ a_{22} \\ a_{32} \end{bmatrix}$ and $C_3 = \begin{bmatrix} a_{13} \\ a_{23} \\ a_{33} \end{bmatrix}$

The columns C_1, C_2, C_3 are linearly dependent if there exist scalars k_1, k_2, k_3 not all zero such that

$$k_1C_1 + k_2C_2 + k_3C_3 = O$$

$$\Rightarrow \quad k_1 \begin{bmatrix} a_{11} \\ a_{21} \\ a_{31} \end{bmatrix} + k_2 \begin{bmatrix} a_{12} \\ a_{22} \\ a_{32} \end{bmatrix} + k_3 \begin{bmatrix} a_{13} \\ a_{23} \\ a_{33} \end{bmatrix} = \begin{bmatrix} 0 \\ 0 \\ 0 \end{bmatrix}$$

$$\therefore \quad \begin{aligned} k_1a_{11} + k_2a_{12} + k_3a_{13} &= 0 \\ k_1a_{21} + k_2a_{22} + k_3a_{23} &= 0 \\ k_1a_{31} + k_2a_{32} + k_3a_{33} &= 0 \end{aligned}$$

i.e., if $$|A| = 0$$

Hence, the columns of A are linearly dependent if $|A| = 0$. Since $|A'| = |A|$, if $|A| = 0$, then $|A'| = 0$. Now if $|A'| = 0$, then the columns of A' are linearly dependent but the columns of A' are the rows of A. Hence if $|A| = 0$, then both the rows and columns of A are linearly dependent. It follows that if $|A| \neq 0$, then its rows and columns are linearly independent and *vice-versa*.

25.39.6 LINEAR DEPENDENCE AND INDEPENDENCE OF ANY MATRIX

Consider a matrix of order $m \times n$, given by

$$A = \begin{bmatrix} a_{11} & a_{12} & \cdots & a_{1n} \\ a_{21} & a_{22} & \cdots & a_{2n} \\ \vdots & \vdots & & \vdots \\ a_{m1} & a_{m2} & \cdots & a_{mn} \end{bmatrix}$$

Let the rank of A be r, then there exists at least one r-minor of A which is non-zero. If A_r be a square submatrix of order $r \times r$ such that $|A_r| \neq 0$, then r rows and columns of A_r are linearly independent, it follows that the matrix A has r rows and columns which are linearly independent. As the rank of A is r so that no set of $(r+1)$ rows and columns of A can be linearly independent. Hence the rank of a matrix A is defined to be the maximum number of linearly independent rows and columns of A.

Since on interchanging rows, the rank of A does not change so without loss of generality we may suppose that the first r rows of A are linearly independent. Let $x_1, x_2, x_3, \ldots, x_r$ denote the r independent vectors and let x_t be one of the remaining $(m - r)$ vectors, then the vectors $x_1, x_2, x_3, \ldots, x_r, x_t$ are linearly independent, therefore there exists scalars $\lambda_1, \lambda_2, \lambda_3, \ldots \lambda_r, \lambda_t$, not all zero such that

$$\lambda_1 x_1 + \lambda_2 x_2 + \ldots + \lambda_r x_r + \lambda_t x_t = O$$

Since $x_1, x_2, \ldots, x_r$ are linearly independent, so we take $\lambda_t \neq 0$, thus

$$x_t = \left(-\frac{\lambda_1}{\lambda_t}\right)x_1 + \left(-\frac{\lambda_2}{\lambda_t}\right)x_2 + \ldots + \left(-\frac{\lambda_r}{\lambda_t}\right)x_r$$

It follows that x_t is a linear combination of $x_1, x_2, \ldots, x_r$.

Hence if the rank of a matrix of order $m \times n$ is r, then it has a set of r linearly independent rows (or columns) and $(m - r)$ linearly dependent rows (or columns).

25.40 HOMOGENEOUS LINEAR EQUATIONS

Let us consider a system of linear homogeneous equations as follows

$$\left.\begin{aligned} a_{11}x_1 + a_{12}x_2 + \ldots + a_{1n}x_n &= 0 \\ a_{21}x_1 + a_{22}x_2 + \ldots + a_{2n}x_n &= 0 \\ \ldots\ldots\ldots\ldots\ldots\ldots\ldots\ldots & \\ a_{m1}x_1 + a_{m2}x_2 + \ldots + a_{mn}x_n &= 0 \end{aligned}\right\} \quad \ldots(1)$$

These equations are m equations in n unknowns. Any set of numbers $x_1, x_2, \ldots, x_n$ that satisfies all the equations (1) is called a solution of (1).

25.40.1 TRIVIAL SOLUTION

The solution $x_1 = 0, x_2 = 0, \ldots x_n = 0$ of the equations (1) is called *trivial* solution.

25.40.2 NON-TRIVIAL SOLUTION

Any other solutions, if exists, is called a *non-trivial* solution of equation (1).

Let the coefficient matrix be
$$A = \begin{bmatrix} a_{11} & a_{12} & \cdots & a_{1n} \\ a_{21} & a_{22} & \cdots & a_{2n} \\ \vdots & \vdots & & \vdots \\ a_{m1} & a_{m2} & \cdots & a_{mn} \end{bmatrix}_{m\times n}$$

and
$$X = \begin{bmatrix} x_1 \\ x_2 \\ x_3 \\ \vdots \\ x_n \end{bmatrix}_{n\times 1}, O = \begin{bmatrix} 0 \\ 0 \\ 0 \\ \vdots \\ 0 \end{bmatrix}_{m\times 1}$$

Then the system of equation (1) can also be written as
$$AX = O \qquad \ldots (2)$$

This equation (2) is called a *matrix equation*.

THEOREM 1. *If X_1 and X_2 are two non-trivial solutions of $AX = O$, then $k_1X_1 + k_2X_2$ is also a solution of $AX = O$, where k_1 and k_2 are any arbitrary numbers.*

Proof. Here $AX = O$ and $AX_1 = O, AX_2 = O$ are given.

Now consider, $A(k_1X_1 + k_2X_2) = k_1(AX_1) + k_2(AX_2) = k_1(O) + k_2(O) = O$

Hence $k_1X_1 + k_2X_2$ is the solution of $AX = O$.

THEOREM 2. *If the rank of A is r, then the number of linearly independent solutions of the equation AX = O which is a system of m homogeneous linear equations in n unknowns is (n – r).*

Proof. Since the equation is
$$AX = O \qquad \ldots (1)$$

where
$$A = \begin{bmatrix} a_{11} & a_{12} & \cdots & a_{1n} \\ a_{21} & a_{22} & \cdots & a_{2n} \\ \vdots & \vdots & & \vdots \\ a_{m1} & a_{m2} & \cdots & a_{mn} \end{bmatrix}_{m\times n} \text{ and } X = \begin{bmatrix} x_1 \\ x_2 \\ x_3 \\ \vdots \\ x_n \end{bmatrix}_{n\times 1}, O = \begin{bmatrix} 0 \\ 0 \\ 0 \\ \vdots \\ 0 \end{bmatrix}_{m\times 1}$$

Since the rank of $A = r$, so A has r linearly independent columns. Suppose the matrix A can be written as
$$A = [c_1\, c_2 \ldots c_r \ldots c_n]_{1\times n}$$

where $c_1, c_2, \ldots c_r, \ldots c_n$ are column vectors of the matrix A. Each $c_1, c_2, \ldots c_n$ has m vectors. Thus the equation (1) can be written as
$$x_1c_1 + x_2c_2 + \ldots + x_rc_r + \ldots + x_nc_n = 0 \qquad \ldots (2)$$

But each $c_{r+1}, c_{r+2}, \ldots c_n$ is a linear combination of $c_1, c_2, \ldots c_r$. Then
$$\left.\begin{aligned} c_{r+1} &= p_{11}c_1 + p_{12}c_2 + \ldots + P_{1r}c_r \\ c_{r+2} &= p_{21}c_1 + p_{22}c_2 + \ldots + P_{2r}c_r \\ &\ldots\ldots\ldots\ldots\ldots\ldots\ldots\ldots \\ C_n &= p_{k1}c_1 + p_{k2}c_2 + \ldots + P_{kr}c_r \end{aligned}\right\} \qquad \ldots (3)$$

where $k = (n - r)$

Now (3) can be written as
$$\left.\begin{aligned} p_{11}c_1 + p_{12}c_2 + \ldots + p_{1r}c_r - 1.c_{r+1} + 0.c_{r+2} + \ldots + 0.c_n &= 0 \\ p_{21}c_1 + p_{22}c_2 + \ldots + p_{2r}c_r + 0.c_{r+1} + 1.c_{r+2} + \ldots + 0.c_n &= 0 \\ \ldots\ldots\ldots\ldots\ldots\ldots\ldots\ldots\ldots\ldots\ldots\ldots \\ p_{k1}c_1 + p_{k2}c_2 + \ldots + P_{kr}c_r + 0.c_{r+1} - 0.c_{r+2} - \ldots + -1.c_n &= 0 \end{aligned}\right\} \qquad \ldots (4)$$

Thus equation (2) and (4) are same, so comparing we get

$$X_1 = \begin{bmatrix} p_{11} \\ p_{12} \\ \vdots \\ p_{1r} \\ -1 \\ 0 \\ \vdots \\ 0 \end{bmatrix}, X_2 = \begin{bmatrix} p_{21} \\ p_{22} \\ \vdots \\ p_{2r} \\ 0 \\ -1 \\ \vdots \\ 0 \end{bmatrix}, \ldots X_{n-r} = \begin{bmatrix} p_{k1} \\ p_{k2} \\ \vdots \\ p_{kr} \\ 0 \\ 0 \\ \vdots \\ -1 \end{bmatrix}$$

where $k = (n - r)$.

Hence, we obtained $(n - r)$ solutions of the equation $AX = O$. Next we have to show that X_1, X_2, X_{n-r} are linearly independent.

For this let us have $l_1X_1 + l_2X_2 + \ldots + l_{n-r}X_{n-r} = O$ ... (5)

Now comparing the $(r+1)^{th}$, $(r+2)^{th}$, n^{th} components on both sides of (5), we get $l_1 = 0 = l_2 = \ldots = l_{n-r}$

Hence $X_1, X_2, X_3, \ldots, X_{n-r}$ are linearly independent. Finally we shall have that every solution of the equation $AX = O$ is a linear combination of $X_1, X_2, \ldots, X_{n-r}$.

Suppose X is any solution of $AX = O$ with components $x_1, x_2, \ldots x_n$. Then

$$X + x_{r+1}X_1 + x_{r+2}X_2 + \ldots + x_nX_{n-r} \quad \ldots (6)$$

is also a solution of $AX = O$.

Obviously, let $(n - r)$ components of the vector (6) be all equal to zero. Let $z_1, z_2, \ldots z_r$ be the first r components of (6). Then $(z_1, z_2, \ldots, z_r, 0, 0, \ldots 0)$ is a solution of $AX = O$. Therefore from (2), we get

$$z_1c_1 + z_2c_2 + \ldots + z_rc_r = O$$

This implies $z_1 = 0 = z_2 = \ldots = z_r$ because $c_1, c_2, \ldots c_r$ are linearly independent, and hence (6) comes out to be zero, then

$$X = -x_{r+1}X_1 - x_{r+2}X_2 - \ldots - x_nX_{n-r}$$

This shows that every solution of $AX = O$ is a linear combination of $X_1, X_2, \ldots X_{n-r}$.

25.41 NATURE OF THE SOLUTION OF THE EQUATION AX = O

Since $AX = O$ is a matrix equation of a system of m homogeneous linear equations in n unknowns and A is a coefficient matrix of order $m \times n$. Let the rank of A be r. Then obviously r cannot be greater than n. So that either r is n or r is less than n. Therefore these are some cases.

Case 1. If $r = n$, then the equation $AX = O$, will have no linearly independent solution. So in this case only trivial solution will exist.

Case 2. If $r < n$, then there will be $(n - r)$ linearly independent solution of $AX = O$ and thus in this case we shall have infinite solutions.

Case 3. Suppose the number of equations is less than number of unknowns, *i.e.*, $m < n$ and since $r \le m$, then obviously $r < n$. Thus in this case a non-zero solution will exist. Therefore, the equation $AX = O$ will have infinite solutions.

WORKING PROCEDURE

In order to determine the solutions of the equation $AX = O$, we proceed to the following steps :

Step 1. *Reduce the matrix A to Echelon form by applying E-row transformations only. The Echelon form gives the rank of A.*

Step 2. *Let A be matrix of order $m \times n$ and let $r(A) = r$. If $r = n$, then $AX = O$ will have zero solution only. If $r < n$, then we will assign $n - r$ arbitrarily chosen values to $n - r$ unknowns.*

Step 3. *Let B be the Echelon form of A, then the equation $AX = O$ is equivalent to the equation $BX = O$. Reduce $BX = O$ to a system of equations and choose $n-r$ unknowns in this system of equations for assigning arbitrary values like $c_1, c_2 \ldots, c_{n-r}$.*

Step 4. *By back substitution of $(n - r)$ unknowns to the system of equations reduced from $BX = O$, we finally obtain the solutions. In case of $r < n$, we get infinite solutions.*

Solved Examples

Example 1. *Find the non-trivial solutions of the equations:*

$$x + y - 6z = 0$$
$$-3x + y + 2z = 0$$
$$x - y + 2z = 0$$

Solution. The given system of equations can be written as

$$AX = O \quad \text{... (1)}$$

where $A = \begin{bmatrix} 1 & 1 & -6 \\ -3 & 1 & 2 \\ 1 & -1 & 2 \end{bmatrix}, X = \begin{bmatrix} x \\ y \\ z \end{bmatrix}$ and $O = \begin{bmatrix} 0 \\ 0 \\ 0 \end{bmatrix}$

Reducing the matrix A into Echelon form, we have

Appling $R_2 \to R_2+3R_1, R_3 \to R_3-R_1$, we get

$$A \sim \begin{bmatrix} 1 & 1 & -6 \\ 0 & 4 & -16 \\ 0 & -2 & 8 \end{bmatrix}$$

Again applying $R_2 \to \frac{1}{4}R_2$, we get

$$A \sim \begin{bmatrix} 1 & 1 & -6 \\ 0 & 1 & -4 \\ 0 & -2 & 8 \end{bmatrix}$$

Again applying $R_3 \to R_3 + 2R_2$ we get

$$A \sim \begin{bmatrix} 1 & 1 & -6 \\ 0 & 1 & -4 \\ 0 & 0 & 0 \end{bmatrix}$$

The last equivalent matrix in Echelon form with two non-zero rows, therefore $\rho(A) = 2$

Thus the given system of equations is equivalent to

$$\begin{bmatrix} 1 & 1 & -6 \\ 0 & 1 & -4 \\ 0 & 0 & 0 \end{bmatrix} \begin{bmatrix} x \\ y \\ z \end{bmatrix} = \begin{bmatrix} 0 \\ 0 \\ 0 \end{bmatrix}$$

$$\Rightarrow \quad x+y-6z = 0 \quad \text{... (2)}$$
$$y - 4z = 0 \quad \text{... (3)}$$

Let us put $z = c$ in (3), we get

$$y = 4c$$

Now putting $y = 4c$ and $z = c$ in (2), we get

$$x = 2c$$

Hence, the non-trival solutions of the given system of equatons are $x=2c$, $y= 4c$, $z = c$, where c is a non-zero arbitrary number.

Example 2. *Show that the only real value of λ for which the following equations have non-zero solutions is 6:*

$$x+2y+3z = \lambda x,\ 3x+y+2z = \lambda y,\ 2x+3y+z = \lambda z$$

Solution. The given system of equations can be rewritten as

$$(1-\lambda)x+2y+3z = 0 \quad \text{...(1)}$$
$$3x+(1-\lambda)y+2z = 0 \quad \text{...(2)}$$
$$2x+3y+(1-\lambda)z = 0 \quad \text{...(3)}$$

This system of equations can be written as

$$AX = O \quad \text{... (4)}$$

where

$$A = \begin{bmatrix} 1-\lambda & 2 & 3 \\ 3 & 1-\lambda & 2 \\ 2 & 3 & 1-\lambda \end{bmatrix}, X = \begin{bmatrix} x \\ y \\ z \end{bmatrix} \text{ and } O = \begin{bmatrix} 0 \\ 0 \\ 0 \end{bmatrix}$$

For non-zero solutions, we must have $|A| = 0$

i.e., $\begin{vmatrix} 1-\lambda & 2 & 3 \\ 3 & 1-\lambda & 2 \\ 2 & 3 & 1-\lambda \end{vmatrix} = 0$

$$\Rightarrow \quad (1-\lambda)(1-\lambda)(1-\lambda)+8+27-6(1-\lambda) -6(1-\lambda)-6(1-\lambda) = 0$$

$$\Rightarrow \quad 1-\lambda^3 \ 3\lambda + 3\lambda^2 + 35 \ 18(1-\lambda) = 0$$

$$\Rightarrow \quad -\lambda^3 + 3\lambda^2 + 15\lambda + 18 = 0$$

$$\Rightarrow \quad \lambda^3 - 3\lambda^2 - 15\lambda - 18 = 0$$

$$\Rightarrow \quad (\lambda-6)(\lambda^2 + 3\lambda + 3) = 0$$

Since $\lambda^2 + 3\lambda + 3 = 0$ given imaginary roots, therefore the only real value of λ for which the system of equations is to have a non-zero solution is 6.

Example 3. *Does the following system of equations possess a common non-zero solution?*

$$x + y + z = 0$$
$$2x - y - 3z = 0$$
$$3x - 5y + 4z = 0$$
$$x + 17y + 4z = 0$$

Solution. The coefficient matrix is

$$A = \begin{bmatrix} 1 & 1 & 1 \\ 2 & -1 & -3 \\ 3 & -5 & 4 \\ 1 & 17 & 4 \end{bmatrix}$$

First reduce A into Echelon form.

Performing

$R_2 \to R_2 - 2R_1, R_3 \to R_3 - 3R_1, R_4 \to R_4 - R_1$

$$\sim \begin{bmatrix} 1 & 1 & 1 \\ 0 & -3 & -5 \\ 0 & -8 & 1 \\ 0 & 16 & 3 \end{bmatrix}$$

Performing $R_2 \to -\frac{1}{3}R_2$

$$\sim \begin{bmatrix} 1 & 1 & 1 \\ 0 & 1 & \frac{5}{3} \\ 0 & -8 & 1 \\ 0 & 16 & 3 \end{bmatrix}$$

Performing $R_3 \to R_3 + 8R_2, R_4 \to R_4 - 16R_4$

$$\sim \begin{bmatrix} 1 & 1 & 1 \\ 0 & 1 & \frac{5}{3} \\ 0 & 0 & \frac{43}{3} \\ 0 & 0 & \frac{71}{3} \end{bmatrix}$$

Performing $R_3 \to \frac{3}{43}R_3$

$$\sim \begin{bmatrix} 1 & 1 & 1 \\ 0 & 1 & \frac{5}{3} \\ 0 & 0 & 1 \\ 0 & 0 & -\frac{71}{3} \end{bmatrix}$$

Performing $R_4 \to R_4 + \frac{71}{3}R_3$

$$\sim \begin{bmatrix} 1 & 1 & 1 \\ 0 & 1 & \frac{5}{3} \\ 0 & 0 & 1 \\ 0 & 0 & 0 \end{bmatrix}$$

This is an Echleon form and having three non-zero rows so A has the rank 3. Since there are 3 number of unknown, hence a trival solution exists here, *i.e.*, $x=1, y=0, z=0$.

Example 4. *Find all the solutions of the following system of linear homogeneous equations.*

$$x - 2y + z - w = 0$$
$$x + y - 2z + 3w = 0$$
$$4x + y - 5z + 8w = 0$$
$$5x - 7y + 2z - w = 0$$

Solution. The coefficient matrix is given by

$$A = \begin{bmatrix} 1 & -2 & 1 & -1 \\ 1 & 1 & -2 & 3 \\ 4 & 1 & -5 & 8 \\ 5 & -7 & 2 & -1 \end{bmatrix}$$

Change this matrix into Echelon form as follows:

Performing $R_2 \to R_2 - R_1, R_3 \to R_3 - 4R_1$ and $R_4 \to R_4 - 5R_1$

$$\sim \begin{bmatrix} 1 & -2 & 1 & -1 \\ 0 & 3 & -3 & 4 \\ 0 & 9 & -9 & 12 \\ 0 & 3 & -3 & 4 \end{bmatrix}$$

Performing $R_2 \to \frac{1}{3}R_2$

$$\sim \begin{bmatrix} 1 & -2 & 1 & -1 \\ 0 & 1 & -1 & \frac{4}{3} \\ 0 & 9 & -9 & 12 \\ 0 & 3 & -3 & 4 \end{bmatrix}$$

Performing $R_3 \to R_3 - 9R_2, R_4 \to R_4 - 3R_2$

$$\sim \begin{bmatrix} 1 & -2 & 1 & -1 \\ 0 & 1 & -1 & \frac{4}{3} \\ 0 & 0 & 0 & 0 \\ 0 & 0 & 0 & 0 \end{bmatrix}$$

This is an Echelon form having two non-zero rows. Hence rank of A=2.

Therefore the given system of equation is equivalent to

$$\begin{bmatrix} 1 & -2 & 1 & -1 \\ 0 & 1 & -1 & \frac{4}{3} \\ 0 & 0 & 0 & 0 \\ 0 & 0 & 0 & 0 \end{bmatrix} \begin{bmatrix} x \\ y \\ z \\ w \end{bmatrix} = 0$$

or
$$x - 2y + z - w = 0 \quad \ldots (1)$$
$$y - z + \frac{4}{3}w = 0 \quad \ldots (2)$$

Let $z = c_1, w = c_2$

From (2) $y = c_1 - \frac{4}{3}c_2$

and from (1) $x = c_1 - \frac{5}{3}c_2$

Hence, solution is

$$x = c_1 - \frac{5}{3}c_2, y = c_1 - \frac{4}{3}c_2, z = c_1, w = c_2$$

where c_1 and c_2 are arbitrary numbers.

EXERCISE 25.6

Find the solution of the following system of linear homogeneous equations:

1. $x + 2y + 3z = 0$, $3x + 4y + 4z = 0$, $7x + 10y + 12z = 0$
2. $x + y - 3z + 2w = 0$, $2x - y + 2z - 3w = 0$, $3x - 2y + z - 4w = 0$, $-4x + y - 3z + w = 0$
3. $x + y + z = 0$, $2x + 5y + 7z = 0$, $2x - 5y + 3z = 0$
4. $3x + 4y - z - 6w = 0$, $2x + 3y + 2z - 3w = 0$, $2x + y + 4z - 9w = 0$, $x + 3y + 13z + 3w = 0$
5. $2x - 3y + z = 0$, $x + 2y - 3z = 0$, $4x - y - 2z = 0$
6. $x + 2y + 3z = 0$, $2x + 3y + 4z = 0$, $7x + 13y + 19z = 0$
7. $x + 3y - 2z = 0$, $2x - y + 4z = 0$, $x - 11y + 14z = 0$
8. $2x - 2y + 5z + 3w = 0$, $4x - y + z + w = 0$, $3x - 2y + 3z + 4w = 0$, $x - 3y + 7z + 6w = 0$

Hint to Selected Problems

1. Performing $R_2 \to R_2 - 3R_1, R_3 \to R_3 - 7R_1$, we get

$$A = \begin{bmatrix} 1 & 2 & 3 \\ 0 & -2 & -5 \\ 0 & -4 & -9 \end{bmatrix} \Rightarrow |A| = 10$$

$\Rightarrow$ Rank of A is 3 which is equal to the number of unknown. Therefore, the only solution is $x = y = z = 0$.

3. Do same as (1).

ANSWERS

1. $x = 0 = y = z$ **2.** $x = 0 = y = z = w$ **3.** $x = 0 = y = z$ **4.** $x = 11c_1 + 6c_2, y = -8c_1 - 3c_2, z = c_1, w = c_2$

5. $x = 0 = y = z$ **6.** $x = c, y = -2c, z = c$ **7.** $x = -\dfrac{10}{7}c, y = \dfrac{8}{7}c, z = c$ **8.** $x = \dfrac{5}{9}c, y = 4c, z = \dfrac{7}{9}c, w = c$

25.42 NON-HOMOGENEOUS EQUATIONS

Let us consider a system of equations which are non-homogeneous as follows:

$$\left.\begin{array}{c} a_{11}x_1 + a_{12}x_2 + \ldots + a_{1n}x_n = b_1 \\ a_{21}x_1 + a_{22}x_2 + \ldots + a_{2n}x_n = b_2 \\ \cdots\cdots\cdots\cdots\cdots\cdots\cdots\cdots \\ a_{m1}x_1 + a_{m2}x_2 + \ldots + a_{mn}x_n = b_m \end{array}\right\} \quad \ldots (1)$$

These are m equations in n unknowns. Let

$$A = \begin{bmatrix} a_{11} & a_{12} & \cdots & a_{1n} \\ a_{21} & a_{22} & \cdots & a_{2n} \\ \vdots & \vdots & & \vdots \\ a_{m1} & a_{m2} & \cdots & a_{mn} \end{bmatrix}_{m \times n}$$

$$X = \begin{bmatrix} x_1 \\ x_2 \\ \vdots \\ x_n \end{bmatrix}_{n \times 1}, B = \begin{bmatrix} b_1 \\ b_1 \\ \vdots \\ b_m \end{bmatrix}_{m \times 1}$$

Then the system of equations (1) can also be written as

$$AX = B \quad \ldots (2)$$

This equation is called a matrix equation. If $x_1, x_2, \ldots x_n$ simultaneously satisfy the equation (2), then $(x_1, x_2, \ldots x_n)$ is called the solution of (2).

25.42.1 CONSISTENCY AND INCONSISTENCY

When there exists one or more than one solution of the equation $AX = B$, then the equations are said to be consistent otherwise they are said to be inconsistent.

25.42.2 AUGMENTED MATRIX

The matrix of the type

$$[A \mid B] = \begin{bmatrix} a_{11} & a_{12} & \cdots & a_{1n} & b_1 \\ a_{21} & a_{22} & \cdots & a_{2n} & b_2 \\ \vdots & \vdots & \cdots & \cdots & \cdots \\ a_{m1} & a_{m2} & \cdots & a_{mn} & b_m \end{bmatrix}$$

is called the augmented matrix of the equations.

25.43 CONDITION FOR CONSISTENCY

THEOREM (Rouche's Theorem). *The equation $AX = B$ is consistent if and only if the rank of A and the rank of the augmented matrix $[A|B]$ are same.*

Proof. Since the equation is $AX = B$

The matrix A can be written as $A = [c_1, c_2, \ldots c_n]$ ... (1)

where $c_1, c_2, \ldots c_n$ are column vectors. Then the equation (1) can be written as

$$[c_1, c_2, \ldots, c_n] \cdot \begin{bmatrix} x_1 \\ x_2 \\ \vdots \\ x_n \end{bmatrix} = B$$

or $$x_1c_1 + x_2c_2 + \ldots + x_nc_n = B \qquad \ldots (2)$$

Suppose the rank of A is r, then A has r linearly independent columns. Let these columns be $c_1, c_2, \ldots c_r$ and these $c_1, c_2, \ldots c_r$ are linearly independent and remaining $(n-r)$ columns are linear combination of $c_1, c_2, \ldots c_r$.

Necessary condition. Suppose the equations are consistent, there must exist $k_1, k_2, \ldots, k_n$ such that

$$k_1c_1 + k_2c_2 + \ldots + k_nc_n = B \qquad \ldots (3)$$

But $c_{r+1}, c_{r+2}, \ldots c_n$ is a linear combination of $c_1, c_2, \ldots c_r$ then from (2) it is obvious that B is also a linear combination of $c_1, c_2, \ldots c_r$ and thus $[A|B]$ has the rank r. Hence, the rank of A is same as the rank of $[A|B]$.

Sufficient condition. Suppose rank A = rank $[A|B] = r$. This implies that $[A|B]$ has r linearly independent columns. But $c_1, c_2, \ldots c_r$ of $[A|B]$ are already linearly independent.

Thus B can be expressed as $\quad B = k_1c_1 + k_2c_2 + \ldots + k_rc_r$; where $k_1, k_2, \ldots k_r$ are scalars. ... (4)

Now, equation (4) becomes $\quad B = k_1c_1 + k_2c_2 + \ldots + k_rc_r + 0.c_{r+1} + \ldots + 0.\,c_n$ ... (5)

Comparing (2) and (5), we get $x_1 = k_1, x_2 = k_2, \ldots x_r = k_r, x_{r+1} = 0, \ldots = x_n = 0$ and these values of $x_1, x_2, \ldots x_n$ are the solution of $AX = B$. Hence, the equations are consistent.

REMARKS

- The n equations in n unknowns have a unique solution.
- If rank of $A <$ rank of $[A|B]$, then there is no solution.
- If $r = n$, then there will be a unique solution.
- If $r < n$, then $(n-r)$ variables can be assigned arbitrary values. Thus there will be infinite solutions and $(n-r+1)$ solutions will be linearly independent.
- If $m<n$ and $r \le m \le n$, then equations will have infinite solutions.

WORKING PROCEDURE

In order to determine the solutions of the equation $AX = B$, we procced the following steps:

Step 1. *Reduce the augmented matrix $[A|B]$ to Echelon form by applying E-row transformations only. The Echelon form gives the rank of A and augmented matrix $[A|B]$.*

Step 2. *(i) If the rank of A is not equal to the rank of $[A|B]$, then the system of equations has no solution, i.e., equations are inconsistent.*

(ii) If the rank of A is equal to the rank of $[A|B]$, then the equations are consistent and they will have unique solution if

rank of A = rank of $[A|B]$ = number of unknowns

and then will have infinite solutions if

rank of A = rank of $[A|B]$ = number of unknowns

Step 3. *Let $[A'|B']$ be the reduced Echelon form of $[A|B]$. Now reduce the equation $A'X = B'$ to* a system of equations, after solving these equations we get the required solution.

SOLVED EXAMPLES

Example 1. *Show that the equations*

$x + 2y - z = 3, 3x - y + 2z = 1,$

$2x - 2y + 3z = 2, x - y + z = -1$

are consistent and solve them.

(Bhilai-2005; Madras-2002)

Solution. The given equations can be written as:

$$\begin{bmatrix} 1 & 2 & -1 \\ 3 & -1 & 2 \\ 2 & -2 & 3 \\ 1 & -1 & 1 \end{bmatrix} \begin{bmatrix} x \\ y \\ z \end{bmatrix} = \begin{bmatrix} 3 \\ 1 \\ 2 \\ -1 \end{bmatrix}, \text{i.e., } AX = B$$

Therefore, augmented matrix is

$$[A|B] = \begin{bmatrix} 1 & 2 & -1 & \vdots & 3 \\ 3 & -1 & 2 & \vdots & 1 \\ 2 & -2 & 3 & \vdots & 2 \\ 1 & -1 & 1 & \vdots & -1 \end{bmatrix}$$

Performing $R_2 \to R_2 - 3R_1$, $R_3 \to R_3 - 2R_1$, $R_4 \to R_4 - R_1$

we get

$$[A \mid B] = \begin{bmatrix} 1 & 2 & -1 & \vdots & 3 \\ 0 & -7 & 5 & \vdots & -8 \\ 0 & -6 & 5 & \vdots & -4 \\ 0 & -3 & 2 & \vdots & -4 \end{bmatrix}$$

Performing $R_2 \to R_2 - R_3$

$$\sim \begin{bmatrix} 1 & 2 & -1 & \vdots & 3 \\ 0 & -1 & 0 & \vdots & -4 \\ 0 & -6 & 5 & \vdots & -4 \\ 0 & -3 & 2 & \vdots & -4 \end{bmatrix}$$

Performing $R_3 \to R_3 - 6R_2, R_4 \to R_4 - 3R_2$

$$\sim \begin{bmatrix} 1 & 2 & -1 & \vdots & 3 \\ 0 & -1 & 0 & \vdots & -4 \\ 0 & 0 & 5 & \vdots & 20 \\ 0 & 0 & 2 & \vdots & 8 \end{bmatrix}$$

Performing $R_3 \to \frac{1}{5}R_3, R_4 \to \frac{1}{2}R_4$

$$\sim \begin{bmatrix} 1 & 2 & -1 & \vdots & 3 \\ 0 & -1 & 0 & \vdots & -4 \\ 0 & 0 & 1 & \vdots & 4 \\ 0 & 0 & 1 & \vdots & 4 \end{bmatrix}$$

Performing $R_4 \to R_4 - R_3$

$$\sim \begin{bmatrix} 1 & 2 & -1 & \vdots & 3 \\ 0 & -1 & 0 & \vdots & -4 \\ 0 & 0 & 1 & \vdots & 4 \\ 0 & 0 & 0 & \vdots & 0 \end{bmatrix}$$

This is an Echelon form and having three non-zero rows. Thus rank A = rank of $[A|B]$ = 3. Therefore the equations are consistent

and $$\begin{bmatrix} 1 & 2 & -1 \\ 0 & -1 & 0 \\ 0 & 0 & 1 \\ 0 & 0 & 0 \end{bmatrix} \begin{bmatrix} x \\ y \\ z \end{bmatrix} = \begin{bmatrix} 3 \\ -4 \\ 4 \\ 0 \end{bmatrix}$$

$$\therefore \quad x + 2y - z = 3, -y = -4, z = 4$$

Hence, the solution is $x = -1$, $y = 4$, $z = 4$.

Example 2. *Solve the following equations by matrix method:*

$$x - 2y + 3z = 6$$
$$3x + y - 4z = -7$$
$$5x - 3y + 2z = 5$$

Solution. The given equations can be written as

$$\begin{bmatrix} 1 & -2 & 3 \\ 3 & 1 & -4 \\ 5 & -3 & 2 \end{bmatrix} \begin{bmatrix} x \\ y \\ z \end{bmatrix} = \begin{bmatrix} 6 \\ -7 \\ 5 \end{bmatrix}$$

i.e., $AX = B$

$\therefore$ Argumented matrix is

$$[A \mid B] = \begin{bmatrix} 1 & -2 & 3 & \vdots & 6 \\ 3 & 1 & -4 & \vdots & -7 \\ 5 & -3 & 2 & \vdots & 5 \end{bmatrix}$$

Performing $R_2 \to R_2 - 3R_1, R_3 \to R_3 - 5R_1$, we get

$$[A \mid B] = \begin{bmatrix} 1 & -2 & 3 & \vdots & 6 \\ 0 & 0 & -13 & \vdots & -25 \\ 0 & 7 & -13 & \vdots & -25 \end{bmatrix}$$

Performing $R_3 \to R_3 - R_2$

$$\sim \begin{bmatrix} 1 & -2 & 3 & \vdots & 6 \\ 0 & 7 & -13 & \vdots & -25 \\ 0 & 0 & 0 & \vdots & 0 \end{bmatrix}$$

This is an Echelon form and having two non-zero rows and rank A = rank $[A|B]$ = 2. Thus the equations are consistent.

$$\begin{bmatrix} 1 & -2 & 3 \\ 0 & 7 & -13 \\ 0 & 0 & 0 \end{bmatrix} \begin{bmatrix} x \\ y \\ z \end{bmatrix} = \begin{bmatrix} 6 \\ -25 \\ 5 \end{bmatrix}$$

i.e., $$x - 2y + 3z = 6$$
$$7y - 13z = -25$$

Let $z = c$, then $$y = -\frac{25}{7} + \frac{13}{7}c$$
$$x = -\frac{8}{7} + \frac{5}{7}c$$

Hence the solution is

$$x = -\frac{8}{7} + \frac{5}{7}c, y = -\frac{25}{7} + \frac{13}{7}c, z = c$$

where c is an arbitrary constant.

Example 3. *Investigate for what values of* λ, μ *the simulataneous equations*

$$x + y + z = 6, x + 2y + 3z = 10, x + 2y + \lambda z = \mu$$

have (i) no solution (ii) a unique solution (iii) infinite solution.

(UKTU-2011, UPTU-2006, 14, Mumbai-2007, Rohtak-2004)

Solution. The given equations can be written as

$$\begin{bmatrix} 1 & 1 & 1 \\ 1 & 2 & 3 \\ 1 & 2 & \lambda \end{bmatrix} \begin{bmatrix} x \\ y \\ z \end{bmatrix} = \begin{bmatrix} 6 \\ 10 \\ \mu \end{bmatrix}$$

i.e., $AX = B$

Therefore, augmented matrix is

$$[A \mid B] = \begin{bmatrix} 1 & 1 & 1 & \vdots & 6 \\ 1 & 2 & 3 & \vdots & 10 \\ 1 & 2 & \lambda & \vdots & \mu \end{bmatrix}$$

Performing $R_2 \to R_2 - R_1, R_3 \to R_3 - R_1$, we get

$$\begin{bmatrix} 1 & 1 & 1 & \vdots & 6 \\ 0 & 1 & 2 & \vdots & 4 \\ 0 & 1 & \lambda - 1 & \vdots & \mu - 6 \end{bmatrix}$$

Performing $R_3 \to R_3 - R_2$

$$\sim \begin{bmatrix} 1 & 1 & 1 & \vdots & 6 \\ 0 & 1 & 2 & \vdots & 4 \\ 0 & 0 & \lambda - 3 & \vdots & \mu - 10 \end{bmatrix}$$

If $\lambda \neq 3$, then rank A = rank $[A|B]$ = 3. Thus in this case a unique solution exists. If $\lambda = 3$ and $\mu_0 \neq 10$, then rank A=2, rank $[A|B]$ is 3. Thus rank $A \neq$ rank $[A|B]$. Hence, in this case equations are inconsistent.

If $\lambda = 3$ and $\mu = 10$, then rank A = rank $[A|B]$= 2. Thus in this case infinite solutions exist.

Example 4. *For what values of η the equations $x+y+z=1$, $x+2y+4z = \eta$, $x+4y+10z = \eta^2$ have a solution? Solve them completely in each case.*

(GBTU-2011, Bhopal-2000, Mumbai-2008, VTU-2006)

Solution. The given system of equations can be written as

$$AX=B \quad \ldots (1)$$

where $A = \begin{bmatrix} 1 & 1 & 1 \\ 1 & 2 & 4 \\ 1 & 4 & 10 \end{bmatrix}, X = \begin{bmatrix} x \\ y \\ z \end{bmatrix}, B = \begin{bmatrix} 1 \\ \eta \\ \eta^2 \end{bmatrix}$

Augmented matrix $[A|B]$ is given by

$$\begin{bmatrix} 1 & 1 & 1 & : & 1 \\ 1 & 2 & 4 & : & \eta \\ 1 & 4 & 10 & : & \eta^2 \end{bmatrix}$$

Applying $R_2 \to R_2 - R_1, R_3 \to R_3 - R_1$, we get

$$\sim \begin{bmatrix} 1 & 1 & 1 & : & 1 \\ 0 & 1 & 3 & : & \eta - 1 \\ 0 & 3 & 9 & : & \eta^2 - 1 \end{bmatrix}$$

Applying $R_3 \to R_3 - 3R_2$, we get

$$\sim \begin{bmatrix} 1 & 1 & 1 & : & 1 \\ 0 & 1 & 3 & : & \eta - 1 \\ 0 & 0 & 0 & : & \eta^2 - 3\eta + 2 \end{bmatrix}$$

This last equivalent matrix is in Echelon form. The given system of equations will have the solutions if

$$\text{rank of } A = \text{rank of } [A|B]$$

For Echelon form, the rank of A is 2 and the augmented matrix $[A|B]$ will have rank 2 if

$$\eta^2 - 3\eta + 2 = 0$$

i.e., if $\quad (\eta - 2)(\eta - 1) = 0$

i.e., if $\quad \eta = 1, 2$

The last equivalent matrix gives the system of equations as follows:

$$\begin{bmatrix} 1 & 1 & 1 \\ 0 & 1 & 3 \\ 0 & 0 & 0 \end{bmatrix} \begin{bmatrix} x \\ y \\ z \end{bmatrix} = \begin{bmatrix} 1 \\ \eta - 1 \\ \eta^2 - 3\eta + 2 \end{bmatrix}$$

$$\Rightarrow \left. \begin{aligned} x + y + z &= 1 \\ y + 3z &= \eta - 1 \end{aligned} \right\} \quad \ldots (2)$$

Since rank of A = rank of $[A|B]$ if $\eta = 1$ and $\eta = 2$

Now we have two cases:

Case I: When $\eta = 1$

From (2), we have

$$\left. \begin{aligned} x + y + z &= 1 \\ y + 3z &= 0 \end{aligned} \right\} \quad \ldots (3)$$

Since rank of A = rank of $[A|B]$=2 and number of unknowns is 3, therefore we will have 3 – 2 =1 unknown to be assigned.

Let us assign z to be c_1, therefore put $z = c_1$ in $y+3z=0$, we get $y = -3c_1$.

Again putting $y = -3c_1$ and $z = c_1$ in $x+y+z=1$, we get $x = 1 + 2c_1$

Thus, in this case the solutions are

$$x = 1 + 2c_1, y = -3c_1, z = c_1$$

where c_1 is an arbitrary number.

Case II : When $\eta = 2$

From (2), we have

$$\left. \begin{aligned} x + y + z &= 1 \\ y + 3z &= 1 \end{aligned} \right\} \quad \ldots (4)$$

Let us assign z to be c_2, therefore, putting $z = c_2$ in $y + 3z = 1$, we get $y = 1 - 3c_2$.

Again, putting $z = c_2, y = 1 - 3c_2$ in $x+y+z = 1$, we get $x = 2c_2$.

Thus, in this case the solutions are

$$x = 2c_2, y = 1 - 3c_2, z = c_2$$

where c_2 is an arbitrary number.

EXERCISE 25.7

1. Use matrix method to solve the equations
   $2x - y + 3z = 9, x + y + z = 6, x - y + z = 2.$
2. Show that the equations $x - 3y - 8z + 10 = 0$, $3x + y - 4z = 0$, $2x + 5y + 6z - 13 = 0$ are consistent and solve them.
3. Examine if the system of equations
   $x + y + 4z = 6, 3x + 2y - 2z = 9, 5x + y + 2z = 13$
   is consistent. Find also the solution if it exists.
4. For what values of λ will the following equations fail to have a unique solution
   $3x - y + \lambda z = 1, 2x + y + z = 2, x + 2y - \lambda z = -1$
   Will the equations have any solution for these values of λ?
5. Solve $2x + 3y + z = 9$, $x + 2y + 3z = 6$, $3x + y + 2z = 8$.

Solve the following equations by matrix method:

6. $5x + 3y + 7z = 4$, $3x + 26y - 2z = 9$, $7x + 2y + 10z = 5$.
   (JNTU-2005, PTU-2005, Bhopal-2008)
7. $5x - 6y + 4z = 15$, $7x + 4y - 3z = 19$, $2x + y + 6z = 46$.
8. $x - y + 2z = 4$, $3x + y + 4z = 6$, $x + y + z = 1$.
9. $x + y + z = 6$, $x + 2y + 3z = 4$, $3x + y - 4z = 0$.
   (UPTU-2007)
10. $2x - y + 3z = 8$, $-x + 2y + z = 4$, $3x + y - 4z = 0$.

11. Show that the following equations are inconsistent

$2x - y + z = 4, 3x - y + z = 6, 4x - y + 2z = 7, -x + y - z = 9.$

12. Show that the equations are inconsistent

$x - 4y + 7z = 14, 3x + 8y - 2z = 13, 7x - 8y + 26z = 5.$

13. Prove that the following system of equations have a unique solution

$5x + 3y + 14z = 4, y + 2z = 1, x - y + 2z = 0.$

14. Solve the following equations by matrix mehod:

$x + y + z = 9, 2x + 5y + 7z = 52, 2x + y - z = 0.$

15. Using matrix method, show that the equations are consistant and hence find the solutions.

$3x+3y+2z = 1, x + 2y = 4, 10y + 3z = -2,$

$2x- 3y -z = 5$ are consistant and hence find the solutions.
(UKTU-2010, GBTU-2010, Nagarjuna-2008)

16. Show that the equations $2x+6y+11 = 0$, $6x+20y-6z+3=0$ and $6y - 18z+1 = 0$ are not consistent.
(UKTU-2011, Rajasthan-2005, 13)

17. Show that the system of equations $2x-3y+7z=5$; $3x+y-3z=13$ and $2x+19y-47z = 32$ is not consistent.
(GBTU-2010)

18. For what value of λ, the system of equations $2x - 2y + z = \lambda x$, $2x - 3y + 2z = \lambda y$, $-x + 2y + 0z = \lambda z$ posses a non-trivial solution. Obtain its general solution
(MTU-2011)

19. Find the value of k so that the equation $x+y+3z=0$, $4x+3y+kz=0$ and $2x+y+2z=0$ have a non-trivial solution.
(UPTU-2008)

20. Show that the system of equations $3x+4y+5z=a$, $4x+5y+6z= b$, $5x+6y+7z=c$ does not have a solution unless $a+c= 2b$.
(UPTU-2008, MTU-2009, Raipur-2004, 14, Nagpur-2001)

Hint to Selected Problems

1. Consider the augmented matrix and perform the following opeartions sequentially

$R_1 \leftrightarrow R_2, R_2 \to R_2 - 2R_1, R_3 \to R_3 - R_1, R_3 \to R_3 - \frac{2}{3}R_2.$

2. Here, Rank $(A) = 2$, which is less than the number of unknowns, therefore, given system of equations have infinite number of solutions.

3. The rank of augmented matrix is equal to the rank of (A).

Therefore, the given system of equation is consistent.

4. The coefficient matrix A is non-singular if $\lambda \neq -\frac{7}{2}$. Thus the given system of equations have a unique solution if $\lambda \neq \frac{7}{2}$.

11. The rank of augmented matrix = 4

Rank of A is 3.

Hence the given system of equations is inconsistent.

ANSWERS

1. $x = 1, y = 2, z = 3$

2. $x = 2c - 1, y = 3 - 2c, z = c$

3. Consistent; $x = 2, y = 2, z = \frac{1}{2}$

4. $\lambda \neq -\frac{7}{2}$ solution is unique; $\lambda = -\frac{7}{2}$, no solution.

5. $x = \frac{35}{18}, y = \frac{29}{18}, z = \frac{5}{18}$

6. $x = \frac{7}{11}, y = \frac{3}{11}, z = 0$

7. $x = 3, y = 4, z = 6$

8. $x = \frac{5}{2} - \frac{3}{2}c, y = -\frac{3}{2} + \frac{1}{2}c, z = c$

9. $x = c - 2, y = 8 - 2c, z = c$

10. $x = 2, y = 2, z = 2$

14. $x = 1, y = 3, z = 4$

15. 2, 1, –4

18. $\lambda = 1, x = 2k_1 - k_2, y = k_1, z = k_2, \lambda = -3, x = -k, y' = -2k, z = k$

19. 8

25.44 GAUSS ELIMINATION METHOD

In this method, the variables from the system of linear equations are eliminated successively and the system of equations is therefore reduced to an upper triangular system from which the variable are determined by back substitution. This method is described as follows: Let us consider a system of linear equation

$$AX = B \quad \ldots(1)$$

Assuming $\det A \neq 0$. Equation (1) has the following form:

$$\left.\begin{aligned} a_{11}x_1 + a_{12}x_2 + \ldots + a_{1n}x_n &= b_1 \\ a_{21}x_1 + a_{22}x_2 + \ldots + a_{2n}x_n &= b_2 \\ \ldots \quad \ldots \quad \ldots \quad &\ldots \\ \ldots \quad \ldots \quad \ldots \quad &\ldots \\ a_{n1}x_1 + a_{n2}x_2 + \ldots + a_{nn}x_n &= b_n \end{aligned}\right\} \quad \ldots(2)$$

Assuming $a_{11} \neq 0$ and divide the first equation by a_{11} and then we subtract this equation multiplied by $a_{21}, a_{31}, \ldots, a_{n1}$ from second, third ... nth equation of (2), we get

$$\left.\begin{aligned} x_1 + a'_{12}x_2 + \ldots + a'_{1n}x_n &= b'_1 \\ a'_{22}x_2 + \ldots + a'_{2n}x_n &= b'_2 \\ \ldots \quad \ldots \quad \ldots \quad \ldots & \quad \ldots \\ \ldots \quad \ldots \quad \ldots \quad \ldots & \quad \ldots \\ a'_{n2}x_2 + \ldots + a'_{n2}x_n &= b'_n \end{aligned}\right\} \quad \ldots(3)$$

Next, we divide second equation of (3) by a'_{22} (assuming $a'_{22} \neq 0$) and subtract this equation multiplied by $a'_{32}, a'_{42}, \ldots, a'_{n2}$ from third, fourth ... nth equation of (3), we get

$$\left.\begin{aligned} x_1 + a'_{12}x_2 + \ldots + a'_{1n}x_n &= b'_1 \\ x_2 + a''_{23}x_3 + \ldots + a''_{2n}x_n &= b''_2 \\ a''_{33}x_3 + \ldots + a''_{3n}x_n &= b''_3 \\ \ldots \quad \ldots \quad \ldots & \quad \ldots \\ a''_{3n}x_3 + \ldots + a''_{nn}x_n &= b''_n \end{aligned}\right\} \quad \ldots(4)$$

Continuing in this way, we get a system of equation as follows:

$$\left.\begin{aligned} x_1 + c_{12}x_2 + c_{13}x_3 + \ldots + c_{1n}x_n &= d_1 \\ x_2 + c_{23}x_3 + \ldots + c_{2n}x_n &= d_2 \\ &\vdots \\ &\vdots \\ c_{nn}x_n &= d_n \end{aligned}\right\} \quad \ldots(5)$$

This is a form of upper triangular system. From back substitution we can find the solution of the system of given equations.

REMARKS

- The coefficient a_{11}, a'_{22} and a''_{33} are called pivots.
- This method will fail if any one of the pivots a_{11}, a'_{22} and a''_{33} becomes zero. In such cases, we rewrite the equations in a different order so that the pivots are non-zero.
- From each of the procedure, the largest coefficient of x is chosen as pivot element.
- This method proposes a systematic astrology for reducing the system of equations to the upper triangular form using the forward elimination approach and then for obtaining values of unknowns using back substitution process.

WORKING PROCEDURE

Let us consider these equations

$$\left.\begin{aligned} a_{11}x_1 + a_{12}x_2 + a_{13}x_3 &= b_1 \\ a_{21}x_1 + a_{22}x_2 + a_{23}x_3 &= b_2 \\ a_{31}x_1 + a_{32}x_2 + a_{33}x_3 &= b_3 \end{aligned}\right\} \quad \ldots(6)$$

Step 1. *First, eliminate x_1 from second and third equations. Assuming $a_{11} \neq 0$, now dividing first equation by a_{11} and then subtract from second and third after multiplied by a_{21} and respectively, we get*

$$\left.\begin{aligned} x_1 + a'_{12}x_2 + a'_{13}x_3 &= b'_1 \\ a'_{22}x_2 + a'_{23}x_3 &= b'_2 \\ a'_{32}x_2 + a'_{33}x_3 &= b'_3 \end{aligned}\right\} \quad \ldots(7)$$

where $a'_{12} = \dfrac{a_{12}}{a_{11}}, a'_{13} = \dfrac{a_{13}}{a_{11}}, \; a'_{22} = a_{22} - a_{21}a'_{12}, a'_{23} = a_{23} - a_{21}a'_{13}$

$a'_{32} = a_{32} - a_{31}a'_{12}, a'_{33} = a_{33} - a_{31}a'_{13}, \; b'_1 = \dfrac{b_1}{a_{11}}, b'_2 = b_2 - a_{21}b'_1, b'_3 = b_3 - a_{31}b'_1$

Step 2. *Now eliminating x_2 from third equation in (7).*

Again assuming $a_{22}' \neq 0$. Dividing second equation in (7) by a_{22}' and then subtract from third equation after multiplied by a_{32}' we get

$$\left.\begin{aligned} x_1 + a'_{12}x_2 + a'_{13}x_3 &= b'_1 \\ x_2 + a''_{23}x_3 &= b''_2 \\ a''_{33}x_3 &= b''_3 \end{aligned}\right\} \quad \ldots(8)$$

where $a''_{23} = \dfrac{a'_{23}}{a'_{22}}, a''_{33} = a'_{33} - a'_{32}a''_{23}, \; b''_2 = \dfrac{b'_2}{a'_{22}}, b''_3 = b_3' - a'_{32}b''_2.$

Step 3. *Evaluating x_1, x_2 and x_3 from (8) by back substitution.*

Solved Examples

Example 1. *Solve the following equations by Gauss's elimination method*

$$6x + 3y + 2z = 6$$
$$6x + 4y + 3z = 0$$
$$20x + 15y + 12z = 0.$$

Solution. Here pivot element is 6. Now Divide first equation by 6, we get

$$x + \frac{1}{2}y + \frac{1}{3}z = 1 \quad ...(1)$$

Now eliminating x from second and third equation with the help of (1). Subtract (1) multiplied by 6 and 20 from second and third equation, respectively we get

$$y + z = -6 \quad ...(2)$$

$$5y + \frac{16}{3}z = -20 \quad ...(3)$$

Now eliminating y from (3) with the help of (2), we get

$$\left(\frac{16}{3} - 5\right)z = -20 + 30$$

$$\frac{1}{3}z = 10 \Rightarrow z = 30$$

Substitute the value of z into (2), we get

$$y = -6 - 30 = -36$$

and again substitute the values of y and z into (1), we get

$$x + \frac{1}{2}(-36) + \frac{1}{3}(30) = 1$$

$$x - 18 + 10 = 1 \Rightarrow x = 9$$

Hence, the solution of the equations are

$$x = 9, y = -36, z = 30.$$

Example 2. *By Gauss's elimination method, solve the following equations*

$$5x - y - 2z = 142$$
$$x - 3y - z = -30$$
$$2x - y - 3z = -50$$

Solution. The largest coefficient in first equation is 5, which is pivot element. So divide first equation by 5, we get

$$x - \frac{1}{5}y - \frac{2}{5}z = \frac{142}{5} \quad ...(1)$$

Now eliminating x from second and third equation with help of (1), we get

$$-\frac{14}{5}y - \frac{3}{5}z = -\frac{292}{5} \quad ...(2)$$

$$-\frac{3}{5}y - \frac{11}{5}z = -\frac{309}{5} \quad ...(3)$$

Eliminating y from (2) and (3), we get

$$-\frac{145}{5}z = -\frac{3450}{5}$$

or

$$z = \frac{3450}{145} = 23.79$$

Substitute the value of z into (3) we get

$$-\frac{3}{5}y - \frac{11}{5}(23.79) = -\frac{309}{5}$$

$$-\frac{3}{5}y = -\frac{309}{5} + \frac{11(23.79)}{5}$$

$$-3y = -309 + 11(23.79)$$

$$-3y = -47.31$$

or $\quad y = 15.77$

Substitute the values of y and z into (1), we get

$$x - \frac{1}{5}(15.77) - \frac{2}{5}(23.79) = \frac{142}{5}$$

$$x = \frac{142}{5} + \frac{15.77}{5} + \frac{2(23.79)}{5} = \frac{205.35}{5}$$

or $\quad x = 41.07$

Hence, the solution are given by

$$x = 41.07, y = 15.77, z = 23.79.$$

Example 3. *Using Gauss's elimination method solve*

$$2x_1 + 4x_2 + x_3 = 3$$
$$3x_1 + 2x_2 - 2x_3 = 2$$
$$x_1 - x_2 + x_3 = 6$$

Solution. Dividing first equation by 2, we get

$$x_1 + 2x_2 + \frac{1}{2}x_3 = \frac{3}{2} \quad ...(1)$$

Multiplying (1) by 3 and subtract from second and also subtract (1) from third of the given equation, we get

$$4x_2 + \frac{7}{2}x_3 = \frac{5}{2} \quad ...(2)$$

$$-3x_2 + \frac{1}{2}x_3 = \frac{9}{2} \quad ...(3)$$

Now dividing (2) by 4 and subtract after multiplies by –3 from (3), we get

or $\quad 25x_3 = 51$

$$x_3 = \frac{51}{25} = 2.04$$

Substitute the value of x_3 into (2), we get

$$+4x_2 + \frac{7}{2}(2.04) = \frac{5}{2}$$

$$4x_2 = \frac{5}{2} - \frac{7(2.04)}{2}$$

$$= \frac{5 - 14.28}{2}$$

$$\Rightarrow \quad x_2 = -\frac{9.28}{8} = -1.16$$

Now substitute the value of x_2 and x_3 into (1), we get

$$x_1 + 2(-1.16) + \frac{1}{2}(2.04) = \frac{3}{2}$$

$$x_1 = \frac{3}{2} + 2(1.16) - \frac{1}{2}(2.04)$$

$$= \frac{3 + 4.64 - 2.04}{2} = \frac{5.6}{2} = 2.8$$

or $\quad x_1 = 2.8$

Hence, the solutions are given by

$$x_1 = 2.8, x_2 = -1.16, x_3 = 2.04.$$

Example 4. *Solve by the Gauss's elimination method.*

$$2x + y + 4z = 12$$
$$8x - 3y + 2z = 23$$
$$4x + 11y - z = 33$$

Solution. Dividing first equation by 2, we get

$$x + \frac{1}{2}y + 2z = 6 \quad ...(1)$$

Now subtract (1) after multiplied by 8 and 4 respectively from second and third equation, we get

$$-7y - 14x = -45 \quad ...(2)$$
$$9y - 9z = 9 \quad ...(3)$$

Now multiplying (4) by 9 and subtract from (3), we get

$$-27z = 9 - \frac{405}{7}$$

or $\quad -27z = -\frac{342}{7} \quad ...(5)$

Hence, the system of equations reduces to upper triangular form as follows:

$$\left.\begin{aligned} x + \frac{1}{2}y + 2z &= 6 \\ y + 2z &= \frac{45}{7} \\ -27z &= -\frac{342}{7} \end{aligned}\right\} \quad ...(6)$$

By back substitution , we get

$$z = \frac{342}{189} = 1.81$$

and $\quad y + 2(1.81) = \frac{45}{7}$

$\Rightarrow \quad y = \frac{45}{7} - 2(1.81) = 6.43 - 3.62$

$$= 2.81$$

and $\quad x + \frac{1}{2}(2.81) + 2(1.81) = 6$

$\therefore \quad x = 6 - \frac{1}{2}(2.81) - 2(1.81)$

$$= 0.975$$

Hence, the solution is

$$x = 0.975, y = 2.81, z = 1.81.$$

Example 5. *Apply Gauss's elimination method to solve the equations*

$$x + 4y - z = -5$$
$$x + y - 6z = -12$$
$$3x - y - z = 4$$

Solution. Eliminating x from second and third equation with the help of first equation. Subtract first equation from second and after multiplied by 3 from third respectively, we get the system of equations as follows:

$$x + 4y - z = -5 \quad ...(1)$$
$$-3y - 5z = -7 \quad ...(2)$$
$$-13y + 2z = 19 \quad ...(3)$$

Elimination y from (3) with help of (2). Divide (2) by -3 and then this equation is subtracted after multiplies by -13 from (3), we get

$$\left.\begin{aligned} x + 4y - z &= -5 \\ y + \frac{5}{3}z &= \frac{7}{3} \\ \frac{71}{3}z &= \frac{148}{3} \end{aligned}\right\} \quad ...(4)$$

By back substitution from (4), we get

$$z = \frac{148}{71}$$

and $\quad y = \frac{7}{3} - \frac{5}{3}z = \frac{7}{3} - \frac{5}{3}\left(\frac{148}{71}\right)$

$\Rightarrow \quad y = -\frac{81}{71}$

and $\quad x = -5 - 4y + z$

$$= -5 - 4\left(-\frac{81}{71}\right) + \left(\frac{148}{71}\right)$$

$$= -5 + \frac{472}{71} + \frac{117}{71}$$

Hence, the solution are

$$x = \frac{117}{71}, y = -\frac{81}{71}, z = \frac{148}{71}.$$

Example 6. *Solve the following system by Gauss's elimination method :*

$$2x + y + z = 10$$
$$3x + 2y + 3z = 18$$
$$x + 4y + 9z = 4$$

Solution. We have

$$2x + y + z = 10 \quad ...(1)$$
$$3x + 2y + 3z = 18 \quad ...(2)$$
$$x + 4y + 9z = 4 \quad ...(3)$$

Divide (1) and 2 and subtract after multiplied by 3 from (2) then subtract from (3), we get

$$x + \frac{1}{2}y + \frac{1}{2}z = 5 \quad ...(4)$$

$$\frac{1}{2}y + \frac{3}{2}z = 3 \quad ...(5)$$

$$\frac{7}{2}y+\frac{17}{2}z=11 \quad \ldots(6)$$

Now divide (5) by $\frac{1}{2}$ and then subtract after multiplied by $\frac{7}{2}$ from (6) we get,

$$x+\frac{1}{2}y+\frac{1}{2}z=5 \quad \ldots(7)$$

$$y+3z=6 \quad \ldots(8)$$

$$-2z=-10 \quad \ldots(9)$$

From back substitution in (9), (8) and (7) we get

$$z=5$$

and $y+3z=6$

$$y+3(5)=6$$

$$y=6-15 \Rightarrow \quad y=-9$$

$$y=-9$$

and $x+\frac{1}{2}(-9)+\frac{1}{2}(5)=5$

$$x=5+\frac{9}{2}-\frac{5}{2}$$

Hence, the solution is $x=7, y=-9, z=5$.

Example 7. *By Gauss's elimination method, solve*

$$4x+11y-z=33$$

$$x+y+4z=12$$

$$8x-3y+2z=20$$

Solution. Given equation are

$$x+y+4z=12 \quad \ldots(1)$$

$$4x+11y-z=33 \quad \ldots(2)$$

$$8x-3y+2z=20 \quad \ldots(3)$$

Eliminating x from (2) and (3) so subtract (1) after multiplied by 4 and 8 from (2) and (3), we get

$$x+y+4z=12 \quad \ldots(4)$$

$$7y-17z=-15 \quad \ldots(5)$$

$$-11y-30z=-76 \quad \ldots(6)$$

Now divide (5) by 7 and then subtract after multiplied by -11 from (6), we get

$$x+y+4z=12 \quad \ldots(7)$$

$$y-\frac{17}{7}z=12 \quad \ldots(8)$$

$$-\frac{397}{7}z=-\frac{697}{7} \quad \ldots(9)$$

By back substitution in (9), (8) and (7) we get

$$z=\frac{697}{397}=1.756$$

From (8) $y=\frac{-15}{7}+\frac{17}{7}z=-\frac{15}{7}+\frac{17}{7}\left(\frac{697}{397}\right)$

$$=\frac{1}{7}\left(\frac{5894}{397}\right)=\frac{5894}{2779}=2.121$$

From (7) $x+y+z=12$

$$\Rightarrow \quad x+\frac{5894}{2779}+4\left(\frac{697}{397}\right)=12$$

$$\Rightarrow \quad x=12-\frac{5894}{2779}+4\left(\frac{697}{397}\right)=\frac{7938}{2779}=2.856$$

Hence, the solution is $x = 2.856$, $y = 2.121$, $z = 1.756$.

Example 8. *Solve by Gauss's elimination method*

$$x+2y+z=3$$

$$2x+3y+3z=10$$

$$3x-y+2z=13$$

Solution. Given equation are

$$x+2y+z=3 \quad \ldots(1)$$

$$2x+3y+3z=10 \quad \ldots(2)$$

$$3x-y+2z=13 \quad \ldots(3)$$

Here pivot element of (1) is 1. Now eliminating x from (2) and (3) by subtracting (1) after multiplied by 2 and 3 respectively from (2) and (3), we get

$$x+2y+z=3 \quad \ldots(4)$$

$$-y+z=4 \quad \ldots(5)$$

$$-7y-z=4 \quad \ldots(6)$$

Now eliminating y from (6) with the help of (5) by subtracting (5) after multiplied by -7 from (6), we get

$$x+2y+z=3 \quad \ldots(7)$$

$$-y+z=4 \quad \ldots(8)$$

$$6z=32 \quad \ldots(9)$$

By back substitution from (7), (8) and (9), we get

From (9) $z=\frac{32}{6}=\frac{16}{3}$

From (8) $-y=4-z$

$$-y=4-\frac{32}{6}=-\frac{8}{6}$$

$$\therefore \quad y=\frac{8}{6}=\frac{4}{3}$$

From (7)

$$x+2y+z=3$$

$$x+2\left(\frac{8}{6}\right)+\frac{32}{6}=3$$

$$x=3-\frac{32}{6}-\frac{16}{6}$$

$$=\frac{18-48}{6}=-\frac{30}{6}$$

$$x=-5$$

Hence, the solution is $x=-5, y=\frac{4}{3}, z=\frac{16}{3}$.

EXERCISE 25.8

1. Solve the following equations by Gauss's elimination method :

(i) $x_1 + x_2 + 2x_3 = 4$
$3x_1 + x_2 - 3x_3 = -4$
$2x_1 - 3x_2 - 5x_3 = -5$

(ii) $2x_1 + x_2 + 4x_3 = 12$
$8x_1 - 3x_2 + 2x_3 = 20$
$4x_1 + 11x_2 - x_3 = 33$

(iii) $x_1 + x_2 + x_3 = 10$
$2x_1 + x_2 + 2x_3 = 17$
$3x_1 + 2x_2 + x_3 = 17$

(iv) $2x + 3y - z = 5$
$4x + 4y - 3z = 3$
$2x - 3y + 2z = 2$

(v) $2x + y + z = 10$
$x + 2y + 3z = 18$
$x + 4y + 9z = 16$

(vi) $2x_1 + 4x_2 + x_3 = 2$
$3x_1 + 2x_2 - 2x_3 = -2$
$x_1 - x_2 + x_3 = 6$

ANSWERS

1.(i) $x_1 = 1, x_2 = -1, x_3 = 2$ (ii) $x_1 = 3, x_2 = 2, x_3 = 1$ (iii) $x_1 = 2, x_2 = 3, x_3 = 5$
(iv) $x = 1, y = 2, z = 3$ (v) $x = 7, y = -9, z = 5$ (vi) $x_1 = 2, x_2 = -1, x_3 = 3$

25.45 EIGENVALUE AND EIGENVECTORS OF A MATRIX

A polynomial in indeterminate λ of the form

$$f(\lambda) = A_0 + A_1\lambda + A_2\lambda^2 + \ldots + A_n\lambda^n$$

where $A_0, A_1, A_2, \ldots, A_n$ are all square matrices of the same order, is called a matric polynomial of degree n if $A_n \neq O$ (null matrix).

From above definition it is clear that every square matrix can be expressed as a matric polynomial of zero degree. If A is a square matrix, then we can write

$$A = \lambda^\circ A$$

REMARK

- Two matric polynomials are said to be equal if and only if the coefficients of like powers of λ are the same.

25.46 THE CHARACTERISTIC EQUATION OF A MATRIX

(UPTU 2008)

Let A be a square matrix of order $n \times n$ and let

$$A = \begin{bmatrix} a_{11} & a_{12} & \cdots & a_{1n} \\ a_{21} & a_{22} & \cdots & a_{2n} \\ \vdots & \vdots & & \vdots \\ a_{n1} & a_{n2} & \cdots & a_{nn} \end{bmatrix}$$

If λ is indeterminate, then the matrix $A - \lambda I$ is called the characteristic matrix of A, where I is the unit matrix of order $n \times n$.

The determinant $|A - \lambda I| = \begin{bmatrix} a_{11} - \lambda & a_{12} & \cdots & a_{1n} \\ a_{21} & a_{22} - \lambda & \cdots & a_{2n} \\ \vdots & \vdots & & \vdots \\ a_{n1} & a_{n2} & \cdots & a_{nn} - \lambda \end{bmatrix}$ is an ordinary polynomial in λ which is called the characteristic polynomial of A and the equation

$$|A - \lambda I| = 0$$

i.e.,

$$\begin{vmatrix} a_{11} - \lambda & a_{12} & \cdots & a_{1n} \\ a_{21} & a_{22} - \lambda & \cdots & a_{2n} \\ \vdots & \vdots & & \vdots \\ a_{n1} & a_{n2} & \cdots & a_{nn} - \lambda \end{vmatrix} = 0$$

is known as the characteristic equation of A. The roots of the equation $|A - \lambda I| = 0$ are called characteristic roots or latent roots or eigenvalues of A. The set of all eigenvalues of a matrix A is called spectrum of A.

25.47 CHARACTERISTIC VECTORS OR EIGENVECTORS OF A MATRIX

(UPTU-2006, 08)

Let $A = [a_{ij}]$ be a matrix of order $n \times n$ and let $X = \begin{bmatrix} x_1 \\ x_2 \\ \vdots \\ x_n \end{bmatrix}$

be a column vector. Consider a vector equation

$$AX = \lambda X \text{; where } \lambda \text{ is a scalar.} \quad ...(1)$$

It is evident that $X = O$ satisfies the equation (1) for every value of λ, thus $X = O$ is a solution of (1). A value of λ for which a non-zero vector, *i.e.*, $X \neq O$ satisfies (1) is called an eigenvalue of the matrix A and the non-zero vector X is called an eigenvector of A corresponding to that eigenvalue λ.

Now equation (1) can be written as $\quad AX = \lambda IX$

or $\quad (A - \lambda I)X = O \quad ...(2)$

where I is the unit matrix of order $n \times n$. Equation (2) represents a matrix equation of a system of n homogeneous equations. The necessary and sufficient condition for the equation (2) to possess a non-zero solution, *i.e.*, $X \neq O$ is that $|A - \lambda I| = 0$, which is a characteristic equation of matrix A.

REMARKS

- The eigenvector is also known as proper vector.
- If X is an eigenvector of a matrix corresponding eigenvalue λ, then for any non-zero scalar kX is also an eigenvector of A corresponding to the same eigenvalue λ.
- Corresponding to an eigenvalue of a matrix A, there will be different eigenvectors of A.
- For a given eigenvector of a matrix A there corresponds one and only one eigenvalue of A.

25.48 RELATION BETWEEN EIGENVALUES AND EIGENVECTORS

THEOREM 1. *λ is an eigenvalue of a matrix A if and only if there exists a non-zero vector X such that $AX = \lambda X$.*

Proof. Suppose that λ is an eigenvalue of A, then

$$|A - \lambda I| = 0$$

$\Rightarrow$ $A - \lambda I$ is a singular matrix.

$\therefore$ The matrix equation $(A - \lambda I)X = O$ has a non-zero solution, thus there exists a non-zero vector X such that

$$(A - \lambda I)X = O \text{ or } AX = \lambda X$$

Conversely, Suppose that there is a non-zero vector X such that $AX = \lambda X$

$\Rightarrow \quad (A - \lambda I)X = O$

Since the matrix equation $(A - \lambda I)X = O$ has a non-zero solution, then the coefficient matrix $A - \lambda I$ is singular, therefore $\quad |A - \lambda I| = 0$

Hence A is an eigenvalue of A.

THEOREM 2. *If X is an eigenvector of a matrix A corresponding to an eigenvalue of A, then kX is also an eigenvector of A corresponding to the same eigenvalue A, where k is any non-zero number.*

Proof. Since X is an eigenvector of a matrix A corresponding to an eigenvalue of A, then we have

$$AX = \lambda X \quad ...(1)$$

Since $k \neq 0$, then multipling both sides of (1) by k, we get $k(AX) = k(\lambda X)$

$\Rightarrow \quad A(kX) = \lambda(kX) \quad ...(2)$

From equation (2), it follows that kX is also an eigenvector of A corresponding to the same eigenvalue λ.

THEOREM 3. *If X is a non-zero eigenvector of a matrix A, then X cannot correspond to more than one eigenvalue of A.*

Proof. If possible, let X be an eigenvector corresponding to eigenvalues λ_1 and λ_2, then

$$AX = \lambda_1 X \quad ...(1)$$

and $\quad AX = \lambda_2 X \quad ...(2)$

From (1) and (2) we have $\quad \lambda_1 X = \lambda_2 X$

$\Rightarrow \quad (\lambda_1 - \lambda_2)X = O$

$\Rightarrow \quad \lambda_1 - \lambda_2 = 0 \quad \because X \neq O$

$\Rightarrow \quad \lambda_1 = \lambda_2$

THEOREM 4. *If X_1 and X_2 be non-zero eigenvectors of a matrix A corresponding to an eigenvalue λ of A, then $k_1X_1 + k_2X_2$ is also an eigenvector of A corresponding to eigenvalue λ, where k_1 and k_2 are non-zero numbers.*

Proof. Since $X_1 \neq 0, X_2 \neq 0$ and $k_1 \neq 0, k_2 \neq 0$, then $k_1X_1 + k_2X_2 \neq O$. Also X_1 and X_2 are eigenvectors of A corresponding to an eigenvalue λ of A, then we have

$$AX_1 = \lambda X_1 \quad ...(1)$$

and
$$AX_2 = \lambda X_2 \quad ...(2)$$

Multiplying (1) by k_1 and (2) by k_2 and then adding, we get

$$Ak_1X_1 + Ak_2X_2 = \lambda(k_1X_1 + k_2X_2)$$

$$\Rightarrow \quad Ak_1X_1 + Ak_2X_2 = \lambda(k_1X_1 + k_2X_2) \quad ...(3)$$

As $k_1X_1 + k_2X_2 \neq O$ then from (3) it follows that $k_1X_1 + k_2X_2$ is also an eigenvector of A corresponding to an eigenvalue of A.

THEOREM 5. *Let A be an $n \times n$ matrix. Then the distinct eigenvectors corresponding to distinct eigenvalues of A are linearly independent.*

Proof. Since A is an $n \times n$ matrix so it will have atmost n eigenvalues. Let $\lambda_1, \lambda_2, \lambda_3,..., \lambda_m$ be m distinct eigenvalues of A out of n eigenvalues and let $X_1, X_2, X_3,...,X_m$ be m distinct eigenvectors corresponding to eigenvalues $\lambda_1, \lambda_2,..., \lambda_m$ respectively. Then we have

$$AX_1 = \lambda_1X_1, AX_2 = \lambda_2X_2,..., AX_m = \lambda_mX_m$$

Let S = $\{X_1, X_2,..., X_m\}$. Then we have to show that S is linearly independent. We shall prove it by induction hypothesis on m.

If $m = 1$, then there is only one non-zero vector, which is obviously linearly independent.

Suppose the result is true for $m = k$, *i.e.*, $\{X_1, X_2,..., X_m\}$ is linearly independent. Let this set be denoted by S_1, then

$$S_1 = \{X_1, X_2, \ldots, X_k\}$$

Finally we shall prove that the set $S_1 \cup \{X_{k+1}\}$ is linearly independent.

For scalars $a_1, a_2, a_3,..., a_k, a_{k+1}$ such that

$$a_1X_1 + a_2X_2 + ... + a_kX_k + a_{k+1}X_{k+1} = O \quad ...(1)$$

$$\Rightarrow \quad A(a_1X_1 + a_2X_2 + ... + a_kX_k + a_{k+1}X_{k+1}) = AO = O$$

$$\Rightarrow \quad a_1AX_1 + a_2AX_2 + \ldots + a_kAX_k + a_{k+1}AX_{k+1} = O$$

$$\Rightarrow \quad a_1\lambda_1X_1 + a_2\lambda_2X_2 + \ldots + a_k\lambda_kX_k + a_{k+1}\lambda_{k+1}X_{k+1} = O \quad ...(2)$$

Multiplying (1) by λ_{k+1} and then substracting it from (2), we get

$$a_1(\lambda_1 - \lambda_{k+1})X_1 + a_2(\lambda_2 - \lambda_{k+1})X_2 + \ldots + a_k(\lambda_k - \lambda_{k+1})X_k = O$$

Since S_1 is linearly independent, hence

$$\Rightarrow \quad a_1(\lambda_1 - \lambda_{k+1}) = a_2(\lambda_2 - \lambda_{k+1}) = \ldots = a_k(\lambda_k - \lambda_{k+1}) = 0$$

$$\Rightarrow \quad a_1 = a_2 = ... = a_k = 0 \quad [\because \lambda_1, \lambda_2,..., \lambda_m \text{ are all distinct.}]$$

Putting $a_1 = a_2 = ... = a_k = 0$ in (1), we get

$$a_{k+1}X_{k+1} = O$$

$$\Rightarrow \quad a_{k+1} = 0 \quad [\because X_{k+1} \neq O]$$

$\therefore$ The set $S_1 \cup \{X_{k+1}\}$ is linearly independent.

Hence, the result is proved by induction.

25.49 EIGENVALUE OF SPECIAL TYPE OF MATRICES

THEOREM 1. *The eigenvalues of a Hermitian matrix are real.* (UKTU - 2011)

Proof. Let A be a Hermitian matrix. Let λ be an eigenvalue of A and let X be its corresponding eigenvector.

Then we have
$$AX = \lambda X \quad ...(1)$$

Pre-multiplying both sides of (1) by X^θ, we have

$$X^\theta AX = X^\theta \lambda X = \lambda X^\theta X \quad ...(2)$$

Taking conjugate transpose of both sides of (2), we have

$$(X^\theta AX)^\theta = (\lambda X^\theta X)^\theta$$

$\Rightarrow$ $$X^\theta A^\theta (X^\theta)^\theta = \bar{\lambda} X^\theta (X^\theta)^\theta$$

$\Rightarrow$ $$X^\theta A^\theta X = \bar{\lambda} X^\theta X \qquad \because (X^\theta)^\theta = X$$

$\Rightarrow$ $$X^\theta AX = \bar{\lambda} X^\theta X \qquad [\because A \text{ is Hermitian} \Rightarrow A^\theta = A]$$

$\Rightarrow$ $$X^\theta \lambda X = \bar{\lambda} X^\theta X \qquad \text{[Using (1)]}$$

or $$\lambda X^\theta X = \bar{\lambda} X^\theta X$$

or $$(\lambda - \bar{\lambda}) X^\theta X = O$$

or $$(\lambda - \bar{\lambda}) = 0 \qquad [\because X \text{ is non-zero} \Rightarrow X^\theta X \neq O]$$

or $$\lambda = \bar{\lambda}$$

Hence, λ is real.

THEOREM 2. *The eigenvalues of a real symmetric matrix are all real.*

Proof. Let A be a real symmetric matrix, then

$$A' = A \qquad \text{...(1)}$$

Let λ be any eigenvalue of A and let X be its corresponding eigenvector, then we have

$$AX = \lambda X \qquad \text{...(2)}$$

Pre-multiplying both sides of (2) by X' we get

$$X'AX = \lambda X'X \qquad \text{...(3)}$$

Taking transpose of both sides of (3), we get

$$(X'AX)' = (\lambda X'X)' \Rightarrow X'A'(X')' = \bar{\lambda} X'(X')'$$

$\Rightarrow$ $$X'A'X = \bar{\lambda} X'X \qquad [\because (X')' = X]$$

$\Rightarrow$ $$X'AX = \bar{\lambda} X'X \qquad [\because A' = A]$$

$\Rightarrow$ $$X'\lambda X = \bar{\lambda} X'X \qquad [\because AX = \lambda X]$$

$\Rightarrow$ $$\lambda X'X = \bar{\lambda} X'X$$

$\Rightarrow$ $$(\lambda - \bar{\lambda}) X'X = O \Rightarrow (\lambda - \bar{\lambda}) = 0 \qquad [\because X \neq O \Rightarrow X'X \neq O]$$

$$\Rightarrow \lambda = \bar{\lambda} \Rightarrow \lambda \text{ is real.}$$

THEOREM 3. *The eigenvalues of a skew-Hermitian matrix are either purely imaginary or zero.*

Proof. Let A be a skew-Hermitian matrix, then

$$A^\theta = -A$$

Now $$(iA)^\theta = -iA^\theta \Rightarrow (iA)^\theta = iA$$

$\Rightarrow$ iA is a Hermitian matrix

Let λ be an eigenvalue of A and X be its corresponding eigenvector, then

$$AX = \lambda X \Rightarrow iAX = i\lambda X$$

$\therefore$ $i\lambda$ is an eigenvalue of a Hermitian matrix. By theorem 1, we can say that $i\lambda$ is real. It follows that either λ is purely real or zero.

Corollary. *The eigenvalues of a real skew-symmetric matrix are either purely imaginary or zero.*

Proof. If the elements of a skew-Hermitian matrix are all real, then it is a real skew-symmetric. Therefore, a real skew-symmetric matrix is skew-Hermitian matrix, hence the result follows from theorem 3.

Theorem 4. *The eigenvalues of a unitary matrix are of unit modulus.* (UKTU-2010, 12)

Proof. Let A be a unitary matrix, then

$$A^\theta A = I \qquad \text{...(1)}$$

Let λ be an eigenvalue of A and let X be its corresponding eigenvector, then

$$AX = \lambda X \quad ...(2)$$

Taking conjugate transpose of both sides of (2), we have

$$(AX)^\theta = (\lambda X)^\theta$$

or $$X^\theta A^\theta = \bar{\lambda} X^\theta \quad ...(3)$$

Now $(X^\theta A^\theta)(AX) = (\bar{\lambda} X^\theta)(\lambda X)$ [Using (2) and (3)]

$\Rightarrow \quad X^\theta (A^\theta A) X = \bar{\lambda}\lambda X^\theta X$

$\Rightarrow \quad X^\theta I X = \bar{\lambda}\lambda X^\theta X$ [Using (1)]

$\Rightarrow \quad X^\theta X = \bar{\lambda}\lambda X^\theta X$

$\Rightarrow \quad (1 - \bar{\lambda}\lambda) X^\theta X = O$

$\Rightarrow \quad \bar{\lambda}\lambda = 1 \quad [\because X \neq O \Rightarrow X^\theta X \neq O]$

$\Rightarrow \quad |\lambda|^2 = 1 \quad \Rightarrow \quad |\lambda| = 1$

Corollary *The eigenvalues of an orthogonal matrix are of unit modulus.*

Proof. We know that if the elements of a unitary matrix are all real, then it is an orthogonal matrix, therefore an orthogonal matrix is a unitary matrix hence, the result follows from theorem 4.

Recapitulations

- $|A - \lambda I| = 0$ is known as characteristic equation and the roots of this equation are called characteristic roots or
- The value of λ for which a non-zero vector, *i.e.*, $X \neq O$ setisfies $AX = \lambda X$ is called an eigenvalues of matrix A and the non-zero vector X is called an eiganvector of A corresponding to that eigenvalue λ.
- λ is an eigenvalue of a matrix A if and only if there exis a non-zero vector X such that $AX = \lambda X$

WORKING PROCEDURE

To find the eigenvalue and eigenvectors

Let A be an $n \times n$ matrix, then it will have n eigenvalues. In order to find the eigenvalues and eigenvectors of A, we use the following steps :

Step 1. *Find the roots of the characteristic equation $|A - \lambda I| = 0$, the roots of λ give the eigenvalues of A.*

Step 2. *Let $X = \begin{bmatrix} x_1 \\ x_2 \\ \vdots \\ x_n \end{bmatrix} \neq O$ be an eigenvector of A corresponding to an eigenvalue λ_1 (say).*

Then X can be determined from the equation $(A - \lambda_1 I) X = O$

which is a system of n homogeneous equations in $x_1, x_2, ..., x_n$. If the rank of $(A - \lambda_1 I)$ is r, then the number of linearly independent solutions is $n - r$.

Solved Examples

Example 1. *If λ is a non-zero eigenvalue of a matrix A, then show that $\frac{1}{\lambda}$ is an eigenvalue of A^{-1}.*

Solution. Let $X \neq O$ be an eigenvector corresponding to the eigenvalue λ of A, then

$$AX = \lambda X$$

$\Rightarrow \quad A^{-1}(AX) = A^{-1}(\lambda X) \quad [\because A^{-1} \text{ exists.}]$

$\Rightarrow \quad (A^{-1}A)X = \lambda(A^{-1}X)$

$\Rightarrow \quad IX = \lambda(A^{-1}X) \quad [\because A^{-1}A = I]$

$\Rightarrow \quad X = \lambda(A^{-1}X) \quad [\because IX = X]$

$\Rightarrow \quad A^{-1}X = \left(\frac{1}{\lambda}\right)X$

Hence, $\frac{1}{\lambda}$ is an eigenvalue of A^{-1}.

Example 2. *Let A be an $n \times n$ matrix. Then show that zero is an eigenvalue of A iff A is singular.*

Solution. Let $X \neq O$ be an eigenvector corresponding to the eigenvalue 0 of A, then

$$AX = 0X = O \quad ...(1)$$

Since (1) represents a system of homogeneous equations, it will have non-zero solution if and only if $\rho(A) < n$

i.e., $\quad$ iff $|A| = 0$

i.e., iff A is singular.

Example 3. *If $\lambda_1, \lambda_2, ..., \lambda_n$ are the eigenvalues of A, then show that $k\lambda_1, k\lambda_2, ..., k\lambda_n$ are eigenvalues of kA, where k is any number.*

Solution. If $k = 0$, then $kA = 0A = O$. Since each eigenvalue of a zero matrix is zero, therefore $0\lambda_1, 0\lambda_2, ..., 0\lambda_n$ are the eigenvalues of kA if $\lambda_1, \lambda_2, ..., \lambda_n$ are eigenvalues of A.

Next, suppose that $k \neq 0$, then we have

$$|kA - k\lambda I| = k^n |A - \lambda I|$$

Now $|kA - k\lambda I| = 0$ iff $|A - \lambda I| = 0$

It follows that $k\lambda$ is an eigenvalue of kA.

Hence, if $\lambda_1, \lambda_2, ..., \lambda_n$ are the eigenvalues of A then $k\lambda_1, k\lambda_2, ..., k\lambda_n$ are the eigenvalues of kA.

Example 4. *If X be a non-zero eigenvector of an $n \times n$ matrix A, then prove that for each positive integer n, X is an eigenvector of A^n corresponding to the eigenvalue λ^n.*

Solution. Since $X \neq O$ is an eigenvector corresponding eigenvalue λ of A, then we have

$$AX = \lambda X \quad ...(1)$$

Now we have to show that $A^n X = \lambda^n X$.

We shall prove this by induction on n.

If $n = 1$, then the result is true by virtue of (1).

Suppose that the result is true for $n = k$, then we have

$$A^k X = \lambda^k X \quad ...(2)$$

Now
$$\begin{aligned} A^{k+1}X &= (A^k A)X \\ &= A^k(AX) \\ &= A^k(\lambda X) \quad \text{[Using (1)]} \\ &= \lambda(A^k X) \\ &= \lambda(\lambda^k X) \quad \text{[Using (2)]} \\ &= \lambda^{k+1} X \\ A^{k+1}X &= \lambda^{k+1}X \end{aligned}$$

Thus, the result is true for $n = k+1$.

Hence by induction the result is true for all positive integers n.

Example 5. *Show that similar matrices have the same eigenvalues.*

Solution. Two matrices A and B of the same order are said to be similar if there exists a non-singular matrix P such that

$$B = P^{-1}AB$$

Let λ be an eigenvalue of A, then X is a root of $|A - \lambda I| = 0$.

Now
$$\begin{aligned} B - \lambda I &= P^{-1}AP - \lambda I \\ &= P^{-1}AP - P^{-1}(\lambda I)P \\ & [\because P^{-1}(\lambda I)P = \lambda P^{-1}P = \lambda I] \\ &= P^{-1}(A - \lambda I)P \end{aligned}$$

$\Rightarrow \quad |B - \lambda I| = |P^{-1}|\,|A - \lambda I|\,|P|$

$\Rightarrow \quad |B - \lambda I| = |A - \lambda I|\,|P^{-1}|\,|P|$

$\Rightarrow \quad |B - \lambda I| = |A - \lambda I|\,|P^{-1}P|$

$\Rightarrow \quad |B - \lambda I| = |A - \lambda I|$

$[\because |P^{-1}P| = |I| = 1]$

Since λ is a root of $|A - \lambda I| = 0$, therefore λ is also a root of $|B - \lambda I| = 0$, it follows that λ is an eigenvalue of B.

Hence, similar matrices have the same eigenvalues.

Example 6. *Let A and B be two matrices of order $n \times n$. Let $X \neq O$ be an eigenvector of A and B corresponding to the eigenvalues λ_1 and λ_2 respectively, then show that X is an eigenvector of AB corresponding to the eigenvalue $\lambda_1\lambda_2$ of AB.*

Solution. Since $X \neq O$ is an eigenvector of A and B corresponding to the eigenvalues λ_1 and λ_2 respectively, then we have

$$AX = \lambda_1 X \quad ...(1)$$

and
$$BX = \lambda_2 X \quad ...(2)$$

Now
$$\begin{aligned} (AB)X &= A(BX) \\ &= A(\lambda_2 X) \quad \text{[Using (2)]} \\ &= \lambda_2(AX) \\ &= \lambda_2(\lambda_1 X) \quad \text{[Using (1)]} \\ &= (\lambda_2\lambda_1)X \\ (AB)X &= (\lambda_1\lambda_2)X \text{ with } X \neq O \end{aligned}$$

It follows that X is an eigenvector of AB corresponding to the eigenvalue $\lambda_1\lambda_2$.

Example 7. *Determine the eigenvalues of the matrix :*

$$A = \begin{bmatrix} 1 & 2 & 3 \\ 0 & -4 & 2 \\ 0 & 0 & 7 \end{bmatrix}$$

Solution. The characteristic equation of A is given by

$$|A - \lambda I| = 0$$

i.e.,
$$\begin{vmatrix} 1-\lambda & 2 & 3 \\ 0 & -4-\lambda & 2 \\ 0 & 0 & 7-\lambda \end{vmatrix} = 0$$

i.e.,
$$(1-\lambda)(-2-\lambda)(7-\lambda) = 0$$

The roots of this characteristic equation are given by $\lambda = 1, -4, 7$.

These are the required eigenvalues of A.

REMARK

- It is clear that the given matrix A is an upper triangular matrix so that the principal diagonal elements 1, – 4, 7 will be the eigenvalues of A.

Example 8. *Determine the eigenvalues of the matrix :*

$$A = \begin{bmatrix} 0 & 1 & 2 \\ 1 & 0 & -1 \\ 2 & -1 & 0 \end{bmatrix}.$$

Solution. The characteristic equation of A is given

$$|A - \lambda I| = 0$$

$$\begin{vmatrix} 0-\lambda & 1 & 2 \\ 1 & 0-\lambda & -1 \\ 2 & -1 & 0-\lambda \end{vmatrix} = 0$$

or
$$-\lambda(\lambda^2 - 1) - 1(-\lambda + 2) + 2(-1 + 2\lambda) = 0$$

or
$$-\lambda^3 + 6\lambda - 4 = 0$$

Solving this equation, we get

$$(\lambda-2)(\lambda^2+2\lambda-2)=0$$

$\Rightarrow$ $\lambda = 2$ and $\lambda = -1 \pm \sqrt{3}$

Hence, the eigenvalues of A are $2, -1 \pm \sqrt{3}$.

Example 9. *Determine the eigenvalues and eigenvectors of the matrix*

$$A = \begin{bmatrix} 5 & 4 \\ 1 & 2 \end{bmatrix}.$$ (Bhopal -2008)

Solution. The characteristic equation of A is given by

$$|A-\lambda I| = 0$$

or $$\begin{vmatrix} 5-\lambda & 4 \\ 1 & 2-\lambda \end{vmatrix} = 0$$

or $$(5-\lambda)(2-\lambda)-4=0$$

or $$\lambda^2-7\lambda+10-4=0$$

or $$\lambda^2-7\lambda+6=0$$

The roots of this equation are $\lambda = 6, 1$.

Thus, the eigenvalues of A are 6, 1.

Eigenvector corresponding to $\lambda_1 = 6$:

Let $X_1 = \begin{bmatrix} x_1 \\ x_2 \end{bmatrix} \neq O$ be an eigenvector of A corresponding to $\lambda_1 = 6$, then we have

$$AX_1 = 6X_1$$

or $$(A-6I)X_1 = O$$

or $$\begin{bmatrix} 5-6 & 4 \\ 1 & 2-6 \end{bmatrix}\begin{bmatrix} x_1 \\ x_2 \end{bmatrix} = \begin{bmatrix} 0 \\ 0 \end{bmatrix}$$

or $$\begin{bmatrix} -1 & 4 \\ 1 & -4 \end{bmatrix}\begin{bmatrix} x_1 \\ x_2 \end{bmatrix} = \begin{bmatrix} 0 \\ 0 \end{bmatrix} \quad \ldots(1)$$

The non-zero solution of (1) will give X_1.

Applying $R_2 \to R_2 + R_1$, we have

$$\begin{bmatrix} -1 & 4 \\ 1 & -4 \end{bmatrix}\begin{bmatrix} x_1 \\ x_2 \end{bmatrix} = \begin{bmatrix} 0 \\ 0 \end{bmatrix} \quad \ldots(2)$$

The coefficient matrix of equation (1) is of rank 1, *i.e.*, $\rho(A-6I) = 1$, therefore the system of equations (1) will have $2 - 1 = 1$ linearly independent solution.

From (2), we have

$$-x_1 + 4x_2 = 0$$

Clearly, $x_1 = 4$ and $x_2 = 1$ satisfy the above equation.

Hence, the eigenvector corresponding to eigenvalue $\lambda_1 = 6$ is

$$X_1 = \begin{bmatrix} 4 \\ 1 \end{bmatrix}$$

Eigenvector corresponding to $\lambda_2 = 1$:

Let $X_2 = \begin{bmatrix} x_1 \\ x_2 \end{bmatrix} \neq O$ be an eigenvector of A corresponding to eigenvalue $\lambda_2 = 1$, then we have

$$AX_2 = \lambda_2 X_2$$

or $$AX_2 = IX_2$$

or $$(A-I)X_2 = O$$

or $$\begin{bmatrix} 5-1 & 4 \\ 1 & 2-1 \end{bmatrix}\begin{bmatrix} x_1 \\ x_2 \end{bmatrix} = \begin{bmatrix} 0 \\ 0 \end{bmatrix}$$

or $$\begin{bmatrix} 4 & 4 \\ 1 & 1 \end{bmatrix}\begin{bmatrix} x_1 \\ x_2 \end{bmatrix} = \begin{bmatrix} 0 \\ 0 \end{bmatrix} \quad \ldots(3)$$

The non-zero solution of (3) will give X_2.

Applying $R_2 \to R_2 - \frac{1}{4}R_1$, we get

$$\begin{bmatrix} 4 & 4 \\ 0 & 0 \end{bmatrix}\begin{bmatrix} x_1 \\ x_2 \end{bmatrix} = \begin{bmatrix} 0 \\ 0 \end{bmatrix} \quad \ldots(4)$$

Clearly, $\rho(A - I) = 1$, therefore the system of equations (3) will have $2 - 1 = 1$ linearly independent solution.

From (4), we get

$$4x_1 + 4x_2 = 0$$

Clearly, $x_1 = 1$ and $x_2 = -1$, satisfy above equation.

Hence, the eigenvector corresponding to eigenvalue $\lambda_2 = 1$ is

$$X_2 = \begin{bmatrix} 1 \\ -1 \end{bmatrix}.$$

Example 10. *Determine the eigenvalues and eigenvectors of the matrix*

$$A = \begin{bmatrix} 8 & -6 & 2 \\ -6 & 7 & -4 \\ 2 & -4 & 3 \end{bmatrix}.$$ (GBTU-2011)

Solution. The characteristic equation of A is given by

$$|A-\lambda I| = 0$$

or $$\begin{vmatrix} 8-\lambda & -6 & 2 \\ -6 & 7-\lambda & -4 \\ 2 & -4 & 3-\lambda \end{vmatrix} = 0$$

or $$(8-\lambda)((7-\lambda)(3-\lambda)-16) + 6(-18+6\lambda+8) + 2(24-14+2\lambda) = 0$$

or $$\lambda^3 - 18\lambda^2 + 45\lambda = 0$$

or $$\lambda(\lambda-3)(\lambda-15) = 0$$

The roots of this equation are $\lambda = 0, 3, 15$.

Thus, the eigenvalues of A are

$$\lambda_1 = 0, \lambda_2 = 3, \lambda_3 = 15.$$

Eigenvector corresponding to $\lambda_1 = 0$:

Let $X_1 = \begin{bmatrix} x_1 \\ x_2 \\ x_3 \end{bmatrix} \neq O$ be an eigenvector corresponding to the eigenvalue $\lambda_1 = 0$, then we have

$$AX_1 = \lambda_1 X1.$$

or $$AX_1 = 0X_1$$

or $$(A - 0I)\, X_1 = O$$

or $$\begin{bmatrix} 8 & -6 & 2 \\ -6 & 7 & -4 \\ 2 & -4 & 3 \end{bmatrix}\begin{bmatrix} x_1 \\ x_2 \\ x_3 \end{bmatrix} = \begin{bmatrix} 0 \\ 0 \\ 0 \end{bmatrix} \qquad \ldots(1)$$

The non-zero solution of (1) will give X_1.

Reducing the coefficient matrix of (1) in Echeleon form by applying elementary row transformations.

Applying $R_1 \leftrightarrow R_3$, we get

$$\begin{bmatrix} 2 & -4 & 3 \\ -6 & 7 & -4 \\ 8 & -6 & 2 \end{bmatrix}\begin{bmatrix} x_1 \\ x_2 \\ x_3 \end{bmatrix} = \begin{bmatrix} 0 \\ 0 \\ 0 \end{bmatrix}$$

Applying $R_2 \to R_2 + 3R_1$, $R_3 \to R_3 - 4R_1$, we get

$$\begin{bmatrix} 2 & -4 & 3 \\ 0 & -5 & 5 \\ 0 & 10 & -10 \end{bmatrix}\begin{bmatrix} x_1 \\ x_2 \\ x_3 \end{bmatrix} = \begin{bmatrix} 0 \\ 0 \\ 0 \end{bmatrix}$$

Applying $R_3 \to R_3 + 2R_2$, we get

$$\begin{bmatrix} 2 & -4 & 3 \\ 0 & -5 & 5 \\ 0 & 0 & 0 \end{bmatrix}\begin{bmatrix} x_1 \\ x_2 \\ x_3 \end{bmatrix} = \begin{bmatrix} 0 \\ 0 \\ 0 \end{bmatrix} \qquad \ldots(2)$$

Clearly $\rho(A - 0.I) = 2$, therefore the system of equations (2) will have $3 - 2 = 1$ (unknowns – rank) linearly independent solution.

From (2), we have

$$2x_1 - 4x_2 + 3x_3 = 0$$
$$-5x_2 + 5x_3 = 0$$

Clearly, $x_1 = \dfrac{1}{2}$, $x_2 = 1$ and $x_3 = 1$ satisfy the above equations.

Hence, the eigenvector corresponding to eigenvalue $\lambda_1 = 0$ is

$$X_1 = \begin{bmatrix} 1/2 \\ 1 \\ 1 \end{bmatrix}$$

Eigenvector corresponding to $\lambda_2 = 3$:

Let $X_2 = \begin{bmatrix} x_1 \\ x_2 \\ x_3 \end{bmatrix} \neq O$ be an eigenvector of A corresponding to $\lambda_2 = 3$, then we have

$$AX_2 = \lambda_2 X_2$$

or $$(A - \lambda_2 I)X_2 = O$$

or $$(A - 3I)X_2 = O$$

or $$\begin{bmatrix} 8-3 & -6 & 2 \\ -6 & 7-3 & -4 \\ 2 & -4 & 3-3 \end{bmatrix}\begin{bmatrix} x_1 \\ x_2 \\ x_3 \end{bmatrix} = O$$

or $$\begin{bmatrix} 5 & -6 & 2 \\ -6 & 4 & -4 \\ 2 & -4 & 0 \end{bmatrix}\begin{bmatrix} x_1 \\ x_2 \\ x_3 \end{bmatrix} = O \qquad \ldots(3)$$

The non-zero solution of (3) will give X_2.

Applying $R_1 \to R_1 + R_2$, we get

$$\begin{bmatrix} -1 & -2 & -2 \\ -6 & 4 & -4 \\ 2 & -4 & 0 \end{bmatrix}\begin{bmatrix} x_1 \\ x_2 \\ x_3 \end{bmatrix} = \begin{bmatrix} 0 \\ 0 \\ 0 \end{bmatrix}$$

Applying $R_2 \to R_2 - 6R_1$, $R_3 \to R_3 + 2R_1$, we get

$$\begin{bmatrix} -1 & -2 & -2 \\ 0 & 16 & 8 \\ 0 & -8 & -4 \end{bmatrix}\begin{bmatrix} x_1 \\ x_2 \\ x_3 \end{bmatrix} = \begin{bmatrix} 0 \\ 0 \\ 0 \end{bmatrix}$$

Applying $R_2 \to \dfrac{1}{8} R_2$, we get

$$\begin{bmatrix} -1 & -2 & -2 \\ 0 & 2 & 1 \\ 0 & -8 & -4 \end{bmatrix}\begin{bmatrix} x_1 \\ x_2 \\ x_3 \end{bmatrix} = \begin{bmatrix} 0 \\ 0 \\ 0 \end{bmatrix}$$

Again applying $R_3 \to R_3 + 4R_2$, we get

$$\begin{bmatrix} -1 & -2 & -2 \\ 0 & 2 & 1 \\ 0 & 0 & 0 \end{bmatrix}\begin{bmatrix} x_1 \\ x_2 \\ x_3 \end{bmatrix} = \begin{bmatrix} 0 \\ 0 \\ 0 \end{bmatrix} \qquad \ldots(4)$$

Clearly $\rho(A - 3I) = 2$, therefore the system of equations (3) will have $3 - 2 = 1$ linearly independent solution.

From (4), we have

$$-x_1 - 2x_2 - 2x_3 = 0$$
$$2x_2 + x_3 = 0$$

Clearly, $x_1 = -2$, $x_2 = -1$ and $x_3 = 2$ satisfy the above equations.

Hence, the eigenvector corresponding to eigenvalue $\lambda_2 = 3$ is

$$X_2 = \begin{bmatrix} -2 \\ -1 \\ 2 \end{bmatrix}$$

Eigenvector corresponding to $\lambda_3 = 15$:

Let $X_3 = \begin{bmatrix} x_1 \\ x_2 \\ x_3 \end{bmatrix} \neq O$ be an eigenvector of A corresponding to $\lambda_3 = 15$, then we have

$$AX_3 = \lambda_3 X_3$$

or $$(A - \lambda_3 I)X_3 = O$$

or $(A-15I)X_3 = O$

or $\begin{bmatrix} 8-15 & -6 & 2 \\ -6 & 7-15 & -4 \\ 2 & -4 & 3-15 \end{bmatrix}\begin{bmatrix} x_1 \\ x_2 \\ x_3 \end{bmatrix} = O$

or $\begin{bmatrix} -7 & -6 & 2 \\ -6 & -8 & -4 \\ 2 & -4 & -12 \end{bmatrix}\begin{bmatrix} x_1 \\ x_2 \\ x_3 \end{bmatrix} = O$ ...(5)

The non-zero solution of (5) will give X_3.

Applying $R_1 \leftrightarrow R_3$, we get

$$\begin{bmatrix} 2 & -4 & -12 \\ -6 & -8 & -4 \\ -7 & -6 & 2 \end{bmatrix}\begin{bmatrix} x_1 \\ x_2 \\ x_3 \end{bmatrix} = \begin{bmatrix} 0 \\ 0 \\ 0 \end{bmatrix}$$

Applying $R_1 \to \frac{1}{2}R_1$, we get

$$\begin{bmatrix} 1 & -2 & -6 \\ -6 & -8 & -4 \\ -7 & -6 & 2 \end{bmatrix}\begin{bmatrix} x_1 \\ x_2 \\ x_3 \end{bmatrix} = \begin{bmatrix} 0 \\ 0 \\ 0 \end{bmatrix}$$

Applying $R_2 \to R_2 + 6R_1, R_3 \to R_3 + 7R_1$, we get

$$\begin{bmatrix} 1 & -2 & -6 \\ 0 & -20 & -40 \\ 0 & -20 & -40 \end{bmatrix}\begin{bmatrix} x_1 \\ x_2 \\ x_3 \end{bmatrix} = \begin{bmatrix} 0 \\ 0 \\ 0 \end{bmatrix}$$

Applying $R_3 \to R_3 - R_2$, we get

$$\begin{bmatrix} 1 & -2 & -6 \\ 0 & -20 & -40 \\ 0 & 0 & 0 \end{bmatrix}\begin{bmatrix} x_1 \\ x_2 \\ x_3 \end{bmatrix} = \begin{bmatrix} 0 \\ 0 \\ 0 \end{bmatrix} \quad \text{...(6)}$$

Clearly $\rho(A - 15I) = 2$, therefore the system of equations (5) will have 3 – 2= 1 linearly independent solution.

From (6), we have

$$x_1 - 2x_2 - 6x_3 = 0$$
$$-20x_2 - 40x_3 = 0$$

Clearly, $x_1 = 2, x_2 = -2$ and $x_3 = 1$ satisfy the above equations.

Hence, the eigenvector corresponding to eigenvalue $\lambda_3 = 15$ is

$$X_3 = \begin{bmatrix} 2 \\ -2 \\ 1 \end{bmatrix}.$$

Example 11. *Determine the eigenvalues and eigenvectors of the matrix*

$$A = \begin{bmatrix} 2 & 1 & 0 \\ 0 & 1 & -1 \\ 0 & 2 & 4 \end{bmatrix}.$$

Solution. The characteristic equation of A is given by

$$|A - \lambda I| = 0$$

or $\begin{vmatrix} 2-\lambda & 1 & 0 \\ 0 & 1-\lambda & -1 \\ 0 & 2 & 4-\lambda \end{vmatrix} = 0$

or $(2-\lambda)\{(1-\lambda)(4-\lambda)+2\} = 0$

or $(2-\lambda)(\lambda^3 - 5\lambda + 6) = 0$

or $(2-\lambda)(2-\lambda)(3-\lambda) = 0$

The roots of this equation are $\lambda = 2, 2, 3$.

Thus the eigenvalues of A are $\lambda_1 = 2, \lambda_2 = 2, \lambda_3 = 3$.

Eigenvector corresponding to the eigenvalue $\lambda_1 = \lambda_2 = 2$:

Let $X_1 = \begin{bmatrix} x_1 \\ x_2 \\ x_3 \end{bmatrix} \neq O$ be an eigenvector corresponding to the eigenvalue 2, then we have

$$AX_1 = 2X_1.$$

or $(A - 2I)\, X_1 = O$

or $\begin{bmatrix} 2-2 & 1 & 0 \\ 0 & 1-2 & -1 \\ 0 & 2 & 4-2 \end{bmatrix}\begin{bmatrix} x_1 \\ x_2 \\ x_3 \end{bmatrix} = \begin{bmatrix} 0 \\ 0 \\ 0 \end{bmatrix}$

or $\begin{bmatrix} 0 & 1 & 0 \\ 0 & -1 & -1 \\ 0 & 2 & 2 \end{bmatrix}\begin{bmatrix} x_1 \\ x_2 \\ x_3 \end{bmatrix} = \begin{bmatrix} 0 \\ 0 \\ 0 \end{bmatrix}$ ...(1)

The non-zero solution of (1) will give X_1.

Applying $R_3 \to R_3 + 2R_2$, we get

$$\begin{bmatrix} 0 & 1 & 0 \\ 0 & -1 & -1 \\ 0 & 0 & 0 \end{bmatrix}\begin{bmatrix} x_1 \\ x_2 \\ x_3 \end{bmatrix} = \begin{bmatrix} 0 \\ 0 \\ 0 \end{bmatrix}$$

Applying $R_2 \to R_2 + R_1$, we get

$$\begin{bmatrix} 0 & 1 & 0 \\ 0 & 0 & -1 \\ 0 & 0 & 0 \end{bmatrix}\begin{bmatrix} x_1 \\ x_2 \\ x_3 \end{bmatrix} = \begin{bmatrix} 0 \\ 0 \\ 0 \end{bmatrix} \quad \text{...(2)}$$

From (2), it is clear that $\rho(A - 2I) = 2$, therefore the system of equations (1) will have 3 – 2 = 1 linearly independent solution.

From (2), we have

$$x_2 = 0 \text{ and } -x_3 = 0$$

Clearly, $x_1 = 1, x_2 = 0, x_3 = 1$ satisfy the above equations.

Hence, the eigenvector corresponding to the eigenvalue $\lambda_1 = \lambda_2 = 2$ is

$$X_1 = \begin{bmatrix} 1 \\ 0 \\ 0 \end{bmatrix}$$

Eigenvector corresponding to the eigenvalue $\lambda_3 = 3$:

Let $X_2 = \begin{bmatrix} x_1 \\ x_2 \\ x_3 \end{bmatrix} \neq O$ be an eigenvector

corresponding to $\lambda_3 = 3$, then we have

$$AX_2 = 3X_2$$

or $$(A - 3I)X_3 = O$$

or $$\begin{bmatrix} 2-3 & 1 & 0 \\ 0 & 1-3 & -1 \\ 0 & 2 & 4-3 \end{bmatrix} \begin{bmatrix} x_1 \\ x_2 \\ x_3 \end{bmatrix} = \begin{bmatrix} 0 \\ 0 \\ 0 \end{bmatrix}$$

or $$\begin{bmatrix} -1 & 1 & 0 \\ 0 & -2 & -1 \\ 0 & 2 & 1 \end{bmatrix} \begin{bmatrix} x_1 \\ x_2 \\ x_3 \end{bmatrix} = \begin{bmatrix} 0 \\ 0 \\ 0 \end{bmatrix} \qquad ...(3)$$

The non-zero solution of (3) will give X_2.

Applying $R_3 \to R_3 + R_2$, we get

$$\begin{bmatrix} -1 & 1 & 0 \\ 0 & -2 & -1 \\ 0 & 0 & 0 \end{bmatrix} \begin{bmatrix} x_1 \\ x_2 \\ x_3 \end{bmatrix} = \begin{bmatrix} 0 \\ 0 \\ 0 \end{bmatrix} \qquad ...(4)$$

Clearly $\rho(A - 3I) = 2$, therefore the system of equations (3) will have 3 – 2= 1 linearly independent solution.

From (4), we have

$$-x_1 + x_2 = 0$$
$$-2x_2 - x_3 = 0$$

Clearly, $x_1 = 1$, $x_2 = 1$ and $x_3 = -2$ satisfy the above equations. Hence, the eigenvector corresponding to eigenvalue $\lambda_3 = 3$ is

$$X_2 = \begin{bmatrix} 1 \\ 1 \\ -2 \end{bmatrix}.$$

Example 12. *Find the eigenvalues and eigenvectors of the matrix*

$$A = \begin{bmatrix} 5 & 4 & 2 \\ 4 & 5 & 2 \\ 2 & 2 & 2 \end{bmatrix}.$$

Solution. The characteristic equation of A is given by

$$|A - \lambda I| = 0$$

or $$\begin{vmatrix} 5-\lambda & 4 & 2 \\ 4 & 5-\lambda & 2 \\ 2 & 2 & 2-\lambda \end{vmatrix} = 0$$

or $$(5-\lambda)\{(5-\lambda)(2-\lambda)-4\} - 4\{4(2-\lambda)-4\} + 2\{8-2(8-\lambda)\} = 0$$

or $$-\lambda^3 + 12\lambda - 21\lambda + 10 = 0$$

or $$-(\lambda-1)^2(\lambda-10) = 0$$

The roots of this equation are 1, 1, 10.

Thus the eigenvalues of A are

$$\lambda_1 = 1, \lambda_2 = 1, \lambda_3 = 10.$$

Eigenvector corresponding to the eigenvalue $\lambda_1 = \lambda_2 = 1$:

Let $X = \begin{bmatrix} x_1 \\ x_2 \\ x_3 \end{bmatrix} \neq O$ be an eigenvector of A

corresponding to the eigenvalue $\lambda_1 = \lambda_2 = 1$, then

we have $AX_1 = IX$ or $(A - I)X_1 = O$

or $$\begin{bmatrix} 5-1 & 4 & 2 \\ 4 & 5-1 & 2 \\ 2 & 2 & 2-1 \end{bmatrix} \begin{bmatrix} x_1 \\ x_2 \\ x_3 \end{bmatrix} = \begin{bmatrix} 0 \\ 0 \\ 0 \end{bmatrix}$$

or $$\begin{bmatrix} 4 & 4 & 2 \\ 4 & 4 & 2 \\ 2 & 2 & 2 \end{bmatrix} \begin{bmatrix} x_1 \\ x_2 \\ x_3 \end{bmatrix} = \begin{bmatrix} 0 \\ 0 \\ 0 \end{bmatrix} \quad ..(1)$$

Applying $R_1 \leftrightarrow R_3$, we get

$$\begin{bmatrix} 2 & 2 & 1 \\ 4 & 4 & 2 \\ 4 & 4 & 2 \end{bmatrix} \begin{bmatrix} x_1 \\ x_2 \\ x_3 \end{bmatrix} = \begin{bmatrix} 0 \\ 0 \\ 0 \end{bmatrix}$$

Applying $R_2 \to R_2 - 2R_1$, we get

$$\begin{bmatrix} 2 & 2 & 1 \\ 0 & 0 & 0 \\ 0 & 0 & 0 \end{bmatrix} \begin{bmatrix} x_1 \\ x_2 \\ x_3 \end{bmatrix} = \begin{bmatrix} 0 \\ 0 \\ 0 \end{bmatrix} \qquad ...(2)$$

From (2), it is clear that $\rho(A - I) = 1$, therefore the system of equation (1) will have 3 – 2 = 1 linearly independent solution.

Now from (2), we have

$$x_2 = 0 \text{ and } -x_3 = 0$$

Clearly $x_1 = 0$, $x_2 = 0$, $x_3 = 0$ satisfy the above equations.

Hence, the eigenvector corresponding to the eigenvalue $\lambda_1 = \lambda_2 = 2$ is

$$X_1 = \begin{bmatrix} 1 \\ 0 \\ 0 \end{bmatrix}.$$

Eigenvector corresponding to the eigenvalue $\lambda_1 = \lambda_2 = 1$

Let $X = \begin{bmatrix} x_1 \\ x_2 \\ x_3 \end{bmatrix} \neq 0$ be an eigenvector corresponding to the eigenvalue $\lambda_1 = \lambda_2 = 1$, then we have

$$AX = IX$$

or $$(A - I)X = O$$

or $$\begin{bmatrix} 5-1 & 4 & 2 \\ 4 & 5-1 & 2 \\ 2 & 2 & 2-1 \end{bmatrix}\begin{bmatrix} x_1 \\ x_2 \\ x_3 \end{bmatrix} = \begin{bmatrix} 0 \\ 0 \\ 0 \end{bmatrix}$$

or $$\begin{bmatrix} 4 & 4 & 2 \\ 4 & 4 & 2 \\ 2 & 2 & 1 \end{bmatrix}\begin{bmatrix} x_1 \\ x_2 \\ x_3 \end{bmatrix} = \begin{bmatrix} 0 \\ 0 \\ 0 \end{bmatrix} \quad ...(1)$$

Applying $R_1 \rightarrow R_3$, we get

$$\begin{bmatrix} 2 & 2 & 1 \\ 4 & 4 & 2 \\ 4 & 4 & 2 \end{bmatrix}\begin{bmatrix} x_1 \\ x_2 \\ x_3 \end{bmatrix} = \begin{bmatrix} 0 \\ 0 \\ 0 \end{bmatrix}$$

Applying $R_2 \rightarrow R_2 - 2R_1$, $R_3 \rightarrow R_3 - 2R_1$, we get

$$\begin{bmatrix} 2 & 2 & 1 \\ 0 & 0 & 0 \\ 0 & 0 & 0 \end{bmatrix}\begin{bmatrix} x_1 \\ x_2 \\ x_3 \end{bmatrix} = \begin{bmatrix} 0 \\ 0 \\ 0 \end{bmatrix} \quad ...(2)$$

From (2), it is clear that $\rho(A - I) = 1$, therefore the equations (1) will have 3–1 =2 linearly independent solutions.

From (2), we have

$$2x_1 + 2x_2 + x_3 = 0$$

Since this equation has two linearly independent solutions so we take $x_2 = c_1$ and $x_3 = c_2$, where c_1 and c_2 are non-zero scalars, then $x_1 = -c - \frac{c_2}{2}$.

Therefore,

$$\begin{bmatrix} x_1 \\ x_2 \\ x_3 \end{bmatrix} = \begin{bmatrix} -c_1 - \frac{c_2}{2} \\ c_1 \\ c_2 \end{bmatrix} = c_1\begin{bmatrix} -1 \\ 1 \\ 0 \end{bmatrix} + c_2\begin{bmatrix} -1/2 \\ 0 \\ 1 \end{bmatrix}$$

Hence, the eigenvectors corresponding to the eigenvalue $\lambda_1 = \lambda_2 = 1$ are

$$X_1 = \begin{bmatrix} -1 \\ 0 \\ 1 \end{bmatrix} \text{ and } X_2 = \begin{bmatrix} -1/2 \\ 0 \\ 1 \end{bmatrix}$$

Eigenvector corresponding to the eigenvalue $\lambda_3 = 10$:

Let $X_3 = \begin{bmatrix} x_1 \\ x_2 \\ x_3 \end{bmatrix} \neq O$ be an eigenvector corresponding to $\lambda_3 = 10$, then we have

$$AX_3 = 10X_3$$

or $$(A - 10I)X_3 = O$$

or $$\begin{bmatrix} 5-10 & 4 & 2 \\ 4 & 5-10 & 2 \\ 2 & 2 & 2-10 \end{bmatrix}\begin{bmatrix} x_1 \\ x_2 \\ x_3 \end{bmatrix} = \begin{bmatrix} 0 \\ 0 \\ 0 \end{bmatrix}$$

or $$\begin{bmatrix} -5 & 4 & 2 \\ 4 & -5 & 2 \\ 2 & 2 & -8 \end{bmatrix}\begin{bmatrix} x_1 \\ x_2 \\ x_3 \end{bmatrix} = \begin{bmatrix} 0 \\ 0 \\ 0 \end{bmatrix} \quad ...(3)$$

Applying $R_1 \rightarrow R_1 + R_2$, we get

$$\begin{bmatrix} -1 & -1 & 4 \\ 4 & -5 & 2 \\ 2 & 2 & -8 \end{bmatrix}\begin{bmatrix} x_1 \\ x_2 \\ x_3 \end{bmatrix} = \begin{bmatrix} 0 \\ 0 \\ 0 \end{bmatrix}$$

Applying $R_2 \rightarrow R_2 + 4R_1$, $R_3 \rightarrow R_3 + 2R_1$, we get

$$\begin{bmatrix} -1 & -1 & 4 \\ 0 & -9 & 18 \\ 0 & 0 & 0 \end{bmatrix}\begin{bmatrix} x_1 \\ x_2 \\ x_3 \end{bmatrix} = \begin{bmatrix} 0 \\ 0 \\ 0 \end{bmatrix} \quad ...(4)$$

From (4), it is clear that $\rho(A - 10I) = 2$, therefore the system of equations (3) will have 3 – 1= 2 linearly independent solutions.

From (4), we have

$$-x_1 - x_2 + 4x_3 = 0$$
$$-9x_2 + 18x_3 = 0$$

Let us take $x_3 = c$, then $x_2 = 2c$ and $x_1 = 2c$.

Therefore,

$$\begin{bmatrix} x_1 \\ x_2 \\ x_3 \end{bmatrix} = \begin{bmatrix} 2c \\ 2c \\ c \end{bmatrix} = c\begin{bmatrix} 2 \\ 2 \\ 1 \end{bmatrix}$$

Hence, the eigenvector corresponding to eigenvalue $\lambda_3 = 10$ is

$$X_3 = \begin{bmatrix} 2 \\ 2 \\ 1 \end{bmatrix}.$$

Example 13. *Find the eigenvalues and eigenvectors of the matrix*

$$A = \begin{bmatrix} 2 & 0 & 1 & -3 \\ 0 & 2 & 10 & 4 \\ 0 & 0 & 2 & 0 \\ 0 & 0 & 0 & 3 \end{bmatrix}.$$

Solution. The characteristic equation of A is given by

$$|A - \lambda I| = 0$$

or $$A = \begin{vmatrix} 2-\lambda & 0 & 1 & -3 \\ 0 & 2-\lambda & 10 & 4 \\ 0 & 0 & 2-\lambda & 0 \\ 0 & 0 & 0 & 3-\lambda \end{vmatrix} = 0$$

or $$(\lambda - 2)^2(\lambda - 3) = 0$$

The roots of characteristic equations are 2, 2, 2, 3.

Thus, the eigenvalues of A are $\lambda_1 = 2, \lambda_2 = 2, \lambda_3 = 2, \lambda_4 = 3$.

Eigenvector corresponding to $\lambda_1 = \lambda_2 = \lambda_3 = 2$:

Let $X = \begin{bmatrix} x_1 \\ x_2 \\ x_3 \\ x_4 \end{bmatrix} \neq O$ be an eigenvector of A corresponding to the eigenvalue 2, then we have

$$AX = 2X.$$

or $$(A - 2I)\, X = O$$

or $$\begin{bmatrix} 2-2 & 0 & 1 & -3 \\ 0 & 2-2 & 10 & 4 \\ 0 & 0 & 2-2 & 0 \\ 0 & 0 & 0 & 3-2 \end{bmatrix}\begin{bmatrix} x_1 \\ x_2 \\ x_3 \\ x_4 \end{bmatrix} = \begin{bmatrix} 0 \\ 0 \\ 0 \\ 0 \end{bmatrix}$$

or $$\begin{bmatrix} 0 & 0 & 1 & -3 \\ 0 & 0 & 10 & 4 \\ 0 & 0 & 0 & 0 \\ 0 & 0 & 0 & 1 \end{bmatrix}\begin{bmatrix} x_1 \\ x_2 \\ x_3 \\ x_4 \end{bmatrix} = \begin{bmatrix} 0 \\ 0 \\ 0 \\ 0 \end{bmatrix} \quad \ldots(1)$$

Applying $R_3 \leftrightarrow R_4$, we get

$$\begin{bmatrix} 0 & 0 & 1 & -3 \\ 0 & 0 & 10 & 4 \\ 0 & 0 & 0 & 1 \\ 0 & 0 & 0 & 0 \end{bmatrix}\begin{bmatrix} x_1 \\ x_2 \\ x_3 \\ x_4 \end{bmatrix} = \begin{bmatrix} 0 \\ 0 \\ 0 \\ 0 \end{bmatrix}$$

Applying $R_2 \to R_2 - 10R_1$, we get

$$\begin{bmatrix} 0 & 0 & 1 & -3 \\ 0 & 0 & 0 & 34 \\ 0 & 0 & 0 & 1 \\ 0 & 0 & 0 & 0 \end{bmatrix}\begin{bmatrix} x_1 \\ x_2 \\ x_3 \\ x_4 \end{bmatrix} = \begin{bmatrix} 0 \\ 0 \\ 0 \\ 0 \end{bmatrix}$$

Applying $R_2 \to \frac{1}{34} R_2$, we get

$$\begin{bmatrix} 0 & 0 & 1 & -3 \\ 0 & 0 & 0 & 1 \\ 0 & 0 & 0 & 1 \\ 0 & 0 & 0 & 0 \end{bmatrix}\begin{bmatrix} x_1 \\ x_2 \\ x_3 \\ x_4 \end{bmatrix} = \begin{bmatrix} 0 \\ 0 \\ 0 \\ 0 \end{bmatrix}$$

Applying $R_2 \to R_3 - R_2$, we get

$$\begin{bmatrix} 0 & 0 & 1 & -3 \\ 0 & 0 & 0 & 1 \\ 0 & 0 & 0 & 0 \\ 0 & 0 & 0 & 0 \end{bmatrix}\begin{bmatrix} x_1 \\ x_2 \\ x_3 \\ x_4 \end{bmatrix} = \begin{bmatrix} 0 \\ 0 \\ 0 \\ 0 \end{bmatrix} \quad \ldots(2)$$

From (2), it is clear that $\rho(A - 2I) = 2$, therefore the equation (1) will have $4 - 2 = 2$ linearly independent solution.

Equation (2) reduces to

$$\left.\begin{aligned} x_3 - 3x_4 &= 0 \\ x_4 &= 0 \end{aligned}\right\} \Rightarrow x_4 = 0, x_3 = 0$$

Let us take $x_1 = c_1, x_2 = c_2$, then we have

$$X = \begin{bmatrix} x_1 \\ x_2 \\ x_3 \\ x_4 \end{bmatrix} = \begin{bmatrix} c_1 \\ c_2 \\ 0 \\ 0 \end{bmatrix} = c_1\begin{bmatrix} 1 \\ 0 \\ 0 \\ 0 \end{bmatrix} + c_2\begin{bmatrix} 0 \\ 1 \\ 0 \\ 0 \end{bmatrix}$$

Hence, the eigenvector corresponding to eigenvalue 2 are

$$X_1 = \begin{bmatrix} 1 \\ 0 \\ 0 \\ 0 \end{bmatrix}, X_2 = \begin{bmatrix} 0 \\ 1 \\ 0 \\ 0 \end{bmatrix}$$

Eigenvector corresponding to the eigenvalue $\lambda_4 = 3$:

Let $X = \begin{bmatrix} x_1 \\ x_2 \\ x_3 \\ x_4 \end{bmatrix} \neq O$ be an eigenvector corresponding to the eigenvalue 3, then we have

$$AX = 3X$$

or $$(A - 3I)X = O$$

or $$\begin{bmatrix} 2-3 & 0 & 1 & -3 \\ 0 & 2-3 & 10 & 4 \\ 0 & 0 & 2-3 & 0 \\ 0 & 0 & 0 & 3-3 \end{bmatrix}\begin{bmatrix} x_1 \\ x_2 \\ x_3 \\ x_4 \end{bmatrix} = \begin{bmatrix} 0 \\ 0 \\ 0 \\ 0 \end{bmatrix}$$

or $$\begin{bmatrix} -1 & 0 & 1 & -3 \\ 0 & -1 & 10 & 4 \\ 0 & 0 & -1 & 0 \\ 0 & 0 & 0 & 0 \end{bmatrix}\begin{bmatrix} x_1 \\ x_2 \\ x_3 \\ x_4 \end{bmatrix} = \begin{bmatrix} 0 \\ 0 \\ 0 \\ 0 \end{bmatrix} \quad \ldots(3)$$

From (3), it is clear that $\rho(A - 3I) = 3$, therefore the equation (3) will have $4 - 3 = 1$ linearly independent solution.

Equation (3) reduces to

$$\begin{aligned} -x_1 + x_3 - 3x_4 &= 0 \\ -x_2 + 10x_3 + 4x_4 &= 0 \\ -x_3 &= 0 \end{aligned}$$

Let us take $x_4 = c$, then from above equaions, we get $x_1 = -3c, x_2 = 4c, x_3 = 0, x_4 = c$

Therefore,

$$\begin{bmatrix} x_1 \\ x_2 \\ x_3 \\ x_4 \end{bmatrix} = \begin{bmatrix} -3c \\ 4c \\ 0 \\ c \end{bmatrix} = c\begin{bmatrix} -3 \\ 4 \\ 0 \\ 1 \end{bmatrix}$$

Hence, the eigenvector corresponding to eigenvalue 3 is

$$X = \begin{bmatrix} -3 \\ 4 \\ 0 \\ 1 \end{bmatrix}.$$

Example 14. *Find the eigenvalues and eigenvectors of the matrix*

$$A = \begin{bmatrix} 2 & -1 \\ 5 & -2 \end{bmatrix}.$$

Solution. The characteristic equation of A is given by

$$|A - \lambda I| = 0$$

or $$\begin{vmatrix} 2-\lambda & -1 \\ 5 & -2-\lambda \end{vmatrix} = 0$$

or $$(2-\lambda)(-2-\lambda) + 5 = 0$$

or $$\lambda^2 + 1 = 0$$

The roots of this equation are $\lambda = \pm i$.

Thus, the eigenvalues of A are $\lambda_1 = -i, \lambda_2 = i$.

[This example confirms that a matrix with real entries may have complex eigenvalues.]

Eigenvector corresponding to the eigenvalue $\lambda_1 = -i$:

Let $X = \begin{bmatrix} x_1 \\ x_2 \end{bmatrix} \neq O$ be an eigenvector of A corresponding to the eigenvalue $\lambda_1 = -i$, then we have

$$AX = \lambda_1 X$$

or $$AX = (-i)X$$

or $$(A - (-i)I)X_1 = O$$

or $$\begin{bmatrix} 2-(-i) & -1 \\ 5 & -2-(-i) \end{bmatrix} \begin{bmatrix} x_1 \\ x_2 \end{bmatrix} = \begin{bmatrix} 0 \\ 0 \end{bmatrix}$$

or $$\begin{bmatrix} 2+i & -1 \\ 5 & -2+i \end{bmatrix} \begin{bmatrix} x_1 \\ x_2 \end{bmatrix} = \begin{bmatrix} 0 \\ 0 \end{bmatrix} \quad \text{...(1)}$$

Applying $R_1 \rightarrow \left(\frac{1}{2+i}\right) R_1$, we get

$$\begin{bmatrix} 1 & -\frac{1}{2+i} \\ 5 & -2+i \end{bmatrix} \begin{bmatrix} x_1 \\ x_2 \end{bmatrix} = \begin{bmatrix} 0 \\ 0 \end{bmatrix}$$

Now $$\frac{-1}{2+i} = \frac{-1(2-i)}{(2+i)(2-i)} = \frac{-2+i}{5} = \frac{-2}{5} + \frac{i}{5}$$

$$\therefore \quad \begin{bmatrix} 1 & -\frac{2}{5} + \frac{i}{5} \\ 5 & -2+i \end{bmatrix} \begin{bmatrix} x_1 \\ x_2 \end{bmatrix} = \begin{bmatrix} 0 \\ 0 \end{bmatrix}$$

Applying $R_2 \rightarrow R_2 - 5R_1$, we get

$$\begin{bmatrix} 1 & -\frac{2}{5} + \frac{i}{5} \\ 0 & 0 \end{bmatrix} \begin{bmatrix} x_1 \\ x_2 \end{bmatrix} = \begin{bmatrix} 0 \\ 0 \end{bmatrix} \quad \text{...(2)}$$

From (2), it is clear that $\rho(A - (-i)I) = 1$, therefore the equation (1) will have $2 - 1 = 1$ linearly independent solution.

Equation (2) reduces to

$$x_1 + \left(-\frac{2}{5} + \frac{i}{5}\right) x_2 = 0$$

Let us take $x_2 = c$, then we get

$$x_1 = -c\left(\frac{-2}{5} + \frac{i}{5}\right) = c\left(\frac{2}{5} - \frac{i}{5}\right)$$

$$\therefore \quad \begin{bmatrix} x_1 \\ x_2 \end{bmatrix} = \begin{bmatrix} c\left(\frac{2}{5} - \frac{i}{5}\right) \\ c \end{bmatrix} = \frac{c}{5} \begin{bmatrix} 2-i \\ 5 \end{bmatrix}$$

Hence, the eigenvector corresponding to eigenvalue $\lambda_1 = -i$ is

$$X_1 = \begin{bmatrix} 2-i \\ 5 \end{bmatrix}$$

Eigenvector corresponding to the eigenvalue $\lambda_2 = i$

Let $X = \begin{bmatrix} x_1 \\ x_2 \end{bmatrix} \neq O$ be an eigenvector corresponding to the eigenvalue $\lambda_2 = i$, then we have

$$AX = \lambda_2 X$$

$$AX = i(X)$$

or $$(A - (i)I)X = O$$

or $$\begin{bmatrix} 2-i & -1 \\ 5 & -2-i \end{bmatrix} \begin{bmatrix} x_1 \\ x_2 \end{bmatrix} = \begin{bmatrix} 0 \\ 0 \end{bmatrix} \quad \text{...(3)}$$

Applying $R_1 \rightarrow \left(\frac{1}{2-i}\right) R_1$, we get

$$\begin{bmatrix} 1 & -\frac{1}{2-i} \\ 5 & -2-i \end{bmatrix} \begin{bmatrix} x_1 \\ x_2 \end{bmatrix} = \begin{bmatrix} 0 \\ 0 \end{bmatrix}$$

Now $$\frac{-1}{2-i} = \frac{-1(2+i)}{(2-i)(2+i)} = \frac{-2-i}{5} = \frac{-2}{5} - \frac{i}{5}$$

$$\therefore \quad \begin{bmatrix} 1 & -\frac{2}{5} - \frac{i}{5} \\ 5 & -2-i \end{bmatrix} \begin{bmatrix} x_1 \\ x_2 \end{bmatrix} = \begin{bmatrix} 0 \\ 0 \end{bmatrix}$$

Applying $R_2 \rightarrow R_2 - 5R_1$, we get

$$\begin{bmatrix} 1 & -\frac{2}{5} - \frac{i}{5} \\ 0 & 0 \end{bmatrix} \begin{bmatrix} x_1 \\ x_2 \end{bmatrix} = \begin{bmatrix} 0 \\ 0 \end{bmatrix} \quad \text{...(4)}$$

From (4) it is clear that $\rho(A - (i)I) = 1$, therefore the system of equation (3) will have $2 - 1 = 1$ linearly independent solution.

Equation (4) reduces to

$$x_1 + \left(-\frac{2}{5} - \frac{i}{5}\right) x_2 = 0$$

Let us take $x_2 = c$, then $x_1 = \left(\frac{2+i}{5}\right) c$

$$\begin{bmatrix} x_1 \\ x_2 \end{bmatrix} = \begin{bmatrix} \left(\frac{2+i}{5}\right) c \\ c \end{bmatrix} = \frac{c}{5} \begin{bmatrix} 2+i \\ 5 \end{bmatrix}$$

Hence, the eigenvector corresponding to eigenvalue $\lambda_2 = i$ is

$$X_1 = \begin{bmatrix} 2+i \\ 5 \end{bmatrix}$$

REMARK

- Let A be an $n \times n$ matrix with real entries. If λ is a complex eigenvalue of A with associated eigenvector X, then $\bar{\lambda}$ is also an eigenvalue of A with associated eigenvector $\bar{X}$.

EXERCISE 25.9

1. Prove that a square matrix A and its transpose A' have the same set of eigenvalues.

2. Let A be an $n \times n$ matrix and let $g(x)$ be any polynomial. If λ is an eigenvalue of A, then prove that $g(\lambda)$ is an eigenvalue of $g(A)$.

3. Show that the eigenvalues of a triangular matrix are just the diagonal elements of the matrix.

4. Let $A = \text{dig. }(\lambda_1, \lambda_2, \ldots, \lambda_n)$ be a diagonal matrix. Prove that each λ_i $(i = 1, 2, 3, \ldots, n)$ is an eigenvalue of A.

5. Let A be an 3×3 matrix. If $\lambda_1, \lambda_2, \lambda_3$ are the eigenvalues of A, then find the eigenvalues of the matrix $(I + aA)^{-1}(1 + bA)$, where a, b are scalars such that $a\lambda_i \neq -1$ for $i = 1, 2, 3$.

6. Let A and B be two $n \times n$ matrices. Let X be an eigenvector of A and B both. Show that X is also an eigenvector of $aA + bB$, where a, b are scalars.

7. Prove that the eigenvectors of a real symmetric matrix corresponding to two distinct eigenvalues are orthogonal.

8. Prove that the eigenvectors of a Hermitian matrix corresponding to two distinct eigenvalues are orthogonal.

9. (i) If λ is an eigenvalue of a matrix A, then show that $k + \lambda$ is an eigenvalue of $A + kI$.

(ii) If the matrix A has characteristic roots $\lambda_1, \lambda_2, \ldots, \lambda_n$ show that the matrix A^2 has such roots as $\lambda_1^2, \lambda_2^2, \ldots, \lambda_n^2$.

10. (i) Find the eigenvalues of a matrix $\begin{bmatrix} 1 & 4 \\ 2 & 3 \end{bmatrix}$.

(ii) Find the eigenvalues of the matrix $A = \begin{bmatrix} a & h & g \\ 0 & b & f \\ 0 & 0 & c \end{bmatrix}$.

11. Find the eigenvalues and eigenvectors of the following matrices :

(i) $\begin{bmatrix} 2 & -4 \\ -1 & -1 \end{bmatrix}$ (ii) $\begin{bmatrix} -1 & 0 \\ 0 & 1 \end{bmatrix}$

(iii) $\begin{bmatrix} 1 & 1 \\ -2 & 4 \end{bmatrix}$ (iv) $\begin{bmatrix} 10 & -18 \\ 6 & -11 \end{bmatrix}$

12. Find the eigenvalues and eigenvectors of the following matrices :

(i) $\begin{bmatrix} 0 & 1 & 0 \\ 0 & 0 & 1 \\ 1 & -3 & 3 \end{bmatrix}$ (ii) $\begin{bmatrix} 5 & 8 & 16 \\ 4 & 1 & 8 \\ -4 & -4 & -11 \end{bmatrix}$

(iii) $\begin{bmatrix} 1 & -1 & -1 \\ -1 & 1 & -1 \\ -1 & -1 & 1 \end{bmatrix}$ (iv) $\begin{bmatrix} 1 & 2 & 2 \\ 1 & 2 & -1 \\ -1 & 1 & 4 \end{bmatrix}$

(v) $\begin{bmatrix} 6 & -2 & 2 \\ -2 & 3 & -1 \\ 2 & -1 & 3 \end{bmatrix}$ (vi) $\begin{bmatrix} -2 & 2 & -3 \\ 2 & 1 & -6 \\ -1 & -2 & 0 \end{bmatrix}$

(UKTU-2011, GBTU-2010)

(vii) $\begin{bmatrix} 1 & 2 & 3 \\ 0 & 2 & 3 \\ 0 & 0 & 2 \end{bmatrix}$ (viii) $\begin{bmatrix} 1 & 1 & 0 \\ 0 & 2 & 2 \\ 0 & 0 & 3 \end{bmatrix}$

(ix) $\begin{bmatrix} 3 & 1 & 1 \\ 2 & 4 & 2 \\ 1 & 1 & 3 \end{bmatrix}$ (x) $\begin{bmatrix} 2 & 1 & 0 \\ 0 & 2 & 1 \\ 0 & 0 & 2 \end{bmatrix}$

13. Find the eigenvalues and eigenvectors of the matrix

$$A = \begin{bmatrix} 1 & 1 & 0 & 0 \\ 0 & 2 & 0 & 0 \\ 0 & 0 & 1 & 1 \\ 0 & 0 & -2 & 4 \end{bmatrix}$$

14. Find all the characteristic roots and the corresponding characteristic vectors of the matrix

$$A = \begin{bmatrix} 2 & 1 & -1 \\ 0 & 3 & -2 \\ 2 & 4 & -3 \end{bmatrix}$$

ANSWERS

5. $\dfrac{1+b\lambda_1}{1+a\lambda_1}, \dfrac{1+b\lambda_2}{1+a\lambda_2}, \dfrac{1+b\lambda_3}{1+a\lambda_3}$

10. (i) $-1, 5$ (ii) a, b, c

11. (i) $\lambda_1 = -2, X_1 = \begin{bmatrix} 1 \\ 1 \end{bmatrix}; \lambda_2 = 3, X_2 = \begin{bmatrix} -4 \\ 1 \end{bmatrix}$

(ii) $\lambda_1 = 1, X_1 = \begin{bmatrix} 0 \\ 1 \end{bmatrix}; \lambda_2 = -1, X_2 = \begin{bmatrix} 1 \\ 0 \end{bmatrix}$ (iii) $\lambda_1 = 2, X_1 = \begin{bmatrix} 1 \\ 1 \end{bmatrix}; \lambda_2 = 3, X_2 = \begin{bmatrix} 1 \\ 2 \end{bmatrix}$ (iv) $\lambda_1 = -2, X_1 = \begin{bmatrix} 3 \\ 2 \end{bmatrix}; \lambda_2 = 1, X_2 = \begin{bmatrix} 2 \\ 1 \end{bmatrix}$

12. (i) $\lambda_1 = \lambda_2 = \lambda_3 = 1, X = \begin{bmatrix} 1 \\ 1 \\ 1 \end{bmatrix}$ (ii) $\lambda_1 = 1, X_1 = \begin{bmatrix} -2 \\ -1 \\ 1 \end{bmatrix}; \lambda_2 = -3, X_2 = \begin{bmatrix} -1 \\ 1 \\ 0 \end{bmatrix}; \lambda_3 = -3, X_3 = \begin{bmatrix} -2 \\ 0 \\ 1 \end{bmatrix}$

(iii) $\lambda_1 = -1, X_1 = \begin{bmatrix} 1 \\ 1 \\ 1 \end{bmatrix}; \lambda_2 = 2, X_2 = \begin{bmatrix} -1 \\ 1 \\ 0 \end{bmatrix}; \lambda_3 = 2, X_3 = \begin{bmatrix} -1 \\ 0 \\ 1 \end{bmatrix}$ (iv) $\lambda_1 = 1, X_1 = \begin{bmatrix} 2 \\ -1 \\ 1 \end{bmatrix}; \lambda_2 = 3, X_2 = \begin{bmatrix} 1 \\ 1 \\ 0 \end{bmatrix}; \lambda_3 = 3, X_3 = \begin{bmatrix} 1 \\ 0 \\ 1 \end{bmatrix}$

(v) $\lambda_1 = 2, X_1 = \begin{bmatrix} -1 \\ 0 \\ 2 \end{bmatrix}; \lambda_2 = 2, X_2 = \begin{bmatrix} 1 \\ 2 \\ 0 \end{bmatrix}; \lambda_3 = 8, X_3 = \begin{bmatrix} 2 \\ -1 \\ 1 \end{bmatrix}$

(vi) $\lambda_1 = -3, X_1 = \begin{bmatrix} -2 \\ 1 \\ 0 \end{bmatrix}; \lambda_2 = -3, X_2 = \begin{bmatrix} 3 \\ 0 \\ 1 \end{bmatrix}; \lambda_3 = 5, X_3 = \begin{bmatrix} 1 \\ 2 \\ 1 \end{bmatrix}$ (vii) $\lambda_1 = \lambda_2 = 1, X = \begin{bmatrix} 1 \\ 0 \\ 0 \end{bmatrix}; \lambda_3 = 2, X_1 = \begin{bmatrix} 2 \\ 1 \\ 0 \end{bmatrix}$

(viii) $\lambda_1 = 1, X_1 = \begin{bmatrix} 1 \\ 0 \\ 0 \end{bmatrix}; \lambda_2 = 2, X_2 = \begin{bmatrix} 2 \\ 1 \\ 0 \end{bmatrix}; \lambda_3 = 3, X_3 = \begin{bmatrix} 1 \\ 2 \\ 1 \end{bmatrix}$ (ix) $\lambda_1 = 2, X_1 = \begin{bmatrix} -1 \\ 1 \\ 0 \end{bmatrix}; \lambda_2 = 2, X_2 = \begin{bmatrix} -1 \\ 0 \\ 1 \end{bmatrix}; \lambda_3 = 6, X_3 = \begin{bmatrix} 1 \\ 2 \\ 1 \end{bmatrix}$

(x) $\lambda_1 = \lambda_2 = \lambda_3 = 2, X = \begin{bmatrix} 1 \\ 0 \\ 0 \end{bmatrix}$ **13.** $\lambda_1 = 1, X_1 = \begin{bmatrix} 1 \\ 0 \\ 0 \\ 0 \end{bmatrix}; \lambda_2 = 2, X_2 = \begin{bmatrix} 1 \\ 1 \\ 0 \\ 0 \end{bmatrix}; \lambda_3 = 2, X_3 = \begin{bmatrix} 0 \\ 0 \\ 1 \\ 1 \end{bmatrix}; \lambda_4 = 3, X_4 = \begin{bmatrix} 0 \\ 0 \\ 0 \\ 1 \end{bmatrix}$

25.50 THE CAYLEY-HAMILTON THEOREM

THEOREM 1. *Every square matrix satisfies its characteristic equation.* (UPTU-2006)

or let A be a square matrix of order n and the characteristic equation of A is

$$|A - \lambda I| = (-1^n)\ [\lambda^n + a_1\lambda^{n-1} + a_2\lambda^{n-2} + \ldots + a_{n-1}\lambda + a_n] = 0$$

then its matrix equation $X^n + a_1X^{n-1} + a_2X^{n-2} + \ldots + a_{n-1}X + a_nI = O$ *is satisfied by the matrix* $X = A$

i.e. $$A^n + a_1A^{n-1} + a_2A^{n-2} + \ldots + a_{n-1}A + a_nI = O$$

where I is a unit matrix of order n and O is null matrix of order n.

Proof. Since A and I are two square matrices of order n and λ is any characteristic root of A, then the matrix $(A - \lambda I)$ is also a square matrix of order n whose elements are at most of degree one in λ. Therefore Adj. $(A - \lambda I)$ will have its elements a polynomials in λ of degree $n-1$ or less and thus Adj. $(A - \lambda I)$ can be expressed as a matrix polynomial in λ as follows :

$$\text{Adj. } (A - \lambda I) = B_0\lambda^{n-1} + B_1\lambda^{n-2} + \ldots + B_{n-2}\lambda + B_{n-1} \qquad \ldots(1)$$

where $B_0, B_1, \ldots, B_{n-1}$ are the square matrices of order n.

Since we know that $\quad A(\text{Adj. } A) = |A|I_n$

$\therefore \quad (A - \lambda I)\text{ Adj. }(A - \lambda I) = |A - \lambda I|I$

or $\quad (A - \lambda I)\text{ Adj. }(A - \lambda I) = (-1^n)\ (\lambda^n + a_1\lambda^{n-1} + a_2\lambda^{n-2} + \ldots + a_{n-1}\lambda + a_n)I \qquad \ldots(2)$

Multiplying both sides of (1) by $(A - \lambda I)$, we get

$$(A - \lambda I)\text{ Adj.}(A - \lambda I) = (A - \lambda I)(B_0\lambda^{n-1} + B_1\lambda^{n-2} + \ldots + B_{n-2}\lambda + B_{n-1}) \qquad \ldots(3)$$

From (2) and (3), we get

$$(A - \lambda I)(B_0\lambda^{n-1} + B_1\lambda^{n-2} + \ldots + B_{n-2}\lambda + B_{n-1}) = (-1^n)\ (\lambda^n + a_1\lambda^{n-1} + a_2\lambda^{n-2} + \ldots + a_{n-1}\lambda + a_n)I$$

Now comparing the coefficients of like powers of λ, we get

$$\left.\begin{aligned} -IB_0 &= (-1)^n\, I \\ AB_0 - IB_1 &= (-1)^n\, a_1I \\ AB_1 - IB_2 &= (-1)^n\, a_2I \\ &\ldots\ldots\ldots\ldots \\ AB_{n-2} - IB_{n-3} &= (-1)^n\, a_{n-1}I \\ AB_{n-1} &= (-1)^n\, a_nI \end{aligned}\right\} \qquad \ldots(4)$$

Premultiplying first, second, third, etc. equations of (4) by A^n, A^{n-1}, A^{n-2}, etc. respectively and then adding, we get

$$-A^nB_0 + A^nB_0 - A^{n-1}B_1 + A^{n-1}B_1 + \ldots = (-1)^n\ (A^n + a_1A^{n-1} + \ldots + a_nI)$$

or $$0 = (-1)^n\ (A^n + a_1A^{n-1} + \ldots + a_nI)$$

Hence $\quad A^n + a_1A^{n-1} + \ldots + a_nI = O$

Corollary 1. *If A be a non-singular matrix of order $n \times n$ and its characteristic polynomial is*

$$|A - \lambda I| = (-1)^n (\lambda^n + a_1\lambda^{n-1} + \ldots + a_{n-1}\lambda + a_n)$$

then $\det(A) = (-1)^n a_n.$

Proof. We have $|A - \lambda I| = (-1)^n (\lambda^n + a_1\lambda^{n-1} + \ldots + a_{n-1}A + a_n)$

Putting $\lambda = 0$, we get $|A - 0I| = (-1)^n a_n$

$\Rightarrow \quad |A| = (-1)^n a_n$

Corollary 2. *If $\lambda_1, \lambda_2, \ldots, \lambda_n$ are eigenvalues of a square matrix of order $n \times n$, then $\det(A) = \lambda_1\lambda_2\lambda_3 \ldots \lambda_n$.*

Proof. If λ is an eigenvalue of A, then it is a root of characteristic equation of A. The characteristic equation of A is given by

$$|A - \lambda I| = (-1)^n (\lambda^n + a_1\lambda^{n-1} + \ldots + a_{n-1}\lambda + a_n) = 0$$

Since $\lambda_1, \lambda_2, \ldots, \lambda_n$ are the eigenvalues of A, hence

$$(-1)^n (\lambda^n + a_1\lambda^{n-2} + \ldots + a_{n-1}\lambda + a_n) = (\lambda_1 - \lambda)(\lambda_2 - \lambda)\ldots(\lambda_n - \lambda) = 0$$

Comparing the constant terms of both sides, we get

$$(-1)^n a_n = \lambda_1\lambda_2\lambda_3 \ldots \lambda_n$$

From Corollary 1, $\det(A) = (-1)^n a_n$

Hence $\det(A) = \lambda_1\lambda_2\lambda_3 \ldots \lambda_n$

Corollary 3. *Let A be an $n \times n$ matrix with characteristic polynomial*

$$f(t) = (-1)^n (t^n + a_1 t^{n-1} + \ldots + a_{n-1}t + a_n)$$

Then A is invertible iff $a_n \neq 0$ and its inverse is

$$A^{-1} = \left(\frac{-1}{a_n}\right)(A^{n-1} + a_1A^{n-2} + \ldots + a_{n-2}A + a_{n-1}I)$$

Proof. By Corollary 1, $|A| = (-1)^n a_n$

$\therefore \quad |A| \neq 0 \Leftrightarrow a_n \neq 0$

i.e., A is invertible iff $a_n \neq 0$

By Cayley-Hamilton theorem, we have

$$f(A) = O$$

$$\Rightarrow \quad (-1)^n (A^n + a_1A^{n-1} + \ldots + a_{n-1}A + a_nI) = O$$

$$\Rightarrow \quad A^n + a_1A^{n-1} + \ldots + a_{n-1}A + a_nI = O$$

$$\Rightarrow \quad A^{-1}(A^n + a_1A^{n-1} + \ldots + a_{n-1}A + a_nI) = A^{-1}O = O$$

$$\Rightarrow \quad A^{n-1} + a_1A^{n-2} + \ldots + a_{n-2}A + a_{n-1}I + a_nA^{-1} = O$$

Hence $A^{-1} = \left(\frac{-1}{a_n}\right)[A^{n-1} + a_1A^{n-2} + \ldots + a_{n-2}A + a_{n-1}I]$

Corollary 4. *If $\lambda_1, \lambda_2, \ldots, \lambda_n$ are the eigenvalue of a matrix A of order $n \times n$, then $Tr(A) = $ Trace of $A = \sum_{i=1}^{n} \lambda_i$*

Proof. Let $A = [a_{ij}]_{n \times n}$. Then $\text{Tr}(A) = \sum_{i=1}^{n} \lambda_i$

Now

$$|A - \lambda I| = \begin{vmatrix} a_{11-\lambda} & a_{12} & \cdots & a_{1n} \\ a_{21} & a_{22-\lambda} & \cdots & a_{2n} \\ \vdots & & & \\ a_{n1} & a_{n2} & \cdots & a_{nn-\lambda} \end{vmatrix}$$

$$= (a_{11} - \lambda)A_{11} - a_{12}A_{12} + a_{13}A_{13} + \ldots + (-1)^n a_{1n}A_n$$

where $A_{11} =$ Minor of $a_{11-\lambda}$, $A_{12} =$ Minor of a_{12} and so on.

Clearly A_{11} is a polynomial of degree $n - 1$ in λ.

Therefore $A_{11} = (a_{22} - \lambda)(a_{33} - \lambda) \ldots (a_{nn} - \lambda) + O(\lambda^{n-3})$

$\Rightarrow \qquad (a_{11}-\lambda)A_{11} = (a_{11}-\lambda)(a_{22}-\lambda)(a_{33}-\lambda)\ ...(a_{nn}-\lambda) + O(\lambda^{n-2})$

Similarly,

$$A_{12} = \text{A polynomial of degree } n-2 \text{ in } \lambda$$
$$A_{13} = \text{A polynomial of degree } n-2 \text{ in } \lambda$$
$$\cdots\cdots\cdots\cdots\cdots\cdots\cdots\cdots\cdots\cdots$$
$$A_{1n} = \text{A polynomial of degree } n-2 \text{ in } \lambda$$

$$\therefore \qquad |A-\lambda I| = (-1)^n\left[\lambda^n - \left(\sum_{i=1}^{n} a_{ii}\right)\lambda^{n-1} + O\left(\lambda^{n-2}\right)\right] \qquad ...(1)$$

Since $\lambda_1, \lambda_2, \ldots, \lambda_n$ are eigenvalues of A, hence

$$|A-\lambda I| = (-1)^n\left[\lambda^n - \left(\sum_{i=1}^{n} \lambda_i\right)\lambda^{n-1} + O\left(\lambda^{n-2}\right)\right] \qquad ...(2)$$

From (1) and (2)

$$(-1)^n\left[\lambda^n - \left(\sum_{i=1}^{n} a_{ii}\right)\lambda^{n-1} + O\left(\lambda^{n-2}\right)\right] = (-1)^n\left[\lambda^n - \left(\sum_{i=1}^{n} \lambda_i\right)\lambda^{n-1} + O\left(\lambda^{n-2}\right)\right]$$

Equating the coefficient of λ^{n-1} of both sides, we get $\sum_{i=1}^{n} a_{ii} = \sum_{i=1}^{n} \lambda_i$

Hence $\qquad \mathrm{Tr}(A) = \sum_{i=1}^{n} a_{ii} = \sum_{i=1}^{n} \lambda_i$

Corollary 5. *If the characteristic equation of a matrix A of order n × n is*

$$|A-\lambda I| = (-1)^n(\lambda^n + a_1\lambda^{n-1} + a_2\lambda^{n-2} + \ldots + a_{n-1}\lambda + a_n) = 0$$

then $Tr(A) = -a_1$

Proof. If $\lambda_1, \lambda_2, \ldots, \lambda_n$ are the eigenvalues of a matrix A, then

$$|A-\lambda I| = (-1)^n(\lambda-\lambda_1)(\lambda-\lambda_2)\ldots(\lambda-\lambda_n) = 0 = (-1)^n\left[\lambda^n - \left(\sum_{i=1}^{n}\lambda_i\right)\lambda^{n-1} + \ldots\right] = 0 \qquad ...(1)$$

But $\qquad |A-\lambda I| = (-1)^n(\lambda^n + a_1\lambda^{n-1} + a_2\lambda^{n-2} + \ldots + a_{n-1}\lambda + a_n) = 0 \qquad ...(2)$

From (1) and (2) we get

$$(-1)^n\left[\lambda^n - \left(\sum_{i=1}^{n}\lambda_i\right)\lambda^{n-1} + \ldots\right] = (-1)^n(\lambda^n + a_1\lambda^{n-1} + a_2\lambda^{n-2} + \ldots + a_{n-1}\lambda + a_n)$$

Taking the coefficient of λ^{n-1} on both sides, we get $\sum_{i=1}^{n}\lambda_i = -a_1$

By Corollary 4, we have $\mathrm{Tr}(A) = \sum_{i=1}^{n}\lambda_i$

Hence $\qquad \mathrm{Tr}(A) = -a_1$

Corollary 6. *Let A be a matrix of order n × n. If m be a positive integer such that m ≥ n, then* A^m *is linearly expressible in terms of those of lower order of A.*

Proof. By Cayley-Hamilton theorem,

$$A^n + a_1A^{n-1} + a_2A^{n-2} + \ldots + a_{n-1}A + a_nI = O \quad ...(1)$$

Multiplying (1) by A^{m-n}, we get

$$A^m + a_1A^{m-1} + a_2A^{m-2} + \ldots + a_nA^{m-n} = O$$

or

$$A^m = (-a_1)A^{m-1} + (-a_2)A^{m-2} + \ldots + (-a_n)A^{m-n}$$

Hence the result.

Recapitulations

- Every square matrix satisfies its characteristic equation.
- If $\lambda_1, \lambda_2 \ldots \lambda_n$ are eigenvalues of a square matrix of order $n \times n$ then $\det(A) = \lambda_1 . \lambda_2 \ldots \lambda_n$.

Solved Examples

Example 1. *Find the characteristic equation of the matrix*

$$A=\begin{bmatrix}1&0&2\\0&2&1\\2&0&3\end{bmatrix}$$

and verify that it is satisfied by A and hence find its inverse.

Solution. The characteristic equation of A is given by

$$|A-\lambda I|=0$$

or
$$\begin{vmatrix}1-\lambda&0&2\\0&2-\lambda&1\\2&0&3-\lambda\end{vmatrix}=0$$

$$\text{or}\,(1-\lambda)\{(2-\lambda)(3-\lambda)-0\}+2\{0-2(2-\lambda)\}=0$$

or
$$-\lambda^3+6\lambda^2-7\lambda-2=0$$

or
$$\lambda^3-6\lambda^2+7\lambda+2=0$$

Next we have to show that

$$A^3-6A^2+7A+2I=O$$

Now $A^2=A.A$

$$=\begin{bmatrix}1&0&2\\0&2&1\\2&0&3\end{bmatrix}\begin{bmatrix}1&0&2\\0&2&1\\2&0&3\end{bmatrix}=\begin{bmatrix}5&0&8\\2&4&5\\8&0&13\end{bmatrix}$$

and $A^3=A^2.A$

$$=\begin{bmatrix}5&0&8\\2&4&5\\8&0&13\end{bmatrix}\begin{bmatrix}1&0&2\\0&2&1\\2&0&3\end{bmatrix}=\begin{bmatrix}21&0&34\\12&8&23\\34&0&55\end{bmatrix}$$

$\therefore\quad A^3-6A^2+7A+2I$

$$=\begin{bmatrix}21&0&34\\12&8&23\\34&0&55\end{bmatrix}-6\begin{bmatrix}5&0&8\\2&4&5\\8&0&13\end{bmatrix}$$

$$+7\begin{bmatrix}1&0&2\\0&2&1\\2&0&3\end{bmatrix}+2\begin{bmatrix}1&0&0\\0&1&0\\0&0&1\end{bmatrix}$$

$$=\begin{bmatrix}21-30+7+2&0-0+0+0&34-48+14+0\\12-12+0+0&8-24+14+2&23-30+7+0\\34-48+14+0&0-0+0+0&55-78+21+2\end{bmatrix}$$

$$=\begin{bmatrix}0&0&0\\0&0&0\\0&0&0\end{bmatrix}=O$$

Hence, $A^3-6A^2+7A+2I=O$ ...(1)

To find A^{-1}:

Since the characteristic equation of A is

$$\lambda^3-6\lambda^2+7\lambda+2=0$$

$\therefore\quad |A|=(-1)^3 2=-2\neq 0\quad [\because\ |A|=(-1)^n a_n]$

$\Rightarrow\quad A^{-1}$ exist.

Premultiplying (1) by A^{-1}, we get

$$A^2-6A+7I+2A^{-1}=O$$

$$\Rightarrow\ A^{-1}=-\frac{1}{2}[A^2-6A+7I]$$

$$\Rightarrow A^{-1}=-\frac{1}{2}\left\{\begin{bmatrix}5&0&8\\2&4&5\\8&0&13\end{bmatrix}-6\begin{bmatrix}1&0&2\\0&2&1\\2&0&3\end{bmatrix}+7\begin{bmatrix}1&0&0\\0&1&0\\0&0&1\end{bmatrix}\right\}$$

$$=-\frac{1}{2}\begin{bmatrix}6&0&-4\\2&-1&-1\\-4&0&2\end{bmatrix}$$

Hence,

$$A^{-1}=-\frac{1}{2}\begin{bmatrix}6&0&-4\\2&-1&-1\\-4&0&2\end{bmatrix}=\frac{1}{2}\begin{bmatrix}-6&0&4\\-2&1&1\\4&0&-2\end{bmatrix}$$

Example 2. *Find the characteristic equation of the matrix* $A=\begin{bmatrix}2&-1&1\\-1&2&-1\\1&-1&2\end{bmatrix}$ *and verify that it is satisfied by A and hence find* A^{-1}.

(UPTU-2006, GBTU-2012, UKTU-2011, Madras-2006)

Solution. The characteristic equation of A is given by

$$|A-\lambda I|=0$$

or
$$\begin{vmatrix}2-\lambda&-1&2\\-1&2-\lambda&-1\\1&-1&2-\lambda\end{vmatrix}=0$$

or $(2-\lambda)\{(2-\lambda)(2-\lambda)-1\}$

$$+1(-2+\lambda+2)+1(1-2+\lambda)=0$$

or
$$-\lambda^3+6\lambda^2-9\lambda+4=0$$

or
$$\lambda^3-6\lambda^2+9\lambda-4=0$$

Next we have to show that

$$A^3-6A^2+9A-4I=O$$

Now $A^2=A.A$

$$=\begin{bmatrix}2&-1&1\\-1&2&-1\\1&-1&2\end{bmatrix}\begin{bmatrix}2&-1&1\\-1&2&-1\\1&-1&2\end{bmatrix}=\begin{bmatrix}6&-5&5\\-5&6&-5\\5&-5&6\end{bmatrix}$$

and $A^3=A^2A$

$$=\begin{bmatrix}6&-5&5\\-5&6&-5\\5&-5&6\end{bmatrix}\begin{bmatrix}2&-1&1\\-1&2&-1\\1&-1&2\end{bmatrix}=\begin{bmatrix}22&-21&21\\-21&22&-21\\21&-21&22\end{bmatrix}$$

Now $A^3-6A^2+9A+2I$

$$=\begin{bmatrix}22&-21&21\\-21&22&-21\\21&-21&22\end{bmatrix}-6\begin{bmatrix}6&-5&5\\-5&6&-5\\5&-5&6\end{bmatrix}$$

$$+9\begin{bmatrix}2&-1&1\\-1&2&-1\\1&-1&2\end{bmatrix}-4\begin{bmatrix}1&0&0\\0&1&0\\0&0&1\end{bmatrix}$$

$$= \begin{bmatrix} 22-36+18-4 & -21+30-9-0 & 21-30+9-0 \\ -21+30-9-0 & 22-36+18-4 & -21+30-9-0 \\ 21-30+9-0 & -21+30-9-0 & 22-36+18-4 \end{bmatrix}$$

$$= \begin{bmatrix} 0 & 0 & 0 \\ 0 & 0 & 0 \\ 0 & 0 & 0 \end{bmatrix} = O$$

Hence, $A^3 - 6A^2 + 9A + 4I = O$ ...(1)

Since $|A| = 2(4-1) + 1(-2+1) + 1(1-2)$

$= 6 - 1 - 1 = 4 \neq 0 \Rightarrow A^{-1}$ exist.

Premultiplying (1) by A^{-1}, we get

$$A^2 - 6A + 9I - 4A^{-1} = O$$

$$\Rightarrow \quad A^{-1} = +\frac{1}{4}[A^2 - 6A + 9I]$$

$$\Rightarrow \quad A^{-1} = \frac{1}{4}\left\{\begin{bmatrix} 6 & -5 & 5 \\ -5 & 6 & -5 \\ 5 & -5 & 6 \end{bmatrix} - 6\begin{bmatrix} 2 & -1 & 1 \\ -1 & 2 & -1 \\ 1 & -1 & 2 \end{bmatrix} + 9\begin{bmatrix} 1 & 0 & 0 \\ 0 & 1 & 0 \\ 0 & 0 & 1 \end{bmatrix}\right\}$$

$$= \frac{1}{4}\begin{bmatrix} 6-12+9 & -5+6+0 & 5-6+0 \\ -5+6+0 & 6-12+9 & -5+6+0 \\ 5-6+0 & -5+6+0 & 6-12+9 \end{bmatrix}$$

$$\therefore \quad A^{-1} = \frac{1}{4}\begin{bmatrix} 3 & 1 & -1 \\ 1 & 3 & 1 \\ -1 & 1 & 3 \end{bmatrix}.$$

Example 3. *Find the characteristic equation of the matrix*

$$A = \begin{bmatrix} 1 & 2 & 0 \\ 2 & -1 & 0 \\ 0 & 0 & -1 \end{bmatrix} \text{ and hence find } A^{-1}.$$

Solution. The characteristic equation of A is given by

$$|A - \lambda I| = 0$$

or
$$\begin{vmatrix} 1-\lambda & 2 & 0 \\ 2 & -1-\lambda & 0 \\ 0 & 0 & -1-\lambda \end{vmatrix} = 0$$

or $(1-\lambda)\{(-1-\lambda)(-1-\lambda)-0\} - 2\{2(-1-\lambda)-0\} = 0$

or $(1-\lambda)(1+\lambda)^2 + 4(1+\lambda) = 0$

or $1+\lambda^2+2\lambda-\lambda-\lambda^3-2\lambda^2+4+4\lambda = 0$

or $-\lambda^3 - \lambda^2 + 5\lambda + 5 = 0$

or $\lambda^3 + \lambda^2 - 5\lambda - 5 = 0$

By Cayley-Hamilton theorem, we have

$$A^3 + A^2 - 5A - 5I = O \quad ...(1)$$

Since $|A - \lambda I| = -\lambda^3 - \lambda^2 + 5\lambda + 5$

$\Rightarrow \quad |A| = 5 \neq 0 \quad$ (Putting $\lambda = 0$)

$\Rightarrow \quad A^{-1}$ exists.

Premultiplying (1) by A^{-1}, we get

$$A^2 + A - 5I - 5A^{-1} = 0$$

$$\Rightarrow \quad A^{-1} = \frac{1}{5}(A^2 + A - 5I) \quad ...(2)$$

Now $A^2 = A.A$

$$= \begin{bmatrix} 1 & 2 & 0 \\ 2 & -1 & 0 \\ 0 & 0 & -1 \end{bmatrix}\begin{bmatrix} 1 & 2 & 0 \\ 2 & -1 & 0 \\ 0 & 0 & -1 \end{bmatrix} = \begin{bmatrix} 5 & 0 & 0 \\ 0 & 5 & 0 \\ 0 & 0 & 5 \end{bmatrix}$$

So

$$A^{-1} = \frac{1}{5}\left\{\begin{bmatrix} 5 & 0 & 0 \\ 0 & 5 & 0 \\ 0 & 0 & 1 \end{bmatrix} + \begin{bmatrix} 1 & 2 & 0 \\ 2 & -1 & 0 \\ 0 & 0 & -1 \end{bmatrix} - 5\begin{bmatrix} 1 & 0 & 0 \\ 0 & 1 & 0 \\ 0 & 0 & 1 \end{bmatrix}\right\}$$

$$= \frac{1}{5}\begin{bmatrix} 1 & 2 & 0 \\ 2 & -1 & 0 \\ 0 & 0 & -5 \end{bmatrix}$$

Example 4. *Show that the matrix* $A = \begin{bmatrix} 0 & c & -b \\ -c & 0 & a \\ b & -a & 0 \end{bmatrix}$ *satisfies Cayley-Hamilton Theorem.*

Solution. The characteristic equation of A is given by

$$|A - \lambda I| = 0$$

or
$$\begin{vmatrix} -\lambda & c & -b \\ -c & -\lambda & a \\ b & -a & -\lambda \end{vmatrix} = 0$$

or $-\lambda(\lambda^2 + a^2) - c(c\lambda - ab) - b(ca + b\lambda) = 0$

or $-\lambda^3 - \lambda(a^2 + b^2 + c^2) = 0$

or $\lambda^3 + \lambda(a^2 + b^2 + c^2) = 0$

We have to show that

$$A^3 + A(a^2 + b^2 + c^2) = O$$

Now $A^2 = A.A$

$$= \begin{bmatrix} 0 & c & -b \\ -c & 0 & a \\ b & -a & 0 \end{bmatrix}\begin{bmatrix} 0 & c & -b \\ -c & 0 & a \\ b & -a & 0 \end{bmatrix}$$

$$= \begin{bmatrix} -(c^2+b^2) & ab & ac \\ ab & -(c^2+a^2) & bc \\ ac & bc & -(a^2+b^2) \end{bmatrix}$$

and $A^3 = A^2 A$

$$= \begin{bmatrix} -(c^2+b^2) & ab & ac \\ ab & -(c^2+a^2) & bc \\ ac & bc & -(a^2+b^2) \end{bmatrix}\begin{bmatrix} 0 & c & -b \\ -c & 0 & a \\ b & -a & 0 \end{bmatrix}$$

$$= \begin{bmatrix} 0 & -c^3-b^2c-a^2c & bc^2+b^2+a^2b \\ c^3+a^2c+b^2c & 0 & -ab^2-ac^2-a^3 \\ -bc^2-b^3-a^2b & ac^2+ab^2+a^3 & 0 \end{bmatrix}$$

$$= \begin{bmatrix} 0 & -c(a^2+b^2+c^2) & b(a^2+b^2+c^2) \\ c(a^2+b^2+c^2) & 0 & -a(a^2+b^2+c^2) \\ -b(a^2+b^2+c^2) & a(a^2+b^2+c^2) & 0 \end{bmatrix}$$

$$A^3 = -(a^2+b^2+c^2)\begin{bmatrix} 0 & c & -b \\ -c & 0 & a \\ b & -a & 0 \end{bmatrix} = -(a^2+b^2+c^2)A$$

Hence, $A^3 + (a^2 + b^2 + c^2)A = O$

Example 5. *Verify Cayley-Hamilton theorem for the matrix*

$$A = \begin{bmatrix} 1 & 1 & 0 & 0 \\ 0 & 2 & 0 & 0 \\ 0 & 0 & -1 & 1 \\ 0 & 0 & -2 & 4 \end{bmatrix}.$$

Solution. The characteristic equation of the matrix A is given by

$$|A - \lambda I| = 0$$

or
$$\begin{vmatrix} 1-\lambda & 1 & 0 & 0 \\ 0 & 2-\lambda & 0 & 0 \\ 0 & 0 & 1-\lambda & 1 \\ 0 & 0 & -2 & 4-\lambda \end{vmatrix} = 0$$

or
$$(1-\lambda)\begin{vmatrix} 2-\lambda & 0 & 0 \\ 0 & 1-\lambda & 1 \\ 0 & -2 & 4-\lambda \end{vmatrix} = 0$$

[Expanding along first column]

or $(1-\lambda)[(2-\lambda)\{(1-\lambda)(4-\lambda)+2\}] = 0$

or $(1-\lambda)[(2-\lambda)(1-\lambda)(4-\lambda)+2(2-\lambda)] = 0$

or $(1-\lambda)(2-\lambda)(\lambda^2 - 5\lambda + 6) = 0$

or $(1-\lambda)(2-\lambda)(2-\lambda)(3-\lambda) = 0$

or $(\lambda-1)(\lambda-2)^2(\lambda-3) = 0$

We have to show that

$$(A - I)(A - 2I)^2(A - 3I) = O$$

Now

$$A - I = \begin{bmatrix} 1 & 1 & 0 & 0 \\ 0 & 2 & 0 & 0 \\ 0 & 0 & 1 & 1 \\ 0 & 0 & -2 & 4 \end{bmatrix} - \begin{bmatrix} 1 & 0 & 0 & 0 \\ 0 & 1 & 0 & 0 \\ 0 & 0 & 1 & 0 \\ 0 & 0 & 0 & 1 \end{bmatrix} = \begin{bmatrix} 0 & 1 & 0 & 0 \\ 0 & 1 & 0 & 0 \\ 0 & 0 & 0 & 1 \\ 0 & 0 & -2 & 3 \end{bmatrix}$$

and

$$A - 2I = \begin{bmatrix} 1 & 1 & 0 & 0 \\ 0 & 2 & 0 & 0 \\ 0 & 0 & 1 & 1 \\ 0 & 0 & -2 & 4 \end{bmatrix} - 2\begin{bmatrix} 1 & 0 & 0 & 0 \\ 0 & 1 & 0 & 0 \\ 0 & 0 & 1 & 0 \\ 0 & 0 & 0 & 1 \end{bmatrix} = \begin{bmatrix} -1 & 1 & 0 & 0 \\ 0 & 0 & 0 & 0 \\ 0 & 0 & -1 & 1 \\ 0 & 0 & -2 & 2 \end{bmatrix}$$

$$\therefore (A - 2I)^2 = \begin{bmatrix} -1 & 1 & 0 & 0 \\ 0 & 0 & 0 & 0 \\ 0 & 0 & -1 & 1 \\ 0 & 0 & -2 & 2 \end{bmatrix}\begin{bmatrix} -1 & 1 & 0 & 0 \\ 0 & 0 & 0 & 0 \\ 0 & 0 & -1 & 1 \\ 0 & 0 & -2 & 2 \end{bmatrix}$$

$$= \begin{bmatrix} 1 & -1 & 0 & 0 \\ 0 & 0 & 0 & 0 \\ 0 & 0 & -1 & 1 \\ 0 & 0 & -2 & 2 \end{bmatrix}$$

$$A - 3I = \begin{bmatrix} 1 & 1 & 0 & 0 \\ 0 & 2 & 0 & 0 \\ 0 & 0 & 1 & 1 \\ 0 & 0 & -2 & 4 \end{bmatrix} - 3\begin{bmatrix} 1 & 0 & 0 & 0 \\ 0 & 1 & 0 & 0 \\ 0 & 0 & 1 & 0 \\ 0 & 0 & 0 & 1 \end{bmatrix}$$

$$= \begin{bmatrix} -2 & 1 & 0 & 0 \\ 0 & -1 & 0 & 0 \\ 0 & 0 & -2 & 1 \\ 0 & 0 & -2 & 1 \end{bmatrix}$$

Now $(A - I)(A - 2I)^2(A - 3I)$

$$= \begin{bmatrix} 0 & 1 & 0 & 0 \\ 0 & 1 & 0 & 0 \\ 0 & 0 & 0 & 1 \\ 0 & 0 & -2 & 3 \end{bmatrix}\begin{bmatrix} 1 & -1 & 0 & 0 \\ 0 & 0 & 0 & 0 \\ 0 & 0 & -1 & 1 \\ 0 & 0 & -2 & 2 \end{bmatrix}\begin{bmatrix} -2 & 1 & 0 & 0 \\ 0 & -1 & 0 & 0 \\ 0 & 0 & -2 & 1 \\ 0 & 0 & -2 & 1 \end{bmatrix}$$

$$= \begin{bmatrix} 0 & 1 & 0 & 0 \\ 0 & 1 & 0 & 0 \\ 0 & 0 & 0 & 1 \\ 0 & 0 & -2 & 3 \end{bmatrix}\begin{bmatrix} -2 & 2 & 0 & 0 \\ 0 & 0 & 0 & 0 \\ 0 & 0 & 0 & 0 \\ 0 & 0 & 0 & 0 \end{bmatrix} = \begin{bmatrix} 0 & 0 & 0 & 0 \\ 0 & 0 & 0 & 0 \\ 0 & 0 & 0 & 0 \\ 0 & 0 & 0 & 0 \end{bmatrix} = O$$

$\therefore \quad (A - I)(A - 2I)^2(A - 3I) = O$

Hence, the Cayley-Hamilton theorem is verified.

Example 6. *Use Cayley-Hamilton theorem to express $2A^5 - 3A^4 + A^2 - 4I$ as a linear polynomial in A, where :*

$$A = \begin{bmatrix} 3 & 1 \\ -1 & 2 \end{bmatrix}.$$

Solution. The characteristic equation of A is given by

$$|A - \lambda I| = 0$$

or
$$\begin{vmatrix} 3-\lambda & 1 \\ -1 & 2-\lambda \end{vmatrix} = 0$$

or $(3-\lambda)(2-\lambda)+1 = 0$

or $\lambda^2 - 5\lambda + 7 = 0$

By Cayley-Hamilton theorem, we have

$$A^2 - 5A + 7I = O \quad \ldots(1)$$

$$\Rightarrow \quad A^2 = 5A - 7I \quad \ldots(2)$$

Now $A^3 = A^2 . A$

$$= (5A - 7I)A = 5A^2 - 7A$$

$$\therefore \quad A^3 = 5A^2 - 7A \quad \ldots(3)$$

Again, $A^4 = A^3 . A = (5A^2 - 7A)A$

$= 5A^3 - 7A^2$

$\Rightarrow \quad A^4 = 5(5A^2 - 7A) - 7(5A - 7I)$ [Using (2) and (3)]

$\Rightarrow \quad A^4 = 25A^2 - 35A - 35A + 49I$

$\Rightarrow \quad A^4 = 25(5A - 7I) - 70A + 49I$ [Using (2)]

$\Rightarrow \quad A^4 = 125 A - 175I - 70A + 49I$

$\Rightarrow \quad A^4 = 55A - 126I \quad \ldots(4)$

Also $\quad A^5 = A^4 . A = (55A - 126I)A$

$A^5 = 55A^2 - 126A$

$\Rightarrow \quad A^5 = 55(5A - 7I) - 126A$ [Using (2)]

$\therefore \quad A^5 = 149A - 385I \quad \ldots(5)$

Now $2A^5 - 3A^4 + A^2 - 4I$

$= 2(149A - 385I) - 3(55A - 126I) + 5A - 7I - 4I$ [Using (2), (4) and (5)]

$= 298A - 770I - 165A + 378I + 5A - 11I$

$= 138A - 403I$

$\therefore \quad 2A^5 - 3A^4 + A^2 - 4I = 138A - 403I$

which is a linear polynomial in A.

EXERCISE 25.10

1. Verify Cayley-Hamilton theorem for the matrix $A = \begin{bmatrix} 1 & 1 \\ 8 & 1 \end{bmatrix}$ and use it to find A^{-1}.
2. Use Cayley-Hamilton theorem to find the inverse of the matrix $A = \begin{bmatrix} 2 & 1 \\ 5 & 3 \end{bmatrix}$.
3. Verify Cayley-Hamilton theorem for the matrix $A = \begin{bmatrix} 0 & 0 & 0 \\ 3 & 1 & 0 \\ -2 & 1 & 4 \end{bmatrix}$ and hence find A^{-1}.
4. Verify Cayley-Hamilton theorem for the following matrix: $A = \begin{bmatrix} 2 & 0 \\ 0 & 1 \end{bmatrix}$.
5. Show that the matrix $A = \begin{bmatrix} 1 & 2 \\ 1 & 1 \end{bmatrix}$ satisfies Cayley-Hamilton theorem.
6. State the Cayley-Hamilton theorem and verify it for the matrix $A = \begin{bmatrix} 1 & 0 & -2 \\ 0 & 0 & 0 \\ -2 & 0 & 4 \end{bmatrix}$
7. Verify Cayley-Hamilton theorem for the matrix $A = \begin{bmatrix} 1 & 4 \\ 2 & 3 \end{bmatrix}$ and hence obtain A^{-1}.
8. Verify Cayley-Hamilton theorem for the matrix $$A = \begin{bmatrix} 1 & 2 & 1 \\ 0 & 1 & -1 \\ 3 & -1 & 1 \end{bmatrix}$$ and hence find A^{-1}.
9. Verify that the matrix $A = \begin{bmatrix} 1 & 2 & 1 \\ -1 & 0 & 3 \\ 2 & -1 & 1 \end{bmatrix}$ satisfies its characteristic equation.
10. Show that the matrix $A = \begin{bmatrix} 2 & 2 & 1 \\ 1 & 3 & 1 \\ 1 & 2 & 2 \end{bmatrix}$ satisfies Cayley-Hamilton theorem.
11. Verify Cayley-Hamilton theorem for the matrix $$A = \begin{bmatrix} 1 & \sqrt{2} & 0 \\ \sqrt{2} & -1 & 0 \\ 0 & 0 & 1 \end{bmatrix}$$ and hence find A^{-1}.
12. Verify Cayley-Hamilton theorem for the matrix $$A = \begin{bmatrix} 1 & 0 & 2 \\ 0 & -1 & 1 \\ 0 & 1 & 0 \end{bmatrix}$$ and hence find A^{-1}.
13. Verify Cayley-Hamilton theorem for the matrix $$A = \begin{bmatrix} 1 & 3 & 7 \\ 4 & 2 & 3 \\ 0 & 2 & 1 \end{bmatrix}$$ and hence find A^{-1}.
14. Verify Cayley-Hamilton theorem for the matrix $$A = \begin{bmatrix} 1 & 2 & 3 \\ 3 & -2 & 1 \\ 4 & 2 & 1 \end{bmatrix}$$ and hence find A^{-1}.
15. Verify Cayley-Hamilton theorem for the matrix $$A = \begin{bmatrix} 1 & 1 & 3 \\ 5 & 2 & 6 \\ -2 & -1 & -3 \end{bmatrix}$$
16. Verify Cayley-Hamilton theorem for the matrix $$A = \begin{bmatrix} 3 & 2 & 4 \\ 4 & 3 & 2 \\ 2 & 4 & 3 \end{bmatrix}$$
17. Verify Cayley-Hamilton theorem for the matrix $$A = \begin{bmatrix} 2 & 3 & -2 \\ 0 & 5 & 4 \\ 1 & 0 & 1 \end{bmatrix}$$
18. If $A = \begin{bmatrix} 1 & 2 \\ -1 & 3 \end{bmatrix}$ express $A^6 - 4A^5 + 8A^4 - 12A^3 + 14A^2$ as a linear polynomial in A.
19. Find the characteristic equation of the matrix $A = \begin{bmatrix} 2 & 1 & 1 \\ 0 & 1 & 0 \\ 1 & 1 & 2 \end{bmatrix}$ and hence compute A^{-1}. Also find the value of $A^8 - 5A^7 + 7A^6 - 3A^5 + A^4 - 5A^3 + 8A^2 - 2A + I$ (UKTU 2010)
20. If $A = \begin{bmatrix} 1 & 0 & 0 \\ 1 & 0 & 1 \\ 0 & 1 & 0 \end{bmatrix}$, show that for every integer $n \geqslant 3$ $A^n = A^{n-2} + A^2 - I$ (UPTU-2009, Mumbai-2006)

21. Verify Cayleg-Hamilton theorem for the following matrices

(i) $\begin{bmatrix} 2 & 2 & 1 \\ 0 & 1 & -1 \\ 3 & -1 & 1 \end{bmatrix}$ (UPTU-2007)

(ii) $\begin{bmatrix} 1 & 0 & -4 \\ 0 & 5 & 4 \\ -4 & 4 & 3 \end{bmatrix}$ (GBTU-2010)

(iii) $\begin{bmatrix} 3 & 0 & 1 \\ 0 & 2 & 0 \\ 0 & 0 & 1 \end{bmatrix}$ (UPTU-2008)

(iv) $\begin{bmatrix} 7 & 2 & -2 \\ -6 & -1 & 2 \\ 6 & 2 & -1 \end{bmatrix}$ (UKTU-2012)

(v) $\begin{bmatrix} 2 & -1 & 1 \\ -1 & 2 & -1 \\ 1 & -1 & 2 \end{bmatrix}$ (SVTV-2008, Anna-2009, Madras-2006)

(vi) $\begin{bmatrix} 3 & 2 & 4 \\ 4 & 3 & 2 \\ 2 & 4 & 3 \end{bmatrix}$ (PTU-2006)

(vii) $\begin{bmatrix} 1 & 3 & 7 \\ 4 & 2 & 3 \\ 1 & 2 & 1 \end{bmatrix}$ (Anna-2005, Bhopal-2008,Kerala 2005)

(viii) $\begin{bmatrix} 1 & 2 & 3 \\ 2 & 4 & 5 \\ 3 & 5 & 6 \end{bmatrix}$ (UPTU-2007)

ANSWERS

1. $A^{-1} = -\frac{1}{7}\begin{bmatrix} 1 & -1 \\ -8 & 1 \end{bmatrix}$

2. $A^{-1} = \begin{bmatrix} 3 & -1 \\ -5 & 2 \end{bmatrix}$

3. $A^{-1} = \frac{1}{5}\begin{bmatrix} 4 & 1 & -1 \\ -12 & 2 & 3 \\ 5 & 0 & 0 \end{bmatrix}$

7. $A^{-1} = -\frac{1}{3}\begin{bmatrix} 3 & -4 \\ -2 & 1 \end{bmatrix}$

8. $A^{-1} = \frac{1}{9}\begin{bmatrix} 0 & 3 & 3 \\ 3 & 2 & -1 \\ 3 & -7 & -1 \end{bmatrix}$

11. $A^{-1} = -\frac{1}{3}\begin{bmatrix} -1 & -\sqrt{2} & 0 \\ -\sqrt{2} & 1 & 0 \\ 0 & 0 & -3 \end{bmatrix}$

13. $A^{-1} = \frac{1}{10}\begin{bmatrix} -4 & 11 & -5 \\ -4 & 1 & 25 \\ 8 & -2 & -10 \end{bmatrix}$

14. $A^{-1} = \frac{1}{36}\begin{bmatrix} -4 & 4 & 8 \\ 1 & -11 & 0 \\ 14 & 6 & -8 \end{bmatrix}$

18. $-4A + 5I$

19. $A^3 - 5A^2 + 7A - 3I = 0;\ A^{-1} = \frac{1}{3}\begin{bmatrix} 2 & -1 & -1 \\ 0 & 3 & 0 \\ -1 & -1 & 2 \end{bmatrix}, \begin{bmatrix} 8 & 5 & 5 \\ 0 & 3 & 0 \\ 5 & 5 & 8 \end{bmatrix}$

25.51 DIAGONALISATION OF A MATRIX

Let A be a square matrix of order n. Then A is said to be diagonalizable iff it is similar to a diagonal matrix. Therefore, A is diagonalizable if there exists a non-singular matrix P such that $P^{-1}AP = D$ where D is a diagonal matrix and the matrix P is said to transform A to diagonal form.

Theorem 1. *An $n \times n$ matrix is diagonalizable if and only if it possesses n linearly independent eigenvectors.*

Proof. Let A be $n \times n$ matrix. Suppose that A is diagonalizable, then it is similar to a diagonal matrix. Let $D=$ diag $(\lambda_1, \lambda_2, \ldots, \lambda_n)$ be that diagonal matrix, therefore there exists a non-singular matrix P (say) such that

$$P^{-1}AP = D \Rightarrow AP = PD \quad \ldots(1)$$

The eigenvalues of D are $\lambda_1, \lambda_2, \ldots, \lambda_n$ and A is similar to D, therefore $\lambda_1, \lambda_2, \ldots, \lambda_n$ are the only eigenvalues of A.

Suppose that $X_1, X_2, \ldots, X_n$ are column vectors of P *i.e.,* $P = [X_1, X_2, \ldots, X_n]$

Since P is invertible so that $X_1, X_2, \ldots, X_n$ are n linearly independent vectors.

Now from (1), we get

$$A[X_1, X_2, \ldots, X_n] = [X_1, X_2, \ldots, X_n]\text{dia.}[\lambda_1, \lambda_2, \ldots, \lambda_n]$$

or
$$[AX_1, AX_2, \ldots, AX_n] = [\lambda_1 X_1, \lambda_2 X_2, \ldots, \lambda_n X_n]$$

$\Rightarrow$
$$AX_1 = \lambda_1 X_1, AX_2 = \lambda_2 X_2, \ldots, AX_n = \lambda_n X_n$$

Therefore, $X_1, X_2, \ldots, X_n$ are the eigenvectors of A corresponding to the eigenvalues of A. Also $X_1, X_2, \ldots, X_n$ are linearly independent, hence A has n linearly independent eigenvectors.

Conversely. Suppose that A has n linearly independent eigenvectors $X_1, X_2, \ldots, X_n$ then there are scalars $\lambda_1, \lambda_2, \ldots, \lambda_n$ (not necessarily distinct) such that

$$AX_1 = \lambda_1 X_1, AX_2 = \lambda_2 X_2, \ldots, AX_n = \lambda_n X_n$$

Let $P = [X_1, X_2, \ldots, X_n]$. Since $X_1, X_2, \ldots, X_n$ are linearly independent eigenvectors, therefore P is invertible.

Let $D = \text{dia.}\ (\lambda_1, \lambda_2, \ldots, \lambda_n)$. Then

$$AP = A\,[X_1, X_2, ..., X_n] = [AX_1, AX_2, ..., AX_n]$$
$$= [\lambda_1 X_1, \lambda_2 X_2, ..., \lambda_n X_n] \quad [X_1, X_2, ..., X_n]\text{dia.}[\lambda_1, \lambda_2, ..., \lambda_n]$$
$$\Rightarrow \quad AP = PD \quad \Rightarrow \quad P^{-1}AP = D \qquad [\because P^{-1} \text{ exists.}]$$

$\Rightarrow$ A is similar to a diagonal matrix.

$\Rightarrow$ A is diagonalizable.

Corollary. *If the eigenvalues of an $n \times n$ matrix are all distinct then it is necessarily diagonalizable.*

Proof. Let A be an $n \times n$ matrix. Since the eigenvectors of A corresponding to the distinct eigenvalues are linearly independent, therefore A has n linearly independent eigenvectors, hence by above theorem, A is diagonalizable.

REMARK

- In view of above theorem, if A is diagonalizable and P diagonalizes A, then

$$P^{-1}AP = \begin{bmatrix} \lambda_1 & 0 & 0 & \cdots & 0 \\ 0 & \lambda_2 & 0 & \cdots & 0 \\ 0 & 0 & \lambda_3 & \cdots & 0 \\ \vdots & & & & \\ 0 & 0 & 0 & \cdots & \lambda_n \end{bmatrix}$$

if and only if the jth column of P is an eigenvector of A corresponding to the eigenvalue λ_i of A for $i = 1, 2, 3, ... n$.

Recapitulations

- A square matrix A is said to be diagonalizable iff it is similiar to a diagonal matrix.
- An $n \times n$ matrix is diagonalizable if and only if it possesses n linearly independent eigenvectors.
- If the eigenvalues of an $n \times n$ matrix are all distinct then it is necessarily diagonalizable.
- If the eigenvalues and corresponding eigenvectors of a matrix A are given, then we can find the matrix A by the relation
$$A = PDP^{-1}$$
where D = dia. $(\lambda_1, \lambda_2, ..., \lambda_n)$ and P is the matrix containing eigenvector X_i corresponding eigenvalues $\lambda_i : i \in$ N.

Solved Examples

Example 1. *Consider the matrix*

$$A = \begin{bmatrix} 3 & 2 & 0 \\ 2 & 0 & 0 \\ 1 & 0 & 2 \end{bmatrix}$$

Find an invertible matrix P such that $P^{-1}AP$ is a diagonal matrix. Also find the diagonal matrix.

Solution. The characteristic equation of A is given by

$$|A - \lambda I| = 0$$

or
$$\begin{vmatrix} 3-\lambda & 2 & 0 \\ 2 & 0-\lambda & 0 \\ 1 & 0 & 2-\lambda \end{vmatrix} = 0$$

or $(3 - \lambda)\{-\lambda(2 - \lambda)\} - 2\{2\,(2 - \lambda)\} = 0$

or $(2 - \lambda)(4 - \lambda)(\lambda + 1) = 0$

$\therefore$ $-1, 2, 4$ are eigenvalues of A.

Since A has distinct eigenvalues so that A is diagonalisable, therefore there exists a non-singular matrix P such that $P^{-1}AP$ is a diagonalisable.

Now to find P we shall find eigenvectors of A.

Eigenvector corresponding to $\lambda_1 = -1$

Let $X_1 = \begin{bmatrix} x_1 \\ x_2 \\ x_3 \end{bmatrix} \neq O$ be an eigenvector corresponding to eigenvalue $\lambda_1 = -1$, then

or $AX_1 = (-1)X_1$

or $(A + I)X_1 = O$

or
$$\begin{bmatrix} 3-(-1) & 2 & 0 \\ 2 & 0-(-1) & 0 \\ 1 & 0 & 2-(-1) \end{bmatrix} \begin{bmatrix} x_1 \\ x_2 \\ x_3 \end{bmatrix} = \begin{bmatrix} 0 \\ 0 \\ 0 \end{bmatrix}$$

or
$$\begin{bmatrix} 4 & 2 & 0 \\ 2 & 1 & 0 \\ 1 & 0 & 3 \end{bmatrix} \begin{bmatrix} x_1 \\ x_2 \\ x_3 \end{bmatrix} = \begin{bmatrix} 0 \\ 0 \\ 0 \end{bmatrix} \qquad ...(1)$$

Applying $R_1 \leftrightarrow R_3$, we get

$$\begin{bmatrix} 1 & 0 & 3 \\ 2 & 1 & 0 \\ 4 & 2 & 0 \end{bmatrix} \begin{bmatrix} x_1 \\ x_2 \\ x_3 \end{bmatrix} = \begin{bmatrix} 0 \\ 0 \\ 0 \end{bmatrix}$$

Applying $R_2 \to R_2 - 2R_1$, $R_3 \to R_3 - 4R_1$, we get

$$\begin{bmatrix} 1 & 0 & 3 \\ 0 & 1 & -6 \\ 0 & 2 & -12 \end{bmatrix} \begin{bmatrix} x_1 \\ x_2 \\ x_3 \end{bmatrix} = \begin{bmatrix} 0 \\ 0 \\ 0 \end{bmatrix}$$

Applying $R_3 \to R_3 - 2R_2$, we get

$$\begin{bmatrix} 1 & 0 & 3 \\ 0 & 1 & -6 \\ 0 & 0 & 0 \end{bmatrix} \begin{bmatrix} x_1 \\ x_2 \\ x_3 \end{bmatrix} = \begin{bmatrix} 0 \\ 0 \\ 0 \end{bmatrix} \qquad ...(2)$$

Clearly $\rho(A + I) = 2$, therefore the equation (1) will have $3 - 2 = 1$ linearly independent solution.

From (2), we have $x_1 + 3x_3 = 0$
$$x_2 - 6x_3 = 0$$

Let us put $x_3 = c$ (an arbitrary constant), then
$$x_2 = c, x_1 = -3c$$

Now $\begin{bmatrix} x_1 \\ x_2 \\ x_3 \end{bmatrix} = \begin{bmatrix} -3c \\ 6c \\ c \end{bmatrix} = c\begin{bmatrix} -3 \\ 6 \\ 1 \end{bmatrix}$

$\therefore \quad X_1 = \begin{bmatrix} -3 \\ 6 \\ 1 \end{bmatrix}$

Eigenvector corresponding to $\lambda_2 = 2$:

Let $X_2 = \begin{bmatrix} x_1 \\ x_2 \\ x_3 \end{bmatrix} \neq O$ be an eigenvector corresponding to eigenvalue $\lambda_2 = 2$, then

or $\quad AX_2 = 2X_2$

or $\quad (A - 2I)X_2 = O$

or $\begin{bmatrix} 3-2 & 2 & 0 \\ 2 & 0-2 & 0 \\ 1 & 0 & 2-2 \end{bmatrix}\begin{bmatrix} x_1 \\ x_2 \\ x_3 \end{bmatrix} = \begin{bmatrix} 0 \\ 0 \\ 0 \end{bmatrix}$

or $\begin{bmatrix} 1 & 2 & 0 \\ 2 & -2 & 0 \\ 1 & 0 & 0 \end{bmatrix}\begin{bmatrix} x_1 \\ x_2 \\ x_3 \end{bmatrix} = \begin{bmatrix} 0 \\ 0 \\ 0 \end{bmatrix}$ ...(3)

Applying $R_2 \to R_2 - 2R_1, R_3 \to R_3 - R_1$, we get

$$\begin{bmatrix} 1 & 2 & 0 \\ 0 & -6 & 0 \\ 0 & -2 & 0 \end{bmatrix}\begin{bmatrix} x_1 \\ x_2 \\ x_3 \end{bmatrix} = \begin{bmatrix} 0 \\ 0 \\ 0 \end{bmatrix}$$

Applying $R_3 \leftrightarrow R_2$ we get

$$\begin{bmatrix} 1 & 2 & 0 \\ 0 & -2 & 0 \\ 0 & -6 & 0 \end{bmatrix}\begin{bmatrix} x_1 \\ x_2 \\ x_3 \end{bmatrix} = \begin{bmatrix} 0 \\ 0 \\ 0 \end{bmatrix}$$

Applying $R_3 \to R_3 - 3R_2$, we get

$$\begin{bmatrix} 1 & 2 & 0 \\ 0 & -2 & 0 \\ 0 & 0 & 0 \end{bmatrix}\begin{bmatrix} x_1 \\ x_2 \\ x_3 \end{bmatrix} = \begin{bmatrix} 0 \\ 0 \\ 0 \end{bmatrix} \quad ...(4)$$

Clearly $\rho(A - 2I) = 2$, therefore the equation (3) will have $3 - 2 = 1$ linearly independent solution.

From (4), we have $\quad x_1 + 2x_2 = 0$

$\quad - 2x_2 = 0$

Let us put $x_3 = c$, we get $x_2 = 0, x_1 = 0$

Now $\begin{bmatrix} x_1 \\ x_2 \\ x_3 \end{bmatrix} = \begin{bmatrix} 0 \\ 0 \\ c \end{bmatrix} = c\begin{bmatrix} 0 \\ 0 \\ 1 \end{bmatrix}$

$\therefore \quad X_2 = \begin{bmatrix} 0 \\ 0 \\ 1 \end{bmatrix}$

Eigenvector corresponding to $\lambda_3 = 4$:

Let $X_3 = \begin{bmatrix} x_1 \\ x_2 \\ x_3 \end{bmatrix} \neq O$ be an eigenvector corresponding to eigenvalue $\lambda_3 = 4$, then

or $\quad AX_3 = 4X_3$

or $\quad (A - 4I)X_3 = O$

or $\begin{bmatrix} 3-4 & 2 & 0 \\ 2 & 0-4 & 0 \\ 1 & 0 & 2-4 \end{bmatrix}\begin{bmatrix} x_1 \\ x_2 \\ x_3 \end{bmatrix} = \begin{bmatrix} 0 \\ 0 \\ 0 \end{bmatrix}$

or $\begin{bmatrix} -1 & 2 & 0 \\ 2 & -4 & 0 \\ 1 & 0 & -2 \end{bmatrix}\begin{bmatrix} x_1 \\ x_2 \\ x_3 \end{bmatrix} = \begin{bmatrix} 0 \\ 0 \\ 0 \end{bmatrix}$ ...(5)

Applying $R_2 \to R_2 + 2R_1, R_3 \to R_3 + R_1$, we get

$$\begin{bmatrix} -1 & 2 & 0 \\ 0 & 0 & 0 \\ 0 & 2 & -2 \end{bmatrix}\begin{bmatrix} x_1 \\ x_2 \\ x_3 \end{bmatrix} = \begin{bmatrix} 0 \\ 0 \\ 0 \end{bmatrix}$$

Applying $R_2 \leftrightarrow R_3$, we get

$$\begin{bmatrix} -1 & 2 & 0 \\ 0 & 2 & -2 \\ 0 & 0 & 0 \end{bmatrix}\begin{bmatrix} x_1 \\ x_2 \\ x_3 \end{bmatrix} = \begin{bmatrix} 0 \\ 0 \\ 0 \end{bmatrix} \quad ...(6)$$

Clearly $\rho(A - 4I) = 2$, therefore the equation (5) will have $3 - 2 = 1$ linearly independent solution.

From (6), we have $\quad -x_1 + 2x_2 = 0$

$\quad 2x_2 - 2x_3 = 0$

Let us put $x_2 = c$, we get $x_2 = 2c, x_3 = c$

Now $\begin{bmatrix} x_1 \\ x_2 \\ x_3 \end{bmatrix} = \begin{bmatrix} 2c \\ c \\ c \end{bmatrix} = c\begin{bmatrix} 2 \\ 1 \\ 1 \end{bmatrix}$

$\therefore \quad X_3 = \begin{bmatrix} 2 \\ 1 \\ 1 \end{bmatrix}$

Since $\quad P = [X_1, X_2, X_3]$

$\therefore \quad = \begin{bmatrix} -3 & 0 & 2 \\ 6 & 0 & 1 \\ 1 & 1 & 1 \end{bmatrix}$

and $\quad P^{-1}AP = \text{dia.}(-1, 2, 4)$

$$= \begin{bmatrix} -1 & 0 & 0 \\ 0 & 2 & 0 \\ 0 & 0 & 4 \end{bmatrix} = D$$

Example 2. *Show that the matrix*

$$A = \begin{bmatrix} -9 & 4 & 4 \\ -8 & 3 & 4 \\ -16 & 8 & 7 \end{bmatrix}$$

is diagonalizable. Also find the diagonal form and a diagonalizing matrix P.

Solution. The characteristic equation of A is given by

$$|A - \lambda I| = 0$$

or $$\begin{vmatrix} -9-\lambda & 4 & 4 \\ -8 & 3-\lambda & 4 \\ -16 & 8 & 7-\lambda \end{vmatrix} = 0$$

Applying $C_1 \to C_1 + C_2 + C_3$, we get

$$\begin{vmatrix} -1-\lambda & 4 & 4 \\ -1-\lambda & 3-\lambda & 4 \\ -1-\lambda & 8 & 7-\lambda \end{vmatrix} = 0$$

or $$-(1+\lambda)\begin{vmatrix} 1 & 4 & 4 \\ 1 & 3-\lambda & 4 \\ 1 & 8 & 7-\lambda \end{vmatrix} = 0$$

Applying $R_1 \to R_2 - R_1$ and $R_3 \to R_3 - R_1$, we get

or $$-(1+\lambda)\begin{vmatrix} 1 & 4 & 4 \\ 1 & -1-\lambda & 0 \\ 1 & 4 & 3-\lambda \end{vmatrix} = 0$$

or $$-(1 + \lambda)(-1 - \lambda)(3 - \lambda) = 0$$

or $$(1 + \lambda)^2 (3 + \lambda) = 0$$

$\therefore$ $-1, -1, 3$ are eigenvalues of A.

Eigenvector corresponding to eigenvalue -1 :

Let $X = \begin{bmatrix} x_1 \\ x_2 \\ x_3 \end{bmatrix} \neq O$ be an eigenvector corresponding to eigenvalue -1, then

or $$AX = (-1)X$$

or $$(A + I)X = O$$

or $$\begin{bmatrix} -9+1 & 4 & 4 \\ -8 & 3+1 & 4 \\ -16 & 8 & 7+1 \end{bmatrix}\begin{bmatrix} x_1 \\ x_2 \\ x_3 \end{bmatrix} = \begin{bmatrix} 0 \\ 0 \\ 0 \end{bmatrix}$$

or $$\begin{bmatrix} -8 & 4 & 4 \\ -8 & 4 & 4 \\ -16 & 8 & 8 \end{bmatrix}\begin{bmatrix} x_1 \\ x_2 \\ x_3 \end{bmatrix} = \begin{bmatrix} 0 \\ 0 \\ 0 \end{bmatrix} \quad \ldots(1)$$

Applying $R_1 \to \frac{1}{4}R_1$' we get

$$\begin{bmatrix} -2 & 1 & 1 \\ -8 & 4 & 4 \\ -16 & 8 & 8 \end{bmatrix}\begin{bmatrix} x_1 \\ x_2 \\ x_3 \end{bmatrix} = \begin{bmatrix} 0 \\ 0 \\ 0 \end{bmatrix}$$

Applying $R_2 \to R_2 - 4R_1$, $R_3 \to R_3 - 8R_1$, we get

$$\begin{bmatrix} -2 & 1 & 1 \\ 0 & 0 & 0 \\ 0 & 0 & 0 \end{bmatrix}\begin{bmatrix} x_1 \\ x_2 \\ x_3 \end{bmatrix} = \begin{bmatrix} 0 \\ 0 \\ 0 \end{bmatrix} \quad \ldots(2)$$

Clearly $\rho(A + I) = 1$, therefore the equation (1) will have $3 - 1 = 2$ linearly independent solutions.

From (2), we have

$$-2x_1 + x_2 + x_3 = 0$$

Let us put $x_1 = c_1$, $x_2 = c_2$ then $x_3 = 2c_1 - c_2$

Now $$\begin{bmatrix} x_1 \\ x_2 \\ x_3 \end{bmatrix} = \begin{bmatrix} c_1 \\ c_2 \\ 2c_1 - c_2 \end{bmatrix} = c_1\begin{bmatrix} 1 \\ 0 \\ 2 \end{bmatrix} + c_2\begin{bmatrix} 0 \\ 1 \\ -1 \end{bmatrix}$$

$\therefore$ $$X_1 = \begin{bmatrix} 1 \\ 0 \\ 2 \end{bmatrix}, X_2 = \begin{bmatrix} 0 \\ 1 \\ -1 \end{bmatrix}$$

Eigenvector corresponding to eigenvalue 3 :

Let $X_3 = \begin{bmatrix} x_1 \\ x_2 \\ x_3 \end{bmatrix} \neq O$ be an eigenvector corresponding to eigenvalue 3, then

or $$AX_3 = 3X_3$$

or $$(A - 3I)X_2 = O$$

or $$\begin{bmatrix} -9-3 & 4 & 4 \\ -8 & 3-3 & 4 \\ -16 & 8 & 7-3 \end{bmatrix}\begin{bmatrix} x_1 \\ x_2 \\ x_3 \end{bmatrix} = \begin{bmatrix} 0 \\ 0 \\ 0 \end{bmatrix}$$

or $$\begin{bmatrix} -12 & 4 & 4 \\ -8 & 0 & 4 \\ -16 & 8 & 4 \end{bmatrix}\begin{bmatrix} x_1 \\ x_2 \\ x_3 \end{bmatrix} = \begin{bmatrix} 0 \\ 0 \\ 0 \end{bmatrix} \quad \ldots(3)$$

Applying $R_2 \leftrightarrow R_1$, we get

$$\begin{bmatrix} -8 & 0 & 4 \\ -12 & 4 & 4 \\ -16 & 8 & 4 \end{bmatrix}\begin{bmatrix} x_1 \\ x_2 \\ x_3 \end{bmatrix} = \begin{bmatrix} 0 \\ 0 \\ 0 \end{bmatrix}$$

Applying $R_1 \to \frac{1}{4}R_1$, we get

$$\begin{bmatrix} -2 & 0 & 1 \\ -12 & 4 & 4 \\ -16 & 8 & 4 \end{bmatrix}\begin{bmatrix} x_1 \\ x_2 \\ x_3 \end{bmatrix} = \begin{bmatrix} 0 \\ 0 \\ 0 \end{bmatrix}$$

Applying $R_2 \to R_2 - 6R_1$, $R_3 \to R_3 - 8R_1$, we get

$$\begin{bmatrix} -2 & 0 & 1 \\ 0 & 4 & -2 \\ 0 & 8 & -4 \end{bmatrix}\begin{bmatrix} x_1 \\ x_2 \\ x_3 \end{bmatrix} = \begin{bmatrix} 0 \\ 0 \\ 0 \end{bmatrix} \quad \ldots(4)$$

Applying $R_3 \to R_3 - 2R_2$, we get

$$\begin{bmatrix} -2 & 0 & 1 \\ 0 & 4 & -2 \\ 0 & 0 & 0 \end{bmatrix}\begin{bmatrix} x_1 \\ x_2 \\ x_3 \end{bmatrix} = \begin{bmatrix} 0 \\ 0 \\ 0 \end{bmatrix}$$

Clearly $\rho(A - 3I) = 2$, therefore the equation (3) will have $3 - 2 = 1$ linearly independent solution.

From (4), we have

$$-2x_1 + x_3 = 0$$
$$4x_2 - 2x_3 = 0$$

Let us put $x_2 = c$, we get $x_3 = 2c$, $x_1 = c$

Now $\begin{bmatrix} x_1 \\ x_2 \\ x_3 \end{bmatrix} = \begin{bmatrix} c \\ c \\ 2c \end{bmatrix} = c\begin{bmatrix} 1 \\ 1 \\ 2 \end{bmatrix}$

$\therefore \quad X_3 = \begin{bmatrix} 1 \\ 1 \\ 2 \end{bmatrix}$

Since $P = [X_1, X_2, X_3]$

$\therefore \quad P = \begin{bmatrix} 1 & 0 & 1 \\ 0 & 1 & 1 \\ 2 & -1 & 2 \end{bmatrix}$

Also $P^{-1}AP = \text{diag.}(-1, -1, 3)$

$$= \begin{bmatrix} -1 & 0 & 0 \\ 0 & -1 & 0 \\ 0 & 0 & 3 \end{bmatrix} = D$$

Example 3. *Show that the matrix* $A = \begin{bmatrix} 8 & -8 & -2 \\ 4 & -3 & -2 \\ 3 & -4 & 1 \end{bmatrix}$ *is diagonalizable.*

Solution. If the matrix A has distinct eigenvalues, then it is essentially diagonalizable, so we shall find its eigenvalues.

The characteristic equation of A is given by

$$|A - \lambda I| = 0$$

or $\begin{vmatrix} 8-\lambda & -8 & -2 \\ 4 & -3-\lambda & -2 \\ 3 & -4 & 1-\lambda \end{vmatrix} = 0$

Applying $R_1 \to R_1 - (R_2 + R_3)$, we get

$$\begin{vmatrix} 1-\lambda & -1+\lambda & -1+\lambda \\ 4 & -3-\lambda & -2 \\ 3 & -4 & 1-\lambda \end{vmatrix} = 0$$

or $(1+\lambda)\begin{vmatrix} 1 & -1 & -1 \\ 4 & -3-\lambda & -2 \\ 3 & -4 & 1-\lambda \end{vmatrix} = 0$

Applying $C_2 \to C_2 + C_1, C_3 \to C_3 + C_1$, we get

or $(1-\lambda)\begin{vmatrix} 1 & 0 & 0 \\ 4 & 1-\lambda & 2 \\ 3 & -1 & 4-\lambda \end{vmatrix} = 0$

or $(1-\lambda)\{(1-\lambda)(4-\lambda) + 2\} = 0$

or $(1-\lambda)(\lambda^2 - 5\lambda + 6) = 0$

or $(1-\lambda)(2-\lambda)(3-\lambda) = 0$

$\therefore$ A has 1, 2, 3 eigenvalues which are all distinct, hence A is diagonalizable.

REMARK

- If some eigenvalues of a matrix are the same, then it need not be non-diagonalizable.

25.52 ALGEBRAIC AND GEOMETRIC MULTIPLICITY OF AN EIGENVALUE

(i) **Algebraic multiplicity :** Let A be an $n \times n$ matrix and let

$$|A - \lambda I_n| = (\lambda - \lambda_1)^{n_1}(\lambda - \lambda_2)^{n_2}(\lambda - \lambda_3)^{n_3} \ldots (\lambda - \lambda_k)^{n_k}$$

where $n_1 + n_2 + n_3 + \ldots + n_k = n$. Then the numbers $n_1, n_2, n_3, \ldots, n_k$ are the algebraic multiplicities of the eigenvalues $\lambda_1, \lambda_2, \lambda_3, \ldots, \lambda_n$ respectively.

(ii) **Eigenspace of a matrix A corresponding to eigenvalue λ:**
Let A be an $n \times n$ matrix, let λ be an eigenvalue of A.

For eigenvalue λ, we find eigenvectors $X = \begin{bmatrix} x_1 \\ x_2 \\ x_3 \\ \vdots \\ x_n \end{bmatrix}$

by solving the linear system $(A - \lambda I)X = O$

The set of all vectors X satisfying $AX = \lambda X$ is called the eigenspace of A corresponding to eigenvalue A, which is denoted by E_k.

(iii) **Geometric multiplicity :**
Let A be an $n \times n$ matrix and λ be one of its eigenvalues, then the dimension of eigenspace E_λ of A corresponding to λ is called the geometric multiplicity of λ.

REMARKS

- Dimension of E_λ = Dimension of the null space $(A - \lambda I)$.
- Geometric multiplicity of $\lambda = n - \rho(A - \lambda I)$.
- The geometric mulitplicity of an eigenvalue cannot exceed its algebraic multiplicity.

Theorem 1. *The necessary and sufficient conditions for a square matrix to be similar to a diagonal matrix is that the geometric multiplicity of each of its eigenvalues coincides with the algebraic multiplicity.*

Proof. Let A be an $n \times n$ matrix.

Necessary Condition : Suppose that A is similar to a diagonal matrix $D =$ dia. $(\lambda_1, \lambda_2, ..., \lambda_n)$, then there exists a non-singular matrix P such that

$$P^{-1}AP = D = \text{dia. } (\lambda_1, \lambda_2, ..., \lambda_n) \quad ...(1)$$

$\therefore$ $\lambda_1, \lambda_2, ..., \lambda_n$ are the eigenvalues of A not necessarily distinct.

Let t be an eigenvalue of A of algebraic multiplicity p, then we have

$$\lambda_1 = t, \ \lambda_2 = t, ..., \lambda_p = t \quad ...(2)$$

If $\rho(A - tI) = m$, then the system of equations

$$(A - tI)X = O$$

will have $n - m$ linearly independent solutions and therefore $n - m$ will be the geometric multiplicity of t.

So, we have to prove that $p = n - m$.

Since the rank of a matrix does not change on premultiplication and postmultiplication by a non-singular matrix.

$\therefore$

$$\rho(A - tI) = \rho\left[P^{-1}(A - tI)P\right] = \rho\left[P^{-1}AP - tI\right] = \rho[D - tI]$$

$$= \rho\left(\text{dia.}\left[\lambda_1 - t, \lambda_2 - t, \ldots, \lambda_p - t, \lambda_{p+1} - t, \lambda_n - t\right]\right)$$

$$= \rho\left(\text{dia.}\left[0, 0, \ldots, 0, \lambda_{p+1} - t, \lambda_n - t\right]\right) \quad \text{[Using (2)]}$$

$$\rho(A - tI) = n - p$$

$$\Rightarrow \quad m = n - p \quad \because \rho(A - tI) = m$$

$$\Rightarrow \quad p = m - n$$

Sufficient Condition : Suppose that the geometric multiplicity of each eigenvalue of A coincides with its algebraic multiplicity. Then we have to show that A is similar to a diagonal matrix, *i.e.*, A is diagonalizable.

Suppose that A has $\lambda_1, \lambda_2, ... \lambda_k$ distinct eigenvalues of multiplicity $n_1, n_2, ... n_k$ respectively, then we have

$$n_1 + n_2 + \ldots + n_k = n$$

Let $\quad X_{11}, X_{12}, \ldots, X_{1n_1}, X_{21}, X_{22}, \ldots, X_{2n_2}; \ldots; X_{k1}, X_{k2}, \ldots, X_{kn_k}\}$

be linearly independent sets of eigenvectors corresponding to the eigenvalues $\lambda_1, \lambda_2, ... \lambda_k$ respectively.

Now we prove that the n vectors given by (3) are linearly independent.

Let

$$(a_{11}X_{11} + a_{12}X_{12} + \ldots + a_{1n_1}X_{1n_1}) + (a_{21}X_{21} + a_{22}X_{22} + \ldots + a_{2n_2}X_{2n_2})$$
$$+ \ldots + (a_{k1}X_{k1} + a_{k2}X_{k2} + \ldots + a_{kn_K}X_{kn_K}) = O \quad ...(4)$$

or

$$X_1 + X_2 + \ldots + X_k = O \quad ...(5)$$

where

$$\left.\begin{aligned} X_1 &= a_{11}X_{11} + a_{12}X_{12} + \ldots + a_{1n_1}X_{1n_1} \\ X_2 &= a_{21}X_{21} + a_{22}X_{22} + \ldots + a_{2n_2}X_{2n_2} \\ &\ldots\ldots\ldots\ldots\ldots\ldots\ldots\ldots \\ X_k &= a_{k1}X_{k1} + a_{k2}X_{k2} + \ldots + a_{kn_k}X_{kn_k} \end{aligned}\right\} \quad ...(6)$$

From (6), we see that X_1 is a linear combination of $X_{11}, X_{12}, ..., X_{1n_1}$ which are eigenvectors of A corresponding to the eigenvalue λ_1. If $X_1 \neq O$ then X_1 is also the eigenvector of A corresponding to eigenvalue λ_1, similarly for other eigenvectors $X_2, X_3, ..., X_k$.

Suppose $X_i \neq O$, then from (3), we see that a system of eigenvectors of A corresponding to distinct eigenvalues of A is linearly dependent which is not possible, hence

$$X_i = O \ \forall \ i = 1, 2, 3, ..., k$$

$$\Rightarrow \quad a_{i1}X_{i1} + a_{i2}X_{i2} + \ldots + a_{in_i}X_{in_i} = O \quad \text{[Using (6)]}$$

But $X_{i1}, X_{i2}, ..., X_{in_1}$ is a set of linearly independent vectors, therefore

$$a_{i1} = a_{i2} = \ldots = a_{in_1} = 0 \ \forall \ i = 1, 2, 3, ..., k$$

It follows that the n vectors given by (3) are linearly independent. Therefore A has n linearly independent eigenvectors, hence A is diagonalizable and hence A is similar to a diagonal matrix.

Solved Examples

Example 1. *Show that the following matrices are not similar to diagonal matrices:*

(i) $\begin{bmatrix} 2 & 3 & 4 \\ 0 & 2 & -1 \\ 0 & 0 & 1 \end{bmatrix}$ (ii) $\begin{bmatrix} 2 & -1 & 1 \\ 2 & 2 & -1 \\ 1 & 2 & -1 \end{bmatrix}$

Solution. (i) Let $A = \begin{bmatrix} 2 & 3 & 4 \\ 0 & 2 & -1 \\ 0 & 0 & 1 \end{bmatrix}$

Clearly, A is an upper triangular matrix therefore its diagonal elements 2, 2, 1 are the eigenvalues of A.

The algebraic multiplicity of eigenvalue is 2. Next, we find the geometric multiplicity of eigenvalue 2.

Let $X = \begin{bmatrix} x_1 \\ x_2 \\ x_3 \end{bmatrix} \neq O$ be an eigenvector corresponding to eigenvalue 2, then

or $$AX_1 = 2X_1$$

or $$(A - 2I)X_1 = O$$

or $$\begin{bmatrix} 2-2 & 3 & 4 \\ 0 & 2-2 & -1 \\ 0 & 0 & 1-2 \end{bmatrix} \begin{bmatrix} x_1 \\ x_2 \\ x_3 \end{bmatrix} = \begin{bmatrix} 0 \\ 0 \\ 0 \end{bmatrix}$$

or $$\begin{bmatrix} 0 & 3 & 4 \\ 0 & 0 & -1 \\ 0 & 0 & -1 \end{bmatrix} \begin{bmatrix} x_1 \\ x_2 \\ x_3 \end{bmatrix} = \begin{bmatrix} 0 \\ 0 \\ 0 \end{bmatrix}$$

Applying $R_3 \to R_3 - R_2$, we get

$$\begin{bmatrix} 0 & 3 & 4 \\ 0 & 0 & -1 \\ 0 & 0 & 0 \end{bmatrix} \begin{bmatrix} x_1 \\ x_2 \\ x_3 \end{bmatrix} = \begin{bmatrix} 0 \\ 0 \\ 0 \end{bmatrix}$$

Clearly $\rho(A - 4I) = 2$, therefore the geometric multiplicity of 2 is 3–2 = 1, which is not equal to the algebraic multiplicity of eigenvalue 2. Hence, A is not similar to the diagonal matrix.

(ii) Let $A = \begin{bmatrix} 2 & -1 & 1 \\ 2 & 2 & -1 \\ 1 & 2 & -1 \end{bmatrix}$

The characteristic equation of A is given by

$$|A - \lambda I| = 0$$

or $$\begin{vmatrix} 2-\lambda & -1 & 1 \\ 2 & 2-\lambda & -1 \\ 1 & 2 & -1-\lambda \end{vmatrix} = 0$$

Applying $C_3 \to C_3 + C_2$, we get

$$\begin{vmatrix} 2-\lambda & -1 & 0 \\ 2 & 2-\lambda & 1-\lambda \\ 1 & 2 & 1-\lambda \end{vmatrix} = 0$$

or $$(1-\lambda)\begin{vmatrix} 2-\lambda & -1 & 0 \\ 2 & 2-\lambda & 1 \\ 1 & 2 & 1 \end{vmatrix} = 0$$

Again applying $R_2 \to R_2 - R_3$, we get

or $$(1-\lambda)\begin{vmatrix} 2-\lambda & -1 & 0 \\ 1 & -\lambda & 0 \\ 1 & 2 & 1 \end{vmatrix} = 0$$

or $$(1 - \lambda)\{-\lambda(2 - \lambda) + 1\} = 0$$

or $$(1 - \lambda)(\lambda^2 - 2\lambda + 1) = 0$$

or $$(1 - \lambda)^3 = 0$$

$\therefore$ 1, 1, 1 are the eigenvalues of A, thus the algebraic multiplicity of 1 is 3.

Next, we find the geometric multiplicity of eigenvalue 1.

Let $X = \begin{bmatrix} x_1 \\ x_2 \\ x_3 \end{bmatrix} \neq O$ be an eigenvector A corresponding to eigenvalue 1, then

or $$AX = (1)X$$

or $$(A - I)X = O$$

or $$\begin{bmatrix} 2-1 & -1 & 1 \\ 2 & 2-1 & -1 \\ 1 & 2 & -1-1 \end{bmatrix} \begin{bmatrix} x_1 \\ x_2 \\ x_3 \end{bmatrix} = \begin{bmatrix} 0 \\ 0 \\ 0 \end{bmatrix}$$

or $$\begin{bmatrix} 1 & -1 & 1 \\ 2 & 1 & -1 \\ 1 & 2 & -2 \end{bmatrix} \begin{bmatrix} x_1 \\ x_2 \\ x_3 \end{bmatrix} = \begin{bmatrix} 0 \\ 0 \\ 0 \end{bmatrix}$$

Applying $R_2 \to R_2 - 2R_1$, $R_3 \to R_3 - R_1$, we get

$$\begin{bmatrix} 1 & -1 & 1 \\ 0 & 3 & -3 \\ 0 & 3 & -3 \end{bmatrix} \begin{bmatrix} x_1 \\ x_2 \\ x_3 \end{bmatrix} = \begin{bmatrix} 0 \\ 0 \\ 0 \end{bmatrix}$$

Again applying $R_3 \to R_3 - R_2$, we get

$$\begin{bmatrix} 1 & -1 & 1 \\ 0 & 3 & -3 \\ 0 & 0 & 0 \end{bmatrix} \begin{bmatrix} x_1 \\ x_2 \\ x_3 \end{bmatrix} = \begin{bmatrix} 0 \\ 0 \\ 0 \end{bmatrix}$$

Clearly $\rho(A - I) = 2$, therefore the geometric multiplicity of 2 is 3 – 2 = 1, which is not equal to the algebraic multiplicity of eigenvalue 2. Hence, A is not similar to the diagonal matrix.

EXERCISE 25.II

1. Prove that the matrix $A=\begin{bmatrix}0 & 1\\-1 & 0\end{bmatrix}$ is not diagonalized over R the set of all real numbers, however A is diagonalizable over C the set of all complex numbers. Find an invertible matrix P over C such that $P^{-1}AP$ is a diagonal matrix.

2. Show that the matrix $A=\begin{bmatrix}1 & -6 & -4\\0 & 4 & 2\\0 & -6 & -3\end{bmatrix}$ is similar to a diagonal matrix. Also find the transforming matrix and diagonal matrix.

3. Transform the matrix $\begin{bmatrix}8 & -12 & 5\\15 & -25 & 11\\24 & -42 & 19\end{bmatrix}$ into diagonal form.

4. Show that each of the following matrices is similar to a diagonal matrix. Also in each case find the diagonal form D and a diagonalizing matrix P :

(i) $\begin{bmatrix}8 & -6 & 2\\-6 & 7 & -4\\2 & -4 & 3\end{bmatrix}$ (ii) $\begin{bmatrix}6 & -2 & 2\\-2 & 3 & -1\\2 & -1 & 3\end{bmatrix}$

(iii) $\begin{bmatrix}17 & 18 & -6\\-18 & 19 & -6\\-9 & 9 & 2\end{bmatrix}$ (iv) $\begin{bmatrix}4 & 2 & -2\\-5 & 3 & 2\\-2 & 4 & 1\end{bmatrix}$

5. Find the non-singualar matrix P such that $P^{-1}AP$ is a diagonal matrix, where $A=\begin{bmatrix}1 & -3 & 3\\3 & -5 & 3\\6 & -6 & 6\end{bmatrix}$.

6. Let $A=\begin{bmatrix}-9 & 4 & 4\\-8 & 3 & 4\\-16 & 8 & 7\end{bmatrix}$. Find an invertible matrix P such that $P^{-1}AP$ is a diagonal matrix.

7. Test the matrix $A=\begin{bmatrix}3 & 1 & 0\\0 & 3 & 0\\0 & 0 & 4\end{bmatrix}$ for diagonalizability.

8. Show that the following matrices are not similar to diagonal matrices :

(i) $\begin{bmatrix}2 & 1 & 0\\0 & 2 & 1\\0 & 0 & 2\end{bmatrix}$ (ii) $\begin{bmatrix}3 & 10 & 5\\-2 & -3 & -4\\3 & 5 & 7\end{bmatrix}$

9. Reduce the matrix $A=\begin{bmatrix}-1 & 2 & -2\\1 & 2 & 1\\-1 & -1 & 0\end{bmatrix}$ to its diagonal form.

(UPTU-2006, UKTU-2010, 11, VTU -2011, Bhopal-2011)

10. The matrix $A=\begin{bmatrix}a & h\\h & b\end{bmatrix}$ is transformed to the diagonal form $D=T^{-1}AT$ where $T=\begin{bmatrix}\cos\theta & \sin\theta\\-\sin\theta & \cos\theta\end{bmatrix}$. Find the value of θ which gives the diagonal transformation.

11. For the matrix $A=\begin{bmatrix}4 & 1 & 0\\1 & 4 & 1\\0 & 1 & 4\end{bmatrix}$, determine a matrix P such that $P^{-1}AP$ is a diagonal matrix. (MTU-2011)

12. Show that the matrix $A=\begin{bmatrix}3 & 1 & -1\\-2 & 1 & 2\\0 & 1 & 2\end{bmatrix}$ is diagonalisable. Hence, find P such that $P^{-1}AP$ is a diagonal matrix.

ANSWERS

1. $P=\begin{bmatrix}-i & i\\1 & 1\end{bmatrix}$ 2. $P=\begin{bmatrix}1 & 2 & 2\\-2 & -2 & 1\\3 & 3 & -2\end{bmatrix}, D=\begin{bmatrix}1 & 0 & 0\\0 & 1 & 0\\0 & 0 & 0\end{bmatrix}$ 4. (i) $D=\begin{bmatrix}0 & 0 & 0\\0 & 3 & 0\\0 & 0 & 15\end{bmatrix}, P=\begin{bmatrix}1 & 2 & 2\\2 & 1 & -2\\2 & -2 & 1\end{bmatrix}$ (ii) $D=\begin{bmatrix}2 & 0 & 0\\0 & 2 & 0\\0 & 0 & 8\end{bmatrix}, P=\begin{bmatrix}-1 & 1 & 2\\0 & 2 & -1\\2 & 0 & 1\end{bmatrix}$

(iii) $D=\begin{bmatrix}-2 & 0 & 0\\0 & 1 & 0\\0 & 0 & 1\end{bmatrix}, P=\begin{bmatrix}2 & 1 & -1\\2 & 1 & 0\\1 & 0 & 3\end{bmatrix}$ (iv) $D=\begin{bmatrix}1 & 0 & 0\\0 & 2 & 0\\0 & 0 & 5\end{bmatrix}, P=\begin{bmatrix}2 & 1 & 0\\1 & 1 & 1\\4 & 2 & 1\end{bmatrix}$ 5. $P=\begin{bmatrix}1 & -1 & 1\\1 & 0 & 1\\0 & 1 & 2\end{bmatrix}$ 6. $P=\begin{bmatrix}1 & 1 & 1\\0 & 2 & 1\\2 & 0 & 2\end{bmatrix}$

7. Not diagonalizable 9. $\begin{bmatrix}1 & 0 & 0\\0 & \sqrt{5} & 0\\0 & 0 & -\sqrt{5}\end{bmatrix}$ 10. $\theta=\frac{1}{2}\tan^{-1}\left(\frac{2h}{b-a}\right)$ 11. $\begin{bmatrix}-1 & 1 & 1\\0 & \sqrt{2} & -\sqrt{2}\\1 & 1 & 1\end{bmatrix}$ 12. $\begin{bmatrix}1 & 1 & 0\\-1 & 0 & 1\\1 & 1 & 1\end{bmatrix}$

25.53 QUADRATIC FORM AND MATRICES

An expression of the form : $\sum_{i=1}^{n}\sum_{j=1}^{n} a_{ij}x_ix_j$ where $a_{ij}\in F$ (a field), is called a quadratic form in n variables $x_1, x_2, ..., x_n$ over a field F, which is denoted by $Q(x_1, x_2, ..., x_n)$ or by Q.

25.53.1 REAL QUADRATIC FORM

The quadratic form $Q=\sum_{i=1}^{n}\sum_{j=1}^{n} a_{ij}x_ix_j$ is called a real quadratic form if a_{ij} are all real numbers *i.e.*, $a_{ij}\in R$ (the field of all real numbers).

For example :

(1) $ax^2 + 2hxy + by^2$ is a real quadratic form in two variables x and y.

(2) $x^2 + y^2 + z^2 + 2yz + 2zx + 2xy$ is a real quadratic form in three variables x, y and z.

(3) $x_1x_2 + x_2x_3 + x_3x_4$ is a real quadratic form in four variables x_1, x_2, x_3 and x_4.

Theorem 1. *Every quadratic form in n variables $x_1, x_2,..., x_n$ over a field F can be expressed in the form $X'BX$, where $X = [x_1, x_2,..., x_n]'$ is a column vector and B is a symmetric matrix of order n over the field F.*

Proof. Let
$$Q = \sum_{i=1}^{n}\sum_{j=1}^{n} a_{ij}x_ix_j \quad ...(1)$$
be a quadratic form in n variables $x_1, x_2, ..., x_n$ over a field F.

Writing the equation (1) such that the terms $a_{ij}x_ix_j$ and $a_{ji}x_jx_i$ are taken together, we get

$$\begin{aligned} Q = a_{11}x_1^2 &+ (a_{12}+a_{21})x_1x_2 + (a_{13}+a_{31})x_1x_3 + \ldots + (a_{1n}+a_{n1})x_1x_n \\ &+ a_{22}x_2^2 + (a_{23}+a_{32})x_2x_3 + \ldots + (a_{2n}+a_{n2})x_2x_n \\ &+ a_{33}x_3^2 + (a_{34}+a_{43})x_3x_4 + \ldots + (a_{3n}+a_{3n})x_3x_n \\ &+ \ldots\ldots + a_{nn}x_n^2 \end{aligned} \quad ...(2)$$

Set $b_{ij} = \frac{1}{2}(a_{ij} + a_{ji})$, then $b_{ij} = b_{ji}$ and $b_{ij} + b_{ji} = a_{ij} + a_{ji}$, using these relations, equations (2) can be written as

$$\begin{aligned} Q = b_{11}x_1^2 + b_{12}x_1x_2 + \ldots + b_{1n}x_1x_n + b_{21}x_2x_1 + b_{22}x_2^2 + \ldots \\ \ldots + b_{2n}x_2x_n + \ldots + b_{n1}x_nx_1 + b_{n2}x_nx_2 + \ldots + b_{nn}x_n^2 \end{aligned}$$

$$\therefore \quad Q = \sum_{i=1}^{n}\sum_{j=1}^{n} b_{ij}x_ix_j \quad ...(3)$$

Let $B = [b_{ij}]$ be a matrix of order $n \times n$. Clearly B is a symmetric matrix.

Let $X = \begin{bmatrix} x_1 \\ x_2 \\ \vdots \\ x_n \end{bmatrix}$, then $X' = [x_1, x_2, ..., x_n]$

Therefore $X'BX$ is a matrix of order 1×1 *i.e.,* $X'BX$ has a single element and this single element is $\sum_{i=1}^{n}\sum_{j=1}^{n} b_{ij}x_ix_j$.

If we regard a matrix of order 1×1 equal to its single element, then we have

$$X'BX = \sum_{i=1}^{n}\sum_{j=1}^{n} b_{ij}x_ix_j$$

But we have $b_{ii} = a_{ii}$ and $b_{ij} = b_{ji}, = \frac{1}{2}(a_{ij} + a_{ji})$, then we have $\sum_{i=1}^{n}\sum_{j=1}^{n} a_{ij}x_ix_j = \sum_{i=1}^{n}\sum_{j=1}^{n} b_{ij}x_ix_j$

Hence,
$$X'BX = \sum_{i=1}^{n}\sum_{j=1}^{n} a_{ij}x_ix_j .$$

25.54 MATRIX OF QUADRATIC FORM $\sum_{i=1}^{n}\sum_{j=1}^{n} a_{ij}x_ix_j$

If $Q = \sum_{i=1}^{n}\sum_{j=1}^{n} a_{ij}x_ix_j$ is a quadratic form in n variables $x_1, x_2, ..., x_n$ over a field, then there exists unique symmetric matrix B such that

$$X'BX = \sum_{i=1}^{n}\sum_{j=1}^{n} a_{ij}x_ix_j$$

The symmetric matrix B is called the matrix of the quadratic form.

In order to find the matrix of the quadratic form $\sum_{i=1}^{n}\sum_{j=1}^{n} a_{ij}x_ix_j$ we shall adjust the coefficients, a_{ij} in such a way that its coefficients form a symmetric matrix.

25.55 CONVERSION OF A SYMMETRIC MATRIX INTO QUADRATIC FORM

Let $A = [a_{ij}]$ be an $n \times n$ symmetric matrix over a field F and let $X = \begin{bmatrix} x_1 \\ x_2 \\ \vdots \\ x_n \end{bmatrix}$, then the quadratic form of A is given by

$$Q = X'AX = [x_1, x_2, ..., x_n]A\begin{bmatrix} x_1 \\ x_2 \\ \vdots \\ x_n \end{bmatrix} = \sum_{i=1}^{n}\sum_{j=1}^{n} a_{ij}x_ix_j$$

Solved Examples

Example 1. *Find the matrix corresponding to the following quadratic form :*

$$Q = 2x_1^2 - 7x_3^2 + 4x_1x_2 - 6x_2x_3$$

Solution. The given quadratic form can be written as

$$Q = 2x_1^2 + (2 + 2)x_1x_2 + (0 + 0)x_1x_3 + 0.x_2^2 + (-3 - 3)x_2x_3 + (0 + 0)x_3x_1 + (-7)x_3^2$$
$$= 2x_1^2 + 2x_1x_2 + 0.x_1x_3 + 2x_2x_1 + 0.x_2^2 + (-3)x_2x_3 + 0x_3x_1 + (-3)x_3x_2 + (-7)x_3^2$$

Therefore, the coefficients of quadratic form will form a matrix as follows :

$$\begin{bmatrix} 2 & 2 & 0 \\ 2 & 0 & -3 \\ 0 & -3 & -7 \end{bmatrix}$$

which is the required matrix of the given quadratic form.

Example 2. *Find the matrices of the following quadratic forms and verify that they can be written as matrix products $X'AX$.*

(i) $x_1^2 - 18x_1x_2 + 5x_2^2$

(ii) $x_1^2 + 2x_2^2 - 5x_3^2 - x_1x_2 + 4x_2x_3 - 3x_3x_1$

Solution. Let $Q = x_1^2 - 18x_1x_2 + 5x_2^2$

The given quadratic form Q can be written as

$$Q = x_1^2 + (-9 - 9)x_1x_2 + 5x_2^2$$
$$= x_1^2 + (-9)x_1x_2 + (-9)x_1x_2 + 5x_2^2$$

Therefore, the matrix corresponding the given quadratic form is

$$A = \begin{bmatrix} 1 & -9 \\ -9 & 5 \end{bmatrix}$$

Let $X = \begin{bmatrix} x_1 \\ x_2 \end{bmatrix}$, then $X' = [x_1 \quad x_2]$

Now $X'AX = [x_1 \quad x_2]\begin{bmatrix} 1 & -9 \\ -9 & 5 \end{bmatrix}\begin{bmatrix} x_1 \\ x_2 \end{bmatrix}$

$$= [x_1 \quad x_2]\begin{bmatrix} x_1 - 9x_2 \\ -9x_1 + 5x_2 \end{bmatrix}$$
$$= x_1(x_1 - 9x_2) + x_2(-9x_1 + 5x_2)$$
$$= x_1^2 - 9x_1x_2 - 9x_2x_1 + 5x_2^2$$
$$= x_1^2 - 18x_1x_2 + 5x_2^2$$

(ii) Let $Q = x_1^2 + 2x_2^2 - 5x_3^2 - x_1x_2 + 4x_2x_3 - 3x_3x_1$

The given quadratic form Q can be written as

$$Q = x_1^2 + \left(-\frac{1}{2} - \frac{1}{2}\right)x_1x_2 + \left(-\frac{3}{2} - \frac{3}{2}\right)x_1x_3 + 2x_2^2 + (2 + 2)x_2x_3 + (-5)x_3^2$$
$$= x_1^2 + \left(-\frac{1}{2}\right)x_1x_2 + \left(-\frac{3}{2}\right)x_1x_3 + \left(-\frac{1}{2}\right)x_2x_1 + 2x_1^2 + 2x_2x_3 + \left(-\frac{3}{2}\right)x_3x_1 + 2x_3x_2 + (-5)x_3^2$$

Therefore, the matrix corresponding to the given quadratic form Q is

$$A = \begin{bmatrix} 1 & -\frac{1}{2} & -\frac{3}{2} \\ -\frac{1}{2} & 2 & 2 \\ -\frac{3}{2} & 2 & -5 \end{bmatrix}$$

Let $X = \begin{bmatrix} x_1 \\ x_2 \\ x_3 \end{bmatrix}$, then $X' = [x_1 \quad x_2 \quad x_3]$

Now $X'AX = [x_1 \quad x_2 \quad x_3]\begin{bmatrix} 1 & -\frac{1}{2} & -\frac{3}{2} \\ -\frac{1}{2} & 2 & 2 \\ -\frac{3}{2} & 2 & -5 \end{bmatrix}\begin{bmatrix} x_1 \\ x_2 \\ x_3 \end{bmatrix}$

$$= [x_1 \quad x_2 \quad x_3]\begin{bmatrix} x_1 - \frac{1}{2}x_2 - \frac{3}{2}x_3 \\ -\frac{1}{2}x_1 + 2x_2 + 2x_3 \\ -\frac{3}{2}x_1 + 2x_2 - 5x_3 \end{bmatrix}$$
$$= x_1\left(x_1 - \frac{1}{2}x_2 - \frac{3}{2}x_3\right) + x_2\left(-\frac{1}{2}x_1 + 2x_2 + 2x_3\right) + x_3\left(-\frac{3}{2}x_1 + 2x_2 - 5x_3\right)$$
$$= x_1^2 - \frac{1}{2}x_1x_2 - \frac{3}{2}x_1x_3 - \frac{1}{2}x_2x_1 + 2x_2^2 + 2x_2x_3 - \frac{3}{2}x_3x_1 + 2x_3x_2 - 5x_3^2$$
$$= x_1^2 + 2x_2^2 - 5x_3^2 - x_1x_2 + 4x_2x_3 - 3x_3x_1.$$

Example 3. *Obtain the matrices corresponding to the following quadratic forms :*

(i) $x^2 + 2y^2 + 3z^2 + 4xy + 5yz + 6zx$

(ii) $ax^2 + by^2 + cz^2 + 2fyz + 2gzx + 2hxy$

(iii) $a_{11}x_1^2 + a_{22}x_2^2 + a_{33}x_3^2 + 2a_{12}x_1x_2 + 2a_{23}x_2x_3 + 2a_{31}x_3x_1$

(iv) $x_1^2 - 2x_2^2 + 4x_3^3 - 4x_4^4 - 2x_1x_2 + 3x_1x_4 + 4x_2x_3 - 5x_3x_4$

(v) $x_1^2 - 2x_2x_3 - x_3x_4$

(vi) $d_1x_1^2 + d_2x_2^2 + d_3x_3^2 + d_4x_4^2$

(vii) $x_1x_2 + x_2x_3 + x_3x_1 + x_1x_4 + x_2x_4 + x_3x_4$

Solution. (i) The given quadratic form can be written as

$$x^2 + (2+2)xy + (3+3)xz + 2y^2 + \left(\frac{5}{2}+\frac{5}{2}\right)yz + 3z^2$$

$$= x^2 + 2xy + 3xz + 2yx + 2y^2 + \frac{5}{2}yz + 3zx + \frac{5}{2}zy + 3z^2$$

Therefore, the matrix corresponding to the given quadratic form is

$$\begin{bmatrix} 1 & 2 & 3 \\ 2 & 2 & \frac{5}{2} \\ 3 & \frac{5}{2} & 3 \end{bmatrix}$$

(ii) The given quadratic form can be written as

$$ax^2 + (h+h)xy + (g+g)xz + by^2 + (f+f)yz + cz^2$$

$$= ax^2 + hxy + gxz + hyx + by^2 + fyz + gzx + fzy + cz^2$$

Therefore, the matrix corresponding to the given quadratic form is

$$\begin{bmatrix} a & h & g \\ h & b & f \\ g & f & c \end{bmatrix}$$

(iii) The given quadratic form can be written as

$$a_{11}x_1^2 + (a_{12}+a_{12})x_1x_2 + (a_{31}+a_{31})x_1x_3 + a_{22}x_2^2 + (a_{23}+a_{23})x_2x_3 + a_{33}x_3^2$$

$$= a_{11}x_1^2 + a_{12}x_1x_2 + a_{31}x_3x_1 + a_{12}x_2x_1 + a_{22}x_2^2 + a_{23}x_2x_3 + a_{31}x_3x_1 + a_{23}x_2x_3 + a_{33}x_3^2$$

Therefore, the matrix corresponding to the given quadratic form is $\begin{bmatrix} a_{11} & a_{12} & a_{31} \\ a_{12} & a_{22} & a_{23} \\ a_{31} & a_{23} & a_{33} \end{bmatrix}$

(iv) The given quadratic form can be written as

$$x_1^2 + (-1-1)x_1x_2 + (0+0)x_1x_3 + \left(\frac{3}{2}+\frac{3}{2}\right)x_1x_4 + (-2)x_2^2 + (2+2)x_2x_3 + (0+0)x_2x_4 + 4x_3^2 + \left(-\frac{5}{2}-\frac{5}{2}\right)x_3x_4 + (-4)x_4^2$$

$$= x_1^2 - x_1x_2 + 0x_1x_3 + \frac{3}{2}x_1x_2 + (-1)x_2x_1 + (-2)x_2^2 + 2x_2x_3 + 0.x_2x_3 + 0.x_3x_1 + 2x_3x_2 + 4x_3^2 + \left(-\frac{5}{2}\right)x_3x_4 + \frac{3}{2}x_4x_1 + 0.x_4x_2 + \left(-\frac{5}{2}\right)x_4x_3 + (-4)x_4^2$$

Therefore, the matrix corresponding to the given quadratic form is

$$\begin{bmatrix} 1 & -1 & 0 & 3/2 \\ -1 & -2 & 2 & 0 \\ 0 & 2 & 4 & -5/2 \\ \frac{3}{2} & 0 & -5/2 & -4 \end{bmatrix}$$

(v) The given quadratic form can be written as

$$x_1^2 + (0+0)x_1x_2 + (0+0)x_1x_3 + (0+0)x_1x_4 + 0.x_2^2 + (-1-1)x_2x_3 + (0+0)x_2x_4 + 0.x_3^2 + \left(-\frac{1}{2}-\frac{1}{2}\right)x_3x_4 + 0.x_4^2$$

$$= x_1^2 + 0.x_1x_2 + 0.x_1x_3 + 0.x_1x_4 + 0.x_2x_1 + 0.x_2^2 + (-1)x_2x_3 + 0.x_2x_4 + 0.x_3x_1 + (-1)x_3x_2 + 0.x_3^2 + \left(-\frac{1}{2}\right)x_3x_4 + 0.x_4x_1 + 0.x_4x_2 + \left(-\frac{1}{2}\right)x_4x_3 + 0.x_4^2$$

Therefore, the matrix corresponding to the given quadratic form is

$$\begin{bmatrix} 1 & 0 & 0 & 0 \\ 0 & 0 & -1 & 0 \\ 0 & -1 & 0 & -\frac{1}{2} \\ 0 & 0 & -\frac{1}{2} & 0 \end{bmatrix}$$

(vi) The given quadratic form can be written as

$$d_1x_1^2 + 0.x_1x_2 + 0.x_1x_3 + 0.x_1x_4 + 0.x_2x_1 + d_2x_2^2 + 0.x_2x_3 + 0.x_2x_4 + 0.x_3x_1 + 0.x_3x_2 + d_3x_3^2 + 0.x_3x_4 + 0.x_4x_1 + 0.x_4x_2 + 0.x_4x_3 + d_4x_4^2$$

Therefore, the matrix corresponding to the given quadratic form is

$$\begin{bmatrix} d_1 & 0 & 0 & 0 \\ 0 & d_2 & 0 & 0 \\ 0 & 0 & d_3 & 0 \\ 0 & 0 & 0 & d_4 \end{bmatrix} = \text{diag.}(d_1, d_2, d_3, d_4)$$

(vii) The given quadratic form can be written as

$$0.x_1^2+\left(\frac{1}{2}+\frac{1}{2}\right)x_1x_2+\left(\frac{1}{2}+\frac{1}{2}\right)x_1x_4$$
$$+0.x_2^2+\left(\frac{1}{2}+\frac{1}{2}\right)x_2x_3+\left(\frac{1}{2}+\frac{1}{2}\right)x_2x_4$$
$$+\left(\frac{1}{2}+\frac{1}{2}\right)x_3x_1+0.x_3^2+\left(\frac{1}{2}+\frac{1}{2}\right)x_3x_4+0.x_4^2$$
$$=0.x_1^2+\frac{1}{2}x_1x_2+\frac{1}{2}x_1x_3+\frac{1}{2}x_1x_4$$
$$+\frac{1}{2}x_2x_1+0.x_2^2+\frac{1}{2}x_2x_3+\frac{1}{2}x_2x_4$$
$$+\frac{1}{2}x_3x_1+\frac{1}{2}x_3x_2+0.x_3^2+\frac{1}{2}x_3x_4+\frac{1}{2}x_4x_1$$
$$+\frac{1}{2}x_4x_2+\frac{1}{2}x_4x_3+0.x_4^2$$

Therefore, the matrix corresponding to the given quadratic form is

$$\begin{bmatrix} 0 & \frac{1}{2} & \frac{1}{2} & \frac{1}{2} \\ \frac{1}{2} & 0 & \frac{1}{2} & \frac{1}{2} \\ \frac{1}{2} & \frac{1}{2} & 0 & \frac{1}{2} \\ \frac{1}{2} & \frac{1}{2} & \frac{1}{2} & 0 \end{bmatrix}$$

Example 4. *Find the matrix of the following quadratic form :*

$$(x_1-x_2+x_3)^2$$

Solution. The given quadratic form can be written as

$$(x_1-x_2+x_3)^2=x_1^2+x_2^2+x_3^2$$
$$-2x_1x_2+2x_1x_3-2x_2x_3$$
$$=x_1^2+(-1-1)x_1x_2+(1+1)$$
$$x_1x_3+x_2^2+(-1-1)x_2x_3+x_3^2$$
$$=x_1^2-x_1x_2+x_1x_3-x_2x_1+x_2^2$$
$$-x_2x_3+x_3x_1-x_3x_2+x_3^2$$

Therefore, the matrix corresponding to the given quadratic form is

$$\begin{bmatrix} 1 & -1 & 1 \\ -1 & 1 & -1 \\ 1 & -1 & 1 \end{bmatrix}$$

Example 5. *Find the quadratic form of the real symmetric matrix*

$$A=\begin{bmatrix} 2 & 2 & 0 \\ 2 & 0 & -3 \\ 0 & -3 & -7 \end{bmatrix}$$

Solution. Let $X=\begin{bmatrix} x_1 \\ x_2 \\ x_3 \end{bmatrix}$, then $X'=[x_1 \quad x_2 \quad x_3]$

Now $X'AX=[x_1 \quad x_2 \quad x_3]\begin{bmatrix} 2 & 2 & 0 \\ 2 & 0 & -3 \\ 0 & -3 & -7 \end{bmatrix}\begin{bmatrix} x_1 \\ x_2 \\ x_3 \end{bmatrix}$

$$=[x_1 \quad x_2 \quad x_3]\begin{bmatrix} 2x_1+2x_2+0.x_3 \\ 2x_1+0.x_2-3x_3 \\ 0.x_1-3x_2-7x_3 \end{bmatrix}$$
$$=x_1(2x_1+2x_2+0.x_3)+x_2(2x_1+0.x_2-3x_3)$$
$$+x_3(0.x_1-3x_2-7x_3)$$
$$=2x_1^2+2x_1x_2+0.x_1x_3+2x_2x_1+0.x_2^2$$
$$-3x_2x_3+0x_3x_1-3x_3x_2-7x_3^2$$
$$=2x_1^2+4x_1x_2-6x_2x_3-7x_3^2$$

$\therefore \quad X'AX=2x_1^2+4x_1x_2-6x_2x_3-7x_3^2$

which is the required quadratic form.

Example 6. *Write down the quadratic forms corresponding to the following symmetric matrices:*

(*i*) $\begin{bmatrix} 1 & 2 & 3 \\ 2 & 0 & 3 \\ 3 & 3 & 1 \end{bmatrix}$ (*ii*) diag.$(\lambda_1, \lambda_2, \ldots, \lambda_n)$

Solution. (i) Let $A=\begin{bmatrix} 1 & 2 & 3 \\ 2 & 0 & 3 \\ 3 & 3 & 1 \end{bmatrix}$ and let $X=\begin{bmatrix} x_1 \\ x_2 \\ x_3 \end{bmatrix}$, then

$X'=[x_1 \quad x_2 \quad x_3]$

Now $X'AX=[x_1 \quad x_2 \quad x_3]\begin{bmatrix} 1 & 2 & 3 \\ 2 & 0 & 3 \\ 3 & 3 & 1 \end{bmatrix}\begin{bmatrix} x_1 \\ x_2 \\ x_3 \end{bmatrix}$

$$=[x_1 \quad x_2 \quad x_3]\begin{bmatrix} x_1+2x_2+3x_3 \\ 2x_1+0.x_2+3x_3 \\ 3x_1+3x_2+x_3 \end{bmatrix}$$
$$=x_1(x_1+2x_2+3x_3)+x_2(2x_1+0.x_2-3x_3)$$
$$+x_3(3x_1+3x_2+x_3)$$
$$=x_1^2+2x_1x_2+3x_1x_3+2x_2x_1+0.x_2^2+3x_2x_3$$
$$+3x_3x_1+3x_3x_2+x_3^2$$
$$=x_1^2+x_3^2+4x_1x_2+6x_1x_3+6x_2x_3$$

which is the required quadratic form.

(ii) Let $A=$ diag.$(\lambda_1, \lambda_2, \ldots, \lambda_n)$. Then A is an $n\times n$ diagonal matrix so let us take

$$X=\begin{bmatrix} x_1 \\ x_2 \\ \vdots \\ x_n \end{bmatrix}$$

then $X'=[x_1 \quad x_2 \quad \cdots \quad x_n]$

$$\text{diag.}(\lambda_1, \lambda_2, \ldots, \lambda_n)\begin{bmatrix} x_1 \\ x_2 \\ \vdots \\ x_n \end{bmatrix}$$
$$=\lambda_1x_1^2+\lambda_2x_2^2+\ldots+\lambda_nx_n^2$$

which is the required quadratic form.

EXERCISE 25.12

1. Find the matrices corresponding to the following quadratic forms as :
   (i) $ax^2 + 2hxy + by^2$
   (ii) $5x_1^2 - 2x_1x_2 + x_2^2$
   (iii) $x_1^2 + 5x_2^2 - 7x_3^2$
   (iv) $4x_1x_3 + 2x_2x_3 + x_3^2$
   (v) $(x_1 + x_2)^2 - x_3^2$
   (vi) $x_1^2 - 2x_2^2 - 3x_3^2 + 4x_1x_2 + 6x_1x_3 - 8x_2x_3$
   (vii) $2x_1x_2 + 6x_1x_3 - 4x_2x_3$
   (viii) $5x_1^2 + 3x_2^2 + 2x_3^2 - x_1x_2 + 8x_2x_3$
   (ix) $8x_1^2 + 7x_2^2 - 3x_3^2 - 6x_1x_2 + 4x_1x_3 - 2x_2x_3$
   (x) $4x_1x_2 + 6x_1x_3 - 8x_2x_3$
   (xi) $5x_1^2 - x_2^2 + 7x_3^2 + 5x_1x_2 - 3x_1x_3$
2. Compute the quadratic form $X'AX$, when
   (i) $A = \begin{bmatrix} 5 & 1/3 \\ 1/3 & 1 \end{bmatrix}$ (ii) $A = \begin{bmatrix} 4 & 0 \\ 0 & 3 \end{bmatrix}$
   (iii) $A = \begin{bmatrix} 3 & -2 \\ -2 & 7 \end{bmatrix}$
3. Write down the quadratic forms corresponding to the following matrices :
   (i) $\begin{bmatrix} 2 & 1 & 5 \\ 1 & 3 & -2 \\ 5 & -2 & 4 \end{bmatrix}$ (ii) $\begin{bmatrix} 1 & 0 & 0 \\ 0 & 2 & 0 \\ 0 & 0 & 3 \end{bmatrix}$
   (iii) $\begin{bmatrix} 1 & 2 & 3 \\ 2 & 2 & 5/2 \\ 3 & 5/2 & 3 \end{bmatrix}$ (iv) $\begin{bmatrix} 0 & 5 & -1 \\ 5 & 1 & 6 \\ -1 & 6 & 2 \end{bmatrix}$
   (v) $\begin{bmatrix} 2 & 2 & 0 \\ 2 & 0 & -3 \\ 0 & -3 & -7 \end{bmatrix}$ (vi) $\begin{bmatrix} 1 & -1/2 & -3/2 \\ -1/2 & 2 & 2 \\ -3/2 & 2 & -5 \end{bmatrix}$
   (vii) $\begin{bmatrix} 0 & 1 & 3 \\ 1 & 0 & -2 \\ 3 & -2 & 0 \end{bmatrix}$ (viii) $\begin{bmatrix} 0 & a & b & c \\ a & 0 & l & m \\ b & l & 0 & p \\ c & m & p & 0 \end{bmatrix}$

ANSWERS

1. (i) $\begin{bmatrix} a & h \\ h & b \end{bmatrix}$ (ii) $\begin{bmatrix} 5 & -1 \\ -1 & 1 \end{bmatrix}$ (iii) $\begin{bmatrix} 1 & 0 & 0 \\ 0 & 5 & 0 \\ 0 & 0 & -7 \end{bmatrix}$ (iv) $\begin{bmatrix} 0 & 0 & 2 \\ 0 & 0 & 1 \\ 2 & 1 & 1 \end{bmatrix}$ (v) $\begin{bmatrix} 1 & 1 & 0 \\ 1 & 1 & 0 \\ 0 & 0 & -1 \end{bmatrix}$

(vi) $\begin{bmatrix} 1 & 2 & 3 \\ 2 & -2 & -4 \\ 3 & -4 & -3 \end{bmatrix}$ (vii) $\begin{bmatrix} 0 & 1 & 3 \\ 1 & 0 & -2 \\ 3 & -2 & 0 \end{bmatrix}$ (viii) $\begin{bmatrix} 5 & -1/2 & 0 \\ -1/2 & 3 & 4 \\ 0 & 4 & 2 \end{bmatrix}$

(ix) $\begin{bmatrix} 8 & -3 & 2 \\ -3 & 7 & -1 \\ 2 & -1 & -3 \end{bmatrix}$ (x) $\begin{bmatrix} 0 & 2 & 3 \\ 2 & 0 & -4 \\ 3 & -4 & 0 \end{bmatrix}$ (xi) $\begin{bmatrix} 5 & 5/2 & -3/2 \\ 5/2 & -1 & 0 \\ -3/2 & 0 & 7 \end{bmatrix}$

2. (i) $5x_1^2 + \frac{2}{3}x_1x_2 + x_2^2$ (ii) $4x_1^2 + 3x_2^2$ (iii) $3x_1^2 + 7x_2^2 - 4x_1x_2$

3. (i) $2x_1^2 + 3x_2^2 + +4x_3^2 + 2x_1x_2 + 10x_1x_3 - 4x_2x_3$ (ii) $x_1^2 + 2x_2^2 + 3x_3^2$ (iii) $x_1^2 + 2x_2^2 + 3x_3^2 + 4x_1x_2 + 6x_1x_3 + 5x_2x_3$

(iv) $x_2^2 + 2x_3^2 + 10x_1x_2 - 2x_1x_3 + 12x_2x_3$ (v) $2x_1^2 - 7x_3^2 + 4x_1x_2 - 6x_2x_3$ (vi) $x_1^2 + 2x_2^2 - 5x_3^2 - x_1x_2 - 3x_1x_3 + 4x_2x_3$

(vii) $2x_1x_2 + 6x_1x_3 - 4x_2x_3$ (viii) $2ax_1x_2 + 2bx_1x_3 + 2cx_1x_4 + 2lx_2x_3 + 2mx_2x_4 + 2px_3x_4$

25.56 RANK OF A QUADRATIC FORM

Let $Q(X) = X'AX$ be a quadratic form over a field F. Then the rank of $Q(X)$ is the rank of the matrix A.

If the rank of $Q(X)$ is r, then there exists a non-singular matrix P which reduces $Q(X)$ to a sum of r square terms.

i.e.,
$$Q(X) = \lambda_1 y_1^2 + \lambda_2 y_2^2 + \ldots \lambda_r y_r^2$$

WORKING PROCEDURE

Let A be $n \times n$ real symmetric matrix. In order to find a non-singular matrix P such that $P'AP$ = diagonal matrix, we use the following steps:

Step 1. *Write $A = I_n A I_n$*

Step 2. *Apply congruent row operations on pre-factor I_n of A on RHS and congruent column operations on post-factor I_n of A. Applying such operations simultaneously on A on LHS till A reduces to a diagonal matrix. When the matrix A reduces to a diagonal matrix, the post-factor I_n ultimately gives the non-singular matrix P such that $P'AP$ = diagonal matrix.*

Solved Examples

Example 1. *Determine a non-singular matrix P such that $P'AP$ is a diagonal matrix, where*

$$A=\begin{bmatrix}0&1&2\\1&0&3\\2&3&0\end{bmatrix}$$

Solution. We have $A = I_3 A I_3$

or
$$\begin{bmatrix}0&1&2\\1&0&3\\2&3&0\end{bmatrix}=\begin{bmatrix}1&0&0\\0&1&0\\0&0&1\end{bmatrix}A\begin{bmatrix}1&0&0\\0&1&0\\0&0&1\end{bmatrix}$$

Applying $R_1 \to R_1 + R_2$, we get

$$\begin{bmatrix}1&1&5\\1&0&3\\2&3&0\end{bmatrix}=\begin{bmatrix}1&1&0\\0&1&0\\0&0&1\end{bmatrix}A\begin{bmatrix}1&0&0\\0&1&0\\0&0&1\end{bmatrix}$$

Applying $C_1 \to C_1 + C_2$, we get

$$\begin{bmatrix}2&1&5\\1&0&3\\5&3&0\end{bmatrix}=\begin{bmatrix}1&1&0\\0&1&0\\0&0&1\end{bmatrix}A\begin{bmatrix}1&0&0\\1&1&0\\0&0&1\end{bmatrix}$$

Applying $R_2 \to R_2-\frac{1}{2}R_1, R_3 \to R_3-\frac{5}{2}R_1$, we get

$$\begin{bmatrix}2&1&5\\0&-1/2&1/2\\0&1/2&-25/2\end{bmatrix}=\begin{bmatrix}1&1&0\\-1/2&1/2&0\\-5/2&-5/2&1\end{bmatrix}A\begin{bmatrix}1&0&0\\1&1&0\\0&0&1\end{bmatrix}$$

Applying $C_2 \to C_2-\frac{1}{2}C_1, C_3 \to C_3-\frac{5}{2}C_1$, we get

$$\begin{bmatrix}2&1&5\\0&-1/2&1/2\\0&1/2&-25/2\end{bmatrix}=\begin{bmatrix}1&1&0\\-1/2&1/2&0\\-5/2&-5/2&1\end{bmatrix}A\begin{bmatrix}1&-1/2&-5/2\\1&1/2&-5/2\\0&0&1\end{bmatrix}$$

Applying $R_3 \to R_3+ R_2$, we get

$$\begin{bmatrix}2&0&0\\0&-1/2&1/2\\0&0&-12\end{bmatrix}=\begin{bmatrix}1&1&0\\-1/2&1/2&0\\-3&-2&1\end{bmatrix}A\begin{bmatrix}1&-1/2&-5/2\\1&1/2&-5/2\\0&0&1\end{bmatrix}$$

Applying $C_3 \to C_3 + C_2$, we get

$$\begin{bmatrix}2&0&0\\0&-1/2&0\\0&0&-12\end{bmatrix}=\begin{bmatrix}1&1&0\\-1/2&1/2&0\\-3&-2&1\end{bmatrix}A\begin{bmatrix}1&-1/2&-3\\1&1/2&-2\\0&0&1\end{bmatrix}$$

$$\Rightarrow \quad \text{diag.}\left(2,-\frac{1}{2},-12\right)=P'AP$$

where
$$P=\begin{bmatrix}1&-1/2&-3\\1&1/2&-2\\0&0&1\end{bmatrix}$$

Example 2. *Determine a non-singular matrix P such that $P'AP$ is a diagonal matrix, where*

$$A=\begin{bmatrix}6&-2&2\\-2&3&-1\\2&-1&3\end{bmatrix}.$$

Interpret the result in terms of quadratic form.

Solution. We have $A = I_3 A I_3$

or
$$\begin{bmatrix}6&-2&2\\-2&3&-1\\2&-1&3\end{bmatrix}=\begin{bmatrix}1&0&0\\0&1&0\\0&0&1\end{bmatrix}A\begin{bmatrix}1&0&0\\0&1&0\\0&0&1\end{bmatrix}$$

Applying $R_2 \to R_2+\frac{1}{3}R_1$, we get

$$\begin{bmatrix}6&-2&2\\0&7/3&-1/3\\2&-1&3\end{bmatrix}=\begin{bmatrix}1&0&0\\1/3&1&0\\0&0&1\end{bmatrix}A\begin{bmatrix}1&0&0\\0&1&0\\0&0&1\end{bmatrix}$$

Applying $C_2 \to C_2+\frac{1}{3}C_1$, we get

$$\begin{bmatrix}6&0&2\\0&7/3&-1/3\\2&-1/3&3\end{bmatrix}=\begin{bmatrix}1&0&0\\1/3&1&0\\0&0&1\end{bmatrix}A\begin{bmatrix}1&1/3&0\\0&1&0\\0&0&1\end{bmatrix}$$

Applying $R_3 \to R_3-\frac{1}{3}R_1$ we get

$$\begin{bmatrix}6&0&2\\0&7/3&-1/3\\0&-1/3&7/3\end{bmatrix}=\begin{bmatrix}1&0&0\\1/3&1&0\\-1/3&0&1\end{bmatrix}A\begin{bmatrix}1&1/3&0\\0&1&0\\0&0&1\end{bmatrix}$$

Applying $C_3 \to C_3-\frac{1}{3}C_1$, we get

$$\begin{bmatrix}6&0&2\\0&7/3&-1/3\\0&-1/3&7/3\end{bmatrix}=\begin{bmatrix}1&0&0\\1/3&1&0\\-1/3&0&1\end{bmatrix}A\begin{bmatrix}1&1/3&-1/3\\0&1&0\\0&0&1\end{bmatrix}$$

Applying $R_3 \to R_3+\frac{1}{7}R_2$, we get

$$\begin{bmatrix}6&0&0\\0&7/3&-1/3\\0&0&16/7\end{bmatrix}=\begin{bmatrix}1&0&0\\1/3&1&0\\-2/7&1/7&1\end{bmatrix}A\begin{bmatrix}1&1/3&-1/3\\0&1&0\\0&0&1\end{bmatrix}$$

Applying $C_3 \to C_3+\frac{1}{7}C_2$, we get

$$\begin{bmatrix}6&0&0\\0&7/3&0\\0&0&16/7\end{bmatrix}=\begin{bmatrix}1&0&0\\1/3&1&0\\-2/7&1/7&1\end{bmatrix}A\begin{bmatrix}1&1/3&-2/7\\0&1&1/7\\0&0&1\end{bmatrix}$$

$$\Rightarrow \quad \text{diag.}\left(6,\frac{7}{3},\frac{16}{7}\right)=P'AP \qquad \ldots(1)$$

where $P = \begin{bmatrix} 1 & 1/3 & -2/7 \\ 0 & 1 & 1/7 \\ 0 & 0 & 1 \end{bmatrix}$.

The quadratic form of the matrix A is given by

$$X'AX = \begin{bmatrix} x_1 & x_2 & x_3 \end{bmatrix} \begin{bmatrix} 6 & -2 & 2 \\ -2 & 3 & -1 \\ 2 & -1 & 3 \end{bmatrix} \begin{bmatrix} x_1 \\ x_2 \\ x_3 \end{bmatrix}$$

$$X'AX = 6x_1^2 + 3x_2^2 + 3x_3^2 - 4x_1x_2 + 4x_1x_3 - 2x_2x_3 \quad \text{...(2)}$$

The non-singular transformation corresponding to the matrix P is given by

$$X = PY \quad \text{...(3)}$$

where $Y = \begin{bmatrix} y_1 \\ y_2 \\ y_3 \end{bmatrix}$ and $X = \begin{bmatrix} x_1 \\ x_2 \\ x_3 \end{bmatrix}$

From (3) we have

$$\begin{bmatrix} x_1 \\ x_2 \\ x_3 \end{bmatrix} = \begin{bmatrix} 1 & 1/3 & -2/7 \\ 0 & 1 & 1/7 \\ 0 & 0 & 1 \end{bmatrix} \begin{bmatrix} y_1 \\ y_2 \\ y_3 \end{bmatrix}$$

i.e.,
$$\left.\begin{aligned} x_1 &= y_1 + \frac{1}{3}y_2 - \frac{2}{7}y_3 \\ x_2 &= y_2 + \frac{1}{7}y_3 \\ x_3 &= y_3 \end{aligned}\right\} \quad \text{...(4)}$$

The transformations given by (4) reduce the quadratic form (2) to the diagonal form

$$X'AX = Y'(P'AP)Y$$

$$= Y'\text{diag.}\left(6, \frac{7}{3}, \frac{16}{7}\right)Y$$

$$= 6y_1^2 + \frac{7}{3}y_2^2 + \frac{16}{7}y_3^2$$

Clearly, $X'AX$ has a sum of 3 square terms, hence the rank of $X'AX$ is 3.

25.57 NORMAL (OR CANONICAL) FORM OF A REAL QUADRATIC MATRIX

Let $X'AX$ be a real quadratic form in n variables over the real field, then there exists a real non-singular linear transformation $X = PY$, which reduces the given quadratic form to the form

$$Y'(P'AP)Y = y_1^2 + y_2^2 + \ldots + y_p^2 - y_{p+1}^2 - y_{p+2}^2 - \ldots - y_r^2$$

This new form is known as the normal (or canonical) form of $X'AX$.

REMARKS

- The number of positive terms in any two normal forms of a real quadratic form is the same.
- The number of negative terms in any two normal forms of a quadratic form is the same.
- The excess of the number of positive terms over the number of negative terms in any two normal forms of a real quadratic form is the same.

25.58 SIGNATURE AND INDEX OF A REAL QUADRATIC FORM

Let $X'AX$ be a real quadratic form of rank r in n variables over the field of reals and its normal form be

$$y_1^2 + y_2^2 + \ldots + y_p^2 - y_{p+1}^2 - y_{p+2}^2 - \ldots - y_r^2$$

Then the number p of positive terms in a normal form is called the index of $X'AX$. The excess of the number of positive terms over the number of negative terms in a normal form, *i.e.*, $p - (r - p) = 2p - r$ is called the signature of $X'AX$.

THEOREM 1. *The signature of a real quadratic form is invariant for its all normal forms.*

Proof. Since the number of positive and negative terms in any two normal forms of a real quadratic form are the same so that their difference are the same, *i.e.*, the signature of given quadratic is the same for its all normal forms.

REMARK

- Two real quadratic forms in n variables over the field of reals are real equivalent if and only if they have the same rank and signature (or index).

25.59 REDUCTION OF A REAL QUADRATIC FORM OVER THE FIELD OF COMPLEX NUMBERS

THEOREM 1. *Let A be an $n \times n$ real symmetric matrix of rank r, then there exists a non-singular matrix P whose elements may be complex numbers such that $P'AP = diag.(1, 1, 1, \ldots 1, 0, 0, \ldots 0)$ where 1 appears r times.*

Proof. Since A is a real symmetric matrix of rank r, then there exists a non-singular real matrix Q such that

$$Q'AQ = \text{diag.}\ (\lambda_1, \lambda_2, ..., \lambda_r, 0, 0, ...\ 0) = D\ (\text{say})$$

where D is a diagonal matrix which has exactly r non-zero elements $\lambda_1, \lambda_2, ..., \lambda_r$ and $\lambda_1, \lambda_2, ..., \lambda_r$ may be positive or negative or both.

Let $$S = \text{diag.}\left(\frac{1}{\sqrt{\lambda_1}}, \frac{1}{\sqrt{\lambda_2}}, \ldots, \frac{1}{\sqrt{\lambda_r}}, 1, 1, \ldots, 1\right)$$

be an $n \times n$ complex diagonal matrix, which is obviously, a complex non-singular diagonal matrix and $S' = S$.

Let us take $P = QS$, clearly P is a complex non-singular matrix.

Now
$$P'AP = (QS)'A(QS)$$
$$= S'(Q'AQ)S = S'DS \qquad [\because Q'AQ = D]$$
$$= SDS \qquad [\because S' = S]$$
$$= \text{diag.}\left(\frac{1}{\sqrt{\lambda_1}}, \frac{1}{\sqrt{\lambda_2}}, \ldots, \frac{1}{\sqrt{\lambda_r}}, 1, 1, \ldots, 1\right) \text{diag.}(\lambda_1, \lambda_2, ..., \lambda_r, 0, 0, ...0)S$$
$$= \text{diag}\left(\sqrt{\lambda_1}, \sqrt{\lambda_2}, \ldots, \sqrt{\lambda_r}, 0, 0, \ldots, 0\right) S$$
$$= \text{diag}\left(\sqrt{\lambda_1}, \sqrt{\lambda_2}, \ldots, \sqrt{\lambda_r}, 0, 0, \ldots, 0\right) \text{diag}\left(\frac{1}{\sqrt{\lambda_1}}, \frac{1}{\sqrt{\lambda_2}}, \ldots, \frac{1}{\sqrt{\lambda_r}}, 1, 1, \ldots, 1\right)$$
$$= \text{diag}(1, 1, 1, \ldots, 1, 0, 0, \ldots, 0)$$

$$\therefore \quad P'AP = \text{diag}(1, 1, 1, \ldots, 1, 0, 0, \ldots, 0)$$

where 1 appears r times.

REMARKS

- Every real quadratic form $X'AX$ is complex equivalent to the form $z_1^2 + z_2^2 + \ldots + z_r^2$ where r is the rank of A.
- Two real quadratic forms in n variables are complex equivalent if and only if they have the same rank.

25.60 ORTHOGONAL REDUCTION OF A REAL QUADRATIC FORM

THEOREM 1. *If $X'AX$ be a real quadratic form of rank r in n variables, then there exists a real orthogonal transformation $X = PY$ which transforms $X'AX$ to the form $\lambda_1 y_1^2 + \lambda_2 y_2^2 + \ldots \lambda_r y_r^2$ where $\lambda_1, \lambda_2, ..., \lambda_r$ are the r non-zero eigenvalues of A and $n - r$ eigenvalues of A being equal to zero.*

Proof. Since A is a real symmetric matrix of order n, then there exists a real orthogonal matrix P such that $P^{-1}AP = D$ where D is a diagonal matrix, whose diagonal elements are the eigenvalues of A. Again, since the rank of A is r, then the rank of $P^{-1}AP$ is also r, therefore D will have exactly r non-zero diagonal elements, hence A has exactly r non-zero eigenvalues and remaining $n - r$ eigenvalues of A are all zero.

So we can take $D = \text{diag.}(\lambda_1, \lambda_2, ..., \lambda_r, 0, 0, .., 0)$ where $\lambda_1, \lambda_2, ..., \lambda_r$ are the r non-zero eigenvalues of A

$$\therefore \quad P^{-1}AP = \text{diag.}(\lambda_1, \lambda_2, ..., \lambda_r, 0, 0, ...\ 0)$$

Since P is an orthogonal matrix, then $P^{-1} = P'$

$$\therefore \quad P^{-1}AP = P'AP = \text{diag.}(\lambda_1, \lambda_2, ..., \lambda_r, 0, 0, ...\ 0) = D$$

It follows that A is congruent to D.

Now, let us take a real orthogonal transformation $X = PY$ such that

$$X'AX = (PY)'A(PY) = Y'(P'AP)Y = Y'DY$$

$\lambda_1 y_1^2 + \lambda_2 y_2^2 + \ldots \lambda_r y_r^2$, if $Y = [y_1, y_2, \ldots y_r, \ldots, y_n]$

THEOREM 2. *Every real quadratic form $X'AX$ in n variables is real equivalent to the form*

$$y_1^2 + y_2^2 + \ldots + y_p^2 - y_{p+1}^2 - y_{p+2}^2 - \ldots - y_r^2$$

where r is the rank of A, and p is the number of positive eigenvalues of A.

Proof. Since A is a real symmetric matrix, then there exists a real orthogonal matrix Q such that

$$Q^{-1}AQ = D$$

or $$Q'AQ = D \qquad [\because Q^{-1} = Q']$$

where D is a diagonal matrix whose diagonal elements are the eigenvalues of A.

Since the rank of A is r so that D is also of rank r, therefore D has exactly r non-zero elements it follows that A has exactly r non-zero eigenvalues and remaining $n - r$ eigenvalues of A are all zero.

If $\lambda_1, \lambda_2, ..., \lambda_r$ be non-zero eigenvalues of A, then we have

$$Q'AQ = D = \text{diag. } (\lambda_1, \lambda_2, ..., \lambda_r, 0, 0, ... 0)$$

Suppose out of r eigenvalues of A, $\lambda_1, \lambda_2, ..., \lambda_p$ are positive eigenvalues and $\lambda_{p+1}, \lambda_{p+2}, ..., \lambda_r$ are negative eigenvalues of A.

Let $$S = \text{diag.}\left(\frac{1}{\sqrt{\lambda_1}}, \frac{1}{\sqrt{\lambda_2}}, ..., \frac{1}{\sqrt{\lambda_p}}, \frac{1}{\sqrt{-\lambda_{p+1}}}, ..., \frac{1}{\sqrt{-\lambda_r}}, 0, ..., 0\right)$$

Then S is a non-singular diagonal matrix, and $S' = S$.

Let us take $P = QS$, clearly P is a real non-singular matrix, then we have

$$\begin{aligned} P'AP &= (QS)'A(QS) = S'(Q'AQ)S \\ &= S'DS && [\because Q'AQ = D] \\ &= SDS && [\because S' = S] \\ &= S \text{ diag. } (\lambda_1, \lambda_2, ..., \lambda_r, 0, 0, ..., 0)S \\ &= \text{diag.}(1, 1, ..., 1, -1, -1, ... -1, 0, 0, ..., 0) \end{aligned}$$

where 1 appears p times and -1 appears $r - p$ times.

Consider a real non-singular linear transformation $X = PY$ which reduces $X'AX$ to the form.

$$Y'(P'AP)Y = y_1^2 + y_2^2 + ... + y_p^2 - y_{p+1}^2 - y_{p+2}^2 - ... - y_r^2$$

REMARKS

- Two real quadratic forms $X'AX$ and $Y'BY$ in the same number of variables are real equivalent if and only if A and B have the same number of positive and negative eigenvalues.
- If $X'AX$ is a real quadratic form, the number of non-zero eigenvalues of A is equal to the rank of $X'AX$ and the number of positive eigenvalues of A is equal to the index of $X'AX$.
- Two real quadratic forms $X'AX$ and $Y'BY$ are orthogonally equivalent if and only if A and B have the same eigenvalues and these occur with the same multiplicities.

Solved Examples

Example 1. *Reduce each of the following quadratic forms in three variables to real canonical form and find its rank and signature. Also write in each case the linear transformation which reduces to normal form.*

(i) $x^2 + 2y^2 + 2z^2 - 2xy - 2yz + zx$

(ii) $x^2 - 2y^2 + 3z^2 - 4yz + 6zx$

(iii) $X'AX = 2x_1^2 + x_2^2 - 3x_3^2 - 8x_2x_3 - 4x_3x_1 + 12x_1x_2$

(iv) $6x_1^2 + 3x_2^2 + 14x_3^2 + 4x_2x_3 + 18x_3x_1 + 4x_1x_2$

Solution. (i) The given quadratic form is

$$X'AX = x^2 + 2y^2 + 2z^2 - 2xy - 2yz + zx$$

$$\therefore \quad A = \begin{bmatrix} 1 & -1 & \frac{1}{2} \\ -1 & 2 & -1 \\ \frac{1}{2} & -1 & 2 \end{bmatrix}$$

We write

$$A = IAI$$

or $$\begin{bmatrix} 1 & -1 & 1/2 \\ -1 & 2 & -1 \\ 1/2 & -1 & 2 \end{bmatrix} = \begin{bmatrix} 1 & 0 & 0 \\ 0 & 1 & 0 \\ 0 & 0 & 1 \end{bmatrix} A \begin{bmatrix} 1 & 0 & 0 \\ 0 & 1 & 0 \\ 0 & 0 & 1 \end{bmatrix}$$

Applying $R_2 \to R_2 + R_1$, we get

$$\begin{bmatrix} 1 & -1 & 1/2 \\ 0 & 1 & -1/2 \\ 1/2 & -1 & 2 \end{bmatrix} = \begin{bmatrix} 1 & 0 & 0 \\ 1 & 1 & 0 \\ 0 & 0 & 1 \end{bmatrix} A \begin{bmatrix} 1 & 0 & 0 \\ 0 & 1 & 0 \\ 0 & 0 & 1 \end{bmatrix}$$

Applying $C_2 \to C_2 + C_1$, we get

$$\begin{bmatrix} 1 & 0 & 1/2 \\ 0 & 1 & -1/2 \\ 1/2 & -1/2 & 2 \end{bmatrix} = \begin{bmatrix} 1 & 0 & 0 \\ 1 & 1 & 0 \\ 0 & 0 & 1 \end{bmatrix} A \begin{bmatrix} 1 & 1 & 0 \\ 0 & 1 & 0 \\ 0 & 0 & 1 \end{bmatrix}$$

Applying $R_3 \to R_3 - \frac{1}{2}R_1$, we get

$$\begin{bmatrix} 1 & 0 & 1/2 \\ 0 & 1 & -1/2 \\ 0 & -1/2 & 7/4 \end{bmatrix} = \begin{bmatrix} 1 & 0 & 0 \\ 1 & 1 & 0 \\ -1/2 & 0 & 1 \end{bmatrix} A \begin{bmatrix} 1 & 1 & 0 \\ 0 & 1 & 0 \\ 0 & 0 & 1 \end{bmatrix}$$

Applying $C_3 \to C_3 - \frac{1}{2}C_1$, we get

$$\begin{bmatrix} 1 & 0 & 0 \\ 0 & 1 & -1/2 \\ 0 & -1/2 & 7/4 \end{bmatrix} = \begin{bmatrix} 1 & 0 & 0 \\ 1 & 1 & 0 \\ -1/2 & 0 & 1 \end{bmatrix} A \begin{bmatrix} 1 & 1 & -1/2 \\ 0 & 1 & 0 \\ 0 & 0 & 1 \end{bmatrix}$$

Applying $R_3 \to R_3 + \frac{1}{2}R_2$, we get

$$\begin{bmatrix}1&0&0\\0&1&-1/2\\0&0&3/2\end{bmatrix}=\begin{bmatrix}1&0&0\\1&1&0\\0&1/2&1\end{bmatrix}A\begin{bmatrix}1&1&-1/2\\0&1&0\\0&0&1\end{bmatrix}$$

Applying $C_3 \to C_3 + \frac{1}{2}C_2$, we get

$$\begin{bmatrix}1&0&0\\0&1&0\\0&0&3/2\end{bmatrix}=\begin{bmatrix}1&0&0\\1&1&0\\0&1/2&1\end{bmatrix}A\begin{bmatrix}1&1&0\\0&1&1/2\\0&0&1\end{bmatrix}$$

Applying $R_3 \to \sqrt{\frac{2}{3}}R_3$, we get

$$\begin{bmatrix}1&0&0\\0&1&0\\0&0&\sqrt{3/2}\end{bmatrix}=\begin{bmatrix}1&0&0\\1&1&0\\0&1/\sqrt{6}&\sqrt{2/3}\end{bmatrix}A\begin{bmatrix}1&1&0\\0&1&1/2\\0&0&1\end{bmatrix}$$

Applying $C_3 \to \sqrt{\frac{2}{3}}C_3$, we get

$$\begin{bmatrix}1&0&0\\0&1&0\\0&0&1\end{bmatrix}=\begin{bmatrix}1&0&0\\1&1&0\\0&1/\sqrt{6}&\sqrt{2/3}\end{bmatrix}A\begin{bmatrix}1&1&0\\0&1&1/\sqrt{6}\\0&0&\sqrt{2/3}\end{bmatrix}$$

$\Rightarrow \quad D = P'AP$

where $P=\begin{bmatrix}1&1&0\\0&1&1/\sqrt{6}\\0&0&\sqrt{2/3}\end{bmatrix}$

Now the real non-singular linear transformation is

$$X = PY$$

i.e., $\begin{bmatrix}x\\y\\z\end{bmatrix}=\begin{bmatrix}1&1&0\\0&1&1/\sqrt{6}\\0&0&\sqrt{2/3}\end{bmatrix}\begin{bmatrix}y_1\\y_2\\y_3\end{bmatrix}$

i.e., $x = y_1 + y_2$

$$y = y_2 + \frac{1}{\sqrt{6}}y_3$$

$$z = \sqrt{\frac{2}{3}}y_3$$

which reduces $X'AX$ to the normal form

$$Y'(P'AP)Y = y_1^2 + y_2^2 + y_3^2$$

The rank of $X'AX$ = number of non-zero terms in normal form = 3

The signature of $X'AX$ = the excess of positive terms over the negative terms

$$= 3 - 0 = 3$$

The index of $X'AX$ = the number of positive terms

$$= 3$$

(ii) The given quadratic form is

$$X'AX = x^2 - 2y^2 + 3z^2 - 4yz + 6zx$$

$$=\begin{bmatrix}x&y&z\end{bmatrix}\begin{bmatrix}1&0&0\\0&-2&-2\\3&-2&3\end{bmatrix}\begin{bmatrix}x\\y\\z\end{bmatrix}$$

$$\therefore \quad A=\begin{bmatrix}1&0&3\\0&-2&-2\\3&-2&3\end{bmatrix}$$

We write

$$A = IAI$$

or

$$\begin{bmatrix}1&0&3\\0&-2&-2\\3&-2&3\end{bmatrix}=\begin{bmatrix}1&0&0\\0&1&0\\0&0&1\end{bmatrix}A\begin{bmatrix}1&0&0\\0&1&0\\0&0&1\end{bmatrix}$$

Applying $R_3 \to R_3 - 3R_1$, we get

$$\begin{bmatrix}1&0&3\\0&-2&-2\\3&-2&-6\end{bmatrix}=\begin{bmatrix}1&0&0\\0&1&0\\-3&0&1\end{bmatrix}A\begin{bmatrix}1&0&0\\0&1&0\\0&0&1\end{bmatrix}$$

Applying $C_3 \to C_3 - 3C_1$, we get

$$\begin{bmatrix}1&0&0\\0&-2&-2\\0&-2&-6\end{bmatrix}=\begin{bmatrix}1&0&0\\0&1&0\\-3&0&1\end{bmatrix}A\begin{bmatrix}1&0&-3\\0&1&0\\0&0&1\end{bmatrix}$$

Applying $R_3 \to R_3 - R_2$, we get

$$\begin{bmatrix}1&0&0\\0&-2&-2\\0&0&-4\end{bmatrix}=\begin{bmatrix}1&0&0\\0&1&0\\-3&-1&1\end{bmatrix}A\begin{bmatrix}1&0&-3\\0&1&0\\0&0&1\end{bmatrix}$$

Applying $C_3 \to C_3 - C_2$, we get

$$\begin{bmatrix}1&0&0\\0&-2&0\\0&0&-4\end{bmatrix}=\begin{bmatrix}1&0&0\\0&1&0\\-3&-1&1\end{bmatrix}A\begin{bmatrix}1&0&-3\\0&1&-1\\0&0&1\end{bmatrix}$$

Applying $R_2 \to \frac{1}{\sqrt{2}}R_2$, we get

$$\begin{bmatrix}1&0&0\\0&-\sqrt{2}&0\\0&0&-4\end{bmatrix}=\begin{bmatrix}1&0&0\\0&1/\sqrt{2}&0\\-3&-1&1\end{bmatrix}A\begin{bmatrix}1&0&-3\\0&1&-1\\0&0&1\end{bmatrix}$$

Applying $C_2 \to \frac{1}{\sqrt{2}}C_2$, we get

$$\begin{bmatrix}1&0&0\\0&-1&0\\0&0&-4\end{bmatrix}=\begin{bmatrix}1&0&0\\0&1/\sqrt{2}&0\\-3&-1&1\end{bmatrix}A\begin{bmatrix}1&0&-3\\0&1/\sqrt{2}&-1\\0&0&1\end{bmatrix}$$

Applying $R_3 \to \frac{1}{\sqrt{4}}R_3$, we get

$$\begin{bmatrix}1&0&0\\0&-1&0\\0&0&-\sqrt{4}\end{bmatrix}$$

$$=\begin{bmatrix}1&0&0\\0&1/\sqrt{2}&0\\-3/\sqrt{4}&-1/\sqrt{4}&1/\sqrt{4}\end{bmatrix}A\begin{bmatrix}1&0&-3\\0&1/\sqrt{2}&-1\\0&0&1\end{bmatrix}$$

Applying $C_3 \to \frac{1}{\sqrt{4}}C_3$, we get

$$\begin{bmatrix}1&0&0\\0&-1&0\\0&0&-1\end{bmatrix}$$

$$=\begin{bmatrix}1&0&0\\0&1/\sqrt{2}&0\\-3/2&-1/2&1/2\end{bmatrix}A\begin{bmatrix}1&0&-3/2\\0&1/\sqrt{2}&-1/2\\0&0&1/2\end{bmatrix}$$

$\Rightarrow \qquad D = P'AP$

where $\quad P = \begin{bmatrix} 1 & 0 & -3/2 \\ 0 & 1/\sqrt{2} & -1/2 \\ 0 & 0 & 1/2 \end{bmatrix}$

Thus the real non-singular linear transformation is $\quad X = PY$

i.e., $\begin{bmatrix} x \\ y \\ z \end{bmatrix} = \begin{bmatrix} 1 & 0 & -3/2 \\ 0 & 1/\sqrt{2} & -1/2 \\ 0 & 0 & 1/2 \end{bmatrix}\begin{bmatrix} y_1 \\ y_2 \\ y_3 \end{bmatrix}$

i.e., $\quad x = y_1 - \frac{3}{2}y_3$

$y = \frac{1}{\sqrt{2}}y_2 - \frac{1}{2}y_3$

$z = \frac{1}{2}y_3$

which reduces $X'AX$ to the normal form

$$Y'(P'AP)Y = \begin{bmatrix} y_1 & y_2 & y_3 \end{bmatrix}\begin{bmatrix} 1 & 0 & 0 \\ 0 & -1 & 0 \\ 0 & 0 & -1 \end{bmatrix}\begin{bmatrix} y_1 \\ y_2 \\ y_3 \end{bmatrix}$$

$$= y_1^2 - y_2^2 - y_3^2$$

The rank of $X'AX$ = number of non-zero terms in normal form

$= 3$

The signature of $X'AX$ = the excess of positive terms over the negative terms

$= 1 - 2 = -1$

The index of $X'AX$ = the number of positive terms in the normal form

$= 1$

(iii) The given quadratic form is

$$X'AX = 2x_1^2 + x_2^2 - 3x_3^2 - 8x_2x_3 - 4x_3x_1 + 12x_1x_2$$

$$= \begin{bmatrix} x_1 & x_2 & x_3 \end{bmatrix}\begin{bmatrix} 2 & 6 & -2 \\ 6 & 1 & -4 \\ -2 & -4 & -3 \end{bmatrix}\begin{bmatrix} x_1 \\ x_2 \\ x_3 \end{bmatrix}$$

$\therefore \qquad A = \begin{bmatrix} 2 & 6 & -2 \\ 6 & 1 & -4 \\ -2 & -4 & -3 \end{bmatrix}$

We write

$A = IAI$

or $\begin{bmatrix} 2 & 6 & -2 \\ 6 & 1 & -4 \\ -2 & -4 & -3 \end{bmatrix} = \begin{bmatrix} 1 & 0 & 0 \\ 0 & 1 & 0 \\ 0 & 0 & 1 \end{bmatrix} A \begin{bmatrix} 1 & 0 & 0 \\ 0 & 1 & 0 \\ 0 & 0 & 1 \end{bmatrix}$

Applying $R_2 \to R_2 - 3R_1, C_2 \to C_2 - 3C_1,$ $R_3 \to R_3 + R_1, C_3 \to C_3 + C_1$ we get

$$\begin{bmatrix} 2 & 0 & 0 \\ 0 & -17 & 2 \\ 0 & 2 & -5 \end{bmatrix} = \begin{bmatrix} 1 & 0 & 0 \\ -3 & 1 & 0 \\ 1 & 0 & 1 \end{bmatrix} A \begin{bmatrix} 1 & -3 & 1 \\ 0 & 1 & 0 \\ 0 & 0 & 1 \end{bmatrix}$$

Applying $R_3 \to R_3 + \frac{2}{17}R_2, C_3 \to C_3 + \frac{2}{17}C_2,$ we get

$$\begin{bmatrix} 2 & 0 & 0 \\ 0 & -17 & 0 \\ 0 & 0 & -81/17 \end{bmatrix} = \begin{bmatrix} 1 & 0 & 0 \\ -3 & 1 & 0 \\ 11/17 & 2/17 & 1 \end{bmatrix} A \begin{bmatrix} 1 & -3 & 11/17 \\ 0 & 1 & 2/17 \\ 0 & 0 & 1 \end{bmatrix}$$

Applying $R_1 \to \frac{1}{\sqrt{2}}R_1, C_1 \to \frac{1}{\sqrt{2}}C_1,$

$R_2 \to \frac{1}{\sqrt{17}}R_2, C_2 \to \frac{1}{\sqrt{17}}C_2, R_3 \to \sqrt{\frac{17}{81}}R_3,$

$C_3 \to \sqrt{\frac{17}{81}}C_3,$ we get

$$\begin{bmatrix} 1 & 0 & 0 \\ 0 & -1 & 0 \\ 0 & 0 & -1 \end{bmatrix} = \begin{bmatrix} 1/\sqrt{2} & 0 & 0 \\ -3/\sqrt{17} & 1/\sqrt{17} & 0 \\ 11\sqrt{17}/9 & 2\sqrt{17}/9 & 1\sqrt{17}/9 \end{bmatrix} A \begin{bmatrix} 1/\sqrt{2} & -3/\sqrt{17} & 11\sqrt{17}/9 \\ 0 & 1/\sqrt{17} & 2\sqrt{17}/9 \\ 0 & 0 & 1\sqrt{17}/9 \end{bmatrix}$$

$\Rightarrow \qquad D = P'AP$

where $\quad P = \begin{bmatrix} 1/\sqrt{2} & -3/\sqrt{17} & 11\sqrt{17}/9 \\ 0 & 1/\sqrt{17} & 2\sqrt{17}/9 \\ 0 & 0 & 1\sqrt{17}/9 \end{bmatrix}$

Thus the real non-singular linear transformation is

$X = PY$

i.e., $\begin{bmatrix} x_1 \\ x_2 \\ x_3 \end{bmatrix} = \begin{bmatrix} 1/\sqrt{2} & -3/\sqrt{17} & 11\sqrt{17}/9 \\ 0 & 1/\sqrt{17} & 2\sqrt{17}/9 \\ 0 & 0 & 1\sqrt{17}/9 \end{bmatrix}\begin{bmatrix} y_1 \\ y_2 \\ y_3 \end{bmatrix}$

i.e., $\quad x_1 = \frac{1}{\sqrt{2}}y_1 - \frac{3}{\sqrt{17}}y_2 + \frac{11}{9}\sqrt{17}y_3$

$x_2 = \frac{1}{\sqrt{17}}y_2 + \frac{2\sqrt{17}}{9}y_3$

$x_3 = \frac{1\sqrt{17}}{9}y_3$

These transformations reduce the given quadratic form $X'AX$ to the normal form

$$Y'(P'AP)Y = Y'DY$$

$$= \begin{bmatrix} y_1 & y_2 & y_3 \end{bmatrix}\begin{bmatrix} 1 & 0 & 0 \\ 0 & -1 & 0 \\ 0 & 0 & -1 \end{bmatrix}\begin{bmatrix} y_1 \\ y_2 \\ y_3 \end{bmatrix}$$

$$= y_1^2 - y_2^2 - y_3^2$$

The rank of $X'AX$ = number of non-zero terms in normal form $= 3$

The signature of $X'AX$ = the excess of the number of positive terms over the negative terms in the normal form $= 1 - 2 = -1$

The index of $X'AX$ = the number of positive terms in the normal form $= 1$

(iv) The given quadratic form is

$$X'AX = 6x_1^2 + 3x_2^2 + 14x_3^2 + 4x_2x_3 + 18x_3x_1 + 4x_1x_2$$

$$= [x_1 \ \ x_2 \ \ x_3] \begin{bmatrix} 6 & 2 & 9 \\ 2 & 3 & 2 \\ 9 & 2 & 14 \end{bmatrix} \begin{bmatrix} x_1 \\ x_2 \\ x_3 \end{bmatrix}$$

$$\therefore \qquad A = \begin{bmatrix} 6 & 2 & 9 \\ 2 & 3 & 2 \\ 9 & 2 & 14 \end{bmatrix}$$

We write $A = IAI$

or
$$\begin{bmatrix} 6 & 2 & 9 \\ 2 & 3 & 2 \\ 9 & 2 & 14 \end{bmatrix} = \begin{bmatrix} 1 & 0 & 0 \\ 0 & 1 & 0 \\ 0 & 0 & 1 \end{bmatrix} A \begin{bmatrix} 1 & 0 & 0 \\ 0 & 1 & 0 \\ 0 & 0 & 1 \end{bmatrix}$$

Applying $R_2 \to R_2 - \frac{1}{3}R_1$, $C_2 \to C_2 - \frac{1}{2}C_1$ and $R_3 \to R_3 - \frac{3}{2}R_1$, $C_3 \to C_3 - \frac{3}{2}C_1$, we get

$$\begin{bmatrix} 6 & 0 & 0 \\ 0 & 7/3 & -1 \\ 0 & -1 & 1/2 \end{bmatrix} = \begin{bmatrix} 1 & 0 & 0 \\ -1/3 & 1 & 0 \\ -3/2 & 0 & 1 \end{bmatrix} A \begin{bmatrix} 1 & -1/3 & -3/2 \\ 0 & 1 & 3/7 \\ 0 & 0 & 1 \end{bmatrix}$$

Applying $R_3 \to R_3 + \frac{3}{7}R_2$, $C_3 \to C_3 + \frac{3}{7}C_1$, we get

$$\begin{bmatrix} 6 & 0 & 0 \\ 0 & 7/3 & 0 \\ 0 & -1 & 1/14 \end{bmatrix} = \begin{bmatrix} 1 & 0 & 0 \\ -1/3 & 1 & 0 \\ -23/14 & 3/7 & 1 \end{bmatrix} A \begin{bmatrix} 1 & -1/3 & -23/14 \\ 0 & 1 & 3/7 \\ 0 & 0 & 1 \end{bmatrix}$$

Applying $R_1 \to \frac{1}{\sqrt{6}}R_1$, $C_1 \to \frac{2}{\sqrt{6}}C_1$,

$R_2 \to \sqrt{\frac{3}{7}}\, R_2$, $C_2 \to \sqrt{\frac{3}{7}}\, C_2$ and $R_3 \to \frac{1}{\sqrt{14}}R_3$,

$C_3 \to \frac{1}{\sqrt{14}}C_3$, we get

$$\begin{bmatrix} 1 & 0 & 0 \\ 0 & 1 & 0 \\ 0 & 0 & 1 \end{bmatrix} = \begin{bmatrix} \frac{1}{\sqrt{6}} & 0 & 0 \\ -\frac{1}{3}\sqrt{\frac{3}{7}} & \sqrt{\frac{3}{7}} & 0 \\ -\frac{23}{14}\frac{1}{\sqrt{14}} & \frac{3}{7}\frac{1}{\sqrt{14}} & 1 \end{bmatrix} A \begin{bmatrix} \frac{1}{\sqrt{6}} & -\frac{1}{3}\sqrt{\frac{3}{7}} & -\frac{23}{14}\frac{1}{\sqrt{14}} \\ 0 & \sqrt{\frac{3}{7}} & \frac{3}{7}\frac{1}{\sqrt{14}} \\ 0 & 0 & 1 \end{bmatrix}$$

$$\Rightarrow \qquad D = P'AP$$

where
$$P = \begin{bmatrix} \frac{1}{\sqrt{6}} & -\frac{1}{3}\sqrt{\frac{3}{7}} & -\frac{23}{14}\frac{1}{\sqrt{14}} \\ 0 & \sqrt{\frac{3}{7}} & \frac{3}{7}\frac{1}{\sqrt{14}} \\ 0 & 0 & 1 \end{bmatrix}$$

Thus the real non-singular linear transformation is

$$X = PY$$

i.e.,
$$\begin{bmatrix} x_1 \\ x_2 \\ x_3 \end{bmatrix} = \begin{bmatrix} \frac{1}{\sqrt{6}} & -\frac{1}{3}\sqrt{\frac{3}{7}} & -\frac{23}{14}\frac{1}{\sqrt{14}} \\ 0 & \sqrt{\frac{3}{7}} & \frac{3}{7}\frac{1}{\sqrt{14}} \\ 0 & 0 & 1 \end{bmatrix} \begin{bmatrix} y_1 \\ y_2 \\ y_3 \end{bmatrix}$$

or
$$x_1 = \frac{1}{\sqrt{6}}y_1 - \frac{1}{3}\sqrt{\frac{3}{7}}y_2 - \frac{23}{14}\frac{1}{\sqrt{14}}y_3$$

$$x_2 = \sqrt{\frac{3}{7}}y_2 + \frac{3}{7}\frac{1}{\sqrt{14}}y_3$$

$$x_3 = y_3$$

These transformations reduce the given quadratic form $X'AX$ to the normal form

$$Y'(P'AP)Y = Y'DY = y_1^2 + y_2^2 + y_3^2$$

The rank of $X'AX$ = number of non-zero terms in normal form

$= 3$

The signature of $X'AX$ = 2(positive terms in normal form) – rank

$= 2(3) - 3$

$= 6 - 3 = 3$

The index of $X'AX$ = the number of positive terms in the normal form

$= 3$

Example 2. *Find an orthogonal matrix P that will diagonalize the real matrix*

$$A = \begin{bmatrix} 0 & 1 & 1 \\ 1 & 0 & -1 \\ 1 & -1 & 0 \end{bmatrix}$$

Interpret the result in terms of quadratic form.

Solution. The characteristic equation of A is given by

$$|A - \lambda I| = 0$$

or
$$\begin{vmatrix} -\lambda & 1 & 1 \\ 1 & -\lambda & -1 \\ 1 & -1 & -\lambda \end{vmatrix} = 0$$

Applying $C_1 \to C_1 + C_2$, we get

$$\begin{vmatrix} 1-\lambda & 1 & 1 \\ 1-\lambda & -\lambda & -1 \\ 0 & -1 & -\lambda \end{vmatrix} = 0$$

or
$$(1-\lambda)\begin{vmatrix} 1 & 1 & 1 \\ 1 & -\lambda & -1 \\ 0 & -1 & -\lambda \end{vmatrix} = 0$$

or $\quad (1-\lambda)(\lambda^2-1+\lambda-1)=0$

or $\quad (1-\lambda)^2(2+\lambda)=0$

Therefore, the eigenvalues of A are 1, 1 and –2.

Eigenvector corresponding to the eigen value 1

Let $X=\begin{bmatrix}x_1\\x_2\\x_3\end{bmatrix}\neq 0$ be an eigenvector corresponding to the eigenvalue 1, then X is a solution of the equation

$$(A-I)X=0$$

or $$\begin{bmatrix}-1&1&1\\1&-1&-1\\1&-1&-1\end{bmatrix}\begin{bmatrix}x_1\\x_2\\x_3\end{bmatrix}=\begin{bmatrix}0\\0\\0\end{bmatrix}$$

Solving these equation, we have

$$-x_1+x_2+x_3=0 \qquad \ldots(1)$$

The two orthogonal solutions, which satisfy (1) are

$$X_1=\begin{bmatrix}1\\0\\1\end{bmatrix} \text{ and } X_2=\begin{bmatrix}1\\2\\-1\end{bmatrix}$$

Thus, the two mutually orthogonal eigenvectors of A are

$$X_1=\begin{bmatrix}1\\0\\1\end{bmatrix} \text{ and } X_2=\begin{bmatrix}1\\2\\-1\end{bmatrix}$$

Eigenvector corresponding to the eigen value –2

Let $X=\begin{bmatrix}x_1\\x_2\\x_3\end{bmatrix}\neq 0$ be an eigenvector corresponding to the eigenvalue –2, then X is a solution of the equation $(A+2I)X=0$

or $$\begin{bmatrix}2&1&1\\1&2&-1\\1&-1&2\end{bmatrix}\begin{bmatrix}x_1\\x_2\\x_3\end{bmatrix}=\begin{bmatrix}0\\0\\0\end{bmatrix}$$

Applying $R_1 \leftrightarrow R_2$, we get

$$\begin{bmatrix}1&2&-1\\2&1&1\\1&-1&2\end{bmatrix}\begin{bmatrix}x_1\\x_2\\x_3\end{bmatrix}=\begin{bmatrix}0\\0\\0\end{bmatrix}$$

Applying $R_2 \to R_2-2R_1$, $R_3 \to R_3-R_1$, we get

$$\begin{bmatrix}1&2&-1\\0&-3&3\\0&-3&3\end{bmatrix}\begin{bmatrix}x_1\\x_2\\x_3\end{bmatrix}=\begin{bmatrix}0\\0\\0\end{bmatrix}$$

Applying $R_3 \to R_3-R_2$, we get

$$\begin{bmatrix}1&2&-1\\0&-3&3\\0&0&0\end{bmatrix}\begin{bmatrix}x_1\\x_2\\x_3\end{bmatrix}=\begin{bmatrix}0\\0\\0\end{bmatrix}$$

Solving these equation, we get

$$x_1+2x_2-x_3=0$$

$$-3x_2+3x_3=0$$

Clearly, $x_1=-1$, $x_2=1$, $x_3=1$ satisfy these equations, therefore, the eigenvector is

$$X_3=\begin{bmatrix}-1\\1\\1\end{bmatrix}$$

Thus, the required matrix P is a matrix whose column vectors are unit vectors which are scalar multiplies of X_1, X_2 and X_3.

$$\therefore \quad P=\begin{bmatrix}1/\sqrt{2}&1/\sqrt{6}&-1/\sqrt{3}\\0&2/\sqrt{16}&1/\sqrt{3}\\1/\sqrt{2}&-1/\sqrt{6}&1/\sqrt{3}\end{bmatrix}$$

Now, we have

$$P'AP=D=\text{diag. }(1,1,-2)$$

The quadratic form of the given symmetric matrix is

$$X'AX=2x_1x_2+2x_1x_3-2x_2x_3$$

Thus, the orthogonal linear transformation $X=PY$ reduces the quadratic form $X'AX$ to the diagonal form

$$X'AX=Y'(P'AP)Y=Y'DY=y_1^2+y_2^2-2y_3^2$$

The rank of $X'AX$ = number of non-zero eigenvalues of A = 3

The signature of $X'AX$ = the excess of the number of positive eigenvalues over the number of negative eigenvalues of A

$$=2-1=1$$

The diagonal form $y_1^2+y_2^2-2y_3^2$ can be reduced to the normal form $z_1^2+z_2^2-z_3^2$.

Example 3. *Reduce the following quadratic form in to canonical form and find its rank and signature* :

$$x^2+4y^2+9z^2+t^2-12yz+6zx-4xy-2xt-6zt$$

Solution. Let $X=\begin{bmatrix}x\\y\\z\\t\end{bmatrix}$, then the given quadratic form can be written as

$$x^2+4y^2+9z^2+t^2-12yz+6zx-4xy-2xt-6zt$$

$$=X'\begin{bmatrix}1&-2&3&-1\\-2&4&-6&0\\3&-6&9&-3\\-1&0&-3&1\end{bmatrix}X=X'AX$$

$$\therefore \quad A=\begin{bmatrix}1&-2&3&-1\\-2&4&-6&0\\3&-6&9&-3\\-1&0&-3&1\end{bmatrix}$$

We write

$$A=IAI$$

$$\begin{bmatrix}1&-2&3&-1\\-2&4&-6&0\\3&-6&9&-3\\-1&0&-3&1\end{bmatrix}=\begin{bmatrix}1&0&0&0\\0&1&0&0\\0&0&1&0\\0&0&0&1\end{bmatrix}A\begin{bmatrix}1&0&0&0\\0&1&0&0\\0&0&1&0\\0&0&0&1\end{bmatrix}$$

Applying $R_2 \to R_2 + 2R_1, C_2 \to C_2 + 2C_1;$
$R_3 \to R_3 + 3R_1, C_3 \to C_3 - 3C_1;$
$R_4 \to R_4 + R_1, C_4 \to C_4 + C_1$

We get

$$\begin{bmatrix} 1 & 0 & 0 & 0 \\ 0 & 0 & 0 & -2 \\ 0 & 0 & 0 & 0 \\ 0 & -2 & 0 & 0 \end{bmatrix} = \begin{bmatrix} 1 & 0 & 0 & 0 \\ 2 & 1 & 0 & 0 \\ -3 & 0 & 1 & 0 \\ 1 & 0 & 0 & 1 \end{bmatrix} A \begin{bmatrix} 1 & 2 & -3 & 1 \\ 0 & 1 & 0 & 0 \\ 0 & 0 & 1 & 0 \\ 0 & 0 & 0 & 1 \end{bmatrix}$$

Applying $R_2 \to R_2 + R_4, C_2 \to C_2 + C_4$, we get

$$\begin{bmatrix} 1 & 0 & 0 & 0 \\ 0 & -4 & 0 & -2 \\ 0 & 0 & 0 & 0 \\ 0 & -2 & 0 & 0 \end{bmatrix} = \begin{bmatrix} 1 & 0 & 0 & 0 \\ 3 & 1 & 0 & 1 \\ -3 & 0 & 1 & 0 \\ 1 & 0 & 0 & 1 \end{bmatrix} A \begin{bmatrix} 1 & 3 & -3 & 1 \\ 0 & 1 & 0 & 0 \\ 0 & 0 & 1 & 0 \\ 0 & 1 & 0 & 1 \end{bmatrix}$$

Applying $R_4 \to R_4 - \frac{1}{2}R_2, C_4 \to C_4 - \frac{1}{2}C_2$, we get

$$\begin{bmatrix} 1 & 0 & 0 & 0 \\ 0 & -4 & 0 & 0 \\ 0 & 0 & 0 & 0 \\ 0 & 0 & 0 & 1 \end{bmatrix} = \begin{bmatrix} 1 & 0 & 0 & 0 \\ 3 & 1 & 0 & 1 \\ -3 & 0 & 1 & 0 \\ -1/2 & -1/2 & 0 & 1/2 \end{bmatrix} \times A \begin{bmatrix} 1 & 3 & -3 & -1/2 \\ 0 & 1 & 0 & -1/2 \\ 0 & 0 & 1 & 0 \\ 0 & 1 & 0 & 1/2 \end{bmatrix}$$

Applying $R_2 \to \frac{1}{2}R_2, C_2 \to \frac{1}{2}C_2$, we get

$$\begin{bmatrix} 1 & 0 & 0 & 0 \\ 0 & -1 & 0 & 0 \\ 0 & 0 & 0 & 0 \\ 0 & 0 & 0 & 1 \end{bmatrix} = \begin{bmatrix} 1 & 0 & 0 & 0 \\ \frac{3}{2} & \frac{1}{2} & 0 & \frac{1}{2} \\ -3 & 0 & 1 & 0 \\ -\frac{1}{2} & -\frac{1}{2} & 0 & \frac{1}{2} \end{bmatrix} A \begin{bmatrix} 1 & \frac{3}{2} & -3 & -\frac{1}{2} \\ 0 & \frac{1}{2} & 0 & -\frac{1}{2} \\ 0 & 0 & 1 & 0 \\ 0 & \frac{1}{2} & 0 & \frac{1}{2} \end{bmatrix}$$

$\Rightarrow \quad D = P'AP$

where $P = \begin{bmatrix} 1 & 3/2 & -3 & -1/2 \\ 0 & 1/2 & 0 & -1/2 \\ 0 & 0 & 1 & 0 \\ 0 & 1/2 & 0 & 1/2 \end{bmatrix}$

Thus, the real non-singular linear transformation $X = PY$ reduces the given quadratic form $X'AX$ to the normal form

$$Y'(P'AP)Y = Y'DY = y_1^2 - y_2^2 + y_4^2$$

The rank of $X'AX$ = the number of non-zero terms in the normal form

$= 3$

The signature of $X'AX$

$= 2(\text{positive terms}) - \text{rank}$

$= 2(2) - 3$

$= 4 - 3 = 1$

EXERCISE 25.13

1. Find the rank of each of the following quadratic forms :
(i) $x^2 - 12xy - 4y^2$ (ii) $3x^2 + 2xy + 3y^2$
(iii) $x^2 - 2xy + y^2$
(iv) $x_1^2 - 2x_1x_2 + 2x_2^2$
(v) $4x_1^2 + x_2^2 - 8x_3^2 + 4x_1x_2 - 4x_1x_3 + 8x_2x_3$
2. Find a real non-singular linear transformation $X = PY$ which reduces the given real quadratic form $x^2 + 2y^2 + 3z^2 + 4xy + 4yz$ to real canonical form. Also find the rank and signature of the given quadratic form.
3. Reduce each of the following quadratic forms to real canonical form and find its rank and signature. Also write in each case the linear transformation which reduce the normal form.
(i) $x_1^2 + 2x_2^2 - 7x_3^2 - 4x_1x_2 + 8x_1x_3$
(ii) $(x_1 + x_2 + x_3)^2 + (x_2 + x_3)^2 + 4x_3^2$
(iii) $x_1^2 + 2x_2^2 + 3x_3^2 + 2x_2x_3 - 2x_3x_1 + 2x_1x_2$
(iv) $4x_1^2 + 9x_2^2 + 2x_3^2 + 8x_2x_3 - 6x_3x_1 + 6x_1x_2$
(v) $3x^2 + 3y^2 + 3z^2 - 2yz + 2zx + 2xy$
(vi) $2x^2 + 9y^2 + 2z^2 - 2yz + 2zx + 6xy$
(vii) $x^2 - 4y^2 + 6z^2 + 2xy - 4xz + 2w^2 - 6zw$
(viii) $x_1x_2 - 4x_1x_4 - 2x_2x_3 + 12x_3x_4$
4. Reduce the quadratic form $7x^2 - 8y^2 - 8z^2 - 2yz - 8zx + 8xy$ to canonical form by an orthogonal transformation and hence find the signature of the quadratic form.

ANSWERS

1. (i) rank = 2 (ii) rank = 2 (iii) rank = 1 (iv) rank = 2 (v) rank = 3

2. $X = \begin{bmatrix} 1 & -1/\sqrt{2} & -2/\sqrt{5} \\ 0 & 1/\sqrt{2} & 1/\sqrt{5} \\ 0 & 0 & 1/\sqrt{5} \end{bmatrix} Y$, canonical form is $y_1^2 - y_2^2 + y_3^2$, rank = 3, signature = 1

3. (i) rank = 3, signature = 1 (ii) rank = 3, signature = 3 (iii) rank = 3, signature = 3 (iv) rank = 3, signature = 1
(v) rank = 3, signature = 3 (vi) rank = 3, signature = 3 (vii) rank = 4, signature = 0 (viii) rank = 4, signature = 0

Objective Evaluations

Fill in the blanks

1. If the rank of a square matrix is not equal to its order, then the matrix is ___________ .
2. The rank of every non-zero matrix is always greater than or equal to ___________.
3. The rank of I_n is ___________ where I_n is a unit matrix of order n .
4. The rank of a matrix is $\le r$. If all $(r+1)$-rowed minors of the matrix ___________.
5. The rank of A and A^T are ___________.
6. The rank of $A=\begin{bmatrix}1 & 2 & 3\\ 2 & 4 & 5\end{bmatrix}$ is ___________.
7. The rank of $A=\begin{bmatrix}1 & 1 & 1 & 1\\ 1 & 1 & 1 & 1\\ 1 & 1 & 1 & 1\\ 1 & 1 & 1 & 1\end{bmatrix}$ is ___________.
8. If A is a non-zero column matrix and B is a non-zero row matrix, then rank of (AB) = ___________.
9. All the elementary matrices are ___________.
10. Elementary transformation does not change the ___________ of the matrix.
11. If $A \sim \begin{pmatrix} I_{10} & 0\\ 0 & 0\end{pmatrix}$, then the rank of A = ___________.
12. The ranks of two equivalent matrices are ___________.
13. The rank of a matrix $\begin{bmatrix}1 & 2 & 3\\ 0 & 1 & 4\\ 0 & 0 & 0\end{bmatrix}$ is ___________.
14. The matrix $\begin{bmatrix}1 & 2 & 3\\ 0 & 1 & 4\\ 0 & 0 & 0\end{bmatrix}$ is an ___________.
15. The rank of the product of two matrices cannot exceed the ___________ of either matrix.
16. The rank of the matrix diag. $[\lambda_1\ \lambda_2\ \lambda_3]$ is ___________.
17. The rank of the matrix $A=\begin{bmatrix}1 & 2 & 0 & 0\\ 0 & 0 & 1 & -1\\ 0 & 0 & 0 & 0\end{bmatrix}$ is ___________.
18. Non-square matrix has ___________ inverse.
19. A matrix A is said to be singular if $|A|$ = ___________.
20. A matrix is said to be ___________ if it is square and non-singular.
21. If $|A| \ne 0$, then matrix is said to be ___________.
22. The inverse of a matrix, if exist is ___________.
23. If A and B be two non-singular matrices of the same order, then AB is ___________.
24. $(AB)^{-1}$ = ___________.
25. The transpose of the matrix of cofactors is known as ___________.
26. The necessary and sufficient condition that a square matrix may possess an inverse is that it be ___________.
27. The inverse of the inverse of a matrix A is equal to ___________.
28. The matrix equation $AX = O$ is a system of linear ___________ equations.
29. If the rank of $A = r$, then the number of linearly independent solutions of m homogeneous equations in n variables is ___________.
30. If X_1 and X_2 are the solutions of $AX = O$, then ___________ is also a solution of $AX = O$.
31. If the rank of A is equal to the number of unknowns in $AX = O$, then there exists only ___________.
32. If the rank of A is less than the number of unknowns in $AX = O$, then there are ___________.
33. The matrix equation $AX = B$ is a system of linear ___________.
34. The equation $AX = B$ is consistent if the rank A = ___________.
35. The rank $A <$ rank$(A|B)$, then the equation $AX = B$ is ___________.
36. If rank A = rank $(A|B)$ and is also equal to the number of unknowns, then there will be ___________.
37. If rank A = rank $(A|B) = r$ and $r < n$ (No. of unknowns), then there will only ___________ linearly independent solutions of $AX = B$.
38. If the number of equations is less than the number of unknowns in $AX = B$, then there will always ___________ solutions.
39. The system of equations $2x-2y+5z+3w=0$, $4x-y+z+w=0$, $3x-2y+3z+4w=0$, $x-3y+7z+6w=0$ has ___________ linearly independent solutions.
40. The equations $3x+4y=5$, $6x+8y=10$ have ___________ solutions.

41. The matrix $[A : B]$ is known as ____________.

42. The system of m non-homogeneous linear equations in n unknowns $AX = B$, is ____________ iff the coefficient matrix A and augmented matrix $[A|B]$ are of the same rank.

43. If A is a square matrix, then its characteristic equation is ____________.

44. If the matrix $(A - \lambda I)$ is singular, then λ is an ____________ of A.

45. If the matrix $(A - \lambda I)$ is singular, then there exists a non-zero vector X such that $AX =$ ____________.

46. If X is an eigenvector corresponding to the eigenvalue of A, the kX is also eigenvector corresponding to ____________.

47. The characteristic vectors corresponding to distinct characteristic roots of a matrix are linearly ____________.

48. The characteristic roots of a Hermitian matrix are ____________.

49. The characteristic roots of a unitary matrix are of ____________.

50. The characteristic roots of the matrix $\begin{bmatrix} 5 & 4 \\ 1 & 2 \end{bmatrix}$ are ____________.

51. The eigenvalues of the matrix A and A^T are ____________.

52. If 0 is the characteristic root of A, then A is ____________.

53. The characteristic roots of a triangular matrix are just the ____________ elements of the same matrix.

54. If $\lambda_1, \lambda_2, \ldots, \lambda_n$ are the characteristic roots of A, then $a\lambda_1, a\lambda_2, \ldots, a\lambda_n$ are the characteristic roots of ____________.

55. If λ is an eigenvalues of A, then the eigenvalue of A^{-1} is ____________.

56. If λ is a characteristic root of a matrix A (non-singular) then the characteristic root of Adj. A is ____________.

57. Every square matrix satisfies its ____________.

58. If $A^2 - 5A + 6I = 0$ for the matrix A, then the eigenvalues of A are ____________.

59. If $A^2 - 4A + 5I = 0$ for the matrix A , then its inverse is ____________.

60. If 1, 2, 3 are eigenvalues of A, then the eigenvalues of A^2 are ____________.

True/False

Write T for True and F for False statement.

1. $\begin{bmatrix} 3 & 5 \\ 7 & 9 \end{bmatrix} = \begin{bmatrix} 3 & 7 \\ 5 & 9 \end{bmatrix}$ **(T/F)**

2. If $A = \begin{bmatrix} 1 & 0 \\ 0 & 1 \end{bmatrix}$, $B = \begin{bmatrix} 3 & 4 \\ 6 & 7 \end{bmatrix}$, then $A + B = \begin{bmatrix} 4 & 4 \\ 6 & 8 \end{bmatrix}$ **(T/F)**

3. If $A = \begin{bmatrix} -1 & 2 & 3 \\ 5 & 6 & 7 \end{bmatrix}$, then $5A = \begin{bmatrix} -5 & -10 & -15 \\ -25 & 30 & 35 \end{bmatrix}$. **(T/F)**

4. Square matrices are always conformable for multiplication. **(T/F)**

5. A matrix A is symmetric if $A' = -A$. **(T/F)**

6. The diagonal elements of a skew-symmetric matrix are all zero. **(T/F)**

7. The rank of zero matrix is 1. **(T/F)**

8. The rank of I_4 is 4. **(T/F)**

9. The rank of a matrix, if it is reduced to an Echelon form, is equal to the number if non-zero rows in Echelon form. **(T/F)**

10. If $A = [a_{ij}]_{m \times n}$ and $|A| = 0$, then the rank of $A \geq n$. **(T/F)**

11. If rank $A = 3$, then the rank of its transpose is 3. **(T/F)**

12. The rank of the matrix $\begin{bmatrix} 0 & 1 & 2 & 3 \\ 0 & 0 & 1 & -1 \\ 0 & 0 & 0 & 0 \end{bmatrix}$ is 3. **(T/F)**

13. The rank of the matrix $\begin{bmatrix} 1 & 1 & 1 & 1 \\ 1 & 1 & 1 & 1 \\ 1 & 1 & 1 & 1 \end{bmatrix}$ is 4. **(T/F)**

14. If $A = \begin{bmatrix} a_{11} \\ a_{21} \\ \vdots \\ a_{m1} \end{bmatrix}$ and $B = [b_{11} \quad b_{12} \quad \ldots \quad b_{1n}]$ then the rank of AB $= 1$. **(T/F)**

15. If the rank of a square matrix of order n is $n - 1$, then $Adj\ A \neq 0$. **(T/F)**

16. The rank of a matrix is always greater than or equal to the rank of its every submatrix. **(T/F)**

17. The rank of $(AB) \geq$ rank of A. **(T/F)**

18. The elementary transofrmation changes the rank of a matrix. **(T/F)**

19. If $A \sim \begin{pmatrix} I_r & O \\ O & O \end{pmatrix}$, then rank of $A = r - 1$. **(T/F)**

20. Then rank of a diag. $[1, 2, 3, \ldots, n]$ is $\frac{n(n+1)}{2}$. **(T/F)**

21. If $A = [a_{ij}]_{m \times n}$ and rank of $A = r$, there exists two singular matrices such that $PAQ = \begin{pmatrix} I_r & O \\ O & O \end{pmatrix}$. **(T/F)**

22. Every square matrix possesses inverse. **(T/F)**

23. Every non-singular matrix possesses inverse. **(T/F)**

24. If A, B are any two $n \times n$ matrices such that $BA = 0$, where 0 is the null matrix. Then at least one of them is non-singular. **(T/F)**

25. The inverse of an orthogonal matrix is not necessarily orthogonal. **(T/F)**

26. $Adj.\ (AB) = Adj.\ (A)\ Adj.\ (B)$ **(T/F)**

27. The inverse of matrix A exists if A is singular. **(T/F)**

28. The matrix equation $AX = B$ represents a system of linearly homogeneous equations. **(T/F)**

29. The matrix equations $AX = O$ represents a system of non-linear homogeneous equations. **(T/F)**

30. If $A = [a_{ij}]_{m\times n}$ in $AX = O$ and rank $A = n$ then there is only zero solution. **(T/F)**

31. If X_1 is a solution of $AX = O$, then $2X_1$ is not a solution of $AX = O$. **(T/F)**

32. If the rank A is less than the number of unknowns, then there will be infinite solutions of $AX = O$. **(T/F)**

33. The eqution $AX = B$ is inconsistent if rank A is equal to the number of unknowns, then there will be unique solutions. **(T/F)**

34. If the equation $AX = B$ is consistent and rank A is equal to the number of unknowns, then there will be unique solution. **(T/F)**

35. The number of equations is less than the number of variables, then there will be infinite solutions. **(T/F)**

36. If the rank of $A = r$ and there are n variables in $AX = B$ and $r < n$, then the number of linearly independent solutions is $n - r + 1$. **(T/F)**

37. The system of equations $x + 2y + 3z = 1$, $3x + y + 3z = 2$, $4x + 5y + 9z = 4$ has only one solution. **(T/F)**

38. If λ is an eigenvalue of A, then $(A - \lambda I)$ is non-singular. **(T/F)**

39. If $\lambda = 0$ is an eigenvalue of A, then A is singular. **(T/F)**

40. If $X \neq 0$ is an eigenvector corresponding to the eigenvalue 3 of A, then $(A - 3I)X = 0$. **(T/F)**

41. The characteristic roots of a skew Hermitian matrix are always real. **(T/F)**

42. The eigenvalues of $A = \begin{bmatrix} 1 & 0 & 0 \\ 0 & 1 & 0 \\ 0 & 0 & 1 \end{bmatrix}$ are 1, 0, 1. **(T/F)**

43. The eigenvector to the eigenvalue 1 of $\begin{bmatrix} 5 & 4 \\ 1 & 2 \end{bmatrix}$ is $X = [1 - \lambda]^T$. **(T/F)**

44. The characteristic roots of $\begin{bmatrix} 2 & 3 & 4 \\ 0 & -2 & 1 \\ 0 & 0 & 1 \end{bmatrix}$ are 2, –2, 1. **(T/F)**

45. If $|A| \neq 0$ and λ is a characteristic root, then $\frac{|A|}{\lambda}$ is a characteristic root of Adj. (A). **(T/F)**

46. If A is a square matrix of order 3, then it has 2 characteristic roots. **(T/F)**

47. If the characteristic equation of A is $\lambda^3 - 6\lambda^2 + 11\lambda + a = 0$. The sum of all its eigenvalues is –6. **(T/F)**

48. If the matrix A satisfies the equation $\lambda^2 - 4\lambda + \mu = 0$ and one of the eigenvalue of A is 3, then the value of μ is 3. **(T/F)**

49. If $A^3 + a_1A^2 + a_2A + a_3I = 0$, then the order A is 3. **(T/F)**

50. If $A^2 - 3A + 2I = 0$, then $|A| = 2$. **(T/F)**

51. Every square matrix satisfies its characteristic equation. **(T/F)**

52. If $\lambda_1, \lambda_2, \ldots, \lambda_n$ are the characteristic roots of A, then the characteristic roots of A^2 are $\lambda_1, \lambda_2, \ldots, \lambda_n$. **(T/F)**

53. The characteristic equation of $A = \begin{bmatrix} 3 & 1 \\ -1 & 2 \end{bmatrix}$ is $\lambda^2 - 3\lambda + 7 = 0$. **(T/F)**

Multiple Choice Questions

Choose the most appropriate one :

1. The unit matrix of order 2 is :

(a) $\begin{bmatrix} 1 & 0 \\ 0 & -1 \end{bmatrix}$ (b) $\begin{bmatrix} 1 & 0 \\ 0 & 1 \end{bmatrix}$

(c) $\begin{bmatrix} -1 & 0 \\ 0 & -1 \end{bmatrix}$ (d) $\begin{bmatrix} 0 & 1 \\ 1 & 0 \end{bmatrix}$

2. A matrix A is skew-symmetric matrix of :

(a) $A' = -A$ (b) $A' = A$

(c) $A' = A^2$ (d) $A' = -A^2$

3. A matrix A is Hermitian if :

(a) $A^\theta = -A$ (b) $A^\theta = 0$

(c) $A^\theta = A$ (d) $A^\theta = A'$

4. $(AB)' = ?$

(a) $A'B'$ (b) $B'A'$

(c) $-A'B'$ (d) $-B'A'$

5. $(AB)^\theta = ?$

(a) $A^\theta B^\theta$ (b) A^θ

(c) B^θ (d) $B^\theta A^\theta$

6. If $A = \begin{bmatrix} 1 & -1 \\ -1 & 1 \end{bmatrix}, B = \begin{bmatrix} -1 & 1 \\ 1 & -1 \end{bmatrix}$, then $A+B = ?$

(a) $\begin{bmatrix} 0 & 0 \\ 0 & 0 \end{bmatrix}$ (b) $\begin{bmatrix} 0 & 0 \\ 0 & 1 \end{bmatrix}$

(c) $\begin{bmatrix} 1 & 0 \\ 0 & 0 \end{bmatrix}$ (d) $\begin{bmatrix} 1 & 1 \\ 1 & 1 \end{bmatrix}$

7. The order of the matrix $\begin{bmatrix} 2 & 3 & 4 & 5 \\ 6 & 7 & 8 & 9 \end{bmatrix}$ is :

(a) 2×3 (b) 2×4

(c) 3×4 (d) 4×2

8. If the rank $A = \begin{bmatrix} a_{11} & a_{12} \\ a_{21} & a_{22} \end{bmatrix}$ is 2, then rank of $\begin{bmatrix} a_{11} & a_{21} \\ a_{12} & a_{22} \end{bmatrix}$ is :

(a) 3 (b) 2

(c) 1 (d) none of these

9. If A is a null matrix, then its rank is :

(a) 0 (b) 1

(c) 2 (d) none of these

10. The rank of I_6 is :

(a) 2 (b) 3

(c) 5 (d) 6

11. If A and B are equivalent, and rank $A = r$, then rank of B is :

(a) $r-1$ (b) $r+1$

(c) r (d) 0

12. The rank of the matrix $\begin{bmatrix} 2 & 3 & 4 & 5 \\ 0 & 3 & 6 & 7 \\ 0 & 0 & -1 & 0 \\ 0 & 0 & 0 & 1 \end{bmatrix}$ is :

(a) 3 (b) 4

(c) 2 (d) 1

13. If $A = \begin{bmatrix} 1 & 1 & 1 \\ 1 & 1 & 1 \\ 1 & 1 & 1 \end{bmatrix}$, then the rank of A^2 is :

(a) 3 (b) 2

(c) 1 (d) 0

14. If $A = \begin{bmatrix} 1 \\ 2 \\ 3 \end{bmatrix}$, $B = [2 \ \ 3 \ \ 4]$, then the rank of AB is :

(a) 2 (b) 3

(c) 1 (d) 0

15. If the rank of a matrix $\geq r$, then there is at least one r-rowed minor of the matrix, whose rank is :

(a) 1 (b) 0

(c) 2 (d) r

16. If the rank of $A = m$ and rank of $B = n$, then :

(a) rank $(AB) \geq$ rank (A)

(b) rank of $(AB) = mn$

(c) rank $(AB) \geq$ rank (B)

(d) rank $(AB) \leq \min\{$rank (A), rank $(B)\}$

17. If A is a matrix such that there exists a square submatrix of order r which is non-singular and every square submatrix of order $r+1$ is singular, then the rank of A is :

(a) $r+1$ (b) r

(c) $r-1$ (d) $r+2$

18. The transpose of the matrix of cofactors is known as :

(a) inverse (b) adjoint

(c) transpose (d) none of these

19. For the inverse of a matrix A it is necessary that A must be :

(a) singular (b) non-singular

(c) diagonal (d) none of these

20. The $\dfrac{(Adj. A)}{|A|}$ is known as :

(a) A^{-1} (b) A^2

(c) A (d) none of these

21. For a matrix A, which one of the following is a number?

(a) A^{-1} (b) *adj. A*

(c) Rank of A (d) none of these

22. If a non-singular matrix A is symmetric, then A^{-1} is :

(a) skew-symmetric (b) Hermitian

(c) diagonal (d) symmetric

23. If A is non-singular matrix, then $(A^{-1})^{-1}$ is :

(a) I (b) A^{-1}

(c) A (d) AA^{-1}

24. The necessary and sufficient condition that a square matrix amy possess an inverse is that it be :

(a) non-singular (b) singular

(c) triangular (d) none of these

25. The diagonal elements of Hermitian matrix are:

(a) complex number (b) real number

(c) matural numbers (d) none of these

26. The diagonal elements of skew-Hermitian matrix are :

(a) pure real numbers or zero

(b) pure imaginary or zero

(c) compelx numbers

(d) none of these

27. Let $A = [a_{ij}]_{m \times n}$, and $\tilde{A} = [a_{ij}, b_i]_{m \times (n+1)}$ for $AX = \bar{b}$, then :

(a) A is coefficient matrix and $\tilde{A}$ the augmented matrix

(b) $\tilde{A}$ is coefficient matrix and A is augmented matrix.

(c) A, $\tilde{A}$ are coefficient matrix

(d) none of these

28. Let $A \neq 0$, and B and C are matrices such that $AB = AC$, then:

(a) $B = C$ (b) $B \neq C$

(c) $B \neq A$ (d) $C \neq A$

29. There are m equations in n variables, then the order of coefficient matrix is :

(a) $n \times m$ (b) $m \times m$ (c) $n \times n$ (d) $m \times n$

30. The matrix equation $AX = O$ represents :

(a) non-homogeneous linear equations

(b) homogeneous linear equations

(c) homogeneous non-linear equations

(d) none of these

31. The equation $AX = B$ is linear and :

(a) homogeneous (b) non-homogeneous

(c) none of these

32. If X_1 and X_2 are the solutions of $AX = O$, then which one is also the solution of $AX = O$?

(a) $X_1^2 + X_2^2$ (b) $(X_1 + X_2)^2$

(c) $X_1 + X_2$ (d) X_1 / X_2

33. If the equation $AX = B$, is consistent and rank of $(A|B) = 4$, then rank of A is :

(a) 4 (b) 8 (c) 3 (d) 2

34. The equations $x + 2y + 3z = \lambda x$, $3x + y + 2z = \lambda y$, $2x + 3y + z = \lambda z$ have non-zero solutions if A equals :

(a) 2 (b) 6 (c) 3 (d) 0

35. If the rank of $A = r$ and there are n variables in $AX = B$, then equation $AX = B$ has $(n - r)$ linearly independent solutions if :
(a) $n = r$ (b) $n > r$
(c) $n < r$ (d) none of these

36. If the rank of A is equal to the number of variables for $AX = O$, then there will be :
(a) unique solution (b) infinite solutions
(c) finite solution (d) zero solution

37. The system of equations $x + 2y + 3z = 1$, $2x + y + 3z = 2$, $5x + 5y + 9z = 4$ has :
(a) only one solution (b) infinitely many solutions
(c) no solution (d) none of these

38. If $\lambda = 0$ is an eigenvalues of A, then det (A) is:
(a) 0 (b) 1
(c) λ (d) none of these

39. If $|A| \neq 0$ and λ is an eigenvalue of A, then the eigenvalue of Adj. (A) is :
(a) λ (b) λ^2 (c) $1/\lambda$ (d) 0

40. If $|A| \neq 0$ and λ is an eigenvalue of A, then the eigenvalue of Adj. (A) is :
(a) $\lambda|A|$ (b) $\frac{1}{\lambda}$ (c) λ^2 (d) $\frac{|A|}{\lambda}$

41. If a matrix A has $(n - 1)$ characteristic roots, then its order is :
(a) n (b) $n - 1$ (c) n^2 (d) $n + 1$

42. If 1, –1, 3 are the characteristic roots of A, then characteristics roots of A^2 are :
(a) 1, 2, 3 (b) 1, –2, 3
(c) 1, 4, 9 (d) –1, 2, –3

43. If $A^3 - 7A^2 + aA + bI = 0$, then the sum of eigenvalues of A is:
(a) –7 (b) 7 (c) a (d) b

44. If $A^3 - 6A^2 + 11A - 6I = 0$, then the det (A) is:
(a) 11 (b) –6 (c) 6 (d) –11

45. If $(-1)^n[A^n + a_1A^{n-1} + \ldots + a_nI] = 0$, then the order of A is :
(a) n (b) $n - 1$
(c) $n + 1$ (d) n^2

46. The eigenvalue of A is λ, then the eigenvalue of A^T is :
(a) λ^2 (b) $\lambda - 1$
(c) $\lambda + 1$ (d) λ

47. If λ is a characteristic root of A, then its characteristic equation is :
(a) $A - \lambda I = 0$ (b) $|A - \lambda I| = 0$
(c) $|A| = 0$ (d) none of these

48. If 1, 2, 3 are the eigenvalues of A, then 10, 20, 30 are the eigenvalues of the matrix :
(a) $5A$ (b) A^2 (c) $10A$ (d) $100A$

49. If λ is a characteristic root of A^{-1}, then the characteristic root of A is :
(a) λ (b) $\frac{1}{\lambda}$ (c) $\lambda - 1$ (d) λ^2

50. If 3, 5, 6 are the characteristic roots of a matrix A, then –2, 0, 1 are the characteristic roots of :
(a) A (b) $A + 5I$
(c) $A - 5I$ (d) $A - 6I$

51. If $A^2 - 2A + I = 0$ for a square matrix A, then the product of its characteristic roots are :
(a) 2 (b) –1 (c) –2 (d) 1

52. If a matrix is singular, then one of its characteristic root is :
(a) 1 (b) –1
(c) 0 (d) none of these

53. The characteristic roots of a real symmetric matrix are :
(a) zero or purely imaginary
(b) zero or real
(c) zero
(d) none of these

ANSWERS

Fill in the Blanks

1. singular **2.** 1 **3.** n **4.** vanish **5.** same **6.** 2 **7.** 1 **8.** 1
9. non-singular **10.** rank **11.** 10 **12.** same **13.** 2 **14.** Echelon form **15.** rank
16. 3 **17.** 2 **18.** no **19.** 0 **20.** invertiable **21.** non-singular
22. unique **23.** non-singular **24.** $B^{-1}A^{-1}$ **25.** adjoint **26.** non-singular **27.** A, itself
28. homogeneous **29.** $n - r$ **30.** $C_1X_1 + C_2X_2$ **31.** zero solution **32.** infinite solution
33. non-homogeneous equations **34.** rank $(A|B)$ **35.** inconsistent **36.** zero solution
37. $n - r$ **38.** infinite solution **39.** one **40.** no solution
41. augmented matrix **42.** consistent. **43.** $|A - \lambda I| = 0$ **44.** eigenvalue
45. λX **46.** λ **47.** independent **48.** real **49.** unit modulus **50.** 6, 1

51. same **52.** singular **53.** diagonal **54.** aA **55.** $\frac{1}{\lambda}$ **56.** $\frac{|A|}{\lambda}$

57. characteristic equation **58.** 2, 3 **59.** $-\frac{1}{5}[A-4I]$ **60.** 1, 4, 9

True/ False

1. T	**2.** T	**3.** F	**4.** T	**5.** F	**6.** T	**7.** F	**8.** T	**9.** T	**10.** F
11. T	**12.** F	**13.** F	**14.** T	**15.** T	**16.** T	**17.** F	**18.** F	**19.** F	**20.** F
21. F	**22.** F	**23.** T	**24.** T	**25.** F	**26.** T	**27.** F	**28.** F	**29.** F	**30.** T
31. F	**32.** T	**33.** T	**34.** T	**35.** T	**36.** T	**37.** T	**38.** F	**39.** T	**40.** T
41. F	**42.** F	**43.** T	**44.** T	**45.** T	**46.** F	**47.** F	**48.** T	**49.** T	**50.** F
51. T	**52.** F	**53.** F							

Multiple choice questions

1. (b)	**2.** (a)	**3.** (c)	**4.** (b)	**5.** (d)	**6.** (a)	**7.** (b)	**8.** (b)	**9.** (a)	**10.** (d)
11. (c)	**12.** (b)	**13.** (c)	**14.** (c)	**15.** (b)	**16.** (d)	**17.** (b)	**18.** (b)	**19.** (b)	**20.** (a)
21. (c)	**22.** (d)	**23.** (c)	**24.** (a)	**25.** (b)	**26.** (b)	**27.** (a)	**28.** (a)	**29.** (d)	**30.** (b)
31. (b)	**32.** (c)	**33.** (a)	**34.** (b)	**35.** (b)	**36.** (d)	**37.** (a)	**38.** (a)	**39.** (c)	**40.** (d)
41. (b)	**42.** (c)	**43.** (b)	**44.** (c)	**45.** (a)	**46.** (d)	**47.** (b)	**48.** (c)	**49.** (b)	**50.** (c)
51. (d)	**52.** (c)	**53.** (a)							

FFFFFF

GATE tutor

KEY Terms and Results

- **Matrix :** A set of mn numbers either real or complex in the form of a rectangular array in which there are m rows and n columns is called a matrix of order $m \times n$.
- **Null matrix :** A matrix of order $m \times n$ is called a null matrix if it contains all mn elements zero.
- **Row matrix :** A matrix having only one row and n column is called a row matrix of order $1 \times n$.
- **Column marix :** A matrix having m rows and one column is called a column matrix of order $m \times 1$.
- **Square matrix :** A matrix having equal number of rows and columns is called square matrix.
- **Determinants :** Let A be a square matrix, then the determinant which is formed by the elements of A usually denoted by $|A|$.
- **Rank of a matrix :** Let A be a matrix of order $m \times n$, then a non-negative integer r is said to be the rank of matrix A if it possesses the following two properties :
  (i) there exists atleast one r-minor of A which is not equal to zero
  (ii) every s-minor of A for all $s > 0$ is zero.
- **Inverse of a matrix :** Let A be a non-singular matrix of order $n \times n$. Then it is said to be invertiable if there exists a non-singular square matrix of order $n \times n$ such that $AB = I_n = BA$.

Results

- Transpose of the matrix of cofactors is called adjoint of the given matrix.
- Sum of two symmetric matrices is again symmetric.
- The rank of a matrix is the order of any highest order of a non-zero minor of the matrix.
- Every non-singular matrix is row equivalent to a unit matrix.
- A square matrix is invertiable if and only if it is non-singular.
- **(Rouche's theorem)** The equation $AX = B$ is consistent if and only if the rank of A and the rank of the augmented matrix $[A/B]$ are same.
- The n equations in n unknowns have a unique solution.
- λ is an eigenvalue of a matrix A if and only if there exists a non-zero vector X such that $AX = \lambda X$.
- If X is a non-zero eigenvector of a matrix A then X cannot correspond to more than one eigenvalue of A.
- The eigenvalues of a Hermitian matrix are real.
- The eigenvalues of a real symmetric matrix are all real.
- The eigenvalue of a skew-Hermitian matrix are either purely imaginary or zero.
- The eigenvalues of a unitary matrix are of unit modulus.
- Similar matrices have the same eigenvalues.
- **(Cayley-Hemilton theorem)** Every square matrix satisfies its characteristic equation.
- An $n \times n$ matrix is diagonalizable if and only if it possesses n linearly independent eigenvectors.

Objective Type Questions

Choose the most appropriate one.

1. Consider the matrices $X_{(4 \times 3)}$, $Y_{(4 \times 3)}$ and $P_{(2 \times 3)}$. The order of $[P(X^TY)^{-1}P^T]^T$ will be : **[GATE(CE)–2005]**

(a) (3×3) (b) (2×2)
(c) (4×3) (d) none of these

2. Multiplication of matrices E and F is G. Matrices E and G are $E = \begin{bmatrix} \cos\theta & -\sin\theta & 0 \\ \sin\theta & \cos\theta & 0 \\ 0 & 0 & 1 \end{bmatrix}$ and $G = \begin{bmatrix} 1 & 0 & 0 \\ 0 & 1 & 0 \\ 0 & 0 & 1 \end{bmatrix}$ where is the matrix F? **[GATE(ME)–2006]**

(a) $\begin{bmatrix} \cos\theta & \cos\theta & 0 \\ -\cos\theta & \sin\theta & 0 \\ 0 & 0 & 1 \end{bmatrix}$ (b) $\begin{bmatrix} \sin\theta & -\cos\theta & 0 \\ \cos\theta & \sin\theta & 0 \\ 0 & 0 & 1 \end{bmatrix}$

(c) $\begin{bmatrix} \cos\theta & -\sin\theta & 0 \\ \sin\theta & \cos\theta & 0 \\ 0 & 0 & 1 \end{bmatrix}$ (d) $\begin{bmatrix} \cos\theta & \sin\theta & 0 \\ -\sin\theta & \cos\theta & 0 \\ 0 & 0 & 1 \end{bmatrix}$

3. The dimension of the null space of the matrix $\begin{bmatrix} 0 & 1 & 1 \\ 1 & -1 & 0 \\ -1 & 0 & -1 \end{bmatrix}$ is : **[GATE(IN)–2013]**

(a) 0 (b) 2
(c) 1 (d) 3

4. A square matrix B is skew-symmetric if : **[GATE(CE)–2009]**

(a) $B^T = B$ (b) $B^T = -B$
(c) $B^{-1} = B$ (d) none of these

5. Real matrix $[A]_{3 \times 1}$, $[B]_{3 \times 3}$, $[C]_{3 \times 5}$, $[D]_{5 \times 3}$, $[E]_{5 \times 5}$ and $[F]_{5 \times 1}$ are given. Matrices $[B]$ and $[E]$ are symmetric.

Following statements are made with respect to these matrices.
(i) Matrix product $[F]^T [C]^T [B] [C] [F]$ is a scalar.
(ii) Matrix product $[D]^T [F] [D]$ is always symmetric.

With reference to above statements, which of the following applies? **[GATE(CE)–2004]**

(a) Statement 1 is false but 2 is true
(b) Statement 1 is true but 2 is false
(c) Both the statements are true
(d) None of these

6. $[A]$ is square matrix which is neither symmetric nor skew–symmetric and $[A]^T$ is its transpose. The sum and difference of these matrices are defined as $[S] = [A] + [A]^T$ and $[D] = [A] - [A]^T$, respectively. Which of the following statements is TRUE ? **[GATE(CE)–2007]**

(a) $[S]$ is symmetric and $[D]$ is skew-symmetric
(b) $[S]$ is skew-symmetric and $[D]$ is symmetric
(c) Both [S] and [D] are symmetric
(d) None of these

7. Given an orthogonal matrix $A = \begin{bmatrix} 1 & 1 & 1 & 1 \\ 1 & 1 & -1 & -1 \\ 1 & -1 & 0 & 0 \\ 0 & 0 & 1 & -1 \end{bmatrix}$, $[AA^T]^{-1}$ is : **[GATE(EC)–2005]**

(a) $\begin{bmatrix} 1 & 0 & 0 & 0 \\ 0 & 1 & 0 & 0 \\ 0 & 0 & 1 & 0 \\ 0 & 0 & 0 & 1 \end{bmatrix}$ (b) $\begin{bmatrix} 1 & -1 & 0 & 0 \\ 0 & 1 & 0 & 0 \\ -1 & 0 & 1 & 0 \\ 0 & 0 & 0 & 1 \end{bmatrix}$

(c) $\begin{bmatrix} 1/2 & 0 & 0 & 0 \\ 0 & 1/2 & 0 & 0 \\ 0 & 0 & 1/2 & 0 \\ 0 & 0 & 0 & 1/2 \end{bmatrix}$ (d) none of these

8. Match List-I with List-II and select the correct answer using the codes given below the lists — **[GATE(ME)–2006]**

List – I	List – II
A. Singular matrix	1. Determinant is not defined.
B. Non-singular matrix	2. Determinant is always one.
C. Real symmetric	3. Determinant is zero.
D. Orthogonal matrix	4. Eigenvalues are always real.
	5. Eigenvalues are not defined.

Codes :

	A	B	C	D
(a)	3	1	4	2
(b)	2	3	4	1
(c)	3	2	5	4
(d)	3	4	2	1

9. The inverse of the 2×2 matrix $\begin{bmatrix} 1 & 2 \\ 5 & 7 \end{bmatrix}$ is : **[GATE(CE)–2007]**

(a) $\frac{1}{3}\begin{bmatrix} 7 & 2 \\ 5 & 1 \end{bmatrix}$ (b) $\frac{1}{3}\begin{bmatrix} -7 & 2 \\ 5 & -1 \end{bmatrix}$

(c) $\frac{1}{3}\begin{bmatrix} 7 & -2 \\ -5 & 1 \end{bmatrix}$ (d) none of these

10. The inverse of the matrix $\begin{bmatrix} 3+2i & i \\ -i & 3-2i \end{bmatrix}$ is : **[GATE(CE)–2010]**

(a) $\frac{1}{12}\begin{bmatrix} 3-2i & -i \\ i & 3+2i \end{bmatrix}$ (b) $\frac{1}{12}\begin{bmatrix} 3+2i & -i \\ i & 3-2i \end{bmatrix}$

(c) $\frac{1}{14}\begin{bmatrix} 3+2i & -i \\ i & 3-2i \end{bmatrix}$ (d) none of these

11. The product of matrices $(PQ)^{-1}P$ is : **[GATE(CE)–2008]**

(a) P^{-1} (b) $P^{-1}Q^{-1}P$
(c) Q^{-1} (d) none of these

12. For which value of X will the matrix given below become singular?

$\begin{bmatrix} 8 & X & 0 \\ 4 & 0 & 2 \\ 12 & 6 & 0 \end{bmatrix}$ **[GATE(ME)–2004]**

(a) 6 (b) 4
(c) 8 (d) 12

13. If $R = \begin{bmatrix} 1 & 0 & -1 \\ 2 & 1 & -1 \\ 2 & 3 & 2 \end{bmatrix}$ then top row of R^{-1} is : **[GATE(EE)–2005]**

(a) [5 –3 1] (b) [5 6 4]
(c) [2 0 –1] (d) none of these

14. For a matrix $[M] = \begin{bmatrix} \frac{3}{5} & \frac{4}{5} \\ X & \frac{3}{5} \end{bmatrix}$, the transpose of the matrix is equal to the inverse of the matrix, $[M]^T = [M]^{-1}$. The value of X is given by : **[GATE(ME)–2009]**

(a) –3/5 (b) –4/5
(c) 3/5 (d) 0

15. Which one of the following does NOT equal to the given determinant

$\begin{vmatrix} 1 & x & x^2 \\ 1 & y & y^2 \\ 1 & z & z^2 \end{vmatrix}$? **[GATE(CS)–2013]**

(a) $\begin{vmatrix} 1 & x+1 & x^2+1 \\ 1 & y+1 & y^2+1 \\ 1 & z+1 & z^2+1 \end{vmatrix}$ (b) $\begin{vmatrix} 1 & x(x+1) & x+1 \\ 1 & y(y+1) & y+1 \\ 1 & z(z+1) & z+1 \end{vmatrix}$

(c) $\begin{vmatrix} 0 & x-y & x^2-y^2 \\ 0 & y-z & y^2-z^2 \\ 1 & z & z^2 \end{vmatrix}$ (d) none of these

16. Let A be an $m \times n$ matrix and B an $n \times m$ matrix. It is given that determinant $[l_m + AB]$ = determinant $(l_n + BA)$ where l_k is the $k \times k$ identity matrix. Using the above property, the determinant of the matrix given below is

$\begin{bmatrix} 2 & 1 & 1 & 1 \\ 1 & 2 & 1 & 1 \\ 1 & 1 & 2 & 1 \\ 1 & 1 & 1 & 2 \end{bmatrix}$: **[GATE(EC)–2013]**

(a) 2 (b) 8
(c) 5 (d) 0

17. A is $m \times n$ full rank matrix with $m > n$ and 1 is an identity matrix. Let $A' = (A^TA)^{-1}A^T$. Then, which one of the following statement is FALSE ? **[GATE(EE)–2008]**

(a) $(AA')^2$ (b) $AA'A = A$
(c) $AA'A = 1$ (d) none of these

18. Let A, B, C, D be $n \times n$ matrices, each with non-zero determinant, if $ABCD = 1$, then B^{-1} is : [GATE(CS)–2004]

(a) ADC (b) CDA
(c) $D^{-1}C^{-1}A^{-1}$ (d) $A^{-1}D^{-1}C^{-1}$

19. Let $A = \begin{bmatrix} 2 & -0.1 \\ 0 & 3 \end{bmatrix}$ and $A^{-1} \begin{bmatrix} \frac{1}{2} & a \\ 0 & b \end{bmatrix}$. Then $(a + b) =$ [GATE(EC)–2005]

(a) 3/20 (b) 7/20
(c) 19/60 (d) 1

20. Given matrix $[A] = \begin{bmatrix} 4 & 2 & 1 & 3 \\ 6 & 3 & 4 & 7 \\ 2 & 1 & 0 & 7 \end{bmatrix}$, the rank of the matrix is : [GATE(CE)–2003]

(a) 4 (b) 2
(c) 3 (d) 1

21. If the A-matrix of the state space model of a SISO linear time invariant system is rank deficient, the transfer function of the system must have : [GATE(IN)–2013]

(a) a pole with positive real part
(b) a pole at the origin
(c) a pole with a negative real part
(d) none of these

22. $X = [x_1, x_2, ..., x_n]$ is an n-tuple non-zero vector. Then $n \times n$ matrix $V = XX^t$: [GATE(EE)–2001]

(a) has rank zero (b) has rank 1
(c) has rank n (d) none of these

23. The rank of the matrix $\begin{bmatrix} 1 & 1 & 1 \\ 1 & -1 & 0 \\ 1 & 1 & 1 \end{bmatrix}$ is : [GATE(EC)–2006]

(a) 0 (b) 2
(c) 1 (d) ∞

24. If the rank of a (5×6) matrix Q is 4, then which one of the following statements is correct ? [GATE(EE)–2008]

(a) Q will have four linearly independent rows and four linearly independent columns
(b) QQ^T will be invertible
(c) Q^TQ will be invertible
(d) none of these

25. Choose the CORRECT set of functions, which are linearly dependent : [GATE(ME)–2013]

(a) $\cos x$, $\sin x$ and $\tan x$ (b) $\cos 2x$, $\sin^2 x$ and $\cos^2 x$
(c) $\cos 2x$, $\sin x$ and $\cos x$ (d) none of these

26. An orthogonal set of vectors having a span that contains P, Q, R is : [GATE(EE)–2006]

(a) $\begin{bmatrix} 4 \\ 3 \\ 11 \end{bmatrix} \begin{bmatrix} 1 \\ 31 \\ 3 \end{bmatrix} \begin{bmatrix} 5 \\ 3 \\ 4 \end{bmatrix}$ (b) $\begin{bmatrix} 6 \\ 7 \\ -1 \end{bmatrix} \begin{bmatrix} -3 \\ 2 \\ -2 \end{bmatrix} \begin{bmatrix} 3 \\ 9 \\ -4 \end{bmatrix}$

(c) $\begin{bmatrix} -4 \\ 2 \\ 4 \end{bmatrix} \begin{bmatrix} 5 \\ 7 \\ -11 \end{bmatrix} \begin{bmatrix} 8 \\ 9 \\ -3 \end{bmatrix}$ (d) $\begin{bmatrix} -6 \\ -3 \\ 6 \end{bmatrix} \begin{bmatrix} 4 \\ -2 \\ 3 \end{bmatrix}$

27. The following vector is linearly dependent upon the solution to the previous problem : [GATE(EE)–2006]

(a) $\begin{bmatrix} 8 \\ 9 \\ 3 \end{bmatrix}$ (b) $\begin{bmatrix} 4 \\ 4 \\ 5 \end{bmatrix}$

(c) $\begin{bmatrix} 0 \\ 0 \\ 1 \end{bmatrix}$ (d) $\begin{bmatrix} -2 \\ -17 \\ 30 \end{bmatrix}$

28. Consider the set of (column) vectors defined by $X = \{x \in R^3 : x_1 + x_2 + x_3 = 0$, where $XT = [x_1, x_2, x_3]\}^T$. Which of the following is TRUE? [GATE(CS)–2007]

(a) X is not a subspace of R^3.
(b) $\{[1, -1, 0]^T, [1, 0, -1]^T\}$ is a basis for the subspace X.
(c) $\{[1, -1, 0]^T, [1, 0, -1]^T\}$ is not a basis of X.
(d) none of these.

29. It is given that $x_1, x_2, ..., x_M$ are M non-zero orthogonal vectors. The dimension of the vector space spanned by $2M$ vectors $x_1, x_2, ..., x_M, -x_1, -x_2, ..., -x_M$ is : [GATE(EC)–2007]

(a) 0 (b) M
(c) $2M$ (d) $M + 1$

30. Consider a non-homogeneous system of linear equations representing mathematically an over-determined system such a system will be : [GATE(CE)–2005]

(a) consistent having a unique solution
(b) consistent having many solution
(c) inconsistent having no solution
(d) all are possible

31. For what values of α and β, the following simultaneous equations have an infinite no. of solutions?

$$x + y + z = 5;$$
$$x + 3y + 3z = 9;$$
$$x + 2y + \alpha z = \beta.$$

[GATE(CE)–2007]

(a) 3, 8 (b) 7, 2
(c) 0, 1 (d) 2, 7

32. Solution for the system defined by the set of equations

$$4y + 3z = 8;$$
$$2x - z = 2$$

and $3x + 2y = 5$ is : [GATE(CE)–2006]

(a) $x = 0; y = 1; z = 4/3$ (b) non-existent
(c) $x = 1; y = 1/2; z = 2$ (d) None of these

33. The following simultaneous equations

$$x + y + z = 3$$
$$x + 2y + 3z = 4$$
$$x + 4y + kz = 6$$

will not have a unique solution for k equal to : [GATE(CE)–2008]

(a) 7 (b) 5
(c) 0 (d) 2

34. Consider the system of simultaneous equations

$$x + 2y + z = 6$$
$$2x + y + 2z = 6$$
$$x + y + z = 5$$

This system has : [GATE(ME)–2003]

(a) unique solution (b) no solution
(c) exactly two solutions (d) none of these

35. For what value of α, if any, will the following system of equations in x, y and z have a solution? [GATE(ME)–2008]

$$2x + 3y = 4$$
$$x + y + z = 4$$
$$x + 2y - z = \alpha$$

(a) 0 (b) 1
(c) any real number (d) none of these

36. A is a 3×4 real matrix and $AX = b$ is a inconsistent system of equations. The highest possible rank of A is : [GATE(ME)–2005]

(a) 0 (b) 1
(c) 2 (d) 3

37. In the matrix equation $Px = q$, which of the following is a necessary condition for the existence of at least one solution for the unknown vector x : [GATE(EE)–2005]

(a) vector q must have only non-zero elements
(b) augmented matrix $[Pq]$ must have the same rank as matrix P
(c) matrix P must be square
(d) none of these

38. Consider the following system of equations

$$2x_1 + x_2 + x_3 = 0$$
$$x_2 - x_3 = 0$$
$$x_1 + x_2 = 0$$

This system has : [GATE(ME)–2011]

(a) no solution (b) infinite number of solutions
(c) two solutions (d) a unique sloution

39. For the set of equations

$$x_1 + 2x_2 + x_3 + 4x_4 = 2$$
$$3x_1 + 6x_2 + 3x_3 + 12x_4 = 6$$

the following statement is true : [GATE(EE)–2013]

(a) no solution
(b) a unique trivial solution exists
(c) multiple non-trivial solutions exist
(d) none of these

40. The system of linear equations

$$4x + 2y = 7$$
$$2x + y = 6$$

has : [GATE(EC)–2008]

(a) no solution (b) a unique solution
(c) zero solution (d) none of these

41. The equation $\begin{bmatrix} 2 & -2 \\ 1 & -1 \end{bmatrix}\begin{bmatrix} x_1 \\ x_2 \end{bmatrix} = \begin{bmatrix} 0 \\ 0 \end{bmatrix}$ has : [GATE(EE)–2013]

(a) no solution (b) non-zero unique solution
(c) multiple solution (d) none of these

42. The system of equations

$$x + y + z = 6$$
$$x + 4y + 6z = 20$$
$$x + 4y + \lambda z = \mu$$

has no solution for value of λ and μ given by : [GATE(EC)–2011]

(a) $\lambda = 6, \mu = 20$ (b) $\lambda \neq 6, \mu = 20$
(c) $\lambda = 6, \mu \neq 20$ (d) none of these

43. The equation $\begin{bmatrix} 2 & -2 \\ 1 & -1 \end{bmatrix}\begin{bmatrix} x_1 \\ x_2 \end{bmatrix} = \begin{bmatrix} 0 \\ 0 \end{bmatrix}$ has : [GATE(EE)–2013]

(a) multiple solution (b) no solution
(c) zero solution (d) only one solution $\begin{bmatrix} x_1 \\ x_2 \end{bmatrix} = \begin{bmatrix} 0 \\ 0 \end{bmatrix}$

44. How many solutions does the following system of linear equations have ?

$$-x + 5y = -1$$
$$x - y = 2$$
$$x + 3y = 3$$

[GATE(CS)–2004]

(a) unique (b) infinitely many
(c) zero solution (d) none of these

45. Consider the following system of linear equations

$$\begin{bmatrix} 2 & 1 & -4 \\ 4 & 3 & -12 \\ 1 & 2 & -8 \end{bmatrix}\begin{bmatrix} x \\ y \\ z \end{bmatrix} = \begin{bmatrix} \alpha \\ 5 \\ 7 \end{bmatrix}$$

Notice that the second and third columns of the coefficient matrix are linearly dependent. For how many values of α, does this system of equations have infinitely many solutions? [GATE(CS)–2003]

(a) 1 (b) 0
(c) 2 (d) none of these

46. Consider the following system of equations in three real variables x_1, x_2 and x_3.

$$2x_1 - x_2 + 3x_3 = 1$$
$$3x_1 - 2x_2 + 5x_3 = 2$$
$$-x_1 - 4x_2 + x_3 = 3$$

This system of equations has : [GATE(CS)–2005]

(a) no solution (b) unique solution
(c) infinite solution (d) none of these

47.
$$x + 2y + z = 4$$
$$2x + y + 2z = 5$$
$$x - y + z = 1$$

The system of algebraic equations given below has : [GATE(ME)–2012]

(a) unique solution (b) zero solution
(c) infinite solution (d) no feasible solution

48. The following system of equations

$$x_1 + x_2 + 2x_3 = 1$$
$$x_1 + 2x_2 + 3x_3 = 2$$
$$x_1 + 4x_2 + ax_3 = 4$$

has a unique solution. The only possible value(s) for a is/are : [GATE(CS)–2008]

(a) 0
(b) 1
(c) 2
(d) any real number other than 5

49. For the matrix $A = \begin{bmatrix} 5 & 3 \\ 1 & 3 \end{bmatrix}$, one of the normalized eigen-vectors is given as : (GATE(ME)–2012)

(a) $\begin{bmatrix} \frac{1}{\sqrt{2}} \\ -\frac{1}{\sqrt{2}} \end{bmatrix}$ (b) $\begin{bmatrix} \frac{3}{\sqrt{10}} \\ -\frac{1}{\sqrt{10}} \end{bmatrix}$

(c) $\begin{bmatrix} \frac{1}{2} \\ \frac{\sqrt{3}}{2} \end{bmatrix}$ (d) $\begin{bmatrix} \frac{1}{\sqrt{5}} \\ \frac{2}{\sqrt{5}} \end{bmatrix}$

50. The eigenvalues of matrix $\begin{bmatrix} 9 & 5 \\ 5 & 8 \end{bmatrix}$ are : [GATE(CE)–2012]

(a) 3.48 and 13.53 (b) 4.70 and 6.86
(c) –2.42 and 6.86 (d) none of these

51. The eigenvalues of the matrix $[P]=\begin{bmatrix} 4 & 5 \\ 2 & -5 \end{bmatrix}$ are : [GATE(CE)–2008]

(a) –6 and 5 (b) 3 and 4
(c) 0 and 1 (d) 1 and 2

52. The eigenvalues of the matrix $\begin{bmatrix} 4 & -2 \\ -2 & 1 \end{bmatrix}$ are : [GATE(CE)–2004]

(a) 1 and 4 (b) 0 and 5
(c) 1 and 2 (d) none of these

53. For the matrix $\begin{bmatrix} 4 & 1 \\ 1 & 4 \end{bmatrix}$ the eigenvalues are : [GATE(ME)–2003]

(a) 3 and 5 (b) 0 and 5
(c) 3 and –3 (d) none of these

54. The number of linearly independent eigenvectors of $\begin{bmatrix} 2 & 1 \\ 0 & 2 \end{bmatrix}$ is : [GATE(ME)–2007]

(a) 0 (b) 2
(c) 3 (d) 1

55. Which one of the following is an eigenvector of the matrix $\begin{bmatrix} 5 & 0 & 0 & 0 \\ 0 & 5 & 5 & 0 \\ 0 & 0 & 2 & 1 \\ 0 & 0 & 3 & 1 \end{bmatrix}$? [GATE(ME)–2005]

(a) $\begin{bmatrix} 0 \\ 0 \\ 1 \\ 0 \end{bmatrix}$ (b) $\begin{bmatrix} 1 \\ -2 \\ 0 \\ 0 \end{bmatrix}$

(c) $\begin{bmatrix} 1 \\ 0 \\ 0 \\ 0 \end{bmatrix}$ (d) none of these

56. The eigenvector of the matrix $\begin{bmatrix} 1 & 2 \\ 0 & 2 \end{bmatrix}$ are written in the form $\begin{bmatrix} 1 \\ a \end{bmatrix}$ and $\begin{bmatrix} 1 \\ b \end{bmatrix}$. What is $a+b$? [GATE(ME)–2008]

(a) 1/2 (b) 0
(c) 1 (d) none of these

57. For the matrix $A=\begin{bmatrix} 3 & -2 & 2 \\ 0 & -2 & 1 \\ 0 & 0 & 1 \end{bmatrix}$, one of the eigenvalues is equal to –2. Which of the following is an eigenvector? [GATE(EE)–2005]

(a) $\begin{bmatrix} 3 \\ -2 \\ 1 \end{bmatrix}$ (b) $\begin{bmatrix} 1 \\ -2 \\ 3 \end{bmatrix}$

(c) $\begin{bmatrix} 2 \\ 5 \\ 0 \end{bmatrix}$ (d) none of these

58. One of the eigenvectors of the matrix $A=\begin{bmatrix} 2 & 2 \\ 1 & 3 \end{bmatrix}$ is : [GATE(ME)–2010]

(a) $\begin{Bmatrix} 2 \\ 1 \end{Bmatrix}$ (b) $\begin{Bmatrix} 0 \\ 1 \end{Bmatrix}$

(c) $\begin{Bmatrix} 1 \\ -1 \end{Bmatrix}$ (d) $\begin{Bmatrix} 2 \\ -1 \end{Bmatrix}$

59. The linear operation $L(x)$ is defined by the cross product $L(x) = b \times x$, where $b = [0\ 1\ 0]^T$ and $x = [x_1\ x_2\ x_3]^T$ are three dimensional vectors. The 3×3 matrix M of this operation satisfies.

$$L(x) = M\begin{bmatrix} x_1 \\ x_2 \\ x_3 \end{bmatrix}$$

Then the eigenvalues of M are : [GATE(EE)–2007]

(a) $i, -i, 0$ (b) 0, 1, –1
(c) 1, –1, 1 (d) none of these

60. An eigenvector of $P=\begin{bmatrix} 1 & 1 & 0 \\ 0 & 2 & 2 \\ 0 & 0 & 3 \end{bmatrix}$ is : [GATE(EE)–2010]

(a) $[1\ 2\ 1]^T$ (b) $[-1\ 1\ 1]^T$
(c) $[1\ -1\ 2]^T$ (d) none of these

61. Given the matrix $\begin{bmatrix} -4 & 2 \\ 4 & 3 \end{bmatrix}$, the eigenvector is : [GATE(EC)–2005]

(a) $\begin{bmatrix} 3 \\ 2 \end{bmatrix}$ (b) $\begin{bmatrix} 2 \\ -1 \end{bmatrix}$

(c) $\begin{bmatrix} 4 \\ 3 \end{bmatrix}$ (d) none of these

62. One pair of eigenvectors corresponding to the two eigenvalues of the matrix $\begin{bmatrix} 0 & -1 \\ 1 & 0 \end{bmatrix}$ is : [GATE(IN)–2013]

(a) $\begin{bmatrix} 1 \\ -j \end{bmatrix}, \begin{bmatrix} j \\ -1 \end{bmatrix}$ (b) $\begin{bmatrix} 0 \\ 1 \end{bmatrix}, \begin{bmatrix} -1 \\ 0 \end{bmatrix}$

(c) $\begin{bmatrix} 1 \\ j \end{bmatrix}, \begin{bmatrix} 0 \\ 1 \end{bmatrix}$ (d) $\begin{bmatrix} 1 \\ j \end{bmatrix}, \begin{bmatrix} j \\ 1 \end{bmatrix}$

63. For the matrix $\begin{bmatrix} 4 & 2 \\ 2 & 4 \end{bmatrix}$ the eigenvalue corresponding to the eigenvector $\begin{bmatrix} 10 & 1 \\ 10 & 1 \end{bmatrix}$ is : [GATE(EC)–2006]

(a) 6 (b) 2
(c) 4 (d) 0

64. How many of the following matrices have an eigenvalue 1?

$\begin{bmatrix} 1 & 0 \\ 0 & 0 \end{bmatrix}, \begin{bmatrix} 0 & 1 \\ 0 & 0 \end{bmatrix}, \begin{bmatrix} 1 & -1 \\ 1 & 1 \end{bmatrix}$ and $\begin{bmatrix} -1 & 0 \\ 1 & -1 \end{bmatrix}$ [GATE(CS)–2008]

(a) 2 (b) 1
(c) 3 (d) 4

65. What are the eigenvalues of the following 2×2 matrix?

$\begin{bmatrix} 2 & -1 \\ -4 & 5 \end{bmatrix}$ [GATE(CS)–2005]

(a) 1 and 6 (b) –1 and 1
(c) 0 and 1 (d) none of these

66. Let A be the 2×2 matrix with elements $a_{11} = a_{12} = a_{21} = +1$ and $a_{22} = -1$. Then the eigenvalues of the matrix A^{19} are : **[GATE(CS)–2012]**

(a) 1024 and –1024 (b) $512\sqrt{2}$ and $-512\sqrt{2}$
(c) $4\sqrt{2}$ and $-4\sqrt{2}$ (d) none of these

67. For a given matrix

$$A = \begin{bmatrix} 2 & -2 & 3 \\ -2 & -1 & 6 \\ 1 & 2 & 0 \end{bmatrix}$$

one of the eigenvalues is 3. The other two eigenvalues are : **[GATE(CE)–2006]**

(a) 3, –5 (b) 2, –5
(c) 0, 1 (d) none of these

68. The minimum and maximum eigenvalues of the matrix $\begin{bmatrix} 1 & 1 & 3 \\ 1 & 5 & 1 \\ 3 & 1 & 1 \end{bmatrix}$ are –2 and 6 respectively. What is the other eigenvalue? **[GATE(CE)–2007]**

(a) 0 (b) 1
(c) 2 (d) 3

69. The sum of the eigenvalues of the matrix given below is :

$$\begin{bmatrix} 1 & 2 & 3 \\ 1 & 5 & 1 \\ 3 & 1 & 1 \end{bmatrix}$$ **[GATE(ME)–2004]**

(a) 7 (b) 5
(c) 0 (d) 1

70. The minimum eigenvalue of the following matrix is :

$$\begin{bmatrix} 3 & 5 & 2 \\ 5 & 12 & 7 \\ 2 & 7 & 5 \end{bmatrix}$$ **[GATE(EC)–2012]**

(a) 1 (b) 0
(c) 2 (d) 3

71. Eigenvalues of the matrix $S = \begin{bmatrix} 3 & 2 \\ 2 & 3 \end{bmatrix}$ are 5 and 1. What are the eigenvalues of the matrix $S^2 = SS$? **[GATE(ME)–2006]**

(a) 6 and 4 (b) 1 and 0
(c) 1 and 25 (d) none of these

72. The eigenvalues of a symmetric matrix are all : **[GATE(ME)–2013]**

(a) pure imaginary (b) real
(c) complex (d) none of these

73. If a square matrix A is real and symmetric, then the eigenvalues are : **[GATE(ME)–2007]**

(a) real and positive (b) always real
(c) real and neagative (d) none of these

74. The matrix $\begin{bmatrix} 1 & 2 & 4 \\ 3 & 0 & 6 \\ 1 & 1 & P \end{bmatrix}$ has one eigenvalue equal to 3. The sum of the other two eigenvalue is : **[GATE(ME)–2008]**

(a) P (b) $P - 3$
(c) $P - 1$ (d) $P - 2$

75. The trace and determinant of a 2×2 matrix are taken to be –2 and –35 respectively. Its eigenvalues are : **[GATE(EE)–2009]**

(a) –7 and 5 (b) 0 and 1
(c) –37 and 1 (d) none of these

76. Eigenvalues of a real symmetric matrix are always : **[GATE(ME)–2011]**

(a) real (b) positive
(c) negative (d) none of these

77. All the four entries of the 2×2 matrix $P = \begin{bmatrix} P_{11} & P_{12} \\ P_{21} & P_{22} \end{bmatrix}$ are non-zero and one of its eigenvalues is zero. Which of the following statement is true? **[GATE(EC)–2008]**

(a) $P_{11}P_{22} - P_{12}P_{21} = -1$ (b) $P_{11}P_{22} - P_{12}P_{21} = 0$
(c) $P_{11}P_{22} - P_{12}P_{21} = -1$ (d) none of these

78. The eigenvalues and the corresponding eigenvectors of a 2×2 matrix are given by :

Eigenvalue	Eigenvector
$\lambda_1 = 8$	$v_1 = \begin{bmatrix} 1 \\ 1 \end{bmatrix}$
$\lambda_2 = 4$	$v_2 = \begin{bmatrix} 1 \\ -1 \end{bmatrix}$

The matrix is : **[GATE(EC)–2006]**

(a) $\begin{bmatrix} 4 & 6 \\ 6 & 4 \end{bmatrix}$ (b) $\begin{bmatrix} 6 & 2 \\ 2 & 6 \end{bmatrix}$

(c) $\begin{bmatrix} 0 & 1 \\ 1 & 0 \end{bmatrix}$ (d) none of these

79. A matrix has eigenvalues –1 and –2. The corresponding eigenvectors are $\begin{bmatrix} 1 \\ -1 \end{bmatrix}$ and $\begin{bmatrix} 1 \\ -2 \end{bmatrix}$ respectively. The matrix is : **[GATE(EE)–2013]**

(a) $\begin{bmatrix} 0 & 1 \\ -2 & -3 \end{bmatrix}$ (b) $\begin{bmatrix} 1 & 2 \\ 2 & 3 \end{bmatrix}$

(c) $\begin{bmatrix} 1 & 1 \\ -1 & -2 \end{bmatrix}$ (d) none of these

80. The eigenvalues of a skew-symmetric matrix are : **[GATE(EC)–2010]**

(a) always zero
(b) either zero or pure imaginary
(c) always real
(d) none of these

81. The eigenvalues of the following matrix are

$$\begin{bmatrix} -1 & 3 & 5 \\ -3 & -1 & 6 \\ 0 & 0 & 3 \end{bmatrix} :$$ **[GATE(EC)–2010]**

(a) $3, -1 + 3j, -1 - 3j$ (b) $3, 3 + 5j, 6 - j$
(c) $3 + j, 3 - j, 5 + j$ (d) none of these

82. Consider the following matrix

$$A = \begin{bmatrix} 2 & 3 \\ x & y \end{bmatrix}$$

If the eigenvalues of A are 4 and 8, then : **[GATE(CS)–2010]**

(a) $x = 4, y = 10$ (b) $x = -4, y = 10$
(c) $x = 5, y = 8$ (d) none of these

83. Consider the matrix $\begin{bmatrix} 1 & 2 & 3 \\ 0 & 4 & 7 \\ 0 & 0 & 3 \end{bmatrix}$.

which one of the following options provides the CORRECT values of the eigenvalues of the matrix? [GATE(CS)–2011]

(a) 3, 7, 3 (b) 0, 1, 2
(c) 1, 4, 3 (d) none of these

84. The characteristic equation of $a(3 \times 3)$ matrix P is defined as

$$a(\lambda) = |\lambda - P| = \lambda^3 + \lambda^2 + 2\lambda + 1 = 0$$

If 1 denotes identity matrix, then the inverse of matrix P will be : [GATE(EE)–2008]

(a) $(P^2 + P + 21)$ (b) $-(P^2 + P + 21)$
(c) $(P^2 + P + 1)$ (d) none of these

85. Given that

$$A = \begin{bmatrix} -5 & -3 \\ 2 & 0 \end{bmatrix} \text{ and } I = \begin{bmatrix} 1 & 0 \\ 0 & 1 \end{bmatrix}$$

the value of A^3 is : [GATE(EC, EE, IN)–2012]

(a) $19A + 30I$ (b) $17A + 15I$
(c) $17A + 21I$ (d) none of these

ANSWERS

1. (b)	**2.** (d)	**3.** (c)	**4.** (b)	**5.** (b)	**6.** (a)	**7.** (a)	**8.** (a)	**9.** (b)	**10.** (a)
11. (c)	**12.** (b)	**13.** (a)	**14.** (b)	**15.** (b)	**16.** (c)	**17.** (b)	**18.** (b)	**19.** (b)	**20.** (b)
21. (b)	**22.** (b)	**23.** (b)	**24.** (a)	**25.** (a)	**26.** (d)	**27.** (d)	**28.** (b)	**29.** (b)	**30.** (d)
31. (d)	**32.** (b)	**33.** (a)	**34.** (b)	**35.** (a)	**36.** (c)	**37.** (b)	**38.** (b)	**39.** (c)	**40.** (a)
41. (c)	**42.** (c)	**43.** (d)	**44.** (a)	**45.** (a)	**46.** (b)	**47.** (c)	**48.** (d)	**49.** (a)	**50.** (a)
51. (a)	**52.** (b)	**53.** (a)	**54.** (d)	**55.** (b)	**56.** (a)	**57.** (c)	**58.** (d)	**59.** (a)	**60.** (a)
61. (b)	**62.** (a, d)	**63.** (a)	**64.** (b)	**65.** (a)	**66.** (b)	**67.** (a)	**68.** (d)	**69.** (a)	**70.** (b)
71. (c)	**72.** (b)	**73.** (b)	**74.** (d)	**75.** (a)	**76.** (a)	**77.** (b)	**78.** (b)	**79.** (a)	**80.** (b)
81. (a)	**82.** (b)	**83.** (c)	**84.** (b)	**85.** (a)					

Hint to Selected Problems

1. With the given order we can say that order of matrices are as follows.

$$X^T \to 3 \times 4$$
$$Y \to 4 \times 3$$
$$X^T Y \to 3 \times 3$$
$$(X^T Y)^{-1} \to 3 \times 3$$
$$P \to 2 \times 3$$
$$P^T \to 3 \times 2$$
$$P(X^T Y)^{-1} P^T \to (2 \times 3)(3 \times 3)(3 \times 2) \to 2 \times 2$$
$$(P(X^T Y)^{-1} P^T)^T \to 2 \times 2$$

2. $E = \begin{bmatrix} \cos\theta & -\sin\theta & 0 \\ \sin\theta & \cos\theta & 0 \\ 0 & 0 & 1 \end{bmatrix}$ and $G = \begin{bmatrix} 1 & 0 & 0 \\ 0 & 1 & 0 \\ 0 & 0 & 1 \end{bmatrix}$

Now $E \times F = G$ or $\begin{bmatrix} \cos\theta & \sin\theta & 0 \\ -\sin\theta & \cos\theta & 0 \\ 0 & 0 & 1 \end{bmatrix} \times F = \begin{bmatrix} 1 & 0 & 0 \\ 0 & 1 & 0 \\ 0 & 0 & 1 \end{bmatrix}$

$$\Rightarrow \quad F = E^{-1} = \frac{\text{Adj}(E)}{|E|} = \begin{bmatrix} \cos\theta & \sin\theta & 0 \\ -\sin\theta & \cos\theta & 0 \\ 0 & 0 & 1 \end{bmatrix}$$

3. $A = \begin{bmatrix} 0 & 1 & 1 \\ 1 & -1 & 0 \\ -1 & 0 & -1 \end{bmatrix}_{3\times3}$

Order of matrix = 3

Rank = 2

$\therefore$ Dimension of null space of $A = 3 - 2 = 1$

5. Statement 1 is true as shown below

$[F]^T$ has a size 1×5

$[C]^T$ has a size 5×3

$[B]$ has a size 3×3

$[C]$ has a size 3×5

$[F]$ has a size 5×1

So $[F]^T[C]^T[B][C][F]$ has a size 1×1. So it is scalar.

So, the statement 1 is true.

Consider statement 2: $D^T FD$ is always symmetric.

Now $D'FD$ does not exist. Since $D'_{3\times5}$, $F_{5\times1}$ and $D_{5\times3}$ are not compatible for multiplication.

Since $D'_{3\times5}F_{5\times1} = X_{3\times1}$ and $X_{3\times1}D_{5\times3}$ does not exist

So, statement 2 is false.

6.
$$S^t = (A + A^t)^t = A^t + (A^t)^t = A^t + A = S$$

i.e., $S^t = S$

$\therefore$ S is symmetric

$\therefore$ $D^t = (A - A^t)^t = A^t - (A^t)^t = A^t - A = -(A - A^t) = -D$

i.e., $D^t = -D$

So D is skew-symmetric.

7. $(AA^T)^{-1} = 1^{-1} = 1$

9. $\begin{bmatrix} a & b \\ c & d \end{bmatrix}^{-1} = \dfrac{1}{(ad - bc)} \begin{bmatrix} d & -b \\ -c & a \end{bmatrix}$

10. $(PQ)^{-1}P = (Q^{-1}P^{-1})P = (Q^{-1})(P^{-1}P) = (Q^{-1})(1) = Q^{-1}$

12. For singularity of matrix $= \begin{vmatrix} 8 & x & 0 \\ 4 & 0 & 2 \\ 12 & 6 & 0 \end{vmatrix} = 0$

$$\Rightarrow \quad 8(0 - 12) - x(0 - 2 \times 12) = 0$$
$$x = 4$$

13. $R=\begin{bmatrix}1 & 0 & -1\\ 2 & 1 & -1\\ 2 & 3 & 2\end{bmatrix}$

$$R^{-1}=\frac{\text{adj}(R)}{|R|}=\frac{[\text{cofactor }(R)]^T}{|R|}$$

$$|R|=\begin{vmatrix}1 & 0 & -1\\ 2 & 1 & -1\\ 2 & 3 & 2\end{vmatrix}=1(2+3)-0(4+2)-1(6-2)=1$$

Now cof. $(1, 1)=+\begin{vmatrix}1 & -1\\ 3 & 2\end{vmatrix}=2+3=5$

$$\text{cof.}(2,1)=+\begin{vmatrix}0 & -1\\ 3 & 2\end{vmatrix}=-3,\ \text{cof}(3,1)=+\begin{vmatrix}0 & -1\\ 1 & -1\end{vmatrix}=+1$$

$$\text{cof}(A)=\begin{bmatrix}5 & \dots & \dots\\ -3 & \dots & \dots\\ 1 & \dots & \dots\end{bmatrix}$$

$$\text{Adj}(A)=[\text{cof. }(A)]=\begin{bmatrix}5 & -3 & 1\\ \dots & \dots & \dots\\ \dots & \dots & \dots\end{bmatrix}$$

Dividing by $|R|=1$ gives

$$R^{-1}=\begin{bmatrix}5 & -3 & 1\\ \dots & \dots & \dots\\ \dots & \dots & \dots\end{bmatrix}$$

top row of $R^{-1}=[5\ -3\ 1]$

14. Given $M^T=M^{-1}$

So $M^TM=1$

$$\begin{bmatrix}\frac{3}{5} & x\\ \frac{4}{5} & \frac{3}{5}\end{bmatrix}\begin{bmatrix}\frac{3}{5} & \frac{4}{5}\\ x & \frac{3}{5}\end{bmatrix}=\begin{bmatrix}1 & 0\\ 0 & 1\end{bmatrix}\quad\Rightarrow\quad x=-4/5$$

15. $\because\begin{vmatrix}1 & x & x^2\\ 1 & y & y^2\\ 1 & z & z^2\end{vmatrix}=-\begin{vmatrix}1 & x(x+1) & x+1\\ 1 & y(y+1) & y+1\\ 1 & z(z+1) & z+1\end{vmatrix}$

Since in a matrix if we interchange two rows of columns then determinant of resultant matrix is multiplied by –1.

So option (b) is correct.

17. $AA'A=A[(A^TA)^{-1}A^T]A=A[(A^TA)^{-1}A^TA]$

Let $A^TA=P$

Then $=A[P^{-1}P]=A.1=A$

18. Given ABCD = 1

$B^{-1}=(A^{-1}D^{-1}C^{-1})^{-1}=(C^{-1})^{-1}.(D^{-1})^{-1}.(A^{-1})^{-1}=CDA$

19. $[AA^{-1}]=1$

$$\begin{bmatrix}2 & -0.1\\ 0 & 3\end{bmatrix}\begin{bmatrix}1/2 & a\\ 0 & b\end{bmatrix}=\begin{bmatrix}1 & 0\\ 0 & 1\end{bmatrix}$$

$$\begin{bmatrix}1 & 2a-0.1b\\ 0 & 3b\end{bmatrix}=\begin{bmatrix}1 & 0\\ 0 & 1\end{bmatrix}$$

$\Rightarrow a=0.1b/2$ ...(1)

$b=1/3$

Now putting (b) in eqn (1), $a=\dfrac{1}{60}$

So $a+b=\dfrac{1}{60}+\dfrac{1}{3}=\dfrac{7}{20}$

20. Consider first 3 × 3 minors, since maximum possible rank is 3.

$$\begin{vmatrix}4 & 2 & 1\\ 6 & 3 & 4\\ 2 & 1 & 0\end{vmatrix}=0,\begin{vmatrix}2 & 1 & 3\\ 3 & 4 & 7\\ 1 & 0 & 1\end{vmatrix}=0,\begin{vmatrix}4 & 1 & 3\\ 6 & 4 & 7\\ 2 & 0 & 1\end{vmatrix}=0$$

and $\begin{vmatrix}4 & 2 & 3\\ 6 & 3 & 7\\ 2 & 1 & 1\end{vmatrix}=0$

Since all 3 × 3 minors are zero, now try 2 × 2 minors.

$$\begin{vmatrix}4 & 2\\ 6 & 3\end{vmatrix}=0,\begin{vmatrix}2 & 1\\ 3 & 4\end{vmatrix}=8-3=5\neq 0$$

so rank = 2

23. Perform, Gauss-elimination

$$\begin{bmatrix}1 & 1 & 1\\ 1 & -1 & 0\\ 0 & 0 & 0\end{bmatrix}\xrightarrow[R_3-R_1]{R_2-R_1}\begin{bmatrix}1 & 1 & 1\\ 0 & -2 & -1\\ 0 & 0 & 0\end{bmatrix}$$

It is in row Echelon form.

So its rank is the number of non-zero rows in this form

i.e., rank = 2

24. If rank of (5 × 6) matrix is 4. Then surely it must have exactly 4 linearly independent rows as well as 4 linearly independent columns.

Since rank = row rank = column rank

31. The augmented matrix for this system is

$$\left[\begin{array}{ccc|c}1 & 1 & 1 & 5\\ 1 & 3 & 3 & 9\\ 1 & 2 & \alpha & \beta\end{array}\right]$$

Using Gauss-elimination method we get

$$\left[\begin{array}{ccc|c}1 & 1 & 1 & 5\\ 1 & 3 & 3 & 9\\ 1 & 2 & \alpha & \beta\end{array}\right]\xrightarrow[R_3-R_1]{R_2-R_1}\left[\begin{array}{ccc|c}1 & 1 & 1 & 5\\ 0 & 2 & 2 & 4\\ 0 & 1 & \alpha-1 & \beta-5\end{array}\right]$$

$$\xrightarrow{R_3-\frac{1}{2}R_2}\left[\begin{array}{ccc|c}1 & 1 & 1 & 5\\ 0 & 2 & 2 & 4\\ 0 & 0 & \alpha-2 & \beta-7\end{array}\right]$$

Now for infinite solution last row must be completely zero.

i.e., $\alpha-2$ and $\beta-7=0$

$\alpha=2$ and $\beta=7$

33. The augmented matrix is

$$\left[\begin{array}{ccc|c}1 & 1 & 1 & 3\\ 1 & 2 & 3 & 4\\ 1 & 4 & K & 6\end{array}\right]$$

Using Gauss-elimination method, we get

$$\left[\begin{array}{ccc|c}1 & 1 & 1 & 3\\ 1 & 2 & 3 & 4\\ 1 & 4 & K & 6\end{array}\right]\xrightarrow[R_3-R_1]{R_2-R_1}\left[\begin{array}{ccc|c}1 & 1 & 1 & 3\\ 0 & 1 & 2 & 1\\ 0 & 3 & K-1 & 3\end{array}\right]$$

$$\xrightarrow{R_3-3R_2}\left[\begin{array}{ccc|c}1 & 1 & 1 & 3\\ 0 & 1 & 2 & 1\\ 0 & 0 & K-7 & 0\end{array}\right]$$

Now if $K\neq 7$

rank (A) = rank $(A|B)=3$

$\therefore$ unique solution

If $K=7$, rank(A) = rank$(A|B)=2$

which is less than number of variables.

$\therefore$ When $K = 7$, unique solution is not possible and only infinite solution is possible.

36. $r(A_{m \times n}) \le \min(m, n)$

So Highest possible rank = Least value of 3 and 4, *i.e.*, highest possible rank (based on size of A) = 3

However if the rank of $A = 3$

Then rank of $[A|B]$ would also be 3.

Which means the system would become consistent.

But it is given the system is inconsistent.

So the maximum rank of A could only be 2.

41. $\begin{bmatrix} 2 & -2 \\ 1 & -1 \end{bmatrix}\begin{bmatrix} x_1 \\ x_2 \end{bmatrix} = \begin{bmatrix} 0 \\ 0 \end{bmatrix}$

$$2x_1 - 2x_2 = 0$$

$$x_1 - x_2 = 0 \quad \Rightarrow \quad x_1 = x_2$$

i.e., x_1 and x_2 are having infinite number of solution

i.e., multiple solution exist.

50. $$A = \begin{bmatrix} 9 & 5 \\ 5 & 8 \end{bmatrix}$$

The characteristic equation is

$$\begin{vmatrix} 9-\lambda & 5 \\ 5 & 8-\lambda \end{vmatrix} = 0$$

$$(9-\lambda)(8-\lambda) - 25 = 0$$

$$\lambda^2 - 17\lambda + 47 = 0$$

So eigenvalues are, $\lambda = 3.48, 13.53$

54. $$A = \begin{bmatrix} 2 & 1 \\ 0 & 2 \end{bmatrix}$$

$$[a - \lambda I] = 0$$

$$\begin{bmatrix} 2-\lambda & 1 \\ 0 & 2-\lambda \end{bmatrix} = 0 \quad \Rightarrow \quad (2-\lambda)^2 = 0$$

$$\Rightarrow \quad \lambda = 2$$

Now, consider the eigenvalue problem

$$[A - \lambda I]\hat{x} = 0$$

$$\begin{bmatrix} 2-\lambda & 1 \\ 0 & 2-\lambda \end{bmatrix}\begin{bmatrix} x_1 \\ x_2 \end{bmatrix} = \begin{bmatrix} 0 \\ 0 \end{bmatrix}$$

Put $\lambda = 2$, we get

$$\begin{bmatrix} 0 & 1 \\ 0 & 0 \end{bmatrix}\begin{bmatrix} x_1 \\ x_2 \end{bmatrix} = \begin{bmatrix} 0 \\ 0 \end{bmatrix}$$

$$x_2 = 0 \quad \text{...(i)}$$

$$0 = 0 \quad \text{...(ii)}$$

$\therefore$ Solution is $x_2 = 0$, x_1 = anything

$$\begin{bmatrix} x_1 \\ x_2 \end{bmatrix} = \begin{bmatrix} K \\ 0 \end{bmatrix}$$

$\because$ There is only one parameter in the infinite solution.

So there is only one linearly independent eigenvector for this problem which may be written as

$$\begin{bmatrix} K \\ 0 \end{bmatrix} \text{ or as } \begin{bmatrix} 1 \\ 0 \end{bmatrix}.$$

Self Assessment Test

1. Find $A+B$ if $A=\begin{bmatrix}2 & 2 & 3\\ 1 & -4 & 1\end{bmatrix}$ and $B=\begin{bmatrix}1 & 3\\ 2 & 4\end{bmatrix}$

2. Show that for any matrix A $-(-A) = A$.

3. Find the size of the product of AB if
$$A = (a_{ij})_{2\times 3} \text{ and } B = (b_{ij})_{3\times 4}.$$

4. If A be an $m\times n$ matrix such that $m > 1$, $n > 1$ find the condition under which $A.u$ and $v.A$ are defined (u and v are vectors.)

5. Show that the matrix AA' and $A'A$ are defined for any matrix A.

6. Evaluate $f(A)$ for the polynomial $f(x) = 2x^2-4x+5$.

7. Show that if $AB=A$ and $BA=B$ then A and B are idempotent.

8. Show that the elementary matrices are invertible and their inverses are also elementary matrices.

9. Show that the matrix
$$A=\begin{pmatrix}1/9 & 8/9 & -4/9\\ 4/9 & -4/9 & -7/9\\ 8/9 & 1/9 & 4/9\end{pmatrix} \text{ is orthogonal.}$$

10. If $A=\begin{pmatrix}1/\sqrt{5} & 2/\sqrt{5}\\ x & y\end{pmatrix}$ is orthogonal. Find the value of x and y.

11. Show that $A=\begin{pmatrix}\frac{1}{3}-\frac{2}{3}i & \frac{2}{3}i\\ -\frac{2}{3}i & -\frac{1}{3}-\frac{2}{3}i\end{pmatrix}$ is unitary.

12. Find the inverse of $A=\begin{bmatrix}3 & 1 & 0\\ -1 & 2 & 2\\ 5 & 0 & -1\end{bmatrix}$.

13. Prove that $A=\begin{bmatrix}3 & 7-4i & -2+5i\\ 7+4i & -2 & 3+i\\ -2-5i & 3-i & 4\end{bmatrix}$ is a Hermitian matrix.

14. Reduce the following matrices to row reduced Echelon form and find their rank

(i) $\begin{bmatrix}1 & 2 & -1 & 0\\ 2 & 1 & -1 & 1\\ 3 & 4 & 2 & 2\end{bmatrix}$ (ii) $\begin{bmatrix}1 & 1 & 1 & 1\\ 1 & 3 & -2 & 1\\ 2 & 0 & -3 & 2\\ 3 & 3 & -3 & 3\end{bmatrix}$

15. Show that the vectors $\begin{pmatrix}1 & 0\\ 0 & 0\end{pmatrix},\begin{pmatrix}1 & 1\\ 0 & 0\end{pmatrix},\begin{pmatrix}1 & 1\\ 1 & 0\end{pmatrix},\begin{pmatrix}0 & 0\\ 0 & 1\end{pmatrix}$ are linearly independent.

16. If $A=\begin{bmatrix}1 & 2 & 3\\ 3 & 2 & 1\\ 1 & 3 & 2\\ 2 & 1 & 3\end{bmatrix}$, then find non-singular matrices P and Q such that PAQ is in the normal form.

17. Reduce the following matrices to their normal forms and find their rank.

(i) $A_1=\begin{bmatrix}1 & 1 & 1 & -1\\ 1 & 2 & 3 & 4\\ 3 & 4 & 5 & 2\end{bmatrix}$

(ii) $A_2=\begin{bmatrix}1 & 4 & 3 & 2\\ 1 & 2 & 3 & 4\\ 2 & 6 & 7 & 5\end{bmatrix}$

18. Find the inverse of the following matrices :

(i) $A_1=\begin{bmatrix}1 & 2 & 1\\ 3 & 2 & 3\\ 1 & 1 & 2\end{bmatrix}$ (ii) $A_2=\begin{bmatrix}1 & 1 & 1\\ 1 & 2 & 3\\ 1 & 3 & 6\end{bmatrix}$

19. Using elementary operations, find the inverse of the matrix
$$A=\begin{bmatrix}1 & -1 & 0 & 2\\ 0 & 1 & 1 & -1\\ 2 & 1 & 2 & 1\\ 3 & -2 & 1 & 6\end{bmatrix}$$

20. If $A=\begin{bmatrix}\cos\theta & \sin\theta & 0\\ \sin\theta & \cos\theta & 0\\ 0 & 0 & 0\end{bmatrix}$ then show that $AA^{-1} = A^{-1}A = I$.

21. Let L be a system of linear equations with more unknown than equations. Show that L cannot have a unique solution.

22. Prove that the following three statements about a system of linear equations are equivalent :
   (i) The system is consistent.
   (ii) No linear combinations of the equations is the equation $0x_1 + 0x_2 +\ldots +0x_n = b \neq 0$.
   (iii) The system is reducible to Echelon form.

23. If u and v are solutions of a non-homogeneous system $AX = B$, then show that difference $w = v - u$ is a solution of the associated homogeneous system $AX = O$.

24. Determine whether the following homogeneous system has a non-zero solution :
$$x_1 - 2x_2 + 3x_3 - 2x_4 = 0$$
$$3x_1 - 7x_2 - 2x_3 + 4x_4 = 0$$
$$4x_1 + 3x_2 + 5x_3 + 2x_4 = 0$$

25. Find a matrix P which transform the matrix $A=\begin{bmatrix}1 & 0 & -1\\ 1 & 2 & 1\\ 2 & 2 & 3\end{bmatrix}$ to diagonal form. Hence, find A^4.

26. Examine for linear dependence [1, 0, 2, 1], [3, 1, 2, 1], [4, 6, 2, –4], [–6, 0, –3, –4] and find the relation between them if possible.

27. Show that the rank of the following $(n+1)\times(n+1)$ matrix where a is real number $\begin{bmatrix}1 & a & a^2 & \cdots & a^n\\ 1 & a & a^2 & \cdots & a^n\\ \vdots & \vdots & \vdots & \cdots & \vdots\\ 1 & a & a^2 & \cdots & a^n\end{bmatrix}$ is 1.

28. Let A be an $m \times n$ matrix where $m < n$. Consider the system of linear equation $AX = B$ where B is $n \times 1$ column vector and $B \neq 0$ then show that the system of equations has a solution if and only if it has infinitely many solutions.

29. In the matrix equation $AX = B$, show that the necessary condition for the existence of at least one solution for the unknown vector X is that augmented matrix $[A|B]$ must have the same rank as A.

30. Show that for the set of equations

$$x_1 + 2x_2 + x_3 + 4x_4 = 2$$
$$3x_1 + 6x_2 + 3x_3 + 12x_4 = 6$$

multiple non-trivial solution exists.

31. Show that the system of simultaneous equations

$x + 2y + z = 6;\ 2x + y + 2z = 6;\ x + y + z = 5$

has no solution.

32. Show that if $A_{2\times 2}$ matrix which satisfy $A^2 - A = 0$ then A must be diagonal.

33. Consider the system of linear equations

$x + y + z = 3;\ x - y - z = 4;\ x - 5y + kz = 6.$

Find the value of k for which the system has an infinite number of solutions. (Ans. $k = -5$)

34. Find the value of λ for which the equations

$$(\lambda - 1)x + (3\lambda + 1)y + 2\lambda z = 0$$
$$(\lambda - 1)x + (4\lambda - 2)y + (\lambda + 3)z = 0$$
$$2x + (3\lambda + 1)y + 3(\lambda - 1)z = 0$$

are consistent.

35. Find the eigenvalue of $A = \begin{bmatrix} 1 & 2 & 3 \\ 0 & -4 & 2 \\ 0 & 0 & 7 \end{bmatrix}$.

36. Show that the characteristic vector corresponding to characteristic root λ of matrix A is also a characteristic vector of every matrix $f(A)$ where $f(x)$ is any scalar polynomial and the corresponding root for $f(A)$ is $f(\lambda)$. In general show that if $g(x) = \dfrac{f_1(x)}{f_2(x)}$ and $|f_2(A)| \neq 0$ and $g(\lambda)$ is a characteristic root of $g(A) = \dfrac{f_1(A)}{f_2(A)}$.

37. Show that any two eigenvectors corresponding to two distinct characteristic roots of a
   (i) Hermitian
   (ii) real symmetric
   (iii) unitary
   matrix are orthogonal.

38. Find the characteristic roots and characteristic vectors for the matrix $A = \begin{bmatrix} 2 & \sqrt{2} \\ \sqrt{2} & 1 \end{bmatrix}$ and show that the matrix A satisfies its characteristic equation.

39. Verify Cayley-Hamilton theorem for the matrix

$$A = \begin{bmatrix} 0 & 0 & 1 \\ 3 & 1 & 0 \\ -2 & 1 & 4 \end{bmatrix}.$$

Hence, evaluate A^{-1}.

40. Prove that the eigenvalues of the matrix $\begin{bmatrix} a_1 & a_2 & a_3 \\ 0 & b_2 & 0 \\ 0 & 0 & c_3 \end{bmatrix}$ are a_1, b_2, c_3.

41. Find all eigenvalues and eigenvectors of the matrix

$$A = \begin{bmatrix} 3 & 1 & 4 \\ 0 & 2 & 6 \\ 0 & 0 & 5 \end{bmatrix}.$$

42. Find a matrix P which transform the matrix $A = \begin{bmatrix} 1 & 0 & -1 \\ 1 & 2 & 1 \\ 2 & 2 & 3 \end{bmatrix}$ to diagonal form. Hence, find A^4.

43. If the characteristic roots of $\begin{bmatrix} 3 & 7 \\ 2 & 5 \end{bmatrix}$ are λ_1 and λ_2, show that the characteristic roots of $\begin{bmatrix} 5 & -7 \\ -2 & 3 \end{bmatrix}$ are $\dfrac{1}{\lambda_1}$ and $\dfrac{1}{\lambda_2}$.

44. If A is any matrix which satisfy $A^3 - A^2 + A - I = 0$ and $A_{3\times 3}$ then show that $A^4 = A = I$.

45. Let $A = \begin{bmatrix} 2 & 3 \\ x & y \end{bmatrix}$. If the eigenvalues of A are 4 and 8 then show that $x = -4$ and $y = 10$.

46. Let A be an $n \times n$ complex matrices whose characteristic polynomial is given by $f(t) = t^n + c_{n-1}t^{n-1} + \ldots + c_1 t + c_0$ then show that $\det(A) = c_0$.

47. If the eigenvalues of a 3×3 matrix A are 1,2 and –3 then show that $A^{-1} = \frac{1}{6}\left[7I - A^2\right]$.

48. If a square matrix of order 10 has exactly 4 distinct eigenvalues, then show that the degree of its minimal polynomial is at least 4.

49. Show that the matrix $M = \begin{bmatrix} 0 & 1 & 2 & 0 \\ 1 & 0 & 1 & 0 \\ 2 & 1 & 0 & 2 \\ 0 & 0 & 2 & 0 \end{bmatrix}$ has both positive and negative real eigenvalues.

50. Show that the characteristic polynomial of $A = \begin{bmatrix} 5 & -6 & -6 \\ -1 & 4 & 2 \\ 3 & -6 & -4 \end{bmatrix}$ is

$$\det|xI - A| = (x-2)^2 (x-1).$$

CCCCCC

SERIES

UNIT OUTLINE

- **Infinite Series**
- **Covergence and Divergence**
- **Tests for Convergence**
- **Alternating Series**
- **Absolute Convergence**
- **Fourier Series**
- **Half Range Series**
- **Parseval's Identity**
- **Complex Form of Fourier Series**

PRE-REQUISITE

- Basic Concepts of Series and Sequence
- Concepts of Sum of the Series
- Concepts of Period of Trigonometric Functions

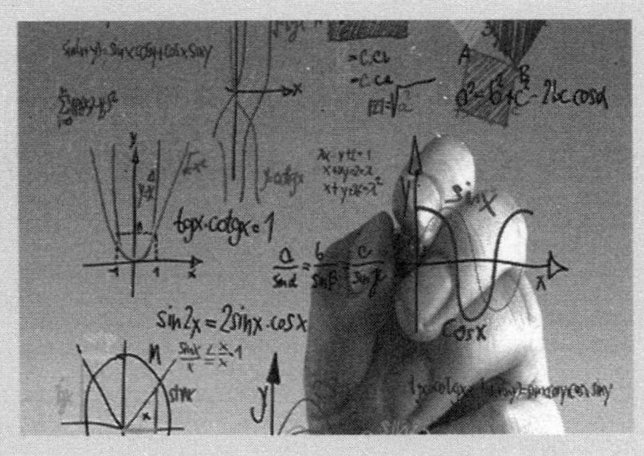

CHAPTER 26 Infinite Series

26.1 INTRODUCTION

Infinite series are essential in the calculation of values of many functions, and can frequently be used for the evaluation of definite integrals. They can also serve to define new and useful functions that are fundamental in many investigations in advanced mathematics and its applications.

Problems connected with the concept of series attracted the Indian mathematician as early as the third century A.D. Their work on series continued till late fourteenth century, but they never took up a critical study of series. In Europe, it was during the sixth century A.D., that the wider significance of finite and infinite series was realized.

The English mathematicians, Brook Taylor (1685-1731) and James Sterling (1692-1770), and the Scotch mathematician Colin Maclaurin (1698-1746), made important contributions to the study of infinite series. The question of convergent of infinite series was first subjected to rigorious investigation by the German Mathematician Carl Friedrich Gauss (1777-1855).

In this chapter, we are going to discuss the convergence behaviour of infinite series of real numbers and shall obtain a few tests for ascertaining the convergence of the infinite series. Some writer use the word Progression instead of word series. But here the word Series, which is due to the writers of the 17lh century and is most commonly used in preferred.

26.2 INFINITE SERIES

Let $<u_n>$ be a sequence of real numbers, then an expression of the form

$$u_1+u_2+ \dots + u_n+\dots \qquad \dots(1)$$

is called an infinite series. In symbols it is generally written as

$$\sum_{n=1}^{\infty} u_n \text{ or } \sum u_n$$

If all the terms of $<u_n>$ after a certain number are zero then the expression

$$u_1 + u_2 + \dots + u_m, \text{ written as } \sum_{n=1}^{m} u_n \text{ is called a finite series.}$$

The term u_n is called the nth term or general term of the series (1). The sum of first n terms of the series is denoted by s_n.

Thus,

$$s_n = u_1+u_2+ \dots +u_n = \sum_{r=1}^{n} u_r$$

26.3 SEQUENCE OF PARTIAL SUM OF AN INFINITE SERIES

An expression of the form $u_1 + u_2 + \dots + u_n \dots$ which involves addition of infinitely many terms has in itself no meaning. In order to give a meaning to the value of such as infinite sum, we form a sequence of partial sums. It is the limit of such a sequence which gives meaning to the infinite series.

Let us associate to the infinite series $u_1 + u_2 + \dots + u_n \dots$; a sequence $<s_n>$ defined by $s_n = u_1 + u_2 + \dots + u_n$.

Then the sequence $<s_n>$ is called the sequence of partial sums of the given series $u_1 + u_2 + \dots + u_n \dots$

26.4 CONVERGENCE, DIVERGENCE OR OSCILLATION OF AN INFINITE SERIES

An infinite series $\sum_{n=1}^{\infty} u_n$ is said to be

(i) convergent if the sequence $<s_n>$ of its partial sums converges to a real number l and in that case l is called the sum of the series $\sum_{n=1}^{\infty} u_n$ and we write $\sum_{n=1}^{m} u_n = l$. In this case, we also say that the series is convergent to l.

(ii) converges absolutely if $\sum_{n=1}^{\infty} |u_n|$ converges.

(iii) converges conditionally if $\sum_{n=1}^{\infty} u_n$ converges but $\sum_{n=1}^{\infty} |u_n|$ does not converge.

(iv) diverges to ∞ (or $-\infty$) if the sequence $<s_n>$ diverges to ∞ (or $-\infty$) and in that case $\sum_{n=1}^{\infty} u_n = \infty \left(\text{or } \sum_{n=1}^{\infty} u_n = -\infty \right)$

(v) oscillate finitely if the sequence $<s_n>$ oscillates finitely.

(vi) oscillate infinitely if the sequence $<s_n>$ oscillates infinitely.

(vii) oscillatory if s_n, the sum of its first n terms, neither tends to a definite finite limit nor to $+\infty$ or $-\infty$ as $n \to \infty$.

REMARKS

- Divergent and oscillatory series are often called non-convergent series.
- The value of s_n of oscillate finitely series fluctuate within a finite range as $n \to \infty$.
- The value of s_n of oscillate infinitely series, tends to infinity as $n \to \infty$ and its sign is alternatively positive and negative.

ILLUSTRATIONS

(1) The series $1 + \frac{2}{3} + \left(\frac{2}{3}\right)^2 + \ldots + \left(\frac{2}{3}\right)^{n-1} + \ldots$ is convergent.

(2) The series $\frac{1}{2} + \frac{1}{2^2} + \frac{1}{2^3} + \ldots$ is convergent.

(3) The series $1 + 2 + 3 + \ldots + n + \ldots$ is divergent.

(4) The series $3 - 3 + 3 - 3 + \ldots$ is oscillatory.

THEOREM 1 (Necessary condition for convergence). *For a series Σu_n to be convergent, it is necessary that* $\lim u_n = 0$.

(PTU–2009)

Proof. Let us suppose, the series Σu_n be convergent. Let s_n denote the sum of n terms of the series

Therefore
$$\left.\begin{aligned} s_n &= u_1 + u_2 + \ldots + u_n \\ \Rightarrow \quad s_{n-1} &= u_1 + u_2 + \ldots + u_{n-1} \end{aligned}\right] \Rightarrow u_n = s_n - s_{n-1} \quad \ldots(1)$$

The series Σu_n is convergent, therefore s_n and s_{n-1} both will tend to the same finite limit, say l as $n \to \infty$.

Now, from (1) $\lim u_n = \lim s_n - \lim s_{n-1} = 1 - 1 = 0$

Hence, for a convergent series, it is necessary that $\lim u_n = 0$.

REMARKS

- The converse of the above theorem is not necessarily true :

  For example, in the series $\sum u_n = 1 + \frac{1}{2} + \frac{1}{3} + \ldots$; $u_n = \frac{1}{n} \to 0$ as $n \to \infty$ but the series is not convergent.
- If a series Σu_n be such that u_n does not tend to zero as $n \to \infty$ then the series does not converge.

THEOREM 2. (Cauchy's General principle of convergence for series). *A necessary and sufficient condition for a series Σu_n to be convergent is that to each $\varepsilon > 0$, there exists a positive integer m such that* $|u_{n+1} + u_{n+2} + \ldots + u_{n+p}| < \varepsilon$ $\forall\ n > m,\ p \geq 1$.

Proof. Let $<s_n>$ be the sequence of partial sums of the series Σu_n. The series Σu_n will converge if and only if the sequence $<s_n>$ of its partial sums converges. But by Cauchy's general principle of convergence for sequences, we know that a necessary and sufficient condition for the convergence of $<s_n>$ is that for each $\varepsilon > 0$, there exists $m \in N$ such that $|u_{n+1} + u_{n+2} + \ldots + u_{n+p}| < \varepsilon$, whenever $n \geq m$ and $p \geq 1$.

$$|s_n - s_m| < \varepsilon, \forall\ n > m$$

$$\Rightarrow \quad |u_{n+1} + u_{n+2} + \ldots + u_{n+p}| < \varepsilon, \forall\ n > m \text{ and } p \geq 1.$$

REMARKS

- The nature of a series remains unaltered if :
  (a) the sign of all term are changed.
  (b) a finite number of terms are added or omitted.
  (c) each term of the series is multiplied or divided by the same fixed number k ($k \neq 0$).
- If Σu_n converges to l and $k \in R$, then Σku_n converges to lk.
- If Σu_n converges to l_1 and Σv_n converges to l_2 then $\Sigma(u_n + v_n)$ converges to $(l_1 + l_2)$.
- If Σu_n diverges and $k \in R$, $k \neq 0$, then Σku_n diverges.
- If Σu_n and Σv_n are two divergent series having all terms positive, then $\Sigma(u_n + v_n)$ also diverges.

THEOREM 3. *A series of positive terms is convergent if s_n, the sum of n terms is less than a fixed number for all values of n.*

Proof. Let $u_1 + u_2 + ... u_n +.....$ be the series of positive terms.
Then $s_n = u_1 + u_2 + ... u_n$.
Obviously if n increases, then s_n increases and may tend to a finite limit or to $+\infty$. The series cannot oscillate. If s_n remains less than a fixed number for all values of n it cannot tend to infinity and so it must tend to a finite limit. Hence, the series is convergent.

26.5 FUNDAMENTAL RESULTS FOR THE CONVERGENCE OF POSITIVE TERM SERIES

THEOREM 1. *A series Σu_n of positive term is convergent if and only if the sequence $<s_n>$ (where $s_n = u_1+u_2+...+u_n$) of its partial sum is bounded above.*

Proof. Since $u_n > 0 \; \forall \; n$, the sequence $<s_n>$ of partial sums of the series is monotonically increasing. Now the series Σu_n is convergent iff the sequence $<s_n>$ is convergent, *i.e.*, iff the sequence $<s_n>$ is bounded above.
($\because$ a monotonically increasing sequence is convergent iff it is bounded above.)

REMARKS

- To show that a series of positive term is convergent it is enough to show the sequence of its partial sum is bounded above and to show that the series of positive term is divergent, we have to show that the sequence of its partial sum is unbounded above.
- A series Σu_n where the terms are not necessarily positive, may fail to the convergent even if the sequence of its partial sums is bounded above.
  For example, consider the series $\sum_{n=1}^{\infty} u_n$ where $u_n = (-1)^n$. Then $s_n = \begin{cases} -1 & \text{if } n \text{ is odd} \\ 0 & \text{if } n \text{ is even.} \end{cases}$
  The sequence $<s_n>$ is bounded above but not convergent and as such, the series is not convergent. Hence, boundedness of the sequence of partial sums of a series Σu_n is only a necessary condition and not a sufficient one. However, it is a sufficient condition for a positive term series.

THEOREM 2. **(Pringsheim theorem).** *If a series Σu_n of positive monotonic decreasing terms converges then not only u_n tends to zero but also nu_n tends to zero as n tends to infinity.*

Proof. By Cauchy's general principle of convergence, we have for a convergent series, that for given $\varepsilon > 0$ there exists a positive integer k such that $|u_{m+1} + u_{m+2} + ... + u_{m+p}| < \varepsilon/2 \; \forall \; m \geq k; \, p \geq 1$

Choose $m+p = n > 2k$ and $m = \left[\frac{n}{2}\right]$ where $\left[\frac{n}{2}\right]$ denote the greatest integer not greater than $\left[\frac{n}{2}\right]$.

Then $$u_{m+1} + u_{m+2} + ... + u_n < \frac{\varepsilon}{2}$$

But Σu_n is monotonic decreasing sequence of positive terms. Therefore,

$$(n-m)u_n < u_{m+1} + u_{m+2} + ... + u_n < \frac{\varepsilon}{2}$$

$$\Rightarrow \quad \frac{1}{2} nu_n < \frac{\varepsilon}{2} \quad \Rightarrow \quad nu_n < \varepsilon, \forall \, n \in N \quad \Rightarrow \quad \lim_{n\to\infty} nu_n = 0.$$

THEOREM 3. *If each term of a series Σu_n of positive terms does not exceed the corresponding terms of a convergent series Σv_n of positive terms, then Σu_n is convergent. On the other hand, if each term of Σu_n exceed (or equals) the corresponding terms of a divergent series of positive terms, then Σu_n is divergent.*

Proof. Let us suppose $u_n < v_n \ \forall n \in N$. Now let $s_n = u_1+u_2+...+u_n$ and $s'_n = v_1+v_2+...+v_n$

Since $u_n \le v_n \ \forall n$, therefore $s_n \le s'_n$.

Now Σv_n is convergent, therefore $\lim s_n \le \lim s'_n = s'$ (a finite quantity).

Thus s_n tends to a finite limit as $n \to \infty$. Hence, the series Σu_n is convergent.

Now, if $u_n > v_n$, $\forall n$ then $s_n > s'_n$.

But Σv_n is divergent, therefore $s'_n \to \infty$ as $n \to \infty$ and hence $s_n \to \infty$ as $n \to \infty$ which gives that Σu_n is divergent.

THEOREM 4 (Convergence of geometric series). *The geometric series* $1 + r + r^2 + ... + r^{n-1} + ...$

(i) *converges to* $\frac{1}{1-r}$ *if* $|r| < 1$ (ii) *diverges to* $+\infty$ *if* $r \ge 1$

(iii) *oscillate finitely if* $r = -1$. (iv) *oscillate infinitely if* $r < -1$.

Proof. Here
$$s_n = 1 + r + r^2 + ... + r^{n-1} = \begin{cases} \frac{1-r^n}{1-r} & \text{if } r \ne 1 \\ n & \text{if } r = 1 \end{cases}$$

Now, there are following cases :

Case (I). If $|r| < 1$.

Then $\lim_{n\to\infty} r^n = 0$ so that $\lim_{n\to\infty} s_n = \frac{1}{1-r} \Rightarrow$ the series is convergent to $\frac{1}{1-r}$.

Case (II). If $r>1$.

Then $\lim_{n\to\infty} r^n = \infty$ so that $s_n = \frac{1-r^n}{1-r} = \frac{1}{1-r} + \frac{r^n}{r-1} \to \infty$ as $n \to \infty$.

Hence, the series diverges to ∞.

if $r=1$, then $s_n = 1 + 1 + ... + 1 + ...$ to n times $= n$

Thus, the sequence $< s_n >$ diverges and hence the series diverges.

Case (III). If $r = -1$.

Then, $s_n = \begin{cases} 0 & \text{if } n \text{ is even} \\ 1 & \text{if } n \text{ is odd} \end{cases}$

Therefore the sequence $<s_n>$ oscillate between 0 and 1. $\Rightarrow$ The series oscillates finitely between 0 and 1.

Case (IV). If $r < -1$

Let $r = -a$ where $a > 1$ Then $s_n = \frac{1}{1+a} - \frac{(-1)^n \cdot a^n}{1+a}$ so that $s_{2n} \to \infty$ and $s_{2n+1} \to \infty$

Therefore, the sequence $<s_n>$ oscillate infinitely between $-\infty$ and $+\infty$.

Hence, the series oscillate infinitely.

THEOREM 5. *A positive terms series Σu_n either converges to a finite limit or diverges to ∞.*

Proof. Let $s_n = u_1 + u_2 + ... + u_n$

$\Rightarrow \quad s_{n+1} = u_1 + u_2 + ... u_{n+1}$

Therefore, $s_{n+1} - s_n = u_{n+1} > 0 \Rightarrow s_{n+1} > s_n, \forall n$

$\Rightarrow$ $<s_n>$ is monotonically increasing sequence.

Since, a monotonically increasing sequence is either convergent to a finite limit or divergent to ∞, the sequence $< s_n >$ of partial sums of the series Σu_n is either convergent a finite limit or divergent to ∞.

Hence, the series Σu_n is either converges or diverges to ∞.

REMARKS

- In view of the above theorem, a positive term series has only two possible behaviours, *i.e.* convergence or divergence while a general term has got five behaviour (*i.e.*, convergent, divergent to ∞, divergent to $-\infty$, oscillate finitely and oscillate infinitely).
- If, in a positive terms series Σu_n, u_n does not tend to 0 as $n \to \infty$, the series is divergent.
- Similarly, it can be proved that a negative term series either converges to a finite limit or diverges to $-\infty$.

THEOREM 6. $\left(\textbf{The Auxillary series} \sum \frac{1}{n^p}\right)$. *The infinite series* $\sum\left(\frac{1}{n^p}\right) = \frac{1}{1^p} + \frac{1}{2^p} + \ldots + \frac{1}{n^p} + \ldots$ *is convergent if* $p > 1$ *and divergent if* $p < 1$. (PTU–2009, VTU–2006, Rohtak–2003)

Proof. **Case (I).** *When* $p > 1$.

Since each term of the given series is positive so that the given series can be written as :

$$\sum\left(\frac{1}{n^p}\right) = \frac{1}{1^p} + \left(\frac{1}{2^p} + \frac{1}{3^p}\right) + \left(\frac{1}{4^p} + \frac{1}{5^p} + \frac{1}{6^p} + \frac{1}{7^p}\right) + \left(\frac{1}{8^p} + \frac{1}{9^p} + \frac{1}{10^p} + \frac{1}{11^p} + \frac{1}{12^p} + \frac{1}{13^p} + \frac{1}{14^p} + \frac{1}{15^p}\right) + \ldots \quad \ldots(1)$$

Since $p > 1$, then

$$3^p > 2^p \Rightarrow \frac{1}{3^p} < \frac{1}{2^p} \qquad \Rightarrow \qquad \frac{1}{2^p} + \frac{1}{3^p} < \frac{1}{2^p} + \frac{1}{2^p} = \frac{2}{2^p} = \frac{1}{2^{p-1}}$$

Also, $5^p > 4^p, 6^p > 4^p, 7^p > 4^p \qquad \Rightarrow \qquad \frac{1}{5^p} < \frac{1}{4^p}, \frac{1}{6^p} < \frac{1}{4^p}, \frac{1}{7^p} < \frac{1}{4^p}$

$$\Rightarrow \frac{1}{4^p} + \frac{1}{5^p} + \frac{1}{6^p} + \frac{1}{7^p} < \frac{1}{4^p} + \frac{1}{4^p} + \frac{1}{4^p} + \frac{1}{4^p} = \left(\frac{1}{2^{p-1}}\right)^2$$

Similarly $\frac{1}{8^p} + \frac{1}{9^p} + \frac{1}{10^p} + \ldots. + \frac{1}{15^p} < \frac{8}{8^p} = \left(\frac{1}{2^{p-1}}\right)^3 \ldots$ and so on.

Now using above inequalities equation (1) becomes

$$\sum\left(\frac{1}{n^p}\right) < 1 + \frac{1}{2^{p-1}} + \left(\frac{1}{2^{p-1}}\right)^2 + \left(\frac{1}{2^{p-1}}\right)^3 + \ldots \quad \ldots(2)$$

The R.H.S. of (2) is a geometric series with common ratio less than 1 as $p > 1$, which is therefore convergent thus the series on L.H.S of (2) is convergent, hence, $\sum\left(\frac{1}{n^p}\right)$ is convergent, when $p > 1$.

Case (II). *When* $p = 1$. Then the given series becomes

$$\sum \frac{1}{n^p} = 1 + \frac{1}{2} + \frac{1}{3} + \ldots$$

Now, this series may be written as follows

$$\sum \frac{1}{n^p} = 1 + \frac{1}{2} + \left(\frac{1}{4} + \frac{1}{4}\right) + \ldots > 1 + \frac{1}{2} + \left(\frac{1}{4} + \frac{1}{4}\right) + \ldots = 1 + \frac{1}{2} + \frac{2}{4} + \ldots. = 1 + \frac{1}{2} + \frac{1}{2} \ldots.$$

Now since $\lim u_n = \frac{1}{2} \neq 0$, the series is divergent.

Case (III). *When* $p < 1$. Then

$$2^p < 2, 3^p < 3, 4^p < 4 \text{ and so on.}$$

Hence, the given series reduces to $\sum \frac{1}{n^p} > 1 + \frac{1}{2} + \frac{1}{3} + \frac{1}{4} + \ldots$

Clearly, the series on the right hand side is divergent. [By case (II)]

Hence, the given series is divergent when $p < 1$.

26.6 COMPARISON TESTS

The most important technique for deciding whether a series is convergent or not is to compare it with another suitable chosen series which is already known to be convergent or divergent.

First form. Let Σu_n and Σv_n be two series of positive terms such that $u_n < k v_n$, $\forall n$

Then,

(i) Σv_n converges $\Rightarrow \Sigma u_n$ converges (ii) Σu_n diverges $\Rightarrow \Sigma v_n$ diverges.

Second form. Let Σu_n and Σv_n be two series of positive terms and let k_1 and k_2 be positive real numbers such that $k_1 v_n \leq u_n \leq k_2 v_n$, $\forall n$

then series Σu_n and Σv_n converge or diverge together.

Third form. If Σu_n and Σv_n be two given positive terms series such that $u_n \le kv_n$, $\forall\, n > m$, $k>0$ and $m \in N$
Then

(i) Σv_n *is* convergent $\Rightarrow \Sigma u_n$ is convergent (ii) Σu_n is divergent $\Rightarrow \Sigma v_n$ is also divergent.

Fourth form. Let Σu_n and Σv_n be two series of positive terms and let k_1, k_2 be positive real numbers such that $k_1 v_n < u_n < k_2 v_n \;\forall\, n > m$, *m* being a fixed positive integer. Then the series Σu_n and Σv_n converge or diverge together.

Fifth form. Let Σu_n and Σv_n be two sereis of positive terms such that

$$\lim_{n\to\infty} \frac{u_n}{v_n} = l \text{ (finite and non-zero)}$$

then both the series converge or diverge together.

REMARKS

- In the above form of the comparison test, the condition $\lim_{n\to\infty} \frac{u_n}{v_n}$ be finite and non-zero cannot be dropped for if $u_n = \frac{1}{n}$ and $v_n = \frac{1}{n^2}$ then $\lim \frac{u_n}{v_n} = +\infty$, Σu_n is divergent and Σv_n is convergent.
  In this case, neither the hypothesis nor the conclusion of the comparison test happens to be true.
- The comparison test is usually applied when the nth term u_n of the given series Σu_n contains the powers of n only which may be positive or negative, integral or fractional.
- v_n can be choosen such that $v_n = \frac{1}{n^{p-q}}$ where *p* and *q* are respectively the highest indices of *n* in the denominator and numerator of u_n when it is in the form of fraction, if u_n can be expanded in ascending powers of $\frac{1}{n}$ then to get v_n we should retain only the lowest power of $\frac{1}{n}$ the numerical factor being disregarded.
- We always denote the given series by Σu_n and the series which is used for comparison by Σv_n
- The series Σv_n is known as auxiliary series. We select the auxiliary series in such a way that $\lim_{n\to\infty}\left(\frac{u_n}{v_n}\right)$ exists finitely and non-zero.

Sixth form. Let Σu_n and Σv_n be two series of positive terms and let $\exists$ a positive integer *m* such that

$$\frac{u_n}{u_{n+1}} \ge \frac{v_n}{v_{n+1}}, \forall n \ge m \text{ then } \Sigma u_n \text{ and } \Sigma v_n \text{ both converge or diverge together.}$$

SOLVED EXAMPLES

Example 1. *If $u_n \ge 0$ for all n and Σu_n converges then show that $\sum \frac{\sqrt{u_n}}{n}$ also converges.*

Solution. Since $u_n \le 0$, then for all $n \in N$

$$\text{or} \quad \frac{u_n}{n^2} < u_n^2 + \frac{2u_n}{n^2} + \frac{1}{n^4}$$

$$\text{or} \quad 0 \le \frac{u_n}{n^2} < \left(u_n + \frac{1}{n^2}\right)^2$$

$$\text{or} \quad 0 \le \frac{\sqrt{u_n}}{n} < u_n + \frac{1}{n^2}$$

Since Σu_n and $\sum \frac{1}{n^2}$ are convergent, therefore $\sum\left(u_n + \frac{1}{n^2}\right)$ is convergent.

Hence, by comparison test $\sum \frac{\sqrt{u_n}}{n}$ is convergent.

Example 2. *If Σu_n is a sereis of positive terms and Σu_n is convergent then show that $\sum \frac{u_n}{1+u_n}$ is convergent.*

Solution. Since $u_n > 0$, then for all $n \in N$, $1 + u_n > 1$ for all $n \in N$

$$\Rightarrow \quad \frac{1}{1+u_n} < 1 \forall n \in N$$

$$\therefore \quad 0 < \frac{u_n}{1+u_n} < u_n \;\; \forall\, n \in N$$

$$\Rightarrow \quad 0 < \frac{u_n}{1+u_n} < \Sigma u_n$$

Since Σu_n is convergent, hence by comparison test $\sum \frac{u_n}{1+u_n}$ is convergent.

Example 3. *If Σu_n is a convergent series of positive terms, prove that $\sum u_n^2$ is also convergent. Is the converse true?*

Solution. Since Σu_n is convergent, then $\lim_{n\to\infty} u_n = 0$

$\therefore$ For given $\in\, > 0$ there exists a positive integer *m* such that

$$| u_n| < \varepsilon \forall\, n \ge m$$

$$\Rightarrow \quad -\varepsilon < u_n < \varepsilon \;\forall\, n \ge m$$

Since $u_n > 0 \; \forall \; n$, choose $\varepsilon < 1$, then

$$0 < u_n < 1 \; \forall \; n$$

$$\Rightarrow \quad 0 < u_n^2 < u_n \; \forall \; n$$

$$[\because u_n < 1 \Rightarrow u_n^2 < u_n]$$

$$\Rightarrow \quad 0 < \Sigma u_n^2 < \Sigma u_n \; \forall \; n$$

Since Σu_n is convergent, hence by comparison test $\sum u_n^2$ is convergent.

Converse is not always true :

For example if $u_n = \frac{1}{n'}$, then $u_n^2 = \frac{1}{n^2}$.

Clearly if $\sum u_n^2 = \Sigma \frac{1}{n^2}$ convergent but $\sum u_n = \Sigma \frac{1}{n}$ is divergent.

Example 4. *If $\sum u_n^2$ and $\sum v_n^2$ are both convergent series, prove that the series $\Sigma u_n v_n$ also convergent.*

Solution. Since $\sum u_n^2$ and $\sum v_n^2$ both are convergent, therefore $\Sigma(u_n^2 + v_n^2)$ is also convergent

$$\Rightarrow \quad \Sigma \frac{1}{2}(u_n^2 + v_n^2) \text{ is also convergent.}$$

We know that G. M. < A. M.

$$\Rightarrow \quad \sqrt{u_n^2 v_n^2} < \frac{1}{2}(u_n^2 + v_n^2)$$

$$\Rightarrow \quad u_n v_n < \frac{1}{2}(u_n^2 + v_n^2)$$

Hence, by comparison test $\Sigma u_n v_n$ is convergent.

Example 5. *Examine the following series for convergence :*

(i) $\Sigma \frac{1}{(\log n)^{\log n}}$ (ii) $\Sigma \frac{1}{(\log \log n)^{\log n}}$

Solution. (i) Since we have $\lim_{n \to \infty} \log(\log n) = \infty$

$\Rightarrow$ There exists a large positive integer n such that log (log n)>2

$$\Rightarrow \quad [\log(\log n)] \log n > 2 \log n$$

$$\Rightarrow \quad (\log n)[\log(\log n] > \log n^2$$

$$\Rightarrow \quad \log[(\log n)^{\log n}] > \log n^2$$

$$\Rightarrow \quad (\log n)^{\log n} > n^2$$

$$\Rightarrow \quad \frac{1}{(\log n)^{\log n}} < \frac{1}{n^2}$$

since $\Sigma \frac{1}{n^2}$ is convergent hence by comparison test $\Sigma(\log n)^{\log n}$ is also convergent.

(ii) Similarly, $\lim_{n \to \infty} \log(\log \log n) = \infty$

$\Rightarrow$there exists a positive integer n is so large such that log (log log n) > 2

$$\Rightarrow \quad \log n \, [\log(\log \log n)] > 2 \log n$$

$$\Rightarrow \quad \log [\{\log (\log n)\}^{\log n}] > \log n^2$$

$$\Rightarrow \quad [\log(\log n)]^{\log n} > n^2$$

$$\Rightarrow \quad \frac{1}{[\log(\log n)]^{\log n}} < \frac{1}{n^2}$$

Since $\Sigma \frac{1}{n^2}$ is convergent, hence by comparison test $\Sigma \frac{1}{[\log(\log n)]^{\log n}}$ is convergent.

Example 6. *Test the convergence of the series*

$$\frac{2}{1} + \frac{3}{4} + \frac{4}{9} + \ldots + \frac{n+1}{n^2} + \ldots$$

Solution. Here $u_n = \frac{n+1}{n^2}$.

Take $v_n = \frac{n}{n^2} = \frac{1}{n}$

Then $\frac{u_n}{v_n} = \frac{\frac{n+1}{n^2}}{\frac{1}{n}} = \frac{n+1}{n^2} \cdot \frac{n}{1} = \frac{n+1}{n}$

Therefore

$$\lim_{n \to \infty} \frac{u_n}{v_n} = \lim_{n \to \infty} \frac{n+1}{n} = \lim_{n \to \infty} \left(1 + \frac{1}{n}\right)$$

= 1, which is finite and non-zero.

Thus, by the comparison test two series are either both convergent or both divergent. But the auxiliary series $\sum v_n = \frac{1}{n}$ is divergent, Hence, the given series Σu_n is also divergent.

Example 7. *Test the convergence of the series*

$$\frac{1}{1 \cdot 2} + \frac{1}{2 \cdot 3} + \ldots + \frac{1}{n(n+1)} + \ldots \quad \text{(VTU–2006)}$$

Solution. Here $u_n = \frac{1}{n(n+1)} = \frac{1}{n} - \frac{1}{n+1}$

If s_n is the partial sum of n terms of the series Σu_n, then

$$s_n = u_1 + u_2 + \ldots + \ldots + u_n$$

$$= \left(1 - \frac{1}{2}\right) + \left(\frac{1}{2} - \frac{1}{3}\right) + \ldots + \left(\frac{1}{n} - \frac{1}{n+1}\right)$$

$$= 1 - \frac{1}{n+1}$$

Now, $\lim_{n \to \infty} s_n = \lim_{n \to \infty} \left[1 - \frac{1}{n+1}\right] = 1$, which is finite and non-zero.

Hence, the given series is convegent.

REMARK

- This type of series is calld "Telescoping series".

Example 8. *Show that the series* $1+\frac{1}{2!}+\frac{1}{3!}+...$ *is convergent.*

Solution. Since, we $\frac{1}{2!}=\frac{1}{2}$

$$\frac{1}{3!}<\frac{1}{2^2}$$

$$\ldots \quad \ldots \quad \ldots$$

$$\ldots \quad \ldots \quad \ldots$$

$$\frac{1}{n!}<\frac{1}{2^{n-1}}$$

Therefore,

$$1+\frac{1}{2!}+\frac{1}{3!}+...+\frac{1}{n!}+...<1+\frac{1}{2}+\frac{1}{2^2}+...$$

The series on the right hand side is a geometric series with common ratio $\frac{1}{2}$ and hence convergent. So the series on the left hand side will also be convergent.

Example 9. *Test the convergence or divergence of*

$$\frac{1}{1\cdot 2\cdot 3}+\frac{3}{2\cdot 3\cdot 4}+\frac{5}{3\cdot 4\cdot 5}+...+\frac{2n-1}{n(n+1)(n+2)}+...$$

(PTU–2009)

Solution. Here, $u_n=\frac{2n-1}{n(n+1)(n+1)}$

Take $v_n=\frac{n}{n(n)(n)}=\frac{1}{n^2}$

Then,

$$\frac{u_n}{v_n}=\frac{2n-1}{n(n+1)(n+2)}\cdot\frac{n^2}{1}=\frac{\left(2-\frac{1}{n}\right)}{\left(1+\frac{1}{n}\right)\cdot\left(1+\frac{2}{n}\right)}$$

$$\Rightarrow \lim_{n\to\infty}\frac{u_n}{v_n}=\lim_{n\to\infty}\frac{\left(2-\frac{1}{n}\right)}{\left(1+\frac{1}{n}\right)\left(1+\frac{2}{n}\right)}=2\text{, which}$$

is finite.

Now, the auxiliary series $\sum v_n=\sum\frac{1}{n^2}$ is convergent ($\because p=2>1$). Hence the given series is convergent.

Example 10. *Test the convergence or divergence of the series*

$$1+\frac{1}{2^2}+\frac{2^2}{3^3}+\frac{3^3}{4^4}+...$$

Solution. Leaving the first term, we get

$$u_n=\frac{n^n}{(n+1)^{n+1}}=\frac{1}{n\left(1+\frac{1}{n}\right)^{n+1}}$$

$$=\frac{1}{n}\left[1+\frac{1}{n}\right]^{-[n+1]}$$

$$=\frac{1}{n}\left[1-\frac{(n+1)}{n}+...\right]=\frac{1}{n}-\left(1+\frac{1}{n}\right)\frac{1}{n}+...$$

Let $\sum v_n=\sum\frac{1}{n}$, where $v_n=\frac{1}{n}$ be the auxiliary series.

$$\text{Then}\quad \lim_{n\to\infty}\frac{u_n}{v_n}=\lim_{n\to\infty}\frac{\frac{1}{n}\left[\frac{1}{(1+1/n)^{n+1}}\right]}{\frac{1}{n}}$$

$$=\lim_{n\to\infty}\left[\frac{\frac{1}{(1+1/n)^n}}{\left(1+\frac{1}{n}\right)}\right]$$

$$=\frac{1}{e}\text{, which is finite and non-zero.}$$

Now, since $\sum v_n=\sum\frac{1}{n}$ is divergent, therefore by comparsion test the given series is also divergent.

Example 11. *Test the convergence of the series whose general term is* $[n^3+1]^{1/3}-n]$. (PTU–2007, Rohtak–2003)

Solution. Here, we have

$$u_n=(n^3+1)^{1/3}-n$$

$$=n\left[\left(1+\frac{1}{n^3}\right)^{1/3}-1\right]$$

$$=n\left[\left(1+\frac{3}{3n^3}+\frac{\frac{1}{3}\left(\frac{1}{3}-1\right)}{2!}\cdot\frac{1}{n^6}+...\right)-1\right]$$

$$=\frac{1}{n^2}\left[\frac{1}{3}-\frac{1}{9n^3}+...\right]$$

Let $v_n=\frac{1}{n^2}$, then the auxiliary series

$$\sum v_n=\sum\frac{1}{n^2}$$

Now, $\lim\frac{u_n}{v_n}=\frac{1}{3}-\frac{1}{9n^3}+...=\frac{1}{3}$ which is finite and non-zero.

Since the series $\sum v_n=\sum\frac{1}{n^2}$ is convergent ($\because p=2>1$), therefore, the given series is also convergent.

Example 12. *Test the convergence of the series whose* n^{th} *term is* $[\sqrt{(n^4+1)}-\sqrt{(n^4-1)}]$.

Solution. Here, we have,

$$u_n=\sqrt{(n^4+1)}-\sqrt{(n^4-1)}$$

$$=n^2\left[\left(1+\frac{1}{n^4}\right)^{1/2}-\left(1-\frac{1}{n^4}\right)^{1/2}\right]$$

$$= n^2\left[\left(1+\frac{1}{2n^4}+\frac{\frac{1}{2}\left(\frac{1}{2}-1\right)}{2!}\cdot\frac{1}{n^8}-\frac{\frac{1}{2}\left(\frac{1}{2}-1\right)\left(\frac{1}{2}-2\right)}{3!}\cdot\frac{1}{n^{12}}+\ldots\right)\right.$$

$$\left.-\left(1-\frac{1}{2n^4}+\frac{\frac{1}{2}\left(\frac{1}{2}-1\right)}{2!}\cdot\frac{1}{n^8}-\frac{\frac{1}{2}\left(\frac{1}{2}-1\right)\left(\frac{1}{2}-2\right)}{3!}\cdot\frac{1}{n^{12}}+\ldots\right)\right]$$

$$= n^2\left[\frac{1}{n^4}+\frac{1}{8n^{12}}+\ldots\right]=\frac{1}{n^2}+\frac{1}{8n^{10}}+\ldots$$

Let $v_n=\frac{1}{n^2}$, then the auxillary series is

$\sum v_n=\sum\frac{1}{n^2}$, which is convergent.

Now $\lim\limits_{n\to\infty}\frac{u_n}{v_n}=\lim\limits_{n\to\infty}\left[\frac{1}{n^2}+\frac{1}{8n^{10}}+\ldots\right]\bigg/\frac{1}{n^2}$

$$=\lim_{n\to\infty}\left[1+\frac{1}{8n^8}+\ldots\right]$$

= 1, which is finite and non-zero.

Therefore, by comparsion test the given series is also divergent.

Example 13. *Test the convergence of the series* $\sum\sin\frac{1}{n}$.

Solution. Here, we have $u_n=\sin\frac{1}{n}$

Let $v_n=\frac{1}{n}$, therefore, the auxiliary series $\sum v_n=\sum\frac{1}{n}$ is divergent.

Now $\lim\limits_{n\to\infty}\frac{u_n}{v_n}=\lim\limits_{n\to\infty}\frac{\sin 1/n}{1/n}=1$, which is finite and non-zero.

Therefore, by comparison test the given series is also divergent.

Example 14. *Test the convergence of the series*

$$\frac{1}{a\cdot1^2+b}+\frac{2}{a\cdot2^2+b}+\frac{3}{a\cdot3^2+b}+\ldots$$

Solution. Here, we have

$$u_n=\frac{n}{a\cdot n^2+b}=\frac{n}{n^2\left[a+\frac{b}{n^2}\right]}=\frac{1}{n}\cdot\frac{1}{\left[a+\frac{b}{n^2}\right]}$$

Let $v_n=\frac{1}{n}$, then the auxiliary series

$\sum v_n=\sum\frac{1}{n}$ is divergent.

Now, $\lim\limits_{n\to\infty}\frac{u_n}{v_n}=\lim\frac{\left[\frac{1}{n}\cdot\frac{1}{(a+b/n^2)}\right]}{\frac{1}{n}}$

$$=\lim_{n\to\infty}\frac{1}{a+\frac{b}{n^2}}=\frac{1}{a},$$

which is finite and non-zero.

Hence, by comparsion test the given series is also divergent.

Example 15. *Test the convergence or divergence of the series*

$$\sum_{n=1}^{\infty}\frac{1}{x^n+x^{-n}}, x>0$$

Solution. **Case (I).** Let $x<1$. Take $v_n=x_n$, then

$$\lim_{n\to\infty}\frac{u_n}{v_n}=\lim_{n\to\infty}\frac{1}{x^n+x^{-n}}\cdot\frac{1}{x^n}=\lim_{n\to\infty}\frac{1}{x^{2n}+1}=1$$

Since, $x^{2n}\to0$ for $x<1$. But the auxiliary series $\Sigma v_n=\Sigma\, x^n$ is convergent (being a G.P. with common ratio $x<1$).

Hence, the given series Σu_n is convergent for $x<1$.

Case (II). Let $x>1$. In this case take $v_n=x^{-n}$. Then

$$\lim_{n\to\infty}\frac{u_n}{v_n}=\lim_{n\to\infty}\frac{1}{x^n+x^{-n}}\cdot\frac{x^n}{1}=\lim_{n\to\infty}\frac{1}{1+x^{-2n}}=1$$

Since, $x^{-2n}\to0$ for $x>1$. But the auxiliary series $\Sigma v_n=\Sigma\, x^{-n}$ is convergent (being a G. P. with common ration $x^{-1}<1$).

Case (III). Let $x=1$. In this case $u_n=\frac{1}{2}$ (for all n) and so

$$s_n=u_1+u_2+\ldots+u_n=\frac{n}{2}\to\infty \text{ as } n\to\infty$$

Therefore, the given series is divergent when $x=1$.

Example 16. *Test the convergence of the series whose* n^{th} *term is* $\sqrt{n^3+1}-\sqrt{n^3}$.

Solution. Here, we have

$$u_n=\sqrt{n^3+1}-\sqrt{n^3}=n^{3/2}\left[1+\frac{1}{n^3}\right]^{1/2}-n^{3/2}$$

$$=n^{3/2}\left[1+\frac{2}{2n^3}-\frac{1}{8n^6}+\ldots\right]-n^{3/2}$$

$$=\frac{1}{2n^{3/2}}-\frac{1}{8n^{9/2}}+\ldots$$

Let us take $v_n = \dfrac{1}{n^{3/2}}$ ($\because$ when u_n is in the form of the series in powers of $1/n$, v_n is taken as the term of lowest power of $1/n$, by ignoring the numerical factor).

Then we have

$$\lim_{n\to\infty}\frac{u_n}{v_n} = \lim_{n\to\infty}\left[\frac{1}{2n^{3/2}} - \frac{1}{8n^{9/2}} + \dots\right] \times \frac{n^{3/2}}{1}$$

$$= \lim_{n\to\infty}\left[\frac{1}{2} - \frac{1}{8n^3} + \dots\right]$$

$= \dfrac{1}{2}$, which is finite and non-zero.

But the auxillary series $\sum v_n = \sum \dfrac{1}{n^{3/2}}$ is convergent ($p = 3/2 > 1$). Hence, the given series is also convergent.

EXERCISE 26.1

Check the convergence of the following series:

1. $\sum u_n = 1 + \frac{1}{3} + \frac{1}{5} + \frac{1}{7} + \dots$
2. $\sum u_n = 1 + \frac{1}{\sqrt{2}} + \frac{1}{\sqrt{3}} + \frac{1}{\sqrt{4}} + \dots$
3. $\sum u_n = 1 + \frac{4}{5} + \frac{6}{10} + \frac{8}{17} + \dots + \frac{2n}{n^2+1} \dots$
4. $\sum u_n = \sqrt{\frac{1}{2^3}} + \sqrt{\frac{2}{3^3}} + \sqrt{\frac{3}{4^3}} + \dots$
5. $\sum u_n = \frac{1}{2} + \frac{\sqrt{2}}{5} + \frac{\sqrt{3}}{10} + \dots \frac{\sqrt{n}}{n^2+1} + \dots$
6. $\sum u_n = \frac{\sqrt{1}}{1+\sqrt{1}} + \frac{\sqrt{2}}{2+\sqrt{2}} + \frac{\sqrt{3}}{3+\sqrt{3}} + \dots$
7. $\sum u_n = \frac{1}{a+b} + \frac{1}{a+2b} + \frac{1}{a+3b} + \dots + \frac{1}{a+nb} + \dots$
8. $\sum u_n = \frac{1}{a(a+b)} + \frac{1}{(a+2b)(a+3b)} + \frac{1}{(a+4b)(a+5b)} + \dots$
9. $\sum u_n = \sum \frac{n}{n^2 + \sqrt{n}}$
10. $\sum u_n = \sum \frac{n}{(a+nb)^2}$
11. $\sum u_n = \sum \frac{\sqrt{n+1} + \sqrt{n-1}}{n}$
12. $\sum u_n = \sum \frac{1}{n} \sin \frac{1}{n}$
13. $\sum u_n = \sum \tan^{-1} \frac{1}{n}$
14. $\sum u_n = \sum \frac{n^p}{(n+1)^q}$
15. $\sum u_n = \sum \frac{n^2-1}{n^2+1}$
16. $\sum u_n = \sum \frac{1}{n}\sqrt{n^2+n+1} - \sqrt{n^2-n+1}$
17. $\frac{1}{4.7.10} + \frac{4}{7.10.13} + \frac{9}{10.13.16} + \dots + \infty$ (VTU–2010)
18. $\sum_{n=1}^{\infty} \frac{1}{\sqrt{n} + \sqrt{(n+1)}}$ (VTU–2008)
19. $\sum_{n=1}^{\infty} \sqrt{\frac{3^n - 1}{2^n + 1}}$ (VTU–2000S)
20. $1 - \frac{1}{3} + \frac{1}{3^2} - \frac{1}{3^3} + \frac{1}{3^4} - \dots \infty$ (JNTU–2000)
21. $\frac{1}{1.2} + \frac{2}{3.4} + \frac{3}{5.6} + \dots \infty$ (Cochin–2001)
22. $\frac{1}{1.3} + \frac{2}{3.5} + \frac{3}{5.7} + \dots \infty$ (PTU–2009)
23. $\sum_{n=1}^{\infty} [\sqrt{(n^2+1)} - n]$ (VTU–2010, PTU–2009)
24. $\frac{1}{1.3.5} + \frac{2}{3.5.7} + \frac{3}{5.7.9} + \dots \infty$ (VTU–2009S)
25. $\sum \frac{\sqrt{n}}{n^2+1}$ (Osmania–2000S)
26. $\sum \frac{(n+1)(n+2)}{n^2\sqrt{n}}$ (JNTU–2006S)
27. $\sum_{n=1}^{\infty} \frac{\sqrt{(n+1)} - 1}{(n+2)^3 - 1}$ (JNTU–2003)

Hint to Selected Problems

1. $\sum u_n = \sum_{n=1}^{\infty}\left(\frac{1}{2n-1}\right) \Rightarrow \lim_{n\to\infty} u_n = \lim_{n\to\infty}\left(\frac{1}{2n-1}\right) = 0$

Now apply comparison test .

4. $u_n = \dfrac{\sqrt{n}}{(n+1)\sqrt{n+1}}$

Then $\lim_{n\to\infty}\frac{u_n}{v_n} = \lim_{n\to\infty}\sqrt{\frac{1}{1+\frac{1}{n}}} \cdot \frac{1}{\left(1+\frac{1}{n}\right)} = n \neq 0$

7. $u_n = \dfrac{1}{a+nb}$ Let $v_n = \dfrac{1}{n}$.

Then $\lim_{n\to\infty}\frac{u_n}{v_n} = \lim_{n\to\infty}\frac{n}{a+nb} = \frac{1}{b} \neq 0.$

8. Here, we have $u_n = \dfrac{1}{[a+(2n-2)b][a+(2n-1)b]}$

Let $v_n = \dfrac{1}{n^2}$

Then $\lim_{n\to\infty}\frac{u_n}{v_n} = \frac{1}{4b^2} \neq 0$

12. $u_n = \frac{1}{n}\sin\frac{1}{n} = \frac{1}{n}\left[\frac{1}{n} - \frac{1}{6n^3} + \frac{1}{120n^5} - ....\right]$

Let $v_n = \frac{1}{n^2}$. Then $\lim_{n\to\infty} \frac{u_n}{v_n} = 1 \neq 0$

14. Here we have $u_n = \frac{n^p}{(n+1)^q}$

Let $v_n = \frac{n^p}{n^{q-p}}$

Then $\lim_{n\to\infty} \frac{v_n}{u_n} = \lim_{n\to\infty} n^{q-p}\left[\frac{n^p}{(n+1)^q}\right] = \lim_{n\to\infty} \frac{1}{\left(1+\frac{1}{n}\right)^q} = 1 \neq 0$

15. $\lim_{n\to\infty} u_n = \lim_{n\to\infty} \frac{\left(1-\frac{1}{n^2}\right)}{\left(1+\frac{1}{n^2}\right)} = 1 \neq 0$

ANSWERS

1. Divergent **2.** Divergent **3.** Divergent **4.** Divergent **5.** Convergent
6. Divergent **7.** Divergent **8.** Convergent **9.** Divergent **10.** Divergent
11. Divergent **12.** Convergent **13.** Divergent
14. Convergent if $p - q + 1 < 0$ and divergent if $p - q + 1 \geq 0$ **15.** Divergent **16.** Divergent **17.** Divergent
18. Divergent **19.** Divergent **20.** Convergent **21.** Divergent **22.** Divergent
23. Divergent **24.** Convergent **25.** Convergent **26.** Divergent **27.** Convergent

26.7 CAUCHY'S ROOT TEST

Let Σu_n be a series of positive terms and let $\lim_{n\to\infty} u_n^{1/n} = l$.
If,
(i) $l < 1$, *then Σu_n converges;* (ii) $l > 1$, *then Σu_n diverges;*
(iii) $l = 1$, *then the test fails and the series may either converge or diverge.*

Proof. Case (I). Let $u_n^{1/n} = l < 1$.
Since $l < 1$, we can choose an $\varepsilon > 0$ such that $l + \varepsilon < 1$.
Let $l + \varepsilon < r$ such that $0 < r < 1$.
Since $\lim_{n\to\infty} u_n^{1/n} = l$, therefore there exists a positive integer m_1 such that.

$$|u_n^{1/n} - l| < \varepsilon, \forall n > m_1 \quad \Rightarrow \quad l - \varepsilon < u_n^{1/n} < l + \varepsilon \,\forall\, n > m_1$$

$$\Rightarrow \quad (l - \varepsilon)^n < u_n < (l + \varepsilon)^n \,\forall\, n > m_1$$

Since $u_n < r^n \,\forall\, n > m_1$ and since Σr^n converges (being a geometric series with common ratio less than one). Then by comparison test Σu_n converges.

Case (II). Let $u_n^{1/n} = l > 1$.
Since $l > 1$, we can choose an $\varepsilon > 0$ such that $l - \varepsilon > 1$.
Let $l - \varepsilon < R$ then $R > 1$.
Since $R^n < u_n \,\forall\, n > m_2$ and since ΣR^n diverges (being a G.P. with common ratio greater than one). Then by comparison test Σu_n diverges.

Case (III). Let $u_n = \frac{1}{n}$ then $\lim_{n\to\infty} u_n^{1/n} = 1$.

Since $\Sigma\left(\frac{1}{n}\right)$ diverges, therefore we find that if $\lim_{n\to\infty} u_n^{1/n} = 1$, then the series Σu_n may diverge.

Again, let $u_n = \frac{1}{n^2}$. In this case also $\lim_{n\to\infty} u_n^{1/n} = 1$ but the series Σu_n converges. Thus we find that if $\lim_{n\to\infty} u_n^{1/n}$, then the series Σu_n may converge. The above two examples show that if

$$\lim_{n\to\infty} (u_n)^{1/n} = 1$$

Then the test fail.

REMARKS

- Cauchy's root test can be applied with advantage to series in which the nth term happens to be an exponential fraction of n.
- In this test, it is understood that $u_n^{1/n}$ stands for the positive n^{th} root of u_n.
- The Cauchy's root test can also be stated as follows :
- "A series Σu_n of positive terms is convergent if for every value of $n \geq m$, m being finite, $(u_n)^{1/n}$ less than a fixed number, which is less than unity, and the series is divergent if $(u_n)^{1/n} \geq 1$ for every value of $n \geq m$.

26.8 D'ALEMBERT RATIO TEST

Let Σu_n be a series of positive terms and let

(a) $\lim_{n\to\infty} \frac{u_n}{u_{n+1}} = l$

Then if,

(i) $l > 1$, *the series converges,* (ii) $l < 1$, *the series diverges*

(iii) $l = 1$, *the series may converge or diverge and therefore the test fails.*

(b) $\frac{u_n}{u_{n+1}} = \infty$ *as* $n \to \infty$. *Then* Σu_n *converges.*

Proof. (a) Case(I). When $l > 1$, Let $\varepsilon > 0$ be a positive number such that $l - \varepsilon > 1$.

Now since $\lim_{n\to\infty} \frac{u_n}{u_{n+1}} = l$, therefore $\exists$ a positive integer m such that $l - \varepsilon < \frac{u_n}{u_{n+1}} < l + \varepsilon$, whenever $n > m$

Now, putting $n = m + 1, m + 2, \ldots p - 1$, in succession in the above inequality, we get

$$l - \varepsilon < \frac{u_{m+1}}{u_{m+2}} < l + \varepsilon$$

$$l - \varepsilon < \frac{u_{m+2}}{u_{m+3}} < l + \varepsilon$$

$$\ldots \quad \ldots \quad \ldots \quad \ldots$$

$$l - \varepsilon < \frac{u_{p-1}}{u_p} < l + \varepsilon$$

Multiplying the corresponding sides of the first part of the above inequalities, we get

$$(l - \varepsilon)^{p-1-m} < \frac{u_{m+1}}{u_{m+2}} \cdot \frac{u_{m+2}}{u_{m+3}} \cdots \frac{u_{p-1}}{u_p} \quad \Rightarrow \quad (l - \varepsilon)^{p-1-m} < \frac{u_{m+1}}{u_p}$$

$$\Rightarrow \quad u_p < u_{m+1}(l - \varepsilon)^{m+1} \cdot (l - \varepsilon)^{-p} \quad \Rightarrow \quad u_p < k(l - \varepsilon)^{-p}, \forall p \geq m + 2 \text{ and } k = u_{m+1}(l - \varepsilon)^{m+1}$$

Since, the series $\Sigma(l - \varepsilon)^{-p}$ converges (being a geometric series with common ratio $(l - \varepsilon)^{-1}$, which is certainly less than unity), then by comparison test it follows that Σu_n converges,

Case (II). When $l < 1$, let $\varepsilon > 0$ be a positive number such that $l + \varepsilon < 1$.

Now since $\lim_{n\to\infty} \frac{u_n}{u_{n+1}} = l$, therefore, $\exists$ a positive intger m such that $l - \varepsilon < \frac{u_n}{u_{n+1}} < l + \varepsilon, \forall n > m$

Putting $n = m + 1, m + 2, \ldots, p - 1$ in succession in the second part of the above inequality, we get

$$\frac{u_{m+1}}{u_{m+2}} < l + \varepsilon, \frac{u_{m+2}}{u_{m+3}} < l + \varepsilon, \ldots \ldots \ldots \frac{u_{p-1}}{u_p} < l + \varepsilon.$$

Multiplying the corresponding sides of the above inequalities, we have

$$\frac{u_{m+1}}{u_p} < (l + \varepsilon)^{p-1-m} \quad \Rightarrow \quad u_p > u_{m+1}(l + \varepsilon)^{m+1}(l + \varepsilon)^{-p}$$

$$\Rightarrow \quad u_p > A(l + e)^{-p} \, \forall \, p \geq m + 2 \text{ and } A = u_{m+1}(l + \varepsilon)^{m+1}$$

Since, $\Sigma(l + \varepsilon)^{-p}$ is a divergent series (being a geometric series with common ratio $(l + \varepsilon)^{-1}$, which is certainly greater than unity), then by comparison test, it follows that Σu_n diverges.

Case (III). Let $l = 1$.

Now, first consider the harmonic series $1 + \frac{1}{2} + \frac{1}{3} + \frac{1}{5} + \ldots + \frac{1}{n} + \ldots$

Then $$\frac{u_n}{u_{n+1}} = \frac{n+1}{n} = 1 + \frac{1}{n} \Rightarrow \lim_{n\to\infty} \frac{u_n}{u_{n+1}} = 1$$

Since, the harmonic series is divergent, we find that if $l = 1$, a series may diverge.

Now, consider the series $\frac{1}{1^2}+\frac{1}{2^2}+...+\frac{1}{n^2}+...$

Then $$\frac{u_n}{u_{n+1}}=\frac{(n+1)^2}{n^2}=\left(1+\frac{1}{n}\right)^2 \Rightarrow \lim_{n\to\infty}\frac{u_n}{u_{n+1}}=1$$

since, the series $\sum\frac{1}{n^2}$ converges, we find that if $l=1$, a series may converge.

(b) Let us suppose $\lim_{n\to\infty}\frac{u_n}{u_{n+1}}=+\infty$ then there exists positive integers m and p such

$$\frac{u_n}{u_{n+1}}>p\,\forall n\geq m, p>1$$

Replacing n by m, $m+1$, $m+2$,..., $n-1$, we have

$$\frac{u_m}{u_{m+1}}>p,\frac{u_{m+1}}{u_{m+2}}>p \quad ... \quad ... \quad \frac{u_{n-1}}{u_n}>p.$$

Multiplying the corresponding sides of the above inequalities, we have

$$\frac{u_m}{u_n}>p^{n-m} \quad\Rightarrow\quad u_n<p^{m-n}\,.\,u_m,$$

$$\Rightarrow \quad u_n<A.p^{-n}\forall n>m \quad\text{and}\quad A=p^m u_m.$$

Since Σp^{-n} is convergent, then by comparison test, the series Σu_n is convergent.

REMARKS

- The ratio test is generally applied when the nth term of the series involves factorials, products of several factors, or combination of powers and factorials.
- The ratio test can also be stated as follows:
"An inifinite series of positive terms is convergent if from and after some terms the ratio of each term to the preceding term is less than a fixed number which is less than unity and series is divergent if the ratio, defined above is greater than or equal to unity,
- The ratio test is easier to apply than the root test. However, the root test is stronger than the ratio test.
- The ratio test does not tell us anything about the convergence of the series Σu_n if we only have
$$\frac{u_n}{u_{n+1}}>1\forall n.$$

SOLVED EXAMPLES

Example 1. *Test the convergence of the series* $x+2x^2+3x^3+4x^4+...$

Solution. Here, we have $u_n=nx^n$

$$\Rightarrow \quad (u_n)^{1/n}=n^{1/n}.x$$

$$\Rightarrow \quad \lim_{n\to\infty}(u_n)^{1/n}=\lim_{n\to\infty}(x.n^{1/n})=x.1=x$$

$$[\because \lim_{n\to\infty}n^{1/n}=1]$$

Then, by Cauchy's root test, Σu_n is convergent if $x<1$ and is divergent if $x>1$. For $x=1$, the Cauchy's root test fails.

In this case, the given series becomes

$$1+2+3+...$$

s_n = sum of n terms of the series $=\frac{1}{2}n(n+1)$, which is finite.

Thus the given series is convergent if $x<1$ and is divergent if $x\geq 1$.

Example 2. *Test the convergence of the series*

$$\frac{1}{2}+\left(\frac{2}{3}\right)x+\left(\frac{3}{4}\right)^2x^2+\left(\frac{4}{5}\right)^3x^3+...\infty,\ x>0.$$

(JNTU–2006)

Solution. Omitting the first term of the series (because it will not effect the convergence or divergence of the series),

we have $$u_n=\left(\frac{n+1}{n+2}\right)^n\cdot x^n$$

Therefore $$\lim_{n\to\infty}u_n^{1/n}=\lim_{n\to\infty}\left[\frac{\left(1+\frac{1}{n}\right)x}{1+\left(\frac{2}{n}\right)}\right]=x.$$

Therefore by Cauchy's root test, the given series Σu_n converges if $x<1$, divegent if $x>1$.

For $x=1$, test fails

$$\therefore \quad \lim_{n\to\infty}u_n=\lim_{n\to\infty}\frac{\left(1+\frac{1}{n}\right)^n}{\left(1+\frac{2}{n}\right)^n}=\frac{e}{e^2}=\frac{1}{e}>0.$$

$\therefore$ The series Σu_n diverges if $x = 1$.

Hence, the given series is convergent if $x < 1$ and divergent if $x \geq 1$.

Example 3. *Test the series for convergence of the series*

$$1+\frac{1}{2^2}+\frac{1}{3^3}+\frac{1}{4^4}+\ldots$$

Solution. Here, we have

$$u_n = \frac{1}{n^n}$$

$$\Rightarrow \quad \lim_{n\to\infty} (u_n)^{1/n} = \lim_{n\to\infty} \frac{1}{n} = 0 < 1$$

Hence by Cauchy's root test the given series is convergent.

Example 4. *Test the convergence of the series*

$$\left(\frac{2^2}{1^2}-\frac{2}{1}\right)^{-1}+\left(\frac{3^3}{2^3}-\frac{3}{2}\right)^{-2}+\left(\frac{4^4}{3^4}-\frac{4}{3}\right)^{-3}+\ldots$$

(VTU–2006)

Solution. Here we have $u_n = \left[\frac{(n+1)^{n+1}}{n^{n+1}}-\frac{(n+1)}{n}\right]^{-n}$

Therefore

$$\lim_{n\to\infty} u_n^{1/n} = \lim_{n\to\infty}\left[\frac{(n+1)^{n+1}}{n^{n+1}}-\frac{n+1}{n}\right]^{-1}$$

$$= \lim_{n\to\infty}\left[\left(1+\frac{1}{n}\right)^{n+1}-\left(1+\frac{1}{n}\right)\right]^{-1}$$

$$= \lim_{n\to\infty}\left(1+\frac{1}{n}\right)^{-1}\left[\left(1+\frac{1}{n}\right)^{n}-1\right]^{-1}$$

$$= (1+0)^{-1}[e-1]^{-1}$$

$$= \frac{1}{e-1} < 1.$$

Hence, by Cauchy's root test the given series is convergent.

Example 5. *Test the convergence of the series* $\Sigma\left(1+\frac{1}{n}\right)^{-n^2}$.

Solution. Here we have

$$u_n = \left(1+\frac{1}{n}\right)^{-n^2} \Rightarrow (u_n)^{1/n} = \left(1+\frac{1}{n}\right)^{-n}$$

$$\Rightarrow \quad \lim_{n\to\infty} (u_n)^{1/n} = \lim_{n\to\infty}\left(1+\frac{1}{n}\right)^{-n} = \frac{1}{e} < 1.$$

Hence, by Cauchy's root test the given series Σu_n is convergent.

Example 6. *Test the convergence of the series*

$$\Sigma\left[\frac{\log n}{\log(n+1)}\right]^{n^2 \log n}.$$

Solution. Here we have

$$u_n = \left[\frac{\log n}{\log(n+1)}\right]^{n^2 \log n}$$

$$\Rightarrow \quad u_n^{1/n} = \left[\frac{\log n}{\log(n+1)}\right]^{n \log n}$$

$$= \left[\frac{\log n\left(1+\frac{1}{n}\right)}{\log n}\right]^{-n\log n}$$

$$= \left[\frac{\log n+\log\left(1+\frac{1}{n}\right)}{\log n}\right]^{-n\log n}$$

$$= \left[\frac{\log n+\frac{1}{n}-\frac{1}{2n^2}+\ldots}{\log n}\right]^{-n\log n}$$

$$= \left[1+\frac{1}{n\log n}-\frac{1}{2n^2\log n}+\ldots\right]^{-n\log n} = k \text{ (say)}$$

Then

$$\log k = \log\left[1+\frac{1}{n\log n}-\frac{1}{2n^2\log n}+\ldots\right]^{-n\log n}$$

$$= (-n\log n)\log\left[\left(1+\frac{1}{n\log n}-\frac{1}{2n^2\log n}+\ldots\right)\ldots\right]$$

$$= -n\log n\left[\left(\frac{1}{n\log n}-\frac{1}{2n^2\log n}+\ldots\right)\ldots\right]$$

$$= -1+\frac{1}{2n}-\ldots$$

$$\Rightarrow \quad \lim_{n\to\infty} \log k = -1$$

$$\Rightarrow \quad \lim_{n\to\infty} k = e^{-1}$$

$$\Rightarrow \quad \lim_{n\to\infty} u_n^{1/n} = \frac{1}{e} < 1. \quad [\because 2 < e < 3]$$

Then, by Cauchy's root test the given series is convergent.

Example 7. *Test the convergence of* $\Sigma\left(\frac{n+1}{n+2}\right)^n x^n, (x > 0)$.

Solution. The nth term of the given series is

$$u_n = \left(\frac{n+1}{n+2}\right)^n x^n$$

Then we have

$$u_n^{1/n} = \left(\frac{n+1}{n+2}\right)x = \left(\frac{1+1/n}{1+2/n}\right)x$$

$$\therefore \quad \lim_{n\to\infty} u_n^{1/n} = \lim_{n\to\infty}\left(\frac{1+\frac{1}{n}}{1+\frac{2}{n}}\right)x = x$$

Now we have the followings cases.

Case I : If $x < 1$ then by Cauhy's root test Σu_n is convergent.

Case II : If $x > 1$ then by Cauchy's root test Σu_n is divergent.

Case III : If $x = 1$, the test fails.

Now, when $x = 1$, we have

$$u_n = \left(\frac{n+1}{n+2}\right)^n = \left(\frac{1+\frac{1}{n}}{1+\frac{2}{n}}\right)^n$$

$$\Rightarrow \quad \lim_{n\to\infty} u_n = \lim_{n\to\infty}\frac{(1+1/n)^n}{(1+2/n)^n} = \frac{e}{e^2} = \frac{1}{e} \neq 0$$

$\Rightarrow \quad \Sigma u_n$ is divergent.

Hence, the given series is convergent if $x < 1$ and divergent if $x \geq 1$.

Example 8. *Test the convergence of the series* $\sum_{n=1}^{\infty} 3^{-n-(-1)^n}$.

Solution. The nth term of the given series is

$$u_n = 3^{-n-(-1)^n}$$

$$= \begin{cases} 3^{-n+1}, & \text{if } n \text{ is odd} \\ 3^{-n-1}, & \text{if } n \text{ is even} \end{cases}$$

$$\therefore \quad \lim_{n\to\infty} u_n^{1/n} = \frac{1}{3}$$

$$\left(\because \lim_{n\to\infty} x^{1/n} = 1 \text{ if } x > 0\right)$$

$$\Rightarrow \quad \lim_{n\to\infty} u_n^{1/n} < 1$$

Hence, by Cauchy's root test the given is convergent.

Example 9. *Examine the convergene of the following series :*

(i) $\Sigma\left(1+\frac{1}{\sqrt{n}}\right)^{-n^{3/2}}$ (PTU–2009, Kurukshetra–2005)

(ii) $\Sigma\frac{(n-\log n)^n}{2^n \cdot n^n}$

(iii) $\Sigma\left(1+\frac{1}{n}\right)^{-n^2}$ (PTU–2010)

(iv) $\sum_{n=2}^{\infty}\frac{1}{(\log n)^n}$ (PTU–2005)

Solution. (i) We have

$$u_n = \left(1+\frac{1}{\sqrt{n}}\right)^{-n^{3/2}}$$

$$\therefore \quad \lim_{n\to\infty} u_n^{1/n} = \lim_{n\to\infty}\left(1+\frac{1}{\sqrt{n}}\right)^{-\sqrt{n}}$$

$$= \lim_{n\to\infty}\left[\left(1+\frac{1}{\sqrt{n}}\right)^{\sqrt{n}}\right]^{-1}$$

$$= e^{-1} = \frac{1}{e} < 1$$

Hence, the given series is convergent.

(ii) We have

$$u_n = \frac{(n-\log n)^n}{2^n \cdot n^n}$$

$$\therefore \quad u_n^{1/n} = \frac{n-\log n}{2n} = \frac{1}{2}\left(1-\frac{\log n}{n}\right)$$

$$\therefore \quad \lim_{n\to\infty} u_n^{1/n} = \lim_{n\to\infty}\frac{1}{2}\left(1-\frac{\log n}{n}\right)$$

$$= \frac{1}{2}(1-0) = \frac{1}{2} < 1$$

$$\left[\because \lim_{n\to\infty}\frac{\log n}{n} = 0\right]$$

Hence, by Cauchy's root test the given series is convergent.

(iii) We have

$$u_n = \left(1+\frac{1}{n}\right)^{-n^2}$$

$$\therefore \quad u_n^{1/n} = \left(1+\frac{1}{n}\right)^{-n}$$

$$\therefore \quad \lim_{n\to\infty} u_n^{1/n} = \lim_{n\to\infty}\left(1+\frac{1}{n}\right)^{-n}$$

$$= \lim_{n\to\infty}\left[\left(1+\frac{1}{n}\right)^n\right]^{-1}$$

$$= e^{-1} = \frac{1}{e} < 1$$

Hence, by Cauchy's root test the given series is convergent.

(iv) We have

$$u_n = \frac{1}{(\log n)^n}$$

$$\therefore \quad u_n^{1/n} = \frac{1}{(\log n)}$$

$$\therefore \quad \lim_{n\to\infty} u_n^{1/n} = \lim_{n\to\infty}\frac{1}{(\log n)} = 0 < 1$$

Hence, by Cauchy's root test the given series is convergent.

Example 10. *Test for convergence the series*

$$1+\frac{2^p}{2!}+\frac{3^p}{3!}+\frac{4^p}{4!}+\ldots$$ (Kurukshetra–2005)

Solution. Here, we have

$$u_n = \frac{n^p}{n!} \Rightarrow u_{n+1} = \frac{(n+1)^p}{(n+1)!}$$

Now

$$\lim_{n\to\infty} \frac{u_{n+1}}{u_n} = \lim_{n\to\infty} \frac{(n+1)^p}{(n+1)!} \frac{n!}{n^p}$$

$$= \lim_{n\to\infty} \left[1+\frac{1}{n}\right]^p \cdot \frac{1}{(n+1)}$$

$$= e^p \times 0 = 0 < 1$$

Hence, by ratio test, the given series is divergent.

Example 11. *Test the series* $x + \frac{x^3}{3!} + \frac{x^5}{5!} + \frac{x^7}{7!} + \ldots$ *for convergence, for all positive value of x.*

Solution. Since x is positive. Hence the given series is of positive term series.

Here

$$u_n = \frac{x^{2n-1}}{(2n-1)!}, u_{n+1} = \frac{x^{2n+1}}{(2n+1)!}$$

$$\Rightarrow \lim_{n\to\infty} \frac{u_n}{u_{n+1}} = \lim_{n\to\infty} \frac{x^{2n-1}}{(2n-1)!} \frac{(2n+1)!}{x^{2n+1}}$$

$$= \lim_{n\to\infty} \frac{2n(2n+1)}{x^2}$$

$$= +\infty, \forall \text{ positive value of } x.$$

Then, by ratio test the given series converges for all positive value of x.

Example 12. *Test for convergence the series*

$$1 + \frac{x}{2^2} + \frac{x^2}{3^2} + \frac{x^3}{4^2} + \ldots$$

Solution. Here we have

$$u_n = \frac{x^{n-1}}{n^2}$$

$$\Rightarrow u_{n+1} = \frac{x^n}{(n+1)^2}$$

Now

$$\frac{u_n}{u_{n+1}} = \frac{x^{n-1}(n+1)^2}{n^2.x^n} = \frac{1}{x}\left(1+\frac{1}{n}\right)^2$$

$$\Rightarrow \lim_{n\to\infty} \frac{u_n}{u_{n+1}} = \lim_{n\to\infty} \frac{1}{x}\left(1+\frac{1}{n}\right)^2 = \frac{1}{x}.$$

Hence, by ratio test the series converges if $\frac{1}{x} > 1$ *i.e.*, $x < 1$, diverges if $x > 1$ and the test fails if $x = 1$.

For $x = 1$, $u_n = \frac{1}{n^2}$. Therefore in the case the series $\Sigma u_n = \Sigma \frac{1}{n^2}$ is convergent.

Example 13. *Test for convergence of the series*

$$\frac{1}{2\sqrt{1}} + \frac{x^2}{3\sqrt{2}} + \frac{x^4}{4\sqrt{3}} + \ldots$$

Solution.

$$u_n = \frac{x^{2n-2}}{(n+1)\sqrt{n}}, u_{n+1} = \frac{x^{2n}}{(n+2)\sqrt{(n+1)}}$$

$$\Rightarrow \frac{u_n}{u_{n+1}} = \frac{x^{2n-2}}{(n+1)\sqrt{n}} \cdot \frac{(n+2)\sqrt{(n+1)}}{x^{2n}}$$

$$= \frac{(1+2/n)}{(1+1/n)} \sqrt{\left(1+\frac{1}{n}\right)} \cdot \frac{1}{x^2}$$

$$\Rightarrow \lim_{n\to\infty} \frac{u_n}{u_{n+1}} = \frac{1}{1}.\sqrt{1}.\frac{1}{x^2} = \frac{1}{x^2}$$

Therefore, by ratio test the given series Σu_n is

(i) convergent if $\frac{1}{x^2} > 1$ *i.e.*, if $x^2 < 1$.

(ii) divergent if $\frac{1}{x^2} < 1$ *i.e.*, if $x^2 > 1$.

and (iii) The test fails if $x^2 = 1$

When $x^2 = 1$, we have $u_n = \frac{1}{(n+1)\sqrt{n}}$. Take $v_n = \frac{1}{n\sqrt{n}}$. Then $\lim_{n\to\infty} \frac{u_n}{v_n} = 1$, which is finite and non-zero. Hence, by comparison test Σu_n and Σv_n are either both convergent or both divergent.

Since $\Sigma v_n = \Sigma \frac{1}{n^{3/2}}$ is convergent as $p = 3/2 > 1$. Hence, the given series Σu_n is also convergent if $x^2 = 1$.

Example 14. *Test for convergence the series*

$$x + \frac{3}{5}x^2 + \frac{8}{10}x^3 + \frac{15}{17}x^4 + \ldots + \frac{n^2-1}{n^2+1}x^n + \ldots.$$

Solution. Here, we have

$$u_n = \frac{n^2-1}{n^2+1}x^n, u_{n+1} = \frac{(n+1)^2-1}{(n+1)^2+1}x^{n+1}$$

$$\Rightarrow \frac{u_n}{u_{n+1}} = \frac{n^2-1}{n^2+1}x^n \cdot \frac{(n+1)^2+1}{(n+1)^2-1} \cdot \frac{1}{x^{n+1}}$$

$$= \frac{1-1/n^2}{1+1/n^2} \cdot \frac{1+2/n+2/n^2}{1+2/n} \cdot \frac{1}{x}$$

$$\Rightarrow \lim_{n\to\infty} \frac{u_n}{u_{n+1}} = \frac{1}{x}$$

Therefore, by ratio test the given series Σu_n is

(i) convergent if $\frac{1}{x} > 1$ *i.e.*, if $x < 1$.

(ii) divergent if $\frac{1}{x} < 1$ *i.e.*, if $x > 1$.

and (iii) test fails if $x = 1$.

When $x = 1$, $u_n = \frac{1-1/n^2}{1+1/n^2} \Rightarrow \lim_{n\to\infty} u_n = 1 > 0$.

The given series Σu_n is divergent if $x = 1$.

Hence, the given series is convergent if $x < 1$ and divergent if $x \geq 1$.

Example 15. *Test the convergence of the series*

$$\sum_{n=1}^{\infty} \frac{1\cdot 3\cdot 5 \ldots\ldots (2n-1)}{2\cdot 4\cdot 6 \ldots\ldots (2n)} (1-x^2)^n, \quad 0 \le x^2 < 1$$

Solution. The nth term of the series is

$$u_n = \frac{1\cdot 3\cdot 5 \ldots\ldots (2n-1)}{2\cdot 4\cdot 6 \ldots\ldots (2n)} (1-x^2)^n$$

Then

$$u_{n+1} = \frac{1\cdot 3\cdot 5 \ldots (2n-1)\cdot (2n+1)}{2\cdot 4\cdot 6 \ldots (2n)(2n+2)} (1-x^2)^{n+1}$$

$$\therefore \quad \frac{u_n}{u_{n+1}} = \frac{2n+2}{2n+1}\cdot\frac{1}{(1-x^2)}$$

or

$$\frac{u_n}{u_{n+1}} = \left(\frac{1+\frac{1}{n}}{1+\frac{1}{2n}}\right)\cdot\frac{1}{(1-x^2)}$$

$$\therefore \quad \lim_{n\to\infty}\frac{u_n}{u_{n+1}} = \lim_{n\to\infty}\frac{\left(1+\frac{1}{n}\right)}{\left(1+\frac{1}{2n}\right)}\cdot\frac{1}{(1-x^2)}$$

$$= \frac{1}{1-x^2}$$

$$\therefore \quad \lim_{n\to\infty}\frac{u_n}{u_{n+1}} = \frac{1}{1-x^2} > 1 \quad \because x^2 < 1$$

Hence, by D'Alembert Ratio test the given series is convergent.

Example 16. *Test the following series for convergence*

(*i*) $1+\frac{2}{5}x+\frac{6}{9}x^2+\frac{14}{17}x^3+\ldots+\frac{2^n-2}{2^n+1}x^{n-2}+\ldots.$

(PTU–2009, VTU–2004)

(*ii*) $\frac{x^2}{2\sqrt{1}}+\frac{x^3}{3\sqrt{2}}+\frac{x^4}{4\sqrt{3}}+\ldots.$

(PTU–2005, VTU–2003, ISM–2001)

Solution. (i) Here, we have

$$u_n = \frac{2^n-2}{2^n+1}x^{n-2}$$

$$\Rightarrow \quad u_{n+1} = \frac{2^{n+1}-2}{2^{n+1}+1}x^{n-1}$$

$$\therefore \quad \frac{u_n}{u_{n+1}} = \frac{2^{n+1}+1}{2^n+1}\cdot\frac{2^n-2}{2^{n+1}-2}\cdot\frac{1}{x}$$

$$= \frac{\left(2+\left(\frac{1}{2}\right)^n\right)}{1+\left(\frac{1}{2}\right)^n}\cdot\frac{\left(1-\frac{2}{2^n}\right)}{\left(2-\frac{2}{2^n}\right)}\cdot\frac{1}{x}$$

so,

$$\lim_{n\to\infty}\frac{u_n}{u_{n+1}} = \lim_{n\to\infty}\left[\frac{\left(2+\left(\frac{1}{2}\right)^n\right)}{\left(1+\left(\frac{1}{2}\right)^n\right)}\cdot\frac{\left(1-\frac{2}{2^n}\right)}{\left(2-\frac{2}{2^n}\right)}\cdot\frac{1}{x}\right]$$

$$= \frac{2+0}{1+0}\cdot\frac{1-0}{2-0}\cdot\frac{1}{x} = \frac{1}{x}$$

Therefore, by D'Alembert's ratio test the given series is convergent if $x < 1$ and is divergent if $x > 1$.

When $x = 1$, we have

$$u_n = \frac{2^n-2}{2^n+1} = \frac{1-\left(\frac{1}{2}\right)^{n-1}}{1+\left(\frac{1}{2}\right)^n}$$

$$\therefore \quad \lim_{n\to\infty} u_n = \lim_{n\to\infty}\frac{1-\left(\frac{1}{2}\right)^{n-1}}{1+\left(\frac{1}{2}\right)^n} = \frac{1-0}{1+0} = 1 \ne 0$$

$\Rightarrow \quad \Sigma u_n$ is divergent.

Hence, the given series is convergent if $x < 1$ and is divergent if $x \ge 1$.

(ii) The nth term of the given series is

$$u_n = \frac{x^{n+1}}{(n+1)\sqrt{n}}$$

$$\Rightarrow \quad u_{n+1} = \frac{x^{n+2}}{(n+2)\sqrt{n+1}}$$

$$\therefore \quad \frac{u_n}{u_{n+1}} = \frac{n+2}{n+1}\cdot\sqrt{\frac{n+1}{n}}\cdot\frac{1}{x}$$

$$= \left(\frac{1+\frac{2}{n}}{1+\frac{1}{n}}\right)\cdot\sqrt{1+\frac{1}{n}}\cdot\frac{1}{x}$$

$$\therefore \quad \lim_{n\to\infty}\frac{u_n}{u_{n+1}} = \lim_{n\to\infty}\left(\frac{1+\frac{2}{n}}{1+\frac{1}{n}}\right)\cdot\sqrt{1+\frac{1}{n}}\cdot\frac{1}{x}$$

$$= \frac{1}{x}$$

Therefore, by D'Alembert ratio test the given series is convergent if $x < 1$ and is divergent if $x > 1$.

When $x = 1$, we have $u_n = \frac{1}{(n+1)\sqrt{n}}$

Then $\quad v_n = \frac{1}{n\sqrt{n}}$

$$\therefore \quad \lim_{n\to\infty}\frac{u_n}{v_n} = \lim_{n\to\infty}\frac{n\sqrt{n}}{(n+1)\sqrt{n}}$$

$$= \lim_{n\to\infty}\frac{1}{\left(1+\frac{1}{n}\right)} = \frac{1}{1+0} = 1 \ne 0$$

$\Rightarrow$ By comparison test, $\Sigma v_n = \Sigma\frac{1}{n^{3/2}}$ is convergent, then Σu_n is convergent.

Hence, the given series is convergent if $x \le 1$ and is divergent if $x > 1$.

EXERCISE 26.2

Based on Cauchy's Root Test :

1. Test the convergence of the following series :

(i) $\sum_{n=1}^{\infty}\left(1+\frac{2}{n}\right)^{-n^2}$ (ii) $\sum_{n=1}^{\infty}\frac{n^{n^2}}{(n+1)^{n^2}}$

(iii) $\sum_{n=1}^{\infty}2^{-n-(-1)^n}$ (iv) $\sum_{n=1}^{\infty}5^{-n-(-1)^n}$

(v) $\sum_{n=1}^{\infty}(n^{1/n}+x)$ for all positive values of x

(vi) $\sum_{n=1}^{\infty}\frac{n^3}{3^n}$ (vii) $\sum_{n=1}^{\infty}\frac{x^n}{n^n},\ x>0$

2. Test the convergence of the following series :

(i) $\sum\left(\frac{n}{n+1}\right)^{n^2}$ (ii) $\sum n^n x^n,\ x>0$

(iii) $\sum\left(\frac{n+1}{3n}\right)^n$ (iv) $\sum\left(\frac{nx}{n+1}\right)^n$

(v) $\sum\frac{(1+nx)^n}{n^n}$ (vi) $\sum(n^{1/n}-1)^n$

3. Test the convergence of the following series :

(i) $\frac{1^3}{3}+\frac{2^3}{3^2}+\frac{3^3}{3^3}+\frac{4^3}{3^4}+....$

(ii) $\frac{2}{1^2}x+\frac{3^2}{2^3}x^2+\frac{4^3}{3^4}x^3+...+\frac{(n+1)^n x^n}{n^{n+1}}+....$ if $x>0$

(iii) $\sum q^{n^2}r^2,\ q,r>0$ (iv) $\sum_{n=2}^{\infty}\frac{1}{[\log(\log n)]^n}$

Based on D'Alembert's Ratio test

4. Test the convergence of the following series :

(i) $\sum_{n=1}^{\infty}\frac{2^{n-1}}{3^n+1}$ (ii) $\sum_{n=1}^{\infty}\frac{n!}{n^n}$

(iii) $\sum_{n=1}^{\infty}\frac{x^n}{n!},x>0$ (iv) $\sum_{n=1}^{\infty}\frac{x^n}{n^n},x>0$

(v) $\sum_{n=1}^{\infty}\frac{2^n n!}{n^n}$ (vi) $\sum_{n=1}^{\infty}\frac{n^n x^n}{n!}$

(vii) $\sum_{n=1}^{\infty}\frac{5^n}{n^2+5}$ (viii) $\sum_{n=1}^{\infty}\frac{n^3+a}{2^n+a}$

(ix) $\sum_{n=1}^{\infty}\frac{\sqrt{n}}{\sqrt{n^2+1}}x^n,\ x>0$ (PTU–2006)

(x) $\sum_{n=1}^{\infty}\sqrt{\frac{n-1}{n^3+1}}x^n,\ x>0$

5. Test the convergence of the series with n^{th} term :

(i) $\frac{1}{x^n+x^{-n}}$ (ii) $\left[\sqrt{n^2+1}-n\right]x^{2n}$

(iii) $\frac{1}{2^n+x},\ x\ge 0$ (iv) $\frac{x^n}{n^2+1}$

(v) $\frac{a^n}{x^n+a^n}$ (vi) $\sqrt{\frac{2^n-1}{3^n-1}}$

6. Test the convergence of the following series :

(i) $\frac{2!}{3}+\frac{3!}{3^2}+\frac{4!}{3^3}+...+\frac{(n+1)!}{3^n}+....$

(ii) $\frac{1^2.2^2}{1!}+\frac{2^2.3^2}{2!}+\frac{3^2.4^2}{3!}+...$

(iii) $\frac{1}{1+2}+\frac{2}{1+2^2}+\frac{3}{1+2^3}+...$

(iv) $1+3x+5x^2+7x^3+...$ (v) $1+\frac{x}{2^2}+\frac{x^2}{3^2}+\frac{x^3}{4^2}+...$

(vi) $2x+\frac{3x^2}{8}+\frac{4x^3}{27}+...+\frac{(n+1)x^n}{n^3}+...$

(vii) $\frac{1}{\sqrt{1}+\sqrt{2}}+\frac{1}{\sqrt{2}+\sqrt{3}}+\frac{1}{\sqrt{3}+\sqrt{4}}+...$

(viii) $\frac{\sqrt{2}-1}{3^3-1}+\frac{\sqrt{3}-1}{4^3-1}+\frac{\sqrt{4}-1}{5^3-1}+...$

(ix) $\frac{1}{2}+\frac{2!}{8}+\frac{3!}{32}+\frac{4!}{128}+...$

(ix) $1+\frac{1}{2.2^{1/100}}+\frac{1}{3.3^{1/100}}+\frac{1}{4.4^{1/100}}+...$

7. Test for convergence the following series :

(i) $\frac{1}{2\cdot3}+\frac{1}{3\cdot4}+\frac{1}{4\cdot5}+\frac{1}{5\cdot6}+...$

(ii) $\frac{1}{1\cdot2\cdot3}+\frac{3}{2\cdot3\cdot4}+\frac{5}{3\cdot4\cdot5}+...$

(iii) $\frac{1\cdot2}{3^2.4^2}+\frac{3\cdot4}{5^2.6^2}+\frac{5\cdot6}{7^2.8^2}+...$

(iv) $\frac{1}{3}+\frac{1\cdot2}{3\cdot5}+\frac{1\cdot2\cdot3}{3\cdot5\cdot7}+\frac{1\cdot2\cdot3\cdot4}{3\cdot5\cdot7\cdot9}+...$

8. Test the series :

$$1+\frac{x^2}{2}+\frac{x^4}{4}+\frac{x^6}{6}+...$$

for convergence for all positive values of x.

9. Test for convergence the series :

$$\frac{x}{1\cdot2}+\frac{x^2}{2\cdot3}+\frac{x^3}{3\cdot4}+\frac{x^4}{4\cdot5}+...,x>0$$

10. Show that the series ($\alpha>0,\ \beta>0$)

$$1+\frac{\alpha+1}{\beta+1}+\frac{(\alpha+1)(2\alpha+1)}{(\beta+1)(2\beta+1)}+\frac{(\alpha+1)(2\alpha+1)(3\alpha+1)}{(\beta+1)(2\beta+1)(3\beta+1)}+...$$

converges if $\beta>\alpha>0$

and diverges if $\alpha\ge\beta>0$

11. Test for convergence the series :

$$\frac{x}{1\cdot3}+\frac{x^2}{2\cdot4}+\frac{x^3}{3\cdot5}+\frac{x^4}{4\cdot6}+...$$

12. Test for convergence the following series :

(i) $1+\frac{x}{2}+\frac{x^2}{3^2}+\frac{x^3}{4^3}+...,x>0$

(ii) $x+2x^2+3x^3+4x^4+...$

(iii) $2+\frac{3}{2}x+\frac{4}{3}x^2+\frac{5}{4}x^3+..., x>0$

(iv) $\frac{(1+a)(1+b)}{1\cdot2\cdot3}+\frac{(2+a)(2+b)}{2\cdot3\cdot4}+\frac{(3+a)(3+b)}{3\cdot4\cdot5}+...$

(v) $x\log x+x^2\log 2x+x^3\log 3x+...+x^n\log nx+...$

(vi) $\sum_{n=1}^{\infty}\frac{n!}{(n^n)^2}$ (PTU–2010)

(vii) $1+\frac{2!}{2^2}+\frac{3!}{3^3}+\frac{4!}{4^4}+...\infty$ (VTU–2008S)

(viii) $\sum_{n=1}^{\infty}\frac{n!3^n}{n^n}$ (Kerala–2005)

(ix) $\frac{2}{3.4}+\frac{2.4}{3.5.6}+\frac{2.4.6}{3.5.7.8}+...$ (VTU–2010)

(x) $\sum_{n=2}^{\infty}\frac{x^n}{n(n-1)(n-2)}$ (JNTU–2006)

(xi) $\sum_{n=1}^{\infty}\left(\frac{n^2}{2^n}+\frac{1}{n^2}\right)$ (Rohtak–2005)

(xii) $\sum_{1}^{\infty}\frac{n^3-n+1}{n!}$ (Madras–2000)

(xiii) $1+\frac{1^2.2^2}{1.3.5}+\frac{1^2.2^2.3^3}{1.3.5.7.9}+...\infty$ (Delhi–2002)

(xiv) $\frac{4}{18}+\frac{4.12}{18.27}+\frac{4.12.20}{18.27.36}+...\infty$ (Madras–2000, 12)

(xv) $\frac{1}{1^P}+\frac{x}{3^P}+\frac{x^2}{5^P}+...\frac{x^{n-1}}{(2n-1)^P}+...\infty$ (JNTU–2006)

(xvi) $\sum_{n=1}^{\infty}\frac{3.6.9...3n}{4.7.10...(3n+1)}\cdot\frac{5^n}{3n+2}$ (VTU–2004)

(xvii) $\frac{3}{4}x+\left(\frac{4}{5}\right)^2x^2+\left(\frac{5}{6}\right)^3x^3+...+\infty\ (x>0)$ (VTU–2007)

13. Test for convergence the series with n^{th} term :

(i) $\frac{n^3-1}{n^3+1}x^n, x>0$ (ii) $\frac{x^n}{a+\sqrt{n}}$

(iii) $\frac{x^n}{x+n}$ (iv) $\frac{3n+1}{4n+3}x^n, x>0$

(v) $\frac{x^n}{(2n+1)^P}$

(vi) $\frac{3^n-2}{3^n+1}x^{n-1}, x>0$ (vii) $\frac{1}{n}\sin\frac{1}{n}$

ANSWERS

1. (i) Convergent (ii) Convergent (iii) Convergent (iv) Convergent (v) Divergent (vi) Convergent (vii) Convergent
2. (i) Convergent (ii) Divergent (iii) Convergent (iv) Convergent if $x<1$, divergent if $x\geq1$
(v) Convergent if $x<1$, divergent if $x\geq1$ (vi) Convergent **3.** (i) Convergent (ii) Convergent if $x<1$ and divergent if $x\geq1$
(iii) Convergent if $0<q<1$ and divergent if $q>1$, Convergent if $0<r<1$, when $q=1$, divergent if $q>1$ or $q=1, r\geq1$
(iv) Convergent
4. (i) Convergent (ii) Convergent (iii) Convergent (iv) Convergent (v) Convergent
(vi) Convergent if $x<1$, divergent if $x\geq1$ (vii) Divergent (viii) Convergent
(ix) Convergent if $x<1$, divergent if $x\geq1$ (x) Convergent if $x<1$, divergent if $x\geq1$
5. (i) Convergent if $x>1$ or $x<1$, and divergent if $x=1$ (ii) Convergent if $x<1$, divergent if $x\geq1$
(iii) Convergent (iv) Convergent if $x\leq1$, divergent if $x>1$ (v) Convergent if $x>a$, divergent if $x\leq a$, (vi) Convergent.
6. (i) Divergent (ii) Convergent (iii) Convergent (iv) Convergent if $x<1$, divergent if $x\geq1$
(v) Convergent if $x\leq1$, divergent if $x>1$ (vi) Convergent if $x\leq1$, divergent if $x>1$
(vii) Divergent (viii) Convergent (ix) Divergent (x) Convergent
7. (i) Convergent (ii) Convergent (iii) Convergent (iv) Convergent
8. Convergent if $x<1$, divergent if $x\geq1$ **9.** Convergent if $x\leq1$, divergent if $x>1$ **11.** Convergent if $x\leq1$, divergent $x>1$
12. (i) Convergent (ii) Convergent if $x<1$, divergent if $x\geq1$ (iii) Convergent if $x<1$, divergent if $x\geq1$ (iv) Divergent
(v) Convergent if $x<1$, divergent if $x\geq1$ (vi) Convergent (vii) Convergent (viii) Convergent (ix) Convergent
(x) Convergent for $x\geq1$, divergent for $x<1$ (xi) Convergent (xii) Convergent (xiii) Divergent (xiv) Convergent
(xv) Convergent $x<1$, divergent for $x>1$; Covergent for $P>1$ and divergent for $P\leq1$
(xvi) Divergent (xvii) Convergent
13. (i) Convergent if $x<1$, divergent if $x\geq1$ (ii) Convergent if $x<1$, divergent if $x\geq1$
(iii) Convergent if $x<1$, divergent if $x\geq1$ (iv) Convergent if $x<1$, divergent if $x\geq1$
(v) Convergent if $x<1$, divergent if $x>1$, when $x=1$, then convergent if $p>1$ and divergent if $p\leq1$
(vi) Convergent if $x<1$, divergent if $x\geq1$ (vii) Convergent.

26.9 RAABE'S TEST

If Σu_n be a series of positive terms such that $\lim_{n\to\infty}\left\{n\left(\frac{u_n}{u_{n+1}}-1\right)\right\}=l.$

Then, if

(i) $l > 1$, *the series converges,* (ii) $l < 1$, *the series diverges,*

(iii) $l = 1$, *the series may either converge or diverge and therefore the test fails.*

Proof. Case (I) When $l > 1$. We can write $l = 1 + r$, where $r > 0$. Choosing $\varepsilon = r/2$, we can find a positive integer m such that

$$l - \varepsilon < n\left(\frac{u_n}{u_{n+1}} - 1\right) < l + \varepsilon \ \forall\ n \geq m$$

Now, from the first part of the above inequality, we have

$$(1+r) - \frac{1}{2}r < n\left(\frac{u_n}{u_{n+1}} - 1\right) \forall\ n \geq m \quad \Rightarrow \quad \frac{1}{2}ru_{n+1} < nu_n - (n+1)u_{n+1} \ \forall n \geq m \qquad ...(1)$$

Putting $n = m+1, m+2, ..., p-1$ in succession in (1), we have

$$\frac{1}{2}ru_{m+2} < (m+1)u_{m+1} - (m+2)u_{m+2}$$

$$\ldots \quad \ldots \quad \ldots \quad \ldots \quad \ldots$$

$$\frac{1}{2}ru_p < (p-1)u_{p-1} - pu_p.$$

Now, adding the corresponding sides of the above inqualities, we have

$$\frac{1}{2}r[u_{m+2} + u_{m+3} + \ldots + u_p] < (m+1)u_{m+1} - pu_p, \quad \Rightarrow \quad \frac{1}{2}r[u_{m+2} + \ldots + u_p] < (m+1)u_{m+1},$$

or $$u_1 + u_2 + \ldots + u_p < \frac{2(m+1)}{r}u_{m+1} < u_1 + u_2 + \ldots + u_{m+1}, \ \forall\ p \geq m+2.$$

The above inequality shows that the sequence $\langle s_n \rangle$ of the partial sums of the series Σu_n is bounded and therefore Σu_n converges.

Case (II) When $l < 1$. Let us choose $\varepsilon = 1 - l$, then we can find a positive integer m such that

$$l - \varepsilon < n\left(\frac{u_n}{u_{n+1}} - 1\right) < 1 (= l + \varepsilon) \forall\ n \geq m \quad \text{or} \quad mu_n < (n+1)\ u_{n+1} \ \forall\ n \geq m$$

Putting $n = m+1, m+2, ..., p-1$ ($p \geq m+2$), in succession, we get

$$(m+1)u_{m+1} < (m+2)u_{m+2},$$

$$(m+1)u_{m+1} < (m+3)u_{m+3},$$

$$\ldots \quad \ldots \quad \ldots \quad \ldots \quad \ldots$$

$$(p-1)\ u_{p-1} < pu_p.$$

From the above inequality, we have by transitivity

$$(m+1)u_{m+1} < pu_p \ \forall\ p \geq m+2 \quad \text{or} \quad u_p > k(1/p) \ \forall\ p \geq m+2 \text{ and } k = (m+1)u_{m+1}.$$

Now, since the series $\Sigma\left(\frac{1}{p}\right)$ diverges, then by comparison test the given series diverges.

Case (III) When $l = 1$. In this case the test fails to give any definite information. For example, consider the series $\Sigma\frac{1}{n}$ and $\Sigma\frac{1}{n(\log n)^2}$ then, we have

$$\lim_{n\to\infty} n\left[\frac{u_n}{u_{n+1}} - 1\right] = 1.$$

But the former sereis is divergent, while the latter is convergent.

REMARKS

- Raabe's test is to be applied when D'Alembert's ratio test fails.
- Raabe's test is stronger than D'Alembert ratio test.
- It can be shown as in the proof of case (I) above that if $\lim_{n\to\infty}\left\{n\left[\frac{u_n}{u_{n+1}} - 1\right]\right\} = +\infty.$, Then, Σu_n converges.
- The case in which $\lim_{n\to\infty}\left\{n\left[\frac{u_n}{u_{n+1}} - 1\right]\right\} = -\infty.$ the given series Σu_n diverges.

26.10 LOGARITHMIC TEST

If Σu_n be a series of positive terms such that $\lim_{n\to\infty}\left(n\log\frac{u_n}{u_{n+1}}\right) = l.$

then Σu_n converges if $l > 1$ and diverges when $l < 1$.

Proof. Case (I) When $l > 1$. In this case, we can choose $\varepsilon > 0$ such that $l - \varepsilon > 1$. Let $l - \varepsilon = p$ (say).

Since $\lim_{n\to\infty}\left(n\log\frac{u_n}{u_{n+1}}\right) = l.$ Therefore, we can find a positive integer m such that

$$l - \varepsilon < n\log\frac{u_n}{u_{n+1}} < l + \varepsilon \ \forall\ n \ge m.$$

Consider the first part of the above inequality, we have

$$n\log\frac{u_n}{u_{n+1}} > p \quad \forall\ n \ge m. \quad \Rightarrow \quad \frac{u_n}{u_{n+1}} > e^{p/n} \quad \forall\ n \ge m. \quad \text{...(1)}$$

Since, $a_n = \left(1+\frac{1}{n}\right)^n$ defines a monotonically increasing sequence converging to e, therefore,

$$e \ge \left(1+\frac{1}{n}\right)^n \quad \forall\ n. \quad \text{...(2)}$$

From (1) and (2), we have

$$\frac{u_n}{u_{n+1}} > \left(1+\frac{1}{n}\right)^p \quad \forall\ n \ge m. \quad \Rightarrow \quad \frac{u_n}{u_{n+1}} > \frac{v_n}{v_{n+1}} \quad \forall\ n \ge m. \quad \text{...(3)}$$

where $\qquad v_n = \frac{1}{n^p}.$

Now since $p > 1$, therefore Σv_n converges and from (3) it then follows by comparison test that Σu_n converges.

Case (II) When $l < 1$. Let the comparison series $\Sigma v_n = \Sigma\frac{1}{n^p}$ be divergent, *i.e.*, $p < 1$.

$\therefore$ $\quad \Sigma u_n$ will be divergent if $\frac{v_n}{v_{n+1}} > \frac{u_n}{u_{n+1}} \quad \Rightarrow \quad \frac{u_n}{u_{n+1}} < \left(1+\frac{1}{n}\right)^p \Rightarrow \log\left(\frac{u_n}{u_{n+1}}\right) < p\log\left(1+\frac{1}{n}\right)$

$$= p\left[\frac{1}{n} - \frac{1}{2n^2} + \frac{1}{3n^3} + \ldots\right]$$

$$\therefore \quad n\log\left(\frac{u_n}{u_{n+1}}\right) = p\left[1 - \frac{1}{2n} + \frac{1}{3n^2} + \ldots\right]$$

$$\therefore \quad \lim_{n\to\infty}\left[n\log\frac{u_n}{u_{n+1}}\right] = p < 1$$

$\therefore$ Σu_n will be divergent if $l < 1$.

REMARKS

- Logarithmic test is to be applied only when :
  (a) ratio test fails
  (b) the ratio test involves the exponent 'e'
- This test is an alternative to Raabe's test.

26.11 SOME MODIFIED FORMS

Various test of convergence, involving limits can be modified in terms of the upper and lower limits. For example, a few modification are given below :

(i) Cauchy's Root test.

The series of non-negative term Σu_n converges or diverges according as

$$\underline{\lim}\, u_n^{1/n} < 1 \quad \text{or} \quad \overline{\lim}\, u_n^{1/n} > 1.$$

(ii) D' Alembert's Ratio test.

The series Σu_n of positive terms converges or diverges according as

$$\underline{\lim}\frac{u_n}{u_{n+1}} > 1 \quad \text{or} \quad \overline{\lim}\frac{u_n}{u_{n+1}} < 1.$$

(iii) Raabe's test.

The series Σu_n of positive terms converges or diverges according as

$$\underline{\lim}\left\{n\left(\frac{u_n}{u_{n+1}}-1\right)\right\}>1 \quad \text{or} \quad \overline{\lim}\left\{n\log\frac{u_n}{u_{n+1}}\right\}<1.$$

(iv) Logarithmic test.

The series Σu_n of positive terms converges or diverges according as

$$\underline{\lim}\left\{n\log\frac{u_n}{u_{n+1}}\right\}>1 \quad \text{or} \quad \overline{\lim}\left\{n\log\frac{u_n}{u_{n+1}}\right\}<1.$$

Some Important Limits

- $\lim_{n\to\infty}\left(1+\frac{x}{n}\right)^n = e^x$
- $\lim_{n\to\infty} n^{1/n} = 1$
- $\lim_{n\to\infty}\frac{\log n}{n} = 0$
- $\lim_{n\to\infty}\left(1+\frac{x}{n}\right)^p = 1$, if p is finite.
- $\lim_{n\to\infty}\left(1+\frac{x}{n}\right)^{n+p} = e^x$, if p is finite.

26.11.1 SOME OTHER IMPORTANT TESTS

(1) De Morgan's and Bertrand's test : The series Σu_n of positive terms is convergent or divergent according as

$$\lim\left[n\left(\frac{u_n}{u_{n+1}}-1\right)-1\right]\log n>1 \quad \text{or} \quad <1.$$

(2) Alternative to Bertrand's test : The series Σu_n of positive terms is convergent or divergent according as

$$\lim\left[\left(n\log\frac{u_n}{u_{n+1}}-1\right)\log n\right]>1 \quad \text{or} \quad <1.$$

Solved Examples

Example 1. *Test the convergence of the series*

$$1+\frac{3}{7}x+\frac{3\cdot 6}{7\cdot 10}x^2+\frac{3\cdot 6\cdot 9}{7\cdot 10\cdot 13}x^3+\ldots$$

Solution. After leaving the first term we have

$$u_n = \frac{3\cdot 6\cdot 9\cdot\ldots\cdot 3n}{7\cdot 10\cdot 13\cdot\ldots\cdot(3n+4)}x^n$$

$$\Rightarrow \quad u_{n+1} = \frac{3\cdot 6\cdot 9\cdot\ldots\cdot 3n(3n+3)}{7\cdot 10\cdot 13\cdot\ldots\cdot(3n+4)(3n+7)}x^{n+1}$$

Now $$\lim_{n\to\infty}\frac{u_{n+1}}{u_n} = \lim_{n\to\infty}\left(\frac{3n+3}{3n+7}\right)x$$

$$= \lim_{n\to\infty}\left(\frac{3+3/n}{3+7/n}\right)x = x$$

Then, by D'Alembert ratio test the series is convergent if $x < 1$, divergent if $x > 1$ and the test fails if $x = 1$.

For $x = 1$, we have $\frac{u_n}{u_{n+1}} = \frac{3n+7}{3n+3}$

or $$n\left(\frac{u_n}{u_{n+1}}-1\right) = n\left(\frac{3n+7}{3n+3}-1\right) = \frac{4n}{3n+3}$$

$$\Rightarrow \lim_{n\to\infty} n\left[\left(\frac{u_n}{u_{n+1}}-1\right)\right] = \lim_{n\to\infty}\frac{4n}{3n+3}$$

$$= \lim_{n\to\infty}\frac{4}{3+3/n}$$

$$= \frac{4}{3} > 1$$

Therefore, by Raabe's test the series is convergent when $x = 1$.

Hence, the given series is convergent when $x \le 1$ and divergent when $x > 1$.

Example 2. *Test the convergence of the following series*

$$\sum_{n=1}^{\infty}\frac{1.3.5....(2n-1)}{2.4.6....(2n)}\cdot\frac{x^{2n}}{2n}, \ (x > 0).$$

Solution. Here, we have

$$u_n = \frac{1.3.5....(2n-1)}{2.4.6....(2n)}\cdot\frac{x^{2n}}{2n}$$

and $$u_{n+1} = \frac{1.3.5....(2n-1)(2n+1)}{2.4.6....(2n)(2n+2)}\cdot\frac{x^{2n+2}}{(2n+2)}$$

$$\Rightarrow \quad \lim_{n\to\infty}\frac{u_n}{u_{n+1}} = \lim_{n\to\infty}\left(\frac{2n+2}{2n+1}\cdot\frac{2n+2}{2n}\cdot\frac{1}{x^2}\right)$$

$$= \frac{1}{x^2}.$$

$\therefore$ By D'Alembert's ratio test, the series is convergent if $x^2 < 1$ and divergent if $x^2 > 1$.

Now since $x > 0$ this gives that the series is convergent if $x < 1$ and divergent if $x > 1$.

If $x = 1$. Then D'Alembert's ratio test fails.

Now consider

$$\lim_{n\to\infty} n\left[\frac{u_n}{u_{n+1}}-1\right] = \lim_{n\to\infty} n\left(\frac{2n+2}{2n+1}\cdot\frac{2n+2}{2n}-1\right)$$

$$= \lim_{n\to\infty}\frac{n(6n+4)}{2n(2n+1)} = \frac{3}{2} > 1.$$

Then by Raabe's test, the series is convergent for $x = 1$.

Hence, the series is convergent if $x \le 1$ and divergent if $x > 1$.

Example 3. *Test the convergence of the series*

$$\frac{a}{b}+\frac{(1+a)}{(1+b)}+\frac{(1+a)(2+a)}{(1+b)(2+b)}+\ldots$$

Solution. Here, we have

$$u_n=\frac{(1+a)(2+a)\ldots(n-1+a)}{(1+b)(2+b)\ldots(n-1+b)}$$

$$\Rightarrow \quad u_{n+1}=\frac{(1+a)(2+a)\ldots(n+a)}{(1+b)(2+b)\ldots(n+b)}$$

$$\therefore \lim_{n\to\infty}\frac{u_n}{u_{n+1}}=\lim_{n\to\infty}\left[\frac{n+b}{n+a}\right]=\lim_{n\to\infty}\left[\frac{1+\frac{b}{n}}{1+\frac{a}{n}}\right]=1.$$

Hence, the D'Alembert's ratio test fails.

Now, consider

$$\lim_{n\to\infty} n\left[\frac{u_n}{u_{n+1}}-1\right]=\lim_{n\to\infty} n\left[\frac{n+b}{n+a}-1\right]$$

$$=\lim_{n\to\infty} n\left[\frac{b-a}{n+b}\right]=\lim_{n\to\infty}\left[\frac{b-a}{1+b/n}\right]$$

$$=(b-a).$$

Then by Raabe's test the given series is convergent if $b-a>1$, *i.e.*, $b>a+1$ and divergent if $b<a+1$.

The test fails for $b=a+1$.

Now, for $b=a+1$, the given series becomes

$$\frac{a}{a+1}+\frac{1+a}{2+a}+\ldots=\Sigma\frac{1+a}{n+a}.$$

Taking $v_n=\frac{1}{n}$, by comparison test, we can easily shown that the series is divergent.

Hence, the given series is convergent if $b>a+1$ and divergent if $b\leq a+1$.

Example 4. *Test the convergence of the series*

$$1+a+\frac{a(a+1)}{1\cdot2}+\frac{a(a+1)(a+2)}{1\cdot2\cdot3}+\ldots$$

Solution. On leaving the first term we have

$$u_n=\frac{a(a+1)(a+2)\ldots(a+n-1)}{1.2\ldots.n}$$

$$\Rightarrow \quad u_{n+1}=\frac{a(a+1)\ldots(a+n)}{1.2\ldots.n(n+1)}$$

$$\therefore \quad \lim_{n\to\infty}\frac{u_n}{u_{n+1}}=\lim_{n\to\infty}\frac{(n+1)}{(a+n)}=\lim_{n\to\infty}\frac{1+\frac{1}{n}}{1+\frac{a}{n}}=1.$$

$\Rightarrow$ The D' Alembert's ratio test fails.

Now $$\lim_{n\to\infty} n\left[\frac{u_n}{u_{n+1}}-1\right]=\lim_{n\to\infty} n\left[\frac{n+1}{a+n}-1\right]$$

$$=\lim_{n\to\infty} n\left[\frac{1-a}{a+n}\right]=\lim_{n\to\infty}\frac{(1-a)}{(1+a/n)}=(1-a).$$

Hence, by Raabe's test the given series is convergent if $1-a>1$, *i.e.*, $a<0$ and divergent if $a>0$ and test fails if $a=0$.

In case $a=0$, the given series becomes $1+0+0+\ldots$

The sum of n terms is always 1. Therefore, the series is convergent if $a=0$. Thus the given series Σu_n is convergent if $a\leq 0$ and divergent if $a>0$.

Example 5. *Test the convergence of the series*

$$\Sigma\frac{n!x^n}{3.5.7\ldots(2n+1)}.$$

Solution. Here, we have

$$u_n=\frac{n!x^n}{3.5.7\ldots(2n+1)}$$

$$\Rightarrow \quad u_{n+1}=\frac{(n+1)!x^{n+1}}{3.5.7\ldots(2n+1)(2n+3)}$$

Now $$\lim_{n\to\infty}\frac{u_n}{u_{n+1}}=\lim_{n\to\infty}\left(\frac{2n+3}{n+1}\right).\frac{1}{x}$$

$$=\lim_{n\to\infty}\left(\frac{2+\frac{3}{n}}{1+\frac{1}{n}}\right).\frac{1}{x}=\frac{2}{x}.$$

Hence, by D'Alembert's ratio test the series is convergent if $2/x>1$, *i.e.*, if $x<2$ and diverges if $2/x<1$, *i.e.*, if $x>2$ and test fails when $2/x=1$, *i.e.*, when $x=2$.

In case $x=2$, apply Raabe's test.

When $x=2$, $$\frac{u_n}{u_{n+1}}=\frac{(2n+3)}{2(n+1)}$$

$$\therefore \quad n\left(\frac{u_n}{u_{n+1}}-1\right)=n\left(\frac{2n+3}{2n+2}-1\right)$$

$$=\frac{n}{2(n+1)}=\frac{1}{2(1+1/n)}$$

$$\therefore \lim_{n\to\infty} n\left(\frac{u_n}{u_{n+1}}-1\right)=\lim_{n\to\infty}\frac{1}{2(1+1/n)}=\frac{1}{2}<1.$$

Hence, by Raabe's test Σu_n is divergent if $x=2$.

Thus, the given series Σu_n is convergent if $x<2$ and divergent if $x\geq 2$.

Example 6. *Test the convergence of the series*

$$1+\frac{1}{2}x+\frac{2!}{3^2}x^2+\frac{3!}{4^3}x^3+\ldots$$

Solution. Here, we have

$$u_n=\frac{(n-1)!}{n^{n-1}}x^{n-1} \quad \Rightarrow \quad u_{n+1}=\frac{n!}{(n+1)^n}x^n$$

$$\therefore \quad \lim_{n\to\infty}\frac{u_n}{u_{n+1}}=\lim_{n\to\infty}\frac{(n+1)^n(n-1)!x^{n-1}}{n!x^n.\,n^{n-1}}$$

$$= \lim_{n\to\infty}\left[1+\frac{1}{n}\right]^n \cdot\frac{1}{x} = \frac{e}{x}.$$

Hence, the given series is convergent if $\frac{e}{x} > 1$, i.e., if $x < e$ and divergent if $x > e$ and the test fails if $x = e$. In this case

$$\lim_{n\to\infty}\left[n\log\frac{u_n}{u_{n+1}}\right] = \lim_{n\to\infty}\left[n\log\frac{\left(1+\frac{1}{n}\right)^n}{e}\right]$$

$$= \lim_{n\to\infty}\left[n^2\left(\frac{1}{n}-\frac{1}{2n^2}+\frac{1}{3n^3}+\ldots\right)-n\right]$$

$$= \lim_{n\to\infty}\left[-\frac{1}{2}+\frac{1}{3n}-\ldots\right] = -\frac{1}{2} < 1.$$

Hence, by log test the series Σu_n is divergent if $x = e$. Thus the given series Σu_n is convergent if $x < e$ and divergent if $x \ge e$.

Example 7. *Test the convergence of the series*

$$x+\frac{2^2x^2}{2!}+\frac{3^3x^3}{3!}+\frac{4^4x^4}{4!}+\ldots$$

(PTU–2008, Cochin–2005, Rohtak–2003)

Solution. Here, we have

$$u_n = \frac{n^n x^n}{n!} \Rightarrow u_{n+1} = \frac{(n+1)^{n+1}.x^{n+1}}{(n+1)!}$$

Therefore, $\lim_{n\to\infty}\frac{u_n}{u_{n+1}} = \lim_{n\to\infty}\frac{(n+1)!n^n x^n}{(n+1)^{n+1}x^{n+1}.n!}$

$$= \lim_{n\to\infty}\frac{1}{\left(1+\frac{1}{n}\right)^n x} = \frac{1}{ex}.$$

Thus, by D'Alembert's ratio test the series is convergent if $ex < 1$ i.e., $x < \frac{1}{e}$, divergent if $x > \frac{1}{e}$ and the test fails if $\frac{1}{ex} = 1$, i.e., $x = \frac{1}{e}$.

In this case

$$\lim_{n\to\infty} n\left[\log\frac{u_n}{u_{n+1}}\right] = \lim_{n\to\infty} n\log\left[\frac{e}{\left(1+\frac{1}{n}\right)^n}\right]$$

$$= \lim_{n\to\infty} n\left[\log e - n\log\left(1+\frac{1}{n}\right)\right]$$

$$= \lim_{n\to\infty} n\left[1-n\left(\frac{1}{n}-\frac{1}{2n^2}+\frac{1}{3n^2}-\ldots\right)\right]$$

$$= \lim_{n\to\infty}\left[\frac{1}{2}-\frac{1}{3n}+\ldots\right] = \frac{1}{2} < 1.$$

Hence, by Logarithmic test, the series is divergent if $x = \frac{1}{e}$.

Thus the given series Σu_n is convergent if $x < \frac{1}{e}$ and divergent if $x \ge \frac{1}{e}$.

Example 8. *Test the convergence of the series*

$$1+\frac{2x}{2!}+\frac{3^2x^2}{3!}+\frac{4^3x^3}{4!}+\ldots$$

Solution. Here, we have

$$u_n = \frac{n^{n-1}x^{n-1}}{n!} \Rightarrow u_{n+1} = \frac{(n+1)^n x^n}{(n+1)!}$$

Now $\frac{u_n}{u_{n+1}} = \frac{(n+1)!n^{n-1}x^{n-1}}{(n+1)^n x^n.n!} = \frac{\left(1+\frac{1}{n}\right)}{\left(1+\frac{1}{n}\right)^n}.\frac{1}{x}$

$\therefore \lim_{n\to\infty}\frac{u_n}{u_{n+1}} = \frac{1}{ex}.$

Hence, by D'Alembert's ratio test the series is convergent if $\frac{1}{ex} > 1$, i.e., $x < \frac{1}{e}$, divergent if $x > \frac{1}{e}$ and the test fails if $x = \frac{1}{e}$.

In this case

$$\lim_{n\to\infty}\left[n\log\frac{u_n}{u_{n+1}}\right] = \lim_{n\to\infty} n\left[\log\frac{\left(1+\frac{1}{n}\right)e}{\left(1+\frac{1}{n}\right)^n}\right]$$

$$= \lim_{n\to\infty} n\left[\log\left(1+\frac{1}{n}\right)+\log e - n\log\left(1+\frac{1}{n}\right)\right]$$

$$= \lim_{n\to\infty} n\left[\left(\frac{1}{n}-\frac{1}{2n^2}+\frac{1}{3n^3}-\ldots\right)+1 - n\left(\frac{1}{n}-\frac{2}{2n^2}+\frac{1}{3n^3}-\ldots\right)\right]$$

$$= \lim_{n\to\infty} n\left[\frac{3}{2}-\frac{5}{6n}+\ldots\right] = \frac{3}{2} > 1.$$

Thus, by Logarithmic test, the series is divergent if $x = \frac{1}{e}$.

Thus the given series Σu_n is convergent if $x \le \frac{1}{e}$ and divergent if $x > \frac{1}{e}$.

Example 9. *Test the convergence of the series*

$$\frac{(a+x)}{1!}+\frac{(a+2x)^2}{2!}+\frac{(a+3x)^3}{3!}+\ldots$$

Solution. Here, we have

$$u_n = \frac{(a+nx)^n}{n!} \Rightarrow u_{n+1} = \frac{[a+(n+1)x]^{n+1}}{(n+1)!}$$

$$\Rightarrow \quad \frac{u_n}{u_{n+1}} = \frac{\left[1+\frac{a/x}{n}\right]^n}{\left[1+\frac{1}{n}\right]^n\left[1+\frac{a/x}{n+1}\right]^{n+1}}\cdot\frac{1}{x}$$

$$\Rightarrow \lim_{n\to\infty}\frac{u_n}{u_{n+1}}$$

$$= \lim_{n\to\infty}\left[\frac{\left[1+\frac{a/x}{n}\right]^n}{\left[1+\frac{1}{n}\right]^n\left[1+\frac{a/x}{n+1}\right]^{n+1}}\cdot\frac{1}{x}\right]$$

$$= \frac{e^{a/x}}{x.e.e^{a/x}} = \frac{1}{ex}.$$

Hence, by D'Alembert's ratio test the series is convergent if $\frac{1}{ex} > 1$, *i.e.*, $x < \frac{1}{e}$, divergent if $x > \frac{1}{e}$ and the test fails if $x = \frac{1}{e}$.

In this case

$$\lim_{n\to\infty} n\log\left(\frac{u_n}{u_{n+1}}\right)$$

$$= \lim_{n\to\infty} n\log\left[\frac{\left[1+\frac{ae}{n}\right]^n}{\left(1+\frac{1}{n}\right)^n\left(1+\frac{ae}{n+1}\right)^{n+1}}\right]$$

$$= \lim_{n\to\infty} n\left[n\log\left(1+\frac{ae}{n}\right)+\log e - n\log\left(1+\frac{1}{n}\right)-(n+1)\log\left(1+\frac{ae}{n+1}\right)\right]$$

$$= \lim_{n\to\infty} n\left[n\left(\frac{ae}{n}-\frac{a^2e^2}{2n^2}+\frac{a^3e^3}{3n^3}\ldots\right)+1 - \left(\frac{1}{n}-\frac{1}{2n^2}+\frac{1}{3n^3}\ldots\right) - (n+1)\left(\frac{ae}{n+1}-\frac{a^2e^2}{2(n+1)^2}+\frac{a^3e^3}{3(n+1)^3}\ldots\right)\right]$$

$$= \lim_{n\to\infty}\left[-\frac{a^2e^2}{2}+\frac{1}{2}+\frac{a^2e^2}{2\left(1+\frac{1}{n}\right)} + \text{terms containing } n \text{ in the denominator}\right]$$

$$= -\frac{a^2e^2}{2}+\frac{1}{2}+\frac{a^2e^2}{2} = \frac{1}{2} < 1.$$

Thus, by Logarithmic test, the series is divergent.

Hence, the given series Σu_n is convergent if $x < \frac{1}{e}$ and divergent if $x \geq \frac{1}{e}$.

Example 10. *Test the convergence of the series*

$$1^p + \left(\frac{1}{2}\right)^p + \left(\frac{1\cdot 3}{2\cdot 4}\right)^p + \left(\frac{1\cdot 3\cdot 5}{2\cdot 4\cdot 6}\right)^p + \ldots$$

Solution. Leaving the first term 1^p, we have

$$u_n = \left[\frac{1\cdot 3\cdot 5\ldots(2n-1)}{2\cdot 4\cdot 6\ldots(2n)}\right]$$

$$\Rightarrow \quad u_{n+1} = \left[\frac{1\cdot 3\cdot 5\ldots(2n-1)(2n+1)}{2\cdot 4\cdot 6\ldots(2n)(2n+2)}\right]^p$$

Now $\frac{u_n}{u_{n+1}} = \left[\frac{(2n+2)}{(2n+1)}\right]^p = \left(\frac{1+\frac{1}{n}}{1+\frac{1}{2n}}\right)^p$

$$\Rightarrow \lim_{n\to\infty}\frac{u_n}{u_{n+1}} = \left(\frac{1}{1}\right)^p \Rightarrow \text{ratio test fails.}$$

Now, applying logarithmic test, we have

$$\log\frac{u_n}{u_{n+1}} = \log\left(\frac{2n+2}{2n+1}\right)^p = \log\left(\frac{1+1/n}{1+1/2n}\right)^p$$

$$= p\left[\log\left(1+\frac{1}{n}\right)-\log\left(1+\frac{1}{2n}\right)\right]$$

$$= p\left[\left(\frac{1}{n}-\frac{1}{2n^2}+\frac{1}{3n^3}-\ldots\right) - \left(\frac{1}{2n}-\frac{1}{2.2^2n^2}+\frac{1}{3.2^3.n^3}-\ldots\right)\right]$$

$$= p\left[\left\{1-\frac{1}{2}\right\}\frac{1}{n}-\frac{1}{2}.\left\{1-\frac{1}{4}\right\}\frac{1}{n^2} + \frac{1}{3}\left\{1-\frac{1}{8}\right\}\frac{1}{n^3}-\ldots\right]$$

$$= p\left[\frac{1}{2n}-\frac{3}{8n^2}+\frac{7}{24n^3}-\ldots\right]$$

$$\therefore \quad n\log\frac{u_n}{u_{n+1}} = p\left[\frac{1}{2}-\frac{3}{8n}+\frac{7}{24n^3}-\ldots\right]$$

Therefore $\lim_{n\to\infty}\left[n\log\frac{u_n}{u_{n+1}}\right] = \frac{p}{2}.$

So that, the series is convergent if $p/2 > 1$, *i.e.*, if $p > 2$, and divergent if $p < 2$ and the test fails if $p = 2$.

EXERCISE 26.3

Test the convergence of the following series

1. $1+\frac{2}{3}\left(\frac{1}{4}\right)+\frac{2.4}{3.5}\left(\frac{1}{6}\right)+\frac{2.4.6}{3.5.7}\left(\frac{1}{8}\right)+\ldots$

2. $\frac{1^2}{4^2}+\frac{1^2.5^2}{4^2.8^2}+\frac{1^2.5^2.9^2}{4^2.8^2.12^2}+\frac{1^2.5^2.9^2.13^2}{4^2.8^2.12^2.16^2}+\ldots$

3. $1+\frac{1}{2}x+\frac{1.3}{2.4}x^2+\frac{1.3.5}{2.4.6}x^3,\ldots,(x>0)$ (Raipur–2005)

4. $x^2+\frac{2^2}{3.4}x^4+\frac{2^2.4^2}{3.4.5.6}x^6+\ldots.$

5. $1+\frac{1}{2}\frac{x^2}{4}+\frac{1.3.5}{2.4.6}\cdot\frac{x^4}{8}+\ldots+\frac{1.3.5.7.9}{2.4.6.8.10}\cdot\frac{x^6}{12}+\ldots$

6. $\sum_{n=1}^{\infty}\frac{n!}{(n+1)^n}x^n, x>0$

7. $\sum_{n=1}^{\infty}\left[\frac{1}{1+\log n}\right]$

8. $1+\frac{2}{3.5}+\frac{2.4}{3.5.7}+\frac{2.4.6}{3.5.7.9}+\ldots$

9. Test the convergence of the series

$x+x^{1+\frac{1}{2}}+x^{1+\frac{1}{2}+\frac{1}{3}}+x^{1+\frac{1}{2}+\frac{1}{3}+\frac{1}{4}}+\ldots$

10. Test the convergence of the following series:

(i) $\frac{1^2}{2^2}+\frac{1^2.3^2}{2^2.4^2}x+\frac{1^2.3^2.5^2}{2^2.4^2.6^2}x^2+\ldots$

(ii) $1+\frac{2^2}{3^2}+\frac{2^2.4^2}{3^2.5^2}+\frac{2^2.4^2.6^2}{3^2.5^2.7^2}+\ldots$

11. Test for convergence, the following series :

(i) $1+\frac{x}{1}+\frac{1}{2}\cdot\frac{x^3}{3}+\frac{1.3}{2.4}\cdot\frac{x^5}{5}+\frac{1.3.5}{2.4.6}\cdot\frac{x^7}{7}+\ldots.$

(ii) $\frac{x}{1}+\frac{1}{2}\cdot\frac{x^2}{3}+\frac{1.3}{2.4}\cdot\frac{x^3}{5}+\frac{1.3.5}{2.4.6}\cdot\frac{x^4}{7}+\ldots(x>0)$

(iii) $\sum_{n=1}^{\infty}\frac{1.3.5\ldots.(4n-5)(4n-3)}{2.4.6\ldots.(4n-4)(4n-2)}\frac{x^{2n}}{4n}, x>0$

(iv) $\sum_{n=1}^{\infty}\frac{2.4.6\ldots.2n}{1.3.5\ldots(2n+1)}$

12. Test for convergence, the following series :

(i) $1+\frac{x}{1!}+\frac{2^2x^2}{2!}+\frac{3^3x^3}{3!}+\ldots$ for $x>0$

(ii) $\frac{1}{2}x+\frac{1.3}{2.4}x^2+\frac{1.3.5}{2.4.6}x^3+\ldots, x>0$

(iii) $1+\frac{2!}{2^2}x+\frac{3!}{3^3}x^2+\ldots, x>0$

13. Test for convergence, the following series :

(i) $1+\frac{a(1-a)}{1^2}+\frac{(1+a)a(1-a)(2-a)}{1^2.2^2}+\frac{(2+a)(1+a)a(1-a)(2-a)(3-a)}{1^2.2^2.3^2}+\ldots$

(ii) $\frac{(1+a)(1+b)}{1.2.3}+\frac{(2+a)(2+b)}{2.3.4}+\frac{(3+a)(3+b)}{3.4.5.}+\ldots$

14. Test for convergence the following series :

(i) $1+\frac{\alpha}{1.\beta}x+\frac{\alpha(\alpha+1)^2}{1.2\beta(\beta+1)^2}x^2+\frac{\alpha(\alpha+1)^2(\alpha+2)^2}{1.2.3\beta(\beta+1)(\beta+2)}x^3+\ldots.$

(ii) $1+\frac{\alpha.\beta}{1.\gamma}x+\frac{\alpha(\alpha+1)\beta(\beta+1)}{1.2.\gamma(\gamma+1)}x^2+\frac{\alpha(\alpha+1)(\alpha+2)\beta(\beta+1)(\beta+2)}{1.2.3.\gamma(\gamma+1)(\gamma+2)}x^3+\ldots$

(Kurukshetra–2005)

15. Test for convergence the following series :

$\frac{a}{a+3}+\frac{a(a+2)}{(a+3)(a+5)}x+\frac{a(a+2)(a+4)}{(a+3)(a+5)(a+7)}x^2+\ldots$

16. Test for convergence the following series :

$\left(\frac{1}{2.4}\right)^{2/3}+\left(\frac{1.3}{2.4.6}\right)^{2/3}+\left(\frac{1.3.5}{2.4.6.8}\right)^{2/3}+\ldots.$

17. Test for convergence the following series :

(i) $\sum_{n=1}^{\infty}\frac{1\cdot3\cdot5\ldots.(2n-1)}{2\cdot4\cdot6\ldots.2n}\cdot\frac{1}{n}$

(ii) $\sum_{n=1}^{\infty}\frac{4\cdot7\cdot10\ldots.(3n+1)}{1\cdot2\cdot3\ldots.n}x^n$ (VTU–2009, PTU–2006S)

(iii) $\sum_{n=1}^{\infty}\frac{3\cdot6\cdot9\ldots.(3n)}{7\cdot10\cdot13\ldots.(3n+4)}x^n, x>0$

(iv) $\sum_{n=1}^{\infty}\frac{(2n)!}{(n!)^2}x^n, x>0$

18. Test for convergence the following series :

$\frac{1^2}{2^2}+\frac{1^2.3^2}{2^2.4^2}+\frac{1^2.3^2.5^2}{2^2.4^2.6^2}+\ldots$

19. Test for convergence the following series :

(i) $\frac{1}{(\log 2)^p}+\frac{1}{(\log 3)^p}+\ldots+\frac{1}{(\log n)^p}+\ldots$

(ii) $x^2(\log 2)^p+x^3(\log 3)^p+x^4(\log 4)^p+\ldots$

20. Test for convergence the following series :

(i) $\frac{x}{1.2}+\frac{x^2}{3.4}+\frac{x^3}{5.6}+\frac{x^4}{7.8}+\ldots\infty\ (x>0)$ (Mumbai–2009)

(ii) $\frac{x}{1.2}+\frac{x^2}{2.3}+\frac{x^3}{3.4}+\frac{x^4}{4.5}+\ldots\infty$ (VTU–2008, JNTU–2003)

(iii) $1+\frac{2}{3}x+\frac{2.3}{3.5}x^2+\frac{2.3.4}{3.5.7}x^3+\ldots\infty$ (VTU–2009S)

(iv) $\frac{x}{1}+\frac{1}{2}\frac{x^3}{3}+\frac{1.3}{2.4}\frac{x^5}{5}+\frac{1.3.5}{2.4.6}\frac{x^7}{7}+\ldots\infty\ (x>0)$

(VTU–2007, Raipur–2005)

(v) $1+\frac{1}{2}\frac{x^2}{4}+\frac{1.3.5}{2.4.6}\frac{x^4}{8}+\frac{1.3.5.7.9}{2.4.6.8.10}\frac{x^6}{12}+\ldots\infty$

(Rohtak–2006S, Roorkee–2000)

(vi) $\frac{1}{1^2}+\frac{1+2}{1^2+2^2}+\frac{1+2+3}{1^2+2^2+3^3}+\ldots$ (VTU–2000)

ANSWERS

1. Convergent **2.** Convergent **3.** $\begin{cases}\text{Convergent if } x<1, \\ \text{Divergent if } x \geq 1\end{cases}$ **4.** Convergent if $x^2 \leq 1$, divergnet if $x^2 > 1$

5. Convergent if $x \leq 1$, divergent if $x > 1$ **6.** Convergent if $x < e$, divergent if $x \geq e$ **7.** Convergent **8.** Convergent

9. Convergent if $x < \frac{1}{e}$, divergent if $x \geq \frac{1}{e}$ **10.** (i)Convergent if $x < 1$, divergent if $x \geq 1$ (ii) Divergent

11. (i) Convergent if $x^2 \leq 1$, divergent if $x^2 > 1$ (ii) Convergent if $0 < x \leq 1$, divergent if $x > 1$
(iii)Convergent if $x \leq 1$, divergent if $x > 1$ (iv) Divergent **12.** (i) Convergent if $x < \frac{1}{e}$, divergent if $x \geq \frac{1}{e}$
(ii) Convergent if $x < 1$, divergent if $x \geq 1$, (iii) Convergent if $x < e$, divergent if $x \geq e$ **13.** (i) Divergent (ii) Divergent

14. (i) Convergent if $x < 1$, divergent if $x > 1$, When $x = 1$, then convergent if $\beta > 2\alpha$, divergent if $\beta \leq 2\alpha$
(ii)Convergent if $x < 1$, divergent if $x > 1$, When $x = 1$, then convergent if $\gamma > \alpha + \beta$, divergent if $\gamma \leq \alpha + \beta$.

15. Convergent if $x \leq 1$, divergent if $x > 1$ **16.** Divergent **17.** (i) Convergent (ii) Convergent if $x < \frac{1}{3}$, divergent if $x \geq \frac{1}{3}$
(iii)Convergent if $x \leq 1$, divergentif $x > 1$ (iv) Convergent if $x < \frac{1}{4}$, divergent if $x \geq \frac{1}{4}$. **18.**Divergent

19. (i) Divergent for all values of p, (ii) Convergent if $x < 1$, divergent if $x \geq 1$

20. (i) Convergent for $x \leq 1$; divergent for $x > 1$ (ii) Convergent for $x \leq 1$; divergent for $x > 1$
(iii) Convergent for $x < 2$; divergent for $x \geq 2$ (iv) Convergent for $x \leq 1$; divergent for $x > 1$
(v) Convergent for $x^2 \leq 1$; divergent for $x^2 > 1$ (vi) Diverges

26.12 GAUSS'S TEST

If Σu_n be a series of positive terms such that

$$\frac{u_n}{u_{n+1}} = \alpha + \frac{\beta}{n} + \frac{\gamma_n}{n^p},$$

where $a > 0$, $p > 1$ and $<\gamma_n>$ is a bounded sequence. Then

(i) *Σu_n converges for $\alpha > 1$, diverges for $\alpha < 1$, whatever β may be.*

(ii) *If $\alpha = 1$, Σu_n converges whenever $\beta > 1$, and diverges whenever $\beta \leq 1$.*

Proof. We have

$$\lim_{n\to\infty} \frac{u_n}{u_{n+1}} = \alpha.$$

Then by D'Alembert's ratio test Σu_n is convergent if $\alpha > 1$ and divergent if $\alpha < 1$.
For $\alpha = 1$, we have

$$n\left[\frac{u_n}{u_{n+1}} - 1\right] = \beta + \frac{\gamma_n}{n^p},$$

where $p > 1$ and $< \gamma_n >$ is a bounded sequence.

$$\therefore \qquad \lim_{n\to\infty} n\left[\frac{u_n}{u_{n+1}} - 1\right] = \beta.$$

Then, by Raabe's test Σu_n is convergent if $\beta > 1$ and divergent if $\beta < 1$.
Now for $\alpha = \beta = 1$, we compare the series with the divergent series Σv_n where $v_n = \frac{1}{n \log n}$.
Now, consider

$$\frac{u_n}{u_{n+1}} - \frac{v_n}{v_{n+1}} = 1 + \frac{1}{n} + \frac{\gamma_n}{n^p} - \frac{(n+1)\log(n+1)}{n \log n} = \frac{\gamma_n}{n^p} - \frac{(n+1)}{n}\left[\frac{\log(n+1)}{\log n} - 1\right]$$

$$= \frac{1}{n^p}\left[\gamma_n - (n+1)\log\left(1 + \frac{1}{n}\right) \cdot \frac{n^{p-1}}{\log n}\right].$$

But $\lim_{n\to\infty}(n+1)\log\left(1+\frac{1}{n}\right) = \lim_{n\to\infty}\left[\log\left(1+\frac{1}{n}\right)^n + \log\left(1+\frac{1}{n}\right)\right]$

$$\lim_{n\to\infty}\frac{n^{p-1}}{\log n} = \infty, p > 1 \text{ and } <\gamma_n> \text{ is bounded.}$$

Therefore, for large value of n, $\gamma_n - (n+1)\log\left(1+\frac{1}{n}\right)\frac{n^{p-1}}{\log n}$ remains negative.

$$\therefore \quad \frac{u_n}{u_{n+1}} - \frac{v_n}{v_{n+1}} < 0 \quad \text{or} \quad \frac{u_n}{u_{n+1}} < \frac{v_n}{v_{n+1}}.$$

Now, since $\Sigma\, v_n$ is divergent, by comparison test Σu_n is divergent.

Hence, the series Σu_n is convergent if $\alpha > 1$ or $\alpha = 1$ and $\beta > 1$ and divergent if $\alpha > 1$ or $\alpha = 1$ and $\beta \leq 1$.

26.13 CAUCHY'S INTEGRAL TEST

Let $f(x)$ be non-negative monotonically decreasing integrable function on $[1, \infty[$ then the series $\sum_{n=1}^{\infty} f(n)$ and the improper integral $\int_1^\infty f(x)\,dx$ converge or diverge together.

26.14 CAUCHY'S CONDENSATION TEST

If $f(n)$ is a monotonically decreasing function of n for all $n \in N$ such that each $f(n)$ is positive,

then two infinite series $\sum_{n=1}^{\infty} f(n)$ and $\sum_{n=1}^{\infty} a^n f(a^n)$ converge or diverge together, where a is a positive integer greater than unity.

26.15 REARRANGEMENT OF TERMS

A series Σv_n is said to be rearrangement of a series Σu_n if there exists one-one correspondence between the terms of the two series and if v_n corresponds to u_n then $v_n = u_n$.

In other words, we can say that a series Σu_n is said to be rearrangement of a series Σv_n if every term of Σu_n is a term of Σv_n and *vice-versa*.

26.16 ALTERNATING SERIES

A series, whose terms are alternatively positive and negative is called an alternating series.

Thus, a series of the form $u_1 - u_2 + u_3 - u_4 + ... + (-1)^{n-1} u_n + ...$ where $u_n > 0\ \forall\ n$, is an alternating series.

26.16.1 ABSOLUTE CONVERGENCE

A series Σu_n is said to be absolutely convergent if the series $\Sigma|u_n|$ is convergent.

26.16.2 UNCONDITIONALLY CONVERGENT SERIES

A series Σu_n is said to be unconditionally convergent if every rearrangement converge to the same sum Σu_n, *i.e*, Σu_n is conditionally convergent iff it is absolutely convergent.

26.16.3 CONDITIONAL CONVERGENCE

A series Σu_n is said to be conditionally convergent if Σu_n is convergent but $\Sigma|u_n|$ is divergent.

REMARK

- The conditional convergence of a series is also known as semi-convergent or non-absolutely convergent.

ILLUSTRATIONS

(1) The series $\Sigma u_n = 1 - \frac{1}{2} + \frac{1}{2^2} - \frac{1}{2^3} + ...$ is absolutely convergent.

(2) The series $\frac{1}{1^2} - \frac{1}{2^2} + \frac{1}{3^2} - \frac{1}{4^2} + ...$ is absolutely convergent.

THEOREM 1. *An absolutely convergent series is convergent.*

Proof. Let us suppose, the series Σu_n is absolutely convergent. Then by definition $|u_n|$ is convergent.

Now $$u_n + |u_n| = \begin{cases} 2u_n, & \text{if } u_n \text{ is positive} \\ 0, & \text{if } u_n \text{ is negative.} \end{cases}$$

Therefore, every term of the series $\Sigma(u_n + |u_n|)$ is ≥ 0 and less than equal to the corresponding term of the convergent series $\Sigma 2|u_n|$.

Hence, $\Sigma(u_n + |u_n|)$ is convergent. Hence Σu_n is convergent.

REMARKS

- The converse of the above theorem is not necessarily true :

  For example : The series $\Sigma u_n = 1 - \frac{1}{2} + \frac{1}{3} - \ldots$ is convergent, but the series $\Sigma|u_n| = 1 + \frac{1}{2} + \frac{1}{3} + \ldots$ is divergent. Hence a convergent series need not be absolutely convergent.

- The usefulness of absolute convergence is partly due to the fact that it is often easier to establish absolute convergence than convergence :

  For example : Consider the series $\Sigma \frac{a^n}{2^n}$, where $a_n = 1$ if n is prime number and $a_n = -1$ otherwise. Here, $\Sigma|a_n| = \Sigma \frac{1}{2^n}$ is convergent. Accordingly $\Sigma a_n/2^n$ is absolutely convergent, and hence convergent.

THEOREM 2. *If the terms of a convergent series of positive terms are rearranged, the series remains convergent and its sum is unaltered.*

Proof. Let us suppose Σu_n be a convergent series, and let the terms be rearranged in any manner. Denote the new series by Σv_n, so that every u is a v and every v is a u.

Let $s_n = u_1 + u_2 + \ldots + u_n$ and $t_n = v_1 + v_2 + \ldots + v_n$.

Then, for any definite value of n, s_n contains n terms each of which occurs, sooner or later, in the v series and so we can find a corresponding m such that t_m contains all the terms of s_n (and possibly other not contained in s_n).

Now, since each term is positive, therefore $s_n \leq t_m$.

Also, suppose that the first m terms of Σv_n are among the first $(n+p)$ terms of Σu_n. Therefore,

$$s_n \leq t_m \leq s_{n+p}.$$

and m tends to infinity with n.

Let Σu_n converges to s, so that $\lim s_n = \lim s_{n+p} = s$

$\therefore$ $\lim t_m = s$.

Hence, Σv_n is convergent and has the same sum as Σu_n.

REMARK

- The arrangement fails for a dearrangement such as $u_1 + u_2 + u_5 + \ldots + u_2 + u_4 + u_6 + \ldots$ where Σu_n is broken up into two (or any finite no. of) infinite series.
  Here, we cannot find an m so that the first n terms of Σu_n occur among the first m terms of Σv_n.
  For instance, u_2 does not occur even if infinitely many of the terms $u_1, u_3, u_5, \ldots$ have been placed.

THEOREM 3. **(Dirichlet's Theorem).** *If the terms of an absolutely convergent series are rearranged, the series remains convergent and its sum is unaltered.*

Proof. Let Σu_n be an absolutely convergent series, and let its terms be rearranged in a different order. Let, the new series be denoted by Σv_n so that every v occurs somewhere in the u series and every u occurs somewhere in the v series.

Now, we have $u_n + |u_n| = 2u_n$ or 0 according as u_n is positive or negative. Now $\Sigma|u_n|$ is a convergent series of positive terms, so also in the series $\Sigma(u_n + |u_n|)$, because its terms are less than equal to be corresponding terms of the series $\Sigma 2|u_n|$.

Let $\Sigma|u_n| = s$ and $\Sigma(u_n + |u_n|) = s'$ so that $\Sigma u_n = s' - s$.

Also, since $\Sigma|u_n|$ and $\Sigma(u_n + |u_n|)$ are convergent series of positive terms, their sum remains unchanged by any rearrangement of term (By theorem 2).

Accordingly, $\Sigma|v_n| = s$ and $\Sigma(v_n + |v_n|) = s'$.

Hence, $\Sigma v_n = s' - s = \Sigma u_n$.

REMARKS

- If we rearrange the order of terms of a semi-convergent series, we may or may not changed the sum of the series.
- The sum will be changed if we interfere too much with the balance between positive and negative terms.
- By a suitable rearrangement of the terms a semi-convergent series may be made to diverge. The reason is that in a semi-convergent series the positive and negative terms taken separately from two divergent series.

THEOREM 4. **(Riemann's Rearrangement theorem).** *By a suitable rearrangement of terms of a conditionally convergent series can be made to converge to any number λ or to diverge to ∞ or $-\infty$ even to oscillate.*

In other words, this theorem can be stated as follows –

To a given conditionally convergent series and to any given number there corresponds a rearrangement of the given series which is convergent and whose sum is the given number.

THEOREM 5. **(Pringsheim theorem).** *Let $f(x)$ be a sequence of positive terms which monotonically converges to zero and let the series $\sum_{n=1}^{\infty}(-1)^{n-1} f(x)$ be rearranged so that in the first p+n terms there are p-positive terms and n negative terms, i.e., $\lim_{n\to\infty} n f(x)=\lambda$ and $\lim_{n\to\infty}\frac{p}{n}=k$ then the sum of the series is increased by $\frac{1}{2}\lambda\log k$.*

26.17 LEIBNITZ'S TEST

If the alternative series $u_1-u_2+u_3-\dots(u_n>0, \forall\, n\in N)$ is such that

(i) $u_{n+1}\le u_n,\ \forall\, n\in N$ (ii) $\lim_{n\to\infty} u_n=0$

Then the series converges.

Proof. Let $s_n=u_1-u_2+u_3-\dots+(-1)^{n-1}u_n$ so that $<s_n>$ is a sequence of partial sums of the given series.

Now for all n

$$s_{2n+2}-s_{2n}=u_{2n+1}-u_{2n+2}\ge 0 \qquad \text{[By (i)]}$$

which gives that s_{2n} is a monotonically increasing sequence.

Further,

$$s_{2n}=u_1-u_2+u_3-\dots u_{2n-1}-u_{2n}=u_1-(u_2-u_3)-(u_4-u_5)-\dots-u_{2n}$$
$$=u_1-[(u_2-u_3)+\dots+u_{2n}]=u_1-\text{some positive number}\le u_1.$$

Therefore, the monotonically increasing sequence $<s_{2n}>$ is bounded above and consequently it is convergent.

Let $\lim_{n\to\infty} s_{2n}=s.$

Now $s_{2n+1}=s_{2n}+u_{2n+1} \Rightarrow \lim_{n\to\infty} s_{2n+1}=\lim_{n\to\infty} s_{2n}+\lim_{n\to\infty} u_{2n+1}$ $\left[\because \lim_{n\to\infty} u_n=0\right]$

$$=s+0=s$$

Thus, the subsequences $<s_{2n}>$ and $<s_{2n+1}>$ both converge to the same limit. Now we shall show that the sequence $<s_n>$ also converges to s.

Let $\varepsilon>0$ be given. Since, the sequences s_{2n} *and* s_{2n+1} both converges to s, there exists positive integers m_1, m_2 such that

$$|s_{2n}-s|<\varepsilon\ \forall\, n\ge m_1,$$

and

$$|s_{2n+1}-s|<\varepsilon\ \forall\, n\ge m_2.$$

Let $m=\max\{m_1,m_2\}.$

Then $|s_n-s|<\varepsilon\ \forall\, n\ge m$

which gives that the sequence $<s_n>$ converges to s.

Hence, the given series $\Sigma(-1)^{n-1}u_n$ converges.

REMARKS

- This test gives us a set of sufficient conditions for the convergence of an alternating series.
- If the test does not show a series to be convergent, we may not immediately say that the series is divergent.

SOLVED EXAMPLES

Example 1. *Show that* $\lim_{n\to\infty}\left[1+\frac{1}{2}+\dots+\frac{1}{n}-\log n\right]$ *exists.*

Solution. Let $f(x)=\frac{1}{x},\ x\in[1,\infty[$.

Then $f(x)>0$ and monotonically decreasing on $[1,\infty[$.

Let $S_n=f(1)+f(2)+\dots+f(n)$

$$=1+\frac{1}{2}+\frac{1}{3}+\dots+\frac{1}{n}$$

and $I_n=\int_1^n f(x)\,dx=\int_1^n\frac{1}{x}dx=[\log x]_1^n=\log n.$

It can be easily shown that

$$f(n) \le S_n - I_n \le f(1) \; \forall \; n \in N$$

or $\quad 0 < \frac{1}{n} \le S_n - I_n \le 1 \; \forall \; n \in N$

which gives that the sequence (u_n), where $u_n = S_n - I_n$, is bounded below.

Now, it can also be shown easily that the sequence $<u_n>$ is a monotonically decreasing. Therefore it converges.

Hence, $\lim_{n\to\infty}\left(1+\frac{1}{2}+...+\frac{1}{n}\right)$ exist.

REMARK

- The limit of the above sequence is called Euler's constant and is denoted by γ.

Example 2. *Show by integral test that* $\Sigma\frac{1}{n^p}$ *converges if* $p > 1$ *and diverges if* $p \le 1$.

Solution. Let $f(x) = \frac{1}{x^p}$, $p > 0$. Then $f(x)$ is positive valued and monotonically decreasing.

Therefore by Cauchy's integral test $\Sigma\frac{1}{n^p}$ and $\int_1^\infty f(x)dx$ converges and diverges together.

Let $\quad I_n = \int_1^n \frac{1}{x^p}dx = \int_1^n x^{-p}dx$

$$= \begin{cases} \left(\frac{n^{1-p}}{1-p} - \frac{1}{1-p}\right), & \text{if } p \ne 1 \\ \log n & , \text{if } p = 1. \end{cases}$$

If $n \to \infty$, $n^{1-p} = \frac{1}{n^{p-1}} \to 0$ as $p > 1$ and tends to ∞ if $p < 1$ and $\log n \to \infty$

$\therefore \quad \lim_{n\to\infty} I_n = -\frac{1}{1-p} = \frac{1}{p-1}$, if $p > 1$

and $\quad \lim_{n\to\infty} I_n = \infty$, if $p \le 1$.

Hence, $\int_1^\infty f(x)dx$ converges if $p>1$ and diverges if $p \le 1$. Then by Cauchy's integral test the series $\Sigma\frac{1}{n^p}$ is convergent if $p > 1$ and divergent if $p \le 1$.

Example 3. *Show by Cauchy's integral test that the series* $\sum_{n=2}^{\infty}\frac{1}{n(\log n)^p}$ *converges if* $p > 1$ *and diverges if* $0 < p \le 1$. (PTU–2010)

Proof. Let us suppose

$$f(x) = \frac{1}{x(\log x)^p}, \; p > 0$$

and $x \in [2, \infty[$; then obviously $f(x)$ is monotonically decreasing in $[2, \infty[$ and positive valued.

Let $\quad I_n = \int_2^n \frac{dx}{x(\log x)^p}$

Then $\quad I_n = \left[\frac{(\log x)^{1-p}}{1-p}\right]_2^n, \; p \ne 1$

$$= \frac{1}{(1-p)}[(\log n)^{1-p} - (\log 2)^{1-p}], p \ne 1$$

and $\quad I_n = [\log\log x]_2^n, \; p = 1$

$$= [\log\log n - \log\log 2], \; p = 1.$$

Therefore, we have

$$\lim_{n\to\infty} I_n = \lim_{n\to\infty}\int_2^n f(x)dx = \infty, \text{ if } p < 1$$

and $\quad \lim_{n\to\infty} I_n = -\frac{1}{(1-p)}(\log 2)^{1-p}$, if $p > 1$.

Thus the integral $\int_2^\infty f(x)dx$ converges if $p > 1$ and diverges if $0 < p \le 1$.

Hence, by Cauchy's integral test, the series

$$\sum_{n=2}^{\infty} f(x) = \sum_{n=2}^{\infty}\frac{1}{n(\log n)^p}$$

converges if $p > 1$ and diverges if $0 < p \le 1$.

Example 4. *Apply the Cauchy's condensation test to discuss the convergence of the series*

$$\sum_{n=2}^{\infty}\frac{1}{(n\log n)(\log\log n)^p}.$$

Solution. Here, we have

$$f(n) = \frac{1}{(n\log n)(\log\log n)^p}$$

$$\therefore \; a^n f(a^n) = \frac{a^n}{(a^n \log a^n)(\log\log a^n)^p}$$

$$= \frac{1}{(n\log a)[\log(n\log a)]^p}$$

Since, a is a positive integer greater than 1 and can be chosen that $\log e \; a > 1$ so that $n \log a > n$.

Then $\quad a^n f(a^n) < \frac{1}{(n\log a)(\log n)^p}$

Since, the series $\frac{1}{\log a}\Sigma\frac{1}{n(\log n)^p}$ is convergent when $p > 1$, therefore $\Sigma a^n f(a^n)$ is also convergent and consequently the given series is convergent when $p > 1$.

Now let $p \le 1$. If we take $a = 2$, then $\log_e a < 1$ so that

$$n\log_e a < n$$

$$\therefore \quad a^n f(a^n) > \frac{1}{(n\log a)(\log n)^p}$$

But the series $\frac{1}{\log a}\Sigma\frac{1}{n(\log n)^p}$ is divergent when $p \le 1$ and therefore $\Sigma a^n f(a^n)$ is also divergent. Then by Cauchy condensation test, the given series is divergent when $p \le 1$.

Example 5. *Test the convergence of the series* $\frac{2^2}{3^2}+\frac{2^2.4^2}{3^2.5^2}+...$

Solution. Here, we have

$$u_n = \frac{2^2.4^2...(2n)^2}{3^2.5^2...(2n+1)^2}..$$

$$\Rightarrow \quad u_{n+1} = \frac{2^2.4^2...(2n+2)^2}{3^2.5^2...(2n+3)^2}$$

$$\therefore \quad \frac{u_n}{u_{n+1}} = \frac{(2n+3)^2}{(2n+2)^2} = \left(1+\frac{3}{2n}\right)^2\left(1+\frac{1}{n}\right)^{-2}$$

$$= \left(1+\frac{3}{n}+\frac{9}{4n^2}\right)\left(1-\frac{2}{n}+\frac{3}{n^2}...\right)$$

(On expanding by binomial expansion)

$$= 1+\frac{1}{n}-\frac{3}{4n^2}+...$$

$$= \alpha+\frac{\beta}{n}+\frac{\gamma_n}{n^2}, \text{where } \gamma_n \to -\frac{3}{4} \text{ as } n \to \infty$$

$\Rightarrow \quad \alpha = 1, \beta = 1$. Then by Gauss test the series Σu_n is divergent.

Example 6. *Test the convergence of the series*

$$\frac{1^2}{2^2}+\frac{1^2.3^2}{2^2.4^2}+\frac{1^2.3^2.5^2}{2^2.4^2.6^2}+....$$

Solution. Here, we have $u_n = \frac{1^2.3^2.5^2...(2n-1)^2}{2^2.4^2.6^2...(2n)^2}$

$$\therefore \quad u_{n+1} = \frac{1^2.3^2...(2n-1)^2(2n+1)^2}{2^2.4^2...(2n)^2(2n+2)^2}$$

$$\therefore \quad \frac{u_n}{u_{n+1}} = \frac{(2n+2)^2}{(2n+1)^2}$$

$$\Rightarrow \quad \lim_{n\to\infty}\frac{u_n}{u_{n+1}} = \lim_{n\to\infty}\frac{\left(2+\frac{2}{n}\right)^2}{\left(2+\frac{1}{n}\right)^2} = 1$$

which gives that, the ratio test is fail.
Now, we can easily see that

$$\lim_{n\to\infty} n\left[\frac{u_n}{u_{n+1}}-1\right] = 1.$$

$\Rightarrow$ Raabe's test also fails.
Now applying Gauss test,
Consider

$$\frac{u_n}{u_{n+1}} = \frac{(2n+2)^2}{(2n+1)^2} = \left(1+\frac{1}{n}\right)^2\left(1+\frac{1}{n}\right)^{-2}$$

$$= \left(1+\frac{2}{n}+\frac{1}{n^2}\right)\left(1-2.\frac{1}{2n}+3.\frac{1}{4n^2}....\right)$$

$$= 1+\frac{1}{n}-\frac{1}{4n^2}+...$$

$$= \alpha+\frac{\beta}{n}+\frac{\gamma_n}{n^2}, \text{ where } \gamma_n \to -\frac{1}{4} \text{ as } n \to \infty.$$

Here, $\alpha = 1$, $\beta = 1$. Therefore by Gauss test the series Σu_n is divergent.

Example 7. *Test the convergence of the series*

$$1+\left(\frac{2}{3}\right)^p+\left(\frac{2.4}{3.5}\right)^p+\left(\frac{2.4.6}{3.5.7}\right)^p+...$$

Solution. Neglecting first term, we have

$$u_n = \left[\frac{2.4.6...(2n)}{3.5.7...(2n+1)}\right]^p$$

$$\Rightarrow \quad u_{n+1} = \left[\frac{2.4.6...(2n)(2n+2)}{3.5.7...(2n+1)(n+3)}\right]^p$$

$$\Rightarrow \quad \frac{u_n}{u_{n+1}} = \left(\frac{2n+3}{2n+2}\right)^p = \frac{\left(1+\frac{3}{2n}\right)^p}{\left(1+\frac{2}{2n}\right)^p}$$

$$= \left(1+\frac{3}{2n}\right)^p\left(1+\frac{1}{n}\right)^{-p}$$

$$= \left[1+p.\frac{3}{2n}+O\left(\frac{1}{n^2}\right)\right]\left[1-\frac{p}{n}+O\left(\frac{1}{n^2}\right)\right]$$

$$= \left[1+\left(\frac{3}{2}-1\right)\frac{p}{n}+O\left(\frac{1}{n^2}\right)\right]$$

$$= 1+\frac{\frac{1}{2}p}{n}+\left(\frac{1}{n^2}\right).$$

Then by Gauss test, the series is convergent if $p/2 > 1$, *i.e.*, $p > 2$ and divergent if $p/2 \le 1$, *i.e*, $p \le 2$.

Example 8. *If x, α, β, γ are all positive, discuss the convergence of hypergeometric series*

$$1+\frac{\alpha.\beta}{1.\gamma}+\frac{\alpha(\alpha+\beta)\beta(\beta+1)}{1.2.\gamma(\gamma+1)}x^2+\frac{\alpha(\alpha+1)(\alpha+2)\beta(\beta+1)(\beta+2)}{1.2.3.\gamma.(\gamma+1).(\gamma+2)}$$

Solution. Since, x, α, β, γ are all positive, the given series is a series of positive terms. Neglecting first term we have,

$$u_n = \frac{\alpha(\alpha+1)...(\alpha+n-1)\beta(\beta+1)...(\beta+n-1)}{1.2...n.\gamma.(\gamma+1)...(\gamma+n-1)}.x^n$$

$$\Rightarrow u_{n+1} = \frac{\alpha(\alpha+1)\ldots(\alpha+n-1)(\alpha+n)\beta(\beta+1)\ldots(\beta+n-1)(\beta+n)}{1.2\ldots n(n+1).\gamma.(\gamma+1)\ldots(\gamma+n-1)(\gamma+n)}$$

$$\therefore \quad u_{n+1} = \frac{(n+1)(\gamma+n)}{(\alpha+n)(\beta+n)}.\frac{1}{x}$$

$$\Rightarrow \quad \lim_{n\to\infty}\frac{u_n}{u_{n+1}} = \frac{(n+1)(\gamma+n)}{(\alpha+n)(\beta+n)}.\frac{1}{x} = \frac{1}{x}$$

$\therefore$ By ratio test, the series is convergent if $\frac{1}{x} > 1$, i.e., $x < 1$ and divergent if $x>1$.
When $x = 1$, the ratio test is fails.
In this case, consider

$$\therefore \quad \frac{u_n}{u_{n+1}} = \left(\frac{(n+1)(n+\gamma)}{(\alpha+n)(\beta+n)}\right) = \frac{\left(1+\frac{1}{n}\right)\left(\frac{\gamma}{n}+1\right)}{\left(1+\frac{\alpha}{n}\right)\left(1+\frac{\beta}{n}\right)}$$

$$= \left[\left(1+\frac{1}{n}\right)\left(1+\frac{\gamma}{n}\right)\right]\left(1+\frac{\alpha}{n}\right)^{-1}\left(1+\frac{\beta}{n}\right)^{-1}$$

$$= \left[1+(1+\gamma)\frac{1}{n}+O\left(\frac{1}{n^2}\right)\right]\left[1-\frac{\alpha}{n}+O\left(\frac{1}{n^2}\right)\right]\left[1-\frac{\beta}{n}+O\left(\frac{1}{n^2}\right)\right]$$

$$= 1+\frac{1+\gamma-\alpha-\beta}{n}+O\left(\frac{1}{n^2}\right)$$

Then by Gauss test, the series is convergent if $1+\gamma-\alpha-\beta > 1$ and divergent if $1+\gamma-\alpha-\beta \le 1$, i.e., the series is convergent if $\gamma > \alpha+\beta$ and divergent if $\gamma \le \alpha+\beta$.

Example 9. *Test the convergence of the series*

$$1-\frac{1}{2^p}+\frac{1}{3^p}-\frac{1}{4^p}+\ldots(p>0).$$

Solution. Since, the given series is an alternating series.

$\therefore$ The nth term

$$t_n = (-1)^{n-1}u_n, \text{ where } u_n = \frac{1}{n^p} > 0, (p>0).$$

$$\text{Now } u_{n+1}-u_n = \frac{1}{(n+1)^p}-\frac{1}{n^p}$$

$$= \frac{n^p-(n+1)^p}{n^p(n+1)^p} < 0 \ \forall n \ge 1.$$

$$\therefore \quad u_{n+1} \le u_n \ \forall n \ge \text{N}.$$

$$\text{Also} \quad \lim_{n\to\infty} u_n = \lim_{n\to\infty}\frac{1}{n^p}, \ p>0.$$

Hence, by Leibnitz test the alternating series $\Sigma(-1)^{n-1}\frac{1}{n^p}$ is convergent.

Example 10. *Test the convergence of the series*

$$\frac{1}{x}-\frac{1}{x+a}+\frac{1}{x+2a}+\ldots, x>0, a>0.$$

Solution. Since, the given series is an alternating series.

$\therefore$ The nth term

$$t_n = (-1)^{n-1}u_n, \text{ where } u_n = \frac{1}{x+(n-1)a} > 0.$$

$$\text{Now } u_{n+1}-u_n = \frac{1}{x+na}-\frac{1}{x+(n-1)a}$$

$$= \frac{[x+(n-1)a]-[x+na]}{[x+na][x+(n-1)a]}$$

$$= \frac{-a}{[x+ma][x+(n-1)a]} < 0$$

$$\therefore \quad u_{n+1} < u_n.$$

$$\text{Also,} \quad \lim_{n\to\infty} u_n = \lim_{n\to\infty}\frac{1}{x+(n-1)a} = 0.$$

Hence, by Leibnitz test, the given series is convergent.

Example 11. *Test the convergence of the series*

$$\frac{\log 2}{2^2}-\frac{\log 3}{3^2}-\frac{\log 4}{4^2}-\ldots$$

Solution. The given series is an alternating series
Here, the nth term

$$t_n = (-1)^n u_n, \text{ where } u_n = \frac{\log(n+1)}{(n+1)^2} > 0$$

$$\lim_{n\to\infty} u_n = \lim_{n\to\infty}\frac{\log(n+1)}{(n+1)^2}$$

$$= \lim_{n\to\infty}\frac{\log(n+1)}{(n+1)}.\frac{1}{(n+1)} = 0.$$

Now, we shall show that $u_{n+1} \le u_n \ \forall n$.

$$\text{Let} \quad f(x) = \frac{\log x}{x^2}$$

$$\text{Then} \quad f'(x) = \frac{x^2.\frac{1}{x}-2x\log x}{x^4}$$

$$= \frac{1-2\log x}{x^3} < 0 \text{ when } x > e^{1/2}.$$

Therefore, the function $f(x)$ is monotonically decreasing for all $x>e^{1/2}$. We know that

$$2 < e < 3 \Rightarrow 2^{1/2} < e^{1/2} < 3^{1/2}$$

$$\Rightarrow 1 < e^{1/2} < 2$$

so $\quad f(n+2) \le f(n+1)$ for all n.

i.e, $\quad u_{n+1} \le u_n \ \forall n$.

Hence, by Leibnitz test the given series is convergent.

Example 12. *Test the convergence of the series* $\sum_{n=1}^{\infty}\frac{(-1)^{n-1}n}{10n-1}$.

Solution. Here, the given series is an alternating series.

The nth term

$$t_n = (-1)^{n-1}.u_n, \text{where } u_n = \frac{n}{10n-1}.$$

Now, $u_{n+1} < u_n \ \forall\ n$ if $\frac{n+1}{10n+9} < \frac{n}{10n-1}$

i.e., if $(n+1)(10n-1) < n(10+9)$

i.e., if $10n^2 + 9n - 1 < 10n^2 + 9n$ which is true.

Now $\lim_{n\to\infty} u_n = \frac{n}{10n-1} \to \frac{1}{10} \neq 0.$

Hence, by Leibnitz test the series does not converge.

Example 13. *Test the absolute convergence of the series*

$$1 - \frac{1}{2\sqrt{2}} + \frac{1}{3\sqrt{3}} - \ldots$$

Solution. Here, $\Sigma u_n = 1 - \frac{1}{2\sqrt{2}} + \frac{1}{3\sqrt{3}} - \ldots$

Then series Σu_n is absolutely convergent if $\Sigma|u_n|$ is convergent.

Now $\Sigma|u_n| = 1 + \frac{1}{2\sqrt{2}} + \frac{1}{3\sqrt{3}} + \ldots$

$$= 1 + \frac{1}{2^{3/2}} + \frac{1}{3^{3/2}}$$

$$= \Sigma \frac{1}{n^{3/2}}.$$

Hence, the series is convergent ($\because p = 3/2 > 1$).

$\Rightarrow$ The given series is absolutely convergent.

Example 14. *Show that the series* $\frac{1}{\sqrt{1}} - \frac{1}{\sqrt{2}} + \frac{1}{\sqrt{3}} - \ldots$ *is conditionally convergent.*

Solution. The given series is an alternating series.

$\therefore$ The nth term

$$t_n = (-1)^{n-1} u_n \text{ where } u_n = \frac{1}{\sqrt{n}} > 0.$$

Now $u_{n+1} - u_n = \frac{1}{\sqrt{n+1}} - \frac{1}{\sqrt{n}}$

$$= \frac{\sqrt{n} - \sqrt{n+1}}{\sqrt{n}\sqrt{n+1}} < 0.$$

$\therefore u_{n+1} < u_n \Rightarrow \lim_{n\to\infty} u_n = \lim_{n\to\infty} \frac{1}{\sqrt{n}} = 0.$

$\therefore$ By Leibnitz test, the given series is convergent.

But the series $\Sigma\left|\frac{(-1)^{n-1}}{\sqrt{n}}\right| = \Sigma\frac{1}{\sqrt{n}}$ is divergent.

$$\left(\because p = \frac{1}{2} < 1\right)$$

Hence, the given series is conditionally convergent.

Example 15. *Discuss the convergence of the series*

$$1 + \frac{x}{1!} + \frac{x^2}{2!} + \ldots \textit{for all values of } x.$$

Solution. Here, we have

$$u_n = \frac{x^{n-1}}{(n-1)!} \Rightarrow u_{n+1} = \frac{x^n}{n!}$$

So, $\frac{|u_n|}{|u_{n+1}|} = \frac{|x|^{n-1}}{(n-1)!}.\frac{n!}{|x|^n} = \frac{n}{|x|}$, for $x \neq 0$

$$\therefore \quad \lim_{n\to\infty}\frac{|u_n|}{|u_{n+1}|} = \lim_{n\to\infty}\frac{n}{|x|} = \infty, \text{for } x \neq 0.$$

$\therefore$ By the ratio test, the series $\sum_{n=1}^{\infty}|u_n|$ is convergent when $x \neq 0$.

Thus $\sum_{n=1}^{\infty} u_n$ is absolutely convergent. If $x=0$, then the series becomes $1+0+0+\ldots$ and so is convergent.

Thus the given series is absolutely convergent.

REMARK

- Since for a convergent series $\sum_{n=1}^{\infty} u_n$, $\lim_{n\to\infty} u_n = 0$. Therefore, $\lim_{n\to\infty}\frac{x^n}{n!} = 0$ is a useful result.

26.17.1 MORE ABOUT CONDITIONAL AND ABSOLUTE CONVERGENCE

(i) If Σu_n is an absolute convergent series, then the series of its positive and the series of its negative terms are both convergent.

(ii) The divergence of $\Sigma|u_n|$ does not imply the divergence of Σu_n. For example, if $u_n = \frac{(-1)^{n-1}}{n}$ then $\Sigma|u_n|$ is divergent, whereas Σu_n is convergent.

(iii) Since, the series $\Sigma|u_n|$ is of positive terms, therefore, all the tests established for testing the convergence of series of positive terms, will also be the tests for determining the absolute convergence of the series Σu_n.

(iv) If Σu_n is conditionally convergent, then the series of its positive terms and the series of its negative terms are both divergent.

(v) A series with mixed signs cannot converge, if the series of its positive terms is convergent (divergent) and the series of its negative terms is divergent (convergent).

26.17.2 SUMMARY OF THE TESTS

For the guidance of the students we given below a working procedure for determining the convergence of a series.

(i) If in a series of positive terms, n^{th} term does not tend to zero, the series is divergent.

(ii) If n^{th} terms tends to zero, then a comparison test may be applied when its nth term neither involves any power of n nor involve factorials.

(iii) If the nth term is the nth power of some expression, then Cauchy's root test may be applied.

(iv) When the series involves increasing power of x or involves factorials, one should start with the ratio test.

(v) If the $\lim\limits_{n\to\infty} \dfrac{u_n}{u_n+1}$ turns out to be 1, then the ratio test fails and Raabe's test or Gauss's test is applied provided $\dfrac{u_n}{u_{n+1}}$ does not involves e, otherwise logarithmic test is applied.

(vi) For an arbitrary terms series, try with the ratio test for absolute convergence. If the limit turns out to be 1, then try some other tests. When the terms have alternating signs, then Leibnitz's test is suggested.

EXERCISE 26.4

1. Test the convergence of the following series

$$1-\frac{1}{2}+\frac{1}{3}-\frac{1}{4}+\ldots$$ (PTU–2009)

2. Prove that the following series is absolute convergent.

$$\left(\frac{\sqrt{2}-1}{1}\right)-\left(\frac{\sqrt{3}-\sqrt{2}}{2}\right)+\left(\frac{\sqrt{4}-\sqrt{3}}{3}\right)-\ldots$$

3. Show that the series $\Sigma \dfrac{\sin n\theta}{n^2}$ is absolutely convergent.

4. Show that the series

$$\frac{1^2}{4^2}+\frac{1^2.5^2}{4^2.8^2}+\frac{1^2.5^2.9^2}{4^2.8^2.12^2}+\ldots$$ is convergent.

5. Examine the convergence of the series

$$1+a+b^2+a^3+b^4+\ldots$$

6. Test the convergence of the series

$$\frac{x}{1}+\frac{1}{2}\cdot\frac{x^2}{3}+\frac{1.3}{2.4}\cdot\frac{x^3}{5}+\ldots$$

7. Show that the series $\Sigma(-1)^{n-1}\sin\dfrac{1}{n}$ is conditionally convergent.

8. Test for convergence the series

$$\Sigma\left(\frac{n^{n-1}.x^{n-1}}{n!}\right).$$

9. Test the convergence of the series

$$1+\frac{a(1-a)}{1^2}+\frac{(1+a)a(1-a)(2-a)}{1^2.2^2}+\ldots$$

10. Show that the series $\dfrac{2}{1^2}-\dfrac{3}{2^2}+\dfrac{4}{3^2}-\dfrac{5}{4^2}+\ldots$ converge conditionally.

11. Show that the series $\dfrac{1}{x+1}-\dfrac{1}{x+2}+\dfrac{1}{x+3}-\ldots$ is convergent except when x is a negative integer.

12. Show that the series $\Sigma(-1)^n[\sqrt{n^2+1}-n]$ is conditionally convergent.

13. Show that the series $\sum\limits_{n=1}^{\infty}\dfrac{(-1)^{n+1}.n}{n^2+1}$ is not absolutely convergent.

14. Show that the binomial series

$$1+nx+\frac{n(n-1)}{2!}x^2+\ldots\frac{n(n-1)\ldots(n-r+1)}{r!}x^r+\ldots$$

is absolutely convergent when $|x|<1$.

15. Test the convergence of the series

$$\sum_{n=1}^{\infty}\left[\frac{1}{n}+\frac{(-1)^{n+1}}{\sqrt{n}}\right].$$

16. Show that the series

$$x+\frac{a-b}{2!}x^2+\frac{(a-b)(a-2b)}{3!}x^3+\frac{(a-b)(a-2b)(a-3b)}{4!}x^4+\ldots$$

is absolutely convergent if $|x|<\dfrac{1}{|b|}$.

17. Show that the series

$$1-\frac{1}{2^3}-\frac{1}{4^3}+\frac{1}{3^3}-\frac{1}{6^3}-\frac{1}{8^3}+\ldots+\frac{1}{(2n-1)^3}-\frac{1}{(4n-2)^3}-\frac{1}{(4n)^3}+\ldots$$

is absolutely convergent.

18. Show that the series

$$2\sin\frac{x}{3}+4\sin\frac{x}{9}+8\sin\frac{x}{27}+\ldots$$

converges absolutely for all finite values of x.

19. Discuss the convergence of the series

$$x^2(\log 2)^q+x^3(\log 3)^q+x^4(\log 4)^q+\ldots$$

20. Discuss the convergence of the following series :

(i) $\dfrac{1}{\log 2}-\dfrac{1}{\log 3}+\dfrac{1}{\log 4}-\dfrac{1}{\log 5}+\ldots$ (PTU–2010)

(ii) $1-\dfrac{1}{5}+\dfrac{1}{9}-\dfrac{1}{13}+\ldots\infty$ (VTU–2010)

(iii) $\sum_{n=0}^{\infty} \frac{(-1)^n}{n!}$ (Delhi–2002)

(iv) $\frac{1}{1.2} - \frac{1}{3.4} + \frac{1}{5.6} - \frac{1}{7.8} + ...\infty$ (Osmania–2003)

(v) $1 - 2x + 3x^2 - 4x^3 + ...\infty \left(x < \frac{1}{2}\right)$ (Cochin–2005)

(vi) $\frac{x}{1+x} - \frac{x^2}{1+x^2} + \frac{x^3}{1+x^3} - \frac{x^4}{1+x^4} + ...\infty \ (0 < x < 1)$

(VTU–2004, Delhi–2002)

Hint to Selected Problems

1. $u_n = \frac{1}{n}, u_{n+1} = \frac{1}{n+1}$

$u_n - u_{n+1} = \frac{1}{n+1} > 0 \ \forall \ n \in N \Rightarrow u_n > u_{n+1}$.

Also $\lim u_n = 0$.

3. Since $\Sigma \left|\frac{\sin n\theta}{n^2}\right| \le \Sigma \frac{1}{n^2}$.

The series $\Sigma \frac{1}{n^2}$ is convergent.

7. $u_n = \sin\left(\frac{1}{n}\right) > 0$

$\sin\left(\frac{1}{n+1}\right) < \frac{1}{\sin n} \Rightarrow u_{n+1} < u_n$.

Also $\lim_{n\to\infty} u_n = \lim_{n\to\infty} \sin \frac{1}{n} = 0$

8. By D'Alembert's test, series is convergent for $x < \frac{1}{e}$ and divergent for $x > \frac{1}{e}$.

Then by logarithmic series, for $x = \frac{1}{e}$ the series is convergent.

18. The given series is convergent if

$$\sum_{n=1}^{\infty} u_n = \sum_{n=1}^{\infty} 2^n \sin\left(\frac{x}{3^n}\right)$$

$$\therefore \quad u_n = 2^n \sin\frac{x}{3^n} > 0, \forall \ n \in N$$

$$\Rightarrow \quad |u_n| = u_n \Rightarrow \Sigma |u_n| = \Sigma u_n$$

$$u_{n+1} = 2^{n+1} \sin\left(\frac{x}{3^{n+1}}\right) \Rightarrow \lim_{n\to\infty}\left|\frac{u_n}{u_{n+1}}\right| = \frac{3}{2} > 1.$$

Then by D'Alembert's ratio test. The given series is absolutely convergent.

ANSWERS

1. Convergent **5.** Convergent. **6.** Convergent if $x \le 1$ and divergent if $x > 1$.

20. (i) Convergent (ii) Convergent (iii) Convergent (iv) Convergent (v) Convergent

26.18 MULTIPLICATION OF SERIES

Consider two convergent infinite series given by

$$\sum_{n=1}^{\infty} a_n = a_1 + a_2 + \cdots + \cdots \text{ and } \quad \sum_{n=1}^{\infty} b_n = b_1 + b_2 + \cdots$$

Then $$\sum_{n=1}^{\infty} c_n = a_1 b_n + a_2 b_{n-1} + \cdots + a_n b_1 = \sum_{r=1}^{n} a_r b_{n-r+1}$$

is called the Cauchy product or simply product of two given series.

THEOREM 1. **(Cauchy's theorem).** *Let Σa_n and Σb_n be two absolutely convergent series then their Cauchy product Σc_n is also absolutely convergent and sum of the cauchy product series is the product of the sums.*

Proof. By definition of Cauchy product of two infinite series Σa_n and Σb_n we can write $\Sigma c_n = \Sigma a_n \cdot \Sigma b_n$ where

$$c_n = \sum_{r=1}^{\infty} a_r b_{n-r+1}$$

Let us suppose $\alpha_n = \sum_{r=1}^{n} a_r \ , \ \beta_n = \sum_{r=1}^{n} a_r \ , \ \sum_{n=1}^{\infty} a_n = l \ $, and $\sum_{n=1}^{\infty} b_n = m$

Then $<\alpha_n>$ and $<\beta_n>$ are the sequences of partial sums of Σa_n and Σb_n respectively. Also l and m are the sums of Σa_n and Σb_n respectively.

Consider $|a_1 b_1| + |a_1 b_2| + |a_2 b_2| + |a_2 b_1| + ...$ to n terms

$$\le (|a_1| + |a_2| + ... |a_n| (|b_1|) + |b_2| + ... + | |b_n|) \le l.m, \forall \ n \in N$$

Thus, the series

$$a_1 b_1 + a_1 b_2 + a_2 b_2 + a_2 b_1 + ... \quad ...(1)$$

must be absolulety convergent. Then by Dirichlet's theorem, (1) is absolutely convergent.

Hence, by grouping, we have

$$\sum_{n=1}^{\infty}\left(\sum_{r=1}^{n} a_r\, b_{n-r+1}\right) \quad i.e. \quad \sum_{n=1}^{\infty} c_n \qquad \ldots(2)$$

is absolutely convergent.

Finally, since the sum of first n^2 terms of (1) is $\alpha_n\,\beta_n$ and $\alpha_n\beta_n \to l.m$ as $n\to\infty$, therefore, the sum of the series (1) is lm. Hence, the series (2) must converge to the same sum.

Hence $\qquad \Sigma c_n = (\Sigma a_n)\,(\Sigma b_n)$

THEOREM 2. (Merten's theorem). *Let $\sum_{n=1}^{\infty} a_n$ and $\sum_{n=1}^{\infty} b_n$ be two convergent series and $\sum_{n=1}^{\infty} a_n$ converges absolutely. Then the Cauchy product series and $\sum_{n=1}^{\infty} c_n$ converges to lm where $l = \sum_{n=1}^{\infty} a_n$ and $m = \sum_{n=1}^{\infty} b_n$.*

Proof. Let $<A_n>$, $<B_n>$ and $<s_n>$ denote the sequence of partial sums of $\sum_{n=1}^{\infty} a_n$, $\sum_{n=1}^{\infty} b_n$ and $\sum_{n=1}^{\infty} c_n$ respectively. As per given, the series $\sum_{n=1}^{\infty} a_n$ and $\sum_{n=1}^{\infty} b_n$ converges to l and m respectively, therefore

$$\lim_{n\to\infty} A_n = l \text{ and } \lim_{n\to\infty} B_n = m \qquad \ldots(1)$$

Suppose that $p_n = s_n - m\ \forall\ n$ so that $\lim_{n\to\infty} p_n = \lim_{n\to\infty}(B_n - m) = m - m = 0 \qquad \ldots(2)$

Now $\qquad s_n = n^{th}$ partial sum of $\sum_{n=1}^{\infty} c_n$

$$\therefore \quad s_n = c_1+c_2+\ldots+c_n = a_1b_1 + a_1b_2+a_2b_1+a_1b_3+a_2b_2+a_3b_1+\ldots+a_1b_n+a_2b_{n-1}+\ldots+a_nb_1$$
$$= a_1(b_1+b_2+\ldots+b_n) + a_2(b_1+b_2+\ldots+b_{n-1}) + a_3(b_1+b_2+\ldots+b_{n-2}) + \ldots + \ldots + a_nb_1$$
$$= a_1B_1 + a_2B_{n-1} + a_2B_{n-2} + \ldots + a_nB_1$$
$$= a_1(p_n+m) + a_2(p_{n-1}+m) + a_3(p_{n-2}+m) + \ldots + a_n(p_1+m)$$
$$= a_1p_n + a_2p_{n-1} + a_3p_{n-2} + \ldots + a_np_1 + m(a_1+a_2+\ldots+a_n)$$
$$= q_n + mA_n \text{ where } q_n = a_1p_n + a_2p_{n-1} + \ldots + a_np_1$$

$$\therefore \quad \lim_{n\to\infty} s_n = \lim_{n\to\infty} q_n + m\lim_{n\to\infty} A_n = \lim_{n\to\infty} q_n + l_m \qquad \ldots(3)$$

Now, we have to show that $q_n \to 0$ as $n\to\infty$.

Here, since $\sum_{n=1}^{\infty} a_n$ converges absolutely $\Rightarrow \sum_{n=1}^{\infty} |a_n|$ converges.

Let $\sum_{n=1}^{\infty} |a_n|$ converges to l'.

From (2) $\lim_{n\to\infty} p_n = 0 \Rightarrow <p_n>$ converges $\Rightarrow <p_n>$ is bounded $\qquad \ldots(4)$

$\Rightarrow \quad \exists\, k > 0$ such that $|p_n| < k\ \forall\ n \qquad \ldots(5)$

By definition of convergence of sequence, for a given $\varepsilon > 0\ \exists$ *a* positive integer m_1 such that

$$|p_n| < \frac{\varepsilon}{2A'+1} \quad \forall n \geq m_1 \qquad \ldots(6)$$

Since, $\sum_{n=1}^{\infty} |a_n|$ converge, so by Cauchy's general principle of convergence there exists a positive integer m_2 such that

$$|a_{m_2}+1| + |a_{m_2}+2| + \ldots + |a_n| < \frac{\varepsilon}{2k+1}\ \forall\ n > m_1 \quad \Rightarrow \quad |a_{m_2}+1| + |a_{m_2}+2| + \ldots + |a_n| < \frac{\varepsilon}{2k+1}\ \forall\ n > m_2 \quad \ldots(7)$$

Let $m^* = \max\{m_1 + m_2\}$

then (6) and (7) are true for $n > m^*$

When $n > 2m^*$ then $n-m^* > m^*$, we have

$$|q_n| = |a_1p_n + a_2p_{n-1} + ... + a_np_1|$$

or $$|q_n| = |a_np_1 + a_{n-1}p_2 + p_{m+1}a_{n-m} + p_{m+2}\,a_{n-m-1} + ... + a_1p_n)$$

$$\leq |p_1||a_n| + |p_2||a_{n-1}| + |p_{m+1}||a_{n-m}| + |p_{m+2}||a_{n-m-1}| + \cdots + |p_n||a_1|$$

$$< k(|a_n| + |a_{n-1}| + \cdots + |a_{n-m}|) + \frac{\varepsilon}{2A'+1}(|a_{n-m-1}| + \cdots + |a_1|)$$

$$< k.\frac{\varepsilon}{2k+1} + \frac{\varepsilon}{2A'+1}(|a_1| + |a_2| + \cdots + |a_{n-m-1}|)$$

$$< k.\frac{\varepsilon}{2k+1} + \frac{\varepsilon}{2A'+1}A' \qquad \left[\therefore \sum_{n=1}^{\infty}|a_n| = A' \Rightarrow \sum_{n=1}^{n-m-1}|a_n| < A'\right]$$

$$< \frac{k\varepsilon}{2k+1} + \frac{\varepsilon}{2A'+1}.A' = \frac{\varepsilon}{2\left(1+\frac{1}{k}\right)} + \frac{\varepsilon}{2\left(1+\frac{1}{A'}\right)}$$

$\Rightarrow \quad |q_n| < \varepsilon$ whenever $n > 2m^*$ $\quad \Rightarrow \lim_{n\to\infty} q_n = 0$

$\Rightarrow \quad \lim_{n\to\infty} S_n = lm$ $\quad \Rightarrow \sum_{n=1}^{\infty} c_n$ converges to $l.m.$

THEOREM 3. (Abel's theorem). *Let $\sum_{n=1}^{\infty} a_n$ and $\sum_{n=1}^{\infty} b_n$ be two convergent series such that $\sum_{n=1}^{\infty} a_n$ and $\sum_{n=1}^{\infty} b_n$ converge to l and m respectively. If their Cauchy product $\sum_{n=1}^{\infty} c_n$ converges, then it converges to l.m.*

Proof. Let $<A_n>$, $<B_n>$ and $<s_n>$ be the sequences of partial sums of $\sum_{n=1}^{\infty} a_n$, $\sum_{n=1}^{\infty} b_n$ and $\sum_{n=1}^{\infty} c_n$ respectively, then

$$\lim_{n\to\infty} A_n = l \text{ and } \lim_{n\to\infty} B_n = m \qquad ...(1)$$

Now $$s_n = c_1 + c_2 + ... + c_n$$
$$= a_1b_1 + a_1b_2 + a_2b_1 + a_1b_3 + a_2b_2 + ... + a_1b_n + a_2b_{n-1} + ... + a_nb_1$$
$$= a_1(b_1+b_2+...+b_n) + a_2(b_1+b_2+...+b_{n-1}) + a_3(b_1+b_2+...+b_{n-2}) + ... + a_nb_1$$
$$= a_1B_n + a_2B_{n-1} + a_3B_{n-2} + ... + a_nB_1 \qquad ...(2)$$

which imples

$$s_{n-1} = a_1B_{n-1} + a_2B_{n-2} + a_3B_{n-3} + ... + a_{n-1}B_1$$
$$s_{n-2} = a_1B_{n-2} + a_2B_{n-3} + a_3B_{n-4} + ... + a_{n-2}B_1$$
....
$$s_1 = a_1B_1$$

On adding all these we get,

$$s_1 + s_2 + ... + s_n = a_1B_n + (a_1+a_2)B_{n-1} + (a_1+a_2+a_3)B_{n-2} + ... + (a_1+a_2+...+a_n)B_1$$
$$= A_1B_n + A_2B_{n-1} + A_3B_{n-2} + ... + A_nB_1$$

$$\Rightarrow \quad \frac{s_1+s_2+...+s_n}{n} = \frac{A_1B_n + A_2B_{n-1} + A_3B_{n-2} + ... + A_nB_n}{n}$$

As per given, $\sum_{n=1}^{\infty} c_n$ converges, Let it converges to s, then $\lim_{n\to\infty} s_n = s$

$\Rightarrow \quad \lim_{n\to\infty} \frac{s_1+s_2+...+s_n}{n} = s$ (By Cauchy's first theorem on limits)

Since $<A_n> \to l$ and $<B_n> \to m$ then by Ceasaro's theorem

$$\lim_{n\to\infty} \frac{A_1B_n + A_2B_{n-1} + ... + A_nB_1}{n} = l.m$$

$\therefore \quad \lim_{n\to\infty} \frac{s_1+s_2+...+s_n}{n} = \lim_{n\to\infty} \frac{A_1B_n + A_2B_{n-1} + ... + A_nB_1}{n} = l.m \quad \Rightarrow \quad s = l.m \Rightarrow \sum_{n=1}^{\infty} c_n$ conveges to $l.m.$

Solved Examples

Example 1. *Show that the Cauchy product of two divergent series given by*

$$\sum_{n=0}^{\infty} a_n = 1-\left(\frac{3}{2}\right)-\left(\frac{3}{2}\right)^2-\left(\frac{3}{2}\right)^3-\ldots$$

and $$\sum_{n=0}^{\infty} b_n = 1+\left(2+\frac{1}{2^2}\right)+\frac{3}{2}\left(2^2+\frac{1}{2^3}\right)+\left(\frac{3}{2}\right)^2\left(2^3+\frac{1}{2^4}\right)+\ldots$$

is convergent.

Solution. Clealy, we can write

$$\sum_{n=0}^{\infty} a_n = 1-\sum_{n=1}^{\infty}\left(\frac{3}{2}\right)^n$$

and $$\sum_{n=0}^{\infty} b_n = 1+\sum_{n=1}^{\infty}\left(\frac{3}{2}\right)^{n-1}\left(2^n+\frac{1}{2^{n+1}}\right)$$

If we leave the first term of both the series, we see that remaining series form G.P. with common ratio greater than 1 and therefore, both are divergent.

If $\sum_{n=0}^{\infty} c_n$ be the Cauchy product, then

$c_0 = a_0b_0=1$

$c_n= a_0b_n+a_1b_{n-1}+a_2b_{n-2}+\ldots+ a_nb_0,\ n\geq 1$

Therefore

$$c_n = 1.\left(\frac{3}{2}\right)^{n-1}\left(2^n+\frac{1}{2^{n+1}}\right)+\left(\frac{-3}{2}\right)\left(\frac{3}{2}\right)^{n-2}\left\{2^{n-1}+\frac{1}{2^n}\right\}+\cdots+\left(\frac{-3}{2}\right)^n\times 1$$

$$=\left(\frac{3}{2}\right)^{n-1}\left[\left(2^n+\frac{1}{2^{n+1}}\right)-\left(2^{n-1}+\frac{1}{2^n}\right)-\left(2^{n-2}+\frac{1}{2^{n-1}}\right)-\left(2+\frac{1}{2^2}\right)\right]-\left(\frac{3}{2}\right)^n$$

$$=\left(\frac{3}{2}\right)^{n-1}\left[2^n+\frac{1}{2^{n+1}}-\{2^{n-1}+2^{n-2}+\ldots+2\}-\left(\frac{1}{2^n}+\frac{1}{2^{n-1}}+\ldots+\frac{1}{2^2}\right)\right]-\left(\frac{3}{2}\right)^n$$

$$=\left(\frac{3}{2}\right)^{n-1}\left[2^n+\frac{1}{2^{n+1}}-2\left(\frac{2^{n-1}-1}{2-1}\right)-\frac{1}{2^2}\frac{\left(1-\frac{1}{2^{n-1}}\right)}{1-\frac{1}{2}}\right]-\left(\frac{3}{2}\right)^n$$

$$=\left(\frac{3}{2}\right)\left[2^n+\frac{1}{2^{n+1}}-2^n+2-\frac{1}{2}+\frac{1}{2^n}\right]-\left(\frac{3}{2}\right)^n$$

$$=\left(\frac{3}{2}\right)^{n-1}\left[\frac{3}{2}+\frac{1}{2^n}+\frac{1}{2^{n+1}}\right]-\left(\frac{3}{2}\right)^n$$

$$=\left(\frac{3}{2}\right)^{n-1}\left[\frac{3}{2}+\frac{3}{2^{n+1}}\right]-\left(\frac{3}{2}\right)^n$$

$$=\left(\frac{3}{2}\right)^n+\left(\frac{3}{2}\right)^{n-1}.\frac{3}{2}.\frac{1}{2^n}-\left(\frac{3}{2}\right)^n$$

$$=\left(\frac{3}{2}\right)^n\frac{1}{2^n}=\frac{3^n}{2^{2n}}=\left(\frac{3}{4}\right)^n$$

$\Rightarrow$ $\sum_{n=1}^{\infty} c_n = \sum_{n=1}^{\infty}\left(\frac{3}{4}\right)^4$ which is a G.P. with common ratio $\frac{3}{4}<1$ and hence convergent.

Example 2. *Prove that*

$$\frac{1}{1-x}\log\frac{1}{1-x}=\sum_{n=1}^{\infty}\left(1+\frac{1}{2}+\frac{1}{3}+\cdots+\frac{1}{n}\right)x^n,$$

for $|x|<1$.

Solution. Consider

$(1+x)^{-1}=1+x+x^2+\cdots=\sum_{n=0}^{\infty}x^n=\sum_{n=0}^{\infty}a_n$

where $a_n= x^n$, $n>0$

and $$\log\frac{1}{1-x}=\log(1-x)^{-1}=-\log(1-x)$$

$$=\frac{x}{1}+\frac{x^2}{2}+\frac{x^3}{3}+\cdots+\frac{x^n}{n}+\cdots$$

$$=\sum_{n=0}^{\infty}\frac{x^{n+1}}{n+1}=\sum_{n=0}^{\infty}b_n \text{ (say)}$$

where $b_n=\dfrac{x^{n+1}}{n+1}, n>0$

Now, $\lim_{n\to\infty}|a_n|^{1/n}=\lim_{n\to\infty}(|x|^n)^{1/n}=|x|$

Hence, by root test $\sum_{n=0}^{\infty}|a_n|$ is convegent for $|x|<1$ and therefore $\sum_{n=0}^{\infty}a_n$ is absolutely convergent for $|x|<1$.

Further

$$\lim_{n\to\infty}|b_{n-1}|^{1/n}=\lim_{n\to\infty}\left(\left|\frac{x^n}{n}\right|\right)^{1/n}=\lim_{n\to\infty}\frac{|x|}{n^{1/n}}=|x|$$

By root test, $\sum_{n=0}^{\infty}|b_n|$ is convergent for $|x|<1$ and therefore $\sum_{n=0}^{\infty}b_n$ is absolutely convergent for $|x|<1$.

If $\sum_{n=0}^{\infty}c_n$ be the Cauchy product of $\sum_{n=0}^{\infty}a_n$ and

$\sum_{n=0}^{\infty} b_n$ then

$$c_n = \sum_{k=0}^{n} a_{n-k}.b_k$$

$$= \sum_{k=0}^{n} x^{n-k}\left(\frac{x^{k+1}}{k+1}\right) = x^{n+1}\sum_{k=0}^{n}\frac{1}{k+1}$$

$$= x^{n+1}\left(1+\frac{1}{2}+\frac{1}{3}+\ldots+\frac{1}{n+1}\right)$$

Now, since both $\sum_{n=0}^{\infty} a_n$ and $\sum_{n=0}^{\infty} b_n$ are absolutely convergent for $|x| < 1$, so their Cauchy product $\sum_{n=0}^{\infty} c_n$ is convergent for $|x| < 1$.

Also, $\sum_{n=0}^{\infty} a_n . \sum_{n=0}^{\infty} b_n = \sum_{n=0}^{\infty} c_n$

$$\Rightarrow \frac{1}{1-x}\log\frac{1}{1-x} = \sum_{n=0}^{\infty} x^{n+1}\left(1+\frac{1}{2}+\frac{1}{3}+\ldots+\frac{1}{n+1}\right)$$

$$= \sum_{n=0}^{\infty} x^{n}\left(1+\frac{1}{2}+\frac{1}{3}+\ldots+\frac{1}{n}\right)$$

EXERCISE 26.5

1. Show that the Cauchy product of the convergent series. $\sum_{n=1}^{\infty}\frac{(-1)^n}{\sqrt{n+1}}$ with itself is divergent.
2. Show that the Cauchy product of the series $\sum_{n=1}^{\infty}\frac{(-1)^n}{(n+1)^p}, p>0$ with itself converges for $p>\frac{1}{2}$ and diverges for $p\le\frac{1}{2}$.
3. Show that the Cauchy product of two series $3+\sum_{n=1}^{\infty}3^n$ and $-2+\sum_{n=1}^{\infty}2^n$ is absolutely convergent, although both the series are divergent.
4. Show that
$$\frac{1}{2}\left(1-\frac{1}{3}+\frac{1}{5}-\frac{1}{7}+\ldots\right)^2 = \frac{1}{2}-\frac{1}{4}\left(1+\frac{1}{3}\right)+\frac{1}{6}\left(1+\frac{1}{3}+\frac{1}{5}\right)\ldots$$
5. Given, $\log 2 = 1-\frac{1}{2}+\frac{1}{3}-\frac{1}{4}+\ldots+\frac{(-1)^n}{n+1}$ prove that
$$\sum_{n=0}^{\infty}(-1)^n\left[\frac{1}{(n+1).1}+\frac{1}{n-2}+\frac{1}{(n-1).3}+\ldots+\frac{1}{1.(n-1)}\right]=(\log 2)^2$$

MISCELLANEOUS EXERCISE

1. Test the series for convergence
$1+\frac{1}{2^2}-\frac{1}{3^2}-\frac{1}{4^2}+\frac{1}{5^2}+\frac{1}{6^2}-\frac{1}{7^2}-\frac{1}{8^2}+\ldots\infty$ (VTU–2006)
2. Test the series $\frac{x}{\sqrt{3}}-\frac{x^2}{\sqrt{5}}+\frac{x^3}{\sqrt{7}}-\ldots$ for absolute convergence and conditional convergence. (VTU–2010)
3. Prove that the series $\frac{\sin x}{1^3}-\frac{\sin 2x}{2^3}+\frac{\sin 3x}{3^3}-\ldots$ converges absolutely. (Rohtak–2006S)
4. Discuss the absolute convergence of $\sum_{n=0}^{\infty}\frac{(-1)^n x^n}{n+1}$. (Hissar–2005S)
5. Find the nature of the series
$\frac{x}{1.2}-\frac{x^2}{2.3}+\frac{x^3}{3.4}-\frac{x^4}{4.5}+\ldots\infty$ (VTU–2009)
6. For what values of x the following series convergent
$x-\frac{x^2}{\sqrt{2}}+\frac{x^3}{\sqrt{3}}-\frac{x^4}{\sqrt{4}}+\ldots\infty$ (PTU–2009S, VTU–2008)
7. Find the readius of convergence of the series $\sum\frac{n!}{n^n}x^n$. (Calicut–2005)
8. Test the series $1-\frac{1}{2\sqrt{2}}+\frac{1}{3\sqrt{3}}-\frac{1}{4\sqrt{4}}$ for :
(i) absolute convergence (ii) conditional convergence (VTU–2007, Rohtak–2005)
9. Test the convergence of the series
$\sum_{n=1}^{\infty}\frac{2^n-2}{2^n+1}x^{n-1}$ $(x>0)$ (Osmania–1999)
10. Examine the following series for uniform convergence :
(i) $\sum_{n=1}^{\infty}\frac{\sin(nx+x^2)}{x(n+2)}$ (PTU–2009)
(ii) $\sum_{n=1}^{\infty}\frac{1}{n^p+n^q x^2}$ (PTU–2005S)
11. Test for uniform convergence the series :
$\frac{\sin x}{1^2}+\frac{\sin 2x}{2^2}+\frac{\sin 3x}{3^2}+\ldots\infty$ (PTU–2003, Andhra–2000)

ANSWERS

1. Convergent
4. Absolute convergent for $0 < x < 1$
5. Convergent for $x \le 1$ and not convergent for $x > 1$
6. $-1 < x \le 1$
7. $-e < x \le e$
8. Absolutely convergent
9. Convergent for $x < 1$; Divergent for $x \ge 1$

Objective Evaluations

Fill in the Blanks

1. A series, which contains __________ no. of terms is called infinite series.
2. If sum of the first n terms of an infinite series tends to a finite limit as n tends to infinity, then series is__________.
3. If sum of first n terms of an infinite series tends to $+\infty$ or $-\infty$ as n tends to infinity then series is said to be__________.
4. If sum of first n terms of an infinite series neither tends to a definite limit nor to $+\infty$ or $-\infty$ as n tends to infinity, then series is said to be__________.
5. For every convergent series, it is necessary that__________.
6. The nature of an infinite series remains ________ by addition or removal of a finite number of terms.
7. Every absolutely convergent series is ________.
8. The sum of an absolutely convergent series is________of the order of terms.
9. A series whose terms are alternatively positive and negative is called an__________.
10. If Σu_n, is convergent, and $\Sigma|u_n|$ is divergent then series Σu_n is said to be__________.

True/ False

Write 'T' for True and 'F' for False statements.

1. For every convergent series, it is necessary that $\lim u_n = 0$. **(T/F)**
2. The series $\sum \frac{1}{n}$ is convergent. **(T/F)**
3. If Σu_n is a series of positive terms then $u_n > 0, \forall n \in N$. **(T/F)**
4. A series of positive terms is divergent if each term after a fixed stage is greater than some fixed positive number. **(T/F)**
5. A series of positive terms is convergent if from and after some terms the ratio of each term to the preceding term is less than a fixed number which is less than unity. **(T/F)**
6. If $\lim u_n > 0$ then series is convergent. **(T/F)**
7. If $\lim u_n = 0$, then the series may or may not be convergent. **(T/F)**
8. If $\lim u_n = 0$, then series is always convergent. **(T/F)**
9. Every convergent series is absolutely convergent. **(T/F)**
10. Every conditionally convergent series is convergent. **(T/F)**

Multiple Choice Questions

Choose the most appropriate one.

1. A series is defined as :
   (a) the sum of the infinite no. of terms of the sequence
   (b) the product of the infinite no. of terms of the sequence
   (c) same as sequence
   (d) none of the above
2. A series is said to be convergent if (when n tends to ∞):
   (a) the sum of the first n terms tends to a finite limit
   (b) the sum of the first n terms tends to a unique limit
   (c) the sum of the first n terms tends to a finite and unique limit
   (d) none of the above
3. A series is said to be infinite if :
   (a) the sum of the first n terms tends to $+\infty$
   (b) the sum of the first n terms tends to $-\infty$
   (c) the no. of terms are infinite
   (d) none of the above
4. A series is said to be divergent if (when $n \to \infty$):
   (a) the sum of the first n terms tends to $+\infty$
   (b) the sum of the first n terms tends to $-\infty$
   (c) both (a) and (b) are true
   (d) none of the above
5. A series is said to be oscillatory if:
   (a) it is not convergent
   (b) it is not divergent
   (c) neither convergent nor divergent
   (d) none of the above
6. If $\lim u_n = 0$ then :
   (a) series is necessarily convergent
   (b) series is necessarily divergent
   (c) may or may not be convergent
   (d) none of the above
7. If Σu_n converges to l_1 and Σv_1 converges to l_2, then $\Sigma(u_n + v_n)$ converges to :
   (a) l_1 (b) l_2
   (c) l_1+l_2 (d) l_1-l_2

8. If Σu_n and Σv_n are two divergent series having all positive terms, then $\Sigma(u_n + v_n)$ is:
(a) convergent (b) divergent
(c) oscillatory (d) none of these

9. The nature of the given series will be change if:
(a) the sign of all terms are changed
(b) a finite no. of terms are added or omitted
(c) each terms of the series is multiplied or divided by a non-zero number
(d) none of the above

10. If Σu_n converges to l and $c \in R$, then Σcu_n:
(a) converges to cl
(b) converges to l
(c) may or may not be converge
(d) none of the above

11. If n^{th} term of the series does not tends to zero as $n \to \infty$ then series is :
(a) necessarily convergent
(b) may or may not be convergent
(c) never convergent
(d) none of the above

12. A geometric series is convergent if the common ratio is :
(a) less than 1 (b) less than equal to 1
(c) greater than equal to 1 (d) none of these

13. The auxiliary series $\sum \frac{1}{n^p}$ is divergent if:
(a) $p = 1$ (b) $p \le 1$
(c) $p \ge 1$ (d) $p = 0$

14. If $\sum u_n$ be a series of positive terms such that $\lim u_n^{1/n} = l$, then series :
(a) converges if $l < 3$
(b) diverges if $l > 1$
(c) both (a) and (b) are true
(d) both (a) and (b) are false.

15. An infinite alternative series is convergent if :
(a) $\lim u_n = 0$
(b) each term is numerically less than the preceding terms
(c) both (a) and (b) are true
(d) none of the above

16. A series Σu_n is said to be absolutely convergent if :
(a) $\lim u_n = 0$ (b) $\lim u_n = +$ ve
(c) $\lim u_n$ exists only (d) Σu_n is convergent.

17. A series Σu_n, is said to be conditionally convergent or semi-convergent if :
(a) Σu_n is convergent and $\Sigma|u_n|$ is not convergent
(b) Σu_n and $\Sigma|u_n|$ both must be convergent
(c) Σu_n and $\Sigma|u_n|$ both must be divergent
(d) none of the above

18. Every absolutely convergent series is :
(a) convergent
(b) divergent
(c) oscillatory
(d) may or may not be convergent

19. Every semi-convergent series is :
(a) convergent (b) absolutely convergent
(c) divergent (d) none of the above

20. If the terms of an absolutely convergent series are rearranged, then the series :
(a) may or may not be convergent
(b) remains convergent
(c) divergent
(d) may or may not be divergent

21. The series $\sum_{n=1}^{\infty} (-1)^n$ is:
(a) convergent (b) divergent
(c) unbounded (d) none of the above

22. The series $\frac{1}{1.3} + \frac{1}{2.4} + \frac{1}{3.5} + \ldots$ is:
(a) convergent (b) divergent
(c) may be convergent (d) none of the above

23. The series $1.x + 1.2x^2 + 1.2.3x^3 + \ldots + n!x^n + \ldots$ is:
(a) convergent everywhere except at $x = 0$
(b) divergent everywhere except at $x = 0$
(c) divergent for $x = 0$
(d) none of the above

24. The series $\sum_{n=1}^{\infty} \frac{n^2}{3^n}$ is:
(a) convergent (b) divergent
(c) unbounded (d) none of these

25. The series for $\sum_{n=1}^{\infty} (-1)^n \frac{x^n}{n}$ for $|x| > 1$ is:
(a) convergent (b) divergent
(c) oscillatory (d) none of the above

26. For infinite series $\sum_{n=1}^{\infty} a_n$ and $\sum_{n=1}^{\infty} b_n$, $b_n \ge 0$ for all n and there is a real number N such that for $n \ge N \Rightarrow |a_n| \le b_n$ if $\sum_{n=1}^{\infty} b_n$ converges, then
(a) Σa_n is absolutely convergent
(b) Σb_n is absolutely convergent
(c) Σa_n is divergent
(d) none of the above

27. The series $1 + \frac{x}{1^2} + \frac{x^2}{2^2} + \frac{x^3}{3^2} + \ldots$:
(a) converges for $|x| < 1$
(b) diverges for $|x| > 1$
(c) diverges for $|x| = 1$
(d) none of the above

28. The series $1 + x + x^2 + \ldots + x^n \ldots +$:
(a) converges for $|x| < 1$
(b) diverges for $|x| > 1$
(c) converges for $|x| = 1$
(d) none of the above

29. Let Σa_n be a convergent series of positive terms and let Σb_n be a divergent series of positive terms converges to 0. Then:
(a) $<a_n>$ converges to 0
(b) $<a_n>$ not converges to 0
(c) $<b_n>$ diverges to 0
(d) none of the above

30. Let Σa_n be a convergent series of positive terms and Σb_n be a divergent series of positive terms. Then:
(a) $\Sigma(a_n+b_n)$ is divergent
(b) $\Sigma(a_n+b_n)$ is convergent
(c) both (a) and (b) may be true
(d) none of the above

31. If for a given series Σu_n
$\frac{u_n}{u_{n+1}} = \alpha + \frac{\beta}{n} + o\left(\frac{1}{n^p}\right), p > 1, \alpha, \beta \in R$. Then
(a) Σu_n converges if $\beta > 1 \; \forall \; \alpha$
(b) Σu_n converges if $\alpha > 1$ for any β
(c) Σu_n diverges to ∞
(d) none of the above

32. If $n^{1/n} \to 1$ as $n \to \infty$, then the series $\sum_{n=1}^{\infty} (n^{1/n} - 1)^n$:
(a) converges
(b) diverges
(c) neither converges nor diverges
(d) none of the above

33. Which of the following inequalities will be used to show the series $\sum_{n=1}^{\infty} e^{-n^2}$ converges ?
(a) $e^n < x$ if $n = 0$ (b) $e^n < x$ if $n > 0$
(c) $e^n > x$ if $n > 0$ (d) None of the above

34. A conditionally convergent series is a series which is
(a) absolutely convergent
(b) convergent but not absolutly convergent
(c) divergent
(d) none of the above

35. For the series $1 + r + r^2 + ...$ $(r > 0)$, which one of the following is true?
(a) It does not converges (b) It does not diverges.
(c) It oscillate. (d) None of these

36. The series $\frac{1}{3^p} + \frac{1}{5^p} + \frac{1}{7^p} + \frac{1}{9^p} + ...$ converges if :
(a) $p < 1$ (b) $p > 1$
(c) $p = 1$ (d) none of the above

37. The series $1 + r + r^2 + ...$ is oscillatory if:
(a) $r < 1$ (b) $r > 1$
(c) $r = 1$ (d) $r = -1$

38. The series $x + \frac{2^2 x^2}{2!} + \frac{3^3 x^3}{3!} + \frac{4^4 . x^4}{4!} + ...$ is convergent if:
(a) $0 < x < \frac{1}{e}$ (b) $x > \frac{1}{e}$
(c) $x = \frac{1}{e}$ (d) none of these

39. Let Σu_n be a series of positive terms. Given that Σu_n is convergent and also $\lim_{n\to\infty} \frac{u_{n+1}}{u_n}$ exists then the limit is:
(a) necessarily equal to 1
(b) necessarily less than 1
(c) necessarily geater than –1
(d) none of the above

40. The series $\sum \frac{n!.2^n}{n^n}$ is:
(a) convergent (b) divergent
(c) oscillatory (d) none of the above

41. Which one of the following test does not give absolute convergence of a series ?
(a) comparison test (b) root test
(c) ratio test (d) none of these

42. If $u_n = \sqrt{n+1} - \sqrt{n}, v_n = \sqrt{n^4+1} - \sqrt{n^4}$ then:
(a) $\sum_{n=1}^{\infty} u_n$ converges but $\sum_{n=1}^{\infty} v_n$ diverges
(b) $\sum_{n=1}^{\infty} u_n$ and $\sum_{n=1}^{\infty} v_n$ both converges
(c) $\sum_{n=1}^{\infty} u_n$ and $\sum_{n=1}^{\infty} v_n$ both diverges
(d) none of the above

43. Ths series:
$\left(\frac{2^2}{1^2} - \frac{2}{1}\right)^{-1} + \left(\frac{3^3}{2^3} - \frac{3}{2}\right)^{-2} + \left(\frac{4^4}{3^4} - \frac{4}{3}\right)^{-3} + ...$ is
(a) convergent (b) divergent
(c) oscillatory (d) none of the above

44. Ths series $\sum \left(1 + \frac{1}{n}\right)^{-n^2}$ is
(a) convergent (b) divergent
(c) oscillatory (d) none of the above

45. Ths series $\sum \frac{1}{\sqrt{n} + \sqrt{n+1} + \sqrt{n+2} + \sqrt{n+3}}$ is:
(a) convergent (b) divergent
(c) oscillatory (d) none of the above

46. Ths series $\sum \frac{1}{n^p + n^{-p}}$ converges of:
(a) $p = 1$
(b) $|p| > 1$
(c) both (a) and (b) are true
(d) none of the above

47. The infinite series $\sum [\sqrt[3]{n^3+1} - n] x^n$ converges if
(a) $|x| \le 1$ (b) $|x| \ge 1$
(c) $|x| = e$ (d) none of the above

48. Which one of the following is divergent
(a) $\sum \sin \frac{1}{n^2}$
(b) $\sum \sin \frac{1}{n}$
(c) both (a) and (b) are true
(d) none of the above

49. If Σu_n is a positive term series such that $\lim_{n\to\infty} u_n \neq 0$ then Σu_n:

(a) converges (b) diverges
(c) oscillatory (d) none of the above

50. The series $\sum_{n=1}^{\infty} (-1)^n . n$ is :

(a) bounded (b) convergent
(c) divergent (d) none of the above

ANSWERS

Fill in the Blanks

1. infinite	**2.** convergent	**3.** divergent	**4.** oscillatory	**5.** $\lim u_n = 0$
6. unchanged	**7.** convergent	**8.** independent	**9.** alternating series	

10. conditionally or semi-convergent.

True/ False

1. T	**2.** F	**3.** T	**4.** T	**5.** T	**6.** F	**7.** T	**8.** F	**9.** F	**10.** T

Multiple Choice Questions

1. (a)	**2.** (c)	**3.** (c)	**4.** (c)	**5.** (c)	**6.** (c)	**7.** (c)	**8.** (b)	**9.** (d)	**10.** (a)
11. (c)	**12.** (a)	**13.** (b)	**14.** (c)	**15.** (c)	**16.** (d)	**17.** (a)	**18.** (a)	**19.** (a)	**20.** (b)
21. (b)	**22.** (a)	**23.** (b)	**24.** (a)	**25.** (b)	**26.** (a)	**27.** (b)	**28.** (a)	**29.** (a)	**30.** (a)
31. (b)	**32.** (a)	**33.** (c)	**34.** (b)	**35.** (a)	**36.** (b)	**37.** (d)	**38.** (a)	**39.** (b)	**40.** (a)
41. (a)	**42.** (b)	**43.** (a)	**44.** (a)	**45.** (b)	**46.** (b)	**47.** (a)	**48.** (a)	**49.** (a)	**50.** (a)

FFFFF

Fourier Series

27.1 INTRODUCTION

In this section, we shall study a special type of functional series extensively studied by Joseph Fourier. Joseph Fourier represented expansions in trigonometrical series in connection with boundary value problem in conduction of heat. Although such expansions had been studied earlier, these series bear the name 'Fourier series'because of the major contributions of Fourier in this field.

27.2 PERIODIC FUNCTIONS

(UPTU–2002)

A function $f(x)$ which satisfies the relation $f(x + T) = f(x)$for all real x and some fixed T is called a periodic function. The smallest positive number T, for which this relation holds, is called the period of $f(x)$.

If T is the period of $T(x)$. Then

$$f(x) = f(x + T) = f(x + 2T) = ... = f(x + nT) = ...$$

Also, $$f(x) = f(x - T) = f(x - 2T) = ... = f(x - nT) = ...$$

$\therefore$ $$f(x) = f(x \pm nT), \text{ where } n \text{ is a positive integer.}$$

For example: Consider the function $f(x) = \sin x$. We have

$$\sin x = \sin (x + 2\pi) = \sin (x + 4\pi) = ...$$

Here, $f(x) = \sin x$ is a periodic function with period 2π. This function is also called sinusoidal periodic function.

We have studied about the Macluarian's theorem which is used to expand a function provided the function's derivative are continuous. Now, the need arise to expand functions which have discontinuities in their derivatives. By Fourier series, we can expand both types of functions under certain conditions as an infinite series of sine and cosine of x and it's integral multiple of a function $f(x)$ is defined in the interval $c < x < c + 2\pi$.

Then, Fourier series of $f(x)$ is given by

$$f(x) = \frac{a_0}{2} + \sum_{n=1}^{\infty} a_n \cos nx + \sum_{n=1}^{\infty} b_n \sin nx \quad ...(1)$$

where a_0, a_n and b_n are called Fourier coefficient of $f(x)$ and their values are given as :

$$a_0 = \frac{1}{\pi} \int_c^{c+2\pi} f(x) dx \quad ...(2)$$

$$a_n = \frac{1}{\pi} \int_c^{c+2\pi} f(x) \cos nx \, dx \quad ...(3)$$

$$b_n = \frac{1}{\pi} \int_c^{c+2\pi} f(x) \sin nx dx \quad ...(4)$$

The series (1) with coefficients a_0, a_n and b_n given by (2), (3) and (4) respectively is called the Fourier series of $f(x)$ and the coefficients a_0, a_n and b_n are called the Fourier coefficients corresponding to $f(x)$.

(i) When $c = 0$, the interval becomes $0 < x < 2\pi$ and formula for a_0, a_n, b_n is obtained by putting $c = 0$.

(ii) When $c = -\pi$, then interval becomes $-\pi < x < \pi$. In this interval, the formula for a_0, a_n and b_n becomes as under :

(a) When $f(x)$ is an odd function, then

$$a_0 = \frac{1}{\pi}\int_{-\pi}^{\pi} f(x)dx = 0 \qquad a_n = \frac{1}{\pi}\int_{-\pi}^{\pi} f(x)\cos nx\, dx = 0 \quad \text{[By property of definite integral]}$$

$$b_n = \frac{1}{\pi}\int_{-\pi}^{\pi} f(x)\sin nx\, dx = \frac{2}{\pi}\int_{0}^{\pi} f(x)\sin x dx$$

Hence, if function $f(x)$ is odd, its Fourier expansion contains only sine series,

i.e., $$f(x) = \sum_{n=1}^{\infty} b_n \sin nx, \text{where } b_n = \frac{2}{\pi}\int_{0}^{\pi} f(x)\sin nx\, dx.$$

(b) When $f(x)$ is even function, then formula for a_0, a_n and b_n are given by

$$a_0 = \frac{1}{\pi}\int_{-\pi}^{\pi} f(x)dx = \frac{2}{\pi}\int_{0}^{\pi} f(x)dx, \quad a_n = \frac{1}{\pi}\int_{-\pi}^{\pi} f(x)\cos nx\, dx = \frac{2}{\pi}\int_{0}^{\pi} f(x)\cos nx\, dx$$

and $$b_n = \frac{1}{\pi}\int_{-\pi}^{\pi} f(x)\sin nx\, dx = 0 \qquad [\because f(x) \sin nx \text{ is odd.}]$$

Hence, if a periodic function $f(x)$ is even, its Fourier expansion contains only cosine terms, *i.e.*, $f(x) = \frac{a_0}{2} + \sum_{n=1}^{\infty}\int_{0}^{\pi} f(x)dx$,

where $$a_0 = \frac{2}{\pi}\int_{0}^{\pi} f(x)dx \text{ and } a_n = \frac{2}{\pi}\int_{0}^{\pi} f(x).\cos nx\, dx$$

27.3 SOME IMPORTANT RESULTS

The following results are useful in the Fourier series :

(i) $\sin n\pi = 0, \cos n\pi = (-1)^n, \cos\left(n + \frac{1}{2}\right)\pi = 0,$ where $n \in Z$.

(ii) $\int uv = uv_1 - u'v_2 + u''v_3 - u'''v_4 + ...,$ where $u' = \frac{du}{dx}, u'' = \frac{d^2u}{dx^2},...$
$v_1 = \int v dx, v_2 = \int v_1 dx,...$

(iv) diver

(iii) $\int_0^{2\pi} \sin nx\, dx = 0$ (iv) $\int_0^{2\pi} \cos nx\, dx = 0$ (v) $\int_0^{2\pi} \sin^2 nx\, dx = \pi$

(vi) $\int_0^{2\pi} \cos^2 nx\, dx = \pi$ (vii) $\int_0^{2\pi} \sin nx.\sin mx\, dx = 0$ (viii) $\int_0^{2\pi} \cos nx.\cos mx\, dx = 0$

(ix) $\int_0^{2\pi} \sin nx.\cos nx\, dx = 0$ (x) $\int_0^{2\pi} \sin nx.\cos nx\, dx = 0$

(xi) $\int e^{ax} \sin bx\, dx = \frac{e^{ax}}{a^2 + b^2}(a \sin bx - b\cos bx) + c$ (xii) $\int e^{ax} \cos bx\, dx = \frac{e^{ax}}{a^2 + b^2}(a\cos bx - b\sin bx) + c$

27.4 DETERMINATION OF FOURIER COEFFICIENTS: EULER'S FORMULAE

The fourier series is given by

$$f(x) = \frac{a_0}{2} + a_1 \cos x + a_2 \cos 2x + ... + a_n \cos nx + b_1 \sin x + ... + b_2 \sin 2x + ... + b_n \sin nx + ... \qquad ...(i)$$

or $$f(x) = \frac{a_0}{2} + \sum_{n=1}^{\infty} a_n \cos nx + \sum_{n=1}^{\infty} b_n \sin nx.$$

To find a_0 : Integrating both sides of equation (1) from $x = c+0, x = c+2\pi$

$$\int_c^{c+2\pi} f(x)dx = \frac{a_0}{2}\int_c^{c+2\pi} dx + \int_c^{c+2\pi}\left(\sum_{n=1}^{\infty} a_n \cos nx\right)dx + \int_c^{c+2\pi}\left(\sum_{n=1}^{\infty} b_n \sin nx\right)dx = \frac{a_0}{2}(c + 2\pi - c) + 0 + 0 = a_0\pi$$

$$\Rightarrow \quad a_0 = \frac{1}{\pi}\int_c^{c+2\pi} f(x)\,dx.$$

To find a_n : Multipling each side of equation (1) by cos nx and integrate w.r.t. x., between the limit c to $c+2\pi$.

$$\int_c^{c+2\pi} f(x)\cos nx\,dx = \frac{a_0}{2}\int_c^{c+2\pi}\cos nx\,dx + \int_c^{c+2\pi}\left(\sum_{n=1}^{\infty} a_n\cos nx\right)\cos nx\,dx + \int_c^{c+2\pi}\left(\sum_{n=1}^{\infty} b_n\sin nx\right)\cos nx\,dx$$

$$= 0 + a_n\pi + 0 = a_n\pi$$

$$\Rightarrow \quad a_n = \frac{1}{\pi}\int_c^{c+2\pi} f(x)\cos nx\,dx.$$

To find b_n : Multiplying each side of equation (1) by sin nx and integrate w.r.t. x between the limit c to $c + 2\pi$.

$$\int_c^{c+2\pi} f(x)\sin nx\,dx = \frac{a_0}{2}\int_c^{c+2\pi}\sin nx dx + \int_c^{c+2\pi}\left(\sum_{n=1}^{\infty} a_n\cos nx\right)\sin nx\,dx + \int_c^{c+2\pi}\left(\sum_{n=1}^{\infty} b_n\sin nx\right)\sin nx\,dx$$

$$= 0 + 0 + b_n\pi = b_n\pi$$

$$\Rightarrow \quad b_n = \frac{1}{\pi}\int_c^{c+2\pi} f(x)\sin nx\,dx$$

These values of a_0, a_n and b_n are called Euler's formulae.

27.5 DIRICHLET'S CONDITIONS

Any function $f(x)$ can be expressed as a Fourier series $\frac{a_0}{2} + \sum_{n=1}^{\infty} a_n\cos nx + \sum_{n=1}^{\infty} b_n\sin nx$, where a_0, a_n and b_n are constants.

(i) $f(x)$ is finite and single valued in the interval $c < x < c + 2\pi$.

(ii) $f(x)$ is periodic with period 2π.

(iii) $f(x)$ and $f'(x)$ are piecewise continuous in the interval $c < x < c + 2\pi$.

The Fourier series with its coefficients converge to

(a) $f(x)$ if x is a point of continuity.

(b) $\frac{f(x+0)+f(x-0)}{2}$, if x is a point of discontinuity.

The conditions (i), (ii) and (iii) imposed on $f(x)$ are sufficient but not necessary, *i.e.*, if the conditions are satisfied, the convergence is guranteed. However, if they are not satisfied the series may or may not converge.

Solved Examples

Example 1. *Expand the function $f(x) = x \sin x$ as a Fourier series in interval $-\pi \le x \le \pi$. Deduce that* $\frac{1}{1.3} - \frac{1}{3.5} + \frac{1}{5.7} - \frac{1}{7.9} + \ldots = \frac{\pi-2}{4}$

(UPTU–2001, 2005, 2008, Q.Bank–2001, SVTU–2009, Bhopal–2009, Rohtak–2006)

Solution. Since $x \sin x$ is an even function of x, so $b_n = 0$, then Fourier series is given by

$$f(x) = x\sin x = \frac{a_0}{2} + \sum_{n=1}^{\infty} a_n\cos nx,$$

$$\text{where } a_0 = \frac{2}{\pi}\int_0^{\pi} x\sin x dx = \frac{2}{\pi}[-x\cos x + \sin x]_0^{\pi}$$

$$= \frac{2}{\pi}(-\pi\cos\pi) = 2$$

$$a_n = \frac{2}{\pi}\int_0^{\pi} x\sin x\cos nx\,dx = \frac{1}{\pi}\int_0^{\pi} x.2\cos nx\sin x\,dx$$

$$= \frac{1}{\pi}\int_0^{\pi} x\{\sin(n+1)x - \sin(n-1)x\}\,dx$$

$$= \frac{1}{\pi}\left[x\left\{\frac{-\cos(n+1)x}{n+1} + \frac{\cos(n+1)x}{n-1}\right\} - 1\left\{\frac{-\sin(n+1)x}{(n+1)^2} + \frac{\sin(n-1)x}{(n+1)^2}\right\}\right]_0^{\pi}$$

$$= \frac{1}{\pi}\left[x\left\{\frac{-\cos(n+1)\pi}{n+1} + \frac{\cos(n-1)\pi}{n-1}\right\}\right]$$

$$= \frac{\cos(n-1)\pi}{n-1} - \frac{\cos(n+1)\pi}{n+1}; n \ne 1$$

$$= \begin{cases} \frac{1}{n-1} - \frac{1}{n+1} = \frac{2}{n^2-1} & \text{if } n \text{ is odd } n \ne 1 \\ \frac{-1}{n-1} + \frac{1}{n+1} = \frac{-2}{n^2-1} & \text{if } n \text{ is even} \end{cases}$$

When $n=1$, then

$$a_1 = \frac{2}{\pi}\int_0^{\pi} x\sin x\cos x dx = \frac{1}{\pi}\int_0^{\pi} x\sin 2x dx$$

$$= \frac{1}{\pi}\left[x\left(\frac{-\cos 2x}{2}\right) - \left(\frac{-\sin 2x}{4}\right)\right]_0^{\pi}$$

$$= \frac{1}{\pi}\left[\frac{-\pi\cos 2\pi}{2}\right] = -\frac{1}{2}$$

$$\therefore \quad x\sin x = 1 - \frac{1}{2}\cos x - 2\left[\frac{\cos 2x}{2^2-1} - \frac{\cos 3x}{3^2-1} + \frac{\cos 4x}{4^2-1} - \frac{\cos 5x}{5^2-1} + \ldots\right]$$

Putting $x = \frac{\pi}{2}$, we get

$$\frac{\pi}{2} = 1 - 2\left(\frac{-1}{2^2-1} + \frac{1}{4^2-1} - \frac{1}{6^2-1} + \ldots\right)$$

$$\Rightarrow \quad \frac{\pi}{2} - 1 = 2\left(\frac{1}{3} - \frac{1}{15} + \frac{1}{35} - \ldots\right)$$

$$\Rightarrow \quad \frac{\pi-2}{4} = \left(\frac{1}{1.3} - \frac{1}{3.5} + \frac{1}{5.7} - \ldots\right).$$

Example 2. *Find the Fourier series to represent e^{ax} in interval $-\pi < x < \pi$.*

Solution. Let

$$f(x) = e^{ax} = \frac{a_0}{2} + \sum_{n=1}^{\infty} a_n \cos nx + \sum_{n=1}^{\infty} b_n \sin nx$$

$$a_0 = \frac{1}{\pi}\int_{-\pi}^{\pi} f(x)dx = \frac{1}{\pi}\int_{-\pi}^{\pi} e^{ax}dx = \frac{1}{\pi}\left[\frac{e^{ax}}{a}\right]_{-\pi}^{\pi}$$

$$= \frac{1}{a\pi}(e^{a\pi} - e^{-a\pi}) = \frac{2\sinh a\pi}{\pi a}$$

$$a_n = \frac{1}{\pi}\int_{-\pi}^{\pi} f(x)\cos nx\,dx = \frac{1}{\pi}\int_{-\pi}^{\pi} e^{ax}\cos nx\,dx$$

$$= \left[\frac{e^{ax}}{\pi(a^2+n^2)} a\cos nx + a\sin nx\right]_{-\pi}^{\pi}$$

$$= \frac{a\cos n\pi(e^{a\pi} - e^{-ia\pi})}{\pi(a^2+n^2)} = \frac{2a(-1)^n \sinh a\pi}{\pi(a^2+n^2)}$$

Similarly, we can set

$$b_n = \frac{2n(-1)^n \sinh a\pi}{\pi(a^2+n^2)}$$

$$\therefore \quad e^{ax} = \frac{\sinh a\pi}{a\pi} + \sum_{n=1}^{\infty} \frac{2a(-1)^n \sinh a\pi}{\pi(a^2+n^2)}\cos nx + \sum_{n=1}^{\infty} \frac{2n(-1)^n \sinh a\pi}{\pi(a^2+n^2)}\sin n\pi$$

$$= \frac{2\sinh a\pi}{\pi}\left[\frac{1}{2a} - a\left(\frac{\cos x}{a^2+1^2} - \frac{\cos 2x}{a^2+2^2} + \frac{\cos 3x}{a^2+3^2} - \ldots\right) - \left(\frac{\sin x}{a^2+1^2} - \frac{2\sin 2x}{a^2+2^2} + \frac{3\sin 3x}{a^2+3^2} - \ldots\right)\right]$$

Example 3. *Obtain the Fourier series for the function $f(x) = x^2$, $-\pi < x < \pi$. Sketch the graph f function $f(x)$. Hence, show that* (PTU–2009, Bhopal–2008, BPTU–2006)

(i) $\frac{1}{1^2} + \frac{1}{2^2} + \frac{1}{3^2} + \frac{1}{4^2} + \ldots = \sum_{n=1}^{\infty} \frac{1}{n^2} = \frac{\pi^2}{6}$

(UPTU(Q. Bank)–2001, Anna–2009, PTU–2009, Osmania–2003, Mumbai–2009, SVTU–2008)

(ii) $\frac{1}{1^2} - \frac{1}{2^2} + \frac{1}{3^2} - \frac{1}{4^2} + \ldots = \frac{\pi^2}{12}$

(UPTU–2004, 08, SVTU–2008)

(iii) $\frac{1}{1^2} + \frac{1}{3^2} + \frac{1}{5^2} + \ldots = \sum_{n=1}^{\infty} \frac{1}{(2n-1)^2} = \frac{\pi^2}{8}$

Solution. $f(x) = x^2$ is an even function, therefore $b_n = 0$

Now $f(x) = x^2 = \frac{a_0}{2} + \sum_{n=1}^{\infty} a_n \cos nx$. Then

$$a_0 = \frac{2}{\pi}\int_0^{\pi} f(x)dx = \frac{2}{\pi}\int_0^{\pi} x^2 dx = \frac{2}{\pi}\left[\frac{x^3}{3}\right]_0^{\pi} = \frac{2}{3}\pi^2$$

$$a_n = \frac{2}{\pi}\int_0^{\pi} f(x)\cos nx dx = \frac{2}{\pi}\int_0^{\pi} x^2 \cos nx\,dx$$

$$= \frac{2}{\pi}\left[x^2\left(\frac{\sin nx}{n}\right) - 2x\left(\frac{-\cos nx}{n^2}\right) + 2\left(\frac{-\sin nx}{n^2}\right)\right]_0^{\pi}$$

$$= \frac{2}{\pi}\left[2x\frac{\cos n\pi}{n^2}\right] = 4\frac{(-1)^n}{n^2}$$

$$\therefore \quad x^2 = \frac{\pi^2}{3} - 4\left(\frac{\cos x}{1^2} - \frac{\cos 2x}{2^2} + \frac{\cos 3x}{3^2} - \frac{\cos 4x}{4^2} + \ldots\right)$$

$$\Rightarrow \quad x^2 = \frac{\pi^2}{3} + 4\sum_{n=1}^{\infty} \frac{(-1)^n}{n^2}\cos nx \quad \ldots (1)$$

Put $x=\pi$ in (1), we get

$$\pi^2 = \frac{\pi^2}{3} - 4\left(-\frac{1}{1^2} - \frac{1}{2^2} - \frac{1}{3^2} - \frac{1}{4^2} - \ldots\right)$$

$$\Rightarrow \quad \frac{2\pi^2}{3} = -4\left(-\frac{1}{1^2} - \frac{1}{2^2} - \frac{1}{3^2} - \frac{1}{4^2} - \ldots\right)$$

$$\therefore \quad \frac{1}{1^2} + \frac{1}{2^2} + \frac{1}{3^2} + \frac{1}{4^2} \ldots = \frac{\pi^2}{6} \quad \ldots (2)$$

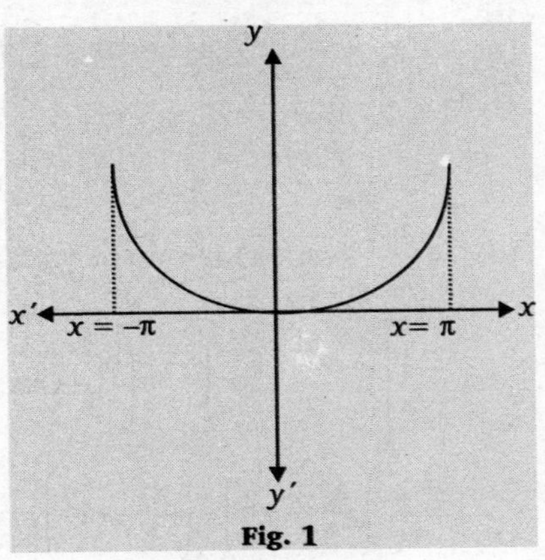

Fig. 1

Put $x = 0$ in (1), we get

$$0=\frac{\pi^2}{3}-4\left(-\frac{1}{1^2}-\frac{1}{2^2}-\frac{1}{3^2}-\frac{1}{4^2}-\ldots\right)$$

$$\therefore \quad \frac{1}{1^2}-\frac{1}{2^2}-\frac{1}{3^2}-\frac{1}{4^2}-\ldots=\frac{\pi^2}{12} \qquad \ldots(3)$$

Adding (2) and (3), we get

$$\frac{\pi^2}{4}=2\left(\frac{1}{1^2}+\frac{1}{3^2}+\frac{1}{5^2}+\ldots\right)$$

$$\therefore \quad \frac{1}{1^2}+\frac{1}{3^2}+\frac{1}{5^2}+\ldots=\frac{\pi^2}{8}.$$

Example 4. *Obtain the Fouries series for $f(x) = e^{-x}$ in the interval $0 < x < 2\pi$.*

(UPTU(Q.Bank)–2001, SVTU–2007)

Solution. Let $f(x) = e^{-x}$. The Fourier series of $f(x)$ can be written as

$$f(x)=e^{-x}=\frac{a_0}{2}+\sum_{n=1}^{\infty}a_n\cos nx+\sum_{n=1}^{\infty}b_n\sin nx$$

Then, $$a_0=\frac{1}{2}\int_0^{2\pi}f(x)dx=\frac{1}{\pi}\int_0^{2\pi}e^{-x}dx$$

$$=\frac{1}{\pi}.\left[-e^{-x}\right]_0^{2\pi}=\frac{1-e^{-2\pi}}{\pi}$$

$$a_n=\frac{1}{\pi}\int_0^{2\pi}f(x)\cos nx\,dx=\frac{1}{\pi}\int_0^{2\pi}e^{-x}\cos nx\,dx$$

$$=\frac{1}{\pi(1+n^2)}[e^{-x}(-\cos nx+n\sin nx)]_0^{2\pi}$$

$$=\frac{1-e^{-2\pi}}{\pi(1+n^2)}$$

$$b_n=\frac{1}{\pi}\int_0^{2\pi}f(x)\sin nx\,dx=\frac{1}{\pi}\int_0^{2\pi}e^{-x}\sin nx\,dx$$

$$=\frac{1}{\pi(1+n^2)}\left[-\sin nx-n\cos nx\right]_0^{2\pi}$$

$$=\frac{1-e^{-2\pi}}{\pi}.\frac{n}{1+n^2}$$

$$\therefore e^{-x}=\frac{1-e^{-2\pi}}{\pi}\left[\frac{1}{2}+\left(\frac{1}{2}\cos x+\frac{1}{5}\cos 2x+\frac{1}{10}\cos 3x+\ldots\right)\right]$$

$$+\left(\frac{1}{2}\sin x+\frac{2}{5}\sin 2x+\frac{3}{10}\sin 3x+\ldots\right)$$

$$=\frac{1-e^{-2\pi}}{2\pi}+\frac{1-e^{-2\pi}}{\pi}\sum_{n=1}^{\infty}\frac{\cos nx}{1+n^2}$$

$$+\frac{1-e^{-2\pi}}{\pi}\sum_{n=1}^{\infty}\frac{n\sin nx}{1+n^2}.$$

Example 5. *Obtain Fourier series for the function $f(x)$, given by*

$$f(x)=\begin{cases}1+\dfrac{2x}{\pi}; & -\pi\le x\le 0\\[2mm] 1-\dfrac{2x}{\pi}; & 0\le x\le\pi\end{cases}$$

(VTU–2010, Mumbai–2007)

Hence, deduce that $\dfrac{1}{1^2}+\dfrac{1}{3^2}+\dfrac{1}{5^2}+\ldots=\dfrac{\pi^2}{8}$.

Solution. When $-\pi\le x\le 0 \Rightarrow 0\le -x\le\pi$

$$\therefore \quad f(-x)=1-\frac{2(-x)}{\pi}=1+\frac{2x}{\pi}=f(x)$$

When $0\le x\le\pi \Rightarrow -\pi\le -x\le 0$

$$\therefore \quad f(-x)=1+\frac{2(-x)}{\pi}=1-\frac{2x}{\pi}=1-\frac{2x}{\pi}=f(x)$$

Therefore, $f(x)$ is an even function of x in the interval $[-\pi, \pi]$. Hence $b_n=0$.

Now, Fourier series of $f(x)$ is given by

$$f(x)=\frac{a_0}{2}+\sum_{n=1}^{\infty}a_n\cos nx$$

Then,

$$a_0=\frac{2}{\pi}\int_0^{\pi}f(x)dx=\frac{2}{\pi}\int_0^{\pi}\left(1-\frac{2x}{\pi}\right)dx=\frac{2}{\pi}\left[x-\frac{x^2}{\pi}\right]_0^{\pi}=0$$

$$a_n=\frac{2}{\pi}\int_0^{\pi}f(x)\cos nx\,dx=\frac{2}{\pi}\int_0^{\pi}\left(1-\frac{2x}{\pi}\right)\cos nx\,dx$$

$$=\frac{2}{\pi}\left[\left(1-\frac{2x}{\pi}\right)\frac{\sin nx}{n}-\left(-\frac{2}{\pi}\right)\left(-\frac{\cos nx}{n^2}\right)\right]_0^{\pi}$$

$$=\frac{2}{\pi}\left[-\frac{2\cos n\pi}{\pi n^2}+\frac{2}{\pi n^2}\right]=\frac{4}{\pi^2n^2}[1-(-1)^n]$$

$$\Rightarrow f(x)=\frac{4}{\pi^2}\sum_{n=1}^{\infty}[1-(-1)^n]\frac{\cos nx}{n^2}$$

$$=\frac{4}{\pi^2}\left(\frac{2\cos x}{1^2}+\frac{2\cos 3x}{3^2}+\frac{2\cos 5x}{5^2}+\ldots\right)$$

$$=\frac{8}{\pi^2}\left(\frac{\cos x}{1^2}+\frac{\cos 3x}{3^2}+\frac{\cos 5x}{5^2}+\ldots\right).$$

Putting $x = 0$, we get $\dfrac{1}{1^2}+\dfrac{1}{3^2}+\dfrac{1}{5^2}+\ldots=\dfrac{\pi^2}{8}$.

[Since $f(0) = 1$)]

Example 6. *Find a Fourier series to represent $x - x^2$ from $x = -\pi$ to $x = \pi$.*

Deduce that $\frac{1}{1^2} - \frac{1}{2^2} + \frac{1}{3^2} - \frac{1}{4^2} + \ldots = \frac{\pi^2}{12}$.

(VTU–2011, Madras–2006)

Solution. The Fourier series for $f(x)$ in $(-\pi, \pi)$ is

$$f(x) = a_0 + \sum_{n=1}^{\infty} a_n \cos nx + \sum_{n=1}^{\infty} b_n \sin nx$$

Here,

$$a_0 = \frac{1}{2\pi}\int_{-\pi}^{\pi}(x - x^2)\,dx = \frac{1}{2\pi}\left[\frac{x^2}{2} - \frac{x^3}{3}\right]_{\pi}^{-\pi} = -\frac{\pi^2}{3}$$

$$a_n = \frac{1}{\pi}\int_{-\pi}^{\pi}(x - x^2)\cos nx\,dx$$

$$= \frac{1}{\pi}\left[(x - x^2)\frac{\sin nx}{n} - (1 - 2x)\left(-\frac{\cos nx}{n^2}\right) + (-2)\left(-\frac{\sin nx}{n^3}\right)\right]_{-\pi}^{\pi}$$

$$= \frac{-4(-1)^n}{n^2}$$

and $b_n = \frac{1}{\pi}\int_{-\pi}^{\pi}(x - x^2)\sin nx\,dx$

$$= \frac{1}{\pi}\left[(x - x^2)\left(-\frac{\cos nx}{n}\right) - (1 - 2x).\left(-\frac{\sin nx}{n^2}\right) + (-2)\left(\frac{\cos nx}{n^3}\right)\right]_{-\pi}^{\pi}$$

$$= \frac{(-2)(-1)^n}{n}$$

$\therefore$ The required Fourier series is

$$x - x^2 = -\frac{\pi^2}{3} + 4\left[\frac{\cos x}{1^2} - \frac{\cos 2x}{2^2} + \frac{\cos 3x}{3^2} - \frac{\cos 4x}{4^2} + \ldots\right]$$

$$+2\left[\frac{\sin x}{1} - \frac{\sin 2x}{2} + \frac{\sin 3x}{3} - \frac{\sin 4x}{4} + \ldots\right] \quad \ldots(1)$$

Deduction. Putting $x = 0$ in (1), we get

$$0 = -\frac{\pi^2}{3} + 4\left(\frac{1}{1^2} - \frac{1}{2^2} + \frac{1}{3^2} - \frac{1}{4^2} + \ldots\right)$$

or $\frac{1}{1^2} - \frac{1}{2^2} + \frac{1}{3^2} - \frac{1}{4^2} + \ldots = \frac{\pi^2}{12}$.

Example 7. *Find the Fourier series of the function defined as*

$$f(x) = \begin{cases} x + \pi & ; \quad 0 \le x \le \pi \\ -x - \pi & ; \quad -\pi \le x \le 0 \end{cases}$$

and $f(x + 2\pi) = f(x)$. (UPTU–2006, Q. Bank–2001)

Solution. Let $f(x) = \frac{a_0}{2} + \sum_{n=1}^{\infty} a_n \cos nx + \sum_{n=1}^{\infty} b_n \sin nx$

Then,

$$a_0 = \frac{1}{\pi}\int_{-\pi}^{\pi} f(x)\,dx = \frac{1}{\pi}\int_{-\pi}^{0} f(x)\,dx + \frac{1}{\pi}\int_{0}^{\pi} f(x)\,dx$$

$$= \frac{1}{\pi}\int_{-\pi}^{0}(-x - \pi)\,dx + \frac{1}{\pi}\int_{0}^{\pi}(x + \pi)\,dx$$

$$= \frac{1}{\pi}\left[\left(-\frac{x^2}{2} - \pi x\right)_{-\pi}^{0} + \left(\frac{x^2}{2} + \pi x\right)_{0}^{\pi}\right]$$

$$= \frac{1}{\pi}\left\{\left(\frac{\pi^2}{2} - \pi^2\right) + \left(\frac{\pi^2}{2} + \pi^2\right)\right\} = \pi$$

$$a_n = \frac{1}{\pi}\int_{-\pi}^{\pi} f(x)\cos nx\,dx$$

$$= \frac{1}{\pi}\int_{-\pi}^{0} f(x).\cos nx\,dx + \frac{1}{\pi}\int_{0}^{\pi} f(x).\cos nx\,dx$$

$$= \frac{1}{\pi}\int_{-\pi}^{0}(-x - \pi)\cos nx\,dx + \frac{1}{\pi}\int_{0}^{\pi}(x + \pi)\cos nx\,dx$$

$$= \frac{1}{\pi}\left[(-x - \pi)\frac{\sin nx}{n} - (-1)\left\{-\frac{\cos nx}{n^2}\right\}\right]_{-\pi}^{0}$$

$$+ \frac{1}{\pi}\left[(x + \pi)\frac{\sin nx}{n} - (-1)\left\{-\frac{\cos nx}{n^2}\right\}\right]_{0}^{\pi}$$

$$= \frac{1}{\pi}\left[-\frac{1}{n^2} + \frac{(-1)^n}{n^2}\right] + \frac{1}{\pi}\left[\frac{(-1)^n}{n^2} - \frac{1}{n^2}\right]$$

$$= \frac{2}{n^2\pi}[(-1)^n - 1] = \begin{cases} -\frac{4}{n^2\pi} & ; \quad \text{if } n \text{ is odd} \\ 0 & ; \quad \text{if } n \text{ is even} \end{cases}$$

Also $b_n = \frac{1}{\pi}\int_{-\pi}^{\pi} f(x)\sin nx\,dx$

$$= \frac{1}{\pi}\left\{\int_{-\pi}^{0} f(x).\sin nx\,dx + \int_{0}^{\pi} f(x).\sin nx\,dx\right\}$$

$$= \frac{1}{\pi}\left\{\int_{-\pi}^{0}(-x - \pi)\sin nx\,dx + \int_{0}^{\pi}(x + \pi)\sin nx\,dx\right\}$$

$$= \frac{1}{\pi}\left[(-x - \pi)\left(-\frac{\cos nx}{n}\right) - (-1)\left\{-\frac{\sin nx}{n^2}\right\}\right]_{-\pi}^{0}$$

$$+ \frac{1}{\pi}\left[(x + \pi)\left(-\frac{\cos nx}{n}\right) - (-1)\left\{-\frac{\sin nx}{n^2}\right\}\right]_{0}^{\pi}$$

$$= \frac{1}{\pi}\left[\frac{\pi}{n}\right] + \frac{1}{\pi}\left[\frac{-2\pi}{n}(-1)^n + \frac{\pi}{n}\right]$$

$$= \frac{1}{n}[1 - 2(-1)^n + 1] = \frac{2}{n}[1 - (-1)^n]$$

$$= \begin{cases} \dfrac{4}{n} & , \text{ if } n \text{ is odd} \\ 0 & , \text{ if } n \text{ is even} \end{cases}$$

The required Fourier series is given by

$$f(x) = \frac{a_0}{2} + a_1 \cos x + a_2 \cos 2x + ... + b_1 \sin x + b_2 \sin 2x + ...$$

$$= \frac{\pi}{2} - \frac{4}{\pi}\left(\frac{\cos x}{1^2} + \frac{\cos 3x}{3^2} + ...\right) + 4\left(\frac{\sin x}{1} + \frac{\sin 3x}{3} + ...\right)$$

Example 8. *Find the Fourier series for the function $f(x) = x + x^2$, $-\pi < x < \pi$. Hence, show that*

(Kurukshetra–2005, UPTU–2003)

(i) $\dfrac{\pi^2}{6} = 1 + \dfrac{1}{2^2} + \dfrac{1}{3^2} + \dfrac{1}{4^2} + ...$ (UPTU–2003)

(ii) $\dfrac{\pi^2}{12} = \dfrac{1}{1^2} - \dfrac{1}{2^2} + \dfrac{1}{3^2} - \dfrac{1}{4^2} + ...$

Solution. Let the Fourier series be

$$x + x^2 = \frac{a_0}{2} + \sum_{n=1}^{\infty} a_n \cos nx + \sum_{n=1}^{\infty} b_n \sin nx \quad ... (1)$$

Here, $a_0 = \dfrac{1}{\pi}\displaystyle\int_{-\pi}^{\pi} (x + x^2)dx$

$$= \frac{1}{\pi}\left[\int_{-\pi}^{\pi} x\,dx + \int_{-\pi}^{\pi} x^2\,dx\right]$$

$$= \frac{2}{\pi}\int_0^{\pi} x^2 dx = \frac{2}{3}\pi^2$$

$$a_n = \frac{1}{\pi}\int_{-\pi}^{\pi}(x + x^2)\cos nx\,dx$$

$$= \frac{1}{\pi}\left[\int_{-\pi}^{\pi} x\cos nx\,dx + \int_{-\pi}^{\pi} x^2 \cos nx\,dx\right]$$

$$= \frac{2}{\pi}\int_0^{\pi} x^2 \cos nx\,dx$$

$$= \frac{2}{\pi}\left[\left(x^2 \frac{\sin nx}{n}\right)_0^{\pi} - \int_0^{\pi} 2x.\frac{\sin nx}{n}dx\right]$$

$$= -\frac{4}{n\pi}\int_0^{\pi} x \sin nx\,dx$$

$$= -\frac{4}{n\pi}\left[\left\{x\left(-\frac{\cos nx}{n}\right)\right\}_0^{\pi} - \int_0^{\pi} 1.\left(-\frac{\cos nx}{n}\right)dx\right]$$

$$= -\frac{4}{n\pi}\left(-\frac{\pi}{n}\cos n\pi\right) = \frac{4}{n^2}\cos n\pi = \frac{4}{n^2}(-1)^n$$

and $b_n = \dfrac{1}{\pi}\displaystyle\int_{-\pi}^{\pi}(x + x^2)\sin nx dx$

$$= \frac{2}{\pi}\int_0^{\pi} x \sin nx\,dx + \frac{2}{\pi}\int_0^{\pi} x^2 \sin nx\,dx$$

$$\left[\because \int_0^{\pi} x^2 \sin nx\,dx = 0\right]$$

$$= \frac{2}{\pi}\left(-\frac{\pi}{n}\cos n\pi\right) = -\frac{2}{n}(-1)^n.$$

From (1), $x + x^2 = \dfrac{\pi^2}{3} + 4\displaystyle\sum_{n=1}^{\infty} \frac{(-1)^n}{n^2} \cos nx - 2\sum_{n=1}^{\infty} \frac{(-1)^n}{n} \sin nx$

$$f(x) = \frac{\pi^2}{3} + 4\left[-\frac{1}{1^2}\cos x + \frac{1}{2^2}\cos 2x - \frac{1}{3^2}\cos 3x + ...\right] - 2\left[-\frac{1}{1}\sin x + \frac{1}{2}\sin 2x - \frac{1}{3}\sin 3x + ...\right].$$

...(2)

We observe that the series on the R.H.S. given by equation (2), always represents $x + x^2$ for all values of x except the end points $-\pi$ or π.

At the point of discontinuity

$$f(-\pi) = \frac{1}{2}(\text{L.H.L.} + \text{R.H.L.})$$

$$= \frac{1}{2}[f(-\pi - 0) + f(-\pi + 0)]$$

$$= \frac{1}{2}[f(\pi - 0) + f(-\pi + 0)]$$

$$= \frac{1}{2}[\pi + \pi^2 + (-\pi) + (-\pi)^2] = \pi^2.$$

Putting $x = -\pi$ in equation (2), we get

$$\pi^2 = \frac{\pi^2}{3} + 4\left[\frac{1}{1^2} + \frac{1}{2^2} + \frac{1}{3^2} + \frac{1}{4^2} + ...\right]$$

Therefore, $\dfrac{\pi^2}{6} = 1 + \dfrac{1}{2^2} + \dfrac{1}{3^2} + \dfrac{1}{4^2} + ...$ ... (3)

Again, putting $x = 0$ in equaton (2), we get

$$0 = \frac{\pi^2}{3} + 4\left[-\frac{1}{1^2} + \frac{1}{2^2} - \frac{1}{3^2} + \frac{1}{4^2} - ...\right]$$

$$\Rightarrow \quad \frac{\pi^2}{12} = \frac{1}{1^2} - \frac{1}{2^2} + \frac{1}{3^2} - \frac{1}{4^2} ...$$

Example 9. *Express $f(x) = |x|$, $-\pi < x < \pi$, as Fourier series. Hence, show that $\dfrac{1}{1^2} + \dfrac{1}{3^2} + \dfrac{1}{5^2} + ... = \dfrac{\pi^2}{8}$.*

(UPTU(Q.Bank)–2001, SVTU–2009, Kerala–2005, PTU–2005)

Solution. Here, $f(-x) = |-x| = |x| = f(x)$

$\therefore$ $f(x)$ is an even function and hence $b_n = 0$.

Let $f(x) = |x| = \dfrac{a_0}{2} + \displaystyle\sum_{n=1}^{\infty} a_n \cos nx$

Then, $a_0 = \dfrac{2}{\pi}\displaystyle\int_0^{\pi} f(x)dx = \frac{2}{\pi}\int_0^{\pi} |x|\,dx$

$$= \frac{2}{\pi}\int_0^{\pi} x\,dx = \frac{2}{\pi}\left[\frac{x^2}{2}\right]_0^{\pi} = \pi$$

and $a_n = \frac{2}{\pi}\int_0^\pi f(x)\cos nx\, dx$

$$= \frac{2}{\pi}\int_0^\pi |x|.\cos nx dx = \frac{2}{\pi}\int_0^\pi x\cos nx dx$$

$$= \frac{2}{\pi}\left[x\left(\frac{\sin nx}{n}\right) - 1\left(-\frac{\cos nx}{n^2}\right)\right]_0^\pi$$

$$= \frac{2}{\pi}\left[\frac{\cos n\pi}{n^2} - \frac{1}{n^2}\right]$$

$$= \frac{2}{\pi n^2}[(-1)^n - 1]$$

$$= \begin{cases} 0 & , \text{ if } n \text{ is even} \\ -\frac{4}{\pi n^2} & , \text{ if } n \text{ is odd} \end{cases}$$

Hence,

$$|x| = \frac{\pi}{2} - \frac{4}{\pi}\left(\cos x + \frac{\cos 3x}{3^2} + \frac{\cos 5x}{5^2} + \ldots\right) \quad \ldots (1)$$

Deduction. Putting $x = 0$, in equation (1), we get $\frac{1}{1^2} + \frac{1}{3^2} + \frac{1}{5^2} + \ldots = \frac{\pi^2}{8}$.

Example 10. *Find the Fourier series expansion of f(x), if*

$$f(x) = \begin{cases} -\pi & , -\pi < x < 0 \\ x & , 0 < x < \pi \end{cases}$$

Deduce that $\frac{1}{1^2} + \frac{1}{3^2} + \frac{1}{5^2} + \ldots = \frac{\pi^2}{8}$.

Solution. Let the Fourier series be

$$f(x) = \frac{a_0}{2} + \sum_{n=1}^{\infty} a_n \cos nx + \sum_{n=1}^{\infty} b_n \sin nx$$

Then, $a_0 = \frac{1}{2\pi}\int_{-\pi}^{\pi} f(x)dx$

$$= \frac{1}{2\pi}\left[\int_{-\pi}^{0}(-\pi)dx + \int_0^\pi x\, dx\right]$$

$$= \frac{1}{\pi}\left[-\pi(x)_{-\pi}^0 + \left(\frac{x^2}{2}\right)_0^\pi\right]$$

$$= \frac{1}{\pi}\left[-\left(\pi^2 + \frac{\pi^2}{2}\right)\right] = -\frac{\pi}{2}$$

$$a_n = \frac{1}{\pi}\int_{-\pi}^{\pi} f(x)\cos nx\, dx$$

$$= \frac{1}{\pi}\left[\int_{-\pi}^{0}(-\pi)\cos nx dx + \int_0^\pi x\cos nx\, dx\right]$$

$$= \frac{1}{\pi}\left[-\pi\left(\frac{\sin nx}{n}\right)_{-\pi}^0 + \left(\frac{x\sin nx}{n} + \frac{\cos nx}{n^2}\right)_0^x\right]$$

$$= \frac{1}{\pi}\left[0 + \frac{1}{n^2}\cos n\pi - \frac{1}{n^2}\right] = \frac{1}{\pi n^2}(\cos n\pi - 1)$$

and

$$b_n = \frac{1}{\pi}\int_{-\pi}^{\pi} f(x)\sin nx\, dx$$

$$= \frac{1}{\pi}\left[\int_{-\pi}^{0}(-\pi)\sin nx\, dx + \int_0^\pi x\sin nx\, dx\right]$$

$$= \frac{1}{\pi}\left[\left(\frac{\pi\cos nx}{n}\right)_{-\pi}^0 + \left(-\frac{\cos nx}{n} + \frac{\sin nx}{n^2}\right)_0^\pi\right]$$

$$= \frac{1}{\pi}\left[\frac{\pi}{n}(0 - \cos n\pi) - \frac{\pi}{n}\cos n\pi\right]$$

$$= \frac{1}{n}(1 - 2\cos n\pi)$$

The required Fourier series is

$$f(x) = -\frac{\pi}{4} - \frac{2}{\pi}\left(\cos x + \frac{\cos 3x}{3^2} + \frac{\cos 5x}{5^2} + \ldots\right)$$
$$+\left(3\sin x - \frac{\sin 2x}{2} + \frac{3\sin 3x}{3} - \frac{\sin 4x}{4} + \ldots\right) \quad \ldots (1)$$

Deduction. Putting $x = 0$ in (1), we get

$$f(0) = \frac{\pi}{4} - \frac{2}{\pi}\left(1 + \frac{1}{3^2} + \frac{1}{5^2} + \ldots\right) \quad \ldots (2)$$

But $f(x)$ is continuous at $x = 0$, and we have $f(0 - 0) = -\pi$ and $f(0 + 0) = 0$

$$\therefore f(0) = \frac{1}{2}[f(0-0) + f(0+0)] = -(\pi/2) \ldots (3)$$

Hence, from (2) and (3), we have

$$-\frac{\pi}{2} = -\frac{\pi}{4} - \frac{2}{\pi}\left[\frac{1}{1^2} + \frac{1}{3^2} + \frac{1}{5^2} + \ldots\right]$$

or $\frac{1}{1^2} + \frac{1}{3^2} + \frac{1}{5^2} + \ldots = \frac{\pi^2}{8}$.

EXERCISE 27.1

1. Express $f(x) = \frac{1}{2}(\pi - x)$ in a Fourier series in the interval $0 < x < 2\pi$.
2. Find the Fourier series to represent the function $f(x) = |\sin x|$, $-\pi < x < \pi$. (UPTU(Q.Bank)–2001)
3. Obtain the Fourier series to represent $f(x) = \frac{1}{4}(\pi - x)^2, 0 < x < 2\pi$. Hence, obtain the following results : (UPTU(Q.Bank)–2001)

   (i) $\frac{1}{1^2} + \frac{1}{2^2} + \frac{1}{3^2} + \frac{1}{4^2} + \ldots = \frac{\pi^2}{6}$

   (ii) $\frac{1}{1^2} - \frac{1}{2^2} + \frac{1}{3^2} - \frac{1}{4^2} + \ldots = \frac{\pi^2}{12}$

   (iii) $\frac{1}{1^2} + \frac{1}{3^2} + \frac{1}{5^2} + \ldots = \frac{\pi^2}{8}$
4. Expand in a Fourier series the function $f(x) = x$ in the interval $0 < x < 2\pi$, sketch its graph from $x = -4\pi$ to $x = 4\pi$. (UPTU(Q.Bank)–2001)

5. Show that for $-\pi < x < \pi$

$$\sin ax = \frac{2\sin a\pi}{\pi}\left(\frac{\sin x}{1^2-a^2} - \frac{2\sin 2x}{2^2-a^2} + \frac{3\sin 3x}{3^2-a^2} - \ldots\right).$$

6. Obtain a Fourier expansion for $\sqrt{1-\cos x}$ in the interval $-\pi < x < \pi$.

7. Obtain a Fourier series to represent e^{-ax} from $x = -\pi$ to $x = \pi$. Hence derive the series for $\frac{\pi}{\sinh \pi}$.

8. Find the Fourier series to represent the periodic function:

$$f(x) = \begin{cases} x & , -\pi/2 < x < \pi/2 \\ \pi - x & , \pi/2 < x < 3\pi/2 \end{cases}$$ (UPTU(Q.Bank)–2001)

9. Find a series of sines and cosines to multiples of x which will represent $\frac{\pi}{\sinh \pi} e^x$ in the interval $-\pi < x < \pi$.

10. Prove that $x^2 = \frac{\pi^2}{3} + 4\sum_{n=1}^{\infty}(-1)^n \frac{\cos nx}{n^2}, -\pi < x < \pi$. (UPTU–2004)

11. Prove that in the interval $x\cos x = -\frac{1}{2}\sin x + 2\sum_{n=2}^{\infty}\frac{n(-1)^n}{n^2-1}\sin nx$ (UPTU(Sp)–2001)

12. If $f(x) = \cos \omega x, -\pi < x < \pi$, where ω is a fraction as a fourier series, prove that $\cot\theta = \frac{1}{\theta} + \frac{2\theta}{\theta^2-\pi^2} + \frac{2\theta}{\theta^2-4\pi^2} + \ldots$ (UPTU(Q.Bank)–2001)

Hint to Selected Problems

3.
$$a_0 = \frac{1}{\pi}\int_0^{2\pi}\frac{1}{4}(\pi-x)^2 dx = \frac{1}{4\pi}\left[\frac{(\pi-3x)^3}{-3}\right]_0^{2\pi}$$

$$= -\frac{1}{12\pi}[-\pi^3 - \pi^3] = \frac{\pi^2}{6}$$

$$a_n = \frac{1}{\pi}\int_0^{2\pi} f(x)\cos nx\, dx = \frac{1}{\pi}\int_0^{2\pi}\frac{1}{4}(\pi-x)^2\cos nx\, dx$$

$$= \frac{1}{4\pi}\left[(\pi-x)^2\frac{\sin nx}{n} - \{-2(\pi-x)\}\left(-\frac{\cos nx}{n^2}\right) + 2\left(\frac{-\sin nx}{n^3}\right)\right]_0^{2\pi}$$

$$= \frac{1}{4\pi}\left[\frac{2\pi}{n^2} + \frac{2\pi}{n^2}\right] = \frac{1}{n^2}$$

and $b_n = \frac{1}{\pi}\int_0^{2\pi}\frac{1}{4}(\pi-x)^2\sin nx\, dx$

$$= \frac{1}{4\pi}\left[(\pi-x)^2\left(-\frac{\cos nx}{n}\right) - \{-2(\pi-x)\}\left(-\frac{\sin nx}{n^2}\right) + 2\left(\frac{\cos nx}{n^3}\right)\right]_0^{2\pi}$$

$$= \frac{1}{4\pi}\left[\left(-\frac{\pi^2}{n} + \frac{2}{n^3}\right) - \left(-\frac{\pi^2}{n} + \frac{2}{n^3}\right)\right] = 0$$

$$\therefore\ f(x) = \frac{\pi^2}{12} + \sum_{n=1}^{\infty}\frac{\cos nx}{n^2}$$

$$= \frac{\pi^2}{12} + \frac{\cos x}{1^2} + \frac{\cos 2x}{2^2} + \frac{\cos 3x}{3^2} + \ldots \quad \ldots(1)$$

(i) Putting $x=0$ in equation (1), we get

$$\frac{\pi^2}{4} = \frac{\pi^2}{12} + \left(\frac{1}{1^2} + \frac{1}{2^2} + \frac{1}{3^2} + \frac{1}{4^2} + \ldots\right)$$

$$\frac{\pi^2}{6} = \frac{1}{1^2} + \frac{1}{2^2} + \frac{1}{3^2} + \frac{1}{4^2} + \ldots \quad \ldots(2)$$

(ii) Putting $x=\pi$ in equation (1), we get

$$0 = \frac{\pi^2}{12} + \left[\left(\frac{-1}{1^2}\right) + \frac{1}{2^2} + \left(-\frac{1}{3^2}\right) + \frac{1}{4^2} + \ldots\right]$$

$$\frac{\pi^2}{12} = \frac{1}{1^2} - \frac{1}{2^2} + \frac{1}{3^2} - \frac{1}{4^2} + \ldots \quad \ldots(3)$$

(iii) Adding equations (2) and (3), we get

$$\Rightarrow \frac{\pi^2}{6} + \frac{\pi^2}{12} = 2\left(\frac{1}{1^2} + \frac{1}{3^2} + \frac{1}{5^2} + \ldots\right)$$

$$\Rightarrow \frac{\pi^2}{4} = 2\left(\frac{1}{1^2} + \frac{1}{3^2} + \frac{1}{5^2} + \ldots\right) \quad \ldots(4)$$

$$\Rightarrow \frac{\pi^2}{8} = \frac{1}{1^2} + \frac{1}{3^2} + \frac{1}{5^2} + \ldots$$

5. Here, $a_0 = 0, a_n = 0$ ($\because f(x)$ is an odd function)

$$b_n = \frac{2}{\pi}\int_0^{\pi}\sin ax \sin nx\, dx = \frac{1}{\pi}\int_0^{\pi}[\cos(n-a)x - \cos(n+a)x]dx$$

$$= \frac{1}{\pi}\left[\frac{\sin(n-a)x}{(n-a)} - \frac{\sin(n+a)x}{n+a}\right]_0^{\pi}$$

$$= \frac{1}{\pi}\left[\frac{\sin(n-a)\pi}{n-a} - \frac{\sin(n+a)\pi}{n+a}\right]$$

$$= \frac{1}{\pi}\left[\frac{(-1)^n(-\sin a\pi)}{n-a} - \frac{(-1)^n\sin a\pi}{n+a}\right]$$

$$= \frac{(-1)^n\sin a\pi}{\pi}\left[\frac{1}{n-a} + \frac{1}{n+a}\right] = (-1)^{n+1}\frac{2n\sin a\pi}{\pi(n^2-a^2)}$$

$$\therefore\ \sin ax = \frac{2\sin a\pi}{\pi}\sum_{n=1}^{\infty}\frac{(-1)^{n+1}n}{n^2-a^2}\sin nx$$

$$= \frac{2\sin a\pi}{\pi}\left[\frac{\sin x}{1^2-a^2} - \frac{2\sin 2x}{2^2-a^2} + \frac{3\sin 3x}{3^2-a^2} - \ldots\right]$$

9.
$$f(n) = \frac{1}{2l}\int_{-l}^{l} f(x)dx + \frac{1}{l}\sum_{n=1}^{\infty}\left[\cos\frac{n\pi x}{l} - \int_{-l}^{l} f(x).\cos\frac{n\pi x}{l}dx + \sin\frac{n\pi x}{l}\int_{-l}^{l} f(x)\sin\frac{n\pi x}{l}dx\right]$$

$$\frac{\pi}{2\sin n\pi}e^x = \frac{1}{2\pi}\int_{-\pi}^{\pi}\frac{\pi}{2\sin n\pi}e^x dx + \frac{1}{\pi}\sum_{n=1}^{\infty}\cos nx\int_{-\pi}^{\pi}\frac{\pi}{2\sin n\pi}e^x\cos nu\, du + \frac{1}{\pi}\sum_{n=1}^{\infty}\sin nx\int_{-\pi}^{\pi}\frac{\pi}{2\sin n\pi}e^x\sin nu\, du.$$

We have $\int_{-\pi}^{\pi} e^u du = \left[e^u\right]_{-\pi}^{\pi} = 2\sin n\pi$

$$\int_{-\pi}^{\pi} e^u \cos nu\, du$$

$$= \left[e^u \frac{\sin nu}{n}\right]_{-\pi}^{\pi} - \frac{1}{n}\int_{-\pi}^{\pi} e^u \sin nu\, du$$

$$= \frac{1}{n^2}\left[e^u \cos nu\right]_{-\pi}^{\pi} - \frac{1}{n^2}\int_{-\pi}^{\pi} e^u \cos nu\, du$$

$$\left(1+\frac{1}{n^2}\right)\int_{-\pi}^{\pi} e^u \cos nu\, du = \frac{1}{n^2}(e^{\pi}-e^{-\pi})\cos n\pi$$

$$\int_{-\pi}^{\pi} e^u \cos nu\, du = \frac{2}{1+n^2}\sin n\pi \cos n\pi$$

$$\int_{-\pi}^{\pi} e^u \sin nu\, du = \left[\frac{-e^u \cos nu}{n}\right]_{-\pi}^{\pi} + \frac{1}{n}\int_{-\pi}^{\pi} e^u \cos nu\, du$$

$$= -\frac{1}{n}(e^{\pi}-e^{\pi})\cos n\pi + \frac{1}{n}\left[\left\{\frac{e^u \sin nu}{n}\right\}_{-\pi}^{\pi} - \int_{-\pi}^{\pi}\frac{e^u \sin nu}{n}du\right]$$

$$\left(1+\frac{1}{n^2}\right)\int_{-\pi}^{\pi} e^u \sin nu\, du = -\frac{1}{n}(e^{\pi}-e^{-\pi})\cos n\pi$$

$$\int_{-\pi}^{\pi} e^u \sin nu\, du = \frac{2}{1+n^2}\sin n\pi \cos n\pi$$

$$\therefore\ \frac{\pi}{2\sin n\pi}e^x = \frac{1}{2} + \sum_{n=1}^{\infty}\frac{\cos n\pi}{1+n^2}\cos nx - \sum_{n=1}^{\infty}\left\{\frac{n}{1+n^2}\cos n\pi \sin nx\right\}$$

$$= \frac{1}{2} - \left(\frac{1}{2}\cos x - \frac{1}{2}\cos 2x + \frac{1}{10}\cos 3x - \frac{1}{17}\cos 4x + \ldots\right)$$

$$+\left(\frac{1}{2}\sin x - \frac{2}{5}\sin 2x + \frac{3}{10}\sin 3x - \frac{4}{17}\sin 4x + \ldots\right)$$

ANSWERS

1. $f(x) = \sum_{n=1}^{\infty}\frac{\sin nx}{n}$

2. $|\sin x| = \frac{2}{\pi} \cdot \frac{4}{\pi}\left(\frac{\cos 2x}{3} + \frac{\cos 4x}{15} + \ldots + \frac{\cos 2nx}{4n^2-1} + \ldots\right)$

3. $f(x) = \frac{\pi^2}{12} + \sum_{n=1}^{\infty}\frac{\cos nx}{n^2} = \frac{\pi^2}{12} + \frac{\cos x}{1^2} + \frac{\cos 2x}{2^2} + \frac{\cos 3x}{3^2} + \ldots$

4. $f(x) = \pi - 2 \cdot \sum_{n=1}^{\infty}\frac{\sin nx}{n}$

5. $\sin ax = \frac{2\sin a\pi}{\pi}\sum_{n=1}^{\infty}\frac{(-1)^{n+1}}{n^2-a^2}\sin nx$

6. $\sqrt{1-\cos x} = \frac{2\sqrt{2}}{\pi} - \frac{4\sqrt{2}}{\pi}\sum_{n=1}^{\infty}\frac{\cos nx}{4n^2-1}$

7. $e^{-ax} = 2\frac{\sin h\, a\pi}{\pi}\left[\left(\frac{1}{2a} - \frac{a\cos x}{1^2+a^2} + \frac{a\cos 2x}{2^2+a^2} - \ldots\right) - \left(\frac{\sin x}{1^2+a^2} - \frac{2\sin 2x}{2^2+a^2} + \frac{3\sin 3x}{3^2+a^2}\ldots\right)\right]$; $\frac{\pi}{\sinh \pi} = 2\left[\frac{1}{2^2+1} - \frac{1}{3^2+1} + \frac{1}{4^2+1} - \ldots\right]$

8. $f(x) = \frac{4}{\pi}\left[\frac{\sin x}{1^2} - \frac{\sin 3x}{3^2} + \frac{\sin 5x}{5^2} - \ldots\right]$

9. $\frac{\pi}{2\sin n\pi}e^x = \frac{1}{2} + \sum_{n=1}^{\infty}\frac{\cos n\pi}{1+n^2}\cos nx - \sum_{n=1}^{\infty}\left\{\frac{n}{1+n^2}\cos nx \sin n\pi\right\} = \frac{1}{2} - \left(\frac{1}{2}\cos x - \frac{1}{2}\cos 2x + \frac{1}{10}\cos 3x - \frac{1}{17}\cos 4x + \ldots\right)$

27.6 FOURIER SERIES FOR DISCONTINUOUS FUNCTIONS

At the point of discontinuity, the value of function for Fourier series is obtained by the average of left hand limit and right hand limit of function at that point of discontinuity.

Solved Examples

Example 1. *Obtain Fourier series for the function*

$$f(x) = \begin{cases} x & ; \ -\pi < x < 0 \\ -x & ; \ 0 < x < \pi \end{cases}$$

and hence show that $\frac{1}{1^2} + \frac{1}{3^2} + \frac{1}{5^2} + \ldots = \frac{\pi^2}{8}$.

(UPTU–2002)

Solution . We know that

$$f(x) = \frac{a_0}{2} + \sum_{n=1}^{\infty} a_n \cos nx + \sum_{n=1}^{\infty} b_n \sin nx \quad \ldots(1)$$

$$a_0 = \frac{1}{\pi}\int_{-\pi}^{\pi} f(x)dx = \frac{1}{\pi}\left[\int_{-\pi}^{0} x\,dx + \int_{0}^{\pi} -x\,dx\right]$$

$$= \frac{1}{\pi}\left[\left(\frac{x^2}{2}\right)_{-\pi}^{0} - \left(\frac{x^2}{2}\right)_{0}^{\pi}\right] = \frac{1}{\pi}\left[0 - \frac{\pi^2}{2} - \frac{\pi^2}{2}\right] = -\pi$$

$$a_n = \frac{1}{\pi}\int_{-\pi}^{\pi} f(x)\cos nx dx$$

$$= \frac{1}{\pi}\left[\int_{-\pi}^{0} x\cos nx dx + \int_{0}^{\pi} -x\cos nx dx\right]$$

$$= \frac{1}{\pi}\left[\left(\frac{x\sin nx}{n}\right)_{-\pi}^{0} - \int_{-\pi}^{0}\frac{\sin nx}{n}dx\right.$$

$$\left. + \left(-x\frac{\sin nx}{n}\right)_{0}^{\pi} - \int_{0}^{\pi}(-1)\frac{\sin nx}{n}dx\right]$$

$$= \frac{1}{\pi}\left[\frac{1}{n^2}(\cos nx)_{-\pi}^{0} - \frac{1}{n^2}(\cos nx)_{0}^{\pi}\right]$$

$$= \frac{1}{\pi}\left[\left\{\frac{1-(-1)^n}{n^2}\right\} - \left\{\frac{(-1)^n - 1}{n^2}\right\}\right]$$

$$= \frac{1}{\pi}\left[\frac{2\{1-(-1)^n\}}{n^2}\right] = \frac{2}{\pi n^2}[1-(-1)^n]$$

$$= \begin{cases} 0 & ; \text{ if } n \text{ is even} \\ \dfrac{4}{\pi n^2} & ; \text{ if } n \text{ is odd} \end{cases}$$

and $b_n = \frac{1}{\pi}\int_{-\pi}^{\pi} f(x)\sin nx dx$

$$= \frac{1}{\pi}\left[\int_{-\pi}^{0} x\sin nx dx + \int_0^{\pi} -x\sin nx\, d\,x\right]$$

$$= \frac{1}{\pi}\left[\left(x\frac{-\cos nx}{n}\right)_{-\pi}^{0} - \int_{-\pi}^{0}\frac{-\cos nx}{n}dx\right.$$

$$\left. + \left(x\frac{\cos nx}{n}\right)_0^{\pi} - \int_0^{\pi}(-1)\frac{-\cos nx}{n}dx\right]$$

$$= \frac{1}{\pi}\left[\frac{-\pi}{n}(-1)^n + \frac{1}{n}(-1)^n\right] = 0$$

From (1)

$$f(x) = -\frac{\pi}{2} + \frac{4}{\pi}\left(\frac{\cos x}{1^2} + \frac{\cos 3x}{3^2} + \frac{\cos 5x}{5^2} + \ldots\right)$$

At the point of discontinuity

$$f(0) = \frac{1}{2}[f(0^-) + f(0^+)] = \frac{1}{2}[0-0] = 0$$

Putting, $x = 0$ in (2), we get

$$0 = -\frac{\pi}{2} + \frac{4}{\pi}\left(\frac{1}{1^2} + \frac{1}{3^2} + \frac{1}{5^2} + \ldots\right)$$

Hence, $\frac{1}{1^2} + \frac{1}{3^2} + \frac{1}{5^2} + \ldots = \frac{\pi^2}{8}$.

Example 2. *Obtain the Fourier series to represent f(x) given as follows :* $f(x) = \begin{cases} x & ; \text{ for } 0 \le x \le \pi \\ 2\pi - x & ; \text{ for } \pi \le x \le 2\pi \end{cases}$

(SVTU–2008, BPTU–2005S)

Solution. Let $f(x) = \frac{a_0}{2} + \sum_{n=1}^{\infty} a_n \cos nx + \sum_{n=1}^{\infty} b_n \sin nx,$

$0 \le x \le 2\pi$ ... (1)

where $a_0 = \frac{1}{\pi}\int_0^{2\pi} f(x)dx$

$$= \frac{1}{\pi}\left[\int_0^{\pi} x\,dx + \int_{\pi}^{2\pi}(2\pi - x)dx\right]$$

$$= \frac{1}{\pi}\left[\left(\frac{x^2}{2}\right)_0^{\pi} + \left(2\pi x - \frac{x^2}{2}\right)_{\pi}^{2\pi}\right]$$

$$= \frac{1}{\pi}\left[\frac{\pi^2}{2} + 2\pi(2\pi - x) - \frac{1}{2}(4\pi^2 - \pi^2)\right]$$

$$= \frac{1}{\pi}(\pi^2) = \pi$$

$$a_n = \frac{1}{\pi}\int_0^{2\pi} f(x)\cos nx dx$$

$$= \frac{1}{\pi}\left[\int_0^{\pi} x\cos nx dx + \int_0^{2\pi}(2\pi - x)\cos nx dx\right]$$

$$= \frac{1}{\pi}\left[\left\{\frac{x\sin nx}{n} + \frac{\cos nx}{n^2}\right\}_0^{\pi}\right.$$

$$\left. + \left\{(2\pi - x)\frac{\sin nx}{n} - \frac{\cos nx}{n^2}\right\}_{\pi}^{2\pi}\right]$$

$$= \frac{1}{\pi}\left[\left(\frac{\cos n\pi - 1}{n^2}\right) - \left(\frac{1-\cos n\pi}{n^2}\right)\right]$$

$$= \frac{2}{n^2\pi}[(-1)^n - 1] = \begin{cases} 0 & , \text{ if } n \text{ is even} \\ -\dfrac{4}{n\pi^2} & , \text{ if } n \text{ is odd} \end{cases}$$

Again $b_n = \frac{1}{\pi}\int_0^{2\pi} f(x).\sin nx\, dx$

$$= \frac{1}{\pi}\left[\int_0^{\pi} x\sin nx\, dx + \int_0^{2\pi}(2\pi - x)\sin nx\, dx\right]$$

$$= \frac{1}{\pi}\left[\left\{-\frac{x\cos nx}{n} + \frac{\sin nx}{n^2}\right\}_0^{\pi}\right.$$

$$\left. + \left\{-(2\pi - x)\frac{\cos nx}{n} - \frac{\sin nx}{n^2}\right\}_{\pi}^{2\pi}\right]$$

$$= \left(\frac{-\pi\cos n\pi}{n} + \frac{\pi\cos n\pi}{n}\right) = 0$$

Therefore, $f(x) = \frac{\pi}{2} - \frac{4}{\pi}\left[\cos x + \frac{\cos 3x}{3^2}\right.$

$$\left. + \frac{\cos 5x}{5^2} + \ldots\right], 0 \le x \le 2\pi$$

which is the required Fourier series for $f(x)$.

Example 3. *If* $f(x) = \begin{cases} 0 & , -\pi \le x \le 0 \\ \sin x & , 0 \le x \le \pi \end{cases}$

Prove that $f(x) = \frac{1}{\pi} + \frac{1}{2}\sin x - \frac{2}{\pi}\sum_{n=1}^{\infty}\frac{\cos 2nx}{4n^2 - 1}$.

(UPTU(Q.Bank)–2001)

Hence, show that

(i) $\frac{1}{1.3} + \frac{1}{3.5} + \frac{1}{5.7} + \ldots = \frac{1}{2}$

(ii) $\frac{1}{1.3} - \frac{1}{3.5} + \frac{1}{5.7} - \ldots = \frac{\pi - 2}{4}$

(Bhopal–2008, Mumbai–2005S, Rohtak–2005)

Solution. Let $f(x) = \frac{a_0}{2} + \sum_{n=1}^{\infty} a_n \cos nx + \sum_{n=1}^{\infty} b_n \sin nx$

Then, $a_0 = \frac{1}{\pi}\int_{-\pi}^{\pi} f(x)dx$

$$= \frac{1}{\pi}\left[\int_{-\pi}^{0} 0.dx + \int_0^{\pi}\sin x dx\right] = \frac{2}{\pi}$$

$$a_n = \frac{1}{\pi}\int_{-\pi}^{\pi} f(x).\cos nx\,dx$$

$$= \frac{1}{\pi}\left[\int_{-\pi}^{0} 0.dx + \int_0^{\pi} \sin x \cos nx\,dx\right]$$

$$= \frac{1}{2\pi}\int_0^{2\pi} 2\cos nx.\sin x dx$$

$$= \frac{1}{2\pi}\int_0^{\pi}[\sin(n+1)x - \sin(n-1)x]dx$$

$$= \frac{1}{2\pi}\left[-\frac{\cos(n+1)x}{n+1} + \frac{\cos(n-1)x}{n-1}\right]_0^{\pi}, n \neq 1$$

$$= \frac{1}{2\pi}\left[-\frac{\cos(n+1)\pi}{n+1} + \frac{\cos(n-1)\pi}{n-1} + \frac{1}{n+1} - \frac{1}{n-1}\right]$$

$$= \frac{1}{2\pi}\left[-\frac{(-1)^{n+1}}{n+1} + \frac{(-1)^{n-1}}{n-1} + \frac{1}{n+1} - \frac{1}{n-1}\right]$$

$$= \begin{cases} \frac{1}{2\pi}\left(-\frac{1}{n+1} + \frac{1}{n-1} + \frac{1}{n+1} - \frac{1}{n-1}\right) &, \text{when } n \text{ is odd} \\ \frac{1}{2\pi}\left(\frac{1}{n+1} - \frac{1}{n-1} + \frac{1}{n+1} - \frac{1}{n-1}\right) &, \text{when } n \text{ is even} \end{cases}$$

$$= \begin{cases} 0 &, \text{when } n \text{ is odd, i.e., } n = 3,5,7,\ldots \\ -\frac{2}{\pi(n^2-1)} &, \text{when } n \text{ is even} \end{cases}$$

When $n=1$, we have

$$a_1 = \frac{1}{\pi}\int_0^{\pi} \sin x \cos x dx = \frac{1}{2\pi}\int_0^{\pi} \sin 2x dx$$

$$= \frac{1}{2\pi}\left[-\frac{\cos 2x}{2}\right]_0^{\pi} = 0$$

and $b_n = \frac{1}{\pi}\int_{-\pi}^{\pi} f(x)\sin nx\,dx$

$$= \frac{1}{\pi}\left[\int_{-\pi}^{0} 0.dx + \int_0^{\pi} \sin x \sin nx\,dx\right]$$

$$= \frac{1}{2\pi}\int_0^{\pi} 2\sin nx \sin x\,dx$$

$$= \frac{1}{2\pi}\int_0^{\pi}[\cos(n-1)x - \cos(n+1)x]dx$$

$$= \frac{1}{2\pi}\left[\frac{\sin(n-1)x}{(n-1)} - \frac{\sin(n+1)x}{(n+1)}\right]_0^{\pi} = 0, n \neq 1$$

When $n = 1$, we have $b_1 = \frac{1}{\pi}\int_0^{\pi} \sin x \sin x\,dx$

$$= \frac{1}{2\pi}\int_0^{\pi}(1-\cos 2x)dx = \frac{1}{2\pi}\left[x - \frac{\sin 2x}{2}\right]_0^{\pi} = \frac{1}{2}$$

$$\therefore\ f(x) = \frac{1}{\pi} - \frac{2}{\pi}\left[\frac{\cos 2x}{2^2-1} + \frac{\cos 4x}{4^2-1} + \frac{\cos 6x}{6^2-1} + \ldots\right] + \frac{1}{2}\sin x$$

$$= \frac{1}{\pi} + \frac{1}{2}\sin x - \frac{2}{\pi}\sum_{n=1}^{\infty}\frac{\cos 2nx}{(2n)^2-1}$$

Putting $x = 0$ in equation (1), we have

$$0 = \frac{1}{\pi} - \frac{2}{\pi}\sum_{n=1}^{\infty}\frac{1}{4n^2-1}$$

$$\frac{1}{2} = \sum_{n=1}^{\infty}\frac{1}{4n^2-1} = \sum_{n=1}^{\infty}\frac{1}{(2n-1)(2n+1)}$$

$$= \frac{1}{1.3} + \frac{1}{3.5} + \frac{1}{5.7} + \ldots$$

Putting $x = \pi/2$ in equation (1), we have,

$$1 = \frac{1}{\pi} + \frac{1}{2} - \frac{2}{\pi}\sum_{n=1}^{\infty}\frac{\cos n\pi}{4n^2-1}$$

$$\Rightarrow\ \frac{1}{2} - \frac{1}{\pi} = -\frac{2}{\pi}\sum_{n=1}^{\infty}\frac{(-1)^n}{4n^2-1}$$

$$\Rightarrow\ \frac{\pi-2}{4} = -\sum_{n=1}^{\infty}\frac{(-1)^n}{(2n-1)(2n+1)}$$

$$= -\left(-\frac{1}{1.3} + \frac{1}{3.5} - \frac{1}{5.7} + \ldots\right)$$

$$\Rightarrow\ \frac{1}{1.3} - \frac{1}{3.5} + \frac{1}{5.7} - \ldots = \frac{\pi-2}{4}$$

EXERCISE 27.2

1. Find the Fourier series for the following function:

$$f(x) = \begin{cases} x^2 &, 0 \le x \le \pi \\ -x^2 &, -\pi \le x \le 0 \end{cases}$$ (Mumbai–2009, Hissar–2007)

2. Find the Fourier series to represent the function:

$$f(x) = \begin{cases} -k &, \text{when } -\pi < x < 0 \\ k &, \text{when } 0 < x < \pi \end{cases}$$

Also deduce that $\frac{\pi}{4} = 1 - \frac{1}{3} + \frac{1}{5} - \frac{1}{7} + \ldots.$

3. Find the Fourier series for the function:

$$f(x) = \begin{cases} -1 &, -\pi < x < -\pi/2 \\ 0 &, -\pi/2 < x < \pi/2 \\ 1 &, \pi/2 < x < \pi \end{cases}$$ (UPTU–2004, 2005)

4. Find the Fourier series expansion for $f(x)$ if

$$f(x) = \begin{cases} -\pi &, -\pi < x < 0 \\ x &, 0 < x < \pi \end{cases}$$ (Bhopal–2008S)

Deduce that $\frac{1}{1^2} + \frac{1}{3^2} + \frac{1}{5^2} + \ldots = \frac{\pi^2}{8}$ (Kottayam–2005)

5. Find the Fourier expansion of the function defined in one period by the relations:

$$f(x)=\begin{cases}1 & , \quad 0<x<\pi \\ 2 & , \quad \pi<x<2\pi\end{cases}$$

and deduce that $\frac{\pi}{4}=1-\frac{1}{3}+\frac{1}{5}-\frac{1}{7}+\dots$.

6. An alternating current after passing through a rectifier has the form $i=\begin{cases}I_0\sin x & \text{for } 0\le x<\pi \\ 0 & \text{for } \pi\le x\le 2\pi\end{cases}$

(VTU–2007, Calicut–2005, UPTU(Q.Bank)–2001)

where I_0 is the maximum current and the period is 2π. Express i as a Fourier series.

Hint to Selected Problems

3. $a_0=\frac{1}{\pi}\int_{-\pi}^{-\pi/2}(-1)dx+\frac{1}{\pi}\int_{-\pi/2}^{\pi/2}0dx+\frac{1}{\pi}\int_{\pi/2}^{\pi}1\,dx=0$

$a_n=\frac{1}{\pi}\int_{-\pi}^{-\pi/2}(-1)\cos nxdx+\frac{1}{\pi}\int_{-\pi/2}^{\pi/2}(0)\cos nxdx$

$+\frac{1}{\pi}\int_{\pi/2}^{\pi}(1)\cos nx\,dx=0$

$b_n=\frac{1}{\pi}\int_{-\pi}^{\pi/2}(-1)\sin nxdx+\frac{1}{\pi}\int_{-\pi}^{\pi/2}(0)\sin nxdx$

$+\frac{1}{\pi}\int_{\pi/2}^{\pi}(1)\sin nx\,dx=0=\frac{2}{n\pi}\left[\cos\frac{n\pi}{2}-\cos n\pi\right]$

$b_1=\frac{2}{\pi}, b_2=-\frac{2}{\pi}, b_3=\frac{2}{3\pi}$

$f(x)=\frac{1}{\pi}\left[2\sin x-2\sin 2x+\frac{2}{3}\sin 3x+\dots\right]$.

ANSWERS

1. $f(x)=2\left(\pi-\frac{4}{\pi}\right)\sin x-\pi\sin 2x+\frac{2}{3}\left(\pi-\frac{4}{9\pi}\right)\sin 3x-\frac{\pi}{2}\sin 4x+..$

2. $f(x)=\frac{4k}{\pi}\left(\sin x+\frac{\sin 3x}{3}+\frac{\sin 5x}{5}+\dots\right)$

3. $f(x)=\frac{2}{\pi}\left[\sin x-\sin 2x+\frac{\sin 3x}{3}+\dots\right]$

4. $f(x)=-\frac{\pi}{4}-\frac{2}{\pi}\left(\cos x+\frac{\cos 3x}{3^2}+\frac{\cos 5x}{5^2}+\dots\right)+\left(3\sin x-\frac{\sin 2x}{2}+\sin 3x-\frac{\sin 4x}{4}+\dots\right)$

5. $f(x)=\frac{3}{2}-\frac{2}{\pi}\left(\sin x+\frac{\sin 3x}{3}+\frac{\sin 5x}{5}+\dots\right)$

6. $i=\frac{I_0}{\pi}+\frac{I_0}{2}\sin x-\frac{2I_0}{\pi}\left(\frac{\cos 2x}{2^2-1}+\frac{\cos 4x}{4^2-1}+\frac{\cos 6x}{6^2-1}+\dots\right)$

27.7 CHANGE OF INTERVAL

In many problems, the interval of Fourier expansion is $2l$ and not 2π. In order to apply this theory, this interval must be transformed into an interval of length 2π.

Consider a periodic function $f(x)$ defined in the interval $c<x<c+2l$. To change the interval into one of length 2π, we put

$$\frac{x}{l}=\frac{z}{\pi} \text{ or } z=\frac{\pi x}{l} \text{ so that at } x=c, z=\frac{\pi c}{l}=d(\text{say})$$

When $\quad x=c+2l, z=\frac{\pi(c+2l)}{l}=\frac{\pi c}{l}+2\pi=d+2\pi$

Thus, the function $f(x)$ of period $2l$ in $(c, c+2l)$ is transformed to the function $f\left(\frac{lz}{\pi}\right)=F(z)$ say, or period in $(d, d+2\pi)$ and then function $F(z)$ can be expressed as a Fourier series such that

$$F(z)=\frac{a_0}{2}+\sum_{n=1}^{\infty}a_n\cos nz+\sum_{n=1}^{\infty}b_n\sin nz \qquad \dots(1)$$

where $\quad a_0=\frac{1}{\pi}\int_d^{d+2\pi}F(z)dz; a_n=\frac{1}{\pi}\int_d^{d+2\pi}F(z)\cos nz\,dz$

and $\quad b_n=\frac{1}{\pi}\int_d^{d+2\pi}F(z)\sin nz\,dz$

Now, making the inverse substitution $z=\frac{\pi x}{l}, dz=\frac{\pi}{l}dx$, when $z=d$, $x=c$ and when $z=d+2\pi, x=c+2l$. The expression (1) becomes

$$F(z)=F\left(\frac{\pi x}{l}\right)=F(x)=\frac{a_0}{2}+\sum_{n=1}^{\infty}a_n\cos\frac{n\pi x}{l}+\sum_{n=1}^{\infty}b_n\sin\frac{n\pi x}{l} \qquad \dots(2)$$

The coefficient a_0, a_n, b_n in (2) becomes

$$a_0=\frac{1}{l}\int_c^{c+2l}f(x)dx,\ a_n=\frac{1}{l}\int_c^{c+2l}f(x)\cos\frac{n\pi x}{l}dx,\ b_n=\frac{1}{l}\int_c^{c+2\pi}f(x)\sin\frac{n\pi x}{l}dx$$

REMARKS

- If $c = 0$, the interval become $0 < x < 2l$ and the a_0, a_n, b_n are given by
$$a_0 = \frac{1}{l}\int_0^{2l} f(x)dx, a_n = \frac{1}{l}\int_0^{2l} f(x)\cos\frac{n\pi x}{l}dx,\ b_n = \frac{1}{l}\int_0^{2l} f(x)\sin\frac{n\pi x}{l}dx.$$
- If $c = -l$, the interval become $-l < x < l$ and a_0, a_n, b_n are given by
$$a_0 = \frac{1}{l}\int_{-l}^{l} f(x)dx,\ a_n = \frac{1}{l}\int_{-l}^{l} f(x)\cos\frac{n\pi x}{l}dx,\ b_n = \frac{1}{l}\int_{-l}^{l} f(x)\sin\frac{n\pi x}{l}dx.$$

Solved Examples

Example 1. *Find the Fourier series to represent $f(x) = x^2 - 2$ when $-2 \le x \le 2$.*

Solution. Here, $b_n = 0$ because $f(x)$ is an even function

Let $f(x) = x^2 - 2 = \dfrac{a_0}{2} + \sum_{n=1}^{\infty} a_n \cos\dfrac{n\pi x}{2}$

$[\because 2l = 4 \Rightarrow l = 2]$

Then,
$$a_0 = \frac{2}{2}\int_0^2 (x^2-2)dx = \left[\frac{x^3}{3} - 2x\right]_0^2 = \frac{8}{3} - 4 = -\frac{4}{3}$$

and
$$a_n = \frac{2}{2}\int_0^2 (x^2 - x)\cos\frac{n\pi x}{2}dx$$
$$= \left[(x^2-2)\frac{\sin n\pi x/2}{(n\pi/2)} - 2x\left(-\frac{\cos\frac{n\pi x}{2}}{(n^2\pi^2/4)}\right) + 2\left(\frac{\sin\frac{n\pi x}{2}}{(n^3\pi^3/8)}\right)\right]_0^2$$
$$= \frac{16\cos n\pi}{n^2\pi^2} = \frac{16(-1)^n}{n^2\pi^2}.$$
$$\therefore\ f(x) = (x^2-2) = -\frac{2}{3} + \frac{16}{\pi^2}\sum\frac{(-1)^n}{n^2}\cos\frac{n\pi x}{2}$$
$$= -\frac{2}{3} - \frac{16}{\pi^2}\left(\cos\frac{\pi x}{2} - \frac{1}{4}\cos\pi x + \frac{1}{9}\cos\frac{3\pi x}{2} - \ldots\right)$$

Example 2. *Obtain the Fourier series for the function*
$$f(x) = \begin{cases} \pi x & ;\ 0 \le x \le 1 \\ \pi(2-x) & ;\ 1 \le x \le 2 \end{cases}$$

(UPTU–2001, VTU–2011, Bhopal–2008, Mumbai–2007)

Solution. Here, $2l = 2 \Rightarrow l = 1$.

Let $f(x) = \dfrac{a_0}{2} + \sum_{n=1}^{\infty} a_n \cos n\pi x + \sum_{n=1}^{\infty} b_n \sin n\pi x$

where
$$a_0 = \int_0^2 f(x)dx = \int_0^1 \pi x dx + \int_1^2 \pi(2-x)dx$$
$$= \pi\left[\frac{x^2}{2}\right]_0^1 + \pi\left[2x - \frac{x^2}{2}\right]_1^2$$
$$= \pi\left(\frac{1}{2}\right) + \pi\left[(4-2) - \left(2 - \frac{1}{2}\right)\right] = \pi$$

$$a_n = \int_0^2 f(x)\cos n\pi x\, dx$$
$$= \int_0^1 \pi x\cos n\pi x\, dx + \int_1^2 \pi(2-x)\cos n\pi x\, dx$$
$$= \left[\pi x\frac{\sin n\pi x}{n\pi} - \pi\left(-\frac{\cos n\pi x}{n^2\pi^2}\right)\right]_0^1 + \left[\pi(2-x)\frac{\sin n\pi x}{n\pi} - (-\pi)\left(-\frac{\cos n\pi x}{n^2\pi^2}\right)\right]_1^2$$
$$= \left(\frac{\cos n\pi}{n^2\pi} - \frac{1}{n^2\pi}\right) + \left[-\frac{\cos 2n\pi}{n^2\pi} + \frac{\cos n\pi}{n^2\pi}\right]$$
$$= \frac{2}{n^2\pi}(\cos n\pi - 1)$$
$$= \frac{2}{n^2\pi}[(-1)^n - 1] = \begin{cases} 0 & ;\ \text{if } n \text{ is even} \\ -\dfrac{4}{n^2\pi} & ;\ \text{if } n \text{ is odd} \end{cases}$$

and
$$b_n = \int_0^2 f(x)\sin n\pi x dx$$
$$= \int_0^1 \pi x\sin n\pi x\, dx + \int_1^2 \pi(2-x)\sin n\pi x\, dx$$
$$= \left[\pi x\left(\frac{-\cos n\pi x}{n\pi}\right) - \pi\left(-\frac{\sin n\pi x}{n^2\pi^2}\right)\right]_0^1 + \left[\pi(2-x)\frac{\cos n\pi x}{n\pi} - (-\pi)\left(-\frac{\sin n\pi x}{n^2\pi^2}\right)\right]_1^2$$
$$= \left[-\frac{\cos n\pi}{n}\right] + \left[\frac{\cos n\pi}{n}\right] = 0$$

Hence,
$$f(x) = \frac{\pi}{2} - \frac{4}{\pi}\left(\frac{\cos\pi x}{1^2} + \frac{\cos 3\pi x}{3^2} + \frac{\cos 5\pi x}{5^2} + \ldots\right)$$

Example 3. *Expand $f(x) = e^{-x}$ as a Fourier series in the interval $(-l, l)$.* (Kerala–2005, VTU–2004)

Solution. Let
$$f(x) = e^{-x} = \frac{a_0}{2} + \sum_{n=1}^{\infty} a_n\frac{\cos n\pi x}{l} + \sum_{n=1}^{\infty} b_n\frac{\sin n\pi x}{l}$$

Then,
$$a_0 = \frac{1}{l}\int_{-l}^{l} e^{-x}dx = \frac{1}{l}\left[-e^{-x}\right]_{-l}^{l}$$
$$= -\frac{1}{l}(e^l - e^{-l}) = \frac{2\sinh l}{l}$$
$$a_n = \frac{1}{l}\int_{-l}^{l} e^{-x}\cos\frac{n\pi x}{l}dx$$

$$= \frac{1}{l}\left[\frac{e^{-x}}{1+\left(\frac{n\pi}{l}\right)^2}\left(-\cos\frac{n\pi x}{l}+\frac{n\pi}{l}\sin\frac{n\pi x}{l}\right)\right]_{-l}^{l}$$

$$= \frac{1}{l^2+(n\pi)^2}[-e^{-l}\cos n\pi + e^{l}\cos n\pi]$$

$$= -\frac{2l\cos n\pi}{l^2+(n\pi)^2}\left(\frac{e^{l}-e^{-l}}{2}\right) = \frac{2l(-1)^n\sinh l}{l^2+(n\pi)^2}$$

$$b_n = \frac{1}{l}\int_{-l}^{l} e^{-x}\sin\frac{n\pi x}{l}dx$$

$$= \frac{1}{l}\left[\frac{e^{-x}}{1+\left(\frac{n\pi}{l}\right)^2}\left(-\sin\frac{n\pi x}{l}-\frac{n\pi}{l}\cos\frac{n\pi x}{l}\right)\right]_{-l}^{l}$$

$$= -\frac{1}{l^2+(n\pi)^2}\left[\frac{n\pi}{l}(e^{-l}-e^{l})\cos n\pi\right]$$

$$= \frac{2n\pi\cos n\pi}{l^2+(n\pi)^2}\left(\frac{e^{l}-e^{-l}}{2}\right) = \frac{2n\pi(-1)^n\sinh l}{l^2+(n\pi)^2}$$

Hence, $e^{-x} = \sinh l\left[\frac{1}{l} - 2l\left(\frac{1}{l^2+\pi^2}\cos\frac{\pi x}{l} - \frac{1}{l^2+2^2\pi^2}\cos\frac{2\pi x}{l} + \frac{1}{l^2+3^2\pi^2}\cos\frac{3\pi x}{l} - \ldots\right)\right.$

$$\left. -2\pi\left(\frac{1}{l^2+\pi^2}\sin\frac{\pi x}{l} - \frac{2}{l^2+2^2\pi^2}\sin\frac{2\pi x}{l} + \frac{3}{l^2+3^2\pi^2}\sin\frac{3\pi x}{l} - \ldots\right)\right].$$

Example 4. *Prove that* $\frac{l}{2} - x = \frac{l}{\pi}\sum_{n=1}^{\infty}\frac{1}{n}\sin\frac{2n\pi x}{l}, 0 < x < l.$

Solution . Let $f(x) = \frac{l}{2} - x, 0 < x < l$

The Fourier series for $f(x)$ in the interval $(0, l)$ is

$$f(x) = \frac{a_0}{2} + \sum_{n=1}^{\infty}\left[a_n\cos\frac{n\pi x}{l/2} + b_n\sin\frac{n\pi x}{l/2}\right]$$

Here,

$$a_0 = \frac{1}{(l/2)}\int_0^l f(x)dx = \frac{2}{l}\int_0^l\left(\frac{1}{2}-x\right)dx$$

$$= \frac{2}{l}\left[\frac{lx}{2} - \frac{x^2}{2}\right]_0^l = 0$$

$$a_n = \frac{1}{(l/2)}\int_0^l f(x)\cos\frac{n\pi x}{(l/2)}dx$$

$$= \frac{2}{l}\int_0^l\left(\frac{l}{2}-x\right)\cos\frac{2n\pi x}{l}dx$$

$$= \frac{2}{l}\left[\left(\frac{l}{2}-x\right)\frac{1}{2n\pi}\sin\frac{2n\pi x}{l} + (-1)\frac{l^2}{4n^2\pi^2}\cos\frac{2n\pi x}{l}\right]_0^l$$

$$= \frac{2}{l}\left[-\frac{l^2}{4n^2\pi^2}\cos 2n\pi + \frac{l^2}{4n^2\pi^2}\right]$$

$$= \frac{2}{l}\cdot\frac{l^2}{4n\pi^2}(-\cos 2n\pi + 1)$$

$$= \frac{1}{2n^2\pi^2}(-1+1) = 0$$

and $b_n = \frac{1}{(l/2)}\int_0^l f(x).\frac{\sin n\pi x}{(l/2)}dx$

$$= \frac{2}{l}\int_0^l\left(\frac{l}{2}-x\right)\sin\frac{2n\pi x}{l}dx$$

$$= \frac{2}{l}\left[\left(\frac{l}{2}-x\right)\left(-\frac{1}{2n\pi}\cos\frac{2n\pi x}{l}\right) - (-1)\left(-\frac{l^2}{4n^2\pi^2}\sin\frac{2n\pi x}{l}\right)\right]_0^l$$

$$= \frac{2}{l}\left[\frac{l}{2}\cdot\frac{1}{2n\pi}\cos 2n\pi + \frac{l}{2}\cdot\frac{1}{2n\pi}(l)\right]$$

$$= \frac{2}{l}\left[\frac{l^2}{2n\pi}\right] = \frac{l}{n\pi}$$

The required Fourier series is

$$f(x) = \sum_{n=1}^{\infty}\frac{1}{n\pi}\frac{\sin 2n\pi x}{l}$$

or $\frac{1}{2} - x = \frac{1}{\pi}\sum_{n=1}^{\infty}\frac{1}{n}\sin\frac{2n\pi x}{l}.$

Example 5. *Find the Fourier expansion for the function* $f(x) = x - x^2; -1 < x < 1.$ (UPTU(Q.Bank)–2001)

Solution . Let $f(x) = \frac{a_0}{2} + \sum_{n=1}^{\infty} a_n\cos n\pi x + \sum_{n=1}^{\infty} b_n\sin n\pi x$

Then, $a_0 = \int_{-1}^{1}(x-x^2)dx = \int_{-1}^{1}xdx - \int_{-1}^{1}x^2dx$

$$= 0 - 2\int_0^1 x^2dx = -2\left[\frac{x^3}{3}\right]_0^1 = -\frac{2}{3}$$

$$a_n = \int_{-1}^{1}(x-x^2)\cos n\pi x\,dx$$

$$= \int_{-1}^{1}x\cos n\pi x\,dx - \int_{-1}^{1}x^2\cos n\pi x\,dx$$

$$= 0 - 2\int_0^1 x^2\cos n\pi x\,dx$$

$$= -2\left[x^2\frac{\sin n\pi x}{n\pi} - 2x\left(-\frac{\cos n\pi x}{n^2\pi^2}\right) + 2\left(-\frac{\sin n\pi x}{n^3\pi^3}\right)\right]_0^1$$

$$= -2\left[\frac{2\cos n\pi}{n^2\pi^2}\right] = -\frac{4(-1)^n}{n^2\pi^2}$$

and

$$b_n = \int_{-1}^{1}(x-x^2)\sin n\pi x\,dx = \int_{-1}^{1}x\sin n\pi x\,dx$$

$$= -1\int_{-1}^{1} x^2 n\pi x\,dx = 2\int_0^1 x\sin n\pi x\,dx - 0$$

$$= 2\left[x\left(-\frac{\cos n\pi x}{n\pi}\right) - 1\left(-\frac{\sin n\pi x}{n^2\pi^2}\right)\right]_0^1$$

$$= 2\left[-\frac{\cos n\pi}{n\pi}\right] = -2\frac{(-1)^n}{n\pi}$$

$$\therefore x - x^2 = -\frac{1}{3} + \frac{4}{\pi^2}\left(\frac{\cos \pi x}{1^2} - \frac{\cos 2\pi x}{2^2} + \frac{\cos 3\pi x}{3^2} - \ldots\right) + \frac{2}{\pi}\left(\frac{\sin \pi x}{1} - \frac{\sin 2\pi x}{2} + \frac{\sin 3\pi x}{3} - \ldots\right).$$

EXERCISE 27.3

1. Develop $f(x)$ in a Fourier series in the interval (0, 2) if
$$f(x) = \begin{cases} x, & 0 < x < 1 \\ 0, & 1 < x < 2 \end{cases}$$

2. Given $f(x) = \begin{cases} 0, & 0 < x < c \\ 1, & c < x < 2c \end{cases}$ expand $f(x)$ in a Fourier series of period $2c$.

3. Expand $f(x)$ in Fourier series in the interval (–2, 2) when
$$f(x) = \begin{cases} 0, & -2 < x < 0 \\ 1, & 0 < x < 2 \end{cases}$$

4. Find a Fourier series for the function given by
$$f(t) = \begin{cases} t, & 0 < t < 1 \\ 1 - t, & 1 < t < 2 \end{cases}$$

5. Find a Fourier series corresponding to the function $f(x)$ defined in (–2, 2) as follows;
$$f(x) = \begin{cases} 2, & \text{if } -2 \le x \le 0 \\ x, & \text{if } 0 < x < 2 \end{cases}$$

6. Find a Fourier series for the function
$$f(x) = \begin{cases} 0, & \text{when } -2 < x < -1 \\ k, & \text{when } -1 < x < 1 \\ 0, & \text{when } 1 < x < 2 \end{cases}$$

7. Find the Fourier series expansion of $f(x) = 2x - x^2$ in (0, 3) and hence show that
$$\frac{1}{1^2} - \frac{1}{2^2} + \frac{1}{3^3} - \frac{1}{4^2} + \ldots - \infty = \frac{\pi}{12}$$
(Mumbai–2005)

Hint to Selected Problems

2. $a_0 = \frac{1}{c}\int_0^{2c} f(x)dx = \frac{1}{c}\int_0^c 0.dx + \frac{1}{c}\int_c^{2c} 1.dx = \frac{1}{c}[x]_c^{2c} = 1$

$$a_n = \frac{1}{c}\int_0^{2c} f(x)\cos\frac{n\pi x}{c}dx$$

$$= \frac{1}{c}\int_0^c 0.\cos\frac{n\pi x}{c}dx + \frac{1}{c}\int_c^{2c} 1.\cos\frac{n\pi x}{c}dx = \frac{1}{c}\left[\frac{c}{n\pi}\sin\frac{n\pi x}{c}\right]_c^{2c}$$

$$= \frac{1}{n\pi}[\sin 2n\pi - \sin n\pi] = 0$$

$$b_n = \frac{1}{c}\int_0^{2c} f(x).\sin\frac{n\pi x}{c}dx$$

$$= \frac{1}{c}\int_0^c 0.\sin\frac{n\pi x}{c}dx + \frac{1}{c}\int_c^{2c} 1.\sin\frac{n\pi x}{c}dx$$

$$= \frac{1}{c}\left[-\frac{c}{n\pi}\cos\frac{n\pi x}{c}\right]_c^{2c}$$

$$= -\frac{1}{n\pi}[\cos 2n\pi - \cos n\pi] = -\frac{1}{n\pi}[1 - (-1)^n]$$

$$= \begin{cases} -\frac{2}{n\pi}, & \text{when } n \text{ is odd} \\ 0, & \text{when } n \text{ is even} \end{cases}$$

Then, $f(x) = \frac{1}{2} - \frac{2}{\pi}\left(\frac{1}{1}\sin\frac{\pi x}{c} + \frac{1}{3}\sin\frac{3\pi x}{c} + \ldots\right)$

5. $a_0 = \frac{1}{l}\int_{-l}^{l} f(x)dx = \frac{1}{2}\left[\int_{-2}^{0} 2dx + \int_0^2 x\,dx\right]$

$$= \frac{1}{2}\left[(2x)_{-2}^{0} + \left(\frac{x^2}{2}\right)_0^2\right] = 3$$

$$a_n = \frac{1}{l}\int_{-l}^{l} f(x)\cos\left(\frac{n\pi x}{l}\right)dx$$

$$= \frac{1}{2}\left[\int_{-2}^{0} 2\cos\frac{n\pi x}{2}dx + \int_0^2 x\cos\frac{n\pi x}{2}dx\right]$$

$$= \frac{1}{2}\left[\frac{4}{n\pi}\left(\sin\frac{n\pi x}{2}\right)_{-2}^{0} + \left(x\frac{2}{n\pi}\sin\frac{n\pi x}{2} + \frac{4}{n^2\pi^2}\cos\frac{n\pi x}{2}\right)_0^2\right]$$

$$= \frac{1}{2}\left[\frac{4}{n^2\pi^2}\cos n\pi - \frac{4}{n^2\pi^2}\right] = \frac{2}{n^2\pi^2}[(-1)^n - 1]$$

$$= \begin{cases} -\frac{4}{n^2\pi^2}, & \text{when } n \text{ is odd} \\ 0, & \text{when } n \text{ is even} \end{cases}$$

$$b_n = \frac{1}{l}\int_{-l}^{l} f(x)\sin\left(\frac{n\pi x}{l}\right)dx$$

$$= \frac{1}{2}\left[2\int_{-2}^{0}\sin\frac{n\pi x}{2}dx + \int_0^2 x\sin\frac{n\pi x}{2}dx\right]$$

$$= \frac{1}{2}\left[2\left(-\frac{2}{n\pi}\cos\frac{n\pi x}{2}\right)\right]_{-2}^{0} + \frac{1}{2}\left[x\left(-\frac{2}{n\pi}\cos\frac{n\pi x}{2}\right) + (1)\frac{4}{n^2\pi^2}\sin\frac{n\pi x}{2}\right]_0^2$$

$$= \frac{1}{2}\left[-\frac{4}{n\pi} + \frac{4}{n\pi}\cos n\pi\right] + \frac{1}{2}\left[-\frac{4}{n\pi}\cos n\pi + \frac{4}{n^2\pi^2}\sin n\pi\right]$$

$$= \frac{1}{2}\left[-\frac{4}{n\pi}\right] = -\frac{2}{n\pi}$$

$$f(x) = \frac{3}{2} - \frac{4}{\pi^2}\left\{\frac{1}{1^2}\cos\frac{\pi x}{2} + \frac{1}{3^2}\cos\frac{3\pi x}{2} + \ldots\right\} - \frac{2}{\pi}\left\{\frac{1}{1}\sin\frac{\pi x}{2} + \frac{1}{2}\sin\frac{2\pi x}{2} + \frac{1}{3}\sin\frac{3\pi x}{3} + \ldots\right\}.$$

ANSWERS

1. $f(x)=\frac{1}{4}-\frac{2}{\pi^2}\left(\cos\pi x+\frac{\cos 3\pi x}{3^2}+\frac{\cos 5\pi x}{5^2}+\ldots\right)+\frac{1}{\pi}\left(\sin\pi x-\frac{\sin 2\pi x}{2}+\frac{\sin 3\pi x}{3}+\ldots\right)$

2. $f(x)=\frac{1}{2}-\frac{2}{\pi}\left\{\sin\frac{\pi x}{c}+\frac{1}{3}\sin\frac{3\pi x}{c}+\ldots\right\}$

3. $f(x)=\frac{1}{2}+\frac{2}{\pi^2}\left(\sin\frac{\pi x}{2}+\frac{1}{3}\sin\frac{3\pi x}{2}+\frac{1}{5}\sin\frac{5\pi x}{2}+\ldots\right)$

4. $f(t)=-\frac{4}{\pi^2}\left(\cos\pi t+\frac{\cos 3\pi t}{3^2}+\frac{\cos 5\pi t}{5^2}+\ldots\right)+\frac{2}{\pi}\left(\sin\pi t+\sin\frac{3\pi t}{3}+\ldots\right)$

5. $f(x)=\frac{3}{2}-\frac{4}{\pi^2}\left\{\frac{1}{1^2}\cos\frac{\pi x}{2}+\frac{1}{3^2}\cos\frac{3\pi x}{2}+\ldots\right\}-\frac{2}{\pi}\left\{\sin\frac{\pi x}{2}+\frac{1}{2}\sin\frac{2\pi x}{2}+\frac{1}{3}\sin\frac{3\pi x}{2}+\ldots\right\}$

6. $f(x)=\frac{k}{2}+\frac{2R}{\pi}\left(\cos\frac{\pi x}{2}-\frac{1}{3}\cos\frac{3\pi x}{2}+\frac{1}{5}\cos\frac{5\pi x}{5}-\ldots\right)$

7. $f(x)=-\sum_{n=1}^{\infty}\frac{9}{n^2\pi^2}\cos\frac{2n\pi x}{3}+\sum_{n=1}^{\infty}\frac{3}{n\pi}\sin\frac{2n\pi x}{3}$

27.8 HALF RANGE SERIES

When we require to expand a function $f(x)$ in the range $(0, \pi)$ in a Fourier series of period 2π or more generally in the range $(0, l)$ in a Fourier series of period $2l$, a function $f(x)$ defined over the interval $0 < x < l$ is capable of two distinct half range series.

The half range cosine series is $f(x)=\frac{a_0}{2l}+\sum_{n=1}^{\infty}a_n\cos\frac{n\pi x}{l}$

where, $a_0=\frac{2}{l}\int_0^l f(x).dx$, where $a_n=\frac{2}{l}\int_0^l f(x)\cos\frac{n\pi x}{l}dx$

The half range sine series is

$$f(x)=\sum_{n=1}^{\infty}b_n\sin\frac{n\pi x}{l}, \text{ where } b_n=\frac{2}{l}\int_0^l f(x)\sin\frac{n\pi x}{l}dx$$

Solved Examples

Example 1. *If* $f(x)=\begin{cases}x & ; \quad 0<x<\pi/2\\ \pi-x & ; \quad \pi/2<x<\pi\end{cases}$ *Show that*

(i) $f(x)=\frac{4}{\pi}\left[\sin x-\frac{\sin 3x}{3^2}+\frac{\sin 5x}{5^2}-\ldots\right]$

(Mumbai–2008, SVTU–2008, VTU–2004)

(ii) $f(x)=\frac{\pi}{4}-\frac{2}{\pi}\left[\frac{\cos 2x}{1^2}+\frac{\cos 6x}{3^2}+\frac{\cos 10x}{5^2}+\ldots\right]$

(VTU–2011)

Solution. (i) Half range sine series, we have $l = \pi$ so

$$f(x)=\sum_{n=1}^{\infty}b_n\sin\frac{n\pi x}{\pi}=\sum_{n=1}^{\infty}b_n\sin nx$$

$$b_n=\frac{2}{\pi}\int_0^{\pi}f(x)\sin nx\,dx$$

$$=\frac{2}{\pi}\left[\int_0^{\pi/2}x\sin nx\,dx+\int_{\pi/2}^{\pi}(\pi-x)\sin nx\,dx\right]$$

$$=\frac{2}{\pi}\left[x\left(-\frac{\cos nx}{n}\right)-1\left(-\frac{\sin nx}{n^2}\right)\right]_0^{\pi/2}$$

$$+\frac{2}{\pi}\left[(\pi-x)\left(-\frac{\cos nx}{nx}\right)-(-1)\left(-\frac{\sin nx}{n^2}\right)\right]_0^{\pi}$$

$$=\frac{2}{\pi}\left[-\frac{\pi}{2n}\cos\frac{n\pi}{2}+\frac{1}{n^2}\sin\frac{n\pi}{2}\right]$$

$$+\frac{2}{\pi}\left[\frac{\pi}{2n}\cos\frac{n\pi}{2}+\frac{1}{n^2}\sin\frac{n\pi}{2}\right]$$

$$=\frac{2}{\pi}\left[\frac{2}{n^2}\sin\frac{n\pi}{2}\right]=\frac{4}{\pi n^2}\sin\frac{n\pi}{2}$$

Hence, $f(x)=\frac{4}{\pi}\left[\sin x-\frac{\sin 3x}{3^2}+\frac{\sin 5x}{5^2}-\ldots\right]$.

(ii) Half range cosine series

Let $f(x)=\frac{a_0}{2}+\sum_{n=1}^{\infty}a_n\cos nx$

Then, $a_0=\frac{2}{\pi}\int_0^{\pi}f(x)dx$

$$=\frac{2}{\pi}\int_0^{\pi/2}x\,dx+\int_{\pi/2}^{\pi}(\pi-x)dx$$

$$=\frac{2}{\pi}\left[\frac{x^2}{2}\right]_0^{\pi/2}+\left[\pi x-\frac{x^2}{2}\right]_{\pi/2}^{\pi}$$

$$=\frac{2}{\pi}\left[\frac{\pi^2}{8}+\left(\pi^2-\frac{\pi^2}{2}\right)-\left(\frac{\pi^2}{2}-\frac{\pi^2}{8}\right)\right]$$

$$=\frac{2}{\pi}\left[\frac{\pi^2}{4}\right]=\frac{\pi}{2}$$

and

$$a_n=\frac{2}{\pi}\int_0^{\pi}f(x)\cos x\,dx$$

$$=\frac{2}{\pi}\left[\int_0^{\pi/2}x\cos nx\,dx+\int_{\pi/2}^{\pi}(\pi-x)\cos nx\,dx\right]$$

$$=\frac{2}{\pi}\left[\frac{x\sin nx}{n}-1\left(-\frac{\cos nx}{n^2}\right)\right]_0^{\pi/2}$$

$$+\frac{2}{\pi}\left[(\pi-x)\frac{\sin nx}{n}-(-1)\left(\frac{\cos nx}{n^2}\right)\right]_{\pi/2}^{\pi}$$

$$=\frac{2}{\pi}\left[\frac{\pi}{2n}\sin\frac{n\pi}{2}+\frac{1}{n^2}\cos\frac{n\pi}{2}-\frac{1}{n^2}\right]$$

$$+\frac{2}{\pi}\left[\frac{\cos n\pi}{n^2}-\frac{\pi}{2n}\sin\frac{n\pi}{2}+\frac{1}{n^2}\cos\frac{n\pi}{2}\right]$$

$$=\frac{2}{\pi}\left[\frac{2}{n^2}\cos\frac{n\pi}{2}-\frac{\cos n\pi}{n^2}-\frac{1}{n^2}\right]$$

$$=\frac{2}{\pi n^2}\left[2\cos\frac{n\pi}{2}-\cos n\pi-1\right]$$

Put $n=0, 1, 2, 3, \ldots$ in equation (1), we get

$$a_1=0, a_2=\frac{2}{\pi.2^2}(2\cos\pi-\cos 2\pi-1)=\frac{-2}{1^2.\pi}$$

$$a_3=0, a_4=0, a_5=0,$$

$$a_6=\frac{2}{6^2\pi}(2\cos 3\pi-\cos 6\pi-1)=\frac{-2}{3^2\pi}$$

$$a_7=a_8=a_9=0,$$

$$a_{10}=\frac{2}{10^2.\pi}(2\cos 5\pi-\cos 10\pi-1)=\frac{-2}{5^2\pi}$$

Hence,

$$f(x)=\frac{\pi}{4}-\frac{2}{\pi}\left[\frac{\cos 2x}{1^2}+\frac{\cos 6x}{3^2}+\frac{\cos 10x}{5^2}+\ldots\right].$$

Example 2. *Develop the* $\sin\frac{\pi x}{l}$ *in half range cosine series in range* $0<x<l$. (UPTU–2001)

Solution . Let $\sin\frac{\pi x}{l}=\frac{a_0}{2}+\sum_{n=1}^{\infty}a_n\cos\frac{n\pi x}{l}$

where, $a_0=\frac{2}{l}\int_0^l\sin\frac{\pi x}{l}dx=\frac{2}{l}\left[-\frac{\cos(\pi x/l)}{\pi/l}\right]_0^l$

$$=\frac{2}{\pi}[\cos\pi-1]=\frac{4}{\pi}$$

and $a_n=\frac{2}{l}\int_0^l\sin\frac{\pi x}{l}\cos\frac{n\pi x}{l}dx$

$$=\frac{1}{l}\int_0^l\left[\sin(n+1)\frac{\pi x}{l}-\sin(n-1)\frac{\pi x}{l}\right]dx$$

$$=\frac{1}{l}\left[-\frac{\cos(n+1)\frac{\pi x}{l}}{(n+1)\pi/l}+\frac{\cos(n+1)\frac{\pi x}{l}}{(n-1)\pi/l}\right]_0^l$$

$$=\frac{1}{\pi}\left[-\frac{(-1)^{n+1}}{n+1}+\frac{(-1)^{n-1}}{n+1}+\frac{1}{n+1}-\frac{1}{n-1}\right]$$

(i) When n is odd

$$a_n=\frac{1}{\pi}\left[-\frac{1}{n+1}+\frac{1}{n-1}+\frac{1}{n-1}-\frac{1}{n-1}\right]=0$$

(ii) When n is even

$$a_n=\frac{1}{\pi}\left[\frac{1}{n+1}-\frac{1}{n-1}+\frac{1}{n-1}-\frac{1}{n-1}\right]$$

$$=\frac{2}{\pi}\left[\frac{1}{n+1}-\frac{1}{n-1}\right]$$

$$=\frac{-4}{\pi(n+1)(n-1)}, n\neq 1$$

$$\therefore\ \sin\frac{\pi x}{l}=\frac{2}{\pi}$$

$$-\frac{4}{\pi}\left[\frac{\cos\frac{2\pi x}{l}}{1.3}+\frac{\cos\frac{4\pi x}{l}}{3.5}+\frac{\cos\frac{6\pi x}{l}}{5.7}+\ldots\right].$$

Example 3. *Obtain the half range sine series for function* $f(x)=x^2$ *in the interval* $0<x<3$.

(UPTU–2001(Sp), 2002)

Solution . The Fourier half range sine series in the interval $(0, c)$ is given by

$$f(x)=\sum_{n=1}^{\infty}b_n\sin nx \quad \ldots(1)$$

where, $b_n=\frac{2}{c}\int_0^c f(x)\sin\frac{n\pi x}{c}dx$

Here, $c=3$ and $f(x)=x^2$

$$\therefore\ b_n=\frac{2}{3}\int_0^3 x^2\sin\frac{n\pi x}{3}dx$$

$$=\frac{2}{3}\left[x^2\left(\frac{-3}{n\pi}\right)\left(\cos\frac{n\pi x}{3}\right)+2x\left(\frac{3}{n\pi}\right)\left(\frac{3}{n\pi}\right)\sin\frac{n\pi x}{3}\right.$$

$$\left.-2\left(\frac{3}{n\pi}\right)\left(\frac{3}{n\pi}\right)\left(\frac{3}{n\pi}\right)\cos\frac{n\pi x}{3}\right]_0^3$$

$$=\frac{2}{3}\left[\left\{-\frac{27}{n\pi}(-1)^n-\frac{54}{n^3\pi^3}(-1)^n\right\}+\frac{54}{n^3\pi^3}\right]$$

$$=\frac{2}{3}\left[\frac{54}{n^3\pi^3}\{1-(-1)^n\}-\frac{27}{n\pi}(-1)^n\right]$$

$$=\begin{cases}\frac{2}{3}\left(\frac{108}{n^3\pi^3}+\frac{27}{n\pi}\right), & \text{if } n \text{ is odd}\\ -\frac{18}{n\pi}, & \text{if } n \text{ is even}\end{cases}$$

Hence, the required half range sine series is given by

$$f(x)=b_1\sin x+b_2\sin 2x+b_3\sin 3x+\ldots$$

$$=\frac{2}{3}\left[\frac{108}{\pi^3}\left(\frac{\sin x}{1^3}+\frac{\sin 3x}{3^3}+\frac{\sin 5x}{5^3}+\ldots\right)\right.$$

$$+\frac{27}{\pi}\left(\frac{\sin x}{1}+\frac{\sin 3x}{3}+\frac{\sin 5x}{5}+\ldots\right)$$

$$\left.-\frac{18}{\pi}\left(\frac{\sin 2x}{2}+\frac{\sin 4x}{4}+\ldots\right)\right].$$

Example 4. *(i) Express $f(x) = x$ as a half range sine series in $0 < x < 2$,* (UPTU–2004)

(ii) Express $f(x) = x$ as a half-range cosine series in $0 < x < 2$. (SVTU–2009, Bhopal–2007, Mumbai–2006)

Solution. (i) The Fourier since series for $F(x)$ in (0, 2) is

$$f(x) = \sum_{n=1}^{\infty} b_n \sin\frac{n\pi x}{2}$$

where

$$b_n = \frac{2}{2}\int_0^2 f(x)\sin\frac{n\pi x}{2}dx = \int_0^2 x\sin\frac{n\pi x}{2}dx$$

$$= \left[-\frac{2x}{n\pi}\cos\frac{n\pi x}{2} + \frac{4}{n^2\pi^2}\sin\frac{n\pi x}{2}\right]_0^2 = \frac{-4(-1)^n}{n\pi}$$

$\Rightarrow b_1 = 4/\pi_1, b_2 = -4/2\pi, b_3 = 4/3\pi,$

$b_4 = -4/4\pi$, etc.

Required half range Fourier sine series is

$$f(x) = \frac{4}{\pi}\left[\sin\frac{\pi x}{2} - \frac{1}{2}\sin\frac{2\pi x}{2} + \frac{1}{3}\sin\frac{3\pi x}{2} - \frac{1}{4}\sin\frac{4\pi x}{2} + \ldots\right]$$

(ii) The Fourier cosine series for $f(x)$ in (0, 2) is

$$f(x) = \frac{a_0}{2} + \sum_{n=1}^{\infty} a_n \cos\frac{n\pi x}{2}$$

where $a_0 = \frac{2}{2}\int_0^2 f(x)dx = \int_0^2 x\,dx = 2$

and $a_n = \frac{2}{2}\int_0^2 f(x)\cos\frac{n\pi x}{2}dx = \int_0^2 x\cos\frac{n\pi x}{2}dx$

$$= \left[\frac{2x}{n\pi}\sin\frac{n\pi x}{2} + \frac{4}{n^2\pi^2}\cos\frac{n\pi x}{2}\right]_0^2$$

$$= \frac{4}{n^2\pi^2}[(-1)^n - 1]$$

$\Rightarrow a_1 = -8/\pi^2, a_2 = 0, a_3 = -8/3^2\pi^2, a_4 = 0,$

$a_5 = -8/5^2\pi^2$

Required half range Fourier series is given by

$$f(x) = 1 - \frac{8}{\pi^2}\left[\frac{\cos \pi x/2}{1^2} + \frac{\cos 3\pi x/2}{3^2} + \frac{\cos 5\pi x/2}{5^2} + \ldots\right].$$

Example 5. *Find a series of cosines of multiples of x which will represent x sin x in the interval (0, π) and show that*

$$\frac{1}{1.3} - \frac{1}{3.5} + \frac{1}{5.7} - \ldots = \frac{\pi - 2}{4}.$$

(UPTU–2002, VTU–2003, Anna–2001)

Solution. Let $x\sin x = \frac{a_0}{2} + \sum_{n=1}^{\infty} a_n \cos nx$

Then

$$a_0 = \frac{2}{\pi}\int_0^\pi x\sin x\,dx = \frac{2}{\pi}[x(-\cos x) - 1.(-\sin x)]_0^\pi$$

$$= \frac{2}{\pi}[-\pi\cos x] = 2$$

and

$$a_n = \frac{2}{\pi}\int_0^\pi x\sin x\cos nx\,dx$$

$$= \frac{1}{\pi}\int_0^\pi x(2\cos nx\sin x)dx$$

$$= \frac{1}{\pi}\int_0^\pi x[\sin(n+1)x - \sin(n-1)x]dx$$

$$= \frac{1}{\pi}\left[x\left\{-\frac{\cos(n+1)x}{n+1} + \frac{\cos(n-1)x}{n-1}\right\} - 1\left\{-\frac{\sin(n+1)x}{(n+1)^2} - \frac{\sin(n-1)\pi}{(n-1)^2}\right\}\right]_0^\pi$$

$$= \frac{1}{\pi}\left[-\frac{\pi\cos(n+1)\pi}{n+1} + \frac{\pi\cos(n-1)\pi}{(n-1)}\right],$$

when $n \neq 1$

$$= \frac{-(-1)^{n+1}}{n+1} + \frac{(-1)^{n-1}}{n-1}$$

$$= (-1)^n\left[\frac{1}{n-1} - \frac{1}{n+1}\right] = \frac{2(-1)^{n-1}}{(n-1)(n+1)}$$

When $n=1$, we have

$$a_1 = \frac{2}{\pi}\int_0^\pi x\sin x\cos x\,dx = \frac{1}{\pi}\int_0^\pi x\sin 2x dx$$

$$= \frac{1}{\pi}\left[x\left(-\frac{\cos 2x}{2}\right) - 1\left(-\frac{\sin 2x}{2^2}\right)\right]_0^\pi$$

$$= \frac{1}{\pi}\left[-\frac{\pi\cos 2x}{2}\right] = -\frac{1}{2}$$

$$\therefore \quad x\sin x = 1 - \frac{1}{2}\cos x - 2\left(\frac{\cos 2x}{1.3} - \frac{\cos 3x}{2.4} + \frac{\cos 4x}{3.5} - \ldots\right)$$

Putting $x = \frac{\pi}{2}$, we get

$$\frac{\pi}{2} = 1 - 2\left(-\frac{1}{1.3} + \frac{1}{3.5} - \frac{1}{5.7} - \ldots\right)$$

$$\therefore \quad 1 + \frac{2}{1.3} - \frac{2}{3.5} + \frac{2}{5.7} - \ldots = \frac{\pi}{2}$$

$$\Rightarrow \quad \frac{2}{1.3} - \frac{2}{3.5} + \frac{2}{5.7} - \ldots = \frac{\pi}{2} - 1$$

Hence, $\frac{1}{1.3} - \frac{1}{3.5} + \frac{1}{5.7} - \ldots = \frac{\pi - 2}{4}.$

Example 6. *Obtain the half range sine series for e^x in $0 < x < 1$.*

Solution. Let $e^x = \sum_{n=1}^{\infty} b_n \sin n\pi x$ $\quad [\because l = 1]$

Then, $b_n = 2\int_0^l e^x \sin n\pi x\,dx$

$$= 2\left[\frac{e^x}{1+(n\pi)^2}(\sin n\pi x - n\pi\cos n\pi x)\right]_0^l$$

$$= 2\left[\frac{e}{1+(n\pi)^2}(-n\pi\cos n\pi x) - \frac{1}{1+(n\pi)^2}(-n\pi)\right]$$

$$= \frac{2}{1+n^2\pi^2}[-en\pi(-1)^n + n\pi]$$

$$= \frac{2n\pi}{1+n^2\pi^2}[1-e(-1)^n]$$

Hence, $e^x = 2\pi\sum_{n=1}^{\infty}\frac{n[1-e(-1)^n]}{1+n^2\pi^2}$

$$= 2\pi\left[\frac{1+e}{1+\pi^2}\sin\pi x + \frac{2(1-e)}{1+4\pi^2}\sin 2\pi x + \frac{3(1+e)}{1+9\pi^2}\sin 3\pi x + \ldots\right].$$

Example 7. *Expand* $f(x) = \begin{cases}\frac{1}{4} - x & ,if \quad 0 < x < \frac{1}{2}\\ x - \frac{3}{4} & ,if \quad \frac{1}{2} < x < 1\end{cases}$ *as the Fourier series of sine terms.*

(UPTU–2001, VTU–2011, Andhra–2000)

Solution . The Fourier sine series for $f(x)$ in (0, 1) is

$$f(x) = \sum_{n=1}^{\infty} b_n \sin n\pi x$$

where, $b_n = \frac{2}{l}\int_0^1 f(x)\sin n\pi x\, dx$

$$= 2\left[\int_0^{1/2}\left(\frac{1}{4} - x\right)\sin n\pi x\, dx + \int_{1/2}^{1}\left(x - \frac{3}{4}\right)\sin n\pi x\, dx\right]$$

$$= 2\left|-\left(\frac{1}{4} - x\right)\frac{\cos n\pi x}{n\pi} - \frac{\sin n\pi x}{n\pi}\right|_0^{1/2} + 2\left|-\left(x - \frac{3}{3}\right)\frac{\cos n\pi x}{n\pi} + \frac{\sin n\pi x}{n^2\pi^2}\right|_{1/2}^{1}$$

$$= 2\left[\frac{1}{4n\pi}\cos\frac{n\pi}{2} + \frac{1}{4n\pi} - \frac{\sin n\pi/2}{n^2\pi^2}\right] + 2\left[-\frac{1}{4n\pi}\cos n\pi - \frac{1}{4n\pi}\cos\frac{n\pi}{2} - \frac{\sin n\pi/2}{n^2\pi^2}\right]$$

$$= \frac{1}{2n\pi}[1-(-1)^n] - \frac{4\sin n\pi/2}{n^2\pi^2}$$

$$\Rightarrow b_1 = \frac{1}{\pi} - \frac{4}{\pi^2}, b_2 = 0, b_3 = \frac{1}{3\pi} + \frac{4}{3^2\pi^2},$$

$$b_4 = 0, b_5 = \frac{1}{5} - \frac{4}{5^2\pi^2}, b_6 = 0 \text{ etc.}$$

Hence, the required Fourier series is

$$f(x) = \left(\frac{1}{\pi} - \frac{4}{\pi^2}\right)\sin\pi x + \left(\frac{1}{3\pi} + \frac{4}{3^2\pi^2}\right)\sin 3\pi x + \left(\frac{1}{5\pi} - \frac{4}{5^2\pi^2}\right)\sin 5\pi x + \ldots$$

Example 8. *Find the half range cosine series of the function*

$$f(t) = \begin{cases}2t & ; \quad 0 < t < 1\\ 2(2-t) & ; \quad 1 < t < 2\end{cases}$$

Solution . The Fourier half range cosine series in interval (0, C) is

$$f(t) = \frac{a_0}{2} + a_1\cos\frac{\pi t}{c} + a_2\cos\frac{2\pi t}{c} + a_3\cos\frac{3\pi t}{c} + \ldots \qquad \ldots(1)$$

Here, $c = 2$ we have

$$a_0 = \frac{2}{c}\int_0^c f(t)dt = \frac{2}{2}\int_0^1 2t\, dt + 2\int_1^2 2(2-t)dt$$

$$= [t^2]_0^1 + \left[2\left(2t - \frac{t^2}{2}\right)\right]_1^2 = 1 + [(4t-t)]_1^2$$

$$= 1 + (8 - 4 - 4 + 1) = 2$$

and

$$a_n = \frac{2}{c}\int_0^c f(t)\cos\frac{n\pi t}{c}dt$$

$$= \frac{2}{2}\int_0^1 2t\cos\frac{n\pi t}{2}dt + \frac{2}{2}\int_1^2 2(2-t)\cos\frac{n\pi t}{2}dt$$

$$= \left[2t\left(\frac{2}{n\pi}\sin\frac{n\pi t}{2}\right) - (2)\left(-\frac{4}{n^2\pi^2}\cos\frac{n\pi t}{2}\right)\right]_0^1 + \left[(4-2t)\left(\frac{2}{n\pi}\sin\frac{n\pi t}{2}\right) - (2)\left(-\frac{4}{n^2\pi^2}\cos\frac{n\pi t}{2}\right)\right]_1^2$$

$$= \left[\frac{4}{n\pi}\sin\frac{n\pi}{2} + \frac{8}{n^2\pi^2}\cos\frac{n\pi}{2} - \frac{8}{n^2 + \pi^2}\right] + \left[0 + \frac{8}{n^2\pi^2}\cos\frac{n\pi}{2} - \frac{4}{n\pi}\sin\frac{n\pi}{2} + \frac{8}{n^2\pi^2}\cos\frac{n\pi}{2}\right]$$

$$= \frac{8}{n^2\pi^2}\cos\frac{n\pi}{2} - \frac{8}{n^2\pi^2} - \frac{4}{n\pi}\sin\frac{n\pi}{2}$$

$$= \frac{8}{n^2\pi^2}\left[\cos\frac{n\pi}{2} - 1 - \frac{n\pi}{2}\sin\frac{n\pi}{2}\right]$$

If $n=1$, $a_1 = \frac{8}{\pi^2}\left[0 - 1 - \frac{\pi}{2}\right] = \frac{-8}{\pi^2} - \frac{4}{\pi}$

If $n=2$, $a_2 = \frac{8}{4\pi^2}[-1-1] = \frac{-16}{4\pi^2} = -\frac{4}{\pi^2}$

If $n=3$, $a_3 = \frac{8}{9\pi^2}\left[0 - 1 + \frac{3\pi}{2}\right] = \frac{-8}{9\pi^2} + \frac{4}{3\pi}$

Putting these values of a_0, a_1, a_2, a_3, ... in equation (1), we get

$$f(1) = 1 - \left(\frac{8}{\pi^2} + \frac{4}{\pi}\right)\cos\frac{\pi t}{2} - \frac{4}{\pi^2}\cos\frac{2\pi t}{2} + \left(-\frac{8}{9\pi^2} + \frac{4}{3\pi}\right)\cos\frac{4\pi t}{2} + \ldots$$

EXERCISE 27.4

1. Find the Fourier half range series expansion of the function

$$f(x) = (-x/l) + 1, 0 \le x \le l.$$

2. Find a series of sines of multiples of x which will represent $f(x)$ in the interval $(0, \pi)$, where

$$f(x) = \begin{cases} \frac{1}{3}\pi & , \quad 0 < x < \frac{1}{3}\pi \\ 0 & , \quad \frac{1}{3}\pi < x < \frac{2}{3}\pi \\ -\frac{1}{3}\pi & , \quad \frac{2}{3}\pi < x < \pi \end{cases}$$

Also, represent this function by a series of cosines of multiples of x as well. Draw graph of these series and find the sine and cosine series where $x = -\frac{1}{3}\pi, -\frac{2}{3}\pi, -\pi$.

3. Find the half range cosine series for function $f(x) = (x - 1)^2$ in the interval $0 < x < 1$. (VTU–2010, JNTU–2006)

Hence show that

(i) $\frac{1}{1^2} + \frac{1}{2^2} + \frac{1}{3^2} + \frac{1}{4^2} + ... = \frac{\pi^2}{6}$,

(ii) $\frac{1}{1^2} - \frac{1}{2^2} + \frac{1}{3^2} - \frac{1}{4^2} + ... = \frac{\pi^2}{12}$,

(iii) $\frac{1}{1^2} + \frac{1}{3^2} + \frac{1}{5^2} + \frac{1}{7^2} + ... = \frac{\pi^2}{8}$. (Anna–2003)

4. If $f(x) = mx, \quad 0 \le x \le \pi/2$

$= m(\pi - x), \quad \pi/2 \le x \le \pi$

Then show that $f(x) = \frac{4m}{\pi}\left[\frac{\sin x}{1^2} - \frac{\sin 3x}{3^2} + \frac{\sin 5x}{5^2} - ...\right]$.

5. If $f(x) = \begin{cases} \frac{hx}{a} & , \quad 0 < x < a \\ \frac{h(l-x)}{l-a}, & \quad a < x < l \end{cases}$

Prove that for all values of x between 0 and l

$$f(x) = \frac{2hl^2}{a(l-a)\pi^2}\left[\sin\frac{\pi a}{l}\sin\frac{\pi x}{l}\sin\frac{\pi x}{l} + \frac{1}{2^2}\sin\frac{2\pi a}{l}\sin\frac{2\pi x}{l} + ...\right]$$

(UPTU(Q.Bank)–2001)

6. Expand $\pi x - x^2$ as a half range sine series in the interval $(0, \pi)$ upto first three terms.

7. Obtain a half range cosine series for

$$f(x) = \begin{cases} kx & , \quad \text{for} \quad 0 \le x \le l/2 \\ k(l-x) & , \quad \text{for} \quad l/2 \le x \le l \end{cases}$$

(Bhopal–2008, VTU–2008)

Show that $f(x) = \frac{4Kl}{\pi^2}\sum_{n=0}^{\infty}\frac{(-1)^n}{(2n+1)^2}\sin\frac{(2n+1)\pi x}{l}$

Deduce the sum of the series $\frac{1}{1^2} + \frac{1}{3^2} + \frac{1}{5^2} + ...$

(UPTU–2003, Rohtak–2006)

8. Find the half range sine series for the function $f(t) = t - t^2$ in the interval $0 < t < 1$. (UPTU–2004)

Hint to Selected Problems

1. $$a_0 = \frac{2}{l}\int_0^l f(x)dx = \frac{2}{l}\int_0^l\left(-\frac{x}{l}+1\right)dx$$

$$= \frac{2}{l}\left[-\frac{x^2}{2l} + x\right]_0^l = \frac{2}{l}\left[-\frac{l^2}{2l} + l\right] = 1$$

$$a_n = \frac{2}{l}\int_0^l f(x)\cos\frac{n\pi x}{l}dx = \frac{2}{l}\int_0^l\left(-\frac{x}{l}+1\right)\cos\frac{n\pi x}{l}dx$$

$$= \frac{2}{l}\left[\left(-\frac{x}{l}+1\right)\left(\frac{l}{n\pi}\sin\frac{n\pi x}{l}\right) - \left(-\frac{1}{l}\right)\left(-\frac{l^2}{n^2\pi^2}\cos\frac{n\pi x}{l}\right)\right]_0^l$$

$$= \frac{2}{l}\left[0 - \frac{l}{n^2\pi^2}\cos n\pi + \frac{l}{n^2\pi^2}\right] = \frac{2}{n^2\pi^2}[1-(-1)^n]$$

$$= \begin{cases} \frac{4}{n^2\pi^2} & , \quad \text{when } n \text{ is odd} \\ 0 & , \quad \text{when } n \text{ is odd} \end{cases}$$

and $b_n = 0$

$$\Rightarrow f(x) = \frac{1}{2} + \frac{4}{\pi^2}\left[\frac{1}{1^2}\cos\frac{\pi x}{l} + \frac{1}{3^2}\cos\frac{3\pi x}{l} + \frac{1}{5^2}\cos\frac{5\pi x}{l} + ...\right]$$

2. $$b_n = \frac{2}{\pi}\int_0^\pi f(v)\sin nv dv$$

$$= \frac{2}{\pi}\left[\int_0^{\pi/2}\frac{\pi}{3}\sin nv dv + \int_{\pi/3}^{2\pi/3} 0.\sin nv\, dv + \int_{2\pi/3}^{\pi} -\frac{\pi}{3}.\sin nv dv\right]$$

$$= \frac{2}{3}\left[-\frac{\cos nv}{n}\right]_0^{\pi/3} - \frac{2}{3}\left[-\frac{\cos nv}{n}\right]_{2\pi/3}^{\pi}$$

$$= \frac{2}{3n}\left[1 - \cos\frac{n\pi}{3} + \cos n\pi - \cos\frac{2n\pi}{3}\right]$$

$$= -\frac{8}{3n}\sin\frac{n\pi}{6}\sin\frac{n\pi}{3}.\cos\frac{n\pi}{2}$$

$$f(x) = -\frac{8}{3}\sum_{n=1}^{\infty}\frac{1}{n}\sin\frac{n\pi}{6}\sin\frac{n\pi}{3}\cos\frac{n\pi}{2}\sin nx$$

$$= \frac{1}{2}\left[\frac{1}{2}\sin 2x + \frac{1}{2}\sin 4x + \frac{1}{3}\sin 3x + \frac{1}{10}\sin 10x + ...\right]$$

$$a_0 = \frac{1}{\pi}\int_0^\pi f(v)dv = \frac{1}{\pi}\int_0^{\pi/3}\frac{\pi}{3}dv + \int_{\pi/3}^{2\pi/3} 0 dv + \int_{2\pi/3}^{\pi} -\frac{\pi}{3}dv = 0$$

$$a_n = \frac{2}{\pi}\int_0^\pi f(v)\cos nv\, dv$$

$$= \frac{2}{\pi}\left[\int_0^{\pi/3} \frac{\pi}{3}.\cos nv dv + \int_{\pi/3}^{2\pi/3} 0.dv + \int_{2\pi/3}^{\pi/3} -\frac{\pi}{3}.\cos nv\, dv\right]$$

$$= \frac{2}{3n}\left[\sin\frac{n\pi}{3} + \sin\frac{2n\pi}{3}\right] = \frac{4}{3n}\sin\frac{n\pi}{2}\cos\frac{n\pi}{6}$$

$$\Rightarrow f(x) = \frac{4}{3}\sum_{n=1}^{\infty}\frac{1}{n}\cos\frac{n\pi}{6}\sin\frac{n\pi}{2}.\cos nx$$

$$= \frac{2}{\sqrt{3}}\left[\cos x - \frac{1}{5}\cos 5x + \frac{1}{7}\cos 7x - \frac{1}{11}\cos 11x + \ldots\right]$$

6. $b_n = \frac{2}{\pi}\int_0^{\pi}(\pi x - x^2)\sin x\, dx$

$$= \frac{2}{\pi}\left[(\pi x - x^2)\left(-\frac{\cos nx}{n}\right) - (\pi - 2x).\left(-\frac{\sin nx}{n^2}\right) + (-2)\left(\frac{\cos nx}{n^3}\right)\right]_0^{\pi}$$

$$= \frac{2}{\pi}\left[-\frac{2\cos n\pi}{n^3} + \frac{2}{n^3}\right] = \frac{4}{\pi n^3}[1-(-1)^n] = 0 \text{ or } \frac{8}{n\pi^3}$$

according as n is even or odd

$$\Rightarrow \pi x - x^2 = \frac{8}{\pi}\left(\sin x + \frac{\sin 3x}{3^3} + \frac{\sin 5x}{5^3} + \ldots\right).$$

7. $a_0 = \frac{2}{l}\int_0^l f(x)\,dx = \frac{2}{l}\left[\int_0^{l/2} kx dx + \int_{l/2}^{l} k(l-x)dx\right] = \frac{kl}{2}$

$$= \frac{2}{l}\left[\int_0^{l/2} k.x\cos\frac{n\pi x}{l}dx + \int_{l/2}^{l} k(l-x)\frac{\cos n\pi x}{l}dx\right]$$

$$= \frac{2kl}{n^2\pi^2}\left[2\cos\frac{n\pi}{2} - 1 - \cos n\pi\right]$$

When n is odd, $\cos\frac{n\pi}{2} = 0$ and $\cos n\pi = -1$,

$$\Rightarrow a_n = 0 \Rightarrow a_1 = a_3 = a_5 = \ldots = 0$$

When n is even,

$$a_2 = \frac{2kl}{2^2\pi^2}[2\cos\pi - 1 - \cos 2\pi] = -\frac{8kl}{2^2\pi^2}$$

$$a_4 = \frac{2kl}{4^2\pi^2}[2\cos 2\pi - 1 - \cos 4\pi] = 0$$

$$a_6 = \frac{2kl}{6^2\pi^2}[2\cos 3\pi - 1 - \cos 6\pi]$$

$$= \frac{2kl}{6^2\pi^2}(-2-1-1) = -\frac{8kl}{6^2\pi^2} \text{ and so on}$$

$$\Rightarrow f(x) = \frac{kl}{4} - \frac{8kl}{\pi^2}\left[\frac{1}{2^2}\cos\frac{2\pi x}{l} + \frac{1}{6^2}\cos\frac{6\pi x}{l} + \ldots\right] \quad \ldots(1)$$

Putting $x = 1, f(x) = 0$

From (1), we have

$$0 = \frac{kl}{4} - \frac{8kl}{\pi^2}\left(\frac{1}{2^2} + \frac{1}{6^2} + \ldots\right)$$

$$\frac{1}{2^2} + \frac{1}{6^2} + \ldots = \frac{\pi^2}{32}$$

$$\Rightarrow \quad \frac{1}{2^2}\left(\frac{1}{1^2} + \frac{1}{3^2} + \ldots\right)\ldots = \frac{\pi^2}{32}.$$

Hence $\quad \frac{1}{1^2} + \frac{1}{3^2} + \ldots = \frac{\pi^2}{8}.$

ANSWERS

1. $f(x) = \frac{1}{2} + \frac{4}{\pi^2}\left[\frac{1}{1^2}\cos\frac{\pi x}{l} + \frac{1}{3^2}\cos\frac{3\pi x}{l} + \frac{1}{5^2}\cos\frac{5\pi x}{l} + \ldots\right]$

2. $f(x) = \frac{1}{2}\left[\frac{1}{2}\sin 2x + \frac{1}{2}\sin 4x + \frac{1}{8}\sin 8x + \frac{1}{10}\sin 10x\ldots\right]$ and $f(x) = \frac{2}{\sqrt{3}}\left[\cos x - \frac{1}{5}\cos 5x + \frac{1}{7}\cos 7x - \frac{1}{11}\cos 11x + \ldots\right]$

3. $\frac{1}{3} + \frac{4}{\pi^2}\left(\cos\pi x + \frac{\cos 2\pi x}{2^2} + \frac{\cos 3\pi x}{3^2} + \ldots\right)$

6. $\pi x - x^2 = \frac{8}{\pi}\left(\sin x + \frac{\sin 3x}{3^2} + \frac{\sin 5x}{5^3} + \ldots\right)$

7. $\frac{2kl}{n^2\pi^2}\left[2\cos\frac{n\pi}{2} - 1 - \cos n\pi\right]$

8. $f(t) = \frac{a}{2} + \frac{2a}{\pi}\left[\sin x + \frac{1}{3}\sin 3x + \frac{1}{5}\sin 5x + \frac{1}{7}\sin 7x + \ldots\right]$

27.9 PARSEVEL'S IDENTITY FOR FOURIER SERIES

Consider the Fourier series $\frac{a_0}{2} + \sum_{n=1}^{\infty}(a_n\cos nx + b_n\sin nx)$ *.If f(x) converges uniformly to f(x) at every point of the interval* (0, 2π), *then*

$$\frac{1}{\pi}\int_0^{2\pi}\{f(x)\}^2 dx = \frac{a_0^2}{2} + \sum_{n=1}^{\infty}(a_n^2 + b_n^2).$$

Proof. Let the series $\frac{a_0}{2} + \sum_{n=1}^{\infty}(a_n\cos nx + b_n\sin nx)$ represents the Fourier series of $f(x)$. Also, let this series converges uniformly to $f(x)$ at every point of the interval (0, 2π) so that

$$f(x) = \frac{a_0}{2} + \sum_{n=1}^{\infty}(a_n\cos nx + b_n\sin nx) \quad \ldots(1)$$

and that term by term integration is possible.

To prove that $\frac{1}{\pi}\int_0^{2\pi}\{f(x)\}^2dx = \frac{a_0^2}{2} + \sum(a_n^2 + b_n^2)$

We have $a_n = \frac{1}{\pi}\int_0^{2\pi} f(x).\cos nx\, dx \qquad (n = 0, 1, 2, 3, ...)$

$b_n = \frac{1}{\pi}\int_0^{2\pi} f(x).\sin nx\, dx \qquad (n = 0, 1, 2, 3 ...)$

Multiplying (1) by $f(x)$ and then integrating from $x=0$ to $x=2\pi$, we get

$$\int_0^{2\pi}\{f(x)\}^2dx = \frac{a_0}{2} + \int_0^{2\pi} f(x)dx + \sum_{n=1}^{\infty}\left(a_n\int_0^{2\pi} f(x).\cos nx dx + b_n\int_0^{2\pi} f(x)\sin nx dx\right) = \frac{a_0}{2}.\pi a_0 + \sum_{n=1}^{\infty}(\pi a_n^2 + \pi b_n^2)$$

Dividing by π, we get $\frac{1}{\pi}\int_0^{2\pi}\{f(x)\}^2dx = \frac{a_0^2}{2} + \sum_{n=1}^{\infty}(a_n^2 + b_n^2)$.

Solved Examples

Example 1. *Obtain the Fourier series expansion of $f(x) = x^2$ in $-\pi < x < \pi$ and prove that $\sum_{n=1}^{\infty}\frac{1}{n^4} = \frac{\pi^4}{90}$ by using Parsevel's theorem.*

Solution. Since $f(x) = x^2$ is even function so $b_n = 0$

Let the Fourier series expansion of $f(x)$ is given by

$$f(x) = x^2 = \frac{a_0}{2} + \sum_{n=1}^{\infty} a_n \cos nx \qquad ...(1)$$

where $a_0 = \frac{2}{\pi}\int_0^{\pi} f(x)dx = \frac{2}{\pi}\int_0^{\pi} x^2dx = \frac{2\pi^2}{3}$

$$a_n = \frac{2}{\pi}\int_0^{\pi} f(x)\cos nx\, dx = \frac{2}{\pi}\int_0^{\pi} x^2 \cos nx\, dx$$

$$\Rightarrow a_n = \frac{2}{\pi}\left[x^2\frac{\sin nx}{n} + 2x.\frac{\cos nx}{n^2} - 2\frac{\sin nx}{n^2}\right]_0^{\pi}$$

$$= \frac{\pi(-1)^2}{n^2}$$

$\therefore$ (1) becomes,

$$x^2 = \frac{\pi^2}{3} + 4\sum_{n=1}^{\infty}\frac{(-1)^n \cos nx}{n^2} \qquad ...(2)$$

which is the required Fourier expansion.

Now, by Parsevel's theorem, we have

$$\int_{-\pi}^{\pi}\{f(x)\}^2dx = \pi\left[\frac{a_0^2}{2} + \sum_{n=1}^{\infty}(a_n^2 + b_n^2)\right]$$

Hence, $\int_{-\pi}^{\pi} x^4dx = \pi\left[\frac{4\pi^4}{2.9} + \sum_{n=1}^{\infty}\frac{16}{n^4}\right]$

$$\Rightarrow \left(\frac{x^5}{5}\right)_{-\pi}^{\pi} = \frac{2\pi^5}{9} + \pi\sum_{n=1}^{\infty}\frac{16}{n^4}$$

or $\frac{2\pi^5}{5} - \frac{2\pi^5}{9} = \pi\sum_{n=1}^{\infty}\frac{16}{n^4} \Rightarrow \frac{\pi^4}{90} = \sum_{n=1}^{\infty}\frac{1}{n^4}$.

Example 2. *By using the sine series for $f(x) = 1$ in $0 < x < \pi$, show that*

$$\frac{\pi^2}{8} = 1 + \frac{1}{3^2} + \frac{1}{5^2} + \frac{1}{7^2} + ...$$

Solution. The Fourier sine series for $f(x) = 1$ in $(0, \pi)$ is

$f(x) = \Sigma\, b_n \sin nx$,

where,

$$b_n = \frac{2}{\pi}\int_0^{\pi} f(x)\sin nx\, dx = \frac{2}{\pi}\int_0^{\pi}(1).\sin nx dx$$

$$= \frac{2}{\pi}\left(-\frac{\cos nx}{n}\right)_0^{\pi}$$

$$= -\frac{2}{n\pi}(\cos n\pi - 1) = -\frac{2}{n\pi}[(-1)^n - 1]$$

$$= \begin{cases}\frac{4}{n\pi}, & \text{if } n \text{ is odd}\\ 0, & \text{if } n \text{ is even}\end{cases}$$

The Fourier sine series is

$$1 = \frac{4}{\pi}\sin x + \frac{4}{3\pi}\sin 3x + \frac{4}{5\pi}\sin 5x + \frac{4}{7\pi}\sin 7x + ...$$

By Parsevel's formula, we get

$$\int_0^{\pi}[f(x)]^2dx = \frac{c}{2}[b_1^2 + b_2^2 + b_3^2 + b_4^2 + b_5^2 + ...]$$

$$\Rightarrow \int_0^{\pi}(1)^2dx = \frac{\pi}{2}\left[\left(\frac{4}{\pi}\right)^2 + \left(\frac{4}{3\pi}\right)^2 + \left(\frac{4}{5\pi}\right)^2 + \left(\frac{4}{7\pi}\right)^2 + ...\right]$$

$$\Rightarrow [x]_0^{\pi} = \left(\frac{\pi}{2}\right)\left(\frac{16}{\pi^2}\right)\left[1 + \frac{1}{3^2} + \frac{1}{5^2} + \frac{1}{7^2} + ...\right]$$

$$\Rightarrow \pi = \frac{\pi}{2}\left(\frac{16}{\pi^2}\right)\left[1 + \frac{1}{3^2} + \frac{1}{5^2} + \frac{1}{7^2} + ...\right]$$

Hence, $\frac{\pi^2}{8} = 1 + \frac{1}{3^2} + \frac{1}{5^2} + \frac{1}{7^2} + ...$

Example 3. *If* $f(x)=\begin{cases}\pi x & , \quad 0<x<1\\ \pi(2-x) & , \quad 1<x<2\end{cases}$

Using half range cosine series, show that

$$\frac{1}{1^4}+\frac{1}{3^4}+\frac{1}{5^4}+\ldots=\frac{\pi^4}{96}.$$

Solution The half range cosine series for $f(x)$ in $(0, c)$ is

$$f(x)=\frac{a_0}{2}+\sum_{n=1}^{\infty}a_n\cos\frac{n\pi x}{c}$$

Here, $a_0=\frac{2}{c}\int_0^c f(x)dx$

$$=\frac{2}{2}\left[\int_0^1 \pi x\,dx+\int_1^2\pi(2-x)dx\right]$$

$$=\pi\left[\frac{x^2}{2}\right]_0^1+\pi\left[2x-\frac{\pi^2}{2}\right]_0^1$$

$$=\frac{\pi}{2}+\pi\left[(4-2)-\left(2-\frac{1}{2}\right)\right]=\pi$$

$$a_n=\frac{2}{c}\int_0^c f(x).\cos\frac{n\pi x}{c}dx$$

$$=\frac{2}{2}\left[\int_0^1 \pi x\cos\frac{n\pi x}{2}dx+\int_1^2\pi(2-x)\cos\frac{n\pi x}{2}dx\right]$$

$$=\pi\left[\frac{x\sin\frac{n\pi x}{2}}{\frac{n\pi}{2}}-\left(-\frac{\cos\frac{n\pi x}{2}}{\frac{n^2\pi^2}{4}}\right)\right]_0^1+\pi\left[(2-x)\frac{\sin\frac{n\pi x}{2}}{\frac{n\pi}{2}}-(-1)\left(-\frac{\cos\frac{n\pi x}{2}}{\frac{n^2\pi^2}{4}}\right)\right]_1^2$$

$$=\pi\left[\frac{2}{n\pi}\sin\frac{n\pi}{2}+\frac{4}{n^2\pi^2}\cos\frac{n\pi}{2}-\frac{4}{n^2\pi^2}\right]$$

$$+\pi\left[0-\frac{4}{n^2\pi^2}\cos n\pi-\frac{2}{n\pi}\sin\frac{n\pi}{2}+\frac{4}{n^2\pi^2}\cos\frac{n\pi}{2}\right]$$

$$=\pi\left[\frac{8}{n^2\pi^2}\cos\frac{n\pi}{2}-\frac{4}{n^2\pi^2}-\frac{4}{n^2\pi^2}\cos n\pi x\right]$$

$$=\frac{4}{n^2\pi}\left[2\cos\frac{n\pi}{2}-1-\cos n\pi\right]$$

Putting $n=1, 2, 3, \ldots$, we get

$$a_1=0, a_2=\frac{-4}{\pi}, a_3=0, a_4=0, a_5=0, a_6=-\frac{4}{9\pi}$$

By Parsevel's formula, we get

$$\int_0^c\{f(x)\}^2dx=\frac{c}{2}\left[\frac{a_0^2}{2}+a_1^2+a_2^2+a_3^2+\ldots\right]$$

$$\int_0^1(\pi x)^2dx+\int_1^2\pi^2(2-x)^2dx=\frac{2}{2}\left[\frac{\pi^2}{2}+\frac{16}{\pi^2}+\frac{16}{81\pi^2}+\ldots\right]$$

$$\pi^2\left[\frac{x^3}{3}\right]_0^1-\pi^2\left[\frac{(2-x)^3}{3}\right]_1^2=\frac{\pi^2}{2}+\frac{16}{\pi^2}+\frac{16}{81\pi^2}+\ldots$$

$$\Rightarrow\quad \frac{\pi^2}{3}-\pi^2\left(0-\frac{1}{3}\right)=\frac{\pi^2}{3^2}+\frac{16}{\pi^2}\left[1+\frac{1}{81}+\ldots\right]$$

$$\Rightarrow\quad \frac{2\pi^2}{3}-\frac{\pi^2}{2}=\frac{16}{\pi^2}\left[1+\frac{1}{3^4}+\frac{1}{5^4}+\ldots\right]$$

$$\Rightarrow\quad \frac{\pi^2}{6}=\frac{16}{\pi^2}\left[1+\frac{1}{3^4}+\frac{1}{5^4}+\ldots\right]$$

$$\Rightarrow\quad \frac{\pi^4}{96}=1+\frac{1}{3^4}+\frac{1}{5^4}+\ldots$$

27.10 COMPLEX FORM OF FOURIER SERIES

In complex notations, the Fourier series is written as

$$f(x)=\sum_{n=-\infty}^{+\infty}C_n\,e^{inx} \qquad (n=0,1,2,\ldots)$$

$$f(x)=C_0+\sum_{n=1}^{\infty}C_ne^{inx}+\sum_{n=1}^{\infty}C_{-n}\,e^{-inx} \qquad \ldots(1)$$

With $C_n=\frac{1}{2\pi}\int_C^{C+2\pi}f(x)\,e^{-inx}dx$ and $C_{-n}=\frac{1}{2\pi}\int_C^{C+2\pi}f(x)\,e^{inx}dx$ $\qquad (n=0,1,2,\ldots)$

$f(x)$ being defined in the interval $[C, C+2\pi]$.

Here, $C_0=\frac{1}{2\pi}\int_C^{C+2\pi}f(x)\,dx=\frac{a_0}{2}.$

$$C_n=\frac{1}{2\pi}\int_C^{C+2\pi}f(x)\,e^{-inx}\,dx=\frac{1}{2\pi}\int_C^{C+2\pi}f(x)(\cos nx-i\sin nx)dx$$

$$=\frac{1}{2\pi}\int_C^{C+2\pi}f(x)\cos nx\,dx-i\frac{1}{2\pi}\int_C^{C+2\pi}f(x)\sin nx\,dx=\frac{a_n-ib_n}{2}$$

Similarly, $C_{-n} = \frac{1}{2\pi}\int_C^{C+2\pi} f(x)\, e^{inx}\, dx = \frac{a_n + ib_n}{2}$.

Now, (1) becomes

$$f(x) = \frac{a_0}{2} + \sum_{n=1}^{\infty}\left\{\left(\frac{a_n - ib_n}{2}\right)\cos nx + \left(\frac{a_n + ib_n}{2}\right)\sin nx\right\} \qquad \ldots(2)$$

with $a_n = \frac{1}{\pi}\int_C^{C+2\pi} f(x).\cos nx\, dx$ and $b_n = \frac{1}{\pi}\int_C^{C+2\pi} f(x)\sin nx\, dx$

Solved Example

Example. *Obtain the complex form of the Fourier series of* $f(x) = e^{-x}$ *in* $-1 \le x \le 1$. (Mumbai–2005S, Madras–2000S)

Solution. The complex form of Fourier series for the given function $f(x)$ is

$$f(x) = \sum_{n=-\infty}^{\infty} C_n\, e^{inx}.$$

Here, $C_n = \int_{-1}^{1} e^{-x}\, e^{-inx\pi} dx = \frac{1}{2}\int_{-1}^{1} e^{-(1+in\pi)x} dx$

$$= \frac{1}{2}\left[\frac{e^{-(1+in\pi)x}}{-(1+in\pi)}\right]_{-1}^{1} = \frac{e^{1+inx} - e^{-(1+inx)}}{2(1+inx)}$$

$$= \frac{e(\cos n\pi + i\sin n\pi) - e^{-1}(\cos n\pi - i\sin n\pi)}{2(1+in\pi)}$$

$$= \frac{e - e^{-1}}{2}(-1)^n\,\frac{1 - in\pi}{1 + n^2\pi^2}$$

$$= \frac{(-1)^n(1 - in\pi)\sinh 1}{1 + n^2\pi^2}$$

Hence, the required complex form of the Fourier series is

$$e^{-x} = \sum_{n=-\infty}^{\infty}\frac{(-1)^n(1 - in\pi)}{1 + n^2\pi^2}\sinh 1.e^{in\pi x}.$$

EXERCISE 27.5

1. Prove that in $0 < x < C$,

$$x = \frac{C}{2} - \frac{4C}{\pi^2}\left(\cos\frac{\pi x}{C} + \frac{1}{3^2}\cos\frac{3\pi x}{C} + \frac{1}{5^2}\cos\frac{5\pi x}{C} + \ldots\right)$$

and deduce that

(i) $\frac{1}{1^4} + \frac{1}{3^4} + \frac{1}{5^4} + \ldots = \frac{\pi^4}{96}$ (ii) $\frac{1}{1^4} + \frac{1}{2^4} + \frac{1}{3^4} + \frac{1}{4^4} + \ldots = \frac{\pi^4}{90}$

2. Find the complex form of the Fourier series of $f(x) = \cos ax,\ -\pi < x < \pi$. (Anna–2009, Mumbai–2009)

3. Find the complex form of the Fourier series of $f(x) = e^{ax},\ -l < x < l$. (Madras–2003)

4. Find the complex form of the Fourier series of $f(x) = \begin{cases} 0, & \text{if } 0 < x < l \\ a, & \text{if } 0 < x < 2l \end{cases}$.

5. Find the complex form of the Fourier series of $f(x) = \cosh 3x + \sinh 3x$ in (–3, 3). (Mumbai–2008)

ANSWERS

2. $\frac{a}{\pi}\sin a\pi \sum_{n=-\infty}^{\infty}\frac{(-1)^n e^{inx}}{a^2 - n^2}$

3. $\frac{2}{\pi} - \frac{2}{\pi}\left[\frac{e^{2it} + e^{-2it}}{1.3} + \frac{e^{4it} + e^{-4it}}{3.5} + \frac{e^{6it} + e^{-6it}}{5.7} + \ldots\right]$

4. $\frac{a}{2} - \frac{a}{\pi}\left[(e^u - e^{-u}) + \frac{1}{3}(e^{3u} - e^{-3u}) + \frac{1}{5}(e^{5u} - e^{-5u}) + \ldots\right]$, where $u = \frac{i\pi x}{l}$

5. $\sinh 9\sum_{-\infty}^{\infty}\frac{(-1)^n(9 + n\pi i)}{81 + (n\pi)^2}e^{n\pi ix/3}$

Objective Evaluations

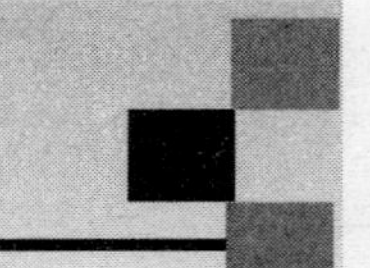

Fill in the Blanks

1. The double sequence $\left\langle \frac{1}{mn} \right\rangle$ converges to ________ .
2. The double sequence $\left\langle (-1)^{m+n} \right\rangle$ oscillates ________ .
3. The double sequence $\left\langle (-1)^{m+n}(m+n) \right\rangle$ oscillates ________ .
4. A necessary condition for convergence of a double series $\sum_{m,n} a_{mn}$ that $\lim_{m,n\to\infty} a_{mn} =$ ________ .
5. Every absolutely convergent double series is ________ .

True / False

Write 'T' for True and 'F' for False statement.

1. $\int_{-\pi}^{\pi} \sin nx\, dx = 0 = \int_{-\pi}^{\pi} \cos nx\, dx \forall n$. **(T/F)**
2. The inequality $\sum_{n=1}^{\infty}(a_n^2 + b_n^2) \le \frac{1}{\pi}\int_{-\pi}^{\pi} f^2 dx$ is called Bessel's inequality. **(T/F)**
3. If f is integrable on $[-\pi, \pi]$ and has period 2π then $\int_{-\pi}^{\pi} f(t)dt = \int_{-\pi}^{\pi} f(t+a)dt$ for any $a \in R$. **(T/F)**
4. $\int_0^{\infty} \frac{\sin t}{T} dt = \pi$ **(T/F)**
5. Fourier series has period 2π. **(T/F)**

Multiple Choice Questions

Choose the most appropriate one

1. A function $f(x)$ is called periodic if it is defined for real $x \in R$ and:
   (a) if there is any positive number p such that $f(x+p) = f(x) + f(p)$
   (b) if there is any positive number p, such that $f(x+p) > f(x)$
   (c) if there is any positive number p such that $f(x+p) = f(r)$,
   (d) none of the above.
2. In Fourier series $f(x) = a_0 + \sum_{n=1}^{\infty}(a_n \cos nx + b_n \sin nx)$, Euler formula is:
   (a) $a_0 = \frac{1}{\pi}\int_{-\pi}^{\pi} f(x)dx$ (b) $a_0 = \frac{1}{2\pi}\int_{-\pi}^{\pi} f(x)dx$
   (c) $a_0 = \frac{1}{2\pi}\int_{-\pi}^{\pi} f(x)\sin x\, dx$ (d) none of the above
3. In Fouries series $f(x) = a_0 + \sum_{n=1}^{\infty}(a_n \cos nx + b_n \sin nx)$, Euler formula is:
   (a) $a_n = \frac{1}{\pi}\int_{-\pi}^{\pi} f(x)dx$ (b) $a_n = \frac{1}{\pi}\int_{-\pi}^{\pi} f(x)\sin nx\, dx$
   (c) $a_n = \frac{1}{\pi}\int_{-\pi}^{\pi} f(x)\cos nx\, dx$ (d) none of the above
4. If a function $f(x)$ of period $p = 2l$ has Fourier series, then in series $f(x) = a_0 + \sum_{n=1}^{\infty}(a_n \cos nx + b_n \sin nx)$, the Fourier coefficients a_n is given by ;
   (a) $\int_{-l}^{l} f(x).\frac{\sin n\pi x}{l} dx$ (b) $\frac{1}{l}\int_{-l}^{l} f(x).\frac{\cos n\pi x}{l} dx$
   (c) $\int_{-l}^{l} f(x).\frac{\cos n\pi x}{l} dx$ (d) none of the above
5. The Fourier series of an even function of period 2π is a:
   (a) Fourier cosine series (b) Fourier sine series
   (c) Fourier complex series (d) none of the above
6. The Fourier series of an odd function of period $2l$ is given by:
   (a) $f(x) = a_0 + \sum_{n=1}^{\infty} a_n \frac{\cos n\pi}{l}$, where $a_0 = \frac{1}{l}\int_0^l f(x)dx,\ a_n = \frac{2}{l}\int_0^l f(x).\frac{\cos n\pi x}{l} dx$
   (b) $f(x) = \sum_{n=1}^{\infty} b_n \sin\frac{n\pi x}{l}$, where $b_0 = \frac{2}{l}\int_0^l f(x)\sin\frac{n\pi x}{l} dx$
   (c) $f(x) = \sum_{n=-\infty}^{\infty} C_n e^{inx}$, where $C_n = \frac{1}{2\pi}\int_{-\pi}^{\pi} f(x)e^{-inx}dx, n = 1, 2, ...$
   (d) none of the above
7. The Fourier series of an odd function of period $2l$ is given by:
   (a) Fourier cosine series (b) Fourier sine series
   (c) Fourier complex series (d) none of the above

ANSWERS

Fill in the Blanks

1. 0 2. finitely 3. infinitely 4. 0 5. convergent

True/False

1. T 2. T 3. T 4. F 5. T

Multiple Choice Questions

1. (c) 2. (c) 3. (a) 4. (b) 5. (a) 6. (b) 7. (b)

FFFFF

GATE tutor

KEY Terms and Results

- **Infinite Series:** An expression of the form
$$\sum_{n=0}^{\infty} a_n = a_0 + a_1 + ... + a_n + ...$$
where each a_n is a real number, is called an infinite series of real numbers.
- **Sequence of Partial Sum:** The sequence $<s_n> = a_1 + a_2 + ... + a_n$ is called the sequence of partial sum of the series $\sum_{n=1}^{\infty} a_n$.
- **Convergent Series:** A series Σa_n is said to be convergent if the sequence $<s_n>$ of partial sums of Σa_n is convergent
- **Divergent Series:** The series Σa_n is said to be divergent, if the sequence $<s_n>$ of partial sums of Σa_n is divergent.
- **Oscillating Series :** The series Σa_n is said to be oscillate if the sequence $<s_n>$ of partial sums of Σa_n oscillate.
- **Positive Series:** An infinite series whose all terms are positive is called a positive term series.
- **Absolute Convergence:** A series Σu_n is said to be absolutely convergent of the series $|\Sigma u_n|$ is convergent.
- **Conditional Convergence:** A series Σu_n is said to be conditionally convergent if (i) Σu_n is convergent and (ii) Σu_n is not absolutely convergent.
- **Fourier Series :** Let f be a real valued bounded and integrable function defined on $[-\pi, \pi]$ such that
$$f(x) = \frac{a_0}{2} + \sum_{n=1}^{\infty} (a_n \cos nx + b_n \sin nx) \quad ... (1)$$
where $a_n = \frac{1}{\pi}\int_{-\pi}^{\pi} f(x)\cos nx\, dx$ and
$b_n = \frac{1}{\pi}\int_{-\pi}^{\pi} f(x)\sin nx\, dx$ then (1) is called Fourier series.
- **Periodic function :** A functon $f(x)$ which satisfies the relation $f(x+\lambda) = f(x)\ \forall\, x$ and some fixed λ, is called periodic function with period λ.
- **Half Range Sine Series :**
$$f(x) = \sum_{n=1}^{\infty} b_n \sin\frac{n\pi x}{l}, \text{where } b_n = \frac{2}{l}\int_0^{2l} f(x)\sin\frac{n\pi x}{l}dx$$
- **Half Range Cosine Series :**
$$f(x) = \frac{a_0}{2l} + \sum_{n=1}^{\infty} a_n \cos\frac{n\pi x}{l}$$
where
$$a_0 = \frac{2}{l}\int_0^l f(x)dx, a_n = \frac{2}{l}\int_0^l f(x)\cos\frac{n\pi x}{l}dx$$
- If Σa_n and Σb_n are two convergent series then $\Sigma(a_n \pm b_n)$ are also convergent.
- If Σa_n converges and Σb_n diverges, then $\Sigma(a_n + b_n)$ diverges.
- A necessary condition for the convergence of the series Σu_n is that $\lim_{n\to\infty} u_n = 0$.
- **(Cauchy's General Principle of Convergence).** A necessary and sufficient condition for a series Σu_n to converge is that to real $\varepsilon > 0\ \exists$ positive integer m such that $|u_{m+1} + u_{m+2} + ... + u_n| < \varepsilon\ \forall\, n \geq m$.
- A positive term series Σu_n is convergent if and only if the sequence $<s_n>$ of partial sums is bounded above.
- A positive term series cannot be oscillate.
- If Σu_n is a positive term series such that $\lim_{n\to\infty} u_n \neq 0$ then Σu_n diverges.
- If Σa_n is a convergent series of positive terms then $\sum a_n^2$ is also convergent.
- **Cauchy's n^th root test:** If Σu_n be a positive term series such that $\lim_{n\to\infty} (u_n)^{1/n} = l$
Then (i) Σu_n converges if $l < 1$ (ii) Σu_n diverges if $l > 1$ (iii) test fails if $l = 1$
- **D'Alembert's ratio test:** Let Σu_n be a positive terms series such that
$$\lim_{n\to\infty}\frac{u_n}{u_{n+1}} = l$$
Then
(i) Σu_n converges if $l > 1$ (ii) Σu_n diverges if $l < 1$
(iii) test fails if $l = 1$.
- **Raabe' test:** Let Σu_n be a positive term series such that
$$\lim_{n\to\infty}\left(\frac{u_n}{u_{n+1}} - 1\right) = l$$
Then
(i) Σu_n converges if $l > 1$
(ii) Σu_n diverges if $l < 1$
(iii) test fails if $l = 1$
- Raabe's test is stronger than ratio test.
- **Cauchy's Integral test :** If $u(x)$ is a non-negative monotonically decreasing and integrable function such that $u(n) = u_n\ \forall\, n \in \text{N}$ then the series $\sum_{n=1}^{\infty} u_n$ is convergent if and only $\int_1^{\infty} u(x)dx$ is convergent.
- **Leibnitz test:** If an alternating series $\sum_{n=1}^{\infty}(-1)^{n-1}u_n$ satisfies
(i) $u_{n+1} \leq u_n\ \forall\, n$ (iii) $\lim_{n\to\infty} u_n = 0$
Then the series $\Sigma(-1)^{n-1} u_n$ converges.
- Every absolutely convergent series is convergent but converse is not necessarily true.

- Any rearrangement of a convergent series of positive terms converges to the same sum.
- Cauchy product of two absolutely convergent series is also absolutely convergent.

Objective Type Questions

1. The series $\sum_{n=1}^{\infty}(-1)^n$ is:
(a) convergent (b) divergent
(c) unbounded (d) none of these

2. The series $\frac{1}{1.3}+\frac{1}{2.4}+\frac{1}{3.5}+\ldots$ is:
(a) convergent (b) divergent
(c) may be convergent (d) none of these

3. The series $1.x + 1.2x^2 + 1.2.3x^3 + \ldots + n!x^n + \ldots$ is:
(a) convergent everywhere except at $x = 0$
(b) divergent everywhere except at $x = 0$
(c) divergent for $x = 0$
(d) none of the above

4. The series $\sum_{n=1}^{\infty}\frac{n^2}{3^n}$ is:
(a) convergent (b) divergent
(c) unbounded (d) none of these

5. The series for $\sum_{n=1}^{\infty}(-1)^n\frac{x^n}{n}$ for $|x| > 1$ is:
(a) convergent (b) divergent
(c) oscillatory (d) none of these

6. For infinite series $\sum_{n=1}^{\infty}a_n$ and $\sum_{n=1}^{\infty}b_n$, $b_n \geq 0$ for all n and there is a real number N such that for $n \geq N \Rightarrow |a_n| \leq b_n$ if $\sum_{n=1}^{\infty}b_n$ converges, then :
(a) Σa_n is absolutely convergent
(b) Σb_n is absolutely convergent
(c) Σa_n is divergent
(d) none of the above

7. The series $1+\frac{x}{1^2}+\frac{x^2}{2^2}+\frac{x^3}{3^2}+\ldots$:
(a) converges for $|x|<1$ (b) diverges for $|x| > 1$
(c) diverges for $|x|=1$ (d) none of the above

8. The series $1 + x + x^2 + \ldots + x^n \ldots +$:
(a) converges for $|x| < 1$ (b) diverges for $|x|>1$
(c) converges for $|x|=1$ (d) none of these

9. Let Σa_n be a convergent series of positive terms and let Σb_n be a divergent series of positive terms converges to 0. Then:
(a) $<a_n>$ converges to 0
(b) $<a_n>$ not converges to 0
(c) $<b_n>$ diverges to 0
(d) none of the above

10. Let Σa_n be a convergent series of positive terms and Σb_n be a divergent series of positive terms. Then:
(a) $\Sigma(a_n+b_n)$ is divergent
(b) $\Sigma(a_n+b_n)$ is convergent
(c) both (a) and (b) may be true
(d) none of the above

11. If for a given series Σu_n
$\frac{u_n}{u_{n+1}}=\alpha+\frac{\beta}{n}+o\left(\frac{1}{n^p}\right), p>1, \alpha,\beta \in R$. Then
(a) Σu_n converges if $\beta > 1\ \forall\ \alpha$
(b) Σu_n converges if $\alpha > 1$ for any β
(c) Σu_n diverges to ∞
(d) none of the above

12. If $n^{1/n} \to 1$ as $n \to \infty$, then the series $\sum_{n=1}^{\infty}(n^{1/n}-1)^n$:
(a) converges
(b) diverges
(c) neither converges nor diverges
(d) none of the above

13. Which of the following inequalities will be used to show the series $\sum_{n=1}^{\infty}e^{-n^2}$ converges ?
(a) $e^n < x$ if $n = 0$ (b) $e^n < x$ if $n > 0$
(c) $e^n > x$ if $n > 0$ (d) none of these

14. A conditionally convergent series is a series which is :
(a) absolutely convergent
(b) convergent but not absolutly convergent
(c) divergent
(d) none of the above

15. For the series $1 + r + r^2 + \ldots$ $(r > 0)$, which one of the following is true?
(a) It does not converges. (b) It does not diverges.
(c) It oscillate. (d) none of these

16. The series $\frac{1}{3^p}+\frac{1}{5^p}+\frac{1}{7^p}+\frac{1}{9^p}+\ldots$ converges if :
(a) $p < 1$ (b) $p >1$
(c) $p = 1$ (d) none of these

17. The series $1+ r + r^2 + \ldots$ is oscillatory if:
(a) $r < 1$ (b) $r > 1$
(c) $r = 1$ (d) $r = -1$

18. The series $x+\frac{2^2x^2}{2!}+\frac{3^3x^3}{3!}+\frac{4^4.x^4}{4!}+\ldots$ is convergent if:
(a) $0 < x < \frac{1}{e}$ (b) $x > \frac{1}{e}$
(c) $x = \frac{1}{e}$ (d) none of these

19. Let Σu_n be a series of positive terms. Given that Σu_n is convergent and also $\lim_{n\to\infty}\frac{u_{n+1}}{u_n}$ exists then the limit is:
(a) necessarily equal to 1
(b) necessarily less than 1
(c) necessarily geater than −1
(d) none of the above

20. The series $\sum \frac{n!.2^n}{n^n}$ is:

(a) convergent (b) divergent
(c) oscillatory (d) none of these

21. Which one of the following test does not give absolute convergence of a series ?

(a) comparison test (b) root test
(c) ratio test (d) none of the above

22. If $u_n = \sqrt{n+1} - \sqrt{n}, v_n = \sqrt{n^4+1} - \sqrt{n^4}$ then:

(a) $\sum_{n=1}^{\infty} u_n$ converges but $\sum_{n=1}^{\infty} v_n$ diverges

(b) $\sum_{n=1}^{\infty} u_n$ and $\sum_{n=1}^{\infty} v_n$ both converges

(c) $\sum_{n=1}^{\infty} u_n$ and $\sum_{n=1}^{\infty} v_n$ both diverges

(d) none of the above

23. Ths series $\left(\frac{2^2}{1^2} - \frac{2}{1}\right)^{-1} + \left(\frac{3^3}{2^3} - \frac{3}{2}\right)^{-2} + \left(\frac{4^4}{3^4} - \frac{4}{3}\right)^{-3} + \ldots$ is :

(a) convergent (b) divergent
(c) oscillatory (d) none of these

24. Ths series $\sum \left(1 + \frac{1}{n}\right)^{-n^2}$ is :

(a) convergent (b) divergent
(c) oscillatory (d) none of these

25. Ths series $\sum \frac{1}{\sqrt{n} + \sqrt{n+1} + \sqrt{n+2} + \sqrt{n+3}}$ is:

(a) convergent (b) divergent
(c) oscillatory (d) none of these

26. Ths series $\sum \frac{1}{n^p + n^{-p}}$ converges if:

(a) $p = 1$
(b) $|p| > 1$
(c) both (a) and (b) are true
(d) none of the above

27. The infinite series $\sum [\sqrt[3]{n^3+1} - n] x^n$ converges if :

(a) $|x| \le 1$ (b) $|x| \ge 1$
(c) $|x| = e$ (d) none of these

28. Which one of the following is divergent ?

(a) $\sum \sin \frac{1}{n^2}$

(b) $\sum \sin \frac{1}{n}$

(c) Both (a) and (b) are true
(d) None of the above

29. If Σu_n is a positive term series such that $\lim_{n \to \infty} u_n \neq 0$ then Σu_n:

(a) converges (b) diverges
(c) oscillation (d) none of these

30. The series $\sum_{n=1}^{\infty} (-1)^n . n$ is :

(a) bounded (b) convergent
(c) divergent (d) none of these

ANSWERS

1. (b)	**2.** (a)	**3.** (b)	**4.** (a)	**5.** (b)	**6.** (a)	**7.** (b)	**8.** (a)	**9.** (a)	**10.** (a)
11. (b)	**12.** (a)	**13.** (c)	**14.** (b)	**15.** (a)	**16.** (b)	**17.** (d)	**18.** (a)	**19.** (b)	**20.** (a)
21. (a)	**22.** (b)	**23.** (a)	**24.** (a)	**25.** (b)	**26.** (b)	**27.** (a)	**28.** (a)	**29.** (a)	**30.** (a)

Self Assessment Test

Verify the following:

1. If $|r| < 1$ then the series $\sum_{n=1}^{\infty} r^n$ converges to $\frac{r}{1-r}$.
2. If Σa_n is convergent then $\Sigma a_n x^n$ is absolututly convergent when $|x| < 1$.
3. If $u_n \geq u_{n+1} \geq 0 \ \forall \ n$ then $\Sigma (u_n - u_{n+1})$ converges.
4. If the series of positive terms Σa_n diverges and $<s_n>$ be its sequence of partial sums then $\sum \frac{a_n}{s_n^2}$ converges.
5. Positive and decreasing terms of convergent series Σu_n implies $nu_n \to 0$.
6. $\sum\left(\sin\frac{1}{n}+\frac{1}{n^2}\right)$ is divergent.
7. If Σu_n is a convergent series of non-negative numbers and v_n is bounded then $\Sigma u_n.v_n$ converges absolutely.
8. The series $\sum a_n^2$ and $\Sigma \mid (1+a_n) e^{-a_n} - 1 \mid$ converge or diverge together.
9. If $f(n)$ be monotonic and non-negative then $\Sigma f(n)$ and $\Sigma 2^n f(2^n)$ are either both convergent or both divergent to ∞.
10. If $u_n > 0$ and $c > 0$ then $\sum\frac{1}{u_n}$ and $\sum\frac{1}{(c+u_n)}$ converge or diverge together.
11. If $u_n > 0 \ \forall \ n$ then Σu_n and $\sum\frac{1}{u_n}$ shall diverge but cannot converge together.
12. If for a positive term series Σu_n, $\lim nu_n = 0$ then Σu_n may or may not converge.
13. If Σa_n, Σb_n be the convergent series of non-negative terms then Σ min [a_n, b_n], Σ max [a_n, b_n] and also $\sum\sqrt{a_n . b_n}$ converges.
14. The series $\sum \frac{n!e^n}{n^n}$ is divergent.
15. The series $\sum_{n=1}^{\infty} \frac{2^2.3^2 \ldots n^2}{(1+2+2^2)(1+3+3^2)\ldots(1+n+n^2)}$ diverges.
16. The series $\sum_{n=2}^{\infty} \frac{(\log n)^p}{n^q}$ converges if $q > 1$ or if $q = 1$ and $p < -1$ and otherwise diverges.
17. For $x > 0$, the series $\sum x^{1+\frac{1}{2}+\ldots+\frac{1}{n}}$ converges if $x < \frac{1}{e}$ and diverges if $x \geq \frac{1}{e}$ .... .
18. The series $1-\frac{1}{2^s}+\frac{1}{3^s}-\frac{1}{4^s}+\ldots$ converges but its rearranged series $1+\frac{1}{3^s}-\frac{1}{2^s}-\frac{1}{5^s}+\frac{1}{7^s}-\frac{1}{4^s}+\ldots$ diverges to $+\infty$ if $0 < s < 1$.
19. The series $1-\frac{1}{2}-\frac{1}{4}+\frac{1}{3}-\frac{1}{6}-\frac{1}{8}+\frac{1}{5}-\frac{1}{10}-\frac{1}{12}+\ldots$ converges to the sum $\frac{1}{2}$ log2.
20. Every conditionally convergent series shall be rearranged so as to converge to a desired sum or diverge or oscillate finitely or infinitety.
21. $\lim \int_0^a g\frac{\sin nx}{\sin x}dx = \lim \int_0^a g\frac{\sin nx}{x}dx \ ; 0 \leq a \leq \pi$
22. $\frac{1}{2}a_0 + \sum_{n=1}^{\infty} a_n = \frac{f(0^-)+f(0^+)}{2}$ where $\int_0^{\infty}\frac{\sin x}{x}dx = \frac{\pi}{2}$
23. $f(x) = |x| = \frac{\pi}{2} - \frac{4}{\pi}\left(\frac{\cos x}{1^2}+\frac{\cos 3x}{3^2}+\frac{\cos 5x}{5^2}+\ldots\right)$
24. The function x^2 is periodic with period $2l$ on the interval $(-l, l)$ and its Fourier series is given by $\frac{l^2}{3}+\frac{4l^2}{\pi^2}\sum_{n=1}^{\infty}\frac{(-1)^n}{n^2}\cos\left(\frac{n\pi x}{l}\right)$.
25. $\left|\cos\left(\frac{\pi x}{l}\right)\right| = \frac{4}{\pi}+\sum_{n=1}^{\infty}\frac{2(-1)^{n+1}}{\pi(4n^2-1)}\cos\left(\frac{2n\pi x}{l}\right)$
26. $x-[x]-\frac{1}{2} = \sum_{n=1}^{\infty}\left(-\frac{1}{n\pi}\right)\sin(2n\pi x) \text{ in} \left[-\frac{1}{2},\frac{1}{2}\right]$
27. For all values of x in $[-\pi, \pi]$, k is not an integer $\cos kx = \frac{\sin k\pi}{\pi}\left[\frac{1}{k}+\sum_{n=1}^{\infty}\frac{(-1)^n 2k\cos nx}{k^2-n^2}\right]$
28. The Fourier series of e^x in $[-l, l]$ is given by $\frac{\sinh l}{l}+2l\sinh l\sum\frac{(-1)^n\cos\frac{n\pi x}{l}}{l^2+n^2\pi^2} - 2\pi\sinh l\sum\frac{(-1)^n n\sin\frac{n\pi x}{l}}{l^2+n^2\pi^2}$

cccccc

DIFFERENTIAL EQUATIONS

UNIT OUTLINES

- Introduction to Ordinary Differential Equations
- Linear Differential Equation with Constant Coefficients
- Homogeneous Linear Differential Equations
- Simultaneous Differential Equations.
- Linear Differential Equation of Second Order with Constant Coefficients
- Series Solution of Second Order Differential Equations.
- Legendre's and Bessel's Functions
- The Laplace Transform
- The Inverse Laplace Transforms
- Introduction to Partial Differential Equations
- Linear and Non-linear Partial Differential Equations
- Partial Differential Equations of Second Order.
- Fourier Transforms
- Z-transforms

PRE-REQUISITE

- Concepts of Differentiation
- Concepts of Integration
- Introductory Concepts of Differential Equations

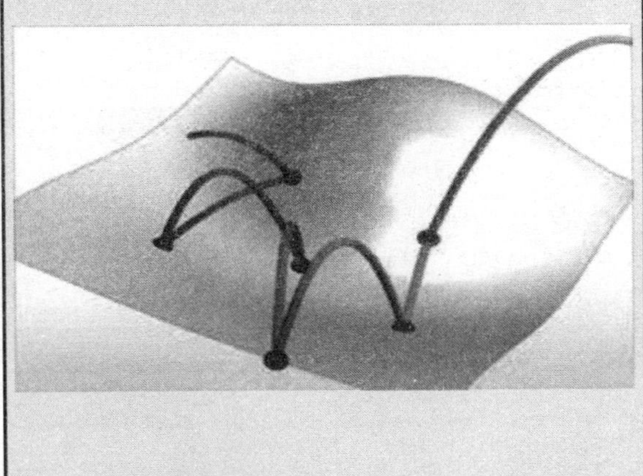

CHAPTER

28 An Introduction to Ordinary Differential Equations

28.1 INTRODUCTION

If $y = f(x)$ be a given function, then its derivative dy/dx can be interpreted as the rate of change of y with respect to x. In any natural process, the variables involved and their rates of changes are connected with one another by means of the basic scientific principles that govern the process. When this connection is expressed in mathematical symbols, the result is often called a differential equation.

Differential equations arise from many problems in Algebra, Geometry, Mechanics, Physics and Chemistry.

28.2 DIFFERENTIAL EQUATIONS

1. **Differential Equation.** An equation involving one dependent variable and its derivatives with respect to one or more independent variables is called a differential equation.

   For example : (i) $e^x dx + e^y dy = 0$, (ii) $y = x\dfrac{dy}{dx}$

   (iii) $\left[1+\left(\dfrac{dy}{dx}\right)^2\right]^{3/2} = c\dfrac{d^2y}{dx^2}$ (iv) $\dfrac{\partial^2 y}{\partial x^2} = \dfrac{1}{c}\dfrac{\partial^2 y}{\partial t^2}$

   (v) $\dfrac{dx}{dt} - \omega y = a\cos pt, \dfrac{dy}{dt} + \omega x = a\sin pt$

2. **Ordinary Differential Equation.** A differential equation which contains only one independent variable is called an ordinary differential equation.

   For example : (i) $xdx + ydy = 0$, (ii) $1+\left(\dfrac{dy}{dx}\right)^2 = \dfrac{d^3y}{dx^3}$ are both ordinary differential equations as they have one independent variable.

3. **Partial Differential Equation.** A differential equation which contains more than one independent variable is called partial differential equation.

   For example : (i) $\dfrac{\partial^2 u}{\partial x^2} + \dfrac{\partial^2 u}{\partial y^2} + \dfrac{\partial^2 u}{\partial z^2} = 0$, (ii) $\dfrac{\partial^2 u}{\partial x^2} = \dfrac{1}{c}\dfrac{\partial^2 u}{\partial y^2}$ are both partial differential equation as they have more than one independent variable such as x, y, z.

4. **Order of Differential Equations.** The order of a differential equation is the order of the highest derivative appearing in it.

5. **Degree of Differential Equations.** The degree of a differential equation is the degree of the highest derivative occurring in it, after the equation has been expressed in a form free from radicals and fractions as far as the derivative are concerned.

   For example :

   (i) $e^x dx + e^y dy = 0$, is of the first order and first degree. (ii) $\dfrac{d^2y}{dx^2} + x^2 y = 0$, is of second order and first degree.

   (iii) The differential equation $\left[1+\left(\dfrac{dy}{dx}\right)^2\right]^{3/2} = c\left(\dfrac{d^2y}{dx^2}\right)$ can be written as $\left[1+\left(\dfrac{dy}{dx}\right)^2\right]^3 = c^2\left(\dfrac{d^2y}{dx^2}\right)^2$, it is of second order and second degree.

Solved Examples

Example 1. *Determine the order and degree of the differential equation*

$$x\left(\frac{d^2y}{dx^2}\right)^3 + y\left(\frac{dy}{dx}\right)^4 + y^2 = 0.$$

Solution. The given differential equation contains second order derivative which is the highest derivatives, so the order of the differential equation is 2. The power of second order derivative is 3, so the degree of the differential equation is 3.

Hence, order = 2 and degree = 3.

Example 2. *Determine the order and degree of the differential equation*

$$\left[1+\left(\frac{dy}{dx}\right)^2\right]^{1/2} = \left(\frac{d^2y}{dx^2}\right)^{1/3}.$$

Solution. Making the differential equation free from radicals, we get

$$\left[1+\left(\frac{dy}{dx}\right)^2\right]^3 = \left(\frac{d^2y}{dx^2}\right)^2$$

Now, in this equation, the order of the highest derivative is 2, so its order is 2. The power of the highest derivative is 2, so its degree is 2.

Hence, the given differential equation is of order 2 and degree 2.

Example 3. *Determine the order and degree of the differential equation*

$$x\left(\frac{dy}{dx}\right) + \frac{2}{(dy/dx)} = y^2.$$

Solution. Making the differential equation free from radicals, we get.

$$x\left(\frac{dy}{dx}\right)^2 + 2 = y^2\left(\frac{dy}{dx}\right).$$

Now, in this equation, the order of the highest derivative is 1 and its power is 2. Hence, the differential equation is of order 1 and degree 2.

Example 4. *Determine the order and degree of the differential equation*

$$y = px + \sqrt{a^2p^2 + b^2}\text{, where } p = dy/dx.$$

Solution. Making the differential equation free from radicals, we get

$$y = px + \sqrt{a^2p^2 + b^2}$$

$$\Rightarrow \quad y - px = \sqrt{a^2p^2 + b^2}$$

$$\Rightarrow \quad (y - px)^2 = a^2p^2 + b^2$$

$$\Rightarrow y^2 + p^2x^2 - 2ypx = a^2p^2 + b^2$$

$$\text{or } (x^2 - a^2)p^2 - 2xyp + y^2 - b^2 = 0$$

$$\text{or } (x^2 - a^2)\left(\frac{dy}{dx}\right)^2 - 2xy\frac{dy}{dx} + y^2 - b^2 = 0.$$

Now, in this equation, the order of the highest derivative is 1 and its power is 2. Hence, the given differential equation is of order 1 and degree 2.

EXERCISE 28.1

Determine the order and degree of each of the following differential equations :

1. $\left(\frac{dy}{dx}\right)^2 + 5y = \sin x$
2. $\frac{d^2y}{dx^2} + 3\left(\frac{dy}{dx}\right)^3 + 2y = 0$
3. $(x^3 - y^3)\,dx + (xy^2 - x^2y)dy = 0$
4. $x^2\left(\frac{dy}{dx}\right) + 2xy - 6x^3 = 0$
5. $\frac{d^2y}{dx^2} + 5xy = -3xe^{-x}$
6. $\left(\frac{d^2y}{dx^2}\right)^3 + 2\left(\frac{dy}{dx}\right)^4 + 9 = \sin x$
7. $(2x - 2y + 5)dx = (x - y + 3)dy$
8. $8y^2 = 4xy\frac{dy}{dx} - \left(\frac{dy}{dx}\right)^3$
9. $y = xp \pm c\sqrt{p^2+1}$ where $p = \frac{dy}{dx}$
10. $\frac{d^2y}{dx^2} = \sqrt{1+\left(\frac{dy}{dx}\right)^2}$
11. $\sqrt{a+x}\left(\frac{dy}{dx}\right) + x = 0$
12. $x\sqrt{1+y^2}\,dx + y\sqrt{1-x^2}\,dy = 0$

Hint to Selected Problems

3. $\frac{dy}{dx} = \frac{x^3 - y^3}{x^2y - xy^2}$

7. $\frac{dy}{dx} = \frac{(2x - 2y + 5)}{(x - y + 3)}$

9. $y = xp \pm c\sqrt{p^2+1} \Rightarrow y - xp = \pm c\sqrt{p^2+1}$ (squaring and then simplifying)

10. Square both sides and then simplify.

11. $\left(\frac{dy}{dx}\right)^2 = \frac{x^2}{a+x}$

12. $\left(\frac{dy}{dx}\right)^2 = \frac{x^2(1-y^2)}{y^2(1-x^2)}$

ANSWERS

1. Order = 1, Degree = 2	**2.** Order = 2, Degree = 1	**3.** Order = 1, Degree = 1
4. Order = 1, Degree = 1	**5.** Order = 2, Degree = 1	**6.** Order = 2, Degree = 3
7. Order = 1, Degree = 1	**8.** Order = 1, Degree = 3	**9.** Order = 1, Degree = 2
10. Order = 2, Degree = 2	**11.** Order = 1, Degree = 2	**12.** Order = 1, Degree = 2

28.3 SOLUTION OF THE DIFFERENTIAL EQUATION

Any relation between the dependent and independent variables *i.e.*, a function of the form $y = f(x) + C$, which satisfies the given differential equation is called its solution or primitive.

For example : Let $\frac{dy}{dx} = \cos x$ be a given differential equation. Then a function $y = \sin x + C$ is its solution.

1. **General Solution :** If the solution of nth order differential equation contains n arbitrary constants, then it is called general solution or complete primitive.
   For example : $y = C_1 \cos x + C_2 \sin x$ (inviolving two arbitrary constants C_1 and C_2) is the general solution of the differential equation $\frac{d^2y}{dx^2} + y = 0$ of second order.
2. **Particular Solution :** A solution obtained from the general solution by giving particular values to the arbitrary constants in the general solution is called particular solution or particular integral.
   For example : $y = C_1 e^x + C_2 e^{-x}$ is the general solution of the differential equation $\frac{d^2y}{dx^2} - y = 0$ whereas $y = e^x - e^{-x}$ or $y = e^x$ are its particular solution.
3. **Singular Solution :** The solution which cannot be obtained from general solution by assigning particular values to the arbitrary constants, is called singular solution.

Solved Examples

Example 1. *Verify that $y = A\cos x - B\sin x$ is a solution of the differential equation $\frac{d^2y}{dx^2} + y = 0$*

Solution. We have $y = A\cos x - B\sin x$. ...(1)

Differentiating (1) w.r.t. x, we get

$$\frac{dy}{dx} = -A\sin x - B\cos x.$$

Differentiating again, we get

$$\frac{d^2y}{dx^2} = -A\cos x + B\sin x$$

$$= -(A\cos x - B\sin x) = -y \qquad \text{(Using (1))}$$

$$\therefore \quad \frac{d^2y}{dx^2} + y = 0$$

Hence, $y = A\cos x - B\sin x$ is a solution of $\frac{d^2y}{dx^2} + y = 0$.

Example 2. *Show that $y = Ae^x + Be^{-x}$ is a solution of the differential equation*

$$\frac{d^2y}{dx^2} - y = 0.$$

Solution. We have $y = Ae^x + Be^{-x}$. ...(1)

Differentiating (1) w.r.t. x, we get

$$\frac{dy}{dx} = Ae^x - Be^{-x}.$$

Differentiating again, we get

$$\frac{d^2y}{dx^2} = Ae^x + Be^{-x} = y \qquad \text{(Using (1))}$$

$$\Rightarrow \quad \frac{d^2y}{dx^2} - y = 0.$$

Hence, $y = Ae^x + Be^{-x}$ is a solution of $\frac{d^2y}{dx^2} - y = 0$.

Example 4. *Show that $y = a\cos(\log x) + b\sin(\log x)$ is a solution of the differential equation*

$$x^2\frac{d^2y}{dx^2} + x\frac{dy}{dx} + y = 0.$$

Solution. We have $y = a\cos(\log x) + b\sin(\log x)$. ...(1)

Differentiating (1) w.r.t. x, we get

$$\frac{dy}{dx} = -\frac{a\sin(\log x)}{x} + \frac{b\cos(\log x)}{x}$$

or $x\frac{dy}{dx} = -a\sin(\log x) + b\cos(\log x)$. ...(2)

Differentiating again, we get

$$x\frac{d^2y}{dx^2} + \frac{dy}{dx} = -\frac{a}{x}\cos(\log x) - \frac{b}{x}\sin(\log x).$$

or $x^2\dfrac{d^2y}{dx^2}+x\dfrac{dy}{dx}$
$= -[a\cos(\log x)+b\sin(\log x)]$ ...(3)

From (1) and (3), we get

$$x^2\frac{d^2y}{dx^2}+x\frac{dy}{dx}=-y.$$

$$\therefore\quad x^2\frac{d^2y}{dx^2}+x\frac{dy}{dx}+y=0$$

Hence,

$y=a\cos(\log x)+b\sin(\log x)$ is a solution of the differential equation $x^2\dfrac{d^2y}{dx^2}+x\dfrac{dy}{dx}+y=0$.

Example 5. *Show that* $y=e^{-x}+Ax+B$ *is a solution of the differential equation* $e^x\dfrac{d^2y}{dx^2}=1$.

Solution. We have $y=e^{-x}+Ax+B$ ...(1)

Differentiating (1) w.r.t. x, we get

$$\frac{dy}{dx}=-e^{-x}+A.$$

Differentiating again, we get

$$\frac{d^2y}{dx^2}=e^{-x}\quad\Rightarrow\quad e^x\frac{d^2y}{dx^2}=1$$

Hence, $y=e^{-x}+Ax+B$ is a solution of the differential equation $e^x\dfrac{d^2y}{dx^2}=1$.

Example 6. *Show that* $Ax^2+By^2=1$ *is a solution of the differential equation*

$$x\left[y\frac{d^2y}{dx^2}+\left(\frac{dy}{dx}\right)^2\right]=y\frac{dy}{dx}.$$

Solution. We have $Ax^2+By^2=1$ ...(1)

Differentiating (1) w.r.t. x, we get

$$2Ax+2By\frac{dy}{dx}=0$$

or $$\frac{A}{B}+\frac{y}{x}\frac{dy}{dx}=0 \qquad ...(2)$$

Differentiating again, we get

$$\frac{d}{dx}\left[\frac{A}{B}+\frac{y}{x}\left(\frac{dy}{dx}\right)\right]=0$$

or $$\frac{d}{dx}\left[\frac{y}{x}\left(\frac{dy}{dx}\right)\right]=0$$

or $$\frac{y}{x}\frac{d^2y}{dx^2}+\frac{dy}{dx}\left[\frac{x\frac{dy}{dx}-y}{x^2}\right]=0$$

$$\therefore\quad x\left[y\frac{d^2y}{dx^2}+\left(\frac{dy}{dx}\right)^2\right]=y\frac{dy}{dx}.$$

Hence, $Ax^2+By^2=1$ is a solution of the differential equation

$$x\left[y\frac{d^2y}{dx^2}+\left(\frac{dy}{dx}\right)^2\right]=y\frac{dy}{dx}.$$

Example 7. *Show that* $y=A\sin(x+B)$ *is a solution of the differential equation*

$$\frac{d^2y}{dx^2}+y=0.$$

Solution. We have $y=A\sin(x+B)$ ...(1)

Differentiating (1) w.r.t. x, we get

$$\frac{dy}{dx}=A\cos(x+B).$$

Differentiating again, we get

$$\therefore\quad\frac{d^2y}{dx^2}=-A\sin(x+B). \qquad ...(2)$$

From (1) and (2), we get $\dfrac{d^2y}{dx^2}+y=0$.

Hence, $y=A\sin(x+B)$ is a solution of the differential equation $\dfrac{d^2y}{dx^2}+y=0$.

EXERCISE 28.2

1. Show that $y=ae^x$ is a solution of the differential equation $\dfrac{dy}{dx}-y=0$
2. Show that $y=ae^{-x}$ is a solution of the differential equation $\dfrac{dy}{dx}+y=0$.
3. Show that $y=c(x-c)^2$ is a solution of the differential equation $8y^2=4xy\dfrac{dy}{dx}-\left(\dfrac{dy}{dx}\right)^3$.
4. Show that $y=4\sin 3x$ is a solution of the differential equation $\dfrac{d^2y}{dx^2}+9y=0$.
5. Verify that $y=A\cos 2x+B\sin 2x$ is a solution of the differential equation $\dfrac{d^2y}{dx^2}+4y=0$.
6. Verify that $y=e^x(A\cos x+B\sin x)$ is a solution of the differential equation $\dfrac{d^2y}{dx^2}-2\dfrac{dy}{dx}+2y=0$. (Andhra–1998)
7. Verify that $y=\dfrac{a}{x}+b$ is a solution of the differential equation $\dfrac{d^2y}{dx^2}+\dfrac{2}{x}\left(\dfrac{dy}{dx}\right)=0$.

8. Verify that $y = Ae^{Bx}$ is a solution of the differential equation $\frac{d^2y}{dx^2} = \frac{1}{y}\left(\frac{dy}{dx}\right)^2$.

9. Show that $y = x^n(A + B\log x)$ is a solution of the differential equation $x^2\frac{d^2y}{dx^2} - (2n-1)x\frac{dy}{dx} + n^2y = 0$.

10. Show that $y = e^{3x}(A + Bx)$ is a solution of the differential equation $\frac{d^2y}{dx^2} - 6\frac{dy}{dx} + 9y = 0$.

11. Verify that $y = ax^3 + bx^2 + c$ is a solution of the differential equation $\frac{d^3y}{dx^3} = 6a$.

12. Verify that $y^2 = 4a(x + a)$ is a solution of the differential equation $y\left[1 - \left(\frac{dy}{dx}\right)^2\right] = 2x\frac{dy}{dx}$.

13. Verify that $y = Ae^{mx} + Be^{nx}$ is a solution of the differential equation $\frac{d^2y}{dx^2} - (m+n)\frac{dy}{dx} + mny = 0$.

14. Show that $x^2 + y^2 = cx$ is a solution of the differential equation $2xy\frac{dy}{dx} = y^2 - x^2$.

15. Show that $y^2 = a(b - x^2)$ is a solution of the differential equation

$$y\frac{dy}{dx} = x\left[y\frac{d^2y}{dx^2} + \left(\frac{dy}{dx}\right)^2\right].$$

16. Show that $y = \log\left[x + \sqrt{x^2 + a^2}\right]^2$ is a solution of the differential equation $(a^2 + x^2)\frac{d^2y}{dx^2} + x\frac{dy}{dx} = 0$.

17. Show that $y = e^{m\cos^{-1}x}$ is a solution of the differential equation $(1 - x^2)\frac{d^2y}{dx^2} - x\frac{dy}{dx} - m^2y = 0$.

18. Show that $y = ae^{\tan^{-1}x}$ is a solution of the differential equation $(1 + x^2)\frac{d^2y}{dx^2} + (2x - 1)\frac{dy}{dx} = 0$.

Hint to Selected Problems

3. $y = C(x - C)^2 \Rightarrow \frac{dy}{dx} = 2C(x - C)$

To prove

$$8y^2 = 4xy\frac{dy}{dx} - \left(\frac{dy}{dx}\right)^3$$

$$\text{R.H.S.} = 4xy.2C(x-C) - 8C^3(x-C)^3$$
$$= 4x.C(x-C)^2.2C(x-C) - 8C^3(x-C)^3$$
$$= 8C^2(x-C)^3(x-C) = 8C^2(x-C)^4$$
$$= 8[C(x-C)^2]^2 = 8y^2 = \text{L.H.S.}$$

10. $y = e^{3x}(A + Bx)$

$$\Rightarrow \frac{dy}{dx} = 3e^{3x}(A + Bx) + e^{3x}\cdot B$$

and $\frac{d^2y}{dx^2} = 3e^{3x}(A+Bx)\cdot 3 + 3e^{3x}\cdot B + 3Be^{3x}$

$$\text{L.H.S.} : \frac{d^2y}{dx^2} - 6\frac{dy}{dx} + 9y$$
$$= \left[9e^{3x}(A+Bx) + 3Be^{3x} + 3Be^{3x}\right]$$
$$-6\left[3e^{3x}(A+Bx) + e^{3x}B\right] + 9e^{3x}(A+Bx)$$

17. $y = e^{m\cos^{-1}x}$

$$\Rightarrow \frac{dy}{dx} = e^{m\cos^{-1}x}.m\left(\frac{-1}{\sqrt{1-x^2}}\right) = \frac{-me^m\cos^{-1}x}{\sqrt{1-x^2}}$$

Let $\frac{dy}{dx} = y_1$ and $\frac{d^2y}{dx^2} = y_2$

$$\Rightarrow y_1 = \frac{-my}{\sqrt{1-x^2}} \Rightarrow y_1\sqrt{1-x^2} = -my$$

Differentiate again

$$y_2\sqrt{1-x^2} + y_1\frac{1}{2}\frac{(-2x)}{\sqrt{1-x^2}} = -my_1$$
$$\Rightarrow y_2\sqrt{1-x^2} - \frac{xy_1}{\sqrt{1-x^2}} = -my_1$$
$$y_2(1-x^2) - xy_1 = -my_1\sqrt{1-x^2}$$

Putting all values of y_1, y_2

$$\frac{d^2y}{dx^2}(1-x^2) - x\frac{dy}{dx} + m(my) = 0$$
$$\Rightarrow \frac{d^2y}{dx^2}(1-x^2) - x\frac{dy}{dx} - m^2y = 0$$
$$(1-x^2)\frac{d^2y}{dx^2} - x\frac{dy}{dx} - m^2y = 0$$

28.4 FORMATION OF A DIFFERENTIAL EQUATION

The general solution of a first order differential equation contains one arbitrary constant, which is called a parameter. If this parameter takes various values, then we get a family of curve of one parameter.

For example : The equation $x^2 + y^2 = c^2$ represents a one parameter family of circles if c takes all real values.

Suppose there is an equation, representing a family of curves, containing n arbitrary constants. Then, in order to find its differential equation, we proceed as follows :

WORKING PROCEDURE

Step 1. *Differentiate the given equation of family of curve n times to get n more equations containing n arbitrary constants and derivatives.*

Step 2. *Eliminate all the n constants from all of these (n + 1) equations to get an equation containing a nth order derivative, which is the required differential equation of the given family of curves.*

Solved Examples

Example 1. *Find the differential equation of family of all straight lines passing through the origin.*

Solution. The general equation of the family of all straight lines passing through the origin is

$$y = mx. \quad \ldots(1)$$

where m is a parameter.

Differentiating (1) w.r.t. x, we get

$$\frac{dy}{dx} = m \cdot \quad \ldots(2)$$

Eliminating m between (1) and (2), we get

$$y = x\frac{dy}{dx}$$

which is the required differential equation.

Example 2. *Find the differential equation of the family of all straight lines making equal intercepts on the co-ordinate axes.*

Solution. The general equation of the family of all straight lines making equal intercepts on the axes is

$$\frac{x}{c} + \frac{y}{c} = 1$$

or $x + y = c$, where c is a parameter. ...(1)

Differentiating (1) w.r.t. x, we get

$$1 + \frac{dy}{dx} = 0$$

which is the required differential equation.

Example 3. *Find the differential equation that will represent the family of all circles having their centres on the x-axis and radii equal to unity.*

Solution. Let $(h, 0)$ be any arbitrary point on the x-axis and suppose $(h, 0)$ be the co-ordinates of the centre of a circle of radius equal to unity.

Then, the general equation of the family of all circles having their centres on x-axis and radii equal to unity is

$$(x - h)^2 + y^2 = 1, \; h \text{ is a parameter.} \quad \ldots(1)$$

Differentiating (1) w.r.t. x, we get

$$2(x - h) + 2y\frac{dy}{dx} = 0$$

or $$(x - h) + y\frac{dy}{dx} = 0 \quad \ldots(2)$$

Eliminating h between (1) and (2), we get

$$(x - h) = -y\frac{dy}{dx}.$$

Putting the value of $(x - h)$ in (1), we get

$$\left[-y\left(\frac{dy}{dx}\right)\right]^2 + y^2 = 1$$

$$y^2\left(\frac{dy}{dx}\right)^2 + y^2 = 1,$$ which is the required differential equation.

Example 4. *Find the differential equation of the family of curves $y = c(x - c)^2$, where c is an arbitrary constant.*

Solution. We have

$$y = c(x - c)^2. \quad \ldots(1)$$

Differentiating (1) w.r.t. x, we get

$$\frac{dy}{dx} = 2c(x - c) \cdot \quad \ldots(2)$$

Dividing (2) by (1), we get

$$\frac{1}{y}\left(\frac{dy}{dx}\right) = \frac{2}{x - c}$$

or $$x - c = \frac{2y}{(dy/dx)} \text{ or } c = x - \frac{2y}{(dy/dx)}$$

Putting the value of c in (1), we get

$$y = \left[x - \frac{2y}{(dy/dx)}\right]\left[\frac{2y}{(dy/dx)}\right]^2.$$

$$y\left(\frac{dy}{dx}\right)^3 = 4y^2\left[x\left(\frac{dy}{dx}\right) - 2y\right]$$

or $$\left(\frac{dy}{dx}\right)^3 = 4xy\frac{dy}{dx} - 8y^2,$$ which is the required differential equation.

Example 5. *Find the differential equation of the family of curves $y^2 - 2ay + x^2 = a^2$, where a is an arbitrary constant.*

Solution. We have

$$y^2 - 2ay + x^2 = a^2. \quad \ldots(1)$$

Differentiating (1) w.r.t. x, we get

$$2y\frac{dy}{dx} - 2a\frac{dy}{dx} + 2x = 0$$

or $$a = y + \frac{x}{(dy/dx)}$$

Putting the value of a in (1), we get

$$y^2 - 2y\left[y + \frac{x}{(dy/dx)}\right] + x^2 = \left[y + \frac{x}{(dy/dx)}\right]^2$$

or $$y^2 - 2y\left[y + \frac{x}{(dy/dx)}\right] + x^2 = y^2 + \frac{x^2}{(dy/dx)^2} + \frac{2xy}{(dy/dx)}$$

or $$-2y\frac{dy}{dx}\left[y\frac{dy}{dx} + x\right] + x^2\left(\frac{dy}{dx}\right)^2 = x^2 + 2xy\frac{dy}{dx}$$

or $$(x^2 - 2y^2)\left(\frac{dy}{dx}\right)^2 - 4xy\frac{dy}{dx} - x^2 = 0$$

which is the required differential equation.

Example 6. *Find the differential equation of the family of curves $y = e^x(a\cos x + b\sin x)$, where a and b are arbitrary constants.*

Solution. We have

$$y = e^x(a\cos x + b\sin x) \qquad \ldots(1)$$

Differentiating (1) w.r.t. x, we get

$$\frac{dy}{dx} = e^x(a\cos x + b\sin x) + e^x(-a\sin x + b\cos x)$$

or $\frac{dy}{dx} = y + e^x(-a\sin x + b\cos x)$. (Using (1)) $\ldots(2)$

Differentiating again, we get

$$\frac{d^2y}{dx^2} = \frac{dy}{dx} + e^x(-a\sin x + b\cos x) + e^x(-a\cos x - b\sin x) \qquad \ldots(3)$$

Eliminating a and b between (1) and (3), the required differential equation is

$$\frac{d^2y}{dx^2} = \frac{dy}{dx} + \left[\frac{dy}{dx} - y\right] - y$$

$\frac{d^2y}{dx^2} - 2\frac{dy}{dx} + 2y = 0$, which is the required differential equation.

Example 7. *Find the differential equation of the family of all circles in the first quadrant which touch the co-ordinate axes.*

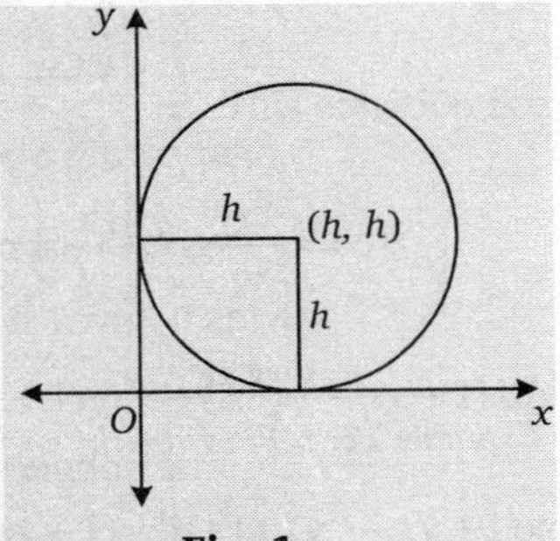

Fig. 1

Solution. The general equation of the family of all circles in the first quadrant which touch the co-ordinate axes is given by

$$(x-h)^2 + (y-h)^2 = h^2 \qquad \ldots(1)$$

Differentiating (1) w.r.t. x, we get

$$2(x-h) + 2(y-h)\frac{dy}{dx} = 0$$

or $2x + 2y\frac{dy}{dx} - 2h - 2h\frac{dy}{dx} = 0$

or $h = \dfrac{\left(x + y\frac{dy}{dx}\right)}{\left(1 + \frac{dy}{dx}\right)} = \left(\dfrac{x+yp}{1+p}\right)$, where $p = \dfrac{dy}{dx}$

Putting the value of h in (1), we get

$$\left[x - \frac{x+yp}{1+p}\right]^2 + \left[y - \frac{x+yp}{1+p}\right]^2 = \left[\frac{x+yp}{1+p}\right]^2$$

or $\left(\dfrac{xp-yp}{1+p}\right)^2 + \left(\dfrac{y-x}{1+p}\right)^2 = \left(\dfrac{x+yp}{1+p}\right)^2$

or $(x-y)^2p^2 + (y-x)^2 = (x+yp)^2$

or $(x-y)^2(p^2+1) = (x+yp)^2$

$$(x-y)^2\left[\left(\frac{dy}{dx}\right)^2 + 1\right] = \left[x + y\frac{dy}{dx}\right]^2$$

which is the required differential equation.

Example 8. *Find the differential equation of the family of all circles having centres on the line $y = x$ and radii equal to unity.*

Solution. The general equation of the family of all circles having centres on the line $y = x$ and radii equal to unity is given by

$$(x-h)^2 + (y-h)^2 = 1 \qquad \ldots(1)$$

Differentiating (1) w.r.t. x, we get

$$2(x-h) + 2(y-h)\frac{dy}{dx} = 0$$

or $h = \dfrac{\left(x + y\frac{dy}{dx}\right)}{\left(1 + \frac{dy}{dx}\right)} = \left(\dfrac{x+yp}{1+p}\right)$, where $p = \dfrac{dy}{dx}$

Putting the value of h in (1), we get

$$\left[x - \frac{x+yp}{1+p}\right]^2 + \left[y - \frac{x+yp}{1+p}\right]^2 = 1$$

or $(x-y)^2p^2 + (y-x)^2 = (1+p)^2$

or $(x-y)^2(p^2+1) = (1+p)^2$

$$\Rightarrow \quad (x-y)^2\left[1 + \left(\frac{dy}{dx}\right)^2\right] = \left[1 + \frac{dy}{dx}\right]^2$$

which is the required differential equation.

Example 9. *Find the differential equation of the family of all circles of radius r.* (Andhra–1999, Mysore–1997, 2009)

Solution. The general equation of the family of all circles with centres (a, b) and radius r is given by

$$(x-a)^2 + (y-b)^2 = r^2 \qquad \ldots(1)$$

Differentiating (1) w.r.t. x, we get

$$2(x-a) + 2(y-b)\frac{dy}{dx} = 0$$

or $(x-a) + (y-b)\frac{dy}{dx} = 0 \qquad \ldots(2)$

Differentiating again, we get

$$1 + (y-b)\frac{d^2y}{dx^2} + \left(\frac{dy}{dx}\right)^2 = 0 \qquad \ldots(3)$$

Eliminating a and b between (1), (2) and (3), we get

$$y - b = \frac{1 + (dy/dx)^2}{-\dfrac{d^2y}{dx^2}}$$

From (2), we get

$$(x-a) = -(y-b)\left(\frac{dy}{dx}\right) = \frac{\frac{dy}{dx}\left[1+\left(\frac{dy}{dx}\right)^2\right]}{\frac{d^2y}{dx^2}}$$

Putting the values of $(x-a)$ and $(y-b)$ in (1), we get

$$\frac{\left(\frac{dy}{dx}\right)^2\left[1+\left(\frac{dy}{dx}\right)^2\right]^2}{\left(\frac{d^2y}{dx^2}\right)^2} + \frac{\left[1+\left(1+\frac{dy}{dx}\right)^2\right]^2}{\left(\frac{d^2y}{dx^2}\right)^2} = r^2$$

or $$\left[1+\left(\frac{dy}{dx}\right)^2\right]^2 = r^2\left(\frac{d^2y}{dx^2}\right)^2$$

which is the required differential equation.

Example 10. *Determine the differential equation whose set of independent solution is* $\{e^x, xe^x, x^2e^x\}$. (UPTU–2002)

Solution. Here, we have

$$y = C_1e^x + C_2xe^x + C_3x^2e^x \quad \ldots(1)$$

Differentiating w.r.t. x, we get

$$y' = C_1e^x + C_2xe^x + C_2e^x + C_3x^2e^x + C_3 2xe^x$$
$$= (C_1e^x + C_2xe^x + C_3x^2e^x) + C_2e^x + C_3 2xe^x$$
$$= y + C_2e^x + C_3 2xe^x \quad \text{(Using (1))}$$
$$\Rightarrow y' - y = C_2e^x + C_3 2xe^x \quad \ldots(2)$$

Again differentiating w.r.t. x, we get

$$y'' - y' = C_2e^x + 2C_3e^x + 2xC_3e^x$$

or $$y'' - y' = (C_2e^x + 2xC_3e^x) + 2C_3e^x \quad \text{(Using (2))}$$

$$\Rightarrow \quad y'' - y' = y' - y + 2C_3e^x$$
$$\Rightarrow \quad y'' - 2y' + y = 2C_3e^x \quad \ldots(3)$$

Differentiating w.r.t. x, we get

$$y''' - 2y'' + y' = 2C_3\, e^x$$
$$\Rightarrow \quad y''' - 2y'' + y' = y'' - 2y' + y \quad \text{(Using (3))}$$
$$\Rightarrow \quad y''' - 3y'' + 3y' - y = 0$$
$$\Rightarrow \quad (D-1)^3 y = 0.$$

Example 11. *A particle falls under gravity in a resisting medium whose resistance varies with velocity. Find the relation between distance and velocity if initially the particle starts from rest.* (UPTU–2003, 04)

Solution. Let v be the velocity when the particle has fallen distance s in time t from rest. If the resistance is mkv, equation of motion is $m\frac{dv}{dt} = mg - mkv$.

Since the forces acting on the particle are the weight mg downwards and resistance mkv upwards.

$$\frac{dv}{dt} = g - kv \quad \Rightarrow \quad \frac{dv}{g-kv} = dt$$

Integration gives $-\frac{1}{k}\log(g-kv) = t + C_1$

$v = 0$ when $t = 0$

$$C_1 = -\frac{1}{k}\log g \quad \Rightarrow \quad \frac{g-kv}{g} = e^{-kt}$$

$$v = \frac{g}{k}(1-e^{-kt}) \quad \ldots(1)$$

$$\frac{ds}{dt} = \frac{g}{k}(1-e^{-kt})$$

Integration yields, $s = \frac{g}{k}t + \frac{g}{k^2}e^{-kt} + C_2$

Since $s = 0$ when $t = 0$

$$C_2 = -\frac{g}{k^2}$$

$$s = \frac{g}{k}t + \frac{g}{k^2}(e^{-kt} - 1) \quad \ldots(2)$$

Eliminating t between (1) and (2), we get

$$s = \frac{g}{k^2}\log\left(\frac{g}{g-kv}\right) - \frac{v}{k}$$

which gives a relation between distance and velocity.

Example 12. *Find the differential equation of the coaxial circles of system* $x^2 + y^2 + 2ax + c^2 = 0$ *where c is a constant and a is a variable.* (JNTU–2003)

Solution. We have, $x^2 + y^2 + 2ax + c^2 = 0 \quad \ldots(1)$

Differentiating w.r.t. x, we get

$$2x + 2y\frac{dy}{dx} + 2a = 0$$

or $$2a = -2\left(x + y\frac{dy}{dx}\right)$$

Substituting in equation (1), we get

$$x^2 + y^2 - 2\left(x + y\frac{dy}{dx}\right)x + c^2 = 0$$

or $$2xy\frac{dy}{dx} = y^2 - x^2 + c^2$$

which is the required differential equation.

EXERCISE 28.3

Find the differential equations of the following families of the curves :

1. $y = 3x + c$, where c is a parameter.
2. $(2x+a)^2 + y^2 = a^2$, where a is a parameter.
3. $x^2 + y^2 = a^2$, where a is a parameter.
4. $x^2 + y^2 = 2ax$, where a is a parameter.
5. $y = a\sin(x+b)$, where a and b are parameters.

6. $y = a\cos(x+b)$, where a and b are parameters.

7. $y = a\cos nx + b\sin nx$, where a and b are parameters.

8. $y = ax + \frac{b}{x}$, where a and b are parameters.

9. $y = ae^{2x} + be^{-x}$, where a and b are parameters.

10. $y = ae^{2x} + be^{-2x}$, where a and b are parameters.

11. $y = ae^{3x} + be^{5x}$, where a and b are parameters.

12. $y^2 = b(a^2 - x^2)$, where a and b are parameters.

13. $(y-b)^2 = 4(x-a)$, where a and b are parameters.

14. $xy = ae^x + be^{-x} + x^2$, where a and b are parameters.

15. Find the differential equation of the family of all circles which passes through the origin and whose centre lie on the y-axis.

Hint to Selected Problems

3. $x^2 + y^2 = a^2 \Rightarrow 2x + 2y\frac{dy}{dx} = 0 \Rightarrow x + y\frac{dy}{dx} = 0$

7. $y = a\cos nx + b\sin nx \Rightarrow \frac{dy}{dx} = -a.n\sin nx + b.n\cos nx$

$$\frac{d^2y}{dx^2} = -a.n^2\cos nx - b.n^2\sin nx = -n^2[a\cos nx + b\sin nx]$$

$$= -n^2y$$

$$\frac{d^2y}{dx^2} + n^2y = 0.$$

11. $y = ae^{3x} + be^{5x}$ ...(1)

$$\frac{dy}{dx} = a.3.e^{3x} + b.5\,e^{5x} \quad ...(2)$$

$$\frac{d^2y}{dx^2} = 9ae^{3x} + 25\,be^{5x} \quad ...(3)$$

From (2) and (3) on removing a and b, we get

$$\frac{d^2y}{dx^2} - 8\frac{dy}{dx} + 15y = 0.$$

ANSWERS

1. $\frac{dy}{dx} = 3$ **2.** $2xy\frac{dy}{dx} - y^2 + 4x^2 = 0$ **3.** $x + y\frac{dy}{dx} = 0$ **4.** $2xy\frac{dy}{dx} + x^2 - y^2 = 0$ **5.** $\frac{d^2y}{dx^2} + y = 0$

6. $\frac{d^2y}{dx^2} + y = 0$ **7.** $\frac{d^2y}{dx^2} + n^2y = 0$ **8.** $x^2\frac{d^2y}{dx^2} + x\frac{dy}{dx} - y = 0$ **9.** $\frac{d^2y}{dx^2} - \frac{dy}{dx} - 2y = 0$ **10.** $\frac{d^2y}{dx^2} - 4y = 0$

11. $\frac{d^2y}{dx^2} - 8\frac{dy}{dx} + 15y = 0$ **12.** $xy\frac{d^2y}{dx^2} + x\left(\frac{dy}{dx}\right)^2 - y\frac{dy}{dx} = 0$ **13.** $2\frac{d^2y}{dx^2} + \left(\frac{dy}{dx}\right)^3 = 0$ **14.** $x\frac{d^2y}{dx^2} + 2\frac{dy}{dx} = xy - x^2 + 2$

15. $(x^2 - y^2)\frac{dy}{dx} - 2xy = 0$

28.5 METHOD OF SOLVING DIFFERENTIAL EQUATION BY SEPARATION OF VARIABLES

If any differential equation can be expressed as

$$f(x)\,dx = g(y)\,dy \quad ...(1)$$

Then we say that variables are separable.

WORKING PROCEDURE

In order to solve the given differential equation, using variable separable,

Step 1. *Separate the variables as $f(x)dx = g(y)dy$.*

Step 2. *Integrate both sides as $\int f(x)dx = \int g(y)dy$.*

Step 3. *Add an arbitrary constant to any of the sides.*

Solved Examples

Example 1. *Solve* $\frac{dy}{dx} = xy + x + y + 1$.

Solution. We have

$$\frac{dy}{dx} = xy + x + y + 1$$

$$\Rightarrow \frac{dy}{dx} = (1+x)(1+y).$$

Separating the variables, we get

$$\frac{dy}{1+y} = (1+x)\,dx.$$

Integrating, we get

$$\log|1+y| = x + \frac{x^2}{2} + C$$

which is the required solution.

Example 2. *Solve* $\frac{dy}{dx} = \log(x+1)$.

Solution. We have

$$\frac{dy}{dx} = \log(x+1)$$

Separating the variables, we get

$dy = \log(x+1)dx$.

Integrating both sides, we get

$$y = \int \log(x+1)dx + C$$

$$= \log(x+1)\int dx - \int\left\{\frac{d}{dx}(\log(x+1)).\int dx\right\}dx + C$$

$$= x\log(x+1) - \int \frac{x}{x+1}dx + C$$

$$= x\log(x+1) - \int \frac{x+1-1}{x+1}dx + C$$

$$= x\log(x+1) - \int dx + \int \frac{dx}{x+1} + C$$

$$\therefore\ y = x\log(x+1) - x + \log|x+1| + C$$

which is the required solution.

Example 3. *Solve* $x\sqrt{1-y^2}\,dx + y\sqrt{1-x^2}\,dy = 0$.

Solution. We have $x\sqrt{1-y^2}\,dx + y\sqrt{1-x^2}\,dy = 0$

Separating the variables, we get

$$\frac{x}{\sqrt{1-x^2}}dx + \frac{y}{\sqrt{1-y^2}}dy = 0$$

Integrating both sides, we get

$$\int \frac{x}{\sqrt{1-x^2}}dx + \int \frac{y}{\sqrt{1-y^2}}dy = C$$

$$\Rightarrow \frac{1}{2}\int \frac{2x}{\sqrt{1-x^2}}dx + \frac{1}{2}\int \frac{2y}{\sqrt{1-y^2}}dy = C$$

$$\Rightarrow \quad -\frac{1}{2}\int \frac{dt}{\sqrt{t}} - \frac{1}{2}\int \frac{du}{\sqrt{u}} = C,$$

where $t = 1-x^2, u = 1-y^2$

$$\Rightarrow \quad -\frac{1}{2}\left[2\sqrt{t}\right] - \frac{1}{2}[2\sqrt{u}] = C$$

$$\Rightarrow \quad -\sqrt{t} - \sqrt{u} = C$$

$$\Rightarrow \quad -\sqrt{1-x^2} - \sqrt{1-y^2} = C$$

$$\therefore\ \sqrt{1-x^2} + \sqrt{1-y^2} = k, \text{ where } k = -C$$

which is the required solution.

Example 4. *Solve* $y - x\frac{dy}{dx} = a\left(y^2 - \frac{dy}{dx}\right)$. (BHOPAL 1991)

Solution. We have $y - x\frac{dy}{dx} = a\left(y^2 + \frac{dy}{dx}\right)$

$$\Rightarrow \quad y - ay^2 = (x+a)\frac{dy}{dx}$$

Separating the variables, we get

$$\int \frac{dx}{x+a} = \frac{1}{(y-ay^2)}dy.$$

Integrating both sides, we get

$$\int \frac{dx}{x+a} = \int \frac{dy}{y-ay^2} + \log C$$

$$\Rightarrow \quad \log|x+a| = \int \frac{dy}{y(1-ay)} + \log C$$

$$= \int \left(\frac{1}{y} + \frac{a}{1-ay}\right)dy + \log C$$

$$= \int \frac{dy}{y} + a\int \frac{dy}{1-ay} + \log C$$

$$\log|x+a| = \log|y| - \log|1-ay| + \log C$$

$$\Rightarrow \quad \log\left|\frac{(x+a)(1-ay)}{y}\right| = \log C$$

$$\Rightarrow \quad \left|\frac{(x+a)(1-ay)}{y}\right| = C$$

$$\Rightarrow \quad \frac{(x+a)(1-ay)}{y} = k, \text{ where } k = \pm C.$$

$\therefore (x+a)(1-ay) = ky$, which is the required solution.

Example 5. *Solve* $\frac{dy}{dx} = e^{x-y} + x^2e^{-y}$.

Solution. We have $\frac{dy}{dx} = e^{x-y} + x^2e^{-y}$

Separating the variables, we get

$$e^y dy = e^x dx + x^2 dx$$

Integrating both sides, we get

$$\int e^y\,dy = \int e^x\,dx + \int x^2dx + C$$

$$\Rightarrow \quad e^y = e^x + \frac{x^3}{3} + C,$$ which is the required solution.

Example 6. *Solve* $(1+x)(1+y^2)dx + (1+y)(1+x^2)dy = 0$.

Solution. We have

$$(1+x)(1+y^2)dx + (1+y)(1+x^2)dy = 0$$

Separating the variables, we get

$$\frac{(1+x)}{(1+x)^2}dx + \frac{(1+y)}{(1+y^2)}dy = 0\cdot$$

Integrating both sides, we get

$$\int \frac{(1+x)}{(1+x^2)}dx + \int \frac{(1+y)}{(1+y^2)}dy = C$$

$$\Rightarrow \quad \int \frac{dx}{1+x^2} + \int \frac{xdx}{1+x^2} + \int \frac{dy}{1+y^2} + \int \frac{ydy}{1+y^2} = C$$

$$\Rightarrow \quad \tan^{-1}x + \frac{1}{2}\int \frac{2x}{1+x^2}dx + \tan^{-1}y + \frac{1}{2}\int \frac{2y}{1+y^2}dy = C$$

$$\Rightarrow \quad \tan^{-1}x + \frac{1}{2}\log|1+x^2| + \tan^{-1}y + \frac{1}{2}\log|1+y^2| = C$$

$$\therefore\ \tan^{-1}x + \tan^{-1}y + \frac{1}{2}[\log(1+x^2) + \log(1+y^2)] = C$$

which is the required solution.

Example 7. *Solve* $(x^2 - yx^2)dy + (y^2 + xy^2)dx = 0\cdot$

Solution. We have

$$(x^2 - yx^2)dy + (y^2 + xy^2)dx = 0$$

Separating the variables, we get

$$x^2(1-y)dy + y^2(1+x)dx = 0.$$

$$\Rightarrow \quad \frac{(1-y)}{y^2}dy + \frac{(1+x)}{x^2}dx = 0.$$

Integrating both sides, we get

$$\int \frac{1-y}{y^2}dy + \int \frac{1+x}{x^2}dx = C$$

$$\Rightarrow \int \frac{1}{y^2}dy - \int \frac{1}{y}dy + \int \frac{1}{x^2}dx + \int \frac{1}{x}dx = C$$

$$\Rightarrow \quad -\frac{1}{y} - \log|y| - \frac{1}{x} + \log|x| = C.$$

$$\Rightarrow \quad \frac{1}{x} + \frac{1}{y} + \log|y| - \log|x| = -C$$

$$\therefore \quad \frac{1}{x} + \frac{1}{y} + \log|y| - \log|x| = k,$$

where $k = -C$, which is the required solution.

Example 8. *Solve* $\log\left(\frac{dy}{dx}\right) = ax + by.$

Solution. We have $\log\left(\frac{dy}{dx}\right) = ax + by$

Taking antilog of both sides, we get

$$\frac{dy}{dx} = e^{ax+by}.$$

Separating the variables, we get

$$e^{-by}dy = e^{ax}dx.$$

Integrating both sides, we get

$$\int e^{-by}dy = \int e^{ax}dx + C$$

$$\Rightarrow \quad \frac{e^{-by}}{-b} = \frac{e^{ax}}{a} + C$$

$\Rightarrow be^{ax} + ae^{-by} = k$, where $k = -abC$, which is the required solution.

Example 9. *Solve* $\cos x(1+\cos y)dx - \sin y(1+\sin x)dy = 0.$

Solution. We have

$$\cos x(1+\cos y)dx - \sin y(1+\sin x)dy = 0$$

Separating the variables, we get

$$\frac{\cos x}{1+\sin x}dx - \frac{\sin y}{1+\cos y}dy = 0.$$

Integrating both sides, we get

$$\int \frac{\cos x}{1+\sin x}dx - \int \frac{\sin y}{1+\cos y}dy = \log C$$

$$\Rightarrow \quad \int \frac{dt}{t} + \int \frac{du}{u} = \log C,$$

where $t = 1+\sin x, u = 1+\cos y$

$$\Rightarrow \quad \log|t| + \log|u| = \log C$$

$$\Rightarrow \log|1+\sin x| + \log|1+\cos y| = \log C$$

$$\therefore \quad |(1+\sin x)(1+\cos y)| = C$$

$$\Rightarrow \quad (1+\sin x)(1+\cos y) = \pm C$$

or $\quad (1+\sin x)(1+\cos y) = k,$

where $k = \pm C$, which is the required solution.

Example 10. *Solve* $\frac{dy}{dx} = \frac{1-\cos x}{1+\cos x}.$

Solution. We have $\frac{dy}{dx} = \frac{1-\cos x}{1+\cos x}$

Separating the variables, we get

$$dy = \left(\frac{1-\cos x}{1+\cos x}\right)dx.$$

Integrating both sides, we get

$$\int dy = \int \frac{1-\cos x}{1+\cos x}dx + C$$

$$\Rightarrow y = \int \frac{2\sin^2\frac{x}{2}}{2\cos^2\frac{x}{2}}dx + C \Rightarrow y = \int \tan^2\frac{x}{2}dx + C$$

$$\Rightarrow \quad y = \int \left(\sec^2\frac{x}{2} - 1\right)dx + C$$

$$\Rightarrow \quad y = \int \sec^2\frac{x}{2}dx - \int dx + C$$

$$\Rightarrow \quad y = 2\tan\frac{x}{2} - x + C$$

$\therefore x + y = 2\tan\frac{x}{2} + C$, which is the required solution.

Example 11. *Solve* $\frac{dy}{dx} = \frac{e^x(\sin^2 x + \sin 2x)}{y(2\log y + 1)}.$

Solution. We have $\frac{dy}{dx} = \frac{e^x(\sin^2 x + \sin 2x)}{y(2\log y + 1)}$

Separating the variables, we get

$$y(2\log y + 1)dy = e^x(\sin^2 x + \sin 2x)dx.$$

Integrating both sides, we get

$$\int y(2\log y + 1)dy = \int e^x(\sin^2 x + \sin 2x)dx + C$$

$$\Rightarrow \quad 2\int y\log y\,dy + \int y\,dy = \int e^x(\sin^2 x + \sin 2x)dx + C$$

$$\Rightarrow \quad 2\left\{\log y\int y\,dy - \int \frac{d}{dy}\left[(\log y).\int y\,dy\right]dy\right\} + \frac{y^2}{2} = \int e^x(\sin^2 x + \sin 2x)dx + C$$

$$\Rightarrow 2\left\{\frac{y^2}{2}\log y - \frac{1}{2}\int y\,dy\right\} + \frac{y^2}{2} = \int e^x(f(x) + f'(x))dx + C,$$

where $f(x) = \sin^2 x$

$$\Rightarrow y^2\log y - \frac{y^2}{2} + \frac{y^2}{2} = e^x f(x) + C$$

$\Rightarrow y^2\log y = e^x\sin^2 x + C$, which is the required solution.

Example 12. *Solve* $(1+y^2)(1+\log x)dx + xdy = 0$, *it being given that* $y = 1$ *when* $x = 1$.

Solution. We have $(1+y^2)(1+\log x)dx + xdy = 0$

Separating the variables, we get

$$\frac{1+\log x}{x}dx + \frac{dy}{1+y^2} = 0.$$

Integrating both sides, we get

$$\int \frac{(1+\log x)}{x}dx + \int \frac{dy}{1+y^2} = C$$

$$\Rightarrow \int t\,dt + \tan^{-1} y = C, \text{ where } t = 1+\log x$$

$$\Rightarrow \frac{t^2}{2} + \tan^{-1} y = C$$

$$\Rightarrow \frac{(1+\log x)^2}{2} + \tan^{-1} y = C.$$

Putting $x = 1$ and $y = 1$, we get

$$\frac{1}{2} + \tan^{-1} 1 = C$$

$$\Rightarrow \quad C = \frac{1}{2} + \frac{\pi}{4}.$$

$$\therefore \quad \frac{(1+\log x)^2}{2} + \tan^{-1} y = \frac{1}{2} + \frac{\pi}{4}$$

$$\text{or } (1+\log x)^2 + 2\tan^{-1} y = 1 + \frac{\pi}{2}.$$

Example 13. *Solve* $(1+e^{2x})dy + (1+y^2)e^x dx = 0$, *it being given that* $y = 1$ *when* $x = 0$.

Solution. We have $(1+e^{2x})dy + (1+y^2)e^x dx = 0$

Separating the variables, we get

$$\frac{dy}{1+y^2} + \frac{e^x}{1+e^{2x}}dx = 0.$$

Integrating both sides, we get

$$\int \frac{dy}{1+y^2} + \int \frac{e^x}{1+e^{2x}}dx = C$$

$$\Rightarrow \tan^{-1} y + \int \frac{dt}{(1+t^2)} = C, \text{ where } t = e^x$$

$$\Rightarrow \tan^{-1} y + \tan^{-1} t = C$$

$$\Rightarrow \tan^{-1} y + \tan^{-1} e^x = C.$$

Putting $x = 0$ and $y = 1$, we get

$$\tan^{-1} 1 + \tan^{-1} e^0 = C$$

$$\Rightarrow \quad C = \frac{\pi}{4} + \frac{\pi}{4} = \frac{\pi}{2}$$

$$\therefore \tan^{-1} y + \tan^{-1} e^x = \frac{\pi}{2}.$$

Example 14. *Solve* $x(1+y^2)dx - y(1+x^2)dy = 0$, *given that* $y = 0$ *when* $x = 1$.

Solution. We have $x(1+y^2)dx - y(1+x^2)dy = 0$

Separating the variables, we get

$$\frac{x}{1+x^2}dx - \frac{y}{1+y^2}dy = 0.$$

Integrating both sides, we get

$$\int \frac{x}{1+x^2}dx - \int \frac{y}{1+y^2}dy = C$$

$$\Rightarrow \frac{1}{2}\int \frac{2x}{1+x^2}dx - \frac{1}{2}\int \frac{2y}{1+y^2}dy = C$$

$$\Rightarrow \frac{1}{2}\log|1+x^2| - \frac{1}{2}\log|1+y^2| = C$$

Putting $y = 0$ and $x = 1$, we get

$$\frac{1}{2}\log 2 = C$$

$$\therefore \frac{1}{2}[\log(1+x^2) - \log(1+y^2)] = \frac{1}{2}\log 2$$

$$\text{or } \log\left|\frac{1+x^2}{1+y^2}\right| = \log 2 \text{ or } (1+x^2) = 2(1+y^2).$$

Example 15. *Solve* $(1+x^2)\sec^2 y\,dy + 2x\tan y\,dx = 0$, *given that* $y = \frac{\pi}{4}$ *when* $x = 1$.

Solution. We have $(1+x^2)\sec^2 y\,dy + 2x\tan y\,dx = 0$

Separating the variables, we get

$$\frac{\sec^2 y}{\tan y}dy + \frac{2x}{1+x^2}dx = 0.$$

Integrating both sides, we get

$$\int \frac{\sec^2 y}{\tan y}dy + \int \frac{2x}{1+x^2}dx = C$$

$$\Rightarrow \int \frac{dt}{t} + \int \frac{du}{u} = C, \text{ where } t = \tan y, u = 1+x^2$$

$$\Rightarrow \log|t| + \log|u| = C$$

$$\Rightarrow \log|\tan y| + \log|1+x^2| = C.$$

Putting $x = 1$ and $y = \frac{\pi}{4}$, we get

$$C = \log\left|\tan\frac{\pi}{4}\right| + \log|1+1|$$

$$\Rightarrow \quad C = \log 1 + \log 2 = \log 2$$

$$\therefore \log|\tan y| + \log(1+x^2) = \log 2$$

$$\therefore \quad \log|(1+x^2)\tan y| = \log 2$$

$$\text{or} \quad 2\cot y = (1+x^2)$$

28.6 DIFFERENTIAL EQUATION REDUCIBLE TO VARIABLE SEPARABLE FORM

Sometimes, we come across with some differential equation, in which the variables cannot be separated. In such type of differential equation, some suitable substitution reduces it to a form in which the variables are separable.

Solved Examples

Example 1. *Solve* $\frac{dy}{dx}=\sin(x+y)$.

Solution. Let $v=x+y$.

Then, $\frac{dv}{dx}=1+\frac{dy}{dx}$ so that $\frac{dy}{dx}=\frac{dv}{dx}-1$.

Now, the given differential equation reduces to

$\frac{dv}{dx}-1=\sin v$ or $\frac{dv}{dx}=1+\sin v$.

Separating the variables, we get

$$\frac{dv}{1+\sin v}=dx.$$

Integrating both sides, we get

$$\int\frac{dv}{1+\sin v}=\int dx+C$$

$$\Rightarrow \int\frac{(1-\sin v)}{(1+\sin v)(1-\sin v)}dv = x+C$$

$$\Rightarrow \int\frac{(1-\sin v)}{1-\sin^2 v}dv=x+C$$

$$\Rightarrow \int\frac{1-\sin v}{\cos^2 v}dv=x+C$$

$$\Rightarrow \int\sec^2 v\,dv-\int\sec v\tan v\,dv=x+C.$$

$\Rightarrow \tan v-\sec v=x+C$

$\therefore \tan(x+y)-\sec(x+y)=x+C$, which is the required solution.

Example 2. *Solve* $\frac{dy}{dx}=(4x+y+1)^2$. (Manglore–1999)

Solution. Let $v=4x+y+1$.

Then, $\frac{dv}{dx}=4+\frac{dy}{dx}$ so that $\frac{dy}{dx}=\frac{dv}{dx}-4$

Now, the given differential equation reduces to

$\frac{dv}{dx}-4=v^2$ or $\frac{dv}{dx}=4+v^2$.

Separating the variables, we get

$$\frac{dv}{4+v^2}=dx.$$

Integrating both sides, we get

$$\int\frac{dv}{4+v^2}=\int dx+C$$

$$\Rightarrow \int\frac{dv}{2^2+v^2}=\int dx+C.$$

$$\Rightarrow \frac{1}{2}\tan^{-1}\frac{v}{2}=x+C$$

$$\therefore \frac{1}{2}\tan^{-1}\left(\frac{4x+y+1}{2}\right)=x+C$$

or $4x+y+1=2\tan(2x+2C)$, which is the required solution.

Example 3. *Solve* $\frac{dy}{dx}=\sin(x+y)+\cos(x+y)$. (VTU–2005)

Solution. Let $x+y=v$.

Then, $1+\frac{dy}{dx}=\frac{dv}{dx}$ so that $\frac{dy}{dx}=\frac{dv}{dx}-1$.

Using above substitution, the given differential equation reduces to

$\frac{dv}{dx}-1=\sin v+\cos v$ or $\frac{dv}{dx}=1+\sin v+\cos v$.

Separating the variables, we get

$$\frac{dv}{1+\sin v+\cos v}=dx.$$

Integrating both sides, we get

$$\int\frac{dv}{1+\sin v+\cos v}=\int dx+C$$

$$\Rightarrow \int\frac{dv}{2\sin\frac{v}{2}\cos\frac{v}{2}+2\cos^2\frac{v}{2}}=x+C$$

$$\Rightarrow \frac{1}{2}\int\frac{\sec^2\frac{v}{2}dv}{\left(\tan\frac{v}{2}+1\right)}=x+C$$

$$\Rightarrow \int\frac{dt}{t}=x+C,\ \text{where } t=1+\tan\frac{v}{2}$$

$$\Rightarrow \log|t|=x+C$$

$$\Rightarrow \log\left|1+\tan\frac{v}{2}\right|=x+C$$

$$\therefore \log\left|1+\tan\left(\frac{x+y}{2}\right)\right|=x+C \quad [\because v=x+y]$$

which is the required solution.

Example 4. *Solve* $\frac{dy}{dx}=\frac{2x+3y+4}{4x+6y+5}$.

Solution. Let $2x+3y=v$.

Then, $2+3\frac{dy}{dx}=\frac{dv}{dx}\Rightarrow\frac{dy}{dx}=\frac{1}{3}\left(\frac{dv}{dx}-2\right)$.

Using above substitution, the given differential equation reduces to

$$\frac{1}{3}\left(\frac{dv}{dx}-2\right)=\frac{v+4}{2v+5}$$

or $$\frac{dv}{dx}-2=\frac{3v+12}{2v+5}$$

or $$\frac{dv}{dx}=2+\frac{3v+12}{2v+5}=\frac{7v+22}{2v+5}.$$

Separating the variables, we get

$$\frac{2v+5}{7v+22}dv=dx.$$

Integrating both sides, we get

$$\int\frac{2v+5}{7v+22}dv=\int dx+C$$

$$\Rightarrow \quad 2\int \frac{v}{7v+22}dv + 5\int \frac{dv}{7v+22} = x + C$$

$$\Rightarrow \quad \frac{2}{7}\int \frac{7v}{7v+22}dv + 5\int \frac{dv}{7v+22} = x + C$$

$$\Rightarrow \frac{2}{7}\int \frac{7v+22-22}{7v+22}dv + 5\int \frac{dv}{7v+22} = x + C$$

$$\Rightarrow \frac{2}{7}\int dv - \frac{44}{7}\int \frac{dv}{7v+22} + 5\int \frac{dv}{7v+22} = x + C$$

$$\Rightarrow \quad \frac{2}{7}v - \frac{9}{7}\int \frac{dv}{7v+22} = x + C$$

$$\Rightarrow \quad \frac{2}{7}v - \frac{9}{49}\log|7v+22| = x + C$$

$$\Rightarrow \quad 14v - 9\log|7v+22| = 49x + 49C$$

$$\Rightarrow 14(2x+3y) - 9\log|14x+21y+22| = 49x + 49C$$

$$\Rightarrow 9\log|14x+21y+22| = 42y - 21x + k, k = 49C,$$

which is the required solution.

Example 5. *Solve the equation*

$$(2x^2+3y^2-7)xdx - (3x^2+2y^2-8)ydy = 0.$$

(UPTU–2005)

Solution. The given equation can be written as

$$\frac{x}{y}\frac{dx}{dy} = \frac{3x^2+2y^2-8}{2x^2+3y^2-7}$$

Applying componendo and dividendo rule, we get

$$\frac{xdx+ydy}{xdx-ydy} = \frac{5x^2+5y^2-15}{x^2-y^2-1}$$

or
$$\frac{xdx+ydy}{x^2+y^2-3} = 5\left(\frac{xdx-ydy}{x^2-y^2-1}\right)$$

Multiplying by 2, we get

$$\left(\frac{2xdx+2ydy}{x^2+y^2-3}\right) = 5\left(\frac{2xdx-2ydy}{x^2-y^2-1}\right)$$

On integrating, we get

$$\log(x^2+y^2-3) = 5\log(x^2-y^2-1) + \log C$$

or
$$x^2+y^2-3 = C(x^2-y^2-1)^5.$$

Exapmple 6. *Solve* $\frac{dy}{dx} = \frac{x(2\log x+1)}{\sin y + y\cos y}$ (VTU–2008)

Solution. Given equation is

$$x(2\log x+1)dx = (\sin y + y\cos y)dy$$

Integrating both sides, we get

$$2\int (\log x.x + x)dx = \int \sin ydy + \int y\cos ydy + c$$

$$2\left[\left(\log x.\frac{x^2}{2} - \int \frac{1}{x}.\frac{x^2}{2}dx\right) + \frac{x^2}{2}\right]$$

$$= -\cos y + [y\sin y - \int \sin y.1dy + c]$$

$$2x^2\log x - \frac{x^2}{2} + \frac{x^2}{2}$$

$$= -\cos y + y\sin y + \cos y + c$$

or
$$2x^2\log x - y\sin y = c.$$

Exapmple 7. *Solve* $\frac{y}{x}\frac{dy}{dx} + \frac{x^2+y^2-1}{2(x^2+y^2)+1} = 0.$ (VTU–2003)

Solution. Putting $x^2 + y^2 = t$, we get

$$2x + 2y\frac{dy}{dx} = \frac{dt}{dx} \quad \text{or} \quad \frac{y}{x}\frac{dy}{dx} = \frac{1}{2x}\frac{dt}{dx} - 1$$

Therefore, the given equation becomes,

$$\frac{1}{2x}\frac{dt}{dx} - 1 + \frac{t-1}{2t+1} = 0 \Rightarrow \quad \frac{1}{2x}\frac{dt}{dx} = 1 - \frac{t-1}{2t+1}$$

or
$$\frac{1}{2x}\frac{dt}{dx} = \frac{t+2}{2t+1} \quad \text{or} \quad 2xdx = \frac{2t+1}{t+2}dt$$

or
$$2xdx = \left(2 - \frac{3}{t+2}\right)dt$$

On integrating, we get

$$x^2 = 2t - 3\log(t+2) + c$$

or
$$x^2 + 2y^2 - 3\log(x^2+y^2+2) + c = 0$$

$$(\because x^2 + y^2 = t)$$

Exapmple 8. *Solve* $\sec^2 x\tan ydx + \sec^2 y\tan xdy = 0.$

(UPTU(B.Pharma)SUM–2009, PTU–2003, VTU–2009)

Solution. Separating the variables, we get

$$\frac{\sec^2 x}{\tan x}dx + \frac{\sec^2 y}{\tan^2 y}dy = 0$$

On integrating, we get

$$\log\tan x + \log\tan y = \log C$$

or
$$\log\tan x\tan y = \log C$$

or
$$\tan x\tan y = C$$

where C is an arbitrary constant.

EXERCISE 28.4

Solve the following differential equations :

1. $x\frac{dy}{dx} = y$
2. $\frac{dy}{dx} + y = 1$
3. $\frac{dy}{dx} = \sqrt{4-y^2}$
4. $\frac{dy}{dx} = e^{x+y}$
5. $\frac{dy}{dx} = (e^x+1)y$
6. $\frac{dy}{dx} = \frac{(1+y^2)}{(1+x^2)}$
7. $\frac{dy}{dx} + \sqrt{\frac{1-y^2}{1-x^2}} = 0$ (Andhra–1999)
8. $(x^2+1)\frac{dy}{dx} = xy$
9. $(x\log x)\frac{dy}{dx} = y$
10. $x\frac{dy}{dx} + y = y^2$
11. $\frac{dy}{dx} = 1 - x + y - xy$
12. $\frac{dy}{dx} = (1+x)(1+y^2)$
13. $(e^x+1)ydy = (y+1)e^x dx$

14. $(x+2)\frac{dy}{dx}=4x^2y$ **15.** $\frac{dy}{dx}+\frac{(xy+y)}{(xy+x)}=0$

16. $x\sqrt{1+y^2}\,dx+y\sqrt{1+x^2}dy=0$

17. $\frac{dy}{dx}+\frac{\cos x\sin y}{\cos y}=0$ **18.** $\frac{dy}{dx}+\frac{(1+\cos 2y)}{(1-\cos 2x)}=0$

19. $(y+xy)dx+(x-xy^2)dy=0$

20. $(x^2y-x^2)dx+(xy^2-y^2)dy=0$

21. $(1+x)ydx+(1-y)x\,dy=0$

22. $(1-x^2)(1-y)dx=xy(1+y)dy$

23. $\sqrt{1+x^2+y^2+x^2y^2}+xy\frac{dy}{dx}=0$

24. $x\cos^2 ydx=y\cos^2 xdy$

25. $3e^x\tan ydx+(1-e^x)\sec^2 ydy=0$

26. $(e^y+1)\cos xdx+e^y\sin xdy=0$ (Manglore–1999)

27. $x(e^{2y}-1)dy+(x^2-1)e^ydx=0$

28. $e^{2x-3y}dx+e^{2y-3x}dy=0$

29. $\sin y\frac{dy}{dx}=\sin^3 x$ **30.** $\sin^2 xdy+\cos^2 ydx=0$

31. $x\cos^2 ydx-y\cos^2 xdy=0$

32. $x\cos ydy=e^x(x\log x+1)dx$

33. $\frac{dy}{dx}=\frac{\sin x+x\cos x}{y(2\log y+1)}$

34. $\frac{dy}{dx}=\frac{x(2\log x+1)}{(\sin y+y\cos y)}$ (VTU–2000)

35. $\tan y\frac{dy}{dx}=\sin(x+y)+\sin(x-y)$

36. $\frac{dy}{dx}=x^2e^{-3y}$, given that $y=0$ when $x=0$.

37. $\frac{dy}{dx}=e^{x+y}$, given that $y=1$ when $x=1$.

38. $e^x\frac{dy}{dx}=3y^3$, given that $y(0)=\frac{1}{2}$.

39. $(x-1)\frac{dy}{dx}=2xy$, given that $y(2)=1$

40. $xy\frac{dy}{dx}=y+2$, given that $y(2)=0$

41. $(1+x^2)\frac{dy}{dx}+(1+y^2)=0$, given that $y=1$ when $x=0$.

42. $\frac{dy}{dx}=\sec y$, given that $y=0$ when $x=0$.

43. $\frac{dy}{dx}=y\sin 2x$, given that $y=0$ when $x=0$.

44. $\frac{dy}{dx}=y\cot 2x$, given that $y\left(\frac{\pi}{4}\right)=2$.

45. Find the equation of the curve that passes through the point (1, 2) and satisfies the differential equation $\frac{dy}{dx}=-\frac{2xy}{(x^2+1)}$.

46. Find the equation of the curve that passes through the point (1, 0) and satisfies the differential equation $(1+y^2)dx-xydy=0$.

47. $\frac{dy}{dx}=(x+y)^2$ **48.** $\frac{dy}{dx}=\sec(x+y)$

49. $(x+y)^2\frac{dy}{dx}=a^2$

50. $(x-y)^2\frac{dy}{dx}=a^2$ (AMIE 1999)

51. $\frac{dy}{dx}=\tan^2(x+y)$ **52.** $\frac{dy}{dx}=\frac{x+y-1}{x+y+1}$

53. $\frac{dy}{dx}=\frac{x-y+3}{2x-2y+5}$ **54.** $\frac{dy}{dx}=\frac{8x+6y+12}{4x+3y+2}$

Hint to Selected Problems

11. $\frac{dy}{dx}=1--x+y-xy=(1-x)(1+y)$

$$\Rightarrow\frac{dy}{(1+y)}=(1-x)dx\Rightarrow\int\frac{dy}{1+y}=\int(1-x)dx+C$$

18. $\frac{dy}{dx}+\left(\frac{1+\cos 2y}{1-\cos 2x}\right)=0\Rightarrow\frac{dy}{dx}+\frac{\cos^2 y}{\sin^2 x}=0$

$$\Rightarrow\sec^2 ydy+\operatorname{cosec}^2 xdx=0$$

23. $\sqrt{1+x^2+y^2+x^2y^2}+xy\frac{dy}{dx}=0$

$$\Rightarrow\sqrt{1+x^2}\sqrt{1+y^2}+xy\frac{dy}{dx}=0$$

$$\Rightarrow\frac{\sqrt{1+x^2}}{x}dx+\frac{y}{\sqrt{1+y^2}}dy=0$$

$$\Rightarrow\int\frac{\sqrt{1+x^2}}{x}dx+\int\frac{y}{\sqrt{1+y^2}}dy=C$$

$$\Rightarrow\int\frac{t^2}{t^2-1}dt+\sqrt{1+y^2}=C,\text{ where }t^2=1+x^2$$

$$\Rightarrow\int dt+\int\frac{dt}{t^2-1}+\sqrt{1+y^2}=C.$$

29. $\sin y\frac{dy}{dx}=\sin^3 x\Rightarrow\sin ydy=\sin^3 xdx$

$$\Rightarrow\sin ydy=\frac{1}{4}(3\sin x-\sin 3x)dx$$

$$\Rightarrow\int\sin ydy=\frac{3}{4}\int\sin xdx-\frac{1}{4}\int\sin 3xdx+C$$

32. $x\cos ydy=e^x(x\log x+1)dx\Rightarrow\cos y\,dy=e^x\left(\log x+\frac{1}{x}\right)dx$

$$\Rightarrow\int\cos y\,dy=\int e^x\left(\log x+\frac{1}{x}\right)dx+C$$

$$\Rightarrow\sin y=\int e^x\left[f(x)+f'(x)\right]dx+C$$

$$\Rightarrow\sin y=e^xf(x)+C=e^x\log x+C$$

35. $\tan y\frac{dy}{dx}=\sin(x+y)+\sin(x-y)$

$$\Rightarrow\tan y\frac{dy}{dx}=2\sin x\cos y\Rightarrow\frac{\tan y}{\cos y}dy=2\sin xdx$$

$$\Rightarrow\int\frac{\tan y}{\cos y}dy=2\int\sin xdx+C$$

45. $\frac{dy}{dx} = \frac{-2xy}{(x^2+1)} \Rightarrow \frac{dy}{y} + \frac{2x}{x^2+1}dx = 0$

$\Rightarrow \int \frac{dy}{y} + \int \frac{2x}{x^2+1}dx = C \Rightarrow \log|y| + \log|x^2+1| = C$

Putting $x = 1$ and $y = 2$, we get $C = 2\log 2 = \log 4$

$\therefore \quad \log|y| + \log|x^2+1| = \log 4 \Rightarrow y(x^2+1) = 4.$

50. Let $x - y = v$ so that $\frac{dy}{dx} = 1 - \frac{dy}{dx}$

Then, $(x-y)^2\frac{dy}{dx} = a^2$

$\Rightarrow v^2\left(1 - \frac{dv}{dx}\right) = a^2 \Rightarrow 1 - \frac{dv}{dx} = \frac{a^2}{v^2}$

$\Rightarrow \frac{dv}{dx} = -\frac{a^2}{v^2} + 1 = \frac{v^2 - a^2}{v^2} \Rightarrow \frac{v^2}{v^2 - a^2}dv = dx$

$\Rightarrow \int \frac{v^2}{v^2 - a^2}dv = \int dx + C \Rightarrow a^2\int \frac{dv}{v^2 - a^2} + \int dv = x + C$

$\Rightarrow a^2\frac{1}{2a}\log\left|\frac{v-a}{v+a}\right| + v = x + C$

$\Rightarrow \frac{a}{2}\log\left|\frac{-a+x-y}{a+x-y}\right| + x - y = x + C$

$\Rightarrow y + C = \frac{a}{2}\log\left|\frac{x-y-a}{x-y+a}\right|$

54. Let $4x + 3y = v$ so that $\frac{dy}{dx} = \frac{1}{3}\left(\frac{dv}{dx} - 4\right)$

Then $\frac{dy}{dx} = \frac{8x+6y+12}{4x+3y+2}$

$\Rightarrow \frac{1}{3}\left(\frac{dv}{dx} - 4\right) = \frac{2v+12}{v+2} \Rightarrow \frac{dv}{dx} = \frac{10v+44}{v+2}$

ANSWERS

1. $y = Cx$ **2.** $\log|1-y| + x = C$ **3.** $y = 2\sin(x+C)$ **4.** $e^x + e^{-y} = C$ **5.** $\log|y| = e^x + x + C$
6. $\tan^{-1}y - \tan^{-1}x = C$ **7.** $\sin^{-1}y + \sin^{-1}x = C$ **8.** $y = C\sqrt{x^2+1}$ **9.** $y = C\log x$
10. $(y-1) = Cxy$ **11.** $\log|1+y| = x - \frac{x^2}{2} + C$ **12.** $\tan^{-1}y = x + \frac{x^2}{2} + C$ **13.** $(e^x+1)(y+1) = Ce^y$
14. $\log|y| = 2x^2 - 8x + 16\log|2+x| + C$ **15.** $x + y + \log|xy| = C$ **16.** $\sqrt{1+y^2} + \sqrt{1+x^2} = C$ **17.** $\log|\sin y| + \sin x = C$
18. $\tan y = \cot x + C$ **19.** $\log|xy| + x - \frac{1}{2}y^2 = C$ **20.** $\frac{1}{2}(x^2+y^2) + (x+y) + \log|(x-1)(y-1)| = C$
21. $\log|xy| + x - y = C$ **22.** $\log|x(1-y)^2| = \frac{x^2}{2} - \frac{y^2}{2} - 2y + C$ **23.** $\log|x| - \log|1+\sqrt{1+x^2}| + \sqrt{1+x^2} + \sqrt{1+y^2} = C$
24. $y\tan y + \log\cos y = x\tan x + \log\cos x + C$ **25.** $\tan y = C(e^x - 1)^3$ **26.** $(1+e^y)\sin x = C$
27. $e^y + e^{-y} + \frac{x^2}{2} - \log|x| = C$ **28.** $e^{5x} + e^{5y} = C$ **29.** $\frac{3}{4}\cos x - \cos y - \frac{\cos 3x}{12} = C$ **30.** $\tan y = \cot x + C$
31. $x\tan x - y\tan y - \log|\sec x| + \log|\sec y| = C$ **32.** $\sin y = e^x\log x + C$ **33.** $y^2\log y = x\sin x + C$ **34.** $y\sin y = x^2\log x + C$
35. $\sec y + 2\cos x = C$ **36.** $e^{3y} = x^3 + 1$ **37.** $e^x + e^{-y} = e + \frac{1}{e}$ **38.** $\frac{1}{6y^2} = e^{-x} - \frac{1}{3}$ **39.** $y = (x-1)^2 e^{2(x-2)}$
40. $y - 2\log(y+2) = \log x - 3\log 2$ **41.** $x + y = 1 - xy$ **42.** $x = \sin y$ **43.** $y^2 e^{\cos 2x} = 0$
44. $y^2 = 4\sin 2x$ **45.** $y(x^2+1) = 4$ **46.** $x^2 = (1+y^2)$ **47.** $x + y = \tan(x+C)$ **48.** $y = \tan\left(\frac{x+y}{2}\right) + C$
49. $y - a\tan^{-1}\left(\frac{x+y}{a}\right) = C$ **50.** $y + C = \frac{a}{2}\log\left|\frac{x-y-a}{x-y+a}\right|$ **51.** $2(y-x) + \sin 2(x+y) + C$
52. $x - y + C = \log|x+y|$ **53.** $x - 2y + \log|x-y+2| = C$ **54.** $15y - 30x - 12\log|20x+15y+22| = C$

28.7 HOMOGENEOUS DIFFERENTIAL EQUATIONS

Definition 1. *A function f(x, y) in x and y is said to be homogeneous function of degree n, if the degree of each term of f(x, y) is n.*

For example : $f(x, y) = ax^2 + 2hxy + by^2$ is a homogeneous function of degree 2.

REMARK

- In general, a homogeneous function of degree n can be expressed as $f(x, y) = x^n F(y/x)$

Definition 2. *A differential equation of the form* $\frac{dy}{dx} = \frac{f_1(x, y)}{f_2(x, y)}$ *is said to be homogeneous, if* $f_1(x, y)$ *and* $f_2(x, y)$ *are homogeneous functions of same degree.*

For example : $\frac{dy}{dx} = \frac{x^2 + y^2}{xy}$ is a homogeneous differential equation.

WORKING PROCEDURE

In order to solve a homogeneous differential equation, we proceed through the following steps :

Step 1. *Substitute* $y = vx$ (*or* $x = vy$).

Step 2. *Reduce the given differential equation in terms of v and x.*

Step 3. *Solve the reduced equation by the method of separation of variables.*

Step 4. *Replace v by* $\frac{y}{x}$ (*or v by* $\frac{x}{y}$) *after integration.*

Solved Examples

Example 1. *Solve* $(x^2 + y^2)\frac{dy}{dx} = xy$. (AMIETE–1999)

Solution. The given differential equation can be expressed as

$$\frac{dy}{dx} = \frac{xy}{x^2 + y^2} \quad \ldots(1)$$

Clearly, (1) is a homogeneous differential equation.

Putting $y = vx$ and $\frac{dy}{dx} = v + x\frac{dv}{dx}$ in (1), we get

$$v + x\frac{dv}{dx} = \frac{vx^2}{x^2 + v^2x^2} = \frac{v}{1+v^2}$$

$$\Rightarrow \quad x\frac{dv}{dx} = \frac{v}{1+v^2} - v = \frac{-v^3}{1+v^2}.$$

Separating the variables, we get

$$-\frac{1+v^2}{v^3}dv = \frac{dx}{x}.$$

Integrating both sides, we get

$$-\int \frac{dv}{v^3} - \int \frac{dv}{v} = \int \frac{dx}{x} + \log|C|$$

$$\Rightarrow \quad \frac{1}{2v^2} - \log|v| = \log|x| + \log|C|$$

$$\Rightarrow \quad \frac{1}{2v^2} = \log|xvC| \Rightarrow xvC = e^{1/2v^2}$$

$$\Rightarrow \quad yC = e^{1/2y^2}$$, which is the required solution.

Example 2. *Solve* $(y^2 - x^2)dy = 3xydx$.

Solution. The given differential equation can be expressed as

$$\frac{dy}{dx} = \frac{3xy}{y^2 - x^2} \quad \ldots(1)$$

Clearly, (1) is a homogeneous differential equation.

Putting $y = vx$ and $\frac{dy}{dx} = v + x\frac{dv}{dx}$ in (1), we get

$$v + x\frac{dv}{dx} = \frac{3x^2v}{v^2x^2 - x^2} = \frac{3v}{v^2 - 1}$$

$$\Rightarrow \quad x\frac{dv}{dx} = \frac{3v}{v^2-1} - v = \frac{4v - v^3}{v^2 - 1}.$$

Separating the variables, we get

$$\frac{v^2-1}{4v - v^3}dv = \frac{dx}{x}$$

or

$$\frac{v^2-1}{v(4-v^2)}dv = \frac{dx}{x} \quad \ldots(2)$$

Let

$$\frac{v^2-1}{v(4-v^2)} = \frac{v^2-1}{v(2+v)(2-v)} = \frac{A}{v} + \frac{B}{2+v} + \frac{C}{2-v}.$$

$$\Rightarrow \quad v^2 - 1 = A(2+v)(2-v) + B(2-v)v + C(2+v)v \quad \ldots(3)$$

Putting $v = 0, 2, -2$ successively in (3), we get

$$A = -\frac{1}{4}, C = \frac{3}{8}, B = -\frac{3}{8}$$

$$\therefore \quad \frac{v^2-1}{v(4-v^2)} = -\frac{1}{4v} - \frac{3}{8(2+v)} + \frac{3}{8(2-v)}.$$

Now, (2) becomes

$$-\frac{dv}{4v} - \frac{3dv}{8(2+v)} + \frac{3dv}{8(2-v)} = \frac{dx}{x}.$$

Integrating both sides, we get

$$-\frac{1}{4}\log|v| - \frac{3}{8}\log|2+v| - \frac{3}{8}\log|2-v| = \log|x| + C,$$

where C is the constant of integration.

$$-\frac{1}{4}\log\left|\frac{y}{x}\right| - \frac{3}{8}\log\left|2+\frac{y}{x}\right| - \frac{3}{8}\log\left|2-\frac{y}{x}\right| = \log|x| + C$$

$$-\frac{1}{4}\log\left|\frac{y}{x}\right| - \frac{3}{8}\log\left|\left(2+\frac{y}{x}\right)\left(2-\frac{y}{x}\right)\right| = \log|x| + C$$

$$-\frac{1}{4}\log\left|\frac{y}{x}\right| - \frac{3}{8}\log\left|\frac{(2x+y)(2x-y)}{x^2}\right| = \log|x| + C$$

$$-\frac{1}{4}\log|y| + \frac{1}{4}\log|x| - \frac{3}{8}\log|(2x+y)(2x-y)| + \frac{3}{4}\log|x| = \log|x| + C$$

$$-\frac{1}{4}\log|y| - \frac{3}{8}\log|(2x+y)(2x-y)| = C$$

or $2\log|y| + 3\log|(2x+y)(2x-y)| = k$,

where $k = -8C$

which is the required solution.

Example 3. *Solve* $(x+y)dy+(x-y)dx=0$.

Solution. The given differential equation can be expressed as

$$\frac{dy}{dx}=\frac{y-x}{y+x} \quad \ldots(1)$$

Clearly, (1) is homogeneous.

Putting $y=vx$ and $\frac{dy}{dx}=v+x\frac{dv}{dx}$ in (1), we get

$$v+x\frac{dv}{dx}=\frac{vx-x}{vx+x}=\frac{v-1}{v+1}$$

$$\Rightarrow \quad v\frac{dv}{dx}=\frac{v-1}{v+1}-v=\frac{-1-v^2}{v+1}.$$

Separating the variables, we get

$$\frac{v+1}{1+v^2}dv+\frac{dx}{x}=0$$

$$\Rightarrow \quad \frac{v}{1+v^2}dv+\frac{1}{1+v^2}dv+\frac{dx}{x}=0$$

Integrating both sides, we get

$$\int\frac{v}{1+v^2}dv+\int\frac{dv}{1+v^2}+\int\frac{dx}{x}=C,$$ where C is the constant of integration.

$$\Rightarrow \quad \frac{1}{2}\log|1+v^2|+\tan^{-1}v+\log|x|=C$$

$$\Rightarrow \quad \frac{1}{2}\log\left(1+\frac{y^2}{x^2}\right)+\tan^{-1}\frac{y}{x}+\log|x|=C$$

$$\Rightarrow \frac{1}{2}\log\left(\frac{x^2+y^2}{x^2}\right)+\tan^{-1}\frac{y}{x}+\log|x|=C$$

$$\Rightarrow \frac{1}{2}\log(x^2+y^2)+\tan^{-1}\frac{y}{x}=C,$$ which is the required solution.

Example 4. *Solve* $xdy-ydx=\sqrt{x^2+y^2}\,dx$.

Solution. The given differential equation can be expressed as

$$\frac{dy}{dx}=\frac{y+\sqrt{x^2+y^2}}{x} \quad \ldots(1)$$

Clearly, (1) is homogeneous as $y+\sqrt{x^2+y^2}$ and x are homogeneous functions of degree 1.

Putting $y=vx$ and $\frac{dy}{dx}=v+x\frac{dv}{dx}$ in (1), we get

$$v+x\frac{dv}{dx}=\frac{vx+\sqrt{x^2+v^2x^2}}{x}=\frac{v+\sqrt{1+v^2}}{1}$$

$$\Rightarrow \quad x\frac{dv}{dx}=\sqrt{1+v^2}.$$

Separating the variables, we get

$$\frac{dv}{\sqrt{1+v^2}}=\frac{dx}{x}.$$

Integrating both sides, we get

$$\int\frac{dv}{\sqrt{1+v^2}}=\int\frac{dx}{x}+\log C$$

$$\log\left|v+\sqrt{1+v^2}\right|=\log|x|+\log C$$

$$\Rightarrow \quad \log\left|\frac{y}{x}+\sqrt{1+\frac{y^2}{x^2}}\right|=\log|x|+\log C$$

$$\Rightarrow \quad \log\left|\frac{y+\sqrt{x^2+y^2}}{x^2}\right|=\log C$$

$$\Rightarrow \quad \left|\frac{y+\sqrt{x^2+y^2}}{x^2}\right|=C$$

$$\Rightarrow \quad y+\sqrt{x^2+y^2}=\pm Cx^2$$

$$\therefore y+\sqrt{x^2+y^2}=kx^2,$$ where $k=\pm C$

Example 5. *Solve* $(1+e^{x/y})dx+e^{x/y}\left(1-\frac{x}{y}\right)dy=0$.

(VTU–2001, 2003, Tirupati–1998, Bhopal–1998, PTU–2006, Rajasthan–2005)

Solution. The given differential equation can be expressed as

$$\frac{dx}{dy}=\frac{e^{x/y}\left(\frac{x}{y}-1\right)}{1+e^{x/y}} \quad \ldots(1)$$

Putting $x=vy$ and $\frac{dx}{dy}=v+y\frac{dv}{dy}$ in (1), we get

$$v+y\frac{dv}{dy}=\frac{e^v(v-1)}{1+e^v}$$

$$\Rightarrow \quad y\frac{dv}{dy}=\frac{e^v(v-1)}{1+e^v}-v=-\frac{v+e^v}{1+e^v}.$$

Separating the variables, we get

$$\frac{(1+e^v)}{(v+e^v)}dv+\frac{dy}{y}=0.$$

Integrating both sides, we get

$$\int\frac{(1+e^v)}{(v+e^v)}dv+\int\frac{dy}{y}=\log C,$$ where C is the constant of integration.

$$\Rightarrow \quad \log|v+e^v|+\log|y|=\log C$$

$$\Rightarrow \quad \log|y(v+e^v)|=\log C$$

$$\Rightarrow \quad |y(v+e^v)|=C$$

$$\Rightarrow \quad y(v+e^v)=\pm C$$

$$\Rightarrow \quad y\left(\frac{x}{y}+e^{x/y}\right)=\pm C$$

$$\Rightarrow \quad x+ye^{x/y}=k,$$ where $k=\pm C$

which is the required solution.

Example 6. *Solve* $(x\sqrt{x^2+y^2}-y^2)dx+xydy=0$.

(AMIETE–2000)

Solution. The given differential equation can be expressed as

$$\frac{dy}{dx}=\frac{y^2-x\sqrt{x^2+y^2}}{xy} \qquad \ldots(1)$$

Clearly (1) is homogeneous.

Putting $y = vx$ and $\frac{dy}{dx}=v+x\frac{dv}{dx}$ in (1), we get

$$v+x\frac{dv}{dx}=\frac{v^2x^2-x\sqrt{x^2+v^2x^2}}{x^2v}=\frac{v^2-\sqrt{1+v^2}}{v}$$

$$\Rightarrow x\frac{dv}{dx}=\frac{v^2-\sqrt{1+v^2}}{v}-v=-\frac{\sqrt{1+v^2}}{v}.$$

Separating the variables, we get

$$\frac{v\,dv}{\sqrt{1+v^2}}+\frac{dx}{x}=0.$$

Integrating both sides, we get

$$\int\frac{v\,dv}{\sqrt{1+v^2}}+\int\frac{dx}{x}=C$$

$$\Rightarrow \sqrt{1+v^2}+\log|x| = C$$

$$\Rightarrow \sqrt{1+\frac{y^2}{x^2}}+\log|x|=C$$

$$\Rightarrow \sqrt{x^2+y^2}+x\log|x|=Cx$$

which is the required solution.

Example 7. *Solve* $(x^3-3xy^2)\,dx=(y^3-3x^2y)dy$.

Solution. The given differential equation can be expressed as

$$\frac{dy}{dx}=\frac{x^3-3xy^2}{y^3-3x^2y} \qquad \ldots(1)$$

Clearly (1) is homogeneous.

Putting $y = vx$ and $\frac{dy}{dx}=v+x\frac{dv}{dx}$ in (1), we get

$$v+x\frac{dv}{dx}=\frac{x^3-3x^3v^2}{x^3v^3-3x^3v}=\frac{1-3v^2}{v^3-3v}$$

$$\Rightarrow x\frac{dv}{dx}=\frac{1-3v^2}{v^3-3v}-v=\frac{1-v^4}{v^3-3v}.$$

Separating the variables, we get

$$\frac{v^3-3v}{1-v^4}dv=\frac{dx}{x}.$$

Integrating both sides

$$\int\frac{v^3-3v}{1-v^4}dv=\int\frac{dx}{x}+\log C$$

$$\Rightarrow\int\left[\frac{1}{2(v+1)}+\frac{1}{2(v-1)}-\frac{2v}{v^2+1}\right]dv=\int\frac{dx}{x}+\log C$$

(Using partial fraction)

$$\Rightarrow \frac{1}{2}\int\frac{dv}{v+1}+\frac{1}{2}\int\frac{dv}{v-1}-\int\frac{2v\,dv}{v^2+1}=\int\frac{dx}{x}+\log C$$

$$\Rightarrow \frac{1}{2}\log|v+1|+\frac{1}{2}\log|v-1|-\log\left|v^2+1\right|=\log|x|+\log C$$

$$\Rightarrow \log\left|\frac{\sqrt{v+1}\,\sqrt{v-1}}{(v^2+1)}\right|=\log|x|+\log C$$

$$\Rightarrow \log\left|\frac{\sqrt{v^2-1}}{x(v^2+1)}\right|=\log C \Rightarrow \left|\frac{\sqrt{v^2-1}}{x(v^2+1)}\right|=C$$

$$\Rightarrow \frac{\sqrt{v^2-1}}{x(v^2+1)}=\pm C \Rightarrow \frac{\sqrt{y^2-x^2}}{(y^2+x^2)}=\pm C$$

$\therefore y^2-x^2=C^2(y^2+x^2)^2$, which is the required solution.

Example 8. *Solve* $x\frac{dy}{dx}=y-x\tan\frac{y}{x}$.

Solution. The given differential equation can be expressed as

$$\frac{dy}{dx}=\frac{y-x\tan\frac{y}{x}}{x}=\frac{y}{x}-\tan\frac{y}{x} \qquad \ldots(1)$$

Clearly (1) is homogeneous, as it is of the form $x^nF\left(\frac{y}{x}\right)$.

Putting $y = vx$ and $\frac{dy}{dx}=v+x\frac{dv}{dx}$ in (1), we get

$$v+x\frac{dv}{dx}=v-\tan v \Rightarrow x\frac{dv}{dx}=-\tan v.$$

Separating the variables, we get

$$\frac{dv}{\tan v}+\frac{dx}{x}=0 \Rightarrow \frac{\cos v}{\sin v}dv+\frac{dx}{x}=0.$$

Integrating both sides, we get

$$\int\frac{\cos v}{\sin v}dv+\int\frac{dx}{x}=\log C$$

$$\Rightarrow \log|\sin v|+\log|x|=\log C$$

$$\Rightarrow \log|x\sin v|=\log C$$

$$\Rightarrow |x\sin v|=C \Rightarrow x\sin v=\pm C$$

$\Rightarrow x\sin\left(\frac{y}{x}\right)=k$, where $k=\pm C$, which is the required solution.

Example 9. *Solve* $\left(x\sin\frac{y}{x}\right)dy=\left(y\sin\frac{y}{x}-x\right)dx$.

Solution. The given differential equation can be expressed as

$$\frac{dy}{dx}=\frac{y\sin\frac{y}{x}-x}{x\sin\frac{y}{x}}=\frac{\frac{y}{x}\sin\frac{y}{x}-1}{\sin\frac{y}{x}} \qquad \ldots(1)$$

Clearly (1) is homogeneous, as it is of the form $x^nF\left(\frac{y}{x}\right)$.

Putting $y = vx$ and $\frac{dy}{dx}=v+x\frac{dv}{dx}$ in (1), we get

$$v+x\frac{dv}{dx}=\frac{v\sin v-1}{\sin v}.$$

$$\Rightarrow x\frac{dv}{dx}=\frac{v\sin v-1}{\sin v}-v=-\frac{1}{\sin v}$$

Separating the variables, we get

$$\sin v\,dv + \frac{dx}{x} = 0.$$

Integrating both sides, we get

$$\int \sin v\,dv + \int \frac{dx}{x} = C$$

$$-\cos v + \log|x| = C \Rightarrow \cos v = \log|x| - C$$

$$\cos v = \log|x| + C_1, \text{ where } C_1 = -C$$

$\cos\left(\frac{y}{x}\right) = \log|x| + C_1$, which is the required solution.

Example 10. *Solve* $\left(x\tan\frac{y}{x} - y\sec^2\frac{y}{x}\right)dx - x\sec^2\frac{y}{x}\,dy = 0.$ (VTU–2006)

Solution. The given differential equation is,

$$\frac{dy}{dx} = \left(\frac{y}{x}\sec^2\frac{y}{x} - \tan\frac{y}{x}\right)\cos^2\frac{y}{x} \quad \text{...(i)}$$

which is a homogeneous equation.

Putting $y = vx$ in equation (i), we get

$$v + x\frac{dv}{dx} = \left(v\sec^2 v - \tan v\right)\cos^2 v$$

or $$x\frac{dv}{dx} = v - \tan v\cos^2 v - v$$

Separating the variables, we get

$$\frac{\sec^2 v}{\tan v}dv = -\frac{dx}{x}$$

Integrating both sides, we get

$$\log\tan v = -\log x + \log c$$

or $x\tan v = c$ or $x\tan\frac{y}{x} = c.$

EXERCISE 28.5

Solve the following differential equations :

1. $\frac{dy}{dx} = \frac{y^2 - x^2}{2xy}$
2. $\frac{dy}{dx} = \frac{x^2 + y^2}{2xy}$
3. $x + y\frac{dy}{dx} = 2y$
4. $x\frac{dy}{dx} = x + y$
5. $\frac{dy}{dx} = \frac{x + y}{x - y}$
6. $(x - y)\frac{dy}{dx} = x + 3y$
7. $\frac{dy}{dx} + \frac{x - 2y}{2x - y} = 0$
8. $\frac{dy}{dx} = \frac{2xy}{x^2 - y^2}$
9. $2xy\frac{dy}{dx} = x^2 + 3y^2$
10. $x^2\frac{dy}{dx} = 2xy + y^2$
11. $y^2 + x^2\frac{dy}{dx} = xy\frac{dy}{dx}$
12. $x^2\frac{dy}{dx} = x^2 + xy + y^2$
13. $y^2dx + (x^2 - xy + y^2)dy = 0$
14. $x^2\left(\frac{dy}{dx}\right) = (x^2 - 2y^2 + xy)$
15. $(x^2 + xy)dy = (x^2 + y^2)dx$
16. $(x^2 + y^2)dx + 3xydy = 0$
17. $x(x - y)dy + y^2dx = 0$
18. $x\frac{dy}{dx} + \frac{y^2}{x} = y$
19. $x\frac{dy}{dx} = y - \sqrt{x^2 + y^2}$
20. $x\frac{dy}{dx} - y = 2\sqrt{y^2 - x^2}$
21. $(x - \sqrt{xy})\,dy = ydx$
22. $y - x\frac{dy}{dx} = x + y\frac{dy}{dx}$
23. $(x^3 + 3xy^2)dx + (y^3 + 3x^2y)dy = 0$ (AMIETE–1997)
24. $x\frac{dy}{dx} = y(\log y - \log x + 1)$
25. $\frac{dy}{dx} = \frac{y}{x} + \sin\frac{y}{x}$ (VTU–2000)
26. $x\frac{dy}{dx} = y - x\cos^2\frac{y}{x}$
27. $\left(x\cos\frac{y}{x}\right)(ydx + xdy) = \left(y\sin\frac{y}{x}\right)(xdy - ydx)$
28. $ydx + \left(x\log\frac{y}{x}\right)dy - 2xdy = 0$
29. $2x^2\frac{dy}{dx} - 2xy + y^2 = 0,\ y(e) = e$
30. $(y + \sqrt{x^2 + y^2})dx - xdy,\ y(1) = 0$ (UPTU–2008)
31. $(x^2y - 2xy^2)dx - (x^3 - 3x^2y)dy = 0$ (Bhopal–2008)
32. $x^2ydx - (x^3 + y^3)dy = 0$ (VTU–2010)
33. $ydx - xdy = \sqrt{x^2 + y^2}dx$ (Raipur–2005)
34. $(3xy - 2ay^2)dx + (x^2 - 2axy)dy = 0$ (SVTU–2009)
35. $\frac{dy}{dx} = \frac{y}{x} + \sin\frac{y}{x}$ (VTU–2005)
36. $ye^{x/y}dx = (xe^{x/y} + y^2)dy$ (VTU–2006)

Hint to Selected Problems

24. $x\frac{dy}{dx} = y(\log y - \log x + 1) \Rightarrow \frac{dy}{dx} = \frac{y}{x}\left(\log\frac{y}{x} + 1\right)$

Putting $y = vx$ and $\frac{dy}{dx} = v + x\frac{dv}{dx}$ in (1), we get

$$v + x\frac{dv}{dx} = v(\log v + 1) \Rightarrow x\frac{dv}{dx} = v\log v$$

$$\Rightarrow \frac{dv}{v\log v} = \frac{dx}{x} \quad \Rightarrow \int\frac{dv}{v\log v} = \int\frac{dx}{x} + \log C$$

28. $ydx + \left(x\log\frac{y}{x}\right)dy - 2xdy = 0$

$$\Rightarrow \frac{dy}{dx} = \frac{-y}{x\log\frac{y}{x} - 2x} = \frac{-y/x}{\log(y/x) - 2}$$

Putting $y = vx$ and $\frac{dy}{dx} = v + x\frac{dv}{dx}$ in (1).

ANSWERS

1. $(x^2+y^2)=Cx$
2. $x=C(x^2-y^2)$
3. $\log(y-x)=C+\dfrac{x}{y-x}$
4. $y=x\log|x|+Cx$
5. $\tan^{-1}\dfrac{y}{x}=\dfrac{1}{2}\log(x^2+y^2)+C$
6. $\log|x+y|+\dfrac{2x}{x+y}=C$
7. $y-x=C(x+y)^3$
8. $y=C(x^2+y^2)$
9. $x^2+y^2=Cx^3$
10. $y=Cx(x+y)$
11. $y=x[C+\log|y|]$
12. $\tan^{-1}\dfrac{y}{x}=\log|x|+C$
13. $|y|=Ce^{\tan^{-1}(y/x)}$
14. $\dfrac{1}{2\sqrt{2}}\log\left|\dfrac{x+\sqrt{2}y}{x-\sqrt{2}y}\right|-\log|x|=C$
15. $C(x-y)^2=xe^{y/x}$
16. $x^2(x^2+4y^2)^3=C$
17. $y=Ce^{y/x}$
18. $\log|x|-\dfrac{x}{y}=C$
19. $y+\sqrt{x^2+y^2}=C$
20. $y+\sqrt{y^2-x^2}=Cx^3$
21. $2\sqrt{\dfrac{x}{y}}+\log|y|=C$
22. $\tan^{-1}\dfrac{y}{x}+\dfrac{1}{2}\log(x^2+y^2)=C$
23. $y^4+6x^2y^2+x^4=C$
24. $y=xe^{Cx}$
25. $\tan\left(\dfrac{y}{2x}\right)=Cx$
26. $\tan\dfrac{y}{x}=C-\log|x|$
27. $xy\cos\left(\dfrac{y}{x}\right)=C$
28. $y=C\left[\log\left(\dfrac{y}{x}-1\right)\right]$
29. $y=\dfrac{2x}{1+\log|x|}$
30. $y+\sqrt{x^2+y^2}=Cx^2$
31. $Cy^3=x^2e^{-x/y}$
32. $\left(\dfrac{x}{y}\right)^3=3\log cy$
33. $y+\sqrt{(x^2+y^2)}=c$
34. $x(c+y)=ay^2$
35. $y=2x\tan^{-1}(cx)$
36. $e^{x/y}=y+c$

28.8 EQUATION REDUCIBLE TO THE HOMOGENEOUS FORM

The differential equation of the type

$$\frac{dy}{dx}=\frac{a_1x+b_1y+c_1}{a_2x+b_2y+c_2} \qquad \ldots(1)$$

can be reduced to the homogeneous form as follows. Put $X=x-h$ and $Y=y-k$.

Therefore, $\dfrac{dY}{dX}=\dfrac{dy}{dx}$

Then, (1) becomes $\dfrac{dY}{dX}=\dfrac{a_1X+b_1Y+(a_1h+b_1k+c_1)}{a_2X+b_2Y+(a_2h+b_2k+c_2)}$

where h and k can be chosen such that $a_1h+b_1k+c_1=0$ and $a_2h+b_2k+c_2=0$.

Then, we get $\dfrac{dY}{dX}=\dfrac{a_1X+b_1Y}{a_2X+b_2Y}$, which is homogeneous differential equation and can be solved by the method discussed earlier.

WORKING PROCEDURE

(A) When $\dfrac{a_1}{a_2}\neq\dfrac{b_1}{b_2}$:

Step 1. *Put $x=X+h$, $y=Y+k$, in the given differential equation.*

Step 2. *Equate the constant terms in the numerator and denominator to zero, and find the value of h and k.*

Step 3. *Solve the resulting homogeneous equation in X and Y.*

Step 4. *Replace X by $x-h$ and Y by $y-k$.*

Step 5. *Put the value of h and k.*

(B) When $\dfrac{a_1}{a_2}\neq\dfrac{b_1}{b_2}$:

Step 1. *Put $a_1x+b_1y=v$.*

Step 2. *Solve the resulting equation.*

Solved Examples

Example 1. *Solve $(2x+y+3)dx=(2y+x+1)dy$.*

Solution. The given equation can be written as

$$\frac{dy}{dx}=\frac{2x+y+3}{2y+x+1}$$

This is the equation reducible to homogeneous form.

To solve it, put $x=X+h, y=Y+k$

We get $\frac{dY}{dX}=\frac{2X+Y+(2h+k+3)}{2Y+X+(h+2k+1)}$

Now putting

$2h+k+3=0$ and $h+2k+1=0$

which gives $h=-5/3, k=1/3$.

We get $\frac{dY}{dX}=\frac{2X+Y}{2Y+X}$ which is the reduced homogeneous equation.

Put $Y=vX \Rightarrow \frac{dY}{dX}=v+X\frac{dv}{dX}$

and solve by the method discussed earlier, we get

$$(X+Y)(X-Y)^3=c$$

Now putting the value of X, Y, h and k, we get

$$\left(x+y+\frac{4}{3}\right)(x-y+2)^3=0.$$

Example 2. *Solve* $(x+y)(dx-dy)=dx+dy$.

Solution. The given equation can be written as

$$\frac{dy}{dx}=\frac{x+y-1}{x+y+1}$$

Put $x+y=v \Rightarrow 1+\frac{dy}{dx}=\frac{dv}{dx}$

$\therefore \frac{dv}{dx}-1=\frac{v-1}{v+1} \Rightarrow \frac{dv}{dx}=\frac{2v}{v+1}$

Now, separating the variables, we get

$$\left(1+\frac{1}{v}\right)dv=2dx.$$

On integrating, we get

$v+\log v=2x+c$ or $x+y+\log(x+y)=2x+c$

$\Rightarrow \log(x+y)=x-y+c$

EXERCISE 28.6

Solve the following ordinary differential equations :

1. $\frac{dy}{dx}=\frac{y-x+1}{y+x-5}$
2. $(2x+y-3)dy=(x+2y-3)dx$ (Madras–2000, VTU–2000, 2009S, Andhra–1998)
3. $\frac{dy}{dx}=\frac{2x+2y-2}{3x+y-5}$
4. $(x+2y-2)dx+(2x-y+3)=0$
5. $\frac{dy}{dx}=\frac{x+y+3}{2x+2y+1}$
6. $\frac{dy}{dx}=\frac{x-y+3}{2x-2y+5}$
7. $(6x+2y-10)\frac{dy}{dx}-2x-9y+20=0$
8. $(x-y)dy=(x+y+1)dx$
9. $(x+y-1)dy=(x+y)dx$
10. $(2x-2y+5)dy-(x-y+3)dx=0$
11. $\frac{dy}{dx}=\frac{x+y+7}{2x+2y+3}$
12. $\frac{dy}{dx}=\frac{2y+x+1}{2x+4y+1}$
13. $\frac{dy}{dx}=\left(\frac{2x-y}{4x-2y+1}\right)^2$
14. $(7y-3)dx+(2x+1)dy=0$
15. $(2x+3y-1)dx+(4x+6y+2)dy=0$
16. $\frac{dy}{dx}=\frac{y+x-2}{y-x-4}$ (Raipur–2005)
17. $(3y+2x+4)dx-(4x+6y+5)dy=0$ (Madras–2000S)

Hint to Selected Problems

1. Put $x=X+h, y=Y+k$ in the given equation, we get $\frac{dY}{dX}=\frac{Y-X}{Y+X}$, where $h=3, k=2$.
4. Proceeding as usual, we get $\frac{dY}{dX}=-\frac{X+2Y}{2X-Y}, h=-\frac{4}{5}, k=\frac{7}{5}$
12. Put $x+2y=v$.
15. Put $2x+3y=v$.

ANSWERS

1. $\tan^{-1}\left(\frac{\beta}{\alpha}\right)+\frac{1}{2}\log(\alpha^2+\beta^2)=c$, where $\beta=y-2, \alpha=x-3$
2. $(x+y-2)=c^2(x-y)^3$
3. $(y-x+3)^4=c(2x+y-3)$
4. $x^2+4xy-y^2-4x+6y=c$
5. $x+c=\frac{2}{3}(x+y)-\frac{5}{9}\log(3x+3y+4)$
6. $x-2y+\log(x-y+2)=c$
7. $(y-2x)^2=c(x+2y-5)$
8. $2\tan^{-1}\{(2y+1)/(2x+1)\}=\log\{c^2(x^2+y^2+x+y+\frac{1}{2})\}$
9. $2(y-x)-\log(2x+2y-1)=c$
10. $x-2y+\log(x-y+2)=c$
11. $\frac{2}{3}(x+y)-\left(\frac{11}{9}\right)\log(3x+3y+10)=x+c$
12. $4(2y-x)-\log(4x+8y+3)=c$
13. $7x-28y+2\log(28x^2-28xy+7y^2-16x+8y+2)+\frac{9}{2\sqrt{2}}\log\frac{14x-7y-4-\sqrt{2}}{14x-7y-4+\sqrt{2}}$
14. $7\log(2x+1)+2\log(7y-3)=c$
15. $x+2y-4\log(2x+3y+7)=c$
16. $x^2+2xy-y^2-4x+8y-14=c$
17. $21x-42y+9\log(14x+21y+22)=c$

28.9 LINEAR DIFFERENTIAL EQUATION

A differential equation in which the dependent variable and all its derivatives are of first degree only and not multiplied together, is called a linear differential equation.

For example : $x\frac{dy}{dx} + y = x$ is a linear differential equation.

28.9.1 GENERAL FORMS OF LINEAR DIFFERENTIAL EQUATION

(i) The equation of the form $\frac{dy}{dx} + Py = Q$ where P and Q are either constants or functions of x only, is most general form of linear differential equation.

WORKING PROCEDURE

To find the solution of such type of equation use the following steps :

Step 1. *Find I.F.* $= e^{\int P\,dx}$.

Step 2. *The solution is* $y\,.\,(I.F.) = \int Q.(I.F.)dx + C$.

Solution of the form $\frac{dx}{dy} + Px = Q$: In order to solve such equation, we proceed as follows :

(ii) The equation of the form $\frac{dx}{dy} + Px = Q$ where P and Q are either constants or functions of y only, is most general form of linear differential equation.

WORKING PROCEDURE

To find the solution of such type of equation use the following steps :

Step 1. *Find I.F.* $= e^{\int P\,dy}$.

Step 2. *The solution is* $x.(I.F.) = \int Q.(I.F.)dy + C$.

Solved Examples

Example 1. *Solve* $(1+x^2)\frac{dy}{dx} + 2xy = \cos x$.

Solution. The given differential equation can be written as

$$\frac{dy}{dx} + \frac{2x}{1+x^2}y = \frac{\cos x}{1+x^2} \qquad ...(1)$$

Comparing (1) with $\frac{dy}{dx} + Py = Q$, we get

$$P = \frac{2x}{1+x^2}, \quad Q = \frac{\cos x}{1+x^2}.$$

$$\text{I.F.} = e^{\int P\,dx} = e^{\int \frac{2x}{1+x^2}dx} = e^{\log(1+x^2)} = 1+x^2.$$

Thus, the solution is

$$y(1+x^2) = \int Q(1+x^2)dx + C$$

$$\Rightarrow \quad y(1+x^2) = \int \frac{\cos x}{(1+x^2)}(1+x^2)dx + C$$

$$\Rightarrow \quad y(1+x^2) = \int \cos x\,dx + C$$

$$\Rightarrow \quad y(1+x^2) = \sin x + C.$$

Example 2. *Solve* $(1+x^2)\frac{dy}{dx} + y = \tan^{-1}x$. (UKTU–2012)

Solution. The given differential equation can be written as

$$\frac{dy}{dx} + \frac{1}{1+x^2}y = \frac{\tan^{-1}x}{1+x^2} \qquad ...(1)$$

Comparing (1) with equation $\frac{dy}{dx} + Py = Q$, we get

$$P = \frac{1}{1+x^2} \text{ and } Q = \frac{\tan^{-1}x}{1+x^2}.$$

So, $\text{I.F.} = e^{\int P\,dx} = e^{\int \frac{1}{1+x^2}dx} = e^{\tan^{-1}x}$.

Thus, the solution of (1) is

$$y.(I.F.) = \int Q.(I.F.)\,dx + C$$

$$\Rightarrow \quad y.(e^{\tan^{-1}x}) = \int \frac{\tan^{-1}x}{1+x^2}e^{\tan^{-1}x}dx + C$$

$$= \int te^t dt + C, \text{ where } t = \tan^{-1}x$$

$$= t\int e^t dt - \int\left\{\frac{d}{dt}(t).\int e^t dt\right\}dt + C$$

(Integrating by parts)

$$= te^t - \int e^t dt + C = te^t - e^t + C$$

$$= (\tan^{-1}x - 1)e^{\tan^{-1}x} + C$$

$\therefore \quad y = (\tan^{-1}x - 1) + Ce^{-\tan^{-1}x}$, which is the required solution.

Example 3. *Solve* $(x^2-1)\frac{dy}{dx}+2xy=\frac{2}{x^2-1}$.

Solution. The given differential equation can be written as

$$\frac{dy}{dx}+\frac{2x}{x^2-1}y=\frac{2}{(x^2-1)^2} \quad ...(1)$$

Comparing (1) with $\frac{dy}{dx}+Py=Q$, we get

$$P=\frac{2x}{x^2-1} \text{ and } Q=\frac{2}{(x^2-1)^2}.$$

$$\text{I.F.}=e^{\int P\,dx}=e^{\int\frac{2x}{x^2-1}dx}=e^{\log(x^2-1)}=x^2-1.$$

Thus, the equation of (1) is

$$y(I.F.)=\int Q.(I.F.)\,dx+C$$

$$\Rightarrow y.(x^2-1)=\int\frac{2}{(x^2-1)^2}(x^2-1)dx+C$$

$$=2\int\frac{1}{x^2-1}dx+C=2.\left[\frac{1}{2}\log\left|\frac{x-1}{x+1}\right|\right]+C$$

$$=\log\left|\frac{x-1}{x+1}\right|+C$$

$$\Rightarrow y(x^2-1)=\log\left|\frac{x-1}{x+1}\right|+C$$

which is the required solution.

Example 4. *Solve* $(x\log x)\frac{dy}{dx}+y=\frac{2}{x}\log x$.

Solution. The given differential equation can be written as

$$\frac{dy}{dx}+\frac{1}{x\log x}y=\frac{2}{x^2} \quad ...(1)$$

Comparing (1) with the equation $\frac{dy}{dx}+Py=Q$, we get

$$P=\frac{1}{x\log x} \text{ and } Q=\frac{2}{x^2}.$$

So, I.F.

$$=e^{\int P\,dx}=e^{\int\frac{1}{x\log x}dx}=e^{\log(\log x)}=\log x.$$

Thus, the solution of (1) is

$$y(I.F.)=\int Q.(I.F.)\,dx+C$$

$$\Rightarrow y.(\log x)=\int\frac{2}{x^2}\log x\,dx+C$$

$$\Rightarrow y\log x$$

$$=2\left\{\log x\int\frac{1}{x^2}dx-\int\left(\frac{d}{dx}(\log x)\int\frac{1}{x^2}dx\right)dx\right\}+C$$

(Integrating by parts)

$$=2\left\{-\frac{1}{x}\log x+\int\frac{1}{x^2}dx\right\}+C$$

$$=2\left[-\frac{1}{x}\log x-\frac{1}{x}\right]+C$$

$$\therefore \quad y\log x+\frac{2}{x}(\log x+1)=C,$$ which is the required solution.

Example 5. *Solve* $x\frac{dy}{dx}+2y=x\cos x$.

Solution. The given differential equation can be written as

$$\frac{dy}{dx}+\frac{2}{x}y=\cos x \quad ...(1)$$

Comparing (1) with the equation $\frac{dy}{dx}+Py=Q$, we get

$$P=\frac{2}{x} \text{ and } Q=\cos x.$$

$$\text{So, I.F.}=e^{\int P\,dx}=e^{\int\frac{2}{x}dx}=e^{\log x^2}=x^2.$$

Thus, the solution of (1) is

$$y(I.F.)=\int Q.(I.F.)\,dx+C$$

$$\Rightarrow yx^2=\int(\cos x).x^2dx+C$$

$$=x^2\int\cos x\,dx-\int\left\{\frac{d}{dx}(x^2)\int\cos x\,dx\right\}dx+C$$

(Integrating by parts)

$$=x^2\sin x-2\int x\sin x\,dx+C$$

$$=x^2\sin x-2\left[x\int\sin x\,dx-\int\left\{\frac{d}{dx}(x)\int\sin x\,dx\right\}dx\right]+C$$

$$=x^2\sin x-2[-x\cos x+\int\cos x\,dx]+C$$

$$=x^2\sin x-2[-x\cos x+\sin x]+C$$

$$\therefore yx^2=x^2\sin x+2x\cos x-2\sin x+C$$

or $yx^2=(x^2-2)\sin x+2x\cos x+C$, which is the required solution.

Example 6. *Solve* $\frac{dy}{dx}+y\tan x=2x+x^2\tan x$.

Solution. The given differential equation is

$$\frac{dy}{dx}+y\tan x=2x+x^2\tan x \quad ...(1)$$

Comparing (1) with the equation $\frac{dy}{dx}+Py=Q$, we get

$$P=\tan x \text{ and } Q=2x+x^2\tan x.$$

$$\text{So, I.F.}=e^{\int P\,dx}=e^{\int\tan x\,dx}=e^{\log\sec x}=\sec x$$

Thus, the solution of (1) is

$$y(I.F.)=\int Q.(I.F.)\,dx+C$$

$$\Rightarrow y\sec x=\int(2x+x^2\tan x)\sec x\,dx+C$$

$$=2\int x\sec x\,dx+x^2\tan x\sec x\,dx+C$$

(Integrating by parts)

$$=2\sec x\int x\,dx-2\int\left\{\frac{d}{dx}(\sec x)\int x\,dx\right\}$$

$$+\int x^2\tan x\sec x\,dx+C.$$

$$=x^2\sec x-\int x^2\sec x\tan x\,dx$$

$$+x^2\tan x\sec x\,dx+C.$$

$\therefore y\sec x=x^2\sec x+C$, which is the required solution.

Example 7. *Solve* $\frac{dy}{dx}+y\cot x = 2x + x^2\cot x$, *given that* $y(0)=0$.

Solution. The given differential equation is

$$\frac{dy}{dx}+y\cot x = 2x + x^2\cot x \quad ...(1)$$

Comparing (1) with the equation $\frac{dy}{dx}+Py=Q$, we get

$P=\cot x$ and $Q=2x+x^2\cot x$.

So, $\text{I.F.}=e^{\int P dx}=e^{\int \cot x dx}$

$=e^{\log\sin x}=\sin x$.

Thus, the solution is

$$y(I.F.)=\int Q(I.F.)dx + C$$

$$\Rightarrow \quad y\sin x = \int (2x+x^2\cot x)\sin x\, dx + C$$

$$= 2\int x\sin x\, dx + \int x^2\cos x\, dx + C$$

$$= 2\sin x\int x\, dx - 2\int\left\{\frac{d}{dx}(\sin x)\int x\, dx\right\}dx + \int x^2\cos x dx + C$$

$$= x^2\sin x - \int x^2\cos x\, dx + \int x^2\cos x\, dx + C$$

$\therefore\ y\sin x = x^2\sin x + C$.

Putting $x=0$ and $y=0$, we get $C=0$.

$\therefore\ y\sin x = x^2\sin x$ or $y=x^2$, which is the required solution.

Example 8. *Solve* $(1+y^2)dx=(\tan^{-1}y-x)dy$.

(Assam–1998, Karnataka–1995, Bhopal–1991, 2008, VTU–2008, UPTU–2005)

Solution. The given differential equation can be written as

$$\frac{dx}{dy}+\frac{1}{1+y^2}x=\frac{\tan^{-1}y}{1+y^2} \quad ...(1)$$

Comparing (1) with the equation $\frac{dx}{dy}+Px=Q$, we get

$$P=\frac{1}{1+y^2} \text{ and } Q=\frac{\tan^{-1}y}{1+y^2}.$$

So, $\text{I.F.}=e^{\int P dy}=e^{\int\frac{1}{1+y^2}dy}=e^{\tan^{-1}y}$.

Thus, the solution of (1) is

$$x.(I.F.)=\int Q.(I.F.)dy + C$$

$$\Rightarrow\ xe^{\tan^{-1}y}=\int\frac{\tan^{-1}y}{1+y^2}e^{\tan^{-1}y}dy+C$$

$$=\int te^t dt + C, \text{ where } t=\tan^{-1}y$$

$$=t\int e^t dt - \int\left\{\frac{d}{dt}(t).\int e^t dt\right\}dt + C$$

$$=te^t-e^t+C=(t-1)e^t+C$$

$$=(\tan^{-1}y-1)e^{\tan^{-1}y}+C$$

$\therefore\ x=(\tan^{-1}y-1)+Ce^{-\tan^{-1}y}$, which is the required solution.

Example 9. *Solve* $(x+y+1)\frac{dy}{dx}=1$.

Solution. The given differential equation can be written as

$$\frac{dx}{dy}-x=1+y. \quad ...(1)$$

Comparing (1) with the equation $\frac{dx}{dy}+Px=Q$,

we get $P=-1$ and $Q=1+y$.

So, $\text{I.F.}=e^{\int P dy}=e^{\int(-1)dy}=e^{-y}$.

Thus, the solution of (1) is

$x.(I.F.)=\int Q.(I.F.)dy + C$, where C is the constant of integration.

$$\Rightarrow\ x(e^{-y})=\int(1+y)e^{-y}dy+C$$

$$=\int e^{-y}dy+\int ye^{-y}dy+C$$

$$=-e^{-y}+y\int e^{-y}dy-\int\left\{\frac{d}{dy}(y).\int e^{-y}dy\right\}dy+C$$

$$=-e^{-y}-ye^{-y}-e^{-y}+C$$

or. $\quad x(e^{-y})=-ye^{-y}-2e^{-y}+C$

$\therefore\ x+y+2=Ce^y$, which is the required solution.

Example 10. *Solve* $y\log y\frac{dx}{dy}+x-\log y=0$.

(UPTU–2000, 04, UPTU(B.Pharm.)–2009)

Solution.

$$y\log y\frac{dx}{dy}+x-\log y=0$$

$$y\log y\frac{dx}{dy}+x=\log y \Rightarrow \frac{dx}{dy}+\frac{x}{y\log y}=\frac{1}{y}$$

$$\text{I.F.}=e^{\int\frac{dy}{y\log y}}=e^{\log(\log y)}=\log y$$

The solution is

$$x.\log y=\int\frac{1}{y}(\log y)\,dy+C$$

$$\Rightarrow\quad x.\log y=\frac{1}{2}(\log y)^2+C.$$

Example 11. *Solve* $\frac{dy}{dx}+\frac{3y}{x}=\frac{1}{x^4}$ (UPTU(CO)–2009)

Solution. Comparing the given differential equation with $\frac{dy}{dx}+Py=Q$, we get $P=\frac{3}{x}$ and $Q=\frac{1}{x^4}$.

So, $\text{I.F.}=e^{\int P dx}=e^{3\int\frac{1}{x}dx}=e^{3\log x}=x^3$.

Thus, the solution is

$$y(I.F.)=\int Q(I.F.)dx+C$$

$$\Rightarrow\quad yx^3=\int\frac{1}{x^4}(x^3)dx+c$$

$$\Rightarrow\quad yx^3=\log x+c$$

where c is a constant of integration.

EXERCISE 28.7

Solve the following differential equations :

1. $\frac{dy}{dx} - y = e^{2x}$
2. $\frac{dy}{dx} + y = e^{x}$
3. $\frac{dy}{dx} - 4y = e^{2x}$
4. $\frac{dy}{dx} + y = e^{-2x}$
5. $\frac{dy}{dx} + 2y = e^{-x}$
6. $\frac{dy}{dx} + 2y = 6e^{x}$
7. $\frac{dy}{dx} - \frac{y}{x} = 2x^2$
8. $\frac{dy}{dx} + 2y = xe^{4x}$
9. $\frac{dy}{dx} + \frac{y}{x} = x^n$
10. $x\frac{dy}{dx} + 3y = x^2$
11. $\frac{dy}{dx} + 2y = 4x$
12. $x\frac{dy}{dx} - y = x+1$
13. $x\frac{dy}{dx} - y = \log x$
14. $(1+x^2)\frac{dy}{dx} + 2xy = \sqrt{x^2+4}$
15. $(1+x^2)\frac{dy}{dx} - 2xy = (x^2+2)(x^2+1)$
16. $\frac{dy}{dx} + \frac{y}{x} = e^{x}$
17. $(1+x^2)\frac{dy}{dx} + 2xy = \cos x$
18. $\frac{dy}{dx} + y = \cos x$
19. $\frac{dy}{dx} + 2y = \sin x$
20. $\frac{dy}{dx} - y\tan x = e^{x}\sec x$
21. $\frac{dy}{dx} + (\sec x)y = \tan x$
22. $(1+x^2)\frac{dy}{dx} + y = e^{\tan^{-1}x}$
23. $\frac{dy}{dx} + y\tan x = \sec x$
24. $\frac{dy}{dx} + y = \cos x - \sin x$
25. $x\frac{dy}{dx} + 2y = \sin x$
26. $\frac{dy}{dx} + y\cot x = x$
27. $\frac{dy}{dx} + y\cot x = \sin 2x$
28. $x\frac{dy}{dx} - y = 2x^2\sec x$
29. $\frac{dy}{dx} + 2y\cot x = 3x^2\operatorname{cosec}^2 x$
30. $(\sec x)\frac{dy}{dx} = y + \tan x$
31. $\frac{dy}{dx} + y\cos x = \sin x\cos x$
32. $\frac{dy}{dx} - y\tan x = -2\sin x$
33. $\frac{dy}{dx} - 2y = \cos 3x$
34. $(1+x^2)\frac{dy}{dx} + 2xy = 4x^2$, given that $y(0) = 0$
35. $\frac{dy}{dx} - 3y\cot x = \sin 2x$, given that $y = 2$ when $x = \frac{\pi}{2}$.
36. $(x+2y^2)\frac{dy}{dx} = y$
37. $(x-y^3)\frac{dy}{dx} + y = 0$
38. $(x-y^2)\frac{dy}{dx} + y = 0$
39. $y^2 + \left(x - \frac{1}{y}\right)\frac{dy}{dx} = 0$
40. $\frac{dx}{dy} + 2xy = e^{-y^2}$
41. $3x(1-x^2)y^2\frac{dy}{dx} + (2x^2-1)y^3 = ax^3$ (Rajasthan–2006)
42. $x\log x\frac{dy}{dx} + y = \log x^2$ (VTU–2011)
43. $(1-x^2)\frac{dy}{dx} - xy = 1$ (VTU–2010)

Hint to Selected Problems

14. $(1+x^2)\frac{dy}{dx} + 2xy = \sqrt{x^2+4} \Rightarrow \frac{dy}{dx} + \frac{2x}{1+x^2}y = \frac{\sqrt{x^2+4}}{1+x^2}$

So, I.F. $= e^{\int \frac{2x}{1+x^2}dx} = e^{\log(1+x^2)} = 1+x^2$

$\therefore\ y(1+x^2) = \int \frac{\sqrt{x^2+4}}{1+x^2}.(1+x^2)dx + C = \int \sqrt{x^2+4}\,dx + C$

30. $\sec x\frac{dy}{dx} = y + \sin x \Rightarrow \frac{dy}{dx} - y\cos x = \sin x\cos x$

So, I.F. $= e^{-\int \cos x\,dx} = e^{-\sin x}$.

$y(e^{-\sin x}) = \int \sin x\cos x\, e^{-\sin x}dx + C = \int te^{-t}dt + C$, where $t = \sin x$.

36. $(x+2y^2)\frac{dy}{dx} = y \Rightarrow \frac{dx}{dy} - \frac{x}{y} = 2y$

So, I.F. $= e^{-\int \frac{1}{y}dy} = e^{-\log y} = \frac{1}{y}$

$\therefore\ x\left\{\frac{1}{y}\right\} = \int 2y.\frac{1}{y}dy + C = 2y + C$

$\Rightarrow\ x = 2y^2 + Cy$

ANSWERS

1. $y = e^{2x} + Ce^{x}$
2. $y = \frac{1}{2}e^{x} + Ce^{-x}$
3. $y = -\frac{1}{2}e^{2x} + Ce^{4x}$
4. $y = -e^{-2x} + Ce^{-x}$
5. $y = e^{-x} + Ce^{-2x}$
6. $y = 2e^{x} + Ce^{-2x}$
7. $y = x^3 + Cx$
8. $y = \frac{1}{6}xe^{4x} - \frac{1}{36}e^{4x} + Ce^{-2x}$
9. $(n+2)xy = x^{n+2} + (n+2)C$
10. $5x^3y - x^5 = C$
11. $y = 2x - 1 + Ce^{-2x}$
12. $y = x\log x - 1 + Cx$
13. $y = Cx - (\log x + 1)$
14. $y(1+x^2) = \frac{x\sqrt{x^2+4}}{2} + 2\log|x + \sqrt{x^2+4}| + C$
15. $y = (1+x^2)(x + \tan^{-1}x + C)$
16. $y = e^{x} - \frac{1}{x}e^{x} + \frac{C}{x}$
17. $y(1+x^2) = \sin x + C$
18. $y = \frac{1}{2}(\cos x + \sin x) + Ce^{-x}$
19. $y = \frac{1}{5}(-4\cos x + 2\sin x) + Ce^{-2x}$
20. $y\cos x = e^{x} + C$
21. $y(\sec x + \tan x) = \sec x + \tan x - x + C$
22. $y = \frac{1}{2}e^{\tan^{-1}x} + Ce^{-\tan^{-1}x}$
23. $y\sec x = \tan x + C$
24. $y = \cos x + Ce^{-x}$
25. $x^2y = (2-x^2)\cos x + 2x\sin x + C$
26. $(y-1)\sin x + x\cos x = C$
27. $y\sin x = \frac{2}{3}\sin^3 x + C$

28. $y = Cx + 2x\log|\sec x + \tan x|$ **29.** $y\sin^2 x = x^3 + C$ **30.** $y = Ce^{\sin x} - (1+\sin x)$ **31.** $y = \sin x - 1 + Ce^{-\sin x}$

32. $2y\cos x = \cos 2x + C$ **33.** $y = \dfrac{3\sin 3x - 2\cos 3x}{13} + Ce^{2x}$ **34.** $3y(1+x^2) = 4x^3$ **35.** $y = 4\sin^3 x - 2\sin^2 x$

36. $x = 2y^2 + Cy$ **37.** $4xy = y^4 + C$ **38.** $3xy = y^3 + C$ **39.** $xy = 1 + y + Cye^{1/y}$ **40.** $xe^{y^2} = y + C$

41. $y^3 = ax + cx\sqrt{1-x^2}$ **42.** $y = \log x + c/\log x$ **43.** $y\sqrt{(1-x^2)} = \sin^{-1} x + c$

28.10 EQUATION REDUCIBLE TO LINEAR FORM (BERNOULLI'S EQUATION)

A differential equation of the form $\dfrac{dy}{dx} + Py = Qy^n$ is called an equation reducible to the linear form or Bernoulli's equation.

Solution of the equation. The given equation can be written as

$$y^{-n}\frac{dy}{dx} + Py^{-n+1} = Q \qquad \ldots(1)$$

Put $\quad y^{-n+1} = v \quad \Rightarrow \quad (-n+1)y^{-n}\dfrac{dy}{dx} = \dfrac{dv}{dx}$.

Then, (1) becomes

$$\frac{-1}{(n-1)}\frac{dv}{dx} + Pv = Q \quad \text{or} \quad \frac{dv}{dx} - (n-1)Pv = Q(1-n)$$

which is a linear equation of first order, and can be solved in a usual manner.

WORKING PROCEDURE

Step 1. *Write the given equation in the form* $\dfrac{dy}{dx} + Py = Qy^n$.

Step 2. *Divide by* y^n*, substitute* $y^{-n+1} = v$ *and get the equation of the form*

$$\frac{dv}{dx} - (n-1)Pv = Q(1-n)$$

Step 3. *Then, apply the method of solution of linear equations.*

Solved Examples

Example 1. *Solve* $x\dfrac{dy}{dx} + y = y^2\log x$.

Solution. The given equation can be written as

$$\frac{1}{y^2}\frac{dy}{dx} + \frac{1}{x}\cdot\frac{1}{y} = \frac{1}{x}\log x$$

Put $\dfrac{1}{y} = v \Rightarrow -\dfrac{1}{y^2}\dfrac{dy}{dx} = \dfrac{dv}{dx}$

Hence, the given equation becomes

$$-\frac{dv}{dx} - \frac{1}{x}v = -\frac{1}{x}\log x, \text{ which is linear in } v.$$

$$\text{I.F.} = e^{\int P\,dx} = e^{-\int \frac{1}{x}dx} = e^{-\log x} = \frac{1}{x}$$

The solution is

$$\frac{1}{x}\cdot v = \int -\frac{1}{x^2}\log x\,dx + c$$

$$= \frac{1}{x}\log x - \int \frac{1}{x}\cdot\frac{1}{x}dx = \frac{1}{x}\log x + \frac{1}{x} + c$$

or $\quad v = \log x + 1 + cx$ or $1 = cyx + y + y\log x$

or $\quad 1 = y(1 + \log x) + cxy$.

Example 2. *Solve* $\dfrac{dy}{dx} + \dfrac{y}{x}\log y = \dfrac{y}{x^2}(\log y)^2$.

(AMIETE 2001, S. PATEL 1997, UKTU–2011)

Solution. The given equation is Bernoulli's type because of presence of non-linear part on the right side, so dividing by $y(\log y)^2$, we get

$$\frac{1}{y(\log y)^2}\cdot\frac{dy}{dx} + \frac{1}{x}\cdot\frac{1}{\log y} = \frac{1}{x^2} \qquad \ldots(1)$$

Now, substituting $u = \dfrac{1}{\log y}$

or $\dfrac{du}{dx} = -\dfrac{1}{y(\log y)^2}\cdot\dfrac{dy}{dx}$ in (1), we get

$$-\frac{du}{dx} + \frac{1}{x}u = \frac{1}{x^2} \Rightarrow \frac{du}{dx} - \frac{1}{x}u = -\frac{1}{x^2}$$

which is a linear equation in u, therefore

$$\text{I.F.} = e^{-\int \frac{1}{x}dx} = \frac{1}{x} \qquad \ldots(2)$$

Multiplying equation (2) by I.F. and after integrating, we get

$\dfrac{u}{x} = \dfrac{1}{2x^2} + c$, where c is the constant of integration.

$\Rightarrow \quad \dfrac{1}{x\log y} = \dfrac{1}{2x^2} + c$, which is the required solution.

Example 3. *Solve* $x^2dy - y(x-y)dx = 0$.

(UPTU(B.Tech.)–2006)

Solution. The given equation can be written as

$$\frac{1}{y^2}\frac{dy}{dx}-\frac{1}{xy}=-\frac{1}{x^2}$$

Put $-\frac{1}{y}=u \Rightarrow \frac{1}{y^2}\frac{dy}{dx}=\frac{du}{dx}$

The given equation reduces to a linear differential equation in u.

$$\frac{du}{dx}-\frac{u}{x}=-\frac{1}{x^2}$$

$$\text{I.F.}=e^{-\int\frac{1}{x}dx}=e^{-\log x}=e^{\log 1/x}=\frac{1}{x}$$

Hence the solution is

$$u.\frac{1}{x}=\int-\frac{1}{x^2}.\frac{1}{x}dx+C=-\int x^{-3}dx+C$$

$$=-\frac{x^{-2}}{-2}+C$$

$\Rightarrow \quad \frac{1}{xy}=-\frac{1}{2x^2}-C$, which is the required solution.

Example 4. *Solve* $xy(1+xy^2)\frac{dy}{dx}=1$. (Nagpur–2006)

Solution. The given differential equation is

$$\frac{dx}{dy}-yx=y^3x^2$$

and dividing by x^2, we have

$$x^{-2}\frac{dx}{dy}-yx^{-1}=y^3 \qquad \ldots(1)$$

Putting $x^{-1}=z$ so that $-x^{-2}\frac{dx}{dy}=\frac{dz}{dy}$

Equation (1) becomes,

$\frac{dz}{dy}+yz=-y^3$ which is linear in z.

Now, I.F. $=e^{\int y dy}=e^{y^2/2}$

$\therefore$ Solution is $z(\text{I.F})=\int(-y^3)(\text{I.F.})dy+c$

or $\quad ze^{y^2/2}=-\int y^2.e^{y^2/2}.ydy+c$

Put $\frac{1}{2}y^2=t$ so that $ydy=dt$

$$=-2\int t.e^t dt+c \quad \text{[Integrate by parts]}$$

$$=-2\left[t.e^t-\int 1.e^t dt\right]+c$$

$$=-2[t.e^t-e^t]+c=(2-y^2)e^{y^2/2}+c$$

or $\quad z=(2-y^2)+ce^{-\frac{1}{2}y^2}$

or $\quad \frac{1}{x}=(2-y^2)+ce^{-\frac{1}{2}y^2}$.

REMARK

- General equation reducible to linear is

$$f'(y)\frac{dy}{dx}+Pf(y)=Q \qquad \ldots(A)$$

where P, Q are functions of x. To solve it, put $f(y)=z$.

Example 5. *Solve* $\frac{dy}{dx}+x\sin 2y=x^3\cos^2 y$.

(VTU–2011, Marathwada–2008, JNTU–2005)

Solution. Dividing by $\cos^2 y$,

$$\sec^2 y\frac{dy}{dx}+2x\frac{\sin y\cos y}{\cos^2 y}=x^3$$

or $\quad \sec^2 y\frac{dy}{dx}+2x\tan y=x^3$

which is of the form (A) above. $\ldots(1)$

$\therefore$ Put $\tan y=z$ so that $\sec^2 y\frac{dy}{dx}=\frac{dz}{dx}$

So equation (1), becomes $\frac{dz}{dx}+2xz=x^3$

This is linear equation in z

$\therefore$ I.F. $=e^{\int 2xdx}=e^{x^2}$

$\therefore$ The solution is

$$ze^{x^2}=\int e^{x^2}x^3dx+c$$

$$=\frac{1}{2}(x^2-1)e^{x^2}+c$$

Replace z by $\tan y$, we get

$$\tan y=\frac{1}{2}(x^2-1)+ce^{-x^2}$$

EXERCISE 28.8

Solve the following ordinary differential equations :

1. $\frac{dy}{dx}+\frac{1}{x}y=x^2y^6$ (Andhra–1998, Delhi–1997)
2. $x\left(\frac{dy}{dx}\right)+y=x^2y^4$
3. $\frac{dy}{dx}+\left(\frac{x}{1-x^2}\right)y=x\sqrt{y}$
4. $\frac{dy}{dx}+\frac{y}{x}=y^3$
5. $y(1+xy)\,dx-xdy=0$
6. $\frac{dy}{dx}+\frac{1}{x}=\frac{e^y}{x^2}$
7. $(x^3y^2+xy)dx=dy$ (BPTU–2005)
8. $\frac{dy}{dx}=x^3y^3-xy$
9. $\frac{dy}{dx}+\left\{\frac{y}{x-1}\right\}=xy^{1/3}$
10. $\frac{dy}{dx}=e^{x-y}(e^x-e^y)$
11. $\frac{dy}{dx}-y\tan x=-y^2\sec x$
12. $x\frac{dy}{dx}+y\log y=xy\,e^x$ (AMIE–2000)
13. $\cos x\,dy=(\sin x-y)\,y\,dx$

14. $\frac{dy}{dx}+\frac{1}{x}\tan y=\frac{1}{x^2}\tan y\sin y$

15. $\sin y\frac{dy}{dx}=\cos y(1-x\cos y)$

16. $\frac{dy}{dx}-\frac{\tan y}{(1+x)}=(1+x)\,e^x\sec y$

(ISM–2001, Marathwada–1993, Bhopal–2009)

17. $\frac{dy}{dx}+y\cos x=y^n\sin 2x$

18. $\frac{dy}{dx}-2y\tan x=y^2\tan^2 x$

19. $nx\frac{dy}{dx}+2y=xy^{n+1}$

20. $x\frac{dy}{dx}+y=y^3$

Hint to Selected Problems

1. Put $\frac{1}{y^6}=v$, we get $\frac{dv}{dx}-\frac{5}{x}v=-5x^2$

2. Put $v=-\frac{1}{y^3}$ **3.** Put $y^{1/2}=v$ **4.** Put $\frac{1}{y^2}=v$

5. Put $v=-\frac{1}{y}$ **6.** Put $v=e^{-y}$ **7.** Put $-\frac{1}{y}=v$

8. Put $\frac{1}{y^2}=v$ **9.** Put $y^{2/3}=v$ **10.** Put $e^y=v$

11. Put $v=-\frac{1}{y}$ **12.** Put $\log y=v$ **13.** Put $1/y=v$

14. Put $-\text{cosec}\, y=v$ **15.** Put $\sec y=v$ **16.** Put $\sin y=v$

17. Put $\frac{1}{y^{n-1}}=v$ **18.** Put $-\frac{1}{y}=v$ **19.** Put $\frac{1}{y^n}=v$

ANSWERS

1. $\frac{1}{x^5y^5}=\frac{5}{2}(1/x^2)+c$ **2.** $\frac{1}{y^3}=x^2(3-cx)$ **3.** $\sqrt{y}(1-x^2)^{1/4}=-\frac{1}{3}(1-x^2)^{3/4}+c$ **4.** $2xy^2+cx^2y^2=1$

5. $-\frac{x}{y}=\frac{1}{2}x^2+c$ **6.** $2x=(2cx^2+1)\,e^y$ **7.** $1=2y\left(1-\frac{x^2}{2}\right)-cye^{-x^2/2}$ **8.** $\frac{1}{y^2}=ce^{x^2}+x^2+1$

9. $y^{2/3}(x-1)^{2/3}=\frac{2}{3}x(x-1)^{5/3}-\frac{3}{20}(x-1)^{8/3}+c$ **10.** $e^y=e^x-1+ce^{-(e^x)}$ **11.** $\sec x=y(\tan x-c)$

12. $x\log y=e^x(x-1)+c$ **13.** $\sec x=y(\tan x-c)$ **14.** $2x=\sin y(1-2cx^2)$ **15.** $\sec y=x+1+ce^x$

16. $\sin y=(1+x)(e^x+c)$ **17.** $\frac{1}{y^{n-1}}=2\sin x-\left\{\frac{2}{1-n}\right\}+ce^{(n-1)\sin x}$ **18.** $\sec^2 x+\frac{1}{3}y\tan^3 x+cy=0$

19. $1/xy^n=cx+1$ **20.** $1/y^2=cx^2+1$

28.11 EXACT DIFFERENTIAL EQUATION

The differential equation which can be derived from its primitive by direct differentiation without any further transformation (such as elimination or reduction), is called an exact differential equation.

For example : The equation $ydx+xdy=0$ is exact, as it is derived from it is primitive $yx=c$.

WORKING PROCEDURE

Step 1. *Compare the given equation with* $Mdx+Ndy=0$ *and find out M and N.*

Step 2. *Show that* $\frac{\partial M}{\partial y}=\frac{\partial N}{\partial x}$, *which conclude the exactness of the given equation.*

Step 3. *Integrate the coefficient of dx (i.e., M) with respect to x, regarding y to be constant.*

Step 4. *Omit the terms containing x in N and find the integral of the coefficient of dy with respect to y.*

Step 5. *Add the above two results and equate this sum to an arbitrary constant i.e.,*

$$\int_{(y\ \text{constant})} Mdx+\int_{\text{those terms which do not contain } x} Ndy=c.$$

REMARKS

- It is clear from the definition that the exact differential equation is formed from its general solution by direct differentiation and without any further operation of elimination or reduction.
- The necessary and sufficient condition for a differential equation (of first order and first degree) $Mdx+Ndy=0$, where M and N are the functions of x and y to be exact equation is that $\frac{\partial M}{\partial y}=\frac{\partial N}{\partial x}$
- The equation $\frac{\partial M}{\partial y}=\frac{\partial N}{\partial x}$ is true whenever both sides exist and are continuous. Here we take the hypothesis that all the functions, which we discuss are sufficiently continuous and differentiable to guarantee the validity of the operations we perform on them.

Solved Examples

Example 1. *Solve* $(ax+hy+g)\,dx+(hx+by+f)dy=0$.

Solution. The given equation

$$(ax+hy+g)\,dx+(hx+by+f)dy=0$$

Compare with $Mdx+Ndy=0$, we get

$M=ax+hy+g$ and $N=hx+by+f$

$$\Rightarrow \frac{\partial M}{\partial y}=h,\ \frac{\partial N}{\partial x}=h \Rightarrow \text{ equation is exact.}$$

Hence, the solution is

$$\int_{(y\text{ constant})} Mdx+\int_{\text{those terms which do not contain }x} Ndy=c$$

$$\Rightarrow \int (ax+hy+g)dx+\int (by+f)dy=c$$

$$\Rightarrow \frac{ax^2}{2}+hxy+gx+\frac{by^2}{2}+fy=c$$

$$\Rightarrow ax^2+2hxy+by^2+2gx+2fy=2c.$$

Example 2. *Solve* $x\,dx+ydy=\dfrac{a^2(xdy-ydx)}{x^2+y^2}$.

(UPTU(B.Tech)–2005, UKTU–2011)

Solution. We have $x\,dx+ydy=\dfrac{a^2(xdy-ydx)}{x^2+y^2}$

$$\Rightarrow \left(x+\frac{a^2y}{x^2+y^2}\right)dx+\left(y-\frac{a^2x}{x^2+y^2}\right)dy=0.$$

Here, $M=x+\dfrac{a^2y}{x^2+y^2}$, $N=y-\dfrac{a^2x}{x^2+y^2}$

$$\frac{\partial M}{\partial y}=\frac{a^2(x^2-y^2)}{(x^2+y^2)^2},\ \frac{\partial N}{\partial x}=\frac{a^2(x^2-y^2)}{(x^2+y^2)^2}$$

$$\Rightarrow \frac{\partial M}{\partial y}=\frac{\partial N}{\partial x}$$

Therefore, equation is exact. Hence,

$$\int\left(x+\frac{a^2y}{x^2+y^2}\right)dx+\int y\,dy=C$$

$$\Rightarrow \frac{x^2}{2}+a^2y\frac{1}{y}\tan^{-1}\left(\frac{x}{y}\right)+\frac{y^2}{2}=C$$

$$\left(\frac{x^2+y^2}{2}\right)+a^2\tan^{-1}\frac{x}{y}=C$$

Example 3. *Solve* $(y^2e^{xy^2}+4x^3)dx+(2xye^{xy^2}-3y^2)dy=0$

(VTU–2006)

Solution. Compare the given equation with $Mdx+Ndy=0$, we get

$$M=y^2e^{xy^2}+4x^3 \text{ and } N=2xye^{xy^2}-3y^2$$

$$\frac{\partial M}{\partial y}=2ye^{xy^2}+y^2e^{xy^2}.2xy=\frac{\partial N}{\partial x}$$

Thus the equation is exact.

Hence, the solution is

$$\int_{(y\text{ constant})} Mdx+\int_{\text{those terms which do not contain }x} Ndy=c$$

$$\Rightarrow \int (y^2e^{xy^2}+4x^3)dx+\int (-3y^2)dy=C$$

$$\Rightarrow e^{xy^2}+x^4-y^3=C$$

Example 4. *Solve* $\left\{y\left(1+\dfrac{1}{x}\right)+\cos y\right\}dx+(x+\log x-x\sin y)dy=0$

(Marathwada–2008S, VTU–2006)

Solution. Compare the given equation with $Mdx+Ndy=0$, we get

$$M=y\left(1+\frac{1}{x}\right)+\cos y$$

and $N=x+\log x-x\sin y$

$$\frac{\partial M}{\partial y}=1+\frac{1}{x}-\sin y=\frac{\partial N}{\partial x}$$

$\Rightarrow$ Equation is exact.

Hence, the solution is

$$\int_{(y\text{ constant})} Mdx+\int_{\text{those terms which do not contain }x} Ndy=c$$

$$\Rightarrow \int\left\{\left(1+\frac{1}{x}\right)y+\cos y\right\}dx=C$$

$$\Rightarrow (x+\log x)y+x\cos y=C$$

Example 5. *Solve* $\dfrac{dy}{dx}+\dfrac{y\cos x+\sin y+y}{\sin x+x\cos y+x}=0$

(Kurukshetra–2005)

Solution. The given equaton is

$$(y\cos x+\sin y+y)dx+(\sin x+x\cos y+x)dy=0$$

Compare the given equation with $Mdx+Ndy=0$, we get $M=y\cos x+\sin y+y$

and $N=\sin x+x\cos y+x$

$$\frac{\partial M}{\partial y}=\cos x+\cos y+1=\frac{\partial N}{\partial x}$$

$\Rightarrow$ Equation is exact.

Hence, the solution is

$$\int_{(y\text{ constant})} Mdx+\int_{\text{those terms which do not contain }x} Ndy=c$$

$$\Rightarrow \int (y\cos x+\sin y+y)dx+\int (0)dy=C$$

or $$y\sin x+(\sin y+y)x=C$$

Example 6. *Solve* $(y^2e^{xy^2}+4x^3)dx+(2xye^{xy^2}-3y^2)dy=0$

(UKTU–2012)

Solution. Compare the given equation with $Mdx+Ndy=0$, we get

$$M=y^2e^{xy^2}+4x^3 \text{ and } N=2xye^{xy^2}-3y^2$$

$$\frac{\partial M}{\partial y}=y^2e^{xy^2}\cdot 2xy+e^{xy^2}.2y=2ye^{xy^2}(1+xy^2)$$

and

$$\frac{\partial N}{\partial x}=2y(xe^{xy^2}\cdot y^2+e^{xy^2})=2ye^{xy^2}(1+xy^2)$$

Since $\frac{\partial M}{\partial y} = \frac{\partial M}{\partial x}$ $\Rightarrow$ Equation is exact.

Hence, the solution is

$$\int_{(y\text{ constant})} Mdx + \int_{\text{those terms which do not contain } x} Ndy = c$$

$$\Rightarrow \quad \int (y^2 e^{xy^2} + 4x^3)dx + \int (-3y^2)dy = C$$

$$\Rightarrow \quad y^2 \frac{e^{xy^2}}{y^2} + x^4 + -y^3 = C$$

$$\Rightarrow \quad e^{xy^2} + x^4 - y^3 = C,$$

where C is an arbitrary constant.

EXERCISE 28.9

Solve the following ordinary differential equations :

1. $(4x+3y+1)\,dx + (3x+2y+1)dy = 0$
2. $\frac{dy}{dx} = \frac{(2x-y)}{x+2y-5}$
3. $xdx + ydy + \frac{xdy - ydx}{x^2+y^2} = 0$
4. $(1+4xy+2y^2)dx + (1+4xy+2x^2)dy = 0$
5. $[1+e^{x/y}]dx + e^{x/y}\,[1-(x/y)]\,dy = 0$
6. $(y\sin 2x)dx - (1+y^2+\cos^2 x)dy = 0$ (VTU–2000)
7. $\frac{xdy}{x^2+y^2} = \left(\frac{y}{x^2+y^2} - 1\right)dx$
8. $(\cos x - x\cos y)dy - (\sin y + y\sin x)dx = 0$ (UPTU–2008)

ANSWERS

1. $2x^2+3xy+y^2+x+y=c$
2. $x^2-y^2-xy+5y=c$
3. $x^2-2\tan^{-1}(x/y)+y^2=c$
4. $x+2x^2y+2xy^2+y=c$
5. $x+ye^{x/y}=c$
6. $y\cos 2x + 2y + \frac{2}{3}y^3 = c$
7. $\tan^{-1}(x/y) - x = c$
8. $-x\sin y - y\cos x = C$

28.12 INTEGRATING FACTOR

If the given equation $Mdx + Ndy = 0$ is not exact, then it can be made exact by multiplying some function of x and y. This multiplier is called an integrating factor (I.F.).

REMARKS

- The number of integrating factor for the equation $Mdx + Ndy = 0$, is infinite.
- If μ is an integrating factor, then $\mu(Mdx + Ndy) = 0$ is an exact differential equation.
- Although an equation of the form $Mdx + Ndy = 0$, always has integrating factor, there is no general method of finding them.

METHODS OF FINDING I.F. : BY INSPECTION

Here, we explain some rules for finding I.F. :

The following list of exact differentials should be noted very carefully.

(1) $d\left(\frac{x}{y}\right) = \frac{ydx - xdy}{y^2}$

(2) $d\left(\frac{y}{x}\right) = \frac{xdy - ydx}{x^2}$

(3) $d\left(\tan^{-1}\frac{y}{x}\right) = \frac{xdy - ydx}{x^2+y^2}$

(4) $d\left(\tan^{-1}\frac{x}{y}\right) = \frac{ydx - xdy}{x^2+y^2}$

(5) $d\left(\log\frac{x}{y}\right) = \frac{ydx - xdy}{xy}$

(6) $d\left(\log\frac{y}{x}\right) = \frac{xdy - ydx}{xy}$

(7) $d(x, y) = xdy + ydx$

(8) $d\left(\frac{1}{xy}\right) = -\left[\frac{xdy + ydx}{x^2y^2}\right]$

(9) $d\left(\frac{x^2}{y}\right) = \frac{2xy\,dx - x^2dy}{y^2}$

(10) $d\left(\frac{y^2}{x}\right) = \frac{2xy\,dy - y^2dx}{x^2}$

(11) $d\left(\frac{y^2}{x^2}\right) = \frac{2x^2ydy - 2y^2x\,dx}{x^4}$

(12) $d\left(\frac{x^2}{y^2}\right) = \frac{2xy^2dx - 2yx^2dy}{y^4}$

(13) $d\left(\frac{e^x}{y}\right) = \frac{ye^x dx - e^x dy}{y^2}$

(14) $d\,[\log(x^2+y^2)] = \frac{2xdx + 2ydy}{x^2+y^2}$

Solved Example

Example. Solve $y(2xy + e^x)dx = e^x\,dy$. (Kurukshetra–2005)

Solution. $y(e^x dx - e^x\,dy) + 2xy^2 dx = 0$

Since the term $2xy^2dx$ should not involve y^2. So, $1/y^2$ may be I.F. Multiplying both side by $1/y^2$, we get

$$\frac{ye^x dx - e^x dy}{y^2} + 2xdx = 0$$

or $$d\left(\frac{e^x}{y}\right) + 2xdx = 0$$

On integrating, we get

$$\frac{e^x}{y} + x^2 = c$$

28.13 RULES FOR FINDING OUT INTEGRATING FACTOR

Rule 1. When $Mx + Ny \neq 0$ and the equation $Mdx + Ndy = 0$ is homogeneous, then an integrating factor of this differential equation is $\dfrac{1}{Mx + Ny}$

Solved Examples

Example 1. Solve $y(y^2 - 2x^2)dx + x(2y^2 - x^2)dy = 0$.

Solution. The given equation is

$$(y^3 - 2x^2y)dx + (2xy^2 - x^3)dy = 0 \quad \ldots(1)$$

Compare with $Mdx + Ndy = 0$, we get

$$M = y^3 - 2x^2y, N = 2xy^2 - x^3$$

$$\therefore \quad \frac{\partial M}{\partial y} = 3y^2 - 2x^2, \frac{\partial N}{\partial x} = 2y^2 - 3x^2$$

$$\because \quad \frac{\partial M}{\partial y} \neq \frac{\partial N}{\partial x} \Rightarrow \text{Eqn. (1) is not exact.}$$

But it is homogeneous in x and y.

Now, $Mx + Ny = x(y^3 - 2x^2y) + y(2xy^2 - x^3)$
$= 3xy(y^2 - x^2)$

$$\therefore \quad \text{I.F.} = \frac{1}{Mx + Ny} = \frac{1}{3xy(y^2 - x^2)}$$

Multiplying eqn. (1) by I.F, we get

$$\frac{y(y^2 - 2x^2)}{3xy(y^2 - x^2)}dx + \frac{x(2y^2 - x^2)}{3xy(y^2 - x^2)}dy = 0$$

$$\Rightarrow \quad \frac{(y^2 - 2x^2)}{x(y^2 - x^2)}dx + \frac{(2y^2 - x^2)}{y(y^2 - x^2)}dy = 0$$

$$\Rightarrow \quad \frac{1}{x}dx - \frac{x}{y^2 - x^2}dx + \frac{1}{y}dy + \frac{y}{y^2 - x^2}dy = 0$$

$$\Rightarrow \quad \frac{1}{x}dx + \frac{1}{y}dy + \frac{1}{2}\left(\frac{2ydy - 2xdx}{y^2 - x^2}\right) = 0$$

$$\Rightarrow \quad d(\log x) + d(\log y) + \frac{1}{2}d\{\log(y^2 - x^2)\} = 0$$

$$\Rightarrow \quad d(\log x) + d(\log y) + d\{\log\sqrt{(y^2 - x^2)}\} = d(\log c)$$

On integrating, we get

$$\log x + \log y + \log\sqrt{(y^2 - x^2)} = \log c$$

$$\Rightarrow \quad xy\sqrt{(y^2 - x^2)} = c$$

where c is an arbitrary constant of integration.

Example 2. Solve $x^2ydx - (x^3 + y^3)dy = 0$.

Solution. The given equation is

$$x^2ydx - (x^3 + y^3)dy = 0 \quad \ldots(1)$$

Compare with $Mdx + Ndy = 0$, we get

$$M = x^2y, N = -(x^3 + y^3)$$

$$\therefore \quad \frac{\partial M}{\partial y} = x^2, \frac{\partial N}{\partial x} = -3x^2$$

$$\because \quad \frac{\partial M}{\partial y} \neq \frac{\partial N}{\partial x} \Rightarrow \text{Eqn. (1) is not exact.}$$

But it is homogeneous in x and y.

Now, $Mx + Ny = (x^2y).x + \{-(x^3 + y^3)\}y$
$= x^3y - x^3y - y^4 = -y^4 \neq 0$

$$\therefore \quad \text{I.F.} = \frac{1}{Mx + Ny} = -\frac{1}{y^4}$$

Multiplying eqn. (1) by $-\dfrac{1}{y^4}$, we get

$$\frac{-x^2}{y^3}dx + \left(\frac{x^3}{y^4} + \frac{1}{y}\right)dy = 0 \quad \ldots(2)$$

Comparing with $Mdx + Ndy = 0$, we get

$$M = -\frac{x^2}{y^3}, N = \frac{x^3}{y^4} + \frac{1}{y}$$

$$\therefore \quad \frac{\partial M}{\partial y} = \frac{3x^2}{y^4}, \frac{\partial N}{\partial x} = \frac{3x^2}{y^4}$$

$$\because \quad \frac{\partial M}{\partial y} = \frac{\partial N}{\partial x} \Rightarrow \text{Eqn. (2) is exact.}$$

Hence, the solution is

$$\int_{(y\text{ constant})} Mdx + \int_{\text{those terms which do not contain } x} Ndy = c$$

$$\Rightarrow \quad \int \left(\frac{-x^2}{y^3}\right)dx + \int \frac{1}{y}dy = c$$

$$\Rightarrow \quad \frac{-x^3}{3y^3} + \log y = c$$

where c is an arbitrary constant of integration.

EXERCISE 28.10

Solve the following ordinary differential equations :

1. $(x^2y - 2xy^2)dx - (x^3 - 3x^2y)dy = 0$ (Osmania–2003S)
2. $xy^2dy - (x^3 + y^3)dx = 0$
3. $(3xy^2 - y^3)dx - (2x^2y - xy^2)dy = 0$

ANSWERS

1. $\frac{x}{y} - 2\log x + 3\log y = c$
2. $\log x - \frac{y^3}{3x^3} = c$
3. $3\log x + \frac{y}{x} - 2\log y = c$

Rule 2. If the equation $Mdx + Ndy = 0$ is of form $f_1(xy)ydx + f_2(xy)xdy = 0$, then $\frac{1}{Mx - Ny}$ is an integrating factor, provided $Mx - Ny \neq 0$.

Solved Example

Example. *Solve* $(y - xy^2)dx - (x + x^2y)dy = 0$.

Solution. The given equation is

$$(y - xy^2)dx - (x + x^2y)dy = 0 \qquad \ldots(1)$$

Comparing with $Mdx + Ndy = 0$, we get

$$M = y - xy^2, N = -(x + x^2y)$$

$$\therefore \quad \frac{\partial M}{\partial y} = 1 - 2xy, \frac{\partial N}{\partial x} = -1 - 2xy$$

$$\because \quad \frac{\partial M}{\partial y} \neq \frac{\partial N}{\partial x} \Rightarrow \text{Eqn. (1) is not exact.}$$

Now, $M = y - xy^2 = y(1 - xy) = yf_1(xy)$

and $N = -(x + x^2y) = -x(1 + xy) = xf_2(xy)$

$$\therefore \quad Mx - Ny = (y - xy^2)x + (x + x^2y)y = 2xy \neq 0$$

$$\therefore \quad \text{I.F.} = \frac{1}{Mx - Ny} = \frac{1}{2xy}$$

Multiplying (1) by $\frac{1}{2xy}$, we get

$$\frac{y - xy^2}{2xy}dx - \frac{x + x^2y}{2xy}dy = 0$$

$$\frac{1}{2}\left(\frac{1}{x} - y\right)dx - \frac{1}{2}\left(\frac{1}{y} + x\right)dy = 0 \qquad \ldots(2)$$

Comparing with $Mdx + Ndy = 0$, we get

$$M = \frac{1}{2}\left(\frac{1}{x} - y\right), N = -\frac{1}{2}\left(\frac{1}{y} + x\right)$$

$$\therefore \quad \frac{\partial M}{\partial y} = -\frac{1}{2}, \frac{\partial N}{\partial x} = -\frac{1}{2}$$

$$\because \quad \frac{\partial M}{\partial y} = \frac{\partial N}{\partial x} \Rightarrow \text{Eqn. (2) is exact.}$$

Hence, the solution is

$$\int_{(y\text{ constant})} Mdx + \int_{\text{those terms which do not contain } x} Ndy = c$$

$$\Rightarrow \quad \int \frac{1}{2}\left(\frac{1}{x} - y\right)dx + \int \frac{-1}{2y}dy = c$$

$$\Rightarrow \quad \frac{1}{2}(\log x - yx) - \frac{1}{2}\log y = c$$

$$\Rightarrow \quad \log x - xy - \log y = 2c$$

$$\Rightarrow \quad \log\left(\frac{x}{y}\right) - xy = A,$$

where $A = 2c$

EXERCISE 28.11

Solve the following ordinary differential equations :

1. $(1 + xy)ydx + (1 - xy)xdy = 0$ (SVTU–2008)
2. $(x^2y^2 + xy + 1)ydx + (x^2y^2 - xy + 1)xdy = 0$
3. $(xy \sin xy + \cos xy)ydx + (xy \sin xy - \cos xy)xdy = 0$
4. $(xy^2 + 2x^2y^3)dx + (x^2y - x^3y^2)dy = 0$

ANSWERS

1. $\log\frac{x}{y} - \frac{1}{xy} = c$
2. $xy - \frac{1}{xy} + \log\frac{x}{y} = c$
3. $x \sec xy = cy$
4. $-\frac{1}{xy} + 2\log x - \log y = c$

Rule 3. If in the equation $Mdx + Ndy = 0$, $\frac{1}{N}\left(\frac{\partial M}{\partial y} - \frac{\partial N}{\partial x}\right)$ is a function of x alone, say $f(x)$, then $e^{\int f(x)dx}$ is an integrating factor of the equation $Mdx + Ndy = 0$.

Rule 4. If in the equation $Mdx + Ndy = 0$, $\frac{1}{M}\left(\frac{\partial N}{\partial x} - \frac{\partial M}{\partial y}\right)$ is a function of y alone, say $f(y)$, then $e^{\int f(y)dy}$ is an integrating factor of the equation $Mdx + Ndy = 0$.

Solved Examples

Example 1. *Solve* $\left(y + \frac{1}{3}y^3 + \frac{1}{2}x^2\right)dx + \frac{1}{4}(1+y^2)xdy = 0.$

Solution. The given equation is

$$\left(y + \frac{1}{3}y^3 + \frac{1}{2}x^2\right)dx + \frac{1}{4}(1+y^2)xdy = 0. \quad \ldots(1)$$

Comparing with $Mdx + Ndy = 0$, we get

$$M = y + \frac{1}{3}y^3 + \frac{1}{2}x^2, N = \frac{1}{4}(1+y^2)x$$

$$\therefore \quad \frac{\partial M}{\partial y} = 1 + y^2, \frac{\partial N}{\partial x} = \frac{1}{4}(1+y^2)$$

$$\because \quad \frac{\partial M}{\partial y} \neq \frac{\partial N}{\partial x} \Rightarrow \text{Eqn. (1) is not exact.}$$

$$\text{Now, } \frac{1}{N}\left(\frac{\partial M}{\partial y} - \frac{\partial N}{\partial x}\right) = \frac{(1+y^2) - \frac{1}{4}(1+y^2)}{\frac{1}{4}(1+y^2)x} = \frac{3}{x} = f(x)$$

$$\therefore \quad \text{I.F.} = e^{\int f(x)dx} = e^{\int \frac{3}{x}dx} = e^{3\log x} = x^3$$

Multiplying (1) by x^3, we get

$$\left(yx^3 + \frac{1}{3}y^3x^3 + \frac{1}{2}x^5\right)dx + \frac{1}{4}(1+y^2)x^4dy = 0 \quad \ldots(2)$$

Comparing with $Mdx + Ndy = 0$, we get

$$M = yx^3 + \frac{1}{3}y^3x^3 + \frac{1}{2}x^5$$

and $$N = \frac{1}{4}(1+y^2)x^4$$

$$\therefore \quad \frac{\partial M}{\partial y} = x^3 + y^2x^3 = x^3(1+y^2),$$

$$\frac{\partial N}{\partial x} = \frac{1}{4}(1+y^2).4x^3 = x^3(1+y^2)$$

$$\because \quad \frac{\partial M}{\partial y} = \frac{\partial N}{\partial x} \Rightarrow \text{Eqn. (2) is exact.}$$

Hence, the solution is

$$\int_{(y\text{ constant})} Mdx + \int_{\text{those terms which do not contain } x} Ndy = c$$

$$\Rightarrow \quad \int\left(yx^3 + \frac{1}{3}y^3x^3 + \frac{1}{2}x^5\right)dx + \int 0.dy = c$$

$$\Rightarrow \quad \frac{yx^4}{4} + \frac{y^3x^4}{12} + \frac{x^6}{12} = c$$

Example 2. *Solve* $2ydx + x(2\log x - y)dy = 0.$

Solution. The given equation is

$$2ydx + x(2\log x - y)dy = 0. \quad \ldots(1)$$

Comparing with $Mdx + Ndy = 0$, we get

$$M = 2y, N = x(2\log x - y)$$

$$\therefore \quad \frac{\partial M}{\partial y} = 2, \frac{\partial N}{\partial x} = 2\left(x.\frac{1}{x} + \log x\right) - y = 2(1+\log x) - y$$

$$\because \quad \frac{\partial M}{\partial y} \neq \frac{\partial N}{\partial x} \Rightarrow \text{Eqn. (1) is not exact.}$$

Now,

$$\frac{1}{N}\left(\frac{\partial M}{\partial y} - \frac{\partial N}{\partial x}\right) = \frac{1}{x(2\log x - y)}(2 - 2 - 2\log x + y) = -\frac{1}{x} = f(x)$$

$$\therefore \quad \text{I.F.} = e^{\int f(x)dx} = e^{-\int \frac{dx}{x}} = e^{-\log x} = \frac{1}{x}$$

Multiplying (1) by $\frac{1}{x}$, we get

$$\frac{2y}{x}dx + (2\log x - y)dy = 0 \quad \ldots(2)$$

Comparing with $Mdx + Ndy = 0$, we get

$$M = \frac{2y}{x}, N = 2\log x - y$$

$$\therefore \quad \frac{\partial M}{\partial y} = \frac{2}{x}, \frac{\partial N}{\partial x} = \frac{2}{x}$$

$$\because \quad \frac{\partial M}{\partial y} = \frac{\partial N}{\partial x} \Rightarrow \text{Eqn. (2) is exact.}$$

Hence, the solution is

$$\int_{(y\text{ constant})} Mdx + \int_{\text{those terms which do not contain } x} Ndy = c$$

$$\Rightarrow \quad \int \frac{2y}{x}dx + \int(-y).dy = c$$

$$\Rightarrow \quad 2y\log x - \frac{y^2}{2} = c$$

Example 3. *Solve* $(y \log y)dx + (x - \log y)dy = 0.$ (UPTU–2004)

Solution. The given equation is

$$(y \log y)dx + (x - \log y)dy = 0. \quad \ldots(1)$$

Comparing with $Mdx + Ndy = 0$, we get

$$M = y \log y, N = x - \log y$$

$$\therefore \quad \frac{\partial M}{\partial y} = y.\frac{1}{y} + \log y = 1 + \log y$$

and $\frac{\partial N}{\partial x} = 1$

$$\because \quad \frac{\partial M}{\partial y} \neq \frac{\partial N}{\partial x} \Rightarrow \text{Eqn. (1) is not exact.}$$

$$\text{Now, } \frac{1}{M}\left(\frac{\partial N}{\partial x} - \frac{\partial M}{\partial y}\right) = \frac{1}{y \log y}(1 - 1 - \log y)$$

$$= -\frac{1}{y} = f(y)$$

$$\therefore \quad \text{I.F.} = e^{\int f(y)dy} = e^{-\int \frac{1}{y}dy} = e^{-\log y} = \frac{1}{y}$$

Multiplying (1) by $\frac{1}{y}$, we get

$$(\log y)dx + \left(\frac{x - \log y}{y}\right)dy = 0 \quad \ldots(2)$$

Comparing with $Mdx + Ndy = 0$, we get

$$M = \log y,\ N = \frac{x}{y} - \frac{\log y}{y}$$

$$\therefore \quad \frac{\partial M}{\partial y} = \frac{1}{y} \text{ and } \frac{\partial N}{\partial x} = \frac{1}{y}$$

$$\because \quad \frac{\partial M}{\partial y} = \frac{\partial N}{\partial x} \Rightarrow \text{Eqn. (2) is exact.}$$

Hence, the solution is

$$\int_{(y\text{ constant})} Mdx + \int_{\text{those terms which do not contain } x} Ndy = c$$

$$\Rightarrow \quad \int \log y dx + \int \left(-\frac{\log y}{y}\right)dy = c$$

$$\Rightarrow \quad x \log y - \frac{1}{2}(\log y)^2 = c$$

Example 4. *Solve* $(xy^3 + y)dx + 2(x^2y^2 + x + y^4)dy = 0.$

Solution. The given equation is

$$(xy^3 + y)dx + 2(x^2y^2 + x + y^4)dy = 0. \quad \ldots(1)$$

Comparing with $Mdx + Ndy = 0$, we get

$$M = xy^3 + y, N = 2(x^2y^2 + x + y^4)$$

$$\therefore \quad \frac{\partial M}{\partial y} = 3xy^2 + 1, \frac{\partial N}{\partial x} = 2(2xy^2 + 1)$$

$$\because \quad \frac{\partial M}{\partial y} \neq \frac{\partial N}{\partial x} \Rightarrow \text{Eqn. (1) is not exact.}$$

$$\text{Now, } \frac{1}{M}\left(\frac{\partial N}{\partial x} - \frac{\partial M}{\partial y}\right) = \frac{2.(2xy^2 + 1) - (3xy^2 + 1)}{xy^3 + y}$$

$$= \frac{xy^2 + 1}{y(xy^2 + 1)} = \frac{1}{y} = f(y)$$

$$\therefore \quad \text{I.F.} = e^{\int f(y)dy} = e^{\int \frac{1}{y}dy}$$

$$= e^{\log y} = y$$

Multiplying (1) by y, we get

$$(xy^4 + y^2)dx + 2(x^2y^3 + xy + y^5)dy = 0 \quad \ldots(2)$$

Comparing with $Mdx + Ndy = 0$, we get

$$M = xy^4 + y^2,\ N = 2(x^2y^3 + xy + y^5)$$

$$\therefore \quad \frac{\partial M}{\partial y} = 4xy^3 + 2y, \frac{\partial N}{\partial x} = 4xy^3 + 2y$$

$$\because \quad \frac{\partial M}{\partial y} = \frac{\partial N}{\partial x} \Rightarrow \text{Eqn. (2) is exact.}$$

Hence, the solution is

$$\int_{(y\text{ constant})} Mdx + \int_{\text{those terms which do not contain } x} Ndy = c$$

$$\Rightarrow \quad \int \frac{2y}{x}dx + \int (-y).dy = c$$

$$\Rightarrow \quad 2y \log x - \frac{y^2}{2} = c$$

EXERCISE 28.12

Solve the following ordinary differential equations :

1. $(xy^2 - e^{1/x^3})dx - x^2ydy = 0$ (SVTU–2009, Mumbai–2007)
2. $(x^2 + y^2 + 2x)dx + 2ydy = 0$
3. $(y^4 + 2y)dx + (xy^3 + 2y^4 - 4x)dy = 0$
4. $(x \sec^2 y - x^2 \cos y)dy = (\tan y - 3x^4)dx$

ANSWERS

1. $\frac{1}{3}e^{x^{-3}} - \frac{1}{2}\frac{y^2}{x^2} = c$
2. $e^x(x^2 + y^2) = c$
3. $\left(y + \frac{2}{y^2}\right)x + y^2 = c$
4. $-\frac{1}{x}\tan y - x^3 + \sin y = c$

Rule 5. If an equation is of the form $x^ay^b[mydx + nxdy] + x^ry^s[pydx + qxdy] = 0$ where a, b, m, n, r, s, p and q are all constants, then the integrating factor will be x^hy^k, where h and k are such that after multiplying by the integrating factor that the condition of exactness is satisfied.

Solved Example

Example. Solve $(2ydx + 3xdy) + 2xy(3ydx + 4xdy) = 0$.

Solution. The given equation is

$$(2ydx + 3xdy) + 2xy(3ydx + 4xdy) = 0. \quad ...(1)$$

$$\Rightarrow \quad (2y + 6xy^2)dx + (3x + 8x^2y)dy = 0. \quad ...(2)$$

Comparing (2) with $Mdx + Ndy = 0$, we get

$$M = 2y + 6xy^2, N = 3x + 8x^2y$$

$$\therefore \quad \frac{\partial M}{\partial y} = 2 + 12xy, \frac{\partial N}{\partial x} = 3 + 16xy$$

$$\because \quad \frac{\partial M}{\partial y} \neq \frac{\partial N}{\partial x} \Rightarrow \text{ Eqn. (2) is not exact.}$$

Eqn. (1) is of the form

$$x^a y^b(mydx + nxdy) + x^r y^s(pydx + qxdy) = 0$$

$\therefore$ $x^h y^k$ is an integrating factor of eqn (1).

Multiplying (2) by $x^h y^k$, we get

$$(2x^h y^{k+1} + 6x^{h+1}y^{k+2})dx + (3x^{h+1}y^k + 8x^{h+2}y^{k+1})dy = 0. \quad ...(3)$$

Comparing (3) with $Mdx + Ndy = 0$, we get

$$M = 2x^h y^{k+1} + 6x^{h+1}y^{k+2}, N = 3x^{h+1}y^k + 8x^{h+2}y^{k+1}$$

Now eqn. (3) must be exact, the condition for which is $\frac{\partial M}{\partial y} = \frac{\partial N}{\partial x}$

$$2(k+1)x^h y^k + 6(k+2)x^{h+1}y^{k+1} = 3(h+1)x^h y^k + 8(h+2)x^{h+1}y^{k+1}$$

Now, equating the coefficients of $x^h y^k$ and $x^{h+1}y^{k+1}$ from both sides, we get

$$2(k+1) = 3(h+1) \Rightarrow 3h - 2k + 1 = 0 \quad ...(4)$$

$$\text{and } 6(k+2) = 8(h+2) \Rightarrow 4h - 3k + 2 = 0 \quad ...(5)$$

On solving (4) and (5), we get $h = 1, k = 2$

Hence, xy^2 is an integrating factor.

Multiplying eqn. (2) by xy^2, we get

$$(2xy^3 + 6x^2y^4)dx + (3x^2y^2 + 8x^3y^3)dy = 0 \quad ...(6)$$

Comparing (6) with $Mdx + Ndy = 0$, we get

$$M = 2xy^3 + 6x^2y^4, N = 3x^2y^2 + 8x^3y^3$$

$$\therefore \quad \frac{\partial M}{\partial y} = 6xy^2 + 24x^2y^3, \frac{\partial N}{\partial x} = 6xy^2 + 24x^2y^3$$

$$\because \quad \frac{\partial M}{\partial y} = \frac{\partial N}{\partial x} \Rightarrow \text{ Eqn. (6) is exact.}$$

Hence, the solution is

$$\int_{(y \text{ constant})} Mdx + \int_{\text{those terms which do not contain } x} Ndy = c$$

$$\Rightarrow \quad \int (2xy^3 + 6x^2y^4)dx + \int 0 \cdot dy = c$$

$$\Rightarrow \quad x^2y^3 + 2x^3y^4 = c$$

EXERCISE 28.13

Solve the following ordinary differential equations :

1. $y(xy + 2x^2y^3)dx + x(xy - x^2y^2)dy = 0$ (Hissar–2005, Kurukshetra–2005)
2. $x(3ydx + 2xdy) + 8y^4(ydx + 3xdy) = 0$

ANSWERS

1. $2\log x - \log y - \frac{1}{xy} = c$
2. $x^3y^2 + 4x^2y^6 = c$

Miscellaneous Solved Examples

Example 1. Solve $(y^2e^x + 2xy)dx - x^2dy = 0$.

Solution. The given equation is

$$(y^2e^x + 2xy)dx - x^2dy = 0$$

Compare with $Mdx + Ndy = 0$, we get

$$M = y^2e^x + 2xy \text{ and } N = -x^2$$

$$\frac{\partial M}{\partial y} = 2ye^x + 2x, \quad \frac{\partial N}{\partial x} = -2x$$

$$\Rightarrow \quad \frac{\partial M}{\partial y} \neq \frac{\partial N}{\partial x}$$

The given equation is not exact.

Now, if in the equation e^x is multiplied by some other function, then it must occur twice in the differential equation. But since it is occurring only once, therefore, we should divide by y^2.

$$\therefore \quad e^x\,dx + \frac{2xydx - x^2dy}{y^2} = 0$$

$$\text{or} \quad \left(e^x + \frac{2x}{y}\right)dx + \left(\frac{-x^2}{y^2}\right)dy = 0$$

Now, $M = e^x + \frac{2x}{y}$ and $N = -\frac{x^2}{y^2}$

$$\Rightarrow \quad \frac{\partial M}{\partial y} = -\frac{2x}{y^2} = \frac{\partial N}{\partial x} \Rightarrow \text{ Equation is exact.}$$

Hence, the solution is

$$\int_{(y \text{ constant})} Mdx + \int_{\text{those terms which do not contain } x} Ndy = c$$

$$\int \left(e^x + \frac{2x}{y}\right)dx = c \Rightarrow e^x + \frac{x^2}{y} = c.$$

Example 2. *Solve* $(x^2y - 2xy^2)dx - (x^3 - 3x^2y)dy = 0$.

Solution. The given equation is

$$(x^2y - 2xy^2)dx - (x^3 - 3x^2y)dy = 0$$

Compare with $Mdx + Ndy = 0$, we get

$$M = x^2y - 2xy^2 \text{ and } N = 3x^2y - x^3$$

$$\Rightarrow \frac{\partial M}{\partial y} = x^2 - 4xy \text{ and } \frac{\partial N}{\partial x} = 6xy - 3x^2$$

$$\Rightarrow \frac{\partial M}{\partial y} \neq \frac{\partial N}{\partial x}$$

The given equation is not exact.

Here, we observe that M and N are the homogeneous functions of x and y and

$$\frac{1}{Mx + Ny} = \frac{1}{x^2y^2} \neq 0, \text{ therefore, I.F.} = \frac{1}{x^2y^2}.$$

Now, multiplying the given equation by I.F., we get

$$\left(\frac{1}{y} - \frac{2}{x}\right)dx + \left(\frac{3}{y} - \frac{x}{y^2}\right)dy = 0$$

In this equation $M = \frac{1}{y} - \frac{2}{x}$ and $N = \frac{3}{y} - \frac{x}{y^2}$

$$\frac{\partial M}{\partial y} = -\frac{1}{y^2} = \frac{\partial N}{\partial x} \Rightarrow \text{This equation is exact.}$$

The solution is $\frac{x}{y} + \log\frac{y^3}{x^2} = c$.

Example 3. *Solve*

$$(x^2y^2 + xy + 1)ydx + (x^2y^2 - xy + 1)xdy = 0.$$

Solution. The given equation is

$$(x^2y^2 + xy + 1)ydx + (x^2y^2 - xy + 1)xdy = 0$$

which is not exact.

M and N are of the form $yf_1(x, y)$ and $xf_2(x, y)$ respectively.

Now,

$$Mx - Ny = x^3y^3 + x^2y^2 + xy - x^3y^3 + x^2y^2 - xy$$
$$= 2x^2y^2 \neq 0$$

$$\therefore \text{ I.F.} = \frac{1}{2x^2y^2}.$$

Multiplying by I.F., the given equation becomes

$$\frac{1}{2}\left(y + \frac{1}{x} + \frac{1}{x^2y}\right)dx + \frac{1}{2}\left(x - \frac{1}{y} + \frac{1}{xy^2}\right)dy = 0$$

which can be shown to be exact and whose solution as usual is given by

$$xy - \frac{1}{xy} + \log\frac{x}{y} = c.$$

Example 4. *Solve* $(xy^3 + y)dx + 2(x^2y^2 + x + y^4)dy = 0$.

Solution. Compare the given equation with $Mdx + Ndy = 0$.

we get $M = xy^3 + y$ and $N = 2(x^2y^2 + x + y^4)$.

$$\Rightarrow \frac{\partial M}{\partial y} = 3xy^2 + 1 \text{ and } \frac{\partial N}{\partial x} = 2(2xy^2 + 1)$$

$$\therefore \frac{1}{M}\left(\frac{\partial N}{\partial x} - \frac{\partial M}{\partial y}\right) = \frac{1}{xy^3 + y}[4xy^2 + 2 - 3xy^2 - 1]$$
$$= \frac{1}{y}$$

which is a function of y only.

$$\text{I.F.} = e^{\int f(y)dy} = e^{\int \frac{1}{y}dy} = e^{\log y} = y.$$

Multiplying the given equation by I.F., we get

$$(xy^4 + y^2)dx + 2(x^2y^3 + xy + y^5)dy = 0.$$

which is an exact differential equation.

Its solution is

$$\int_{(y \text{ constant})}(xy^4 + y^2)dx + \int_{\text{those terms which do not contain } x} 2y^5dy = c$$

$$\Rightarrow \frac{1}{2}x^2y^4 + xy^2 + \frac{1}{3}y^6 = c.$$

Example 5. *Solve* $(y^2 + 2x^2y)dx + (2x^3 - xy)dy = 0$. (Rajasthan–2005)

Solution. Here, the given equation is

$$(y^2 + 2x^2y)dx + (2x^3 - xy)dy = 0 \quad \ldots(1)$$

which can be written as

$$y(y + 2x^2)dx + 2x^2(ydx + xdy) = 0.$$

Let us suppose that the possible integrating factor be $x^h y^k$.

Multiplying (1) by I.F. (*i.e.*, $x^h y^k$), we have

$$(x^h y^{k+2} + 2x^{h+2}y^{k+1})dx + (2x^{h+3}y^k - x^{h+1}y^{k+1})dy = 0 \quad \ldots(2)$$

Now, $M = x^h y^{k+2} + 2x^{h+2}y^{k+1}$

and $N = 2x^{h+3}y^k - x^{h+1}y^{k+1}$

$$\Rightarrow \frac{\partial M}{\partial y} = (k+2)x^h y^{k+1} + 2(k+1)x^{h+2}y^k$$

and $\frac{\partial N}{\partial x} = 2(h+3)x^{h+2}y^k - (h+1)x^h y^{k+1}$.

If the equation (2) is exact, then $\frac{\partial M}{\partial y} = \frac{\partial N}{\partial x}$.

$\therefore$ Equating the coefficients of $x^h y^{k+1}$ and $x^{h+2}y^k$ on both sides, we have

$$k + 2 = -h - 1 \Rightarrow h + k + 3 = 0$$

and $2k + 2 = 2h + 6 \Rightarrow h - k + 2 = 0$

Solving these equations, we get $h = -5/2$ and $k = -1/2$

The I.F. $= x^h y^k = x^{-5/2} y^{-1/2}$

Now, equation (2) becomes

$$(x^{-5/2}y^{3/2} + 2x^{-1/2}y^{1/2})dx$$
$$+ (2x^{1/2}y^{-1/2} - x^{-3/2}y^{1/2})dy = 0.$$

We can easily verify that this equation is exact by condition $\frac{\partial M}{\partial y} = \frac{\partial N}{\partial x}$.

Hence, the solution is

$$\int_{(y\text{ constant})} M dx + \int_{\text{those terms which do not contain } x} N dy = c$$

$$\Rightarrow \quad -\frac{2}{3}x^{-3/2}y^{3/2} + 4x^{1/2}y^{1/2} = c$$

is required solution.

MISCELLANEOUS EXERCISE

Solve the following ordinary differential equations :

1. $(1+xy)\, y\, dx + (1-xy)x\, dy = 0$ (AMIE–1996, Madurai–1990, MDU–2000)
2. $ydx - xdy + (1+x^2)dx + x^2 \sin y\, dy = 0$
3. $(x^3y^3 + x^2y^2 + xy + 1)y\, dx + (x^3y^3 - x^2y^2 - xy + 1)x\, dy = 0$
4. $x^2ydx - (x^3 + y^3)dy = 0$
5. $y(2x^2y + e^x)dx - (e^x + y^3)dy = 0$ (Bangalore–1990)
6. $xdy - ydx = -2x^3dx$
7. $xdx + ydy + (x^2 + y^2)dy = 0$
8. $(x^2 + y^2)dx - 2xy\, dy = 0$
9. $x(3y\, dx + 2x\, dy) + 8y^4(y\, dx + 3x\, dy) = 0$
10. $(xy \sin xy + \cos xy)\, y\, dx + (xy \sin xy - \cos xy)x\, dy = 0$
11. $(3x^2y^4 + 2xy)dx + (2x^3y^3 - x^2)dy = 0$
12. $(2ydx + 3x\, dy) + 2xy\,(3y\, dx + 4x\, dy) = 0$
13. $(xy^2 + 2x^2y^3)dx + (x^2y - x^3y^2)dy = 0$
14. $(xy^2 - x^2)dx + (3x^2y^2 + x^2y - 2x^3 + y^2)dy = 0$
15. $xdx + ydy = \frac{a^2(xdy - ydx)}{x^2 + y^2}$ (UPTU–2005)
16. $(4xy + 3y^2 - x)dx + x(x + 2y)dy = 0$ (Mumbai–2006)
17. $(y - xy^2)dx + (x + x^2y)dy = 0$ (Mumbai–2006)
18. $ydx - xdy + 3x^2y^2e^{x^3}\, dx = 0$ (Kurukshetra–2006)
19. $2ydx + x(2\log x - y)dy = 0$ (PTU–2005)

Hint to Selected Problems

1. The given equation can be written as $d(yx) + xy^2dx - x^2ydy = 0$ Divide by x^2y^2 and then put $xy = v$.
2. The given equation can be written as $d(e^x) + d\left(\frac{x^2}{y}\right) = 0$. Now integrate it.
4. I.F. $= -\frac{1}{y^4}$
5. I.F. $= \frac{1}{y^2}$
6. Dividing throughout by x^6, we get $d\left(\frac{y}{x}\right) + 2x\, dx = 0$. Then integrate it.
7. The given equation can be written as $\frac{d(x^2 + y^2)}{x^2 + y^2} + 2dy = 0$.
8. I.F. $= \frac{1}{N}\left(\frac{\partial M}{\partial y} - \frac{\partial N}{\partial x}\right) = \frac{1}{x^2}$
9. I.F. $= xy$
10. I.F. $= \frac{1}{Mx - Ny} = \frac{1}{2yx \cos xy}$
11. I.F. $= \frac{1}{M}\left(\frac{\partial N}{\partial x} - \frac{\partial M}{\partial y}\right) = \frac{1}{y^2}$
12. I.F. $= xy^2$
13. I.F. $= \frac{1}{Mx - Ny} = \frac{1}{3x^3y^3}$
14. I.F. $= \frac{1}{M}\left(\frac{\partial N}{\partial x} - \frac{\partial M}{\partial y}\right) = e^{6y}$

ANSWERS

1. $\log(x/y) = c + (1/xy)$
2. $\frac{y}{x} + \frac{1}{x} - x + \cos y = c$
3. $xy + (-1/xy) - 2\log y = c$
4. $y = ce^{x^3/3y^3}$
5. $\frac{2}{3}x^3 - \frac{1}{2}y^2 + \frac{e^x}{y} = c$
6. $y + x^3 = cx$
7. $x^2 + y^2 = e^{c-2y}$
8. $x^2 - y^2 = cx$
9. $x^3y^2 + 4y^6x^2 = c$
10. $x \sec(xy) = cy$
11. $x^3y^2 + (x^2/y) = c$
12. $x^2y^3 + 2x^3y^2 = c$
13. $x^2 = cy\, e^{1/xy}$
14. $e^{6y}\left[x^2\left(\frac{y^2}{2} - \frac{x}{3}\right) + \left(\frac{y^2}{6} - \frac{y}{18} + \frac{1}{108}\right)\right] = c$
15. $x^2 + y^2 - 2a^2 \tan^{-1}\left(\frac{y}{x}\right) = c$
16. $4x^4y + 4x^3y^2 - x^4 = c$
17. $\log\left(\frac{x}{y}\right) = c + xy$
18. $\left(\frac{x}{y}\right) + e^{x^3} = c$
19. $4y \log x = y^2 + c$

Objective Evaluations

Fill in the Blanks

1. An equation, which contains, dependent, independent variables and different derivatives is called ________.
2. A differential equation, which involves only one independent variable is called ________.
3. A differential equation, which involve two or more independent variables and partial differential coefficients, is called ________.
4. The order of the differential equation $\left[1+\left(\frac{dy}{dx}\right)^2\right]^{3/2} = \frac{d^2y}{dx^2}$ is ________.
5. The degree of a differential equation is the degree of its ________ derivative which appears in it.
6. The solution of the differential equation, which contains the full number of arbitrary constant is known as ________.
7. The solution, derived from a general solution is known as ________.
8. The ________ solution does not contain any arbitrary constants.
9. To solve a homogeneous equation, we put $y =$ ________.
10. An equation, which can be derived by its primitive by direct differentiation is known as ________ differential equation.

True/False

Write 'T' for True and 'F' for False statement.

1. An exact differential equation can be derived from its primitive by direct differentiation. **(T/F)**
2. If μ is an integrating factor, then $\mu(Mdx + Ndy) = 0$ is not necessarily exact. **(T/F)**
3. If the given equation is homogeneous and $Mx + Ny \neq 0$, then $1/Mx + Ny$ is the I.F. of the given equation. **(T/F)**
4. A differential equation always having infinite number of integrating factor (I.F.). **(T/F)**
5. An equation $M\,dx + N\,dy = 0$ always has integrating factor. **(T/F)**
6. The equation of the type $\frac{dy}{dx} + Py = Q$ where P and Q are functions of x and y is necessarily linear. **(T/F)**
7. A homogeneous equation of the first order and first degree, can be written as $\frac{dy}{dx} = f(y/x)$. **(T/F)**
8. The general solution of the differential equation of the first order may contain two arbitrary constant. **(T/F)**
9. In a linear equation, no product of dependent variable or derivative occur. **(T/F)**
10. The solution of the differential equation does not involve the derivatives of the dependent variable with respect to the independent variable or variables. **(T/F)**

Multiple Choice Questions

Choose the most appropriate one.

1. The general solution of a first degree equation contains :
   (a) one arbitrary constant (b) two arbitrary constants
   (c) three arbitrary constants (d) none of these
2. The equation $(x^2 + y^2)\frac{dy}{dx} = xy$ is :
   (a) homogeneous (b) linear
   (c) reducible to linear (d) none of these
3. The number of integrating factor for a differential equation of the type $M\,dx + N\,dy = 0$ is :
   (a) one (b) finite
   (c) infinite (d) none of these
4. If the given equation of the type $f_1(x, y)\, y\, dx + f_2(x, y)\, x\, dy = 0$, then I.F. is :
   (a) $Mx - Ny$ (b) $1/(Mx - Ny)$
   (c) $Mx + Ny$ (d) $1/(Mx + Ny)$
5. An equation of the type $(dy/dx) + Py = Qy^n$ is said to be :
   (a) linear
   (b) homogeneous
   (c) exact
   (d) Bernoulli's differential equation
6. The solution, which is the envelope of the family of lines represented by the complete primitive is said to be :
   (a) singular (b) non-singular
   (c) particular (d) none of these
7. The ordinary differential equation always involve :
   (a) one independent variable
   (b) 2 I.V. or two independent variables
   (c) more than 2 P.V.
   (d) none of these

8. If the given differential equation is of the form $\left(d^n y / dx^n\right)^m = c$ then its :
(a) order is n
(b) degree is m
(c) both (a) and (b) are true
(d) none of these

9. The necessary and sufficient condition of $M\,dx + N\,dy = 0$ is to be exact if :
(a) $\frac{\partial M}{\partial y}$ and $\frac{\partial N}{\partial x}$ must exist (b) $\frac{\partial M}{\partial y} = \frac{\partial N}{\partial x}$
(c) $\frac{\partial M}{\partial y} + \frac{\partial N}{\partial x} = 0$ (d) $\frac{\partial M}{\partial x} = \frac{\partial M}{\partial y}$

10. If $\frac{1}{M}\left(\frac{\partial N}{\partial x} - \frac{\partial M}{\partial y}\right)$ is a function of y alone, say $f(y)$, then the required factor is :
(a) $f(y)$ (b) $e^{\int f(y)dy}$
(c) $\int f(y)dy$ (d) none of these

ANSWERS

Fill in the Blanks

1. differential equation	**2.** ODE	**3.** PDE	**4.** 2	**5.** higher
6. general solution	**7.** P.I.	**8.** singular	**9.** vx	**10.** exact

True/False

1. T	**2.** F	**3.** T	**4.** T	**5.** T	**6.** F	**7.** T	**8.** F	**9.** T	**10.** T

Multiple Choice Questions

1. (a)	**2.** (a)	**3.** (c)	**4.** (b)	**5.** (d)	**6.** (a)	**7.** (a)	**8.** (c)	**9.** (b)	**10.** (b)

FFFFF

CHAPTER 29 Linear Differential Equation with Constant Coefficients

29.1 INTRODUCTION

The equation $\frac{d^n y}{dx^n} + A_1 \frac{d^{n-1} y}{dx^{n-1}} + A_2 \frac{d^{n-2} y}{dx^{n-2}} + ... + A_n y = B$ ...(1)

having $A_1, ..., A_n$ and B either constant or function of x, is called the linear differential equation of nth order.

If $A_1, A_2, ..., A_n$ are all constants and B may not be constant, then equation (1) is said to be linear differential equation of nth degree with constant coefficients.

If we take $B = 0$, then the corresponding equation is called homogeneous equation.

Using the symbols $D, D^2, ..., D^n$ for $\frac{d}{dx}, \frac{d^2}{dx^2}, ..., \frac{d^n}{dx^n}$ respectively in (1), then we get

$$D^n y + A_1 D^{n-1} y + A_2 D^{n-2} y + ... + A_n y = B$$

$$\Rightarrow \quad (D^n + A_1 D^{n-1} + A_2 D^{n-2} + ... + A_n) y = B \Rightarrow f(D) y = B \qquad ...(2)$$

where, $f(D) = D^n + A_1 D^{n-1} + A_2 D^{n-2} + ... + A_n$.

Now, consider the homogeneous differential equation

$$f(D) y = 0 \qquad ...(3)$$

(Obtained by putting right hand side, *i.e.*, B equal to zero).

Now, we shall show that if $y_1, y_2, ..., y_n$ are n linearly independent solutions of (3), then $(C_1 y_1 + C_2 y_2 + ... + C_n y_n)$ is also a solution of (3), where $C_1, C_2, ..., C_n$ are arbitrary constants.

Since, we assumed that $y_1, y_2, ..., y_n$ are solution of (3) $\Rightarrow y_1, y_2, ..., y_n$ must satisfy (3).

which gives
$$\left.\begin{array}{l} f(D) y_1 = 0 \\ f(D) y_2 = 0 \\ \ldots\ldots\ldots\ldots\ldots \\ f(D) y_n = 0 \end{array}\right] \qquad ...(4)$$

Now consider

$$f(D)(C_1 y_1 + C_2 y_2 + ... + C_n y_n) = f(D)(C_1 y_1) + f(D)(C_2 y_2) + ... + f(D)(C_n y_n)$$
$$= C_1 f(D) y_1 + C_2 f(D) y_2 + ... + C_n f(D) y_n$$
$$= C_1.0 + C_2.0 + ... + C_n.0 \qquad \text{(By using (4))}$$

Therefore, we have $f(D)\,[C_1 y_1 + C_2 y_2 + ... C_n y_n] = 0$ ...(5)

$\Rightarrow$ $(C_1 y_1 + C_2 y_2 + ... + C_n y_n)$ satisfies (3).

$\Rightarrow$ $(C_1 y_1 + C_2 y_2 + ... + C_n y_n)$ is also a solution of (3).

Hence, we can say that if $y_1, y_2, ..., y_n$ are n linearly independent solution of (3), then $(C_1 y_1 + C_2 y_2 + ... + C_n y_n)$ is also a solution of (3) known as complete or general solution of (3), containing n arbitrary constants $C_1, C_2, ..., C_n$.

Now, let us suppose $(C_1 y_1 + C_2 y_2 + ... + C_n y_n) = u$ (say).

Then, from (5), we have

$$f(D)\,u = 0 \qquad ...(6)$$

Again, let v be any particular solution of (2). Therefore, we have

$$f(D)\,v = B \qquad ...(7)$$

Now, $$f(D)(u+v) = f(D)u + f(D)v \qquad \text{(Using (6) and (7))}$$

which shows that $(u+v)$, *i.e.*, $\{(C_1y_1 + C_2y_2 + ... + C_ny_n) + v\}$ is the general solution of (2).

WORKING PROCEDURE

Step 1. *Firstly, we find the general solution of (2), which is called the complimentary function (C.F.), contains as many arbitrary constants as is the order of the given differential equation.*

Step 2. *Next, find the solution of (1), with no arbitrary constant which is called the particular integral (P.I.).*

Step 3. *To find the general solution of (1), add C.F. and P.I. obtained in (1) and (2), i.e.,* $y = u + v =$ *C.F. + P.I.*

REMARKS

- Here, the operator D stands for d/dx, D^2 for d^2/dx^2 and so on.
- The operator D^{-1} stands for integration.
- Since, the symbol D satisfies the fundamental laws of algebra, therefore it can be regarded as an algebraic quantity.
- The general solution of (1) is $y = C.F. + P.I.$, where C.F. involves n arbitrary constants and P.I. does not involve any arbitrary constant.
- Since P.I. appears due to B in (1), therefore, if a linear differential equation with constant coefficients is given with $B = 0$, then its general solution will not involve P.I. and hence the general solution of the differential equation is given by $y =$ C.F.
- The equations, discussed in this chapter are most important in the study of vibrations of all kinds, mechanical, acoustical and electrical.
- The method (discussed above) of solving these type of equations, is given by Euler and D'Alembert.

29.1.1 AUXILIARY EQUATION

Consider the differential equation (1) with $B = 0$, *i.e.*,

$$(D^n + A_1D^{n-1} + A_2D^{n-2} + ... + A_n)y = 0 \quad \text{or} \quad f(D)y = 0. \qquad ...(1)$$

Substitute $y = e^{mx}$ on the trial basis, then we get $e^{mx}(m^n + A_1m^{n-1} + A_2m^{n-2} + ... + A_n) = 0$

which holds if

$$m^n + A_1m^{n-1} + A_2m^{n-2} + ... + A_n = 0 \quad \text{or} \quad f(m) = 0. \qquad ...(2)$$

Equation (2) is called the auxiliary equation.

From (1) and (2), we observe that the auxiliary equation $f(m) = 0$ will give the same value of m as the equation $f(D) = 0$ gives the value of D.

29.2 METHOD OF FINDING THE COMPLEMENTARY FUNCTION (C.F.)

To find the C.F., the roots of the auxiliary equation (2) are to be considered. Three different cases arise :

(i) The roots of auxiliary equation (2) are real.

(ii) The roots of auxiliary equation (2) are complex, *i.e.*, $\alpha \pm i\beta$ type.

(iii) The roots of auxiliary equation (2) are surds, *i.e.*, $\alpha \pm \sqrt{\beta}$ type.

Case (i) : (a) Suppose that the auxiliary equation (2) has n distinct roots $m_1, m_2, ..., m_n$, then C.F. is given by

$$C_1e^{m_1x} + C_2e^{m_2x} + ... + C_ne^{m_nx}$$

where $C_1, C_2, ..., C_n$ are arbitrary constants.

(b) If the auxiliary equation having r roots are equal to m_1(say) and remaining roots are distinct, then the C.F. is given by

$$[C_1 + C_2x + C_3x^2 + ... + C_rx^{r-1}]\,e^{m_1x} + C_{r+1}e^{m_{r+1}x} + ... + C_ne^{m_nx}.$$

Case (ii): If some of the roots of the auxiliary equation are complex, then we shall use the following procedure.

Let $\alpha \pm i\beta$ be the roots of the auxiliary equation, then the corresponding part becomes

$$= C_1e^{(\alpha+i\beta)x} + C_2e^{(\alpha-i\beta)x} = C_1e^{\alpha x}.e^{i\beta x} + C_2e^{\alpha x}.e^{-i\beta x}$$

$$= e^{\alpha x}[C_1\cos\beta x + iC_1\sin\beta x] + e^{\alpha x}[C_2\cos\beta x - iC_2\sin\beta x]$$

$$= e^{\alpha x}[(C_1 + C_2)\cos\beta x + (iC_1 - iC_2)\sin\beta x]$$

$$\text{C.F.} = e^{\alpha x}\,[B_1 \cos\beta x + B_2 \sin\beta x] \qquad ...(1)$$

where, B_1, B_2 are arbitrary constants.

The expression (1) can also be written as

(a) $B_1 e^{\alpha x}\cos(\beta x + B_2)$

(b) $B_1 e^{\alpha x}\sin(\beta x + B_2)$.

If, the equation has two equal pair of complex roots $\alpha + i\beta$ and $\alpha - i\beta$, say, occur twice, then the corresponding part of C.F. is written as

$$e^{\alpha x}[(B_1 + B_2 x)\cos\beta x + (B_3 + B_4 x)\sin\beta x].$$

In general, if $\alpha \pm i\beta$ occur k times, then the corresponding part of the C.F. can be written as

$$e^{\alpha x}\{(B_1 + B_2 x + ... + B_k x^{k-1})\cos\beta x + (B_{k+1} + B_{k+2}x + ... + B_{2k}x^{k-1})\}\,\sin\beta x$$

where $B_1, B_2, ..., B_k, B_{k+1}, ..., B_{2k}$ are arbitrary constants.

Case (iii) : If a pair of the roots of the auxiliary equation involves surds, say $\alpha \pm \sqrt{\beta}$, where $\beta > 0$, then the corresponding part of C.F. in one of the following three forms

(a) $e^{\alpha x}[B_1 \cosh(x\sqrt{\beta}) + B_2 \sinh(x\sqrt{\beta})]$ (b) $B_1 e^{\alpha x}\cosh(x\sqrt{\beta} + B_2)$ (c) $B_1 e^{\alpha x}\sinh(x\sqrt{\beta} + B_2)$

Remarks

- The results obtained in case (iii), are exactly similar to those of case (ii) except that sin and cos replaced by sinh and cosh respectively.
- The method of finding the complimentary function (C.F.) of the following differential equation of the form

$$(D^n + A_1 D^{n-1} + A_2 D^{n-2} + ... + A_n)y = 0$$

can be concluded as follows :

S. No.	Nature of the Roots	Solution
1.	Real and distinct say $m_1, m_2, ..., m_n$	$y = B_1 e^{m_1 x} + B_2 e^{m_2 x} ... + B_n e^{m_n x}$
2.	Real and equal, say m_1	$y = (B_1 + B_2 x + B_3 x^2 + ... + B_n x^{n-1})e^{m_1 x}$
3.	Non-repeated roots : $\alpha \pm i\beta$	(a) $y = (B_1 \cos\beta x + B_2 \sin\beta x)e^{\alpha x}$ (b) $y = B_1 e^{\alpha x}\cos(\beta x + B_2)$
4.	Repeated roots : $\alpha \pm i\beta$, r times	$y = \{(B_1 + B_2 x + ... + B_r x^{r-1})\cos\beta x$ $+ (B_1' + B_2' x + ... + B_r' x^{r-1})\sin\beta x\}e^{\alpha x}$
5.	Irrational roots : $\alpha \pm \sqrt{\beta}$	(a) $y = B_1 e^{\alpha x}\cosh(x\sqrt{\beta} + B_2)$ (b) $y = B_1 e^{\alpha x}\sinh(x\sqrt{\beta} + B_2)$

Solved Examples

Example 1. *Solve* $[D^3 + 6D^2 + 11D + 6]\,y = 0$.

Solution. Here, the given differential equation is

$$[D^3 + 6D^2 + 11D + 6]\,y = 0$$

To find the auxiliary equation, replace D by m, then (1) becomes,

$$m^3 + 6m^2 + 11m + 6 = 0$$

$$\Rightarrow \quad (m+1)(m^2 + 5m + 6) = 0$$

$$\Rightarrow \quad (m+1)(m+2)(m+3) = 0$$

$$\Rightarrow \quad m = -1, -2, -3$$

i.e., Roots are real and unequal. Hence, the general solution is

$$y = C_1 e^{-x} + C_2 e^{-2x} + C_3 e^{-3x}.$$

Example 2. *Solve* $[D^4 + 2D^3 - 3D^2 - 4D + 4]\,y = 0$.

Solution. Here, the auxiliary equation is

$$m^4 + 2m^3 - 3m^2 - 4m + 4 = 0$$

or $$(m-1)(m^3 + 3m^2 - 4) = 0$$

$$\Rightarrow (m-1)(m-1)(m^2 + 4m + 4) = 0$$

$$\Rightarrow \quad (m-1)(m-1)(m+2)^2 = 0$$

$$\Rightarrow \quad m = +1, +1, -2, -2$$

$\Rightarrow$ Repeated real roots exist.

Hence, general solution is

$$y_1 = (C_1 + C_2 x)e^{x} + (C_3 + C_4 x)\,e^{-2x}.$$

Example 3. *Solve* $(D^4 + k^4)y = 0$. [UPTU(Q. Bank)–2001]

Solution. Here, the auxiliary equation is

$$m^4 + k^4 = 0 \text{ or } (m^2 + k^2)^2 - 2k^2m^2 = 0$$

$$\Rightarrow (m^2 + k^2)^2 - (\sqrt{2}\,.\,km)^2 = 0$$

$$\Rightarrow (m^2 + k^2 - \sqrt{2}.km)(m^2 + k^2 + \sqrt{2}.km) = 0$$

$$\Rightarrow m^2 - \sqrt{2}.km + k^2 = 0$$

and $$m^2 + k^2 + \sqrt{2}\,.\,km = 0$$

$$\Rightarrow m = \frac{\sqrt{2}\,k \pm \sqrt{(2k^2 - 4k^2)}}{2}$$

and $$m = \frac{-\sqrt{2}\,k \pm \sqrt{(2k^2 - 4k^2)}}{2}$$

$$\Rightarrow m = \frac{k}{\sqrt{2}} \pm i\frac{k}{\sqrt{2}} \text{ and } m = -\frac{k}{\sqrt{2}} \pm i\frac{k}{\sqrt{2}}$$

Hence, the solution is

$$y = e^{kx/\sqrt{2}}\{C_1 \cos(kx/\sqrt{2}) + C_2 \sin(kx/\sqrt{2})$$

$$+ e^{-kx/\sqrt{2}}\{C_3 \cos(kx/\sqrt{2}) + C_4 \sin(kx/\sqrt{2})\}$$

Example 4. *Solve* $[D^4 - 4D^3 + 8D^2 - 8D + 4]\,y = 0$.

Solution. Here, the auxiliary equation is

$$m^4 - 4m^3 + 8m^2 - 8m + 4 = 0$$

$$\Rightarrow (m^2 - 2m + 2)^2 = 0 \Rightarrow m = 1 \pm i,\ 1 \pm i$$

$\Rightarrow$ Repeated complex roots exist.

Hence, the solution of the given equation is

$$y = e^x\ \{(C_1 + C_2x)\cos x + (C_3 + C_4x)\sin x\}.$$

Example 5. *Solve* $(D^2 + 6D + 4)y = 0$.

Solution. Here, the auxiliary equation is

$$m^2 + 6m + 4 = 0 \ \Rightarrow\ m = -3 \pm \sqrt{5}.$$

$\Rightarrow$ Irrational roots exist.

Hence, the solution of the given equation is

$$y = e^{-3x}\ (C_1 \cosh x\sqrt{5} + C_2 \sinh x\sqrt{5}).$$

Example 6. *Solve* $(D^4 - n^4)y = 0$, *where* $D \equiv \frac{d}{dx}$ [UPTU(Q. Bank)–2001]

Solution. The auxiliary equation is

$$m^4 - n^4 = 0 \quad \Rightarrow \quad (m^2 - n^2)(m^2 + n^2) = 0$$

$$\Rightarrow \quad m = \pm n,\ \pm ni$$

$$\text{C.F.} = C_1e^{nx} + C_2e^{-nx} + e^x(C_3 \cos nx + C_4 \sin nx)$$

$$= C_1e^{nx} + C_2e^{-nx} + C_3 \cos nx + C_4 \sin nx$$

Hence, the solution is

$$y = C_1e^{mx} + C_2e^{-nx} + C_3 \cos nx + C_4 \sin nx.$$

EXERCISE 29.1

Solve the following equations :

1. $\frac{d^2y}{dx^2} + 3\frac{dy}{dx} + 2y = 0$
2. $(D^3 - 9D^2 + 23D - 15)y = 0$
3. $(D^4 - D^3 - 9D^2 - 11D - 4)y = 0$
4. $(D^2 + 1)^2\,(D - 1)^2 y = 0$
5. $(D^3 - D^2 - 12D)y = 0$
6. $(D^4 + 2n^2D^2 + n^4)y = 0$
7. $[D^2 - 2\lambda D + (\lambda^2 + \mu^2)]y = 0$
8. $(D^4 + D^3 + 2D^2 - D + 3)\ y = 0$
9. $(D^5 - 13D^3 + 26D^2 + 82D + 104)y = 0$
10. $\frac{d^4y}{dx^4} + y = 0$
11. $(D^3 - 3D^2 + 4)y = 0$ [UPTU(Q. Bank)–2001]
12. $(D^2 - 2D + 4)^2 y = 0$ [UPTU(Q. Bank)–2001]
13. $\frac{d^2x}{dt^2} + 5\frac{dx}{dt} + 6x = 0$, given $x(0) = 0$, $\frac{dx(0)}{dt} = 15$ (VTU–2010)
14. $(D^4 - 4D + 4)y = 0$ (Bhopal–2008)
15. $(D^2 + 1)^3 y = 0$, where $D \equiv d/dx$
16. $\frac{d^2x}{dt^2} - 4\frac{dx}{dt} + 13x = 0, x(0), \frac{dx(0)}{dt} = 2$ (VTU–2008)
17. $\frac{d^3y}{dx^3} + y = 0$ (VTU–2000S)
18. $\frac{d^4y}{dx^4} + 8\frac{d^2y}{dx^2} + 16y = 0$ (JNTU–2005)
19. $(4D^4 - 8D^3 - 7D^2 + 11D + 6)y = 0$ (VTU–2008)

Hint to Selected Problems

1. $m = -1, -2$
2. $m = 1, 5, 3$
3. $m = -1, -1, -1, 4$
4. $m = \pm i, \pm i, 1, 1$
5. $m = 0, 4, 3$
6. $m = \pm ni,\ \pm ni$
7. $m = \lambda \pm \mu i$
8. $m = -1 \pm i\sqrt{2},\ \frac{1}{2} \pm i\frac{\sqrt{3}}{2}$
9. $m = -1 \pm i,\ -3 \pm 2i,\ -4$
10. $m = \frac{-1 \pm i}{\sqrt{2}},\ \frac{1 \pm i}{\sqrt{2}}$
11. $m = -1, 2, 2$
12. $m = 1 \pm \sqrt{3}i,\ 1 \pm \sqrt{3}i$

ANSWERS

1. $y = C_1e^{-x} + C_2e^{-2x}$
2. $y = C_1e^{x} + C_2e^{3x} + C_3e^{5x}$
3. $y = e^{-x}(C_1 + C_2x + C_3x^2) + C_4e^{4x}$

4. $y=(C_1+C_2x)\sin x+(C_3+C_4x)\cos x+(C_5+C_6x)e^x$

5. $y=C_1+C_2e^{4x}+C_3e^{-3x}$

6. $y=(C_1-C_2x)\cos nx+(C_3+C_4x)\sin nx$

7. $y=e^{\lambda x}(C_1\cos\mu x+C_2\sin\mu x)$

8. $y=e^{-x}\left[C_1\cos(\sqrt{2}x)+C_2\sin(\sqrt{2}x)+e^{x/2}\left(C_3\cos\frac{\sqrt{3}}{2}x+C_4\sin\frac{\sqrt{3}}{2}x\right)\right]$

9. $y=C_1e^{-x}\cos(x+\alpha)+C_2e^{-3x}\cos(2x+\beta)+C_3e^{-4x}$

10. $y=C_1e^{x/\sqrt{2}}\cos\left(\frac{x}{\sqrt{2}}+C_2\right)+C_3e^{-x/\sqrt{2}}\cos\left(\frac{x}{\sqrt{2}}+C_4\right)$

11. $y=C_1e^{-x}+(C_2+C_3x)e^{2x}$

12. $y=e^x[(C_1+C_2x)\cos\sqrt{3}x+(C_3+C_4x)\sin\sqrt{3}x]$

13. $x=15(e^{-2t}-e^{-3t})$

14. $y=((C_1+C_2x)e^{\sqrt{2}x}+(C_3+C_4x)e^{-\sqrt{2}x})$

15. $y=(C_1+C_2x+C_3x^2)\cos x+(C_4+C_5x+C_6x^2)\sin x$

16. $\frac{2}{3}e^{2t}\sin 3t$

17. $y=C_1e^{-x}+e^{x/2}\left(C_2\cos\frac{\sqrt{3x}}{2}+C_3\sin\frac{\sqrt{3x}}{2}\right)$

18. $y=(C_1+C_2x)\cos 2x+(C_3+C_4x)\sin 2x$

19. $y=C_1e^{-x}+C_2e^{2x}+e^{x/2}\left(C_3\cos\frac{x}{\sqrt{2}}+C_4\sin\frac{x}{\sqrt{2}}\right)$

29.3 PARTICULAR INTEGRAL

Consider the differential equation $\quad f(D)y=B \Rightarrow y=\frac{1}{f(D)}.B$

Let $\frac{1}{f(D)}.B$ denote some function of x, which operated upon by $f(D)$ produces B. Hence, P.I. $=\frac{1}{f(D)}.B$.

29.3.1 GENERAL METHOD OF FINDING P.I.

THEOREM 1. *If B is a function of x, then* $\frac{1}{D-a}B=e^{ax}\int Be^{-ax}\,dx$.

Proof. Le t $y=\frac{1}{D-a}B \Rightarrow (D-a)y=B \Rightarrow \left(\frac{d}{dx}-a\right)y=B \Rightarrow \frac{dy}{dx}-ay=B$

which is the linear differential equation. I.F. $=e^{-\int a dx}=e^{-ax}$.

Hence, solution is given by $ye^{-ax}=\int Be^{-ax}dx$.

(Since we find the P.I., therefore we omit the constant of integration.)

$\therefore \quad y=e^{ax}\int Be^{-ax}\,dx \quad \Rightarrow \quad \frac{1}{D-a}.B=e^{ax}\int Be^{-ax}dx$

REMARKS

- P.I. never contains any arbitrary constant.
- The method discussed above can be used to evaluate P.I. in any problem. It does not depend upon the form of B.
- The method discussed above must be used when B is of the form $\sec ax$, $\operatorname{cosec} ax$, $\tan ax$, etc.
- Here, the operator $\frac{1}{f(D)}$ (known as increase operator) having the following properties :

(a) If $B=u_1+u_2+\ldots+u_n$, then $\frac{1}{f(D)}.B=\frac{1}{f(D)}.u_1+\frac{1}{f(D)}.u_2+\ldots\frac{1}{f(D)}.u_n$

(b) $\frac{1}{f(D)}(KB)=\frac{K}{f(D)}.B$

(c) $\frac{1}{f(D)}$ can be resolved into factors.

(d) $\frac{1}{f(D)}$ can be broken into partial fractions.

(e) $\frac{1}{f(D)}.B$ is a particular integration.

Solved Examples

Example 1. *Solve* $D^2-5D+6=e^{3x}$.

Solution. The given equation can be written as

$(D-3)(D-2)y=e^{3x}$

$$\text{C.F.}=C_1e^{3x}+C_2e^{2x}$$

and

$$\text{P.I.}=\frac{1}{D-3}\cdot\frac{1}{D-2}e^{3x}=\frac{1}{D-3}e^{2x}\int e^{3x}e^{-2x}dx$$

$$=\frac{1}{D-3}e^{2x}.e^{x}=e^{3x}\int e^{3x}.e^{-3x}dx=xe^{3x}.$$

Now, general solution = C.F. + P.I.

$$\Rightarrow\quad y=C_1e^{3x}+C_2e^{2x}+xe^{3x}.$$

Example 2. *Solve* $(D^2+1)y=\sec^2 x$.

Solution. Here, the given equation is

$$(D^2+1)y=\sec^2 x \qquad ...(1)$$

To find the C.F. of (1).

The auxiliary equation of (1) is given by

$m^2+1=0 \Rightarrow m=\pm i$

$\Rightarrow$ C.F. $=C_1\cos x+C_2\sin x$

$$\text{P.I.}=\frac{1}{D^2+1}\sec^2 x$$

$$=\frac{1}{(D+i)(D-i)}\sec^2 x=\frac{1}{2i}\left[\frac{1}{D-i}-\frac{1}{D+i}\right]\sec^2 x$$

$$=\frac{1}{2i}\left[e^{xi}\int e^{-ix}\sec^2 x\,dx-e^{-ix}\int e^{ix}\sec^2 x\,dx\right]$$

$$=\frac{1}{2i}\left\{e^{ix}\int\frac{\cos x-i\sin x}{\cos^2 x}dx-e^{-ix}\int\frac{\cos x+i\sin x}{\cos^2 x}dx\right\}$$

$$=\frac{1}{2i}\left\{e^{ix}\int(\sec x-i\sec x\tan x)dx\right.$$
$$\left.-e^{-ix}\int(\sec x+i\sec x\tan x)dx\right\}$$

$$=\frac{1}{2i}\left\{(e^{ix}-e^{-ix})\int\sec x dx-i(e^{ix}+e^{-ix})\right.$$
$$\left.\int\tan x\sec x dx\right\}$$

$$=\frac{1}{2i}\{2i\sin x\log(\sec x+\tan x)-2i\cos x\sec x\}$$

$$=\sin x\log(\sec x+\tan x)-1.$$

Hence, the general solution is y = C.F. + P.I.

$$\Rightarrow\quad y=C_1\cos x+C_2\sin x$$
$$+\sin x\log(\sec x+\tan x)-1.$$

Example 3. *Solve* $(D^2+9)y=\sec 3x$.

Solution. Auxiliary equation is $m^2+9=0 \Rightarrow m=\pm 3i$

$\therefore$ C.F. $=c_1\cos 3x+c_2\sin 3x$

$$\text{P.I.}=\frac{\sec 3x}{D^2+9}=\frac{\sec 3x}{(D+3i)(D-3i)}$$

$$=\frac{1}{6i}\left[\frac{1}{D-3i}-\frac{1}{D+3i}\right]\sec 3x$$

$$=\frac{1}{6i}\left[e^{3ix}\int e^{-3ix}\sec 3x dx-e^{-3ix}\int e^{3ix}\sec 3x dx\right]$$

$$=\frac{1}{6i}\left[e^{3ix}\left\{\int\left(1-i\frac{\sin 3x}{\cos 3x}\right)dx\right\}\right.$$
$$\left.-e^{-3ix}\left\{\int\left(1+i\frac{\sin 3x}{\cos 3x}\right)dx\right\}\right]$$

$$=\frac{1}{6i}\left[e^{3ix}\left\{x+\frac{i}{3}\log\cos 3x\right\}\right.$$
$$\left.-e^{-3ix}\left\{x-\frac{i}{3}\log\cos 3x\right\}\right]$$

$$=\frac{1}{6i}\left[(\cos 3x+i\sin 3x)\left(x+\frac{i}{3}\log\cos 3x\right)\right.$$
$$\left.-(\cos 3x-i\sin 3x)\left(x-\frac{i}{3}\log\cos 3x\right)\right]$$

$$=\frac{1}{6i}\left[\frac{2i}{3}\cos 3x\log\cos 3x+2i\sin 3x\right].$$

$$=\frac{1}{9}\cos 3x\log\cos 3x+\frac{x}{3}\sin 3x$$

Hence, y = C.F. + P.I.

$$=c_1\cos 3x+c_2\sin 3x$$
$$+\frac{x}{3}\sin 3x+\frac{1}{9}\cos 3x\log\cos 3x.$$

EXERCISE 29.2

Solve the following differential equations :

1. $(D^2+a^2)y=\sec ax$ (MTU(B.Pharma.)–2011, UKTU–2011)
2. $(D^2+a^2)y=\tan ax$
3. $(D^2+1)y=\text{cosec } x$
4. $(D^2+n^2)y=\cot nx$
5. $(D^2+n^2)y=\tan nx$

Hint to Selected Problems

1. $m=\pm ai \Rightarrow$ C.F. $=C_1\cos ax+C_2\sin ax$

$$\text{P.I.}=\frac{1}{D^2+a^2}\sec ax=\frac{1}{(D+ai)(D-ai)}\sec ax$$

$$=\frac{1}{2ai}\left[\frac{1}{(D-ai)}-\frac{1}{(D+ai)}\right]\sec ax$$

$$=\frac{1}{2ai}\left\{e^{iax}\int e^{-iax}\sec ax\,dx-e^{-iax}\int e^{iax}\sec ax\,dx\right\}$$

2. P.I. $= \frac{1}{D^2+a^2}\tan ax = \frac{1}{2ai}\left[\frac{1}{D-ia} - \frac{1}{D+ia}\right]\tan ax$.

Then proceed as above.

3. P.I. $= \frac{1}{D^2+1}\operatorname{cosec} x$

$= \frac{1}{(D+i)(D-i)}\operatorname{cosec} x = \frac{1}{2i}\left[\frac{1}{D-i} + \frac{1}{D+i}\right]\operatorname{cosec} x$.

Now proceed as above.

4. C.F. $= C_1\cos nx + C_2 \sin nx$

P.I. $= \frac{1}{D^2+n^2}\cot nx = \frac{1}{(D-in)(D+in)}\cot nx$

$= \frac{1}{2in}\left[\frac{1}{D-in}\cot nx - \frac{1}{D+in}\cot nx\right]$

ANSWERS

1. $y = C_1\cos ax + C_2 \sin ax + \frac{x}{a}\sin ax + \frac{1}{a^2}\cos ax \log \cos ax$

2. $y = C_1\cos ax + C_2 \sin ax - \frac{1}{a^2}\cos ax \log \tan\left(\frac{\pi}{4} + \frac{ax}{2}\right)$

3. $y = C_1\cos x + C_2 \sin x + \sin x \log \sin x - x\cos x$

4. $y = C_1\cos nx + C_2 \sin nx + \frac{1}{n^2}\sin nx \log(\operatorname{cosec} nx - \cot nx)$

5. $y = C_1\cos nx + C_2 \sin nx - \frac{1}{n^2}\cos nx \log(\sec nx + \tan nx)$

29.3.2 SHORT METHODS OF GETTING P.I.

The general method for getting P.I. discussed above requires lot of calculations. In certain cases, the P.I. can be obtained by methods which are shorter than the general method.

(1) To evaluate P.I., when B is of the form e^{ax} :

Here, we want to evaluate $\frac{1}{f(D)}e^{ax}$ where, $f(D) = A_0D^n + A_1D^{n-1} + ... + A_n$ with $f(a) \neq 0$.

Here, $B = e^{ax}$, we have

$$D(e^{ax}) = ae^{ax}$$

$$D^2(e^{ax}) = a^2e^{ax}$$

....

....

$$D^n(e^{ax}) = a^ne^{ax}$$

$$\Rightarrow \quad f(D)e^{ax} = (A_0D^n + A_1D^{n-1} + ... + A_n)e^{ax} = A_0D^ne^{ax} + A_1D^{n-1}e^{ax} + ... + A_ne^{ax}$$

$$= A_0a^ne^{ax} + A_1a^{n-1}e^{ax} + ... + A_ne^{ax} = (A_0a^n + A_1a^{n-1} + ... + A_n)\, e^{ax}$$

$$\Rightarrow \quad f(D)\, e^{ax} = f(a)\, e^{ax}.$$

Operating upon both sides with $\frac{1}{f(D)}$, we get $\frac{1}{f(D)} \cdot f(D).e^{ax} = \frac{1}{f(D)} \cdot f(a)\, e^{ax} \Rightarrow e^{ax} = f(a)\frac{1}{f(D)}e^{ax}$

$$\Rightarrow \quad \frac{1}{f(D)}e^{ax} = \frac{e^{ax}}{f(a)}, \text{ provided } f(a) \neq 0.$$

SOLVED EXAMPLES

Example 1. *Solve* $(D^2 - 3D + 2)y = e^{5x}$.

Solution. The given equation is

$$(D^2 - 3D + 2)y = e^{5x}$$

Auxiliary equation is $m^2 - 3m + 2 = 0$

$\Rightarrow \quad (m-1)(m-2) = 0 \Rightarrow m = 1, 2$.

$\therefore$ C.F. $= C_1e^x + C_2e^{2x}$

Now, P.I. $= \frac{1}{D^2 - 3D + 2}.e^{5x} = \frac{1}{25 - 3\times 5 + 2}e^{5x}$

$= \frac{1}{12}e^{5x}$

Hence, the general solution is

$y =$ C.F. + P.I.

$\Rightarrow \quad y = C_1e^x + C_2e^{2x} + \frac{1}{12} \cdot e^{5x}$.

Example 2. *Solve* $(D^3 + 1)y = (e^x + 1)^2$.

Solution. The given equation is

$$(D^3 + 1)y = (e^x + 1)^2$$

The auxiliary equation is $m^3 + 1 = 0$

$\Rightarrow (m+1)(m^2 - m + 1) = 0 \Rightarrow m = -1, \frac{1}{2} \pm \frac{i\sqrt{3}}{2}$

Therefore,

$$\text{C.F.} = C_1e^{-x} + e^{x/2}\left[C_2\cos\left(\frac{x\sqrt{3}}{2}\right) + C_3\sin\left(\frac{x\sqrt{3}}{2}\right)\right].$$

Now,

$$\text{P.I.} = \frac{1}{(D^3+1)}[e^x+1]^2 = \frac{1}{(D^3+1)}(e^{2x}+2e^x+1)$$

$$= \frac{1}{D^3+1}(e^{2x}+2e^x+e^{0x})$$

$$= \frac{1}{D^3+1}e^{2x} + 2\frac{1}{D^3+1}e^x + \frac{1}{D^3+1}e^{0x}.$$

$$= \frac{1}{2^3+1}e^{2x} + 2\frac{1}{1^3+1}e^x + \frac{1}{0+1}e^{0x} = \frac{1}{9}e^{2x} + e^x + 1$$

Here, the general solution is

$y = \text{C.F.} + \text{P.I.}$

$$\Rightarrow\quad y = C_1e^{-x} + e^{x/2}\left[C_2\cos\left(\frac{x\sqrt{3}}{2}\right) + C_3\sin\left(\frac{x\sqrt{3}}{2}\right)\right] + \frac{1}{9}e^{2x} + e^x + 1$$

Example 3. *Solve* $(D^3 - 2D^2 + 4D - 8)y = 8.$ (UPTU(B.Pharm)SUM–2009)

Solution. The given equation is

$$(D^3 - 2D^2 + 4D - 8)y = 8$$

Auxiliary equation is

$$m^3 - 2m^2 + 4m - 8 = 0$$

$$\Rightarrow\quad (m^2 + 4)(m - 2) = 0$$

$$\Rightarrow\quad m = 2, \pm 2i$$

$$\therefore\quad \text{C.F.} = C_1e^{2x} + C_2\cos 2x + C_3\sin 2x$$

$$\text{Now P.I.} = \frac{1}{D^3 - 2D^2 + 4D - 8}(8e^{0x})\;(\because e^{0x} = 1)$$

$$= \frac{1}{(0)^3 - 2(0)^2 + 4(0) - 8}(8e^{0x}) = -1$$

Hence, the general solution is

$$y = \text{C.F.} + \text{P.I.}$$

$$\Rightarrow\quad y = C_1e^{2x} + C_2\cos 2x + C_3\sin 2x - 1$$

Example 4. *Solve* $(D-2)^3y = 17e^{2x}$ (MTU–2011)

Solution. The given equation is

$$(D-2)^3y = 17e^{2x}$$

Auxiliary equation is

$$(m-2)^3 = 0$$

$$\Rightarrow\quad m = 2, 2, 2$$

$$\therefore\quad \text{C.F.} = (C_1 + C_2x + C_3x^2)e^{2x}$$

$$\text{Now P.I.} = \frac{1}{(D-2)^3}17e^{2x} \qquad |\text{Case of failure}$$

$$= 17x\left[\frac{1}{3(D-2)^2}e^{2x}\right] \qquad |\text{Again case of failure}$$

$$= \frac{17}{3}x^2\left[\frac{1}{2(D-2)}e^{2x}\right] \qquad |\text{Again case of failure}$$

$$= \frac{17}{6}x^3e^{2x}$$

Example 5. *Solve* $2\frac{d^3y}{dx^3} - \frac{d^2y}{dx^2} + 4\frac{dy}{dx} - 2y = e^x$ (UPTU–2007)

Solution. The given equation can be written as

$$(2D^3 - D^2 + 4D - 2)y = e^x$$

Auxiliary equation is

$$2m^3 - m^2 + 4m - 2 = 0$$

$$\Rightarrow\quad (2m-1)(m^2+2) = 0$$

$$\Rightarrow\quad m = \frac{1}{2}, \pm\sqrt{2}i$$

$$\therefore\quad \text{C.F.} = C_1e^{x/2} + C_2\cos\sqrt{2}x + C_3\sin\sqrt{2}x$$

$$\text{Now P.I.} = \frac{1}{2D^3 - D^2 + 4D - 2}e^x$$

$$= \frac{1}{2(1)^3 - (1)^2 + 4(1) - 2}e^x = \frac{1}{3}e^x$$

Hence, the complete solution is

$$y = \text{C.F.} + \text{P.I.}$$

$$= C_1e^{x/2} + C_2\cos\sqrt{2}x + C_3\sin\sqrt{2}x + \frac{1}{3}e^x$$

EXERCISE 29.3

Solve the following differential equations :

1. $(D^2 - 4D + 1)y = e^{2x} - e^{-x}$
2. $(D^2 + 5D + 6)y = e^{2x}$
3. $(4D^2 + 4D - 3)y = e^{2x}$
4. $(D^2 - 2D + 1)y = 2e^{5x/2}$
5. $(D^2 + D + 1)y = e^{-x}$
6. $D^2(D+1)^2(D^2+D+1)^2y = e^x$
7. $[D^2 + 2pD + (p^2+q^2)]\,y = e^{ax}$
8. $(4D^2 + 12D + 9)y = 144e^{-3x}$
9. $(D^2 - 4D + 3)y = e^{3x}$ [UPTU(Q. Bank)–2001]
10. $(D^2 - a^2)y = e^{ax} - e^{-ax}$ [UPTU(Q. Bank)–2001]
11. $(D^2 + D + 1)y = (1 + e^x)^2$ [UPTU(Q. Bank)–2001]
12. $\frac{d^3y}{dx^3} - 3\frac{d^2y}{dx^2} + 3\frac{dy}{dx} - y = e^x + 2$ [UPTU(Q. Bank)–2001]

Hint to Selected Problems

1. $m = 2 \pm \sqrt{3} \Rightarrow$ C.F. $= e^{2x}[C_1 \cosh x\sqrt{3} + C_2 \sinh x\sqrt{3}]$

$$\text{P.I.} = \frac{1}{D^2 - 4D + 1}[e^{2x} - e^{-x}]$$

$$= \frac{1}{D^2 - 4D + 1} e^{2x} - \frac{1}{(D^2 - 4D + 1)} e^{-x}$$

$$= \frac{e^{2x}}{2^2 - 4 \times 2 + 1} - \frac{e^{-x}}{(-1)^2 - 4(-1) + 1} = -\frac{e^{2x}}{3} - \frac{e^{-x}}{6}$$

4. $m = 1, 1 \Rightarrow$ C.F. $= (C_1 + C_2 x)e^x$

$$\text{P.I.} = \frac{1}{D^2 - 2D + 1}(2e^{5x/2}) = 2.\frac{e^{5x/2}}{\frac{25}{4} - 4}.$$

6. $m = 0, 0, -1, -1, -\frac{1}{2} \pm \frac{i\sqrt{3}}{2}, \ -\frac{1}{2} \pm \frac{i\sqrt{3}}{2}.$

7. $m = -p \pm iq$.

8. $m = -\frac{3}{2}, -\frac{3}{2} \Rightarrow$ C.F. $= (C_1 + C_2 x)\, e^{-3x/2}$

$$\text{P.I.} = 144\left(\frac{1}{4D^2 + 12D + 9}\right)e^{-3x} = \frac{144\, e^{-3x}}{9}$$

9. $\text{P.I.} = \frac{1}{D^2 - 4D + 3} e^{3x} = \frac{1}{2D - 4} e^{3x} = \frac{x}{2} . e^{3x}$

10. $m = \pm a$

11. $m = -\frac{1}{2} \pm \frac{\sqrt{3}}{2} i$

12. $m = 1, 1, 1$

ANSWERS

1. $y = e^{2x}(C_1 \cosh x\sqrt{3} + C_2 \sinh x\sqrt{3}) - \frac{1}{3}e^{2x} - \frac{1}{6}e^{-x}$

2. $y = C_1 e^{-2x} + C_2 e^{-3x} + \frac{1}{20}e^{2x}$

3. $y = C_1 e^{x/2} + C_2 e^{-3x/2} + \frac{1}{21}e^{2x}$

4. $y = (C_1 + C_2 x)\, e^x + \frac{8}{9}e^{5x/2}$

5. $y = e^{-x/2}\left[C_1 \cos\left(\frac{1}{2}x\sqrt{3}\right) + C_2 \sin\left(\frac{1}{2}x\sqrt{3}\right)\right] + e^{-x}$

6. $y = (C_1 + C_2 x)e^{0x} + (C_3 + C_4 x)e^{-x} + e^{-x/2}\left[(C_5 + C_6 x)\cos\left(\frac{1}{2}\sqrt{3}x\right) + (C_7 + C_8 x)\sin\left(\frac{1}{2}\sqrt{3}x\right)\right] + \frac{1}{36}e^x$

7. $y = e^{-px}(C_1 \cos qx + C_2 \sin qx) + \frac{e^{ax}}{[(p+a)^2 + q^2]}$

8. $y = (C_1 + C_2 x)\, e^{-3x/2} + 16e^{-3x}$

9. $y = C_1 e^x + C_2 e^{3x} + \frac{x}{2}e^{3x}$

10. $y = C_1 e^{ax} + C_2 e^{-ax} + \frac{x}{9}\cosh ax$

11. $y = e^{-x/2}\left[C_1 \cos\frac{\sqrt{3}}{2}x + C_2 \sin\frac{\sqrt{3}}{2}x\right] + 1 + \frac{1}{7}e^{2x} + \frac{2}{3}e^x$

12. $y = (C_1 + C_2 x + C_2 x^2)e^x + \frac{x^3}{6}e^x - 2$

(2) To evaluate P.I., when B is of the form $\sin ax$ or $\cos ax$:

Case (I) : If $f(D)$ contains even power of D :

Let us suppose

$$f(D^2) = A_0(D^2)^n + A_1(D^2)^{n-1} + \ldots + A_n .$$

Here, we observe that

$$D^2 \sin ax = -a^2 \sin ax$$
$$D^4 \sin ax = (-a^2)^2 \sin ax$$
$$D^6 \sin ax = (-a^2)^3 \sin ax$$
$$\ldots\ldots\ldots\ldots\ldots\ldots$$
$$(D^2)^n \sin ax = (-a^2)^n \sin ax$$

Consider
$$f(D^2)\sin ax = [A_0(D^{2n}) + A_1(D^{2n-2}) + \ldots + A_n]\sin ax$$
$$= A_0 D^{2n} \sin ax + A_1 D^{2n-2} \sin ax + \ldots + A_n \sin ax$$
$$= A_0(-a^2)^n \sin ax + A_1(-a^2)^{n-1} \sin ax + \ldots + A_n \sin ax = f(-a^2)\sin ax$$

Now, operating on both sides with $\frac{1}{f(D^2)}$, we get $\frac{1}{f(D^2)} . f(D^2)\sin ax = f(-a^2)\frac{1}{f(D^2)}\sin ax$

$$\Rightarrow \qquad \sin ax = f(-a^2)\left[\frac{1}{f(D^2)}\sin ax\right] \Rightarrow \frac{1}{f(D^2)}\sin ax = \frac{1}{f(-a^2)}\sin ax .$$

Case (II) : If $f(D)$ contains odd power of D :

Let us suppose, it be put in the form $f_1(D^2)+f_2(D^2)D$, then

$$\frac{1}{f(D)}\sin ax=\frac{1}{f_1(D^2)+f_2(D^2)D}\sin ax=\frac{1}{f_1(-a^2)+f_2(-a^2)D}\sin ax$$

$$=\frac{1}{p+qD}\sin ax \text{ (say)} \qquad \text{(Where } p=f_1(-a^2),\ q=f_2(-a^2)]$$

$$=(p-qD)\left[\frac{1}{(p-qD)(p+qD)}\sin ax\right]=(p-qD)\left[\frac{1}{p^2-q^2D^2}\sin ax\right]$$

$$=(p-qD)\left[\frac{1}{p^2+q^2a^2}\sin ax\right] \qquad \text{(By putting } D^2=-a^2)$$

$$=\frac{(p-qD)\sin ax}{(p^2+a^2q^2)}=\frac{p\sin ax-qa\cos ax}{p^2+a^2q^2}$$

$$\Rightarrow \quad \frac{1}{f(D)}\sin ax=\frac{f_1(-a^2)\sin ax-f_2(-a^2)a\cos ax}{\{f_1(-a^2)\}^2+a^2\{f_2(-a^2)\}^2}$$

REMARKS

- To find P.I. $=\frac{1}{f(D)}\sin ax$, replace D^2 by $-a^2$ provided $f(-a^2)\neq 0$.
- If the linear factors of D contains the odd powers of D, then first multiplying the numerator and denominator by the conjugate factors $(P\pm qD)$ and then putting D^2 by$(-a^2)$.
- Similar results are true for $\frac{1}{f(D)}\cos ax$.

Solved Examples

Example 1. Solve $\frac{d^2y}{dx^2}-3\frac{dy}{dx}+2y=\cos 3x$.

Solution. The given differential equation can be written as

$$(D^2-3D+2)y=\cos 3x \qquad \ldots(1)$$

To find C.F., the auxiliary equation is

$$m^2-3m+2=0$$

$$\Rightarrow (m-1)(m-2)=0$$

which gives $m=1$ and $m=2$.

Therefore, C.F. $=C_1e^x+C_2e^{2x}$.

Now,

$$\text{P.I.}=\frac{1}{D^2-3D+2}\cos 3x=\frac{1}{-9-3D+2}\cos 3x \qquad [\because D^2=-a^2=-9]$$

$$=\frac{1}{-7-3D}\cos 3x=-\frac{(7-3D)}{(7^2-9D^2)}\cos 3x$$

$$=-\frac{(7-3D)}{7^2-9(-9)}\cos 3x=-\frac{1}{130}[7\cos 3x-3D\cos 3x]$$

$$=-\frac{7}{130}\cos 3x-\frac{9}{130}\sin 3x=-\frac{1}{130}(7\cos 3x+9\sin 3x).$$

Hence, the general solution of (1) is given by

$$y=\text{C.F.}+\text{P.I.}$$

$$\Rightarrow y=C_1e^x+C_2e^{2x}-\frac{1}{130}[7\cos 3x+9\sin 3x].$$

Example 2. Solve $\frac{d^2y}{dx^2}-\frac{dy}{dx}-2y=\sin 2x$.

Solution. Here, the given equation can be written as

$$(D^2-D-2)y=\sin 2x \qquad \ldots(1)$$

To find the C.F. of (1), the auxiliary equation is

$$m^2-m-2=0$$

which gives, $(m+1)(m-2)=0 \Rightarrow m=-1,2$

C.F. $=C_1e^{-x}+C_2e^{2x}$.

$$\text{Now, P.I.}=\frac{1}{D^2-D-2}\sin 2x=\frac{1}{-4-D-2}\sin 2x \qquad [\because D^2=-a^2=-4]$$

$$=-\frac{1}{D+6}\sin 2x=-\frac{(D-6)}{(D+6)(D-6)}\sin 2x$$

$$=-\frac{(D-6)}{D^2-36}\sin 2x=-\frac{(D-6)}{-4-36}\sin 2x$$

$$=\frac{1}{40}[(D-6)\sin 2x]$$

$$=\frac{1}{40}[D\sin 2x-6\sin 2x]$$

$$=\frac{1}{40}[2\cos 2x-6\sin 2x]$$

$$=\frac{1}{20}\cos 2x-\frac{3}{20}\sin 2x$$

Hence, the complete solution is given by

$y = \text{C.F.} + \text{P.I.}$

$$\Rightarrow\ y = C_1e^{-x} + C_2e^{2x} + \frac{1}{20}\cos 2x - \frac{3}{20}\sin 2x.$$

Example 3. *Solve* $(D^2+4)y = \sin 3x + \cos 2x$.

(UPTU(SUM)–2008)

Solution. The given differential equation is

$$(D^2+4)y = \sin 3x + \cos 2x \qquad ...(1)$$

To find the C.F. of (1), the auxiliary equation is

$$m^2+4=0$$

$$\Rightarrow \quad m = \pm 2i$$

Therefore, C.F. $= C_1\cos 2x + C_2 \sin 2x$.

$$\text{Now, P.I.} = \frac{1}{D^2+4}\sin 3x + \frac{1}{D^2+4}(\cos 2x)$$

$$= \frac{1}{-(3)^2+4}\sin 3x + x.\frac{1}{2D}(\cos 2x)$$

$$= -\frac{1}{5}\sin 3x + \frac{x}{2}\left(\frac{\sin 2x}{2}\right)$$

$$= -\frac{1}{5}\sin 3x + \frac{x}{4}\sin 2x$$

Hence, the general solution of (1) is given by

$y = \text{C.F.} + \text{P.I.}$

$$= C_1\cos 2x + C_2\sin 2x - \frac{1}{5}\sin 3x + \frac{x}{4}\sin 2x.$$

Example 4. *Solve* $\dfrac{d^2y}{dx^2} + a^2y = \sin ax$. (UPTU–2008)

Solution. Here, the given equation can be written as

$$(D^2+a^2)y = \sin ax \qquad ...(1)$$

To find the C.F. of (1), the auxiliary equation is

$$m^2+a^2=0 \ \Rightarrow\ m = \pm ai$$

Therefore C.F. $= C_1\cos ax + C_2\sin ax$.

$$\text{Now,}\quad \text{P.I.} = \frac{1}{D^2+a^2}\sin ax = \frac{x}{2D}\sin ax$$

$$= \frac{x}{2}\left[\frac{-\cos ax}{a}\right] = -\frac{x}{2a}\cos ax$$

Hence, the general solution of (1) is given by

$y = \text{C.F.} + \text{P.I.}$

$$= C_1\cos ax + C_2\sin ax - \frac{x}{2a}\cos ax.$$

Example 5. *Solve* $(D^2+4)y = \cos^2 x$. (MTU(B.Pharm)–2011)

Solution. Here, the given differential equation is

$$(D^2+4)y = \cos^2 x \qquad ...(1)$$

To find the C.F. of (1), the auxiliary equation is

$$m^2+4=0 \quad \Rightarrow \quad m = \pm 2i$$

$\therefore$ C.F. $= C_1\cos 2x + C_2\sin 2x$.

$$\text{Now, P.I.} = \frac{1}{D^2+4}\cos^2 x$$

$$= \frac{1}{2}\left[\frac{1}{D^2+4}(1+\cos 2x)\right]$$

$$= \frac{1}{2}\left[\frac{1}{D^2+4}(e^{0x}) + \frac{1}{D^2+4}(\cos 2x)\right]$$

$$= \frac{1}{2}\left[\frac{1}{4} + x.\frac{1}{2D}(\cos 2x)\right]$$

$$= \frac{1}{2}\left[\frac{1}{4} + \frac{x}{4}\sin 2x\right] = \frac{1}{8}(1+x+\sin 2x)$$

Hence, the general solution of (1) is given by

$y = \text{C.F.} + \text{P.I.}$

$$= C_1\cos 2x + C_2\sin 2x + \frac{1}{8}(1+x+\sin 2x)$$

EXERCISE 29.4

Solve the following differential equations :

1. $(D^2+9)y = \cos 4x$
2. $(D^2-2D+5)y = \sin 3x$
3. $(D^2-3D+2)y = \sin 3x$
4. $(D^4+2D^3-3D^2)y = 3e^{2x} + 4\sin x$
5. $(D^3-2D^2+3)y = \cos x$
6. $(D^2+16)y = \sin 2x$, given that $y = 0$ and $\dfrac{dy}{dx} = \dfrac{5}{6}$ when $x = 0$.
7. $(D^4-2D^2+1)y = \cos x$
8. $(D^2+2D+2)y = \cos 2x$
9. $(D^2-9)y = \sin x + \cos x$
10. Solve $\dfrac{d^2y}{dx^2} + 2\dfrac{dy}{dx} + 10y + 37\sin 3x = 0$ and find the value of y when $x = \dfrac{\pi}{2}$ being given that $y = 3$, $\dfrac{dy}{dx} = 0$ when $x = 0$. (GBTU–2011)
11. $\dfrac{d^2y}{dx^2} + 4y = e^x + \sin 2x$ (UPTU(B.Pharm)–2009, 2010)
12. $(D^2+5D-6)y = \sin 3x + \cos 2x$ (GBTU–2010, GBTU(CO)–2011)
13. $(D^2+5D-6)y = \sin 4x \sin x$ (UPTU(SUM)–2009)

Hint to Selected Problems

1. $m = \pm 3i \Rightarrow \text{C.F.} = C_1\cos 3x + C_2\sin 3x$

$$\text{P.I.} = \frac{1}{D^2+9}\cos 4x = \frac{1}{-4^2+9}\cos 4x = -\frac{1}{7}\cos 4x.$$

5. $m=-1, \frac{3}{2}\pm\frac{i\sqrt{3}}{2}$

$\Rightarrow$ C.F. $= C_1e^{-x} + e^{3/2.x}\left[C_2\cos\left(\frac{\sqrt{3}}{2}x\right)+C_3\sin\left(\frac{\sqrt{3}}{2}x\right)\right]$

P.I. $= \frac{1}{(D+1)(D^2-3D+3)}\cos x$.

9. C.F. $= C_1e^{3x} + C_2e^{-3x}$

P.I. $= \frac{1}{D^2-9}(\sin x+\cos x) = \frac{1}{D^2-9}\sin x + \frac{1}{D^2-9}\cos x$

$= \frac{1}{-1-9}\sin x + \frac{1}{-1-9}\cos x = -\frac{1}{10}\sin x - \frac{1}{10}\cos x$

ANSWERS

1. $y = C_1\cos 3x + C_2\sin 3x - \frac{1}{7}\cos 4x$

2. $y = e^x[C_1\cos 2x + C_2\sin 2x] + \frac{1}{26}(3\cos 3x - 2\sin 3x)$

3. $y = C_1e^x + C_2e^{2x} + \frac{1}{130}(9\cos 3x - 7\sin 3x)$

4. $y = (C_1 + C_2x) + C_3e^x + C_4e^{-3x} + \frac{3}{20}e^{2x} + \frac{4}{5}\sin x + \frac{2}{5}\cos x$

5. $y = C_1e^{-x} + \left\{C_2\cos\left(\frac{x\sqrt{3}}{2}\right) + C_3\sin\left(\frac{x\sqrt{3}}{2}\right)\right\}e^{3x/2} + \frac{1}{26}[5\cos x - \sin x]$

6. $y = \frac{1}{6}\sin 4x + \frac{1}{12}\sin 2x$

7. $y = (C_1 + C_2x)\,e^x + (C_3 + C_4x)e^{-x} + \frac{1}{4}\cos x$

8. $y = e^{-x}[C_1\cos x + C_2\sin x] - \frac{1}{10}(\cos 2x - 2\sin 2x)$

9. $y = C_1e^{3x} + C_2e^{-3x} - \frac{1}{10}[\sin x + \cos x]$

10. $y = e^{-x}(C_1\cos 3x + C_2\sin 3x) + 6\cos 3x - \sin 3x$ and $y = 1$ at $x = \pi/2$

11. $y = C_1\cos 2x + C_2\sin 2x + \frac{1}{5}e^x - \frac{x}{4}\cos 2x$

12. $y = C_1e^x + C_2e^{-6x} - \frac{1}{30}(\cos 3x + \sin 3x) + \frac{1}{20}(\sin 2x - \cos 2x)$

13. $y = C_1e^x + C_2e^{-6x} + \frac{1}{2}\left[\frac{\sin 3x - \cos 3x}{30} + \frac{31\cos 5x - 25\sin 5x}{1586}\right]$

(3) To evaluate P.I., when B is of the form x^m, when m is positive integer :

i.e., to evaluate $\frac{1}{f(D)}x^m,\ m \in Z^+$ and $f(D) = A_0D^n + A_1D^{n-1} + \ldots + A_n$

Let us consider $\frac{1}{D-a}x^m$

i.e., $\frac{1}{(D-a)}x^m = e^{ax}\int e^{-ax}x^m\,dx = e^{ax}\left\{\frac{e^{-ax}x^m}{a} - \frac{mx^{m-1}e^{-ax}}{a^2} - \frac{m(m-1)x^{m-2}e^{-ax}}{a^3} \ldots - \frac{m(m-1)\ldots 2.1\,e^{-ax}}{a^{m+1}}\right\}$

If we expand $\frac{1}{D-a}$ in powers of D, we get ...(1)

$$\frac{1}{(D-a)}x^m = -\frac{1}{a(1-D/a)}x^m = -\frac{1}{a}\left[1+\frac{D}{a}+\frac{D^2}{a^2}+\ldots\right]x^m$$

$$\frac{1}{D-a}x^m = -\frac{1}{a}\left[x^m + \frac{mx^{m-1}}{a} + \frac{m(m-1)x^{m-2}}{a^2} + \ldots + \frac{m(m-1)\ldots 2.1}{a^m}\right] \quad \ldots(2)$$

Here, we observe that (1) and (2) are the same.

WORKING-PROCEDURE

Take the lowest degree term from $f(D)$ and remaining factor will be of the form $[1+f(D)]$ or $[1-f(D)]$. Now, this factor can be taken in the numerator with a negative index, which can be expanded by Binomial theorem. Here, it should be noted that the expansion is to be carried upto the term D^m, since we always have $D^{m+1}x^m = 0,\ D^{m+2}x^m = 0$ and all other higher differential coefficients of x^m are zero.

SOME IMPORTANT EXPANSIONS (TO BE USED DIRECTLY)

1. $[1+x]^n = 1 + nx + \frac{n(n-1)}{2!}x^2 + \frac{n(n-1)(n-2)}{3!}x^3 + \ldots$

2. $(1+x)^{-1} = 1 - x + x^2 - x^3 + x^4 - x^5 + \ldots$

3. $(1-x)^{-1} = 1 + x + x^2 + x^3 + x^4 + \ldots$

4. $(1-x)^{-2} = 1 + 2x + 3x^2 + 4x^3 + \ldots$

5. $(1+x)^{-2} = 1 - 2x + 3x^2 - 4x^3 + \ldots$

Solved Examples

Example 1. *Solve* $(D^2+D-2)y = x+\sin x$.

Solution. The given equation is

$$((D^2+D-2)y = x+\sin x \quad \ldots(1)$$

To find C.F., the auxiliary equation is

$$m^2+m-2=0$$

$$\Rightarrow \quad (m-1)(m+2)=0 \Rightarrow m=1,-2$$

$$\therefore \quad \text{C.F.} = C_1e^x + C_2e^{-2x}$$

Now, $$\text{P.I.} = \frac{1}{(D^2+D-2)}(x+\sin x)$$

$$= \frac{1}{(D^2+D-2)}x + \frac{1}{(D^2+D-2)}\sin x$$

$$= \frac{1}{-2\left(1-\frac{1}{2}D-\frac{1}{2}D^2\right)}x + \frac{1}{-1+D-2}\sin x$$

$$= -\frac{1}{2}\left[1-\left(\frac{1}{2}D+\frac{1}{2}D^2\right)\right]^{-1} x + \frac{(D+3)}{(D-3)(D+3)}\sin x$$

$$= -\frac{1}{2}\left(1+\frac{1}{2}D+\ldots\right)x + \frac{(D+3)}{D^2-9}\sin x$$

$$= -\frac{1}{2}\left(x+\frac{1}{2}\right) + \frac{D+3}{-1-9}\sin x$$

$$= -\frac{1}{2}\left(x+\frac{1}{2}\right) - \left(\frac{1}{10}\right)[D(\sin x)+3\sin x]$$

$$= -\frac{1}{2}x - \frac{1}{4} - \frac{1}{10}(\cos x + 3\sin x).$$

Hence, the complete solution is given by

$$y = \text{C.F.} + \text{P.I.}$$

$$\therefore y = C_1e^x + C_2e^{-2x} - \frac{1}{2}x - \frac{1}{4} - \frac{1}{10}(\cos x + 3\sin x)$$

Example 2. *Solve* $(D^2-4)y = x^2$.

Solution. The differential equation is

$$(D^2-4)y = x^2 \quad \ldots(1)$$

To find the C.F. of (1), the auxiliary equation is

$$m^2-4=0 \Rightarrow \quad m=\pm 2$$

$$\therefore \text{ C.F.} = C_1e^{2x} + C_2e^{-2x}$$

Now, $$\text{P.I.} = \frac{1}{D^2-4}x^2 = \frac{1}{-4\left[1-\frac{1}{4}D^2\right]}x^2$$

$$= -\frac{1}{4}\left[1-\frac{1}{4}D^2\right]^{-1}x^2$$

$$= -\frac{1}{4}\left[1+\frac{1}{4}D^2+\ldots\right]x^2$$

$$= -\frac{1}{4}\left[x^2+\frac{1}{4}D^2(x^2)\right] = -\frac{1}{4}\left[x^2+\frac{1}{2}\right]$$

Hence, the complete solution is given by

$$y = \text{C.F.} + \text{P.I.}$$

$$\Rightarrow \quad y = C_1e^{2x} + C_2e^{-2x} - \frac{1}{4}\left[x^2+\frac{1}{2}\right].$$

Example 3. *Solve* $(D^2-4D+4)y = x^2+e^x+\cos 2x$.

(UPTU(B.Pharma)SUM–2009)

Solution. The given differential equation is

$$(D^2-4D+4)y = x^2+e^x+\cos 2x \quad \ldots(1)$$

To find C.F., the auxiliary equation is given by

$$m^2-4m+4=0 \Rightarrow (m-2)^2=0 \Rightarrow m=2,2$$

$$\therefore \quad \text{C.F.} = (C_1+C_2x)\,e^{2x}$$

Now, $$\text{P.I.} = \frac{1}{(D^2-4D+4)}(x^2+e^x+\cos 2x)$$

$$= \frac{1}{(D-2)^2}x^2 + \frac{1}{(D-2)^2}e^x + \frac{1}{(D^2-4D+4)}\cos 2x$$

$$= \frac{1}{4\left(1-\frac{D}{2}\right)^2}x^2 + \frac{1}{(1-2)^2}e^x + \frac{1}{(-2^2-4D+4)}\cos 2x$$

$$= \frac{1}{4}\left(1-\frac{D}{2}\right)^{-2}x^2 + \frac{e^x}{1} - \frac{1}{4D}\cos 2x$$

$$= \frac{1}{4}\left[1+D+\frac{3}{4}D^2+\ldots\right]x^2 + e^x - \frac{1}{4}\int \cos 2x\,dx$$

$$= \frac{1}{4}\left(x^2+D(x^2)+\frac{3}{4}D^2(x^2)\right) + e^x - \frac{1}{4}\cdot\frac{1}{2}\sin 2x$$

$$= \frac{1}{4}\left[x^2+2x+\frac{3}{2}\right] + e^x - \frac{1}{8}\sin 2x\,.$$

Hence, the complete solution is given by

$$y = \text{C.F.} + \text{P.I.}$$

$$\Rightarrow y = (C_1+C_2x)e^{2x} + \frac{1}{4}\left(x^2+2x+\frac{3}{2}\right) + e^x - \frac{1}{8}\sin 2x\,.$$

Example 4. *Find the solution of the following differential equation :*

$$(D^2-4D-5)y = e^{2x}+3\cos(4x+3),$$

where $D \equiv \frac{d}{dx}$. (UPTU–2008)

Solution. Here, we have

$$(D^2-4D-5)y = e^{2x}+3\cos(4x+3)$$

$$\text{A.E. is } (m^2-4m-5)=0$$

$$\Rightarrow m^2-5m+m-5=0$$

$$\Rightarrow \quad (m-5)(m+1)=0 \Rightarrow m=-1,5.$$

$$\text{C.F.} = C_1e^{-x} + C_2e^{5x}$$

$$= \frac{1}{(D^2-4D-5)}\left[e^{2x}+3\cos(4x+3)\right]$$

$$= \frac{1}{D^2-4D-5}e^{2x} + 3\frac{1}{D^2-4D-5}\cos(4x+3)$$

$$= \frac{e^{2x}}{2^2 - 4(2) - 5} + 3.\frac{1}{-16 - 4D - 5}\cos(4x+3)$$

$$= \frac{e^{2x}}{4 - 8 - 5} + 3\frac{1}{-4D - 21}\cos(4x+3)$$

$$= -\frac{e^{2x}}{9} - 3\frac{1}{4D + 21}\cos(4x+3)$$

$$= -\frac{e^{2x}}{9} - 3\frac{4D - 21}{16D^2 - 441}\cos(4x+3)$$

$$= -\frac{e^{2x}}{9} - 3\frac{(4D - 21)}{16(-16) - 441}\cos(4x+3)$$

$$= \frac{e^{2x}}{9} + \frac{3}{697}(4D - 21)\cos(4x+3)$$

$$= -\frac{e^{2x}}{9} + \frac{3}{697}\left[4D\cos(4x+3) - 21\cos(4x+3)\right]$$

$$= -\frac{e^{2x}}{9} + \frac{3}{697}\left[-16\sin(4x+3) - 21\cos(4x+3)\right]$$

$$= -\frac{e^{2x}}{9} - \frac{3}{697}\left[16\sin(4x+3) + 21\cos(4x+3)\right]$$

Complete solution is given by

y = C.F. + P.I.

$$= C_1e^{-x} + C_2e^{5x} - \frac{e^{2x}}{9} - \frac{3}{697}\left[16\sin(4x+3) + 21\cos(4x+3)\right]$$

Example 5. *Solve* $\frac{d^3y}{dx^3} - 3\frac{d^2y}{dx^2} + 4\frac{dy}{dx} - 2y = e^x + \cos x.$ (UPTU–2001, 2006)

Solution. The given equation can be written as

$$(D^3 - 3D^2 + 4D - 2)y = e^x + \cos x$$

The auxiliary equation is $m^3 - 3m^2 + 4m - 2 = 0$

or $(m-1)(m^2 - 2m + 2) = 0$, *i.e.*, $m = 1,\ 1 \pm i$

$\therefore$ C.F. $= C_1e^x + e^x(C_2\cos x + C_3\sin x)$

$$\text{P.I.} = \frac{1}{(D-1)(D^2 - 2D + 2)}e^x + \frac{1}{(D^3 - 3D^2 + 4D - 2)}\cos x$$

$$= \frac{1}{(D-1)(1 - 2 + 2)}e^x + \frac{1}{(-1)D - 3(-1) + 4D - 2}\cos x$$

$$= \frac{1}{(D-1)}e^x + \frac{1}{3D+1}\cos x$$

$$= e^x\frac{1}{D}.1 + \frac{(-3\sin x - \cos x)}{-9-1}$$

$$= e^x.x + \frac{1}{10}(3\sin x + \cos x)$$

Hence, complete solution is

y = C.F. + P.I.

$$\therefore \quad y = C_1e^x + e^x(C_2\cos x + C_3\sin x) + xe^x + \frac{1}{10}(3\sin x + \cos x)$$

Example 6. *Determine the general solution of the differential equation*

$$y'' - 2y' + 2y = x + e^x\cos x \qquad \text{(UPTU–2002)}$$

Solution. We have $y'' - 2y' + 2y = x + e^x\cos x$

The auxiliary equation is

$$m^2 - 2m + 2 = 0 \Rightarrow m = 1 \pm i$$

C.F. $= e^x\,(A\cos x + B\sin x)$

$$\therefore \text{P.I.} = \frac{1}{(D^2 - 2D + 2)}.x + \frac{1}{D^2 - 2D + 2}e^x\cos x$$

Let $I_1 = \frac{1}{(D^2 - 2D + 2)}x$

and $I_2 = \frac{1}{(D^2 - 2D + 2)}.e^x\cos x$

$$\therefore \quad I_1 = \frac{1}{(D^2 - 2D + 2)}x = \frac{1}{2\left[1 - D + \frac{D^2}{2}\right]}x$$

$$= \frac{1}{2\left[1 - \left(D - \frac{D^2}{2}\right)\right]}.x$$

$$= \frac{1}{2}\left[1 - \left(D - \frac{D^2}{2}\right)\right]^{-1}x$$

$$= \frac{1}{2}\left[1 + \left(D - \frac{D^2}{2}\right) + \ldots\right]x$$

$$= \frac{1}{2}\left[x + Dx - \frac{D^2}{2}x + \ldots\right] = \frac{1}{2}(x+1)$$

Now, $I_2 = \frac{1}{(D^2 - 2D + 2)}e^x\cos x$

$$= e^x.\frac{1}{(D+1)^2 - 2(D+1) + 2}\cos x$$

$$= e^x\frac{1}{D^2 + 1}\cos x = e^x.x\frac{1}{2D}\cos x$$

$$= \frac{1}{2}xe^x\sin x$$

Hence, the solution is y = C.F. + P.I.

$$\therefore \quad y = e^x(A\cos x + B\sin x) + \frac{1}{2}(x+1) + \frac{1}{2}xe^x\sin x.$$

Example 7. *Find the complete solution of*

$$\frac{d^2y}{dx^2} - 3\frac{dy}{dx} + 2y = xe^{3x} + \sin 2x.$$

(UPTU–2003, VTU–2008, Kottayam–2005, GBTU–2010)

Solution. The auxiliary equation is $m^2 - 3m + 2 = 0$

$\Rightarrow \quad m^2 - 2m - m + 2 = 0$

$\Rightarrow \quad m(m-2) - 1(m-2) = 0$

$\Rightarrow \quad (m-2)(m-1) = 0 \quad \Rightarrow m = 1, 2$

C.F. $= C_1e^x + C_2e^{2x}$.

$$\text{P.I.} = \frac{1}{D^2 - 3D + 2}.(xe^{3x} + \sin 2x)$$

$$= \frac{1}{D^2 - 3D + 2}xe^{3x} + \frac{1}{D^2 - 3D + 2}\sin 2x$$

$$= e^{3x}\frac{1}{(D+3)^2 - 3(D+3) + 2}x + \frac{1}{-4 - 3D + 2}\sin 2x$$

$$= e^{3x}.\frac{1}{D^2 + 6D + 9 - 3D - 9 + 2}x + \frac{1}{-3D - 2}\sin 2x$$

$$= e^{3x}\frac{1}{D^2 + 3D - 2}x - \frac{1}{3D + 2}\sin 2x$$

$$= \frac{e^{3x}}{2}\left[1 + \left(\frac{3D + D^2}{2}\right)\right]^{-1}x - \frac{(3D-2)}{9D^2 - 4}\sin 2x$$

$$= \frac{e^{3x}}{2}\left[1 - \left(\frac{3D + D^2}{2}\right) + \dots\right]x - \frac{3D - 2}{9(-4) - 4}\sin 2x$$

$$= \frac{e^{3x}}{2}\left[x - \left(\frac{3D + D^2}{2}\right)x + \dots\right] - \frac{3D - 2}{-36 - 4}\sin 2x$$

$$\text{P.I.} = \frac{e^{3x}}{4}(2x - 3) + \frac{1}{40}(6\cos 2x - 2\sin 2x)$$

$$= \frac{e^{3x}}{4}(2x - 3) + \frac{3}{20}\cos 2x - \frac{1}{20}\sin 2x$$

The complete solution is y = C.F. + P.I.

$$y = C_1e^x + C_2e^{2x} + \frac{e^{3x}}{4}(2x - 3) + \frac{3}{20}\cos 2x - \frac{1}{20}\sin 2x$$

Example 8. *A body executes damped forced vibrations given by the equation*

$$\frac{d^2x}{dt^2} + 2k\frac{dx}{dt} + b^2x = e^{-kt}\sin wt.$$

Solve the differential equation for both the cases where $w^2 \neq b^2 - k^2$ and when $w^2 = b^2 - k^2$. (UPTU–2002)

Solution. The given equation can be written as

$$(D^2 + 2kD + b^2)x = e^{-kt}\sin wt$$

The auxiliary equation is $m^2 + 2km + b^2 = 0$

$$\Rightarrow \quad m = \frac{-2k \pm \sqrt{4k^2 - 4b^2}}{2} = -k \pm \sqrt{k^2 - b^2}$$

As the given problem is on vibration, we must have $k^2 < b^2$.

$$\therefore \quad m = -k \pm \sqrt{-(b^2 - k^2)} = -k \pm i\sqrt{(b^2 - k^2)}$$

$$\therefore \text{C.F.} = e^{-kt}\left\{C_1\cos\sqrt{(b^2 - k^2)}t + C_2\sin\sqrt{(b^2 - k^2)t}\right\}$$

$$\therefore \text{P.I.} = \frac{1}{D^2 + 2kD + b^2}e^{-kt}\sin wt$$

$$= e^{-kt}\frac{1}{(D-k)^2 + 2k(D-k) + b^2}\sin wt$$

$$= e^{-kt}\frac{1}{b^2 + (b^2 - k^2)}\sin wt$$

$$= e^{-kt}\frac{1}{-w^2 + (b^2 - k^2)}\sin wt, \text{ if } w^2 \neq b^2 - k^2$$

$$= e^{-kt}\, t\frac{1}{2D}\sin wt = e^{-kt}\left(-\frac{t}{2w}\cos wt\right) \text{ if } w^2 = b^2 - k^2$$

EXERCISE 29.5

Solve the following differential equations :

1. $(D^3 - D^2 - 6D)y = x^2 + 1$ (UPTU(Q. Bank)–2001)
2. $(D^4 - a^4)y = x^4$
3. $(D^3 + 2D^2 + D)y = e^{2x} + x^2 + x$
4. $(D^3 - 3D - 2)\ y = x^3$
5. $(D^3 - 3D^2 + 2D)\ y = 4 + 60e^{5x}$
6. $(D^3 + 1)\ y = \sin 3x - \cos^2\frac{x}{2}$
7. $(D^2 - 2D + 3)y = \cos x + x^2$
8. $(D^2 - 5D + 6)y = x + e^{mx}$
9. $(D^2 + 16)y = \cos 3x + e^{3x} + x^4$.
10. $(D^2 + 4)y = \sin 3x + x^2$
11. $\frac{d^2y}{dx^2} - \frac{dy}{dx} + 4y = x^2 + e^x$ (UPTU(B.Pharma)SUM–2010)
12. If $\frac{d^2x}{dt^2} + \frac{g}{b}(x - a) = 0$; a, b and g are positive numbers and $x = a'$, $\frac{dx}{dt} = 0$ when t = 0, show that

$$x = a + (a' - a)\cos\sqrt{\frac{g}{b}}t$$ (UPTU(SUM)–2007)

Hint to Selected Problems

1. $m = 0, -2, 3 \Rightarrow$ C.F. $= C_1e^{0x} + C_2e^{-2x} + C_3e^{3x}$

$$\text{P.I.} = \frac{1}{D^3 - D^2 - 6D}(1+x^2) = -\frac{1}{6D}\left[1 + \frac{D}{6} - \frac{D^2}{6}\right]^{-1}(1+x^2)$$

$$= -\frac{1}{6D}\left[1 - \frac{1}{6}(-D + D^2)\right]^{-1}(1+x^2)$$

Now expand by binomial theorem and D for differentiation and $1/D$ for integration.

7. $m = 1 \pm i\sqrt{2} \Rightarrow$ C.F. $= e^x\,[C_1 \cos\sqrt{2}x + C_2 \sin\sqrt{2}\,x]$

$$\text{P.I.} = \frac{1}{D^2 - 2D + 3}(\cos x + x^2)$$

$$= \frac{1}{D^2 - 2D + 3}\cos x + \frac{1}{D^2 - 2D + 3}x^2$$

$$= \frac{1}{-1 - 2D + 3}\cos x + \frac{1}{3}\left[1 - \left(\frac{2D}{3} - \frac{D^2}{3}\right)\right]^{-1} x^2.$$

10. $m = \pm 2i \Rightarrow$ C.F. $= C_1 \cos 2x + C_2 \sin 2x$

$$\text{P.I.} = \frac{1}{D^2 + 4}(\sin 3x + x^2) = \frac{1}{D^2 + 4}\sin 3x + \frac{1}{D^2 + 4}.x^2$$

$$= \frac{1}{-9 + 4}\sin 3x + \frac{1}{4\left(1 + \frac{D^2}{4}\right)}.x^2$$

$$= -\frac{1}{5}\sin 3x + \frac{1}{4}\left(1 + \frac{D^2}{4}\right)^{-1}.x^2$$

Now expand by Binomial expansion.

ANSWERS

1. $y = C_1 + C_2e^{3x} + C_3e^{-2x} - \frac{25}{108}x - \frac{1}{18}x^3 + \frac{1}{36}x^2$

2. $y = C_1e^{ax} + C_2e^{-ax} + C_3 \cos ax + C_4 \sin ax - \frac{x^4}{a^4} - \frac{24}{a^8}$

3. $y = C_1 + (C_2 + C_3x)e^{-x} + \frac{1}{18}e^{2x} + \frac{1}{3}x^3 - \frac{3}{2}x^2 + 4x$

4. $y = (C_1 + C_2x)e^{-x} + C_3e^{2x} - \frac{1}{2}x^3 + \frac{9}{4}x^2 - \frac{27}{4}x + 15$

5. $y = C_1 + C_2e^x + C_3e^{2x} + 2x + e^{5x}$

6. $y = C_1e^{-x} + e^{x/2}\left\{C_2 \cos\frac{x\sqrt{3}}{2} + C_3 \sin\frac{x\sqrt{3}}{2}\right\} + \frac{1}{730}[\sin 3x + 27\cos 3x] - \frac{1}{2} - \frac{1}{4}(\cos x - \sin x)$

7. $y = e^x[C_1 \cos(x\sqrt{2}) + C_2 \sin(x\sqrt{2})] + \frac{1}{4}(\cos x - \sin x) + \frac{x^2}{3} + \frac{4}{9}x + \frac{2}{27}$

8. $y = C_1e^{2x} + C_2e^{3x} + \frac{1}{6}\left[x + \frac{5}{6}\right] + [e^{mx}/m^2 - 5m + 6)]$

9. $y = C_1 \cos 4x + C_2 \sin 4x + \frac{1}{7}\cos 3x + \frac{1}{25}e^{3x} + \frac{1}{16}x^4 - \frac{3}{64}x^2 + \frac{3}{512}$

10. $y = C_1 \cos 2x + C_2 \sin 2x - \frac{1}{5}\sin 3x + \frac{1}{4}x^2 - \frac{1}{8}$

11. $y = e^{x/2}\left(C_1 \cos\frac{\sqrt{15}}{2}x + C_2 \sin\frac{\sqrt{15}}{2}x\right) + \frac{1}{4}\left(e^x + x^2 + \frac{x}{2} - \frac{3}{8}\right)$

12. $x = (a' - a)\cos\sqrt{\frac{g}{b}}\,t + a$

(4) To evaluate $\frac{1}{f(D)}e^{ax}.X$, where X is any function of x :

Let us consider any function X_1 of x. Then, by simple differentiation, we get

$$D(e^{ax}.X_1) = e^{ax}D(X_1) + X_1ae^{ax} = e^{ax}(D + a)X_1. \quad \text{...(1)}$$

Now, let us assume

$$D^n[e^{ax}.X_1] = e^{ax}(D + a)^n.X_1 \quad \text{...(2)}$$

Then, consider $D^{n+1}[e^{ax}.X_1] = D[D^n(e^{ax}.X_1)] = D[e^{ax}(D + a)^n.X_1] = ae^{ax}(D + a)^n.X_1 + e^{ax}.D(D + a)^n.X_1$

$$= e^{ax}(D + a)^{n+1}.X_1$$

Therefore, by the method of induction, we have $D^n[e^{ax}.X_1] = e^{ax}(D + a)^n X_1$, for all positive integer n

$$\therefore \quad f(D)e^{ax}.X_1 = e^{ax}f(D + a)X_1. \quad \text{...(3)}$$

Now, operating on equation (3) with $\frac{1}{f(D)}$, we get

$$\frac{1}{f(D)}.f(D)e^{ax}.X_1 = \frac{1}{f(D)}e^{ax}f(D + a).X_1 \quad \Rightarrow \quad e^{ax}.X_1 = \frac{1}{f(D)}e^{ax}f(D + a).X_1. \quad \text{...(4)}$$

Let $\quad X = f(D + a).X_1 \Rightarrow X_1 = \frac{X}{f(D + a)}.$

Now, (4) becomes

$$e^{ax}.\frac{X}{f(D+a)}=\frac{1}{f(D)}e^{ax}.\frac{X}{f(D+a)}.f(D+a) \Rightarrow \quad \frac{1}{f(D)}[e^{ax}.X] = e^{ax}\left[\frac{1}{f(D+a)}.X\right]$$

REMARKS

- Here, we observe that if e^{ax} is brought to the left from the right of $\frac{1}{f(D)}$, then D should be replaced by $(D+a)$.
- This method will be used if X is $\cos ax$, $\sin ax$ or x^m or a polynomial of degree m.
- This method is also capable to find $\left\{\frac{1}{f(D)}e^{ax}\right\}$, when $f(a) = 0$.

WORKING PROCEDURE

Replace D by $(D+a)$ and brought e^{ax} before the operator $\frac{1}{f(D)}$. After that, determine $\frac{1}{f(D+a)}.X$ as usual.

Solved Examples

Example 1. *Solve* $(D^2+4D-12)y=(x-1)e^{2x}$.

Solution. The given differential equation is

$$(D^2+4D-12)y=(x-1)e^{2x} \qquad \ldots(1)$$

To find C.F. of (1), the auxiliary equation is

$$m^2+4m-12=0 \Rightarrow (m-2)(m+6)=0$$

which gives $m = 2$ and $m = -6$.

$\therefore$ C.F. $= C_1e^{2x}+C_2e^{-6x}$

Now, P.I. $=\frac{1}{(D^2+4D-12)}e^{2x}(x-1)$

$$= e^{2x}\frac{1}{[(D+2)^2+4(D+2)-12]}(x-1)$$

$$= e^{2x}\frac{1}{(D^2+8D)}(x-1) = e^{2x}\frac{1}{8D\left(1+\frac{D}{8}\right)}(x-1)$$

$$=\frac{1}{8}e^{2x}\frac{1}{D}\left(1+\frac{1}{8}D\right)^{-1}(x-1)$$

$$=\frac{1}{8}e^{2x}\frac{1}{D}\left(1-\frac{1}{8}D+\ldots\right)(x-1)$$

$$=\frac{1}{8}e^{2x}\frac{1}{D}\left(x-1-\frac{1}{8}\right)=\frac{1}{8}e^{2x}\frac{1}{D}\left(x-\frac{9}{8}\right)$$

$$=\frac{1}{8}e^{2x}\int\left(x-\frac{9}{8}\right)dx$$

$$=\frac{1}{8}e^{2x}\left(\frac{x^2}{2}-\frac{9}{8}x\right).$$

Hence, the general solution of (1), is given by

$y =$ C.F. + P.I.

$$\Rightarrow \quad y=C_1e^{2x}+C_2e^{-6x}+\frac{1}{8}e^{2x}\left[\frac{x^2}{2}-\frac{9}{8}x\right].$$

Example 2. *Solve* $(D^2-2D+4)y=e^x\cos x$.

Solution. The differential equation is

$$(D^2-2D+4)y=e^x\cos x \qquad \ldots(1)$$

To find the C.F. of (1), the auxiliary equation is

$$m^2-2m+4=0 \Rightarrow m=1\pm i\sqrt{3}.$$

Therefore, C.F. $= e^x(C_1\cos\sqrt{3}.x + C_2\sin\sqrt{3}.x)$

Now, P.I. $=\frac{1}{(D^2-2D+4)}e^x\cos x$

$$= e^x\frac{1}{[(D+1)^2-2(D+1)+4]}\cos x$$

$$= e^x\frac{1}{(D^2+3)}\cos x$$

$$= e^x\frac{1}{-1^2+3}\cos x=\frac{1}{2}e^x\cos x.$$

Hence, the complete solution of (1) is given by

$y =$ C.F. + P.I.

$$\Rightarrow \; y=e^x\,[C_1\cos\sqrt{3}.x+C_2\sin\sqrt{3}.x]+\frac{1}{2}e^x\cos x.$$

Example 3. *Solve* $(D^2-5D+6)y=e^{2x}\sin 2x$.

Solution. The given differential equation is

$$(D^2-5D+6)y=e^{2x}\sin 2x \qquad \ldots(1)$$

To find the C.F. of (1), the auxiliary equation is given by

$$m^2-5m+6=0 \Rightarrow (m-2)(m-3)=0$$

which gives, $m = 2$ and $m = 3$.

$\therefore$ C.F. $= C_1e^{2x}+C_2e^{3x}$

Now, P.I. $=\frac{1}{D^2-5D+6}e^{2x}\sin 2x$

$$= e^{2x}\frac{1}{[(D+2)^2-5(D+2)+6]}\sin x$$

$$= e^{2x}\frac{1}{D^2-D}\sin 2x = e^{2x}\frac{1}{-2^2-D}\sin 2x$$

$$= e^{2x}\frac{1}{-4-D}\sin 2x = -e^{2x}\frac{1}{(4+D)}\sin 2x$$

$$= -e^{2x}\frac{(D-4)}{(D+4)(D-4)}\sin 2x$$

$$= -e^{-2x}\left[\frac{D-4}{D^2-16}\right]\sin 2x$$

$$= -e^{2x}\left[\frac{D-4}{-4-16}\right]\sin 2x$$

$$= \frac{e^{2x}}{20}(D-4)\sin 2x$$

$$= \frac{e^{2x}}{20}[D\sin 2x - 4\sin 2x]$$

$$= \frac{e^{2x}}{20}[2\cos 2x - 4\sin 2x]$$

Hence, the complete solution of (1) is given by

$$y = \text{C.F.} + \text{P.I.}$$

$$\Rightarrow \quad y = C_1e^{2x} + C_2e^{3x} + \frac{e^{2x}}{20}[2\cos 2x - 4\sin 2x]$$

Example 4. *Solve* $\frac{d^2y}{dx^2} - 2\frac{dy}{dx} + y = xe^x\cos x$. (UPTU-2009)

Solution. The differential equation can be written as

$$(D^2 - 2D + 1)y = xe^x\cos x \qquad \ldots(1)$$

To find the C.F., the auxiliary equation is

$$m^2 - 2m + 1 = 0 \Rightarrow m = 1, 1$$

Therefore, C.F. $= (C_1 + C_2x)e^x$

$$\text{Now, P.I.} = \frac{1}{D^2-2D+1}xe^x\cos x$$

$$= \frac{1}{(D-1)^2}xe^x\cos x$$

$$= e^x\frac{1}{(D+1-1)^2}x\cos x = e^x\frac{1}{D^2}x\cos x$$

$$= e^x\frac{1}{D}(x\sin x + \cos x)$$

$$= e^x(-x\cos x + 2\sin x)$$

Hence, the complete solution of (1) is given by

$$y = \text{C.F.} + \text{P.I.}$$

$$= (C_1 + C_2x)e^x + e^x(-x\cos x + 2\sin x)$$

EXERCISE 29.6

Solve the following differential equations :

1. $(D^2 - 2D + 1)y = e^x.x^2$
2. $(D^2 - 5D + 6)y = x^3.e^{2x}$
3. $(D^2 - 1)y = e^x(1 + x^2)$
4. $(D^2 - 4D + 1)y = e^{2x}\sin x$
5. $(D^2 - 2D + 1)y = x^2e^{3x}$
6. $(D^2 - 1)y = e^x\cos x$
7. $(D^2 - 2D + 5)y = e^{2x}\sin x$ (MTU(AG)–2011)
8. $(D^2 - 2D + 6)y = e^x\cos x$
9. $(D^2 - 1)y = \cosh x\cos x + a^x$
10. $(D^2 - 4D - 5)y = xe^{-x}$ given that $y = 0$ and $\frac{dy}{dx} = 0$ at $x = 0$.
11. $(D^2 - 4D + 4)y = e^x\cos x$ (GBTU(CO)–2010)
12. $\frac{d^2y}{dx^2} - 2\frac{dy}{dx} + 4y = e^{2x}\cos x$ (MTU(B.Pharm.)–2011)
13. $(D^2 - 3D + 2)y = xe^x + \sin 2x$ (UPTU–2008)
14. $(D^2 - 1)y = xe^x + \cos^2 x$ (UPTU(SUM)–2007)
15. $(D^2 - 1)y = x\sin x + x^2e^x$ (UPTU(SUM)–2009)
16. $(D^2 - 2D + 1)y = x\sin x$ (UKTU–2012)
17. $\frac{d^2y}{dx^2} + 2\frac{dy}{dx} + y = x^2e^{-x}\cos x$ (GBTU–2012)

Hint to Selected Problems

1. $m = 1, 1, \therefore$ C.F. $= (C_1 + C_2x)e^x$

$$\text{P.I.} = \frac{1}{D^2-2D+1}e^x.x^2 = \left[\frac{1}{(D-1)^2}e^x.x^2\right]$$

$$= e^x\left[\frac{1}{[(D+1)-1]^2}.x^2\right] = e^x.\frac{1}{D^2}.x^2 = \frac{e^x.x^4}{12}$$

3. $m = \pm 1, \therefore$ C.F. $= C_1e^x + C_2e^{-x}$

$$\text{P.I.} = \frac{1}{D^2-1}e^x(1+x^2) = e^x\frac{1}{(D+1)^2-1}(1+x^2)$$

$$= e^x\left[\frac{1}{D^2+2D+1-1}\right].(1+x^2)$$

$$= e^x.\frac{1}{D^2+2D}(1+x^2) = \frac{e^x}{2D}\left[1+\frac{D}{2}\right]^{-1}[1+x^2]$$

Expand by binomial expansion.

7. $m = 1 \pm 2i, \therefore$ C.F. $= e^x(C_1\cos 2x + C_2\sin 2x)$

$$\text{P.I.} = \frac{1}{D^2-2D+5}e^{2x}\sin x = e^{2x}\frac{1}{(D+2)^2-2(D+2)+5}.\sin x$$

$$= e^{2x}.\frac{1}{D^2+2D+5}\sin x.$$

9. C.F. $= C_1e^x + C_2e^{-x}$

$$\text{P.I.} = \frac{1}{D^2-1}\cosh x\cos x + \frac{1}{D^2-1}a^x$$

$$= \frac{1}{D^2-1}\left(\frac{e^x+e^{-x}}{2}\right)\cos x + \frac{1}{(D^2-1)}e^{\log a^x}$$

$$= \frac{1}{2}e^x\left\{\frac{1}{(D+1)^2-1}\cos x + \frac{1}{2}e^{-x}\frac{1}{(D-1)^2-1}\cos x + \frac{1}{(\log a)^2-1}e^{x\log a}\right\}.$$

ANSWERS

1. $y=(C_1+C_2x)e^x+\frac{1}{12}e^x.x^4$
2. $y=C_1e^{2x}+C_2e^{3x}-e^{2x}\left[\frac{x^4}{4}+x^3+3x^2+6x\right]$
3. $y=C_1e^x+C_2e^{-x}+\frac{1}{12}e^x[9x+2x^3-3x^2]$
4. $y=C_1e^{(2+\sqrt{3})x}+C_2e^{(2-\sqrt{3})x}-\frac{1}{4}e^{2x}\sin x$
5. $y=(C_1+C_2x)e^x+\frac{1}{8}e^{3x}(2x^2-4x+3)$
6. $y=C_1e^x+C_2e^{-x}-\frac{1}{5}e^x(\cos x-2\sin x)$
7. $y=e^x[C_1\cos 2x+C_2\sin 2x]-\frac{1}{10}e^{2x}(\cos x-2\sin x)$
8. $y=e^x[C_1\cos\sqrt{5}.x+C_2\sin\sqrt{5}.x]+\frac{1}{4}e^x\cos x$
9. $y=C_1e^x+C_2e^{-x}+\frac{1}{10}e^x[2\sin x-\cos x]-\frac{1}{10}e^{-x}(2\sin x+\cos x)+\frac{a^x}{(\log a)^2-1}$
10. $y=-\frac{1}{216}e^{-x}+\frac{1}{216}e^{5x}-\frac{1}{36}xe^{-x}-\frac{1}{12}x^2e^{-x}$
11. $y=(C_1+C_2x)e^{2x}-\frac{e^x}{2}\sin x$
12. $y=e^x(C_1\cos\sqrt{3}x+C_2\sin\sqrt{3}x)+\frac{1}{13}e^{2x}(2\sin x+3\cos x)$
13. $y=C_1e^x+C_2e^{2x}-e^x\left(\frac{x^2}{2}+x\right)+\frac{1}{20}(3\cos 2x-\sin 2x)$
14. $y=C_1e^x+C_2e^{-x}+\frac{1}{4}e^x(x^2-x)-\frac{1}{2}-\frac{1}{10}\cos 2x$
15. $y=C_1e^x+C_2e^{-x}-\frac{1}{2}(x\sin x+\cos x)+\frac{xe^x}{12}(2x^2-3x+3)$
16. $y=(C_1+C_2x)e^x+\frac{1}{2}[(x+1)\cos x-\sin x]$
17. $y=(C_1+C_2x)e^{-x}+e^{-x}(-x^2\cos x+4x\sin x+6\cos x)$

(5) To evaluate $\frac{1}{f(D)}e^{ax}.X$, when $f(a)=0$:

Let us suppose $f(a)=0$. In this case $(D-a)$ is at least one factor of $f(D)$.

Let $f(D)=(D-a)^r\ g(D)$, where $g(a)\neq 0$.

Then, $\frac{1}{f(D)}e^{ax}=\frac{1}{(D-a)^r}\cdot\frac{1}{g(a)}e^{ax}=\frac{1}{g(a)}\cdot\frac{1}{(D-a)^r}e^{ax}=\frac{1}{g(a)}\cdot\frac{1}{(D-a)^{r-1}}e^{ax}\int e^{ax}.e^{-ax}dx$

$$=\frac{1}{g(a)}\cdot\frac{1}{(D-a)^{r-1}}xe^{ax}=\frac{1}{g(a)}\cdot\frac{1}{(D-a)^{r-2}}e^{ax}\int xe^{ax}.e^{-ax}dx=\frac{1}{g(a)}\cdot\frac{1}{(D-a)^{r-2}}\cdot\frac{x^2}{2!}e^{ax}.$$

Proceeding in the same way, finally, we get $\frac{1}{f(D)}e^{ax}=\frac{1}{g(a)}\cdot\frac{x^r}{r!}e^{ax}$.

REMARKS

- Substitute $D=a$ in those factors of $f(D)$ which do not vanish for $D=a$ and then make the question as P.I. of a product of e^{ax} and 1, which is calculated by previous section and reduce to the calculation of $\frac{1}{D}.1$ or $\frac{1}{D^2}.1$ or $\frac{1}{D^3}.1$ and so on.
- Here, $\frac{1}{D^n}$ implies n times integral of 1, with respect to x.

SOLVED EXAMPLES

Example 1. *Solve* $(D^2+D-6)y=e^{2x}$.

Solution. The given equation is

$$(D^2+D-6)y=e^{2x} \quad \ldots(1)$$

To find C.F. of (1), the auxiliary equation is

$$m^2+m-6=0$$

$\Rightarrow\ (m+3)(m-2)=0\ \Rightarrow\ m=2,-3$

$\therefore\quad$ C.F. $=C_1e^{2x}+C_2e^{-3x}$

Now, P.I. $=\frac{1}{D^2+D-6}e^{2x}=\frac{1}{(D+3)(D-2)}e^{2x}$

$$=\frac{1}{(2+3)(D-2)}e^{2x}=\frac{1}{5(D-2)}e^{2x}.1$$

$$=\frac{1}{5}e^{2x}\frac{1}{(D+2)-2}.1=\frac{1}{5}e^{2x}\frac{1}{D}.1=\frac{1}{5}xe^{2x}.$$

Hence, the complete solution of (1) is given by

$$y=\text{C.F.}+\text{P.I.}$$

$$\Rightarrow\quad y=C_1e^{2x}+C_2e^{-3x}+\frac{1}{5}xe^{2x}.$$

Example 2. Solve $\dfrac{d^2y}{dx^2} - 3\dfrac{dy}{dx} + 2y = e^x$.

Solution. The given differential equation can be written as

$$(D^2 - 3D + 2)y = e^x \quad \ldots(1)$$

To find the C.F. of (1), the auxiliary equation is

$$m^2 - 3m + 2 = 0$$

$$\Rightarrow \quad (m-1)(m-2) = 0 \Rightarrow m = 1, 2$$

$$\therefore \quad \text{C.F.} = C_1e^x + C_2e^{2x}$$

Now, $$\text{P.I.} = \frac{1}{(D^2 - 3D + 2)}e^x$$

$$= \frac{1}{(D-2)(D-1)}e^x = \frac{1}{(1-2)(D-1)}e^x$$

(By putting 1 for D in $(D-2)$, because at $D = 1$ $(D-2) \neq -1$)

$$= -\frac{1}{D-1}e^x = -\frac{1}{D-1}e^x.1$$

$$= -e^x\frac{1}{(D+1)-1}.1 = -e^x.\frac{1}{D}.1 = -e^x.x$$

Hence, the complete solution of (1) is given by

y = C.F. + P.I. $\Rightarrow y = C_1e^x + C_2e^{2x} - xe^x$.

Example 3. Solve $(D^3 + 3D^2 + 3D + 1)\, y = e^{-x}$.

Solution. The given differential equation is

$$(D^3 + 3D^2 + 3D + 1)\, y = e^{-x} \quad \ldots(1)$$

To find the C.F. of (1), the auxiliary equation is given by $(m+1)^3 = 0 \Rightarrow m = -1, -1, -1$

$$\therefore \quad \text{C.F.} = (C_1 + C_2x + C_3x^2)\, e^{-x}$$

Now, $$\text{P.I.} = \frac{1}{(D+1)^3}e^{-x} = e^{-x}\frac{1}{(D-1+1)^3}.1$$

$$= e^{-x}.\frac{1}{D^3}.1 = e^{-x}.\frac{x^3}{3!}.$$

Hence, the complete solution of (1) is given by

y = C.F. + P.I.

$$\Rightarrow \quad y = (C_1 + C_2x + C_3x^2)e^{-x} + e^{-x}.\frac{x^3}{3!}.$$

EXERCISE 29.7

Solve the following differential equations :

1. $(D^2 + 4D + 3)\, y = e^{-3x}$
2. $(D^2 + 6D + 9)y = 2e^{-3x}$
3. $(D^4 + D^3 + D^2 - D - 2)y = e^x$
4. $(D^2 - 9D + 18)y = \cosh 3x$
5. $(D-1)^2(D^2+1)^2 y = e^x$
6. $(D^2 - 3D + 2)y = e^x$ when $y = 3$, $\dfrac{dy}{dx} = 3$ at $x = 0$
7. $(D-1)^3(D+1)y = e^x + e^{-x}$
8. $(D^2 - 6D + 9)y = 4e^{3x}$
9. $(D^2 - 1)y = \cosh x$
10. $(D^2 - 4D + 4)y = 8(x^2 + e^{2x} + \sin 2x)$

Hint to Selected Problems

1. C.F. $= C_1e^{-x} + C_2e^{-3x}$

$$\text{P.I.} = \frac{1}{D^2 + 4D + 3}e^{-3x} = \frac{1}{(D+1)(D+3)}e^{-3x}$$

$$= \frac{1}{(-3+1)(D+3)}e^{-3x} = -\frac{1}{2(D+3)}e^{-3x}.1$$

$$= -\frac{1}{2}e^{-3x}\frac{1}{[(D-3)+3]}.1 = -\frac{1}{2}e^{-3x}.\frac{1}{D}.1 = -\frac{1}{2}e^{-3x}.x$$

4. C.F. $= C_1e^{3x} + C_2e^{6x}$

$$\text{P.I.} = \frac{1}{D^2 - 9D + 18}\cosh 3x = \frac{1}{D^2 - 9D + 18}\left(\frac{e^{3x} + e^{-3x}}{2}\right)$$

$$= \frac{1}{2(D-3)(D-6)}(e^{3x} + e^{-3x})$$

7. C.F. $= (C_1 + C_2x + C_3x^2)\, e^x + C_4e^{-x}$

$$\text{P.I.} = \frac{1}{(D-1)^3(D+1)}(e^x + e^{-x})$$

$$= \frac{1}{2}\frac{1}{(D-1)^3}e^x.1 - \frac{1}{8}\frac{1}{(D+1)}e^{-x}.1.$$

10. $$\text{P.I.} = \frac{1}{(D^2 - 4D + 4)}(8x^2 + 8e^{2x} + 8\sin 2x)$$

$$= \frac{1}{(D-2)^2}8x^2 + \frac{1}{(D-2)^2}8e^{2x} + \frac{1}{(D-2)^2}.8\sin 2x.$$

ANSWERS

1. $y = C_1e^{-x} + C_2e^{-3x} - \dfrac{x}{2}e^{-3x}$
2. $y = (C_1 + C_2x)\, e^{-3x} + x^2e^{-3x}$
3. $y = C_1e^x + C_2e^{-x} + e^{-x/2}\left[C_3\cos\left(\dfrac{\sqrt{7}}{2}x\right) + C_4\sin\left(\dfrac{\sqrt{7}}{2}x\right)\right] + \dfrac{1}{8}xe^x$
4. $y = C_1e^{3x} + C_2e^{6x} - \dfrac{1}{6}xe^{3x} + \dfrac{1}{108}e^{-3x}$
5. $y = (C_1 + C_2x)e^x + (C_3 + C_4x)\cos x + (C_5 + C_6x)\sin x + \dfrac{1}{8}x^2e^x$
6. $y = 2e^x + e^{2x} - xe^x$

7. $y=(C_1+C_2x+C_3x^2)e^x+C_4e^{-x}+\frac{1}{12}x^3e^x-\frac{x}{8}e^{-x}$

8. $y=(C_1+C_2x)\,e^{3x}+2x^2e^{3x}$

9. $y=C_1e^x+C_2e^{-x}+\frac{1}{2}x\sinh x$

10. $(C_1+C_2x)\,e^{2x}+2x^2+3+4x+4x^2e^{2x}+\cos 2x$

(6) To evaluate $\frac{1}{f(D^2)}\sin ax$ or $\cos ax$, when $f(-a^2)=0$:

To find the particular integral of such cases, we shall calculate P.I. for e^{iax} instead of $\sin ax$ or $\cos ax$.

Here, we have $e^{iax}=\cos ax+i\sin ax$.

Thus, P.I. for e^{iax} = P.I. for $(\cos ax+i\sin ax)$

$\Rightarrow$ P.I. for $\cos ax$ = Real part of P.I. for e^{iax} and P.I. for $\sin ax$ = imaginary part of P.I. for e^{iax}.

Therefore, $\frac{\cos ax}{D^2+a^2}$ and $\frac{\sin ax}{D^2+a^2}$ are respectively, real and imaginary part of $\frac{e^{iax}}{D^2+a^2}$

$$=\frac{e^{iax}}{(D+ai)\,(D-ai)}=\frac{e^{iax}}{(ai+ai)\,(D-ai)}$$

(By putting ai for in $(D+ai)$ because at $D=ai$ it does not vanish.)

$$=\frac{e^{iax}}{2ai}\left[\frac{1}{D+ai-ai}.1\right]=\frac{e^{iax}}{2ai}.\frac{1}{D}.1=\frac{x}{2ai}(e^{aix})=-\frac{ix(\cos ax+i\sin ax)}{2a}=-\frac{ix}{2a}\cos ax+\frac{x}{2a}\sin ax$$

$$\Rightarrow\quad \frac{1}{D^2+a^2}\sin ax=-\frac{x}{2a}\cos ax=\frac{x}{2}\int\sin ax\,dx\quad\text{and}\quad\frac{1}{D^2+a^2}\cos ax=\frac{x}{2a}\sin ax=\frac{x}{2}\int\cos ax\,dx.$$

Solved Examples

Example 1. *Solve* $(D^2+a^2)y=\sin ax$.

Solution. The given equation is

$$(D^2+a^2)y=\sin ax \qquad \ldots(1)$$

To find the C.F. of (1), the auxiliary equation is

$$m^2+a^2=0\Rightarrow m=0\pm ai$$

$$\therefore\quad \text{C.F.}=e^{0x}[C_1\cos ax+C_2\sin ax]=[C_1\cos ax+C_2\sin ax]$$

Now, P.I. $=\frac{1}{D^2+a^2}\sin ax$

$=$ Imaginary part of $\left[\frac{1}{D^2+a^2}(\cos ax+i\sin ax)\right]$

$=$ Imaginary part of $\left[\frac{1}{D^2+a^2}e^{iax}\right]$

$=$ Imaginary part of $\left[\frac{1}{(D+ai)(D-ai)}e^{iax}\right]$

$=$ Imaginary part of $\left[\frac{1}{(ai+ai)(D-ai)}e^{iax}\right]$

$=$ Imaginary part of $\left[\frac{1}{2ai}.\frac{1}{(D-ai)}e^{iax}\right]$

$=$ Imaginary part of $\frac{1}{2ai}\frac{1}{(D-ai)}e^{iax}.1$

$=$ Imaginary part of $\frac{1}{2ai}e^{iax}\frac{1}{[(D+ia)-ia]}.1$

$=$ Imaginary part of $\frac{1}{2ai}e^{iax}\frac{1}{D}.1$

$=$ Imaginary part of $\frac{1}{2ai}e^{iax}.x$

$=$ Imaginary part of $\frac{1}{2ai}.x\,(\cos ax+i\sin ax)$

$=$ Imaginary part of $\frac{1}{2a}x\left[\frac{1}{i}\cos ax+\sin ax\right]$

$=$ Imaginary part of $\frac{1}{2a}x\left[\frac{i}{i^2}\cos ax+\sin ax\right]$

$=$ Imaginary part of $\frac{1}{2a}x[-i\cos ax+\sin ax]$

$=-\frac{x}{2a}\cos ax$.

Hence, the complete solution of (1) is given by

$$y=\text{C.F.}+\text{P.I.}$$

$$\Rightarrow\quad y=C_1\cos ax+C_2\sin ax-\frac{x}{2a}\cos ax.$$

Example 2. *Solve* $(D^2+1)y=\sin x\sin 2x$.

Solution. The given equation is

$$(D^2+1)y=\sin x\sin 2x \qquad \ldots(1)$$

To find the C.F. of (1), the auxiliary equation is

$$(m^2+1)=0\Rightarrow m=\pm i$$

$$\therefore\quad \text{C.F.}=C_1\cos x+C_2\sin x$$

Now, P.I. $=\frac{1}{(D^2+1)}(\sin x\sin 2x)$

$$= \frac{1}{(D^2+1)}\cdot\frac{1}{2}[2\sin x \sin 2x]$$

$$= \frac{1}{2}\cdot\frac{1}{D^2+1}[\cos x - \cos 3x]$$

$$= \frac{1}{2}\left[\frac{1}{D^2+1}\cos x - \frac{1}{D^2+1}\cos 3x\right].$$

Now,

$$\frac{1}{D^2+1}\cos 3x = \frac{1}{-3^2+1}\cos 3x = -\frac{1}{8}\cos 3x$$

Again, $\frac{1}{D^2+1}\cos x = \text{Real part of}\left[\frac{1}{D^2+1}e^{ix}\right]$

$$= \text{Real part of}\left[\frac{1}{(D+i)(D-i)}e^{ix}\right]$$

$$= \text{Real part of}\left[\frac{1}{(i+i)(D-i)}e^{ix}\right]$$

$$= \text{Real part of}\left[\frac{1}{2i}\cdot\frac{1}{(D-i)}e^{ix}\right]$$

$$= \text{Real part of}\left[\frac{1}{2i}\,\frac{1}{D-i}e^{ix}.1\right]$$

$$= \text{Real part of}\left[\frac{1}{2i}\cdot\frac{1}{(D+i-i)}.1\right]$$

$$= \text{Real part of}\left[\frac{1}{2i}e^{ix}\,\frac{1}{D}.1\right]$$

$$= \text{Real part of}\left[\frac{1}{2i}.e^{ix}.x\right]$$

$$= \text{Real part of}\left[\frac{x}{2i}(\cos x + i\sin x)\right]$$

$$= \text{Real part of}\left[-i.\frac{1}{2}x\cos x + \frac{1}{2}x\sin x\right]$$

$$= \frac{1}{2}x\sin x$$

Hence, the complete solution of (1) is given by

y = C.F. + P.I.

$$\Rightarrow y = C_1\cos x + C_2\sin x + \frac{x}{4}\sin x - \frac{1}{16}\cos 3x.$$

Example 3. *Find the solution of the equation*

$$\frac{d^2y}{dx^2} + 4y = 8\cos 2x$$

given that $y = 0$ *and* $\frac{dy}{dx} = 2$ *when* $x = 0$.

Solution. The given differential equation can be written as

$(D^2+4)y = 8\cos 2x$ ...(1)

To find the C.F. of (1), the auxiliary equation is given by $m^2 + 4 = 0 \Rightarrow m = \pm 2i$

$\therefore$ C.F. $= C_1\cos 2x + C_2\sin 2x$

Now, P.I. $= 8.\frac{1}{D^2+4}\cos 2x$

$$= 8.\text{Real part of}\left[\frac{1}{D^2+4}(\cos 2x + i\sin 2x)\right]$$

$$= 8.\text{Real part of}\left[\frac{1}{D^2+4}e^{2ix}\right]$$

$$= 8.\text{Real part of}\left[\frac{1}{(D+2i)(D-2i)}e^{2ix}\right]$$

$$= 8.\text{Real part of}\left[\frac{1}{(2i+2i)(D-2i)}e^{2ix}\right]$$

$$= 8.\text{Real part of}\left[\frac{1}{4i}\,\frac{1}{(D-2i)}e^{2ix}.1\right]$$

$$= 8.\text{Real part of}\left[\frac{1}{4i}\,\frac{1}{D-2i}e^{2ix}.1\right]$$

$$= 8.\text{Real part of}\left[\frac{1}{4i}e^{2ix}\frac{1}{D}.1\right]$$

$$= 8.\text{Real part of}\left[\frac{1}{4i}e^{2ix}.x\right]$$

$$= 8.\text{Real part of}\left[\frac{1}{4i}(\cos 2x + i\sin 2x).x\right]$$

$$= 8.\text{Real part of}\frac{1}{-4}[i\cos 2x - \sin 2x].x$$

$$= 8.\frac{x}{4}\sin 2x = 2x\sin 2x$$

Hence, the complete solution of (1) is given by

y = C.F. + P.I.

$$\Rightarrow \quad y = C_1\cos 2x + C_2\sin 2x + 2x\sin 2x$$

$$\Rightarrow \quad \frac{dy}{dx} = -2C_1\sin 2x + 2C_2\cos 2x + 2\sin 2x + 4x\cos 2x.$$

Now, using the given conditions

$y = 0 = \frac{dy}{dx}$ at $x = 0$ gives $C_1 = 0$ and $C_2 = 1$

$\therefore$ General solution is $y = \sin 2x + 2x\sin 2x$

Example 4. *Solve the given differential equation*

$$\frac{d^2y}{dx^2} + 9y = 2\sin 3x + \cos 3x\cdot$$

Solution. The given differential equation can be written as

$(D^2+9)y = 2\sin 3x + \cos 3x$ ...(1)

To find the C.F. of (1), the auxiliary equation is given by $m^2 + 9 = 0 \Rightarrow m = \pm 3i$

$\therefore$ C.F. $= C_1\cos 3x + C_2\sin 3x$

Now, P.I. $= \frac{1}{D^2+9}(2\sin 3x + \cos 3x)$

$$= 2\frac{1}{D^2+9}\sin 3x + \frac{1}{D^2+9}\cos 3x.$$

Now, $\frac{1}{D^2+9}\cos 3x + i.\frac{1}{D^2+9}\sin 3x$

$= \frac{1}{D^2+9}(\cos 3x + i\sin 3x)$

$= \frac{1}{D^2+9}e^{i3x}$

$= \frac{1}{(D+3i)(D-3i)}e^{i3x}$

$= \frac{1}{(3i+3i)(D-3i)}e^{3ix}.1$

$= \frac{e^{i3x}}{6i}.\frac{1}{[(D+3i)-3i]}.1$

$= \frac{1}{6i}.e^{i3x}.\frac{1}{D}.1 = \frac{1}{6i}e^{i3x}.x$

$= \frac{x}{6i}[\cos 3x + i\sin 3x]$

$= \frac{x}{6}\sin 3x - i\frac{x}{6}\cos 3x$

Now, equating real and imaginary part of both sides, we get

$$\frac{1}{D^2+9}\cos 3x = \frac{x}{6}\sin 3x$$

and $\frac{1}{D^2+9}\sin 3x = -\frac{x}{6}\cos 3x$.

$\therefore$ Required P.I. $= 2\left[-\frac{x}{6}\cos 3x\right] + \frac{x}{6}\sin 3x$

$= -\frac{x}{3}\cos 3x + \frac{x}{6}\sin 3x.$

Hence, the complete solution of (1) is given by

$y =$ C.F. + P.I.

$$\Rightarrow \quad y = C_1\cos 3x + C_2\sin 3x - \frac{1}{3}x\cos x + \frac{1}{6}x\sin 3x$$

EXERCISE 29.8

Solve the following differential equations :

1. $(D^2+a^2)y = \cos ax$
2. $(D^2+4)y = \cos 2x$
3. $(D^2+4)y = e^x + \sin 2x$
4. $(D^3+a^2D)y = \sin ax$
5. $(D^3+1)y = \cos 2x$
6. $(D^4+D^2+1)y = e^{-x/2}\cos\frac{x\sqrt{3}}{2}$ (Rajasthan–2006)
7. $(D^2+2D+2)y = 2e^{-x}\sin x$
8. $(D^4+2D^2+1)y = \cos x$
9. $(D^2+4)y = 4+\sin 2x$

Hint to Selected Problems

1. $m = \pm ai \Rightarrow$ C.F. $= C_1\cos ax + C_2\sin ax$

P.I. $= \frac{1}{D^2+a^2}\cos ax =$ Real part of $\left[\frac{1}{D^2+a^2}e^{iax}\right]$.

3. $m = \pm 2i$, C.F. $= C_1\cos 2x + C_2\sin 2x$

P.I. $= \frac{1}{(D^2+4)}(e^x+\sin 2x) = \frac{e^x}{D^2+4} + \frac{1}{D^2+4}.\sin 2x$

$= \frac{e^x}{1+4} +$ Imag. part of $\left[\frac{1}{D^2+4}.e^{2ix}\right]$

4. P.I. $= \frac{1}{D^2+a^2}\sin ax =$ Imaginary part of $\left[\frac{1}{(D+ai)(D-ai)}e^{iax}\right]$

ANSWERS

1. $y = C_1\cos ax + C_2\sin ax + \frac{x}{2a}\sin ax$
2. $y = C_1\cos 2x + C_2\sin 2x + \frac{x}{4}\sin 2x$
3. $y = C_1\cos 2x + C_2\sin 2x + \frac{1}{5}e^x - \frac{1}{4}x\cos 2x$
4. $y = C_1 + C_2\cos ax + C_3\sin ax - \frac{1}{2a^2}x\sin ax$
5. $y = C_1e^{-x} + e^{x/2}\left[C_2\cos\frac{\sqrt{3}}{2}x + C_3\sin\frac{\sqrt{3}}{2}x\right] + \frac{1}{65}(\cos 2x - 8\sin 2x)$
6. $y = e^{-x/2}\left[C_1\cos\frac{1}{2}x\sqrt{3} + C_2\sin\frac{1}{2}x\sqrt{3}\right] + e^{x/2}\left[C_3\cos\frac{1}{2}x\sqrt{3} + C_4\sin\frac{1}{2}x\sqrt{3}\right] - \frac{1}{12}\sqrt{3}\,x\,e^{-x/2}\sin\frac{\sqrt{3}x}{2} + \frac{x}{4}e^{-x/2}\cos\frac{\sqrt{3}}{2}x$
7. $y = e^{-x}(C_1\cos x + C_2\sin x) - xe^{-x}\cos x$
8. $y = (C_1+C_2x)\cos x + (C_3+C_4x)\sin x - \frac{1}{8}x^2\cos x$
9. $y = C_1\cos 2x + C_2\sin 2x + 1 - \frac{x}{4}\cos 2x$

(7) To evaluate $\frac{1}{f(D)}x.X$, where X is any function of x (except e^{ax}) :

Consider $D^n(x.X) = xD^n.X + {}^nC_1\,D^{n-1}.X$ (By Leibnitz's theorem)

We have $f(D)(xX) = x\,f(D)X + f'(D).X$

Now, taking the inverse operator, we have $\frac{1}{f(D)}(xX) = x.\frac{1}{f(D)}X + \left[\frac{d}{dD}\frac{1}{f(D)}\right]X$

But we have $\frac{d}{dD}\left[\frac{1}{f(D)}\right] = -\frac{f'(D)}{\{f(D)\}^2}$. Therefore, $\frac{1}{f(D)}(x.X) = x.\frac{1}{f(D)}X - \frac{f'(D)}{\{f(D)\}^2}X$.

REMARK

- If we want to find P.I. when B is of the form $x^m.X$, where X is any function of x, then there are two cases
  (a) If $X = x^n$, then $x^m.X = x^{m+n}$.
  Then B is of the form x^{m+n} (Polynomial). Here, we should apply the method of finding P.I. for polynomial discussed earlier.
  (b) If $X = e^{ax}$, then $x^m.X = x^m.e^{ax}$ and we should apply the method, discussed earlier.
  (c) If $X = \cos ax$, then $x^m.X = x^m \cos ax$
  Then P.I. $= \frac{1}{f(D)} x^m \cos ax = \frac{1}{f(D)}$ (Real part of $x^m.e^{iax}$) = Real part of $\frac{1}{f(D)} x^m.e^{iax}$, which can be easily calculated.
  Similar results hold if $X = \sin ax$, then taking imaginary part.

Solved Examples

Example 1. *Solve* $(D^2+2D+1)y = x\cos x$. (Rajasthan–2006)

Solution. The given equation is

$$((D^2+2D+1)y = x\cos x \quad \ldots(1)$$

To find the C.F. of (1), the auxiliary equation is

$$m^2+2m+1=0 \Rightarrow m=-1,-1$$

$$\therefore \quad \text{C.F.} = (C_1+C_2x)\,e^{-x}$$

Now, $\text{P.I.} = \frac{1}{(D^2+2D+1)}.x\cos x$

$$= x.\frac{1}{D^2+2D+1}\cos x - \frac{2D+2}{(D^2+2D+1)^2}\cos x$$

$$= x.\frac{1}{2D}\cos x - \frac{2D+2}{4D^2}\cos x$$

$$= \frac{x}{2}\sin x + \frac{(D+1)}{2}\cos x$$

$$= \frac{x}{2}\sin x + \frac{\cos x}{2} - \frac{\sin x}{2}$$

Hence, the complete solution of (1) is given by

$$y = \text{C.F.} + \text{P.I.}$$

$$\Rightarrow \quad y = (C_1+C_2x)e^{-x} + \frac{x}{2}\sin x + \frac{\cos x}{2} - \frac{\sin x}{2}.$$

Example 2. *Solve* $(D^2-4D+4) = 8x^2e^{2x}\sin 2x$.
(JNTU–2006, UPTU 2004, 05, 09)

Solution. The given equation is

$$(D^2-4D+4) = 8x^2e^{2x}\sin 2x \quad \ldots(1)$$

To find the C.F. of (1), the auxiliary equation is

$$m^2-4m+4=0 \Rightarrow m=2,2$$

$$\therefore \quad \text{C.F.} = (C_1+C_2x)\,e^{2x}$$

Now, $\text{P.I.} = 8.\frac{1}{(D-2)^2}e^{2x}(x^2\sin 2x)$

$$= 8e^{2x}.\frac{1}{(D+2-2)^2}.(x^2\sin 2x)$$

$$= 8e^{2x}.\frac{1}{D^2}(x^2\sin 2x) = 8\,e^{2x}.I_1$$

where, $I_1 = \frac{1}{D^2}(x^2\sin 2x)$

$$= \text{Imaginary part of} \frac{1}{D^2}x^2e^{2ix}$$

$$= \text{Imaginary part of}\, e^{2ix}\frac{1}{(D+2i)^2}x^2$$

$$= \text{Imaginary part of} \frac{e^{2ix}}{4i^2}\left(1+\frac{D}{2i}\right)^{-2}x^2$$

$$= \text{Imaginary part of} \frac{e^{2ix}}{-4}\left(1-\frac{iD}{2}\right)^{-2}x^2$$

$$= \text{Imaginary part of} \frac{e^{2ix}}{-4}\left[1+2\left(\frac{iD}{2}\right)+3\left(\frac{iD}{2}\right)^2+\ldots\right]x^2$$

$$= \text{Imaginary part of} \frac{e^{2ix}}{-4}\left[1+Di-\frac{3}{4}D^2+\ldots\right]x^2$$

$$= \text{Imaginary part of} \frac{e^{2ix}}{-4}\left[x^2+2ix-\frac{3}{2}\right]$$

$$= \text{Imaginary part of}$$

$$\left\{-\frac{1}{4}(\cos 2x + i\sin 2x)\left(x^2+2ix-\frac{3}{2}\right)\right\}$$

$$= -\frac{1}{4}\left[\left(x^2-\frac{3}{2}\right)\sin 2x + 2x\cos 2x\right]$$

$$= -\frac{1}{8}[(2x^2-3)\sin 2x + 4x\cos 2x$$

$$\therefore \quad \text{P.I.} = 8e^{2x}.I_1 = 8e^{2x}\left[-\frac{1}{8}\{(2x^2-3)\sin 2x + 4x\cos 2x\}\right]$$

$$= -e^{2x}[(2x^2-3)\sin 2x + 4x\cos 2x].$$

Hence, the complete solution of (1) is given by

$$y = \text{C.F.} + \text{P.I.}$$

$$\Rightarrow \quad y = e^{2x}[C_1 + C_2x + 3\sin 2x - 2x^2\sin 2x - 4x\cos 2x].$$

Example 3. *Solve* $(D^2-2D+1)\,y = xe^x\sin x$. (SVTU–2007, JNTU–2006, UPTU–2005, MTU(B.Pharm.)–2011)

Solution. The given differential equation can be written as

$$(D^2-2D+1)\,y = xe^x\sin x \quad \ldots(1)$$

To find the C.F. of (1), the auxiliary equation is given by $m^2-2m+1=0 \Rightarrow m=1,1$

$$\therefore \quad \text{C.F.} = (C_1+C_2x)\,e^x$$

Now, $\text{P.I.} = \frac{1}{(D^2 - 2D + 1)} xe^x \sin x$

$$= \frac{1}{(D-1)^2} e^x (x \sin x)$$

$$= e^x \frac{1}{(D+1)^2 - 2(D+1) + 1} x \sin x$$

$$= e^x \frac{1}{D^2} (x \sin x) = e^x . \frac{1}{D} \int x \sin x \, dx$$

$$= e^x \left(\frac{1}{D}\right)(-x \cos x + \sin x)$$

$$= e^x \left[\int -x \cos x dx + \int \sin x dx\right]$$

$$= e^x [-x \sin x - 2 \cos x].$$

Hence, the complete solution of (1) is given by

$$y = \text{C.F.} + \text{P.I.}$$

$$\Rightarrow \quad y = (C_1 + C_2 x) e^x - e^x (x \sin x + 2 \cos x).$$

Example 4. *Solve* $\frac{d^2y}{dx^2} + 4y = x \sin x$. (Madras–2004)

Solution. The given differential equation can be written as

$$(D^2 + 4)y = x \sin x \quad \ldots(1)$$

To find the C.F. of (1), the auxiliary equation is given by $m^2 + 4 = 0 \Rightarrow m = \pm 2i$

$\therefore$ C.F. $= C_1 \cos 2x + C_2 \sin 2x$

Now, $\text{P.I.} = \frac{1}{D^2 + 4} x \sin x$

$$= x \frac{1}{D^2 + 4} \sin x - \frac{2D}{(D^2 + 4)^2} \sin x$$

$$= \frac{x \sin x}{-1^2 + 4} - \frac{2D}{(-1^2 + 4)^2} \sin x$$

$$= \frac{1}{3} x \sin x - \frac{2}{9} D(\sin x)$$

$$= \frac{1}{3} x \sin x - \frac{2}{9} \cos x$$

Hence, the complete solution of (1) is given by

$$y = \text{C.F.} + \text{P.I.}$$

$$\Rightarrow y = C_1 \cos 2x + C_2 \sin 2x + \frac{x}{3} \sin x - \frac{2}{9} \cos x.$$

EXERCISE 29.9

Solve the following differential equations :

1. $(D^2 - 2D + 1)y = x \sin x$
2. $(D^2 + m^2)y = x \cos mx$
3. $(D^2 - 1) = x^2 \sin x$
4. $(D^2 + a^2)^2 y = \sin ax$ (SRM–2013)
5. $(D^4 - 1)y = x \sin x$
6. $(D^4 - 1)y = e^x \cos x$
7. $(D^4 + 2D^2 + 1)y = x^2 \cos x$ (Nagarjuna–2008, Rajasthan–2005)
8. $(D^2 + 1)y = x^2 \sin 2x$
9. $(D^2 + 1)^2 y = 24x \cos x$, given that $x = 0, y = 0, Dy = 0, D^2 y = 0, D^3 y = 0$.

Hint to Selected Problems

1. C.F. $= (C_1 + C_2 x) e^x$

$$\text{P.I.} = \frac{1}{D^2 - 2D + 1} x \sin x$$

$$= x \frac{1}{D^2 - 2D + 1} \sin x - \frac{(2D-2)}{(D^2 - 2D + 1)^2} \sin x$$

$$= x . \frac{1}{-1 - 2D + 1} \sin x - \frac{(2D-2)}{(-1 - 2D + 1)^2} \sin x$$

$$= x . \frac{1}{-2D} \sin x - \frac{(2D-2)}{4D^2} \sin x.$$

2. $\text{P.I.} = \frac{1}{D^2 + m^2} x \cos mx = \text{Real part of} \frac{1}{(D^2 + m^2)} x e^{imx}$

$$= \text{Real part of } e^{imx} \left[\frac{1}{(D + im)^2 + m^2} . x\right]$$

8. $\text{P.I.} = \frac{1}{D^2 + 1} x^2 \sin 2x = \text{Imaginary part of} \frac{1}{D^2 + 1} . xe^{2ix}$

$$= \text{Imag. part of } e^{2ix} \left[\frac{1}{(D + 2i)^2 + 1}\right] . x.$$

10. $\text{P.I.} = \frac{1}{(D^2 + 1)^2} 24x \cos x = 24 \frac{1}{(D^2 + 1)^2} x \cos x$

$$= \text{Real part of } 24e^{ix} \frac{1}{[(D + i)^2 + 1]^2} . x$$

ANSWERS

1. $y = (C_1 + C_2 x) e^x + \frac{1}{2}(x \cos x + \cos x - \sin x)$
2. $y = C_1 \cos mx + C_2 \sin mx + \frac{x^2}{4m} \sin mx + \frac{x}{4m^2} \cos mx$
3. $y = C_1 e^x + C_2 e^{-x} - x \cos x - \frac{1}{2}(x^2 - 1) \sin x$
4. $y = (C_1 + C_2 x) \cos ax + (C_3 + C_4 x) \sin ax - \frac{1}{8a^2}(x^2 \sin ax)$
5. $y = C_1 e^x + C_2 e^{-x} + C_3 \cos x + C_4 \sin x + \frac{1}{8}(x^2 \cos x - 3x \sin x)$
6. $y = C_1 e^x + C_2 e^{-x} + C_3 \cos x + C_4 \sin x - \frac{1}{5} e^x \cos x$
7. $y = (C_1 + C_2 x) \cos x + (C_3 + C_4 x) \sin x - \frac{1}{48}(x^4 - 9x^2) \cos x + \frac{1}{12} x^3 \sin x$
8. $y = C_1 \cos x + C_2 \sin x - \frac{1}{27}[24x \cos 2x + (9x^2 - 26) \sin 2x]$
9. $y = 3x^2 \sin x - x^3 \cos x$

Objective Evaluations

Fill in the Blanks

1. If the order of the given differential equation is n, then C.F. of this equation contains ________ arbitrary constant.
2. If the order of the given differential equation is n, then ________ of this equation does not contain any arbitrary constants.
3. The ________ can be obtained by adding complementary function and particular integral.
4. The auxiliary equation can be obtained by replacing D by ________.
5. If the auxiliary equation having two roots namely m_1 and m_2 such that $m_1 \neq m_2$, then C.F. is ________.
6. If the auxiliary equation having two equal roots namely $m_1 = m_2 = m$, then C.F. is ________.
7. The imaginary roots of an equation always occurs in ________.
8. If the roots of auxiliary equation are $\alpha \pm i\beta$, then C.F. is ________.
9. The value of $\frac{1}{f(D)} e^{ax} = \frac{1}{f(a)} e^{ax}$ provided ________.
10. The particular integral of the equation is $(D^2 - 2D + 5)y = e^{-x}$ is ________.

True/False

Write 'T' for True and 'F' for False statement.

1. The particular integral of an nth order differential equation contains n independent arbitrary constants. **(T/F)**
2. The complementary functions of a differential equation of order n contains n independent arbitrary constant. **(T/F)**
3. If $f(D) = 0$, then we cannot find the complementary function. **(T/F)**
4. If P.I. $= \frac{1}{f(D)} e^{ax}$, then we put a for D in $f(D)$ and we get the required P.I. provided $F(a) \neq 0$. **(T/F)**
5. The C.F. of the differential equation $(D^2 - 3D + 2)y = e^{\sqrt{x}}$ is $C_1 e^x + C_2 e^{2x}$. **(T/F)**
6. To find the P.I. when $Q = e^{ax}.X$, where X is a function of x, then we replace D by $(D + a)$ and bring e^{ax} before the operator $1/f(D)$. **(T/F)**
7. The method discussed in (6) cannot be used to find $\frac{1}{f(D)} e^{ax}$ when $f(a) = 0$. **(T/F)**
8. The general solution of a differential equation cannot be found with the particular integral and complimentary function. **(T/F)**

Multiple Choice Questions

Choose the most appropriate one.

1. The solution of $(D^2 + 1)y = 0$ is $y =$:
   (a) $A\cos x - B\sin x$ (b) $-A\cos x - B\sin x$
   (c) $A\cos x + B\sin x$ (d) $-A\cos x + B\sin x$
2. The general solution of the differential equation $(D^2 + 1)y = 0$ is :
   (a) $y = \cos x$ (b) $y = C\cos x$
   (c) $y = C_1\cos(x + C_2)$ (d) $C_1\cos(C_2 + C_3 x)$
3. The complete solution of $(D^2 - 3D + 4)y = 0$ is :
   (a) $y = C_1 e^{-x} + C_2 e^{4x}$ (b) $y = C_1 x + C_2 + x$
   (c) $y = (C_1 + C_2 x)e^x$ (d) none of these
4. The C.F. of $(D^2 + 2D + 1)\ y = (x - 1)$ is :
   (a) $y = (C_1 + C_2 x)e^x$ (b) $y = (C_1 + C_2 x)e^{-x}$
   (c) $y = C_1 e^x + C_2 e^{-x}$ (d) none of these
5. The particular integral of n th order differential equation contains :
   (a) $(n + 1)$ arbitrary constants
   (b) n arbitrary constants
   (c) one arbitrary constant
   (d) none of these
6. The complete primitive can be obtained by :
   (a) C.F (b) P.I.
   (c) C.F. + P.I. (d) none of these
7. The P.I. of $(D^2 - 5D + 6)y = e^{mx}$ is :
   (a) e^{mx} (b) $\frac{e^{mx}}{m^2 - 5m + 6}$
   (c) $m^2 - 5m + 6$ (d) none of these
8. To find $\frac{1}{f(D)} e^{ax}.X$, we bring e^{ax} to the left from right of $\frac{1}{f(D)}$, then D must be replaced by :
   (a) $D - a$ (b) a (c) m (d) $D + a$
9. The general solution of $(2D + 1)^2 y = 4e^{-x/2}$ is $y =$:
   (a) $(C_1 + C_2 x)\,e^{-x/2}$ (b) $e^{-x/2}\left(\frac{x^2}{2}\right)$
   (c) (a) + (b) (d) none of these
10. The general solution of $(D^2 + a^2)y = \cos ax$ is $y =$:
   (a) $C_1\cos ax + C_2\sin ax$ (b) $\cos ax$
   (c) $\sin ax$ (d) (a) + (c)

Answers

Fill in the Blanks

1. n	**2.** particular integral	**3.** general solution	**4.** m	**5.** $C_1 e^{m_1 x} + C_2 e^{m_2 x}$
6. $(C_1 + C_2 x)\,e^{mx}$	**7.** pairs	**8.** $e^{\alpha x}[C_1\cos\beta x + C_2\sin\beta x]$	**9.** $f(a) \neq 0$	**10.** $\frac{1}{8}e^{-x}$

True/False

1. T	**2.** F	**3.** F	**4.** T	**5.** T	**6.** T	**7.** F	**8.** F

Multiple Choice Questions

1. (c)	**2.** (c)	**3.** (a)	**4.** (b)	**5.** (d)	**6.** (c)	**7.** (b)	**8.** (d)	**9.** (c)	**10.** (d)

FFFFF

CHAPTER 30 Homogeneous Linear Differential Equations

30.1 INTRODUCTION

Any differential equation of the form

$$x^n \frac{d^n y}{dx^n} + A_1 \frac{d^{n-1} y}{dx^{n-1}} + \dots + A_n y = X$$

is called a homogeneous linear differential equation of nth order, where $A_1, A_2, \dots, A_n$ are constants and X is a function of x or a constant.

For example : Consider the following differential equations :

(i) $x^2 \frac{d^2 y}{dx^2} + 3x \frac{dy}{dx} + 4y = e^x$ (ii) $x^2 \frac{d^2 y}{dx^2} + 2x \frac{dy}{dx} + 2y = e^x$

The above two differential equations are linear as the dependent variable y and its derivatives appear in their first degree and are not multiplied together.

REMARKS

- The differential equation in which the powers of x in the coefficients are equal to the orders of the derivative associated with them, is called the homogeneous linear differential equation.
- In linear homogeneous differential equation, the dependent variable y and its derivatives with respect to independent variable x, appears in their first degree and are not multiplied together.
- The homogeneous linear equations, discussed above, are also known as Cauchy-Euler equation.

30.2 SOLUTION OF HOMOGENEOUS LINEAR EQUATION

Consider the homogeneous linear differential equation

$$x^n \frac{d^n y}{dx^n} + A_1 x^{n-1} \frac{d^{n-1} y}{dx^{n-1}} + A_2 x^{n-2} \frac{d^{n-2} y}{dx^{n-2}} + \dots + A_n y = X \qquad \dots(1)$$

where, $A_1, A_2, \dots, A_n$ are constants and X is a function of x or constant.

Now, equation (1) can be transformed to an equivalent equation (linear differential equation) with constant coefficients by changing the independent variable by the relation

$$x = e^z, \text{ i.e., } z = \log x$$

$$\Rightarrow \quad \frac{dz}{dx} = \frac{1}{x}$$

Now $$\frac{dy}{dx} = \frac{dy}{dz}\frac{dz}{dx} = \frac{1}{x}.\frac{dy}{dz} \quad \Rightarrow \quad x\frac{dy}{dx} = \frac{dy}{dz}$$

Again, $$\frac{d^2 y}{dx^2} = \frac{d}{dx}\left(\frac{dy}{dx}\right) = \frac{d}{dx}\left(\frac{1}{x}.\frac{dy}{dz}\right)$$

$$= \frac{1}{x}\frac{d}{dx}\left(\frac{dy}{dx}\right) + \frac{d}{dx}\left(\frac{1}{x}\right)\frac{dy}{dz}$$

$$= \frac{1}{x}\frac{d^2y}{dz^2}.\frac{dz}{dx} - \frac{1}{x^2}\frac{dy}{dz} = \frac{1}{x^2}.\frac{d^2y}{dz^2} - \frac{1}{x^2}.\frac{dy}{dz}$$

$$\Rightarrow \quad x^2\frac{d^2y}{dx^2} = \frac{d^2y}{dz^2} - \frac{dy}{dz}$$

Proceeding likewise, we get

$$x^n\frac{d^ny}{dx^n} = \left[\frac{d^ny}{dz^n} - \frac{n(n-1)}{2!}.\frac{d^{n-1}y}{dz^{n-1}} + ... + (-1)^{n-1}n!\frac{dy}{dz}\right]$$

If we write $D \equiv \frac{d}{dz}$, then we get

$$x\frac{dy}{dx} = Dy, x^2\frac{d^2y}{dx^2} = D(D-1)y, x^3\frac{d^3y}{dx^3} = D(D-1)(D-2)y$$

and so on

$$x^n\frac{d^ny}{dx^n} = D(D-1)(D-2)...(D-n+1)y$$

Let us consider, the transformed equation

$$f(D)y = X \qquad ...(2)$$

The general solution of (2) is the sum of a particular solution of (2) and complementary function of (2). Now, it can be easily solved by the usual method given in Chapter 29.

WORKING PROCEDURE

Step 1. *Put* $x = e^z$, $x\frac{d}{dx} = D = \frac{d}{dz}$, $x^2\frac{d^2}{dx^2} = D(D-1)$ *and so on.*

Step 2. *Obtain the equation in terms of D (linear equation).*

Step 3. *To find the C.F. and P.I. used the usual method given in Chapter 29.*

Step 4. *Find general solution by adding C.F. and P.I.*

Step 5. *Finally, substitute* $z = \log x$.

REMARK

- To solve the homogeneous linear differential equation, we change the independent variable x to z by substitution $x = e^z$. The substitution can be easily justified as follows : "Since the exponential function e^{mz} has the property that its derivative are all constant multiples of the function itself. This leads us to consider $x = e^{mz}$ as a possible solution of the given equation."

Solved Examples

Example 1. *Solve* $x^2\frac{d^2y}{dx^2} - 4x\frac{dy}{dx} + 6y = x \qquad ...(1)$

Solution. Putting $x = e^z$

$\Rightarrow \quad z = \log x$ and $D \equiv \frac{d}{dz}$

Thus, the given equation becomes

$$[D(D-1) - 4D + 6]\,y = e^z$$

$$\Rightarrow \quad (D^2 - 5D + 6)y = e^z \qquad ...(2)$$

which is a linear equation in y.

To find the C.F. of (2), the auxiliary equation is

$$m^2 - 5m + 6 = 0$$

$\Rightarrow \quad (m-2)(m-3) = 0$ which gives $m = 2, 3$

$$\therefore \quad \text{C.F.} = C_1e^{2z} + C_2e^{3z}$$

Now, $\text{P.I.} = \frac{1}{D^2 - 5D + 6}e^z = \frac{1}{1-5+6}e^z = \frac{1}{2}e^z$

Hence, the general solution is given by

$$y = \text{C.F.} + \text{P.I.}$$

$$\Rightarrow \quad y = C_1e^{2z} + C_2e^{3z} + \frac{1}{2}e^z$$

Now, put $\quad e^z = x.$

$$y = C_1x^2 + C_2x^3 + \frac{1}{2}x$$

Example 2. *Solve* $x^2\frac{d^2y}{dx^2} - 3x\frac{dy}{dx} + 4y = 2x^2$.

Solution. Here, the given equation is

$$x^2\frac{d^2y}{dx^2} - 3x\frac{dy}{dx} + 4y = 2x^2 \qquad ...(1)$$

Let $x = e^z$

then

$$z = \log x,\ x\frac{dy}{dx} = Dy,\ x^2\frac{d^2y}{dx^2} = D(D-1)y$$

Put in (1), we get

$$(D(D-1) - 3D + 4)y = 2e^{2z}$$

$$(D^2 - 4D + 4)\ y = 2e^{2z} \quad ...(2)$$

To find the C.F. of (1), the auxiliary equation is

$$m^2 - 4m + 4 = 0$$

$$\Rightarrow \quad (m-2)^2 = 0 \Rightarrow m = 2, 2$$

$$\therefore \quad \text{C.F.} = (C_1 + C_2 z)e^{2z} = (C_1 + C_2 \log x)\, x^2$$

Now, $\text{P.I.} = \dfrac{1}{(D-2)^2} 2e^{2z}$

$$= 2e^{2z}.\frac{1}{D^2}.1 = 2\frac{1}{(D-2)^2}e^{2z}$$

$$= 2e^{2z}\frac{1}{[(D+2)-2]^2}.1 = 2e^{2z}.\frac{1}{D^2}.1$$

$$= 2e^{2z}.\frac{z^2}{2} = z^2 e^{2z} = (\log x)^2 . x^2$$

Hence, the complete solution of (1) is given by

$$y = \text{C.F.} + \text{P.I.}$$

$$\Rightarrow \quad y = (C_1 + C_2 \log x)\, x^2 + x^2 (\log x)^2$$

Example 3. Solve $x^2\dfrac{d^2y}{dx^2} + 2x\dfrac{dy}{dx} - 20y = (x+1)^2$

Solution. The given equation is

$$x^2\frac{d^2y}{dx^2} + 2x\frac{dy}{dx} - 20y = (x+1)^2 \quad ...(1)$$

Putting $x = e^x,\ D \equiv \dfrac{d}{dz}$ in (1), we get

$$[D(D-1) + 2D - 20]\ y = (e^z + 1)^2$$

$$\Rightarrow \quad [D^2 + D - 20]y = (e^z + 1)^2$$

$$\text{or} \quad (D+5)\,(D-4)y = e^{2z} + 2e^z + 1 \quad ...(2)$$

To find the C.F. of (1), the auxiliary equation is

$$(m+5)(m-4) = 0 \ \Rightarrow\ m = 4, -5$$

$$\therefore \quad \text{C.F.} = C_1 e^{4z} + C_2 e^{-5z} = C_1 x^4 + C_2 x^{-5}$$

Now, $\text{P.I.} = \dfrac{1}{(D+5)(D-4)}e^{2z}$

$$+ \frac{1}{(D+5)(D-4)}2e^z + \frac{1}{(D+5)(D-4)}e^{0z}$$

$$= \frac{e^{2z}}{(2+5)(2-4)} + 2\frac{1.e^z}{(1+5)(1-4)} + \frac{1}{5(-4)}e^{0z}$$

$$= \frac{e^{2z}}{-14} + \frac{2e^z}{-18} - \frac{1}{20}$$

$$= \frac{e^{2z}}{-14} - \frac{1}{9}e^z - \frac{1}{20} = -\frac{1}{14}x^2 - \frac{1}{9}x - \frac{1}{20}$$

Hence, the general solution is given by

$$y = \text{C.F.} + \text{P.I.}$$

$$\Rightarrow \quad y = C_1 x^4 + C_2 x^{-5} - \frac{1}{14}x^2 - \frac{1}{9}x - \frac{1}{20}$$

Example 4. Solve $x^2\dfrac{d^2y}{dx^2} + 7x\dfrac{dy}{dx} + 13y = \log x$.

Solution. The given equation is

$$x^2\frac{d^2y}{dx^2} + 7x\frac{dy}{dx} + 13y = \log x \quad ...(1)$$

Putting $x = e^z$ and denoting $\dfrac{d}{dz} \equiv D$ in (1), we get

$$D(D-1)y + 7Dy + 13y = z$$

$$\Rightarrow \quad [D^2 + 6D + 13]\ y = z \quad ...(2)$$

To find the C.F. of (1), the auxiliary equation is

$$m^2 + 6m + 13 = 0$$

which gives $m = \dfrac{-6 \pm \sqrt{(36-52)}}{2} = -3 \pm 2i$

$$\therefore \quad \text{C.F.} = C_1 e^{-3z}\cos(2z + C_2)$$

$$= C_1 x^{-3}\cos(2\log x + C_2)$$

Now, $\text{P.I.} = \dfrac{1}{D^2 + 6D + 13}z$

$$= \frac{1}{13\left[1 + \left(\frac{6}{13}D + \frac{1}{13}D^2\right)\right]}z$$

$$= \frac{1}{13}\left[1 + \left(\frac{6}{13}D + \frac{1}{13}D^2\right)\right]^{-1} z$$

$$= \frac{1}{13}\left[1 - \frac{6}{13}D - \frac{1}{13}D^2 + ...\right]z$$

$$= \frac{1}{13}\left(z - \frac{6}{13}\right) = \frac{1}{13}\left[\log x - \frac{6}{13}\right]$$

Hence, the general solution is given by

$$y = \text{C.F.} + \text{P.I.}$$

$$\Rightarrow \quad y = \frac{C_1}{x^3}\cos[2\log x + C_2] + \frac{1}{13}\log x - \frac{6}{169}$$

Example 5. Solve $x^3\dfrac{d^3y}{dx^3} + 3x^2\dfrac{d^2y}{dx^2} + x\dfrac{dy}{dx} + y = x + \log x$.

(UPTU–2001, Bhopal–2008)

Solution. Putting $x = e^z$, in the given equation, we get

$$(D(D-1)(D-2) + 3D(D-1) + (D+1)y$$

$$= e^z + z$$

where $D \equiv \dfrac{d}{dz}$

$$\Rightarrow \quad (D^3 + 1)y = e^z + z \quad ...(1)$$

To find the C.F. of (1), the auxiliary equation is

$$m^3 + 1 = 0 \Rightarrow (m+1)(m^2 - m + 1) = 0$$

$$\Rightarrow \quad m = -1, \frac{1 \pm i\sqrt{3}}{2}$$

$\therefore$ C.F. $= C_1 e^{-z} + e^{z/2}\left[C_2 \cos\left(\frac{\sqrt{3}}{2}\right)z + C_3 \sin\left(\frac{\sqrt{3}}{2}\right)z\right]$

$$= C_1 x^{-1} + x^{1/2}\left[C_2 \cos\left(\frac{\sqrt{3}}{2}\log x\right) + C_3 \sin\left(\frac{\sqrt{3}}{2}\log x\right)\right]$$

Now, P.I. $= \frac{1}{D^3+1}e^z + z = \frac{1}{D^3+1}e^z + \frac{1}{D^3+1}z$

$$= \frac{e^z}{1^3+1} + (1+D^3)^{-1}z = \frac{1}{2}e^z + (1-D^3+\ldots)z$$

$$= \frac{1}{2}e^z + z = \frac{1}{2}x + \log x$$

Hence, the general solution of (1) is given by

$y =$ C.F. + P.I.

$$\Rightarrow \quad y = \frac{C_1}{x} + \sqrt{x}\left[C_2 \cos\left(\frac{\sqrt{3}}{2}\log x\right) + C_3 \sin\left(\frac{\sqrt{3}}{2}\log x\right)\right] + \frac{1}{2}x + \log x$$

Example 6. *Solve* $x^2\frac{d^2y}{dx^2} - x\frac{dy}{dx} + 4y = \cos(\log x) + x\sin(\log x)$

Solution. Putting, $x = e^z \Rightarrow z = \log x$ in the given equation, we get

$$[D(D-1) - D + 4]\, y = \cos z + e^z \sin z$$

where $D \equiv \frac{d}{dz}$

$$\Rightarrow \quad (D^2 - 2D + 4)\, y = \cos z + e^z \sin z \quad \ldots(1)$$

To find the C.F. of (1), the auxiliary equation is

$m^2 - 2m + 4 = 0 \Rightarrow m = 1 \pm i\sqrt{3}$

C.F. $= e^z\,[C_1 \cos\sqrt{3}z + C_2 \sin\sqrt{3}z]$

$$= x\,[C_1 \cos(\sqrt{3}\log x) + C_2 \sin(\sqrt{3}\log x)]$$

Now, P.I. $= \frac{1}{(D^2-2D+4)}\cos z + \frac{1}{(D^2-2D+4)}e^z.\sin z$

$$= \frac{1}{-1^2-2D+4}\cos z + e^z\frac{1}{(D+1)^2 - 2(D+1)+4}\sin z$$

$$= \frac{1}{(3-2D)}\cos z + e^z\frac{1}{D^2+3}\sin z$$

$$= \frac{(3+2D)}{9-4D^2}\cos z + e^z.\frac{1}{-1^2+3}\sin z$$

$$= \frac{3\cos z - 2\sin z}{9+4} + \frac{e^z \sin z}{2}$$

$$= \frac{1}{13}[3\cos(\log x) - 2\sin(\log x)] + \frac{1}{2}x\sin(\log x)$$

Hence, the general solution of the given equation is

$y =$ C.F. + P.I.

$$\Rightarrow \quad y = x\,[C_1 \cos(\sqrt{3}\log x) + C_2 \sin(\sqrt{3}\log x)] + \frac{1}{13}[3\cos(\log x) - 2\sin(\log x)] + \frac{1}{2}x\sin(\log x)$$

Example 7. *Solve* $x^3\frac{d^3y}{dx^3} + 2x^2\frac{d^2y}{dx^2} + 2y = 10\left[x + \frac{1}{x}\right]$.

(SAMBHALPUR–1998, PTU–2003, SVTU–2006, UPTU(Co)–2009)

Solution. Putting $x = e^z$, in the given equation, we get

$$[D(D-1)(D-2) + 2D(D-1) + 2]\, y = 10(e^z + e^{-z}), \text{ where } \frac{d}{dz} \equiv D$$

$$\Rightarrow \quad (D^3 - D^2 + 2)\, y = 10(e^z + e^{-z}) \quad \ldots(1)$$

To find the C.F. of (1), the auxiliary equation is

$m^3 - m^2 + 2 = 0$

$\Rightarrow \quad (m+1)(m^2 - 2m + 2) = 0 \Rightarrow m = -1, 1 \pm i$

$\therefore$ C.F. $= C_1 e^{-z} + e^z(C_2 \cos z + C_3 \sin z)$

$$= \frac{C_1}{x} + x[C_2 \cos(\log x) + C_3 \sin(\log x)]$$

Now, P.I. $= \frac{10}{(D+1)(D^2-2D+2)}e^z + \frac{10}{(D+1)(D^2-2D+2)}e^{-z}$

$$= \frac{10e^z}{(1+1)(1-2+2)} + \frac{10}{(1+2+2)}.\frac{1}{(D+1)}e^{-z}.1$$

$$= 5e^z + 2e^{-z}\frac{1}{D-1+1}.1$$

$$= 5e^z + 2e^{-z}.z = 5x + \frac{2}{x}\log x$$

Hence, the general solution of the given equation is

$y =$ C.F. + P.I.

$$\Rightarrow \quad y = \frac{C_1}{x} + x\,[C_2 \cos(\log x) + C_3 \sin(\log x)] + 5x + \frac{2}{x}\log x$$

Example 8. *Solve* $x^2\frac{d^2y}{dx^2} + 4x\frac{dy}{dx} + 2y = e^x$.

[UPTU-2005, Kurukshetra-2005]

Solution. Here, the given equation is

$$x^2\frac{d^2y}{dx^2}+4x\frac{dy}{dx}+2y=e^x \quad \text{...(1)}$$

Putting $x=e^z,\ z=\log x$ and $\frac{d}{dz}\equiv D$, we get

$$[D(D-1)+4D+2]\,y=e^{e^z}$$

or $$(D^2+3D+2)\,y=e^{e^z} \quad \text{...(2)}$$

To find the C.F. of (2), the auxiliary equation is

$$m^2+3m+2=0 \Rightarrow m=-1,-2$$

$$\therefore \quad \text{C.F.}=C_1e^{-z}+C_2e^{-2z}=\frac{C_1}{x}+\frac{C_2}{x^2} \quad \text{...(3)}$$

Now, $$\text{P.I.}=\frac{1}{(D+2)}\left[\frac{1}{(D+1)}e^{e^z}\right]$$

Let $$\frac{1}{(D+1)}e^{e^z}=u$$

$$\therefore \quad \frac{du}{dz}+u=e^{e^z}$$

This is a linear equation, with I.F. $=e^z$.

$$\therefore \quad ue^z=\int e^{e^z}.e^z\,dz=e^{e^z}$$

$$\therefore \quad u=e^{e^z}.e^{-z}$$

$$\therefore \quad \text{P.I.}=\frac{1}{(D+2)}(e^{e^z}.e^{-z})=v \text{ (say)}$$

$$\therefore \quad \frac{dv}{dz}+2v=e^{e^z}.e^{-z}$$

This is again a linear equation with I.F. $=e^{2z}$

$$\therefore \quad v.e^{2z}=\int e^{e^z}.e^{-z}.e^{2x}dz$$

$$=\int e^{e^z}.e^z dz=e^{e^z}$$

$$\Rightarrow \quad v=e^{e^z}.e^{-2z}=\frac{e^x}{x^2}$$

Hence, the general solution of the given equation is given by

$$y=\frac{C_1}{x}+\frac{C_2}{x^2}+\frac{e^x}{x^2}$$

Example 9. *Solve the homogeneous linear differential equation*

$$x^2\frac{d^2y}{dx^2}+x\frac{dy}{dx}+y=(\log x)\sin(\log x)$$

[UPTU-2002, Kurukshetra-2006, Madras-2006, Kerala-2005]

Solution. Since given equation is homogeneous

Put $x=e^z \Rightarrow \log x=z$

Also, $$x\frac{dy}{dx}=Dy$$

$$x^2\frac{d^2y}{dx^2}=D(D-1)y \qquad \left[\because D\equiv\frac{d}{dz}\right]$$

The transformed equation is

$$D(D-1)y+Dy+y=z\sin z$$

$$\Rightarrow \quad (D^2-D+D+1)y=z\sin z$$

$$\Rightarrow \quad (D^2+1)y=z\sin z$$

A.E. is $m^2+1=0 \Rightarrow m=\pm i$

C.F. $=C_1\cos z+C_2\sin z$

$$\text{P.I.}=\frac{1}{D^2+1}z\sin z$$

$$=\text{Imaginary part of }\frac{1}{D^2+1}ze^{iz}$$

$$=\text{Imaginary part of }e^{iz}\frac{1}{(D+i)^2+1}z.$$

$$=\text{Imaginary part of }e^{iz}\frac{1}{D^2+2iD-1+1}z$$

$$=\text{Imaginary part of }e^{iz}\frac{1}{D^2+2iD}z$$

$$=\text{Imaginary part of }e^{iz}\frac{1}{2iD}\frac{1}{\left(1+\frac{D}{2i}\right)}z$$

$$=\text{Imaginary part of }e^{iz}\frac{1}{2iD}\left(1-\frac{D}{2i}\right)z$$

$$=\text{Imaginary part of }e^{ir}\frac{1}{2iD}\left(z-\frac{1}{2i}\right)$$

$$=\text{Imaginary part of }e^{iz}\frac{1}{2i}\left(\frac{z^2}{2}-\frac{z}{2i}\right)$$

$$=\text{Imaginary part of }\frac{1}{2i}(\cos z+i\sin z)\left(\frac{z^2}{2}-\frac{z}{2i}\right)$$

$$=\text{Imaginary part of }(\cos z+i\sin z)\left(\frac{z^2}{4i}+\frac{z}{4}\right)$$

$$=-\frac{z^2}{4}\cos z+\frac{z}{4}\sin z$$

Complete solution is

$$y=\text{C.F.}+\text{P.I.}$$

$$y=C_1\cos z+C_2\sin z-\frac{z^2}{4}\cos z+\frac{z}{4}\sin z$$

$$y=C_1\cos(\log x)+C_2\sin(\log x)$$
$$-\frac{1}{4}(\log x)^2\cos(\log x)+\frac{1}{4}(\log x)\sin(\log x)$$

Example 10. *Solve* $\frac{x^2d^2y}{dx^2}+x\frac{dy}{dx}-\lambda^2y=0$ (UPTU-2007)

Solution. Put $x=e^z \Rightarrow \log x=Z$

Also $$x\frac{dy}{dx}=Dy$$

$$x^2\frac{d^2y}{dx^2} = D(D-1)y$$

The transformed equation is

$$D(D-1)y + Dy - \lambda^2 y = 0$$

$$\Rightarrow \quad [D(D-1)+D-\lambda^2]y = 0$$

$$\Rightarrow \quad (D^2-\lambda^2)y = 0$$

A.E. is $\quad m^2 - \lambda^2 = 0 \Rightarrow m = \pm\lambda$

C.F. $= C_1e^{\lambda z} + C_2e^{-\lambda z}$

P.I. $=0$

Complete solution is

$$y = \text{C.F.} + \text{P.I.}$$

$$= C_1e^{\lambda z} + C_2e^{-\lambda z} = C_1x^{\lambda} + C_2x^{-\lambda}$$

Example 11. *Solve*

$$(x+1)^2\frac{d^2y}{dx^2} + (x+1)\frac{dy}{dx} = (2x+3)(2x+4)$$

[MTU (Sum)-2011, GBTU (CO)-2011]

Solution. Put $x+1 = e^z \Rightarrow z = \log(x+1)$

Also $D \equiv \dfrac{d}{dz}$

The transformed equation is

$$[D(D-1)+D]y = (2e^z+1)(2e^z+2)$$

$$D^2y = 4e^{2z} + 6e^z + 2$$

A.E. is $m^2 = 0 \Rightarrow m = 0, 0$

$\therefore$ C.F. $= C_1 + C_2z$

P.I. $= \dfrac{1}{D^2}(4e^{2z}+6e^z+2) = e^{2z} + 6e^z + z^2$

Complete solution is

$$y = \text{C.F.} + \text{P.I.}$$

$$= C_1 + C_2z + e^{2z} + 6e^z + z^2$$

$$= C_1 + C_2\log(x+1) + (x+1)^2 + 6(x+1) + [\log(x+1)]^2$$

EXERCISE 30.1

Solve the following differential equations :

1. $x^2\dfrac{d^2y}{dx^2} - x\dfrac{dy}{dx} + y = x$
2. $x^2\dfrac{d^2y}{dx^2} - 4x\dfrac{dy}{dx} + 6y = x^4$
3. $x^2\dfrac{d^2y}{dx^2} + x\dfrac{dy}{dx} - 4y = x^2$
4. $x^4\dfrac{d^3y}{dx^3} + 2x^3\dfrac{d^2y}{dx^2} - x^2\dfrac{dy}{dx} + xy = 1$
5. $x^3\dfrac{d^3y}{dx^3} + 3x^2\dfrac{d^2y}{dx^2} + x\dfrac{dy}{dx} + y = x\log x$
6. $x^3\dfrac{d^3y}{dx^3} + 2x^2\dfrac{d^2y}{dx^2} + 3x\dfrac{dy}{dx} - 3y = x^2 + x$
7. $x^3\dfrac{d^3y}{dx^3} - x^2\dfrac{d^2y}{dx^2} + 2x\dfrac{dy}{dx} - 2y = x^3 + 3x$
8. $x^2\dfrac{d^2y}{dx^2} + x\dfrac{dy}{dx} - y = x^m$
9. $(x^3D^3 + 3x^2D^2 - 2xD + 2)y = 0$
10. $x^2\dfrac{d^2y}{dx^2} - x\dfrac{dy}{dx} + y = 2\log x$
11. $(x^4D^4 + 6x^3D^3 + 9x^2D^2 + 3xD + 1)y = (1+\log x)^2$
12. $x^3\dfrac{d^3y}{dx^3} + 2x^2\dfrac{d^2y}{dx^2} + x\dfrac{dy}{dx} - y = \cos(\log x)$
13. $x^2\dfrac{d^2y}{dx^2} - 2x\dfrac{dy}{dx} - 4y = x^4$ [AMIE-2000]
14. $x^2\dfrac{d^2y}{dx^2} + 6x\dfrac{dy}{dx} + 6y = (\log x)^2$
15. $x^3\dfrac{d^3y}{dx^3} + 2x^2\dfrac{d^2y}{dx^2} + 3x\dfrac{dy}{dx} - 3y = 0$
16. $(3x+2)^2\dfrac{d^2y}{dx^2} + (3x+2)\dfrac{dy}{dx} - 12y = 6x$ (GBTU-2011)
17. $x^2\dfrac{d^2y}{dx^2} - x\dfrac{dy}{dx} + y = \log x$ (VTU-2010)
18. $x^2\dfrac{d^2y}{dx^2} - 3x\dfrac{dy}{dx} + y = \log x\dfrac{\sin(\log x + 1)}{x}$ (ISM-2001)

Hint to Selected Problems

1. Put $x = e^x$, $D \equiv \dfrac{d}{dz}$.

Then, given equation reduces to $(D^2 - 2D + 1)y = e^z$

C.F. $= (C_1 + C_2z)e^z$

P.I. $= \dfrac{1}{(D-1)^2}e^z = e^z\left(\dfrac{1}{D^2}\right).1$

3. C.F. $= C_1e^{2z} + C_2e^{-2z}$

P.I. $= \dfrac{1}{(D^2-4)}e^{2z} = \dfrac{1}{(D+2)(D-2)}e^{2z} = \dfrac{1}{4(D-2)}e^{2z}$

$= \dfrac{1}{4}e^{2z}\dfrac{1}{(D+2)-2}.1 = \dfrac{1}{4}e^{2z}.\dfrac{1}{D}.1 = \dfrac{1}{4}ze^{2z}$

5. C.F. $= C_1e^{-z} + C_2e^{z/2}\cos\left(\dfrac{\sqrt{3}}{2}z + C_3\right)$

P.I. $= \dfrac{1}{(D^3+1)}ze^z = e^z.\dfrac{1}{(D^3+3D^2+3D+2)}.z$

12. C.F. $= C_1x + C_2\cos(\log x) + C_3\sin(\log x)$

P.I. $= \dfrac{1}{(D-1)(D^2+1)}.\cos z = \dfrac{D+1}{D^4-1}\cos z$

$$= \frac{1}{D^2+1}\left[\frac{(D+1)\cos z}{(D^2-1)}\right] = \frac{1}{D^2+1}\left[\frac{-\sin z+\cos z}{-1-1}\right]$$

ANSWERS

1. $y = x(C_1 + C_2\log x) + \frac{x}{2}.(\log x)^2$ **2.** $y = C_1x^2 + C_2x^3 + \frac{1}{2}x^4$ **3.** $y = C_1x^2 + \frac{C_2}{x^2} + \frac{1}{4}x^2\log x$

4. $y = (C_1 + C_2\log x)x + C_3x^{-1} + \frac{1}{4x}\log x$ **5.** $y = C_1x^{-1} + C_2\sqrt{x}\cos\{(\sqrt{3}/2)\log x + C_3\} + \frac{1}{2}x\log x - \frac{3}{4}x$

6. $y = C_1x + C_2\cos\{(\sqrt{3}\log x) + C_3\sin(\sqrt{3}\log x)\} + \frac{1}{7}x^2 + \frac{1}{4}x\log x$

7. $y = (C_1 + C_2\log x)x + C_3x^2 + \frac{x^3}{4} - \frac{3}{2}x(\log x)^2$ **8.** $y = C_1x + C_2x^{-1} + \frac{x^m}{(m^2-1)}$ **9.** $y = x[C_1 + C_2\log x] + C_3x^{-2}$

10. $y = x[C_1 + C_2\log x] + 2\log x + 4$ **11.** $y = (C_1 + C_2\log x)\cos(\log x) + (C_3 + C_4\log x)\sin(\log x) + (\log x)^2 + 2(\log x) - 3$

12. $y = C_1x + C_2\cos(\log x) + C_3\sin(\log x) - \frac{1}{4}\log x[\cos(\log x) + \sin(\log x)]$ **13.** $y = C_1x^4 + \frac{C_2}{x} + \frac{x^4\log x}{5}$

14. $y = C_1x^{-2} + C_2x^{-3} + \frac{1}{108}\{18(\log x)^2 - 30(\log x) + 19\}$

15. $y = C_1x + C_2\cos(\sqrt{3}\log x) + C_3\sin(\sqrt{3}\log x)$ **16.** $y = C_1(3x+2)^2 + C_2(3x+2)^{-2/3} - \frac{2}{15}(3x+2) + \frac{1}{3}$

17. $y = (C_1 + C_2\log x)x + \log x + 2$

18. $y = x^2(C_1x\sqrt{3} + C_2x^{-\sqrt{3}}) + \frac{1}{x}\left[\frac{1}{6}(\log x+1) + \frac{\log x}{61}\{5\sin(\log x) + 6\cos(\log x) + \frac{2}{3721}[27\sin(\log x) + |9|\cos(\log x)\}\right]$

30.3 AN ALTERNATIVE APPROACH FOR GETTING P.I. WHEN THE R.H.S. IS KEPT UNCHANGED

If the given equation can be written as $F(D_1)y = f(x)$ ...(1)

Then, use the following results :

(a) $\dfrac{1}{D_1-\alpha}f(x) = x^{\alpha}\int x^{-\alpha-1}f(x)\,dx$ (b) $\dfrac{1}{D_1+\alpha}f(x) = x^{-\alpha}\int x^{\alpha-1}f(x)\,dx$

Here, to find the P.I., first factorize $F(D_1)$ into linear factors and then use any of the following method :

(i) The operator $\dfrac{1}{F(D_1)}$ can be broken up into partial fractions.

Then, P.I. $= \dfrac{1}{f(D_1)}f(x) = \left[\dfrac{A_1}{D_1-\alpha_1} + \dfrac{A_2}{D_1-\alpha_2} + \ldots + \dfrac{A_n}{D_1-\alpha_n}\right]f(x)$

which can be obtained easily, by using (a) and (b).

(ii) P.I. $= \dfrac{1}{(D_1-\alpha_1)(D_1-\alpha_2)\ldots(D_1-\alpha_n)}f(x)$

Now, using (a) and (b), by taking the factors in succession beginning with the first on the right.

Solved Examples

Example 1. *Solve* $x^2\dfrac{d^2y}{dx^2} + 4x\dfrac{dy}{dx} + 2y = e^x$.

Solution. Putting $x = e^z \Rightarrow z = \log x$ and $\dfrac{d}{dz} \equiv D$, in given equation, we get (without changing R.H.S.)

$[D(D-1) + 4D + 2]\,y = e^x$

$\Rightarrow (D^2 + 3D + 2)y = e^x$ ... (1)

To find the C.F. of (1), the auxiliary equation is

$m^2 + 3m + 2 = 0 \Rightarrow m = -1, -2$

$\therefore$ C.F. $= C_1e^{-z} + C_2e^{-2z} = C_1x^{-1} + C_2x^{-2}$

Now, P.I. $= \dfrac{1}{(D+2)(D+1)}e^x$

$= \left[\dfrac{1}{D+1} - \dfrac{1}{D+2}\right]e^x$

[By breaking up into partial fractions]

$= \dfrac{1}{D+1}e^x - \dfrac{1}{D+2}e^x$

$= x^{-1}\int x^{1-1}e^x dx - x^{-2}\int x^{2-1}e^x dx$

[By using (a)]

$$= x^{-1}\int e^x dx - x^{-2}\int x e^x\, dx$$
$$= x^{-1}e^x - x^{-2}[xe^x - \int 1.e^x dx]$$
$$= x^{-1}e^x - x^{-2}[xe^x - e^x] = x^{-2}e^x$$

Hence, the general solution is given by

$$y = \text{C.F.} + \text{P.I.}$$
$$\Rightarrow \quad y = C_1x^{-1} + C_2x^{-2} + x^{-2}e^x$$

Example 2. *Solve* $x^2\dfrac{d^2y}{dx^2} + 3x\dfrac{dy}{dx} + y = \dfrac{1}{(1-x)^2}$.

(PTU-2003)

Solution. Putting $x = e^z$, *i.e.*, $z = \log x$, we get

$$[D(D-1)+(3D+1)]y = \frac{1}{(1-x)^2}$$

[Without changing RHS]

$$\Rightarrow \quad (D+1)^2 y = \frac{1}{(1-x)^2} \quad ...(1)$$

To find the C.F. of (1), the auxiliary equation is

$$(m+1)^2 = 0 \Rightarrow m = -1, -1$$

$$\therefore \quad \text{C.F.} = (C_1 + C_2 z)e^{-z} = (C_1 + C_2 \log x)\, x^{-1}$$

$$\text{Now, P.I.} = \frac{1}{D+1}\cdot\frac{1}{D+1}(1-x)^{-2}$$
$$= \frac{1}{D+1}x^{-1}\int x^{1-1}(1-x)^{-2}dx$$
$$= \frac{1}{D+1}x^{-1}(1-x)^{-1}$$
$$= x^{-1}\int x^{1-1}x^{-1}(1-x)^{-1}dx$$
$$= x^{-1}\int\frac{dx}{x(1-x)} = x^{-1}\int\left(\frac{1}{x}+\frac{1}{1-x}\right)dx$$
$$= x^{-1}\,[\log x - \log(1-x)]$$
$$= x^{-1}\log\left(\frac{x}{1-x}\right)$$

Hence, the required general solution is given by

$$y = \text{C.F.} + \text{P.I.}$$
$$\Rightarrow \quad y = (C_1 + C_2\log x)\,x^{-1} + x^{-1}\log\left(\frac{x}{1-x}\right)$$

30.4 EQUATION REDUCIBLE TO HOMOGENEOUS FORM

Any differential equation of the form

$$(a+bx)^n\frac{d^ny}{dx^n} + A_1(a+bx)^{n-1}\frac{d^{n-1}y}{dx^{n-1}} + \ldots + A_{n-1}(a+bx)\frac{dy}{dx} + A_n y = X(x) \quad ...(1)$$

where the coefficients $A_1, A_2, \ldots, A_n$ are constants, can be transformed into homogeneous linear equation with constant coefficients by changing independent variable from x to z, by a suitable substitution $z = a+bx$.

Let $\quad z = a+bx \Rightarrow z = a+bx \Rightarrow \dfrac{dz}{dx} = b$

Now we have

$$\frac{dy}{dx} = \frac{dy}{dz}\cdot\frac{dz}{dx} = \frac{dy}{dz}.b$$
$$\frac{d^2y}{dx^2} = \frac{d}{dx}\left(\frac{dy}{dx}\right) = \frac{d}{dz}\left(b\frac{dy}{dz}\right)\frac{dz}{dx} = b^2\frac{d^2y}{dz^2}$$

....

....

$$\frac{d^ny}{dx^n} = b^n\frac{d^ny}{dz^n}$$

Putting all these values in equation (1), and dividing throughout by b^n, we get

$$z^n\frac{d^ny}{dz^n} + \frac{A_1}{b}z^{n-1}\frac{d^{n-1}y}{dz^{n-1}} + \frac{A_2}{b^2}z^{n-2}\frac{d^{n-2}y}{dz^{n-2}} + \ldots + \frac{A_{n-1}}{b^{n-1}}z\frac{dy}{dz} + \frac{A_n}{b^n}y = \frac{1}{b^n}X\left(\frac{z-a}{b}\right) \quad ...(2)$$

Now, this is the standard homogeneous equation.

REMARK

- Sometimes, the given equation can be solved easily by making the substitution $e^z = a+bx \Rightarrow z = \log(a+bx)$

Solved Examples

Example 1. *Solve* $(1+x)^2\dfrac{d^2y}{dx^2} + (1+x)\dfrac{dy}{dx} + y$

$$= 4\cos\log(1+x)$$

Solution. Let $(1+x) = e^z$ and $\dfrac{d}{dz} \equiv D$.

Then, the given equation becomes

$$[D(D-1) + D+1]\; y = 4\cos z$$
$$\Rightarrow \quad (D^2+1)\; y = 4\cos z \quad ...(1)$$

To find the C.F. of (1), the auxiliary equation is

$$m^2 + 1 = 0 \Rightarrow m = \pm i$$

$\therefore$ C.F. $= e^{0z}(C_1 \cos z + C_2 \sin z)$
$= C_1 \cos\log(1+x) + C_2 \sin\log(1+x)$

Now, P.I. $= \dfrac{1}{D^2+1} 4 \cos z = 4\dfrac{1}{D^2+1}\cos z$

$= \text{Real part of } 4.\dfrac{1}{D^2+1}(\cos z + i \sin z)$

$= \text{Real part of } 4.\dfrac{1}{D^2+1}.e^{iz}$

Consider $\dfrac{1}{D^2+1}e^{iz} = \dfrac{1}{(D+i)(D-i)}e^{iz}$

$= \dfrac{1}{(i+i)(D-i)}e^{iz}$

$= \dfrac{1}{2i(D-i)}e^{iz}.1 = \dfrac{1}{2i}e^{iz}\dfrac{1}{D+i-i}.1$

$= \dfrac{e^{iz}}{2i}.\dfrac{1}{D}.1 = \dfrac{ze^{iz}}{2i}$

$= \dfrac{z}{2i}(\cos z + i \sin z)$

$= -\dfrac{iz}{2}(\cos z + i \sin z)$

$= -i\dfrac{z}{2}\cos z + \dfrac{z}{2}\sin z$

$\therefore$ P.I. $= \text{Real part of } 4\left(-\dfrac{iz}{2}\cos z + \dfrac{z}{2}\sin z\right)$

$= 2z \sin z = 2\log(1+x)\sin\log(1+x)$

Hence, the general solution is given by

$y =$ C.F. + P.I.

$\Rightarrow \quad y = C_1 \cos\log(1+x) + C_2 \sin\log(1+x) + 2\log(1+x)\sin\log(1+x)$

Example 2. *Solve* $2x^2y\dfrac{dy}{dx} + 4y^2 = x^2\left(\dfrac{dy}{dx}\right)^2 + 2xy\dfrac{dy}{dx}$ *after making it homogeneous by the substitution* $y = z^2$.

Solution. Let $y = z^2 \Rightarrow \dfrac{dy}{dx} = 2z.\dfrac{dz}{dx}$

and $\dfrac{d^2y}{dx^2} = 2z\dfrac{d^2y}{dx^2} + 2\left(\dfrac{dz}{dx}\right)^2$

Putting all these values in the given equation, we get

$$2x^2z^2\left[2z\frac{d^2z}{dx^2} + 2\left(\frac{dz}{dx}\right)^2\right] + 4z^2$$
$$= x^2\left(2z\frac{dz}{dx}\right)^2 + 2xz^2.2z\frac{dz}{dx}$$

$\Rightarrow \quad x^2\dfrac{d^2z}{dx^2} - x\dfrac{dz}{dx} + z = 0$ ...(1)

which is standard homogeneous equation.

Putting $x = e^t$ or $\log x = t$ in (1), we get

$(D(D-1) - D + 1)z = 0$, where $D \equiv \dfrac{d}{dt}$

$\Rightarrow \quad (D-1)^2 z = 0$ ...(2)

To find the C.F. of (2), the auxiliary equation is

$(m-1)^2 = 0 \Rightarrow m = 1,1$

C.F. $= (C_1 + C_2t)e^t = (C_1 + C_2 \log x)\, x$

$\Rightarrow$ the general solution is $z =$ C.F.

[R.H.S. of (1) is zero]

$\Rightarrow \quad z = (C_1 + C_2 \log x)x$

Squaring both sides, we get ,

$z^2 = x^2(C_1 + C_2 \log x)$

$\Rightarrow \quad y = x^2(C_1 + C_2 \log x)$

which is the required complete solution.

Example 3. *Solve* $(x+a)^2\dfrac{d^2y}{dx^2} - 4(x+a)\dfrac{dy}{dx} + 6y = x$

Solution. Let $(x+a) = e^z$ and $d/dz \equiv D$

Then, $(x+a)\dfrac{dy}{dx} = Dy$ and

$(x+a)^2\dfrac{d^2y}{dx^2} = D(D-1)y$.

$\therefore \quad (x+a)^2\dfrac{d^2y}{dx^2} - 4(x+a)\dfrac{dy}{dx} + 6y = x$

$\Rightarrow \quad D(D-1)y - 4Dy + 6y = e^z - a$

$\Rightarrow \quad (D^2 - D - 4D + 6)y = e^z - a$

$\Rightarrow \quad (D^2 - 5D + 6)y = e^z - a$

Auxiliary equation is

$m^2 - 5m + 6 = 0 \Rightarrow (m-3)(m-2) = 0$

or $m = 2,3$

$\therefore$ C.F. $= C_1e^{2z} + C_2e^{3z}$

P.I. $= \dfrac{e^z - a}{D^2 - 5D + 6}$

$= \dfrac{e^z}{D^2-5D+6} - \dfrac{a}{D^2-5D+6}$

$= \dfrac{e^z}{1-5+6} - \dfrac{a}{6} = \dfrac{e^z}{2} - \dfrac{a}{6}$

$\therefore \quad y =$ C.F. + P.I.

$\Rightarrow \quad y = C_1e^{2z} + C_2e^{3z} + \dfrac{e^z}{2} - \dfrac{a}{6}$

$= C_1(x+a)^2 + C_2(x+a)^3 + \dfrac{1}{2}(x+a) - \dfrac{a}{6}$

Example 4. *Solve* $(1+x)^2\dfrac{d^2y}{dx^2} + (1+x)\dfrac{dy}{dx} + y = 2\sin[\log(1+x)]$

[(VTU)-2009 JNTU-2005, Kewals-2005)

Solution. Let $1 + x = e^z$ *i.e.*, $z = \log(1+x)$

so that $(1+x)\frac{dy}{dx} = Dy$

and $(1+x)^2\frac{d^2y}{dx^2} = D(D-1)y$,

where $D \equiv \frac{d}{dz}$

Then, the given equation becomes

$$[D(D-1)+D+1]y = 2\sin z$$

$$(D^2+1)y = 2\sin Z \quad \text{... (i)}$$

To find the C.F. of (1)

A.F. is $m^2+1=0 \Rightarrow m = \pm i$

$\therefore$ C.F. $= e^{oz}(c_1\cos z + c_2\sin z)$

$= c_1\cos\log(1+x) + c_2\sin\log(1+x)$

Now, P.I. $= 2\frac{1}{D^2+1}\sin z = 2z\cdot\frac{1}{2D}\sin z$

$= z\int \sin z dz = -z\cos z$

$= -\log(1+x)\cos\log(1+x)$

($\because$ On replacing D^2 by -1^2, $D^2+1=0$)

Hence, the general solution is given by

$y =$ C.F. +P.I.

$\Rightarrow$ $y = c_1\cos\log(1+x) + c_2\sin\log(1+x)$

$-\log(1+x)\cos\log(1+x)$

Example 5. *Solve* $(2x-1)^2\frac{d^2y}{dx^2} + (2x-1)\frac{dy}{dx} - 2y$

$= 8x^2 - 2x + 3$ (VTU 2006)

Solution. Let $2x-1 = e^2$, *i.e.*, $z = \log(2x-1)$

So that $(2x-1)\frac{dy}{dx} = 2Dy$

and $(2x-1)^2\frac{d^2y}{dx^2} = 4D(D-1)y$,

where $D \equiv \frac{d}{dz}$

Then, the given equation becomes.

$4D(D-1)y + 2Dy - 2y$

$$= 8\left(\frac{1+e^z}{2}\right)^2 - 2\left(\frac{1+e^z}{2}\right) + 3$$

or $2D^2y - Dy - y = e^{2z} + \frac{3}{2}e^z + 2$ ... (i)

To find the C.F of (1)

A. E. is $2D^2 - D - 1 = 0 \Rightarrow D = 1, -1/2$

$\therefore$ C.F. $= c_1e^z + c_2e^{-z/2}$

Now, P.I. $= \frac{1}{2D^2-D-1}\left(e^{2z} + \frac{3}{2}e^z + 2\right)$

$$= \frac{1}{2\cdot4-2-1}e^{2z} + \frac{3}{2}\frac{z}{4D-1}e^z$$

$$+2\cdot\frac{1}{2\cdot0^2-0-1}e^{oz}$$

[$\because$ On putting $z=1$, $2D^2-D-1=0$]

$$= \frac{1}{5}e^{2z} + \frac{3z}{2}\cdot\frac{1}{4-1}e^z - 2$$

$$= \frac{1}{5}e^{2z} + \frac{z}{2}e^z - 2$$

Hence, the general solution is given by

$y =$ C.F.+P. I.

$= C_1e^z + C_2e^{-z/2} + \frac{1}{5}e^{2z} + \frac{z}{2}e^z - 2$

$= C_1(2x-1) + C_2(2x-1)^{-1/2}$

$+\frac{1}{5}(2x-1)^2 + \frac{1}{2}(2x-1)\log(2x-1) - 2$

[On replacing z by $\log(2x-1)$]

EXERCISE 30.2

Solve the following differential equations :

1. $(3x+2)^2\frac{d^2y}{dx^2} + 3(3x+2)\frac{dy}{dx} - 36y = 3x^2 + 4x + 1$
2. $(3x+2)^2\frac{d^2y}{dx^2} + 5(3x+2)\frac{dy}{dx} - 3y = x^2 + x + 1$ (Mumbai-2006)
3. $(5+2x)^2\frac{d^2y}{dx^2} - 6(5+2x)\frac{dy}{dx} + 8y = 0$
4. $(1+2x)^2\frac{d^2y}{dx^2} - 6(1+2x)\frac{dy}{dx} + 16y = 8(1+2x)^2$
5. $16(x+1)^4\frac{d^4y}{dx^4} + 96(x+1)^3\frac{d^3y}{dx^3} + 104(x+1)^2\frac{d^2y}{dx^2} + 8(x+1)\frac{dy}{dx} + y = x^2 + 4x + 3$
6. $(2x-1)^3\frac{d^3y}{dx^3} + (2x-1)\frac{dy}{dx} - 2y = 0$
7. $(x+1)^2\frac{d^2y}{dx^2} + (x+1)\frac{dy}{dx} = 4x^2 + 14x + 12$

Hint to Selected Problems

2. Putting $3x+2=e^z$

$$\text{C.F.} = C_1e^{-z}+C_2e^{z/3}$$

$$\text{P.I.} = \frac{1}{27}\frac{1}{(D+1)(3D-1)}(e^{2z}-e^z+7)$$

$$=\frac{1}{27}\left[\frac{1}{(D+1)(3D-1)}e^{2z}-\frac{1}{(D+1)(3D-1)}e^z + \frac{1}{(D+1)(3D-1)}.7e^{0z}\right]$$

5. Putting $x+1=e^z$

$$\text{C.F.} = [C_1+C_2\{\log(1+x)\}]\,e^{\log(1+x)/z}+[C_3+C_4\{\log(1+x)\}]\,e^{-\log\frac{(1-x)}{2}}$$

$$\text{P.I.} = \frac{1}{(4D^2-1)^2}[e^{2z}+2e^z]$$

6. $\text{C.F.} = C_1e^z+e^z\left[C_2\cosh\frac{\sqrt{3}}{2}z+C_3\sinh\frac{\sqrt{3}}{2}z\right]$,

where $2x-1=e^z$

ANSWERS

1. $y=C_1(3x+2)^2+C_2(3x+2)^{-2}+\frac{1}{108}[(3x+2)^2\log(3x+2)+1]$

2. $y=C_1(3x+2)^{1/3}+C_2(3x+2)^{-1}+\frac{(3x+2)^2}{405}-\frac{(3x+2)}{108}-\frac{7}{27}$

3. $y=(5+2x)^2\,[C_1(5+2x)^{\sqrt{2}}+C_2(5+2x)^{-\sqrt{2}}]$

4. $y=[C_1+C_2\log(1+2x)]\,(1+2x)^2+(1+2x)^2\,[\log(1+2x)]^2$

5. $y=(C_1+C_2t)\,e^{t/2}+(C_3+C_4t)e^{-t/2}+\frac{e^{2t}}{225}+\frac{2}{9}e^t$, where $t=\log(1+x)$

6. $y=C_1(2x-1)+(2x-1)C_2\cosh\left\{\left(\frac{\sqrt{3}}{2}\right)\log(2x-1)\right\}+C_3\sinh\left\{\left(\frac{\sqrt{3}}{2}\right)\log(2x-1)\right\}$

7. $y=C_1+C_2\log(1+x)+x^2+8x+7+[\log(1+x)]^2$

Objective Evaluations

Fill in the Blanks

1. In any homogeneous equation $x^2\frac{d^2y}{dx^2}+x\frac{dy}{dx}+y=X$, the function X is either constant or a function of __________only.
2. The homogeneous linear equation can be reduced to a linear differential equation with constant coefficients by putting $x=$ __________ .
3. In linear homogeneous differential equation, the dependent variable y and its derivative occurs in the __________ degree only.
4. In homogeneous linear differential equation, the powers of x in the coefficients are __________ to the order of associated derivative.
5. If we replace $x = e^z$ in the given differential equation, then the value of $x^3\frac{d^3y}{dx^3}$ is __________, where $D \equiv d/dz$.

True / False

Write 'T' for True and 'F' for False statement :

1. The homogeneous linear differential equation can be reduced to linear differential equation with constant coefficients. **(T/F)**
2. The differential equation, in which the power of x in the coefficients are greater than the order of its derivative, associated with them, is called homogeneous linear equation. **(T/F)**
3. The differential equation in which the power of x in the coefficients are equal to the order of its derivative, associated with them is called the homogeneous linear equation. **(T/F)**
4. The homogeneous linear equations are also known as Cauchy-Euler equations. **(T/F)**
5. The equation $(D^2+1)\,y=\cos z$ is not linear. **(T/F)**
6. $x^2\frac{d^2y}{dx^2}-x\frac{dy}{dx}-3y=x^2\log x$ is a linear homogeneous differential equation of order 2. **(T/F)**

Multiple Choice Questions

Choose the most appropriate one;

1. The particular integral of $(x^2D^2-2xD)y=\log x$ is :
   (a) $\log x$ (b) $-\frac{1}{9}\log x-\frac{1}{6}(\log x)^2$
   (c) $A\log x$ (d) none of these
2. The P.I. is $(x^2D^2+5xD+4)y=x\log x$ is given by :
   (a) $\frac{1}{9}x\log x$ (b) $\frac{2}{27}x$
   (c) $\frac{1}{9}x\log x-\frac{2}{27}x$ (d) none of these
3. In homogeneous linear equation, the powers of x in the coefficients and the order of its associated derivatives are :
   (a) equal
   (b) not equal
   (c) order of the derivative is greater
   (d) none of these
4. In homogeneous linear equation the dependent variable y and its derivative occurs in :
   (a) first order only (b) first degree only
   (c) first order and first degree (d) none of these
5. The complete solution of $x^2\frac{d^2y}{dx^2}-x\frac{dy}{dx}+2y=x\log x$:
   (a) $x[C_1\cos(\log x)+C_2\sin(\log x)]$ (b) $x\log x$
   (c) (a) + (b) (d) none of these

ANSWERS

Fill in the Blanks

1. x 2. e^x 3. first 4. equal 5. $D(D-1)(D-2)$

True / False

1. T 2. F 3. T 4. T 5. F 6. T

Multiple Choice Questions

1. (b) 2. (c) 3. (a) 4. (b) 5. (c)

FFFFF

CHAPTER 31 Ordinary Simultaneous Linear Differential Equations

31.1 INTRODUCTION

In this chapter, we shall discuss the ordinary differential equations involving two or more dependent variables. Here, we shall discuss the case when there are as many simultaneous equations as there are dependent variables. For solving such equations, obtain an equation involving one dependent variable with one independent variable, by the process of elimination. After solving the derived equation, we substitute back to get the other dependent variable.

Consider the simultaneous equation as

$$\left.\begin{aligned} f_1(D)x + f_2(D)y &= f(t) \\ \text{and} \quad g_1(D)x + g_2(D)y &= g(t) \end{aligned}\right] \quad \text{with} \quad D \equiv \frac{d}{dt} \qquad \ldots (1)$$

where, x and y are functions of t and $f_1(D), f_2(D), g_1(D)$ and $g_2(D)$ are rational integral functions with constant coefficients, $f(t)$ and $g(t)$ are the functions of the independent variable t. Now define the determinant Δ such as

$$\Delta = \begin{vmatrix} f_1(D) & f_2(D) \\ g_1(D) & g_2(D) \end{vmatrix} \qquad \ldots (2)$$

then we can say Δ involves the operator coefficients of x and y in (1).

The equation (2) can be solved by the usual methods.

REMARKS

- To solve the simultaneous equations completely, we always require as many simultaneous equations as are the number of dependent variables.
- The method of solving the simultaneous differential equations with constant coefficients is similar to that of solving a set of simultaneous equations in Algebra.
- The number of arbitrary constants appearing in the general solution of the system (1) is equal to the degree in D of the determinant Δ given by (2), provided determinant is non-zero.
- The determinant Δ, defined by (2), involves the operator coefficients of x and y in (1).

31.2 METHOD OF SOLVING SIMULTANEOUS LINEAR DIFFERENTIAL EQUATION WITH CONSTANT COEFFICIENTS

Let x and y be the dependent variables and t be the independent variable. Generally, there are two methods for the solution of simultaneous linear differential equations with constant coefficients.

METHOD-1: SYMBOLIC METHOD WITH USE OF D

Consider the simultaneous equation such as

$$f_1(D)x + g_1(D)y = T_1 \qquad \ldots (1)$$

and

$$f_2(D)x + g_2(D)y = T_2 \qquad \ldots (2)$$

where T_1 and T_2 are the functions of independent variable t and f_1, f_2, g_1, g_2 are polynomial functions with constant coefficients.

Operate on both sides of equation (1) by $g_2(D)$ and equation (2) by $g_1(D)$, we get

$$g_2(D)f_1(D)x + g_2(D)g_1(D)y = g_2(D)T_1 \qquad \ldots (3)$$

and $\qquad g_1(D)\, f_2(D)x + g_1(D)g_2(D)y = g_1(D)\, T_2$ ... (4)

Now, since $g_1(D)$ and $g_2(D)$ both have the constant coefficients then

$$g_1(D)\, g_2(D)y = g_2(D)\, g_1(D)y$$

therefore, from (3) and (4)

$$[g_2(D)f_1(D) - g_1(D)f_2(D)]x = g_2(D)T_1 - g_1(D)T_2 \quad \text{... (5)}$$

Equation (5) is an ordinary differential equation with one dependent variable and can be solved by the usual methods. Thus, x can be obtained as a function of t. The value of y is then obtained by substituting the value of x in any of the given equations and integrating the resulting equation, if necessary. If however, y is obtained by an independent elimination as in the case of x, the values of x and y are to be substituted in given equation (1) and (2) and the arbitrary constants in x and y are to be so adjusted that the given equations are satisfied. Here, the number of independent arbitrary constants entering in the general solution is the index of the highest power of D in

$$[g_2(D)\, f_1(D) - g_1(D)\, f_2(D)]$$

METHOD-2: USE OF DIFFERENTIATION

If two equations containing $x, y, \frac{dx}{dt}$ and $\frac{dy}{dt}$ are given. Then we can obtain more equations containing $x, y, \frac{dx}{dt}, \frac{dy}{dt}, \frac{d^2x}{dt^2}$ and $\frac{d^2y}{dt^2}$ by differentiating the given equations with respect to t.

From these equations, we can obtain an equation containing x (or y) and its derivative, by eliminating x (or y) and its derivatives. Now, solve this new equation for x (or y) and substituting the value of x (or y) in any of the given equation and if necessary, solve the resulting equation.

REMARKS

- The method of differentiation will be used when found very necessary.
- Generally t will be the independent variable and x and y will be dependent variables. In some problems any other variable, say x, will be given as the independent variable and y and z as the dependent variables.

Solved Examples

Example 1. *Solve* $\frac{dx}{dt} - 7x + y = 0$

$\frac{dy}{dt} - 2x - 5y = 0.$

Solution. The given equation can be written as

$(D-7)x + y = 0$ ... (1)

and $\quad (D-5)y - 2x = 0$ ... (2)

Now, eliminating y, we get

$(D-7)(D-5)x + 2x = 0$

$\Rightarrow \quad \frac{d^2x}{dt^2} - 12\frac{dx}{dt} + 37x = 0$ ... (3)

The auxiliary equation is

$m^2 - 12m + 37 = 0 \;\Rightarrow\; m = 6 \pm i$

$\therefore \quad x = e^{6t}\,[C_1 \cos t + C_2 \sin t]$

Putting this value of x in (1), we get

$e^{6t}[-C_1 \sin t + C_2 \cos t] + 6e^{6t}[C_1 \cos t + C_2 \sin t] - 7e^{6t}(C_1 \cos t + C_2 \sin t) + y = 0$

$\Rightarrow \quad y = e^{6t}(C_1 - C_2)\cos t + (C_1 + C_2)\sin t$

Hence, the solution is

$$x = e^{6t}\,[C_1 \cos t + C_2 \sin t]$$

$$y = e^{6t}\,[(C_1 - C_2)\cos t + (C_1 + C_2)\sin t]$$

Example 2. *Solve* $\frac{dx}{dt} = -\omega y$ ... (1)

$\frac{dy}{dt} = \omega x$ ... (2)

[UPTU (SUM)-2007, UPTU-2008]

Solution. Differentiating (1) with respect to t, we get

$\frac{d^2x}{dt^2} + \omega\frac{dy}{dt} = 0 \;\Rightarrow\; \frac{d^2x}{dt^2} + \omega^2 x = 0$ [By using (2)]

$\therefore \quad x = C_1 \cos \omega t + C_2 \sin \omega t$

Putting this value of x in (1), we get

$y = -(1/\omega)[-C_1\omega \sin \omega t + C_2\omega \cos \omega t]$

$= -C_2 \cos \omega t + C_1 \sin \omega t$

Example 3. *Solve* $\frac{dx}{dt} + 2\frac{dy}{dt} - 2x + 2y = 3e^t$

$3\frac{dx}{dt} + \frac{dy}{dt} + 2x + y = 4e^{2t}.$

Solution. The given equation can be written as

$(D-2)x + 2(D+1)y = 3e^t$ ... (1)

and $\quad (3D+2)x + (D+1)y = 4e^{2t}$ ... (2)

Eliminating y between (1) and (2), we obtain

$[2(3D+2) - (D-2)]\,x = 8e^{2t} - 3e^t$

or $\quad (5D+6)x = 8e^{2t} - 3e^t$

or $$\frac{dx}{dt}+\frac{6}{5}x=\frac{8}{5}e^{2t}-\frac{3}{5}e^{t}$$
which is a linear differential equation with

$$\text{I.F.} = e^{\int 6/5 dt} = e^{6t/5}$$

Now, solution becomes

$$x.e^{6t/5}=\int e^{6t/5}\left[\frac{8}{5}e^{2t}-\frac{3}{5}e^{t}\right]dt + C_1$$

$$=\frac{8}{5}\int e^{16t/5}-\frac{3}{5}\int e^{11t/5}dt+C_1$$

$$=\frac{1}{2}e^{16t/5}-\frac{3}{11}e^{11t/5}+C_1$$

$$\therefore \quad x=\frac{1}{2}e^{2t}-\frac{3}{11}e^{t}+C_1e^{-6t/5}$$

Now, $$\frac{dx}{dt}=e^{2t}-\frac{3}{11}e^{t}-\frac{6}{5}C_1e^{-6t/5}$$

Putting this value in (1), we get

$$2Dy+2y+e^{2t}-\frac{3}{11}e^{t}-\frac{6}{5}C_1e^{-6t/5}$$
$$-e^{2t}+\frac{6}{11}e^{t}-2C_1e^{-6t/5}=3e^{t}$$

or $$2\frac{dy}{dt}+2y=\frac{30}{11}e^{t}+\frac{16}{5}C_1e^{-6t/5}$$

or $$\frac{dy}{dt}+y=\frac{15}{11}e^{t}+\frac{8}{5}C_1e^{-6t/5}$$

which is a linear differential equation with

$$\text{I.F.} = e^{\int dt}=e^{t}$$

The solution is

$$\therefore \quad y.e^{t}=\frac{15}{11}\int e^{2t}dt+\frac{8}{5}C_1\int e^{-t/5}dt$$
$$+C_2=\frac{15}{22}e^{2t}-8C_1e^{-t/5}+C_2$$

Hence, the solution is given by

$$x=\frac{1}{2}e^{2t}-\frac{3}{11}e^{t}+C_1e^{-6t/5}$$

$$y=\frac{15}{22}e^{t}-8C_1e^{-6t/5}+C_1e^{-t}$$

Example 4. *Solve* $\frac{dx}{dt}+\frac{dy}{dt}-2y=2\cos t-7\sin t$

$$\frac{dx}{dt}-\frac{dy}{dt}+2x=4\cos t-3\sin t$$ [UKTU-2001]

Solution. The given equation can be written as

$$Dx+(D-2)y=2\cos t-7\sin t \quad \text{... (1)}$$

and $$(D+2)x-Dy=4\cos t-3\sin t \quad \text{... (2)}$$

Eliminating y between (1) and (2), we obtain

$$[D^2+(D-2)(D+2)]x$$
$$=D(2\cos t-7\sin t)+(D-2)(4\cos t-3\sin t)$$

or $$(D^2-2)x=-9\cos t$$

Auxiliary equation is $m^2-2=0 \Rightarrow m=\pm\sqrt{2}$

$$\text{C.F.} = C_1e^{\sqrt{2}t}+C_2e^{-\sqrt{2}t}$$

and $$\text{P.I.} = -9\frac{1}{D^2-2}\cos t=\frac{-9\cos t}{D^2-2}=3\cos t$$

$$x=C_1e^{\sqrt{2}t}+C_2e^{-\sqrt{2}t}+3\cos t$$

$$\frac{dx}{dt}=\sqrt{2}\,C_1\,e^{\sqrt{2}t}-C_2\sqrt{2}\,e^{-\sqrt{2}t}-3\sin t$$

Now adding (1) and (2), we get

$$2Dx+2x-2y=6\cos t-10\sin t$$

or $$y=\frac{dx}{dt}+x-3\cos t+5\sin t$$

$$=\sqrt{2}C_1e^{\sqrt{2}t}-C_2\sqrt{2}e^{-\sqrt{2}t}-3\sin t+C_1e^{\sqrt{2}t}$$
$$+C_2e^{-\sqrt{2}t}+3\cos t-3\cos t+5\sin t$$

$$=(\sqrt{2}+1)C_1\,e^{\sqrt{2}t}+(1-\sqrt{2})C_2e^{-\sqrt{2}t}+2\sin t$$

Hence, the solution is given as

$$x=C_1e^{\sqrt{2}t}+C_2e^{-\sqrt{2}t}+3\cos t$$

$$y=(\sqrt{2}+1)C_1\,e^{\sqrt{2}t}+(1-\sqrt{2})C_2e^{-2t}+2\sin t$$

Example 5. *Solve* $\frac{dx}{dt}=ax+by,\ \frac{dy}{dt}=a'x+b'y.$

Solution. The given equation can be written as

$$(D-a)x-by=0 \quad \text{... (1)}$$

and $$-a'x+(D-b')y=0 \quad \text{... (2)}$$

Eliminating y between (1) and (2), we get

$$[(D-a)(D-b')-a'b]\,x=0$$

or $$[D^2-(a+b')D+(ab'-a'b)]\,x=0$$

$\therefore$ Auxiliary equation is

$$m^2-(a+b')m+(ab'-a'b)=0$$

$$m=\frac{(a+b')\pm\sqrt{[(a+b')^2-4(ab'-a'b)]}}{2}$$

$$=\frac{(a+b')\pm\sqrt{(a-b')^2+4a'b}}{2}$$

where roots m_1 and m_2 is

$$\left.\begin{aligned} m_1&=\frac{(a+b')+\sqrt{(a-b')^2+4a'b}}{2}\\ m_2&=\frac{(a+b')-\sqrt{(a-b')^2+4a'b}}{2}\end{aligned}\right] \quad \text{... (3)}$$

$$\therefore \quad x=C_1e^{m_1t}+C_2e^{m_2t}$$

$$\frac{dx}{dt}=C_1m_1e^{m_1t}+C_2m_2e^{m_2t}$$

$\therefore$ From (1), we get

$$y=\frac{1}{b}\left[\frac{dx}{dt}-ax\right]$$

$$=\frac{1}{b}[(m_1-a)C_1e^{m_1t}+(m_2-a)\,C_2e^{m_2t}]$$

Hence, the solution is

$$x = C_1 e^{m_1 t} + C_2 e^{m_2 t}$$

$$y = \frac{1}{b}[(m_1 - a)C_1 e^{m_1 t} + (m_2 - a)C_2 e^{m_2 t}]$$

Example 6. *Solve* $\frac{dx}{dt} + 4x + 3y = t, \quad \frac{dy}{dt} + 2x + 5y = e^t$

[UPTU-2006, UPTU Q.Bank-2001]

Solution. The given equation can be written as

$$(D+4)x + 3y = t \qquad \text{... (1)}$$

and $2x + (D+5)y = e^t$ ... (2)

Eliminating y between (1) and(2), we get

$$[(D+4)(D+5) - 6]x = (D+5)t - 3e^t$$

or $(D^2 + 9D + 14)x = 1 + 5t - 3e^t$

$\therefore$ Auxiliary equation is $m^2 + 9m + 14 = 0$

$\therefore$ $m = -2, -7$

Complementary function $= C_1 e^{-2t} + C_2 e^{-7t}$

Particular integral

$$= \frac{1}{14 + 9D + D^2}(1+5t) - \frac{1}{14 + 9D + D^2} 3e^t$$

$$= \frac{1}{14}\left[1 + \frac{9}{14}D + \frac{1}{14}D^2\right]^{-1}(1+5t) - \frac{3e^t}{14+9+1}$$

[By replacing D by 1]

$$= \frac{1}{14}\left[1 - \frac{9}{14}D - \frac{1}{14}D^2 + ...\right](1+5t) - \frac{1}{8}e^t$$

$$= \frac{1}{14}\left[1 + 5t - \frac{9}{14}.5\right] - \frac{1}{8}e^t$$

$$= \frac{1}{14}\left[5t - \frac{31}{14}\right] - \frac{1}{8}e^t$$

$$\therefore \quad x = C_1 e^{-2t} + C_2 e^{-7t} + \frac{5}{14}t - \frac{1}{8}e^t - \frac{31}{196}$$

$$\frac{dx}{dt} = -2C_1 e^{-2t} - 7C_2 e^{-7t} + \frac{5}{14} - \frac{1}{8}e^t$$

Putting above value in (1), we get

$$3y = -\frac{dx}{dt} - 4x + t = -2C_1 e^{-2t}$$

$$+3C_2 e^{-7t} - \frac{10}{7}t + t - \frac{5}{14} + \frac{31}{49} + \frac{1}{8}e^t + \frac{1}{2}e^t$$

$$\text{or } y = \frac{1}{3}\left[-2C_1 e^{-2t} + 3C_2 e^{-7t} - \frac{3}{7}t + \frac{27}{98} + \frac{5}{8}e^t\right]$$

$$\text{or } x = C_1 e^{-2t} + C_2 e^{-7t} + \frac{5}{14}t - \frac{31}{196} - \frac{1}{8}e^t$$

Example 7. *Solve* $t dx = (t - 2x)dt = 0$ *and* $t\, dy = (tx + ty + 2x - t)dt$.

Solution. The given equation can be written as

$$t\, dx - (t - 2x)dt = 0 \qquad \text{... (1)}$$

and $t\, dy - (tx + ty + 2x - t)dt = 0$ ... (2)

From (1), we get $\frac{dx}{dt} + \frac{2}{t}x = 1$

which is a linear equation

$$\text{I.F.} = e^{\int 2/t\,dt} = e^{2\log t} = t^2 \qquad \text{...(3)}$$

$$\therefore \quad xt^2 = \int t^2.1\, dt + C_1 = \frac{1}{3}t^3 + C_1$$

$$\therefore \quad x = \frac{1}{3}t + C_1 t^{-2}$$

Now adding (1) and (2), we get

$$t(dx + dy) = t(x+y)dt \text{ or } = \frac{dx + dy}{x+y} = dt$$

Integrating, $\log(x+y) = t + \log C_2$

$$y = C_2 e^t - C_1 t^{-2} - \frac{1}{3}t$$

Hence, solution is

$$x = \frac{1}{3}t + C_1 t^{-2}; \; y = C_2 e^t - C_1 t^{-2} - \frac{1}{3}t$$

Example 8. *Solve* $x\frac{dy}{dx} + z = 0,\; x\frac{dz}{dx} + y = 0$, *both equations be simultaneous differential equations.*

Solution. Differentiating first equation w.r.t. x, we get

$$x\frac{d^2y}{dx^2} + \frac{dy}{dx} + \frac{dz}{dx} = 0$$

or $x^2\frac{d^2y}{dx^2} + x\frac{dy}{dx} + x\frac{dz}{dx} = 0$ ... (A)

and from second equation,

$$x\frac{dz}{dx} = -y$$

Put this value of $x\frac{dz}{dx}$ in equation (A)

$$x^2\frac{d^2y}{dx^2} + x\frac{dy}{dx} - y = 0 \qquad \text{... (B)}$$

This equation is a homogeneous equation.

Put $x = e^t$ so that $\frac{dy}{dx} = \frac{dy}{dt}.\frac{dt}{dx} = \frac{1}{x}\frac{dy}{dt}$

$$\therefore \quad x\frac{dy}{dx} = \frac{dy}{dt} \text{ or } x\frac{d}{dx} \equiv \frac{d}{dt}$$

$$x\frac{d}{dx}\left[x\frac{dy}{dx}\right] = x^2\frac{d^2y}{dx^2} + x\frac{dy}{dx}$$

or $x^2\frac{d^2y}{dx^2} = \left(x\frac{d}{dx} - 1\right)x\frac{dy}{dx} = (D-1)Dy$

$$\left[\because \frac{d}{dt} \equiv D\right]$$

$\therefore$ From (B), we get

$\{(D-1)D + (D-1)\}y = 0$ or $(D^2 - 1)y = 0$

$\therefore$ $y = C_1 e^t + C_2 e^{-t}$ or $y = C_1 x + C_2 x^{-1}$ ... (1)

so that $\frac{dy}{dx} = C_1 - \frac{C_2}{x^2}$

$\therefore$ From first equation, we get $z = -x\frac{dy}{dx}$

or Desired solution is

$$y = C_1 x + C_2 x^{-1},\; z = -C_1 x + C_2 x^{-1}$$

Example 9. *Solve* $\frac{d^2x}{dt^2}+m^2y=0,\ \frac{d^2y}{dt^2}-m^2x=0.$

Solution. The given equation can be written as

$$D^2x+m^2y=0 \quad \text{... (1)}$$

$$-m^2x+D^2y=0 \quad \text{... (2)}$$

Eliminating y between (1) and (2),

$(D^4+m^4)x=0$.

Auxiliary equation is $M^4+m^4=0$

or $(M^2+m^2)^2-2M^2m^2=0$

or $(M^2-\sqrt{2}Mm+m^2)(M^2+\sqrt{2}Mm+m^2)=0$

$\therefore$ $M^2-\sqrt{2}mM+m^2=0$

or $M^2+\sqrt{2}mM+m^2=0$

$$\therefore \quad M=\frac{\sqrt{2}m\pm\sqrt{(2m^2-4m^2)}}{2},$$

$$M=\frac{-\sqrt{2}m\pm\sqrt{(2m^2-4m^2)}}{2}$$

$$=\frac{m}{\sqrt{2}}\pm\frac{m}{\sqrt{2}}i,\ =-\frac{m}{\sqrt{2}}\pm\frac{m}{\sqrt{2}}i$$

$$\therefore \quad x=e^{mt/\sqrt{2}}\left[C_1\cos\frac{mt}{\sqrt{2}}+C_2\sin\frac{mt}{\sqrt{2}}\right]$$

$$+e^{-mt/\sqrt{2}}\left[C_3\cos\frac{mt}{\sqrt{2}}+C_4\sin\frac{mt}{\sqrt{2}}\right] \quad \text{... (3)}$$

so that

$$\frac{dx}{dt}=\frac{m}{\sqrt{2}}e^{mt/\sqrt{2}}\left[C_1\cos\frac{mt}{\sqrt{2}}+C_2\sin\frac{mt}{\sqrt{2}}\right]$$

$$+e^{-mt/\sqrt{2}}\frac{m}{\sqrt{2}}\left[-C_1\sin\frac{mt}{\sqrt{2}}+C_2\cos\frac{mt}{\sqrt{2}}\right]$$

$$+\left(-\frac{m}{\sqrt{2}}\right)e^{-mt/\sqrt{2}}\left[C_3\cos\frac{mt}{\sqrt{2}}+C_4\sin\frac{mt}{\sqrt{2}}\right]$$

$$+e^{-mt/\sqrt{2}}\frac{m}{\sqrt{2}}\left[-C_3\sin\frac{mt}{\sqrt{2}}+C_4\cos\frac{mt}{\sqrt{2}}\right]$$

and
$$\frac{d^2x}{dt^2}=\frac{m^2}{2}e^{mt/\sqrt{2}}\left[C_1\cos\frac{mt}{\sqrt{2}}+C_2\sin\frac{mt}{\sqrt{2}}\right]$$

$$+\frac{m^2}{2}e^{mt/\sqrt{2}}\left[-C_1\sin\frac{mt}{\sqrt{2}}+C_2\cos\frac{mt}{\sqrt{2}}\right]$$

$$+\frac{m^2}{2}e^{mt/\sqrt{2}}\left[-C_1\sin\frac{mt}{\sqrt{2}}+C_2\cos\frac{mt}{\sqrt{2}}\right]$$

$$-\frac{m^2}{2}e^{mt/\sqrt{2}}\left[C_1\cos\frac{mt}{\sqrt{2}}+C_2\sin\frac{mt}{\sqrt{2}}\right]$$

$$+\frac{m^2}{2}e^{-mt/\sqrt{2}}\left[C_3\cos\frac{mt}{\sqrt{2}}+C_4\sin\frac{mt}{\sqrt{2}}\right]$$

$$-\frac{m^2}{2}e^{-mt/\sqrt{2}}\left[-C_3\sin\frac{mt}{\sqrt{2}}+C_4\cos\frac{mt}{\sqrt{2}}\right]$$

$$-\frac{m^2}{2}e^{-mt/\sqrt{2}}\left[-C_3\sin\frac{mt}{\sqrt{2}}+C_4\cos\frac{mt}{\sqrt{2}}\right]$$

$$-\frac{m^2}{2}e^{-mt/\sqrt{2}}\left[C_3\cos\frac{mt}{\sqrt{2}}+C_4\sin\frac{mt}{\sqrt{2}}\right]$$

$$=m^2e^{mt/\sqrt{2}}\left[-C_1\sin\frac{mt}{\sqrt{2}}+C_2\cos\frac{mt}{\sqrt{2}}\right]$$

$$-m^2e^{-mt/\sqrt{2}}\left[-C_3\sin\frac{mt}{\sqrt{2}}+C_4\cos\frac{mt}{\sqrt{2}}\right]$$

$\therefore$ From (1)

$$y=-\frac{1}{m^2}\frac{d^2x}{dt^2}$$

$$=e^{mt/\sqrt{2}}\left[C_1\sin\frac{mt}{\sqrt{2}}-C_2\cos\frac{mt}{\sqrt{2}}\right]$$

$$+e^{-mt/\sqrt{2}}\left[-C_3\sin\frac{mt}{\sqrt{2}}+C_4\cos\frac{mt}{\sqrt{2}}\right] \quad \text{... (4)}$$

Hence, (3) and (4) be complete solution.

Example 10. *Solve the following system of differential equations* $Dx+Dy+3x=\sin t$ *and* $Dx+y-x=\cos t$. [UPTU-2003]

Solution. We have

$$Dx+Dy+3x=\sin t \quad \text{... (1)}$$

and
$$Dx+y-x=\cos t \quad \text{... (2)}$$

Differentiating equation (2), we get

$$D(D-1)x+Dy=-\sin t \quad \text{... (3)}$$

Subtracting (1) from (3), we get

$$\{D(D-1)-(D+3)\}\,x=-2\sin t$$

or $\{D^2-D-D-3\}\,x=-2\sin t$

or $(D^2-2D-3)x=-2\sin t$

The auxiliary equation is

$$m^2-2m-3=0 \text{ or } (m-3)(m+1)=0,$$

i.e., $m=3,-1$

$\therefore$ C.F. $=C_1e^{3t}+C_2e^{-t}$

$$\text{P.I.}=\frac{1}{D^2-2D-3}(-2\sin t)=-2\frac{1}{(-1)-2D-3}\sin t$$

$$=2.\frac{1}{2(D+2)}\sin t=\frac{(D-2)}{D^2-4}\sin t$$

$$=\frac{(D-2)}{-1-4}\sin t=\frac{\cos t-2\sin t}{-5}$$

$$=\frac{1}{5}(2\sin t-\cos t)$$

Hence, the general solution is

$y=$ C.F. + P.I.

$$x=C_1e^{3t}+C_2e^{-t}+\frac{1}{5}(2\sin t-\cos t) \quad \text{... (4)}$$

From (2), we get

$(D-1)x+y=\cos t$

or
$$(D-1)\left\{C_1e^{3t}+C_2e^{-t}+\frac{1}{5}(2\sin t-\cos t)\right\}+y=\cos t$$

or $y = \cos t$

$$-D\left\{C_1e^{3t} + C_2e^{-t} + \frac{1}{5}(2\sin t - \cos t)\right\}$$

$$+\left\{C_1e^{3t} + C_2e^{-t} + \frac{1}{5}(2\sin t - \cos t)\right\}$$

$$= \cos t - 3C_1e^{3t} + C_2e^{-t} - \frac{1}{5}(2\cos t + \sin t)$$

$$+C_1e^{3t} + C_2e^{-t} + \frac{1}{5}(2\sin t - \cos t)$$

$$= \cos t - 2C_1e^{3t} + 2C_2e^{-t} - \frac{1}{5}(3\cos t - \sin t)$$

$$= \frac{2}{5}\cos t + \frac{1}{5}\sin t + 2C_2e^{-t} - 2C_1e^{3t}$$

$$\therefore\ y = \frac{1}{5}(2\cos t + \sin t) - 2C_1e^{3t} + 2C_2e^{-t} \quad \text{... (5)}$$

and $x = C_1e^{3t} + C_2e^{-t} + \frac{1}{5}(2\sin t - \cos t)$ ... (6)

which is required solution.

Example 11. *Solve* $\frac{dx}{dt} = 2y,\ \frac{dy}{dt} = 2z,\ \frac{dz}{dt} = 2x.$

[UPTU-2004, 07, SVTU-2006 S]

Solution. We have

$$\frac{dx}{dt} = 2y \Rightarrow Dx = 2y \quad \text{... (1)}$$

$$\frac{dy}{dt} = 2z \Rightarrow Dy = 2z \quad \text{... (2)}$$

and $\frac{dz}{dt} = 2x \Rightarrow Dz = 2x$ ... (3)

From (1), $\frac{dx}{dt} = 2y$

or $\frac{d^2x}{dt^2} = 2\frac{dy}{dt} = 2(2z) = 4z$

$$\frac{d^3x}{dt^3} = 4\frac{dz}{dt} = 4(2x) = 8x$$

$$\frac{d^3x}{dt^3} - 8x = 0 \Rightarrow (D^3 - 8)x = 0$$

The auxiliary equation is $m^3 - 8 = 0$

or $(m-2)(m^2 + 2m + 4) = 0$

$\Rightarrow$ $m - 2 = 0 \Rightarrow m = 2$

and $m^2 + 2m + 4 = 0 \Rightarrow m = \frac{-2 \pm \sqrt{4-16}}{2}$

$$= \frac{-2 \pm i\sqrt{12}}{2} = -1 \pm i\sqrt{3}$$

$\therefore$ The general solution is

$$x = C_1e^{2t} + e^{-t}(A\cos\sqrt{3}t + B\sin\sqrt{3}t)$$

$$\left[A = C_2\cos\alpha,\ B = C_2\sin\alpha,\ \tan\alpha = \frac{B}{A}\right.$$

or $\left.\alpha = \tan^{-1}\left(\frac{B}{A}\right)\right]$

$$x = C_1e^{2t} + e^{-t}\,[C_2\cos\alpha\cos\sqrt{3}t + C_2\sin\alpha\sin\sqrt{3}t\,]$$

$$x = C_1e^{2t} + e^{-t}C_2\cos(\sqrt{3}t - \alpha)$$

$$= C_1e^{2t} + C_2e^{-t}\cos(\sqrt{3}t - \alpha)$$

From (3), we have $\frac{dz}{dt} = 2x$

$$\Rightarrow\ \frac{dz}{dt} = 2C_1e^{2t} + 2C_2e^{-t}\cos(\sqrt{3}t - \alpha)$$

$$z = C_1e^{2t} + 2C_2\frac{e^{-t}}{\sqrt{1+3}}\cos(\sqrt{3}t - \alpha - \beta)$$

$$\left[\because \int e^{ax}\cos bx\,dx = \frac{e^{ax}}{\sqrt{a^2+b^2}}\cos(bx - \beta)\right.$$

where $\left.\beta = \tan^{-1}\frac{\sqrt{3}}{-1} = \frac{2\pi}{3} \text{ and } -\frac{2\pi}{3} = \frac{4\pi}{3}\right]$

$$z = C_1e^{2t} + C_2e^{-t}\cos\left(\sqrt{3}t - \alpha + \frac{4\pi}{3}\right)$$

From (2), $\frac{dy}{dt} = 2z$

$$\Rightarrow\ \frac{dy}{dt} = 2C_1e^{2t} + 2C_2e^{-t}\cos\left(\sqrt{3}t - \alpha + \frac{4\pi}{3}\right)$$

$$y = \int 2C_1e^{2t}dt + 2C_2\int e^{-t}\cos(\sqrt{3}t - \alpha + \frac{4\pi}{3})\,dt$$

$$\left(\because \gamma = \tan^{-1}\frac{\sqrt{3}}{-1} = \frac{2\pi}{3}\right)$$

$$= C_1e^{2t} + 2C_2\frac{e^{-x}}{\sqrt{1+3}}\cos\left(\sqrt{3}t - \alpha + \frac{4\pi}{3} - \gamma\right)$$

$$= C_1e^{2t} + 2C_2\frac{e^{-t}}{\sqrt{1+3}}\cos\left(\sqrt{3}t - \alpha + \frac{4\pi}{3} - \frac{2\pi}{3}\right)$$

$$\therefore\ y = C_1e^{2t} + C_2e^{-t}\cos\left(\sqrt{3}t - \alpha + \frac{2\pi}{3}\right)$$

Example 12. *Solve* $\frac{d^2x}{dt^2} + \frac{dy}{dt} + 3y = e^{-t},$

$$\frac{d^2y}{dt^2} - 4\frac{dy}{dt} + 3y = \sin 2t.$$

[UPTU-2007, MTU (SUM)-2011]

Solution. The given equation can be written as

$$D^2x + Dy + 3x = e^{-t}$$

$$\Rightarrow\ (D^2 + 3)x + Dy = e^{-t} \quad \text{... (1)}$$

and $D^2y - 4Dx + 3y = \sin 2t$

$$\Rightarrow -4Dx + (D^2 + 3)y = \sin 2t \quad \text{... (2)}$$

From (1) and (2), we get

$$(D^4 + 6D^2 + 9)x + D(D^2 + 3)y = e^{-t} + 3e^{-t} \quad \text{...(3)}$$

$$-4D^2x + D(D^2 + 3)y = 2\cos 2t \quad \text{... (4)}$$

Subtracting (4) from (3), we get

$$(D^4 + 10D^2 + 9) = 4e^{-t} - 2\cos 2t$$

The auxiliary equation is $m^4 + 10m^2 + 9 = 0$

$\Rightarrow (m^2 + 1)(m^2 + 9) = 0$, *i.e.*, $m = \pm i,\ \pm 3i$

$$\text{C.F.} = C_1 \cos t + C_2 \sin t + C_3 \cos 3t + C_4 \sin 3t$$

$$\text{P.I.} = \frac{1}{D^4 + 10D^2 + 9} 4e^{-t} - \frac{1}{D^4 + 10D^2 + 9}(2\cos 2t)$$

$$= \frac{4}{1+10+9} e^{-t} - \frac{1}{(-4)^2 + 10(-4) + 9}(2\cos 2t)$$

$$= \frac{e^{-t}}{5} + \frac{2}{15}\cos 2t$$

$$x = C_1 \cos t + C_2 \sin t + C_3 \cos 3t + C_4 \sin 3t + \frac{e^{-t}}{5} + \frac{2}{15}\cos 2t$$

Putting the value of x in (2), we get

$$-4D\,[C_1 \cos t + C_2 \sin t + C_3 \cos 3t + C_4 \sin 3t + \frac{e^{-t}}{5} + \frac{2}{15}\cos 2t] + (D^2 + 3)y = \sin 2t$$

or $$-4[-C_1 \sin t + C_2 \cos t - 3C_3 \sin 3t + 3C_4 \cos 3t - \frac{e^{t}}{5} - \frac{4}{15}\sin 2t] + (D^3 + 3)y = \sin 2t$$

or $$(D^3 + 3)y = \sin 2t - 4C_1 \sin t + 4C_2 \cos t - 12C_3 \sin 3t + 12C_4 \cos 3t - \frac{4}{5}e^{-t} - \frac{16}{15}\sin 2t$$

The auxiliary equation is $m^2 + 3 = 0 \Rightarrow \pm i\sqrt{3}$

$$\therefore \text{ C.F.} = C_1 \cos\sqrt{3}t + C_2 \sin\sqrt{3}t$$

$$\text{P.I.} = \frac{1}{D^2+3}[\sin 2t - 4C_1 \sin t + 4C_2 \cos t - 12C_3 \sin 3t + 12C_4 \cos 3t - \frac{4}{5}e^{-t} - \frac{16}{15}\sin 2t]$$

$$= \frac{1}{D^2+3}\left(-\frac{1}{15}\sin 2t\right) + \frac{1}{D^2+3}(-4C_1 \sin t) + \frac{1}{D^2+3}4C_2 \cos t + \frac{1}{D^2+3}(-12C_3 \sin 3t) + \frac{1}{D^2+3}(12C_4 \cos 3t) + \frac{1}{D^2+3}\left(-\frac{4}{5}e^{-t}\right)$$

$$= \frac{1}{-4+3}\left(-\frac{1}{15}\sin 2t\right) + \frac{1}{-1+3}(-4C_1 \sin t) + \frac{1}{-1+3}4C_2 \cos t + \frac{1}{-9+3}(-12C_3 \sin 3t) + \frac{1}{-9+3}(12C_4 \cos 3t) + \frac{1}{1+3}(-\frac{4}{5}e^{-t})$$

$$= \frac{1}{15}\sin 2t - 2C_1 \sin t + 2C_2 \cos t + 2C_3 \sin 3t - 2C_4 \cos 3t - \frac{1}{5}e^{-t}$$

$$\therefore \quad y = \text{C.F.} + \text{P.I.}$$

$$\therefore \quad y = C_1 \cos\sqrt{3}t + C_2 \sin\sqrt{3}t + \frac{1}{15}\sin 2t - 2C_1 \sin t + 2C_2 \cos t + 2C_3 \sin 3t - 2C_4 \cos 3t - \frac{1}{5}e^{-t}$$

and

$$x = C_1 \cos 3t + C_2 \sin t + C_3 \cos 3t + C_4 \sin 3t + \frac{e^{-t}}{5} + \frac{2}{15}\cos 2t$$

EXERCISE 31.1

Solve the following simultaneous differential equations :

1. $\dfrac{d^2x}{dt^2} - 3x - 4y = 0, \quad \dfrac{d^2y}{dt^2} + x + y = 0$ [UPTU–2005]
2. $\dfrac{dx}{dt} = 3x + 2y, \quad \dfrac{dy}{dt} = 5x + 3y$ [MTU (B. Pharma)–2011, UPTU (SUM)–2008]
3. $\dfrac{d^2x}{dt^2} + 4x + y = te^{3t}, \quad \dfrac{d^2y}{dt^2} + y - 2x = \cos^2 t$
4. $\dfrac{dx}{dt} + 5x + y = e^t, \quad \dfrac{dy}{dt} - x + 3y = 0$
5. $\dfrac{dx}{dt} = 3x + 2y, \quad \dfrac{dy}{dt} + 5x + 3y = 0$
6. $\dfrac{dx}{dt} + 2x - 3y = t, \quad \dfrac{dy}{dt} - 3x + 2y = e^{2t}$ [Nagpur–2009, UKTU-2012]
7. $(D - 17)y + (2D - 8)z = 0, (13D - 53)y - 2z = 0$
8. $2\dfrac{d^2y}{dx^2} - \dfrac{dz}{dx} - 4y = 2x, \quad 2\dfrac{dy}{dx} + 4\dfrac{dz}{dx} - 3z = 0$
9. $\dfrac{dx}{dt} + \dfrac{2}{t}(x - y) = 1, \quad \dfrac{dy}{dt} + \dfrac{1}{t}(x + 5y) = t$ [UPTU–2005]
10. $\dfrac{dx}{dt} + 5x - 2y = t, \quad \dfrac{dy}{dt} + 2x + y = 0$ [UPTU-2008, SVTU-2009, Kurukshetra-2005]
11. $\dfrac{d^2x}{dt^2} + y = \sin t, \quad \dfrac{d^2y}{dt^2} + x = \cos t$ [UPTU-2004]
12. $(D^2 - 1)x + 8Dy = 16e^t, \; Dx + 3(D^2 + 1)y = 0$ [UPTU-2001]
13. $\dfrac{dx}{dt} + 7x - y = 0, \dfrac{dy}{dt} + 2x + 5y = 0\,)$ [GBTU (AG)SUM-2010]
14. $\dfrac{dx}{dt} + x - 2y = 0, \dfrac{dy}{dt} + x + 4y = 0\,; \; x(0) = y(0) = 1$ (MTU-2011)
15. $\dfrac{dx}{dt} = 3x + 8y, \dfrac{dy}{dt} = -x - 3y\,; \; x(0) = 6, y(0) = -2$ (GBTU-2010)
16. $\dfrac{dx}{dt} = y + 1, \dfrac{dy}{dt} = x + 1$ [UPTU (SUM)-2009]
17. $\dfrac{dx}{dt} - y = e^t, \dfrac{dy}{dt} + x = \sin t\,; \; x(0) = 1, y(0) = 0$ [GBTU (SUM)-2010, GBTU (CO)-2011]
18. $\dfrac{dx}{dt} + 5x + y = e^t, \dfrac{dy}{dt} + x + 5y = e^{5t}$ [UPTU (CO)-2009]
19. $\dfrac{dx}{dt} + \dfrac{dy}{dt} + 2x + y = 0, \dfrac{dy}{dt} + 5x + 3y = 0$ [GBTU (CO)-2011]

20. $\frac{dx}{dt}=-4(x+y), \frac{dx}{dt}+4\frac{dy}{dt}=-4y$ with conditions

$x(0)=1, y(0)=0$ (GBTU-2011)

21. $\frac{dx}{dt}+y=\sin t, \frac{dy}{dt}+x=\cos t$ given that $x = 2$ and $y = 0$ when $t = 0$ (Bhopal-2009, JNTU-2006, Kerala-2005)

22. $\frac{dx}{dt}+2x+3y=0,\ 3x+\frac{dy}{dt}+2y=2e^{2t}$ (Delhi-2002)

23. $\frac{dx}{dt}+2y=e^t, \frac{dy}{dt}-2x=e^{-t}$ (Bhopal-2005)

24. $\frac{d^2x}{dt^2}-3x-4y=0, \frac{d^2y}{dt^2}+x+y=0$ (UPTU-2005)

25. $\frac{d^2x}{dt^2}+y=\sin t, \frac{d^2y}{dt^2}+x=\cos t$ (UPTU-2004)

Hint to Selected Problems

1. Eliminating y, we get

$$[(D^2+1)(D^2-3)+4]x=0$$

$$\Rightarrow \quad (D^2-1)^2x=0$$

The auxiliary equation is $(m^2-1)^2=0 \Rightarrow m=\pm1,\pm1$

$$x=(C_1+C_2t)e^{-t}+(C_3+C_4t)e^t$$

$$\frac{dx}{dt}=-(C_1+C_2t)e^{-t}+C_2e^{-t}+(C_3+C_4t)e^t+C_4e^t$$

$$\frac{d^2x}{dt^2}=(C_1+C_2t)e^{-t}-2C_2e^{-t}+(C_3+C_4t)^t+2C_4e^t$$

Now, for given equation

$$4y=D^2x-3x$$
$$=(C_1+C_2t)\,e^{-t}-2C_2e^{-t}+(C_3+C_4t)\,e^t$$
$$+2C_4e^t-3(C_1+C_2t)e^{-t}-3(C_2t+C_4t)\,e^t$$
$$=-(2C_1+2C_2+2C_2t)\,e^{-t}+(-2C_3+2C_4-2C_4t)\,e^t$$
$$=-\frac{1}{2}(C_1+C_2+C_2t)\,e^{-t}+\frac{1}{2}(C_4+C_3-C_5t)\,e^t$$

3. The given equation can be written as

$$(D^4+4)x+y=t\,e^{3t} \quad \text{... (1)}$$

$$-2x+(D^2+1)y=\cos^2 t \quad \text{... (2)}$$

Eliminating y, we get

$$\left[(D^2+1)\,(D^2+4)+2\right]x=(D^2+1)t\,e^{3t}-\cos^2 t$$

$$(D^4+5D^2+6)x=10t\,e^{3t}-\cos^2 t+6e^{3t}$$

$$m^4+5m^2+6=0$$

$$(m^3+3)\,(m^2+2)=0$$

$$m=\pm\sqrt{3}\,i,\ \pm\sqrt{2}\,i$$

C.F. $=(C_1\cos\sqrt{3}\,t+C_2\sin\sqrt{3}\,t)+(C_3\cos\sqrt{2}\,t+C_4\sin\sqrt{2}\,t)$

$$\text{P.I.}=\frac{10}{D^4+5D^2+6}t\,e^{3t}+\frac{6}{D^4+5D^2+6}e^{3t}-\frac{1}{D^4-5D^2+6}\cos^2 t$$

$$=10\,e^{3t}.\frac{1}{(D+3)^4+5(D+3)^2+6}t+\frac{6e^{3t}}{3^4+5.3^2+6}-\frac{1}{(D^4+5D^2+6)}.\frac{1}{2}(1+\cos 2t)$$

$$=10e^{3t}.\frac{1}{131+138D+59D^2+\ldots}t+\frac{1}{22}e^{3t}-\frac{1}{6+5D^2+D^4}.\frac{1}{2}-\frac{1}{D^4+5D^2+6}\left(\frac{1}{2}\cos 2t\right)$$

$$=10^{3t}\frac{1}{132}\left(1+\frac{23}{22}D+\frac{59}{132}D^2+\ldots\right)^{-1}t$$

$$+\frac{e^{3t}}{22}-\frac{1}{6}\left(1+\frac{5D^2}{6}+\frac{D^4}{6}\right)^{-1}\frac{1}{2}-\frac{\frac{1}{2}\cos 2t}{(-2)^2+5(-2)^2+6}$$

$$=\frac{5}{66}te^{3t}-\frac{49}{1452}e^{3t}-\frac{1}{12}-\frac{1}{4}\cos 2t$$

$$x=(C_1\cos\sqrt{3}\,t+C_2\sin\sqrt{3}\,t)\ +(C_3\cos\sqrt{2}\,t+C_4\sin\sqrt{2}\,t)$$
$$+\frac{5}{66}t\,e^{3t}-\frac{49}{1452}e^{3t}-\frac{1}{12}-\frac{1}{4}\cos 2t \quad \text{... (3)}$$

$$\frac{dx}{dt}=\left(-C_1\sqrt{3}\sin\sqrt{3}\,t+C_2\sqrt{3}\cos\sqrt{3}\,t\right)$$
$$+\left(-C_3\sqrt{3}\sin\sqrt{2}\,t+C_4\sqrt{2}\cos\sqrt{2}\,t\right)$$
$$+\frac{5}{66}\left(3t\,e^{3t}+e^{3t}\right)-\frac{49}{1452}3e^{3t}+\frac{1}{2}\sin 2t$$

$$\frac{d^2x}{dt^2}=-3(C_1\cos\sqrt{3}\,t+C_2\sin\sqrt{3}\,t)-2(C_3\cos\sqrt{2}\,t+C_4\sin\sqrt{2}t)$$
$$+\frac{5}{66}(9te^{3t}+6e^{3t})-\frac{49}{1452}9e^{3t}+\cos 2t$$

Substituting in (1), we get

$$y=-\frac{d^2x}{dt^2}-4x+te^{3t}$$

$$y=\left(C_1\cos\sqrt{3}\,t+C_2\sin\sqrt{3}\,t\right)$$
$$-2(C_3\cos\sqrt{2}\,t+C_4\sin\sqrt{2}\,t)+\frac{1}{66}te^{2t}-\frac{23}{1452}e^{3t}+\frac{1}{3} \quad \text{... (4)}$$

6. The given equation can be written as

$$(D+2)x-3y=t \quad \text{... (1)}$$
$$-3x+(D+2)y=e^{2t} \quad \text{... (2)}$$

Eliminating y, we get

$$[(D+2)^2-9]x=(D+2)t+3e^{2t}$$

$$(D^2+4+4D-9)x=(D+2)t+3e^{2t}=(1+2t)+3e^{2t}$$

A.E. $m^2+4m-5=0 \quad m=1,-5$

C.F. $=C_1e^{-5t}+C_2e^t$

$$\text{P.I.}=\frac{1+2t}{(D-1)(D+5)}+3\frac{e^{2t}}{(D-1)(D+5)}$$

$$=-\frac{1}{5}\left[1-\frac{4}{5}D-\frac{1}{5}D^2\right]^{-1}(1+2t)+\frac{3e^{2t}}{(2-1)(2+5)}$$

$$=-\frac{1}{5}\left[1+\frac{4}{5}D+\frac{1}{5}D^2+\ldots\right](1+2t)+\frac{3}{7}e^{2t}$$

$$=-\frac{13}{25}-\frac{2}{5}t+\frac{3}{7}e^{2t}$$

$$x=C_1e^{-5t}+C_2e^t+\frac{3}{7}e^{2t}-\frac{2}{5}t-\frac{13}{25}$$

$$\frac{dx}{dt}=-5C_1e^{-5t}+C_2e^t+\frac{6}{7}e^{2t}-\frac{2}{5}$$

From (1) : $3y=-5C_1e^{-5t}+C_2e^t+\frac{6}{7}e^{2t}-\frac{2}{5}+2C_1e^{-5t}$
$$+2C_2e^t+\frac{6}{7}e^{2t}-\frac{4}{5}t-\frac{26}{25}-t$$

$$y=-C_1e^{-5t}+C_2e^t+\frac{4}{7}e^{2t}-\frac{12}{25}-\frac{3}{5}t$$

9. The given equation can be written as

$$t\frac{dx}{dt}+2(x-y)=t \quad \ldots (1)$$

$$t\frac{dy}{dt}+x+5y=t^2 \quad \ldots (2)$$

Differentiating (1) w.r.t. t, we have

$$t\frac{d^2x}{dt^2}+\frac{dx}{dt}+2\frac{dx}{dt}-2\frac{dy}{dt}=1$$

$$t^2\frac{d^2x}{dt^2}+t\frac{dx}{dt}+2t\frac{dx}{dt}-2t\frac{dy}{dt}=t \quad \ldots (3)$$

Substituting the value of $t\frac{dy}{dt}$ from (2) in (1), we get

$$t^2\frac{d^2x}{dt^2}+3t\frac{dx}{dt}+2x+5\left(t\frac{dx}{dt}+2x-t\right)-2t^2=t$$

$$\Rightarrow \quad t\frac{d^2x}{dt^2}+8t\frac{dx}{dt}+12x=2t^2+6t \quad \ldots (4)$$

which is a homogeneous linear equation.

Put $t=e^z$

$$\frac{dx}{dt}=\frac{dx}{dz}\cdot\frac{dz}{dt}=\frac{dx}{dz}\cdot\frac{1}{t}$$

$$t\frac{d}{dt}\left(t\frac{dx}{dt}\right)=t^2\frac{d^2x}{dt^2}+t\frac{dx}{dt}$$

$$t^2D^2x=(D-1)Dx$$

Equation (4) gives

$$[(D-1)D+8D+12]x = 2e^{2z}+6e^z$$

$$(D^2+7D+12)x=2e^{2z}+6e^z$$

A.E. is $m^2+7m+12=0 \Rightarrow m=-3,-4$

$$\text{C.F.} = C_1e^{-3z}+C_2e^{-4z}$$

$$\text{P. I.} = \frac{2}{(D^2+7D+12)}e^{2z}+\frac{6}{(D^2+7D+12)}e^z$$

$$=\frac{2}{30}e^{2z}+\frac{6}{20}e^z=\frac{1}{15}e^{z}+\frac{3}{10}e^z$$

$$x=C_1e^{-3z}+C_2e^{-4z}+\frac{1}{15}e^{2z}+\frac{3}{10}e^z$$

$$x=\frac{C_1}{t^3}+\frac{C_2}{t^4}+\frac{t^2}{15}+\frac{3}{10}t$$

$$\frac{dx}{dt}=-\frac{3C_1}{t^4}-\frac{4C_2}{t^5}+\frac{2t}{15}+\frac{3}{10}$$

From (1) :

$$2y=-\frac{3C_1}{t^3}-\frac{4C_2}{t^4}+\frac{2t^2}{15}+\frac{3t}{10}+\frac{2C_1}{t^3}+\frac{2C_2}{t^3}+\frac{2t^2}{15}+\frac{6}{10}t-t$$

$$y=-\frac{C_1}{2t^3}-\frac{C_2}{t^4}+\frac{2t^2}{15}+\frac{3}{10}$$

ANSWERS

1. $x=(C_1+C_2t)e^{-t}+(C_3+C_4t)\,e^t$, $y=-\frac{1}{2}(C_1+C_2+C_2t)e^{-t}+\frac{1}{2}(C_4-C_3-C_4t)e^t$

2. $x=C_1e^{(3+\sqrt{10})t}+C_2e^{(3-\sqrt{10})t}$, $y=\frac{1}{2}\sqrt{10}\,[C_1e^{(3+\sqrt{10})t}-C_2e^{(3-\sqrt{10})t}]$

3. $x=(C_1\cos\sqrt{3}t+C_2\sin\sqrt{3}t)+(C_3\cos\sqrt{2}t+C_4\sin\sqrt{2}t)+\frac{5}{66}te^{3t}-\frac{49}{1452}e^{3t}-\frac{1}{12}-\frac{1}{4}\cos 2t$

$y=(C_1\cos\sqrt{3}t+C_2\sin\sqrt{3}t)-(C_3\cos\sqrt{2}t+C_4\sin\sqrt{2}t)+\frac{1}{60}te^{3t}-\frac{23}{1452}e^{3t}+\frac{1}{3}$

4. $x=(C_1+C_2t)e^{-4t}+\frac{4}{25}e^t-\frac{1}{36}e^{2t}$, $y=-(C_1+C_2+C_2t)\,e^{-4t}+\frac{7}{36}e^{2t}+\frac{1}{25}e^t$

5. $x=C_1\cos t+C_2\sin t$, $y=\frac{1}{2}(C_2-3C_1)\cos t-\frac{1}{2}(C_1+3C_2)\sin t$

6. $x=C_1e^{-5t}+C_2e^t+\frac{3}{7}e^{2t}-\frac{2}{5}t-\frac{13}{25}$, $y=-C_1e^{-5t}+C_2e^t+\frac{4}{7}e^{2t}-\frac{3}{5}t-\frac{12}{25}$

7. $y=C_1e^{3x}+C_2e^{5x}$, $z=-7C_1e^{3x}+6C_2e^{5x}$

8. $y=(C_1+C_2x)\,e^x+C_3e^{-3x/2}-\frac{1}{2}x$, $z=-2(C_1+C_2x-3C_2)e^x-\frac{1}{3}C_3e^{-3x/2}-\frac{1}{3}$

9. $x=\frac{C_1}{t^3}+\frac{C_2}{t^4}+\frac{t^2}{15}+\frac{3}{10}t$, $y=-\frac{C_1}{2t^3}-\frac{C_2}{t^4}+\frac{2t^2}{15}-\frac{t}{20}$

10. $x=-\frac{1}{27}(1+6t)e^{-3t}+\frac{1}{27}(1+3t)$, $y=-\frac{2}{27}(2+3t)e^{-3t}+\frac{2}{27}(2-3t)$

11. $x=C_1e^t+C_2e^{-t}+C_3\cos t+C_4\sin t+\frac{t}{4}(\sin t-\cos t)$, $y=-C_1e^t-C_2e^{-t}+C_3\cos t+C_4\sin t+\frac{1}{4}(2+t)(\sin t-\cos t)$

12. $y=C_1\cos\frac{t}{\sqrt{3}}+C_2\sin\frac{t}{\sqrt{3}}+C_3\cosh\sqrt{3}.t+C_4\sinh\sqrt{3}t+2e^t$; $x=\sqrt{3}C_1\sin\frac{t}{\sqrt{3}}-\sqrt{3}C_2\cos\frac{t}{\sqrt{3}}-3\sqrt{3}C_3\sinh\sqrt{3}t-3\sqrt{3}C_4\cosh\sqrt{3}t-6e^t-3t$

13. $x=e^{-6t}(A\cos t+B\sin t), y=e^{-6t}[(A+B)\cos t-(A-B)\sin t]$

14. $x=4e^{-2t}-3e^{-3t}, y=-2e^{-2t}+3e^{-3t}$

15. $x=4e^t+2e^{-t}, y=-e^t-e^{-t}$

16. $x=C_1e^t+C_2e^{-t}-1, y=C_1e^t-C_2e^{-t}-1$

17. $x=2\sin t+\frac{3}{2}\cos t+\frac{t}{2}\cos t-\frac{1}{2}e^t, y=\frac{1}{2}\cos t-\frac{3}{2}\sin t+\frac{t}{2}\sin t-\frac{1}{2}e^t$

18. $x=C_1e^{-6t}+C_2e^{-4t}+\frac{6e^t}{35}-\frac{e^{5t}}{99}$, $y=C_1e^{-6t}-C_2e^{-4t}-\frac{1}{35}e^t+\frac{10}{99}e^{5t}$

19. $x=\left(\frac{C_1-3C_2}{5}\right)\sin t-\left(\frac{C_2+3C_1}{5}\right)\cos t$, $y=C_1\cos t+C_2\sin t$

20. $x=(1-2t)e^{-2t}, y=te^{-2t}$

21. $x=e^t+e^{-t}, y=e^{-t}-e^t+\sin t$

22. $x=C_1e^t+C_2e^{-st}+\frac{6}{7}e^{2t}$; $y=C_2e^{-st}-C_1e^t+\frac{8}{7}e^{2t}$

23. $x=\frac{1}{5}e^t+\frac{2}{5}e^{-t}-C_1\sin 2t+C_2\cos 2t$, $y=\frac{2}{5}e^t+\frac{1}{5}e^{-t}+C_1\cos 2t+C_2\sin 2t$

24. $x=(C_1+C_2t)e^{-t}+(C_3+C_4t)e^t$, $y=-\frac{1}{2}[C_1+C_2(1+t)]e^{-t}+\frac{1}{2}[C_4(1-t)-C_3]e^t$

25. $x=C_1e^t+C_2e^{-t}+C_3\cos t+C_4\sin t-\frac{t}{4}\cos t+\frac{t}{4}\sin t$, $y=-C_1e^t-C_2e^{-t}+C_3\cos t+C_4\sin t+\frac{1}{4}(2+t)(\sin t-\cos t)$

31.3 SIMULTANEOUS EQUATIONS IN DIFFERENT FORM

Consider the equations of the type $P_1dx + Q_1dy + R_1dz = 0$

$$P_2dx + Q_2dy + R_2dz = 0 \quad \text{... (1)}$$

where P_1, P_2, Q_1, Q_2, R_1 and R_2 are functions of x, y, z.

Equation (1) can be written as

$$P_1\frac{dx}{dz} + Q_1\frac{dy}{dt} + R_1 = 0, \; P_2\frac{dx}{dz} + Q_2\frac{dy}{dz} + R_2 = 0.$$

Solving the above equations for $\frac{dx}{dz}, \frac{dy}{dz}$, we get $\frac{dx}{dz} = \frac{Q_1R_2 - Q_2R_1}{P_1Q_2 - Q_1P_2}, \; \frac{dy}{dz} = \frac{R_1P_2 - P_1R_2}{P_1Q_2 - Q_1P_2}$

Hence,

$$\frac{dx}{Q_1R_2 - Q_2R_1} = \frac{dy}{R_1P_2 - R_2P_1} = \frac{dz}{P_1Q_2 - P_2Q_1}$$

i.e., the equation can be put in the form $\frac{dx}{P} = \frac{dy}{Q} = \frac{dz}{R}$

where P, Q and R the functions of x, y and z.

WORKING PROCEDURE

Method-I

Step (1). *Take any two member of an equation (1) say* $\frac{dx}{P} = \frac{dy}{Q}$ *(say)*
After integrating it, we may get an equation.

Step (2). *Again take two member of equation (1) say* $\frac{dy}{Q} = \frac{dz}{R}$ *(say)*
After integrating it, we also get an equation.

Step (3). *The solution obtained from* (i) *and* (ii) *give the required general solution.*

Method-II

Step (1). *The given equation is* $\frac{dx}{P} = \frac{dy}{Q} = \frac{dz}{R}$

If we choose l, m *and* n *such that* $\frac{dx}{P} = \frac{dy}{Q} = \frac{dz}{R} = \frac{ldx + mdy + ndz}{lP + mQ + nR}$

If $lP + mQ + nR = 0$, *then,* $ldx + mdy + ndz = 0$

If it is an exact differential, say du, then $u = a$ *is one equation of the complete solution.*

REMARKS

- To find a solution of the given equation, we choose l, m, n such that $ldx+mdy+ndz$ is differential of $lP+mQ+nR$.
- If we have obtained one solution, then this solution can be used to simplify the other differential equations in the integrable form.
- Sometimes, we use only one set of multiples, but in some cases, we have a need of more than one set of multipliers.
- We can obtain one relation, say $u = a$ by the first method and the second relation by the second method.

Geometrical meaning of $\frac{dx}{P} = \frac{dy}{Q} = \frac{dz}{R}$

Since, we know that the direction cosines of the tangent to a curve at any point (x, y, z) are $\frac{dx}{ds}, \frac{dy}{ds}, \frac{dz}{ds}$ or proportional to dx, dy, dz. Therefore, geometrically the above situations represents a system of curves in such a way that the direction-ratios of the tangent from it at any point $A(x, y, z)$ are proportional to P, Q and R. If $u = a$ and $v = b$ are the complete solutions $\frac{dx}{P} = \frac{dy}{Q} = \frac{dz}{R}$, then system of curves is intersection of the surfaces $u = a$, $v = b$. It is also clear that since a, b are arbitrary constants, the system of curves represented by the equations is doubly infinite.

Solved Examples

Example 1. *Solve the simultaneous equations.*

$$\frac{a\,dx}{(b-c)\,yz} = \frac{b\,dy}{(c-a)zx} = \frac{c\,dz}{(a-b)\,xy}.$$

Solution. Let us take the x, y, z as multipliers.

$$\text{Each fraction} = \frac{ax\,dx + by\,dy + cz\,dz}{0}$$

$\therefore \quad axdx + bydy + cz\,dz = 0$

Integrating $ax^2 + by^2 + cz^2 = C_1 \quad \text{... (1)}$

Now taking ax, by, cz as multipliers.

$$\text{Each fraction} = \frac{a^2xdx + b^2ydy + c^2zdz}{0}$$

$\therefore \quad a^2xdx + b^2ydy + c^2zdz = 0$

On integrating,

$$a^2x^2 + b^2y^2 + c^2z^2 = 0 \quad \text{... (2)}$$

Hence, complete solution is

$$\phi(ax^2 + by^2 + cz^2,\; a^2x^2 + b^2y^2 + c^2z^2) = 0$$

Example 2. *Solve* $\dfrac{xdx}{z^2-2yz-y^2}=\dfrac{dy}{y+z}=\dfrac{dz}{y-z}$.

Solution. Let us take $1, y, z$ as multiplier, we get

$$\text{Each fraction} = \frac{xdx+ydy+zdz}{0}$$

$$xdx+ydy+zdz=0$$

Integrating, $x^2+y^2+z^2=C_1$ ... (1)

Again, last two members, we get

$$\frac{dy}{y+z}=\frac{dz}{y-z}$$

$$ydy-zdy=ydz+zdz$$

or $ydy-(ydz+zdy)-zdz=0$

Integrating $y^2-2yz-z^2=C_2$ ... (2)

Complete solution is

$$x^2+y^2+z^2=C_1$$
$$y^2-2yz-z^2=C_2$$

Example 3. *Solve the simultaneous equation*

$$\frac{dx}{y^2+z^2-x^2}=\frac{dy}{-2xy}=\frac{dz}{-2xz}.$$

Solution. From last two members, we get

$$\frac{dy}{y}=\frac{dz}{z}$$

$\therefore$ $y=C_1z$ ... (1)

Now, taking x,y,z as multiplier, we get

$$\text{Each fraction} = \frac{dz}{-2xz}=\frac{xdx+ydy+zdz}{-x(x^2+y^2+z^2)}$$

$$\frac{dz}{z}=\frac{2xdx+2ydy+2zdz}{x^2+y^2+z^2}$$

Integrating $\log z+\log C_2=\log(x^2+y^2+z^2)$

$x^2+y^2+z^2=C_2z$ ... (2)

Complete solution is

$$y=C_1z$$
$$x^2+y^2+z^2=C_2z$$

Example 4. *Solve* $\dfrac{dx}{x^2+y^2+yz}=\dfrac{dy}{x^2+y^2-xz}=\dfrac{dz}{z(x+y)}$.

Solution. Given equation can change to be new form as

$\dfrac{dx-dy}{z(x+y)}=\dfrac{dz}{z(x+y)}$ or $dx-dy=dz$

Integrating $x-y-z=C_1$ ... (1)

Again from the given equation, we get

$$\frac{xdx+ydy}{(x+y)(x^2+y^2)}=\frac{dz}{z(x+y)}$$

or $\dfrac{xdx+ydy}{x^2+y^2}=\dfrac{dz}{z}$.

Integrating, $\log(x^2+y^2)=2\log z+\log C_2$

$\therefore$ $x^2+y^2=z^2C_2$ ... (2)

From (1) and (2), we get complete solution

$$x-y-z=C_1$$
$$x^2+y^2=z^2C_2$$

Example 5. *Solve* $\dfrac{dx}{x(y^2-z^2)}=\dfrac{dy}{-y(z^2+x^2)}=\dfrac{dz}{z(x^2+y^2)}$.

Solution. Taking $\dfrac{1}{x},-\dfrac{1}{y},-\dfrac{1}{z}$ as multipliers, we get

$$\text{Each fraction} = \frac{\frac{dx}{x}-\frac{dy}{y}-\frac{dz}{z}}{0}$$

$\therefore$ $\dfrac{dx}{x}-\dfrac{dy}{y}-\dfrac{dz}{z}=0$ or $\dfrac{dy}{y}+\dfrac{dz}{z}=\dfrac{dx}{x}$

Integrating, $\log y+\log z=\log x+\log C_1$.

$\therefore$ $yz=C_1x$ ... (A)

Again using x,y,z as multipliers, we get

$$\text{Each fraction} = \frac{xdx+ydy+zdz}{0}$$

$\therefore$ $xdx+ydy+zdz=0$

Integrating $x^2+y^2+z^2=C_2$. ... (B)

From (A) and (B), we obtain complete solution.

EXERCISE 31.2

Solve the following simultaneous differential equations :

1. $\dfrac{dx}{xy}=\dfrac{dy}{y^2}=\dfrac{dz}{zyx-2x^2}$
2. $\dfrac{dx}{x^2+y^2}=\dfrac{dy}{2xy}=\dfrac{dz}{(x+y).z}$
3. $\dfrac{dx}{(x^2-yz)}=\dfrac{dy}{y^2-zx}=\dfrac{dz}{z^2-xy}$
4. $\dfrac{dx}{yz}=\dfrac{dy}{zx}=\dfrac{dz}{xy}$
5. $\dfrac{dx}{y+z}=\dfrac{dy}{z+x}=\dfrac{dz}{x+y}$
6. $\dfrac{dx}{mz-ny}=\dfrac{dy}{nx-lz}=\dfrac{dz}{ly-mx}$
7. $\dfrac{dx}{z(x+y)}=\dfrac{dy}{z(x-y)}=\dfrac{dz}{x^2+y^2}$
8. $\dfrac{dx}{z}=\dfrac{dy}{-z}=\dfrac{dz}{z^2+(x+y)^2}$
9. $\dfrac{dx}{x(y-z)}=\dfrac{dy}{y(z-x)}=\dfrac{dz}{z(x-y)}$
10. $\dfrac{dx}{x^2+y^2}=\dfrac{dy}{2xy}=\dfrac{dz}{(x+y)^2}$
11. $\dfrac{dx}{y^2+yz+z^2}=\dfrac{dy}{z^2+zx+x^2}=\dfrac{dz}{x^2+xy+y^2}$
12. $\dfrac{dx}{\cos(x+y)}=\dfrac{dy}{\sin(x+y)}=\dfrac{dz}{z}$

Hint to Selected Problems

1. Taking the first two members, we get

$$\frac{dx}{xy}=\frac{dy}{y^2} \quad\text{or}\quad \frac{dx}{x}=\frac{dy}{y}$$

Integrating $\log x=\log y+\log C_1$

$x=C_1y$ ... (1)

Again taking the last two members, we have

$$\frac{dy}{y^2}=\frac{dz}{zxy-2x^2}$$

$\dfrac{dy}{y^2}=\dfrac{dz}{zC_1y^2-2C_1^2y^2}$ from (1)

$$dy = \frac{dz}{zC_1 - 2C_1^2} \quad \text{or} \quad C_1 dy = \frac{dz}{z - 2C_1}$$

Integrating, we get

$$C_1 y = \log(z - 2C_1) + C_2$$

$$x = \log\left(z - \frac{2x}{y}\right) + C_2 \qquad [\because C_1 y = x]$$

$$x = \log(zy^{-2}) - \log y + C_3 \qquad \text{... (2)}$$

3. Obviously, each of the given ratios

$$= \frac{dx - dy}{x^2yz - y^2 + zx} \qquad \text{... (1)}$$

$$= \frac{dy - dz}{y^2 - zx - z^2 + xy} \qquad \text{... (2)}$$

$$= \frac{dz - dx}{x^2 - xy - x^2 + yz} \qquad \text{... (3)}$$

From (1) and (2)

$$\frac{dx - dy}{(x-y)(x+y+z)} = \frac{dy - dz}{(y-z)(x+y+z)}$$

$$\frac{dx - dy}{x - y} = \frac{dy - dz}{y - z}$$

Integrating, we get

$$\log(x - y) = \log(y - z) + \log a$$

$$\frac{x - y}{y - z} = a \qquad \text{... (4)}$$

From (2) and (3), similarly

$$\frac{y - z}{z - x} = b \qquad \text{... (5)}$$

From (4) and (5), the complete solution of the equation is

$$\phi\left(\frac{x - y}{y - z}, \frac{y - z}{z - x}\right) = 0$$

7. Using $x, -y, -z$ as multipliers, we have

$$\text{Each fraction } = \frac{xdx - ydy - zdz}{xz(x+y) - yz(x-y) - z(x^2 + y^2)}$$

$$= \frac{xdx + ydy - zdz}{0}$$

$$xdx - ydy - zdz = 0$$

Integrating, $\quad x^2 - y^2 - z^2 = C_1 \qquad$ ... (1)

Similarly, using $y, x, -z$ as multipliers, we get

$$\text{Each fraction } = \frac{ydx + xdy - zdz}{yz(x+y) + xz(x-y) - z(x^2 + y^2)}$$

$$= \frac{ydx + xdy - zdz}{0}$$

$$\therefore \quad ydx + xdy - zdz = 0$$

Integrating, $\quad 2xy - z^2 = C_2$

9. Obviously, $\dfrac{dx}{x(y-z)} = \dfrac{dy}{y(z-x)} = \dfrac{dz}{z(x-y)}$

$$= \frac{dx + dy + dz}{xy - xz + yz - yx + zx - zy}$$

$$dx + dy + dz = 0$$

Integrating, $\quad x + y + z = C_1 \qquad$ ... (1)

Now using $\frac{1}{x}, \frac{1}{y}, \frac{1}{z}$ as multipliers, we have

$$\underset{I}{\frac{\frac{1}{x}dx}{y - z}} = \underset{II}{\frac{\frac{1}{y}dy}{z - x}} = \underset{III}{\frac{\frac{1}{z}dz}{x - y}}$$

$$I = II = III = \frac{\frac{1}{x}dx + \frac{1}{y}dy + \frac{1}{z}dz}{y - z + z - x + x - y}$$

$$\frac{1}{x}dx + \frac{1}{y}dy + \frac{1}{z}dz = 0$$

On integrating, we get

$$\log x + \log y + \log z = \log C_2$$

$$xyz = C_2 \qquad \text{... (2)}$$

10. Obviously, $\dfrac{dx + dy}{x^2 + y^2 + 2xy} = \dfrac{dx - dy}{x^2 + y^2 - 2xy} = \dfrac{dz}{(x+y)^2}$

$$\underset{I}{\frac{dx + dy}{(x+y)^2}} = \underset{II}{\frac{dx - dy}{(x-y)^2}} = \underset{III}{\frac{dz}{(x+y)^2}}$$

Taking first two members

$$\frac{dx + dy}{(x+y)^2} = \frac{dx - dy}{(x-y)^2}$$

Integrating, we get

$$-(x+y)^{-1} = -(x-y)^{-1} + C_1$$

$$\frac{1}{x - y} - \frac{1}{x + y} = C_1$$

$$\frac{2y}{x^2 - y^2} = C_1 \qquad \text{... (1)}$$

Now, taking first and last members

$$\frac{dx + dy}{(x+y)^2} = \frac{dz}{(x+y)^2}$$

$$dx + dy - dz = 0$$

Integrating, we get

$$x + y - z = C_2 \qquad \text{... (2)}$$

From equation (1) and (2), the complete solution is given by

$$\phi\left(\frac{2y}{x^2 - y^2}, x + y - z\right) = 0$$

ANSWERS

1. $\phi\left[\left(\frac{x}{y}\right), x - \log(zy - 2) + \log y\right] = 0$ **2.** $\phi\left(\frac{x+y}{z}, \frac{2y}{y^2 - x^2}\right) = 0$ **3.** $\phi\left(\frac{x-y}{y-z}, \frac{y-z}{z-x}\right) = 0$ **4.** $\phi(x^2 - y^2, x^2 - z^2) = 0$

5. $\phi\left[\left(\frac{y-x}{z-y}\right), (x-y)^2(x+y+z)\right] = 0$ **6.** $\phi(lx + my + nz, x^2 + y^2 + z^2) = 0$ **7.** $\phi(x^2 - y^2 - z^2, 2xy - z^2) = 0$

8. $\phi[x + y, \log\{z^2 + (x+y)^2\} - 2x] = 0$ **9.** $\phi(x + y + z, xyz) = 0$ **10.** $\phi\left(\frac{2y}{x^2 - y^2}, x + y - z\right) = 0$ **11.** $\phi\left(\frac{y-x}{z-x}, \frac{y-x}{z-y}\right) = 0$

12. $f\{(\cos(x+y) + \sin(x+y)\} e^{y-x} z^{\sqrt{2}} \cot\left(\frac{x+y}{2} + \frac{\pi}{8}\right) = 0$

FFFFF

CHAPTER

32 Linear Differential Equation of Second Order with Variable Coefficients

32.1 INTRODUCTION

A differential equation of the form

$$\frac{d^2y}{dx^2}+P\frac{dy}{dx}+Qy=X \qquad \ldots(1)$$

where P, Q and X are function of x alone, are called linear equation of second order.

We have several methods of solving the equation (1).

For example: Linear equations with constant coefficients, homogeneous equations and exact equations.

When these methods are not applicable, we shall try the method, discussed in this chapter.

32.2 THE COMPLETE SOLUTION IN TERMS OF A KNOWN SOLUTION

Consider the differential equation $\frac{d^2y}{dx^2}+P\frac{dy}{dx}+Qy=0$

Let $y=u$ be a known solution of the complementary function of (1).

$\Rightarrow$ $y=u$ be a solution of $\frac{d^2y}{dx^2}+P\frac{dy}{dx}+Qy=0$. Therefore,

$$\frac{d^2u}{dx^2}+P\frac{du}{dx}+Qu=0. \qquad \ldots(2)$$

On substituting $y=uv$, we get $\frac{dy}{dx}=v\frac{du}{dx}+u\frac{dv}{dx}$ and $\frac{d^2y}{dx^2}=v\frac{d^2u}{dx^2}+2\frac{du}{dx}.\frac{dv}{dx}+u\frac{d^2v}{dx^2}$

Putting all these values in equation (1), we get

$$\left(v\frac{d^2u}{dx^2}+2\frac{du}{dx}.\frac{dv}{dx}+u\frac{d^2v}{dx^2}\right)+P\left(v\frac{du}{dx}+u\frac{dv}{dx}\right)+Qu.v=X \Rightarrow u\frac{d^2v}{dx^2}+\frac{dv}{dx}\left(2\frac{du}{dx}+Pu\right)+v\left(\frac{d^2u}{dx^2}+P\frac{du}{dx}+Qu\right)=X$$

$$\Rightarrow \quad u\frac{d^2v}{dx^2}+\frac{dv}{dx}\left(2\frac{du}{dx}+Pu\right)=X \qquad \text{(By using (2))}$$

$$\Rightarrow \quad \frac{d^2v}{dx^2}+\left(P+\frac{2}{u}\frac{du}{dx}\right)\frac{dv}{dx}=\frac{X}{u}. \qquad \ldots(3)$$

Putting $\frac{dv}{dx}=p \Rightarrow \frac{d^2v}{dx^2}=\frac{dp}{dx}$ in (3), we get $\frac{dp}{dx}+\left(P+\frac{2}{u}\frac{du}{dx}\right)p=\frac{X}{u}$. $\qquad \ldots(4)$

Equation (4) is a linear differential equation with p as dependent variable.

$$\text{I.F.}=e^{\int\left(P+\frac{2}{u}\frac{du}{dx}\right)dx}=e^{\{2\log u+\int P dx\}}=u^2e^{\int P dx}.$$

The solution is $\quad pu^2e^{\int P dx}=\int\left[\frac{X}{u}.u^2e^{\int P dx}\right]dx+C_1 \qquad \ldots(5)$

$$\Rightarrow \qquad p = \frac{dv}{dx} = \frac{C_1 e^{-\int P dx}}{u^2} + \frac{e^{-\int P dx}}{u^2} \int uX e^{\int P dx} dx$$

On integrating we get $v = C_2 + C_1 \int \frac{e^{-\int P dx}}{u^2} dx + \int \left[\frac{e^{-\int P dx}}{u^2} \int uX e^{\int P dx} . dx \right] dx.$

Hence, the solution of (1) is $y = u.v = C_2 + C_1 u \int \frac{e^{-\int P dx}}{u^2} dx + u \int \left[\frac{e^{-\int P dx}}{u^2} \int uX e^{\int P dx} . dx \right] dx.$ ...(6)

WORKING PROCEDURE

Step 1. *Put in given equation into standard form* $\frac{d^2y}{dx^2} + P\frac{dy}{dx} + Qy = X$. *The coefficient of* $\frac{d^2y}{dx^2}$ *must be unity.*

Step 2. *Find an integral u.*

Step 3. *Assume that the complete solution is given by* $y = uv$. *Then the given equation reduces to*

$$\frac{d^2v}{dx^2} + \left(P + \frac{2}{u}\frac{du}{dx}\right)\frac{dv}{dx} = \frac{X}{u} \qquad ...(1)$$

Step 4. *Put* $\frac{dv}{dx} = p, \frac{d^2v}{dx^2} = \frac{dp}{dx}$ *in (1) and then solve.*

Step 5. *Put the value of v in the assumed solution* $y = uv$ *and get the desired complete solution.*

REMARKS

- The solution, given by (6) contains two arbitrary constants, hence it is the complete solution of the given equation in terms of the known integral.
- To find one integral belonging to the complementary function by inspection, the following points must be kept into mind.

(i) $u = e^x$ is a part of C.F if $P + Q + 1 = 0$

(ii) $u = e^x$ is a part of C.F. if $1 - P + Q = 0$

(iii) $u = e^{ax}$ is a part of C.F. if $1 + \frac{P}{a} + \frac{Q}{a^2} = 0$

(iv) $u = x$ is a part of C.F. if $P + Q.x = 0$

(v) $u = x^2$ is a part of C.F. if $2 + 2Px + Qx^2 = 0$

SOLVED EXAMPLES

Example 1. *Solve* $x^2\frac{d^2y}{dx^2} - 2x(1+x)\frac{dy}{dx} + 2(1+x)y = x^3$.

Solution. The given equation can be written as

$$\frac{d^2y}{dx^2} - 2\left(1 + \frac{1}{x}\right)\frac{dy}{dx} + 2\left(\frac{1}{x^2} + \frac{1}{x}\right)y = x. \quad ...(1)$$

Compare with the standard equation, we get

$$P = -2\left(1 + \frac{1}{x}\right),\ Q = 2\left(\frac{1}{x} + \frac{1}{x^2}\right),\ X = x$$

Here, we observe that

$$P + Qx = -2\left(\frac{1}{x} + 1\right) + 2x\left(\frac{1}{x} + \frac{1}{x^2}\right) = 0$$

$\Rightarrow \quad u = x$ is a part of C.F.

Putting $y = vx$

$$\therefore \quad \frac{dy}{dx} = x\frac{dv}{dx} + v,\ \frac{d^2y}{dx^2} = x\frac{d^2v}{dx^2} + 2\frac{dv}{dx}$$

Now, putting all those values in equation (1), we get

$$\frac{d^2v}{dx^2} - 2\frac{dv}{dx} = 1. \quad ...(2)$$

Let $p = \frac{dv}{dx}$, then (2) gives $\frac{dp}{dx} - 2p = 1$

which is a linear equation

$$\text{I.F.} = e^{-2\int dx} = e^{-2x}.$$

$\therefore$ Solution is

$$p\,e^{-2x} = \int 1 . e^{-2x} dx + C_1 = -\frac{1}{2}e^{-2x} + C_1$$

$$\therefore \quad p = \frac{dv}{dx} = -\frac{1}{2} + C_1 e^{2x}.$$

On integrating, we get $v = -\frac{1}{2}x + \frac{C_1}{2}e^{2x} + C_2$.

Hence, the complete solution is given by

$$y = vx = -\frac{1}{2}x^2 + \frac{C_1}{2}x e^{2x} + C_2 x.$$

Example 2. *Solve* $(x+2)\frac{d^2y}{dx^2} - (2x+5)\frac{dy}{dx} + 2y = (x+1)e^x$.

Solution. The given equation can be written as

$$\frac{d^2y}{dx^2} - \left(\frac{2x+5}{x+2}\right)\frac{dy}{dx} + \frac{2}{(x+2)}y = \left(\frac{x+1}{x+2}\right)e^x. \quad ...(1)$$

Compare with the standard equation, we get

$$P = -\left(\frac{2x+5}{x+2}\right), Q = \frac{2}{x+2} \text{ and } X = \left(\frac{x+1}{x+2}\right)e^x.$$

Here, we observe that

$$2^2+2P+Q=4-2\left(\frac{2x+5}{x+2}\right)+\frac{2}{x+2}=0$$

which implies $u=e^{2x}$ is a part of C.F. of (1).
Suppose the complete solution of (1) is given by

$$y=uv.$$

Then (1) reduces to $\frac{d^2v}{dx^2}+\left(P+\frac{2}{u}\frac{du}{dx}\right)\frac{dv}{dx}=\frac{X}{u}$

$$\Rightarrow \frac{d^2v}{dx^2}+\left(-\frac{2x+5}{x+2}+\frac{2}{e^{2x}}.2e^{2x}\right)\frac{dv}{dx}=\frac{x+1}{x+2}\cdot\frac{e^x}{e^{2x}} \qquad ...(2)$$

$$\Rightarrow \frac{d^2v}{dx^2}+\frac{2x+3}{x+2}\frac{dv}{dx}=\frac{x+1}{x+2}e^{-x}$$

Let $\frac{dv}{dx}=p$ so that $\frac{d^2v}{dx^2}=\frac{dp}{dx}$, then (2) gives

$$\frac{dp}{dx}+\frac{2x+3}{x+2}p=\frac{x+1}{x+2}e^{-x}. \qquad ...(3)$$

Now, $\int\frac{2x+3}{x+2}dx=\int\left[2-\frac{1}{x+2}\right]dx$

$$=2x-\log(x+2)$$

I.F. of (3) $=e^{2x-\log(x+2)}=e^{2x}(x+2)^{-1}$.
Therefore, the solution of (3) is given by

$$pe^{2x}\frac{1}{x+2}=\int\frac{x+1}{(x+2)^2}e^x dx+C_1$$

$$=C_1+\int e^x\frac{x+2-1}{(x+2)^2}dx=C_1+\frac{e^x}{(x+2)}$$

(By using $\int e^x[f(x)+f'(x)]dx=e^x f(x)$)

$$\therefore \quad p=e^{-x}+C_1(x+2)e^{-2x}$$

or $$\frac{dv}{dx}=e^{-x}+C_1(x+2)e^{-2x}$$

$$\therefore \quad v=C_2+\int e^{-x}dx+C_1\int(x+2)e^{-2x}dx$$

$$\Rightarrow \quad v=C_2-e^{-x}+C_1\left[(x+2)(-\frac{1}{2}e^{-2x})-1.\left(\frac{1}{4}e^{-2x}\right)\right]$$

$$=C_2-e^{-x}-\frac{1}{4}C_1e^{-2x}(2x+5) \qquad ...(4)$$

Hence, the complete solution is given by

$$y=uv=e^{2x}\left[C_2-e^{-x}-\frac{1}{4}C_1e^{-2x}(2x+5)\right]$$

$$y=C_2e^{2x}-\frac{1}{4}C_1(2x+5)-e^x.$$

Example 3. *Solve* $\sin^2x\frac{d^2y}{dx^2}=2y$, *given* $y=\cot x$ *is a solution.* (Bhopal–2007)

Solution. The given equation is

$$\sin^2x\frac{d^2y}{dx^2}=2y. \qquad ...(1)$$

Put $y=v\cot x$

$$\frac{dy}{dx}=\frac{dv}{dx}.\cot x-v\,\text{cosec}^2x$$

$$\frac{d^2y}{dx^2}=\frac{d^2v}{dx^2}\cot x-2\,\text{cosec}^2x\frac{dv}{dx}+2v\,\text{cosec}^2x\cot x.$$

In (1), we get

$$\cos x\sin x\frac{d^2v}{dx^2}-2\frac{dv}{dx}=0$$

or $$\frac{d^2v}{dx^2}-\frac{2}{\sin x\cos x}\frac{dv}{dx}=0$$

or $$\frac{dp}{dx}-\frac{2}{\sin x\cos x}.p=0 \qquad \left(p=\frac{dv}{dx}\right)$$

On separating the variables, we get

$$\frac{dp}{p}=\frac{2}{\sin x\cos x}dx=\frac{2\sec^2x}{\tan x}dx.$$

On integrating, we get

$$\log p=2\log\tan x+\log c$$

$$p=C_1\tan^2x$$

$$\Rightarrow \quad \frac{dv}{dx}=C_1\tan^2x=C_1(\sec^2x-1).$$

Integrating, we get $v=C_1(\tan x-x)+C_2$.
Hence, the complete solution is given by

$$y=v\cot x=C_1[1-x\cot x]+C_2\cot x.$$

Example 4. *Solve* $x\frac{d^2y}{dx^2}-(2x+1)\frac{dy}{dx}+(x+1)y=(x^2+x-1)e^{2x}$.

Solution. The given equation can be written as

$$\frac{d^2y}{dx^2}-\left(\frac{2x+1}{x}\right)\frac{dy}{dx}+\left(1+\frac{1}{x}\right)y=\left(x+1-\frac{1}{x}\right)e^{2x}. \qquad ...(1)$$

Compare with the standard equation, we get

$$P=-\left(\frac{2x+1}{x}\right), Q=\left(1+\frac{1}{x}\right)$$

and $$X=\left(x+1-\frac{1}{x}\right)e^{2x}.$$

Here, we observe that
$1+P+Q=0\Rightarrow y=e^x$ is a part of C.F.
Now putting $y=ve^x$ in (1), we get

$$\frac{d^2v}{dx^2}-\frac{1}{x}\frac{dv}{dx}=\left(x+1-\frac{1}{x}\right)e^x.$$

$$\Rightarrow \quad \frac{dp}{dx}-\frac{1}{x}p=\left(x+1-\frac{1}{x}\right)e^x \qquad \left(\frac{dv}{dx}=p\right)$$

which is the equation I.F $=e^{\int\frac{1}{x}dx}=e^{-\log x}=\frac{1}{x}$.

$\therefore$ Solution is

$$p.\frac{1}{x} = \int\left(x+1-\frac{1}{x}\right)e^x \cdot \frac{1}{x}\,dx + C$$

$$= \int\left(e^x + \frac{1}{x}e^x - \frac{1}{x^2}e^x\right)dx + C$$

$$= e^x + \frac{e^x}{x} + C$$

$$\Rightarrow \quad p = \frac{dv}{dx} = xe^x + e^x + C.$$

On integrating, we get

$$v = xe^x + \frac{C}{2}x^2 + C_2 \Rightarrow v = xe^x + C_1x^2 + C_2.$$

Hence, the complete solution is given by

$$y = v\,e^x = x\,e^{2x} + C_1\,x^2\,e^x + C_2\,e^x.$$

Example 5. *Solve* $y'' - 4xy' + (4x^2 - 2)y = 0$ *given that* $y = e^{x^2}$ *is an integral included in the complementary function.* (UPTU–2004)

Solution. We have $y'' - 4xy' + (4x^2 - 2)y = 0$ ...(1)

On putting $y = ve^{x^2}$ in (1), then

$$\frac{d^2v}{dx^2} + \left(P + \frac{2}{u}\right)\frac{dv}{dx} = 0$$

$$[\because P = -4x, Q = 4x^2 - 2, X = 0]$$

$$\text{or} \quad \frac{d^2v}{dx^2} + \left[-4x + \frac{2}{e^{x^2}}(2x\,e^{x^2})\right]\frac{dv}{dx} = 0$$

$$\text{or} \quad \frac{d^2v}{dx^2} + [-4x + 4x]\frac{dv}{dx} = 0$$

$$\text{or} \quad \frac{d^2v}{dx^2} = 0 \Rightarrow \frac{dv}{dx} = C_1 \Rightarrow v = C_1x + C_2$$

$$(\because u = e^{x^2})$$

then $y = uv$

$$\therefore \quad y = e^{x^2}(C_1x + C_2).$$

EXERCISE 32.1

Solve the following equations :

1. $x^2\frac{d^2y}{dx^2} - (x^2 + 2x)\frac{dy}{dx} + (x+2)y = x^3 e^x.$
2. $(x\sin x + \cos x)\frac{d^2y}{dx^2} - x\cos x\frac{dy}{dx} + y\cos x = \sin x(x\sin x + \cos x)^2.$
3. $(1-x^2)\frac{d^2y}{dx^2} + x\frac{dy}{dx} - y = x(1-x^2)^{3/2}.$
4. $\frac{d^2y}{dx^2} + \left(1 + \frac{2}{x}\cot x - \frac{2}{x^2}\right)y = x\cos x$ given that $\frac{\sin x}{x}$ is a C.F.
5. $x\frac{d^2y}{dx^2} - (2x-1)\frac{dy}{dx} + (x-1)\,y = 0.$ (Bhopal–2008S)
6. $\frac{d^2y}{dx^2} - 2(x+1)\frac{dy}{dx} + (x+2)\,y = (x-2)e^{2x}.$
7. $(x+1)\frac{d^2y}{dx^2} - 2(x+3)\frac{dy}{dx} + (x+5)\;y = e^x$
8. $x\frac{d^2y}{dx^2} + (x-2)\frac{dy}{dx} - 2y = x^3.$
9. $x\frac{d^2y}{dx^2} - (x+2)\frac{dy}{dx} + 2y = x^3.$
10. $x\frac{d^2y}{dx^2} + (1-x)\frac{dy}{dx} - y = e^x.$
11. $x\frac{d^2y}{dx^2} + (x-1)\frac{dy}{dx} - y = x^2.$
12. $x^2\frac{d^2y}{dx^2} + x\frac{dy}{dx} - 9y = 0,$ given that $y = x^3$ is a C.F.
13. $x\frac{dy}{dx} - y = (x-1)\left(\frac{d^2y}{dx^2} - x + 1\right).$
14. Solve $x\frac{d^2y}{dx^2} - \frac{dy}{dx} - 4x^3y = -4x^5$. Given that $y = e^{x^2}$ is a solution is the left hand side is equated to zero.
15. Verify that $f_1(x) = x^2$ is solution of the differential equation $\frac{d^2y}{dx^2} - \frac{2}{x^2}y = 0,\ 0 < x < \infty$ and find a second independent solution. Also obtain the solution of the given equation.
16. Solve $\frac{d^2y}{dx^2} - \cot x\frac{dy}{dx} - (1 - \cot x)y = e^x \sin x.$
17. Solve $(1-x^2)y'' - 2xy' + 2y = 0$ given that $y = x$ is a solution. (BPTU–2005S)
18. Solve $x\frac{d^2y}{dx^2} - (2x-1)\frac{dy}{dx} + (x-1)y = e^x$ given that $y = e^x$ is one integral. (Bhopal–2007S)

Hint to Selected Problems

1. $P + Qx = 0 \Rightarrow y = x$ is a part of C.F. Take $y = vx$.
2. $P + Qx = 0 \Rightarrow y = x$ is a part of C.F.
4. Take $y = v.\frac{\sin x}{x}$.
6. Here $1 + P + Q = 0 \Rightarrow y = e^x$ is a part of C.F.
8. Here $1 - P + Q = 0 \Rightarrow y = e^{-x}$ is a part of C.F.
10. Here $1 + P + Q = 0 \Rightarrow y = e^x$ is a part of C.F.
11. Here $1 - P + Q = 0 \Rightarrow y = e^{-x}$ is a part of C.F.
13. Here $P + Qx = 0 \Rightarrow y = x$ is a part of C.F.
15. $y = x^2$ satisfying the given equation, therefore we can take $y = vx^2$.

ANSWERS

1. $y = x^2e^x - xe^x + C_1xe^x + C_2x.$
2. $y = \frac{1}{4}x\cos 2x - \frac{1}{2}\sin 2x - C_1\frac{\cos x}{x} + C_x$
3. $y = -\frac{1}{9}x(1-x^2)^{3/2} - C_1\{x\sin^{-1}x + \sqrt{(1-x^2)}\} + C_2x.$
4. $y = \frac{x^2\sin x}{6} + C_1\left[-x\cos x + 2\sin x\log\sin x - \frac{2\sin x}{x}\int \log\sin x\, dx\right] + C_2\frac{\sin x}{x}$
5. $y = e^x(C_1\log x + C_2).$
6. $y = \frac{1}{3}C_1x^3e^x + C_2e^x + e^{2x}.$
7. $y = \frac{1}{5}C_1e^x(x+1)^5 - \frac{1}{4}xe^x + C_2e^x.$
8. $y = x^3 + (C_1-3)(x^2-2x+2) + C_2e^{-x}.$
9. $y = -x^3 - (C_1-3)(x^2+2x+2) + C_2e^x.$
10. $y = e^x\log x + C_1e^x\int\frac{e^{-x}}{x}dx + C_2e^x.$
11. $y = C_1(x-1) + C_2e^{-x} + x^2 - 2x + 2.$
12. $\frac{-4}{6}x^{-3} + C_2 + x^{-3}.$
13. $y = C_1e^x + C_2x - x^2 - x^2 - 1.$
14. $y = e^{x^2}x^2e^{-x^2} - \frac{C_1}{4}e^{-2x^2} + C_2.$
15. $y = -\frac{1}{3}\left(\frac{C_1}{x}\right) + C_2x^2.$
16. $y = -\frac{C_1}{2x} + C_2\left(x + \frac{1}{x}\right).$
17. $y = x\left[C_1\left\{\log\left(\frac{x}{1-x}\right)^2 - \frac{1}{x}\right\} + C_2\right]$
18. $y = e^x(C_1\log x + x + C_2)$

32.3 METHOD OF REMOVAL OF THE FIRST DERIVATIVE

Transformation of the Equation into Normal Form

To obtain a suitable substitution for the dependent variable which transforms the equation $\frac{d^2y}{dx^2} + P\frac{dy}{dx} + Qy = X$ *into normal form, i.e., the form where the first derivative is absent :*

Consider the equation

$$\frac{d^2y}{dx^2} + P\frac{dy}{dx} + Qy = X. \qquad \ldots(1)$$

Let us suppose $y = uv$ is the general solution of (1), where u is a function of x and is not a part of C.F.

Now, $y = uv \Rightarrow \frac{dy}{dx} = v\frac{du}{dx} + u\frac{dv}{dx}$ and $\frac{d^2y}{dx^2} = v\frac{d^2u}{dx^2} + 2\frac{dv}{dx}\cdot\frac{du}{dx} + u\frac{d^2v}{dx^2}$

Putting all these values in equation (1), we get

$$\left\{v\frac{d^2u}{dx^2} + 2\frac{dv}{dx}\cdot\frac{du}{dx} + u\frac{d^2v}{dx^2}\right\} + P\left(v\frac{du}{dx} + u\frac{dv}{dx}\right) + Q\,uv = X \Rightarrow u\frac{d^2v}{dx^2} + u\frac{dv}{dx}\left[P + \frac{2}{u}\frac{du}{dx}\right] + v\left[\frac{d^2u}{dx^2} + P\frac{du}{dx} + Qu\right] = X \qquad \ldots(2)$$

To remove the term of first derivative, we shall choose u such that

$$P + \frac{2}{u}\frac{du}{dx} = 0 \qquad \Rightarrow \qquad \frac{du}{u} = -\frac{P}{2}dx$$

On integrating, we get

$$\log u = -\int\frac{P}{2}dx \qquad \text{or} \qquad u = e^{\left\{-\int\frac{P}{2}dx\right\}}. \qquad \ldots(3)$$

Now equation (2) becomes

$$u\frac{d^2v}{dx^2} + v\left[\frac{d^2u}{dx^2} + P\frac{du}{dx} + Qu\right] = X \qquad \Rightarrow \qquad \frac{d^2v}{dx^2} + \frac{v}{u}\left[\frac{d^2u}{dx^2} + P\frac{du}{dx} + Qu\right] = \frac{X}{u} \qquad \ldots(4)$$

From (3), we get

$$u = e^{-\int P/2\, dx} \qquad \Rightarrow \qquad \frac{du}{dx} = -\frac{P}{2}u$$

and

$$\frac{d^2u}{dx^2} = -\frac{1}{2}\left(P\frac{du}{dx} + u\frac{dP}{dx}\right) = -\frac{1}{2}\left[P\left(-\frac{P}{2}u\right) + u\frac{dP}{dx}\right] = \frac{1}{4}P^2u - \frac{u}{2}\frac{dP}{dx}.$$

Putting these values in (4), we get

$$\frac{d^2v}{dx^2} + v\left[\frac{1}{4}P^2 - \frac{P^2}{2} + Q - \frac{1}{2}\frac{dP}{dx}\right] = X_1e^{\left\{\int\frac{1}{2}P\,dx\right\}} \qquad \text{or} \qquad \frac{d^2v}{dx^2} + v\left[Q - \frac{1}{4}P^2 - \frac{1}{2}\frac{dP}{dx}\right] = X_1e^{\left\{\int\frac{1}{2}Pdx\right\}}$$

which is known as the normal form of the equation (1) and be easily solved.

WORKING PROCEDURE

Step 1. *Put the given equation into standard form* $\frac{d^2y}{dx^2}+P\frac{dy}{dx}+Qy=X$ *With coefficient of* $\frac{d^2y}{dx^2}$ *is unity.*

Step 2. *To remove the first derivative, choose* $u=e^{-\int\frac{1}{2}P\,dx}$.

Step 3. *Assume that the complete solution of the given equation is $y=uv$. Then the equation reduces to normal form*

$$\frac{d^2v}{dx^2}+P_1\,v=X_1,\text{ where } P_1=Q-\frac{1}{4}P^2-\frac{1}{2}\frac{dP}{dx}\text{ and } X_1=\frac{X}{u}.$$

Step 4. *Solve the equation (obtained in step 3) for v, then the complete solution is $y=uv$.*

REMARK

- Students are advised to remember the equation form, so that it may be written directly.

To solve the equation $\frac{d^2v}{dx^2}+P_1v=X_1$, we have following two cases :

(a) If P_1 is constant, then equation being constant coefficient and can be solved by usual methods.

(b) If $P_1=\frac{\text{constant}}{x^2}$, then the resulting equation reduces to homogeneous form.

Solved Examples

Example 1. *Solve*
$$\frac{d^2y}{dx^2}+\frac{1}{x^{1/3}}\frac{dy}{dx}+\left(\frac{1}{4x^{2/3}}-\frac{1}{6x^{4/3}}-\frac{6}{x^2}\right)y=0.$$

Solution. The given equation is
$$\frac{d^2y}{dx^2}+\frac{1}{x^{1/3}}\frac{dy}{dx}+\left(\frac{1}{4x^{2/3}}-\frac{1}{6x^{4/3}}-\frac{6}{x^2}\right)y=0.$$

Comparing with the standard equation, we get
$$P=\frac{1}{x^{1/3}},Q=\frac{1}{4x^{2/3}}-\frac{1}{6x^{4/3}}-\frac{6}{x^2}\text{ and } X=0.$$

Let us take
$$u=e^{-\int\frac{1}{2}Pdx}=e^{-\int\frac{1}{2}x^{-1/3}dx}=e^{\left(-\frac{3}{4}x^{2/3}\right)}.$$

Now putting all these values, into the normal form
$$\frac{d^2v}{dx^2}+v\left[Q-\frac{1}{4}P^2-\frac{1}{2}\frac{dP}{dx}\right]=Xe^{\int\frac{1}{2}P\,dx}$$

We get
$$\frac{d^2v}{dx^2}+v\left[\left(\frac{1}{4x^{2/3}}-\frac{1}{6x^{4/3}}-\frac{6}{x^2}\right)-\frac{1}{4x^{2/3}}-\frac{1}{2}\left(-\frac{1}{3x^{4/3}}\right)\right]=0$$

$$\Rightarrow\quad\frac{d^2v}{dx^2}-\frac{6}{x^2}v=0\Rightarrow x^2\frac{d^2v}{dx^2}-6v=0$$

This is a homogeneous equation. To solve, put $x=e^z$ and let $\frac{d}{dz}\equiv D$, then we get

$$D(D-1)v-6v=0\quad\text{or}\quad(D+2)(D+3)v=0.$$

Auxiliary equation is
$$(m+2)(m-3)=0\Rightarrow m=-2,+3.$$

$$\Rightarrow\quad v=C_1e^{-2z}+C_2e^{3z}=C_1\cdot\frac{1}{x^2}+C_2\cdot x^3.$$

Hence, the complete solution is given by
$$y=uv=\left(\frac{C_1}{x^2}+C_2x^3\right)e^{\left(\frac{-3}{4}x^{2/3}\right)}.$$

Example 2. *Solve*
$$x^2\frac{d^2y}{dx^2}-2(x^2+x)\frac{dy}{dx}+(x^2+2x+2)y=0.$$

Solution. The given equation can be written as
$$\frac{d^2y}{dx^2}-2\left(1+\frac{1}{x}\right)\frac{dy}{dx}+\left(1+\frac{2}{x}+\frac{2}{x^2}\right)y=0$$

Here, $P=-2\left(1+\frac{1}{x}\right),Q=1+\frac{2}{x}+\frac{2}{x^2},X=0$

To remove the first derivative, let us choose
$$u=e^{-\frac{1}{2}\int Pdx}=e^{\int\left(1+\frac{1}{x}\right)dx}=e^{x+\log x}$$
$$=e^x.e^{\log x}=xe^x.$$

Now, putting all these values, into the normal form, we get
$$\frac{d^2v}{dx^2}=0\;\Rightarrow\;v=(C_1+C_2x)e^{0.x}=(C_1+C_2x)$$

Hence, the complete solution is given by
$$y=uv\;\Rightarrow\;y=x\,e^x(C_1+C_2x).$$

Example 3. *Solve* $\frac{d^2y}{dx^2}-2bx\frac{dy}{dx}+b^2x^2y=x.$

Solution. We observe that
$$P=-2bx,Q=b^2x^2,X=x.$$

To remove the first derivative, let us choose
$$u=e^{-\frac{1}{2}\int Pdx}=e^{b\int xdx}=e^{\frac{1}{2}bx^2}.$$

Putting all these values into normal form, we get
$$\frac{d^2v}{dx^2}+bv=xc-\frac{1}{2}bx^2$$

A.E. is $m^2+b=0\Rightarrow m=\pm i\sqrt{b}$

$\therefore \quad C.F. = C_1 \cos(x\sqrt{b}) + C_2 \sin(x\sqrt{b})$

$\therefore \quad P.I. = \dfrac{1}{D^2+b}\left(xe^{\frac{1}{2}bx^2}\right)$

Therefore, $v = C_1 \cos(x\sqrt{b}) + C_2 \sin(x\sqrt{b})$

$$+\frac{1}{D^2+b}(xe^{-1/2bx^2}).$$

Hence, the complete solution is given by $y = uv$

$\Rightarrow \quad y = e^{1/2bx^2}\left[C_1 \cos(x\sqrt{b}) C_2 \sin(x\sqrt{b})\right.$

$$\left. + \frac{1}{D^2+b}(xe^{-1/2bx^2})\right].$$

Example 4. *Remove the second term from the given equation and hence solve*

$$\frac{d^2y}{dx^2} - 4x\frac{dy}{dx} + (4x^2-1)y = -3e^{x^2}\sin 2x.$$

(UPTU–2004)

Solution. The given equation is

$$\frac{d^2y}{dx^2} - 4x\frac{dy}{dx} + (4x^2-1)y = -3e^{x^2}\sin 2x.$$

Comparing with the standard form, we get

$$P = -4x,\ Q = 4x^2-1,\ X = -3e^{x^2}\sin 2x$$

To remove second term, let us choose

$$u = e^{-1/2\int P\,dx} = e^{-1/2\int(-4x)\,dx} = e^{x^2}.$$

Putting all those values in the normal form, we get $\dfrac{d^2v}{dx^2} + v = -3\sin 2x.$

The auxiliary equation is $m = \pm i$

$\therefore \quad C.F. = C_1 \cos x + C_2 \sin x$

$$P.I. = -3\frac{1}{D^2+1}\sin 2x$$

$$= -3.\frac{1}{-2^2+1}\sin 2x = \sin 2x$$

$\Rightarrow \quad v = C.F. + P.I = C_1 \cos x + C_2 \sin x + \sin 2x$

Hence, the required complete solution of (1) is given by $y = uv$

$\Rightarrow \quad y = e^{x^2}(C_1 \cos x + C_2 \sin x + \sin 2x).$

Example 5. *Solve* $\dfrac{d^2y}{dx^2} - 2\tan x.\dfrac{dy}{dx} + 5y = 0.$

Solution. Comparing the given equation with standard equation, we get $P = -2\tan x, Q = 5y, X = 0.$ To remove the term of first derivative, let us choose

$$u = e^{-1/2\int P dx} = e^{-1/2\int(-2\tan x)}$$

$$= e^{\log \sec x} = \sec x.$$

Put all these values into normal form, we get

$$\frac{d^2v}{dx^2} + 6v = 0.$$

Auxiliary equation is $m^2 + 6 = 0 \Rightarrow m \pm i\sqrt{6}$

$\therefore \quad v = C_1 \cos(x\sqrt{6}) + C_2 \sin(x\sqrt{6})$

Hence, the complete solution is given by $y = uv$

$\Rightarrow \quad y = \sec x\,[C_1 \cos(x\sqrt{6}) + C_2 \sin(x\sqrt{6})].$

EXERCISE 32.2

Solve the following differential equations :

1. $x\dfrac{d}{dx}\left(x\dfrac{dy}{dx} - y\right) - 2x\dfrac{dy}{dx} + 2y + x^2y = 0.$
2. $\dfrac{d^2y}{dx^2} - 4x\dfrac{dy}{dx} + (4x^2-3)y = e^{x^2}.$
3. $\dfrac{d^2y}{dx^2} - \dfrac{2}{x}\dfrac{dy}{dx} + \left(1+\dfrac{2}{x^2}\right)y = xe^x$
4. $\dfrac{d^2y}{dx^2} - \dfrac{1}{x^{1/2}}\dfrac{dy}{dx} + \dfrac{1}{4x^2}(-8+x^{1/2}+x)y = 0$
5. $\dfrac{d^2y}{dx^2} - 2\tan x\dfrac{dy}{dx} + 5y = \sec x.e^x$
6. $\left(\dfrac{d^2y}{dx^2} + y\right)\cot x + 2\left(\dfrac{dy}{dx} + y\tan x\right) = \sec x$
7. $\dfrac{d^2y}{dx^2} - 4x\dfrac{dy}{dx} + (4x^2-1)y = -3e^{x^2}[\sin 2x + 5e^{-2x} + 6]$
8. $\dfrac{d^2y}{dx^2} + 2x\dfrac{dy}{dx} + (x^2+1)y = x^3 + 3x.$
9. $\dfrac{d}{dx}\left(\cos^2 x\dfrac{dy}{dx}\right) + y\cos^2 x = 0$
10. $x^2(\log x)^2\dfrac{d^2y}{dx^2} - 2x\log x\dfrac{dy}{dx} + [2 + \log x - 2(\log x)^2]y = (\log x)^3 x^2$
11. $\dfrac{d^2y}{dx^2} + 2x\dfrac{dy}{dx} + (x^2+5)y = xe^{-1/2x^2}$
12. $\dfrac{d^2y}{dx^2} - 2x\dfrac{dy}{dx} + (x^2+2)y = e^{\frac{1}{2}(x^2+2x)}$ (GBTU(CO)–2011)
13. $\dfrac{d^2y}{dx^2} + 2x\dfrac{dy}{dx} + (x^2-8)y = x^2e^{-x^2/2}$ (GBTU–2012)

Hint to Selected Problems

1. $P = \dfrac{-2}{x},\ Q = 2 + \dfrac{1}{x^2},\ X = 0,$

   $u = e^{-(1/2)\int Pdx} = e^{\int(1/x)dx} = e^{\log x} = x$

3. $P = -\dfrac{2}{x},\ Q = 1 + \dfrac{2}{x},\ X = xe^x,$

   $u = e^{(-1/2)\int Pdx} = e^{\int(1/x)dx} = e^{\log x} = x$

4. $P = -\dfrac{1}{x^{-1/2}},\ Q = \dfrac{1}{4x^2}(x + x^{1/2} - 8),\ X = 0,$

   $u = e^{(-1/2)\int Pdx} = e^{-1/2}\int x^{-1/2}dx = e^{x^{1/2}} = e^{\sqrt{x}}$

7. $P = -4x,\ Q = 4x^2 - 1,\ X = -3e^{x^2}[\sin 2x + 5e^{-2x} + 6],$

   $u = e^{-1/2\int Pdx} = e^{-1/2\int -4xdx} = e^{x^2}$

8. $P = 2x,\ Q = x^2+1,\ X = x(x^2+3)\ ,u = e^{-1/2\int Pdx} = e^{-x^2/2}$

9. $P = -2\tan x,\ Q = 1,\ X = 0\ ,$

$u = e^{-1/2\int Pdx = e^{\int \tan x dx} = e^{\log \sec x}} = \sec x$

ANSWERS

1. $y = x[C_1\cos x + C_2 \sin x]$

2. $y = e^{x^2}[C_1e^x + C_2e^{-x} - 1]$

3. $y = x\left[C_1\cos x + C_2\sin x + \dfrac{e^x}{2}\right]$

4. $y = e^{\sqrt{x}}[C_1x^2 + C_2x^{-1}]$

5. $y = \sec x\left[C_1\cos(\sqrt{6}x) + C_2\sin(\sqrt{6}x) + \dfrac{e^x}{7}\right]$

6. $y = \dfrac{1}{2}\sin x + (C_1x + C_2)\cos x$

7. $y = e^{x^2}(C_1\cos x + C_2\sin x + \sin 2x - 3e^{-2x} - 18)$

8. $y = x + (C_1x + C_2)e^{-x^2/2}$

9. $y = \sec x(C_1\cos\sqrt{2x} + C_2\sin\sqrt{2x})$

10. $y = (\log x)(C_1x^2 + C_2x^{-1} + \dfrac{1}{3}x^2\log x)$

11. $y = e^{-x^2/2}\left[C_1\cos(2x + C_2) + \dfrac{1}{4}x\right]$

12. $y = e^{x^2/2}(C_1\cos\sqrt{3}x + C_2\sin\sqrt{3}x) + \dfrac{1}{4}e^{\frac{1}{2}(x^2+2x)}$

13. $y = e^{-\frac{x^2}{2}}\left[C_1e^{3x} + C_2e^{-3x} - \dfrac{1}{9}\left(x^2 + \dfrac{2}{9}\right)\right]$

32.4 TRANSFORMATION OF THE EQUATION BY CHANGING THE INDEPENDENT VARIABLE

Consider the equation

$$\frac{d^2y}{dx^2} + P\frac{dy}{dx} + Qy = X \qquad \text{...(1)}$$

where P, Q and X are functions of z. Let the independent variable be changed from x to z, where $z = f(x)$ (say), we know that

$$\frac{df}{dx} = \frac{df}{dz}\cdot\frac{dz}{dx} \Rightarrow \frac{dy}{dx} = \frac{dy}{dz}\cdot\frac{dz}{dx} \qquad \frac{d^2y}{dx^2} = \frac{d}{dx}\left(\frac{dy}{dz}\cdot\frac{dz}{dx}\right) = \frac{d}{dz}\left(\frac{dy}{dz}\cdot\frac{dz}{dx}\right)\cdot\frac{dz}{dx} = \frac{d^2y}{dz^2}\left(\frac{dz}{dx}\right)^2 + \frac{dy}{dz}\cdot\frac{d^2z}{dx^2}$$

Putting all those values in equation (1), we get

$$\left\{\frac{d^2y}{dz^2}\left(\frac{dz}{dx}\right)^2 + \frac{dy}{dz}\cdot\frac{d^2z}{dx^2}\right\} + P\left\{\frac{dy}{dz}\cdot\frac{dz}{dx}\right\} + Q.y = X$$

$$\Rightarrow \quad \frac{d^2y}{dz^2} + \left\{\frac{\dfrac{d^2z}{dx^2} + P\dfrac{dz}{dx}}{\left(\dfrac{dz}{dx}\right)^2}\right\}\frac{dy}{dz} + \frac{Q}{\left(\dfrac{dz}{dx}\right)^2}y = \frac{X}{\left(\dfrac{dz}{dx}\right)^2} \quad \Rightarrow \quad \frac{d^2y}{dz^2} + P_1\frac{dy}{dz} + Q_1y = X_1 \qquad \text{...(2)}$$

where,
$$P_1 = \left\{\frac{\dfrac{d^2z}{dx^2} + P\dfrac{dz}{dx}}{\left(\dfrac{dz}{dx}\right)^2}\right\},\ Q_1 = \left\{\frac{Q}{\left(\dfrac{dz}{dx}\right)^2}\right\},\ \text{and } X_1 = \left\{\frac{X}{\left(\dfrac{dz}{dx}\right)^2}\right\}.$$

Here, P_1, Q_1, X_1 are functions of x but can be expressed in terms of z by the given relation between z and x.

WORKING PROCEDURE

Step 1. *Put the given equation into standard form, and find the value of P, Q and X.*

Step 2. *Suppose* $Q = \pm C\ f(x)$, *then assume a relation between z and x given by* $\left(\dfrac{dz}{dx}\right)^2 = Cf(x)$. *(Here, we always omit –ve sign of Q).*

Step 3. *Now solve*

$$\left(\frac{dz}{dx}\right)^2 = Cf(x) \Rightarrow \frac{dz}{dx} = \sqrt{C\ f(x)} \qquad \text{...(1)}$$

(Omit the negative sign, to get a relation between z and x).
After solving (1), we get a relation between z and x (Don't use the constant of integration).

Step 4. *Transform the equation into the form*

$$\frac{d^2y}{dz^2} + P_1\frac{dy}{dz} + Q_1y = X_1 \qquad \text{...(2)}$$

Step 5. *Now solve (2) for z and then replace z by x by the relation between x and z.*

REMARKS

- If we equate Q_1 to a constant quantity, then P_1 also becomes constant, then the equation (2) can be solved easily (since, then it will be linear equation with constant coefficients).
- Equation (2) can be solved in following two ways :

(i) We choose z in such a way that $P_1 = 0$, *i.e.*, $\dfrac{d^2z}{dx^2} + P\dfrac{dz}{dx} = 0 \Rightarrow z = \int e^{-\int P dx} dx$

Then, the given equation is changed into $\dfrac{d^2y}{dz^2} + Q_1 y = X_1$

This is easily integrable, provided Q_1 comes out to be constant or of the form $\dfrac{\text{constant}}{z^2}$.

(ii) Choose z in such a way that $Q_1 = \left\{\dfrac{Q}{(dz/dx)^2}\right\}$ is a constant, say C^2, then $\dfrac{Q}{(dz/dx)^2} = C^2 \Rightarrow C\dfrac{dz}{dx} = \sqrt{Q}$

$\therefore\ Cz = \int \sqrt{Q}\,.\,dz$

Put this value in the given equation, we get $\dfrac{d^2y}{dz^2} + P_1\dfrac{dy}{dz} + C^2 y = X_1$

This differential equation can be integrated provided P_1 also comes out to be a constant.

- The value of P_1, Q_1 and X_1 must be remembered for its direct use.

Solved Examples

Example 1. *Solve* $x\dfrac{d^2y}{dx^2} - \dfrac{dy}{dx} + 4x^3y = x^5$.

(UPTU–2002, 03, 05)

Solution. The given equation can be written as

$$\frac{d^2y}{dx^2} - \frac{1}{x}\frac{dy}{dx} + 4x^2y = x^4 \qquad \ldots(1)$$

Comparing with the standard equation, we get

$$P = -\frac{1}{x},\ Q = 4x^2,\ X = x^4.$$

Let us choose z such that

$$\left(\frac{dz}{dx}\right)^2 = 4x^2 \Rightarrow \frac{dz}{dx} = 2x.$$

$$\therefore z = x^2$$

Then, the transformed equation is

$$\frac{d^2y}{dz^2} + P_1\frac{dy}{dz} + Q_1 y = X_1 \qquad \ldots(2)$$

where, $P = \dfrac{\dfrac{d^2z}{dx^2} + P\dfrac{dz}{dx}}{(dz/dx)^2} = \dfrac{2 + \left(-\dfrac{1}{x}\right)2x}{4x^2} = 0,$

$$Q_1 = \frac{Q}{(dz/dx)^2} = 1$$

and $X_1 = \dfrac{X}{(dz/dx)^2} = \dfrac{x^4}{4x^2} = \dfrac{x^2}{4} = \dfrac{z}{4}$.

Put all these values in (2), we get $\dfrac{d^2y}{dz^2} + y = \dfrac{z}{4}$.

Auxiliary equation is $m^2 + 1 = 0$

$\Rightarrow$ C.F. $= C_1 \cos z + C_2 \sin z$

$$\text{P.I.} = \frac{1}{D_1^2+1}\cdot\frac{z}{4} = \frac{z}{4}(D_1+1)^{-1} = \frac{z}{4} \qquad \left(D_1 \equiv \frac{d}{dz}\right)$$

$$\therefore y = \text{C.F.} + \text{P.I.} = C_1 \cos z + C_2 \sin z + \frac{z}{4}$$

$$\Rightarrow y = C_1 \cos x^2 + C_2 \sin x^2 + \frac{x^2}{4}.$$

Example 2. *Solve* $(1+x^2)^2\dfrac{d^2y}{dx^2} + 2x(1+x^2)\dfrac{dy}{dx} + 4y = 0$.

(UPTU(B.Tech.)(Q. Bank)–2001)

Solution. The given equation can be written as

$$\frac{d^2y}{dx^2} + \frac{2x}{(1+x^2)}\frac{dy}{dx} + \frac{4}{(1+x^2)^2}y = 0 \qquad \ldots(1)$$

Comparing with the standard equation, we get

$$P = \frac{2x}{(1+x^2)},\quad Q = \frac{4}{(1+x^2)^2},\quad X = 0.$$

Let us choose z such that

$$\left(\frac{dz}{dx}\right)^2 = \frac{4}{(1+x^2)^2} \Rightarrow \frac{dz}{dx} = \frac{2}{1+x^2}.$$

On integrating, we get

$$z = 2\int \frac{dx}{1+x^2} \Rightarrow z = 2\tan^{-1}x.$$

Thus, the transformed equation is

$$\frac{d^2y}{dz^2} + P_1\frac{dy}{dz} + Q_1 y = X_1 \qquad \ldots(2)$$

where,

$$P_1 = \frac{\dfrac{d^2z}{dx^2} + P\dfrac{dz}{dx}}{(dz/dx)^2} = \frac{-\dfrac{4x}{(1+x^2)^2} + \dfrac{2x}{1+x^2}\cdot\dfrac{2}{1+x^2}}{4/(1+x^2)^2} = 0$$

$$Q_1 = \frac{Q}{(dz/dx)^2} = 1 \text{ and } X_1 = \frac{X}{(dz/dx)^2} = 0.$$

Putting all these in (2), we get $\dfrac{d^2y}{dz^2} + y = 0$

Auxiliary equation is $m^2+1=0 \Rightarrow m=\pm i$

$\therefore$ C.F. $=C_1\cos z+C_2\sin z$.

Hence, the complete solution is

$\Rightarrow y=C_1\cos(2\tan^{-1}x)+C_2\sin(2\tan^{-1}x)$.

Example 3. *Solve* $x\dfrac{d^2y}{dx^2}-\dfrac{dy}{dx}-4x^3y=8x^3\sin x^2$.

Solution. The given equation can be written as

$$\frac{d^2y}{dx^2}-\frac{1}{x}\frac{dy}{dx}-4x^2y=8x^2\sin x^2. \quad ...(1)$$

Comparing with the standard equation, we get

$$P=-\frac{1}{x},\ Q=-4x^2 \text{ and } X=8x^2\sin x^2.$$

Let us choose z such that

$$\left(\frac{dz}{dx}\right)^2=4x^2 \text{ or } \frac{dz}{dx}=2x \Rightarrow z=x^2.$$

Then the transformed equation is

$$\frac{d^2y}{dz^2}+P_1\frac{dy}{dz}+Q_1y=X_1 \quad ...(2)$$

where, $P_1=\dfrac{\dfrac{d^2z}{dx^2}+P\dfrac{dz}{dx}}{(dz/dx)^2}=0$ and $Q_1=-1$

and $X_1=2\sin z$.

Putting all these in (2), we get

$$(D_1^2-1)y=2\sin z \qquad \left(D_1\equiv\frac{d}{dz}\right)$$

$\Rightarrow$ C.F. $=C_1e^{z}+C_2e^{-z}$

$$\text{P.I.}=\frac{1}{D_1^2-1}.2\sin z=2.\frac{1}{-1^2-1}\sin z$$

$$=-\sin z.$$

Hence, the complete solution is

$$y=C_1e^{z}+C_2e^{-z}-\sin z$$

$$\Rightarrow y=C_1e^{x^2}+C_2e^{-x^2}-\sin x^2.$$

Example 4. *Transform the differential equation*

$$\cos x\frac{d^2y}{dx^2}+\sin x\frac{dy}{dx}-2y\cos^3x=2\cos^5x$$

into one having z as independent variable, where $z=\sin x$ and solve it.

(UPTU(Q. Bank)–2001, Bhopal–2006S)

Solution. Given that

$$z=\sin x\Rightarrow\frac{dz}{dx}=\cos x \quad ...(1)$$

Now, $$\frac{dy}{dx}=\frac{dy}{dz}.\frac{dz}{dx}=\cos x\frac{dy}{dz} \quad ...(2)$$

and $$\frac{d^2y}{dx^2}=\frac{d}{dx}\left(\frac{dy}{dz}\right)=\frac{d}{dx}\left(\cos x\frac{dy}{dz}\right)$$

$$=-\sin x\frac{dy}{dz}+\cos x\frac{d}{dx}\left(\frac{dy}{dz}\right)$$

$$=-\sin x\frac{dy}{dz}+\cos x\frac{d}{dz}\left(\frac{dy}{dz}\right).\frac{dz}{dx}$$

$$\Rightarrow \frac{d^2y}{dx^2}=-\sin x\frac{dy}{dz}+\cos^2x\frac{d^2y}{dz^2}$$

(By using (1))

Now, using (2) and (3), given equation reduces to

$$\cos x\left[-\sin x\frac{dy}{dz}+\cos^2x\frac{d^2y}{dz^2}\right]$$

$$+\sin x\cos x\frac{dy}{dz}-2\cos x^2x.y=2\cos^5x$$

$$\Rightarrow \frac{d^2y}{dz^2}-2y=2\cos^2x$$

$$\Rightarrow \frac{d^2y}{dz^2}-2y=2(1-\sin^2x)$$

$$\Rightarrow (D_1^2-2)y=2(1-z^2) \qquad \left(D_1\equiv\frac{d}{dz}\right)$$

Now, the auxiliary equation is

$m^2-2=0 \Rightarrow m=\pm\sqrt{2}$

$\Rightarrow$ C.F. $=C_1e^{\sqrt{2}.z}+C_2e^{-\sqrt{2}.z}$

$$\text{P.I.}=2.\frac{1}{D_1^2-2}(1-z^2)$$

$$=2.\frac{1}{-2\left(1-\dfrac{D_1^2}{2}\right)}(1-z^2).$$

$$=-\left(1-\frac{D_1^2}{2}\right)^{-1}(1-z^2)$$

$$=-\left(1+\frac{D_1^2}{2}+...\right)(1-z^2)$$

$$=-(1-z^2-1)=z^2.$$

Therefore, the complete solution is given by

$$y=C_1e^{z\sqrt{2}}+C_2e^{-z\sqrt{2}}+z^2.$$

Putting $z=\sin x$, we get the required general solution as

$$y=C_1e^{\sqrt{2}.\sin x}+C_2e^{-\sqrt{2}\sin x}+\sin^2x.$$

EXERCISE 32.3

Solve the following differential equations :

1. $x^6\dfrac{d^2y}{dx^2}+3x^5\dfrac{dy}{dx}+a^2y=\dfrac{1}{x^2}$
2. $\sin^2x\dfrac{d^2y}{dx^2}+\sin x\cos x\dfrac{dy}{dx}+4y=0$
3. $\dfrac{d^2y}{dx^2}+\cot x\dfrac{dy}{dx}+4\,\text{cosec}^2x.y=0$
4. $\dfrac{d^2y}{dx^2}+\dfrac{2}{x}\dfrac{dy}{dx}+\dfrac{a^2}{x^4}y=0$
5. $\dfrac{d^2y}{dx^2}+\tan x\dfrac{dy}{dx}+y\cos^2x=0$ (Bhopal–2005)

6. $\frac{d^2y}{dx^2} - \cot x \frac{dy}{dx} - y\sin^2 x = \cos x - \cos^3 x$ (GBTU–2011)

7. $x\frac{d^2y}{dx^2} + (4x^2-1)\frac{dy}{dx} + 4x^3y = 2x^3$ (UPTU–2006)

8. $\frac{d^2y}{dx^2} - (8e^{2x}+2)\frac{dy}{dx} + 4e^{4x}.y = e^{6x}$

9. $\frac{d^2y}{dx^2}\left(1-\frac{1}{x}\right)\frac{dy}{dx} + 4x^2e^{-2x}y = 4(x^2+x^3)\,e^{-3x}$

10. $(x^3-x)\frac{d^2y}{dx^2} + \frac{dy}{dx} + n^2x^3y = 0$

11. $\frac{d^2y}{dx^2} - \frac{1}{x}\frac{dy}{dx} + 4x^2y = x^4$ (UPTU–2003, Special Exam–2001)

12. $x^4\frac{d^2y}{dx^2} + 2x^3\frac{dy}{dx} + n^2y = 0$ (UKTU–2011)

Hint to Selected Problems

1. $P = \frac{3}{x},\ Q = \frac{a^2}{x^6},\ X = \frac{1}{x^8}$. Then take

$$\frac{Q}{(dz/dx)^2} = \frac{a^2/x^6}{(dz/dx)^2} = a^2 \text{ (say)}$$

$$\Rightarrow \left(\frac{dz}{dx}\right)^2 = \frac{1}{x^6} \Rightarrow \frac{dz}{dx} = \frac{1}{x^3} \Rightarrow z = -\frac{1}{2x^2}$$

2. $P = \cot x,\ \ Q = 4\,\mathrm{cosec}^2 x,\ X = 0,$

$$P_1 = \frac{\frac{d^2z}{dx^2} + P\frac{dz}{dx}}{(dz/dx)^2}, Q = \frac{Q}{(dz/dx)^2}, X = \frac{X}{(dz/dx)^2}$$

$\Rightarrow Q_1 = 1, P_1 = 0$.

Then transformed equation is $(D^2+1)y = 0$.

4. $P = \frac{2}{x}, Q = \frac{a^2}{x^4}, X = 0$.

Now, $\frac{Q}{(dz/dx)^2} = \frac{a^2/x^4}{(dz/dx)^2} = a^2 \Rightarrow \frac{dz}{dx} = -\frac{1}{x^2} \Rightarrow z = \frac{1}{x}$

6. $P = -\cot x, Q = -\sin^2 x, X = \cos x - \cos^3 x$.

Now, $\frac{Q}{(dz/dx)^2} = -\frac{\sin^2 x}{(dz/dx)^2} = -1$ (say) $\Rightarrow\ z = -\cos x$.

$\therefore\ P_1 = 0,\ Q_1 = -1,\ X_1 = -z$.

Transformed equation is given by $(D^2-1)y = -z$.

10. $P = \frac{1}{x^3-x},\ \ Q = \frac{n^2x^2}{x^2-1},\ \ X = 0$.

Proceed as usual, we get $z = \sqrt{x^2-1}$, $P_1 = 0, Q = n^2, X_1 = 0$.

Then transformed equation is $\frac{d^2y}{dx^2} + n^2y = 0$.

ANSWERS

1. $y = C_1\cos\left(\frac{a}{2x^2}\right) - C_2\sin\left(\frac{a}{2x^2}\right) + \frac{1}{a^2x^2}$

2. $y = C_1\cos\left(2\log\tan\frac{x}{2}\right) + C_2\sin\left(2\log\tan\frac{x}{2}\right)$

3. Same as (2)

4. $y = C_1\cos\frac{a}{x} + C_2\sin\frac{a}{x}$

5. $y = C_1\cos(\sin x) + C_2\sin(\sin x)$

6. $y = C_1e^{-\cos x} + C_2e^{\cos x} - \cos x$

7. $y = e^{-x^2}[C_1x^2 + C_2] + \frac{1}{2}$

8. $y = C_1e^{(2+\sqrt{3})e^{2x}} + C_2e^{(2-\sqrt{3})e^{2x}} + \frac{1}{4}e^{2x} + 1$

9. $y = C_1\cos\{2(x+1)e^{-x}\} - C_2\sin[2(x+1)e^{-x}\} + (x+1)e^{-x}$

10. $y = C_1\sin\left[n\sqrt{(x^2-1)} + C_2\right]$

11. $y = C_1\cos(x^2) + C_2\sin(x^2) + \frac{x^2}{4}$

12. $y = C_1\cos\left(-\frac{n}{x}\right) + C_2\sin\left(-\frac{n}{x}\right)$

32.5 METHOD OF VARIATION OF PARAMETERS

Consider the differential equation

$$\frac{d^2y}{dx^2} + P\frac{dy}{dx} + Qy = X \quad \ldots(1)$$

Let us suppose the C.F. of (1) is given by

$$y = A_1u + B_1v \quad \ldots(2)$$

where A_1 and B_1 are constants and u, v are functions of x.

Now, since (2) is a C.F. of (1), therefore, u and v must be the solution of

$$\frac{d^2y}{dx^2} + P\frac{dy}{dx} + Qy = 0 \quad \ldots(3)$$

$$\Rightarrow \quad \frac{d^2u}{dx^2} + P\frac{du}{dx} + Qu = 0 \text{ and } \frac{d^2v}{dx^2} + P\frac{dv}{dx} + Qv = 0 \quad \ldots(4)$$

Now, let us assume $$y = A_2u + B_2v \qquad \ldots(5)$$
is the complete solution of (1), where A_2, B_2 are functions of x and to be so chosen that (1) shall be satisfied.
Now, we have two unknown quantities, A_2 and B_2 in terms of which y has been expressed by (5).

To determine A_2 and B_2, let us take $$u\frac{dA}{dx} + v\frac{dB}{dx} = 0 \qquad \ldots(6)$$

Differentiating (5) and using (6), we get $$\frac{dy}{dx} = A_1\frac{du}{dx} + B_1\frac{dv}{dx}. \qquad \ldots(7)$$

Now, differentiating (6), we get $$\frac{d^2y}{dx^2} = A_1\frac{d^2u}{dx^2} + \frac{dA_1}{dx}.\frac{du}{dx} + B_1\frac{d^2v}{dx^2} + \frac{dB_1}{dx}.\frac{dv}{dx}. \qquad \ldots(8)$$

Substituting all these values in (1), we get
$$A_1\left[\frac{d^2u}{dx^2} + P\frac{du}{dx} + Qu\right] + B_1\left[\frac{d^2v}{dx^2} + P\frac{dv}{dx} + Q.v\right] + \frac{dA_1}{dx}.\frac{du}{dx} + \frac{dB_1}{dx}.\frac{dv}{dx} = X.$$

Using (4), we get $$A_1.0 + B_1.0 + \frac{dA_1}{dx}.\frac{du}{dx} + \frac{dB_1}{dx}.\frac{dv}{dx} = X$$
$$\frac{dA_1}{dx}.\frac{du}{dx} + \frac{dB_1}{dx}.\frac{dv}{dx} = X. \qquad \ldots(9)$$

Solving (6) and (9), we get $$\frac{dA_1}{dx} = -\frac{vX}{u\frac{dv}{dx} - v\frac{du}{dx}} \quad \text{and} \quad \frac{dB_1}{dx} = \frac{uX}{u\frac{dv}{dx} - v\frac{dy}{dx}}.$$

On integrating, we get $$A_1 = f(x) + C_1 \quad \text{and} \quad B_1 = g(x) + C_2.$$
Hence, from (5), the required complete solution is given by $y = C_1u + C_2v + u\ f(x) + v\ g(x)$ with C_1 and C_2 arbitrary constants.

WORKING PROCEDURE

Step 1. *Reduce the given equation into normal form.*

Step 2. *Consider* $\frac{d^2y}{dx^2} + P\frac{dy}{dx} + Qy = 0$ *which can be solved by any method.* ...(1)

Step 3. *Let* $y = Au + Bv$ ...(2)
be the solution of (1), *where A and B are arbitrary constants and u and v are known functions of x. Then* $Au + Bv$ *is complementary function.*

Step 4. *Choose* A_1 *and* B_1 *such that* $u\frac{dA_1}{dx} + v\frac{dB_1}{dx} = 0$ ...(3)

Step 5. *Let* $y = A_1u + B_1v$
be the general solution of the given equation. Then A_1 *and* B_1 *are functions of x, which are to be determined.*

Step 6. *Differentiating* (2) *and using* (3), *we get* $\frac{dy}{dx} = A_1\frac{du}{dx} + B_1\frac{dv}{dx}$ ...(4)

Now, differentiating (4), *we get* $\frac{d^2y}{dx^2} = \frac{dA_1}{dx}\frac{du}{dx} + A_1\frac{d^2u}{dx^2} + \frac{dB_1}{dx}.\frac{dv}{dx} + B_1.\frac{d^2v}{dx^2}$ ...(5)

Substituting all these values in the standard form of the given equation and after simplification, we get
$$\frac{dA_1}{dx}.\frac{du}{dx} + \frac{dB_1}{dx}.\frac{dv}{dx_1} = X \qquad \ldots(6)$$

Step 7. *Solve* (3) *and* (6) *and get* $\frac{dA_1}{dx}$ *and* $\frac{dB_1}{dx}$, *integrate these to get* A_1 *and* B_1. *Then putting the values of* A_1 *and* B_1 *in* (2) *and get the required general solution.*

REMARKS

- Since, the form of y is the same for the equations (3) and (1), but the constants, which will occur, are different, therefore this method is known as variation of parameters.
- In this method, we must require a complete knowledge of the complementary functions.
- This method must be used only if instructed to do so.
- This method is more useful when the C.F. is known easily but particular integral of the differential equation cannot be obtained by any previous method (discussed in this chapter).

Solved Examples

Example 1. *Solve the following equation by variation of parameters* $\frac{d^2y}{dx^2}+a^2y=\sec ax$

(UPTU(SUM)–2002, 08 Q. Bank, AMIE–2004, UKTU–2012)

Solution. The solution of the given equation $\frac{d^2y}{dx^2}+a^2y=0$ is given by

$$y=A\cos ax+B\sin ax \qquad \ldots(1)$$

Now, let us suppose that A and B are functions of x and let (1) satisfies the given equation.

Therefore, $\frac{dy}{dx}=-Aa\sin ax+B.a\cos ax$

$$+\cos ax\frac{dA}{dx}+\sin ax\frac{dB}{dx}$$

Now, let us choose A and B such that

$$\cos ax\frac{dA}{dx}+\sin ax\frac{dB}{dx}=0. \qquad \ldots(2)$$

Then, $\frac{dy}{dx}=-A.a\sin ax+B.a\cos ax$

$$\Rightarrow \frac{d^2y}{dx^2}=-Aa^2\cos ax-Ba^2\sin ax$$

$$-\frac{dA}{dx}.a\sin ax+\frac{dB}{dx}.a\cos ax.$$

If equation (1) satisfy the given equation, when A and B are functions of x, we have

$$-\frac{dA}{dx}.a\sin ax+a\cos ax\frac{dB}{dx}=\sec ax \qquad \ldots(3)$$

Solving (2) and (3), we get

$$a\frac{dB}{dx}=1 \text{ and } a\frac{dA}{dx}=-\tan ax$$

$$\Rightarrow A=\frac{1}{a^2}\log\cos ax+C_1 \text{ and } B=\frac{x}{a}+C_2.$$

Hence, the complete solution of the given equation is $y=C_1\cos ax+C_2\sin ax$

$$+\frac{1}{a^2}(\log\cos ax)\cos ax+\frac{x}{a}\sin ax.$$

Example 2. *Solve by the method of variation of parameters*

(a) $\frac{d^2y}{dx^2}+4y=4\tan 2x$.

Solution. Here, the given equation is

$$\frac{d^2y}{dx^2}+4y=4\tan 2x \qquad \ldots(1)$$

To find the C.F., the auxiliary equation is

$$m^2+4=0 \Rightarrow m=\pm 2i$$

$$\Rightarrow y=A\cos 2x+B\sin 2x.$$

If A and B are functions of x, then

$$\frac{dy}{dx}=-2A\sin 2x+2B\cos 2x$$

$$+\cos 2x\frac{dA}{dx}+\sin 2x\frac{dB}{dx}$$

Now, if $\cos 2x\frac{dA}{dx}+\sin 2x\frac{dB}{dx}=0 \qquad \ldots(2)$

Then, $\frac{d^2y}{dx^2}=-4A\cos 2x-4B\sin 2x$

$$-2\sin 2x\frac{dA}{dx}+2\cos 2x\frac{dB}{dx}.$$

Substituting these values in the given equation, we have

$$\cos 2x\frac{dB}{dx}-\sin 2x\frac{dA}{dx}=2\tan 2x \qquad \ldots(3)$$

Solving (2) and (3), we get

$$\frac{dB}{dx}=2\sin 2x \Rightarrow B=-\cos 2x+C_1$$

and $\frac{dA}{dx}=-\frac{2\sin^2 2x}{\cos 2x}=-2\frac{(1-\cos^2 2x)}{\cos 2x}$

$$=2\cos 2x-2\sec 2x$$

$$\therefore \quad A=\sin 2x-\log(\sec 2x+\tan 2x)+C_2.$$

Hence, the solution of the given equation is

$$y=[\sin 2x-\log(\sec 2x+\tan 2x)+C_2]\cos 2x+(C_1-\cos 2x)\sin 2x$$

$$\Rightarrow y=C_1\sin 2x+C_2\cos 2x-\log[\sec 2x+\tan 2x]\cos 2x.$$

(b) $\frac{d^2y}{dx^2}+y=\tan x$ (UPTU–2009, UKTU–2011)

Solution. Proceed same as above, we get

Parts of C.F. are $u=\cos x$ and $v=\sin x$

$\therefore \quad y=A\cos x+B\sin x$

where A and B are determined as :

$$A=-\int\frac{Rv}{uv_1-u_1v}dx+C_1$$

$$=-\int\frac{\tan x.\sin x}{\cos^2 x+\sin^2 x}dx+C_1$$

$[\because R=\tan x]$

$$=-\int\frac{1-\cos^2 x}{\cos x}dx+C_1$$

$$=-\int(\sec x-\cos x)dx+C_1$$

$$\Rightarrow A=\sin x-\log(\sec x+\tan x)+C_1$$

and $B=\int\frac{Ru}{uv_1-u_1v}dx+C_2$

$$=\int\frac{\tan x.\cos x}{1}dx+C_2$$

$$\Rightarrow \quad B=-\cos x+C_2$$

Hence, the complete solution is

$$y=A\cos x+B\sin x$$

$$=[\sin x-\log(\sec x+\tan x)+C_1]\cos x+(-\cos x+C_2)\sin x$$

$$\Rightarrow y=C_1\cos x+C_2\sin x-\cos x\log(\sec x+\tan x).$$

Example 3. *Apply the method of variation of parameters to solve*

$$x^2\frac{d^2y}{dx^2}+x\frac{dy}{dx}-y=x^2e^x$$ (UPTU–2004, 06)

Solution. Clearly, the given equation is homogeneous equation.

To solve it, put $x = e^x$ and $D \equiv \frac{d}{dz}$.

$$[D(D-1)+D-1]y = 0 \Rightarrow (D^2-1)y = 0 \quad ...(1)$$

Therefore, the solution is $y = Ae^z + Be^{-z}$

i.e., The complementary function of the given equation is

$$y = Ax + B(1/x) \quad ...(2)$$

Now, if A and B are the functions of x, therefore

$$\frac{dy}{dx} = A - \frac{B}{x^2} + x\frac{dA}{dx} + \frac{1}{x}\frac{dB}{dx}.$$

Now, choosing A and B such that

$$x\frac{dA}{dx} + \frac{1}{x}\frac{dB}{dx} = 0. \quad ...(3)$$

Then, $\frac{d^2y}{dx^2} = \frac{dA}{dx} - \frac{1}{x^2}\frac{dB}{dx} + \frac{2B}{x^3}$.

Putting all these values in the given equation, we get

$$x^2\frac{dA}{dx} - \frac{dB}{dx} = x^2e^x \quad ...(4)$$

Now, solving (3) and (4), we get

$$2x^2\frac{dA}{dx} = x^2e^x \Rightarrow A = \frac{1}{2}e^x + C_1$$

and $2\frac{dB}{dx} = -x^2e^x$

$$\Rightarrow B = -\frac{1}{2}(x^2e^x - 2xe^x + 2e^x + C_2).$$

Hence, the solution of the given equation is

$$y = x\left(C_1 + \frac{1}{2}e^x\right) + \frac{1}{x}\left(-\frac{1}{2}x^2e^x + xe^x - e^x + \frac{1}{2}C_2\right)$$

Example 4. *Solve by method of variation of parameters*

$$\frac{d^2y}{dx^2} - y = \frac{2}{1+e^x}.$$

(UPTU–2001, AMIETE–2001, GBTU(SUM)–2010)

Solution. The C.F. of the given equation is

$$y = C_1e^x + C_2e^{-x} \quad ...(1)$$

Now, if A and B are the functions of x, therefore

$$\frac{dy}{dx} = Ae^x - Be^{-x} + \frac{dA}{dx}e^x + \frac{dB}{dx}e^{-x}.$$

Now choosing A and B such that

$$e^x\frac{dA}{dx} + e^{-x}\frac{dB}{dx} = 0 \quad ...(2)$$

$$\Rightarrow \frac{dy}{dx} = Ae^x - Be^{-x}$$

$$\Rightarrow \frac{d^2y}{dx^2} = \frac{dA}{dx}.e^x - \frac{dB}{dx}e^{-x} + Ae^x + Be^{-x}.$$

Putting all these values in the given equation, we get

$$e^x\frac{dA}{dx} - e^{-x}\frac{dB}{dx} = \frac{2}{1+e^x}. \quad ...(3)$$

Solving (2) and (3), we get

$$2e^x\frac{dA}{dx} = \frac{2}{1+e^x} \Rightarrow \frac{dA}{dx} = \frac{e^{-x}}{1+e^x}$$

$$\Rightarrow A = \int\frac{e^{-x}dx}{1+e^x} = \int\frac{dz}{z^2(1+z)}$$

(By putting $e^x = z$ and $e^x\,dx = dz$)

$$= \int\left(\frac{1}{z^2} - \frac{1}{z} + \frac{1}{1+z}\right)dz$$

$$= -\frac{1}{z} - \log z + \log(1+z) + C_1$$

$$= \log\frac{(1+e^x)}{e^x} - e^{-x} - C_1.$$

Similarly, we can find

$$B = -\log(1+e^x) + C_2.$$

Hence, the complete solution of the given equation is

$$y = C_1e^x + C_2e^{-x} + e^x\log\frac{1+e^x}{e^x} - 1 - e^{-x}\log(1+e^x).$$

Example 5. *By the method of variation of parameters, solve the differential equation*

$$\frac{d^2y}{dx^2} + (1-\cot x)\frac{dy}{dx} - y\cot x = \sin^2 x.$$

(UPTU–2002, Special Exam.–2001)

Solution. Compare the given equation to standard form

$$\frac{d^2y}{dx^2} + P\frac{dy}{dx} + Qy = 0$$

Here, $P = 1-\cot x,\ Q = -\cot x$

$\therefore\ 1-P+Q = 1-(1-\cot x)-\cot x = 0$,

$\therefore\quad y = e^{-x}$ is a part of C.F.

Putting all these above values in given equation, we get

$$\frac{d^2v}{dx^2} - (1+\cot x)\frac{dv}{dx} = 0$$

or $\frac{dP}{dx} - (1+\cot x)P = 0$, where $p = \frac{dv}{dx}$

or $\frac{dP}{P} = (1+\cot x)dx$

On integrating, we get

$$\log P = x + \log\sin x + \log C_1$$

$$\Rightarrow \log\left(\frac{P}{C_1\sin x}\right) = x$$

$$\Rightarrow \frac{P}{C_1\sin x} = e^x \Rightarrow P = C_1e^x\sin x$$

$$\Rightarrow \quad P = \frac{dv}{dx} = C_1 e^x \sin x \Rightarrow v = C_1 \int e^x \sin x$$

$$\Rightarrow \quad v = C_1 \frac{1}{2} e^x (\sin x - \cos x) + C_2$$

Hence, $y = v e^{-x}$

$$\therefore \quad y = C_1 . \frac{1}{2} (\sin x - \cos x) + C_2 e^{-x}$$

Let $y = A(\sin x - \cos x) + Be^{-t}$ be the complete solution of the given equation, where A and B are functions of x.

Now, $\dfrac{dy}{dx} = A(\cos x + \sin x) - Be^{-x}$

$$+ \frac{dA}{dx}(\sin x - \cos x) + \frac{dB}{dx} e^x$$

Let us suppose A and B such that

$$\frac{dA}{dx}(\sin x - \cos x) + \frac{dB}{dx} e^{-x} = 0 \qquad \ldots(3)$$

$$\frac{dy}{dx} = A(\cos x + \sin x) - Be^{-x}$$

and $\dfrac{d^2y}{dx^2} = \dfrac{dA}{dx}(\cos x + \sin x) - \dfrac{dB}{dx} e^{-x}$

$$+ A(-\sin x + \cos x) + Be^{-x}$$

Putting these above values in the given equation, we get

$$\frac{dA}{dx}(\cos x + \sin x) - \frac{dB}{dx} e^{-x} = \sin^2 x \qquad \ldots(4)$$

From (3) and (4), we get

$$\frac{dA}{dx} = \frac{1}{2} \sin x$$

and $\dfrac{dB}{dx} = \dfrac{1}{2} e^x (\sin x \cos x - \sin^2 x)$

$$= \frac{e^x}{4} (\sin 2x + \cos 2x - 1)$$

On integrating, we get

$$A = -\frac{1}{2} \cos x + C_1$$

and $B = \dfrac{1}{4} \int e^x (\sin 2x - 1 + \cos 2x) dx + C_2$

$$= \frac{1}{4} \int e^x \sin 2x dx - \frac{1}{4} \int e^x dx$$

$$+ \frac{1}{4} \int e^x \cos 2x + C_2$$

$$= \frac{1}{4} . \frac{e^x}{5} (\sin 2x - 2\cos 2x) - \frac{e^x}{4}$$

$$+ \frac{1}{4} . \frac{e^x}{5} (\cos 2x + 2\sin 2x) + C_2$$

$$= \frac{e^x}{20} (3\sin 2x - \cos 2x) - \frac{e^{-x}}{4} + C_2$$

Putting the values of A and B in (2), the general solution is

$$y = \left(-\frac{1}{2} \cos x + C_1\right) (\sin x - \cos x)$$

$$+ \left\{\frac{e^x}{20} (3\sin 2x - \cos 2x) - \frac{e^x}{2} + C_2\right\} e^{-x}$$

$$\therefore \quad y = C_1 (\sin x - \cos x) + C_2 e^{-x}$$

$$- \frac{1}{10} (\sin 2x - 2\cos 2x).$$

Example 6. *Solve by method of variation of parameters*

$$\frac{d^2y}{dx^2} + 2\frac{dy}{dx} + y = e^{-x} \log x. \qquad \text{(UPTU–2008)}$$

Solution. We have $\dfrac{d^2y}{dx^2} + 2\dfrac{dy}{dx} + y = e^{-x} \log x$

A.E. is $m^2 + 2m + 1 = 0$

$$\Rightarrow \quad (m+1)^2 = 0 \Rightarrow m = -1, -1.$$

C.F. $= C_1 e^{-x} + C_2 x e^{-x} = C_1 y_1 + C_2 y_2$,

where $y_1 = e^{-x}$ and $y_2 = xe^{-x}$

P.I. $= uy_1 + vy_2 \Rightarrow$ P.I. $= ue^{-x} + vxe^{-x}$.

where, $u = -\int \dfrac{y_2 X}{y_1 y_2' - y_1' y_2} dx$

$$= -\int \frac{xe^{-x} (e^{-x} \log x)}{e^{-x}(e^{-x} - xe^{-x}) - (e^{-x})(xe^{-x})} dx$$

$$= -\int \frac{x\, e^{-2x} \log x}{e^{-2x} - xe^{-2x} + xe^{-2x}} dx$$

$$= -\int \frac{e^{-2x} x \log x}{e^{-2x}} dx$$

$$= -\int x \log x \, dx$$

$$= -\left[\log x \frac{x^2}{2} - \int \left(\frac{1}{x}\right)\left(\frac{x^2}{2}\right) dx\right]$$

$$= -\frac{x^2}{2} \log x + \int \frac{x}{2} dx$$

$$= -\frac{x^2}{2} \log x + \frac{x^2}{4}.$$

$$v = \int \frac{y_1 X}{y_1 y_2' - y_1' y_2} dx$$

$$= \int \frac{e^{-x} e^{-x} \log x}{e^{-x}(e^{-x} - xe^{-x}) - (e^{-x})(xe^{-x})} dx$$

$$= \int \frac{e^{-2x} \log x}{e^{-2x}} dx$$

$$= \int \log x \, dx = (\log x) x - \int \frac{1}{x} . x \, dx$$

$$= x \log x - \int dx = x \log x - x.$$

P.I. $= uy_1 + vy_2$

$$= \left(-\frac{x^2}{2} \log x + \frac{x^2}{4}\right) e^{-x} + (x \log x - x) xe^{-x}$$

$$= xe^{-x} \left[\left(-\frac{x}{2} \log x + \frac{x}{4}\right) + x \log x - x\right]$$

$$= xe^{-x} \left[\frac{x}{2} \log x - \frac{3x}{4}\right]$$

Complete solution is y = C.F. + P.I.

$$y = C_1 e^{-x} + C_2 x e^{-x} + x e^{-x}\left[\frac{x}{2}\log x - \frac{3x}{4}\right].$$

Example 7. *Using variation of parameters method, solve*

$$x^2\frac{d^2y}{dx^2} + 2x\frac{dy}{dx} - 12y = x^3\log x. \quad \text{(UPTU–2004)}$$

Solution. The given equation is

$$x^2\frac{d^2y}{dx^2} + 2x\frac{dy}{dx} - 12y = x^3\log x \qquad ...(1)$$

Put $x = e^z$, so that $z = \log x$ and let $D \equiv \frac{d}{dz}$, then the given equation (1) reduces to

$$[D(D-1)+2D-12]\,y = ze^{3z}$$

$$\Rightarrow \qquad (D^2 + D - 12)y = z\,e^{3z}.$$

Auxiliary equation is

$$m^2 + m - 12 = 0 \Rightarrow m = 3, -4$$

$$\text{C.F.} = C_1e^{3z} + C_2e^{-4z} = C_1x^3 + C_2x^{-4}.$$

Hence, parts of C.F. are x^3 and x^{-4}

Let, $u = x^3$ and $v = x^{-4}$. Also, $R = x\log x$.

Let $y = Au + Bv$ be the complete solution, where A and B are some suitable functions of x. A and B are determined as follows :

$$A = -\int\frac{Rv}{uv_1 - u_1v}dx + C_1$$

$$= -\int\frac{x\log x . x^{-4}}{x^3(-4x^{-5}) - 3x^2(x^{-4})}dx + C_1$$

$$= -\int\frac{x^{-3}\log x}{-7x^{-2}}dx + C_1$$

$$= \frac{1}{7}\int\frac{\log x}{x}dx + C_1$$

$$= \frac{1}{7}\frac{(\log x)^2}{2} + C_1 = \frac{1}{14}(\log x)^2 + C_1$$

$$B = \int\frac{Ru}{uv_1 - u_1v}dx + C_2$$

$$= \int\frac{x\log x . x^3}{-7x^{-2}}dx + C_2$$

$$= -\frac{1}{7}\int x^6\log x\,dx + C_2$$

$$= -\frac{1}{7}\left[\log x . \frac{x^7}{7} - \int\frac{1}{x}.\frac{x^7}{7}dx\right] + C_2$$

$$= -\frac{1}{7}\left[\frac{x^7\log x}{7} - \frac{1}{7}\left(\frac{x^7}{7}\right)\right] + C_2$$

$$= \frac{x^7}{49}\left(\frac{1}{7} - \log x\right) + C_2$$

Hence, the complete solution is given by

$$y = Ax^3 + Bx^{-4}$$

$$= \left[\frac{1}{14}(\log x)^2 + C_1\right]x^3 + \left[\frac{x^7}{49}\left(\frac{1}{7} - \log x\right) + C_2\right]x^{-4}$$

$$= C_1x^3 + C_2x^{-4} + \frac{x^3}{98}\log x(7\log x - 2) + \frac{x^3}{343}$$

$$y = \left(C_1 + \frac{1}{343}\right)x^3 + C_2x^{-4} + \frac{x^3}{98}\log x(7\log x - 2)$$

Example 8. *Solve by method of variation of parameters*

$$(D^2 - 1)y = 2(1 - e^{-2x})^{-1/2}.$$

(MTU–2011, UPTU(SUM)–2009)

Solution. Proceeding same as above, we get

Parts of C.F. are $u = e^x$ and $v = e^{-x}$

$\therefore \quad y = A\,e^x + B\,e^{-x}$

where A and B are determined as follows :

$$A = -\int\frac{Rv}{uv_1 - u_1v}dx + C_1$$

$$= -\int\frac{2(1-e^{-2x})^{-1/2}.e^{-x}}{e^x(-e^{-x}) - e^x.e^{-x}}dx + C_1$$

$$= -2\int\frac{e^{-x}}{-2\sqrt{1-e^{-2x}}}dx + C_1$$

$$= \int\frac{e^{-x}}{\sqrt{1-e^{-2x}}}dx + C_1$$

Put $e^{-x} = t \therefore e^{-x}dx = -dt$

$$= -\int\frac{dt}{\sqrt{1-t^2}} + C_1$$

$$\Rightarrow \quad A = -\sin^{-1}(e^{-x}) + C_1$$

and $B = \int\frac{Ru}{uv_1 - u_1v}dx + C_2$

$$= -\int\frac{2(1-e^{-2x})^{-1/2}.e^{-x}}{(-2)}dx + C_2$$

$$= -\int\frac{e^{-x}}{\sqrt{1-e^{-2x}}}dx + C_2$$

$$= -\int\frac{e^{2x}}{\sqrt{e^{2x}-1}}dx + C_2$$

Put $e^{2x} = t \therefore e^{2x}dx = dt/2$

$$= -\frac{1}{2}\int\frac{dt}{\sqrt{t-1}} + C_2 = -\frac{1}{2}\frac{(t-1)^{1/2}}{\left(\frac{1}{2}\right)} + C_2$$

$$\Rightarrow \quad B = -(e^{2x}-1)^{1/2} + C_2$$

Hence, the complete solution is

$$y = A\,e^x + B\,e^{-x}$$
$$y = [-\sin^{-1}(e^{-x}) + C_1]e^x + [-(e^{2x}-1)^{1/2} + C_2]e^{-x}$$

$$\Rightarrow \quad y = C_1e^x + C_2e^{-x} - e^x \sin^{-1}(e^{-x}) - e^{-x}(e^{2x}-1)^{1/2}.$$

EXERCISE 32.4

Apply method of variation of parameters solve the following equations :

1. $x^2\dfrac{d^2y}{dx^2} - 2x(1+x)\dfrac{dy}{dx} + 2(x+1)y = x^3$
2. $\dfrac{d^2y}{dx^2} + a^2y = \operatorname{cosec} ax$
3. $(1-x)\dfrac{d^2y}{dx^2} + x\dfrac{dy}{dx} - y = (1-x)^2$
4. $(x+2)\dfrac{d^2y}{dx^2} - (2x+5)\dfrac{dy}{dx} + 2y = (x+1)e^x$
5. $\dfrac{d^2y}{dx^2} + y = \operatorname{cosec} x$
6. $\dfrac{d^2y}{dx^2} + 9y = \sec 3x$
7. $\dfrac{d^2y}{dx^2} + (\tan x - 3\cos x)\dfrac{dy}{dx} + 2y\cos^2 x = \cos^4 x$
8. $\dfrac{d^3y}{dx^3} - 6\dfrac{d^2y}{dx^2} + 11\dfrac{dy}{dx} - 6y = e^{2x}$
9. If $y = x, y = x^2 - 1$ are linearly independent solution of
$$(x^2+1)\frac{d^2y}{dx^2} - 2x\frac{dy}{dx} + 2y = 0$$
find the general solution of
$$(x^2+1)\frac{d^2y}{dx^2} - 2x\frac{dy}{dx} + 2y = 6(x^2+1)^2.$$
10. $(1-x^2)\dfrac{d^2y}{dx^2} + x\dfrac{dy}{dx} - y = x(1-x^2)^{3/2}$ (UPTU(Q.Bank)–2001)
11. $\dfrac{d^2y}{dx^2} - 2\dfrac{dy}{dx} = e^x \sin x$ (UPTU(Q.Bank)–2001)
12. $\dfrac{d^2y}{dx^2} - 3\dfrac{dy}{dx} + 2y = \dfrac{e^x}{1+e^x}$ (MTU(SUM)–2011)
13. $\dfrac{d^2y}{dx^2} - 3\dfrac{dy}{dx} + 2y = \sin e^{-x}$ (MTU–2012)
14. $x^2\dfrac{d^2y}{dx^2} + 4x\dfrac{dy}{dx} + 2y = e^x$ (GBTU–2012)
15. $x^2\dfrac{d^2y}{dx^2} + x\dfrac{dy}{dx} - 9y = 48x^5$ (GBTU(CO)–2010)

Hint to Selected Problems

1. We have $P + Q.x = 0 \quad \Rightarrow \quad y = x$ is a part of C.F.
On solving by usual procedure, we get $y = Ax + Bxe^{2x}$ is the general solution. Now, by letting A and B also the functions of x, use the technique of variation of parameters.
2. C.F. of the equation is given by $y = A\cos ax + B\sin ax$,
where A and B both are functions of x. Now use method of variation of parameters.
4. Here, $y = e^{2x}$ is a part of C.F. Putting $y = ve^{2x}$ and on solving, we get $y = A(2x+5) + Be^{2x}$ is the solution.
Now by letting A and B, both functions of x, use the method of variation of parameters.
9. It is given that $y = x$ and $y = x^2 - 1$ are two linearly independent solutions of the given differential equation.
Let $y = A(x^2-1) + Bx$ be the general solution of given equation. Then use method of variation of parameters by assuming A and B both are the functions of x.

ANSWERS

1. $y = C_1x + C_2xe^{2x} - \dfrac{x^2}{2} - \dfrac{x}{4}$
2. $y = C_1\cos ax + C_2\sin ax - \dfrac{x}{a}\cos ax + \dfrac{1}{a^2}\sin ax.\log\sin ax$
3. $y = C_1e^x + C_2x + x + 1 + x^2$
4. $y = C_1(2x+5) + C_2e^{2x} - e^x$
5. $y = C_1\cos x + C_2\sin x - x\cos x + \sin x\log\sin x$
6. $y = C_1\cos 3x + C_2\sin 3x + \dfrac{1}{9}\cos 3x\log\cos 3x + \dfrac{x}{3}\sin 3x$
7. $y = C_1e^{\sin x} + C_2e^{2\sin x} - \dfrac{5}{4} - \dfrac{3}{2}\sin x - \dfrac{1}{2}\sin^2 x$
8. $y = C_1e^x + (C_2 - x)e^{2x} + C_3e^{3x}$
9. $y = C_1(x^2-1) + C_2x + x^4 + 3x^2$
10. $y = C_1(\sqrt{1-x^2} + x\sin^{-1}x) + C_2x - \dfrac{x}{9}(1-x^3)^{3/2}$
11. $y = C_1 + C_2e^{2x} - \dfrac{e^x}{2}\sin x$
12. $y = [\log(e^{-x}+1) + C_1]e^x + [\log(1+e^{-x}) - (1+e^{-x}) + C_2]e^{2x}$
13. $y = C_1e^x + C_2e^{2x} - e^{2x}\sin e^{-x}$
14. $y = (e^x + C_1)\dfrac{1}{x} + [(1-x)e^x + C_2]\dfrac{1}{x^2}$
15. $y = (4x^2 + C_1)x^3 + (C_2 - x^8)x^{-3}$

Objective Evaluations

Fill in the Blanks

1. For a standard linear equation of second order $y = e^x$ is a part of solution if ________.
2. For a standard linear equation of second order $y = x^2$ is a part of solution if ________.
3. The normal for of the equation does not contain the terms of ________.
4. To remove the term of first derivative, we obtain ________ form.
5. A method in which the solution is obtained by varying the arbitrary constants of the complimentary function is known as of ________.

True/False

Write 'T' for True and 'F' for False statement.

1. The equation $\frac{d^2y}{dx^2} + y = \text{cosec } x$ is a linear equation of second order. **(T/F)**
2. The normal form of the given equation of second order must contain the term of first derivative. **(T/F)**
3. The normal form of the given equation of second order does not contain the term of second derivative. **(T/F)**
4. In the method of variation of parameter, we take A and B are the functions of x. **(T/F)**
5. In the method of variation of parameter, there is a need of complete knowledge of complimentary function. **(T/F)**

Multiple Choice Questions

Choose the most appropriate one.

1. If $1 + P + Q = 0$, then the one solution of the linear equation of second order is :
   (a) $y = e^x$ (b) $y = e^{2x}$
   (c) $y = ae^x$ (d) $y = x$
2. If $P + Qx = 0$, then the part of a complimentary function is :
   (a) $y = x$ (b) $y = x^2$
   (c) $y = x^3$ (d) $y = x^4$
3. In the method of variation of parameter, we choose A and B of :
   (a) functions of x (b) functions of y
   (c) constants (d) none of these
4. $y = x^2$ is a part of C.F. if :
   (a) $1 + P + Q = 0$ (b) $2 + 2Px + Qx^2 = 0$
   (c) $1 - P + Q = 0$ (d) $P + Qx = 0$
5. If $y = e^{mx}$ is a solution of linear equation of second order then :
   (a) $m^2 + Pm + Q = 0$ (b) $m^2 + Pm + Q = 1$
   (c) $m^2 + Pm + Q \neq 0$ (d) none of these

ANSWERS

Fill in the Blanks

1. $1 + P + Q = 0$	2. $2 + 2Px + Qx^2 = 0$	3. first derivative	4. normal	5. variation of parameters

True/False

1. T	2. F	3. F	4. T	5. T

Multiple Choice Questions

1. (a)	2. (a)	3. (a)	4. (b)	5. (a)

FFFFF

CHAPTER 33 Applications of Ordinary Differential Equation to Engineering Problems

33.1 INTRODUCTION

Differential equations are widely used in solving engineering problems. The physical principles forming the background of the problems are expressed mathematically by the formulation of one or more differential equations. Here, we study the applications concerned with the motion of a particle in a resistance medium, SHM, bending of beams, electric circuits, etc.

33.2 APPLICATION TO ELECTRIC CIRCUITS

If q be the electrical charge on a conductor of capacity C and i be the current, then

(a) $i = \frac{dq}{dt}$ or $q = \int i\, dt$

(b) the potential drop across the resistance R is Ri .

(c) the potential drop across the inductance L is $L\frac{di}{dt}$.

(d) the potential drop across the capacitance C is $\frac{q}{C}$.

33.2.1 KIRCHOFF'S LAW

1. **Voltage Law :** The algebraic sum of the voltage drop around any closed circuit is equal to the resultant electromotive force in the circuit.
2. **Current Law :** At a junction or node, current coming is equal to the current going.

(i) L-C Electrical Circuit (without e.m.f): Let q be the electrical charge on the condenser plate and i be the current in an electrical circuit containing an inductance L and capacitance C. Then

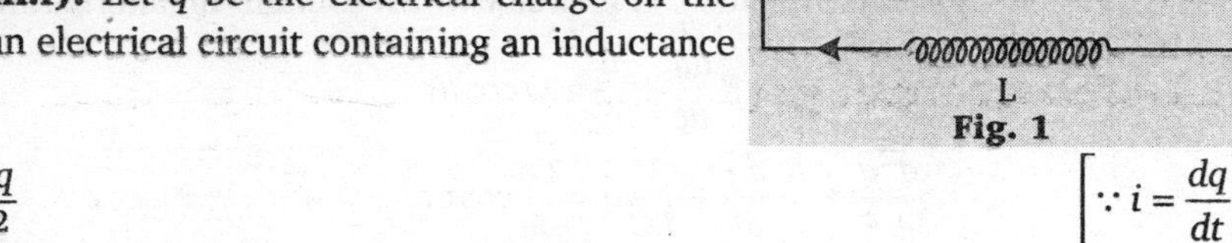

Fig. 1

(a) The voltage drop across L is

$$L\frac{di}{dt} = L\frac{d^2q}{dt^2} \qquad \left[\because i = \frac{dq}{dt}\right]$$

(b) The voltage drop across C is $\frac{q}{C}$.

By Kirchoff's law, we have $L\frac{d^2q}{dt^2} + \frac{q}{C} = 0$ [$\because$ No e.m.f. is applied.]

$$\frac{d^2q}{dt^2} + \frac{1}{LC}q = 0$$

$$\frac{d^2q}{dt^2} + \omega^2 q = 0 \qquad \left[\because \omega^2 = \frac{1}{LC}\right]$$

which is an equation of S.H.M. Thus it represents free electrical oscillations of current having period

$$T = \frac{2\pi}{\omega} = 2\pi\sqrt{LC}$$

Hence, the discharging of a condenser through an inductance L is the same as the motion of the mass m attached at the end of a spring.

(ii) L-R Series Circuit : Let i be the current flowing in the circuit containing resistance R and inductance L in series, with voltage source E at any time t.

By voltage law, $$Ri + L\frac{di}{dt} = E$$

$$\frac{di}{dt} + \frac{R}{L}i = \frac{E}{L} \qquad \text{... (1)}$$

This is the linear differential equation

$$\text{I.F.} = e^{\int \frac{R}{L}dt} = e^{\frac{R}{L}t}$$

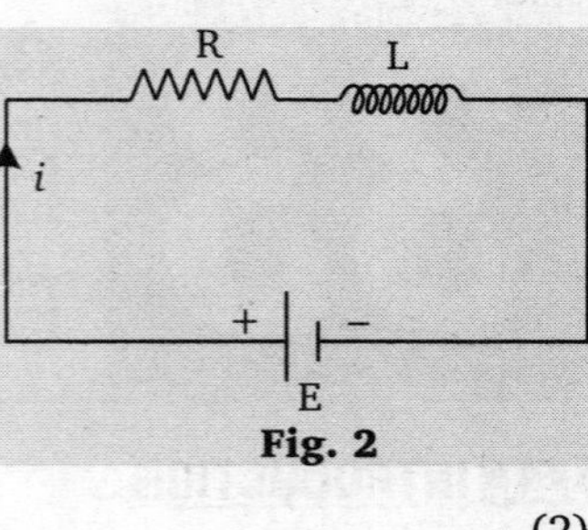

Fig. 2

Its solution is given by $$i\,e^{\frac{R}{L}t} = \int \frac{E}{L}e^{\frac{R}{L}t}\,dt + C$$

$$\text{i.e.}\; e^{\frac{R}{L}t} = \frac{E}{L} \times \frac{L}{R}e^{\frac{R}{L}t} + C$$

$$i = \frac{E}{R} + Ce^{-\frac{R}{L}t} + C \qquad \text{... (2)}$$

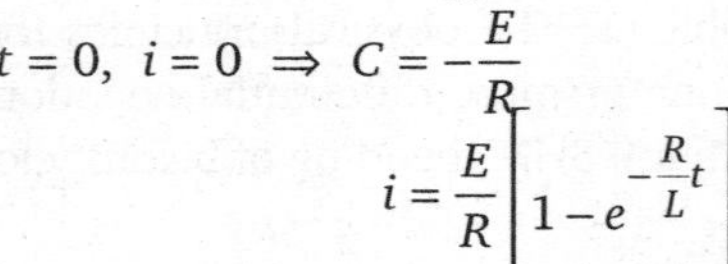

At $t = 0,\ i = 0 \Rightarrow C = -\frac{E}{R}$

From (2), $$i = \frac{E}{R}\left[1 - e^{-\frac{R}{L}t}\right]$$

(iii) L-C Electrical Circuit with e.m.f. $Q \cos nt$: The equation for an L-C circuit with e.m.f. $Q \cos nt$ is

$$L\frac{d^2q}{dt^2} + \frac{q}{C} = Q\cos nt \;\Rightarrow\; \frac{d^2q}{dt^2} + \frac{1}{LC}q = \frac{Q}{L}\cos nt$$

$$\frac{d^2q}{dt^2} + \omega^2 q = E\cos nt\text{, where, } \omega^2 = \frac{1}{LC} \text{ and } E = \frac{Q}{L}.$$

REMARK

- Practically we see that we tune in a radio station. By changing C, the natural frequency of the tuning L-C circuit is made equal to the frequency of the desired receiving radio station, and so the output of the receiver becomes large at the desired receiving station.

(iv) L-C-R Electrical Circuit with e.m.f. $Q \cos nt$:

The differential equation of the L-C-R circuit which contains an alternating e.m.f. $Q \cos nt$ is

$$L\frac{d^2q}{dt^2} + R\frac{dq}{dt} + \frac{q}{C} = Q\cos nt$$

$$\frac{d^2q}{dt^2} + \frac{R}{L}\frac{dq}{dt} + \frac{1}{LC}q = \frac{Q}{L}\cos nt$$

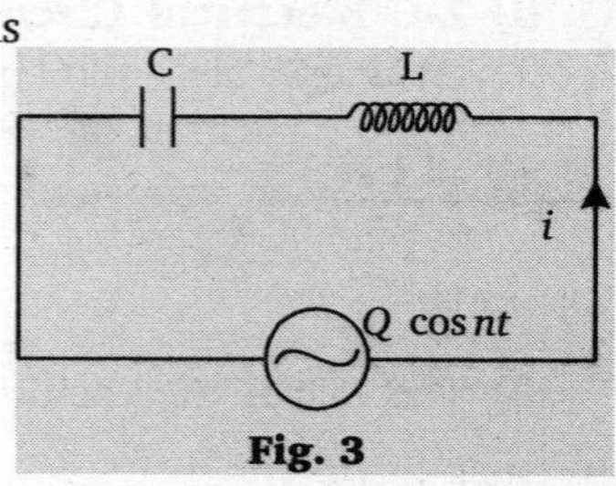

Fig. 3

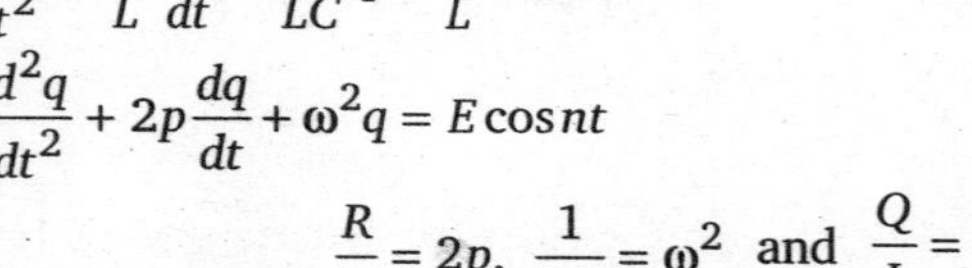

$$\frac{d^2q}{dt^2} + 2p\frac{dq}{dt} + \omega^2 q = E\cos nt$$

where, $$\frac{R}{L} = 2p,\ \frac{1}{LC} = \omega^2 \text{ and } \frac{Q}{L} = E.$$

Solved Examples

Example 1. *A condenser of capacity C Farades with V_0 is discharged through a resistance R Ohms. Show that if q Coulomb is the charge on the condenser, i Ampere the current and V the voltage at time t.*

$$q = CV,\ V = Ri \text{ and } i = \frac{dq}{dt}$$

Hence show that $V = V_0\, e^{-\frac{t}{RC}}$.

Solution. Voltage across resistance $R = Ri$.

Voltage drop across capacitance $= \frac{q}{C}$.

The equation of discharge of condenser can be written, when after release of key the condenser gets discharged and at that time, voltage across the battery gets zero so that $V_0 = 0$.

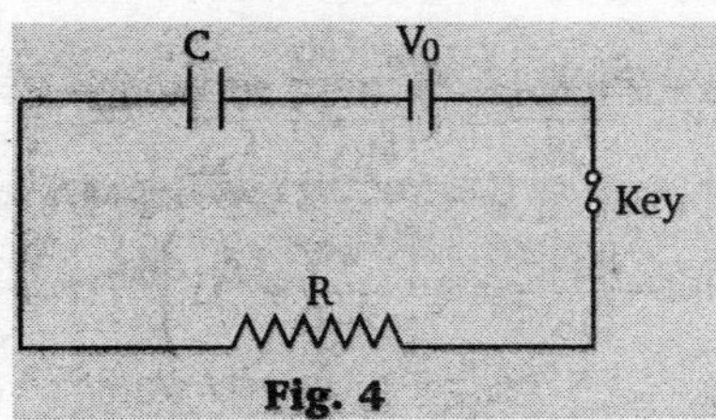

Fig. 4

The differential equation of the above circuit is

$$Ri + \frac{q}{C} = 0$$

$$R\frac{dq}{dt} + \frac{q}{C} = 0$$

$$\frac{dq}{dt} + \frac{q}{RC} = 0$$

or
$$\frac{dq}{dt} = -\frac{q}{RC}.$$

$$\frac{dq}{q} = -\frac{1}{RC}dt$$

Integrating both sides, we get

$$\int \frac{dq}{q} = -\frac{1}{RC}\int dt$$

$$\log q = -\frac{1}{RC}t + a \qquad \text{... (1)}$$

At $t = 0$, the change at the condenser is q_0 such that

$$\log q_0 = -\frac{1}{RC}(0) + a \qquad \text{... (2)}$$

$$\Rightarrow \quad a = \log q_0$$

Now, from (1), we have

$$\log q = -\frac{1}{RC}t + \log q_0$$

$$\log q - \log q_0 = -\frac{1}{RC}t$$

$$\Rightarrow \quad \log\frac{q}{q_0} = -\frac{1}{RC}t \quad \Rightarrow \quad \frac{q}{q_0} = e^{-t/RC}$$

$$q = q_0\, e^{-t/RC} \qquad \text{... (3)}$$

Dividing both sides of (3) by C, we get

$$\Rightarrow \quad \frac{q}{C} = \frac{q_0}{C}\, e^{-t/RC}$$

Hence, $\quad V = V_0\, e^{-t/RC}$

Example 2. *Show that the frequency of free vibrations in a closed electrical circuit with inductance L and capacity C in series is* $\dfrac{30}{\pi\sqrt{LC}}$ *per minute.*

[UPTU(Q. Bank)–2002]

Solution. Let i be the current and q be the charge in the condenser plate at any time t.

The voltage drop across $L = L\dfrac{di}{dt} = L\dfrac{d^2q}{dt^2}$

The voltage drop across $C = \dfrac{q}{C}$.

Since, there is no applied e.m.f. in the circuit, we have by Kirchoff's law :

$$L\frac{d^2q}{dt^2} + \frac{q}{C} = 0 \text{ or } \frac{d^2q}{dt^2} + \frac{q}{CL} = 0.$$

$$\Rightarrow \quad \frac{d^2q}{dt^2} = -\frac{1}{LC}q$$

Writing $\dfrac{1}{LC} = \omega^2$, it becomes $\dfrac{d^2q}{dt^2} = -\omega^2 q$

It represents oscillatory current with period

$$\frac{2\pi}{\omega} = 2\pi\sqrt{LC}.$$

$$\text{Frequency} = \frac{1}{T} \text{ per second}$$

$$= \frac{60}{2\pi\sqrt{LC}} \text{ per minute}$$

$$= \frac{30}{\pi\sqrt{LC}} \text{ per minute.}$$

Example 3. *A condenser of capacity C is discharged through the inductance L and a resistance R in series and the charge q at any time t satisfies the equation*

$$L\frac{d^2q}{dt^2} + R\frac{dq}{dt} + \frac{q}{C} = 0.$$

Given that L=0.25 Henry, R = 250 Ohms, $C = 2\times10^{-6}$ Farad and that when t = 0, the charge q is 0.02 Coulomb and the current $\dfrac{dq}{dt} = 0$. Obtain the value of q in terms of t.

Solution. The given equation of charge is

$$L\frac{d^2q}{dt^2} + R\frac{dq}{dt} + \frac{q}{C} = 0 \qquad \text{... (1)}$$

Substituting the given value in (1), we get

$$\frac{d^2q}{dt^2} + \frac{250}{0.25}\frac{dq}{dt} + \frac{q}{2\times10^{-6}\times 0.25} = 0 \qquad \text{... (2)}$$

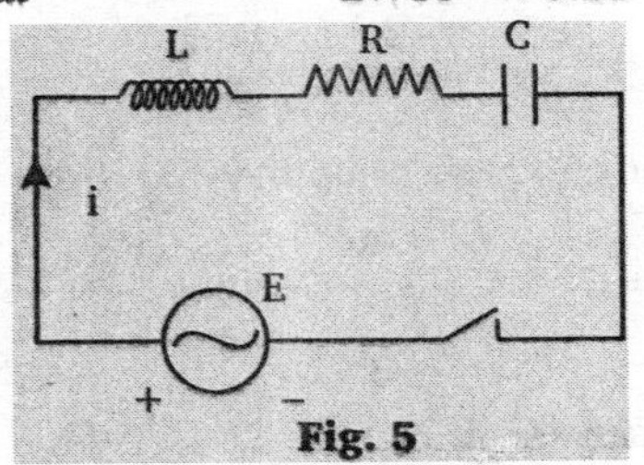

Fig. 5

$$\frac{d^2q}{dt^2} + 1000\frac{dq}{dt} + 2\times10^6 q = 0$$

A.E. is $\quad m^2 + 1000m + 2\times10^6 = 0$

$$\Rightarrow \quad m = -500 \pm 1323i$$

The solution is

$$q = e^{-500t}\,[A\cos 1323t + B\sin 1323t] \qquad \text{... (3)}$$

Putting $t = 0,\ q = 0.002$ in equation (3), we get $A = 0.002$.

From (3), we have

$$\frac{dq}{dt} = -500\, e^{-500t}\,[A\cos 1323t + B\sin 1323t]$$
$$+1323e^{-500t}\,[-A\sin 1323t + B\cos 1323t] \quad \ldots (4)$$

Putting $t = 0,\ \frac{dq}{dt} = 0$ in (4), we get $B = 0.0008$.

Hence, the required value of q is

$$q = e^{-0.500t}[0.002\cos 1323t + 0.0003\sin 1323t]$$

Example 4. *An uncharged condenser of capacity C is charged by applying an e.m.f., $E\sin\frac{t}{\sqrt{LC}}$ through leads of self-inductance L and negligible resistance. Prove that at time t, the charge on one of the plate is*

$$\frac{EC}{2}\left[\sin\frac{t}{\sqrt{LC}} - \frac{t}{\sqrt{LC}}\cos\frac{t}{\sqrt{LC}}\right]$$

(UPTU–2003)

Solution. Let q be the charge on the condenser at any time t.

The differential equation for the circuit is

$$L\frac{d^2q}{dt^2} + \frac{q}{C} = E\sin\frac{t}{\sqrt{LC}} \quad \ldots (1)$$

It's A.E. is $Lm^2 + \frac{1}{C} = 0$ or $m^2 = -\frac{1}{LC}$

so that $m = \pm\frac{i}{\sqrt{LC}}$

$$\text{C.F.} = C_1\cos\frac{t}{\sqrt{LC}} + C_2\sin\frac{t}{\sqrt{LC}}$$

$$\text{P.I.} = \frac{1}{LD^2 + \frac{1}{C}}E\sin\frac{t}{\sqrt{LC}} = Et\,.\,\frac{1}{2LD}\sin\frac{t}{\sqrt{LC}}$$

$$= \frac{Et}{2L}\left[-\sqrt{LC}\cos\frac{t}{\sqrt{LC}}\right] = -\frac{Et}{2}\sqrt{\frac{C}{L}}\cos\frac{t}{\sqrt{LC}}$$

The complete solution is

$$q = C_1\cos\frac{t}{\sqrt{LC}} + C_2\sin\frac{t}{\sqrt{LC}} - \frac{Et}{2}\sqrt{\frac{C}{L}}\cos\frac{t}{\sqrt{LC}} \quad \ldots (2)$$

When $t = 0,\ q = 0,\ \therefore C_1 = 0$.

Differentiating (2) w.r.t. t, we get

$$\frac{dq}{dt} = -\frac{q}{\sqrt{LC}}\sin\frac{t}{\sqrt{LC}} + \frac{C_2}{\sqrt{LC}}\cos\frac{t}{\sqrt{LC}}$$
$$-\frac{E}{2}\sqrt{\frac{C}{L}}\left[\cos\frac{t}{\sqrt{LC}} - \frac{t}{\sqrt{LC}}\sin\frac{t}{\sqrt{LC}}\right]$$

$\frac{dq}{dt} = i = 0$, when $t = 0$

$$\Rightarrow \frac{C_2}{\sqrt{LC}} - \frac{E}{2}\sqrt{\frac{C}{L}} = 0 \Rightarrow C_2 = \frac{EC}{2}$$

Now, from (2), we get

$$q = \frac{EC}{2}\sin\frac{t}{\sqrt{LC}} - \frac{Et}{2}\sqrt{\frac{C}{L}}\cos\frac{t}{\sqrt{LC}}$$

or $$q = \frac{EC}{2}\left[\sin\frac{t}{\sqrt{LC}} - \frac{t}{\sqrt{LC}}\cos\frac{t}{\sqrt{LC}}\right]$$

Example 5. *In an L-C-R circuit, the charge q on a plate of a condenser is given by $L\frac{d^2q}{dt^2} + R\frac{dq}{dt} + \frac{q}{C} = E\sin pt$. The circuit is tuned to resonance so that $p^2 = 1/LC$. If initially the current i and the charge q be zero, show that for small values of R/L, the current in the circuit at time t is given by*

$$\left\{\frac{Et}{2L}\right\}\sin pt$$

(UPTU–2004, 05)

Solution. The given equation is

$$\left(LD^2 + RD + \frac{1}{C}\right)q = E\sin pt;\quad D \equiv \frac{d}{dt} \quad \ldots (1)$$

A.E. is $Lm^2 + Rm + \frac{1}{C} = 0$

$$m = \frac{-R \pm \sqrt{R^2 - \frac{4L}{C}}}{2L} = -\frac{R}{2L} \pm \frac{1}{2}\sqrt{\frac{R^2}{L^2} - \frac{4}{LC}}$$
$$= -\frac{R}{2L} \pm \frac{1}{2}\sqrt{\frac{-4}{LC}}$$

$$m = -\frac{R}{2L} \pm \frac{i}{\sqrt{LC}} = -\frac{R}{2L} \pm pi$$

$$\text{C.F.} = e^{-\frac{Rt}{2L}}(C_1\cos pt + C_2\sin pt)$$
$$= \left(1 - \frac{Rt}{2L}\right)(C_1\cos pt + C_2\sin pt)$$

$$\text{P.I.} = \frac{1}{LD^2 + RD + \frac{1}{C}}(E\sin pt)$$
$$= E.\frac{1}{-Lp^2 + RD + \frac{1}{C}}\sin pt$$
$$= E\,.\,\frac{1}{RD}\sin pt = -\frac{E}{Rp}\cos pt$$

Complete solution is given by

$$q = \left(1 - \frac{Rt}{2L}\right)(C_1\cos pt + C_2\sin pt) - \frac{E}{Rp}\cos pt \quad \ldots (2)$$

$$i = \frac{dq}{dt} = \left(1 - \frac{Rt}{2L}\right)(-pC_1\sin pt + pC_2\cos pt)$$
$$-\frac{R}{2L}(C_1\cos pt + C_2\sin pt) + \frac{E}{R}\sin pt \quad \ldots (3)$$

When $t = 0$, $q = 0$ and $i = 0$, then

from (2), $0 = C_1 - \frac{E}{Rp} \Rightarrow C_1 = \frac{E}{Rp}$

from (3), $0 = C_2 p - \frac{R}{2L} C_1 \Rightarrow C_2 = \frac{E}{2Lp^2}$

Now, from (3),

$$i = \left(1 - \frac{Rt}{2L}\right)\left(-\frac{E}{R}\sin pt + \frac{E}{2Lp}\cos pt\right)$$

$$-\frac{R}{2L}\left(\frac{E}{Rp}\cos pt + \frac{E}{2Lp^2}\sin pt\right) + \frac{E}{R}\sin pt$$

$$= \frac{Et}{2L}\sin pt - \frac{ERt}{4pL^2}\cos pt - \frac{ER}{4L^2p^2}\sin pt$$

$$= \frac{Et}{2L}\sin pt; \quad \frac{R}{L} \text{ being small.}$$

Example 6. Solve the equation $L\frac{di}{dt} + Ri = E_0 \sin \omega t$, where L, R and E_0 are constants and discuss the case when t increases indefinitely.

Solution. $L\frac{di}{dt} + Ri = E_0 \sin \omega t$ or $\frac{di}{dt} + \frac{R}{L}i = \frac{E_0}{L}\sin \omega t$

$$\text{I.F.} = e^{\int \frac{R}{L}dt} = e^{\frac{R}{L}t}$$

Solution is given by

$$ie^{\frac{R}{L}t} = \frac{E_0}{L}\int e^{\frac{R}{L}t}\sin \omega t \, dt + A$$

$$i\,e^{\frac{R}{L}t} = E_0 \frac{e^{\frac{R}{L}t}}{\sqrt{\frac{R^2}{L^2} + \omega^2}} \sin\left(\omega t - \tan^{-1}\frac{L\omega}{R}\right) + A$$

$$i = \frac{E_0 L}{\sqrt{R^2 + L^2\omega^2}} \sin\left(\omega t - \tan^{-1}\frac{L\omega}{R}\right) + Ae^{-\frac{R}{L}t}$$

As t increases indefinitely, then $Ae^{-\frac{R}{L}t}$ tends to zero

$$i = \frac{E_0 L}{\sqrt{R^2 + L^2\omega^2}} \sin\left(\omega t - \tan^{-1}\frac{L\omega}{R}\right).$$

Example 7. *For an electric circuit constants L, R, C, the charge q on a plate condenser is given by*

$$L\frac{d^2q}{dt^2} + R\frac{dq}{dt} + \frac{q}{C} = E \text{ and the current by } i = \frac{dq}{dt}.$$

Let $L = 1$ Henry, $C = 10^{-4}$ Farad, $R = 100$ Ohms, $E = 100$ Volts.

Suppose that no charge is present and no current is flowing at time $t = 0$, when the e.m.f. is applied. Determine q and i at time t.

Solution. The differential equation is

$$L\frac{d^2q}{dt^2} + R\frac{dq}{dt} + \frac{q}{C} = E$$

or
$$\frac{d^2q}{dt^2} + \frac{R}{L}\frac{dq}{dt} + \frac{q}{LC} = \frac{E}{L}$$

Put $\frac{R}{L} = 2b$ and $\frac{1}{LC} = k^2$ in above equation, we have

$$\frac{d^2q}{dt^2} + 2b\frac{dq}{dt} + k^2q = \frac{E}{L} \quad \text{... (1)}$$

The solution is given by

$$q = \frac{E}{k^2L} + e^{-bt}\left[A\cos\sqrt{k^2-b^2}\,t + B\sin\sqrt{k^2-b^2}\,t\right] \quad \text{... (2)}$$

Putting $q = 0$ and $t = 0$, we get

$$0 = \frac{E}{k^2L} + A \text{ or } A = -\frac{E}{k^2L}.$$

Differentiating (2), we have

$$\frac{dq}{dt} = -be^{-bt}\left[A\cos\sqrt{k^2-b^2}\,t + B\sin\sqrt{k^2-b^2}\,t\right] \quad \text{... (3)}$$

$$+ e^{-bt}\left[A\sqrt{k^2-b^2}\sin\sqrt{k^2-b^2}\,t + B\sqrt{k^2-b^2}\cos\sqrt{k^2-b^2}\,t\right]$$

Putting $\frac{dq}{dt} = 0$ and $t = 0$ in (3), we have

$$0 = -bA + B\sqrt{k^2-b^2}$$

$$B = \frac{bA}{\sqrt{k^2-b^2}} = \frac{-\frac{bE}{k^2L}}{\sqrt{k^2-b^2}} = \frac{-bE}{k^2L\sqrt{k^2-b^2}}$$

Putting the values of A and B in equation (2), we get

$$q = \frac{E}{k^2L} + e^{-bt}\left[-\frac{E}{k^2L}\cos\sqrt{k^2-b^2}t - \frac{bE}{k^2L\sqrt{k^2-b^2}}\sin\sqrt{k^2-b^2}t\right]$$

$$q = \frac{E}{k^2L}\left[1 - e^{-bt}\left(\cos\sqrt{k^2-b^2}t + \frac{b}{\sqrt{k^2-b^2}}\sin\sqrt{k^2-b^2}t\right)\right] \quad \text{... (4)}$$

$$\frac{E}{k^2 L} = \frac{E}{\frac{1}{LC}.L} = EC = 100 \times 10^{-4} = \frac{1}{100}$$

$$b = \frac{R}{2L} = \frac{100}{2 \times 1} = 50$$

$$\sqrt{k^2 - b^2} = \sqrt{\frac{1}{LC} - (50)^2} = \sqrt{\frac{1}{10^{-4}} - (50)^2}$$

$$= \sqrt{7500} = 50\sqrt{3}$$

Now, from (4), we get

$$q = \frac{1}{100}\left[1 - e^{-50t}(\cos 50\sqrt{3}t + \frac{1}{\sqrt{3}}\sin 50\sqrt{3}t)\right].$$

Example 8. *The differential equation for a circuit in which self-inductance neutralize each other is*

$$L\frac{d^2 i}{dt^2} + \frac{i}{C} = 0$$

Find the current i as a function of t, given that I is the maximum current and $i = 0$ when $t = 0$.

Solution. The differential equation is

$$L\frac{d^2 i}{dt^2} + \frac{i}{C} = 0$$

or $$\frac{d^2 i}{dt^2} + \frac{i}{LC} = 0$$

A.E. is $$D^2 + \frac{1}{LC} = 0$$

$$D^2 = -\frac{1}{LC} = \frac{j^2}{LC} \text{ or } D = \pm\frac{j}{\sqrt{LC}},$$

where $j^2 = -1$

Solution is given by

$$i = C_1 \sin\frac{1}{\sqrt{LC}}t + C_2 \cos\frac{1}{\sqrt{LC}}t \quad \text{... (1)}$$

Putting $t = 0$, $i = 0$ in equation (1), we get $C_2 = 0$.

Thus, (1) reduces to $i = C_1 \sin\frac{t}{\sqrt{LC}}$ ... (2)

From (2), maximum value of current i is obtained by putting $\sin\frac{t}{\sqrt{LC}} = 1$.

The maximum current, $I = C_1$.

Required solution is $i = I\sin\frac{t}{\sqrt{LC}}$.

Example 9. *A 20 Ohms resistor is connected in series with a capacitor of 0.01 Farad and e.m.f. E Volts given by $40e^{-3t} + 20e^{-6t}$. If $q = 0$, at $t = 0$. Show that the maximum charge on the capacitor is 0.25 Coulomb.*

Solution. Equation of charge and discharge can be written as follows :

$$R\frac{dQ}{dt} + \frac{Q}{C} = 40e^{-3t} + 20e^{-6t}$$

$$\Rightarrow \frac{dQ}{dt} + \frac{Q}{RC} = \frac{40}{R}e^{-3t} + \frac{20}{R}e^{-6t}$$

$$\Rightarrow \frac{dQ}{dt} + \frac{Q}{RC} = \frac{40}{20}e^{-3t} + \frac{30}{20}e^{-6t}$$

$$\Rightarrow \frac{dQ}{dt} + \frac{Q}{20 \times 0.01} = 2e^{-3t} + e^{-6t}$$

$$\Rightarrow \frac{dQ}{dt} + 5Q = 2e^{-3t} + e^{-6t}$$

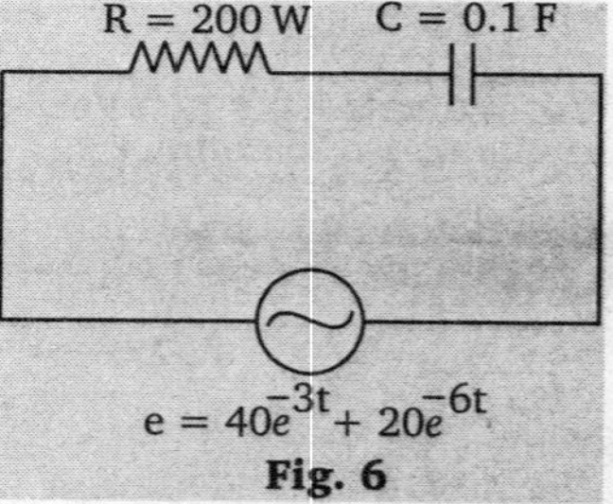

Fig. 6

$$\text{I.F.} = e^{5\int dt} = e^{5t}.$$

Its solution is

$$Qe^{5t} = \int e^{5t}(2e^{-3t} + e^{-6t})\,dt + a.$$

$$= \int (2e^{2t} + e^{-t})\,dt + a = e^{2t} - e^{-t} + a.$$

$$\Rightarrow Q = e^{-3t} - e^{-6t} + ae^{-5t} \quad \text{... (1)}$$

Putting $t = 0$, $Q = 0$ in (1), we get

$$0 = ae^{-3\times 0} + e^{-6\times 0} + ae^{-5\times 0}$$

$$\Rightarrow 0 = 1 - 1 + a \Rightarrow a = 0$$

Putting value of a in (1)

$$Q = e^{-3t} - e^{-6t}$$

For maximum value $\frac{dQ}{dt} = 0$.

$$-3e^{-3t} - (-6)e^{-6t} = 0 \text{ or } 3e^{-3t} = 6e^{-6t}.$$

$$e^{-3t} = 2e^{-6t} \text{ or } \frac{1}{2} = e^{-3t}$$

or $$2 = e^{3t}$$

$$\log 2 = 3t$$

$$t = \frac{1}{3}\log 2 = \log(2^{1/3})$$

Maximum charge (by putting the value of t)

$$Q = e^{-3t} - e^{-6t} = e^{-3(\log(2^{1/3}))} - e^{-6(\log 2^{1/3})}$$

$$= e^{\log 2^{(-3/3)}} - e^{\log 2^{(-6/2)}}$$

$$= e^{\log(1/2)} - e^{\log(1/4)} = \frac{1}{2} - \frac{1}{4} = \frac{2-1}{4}$$

$$= \frac{1}{4} = 0.25 \text{ Amp.}$$

Example 10. *The voltage V and the current i at a distance x from the sending end of the transmission line satisfy the equation :*

$$-\frac{dV}{dx} = Ri, \quad -\frac{di}{dx} = GV$$

where R and G are constants. If $X = V_0$ at the sending end ($x = 0$) and $V = 0$ at receiving end ($x = l$). Show that $V = V_0 \left\{ \frac{\sinh n(l-x)}{\sinh nl} \right\}$, *when* $n^2 = RG$.

Solution. We have $-\frac{dV}{dx} = Ri$ ... (1)

$$-\frac{di}{dx} = GV \quad \text{... (2)}$$

When $x = 0$, $V = V_0$, when $x = l$, $V = 0$

putting the value of i from (1) into (2), we get

$$-\frac{d}{dx}\left(-\frac{dV}{dx}.\frac{1}{R}\right) = GV$$

or
$$\frac{d^2V}{dx^2} = RGV$$

$$\frac{d^2V}{dx^2} - (RG)V = 0$$

$$(D^2 - RG)V = 0$$

A.E. $(D^2 - n^2) = 0 \Rightarrow D = \pm n$

$$V = A\,e^{+nx} + Be^{-nx} \quad \text{... (3)}$$

On putting $x = 0$ and $V = V_0$ in (3), we get

$$V_0 = A + B \quad \text{... (4)}$$

On putting $x = l$ and $V = 0$ in (3), we get

$$0 = Ae^{+nl} + Be^{-nl} \quad \text{... (5)}$$

From (4) and (5), we have

$$A = \frac{V_0}{1-e^{2nl}}, \quad B = -\frac{V_0 e^{2nl}}{1-e^{2nl}}$$

Now, from (3), we have

$$V = \frac{V_0 e^{nx}}{1-e^{2nl}} - V_0 \frac{e^{2nl}e^{-nx}}{1-e^{2nl}}$$

$$= \frac{V_0[e^{nx} - e^{2nl-nx}]}{1-e^{2nl}}$$

$$= \frac{V_0[e^{(nl-nx)} - e^{-(nl-nx)}]}{e^{nl} - e^{-nl}}$$

[Dividing numerator and denominator by e^{nl}]

$$= V_0 \left\{ \frac{\sinh n(l-x)}{\sinh nl} \right\}.$$

EXERCISE 33.1

1. In a condenser discharging electricity the voltage V satisfies the equation $k\frac{dV}{dt} + V = 0$, where k is a constant and t is time measured in seconds. Given $k = 50$, find the time t, in which V decreases to one tenth of its original value.

2. A resistance of 100 Ohms, an inductance of 0.5 Henry are connected in series with a battery of 20 Volts. Find the current in the circuit as a function of time.

3. In an LCR circuit, the charge q on a plate of the condenser is given by $L\frac{d^2q}{dt^2} + R\frac{dq}{dt} + \frac{q}{C} = E\sin\omega t$, where $i = \frac{dq}{dt}$. The circuit is tuned to resonance so that $\omega^2 = \frac{1}{LC}$. If $R^2 < \frac{4L}{C}$ and $q = 0 = i$, when $t = 0$, show that

$$q = \frac{E}{R\omega}\left[-\cos\omega t + e^{-\frac{Rt}{2L}}\left(\cos pt + \frac{R}{2Lp}\sin pt\right)\right]$$

and $i = \frac{E}{R}\left[\sin\omega t - \frac{1}{p\sqrt{LC}}e^{-\frac{Rt}{2L}}\sin pt\right]$, where $p^2 = \frac{1}{LC} - \frac{R^2}{4L^2}$.

(UPTU(B.Tech)–2003)

4. The equations of electromotive force in terms of current i for an electrical circuit having resistance R and a condenser of capacity C_1 in series is $E = Ri + \int \frac{i}{C}dt$. Find the current i at any time t, when $E = E_0 \sin\omega t$. (UPTU–2006)

5. The damped LCR circuit is governed by the equation

$$L\frac{d^2q}{dt^2} + R\frac{dq}{dt} + \left(\frac{1}{C}\right)q = 0$$

where L, C, R are positive constants. Find the conditions under which the circuit is overdamped, underdamped and critically damped. Find also the critical resistance.

(UPTU–2005, (CO), 2008)

6. A condenser of capacity C is discharged through an inductance L and resistance R in series and the charge q at time t satisfies the equation

$$L\frac{d^2q}{dt^2} + R\frac{dq}{dt} + \frac{q}{C} = 0$$

given that $L = 0.25$ Henry, $R = 250$ Ohm, $C = 2\times10^{-6}$ Farad and that when $t = 0$, charge q is 0.002 Coulomb and current $i = 0$. Obtain the value of q in terms of t.

(UPTU(Q. Bank)–2002)

7. A constant e.m.f. E at $t = 0$ is applied to a circuit consisting of an inductance L, resistance R and capacitance C in series. The initial values of the current and the charge being zero, find the current at any time t, if $CR^2 < 4L$. Show that the amplitudes of the successive vibrations are in geometrical progression. (UPTU(Q. Bank)–2001)

8. The differential equation $\frac{d^2x}{dt^2} + 2k\frac{dx}{dt} + n^2x = 0$; $(k < n)$, represents the damped harmonic oscillations of apertures. Solve this equation and show that the ratio of the amplitude of any oscillation to that of the preceding one is constant, *i.e.*, its amplitude forms a G.P.

9. An inductance (L) of 2.0 H and a resistance (R) of 20 Ω are connected in series with an e.m.f. E Volt. If the current i is zero, when $t=0$, find the current (i) at the end of 0.01 seconds if $E=100\,V$, using the following differential equation

$$L\frac{di}{dt}+iR=E$$ [UPTU–2008]

10. A resistance R of 5 Ω and an inductance L of 0.1 H are conneated in series with a battery of 12 V. Find the current i in the circuit as a funation of time using the following differential equation .

$$Ri+L\frac{di}{dt}=E$$ [UPTU(SUM)–2008]

11. When a resistance R ohms is connected in series with an induatance L henries, an e.m.f. of E volts, the current i amperes at time t is given by

$$L\frac{di}{dt}+Ri=E$$

If E = 10 sin t volts and $i=0$ when $t=0$, find i as a function of t.

12. An R-L circuit has an e.m.f. given (in volts) by 4 sin t, a resistance of 100 ohms, an inductance of 4 henries with no initial current. Find the current at any time t.

[GBTU–2012, GBTU(CO)–2011, MTU(SUM)–2011]

13. An inductance L of 5.0 H and a resistance R of 25 Ω are connected in series with an e.m.f. E volt. If the current I is zero when t = 0, find the current I at the end of 1 second if E = 100 V using the differential equation

$$L\frac{dI}{dt}+IR=E$$ [GBTU(CO)–2010]

14. Find the steady state solution in R-L-C circuit equation consisting of inductance L = 0.05 H, resistance R = 5 ohms and a condenser of capacitance 4×10^{-4} Farad, if $Q=I=0$ when t = 0 and there is an alternating e.m.f. of 200 cos 100t. Find $Q(t)$ and $I(t)$. [MTU–2011]

15. An R-L-C circuit connected in series has $R=90\Omega, C=\frac{1}{140}$ Farad, L = 10 henries and an applied voltage $E(t)$ = 10 cos t. Assuming no initial charge on the capacitor but an initial current of 1 ampere at t = 0, when the voltage is first applied, find the subsequent charge on the capacitor and the amlplitude of the steady state charge. [M.T.U. 2012]

ANSWERS

1. 115.13 secs **2.** $\frac{1}{5}[1-e^{-200t}]$ **4.** $i=E_0\omega\frac{C}{1+\omega^2R^2C^2}[\cos\omega t+\omega RC\sin\omega t]+C\,e^{-t/RC}$

5. $R>2\sqrt{\frac{L}{C}}$ [over damping], $R=2\sqrt{\frac{L}{C}}$ [critically damped] **6.** $q=e^{-500t}[0.002\cos 1323t+0.0008\sin 1323t]$

10. $i=2.4(1-e^{-50t})$ ampere **11.** $i=\frac{10}{L^2+R^2}(R\sin t-L\cos t+Le^{-Rt/L})$ **12.** $i(t)=\frac{1}{626}(e^{-25t}-\cos t+25\sin t)$ **3.** 3.973 ampere

14. $Q(t)=\frac{2}{85}(\sin 100t+4\cos 100t)$; $I(t)=\frac{40}{17}(\cos 100t-4\sin 100t)$ **15.** $q=\frac{3}{25}e^{-2t}-\frac{43}{250}e^{-7t}+\frac{1}{250}(9\sin t+13\cos t)$; $\frac{\sqrt{10}}{50}$

33.3 APPLICATIONS AS RATE OF COOLING

Example 1. *According to Newton's law of cooling, the rate at which a substance cools in moving air is proportional to the difference between the temperature of the substance and that of the air. If the temperature of the air is 30°C and the substance cools from 100°C to 70°C in 15 minutes, find when the temperature will be 40°C.*

Solution. By Newton's law of cooling, we have

$$\frac{dT}{dt}=-k(T-30) \text{ or } \frac{dT}{T-30}=-kdt$$

On integrating $\log(T-30)=-kt+C$ ... (1)

Initially, when t = 0, $T=100$

$$C=\log 70$$

Substituting the value of C in equation (1), we get

$$\log(T-30)=-kt+\log 70 \quad \text{... (2)}$$

Also, when $t=15$, $T=70$

$$15k=\log 70-\log 40 \quad \text{... (3)}$$

Dividing (2) by (3), we get

$$\frac{t}{15}=\frac{\log 70-\log(T-30)}{\log(70)-\log 40} \quad \text{... (4)}$$

Now, when $T=40$, we have from (4)

$$\frac{t}{15}=\frac{\log 70-\log 10}{\log 70-\log 40}=\frac{\log_e 7}{\log_e(7/4)}$$

$$=\frac{\log_{10}7}{\log_{10}(7/4)}=3.48$$

$$t=15\times3.48=52.20$$

Hence, the temperature will be 40° after 52.20 minutes.

Example 2. *The rate at which the ice melts is proportional to the amount of ice at the instant. Find the amount of ice after 2 hours if half the quantity melts in 30 minutes.*

Solution. Let A be the amount of ice at any time t.

$$\frac{dA}{dt} = kA \quad \text{or} \quad \frac{dA}{A} = kdt$$

$$\int \frac{dA}{A} = k\int dt + C$$

$$\log A = kt + C \qquad \text{... (1)}$$

Putting $t = 0$, $A = A_0$ in equation (1),

we get $\log A_0 = 0 + C \Rightarrow C = \log A_0$

Substituting the value of c in equation (1), we get

$$\log A = kt + \log A_0 \qquad \text{... (2)}$$

$$A = A_0 / 2, \text{ when } t = 1/2 \text{ hour}$$

$$\log \frac{A_0}{2} = \frac{k}{2} + \log A_0 \quad \Rightarrow \log \frac{A_0}{2A_0} = \frac{k}{2}$$

$$\Rightarrow \quad k = 2.\log \frac{1}{2}$$

On putting the value of k in equation (2), we get

$$\log A = \left(2\log \frac{1}{2}\right)t + \log A_0 \qquad \text{... (3)}$$

On putting $t = 2$ hours in equation (3), we get

$$\log A = 4\log \frac{1}{2} + \log A_0$$

$$\log \frac{A}{A_0} = \log\left(\frac{1}{2}\right)^4 \Rightarrow \frac{A}{A_0} = \left(\frac{1}{2}\right)^4$$

$$\frac{A}{A_0} = \frac{1}{16} \qquad \Rightarrow \qquad A = \frac{A_0}{16}$$

After 2 hours, amount of ice left $= \frac{1}{16}$ of amount of ice at the beginning.

33.4 CHEMICAL ACTION

Example 1. *Uranium disintegrates at a rate proportional to the amount present at any instant. If M_1 and M_2 grams of uranium are present at times T_1 and T_2 respectively, show that the half life of Uranium is* $\dfrac{(T_2 - T_1)\log 2}{\log(M_1 / M_2)}$.

Solution. Let M grams of Uranium be present at any time t. Then the equation of disintegration of Uranium is

$$\frac{dM}{dt} = -kM_1 \text{, where } k \text{ is a constant.}$$

$$\frac{dM}{dt} = -Rdt \qquad \text{... (1)}$$

Integrating (1) as

$$\int_{M_1}^{M_2} \frac{dM}{M} = -k\int_{T_1}^{T_2} dt$$

$$\Rightarrow \quad \log \frac{M_2}{M_1} = -k(T_2 - T_1)$$

$$\Rightarrow \quad k = \frac{\log(M_1 / M_2)}{T_2 - T_1} \qquad \text{... (2)}$$

Let the half life of Uranium, *i.e.*, $t = T$, $M = \frac{1}{2}M_0$

From (1) again

$$\int_M^{M/2} \frac{dM}{M} = -k\int T\,dt$$

$$\log \frac{1}{2} = kT \quad \Rightarrow \quad T = \frac{\log 2}{k}$$

Hence, $$T = \frac{(T_2 - T_1)\log 2}{\log(M_1 / M_2)}.$$

EXERCISE 33.2

1. If the temperature of the air is 30°C and the substance cools from 100°C to 70°C in 15 minutes, find when the temperature will be 40°C.
2. Water at temperature 100 °C cools in 10 minutes to 88 °C in a room of temperature 25°C. Find the temperature of water after 20 minutes.
3. A body originally at 80 °C cools down to 60 °C in 20 minutes, the temperature of the air being 40 °C. What will be the temperature of the body after 40 minutes from the original.
4. The rate at which a body cools in proportional to the difference between the temperature of the body and that of the surrounding air. If a body in air at 25°C will cool from 100°C to 75°C in one minutes. Find its temperature at the end of three minutes.
5. If a thermometer is taken outdoors where the temperature is 0°C from a room in which the temperature is 21°C and the reading drops at 10°C in 1 minute. How long after its removal will the reading be 5°C ?
6. Radium decomposes at a rate proportional to the amount present. If a fraction P of the original amount disappears in one year. How much will remain at the end of 21 years?
7. Under certain conditions cane sugar is converted into dextrose at a rate, which is proportional to the amount unconverted at any time. If out of 75 grams of sugar at $t = 0$, 8 grams are converted during the first 3 minutes, find the amount converted in 1½ hours.
8. A radioactive substance decomposes at a rate proportional to its mass. When the mass is 10 milligrams, the rate of disintegration is 0.051 milligram per day. How long will it take for the mass to be reduced from 10 milligrams to 5 milligrams?

ANSWERS

1. 52.5 minutes	**2.** 77.9°C	**3.** 50 °C	**4.** 47.22	**5.** 2 minutes, 13 seconds
6. $\left(1-\frac{1}{p}\right)^{21}$ times of the original amount		**7.** 21.53 grams	**8.** 136 days	

33.5 SIMPLE HARMONIC MOTION (S.H.M.)

A motion in which a particle moves in a straight line in such a way that its acceleration is always directed towards a fixed point on the line (called centre of the force) and varies as the distance of the particle from the fixed point, is called simple harmonic motion.

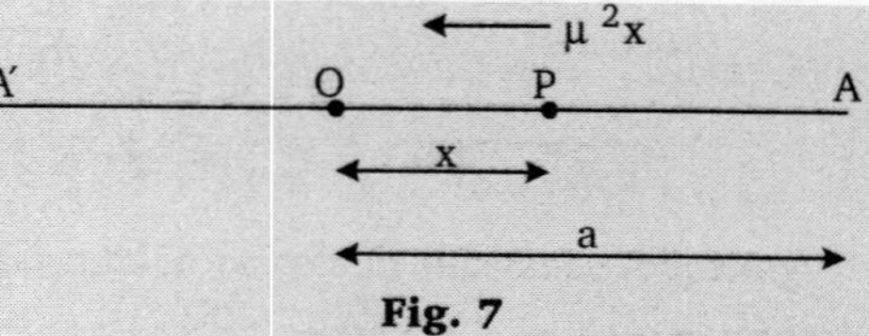

Fig. 7

Let O be the fixed point in the line $A'A$. Let P be the position of the particle at any time t, where $OP = x$.

Since the acceleration is always directed towards O, *i.e.*, the acceleration is in the direction opposite to that in which x increases, the equation of motion of the particle is $\frac{d^2x}{dt^2} = -\mu^2 x$.

$$(D^2 + \mu^2)x = 0, \text{ where } D = d/dx. \quad \text{... (1)}$$

It is a linear differential equation with constant coefficients. Its auxiliary equation is $m^2 + \mu^2 = 0$ so that $m = \pm i\mu$

The solution of (1) is $x = C_1 \cos \mu t + C_2 \sin \mu t$ ... (2)

Velocity of particle at $P = \frac{dx}{dt} = -C_1\mu \sin \mu t + C_2 \mu \cos \mu t$ ... (3)

If the particle starts from rest at A, where $OA = a$,

then from (2), (at $t = 0$, $x = a$), $C_1 = a$ and from (3), (at $t = 0$, $\frac{dx}{dt} = 0$); $C_2 = 0$

$$x = a \cos \mu t \quad \text{... (4)}$$

$$\frac{dx}{dt} = -a\mu \sin \mu t \quad \text{... (5)}$$

$$= -a\mu\sqrt{1-\cos^2 \mu t} = -a\mu\sqrt{1-\frac{x^2}{a^2}}$$

$$= -\mu\sqrt{a^2 - x^2} \quad \text{... (6)}$$

The equation (6) gives the velocity of the particle at any point P.

REMARKS

- We observe that velocity is maximum when $x = 0$. Thus in SHM, the velocity is maximum at the centre of force O.
- We observe from equation (6), that velocity $= 0$ when $x = a$.
  Thus, in S.H.M., the velocity is zero at points equidistant from the centre of force.

33.5.1 NATURE OF MOTION

The particle starts from rest at A where its acceleration is maximum and is μa towards O. It begins to move towards the centre of attraction O and as it approaches the centre of force O, its velocity goes on increasing. When the particle reaches O its acceleration is zero and its velocity is maximum and is $a\mu$ in the direction OA'. Due to this velocity gained at O the particle moves towards the left of O. But on account of the centre of attraction at O, a force begins to act upon the particle against its direction of motion. So, its velocity goes on decreasing and it comes to instantaneous rest at A' where $OA' = OA$. The rest at A' is only instantaneous. The particle at once begins to move towards the centre of attraction O and retracing its path, it again comes to instantaneous rest at A. Thus the motion of particle is oscillatory and it continues to oscillate between A and A'. To start from A and to come back to A is called one complete oscillation.

33.5.2 AMPLITUDE

In S.H.M., the distance from the centre to the position of maximum displacement is called the amplitude of the motion. OA is maximum distance and is called the amplitude.

From (6),
$$-\frac{dx}{\sqrt{a^2 - x^2}} = \mu dt$$

Integrating, we get
$$\cos^{-1}\frac{x}{a} = \mu t + A \quad \text{... (7)}$$

Putting $t = 0,\ x = a$ in equation (7), we get $\quad 0 = 0 + A \quad \Rightarrow \quad A = 0.$

On putting the value of A, equation (7) becomes

$$\cos^{-1}\frac{x}{a} = \mu t \quad \Rightarrow \quad x = a\cos\mu t$$

Particle will reach O in time t_1, therefore,

$$0 = a\cos\mu t_1 \Rightarrow \quad 0 = \cos\mu t_1$$

$$\cos\frac{\pi}{2} = \cos\mu t_1 \quad \Rightarrow \quad \frac{\pi}{2} = \mu t_1$$

Hence,
$$t_1 = \frac{\pi}{2\mu}$$

33.5.3 TIME PERIOD

In a S.H.M., the time taken to make a complete oscillation is called time period.

33.5.4 FREQUENCY

The number of complete oscillations per second is called the frequency of motion. If n is the frequency and t is the time period.

$$\text{Time period } T = 4\left(\frac{\pi}{2}\mu\right) = \frac{2\pi}{\mu}$$

$$\text{Frequency } n = \frac{1}{T} = \frac{\mu}{2\pi}$$

REMARKS

- The equation of S.H.M. is $\dfrac{d^2x}{dt^2} = -\omega^2 x$
- The velocity V at a distance x from the centre at time t is $V^2 = \omega^2(a^2 - x^2)$.
  $x = a\cos\omega t$ where a is the amplitude and ω is the angular velocity (at the extreme point).
- Maximum acceleration $= \omega^2 a$.
- Maximum velocity $= \omega a$.
- Time period $T = \dfrac{2\pi}{\omega}$.

33.6 GEOMETRICAL REPRESENTATION OF S.H.M.

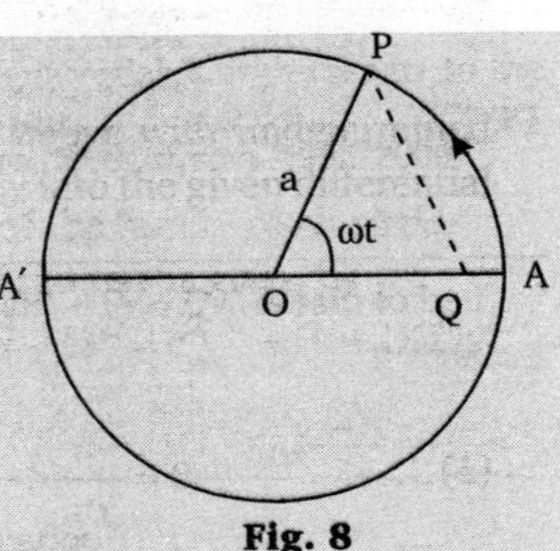

Fig. 8

Let a particle moves with a uniform angular velocity ω round the circumference of a circle of radius a. Suppose AA' is a fixed diameter of the circle. If the particle starts from A and P is its position at time t, then $\angle AOP = \omega t$. Draw PQ perpendicular to the diameter AA'.

If $OQ = x$, then $\quad x = a\cos\omega t \quad \ldots (1)$

As the particle P moves round the circumference, the foot Q of the perpendicular on the diameter AA' oscillates on AA' from A to A' and from A' to A back. Thus the motion of the point Q is periodic.

From (1), we have
$$\frac{dx}{dt} = -a\omega\sin\omega t \quad \ldots (2)$$

and
$$\frac{d^2x}{dt^2} = -a\omega^2\cos\omega t = -\omega^2 x \quad \ldots (3)$$

The equations (2) and (3) gives the velocity and acceleration of Q at any time t.

The equation (3) shows that Q executes a simple harmonic motion with centre at the origin O. From equation (1), we see that the amplitude of this S.H.M. is a because the maximum value of x is a.

The periodic time of Q = The time required by P to turn through an angle 2π with a uniform angular velocity ω .

$$= \frac{2\pi}{\omega}.$$

Thus, if a particle describes a circle with constant angular velocity, the foot of the perpendicular from it on any diameter executes a S.H.M.

Solved Examples

Example 1. *A point moving in a straight line with S.H.M. has velocities V_1 and V_2 when its distances from the centre are x_1 and x_2. Show that the period of motion is*

$$2\pi\sqrt{\frac{x_1^2 - x_2^2}{V_2^2 - V_1^2}}.$$

Solution. Let the equation of S.H.M. with centre O as origin be $\frac{d^2x}{dt^2} = -\mu x^2$. Then the time period $T = \frac{2\pi}{\mu}$. If a be the amplitude of the motion, we have

$$V^2 = \mu^2(a^2 - x^2)$$

where V is the velocity at a distance x from the centre.

But when $x = x_1$, $V = V_1$ and

when $x = x_2, V = V_2$.

Therefore, from (1), we have

$$V_1^2 = \mu^2(a^2 - x_1^2)$$

and $V_2^2 = \mu^2(a^2 - x_2^2)$.

These gives

$$V_2^2 - V_1^2 = \mu^2\{(a^2 - x_2^2) - (a^2 - x_1^2)\}.$$

$$= \mu^2 (x_1^2 - x_2^2)$$

$$\Rightarrow \quad \mu^2 = \frac{V_2^2 - V_1^2}{x_1^2 - x_2^2}$$

Hence, the time period

$$T = \frac{2\pi}{\mu} = 2\pi\sqrt{\frac{x_1^2 - x_2^2}{V_2^2 - V_1^2}}.$$

Example 2. *At the end of the three successive seconds, the distance of a point moving with S.H.M. from its mean position are x_1, x_2, x_3 respectively. Show that the time of a complete oscillation is*

$$\frac{2\pi}{\cos^{-1}\left(\frac{x_1 + x_3}{2x_2}\right)}.$$

Solution. Let the moving point be at distances x_1, x_2, x_3 from the mean position at the end of $t, t+1, t+2$ seconds respectively.

Using $x = a\cos\mu t$

$$x_1 = a\cos\mu t \quad \text{... (1)}$$

$$x_2 = a\cos\mu(t+1) \quad \text{... (2)}$$

$$x_3 = a\cos\mu(t+2) \quad \text{... (3)}$$

Adding (1) and (3), we get

$$x_1 + x_3 = a\,[\cos\mu(t+2) + \cos\mu t]$$

$$= a.2\cos\left\{\frac{\mu(t+2)+\mu t}{2}\right\}\cos\left\{\frac{\mu(t+2)-\mu t}{2}\right\}$$

$$= 2a\cos\mu(t+1)\cos\mu = 2x_2\cos\mu$$

[Using (2)]

$$\mu = \cos^{-1}\left(\frac{x_1 + x_3}{2x_2}\right)$$

Hence, the time of a complete oscillation

$$= \frac{2\pi}{\mu} = \frac{2\pi}{\cos^{-1}\left(\frac{x_1 + x_3}{2x_2}\right)}.$$

Example 3. *A particle moves with S.H.M. of period 12 seconds, travels 8 cm from the position of rest in 2 seconds. Find the amplitude, the maximum velocity and the velocity at the end of 2 seconds.*

Solution. $T = \frac{2\pi}{\sqrt{\mu}} = 12 \Rightarrow \sqrt{\mu} = \frac{\pi}{6}$

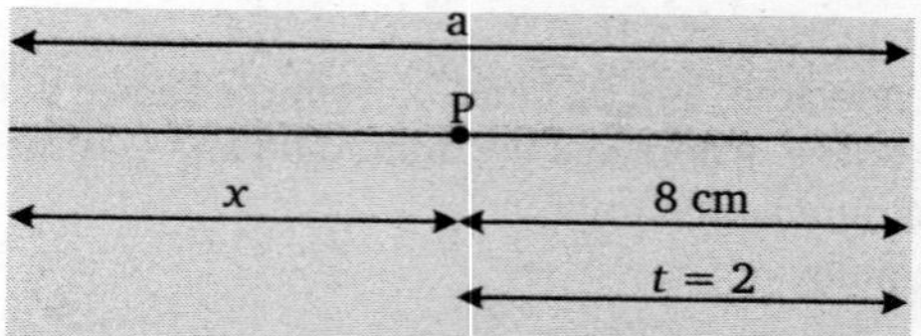

a
P
x
8 cm
t = 2

Fig. 9

Let a be the amplitude OA

$AP = 8$ cm,

$OP = x = a - 8$, $t = 2$ seconds

We know that $x = a\cos\sqrt{\mu}\,t$

$$a - 8 = a\cos 2\sqrt{\mu} = a\cos\frac{\pi}{3} = \frac{a}{2} \Rightarrow a = 16$$

Maximum velocity

$$= a\sqrt{\mu} = \frac{\pi}{6}\times 16 = \frac{8\pi}{3} = 4.619 \text{ cm/sec}$$

Velocity V at the end of two seconds

$$= \sqrt{\mu}\,\sqrt{a^2 - x^2} = \frac{\pi}{6}\sqrt{256 - 64}$$

$$(\because x = a - 8 = 16 - 8 = 8)$$

$$= \frac{\pi}{6}\times\sqrt{192} = \frac{4\pi\sqrt{3}}{3} \text{ cm/sec.}$$

Example 4. *A particle is performing a simple harmonic motion of period T about a centre O and it passes through a point P where $OP = b$ with velocity V in the direction OP; prove that the time which elapses before it returns to P is*

$$\frac{T}{\pi}\tan^{-1}\left(\frac{VT}{2\pi b}\right).$$

Solution. Let the equation of S.H.M. with centre O as origin as $\dfrac{d^2x}{dt^2} = -\mu x$.

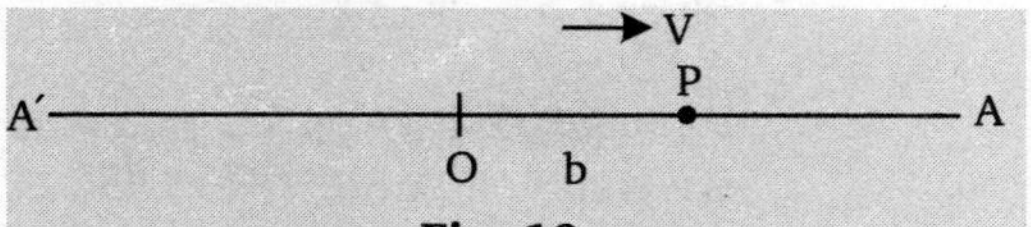

Fig. 10

The time period, $T = \dfrac{2\pi}{\sqrt{\mu}}$.

Let the amplitude be a.

Then, $\left(\dfrac{dx}{dt}\right)^2 = \mu\,(a^2 - x^2)$.

When the particle passes through P, its velocity is given to be V in the direction OP. Also, $OP= b$. So putting $x = b$ and $dx/dt = V$ in (1), we get

$$V^2 = \mu(a^2 - b^2) \qquad \text{... (2)}$$

Let A be an extremity of the motion. From P the particle comes to instantaneous rest at A and then returns back to P. In S.H.M., the time from P to A is equal to the time from A to P.

$\therefore$ The required time $= 2 \times$ time from A to P.

Now, for the motion from A to P, we have

$$\frac{dx}{dt} = -\sqrt{\mu}\,\sqrt{a^2 - x^2}$$

$$\Rightarrow \quad dt = -\frac{1}{\sqrt{\mu}}\frac{dx}{\sqrt{a^2 - x^2}}$$

Let t_1 be the time from A to P. Then at $A, t = 0, x = a$ and at $P, t = t_1$ and $x = b$. Therefore, integrating (3), we get

$$\int_0^{t_1} dt = \frac{1}{\sqrt{\mu}}\int_a^b \frac{-dx}{\sqrt{a^2 - x^2}}$$

$$t_1 = \frac{1}{\sqrt{\mu}}\left[\cos^{-1}\frac{x}{a}\right]_a^b$$

$$= \frac{1}{\sqrt{\mu}}\left[\cos^{-1}\frac{b}{a} - \cos^{-1}1\right]$$

$$= \frac{1}{\sqrt{\mu}}\cos^{-1}\left(\frac{b}{a}\right).$$

Hence, the required time

$$= 2t_1 = \frac{2}{\sqrt{\mu}}\cos^{-1}\frac{b}{a} = \frac{2}{\sqrt{\mu}}\tan^{-1}\left\{\frac{\sqrt{a^2 - b^2}}{b}\right\}$$

$$= \frac{2}{\sqrt{\mu}}\tan^{-1}\left(\frac{V}{b\sqrt{\mu}}\right)$$

$$\left[\text{From (2), } \sqrt{a^2 - b^2} = \frac{V}{\sqrt{\mu}}\right]$$

$$= \frac{2}{2\pi/T}\tan^{-1}\left\{\frac{V}{b\,(2\pi/T)}\right\}$$

$$\left(\because T = \frac{2\pi}{\sqrt{l}} \text{ so that } \sqrt{\mu} = \frac{2\pi}{T}\right)$$

$$= \frac{T}{\pi}\tan^{-1}\left(\frac{VT}{2\pi b}\right).$$

Example 5. *A point moves in a straight line towards a centre of force $\mu/(\text{distance})^3$ starting from rest at a distance 'a' from the centre of force. Show that the time of reaching a point distant 'b' from the centre of force is $(a/\sqrt{\mu})\sqrt{a^2 - b^2}$, and that its velocity is $\dfrac{\sqrt{\mu}}{ab}\sqrt{a^2 - b^2}$.* [UPTU–2001]

Solution. Let a point moves from A towards the centre of force O.

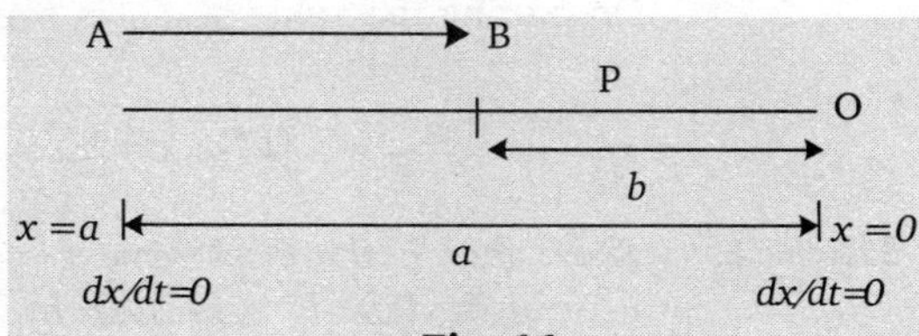

Fig. 11

$$\frac{m\mu}{x^3} = -m\frac{d^2x}{dt^2} \Rightarrow \frac{d^2x}{dt^2} = -\frac{\mu}{x^3}.$$

$$2\frac{dx}{dt}\cdot\frac{d^2x}{dt^2} = -2\frac{\mu}{x^3}\cdot\frac{dx}{dt}$$

Integrating, we get

$$\left(\frac{dx}{dt}\right)^2 = -2\mu\frac{x^{-2}}{-2} + C$$

$$\Rightarrow \quad V^2 = \frac{\mu}{x^2} + C \qquad \text{... (1)}$$

At A, $V = 0$, and $x = a$,

$$\therefore\ 0 = \frac{\mu}{a^2} + C \Rightarrow C = -\frac{\mu}{a^2}$$

On putting the value of C, equation (1) becomes

$$V^2 = \frac{\mu}{x^2} - \frac{\mu}{a^2} = \mu\left(\frac{a^2 - x^2}{x^2a^2}\right) \qquad \text{... (2)}$$

Therefore, velocity when $x = b$ is given by

$$V^2 = \mu\left(\frac{a^2 - b^2}{a^2b^2}\right) \Rightarrow V = \pm\sqrt{\mu}\frac{\sqrt{a^2 - b^2}}{ab}$$

$$\Rightarrow V = \sqrt{\mu}\frac{\sqrt{a^2 - b^2}}{ab} \quad \text{(Numerical value)}$$

From (2), $\left(\dfrac{dx}{dt}\right)^2 = \mu\dfrac{(a^2 - x^2)}{x^2a^2} \Rightarrow \dfrac{dx}{dt}$

$$= -\sqrt{\mu}\frac{\sqrt{a^2 - x^2}}{xa}$$

$$dt = -\frac{1}{\sqrt{\mu}}\frac{xa}{\sqrt{a^2-x^2}}dx$$

Integrating, we get

$$t = -\frac{1}{\sqrt{\mu}}\int\frac{xa\,dx}{\sqrt{a^2-x^2}}$$

(As the point 'A' is moving towards O)

Let $a^2 - x^2 = z^2 \Rightarrow \quad -2xdx = 2zdz$

$$t = \frac{a}{\sqrt{\mu}}\int\frac{zdz}{z} = \frac{a}{\sqrt{\mu}}\int dz = \frac{a}{\sqrt{\mu}}z + C_1 \qquad \text{... (3)}$$

$$= \frac{a}{\sqrt{\mu}}\sqrt{a^2-x^2} + C_1$$

At $A_1,\ t = 0,\ x = a$

On putting $t = 0,\ x = a$ in equation (3), we get $0 = 0 + C_1 \quad \Rightarrow C_1 = 0$

On putting the value of C_1 in equation (3), we get

$t = \frac{a}{\sqrt{\mu}}\sqrt{a^2-x^2}$. At B_1, $x = b, t = \frac{a}{\sqrt{\mu}}\sqrt{a^2-b^2}$

Example 6. *Show that if the displacement of a particle in a straight line is expressed by the equation $x = a\cos\mu t + b\sin\mu t$, it describes a simple harmonic motion whose amplitude is $\sqrt{a^2+b^2}$ and time period is $2\pi/\mu$.*

Solution. We have $\quad x = a\cos\mu t + b\sin\mu t \qquad \text{... (1)}$

$$\therefore \quad \frac{dx}{dt} = -a\mu\cos\mu t + b\mu\cos\mu t \qquad \text{... (2)}$$

and $\quad \frac{d^2x}{dt^2} = -a\mu^2\cos\mu t - b\mu^2\sin\mu t = -\mu^2 x$

which represents simple harmonic motion with centre at origin.

Time period, $T = \frac{2\pi}{\omega} = \frac{2\pi}{\mu}$

Also, amplitude is the value of x, when $\frac{dx}{dt} = 0$.

From (2), $0 = -a\mu\sin\mu t + b\mu\cos\mu t$

$\Rightarrow \quad \tan\mu = \frac{b}{a}$ and $\sin\mu t = \frac{b}{\sqrt{a^2+b^2}}$

and $\quad \cos\mu t = \frac{a}{\sqrt{a^2+b^2}}$

From (1), $x = a.\frac{a}{\sqrt{a^2+b^2}} + b.\frac{b}{\sqrt{a^2+b^2}}$

$\Rightarrow \quad x = \sqrt{a^2+b^2}$

Example 7. *A moving body opposed by a force per unit mass of value Cx and resistance per unit mass of value bV^2 where ax and V are the displacement and velocity of the particle in terms of x, if it starts from rest.*

Solution. By Newton's second law of motion, the equation of motion of the body is

$$V\frac{dV}{dx} = -Cx - bV^2$$

$$v\frac{dV}{dx} + bV^2 = -Cx \qquad \text{... (1)}$$

Putting $V^2 = z,\ 2V\frac{dV}{dx} = \frac{dz}{dx}$ in equation (1), we get

$$\frac{1}{2}\frac{dz}{dx} + bz = -Cx$$

$$\Rightarrow \frac{dz}{dx} + 2bz = -2Cx$$

$$\text{I.F.} = e^{\int 2b\,dx} = e^{2bx}$$

Solution is given by

$$z.e^{2bx} = \int -2Cx\,.\,e^{2bx}dx + C'$$

$$= -2C\left[\frac{xe^{2bx}}{2b} - \int\frac{e^{2bx}}{2b}dx\right] + C'$$

$$= -\frac{C}{b}x\,e^{2bx} + \frac{C}{b}\frac{e^{2bx}}{2b} + C'$$

$$= -\frac{C}{b}x + \frac{C}{2b^2} + C'\,e^{-2bx}$$

$$V^2 = -\frac{Cx}{b} + \frac{C}{2b^2} + C'e^{-2bx}$$

Putting $V = 0$ and $x = 0$ in equation (2), we have

$$0 = \frac{C}{2b^2} + C' \Rightarrow C' = -\frac{C}{2b^2}$$

Equation (2) becomes

$$V^2 = -\frac{Cx}{b} + \frac{C}{2b^2} - \frac{C}{2b^2}e^{-2bx}.$$

Example 8. *A point executes S.H.M. such that in two of its position velocities are u, v and the two corresponding accelerations are $\alpha.\beta$; show that the distance between the two positions is $\frac{v^2-u^2}{\alpha+\beta}$ and the amplitude of motion is $\frac{\{(v^2-u^2)(\alpha^2v^2-\beta^2u^2)\}^{1/2}}{\alpha^2-\beta^2}$.*

Solution. Let the equation of S.H.M. with centre at origin be $\frac{d^2x}{dt^2} = -\mu x$

If a be the amplitude of the motion, we have

$\left(\frac{dx}{dt}\right)^2 = \mu(a^2 - x^2)$, where dx/dt is the velocity at a distance x from the centre.

Let x_1 and x_2 be the distance from the centre of two positions where u and v are the velocities and α and β are the accelerations respectively.

Then $\alpha = \mu\,x_1 \qquad \text{... (1)}$

$\beta = \mu x_2$ ... (2)

$v^2 = \mu(a^2 - x_1^2)$ ... (3)

and $v^2 = \mu(a^2 - x_2^2)$ ... (4)

Adding (1) and (2), we get

$\alpha + \beta = \mu(x_1 + x_2)$ ... (5)

Also subtracting (3) from (4), we get

$$v^2 - u^2 = \mu(x_1^2 - x_2^2)$$
$$= \mu(x_1 - x_2)(x_1 + x_2) = (\alpha + \beta)(x_1 - x_2) \quad \text{[From (5)]}$$

$$(x_1 - x_2) = (v^2 - u^2)/(\alpha + \beta)$$

This gives the distance between the two positions.

Now to get the amplitude a, it is obvious that we have to eliminate x_1, x_2 and μ from the equations (1), (2), (3) and (4). Substituting for x_1 and x_2 from (1) and (2) in (3) and (4), we have

$$u^2 = \mu\left(a^2 - \frac{\alpha^2}{\mu^2}\right), \text{ i.e., } a^2\mu^2 - u^2\mu - \alpha^2 = 0 \quad \text{... (6)}$$

$$\text{and } v^2 = \mu\left(a^2 - \frac{\beta^2}{\mu^2}\right), \text{ i.e., } a^2\mu^2 - v^2\mu - \beta^2 = 0 \quad \text{... (7)}$$

By method of cross multiplication, we get

$$\frac{\mu^2}{u^2\beta^2 - v^2\alpha^2} = \frac{\mu}{-a^2\alpha^2 + a^2\beta^2} = \frac{1}{a^2u^2 - a^2v^2}$$

Equating the two values of μ^2 found from the above equations, we get

$$\frac{\alpha^2v^2 - u^2\beta^2}{a^2(v^2 - u^2)} = \left[\frac{a^2(\alpha^2 - \beta^2)}{a^2(v^2 - u^2)}\right]^2$$

or
$$\frac{\alpha^2v^2 - u^2\beta^2}{a^2(v^2 - u^2)} = \frac{(\alpha^2 - \beta^2)^2}{(v^2 - u^2)^2}$$

$$a^2 = \frac{(\alpha^2v^2 - \beta^2u^2)(v^2 - u^2)}{(\alpha^2 - \beta^2)^2}$$

$$\Rightarrow \quad a = \frac{\{(v^2 - u^2)(\alpha^2v^2 - \beta^2u^2)\}^{1/2}}{\alpha^2 - \beta^2}.$$

Example 9. *In a system, the amplitude of motion is 5 metres and the period is 4 seconds. Find the time required by the particle in passing between the points which are at distances of 4 metres and 2 metres from the centre of force and are on the same side of it. Also find the velocities at these points.*

Solution. Equation of S.H.M. is

$$\frac{d^2x}{dt^2} = -\mu^2 x \quad \text{... (1)}$$

Time period, $T = \frac{2\pi}{\mu} = 4 \Rightarrow \mu = \pi/2$

Solution to (1) is $x = a\cos\mu t = 5\cos\frac{\pi}{2}t$ ... (2)

Let t_1 sec and t_2 sec be the times when the particle is at a distance of 4 metres and 2 metres respectively from the centre of force. Then

From (2), $4 = 5\cos\frac{\pi}{2}t_1 \Rightarrow t_1 = \frac{2}{\pi}\cos^{-1}\left(\frac{4}{5}\right)$

and $2 = 5\cos\frac{\pi}{2}t_2 \Rightarrow t_2 = \frac{2}{\pi}\cos^{-1}\left(\frac{2}{5}\right)$

$\therefore$ Time required in passing through these points

$$= t_2 - t_1 = \frac{2}{\pi}\left[\cos^{-1}\left(\frac{2}{5}\right) - \cos^{-1}\left(\frac{4}{5}\right)\right]$$
$$= 0.33 \text{ sec}$$

Differentiating equation (2) w.r.t. 't', we get

$$\frac{dx}{dt} = -\frac{5\pi}{2}\sin\frac{\pi t}{2}$$
$$= -\frac{5\pi}{2}\sqrt{1 - \cos^2\frac{\pi t}{2}}$$
$$= -\frac{5\pi}{2}\sqrt{1 - \frac{x^2}{25}}$$
$$= -\frac{\pi}{2}\sqrt{25 - x^2}$$

When $x = 4m$,

$$V = -\frac{\pi}{2}\sqrt{25 - 16} = -\frac{3\pi}{2} \text{ metre}$$

When $x = 2m$,

$$V = -\frac{\pi}{2}\sqrt{25 - 4} = -\frac{\pi\sqrt{21}}{2} \text{ metre}$$

Negative sign indicates that it is directed towards centre of force.

EXERCISE 33.3

1. A particle is executing simple harmonic motion with amplitude 20 cm and time 4 seconds. Find the time required by the particle in passing between points which are at distances 15 cm and 5 cm from the centre of force and are on the same side of it. [UPTU Q. Bank–2002]

2. A particle moves with S.H.M. if, when at a distance of 3 cm and 4 cm from the centre of the path, its velocities are 8 cm and 6 cm per sec respectively. Find its period, maximum velocity and acceleration when at its greatest distance from the centre.

3. A point executes S.H.M. such that in two of its positions, the velocities are u, v and the corresponding accelerations α, β. Show that the distance between the positions is $\dfrac{v^2-u^2}{\alpha+\beta}$ and find the amplitude of the motion.

4. A particle moves with S.H.M. in a straight line under the action of force which is proportional to the distance of the particle from $x=0$. If it starts at $x=5$ cm with a velocity of 10 cm/sec and it reaches an extreme positions x = 10 cm, at what speed does it passes through the origin?

5. A particle whose mass is m, is acted upon by a force $m\mu(x+a^4x^{-3})$ towards the origin. If it starts from rest at a distance a, show that it will arrive at the origin in time $\dfrac{\pi}{4\sqrt{\mu}}$.

6. At what distance from the centre, the velocity in a S.H.M. will be one forth of the maximum?

7. A particle begins to move from a distance 'a' towards a fixed centre which repels it with retardation μx. If its initial velocity is $a\sqrt{\mu}$, show that it continually approach the fixed centre but will never reach it. [UPTU(SUM)–2007]

8. A particle of mass m moves in a straight line under the action of force mn^2x which is always directed towards a fixed point O on the line. Determine the displacement $x(t)$ if the resistance to the motion is $2\lambda mnv$ given that initially x = 0, $\dfrac{dx}{dt}=0, (0<\lambda<1)$ (GBTU–2010)

Hint to Selected Problems

7. $\ddot{x}=\mu x; t=0, x=a, \dot{x}=a\sqrt{\mu}$ so that $x=ae^{\sqrt{\mu}t}$

8. Equation of motion is $m\ddot{x}=-2\lambda mn\dot{x}-mn^2x$

ANSWERS

1. 0.38 sec **2.** π sec, 10 cm/sec, 20 cm/sec^2 **4.** 11.546 cm/sec

6. $x=\pm\dfrac{\sqrt{15}\,a}{4}$ **8.** $x(t)=\dfrac{x_0}{n\omega}e^{-\lambda nt}\sin n\omega t$, where $\omega=\sqrt{1-\lambda^2}$

33.7 SIMPLE PENDULUM

A light inextensible string and a heavy particle of negligible size tied to one end of the string whose other end is attached to a fixed point and oscillating in a vertical plane under gravity through a small angle, are said to form a simple pendulum.

33.7.1 OSCILLATING OF A SIMPLE PENDULUM

Let O be the fixed point, l the length of the string and m the mass of the bob (heavy mass).

Let P be the positions of the bob at any time t. Let arc $AP=s$ and $\angle AOP=\theta$, where OA is the vertical line through O.

The forces acting on the bob are

(i) Its weight mg acting vertically downward

(ii) The tension T in the string.

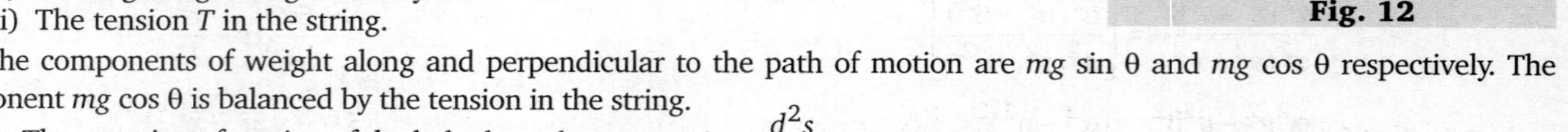

Fig. 12

The components of weight along and perpendicular to the path of motion are $mg\sin\theta$ and $mg\cos\theta$ respectively. The component $mg\cos\theta$ is balanced by the tension in the string.

$\therefore$ The equation of motion of the bob along the tangent is $m\dfrac{d^2s}{dt^2}=-mg\sin\theta$

$$\frac{d^2(l\theta)}{dt^2}=-g\sin\theta \qquad (\because s=l\theta)$$

$$\frac{d^2\theta}{dt^2}=-\frac{g}{l}\left(\theta-\frac{\theta^3}{3!}+\ldots\right)=-\frac{g}{l}\theta \text{ to a first approximation.}$$

or
$$\frac{d^2\theta}{dt^2}+\omega^2\theta=0,\ \text{where, } \omega^2=\frac{g}{l}.$$

Its auxiliary equation is

$$m^2+\omega^2=0 \Rightarrow m=\pm i\omega$$

Solution is given by
$$\theta=C_1\cos\omega t+C_2\sin\omega t$$

$$\theta=C_1\cos\sqrt{\frac{g}{l}}\,t+C_2\sin\sqrt{\frac{g}{l}}\,t$$

The motion of the bob is simple harmonic and the time of an oscillation is $\dfrac{2\pi}{\omega}=2\pi\sqrt{\dfrac{l}{g}}$.

33.7.2 BEAT OF A PENDULUM

A beat of a pendulum means its going from one extreme position of rest to the other position of rest, *i.e.*, half of the complete oscillation.

$\therefore$ The time of a beat $= \frac{1}{2}T = \pi\sqrt{\frac{l}{g}}$.

33.7.3 THE SECOND'S PENDULUM

If a simple pendulum oscillates from rest in one second, *i.e.*, if the time of one beat of a simple pendulum is one second, then it is called a second's pendulum and such a clock is said to be a correct clock.

Thus for a second pendulum,

$$1 = \pi\sqrt{\frac{l}{g}}, \quad l = \frac{g}{\pi^2} = \frac{g}{(3.1416)^2}$$

In F.P.S. system, $g = 32.2$, then $l = \frac{32.2}{(3.1416)^2} = 39.14$ inches (approx.)

and in C.G.S. system, $g = 981$, then $l = \frac{981}{(3.1416)^2} = 99.4$ cm (approx.)

33.7.4 GAIN OR LOSS OF BEATS BY A CLOCK

Let t be the time of n beat.

$$t = n\pi\sqrt{\frac{l}{g}} \text{ or } n = \frac{t}{\pi}\sqrt{\frac{g}{l}} \quad \text{... (1)}$$

Taking log on both sides of (1), we get

$$\log n = \log t - \log \pi + \frac{1}{2}\log g - \frac{1}{2}\log l \quad \text{... (2)}$$

Taking differential of (2), we have

$$\frac{dn}{n} = \frac{1}{2}\frac{dg}{g} - \frac{1}{2}\frac{dl}{l} \quad \text{... (3)}$$

If l changes, g remains constant, then $\frac{dn}{n} = -\frac{1}{2}\frac{dl}{l}$

If g changes, l remains constant, then $\frac{dn}{n} = \frac{1}{2}\frac{dg}{g}$

Solved Examples

Example 1. *A clock is provided with a seconds pendulum is gaining 5 minutes a day. Find how much the length of the pendulum should be increased so as to correct the clock.*

Solution. Here, n = Number of beats in a second pendulum $= 86400$.

δn = Change in the number of beats $= 5 \times 60 = 300$

The length of second's pendulum = 99.4 cm

We have, $dn = -\frac{n}{2}\frac{\delta l}{l}$ ... (1)

$$300 = -\frac{86400}{2}\frac{\delta l}{99.4}$$

$$\delta l = -\frac{300 \times 2 \times 99.4}{86400} = -0.690$$

As the pendulum gains 5 minutes a day, its length is less than its correct length by 0.690 cm. Hence, the length of the pendulum should be increased by 0.690 cm in order to correct the clock.

Example 2. *A second's pendulum which gains 10 seconds per day at one place loses 10 seconds per day at another. Compare the accelerations due to gravity at the two places.* (Kurukshetra–2005)

Solution. Let g be the acceleration due to gravity when the pendulum beats seconds.

Let $g + g_1$ be the acceleration due to gravity at the place, where it gains 10 seconds per day, then :

$$dn = 10, \quad n = 86400$$

Since $dn = \frac{n}{2} \times \frac{g_1}{g}$

(Here dg has been replaced by g_1)

$$10 = \frac{86400}{2} \cdot \frac{g_1}{g}$$

$$\frac{g_1}{g} = \frac{1}{4320}$$

Adding (1) to both sides

$$\frac{g+g_1}{g} = \frac{4321}{4320} \quad \text{... (1)}$$

Let $g+g_2$ be the acceleration due to gravity at the places where it loses 10 seconds per day, then

$$dn = -10$$

$$\Rightarrow \quad -10 = \frac{86400}{2} \cdot \frac{g_2}{g}$$

$$\Rightarrow \quad \frac{g_2}{g} = -\frac{1}{4320}$$

$$\Rightarrow \quad \frac{g+g_2}{g} = \frac{4319}{4320} \quad \text{... (2)}$$

Dividing (1) by (2), we get $\dfrac{g+g_1}{g+g_2} = \dfrac{4321}{4319}$, which is the required ratio.

Example 3. *If a pendulum of length l makes n complete oscillations in a given time, show that if g is changed to (g+g′), the number of oscillations gained is ng′/(2g).*

Solution. For a pendulum of length l, the time of one complete oscillation T is given by

$$T = 2\pi\sqrt{\frac{l}{g}}$$

n = the number of complete oscillation in a given time t

$$= \frac{t}{T} = \frac{t}{2\pi}\sqrt{\frac{g}{l}}$$

$$\log n = \log\frac{t}{2\pi} + \frac{1}{2}\log g - \frac{1}{2}\log l$$

Differentiating, we get

$$\frac{1}{n}\delta n = \frac{1}{2g}\delta g - \frac{1}{2l}\delta l \quad \text{... (1)}$$

$$(\because t/2\pi \text{ is constant})$$

If l is fixed then $\delta l = 0$ and if g is changed to $(g+g')$, then $\delta g - g'$

from (1), we get

$$\frac{1}{n}\delta n = \frac{1}{2g}g'$$

Hence, the number of oscillations gained

$$= \delta n = \frac{ng'}{2g}$$

Example 4. *Find how many seconds a clock would lose per day, if the length of its pendulum were increased in the ratio 900 : 901.*

Solution. Let the original length and $l + dl$, the increased length of the pendulum, then

$$\frac{l}{d+dl} = \frac{900}{901}$$

$$\frac{dl}{l} = \frac{901}{900} - 1 = \frac{1}{900}$$

Let n be the number of beats per day, then

$$n = 8.6400.$$

If dn is the change in the number of beats, then

$$dn = -\frac{n}{2}\frac{dl}{l} = -\frac{86400}{2} \times \frac{1}{900} = -48$$

Since dn is negative, the clock will lose 48 seconds per day.

Example 5. *If a pendulum of length l makes n complete oscillations in a given time, show that, if the length be changed to l + l′, the number of oscillations lost is nl′/(2l).*

Solution. For a pendulum of length l, the time of one complete oscillation T is given by $T = 2\pi\sqrt{\dfrac{l}{g}}$.

$\therefore$ n = the number of complete oscillations in a given time $t = \dfrac{t}{T} = \dfrac{t}{2\pi}\sqrt{\dfrac{g}{l}}$

$$\log n = \log\left(\frac{t}{2\pi}\right) + \frac{1}{2}\log g - \frac{1}{2}\log l$$

Differentiating, we get

$$\frac{1}{n}\delta n = \frac{1}{2g}\delta g - \frac{1}{2l}\delta l$$

$$(\because t/2\pi \text{ is constant})$$

If g is fixed, then $\delta g = 0$ and if l is changed to $l+l'$, then

$$\delta l = l'$$

From (1), we get

$$\frac{1}{n}\delta n = 0 - \frac{1}{2l}l'$$

$$\delta n = -\frac{n}{2l}l'$$

Hence, the number of oscillations lost in time t

$$= -\delta n = \frac{nl'}{2l}.$$

EXERCISE 33.4

1. Find how many seconds a clock would lose per day if the length of its pendulum were increased in the ratio 450 : 451.
2. A clock with a second's pendulum loses 10 seconds per day at a place where g = 980 cm/sec^2. What change in the gravity is necessary to make it accurate.
3. If a pendulum clock loses 9 minutes per week, find in mm, what change is required in the length of the pendulum in order that the clock may keep correct time.
4. A pendulum of length l hangs against a wall inclined at an angle θ to be horizontal. Show that the time of complete oscillation is $2\pi\sqrt{\dfrac{l}{g\sin\theta}}$.

5. If l_1 be the length of an imperfectly adjusted second's pendulum which gains n seconds in one hour and l_2 the length of one which loses n seconds in one hour, at the same place, show that the true length of second's pendulum is $\dfrac{4l_1l_2}{l_1+l_2+2\sqrt{l_1l_2}}$.

6. A pendulum of length l has one end of the string fastned to a peg on a smooth plane inclined to the horizon at an angle α. With the string and the weight on the plane, its time of oscillation in t sec. If the pendulum of length l' oscillates in one second when suspended vertically, prove that $\alpha = \sin^{-1}\left(\dfrac{1}{l't^2}\right)$ (Kurukshetra–2006)

7. It $I\dfrac{d^2\theta}{dt^2} = -mgl\sin\theta$, where I, m, g, l are constants, given that at $t=0$, $\theta = 0$ and $\dfrac{d\theta}{dt} = \omega_0 = m\sqrt{mgl/I}$ then show that $t = \dfrac{2}{\omega_0}\log\dfrac{\pi+\theta}{4}$ (Nagpur–2009)

8. A point moves in a straight line towards the centre of force $\mu/(\text{distance})^2$ starting from rest at a distance a from the centre of force, show that the time of reaching a point b from the centre of force is $a\sqrt{a^2-b^2}/\sqrt{\mu}$ and that its velocity then is $\dfrac{\sqrt{\mu}}{ab}\sqrt{a^2-b^2}$ (UPTU–2001)

ANSWERS

1. 96 seconds per day 2. to be increased by 0.227 cm/sec^2

33.8 LINEAR MOTION OF A PARTICLE IN A RESISTING MEDIUM

When a body is moving vertically in a resisting medium, there are two forces acting on the body *i.e.*, its weight downward and the resistance of the air opposite to the direction of the motion. If it is moving horizontally, the only force acting on it is the resistance.

If the particle is falling under gravity in a resisting medium, the velocity will never exceed some definite quantity. When the resistance becomes equal to the weight, the acceleration is then zero, and the particle will continue to move with a constant velocity. This velocity for which the acceleration becomes zero is called the limiting or terminal velocity.

Solved Examples

Example 1. *A particle of mass m is projected vertically upward under gravity, the resistance of the air being mk times the velocity. Show that the greatest height attained by the particle is* $\dfrac{V^2}{g}[\lambda - \log(1+\lambda)]$, *where V is the greatest velocity which the above mass will attain when it falls freely and* λV *is the initial velocity,*

Solution. Let V be the velocity of the particle at time t. The forces acting on the particle are

(i) its weight mg acting vertically downwards.

(ii) the resistance mkV of the air acting vertically downwards.

Accelerating force on the particle $= -mg - mkV$

$\therefore$ By Newton's second law, the equation of motion of the particle is $mV\dfrac{dV}{dx} = -mg - mkV$.

or $$V\frac{dV}{dx} = -g - kV \quad \text{... (1)}$$

When the particle falls freely (under gravity), equation (1) becomes (changing g to $-g$)

$$V\frac{dV}{dx} = g - kV \quad \text{... (2)}$$

When the particle attains the greatest velocity V, its acceleration is zero.

From (2), $0 = g - kV$

$\Rightarrow$ $k = g/V$

Putting this value of k in equation (1), we have

$$V\frac{dV}{dx} = -g - \frac{g}{V}v = -\frac{g}{V}(V+v)$$

$$\frac{v}{V+v}dv = -\frac{g}{V}dx$$

Integrating $\displaystyle\int \frac{v}{V+v}dv = -\frac{g}{V}\int dx + C$

or $\displaystyle\int\left(1 - \frac{V}{V+v}\right)dv = -\frac{g}{V}x + C$

or $$v - V\log(V+v) = -\frac{g}{V}x + C \quad \text{... (3)}$$

Initially, when $x = 0$, $v = \lambda V$

From (3), we have $\lambda V - V\log(V+\lambda V) = C$

or $C = V[\lambda - \log V(1+\lambda)]$

Substituting the value of C in (3), we get

$$v - V\log(V+v) = -\frac{g}{V}x + V[\lambda - \log V(1+\lambda)] \quad \text{... (4)}$$

Let h be the greatest height attained by the particle, then $x = h$ when $V = 0$.

From (4), we have

$$-V\log V = -\frac{g}{h}h + V[\lambda - \log V(1+\lambda)]$$

$$\frac{g}{V}h = V\lambda - V[\log V(1+\lambda) - \log V]$$

$$= V\lambda - V\log\frac{V(1+\lambda)}{V}$$

Hence, $$h = \frac{V^2}{g}[\lambda - \log(1+\lambda)]$$

Example 2. *A particle falls under gravity in a resisting medium whose resistance varies with velocity. Find the relation between distance and velocity if initially the particle starts from rest.* [UPTU–2003]

Solution. By Newton's second law of motion, the equation of motion of the body is

$$mV\frac{dV}{dx} = mg - mkV$$

$$\Rightarrow V\frac{dV}{dx} = g - kV \quad \Rightarrow \quad \frac{V\,dV}{g - kV} = dx$$

[On separating the variables]

$$\Rightarrow \quad -\frac{dV}{k} + \frac{g}{k}\frac{dV}{g - kV} = dx$$

Integrating, we get

$$-\frac{V}{k} = \frac{g}{k}\left(-\frac{1}{k}\right)\log(g - kV) = x + C \quad \text{... (1)}$$

$$-\frac{V}{k} - \frac{g}{k^2}\log(g - kV) = x + C$$

Initially, $x = 0,\ V = 0$

so, $-\frac{g}{k^2}\log g = C$

Equation (1) becomes

$$-\frac{V}{k} - \frac{g}{k^2}\log(g - kV) = x - \frac{g}{k^2}\log g$$

$$\Rightarrow -\frac{V}{k} - \frac{g}{k^2}\log\frac{g - kV}{g} = x$$

Example 3. *A particle is projected with velocity V along a smooth horizontal plane in a medium whose resistance per unit mass is μ times the cube of the velocity. Show that the distance it was described in time t in* $(1/\mu V)\left[\sqrt{1 + 2\mu V^2 t} - 1\right]$ *and that its velocity then is* $V/\sqrt{(1 + 2\mu V^2 t)}$.

Solution. Take the point of projection O as origin. Let v be the velocity of the particle at time t at a point distant x from the fixed point O. Then the resistance at this point will be $m\mu v^3$, acting in the direction of x decreasing. Here, the resistance is the only force acting on the particle during its motion.

∴ The equation of motion of the particle is

$$m\frac{dv}{dt} = -m\mu v^3$$

$$\Rightarrow \quad \frac{dv}{v^3} = -\mu\,dt$$

Integrating, $-\frac{1}{2v^2} = -\mu t + A$, where A is a constant.

But initially, when $t = 0,\ v = V$ and

so $$A = -\frac{1}{2V^2}$$

$$-\frac{1}{2v^2} = -\mu t - \frac{1}{2V^2} \Rightarrow \frac{1}{v^2} = (2\mu V^2 t + 1)/V^2$$

$$v = V/\sqrt{(1 + 2\mu V^2 t)} \quad \text{... (1)}$$

which gives the velocity of the particle at time t. Since the particle is moving in the direction of x increasing, therefore from the equation (1), we have

$$\frac{dx}{dt} = v = V/\sqrt{(1 + 2\mu V^2 t}$$

$$\Rightarrow \quad dx = V(1 + 2\mu V^2 t)^{-1/2}\,dt$$

Integrating, $x = \frac{1}{\mu V}(1 + 2\mu V^2 t)^{1/2} + B$,

where B is a constant.

But initially, when $t = 0,\ x = 0$, $\therefore B = -\frac{1}{\mu V}$

$$x = \frac{1}{\mu V}(1 + 2\mu V^2 t)^{1/2} - \frac{1}{\mu V}$$

Hence, $x = \frac{1}{\mu V}\left[\sqrt{(1 + 2\mu V^2 t)} - 1\right]$

which gives the distance described in time t.

Example 4. *A moving body is opposed by a force per unit mass of value Cx and resistance per unit mass of value* bv^2, *where x and v are the displacement and velocity of the particle at that instant. Show that the velocity of the particle, if it starts from rest, is given by* $v^2 = \frac{C}{2b^2}(1 - e^{-2bx}) - \frac{Cx}{b}$.

Solution. By Newton's second law of motion, the equation of motion of the body is

$$v\frac{dv}{dx} = -Cx - bv^2$$

$$\Rightarrow \quad v\frac{dv}{dx} + bv^2 = -Cx \quad \text{... (1)}$$

Putting $v^2 = z$ and $2v\frac{dv}{dx} = \frac{dz}{dx}$

Equation (1) becomes $\frac{1}{2}\frac{dz}{dx} + bz = -Cx$.

$$\frac{dz}{dx} + 2bz = -2Cx \quad \text{... (2)}$$

which is Leibnitz's linear equation

$$\text{I.F.} = e^{\int 2b\,dx} = e^{2bx}.$$

The solution of (2) is

$$z.e^{2bx} = \int -2Cx.e^{2bx}\,dx + C_1$$

$$= -2C\left[x.\frac{e^{2bx}}{2b} - \int 1.\frac{e^{2bx}}{2b}dx\right] + C_1$$

$$= -\frac{Cx}{b}e^{2bx} + \frac{C}{2b^2}e^{2bx} + C_1$$

$$\Rightarrow v^2.e^{2bx} = -\frac{Cx}{b}e^{2bx} + \frac{C}{2b^2}e^{2bx} + C_1$$

$$\Rightarrow \qquad v^2 = -\frac{Cx}{b} + \frac{C}{2b^2} + C_1 e^{-2bx} \qquad \text{... (3)}$$

Initially when $x = 0,\ v = 0$

$$\frac{C}{2b^2} + C_1 = 0 \Rightarrow C_1 = -\frac{C}{2b^2}$$

Substituting the value of C_1 in (3), we have

$$v^2 = -\frac{Cx}{b} + \frac{C}{2b^2} - \frac{C}{2b}e^{-2bx}$$

Hence, $v^2 = \dfrac{C}{2b^2}(1 - e^{-2bx}) - \dfrac{Cx}{b}$

Example 5. *The accelertion and velocity of a body falling in the air approximately satisfy the equtation: Acceleration = $g - kV^2$, where V is the velocity of the body at any time t, and g, k are constants. Find the distance traversed as a function of the time, if the body falls from rest. Show that value of v will never exceed* $\sqrt{\dfrac{g}{k}}$

Solution. Acceleration = $g - kV^2$

$$\frac{dV}{dt} = g - kV^2 \Rightarrow \qquad \frac{dV}{g - k^2} = dt$$

$$\Rightarrow \quad \frac{1}{2\sqrt{g}}\left[\frac{1}{\sqrt{g} + \sqrt{k}V} + \frac{1}{\sqrt{g} - \sqrt{k}\,V}\right] dV = dt$$

On integrating, we get

$$\frac{1}{2\sqrt{g}}\frac{1}{\sqrt{k}}\log(\sqrt{g} + \sqrt{k}V)$$

$$-\frac{1}{2\sqrt{gk}}\log(\sqrt{g} - \sqrt{k}V) = t + C$$

$$\Rightarrow \quad \frac{1}{2\sqrt{gk}}\log\frac{\sqrt{g} + \sqrt{k}.V}{\sqrt{g} - \sqrt{k}.V} = t + C \qquad \text{... (1)}$$

On putting $t = 0$, $V = 0$ in equation (1), we get

$$\frac{1}{2\sqrt{gk}}\log 1 = 0 + C \quad \Rightarrow \quad C = 0$$

Equation (1) becomes

$$\frac{1}{2\sqrt{gk}}\log\frac{\sqrt{g} + \sqrt{k}.V}{\sqrt{g} - \sqrt{k}.V} = t$$

$$\Rightarrow \qquad \log\frac{\sqrt{g} + \sqrt{k}.V}{\sqrt{g} - \sqrt{k}.V} = 2\sqrt{gk}\ t$$

$$\frac{\sqrt{g} + \sqrt{k}.V}{\sqrt{g} - \sqrt{k}.V} = e^{2\sqrt{gk}\ t}$$

By componendo and dividendo, we have

$$\frac{\sqrt{k}\,V}{\sqrt{g}} = \frac{e^{2\sqrt{gk}\,t} - 1}{e^{2\sqrt{gk}\,t} + 1} = \frac{e^{\sqrt{gk}\,t} - e^{\sqrt{gk}\,t}}{e^{\sqrt{gk}\,t} + e^{\sqrt{gk}\,t}}$$

$$= \tanh\sqrt{gk}\ t$$

$$\Rightarrow \qquad V = \sqrt{\frac{g}{k}}\tanh\sqrt{gk}\ t$$

Whenever the value of t may be

$$\tanh\sqrt{gk}\ t \le 1.$$

Hence, the value of V will never exceed $\sqrt{\dfrac{g}{k}}$.

$$\therefore \qquad \frac{dx}{dt} = \sqrt{\frac{g}{k}}\tanh\sqrt{gk}\ t$$

Integrating again, $x = \sqrt{\dfrac{g}{k}}\displaystyle\int \tanh\sqrt{gk}\,t + B$

$$\Rightarrow x = \frac{1}{k}\log\cosh\sqrt{gk}.t + B$$

When $t = 0,\ x = 0$, then $B = 0$.

Hence, $x = \dfrac{1}{k}\log\cosh\sqrt{gk}\ t$

Example 6. *A particle of mass m is falling under the influence of gravity through a medium whose resistance equals μ times the velocity. If the particle were released from rest, show that the distance fallen through in time t is* $\dfrac{gm^2}{\mu^2}\left[e^{-(\mu/m)t} - 1 + \dfrac{\mu t}{m}\right]$.

Solution. Let a particle of mass m falling under gravity be at a distance x from the starting point after time t. If v is its velocity at this point, then the resistance on the particle is μv acting vertically upwards, *i.e.*, in the direction of x decreasing. The weight mg of the particle acts vertically downwards, *i.e.*, in the direction of x increasing.

$\therefore$ The equation of motion of the particle is

$$m\frac{d^2x}{dt^2} = mg - \mu v$$

$$\frac{dv}{dt} = g - \frac{\mu}{m}v \Rightarrow dt = \frac{dv}{g - (\mu/m)v}$$

Integrating, we get

$$t = -\frac{m}{\mu}\log\left(g - \frac{\mu}{m}v\right) + A,$$

where A is a constant.

But initially, when $t = 0$, $v = 0$ and

so $\qquad A = (m/\mu)\log g$

$$t = -\frac{m}{\mu}\log\left(g - \frac{\mu}{m}v\right) + \frac{m}{\mu}\log g$$

$$\Rightarrow \qquad t = -\frac{m}{\mu}\log\left\{\frac{g - (\mu/m)v}{g}\right\}$$

$$-\frac{\mu t}{m} = \log\left(1-\frac{\mu}{gm}v\right)$$

$$1-\frac{\mu v}{gm} = e^{-\mu t/m}$$

$$\Rightarrow \quad v = \frac{dx}{dt} = \frac{gm}{\mu}\left(1-e^{-\mu t/m}\right)$$

$$\Rightarrow \quad dx = \frac{gm}{\mu}(1-e^{-\mu t/m})\,dt$$

Integrating, we get

$$x = \frac{gm}{\mu}\left[t+\frac{m}{\mu}e^{-\mu t/m}\right]+B \qquad \text{... (1)}$$

where B is a constant.

But initially, when $t = 0$, $x = 0$. Therefore,

$$0 = \frac{gm}{\mu}\left[\frac{m}{\mu}\right]+B \qquad \text{... (2)}$$

Subtracting (2) from (1), we have

$$x = \frac{gm}{\mu}\left[\frac{m}{\mu}e^{-\mu t/m}-\frac{m}{\mu}+t\right]$$

$$= \frac{gm^2}{\mu^2}\left\{e^{-(\mu t/m)}-1+\frac{\mu t}{m}\right\}$$

Example 7. *A body falling vertically under gravity encounters resistance of the atmosphere. If the resistance varies as the velocity, show that the equation of motion is given by*

$$\frac{du}{dt} = g-ku$$

where u is the velocity, k is a constant and g is the acceleration due to gravity. Show that as t increases, u approaches the value g/k. Also, if $u = dx/dt$, where x is the distance fallen by the body from rest in time t, show that $x = \frac{gt}{k}-\frac{g}{k^2}(1-e^{-kt})$.

Solution. Let the mass of the falling body be unity.

Acceleration $= \frac{du}{dt}$.

Force acting downward $= 1.\frac{du}{dt} = \frac{du}{dt}$

Force of resistance $= -ku$

Net force acting downward $= g-ku$.

$$\frac{du}{dt} = g-ku \qquad \text{... (1)}$$

$$\Rightarrow \quad \frac{du}{g-ku} = dt$$

Integrating $\int \frac{du}{g-ku} = \int dt$

$$\Rightarrow \quad t = -\frac{1}{k}\log(g-ku)+\log a$$

$$= \log(g-ku)^{-t/k}a$$

$$a(g-ku)^{-1/k} = e^t \Rightarrow (g-ku) = a^k e^{-kt}$$

$$\Rightarrow \quad u = \frac{g}{k}-\frac{a^k}{k}e^{-kt}$$

If f increases very large, then $\frac{a^k}{k}e^{-kt} = 0$

$u = \frac{g}{k}$, when $t \to \infty$

Now, $\quad u = \frac{dx}{dt} \Rightarrow \frac{du}{dt} = \frac{d^2x}{dt^2}$.

Putting the value of $\frac{d^2u}{dt^2}$ and u in (1), we get

$$\frac{d^2x}{dt^2}+k\frac{dx}{dt} = g \text{ or } (D^2+kD)x = g.$$

Auxiliary equation is $D(D+k) = 0$

$$\Rightarrow \quad D = 0,\ D = -k$$

$$\text{C.F.} = C_1+C_2e^{-kt}.$$

$$\text{P.I.} = \frac{1}{D^2+kD}g = t\frac{1}{2D+k}g = \frac{tg}{k}.$$

$$\Rightarrow \quad x = C_1+C_2e^{-kt}+\frac{gt}{k} \qquad \text{... (2)}$$

Putting the value of $t = 0$ and $x = 0$ in equation (2), we get

$$0 = C_1+C_2 \text{ or } C_2 = -C_1.$$

Equation (2) becomes

$$x = C_1-C_1e^{-kt}+\frac{gt}{k} \qquad \text{... (3)}$$

On differentiating (3), we get

$$\frac{dx}{dt} = C_1ke^{-kt}+\frac{g}{k} \qquad \text{... (4)}$$

On putting $\frac{dx}{dt} = 0$, when $t = 0$ in equation (4), we get

$$0 = C_1k+\frac{g}{k} \Rightarrow C_1 = -\frac{g}{k^2}$$

Hence, $x = -\frac{g}{k^2}+\frac{g}{k^2}e^{-kt}+\frac{gt}{k}$

$$= \frac{gt}{k}-\frac{g}{k^2}(1-e^{-kt}).$$

Example 8. *A body of mass m, falling from rest, is subject to the force of gravity and an air resistance proportional to the square of the velocity (i.e., kV^2). If it falls through a distance x and possesses a velocity v at that instant, prove that $\frac{2kx}{m} = \log\frac{a^2}{a^2-V^2}$, where $mg = ka^2$.*

Solution. The forces acting on the body are

(i) its weight mg acting vertically downwards.

(ii) the resistance kv^2 of the air acting vertically upwards.

Accelerating force on the body $= mg - kV^2$

$$= ka^2 - kV^2 = k(a^2 - V^2).$$

By Newton's second law, the equation of motion of the body is

$$mv\frac{dv}{dx} = k(a^2 - v^2)$$

$$\frac{v}{a^2 - v^2}dv = \frac{k}{m}dx.$$

Integrating, $\int \frac{v}{a^2 - v^2}dv = \frac{k}{m}\int dx + C$

$$-\frac{1}{2}\log(a^2 - v^2) = \frac{k}{m}x + C \quad \dots (1)$$

Initially, when $x = 0,\ v = 0,\ \therefore\ C = -\frac{1}{2}\log a^2$

From (1),

we have $-\frac{1}{2}\log(a^2 - v^2) = \frac{k}{m}x - \frac{1}{2}\log a^2$

$$\Rightarrow \frac{2kx}{m} = \log a^2 - \log(a^2 - v^2)$$

$$= \log \frac{a^2}{a^2 - v^2}$$

EXERCISE 33.5

1. A particle falls in a vertical line under gravity (supposed constant) and the force of air resistance to its motion is proportional to its velocity. Show that its velocity cannot exceed a particular limit.

2. A body falling from rest is subjected to a force of gravity and an air resistance of n^2/g times of the square of velocity. Show that the distance travelled by the body in t seconds is $\frac{g}{n^2}\log\cosh nt$.

3. A particle of unit mass moves in a straight line under retardation which is k times its velocity. Initially, the particle is at a distance a from a given point O in the line and is moving towards O with velocity $u(>ak)$. Prove that it will reach O in time $\frac{1}{k}\log\frac{u}{u - ak}$.

4. A particle of unit mass is projected vertically upward with velocity V and the resistance of the air produces a retardation kv^2, where v is the velocity. Show that the velocity V', with which the particle will return to the point of projection, is given by $\frac{1}{V'^2} = \frac{1}{V^2} + \frac{k}{g}$.

5. A moving body is opposed by a force proportional to the displacement and by a resistance proportional to the square of the velocity. Prove that the velocity is given by

$$V^2 = aC - \frac{Cx}{b} + \frac{C}{ab^2}$$

ANSWERS

1. $V = \frac{g}{k}$

33.9 HOOKE'S LAW

The tension of an elastic string is proportional to the extension of the string beyond its natural length.

If x is the stretched length of a string of natural length l, then by Hooke's law the tension T in the string is given by $T = E\frac{x - l}{l}$, where E is called the modulus of elasticity of the string that the direction of the tension is always opposite to the extension.

33.9.1 PARTICLE ATTACHED TO ONE END OF A HORIZONTAL ELASTIC STRING

If one end of the elastic string be fixed at O on a table. The other end A of the elastic string of length l is attached to a particle of mass m.

The string is stretched to a point B and then released. The particle comes into motion. Let the particle be at a distance x from A at any time t. The weight mg of the particle is acting downward and is balanced by the reaction R of the table.

The only force acting upon the particle is the tension of the string

By Hooke's law, $\frac{\text{Stress}}{\text{Strain}}$ = Constant of elasticity

$$T = E\frac{x}{l}$$

where, x = Extension of the length of the string,

E = Modulus of elasticity.

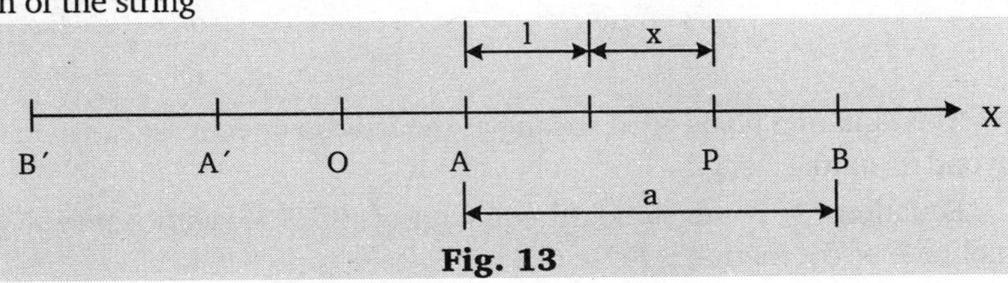

Fig. 13

Equation of motion is

$$m\frac{d^2x}{dt^2} = -\frac{Ex}{l}$$

$$\frac{d^2x}{dt^2} = -\left(\frac{E}{ml}\right)x \quad \text{... (1)}$$

The motion of the equation is S.H.M.

On multiplying (1) by $2\frac{dx}{dt}$, we get $2\frac{d^2x}{dt^2}\cdot\frac{dx}{dt} = -2\frac{Ex}{lm}\cdot\frac{dx}{dt}$

On integrating, we get $\left(\frac{dx}{dt}\right)^2 = \frac{-Ex^2}{lm} + C \quad \Rightarrow \quad V^2 = -\frac{Ex^2}{ml} + C$ ... (2)

If the velocity of the particle is zero at B, amplitude $AB = a$. The particle moves from A to B and back B to A will be a S.H.M. The particle moves towards O. Then it moves with uniform velocity upto A'.

On putting $V = 0,\ x = a$ we get $0 = -\frac{Ea^2}{ml} + C$ or $C = \frac{Ea^2}{ml}$

$$\Rightarrow \quad V^2 = \frac{E}{lm}(a^2 - x^2)$$

At $A,\ x = 0,\ V = \sqrt{\frac{E}{l}}\,a$. This is the maximum velocity. The particle moves from A to A' with this velocity. After that the string again stretches and motion becomes S.H.M.

Periodic time of S.H.M. (from A to B, B to A, A' to B', B to A') + time taken by particle from A to A' and A' to A with constant velocity $\sqrt{\frac{E}{l}}\,a$.

$$\frac{2\pi}{\sqrt{E/l}} + \frac{4l}{\sqrt{(E/lm)}\,a} = \sqrt{\left(\frac{lm}{E}\right)}\cdot\left(\pi + \frac{2l}{a}\right)$$

33.9.2 PARTICLE SUSPENDED BY AN ELASTIC STRING

Let one end of the string OA of natural length l be attached to the fixed point O and a particle of mass m be attached to the other end A. Due to the weight mg of the particle the string OA is stretched and if B is the position of equilibrium of the particle such that $AB = d$, then the tension T_B in the string will balance the weight of the particle, *i.e.*,

$$mg = T_B \quad \text{or} \quad mg = E\frac{AB}{OA} = E\frac{d}{l} \quad \text{... (1)}$$

The particle is pulled down to a point C such that $BC = C$ and then released. At the point C, the tension in the string is greater than weight of the particle and so the particle starts moving vertically upwards with velocity zero at C. Let P be the position of the particle at any time t, where $BP = x$. The tension in the string when the particle is at P is $T_P = E\frac{d+x}{l}$, acting vertically upwards.

The resultant force acting on the particle at P is the vertically upwards direction.

$$= T_P - mg = E\left(\frac{d+x}{l}\right) - mg = \frac{Ed}{l} + \frac{Ex}{l} - mg = \frac{Ex}{a} \quad \text{[Using (1)]}$$

Also, the acceleration of the particle at P is d^2x/dx^2 in the direction of x increasing, *i.e.*, in the vertically downwards direction.

$\therefore$ By Newton's law, the equation of motion of P is given by

$$m\frac{d^2x}{dt^2} = -\frac{Ex}{l} \quad \text{or} \quad \frac{d^2x}{dt^2} = -\frac{E}{lm}x \quad \text{... (2)}$$

This equation holds good so long as the tension operates, *i.e.*, when the string is extended beyond its natural length.

Equation (2) is the standard equation of a S.H.M. with centre at the origin B and the amplitude of the motion is $BC = C$.

The periodic time T of the S.H.M. represented by the equation (2) is given by

$$T = \frac{2\pi}{\sqrt{E/ml}} = 2\pi\sqrt{\frac{lm}{E}} \quad \text{... (3)}$$

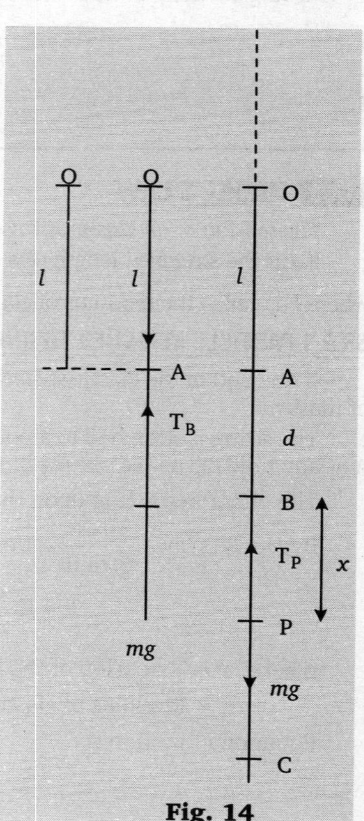

Fig. 14

Solved Examples

Example 1. *Two particles of mass m_1 and m_2 are tied to the ends of an elastic string of natural length a and modulus λ. They are placed on a smooth table so that the string is just taut and m_2 is projected with any velocity directly away from m_1. Prove that the string will become slack after the lapse of time :*

$$\pi\left[am_1m_2 / \lambda(m_1+m_2)\right]^{1/2}.$$

Solution. Let initially at $t = 0$, the particle m_1 be placed at O, the fixed point of reference. Given that the particle m_2 is at a distance a from m_1. Let at time t the particle m_1 be at a distance x from O and the particle m_2 at a distance y from O.

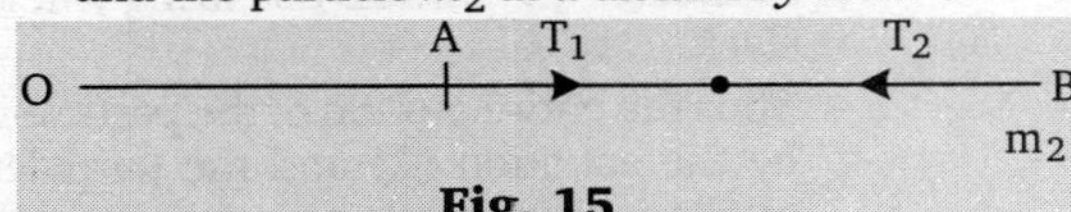

Fig. 15

Let T_1 be the tension in the string.

The equation of motion of m_1 is

$$m_1\frac{d^2x}{dt^2} = +T_1 \quad \ldots (1)$$

The equation of motion of m_2 is

$$m_2\frac{d^2y}{dt^2} = -T_1 \quad \ldots (2)$$

and
$$T_1 = \lambda\,\frac{y-x}{a} \quad \ldots (3)$$

From (1) and (2), we get

$$\frac{d^2y}{dt^2}-\frac{d^2x}{dt^2} = -\frac{T_1}{m_2}-\frac{T_1}{m_1}$$

$$\Rightarrow \quad \frac{d^2(y-x)}{dt^2} = -\left(\frac{1}{m_1}+\frac{1}{m_2}\right)\lambda\,\frac{y-x}{a}$$

or
$$\frac{d^2z}{dt^2} = \frac{-\lambda(m_1+m_2)}{m_1m_2a}\,z,$$

taking $y-x=z$.

This is the equation of motion of S.H.M. with periodic time.

$$T = \frac{2\pi}{\sqrt{\dfrac{\lambda(m_1+m_2)}{a(m_1\,.\,m_2)}}}.$$

Now, when m_2 starts coming towards m_1 and when the string acquires its original length, the string becomes slack after this time. This required time

$$=\frac{T}{4}+\frac{T}{4}=\frac{T}{2}.$$

Hence, the string becomes slack after time

$$\frac{T}{2} = \pi\left[am_1m_2 / \lambda(m_1+m_2)\right]^{1/2}$$

Example 2. *A light elastic string of original length l is hung by one end to the other end are tied successively particles of masses m, m′. If t_1 and t_2 be the periods of small oscillations corresponding to these weights are C_1 and C_2 the statical extensions, prove that $g(t_1^2-t_2^2) = 4\pi^2(C_1-C_2)$.*

Solution. We have

$$mg = T_1 \quad \ldots (1)$$

$$mg = E\,\frac{C_1}{l} \quad \ldots (2)$$

$$m'g = E\,\frac{C_2}{l} \quad \ldots (3)$$

Equation of motion of first particle,

$$m\frac{d^2x}{dt^2} = mg - E\frac{(x+C_1)}{l}$$

$$\Rightarrow \quad m\frac{d^2x}{dt^2} = mg - E\frac{C_1}{l}-\frac{Ex}{l} = \frac{-Ex}{l}$$

[Using (2)]

Motion of S.H.M. with $t_1 = \dfrac{2\pi}{\sqrt{E/lm}}$

Similarly, $t_2 = \dfrac{2\pi}{\sqrt{E/lm'}}$

Now, $$t_1^2-t_2^2 = 4\pi^2\,\frac{l}{E}(m-m')$$

$$= 4\pi^2\left(\frac{C_1}{g}-\frac{C_2}{g}\right)$$

[Using (2) and (3)]

Hence, $g(t_1^2-t_2^2) = 4\pi^2(C_1-C_2)$.

Example 3. *An elastic string without weight of which the unstretched length is l and modulus of elasticity is the weight of n kg is suspended by one end and a mass m kg is attached to the other end. Show that the time of a small vertical oscillation is $2\pi\sqrt{\dfrac{ml}{ng}}$.*

Solution. Let $OA = l$ be the natural length of a string whose one end is fixed at O. B is the position of equilibrium of a particle of mass m kg attached to the other end of the string. Considering the equilibrium of the particle at B, we have $mg =$ the tension T_B in the string OB.

$$\therefore \quad mg = ng\,\frac{AB}{l} \quad \ldots (1)$$

because modulus of elasticity of the string is given to be ng.

Now, suppose the particle is pulled slightly upto C (so that $BC < AB$ and then let go. It starts moving vertically upwards with velocity zero at C. Let P be its position at any point t, where $BP = x$.

The direction BP is that of x increasing and the direction PB is that of x decreasing. At P, there are two forces acting on the particle :

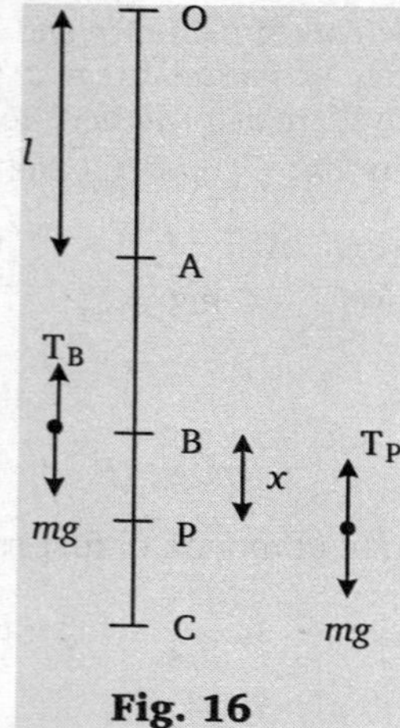

Fig. 16

(i) The weight mg acting vertically downwards, *i.e.*, in the direction of x increasing, and

(ii) the tension $T_P = ng\frac{AB + x}{l}$ in the string OP, acting vertically upwards, *i.e.*, in the direction of x decreasing.

Hence, by Newton's second law of motion, the equation of motion of the particle at P is

$$m\frac{d^2x}{dt^2} = mg - ng\frac{AB + x}{l}$$

$$= mg - ng\frac{AB}{l} - ng\frac{x}{l} = -ng\frac{x}{l}$$

[Using (1)]

$$\therefore \quad \frac{d^2x}{dt^2} = -\frac{ng}{ml}x \qquad \text{... (2)}$$

which is the equation of a simple harmonic motion with centre at the origin B and amplitude BC.

Since $BC < AB$, therefore during the entire motion of the particle the string will not become slack.

Thus the entire motion of the particle is governed by the equation (2) and the particle will make oscillations in simple harmonic motion about the centre B.

The time of oscillation

$$= \frac{2\pi}{\sqrt{\mu}} = \frac{2\pi}{\sqrt{ng / lm}} = 2\pi\sqrt{\frac{lm}{ng}}.$$

EXERCISE 33.6

1. Two light elastic strings are fastened to a particle of mass m and their other ends to fixed points so that the strings are taut. The modulus of each is λ, the tension T, and length a and b. Show that the period of an oscillation along the line of the strings is $2\pi\left[\frac{mab}{(T+\lambda)(a+b)}\right]^{1/2}$.
2. An elastic string of natural length $(a+b)$, where $a > b$ and modulus of elasticity λ has a particle of mass m attached to it at a distance a from one end, which is fixed to a point A of a smooth horizontal plane. The other end of the string is fixed to a point B so that the string is just unstretched. If the particle will be held at B and then released, show that it will oscillate to and fro through a distance $\frac{b(\sqrt{a} + \sqrt{b})}{\sqrt{a}}$ in a periodic time $\pi(\sqrt{a} + \sqrt{b})\sqrt{m / \lambda}$.
3. A particle is performing S.H.M. in the line joining two points A and B on a smooth plane and is connected with these points by elastic strings of natural lengths a and a', the moduli of elasticity being λ and λ' respectively. Show that the periodic time is $2\pi\sqrt{m\left(\frac{\lambda}{a} + \frac{\lambda'}{a'}\right)^{-1}}$.
4. A mass m hangs from a light spring and is given a small vertical displacement. If l is the length of the spring when the system is in equilibrium and n the number of oscillations per second, show that the natural length of the spring is $l - (g / 4\pi^2 n^2)$.
5. A heavy particle is attached to one point of a uniform elastic string. The ends of the string are attached to two points in a vertical line. Show that the period of a vertical oscillation in which the string remains taut is $2\pi\sqrt{mh / 2\pi}$, where λ is the coefficient of elasticity of the string and h the harmonic mean of the unstretched lengths of the two parts of the string.
6. A light elastic string of natural length l has one extremity fixed at point O and the other attached to a stone, the weight of which in equilibrium would extend the string to a length l_1. Show that if the stone be dropped from rest at O, it will come to instantaneous rest at a depth $\sqrt{l_1^2 - l^2}$ below equilibrium position.

33.10 MECHANICAL OSCILLATORY SYSTEMS

33.10.1 FREE OSCILLATIONS

Consider a spring OA suspended vertically from a fixed support at O. Let a body of mass m be suspended from the end A, the mass of the body being so large in comparison with the mass of the spring that the latter may be neglected. Let e (= AB) be the elongation produced by the mass m hanging in equilibrium, then B is called the position of static equilibrium and e is called the static extension. Let k be the stiffness of the spring k, the restoring force per unit stretch of the spring due to elasticity.

For equilibrium at B, $mg = T = ke$ ... (1)

Let the mass be displaced through a further distance x from the equilibrium position. The acceleration of the mass m at this position is $\frac{d^2x}{dt^2}$ and the forces acting upon it are weight mg downwards and the restoring force $k(e + x)$ upwards

$\therefore$ The equation of motion of mass m is

$$m\frac{d^2x}{dt^2} = mg - k(e+x) = -kx \qquad \text{[Using (1)]}$$

$$\Rightarrow \qquad \frac{d^2x}{dt^2} + \frac{k}{m}x = 0$$

Writing $\omega^2 = \frac{k}{m}$, it becomes $\frac{d^2x}{dt^2} + \omega^2 x = 0$. ... (2)

The solution of equation (2) is $x = C_1 \cos \omega t + C_2 \sin \omega t$, which can also be written as

$$x = A\cos(\omega t + B) \qquad ... (3)$$

where C_1, C_2 or A, B are constants to be determined from the initial conditions of the problem.

Equation (3) represents oscillatory variations of the variable x with amplitude A and period $T = \frac{2\pi}{\omega}$. It represents the S.H.M. of the mass m with period

$$2\pi\sqrt{\frac{m}{k}} = 2\pi\sqrt{\frac{e}{g}} \qquad \left(\text{Since } \omega^2 = \frac{k}{m} \text{ and } mg = ke\right)$$

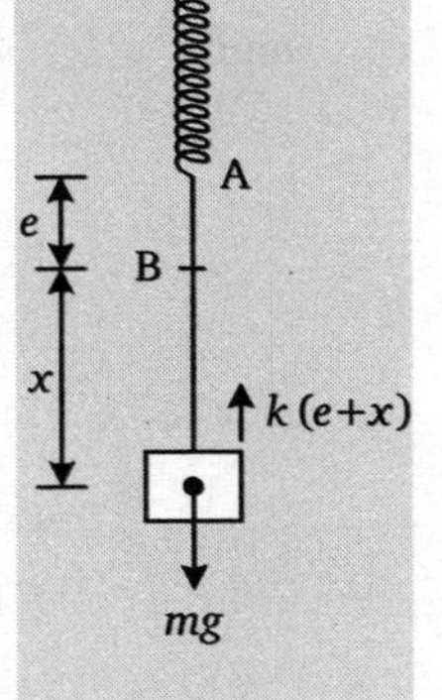

Fig. 17

33.10.2 DAMPED FREE OSCILLATIONS

We have assumed so far that no frictional forces acting on the oscillatory body. This situation is impractical, for if it were true, a pendulum or a weight on a string would continue to oscillate indefinitely. Due to friction, the amplitude gradually decreases to zero. This motion is called damped harmonic motion. The damping force generally arises due to air resistance or internal forces. Its magnitude depends generally on the speed of the body. We will consider the case when the damping force is proportional to the velocity or equal to $\lambda\frac{dx}{dt}$ of the body and is directed opposite to it, where λ is damping constant.

Equation of damped harmonic motion is

$$m\frac{d^2x}{dt^2} = mg - ke - kx - \lambda\frac{dx}{dt} = -kx - \lambda\frac{dx}{dt} \qquad [\because mg = ke]$$

$$\Rightarrow \qquad m\frac{d^2x}{dt^2} + kx + \lambda\frac{dx}{dt} = 0 \qquad \text{or} \qquad \frac{d^2x}{dt^2} + \left(\frac{\lambda}{m}\right)\frac{dx}{dt} + \left(\frac{k}{m}\right)x = C$$

Let $\frac{\lambda}{m} = 2p$ and $\frac{k}{m} = \omega^2$, then $\frac{d^2x}{dt^2} + 2p\frac{dx}{dt} + \omega^2 x = 0$ ... (1)

which is a linear differential equation with constant coefficients.

A.E. is $m^2 + 2km + \omega^2 = 0 \quad \Rightarrow \quad m = -k \pm \sqrt{k^2 - \omega^2}$

Hence, the general solution $x = A\,e^{(-p+\sqrt{p^2-\omega^2})\,t} + Be^{(-p-\sqrt{p^2-\omega^2})t}$

At $t = 0$, $x = x_0$ = maximum and so $A + B = x_0$ ... (2)

At $t = 0$, $\frac{dx}{dt} = 0$ and so $A - B = \frac{px_0}{\sqrt{p^2 - \omega^2}}$ ... (3)

Solving (2) and (3), we get $A = \frac{x_0}{2}\left[1 + \frac{p}{\sqrt{p^2 - \omega^2}}\right]$, $B = \frac{x_0}{2}\left[1 - \frac{p}{\sqrt{p^2 - \omega^2}}\right]$

In damped harmonic motion there arise three cases :

Case I: When $\frac{\lambda}{m} < \sqrt{\frac{p}{m}}$, *i.e.*, $2p < \omega$, *i.e.*, $\sqrt{p^2 - \omega^2}$ is imaginary.

Hence, $$A = \frac{x_0}{2}\left[1 + \frac{p}{\sqrt{p^2 - \omega^2}}\right] \text{ or } A = \frac{x_0}{2}\left[1 + \frac{p}{i\sqrt{\omega^2 - p^2}}\right]$$

and $$B = \frac{x_0}{2}\left[1 - \frac{p}{i\sqrt{\omega^2 - p^2}}\right]$$

$\therefore$ $$x = \frac{x_0}{2}\left[1 + \frac{p}{i\sqrt{\omega^2 - p^2}}\right] e^{(-p + i\sqrt{\omega^2 - p^2})t} + \left[1 - \frac{p}{i\sqrt{\omega^2 - p^2}}\right] e^{(-p - i\sqrt{\omega^2 - p^2})t}$$

$$x = \frac{x_0 e^{-pt}}{\sqrt{\omega^2 - p^2}}\left[\sqrt{\omega^2 - p^2} \sin\sqrt{(\omega^2 - p^2)}\, t + k\cos\sqrt{\omega^2 - k^2 t}\right]$$

$$x = \frac{x_0 e^{-pt}\omega}{\sqrt{\omega^2 - p^2}} \sin\left[\sqrt{\omega^2 - p^2}\, t + p\right], \text{ where, } \tan\phi = \frac{p}{\sqrt{\omega^2 - k^2}}.$$

$$x = Ae^{(-\lambda t/2m)} \sin\left[\sqrt{\left(\omega^2 - \frac{\lambda^2}{4m^2}\right)} t + \phi\right], \text{ where, } A = \frac{x_0\omega}{\sqrt{\omega^2 - \frac{\lambda^2}{4m^2}}}.$$

INTERPRETATION

This equation represents damped harmonic motion with period $T = \dfrac{2\pi}{\sqrt{\omega^2 - \dfrac{\lambda^2}{4m^2}}}$

Period is longer and frequency is smaller when friction is present. The amplitude is not constant but it decays exponentially.

Case I : When $k = \omega$. Here $x = 0$, $\lambda = -k$. Hence, $x = x e^{-kt}$, $x = x_0 e^{-\lambda t/2m}$

This is critical damping. There is no oscillatory term in the solution, hence the motion becomes non-oscillatory.

Case II : When $k > \omega$

In this case $\sqrt{k^2 - \omega^2}$ is real and $-k + \sqrt{k^2 - \omega^2}$ and $-k - \sqrt{k^2 - \omega^2}$ are both negative.

The displacement falls exponentially. Hence, in this case a body merely returns to the equilibrium position when released from its initial displacement A.

33.10.3 FORCED OSCILLATIONS (WITHOUT DAMPING) AND RESONANCE

If an external force is applied on the point of support of the spring, it oscillates. The motion is called the forced oscillatory motion.

Let the external force be $q \cos nt$.

Equation of motion is $$m\frac{d^2x}{dt^2} = mg - me - kx + q\cos nt \qquad [\because mg = ke]$$

$$\Rightarrow \quad m\frac{d^2x}{dt^2} = -kx + q\cos nt$$

$$\Rightarrow \quad \frac{d^2x}{dt^2} = -\frac{k}{m} + \frac{q}{m}\cos nt \qquad \text{... (1)}$$

Let $\dfrac{k}{m} = \omega^2$ and $\dfrac{q}{m} = E$, then (1) becomes $\dfrac{d^2x}{dt^2} = -\omega^2 x + E\cos nt \Rightarrow \dfrac{d^2x}{dt^2} + \omega^2 x = E\cos nt$

$$\Rightarrow \quad (p^2 + \omega^2)x = E\cos nt \qquad \text{... (2)}$$

A.E. is $D^2 + \omega^2 = 0$ or $D = \pm i\omega$

$$\text{C.F.} = C_1\cos\omega t + C_2\sin\omega t$$

and $$\text{P.I.} = \frac{1}{D^2 + \omega^2} E\cos nt$$

Case I : If $\omega \neq n$, P.I. $= E\dfrac{1}{-n^2 + \omega^2}\cos nt$

Complete solution of (1) is $$x = C_1 \cos \omega t + C_2 \sin \omega t + \frac{E}{\omega^2 - n^2} \cos nt$$

or $$x = A\cos(\omega t + \alpha) + \frac{E}{\omega^2 - n^2} \cos nt \qquad \text{... (2)}$$

Equation (2) shows that the motion is the resultant of two oscillatory motions, *i.e.*, the first due to $A\cos(\omega t + \alpha)$ gives free oscillation of period $2\pi / \omega$ and the second due to $\frac{E}{\omega^2 - n^2} \cos nt$ gives forced oscillations of period $\frac{2\pi}{n}$. If ω is large, then the frequency of free oscillations is very high, then the amplitude $\frac{E}{\omega^2 - n^2}$ of forced oscillations is small.

Case II : If $\omega = n$,
$$\text{P.I.} = \frac{1}{D^2 + \omega^2} E \cos nt$$
$$= E.t \frac{1}{2D} \cos nt = \frac{Et}{2} \int \cos nt \, dt$$

Now, $$\text{P.I.} = \frac{Et}{2} \int \cos nt \, dt$$
$$= \frac{Et}{2} \left(\frac{\sin nt}{n} \right)$$

So, $$x = C_1 \cos \omega t + C_2 \sin \omega t + E\,t \frac{\sin nt}{2n}$$

$\Rightarrow$ $$x = C_1 \cos \omega t + C_2 \sin \omega t + E \frac{t}{2\omega} \sin \omega t \qquad [n = \omega]$$

or $$x = C_1 \cos \omega t + \left(C_2 + \frac{Et}{2\omega} \right) \sin \omega t$$

Let $$C_1 = r \sin \phi \text{ and } \left(C_2 + \frac{Et}{2\omega} \right) = r \cos \phi$$

Then, $$x = r \sin \phi \cos \omega t + r \cos \phi \sin \omega t$$

$\Rightarrow$ $$x = r \sin (\omega t + \phi)$$

The period of oscillations $= \frac{2\pi}{\omega}$

and amplitude $= \sqrt{C_1^2 + \left(C_2 + \frac{Et}{2\omega} \right)}$ and it increases as t increases.

After long time, the amplitude of the oscillation may become abnormally large causing over strain and consequently break down the system. But it does not happen as there is always some resistance in the system.

33.10.4 RESONANCE

If the frequency due to external periodic force becomes equal to the natural frequency of the system, the phenomenon is known as resonance.

In designing a machine or structure, occurrence of the resonance is always to be avoided so that the system may not breakdown. While marching over a bridge, the soldiers avoid their steps may not be in rhythm with the natural frequency of the bridge. Resonance may cause the bridge to collapse.

33.10.5 FORCED OSCILLATIONS (WITH DAMPING)

In the above case, if in addition, there is a damping force which is proportional to the instantaneous velocity of the mass, say $\lambda \frac{dx}{dt}$, then the equation of motion of the mass m is
$$m\frac{d^2x}{dt^2} = mg - k(e + x) - \lambda \frac{dx}{dt} + q \cos nt$$

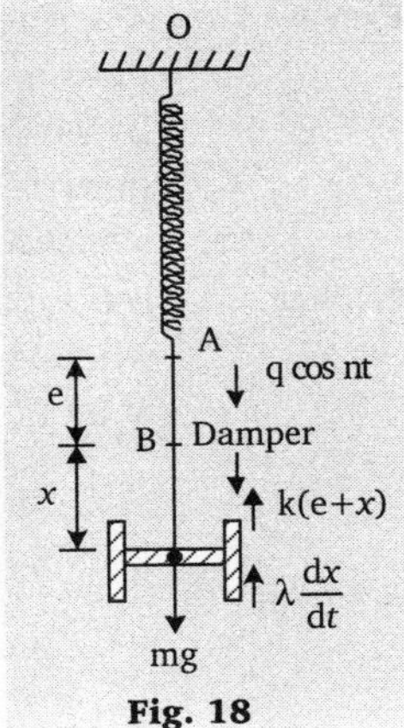

Fig. 18

Since $mg = ke$, it becomes $$m\frac{d^2x}{dt^2} = -kx - \lambda \frac{dx}{dt} + q \cos nt$$

$\Rightarrow$ $$\frac{d^2x}{dt^2} + \frac{\lambda}{m}\frac{dx}{dt} + \frac{k}{m} x = \frac{q}{m} \cos nt$$

Writing $\frac{\lambda}{m} = 2p, \frac{k}{m} = \omega^2$ and $\frac{q}{m} = E$, it becomes

$$\frac{d^2x}{dt^2} + 2p\frac{dx}{dt} + \omega^2 x = E\cos nt \quad \text{... (1)}$$

Equation (1) is a linear differential equation with constant coefficients.

It's A.E. is $m^2 + 2pm + \omega^2 = 0 \quad \Rightarrow \quad m = -p \pm \sqrt{p^2 - \omega^2}$

$\therefore$

$$\text{C.F.} = e^{-pt}\left(C_1 e^{\sqrt{p^2-\omega^2}t} + C_2 e^{-\sqrt{p^2-\omega^2}t}\right)$$

$$\text{P.I.} = E\frac{1}{D^2 + 2pD + \omega^2}\cos nt = E.\frac{1}{-n^2 + 2pD + \omega^2}\cos nt$$

$$= E.\frac{1}{(\omega^2 - n^2) + 2pD}\cos nt = E.\frac{(\omega^2 - n^2) - 2pD}{(\omega^2 - n^2)^2 - 4p^2D^2}\cos nt$$

$$= \frac{E(\omega^2 - n^2)\cos nt + 2pn\sin nt}{(\omega^2 - n^2) + 4p^2n^2}$$

Now, let $\omega^2 - n^2 = r\cos\phi,\ 2pn = r\sin\phi$, so that $r = \sqrt{(\omega^2 - n^2)^2 + 4p^2n^2}$ and $\phi = \tan^{-1}\left(\frac{2pn}{\omega^2 - n^2}\right)$

$\therefore$

$$\text{P.I.} = E\frac{r\cos(nt - \phi)}{r^2} = \frac{E}{r}.\cos(nt - \phi)$$

The complete solution of equation (1) is given by

$$x = e^{-pt}\left(C_1 e^{\sqrt{p^2-\omega^2}\,t} + C_2 e^{-\sqrt{p^2-\omega^2}\,t}\right) + \frac{E\cos\left\{nt - \tan^{-1}\left(\frac{2pn}{\omega^2 - n^2}\right)\right\}}{\sqrt{(\omega^2 - n^2)^2 + 4p^2n^2}}$$

The C.F. represents free oscillations of the system which die out as $t \to \infty$ due to the presence of the factory e^{-pt} .

The P.I. represents the forced oscillations of the system having a constant amplitude.

$$= \frac{E}{\sqrt{(\omega^2 - n^2)^2 + 4p^2n^2}}$$

and the period $= \frac{2\pi}{n}$, which is the same as that of the impressed force.

Thus, as t increases, the free oscillations (given by the C.F.) die out while the forced oscillations (given by the P.I.) persist giving the steady state of motion.

Solved Examples

Example 1. *A spring of negligible weight hangs vertically. A mass m is attached to the other end. If the mass is moving with velocity u when the spring is unstretched, find the velocity v as a function of the stretch x.*

Solution. Let x be the increase in length of the spring when velocity of the mass m is v. The equation of motion is

$$mv\frac{dv}{dx} = mg - T$$

$$mv\frac{dv}{dx} = mg - kx$$

$$mv\,dv = (mg - kx)\,dx$$

Integrating, we get $\frac{mv^2}{2} = mgx - k\frac{x^2}{2} + C$.

When $x = 0$, $v = u$ we have and so $C = \frac{mu^2}{2}$

$$\frac{mv^2}{2} = mgx - \frac{kx^2}{2} + \frac{mu^2}{2}$$

$$mv^2 = 2mgx - kx^2 + mu^2$$

$v^2 = 2gx + u^2 - \frac{k}{m}x^2$, which gives the required velocity.

Example 2. *A mass M suspended from the end of a helical spring is subjected to a periodic force $f = F\sin\omega t$ in the direction of its length. The force f is measured positive vertically downwards and at zero time M is at rest. If the spring stiffness is S, prove that the displacement of M at time t from the commencement of motion is given by*

$$x = \frac{F}{M(p^2 - \omega^2)}\left[\sin\omega t - \frac{\omega}{p}\sin pt\right]$$ *,where*

$p^2 = \frac{S}{M}$ *and damping effects are neglected.*

[UPTU–2002]

Solution. Let x be the displacement from the equilibrium position, the equation of motion is

$$M\frac{d^2x}{dt^2} = -Sx + F\sin\omega t$$

$$\frac{d^2x}{dt^2} + \frac{S}{M}x = \frac{F}{M}\sin\omega t$$

$$\Rightarrow \quad \frac{d^2x}{dt^2} + P^2x = \frac{F}{M}\sin\omega t \qquad \left(\because \frac{S}{M} = p^2\right)$$

$$(D^2 + P^2)x = \frac{F}{M}\sin\omega t$$

A.E. is $D^2 + P^2 = 0,\ D = \pm iP$.

C.F. $= (C_1\cos pt + C_2\sin pt)$

$$\text{P. I.} = \frac{1}{D^2 + P^2}\frac{F}{M}\sin\omega t = \frac{F}{M}\frac{1}{-\omega^2 + P^2}\sin\omega t$$

$$\therefore\ x = C_1\cos pt + C_2\sin pt + \frac{F}{M}\frac{1}{p^2 - \omega^2}\sin\omega t \qquad \ldots (1)$$

Putting $t = 0$ and $x = 0$ in (1), we get $0 = C_1$

Equation (1) becomes

$$x = C_2\sin pt + \frac{F}{M}\frac{1}{p^2 - \omega^2}\sin\omega t \qquad \ldots (2)$$

Differentiating (2), we obtain

$$\frac{dx}{dt} = C_2 p\cos pt + \frac{F}{M}\frac{\omega}{p^2 - \omega^2}\cos\omega t \qquad \ldots (3)$$

Putting $\frac{dx}{dt} = 0$ and $t = 0$ in (3), we have

$$0 = C_2P = \frac{F}{M}\frac{\omega}{p^2 - \omega^2}$$

$$\Rightarrow \quad C_2 = -\frac{F\omega}{PM\,(p^2 - \omega^2)}.$$

On substituting the value of C_2 in (2), we get

$$x = -\frac{\omega}{p}\frac{F}{PM\,(p^2 - \omega^2)}\sin pt + \frac{F}{M}\frac{1}{p^2 - \omega^2}\sin\omega t$$

$$= \frac{F}{M(p^2 - \omega^2)}\left[\sin\omega t - \frac{\omega}{p}\sin pt\right].$$

Example 3. *In a spring of T_1 and T_2 be the periods corresponding to two different weights attached and s_1 and s_2 the statical extensions due to these weights, prove that*

$$g = \frac{4\pi^2(s_1 - s_2)}{T_1^2 - T_2^2}.$$

Solution. Let k be the stiffness of the spring and m_1, m_2 be the masses attached to the spring for which the periods are T_1 and T_2 respectively, we have

$$T_1 = 2\pi\sqrt{\frac{m_1}{k}} \Rightarrow T_1^2 = 4\pi^2 m_1/k \qquad \ldots (1)$$

and $$T_2 = 2\pi\sqrt{\frac{m_2}{k}} \Rightarrow T_2^2 = \frac{4\pi^2 m_2}{k} \qquad \ldots (2)$$

Subtracting (2) from (1), we get

$$T_1^2 - T_2^2 = \frac{4\pi^2}{k}(m_1 - m_2) \qquad \ldots (3)$$

During static equilibrium spring force is equal to gravitational force, *i.e.*,

$$ks = mg$$

$$ks_1 = m_1 g \qquad \ldots (4)$$

and $$ks_2 = m_2 g \qquad \ldots (5)$$

Subtracting (4) from (5), we get

$$k(s_1 - s_2) = (m_1 - m_2)\,g$$

$$k^{-1}(m_1 - m_2) = \frac{1}{g}(s_1 - s_2)$$

Substituting the value of $(m_1 - m_2)$ in (2), we get $T_1^2 - T_2^2 = 4\pi^2\frac{1}{g}(s_1 - s_2)$

$$g = \frac{4\pi^2(s_1 - s_2)}{T_1^2 - T_2^2}.$$

Example 4. *A spring for which the spring constant $k = 700\ Nm^{-1}$ hangs in a vertical position with its upper end fixed to a support. A mass of 20 kg is attached to the lower end and system brought to rest. Find the position of the mass at time t, if a force 70 sin 2tN is applied to the support.* [UPTU Q. Bank–2002]

Solution. Equation of motion is

$$m\frac{d^2x}{dt^2} = -kx + 70\sin 2t$$

$$\Rightarrow \quad 20\frac{d^2x}{dt^2} = -700x + 70\sin 2t$$

$$\Rightarrow \quad \frac{d^2x}{dt^2} + 35x = \frac{7}{2}\sin 2t$$

Auxiliary equation is

$$m^2 + 35 = 0$$

$$m = \pm i\sqrt{35}$$

C.F. $= C_1\cos\sqrt{35}\,t + C_2\sin\sqrt{35}\,t$

$$\text{P.I.} = \frac{1}{D^2 + 35}\left(\frac{7}{2}\sin 2t\right) = \frac{7}{62}\sin 2t$$

$$x = C_1\cos\sqrt{35}\,t + C_2\sin\sqrt{35}\,t + \frac{7}{62}\sin 2t \qquad \ldots (1)$$

At $t = 0$, $x = 0$, $\therefore$ from (1), $C_1 = 0$.

From (1), $x = C_2 \sin\sqrt{35}\,t + \frac{7}{62}\sin 2t$ ... (2)

Differentiating w.r.t. t, we get

$$\frac{dx}{dt} = \sqrt{35}\,C_2 \cos\sqrt{35}\,t + \frac{7}{31}\cos 2t \quad \text{... (3)}$$

Also, when $t = 0$, $v = \frac{dx}{dt} = 0$ in (3), we get

$$0 = \sqrt{35}\,C_2 + \frac{7}{31} \Rightarrow C_2 = -\frac{7}{31\sqrt{35}}$$

$\therefore$ From (2), $x = -\frac{7}{31\sqrt{35}}\sin\sqrt{35}\,t + \frac{7}{62}\sin 2t$.

Example 5. *A spring fixed at the upper end supports a weight of 980 cm at its lower end. The spring stretches 1/2 cm under a load of 10 gm and the resistance (in gm weight) to the motion of the weight is numerically equal to 1/10 of the speed of the weight in cm/sec. The weight is pulled down 1/4 cm below its equilibrium position and then released. Find the expression for the distance of weight from its equilibrium position at time t during its first upward motion. Also find the time t it takes the damping factor to drop to 1/10 of its initial value.*

Solution. Let OA be a spring fixed at O and a load of 10 gm is attached at A. The spring is stretched by $\frac{1}{2}$ cm.

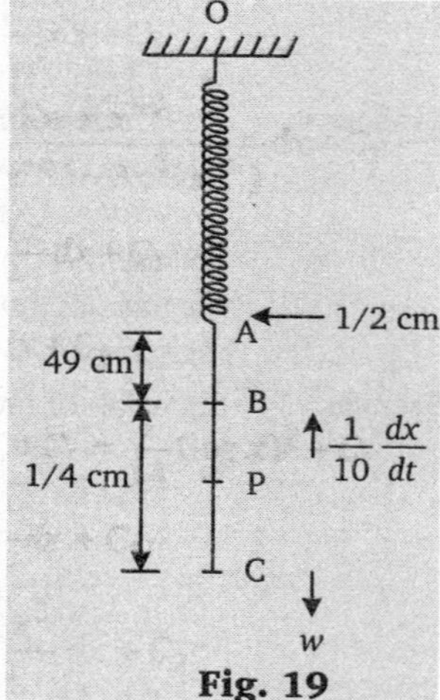

Fig. 19

$$mg = T_0$$

or

$$10 = T_0 = k.\frac{1}{2}$$

$\Rightarrow k = 20$ g/cm

Let B be the equilibrium after attaching a weight 980 gm at A.

$$mg = kx,$$

where $x = AB$

$$980 = 20x \Rightarrow x = AB = \frac{980}{20} = 49 \text{ cm.}$$

After static equilibrium, the weight is pulled down to C and released ($BC = \frac{1}{4}$ cm). After release, the weight be at P after time t.

$$BP = x$$

$$T = k.AP = 20\,(49 + x) = 980 + 20x$$

Equation of motion is

$$m\frac{d^2x}{dt^2} = \omega - T - \frac{1}{10}\frac{dx}{dt}$$

(Resistance $= \frac{1}{10}\frac{dx}{dt}$)

$$\frac{980}{g}\frac{d^2x}{dt^2} = 980 - (980 + 2x) - \frac{1}{10}\frac{dx}{dt}$$

$[g = 980 \text{ cm/sec}^2]$

$$\Rightarrow 10\frac{d^2x}{dt^2} + \frac{dx}{dt} + 200x = 0$$

$$\Rightarrow 10D^2x + Dx + 200x = 0$$

A.E. is $10m^2 + m + 200 = 0$

$$m = \frac{-1 \pm \sqrt{1 - 8000}}{20} = \frac{-1 \pm i\,89.4}{20}$$

$\Rightarrow m = -0.05 \pm i.45$

C.F. $\Rightarrow x = e^{-0.05t}[C_1 \cos 4.5t + C_2 \sin 4.5t]$ ... (1)

On putting $t = 0$ and $x = \frac{1}{4}$ in (1),

we get $\frac{1}{4} = C_1$

On differentiating (1), we get

$$\frac{dx}{dt} = -0.05\,e^{-0.05t}[C_1 \cos 4.5t + C_2 \sin 4.5t]$$
$$+ e^{-0.05t}[-4.5C_1 \sin 4.5t + 4.5\,C_2 \cos 4.5t] \quad \text{... (2)}$$

On putting $\frac{dx}{dt} = 0$ and $t = 0$ in (2), we have

$$0 = -0.05C_1 + 4.5\,C_2$$

$$\Rightarrow \quad C_2 = \frac{0.05}{4.5}C_1 = \frac{0.05}{4.5}\left(\frac{1}{4}\right) = 0.0028$$

On substituting the values of C_1 and C_2 in (1), we obtain

$$x = e^{-0.05t}[0.25 \cos 4.5t + 0.0028 \sin 4.5t]$$

Damping factor $= be^{-0.05t}$

(b = constant of proportionality)

Initial value of damping factor $= be^0 = b$

Let damping factor after time t be $\frac{b}{10}$. Then,

$$\frac{b}{10} = b\,e^{-0.05t} \text{ or } e^{t/20} = 10 \Rightarrow \frac{t}{20} = \log_e 10.$$

$$t = 20\log_e 10 \Rightarrow t = 20\frac{\log_e 10}{\log_{10} e}$$

$$= \frac{20}{\log_{20} e} = 20 \times 2.3 = 46 \text{ sec}$$

EXERCISE 33.7

1. A body weighing 4.9 kg is hung from a spring. A pull of 10 kg will stretch the spring to 5 cm. The body is pulled down 6 cm below the static equilibrium position and then released. Find the displacement of the body from its equilibrium position at time t seconds, the maximum velocity and the period of oscillation. [UPTU Q. Bank–2002]
2. A mass of 200 gm is toed at the end of a spring which extends to 4 cm under a force 196,000 dynes. The spring is pulled 5 cm and released. Find the displacement, t seconds after release, if there be a damping force of 2000 dynes per cm per second. [UPTU Q. Bank–2002]
3. A spring for which stiffness K=700 Newton/m hangs in a vertical position with its upper end fixed. A mass of 7 kg is attached to the lower end. After coming to rest, the mass is pulled down 0.05 m and released. Discuss the resulting motion of the mass, neglecting air resistance.
4. A body executes damped forced vibrations given by the equation $\frac{d^2x}{dt^2}+2K\frac{dx}{dt}+b^2x=e^{-kt}\sin\omega t$. Solve the equation for both the cases, when $\omega^2 \neq b^2-K^2$ and when $\omega^2 = b^2-K^2$.
5. A spring of negligible weight which stretches l inch under tension of 2 lb is fixed at one end is attached to a weight of w lb at the other. It is found that resonance occurs when an axial periodic force 2 cos $2t$ lb acts on the weight. Show that when the free vibrations have dies out, the forced vibrations are given by $x = Ct \sin 2t$, and find values of w and C.
6. A spring which stretches by an amount e under a force mk^2e is suspended from a support P and has a mass m at its lower end. At time $t=0$, the mass is at rest in its equilibrium position at a point A below P. A vertical oscillation is now given to the support P such that at any time $t(>0)$, its displacement below its initial position is $a \sin nt$. Show that the displacement x of the mass below A is given by

$$\frac{d^2x}{dt^2}+K^2x=K^2a\sin nt$$

Hence, show that if $n \neq k$, the displacement is given by

$$x=\frac{Ka}{K^2-n^2}(K\sin nt-n\sin nt)$$

What happens, when n=k? [UPTU Q. Bank–2002]

33.11 BENDING OF BEAMS

Beam or Bar : A rod of a circular or rectangular cross-section with its length very much greater than its thickness is called a beam.

Supported Beam : If a beam may just rest on a support like a knife edge is called a supported beam.

Fixed Beam : If one of the ends of a beam is firmly fixed, then it is called a fixed beam.

Cantilever : If one end of the beam is fixed and the other end is loaded, it is called a cantilever.

Strut : A beam of homogeneous isotropic material subject to compressive stress is called strut.

33.11.1 DEFLECTION OF BEAMS

A beam can be considered to be made up of fibres running lengthwise. When a beam is bent under the given loading, it is observed that the fibres in the upper of its face are compressed while those in the opposite face to the lower half are stretched. In between these surfaces of compression and tension, there must be a transition surface where the fibres are neither compressed nor stretched. This surface is called the neutral surface of the beam. The curve defined by any fibre in this surface known as the deflection curve or the elastic curve of the beam. The line in which any plane section of the beam cuts the neutral surface is called the neutral axis of that section.

Before we derive the equation of the deflection curve, we make the following assumptions :

(i) The beam is of uniform cross-section and its breadth and thickness are small compared to its length.

(ii) The axis of the beam is a horizontal line through the centre of the gravity of the cross-section.

(iii) The beam is perfectly elastic and obeys Hooke's law.

(iv) The deflections of the beam are small so that the slope at any point of the elastic curve is assumed small.

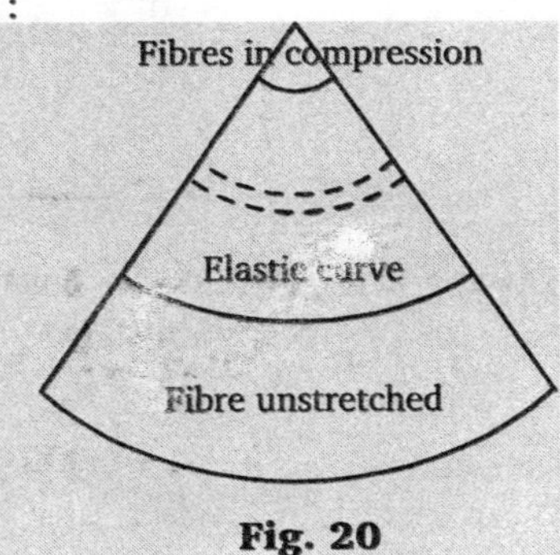

Fig. 20

Consider a cross-section of the beam which cuts the neutral surface in the point P at a distance x from O, where OX and OY are the axes of coordinates. Let CD be the neutral axis of this section. We know from mechanics that in the equilibrium of any cross-section of the beam, the external bending moment M about CD, due to different loads acting on the beam, must balance the internal moment $\frac{EI}{R}$ which is the sum of the moment of the force about CD.

Hence, we get $$M = \frac{EI}{R} \quad \text{... (1)}$$

where, E is the modulus of elasticity of the material of the beam I is the moment of inertia of the area of cross-section of the beam and R is the radius of curvature of the elastic curve at the point P.

Since $R = \dfrac{\left\{1+(dy/dx)^2\right\}^{3/2}}{d^2y/dx^2}$ and $\dfrac{dy}{dx}$ is small; $\left(\dfrac{dy}{dx}\right)^2$ can be ignored.

Thus, (1) gives bending moment

$$M = EI\frac{d^2y}{dx^2} \quad \text{... (2)}$$

If the beam is subjected to transverse load only, we further obtain from equation (2) on differentiation

$$\frac{dM}{dx} = \text{Shear force} = EI\frac{d^3y}{dx^3} = F \quad \text{... (3)}$$

and $$\frac{d^2M}{dx^2} = \text{Intensity of loading} = \frac{dF}{dx} = EI\frac{d^4y}{dx^4} \quad \text{... (4)}$$

REMARKS

- Equation (1) is known as Bernoulli's-Euler formula.
- The product EI is called the flexural rigidity of the beam.
- The moment M about the neutral axis of all external forces acting on either side of the two portions of the beam separated by the cross-section is independent of the portion considered.
- The moment M is the algebraic sum of the moments of the external forces acting on the portion of the beam about the neutral axis. The upward forces (in anti-clockwise direction) gives negative moments.

BOUNDARY CONDITIONS

(i) End Freely Supported :

At the freely supported end O, there will be no deflection and no bending moment.

$$y = 0, \quad \frac{d^2y}{dx^2} = 0.$$

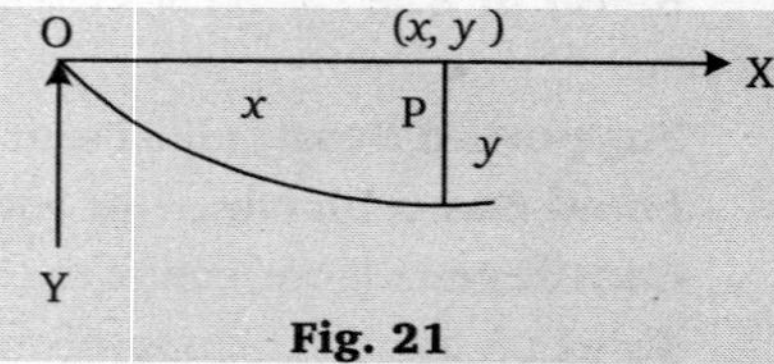

Fig. 21

(ii) Fixed End Horizontally :

Deflection and slope of the beam are zero

$$y = 0, \quad \frac{dy}{dx} = 0.$$

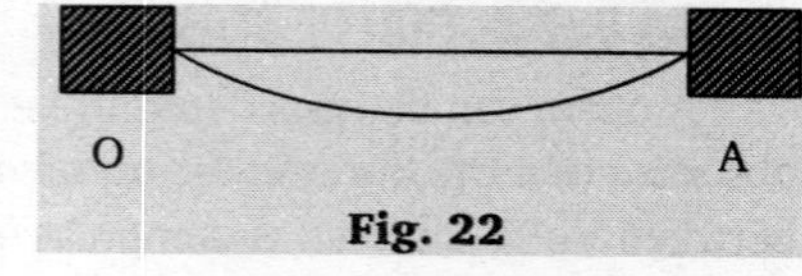

Fig. 22

(iii) Perfectly Free End :

At the free end, there is no bending moment of shear force.

$$\frac{d^2y}{dx^2} = 0, \quad \frac{d^3y}{dx^3} = 0.$$

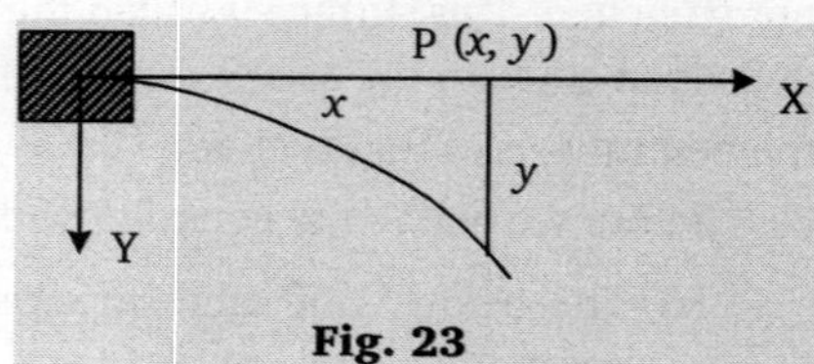

Fig. 23

Solved Examples

Example 1. *For a strut of length l, freely hinged at each end, prove that the bending moment at the centre is $\dfrac{-Wl^2}{8}\dfrac{Q}{Q-P}$, where $Q = \dfrac{EI\pi^2}{l^2}$. The differential equation for a strut is*

$$EI\frac{d^2y}{dx^2} + Py = -\frac{Wl^2}{8}\sin\frac{\pi x}{l}$$

Solution. We have

$$EI\frac{d^2y}{dx^2} + Py = -\frac{Wl^2}{8}\sin\frac{\pi x}{l} \quad \text{... (1)}$$

$$\Rightarrow \quad \frac{d^2y}{dx^2} + \frac{P}{EI}y = -\frac{W\pi^2}{8Q}\sin\frac{\pi x}{l}$$

$$\left[\because EI\frac{\pi^2}{l^2} \Rightarrow \frac{1}{EI} = \frac{\pi^2}{l^2Q}\right]$$

$$\Rightarrow \quad \frac{d^2y}{dx^2} + \frac{P\pi^2}{Ql^2} y = \frac{-W\pi^2}{8Q} \sin\frac{\pi x}{l}$$

$$\left(D^2 + \frac{P\pi^2}{Ql^2}\right) y = \frac{-W\pi^2}{8Q} \sin\frac{\pi x}{l}$$

It's A.E. is $D^2 + \frac{P\pi^2}{Ql^2} = 0$

$$\Rightarrow \quad D = \pm\frac{\pi}{l}\sqrt{\frac{P}{Q}}$$

$$\text{C.F.} = C_1 \cos\left(\frac{\pi}{l}\sqrt{\frac{P}{Q}}\right)x + C_2 \sin\left(\frac{\pi}{2}\sqrt{\frac{P}{Q}}\right)x$$

$$\text{P.I.} = \frac{1}{D^2 + \frac{P\pi^2}{Ql^2}}\left(-\frac{W\pi^2}{8Q}\sin\frac{\pi x}{l}\right)$$

$$= -\frac{W\pi^2}{8Q} \frac{1}{-\frac{\pi^2}{l^2} + \frac{P\,\pi^2}{Q\,l^2}} \sin\frac{\pi x}{l}$$

$$= \frac{W\pi^2}{8Q} \frac{l^2}{\pi^2} \cdot \frac{1}{1 - \frac{P}{Q}} \sin\frac{\pi x}{l}$$

$$= \frac{Wl^2}{8} \cdot \frac{1}{Q - P} \sin\frac{\pi x}{l}$$

The general solution of (1) is

$$y = C_1 \cos\left(\frac{\pi}{l}\sqrt{\frac{P}{Q}}\right)x \quad \text{... (2)}$$

$$+ C_2 \sin\left(\frac{\pi}{l}\sqrt{\frac{P}{Q}}\right)x + \frac{Wl^2}{8} \frac{1}{Q-P} \sin\frac{\pi x}{l}$$

Since the ends are freely hinged.

$$y = 0, \quad \frac{d^2y}{dx^2} \text{ at } x = 0$$

$$\text{and } y = 0, \quad \frac{d^2y}{dx^2} \text{ at } x = l.$$

On putting $x = 0$, $y = 0$ in (2), we get $0 = C_1$.
Then, (2) becomes

$$y = C_2 \sin\left(\frac{\pi}{l}\sqrt{\frac{P}{Q}}\right)x + \frac{Wl^2}{8} \frac{1}{Q-P} \sin\frac{\pi x}{l} \quad \text{... (3)}$$

On putting $x = l$ and $y = 0$ in (3), we get

$$0 = C_2 . \sin \pi\sqrt{\frac{P}{Q}} \quad \Rightarrow \quad C_2 = 0$$

(3) reduces to

$$y = \frac{Wl^2}{8} \frac{1}{Q-P} \sin\frac{\pi x}{l}$$

But bending moment is given by

$$M = EI \frac{d^2y}{dx^2}$$

$$M = EI\left[-\frac{Wl^2}{8} \frac{1}{Q-P} \frac{\pi^2}{l^2} \sin\frac{\pi x}{l}\right]$$

$$= EI\left[-\frac{W\pi^2}{8} \frac{1}{Q-P} \sin\frac{\pi x}{l}\right]$$

$$M = -\frac{Wl^2}{8} \frac{Q}{Q-P} \sin\frac{\pi x}{l} \quad \left(Q = \frac{EI\,\pi^2}{l^2}\right)$$

Hence, M at centre (at $x = l/2$)

$$= -\frac{Wl^2}{8} \frac{Q}{Q-P} \sin\frac{\pi}{2}$$

$$= -\frac{Wl^2}{8} \frac{Q}{Q-P}.$$

Example 2. *A beam of length l is clamped horizontally at its end $x = 0$ and is free at the end $x = l$. A point load W is applied at the end $x = l$, in addition to a uniform load w per unit length from $x = 0$ to $x = l/2$. Find the deflections at any point.*

[UPTU B.Tech–2002]

Solution. From the equation of balance, we get

$$R = W + \frac{wl}{2} \quad \text{... (1)}$$

Choose a random axis NN'. Let (x, y) be the coordinates of N. Taking moment about N, we get

$$EI \frac{d^2y}{dx^2} = \frac{wl}{2}\left(x - \frac{l}{4}\right) - Rx$$

$$= \frac{wl}{2}\left(x - \frac{l}{4}\right) - \left(W - \frac{wl}{2}\right)x$$

[From (1)]

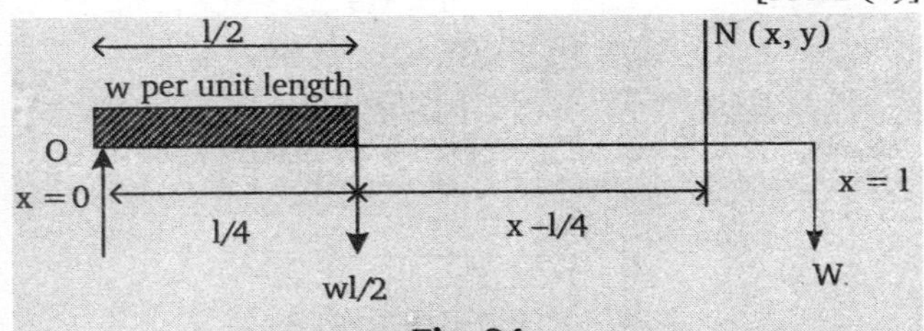

Fig. 24

$$\text{or } EI \frac{d^2y}{dx^2} = -\frac{wl^2}{8} - Wx \quad \text{... (2)}$$

Integrating (2) w.r.t. x, we get

$$EI\frac{dy}{dx} = -\frac{wl^2}{8}x - \frac{Wx^2}{2} + C.$$

Applying boundary conditions, we have $x = 0$

$$\Rightarrow \text{slope} = \frac{dy}{dx} = 0, \quad \therefore C = 0$$

$$EI\frac{dy}{dx} = -\frac{wl^2}{8}x - \frac{Wx^2}{2} \quad \text{... (3)}$$

Again integrating w.r.t. x, we get

$$Ely = \frac{-wl^2x^2}{16} - \frac{Wx^3}{6} + C_1$$

Applying boundary conditions, we have

$$x = 0,\ y = 0 \Rightarrow C_1 = 0.$$

$$Ely = \frac{-wl^2x^2}{16} - \frac{Wx^3}{6}$$

or $$y = -\frac{1}{El}\left[\frac{Wx^3}{6} + \frac{wl^2x^2}{16}\right].$$

Example 3. *The deflection of a strut of length l with one end $(x = 0)$ built in and the other supported and subjected to end thrust P, satisfies the equation $\frac{d^2y}{dx^2} + a^2y = \frac{a^2R}{P}(l - x)$. Prove that deflection curve in $y = \frac{R}{P}\left(\frac{\sin ax}{a} - l\cos ax + l - x\right)$, where $al = \tan al$.* [UPTU–2001]

Solution. The given equation is $\frac{d^2y}{dx^2} + a^2y = \frac{a^2R}{P}(l - x)$

$$(D^2 + a^2)y = \frac{a^2R}{P}(l - x) \quad \text{... (1)}$$

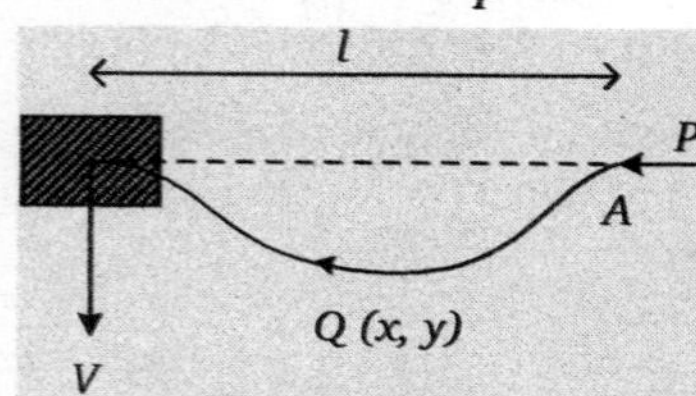

Fig. 25

It's A.E. is $m^2 + a^2 = 0$ so that

$$m^2 + a^2 = 0 \text{ so that}$$

$$m = \pm ia$$

$\therefore$ C.F. $= C_1 \cos ax + C_2 \sin ax$.

$$\text{P. I.} = \frac{1}{D^2 + a^2}\left\{\frac{a^2R}{P}(l - x)\right\}$$

$$= \frac{a^2R}{P}\frac{1}{a^2\left(1 + \frac{D^2}{a^2}\right)}(l - x).$$

$$= \frac{R}{P}\left(1 + \frac{D^2}{a^2}\right)^{-1}(l - x)$$

$$= \frac{R}{P}\left(1 - \frac{D^2}{a^2} + ...\right)(l - x)$$

$$= \frac{R}{P}(l - x)$$

$\therefore$ The complete solution of (1) is

$$y = C_1 \cos ax + C_2 \sin ax + \frac{R}{P}(l - x) \quad \text{... (2)}$$

Differentiating (2) w.r.t. x,

$$\frac{dy}{dx} = -aC_1 \sin ax + aC_2 \cos ax - \frac{R}{P} \quad \text{... (3)}$$

Since the end O is build in, $y = 0$ and $\frac{dy}{dx} = 0$ at $x = 0$

$\therefore$ From (2), $0 = C_1 + \frac{Rl}{P} \Rightarrow C_1 = -\frac{Rl}{P}$.

From (4), $0 = aC_2 - \frac{R}{P} \Rightarrow C_2 = \frac{R}{aP}$.

Substituting the values of C_1 and C_2 in (2), we have

$$y = \frac{R}{P}\left(\frac{\sin ax}{a} - l\cos ax + l - x\right) \quad \text{...(4)}$$

which is the equation of the deflection curve.

Also, at the end A, $y = 0$, when $x = l$

$\therefore$ From (1), $0 = \frac{R}{P}\left(\frac{\sin al}{a} - l\cos al\right)$

or $$\frac{\sin al}{a} = l\cos al$$

$\therefore$ $$al = \tan al.$$

Example 4. *The differential equation satisfied by a beam uniformly loaded (W kg/metre) with one end fixed and the second end subjected to tensile force P, is given by :*

$$EI\frac{d^2y}{dx^2} = Py - \frac{1}{2}Wx^2$$

Show that the elastic curve for the beam with conditions $y = 0 = \frac{dy}{dx}$ at $x = 0$, is given by

$$y = \frac{W}{Pn^2}(1 - \cosh nx) + \frac{Wx^2}{2P}, \text{ where } n^2 = \frac{P}{EI}.$$

Solution. We have

$$EI\frac{d^2y}{dx^2} = Py - \frac{1}{2}Wx^2 \quad \text{... (1)}$$

$$\Rightarrow \quad \frac{d^2y}{dx^2} - \frac{P}{EI}y = -\frac{W}{2EI}x^2$$

$$\left(D^2 - \frac{P}{EI}\right)y = -\frac{W}{2EI}x^2$$

A.E. is $D^2 - \frac{P}{EI} = 0$ or $D^2 = \frac{P}{EI} = n^2$

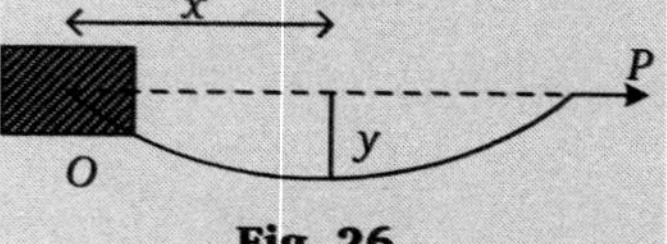

Fig. 26

or $D = \pm n$ $\qquad \left(n^2 = \frac{P}{EI}\right)$

C.F. $= C_1e^{nx} + C_2e^{-nx}$

$$\text{P.I.} = \frac{1}{D^2 - \frac{P}{EI}}\left(-\frac{W}{2EI}\right)x^2 = -\frac{W}{2EI}\cdot\frac{1}{D^2 - n^2}.x^2$$

$$= \frac{W}{2n^2.EI}\left(1 - \frac{D^2}{n^2}\right)^{-1} x^2$$

$$= \frac{W}{2n^2EI}\left(1 + \frac{D^2}{n^2}\right)x^2$$

[By neglecting higher order derivatives]

$$= \frac{W}{2n^2EI}\left(x^2 + \frac{2}{n^2}\right)$$

$$y = C_1e^{nx} + C_2e^{-nx} + \frac{W}{2n^2EI}\left(x^2 + \frac{2}{n^2}\right) \quad \text{... (2)}$$

Differentiating (2) w.r.t. x, we get

$$\frac{dy}{dx} = nC_1e^{nx} - nC_2e^{-nx} + \frac{W}{2n^2EI}(2x) \quad \text{... (3)}$$

Putting $x = 0$, $\frac{dy}{dx} = 0$ in (3), we get

$$0 = nC_1 - nC_2 \text{ or } C_1 = C_2.$$

Putting $x = 0$, $y = 0$ in (2), we get

$$0 = C_1 + C_2 + \frac{W}{2n^2EI}\frac{2}{n^2}$$

$$0 = C_1 + C_2 + \frac{W}{n^4EI} \quad \text{... (4)}$$

Putting $C_1 = C_2$ in (4), we get $0 = 2C_1 + \frac{W}{n^4EI}$

$$\Rightarrow \quad C_1 = -\frac{W}{2n^4\,EI}$$

Now $n^2 = \frac{P}{EI}$ or $n^2EI = P$

$$C_1 = C_2 = -\frac{W}{2n^2P}$$

Putting the values of C_1 and C_2 in (2), we get

$$y = -\frac{W}{2n^2P}(e^{nx} + e^{-nx}) + \frac{W}{2P}\left(x^2 + \frac{2}{n^2}\right)$$

$$\Rightarrow y = -\frac{W}{n^2P}\cosh nx + \frac{W}{2P}x^2 + \frac{W}{Pn^2}$$

$$\Rightarrow y = \frac{W}{Pn^2}(1 - \cosh nx) + \frac{Wx^2}{2P}.$$

33.12 WHIRLING OF SHAFTS

Sometimes, rotating shafts rotate about its geometrical axes even in the absence of external load. The magnitude of the deflection depends upon the

(i) Stiffness of shaft and its support
(ii) The total mass of shaft and attached parts
(iii) The imbalance of the mass with respect to the axis of rotation.
(iv) The amount of damping in the system.

At certain speeds the deflection of a rotating shaft tends to become large and the shaft will fracture unless the speed is lowered and the deflection becomes normal again. These dangerous speeds are called whirling speeds or critical speeds of the shaft.

The differential equation for the whirling of the shaft of weight W per unit length which is rotating with angular velocity ω (radian/sec) is

$$(D^4 - a^4)\,y = 0, \text{ where } a^4 = \frac{W\omega^2}{gEI}$$

A.E. is $\quad m^4 - a^4 = 0 \Rightarrow m = \pm a, \pm ai$

Hence, its complete solution is

$$y = C_1e^{ax} + C_2e^{-ax} + C_3\cos ax + C_4\sin ax$$

$$y = A\cosh ax + B\sinh ax + C\cos ax + D\sin ax.$$

To determine arbitrary constants, boundary conditions are

(1) In case of short or flexible bearings : There is no deflection and no bearing moment. Boundary conditions are

$$y = 0, \quad \frac{d^2y}{dx^2} = 0$$

(2) In case of load bearings : The deflection and the slope of the shaft both will be zero. Boundary conditions are :

$$y = 0, \quad \frac{dy}{dx} = 0$$

(3) When end of shaft is perfectly free : There is no bending moment and no shear force. Boundary conditions are

$$\frac{d^2y}{dx^2} = 0 \text{ and } \frac{d^3y}{dx^3} = 0$$

Solved Examples

Example 1. *The whirling speed of a shaft of length l is given by* $\frac{d^4y}{dx^4} - a^4y = 0$ *, where* $a^4 = \frac{Ww^2}{gEI}$ *and y is the displacement. If* $\cos al.\cosh al = 1$*, find the whirling speed of the shaft.*

Solution. The given differential equation is

$$\frac{d^4y}{dx^4} - a^4y = 0$$

C.F. is $y = A\cosh ax + B\sinh ax$

$$+ C\cos ax + D\sin ax \quad \text{... (1)}$$

Differentiating (1) w.r.t. x, we get

$$\frac{dy}{dx} = Aa\sinh ax + Ba\cosh ax$$

$$-Ca\sin ax + Da\cos ax$$

$$\frac{1}{a}\frac{dy}{dx} = A\sinh ax + B\cosh ax$$

$$- C\sin ax + D\cos ax \quad \text{... (2)}$$

As the ends of the shaft are fixed, the boundary conditions are

$$y = 0, \frac{dy}{dx} = 0, \text{ when } x = 0 \quad \text{... (3)}$$

and $y = 0, \frac{dy}{dx} = 0$, when $x = 2$ ... (4)

Using (3), from (1), we have $0 = A + C$,

or $C = -A$ ... (5)

and from (2), we have $0 = B + D$

$D = -B$ ... (6)

From (4) and (1), we have

$$0 = A\cosh al + B\sinh al + C\cos al + D\sin al \quad \text{... (7)}$$

and from (2) and (1), we have

$$0 = A\sin al + B\cosh al - C\sin al + D\cos al \quad \text{... (8)}$$

Substituting the values of C and D from (5) and (6) in (7) and (8), we get

$$A(\cosh al - \cos al) + B(\sinh al - \sin al) = 0$$

and $A(\sinh al + \sin al) + B(\cosh al - \cos al) = 0$

Solving for A and B, we get

$$\frac{\cosh al - \cos al}{\sinh al - \sin al} = -\frac{B}{A} = \frac{\sinh al + \sin al}{\cosh al - \cos al}$$

$$\cosh^2 al - 2\cosh al \cos al + \cos^2 al$$

$$= \sinh^2 al - \sin^2 al$$

$$-2\cosh al \cos al + 2 = 0$$

$$\cos al \cosh al = 1$$

When the shaft whirls, this equation must satisfy. The solution of the equation gives

$$al = 4.73 = \frac{3\pi}{2} \text{ radians approximately.}$$

Now, $a^4 = \frac{W\omega^2}{gEI}$ or $\left(\frac{W\omega^2}{gEI}\right)^{1/2} = a^2$

$$\Rightarrow \quad \omega\sqrt{\left(\frac{W}{gEI}\right)}l^2 = a^2l^2 = \frac{9\pi^2}{4}$$

$\therefore$ Whirling speed of a shaft with ends in long bearings

$$\omega = \frac{9\pi^2}{4}\sqrt{\left(\frac{gEI}{W}\right)} \text{ approximately.}$$

Example 2. *If the shaft has one long bearing, find the whirling speed of the shaft.*

Solution. $y = A\cosh ax + B\sinh ax + C\cos ax + D\sin ax$

Conditions : $y = 0, \frac{du}{dx} = 0$

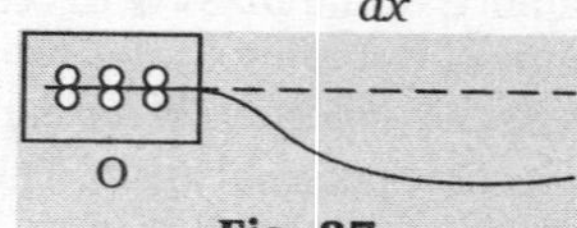

Fig. 27

Then $\cosh al \cos al = -1$

$$\Rightarrow \quad al = 1.865$$

$$\omega\sqrt{\frac{W}{gEI}}\,l^2 = a^2l^2$$

$$\omega\sqrt{\frac{W}{gEI}}\,l^2 = (1.865)^2$$

$$\Rightarrow \quad \omega = \frac{3.5}{l^2}\sqrt{\frac{gEI}{W}}.$$

EXERCISE 33.8

1. A light horizontal strut AB of length l is freely pinned at A and B and is under the action of equal and opposite compressive forces P at each of its ends and carries a load W at its centre. Prove that the deflection at the centre is $\frac{W}{2P}\left(\frac{l}{n}\tan\frac{nl}{2} - \frac{l}{2}\right)$, where $n^2 = \frac{P}{EI}$.

2. Solve the differential equation

$$EI\frac{d^2y}{dx^2} + Py = -\frac{Wl^2}{8}\sin\frac{\pi x}{l}$$

for a strut of length l, freely hinged at each end, prove that bending moment at the centre is $-\frac{Wl^2}{8}\left(\frac{Q}{Q-P}\right)$, where $Q = EI\frac{\pi^2}{l^2}$.

3. A beam of length l metres is fixed horizontally at one end. Find the equation of the elastic curve and the maximum deflection, if there is uniform load W Newton/m along the length and load W Newton at the free end.

[UPTU Q. Bank–2001]

4. The differential equation for the displacement y of a whirling shaft when the weight of the shaft is taken into account is

$$EI\frac{d^4y}{dx^4} - \frac{Ww^2}{g}y = W\cdot$$

Taking the shaft of length $2l$ with the origin at the centre and short bearings at both ends, show that the maximum deflection of the shaft is

$$\frac{g}{2w^2}(\operatorname{sech} al + \sec al - 2)$$

5. A horizontal tie rod of length $2l$ with concentrated load W at the centre and ends freely hinged, satisfies the differential equation

$EI\frac{d^2y}{dx^2} = Py - \frac{W}{2}x$ with conditions $x = 0,\ y = 0$ and $x = l,\ \frac{dy}{dx} = 0$. Prove that the deflection δ and bending moment M at the centre $(x = l)$ are $\delta = \frac{W}{Pn}(nl - \tanh nl)$ and $M = -\frac{W}{2n}\tanh nl$, where $n^2EI = P$.

6. The differential equation for the displacement of a heavy whirling shaft is $\frac{d^4y}{dx^4} = a^4\left(y + \frac{g}{w^2}\right)$, where $a^4 = \frac{Ww^2}{gEI}$.

If both ends are short bearings, then ends being $x = 0$ and $x = l$, find the bending moment of the centre of the shaft.

ANSWERS

2. $y = C_1\cos\frac{\pi}{l}\sqrt{\frac{P}{Q}}x + C_2\sin\frac{\pi}{l}\sqrt{\frac{P}{Q}}x + \frac{Wl^2}{8(Q-P)}\sin\frac{\pi x}{l}$ **3.** $y = \frac{W}{24EI}(-3l^4 - 8l^3) = \frac{Wl^3}{24EI}(3l + 8)$ **6.** $\frac{W}{2a^2}\left(\operatorname{sech}\frac{al}{2} - \sec al\right)$

33.13 APPLICATIONS OF SIMULTANEOUS LINEAR DIFFERENTIAL EQUATIONS

Example 1. *A particle is projected with velocity u making an angle α with the horizontal. Neglecting air resistance, show that the equation of its path is the parabola*

$$y = x\tan\alpha - \frac{9x^2}{2u^2\cos^2\alpha}$$

Find the time of flight, the greatest height attained and range on the horizontal plane.

Solution. Let a particle of mass m be projected from a point O with velocity u in a direction making an angle α with the horizontal. Let the horizontal and the vertical lines through O is the plane of motion of the particle be taken as the axes of x and y respectively. Let $P(x, y)$ be the position of the particle at time t.

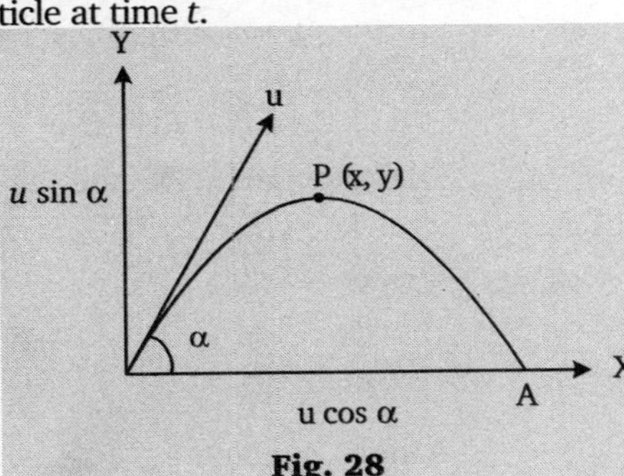

Fig. 28

The horizontal and vertical components of u are y cos a and u sin α respectively.

The only force acting on the particle is its weight mg acting vertically downwards.

$\therefore$ The equation of motion are

Parallel to x-axis

$$m\frac{d^2x}{dt^2} = 0 \Rightarrow \frac{d^2x}{dt^2} = 0 \qquad \text{... (1)}$$

Parallel to y-axis

$$m\frac{d^2x}{dt^2} = -mg \Rightarrow \frac{d^2y}{dt^2} \qquad \text{... (2)}$$

Integrating (1) w.r.t. t, we get $\frac{dx}{dt} = C_1$

Initially, when $t = 0$, $dx/dt = u\cos\alpha$,

$\therefore \quad C_1 = u\cos\alpha$

$$\therefore \quad \frac{dx}{dt} = u\cos\alpha \qquad \text{... (3)}$$

Integrating w.r.t. t, $x = (u\cos\alpha)t + C_2$

Initially, when $t = 0, x = 0,\ \therefore C_2 = 0$

$$x = (u\cos\alpha)\,t \qquad \text{... (4)}$$

Integrating (2) w.r.t. t, $\frac{dy}{dt} = -gt + C_3$

Initially, when $t = 0$, $\frac{dy}{dt} = u\sin\alpha$

so that $\quad C_3 = u\sin\alpha$

$$\frac{dy}{dt} = u\sin\alpha - gt \qquad \text{... (5)}$$

Integrating w.r.t. t, we get

$$y = (u\sin\alpha)t - \frac{1}{2}gt^2 + C_4$$

Initially, when $t = 0,\ y = 0,\ \therefore C_4 = 0$

$$y = (u\sin\alpha)t - \frac{1}{2}gt^2 \qquad \text{... (6)}$$

Equation (4) and (6) give the position of the particle at any time t. The equation of the path described by the particle is obtained by eliminating the parameter t between equations (4) and (6).

From (4), $t = \dfrac{x}{u\cos\alpha}$

Substituting this value of t in equation (6), we get

$$y = (u\sin\alpha).\frac{x}{u\cos\alpha} - \frac{1}{2}g\left(\frac{x^2}{u^2\cos^2\alpha}\right)$$

$$y = x\tan\alpha - \frac{gx^2}{2u^2\cos^2\alpha} \quad \text{... (7)}$$

which is the equation of the path of the projectile. Clearly, the path is a parabola.

The time of flight of the projectile is the time taken by the particle to reach the horizontal plane through O.

At the point A, $y = 0$

From (6), $0 = (u\sin\alpha)t - \frac{1}{2}gt^2$ or $t = \dfrac{2u\sin\alpha}{g}$

At the highest point, the vertical component of velocity vanishes, *i.e.*,

$$dy/dt = 0$$

so that $u\sin\alpha - gt = 0$ or $t = \dfrac{u\sin\alpha}{g}$.

$\therefore$ The greatest height $= y$ (when $\dfrac{dy}{dx} = 0$).

$$= (u\sin\alpha)\,t - \frac{1}{2}gt^2 \text{ (when } t = \frac{u\sin\alpha}{g})$$

$$= u\sin\alpha.\frac{u\sin\alpha}{g} - \frac{1}{2}g\frac{u^2\sin^2\alpha}{g}$$

$$= \frac{u^2\sin^2\alpha}{g} - \frac{u^2\sin^2\alpha}{2g} = \frac{u^2\sin^2\alpha}{2g}.$$

The range $R\,(= OA)$ is the horizontal distance covered by the particle in the time of flight.

$$R = u\cos\alpha \times \frac{2u\sin\alpha}{g} = \frac{u^2\sin 2\alpha}{g}.$$

Example 2. *The equation of motion under certain conditions are*

$$m\frac{d^2x}{dt^2} + eh\frac{dy}{dt} = eE \quad \text{... (1)}$$

$$m\frac{d^2y}{dt^2} - eh\frac{dx}{dt} = 0 \quad \text{... (2)}$$

with condition $x = \dfrac{dx}{dt} = y = \dfrac{dy}{dt} = 0$, *when* $t = 0$, *find the path of electron.*

Solution. Multiplying (2) by k and adding to (1), we get

$$m\frac{d^2x}{dt^2} + mk\frac{d^2y}{dt^2} + eh\frac{dy}{dt} - ehk\frac{dx}{dt} = eE$$

$$\Rightarrow \quad m\frac{d^2}{dt^2}(x+ky) - ehk\frac{d}{dt}\left(-\frac{y}{k} + x\right) = e.E \quad \text{... (3)}$$

Let us choose k such that $x + ky = x - \dfrac{y}{x}$.

$$k = -\frac{1}{k} \Rightarrow k^2 = -1 \Rightarrow k = \pm i$$

Putting $x + ky = u$ in (3), we have

$$m\frac{d^2u}{dt^2} - ehk\frac{du}{dt} = eE$$

$$\Rightarrow \quad \frac{d^2u}{dt^2} - \omega k\frac{du}{dt} = \frac{eE}{m} \quad \left(\omega = \frac{eh}{m}\right)$$

$$\Rightarrow \quad D^2u - \omega k\,Du = \frac{eE}{m}$$

A.E. is $\quad m^2 - \omega km = 0$

$$m(m - \omega k) = 0$$

$$m = 0 \text{ or } m = \omega k$$

$$\text{C.F.} = C_1 + C_2e^{\omega kt}$$

$$\text{P.I.} = \frac{1}{D^2 - \omega kD}\frac{eE}{m} = \frac{eE}{m}\frac{1}{D^2 - \omega kD}e^{0t}$$

$$= \frac{eEt}{m}\frac{1}{2D - \omega k}e^{0t} = \frac{\omega Et}{h}\frac{1}{0 - \omega k} = \frac{-Et}{hk} \quad \left(\omega = \frac{eh}{m}\right)$$

The complete solution is

$$v = C_1 + C_2e^{\omega kt} - \frac{Et}{hk}$$

or $\quad x + ky = C_1 + C_2e^{\omega kt} - \dfrac{Et}{hk}. \quad \text{... (4)}$

Putting the value of k, *i.e.*, $i, -i$ in (4), we get

$$x + iy = C_1 + C_2e^{i\omega t} - \frac{Et}{ih} \quad \text{... (5)}$$

$$x - iy = C_3 + C_4e^{-i\omega t} + \frac{Et}{ih} \quad \text{... (6)}$$

Differentiating (5) and (6), we get

$$\frac{dx}{dt} + i\frac{dy}{dt} = C_2i\omega e^{i\omega t} + \frac{iE}{h} \quad \text{... (7)}$$

$$\frac{dx}{dt} - i\frac{dy}{dt} = -i\omega C_4e^{-i\omega t} - \frac{iE}{h} \quad \text{... (8)}$$

Initial conditions, $x = y = \dfrac{dx}{dt} = \dfrac{dy}{dt} = 0$, when $t = 0$.

Putting these values in (5), (6), (7) and (8), we get

$$0 = C_1 + C_2 \Rightarrow C_2 = -C_1$$

$$0 = C_3 + C_4 \Rightarrow C_4 = -C_3$$

$$0 = i\omega C_2 + \frac{iE}{h} \Rightarrow C_2 = -\frac{E}{\omega h}$$

$$0 = -i\omega C_4 - \frac{iE}{h} \Rightarrow C_4 = \frac{-E}{\omega h}$$

On substituting the values of C_1, C_2, C_3 and C_4 in (5) and (6), we get

$$x + iy = \frac{E}{\omega h} - \frac{E}{h\omega} e^{i\omega t} + i\frac{Et}{h} \quad \text{... (9)}$$

$$x - iy = \frac{E}{\omega h} - \frac{E}{\omega h} e^{-i\omega t} - i\frac{Et}{h} \quad \text{... (10)}$$

On adding (9) and (10), we get

$$2x = \frac{2E}{h\omega} - \frac{E}{h\omega}(e^{i\omega t} + e^{-i\omega t})$$

$$x = \frac{E}{h\omega} - \frac{E}{h\omega}\cos\omega t$$

$$x = \frac{E}{h\omega}(1 - \cos\omega t)$$

Substituting (10) from (9), we obtain

$$2iy = -\frac{E}{h\omega}(e^{i\omega t} - e^{-i\omega t}) + \frac{2iEt}{h}$$

$$y = -\frac{-E}{\omega h}\left(\frac{e^{i\omega t} - e^{-i\omega t}}{2i}\right) + \frac{Et}{h}$$

$$= -\frac{E}{\omega h}\sin\omega t + \frac{Et}{h}$$

$$y = \frac{E}{h\omega}(\omega t - \sin\omega t).$$

Example 3. *Two particles each of mass m gm are suspended from two strings of same stiffness k. After the system comes to rest, the lower mass is pulled l cm downwards and released. Discuss their motion.*

Solution. Let the displacements of the upper and lower masses at time t from their respective positions of equilibrium be denoted by x and y. The stretch of the upper string is x and the stretch of the lower spring is $(y - x)$.

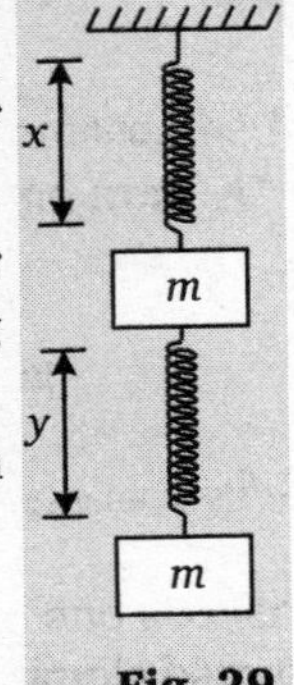

Fig. 29

Thus the restoring force acting on the upper mass

$$= -kx + k(y - x)$$
$$= k(y - 2x)$$

and the restoring force acting on the lower mass $= -k(y - x)$.

Thus the equations of motion are

$$m\frac{d^2x}{dt^2} = k(y - 2x) \text{ and } m\frac{d^2y}{dt^2} = -k(y - x).$$

$$(mD^2 + 2k)x - ky = 0 \quad \text{... (1)}$$

and $$(mD^2 + k)y - kx = 0 \quad \text{... (2)}$$

Operating (1) by $(mD^2 + k)$ and multiplying (2) by k and then adding, we get

$$[(mD^2 + k)(mD^2 + 2k) - k^2]x = 0$$

$$(D^4 + 3hD^2 + h^2)x = 0, \text{ where } h^2 = k/m.$$

The auxiliary equation is $m^4 + 3hm^2 + h^2 = 0$

$$m^2 = \frac{-3h \pm \sqrt{9h^2 - 4h^2}}{2} = -2.62h$$

or $-0.38h = -a^2, -b^2$ (say)

and so $m = \pm ia, \pm ib$.

Thus, we have

$$x = C_1\cos at + C_2\sin at + C_3\cos bt + C_4\sin bt \quad \text{... (3)}$$

From (1), we have

$$y = \left(\frac{D^2}{h} + 2\right)x$$

$$= \left(2 - \frac{a^2}{h}\right)(C_1\cos at + C_2\sin at)$$
$$+ (2 - b^2/h)(C_3\cos bt + C_4\sin bt) \quad \text{... (4)}$$

Initially, we have $t = 0,\ x = y = l$

and $\frac{dx}{dt} = \frac{dy}{dt} = 0$.

Therefore, from (3), we have

$l = C_1 + C_3;\ 0 = aC_2 + bC_4$ and from (4), we have

$$l = (2 - a^2/h)C_1 + (2 - b^2/h)C_3$$
$$0 = (2 - a^2/h)C_2 + (2 - b^2/h)bC_4$$

Thus, $$C_1 = \frac{l(h - b^2)}{a^2 - b^2},$$

$$C_2 = \frac{l(h - a^2)}{b^2 - a^2},\ C_3 = C_4 = 0$$

Putting these values of C_1, C_2, C_3 and C_4 in (3) and (4), we get x and y. From the values of x and y, it is clear that the motion of the given system of springs in a combination of springs is a combination of two simple harmonic motion of periods $2\pi/a$ and $2\pi/b$.

Example 4. *Assuming that a spherical rain drop evaporates at rate proportional to its surface area and if its radius originally is 3 mm and one hour later has been reduced to 2 mm, find an expression for the radius of the rain drop at any time.*

Solution. Evaporation $\propto$ surface area

$$\frac{dV}{dt} \propto S$$

$$\Rightarrow \quad \frac{dV}{dt} = kS \quad \text{... (1)}$$

$$V = \frac{4}{3}\pi r^3 \quad \text{... (2)}$$

$$\frac{dV}{dt} = 4\pi r^2 \frac{dr}{dt}$$

Putting the value of $\frac{dV}{dt}$ from (1) in (2), we have

$$4\pi r^2 \frac{dr}{dt} = kS$$

$$S\frac{dr}{dt} = kS \text{ or } \frac{dr}{dt} = k.$$

$$r = kt + C \quad \text{... (3)}$$

Putting $t = 0, r = 3$ in (3), we get $3 = C$.

(3) becomes $r = kt + 3$ ... (4)

Putting $t = 1$ and $r = 2$ in (4),

we get $2 = k + 3 \Rightarrow k = -1$

(4) becomes

$$r = -t + 3.$$

EXERCISE 33.9

1. A particle moving in a plane is subjected to a force directed towards a fixed point O and proportional to the distance of the particle from O. Show that the differential equations of motion are of the form

$$\frac{d^2x}{dt^2} = -K^2x, \quad \frac{d^2y}{dt^2} = -K^2y.$$

Find the cartesian equation of path of the particle if $x = 1, \ y = 0, \ \frac{dx}{dt} = 0$ and $\frac{dy}{dt} = 2$, when $t = 0$.

2. A particle of unit mass is projected with velocity u at an inclination a above the horizon in a medium whose resistances is k times the velocity. Show that its direction will again make an angle a with horizontal after a time

$$\frac{1}{k}\log\left(1 + \frac{2ku}{g}\sin\alpha\right).$$

3. In a chemical transformation of certain substance following equation occur :

$$\frac{dx}{dt} = ax = 0, \quad \frac{dz}{dt} = by, \quad x + y + z = c$$

where a, b, c are constants. Obtain a differential equation for z.

Hence, prove that if $z = -\frac{dz}{dt} = 0$ when $t = 0$, then

$$z = c + \frac{c}{a - b}(be^{-at} - ae^{-bt})$$

4. Two particles each of mass m gm are suspended from two springs of same stiffness k. After the system comes to rest, the lower mass is pulled l cm downwards and released. Discuss their motion.

ANSWERS

1. $4x^2 + k^2y^2 = 4$

4. $x = \frac{l(\lambda - \beta^2)}{\alpha^2 - \beta^2}\cos\alpha t + \frac{l(\lambda - \alpha^2)}{\beta^2 - \alpha^2}\sin\alpha t$

FFFFFF

CHAPTER

34 Series Solution of Second Order Differential Equations

34.1 INTRODUCTION

We have observed that the linear homogeneous differential equation with constant coefficient can be solved by some algebraic methods, which give the solution as elementary function. If we consider the differential equations having variable coefficients, we get a more complicated situation and the solutions thus obtained may be non-elementary functions. Therefore, Legendre's, Bessel's and the hypergeometric differential equations and their solutions play an important role in the field of pure and applied mathematics and in engineering mathematics as well. We shall now discuss in this chapter a method for solving such type of differential equation of order two having their solutions in the forms of power series, and the method is therefore called a power series method. Besides the solutions, we shall also discuss some properties of the solutions.

34.2 POWER SERIES METHOD

This method is very effective with respect to linear homogeneous differential equation with variable coefficients. This method gives the solution of the differential equations in the form of a power series. Therefore, an infinite series of the form

$$\sum_{m=0}^{\infty} a_m x^m = a_0 + a_1 x + a_2 x^2 + \ldots + a_m x^m + \ldots$$

is called a power series. This power series is said to be convergent at a point x if $\lim_{n\to\infty} \sum_{m=0}^{n} a_m x^m$ exists. It is clear that the above series is always convergent at $x = 0$. To explain this method clear, let us consider a general homogeneous differential equation of second order $y'' + P(x)y' + Q(x)y = 0$.

The solution y of this given differential equation is assumed in the form of a power series as above with undetermined coefficient and these coefficients are determined by putting that series and the series for derivatives of y into the given differential equation.

34.2.1 ORDINARY AND SINGULAR POINTS

Consider a general homogeneous linear differential equation of order two :

$$\frac{d^2y}{dx^2} + P(x)\frac{dy}{dx} + Qxy = 0 \quad \text{or} \quad y'' + P(x)y' + Q(x)y = 0. \qquad \ldots(1)$$

The main concept about the solution of (1) is that the behaviour of the solutions near a point $x = x_0$ depends on the behaviour of $P(x)$ and $Q(x)$ near this point x_0. If $P(x)$ and $Q(x)$ are analytic at this point x_0, then power series method is applicable in some neighbourhood of x_0. Then this point x_0 is called an ordinary point of the differential equation (1). Thus we can say that every solution of (1) is analytic at x_0. If x_0 is not an ordinary point, then this point x_0 is called a singular point.

34.2.2 REGULAR SINGULAR POINTS

In the above section, we have seen that if one of the coefficient functions $P(x)$ and $Q(x)$ is not differentiable at x_0 then this point is called a singular point. Thus a point x_0 of the differential equation (1) is called regular if the functions $(x - x_0)P(x)$ and $(x - x_0)^2 Q(x)$ are analytic at $x = x_0$.

If a singular point x_0 is located at the origin, then the general form of an analytic function at $x = x_0$ is $\sum_{m=0}^{\infty} a_m x^m$.

This implies that the origin will definitely be a singular point of (1) of $P(x)$ and $Q(x)$ have at least one of the coefficients with negative subscripts non-zero. In this case we assume the solution of the differential equation (1) of the form

$$y = x^n \sum_{m=0}^{\infty} a_m x^m = \sum_{m=0}^{\infty} a_m x^{m+n}$$

where n may be a negative integer or may be a fraction or even an irrational number.

REMARKS

- The solution of the form $y = \sum_{m=0}^{\infty} a_m x^m$ must be valid at least on the interval $|x| < 1$.
- The power series of the type $y = \sum_{m=0}^{\infty} a_m x^{m+n}$ is known as "quasi power series".

34.3 POWER SERIES SOLUTION

34.3.1 SOLUTION NEAR AN ORDINARY POINT

Consider the differential equation $\dfrac{d^2y}{dx^2} + P(x)\dfrac{dy}{dx} + Q(x)y = 0$ ...(1)

Let us take a trial solution of the form $y = \sum_{n=0}^{\infty} C_n x^n$ ...(2)

$$\Rightarrow \left.\begin{aligned} \frac{dy}{dx} = \sum_{n=0}^{\infty} n\, C_n\, x^{n-1} \text{ and } \frac{d^2y}{dx^2} = \sum_{n=0}^{\infty} n(n-1)\, C_n\, x^{n-2} \end{aligned}\right\} \quad ...(3)$$

Also, by letting $P(x)$ and $Q(x)$ are not polynomial in x, we can expand such that

$$P(x) = \sum_{n=1}^{\infty} p_n x^n \quad \text{and} \quad Q(x) = \sum_{n=0}^{\infty} q_n . x^n \quad ...(4)$$

Now, putting all these values in equation (1), we get the required solution.

34.3.2 SOLUTION NEAR A REGULAR SINGULAR POINT

Here, we assume a trial series solution of the type

$$y = x^m\,(C_0 + C_1 x + C_2 x^2 + ...) \quad ...(1)$$

$$= x^m . \sum_{n=0}^{\infty} C_n x^n \text{, where all } C_n\text{'s constant with } C_n \neq 0.$$

WORKING PROCEDURE

To find the values of m and C's, we proceed as follows :

Step 1. *Put the values of* $\dfrac{dy}{dx}$ *and* $\dfrac{d^2y}{dx^2}$ *in the given differential equation.*

Step 2. *By equating to zero the coefficients of the lowest power of* x, *get a quadratic equation in m, which is called indicial equation.*

Step 3. *To find the values of the equations* $C_1, C_2, ...,$ *etc. in terms of* C_0 *equating to zero the coefficients of other powers of* x.

Step 4. *The nature of the root can be determined as follows :*

(A) If roots of the indicial equation are equal :

Let $m = n$ *be two equal roots. Then putting* $m = n$ *in* y *and in* $\dfrac{\partial y}{\partial m}$ *we may get the two independent solutions.*

(B) If roots of the indicial equation unequal and not differing by an integer

If the indicial equation has two unequal roots $m = m_1$ *and* m_2 *which do not differ by an integer, then by putting* $m = m_1$ *and* m_2 *in the series, we get two independent solutions.*

(C) If the roots of the indicial equation differing by an integer and making the coefficients of some powers of x in the series for y infinity :

Let $m = m_1$ *and* m_2 *be two roots of the indicial equation which differ by an integer and some of the coefficients of powers of* k *in the series for* y *infinity for* $m = m_2$.

Here, put $C(m - m_2)$ *for* C_0, *then we get two independent solutions for* $m = m_2$. *Then proceed as in case 1.*

(D) If the roots of the indicial equation differing by an integer and making the coefficients of the series for y indeterminate :

If $m = m_1$ *and* $m_2 (m_1 > m_2)$ *are two roots of the indicial equation which differ by an integer. If one of the coefficients of the series for y becomes indeterminate when* $m = m_2$, *the complete solution is given by putting* $m = m_2$ *in y, which have two arbitrary constants.*

REMARKS

- The second solution always consists of a numerical multiple of the product of the first solution and log x added to another series.
- The series obtained by putting $m = m_1$ in y is merely a numerical multiple of one of the series contained in the first solution.

Solved Examples

Example 1. Solve $x\dfrac{d^2y}{dx^2} + \dfrac{dy}{dx} + xy = 0$.

(UPTU-2003, VTU-2010, SVTU-2007)

Solution. The given equation is

$$x\frac{d^2y}{dx^2} + \frac{dy}{dx} + xy = 0 \quad \ldots(1)$$

Putting $y = x^m$ in the LHS of (1), we get

$$xm(m-1)x^{m-2} + mx^{m-1} + x.x^m$$
$$= x^{m+1} + m^2x^{m-1}.$$

Clearly, the common difference of the powers is $(m+1)-(m-1)$, *i.e.*, 2.

Let

$$y = \sum_{r=0}^{\infty} C_r x^{m+2r}$$
$$= C_0x^m + C_1x^{m+2} + C_2x^{m+4} + \ldots \quad \ldots(2)$$

is the solution of (1).

Then, we have

$$\frac{dy}{dx} = \sum_{r=0}^{\infty} C_r(m+2r)\,x^{m+2r-1}$$

$$\frac{d^2y}{dx^2} = \sum_{r=0}^{\infty} C_r(m+2r)(m+2r-1)^{m+2r-2}.$$

Put all these values in (1), we get

$$\sum_{r=0}^{\infty} C_r[(m+2r)(m+2r-1)x^{m+2r-1}$$
$$+ (m+2r)x^{m+2r-1} + x^{m+2r+1}] = 0$$

$$\Rightarrow \sum_{r=0}^{\infty} C_r[x^{m+2r+1} + (m+2r)^2\,x^{m+2r-1}] = 0.$$

Equating to zero, the coefficient of the lowest power of x, *i.e.*, of x^{m-1}, we have

$$C_0m^2 = 0$$

which is the required indicial equation.

Since $C_0 \neq 0$, therefore $m = 0$, 0 are two equal roots.

Now, equating to zero the coefficient of the general term, *i.e.*, of x^{m+2p+1}, we get

$$C_p + (m+2p+2)^2C_{p+1} = 0$$

$$\Rightarrow C_{p+1} = -\frac{1}{(m+2p+2)^2}C_p \quad \ldots(3)$$

Putting $p = 0, 1, 2, \ldots$ in (3), we get

$$C_1 = -\frac{1}{(m+2)^2}C_0,$$

$$C_2 = -\frac{1}{(m+4)^2}C_1$$
$$= (-1)^2\frac{1}{(m+2)^2(m+4)^2}C_0$$

$$C_3 = -\frac{1}{(m+6)^2}C_2$$
$$= (-1)^3\frac{1}{(m+2)^2(m+4)^2(m+6)^2}C_0 \ldots$$

and so on.

Put all these values in (2), we get

$$y = C_0x^m\left[1 - \frac{x^2}{(m+2)^2} + \frac{x^4}{(m+2)^2(m+4)^2} - \ldots\right] \quad \ldots(4)$$

Putting $m = 0$, we get

$$y = C_0\left[1 - \frac{x^2}{2^2} + \frac{x^4}{2^2.4^2} - \frac{x^6}{2^2.4^2.6^2}\right] + \ldots \quad \ldots(5)$$

$= C_0.u$ (say),

which is the first solution of the given equation (1).

when $u = 1 - \dfrac{x^2}{2^2} + \dfrac{x^2}{2^2.4^2} - \dfrac{x^6}{2^2.4^2.6^2} + \ldots$

Since, there are two equal values of m, therefore, second solution be obtained from (4).

Now, from (4)

$$\frac{dy}{dx} = C_0\left[mx^{m-1} - \frac{(m+2)x^{m+1}}{(m+2)^2} + \frac{(m+4)x^{m+3}}{(m+2)^2(m+4)^2} - \ldots\right]$$

$$\Rightarrow \frac{d^2y}{dx^2} = C_0\left[m(m-1)x^{m-2} - \frac{(m+2)(m+1)}{(m+2)^2}x^m + \frac{(m+4)(m+3)x^{m+2}}{(m+2)^2(m+4)^2} - \ldots\right]$$

Put above two values in (1), we get

$$\text{LHS} = xC_0\left[m(m-1)x^{m-2} - \frac{(m+2)(m+1)}{(m+2)^2}x^m + \frac{(m+4)(m+3)x^{m+2}}{(m+2)^2(m+4)^2} - \ldots\right]$$

$$+C_0\left[mx^{m-1}-\frac{(m+2)x^{m+1}}{(m+2)^2}+\frac{(m+4)x^{m+3}}{(m+2)^2(m+4)^2}-\ldots\right]$$

$$+xC_0\left[x^m-\frac{x^{m+2}}{(m+2)^2}+\frac{x^{m+4}}{(m+2)^2(m+4)^2}-\ldots\right]$$

$$=C_0m^2x^{m-1}$$

$$\therefore\left[x\frac{d^2}{dx^2}+\frac{d}{dx}+x\right]y = C_0m^2x^{m-1}.$$

Differentiating both sides partially, w.r.t. m, we get

$$\frac{\partial}{\partial m}\left[x\frac{d^2}{dx^2}+\frac{d}{dx}+x\right]y=\frac{\partial}{\partial m}(C_0m^2x^{m-1})$$

$$\Rightarrow\quad\left[x\frac{d^2}{dx^2}+\frac{d}{dx}+x\right]\left(\frac{\partial y}{\partial m}\right)$$

$$=C_0.2mx^{m-1}+C_0m^2x^{m-1}\log x$$

Putting $m=0$, we get

$$\left[x\frac{d^2}{dx^2}+\frac{d}{dx}+x\right]\left[\frac{\partial y}{\partial m}\right]=0$$

$\Rightarrow\left(\frac{\partial y}{\partial m}\right)_{m=0}$ satisfy the equation (1), therefore it is also a solution of (1).

Differentiating (4) partially w.r.t. m, we get

$$\frac{\partial y}{\partial m}=C_0x^m\log x\left[1-\frac{x^2}{(m+2)^2}+\frac{x^4}{(m+2)^2(m+4)^2}-\ldots\right]$$

$$+C_0x^m\left[\frac{2x^2}{(m+2)^3}+\left\{\frac{-2}{(m+2)^3(m+4)^2}+\frac{-2}{(m+2)^2(m+4)^3}\right\}x^4+\ldots\right]$$

Putting $m=0$, we get

$$\left(\frac{\partial y}{\partial m}\right)_{m=0}=C_0\log x\left[1-\frac{x^2}{2^2}+\frac{x^4}{2^2.4^2}-\ldots\right]$$

$$+C_0\left[\frac{x^2}{2^2}+\left\{\frac{-2}{2^3.4^2}+\frac{-2}{2^2.4^3}\right\}x^4\ldots\right]$$

$$=bu\log x+b\left[\frac{x^2}{2^2}-\frac{3}{2^3.4^2}x^4+\ldots\right]$$

$$=bv\text{(say)}$$

where, $v=u\log x+\left[\frac{x^2}{2^2}-\frac{3}{2^3.4^2}x^4+\ldots\right]$ and b is any arbitrary constant which replaces C_0. Hence, the required general solution of (1) is given by

$y=au+bv$, where a and b are arbitrary constants.

Example 2. *Solve the following Legendre's equation*

$$(1-x^2)y''-2xy'+p(p+1)y=0$$

in descending powers of x.

(UPTU(Q. Bank)–2002, GBTU–2010)

Solution. The given equation can be written as

$$(1-x^2)\frac{d^2y}{dx^2}-2x\frac{dy}{dx}+p(p+1)y=0 \qquad \ldots(1)$$

Putting $y=x^m$ in the LHS of (1), we get

$$(1-x^2)m(m-1)x^{m-2}-2x.mx^{m-1}+p(p+1)x^m$$

$$\text{or}\,(-m^2-m+p^2+p)x^m+m(m-1)x^{m-2}.$$

Clearly, the common difference of the powers is $m-(m-2)$, *i.e.*, 2.

Let the solution of (1) in descending powers of x be

$$y=C_0x^m+C_1x^{m-2}+C_2x^{m-4}+\ldots$$

$$=\sum_{r=0}^{\infty}C_rx^{m-2r} \qquad \ldots(2)$$

$$\Rightarrow\quad\frac{dy}{dx}=\sum_{r=0}^{\infty}C_r(m-2r)\,x^{m-2r-1}$$

$$\text{and}\ \frac{d^2y}{dx^2}=\sum_{r=0}^{\infty}C_r(m-2r)(m-2r-1)x^{m-2r-2}$$

Put all these values in (1), we get

$$\sum_{r=0}^{\infty}C_r[(1-x^2)(m-2r)(m-2r-1)x^{m-2r-2}-2x(m-2r)x^{m-2r-1}+p(p+1)x^{m-2r}]=0$$

$$\Rightarrow\quad\sum_{r=0}^{\infty}C_r[(-(m-2r)(m-2r-1)-2(m-2r)+p(p+1)x^{m-2r}+(m-2r)(m-2r-1)x^{m-2r-2}]=0$$

$$\Rightarrow\sum_{r=0}^{\infty}C_r\ [\{p^2-(m-2r)^2+(p-m+2r)\}\ x^{m-2r}+(m-2r)(m-2r-1)x^{m-2r-2}]=0$$

$$\Rightarrow\quad\sum_{r=0}^{\infty}C_r\ [(p-m+2r)(p-m-2r+1)\ x^{m-2r}+(m-2r)(m-2r-1)x^{m-2r-2}]=0$$

Equating to zero, the coefficients of the highest power of x, *i.e.*, x^m, we get the initial equation as $C_0(p-m)(p+m+1)=0$.

Since $C_0 \neq 0$, therefore, we get $m = p, -(p+1)$.

Now, equating to zero the coefficients of x^{m-2r} we get

$$C_r(p-m+2r)(p+m-2r+1) + (m-2r+2)(m-2r+1)C_{r-1} = 0$$

$$\Rightarrow \quad C_r = \frac{(m-2r+2)(m-2r+1)}{(p-m+2r)(p+m-2r+1)} C_{r-1}.$$

Putting $r = 1, 2, \ldots$, we get

$$C_1 = -\frac{m(m-1)}{(p-m+2)(p-m-1)} C_0$$

$$C_2 = -\frac{(m-2)(m-3)}{(p-m+4)(p+m-3)} C_1$$

$$= (-1)^2 \frac{m(m-1)(m-2)(m-3)}{(p-m+2)(p-m+4)(p+m-1)(p+m-3)} C_0$$

... ... and so on.

Put all these values in (2), we get

$$y = C_0 \left[x^m - \frac{m(m-1)}{(p-m+2)(p+m-1)} x^{m-2} + \frac{m(m-1)(m-2)(m-3)}{(p-m+2)(p-m+4)(p+m-1)(p+m-3)} x^{m-4} - \ldots \right]$$

Now putting $m = p, -(p+1)$ successively, we get

$$y = C_0 \left[x^p - \frac{p(p-1)}{2(2p-1)} x^{p-2} + \frac{p(p-1)(p-2)(p-3)}{2.4.(2p-1)(2p-3)} x^{p-4} - \ldots \right]$$

$$= au \text{ (say)}$$

which is one solution of the given equation.

Also,

$$y = C_0 \left[x^{-p-1} + \frac{(p+1)(p+2)}{2(2p+3)} x^{-p-3} + \frac{(p+1)(p+2)(p+3)(p+4)}{2.4(2p+3)(2p+5)} x^{-p-5} + \ldots \right]$$

$$= bv \text{ (say)}$$

Here, the required solution of the given equation is $y = au + bv$, where a and b are arbitrary constants.

Example 3. *Find a general solution in series of powers of x of the equation*

$$x^2 \frac{d^2y}{dx^2} + x\frac{dy}{dx} + (x^2-1)y = 0.$$

Solution. The given equation is

$$x^2 \frac{d^2y}{dx^2} + x\frac{dy}{dx} + (x^2-1)y = 0 \qquad \ldots(1)$$

Putting $y = x^m$ in the LHS of (1), we get

$$(m^2-1)x^m + x^{m+2}.$$

Here, clearly the common difference of the powers is 2.

Let us assume the solution of (1) is given by

$$y = \sum_{r=0}^{\infty} C_r x^{m+2r} \qquad \ldots(2)$$

Now, we have

$$\frac{dy}{dx} = \sum_{r=0}^{\infty} C_r (m+2r)\, x^{m+2r-1}.$$

Also, $$\frac{d^2y}{dx^2} = \sum_{r=0}^{\infty} C_r (m+2r)(m+2r-1).x^{m+2r-2}.$$

Putting all these values in equation (1), we get

$$\sum_{r=0}^{\infty} C_r [x^2(m+2r)(m+2r-1)x^{m+2r-2} + x(m+2r)x^{m+2r-1} + (x^2-1)x^{m+2r}] = 0$$

$$\Rightarrow \sum_{r=0}^{\infty} C_r [\{(m+2r)^2 - 1\}\, x^{m+2r} + x^{m+2r+2}] = 0$$

$$\Rightarrow \sum_{r=0}^{\infty} C_r [(m+2r+1)(m+2r-1)x^{m+2r} + x^{m+2r+2}] = 0.$$

Equating to zero the coefficients of the lowest power of x, *i.e.*, of x^m, we get

$$C_0 (m+1)\,(m-1) = 0$$

Now, $C_0 \neq 0 \Rightarrow m = 1, -1$.

$\Rightarrow$ Roots are unequal and differ by an integer (*i.e.*, 2).

Therefore, equating to zero the coefficients of the general term, *i.e.*, of x^{m+2p}, we get

$$C_p (m+2p+1)(m+2p-1) + C_{p-1} = 0$$

$$\Rightarrow \quad C_p = -\frac{1}{(m+2p+1)\,(m+2p-1)} C_{p-1}.$$

Putting $p = 1, 2, 3, \ldots$ successively, we get

$$C_1 = -\frac{1}{(m+3)(m+1)} C_0,$$

$$C_2 = -\frac{1}{(m+5)(m+3)} C_1$$

$$= (-1)^2 \frac{1}{(m+5)(m+3)^2(m+1)} C_0.$$

$$C_3 = -\frac{1}{(m+7)(m+5)} C_2$$

$$= (-1)^3 \frac{1}{(m+7)(m+5)^2(m+3)^2(m+1)} C_0$$

... ... and so on.

Putting all these values in (2), we get

$$y = C_0 x^m \left[1 - \frac{x^2}{(m+1)(m+3)} + \frac{x^4}{(m+1)(m+3)^2(m+5)} - \frac{x^6}{(m+1)(m+3)^2(m+5)^2(m+7)} + \ldots \right] \qquad \ldots(3)$$

If we put $m=-1$ in the above series, the coefficients of x^2 and higher order becomes infinite. To remove the difficulty, put $C(m+1)$ for C_0 in (3).

$$\therefore y = Cx^m\left[(m+1) - \frac{1}{(m+3)}x^2 + \frac{1}{(m+3)^2(m+5)}x^4 - \frac{1}{(m+3)^2(m+5)^2(m+7)}x^6 + \ldots\right] \quad \ldots(4)$$

Now, putting $m=-1$ in (4), we get

$$y = Cx^{-1}\left[-\frac{1}{2}x^2 + \frac{1}{2^2.4}x^4 - \frac{1}{2^2.4^2.6}.x^6 + \ldots\right] \quad \ldots(5)$$

$= au$ (say)

which is the one solution of the given equation (1).

Now, differentiating (4) partially w.r.t. m, we get

$$\frac{\partial y}{\partial m} = Cx^m \log x\left[(m+1) - \frac{x^2}{(m+3)} + \frac{x^4}{(m+3)^2(m+5)} - \ldots\right]$$

$$+ Cx^m\left[1 + \frac{x^2}{(m+3)^2} - \left\{\frac{2}{(m+3)^2(m+5)} + \frac{1}{(m+3)^2(m+5)^5}\right\}x^4 + \ldots\right].$$

Putting $m=-1$, we get

$$\left(\frac{\partial y}{\partial m}\right)_{m=-1} = Cu\log x + Cx^{-1}\left[1 + \frac{x^2}{4} - \frac{5}{64}x^4 + \ldots\right]$$

$= bv$ (say).

Hence, the required general solution of the given equation (1) is given by $y = au + bv$.

Example 4. *Solve the following differential equation in series*

$$2x^2\frac{d^2y}{dx^2} - x\frac{dy}{dx} + (x-5)y = 0. \quad \text{(UPTU–2008)}$$

Solution. We have

$$2x^2\frac{d^2y}{dx^2} - x\frac{dy}{dx} + (x-5)y = 0 \quad \ldots(1)$$

Here $xP_1(x)$ and $x^2P_2(x)$ are analytic (not infinite at $x=0$). So x_0 is regular singular point.

We assume the solution in the form :

$$y = \sum_{k=0}^{\infty} a_k\, x^{m+k},\ \frac{dy}{dx} = \Sigma a_k\,(m+k)\,x^{m+k-1}$$

$$\frac{d^2y}{dx^2} = \Sigma a_k(m+k)(m+k-1)\,x^{m+k-2}$$

Putting the values of y, $\frac{dy}{dx}$ and $\frac{d^2y}{dx^2}$ in equation (1), we get

$$2x^2\Sigma a^k(m+k)(m+k-1)x^{m+k-2} - x\Sigma a_k(m+k)x^{m+k-1} + (x-5)\Sigma a_k x^{m+k} = 0$$

$$\Rightarrow \Sigma a_k\left[2(m+k)(m+k-1) - (m+k) - 5\right]x^{m+k} + \Sigma a_k x^{m+k+1} = 0$$

$$\Sigma a_k\left[2(m+k)^2 - 3(m+k) - 5\right]x^{m+k} + \Sigma a_k x^{m+k+1} = 0$$

Equating the coefficient of the lowest degree term x^n to zero by putting $k=0$ in first summation of (2), we get

$$a_0(m+1)(2m-5) = 0 \Rightarrow (m+1) = 0$$

or $2m-5=0$

$\Rightarrow$ $m=-1$ or $m=5/2$

Now, equating the coefficient of next lower degree term x^{m+1} to zero by putting $k=1$ in first summation and $k=0$ in second summation, we get

$$a_1(m+2)(2m-3) + a_0 = 0$$

$$\Rightarrow \quad a_1(m+2)(2m-3) = a_0$$

$$a_1 = \frac{a_0}{(m+2)(2m-3)}$$

Now, equating the coefficient of x^{m+k+1} to zero by putting $k = k+1$ in first summation

$$a_{k+1}(m+k+2)(2m+2k-3) + a_k = 0$$

$$a_{k+1} = \frac{-a_k}{(m+k+2)(2m+2k-3)}$$

If $k=0$, $a_1 = \frac{-a_0}{(m+2)(2m-3)}$

If $k=1$, $a_2 = \frac{-a_1}{(m+3)(2m-1)}$

$$= \frac{a_0}{(m+2)(m+3)(2m-1)(2m-3)}$$

If $k=2$, $a_3 = \frac{-a_2}{(m+4)(2m+1)}$

$$= -\frac{a_0}{(m+2)(m+3)(m+4)(2m+1)(2m-1)(2m-3)}$$

$m=-1$	$m=5/2$
$a_1 = a_0/5$	$a_1 = -a_0/9$
$a_2 = a_0/30$	$a_2 = a_0/198$
$a_3 = -a_0/90$	$a_3 = -a_0/7722$

We have, $y = x^m(a_0 + a_1x + a_2x^2 + a_3x^3 + ...)$

For $m=-1$,

$$y_1 = x^{-1}\left(a_0 + \frac{a_0}{5}x + \frac{a_0}{30}x^2 - \frac{a_0}{90}x^3 + ...\right)$$

$$y_1 = a_0x^{-1}\left(1 + \frac{x}{5} + \frac{x^2}{30} - \frac{x^3}{90} + ...\right)$$

For $m = 5/2$,

$$y_2 = x^{5/2}\left(a_0 - \frac{a_0}{9}x + \frac{a_0}{198}x^2 - \frac{a_0}{7722}x^2 + ...\right)$$

$$\Rightarrow y_2 = a_0x^{5/2}\left(1 - \frac{x}{9} + \frac{x^2}{198} - \frac{x^3}{7722} + ...\right)$$

Since two solutions are linearly independent, so the general solution of (1) may be represented as $y = Ay_1 + By_2$.

Example 5. *Solve* $(1-x^2)\dfrac{d^2y}{dx^2} + 2x\dfrac{dy}{dx} + y = 0$.

(UPTU (Q. Bank) 2002, AMIETE-2000)

Solution. The given equation is

$$(1-x^2)\frac{d^2y}{dx^2} + 2x\frac{dy}{dx} + y = 0 \qquad ...(1)$$

Let us assume the series solution of the equation (1) is

$$y = \sum_{r=0}^{\infty} C_r x^{m+r}$$

$$\therefore \quad \frac{dy}{dt} = \sum_{r=0}^{\infty} C_r(m+r)\, x^{m+r-1}$$

and $$\frac{d^2y}{dx^2} = \sum_{r=0}^{\infty} C_r(m+r)(m+r-1)x^{m+r-2}.$$

Putting all these values in (1), we get

$$\sum_{r=0}^{\infty} C_r[(1-x^2)(m+r)(m+r-1)x^{m+r-2} + 2x(m+r)x^{m+r-1} + x^{m+r}] = 0$$

$$\Rightarrow \sum_{r=0}^{\infty} C_r\, [-\{(m+r)(m+r-3)-1\}\, x^{m+r} + (m+r)(m+r-1)x^{m+r-2}] = 0.$$

Now, equating to zero the coefficients of the lower power of x, *i.e.*, of x^{m-2}, we get the indicial equation is $C_0m(m-1)=0$

Now, $\quad C_0 \neq 0 \Rightarrow m = 0, 1$

$\Rightarrow$ Roots are real and unequal differing by an integer.

Now equating to zero the coefficients of the next higher power of x, *i.e.*, of x^{m-1}, we get

$$C_1(m+1)m = 0.$$

If $m=0$, we get $C_1 = 0$.

Now, equating to zero the coefficients of the general term, *i.e.*, of x^{m+p}, we get

$$C_{p+2} = \frac{(m+p)(m+p-3)-1}{(m+p+1)(m+p+2)}C_p.$$

Putting $p = 0, 1, 2, ...$, successively, we get

$$C_2 = \frac{m(m-3)-1}{(m+1)(m+2)}C_0.$$

$$C_3 = \frac{(m+1)(m-2)-1}{(m+2)(m+3)}C_1,$$

$$C_4 = \frac{(m+2)(m-1)-1}{(m+3)(m+4)}C_2 = \frac{\{(m+2)(m-1)-1\}\{m(m-3)-1\}}{(m+1)(m+2)(m+3)(m+4)}C_0,$$

$$C_5 = \frac{(m+3)m-1}{(m+4)(m+5)}C_3 = \frac{\{m(m+3)-1\}\{(m+1)(m-2)-1\}}{(m+2)(m+3)(m+4)(m+5)}C_1$$

... ... and so on.

Put all these values in (2), we get

$$y = C_0x^m\left[1 + \frac{m(m-3)-1}{(m+1)(m+2)}x^2 + \frac{\{(m+2)(m-1)-1\}\,\{m(m-3)-1\}}{(m+1)(m+2)(m+3)(m+4)}x^4 + ...\right]$$

$$+ C_1x^m\left[x + \frac{(m+1)(m-2)-1}{(m+2)(m+3)}x^3 + \frac{\{m(m+3)-1\}\,\{(m+1)(m-2)-1\}}{(m+2)(m+3)(m+4)(m+5)}x^5 + ...\right].$$

Putting $m=0$, we get

$$y = C_0\left[1 - \frac{1}{2}x^2 + \frac{1}{8}x^4 + ...\right] + C_1\left[x - \frac{1}{2}x^3 + \frac{1}{40}x^5 + ...\right]$$

$$= a\left[1 - \frac{x^2}{2} + \frac{x^4}{8} + ...\right] + b\left[x - \frac{x^3}{2} + \frac{x^5}{40} + ...\right]$$

where a and b are arbitrary constants.

This is the required general solution of the given equation.

Example 6. *Solve* $(x-x^2)\dfrac{d^2y}{dx^2}+(1-5x)\dfrac{dy}{dx}-4y=0$.

Solution. The given equation is

$$(x-x^2)\frac{d^2y}{dx^2}+(1-5x)\frac{dy}{dx}-4y=0 \quad \ldots(1)$$

Putting $y=x^m$ on the LHS of (1), we get

$$(m^2+4m+4)x^m - m^2x^{m-1}$$

Clearly, the common difference of the powers is $m-(m-1)=1$.

Now, let us assume that the solution of the given equation (1) is

$$y=\sum_{r=0}^{\infty} C_r x^{m+r}$$

$$\therefore \quad \frac{dy}{dx}=\sum_{r=0}^{\infty} C_r (m+r)x^{m+r-1}$$

$$\text{and} \quad \frac{d^2y}{dx^2}=\sum_{r=0}^{\infty} C_r (m+r)(m+r-1)x^{m+r-2}.$$

Put all these values in (1), we get

$$\sum_{r=0}^{\infty} C_r[(x-x^2)(m+r)(m+r-1)x^{m+r-2}$$
$$+(1-5x)(m+r)x^{m+r-1}-4x^{m+r}]=0$$

$$\Rightarrow \quad \sum_{r=0}^{\infty} C_r[-(m+r+2)^2x^{m+r}+(m+r)^2x^{(m+r)}] \quad \ldots(2)$$

Equating to zero the coefficients of lowest power of x, *i.e.*, of x^{m-1}, we get

$$m^2C_0=0 \Rightarrow m=0,0.$$

Now equating to zero the coefficients of the next higher powers of x, *i.e.*, of x^m, we get

$$-C_0(m+2)^2+C_1(m+1)^2=0$$

$$\Rightarrow \quad C_1=\frac{(m+2)^2}{(m+1)^2}C_0.$$

Again equating to zero the coefficients of the general term, *i.e.*, of x^{m+p}, we have

$$C_{p+1}=\frac{(m+p+2)^2}{(m+p+1)^2}C_p. \quad \ldots(3)$$

Putting $p=1,2,\ldots$ in (3) successively we get

$$C_2=\frac{(m+3)^2}{(m+2)^2}C_1=\frac{(m+3)^2}{(m+1)^2}C_0$$

$$C_3=\frac{(m+4)^2}{(m+3)^2}C_2=\frac{(m+4)^2}{(m+1)^2}C_0\ldots$$

and so on.

Now put all these values in equation (2), we get

$$y=C_0x^m\left[1+\left(\frac{m+2}{m+1}\right)^2x+\left(\frac{m+3}{m+1}\right)^2x^2+\left(\frac{m+4}{m+1}\right)^2x^3+\ldots\right] \quad \ldots(4)$$

Putting $m=0$, we get

$$y=C_0[1+2^2x+3^2x^2+4^2x^3+\ldots]$$

which is the one solution of the given equation (1).

Now, to find the second solution of the given equation, we proceed as follows.

Differentiating (4) partially w.r.t. m, we get

$$\frac{\partial y}{\partial m}=C_0x^m\log x\left[1+\left(\frac{m+2}{m+1}\right)^2x+\left(\frac{m+3}{m+1}\right)^2x^2+\left(\frac{m+4}{m+1}\right)^2x^3+\ldots\right]$$
$$+C_0x^m\left[2\left(\frac{m+2}{m+1}\right)\left\{\frac{1}{m+1}-\frac{m+2}{(m+1)^2}\right\}x+2\left(\frac{m+3}{m+1}\right)\left\{\frac{1}{m+1}-\frac{m+3}{(m+1)^2}\right\}x^2+\ldots\right]$$
$$=y\log x+C_0x^m\left[2\left(\frac{m+2}{m+1}\right)\left\{\frac{1}{m+1}-\frac{m+2}{(m+1)^2}\right\}x+2\left(\frac{m+3}{m+1}\right)\left\{\frac{1}{m+1}-\frac{m+3}{(m+1)^2}\right\}x^2+\ldots\right]$$

Now putting $m=0$ and $C_0=a$ and b respectively in the two series, we get

$$y=a[1+2^2x+3^2x^2+4^2x^3+\ldots]=au \text{ (say)}$$

$$\text{and} \quad \left(\frac{\partial y}{\partial m}\right)_{m=0}=bu\log x$$
$$+b[1.2(1-2)x+1.3(1-3)x^2+\ldots]$$
$$=bu\log x-2b[1.2x+2.3x^2+3.4x^3+\ldots]$$
$$=bv \text{ (say)}.$$

Hence, the complete solution of the given equation is $y=au+bv$, where a and b are arbitrary constants.

Example 7. *Solved* $9x(1-x)\dfrac{d^2y}{dx^2}-12\dfrac{dy}{dx}+4y=0$.

(Roorkee-2000, Madras-2006)

Solution. Here, the given equation is

$$9x(1-x)\frac{d^2y}{dx^2}-12\frac{dy}{dx}+4y=0 \quad \ldots(1)$$

Now, putting $y=x^m$ on the LHS of (1), we get

$$\{-9m(m-1)+4\}\,x^m+\{9m(m-1)-12m\}\,x^{m-1}.$$

Clearly, the common difference of the powers is one.

Let us assume $y = \sum_{r=0}^{\infty} C_r x^{m+r}$ be the solution of (1).

Therefore, $\frac{dy}{dx} = \sum_{r=0}^{\infty} C_r (m+r)\, x^{m+r-1}$

and $\frac{d^2y}{dx^2} = \sum_{r=0}^{\infty} C_r (m+r)(m+r-1)\, x^{m+r-2}$.

Putting all these values in (1) and after simplification, we get

$$\sum_{r=0}^{\infty} C_r[(3m+3r-4)(3m+3r+1)\, x^{m+r} - 3(m+r)(3m+3r-7)x^{m+r-1}]. \quad \ldots(2)$$

Equating to zero the coefficient of lower power of x, *i.e.*, of x^{m-1}, we have $-3C_0 m(3m-7) = 0$.

Since $C_0 \neq 0$. Therefore, $m = 0, 7/3$.

Now equating to zero the coefficient of general term, *i.e.* x^{m+p}, we have

$$C_p[(3m+3p-4)(3m+3p+1) - 3(m+p+1)(3m+3p+3-7)]C_{p+1} = 0$$

$$\Rightarrow \quad C_{p+1} = \frac{(3m+3p+1)}{3(m+p+1)} C_p.$$

Putting $p = 0, 1, 2, \ldots$ successively, we have

$$C_1 = \frac{(3m+1)}{3(m+1)} C_0$$

$$C_2 = \frac{(3m+4)}{3(m+2)} C_1 = \frac{(3m+4)(3m+1)}{3^2(m+2)(m+1)} C_0$$

$$C_3 = \frac{(3m+7)}{3(m+3)} C_2 = \frac{(3m+7)(3m+4)(3m+1)}{3(m+3)(m+2)(m+1)} C_0 \ldots$$

and so on.

Therefore, we have

$$y = \sum_{r=0}^{\infty} C_r x^{m+r} = C_0 x^m + C_1 x^{m+1} + C_2 x^{m+2} + \ldots$$

$$= C_0 x^m \left[1 + \frac{3m+1}{3(m+1)} x + \frac{(3m+1)(3m+4)}{3^2(m+1)(m+2)} x^2 + \frac{(3m+1)(3m+4)(3m+7)}{3^3(m+1)(m+2)(m+3)} x^3 + \ldots\right].$$

If $m = 0$, then taking $C_0 = a$.

$$\therefore \quad y = a\left[1 + \frac{1}{3}x + \frac{1.4}{3.6}x^2 + \frac{1.4.7}{3.6.9}x^3 + \ldots\right]$$

$= au$ (say)

which is one solution of the given equation.

Also, if $m = \frac{7}{3}$, taking $C_0 = b$, we get

$$y = bx^{7/3}\left[1 + \frac{8}{10}x + \frac{8.11}{10.13}x^2 + \frac{8.11.14}{10.13.16}x^3 + \ldots\right]$$

$= bv$ (say).

Hence, the complete solution of the given equation is $y = au + bv$, where a and b are arbitrary constants.

34.3.3 METHOD OF DIFFERENTIATION

Example 8. *Solve* $(1+x^2)\frac{d^2y}{dx^2} + 2x\frac{dy}{dx} = 0$,

where $x = 0, \frac{dy}{dx} = 1, y = 0$.

Solution. The given equation can be written as

$$(1+x^2)y_2 + 2xy_1 = 0 \quad \ldots(1)$$

Differentiating the equation (1) n times by Leibnitz's theorem, we have

$$(1+x^2)y_{n+2} + 2(n+1)xy_{n+1} + n(n+1)y_n = 0$$

when $x = 0$, we have

$$(y_{n+2})_0 = -n(n+1)(y_n)_0. \quad \ldots(2)$$

Using the given equation, when $x = 0$, we have

$$(y_2)_0 = 0.$$

Putting $n = 2, 4, \ldots$ in (2), we have

$$(y_2)_0 = (y_4)_0 = \ldots = 0.$$

Again putting $n = 1, 2, \ldots$, in (2), we have

$$(y_3)_0 = -1.2.(y_1)_0 = -2!$$

$$(y_5)_0 = -3.4(y_3)_0 = -4!$$

$$(y_7)_0 = -5.6(y_5)_0 = -6!$$

... ... etc.

Hence, $y = (y)_0 + (y_1)_0 x + (y_2)_0 \frac{x^2}{2!} + \ldots$

$$= x - 2!\frac{x^3}{3!} + 4!\frac{x^5}{5!} - 6!\frac{x^7}{7!} \ldots$$

$$= x - \frac{x^3}{3} + \frac{x^5}{5} - \frac{x^7}{7} + \ldots$$

EXERCISE 34.1

1. Solve $\frac{d^2y}{dx^2} - 2x^2\frac{dy}{dx} + 4xy = x^2 + 2x + 2$ in powers of x.

2. Solve $x\frac{d^2y}{dx^2} + \frac{dy}{dx} + xy = 0$.

3. Solve $x\frac{d^2y}{dx^2} + (1+x)\frac{dy}{dx} + 2y = 0$.

4. Transform the equation $\frac{d^2y}{dx^2} - y = 0$ by the substitution $x = \frac{1}{z}$ and show that it has no integrals that are regular in descending powers of x.

5. Solve completely in series the equation
$$x\frac{d^2y}{dx^2} + (x+n)\frac{dy}{dx} + (n+1)y = 0$$

6. Solve $(2x + x^3)\frac{d^2y}{dx^2} - \frac{dy}{dx} - 6xy = 0$. (Bhopal-2006)

7. Solve $2x^2\frac{d^2y}{dx^2} - x\frac{dy}{dx} + (1-x^2)y = x^2$.

8. Solve $(x - x^2)\frac{d^2y}{dx^2} + (1-x)\frac{dy}{dx} - y = 0$.

9. Solve $x^4\frac{d^2y}{dx^2} + x\frac{dy}{dx} + y = \frac{1}{x}$.

10. Solve $\frac{d^2y}{dx^2} - y = x$.

11. Solve $\frac{d^2y}{dx^2} + x^2y = 0$. (AMIETE-1995)

12. Solve $(1+x^2)\frac{d^2y}{dx^2} + x\frac{dy}{dx} - y = 0$. (AMIETE-1997)

13. Solve $x^2\frac{d^2y}{dx^2} + x\frac{dy}{dx} + (x^2 - 4)y = 0$.
(AMIETE-1997, Bhopal-2008, Rajasthan-2003)

14. Solve $x^2\frac{d^2y}{dx^2} + 4x\frac{dy}{dx} + (x^2 + 2)y = 0$. (AMIETE-1998)

15. Solve $x^2\frac{d^2y}{dx^2} + 6x\frac{dy}{dx} + (6 - x^2)y = 0$. (AMIETE-1997)

16. Solve $2x(1-x)\frac{d^2y}{dx^2} + (1-x)\frac{dy}{dx} + 3y = 0$.
(UPTU(B.Tech.)-2004, GBTU-2010)

17. Solve $2x^2\frac{d^2y}{dx^2} - x\frac{dy}{dx} + (x^2 + 1)y = 0$.
(AMIETE-1999, GBTU-2011)

18. Solve $x^2\frac{d^2y}{dx^2} + (x-1)\frac{dy}{dx} - y = 0$. (AMIETE-1996, 98)

19. Solve $x^2\frac{d^2y}{dx^2} + \frac{dy}{dx} + xy = 0$. (AMIETE-1997, 2000, 01)

20. Solve $x^2\frac{d^2y}{dx^2} + 5x\frac{dy}{dx} + x^2y = 0$. (UPTU-2002)

21. Solve $(1 - x^2)y'' - xy' + 4y = 0$ in series.
(UPTU(Q. Bank)-2002, Bhopal-2008, UPTU-2006, GBTU-2012)

22. Find the power series solution of the following differential equation about $x = 0$
$$(1 - x^2)\frac{d^2y}{dx^2} - 2x\frac{dy}{dx} + 2y = 0$$ (UPTU-2004)

23. Solve $y'' + xy' + x^2y = 0$. (UPTU-2005, Q. Bank-2002)

24. Solve in series $x(2 + x^2)\frac{d^2y}{dx^2} - \frac{dy}{dx} - 6xy = 0$
(UPTU(Q. Bank)–2002)

25. Solve in series the equation $\frac{d^2y}{dx^2} + xy = 0$. (VTU-2010)

26. Solve in series the equation $\frac{d^2y}{dx^2} + y = 0, y(0) = 0$.
(BPTU-2005)

27. Solve in series the equation $(1+x^2)\frac{d^2y}{dx^2} + x\frac{dy}{dx} - y = 0$.
(SVTU-2008)

28. $4x\frac{d^2y}{dx^2} + 2\frac{dy}{dx} + y = 0$ (PTU-2005)

29. $8x^2\frac{d^2y}{dx^2} + 10x\frac{dy}{dx} - (1+x)y = 0$ (PTU-2009)

30. $3x\frac{d^2y}{dx^2} + (1-x)\frac{dy}{dx} - y = 0$ (SVTU-2008)

Hint to Selected Problems

1. Let a trial solution of the given equation is
$$y = C_0 + C_1x + C_2x^2 + \ldots \Rightarrow \frac{dy}{dx} = C_1 + 2C_2x + 3C_3x + \ldots;$$
$$\frac{d^2y}{dx^2} = 2C_2 + 6C_3x + \ldots$$
Using these values in the given equation.

2. Let the series solution be
$$y = \sum_{r=0}^{\infty} C_r\, x^{m+2r} \Rightarrow \frac{dy}{dx} = \sum_{r=0}^{\infty} C_r(m+2r)\, x^{m+2r-1}$$
and $\frac{d^2y}{dx^2} = \sum_{r=0}^{\infty} C_r(m+2r)(m+2r-1)\, x^{m+2r-2}$. Put all these values in the given equation.

3. Indicial equation has two equal roots, *i.e.*, 0, 0.

20. $y = \Sigma a_k (m+k)(m+k+4)x^{m+k} + \Sigma a_k x^{m+k-2}$.
Indical equation is $a_0 m(m-4) = 0$.

ANSWERS

1. $y = C_0\left(1 - \frac{2}{3}x^3 - \frac{2}{45}x^6 \ldots\right) + C_1\left(x - \frac{1}{6}x^4 - \frac{1}{63}x^7\right) + x^2 + \frac{1}{3}x^3 + \frac{1}{12}x^4 + \frac{1}{45}x^6 + \ldots$

2. $y = au + bv$, where $u = 1 - \frac{x^2}{2^2} + \frac{x^4}{2^2.4^2} - \frac{x^6}{2^2.4^2.6^2} + \ldots$ and $v = u \log x + \left[\frac{x^2}{2^2} - \frac{3}{2^3.4^2}x^4 + \ldots\right]$

3. $u = 1 - 2x + \frac{3}{2!}x^2 - \frac{4}{3!}x^3 + \ldots$; $v = bu \log x + b\left[2\left(2 - \frac{1}{2}\right)x - \frac{3}{2!}\left(-\frac{1}{3} + 2 + \frac{1}{2}\right)x^2 + \ldots\right]$

4. Transformed equation is $z^4 \frac{d^2y}{dz^2} + 2z^2 \frac{dy}{dz} - y = 0$

5. $y = a\left[n - (n+1)x + (n+2)\frac{x^2}{2!} - (n+3)\frac{x^3}{3!} + \ldots\right] + bx^{1-n}\left[1 + \frac{2}{n-2}x + \frac{3}{(n-2)(n-3)}x^2 + \frac{4}{(n-2)(n-3)(n-4)}x^3 + \ldots\right]$

6. $y = a\left[1 + 3x^2 + \frac{3}{5}x^4 - \frac{1}{15}x^6 + \ldots\right] + bx^{3/2}\left[1 + \frac{3}{8}x^2 - \frac{1}{8}.\frac{3}{16}x^4 + \frac{1.3.5}{8.16.24}x^6 \ldots\right]$

7. $y = ax\left[1 + \frac{x^2}{2.5} + \frac{x^4}{2.4.5.9} + \frac{x^6}{2.4.6.5.9.15} + \ldots\right] + bx^{-1/2}\left[1 + \frac{x^2}{2.3} + \frac{x^4}{2.4.3.7} + \frac{x^6}{2.4.6.3.7.11} + \ldots\right]$

$+ \frac{1}{3}x^2 + \frac{1}{3}.\frac{1}{3.7}x^4 + \frac{1}{3}.\frac{1}{3.5.7.11}x^6 + \ldots$

8. $y = a\left[1 + x + \frac{2}{4}x^2 + \frac{2.5}{4.9}x^3 + \ldots\right]\left[4 \log x + \left(-2x - x^2 - \frac{14}{27}x^3 - \ldots\right)\right]$

9. $y = a\left[1 - \frac{1}{3!}.x^{-2} - \frac{1}{5!}x^{-4} - \frac{1.3}{7!}x^{-6} - \ldots\right] + b\left[x - \frac{1}{x}\right] + 2x^{-3}\left(\frac{1}{4!} + \frac{2}{6!}x^{-2} + \frac{2.4}{8!}x^{-4} + \ldots\right)$

10. $y = a\left[1 + \frac{1}{2}x^2 + \frac{1}{24}x^4 + \ldots\right] + b\left[x + \frac{1}{6}x^3 + \frac{1}{120}x^5 + \ldots\right] + \left(\frac{1}{6}x^3 + \frac{1}{120}x^5 + \ldots\right)$

11. $y = a\left(1 - \frac{1}{12}x^4 + \frac{x^8}{12\times7\times8} - \frac{x^{12}}{12\times8\times7\times11\times12} + \ldots\right) + b\left(x - \frac{x^5}{20} + \frac{x^9}{20\times8\times1} + \ldots\right)$

12. $y = a\left(1 + \frac{x^2}{2} - \frac{x^4}{8} \ldots\right) + bx$

13. $y = ax^2\left[1 - \frac{x^2}{2\times6} + \frac{x^4}{2.4.6.8} - \frac{x^6}{2.4.6.8.10} + \ldots\right] + b\left[x^2 \log x\left(-\frac{1}{2^2.4} + \frac{x^2}{2^3.4.6} - \frac{x^4}{2^3.4^3.6.8} + \ldots\right)\right] + x^{-2}\left(1 + \frac{x^2}{2^2} + \frac{x^4}{2^2.4^2} + \ldots\right)$

14. $y_1 = \frac{a}{x}\left[1 - \frac{x^2}{6} + \frac{x^4}{12} + \ldots\right] + b\left[1 - \frac{x^2}{12} + \frac{x^4}{360} + \ldots\right]$, $y_2 = x^{-2}(a \cos x + b \sin x)$

15. $y = Ay_1 + By_2$,
when $y_1 = ax^{-2}\left(1 + \frac{x^2}{3!} + \frac{x^4}{5!} + \ldots\right) + bx^{-2}\left(x + \frac{x^3}{3.4} + \frac{x^5}{3.4.5.6} + \ldots\right)$, $y_2 = ax^{-3}\left(1 + \frac{x^2}{2!} + \frac{x^4}{4!} + \ldots\right) + bx^{-3}\left(x + \frac{x^3}{3!} + \frac{x^5}{5!} + \ldots\right)$

16. $y = a\sqrt{x}.(1-x) + b\left(1 - 3x + \frac{3x^2}{1.3} + \frac{3.x^3}{3.5} + \frac{3x^4}{5.7} + \ldots\right)$.

17. $y = ax\left(1 - \frac{x^2}{10} + \frac{x^4}{360} + \ldots\right) + bx^{1/2}\left(1 + \frac{x^2}{6} + \frac{x^4}{168} + \ldots\right)$.

18. $y = a\left(1 - x + \frac{x^2}{2!} - \frac{x^3}{3!} + \ldots\right) + b\left(x^2 - \frac{2x^3}{3!} + \frac{2x^4}{4!} - \frac{2x^5}{5!} + \ldots\right)$

19. $y = a\left(1 - \frac{x^2}{2^2} + \frac{x^4}{2^2.4^2} - \frac{x^6}{2^2.4^2.6^2} + \ldots\right) + b\left(y_1 \log x + C_0\left\{\frac{x^2}{2^2} + \frac{1}{2^2.4^2}\left(1 + \frac{1}{x}\right)x^4 + \frac{1}{2^2.4^2.6^2}\left(1 + \frac{1}{2} + \frac{1}{3}\right)x^6 + \ldots\right\}\right)$

20. $y = a\left(1 - \frac{x^2}{12} + \frac{x^4}{364} - \ldots\right) + bx^{-4}\log x\left(1 - \frac{x^4}{16} - \frac{x^6}{16}\ldots\right) + 6x^{-2}\left(\frac{1}{4} + \frac{x^2}{64}\right) + \ldots.$

21. $y = a_0(1 - 2x^2) + a_1 x\left(1 - \frac{x^2}{2} - \frac{x^4}{8} - \ldots\right)$ **22.** $y = a_0\left(1 - x^2 - \frac{x^4}{3} - \frac{x^6}{5} - \ldots\right) + a_1 x$ **23.** $y = a_0\left(1 - \frac{x^4}{12} + \frac{x^6}{90} - \ldots\right) + a_1\left(x - \frac{x^3}{6} - \frac{x^5}{40} - \ldots\right)$

24. $y = A\left(1 + 3x^2 + \frac{3}{5}x^4 - \frac{1}{15}x^6 + \ldots\right) + Bx^{3/2}\left(1 + \frac{3}{8}x^2 - \frac{3.1}{8.16}x^4 + \frac{5.3.1}{8.16.24}x^6 - \ldots\right)$

25. $y = a_0\left(1 - \frac{x^3}{3!} + \frac{1.4.x^6}{6!} - \frac{1.4.7.x^9}{9!} + \ldots\right) + a_1\left(x - \frac{2.x^4}{4!} + \frac{2.5x^7}{7!} - \ldots\right)$ **26.** $y = a_1\left(x - \frac{x^3}{3!} + \frac{x^5}{5!} - \ldots\right)$

27. $y = a_0\left(1 + \frac{x^2}{2} - \frac{x^4}{8} + \frac{x^6}{16} - \frac{5x^8}{128} + \ldots\right) + a_1 x$ **28.** $y = C_1\cos\sqrt{x} + C_2\sin\sqrt{x}$

29. $y = C_1 x^{-1/2}\left(1 + \frac{x}{2} + \frac{x^2}{40} + \ldots\right) + C_2 x^{1/4}\left(1 + \frac{x}{14} + \frac{x^2}{616} + \ldots\right) + C_2\sqrt{x}\left(x + \frac{x^2}{2.3} + \frac{x^4}{2.4.3.7} + \frac{x^6}{2.4.6.3.7.11} + \ldots\right)$

30. $y = C_1\left(1 + x + \frac{x^2}{4} + \frac{x^3}{4}.7 + \ldots\right) + C_2 x^{2/3}\left(1 + \frac{1}{3}x + \frac{x^2}{3.6} + \frac{x^3}{3.6.9} + \ldots\right)$

FFFFFF

CHAPTER

35 Legendre and Bessel's Functions

35.1 INTRODUCTION

It is known that the functions like e^x, $\sin x$ and $\cos x$ are categorised as elementary transcendental functions. The theory of higher transcendental functions constitutes the special functions of Mathematics. We can express the variety of the properties of these functions in terms of differential equation satisfied by them. The special function constitute the complex exponential functions, hypergeometric functions, Jacobi function, Legendre functions, Bessel's and Spherical Bessel's functions, etc.

35.2 LEGENDRE'S FUNCTIONS

Consider a homogeneous linear differential equation of order two of the form

$$(1-x^2)\frac{d^2y}{dx^2}-2x\frac{dy}{dx}+n(n+1)y=0 \qquad \ldots(1)$$

where n is a real number. This differential equation is known as Legendre's differential equation, and any solution of (1) is called a Legendre function.

35.2.1 SOLUTION OF LEGENDRE'S EQUATION

Dividing (1) by $(1-x^2)$, we get $\frac{d^2y}{dx^2}-\frac{2x}{1-x^2}\frac{dy}{dx}+n(n+1).\frac{1}{1-x^2}y=0$

Now compare this equation with the standard form $\frac{d^2y}{dx^2}+P(x)\frac{dy}{dx}+Q(x)y=0$

we get $$P(x)=-\frac{2x}{1-x^2}, Q(x)=\frac{n(n+1)}{1-x^2}.$$

It is trivially obtained that $P(x)$ and $Q(x)$ are analytic at $x=0$, so for finding the solution of (1) we apply the power series method. Let us assume the solution of (1)

$$y=\sum_{m=0}^{\infty} a_m x^m \qquad \ldots(2)$$

Now differentiating (2) w.r.t. x one time and then two times, we get

$$\frac{dy}{dx}=\sum_{m=1}^{\infty} m a_m x^{m-1} \qquad \ldots(3)$$

and $$\frac{d^2y}{dx^2}=\sum_{m=2}^{\infty} m(m-1)a_m x^{m-2} \qquad \ldots(4)$$

Substitute the values of $y, \frac{dy}{dx}$ and $\frac{d^2y}{dx^2}$ from (2), (3) and (4) into eq. (1), we get

$$(1-x^2)\sum_{m=2}^{\infty} m(m-1)a_m x^{m-2}-2x\sum_{m=1}^{\infty} m a_m x^{m-1}+n(n+1)\sum_{m=0}^{\infty} a_m x^m=0$$

or $$\sum_{m=2}^{\infty} m(m-1)a_m x^{m-2}-\sum_{m=2}^{\infty} m(m-1)a_m x^m-2\sum_{m=1}^{\infty} m a_m x^m+n(n+1)\sum_{m=0}^{\infty} a_m x^m=0$$

or $\{2.1\,a_2 + 3.2a_3x + 4.3a_4x^2 + ... + (r+2)(r+1)a_{r+2}x^r + ...\}$

$$-\{2.1a_2x^2 + 3.2a_3x^3 + ... + r(r-1)a_rx^r + ...\} - 2\{a_1x + 2a_2x^2 + ... + ra_rx^r + ...\} + n(n+1)\{a_0 + a_1x + ... + a_rx^r + ...\} = 0. \quad ...(5)$$

If equation (2) is a solution of (1), then equation (5) must be an identity in x. Thus in (5) the sum of the coefficients of each power of x must be zero. We therefore obtain

$$2a_2 + n(n+1)a_0 = 0.$$

$$6a_3 + \{-2 + n(n+1)\}a_1 = 0,\ 12a_4 + \{-2a_2 - 4a_2 + n(n+1)a_2\} = 0$$

$$\therefore \quad a_{n-2} = -\frac{n(n-1)}{2(2n-1)} \cdot \frac{(2n)!}{2^n(n!)^2} \qquad \left(\because a_n = \frac{(2n!)}{2^n(n!)^2}\right)$$

$$= -\frac{n(n-1)2n.(2n-1).(2n-2)!}{2(2n-1).2^n.n!.n(n-1).(n-2)!} = -\frac{n(n-1)2n.(2n-1).(2n-2)!}{2(2n-1).2^n n.(n-1)!.n(n-1).(n-2)!} = -\frac{(2n-2)!}{2^n(n-1)!(n-2)!}.$$

Similarly, $a_{n-4} = -\dfrac{(n-2)(n-3)}{4(2n-3)}a_{n-2} = -\dfrac{(n-2)(n-3)}{4(2n-3)} \cdot \dfrac{-(2n-2)!}{2^n(n-1)!(n-2)!}$

$$= \frac{(n-2)(n-3).(2n-2)(2n-3)(2n-4)!}{4(2n-3)2^n(n-1).(n-2)!(n-2)(n-3).(n-4)!} = \frac{(2n-4)!}{2^n.(2)!(n-2)!(n-4)!}$$

Continuing in this way, we get in general,

$$a_{n-2m} = \frac{(-1)^m(2n-2m)!}{2^n(m)!(n-m)!(n-2m)!},\quad n-2m \ge 0$$

Thus we obtain the first kind of Legendre polynomial of degree n and it is denoted by $P_n(x)$ which is given as

$$P_n(x) = \sum_{m=0}^{N} a_{n-2m}x^{n-2m}$$

In general, we obtain

$$(r+2)(r+1)a_{r+2} + \{-r(r-1) - 2r + n(n+1)\}a_r = 0 \text{ for } r = 2,3,4,...$$

or $$(r+2)(r+1)a_{r+2} + (n-r)(n+r+1)a_r = 0$$

or $$a_{r+2} = -\frac{(n-r)(n+r+1)}{(r+2)(r+1)}a_r,\ r = (0,1,2,...) \quad ...(6)$$

This equation (6) is known as recursion formula. Now finding the coefficients successively for $r = 0,1,2,3,...$

$$a_2 = -\frac{n(n+1)}{2.1}a_0 = -\frac{n(n+1)}{(2)!}a_0;\ a_3 = -\frac{(n-1)(n+2)}{3.2}a_1 = -\frac{(n-1)(n+2)}{(3)!}a_1$$

$$a_4 = -\frac{(n-2)(n+3)}{4.3}a_2 = -\frac{(n-2)(n+3)}{4.3} \cdot \frac{-n(n+1)}{2.1}a_0 = \frac{(n-2)n(n+1)(n+3)}{(4)!}a_0$$

$$a_5 = -\frac{(n-3)(n+4)}{5.4}a_3 = \frac{(n-3)(n-1)(n+2)(n+4)}{(5)!}a_1$$

$$\vdots$$

etc.

We observed from above coefficients that all the even numbered coefficients are obtained in terms of a_0 while all odd numbered coefficients are obtained in terms of a_1. Thus we obtain the solution as

$$y = a_0y_1(x) + a_1y_2(x) \text{ where } y_1(x) = 1 - \frac{n(n+1)}{(2)!}x^2 + \frac{(n-2)n(n+1)(n+3)}{(4)!}x^4 - ...$$

and $$y_2(x) = x - \frac{(n-1)(n+2)}{(3)!}x^3 + \frac{(n-3)(n-1)n(n+2)(n+4)}{(5)!}x^5 - .....$$

These both series are convergent if $|x| < 1$. Sometimes, we have observed that the parameter n in the Legendre's differential equation will be non-negative. Then, from recursion formula (6) we obtain

$a_{r+2} = 0$, when $r = n$ i.e., $a_{n+2} = a_{n+4} = ... = 0$

Hence we can say that if n is even. $y_1(x)$ becomes a polynomial of degree n whereas n is odd $y_2(x)$ becomes a polynomial of degree n. Therefore if $y_1(x)$ is multiplied by some constant, then this polynomial is called Legendre's polynomial of first kind and if $y_2(x)$ is multiplied by some constant, then $y_2(x)$ is called Legendre's polynomial of second kind. Now to obtain first kind of Legendre's polynomial we proceed as follows :

The recursion formula given in (6) may be written as

$$a_r = -\frac{(r+2)(r+1)}{(n-r)(n+r+1)}a_{r+2} \text{ for } r \le n-2$$

Also, all *a*'s may express in terms of the coefficient a_n which is the coefficient of the highest power of x of the polynomial. This a_n is an arbitrary so choose $a_n = 1$ when $n = 0$ and $a_n = \frac{1.3.5...(2n-1)}{n!} = \frac{2n!}{2^n.(n)^2!}$ for all $n = 1,2,3...$. This a_n is chosen in such a way that the values of all those polynomial will be 1 when $x = 1$. Now finding the coefficients as follows :

$$a_{n-2} = -\frac{n(n-1)}{2(2n-1)}a_n \quad \text{or} \quad P_n(x) = \sum_{m=0}^{N} \frac{(-1)^m(2n-2m)!}{2^n(m)!(n-m)!(n-2m)!}.x^{n-2m}$$

where $N = \begin{cases} n/2 & ; \text{ if } n \text{ is even} \\ (n-1)/2 & ; \text{ if } n \text{ is odd.} \end{cases}$

Similarly, we can obtain second kind of Legendre polynomial. From recursion formula (6) we also have

$$a_{r+2} = 0 \text{ if } r = -(n+1) \text{ i.e., } a_{-n+3} = a_{-n+5} = ... = 0$$

In this case we obtain second kind of Legendre polynomial of an infinite series. It is denoted by $Q_n(x)$. If we choose

$$a_{-n-1} = \frac{(n)!}{1.3.5...(2n+1)}$$

Then from the following relation

$$a_r = -\frac{(r+2)(r+1)}{(n-r)(n+r+1)}a_{r+2}, \text{ we obtain } a_{-n-3} = \frac{(n+1)((n+2)}{2(2n+3)}.a_{-n-1}$$

and
$$a_{-n-5} = \frac{(n+3)(n+4)}{4(2n+5)}.a_{-n-3} = \frac{(n+1)(n+2)(n+3)(n+4)}{2.4(2n+3)(2n+5)}a_{-n-1} \text{ and so on.}$$

Hence $Q_n(x)$ is given as

$$Q_n(x) = a_{-n-1}\left[x^{-n-1} + \frac{(n+1)(n+2)}{2(2n+3)}x^{-n-3} + \frac{(n+1)(n+2)(n+3)(n+4)}{2.4(2n+3)(2n+5)}x^{-n-5} + ...\right]$$

$$\Rightarrow \quad Q = \frac{(n)!}{1.3.5...(2n+1)}\left[x^{-n-1} + \frac{(n+1)(n+2)}{2(2n+3)}x^{-n-3} + \frac{(n+1)(n+2)(n+3)(n+4)}{2.4(2n+3)(2n+5)}x^{-n-5} + ...\right] \quad \text{(VTU-2006)}$$

35.3 GENERATING FUNCTIONS OF LEGENDRE'S POLYNOMIAL $P_n(X)$

(UPTU-2005, Kerala-2005)

The functions of the type $\frac{1}{\sqrt{1-2xt+t^2}}$ *generates Legendre polynomial* $P_n(x)$, *is called generating functions. Thus we obtain* $\frac{1}{\sqrt{1-2xt+t^2}} = \sum_{n=0}^{\infty} P_n(x)t^n$.

Proof. Consider R.H.S. $= \frac{1}{\sqrt{1-2xt+t^2}} = \frac{1}{\sqrt{1-s}}$ $\quad [\because s = 2xt - t^2]$

$$= (1-s)^{-1/2} = 1 + \frac{1}{2}s + \frac{1.3}{2.4}s^2 + \frac{1.3.5}{2.4.6}s^3 + ... + \frac{1.3.5...(2n-3)}{2.4.6...(2n-2)}s^{n-1} + \frac{1.3.5...(2n-1)}{2.4.6...(2n)}s^n + ...$$

(Expand by binomial theorem) ...(1)

Since $s = 2xt - t^2$

$\therefore$ $s^n = (2xt - t^2)^n = t^n(2x-t)^n = t^n[{}^nC_0(2x)^n - {}^nC_1(2x)^{n-1}t + ...]$.

Similarly $s^{n-1} = t^{n-1}\left[{}^{n-1}C_0(2x)^{n-1} - {}^{n-1}C_1(2x)^{n-2}t + ...\right]$

and $s^{n-2} = t^{n-2}\left[{}^{n-2}C_0(2x)^{n-2} - {}^{n-2}C_1(2x)^{n-3}t + {}^{n-2}C_2(2x)^{n-4}t^2 ...\right]$

$\vdots$

etc.

Substitute these value in the above equation (1), we get

$$\text{L.H.S.} = 1+\frac{1}{2}t(2x-t)+\frac{1.3}{2.4}t^2(2x-t)^2+\frac{1.3.5}{2.4.6}t^3(2x-t)^3+...$$

$$...+\frac{1.3.5...(2x-5)}{2.4.6...(2x-4)}t^{n-2}\left[{}^{n-2}C_0(2x)^{n-2}-{}^{n-2}C_1(2x)^{n-3}t+{}^{n-2}C_2(2x)^{n-4}t^2+...\right]$$

$$+\frac{1.3.5...(2n-3)}{2.4.6...(2n-2)}t^{n-1}\left[{}^{n-1}C_0(2x)^{n-1}-{}^{n-1}C_1(2x)^{n-2}t+...\right]$$

$$+\frac{1.3.5...(2n-1)}{2.4.6...(2n)}t^{n}\left[{}^{n}C_0(2x)^{n}-{}^{n}C_1(2x)^{n-1}t+...\right]$$

Now collecting the coefficients of t^n, we get

$$=\frac{1.3.5...(2n-1)}{2.4.6...(2n)}{}^nC_0(2x)^n-\frac{1.3.5...(2n-3)}{2.4.6...(2n-2)}{}^{n-1}C_1(2x)^{n-2}+\frac{1.3.5...(2n-5)}{2.4.6...(2n-4)}{}^{n-2}C_2(2x)^{n-4}-...$$

$$=\frac{1.3.5...(2n-1)}{2.4.6...(2n)}.2^n x^n-\frac{1.3.5...(2n-3)}{2.4.6...(2n-2)}.\frac{(n-1)}{(1)!}2^{n-2}.x^{n-2}$$

$$+\frac{1.3.5...(2n-5)}{2.4.6...(2n-4)}.\frac{(n-2)(n-3)}{(2)!}2^{n-4}.x^{n-4}-...$$

$$=\frac{1.3.5...(2n-1)}{2.4.6...(2n)}2^n\left[x^n-\frac{2n(n-1)}{(2n-1)2^2}x^{n-2}+\frac{2n(2n-2)(n-2)(n-3)}{(2n-1)(2n-3)(2)!\,.\,2^4}.x^{n-4}-...\right]$$

$$=\frac{1.3.5...(2n-1)}{(n)!}\left[x^n-\frac{n(n-1)}{2(2n-1)}x^{n-2}+\frac{n(n-1)(n-2)(n-3)}{2.4(2n-1)(2n-3)}.x^{n-4}-...\right]$$

Hence we obtain $\frac{1}{\sqrt{1-2xt+t^2}}=\sum_{n=0}^{\infty}P_n(x)t^n$

35.3.1 RODRIGUE'S FORMULA

(UPTU-2004, 07, VTU-2008, Bhopal-2007, Madras-2003)

The expression for $P_n(x)$, given by $P_n(x)=\frac{1}{2^n(n)!}\frac{d^n}{dx^n}(x^2-1)^n$ is called Rodrigue's Formula.

Proof. Since $P_n(x)$ is a Legendre polynomial whose expression is given as

$$P_n(x)=\sum_{m=0}^{[n/2]}\frac{(-1)^m(2n-2m)!}{2^n(m)!(n-m)!(n-2m)!}.x^{n-2m} \quad ...(1)$$

where $[n/2]$ is an integral value of $n/2$ not exceed $n/2$. Rearrange (1), we get

$$P_n(x)=\sum_{m=0}^{[n/2]}\frac{(-1)^m}{2^n(m)!(n-m)!}\left\{\frac{(2n-2m)!}{(n-2m)!}.x^{n-2m}\right\}=\sum_{m=0}^{[n/2]}\frac{(-1)^m}{2^n(m)!(n-m)!}.\frac{d^n}{dx^n}x^{2n-2m}$$

$$\left(\because \frac{d^r}{dx^r}x^{2n-2m}=\frac{(2n-2m)!}{(2n-2m-r)!}.x^{2n-2m-r}\right)$$

$$=\frac{1}{2^n(n)!}\sum_{m=0}^{[n/2]}\frac{(n)!}{(m)!(n-m)!}.\frac{d^n}{dx^n}(x^2)^{n-m}.(-1)^m$$

Now extending the range of m from 0 to n. To do so no change will occur in the above expression, because n^{th} derivatives of those terms whose degree are less than n will be zero. Thus above expression can be written as

$$=\frac{1}{2^n(n)!}\frac{d^n}{dx^n}\sum_{m=0}^{n}\frac{(n)!}{(m)!(n-m)!}(x^2)^{n-m}(-1)^m=\frac{1}{2^n(n)!}\frac{d^n}{dx^n}\sum_{m=0}^{n}{}^nC_m(x^2)^{n-m}(-1)^m$$

$$\left(\because {}^nC_m=\frac{(n)!}{(m)!(n-m)!}\right)$$

$$= \frac{1}{2^n (n)!} \frac{d^n}{dx^n}\left[{}^nC_0(x^2)^n - {}^nC_1(x^2)^{n-1} + {}^nC_2(x^2)^{n-2} + \ldots + {}^nC_n(-1)^n\right] = \frac{1}{2^n (n)!} \frac{d^n}{dx^n}(x^2-1)^n$$

(By Binomial theorem)

Hence $P_n(x) = \frac{1}{2^n (n)!} \frac{d^n}{dx^n}(x^2-1)^n$.

35.3.2 LAPLACE INTEGRAL FOR $P_n(x)$

(i) *Laplace's First Integral for $P_n(x)$:*

$$P_n(x) = \frac{1}{\pi}\int_0^{\pi} [x \pm \sqrt{(x^2-1)}\cos\theta]^n\, d\theta,$$ *where n is any positive integer.*

Proof. Since we know that $\int_0^{\pi} \frac{d\theta}{a \pm b\cos\theta} = \frac{\pi}{\sqrt{a^2-b^2}}$, where $a^2 > b^2$ ...(1)

Let us take $a = 1 - tx$ and $b = t\sqrt{x^2-1}$,

Then, $a^2 - b^2 = (1-tx)^2 - t^2(x^2-1) = 1 + t^2x^2 - 2tx - t^2x^2 + t^2 = 1 - 2tx + t^2$.

Thus (1) becomes $\int_0^{\pi} \frac{d\theta}{(1-tx) \pm t\sqrt{x^2-1}\cos\theta} = \frac{\pi}{\sqrt{1-2tx+t^2}}$ ...(2)

Since generating function gives $\frac{1}{\sqrt{1-2xt+t^2}} = \sum_{n=0}^{\infty} P_n(x)t^n$

$\therefore$ (2) becomes

$$\pi \sum_{n=0}^{\infty} P_n(x)t^n = \int_0^{\pi} \frac{d\theta}{1 - tx \pm t\sqrt{(x^2-1)}\cos\theta} = \int_0^{\pi} \frac{d\theta}{[1 - t\{x \mp \sqrt{(x^2-1)}\cos\theta\}]}$$

$$= \int_0^{\pi} [1 - t\{x \pm \sqrt{(x^2-1)}\cos\theta\}]^{-1} d\theta = \int_0^{\pi} (1-ts)^{-1} d\theta, \quad \text{where} \quad s = x \pm \sqrt{(x^2-1)}\cos\theta$$

$$= \int_0^{\pi} (1 + ts + t^2s^2 + \ldots + t^n s^n + \ldots) d\theta = \int_0^{\pi} \sum_{n=0}^{\infty} t^n s^n d\theta = \sum_{n=0}^{\infty} \int_0^{\pi} s^n t^n d\theta$$

$$= \sum_{n=0}^{\infty} t^n \int_0^{\pi} [x \mp \sqrt{(x^2-1)}\cos\theta]^n d\theta$$

$$\therefore \pi \sum_{n=0}^{\infty} P_n(x)t^n = \sum_{n=0}^{\infty} t^n \int_0^{\pi} [x \mp \sqrt{(x^2-1)}\cos\theta]^n\, d\theta$$

$$\therefore \quad \pi P_n(x) = \int_0^{\pi} [x \pm \sqrt{(x^2-1)}\cos\theta]^n\, d\theta$$

Hence, $P_n(x) = \frac{1}{\pi}\int_0^{\pi} [x \pm \sqrt{(x^2-1)}\cos\theta]^n\, d\theta$. $\because$

(ii) *Laplace's Second integral for $P_n(x)$:*

$$P_n(x) = \frac{1}{\pi}\int_0^{\pi} \frac{d\theta}{[x \pm \sqrt{(x^2-1)}\cos\theta]^{n+1}},$$ *where n is any positive integer.*

Proof. Since we know that $\int_0^{\pi} \frac{d\theta}{a \pm b\cos\theta} = \frac{\pi}{\sqrt{a^2-b^2}}$, where $a^2 > b^2$ ...(1)

Here taking $a = xt - 1$, and $b = t\sqrt{(x^2-1)}$, then $a^2 - b^2 = 1 - 2xt + t^2$

$\therefore$ (1) becomes $\int_0^{\pi} \frac{d\theta}{(xt-1) \pm t\sqrt{(x^2-1)}\cos\theta} = \frac{\pi}{\sqrt{1-2tx+t^2}}$ ...(2)

Since $\sum_{n=0}^{\infty} P_n(x)\, t^n = \frac{1}{\sqrt{1-2xt+t^2}}$

$\therefore$ (2) becomes

$$\pi \sum_{n=0}^{\infty} P_n(x)\, t^n = \int_0^\pi \frac{d\theta}{[-1+t\{x \pm \sqrt{(x^2-1)}\cos\theta\}]} = \int_0^\pi [t\{x \pm \sqrt{(x^2-1)}\cos\theta\}-1]^{-1} d\theta$$

$$= \int_0^\pi (ts-1)^{-1} d\theta,\ s = x \pm \sqrt{(x^2-1)}\cos\theta$$

$$= \int_0^\pi \frac{1}{ts}\left(1-\frac{1}{ts}\right)^{-1} d\theta = \int_0^\pi \frac{1}{ts}\left(1+\frac{1}{ts}+\frac{1}{t^2s^2}+\ldots+\frac{1}{t^n s^n}+\ldots\right) d\theta$$

$$= \int_0^\pi \left(\frac{1}{ts}+\frac{1}{t^2s^2}+\ldots+\frac{1}{t^{n+1}s^{n+1}}+\ldots\right) d\theta = \int_0^\pi \sum_{n=0}^{\infty} \frac{1}{t^{n+1}s^{n+1}} d\theta$$

$$\therefore \quad \pi \sum_{n=0}^{\infty} P_n(x) t^n = \sum_{n=0}^{\infty} \frac{1}{t^{n+1}} \int_0^\pi \frac{d\theta}{[x \pm \sqrt{(x^2-1)}\cos\theta]^{n+1}}$$

or
$$\pi \cdot \frac{1}{\sqrt{1-2xt+t^2}} = \sum_{n=0}^{\infty} \frac{1}{t^{n+1}} \int_0^\pi \frac{d\theta}{[x \pm \sqrt{(x^2-1)}\cos\theta]^{n+1}}$$

or
$$\frac{\pi}{t} \cdot \frac{1}{\sqrt{1-2x \cdot \frac{1}{t}+\frac{1}{t^2}}} = \sum_{n=0}^{\infty} \frac{1}{t^{n+1}} \int_0^\pi \frac{d\theta}{[x \pm \sqrt{(x^2-1)}\cos\theta]^{n+1}}$$

or
$$\frac{\pi}{t} \sum_{n=0}^{\infty} \frac{1}{t^n} P_n(x) = \sum_{n=0}^{\infty} \frac{1}{t^{n+1}} \int_0^\pi \frac{d\theta}{[x \pm \sqrt{(x^2-1)}\cos\theta]^{n+1}}$$

or
$$\pi \sum_{n=0}^{\infty} \frac{1}{t^{n+1}} P_n(x) = \sum_{n=0}^{\infty} \frac{1}{t^{n+1}} \int_0^\pi \frac{d\theta}{[x \pm \sqrt{(x^2-1)}\cos\theta]^{n+1}}$$

$$\therefore \quad \pi P_n(x) = \int_0^\pi \frac{d\theta}{[x \pm \sqrt{(x^2-1)}\cos\theta]^{n+1}}$$

Hence
$$P_n(x) = \frac{1}{\pi} \int_0^\pi \frac{d\theta}{[x \pm \sqrt{(x^2-1)}\cos\theta]^{n+1}}.$$

35.4 ORTHOGONAL PROPERTIES OF LEGENDRE'S POLYNOMIALS

(i) $\int_{-1}^{1} P_m(x) P_n(x)\, dx = 0,$ *when* $m \neq n.$

(ii) $\int_{-1}^{1} [P_n(x)]^2\, dx = \frac{2}{2n+1},$ *when* $m = n.$ (UPTU-2001, 02, 04, 08, SVTU-2008, Madras-2006, VTU-2006)

Proof. (i) Legendre differential equation is given by

$$(1-x^2)\frac{d^2y}{dx^2} - 2x\frac{dy}{dx} + n(n+1)y = 0 \quad \text{or} \quad \frac{d}{dx}\left\{(1-x^2)\frac{dy}{dx}\right\} + n(n+1)y = 0 \qquad \ldots(1)$$

Since $P_m(x)$ and $P_n(x)$ are the solution of (1), so we have

$$\frac{d}{dx}\left\{(1-x^2)\frac{dP_m(x)}{dx}\right\} + m(m+1)P_m(x) = 0 \qquad \ldots(2)$$

and
$$\frac{d}{dx}\left\{(1-x^2)\frac{dP_n(x)}{dx}\right\} + n(n+1)P_n(x) = 0 \qquad \ldots(3)$$

Now, multiplying (2) by $P_n(x)$ and (3) by $P_m(x)$ and then subtract, we get

$$\frac{d}{dx}\left\{(1-x^2)\frac{dP_m(x)}{dx}\right\}P_n(x) - \frac{d}{dx}\left\{(1-x^2)\frac{dP_n(x)}{dx}\right\}P_m(x) + [m(m+1)-n(n+1)]P_m(x)P_n(x) = 0 \qquad \ldots(4)$$

Integrating (4) w.r.t. x from $x = -1$ to $x = 1$, , we get

$$\int_{-1}^{1} \frac{d}{dx}\left\{(1-x^2)\frac{dP_m(x)}{dx}\right\}P_n(x)dx - \int_{-1}^{1} \frac{d}{dx}\left\{(1-x^2)\frac{dP_n(x)}{dx}\right\}P_m(x)dx$$

$$+(m-n)(m+n+1)\int_{-1}^{1} P_m(x)P_n(x)\, dx = 0 \qquad \ldots(5)$$

Let $I_1 = \int_{-1}^{1} \frac{d}{dx}\left\{(1-x^2)\frac{dP_m(x)}{dx}\right\}P_n(x)dx$ and $I_2 = \int_{-1}^{1} \frac{d}{dx}\left\{(1-x^2)\frac{dP_n(x)}{dx}\right\}P_m(x)dx$.

$\therefore$ (5) becomes $I_1 - I_2 + (m-n)(m+n+1)\int_{-1}^{1} P_m(x)P_n(x)\,dx = 0$ ...(6)

Now solving I_1 and I_2, we get

$$I_1 = \int_{-1}^{1} \frac{d}{dx}\left\{(1-x^2)\frac{dP_m(x)}{dx}\right\}P_n(x)dx = P_n(x)\left[(1-x^2)\frac{dP_m(x)}{dx}\right]_{-1}^{1} - \int_{-1}^{1}\frac{dP_n(x)}{dx}(1-x^2)\frac{dP_m(x)}{dx}\,dx$$

(Integration by parts)

$$= 0 - \int_{-1}^{1} (1-x^2)\frac{dP_n(x)}{dx}\cdot\frac{dP_m(x)}{dx}dx$$

$$\therefore \quad I_1 = -\int_{-1}^{1} (1-x^2)\frac{dP_n(x)}{dx}\cdot\frac{dP_m(x)}{dx}dx.$$

Also,

$$I_2 = \int_{-1}^{1} \frac{d}{dx}\left\{(1-x^2)\frac{dP_n(x)}{dx}\right\}P_m(x)dx = P_m(x)\left[(1-x^2)\frac{dP_n(x)}{dx}\right]_{-1}^{1} - \int_{-1}^{1}(1-x^2)\frac{dP_m(x)}{dx}\cdot\frac{dP_n(x)}{dx}\,dx$$

$$= 0 - \int_{-1}^{1} (1-x^2)\frac{dP_m(x)}{dx}\cdot\frac{dP_n(x)}{dx}\,dx$$

$$\therefore \quad I_2 = -\int_{-1}^{1} (1-x^2)\frac{dP_m(x)}{dx}\cdot\frac{dP_n(x)}{dx}\,dx.$$

Thus $I_1 - I_2 = 0$. Now (6) becomes $0 + (m-n)(m+n+1)\int_{-1}^{1} P_m(x)P_n(x)\,dx = 0$

If $m \neq n$, then $\int_{-1}^{1} P_m(x)P_n(x)\,dx = 0$.

Proof. (ii) $\int_{-1}^{1} [P_n(x)]^2\,dx = \frac{2}{2n+1}$, if $m = n$

Since we know that $\frac{1}{\sqrt{1-2xt+t^2}} = \sum_{n=0}^{\infty} P_n(x)t^n$

or $$\frac{1}{\sqrt{1-2xt+t^2}} = P_0(x) + t\,P_1(x) + t^2\,P_2(x) + \ldots + t^n\,P_n(x) + \ldots$$

Squaring of both sides, we get

$$\frac{1}{1-2xt+t^2} = [P_0(x) + t\,P_1(x) + t^2\,P_2(x) + \ldots + t^n\,P_n(x) + \ldots]^2$$

$$= [P_0(x)]^2 + [t\,P_1(x)]^2 + [t^2\,P_2(x)]^2 + \ldots + [t^n\,P_n(x)]^2 + \ldots$$
$$+ 2[tP_0(x)\,P_1(x) + t^2\,P_0(x)P_2(x) + \ldots + t^n\,P_0(x)P_n(x) + \ldots$$
$$\ldots + t^3\,P_1(x)\,P_2(x) + t^4\,P_1(x)P_3(x) + \ldots + t^{n+1}\,P_1(x)P_n(x) + \ldots]$$

$$= \sum_{n=0}^{\infty} t^{2n}[P_n(x)]^2 + 2\sum_{\substack{m,n=0\\ m\neq n}}^{\infty} t^{m+n}P_m(x)P_n(x)$$

$$\therefore \quad \frac{1}{1-2xt+t^2} = \sum_{n=0}^{\infty} t^{2n}[P_n(x)]^2 + 2\sum_{\substack{m,n=0\\ m\neq n}}^{\infty} t^{m+n}P_m(x)P_n(x).$$

Integrating both sides w.r.t. x from $x = -1$ to 1, we get

$$\int_{-1}^{1} \frac{1}{1-2xt+t^2} = \int_{-1}^{1}\sum_{n=0}^{\infty} t^{2n}[P_n(x)]^2\,dx + 2\int_{-1}^{1}\sum_{\substack{m,n=0\\ m\neq n}}^{\infty} t^{m+n}P_m(x)P_n(x)dx$$

$$= \sum_{n=0}^{\infty} t^{2n}\int_{-1}^{1}[P_n(x)]^2\,dx + 2\sum_{\substack{m,n=0\\ m\neq n}}^{\infty} t^{m+n}\int_{-1}^{1}P_m(x)P_n(x)dx$$

$$= \sum_{n=0}^{\infty} t^{2n}\int_{-1}^{1}[P_n(x)]^2\,dx + 0 \qquad \left[\because \int_{-1}^{1}P_m(x)P_n(x)dx = 0 \text{ when } m \neq n\right]$$

$$= \sum_{n=0}^{\infty} t^{2n} \int_{-1}^{1} [P_n(x)]^2 \, dx = \int_{-1}^{1} \frac{dx}{1-2xt+t^2}$$

$$= -\frac{1}{2t}\Big[\log(1-2xt+t^2)\Big]_{-1}^{1} = -\frac{1}{2t}[\log(1-2t+t^2) - \log(1+2t+t^2)]$$

$$= -\frac{1}{2t}[\log(1-t)^2 - \log(1+t)^2] = -\frac{1}{2t}\left[\log\left(\frac{1-t}{1+t}\right)^2\right] = -\frac{1}{t}\left[\log\frac{1-t}{1+t}\right]$$

$$= \frac{1}{t}\left[\log\frac{1+t}{1-t}\right] = \frac{1}{t}[\log(1+t) - \log(1-t)]$$

$$= \frac{1}{t}\left[\left\{t - \frac{t^2}{2} + \frac{t^3}{3} - \frac{t^4}{4} + \ldots\right\} - \left\{-t - \frac{t^2}{2} - \frac{t^3}{3} - \frac{t^4}{4} - \ldots\right\}\right]$$

$$= \frac{1}{t}\left[2t + \frac{2t^3}{3} + \frac{2t^5}{5} + \ldots\right] = 2\left[1 + \frac{t^2}{3} + \frac{t^4}{5} + \ldots\right] = 2\sum_{n=0}^{\infty} \frac{t^{2n}}{2n+1}$$

$$\therefore \quad \sum_{n=0}^{\infty} t^{2n} \int_{-1}^{1} [P_n(x)]^2 \, dx = \sum_{n=0}^{\infty} \frac{2}{2n+1} . t^{2n}$$

Hence, $\int_{-1}^{1} [P_n(x)]^2 \, dx = \dfrac{2}{2n+1}$.

35.5 RECURRENCE RELATIONS FOR LEGENDRE'S FUNCTION

(I) $\boldsymbol{(2n+1)xP_n = (n+1)P_{n+1} + nP_{n-1}}$ (Madras-2006)

Proof. Since we know that $(1-2xt+t^2)^{-1/2} = \sum_{n=0}^{\infty} t^n P_n(x)$ ...(1)

Differentiating (1) both sides w.r.t. 't', we get

$$-\frac{1}{2}(1-2xt+t^2)^{-3/2} . (-2x+2t) = \sum_{n=1}^{\infty} n t^{n-1} P_n(x) \quad \text{or} \quad \frac{(x-t)(1-2xt+t^2)^{-1/2}}{(1-2xt+t^2)} = \sum_{n=1}^{\infty} n t^{n-1} P_n(x)$$

or $\quad (x-t)(1-2xt+t^2)^{-1/2} = (1-2xt+t^2) \sum_{n=1}^{\infty} nt^{n-1} P_n(x)$

or $\quad (x-t)\sum_{n=0}^{\infty} t^n P_n(x) = (1-2xt+t^2)\sum_{n=1}^{\infty} nt^{n-1} P_n(x)$ (From (1))

or $\quad x\sum_{n=0}^{\infty} t^n P_n(x) - \sum_{n=0}^{\infty} t^{n+1} P_n(x) = \sum_{n=1}^{\infty} n t^{n-1} P_n(x) - 2x\sum_{n=1}^{\infty} nt^n P_n(x) + \sum_{n=1}^{\infty} nt^{n+1} P_n(x)$

$$x[P_0(x) + tP_1(x) + \ldots + t^n P_n(x) + \ldots] - [tP_0(x) + t^2 P_1(x) + \ldots + t^n P_{n-1}(x) + \ldots]$$

$$= [P_1(x) + 2tP_2(x) + \ldots + (n+1)\, t^n P_{n+1}(x) + \ldots] - 2x[tP_1(x) + 2t^2 P_2(x) + \ldots + nt^n P_n(x) + \ldots]$$
$$+ [t^2 P_1(x) + 2t^3 P_2(x) + \ldots + (n-1)\, t^n P_{n-1}(x) + \ldots]$$

Taking the coefficient of t^n both sides, we get

$$xP_n(x) - P_{n-1}(x) = (n+1)\, P_{n+1}(x) - 2nxP_n(x) + (n-1)\, P_{n-1}(x)$$

or $\quad (2n+1)\, xP_n(x) = (n+1)\, P_{n+1}(x) + nP_{n-1}(x)$

or $\quad (2n+1)xP_n = (n+1)P_{n+1} + nP_{n-1}$

REMARK

- Putting $(n-1)$ in place of n, we get $(2n-1)xP_{n-1} = nP_n + (n-1)\, P_{n-2}$

(II) $\boldsymbol{nP_n = xP_n' - P_{n-1}'}$ where $P_n' \equiv \dfrac{dP_n}{dx}$. (UPTU-2006)

Proof. Since we have $\dfrac{1}{\sqrt{1-2xt+t^2}} = \sum_{n=0}^{\infty} t^n P_n(x)$ ...(1)

Differentiating (1) both sides w.r.t. 't' and w.r.t. x, respectively, we get

$$(x-t)(1-2xt+t^2)^{-3/2} = \sum_{n=1}^{\infty} nt^{n-1}P_n(x) \qquad \text{...(2)}$$

and $$t(1-2xt+t^2)^{-3/2} = \sum_{n=0}^{\infty} t^n P_n'(x) \qquad \text{...(3)}$$

From (2) and (3), we get

$$(x-t)\sum_{n=0}^{\infty} t^n P_n'(x) = t\sum_{n=1}^{\infty} nt^{n-1}P_n(x) \quad \text{or} \quad x\sum_{n=0}^{\infty} t^n P_n'(x) - \sum_{n=0}^{\infty} t^{n+1}P_n'(x) = \sum_{n=1}^{\infty} nt^n P_n(x)$$

or $$x[P_0'(x)+tP_1'(x)+\ldots+t^nP_n'(x)+\ldots]-[tP_0'(x)+t^2P_1'(x)+\ldots+t^nP_{n-1}'(x)+\ldots]$$
$$= tP_1(x)+2t^2P_2(x)+\ldots+nt^nP_n(x)+\ldots$$

Taking the coefficients of t^n of both sides, we get

$$xP_n'(x) - P_{n-1}'(x) = nP_n(x)$$

$$\therefore \quad nP_n = xP_n' - P_{n-1}'$$

(III) $\boldsymbol{(2n+1)P_n = P_{n+1}' - P_{n-1}'}$.

Proof. From recurrence relation (I), we have

$$(2n+1)\,xP_n = (n+1)\,P_{n+1} + nP_{n-1}.$$

Differentiating this w.r.t. 'x' of both sides, we get

$$(2n+1)\,P_n + (2n+1)\,xP_n' = (n+1)\,P_{n+1}' + nP_{n-1}' \qquad \text{...(1)}$$

From recurrence relation (II), we have $nP_n = xP_n' - P_{n-1}'$

or $$xP_n' = nP_n + P_{n-1}'$$

Substitute this value of xP_n' into (1), we get

$$(2n+1)P_n + (2n+1)(nP_n + P_{n-1}') = (n+1)\,P_{n+1}' + nP_{n-1}'$$

or $$(n+1)(2n+1)\,P_n = (n+1)P_{n+1}' - (2n+1)P_{n-1}' + nP_{n-1}' = (n+1)\,P_{n+1}' - (n+1)\,P_{n-1}'$$

$$\therefore \quad (2n+1)\,P_n = P_{n+1}' - P_{n-1}'$$

(IV) $\boldsymbol{(n+1)P_n = P_{n+1}' - xP_n'}$.

Proof. From recurrence relations (II) and (III), we have

$$nP_n = x\,P_n' - P_{n-1}' \qquad \text{...(1)}$$

and $$(2n+1)\,P_n = P_{n+1}' - P_{n-1}' \qquad \text{...(2)}$$

Subtract (1) from (2), we get

$$(2n+1)\,P_n - nP_n = P_{n+1}' - xP_n' \quad \text{or} \quad (n+1)\,P_n = P_{n+1}' - xP_n'.$$

(V) $\boldsymbol{(1-x^2)P_n' = n(P_{n-1} - xP_n)}$.

Proof. From recurrence relations (II) and (IV), we have

$$nP_n = x\,P_n' - P_{n-1}' \qquad \text{...(1)}$$

and $$(n+1)\,P_n = P_{n+1}' - xP_n' \qquad \text{...(2)}$$

Putting $(n-1)$ in place of n in (2), we get

$$nP_{n-1} = P_n' - xP_{n-1}' \qquad \text{...(3)}$$

Now, multiplying (1) by x and subtract from (3), we get

$$nP_{n-1} - nxP_n = P_n' - x^2P_n' \quad \text{or} \quad n(P_{n-1} - xP_n) = (1-x^2)\,P_n'$$

$$\therefore \quad (1-x^2)\,P_n' = n(P_{n-1} - xP_n)$$

(VI) $\boldsymbol{(1-x^2)P_n' = (n+1)(xP_n - P_{n+1})}$. (UPTU-2007)

Proof. From recurrence relations (I) and (V), we have

$$(2n+1)\,xP_n = (n+1)\,P_{n+1} + nP_{n-1} \qquad \text{...(1)}$$

and $$(1-x^2)\,P_n' = n(P_{n-1} - xP_n) \qquad \text{...(2)}$$

Substitute the value of nP_{n-1} from (1) into (2), we get

$$(1-x^2)\,P_n' = (2n+1)\,xP_n - (n+1)P_{n+1} - nxP_n = (n+1)xP_n - (n+1)\,P_{n+1}$$

$$(1-x^2)\,P_n' = (2n+1)\,xP_n - (n+1)P_{n+1} - nxP_n \Rightarrow (1-x^2)\,P_n' = (n+1)\,(xP_n - P_{n+1})$$

BELTRAMI'S RELATION

The following relation $(2n+1)(x^2-1)\,P_n' = n(n+1)\,)\,(P_{n+1} - P_{n-1})$ *is known as Beltrami's Relation.*

Proof. From recurrence relations (V) and (VI), we have

$$(1-x^2)\,P_n' = n(P_{n-1} - xP_n) \qquad \dots(1)$$

and

$$(1-x^2)\,P_n' = (n+1)\,(xP_n - P_{n+1}) \qquad \dots(2)$$

Eliminating xP_n from (1) and (2), we get

$$\frac{(1-x^2)P_n'}{n} + \frac{(1-x^2)P_n'}{n+1} = P_{n-1} - P_{n+1} \quad \text{or} \quad \frac{(n+1)(1-x^2)P_n' + n(1-x^2)P_n'}{n(n+1)} = P_{n-1} - P_{n+1}$$

or

$$(2n+1)(1-x^2)\,P_n' = n(n+1)\,)\,(P_{n-1} - P_{n+1})$$

$$\therefore \quad (2n+1)(x^2-1)\,P_n' = n(n+1)\,)\,(P_{n+1} - P_{n-1})$$

35.6 CHRISTOFFEL'S EXPANSION

The following series $P_n' = (2n-1)\,P_{n-1} + (2n-5)\,P_{n-3} + (2n-9)P_{n-5} + \dots + l$

here, $l = \begin{cases} 3P_1, & \text{if } n \text{ is even} \\ P_0, & \text{if } n \text{ is odd} \end{cases}$ *is known as Christoffel's Expansion.*

Proof. From recurrence relation (III), we have

$$(2n+1)\,P_n = P_{n+1}' - P_{n-1}'$$

$$\therefore \quad P_{n+1}' = (2n+1)\,P_n + P_{n-1}' \qquad \dots(1)$$

Now, putting $(n-1)$ in place of n in (1), we get

$$P_n' = (2n-1)\,P_{n-1} + P_{n-2}' \qquad \dots(2)$$

Now putting $(n-2)$, $(n-4)$, $(n-6)$ in place of n in (2), we get

$$P_{n-2}' = (2n-5)\,P_{n-3} + P_{n-4}' \qquad \dots(3)$$

$$P_{n-4}' = (2n-9)\,P_{n-5} + P_{n-6}' \qquad \dots(4)$$

$$P_{n-6}' = (2n-13)\,P_{n-7} + P_{n-8}' \qquad \dots(5)$$

$$\vdots$$

$$P_2' = 3P_1 + P_0', \text{ if } n \text{ is even.}$$

Adding (2), (3), (4), (5), ..., we get

$$P_n' = (2n-1)\,P_{n-1} + (2n-5)P_{n-3} + (2n-9)P_{n-5} + (2n-13)P_{n-7} + \dots + 3P_1 + P_0'$$

$$= (2n-1)\,P_{n-1} + (2n-5)P_{n-3} + (2n-9)P_{n-5} + \dots + 3P_1 \qquad \left[\because P_0' = 0\right]$$

If n is odd, then

$$P_n' = (2n-1)\,P_{n-1} + (2n-5)P_{n-3} + (2n-9)P_{n-5} + \dots + 5P_2 + P_1'$$

$$= (2n-1)\,P_{n-1} + (2n-5)P_{n-3} + (2n-9)P_{n-5} + \dots + 5P_2 + P_0 \qquad \left[\because P_0 = 1 = P_1'\right]$$

Hence, we obtained Christoffel's Expansion.

35.6.1 CHRISTOFFEL'S SUMMATION FORMULA

The following summation

$$\sum_{K=0}^{n} (2k+1)P_k(x)P_k(y) = (n+1)\left[\frac{P_{n+1}(x)\,P_n(y) - P_{n+1}(y)P_n(x)}{(x-y)}\right]$$

is known as a Christoffel's summation.

Proof. From recurrence relation I, we have

$$(2k+1)\,xP_k(x) = (k+1)P_{k+1}(x) + kP_{k-1}(x) \qquad \dots(1)$$

and

$$(2k+1)\,yP_k(y) = (k+1)P_{k+1}(y) + kP_{k-1}(y) \qquad \dots(2)$$

Multiplying (1) by $P_k(y)$ and (2) by $P_k(x)$ and then subtract, we get

$$(2k+1)(x-y)P_k(x)P_k(y) = (k+1)[P_{k+1}(x)P_k(y) - P_k(x)P_{k+1}(y)] + k[P_{k-1}(x)P_k(y) - P_k(x)P_{k-1}(y)]$$

Taking summation from $k = 0$ to $k = n$, we get

$$(x-y)\sum_{K=0}^{n}(2k+1)\,P_k(x)\,P_k(y)$$

$$= \sum_{k=0}^{n}(k+1)\,[P_{k+1}(x)\,P_k(y) - P_k(x)P_{k+1}(y)] + \sum_{k=0}^{n} k\,[P_{k-1}(x)\,P_k(y) - P_k(x)P_{k-1}(y)]$$

$$= \{[P_1(x)P_0(y) - P_0(x)P_1(y)] + 2[P_2(x)P_1(y) - P_1(x)P_2(y)] + 3[P_3(x)P_2(y) - P_2(x)P_3(y)] + \ldots$$
$$+ n[P_n(x)P_{n-1}(y) - P_{n-1}(x)P_n(y)] + (n+1)\,[P_{n+1}(x)P_n(y) - P_n(x)P_{n+1}(y)]\}$$
$$+ \{[P_0(x)P_1(y) - P_1(x)P_0(y)] + 2[P_1(x)P_2(y) - P_2(x)P_1(y)] + 3[P_2(x)P_3(y) - P_3(x)P_2(y)] + \ldots$$
$$+ (n-1)[P_{n-2}(x)P_{n-1}(y) - P_{n-1}(x)P_{n-2}(y)] + n\,[P_{n-1}(x)P_n(y) - P_n(x)P_{n-1}(y)]\}$$

$$= (n+1)\,[P_{n+1}(x)\,P_n(y) - P_n(x)P_{n+1}(y)]$$ (All the terms cancel except above)

$$\therefore \quad \sum_{k=0}^{n}(2k+1)\,P_k(x)\,P_k(y) = (n+1)\left[\frac{P_{n+1}P_n(y) - P_n(x)P_{n+1}(y)}{(x-y)}\right]$$

Solved Examples

Example 1. *Prove that* $|P_n(x)| < 1$, *when* $-1 < x < 1$.

Solution. From Laplace first integral for $P_n(x)$, we have

$$P_n(x) = \frac{1}{\pi}\int_0^{\pi}[x \pm \sqrt{(x^2-1)}\cos\theta]^n\,d\theta. \quad \ldots(1)$$

Now taking

$$\left|[x \pm \sqrt{(x^2-1)}\cos\theta]\right| = \left|x \pm i\sqrt{(1-x^2)}\cos\theta\right|$$

$$= \sqrt{x^2 + (1-x^2)\cos^2\theta} = \sqrt{1-(1-x^2)\sin^2\theta}$$

$\therefore |x \pm \sqrt{(x^2-1)}\cos\theta| < 1$ except $\theta = 0$ and $\theta = \pi$.

From (1), we have

$$|P_n(x)| = \left|\frac{1}{\pi}\int_0^{\pi}[x \pm \sqrt{(x^2-1)}\cos\theta]^n\,d\theta\right|$$

$$\le \frac{1}{\pi}\int_0^{\pi}|x \pm \sqrt{(x^2-1)}\cos\theta|^n\,d\theta$$

$$< \frac{1}{\pi}\int_0^{\pi}1.d\theta = \frac{1}{\pi}.\pi = 1$$

$\therefore |P_n(x)| < 1.$

Example 2. *Show that*

(i) $P_n(-x) = (-1)^n P_n(x)$ (Bhopal-2008, VTU-2003)

and (ii) $P_n'(-x) = (-1)^{n+1}P_n'(x)$.

Solution. (i) Since we have

$$P_n(x) = \sum_{m=0}^{[n/2]}\frac{(-1)^m(2n-2m)!}{2^n(m)!(n-m)!(n-2m)!}.x^{n-2m}$$

Putting $-x$ in place of x, we get

$$P_n(-x) = \sum_{m=0}^{[n/2]}\frac{(-1)^m(2n-2m)!}{2^n(m)!(n-m)!(n-2m)!}.(-x)^{n-2m}$$

$$= \sum_{m=0}^{[n/2]}\frac{(-1)^m(2n-2m)!}{2^n(m)!(n-m)!(n-2m)!}.(-1)^{n-2m}.x^{n-2m}$$

$$= (-1)^n\sum_{m=0}^{[n/2]}\frac{(-1)^m(2n-2m)!}{2^n(m)!(n-m)!(n-2m)!}.x^{n-2m}$$

$[\because (-1)^{-2m} = 1]$

$$= (-1)^n P_n(x).$$

Hence $P_n(-x) = (-1)^n P_n(x)$.

(ii) To show $P_n'(-x) = (-1)^{n+1}P_n'(x)$.

From above result we have

$$P_n(-x) = (-1)^n P_n(x).$$

Differential both sides w.r.t. x we get

$$-P_n'(-x) = (-1)^n P_n'(x)$$

or $P_n'(-x) = (-1)^{n+1}P_n'(x)$.

Hence proved the result.

Example 3. *Show that* $P_n(1) = 1$ *and* $P_n(-1) = (-1)^n$.

(VTU-2003, Delhi-2002, BPTU-2005)

Solution. Since we have

$$\sum_{n=0}^{\infty}P_n(x)t^n = (1-2xt+t^2)^{-1/2} \quad \ldots(1)$$

putting $x = 1$ in (1), we get

$$\sum_{n=0}^{\infty}P_n(1)t^n = (1-2t+t^2)^{-1/2} = (1-t)^{-1}$$

or $[P_0(1) + tP_1(1) + \ldots + t^nP_n(1) + \ldots]$
$= [1 + t + t^2 + \ldots + t^n + \ldots].$

Taking the coefficient of t^n of both sides, we get $P_n(1) = 1$. Hence proved.

Next putting $x = -1$ in (1), we get

$$\sum_{n=0}^{\infty} P_n(-1)t^n = (1+2t+t^2)^{-1/2} = (1+t)^{-1}$$

or $$[P_0(-1) + tP_1(-1) + ... + t^n P_n(-1) + ...]$$
$$= [1 - t + t^2 - ... + (-1)^n t^n + ...].$$

Comparing the coefficient of t^n of both sides, we get $P_n(-1) = (-1)^n$.

Example 4. *Show that*

(i) $P_n(0) = 0$, *if n is odd.*

(ii) $P_n(0) = \dfrac{(-1)^{n/2}(n)!}{2^n((n/2)!)^2}$, *if n is even.*

(UPTU-2008, SVTU-2008)

Solution. Since we have

$$\sum_{n=0}^{\infty} P_n(x)t^n = (1-2xt+t^2)^{-1/2} \quad ...(1)$$

Putting $x = 0$ in (1), we get

$$\sum_{n=0}^{\infty} P_n(0)t^n = (1+t^2)^{-1/2}$$

or $$[P_0(0) + t P_1(0) + ... + t^n P_n(0) + ...]$$
$$= \left[1 - \frac{1}{2}t^2 + \frac{1.3}{2.4}t^4 - ... + \frac{(-1)^n 1.3.5...(2n-1)}{2^n(n)!}t^{2n} + ...\right].$$

(i) If n is odd, then comparing the coefficient of t^n, we get $P_n(0) = 0$

(ii) If n is even, then comparing the coefficient of t^n of both sides, *i.e.*, let $n = 2m$ then comparing the coefficient of t^{2m}, we get

$$P_{2m}(0) = \frac{(-1)^m 1.3.5...(2m-1)}{2.4.6...2m}$$

$$\therefore \quad P_{2m}(0) = \frac{(-1)^m(2m)!}{2^{2m}((m)!)^2}$$

or $$P_n(0) = \frac{(-1)^{n/2}(n)!}{2^n((n/2)!)^2} \qquad [\because n = 2m]$$

Example 5. *Show that*

(i) $P_{2n}(0) = (-1)^n \dfrac{1.3.5...(2n-1)}{2.4.6...2n}$

(ii) $P_{2n+1}(0) = 0.$ (UPTU-2005)

Solution. We know that $\sum z^{2n} P_{2n}(x) = (1-2xz+z^2)^{-1/2}$

On putting $x = 0$, $\Sigma z^{2n} P_{2n}(0) = (1+z^2)^{-1/2}$

$$= 1 + \left(-\frac{1}{2}\right)z^2 + \frac{\left(-\frac{1}{2}\right)\left(-\frac{3}{2}\right)}{2!}(z^2)^2 + \frac{\left(-\frac{1}{2}\right)\left(-\frac{3}{2}\right)\left(-\frac{5}{2}\right)}{3!}(z^2)^3$$
$$+ ... + \frac{\left(-\frac{1}{2}\right)\left(-\frac{3}{2}\right)\left(-\frac{5}{2}\right)...\left(-\frac{1}{2}-n+1\right)}{n!}(z^2)^n + ... \quad ...(1)$$

Equating the coefficient of z^{2n} both sides, we get

$$P_{2n}(0) = \frac{\left(-\frac{1}{2}\right)\left(-\frac{3}{2}\right)\left(-\frac{5}{2}\right)...\left(-\frac{1}{2}-n+1\right)}{n!}$$
$$= (-1)^n \frac{1.3.5...(2n-1)}{2^n.n!}$$
$$= (-1)^n \frac{1.3.5...(2n-1)}{2.4.6...2n}$$

Coefficient of $z^{2n+1} = P_{2n+1}(0) = 0$.

Example 6. *Show that any polynomial f(x) of degree n can be represented in the form*

$$f(x) = C_0P_0(x) + C_1P_1(x) + ... + C_nP_n(x),$$

where, $C_n = \dfrac{2n+1}{2}\int_{-1}^{1} f(x)\, P_n(x)\, dx$.

Solution. Since we have

$$P_n(x) = \sum_{m=0}^{[n/2]} \frac{(-1)^m(2n-2m)!}{2^n(m)!(n-m)!(n-2m)!}.x^{n-2m}$$

Putting $n = 0, 1, 2, 3, ...$, we get

$$P_0(x) = 1, P_1(x) = x, P_2(x) = \frac{1}{2}(3x^2-1),$$
$$P_3(x) = \frac{1}{2}(5x^3 - 3x)$$
$$P_4(x) = \frac{1}{8}(35x^4 - 30x^2 - 3) ...$$

From above Legendre's function, we get

$$1 = P_0(x), x = P_1(x),$$
$$x^2 = \frac{1}{3} + \frac{2}{3}P_2(x) = \frac{1}{3}P_0(x) + \frac{2}{3}P_2(x)$$
$$x^3 = \frac{3}{5}x + \frac{2}{5}P_3(x) = \frac{3}{5}P_1(x) + \frac{2}{5}P_3(x)...$$

Suppose $f(x)$ is a polynomial of degree 3, *i.e.*,

$$f(x) = a_0 + a_1x + a_2x^2 + a_3x^3$$

Substitute the values of $x^3, x^2, x, 1$, we get

$$f(x) = a_0P_0(x) + a_1P_1(x) + a_2\left[\frac{1}{3}P_0(x) + \frac{2}{3}P_2(x)\right] + a_3\left[\frac{3}{5}P_1(x) + \frac{2}{5}P_3(x)\right]$$
$$= \left(a_0 + \frac{a_2}{3}\right)P_0(x) + \left(a_1 + \frac{3}{5}a_3\right)P_1(x) + \frac{2}{3}a_2P_2(x) + \frac{2}{5}a_3P_3(x)$$

$$= C_0P_0(x) + C_1P_1(x) + C_2P_2(x) + C_3P_3(x)$$

$$\therefore \quad f(x) = \sum_{m=0}^{3} C_m P_m(x)$$

Further, since $P_n(x)$ is a polynomial of degree n for every positive integer n, therefore x^n can always be expressed as a linear combination of $P_0(x), P_1(x), \ldots, P_n(x)$. Thus, if $f(x)$ is a polynomial of degree n, then

$$f(x) = C_0P_0(x) + C_1P_1(x) + C_2P_2(x) + \ldots + C_nP_n(x)$$

Next, multiplying above equation by $P_n(x)$ and then integrate from $x = -1$ and $x = 1$, we get

$$\int_{-1}^{1} f(x)P_n(x)dx = C_0\int_{-1}^{1} P_0(x)P_n(x)dx + C_1\int_{-1}^{1} P_1(x)P_n(x)dx + \ldots + C_n\int_{-1}^{1}[P_n(x)]^2 dx$$

$$= 0 + 0 + \ldots + C_n\left[\frac{2}{2n+1}\right]$$

$$= C_n\left(\frac{2}{2n+1}\right)$$

(By orthogonal properties of Legendre polynomial)

$$\therefore \quad C_n = \frac{2n+1}{2}\int_{-1}^{1} f(x)\, P_n(x)\, dx$$

Example 7. *Express $f(x) = x^4 + 2x^3 + 2x^2 - x - 3$ in terms of Legendre's polynomials.* (UPTU(Q. Bank)-2002)

Solution. Degree of $f(x)$ is four. Therefore, we take Legendre's polynomials $P_0(x)$, $P_1(x)$, $P_2(x)$, $P_3(x)$, $P_4(x)$ as follows :

$$P_0(x) = 1, P_1(x) = x, P_2(x) = \frac{1}{2}(3x^2 - 1),$$

$$P_3(x) = \frac{1}{2}(5x^3 - 3x)$$

$$P_4(x) = \frac{1}{8}(35x^4 - 30x^2 - 3).$$

From Legendre's polynomial, we have

$$1 = P_0(x), x = P_1(x),$$

$$x^2 = \frac{1}{3} + \frac{2}{3}P_2(x) = \frac{1}{3}P_0(x) + \frac{2}{3}P_2(x)$$

$$x^3 = \frac{3}{5}x + \frac{2}{5}P_3(x) = \frac{3}{5}P_1(x) + \frac{2}{5}P_3(x)$$

and
$$x^4 = -\frac{3}{35} + \frac{30}{35}x^2 + \frac{8}{35}P_4(x)$$

$$= -\frac{3}{35}P_0(x) + \frac{30}{35}\left(\frac{1}{3}P_0(x) + \frac{2}{5}P_2(x)\right) + \frac{8}{35}P_4(x)$$

$$= \frac{1}{5}P_0(x) + \frac{20}{35}P_2(x) + \frac{8}{35}P_4(x).$$

Now, substitute these values of $1, x, x^2, x^3$ and x^4 into

$$f(x) = x^4 + 2x^3 + 2x^2 - x - 3$$

$$= \left[\frac{1}{5}P_0(x) + \frac{20}{35}P_2(x) + \frac{8}{35}P_4(x)\right] + 2\left[\frac{3}{5}P_1(x) + \frac{2}{5}P_3(x)\right]$$

$$+ 2\left[\frac{1}{3}P_0(x) + \frac{2}{3}P_2(x)\right] - [P_1(x)] - 3P_0(x)$$

$$\therefore \quad f(x) = \frac{8}{35}P_4(x) + \frac{4}{5}P_3(x) + \frac{40}{21}P_2(x) + \frac{1}{5}P_1(x) - \frac{32}{15}P_0(x).$$

Example 8. *Prove that*

$$\int_{-1}^{1} x^2 P_{n+1}P_{n-1}\, dx = \frac{2n(n+1)}{(2n-1)(2n+1)(2n+3)}.$$

(UPTU(Q. Bank)-2002, JNTU-2006, Kerala(M.Tech)-2005)

Solution. From Recurrence relation I, we have

$$(2n+1)x P_n = (n+1)P_{n+1} + nP_{n-1} \quad \ldots(1)$$

Putting $(n-1)$ and $(n+1)$ in place of n respectively, we get

$$(2n-1)xP_{n-1} = nP_n + (n-1)P_{n-2} \quad \ldots(2)$$

$$(2n+3)xP_{n+1} = (n+2)P_{n+2} + (n+1)P_n \quad \ldots(3)$$

Multiplying (2) and (3), we get

$$(2n-1)(2n+3)x^2P_{n+1}P_{n-1}$$

$$= [nP_n + (n-1)P_{n-2}][(n+2)P_{n+2} + (n+1)P_n]$$

$$= n(n+2)P_nP_{n+2} + n(n+1)(P_n)^2 + (n-1)(n+2)P_{n-2}P_{n+2} + (n^2-1)P_{n-2}P_n.$$

Now integrating from $x = -1$ to $x = 1$ w.r.t. x we get

$$(2n-1)(2n+3)\int_{-1}^{1} x^2P_{n+1}P_{n-1}dx$$

$$= n(n+2)\int_{-1}^{1} P_nP_{n+2}dx + n(n+1)\int_{-1}^{1}[P_n]^2 dx + (n-1)(n+2)\int_{-1}^{1} P_{n-2}P_{n+2}dx + (n^2-1)\int_{-1}^{1} P_{n-2}P_n\, dx$$

$$= n(n+1)\int_{-1}^{1}[P_n]^2\, dx + 0 + 0 + 0$$

$$= n(n+1)\left[\frac{2}{2n+1}\right] \quad \text{(By orthogonal property)}$$

$$\therefore \int_{-1}^{1} x^2P_{n+1}P_{n-1}\, dx = \frac{2n(n+1)}{(2n-1)(2n+1)(2n+3)}.$$

Example 9. *Prove that*

$$\int_{-1}^{1} (x^2-1)\, P_{n+1}\, P_n'\, dx = \frac{2n(n+1)}{(2n+1)(2n+3)}.$$

Solution. Since we have

$$(2n+1)(x^2-1)\, P_n' = n(n+1)\,(P_{n+1} - P_{n-1})$$

(Beltrami's result)

Now, multiplying by P_{n+1} and then integrating from $x = -1$ to 1, we get

$$(2n+1)\int_{-1}^{1}(x^2-1)\,P_{n+1}\,P_n' dx$$

$$= n(n+1)\int_{-1}^{1}[P_{n+1}]^2 dx - n(n+1)\int_{-1}^{1}P_{n+1}P_{n-1}dx$$

$$= n(n+1)\left[\frac{2}{2n+3}\right] - 0$$

(By orthogonal properties)

$$(2n+1)\int_{-1}^{1}(x^2-1)\,P_{n+1}\,P_n'\,dx = \frac{2n(n+1)}{(2n+3)}$$

$$\therefore \int_{-1}^{1}(x^2-1)\,P_{n+1}P_n'\,dx = \frac{2n(n+1)}{(2n+1)(2n+3)}.$$

Example 10. *Show that* $\int_{-1}^{1} xP_nP_{n-1}dx = \dfrac{2n}{4n^2-1}$.

Solution. From recurrence relation I, we have

$$(2n+1)x\,P_n = (n+1)\,P_{n+1} + n\,P_{n-1} \quad \ldots(1)$$

Multiplying (1) by P_{n-1} and then integrating from $x = -1$ to 1,

$$(2n+1)\int_{-1}^{1}xP_nP_{n-1}dx$$

$$= (n+1)\int_{-1}^{1}P_{n+1}P_{n-1}dx + n\int_{-1}^{1}[P_{n-1}]^2 dx$$

$$= 0 + n\left[\frac{2}{2n-1}\right] = \frac{2n}{2n-1}$$

$$\therefore \int_{-1}^{1}xP_nP_{n-1}dx = \frac{2n}{(2n+1)(2n-1)} = \frac{2n}{4n^2-1}.$$

Example 11. *Show that* $\int_{-1}^{1}\dfrac{P_n(x)}{\sqrt{1-2xt+t^2}}dx = \dfrac{2t^n}{2n+1}$.

Solution. Since we have $\dfrac{1}{\sqrt{1-2xt+t^2}} = \sum_{n=0}^{\infty} P_n(x)\,t^n$

or $\dfrac{1}{\sqrt{1-2xt+t^2}} = P_0(x) + tP_1(x) + \ldots$

$$+ t^nP_n(x) + t^{n+1}P_{n+1}(x) + \ldots$$

Now, multiplying this equation by $P_n(x)$ and then integrating from $x = -1$ to 1, we get

$$\int_{-1}^{1}\frac{P_n(x)}{\sqrt{1-2xt+t^2}}dx = \int_{-1}^{1}P_0(x)P_n(x)dx$$

$$+ t\int_{-1}^{1}P_1(x)\,P_n(x)\,dx + \ldots + t^n\int_{-1}^{1}[P_n(x)]^2 dx$$

$$+ t^{n+1}\int_{-1}^{1}P_{n+1}(x)\,P_n(x)\,dx + \ldots$$

$$= t^n\int_{-1}^{1}[P_n(x)]^2\,dx$$

$$= t^n\left[\frac{2}{2n+1}\right]$$

(All integral except one is zero)

$$\therefore \int_{-1}^{1}\frac{P_n(x)}{\sqrt{1-2xt+t^2}}dx = \frac{2t^n}{2n+1}.$$

Example 12. *Prove that*

$$\frac{1+t}{t\sqrt{1-2xt+t^2}} - \frac{1}{t} = \sum_{n=0}^{\infty}[P_n(x)+P_{n+1}(x)]t^n.$$

Solution.

$$\text{RHS} = \sum_{n=0}^{\infty}[P_n(x)+P_{n+1}(x)]\,t^n$$

$$= \sum_{n=0}^{\infty} t^n\,P_n(x) + \sum_{n=0}^{\infty} t^nP_{n+1}(x)$$

$$= \frac{1}{\sqrt{1-2xt+t^2}} + \frac{1}{t}\sum_{n=0}^{\infty} t^{n+1}\,P_{n+1}(x)$$

$$= \frac{1}{\sqrt{1-2xt+t^2}} + \frac{1}{t}\left[\sum_{n=0}^{\infty} t^nP_n(x) - P_0(x)\right]$$

$$= \frac{1}{\sqrt{1-2xt+t^2}} + \frac{1}{t}\left[\sum_{n=0}^{\infty} t^nP_n(x) - P_0(x)\right]$$

$$= \frac{1}{\sqrt{1-2xt+t^2}} + \frac{1}{t}\left[\sum_{n=0}^{\infty} t^n\,P_n(x) - 1\right]$$

$$[\because P_0(x) = 1]$$

$$= \frac{1}{\sqrt{1-2xt+t^2}} + \frac{1}{t\sqrt{1-2xt+t^2}} - \frac{1}{t}$$

$$= \frac{1+t}{t\sqrt{1-2xt+t^2}} - \frac{1}{t} = \text{LHS}.$$

Example 13. *Prove that* $\int_{-1}^{1}(1-x^2)P_m'\,P_n'\,dx = 0, m \neq n$.

(UPTU-2006)

Solution. From Legendre's differential equation, we have

$$(1-x^2)\frac{d^2y}{dx^2} - 2x\frac{dy}{dx} + n(n+1)y = 0 \quad \ldots(1)$$

Since $P_n(x)$ and $P_m(x)$ are the solution of (1), so

$$(1-x^2)\frac{d^2P_m}{dx^2} - 2x\frac{dP_m}{dx} + m(m+1)P_m = 0$$

or $$\frac{d}{dx}\left\{(1-x^2)\frac{dP_m}{dx}\right\} + m(m+1)\,P_m = 0 \quad \ldots(2)$$

Multiplying (2) by P_n and then integrating from $x = -1$ to 1, we get

$$\int_{-1}^{1}\frac{d}{dx}\left\{(1-x^2)\frac{dP_m}{dx}\right\}P_n dx$$

$$+ m(m+1)\int_{-1}^{1}P_m\,P_n dx = 0$$

or $$\left[(1-x^2)\frac{dP_m}{dx}P_n\right]_{-1}^{1} - \int_{-1}^{1}(1-x^2)\frac{dP_m}{dx}.\frac{dP_n}{dx}dx$$

$$+ m(m+1)\int_{-1}^{1}P_mP_ndx = 0$$

or $$0 - \int_{-1}^{1}(1-x^2)P_m'\,P_n'\,dx$$

$$+ m(m+1)\int_{-1}^{1}P_mP_ndx = 0$$

or $$\int_{-1}^{1}(1-x^2)P_m'\,P_n'\,dx = m(m+1)\int_{-1}^{1}P_mP_ndx \quad \ldots(3)$$

Since $m \neq n$, then $\int_{-1}^{1}P_m\,P_n\,dx = 0$

$$\Rightarrow \int_{-1}^{1}(1-x^2)P_m'\,P_n'\,dx = 0.$$

Example 14. *Prove that* $\int_{-1}^{1} (1-x^2)(P_n')^2\, dx = \frac{2n(n+1)}{2n+1}$.

(UPTU(B.Tech.)2006, SVTU-2008, Kerala(ME)-2005)

Solution. From equation (3) in Ex. 13, we have

$$\int_{-1}^{1}(1-x^2)P_m' P_n'\, dx = m(m+1)\int_{-1}^{1} P_m P_n dx$$

If $m = n$, then

$$\int_{-1}^{1}(1-x^2)(P_n')^2\, dx = n(n+1)\int_{-1}^{1}(P_n)^2 dx$$

$$= n(n+1)\,.\frac{2}{2n+1} = \frac{2n(n+1)}{2n+1}.$$

Example 15. *Prove that* $\int_{-1}^{1} x^m P_n(x) = 0$ if $m < n$.

Solution. From Rodrigue's formula, we have

$$P_n(x) = \frac{1}{2^n (n)!}\frac{d^n}{dx^n}(x^2-1)^n \quad \ldots(1)$$

Multiplying (1) by x^m and then integrating from $x=-1$ to 1, we get

$$\int_{-1}^{1} x^m P_n(x)dx$$

$$= \frac{1}{2^n (n)!}\int_{-1}^{1} x^n \frac{d^n}{dx^n}(x^2-1)^n dx$$

(Integration by part)

$$= \frac{1}{2^n (n)!}\left[\left\{x^m.\frac{d^{n-1}}{dx^{n-1}}(x^2-1)^n\right\}_{-1}^{1} - \int_{-1}^{1} m x^{m-1}\frac{d^{n-1}}{dx^{n-1}}(x^2-1)^n dx\right]$$

$$= -\frac{m}{2^n (n)!}\int_{-1}^{1} x^{m-1}\frac{d^{n-1}}{dx^{n-1}}(x^2-1)^n dx$$

$$\left[\because \left[\frac{d^{n-1}}{dx^{n-1}}(x^2-1)^n\right]_{-1}^{1} = 0\right]$$

Continuing the integration upto m times, we get

$$= \frac{(-1)^m (m)!}{2^n (n)!}\int_{-1}^{1} x^0 \frac{d^{n-m}}{dx^{n-m}}(x^2-1)^n dx$$

$$= \frac{(-1)^m (m)!}{2^n (n)!}\left[\frac{d^{n-m-1}}{dx^{n-m-1}}(x^2-1)^n\right]_{-1}^{1} = 0$$

$$\therefore \int_{-1}^{1} x^m P_n(x) = 0 \quad (m<n).$$

EXERCISE 34.1

1. Express the following function in terms of Legendre's polynomial.
   (i) $2x+10x^3$
   (ii) x^3+2x^2-x-3 (Osmania-2003)
   (iii) $x^4+3x^3-x^2+5x-2$ (VTU-2010, SVTU-2007, Bhopal-2008, Madras-2006)
   (iv) $4x^3+6x^2+7x+2$ (SVTU-2008)
2. Find $P_6(t)$.
3. Show that $\frac{1-t^2}{(1-2xt+t^2)^{3/2}} = \sum_{n=0}^{\infty}(2n+1)P_n(x)t^n$.
4. Prove that $\int_{-1}^{1}(P_n')^2 dx = n(n+1)$.
5. Show that $2P_2(x) - 3P_1(x)P_1(x) + 1 = 0$.
6. Prove that $P_{n+1}' + P_n' = \sum_{r=0}^{n}(2r+1)P_r(x)$.
7. Prove that $(1-2xt+t^2)^{-1/2}$ is a solution of the equation
   $$t\frac{\partial^2(ts)}{\partial t^2} + \frac{\partial}{\partial x}\left\{(1-x^2)\frac{\partial s}{\partial x}\right\} = 0$$
8. Prove that $\int_{-1}^{1}(1-x^2)P_m' P_n'\, dx = \frac{2n(n+1)}{(2n+1)}\delta_{mn},\ \delta_{mn} = \begin{cases}1, m=n\\ 0, m\neq n\end{cases}$.
9. Prove that
   (i) $\int P_n(x)\,dx = \left[\frac{P_{n+1}(x)-P_{n-1}(x)}{2n+1}\right]+C$
   (ii) $\int_x^1 P_n(x)\,dx = \frac{P_{n-1}(x)-P_{n+1}(x)}{2n+1}$
10. Prove that $\int_x^1 x^n P_n(x)\,dx = \frac{2^{n+1}[(n!)]^2}{(2n+1)!}$
11. Prove that
    (i) $\int_{-1}^{1} P_n(x)\,dx = 0,\ n\neq 0$,
    (ii) $\int_{-1}^{1} P_0(x)\,dx = 2$
12. Find the value of integrals
    (i) $\int_{-1}^{1} x^{99} P_{100}(x)\,dx$, (ii) $\int_{-1}^{1} x^2 P_2(x)\,dx$
13. Prove that
    (i) $P_n'(1) = \frac{1}{2}n(n+1)$, (ii) $P_n'(-1) = (-1)^{n-1}.\frac{1}{2}n(n+1)$
14. Prove that if $r_2 > 0$
    $$\frac{1}{r} = \frac{1}{r_2}\left[P_0 + P_1(\cos\theta)\left(\frac{r_1}{r_2}\right) + \left(\frac{r_1}{r_2}\right)^2 P_2(\cos\theta)+\ldots\right],$$

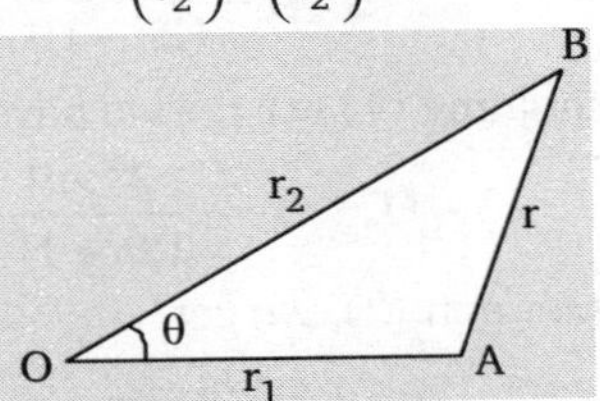

where in ΔOAB

15. Prove that $|P_n(x)| < \sqrt{\frac{\pi}{2n(1-x^2)}}$, when $-1<x<1$.
16. If $f(x) = \begin{cases}0 & -1<x<0\\ x & 0<x<1\end{cases}$, show that
    $$f(x) = \frac{1}{4}P_0(x) + \frac{1}{2}P_1(x) + \frac{5}{16}P_2(x) - \frac{3}{32}P_4(x)+\ldots$$
    (UPTU-2003)

Hint to Selected Problems

1. (i) Use $P_n(x)=\sum_{k=0}^{[n/2]}\frac{(-1)(2n-2k)!}{2^n.k!(n-2k)!(n-k)!}x^{n-2k}$, i.e.,

$P_0(x)=1, P_1(x)=x, P_2(x)=\frac{1}{2}(3x^2-1), P_3(x)=\frac{1}{2}(5x^3-3x)$

3. Since we have $\frac{1}{\sqrt{1-2xt+t^2}}=\sum_{n=0}^{\infty}P_n(x)t^n$ ...(1)

On differentiating, we get (after simplification)

$$2t(x-t)(1-2xt-t^2)^{-3/2}=\sum_{n=0}^{\infty}2nP_n(x)t^n \quad ...(2)$$

Then adding (1) and (2).

4. Use Christoffel's expansion.

5. Use $P_1(x)=x,\ P_2(x)=\frac{1}{2}(3x^2-1)$.

6. Put $n=1,2,3,...,n-1$, n in the recurrence relation $P'_{n+1}-P_{n-1}=(2n+1)P_n$ and then adding all these expression.

7. The given equation can be written as

$$t^2\frac{\partial^2 s}{\partial t^2}+2t\frac{\partial s}{\partial t}+(1-x^2)\frac{\partial^2 s}{\partial x^2}-2x\frac{\partial s}{\partial x}=0.$$

Then, find the values of different derivatives by using $s=(1-2xt+t^2)^{-1/2}$.

8. Use Beltrami's result and Christoffel's expansion.

9. (i) We have $(2n+1)P_n=P'_{n+1}-P'_{n-1}$

$\Rightarrow P_n(x)=\frac{P'_{n+1}(x)-P_{n-1}(x)}{(2n+1)}$. Then integrate.

10. From Rodrigue's formula, we have

$$P_n(x)=\frac{1}{2^n.n!}\frac{d^n}{dx^n}(x^2-1)^n \Rightarrow x^nP_n(x)=\frac{1}{2^n.n!}x^n\frac{d^n}{dx^n}(x^2-1)^n.$$

Now, integrating w.r.t. x from $x=-1$ to $x=+1$.

11. On integrating the Rodrigue's formula from $x=-1$ to $x=+1$.

12. Use $\int_{-1}^{1}x^m P_n(x_0)=0$, if $m<n$.

13. Differentiating the generating function

$\sum_{n=0}^{\infty}t^nP_n(x)=(1-2xt+t^2)^{-1/2}$ and then putting $x=1$.

14. We have $\cos\theta=\frac{r_1^2+r_2^2-r^2}{2r_1r_2}$

$$\Rightarrow \frac{1}{r}=\frac{1}{r_2}\left[\frac{1}{\sqrt{1-2\left(\frac{r_1}{r_2}\right)\cos\theta+\left(\frac{r_1}{r_2}\right)^2}}\right]$$

Then use the generating function $\frac{1}{\sqrt{1-2xt+t^2}}=\sum_{n=0}^{\infty}t^nP_n(x)$

By putting $t=\frac{r_1}{r_2}$ and $x=\cos\theta$.

ANSWERS

1. (i) $8P_1(x)+4P_3(x)$ (ii) $\frac{2}{5}P_3+\frac{4}{3}P_2-\frac{2}{5}P_1-\frac{7}{3}P_0$ (iii) $\frac{8}{35}P_4(x)+\frac{6}{5}P_3(x)-\frac{2}{21}P_2(x)+\frac{34}{5}P_1(x)-\frac{224}{105}P_0(x)$

(iv) $\frac{8}{5}P_3-4P_2+\frac{47}{5}P_1+4$ **2.** $\frac{1}{16}(231x^6-315x^4+105x^2-5)$ **12.** (i) 0 (ii) 4/15

35.7 LEGENDRE'S FUNCTION OF THE SECOND KIND

Here we have $Q_n(x)=\frac{n!}{1.3...(2n+1)}\left[x^{-n-1}+\frac{(n+1)(n+2)}{2(2n+3)}x^{-n-3}+\frac{(n+1)(n+2)(n+3)(n+4)}{2.4(2n+3)(2n+5)}x^{-n-5}+...\right]$

$$=\frac{1}{1.3...(2n+1)}\sum_{r=0}^{\infty}\frac{(n+2r)!x^{-(n+2r+1)}}{2.4...2r(2n+3)(2n+5)...(2n+2r+1)}$$

$$\Rightarrow \quad Q_n(x)=\frac{2^n.n!}{(2n+1)!}\sum_{r=0}^{\infty}\frac{(n+2r)!\,x^{-(n+2r+1)}}{2^r\,r!\,(2n+3)(2n+5)...(2n+2r+1)}. \quad ...(1)$$

Now, differentiating (1) w.r.t. x we have

$$Q'_n(x)=-\frac{2^n.n!}{(2n+1)!}\sum_{r=0}^{\infty}\frac{(n+2r+1)!\,x^{-(n+2r+2)}}{2^r.r!(2n+3)(2n+5)...(2n+2r+1)} \quad ...(2)$$

Putting $n-1$ for n in (2), we get

$$Q'_{n-1}(x)=-\frac{-2^{n-1}.(n-1)!}{(2n-1)!}\sum_{r=0}^{\infty}\frac{(n+2r)!\,x^{-(n+2r+1)}}{2^r.r!(2n+1)(2n+3)...(2n+2r+1)} \quad ...(3)$$

Also, putting $n+1$ for n in (2), we get

$$Q'_{n+1}(x)=-\frac{2^{n+1}.(n+1)!}{(2n+3)!}\sum_{r=0}^{\infty}\frac{(n+2r+2)!\,x^{-(n+2r+3)}}{2^r.r!(2n+5)...(2n+2r+1)(2n+2r+3)}$$

$$=-\frac{2^n.n!(2n+2)}{(2n+3)(2n+2)(2n+1)(2n)!}.\sum_{r=0}^{\infty}\frac{(n+2r+2)!\,x^{-(n+2r+3)}}{2^r.r!(2n+5)...(2n+2r+3)}$$

$$=-\frac{2^n.n!}{(2n)!}\sum_{r=0}^{\infty}\frac{(n+2r+2)!\,x^{-(n+2r+3)}}{2^r.(r!)(2n+1)(2n+3)...(2n+2r+3)} \quad ...(4)$$

35.8 IMPORTANT RECURRENCE RELATIONS

(i) $Q'_{n+1} - Q'_{n-1} = (2n+1)\, Q_n$

(ii) $nQ'_{n+1} + (n+1)\, Q'_{n-1} = (2n+1)\, xQ'_n$

(iii) $(2n+1)\, xQ_n = (n+1)\, Q_{n+1} + nQ_{n-1}$

(iv) $(2n+1)(1-x^2)\, Q'_n = n\,(n+1)\,(Q_{n-1} - nQ_{n+1})$

(v) $x\, Q'_n - Q'_{n-1} = n\, Q_n$

(vi) $Q'_n - xQ'_{n-1} = n\, Q_{n-1}$

(vii) $(x^2-1)\, Q'_n = nxQ_n - nQ_{n-1}$

(viii) $(x^2-1)\, Q'_n = (n+1) Q_{n+1} - (n+1)\, xQ_n$

(ix) $(x^2-1)\,(Q_n P'_n - P_n Q'_n) = c$, a constant.

(x) $\dfrac{Q_n(x)}{P_n(x)} = \int_x^\infty \dfrac{dx}{(x^2-1)\, P_n^2(x)}$

(xi) $n[P_n Q_{n-1} - Q_n P_{n-1}] = 1$.

(xii) $P_n Q_{n-2} - Q_n P_{n-2} = \dfrac{(2n-1)}{n(n-1)} x$.

THEOREM 1. *If P_n is a solution of Legendre's equation, then the complete solution of the Legendre's equation is given by $aP_n + bQ_n$, where $Q_n = cP_n \int \dfrac{dx}{(1-x^2) P_n^{\,2}}$, c being a constant.*

Proof. Consider the Legendre's equation

$$(1-x^2)\frac{d^2y}{dx^2} - 2x\frac{dy}{dx} + n(n+1)y = 0. \qquad \text{...(1)}$$

Let $y = uP_n$ be the complete solution of (1), where u is a function of x.

$$\text{Therefore, } \frac{dy}{dx} = u\frac{dP_n}{dx} + P_n\frac{du}{dx},\quad \frac{d^2y}{dx^2} = u\frac{d^2P_n}{dx^2} + 2\frac{dP_n}{dx}.\frac{du}{dx} + P_n\frac{d^2u}{dx^2} \qquad \text{...(2)}$$

Putting all these values in (1), we get

$$(1-x^2)\left\{u\frac{d^2P_n}{dx^2} + 2\frac{dP_n}{dx}.\frac{du}{dx} + P_n\frac{d^2u}{dx^2}\right\} - 2x\left\{u\frac{dP_n}{dx} + P_n\frac{du}{dx}\right\} + n(n+1)uP_n = 0$$

$$\Rightarrow \quad (1-x^2)\left\{P_n\frac{d^2u}{dx^2} + 2\frac{dP_n}{dx}\frac{du}{dx}\right\} + (1-x^2)u\frac{d^2P_n}{dx^2} - 2xu\frac{dP_n}{dx} - 2u\frac{dP_n}{dx} + n(n+1)uP_n = 0$$

$$\Rightarrow (1-x^2)\left\{P_n\frac{d^2u}{dx^2} + 2\frac{du}{dx}\frac{dP_n}{dx}\right\} + u\left\{(1-x^2)\frac{d^2P_n}{dx^2} - 2x\frac{dP_n}{dx} + n(n+1)P_n\right\} - 2xP_n\frac{du}{dx} = 0. \qquad \text{...(3)}$$

Now since $P_n(x)$ is the solution of (1), therefore, we have

$$(1-x^2)\frac{d^2P_n}{dx^2} - 2x\frac{dP_n}{dx} + n(n+1)P_n = 0.$$

Using this in (3), we get

$$(1-x^2)\left\{P_n\frac{d^2u}{dx^2} + 2\frac{du}{dx}\frac{dP_n}{dx}\right\} - 2xP_n\frac{du}{dx} = 0 \quad \Rightarrow \quad \frac{d^2u/dx^2}{du/dx} + 2\frac{dP_n/dx}{P_n} - \frac{2x}{1-x^2} = 0.$$

On integrating, we have

$$\log\frac{du}{dx} + 2\log P_n + \log(1-x^2) = \log k \quad \Rightarrow \quad \log\left\{\frac{du}{dx}.P_n^{\,2}(1-x^2)\right\} = \log k$$

$$\Rightarrow \quad \frac{du}{dx}.P_n^{\,2}(1-x^2) = k \quad \Rightarrow \quad \frac{du}{dx} = \frac{k}{(1-x^2)P_n^{\,2}}.$$

On integrating, we have $u = k\int \dfrac{1}{(1-x^2)P_n^{\,2}} dx + A$, where A and k are arbitrary constants of integration. Hence, the complete solution of Legendre's equation is

$$y = uP_n = \left[k\int \frac{dx}{(1-x^2)P_n^{\,2}} + A\right]P_n = aP_n + kP_n\int\frac{dx}{(1-x^2)P_n^{\,2}} = aP_n + \frac{k}{c}.cP_n\int\frac{dx}{(1-x^2)P_n^{\,2}} = aP_n + bQ_n$$

where $b = k/c$.

Solved Examples

Example 1. *Show that $Q_2(x) = \dfrac{1}{2}P_2(x)\log\dfrac{x+1}{x-1} - \dfrac{3}{2}x$.*

Solution. Since we know that

$$(n+1)Q_{n+1} = (2n+1)xQ_n - nQ_{n-1}.$$

Putting $n = 1$, we get $2Q_2 = 3xQ_1 - Q_0$

$$= 3x\left[\frac{x}{2}\log\frac{x+1}{x-1} - 1\right] - \frac{1}{2}\log\frac{x+1}{x-1}$$

$$= \frac{1}{2}(3x^2 - 1)\log\frac{x+1}{x-1} - 3x$$

$$= P_2(x)\log\frac{x+1}{x-1} - 3x.$$

Hence, $Q_2(x) = \frac{1}{2}P_2(x)\log\frac{x+1}{x-1} - \frac{3}{2}x.$

Example 2. *Show that*

(i) $n(Q_n P_{n-1} - Q_{n-1}P_n) = (n-1)(Q_{n-1}P_{n-2} - Q_{n-2}P_{n-1})$

(ii) $n(Q_n P_{n-1} - Q_{n-1}P_n) = -1.$

Solution. (i) Using recurrence relation of $P_n(x)$ and $Q_n(x)$, we have

$$(2n+1)xP_n = (n+1)P_{n+1} + nP_{n-1} \quad \ldots(1)$$

$$(2n+1)xQ_n = (n+1)Q_{n+1} + nQ_{n-1}. \quad \ldots(2)$$

Replacing n by $n-1$ in (1) and (2), we get

$$(2n-1)xP_{n-1} = nP_n + (n-1)P_{n-2} \quad \ldots(3)$$

$$(2n-1)xQ_{n-1} = nQ_n + (n-1)Q_{n-2}. \quad \ldots(4)$$

Multiplying (3) by Q_{n-1} and (4) by P_{n-1} and then subtracting, we have

$$0 = n(P_n Q_{n-1} - P_{n-1}Q_n) + (n-1)(P_{n-2}Q_{n-1} - P_{n-1}Q_{n-2})$$

$$\Rightarrow n(Q_n P_{n-1} - Q_{n-1}P_n) = (n-1)(Q_{n-1}P_{n-2} - Q_{n-2}P_{n-1}). \ldots(5)$$

(ii) Let $u_n = n(Q_n P_{n-1} - Q_{n-1}P_n)$. Then (5) can be written as $u_n = u_{n-1}$

$$\Rightarrow u_{n-1} = u_{n-2} = u_{n-3} = \ldots = u_3 = u_2 = u_1$$

$$\therefore \quad u_n = u_1$$

$$\Rightarrow n(Q_n P_{n-1} - Q_{n-1}P_n) = Q_1 P_0 - Q_0 P_1$$

Hence, $n(Q_n P_{n-1} - Q_{n-1}P_n) = Q_1 - xQ_0$

$$= \frac{x}{2}\log\left(\frac{x+1}{x-1}\right) - 1 - \frac{x}{2}\log\frac{x+1}{x-1} = -1.$$

EXERCISE 35.2

1. Show that $\frac{Q_n}{P_n} = \int_x^\infty \frac{dx}{(x^2-1)\,P_n^2}$.

2. Show that $Q_1(x) = \frac{x}{2}\log\frac{x+1}{x-1} - 1$.

3. Show that

   (a) $P_n Q_{n-1} - Q_n P_{n-1} = \frac{1}{n}$.

   (b) $P_n Q_{n-2} - Q_n P_{n-2} = \frac{(2n-1)\,x}{n(n-1)}$.

4. Show that $Q_n(x) = 2^n n! \int_x^\infty dx \int_x^\infty dx \ldots \int_x^\infty (x^2-1)^{-n-1}dx$.

5. Show that

$$(2n+1)(1-x^2)\,Q_n'(x) = n(n+1)\,[Q_{n-1}(x) - Q_{n+1}(x)].$$

Hint to Selected Problems

1. Putting the values of P_n, Q_n, P_n' and Q_n' in the following relation

$$Q_n P_n' - P_n Q_n' = \frac{c}{x^2-1} = \frac{c}{x^2}\left(1 - \frac{1}{x^2}\right)^{-1} = c\left(\frac{1}{x^2} + \frac{1}{x^4} + \frac{1}{x^6} + \ldots\right)$$

Then equating the coefficient of $\frac{1}{x^2}$ of both the sides, we get $c = 1$.

Then write $\frac{Q_n P_n' - P_n Q_n'}{P_n^2} = -\frac{d}{dx}\left(\frac{Q_n}{P_n}\right) = \frac{1}{(x^2-1)P_n^2}$

Then taking the limit after integrating from x to ∞.

2. Putting $n = 1$ in eq. (4) of example 1, we get

$$\frac{Q_1(x)}{P_1(x)} = \int_x^\infty \frac{dx}{(x^2-1)P_1^2(x)}$$

$\Rightarrow \; Q_1(x) = x\int_0^\infty \frac{dx}{(x^2-1)x^2}$. Now, integrating after breaking using partial fraction.

3. Using the following recurrence relation

$$(2n+1)xP_n = (n+1)P_{n+1} + n\,P_{n-1}$$

and $(2n+1)xQ_n = (n+1)Q_{n+1} + nQ_{n-1}$.

4. We have

$$(x^2-1)^{-n-1} = x^{-2n-1}\left(1 - \frac{1}{x^2}\right)^{-n-1}$$

$$= x^{-2n-2}\left[1 + (n+1)\frac{1}{x^2} + \frac{(n+1)(n+2)}{2!}\cdot\frac{1}{2^4}\ldots\right]$$

$$= x^{-(2n+2)} + (n+1)x^{-(2n+4)} + \frac{(n+1)(n+2)}{2!}x^{-(2n+6)} + \ldots$$

Now integrating both sides w.r.t. x between x to ∞, n times.

5. Since $Q_n(x)$ is the solution of $\frac{d}{dx}\left[(1-x^2)\frac{dy}{dx}\right] + n(n+1)\ldots$

So, $\frac{d}{dx}\left((1-x^2)Q_n'\right) = -n(n+1)Q_n$. Then, integrating both sides from x to ∞.

35.9 BESSEL'S FUNCTION

The homogeneous linear differential equation of the form

$$x^2\frac{d^2y}{dx^2} + x\frac{dy}{dx} + \left(x^2 - n^2\right)y = 0 \quad \ldots(1)$$

is known as Bessel's differential equation, where n is a non-negative real number.

35.9.1 SOLUTION OF THE BESSEL'S FUNCTION

Change the differential equation (1) into standard form by dividing (1) by x^2.

$$\frac{d^2y}{dx^2}+\frac{1}{x}\frac{dy}{dx}+\left(1-\frac{n^2}{x^2}\right)y=0 \qquad \ldots(2)$$

Now compare this differential equation with following equation

$$\frac{d^2y}{dx^2}+P(x)\frac{dy}{dx}+Q(x)y=0$$

We get $$P(x)=\frac{1}{x},\ Q(x)=\left(1-\frac{n^2}{x^2}\right).$$

It is obvious from $P(x)$ and $Q(x)$, that $x=0$ is a singular point which is located at the origin. Therefore we assume the solution of (1) in the form of a power series of the following type

$$y=\sum_{m=0}^{\infty} a_m x^{m+r}\ (a_0\neq 0) \qquad \ldots(3)$$

Differentiating (3) w.r.t. x we get

$$\frac{dy}{dx}=\sum_{m=0}^{\infty} a_m(m+r)x^{m+r-1} \qquad \ldots(4)$$

Again differentiating (4) w.r.t. x we get

$$\frac{d^2y}{dx^2}=\sum_{m=0}^{\infty} a_m(m+r)(m+r-1)x^{m+r-2}. \qquad \ldots(5)$$

Now substitute the values of $y, \frac{dy}{dx}, \frac{d^2y}{dx^2}$ from (3), (4) and (5) into (1), we have

$$x^2\sum_{m=0}^{\infty} a_m(m+r)(m+r-1)x^{m+r-2}+x\sum_{m=0}^{\infty} a_m(m+r)x^{m+r-1}+(x^2-n^2)\sum_{m=0}^{\infty} a_m x^{m+r}=0$$

$$\sum_{m=0}^{\infty} a_m(m+r)(m+r-1)x^{m+r}+\sum_{m=0}^{\infty} a_m(m+r)x^{m+r}-\sum_{m=0}^{\infty} a_m n^2 x^{m+r}+\sum_{m=0}^{\infty} a_m x^{m+r+2}=0 \qquad \ldots(6)$$

Equation (6) will be an identity if the equation (3) is a solution of (1), then coefficient of each terms in (6) will be zero. Thus taking the coefficient of x^r, x^{r+1}

$$a_0 r(r-1)+a_0 r-n^2 a_0=0 \qquad \ldots(7)$$

$$a_1(r+1)r+a_1(r+1)-n^2 a_1=0. \qquad \ldots(8)$$

In general, taking the coefficients of x^{s+r}

$$a_s(s+r)(s+r-1)+a_s(s+r)-n^2 a_s+a_{s-2}=0 \text{ for } s=2,3,4,\ldots \qquad \ldots(9)$$

From (7), we have $r(r-1)+r-n^2=0$ $[\because a_0\neq 0]$

or $r^2-n^2=0$ or $r=n,-n.$

From (8), we have $[(r+1)r+(r+1)-n^2]a_1=0$.

For any value of $r=n,-n$, we get $a_1=0$.

From (9), we have

$$a_s[(s+r)(s+r-1)+s+r-n^2]+a_{s-2}=0 \quad \text{or} \quad a_s[(s+r)^2-n^2]+a_{s-2}=0$$

or $$a_s(s+r-n)(s+r+n)+a_{s-2}=0 \qquad \ldots(10)$$

For case if $r=n$, then (1) becomes

$$a_s(s)(s+2n)+a_{s-2}=0 \quad \text{or} \quad a_s=-\frac{1}{s(s+2n)}.a_{s-2}$$

Putting $s=2,3,4,5,\ldots$

$$a_2=-\frac{1}{2(2+2n)}a_0;\ a_3=-\frac{1}{3(3+2n)}a_1=0 \qquad [\because a_1=0]$$

$$a_4=-\frac{1}{4(4+2n)}a_2=-\frac{1}{4(4+2n)}\left(-\frac{1}{2(2+2n)}\right)a_0=(-1)^2\frac{1}{2.4(2+2n)(4+2n)}a_0$$

⋮

etc.

We observe that $a_1 = a_3 = a_5 = ... = 0$. Since a_0 is arbitrary. Let us choose $a_0 = \frac{1}{2^n \Gamma(n+1)}$ where $\Gamma(n+1)$ is a gamma function, therefore, we know that $\Gamma(n+1) = n\Gamma(n)$ and if n is positive integer and $\Gamma(n+1) = (n)!$. Thus

$$a_2 = -\frac{1}{2(2+2n)}a_0 = -\frac{1}{2^2(1+n)} \cdot \frac{1}{2^n \Gamma(n+1)} \qquad \left(\because a_0 = \frac{1}{2^n \Gamma(n+1)}\right)$$

$$= -\frac{1}{2^{n+2}\Gamma(n+2)}$$

and

$$a_4 = (-1)^2 \frac{1}{2^4.(2)!(1+n)(2+n)} \cdot \frac{1}{2^n \Gamma(n+1)} = (-1)^2 \frac{1}{2^{n+4}.(2)!\,\Gamma(n+3)}$$

and so on. Now From (3), we have

$$y = \sum_{m=0}^{\infty} a_m x^{m+r} = a_0 x^r + a_1 x^{1+r} + a_2 x^{2+r} + a_3 x^{3+r} + a_4 x^{4+r} + ...$$

$$= a_0 x^r + a_2 x^{2+r} + a_4 x^{4+r} + ... = a_0 x^n + \left\{-\frac{1}{2^{n+2}\Gamma(n+2)} x^{n+2}\right\} + \frac{1}{2^{n+4}(2)!\Gamma(n+3)} x^{n+4} + ...$$

$$= \frac{1}{2^n \Gamma(n+1)} x^n - \frac{1}{2^{n+2}(1)!\,\Gamma(n+2)} x^{n+2} + \frac{1}{2^{n+4}(2)!\,\Gamma(n+3)} x^{n+4} + ...$$

$$= \sum_{m=0}^{\infty} \frac{(-1)^m x^{n+2m}}{2^{n+2m}(m)!\,\Gamma(n+m+1)}.$$

This solution is known as Bessel's function, which is denoted by $J_n(x)$. This function is also known as Bessel's function of first kind.

$$\therefore \quad J_n(x) = \sum_{m=0}^{\infty} (-1)^m \frac{x^{n+2m}}{2^{n+2m}(m)!\,\Gamma(n+m+1)} \qquad ...(11)$$

For case if $r = -n$, we have

$$J_{-n}(x) = \sum_{m=0}^{\infty} \frac{(-1)^m x^{-n+2m}}{2^{-n+2m}(m)!\,\Gamma(-n+m+1)} \qquad ...(12)$$

35.9.2 GENERAL SOLUTION

The solution of the Bessel's differential equation of the type $y(x) = A J_n(x) + B J_{-n}(x)$ where A and B are arbitrary constants, is called general solution. (VTU-2006)

35.9.3 LINEAR DEPENDENCE

For an integer $r = n$, the Bessel's function $J_n(x)$ and $J_{-n}(x)$ are linearly dependent because $J_{-n}(x) = (-1)^n J_n(x)$ for $n = 1, 2, ...$

Proof. Since
$$J_{-n}(x) = \sum_{m=0}^{\infty} \frac{(-1)^m x^{-n+2m}}{2^{-n+2m}(m)!\,\Gamma(-n+m+1)}, \qquad ...(1)$$

if n is a positive integer, then the gamma functions in the coefficients of first n terms becomes infinite and coefficients of (1) becomes zero. Thus the summation will start at $m = n$ and in this case $\Gamma(-n+m+1) = (m-n)!$.

From (1), we now have,

$$J_{-n}(x) = \sum_{m=n}^{\infty} \frac{(-1)^m x^{-n+2m}}{2^{-n+2m}(m)!\,(m-n)!} = \sum_{k=0}^{\infty} \frac{(-1)^{n+k} x^{n+2k}}{2^{n+2k}(k)!\,(n+k)!} \qquad [\because m = n+k]$$

$$= (-1)^n \sum_{k=0}^{\infty} \frac{(-1)^k x^{n+2k}}{2^{n+2k}(k)!\,\Gamma(n+k+1)}$$

$$\therefore \quad J_{-n}(x) = (-1)^n J_n(x)$$

35.9.4 DEFINITION OF $J_n(x)$, WHEN $n = 0$

Putting $n = 0$ in the Bessel's differential equation, we get

$$x\frac{d^2y}{dx^2} + \frac{dy}{dx} + xy = 0 \qquad ...(1)$$

Let us assume the solution

$$y = \sum_{m=0}^{\infty} a_m x^{m+r} \quad (a_0 \neq 0) \qquad ...(2)$$

$$\therefore \quad \frac{dy}{dx} = \sum_{m=0}^{\infty} a_m (m+r) x^{m+r-1} \quad \text{and} \quad \frac{d^2y}{dx^2} = \sum_{m=0}^{\infty} a_m (m+r)(m+r-1) x^{m+r-2}.$$

Substitute these values in (1), we get

$$x \sum_{m=0}^{\infty} a_m (m+r)(m+r-1) x^{m+r-2} + \sum_{m=0}^{\infty} a_m (m+r) x^{m+r-1} + x \sum_{m=0}^{\infty} a_m x^{m+r} = 0$$

or
$$\sum_{m=0}^{\infty} a_m (m+r)(m+r-1) x^{m+r-1} + \sum_{m=0}^{\infty} a_m (m+r) x^{m+r-1} + \sum_{m=0}^{\infty} a_m x^{m+r+1} = 0 \quad \ldots(3)$$

If (2) is the solution of (1), then (3) will be an identity, Thus coefficients of each terms will be zero. So that taking the coefficients of x^{r-1}, we get

$$a_0 r(r-1) + a_0 r = 0 \quad \text{or} \quad r^2 a_0 = 0 \quad \text{or} \quad r = 0 \quad [\because a_0 \neq 0]$$

Now taking the coefficient of x^r, we have

$$a_1(1+r)r + a_1(1+r) = 0 \quad \text{or} \quad a_1(1+r)^2 = 0 \quad \text{or} \quad a_1 = 0 \quad [\because r = 0]$$

In general, taking the coefficient of x^{m+r}

$$a_{m+1}(m+r+1)(m+r) + a_{m+1}(m+r+1) + a_{m-1} = 0$$

or
$$a_{m+1}(m+r+1)^2 + a_{m-1} = 0 \quad \text{or} \quad a_{m+1} = -\frac{a_{m-1}}{(m+r+1)^2}$$

For the case $r = 0$, $a_{m+1} = -\frac{a_{m-1}}{(m+1)^2}$.

Putting $m = 1,2,3,4,5,\ldots$, $a_3 = -\frac{a_1}{9} = 0$ $\quad [\because a_1 = 0]$

$$a_2 = -\frac{a_0}{2^2};\ a_4 = -\frac{a_2}{4^2} = \frac{(-1)^2 a_0}{2^2 \,.\, 4^2};\ a_5 = 0 \ldots \text{ etc.}$$

Thus we obtained $a_1 = a_3 = a_5 = \ldots = 0$. Hence, $y = a_0\left(1 - \frac{x^2}{2^2} + \frac{x^4}{2^2 \,.\, 4^2} - \frac{x^6}{2^2 \,.\, 4^2 \,.\, 6^2} + \ldots\right)$

If $a_0 = 1$, then $y = J_0(x)$.

$$\therefore \quad J_0(x) = 1 - \frac{x^2}{2^2} + \frac{x^4}{2^2 \,.\, 4^2} - \frac{x^6}{2^2 \,.\, 4^2 \,.\, 6^2} + \ldots$$

$J_0(x)$ is also known as Bessel's function of order zero.

35.9.5 GENERATING FUNCTION FOR $J_n(x)$

The function of the form $e^{\left[\frac{1}{2}x\left(t-\frac{1}{t}\right)\right]}$ *generates* $J_n(x)$, *if taking coefficient of* t^n. *Thus this function is known as Generating function for* $J_n(x)$. (VTU-2007)

Proof. Expand $e^{\left[\frac{1}{2}x\left(t-\frac{1}{t}\right)\right]} = e^{\frac{xt}{2}} \cdot e^{-\frac{x}{2t}}$

$$= \left[1 + \frac{xt}{2} + \frac{1}{(2)!}\left(\frac{xt}{2}\right)^2 + \ldots + \frac{1}{(n)!}\left(\frac{xt}{2}\right)^n + \frac{1}{(n+1)!}\left(\frac{xt}{2}\right)^{n+1} + \frac{1}{(n+2)!}\left(\frac{xt}{2}\right)^{n+2} + \ldots\right]$$

$$\cdot\left[1 - \frac{x}{2t} + \frac{1}{(2)!}\left(\frac{x}{2t}\right)^2 + \ldots + \frac{(-1)^n}{(n)!}\left(\frac{x}{2t}\right)^n + \frac{(-1)^{n+1}}{(n+1)!}\left(\frac{x}{2t}\right)^{n+1} + \frac{(-1)^{n+2}}{(n+2)}\left(\frac{x}{2t}\right)^{n+2} + \ldots\right]$$

Now collecting the coefficient of t^n, in above expression (obtained after multiplication) is given by

$$= \frac{1}{(n)!}\left(\frac{x}{2}\right)^n - \frac{1}{(n+1)!}\left(\frac{x}{2}\right)^{n+2} + \frac{1}{(n+2)!} \cdot \frac{1}{(2)!}\left(\frac{x}{2}\right)^{n+4} + \ldots$$

$$= \sum_{m=0}^{\infty} (-1)^m \cdot \frac{1}{(m)!(n+m)!} \cdot \left(\frac{x}{2}\right)^{n+2m} = \sum_{m=0}^{\infty} \frac{(-1)^m x^{n+2m}}{2^{n+2m}(m)!\Gamma(m+n+1)} = J_n(x)$$

$$[\because \Gamma(m+n+1) = (m+n!)]$$

$$\therefore \quad e^{\left[\frac{1}{2}x\left(t-\frac{1}{t}\right)\right]} = \sum_{n=0}^{\infty} t^n J_n(x).$$

If taking the coefficient of t^{-n}, we get

$$= \frac{(-1)^n}{(n)!}\left(\frac{x}{2}\right)^n + \frac{(-1)^{n+1}}{(n+1)!}\left(\frac{x}{2}\right)^{n+2} + \frac{(-1)^{n+2}}{(n+2)!}\cdot\frac{1}{(2)!}\left(\frac{x}{2}\right)^{n+4} + \ldots$$

$$= (-1)^n\left[\frac{1}{(n)!}\left(\frac{x}{2}\right)^n - \frac{1}{(n+1)!}\left(\frac{x}{2}\right)^{n+2} + \frac{1}{(n+2)!}\cdot\frac{1}{(2)!}\left(\frac{x}{2}\right)^{n+4} - \ldots\right]$$

$$= (-1)^n \sum_{m=0}^{\infty} \frac{(-1)^m x^{n+2m}}{2^{n+2m}(m)!\Gamma(n+m+1)} = (-1)^n J_n(x) = J_{-n}(x) \qquad [\because J_{-n}(x) = (-1)^n J_n(x)]$$

Hence we obtain $e^{\left[\frac{1}{2}x\left(t-\frac{1}{t}\right)\right]} = \sum_{n=-\infty}^{\infty} t^n J_n(x)$.

35.10 ORTHOGONAL PROPERTIES OF BESSEL'S FUNCTION

$\int_0^1 xJ_n(\alpha x)\, J_n(\beta x)\, dx = 0$, *where* α, β *are the roots of* $J_n(x) = 0$. (VTU-2006, 13, UPTU(Q. BANK)-2002)

Proof. We know that $$x^2\frac{d^2y}{dx^2} + x\frac{dy}{dx} + (\alpha^2x^2 - n^2)y = 0 \qquad \ldots(1)$$

and $$x^2\frac{d^2z}{dx^2} + x\frac{dz}{dx} + (\beta^2x^2 - n^2)z = 0 \qquad \ldots(2)$$

where, $y = J_n(\alpha x)$ and $z = J_n(\beta x)$

Multiplying (1) by z/x and (2) by $-y/x$ and then add, we get

$$x\left[z\frac{d^2y}{dx^2} - y\frac{d^2z}{dx^2}\right] + \left[z\frac{dy}{dx} - y\frac{dz}{dx}\right] + (\alpha^2 - \beta^2)xyz = 0 \;\Rightarrow\; \frac{d}{dx}\left[x\left(z\frac{dy}{dx} - y\frac{dz}{dx}\right)\right] + (\alpha^2 - \beta^2)xyz = 0 \qquad \ldots(3)$$

On integrating from 0 to 1, we get

$$\left[x\left(z\frac{dy}{dx} - y\frac{dz}{dx}\right)\right]_0^1 + (\alpha^2 - \beta^2)\int_0^1 xyz\, dx = 0$$

$$\Rightarrow \quad (\beta^2 - \alpha^2)\int_0^1 xyz\, dx = \left[x\left(z\frac{dy}{dx} - y\frac{dz}{dx}\right)\right]_0^1 = \left[z\frac{dy}{dx} - y\frac{dz}{dx}\right]_{x=1} \qquad \ldots(4)$$

Now, $y = J_n(\alpha x) \Rightarrow \dfrac{dy}{dx} = \alpha J_n'(\alpha x)$, $z = J_n(\beta x) \Rightarrow \dfrac{dz}{dx} = \beta J_n'(\beta x)$

Putting these values in (4), we get

$$(\beta^2 - \alpha^2)\int_0^1 x\, J_n(\alpha x)\, J_n(\beta x) dx = \left[\alpha J_n'(\alpha x)\, J_n(\beta x) - \beta J_n'(\beta x)\, J_n(\alpha x)\right]_{x=1}$$

$$= \alpha J_n'(\alpha)\, J_n(\beta) \ldots \beta J_n'(\beta)\, J_n(\alpha) \qquad \ldots(5)$$

Now, since α, β are the roots of $J_n(x) = 0 \Rightarrow J_n(\alpha) = J_n(\beta) = 0$.

Putting these values in (5), we get

$$(\beta^2 - \alpha^2)\int_0^1 x\, J_n(\alpha x).J_n(\beta x)\, dx = 0.$$

Hence, $\int_0^1 x\, J_n(\alpha x)\, J_n(\beta x)\, dx = 0$.

35.11 RECURRENCE RELATIONS FOR $J_n(X)$

(I) $\boldsymbol{x J_n'(x) = nJ_n(x) - xJ_{n+1}(x)}$**, where** $\boldsymbol{J_n'(x) = \dfrac{dJ_n(x)}{dx}}$

Proof. Since we have $$J_n(x) = \sum_{m=0}^{\infty} (-1)^m \frac{1}{(m)!\,\Gamma(n+m+1)}\cdot\left(\frac{x}{2}\right)^{n+2m} \qquad \ldots(1)$$

Differentiating (1) w.r.t. x, we get

$$J_n'(x) = \sum_{m=0}^{\infty} \frac{(-1)^m\,(n+2m)}{(m)!\,\Gamma(n+m+1)}\cdot\frac{1}{2}\cdot\left(\frac{x}{2}\right)^{n+2m-1}$$

or $$xJ_n'(x) = \sum_{m=0}^{\infty} \frac{(-1)^m (n+2m)}{(m)!\Gamma(n+m+1)} \cdot \left(\frac{x}{2}\right)^{n+2m} = \sum_{m=0}^{\infty} \frac{(-1)^m (n+2m)}{(m)!\Gamma(n+m+1)} \cdot \left(\frac{x}{2}\right)^{n+2m}$$

$$= \sum_{m=0}^{\infty} \frac{(-1)^m \, n}{(m)!\,\Gamma(n+m+1)} \cdot \left(\frac{x}{2}\right)^{n+2m} + \sum_{m=0}^{\infty} \frac{(-1)^m \, 2m}{(m)!\,\Gamma(n+m+1)} \cdot \left(\frac{x}{2}\right)^{n+2m}$$

$$= n\,J_n(x) + \sum_{m=0}^{\infty} \frac{(-1)^m \,.\, 2}{(m-1)!\,\Gamma(n+m+1)} \cdot \frac{x}{2} \cdot \left(\frac{x}{2}\right)^{n-1+2m}$$

$$= n\,J_n(x) + x \sum_{m=0}^{\infty} \frac{(-1)^m}{(m-1)!\,\Gamma(n+m+1)} \cdot \left(\frac{x}{2}\right)^{n-1+2m}$$

$$= n\,J_n(x) + x \sum_{m=1}^{\infty} \frac{(-1)^m}{(m-1)!\,\Gamma(n+m+1)} \cdot \left(\frac{x}{2}\right)^{n-1+2m} \qquad \left(\because \frac{1}{(-1)!} = 0\right)$$

$$= n\,J_n(x) + x \sum_{k=0}^{\infty} \frac{(-1)^{k+1}}{(k)!\,\Gamma(n+1+k+1)} \cdot \left(\frac{x}{2}\right)^{n+1+2k} = nJ_n(x) - xJ_{n+1}(x)$$

$$\therefore \quad x\,J_n'(x) = n\,J_n(x) - x\,J_{n+1}(x)$$

REMARK

- $\frac{d}{dx}(x^{-n} J_n) = -x^{-n} J_{n+1}$

(II) $xJ_n'(x) = -nJ_n(x) + xJ_{n-1}(x)$. (UPTU-2004, 06)

Proof. Since we have $J_n(x) = \sum_{m=0}^{\infty} \frac{(-1)^m}{(m)!\,\Gamma(n+m+1)} \cdot \left(\frac{x}{2}\right)^{n+2m}$ ...(1)

Differentiating (1) w.r.t. x, we get

$$J_n'(x) = \sum_{m=0}^{\infty} \frac{(-1)^m (n+2m)}{(m)!\Gamma(n+m+1)} \cdot \frac{1}{2} \cdot \left(\frac{x}{2}\right)^{n+2m-1}$$

or $$xJ_n'(x) = \sum_{m=0}^{\infty} \frac{(-1)^m (n+2m)}{(m)!\,\Gamma(n+m+1)} \cdot \left(\frac{x}{2}\right)^{n+2m} = \sum_{m=0}^{\infty} \frac{(-1)^m \,(2n+2m-n)}{(m)!\,\Gamma(n+m+1)} \cdot \left(\frac{x}{2}\right)^{n+2m}$$

$$= -n \sum_{m=0}^{\infty} \frac{(-1)^m}{(m)!\,\Gamma(n+m+1)} \cdot \left(\frac{x}{2}\right)^{n+2m} + \sum_{m=0}^{\infty} \frac{(-1)^m \, 2(n+m)}{(m)!\,\Gamma(n+m+1)} \cdot \left(\frac{x}{2}\right)^{n+2m}$$

$$= -n\,J_n(x) + \sum_{m=0}^{\infty} \frac{(-1)^m \,.\, 2}{(m)!\,\Gamma(n+m)} \cdot \frac{x}{2} \cdot \left(\frac{x}{2}\right)^{n+2m-1}$$

$$= -n\,J_n(x) + x \sum_{m=0}^{\infty} \frac{(-1)^m}{(m)!\,\Gamma(n-1+m+1)} \cdot \left(\frac{x}{2}\right)^{n-1+2m} = -n\,J_n(x) + xJ_{n-1}(x)$$

$$\therefore \quad xJ_n'(x) = -nJ_n(x) + xJ_{n-1}(x).$$

REMARK

- $\frac{d}{dx}(x^n J_n) = x^n J_{n-1}$

(III) $2J_n'(x) = J_{n-1}(x) - J_{n+1}(x)$. (SVTU-2007, Madras-2006, JNTU-2006, PTU-2005, Anna-2005, VTU-2005)

Proof. From recurrence relations I and II, we have

$$x\,J_n'(x) = n\,J_n(x) - x\,J_{n+1}(x) \qquad ...(1)$$

$$x\,J_n'(x) = -n\,J_n(x) + x\,J_{n-1}(x) \qquad ...(2)$$

Adding (1) and (2), we get

$$2x\,J_n'(x) = xJ_{n-1}(x) - xJ_{n+1}(x)$$

$$\therefore \quad 2J_n'(x) = J_{n-1}(x) - J_{n+1}(x)$$

(IV) $2nJ_n(x) = x[J_{n-1}(x) + J_{n+1}(x)]$.

Proof. From recurrence relations I and II, we have

$$x J_n'(x) = n J_n(x) - x J_{n+1}(x) \quad \ldots(1)$$

$$x J_n'(x) = -n J_n(x) + x J_{n-1}(x) \quad \ldots(2)$$

From (1) and (2), we get

$$nJ_n(x) - xJ_{n+1}(x) = -nJ_n(x) + xJ_{n-1}(x)$$

$$\therefore \quad 2n J_n(x) = x[J_{n-1}(x) + J_{n+1}(x)].$$

(V) $\frac{d}{dx}[x^{-n}J_n(x)] = -x^{-n}J_{n+1}(x)$. (UPTU-2003, BPTU-2005, PTU-2006)

Proof.

$$\text{LHS} = \frac{d}{dx}[x^{-n}J_n(x)] = x^{-n}J_n'(x) - nx^{-n-1}J_n(x) = x^{-n-1}[xJ_n'(x) - nJ_n(x)]$$

$$= x^{-n-1}[-x J_{n+1}(x)] \quad \text{(from recurrence relation I)}$$

$$= -x^{-n} J_{n+1}(x) = \text{RHS}$$

$$\therefore \quad \frac{d}{dx}[x^{-n}J_n(x)] = -x^{-n} J_{n+1}(x)$$

(VI) $\frac{d}{dx}[x^n J_n(x)] = x^n J_{n-1}(x)$. (UPTU-2005, VTU-2005, Bhopal-2008,)

Proof. Consider

$$\text{LHS} = \frac{d}{dx}[x^n J_n(x)] = x^n J_n'(x) + nx^{n-1}J_n(x) = x^{n-1}[xJ_n'(x) + nJ_n(x)]$$

$$= x^{n-1}[x J_{n-1}(x)] \quad \text{(From recurrence relation II)}$$

$$= x^n J_{n-1}(x) = \text{RHS}$$

$$\therefore \quad \frac{d}{dx}[x^n J_n(x)] = x^n J_{n-1}(x).$$

Solved Examples

Example 1. *Show that $J_n(x)$ is even and odd function for even n and for odd n respectively.*

Solution. Since we have

$$J_n(x) = \sum_{m=0}^{\infty} \frac{(-1)^m}{(m)!\Gamma(m+n+1)} \cdot \left(\frac{x}{2}\right)^{n+2m} \quad \ldots(1)$$

Putting $-x$ in place of x, we get

$$J_n(-x) = \sum_{m=0}^{\infty} \frac{(-1)^m}{(m)!\Gamma(m+n+1)} \cdot \left(-\frac{x}{2}\right)^{n+2m}$$

$$= \sum_{m=0}^{\infty} \frac{(-1)^m}{(m)!\Gamma(n+m+1)} \cdot (-1)^{n+2m} \cdot \left(\frac{x}{2}\right)^{n+2m}$$

$$= (-1)^n \sum_{m=0}^{\infty} \frac{(-1)^m}{(m)!\Gamma(n+m+1)} \cdot \left(\frac{x}{2}\right)^{n+2m}$$

$$= (-1)^n J_n(x).$$

(i) If n is even, then $(-1)^n = 1$

$\therefore$ $J_n(-x) = J_n(x)$

$\therefore$ $J_n(x)$ is even.

(ii) If n is odd, then $(-1)^n = -1$

$\therefore$ $J_n(-x) = -J_n(x)$

$\therefore$ $J_n(x)$ is odd.

Example 2. *Show that $J_0'(x) = -J_1(x)$.*

Solution. From recurrence relation (1). we have

$$x J_n'(x) = nJ_n(x) - x J_{n+1}(x).$$

Putting $n = 0$, we get $x J_0'(x) = -x J_1(x)$.

$\therefore$ $J_0'(x) = -J_1(x)$.

Example 3. *Prove the following using Generating function for $J_n(x)$.*

(i) $\cos(x\sin\theta) = J_0 + 2J_2\cos 2\theta + 2J_4\cos 4\theta + \ldots$ (Kerala(M.Tech)-2005)

(ii) $\sin(x\sin\theta) = 2J_1\sin\theta + 2J_3\sin 3\theta + \ldots$ (Anna-2005)

(iii) $\cos(x\cos\theta) = J_0 - 2J_2\cos 2\theta + 2J_4\cos 4\theta - \ldots$

(iv) $\sin(x\cos\theta) = 2J_1\cos\theta - 2J_3\cos 3\theta + \ldots$

(v) $\cos x = J_0 - 2J_2 + 2J_4 - \ldots$

(vi) $\sin x = 2J_1 - 2J_3 + 2J_5 - \ldots$

Solution. Since we have $e^{\left[\frac{1}{2}x\left(t-\frac{1}{t}\right)\right]} = \sum_{n=-\infty}^{\infty} t^n J_n(x)$

$$= J_0 + \left(t - \frac{1}{t}\right)J_1 + \left(t^2 + \frac{1}{t^2}\right)J_2 + \left(t^3 - \frac{1}{t^3}\right)J_3 + \ldots \quad \ldots(1)$$

Let us put $t = e^{i\theta}$, then

$$t^n = e^{in\theta} = \cos n\theta + i\sin n\theta$$

$$t^{-n} = e^{-in\theta} = \cos n\theta - i\sin n\theta$$

$$\therefore\ t^n - \frac{1}{t^n} = 2i\sin n\theta \text{ and } t^n + \frac{1}{t^n} = 2\cos n\theta,$$

$n = 1, 2, 3, \ldots$

From (1), we have

$$e^{i(x\sin\theta)} = J_0 + (2i\sin\theta)J_1 + (2\cos 2\theta)J_2 + (2i\sin 3\theta)J_3 + \ldots$$

or $\cos(x\sin\theta) + i\sin(x\sin\theta)$

$$= (J_0 + 2J_2\cos 2\theta + 2J_4\cos 4\theta + \ldots) + i(2J_1\sin\theta + 2J_3\sin 3\theta + \ldots)$$

Separate real and imaginary parts, we get

(i) $\cos(x\sin\theta) = J_0 + 2J_2\cos 2\theta + 2J_4\cos 4\theta + \ldots$

and (ii) $\sin(x\sin\theta) = 2J_1\sin\theta + 2J_3\sin 3\theta + \ldots$

Now putting $\frac{\pi}{2} - \theta$ in place of θ in (i) and (ii), we get

(iii) $\cos\left(x\sin\left(\frac{\pi}{2} - \theta\right)\right) = J_0 + 2J_2\cos 2\left(\frac{\pi}{2} - \theta\right) + 2J_4\cos 4\left(\frac{\pi}{2} - \theta\right) + \ldots$

$$\therefore \cos(x\cos\theta) = J_0 + 2J_2\cos(\pi - 2\theta) + 2J_4\cos(2\pi - 4\theta) + \ldots$$

$$= J_0 - 2J_2\cos 2\theta + 2J_4\cos 4\theta + \ldots$$

(iv) $\sin\left(x\sin\left(\frac{\pi}{2} - \theta\right)\right)$

$$= 2J_1\sin\left(\frac{\pi}{2} - \theta\right) + 2J_3\sin 3\left(\frac{\pi}{2} - \theta\right) + \ldots$$

$$\therefore\ \sin(x\cos\theta) = 2J_1\cos\theta - 2J_3\cos 3\theta + \ldots$$

Now putting $\theta = \pi/2$ in (i) and (ii), we get

(v) $\cos\left(x\sin\frac{\pi}{2}\right)$

$$= J_0 + 2J_2\cos\pi + 2J_4\cos 2\pi + \ldots$$

$$\therefore\ \cos x = J_0 - 2J_2 + 2J_4 - \ldots$$

(vi) and

$$\sin\left(x\sin\frac{\pi}{2}\right) = 2J_1\sin\frac{\pi}{2} + 2J_3\sin\frac{3\pi}{2} + \ldots$$

$$\sin x = 2J_1 - 2J_3 + \ldots$$

Example 4. *Prove that* $J_2'(x) = \left(1 - \frac{4}{x^2}\right)J_1(x) + \frac{2}{x}J_0(x)$, *where* $J_n(x)$ *is the Bessel's function of first kind.*

(UPTU-2001, 08)

Solution. $xJ_n' = -nJ_n + xJ_{n-1}$ (Recurrence Formula) ...(1)

On putting $n = 2$ in (1), we have

$$xJ_2' = -2J_2 + xJ_1$$

$$\Rightarrow\quad J_2' = -\frac{2}{x}J_2 + J_1 \quad \ldots(2)$$

$$xJ_n' = n\,J_n - xJ_{n+1} \quad \text{(Recurrence relation)} \quad \ldots(3)$$

From (1) and (3), we have

$$-nJ_n + xJ_{n-1} = nJ_n - xJ_{n+1}$$

On putting $n = 1$,

$$-J_1 + xJ_0 = J_1 - xJ_2$$

$$-\frac{1}{x}J_1 + J_0 = \frac{1}{x}J_1 - J_2$$

$$\Rightarrow\quad J_2 = \frac{2}{x}J_1 - J_0 \quad \ldots(4)$$

Putting the value of J_2 from (4) in (2), we get

$$J_2' = -\frac{2}{x}\left(\frac{2}{x}J_1 - J_0\right) + J_1$$

$$= -\frac{4}{x^2}J_1 + \frac{2}{x}J_0 + J_1$$

$$= \left(1 - \frac{4}{x^2}\right)J_1 + \frac{2}{x}J_0$$

Example 5. *Prove that*

$$\frac{d}{dx}[J_n^2 + J_{n+1}^2] = 2\left(\frac{n}{x}J_n^2 - \frac{n+1}{x}J_{n+1}^2\right).$$

(UPTU-2005, Q. Bank-2002, VTU-2000, 06)

Solution. LHS $= \frac{d}{dx}[J_n^2 + J_{n+1}^2]$

$$= 2J_n\,J_n' + 2J_{n+1}\,J_{n+1}' \quad \ldots(1)$$

From recurrence relation (1), we have

$$x\,J_n' = n\,J_n - xJ_{n+1}$$

$$\therefore\quad J_n' = \frac{n}{x}J_n - J_{n+1} \quad \ldots(2)$$

From recurrence relation (2), we have

$x\,J_n' = -n\,J_n + xJ_{n-1}$ or $J_n' = -\frac{n}{x}J_n + J_{n-1}$.

Putting $(n+1)$ in place of n, we get

$$J_{n+1}' = -\frac{n+1}{x}J_{n+1} + J_n \quad \ldots(3)$$

Substitute the value of J_n' and J_{n+1}' from (2) and (3) into (1), we get

$$\text{LHS} = 2J_n\left[\frac{n}{x}J_n - J_{n+1}\right] + 2\left[-\frac{n+1}{x}J_{n+1} + J_n\right]J_{n+1}$$

$$= 2\frac{n}{x}J_n^2 - 2J_n\,J_{n+1} - 2\frac{(n+1)}{x}J_{n+1}^2 + 2J_nJ_{n+1}$$

$$= 2\left(\frac{n}{x}J_n^2 - \frac{n+1}{x}J_{n+1}^2\right) = \text{R.H.S.}$$

Hence $\frac{d}{dx}[J_n^2 + J_{n+1}^2] = 2\left(\frac{n}{x}J_n^2 - \frac{n+1}{x}J_{n+1}^2\right).$

Example 6. *Prove that* $\frac{d}{dx}[x J_n J_{n+1}] = x\left(J_n^2 - J_{n+1}^2\right).$

Solution. LHS $= \frac{d}{dx}(xJ_nJ_{n+1})$

$$= xJ_nJ'_{n+1} + xJ'_n J_{n+1} + J_nJ_{n+1} \quad \ldots(1)$$

From recurrence relations I and II, we have

$$xJ'_n = nJ_n - xJ_{n+1} \quad \ldots(2)$$

and $xJ'_n = -nJ_n + xJ_{n-1} \quad \ldots(3)$

Putting $(n + 1)$ in place of n in (3), we get

$$xJ'_{n+1} = -(n+1)J_{n+1} + xJ_n \quad \ldots(4)$$

Substitute the values of xJ'_n and xJ'_{n+1} from (2) and (4) into (1), we get

$$\text{LHS} = J_n[-(n+1)J_{n+1} + xJ_n] + J_{n+1}[nJ_n - xJ_{n+1}] + J_nJ_{n+1}$$

$$= -nJ_nJ_{n+1} - J_nJ_{n+1} + xJ_n^2 + nJ_nJ_{n+1} - xJ_{n+1}^2 + J_nJ_{n+1}$$

$$= xJ_n^2 - xJ_{n+1}^2 = x(J_n^2 - J_{n+1}^2) = \text{RHS}$$

Hence, $\frac{d}{dx}[xJ_nJ_{n+1}] = x[J_n^2 - J_{n+1}^2].$

Example 7. *Prove the following relation :*

$$x^2J''_n(x) = (n^2 - n - x^2)J_n(x) + xJ_{n+1}(x)$$

(UPTU-2006, 07, Q. Bank-2002)

Solution.
$$x^2\frac{d^2y}{dx^2} + x\frac{dy}{dx} + (x^2 - n^2)y = 0 \quad \ldots(1)$$

(Bessel's equation)

Clearly, $J_n(x)$ is the solution of (1)

So, $x^2J''_n + xJ'_n + (x^2 - n^2)J_n = 0 \quad \ldots(2)$

We know that

$$xJ'_n = nJ_n - xJ_{n+1} \quad \text{(Recurrence relation)} \quad \ldots(3)$$

Putting the value of xJ'_n from (3) in (2), we get

$$x^2J''_0 = -nJ_n + xJ_{n+1} + (n^2 - x^2)J_n$$

$$x^2J'' = (n^2 - n - x^2)J_n + xJ_{n+1}.$$

Example 8. *Prove the followings :*

(i) $J_{1/2}(x) = \sqrt{\frac{2}{\pi x}}.\sin x$ (JNTU-2003, VTU-2009)

(ii) $J_{-1/2}(x) = \sqrt{\frac{2}{\pi x}}.\cos x$

(VTU-2003, Anna-2005, WBTU-2005)

Solution. (i) Since we have

$$J_n(x) = \frac{x^n}{2^n\Gamma(n+1)}\left[1 - \frac{x^2}{2(2n+2)} + \frac{x^4}{2.4(2n+2)(2n+4)}\ldots\right] \quad \ldots(1)$$

Putting $n = 1/2$ in (1), we get

$$J_{1/2}(x) = \frac{x^{1/2}}{2^{1/2}\Gamma\left(1+\frac{1}{2}\right)}\left[1 - \frac{x^2}{2.3} + \frac{x^4}{2.4.3.5} - \ldots\right]$$

$$= \sqrt{\frac{x}{2}}.\frac{1}{\frac{1}{2}\Gamma(1/2)}\left[1 - \frac{x^2}{(3)!} + \frac{x^4}{(5)!} - \ldots\right]$$

$$= \sqrt{\frac{2}{x}}.\frac{1}{\Gamma(1/2)}\left[x - \frac{x^3}{(3)!} + \frac{x^5}{(5)!} - \ldots\right]$$

$$= \sqrt{\frac{2}{\pi x}}.\sin x$$

$$\left(\because \Gamma\left(\frac{1}{2}\right) = \sqrt{\pi} \text{ and } \sin\theta = \theta - \frac{\theta^3}{(3)!} + \frac{\theta^5}{(5)!} - \ldots\right)$$

(ii) Putting $n = -1/2$ in (1), we get

$$J_{-1/2}(x) = \frac{x^{-1/2}}{2^{-1/2}\Gamma\left(1-\frac{1}{2}\right)}\left[1 - \frac{x^2}{1.2} + \frac{x^4}{1.2.3.4} - \ldots\right]$$

$$= \sqrt{\frac{2}{x}}.\frac{1}{\Gamma\left(\frac{1}{2}\right)}\left[1 - \frac{x^2}{(2)!} + \frac{x^4}{(4)!} - \ldots\right]$$

$$= \sqrt{\frac{2}{\pi x}}\left[1 - \frac{x^2}{(2)!} + \frac{x^4}{(4)!} - \ldots\right]$$

$$= \sqrt{\frac{2}{\pi x}}.\cos x$$

$$\left(\because \cos\theta = 1 - \frac{\theta^2}{(2)!} + \frac{\theta^4}{(4)!} - \ldots\right)$$

$$\therefore J_{-1/2}(x) = \sqrt{\frac{2}{\pi x}}.\cos x.$$

Example 9. *Prove the following :*

$$J_3(x) + 3J_0(x) + 4J_0'''(x) = 0.$$

(UPTU(B.Tech.)-2001, Q. Bank-2002)

Solution. We know that $2J'_n = J_{n-1} - J_{n+1}.$

Differentiating and multiplying by 2, we get

$$\begin{aligned} 2^2 J_n'' &= 2J_{n-1}' - 2J_{n+1}' \\ &= (J_{n-2} - J_n) - (J_n - J_{n+2}) \\ &= J_{n-2} - 2J_n + J_{n+2}. \end{aligned}$$

Differentiating again and multiplying by 2, we get

$$\begin{aligned} 2^3 J_n''' &= 2J_{n-2}' - 4J_n' + 2J_{n+2}' \\ &= (J_{n-3} - J_{n-1}) - 2(J_{n-1} - J_{n+1}) + (J_{n+1} - J_{n+3}) \\ &= J_{n-3} - 3J_{n-1} + 3J_{n+1} - J_{n+3}. \end{aligned}$$

Putting $n = 0$, we get

$$\begin{aligned} 2^3 J_0''' &= J_{-3} - 3J_{-1} + 3J_1 - J_3 \\ &= (-1)^3 J_3 - 3(-1)J_1 + 3J_1 - J_3 \\ &= -2J_3 + 6J_1 \\ 4J_0''' &= -J_3 - 3J_1 = -J_3 + 3(-J_0') \\ &= J_3 + 3J_0' + 4J_0''' = 0 \end{aligned}$$

$$\Rightarrow \quad J_3(x) + 3J_0'(x) + 4J_0'''(x) = 0$$

Example 10. *Prove that*

(*i*) $[J_{1/2}(x)]^2 + [J_{-1/2}(x)]^2 = \dfrac{2}{\pi x}$ (Delhi-2002)

(*ii*) $J_{-3/2}(x) = -\sqrt{\dfrac{2}{\pi x}}\left(\dfrac{1}{x}\cos x + \sin x\right)$

Solution. (i) In Example 6, we have proved that

$$J_{1/2}(x) = \sqrt{\frac{2}{\pi x}}\,.\sin x$$

and $J_{-1/2}(x) = \sqrt{\dfrac{2}{\pi x}}\,.\cos x$

Squaring these and add, we get

$$[J_{1/2}(x)]^2 + [J_{-1/2}(x)]^2 = \frac{2}{\pi x}(\sin^2 x + \cos^2 x) = \frac{2}{\pi x}$$

(ii) Since we know that

$$2nJ_n(x) = x\,[J_{n-1}(x) + J_{n+1}(x)]$$

or $J_{n-1}(x) = \dfrac{2n}{x} J_n(x) - J_{n+1}(x)$

Now, putting $n = -1/2$, we get

$$\begin{aligned} J_{-3/2}(x) &= \frac{2\left(-\frac{1}{2}\right)}{x} J_{-1/2} - J_{1/2}(x) \\ &= -\frac{1}{x} J_{-1/2}(x) - J_{1/2}(x) \quad \ldots(1) \end{aligned}$$

Putting the value of $J_{1/2}(x) = \sqrt{\dfrac{2}{\pi x}}\,.\sin x$

and $J_{-1/2}(x) = \sqrt{\dfrac{2}{\pi x}}\,.\cos x$ into (1),

we get $J_{-3/2}(x) = -\sqrt{\dfrac{2}{\pi x}}\left(\dfrac{1}{x}\cos x + \sin x\right)$

Example 11. *Prove that* $\lim\limits_{x\to 0} \dfrac{J_n(x)}{x^n} = \dfrac{1}{2^n \Gamma(n+1)}, n > -1.$

Solution. Since we know that

$$J_n(x) = \frac{x^n}{2^n \Gamma(n+1)}\left[1 - \frac{x^2}{2(2n+2)} + \frac{x^4}{2.4(2n+2)(2n+4)} \ldots\right]$$

or

$$\frac{J_n(x)}{x^n} = \frac{1}{2^n \Gamma(n+1)}\left[1 - \frac{x^2}{2(2n+2)} + \frac{x^4}{2.4(2n+2)(2n+4)} \ldots\right]$$

Taking limit of both sides as $x \to 0$

$$\therefore \lim_{x\to 0} \frac{J_n(x)}{x^n} = \frac{1}{2^n \Gamma(n+1)} \lim_{x\to 0}\left[1 - \frac{x^2}{2(2n+2)} + \frac{x^4}{2.4(2n+2)(2n+4)} \ldots\right]$$

$$= \frac{1}{2^n \Gamma(n+1)}\,.1 = \frac{1}{2^n \Gamma(n+1)}$$

Example 12. *Prove that* $J_0^2 + 2(J_1^2 + J_2^2 + J_3^2 + \ldots) = 1$ *and deduce that* $|J_0(x)| \le 1, |J_n(x)| \le \dfrac{1}{\sqrt{2}}, n \ge 1.$

(UPTU-2003, 08, Kerala(M.Tech)-2005, VTU-2003)

Solution. Since we know that

$$e^{\left[\frac{1}{2}x\left(t-\frac{1}{t}\right)\right]}.e^{-\left[\frac{1}{2}x\left(t-\frac{1}{t}\right)\right]} = 1 \quad \ldots(1)$$

$$\sum_{n=\infty}^{\infty} t^n J_n(x).\sum_{n=-\infty}^{\infty} J_n(-x).t^n = 1$$

(by generating function)

$$\therefore \left[\sum_{n=-\infty}^{\infty} t^n J_n(x)\right].\left[\sum_{n=-\infty}^{\infty} (-1)^n J_n(x).t^n\right] = 1$$

$$[\because J_n(-x) = (-1)^n J_n(x)]$$

or

$$\begin{aligned} &[\ldots + J_2(x).t^{-2} - J_1(x).t^{-1} + J_0(x) + tJ_1(x) + t^2.J_2(x)\ldots] \\ &\times[\ldots + t^{-2}J_2(x) + t^{-1}J_1(x) + J_0(x) - tJ_1(x) + t^2 J_2(x) - \ldots] = 1 \end{aligned}$$

Comparing the common terms of both sides, we get

$$J_0^2(x) + 2J_1^2(x) + 2J_2^2(x) + \ldots = 1$$

or $\qquad J_0^2 + 2(J_1^2 + J_2^2 + J_3^2 + \ldots) = 1$

Further, since, if x is real and we have all the terms in the left hand side of above obtained result. Then,

$|J_0(x)| \le 1$ and $2|J_n(x)|^2 \le 1$ or $|J_n(x)| \le \dfrac{1}{\sqrt{2}}.$

Example 13. *Prove that*

$$x = 2J_0J_1 + 6J_1J_2 + \ldots + 2(2n+1)\, J_nJ_{n+1} + \ldots$$

Solution. Since we have a result

$$\frac{d}{dx}(xJ_nJ_{n+1}) = x(J_n^2 - J_{n+1}^2) \qquad \ldots(1)$$

Putting $n = 0, 1, 2, 3, 4, \ldots$, we get

$$\frac{d}{dx}(xJ_0J_1) = x(J_0^2 - J_1^2) \qquad \ldots(2)$$

$$\frac{d}{dx}(xJ_1J_2) = x(J_1^2 - J_2^2) \qquad \ldots(3)$$

$$\frac{d}{dx}(xJ_2J_3) = x(J_2^2 - J_3^2) \qquad \ldots(4)$$

$$\frac{d}{dx}(xJ_3J_4) = x(J_3^2 - J_4^2)$$

and so on.

Now, multiplying (2), (3), (4), ... by 1, 3, 5,... respectively and adding, we get

$$\frac{d}{dx}[x(J_0J_1 + 3J_1J_2 + 5J_2J_3 + \ldots)]$$

$$= x[J_0^2 + 2(J_1^2 + J_2^2 + \ldots)] = x[1]$$

$$[\because J_0^2 + 2(J_1^2 + J_2^2 + \ldots) = 1]$$

$$\therefore \frac{d}{dx}[x(J_0J_1 + 3J_1J_2 + 5J_2J_3 + \ldots)] = x\,.$$

Integrating both sides, we get

$$x(J_0J_1 + 3J_1J_2 + 5J_2J_3 + \ldots) = \frac{x^2}{2} + A$$

When $x = 0 \Rightarrow A = 0$

$$\therefore x(J_0J_1 + 3J_1J_2 + 5J_2J_3 + \ldots) = \frac{x^2}{2}$$

$$\therefore x = 2J_0J_1 + 6J_1J_2 + 10J_2J_3 + \ldots$$

or $x = 2J_0J_1 + 6J_1J_2 + \ldots + 2(2n+1)J_nJ_{n+1} + \ldots$

35.12 BESSEL'S INTEGRAL

We have

(a) $J_0(x) = \frac{1}{\pi}\int_0^\pi \cos(x\sin\theta)d\theta$ (Madras-2006) (b) $J_n(x) = \frac{1}{\pi}\int_0^\pi \cos(n\theta - x\sin\theta)d\theta$ (VTU-2006)

Proof. It is known that $\cos(x\sin\theta) = J_0 + 2J_2\cos 2\theta + 2J_4\cos 4\theta + \ldots$ ...(1)

and $\sin(x\sin\theta) = 2J_1\sin\theta + 2J_3\sin 3\theta + 2J_5\sin 5\theta + \ldots$ ...(2)

On integrating (1) between 0 to π, we get

$$\int_0^\pi \cos(x\sin\theta)d\theta = \int_0^\pi (J_0 + 2J_2\cos 2\theta + 2J_4\cos 4\theta + \ldots)d\theta = J_0\int_0^\pi d\theta + 2J_2\int_0^\pi \cos 2\theta d\theta + 2J_4\int_0^\pi \cos 4\theta d\theta + \ldots$$

$$= J_0(\pi) + 0 + 0 + \ldots$$

Hence, $J_0 = \frac{1}{\pi}\int_0^\pi \cos(x\sin\theta)d\theta$

Similarly, multiplying (1) by $\cos n\theta$ and integrating between 0 to π, we get

$$\int_0^\pi \cos(x\sin\theta)\cos n\theta d\theta = \int_0^\pi (J_0\cos n\theta + 2J_2\cos 2\theta + 2J_4\cos 4\theta\cos n\theta + \ldots)d\theta$$

$$= 2J_0\int_0^\pi \cos n\theta d\theta + 2J_2\int_0^\pi \cos 2\theta\cos n\theta d\theta + \ldots = \begin{cases} 0; & \text{if } n \text{ is odd} \\ \pi J_n; & \text{if } n \text{ is even} \end{cases} \qquad \ldots(3)$$

Further, multiplying (2) by $\sin n\theta$ and integrating between 0 to π, we get

$$\int_0^\pi \sin(x\sin\theta)\sin n\theta d\theta = \int_0^\pi (2J_1\sin\theta\sin n\theta + 2J_3\sin 3\theta\sin n\theta + \ldots)d\theta$$

$$= 2J_1\int_0^\pi \sin\theta\sin n\theta d\theta + 2J_3\int_0^\pi \sin 3\theta\,\sin n\theta\, d\theta + \ldots = \begin{cases} 0; & \text{if } n \text{ is even} \\ \pi J_n; & \text{if } n \text{ is odd} \end{cases} \qquad \ldots(4)$$

Using (3) and (4), we get

$$\int_0^\pi (\cos(x\sin\theta)\cos n\theta + \sin(x\sin\theta)\sin n\theta)d\theta = \pi J_n$$

or $\int_0^\pi \cos(n\theta - x\sin\theta)d\theta = \pi J_n$ or $J_n = \frac{1}{\pi}\int_0^\pi \cos(n\theta - x\sin\theta)\, d\theta$

35.13 FOURIER-BESSEL EXPANSION

Let $f(x)$ be a function, which is continuous and has a finite number of oscillations in the interval $0 \le x \le a$, then $f(x)$ can be expanded as follows

$$f(x) = C_1J_n(\alpha_1 x) + C_2J_n(\alpha_2 x) + C_3J_n(\alpha_3 x) + \ldots + C_nJ_n(\alpha_n x) + \ldots$$

or

$$f(x) = \sum_{i=1}^{\infty} C_i\, J_n(\alpha_i x)$$

where, $\alpha_1, \alpha_2, \ldots$ are the roots of the equation $J_n(x) = 0$.

Proof. It is given that $f(x) = \sum_{i=1}^{\infty} C_iJ_n(\alpha_i x)$ ...(1)

Multiplying (1) by $xJ_n(\alpha_j x)$, we get $f(x)\, J_n(\alpha_j x) = \sum_{i=1}^{\infty} C_i x\, J_n(\alpha_j x)\, J_n(\alpha_i x)$ ...(2)

On integrating from $x = 0$ to a, we get

$$\int_0^a xJ_n(\alpha_i x)\, J_n(\alpha_j x)\, dx = \begin{cases} 0\ ; & \text{if } i \neq j \\ \dfrac{a^2}{2} J_{n+1}^2(\alpha_i a)\ ; & \text{if } i = j \end{cases} \quad \text{...(4)}$$

Using (4) in (3), we get

$$\int_0^a x\, f(x)\, J_n(\alpha_i x)\, dx = C_i . \frac{a^2}{2}\, J_{n+1}^2(\alpha_i a) \qquad \Rightarrow C_i = \frac{2\int_0^a x\, f(x)\, J_n(\alpha_i x)\, dx}{a^2 . J_{n+1}^2\, (\alpha_i a)}$$

Putting these values in (1), we get $f(x) = C_1 J_n(\alpha_1 x) + C_2 J_n(\alpha_2 x) + C_3 J_n(\alpha_3 x) + ... + C_n J_n(\alpha_n x) + ...$

35.14 BER AND BEI FUNCTION

Consider the differential equation

$$x\frac{d^2y}{dx^2} + \frac{dy}{dx} - ix\, y = 0 \quad \text{...(1)}$$

which occurs in certain problems of electrical engineering.

We have $y = J_0(kx) = J_0[(-i)^{1/2} x] = J_0(i^{3/2} x)$

Replacing $i^{3/2}$ in the series for $J_0(x)$, we get

$$y = 1 - \frac{i^3 x^2}{2^2} + \frac{i^6 x^4}{(2!)^2 2^4} - \frac{i^9 x^6}{(3!)^2 2^6} + \frac{i^{12} x^8}{(4!)^2 2^8} - ...$$

$$= \left[1 - \frac{x^4}{2^2 . 4^2} + \frac{x^8}{2^2 . 4^2 . 6^2 . 8^2} - ...\right] + i\left[\frac{x^2}{2^2} - \frac{x^6}{2^2 . 4^2 . 6^2} + \frac{x^{10}}{2^2 . 4^2 . 6^2 . 8^2 . 10^2} - ...\right] \quad \text{...(2)}$$

which is complex for x real. The series in the above brackets are taken to define Bessel-real (or ber) and Bessel-imaginary (or bei) functions.

Thus $$\text{ber } x = 1 + \sum_{m=1}^{\infty} (-1)^m \frac{x^{4m}}{2^2 . 4^2 . 6^2 ... (4m)^2} \quad \text{...(3)}$$

and $$\text{bei } x = -\sum_{m=1}^{\infty} (-1)^m \frac{x^{4m-2}}{2^2 . 4^2 . 6^2 ...... (4m-2)^2}$$

So that $y = \text{ber } x + i \text{ bei } x$ is a solution of (1).

Solved Examples

Example 1. *Prove that* $\sum_{i=1}^{\infty} \frac{2J_0(a_i x)}{a_i J_1(a_i i)} = 1$, *where* a_i *are the roots of* $J_0(x)$. (UPTU(Q. Bank)-2002)

Solution. We have

$$f(x) = \sum_{i=1}^{\infty} C_i J_n(a_i x) \quad \text{...(1)}$$

$$C_i = \frac{2}{J_{n+1}^2(a_i)} \int_0^1 xJ_n(a_i x)\, f(x)\, dx \quad \text{...(2)}$$

Putting $f(x) = 1$ and $n = 0$ in (1), we get

$$1 = \sum_{i=1}^{\infty} C_i\, J_0(a_i x)$$

$$\therefore\ C_i = \frac{2}{J_1^2(a_i)} \int_0^1 xJ_0(a_i x) dx$$

$$= \frac{2}{J_1^2(a_i)} \left(\frac{J_1(a_i)}{a_i}\right) = \frac{2}{a_i J_1(a_i)}$$

Then using these values in (1), we get

$$1 = \sum_{i=1}^{\infty} \frac{2}{a_i\, J_1(a_i)}\, J_0(a_i x).$$

Hence, $\sum \frac{2J_0(a_i x)}{a_i J_1(a_i)} = 1$

Example 2. *Expand* $f(x) = x^2$ *in the interval* $0 < x < 2$ *in terms of* $J_\alpha(\alpha_n x)$, *where* α_n *are the roots of* $J_2(2\alpha_n) = 0$.

Solution. Here, we have

$$f(x) = x^2$$

i.e., $$x^2 = \sum_{i=1}^{\infty} C_i\, J_2(\alpha_i x) \quad \text{...(1)}$$

Multiplying (1) by $nJ_2(\alpha_j x)$, we get

$$x^3 J_2(\alpha_j x) = \sum_{i=1}^{\infty} xJ_2(\alpha_i x) J_2(\alpha_j x) \quad \text{...(2)}$$

On integrating from $x = 0$ to 2, we get

$$\int_0^2 x^3 J_2(\alpha_j x) dx = \sum_{i=1}^{\infty} C_i \int_0^2 xJ_2(\alpha_i x) J_2(\alpha_j x) dx$$

$$\Rightarrow \left[\frac{x^3 J_3(\alpha_i x)}{\alpha_i}\right]_0^2 = C_i \int_0^2 x J_2^2(\alpha_i x)dx$$

$$\Rightarrow \frac{8J_3(2\alpha_i)}{\alpha_i} = C_i \frac{2^2}{2} J_3^2(2\alpha_i)$$

$$\Rightarrow C_i = \frac{8J_3(2\alpha_i)}{\alpha_i} \cdot \frac{2}{4J_3^2(2\alpha_i)} = \frac{4}{\alpha_i J_3(2\alpha_i)}$$

Putting this value in (1), we get

$$x^2 = \sum_{i=1}^{\infty} \frac{4J_\alpha(\alpha_i x)}{\alpha_i J_3(2\alpha_i)}$$

EXERCISE 35.3

1. Prove that
   (i) $\int_0^x x^n J_{n-1}(x)dx = x^n J_n(x)$
   (ii) $\int_0^x x^{n+1} J_n(x)\,dx = x^{n+1} J_{n+1}(x)$
2. Prove that $\int_0^{\pi/2} \sqrt{\pi x}\, J_{1/2}(2x)\,dx = 1$.
3. Prove that $J_0(x) = \frac{1}{\pi}\int_0^\pi \cos(x\sin\phi)\,d\phi$.
4. Show that if $n > -1$, $\int_0^x x^{-n} J_{n+1}(x)\,dx = \frac{1}{2^n \Gamma(n+1)} - x^{-n} J_n(x)$.
5. Prove that $4J_n''(x) = J_{n-2}(x) - 2J_n(x) + J_{n+2}(x)$.
6. Prove that $J_n J_{-n}' - J_{-n} J_n' = -\frac{2\sin n\pi}{\pi x}$.

   Hence deduce that $\frac{d}{dx}\left[\frac{J_{-n}}{J_n}\right] = -\frac{2\sin n\pi}{\pi x J_n^2}$.
7. Prove that
   (i) $J_2 = J_0'' - \frac{1}{x} J_0'$ (ii) $J_2 - J_0 = 2J_0''$
8. Prove that $J_{n+3} + J_{n+5} = \frac{2}{x}(n+4)\,J_{n+4}$.
9. Prove that $J_n' = \frac{2}{x}\left[\frac{n}{2} J_n - (n+2)J_{n+2} + (n+4)\,J_{n+4} - \ldots\right]$.
10. Prove that $\int J_{n+1}(x)\,dx = \int J_{n-1}(x)\,dx - 2J_n(x) + A$.
11. Prove that $J_{n-1} = \frac{2}{x}[nJ_n - (n+2)\,J_{n+2} + (n+4)\,J_{n+4} - \ldots]$ and hence deduce that

    $\frac{1}{2} x J_n = (n+1)\,J_{n+1} - (n+3)J_{n+3} + (n+5)\,J_{n+5} - \ldots$
12. Prove that
    (i) $J_{3/2}(x) = \sqrt{\frac{2}{\pi x}}\left[\frac{1}{x}\sin x - \cos x\right]$
    (ii) $J_{-5/2}(x) = \sqrt{\frac{2}{\pi x}}\left[\left(\frac{3-x^2}{x^2}\right)\cos x + \frac{3}{x}\sin x\right]$ (UPTU(Q. Bank)-2002)
    (iii) $J_{5/2}(x) = \sqrt{\frac{2}{\pi x}}\left[\left(\frac{3-x^2}{x^2}\right)\sin x - \frac{3}{x}\cos x\right]$
13. Show that $J_n(x) = \frac{1}{\pi}\int_0^\pi \cos(n\theta - x\sin\theta)\,d\theta$, where n is positive integer.
14. Prove that $\int_0^\infty e^{-ax} J_0(bx)dx = \frac{1}{\sqrt{a^2+b^2}}, a > 0$.
15. Show that $\int xJ_0^2(x)dx = \frac{1}{2}x^2\left[J_0^2(x) + J_1^2(x)\right] + C$. (UPTU-2003, 04)
16. Show that $4J_0'''(x) + 3J_0'(x) + J_3(x) = 0$. (UPTU-2002, Osmania-2003, 11)
17. Show that $J_0(x) = \frac{1}{\pi}\int_0^\pi \cos(x\cos\phi)\,d\phi$. (UPTU(Q. Bank)-2002)
18. Show that the solution of the differential equation

    $$\frac{d^2y}{dx^2} + \left(9x - \frac{20}{x^2}\right)y = 0$$

    in terms of Bessel's function is given by

    $y = \left[\sqrt{x}\,(C_1 J_3(2x^{3/2}) + C_2 Y_3(2x^{3/2})\right]$ (UPTU-2002)
19. Show that the solution of the differential equation

    $y'' + \frac{y'}{x}\left(8 - \frac{1}{x^2}\right)y = 0$ is given by

    $y = C_1 J_2(4\sqrt{2}x) + C_2 Y_2(4\sqrt{2}.x)$ (UPTU(Q. Bank)-2002)
20. Show that the solution of the differential equation

    $$y'' + \frac{y'}{x} + 4\left(x^2 - \frac{n^2}{x^2}\right)y = 0$$

    is given by $y = C_1 J_n(x^2) + C_2 Y_n(x^2)$.

Hint to Selected Problems

1. Integrate both sides from 0 to x of the following relation

   $\frac{d}{dx}\int x^n J_n(x) = x^n J_{n-1}(x)$.
2. $J_{1/2}(x) = \sqrt{\frac{2}{\pi x}}.\sin x \Rightarrow J_{1/2}(2x) = \sqrt{\frac{2}{2\pi x}}.\sin 2x$

   $\Rightarrow \sqrt{\pi x}.\,J_{1/2}(2x) = \sin 2x$. Now solve.
3. Integrating both sides w.r.t. ϕ from $\phi = 0$ to π of the following relation $\cos(x\sin\phi) = J_0 + 2J_2\cos 2\phi + 2J_4\cos 4\phi + \ldots$
4. Integrating the relation $\frac{d}{dx}[x^{-n}J_n(x)] = -x^{-n}J_{n+1}^{(x)}$
5. Differentiating $2J_n'(x) = J_{n-1}(x) - J_{n+1}(x)$ and then replace $n-1$ and $n+1$ in place of n.
7. Put $n = 0$ in the recurrence relation $xJ_n' = nJ_n - xJ_{n+1}$.
8. Use recurrence relation $2n\,J_n = x(J_{n-1} + J_{n+1})$.
9. Replacing n by $n+2, n+4, n+6, \ldots$ in $J_{n-1} + J_{n+1} = \frac{2n}{x} J_n$.
13. Using $\cos(x\sin\theta) = J_0 + 2J_2\cos 2\theta + \ldots + 2J_{2m}\cos 2m\theta \ldots$ and $\sin(x\sin\theta) = 2\sin\theta.J_1 + 2\sin 3\theta J_3 + \ldots + 2J_{2m+1}\sin(2m+1)\theta + \ldots$
14. We have $J_0(x) = \frac{1}{\pi}\int_0^\pi \cos(x\sin\theta)d\phi \Rightarrow J_0(6x) = \frac{1}{\pi}$

    Now multiplying by e^{-9x} and then integrate from $x = 0$ to ∞.

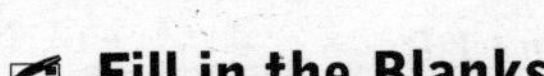

Objective Evaluations

Fill in the Blanks

1. The solution of Legendre's differential equation is known as ________.

2. $x =$ ________ is a regular point of Legendre's differential equation.

3. $P_n(x)$, the Legendre polynomial has a degree ________ if n is even.

4. $\dfrac{1}{\sqrt{1-2xt+t^2}} = \sum_{n=0}^{\infty} t^n$ ________.

5. If $\int_{-1}^{1} P_m\, P_n\, dx = \begin{cases} 0, & m \neq n \\ \dfrac{2}{2n+1}, & m = n \end{cases}$, then property is known as ________.

6. $|P_n(x)| <$ ________ if $-1 < x < 1$.

7. $\int_{-1}^{1} P_n(x) dx =$ ________, if $n \neq 0$.

8. $\int_{-1}^{1} P_0(x) dx =$ ________.

9. $P_n(-1) =$ ________.

10. $P_n(-x) =$ ________.

11. $P_n(1) =$ ________.

12. $\int_{-1}^{1} x^{10} P_{100}(x) dx =$ ________.

13. All the roots of $P_n(x) = 0$ are ________.

14. All the roots of $P_n(x) = 0$ lie between ________.

15. The differential equation of the type

$$x^2 \frac{d^2y}{dx^2} + x\frac{dy}{dx} + (x^2 - n^2)y = 0 \text{ is known as ________.}$$

16. $x =$ ________ is a singular point for Bessel's differential equation.

17. The solution of Bessel's differential equation is denoted by ________.

18. $\dfrac{d}{dx}[x^n\, J_n] = x^n$ ________.

19. $\text{Exp}\left[\dfrac{x\left(t - \dfrac{1}{t}\right)}{2}\right] = \sum_{n=-\infty}^{\infty} t^n$ ________.

20. $J_{-n}(x)(x) = (-1)^n$ ________.

21. $J_n(x)$ is even function of n is ________.

22. $J_0'(x) =$ ________

True/False

Write 'T' for True and 'F' for False statement.

1. The differential equation $\dfrac{d^2y}{dx^2} + P(x)\dfrac{dy}{dx} + Q(x)y = 0$ is homogeneous. **(T/F)**

2. $x = -1$ is an ordinary point for Legendre's differential equation. **(T/F)**

3. $x = 1$ is a singular point of Legendre's differential equation. **(T/F)**

4. $J_0'(x) = -J_1'(x)$. **(T/F)**

5. $P_n(x)$ is not a solution of Legendre's differential equation. **(T/F)**

6. The equation $P_n(x) = 0$ has all roots real. **(T/F)**

7. $P_n(x)$ is a polynomial of degree n. **(T/F)**

8. $P_n(1) = 0$. **(T/F)**

9. $P_n(x)$ is an odd function if n is even. **(T/F)**

10. $P_2(-x) = P_2(x)$. **(T/F)**

11. $P_n(0) = 0$ if n is odd. **(T/F)**

12. $P_n(x) = \dfrac{1}{(n)!\, 2^n} \dfrac{d^n}{dx^n}(x^2 - 1)^n$. **(T/F)**

13. $\int_{-1}^{1} P_m P_n dx = 0$, if $m \neq n$. **(T/F)**

14. $P_3(x) = \dfrac{5x^3 - 3x^2}{2}$ **(T/F)**

15. $2P_2(x) = 3[P_1(x)]^2 - 1$ **(T/F)**

16. $\int_{-1}^{1} P_n(x) dx = 0, n \neq 0$ **(T/F)**

17. $\int_{-1}^{1} x^m P_n(x) = 0$ if $m < n$. **(T/F)**

18. All the roots of $P_n(x) = 0$ lie between -1 and $+1$. **(T/F)**

19. $\dfrac{d}{dx}[x^{-n} J_n] = x^{-n} J_{n+1}$ **(T/F)**

20. $J_{-n}(x) = (-1)^n J_{n+1}$ **(T/F)**

21. $[J_{1/2}(x)]^2 + [J_{-1/2}(x)]^2 = \dfrac{2}{\pi x}$ **(T/F)**

22. $|J_0(x)| \leq 1,\ n \geq 1$ **(T/F)**

23. $|J_n(x)| \geq 2^{-1/2},\ n \geq 1$ **(T/F)**

24. $\lim_{x \to 0} \dfrac{J_n(x)}{x^n} = \dfrac{1}{2^n\, \Gamma(n+1)}$, $n > -1$. **(T/F)**

Multiple Choice Questions

Choose the most appropriate one.

1. The ordinary point for differential equation $\dfrac{d^2y}{dx^2} + y = 0$:

(a) $[-1,1]$ (b) R (set of real nos.)

(c) 0 (d) 1

2. Singular point of the Legendre's differential equation is :

(a) 0 (b) 2

(c) 1 (d) none of these

3. Singular point of the Bessel's differential equation is
(a) 0 (b) −1
(c) 1 (d) 2

4. All the roots of $P_n(x) = 0$ lies between :
(a) −2 and −1 (b) −1 and 0
(c) −1 and +1 (d) 0 and 1

5. All the roots of $P_n(x) = 0$ are :
(a) real (b) some reals
(c) 0 (d) complex

6. $P_n(x)$ is an even function if n equals :
(a) −1 (b) 0
(c) 3 (d) 4

7. $P_n(x)$ is a polynomial of odd degree if n equals :
(a) 3 (b) 2
(c) 6 (d) none of these

8. $P_1(x)$ equals :
(a) $\dfrac{x^2}{2}$ (b) x
(c) 1 (d) $-x$

9. $P_n'(x) - P_{n-2}'(x)$ equals :
(a) $(2n+1)P_n(x)$ (b) $2nP_{n-1}(x)$
(c) $(2n-1)P_{n-1}(x)$ (d) $P_n(x)$

10. $(xP_9'(x) - P_8'(x) / P_9(x)$ equals :
(a) 9 (b) 8
(c) $P_8(x)$ (d) x

11. The value of the integral $\int_{-1}^{1}(1-x^2)P_m' P_n' \, dx$:
(a) 1 (b) 0
(c) −1 (d) 2

12. The value of the integral $\int_{-1}^{1}[P_n(x)]^2 \, dx$:
(a) $\dfrac{1}{2n+1}$ (b) $\dfrac{1}{n}$
(c) $\dfrac{2}{2n+1}$ (d) $\dfrac{2}{2n-1}$

13. The value of the integral $\int_{-1}^{1} x^3 P_4(x) \, dx$:
(a) −1 (b) 1
(c) 2 (d) 0

14. $\dfrac{d}{dx}\left[(1-x^2)\dfrac{dP_n}{dx}\right]$ equals :
(a) $-n(n+1)P_n$ (b) $n(n+1)P_n$
(c) $(n+1)P_n$ (d) nP_n

15. $J_0(x)$ is a Bessel's function of order :
(a) 1 (b) n
(c) 0 (d) 2

16. $P_n(0)$ equals, if n is odd :
(a) 0 (b) 1
(c) −1 (d) none of these

17. $x[J_{n-1} + J_{n+1}]$ equals :
(a) $2nJ_{n-1}$ (b) nJ_n
(c) $2nJ_n$ (d) $2nJ_{n+1}$

18. $(-1)^n J_n(x)$ equals :
(a) $J_n(x)$ (b) $J_{-n}(x)$
(c) $J_{n-1}(x)$ (d) $J_{n+1}(x)$

19. Magnitude of $J_n(x)$ is less than or equals to :
(a) $\sqrt{2}$ (b) $1/\sqrt{2}$
(c) 2 (d) 1/2

20. The value of the integral $\int_0^x x^n J_{n-1}(x) \, dx$:
(a) $x^n J_n(x)$ (b) $J_n(x)$
(c) x^n (d) $x^{-n}J_n(x)$

21. $J_2(x) - J_0(x)$ equals :
(a) $J_0''(x)$ (b) $2J_0'(x)$
(c) $2J_0''(x)$ (d) $2J_0(x)$

22. The value of the integral $\int_0^{\pi/2} \sqrt{\pi x}\, J_{1/2}(2x) \, dx$:
(a) 0 (b) −1
(c) 1 (d) none of these

ANSWERS

Fill in the Blanks

1. Legendre's function	**2.** −1, 1	**3.** even	**4.** $P_n(x)$	**5.** orthogonal	**6.** 1
7. 0	**8.** 2	**9.** $(-1)^n$	**10.** $(-1)^n P_n(x)$		**11.** 1
12. 0	**13.** real	**14.** −1 and +1	**15.** Bessel's differential equation		**16.** 0
17. $J_n(x)$	**18.** $J_{n-1}(x)$	**19.** $J_n(x)$	**20.** $J_n(x)$	**21.** even	**22.** $J_1(x)$

True/False

1. T	**2.** F	**3.** T	**4.** T	**5.** F	**6.** T	**7.** T	**8.** F	**9.** F	**10.** T
11. T	**12.** T	**13.** T	**14.** F	**15.** T	**16.** T	**17.** T	**18.** T	**19.** T	**20.** T
21. F	**22.** F	**23.** T	**24.** T	**25.** F	**26.** T				

Multiple Choice Questions

1. (b)	**2.** (c)	**3.** (a)	**4.** (c)	**5.** (a)	**6.** (d)	**7.** (a)	**8.** (b)	**9.** (c)	**10.** (a)
11. (b)	**12.** (c)	**13.** (d)	**14.** (a)	**15.** (c)	**16.** (a)	**17.** (c)	**18.** (b)	**19.** (c)	**20.** (b)
21. (a)	**22.** (c)	**23.** (c)	**24.** (a)						

FFFFFF

36 The Laplace Transform

36.1 INTRODUCTION

An integral of the type $\int_{-\infty}^{\infty} k(p,t)F(t)dt$ is defined as the integral transform of $F(t)$, provided it is convergent. It is denoted by $f(p)$ or $T[F(t)]$.

$$f(p) = TF(t) = \int_{-\infty}^{\infty} k(p,t)F(t)dt$$

REMARK

- The function $k(p,t)$ appearing in the integral is called kernel of the transform. Here p is a parameter and is independent of $t.p$ may be real or complex number.

Definition 1. *If F(t) be a function of t defined for all values of t, then Laplace transform of f(t), denoted by L* $\{F(t)\}$ *or f(p) is defined by*

$$L\{F(t)\} = f(p) = \int_0^{\infty} e^{-pt}\ F(t)\ dt \qquad \ldots(1)$$

REMARKS

- If the integral (1) converges for some value of p, then only the Laplace transform of $f(t)$ exists otherwise not.
- L is called Laplace transform operator.

Definition 2. *A function f(x) is said to be exponential order a as* $x \to \infty$ *if* $\lim_{x\to\infty} e^{-ax} f(x) = a$ *finite quantity, i.e., for a given positive integer n, if a real number M such that* $|e^{-ax}\ f(x)| < M,\ \forall\ x \geq n$ *which can be written as* $f(x) = O(e^{-ax}),\ x \to \infty$.

Definition 3. *A function* $f(x)$ *is called sectionally continuous over the closed interval* $x_1 \leq x \leq x_2$ *if the closed interval can be divided into a finite number of subintervals* $a \leq x \leq b$ *such that*

(i) $f(x)$ *is continuous in the closed interval (a, b).*

(ii) $\lim_{x\to a+0} f(x)$ *and* $\lim_{x\to b-0} f(x)$ *both exist.*

Definition 4. *A function which is sectionally (or piecewise) continuous over every finite interval in the range* $t \geq 0$ *and* ω *of exponential order as* $t \to \infty$ *is called a function of class A.*

36.2 LINEARITY PROPERTY

THEOREM. *The Laplace transformation is a linear transformation* $L\{a_1 F_1(t) + a_2 F_2(t)\} = a_1\ L\{F_1(t)\} + a_2\ L\{F_2(t)\}$ *if* a_1 *and* a_2 *be constants.*

Proof. We know that $\quad L = \{F(t)\} = \int_0^{\infty} e^{-pt} F(t)\ dt$.

Therefore,

$$L\{a_1F_1(t) + a_2F_1(t)\} = \int_0^{\infty} e^{-pt}\ \{a_1F_1(t) + a_2F_1(t)\}\ dt = a_1 \int_0^{\infty} e^{-pt}\ F_1(t)\ dt + a_2 \int_0^{\infty} e^{-pt} F_2(t)dt$$
$$= a_1\ L\{F_1(t)\} + a_2 L\{F_2(t)\}.$$

36.3 EXISTENCE OF LAPLACE TRANSFORM

THEOREM. *If F(t) is a function which is piecewise continuous on every finite interval in the range* $t \geq 0$ *and satisfies* $|F(t)| \leq M\ e^{at}$ *for all* $t \geq 0$ *and for constant a and M then the Laplace transform of f(t) exists for all* $p > a$.

Proof. We know that $L\{F(t)\} = \int_0^\infty e^{-pt} F(t)\,dt = \int_0^{t_0} F(t)e^{-pt}dt + \int_{t_0}^\infty F(t)e^{-pt}dt$ ...(1)

Now, $\int_0^{t_0} F(t)\,e^{-pt}dt$ exists since $F(t)$ is sectionally continuous on every finite interval $0 \le t \le t_0$

and $\left|\int_{t_0}^\infty F(t)\,e^{-pt}\,dt\right| \le \int_{t_0}^\infty |F(t)\,e^{-pt}|\,dt \le \int_{t_0}^\infty e^{-pt}\,M\,e^{at}\,dt$ $[\because |F(t)| \le Me^{at}]$

$$= \int_{t_0}^\infty e^{(a-p)t}\,M\,dt = M\left[\frac{e^{-(p-a)t}}{-(p-a)}\right]_{t_0}^\infty = \frac{M}{p-a}\,e^{-(p-a)t_0}\text{, if } p > a$$

$$\Rightarrow \left|\int_{t_0}^\infty F(t)\,e^{-pt}\,dt\right| \le \frac{M}{p-a}\,e^{-(p-a)t_0}\text{, if } p > a.$$

Now, $\dfrac{Me^{-(p-a)t_0}}{p-a}$ can be made small as we please by taking t_0 sufficiently large. Hence, from (1), we conclude that $L\{f(t)\}$ exists for all $p < a$.

REMARK

- The above conditions are sufficient but not necessary for the existance of the Laplace transform. If these conditions are satisfied, the Laplsce transform must exist. If these conditions are not satisfied, the Laplace transform may or may not exist.

36.4 LAPLACE TRANSFORM OF SOME ELEMENTARY FUNCTIONS

(i) $F(t) = 1$

Solution. We have $L\{F(t)\} = \int_0^\infty e^{-pt} F(t)\,dt$ ...(1)

Here $F(t) = 1$.

Therefore, from (1) $L\{1\} = \int_0^\infty e^{-pt}\,.\,1\,dt = \left[\frac{e^{-pt}}{p}\right]_0^\infty = \frac{1}{p},\ p > 0$

Hence, $L\{1\} = \dfrac{1}{p}$.

(ii) $F(t) = t^n$

Solution. We have $L\{F(t)\} = \int_0^\infty e^{-pt} F(t)\,dt$

$$\Rightarrow L\{t^n\} = \int_0^\infty e^{-pt}\,t^n\,dt = \int_0^\infty e^{-pt}\,.\,t^{(n+1)-1}\,dt$$

$$= \frac{\Gamma(n+1)}{p^{n+1}} = \frac{n!}{p^{n+1}},\ p > 0 \qquad \left[\because \int_0^\infty e^{-u}\,u^n du = \Gamma(n+1)\right]$$

Hence, $L\{t^n\} = \dfrac{n!}{p^{n+1}}$.

(iii) $F(t) = t$

Solution. We have $L\{F(t)\} = \int_0^\infty e^{-pt}\,.\,t\,dt = \left[-\frac{1}{p}\,t\,e^{-pt}\right]_0^\infty + \frac{1}{p}\int_0^\infty e^{-pt}dt = \frac{1}{p^2},\ p > 0$

(iv) $F(t) = e^{at}$

Solution. We have $L\{F(t)\} = \int_0^\infty e^{-pt}e^{at}\,dt = \int_0^\infty e^{-(p-a)t}\,dt$.

If $p \le a$, integral diverges. For $p > a$, the integral converges. Hence, for $p > a$,

$$L\{e^{at}\} = \int_0^\infty e^{-(p-a)t}\,dt = \left[-\frac{e^{-(p-a)t}}{p-a}\right]_0^\infty = 0 + \frac{1}{p-a} = \frac{1}{p-a},\ p > a.$$

(v) $F(t) = \sin at$

Solution. $L\{\sin at\} = \int_0^\infty e^{-pt}\,\sin at\,dt = \left[\frac{e^{-pt}(-p\sin at - a\cos at)}{p^2+a^2}\right]_0^\infty$ $\because \int e^{ax}\sin bx dx = e^{ax}\dfrac{[a\sin bx - b\cos bx]}{a^2+b^2}$

$$= \frac{a}{p^2+a^2},\ p > a$$

Hence, $L\{\sin at\} = \dfrac{a}{p^2+a^2}$.

(vi) $F(t) = \cos at$

Solution. We know that

$$\int e^{ax} \cos bx \, dx = \frac{e^{ax}(a\cos bx + b\sin bx)}{a^2 + b^2}$$

Therefore, we have

$$L\{\cos at\} = \int_0^\infty e^{-pt} \cos at \, dt = \left[\frac{e^{-pt}(-p\cos at + a\sin at)}{a^2 + p^2}\right]_0^\infty = \frac{p}{p^2 + a^2}, \; p > 0.$$

(vii) $F(t) = \sinh at$

Solution. Consider $L\{\sinh at\} = L\left\{\frac{e^{at} - e^{-at}}{2}\right\} = \frac{1}{2}L\{e^{at}\} - \frac{1}{2}L\{e^{-at}\}$ (Using (iv))

$$= \frac{1}{2}\cdot\frac{1}{p-a} - \frac{1}{2}\cdot\frac{1}{p+a} = \frac{a}{p^2 - a^2}$$

Hence, $L\{\sinh at\} = \dfrac{a}{p^2 - a^2}$.

(viii) $F(t) = \cosh at$

Solution. Consider

$$L\{\cosh at\} = L\left[\frac{1}{2}(e^{at} + e^{-at})\right] = \frac{1}{2}L\{e^{at}\} + \frac{1}{2}L\{e^{-at}\} = \frac{1}{2}\cdot\frac{1}{p-a} + \frac{1}{2}\cdot\frac{1}{p+a}, \; p > a \text{ and } p > -a$$

$$= \frac{p}{p^2 - a^2}, \; p > |a|$$

Hence, $L\{\cosh at\} = \dfrac{p}{p^2 - a^2}$.

Solved Examples

Example 1. *Find the Laplace transform of the function* $F(t) = \dfrac{e^{at} - 1}{a}$.

Solution. We have

$$L\{F(t)\} = L\left\{\frac{e^{at} - 1}{a}\right\} = L\left\{\frac{1}{a}e^{at} - \frac{1}{a}\right\}$$

$$= \frac{1}{a}L\{e^{at}\} - \frac{1}{a}L\{1\}$$

$$= \frac{1}{a}\left(\frac{1}{p-a}\right) - \frac{1}{a}\left(\frac{1}{p}\right) = \frac{1}{p(p-a)}.$$

Example 2. *Find* $L\{(t^2 + 1)^2\}$.

Solution.

$$L\{(t^2 + 1)^2\} = L\{t^4 + 2t^2 + 1\}$$

$$= L\{t^4\} + 2L\{t^2\} + L(1)$$

(By linearity property)

$$= \frac{4!}{p^5} + 2.\frac{2!}{p^3} + \frac{1}{p}$$

$$= \frac{24 + 4p^2 + p^4}{p^5}, \; p > 0.$$

Example 3. *Find* $L\{F(t)\}$ *where* $F(t) = (\sin t - \cos t)^2$.

Solution. Consider

$$L\{(\sin t - \cos t)^2\}$$

$$= L\{\sin^2 t + \cos^2 t - 2\sin t \cos t\}$$

$$= L\{1 - \sin 2t\} = L\{1\} - L\{\sin 2t\}$$

$$= \frac{1}{p} - \frac{2}{p^2 + 2^2}, \; p > 0 = \frac{p^2 - 2p + 4}{p(p^2 + 4)}, \; p > 0$$

Example 4. *Find* $L\{6\sin 2t - 5\cos 2t\}$.

Solution.

$$L\{6\sin 2t - 5\cos 2t\} = 6L\{\sin 2t\} - 5L\{\cos 2t\}$$

$$= 6.\frac{2}{p^2 + 2^2} - 5.\frac{p}{p^2 + 2^2}, \; p > 0$$

$$= \frac{12 - 5p}{p^2 + 4}, \; p > 0.$$

Example 5. *Find*

$$L\{7e^{2t} + 9e^{-2t} + 5\cos t + 7t^3 + 5\sin 3t + 2\}.$$

Solution.

$$L\{7e^{2t} + 9e^{-2t} + 5\cos t + 7t^3 + 5\sin 3t + 2\}$$

$$= 7L\{e^{2t}\} + 9L\{e^{-2t}\} + 5L\{\cos t\} + 7.L\{t^3\} + 5.L\{\sin 3t\} + 2.L\{1\}$$

$$= \frac{7}{p-2} + \frac{9}{p+2} + \frac{5p}{p^2+1} + \frac{4^2}{p^4} + \frac{15}{p^2+9} + \frac{q}{p}$$

$$= \frac{4(4p-1)}{p^2 - 4} + \frac{5p}{p^2+1} + \frac{42}{p^4} + \frac{15}{p^2+4} + \frac{q}{p}.$$

Example 6. *Find* $L\{2e^{3t} - e^{-3t}\}$.

Solution. $L\{2e^{3t} - e^{-3t}\} = 2L\{e^{3t}\} - L\{e^{-3t}\}$

$$= 2.\frac{1}{p-3} - \frac{1}{p+3},\ p > 3 \text{ and } p > -3$$

$$= \frac{p+9}{p^2-9},\ p > |3|.$$

Example 7. *Find* $L\{F(t)\}$, *if* $F(t) = \begin{cases} e^t, & 0 < t \le 1 \\ 0, & t > 1 \end{cases}$.

Solution. $L\{f(t)\} = \int_0^\infty e^{-pt} F(t)\, dt$

$$= \int_0^1 e^{-pt}.e^t\, dt + \int_1^\infty e^{-pt}.0\, dt$$

$$= \int_0^1 e^{-(p-1)t}.dt = \left[-\frac{e^{-(p-1)t}}{p-1}\right]_0^1$$

$$= \frac{1}{(p-1)}[1 - e^{-(p-1)}],\ p \ne 1.$$

Example 8. *Find* $L\{F(t)\}$, *where* $F(t) = \begin{cases} 0, & 0 < t < 1 \\ t, & 1 < t < 2 \\ 0, & t > 2 \end{cases}$.

(JNTU–2006, WBTU–2005)

Solution. We have that $F(t)$ is not defined at $t = 0, 1$ and 2.

$$\therefore L\{F(t)\} = \int_0^\infty e^{-pt} F(t)\, dt$$

$$= \int_0^1 e^{-pt}.0\, dt + \int_1^2 e^{-pt}.t\, dt + \int_2^\infty e^{-pt}.0\, dt$$

$$= \int_1^2 e^{-pt}.tdt = \left[-t\frac{e^{-pt}}{p} - \frac{e^{-pt}}{p^2}\right]_1^2, p \ne 0$$

$$= -\left(\frac{2}{p} + \frac{1}{p^2}\right)e^{-2p} + \left(\frac{1}{p} + \frac{1}{p^2}\right)e^{-p}, p \ne 0.$$

Example 9. *Find the Laplace transform of*

$$F(t) = \begin{cases} t^2, & 0 < t < 2 \\ t-1, & 2 < t < 3 \\ 7, & t > 3 \end{cases}$$

(UPTU–2007, Mumbai–2007)

Solution. $L[F(t)] = \int_0^\infty e^{-pt} F(t) dt$

$$= \int_0^2 t^2 e^{-pt} dt + \int_2^3 (t-1)e^{-pt} dt + \int_3^\infty 7e^{-pt} dt$$

$$= \left[t^2\frac{e^{-pt}}{(-p)} - 2t\frac{e^{-pt}}{(-p)^2} + 2\frac{e^{-pt}}{(-p)^3}\right]_0^2$$

$$+ \left[(t-1)\left(\frac{e^{-pt}}{(-p)}\right) - \frac{e^{-pt}}{(-p)^2}\right]_2^3 + 7\left[\frac{e^{-pt}}{-p}\right]_3^\infty$$

$$= \left[-4\left(\frac{e^{-2p}}{p}\right) - 4\left(\frac{e^{-2p}}{p^2}\right) + \frac{2}{p^3}\right]$$

$$+ \left[2\left(\frac{e^{-3p}}{-p}\right) - \left(\frac{e^{-3p}}{p^2}\right) + \left(\frac{e^{-2p}}{p}\right) + \frac{e^{-2p}}{p^2}\right]$$

$$+ 7\left[0 + \frac{e^{-3p}}{p}\right]$$

$$= \frac{2}{p^3} + e^{-2p}\left[-\frac{4}{p} - \frac{4}{p^2} - \frac{2}{p^3}\right]$$

$$+ e^{-3p}\left[-\frac{2}{p} - \frac{1}{p^2}\right]$$

$$+ e^{-2p}\left[\frac{1}{p} + \frac{1}{p^2}\right] + e^{-3p}\left[\frac{7}{p}\right]$$

$$= \frac{2}{p^3} + e^{-2p}\left[-\frac{4}{p} - \frac{4}{p^2} - \frac{2}{p^3} + \frac{1}{p} + \frac{1}{p^2}\right]$$

$$+ e^{-3p}\left[-\frac{2}{p} - \frac{1}{p^2} + \frac{7}{p}\right]$$

$$= \frac{2}{p^3} - \frac{e^{-2p}}{p^3}(2 + 3p + 3p^2) + \frac{e^{-3p}}{p^2}(5p - 1)$$

Example 10. *Find* $L\{F(t)\}$, *if* $F(t) = \begin{cases} 1, & 0 < t < 2 \\ t, & t > 2 \end{cases}$.

Solution. $L\{F(t)\} = \int_0^\infty F(t)\, e^{-pt}\, dt$

$$= \int_0^2 1.e^{-pt} dt + \int_2^\infty t\, e^{-pt}\, dt$$

$$= \left[\frac{e^{-pt}}{-p}\right]_0^2 + \left[\left(\frac{e^{-pt}}{-p}\right)t - \left(\frac{e^{-pt}}{p^2}\right).1\right]_2^\infty$$

$$= -\frac{e^{-2p}}{p} + \frac{1}{p} - \frac{1}{p}\lim_{t\to\infty}\frac{t}{e^{pt}} + \frac{2}{p}e^{-2p}$$

$$- \frac{1}{p^2}\lim_{t\to\infty} e^{-pt} + \frac{e^{-2p}}{p^2}$$

$$= \frac{1}{p}(1 + e^{-2p}) + \frac{1}{p^2}e^{-2p}$$

$$- \frac{1}{p}\lim_{t\to\infty}\frac{t}{e^{pt}} - \frac{1}{p^2}\lim_{t\to\infty} e^{-pt}.$$

Now, if $p > 0$, we have

$$\lim_{t\to\infty} e^{-pt} = 0 \text{ and } \lim_{t\to\infty}\frac{t}{e^{pt}} \qquad (\text{Form } \infty/\infty)$$

$$= \lim_{t\to\infty}\frac{1}{pe^{pt}} = 0$$

Hence, $L\{F(t)\} = \frac{1}{p}[1 + e^{-2p}] + \frac{1}{p^2}e^{-2p}, p > 0.$

Example 11. *Show that* $L\left\{\frac{1}{\sqrt{\pi t}}\right\} = \frac{1}{\sqrt{p}}$.

Solution. We have $L\left\{\frac{1}{\sqrt{\pi t}}\right\} = \int_0^\infty e^{-pt}.\frac{1}{\sqrt{\pi t}}.dt$

$$= \frac{1}{\sqrt{\pi}}\int_0^\infty e^{-pt}.\frac{1}{\sqrt{t}} dt$$

$$= \frac{1}{\sqrt{\pi}}\int_0^\infty e^{-pt}.t^{-1/2}\, dt$$

$$= \frac{1}{\sqrt{\pi}} \int_0^\infty e^{-pt} . t^{-1/2-1}\, dt$$

$$= \frac{1}{\sqrt{\pi}} \frac{\boxed{1/2}}{p^{1/2}}, \; p > 0$$

(Using gamma function)

$$= \frac{1}{\sqrt{\pi}} \cdot \frac{\sqrt{\pi}}{p^{1/2}} \quad \left[\because \Gamma(\tfrac{1}{2}) = \sqrt{\pi}\right]$$

$$= \frac{1}{\sqrt{p}}.$$

Example 12. *Show that* $L\left\{\frac{\cos\sqrt{t}}{\sqrt{t}}\right\} = \sqrt{\left(\frac{\pi}{p}\right)}\, e^{-1/4p}$. (Mumbai–2009)

Solution. Here, we have

$$\frac{\cos\sqrt{t}}{\sqrt{t}} = \frac{1}{\sqrt{t}}\left\{1 - \frac{1}{2!}(\sqrt{t})^2 + \frac{1}{4!}(\sqrt{t})^4 - \frac{1}{6!}(\sqrt{t})^6 + \ldots\right\}$$

$$= t^{-1/2} - \frac{1}{2!} t^{1/2} + \frac{1}{4!} t^{3/2} - \frac{1}{6!} t^{5/2} + \ldots$$

Therefore,

$$L\left\{\frac{\cos\sqrt{t}}{\sqrt{t}}\right\} = L\{t^{-1/2}\} - \frac{1}{2!} L\{t^{1/2}\} + \frac{1}{4!} L\{t^{3/2}\} - \frac{1}{6!} L\{t^{5/2}\} + \ldots$$

$$= \frac{\Gamma(\frac{1}{2})}{p^{1/2}} - \frac{1}{2!}\frac{\Gamma(\frac{3}{2})}{p^{3/2}} + \frac{1}{4!}\frac{\Gamma(\frac{5}{2})}{p^{5/2}} - \frac{1}{6!}\frac{\Gamma(\frac{7}{2})}{p^{7/2}} + \ldots, p > 0$$

$$= \frac{\sqrt{\pi}}{p^{1/2}} - \frac{1}{1.2} \cdot \frac{\frac{1}{2}.\sqrt{\pi}}{p^{3/2}} + \frac{\frac{3}{2}.\frac{1}{2}.\sqrt{\pi}}{1.2.3.4} \cdot \frac{1}{p^{5/2}} - \frac{\frac{5}{2}.\frac{3}{2}.\frac{1}{2}.\sqrt{\pi}}{1.2.3.4.5.6} \cdot \frac{1}{p^{7/2}} + \ldots$$

$$= \sqrt{\left(\frac{\pi}{p}\right)}\left[1 - \frac{1}{1!}\left(\frac{1}{4p}\right) + \frac{1}{2!}\left(\frac{1}{4p}\right)^2 - \frac{1}{3!}\left(\frac{1}{4p}\right)^3 + \ldots\right]$$

$$= \sqrt{\left(\frac{\pi}{p}\right)} . e^{-1/4p}.$$

EXERCISE 36.1

Find the Laplace transform of the following functions : (Ques. 1 to 11)

1. $\sin t \cos t$
2. $4\cos^2 2t$
3. $\sin^2 at$
4. $3\cosh 5t - 4\sinh 5t$ (Nagarjuna–2006)
5. $3t^4 - 2t^3 + 4e^{-3t} - 2\sin 5t + 3\cos 2t$
6. $e^{-2t} - e^{-3t}$
7. $F(t) = \begin{cases} \sin t, & 0 < t < \pi \\ 0, & t > \pi \end{cases}$ (Madras–2000S)
8. $F(t) = \begin{cases} (t-1)^2, & t > 1 \\ 0, & 0 < t < 1 \end{cases}$
9. $F(t) = \begin{cases} e^t, & 0 < t < 5 \\ 3, & t > 5 \end{cases}$
10. $F(t) = \begin{cases} t, & 0 < t < 4 \\ 5, & t > 4 \end{cases}$
11. $F(t) = \sin\sqrt{t}$
12. Show that t^2 is of exponential order 3.
13. Show that the function e^{t^2} is not of exponential order as $t \to \infty$.
14. Show that the Laplace transforms of the function $F(t) = t^n, -1 < n < 0$ exists, although it is not a function of class A.
15. Find the Laplace transform of $F(t) = \begin{cases} e^t, & 0 < t < 1 \\ 0, & t > 1 \end{cases}$. (UPTU–2004)

Hint to Selected Problems

1. The given function can be written as $F(t) = \sin t . \cos t = \frac{1}{2}\sin 2t$.
2. Lüüüüfüütüü±üLüüüü + t = L + L t .
7. $L\{F(t)\} = \int_0^\infty e^{-pt} F(t) dt = \int_0^\pi e^{-pt} . \sin t dt + \int_0^\infty e^{-pt} .0 dt$
8. $L\{F(t)\} = \int_0^\infty F(t) e^{-pt} dt = \int_0^1 0.e^{-pt} dt + \int_0^\infty (t-1)^2 e^{-pt} dt$.
11. $L\{\sin\sqrt{t}\} = L\left[\sqrt{t} - \frac{(\sqrt{t})^3}{3!} + \frac{(\sqrt{t})^5}{5!} - \frac{(\sqrt{t})^7}{7!} + \ldots\right]$

ANSWERS

1. $\frac{1}{p^2+4}, \; p > 0$
2. $\frac{4(p^2+8)}{p(p^2+16)}, \; p > 0$
3. $\frac{2a^2}{p(p^2+4a^2)}, \; p > 0$
4. $\frac{3p-20}{p^2-25}, \; p > 5$
5. $\frac{72}{p^5} - \frac{12}{p^4} + \frac{4}{p+3} - \frac{10}{p^2+25} + \frac{3p}{p^2+4}, \; p > 0$
6. $\frac{1}{p^2+5p+6}, \; p > -2$
7. $\frac{e^{-p\pi}+1}{p^2+1}$
8. $\frac{2e^{-p}}{p^3}, \; p > 0$
9. $\frac{1-e^{-5(p-1)}}{p-1} + \frac{3}{p}e^{-5p}, \; p > 0$
10. $\frac{1+(p-1)\,e^{-4p}}{p^2}, \; p > 0$
11. $\frac{\sqrt{\pi}}{2p^{3/2}}\, e^{-1/4p}$
15. $\frac{e^{1-p}-1}{1-p}$

36.5 TRANSLATION OR SHIFTING THEOREMS

THEOREM 1. **(First Translation or Shifting Theorem).** *If $f(p)$ is the Laplace transform of $F(t)$, then $f(p-a)$ is the Laplace transforms of e^{at} $F(t)$, i.e., if $L\{F(t)\} = f(P)$, when $p > a$, then $L\{e^{at} F(t)\} = f(p-a),\ p > a$.*

Proof. We have, by definition of Laplace transform

$$L\{F(t)\} = f(p) = \int_0^\infty e^{-pt} F(t)\, dt$$

Therefore, $$L\{e^{at} F(t)\} = \int_0^\infty e^{-pt}.e^{at} F(t)\, dt = \int_0^\infty e^{-(p-a)t}.F(t)\, dt = \int_0^\infty e^{-ut} F(t)\, dt,$$

where $$u = p - a > 0$$

$$= f(u) \qquad \text{(By definition)}$$

$$= f(p-a).$$

THEOREM 2. **(Second Translation or Heaviside's Shifting Theorem).**

If $L\{F(t)\} = f(p)$ and $G(t) = \begin{cases} F(t-a), & t > a \\ 0, & t < a \end{cases}$ then, $L\{G(t)\} = e^{-ap} f(p)$. (UPTU–2006, 08)

Proof. Let $L\{F(t)\} = f(p)$ and $G(t) = \begin{cases} F(t-a), & \text{if } t > a \\ 0, & \text{if } t < a \end{cases}$

Then, $$L\{G(t)\} = \int_0^\infty e^{-pt} G(t)\, dt$$

$$= \int_0^a e^{-pt} G(t)\, dt + \int_a^\infty e^{-pt} G(t)\, dt = \int_0^a e^{-pt}.0\, dt + \int_a^\infty e^{-pt} F(t-a)\, dt$$

$$= 0 + \int_a^\infty e^{-pt} F(t-a)\, dt.$$

Let $t - a = u$, therefore $dt = du$.

If $t = a$, then $u = t - a = a - a = 0$ and if $t = \infty$, then $u = \infty - a = \infty$.

Hence, $L\{G(t)\} = \int_0^\infty e^{-p(u+a)} F(u)\, du = e^{-pa} \int_0^\infty e^{-pu} F(u)\, du = e^{-pa} f(p)$.

THEOREM 3. **(Change of Scale Property).** *If $L\{F(t)\} = f(p)$, then $L\{F(at)\} = \frac{1}{a} f\left(\frac{p}{a}\right)$.*

Proof. By definition

$$L\{F(at)\} = \int_0^\infty e^{-pt} F(at)\, dt = \int_0^\infty e^{-pu/a} F(u) \frac{du}{a} \qquad (\text{where } at = u)$$

$$= \frac{1}{a}\int_0^\infty e^{-pu/a} F(u)\, du = \frac{1}{a}\int_0^\infty e^{-su} F(u)\, du, \text{ where } s = \frac{p}{a}$$

$$= \frac{1}{a} f(s) = \frac{1}{a} f\left(\frac{p}{a}\right).$$

Solved Examples

Example 1. *Find* $L\left\{\frac{e^{-at}\, t^{n-1}}{(n-1)!}\right\}$.

Solution. We have $L\left\{\frac{t^{n-1}}{(n-1)!}\right\} = \frac{1}{(n-1)!}.\frac{(n-1)!}{p^n} = \frac{1}{p^n}$.

Therefore, using first shifting theorem, we have

$$L\left\{e^{-at}\frac{t^{n-1}}{(n-1)!}\right\} = f(p+a) = \frac{1}{(p+a)^n}.$$

Example 2. *Find* $L\{e^t \cos^2 t\}$.

Solution. We have

$$L\{\cos^2 t\} = L\left\{\frac{1}{2}(1+\cos 2t)\right\}$$

$$= \frac{1}{2}[L\{1\} + L\{\cos 2t\}]$$

$$= \frac{1}{2}\left\{\frac{1}{p} + \frac{p}{p^2+2^2}\right\}$$

$$= \frac{p^2+2}{p(p^2+4)} = f(p) \text{ (say)}$$

Using first shifting theorem, we have

$$L\{e^t \cos^2 t\} = f(p-1) = \frac{(p-1)^2+2}{(p-1)(p-1)^2+4}$$

$$= \frac{p^2-2p+3}{(p-1)(p^2-2p+5)}$$

Example 3. *Find* $L\{e^{-t}(3\sin 2t - 5\cosh 2t)\}$.

Solution. We have

$$L\{3\sin 2t - 5\cosh 2t\}$$

$$= 3.\frac{2}{p^2+2^2} - \frac{5p}{p^2-2^2} = f(p) \text{ (say)}.$$

Using first shifting theorem, we have

$$L\{e^{-t}(3\sin 2t - 5\cosh 2t)\} = f(p+1)$$
$$= \frac{6}{(p+1)^2+4} - \frac{5(p+1)}{(p+1)^2-4}$$
$$= \frac{6}{p^2+2p+5} - \frac{5(p+1)}{p^2+2p-3}.$$

Example 4. *Find* $L\{e^{-t}(3\sinh 2t - 5\cosh 2t)\}$.

Solution. We have

$$L\{3\sinh 2t - 5\cosh 2t\}$$
$$= 3.\frac{2}{p^2-2^2} - 5\frac{p}{p^2-2^2}$$
$$= \frac{6-5p}{p^2-4} = f(p) \quad \text{(say)}$$

Using first shifting theorem, we have

$$L\{e^{-t}(3\sinh 2t - 5\cosh 2t)\} = f(p+1)$$
$$= \frac{6-5(p+1)}{(p+1)^2-4} = \frac{1-5p}{p^2+2p-3}.$$

Example 5. *Find* $L\{e^t(t+3)^2\}$.

Solution. We have

$$L\{(t+3)^2\} = L\{t^2+6t+9\}$$
$$= \frac{2!}{p^3} + 6.\frac{1!}{p^2} + \frac{9}{p}$$
$$= \frac{2+6p+9p^2}{p^3} = f(p) \quad \text{(say)}$$

Using first shifting theorem, we have

$$L\{(t+3)^2 e^t\} = f(p-1)$$
$$= \frac{2+6(p-1)+9(p-1)^2}{(p-1)^3}$$
$$= \frac{9p^2-12p+5}{(p-1)^3}.$$

Example 6. *If* $L\{\cos^2 t\} = \dfrac{p^2+2}{p(p^2+4)}$, *find* $L[\cos^2 at]$.

(UPTU–2006)

Solution. We have $L\{\cos^2 t\} = \dfrac{p^2+2}{p(p^2+4)}$

By change of scale property, we have

$$L\{\cos^2 at\} = \frac{1}{a}.\frac{\left(\frac{p}{a}\right)^2}{\left(\frac{p}{a}\right)\left[\left(\frac{p}{a}\right)^2+4\right]}$$

$$= \frac{1}{p}\left[\frac{p^2+2a^2}{\frac{p}{a}(p^2+4a^2)}\right] = \frac{p^2+2a^2}{p(p^2+4a^2)}.$$

Example 7. *Given* $L\{F(t)\} = \dfrac{p^2-p+1}{(2p+1)^2(p-1)}$.

Applying the change of scale property, show that

$$L\{F(2t)\} = \frac{p^2-2p+4}{4(p+1)(p-2)}.$$

Solution. Given that

$$L\{F(t)\} = \frac{p^2-p+1}{(2p+1)^2(p-1)} = f(p) \text{ (say)}$$

By using change of scale property, we have

$$L\{F(2t)\} = \frac{1}{2}f\left(\frac{p}{2}\right)$$

$$= \frac{1}{2}.\frac{\left(\frac{p}{2}\right)^2 - \left(\frac{p}{2}\right) + 1}{\left[2.\left(\frac{p}{2}\right)+1\right]^2\left(\frac{p}{2}-1\right)}$$

$$= \frac{p^2-2p+4}{4(p+1)^2(p-2)}.$$

Example 8. *Find* $L\{F(t)\}$, *where*

$$F(t) = \begin{cases} \cos\left(t - \frac{2}{3}\pi\right), & t > \frac{2\pi}{3} \\ 0, & t < \frac{2\pi}{3} \end{cases}$$

Solution. Let $F(t) = \cos t$

Then, $G(t) = \begin{cases} F\left(t - \frac{2\pi}{3}\right), & t > 2\pi/3 \\ 0, & t < 2\pi/3 \end{cases}$

We have $L\{F(t)\} = L\{\cos t\} = \dfrac{p}{p^2+1} = f(p)$ (say)

Using second shifting theorem, we have

$$L\{G(t)\} = e^{\left(-\frac{2\pi}{3}\right)p}.f(p) = e^{-2\pi p/3}.\frac{p}{p^2+1}.$$

Example 9. *Find* $L\{G(t)\}$, *where* $G(t) = \begin{cases} e^{t-a}, & t > a \\ 0, & t < a \end{cases}$.

(UPTU–2008)

Solution. By second shifting theorem, we have

$L\{F(t)\} = f(p)$ and $G(t) = \begin{cases} F(t-a), & t > a \\ 0, & t < a \end{cases}$

Then, $L\{G(t)\} = e^{-ap} f(p)$

Let $F(t) = e^t$

Then, $L\{F(t)\} = L\{e^t\} = \int_0^\infty e^{-pt}.e^t\,dt$

$$= \frac{1}{p-1}, p > 1 = f(p) \text{ (say)}$$

Now, let $G(t) = \begin{cases} F(t-a) = e^{t-a}, & t > a \\ 0, & t < a \end{cases}$

Then, $L\{G(t)\} = e^{-ap} f(p) = \dfrac{e^{-ap}}{p-1}, \; p > 1.$

EXERCISE 36.2

1. Find $L\{t^3 e^{-3t}\}$.
2. Find $L\{e^{3t}\cos 5t\}$.
3. Find $L\{e^{-t}\sin^2 t\}$. (Mumbai–2009)
4. Find $L\{e^{t}\sin^2 t\}$
5. Find $L\{e^{-4t}\cosh 2t\}$
6. Find $L\{e^{-2t}(3\cos 6t - 5\sin 6t)\}$
7. Using first shifting theorem, find the value of $L\{e^{6t}(t+2)^2\}$
8. If $L\{F(t)\} = f(p)$, find $L\{F(t)\cos\omega t\}$
9. Applying change of scale property, find
   (i) $L\{\sinh 3t\}$, (ii) $L\{\cos 5t\}$
10. Find $L\{F(t)\}$, where $F(t) = \begin{cases} \sin\left(t - \dfrac{\pi}{3}\right), & t > \pi/3 \\ 0, & t < \pi/3 \end{cases}$
11. Find $L\{F(t)\}$, where $F(t) = \begin{cases} \sin\left(t - \dfrac{2}{3}\pi\right), & t > 2\pi/3 \\ 0, & t < 2\pi/3 \end{cases}$
12. If $\{F(t)\} = \dfrac{1}{p}e^{-1/p}$, show that $L\{e^{-t}F(3t)\} = \dfrac{e^{-3/(p+1)}}{p+1}$.
13. Find the Laplace transform of $e^{t}\, t^{-1/2}$. (UPTU(Q. Bank)–2001)
14. Find the Laplace transform of :
    (i) $t^2 e^t \sin 4t$ (UPTU(Sp.)–2001)
    (ii) $t\, e^{-t}\sin 2t$ (UPTU(Sp.)–2002)
15. Find the Laplace transform of
$$F(t) = \begin{cases} 1, & 0 \le t < 1 \\ t, & 1 \le t < 2 \\ t^2, & 2 \le t < \infty \end{cases}$$ (UPTU(Q. Bank)–2001)

Hint to Selected Problems

1. $L\{t^3\} = \dfrac{3!}{p^4}$, then $L\{t^3 e^{-3t}\} = f(p+3) = \dfrac{6}{(p+3)^4}$

5. $L(\cosh 2t) = \dfrac{p}{p^2 - 2^2} = \dfrac{p}{p^2 - 4} = f(p)$, then apply first shifting theorem.

7. $L\{(t+3)^2\} = L\{t^2 + 6t + 9\} = \dfrac{2!}{p^3} + 6.\dfrac{1}{p^2} + \dfrac{9}{p} = f(p)$ (say),

then applying first shifting theorem.

9. (i) $L\{\sinh t\} = \dfrac{1}{(p^2-1)} = f(p)$, then by change of scale property
$$L\{\sinh 3t\} = \frac{1}{3} f\left(\frac{p}{3}\right) = \frac{1}{3}.\frac{1}{(p/3)^2 - 1} = \frac{3}{p^2 - 9}$$

10. Let $G(t) = \sin t$, then $F(t) = \begin{cases} G\left(t - \dfrac{\pi}{3}\right), & t > \pi/3 \\ 0, & t < \pi/3 \end{cases}$

Then, $L\{G(t)\} = L\{\sin t\} = \dfrac{1}{p^2+1} = f(p)$ (say) then apply second shifting theorem.

ANSWERS

1. $\dfrac{6}{(p+3)^4}$ 2. $\dfrac{p-3}{p^2 - 6p + 25}$ 3. $\dfrac{2}{(p+1)(p^2+2p+5)}$ 4. $\dfrac{2}{(p-1)(p^2-2p+5)}$ 5. $\dfrac{p+4}{p^2+8p+12}$

6. $\dfrac{3p-24}{p^2+4p+40}$ 7. $\dfrac{4p^2 - 44p + 122}{(p-6)^3}$ 8. $\dfrac{1}{2}[f(p - i\omega) + f(p + i\omega)]$ 9. (i) $\dfrac{3}{p^2 - 9}$ (ii) $\dfrac{p}{p^2+25}$, $p > 0$

10. $e^{-\pi p/3}.\dfrac{1}{p^2+1}$, $p > 0$ 11. $\dfrac{e^{-2\pi p/3}}{p^2+1}$, $p > 0$ 13. $\dfrac{\sqrt{\pi}}{\sqrt{p-1}}$ 14. (i) $\dfrac{8(3p^2 - 6p - 13)}{(p^2 - 2p + 17)^3}$ (ii) $\dfrac{4p+4}{(p^2+2p+5)^2}$

15. $\dfrac{1}{p} + \dfrac{2}{p}e^{-2p} + \dfrac{e^{-p}}{p^2} + \dfrac{3}{p^2}e^{-2p} + \dfrac{2}{p^3}e^{-2p}$

36.6 LAPLACE TRANSFORM OF DERIVATIVES

THEOREM 1. *Let $F(t)$ be continuous for all $t \ge 0$ and be of exponential order as $t \to \infty$ and if $F'(t)$ is of class A, the Laplace transforms of derivatives $F'(t)$ exists when $p > a$ and $L\{F'(t)\} = p\, L\{F(t)\} - F(0)$.*

Proof. By definition, we have

$$L\{F'(t)\} = \int_0^\infty e^{-pt} F'(t)\, dt = \left[e^{-pt}F(t)\right]_0^\infty + p\int_0^\infty e^{-pt}\, F(t)\, dt \quad \text{(On integrating by parts)}$$

$$= -F(0) + pL\{F(t)\} \qquad \left[\because \lim_{t\to\infty} e^{-pt}\, F(t) = 0\right]$$

$$= pL\{F(t)\} - F(0).$$

REMARK

- Proceeding same as above, we get
$L\{F''(t)\} = pL\{F'(t)\} - F'(0) = p[p\,L\{F(t)\} - F(0)] - F'(0) = p^2 L\{F(t)\} - p\,F(0) - F'(0) = p^2 f(p) - p\,F(0) - F'(0)$.

THEOREM 2. *If $F(t), F'(t), \ldots, F^{n-1}(t)$ are continuous for $t \geq 0$ and be of exponential order as $t \to \infty$ and if $F^n(t)$ is of class A and if $L\{F(t)\} = f(p)$, then*

$$L\{F^n(t)\} = p^n f(p) - p^{n-1}F(0) - p^{n-2}F(0) \ldots pF^{(n-2)}(0) - F^{(n-1)}(0) = p^n f(p) - \sum_{r=0}^{n-1} p^{n-1-r} F^r(0)$$

Proof. Using above theorem, we have

$$L\{F'(t)\} = pL\{F(t)\} - F(0) \quad \text{and} \quad L\{F''(t)\} = p^2 L\{F(t)\} - pF(0) - F'(0)$$

Similarly, we can find

$$L\{F'''(t)\} = pL\{F''(t)\} - F''(0) = p[p^2 L\{F(t)\} - p\,F(0) - F'(0)] - F''(0)$$
$$= p^3 L\{F(t)\} - p^2 F(0) - pF'(0) - F''(0).$$

Proceeding, similarly, we get

$$L\{F^n(t)\} = p^n L\{F(t)\} - p^{n-1}F(0) - p^{n-2}F'(0) - \ldots - F^{n-1}(0) = p^n L\{F(t)\} - \sum_{r=0}^{n-1} p^{n-1-r} F^r(0).$$

THEOREM 3. *If $F(t)$ is a function of class A and if $L\{F(t)\} = f(p)$, then $L\{t\,.\,F(t)\} = -f'(p)$*

Proof. We know that

$$f(p) = L\{F(t)\} = \int_0^\infty e^{-pt} F(t)\,dt.$$

Therefore,

$$f'(p) = \frac{d}{dp}\int_0^\infty e^{-pt} F(t) dt = \int_0^\infty \frac{\partial}{\partial p}\{e^{-pt}F(t)\}\,dt$$

(By Leibnitz's rule of differentiation under the sign of integral)

$$= -\int_0^\infty t\,e^{-pt} F(t)\,dt = -\int_0^\infty e^{-pt}\{t\,F(t)\}\,dt = -L\{t\,F(t)\}$$

$$\Rightarrow \quad L\{t\,F(t)\} = -f'(p).$$

THEOREM 4. *If $F(t)$ is a function of class A and if $L\{F(t)\} = f(p)$ Then, $L\{t^n F(t)\} = (-1)^n \dfrac{d^n}{dp^n} f(p)$.* (UPTU–2005)

Proof. We shall prove this theorem by the Principle of Mathematical induction.

Step 1. Using previous theorem, we have

$$L\{t\,F(t)\} = (-1)^1 \frac{d}{dp} f(p) \quad \Rightarrow \quad \text{Theorem is true for } n = 1.$$

Step 2. Assume that the theorem is true for a particular value of n say k. Then, we have

$$L\{t^k F(t)\} = (-1)^k \frac{d^k}{dp^k} f(p) \quad \Rightarrow \quad \int_0^\infty e^{-pt} t^k F(t)\,dt = (-1)^k \frac{d^k}{dp^k} f(p).$$

Step 3. Differentiating both sides w.r.t. p, we have

$$\frac{d}{dp}\int_0^\infty e^{-pt} t^k F(t)\,dt = (-1)^k \frac{d^{k+1}}{dp^{k+1}} f(p).$$

Applying, Leibnitz's rule for differentiation under the sign of integration, we have

$$-\int_0^\infty e^{-pt} t^{k+1} F(t)\,dt = (-1)^{k+2} \frac{d^{k+1}}{dp^{k+1}} f(p) \quad \Rightarrow \quad \int_0^\infty e^{-pt}\{t^{k+1} F(t)\}\,dt = (-1)^{k+1} \frac{d^{k+1}}{dp^{k+1}} f(p)$$

$$\Rightarrow \quad L\{t^{k+1} F(t)\} = (-1)^{k+1} \frac{d^{k+1}}{dp^{k+1}} f(p) \quad \Rightarrow \quad \text{Theorem is true for } n = k+1.$$

Hence, by the principle of mathematical induction, it is true for every positive integral value of n.

THEOREM 5. *Let a function $F(t)$ be periodic with period w, so that $F(t + nw) = F(t)$ for $n = 1, 2, 3, \ldots,$*

then
$$L\{F(t)\} = \frac{1}{1 - e^{-pw}} \int_0^w e^{-pt} F(t)\,dt.$$

Proof. We know that

$$L\{F(t)\} = \int_0^\infty e^{-pt}\, F(t)\, dt$$

$$= \int_0^w e^{-pt}\, F(t)\, dt + \int_w^{2w} e^{-pt}\, F(t)\, dt + \ldots + \int_{nw}^{(n+1)w} e^{-pt}\, F(t)\, dt + \ldots$$

$$= \sum_{n=0}^{\infty} \int_{nw}^{(n+1)w} e^{-pt}\, F(t)\, dt = \sum_{n=0}^{\infty} \int_0^w e^{-p(x+nw)}\, F(x+nw)\, dx \quad [t = x + nw]$$

$$= \sum_{n=0}^{\infty} \int_0^w e^{-px}\, e^{-npw}\, F(x)\, dx \quad [\because \text{By definition of periodic function } f(x+nw) = f(x)]$$

$$= \sum_{n=0}^{\infty} e^{-npw} \int_0^w e^{-px}\, F(x)\, dx = (1 + e^{-pw} + e^{-2pw} + \ldots) \int_0^w e^{-px}\, F(x)\, dx$$

$$= \frac{1}{1 - e^{-pw}} \int_0^w e^{-px}\, F(x)\, dx \qquad \left[\because e^{-pw} < 1\right]$$

THEOREM 6 **(Initial Value Theorem).** *Let $F(t)$ be continuous for all $t \geq 0$ and be of exponential order as $t \to \infty$ and if $F'(t)$ is of class A, then* $\lim_{t\to 0} F(t) = \lim_{p\to\infty} pL\{F(t)\}$.

Proof. We know that

$$L\{F'(t)\} = \int_0^\infty e^{-pt}\, F'(t) dt = pL\,\{F(t)\} - F(0). \qquad \ldots(1)$$

Since $F'(t)$ is sectionally continuous and of exponential order.

Therefore, $\lim_{p\to\infty} \int_0^\infty e^{-pt}\, F'(t)\, dt = 0$

Now, taking limit as $p \to \infty$ in (1), we have

$$0 = \lim_{p\to\infty} pL\,\{F(t)\} - F(0) \quad \Rightarrow \quad F(0) = \lim_{p\to\infty} pL\,\{F(t)\} \quad \Rightarrow \quad \lim_{t\to 0} F(t) = \lim_{p\to\infty} pL\,\{F(t)\}.$$

THEOREM 7 **(Final Value Theorem).** *Let $F(t)$ be continuous for all $t \geq 0$ and be of exponential order as $t \to \infty$ and if $F'(t)$ is of class A, then* $\lim_{t\to\infty} F(t) = \lim_{p\to 0} pL\,\{F(t)\}$.

Proof. We know that

$$L\{F'(t)\} = \int_0^\infty e^{-pt}\, F'(t)\, dt = pL\,\{F(t)\} - F(0). \qquad \ldots(1)$$

Taking limit as $p \to 0$ in (1), we get

$$\lim_{p\to 0} \int_0^\infty e^{-pt}\, F'(t) dt = \lim_{p\to 0} [pL\,\{F(t)\} - F(0)] \quad \Rightarrow \quad \int_0^\infty F'(t)\, dt = \lim_{p\to 0} pL\,\{F(t)\} - F(0)$$

$$\Rightarrow \quad \left[F(t)\right]_0^\infty = \lim_{p\to 0} pL\,\{F(t)\} - F(0) \quad \Rightarrow \quad \lim_{t\to\infty} F(t) - F(0) = \lim_{p\to 0} pL\,\{F(t)\} - F(0)$$

$$\Rightarrow \quad \lim_{t\to\infty} F(t) = \lim_{p\to 0} pL\,\{F(t)\}.$$

THEOREM 8 **(Laplace Transform of the Laplace Transform).** *We have*

$$L\,[L\,\{F(t)\}] = L\left\{\int_0^\infty e^{-pt} F(t)\, dt\right\} = \int_0^\infty e^{-up} \left\{\int_0^\infty e^{-pt}\, F(t)\, dt\right\} dp.$$

Proof. The area of integration being the whole positive quadrant. Now, changing the order of integration, we get

$$L\,[L\,\{F(t)\}] = \int_0^\infty F(t) \left\{\int_0^\infty e^{-p(t+u)} dp\right\} dt = \int_0^\infty F(t) \left\{\left[\frac{e^{-p(t+u)}}{-(t+u)}\right]_{p=0}^{\infty}\right\} dt = \int_0^\infty \frac{F(t)}{t+u}\, dt.$$

THEOREM 9 **(Laplace Transforms of Integrals).** *If $F(t)$ is piecewise continuous and satisfies $|F(t)| \leq Me^{at}, \forall t \geq 0$ for some constant a and M, then*

$$L\left\{\int_0^t F(x)\, dx\right\} = \frac{1}{p} L\,\{F(t)\} \qquad (p > 0,\ p > a)$$

Proof. Let $F(t)$ be piecewise continuous such that

$$|\, F(t) \,| \leq M\, e^{at} \qquad \ldots(1)$$

for some constants a and M.

If (1) holds for some negative value of a, then it also holds for positive value of a. Therefore, suppose that a is positive.

Let $$G(t) = \int_0^t F(x)\,dx.$$

Then $G(t)$ is continuous. ($\because$ Integral of an integrable function is continuous)

Now, $|G(t)| \le \int_0^t |F(x)|\,dx \le \int_0^t Me^{ax}\,dx \Rightarrow |G(t)| \le \frac{M}{a}(e^{at}-1),\ a>0$ ...(2)

Further, $G'(t) = F(t)$, except for points at which $F(t)$ is discontinuous. Therefore, $G'(t)$ is piecewise continuous on each finite interval.

$\therefore$ $$L\{G'(t)\} = pL\{G(t)\} - G(0) = pL\{G(t)\} \qquad [\because G(0)=0]$$

$$\Rightarrow L\{G(t)\} = \frac{1}{p}L\{G'(t)\} \Rightarrow L\left\{\int_0^t F(x)\,dx\right\} = \frac{1}{p}L\{F(t)\}.$$

THEOREM 10 **(Division by t).** *If* $L\{F(t)\} = f(p)$, *then* $L\left\{\frac{1}{t}F(t)\right\} = \int_p^\infty f(x)\,dx$ *provided* $\lim_{t\to 0}\left\{\frac{1}{t}F(t)\right\}$ *exists.* (UPTU–2005, 07)

Proof. Let $$G(t) = \frac{1}{t}F(t),\ i.e.,\ F(t) = t\,G(t)$$

Therefore, $$L\{F(t)\} = L\{t\,G(t)\} = -\frac{d}{dp}L\{G(t)\} \Rightarrow f(p) = -\frac{d}{dp}L\{G(t)\}.$$

On integrating both sides with respect to p to ∞, we get

$$-\left[L\{G(t)\}\right]_p^\infty = \int_p^\infty f(p)\,dp \Rightarrow -\lim_{p\to\infty} L\{G(t)\} + L\{G(t)\} = \int_p^\infty f(p)\,dp$$

$$\Rightarrow 0 + L\{G(t)\} = \int_p^\infty f(p)\,dp,$$ (By using $\lim_{p\to\infty} L\{G(t)\} = \lim_{p\to\infty}\int_0^\infty e^{-pt}G(t)dt = 0$)

$$\Rightarrow L\left\{\frac{1}{t}F(t)\right\} = \int_p^\infty f(x)\,dx.$$

Solved Examples

Example 1. *Find* $L\{t\cos at\}$. (Raipur–2005)

Solution. We know that

$$L\{\cos at\} = \frac{p}{p^2+a^2},\ p>0.$$

Therefore,

$$L\{t\cos at\} = -\frac{d}{dp}L\{\cos at\} = -\frac{d}{dp}\left(\frac{p}{p^2+a^2}\right).$$

$$= \frac{p^2-a^2}{(p^2+a^2)^2}.$$

Example 2. *Find* $L\{t^2\sin at\}$. (UPTU(Q. Bank)–2001)

Solution. We know that $L\{\sin at\} = \frac{a}{p^2+a^2}$

Therefore, $$L\{t^2\sin at\} = (-1)^2\frac{d^2}{dp^2}L\{\sin at\}$$

$$= \frac{d^2}{dp^2}\left\{\frac{a}{p^2+a^2}\right\}$$

$$= \frac{d}{dp}\left\{\frac{-2ap}{(p^2+a^2)^2}\right\}$$

$$= \frac{2a(3p^2-a^2)}{(p^2+a^2)^3},\ p>0.$$

Example 3. *Find* $L\{(\sin at - at\cos at)\}$.

Solution. Consider $L\{\sin at - at\cos at\}$

$$= L\{\sin at\} - aL\{t\cos at\}$$

$$= \frac{a}{p^2+a^2} - a.(-1)\frac{d}{dp}[L\{\cos at\}]$$

$$= \frac{a}{p^2+a^2} + a\frac{d}{dp}\left(\frac{p}{p^2+a^2}\right)$$

$$= \frac{a}{p^2+a^2} + \frac{a(a^2-p^2)}{(p^2+a^2)^2}$$

$$= \frac{2a^3}{(p^2+a^2)^2}.$$

Example 4. *Find the Laplace transform of the function*

$$F(t) = t\,e^{-t}\sin 2t$$

(UPTU–2002, Kurukshetra–2005, 13)

Solution. $$L\{\sin 2t\} = \frac{2}{p^2+4},$$

$$L\{e^{-t}\sin 2t\} = \frac{2}{(p+1)^2+4} = f(p)\quad \text{(say)}$$

$$L\{te^{-t}\sin 2t\} = f'(p) = -\frac{d}{dp}\left[\frac{2}{(p+1)^2+4}\right]$$

$$= \frac{-2.2(p+1)}{[(p+1)^2+4]^2} = \frac{4(p+1)}{[(p+1)^2+4]^2}.$$

Example 5. *Obtain the Laplace transform of* $t^2 e^t \sinh t$. (UPTU–2002)

Solution.
$$L\{\sin 4t\} = \frac{4}{p^2+16},$$
$$L\{e^t \sin 4t\} = \frac{4}{(p-1)^2+16}$$
$$L\left\{te^t \sin 4t\right\} = -\frac{d}{dp}\left(\frac{4}{p^2-2p+17}\right) = \frac{4(2p-2)}{(p^2-2p+17)^2}$$
$$L\left\{t^2e^t \sin 4t\right\} = -\frac{d}{dp}\left(\frac{4(2p-2)}{(p^2-2p+17)^2}\right)$$
$$= \frac{-4(2p^2-4p+34-8p^2+16p-8)}{(p^2-2p+17)^3}$$
$$= \frac{-4(-6p^2+12p+26)}{(p^2-2p+17)^3}$$
$$= \frac{8\,(3p^2-6p-13)}{(p^2-2p+17)^3}.$$

Example 6. *Given* $L\{\sin\sqrt{t}\} = \frac{\sqrt{\pi}}{2p^{3/2}} e^{-1/4p}$, *show that*
$$L\left\{\frac{\cos\sqrt{t}}{\sqrt{t}}\right\} = \sqrt{\left(\frac{\pi}{p}\right)}.\, e^{-1/4p}.$$ (Mumbai–2009)

Solution. Let $F(t) = \sin\sqrt{t}$

Then we have $F'(t) = \cos\frac{\sqrt{t}}{2\sqrt{t}}$ and $F(0) = 0$

Put all these values in
$$L\{F'(t)\} = pL\{F(t)\} - F(0)$$
We get $L\left\{\frac{\cos\sqrt{t}}{2\sqrt{t}}\right\} = pL\{\sin\sqrt{t}\}$
$$= p\left[\frac{\sqrt{\pi}}{2p^{3/2}} e^{-1/4p}\right] = \frac{1}{2}\sqrt{\left(\frac{\pi}{p}\right)} e^{-1/4p}.$$
Hence, $L\left\{\frac{\cos\sqrt{t}}{\sqrt{t}}\right\} = \sqrt{\left(\frac{\pi}{p}\right)}.\, e^{-1/4p}.$

Example 7. *Show that* $L\left\{\frac{\sin t}{t}\right\} = \tan^{-1}\frac{1}{p}$ *and hence find* $L\left\{\frac{\sin at}{t}\right\}$. *Does the Laplace transform of* $\frac{\cos at}{t}$ *exists?* (UPTU–2005, PTU–2010)

Solution. Let $F(t) = \sin t$

Then, $\lim_{t\to 0}\frac{F(t)}{t} = \lim_{t\to 0}\frac{\sin t}{t} = 1.$

We know that $L\{\sin t\} = \frac{1}{p^2+1} = f(p)$ (say)

Then we have
$$L\left\{\frac{\sin t}{t}\right\} = \int_p^\infty f(x)\,dx$$
$$= \int_p^\infty \frac{dx}{x^2+1} = \left(\tan^{-1} x\right)_p^\infty$$
$$= \frac{\pi}{2} - \tan^{-1} p = \cot^{-1} p = \tan^{-1}\left(\frac{1}{p}\right).$$
Now, $L\left\{\frac{\sin at}{t}\right\} = aL\left\{\frac{\sin at}{at}\right\}$
$$= a\,.\,\frac{1}{a}\tan^{-1}\left(\frac{1}{p/a}\right)$$
$$\left[\because L\{f(at)\} = \frac{1}{a} f\left(\frac{p}{a}\right)\right]$$
$$= \tan^{-1}\left(\frac{a}{p}\right).$$
Also, since $L\{\cos at\} = \frac{p}{p^2+a^2} = f(p)$ (say)

Then, $L\left\{\frac{\cos at}{t}\right\} = \int_p^\infty \frac{x}{x^2+a^2}\,dx$
$$= \left[\frac{1}{2}\log(x^2+a^2)\right]_p^\infty$$
$$= \frac{1}{2}\lim_{x\to\infty}\log(x^2+a^2) - \frac{1}{2}\log(p^2+a^2)$$
which does not exist since $\lim_{x\to\infty}\log(x^2+a^2)$ is infinite.

Therefore, $L\left\{\frac{\cos at}{t}\right\}$ does not exist.

Example 8. *Find* $L\left\{\frac{\sinh t}{t}\right\}$.

Solution. Let us assume $F(t) = \sinh t$.

Now, $\lim_{t\to 0}\frac{F(t)}{t} = \lim_{t\to 0}\frac{\sinh t}{t} = 1.$

Since $L\{\sinh t\} = \frac{1}{p^2-1} = f(p)$ (say)
$$\therefore\quad L\left\{\frac{\sinh t}{t}\right\} = \int_p^\infty f(x)\,dx = \int_p^\infty \frac{dx}{x^2-1}$$
$$= \left[\frac{1}{2}\log\frac{x-1}{x+1}\right]_p^\infty = \left[\frac{1}{2}\log\frac{1-1/x}{1+1/x}\right]_p^\infty$$
$$= 0 - \frac{1}{2}\log\frac{p-1}{p+1} = \frac{1}{2}\log\frac{p+1}{p-1}$$

Example 9. *If* $F(t) = \frac{e^{at}-\cos bt}{t}$, *find the Laplace transform of* $F(t)$. (UPTU–2003)

Solution. $F(t) = \frac{e^{at}-\cos bt}{t} = \frac{e^{at}}{t} - \frac{\cos bt}{t}$

We know that

$$L\left(e^{at}-\cos bt\right)=\left(\frac{1}{p-a}-\frac{p}{p^2+b^2}\right)$$

$$\therefore L\left(\frac{e^{at}-\cos bt}{t}\right)=\int_p^{\infty}\left(\frac{1}{p-a}-\frac{p}{p^2+b^2}\right)ds$$

$$=\left[\log(p-a)-\frac{1}{2}\log(p^2+b^2)\right]_p^{\infty}$$

$$=\left[\frac{2\log(p-a)-\log(p^2+b^2)}{2}\right]_p^{\infty}$$

$$=\frac{1}{2}\left[\log(p-a)^2-\log(p^2+b^2)\right]_p^{\infty}$$

$$=\frac{1}{2}\left[\log\frac{(p-a)^2}{p^2+b^2}\right]_p^{\infty}=\frac{1}{2}\left[\log\left[\frac{(1-(a/p))}{(1+(b^2/p^2))}\right]\right]_p^{\infty}$$

$$=\frac{1}{2}\left[0-\log\frac{(1-(1/p))^2}{(1+(b^2/p^2))}\right]=\frac{1}{2}\left[\log\frac{p^2+b^2}{(p-a)^2}\right]$$

EXERCISE 36.3

1. Show that $L\{-a\sin at\}=-\dfrac{a^2}{p^2+a^2}$.
2. Evaluate
   (i) $L\{t\cosh 3t\}$, (ii) $L\{t\sinh at\}$
3. Show that $L\{t^2\cos at\}=\dfrac{2p(p^2-3a^2)}{(p^2+a^2)^3},\ p>0$.
4. Show that $L(t^n e^{at})=\dfrac{n!}{(p-a)^{n+1}},\ p>a$.
5. Show that $L\{t(3\sin 2t-2\cos 2t)\}=\dfrac{8+12p-2p^2}{(p^2+4)^2}$.
6. Show that $L\{\sin\alpha t+t\cos\alpha t\}=\dfrac{(\alpha+1)p^2+(\alpha-1)\alpha^2}{(p^2+\alpha^2)^2}$.
7. Show that
   $L\{t^2-3t+2\}\sin 3t=\dfrac{6p^4-18p^3+126p^2-162p+432}{(p^2+9)^3}$.
8. If $L\{F(t),t\to p\}=f(p)$, show that
   $L\left\{\int_0^t\frac{F(u)}{u}du,\ t\to p\right\}=\frac{1}{p}\int_p^{\infty}f(y)\,dy$.
   Hence, show that $L\left\{\int_0^t\frac{\sin u}{u}du,\ t\to p\right\}=\frac{\cot^{-1}p}{p}$.
9. Show that if $L\{F(t)\}=f(p)$, then
   $\int_0^{\infty}\frac{F(t)}{t}dt=\int_0^{\infty}f(x)\,dx$, provided that the integral converges.
10. If $L\{t\sin wt\}=\dfrac{2wp}{(p^2+w^2)^2}$, evaluate $L\{wt\cos wt+\sin wt\}$. (UPTU(Q. Bank)–2001)
11. Find the Laplace transform of
    (i) $\int_0^t e^{-t}\cos t\,dt$,
    (ii) $\int_0^t\frac{\sin t}{t}dt$ (UPTU(Q. Bank)–2001, JNTU–2005)
12. Find the Laplace transform of
    (i) $te^{-t}\sin 2t$, (UPTU–2002)
    (ii) $t^2e^t\sin 4t$ (UPTU(Sp)–2001)

Hint to Selected Problems

1. $F(t)=-a\sin at, F'(t)=-a^2\cos at, F''(t)=a^3\sin at$
   $F'(0)=-a^2$ and $F(0)=0$, then using
   $L\{F''(t)\}=p^2\,L\,F\{t\}-pF(0)-F'(0)$
9. Use Theorem 10.

ANSWERS

2. (a) $\dfrac{p^2+9}{(p^2-9)^2}$ (b) $\dfrac{2ap}{(p^2-a^2)^2}$ 10. $\dfrac{2wp^2}{(p^2+w^2)^2}$ 11. (i) $\dfrac{p+1}{p(p^2+2p+2)}$, (ii) $\dfrac{1}{p}\cot^{-1}p$

12. (i) $\dfrac{4p+4}{(p^2+2p+5)^2}$, (ii) $\dfrac{8(3p^2-6p-13)}{(p^2-2p+17)^3}$

36.7 EVALUATION OF INTEGRALS

If $L\{F(t)\}=f(p)$, i.e., $\int_0^{\infty}e^{-pt}F(t)\,dt=f(p)$

By taking limit as $p\to 0$, we have $\int_0^{\infty}F(t)\,dt=f(0)$, provided the integral is convergent.

36.8 SOME IMPORTANT SPECIAL FUNCTIONS

(i) The sine and cosine integrals. The sine the cosine integrals, which are denoted by $S_i(t)$ and $C_i(t)$ respectively are defined by $S_i(t)=\int_0^t\frac{\sin u}{u}du$ and $C_i(t)=\int_t^{\infty}\frac{\cos u}{u}du$.

(ii) Error Function. The error function denoted by $erf\ (t)$, is defined by $erf\ (t) = \frac{2}{\sqrt{\pi}} \int_0^t e^{-u^2}\, du$.

(iii) The Gamma function. If $n > 0$, the gamma function is defined by $\Gamma(n) = \int_0^\infty u^{n-1} e^{-u}\, du$.

(iv) Heaviside's unit function. The unit step function or heaviside's unit function denoted by $H(t-a)$ is defined by

$$H(t-a) = \begin{cases} 0, & t < a \\ 1, & t \geq a \end{cases}.$$

(v) Bessel's functions. $J_n(t) = \frac{t^n}{2^n\, \Gamma(n+1)} \left[1 - \frac{t^2}{2(2n+2)} + \frac{t^4}{2.4(2n+2)(2n+4)} \dots\right]$.

Solved Examples

Example 1. *Find* $\int_0^\infty \frac{(e^{-at} - e^{-bt})}{t}\, dt$.

(SVTU–2009, Mumbai–2007, JNTU–2006)

Solution. Let $F(t) = e^{-at} - e^{-bt}$.

Thus, we have

$$L\{F(t)\} = L\{e^{-at}\} - L\{e^{-bt}\}$$

$$= \frac{1}{p+a} - \frac{1}{p+b} = f(p) \text{ (say)}.$$

Therefore,

$$L\left\{\frac{F(t)}{t}\right\} = \int_p^\infty f(x)\, dx = \int_p^\infty \left(\frac{1}{x+a} - \frac{1}{x+b}\right) dx$$

$$= \left[\log\left(\frac{x+a}{x+b}\right)\right]_p^\infty = \lim_{x\to\infty} \log\frac{x+a}{x+b} - \log\frac{p+a}{p+b}$$

$$= \lim_{x\to\infty} \log\frac{1+a/x}{1+b/x} - \log\frac{p+a}{p+b}$$

$$= 0 - \log\frac{p+a}{p+b} = \log\frac{p+b}{p+a}$$

Therefore, $L\left\{\frac{F(t)}{t}\right\} = \int_0^\infty e^{-pt} . \frac{e^{-at} - e^{-bt}}{t} dt$

$$= \log\frac{p+b}{p+a}$$

Hence, taking limit as $p \to 0$, we have

$$\int_0^\infty \frac{e^{-at} - e^{-bt}}{t}\, dt = \log\frac{b}{a}.$$

Example 2. *Show that* $\int_0^\infty t\, e^{-2t} \cos t\, dt = \frac{3}{25}$.

Solution. We have $L\{t \cos t\} = -\frac{d}{dp} L\{\cos t\}$

or $\int_0^\infty e^{-pt} . t \cos t dt = -\frac{d}{dp}\left(\frac{p}{p^2+1}\right) = \frac{p^2-1}{(p^2+1)^2}$

Taking $p = 2$, we get

$$\int_0^\infty t\, e^{-2t} \cos t\, dt = \frac{3}{25}.$$

Example 3. *Show that*

(i) $L\{\sinh at \cos at\} = \frac{a(p^2 - 2a^2)}{p^4 + 4a^4}$

(ii) $L\{\sinh at \sin at\} = \frac{2a^2 p}{p^4 + 4a^4}$.

Solution. (i) We know that

$$L\{\sinh at\} = \frac{a}{p^2 - a^2} = f(p) \text{ (say)}$$

Therefore, $L\{e^{iat} \sin at\} = f(p - ia)$

$$= \frac{a}{(p-ia)^2 - a^2} = \frac{a}{(p^2 - 2a^2) - 2iap}$$

$$= \frac{a\{(p^2 - 2a^2) + 2iap\}}{(p^2 - 2a^2)^2 - (2ipa)^2}$$

$\Rightarrow$ $L\{\sinh at(\cos at + i \sin at)\}$

$$= \frac{a(p^2 - 2a^2) + 2ia^2 p}{p^4 + 4a^4}$$

$\Rightarrow$ $L\{\sinh at \cos at\} + iL\{\sinh at \sin at\}$

$$= \frac{a(p^2 - 2a^2)}{p^4 + 4a^4} + i\frac{2a^2 p}{p^4 + 4a^4}$$

Equating real and imaginary parts of both the sides, we get

$$L\{\sinh at \cos at\} = \frac{a(p^2 - 2a^2)}{p^4 + 4a^4}$$

and $L\{\sinh at \sin at\} = \frac{2a^2 p}{p^4 + 4a^4}$.

Example 4. *Find* $L\{erf \sqrt{t}\}$ *and hence prove that*

$$L\{t . erf(2\sqrt{t})\} = \frac{3p+8}{p^2(p+4)^{3/2}}.$$

(UPTU–2001)

Solution. We know that

$$erf \sqrt{t} = \frac{2}{\sqrt{\pi}} \int_0^{\sqrt{t}} e^{-x^2} dx$$

$$= \frac{2}{\sqrt{\pi}} \int_0^{\sqrt{t}} \left(1 - x^2 + \frac{x^4}{2!} - \frac{x^6}{6!} + \dots\right) dx$$

$$= \frac{2}{\sqrt{\pi}} \left[x - \frac{x^3}{3} + \frac{x^5}{10} - \frac{x^7}{42} - \dots\right]_0^{\sqrt{t}}$$

$$= \frac{2}{\sqrt{\pi}}\left[\sqrt{t} - \frac{t^{3/2}}{3} + \frac{t^{3/2}}{10} - \frac{t^{7/2}}{42} + \ldots\right]$$

$$L\{erf\sqrt{t}\} = \frac{2}{\sqrt{\pi}}\left[\frac{\Gamma(3/2)}{p^{3/2}} - \frac{\Gamma(5/2)}{3p^{5/2}} + \frac{\Gamma(7/2)}{10p^{7/2}} - \frac{\Gamma(9/2)}{42p^{9/2}} + \ldots\right]$$

$$= \frac{2}{\sqrt{\pi}}\left[\frac{\frac{1}{2}\Gamma(1/2)}{p^{3/2}} - \frac{\frac{3}{2}.\frac{1}{2}\Gamma(1/2)}{3p^{5/2}} + \frac{\frac{5}{2}.\frac{3}{2}.\frac{1}{2}\Gamma(1/2)}{10p^{7/2}} - \frac{\frac{7}{2}.\frac{5}{2}.\frac{3}{2}.\frac{1}{2}\Gamma(1/2)}{42p^{9/2}} + \ldots\right]$$

$$= \frac{1}{p^{3/2}} - \frac{1}{2}\frac{1}{p^{5/2}} + \frac{1.3}{2.4}.\frac{1}{p^{7/2}} - \frac{1}{2}.\frac{3}{4}.\frac{5}{6}.\frac{1}{p^{9/2}} + \ldots$$

$$= \frac{1}{p^{3/2}}\left[1 - \frac{1}{2}.\frac{1}{p} + \frac{1}{2}.\frac{3}{4}.\frac{1}{p^2} - \frac{1}{2}.\frac{3}{4}.\frac{5}{6}.\frac{1}{p^3} + \ldots\right]$$

$$= \frac{1}{p^{3/2}}\left[1 - \frac{1}{2}.\frac{1}{p} + \frac{\left(-\frac{1}{2}\right)\left(-\frac{3}{2}\right)}{2!}\frac{1}{p^2} + \frac{\left(-\frac{1}{2}\right)\left(-\frac{3}{2}\right)\left(-\frac{5}{2}\right)}{3!}\frac{1}{p^3} + \ldots\right]$$

$$= \frac{1}{p^{3/2}}\left[1 + \frac{1}{p}\right]^{-1/2}$$

$$= \frac{1}{p^{3/2}}\left[\frac{p}{p+1}\right]^{1/2} = \frac{1}{p\sqrt{p+1}}$$

Now, $L\{erf\ 2\sqrt{t}\} = L\{erf\ \sqrt{4t}\}$

$$= \frac{1}{4}\frac{1}{\frac{p}{4}\sqrt{\frac{p}{4}+1}} = \frac{2}{p\sqrt{p+4}} = \frac{2}{p\sqrt{p+4}}$$

$$L\{t.erf\ (2\sqrt{t})\} = -\frac{d}{dp}\frac{2}{\sqrt{p^3+4p^2}}$$

$$= -2\left(-\frac{1}{2}\right)\left(p^3+4p^2\right)^{-3/2}(3p^2+8p)$$

$$= \frac{3p^2+8p}{(p^3+4p^2)^{3/2}} = \frac{3p+8}{p^2(p+4)^{3/2}}.$$

Example 5. *Show that* $L\left\{\frac{\sin^2 t}{t}\right\} = \frac{1}{4}\log\left(\frac{p^2+4}{p^2}\right)$.

Solution. We know that $\sin^2 t = \frac{1}{2}(1-\cos 2t)$.

Now let $F(t) = \sin^2 t = \frac{1}{2}(1-\cos 2t)$

$$\Rightarrow L\{F(t)\} = \frac{1}{2}[L\{1\} - L\{\cos 2t\}]$$

$$= \frac{1}{2}\left[\frac{1}{p} - \frac{p}{p^2+4}\right] = f(p) \text{ (say)}$$

Now, $\lim_{t\to 0}\left\{\frac{1}{t}F(t)\right\} = \lim_{t\to 0}\left(\frac{\sin t}{t}\right).\sin t$
$= 1.0 = 0 \Rightarrow$ limit exists.

Therefore, $L\left\{\frac{1}{t}F(t)\right\} = \int_p^\infty f(x)\,dx$

$$= \frac{1}{2}\int_p^\infty\left(\frac{1}{x} - \frac{x}{x^2+4}\right)dx$$

$$= \frac{1}{2}\left[\log x - \frac{1}{2}\log(x^2+4)\right]_p^\infty$$

$$= \frac{1}{4}\left[\log\frac{x^2}{x^2+4}\right]_p^\infty$$

$$= \frac{1}{4}\left[\lim_{x\to\infty}\log\frac{x^2}{x^2+4} - \log\frac{p^2}{p^2+4}\right]$$

$$= \frac{1}{4}\left[\lim_{x\to\infty}\log\frac{1}{1+(4/x^2)} - \log\frac{p^2}{p^2+4}\right]$$

$$= \frac{1}{4}\left[\log 1 - \log\frac{p^2}{p^2+4}\right]$$

$$= \frac{1}{4}\left[0 - \log\frac{p^2}{p^2+4}\right]$$

$$= -\frac{1}{4}\log\frac{p^2}{p^2+4}$$

$$= \frac{1}{4}\log\frac{p^2+4}{p^2}.$$

Example 6. *Show that*

$$\int_0^\infty \frac{\cos at - \cos bt}{t}dt = \frac{1}{2}\log\left(\frac{p^2+b^2}{p^2+a^2}\right).$$

(UPTU–2004)

Solution. Let $F(t) = \cos at - \cos bt$

$$\Rightarrow L\{F(t)\} = L\{\cos at\} - L\{\cos bt\}$$

$$= \frac{p}{p^2+a^2} - \frac{p}{p^2+b^2} = f(p) \text{ (say)}$$

Now, $\lim_{t\to 0} \frac{F(t)}{t} = \lim_{t\to 0} \frac{\cos at - \cos bt}{t}$

$$= \lim_{t\to 0} \frac{-a\sin at + b\sin bt}{1}$$

(By L-Hospital's rule)

$= 0 \Rightarrow$ limit exist.

Therefore,

$$L\left\{\frac{F(t)}{t}\right\} = \int_p^\infty F(x)\,dx$$

$$= \int_p^\infty \left[\frac{x}{x^2+a^2} - \frac{x}{x^2+b^2}\right] dx$$

$$= \frac{1}{2}\left[\log(x^2+a^2) - \log(x^2+b^2)\right]_p^\infty$$

$$= \frac{1}{2}\left[\log\frac{x^2+a^2}{x^2+b^2}\right]_p^\infty$$

$$= \frac{1}{2}\lim_{x\to\infty} \log\frac{x^2+a^2}{x^2+b^2} - \frac{1}{2}\log\frac{p^2+a^2}{p^2+b^2}$$

$$= \frac{1}{2}\lim_{x\to\infty} \log\frac{1+a^2/x^2}{1+b^2/x^2} - \frac{1}{2}\log\frac{p^2+a^2}{p^2+b^2}$$

$$= 0 - \frac{1}{2}\log\frac{p^2+a^2}{p^2+b^2} = -\frac{1}{2}\log\frac{p^2+a^2}{p^2+b^2}$$

$$= \frac{1}{2}\log\frac{p^2+b^2}{p^2+a^2}.$$

Example 7. *Find the Laplace transform of $S_i(t)$.*

Solution. By definition, we have

$$S_i(t) = \int_0^t \frac{\sin u}{u}\,du$$

$$= \int_0^t \left(1 - \frac{u^2}{3!} + \frac{u^4}{5!} - \frac{u^6}{7!} + \ldots\right) du$$

$$= t - \frac{t^3}{3.3!} + \frac{t^5}{5.5!} - \frac{t^7}{7.7!} + \ldots$$

Therefore,

$$L\{S_i(t)\} = L(t) - \frac{1}{3.3!}L\{t^3\} + \frac{1}{5.5!}L\{t^5\} - \frac{1}{7.7!}L\{t^7\} + \ldots$$

$$= \frac{1!}{p^2} - \frac{1}{3.3!}\cdot\frac{3!}{p^4} + \frac{1}{5.5!}\cdot\frac{5!}{p^6} - \frac{1}{7.7!}\cdot\frac{7!}{p^8} + \ldots$$

$$= \frac{1}{p}\left[\frac{1}{p} - \frac{1}{3}\cdot\frac{1}{p^3} + \frac{1}{5}\cdot\frac{1}{p^5} - \frac{1}{7}\cdot\frac{1}{p^7} + \ldots\right]$$

$$= \frac{1}{p}\tan^{-1}\frac{1}{p}.$$

Example 8. *Evaluate L {F(t)} if, wherehas period 2π.*

Solution. Sinceis a function with period $T = 2\pi$. Therefore, we have

$$L\{F(t)\} = \frac{\int_0^T e^{-pt}F(t)\,dt}{1-e^{-pT}}$$

$$= \frac{\int_0^\pi e^{-pt}\sin t\,dt + \int_\pi^{2\pi} 0.e^{-pt}\,dt}{1-e^{-2\pi p}}$$

$$= \frac{1}{1-e^{-2\pi p}}\int_0^\pi e^{-pt}\sin t\,dt$$

$$= \frac{1}{1-e^{-2\pi p}}\left[\frac{e^{-pt}}{p^2+1}(-p\sin t - \cos t)\right]_0^\pi$$

$$= \frac{1}{1-e^{-2\pi p}}\left[\frac{e^{-p\pi}}{p^2+1} + \frac{1}{p^2+1}\right]$$

$$= \frac{1+e^{-p\pi}}{(p^2+1)[1-(e^{-p\pi})^2]}$$

$$= \frac{1+e^{-p\pi}}{(p^2+1)(1-e^{-p\pi})(1+e^{-p\pi})}$$

$$= \frac{1}{(p^2+1)(1-e^{-p\pi})}$$

EXERCISE 36.4

1. Show that $L\{(1+te^{-t})^3\} = \frac{1}{p} + \frac{3}{(p+1)^2} + \frac{6}{(p+2)^3} + \frac{6}{(p+3)^4}$.

2. If $L\left\{2\sqrt{\frac{t}{\pi}}\right\} = \frac{1}{p^{3/2}}$, then show that $\frac{1}{p^{1/2}} = L\left\{\frac{1}{\sqrt{\pi t}}\right\}$.

   (UPTU–2005, Madras–2003)

3. Find (i) $L\{F(t)\}$ and (ii) $L\{F'(t)\}$ for the function defined by

   $F(t) = \begin{cases} 2t, & 0 \le t \le 1 \\ t, & t > 1 \end{cases}$.

4. Show that $\int_0^\infty \frac{\sin^2 t}{t^2}\,dt = \frac{\pi}{2}$.

5. Show that $\int_0^\infty \frac{\cos 6t - \cos 4t}{t}\,dt = \log\left(\frac{2}{3}\right)$.

   (Mumbai–2008, PTU–2006)

6. Show that

   (i) $L\{J_0(t)\} = \frac{1}{\sqrt{1+p^2}}$ (ii) $L\{t\,J_0(at)\} = \frac{p}{(p^2+a^2)^{3/2}}$.

7. Show that $L\{J_1(t)\} = 1 - \frac{p}{\sqrt{p^2+1}}$, where $J_1(t)$ is the Bessel function of order one and hence deduce that $L\{tJ_1(t)\} = \frac{1}{(p^2+1)^{3/2}}$.

8. Prove that $L\{J_0(a\sqrt{t})\} = \frac{1}{p}e^{-a^2/4p}$.

9. Show that $L\{t.erf(2\sqrt{t})\} = \frac{3p+8}{p^2(p+4)^{3/2}}$.

10. Show that $L\{e^{3t}.erf\sqrt{t}\} = \frac{1}{(p-3)\sqrt{p-2}}$.

11. Show that $L\{c_i(t)\} = \frac{1}{2p}.\log(p^2+1)$, where $c_i(t) = \int_0^\infty \frac{\cos u}{u}du$.

12. If $F(t) = t^2,\ 0 < t < 2$ and $F(t+2) = F(t)$. Then show that

$$L\{F(t)\} = \frac{-(4p^2+4p+2)e^{-2p+2}}{p^3(1-e^{-2p})}$$

13. If $F(t) = \begin{cases} 3t, & 0 < t < 2 \\ 6, & 2 < t < 4 \end{cases}$ and $F(t)$ is a periodic function of period 4, then show that $L\{F(t)\} = \frac{3-3e^{-2p}-6pe^{-4p}}{p^2(1-e^{-4p})}$

14. Find the Laplace transform of the Heaviside's unit step function $H(t-a)$.

15. Show that $\int_0^\infty \frac{\sin t}{t}dt = \frac{\pi}{2}$.

16. Show that $\int_0^\infty \frac{e^{-t}-e^{-3t}}{t}dt = \log 3$.

17. Show that $\int_0^\infty t^3 e^{-t}\sin t\, dt = 0$.

18. Show that $\int_0^\infty e^{-t}\frac{\sin t}{t}dt = \frac{1}{4}\log\frac{p^2+4}{p^2}$.

(UPTU–2008, VTU–2009S)

Hint to Selected Problems

1. Let $F(t) = 2\sqrt{\left(\frac{t}{\pi}\right)} \Rightarrow F'(t) = \frac{1}{\sqrt{\pi t}}$.

Then use $L\{F'(t)\} = pLF(t) - F(0)$.

5. (i) Use $J_0(t) = 1 - \frac{t^2}{2^2} + \frac{t^4}{2^2.4^2} - \frac{t^6}{2^2.4^2.6^2}$,

(ii) $L\{tJ_0(t)\} = -\frac{d}{dp}L\{J_0(t)\}$.

6. Since $J_0'(t) = -J_1(t)$. Now using $L\{f'(t)\} = pL\{f(t)\} - f(0)$

$L\{J_1(t)\} = L\{-J_0'(t)\} = -L\{J_0'(t)\}$.

7. Use $J_0(t) = 1 - \frac{t^2}{2^2} + \frac{t^4}{2^2.4^2} - \frac{t^6}{2^2.4^2.6^2} + \ldots$

$\Rightarrow J_0(a\sqrt{t}) = 1 - \frac{a^2t}{2^2} + \frac{a^4t^2}{2^2.4^2} - \frac{a^6.t^3}{2^2.4^2.6^2} + \ldots$

8. Use $erf(\sqrt{t}) = \frac{2}{\sqrt{\pi}}\int_0^{\sqrt{t}} e^{-u^2}du$

$$= \frac{2}{\sqrt{\pi}}\int_0^{\sqrt{t}}\left(1 - u^2 + \frac{u^4}{2!} - \frac{u^6}{3!} + \ldots\right)du.$$

10. Using $L\{c_i(t)\} = L\left\{\int_t^\infty \frac{\cos u}{u}du\right\}$.

11. Using $L\{F(t)\} = \frac{\int_0^T e^{-pt}F(t)\,dt}{1-e^{-pT}} = \frac{\int_0^2 t^2e^{-pt}dt}{1-e^{-2p}}$.

ANSWERS

2. (i) $\frac{2}{p^2} - \left(\frac{1}{p} + \frac{1}{p^2}\right)e^{-p},\ p > 0$ (ii) $\frac{1}{p}(2-e^{-p})$

36.9 THE UNIT STEP FUNCTION

The unit step function, denoted by $H(t-a)$ is defined by $H(t-a) = \begin{cases} 0, & t < a \\ 1, & t > a \end{cases}$

$$\therefore \quad L\{H(t-a)\} = \int_0^\infty e^{-pt}H(t-a)dt = \int_a^\infty e^{-pt}.1\,dt = \left[\frac{e^{-pt}}{-p}\right]_a^\infty$$

$$= \left[\lim_{t\to\infty}\frac{e^{-pt}}{-p}\right] + \frac{e^{-ap}}{p} = 0 + \frac{e^{-ap}}{p},\ p > 0 = \frac{e^{-ap}}{p},\ p > 0$$

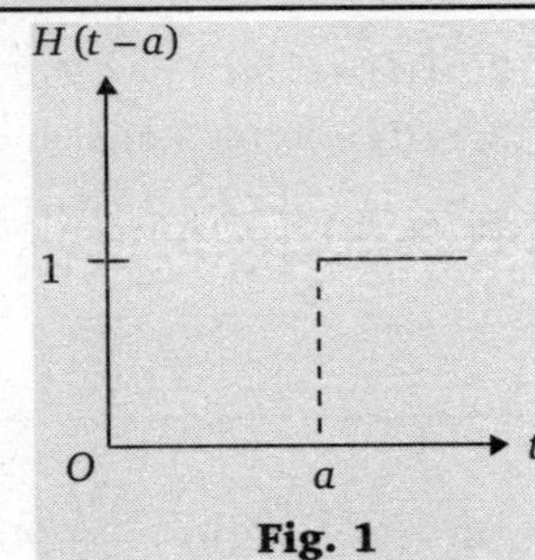

Fig. 1

THEOREM 1. *If $L\{F(t)\} = f(p)$, then $L\{F(t-a).H(t-a)\} = e^{-ap}f(p)$.*

Proof. We have $L\{F(t-a).H(t-a)\} = \int_0^\infty e^{-pt}[F(t-a).H(t-a)]\,dt = \int_0^a e^{-pt}F(t-a).0\,dt + \int_a^\infty e^{-pt}F(t-a).(1)dt$

$$= \int_a^\infty e^{-pt}F(t-a)dt = \int_0^\infty e^{-p(u+a)}F(u)du \qquad [\text{Putting } t-a = u]$$

$$= e^{-ap}\int_0^\infty e^{-pu}F(u)\,du = e^{-ap}f(p).$$

THEOREM 2. $L\{F(t).H(t-a)\} = e^{-ap}L\{F(t+a)\}dt$.

Proof. We have $L\{F(t).H(t-a)\} = \int_0^\infty e^{-pt}[F(t).H(t-a)]\,dt = \int_0^a e^{-pt}[F(t).H(t-a)]\,dt + \int_a^\infty e^{-pt}[F(t).H(t-a)]\,dt$

$$= 0 + \int_0^\infty e^{-pt}.F(t).(1)dt = \int_0^\infty e^{-p(u+a)}F(u+a)du, \qquad [\text{Putting } t-a = u]$$

$$= e^{-ap}\int_0^\infty e^{-pu}F(u+a)\,du = e^{-ap}\int_0^\infty e^{-pt}F(t+a)dt = e^{-ap}L\{F(t+a)\}$$

Solved Examples

Example 1. *Find the Laplace transform of $t^2H(t-3)$.*

Solution. We have $L\{t^2H(t-3)\} = e^{-3p}L\{(t+3)^2\}$

$$= e^{-3p}L\{t^2+6t+9\}$$

$$= e^{3p}\left[\frac{2}{p^3}+\frac{6}{p^2}+\frac{9}{p}\right]$$

Example 2. *Express the following function in terms of unit step function and find its Laplace transform :*

$$F(t) = \begin{cases} t-1, & 1<t<2 \\ 3-t, & 2<t<3 \end{cases}.$$

Solution. We have

$$F(t) = \begin{cases} t-1, & 1<t<2 \\ 3-t, & 2<t<3 \end{cases}$$

$$= (t-1)[H(t-1)-H(t-2)] + (3-t)[H(t-2)-H(t-3)]$$

$$= (t-1)H(t-1)-(t-1)H(t-2) + (3-t)H(t-2)+(t-3)H(t-3)$$

$$= (t-1)H(t-1)-2(t-2)H(t-2) + (t-3)H(t-3).$$

$$\therefore\ L\{F(t)\} = \frac{e^{-p}}{p^2}-2\frac{e^{-2p}}{p^2}+\frac{e^{-3p}}{p^2}.$$

Example 3. *Express the following function in terms of unit step function and find its Laplace transform :* (UPTU–2002)

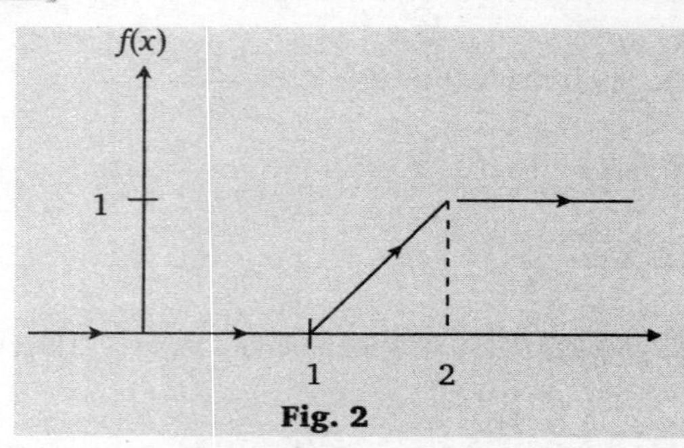

Fig. 2

Solution. The algebraic form of the function in the figure is

$$F(t) = \begin{cases} 0, & 0<t<1 \\ t-1, & 1<t<2 \\ 1, & 2<t \end{cases} \quad \ldots(1)$$

$$= (t-1)[H(t-1)-H(t-2)]+H(t-2)$$

$$= (t-1)H(t-1)-(t-1-1)H(t-2)$$

$$= (t-1)H(t-1)-(t-2)H(t-2)$$

$$\therefore\ L\{F(t)\} = L\{(t-1)H(t-1)\}-L\{(t-2)H(t-2)\}$$

$$= \frac{e^{-p}}{p^2}-L\{(t-2)H(t-2)\}$$

$$= \frac{e^{-p}}{p^2}-\frac{e^{-2p}}{p^2}.$$

EXERCISE 36.5

Express the following functions in terms of unit step function. Also find their Laplace transforms :

1. $F(t)=2t$ for $0<t<\pi$, $F(t)=1$ for $t>\pi$.
2. $F(t)=t^2$ for $0<t\le 2$, $F(t)=0$ for $t>0$.
3. $F(t)=\cos(\omega t+\phi)$ for $0<t<T$, $F(t)=0$ for $t>T$.
4. $f(t) = \begin{cases} t, & 0<t<2 \\ 0, & t>2 \end{cases}.$
5. $f(t) = \begin{cases} \sin t, & 0<t<\pi \\ t, & t>\pi \end{cases}$
6. $f(t) = \begin{cases} 4, & 0<t<1 \\ -2, & 1<t<3 \\ 5, & t>3 \end{cases}$
7. Find Laplace transform of $tU_2(t)$.
8. Using unit step function, find the Laplace transform of
   (i) $(t-1)^2u(t-1)$ (ii) $\sin t\ u(t-4)$ (UPTU–2008)

Answers

8. (i) $\frac{2e^{-p}}{p^3}$ (ii) $\frac{e^{-4p}}{p^3}(4p^2+4p+2)$

Miscellaneous Solved Examples

Example 1. *Find $L\left\{\sqrt{t}e^{3t}\right\}$.* (PTU–2009)

Solution. $$L\{\sqrt{t}\} = \frac{\boxed{3/2}}{p^{3/2}} = \frac{(1/2).\sqrt{\pi}}{p^{3/2}}$$

$\therefore$ By shifting property, we get

$$L(e^{3t}\sqrt{t}) = \frac{\sqrt{\pi}}{2} = \frac{1}{(p-3)^{3/2}}.$$

Example 2. *Find the Laplace transform of the function*

$$f(t)=|t-1|+|t+1|, t\ge 0.$$ (SVTU–2009)

Solution. Given function is equivalent to

$$f(t) = \begin{cases} 2 & 0\le t\le 1 \\ 2t & t\ge 1 \end{cases}$$

$$\therefore\ Lf(t) = \int_0^1 e^{-pt}(2)dt + \int_1^{\infty} e^{-pt}(2t)dt$$

$$= 2\left[\left|\frac{e^{-pt}}{-p}\right|_0^1 + 2\left|\frac{t.e^{-pt}}{-p}\right|_1^\infty - \left|\frac{e^{-pt}}{(-p)^2}\right|_1^\infty\right]$$

$$= 2\left(\frac{e^{-p}}{-p} + \frac{1}{p}\right) + 2\left(\frac{0-e^{-p}}{-p} - \frac{0-e^{-p}}{p^2}\right)$$

$$= \frac{2}{p}\left(1 + \frac{e^{-p}}{p}\right).$$

Example 3. *Find the Laplace trransform of the function $f(t) = [t]$ where [ ] stands for the greatest integer function.* (PTU–2010)

Solution. Given function is equivalent to

$[t] = 0$ in $(0, 1) + 1$ in $(1, 2) + 2$ in $(2, 3) + 3$ in $(3, 4) + ...$

$$\therefore \quad L[F(t)] = \int_0^\infty e^{-pt}[f(t)]dt = \int_0^\infty e^{-pt}[t]dt$$

$$= \int_0^1 e^{-pt}(0)dt + \int_1^2 e^{-pt}(1)dt + \int_2^3 e^{-pt}(2)dt + \int_3^4 e^{-pt}(3)dt + ... + \infty$$

$$= 0 + \left|\frac{e^{-pt}}{-p}\right|_1^2 + 2\left|\frac{e^{-pt}}{-p}\right|_2^3 + 3\left|\frac{e^{-pt}}{-p}\right|_3^4 + ... + \infty$$

$$= \frac{-1}{p}[(e^{-2p} - e^{-p}) + 2(e^{-3p} - e^{-2p}) + 3(e^{-4p} - e^{-3p}) + ... + \infty]$$

$$= \frac{1}{p}(e^{-p} + e^{-2p} + e^{-3p} ... + \infty)$$

$$= \frac{1}{p}\left(\frac{e^{-p}}{1-e^{-p}}\right) = \frac{1}{p(e^{p}-1)}.$$

Example 4. *Find $L(erf\, 2\sqrt{t})$.* (Mumbai–2006)

Solution. Since we know that $L(erf\sqrt{t}) = \dfrac{1}{p(p+1)}$

$$\therefore L(erf\, 2\sqrt{t}) = L[erf\sqrt{(4t)}]$$

$$= \frac{1}{4}.\frac{1}{\frac{p}{4}\sqrt{\left(\frac{p}{4}+1\right)}} = \frac{2}{p\sqrt{(p+4)}}.$$

Example 5. *Find $L\{t^3 e^{-3t}\}$.* (Kottayam–2005)

Solution. Since $L(e^{-3t}) = \dfrac{1}{p+3}$

$$\therefore \quad L(t^3 e^{-3t}) = (-1)^3 \frac{d^3}{dp^3}\left(\frac{1}{p+3}\right) = -\frac{(-1)^3.3!}{(p+3)^{3+1}}$$

$$= \frac{6}{(p+3)^4}.$$

Example 6. *Find $L\{(1-e^t)/t\}$.* (Madras–2000)

Solution. $\because \quad L(1-e^t) = L(1) - L(e^t) = \dfrac{1}{p} - \dfrac{1}{p-1}$

$$\therefore L\left(\frac{1-e^t}{t}\right) = \int_p^\infty \left(\frac{1}{p} - \frac{1}{p-1}\right)dp$$

$$= |\log p - \log(p-1)|_p^\infty = \left|\log\left(\frac{p}{p-1}\right)\right|_p^\infty$$

$$= -\log\left[\frac{1}{1-1/p}\right] = \log\left(\frac{p-1}{p}\right).$$

Example 7. *Find $L\left\{\dfrac{\cos at - \cos bt}{t} + t\sin at\right\}$.* (VTU–2010)

Solution. $\because \quad L(\cos at - \cos bt) = \dfrac{p}{p^2+a^2} - \dfrac{p}{p^2+b^2}$

and $L(\sin at) = \dfrac{a}{p^2+a^2}$

$$\therefore L\left(\frac{\cos at - \cos bt}{t}\right) + L(t\sin at)$$

$$= \int_p^\infty \left(\frac{p}{p^2+a^2} - \frac{p}{p^2+b^2}\right)dp - \frac{d}{dp}\left(\frac{a}{p^2+a^2}\right)$$

$$= \left|\frac{1}{2}\log(p^2+a^2) - \frac{1}{2}\log(p^2+b^2)\right|_p^\infty - a\frac{(-2p)}{(p^2+a^2)^2}.$$

$$= \frac{1}{2}\lim_{p\to\infty}\log\frac{p^2+a^2}{p^2+b^2} - \frac{1}{2}\log\frac{p^2+a^2}{p^2+b^2} + \frac{2ap}{(p^2+a^2)^2}$$

$$= \frac{1}{2}\log\left(\frac{1+0}{1+0}\right) - \frac{1}{2}\log\left(\frac{p^2+a^2}{p^2+b^2}\right) + \frac{2ap}{(p^2+a^2)^2}$$

$$= \log\left(\frac{p^2+b^2}{p^2+a^2}\right)^{1/2} + \frac{2ap}{(p^2+a^2)^2}. \qquad [\because \log 1 = 0]$$

Example 8. *Find $L\left\{e^{-t}\int_0^t \frac{\sin t}{t}dt\right\}$.* (Madras–2006)

Solution. $\because \quad L\{\sin t\} = \dfrac{1}{p^2+1}$

$$L\left(\frac{\sin t}{t}\right) = \int_0^\infty \frac{1}{p^2+1}dp$$

$$= \frac{\pi}{2} - \tan^{-1} p = \cot^{-1} p$$

$$\therefore \quad L\left(\int_0^t \frac{\sin t}{t}dt\right) = \frac{1}{p}\cot^{-1} p.$$

Now by shiefting property,

$$L\left\{e^{-t}\left(\int_0^t \frac{\sin t}{t}dt\right)\right\} = \frac{1}{p+1}\cot^{-1}(p+1).$$

Example 9. *Find $L\left\{t\int_0^t \frac{e^{-t}\sin t}{t}dt\right\}$.* (PTU–2005)

Solution. $\because \quad L\left(\dfrac{\sin t}{t}\right) = \cot^{-1} p$

$$\therefore L\left(e^{-t}.\frac{\sin t}{t}\right) = \cot^{-1}(p+1)$$

and $L\left\{\int_0^t e^{-t}.\frac{\sin t}{t}dt\right\} = \frac{1}{p}\cot^{-1}(p+1)$

Hence, $L\left\{t\int_0^t e^{-t}.\frac{\sin t}{t}dt\right\} = -\frac{d}{dp}\left[\frac{\cot^{-1}(p+1)}{p}\right]$

$$= -\frac{p\left[\frac{-1}{1+(p+1)^2}\right] - \cot^{-1}(p+1)}{p^2}$$

$$= \frac{p+(p^2+2p+2)\cot^{-1}(p+1)}{p^2(p^2+2p+2)}.$$

Example 10. Find $L\left\{\int_0^t\int_0^t\int_0^t (t\sin t)dtdtdt\right\}$. (Mumbai–2006)

Solution. $\because \quad L(\sin t) = \frac{1}{p^2+1}$

$$\therefore \quad L(t\sin t) = -\frac{d}{dp}\frac{1}{(p^2+1)} = \frac{2p}{(p^2+1)^2}$$

Thus, $L\left\{\int_0^t\int_0^t\int_0^t (t\sin t)dtdtdt\right\} = \frac{1}{p^3}L(t\sin t)$

$$= \frac{1}{p^3}.\frac{2p}{(p^2+1)^2} = \frac{2}{p^2(p^2+1)^2}$$

Example 11. Find $L\int_0^\infty te^{-3t}\sin t dt$. (VTU–2007)

Solution. $\int_0^\infty te^{-3t}\sin t dt = \int_0^\infty e^{-pt}(t\sin t)dt$, where $p = 3$

$= L(t\sin t)$, By def.

$$= (-1)\frac{d}{dp}\left(\frac{1}{p^2+1}\right)$$

$$= \frac{2p}{(p^2+1)^2} = \frac{2\times 3}{(3^2+1)^2} = \frac{3}{50}.$$

Example 12. Find $\int_0^\infty e^{-t}\left(\frac{\cos at - \cos bt}{t}\right)dt$. (Mumbai–2009)

Solution. $\because L(\cos at) = \frac{p}{p^2+a^2}$ and $L(\cos bt) = \frac{p}{p^2+b^2}$

$$\therefore L\frac{\cos at - \cos bt}{t} = \int_p^\infty\left(\frac{p}{p^2+a^2} - \frac{p}{p^2+b^2}\right)dp$$

$$= \frac{1}{2}\left\{\log\left(\frac{p^2+a^2}{p^2+b^2}\right)\right\}_p^\infty$$

$$= \frac{1}{2}\log\left(\frac{p^2+b^2}{p^2+a^2}\right)$$

$$\Rightarrow \int_0^\infty e^{-pt}\left(\frac{\cos at - \cos bt}{t}\right)dt = \frac{1}{2}\log\left(\frac{p^2+b^2}{p^2+a^2}\right)$$

Taking $p = 1$, we get

$$\int_0^\infty e^{-t}\left(\frac{\cos at - \cos bt}{t}\right)dt = \frac{1}{2}\log\left(\frac{1+b^2}{1+a^2}\right).$$

MISCELLANEOUS EXERCISE

Find the Laplace transform of :

1. $e^{2t} + 4t^3 - 2\sin 3t + 3\cos 3t$ (JNTU–2003)
2. $\sin 2t \cos 3t$ (Kottayam–2005)
3. $\sin^5 t$ (Mumbai–2007)
4. $e^{2t}(3t^5 - \cos 4t)$ (PTU–2007)
5. $e^{-3t}\sin 5t \sin 3t$ (VTU–2006)
6. $e^{2t}\sin^4 t$ (Mumbai–2007)
7. $\cosh at \sin at$ (Delhi–2002)
8. $\sinh 3t \cos^2 t$ (Madras–2000)
9. $t^2 e^{2t}$ (VTU–2008S)
10. $t\sqrt{(1+\sin t)}$ (Mumbai–2007)
11. $f(t) = \begin{cases} 4, & 0 \le t \le 1 \\ 3, & t > 1 \end{cases}$ (UPTU–2009)
12. $f(x) = \begin{cases} \sin(x-\pi/3), & x > \pi/3 \\ 0, & x < \pi/3 \end{cases}$ (Rajasthan–2006)
13. Find the Laplace transform of the saw-toothed wave of period T, given $f(t) = t/T$ for $0 < t < T$. (VTU–2007)
14. Find the Laplace transform of the square wave function of period a defined as
 $f(t) = k$ when $0 < t < a$
 $= -k$ when $a < t < 2a$. (VTU–2011)
15. Find the Laplace transform of the triangular wave of period $2a$ given by
 $f(t) = t, 0 < t < a$
 $= 2a - t, a < t < 2a$.
 (Nagarjuna–2008, VTU–2008S, UPTU–2002)
16. $t\sin^2 t$ (Nagarjuna–2008)
17. $\sin 2t - 2t\cos 2t$ (Anna–2003)
18. $e^{2t}\sin 3t$ (Madras–2003)
19. $te^{-2t}\sin 4t$ (VTU–2008)
20. $t^2 e^{-3t}\sin 2t$ (Madras–2000S)
21. $(e^{-at} - e^{-bt})/t$ (Anna–2005S)
22. $\frac{(\sin t \sin 5t)}{t}$ (Mumbai–2008)
23. $(1-\cos 3t)/t$ (VTU–2006)
24. $(1-\cos t)/t^2$ (Hazaribag–2008)
25. $2^t + \frac{\cos 2t - \cos 3t}{t} + t\sin t$ (VTU–2004)
26. $\int_0^\infty \frac{e^{-\sqrt{2}t}\sin nt \sin t}{t}dt$ (Mumbai–2005)

27. $\int_0^\infty te^{-2t}\sin 3t\,dt$ (VTU–2008)

28. Prove that $\int_0^\infty \frac{e^{-2t}\sin nt}{t}dt = \frac{1}{2}\log 3$. (Mumbai–2008)

29. Prove that $\int_0^\infty \frac{e^{-t}\sin^2 t}{t}dt = \frac{1}{4}\log 5$. (Kurukshetra–2006)

30. $L\int_0^t \frac{e^t \sin t}{t}dt$ (PTU–2009S, SVTU–2009, Bhopal–2008)

ANSWERS

1. $\frac{1}{p-2}+\frac{24}{p^4}+\frac{3(p-2)}{p^4+9}$ **2.** $\frac{2(p^2-5)}{(p^2+1)(p^2+25)}$ **3.** $\frac{5}{4}\left\{\frac{1}{p^2+1}-\frac{3/2}{p^2+9}+\frac{1/2}{p^2+25}\right\}$ **4.** $\frac{60}{p-2}-\frac{p-2}{p^2-4p+20}$

5. $\frac{30(p+3)}{(p^2+6p+13)(p^2+6p+73)}$ **6.** $\frac{1}{8}\left\{\frac{3}{(p-2}-\frac{4(p-2)}{p^2-4p+8}+\frac{p-4}{p^2-8p+32}\right\}$ **7.** $\frac{a(p^2+2a^2)}{p^4+4a^4}$

8. $\frac{3}{2}\left[\frac{1}{p^2-9}+\frac{p^2-13}{p^4-10p^2+169}\right]$ **9.** $\frac{2}{(p+2)^3}$ **10.** $\frac{4(4p^2+4p-1)}{(4p^2+1)^2}$ **11.** $\frac{4}{p}-\frac{e^{-p}}{p}$ **12.** $\frac{e^{-\pi p/3}}{p^2+1}$

13. $(1/p^2T)-e^{-pT}/p(1-e^{pT})$ **14.** $(a/p)\tanh(ap/2)$ **15.** $(1/p^2)\tanh\frac{1}{2}ap$ **16.** $\frac{2(3p^2+4)}{p^2(p^2+4)^2}$ **17.** $\frac{16}{(p^2+4)^2}$

18. $\frac{6(p-2)}{(p^2-4p+13)^2}$ **19.** $\frac{8(p+2)}{p^2+4p+20}$ **20.** $\frac{2(p^3+6p^2+9p+2)}{(p^2+4p+5)^3}$ **21.** $\log\{(p+b)/(p+a)\}$

22. $\frac{1}{2}\log\{(p^2+36)/(p^2+16)\}$ **23.** $\frac{1}{2}\log\left(\frac{p^2+9}{p^2}\right)$ **24.** $\cot^{-1}p-\frac{1}{2}p\log(1+p^{-2})$

25. $\frac{1}{p-\log 2}+\frac{2p}{(p^2+1)^2}+\frac{1}{2}\log\left(\frac{p^2+9}{p^2+4}\right)$ **26.** $\pi/8$ **27.** 12/169 **30.** $\frac{\cot^{-1}(p-1)}{p}$

Objective Evaluations

Fill in the Blanks

1. An integral of the type $\int_{-\infty}^{\infty} k(p,t)\, F(t)\, dt$ is called __________ of $F(t)$.
2. The function $k(p,t)$ is known as __________ of the transform.
3. If the integral $\int_0^{\infty} e^{-pt} F(t)\, dt$ __________ for some value of p, then only the Laplace transform of $f(t)$ exists.
4. $L\{1\} =$ __________
5. $L\{t^n\} = \frac{\_\_\_}{p^{n+1}}$

True/False

Write 'T' for True and 'F' for False statement.

1. $L\{\cos at\} = \frac{p}{p^2+a^2}$. **(T/F)**
2. $L\{\sin at\} = \frac{a}{p^2+a^2}$. **(T/F)**
3. $L\{e^{at}\} = \frac{1}{p-a}$. **(T/F)**
4. $L\{\sinh at\} = \frac{p}{p^2-a^2}$. **(T/F)**
5. $L\{\cosh at\} = \frac{p}{p^2-a^2}$. **(T/F)**

Multiple Choice Questions

Choose the most appropriate one.

1. The Laplace transform of 1 is :
   (a) $1/p$ (b) $1/p^2$ (c) $1/\sqrt{p}$ (d) none of these
2. The Laplace transform of t is :
   (a) $1/p$ (b) $1/p^2$ (c) $1/\sqrt{p}$ (d) none of these
3. The Laplace transform of $t^{n-1}/(n-1)!$ is :
   (a) $\frac{1}{p^{n-1}}$ (b) $1/p^n$ (c) $\frac{1}{p^{n+1}}$ (d) none of these
4. The Laplace transform of $\frac{t^{n-1}}{\Gamma(a)}$ is :
   (a) $\frac{1}{p^{n-1}}$ (b) $1/p^n$ (c) $\frac{1}{p^{n+2}}$ (d) none of these
5. The Laplace transform of e^{at} is :
   (a) $\frac{1}{p-a}$ (b) $\frac{1}{(p-a)^2}$ (c) $\frac{1}{(p-a)^n}$ (d) none of these
6. The Laplace transform of te^{at} is :
   (a) $\frac{1}{p-a}$ (b) $\frac{1}{(p-a)^2}$ (c) $\frac{1}{(p-a)^n}$ (d) none of these
7. The Laplace transform of $\frac{1}{(n-1)!} t^{n-1} e^{at}$ is :
   (a) $\frac{1}{p-a}$ (b) $\frac{1}{(p-a)^2}$
   (c) $\frac{1}{(p-a)^n}, n = 1, 2, 3, \ldots$ (d) none of these
8. The Laplace transform of $\frac{1}{a-b}(e^{at} - e^{bt})$ is :
   (a) $\frac{1}{(p-a)(p-b)}, (a \neq b)$ (b) $\frac{p}{(p-a)(p-b)}, (a \neq b)$
   (c) $\frac{p}{(p+a)(p+b)}, (a \neq b)$ (d) none of these
9. If $L\{F(t)\} = f(p)$, then $L\{F'(t)\}$ is :
   (a) $L\{f'(t)\} = f(p)$ (b) $L\{f'(t)\} = f(b) + f(0)$
   (c) $L\{f'(t)\} = pf(p) + f(0)$ (d) $L\{f'(t)\} = pf(p) - f(0)$
10. If $L\{F(t)\} = f(p)$, then $L\{F''(t)\}$ is :
    (a) $p^2 f(p) - pF(0) - F'(0)$ (b) $pf(p) - pF(0)$
    (c) $f''(p)$ (d) none of these
11. The Laplace transform of $f(t)$ is $f(p)$, then :
    (a) $L\{t\, F(t)\} = f(p)$ (b) $L\{tF(t)\} = -f(p)$
    (c) $L\{tF(t)\} = f'(p)$ (d) none of these
12. If $u(x,t)$ is a function of two variables x and t and $L\{u(x,t)\} = U(x,p)$, then $L\left(\frac{\partial u}{\partial t}\right) =$:
    (a) $pU(x,p) - u(x,0)$ (b) $pU(x,p)$
    (c) $u(x,0)$ (d) none of these

Answers

Fill in the Blanks

1. integral transform 2. kernel 3. exist 4. $1/p$ 5. $n!$

True/False

1. T 2. T 3. T 4. F 5. T

Multiple Choice Questions

1. (a) 2. (b) 3. (b) 4. (b) 5. (a) 6. (b) 7. (c) 8. (a) 9. (d) 10. (a)
11. (b) 12. (a)

FFFFFF

CHAPTER 37 The Inverse Laplace Transform

37.1 INTRODUCTION

If the Laplace transform of a function $F(t)$ is $f(p)$, i,e., if $L\{F(t)\} = f(p)$. Then, $F(t)$ is known as inverse Laplace transform of $F(p)$. Symbolically, $F(t) = L^{-1}\{f(p)\}$. where, L^{-1} is called the inverse Laplace transformation operator.

For example, If $L\{e^{-2t}\} = \dfrac{1}{p+2}$. Then we can write $L^{-1}\left(\dfrac{1}{p+2}\right) = e^{-2t}$

37.1.1 NULL FUNCTION

A function $N(t)$ of t such that $\int_0^t N(t)\,dt = 0, \quad \forall t > 0$ is called null function.

37.1.2 UNIQUENESS OF INVERSE LAPLACE TRANSFORMS

Since, we know that the Laplace transform of a null function $N(t)$ is zero. Also, it is clearly that if $L(F(t)\} = f(p)$, then also

$$L\{F(t) + N(t)\} = f(p)$$

It follows that we can have two different functions with same Laplace transform.

If we allow null functions, we see that the inverse Laplace transform is nol unique. It is unique, however, if we disallow null functions.

37.1.3 LEARCH THEOREM

If we restrict ourselves to functions $F(t)$ which are sectionally continuous in every finite interval $0 \le t \le N$ and of exponential order for $t > N$, then the inverse Laplace transform of $f(p)$.

i.e., $L^{-1}\{f(p)\} = F(t)$, is unique.

37.2 SOME INVERSE LAPLACE TRANSFORMS

	$f(p)$	$L^{-1}[f(p)] = F(t)$		$f(p)$	$L^{-1}[f(p)] = F(t)$
(i)	$\dfrac{1}{p}$	1	**(vi)**	$\dfrac{1}{p^2+a^2}$	$\cos at$
(ii)	$\dfrac{1}{p^2}$	t	**(vii)**	$\dfrac{1}{p^2-a^2}$	$\dfrac{\sin h\, at}{a}$
(iii)	$\dfrac{1}{p^{n+1}}, n = 0, 1, 2...$	$\dfrac{t^n}{n!}$	**(viii)**	$\dfrac{p}{p^2-a^2}$	$\cos h\, at$
(iv)	$\dfrac{1}{p-a}$	e^{at}	**(ix)**	$\dfrac{\Gamma(a+1)}{p^n}$	$t^a,\ a > -1$
(v)	$\dfrac{1}{p^2+a^2}$	$\dfrac{\sin at}{a}$			

37.3 IMPORTANT PROPERTIES OF INVERSE LAPLACE TRANSFORM

(i) Linearity Property.

If C_1 and C_2 are any constants while $f_1(p)$ and $f_2(p)$ are the Laplace transform $F_1(t)$ and $F_2(t)$ respectively, then

$$L^{-1}\{C_1 f_1(p) + C_2 f_2(p)\} = C_1 L^{-1}\{f_1(p)\} + C_2 L^{-1}\{f_2(p)\}$$

Proof. We have

$$L\{C_1 F_1(t) + C_2 F_2(t)\} = C_1 L\{F_1(t)\} + C_2 L\{F_2(t)\} = C_1 f_1(p) + C_2 f_2(p)$$

$$\Rightarrow \quad L^{-1}\{C_1 f_1(p) + C_2 f_2(p)\} = C_1 F_1(t) + C_2 F_2(t) = C_1 L^{-1}\{f_1(p)\} + C_2 L^{-1}\{f_2(p)\}.$$

(ii) First Translation or Shifting Theorem.

If $\quad L^{-1}\{f(p)\} = F(t)$, then

$$L^{-1}\{f(p-a)\} = e^{at}\ F(t) = e^{at} L^{-1}\{f(p)\}$$

Proof. We have $\quad f(p) = \int_0^\infty e^{-pt}\ F(t)\,dt$

$$\Rightarrow \quad f(p-a) = \int_0^\infty e^{-(p-a)t}\ F(t)\,dt = \int_0^\infty e^{-pt}\{e^{at}\ F(t)\}\,dt = L\,\{e^{at}\ F(t)\}$$

Hence, $\quad L^{-1}\{f(p-a)\} = e^{at} F(t) = e^{at}\ L^{-1}\ \{f(p)\}.$

(iii) Second Translation or Shifting Theorem. If $L^{-1}\{f(p)\} = F(t)$, then $L^{-1}\{e^{-ap} f(p)\} = G(t)$, where

$$G(t) = \begin{cases} F(t-a), & t > a \\ 0, & t < a \end{cases}$$

Proof. We know that $\quad f(p) = \int_0^\infty e^{-pt}\ F(t)\,dt$

Therefore, $\quad e^{-ap} f(p) = \int_0^\infty e^{-p(t+a)}\ F(t)\,dt$

$$= \int_0^\infty e^{-px} F(x-a)dx, \text{ putting } t+a = x \Rightarrow dt = dx$$

$$= \int_0^a e^{-px}.0\,dx + \int_a^\infty e^{-px} F(x-a)\,dx = \int_0^a e^{-pt}.0\,dt + \int_a^\infty e^{-pt} F(t-a)\,dt$$

$$= \int_0^\infty e^{-pt} G(t)\,dt \;=\; L\{G(t)\}$$

where, $G(t) = \begin{cases} F(t-a), & t > a \\ 0, & t < a \end{cases}$ shows, $L^{-1}\{e^{ap}\ f(p)\} = G(t)$

(iv) Change of Scale Property.

If $L^{-1}\{f(p)\} = F(t)$, then $L^{-1}\{f(ap)\} = \dfrac{1}{a} F\left(\dfrac{t}{a}\right)$

Proof. We know that $\quad f(ap) = \dfrac{1}{a}\int_0^\infty e^{-px}\ F\left(\dfrac{x}{a}\right)dx \quad \Rightarrow \quad f(ap) = \int_0^\infty e^{-apt} F(t)\,dt$

Putting $\quad at = x \quad \Rightarrow dt = \dfrac{1}{a}dx$, we get

$$f(ap) = \frac{1}{a}\int_0^\infty e^{-px} F\left(\frac{x}{a}\right)dx = \frac{1}{a}\int_0^\infty e^{-pt} F\left(\frac{t}{a}\right)dt \quad \text{[By the property of definite integral]}$$

$$= \frac{1}{a} L\left\{F\left(\frac{t}{a}\right)\right\} = L\left\{\frac{1}{a} F\left(\frac{t}{a}\right)\right\}$$

Hence, $\quad L^{-1}\{f(ap)\} = \dfrac{1}{a} F\left(\dfrac{t}{a}\right)$

Solved Examples

Example 1. *Find the inverse Laplace transforms of the following functions :*

(i) $\dfrac{2p+1}{p(p+1)}$ $\qquad$ (ii) $\dfrac{3p-8}{4p^2+25}$

Solution. (i) We have

$$L^{-1}\left\{\frac{2p+1}{p(p+1)}\right\} = L^{-1}\left\{\frac{p+(p+1)}{p(p+1)}\right\}$$

$$= L^{-1}\left\{\frac{1}{p+1}\right\} + L^{-1}\left\{\frac{1}{p}\right\} = e^{-t}+1$$

(ii) Here, we have

$$L^{-1}\left\{\frac{3p-8}{4p^2+25}\right\} = \frac{3}{4}L^{-1}\left\{\frac{p}{p^2+\left(\frac{5}{2}\right)^2}\right\} - 2L^{-1}\left\{\frac{1}{p^2+\left(\frac{5}{2}\right)^2}\right\}$$

$$= \frac{3}{4}\cos\left(\frac{5}{2}t\right) - 2.\frac{2}{5}\sin\left(\frac{5}{2}t\right)$$

$$= \frac{3}{4}\cos\left(\frac{5}{2}t\right) - \frac{4}{5}\sin\left(\frac{5}{2}t\right)$$

Example. 2 *Show that* $\dfrac{1}{p^{1/2}} = L\left[\dfrac{1}{\sqrt{\pi t}}\right]$ [UPTU 2005]

Solution. We have to show that $\dfrac{1}{p^{1/2}} = L\left[\dfrac{1}{\sqrt{\pi t}}\right]$

Now, $L^{-1}\left[\dfrac{1}{p^n}\right] = \dfrac{t^{n-1}}{(n-1)!} = \dfrac{t^{n-1}}{\Gamma(n)}$

So, $L^{-1}\left[\dfrac{1}{p^{1/2}}\right] = \dfrac{t^{\frac{1}{2}-1}}{\Gamma(1/2)} = \dfrac{t^{-1/2}}{\Gamma(1/2)} = \dfrac{t^{-1/2}}{\sqrt{\pi}}$

$$L^{-1}\left[\frac{1}{p^{1/2}}\right] = \frac{1}{\sqrt{\pi t}} \Rightarrow \frac{1}{p^{1/2}} = L\left[\frac{1}{\sqrt{\pi t}}\right].$$

Example 3. *Find* $L^{-1}\left\{\dfrac{3p-2}{p^{5/2}} - \dfrac{7}{3p+2}\right\}$

Solution. Here, we have

$$L^{-1}\left\{\frac{3p-2}{p^{5/2}} - \frac{7}{3p+2}\right\} = 3L^{-1}\left\{\frac{1}{p^{3/2}}\right\}$$

$$-2L^{-1}\left\{\frac{1}{p^{5/2}}\right\} - \frac{7}{3}L^{-1}\left\{\frac{1}{p+(2/3)}\right\}$$

$$= 3\frac{t^{1/2}}{\Gamma\left(\frac{3}{2}\right)} - 2\frac{t^{3/2}}{\Gamma\left(\frac{5}{2}\right)} - \frac{7}{3}e^{\left(-\frac{2}{3}\right)t}$$

$$= 6\sqrt{\left(\frac{t}{\pi}\right)} - \frac{8}{3}t\sqrt{\left(\frac{t}{\pi}\right)} - \frac{7}{3}e^{-2t/3}.$$

Example 4. *Find*

$$L^{-1}\left\{\frac{3}{p^2-3} + \frac{3p+2}{p^3} - \frac{3p-27}{p^2+9} + \frac{6-30\sqrt{p}}{p^4}\right\}.$$

Solution. We have

$$L^{-1}\left\{\frac{3}{p^2-3} + \frac{3p+2}{p^3} - \frac{3p-27}{p^2+9} + \frac{6-30\sqrt{p}}{p^4}\right\}$$

$$= L^{-1}\left\{\frac{3}{p^2-3} + \frac{3}{p^2} + \frac{2}{p^3} - \frac{3p}{p^2+9} + \frac{27}{p^2+9} + \frac{6}{p^4} - \frac{30}{p^{7/2}}\right\}$$

$$= 3L^{-1}\left\{\frac{1}{p^2-(\sqrt{3})^2}\right\} + 3L^{-1}\left\{\frac{1}{p^2}\right\}$$

$$+ 2L^{-1}\left\{\frac{1}{p^3}\right\} - 3L^{-1}\left\{\frac{p}{p^2+3^2}\right\}$$

$$+ 27L^{-1}\left\{\frac{1}{p^2+3^2}\right\}$$

$$+ 6L^{-1}\left\{\frac{1}{p^4}\right\} - 30L^{-1}\left\{\frac{1}{p^{7/2}}\right\}$$

$$= 3.\frac{1}{\sqrt{3}}\sinh\sqrt{3}.t + 3.\frac{t^{2-1}}{(2-1)!}$$

$$+ 2.\frac{t^{3-1}}{(3-1)!} - 3\cos 3t + \frac{27}{3}\sin 3t$$

$$+ 6\frac{t^4-1}{(4-1)!} - 30\frac{t^{7/2-1}}{\Gamma\left(\frac{7}{2}\right)}$$

$$= \sqrt{3}\sinh\sqrt{3}t + 3t + t^2 - 3\cos 3t$$

$$+9\sin 3t + t^3 - 16t^2\sqrt{\left(\frac{t}{\pi}\right)}.$$

Example 5. *A function* $f(t)$ *obey the equation* $f(t) + 2\int_0^t f(t)\,dt = \cosh 2t$. *Find the Laplace transformation of* $f(t)$. [UPTU 2006]

Solution. We have $f(t) + 2\int_0^t f(t)\,dt = \cosh 2t$

Taking Laplace transformation of both the sides, we get

$$L\{f(t)\} + 2L\int_0^t f(t)\,dt = L(\cosh 2t)$$

$$\Rightarrow \quad F(p) + 2.\frac{1}{p}F(p) = \frac{p}{p^2-4}$$

$$\Rightarrow \quad F(p)\left[1+\frac{2}{p}\right] = \frac{p}{p^2-4}$$

$$\Rightarrow \quad F(p).\left[\frac{p+2}{p}\right] = \frac{p}{p^2-4}$$

$$\Rightarrow \quad F(p) = \left(\frac{p}{p^2-4}\right).\left(\frac{p}{p+2}\right)$$

$$\Rightarrow \quad F(p) = \frac{p^2}{(p^2-4)(p+2)}$$

Example 6. *Show that* $L^{-1}\left\{\dfrac{1}{p}\cos\dfrac{1}{p}\right\} = 1 - \dfrac{t^2}{(2!)^2} + \dfrac{t^4}{(4!)^2} - \dfrac{t^6}{(6!)^2} + \ldots$

Solution.

$$L^{-1}\left\{\frac{1}{p}\cos\frac{1}{p}\right\} = L^{-1}\left\{\frac{1}{p}\left(1 - \frac{(1/p)^2}{2!} + \frac{(1/p)^4}{4!} - \frac{(1/p)^6}{6!} + \ldots\right)\right\}$$

$$= L^{-1}\left\{\frac{1}{p}\right\} - \frac{1}{2!}L^{-1}\left\{\frac{1}{p^3}\right\} + \frac{1}{4!}L^{-1}\left\{\frac{1}{p^5}\right\} - \frac{1}{6!}L^{-1}\left\{\frac{1}{p^7}\right\} + \ldots$$

$$= 1 - \frac{t^2}{(2!)^2} + \frac{t^4}{(4!)^2} - \frac{t^6}{(6!)^2} + \ldots$$

Example. 7 *Evaluate* $L^{-1}\left\{\dfrac{3p-2}{p^2-4p+20}\right\}$.

Solution. We have

$$L^{-1}\left\{\frac{3p-2}{p^2-4p+20}\right\}=L^{-1}\left\{\frac{3(p-2)+4}{(p-2)^2+16}\right\}$$

$$=L^{-1}\left\{\frac{3(p-2)}{(p-2)^2+16}+\frac{4}{(p-2)^2+16}\right\}$$

$$=3L^{-1}\left\{\frac{p-2}{(p-2)^2+4^2}\right\}+4L^{-1}\left\{\frac{1}{(p-2)^2+4^2}\right\}$$

$$=3e^{2t}L^{-1}\left\{\frac{p}{p^2+4^2}\right\}+4e^{2t}L^{-1}\left\{\frac{1}{p^2+4^2}\right\}$$

$$=3e^{2t}\cos 4t+e^{2t}\sin 4t$$

Example 8. *Evaluate the inverse Laplace transform*

$$L^{-1}\left\{\frac{p}{(p+1)^{5/2}}\right\}$$

Solution.

$$L^{-1}\left\{\frac{p}{(p+1)^{5/2}}\right\}=L^{-1}\left\{\frac{(p+1)-1}{(p+1)^{5/2}}\right\}$$

$$=e^{-t}L^{-1}\left\{\frac{p-1}{p^{5/2}}\right\}$$

$$=e^{-t}L^{-1}\left[\left\{\frac{1}{p^{3/2}}\right\}-\frac{1}{p^{5/2}}\right]$$

$$=e^{-t}L^{-1}\left\{\frac{1}{p^{3/2}}\right\}-e^{-t}L^{-1}\left\{\frac{1}{p^{5/2}}\right\}$$

$$=e^{-t}\frac{t^{(3/2)-1}}{\Gamma\left(\frac{3}{2}\right)}-e^{-t}\frac{t^{(5/2)-1}}{\Gamma(5/2)}$$

$$=2e^{-t}\sqrt{\left(\frac{t}{\pi}\right)}-\frac{4}{3}e^{-t}.t\sqrt{\left(\frac{t}{\pi}\right)}$$

$$=\frac{2}{3}e^{-t}\sqrt{\left(\frac{t}{\pi}\right)}(3-2t)$$

Example 9. *Evaluate* $L^{-1}\left\{\frac{1}{(p+2)(p-1)^2}\right\}$

Solution.

$$L^{-1}\left\{\frac{1}{(p+2)(p-1)^2}\right\}$$

$$=L^{-1}\left\{\frac{1}{(p-1+3)(p-1)^2}\right\}$$

$$=e^tL^{-1}\left\{\frac{1}{p+3}\cdot\frac{1}{p^2}\right\}$$

$$=e^tL^{-1}\left\{\frac{1}{p^2}\left(\frac{1}{3}-\frac{1}{9}p+\frac{1}{9}\frac{p^2}{p+3}\right)\right\}$$

(Dividing 1 by 3 + p till p^2 is a common factor in the remainder)

$$=e^tL^{-1}\left\{\frac{1}{3}.\frac{1}{p^2}-\frac{1}{9}.\frac{1}{p}+\frac{1}{9}.\frac{1}{(p+3)}\right\}$$

$$=e^t\left(\frac{1}{3}t-\frac{1}{9}+\frac{1}{9}e^{-3t}\right)$$

$$=\frac{1}{9}[(3t-1)e^t+e^{-2t}]$$

Example 10. *If* $L^{-1}\left\{\frac{p}{(p^2+1)^2}\right\}=\frac{1}{2}t.\sin t$

find $L^{-1}\left\{\frac{32p}{(16p^2+1)^2}\right\}$

Solution. We have

$$L^{-1}\left\{\frac{p}{(p^2+1)^2}\right\}=\frac{1}{2}t\sin t$$

$$\therefore\quad L^{-1}\left\{\frac{ap}{(a^2p^2+1)^2}\right\}=\frac{1}{2}.\frac{1}{a}.\frac{t}{a}.\sin\frac{t}{a}$$

$$\Rightarrow\quad L^{-1}\left\{\frac{2a^2p}{(a^2p^2+1)^2}\right\}=\frac{t}{a}\sin\frac{t}{a}$$

Now putting $a=4$, we get

$$L^{-1}\left\{\frac{32p}{(16p^2+1)^2}\right\}=\frac{t}{4}\sin\frac{t}{4}$$

Example 11. *Evaluate* $L^{-1}\left\{\frac{e^{4-3p}}{(p+4)^{5/2}}\right\}$

Solution. We have

$$L^{-1}\left\{\frac{1}{(p+4)^{5/2}}\right\}=e^{-4t}L^{-1}\left\{\frac{1}{p^{5/2}}\right\}$$

$$=e^{-4t}\frac{t^{(5/2)-1}}{\Gamma\left(\frac{5}{2}\right)}=\frac{4t^{3/2}e^{-4t}}{3\sqrt{\pi}}$$

Therefore,

$$L^{-1}\left\{\frac{e^{4-3p}}{(p+4)^{5/2}}\right\}=e^4\,L^{-1}\left\{\frac{e^{-3p}}{(p+4)^{5/2}}\right\}$$

$$=\begin{cases}e^4.\dfrac{4}{3\sqrt{\pi}}(t-3)^{3/2}e^{-4(t-3)}, & t>3\\ 0, & t<3\end{cases}$$

$$=\begin{cases}\dfrac{4}{3\sqrt{\pi}}(t-3)^{3/2}\,e^{-4(t-4)}, & t>3\\ 0, & t<3\end{cases}$$

$$=\frac{4}{3\sqrt{\pi}}(t-3)^{3/2}e^{-4(t-4)}.H(t-3).$$

Example 12. *Find the inverse Laplace transform of* $\frac{e^{-cp}}{p^2(p+a)}$, $c>0$ [UPTU-2001 Sp., 2002]

Solution. We have

$$L^{-1}\left[\frac{e^{-cp}}{p^2(p+a)}\right]$$

$$= L^{-1}\left[-\frac{e^{-cp}}{a^2 p}+\frac{e^{-cp}}{ap^2}+\frac{e^{-cp}}{a^2(p+a)}\right]$$

[By Partial Fractions]

$$= L^{-1}\left[-\frac{1}{a^2}\frac{e^{-cp}}{p}+\left(\frac{1}{a}\right)\frac{e^{-cp}}{p^2}+\frac{1}{a^2}\cdot\frac{e^{-c(p+a)}}{e^{-ca}(p+a)}\right]$$

$$= -\frac{1}{a^2}u(t-c)+\frac{1}{a}(t-c)u(t-c)$$

$$+\frac{1}{a^2 e^{-ca}}.e^{-at}u(t-c)$$

$$= u(t-c)\left[-\frac{1}{a^2}+\frac{1}{a}(t-c)+\frac{1}{a^2}e^{-a(c+t)}\right]$$

Where $u(t-c)$ = unit step function.

Example 13. *Find a function for which*

$$F(t) = L^{-1}\left\{\frac{3}{p}-\frac{4e^{-p}}{p^2}+\frac{4e^{-3p}}{p^2}\right\}$$

Solution. We have

$$F(t) = L^{-1}\left\{\frac{3}{p}-\frac{4e^{-p}}{p^2}+\frac{4e^{-3p}}{p^2}\right\}$$

$$= 3L^{-1}\left\{\frac{1}{p}\right\} - 4L^{-1}\left[\frac{e^{-p}}{p^2}\right] + 4L^{-1}\left[\frac{e^{-3p}}{p^2}\right] \quad \dots (1)$$

Now, $L^{-1}\left[\frac{1}{p}\right] = 1, \quad L^{-1}\left\{\frac{1}{p^2}\right\} = t$

Therefore,

$$L^{-1}\left\{\frac{e^{-p}}{p^2}\right\} = (t-1)H(t-1)$$

and $L^{-1}\left\{\frac{e^{-3p}}{p^2}\right\} = (t-3)\,H(t-3)$.

Putting all these values in (1), we get

$$F(t) = 3-4(t-1)\,H(t-1)+4(t-3)H(t-3)$$

Example 14. *Evalute* $L^{-1}\left\{\frac{p+1}{p^2+6p+25}\right\}$

Solution. We have

$$L^{-1}\left\{\frac{p+1}{p^2+6p+25}\right\} = L^{-1}\left\{\frac{(p+3)-2}{(p+3)^2+16}\right\}$$

$$= e^{-3t}L^{-1}\left\{\frac{p-2}{p^2+16}\right\}$$

$$= e^{-3t}\left[L^{-1}\left\{\frac{p}{p^2+4^2}\right\}-2L^{-1}\left\{\frac{1}{p^2+4^2}\right\}\right]$$

$$= e^{-3t}\left[\cos 4t-\frac{1}{2}\sin 4t\right].$$

Example 15. *Evalute* $L^{-1}\left[\frac{e^{-p}-3e^{-3p}}{p^2}\right]$

Solution. We have

$$L^{-1}\left[\frac{e^{-p}-3e^{-3p}}{p^2}\right] = L^{-1}\left[\frac{e^{-p}}{p^2}-\frac{3e^{-3p}}{p^2}\right] \quad \dots (1)$$

We know that $Lu(t-a) = \frac{e^{-ap}}{p}$

and $L[(t-a)\,u(t-a)] = \frac{e^{-ap}}{p^2}$

Using these results in (1), we get

$$L^{-1}\left[\frac{e^{-p}-3e^{-3p}}{p^2}\right]$$

$$= (t-1)u(t-1)-3(t-3)u(t-3).$$

Example 16. *Evalute* $L^{-1}\left\{\frac{p+5}{(p+2)(p^2+4)}\right\}$

Solution. We have $L^{-1}\left\{\frac{p+5}{(p+2)(p^2+4)}\right\}$

$$= L^{-1}\left\{\frac{1}{8}\left(\frac{3}{p+2}-\frac{3p-14}{p^2+4}\right)\right\}$$

$$= \frac{1}{8}\left[3L^{-1}\left\{\frac{1}{p+2}\right\}-3L^{-1}\left\{\frac{p}{p^2+4}\right\} +14L^{-1}\left\{\frac{1}{p^2+4}\right\}\right]$$

$$= \frac{1}{8}(3e^{-2t}-3\cos 2t+7\sin 2t)$$

Example 17. *Show that*

$$L^{-1}\left\{\frac{p}{p^4+p^2+1}\right\} = \frac{2}{\sqrt{3}}\sinh\frac{t}{2}.\sin\frac{1}{2}\sqrt{3}\,t$$

(Raipur-2005)

Solution. We have

$$L^{-1}\left\{\frac{p}{p^4+p^2+1}\right\} = L^{-1}\left\{\frac{p}{(p^2+1)^2-p^2}\right\}$$

$$= L^{-1}\left\{\frac{p}{(p^2+p+1)(p^2-p+1)}\right\}$$

$$= L^{-1}\left\{\frac{1}{2}\frac{(p^2+p+1)-(p^2-p+1)}{(p^2-p+1)(p^2+p+1)}\right\}$$

$$= L^{-1}\left\{\frac{1}{2(p^2-p+1)}-\frac{1}{2(p^2+p+1)}\right\}$$

$$= \frac{1}{2}L^{-1}\left[\frac{1}{\left\{\left(p-\frac{1}{2}\right)^2+\frac{3}{4}\right\}}\right]-\frac{1}{2}L^{-1}\left[\frac{1}{\left\{\left(p+\frac{1}{2}\right)^2+\frac{3}{4}\right\}}\right]$$

$$= \frac{1}{2}e^{t/2}L^{-1}\left[\frac{1}{\left\{p^2+\left(\frac{1}{2}\sqrt{3}\right)^2\right\}}\right]-\frac{1}{2}e^{-t/2}L^{-1}\left[\frac{1}{\left\{p^2+\left(\frac{1}{2}\sqrt{3}\right)^2\right\}}\right]$$

$$= \frac{1}{2}e^{t/2}\frac{2}{\sqrt{3}}\sin\left(\sqrt{3}\,.\frac{t}{2}\right)$$

$$-\frac{1}{2}e^{-t/2}\frac{2}{\sqrt{3}}\sin\left(\sqrt{3}.\frac{t}{2}\right)$$

$$= \frac{1}{\sqrt{3}}(e^{t/2} - e^{-t/2})\sin\left(\sqrt{3}\,.\frac{t}{2}\right)$$

$$= \frac{2}{\sqrt{3}}\sinh\frac{t}{2}\sin\left(\sqrt{3}.\frac{t}{2}\right).$$

Example 18. *Find* $L^{-1}\left\{\dfrac{2p^3+2p^2+4p+1}{(p^2+1)(p^2+p+1)}\right\}$

Solution. We have $L^{-1}\left\{\dfrac{2p^3+2p^2+4p+1}{(p^2+1)(p^2+p+1)}\right\}$

$$= L^{-1}\left\{\frac{p+2}{p^2+1} + \frac{p-1}{p^2+p+1}\right\}$$

$$= L^{-1}\left\{\frac{p}{p^2+1}\right\} + 2L^{-1}\left\{\frac{1}{p^2+1}\right\}$$

$$+L^{-1}\left\{\frac{\left(p+\frac{1}{2}\right)-\frac{3}{2}}{\left(p+\frac{1}{2}\right)^2+\frac{3}{4}}\right\}$$

$$= \cos t + 2\sin t + e^{-t/2}L^{-1}\left[\frac{p-3/2}{p^2+3/4}\right]$$

$$= \cos t + 2\sin t + e^{-t/2}\left[L^{-1}\frac{p}{p^2+\left(\frac{\sqrt{3}}{2}\right)^2}\right.$$

$$\left. -\frac{3}{2}L^{-1}\left[\frac{1}{p^2+\left(\frac{\sqrt{3}}{2}\right)^2}\right]\right]$$

$$= \cos t + 2\sin t + e^{-t/2}$$

$$\left\{\cos\left(\frac{1}{2}\sqrt{3}t\right) - \frac{3}{2}.\frac{2}{\sqrt{3}}\sin\left(\frac{1}{2}\sqrt{3}t\right)\right\}$$

$$= \cos t + 2\sin t + e^{-t/2}$$

$$\left[\cos\left(\frac{1}{2}.\sqrt{3}t\right) - \sqrt{3}\sin\left(\frac{1}{2}\sqrt{3}t\right)\right]$$

Example 19. *Evaluate* $L^{-1}\left\{(p+1)\dfrac{e^{-\pi p}}{p^2+p+1}\right\}.$

Solution. We get

$$L^{-1}\left\{\frac{p+1}{p^2+p+1}\right\} = L^{-1}\left\{\frac{\left(p+\frac{1}{2}\right)+\frac{1}{2}}{\left(p+\frac{1}{2}\right)^2+\frac{3}{4}}\right\}$$

$$= e^{-t/2}L^{-1}\left\{\frac{p+\frac{1}{2}}{p^2+\frac{3}{4}}\right\}$$

$$= e^{-t/2}L^{-1}\left[\frac{p}{p^2+\left(\frac{\sqrt{3}}{2}\right)^2}\right] + \frac{1}{2}e^{-t/2}L^{-1}\left[\frac{1}{p^2+\left(\frac{\sqrt{3}}{2}\right)^2}\right]$$

$$= e^{-t/2}\cos\left(\frac{\sqrt{3}t}{2}\right) + \frac{1}{2}e^{-t/2}\left(\frac{2}{\sqrt{3}}\right)\sin\left(\frac{\sqrt{3}t}{2}\right)$$

$$\therefore \quad L^{-1}\left\{\frac{(p+1)e^{-\pi p}}{p^2+p+1}\right\}$$

$$= \begin{cases} \dfrac{e^{-(t-\pi)/2}}{\sqrt{3}}\left[\sqrt{3}\cos\dfrac{\sqrt{3}}{2}(t-\pi) + \sin\dfrac{\sqrt{3}}{2}(t-\pi)\right], & t > \pi \\ 0, & t < \pi \end{cases}$$

$$= \frac{e^{-(t-\pi)/2}}{\sqrt{3}}$$

$$\left[\sqrt{3}\cos\frac{\sqrt{3}}{2}(t-\pi) + \sin\frac{\sqrt{3}}{2}(t-\pi)\,H(t-\pi)\right].$$

Example 20. *Find* $L^{-1}\left\{\dfrac{5p+3}{(p-1)(p^2+2p+5)}\right\}$

[UPTU-2005, Rohtak-2009]

Solution. We have

$$\frac{5p+3}{(p-1)(p^2+2p+5)} = \frac{1}{p-1} + \frac{2-p}{p^2+2p+5}$$

[By partial fractions]

$$\therefore L^{-1}\left\{\frac{5p+3}{(p-1)(p^2+2p+5)}\right\}$$

$$= L^{-1}\left\{\frac{1}{p-1} + \frac{2-p}{p^2+2p+5}\right\}$$

$$= L^{-1}\left\{\frac{1}{p-1} - \frac{p+1}{(p+1)^2+4} + \frac{3}{(p+1)^2+4}\right\}$$

$$= L^{-1}\left\{\frac{1}{p-1}\right\} - L^{-1}\left\{\frac{p+1}{(p+1)^2+4}\right\}$$

$$+3L^{-1}\left\{\frac{1}{(p+1)^2+4}\right\}$$

$$= e^t - e^{-t}L^{-1}\left\{\frac{p}{p^2+4}\right\} + 3e^{-t}L^{-1}\left\{\frac{1}{p^2+4}\right\}$$

$$= e^t - e^{-t}\cos 2t + \frac{3}{2}e^{-t}\sin 2t$$

$$= e^t - e^{-t}\left(\cos 2t - \frac{3}{2}\sin 2t\right).$$

EXERCISE 37.1

1. Find the inverse Laplace transform of the following functions :

(a) $\frac{1}{p^4}$ (b) $\frac{1}{p^2+4}$ (c) $\frac{4}{p-2}$ (d) $\frac{1}{\sqrt{p}}$

(e) $\frac{p}{p^2+2}+\frac{6p}{p^2-16}+\frac{3}{p-3}$ (f) $\frac{2p-5}{p^2-9}$

(g) $\frac{6}{2p-3}-\frac{3+4p}{9p^2-16}+\frac{8-6p}{16p^2+9}$ [UPTU-2001]

(h) $\frac{3(p^2-1)^2}{2p^5}+\frac{4p-18}{9-p^2}+\frac{(p+1)(2-\sqrt{p})}{p^{5/2}}$

(i) $\frac{1}{p}\sin\frac{1}{p}$

2. Find the inverse Laplace transform of the following functions :

(a) $\frac{1}{p^2-6p+10}$ (b) $\frac{p+b}{(p+b)^2+a^2}$

(c) $\frac{3p+7}{p^2-2p-3}$ (d) $\frac{1}{(p+a)^n}$

(e) $\frac{p}{(p+1)^5}$ (f) $\frac{p^2-2p+3}{(p-1)^2(p+1)}$

(g) $\frac{2p^2-1}{(p^2+1)(p^2+4)}$ [UPTU-2004]

(h) $\frac{2p^2-6p+5}{p^3-6p^2+11p-6}$ [UPTU-2004; VTU-2007]

3. If $L^{-1}\left\{\frac{e^{-1/p}}{p^{1/2}}\right\}=\frac{\cos 2\sqrt{t}}{\sqrt{\pi t}}$, evaluate $L^{-1}\left\{\frac{e^{-a/p}}{p^{1/2}}\right\},\ a>0$

4. Find the inverse Laplace transforms of the following functions

(a) $\frac{e^{-5p}}{(p-2)^4}$ (b) $\frac{e^{-4p}}{(p-3)^4}$

(c) $\frac{e^{-3p}}{p^3}$ (d) $\frac{p+8}{p^2+8p+5}$

5. Show that

(a) $L^{-1}\left\{\frac{pe^{-ap}}{p^2-w^2}\right\}=\cosh w(t-a)\,H(t-a),\ a>0$

(b) $L^{-1}\left\{\frac{p+2}{p^2-2p+5}\right\}=e^t[\cos 2t+\frac{3}{2}\sin 2t]$

(c) $L^{-1}\left\{\frac{6p^2+22p+18}{p^3+6p^2+11p+6}\right\}=e^{-t}+2e^{-2t}+3e^{-3t}$

(d) $L^{-1}\left\{\frac{4p+5}{(p-1)^2(p+2)}\right\}=3te^t+\frac{1}{3}e^t-\frac{1}{3}e^{-2t}$ (Kurukshetra-2005)

(e) $L^{-1}\left\{\frac{4p+5}{(p-4)^2(p+3)}\right\}=-\frac{1}{7}e^{-3t}+\frac{1}{7}e^{4t}+3te^{4t}$

(f) $L^{-1}\left\{\frac{5p^2-15p-11}{(p+1)(p-2)^3}\right\}$
$=-\frac{1}{3}e^{-t}+\frac{1}{3}e^{2t}+4te^{2t}+4te^{2t}-\frac{7}{2}t^2e^{2t}$

(g) $L^{-1}\left\{\frac{2p+1}{(p+2)^2(p-1)^2}\right\}=\frac{1}{3}t(e^t-e^{-2t})$

(h) $L^{-1}\left\{\frac{p}{(p^2-2p+2)(p^2+2p+2)}\right\}=\frac{1}{2}\sin t\sinh t$

6. Show that

(a) $L^{-1}\left\{\frac{p^2}{p^4+4a^4}\right\}=\frac{1}{2a}(\cosh at\sin at+\sinh at\cos at)$

(b) $L^{-1}\left\{\frac{p^2}{p^4+4a^4}\right\}=\frac{1}{2a^2}\sin at\sinh at$ (Mumbai-2008)

(c) $L^{-1}\left\{\frac{1}{(p^2+4)(p+1)^2}\right\}=\frac{1}{25}\left\{e^{-t}(2+5t)-2\cos 2t-\frac{3}{2}\sin 2t\right\}$

(d) $L^{-1}\left\{\frac{3p^3-3p^2-40p+36}{(p^2-4)^2}\right\}=(5t+3)e^{-2t}-2te^{2t}$

Hint to Selected Problems

1. (g) $L^{-1}\left[\frac{6}{2p-3}-\frac{3+4p}{9p^2-16}+\frac{8-6p}{16p^2+9}\right]$

$$=3L^{-1}\left(\frac{1}{p-\left(\frac{3}{2}\right)}\right)-\frac{1}{3}L^{-1}\left(\frac{1}{p^2-\left(\frac{4}{3}\right)^2}\right)-\frac{4}{9}L^{-1}\left(\frac{p}{p^2-\left(\frac{4}{3}\right)^2}\right)+\frac{1}{2}L^{-1}\left(\frac{1}{p^2+\left(\frac{3}{4}\right)^2}\right)-\frac{3}{8}L^{-1}\left(\frac{p}{p^2+\left(\frac{3}{4}\right)^2}\right)$$

$$=3e^{3t/2}-\frac{1}{4}\sinh\frac{4t}{3}-\frac{4}{9}\cosh\frac{4t}{3}+\frac{2}{3}\sin\frac{3t}{4}-\frac{3}{8}\cos\frac{3}{4}t$$

2. (a) $L^{-1}\left\{\frac{1}{p^2-6p+10}\right\}=L^{-1}\left[\frac{1}{(p-3)^2+1}\right]$

$$=e^{3t}L^{-1}\left[\frac{1}{p^2+1}\right]=e^{3t}.\sin t$$

3. $L^{-1}\left\{\frac{e^{-1/p}}{p^{1/2}}\right\}=\frac{\cos 2\sqrt{t}}{\sqrt{\pi t}}\Rightarrow L^{-1}\left(\frac{e^{-1/pk}}{(pk)^{1/2}}\right)=\frac{1}{k}\frac{\cos 2\sqrt{t/k}}{\sqrt{\pi t/k}}$.

Now taking $k=1/a$

5. (a) Since $L^{-1}\left(\frac{p}{p^2-w^2}\right)=\cosh wt$

$$\Rightarrow\quad L^{-1}\left[\frac{pe^{-ap}}{p^2-w^2}\right]=\cosh w(t-a)\,H(t-a)$$

6. (a) $L^{-1}\left[\frac{p^2}{p^4+4a^4}\right]=L^{-1}\left[\frac{p^2}{(p^2+2a^2)^2-4a^2p^2}\right]$

$$=L^{-1}\left[\frac{p^2}{(p^2-2ap+2a^2)(p^2+2ap+2a^2)}\right]$$

$$= L^{-1}\left[\frac{1}{4a}\frac{p(p^2+2ap+2a^2)-p(p^2-2ap+2a^2)}{(p^2+2ap+2a^2)(p^2-2ap+2a^2)}\right] = \frac{1}{4a}L^{-1}\left[\frac{(p-a)+a}{(p-a)^2+a^2}\right]-\frac{1}{4a}L^{-1}\left[\frac{(p+a)-a}{(p+a)^2+a^2}\right]$$

$$= L^{-1}\left[\frac{p}{4a(p^2-2ap+2a^2)}-\frac{p}{4a(p^2+2ap+2a^2)}\right] = \frac{e^{at}}{4a}L^{-1}\left[\frac{p+a}{p^2+a^2}\right]-\frac{1}{4a}e^{-at}L^{-1}\left[\frac{p-a}{p^2+a^2}\right]$$

ANSWERS

1. (a) $\frac{t^3}{6}$; (b) $\frac{1}{2}\sin 2t$; (c) $4e^{2t}$; (d) $\frac{1}{\sqrt{\pi t}}$; (e) $\cos\sqrt{2}\,t+6\cosh 4t+3e^{3t}$; (f) $2\cosh 3t-\frac{5}{3}\sinh 3t$;

(g) $3e^{3t/2}-\frac{1}{4}\sinh\frac{4t}{3}-\frac{4}{9}\cosh\frac{4t}{3}+\frac{2}{3}\sin\frac{3t}{4}-\frac{3}{8}\cos\frac{3}{4}t$; (h) $\frac{1}{2}-\frac{3}{2}t^2+\frac{1}{16}t^4-4\cosh 3t+6\sinh 3t+4\sqrt{\frac{t}{\pi}}+\frac{8}{3}t\sqrt{\frac{t}{\pi}}-t$;

(i) $t-\frac{t^3}{(3!)^2}+\frac{t^5}{(5!)^2}-\frac{t^7}{(7!)^2}+\ldots$

2. (a) $e^{3t}\sin t$; (b) $e^{-bt}\cos at$; (c) $4e^{3t}-e^{-t}$; (d) $e^{-at}\frac{t^{n-1}}{(n-1)!}, n\in\mathbf{Z}^+$; (e) $e^{-t}(4t^3-t^4)/24$;

(f) $\left(t-\frac{1}{2}\right)e^t+\frac{3}{2}e^{-t}$ (g) $-\sin t+\frac{3}{2}\sin 2t$ (h) $\frac{1}{2}e^t-e^{2t}+\frac{5}{2}e^{3t}$

3. $\cos\frac{2\sqrt{at}}{\sqrt{\pi t}}$ **4.** (a) $\frac{1}{6}(t-5)^3e^{2(t-5)}H(t-5)$ (b) $\frac{1}{6}(t-4)^3e^{3(t-4)}.H(t-4)$

(c) $\frac{1}{2}(t-3)^2\,H(t-3)$ (d) $e^{-4t}\left[\cosh(\sqrt{11}\,t)+\left(\frac{4}{\sqrt{11}}\right)\sinh(\sqrt{11}.t)\right]$

37.4 INVERSE LAPLACE TRANSFORMS OF DERIVATIVES

Theorem 1. *If* $L^{-1}\{f(p)\}=F(t)$, *then* $L^{-1}\{f(n)(p)\}=(-1)^n.t^n.F(t)$

Proof. Since we know that $L\{t^n\,F(t)\}=(-1)^n f^{(n)}(p)$

Therefore, $t^n\,F(t)=L^{-1}\{(-1)^n f^{(n)}(p)\}=(-1)^n\,L^{-1}\{f^{(n)}(p)\}$

Hence, $L^{-1}\{f^{(n)}(p)\}=(-1)^n\,t^n\,F(t)$

37.5 DIVISION BY p

Theorem 1. *If* $L^{-1}\{f(p)\}=F(t)$, *then* $L^{-1}\left\{\frac{f(p)}{p}\right\}=\int_0^t F(u)\,du$

Proof. Since we know that $\frac{f(p)}{p}=L\left\{\int_0^t F(u)\,du\right\} \Rightarrow L^{-1}\left\{\frac{f(p)}{p}\right\}=\int_0^t F(u)\,du$

37.6 MULTIPLICATION BY POWERS OF p

Theorem 1. *If* $L^{-1}\{f(p)\}=F(t)$ *and* $F(0)=0$, *then* $L^{-1}\{pf(p)\}=F'(t)$

Proof. We know that

$$L\{F'(t)\}=pL\{F(t)\}-F(0)=pL[F(t)]=p\,f(p) \qquad [\because f(0)=0].$$

Hence, $L^{-1}\{p\,f(p)\}=F'(t)$

37.7 INVERSE LAPLACE TRANSFORMS OF INTEGRALS

Theorem 1. *If* $L^{-1}\{f(p)\}=F(t)$, *then* $L^{-1}\left[\int_p^\infty f(x)\,dx\right]=\frac{F(t)}{t}$.

Proof. We know that

$$L\left\{\frac{1}{t}F(t)\right\}=\int_p^\infty f(x)\,dx \text{ provided } \lim_{t\to 0}\left\{\frac{F(t)}{t}\right\} \text{ exists.}$$

Hence, $L^{-1}\left\{\int_p^\infty f(x)\,dx\right\}=\frac{F(t)}{t}$.

Solved Examples

Example 1. *Find* $L^{-1}\left\{\dfrac{p}{(p^2+a^2)^2}\right\}$. (SVTU-2009, VTU-2010)

Solution. We have

$$L^{-1}\left\{\frac{p}{(p^2+a^2)^2}\right\}=L^{-1}\left\{-\frac{1}{2}\frac{d}{dp}\left(\frac{1}{p^2+a^2}\right)\right\}$$

$$=-\frac{1}{2}L^{-1}\left\{\frac{d}{dp}\left(\frac{1}{p^2+a^2}\right)\right\}$$

$$=-\frac{1}{2}t(-1)L^{-1}\left\{\frac{1}{p^2+a^2}\right\}=\frac{t}{2a}\sin at$$

Example 2. *Evaluate* $L^{-1}\left\{\log\left(1-\dfrac{1}{p^2}\right)\right\}$.

Solution. Let us suppose $f(p)=\log\left(1-\dfrac{1}{p^2}\right)$.

$$=\log\left(\frac{p^2-1}{p^2}\right)=-2\log p+\log(p^2-1)$$

$$\Rightarrow \quad f'(p)=-2\left(\frac{1}{p}-\frac{p}{p^2-1}\right)$$

$$\Rightarrow \quad L^{-1}\{f'(p)\}=-2(1-\cosh t)$$

$$\Rightarrow \quad -tL^{-1}\{f(p)\}=-2(1-\cosh t)$$

$$\Rightarrow \quad L^{-1}\left\{\log\left(1-\frac{1}{p^2}\right)\right\}=\frac{2}{t}(1-\cosh t).$$

Example 3. *Find the function whose Laplace transform is* $\log\left(1+\dfrac{1}{p}\right)$ [UPTU-2007]

Solution.

$$L^{-1}\left[\log\left(1+\frac{1}{p}\right)\right]=-\frac{1}{t}L^{-1}\left[\frac{d}{dp}\log\left(\frac{p+1}{p}\right)\right]$$

$$=-\frac{1}{t}L^{-1}\left[\left(\frac{p}{p+1}\right)\left(-\frac{1}{p^2}\right)\right]$$

$$=-\frac{1}{t}L^{-1}\left[-\frac{1}{p(p+1)}\right]=-\frac{1}{t}L^{-1}\left[\frac{1}{p+1}-\frac{1}{p}\right]$$

$$=-\frac{1}{t}[e^{-t}-1]=\frac{1}{t}[1-e^{-t}].$$

Example 4. *Evaluate* $L^{-1}\left\{\dfrac{p+2}{p^2(p+3)}\right\}$

Solution. Consider $L^{-1}\left\{\dfrac{p+2}{p^2(p+3)}\right\}=L^{-1}\left\{\dfrac{(p+3)-1}{p^2(p+3)}\right\}$

$$=L^{-1}\left\{\frac{1}{p^2}-\frac{1}{p^2(p+3)}\right\}$$

$$=L^{-1}\left\{\frac{1}{p^2}\right\}-L^{-1}\left\{\frac{1}{p^2(p+3)}\right\} \quad \text{... (1)}$$

Since, $L^{-1}\left\{\dfrac{1}{p^2}\right\}=t$

and $L^{-1}\left\{\dfrac{1}{p+3}\right\}=e^{-3t}=F(t)$ (say)

Then, we have

$$L^{-1}\left\{\frac{1}{p(p+3)}\right\}=\int_0^t F(x)\,dx \quad \text{...(1)}$$

$$=\int_0^t e^{-3x}dx=\frac{1}{3}(1-e^{-3t})=F_1(t) \quad \text{(say)}$$

Therefore, $L^{-1}\left\{\dfrac{1}{p^2(p+3)}\right\}=\int_0^1 F_1(x)\,dx$

$$=\frac{1}{3}\int_0^t(1-e^{-3x})\,dx$$

$$=\frac{1}{3}t+\frac{1}{9}(e^{-3t}-1)$$

Hence, from (1), we have

$$L^{-1}\left\{\frac{p+2}{p^2(p+3)}\right\}=t-\frac{t}{3}-\frac{1}{9}(e^{-3t}-1)$$

$$=\frac{2}{3}t-\frac{1}{9}e^{-3t}+\frac{1}{9}$$

Example 5. *Find the inverse Laplace transform of* $f(p)=\log\dfrac{p+a}{p+b}$ [UPTU-2003, Anna-2003]

Solution.

$$L^{-1}\log\left[\frac{p+a}{p+b}\right]=-\frac{1}{t}L^{-1}\left[\frac{d}{dp}\log\frac{p+a}{p+b}\right]$$

$$=-\frac{1}{t}L^{-1}\left[\frac{d}{dp}\log(p+a)-\frac{d}{dp}\log(p+b)\right]$$

$$=-\frac{1}{t}L^{-1}\left[\frac{1}{p+a}-\frac{1}{p+b}\right]$$

$$=-\frac{1}{t}\left[e^{-at}-e^{-bt}\right]=\frac{1}{t}\left[e^{-bt}-e^{-at}\right].$$

Example 6. *Evaluate* $L^{-1}\left\{\dfrac{1}{p^4(p^2+1)}\right\}$

Solution. Since we know that $L^{-1}\left\{\dfrac{1}{p^2+1}\right\}=\sin t$

Therefore

$$L^{-1}\left\{\frac{1}{p(p^2+1)}\right\}=\int_0^t \sin x\,dx=1-\cos t$$

Also,

$$L^{-1}\left\{\frac{1}{p^2(p^2+1)}\right\}=\int_0^t(1-\cos x)dx=t-\sin t$$

and $L^{-1}\left\{\dfrac{1}{p^3(p^2+1)}\right\}=\int_0^t(x-\sin x)dx$

$$=\frac{t^2}{2}+\cos t-1$$

Hence, $L^{-1}\left\{\dfrac{1}{p^4(p^2+1)}\right\}$

$$= \int_0^t \left\{\frac{1}{2}x^2 + \cos x - 1\right\} dx = \frac{1}{6}t^3 + \sin t - t$$

Example 7. *Evaluate* $L^{-1}\left\{\dfrac{1}{p(p+1)^3}\right\}$

Solution. We know that

$$L^{-1}\left\{\frac{1}{p(p+1)^3}\right\} = L^{-1}\left\{\frac{1}{(p+1-1)(p+1)^3}\right\}$$

$$= e^{-t}\, L^{-1}\left\{\frac{1}{(p-1)\,p^3}\right\} \quad \text{... (1)}$$

Also, since $L^{-1}\left\{\dfrac{1}{p-1}\right\} = e^t$

Therefore, $L^{-1}\left\{\dfrac{1}{p(p-1)}\right\} = \int_0^t e^x dx = (e^t - 1)$

and

$$L^{-1}\left\{\frac{1}{p^2(p-1)}\right\} = \int_0^t (e^x - 1)dx = e^t - t - 1$$

and $L^{-1}\left\{\dfrac{1}{p^3(p-1)}\right\} = \int_0^t (e^x - x - 1)dx$

$$= e^t - \frac{1}{2}t^2 - t - 1$$

Therefore, from (1), we get

$$L^{-1}\left\{\frac{1}{p(p+1)^3}\right\} = e^{-t}\left\{e^t - \frac{1}{2}t^2 - 1 - t\right\}$$

$$= 1 - e^{-t}\left(1 + t + \frac{t^2}{2}\right)$$

Example 8. *Obtain the inverse Laplace transformation*

$\cot^{-1}\left(\dfrac{p+3}{2}\right)$ [UPTU-2001 (Sp.), 2002]

Solution. We know that $L^{-1}[f(p)] = -\dfrac{1}{t}L^{-1}\left[\dfrac{d}{dp}f(p)\right]$

$$\therefore \left[\cot^{-1}\left(\frac{p+3}{2}\right)\right] = -\frac{1}{t}L^{-1}\left[\frac{d}{dp}\cot^{-1}\left(\frac{p+3}{2}\right)\right]$$

$$= -\frac{1}{t}L^{-1}\left[\frac{-\frac{1}{2}}{1+\left(\frac{p+3}{2}\right)^2}\right] = \frac{1}{2t}L^{-1}\left[\frac{4}{4+(p+3)^2}\right]$$

$$= \frac{1}{t}L^{-1}\left[\frac{2}{2^2+(p+3)^2}\right] = \frac{1}{t}e^{-3t}L^{-1}\left[\frac{2}{2^2+p^2}\right]$$

$$= \frac{e^{-3t}}{t}\sin 2t \,.$$

Example 9. *Evaluate*

(i) $L^{-1}\left\{\log\left(1+\dfrac{1}{p^2}\right)\right\}$ [UPTU Q.Bank-2001]

(ii) $L^{-1}\left\{\dfrac{1}{p}\log\left(1+\dfrac{1}{p^2}\right)\right\}$

Solution. (i) Let $f(p) = \log\left(1+\dfrac{1}{p^2}\right) = -\log\left(\dfrac{p^2}{p^2+1}\right)$

$$= -2\log p + \log(p^2+1)$$

Therefore, $f'(p) = -\dfrac{2}{p} + \dfrac{2p}{p^2+1}$

$\Rightarrow \quad L^{-1}\{f'(p)\} = -2 + 2\cos t$

$\Rightarrow \quad -tL^{-1}\{f(p)\} = -2(1-\cos t)$

Hence, $L^{-1}\left\{\log\left(1+\dfrac{1}{p^2}\right)\right\} = \dfrac{2(1-\cos t)}{t}$

(ii) Since

$$L^{-1}\left\{\log\left(1+\frac{1}{p^2}\right)\right\} = \frac{2(1-\cos t)}{t}$$

Therefore,

$$L^{-1}\left\{\frac{1}{p}\log\left(1+\frac{1}{p^2}\right)\right\} = L^{-1}\left\{\frac{1}{p}\,f(p)\right\}$$

$$= \int_0^t F(x)dx = \int_0^t \frac{2}{x}(1-\cos x)\,dx\,.$$

EXERCISE 37.2

1. Evaluate the following inverse Laplace transforms :

(a) $L^{-1}\left\{\dfrac{p}{(p^2-a^2)^2}\right\}$ (b) $L^{-1}\left\{\dfrac{p}{(p^2-16)^2}\right\}$

(c) $L^{-1}\left\{\dfrac{1}{(p-a)^3}\right\}$ (d) $L^{-1}\left\{\dfrac{p+1}{(p^2+2p+2)^2}\right\}$

(e) $L^{-1}\left\{\dfrac{p^2}{(p^2+4)^2}\right\}$ [UPTU Q. Bank -2001]

2. Show that :

(a) $L^{-1}\left\{\dfrac{1}{p^3(p+1)}\right\} = 1 - t + \dfrac{t^2}{2} - e^{-t}$

(b) $L^{-1}\left\{\dfrac{1}{p^3(p^2+1)}\right\} = \dfrac{t^2}{2} + \cos t - 1$ (GBTU-2012)

(c) $L^{-1}\left\{\log\dfrac{p+2}{p+1}\right\} = \dfrac{1}{t}(e^{-t} - e^{-2t})$

(d) $L^{-1}\left\{\dfrac{1}{p}\log\dfrac{p+2}{p+1}\right\} = \int_0^t \dfrac{1}{x}(e^{-x} - e^{-2x})dx$

(e) $L^{-1}\left\{\dfrac{1}{p}\log\dfrac{p+3}{p+2}\right\} = \int_0^t \dfrac{1}{x}(e^{-2x} - e^{-3x})\,dx$

(f) $L^{-1}\left(\tan^{-1}\left(\frac{2}{p^2}\right)\right)=\frac{2}{t}\sin t\sinh t$

[UPTU Q. Bank–2001, Mumbai–2005; VTU–2011]

3. If $L^{-1}\left\{\frac{p}{(p^2+1)^2}\right\}=\frac{1}{2}t\sin t$, then show that

$$L^{-1}\left\{\frac{1}{(p^2+1)^2}\right\}=\frac{1}{2}(\sin t-t\cos t).$$

Hint to Selected Problems

1. (a) Since $\frac{d}{dp}(p^2-a^2)^{-1}=-\frac{2p}{(p^2-a^2)^2}$,

therefore $\frac{p}{(p^2-a^2)^2}=-\frac{1}{2}\frac{d}{dp}\left(\frac{1}{p^2-a^2}\right)$

$$\Rightarrow L^{-1}\left(\frac{p}{(p^2-a^2)^2}\right)=-\frac{1}{2}L^{-1}\left[\frac{d}{dp}\left(\frac{1}{p^2-a^2}\right)\right]$$

$$=-\frac{1}{2}(-1)^t t\,L^{-1}\left(\frac{1}{p^2-a^2}\right)=\frac{1}{2}t\frac{1}{a}\sinh at$$

2. (a) $L^{-1}\left[\frac{1}{p+1}\right]=e^{-t}=F(t)$ (say)

$$\therefore L^{-1}\left[\frac{1}{p(p+1)}\right]=\int_0^t F(x)\,dx=\int_0^t e^{-x}dx=1-e^{-t}$$

$$\Rightarrow L^{-1}\left[\frac{1}{p^2(p+1)}\right]=\int_0^t(1-e^{-x})dx=t+e^{-t}-1$$

and $L^{-1}\left[\frac{1}{p^3(p+1)}\right]=\int_0^t(x+e^{-x}-1)dx=1-t+\frac{t^2}{2}-e^{-t}$

3. We have $L^{-1}\left[\frac{p}{(p^2+1)^2}\right]=\frac{1}{2}t\sin t=F(t)$ (say)

$$\Rightarrow L^{-1}\left[\frac{1}{(p^2+1)^2}\right]=L^{-1}\left[\frac{1}{p}\cdot\frac{p}{(p^2+1)^2}\right]=\int_0^t F(x)dx$$

$$=\frac{1}{2}\int_0^t x.\sin x\,dx=\frac{1}{2}(\sin t-t\cos t)$$

ANSWERS

1. (a) $\frac{t}{2a}\sinh at$; (b) $\frac{t}{8}\sinh 4t$; (c) $\frac{1}{2}t^2e^{at}$; (d) $\frac{t}{2}e^{-t}\sin t$; (e) $\frac{1}{4}(\sin 2t+2t\cos 2t)$

37.8 CONVOLUTION

If $L^{-1}\{f(p)\}=F(t)$ and $L^{-1}\{g(p)\}=G(t)$, where F(t) and G(t) are two functions of class A, then [UPTU-2002]

$$L^{-1}\{f(p).g(p)\}=\int_0^t F(u)\,G(t-u)\,du=F*G$$

*We call F * G the convolution or falting of F and G.*

Proof. $\int_0^t F(x)\,G(t-x)\,dx=H(t)$

Then, $L\{H(t)\}=\int_0^\infty e^{-pt}H(t)\,dt$

$$=\int_0^\infty e^{-pt}\left[\int_0^t F(x)G(t-x)dx\right]dt$$

$$=\int_0^\infty\left[\int_0^t e^{-pt}F(x)G(t-x)dx\right]dt \quad ...(1)$$

The integration being first with respect to x and then t.

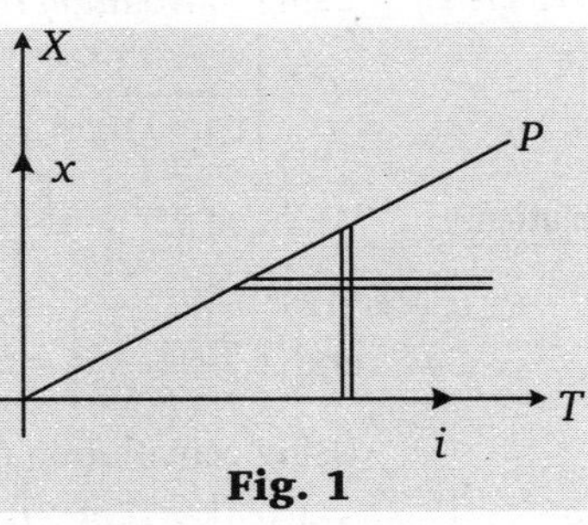

Fig. 1

The integration (1) is within the region lying below the line OP whose equation is $x=t$ and above OT, t being taken along OT and x along OX, with O is the origin the axes being perpendicular to each other. If the order of integration is changed, the strip will be taken parallel to OT, so that the limits of t are from x to ∞ and of x from 0 to ∞.

Therefore, $L\{H(t)\}=\int_0^\infty dx\int_x^\infty e^{-pt}F(x)\,G(t-x)\,dt$

$$=\int_0^\infty e^{-px}F(x)\,dx\int_x^\infty e^{-p(t-x)}G(t-x)\,dt$$

Putting $t-x=\theta\Rightarrow dt=d\theta$

$$\Rightarrow L\{H(t)\}=\int_0^\infty e^{-px}F(x)\left\{\int_0^\infty e^{-p\theta}G(\theta)\,d\theta\right\}dx$$

$$=\int_0^\infty e^{-px}F(x)\,g(p)\,dx=f(p)\,g(p)$$

$$\Rightarrow L\left\{\int_0^t F(x)\,G(t-x)dx\right\}=f(p)\,g(p)$$

$$\Rightarrow \int_0^t F(x)\,G(t-x)\,dx=L^{-1}\{f(p)\,g(p)\}=F*G.$$

REMARKS

- $F * G$ is commutative, *i.e.*, $F * G = G * F$.
- $F * G$ is associative.
- $F * G$ is distributive over addition.

37.9 THE HEAVISIDE EXPANSION FORMULA

Theorem. *If F(P) and G(P) are polynomials in P, the degree of F(P) being less than that of G(P) and if*

$$G(p) = (p-\alpha_1)(p-\alpha_2)\ldots(p-\alpha_n)$$

where, $\alpha_1, \alpha_2, \ldots, \alpha_n$ are distinct constants, real or complex, then

$$L^{-1}\left\{\frac{F(p)}{G(p)}\right\} = \sum_{r=1}^{n} \frac{F(\alpha_r)}{G'(\alpha_r)} e^{\alpha_r . t}$$

Proof. By the method of partial fractions, let

$$\frac{f(p)}{G(p)} = \frac{A_1}{p-\alpha_1} + \frac{A_2}{p-\alpha_2} + \ldots + \frac{A_r}{p-\alpha_r} + \ldots + \frac{A_n}{p-\alpha_n}$$

Multiplying both sides by $(p - \alpha_r)$ and taking the limit $(p-\alpha_r)$, we get

$$A_r = \lim_{p\to\alpha_r} \frac{F(p)\,(p-\alpha_r)}{G(p)} = \lim_{p\to\alpha_r} F(p) \,.\, \lim_{p\to\alpha_r} \frac{p-\alpha_r}{G(p)}$$

$$= \lim_{p\to\alpha_r} F(p) \,.\, \lim_{p\to\alpha_r} \frac{1}{G'(p)} = \frac{F(\alpha_r)}{G'(\alpha_r)}$$

[By L'Hospital's rule]

Therefore,

$$\frac{F(p)}{G(p)} = \frac{F(\alpha_1)}{G'(\alpha_1)} \cdot \frac{1}{p-\alpha_1} + \frac{F(\alpha_2)}{G'(\alpha_2)} \cdot \frac{1}{p-\alpha_2} \ldots + \frac{F(\alpha_r)}{G'(\alpha_r)} \cdot \frac{1}{p-\alpha_r} + \ldots + \frac{F(\alpha_n)}{G'(\alpha_n)} \cdot \frac{1}{p-\alpha_n}$$

Hence, $$L^{-1}\left[\frac{F(p)}{G(p)}\right] = \frac{F(\alpha_1)}{G'(\alpha_1)} \,.\, e^{\alpha_1 t} + \frac{F(\alpha_2)}{G'(\alpha_2)} e^{\alpha_2 t} + \ldots + \frac{F(\alpha_r)}{G'(\alpha_r)} e^{\alpha_r t} + \ldots + \frac{F(\alpha_n)}{G'(\alpha_n)} e^{\alpha_n t}$$

$$= \sum_{r=1}^{n} \frac{F(\alpha_r)}{G'(\alpha_r)} . e^{\alpha_r . t}$$

Solved Examples

Example 1. *Using convolution theorem, evolute*

$$L^{-1}\left\{\frac{1}{(p-1)(p+2)}\right\}$$

Solution. We have $L^{-1}\left\{\frac{1}{p-1}\right\} = e^t = F_1(t)$ (say)

and $L^{-1}\left\{\frac{1}{p+2}\right\} = e^{-2t} = F_2(t)$ (say)

Using convolution theorem, we have

$$L^{-1}\left\{\frac{1}{p-1} \cdot \frac{1}{p+2}\right\} = F_1 * F_2$$

$$= \int_0^t F_1(x)\, F_2(t-x)\, dx$$

$$= \int_0^t e^x e^{-2(t-x)} dx$$

$$= e^{-2t} \int_0^t e^{3x} dx = \frac{1}{3}(e^t - e^{-2t})$$

Example 2. *Use the convolution theorem to find*

$$L^{-1}\left\{\frac{p^2}{(p^2+a^2)^2}\right\}$$ (Hazaribag-2009)

Solution. We know that $L^{-1}\left\{\frac{p}{(p^2+a^2)}\right\} = \cos at$

Therefore, by convolution theorem, we have

$$L^{-1}\left\{\frac{p^2}{(p^2+a^2)^2}\right\} = L^{-1}\left\{\frac{p}{p^2+a^2} \cdot \frac{p}{p^2+a^2}\right\}$$

$$= \int_0^t \cos ax \cos a(t-x)\, dx$$

$$= \int_0^t \cos ax\, (\cos at \cos ax + \sin at \sin ax)\, dx$$

$$= \cos at \int_0^t \cos^2 ax dx + \sin at \int_0^t \cos ax \sin ax\, dx$$

$$= \frac{1}{2}\cos at \int_0^t (1+\cos 2ax)\, dx + \frac{1}{2}\sin at \int_0^t \sin 2ax\, dx$$

$$= \frac{1}{2}\cos at \left[x + \frac{1}{2a}\sin 2ax\right]_0^t + \frac{1}{2}\sin at\left[-\frac{1}{2a}\cos 2ax\right]_0^t$$

$$= \frac{1}{2}\cos at\left[t + \frac{1}{2a}\sin 2a\right] + \frac{1}{4a}\sin at(1-\cos 2at)$$

$$= \frac{1}{2} t \cos at + \frac{1}{4a} \sin at$$
$$+ \frac{1}{4a} (\sin 2at \cos at - \sin at \cos 2at)$$
$$= \frac{1}{2} t \cos at + \frac{1}{4a} [(\sin at + \sin(2at - at)]$$
$$= \frac{1}{2a} [at \cos at + \sin at]$$

Example 3. *Evaluate* $L^{-1}\left[\dfrac{p}{(p^2+1)(p^2+4)}\right]$ [UPTU-2002]

Solution. We know that

$$L^{-1} \frac{p}{p^2+1} = \cos x \text{ and } L^{-1} \frac{2}{p^2+2^2} = \sin 2x$$

$$L^{-1}\left[\frac{p}{(p^2+1)(p^2+4)}\right] = \frac{1}{2} L^{-1}\left[\left(\frac{p}{p^2+1}\right) \cdot \left(\frac{2}{p^2+4}\right)\right]$$
$$= \frac{1}{2} \int_0^t \sin 2x \cos(t-x) dx$$
$$= \int_0^t \sin x \cos x \{\cos t \cos x + \sin t \sin x\}\, dx$$
$$= \int_0^t (\sin x \cos^2 x \cos t + \sin^2 x \cos x \sin t)\, dx$$
$$= \left[-\frac{\cos^3 x}{3} \cos t + \frac{\sin^3 x}{3} \sin x\right]_0^t$$
$$= -\frac{\cos^4 t}{3} + \frac{\sin^4 t}{3} + \frac{\cos t}{3}$$
$$= \frac{1}{3}\left[\sin^4 t - \cos^4 t\right] + \frac{\cos t}{3}$$
$$= \frac{1}{3} (\sin^2 t + \cos^2 t)(\sin^2 t - \cos^2 t) + \frac{\cos t}{3}$$
$$= \frac{1}{3} (\sin^2 t - \cos^2 t) + \frac{\cos t}{3} = -\frac{1}{3} \cos 2t + \frac{\cos t}{3}$$
$$= \frac{1}{3} (\cos t - \cos 2t)$$

Example 4. *Using convolution theorem, prove that*

$$L^{-1}\left[\frac{1}{p^3(p^2+1)}\right] = \frac{t^2}{2} + \cos t - 1$$

[UPTU-2005, VTU-2007, GBTU-2012]

Solution. $L^{-1}\left[\dfrac{1}{p^3(p^2+1)}\right]$ by convolution theorem.

$$L^{-1}\left\{\frac{1}{p^3}\right\} = \frac{t^2}{2!}$$

$$L^{-1}\left\{\frac{1}{p^2+1}\right\} = \sin t$$

Using convolution theorem

$$L^{-1}\left[\frac{1}{p^3(p^2+1)}\right] = \int_0^t \frac{(t-x)^2}{2!} \sin x\, dx$$
$$= \frac{1}{2} \int_0^t (t^2 + x^2 - 2tx) \sin x\, dx$$
$$= \frac{1}{2}\Big[(t^2 + x^2 - 2tx)(-\cos x) - \int (2x - 2t)(-\cos x) dx\Big]$$
$$= \frac{1}{2}\Big[-\cos x(t^2 + x^2 - 2tx) + 2\int (x-t) \cos x dx\Big]_0^t$$
$$= \frac{1}{2}\Big[-\cos x(t^2 + x^2 - 2tx) + 2(x-t)\sin x + 2\cos x\Big]_0^t$$
$$= \frac{1}{2}\Big[-\cos x(t^2 + x^2 - 2tx) + 2(x-t)\sin x + 2\cos x\Big]_0^t$$
$$= \frac{1}{2}\Big[-\cos t(t^2 + t^2 - 2t^2) + 0 + 2\cos t + t^2 \cos 0 - 2\cos 0\Big]_0^t$$
$$= \cos t + \frac{t^2}{2} - 1 = \frac{t^2}{2} + \cos t - 1.$$

Example 5. *Using the convolution theorem, find*

$$L^{-1}\left[\frac{p^2}{(p^2+a^2)(p^2+b^2)}\right], \ a \neq b$$

[UPTU-2004, Mumbai-2007, Bhopal-2008, UKTU-2011, VTU-2011S]

Solution. We have $L(\cos at) = \dfrac{p}{p^2+a^2}$

$$L(\cos bt) = \frac{p}{p^2+b^2}$$

Hence, by convolution theorem,

$$L\left[\int_0^t \cos ax \cos(bt - x) dx\right] = \frac{p^2}{(p^2+a^2)(p^2+b^2)}$$

Therefore $L^{-1}\left[\dfrac{p^2}{(p^2+a^2)(p^2+b^2)}\right]$

$$= \int_0^t \cos ax \cos b(t-x)\, dx$$
$$= \frac{1}{2} \int_0^t \{\cos(ax + bt - bx) + \cos(ax - bt + bx)\} dx$$
$$= \frac{1}{2} \int_0^t \cos\{(a-b)x + bt\} dx + \frac{1}{2} \int_0^t \cos[(a+b)x - bt] dx$$
$$= \frac{\sin at - \sin bt}{2(a-b)} + \frac{\sin at + \sin bt}{2(a+b)}$$
$$= \frac{a \sin at - b \sin bt}{a^2 - b^2}.$$

Example 6. *Evaluate* $L^{-1}\left\{\frac{1}{\sqrt{p}}\frac{1}{(p-a)}\right\}$ *by the convolution theorem.*

Solution. We know that

$$L^{-1}\left\{\frac{1}{\sqrt{p}}\right\} = L^{-1}\left\{\frac{1}{p^{1/2}}\right\} = \frac{t^{(1/2)-1}}{\Gamma(1/2)}$$

$$= \frac{1}{\sqrt{\pi}.\sqrt{t}} = F_1(t) \text{ (say)}$$

Also, $L^{-1}\left\{\frac{1}{p-a}\right\} = e^{at} = F_2(t)$

Then, by convolution theorem, we have

$$L^{-1}\left\{\frac{1}{\sqrt{p}\,(p-a)}\right\} = F_1(t) * F_2(t)$$

$$= \int_0^t F_1(x).F_2(t-x)\,dx$$

$$= \int_0^t \frac{1}{\sqrt{\pi}}\cdot\frac{1}{\sqrt{x}}\,e^{a(t-x)}\,dx$$

$$= \frac{e^{at}}{\sqrt{\pi}}\int_0^{\sqrt{at}} \frac{\sqrt{a}}{u}\,e^{-u^2}.\frac{2u}{a}.du$$

[By Putting] $ax = u^2 \Rightarrow dx = \frac{2u\,du}{a}$

$$= \frac{e^{at}}{\sqrt{a}}\cdot\frac{2}{\sqrt{\pi}}\int_0^{\sqrt{(at)}} e^{-u^2}\,du = \frac{e^{at}}{\sqrt{a}}\,erf\,[\sqrt{at})]$$

Example 7. *Using convolution theorem, show that*

$$B(m, n) = \int_0^1 x^{m-1}(1-x)^{n-1}dx$$

$$= \frac{\Gamma(m)\,\Gamma(n)}{\Gamma(m+n)},\quad m > 0,\ n > 0$$

Solution. Let $F(t) = \int_0^t x^{m-1}(t-x)^{n-1}dx$

$$= \int_0^t F_1(x).F_2(t-x)\,dx$$

where, $F_1(t) = t^{m-1} = F_1 * F_2$ and $F_2(t) = t^{n-1}$

Therefore $L(F(t)\} = L\{F_1 * F_2\}$

$$= L\{F_1(t)\}\,.\,L\{F_2(t)\} = L\{t^{m-1}\}.L\{t^{n-1}\}$$

$$= \frac{\Gamma(m)}{p^m}.\frac{\Gamma(n)}{p^n} = \frac{\Gamma(m)\,\Gamma(n)}{p^{m+n}}$$

$$\Rightarrow\quad F(t) = L^{-1}\left\{\frac{\Gamma(m)\,.\,\Gamma(n)}{p^{m+n}}\right\}$$

$$\Rightarrow\quad F(t) = \int_0^t x^{m-1}(t-x)^{n-1}dx$$

$$= L^{-1}\left\{\frac{\Gamma(m)\,\Gamma(n)}{p^{m+n}}\right\}$$

$$= \Gamma(m)\,.\,\Gamma(n)\,.\,L^{-1}\left\{\frac{1}{p^{m+n}}\right\}$$

$$= \frac{\Gamma(m)\,\Gamma(n)}{\Gamma(m+n)}\,t^{m+n-1}$$

Let $t = 1$, then we have

$$B(m, n) = \int_0^1 x^{m-1}(1-x)^{n-1}dx = \frac{\Gamma(m)\,\Gamma(n)}{\Gamma(m+n)}.$$

Example 8. *Using Heaviside's expansion formula, evaluate* $L^{-1}\left\{\frac{3p+1}{(p-1)(p^2+1)}\right\}$

Solution. We have $F(p) = 3p+1$

and $G(p) = (p-1)\,(p^2+1)$

$$= (p-1)\,(p+i)(p-i)$$

Clearly, $G(p)$ has 3 distinct zeroes $\alpha_1 = 1, \alpha_2 = i$ and $\alpha = -i$

Also, $G'(p) = 3p^2 - 2p + 1$

Using Heaviside's expansion formula, we have

$$L^{-1}\left\{\frac{3p+1}{(p-1)(p^2+1)}\right\} = \frac{F(1)}{G'(1)}e^t + \frac{F(i)}{G'(i)}e^{it} + \frac{F(-i)}{G'(-i)}e^{-it}$$

$$= \frac{4e^t}{2} + \frac{3i+1}{-(2+2i)}e^{it} + \frac{(-3i+1)}{(-2+2i)}e^{-it}$$

$$= 2e^t - \frac{(3i+1)(1-i)}{2(1+i)(1-i)}e^{i.t} + \frac{(3i-1)\,(1+i)}{2(1-i)(1+i)}e^{-i.t}$$

$$= 2e^t - \frac{1}{2}(i+2)e^{it} + \frac{1}{2}(i-2)e^{-it}$$

$$= 2e^t - \frac{1}{2}i(e^{it} - e^{-it}) - (e^{it} + e^{-it})$$

$$= 2e^t - \frac{1}{2}.i.2i\,\sin t - 2\cos t$$

$$= 2e^t + \sin t - 2\cos t\,.$$

Example 9. *Find* $L^{-1}\left[\frac{2p^2-6p+5}{p^3-6p^2+11p-6}\right]$. [UPTU-2004]

Solution. Let $f(p) = 2p^2 - 6p + 5$

and $G(p) = p^3 - 6p^2 + 11p - 6$

$$= (p-1)(p-2)(p-3)$$

$$G'(p) = 3p^2 - 12p + 11$$

$G(p) = 0$ has three roots 1, 2, 3.

$\alpha_1 = 1, \alpha_2 = 2, \alpha_3 = 3$

By Heaviside's inverse formula, we have

$$L^{-1}\left[\frac{F(p)}{G(p)}\right] = \sum_{i=1}^{n}\frac{F(\alpha_i)}{G'(\alpha_i)}e^{t\alpha_i}$$

$$L^{-1}\left[\frac{2p^2-6p+5}{p^3-6p^2+11p-6}\right] = \frac{F(\alpha_1)}{G'(\alpha_1)}e^{t\alpha_1} + \frac{F(\alpha_2)}{G'(\alpha_2)}e^{t\alpha_2} + \frac{F(\alpha_3)}{G'(\alpha_3)}e^{t\alpha_3}$$

$$= \frac{F(1)}{G'(1)}e^t + \frac{F(2)}{G'(2)}e^{2t} + \frac{F(3)}{G'(3)}e^{3t}$$

$$= \frac{1}{2}e^t + \frac{1}{-1}e^{2t} + \frac{5}{2}e^{3t} = \frac{1}{2}e^t - e^{2t} + \frac{5}{2}e^{3t}.$$

Example 10. *Show that* $L^{-1}\left\{\dfrac{1}{\sqrt{p^2+a^2}}\right\}=\int_0^t J_0(ak)\,dx$.

Solution. Here, we have

$$L^{-1}\left\{\frac{1}{\sqrt{p^2+a^2}}\right\}=L^{-1}\left\{\frac{1}{p}\left(1+\frac{a^2}{p^2}\right)^{-1/2}\right\}$$

$$=L^{-1}\left\{\frac{1}{p}\left(1-\frac{1}{2}\frac{a^2}{p^2}+\frac{1.3}{2.4}\frac{a^4}{p^4}-\frac{1.3.5}{2.4.6}\frac{a^6}{p^6}+\ldots\right)\right\}$$

$$=L^{-1}\left\{\frac{1}{p}-\frac{a^2}{2}.\frac{1}{p^3}+\frac{1.3}{2.4}\frac{a^4}{p^5}-\frac{1.3.5}{2.4.6}\frac{a^6}{p^7}+\ldots\right\}$$

$$=1-\frac{a^2}{2}.\frac{t^2}{2!}+\frac{1.3}{2.4}\frac{a^4t^4}{4!}-\frac{1.3.5}{2.4.6}\frac{a^6t^6}{6!}+\ldots$$

$$=1-\frac{(at)^2}{2^2}+\frac{(at)^4}{2^2.4^2}-\frac{(at)^6}{2^2.4^2.6^2}+\ldots$$

$$=J_0(at).$$

Example 11. *Evaluate* $L^{-1}\left\{\dfrac{e^{-\sqrt{p}}}{p}\right\}$ *and hence deduce that*

$$L^{-1}\left\{\frac{e^{x\sqrt{p}}}{p}\right\}=erfc\left(\frac{x}{2\sqrt{t}}\right)$$

Solution. Let $f(p)=e^{-\sqrt{p}}$

Therefore, $F(t)=L^{-1}\{e^{-\sqrt{p}}\}$

$$=L^{-1}\left\{1-\sqrt{p}+\frac{p}{2!}-\frac{p^{3/2}}{3!}+\frac{p^2}{4!}-\frac{p^{5/2}}{5!}+\ldots\right\}$$

$$=L^{-1}\{1\}-L^{-1}\{p^{1/2}\}+\frac{1}{2!}L^{-1}\{p\}$$

$$-\frac{1}{3!}L^{-1}\{p^{3/2}\}+\frac{1}{4!}L^{-1}\{p^2\}$$

$$-\frac{1}{5!}L^{-1}\{p^{5/2}\}+\ldots \qquad \ldots (1)$$

Now, $L^{-1}\{p^{n+(1/2)}\}=L^{-1}\left\{\dfrac{1}{p^{-n-(1/2)}}\right\}$

$$=\frac{t^{-n-(3/2)}}{\Gamma\left(-n-\frac{1}{2}\right)},\ n\in\mathbf{Z}$$

$$=\frac{(-1)^{n+1}}{\sqrt{n}}\left(\frac{1}{2}\right)\left(\frac{3}{2}\right)\left(\frac{5}{2}\right)\ldots\left(\frac{2n+1}{2}\right)t^{-n-(3/2)}$$

Now, from (1), we have

$$F(t)=-\frac{(-1)t^{-3/2}}{\sqrt{\pi}}.\frac{1}{2}-\frac{1}{3!}.\frac{(-1)^2}{\sqrt{\pi}}\left(\frac{1}{2}\right)\left(\frac{3}{2}\right)t^{-5/2}$$

$$-\frac{1}{5!}\frac{(-1)^3}{\sqrt{\pi}}\left(\frac{1}{2}\right)\left(\frac{3}{2}\right)\left(\frac{5}{2}\right)t^{-7/2}+\ldots$$

$$=\frac{1}{2\sqrt{\pi}.t^{3/2}}$$

$$\left[1-\frac{1}{4t}+\frac{(1/4t)^2}{2!}-\frac{(1/4t)^3}{3!}+\ldots\right]$$

$$=\frac{1}{2\sqrt{\pi}\,.\,t^{3/2}}e^{-1/4t}.$$

Since $L^{-1}\left\{\dfrac{f(p)}{p}\right\}=\int_0^t F(x)\,dx$

where $F(t)=L^{-1}\{f(p)\}$

Therefore,

$$L^{-1}\left\{\frac{e^{-\sqrt{p}}}{p}\right\}=\int_0^t\frac{1}{2\sqrt{\pi}\,x^{3/2}}e^{-1/(4x)}.dx$$

$$=-\frac{2}{\sqrt{\pi}}\int_\infty^{1/2\sqrt{t}}e^{-y^2}\,dy$$

(where $x=\dfrac{1}{4y^2}\Rightarrow dx=-\dfrac{dy}{2y^3}$)

$$=\frac{2}{\sqrt{\pi}}\int_{1/(2\sqrt{t})}^\infty e^{-y^2}\,dy=erfc\left(\frac{1}{2\sqrt{t}}\right)$$

Deduction. We have

$$\because\quad L^{-1}\left\{\frac{e^{-\sqrt{p}}}{p}\right\}=erfc\left(\frac{1}{2\sqrt{t}}\right)$$

$$\therefore\quad L^{-1}\left\{\frac{e^{-\sqrt{(x^2p)}}}{x^2p}\right\}=\frac{1}{x^2}erfc\left(\frac{1}{2\sqrt{t/x^2}}\right)$$

(By change of scale property)

or $L^{-1}\left\{\dfrac{e^{-x\sqrt{p}}}{p}\right\}=erfc\left(\dfrac{x}{2\sqrt{t}}\right)$

Example 12. *Show that*

$$L^{-1}\left\{\frac{1}{p^3+1}\right\}=\frac{t^2}{2!}-\frac{t^5}{5!}+\frac{t^8}{8!}-\frac{t^{11}}{11!}+\ldots$$

Solution. We have $\dfrac{1}{p^3+1}=\dfrac{1}{p^3}\left(1+\dfrac{1}{p^3}\right)^{-1}$

$$=\frac{1}{p^3}\left[1-\frac{1}{p^3}+\frac{1}{p^6}-\frac{1}{p^9}+\frac{1}{p^{12}}-\ldots\right]$$

$$=\frac{1}{p^3}-\frac{1}{p^6}+\frac{1}{p^9}-\frac{1}{p^{12}}+\ldots$$

Hence, $L^{-1}\left\{\dfrac{1}{p^3+1}\right\}=\dfrac{t^2}{2!}-\dfrac{t^5}{5!}+\dfrac{t^8}{8!}-\dfrac{t^{11}}{11!}+\ldots$

Example 13. *Show that* $\int_0^\infty \cos x^2dx=\dfrac{1}{2}\sqrt{\dfrac{\pi}{2}}$

Solution. Let, $F(t)=\int_0^\infty \cos t\,.\,x^2\,dx$

Therefore, $L\{F(t)\} = \int_0^\infty e^{-pt} F(t)\, dt$

$$= \int_0^\infty e^{-pt}\left[\int_0^\infty \cos t.x^2 dx\right] dt$$

$$= \int_0^\infty \left[e^{-pt}\int_0^\infty \cos t.x^2 dt\right] dx$$

$$= \int_0^\infty L\{\cos tx^2\}\, dx = \int_0^\infty \frac{p}{p^2+x^4} dx$$

$$= \frac{1}{2\sqrt{p}}\int_0^{\pi/2} \frac{d\theta}{\sqrt{(\tan\theta)}}$$

where $x = \sqrt{p\tan\theta} \Rightarrow dx = \dfrac{p\sec^2\theta\, d\theta}{2\sqrt{p\tan\theta}}$

$$= \frac{1}{2\sqrt{p}}\int_0^{\pi/2} \sin^{-1/2}\theta \cos^{1/2}\theta\, d\theta$$

$$= \frac{1}{2\sqrt{p}} \frac{\Gamma\left(\frac{1}{4}\right)\Gamma\left(\frac{3}{4}\right)}{2\,\Gamma(1)} = \frac{1}{2\sqrt{p}} \frac{\Gamma\left(\frac{1}{4}\right)\Gamma\left(1-\frac{1}{4}\right)}{2}$$

$$= \frac{1}{4\sqrt{p}} \cdot \frac{\pi}{\sin\frac{\pi}{4}} = \frac{\pi}{2\sqrt{2p}}$$

[Using $\Gamma(p)\,\Gamma(1-p) = \dfrac{\pi}{\sin p\pi}$]

Hence, $F(t) = \dfrac{\pi}{2\sqrt{2}} L^{-1}\left\{\dfrac{1}{p^{1/2}}\right\}$

$$= \frac{\pi}{2\sqrt{2}} \frac{t^{(1/2)-1}}{\Gamma(1/2)} = \frac{1}{2}\sqrt{\frac{\pi}{2t}}$$

$$= \int_0^\infty \cos t\,.\, x^2\, dx = \frac{1}{2}\sqrt{\frac{\pi}{2t}}$$

Let $t=1$, then we have $\int_0^\infty \cos x^2 dx = \dfrac{1}{2}\sqrt{\dfrac{\pi}{2}}$.

Example 14. *Show that* $\int_0^\infty e^{-x^2} dx = \dfrac{\sqrt{\pi}}{2}$

Solution. Let, $F(t) = \int_0^\infty e^{-tx^2} dx$

Then proceed as in example (13), we get

$$L\{F(t)\} = \int_0^\infty L\{e^{-tx^2}\}\, dx = \int_0^\infty \frac{dx}{p+x^2}$$

$$= \left(\frac{1}{\sqrt{p}}.\tan^{-1}\frac{x}{\sqrt{p}}\right)_0^\infty = \frac{\pi}{2\sqrt{p}}$$

$$\Rightarrow F(t) = \frac{\pi}{2}L^{-1}\left(\frac{1}{\sqrt{p}}\right) = \frac{\pi}{2}.\frac{1}{\sqrt{(\pi t)}} = \frac{1}{2}\sqrt{\frac{\pi}{t}}$$

$$\Rightarrow \int_0^\infty e^{-tx^2} dx = \frac{1}{2}\sqrt{\left(\frac{\pi}{t}\right)}$$

Taking $t=1$, then we have $\int_0^\infty e^{-x^2} dx = \dfrac{1}{2}\sqrt{\pi}$

Example 15. *Show that*

$$L^{-1}\left\{\frac{8}{(p^2+1)^3}\right\} = (3-t^2)\sin t - 3t\cos t$$

Solution. We know that $L^{-1}\left\{\dfrac{1}{p^2+1}\right\} = \sin t$

$$L^{-1}\left\{\frac{1}{(p^2+1)}\cdot\frac{1}{(p^2+1)}\right\} = \int_0^t \sin x \sin(t-x)\, dx$$

$$= \int_0^t \sin x\,(\sin t\cos x - \cos t\sin x)\, dx$$

$$= \sin t\int_0^t \sin x\cos x\, dx - \cos t\int_0^t \sin^2 x\, dx$$

$$= \frac{1}{2}\sin t\int_0^t \sin 2x\, dx - \frac{1}{2}\cos t\int_0^t (1-\cos 2x) dx$$

$$= \frac{1}{2}\sin t.\frac{1}{2}(1-\cos 2t) - \frac{1}{2}\cos t\left(t - \frac{1}{2}\sin 2t\right)$$

$$= \frac{1}{2}\sin t\sin^2 t - \frac{1}{2}t\cos t + \frac{1}{2}\cos t\sin t\cos t$$

$$\Rightarrow L^{-1}\left\{\frac{1}{(p^2+1)^2}\right\} = \frac{1}{2}\sin t - \frac{t}{2}\cos t$$

$$\Rightarrow L^{-1}\left\{\frac{8}{(p^2+1)^3}\right\} = 8L^{-1}\left\{\frac{1}{(p^2+1)^2}\cdot\frac{1}{(p^2+1)}\right\}$$

$$= 8\int_0^t \left(\frac{1}{2}\sin x - \frac{x}{2}\cos x\right)\sin(t-x)\, dx$$

[By convolution theorem]

$$= 4\int_0^t (\sin x - x\cos x)(\sin t\cos x - \cos t\sin x)\, dx$$

$$= 4\sin t\int_0^t (\sin x\cos x - x\cos^2 x) dx - 4\cos t\int_0^t (\sin^2 x - x\sin x.\cos x) dx$$

$$= 2\sin t\int_0^t \{\sin 2x - x(1+\cos 2x)\}\, dx - 2\cos t\int_0^t \{(1-\cos 2x) - (x\sin 2x)\}\, dx$$

$$= 2\sin t\left[-\frac{t^2}{2} + \frac{1-\cos 2t}{2} - \frac{t}{2}\sin 2t + \frac{1-\cos 2t}{4}\right] - 2\cos t\left[t - \frac{1}{2}\sin 2t + \frac{t\cos 2t}{2} - \frac{\sin 2t}{4}\right]$$

$$= -t^2\sin t + \frac{3}{2}\sin t - \frac{3}{2}\sin t\cos 2t - t\sin t\sin 2t - 2t\cos t + \frac{3}{2}\sin 2t\cos t - t\cos t\cos 2t$$

$$= -t^2\sin t + \frac{3}{2}\sin t + \frac{3}{2}(-\sin t\cos 2t + \sin 2t\cos t) - t(\cos 2t\cos t + \sin t\sin 2t) - 2t\cos t$$

$$= (3-t^2)\sin t - 3t\cos t$$

Example 16. Show that $1*1*1*\ldots*1$ (n times) $= \dfrac{t^{n-1}}{(n-1)!}$ where $n = 1, 2, 3, \ldots$

Solution. Since we know that

$$F*G = \int_0^t F(x)\,G(t-x)\, dx$$

$$\therefore \quad 1*1 = \int_0^1 1.1\, dx = t$$

Again, $1*1*1 = t*1 = \int_0^t x.1\,dx = \frac{t^2}{2}$

$$1*1*1 = \left(\frac{t^2}{2}\right)*1 = \int_0^t \frac{x^2}{2}.1\,dx = \left(\frac{x^3}{2.3}\right)_0^t = \frac{t^3}{3!}$$

Proceeding in the similar way, we get

$$1*1*1*...*1 \text{ (}n\text{ times)} = \frac{t^{n-1}}{(n-1)!}$$

where $n = 1, 2, 3, ...$

EXERCISE 37.3

1. Using the convolution theorem, show that

(a) $L^{-1}\left\{\frac{1}{(p+1)(p-2)}\right\} = \frac{1}{3}[e^{2t} - e^{-t}]$

(b) $L^{-1}\left\{\frac{p}{(p^2+a^2)^2}\right\} = \frac{1}{2a} t \sin at$ [UPTU-2008]

(c) $L^{-1}\left\{\frac{1}{p(p^2+4)^2}\right\} = \frac{1}{16}(1 - t\sin 2t - \cos 2t)$

(d) $L^{-1}\left\{\frac{1}{(p-2)(p^2+1)}\right\} = \frac{1}{5}[e^{2t} - 2\sin t - \cos t]$

(e) $\int_0^t \sin u \cos(t-u)\,du = \frac{t}{2}\sin t$

(f) $\int_0^t J_0(u) J_0(t-u)\,du = \sin t$

(g) $L^{-1}\left(\frac{p}{(p^2+4)^2}\right) = \frac{t}{4}\sin 2t$ [UPTU-2004, GBTU-2010]

(h) $L^{-}\left\{\frac{1}{(p^2+1)(p^2+q)}\right\} = \frac{1}{8}\left(\sin t - \frac{1}{3}\sin 3t\right)$ (Mumbai-2005S)

(i) $L^{-}\left\{\frac{p}{(p^2+1)(p^2+4)(p^2+q)}\right\}$

$= \frac{1}{12}\cos t - \frac{1}{10}\cos 2t + \frac{1}{60}\cos 3t$ (Mumbai-2006)

(j) $L^{-}\left\{\frac{1}{(p-2)(p+2)^2}\right\} = \frac{1}{16}(e^{2t} - e^{-2t} - 4te^{-2t})$ (Mumbai-2009)

(k) $L^{-}\left\{\frac{P}{(P+2)(P^2+q)}\right\} = \frac{1}{13}(3\sin 3t + 2\cos 2t - 2e^{-2t})$ (VTU-2008 S)

(l) $L^{-}\left\{\frac{1}{(P^2+4P+13)^2}\right\} = \frac{e^{-2t}}{54}(\sin 3t - 3t\cos 3t)$ (Mumbai-2008)

2. Using the Heaviside formula, show that

(a) $L^{-1}\left\{\frac{p^2-6}{p^3+4p^2+3p}\right\} = -2 + \frac{5}{2}e^{-t} + \frac{1}{2}e^{-3t}$

(b) $L^{-1}\left\{\frac{19p+37}{(p+1)(p-2)(p+3)}\right\} = -3e^{-t} + 5e^{2t} - 2e^{-3t}$

(c) $L^{-1}\left(\frac{1}{p^3-1}\right) = \frac{1}{3}\left[e^t - e^{-t/2}\left\{\cos\left(\frac{1}{2}\sqrt{3}t\right) + \sqrt{3}\sin\left(\frac{1}{2}\sqrt{3}t\right)\right\}\right]$

(d) $L^{-1}\left(\frac{1}{p^3+1}\right) = \frac{1}{3}\left[e^{-t} - e^{t/2}\left\{\cos\left(\frac{1}{2}\sqrt{3}t\right) - \sqrt{3}\sin\left(\frac{1}{2}\sqrt{3}t\right)\right\}\right]$

Hint to Selected Problems

1. (a) $L^{-1}\left(\frac{1}{p+1}\right) = e^{-t} = F_1(t)$ (say)

and $L^{-1}\left(\frac{1}{p+2}\right) = e^{2t} = F_2(t)$ (say)

Then using convolution theorem.

2. (a) $f(p) = 2p^2 - 6p + 5$

and $G(p) = p^3 - 6p^2 + 11p - 6 = (p-1)(p-2)(p-3)$

$\Rightarrow G'(p) = 3p^2 - 12p + 11$

$\Rightarrow G(p)$ has 3 distinct roots $\alpha_1 = 1,\ \alpha_2 = 2$ and $\alpha_3 = 3$.

Then using Heavyside expansion formula.

2. (c) $f(p) = 19p + 37$

$G(p) = (p+1)(p-2)(p+3)$

$\Rightarrow G(p)$ has three distinct roots $\alpha_1 = -1,\ \alpha_2 = 2,\ \alpha_3 = -3$.

Then by Heavyside's formula, we have

$$L^{-1}\left[\frac{19p+37}{(p+1)(p-2)(p+3)}\right] = \frac{f(-1)}{G'(-1)}e^{-t} + \frac{f(2)}{G'(2)}e^{2t} + \frac{f(-3)}{G'(-3)}e^{-3t} = -3e^{-t} + 5e^{2t} - 2e^{-3t}$$

Miscellaneous Solved Examples

Example 1. *Find the inverse transforms of* $\frac{p+2}{p^2-4p+13}$. (VTU 2008)

Solution.
$$L^{-1}\left(\frac{p+2}{p^2-4p+13}\right) = L^{-1}\left(\frac{p+2}{(p-2)^2+9}\right)$$

$$= L^{-1}\left[\frac{p-2+4}{(p-2)^2+3^2}\right]$$

$$= L^{-1}\left[\frac{p-2}{(p-2)^2+3^2}\right] + 4L^{-1}\left(\frac{1}{(p-2)^2+3^2}\right)$$

$$= e^{2t}\cos 3t + \frac{4}{3}e^{2t}\sin 3t.$$

Example 2. *Find* $L^{-1}\left\{\frac{(p+2)^2}{(p^2+4p+8)^2}\right\}$ (Mumbai–2005)

Solution. $L^{-1}\left\{\frac{(p+2)^2}{(p^2+4p+8)^2}\right\} = L^{-1}\left\{\frac{(p+2)^2}{(p^2+4p+4+4)^2}\right\}$

$$= L^{-1}\left\{\frac{(p+2)^2}{[(p+2)^2+4]^2}\right\}$$

$$= e^{-2t}L^{-1}\left\{\frac{p^2}{(p^2+4)^2}\right\} = e^{-2t}L^{-1}\left\{\frac{p^2+4-4}{(p^2+4)^2}\right\}$$

$$= e^{-2t}L^{-1}\left\{\frac{1}{p^2+4} - \frac{4}{(p^2+4)^2}\right\}$$

$$= \frac{e^{-2t}\sin 2t}{2} - 4e^{-2t}L^{-1}\left\{\frac{1}{(p^2+4)^2}\right\}$$

$$= \frac{e^{-2t}\sin 2t}{2} - 4e^{-2t}\left\{\frac{1}{4}\left(\frac{\sin 2t}{4} - \frac{t\cos 2t}{2}\right)\right\}$$

$$= e^{-2t}\left\{\frac{\sin 2t}{2} - \frac{\sin 2t}{4} + \frac{t\cos 2t}{2}\right\}$$

$$= e^{-2t}\left\{\frac{\sin 2t}{4} + \frac{t\cos 2t}{2}\right\} = e^{-2t}\left\{\frac{\sin 2t}{4} + \frac{t\cos 2t}{2}\right\}$$

Example 3. *Find* $L^{-1}\left\{\frac{1}{p(p^2+a^2)}\right\}$ (PTU–2003)

Solution. $\because$ $L^{-1}\left(\frac{1}{p^2+a^2}\right) = \frac{1}{a}\sin at$

$\therefore$ $L^{-1}\left\{\frac{1}{p(p^2+a^2)}\right\} = \int_0^t \frac{1}{a}\sin at\,dt$

$$= \frac{1}{a^2}[-\cos at]_0^t = \frac{(1-\cos at)}{a^2}.$$

Example 4. *Find* $L^{-1}\left\{\frac{p+2}{p^2(p+1)(p-2)}\right\}$ (VTU-2003)

Solution. $L^{-1}\left\{\frac{p+2}{(p+1)(p-2)}\right\} = \frac{4}{3}L^{-1}\left(\frac{1}{p-2}\right) - \frac{1}{3}L^{-1}\left(\frac{1}{p+1}\right)$

$$= \frac{4}{3}e^{2t} - \frac{1}{3}e^{-t}$$

Now, $L^{-1}\left\{\frac{p+2}{p(p+1)(p-2)}\right\}$

$$= \int_0^t L^{-1}\left(\frac{p+2}{(p+1)(p-2)}\right)dt$$

$$= \int_0^t\left(\frac{4}{3}e^{2t} - \frac{1}{3}e^{-t}\right)dt = \frac{2}{3}e^{2t} + \frac{1}{3}e^{-t} - 1$$

Now, $L^{-1}\left\{\frac{p+2}{p^2(p+1)(p-2)}\right\}$

$$= \int_0^t L^{-1}\left\{\frac{p+2}{p(p+1)(p-2)}\right\}dt$$

$$= \int_0^t\left(\frac{2}{3}e^{2t} + \frac{1}{3}e^{-t} - 1\right)dt = \frac{1}{3}(e^{2t} - e^{-t} - t)$$

Example 5. *Find* $L^{-1}\left\{\frac{p+2}{(p^2+4p+5)^2}\right\}$. (SVTU-2009, PTU-2005)

Solution. $L^{-1}\left\{\frac{1}{p^2+4p+5}\right\}$

$$= L^{-1}\left\{\frac{1}{(p+2)^2+1}\right\} = e^{-2t}\sin t$$

Now, $L^{-1}\left\{\frac{d}{dp}\left(\frac{1}{p^2+4p+5}\right)\right\} = (-1)t\cdot e^{-2t}\sin t$

$$L^{-1}\left\{\frac{-(2p+4)}{(p^2+4p+5)^2}\right\} = -t\cdot e^{-2t}\sin t$$

$$L^{-1}\left\{\frac{p+2}{(p^2+4p+5)^2}\right\} = \frac{1}{2}t\cdot e^{-2t}\sin t$$

Example 6. *Find* $L^{-1}\left\{\log\frac{p+1}{p-1}\right\}$

(Bhopal–2008, SVTU–209, UPTU–2009, UKTU–2012)

Solution. It $f(t) = L^{-1}\log\frac{p+1}{p-1}$

then, $tf(t) = L^{-1}\left\{\frac{-d}{dp}\log\left(\frac{p+1}{p-1}\right)\right\}$

$$= -L^{-1}\left\{\frac{d}{dp}\log(p+1)\right\} + L^{-1}\left\{\frac{d}{dp}\log(p-1)\right\}$$

$$= -L^{-1}\left(\frac{1}{p+1}\right) + L^{-1}\left(\frac{1}{p-1}\right)$$

$$= -e^{-t} + e^{t} = 2\sinh t$$

Thus $f(t) = (2\sinh t)/t$

Example 7. *Find* $L^{-1}\left\{\log\frac{p^2+1}{p(p+1)}\right\}$ (SVTU-2009, VTU-2008)

Solution. If $f(t) = L^{-1}\log\frac{p^2+1}{p(p+1)}$

Then $t\,f(t) = L^{-1}\left\{-\frac{d}{dp}\log\left(\frac{p^2+1}{p(p+1)}\right)\right\}$

$$= L^{-1}\left\{\frac{d}{dp}\log(p^2+1)\right\} + L^{-1}\left\{\frac{d}{dp}\log p\right\} + L^{-1}\left\{\frac{d}{dp}\log(p+1)\right\}$$

$$= -L^{-1}\left(\frac{2p}{p^2+1}\right) + L^{-1}\left(\frac{1}{p}\right) + L^{-1}\left(\frac{1}{p+1}\right)$$

$$= -2\cos t + 1 + e^{-t}$$

Thus, $f(t) = \frac{1}{t}(1 + e^{-t} - 2\cos t)$

MISCELLANEOUS EXERCISE

(A) Find the inverse laplace transform of the following functions.

1. $\frac{p}{(2p-1)(3p-1)}$ (VTU-2010)
2. $\frac{1}{p^2-5p+6}$ (SVTU-2008)
3. $\frac{3p+2}{p^2-p-2}$ (VTU-2010 S)
4. $\frac{1}{p(p^2-1)}$ (Nagarjuna-2008)
5. $\frac{1-7p}{(p-3)(p-1)(p+2)}$ (BPTU-2005 S)
6. $\frac{p}{(p^2-1)^2}$ (Kurukshetra-2005)
7. $\frac{p^3}{p^4-a^4}$ (Kurukshetra-2005)
8. $\frac{p^2+2p+3}{(p^2+2p+2)(p^2+2p+5)}$ (Mumbai-2008)
9. $\frac{a(p^2-2a^2)}{p^4+4a^4}$ (Mumbai-2009)
10. $\frac{1}{p^2(p+5)}$ (Madras–2003 S)
11. $\frac{p}{a^2p^2+b^2}$ (Madras-2000 S)
12. $\frac{p+2}{(p^2+4p+8)^2}$ (Mumbai-2006)
13. $\frac{1}{2}\log\left(\frac{p^2+b^2}{p^2+a^2}\right)$ (Mumbai-2008, VTU-2008)
14. $\log\frac{p^2+1}{(p-1)^2}$ (Madras-2000 S)
15. $\tan^{-1}\left(\frac{2}{p}\right)$ (Mumbai-2007, PTU-2005)
16. $\cot^{-1}(p)$ (VTU-2005)
17. $p\log\frac{p-1}{p+1}$ (Madras-1999)
18. $\frac{2p+1}{p^2-4}$ (UPTU (SUM)-2009)
19. $\frac{e^{-2\pi p}}{p(p^2+1)}$ (GBTU-2011)
20. $\frac{p-1}{p^2(p-7)}$ (GBTU (CO)-2011)
21. $\frac{14p+10}{49p^2+28p+13}$ (UPTU (SUM)-2007)
22. $\frac{p}{p^2+6p+25}$ (GBTU-2011)
23. $\frac{p+1}{p^2-6p+25}$ (GBTU-2010)
24. $\log\left(\frac{p^2+4p+5}{p^2+2p+5}\right)$ (MTU (SUM)-2011)
25. $\frac{1}{p^4+4}$ (UPTU (SUM)-2007)
26. $\frac{1}{p(p+1)(p+2)}$ (UPTU (SUM)-2008)
27. $\log\left(1-\frac{a^2}{p^2}\right)$ (UPTU (CO)-2009)
28. $\frac{1}{p(p+a)^3}$ (UKTU-2012)
29. $\frac{1}{p^2(p^2+1)}$ (UPTU (SUM)-2009)
30. $\frac{p^2}{(p^2+w^2)^2}$ (GBTU(CO) 2010)
31. $\frac{8p}{(p^2+16)(p^2+1)^2}$ (UPTU (CO)-2009)

(B) Use convolution theorm to find:

1. $L^{-1}\left[\frac{1}{(p^2+a^2)^2}\right]$ (UPTU (SUM)-2007, UPTU-2009)
2. $L^{-1}\left[\frac{16}{(p-2)(p+2)^2}\right]$ (MTU-2011)
3. $L^{-1}\left[\frac{p}{(p^2+a^2)^3}\right]$ (UPTU-2008, MTU-2012)

ANSWERS

(A)

1. $3e^{t/2}+2e^{t/3}$
2. $e^{3t}-e^{2t}$
3. $\frac{1}{3}(8e^{2t}-e^{-t})$
4. $\cosh t$
5. $e^t+e^{-2t}-2e^{3t}$
6. $\frac{1}{2}t\sinh t$
7. $\frac{1}{2}[\cos at+\cosh at]$
8. $\frac{1}{3}e^{-t}(\sin t+\sin 2t)$
9. $\cos at\sinh at$
10. $\frac{1}{25}(e^{-5t}+5t-1)$
11. $\frac{1}{a^2}\cos\left(\frac{bt}{a}\right)$
12. $\frac{1}{2}te^{-2t}\sin 2t$
13. $\frac{1}{t}(\cos at-\cos bt)$
14. $\frac{2}{t}(e^t-\cos t)$
15. $\frac{\sin 2t}{t}$
16. $\frac{\sin t}{t}$
17. $\frac{2(\sinh t-t\cosh t)}{t^2}$
18. $2\cosh 2t+\frac{1}{2}\sinh 2t$
19. $1-\cos t u(t-2\pi)$
20. $-\frac{6}{49}+\frac{1}{7}t+\frac{6}{49}e^{7t}$
21. $\frac{2}{7}e^{-2t/7}\left(\cos\frac{3}{7}t+\sin\frac{3}{7}t\right)$
22. $e^{-3t}\left(\cos 4t-\frac{3}{4}\sin 4t\right)$
23. $e^{3t}(\cos 4t+\sin 4t)$
24. $\frac{2}{t}(e^{-t}\cos 2t-e^{-2t}\cos t)$
25. $\frac{1}{4}(\sin t\cosh t-\cos t\sinh t)$
26. $\frac{1}{2}-e^{-t}+\frac{1}{2}e^{-2t}$
27. $\frac{2}{t}(1-\cosh at)$
28. $\frac{1}{a^3}-\frac{1}{a^3}e^{-at}\left(1+at+\frac{a^2t^2}{2}\right)$
29. $t-\sin t$
30. $\frac{1}{2\omega}\sin\omega t+\frac{t}{2}\cos\omega t$
31. $\frac{60t\sin t-8\cos t+8\cos 4t}{225}$

(B)

1. $\frac{1}{2a^3}(\sin at-at\cos at)$
2. $e^{2t}-e^{-2t}(1+4t)$
3. $\frac{t}{8a^3}(\sin at-at\cos at)$

Objective Evaluations

Fill in the Blanks

1. A function $N(t)$ of t such that $\int_0^t N(t) = 0\ \forall t$ is called .... function.
2. If $L^{-1}[f(p)] = F(t)$, then $L^{-1}\{f(p-a)\} =$ ....
3. If $L^{-1}\{f(p)\} = F(t)$, then $L^{-1}\{f(ap)\} =$ .... $F\left(\frac{t}{a}\right)$.
4. $L^{-1}\left[\frac{1}{p}\right] =$ ....
5. $L^{-1}\left[\frac{p}{p^2+a^2}\right] =$ ....

True or False

Write 'T' for True and 'F' for False statement.

1. $L^{-1}\left(\frac{1}{p^2}\right) = t$ **(T/F)**
2. $L^{-1}\left(\frac{1}{p^{n+1}}\right) = \frac{t^n}{n!}$ **(T/F)**
3. $L^{-1}\left(\frac{1}{p^2-a^2}\right) = \frac{\sinh at}{a}$ **(T/F)**
4. $L^{-1}\left(\frac{p}{p^2-a^2}\right) = \cos at$ **(T/F)**
5. $L^{-1}\left(\frac{1}{p-a}\right) = e^{at}$ **(T/F)**

Multiple Choice Questions

Choose the most appropriate one.

1. The function whose Laplace transform is $\frac{1}{p^2+w^2}$ is :
   (a) $\frac{1}{w}\sin wt$ (b) $\cos wt$
   (c) $\sin wt$ (d) none of these
2. The funciton whose Laplace transform is $\frac{p}{p^2+a^2}$ is:
   (a) $\cos wt$ (b) $\sin wt$
   (c) $\frac{1}{w}\cos wt$ (d) none of these
3. The funciton whose Laplace transform is $\frac{1}{p^2-a^2}$ is:
   (a) $\frac{1}{a}\sinh at$ (b) $\frac{1}{a}\cosh at$
   (c) $\sinh at$ (d) none of these
4. The funciton whose Laplace transform is $\frac{p-a}{(p-a)^2+w^2}$ is:
   (a) $e^{at}\cos wt$ (b) $\cos wt$
   (c) $e^{at}\sin wt$ (d) none of these
5. The funciton whose Laplace transform is $\frac{1}{p(p^2+w^2)}$ is:
   (a) $\frac{1}{w^3}(wt - \sin wt)$ (b) $\frac{1}{2w^2}(\sin wt - wt\cos wt)$
   (c) $\frac{1}{2w}\sin wt$ (d) none of these
6. The funciton whose Laplace transform is $\frac{p}{(p^2+w^2)^2}$ is:
   (a) $\frac{1}{w}(w - \sin wt)$ (b) $\frac{1}{2w}\sin wt$
   (c) $\frac{1}{w}\sin wt$ (d) none of these
7. The funciton whose Laplace transform is $\frac{p^2}{(p^2+w^2)^2}$ is:
   (a) $\frac{1}{2w}(\sin wt + wt + \cos wt)$ (b) $\sin wt + wt\cos wt$
   (c) $\sin wt$ (d) none of these
8. The funciton whose Laplace transform is $\frac{1}{p^4-a^4}$ is:
   (a) $\sinh at$ (b) $\cosh at$
   (c) $\sinh at - \cosh at$ (d) none of these

ANSWERS

Fill in the Blanks

1. null 2. $e^{at}F(t)$ 3. $1/a$ 4. 1 5. cos at

True or False

1. (T) 2. (T) 3. (T) 4. (F) 5. (T)

Multiple Choice Questions

1. (a) 2. (a) 3. (a) 4. (a) 5. (b)
6. (b) 7. (a) 8. (d)

FFFFFF

CHAPTER 38 Applications of Laplace Transform to Solution of Ordinary Differential Equation

38.1 SOLUTION OF ORDINARY DIFFERENTIAL EQUATION WITH CONSTANT COEFFICIENTS

Consider a linear differential equation with constant coefficients

$$\frac{d^n y}{dt^n} + A_1 \frac{d^{n-1} y}{dt^{n-1}} + \ldots + A_{n-1} \frac{dy}{dt} + A_n y = F(t) \qquad \ldots (1)$$

where t is the independent variable and $F(t)$ is a function of t.

Let $\qquad y(0) = C_1,\ y'(0) = C_2, \ldots, y^{n-1}(0) = C_{n-1} \qquad \ldots (2)$

be the given initial or boundary conditions, where $C_1, C_2, \ldots, C_{n-1}$ are constants. Now, taking the Laplace transform of both sides of (1) and using the conditions given by (2), we get an algebraic equation from which $\bar{y}(p) = L\{y(t)\}$ is determined. The required solution is then obtained by finding the inverse Laplace transform of $\bar{y}(p)$.

REMARKS

- The algebraic equation, obtained above is known as subsidiary equation.
- The above method is easily extended to higher order differential equation.

WORKING PROCEDURE

Step 1. *Taking Laplace transform of both the sides of the given differential equation and use given initial conditions.*

Step 2. *Solve the equation obtained in step (1) for L{y}.*

Step 3. *Taking inverse Laplace transform to find y.*

Solved Examples

Example 1. *Solve* $\frac{d^2y}{dt^2} + y = 0$ *under the condition that* $y = 1, \frac{dy}{dt} = 0$ *when* $t = 0$.

Solution. Here, the given equation is

$$\frac{d^2y}{dt^2} + y = 0. \qquad \ldots (1)$$

Taking the Laplace transform of both sides of the given differential equation, we get

$$L\{y''\} + L\{y\} = 0$$

$$\Rightarrow \quad p^2L\{y\} - py(0) - y'(0) + L\{y\} = 0$$

$$\Rightarrow \quad (p^2+1)\, L\{y\} - p.1 - 0 = 0$$

[Using the given conditions]

$$\Rightarrow \quad L\{y\} = \frac{p}{p^2+1}$$

Therefore, $y = L^{-1}\left\{\frac{p}{p^2+1}\right\} = \cos t$.

Example 2. *Solve* $(D^2+1)y = 6\cos 2t$ *if* $y = 3, Dy = 1$ *when* $t = 0$.

Solution. The given equation can be written as

$$y'' + y = 6\cos 2t$$

Taking the Laplace transform of both the sides of the given differential equation, we get

$$L\{y''\} + L\{y\} = 6L\{\cos(2t)\}$$

$$\Rightarrow \quad p^2L\{y\} - py(0) - y'(0) + L\{y\} = 6\frac{p}{p^2+2^2}$$

$$\Rightarrow \quad (p^2+1)\, L\{y\} - 3p - 1 = \frac{6p}{p^2+4}$$

[Using the given conditions]

$$\Rightarrow L\{y\} = \frac{3p}{p^2+1} + \frac{1}{p^2+1} + \frac{6p}{(p^2+1)(p^2+4)}$$

$$= \frac{3p}{p^2+1} + \frac{1}{p^2+1} + \frac{2p[(p^2+4)-(p^2+1)]}{(p^2+1)(p^2+4)}$$

$$= \frac{3p}{p^2+1} + \frac{1}{p^2+1} + 2p\left\{\frac{1}{p^2+1} - \frac{1}{p^2+4}\right\}$$

$$= \frac{5p}{p^2+1} + \frac{1}{p^2+1} - \frac{2p}{p^2+4}$$

Therefore,

$$y = 5L^{-1}\left\{\frac{p}{p^2+1}\right\} + L^{-1}\left\{\frac{1}{p^2+1}\right\} - 2L^{-1}\left\{\frac{p}{p^2+4}\right\}$$

$$\Rightarrow \quad y = 5\cos t + \sin t - 2\cos 2t$$

Example 3. *Using Laplace transforms, find the solution of the initial value problem* :

$$y'' + 9y = 6\cos 3t,\ y(0) = 2,\ y'(0) = 0$$

[UPTU–2006]

Solution. The given equation can be written as

$$y'' + 9y = 6\cos 3t,\ y(0) = 2,\ y'(0) = 0 \quad \text{... (1)}$$

Taking Laplace transform of (1), we get

$$[p^2L\{y\} - py(0) - y'(0)] + 9L\{y\} = 6\frac{p}{p^2+9}$$

Putting the value of $y(0)$ and $y'(0)$ in (2), we have

$$p^2L\{y\} - 2p + 9L\{y\} = \frac{6p}{p^2+9}$$

$$(p^2+9)L\{y\} = 2p + \frac{6p}{p^2+9}$$

$$L\{y\} = \frac{2p}{p^2+9} + \frac{6p}{(p^2+9)^2}$$

$$\Rightarrow \quad y = L^{-1}\left\{\frac{2p}{p^2+9}\right\} + L^{-1}\left\{\frac{6p}{(p^2+9)^2}\right\}$$

$$= 2\cos 3t + 3L^{-1}\frac{d}{dp}\left[-\frac{3}{p^2+9}\right]$$

$$= 2\cos 3t + t\sin 3t$$

Example 4. *Solve* $(D^2+9)y = \cos 2t$ *if* $y(0) = 1,\ y\left(\frac{\pi}{2}\right) = -1.$

[UPTU–2002, 06, Bhopal–2008]

Solution. The given equation can be written as

$$y'' + 9y = \cos 2t \quad \text{... (1)}$$

Taking the Laplace transform of both the sides of (1), we get

$$L\{y''\} + 9L\{y\} = L\{\cos 2t\}$$

$$\Rightarrow \quad p^2L\{y\} - py(0) - y'(0) + 9L\{y\} = \frac{p}{p^2+4}$$

$$\Rightarrow \quad (p^2+9)\,L\{y\} - p - C = \frac{p}{p^2+4},\ \text{where}$$

$$C = y'(0)$$

$$\therefore \quad L\{y\} = \frac{p+C}{p^2+9} + \frac{p}{(p^2+9)(p^2+4)}$$

$$= \frac{p}{p^2+9} + \frac{C}{p^2+9} + \frac{p}{5(p^2+4)} - \frac{p}{5(p^2+9)}$$

Therefore,

$$y = L^{-1}\left\{\frac{p}{p^2+9}\right\} + CL^{-1}\left\{\frac{1}{p^2+9}\right\}$$

$$+ \frac{1}{5}L^{-1}\left\{\frac{p}{p^2+4}\right\} - \frac{1}{5}L^{-1}\left\{\frac{p}{p^2+9}\right\}$$

$$= \cos 3t + \frac{1}{3}C\sin 3t + \frac{1}{5}\cos 2t - \frac{1}{5}\cos 3t$$

$$= \frac{4}{5}\cos 3t + \frac{1}{3}C\sin 3t + \frac{1}{5}\cos 2t \quad \text{... (2)}$$

Now, since $y\left(\frac{\pi}{2}\right) = -1$, therefore, from (1), we have $-1 = \frac{4}{5}\cos\frac{3\pi}{2} + \frac{1}{3}C\sin\frac{3\pi}{2} + \frac{1}{5}\cos\pi$

On solving, we get $C = \frac{12}{5}$

Put this value in (2), we get

$$y = \frac{4}{5}\cos 3t + \frac{4}{5}\sin 3t + \frac{1}{5}\cos 2t.$$

Example 5. *Solve using Laplace transform method*

$$y''(t) + 4y'(t) + 4y(t) = 6e^{-t}, \textit{with}$$

$$y(0) = -2,\ y'(0) = 8.$$

[UPTU–2007]

Solution. The given equation can be written as

$$y''(t) + 4y'(t) + 4y(t) = 6e^{-t}$$

Taking Laplace transform on both sides of the given equation, we get

$$[p^2L\{y\} - py(0) - y'(0)] + 4[pL\{y\}$$

$$-y(0)] + 4L\{y\} = \frac{6}{p+1} \quad \text{... (1)}$$

Putting $y(0) = -2$ and $y'(0) = 8$ in (1), we get

$$[p^2L\{y\} - p(-2) - 8] + 4[pL\{y\}$$

$$+2] + 4L\{y\} = \frac{6}{p+1}$$

$$\Rightarrow \quad (p^2+4p+4)L\{y\} + 2p = \frac{6}{p+1}$$

$$\Rightarrow \quad (p^2+4p+4)L\{y\} = -2p + \frac{6}{p+1}$$

$$\Rightarrow \quad (p+2)^2L\{y\} = \frac{-2p^2-2p+6}{(p+1)}$$

$$\Rightarrow \quad L\{y\} = \frac{-2p^2-2p+6}{(p+1)(p+2)^2}$$

Let $\frac{-2p^2-2p+6}{(p+1)(p+2)^2} = \frac{A}{p+1} + \frac{B}{p+2} + \frac{C}{(p+2)^2}$

$$-2p^2 - 2p + 6 = A(p+2)^2$$

$$+B(p+1)(p+2) + C(p+1)$$

$$-2 + 2 + 6 = A(-1+2)^2 \Rightarrow A = 6$$

[Putting $p=-1$]

$$-8+4+6 = C(-2+1) \Rightarrow C = -2$$ [Putting $p = -2$]

Comparing the coefficients of p^2 on both sides, we get

$$-2 = A+B \Rightarrow -2 = 6+B \Rightarrow B = -8$$

$$L\{y\} = \frac{6}{p+1} + \frac{-8}{p+2} + \frac{2}{(p+2)^2}$$

$$y = L^{-1}\left[\frac{6}{p+1} - \frac{8}{p+2} - \frac{2}{(p+2)^2}\right]$$

Hence, $y = 6e^{-t} - 8e^{-2t} - 2e^{-2t}t$

Example 6. *Solve* $(D^3 - 2D^2 + 5D)y = 0$ *given that* $y(0) = 0,\ y'(0) = 1,\ y\left(\frac{\pi}{8}\right) = 1$.

[UPTU Q. Bank–2001]

Solution. The given equation can be written as

$$y''' - 2y'' + 5y' = 0 \qquad \text{... (1)}$$

Taking the Laplace transforms of both sides of (1), we get

$$L\{y'''\} - 2L\{y''\} + 5L\{y'\} = 0$$

$$\Rightarrow \quad p^3L\{y\} - p^2y(0) - py'(0) - y''(0)$$
$$-2[p^2L\{y\} - py(0) - y'(0)]$$
$$+5[pL\{y\} - y(0)] = 0$$

$$\Rightarrow \quad [p^3 - 2p^2 + 5p]\,L\{y\} - p$$
$$-C - 2(-1) + 5.0 = 0,$$

where $y''(0) = C$

$$L\{y\} = \frac{C-2+p}{p^3-2p^2+5p} = \frac{C-2}{p(p^2-2p+5)} + \frac{1}{p^2-2p+5}$$

$$= \frac{C-2}{5p} - \frac{C-2}{5}\cdot\frac{p-2}{p^2-2p+5} + \frac{1}{p^2-2p+5}$$

$$= \frac{C-2}{5p} - \frac{C-2}{5}\cdot\frac{(p-1)-1}{(p-1)^2+4} + \frac{1}{(p-1)^2+4}$$

$$= \frac{C-2}{5p} - \frac{C-2}{5}\cdot\frac{(p-1)}{(p-1)^2+4} + \frac{C+3}{10}\cdot\frac{2}{(p-1)^2+4}$$

Therefore,

$$y = \frac{C-2}{5}.L^{-1}\left\{\frac{1}{p}\right\} - \frac{C-2}{5}L^{-1}\left\{\frac{p-1}{(p-1)^2+4}\right\} + \frac{C+3}{10}L^{-1}\left\{\frac{2}{(p-1)^2+4}\right\}$$

$$= \frac{C-2}{5} - \frac{C-2}{5}e^t\cos 2t + \frac{C+3}{10}e^t\sin 2t \qquad \text{... (2)}$$

Now, since $y\left(\frac{\pi}{8}\right) = 1$, therefore

$$1 = \frac{C-2}{5} - \frac{C-2}{5}e^{\pi/8}\frac{1}{\sqrt{2}} + \frac{C+3}{10}e^{\pi/8}.\frac{1}{\sqrt{2}}$$

$$\Rightarrow \quad \frac{7-C}{5} = \frac{e^{\pi/8}}{10\sqrt{2}}(-2C+4+C+3)$$

$$\Rightarrow \quad \left(\frac{7-C}{5}\right)\cdot\left(1 - \frac{e^{\pi/8}}{2\sqrt{2}}\right) = 0$$

$$\Rightarrow \quad C = 7$$

Put this value of C in (2), we get

$$y = 1 + e^t(\sin 2t - \cos 2t).$$

Example 7. *Solve* $(D^2 - 3D + 2)y = 1 - e^{2t}$, $y = 1$, $Dy = 0$ *when* $t = 0$.

Solution. The given equation can be written as

$$y'' - 3y' + 2y = 1 - e^{2t} \qquad \text{... (1)}$$

Taking Laplace transform of both the sides of (1), we get

$$L\{y''\} - 3L\{y'\} + 2L\{y\} = L\{1\} - L\{e^{2t}\}$$

$$p^2L\{y\} - py(0) - y'(0) - 3[pL\{y\} - y(0)] + 2L\{y\} = \frac{1}{p} - \frac{1}{p-2}$$

$$\Rightarrow \quad (p^2 - 3p + 2)L\{y\} - p + 3 = -\frac{2}{p(p-2)}$$

$$\Rightarrow \quad (p-1)(p-2)L\{y\} = -\frac{2}{p(p-2)} + (p-3)$$

$$\Rightarrow \quad L\{y\} = \frac{p^3 - 5p^2 + 6p - 2}{p(p-1)(p-2)^2} = \frac{p^2 - 4p + 2}{p(p-2)^2}$$

$$= \frac{1}{2p} + \frac{1}{2(p-2)} - \frac{1}{(p-2)^2}$$

Therefore, $y = \frac{1}{2}L^{-1}\left\{\frac{1}{p}\right\} + \frac{1}{2}L^{-1}\left\{\frac{1}{p-2}\right\} - L^{-1}\left\{\frac{1}{(p-2)^2}\right\}$

$$= \frac{1}{2} + \frac{1}{2}e^{2t} - te^{2t}$$

EXERCISE 38.1

1. Solve $\frac{dy}{dt} + y = 1$ if $y = 2$ when $t = 0$.
2. Show that the general solution of the equation $(D^2 + k^2)y = 0$ is
   $$y = C_1\cos kt + C_2\sin kt$$
3. Solve $y''(t) + y(t) = t$ if $y'(0) = 1,\ y(\pi) = 0$.
4. Solve $(D^2 - 1)y = a\cosh nt$ if $y = Dy = 0$, when $t = 0$.
5. Solve $(D^2 + m^2)x = a\cos nt$, $t \neq 0$, where x, Dx equal to x_0 and x_1, when $t = 0, n \neq m$.
6. Solve $(D^2 + m^2)y = a\cos nt, t > 0$ if $y = 0 = Dy$, when $t = 0$.
7. Solve $(D^2 + m^2)x = a\sin nt, t > 0$, where x, Dx equal to x_0 and x_1, when $t = 0, n \neq m$.

8. Solve $(D+2)^2 y = 4e^{-2t}$, $y(0) = -1$ and $y'(0) = 4$.
9. Solve $(D^2+6D+9)y = \sin x$, where $y(0)=1$, $y'(0)=0$.
10. Solve $(D^2+4D+4)x = \sin \omega t$, $t > 0$, where x_0 and x_1 are the values of x and Dx, when $t = 0$.
11. Solve $(D^2+3D+2)y = 0$, where $y = y_0$ and $Dy = y_1$ at $t = 0$.
12. Solve $(D^2+9)y = 18t$, if $y(0)=0$, $y\left(\frac{\pi}{2}\right)=0$.
13. Solve $(D^2+2D+1)y = 3te^{-t}$, $t>0$ subject to the conditions $y=4$, $Dy=2$, when $t=0$.
14. Solve $(D^2+1)y = \sin t \sin 2t, t>0$ if $y=1, Dy=0$, when $t=0$. [UPTU–2001(Sp.), 02]
15. Solve $(D^2+n^2)y = a\sin(nt+\alpha)$ if $y = Dy = 0$, when $t=0$. (GBTU(CO)–2010)
16. Solve $(D^3+1)y = 1, t>0$ if $y = Dy = D^2y = 0$, when $t=0$.
17. Solve $(D^3-D)y = 2\cos t$, $y=3$, $Dy=2$, $D^2y=1$, when $t=0$.
18. Solve $(D^3+D)y = e^{2t}$, $y(0) = y'(0) - y''(0) = 0$.
19. Solve $(D^4-1)y = 1$ if $y = Dy = D^2y = D^3y = 0$ at $t=0$.
20. Solve $(D^4+2D^2+1)y = 0$ if $y(0)=0$, $y'(0)=1$, $y''(0)=2$ and $y'''(0)=-3$.
21. Solve $(D^2+D)y = t^2+2t$ if $y(0)=4$, $y'(0)=-2$.
22. Solve $\frac{d^3y}{dt^3} - 3\frac{d^2y}{dt^2} + 3\frac{dy}{dt} - y = t^2e^t$ where $y(0) = 1$,

$$y(0)=1, \left(\frac{dy}{dt}\right)_{t=0} = 0, \left(\frac{d^2y}{dt^2}\right)_{t=0} = -2$$

(UPTU (SUM)–2008, SVTU–2009)

23. Voltage Ee^{-at} is applied at $t=0$ to a circuit of inductance L and resistance R. Show that the current at time t is $\frac{E}{R-aL}(e^{-at} - e^{-Rt/L})$ (UPTU (SUM)–2007, VTU–2000)
24. Solve $y'' + 4y' + 3y = e^{-t}, y(0) = y'(0) = 1$ (VTU–2008 S, Kurukshetra–2005)
25. Solve $y'' + y = t, y(0) = 1, y'(0) = 0$ (Mumbai–2009)
26. Solve $y'' - 3y' + 2y = e^{3t}$ when $y(0)=1$ and $y'(0)=0$ (VTU–2010)
27. Solve $(D^2-3D+2)y = 4e^{2t}$ with $y(0)=-3, y(0)=5$ (Mumbai–2008)
28. Solve $y''+25y = 10\cos 5t$ given that $y(0)=2, y''(0)=0$ (SVTU–2008)
29. Solve $\frac{d^2y}{dt^2} + 2\frac{dy}{dt} - 3y = \sin t, y = \frac{dy}{dt} = 0$ when $t=0$ (Kurukshetra–2005, Madras–2003)
30. Solve $y''+2y'+5y = 5(t-2), y(0)=0, y'(0)=0$ (PTU–2005 S)
31. Solve $\frac{d^2x}{dt^2} + 9x = \sin 2t, x(0)=1, x'(0)=0$ (GBTU (CO)–2011)
32. Solve $\frac{d^2y}{dt^2} + 9x = \sin 3t$, given $y=0, \frac{dy}{dt} = 0$ at $t=0$ (MTU–2012)
33. Solve $\frac{d^2x}{dt^2} + 6\frac{dx}{dt} + 8x = e^{-3t} - e^{-5t}$; $x(0)=0, x'(0)=0$ (UPTU(CO)–2009)
34. Solve $y''+3y'+2y = te^{-t}; y(0)=1, y'(0)=0$ (GBTU–2012)
35. Solve $y''+2y'+y = te^{-t}; y(0)=1, y'(0)=2$ (MTU (SUM)–2011)
36. Solve $\frac{d^2x}{dt^2} + 3\frac{dx}{dt} + 2x = r(t)$

where $r(t) = \begin{cases} e^t, & 0<t<2 \\ 0, & t>2 \end{cases}$ and $x(0)=1, x'(0)=-2$ (GBTU(CO)–2010)

37. Solve $\frac{d^2y}{dt^2} + 9y = r(t)$ with intial conditions $y(0)=0$ and $y'(0)=4$ where $r(t) = \begin{cases} 8\sin t & 0<t<\pi \\ 0 & t>\pi \end{cases}$ (GBTU–2011)
38. A particle moves in a line so that its displacement x from a fixed point O at any time t, is given by

$$\frac{d^2x}{dt^2} + 4\frac{dx}{dt} + 5x = 80\sin 5t$$

Using Laplace transform, find its displacemnt at any time t if initially particle is at rest at x=0. (UPTU (CO)–2009)

39. An alternating $e.m.f$ $E\sin \omega t$ is applied to circuit with an inductance L and a capacitance C in series. Show that the current in the circuit is $\frac{E\omega}{(n^2-\omega^2)L}(\cos\omega t - \cos nt)$ where $n^2 = \frac{1}{LC}$. (GBTU–2010)

Hint to Selected Problems

1. Taking the Laplace transform of the given equation, we get $L(y') + L(y) = L(1)$

$$\Rightarrow \quad pL\{y\} - y\{0\} + L\{y\} = \frac{1}{p}$$

$$\Rightarrow \quad L\{y\} = \frac{2p+1}{p(p+1)} = \frac{1}{p+1} + \frac{1}{p}.$$

Now taking inverse Laplace transform.

3. $L\{y''\} + L\{y\} = L\{t\}$

$$\Rightarrow p^2L\{y\} - py(0) - y'(0) + L\{y\} = \frac{1}{p^2}$$

$$\Rightarrow \quad (p^2+1)L\{y\} - pA - 1 = \frac{1}{p^2}. \quad [\because A = y(0),\ y'(0) = 1]$$

$$\Rightarrow \quad L\{y\} = \frac{pA}{p^2+1} + \frac{1}{p^2}$$

Now taking inverse Laplace transform.

4. Taking Laplace transform of the given equation and after simplification, we get

$$L\{y\} = \frac{ap}{(p^2-1)(p^2-n^2)}$$

$$= \frac{ap}{(n^2-1)}\left[\frac{1}{p^2-n^2} - \frac{1}{p^2-1}\right] = \frac{1}{(n^2-1)}\left[\frac{p}{p^2-n^2} - \frac{p}{p^2-1}\right]$$

Now taking inverse Laplace transform of both the sides.

7. Proceeding as usual, we get

$$L\{x\} = \frac{px_0}{(p^2+m^2)} + \frac{x_1}{(p^2+m^2)}$$

$$+\frac{a_n}{(p^2+m^2)(p^2+n^2)}=\frac{px_0}{(p^2+m^2)}+\frac{x_1}{(p^2+m^2)}+\frac{a}{(m^2-n^2)}\left[\frac{n}{p^2+n^2}-\frac{n}{p^2+m^2}\right]$$

Now taking inverse Laplace transform of both the sides.

9. We have $L\{y\}=\frac{(p+6)}{(p+3)^2}+\frac{1}{(p^2+1)(p+3)^2}$

$$=\frac{1}{(p+3)}+\frac{3}{(p+3)^2}+\left[\frac{3}{50(p+3)}+\frac{1}{10(p+3)^2}-\frac{3p-4}{50(p^2+1)}\right]$$

$$=\frac{1}{50}\left[\frac{53}{p+3}+\frac{155}{(p+3)^2}-\frac{3p}{(p^2+1)}+\frac{4}{(p^2+1)}\right]$$

Now taking inverse Laplace transform of both the sides.

11. Proceeding as usual, we get

$$L\{y\}=\frac{p+3}{(p^2+3p+2)}y_0+\frac{y_1}{(p^2+3p+2)}$$

$$=\frac{(p+3)}{(p+1)(p+2)}y_0+\frac{y_1}{(p+1)(p+2)}$$

$$=\left[\frac{2}{p+1}-\frac{1}{p+2}\right]y_0+\left[\frac{1}{p+1}-\frac{1}{p+2}\right]y_1$$

$$=\frac{(2y_0+y_1)}{(p+1)}-\frac{(y_0+y_1)}{(p+2)}$$

Now taking inverse Laplace transform of both the sides.

14. $(D^2+1)y=\sin t\sin 2t=\frac{1}{2}[\cos t-\cos 3t]$

$$\Rightarrow\quad L\{y''\}+L\{y\}=\frac{1}{2}[L\{\cos t\}-L\{\cos 3t\}]$$

After simplification, we get

$$L\{y\}=\frac{p}{p^2+1}+\frac{p}{2(p^2+1)^2}-\frac{p}{16}\left[\frac{(p^2+9)-(p^2+1)}{(p^2+9)(p^2+1)}\right]$$

$$=\frac{p}{p^2+1}-\frac{1}{4}\left[\frac{d}{dp}\left(\frac{1}{p^2+1}\right)\right]-\frac{p}{16(p^2+1)}+\frac{p}{16(p^2+9)}$$

Now taking inverse Laplace transform of both the sides.

15. The given equation can be written as

$$y''+n^2y=a[\sin nt\cos\alpha+\cos nt\sin\alpha]$$

Taking Laplace transform and simplifying, we get

$$L\{y\}=a\cos\alpha.\frac{n}{(p^2+n^2)^2}+a\sin\alpha\frac{p}{(p^2+n^2)^2}$$

$$\Rightarrow\ y=a.n\cos\alpha L^{-1}\left\{\frac{1}{(p^2+n^2)^2}\right\}+a\sin\alpha L^{-1}\left\{\frac{p}{(p^2+n^2)^2}\right\}$$

$$=a.n\cos\alpha\int_0^t\left(\frac{1}{n}\sin nx\right)\frac{1}{n}\sin n(t-x)dx-\frac{a\sin\alpha}{2}L^{-1}\left\{\frac{d}{dp}\frac{1}{(p^2+n^2)^2}\right\}$$

(Using convolution theorem)

$$=a\frac{\cos\alpha}{2n}\int_0^t[\cos n(t-2x)-\cos nt]dx+\frac{a\sin\alpha}{2}t.L^{-1}\left\{\frac{1}{p^2+n^2}\right\}$$

16. Proceeding as usual, we get

$$L\{y\}=\frac{1}{p}-\frac{1}{3(p+1)}-\frac{2\left(p-\frac{1}{2}\right)}{3\left[\left(p-\frac{1}{2}\right)^2+\frac{3}{4}\right]}$$

$$\Rightarrow\quad y=1-\frac{e^{-t}}{3}-\frac{2}{3}e^{t/2}L^{-1}\left\{\frac{p}{p^2+(\sqrt{3}/2)^2}\right\}$$

ANSWERS

1. $y=e^{-t}+1$ **3.** $y=\pi\cos t+t$ **4.** $y=\frac{a}{n^2-1}(\cosh nt-\cosh t)$ **5.** $x=x_0\cos mt+\frac{x_1}{m}\sin mt+\frac{a}{m^2-n^2}(\cos nt-\cos mt)$

6. $y=\frac{a}{m^2-n^2}(\cos nt-\cos mt)$ **7.** $x=x_0\cos mt+\frac{x_1}{m}\sin mt+\frac{a}{m^2-n^2}\left(\sin nt-\frac{n}{m}\sin mt\right)$ **8.** $y=e^{-2t}(2t^2+2t-1)$

9. $\frac{1}{50}[(53+155x)e^{-3x}-(3\cos x-4\sin x)]$

10. $x=e^{-2t}\left[x_0(1-2t)+(x_1+4x_0)+\frac{w}{(4+w^2)}t+\frac{4w}{(4+w^2)^2}\right]-\frac{4w}{(4+w^2)^2}\cos wt+\frac{(4-w^2)}{(4+w^2)^2}\sin wt$

11. $y=(2y_0+y_1)e^{-t}-(y_0+y_1)e^{-2t}$ **12.** $y=\pi\sin 3t+2t$ **13.** $y=\frac{1}{2}e^{-t}.t^3+4e^{-t}+6te^{-t}$

14. $y=\frac{15}{16}\cos t+\frac{1}{4}t\sin t+\frac{1}{16}\cos 3t$ **15.** $y=\frac{a}{2n^2}[\cos\alpha\sin nt-nt\cos(\alpha+nt)]$ **16.** $y=1-\frac{1}{3}e^{-t}-\frac{2}{3}e^{t/2}\cos\left(\frac{\sqrt{3}t}{2}\right)$

17. $y=3\sinh t-\sin t+\cosh t+2$ **18.** $y=\frac{1}{3}e^{-t}(\sin 2t+\sin t)$ **19.** $y=-1+\frac{1}{2}\cosh t+\frac{1}{2}\cos t$

20. $y=t(\sin t+\cos t)$ **21.** $y=\frac{1}{3}t^3+2+2e^{-t}$ **22.** $y=\left(1-t-\frac{t^2}{2}+\frac{t^5}{60}\right)e^t$

24. $y=\frac{7}{4}e^{-t}-\frac{3}{4}e^{-3t}-\frac{1}{2}te^{-t}$ **25.** $y=t-3\sin t+\cos t$ **26.** $y=2t+3+\frac{1}{2}(e^{3t}-e^t)-2e^{2t}$

27. $y=4e^{2t}(1+t)-7e^t$ **28.** $y=2\cos 5t+t\sin 5t$ **29.** $y=\frac{1}{8}e^t-\frac{1}{40}e^{-3t}-\frac{1}{10}(2\sin t+\cos t)$

30. $y=\frac{-12}{5}+\frac{12}{5}e^{-t}\cos 2t+\frac{7}{10}e^{-t}\sin 2t$ **31.** $x=\cos 3t+\frac{1}{5}\sin 2t-\frac{2}{15}\sin 3t$ **32.** $y=\frac{1}{18}(\sin 3t-3t\cos 3t)$

33. $x=\frac{1}{3}(e^{-2t}-e^{-5t})-e^{-3t}+e^{-4t}$ **34.** $y=3e^{-t}-2e^{-2t}+e^{-t}\left(\frac{t^2}{2}-t\right)$

35. $y = e^{-t}\left(1+3t+\frac{t^3}{6}\right)$ **36.** $x = \frac{4}{3}e^{-2t} + \frac{1}{6}e^{t}[1-u(t-2)] - \frac{1}{2}e^{-t} + \frac{1}{2}e^{4-t}u(t-2) - \frac{1}{3}e^{6-2t}u(t-2)$

37. $y = \sin 3t + \sin t + [\sin(t-\pi) - \frac{1}{3}\sin 3(t-\pi)]u(t-\pi)$ **38.** $x = e^{-2t}(2\cos t + 14\sin t) - 2\cos 5t - 2\sin 5t$

38.2 SOLUTION OF ORDINARY DIFFERENTIAL EQUATION WITH VARIABLE COEFFICIENTS

The Laplace transform can also be used in solving some ordinary differential equations in which the coefficients are variable. A particular differential equation when the method proves useful is one in which the terms have the form $t^m\, y^n(t)$ whose Laplace transform is

$$(-1)^m \frac{d^m}{dp^m}[L\{y^n(t)\}]$$

Solved Examples

Example 1. *Solve* $(tD^2 + D + 4t)y = 0$ *if* $y(0) = 3$, $y'(0) = 0$.

Solution. The given equation can be written as

$$ty'' + y' + 4ty = 0 \quad \text{... (1)}$$

Taking the Laplace transform of both sides of (1), we get

$$L\{ty''\} + L\{y'\} + 4L\{ty\} = 0$$

$$\Rightarrow \quad -\frac{d}{dp}L\{y''\} + L\{y'\} + 4(-1)\frac{d}{dp}L\{y\} = 0$$

$$\Rightarrow \quad -\frac{d}{dp}[p^2L\{y\} - py(0) - y'(0)] + [pL\{y\} - y(0)] - 4\frac{d}{dp}L\{y\} = 0$$

$$\Rightarrow \quad -\frac{d}{dp}[p^2L\{y\} - 3p] + (pL\{y\} - 3) - \frac{4d[L\{y\}]}{dp} = 0$$

$$\Rightarrow \quad -(p^2+4)\frac{d[L\{y\}]}{dp} - pL\{y\} = 0$$

$$\Rightarrow \quad \frac{d[L\{y\}]}{L\{y\}} + \frac{p}{p^2+4}dp = 0$$

On integrating, we get

$$\log[L\{y\}] + \frac{1}{2}\log(p^2+4) = \log C_1$$

$$\Rightarrow \quad L\{y\} = \frac{C_1}{\sqrt{p^2+4}}$$

Therefore, $y = L^{-1}\left\{\frac{C_1}{\sqrt{p^2+4}}\right\} = C_1J_0(2t)$.

Example 2. *Solve* $[tD^2 + (t-1)D - 1]y = 0$ *if* $y(0) = 5$, $y(\infty) = 0$.

Solution. The given equation can be written as

$$ty'' + ty' - y' - y = 0 \quad \text{... (1)}$$

Taking the Laplace transforms of both sides of (1), we get

$$L\{ty''\} + L\{ty'\} - L\{y'\} - L\{y\} = 0$$

$$\Rightarrow \quad -\frac{d}{dp}[L\{y''\}] - \frac{d}{dp}[L\{y'\}] - [pL\{y\} - y(0) - L\{y\}] = 0$$

$$\Rightarrow \quad -\frac{d}{dp}[p^2L\{y\} - py(0) - y'(0)] - \frac{d}{dp}[pL\{y\} - y(0)] - pL\{y\} + 5 - L\{y\} = 0$$

$$\Rightarrow \quad -\frac{d}{dp}[p^2L\{y\} - 5p - A] - \frac{d}{dp}[pL\{y\} - 5] - (p+1)L\{y\} + 5 = 0, \text{where } A = y'\{0\}$$

$$\Rightarrow \quad \frac{d[L\{y\}]}{dp} + \frac{3p+2}{p^2+p}L\{y\} = \frac{10}{p^2+p} \quad \text{... (2)}$$

which is a linear differential equation in $L\{y\}$. Therefore,

$$\text{I.F.} = e^{\int\left\{\frac{3p+2}{(p^2+p)}\right\}dp} = e^{\int\left(\frac{2}{p}+\frac{1}{p+1}\right)dp} = e^{[2\log p + \log(p+1)]} = p^2(p+1).$$

Hence, the solution of equation (2) is given by

$$L\{y\}.p^2(p+1) = C_1 + \int \frac{10}{p^2+p}.p^2(p+1)dp = C_1 + 10\int p\,dp = C_1 + 5p^2$$

$$\Rightarrow \quad L\{y\} = \frac{C_1}{p^2(p+1)} + \frac{5}{p+1} = C_1\left\{\frac{1}{p^2} - \frac{1}{p} + \frac{1}{p+1}\right\} + \frac{5}{p+1}$$

Therefore, $y = C_1L^{-1}\left\{\frac{1}{p^2} - \frac{1}{p} + \frac{1}{p+1}\right\} + 5L^{-1}\left\{\frac{1}{p+1}\right\} = C_1(t - 1 + e^{-t}) + 5e^{-t}$.

Now, using the given conditions $y(\infty) = 0$. We must have $C_1 = 0$. Hence, $y = 5e^{-t}$ is the required solution.

Example 3. *Solve* $\dfrac{d^2y}{dx^2}+2\dfrac{dy}{dx}+5y=e^{-x}\sin x$, *where* $y(0)=0,\ y'(0)=1$.

(UPTU–2004, 08, (SUM)–2009, PTU–2010, MTU–2011)

Solution. $\dfrac{d^2y}{dx^2}+2\dfrac{dy}{dx}+5y=e^{-x}\sin x$

Taking the Laplace transform on both the sides, we get $[p^2L\{y\}-py(0)-y'(0)]+2[pL\{y\}-y(0)]+5L\{y\}=L\{e^{-x}\sin x\}$

$$[p^2L\{y\}-py(0)-y'(0)]+2[pL\{y\}-y(0)]+5L\{y\}=\frac{1}{(p+1)^2+1} \quad \text{... (1)}$$

On substituting the values of $y(0)$ and $y'(0)$ in (1), we get

$$(p^2L\{y\}-1)+2pL\{y\}+5L\{y\}=\frac{1}{p^2+2p+2}$$

$$(p^2+2p+5)L\{y\}=1+\frac{1}{p^2+2p+2}=\frac{p^2+2p+3}{p^2+2p+2}$$

$$L\{y\}=\frac{p^2+2p+3}{(p^2+2p+5)(p^2+2p+2)}$$

On resolving R.H.S. into partial fractions, we get

$$L\{y\}=\frac{2}{3}\cdot\frac{1}{p^2+2p+5}+\frac{1}{3}\cdot\frac{1}{p^2+2p+2}$$

On inversion, we obtain

$$y=\frac{2}{3}L^{-1}\frac{1}{p^2+2p+5}+\frac{1}{3}L^{-1}\frac{1}{p^2+2p+2}$$

$$y=\frac{1}{3}L^{-1}\frac{2}{(p+1)^2+(2)^2}+\frac{1}{3}L^{-1}\frac{1}{(p+1)^2+(1)^2}$$

$$\Rightarrow\quad y=\frac{1}{3}e^{-x}\sin 2x+\frac{1}{3}e^{-x}\sin x$$

$$y=\frac{1}{3}.e^{-x}(\sin x+\sin 2x)$$

Example 4. *Solve* $(D^2+1)y=t\cos 2t$ *subject to the condition* $y=0,\ \dfrac{dy}{dt}=0$ *when* $t=0$.

[UPTU–2005, UKTU–2012, Raipur–2005]

Solution. The given equation can be written as

$$y''+y=t\cos 2t \quad \text{... (1)}$$

Taking the Laplace transform of both sides of (1), we get

$$L\{y''\}+L\{y\}=L\{t\cos 2t\}$$

$$\Rightarrow\quad p^2L\{y\}-py(0)-y'(0)+L\{y\}=-\frac{d}{dp}[L\{\cos 2t\}]$$

$$\Rightarrow\quad (p^2+1)L\{y\}=-\frac{d}{dp}\left(\frac{p}{p^2+4}\right)=-\frac{1}{p^2+4}+\frac{2p^2}{(p^2+4)^2}$$

$$\therefore\quad L\{y\}=\frac{p^2-4}{(p^2+1)(p^2+4)^2}$$

$$=-\frac{5}{9(p^2+1)}+\frac{5}{9(p^2+4)}+\frac{8}{3(p^2+4)^2}$$

[Resolving into partial fractions]

$$\Rightarrow\quad y=-\frac{5}{9}L^{-1}\left\{\frac{1}{p^2+1}\right\}+\frac{5}{9}L^{-1}\left\{\frac{1}{p^2+4}\right\}+\frac{8}{3}L^{-1}\left\{\frac{1}{(p^2+4)^2}\right\}$$

$$=-\frac{5}{9}\sin t+\frac{5}{18}\sin 2t+\frac{8}{3}\int_0^t\frac{1}{2}\sin 2x.\frac{1}{2}\sin 2(t-x)dx$$

[By convolution theorem and using $L^{-1}\left\{\dfrac{1}{p^2+4}\right\}=\dfrac{1}{2}\sin 2t$]

$$=-\frac{5}{9}\sin t+\frac{5}{18}\sin 2t+\frac{1}{3}\int_0^t\{\cos(2t-4x)-\cos 2t\}dx$$

$$=-\frac{5}{9}\sin t+\frac{5}{18}\sin 2t+\frac{1}{3}\left[-\frac{1}{4}\sin(2t-4x)-x\cos 2t\right]_0^t$$

$$=-\frac{5}{9}\sin t+\frac{5}{18}\sin 2t+\frac{1}{12}\sin 2t-\frac{1}{3}t\cos 2t+\frac{1}{12}\sin 2t$$

$$=-\frac{5}{9}\sin t+\frac{4}{9}\sin 2t-\frac{1}{3}t\cos 2t$$

Example 5. *Solve* $(D^3-D^2-D+1)y=8te^{-t}$ *if* $y=D^2y=0,\ Dy=0$ *when* $t=0$.

Solution. The given equation can be written as

$$y'''-y''-y'+y=8te^{-t} \quad \text{... (1)}$$

Taking the Laplace transforms of both sides of (1), we get

$$L\{y'''\}-L\{y''\}-L\{y'\}+L\{y\}=8L\{te^{-t}\}$$

$$\Rightarrow\quad p^3L\{y\}-p^2y(0)-py'(0)-y''(0)-[p^2L\{y\}-py(0)-y'(0)]-[pL\{y\}-y(0)]+L\{y\}=-8\frac{d}{dp}L\{e^{-t}\}$$

$$\text{or } (p^3-p^2-p+1)L\{y\}-p+1=-8\frac{d}{dp}\left[\frac{1}{p+1}\right]$$

$$\Rightarrow\quad (p-1)^2(p+1)L\{y\}=p-1+\frac{8}{(p+1)^2}$$

$$\Rightarrow\quad L\{y\}=\frac{1}{(p-1)(p+1)}+\frac{8}{(p-1)^2(p+1)^3}$$

$$= \frac{1}{2}\left(\frac{1}{p-1} - \frac{1}{p+1}\right) - \frac{3}{2(p-1)} + \frac{1}{(p-1)^2} + \frac{3}{2(p+1)} + \frac{2}{(p+1)^2} + \frac{2}{(p+1)^3}$$

$$= -\frac{1}{p-1} + \frac{1}{p+1} + \frac{1}{(p-1)^2} + \frac{2}{(p+1)^2} + \frac{2}{(p+1)^3}$$

Therefore,

$$y = -L^{-1}\left\{\frac{1}{p-1}\right\} + L^{-1}\left\{\frac{1}{p+1}\right\} + L^{-1}\left\{\frac{1}{(p-1)^2}\right\} + 2L^{-1}\left\{\frac{1}{(p+1)^2}\right\} + 2L^{-1}\left\{\frac{1}{(p+1)^3}\right\}$$

$$= -e^t + e^{-t} + e^t L^{-1}\left\{\frac{1}{p^2}\right\} + 2e^{-t}L^{-1}\left\{\frac{1}{p^2}\right\} + 2e^{-t}L^{-1}\left\{\frac{1}{p^3}\right\}$$

$$= -e^t + e^{-t} + e^t.t + 2e^{-t}.t + 2e^{-t}\left(\frac{t^2}{2!}\right)$$

$$= (1 + 2t + t^2)\,e^{-t} - (1-t)\,e^t.$$

Example 6. *Solve* $[tD^2 + (1-2t)D - 2]y = 0$, *where* $y(0) = 1,\ y'(0) = 2$.

[UPTU–2002, PTU–2002]

Solution. Here, $tD^2y + (1-2t)Dy - 2y = 0$

$$\Rightarrow \quad ty'' + y' - 2ty' - 2y = 0$$

Taking Laplace transform of given differential equation, we get

$$L\{ty''\} + L\{y'\} - 2L\{ty'\} - 2L\{y\} = 0$$

$$\Rightarrow \quad -\frac{d}{dp}L\{y''\} + L\{y'\} + 2\frac{d}{dp}L\{y'\} - 2L\{y\} = 0$$

$$-\frac{d}{dp}\left[p^2L\{y\} - py(0) - y'(0)\right] + [pL\{y\} - y(0)] + 2[pL\{y\} - y(0)] - 2L\{y\} = 0$$

Putting the values of $y(0)$ and $y'(0)$, we get

$$-\frac{d}{dp}(p^2L\{y\} - p - 2) + (pL\{y\} - 1) + 2\frac{d}{dp}(pL\{y\} - 1) - 2L\{y\} = 0$$

$$[\because y(0) = 1, y'(0) = 2]$$

$$\Rightarrow \quad -p^2\frac{dL\{y\}}{dp} - 2pL\{y\} + 1 + pL\{y\} - 1 + 2\left(p\frac{dL\{y\}}{dp} + L\{y\}\right) - 2L\{y\} = 0$$

$$\Rightarrow \quad -(p^2 - 2p)\frac{dL\{y\}}{dp} - pL\{y\} = 0$$

$$\Rightarrow \quad -\frac{dL\{y\}}{y} - \frac{1}{p-2}dp = 0$$

[Separating the variables]

$$\Rightarrow \quad \int \frac{dL\{y\}}{y} + \int \frac{dp}{p-2} = 0$$

$$\Rightarrow \quad \log L\{y\} + \log(p-2) = \log C$$

$$\Rightarrow \quad \log L\{y\}(p-2) = \log C$$

$$\Rightarrow \quad L\{y\}(p-2) = C$$

$$\Rightarrow \quad L\{y\} = \frac{C}{p-2}$$

$$\Rightarrow \quad y = CL^{-1}\left\{\frac{1}{p-2}\right\} \Rightarrow y = Ce^{2t} \qquad \ldots(1)$$

At $\quad x = 0,\ y(0) = Ce^0 \qquad \ldots(2)$

Putting $y(0) = 1$, in (2), we get

$$1 = Ce^0 \Rightarrow C = 1$$

Putting $C = 1$ in (1), we get $y = e^{2t}$. This is the required solution.

Example 7. *Solve* $y'' - ty' + y = 1$ *if* $y(0) = 1,\ y'(0) = 2$.

Solution. Taking the Laplace transforms of both sides of the given equation, we get

$$L\{y''\} - L\{ty'\} + L\{y\} = L\{1\}$$

$$\Rightarrow \quad p^2L\{y\} - py(0) - y'(0) + \frac{d}{dp}[L\{y'\}] + L\{y\}] = \frac{1}{p}$$

$$\Rightarrow \quad p^2L\{y\} - p - 2 + \frac{d}{dp}[pL\{y\} - y(0)] + L\{y\} = \frac{1}{p}$$

$$\Rightarrow \quad p^2L\{y\} - p - 2 + \frac{d}{dp}[pL\{y\} - 1] + L\{y\} = \frac{1}{p}$$

$$\Rightarrow \quad p\frac{d[L\{y\}]}{dp} + (p^2 + 2)L\{y\} = p + 2 + \frac{1}{p}$$

$$\Rightarrow d\frac{[L\{y\}]}{dp} + \left(p + \frac{2}{p}\right)L\{y\} = 1 + \frac{2}{p} + \frac{1}{p^2} \ldots (1)$$

which is a linear differential equation in $L(y)$.

$$\therefore \quad \text{I.F.} = e^{\int\left(p+\frac{2}{p}\right)dp} = e^{\frac{p^2}{2}+2\log p} = p^2e^{p^2/2}.$$

Therefore, solution of (1) is given by

$$p^2e^{p^2/2}L\{y\} = C_1 + \int\left(1 + \frac{2}{p} + \frac{1}{p^2}\right)p^2e^{p^2/2}dp$$

Hence, the solution of (1) is given by

$$p^2e^{p^2/2}L\{y\} = C_1 + \int\left(1 + \frac{2}{p} + \frac{1}{p^2}\right)p^2e^{p^2/2}.dp$$

$$= C_1 + \int (p^2 + 2p + 1)\,e^{p^2/2}dp$$

$$= C_1 + \int (p^2 + 1)e^{p^2/2}dp + 2\int pe^{p^2/2}dp$$

$$= C_1 + \int (2v + 1)e^v.\frac{dv}{\sqrt{2v}} + 2\int \sqrt{2v}.e^v.\frac{dv}{\sqrt{2v}}$$

[where $\frac{p^2}{2} = v \Rightarrow pdp = dv$, *i.e.*, $dp = \frac{dv}{\sqrt{2v}}$]

$$= C_1 + \sqrt{2v}.e^v - \int \frac{e^v}{\sqrt{2v}} dv + \int \frac{e^v}{\sqrt{2v}} dv + 2\int e^v dv$$

$$= C_1 + \sqrt{2v}\, e^v + 2e^v = C_1 + pe^{p^2/2} + 2e^{p^2/2}.$$

Therefore,

$$L\{y\} = \frac{C_1}{p^2} e^{-p^2/2} + \frac{1}{p} + \frac{2}{p^2}$$

$$= \frac{C_1}{p^2}\left(1 - \frac{p^2}{2} + \frac{p^4}{4\,.\,2!} - ...\right) + \frac{1}{p} + \frac{2}{p^2}$$

$$= \frac{(2+C_1)}{p^2} - \frac{C_1}{2} + \frac{C_1}{8} p^2 ... + \frac{1}{p}$$

[On expanding the exponential function]

Hence,

$$y = (2+C_1)L^{-1}\left\{\frac{1}{p^2}\right\} - \frac{1}{2}C_1 L^{-1}\{1\} + \frac{1}{8}C_1 L^{-1}\{p^2\} + ... + L^{-1}\left\{\frac{1}{p}\right\}$$

$$= (2+C_1)t + 1$$

$[\because L^{-1}\{p^n\} = 0, \text{ for } n = 0,1,2,...]$

Also, given that

$y'(0) = 2$

$\therefore \quad 2 = 2 + C_1 \Rightarrow C_1 = 0$

which gives $y = 2t + 1$ is the required solution.

Example 8. *Using Laplace transform, solve the following differential equation*

$$y'' + 2ty' - y = t$$

where, $y(0) = 0$ and $y'(0) = 1$. [UPTU–2003]

Solution. We have $\quad y'' + 2ty' - y = t \qquad ... (1)$

Taking Laplace transform of (1), we get

$$[p^2 L\{y\} - py(0) - y'(0)] - 2\frac{d}{dp}[pL\{y\} - y(0)] - L\{y\} = \frac{1}{p^2} \qquad ... (2)$$

On putting $y(0) = 0$ and $y'(0) = 1$ in (2), we get

$$(p^2 L\{y\} - 1) - 2\frac{d}{dp}(pL\{y\} - 0) - L\{y\} = \frac{1}{p^2}$$

$$\Rightarrow \quad (p^2 L\{y\} - 1) - 2L\{y\} - 2p\frac{dL\{y\}}{dp} - L\{y\} = \frac{1}{p^2}$$

$$\Rightarrow -2p\frac{dL\{y\}}{dp} + (p^2 - 3)L\{y\} = \frac{1}{p^2} + 1 = \frac{1+p^2}{p^2}$$

$$\Rightarrow \quad \frac{dL\{y\}}{dp} - \frac{p^2-3}{2p} L\{y\} = \frac{1+p^2}{-2p^3}$$

$$\Rightarrow \frac{dL\{y\}}{dp} - \left(\frac{p}{2} - \frac{3}{2p}\right)L\{y\} = -\frac{1}{2p^3} - \frac{1}{2p} \qquad ... (3)$$

Thus, (3) is a linear differential equation

$$\text{I.F.} = e^{\frac{1}{2}\int\left(\frac{3}{p} - p\right)dp} = e^{\frac{1}{2}\left(3\log p - \frac{p^2}{2}\right)} = e^{-\frac{p^2}{4}}.p^{3/2}$$

Solution of differential equation (3) is

$$L\{y\}\, e^{-p^2/4}.p^{3/2}$$

$$= -\frac{1}{2}\int\left(\frac{1}{p^3} + \frac{1}{p}\right)p^{3/2}.\, e^{-p^2/4} dp$$

$$= -\frac{1}{2}\int\left(\sqrt{p} + \frac{1}{p^{3/2}}\right)e^{-p^2/4} dp$$

Put $p^2 = ut \Rightarrow p = 2\sqrt{t}$ so that $dp = \dfrac{dt}{\sqrt{t}}$.

Then we have

$$L\{y\}p^{3/2}e^{-p^2/4}$$

$$= -\frac{1}{2}\int\left(\sqrt{2}\,t^{1/4} + \frac{1}{2\sqrt{2}}t^{-3/4}\right)e^{-t}\frac{dt}{\sqrt{t}}$$

$$= -\frac{1}{\sqrt{2}}\int\left(t^{-1/4} + \frac{1}{4}t^{-5/4}\right)e^{-t}dt$$

$$= -\frac{1}{\sqrt{2}}\int t^{-1/4}e^{-t}dt - \frac{1}{4\sqrt{2}}\int t^{-5/4}e^{-t}dt$$

$$= -\frac{1}{\sqrt{2}}\left[t^{-1/4}\frac{e^{-t}}{-1} + \int\left(-\frac{1}{4}\right)t^{-5/4}e^{-t}dt\right] + \frac{1}{4\sqrt{2}}\int t^{-5/4}e^{-t}dt$$

$$= \frac{1}{\sqrt{2}}e^{-t}.t^{-1/4} = \frac{1}{\sqrt{2}}e^{-p^2/4}\left(\frac{p^2}{4}\right)^{-1/4}$$

$$= \frac{1}{\sqrt{p}}e^{-p^2/4}$$

$$\Rightarrow L\{y\} = \frac{1}{p^2}$$

$$\Rightarrow L\{y\} = \frac{1}{p^2} + C \Rightarrow y = L^{-1}\left\{\frac{1}{p^2} + C\right\} = t + C.$$

38.3 SOLUTION OF SIMULTANEOUS ORDINARY DIFFERENTIAL EQUATIONS

The Laplace transform can be used to solve two or more simultaneous ordinary differential equations. The procedure is essentially the same as that described in previous sections.

Solved Examples

Example 1. *Solve* $(D^2+2)x - Dy = 1$, $Dx + (D^2+2)y = 0$, *if* $x = 0 = Dx = y = Dy$, *when* $t = 0$.

Solution. Taking Laplace transforms of both sides of the given equations, we have

$$L\{x''\} + 2L\{x\} - L\{y'\} = L\{1\}$$

and $L\{x'\} + 2L\{y''\} + 2L\{y\} = 0$

$$\Rightarrow \quad p^2L\{x\} - px(0) - x'(0) + 2L\{x\} - [pL\{y\} - y(0)] = \frac{1}{p}$$

and $pL\{x\} - x(0) + p^2L\{y\} - py(0) - y'(0) + 2L\{y\} = 0$

which gives $\quad (p^2+2)L(x) - pL\{y\} = \frac{1}{p}$

and $pL\{x\} + (p^2+2)L\{y\} = 0$

Solving for $L\{x\}$ and $L\{y\}$, we have

$$L\{x\} = \frac{p^2+2}{p(p^4+5p^2+4)} = \frac{1}{2p} - \frac{1}{6}\left[\frac{2p}{p^2+1} + \frac{p}{p^2+4}\right]$$

and $L\{y\} = \dfrac{-1}{p^4+5p^2+4} = \dfrac{1}{3}\left[\dfrac{1}{p^2+4} - \dfrac{1}{p^2+1}\right]$.

Therefore,

$$x = \frac{1}{2}L^{-1}\left\{\frac{1}{p}\right\} - \frac{1}{6}\left[2L^{-1}\left\{\frac{p}{p^2+1}\right\} + L^{-1}\left\{\frac{p}{p^2+4}\right\}\right] = \frac{1}{2} - \frac{1}{6}[2\cos t + \cos 2t]$$

and $\quad y = \dfrac{1}{3}\left[\dfrac{1}{2}\sin 2t - \sin t\right] = \dfrac{1}{6}[\sin 2t - 2\sin t]$

Example 2. *Solve the simultaneous equation* $\dfrac{dx}{dt} - y = e^t$, $\dfrac{dy}{dt} + x = \sin t$, *given* $x(0) = 1$, $y(0) = 0$.

[UPTU–2006, Q.Bank–2001, GBTU (SUM)–2010, UKTU–2011, Delhi–2002]

Solution. Taking Laplace transforms of the given equations, we get

$$[p\bar{x} - x(0)] - \bar{y} = \frac{1}{p-1},$$

where $\quad \bar{x} = L(x), \ \bar{y} = L(y)$

i.e., $\quad p\bar{x} - 1 - \bar{y} = \dfrac{1}{p-1} \qquad [\because x(0) = 1]$

$$p\bar{x} - \bar{y} = \frac{p}{p-1} \text{ and } [p\bar{y} - y(0)] + \bar{x} = \frac{1}{p^2+1}$$

i.e., $\quad \bar{x} + p\bar{y} = \dfrac{1}{p^2+1} \qquad [\because y(0) = 0] \qquad \text{... (2)}$

Solving (1) and (2) for $\bar{x}$ and $\bar{y}$, we have

$$\bar{x} = \frac{p^2}{(p-1)(p^2+1)} + \frac{1}{(p^2+1)^2} = \frac{1}{2}\left[\frac{1}{p-1} + \frac{p}{p^2+1} + \frac{1}{p^2+1}\right] + \frac{1}{(p^2+1)^2}$$

$$\bar{y} = \frac{p}{(p^2+1)^2} - \frac{p}{(p-1)(p^2+1)} = \frac{p}{(p^2+1)^2} - \frac{1}{2}\left[\frac{1}{p-1} - \frac{p}{p^2+1} + \frac{1}{p^2+1}\right]$$

Taking inverse Laplace transform of both sides, we get

$$x = \frac{1}{2}L^{-1}\left\{\frac{1}{p-1} + \frac{p}{p^2+1} + \frac{1}{p^2+1}\right\} + L^{-1}\left\{\frac{1}{(p^2+1)^2}\right\}$$

$$= \frac{1}{2}\left[e^t + \cos t + \sin t\right] + \frac{1}{2}(\sin t - t\cos t)$$

$$= \frac{1}{2}\left[e^t + \cos t + 2\sin t - t\cos t\right]$$

$$y = L^{-1}\left\{\frac{p}{(p^2+1)^2}\right\} - \frac{1}{2}L^{-1}\left\{\frac{1}{p-1} - \frac{p}{p^2+1} + \frac{1}{p^2+1}\right\}$$

$$= \frac{1}{2}t\sin t - \frac{1}{2}\left[e^t - \cos t + \sin t\right]$$

$$= \frac{1}{2}\left[t\sin t - e^t + \cos t - \sin t\right]$$

Hence, $x = \dfrac{1}{2}(e^t + \cos t + 2\sin t - t\cos t)$

$$y = \frac{1}{2}(t\sin t - e^t + \cos t - \sin t)$$

Example 3. *Using Laplace transformation, solve*

$(D-2)x - (D+1)y = 6e^{3t}$

$(2D-3)x + (D-3)y = 6e^{3t}$

Given $x = 3$, $y = 0$ when $t = 0$. [UPTU–2001]

Solution. Taking Laplace transformation of the given equations, we get

$$\left.\begin{aligned} LDx - 2Lx - LDy - Ly &= 6Le^{3t} \\ 2LDx - 3Lx + LDy - 3Ly &= 6Le^{3t} \end{aligned}\right\}$$

$$\Rightarrow \left.\begin{aligned} p\bar{x} - x(0) - 2\bar{x} - p\bar{y} + y(0) - \bar{y} &= 6\frac{1}{p-3} \\ 2p\bar{x} - 2x(0) - 3\bar{x} + p\bar{y} - y(0) - 3\bar{y} &= \frac{6}{p-3} \end{aligned}\right\},$$

where $\bar{x} = L(x)$

$$\Rightarrow \left.\begin{aligned}(p-2)\bar{x}-(p+1)\bar{y}-3 &= \frac{6}{p-3}\\ (2p-3)\bar{x}+(p-3)\bar{y}-6 &= \frac{6}{p-3}\end{aligned}\right\}$$

$$\Rightarrow \left.\begin{aligned}(p-2)\bar{x}-(p+1)\bar{y} &= \frac{3p-3}{p-3}\\ (2p-3)\bar{x}+(p-3)\bar{y} &= \frac{6p-12}{p-3}\end{aligned}\right\}$$

$$\Rightarrow \left.\begin{aligned}&(p-3)(p-2)\bar{x}-(p-3)(p+1)\bar{y} = 3p-3\\ &(p+1)(2p-3)\bar{x}+(p+1)(p-3)\bar{y} = \frac{(p+1)(6p-12)}{p-3}\end{aligned}\right\}$$

On adding, we get

$$(3p^2-6p+3)\bar{x} = 3(p-1)+\frac{6(p^2-p-2)}{p-3}$$

$$\Rightarrow \quad \bar{x} = \frac{3(p-1)}{3(p-1)^2}+\frac{6(p^2-p-2)}{3(p-1)^2(p-3)}$$

$$x = L^{-1}\left\{\frac{1}{p-1}+\frac{2}{(p-1)^2}+\frac{2}{p-3}\right\}$$

$$= e^t+2te^t+2e^{3t}$$

Putting the value of x in (1), we get

$$(D-2)(e^t+2te^t+2e^{3t})-(D+1)y = 6e^{3t}$$

$$\Rightarrow \quad e^t+2te^t+2e^t+6e^{3t}-2e^t-4te^t-4e^{3t}-(D+1)y = 6e^{3t}$$

$$\Rightarrow \quad (D+1)y = e^t-2te^t-4e^{3t} \qquad \dots (2)$$

Taking Laplace transform of (2), we get

$$p\bar{y}-y(0)+\bar{y} = \frac{1}{p-1}-\frac{2}{(p-1)^2}-\frac{4}{p-3}$$

$$\Rightarrow \quad (p+1)\bar{y} = \frac{1}{p-1}-\frac{2}{(p-1)^2}-\frac{y}{p-3}$$

$$\bar{y} = \frac{1}{p^2-1}-\frac{2}{(p+1)(p-1)^2}-\frac{4}{(p+1)(p-3)}$$

$$\bar{y} = \frac{1}{p^2-1}-\frac{1/2}{p+1}+\frac{1/2}{p-1}-\frac{1}{(p-1)^2}+\frac{1}{p+1}-\frac{1}{p-3}$$

$$\bar{y} = \frac{1}{p^2-1}+\frac{1/2}{p+1}+\frac{1/2}{p-1}-\frac{1}{(p-1)^2}-\frac{1}{p-3}$$

$$\Rightarrow \quad y = L^{-1}\left\{\frac{1}{p^2-1}+\frac{1}{2}\frac{1}{p+1}+\frac{1}{2}\frac{1}{p-1}-\frac{1}{p-3}-\frac{1}{(p-1)^2}\right\}$$

$$\Rightarrow \quad y = \sinh t+\frac{1}{2}e^{-t}+\frac{1}{2}e^t-e^{3t}-te^t$$

$$\Rightarrow \quad y = \sinh t+\cosh t-e^{-3t}-te^t$$

Example 4. *Solve* $(D^2-1)x+5Dy = t, -Dx+(D^2-4)y = -2$, *when* $x = 0 = Dx = y = Dy, dt = 0$.

Solution. Taking the Laplace transforms of both sides of the given equations, we have

$$L\{x''\}-L\{x\}+5L\{y'\} = L\{t\}$$

and $-2L\{x'\}+L\{y''\}-4L\{y\} = -2L\{1\}$

or $p^2L\{x\}-px(0)-x'(0)-L\{x\}+5[pL\{y\}-y(0)] = 1/p^2$

and $-2[pL\{x\}-x(0)]+p^2L\{y\}-py(0)-y'(0)-4L\{y\} = -2/p$

which gives

$$(p^2-1)L\{x\}+5pL\{y\} = 1/p^2 \qquad \dots (1)$$

and $-2pL\{x\}+(p^2-4)L\{y\} = -2/p \qquad \dots (2)$

On solving (1) and (2) for $L(x)$ and $L(y)$, we get

$$L(x) = \frac{11p^2-4}{p^2(p^2+1)(p^2+4)}$$

$$= -\frac{1}{p^2}+\frac{5}{p^2+1}-\frac{4}{p^2+4}$$

and $L\{y\} = \dfrac{-2p^2+4}{p(p^2+1)(p^2+4)}$

$$= \frac{1}{p}-\frac{2p}{p^2+1}+\frac{p}{p^2+4}$$

Therefore, we get

$$x = -L^{-1}\left\{\frac{1}{p^2}\right\}+5L^{-1}\left\{\frac{1}{p^2+1}\right\}-4L^{-1}\left\{\frac{1}{p^2+4}\right\}$$

$$= -t+5\sin t-2\sin 2t$$

and $y = L^{-1}\left\{\dfrac{1}{p}\right\}-2L^{-1}\left\{\dfrac{p}{p^2+1}\right\}+L^{-1}\left\{\dfrac{p}{p^2+4}\right\}$

$$= 1-2\cos t+\cos 2t.$$

Example 5. *Solve* $Dx+Dy = t$; $D^2x-y = e^{-t}$, *when* $x(0) = 3,\ x'(0) = -2,\ y(0) = 0$.

Solution. Taking the Laplace transforms of both the sides of the given equations, we get

$$L\{x'\}+L\{y'\} = L\{t\}$$

and $L\{x''\}-L\{y\} = L\{e^{-t}\}$

which gives

$$pL\{x\}-x(0)+pL\{y\}-y(0) = 1/p^2$$

and $p^2L\{x\}-px(0)-x'(0)-L\{y\} = \dfrac{1}{p+1}$

or $pL\{x\}+pL\{y\} = 3+\dfrac{1}{p^2} \qquad \dots (1)$

and $p^2L\{x\}-L\{y\} = 3p-2+\dfrac{1}{p+1} \qquad \dots (2)$

Solving (1) and (2) for $L\{x\}$ and $L\{y\}$, we get

$$L\{x\} = \frac{2}{p} + \frac{1}{p^3} + \frac{1}{2(p+1)} + \frac{p}{2(1+p^2)} - \frac{3}{2(p^2+1)}$$

and
$$L\{y\} = \frac{1}{p(p+1)(p^2+1)} + \frac{2}{p^2+1}$$
$$= \frac{1}{p} - \frac{p}{2(p+1)} - \frac{p}{2(p^2+1)} - \frac{1}{2(p^2+1)} + \frac{2}{p^2+1}.$$
$$= \frac{1}{p} - \frac{1}{2(p+1)} - \frac{p}{2(p^2+1)} + \frac{3}{2(p^2+1)}$$

Therefore,
$$x = 2L^{-1}\left\{\frac{1}{p}\right\} + L^{-1}\left\{\frac{1}{p^3}\right\} + \frac{1}{2}L^{-1}\left\{\frac{1}{p+1}\right\} + \frac{1}{2}L^{-1}\left\{\frac{p}{p^2+1}\right\} - \frac{3}{2}L^{-1}\left\{\frac{1}{p^2+1}\right\}$$
$$= 2 + \frac{1}{2}t^2 + \frac{1}{2}e^{-t} + \frac{1}{2}\cos t - \frac{3}{2}\sin t$$

and
$$y = L^{-1}\left\{\frac{1}{p}\right\} - \frac{1}{2}L^{-1}\left\{\frac{1}{p+1}\right\} - \frac{1}{2}L^{-1}\left\{\frac{p}{p^2+1}\right\} + \frac{3}{2}L^{-1}\left\{\frac{1}{p^2+1}\right\}$$
$$= 1 - \frac{1}{2}e^{-t} - \frac{1}{2}\cos t + \frac{3}{2}\sin t.$$

EXERCISE 38.2

Solve the following simultaneous equations :

1. $3\frac{dx}{dt} - y = 2t, \frac{dx}{dt} + \frac{dy}{dt} - y = 0$ with the conditions $x(0) = y(0) = 0$ [UPTU (SUM)–2008]

2. $\frac{dx}{dt} + \frac{dy}{dt} + x + y = 1, \frac{dy}{dt} = 2x + y;\ x(0) = 0, y(0) = 1$ [MTU (SUM)–2011]

3. $\frac{d^2x}{dt^2} - x = y, \frac{d^2y}{dt^2} + y = -x$, given that $t = 0; x = 2, y = -1, \frac{dx}{dt} = 0$ and $\frac{dy}{dt} = 0$ (PTU–2009S)

4. $3\frac{dx}{dt} + \frac{dy}{dt} + 2x = 1, \frac{dx}{dt} + 4\frac{dy}{dt} + 3y = 0;$ given $x = 0, y = 0$ when $t = 0$. (Madras–2003S)

ANSWERS

1. $x = \frac{t^2}{2} + \frac{t}{2} - \frac{3}{4}e^{2t/3} + \frac{3}{4}$

2. $x = e^{-t} - 1, y = 2 - e^{-t}$

3. $x = 2 + t^2/2, y = -1 - t^2/2$

4. $x = \frac{1}{10}(5 - 2e^{-t} - 3e^{-6t/11})$, $y = \frac{1}{5}(e^{-t} - e^{-6t/11})$

38.4 APPLICATION OF LAPLACE TRANSFORM TO ENGINEERING PROBLEMS

Example 1. *A mass m moves along the x-axis under the influence of a force which is proportional to its instantaneous speed and in a direction opposite to the direction of motion. Assuming that at $t = 0$, the particle is located at $x = 0$ and moving to the right with speed V_0. Find the position where the mass comes to rest.*

Solution. The motion of the particle is described as below :

P

Fig. 1

By the Newton's second law of motion, we get the equation of motion of the particle, given by

$$m\frac{d^2x}{dt^2} = -\mu\frac{dx}{dt} \quad \text{... (1)}$$

with initial conditions $x(0) = a$ and $x'(0) = V_0$.

Taking the Laplace transform of both sides of (1), we get

$$mL\left\{\frac{d^2x}{dt^2}\right\} = -\mu L\left\{\frac{dx}{dt}\right\}$$

$$\Rightarrow \quad m[p^2 L\{x\} - px(0) - x'(0)] = -\mu[pL\{x\} - x(0)]$$

$$\Rightarrow \quad (mp^2 + \mu p)L\{x\} = m(ap + V_0) + a\mu$$

$$\Rightarrow \quad L\{x\} = \frac{m(ap + V_0) + a\mu}{p(mp + \mu)}$$
$$= \frac{mV_0 + \mu a}{\mu p} - \frac{mV_0}{\mu\left(p + \frac{\mu}{m}\right)}$$

Therefore,
$$x = \left(\frac{mV_0}{\mu} + a\right)L^{-1}\left\{\frac{1}{p}\right\} - \frac{mV_0}{\mu}L^{-1}\left\{\frac{1}{p + \frac{\mu}{m}}\right\}$$
$$= \left(\frac{mV_0}{\mu} + a\right) - \frac{mV_0}{\mu}.e^{-\mu t/m}. \quad \text{... (2)}$$

If $\frac{dx}{dt} = V_0 e^{\mu t/m} = 0$.

Then, from (2), we have $x = \frac{mV_0}{\mu} + a$.

Hence, the mass m comes to rest at a distance $\frac{mV_0}{\mu} + a$ from the centre.

Example 2. *A particle moves along a line so that its displacement X from a fixed point at any time x is given by*

$$X''(t)+4X'(t)+5X(t)=80\sin 5t.$$

Find its displacement at any time $x>0$, if at $t=0$, the particle is at rest at $X=0$.

Solution. Here, the displacement of the particle is given by the differential equation

$$X''(t)+4X'(t)+5X(t)=80\sin 5t \quad \text{... (1)}$$

where $X(0)=0$ and $X'(0)=0$.

Taking Laplace transform of both sides of (1), we get

$$L\{X''(t)\}+4L\{X'(t)\}+5L\{X(t)\}=80L\{\sin 5t\}$$

$$\Rightarrow \quad p^2L\{X(t)\}-pX(0)-X'(0)$$

$$+4[pL\{X(t)\}-X(0)]+5L\{X(t)\}=80\times\frac{5}{p^2+25}$$

$$\Rightarrow \quad (p^2+4p+5)\,L\{X(t)\}=\frac{400}{p^2+25}$$

$$\Rightarrow \quad L\{X(t)\}=\frac{400}{(p^2+4p+5)(p^2+25)}$$

$$=\frac{-2(p+5)}{p^2+25}+\frac{2(p+9)}{p^2+4p+5}.$$

Therefore, $X(t)=-2L^{-1}\left\{\frac{p}{p^2+25}+\frac{5}{p^2+25}\right\}$

$$+2L^{-1}\left\{\frac{(p+2)+7}{(p+2)^2+1}\right\}$$

$$=-2(\cos 5t+\sin 5t)+2e^{-2t}L^{-1}\left\{\frac{p+7}{p^2+1}\right\}$$

$$\Rightarrow X(t)=-2(\cos 5t+\sin 5t)$$

$$+2e^{-2t}(\cos t+7\sin t).$$

Example 3. *A resistance R in series with inductance L is connected with emf $E(t)$. The current i is given by*

$$L\frac{di}{dt}+Ri=E(t)$$

If the switch is connected at $t=0$ and disconnected at $t=a$, find the current i in terms of t. [UPTU–2001]

Solution. Conditions under which current i flows are $i=0$ at $t=0$.

$$E(t)=\begin{cases}E, & 0<t<a\\ 0, & t>a\end{cases}$$

Given equation is

$$L\frac{di}{dt}+Ri=E(t) \quad \text{... (1)}$$

Taking Laplace transform of (1), we get

$$L[pL\{i\}-i(0)]+Ri=\int_0^\infty e^{-pt}E(t)\,dt$$

$$LpL\{i\}+Ri=\int_0^\infty e^{-pt}E(t)\,dt \qquad [i(0)=0]$$

$$(Lp+R)L\{i\}=\int_0^\infty e^{-pt}E(t)dt$$

$$=\int_0^a e^{-pt}Edt+\int_0^\infty e^{-pt}Edt$$

$$=E\left[\frac{e^{-pt}}{-p}\right]_0^a+0=\frac{E}{p}\left[1-e^{-ap}\right]=\frac{E}{p}-\frac{E}{p}e^{-ap}$$

$$\Rightarrow \quad L\{i\}=\frac{E}{p(Lp+R)}-\frac{Ee^{-ap}}{p(Lp+R)}$$

Taking inverse Laplace transform, we obtain

$$i=L^{-1}\left\{\frac{E}{p(Lp+R)}\right\}-L^{-1}\left\{\frac{Ee^{-ap}}{p(Lp+R}\right\} \quad \text{... (2)}$$

Now, we have to find the value of $L^{-1}\left\{\frac{E}{p(Lp+R)}\right\}$

$$L^{-1}\left\{\frac{E}{p(Lp+R)}\right\}=\frac{E}{L}L^{-1}\left\{\frac{1}{p\left(p+\frac{R}{L}\right)}\right\}$$

$$=\frac{E}{L}\cdot\frac{L}{R}L^{-1}\left\{\frac{1}{p}-\frac{1}{p+\frac{R}{L}}\right\}=\frac{E}{R}\left[1-e^{-R/Lt}\right]$$

and

$$L^{-1}\left\{\frac{Ee^{-ap}}{p(Lp+R}\right\}=\frac{E}{R}\left[1-e^{-\frac{R}{L}(t-a)}\right]u(t-a)$$

[By the second shifting theorem]

On substituting the values of the inverse transforms in (2), we get

$$i=\frac{E}{R}\left[1-e^{-\frac{R}{L}t}\right]-\frac{E}{R}\left[1-e^{-\frac{R}{L}(t-a)}\right]u(t-a)$$

Hence, $i=\frac{E}{R}\left[1-e^{-\frac{R}{L}t}\right]$,

for $0<t<a$, $[u(t-a)=0]$

$$i=\frac{E}{R}\left[1-e^{-\frac{R}{L}t}\right]-\frac{E}{R}\left[1-e^{-\frac{R}{L}(t-a)}\right], \text{ for } t>a,$$

$[u(t-a)=1]$.

$$=\frac{E}{R}\left[e^{-\frac{R}{L}(t-a)}-e^{-\frac{R}{L}t}\right]=\frac{E}{R}e^{-\frac{R}{L}t}\left[e^{\frac{Ra}{L}}-1\right]$$

EXERCISE 38.3

1. Solve $[tD^2+(1-2t)D-2]y=0$ if $y(0)=1,\ y'(0)=2$.
2. Solve $y''+ty'-y=0$ if $y(0)=0,\ y'(0)=1$.
3. Solve $y''(t)+aty'(t)-2ay(t)=1$ if $y(0)=y'(0)=0,\ a>0$.
4. Solve $(D-2)x+3y=0;\ 2x+(D-1)y=0$, if $x(0)=8$ and $y(0)=3$.
5. Solve $(D^2-3)x-4y=0$
   $x+(D^2+1)y=0,\ t>0$,

if $x = y = Dy = 0, Dx = 2$, when $t = 0$. [UPTU–2004]

6. Solve $(D-2)x-(D+1)y = 6e^{3t}$, $(2D-3)x+(D-3)y = 6e^{3t}$, if $x = 3$, $y = 0$ when $t = 0$.

7. Solve $(D-2)x-(D-2)y = \sin t$
$(D^2+1)x+2Dy = 0$ if $x = 0 = x'(0) = y(0)$.

8. Using $\frac{dy}{dt}+2x = \sin 2t$, $\frac{dx}{dt}-2y = \cos 2t$ $(t > 0)$ such that $t = 0, x = 1$ and $y = 0$. Show that the particle moves along the curve $4x^2+4xy+5y^2 = 4$. [UPTU–2003, MTU(SUM)–2011]

9. Solve $\frac{dx}{dt}+y = \sin t$, $\frac{dy}{dt}+x = \cos t$, using $x = 2$, $y = 0$ at $t = 0$. [UPTU–2004, GBTU–2012, Kerala–2005]

10. Solve $\frac{dx}{dt}+4\frac{dy}{dt}-y = 0$; $\frac{dx}{dt}+2y = e^{-t}$ with $x(0) = y(0) = 0$. [UPTU 2008]

11. A body falls from rest in a liquid whose density is one fourth that of the body. If the liquid offers resistance proportional to the velocity and the velocity approaches a limiting value of 9 meter/sec, find the distance fallen in 5 seconds. [UPTU Q. Bank–2001]

12. A beam of length l is clamped horizontally at its ends $x = 0$ and is free at the end $x = l$. A point load w is applied at the end $x = l$ in addition to a uniform load w per unit length from $x = 0$ to $x = l/2$. Find the deflection of the beam. [UPTU(Sp)–2001]

13. Solve $y^{iv}(t)+y'''(t) = \cos t$, $y(0) = y'(0) = y'''(0) = 0$, $y''(0)$ is arbitrary.

14. Solve $y'''(t)+y''(t)-4y'(t)-4y(t) = F(t)$ if $y(0) = y''(0) = 0$ and $y'(0) = 2$.

15. Solve $(D^3-D^2+4D-4)y = 68e^t \sin 2t$, $y = 1$, $Dy = -19$, $D^2y = -37$ at $t = 0$.

16. Solve $y''-4y'+3y = F(t)$ if $y(0) = 1$, $y'(0) = 0$.

17. A particle P of mass 2 grams moves on the x-axis and is attached towards origin 0 with a force numerically about to $8x$. If it is initially at rest at $x = 10$, find its position at any subsequent time assuming :

(a) no other force acts

(b) a damping force numerically equal to 8 times the instantaneous velocity acts. [UPTU Q. Bank–2001]

Hint to Selected Problems

1. The given equation can be written as

$$ty''+y'-2ty'-2y = 0 \Rightarrow L\{ty''\}+L\{y'\}-2L\{ty'\}-2L\{y\} = 0$$

$$\Rightarrow -\frac{d}{dp}L\{y''\}+L\{y'\}+2\frac{d}{dp}L\{y'\}-2L\{y\} = 0$$

$$\Rightarrow \frac{dL\{y\}}{L\{y\}}+\frac{1}{(p-2)}dp = 0 \quad \text{Now integrate.}$$

2. Proceeding as usual, we get $\frac{d}{dp}[L\{y\}]-\left(p-\frac{2}{p}\right)L\{y\} = -\frac{1}{p}$

which is a linear differential equation of first order with

$$\text{I.F.} = e^{-\int\left(p-\frac{2}{p}\right)dp} = p^2e^{-p^2/2}$$

The solution of (1) is given by

$$L\{y\} = \frac{C_1}{p^2}e^{p^2/2}+\frac{1}{p^2} = \frac{C_1}{p^2}\left[1+\frac{p^2}{2}+\frac{1}{2!}\cdot\frac{1}{4}p^4+\ldots\right]+\frac{1}{p^2}$$

Now taking inverse Laplace transform.

3. Proceeding same as above, we get

$$\frac{d}{dp}L\{y\}+\left[-\frac{p}{a}+\frac{3}{p}\right]L\{y\} = -\frac{1}{ap^2}, \text{ which is linear in } L\{y\}.$$

7. $L\{x\} = \frac{1}{9(p+1)}+\frac{1}{3(p+1)^2}+\frac{4}{45(p-2)}-\frac{(p+2)}{5(p^2+1)}$

and $L\{y\} = \frac{1}{p^3-3p^{-2}} = \frac{1}{9(p+1)}+\frac{1}{3(p+1)^2}-\frac{1}{9(p-2)}$

ANSWERS

1. $y = e^{2t}$ **2.** $y = t$ **3.** $y = \frac{t^2}{2}$ **4.** $y = 5e^{-t}-2e^{4t}$ **5.** $y = \frac{1}{2}(1-t)e^t-\frac{1}{2}(1+t)e^{-t}$ **6.** $x = e^t+2te^t+2e^{3t}$ and $y = e^t-te^t-e^{3t}$

7. $x = \frac{1}{9}e^{-t}+\frac{4}{45}e^{2t}-\frac{1}{5}\cos t-\frac{2}{5}\sin t+\frac{1}{3}te^{-t}$, $y = \frac{1}{9}e^{-t}-\frac{1}{9}e^{2t}+\frac{1}{3}e^{-t}.t$ **9.** $x = e^{-t}+e^t$; $y = \sin t+e^{-t}-e^{-t}$

10. $x = -\frac{5}{7}e^{-t}+\frac{8}{21}e^{3t/4}+\frac{1}{4}$; $y = \frac{1}{7}e^{-t}-\frac{1}{7}e^{\frac{3}{4}t}$ **11.** $x = 34.17$ meters

12. $y = \frac{1}{24EI}\left[wx^2\left(x^2-2lx+\frac{3l^2}{2}\right)-w\left(x-\frac{l}{2}\right)^4 u\left(x-\frac{l}{2}\right)-4wx^2(3l-x)\right]$

13. $y = -1+t+Ct^2+\frac{1}{2}(e^{-t}+\cos t-\sin t)$ **14.** $y = \sinh 2t+\frac{1}{12}F(t)(-4e^{-t}+e^{2t}+3e^{-2t})$

15. $y = \frac{1}{5}(e^t+14\cos 2t-3\sin 2t)$ $-2e^t(\cos 2t+4\sin 2t)$ **16.** $y = \frac{3}{2}e^t-\frac{1}{2}e^{3t}+\frac{1}{2}\int_0^t(e^{3x}-e^x)\,F(t-x)dx$

17. (a) $X = 10\cos 2t$, (b) $X = 10e^{-2t}+20te^{-2t}$.

FFFFFF

39 An Introduction to Partial Differential Equations

39.1 INTRODUCTION

In this chapter, we shall discuss the differential equations, with number of independent variables are two or more. In such cases, any dependent variable is likely to be a function of more than one variables, so that it possesses not ordinary derivatives with respect to a single variable but partial derivatives with respect to several variables. The partial differential equation implies necessarily the existence of more than one independent variables. We shall usually take z as dependent variable and x, y as independent variable and throughout the chapter we shall denote the partial derivatives

$\frac{\partial z}{\partial x}, \frac{\partial z}{\partial y}, \frac{\partial^2 z}{\partial x^2}, \frac{\partial^2 z}{\partial x \partial y}$ and $\frac{\partial^2 z}{\partial y^2}$ by p, q, r, s and t respectively.

Definition. *The equation of the type* $F\left(\frac{\partial z}{\partial x}, \ldots, \frac{\partial^2 z}{\partial x^2}, \ldots, \frac{\partial^2 z}{\partial x \partial y}, \ldots\right) = 0$ *is called a partial differential equation.*

39.2 ORDER AND DEGREE

39.2.1 ORDER OF PDE

The order of the partial differential equation is the order of its highest derivative.

(i) First order PDE. A first order partial differential equation for a function $z = f(x, y)$ contains at least one of the partial derivatives $\frac{\partial z}{\partial x}$ or $\frac{\partial z}{\partial y}$. But no partial derivative of order higher then one.

For example : $x\frac{\partial z}{\partial x} + y\frac{\partial z}{\partial y} = 0.$

(ii) Second order PDE. A second order partial differential equation for $z = f(x, y)$ contains at least one of the partial derivatives $\frac{\partial^2 z}{\partial x^2}, \frac{\partial^2 z}{\partial y^2}, \frac{\partial^2 z}{\partial x \partial y}$ but no partial derivatives of order higher than two.

For example : (i) $\frac{\partial^2 \phi}{\partial x^2} + \frac{\partial^2 \phi}{\partial y^2} + \frac{\partial^2 \phi}{\partial z^2} = 0$ (ii) $\frac{\partial z}{\partial t} - C\frac{\partial^2 z}{\partial x^2} = 0.$

REMARK

- The second order partial differential equation may also contain first order term, like $\frac{\partial z}{\partial x}, \frac{\partial z}{\partial y}$ etc.

39.3.2 DEGREE OF PDE

The degree of partial differential equation is the power of the highest derivative in the equation.

For example :

(i) $\frac{\partial^2 \phi}{\partial x^2} + \frac{\partial^2 \phi}{\partial y^2} + \frac{\partial^2 \phi}{\partial z^2} = 0$ (ii) $\frac{\partial z}{\partial t} - C\frac{\partial^2 z}{\partial x^2} = 0$ (iii) $x\frac{\partial z}{\partial x} + y\frac{\partial z}{\partial y} = 0$

(iv) $\frac{\partial^2 z}{\partial t^2} = C^2 \frac{\partial^2 z}{\partial y^2}$ (v) $\left(\frac{\partial z}{\partial x}\right)^3 + \frac{\partial z}{\partial x} = 0.$

Equation (i), (ii), (iii) and (iv) are PDEs of degree one, and the equation (v) is a PDE of degree 3.

39.3 CLASSIFICATION OF PARTIAL DIFFERENTIAL EQUATIONS

39.3.1 LINEAR AND NON-LINEAR PARTIAL DIFFERENTIAL EQUATION

A partial differential equation is said to be linear if :

(i) it is of the first degree in the dependent variable and its partial derivatives.

(ii) it does not contain the product of dependent variables and either of its partial derivatives, and

(iii) it does not contain any transcendental function.

For example : (i) $\frac{\partial^2 \phi}{\partial x^2} + \frac{\partial^2 \phi}{\partial y^2} + \frac{\partial^2 \phi}{\partial z^2} = 0$ (ii) $\frac{\partial T}{\partial t} - K\frac{\partial^2 T}{\partial t^2} = 0$

(iii) $\frac{\partial^2 u}{\partial t^2} = C^2 \frac{\partial^2 u}{\partial y^2}$ (iv) $\frac{\partial^2 u}{\partial x^2} + \frac{\partial^2 u}{\partial y^2} = f(x, y)$

The above all equations are linear.

39.3.2 NON-LINEAR PDE

A partial differential equation, which is not linear is called non-linear equation.

For example : $\left(\frac{\partial f}{\partial x}\right)^3 + \frac{\partial f}{\partial t} = 0.$

39.3.3 QUASI-LINEAR

Consider a non-linear equation $R_1 r + S_1 s + T_1 t = V_1$, where R_1, S_1, T_1 and V_1 are the functions of p and q as well as of x, y and z. Then, we observe that, it has a certain formal resemblance to a linear equation. Due to this resemblance with linear equation, equation (1) is said to be quasi-equation.

REMARKS

- Quasi-linear equation is also called the uniform non-linear equation.
- Quasi-linear equation can be easily solved by Monge's method.

39.3.4 HOMOGENEOUS AND NON-HOMOGENEOUS EQUATIONS

A linear partial differential equation can be classified as follows :

(i) Homogeneous linear equation.

(ii) Non-homogeneous linear equation.

(i) Homogeneous linear equation :

If each term of a partial differential equation contains either the dependent variable (or unknown function) or one of its partial derivatives, it is said to be homogeneous.

For example : (i) $\frac{\partial^2 \phi}{\partial x^2} + \frac{\partial^2 \phi}{\partial y^2} + \frac{\partial^2 \phi}{\partial z^2} = 0$ (ii) $\frac{\partial^2 z}{\partial t^2} = c^2 \frac{\partial^2 z}{\partial y^2}.$

(ii) Non-homogeneous linear equation :

An equation, which is not homogeneous is called non-homogeneous linear equation.

For example : (i) $\frac{\partial^2 z}{\partial x^2} + \frac{\partial^2 z}{\partial y^2} = f(x, y)$ (ii) $\frac{\partial^3 u}{\partial x^3} + 2\frac{\partial^3 u}{\partial x \partial y^2} - 6\left(\frac{\partial u}{\partial y}\right)^4 = 0.$

39.4 SOLUTION OF PARTIAL DIFFERENTIAL EQUATIONS

A solution of PDE in some region R of the space of independent variables is a function all of whose partial derivatives appearing in the equation exist in some domain containing R and which satisfies the equation everywhere in R.

REMARKS

- The solution of ODE involves arbitrary constants, the solution of a PDE involves arbitrary functions.
- As we increase the order of partial derivation in a PDE, we must introduce more arbitrary functions.
- If u_1 and u_2 are any linearly independent solution of linear homogenous PDE, then $u = c_1u_1 + c_2u_2$, where c_1 and c_2 are arbitrary constants, is also a solution of that equation. This is known as Principle of superposition and can by extended to the case where n solution of a PDE exists.

EXERCISE 39.1

Classify the given PDE's by a way order and degree linearity (L)/ non-linearity (NL), homogeneity (H)/ non-homogeneity (NH).

1. $\dfrac{\partial^2 u}{\partial t^2} - c^2\dfrac{\partial^2 u}{\partial x^2} + 2\beta\dfrac{\partial u}{\partial t} + \alpha u = 0.$

2. $\dfrac{\partial^2 u}{\partial t^2} - c^2\dfrac{\partial^2 u}{\partial x^2} = 0.$

3. $-\dfrac{h^2}{2m}\left(\dfrac{\partial^2\phi}{\partial x^2} + \dfrac{\partial^2\phi}{\partial y^2} + \dfrac{\partial^2\phi}{\partial z^2}\right) + v(r)d = ih\dfrac{\partial\phi}{\partial t}.$

4. $\dfrac{\partial\rho}{\partial t} + \rho\left(\dfrac{\partial u}{\partial x} + \dfrac{\partial v}{\partial y} + \dfrac{\partial w}{\partial z}\right) = 0.$

5. $\dfrac{\partial u}{\partial t} - K\left(\dfrac{\partial^2 u}{\partial x^2} + \dfrac{\partial^2 u}{\partial y^2}\right) = 0.$

6. $\dfrac{\partial^2 u}{\partial x^2} + \dfrac{\partial^2 u}{\partial y^2} + \dfrac{\partial^2 u}{\partial z^2} = \dfrac{1}{\varepsilon}\rho(x,y,z).$

7. $x^2\dfrac{\partial^2 f}{\partial x^2} + y^2\dfrac{\partial^2 f}{\partial y^2} = 0.$

8. $xy\dfrac{\partial^2 f}{\partial x^2} + x\dfrac{\partial f}{\partial x}\dfrac{\partial f}{\partial y} + y\dfrac{\partial^2 f}{\partial y^2} = x^2 + y^2.$

ANSWERS

1. 2, 1, L, H **2.** 2, 1, L, H **3.** 2, 1, L, H **4.** 1,1, L, H **5.** 2, 1, L, H
6. 2, 1, L, NH **7.** 2, 1, L, H **8.** 2, 1, L, NH

39.5 LINEAR PARTIAL DIFFERENTIAL EQUATION OF FIRST ORDER

A differential equation involving partial derivatives p and q only, no higher derivative is called of order 1. If the degree of p and q is one, then it is called a linear partial differential equation of order one.

39.5.1 COMPLETE INTEGRAL

Let us consider the partial differential equation $f(x,y,z,p,q) = 0$ where x, y are independent variable, and z is dependent, while $p = \dfrac{\partial z}{\partial x}, q = \dfrac{\partial z}{\partial y}$, then a relation of type $F(x,y,z,a,b) = 0$ containing as many arbitrary constants as the number of independent variable in the above partial differential equation is called complete integral.

39.5.2 PARTICULAR INTEGRAL

In the complete integral $F(x,y,z,a,b) = 0$ giving the particular values to the constants a and b, we get the particular integral.

39.5.3 SINGULAR INTEGRAL

The envelope of the surfaces given by the complete integral $F(x,y,z,a,b) = 0$ is called singular integral. Therefore the singular integral is obtained by eliminating a and b from $F(x,y,z,a,b) = 0, \dfrac{\partial F}{\partial a} = 0$ and $\dfrac{\partial F}{\partial b} = 0.$

39.5.4 GENERAL INTEGRAL

Let $u = u(x,y,z)$ and $v = v(x,y,z)$ be two functions of x, y and z then the solution of the differential equation $pP + qQ = R$ of the types $f(u,v) = 0$ is called the general integral. This also can be taken as $u = f(v)$ or $v = f(u)$.

39.6 DERIVATION OF PARTIAL DIFFERENTIAL EQUATION BY THE ELIMINATION OF ARBITRARY CONSTANTS

Consider the equation

$$F(x,y,z,a,b) = 0 \qquad \ldots(1)$$

where a and b are arbitrary constants. Differentiating (1) partially with respect to x, regarding z as a function of two independent variables x and y, we get

$$\frac{\partial F}{\partial x} + p\frac{\partial F}{\partial z} = 0 \quad \text{and} \quad \frac{\partial F}{\partial y} + q\frac{\partial F}{\partial z} = 0 \qquad \ldots(2)$$

By the elimination of a and b from (1) and (2), we shall get an equation of the type

$$F(x,y,z,p,q)=0 \qquad \ldots(3)$$

which is the required partial differential equation of the first order.

REMARK

- It can be easily shown that if there are more arbitrary constants, then the number of independent variables, the above procedure of elimination will gives rise to partial differential equation of higher order than first.

Solved Examples

Example 1. *Construct a partial differential equation, by eliminating a, b and c form*

$$z=a(x+y)+b(x-y)+abt+c. \quad \text{(IAS–1998)}$$

Solution. The given equation is

$$z=a(x+y)+b(x-y)+abt+c \qquad \ldots(1)$$

Now, differentiating (1) partially with respect to x, y and z, we get

$$\frac{\partial z}{\partial x}=a+b, \frac{\partial z}{\partial y}=a-b, \frac{\partial z}{\partial t}=ab \qquad \ldots(2)$$

Now, using

$$(a+b)^2-(a-b)^2=4ab$$

$$\Rightarrow \quad \left(\frac{\partial z}{\partial x}\right)^2-\left(\frac{\partial z}{\partial y}\right)^2=4\frac{\partial z}{\partial t}$$

which is the required partial differential equation.

Example 2. *Form a partial differential equation by eliminating a, b, c from* $\frac{x^2}{a^2}+\frac{y^2}{b^2}+\frac{z^2}{c^2}=1.$

Solution. The given equation is

$$\frac{x^2}{a^2}+\frac{y^2}{b^2}+\frac{z^2}{c^2}=1 \qquad \ldots(1)$$

Differentiating (1) partially with respect to x and y, we get

$$\frac{2x}{a^2}+\frac{2z}{c^2}\cdot\frac{\partial z}{\partial x}=0 \qquad \ldots(2)$$

and

$$\frac{2y}{b^2}+\frac{2z}{c^2}\cdot\frac{\partial z}{\partial y}=0 \qquad \ldots(3)$$

Now, differentiating (2), w.r.t. x and (3) w.r.t. y, we get

$$\frac{2}{a^2}+\frac{2}{c^2}\left\{\left(z\frac{\partial^2 z}{\partial x^2}\right)+\left(\frac{\partial z}{\partial x}\right)^2\right\}=0$$

$$\Rightarrow \quad c^2+a^2\left(\frac{\partial z}{\partial x}\right)^2+a^2z\frac{\partial^2 z}{\partial x^2}=0 \qquad \ldots(4)$$

and

$$c^2+b^2\left(\frac{\partial z}{\partial x}\right)^2+b^2z\frac{\partial^2 z}{\partial y^2}=0 \qquad \ldots(5)$$

Putting the value of c^2 from (2) in (4), we get

$$\left(-a^2\frac{z}{x}\cdot\frac{\partial z}{\partial x}\right)+a^2\left(\frac{\partial z}{\partial x}\right)^2+a^2z\frac{\partial^2 z}{\partial x^2}=0$$

$$\Rightarrow \quad -\frac{z}{x}\frac{\partial z}{\partial x}+\left(\frac{\partial z}{\partial x}\right)^2+z\frac{\partial^2 z}{\partial x^2}=0$$

$$\Rightarrow \quad -z\frac{\partial z}{\partial x}+x\left(\frac{\partial z}{\partial x}\right)^2+xz\frac{\partial^2 z}{\partial x^2}=0.$$

Similarly, from (3) and (5), we get

$$-z\frac{\partial z}{\partial y}+y\left(\frac{\partial z}{\partial y}\right)^2+yz\frac{\partial^2 z}{\partial y^2}=0.$$

39.7 DERIVATION OF A PARTIAL DIFFERENTIAL EQUATION BY THE ELIMINATION OF AN ARBITRARY FUNCTION

Let u and v be any two functions of x, y, z connected by the relation

$$\phi(u,v)=0 \qquad \ldots(1)$$

Now, it is to be shown on the elimination of the arbitrary function ϕ from (1), a partial differential equation will be formed and moreover, this equation will be linear. Differentiating (1) partially with respect to x and y regarding z as independent variables, we have

$$\frac{\partial\phi}{\partial u}\left(\frac{\partial u}{\partial x}+\frac{\partial u}{\partial z}\cdot\frac{\partial z}{\partial x}\right)+\frac{\partial\phi}{\partial v}\left(\frac{\partial v}{\partial x}+\frac{\partial v}{\partial z}\cdot\frac{\partial z}{\partial x}\right)=0 \quad \Rightarrow \quad \frac{\partial\phi}{\partial u}\left(\frac{\partial u}{\partial x}+p\frac{\partial u}{\partial z}\right)+\frac{\partial\phi}{\partial v}\left(\frac{\partial v}{\partial x}+p\frac{\partial v}{\partial z}\right)=0 \qquad \ldots(2)$$

and

$$\frac{\partial\phi}{\partial u}\left(\frac{\partial u}{\partial y}+\frac{\partial u}{\partial z}\cdot\frac{\partial z}{\partial y}\right)+\frac{\partial\phi}{\partial v}\left(\frac{\partial v}{\partial y}+\frac{\partial v}{\partial z}\cdot\frac{\partial z}{\partial y}\right)=0 \quad \Rightarrow \quad \frac{\partial\phi}{\partial u}\left(\frac{\partial u}{\partial y}+q\frac{\partial u}{\partial z}\right)+\frac{\partial\phi}{\partial v}\left(\frac{\partial v}{\partial y}+q\frac{\partial v}{\partial z}\right)=0 \qquad \ldots(3)$$

Now eliminating $\frac{\partial\phi}{\partial u}, \frac{\partial\phi}{\partial v}$ between (2) and (3) by the method of determinant, we get

$$\begin{vmatrix} \left(\frac{\partial u}{\partial x}+p\frac{\partial u}{\partial z}\right) & \left(\frac{\partial v}{\partial x}+p\frac{\partial v}{\partial z}\right) \\ \left(\frac{\partial u}{\partial y}+q\frac{\partial u}{\partial z}\right) & \left(\frac{\partial v}{\partial y}+q\frac{\partial v}{\partial z}\right) \end{vmatrix} = 0 \quad \Rightarrow \quad \left(\frac{\partial u}{\partial y}\cdot\frac{\partial v}{\partial z}-\frac{\partial u}{\partial z}\cdot\frac{\partial v}{\partial y}\right)p+\left(\frac{\partial u}{\partial z}\cdot\frac{\partial v}{\partial x}-\frac{\partial u}{\partial x}\cdot\frac{\partial v}{\partial z}\right)q=\frac{\partial u}{\partial x}\cdot\frac{\partial v}{\partial y}-\frac{\partial u}{\partial y}\cdot\frac{\partial v}{\partial x}$$

$$\Rightarrow \quad \frac{\partial(u,v)}{\partial(y,z)}p+\frac{\partial(u,v)}{\partial(z,x)}q=\frac{\partial(u,v)}{\partial(x,y)}$$

which is the linear PDE of first order and first degree in p and q which can also be written as $Pp+Qq=R$

where, $P=\frac{\partial(u,v)}{\partial(y,z)}, Q\frac{\partial(u,v)}{\partial(z,x)}$ and $R=\frac{\partial(u,v)}{\partial(x,y)}$.

Solved Examples

Example 1. *By means of a partial differential equation, eliminate the arbitrary function from the equation*

$$x+y+z=f(x^2+y^2+z^2) \quad \ldots(1)$$

(SVTU–2007)

Solution. Differentiating (1) partially w.r.t. x and y, we get

$$(1+p)=f'(x^2+y^2+z^2).(2x+2zp) \quad \ldots(2)$$

and $(1+q)=f'(x^2+y^2+z^2).(2y+2zq) \quad \ldots(3)$

From (2) and (3) we have

$$\frac{(1+p)}{(2x+2zp)}=\frac{(1+q)}{(2y+2zq)}$$

$\Rightarrow (1+p)(y+zq)=(1+q)(x+zp)$

$\Rightarrow (y-z)p+(z-x)q=(x-y)$, which is the required PDE.

Example 2. *Eliminate the arbitrary functions f and g from* $y=(f(x-at)+g(x+at)$. (VTU–2009)

Solution. The given equation is

$$y=f(x-at)+g(x+at) \quad \ldots(1)$$

$$\Rightarrow \quad \frac{\partial y}{\partial x}=f'(x-at)+g'(x+at)$$

and $\frac{\partial^2 y}{\partial x^2}=f''(x-at)+g''(x+at) \quad \ldots(2)$

Now $\frac{\partial y}{\partial t}=f'(x-at).(-a)+g'(x+at)(a)$

$$\Rightarrow \quad \frac{\partial^2 y}{\partial t^2}=f''(x-at)(-a)^2+g''(x+at)(a)^2$$

$$=a^2[f''(x-at)+g''(x+at)]=a^2\frac{\partial^2 y}{\partial x^2}$$

(Using (2))

$\Rightarrow \frac{\partial^2 y}{\partial t^2}=a^2\frac{\partial^2 y}{\partial x^2}$, which is the required PDE.

Example 3. *Form the partial differential equation by eliminating the arbitrary function from* $z=f(x+it)+g(x-it)$. (UKTU–2011)

Solution. Let $z=f(x+it)+g(x-it)$

Differentiating z partially w.r.t. z and t we get

$$\frac{\partial z}{\partial x}=f'(x+it)+g'(x-it)$$

$$\frac{\partial^2 z}{\partial x^2}=f''(x+it)+g''(x-it) \quad \ldots(1)$$

and $\frac{\partial z}{\partial t}=if'(x+it)-ig(x-it)$

$$\frac{\partial^2 z}{\partial t^2}=i^2f''(x+it)+i^2g''(x-it)$$

$$=-f''(x+it)-g(x-it) \quad \ldots(2)$$

On adding (1) and (2) we get

$\frac{\partial^2 z}{\partial x^2}+\frac{\partial^2 z}{\partial t^2}=0$, which is the required partial differential equation.

EXERCISE 39.2

A. Form a PDE, by eliminating arbitrary constants for the following equations :

1. $z=(x+a)(y+b)$. 2. $z=ax+by+ab$.
3. $z=ax+a^2y^2+b^2$
4. $(x-h)^2+(y-k)^2+z^2=c^2$ (IAS–1996, Kottayam–2005)
5. $z=axe^y+\frac{1}{2}a^2e^{2y}+b$. 6. $z=(x-a)^2+(y-b)^2$
7. Find the differential of all spheres of radius λ, having centre in the xy-plane. (IAS–1996)
8. Find the differential equation by eliminating a and b from $z=(x^2+a)(y^2+b)$. (IAS–1997)
9. Find the differential equation of the set of all right circular cone whose axes coincide with z-axis (IAS–1998)

B. Form a PDE by eliminating the arbitrary function form the following equations:

1. $z=y^2+2f\left(\frac{1}{x}+\log y\right)$

(VTU–2010, JNTU–2010, Madras–2000)

2. $z = f(y/x)$;
3. $f(x+y+z)+(x^2+y^2+z^2)=0$.
4. $lx+my+nz = f(x^2+y^2+z^2)$
5. $z = f(x)+xg(y)$ (VTU–2004)
6. $z = f(x^2-y^2)$ (SVTU–2008)
7. $f(xy+z^2, x+y+z)=0$ (PTU–2006)
8. $z = x^2f(y)+y^2g(x)$ (Anna–2003)
9. $z = f(x^2+y^2)+x+y$ (Anna–2009)
10. $z = f_1(y+2x)+f_2(y-3x)$ (Kurukshetra–2005)
11. $z = e^{my}\phi(x-y)$ (PTU–2002)

C. Find the differential equation of all surfaces of revolution having z-axis as the axis of rotation (IAS–1997)

D. Find a partial differential equation by eliminating the arbitrary functions f and g from $z = yf(x)+xg(x)$

E. Find the differential equation of the surface which are the envelope of a one parametric family of planes. (IAS–1995)

F. Form a partial differential equation by eliminating the arbitrary function ϕ from $\phi(x^2+y^2+z^2, z^2-2xy)=0$

G. Form a PDE by eliminating f and g from

$$z = f(x^2-y)+g(x^2+y)$$ (IAS–1996)

H. Find the differenmtial equation of all planes which are at a constant distance a from the origin.

(VTU–2009, Kurukshetra–2006)

ANSWERS

A. 1. $z = pq$ 2. $z = px+qy+pq$ 3. $q = 2yp^2$ 4. $z^2(p^2+q^2+1)=c^2$ 5. $q = px+p^2$
6. $p^2+q^2=z^2$ 8. $4xyz = pq$ 9. $py = xq$

B. 1. $x^2p+yq=2y^2$ 2. $px+qy=0$ 3. $(y+z)p-(z+x)q = x-y$ 4. $(l+np)y+z(lq-mp)=(m+nq)x$
5. $\frac{\partial z}{\partial y} = x\frac{\partial^2 z}{\partial x\partial y}$ 6. $py+qx=0$ 7. $xys = px+py-z$ 8. $xyr = 2(px+qy-2z)$ 9. $qx-py = x+y$
10. $\frac{\partial^2 z}{\partial x^2}+\frac{\partial^2 z}{\partial x\partial y}-6\frac{\partial^2 z}{\partial y^2}=0$ 11. $p+q = mz$

C. $py = xq$ **D.** $xyr = xp+yq-z$ **F.** $(p-q)z = y-x$. **G.** $x\frac{\partial^2 z}{\partial x^2} = p+4x^3\frac{\partial^2 z}{\partial y^2}$ **H.** $z = px+qy+a\sqrt{1+p^2+q^2}$

39.8 LAGRANGE'S LINEAR EQUATION

The partial differential equation of the form $Pp+Qq=R$, where P,Q,R are functions of x,y,z in the standard form of the linear partial differential equation of the order one is called Lagrange's linear equation.

39.9 SOLUTION OF STANDARD FORMS (NON-LINEAR EQUATION)

In this section, we shall deal with some special types of equations which can be solved easily by some special methods, other than the general method.

39.9.1 STANDARD FORM (1) : EQUATION INVOLVING ONLY p AND q AND NO x, y, z :

The complete integral of equation of the type $f(p,q)=0$, *i.e.*, in which x,y,z do not occur, is

$$z = ax+by+c \quad ...(1)$$

where a and b are connected by the relation

$$f(a,b)=0 \quad ...(2)$$

Since, we have $p = \frac{\partial z}{\partial x} = a$ and $q = \frac{\partial z}{\partial y} = b$, which on substitution in (2) becomes the given equation.

Let us suppose from (2), $b = g(a)$ and replacing c by $\phi(a)$, the general solution is obtained by eliminating 'a' from the following equation

$$z = ax+g(a)y+\phi(a) \quad ...(3)$$

Differentiating (3) with respect to a, we get

$$0 = x+yg'(a)+\phi'(a) \quad ...(4)$$

Now, to find the singular integral, differentiating $z = ax+g(a)y+c$ with respect to a and c, we get

$0 = x+yg'(a)$ and $0 = 1 \Rightarrow$ There is no singular solution.

39.9.2 STANDARD FORM (II) : EQUATION INVOLVING ONLY P, Q AND Z

The equation which do not contain x and y, *i.e.*, which are of the form

$$f(z,p,q)=0 \quad ...(1)$$

Equation (1) can be solved in the following way :

Write $X = x + ay$, where a is an arbitrary constant and assume z to be function of $(x + ay)$, i.e., of X alone

$$\therefore \qquad z = f(X) = f(x+ay)$$

$$\Rightarrow \qquad p = \frac{\partial z}{\partial x} = \frac{dz}{dX}\cdot\frac{\partial X}{\partial x} = \frac{dz}{dX}.1 \qquad \text{and} \qquad q = \frac{\partial z}{\partial y} = \frac{dz}{dX}\cdot\frac{\partial X}{\partial y} = a\frac{dz}{dX}.$$

Now, the equation (1), becomes $F\left(z, \frac{dz}{dX}, a\frac{dz}{dX}\right) = 0$, which is an ordinary differential equation of the first order and can be integrated. So the complete integral will be known. If $f = 0$ is the complete integral involving two constants a and b, then replacing b by $f(a)$, the general integral is obtained by eliminating a form $f = 0, \frac{df}{da} = 0$. The singular integral is obtained by eliminating a and b, from $f = 0, \partial f/\partial a = 0$ and $\partial f/\partial b = 0$.

Solved Examples

Example 1. *Solve $p^2 + q^2 = 1$.* (Osmania–2005)

Solution. The given equation is of the form $f(p,q) = 0$.

The solution is given by $z = ax + by + c$, where a and b are related by $f(a,b) = 0$

$$\Rightarrow \quad a^2 + b^2 = 1 \quad \Rightarrow \quad b = \sqrt{(1-a^2)}$$

Hence, the complete integral is

$$z = ax + \sqrt{(1-a^2)}\, y + c.$$

For the general integral write $c = \phi(a)$.

Then it is obtained by eliminating a from

$$z = ax + \sqrt{(1-a^2)}\, y + \phi(a)$$

$$\text{and} \quad 0 = x + \frac{-a}{\sqrt{(1-a^2)}}y + \phi'(a).$$

Example 2. *Solve $x^2p^2 + y^2q^2 = z^2$.*

Solution. The given equation can be written as

$$\left(\frac{x}{z}\cdot\frac{\partial z}{\partial x}\right)^2 + \left(\frac{y}{z}\cdot\frac{\partial z}{\partial y}\right)^2 = 1 \qquad \ldots(1)$$

Putting $\frac{1}{z}dz = dZ$, i.e., $z = e^Z$;

$\frac{1}{x}dx = dX$,i.e., $x = e^X$

and $\frac{1}{y}dy = dY$ i.e., $y = e^Y$

in (1), we get $\left[\frac{\partial Z}{\partial X}\right]^2 + \left[\frac{\partial Z}{\partial Y}\right]^2 = 1$, which is of the type $f(p,q) = 0$.

Therefore, the complete integral is given by $Z = aX + bY + c_1$, where a and b are related by $a^2 + b^2 = 1$.

$$\Rightarrow \; b = \sqrt{(1-a^2)} \quad \Rightarrow z = aX + \sqrt{(1-a^2)}Y + c_1$$

$$\Rightarrow \log z = a\log x + \sqrt{(1-a^2)}\log y + c_1.$$

To find the general solution put $a = \cos\theta$.

$$\Rightarrow \quad \log z = \cos\theta \log x + \sin\theta \log y + \log c$$

$$\Rightarrow \quad z = cx^{\cos\theta}.y^{\sin\theta}.$$

Now, we eliminate θ from

$$z = g(\theta)x^{\cos\theta}y^{\sin\theta} \qquad (\text{take } c = g(\theta))$$

$$\text{and} \quad 0 = g'(\theta)x^{\cos\theta}y^{\sin\theta} + g(\theta)x^{\cos\theta}y^{\sin\theta}(-\sin\theta)\log_e x + g(\theta)x^{\cos\theta}y^{\sin\theta}\cos\theta\log_e y$$

which is the required general solution.

To find singular integral, we eliminate θ and c, from $z = cx^{\cos\theta}.y^{\sin\theta}$

$$\Rightarrow \quad \frac{\partial z}{\partial \theta} = -c\sin\theta x^{\cos\theta}y^{\sin\theta}\log_e x + c\cos\theta.x^{\cos\theta}.y^{\sin\theta}\log_e y = 0$$

$$\text{and} \quad \frac{\partial z}{\partial c} = x^{\cos\theta}.y^{\sin\theta} = 0$$

$\Rightarrow$ $z = 0$ is the singular integral of the given equation.

Example 3. *Find the complete integral of*

$$(y-x)(qy-px) = (p-q)^2.$$

Solution. Substitute $X = x + y$ and $Y = xy$

so that $p = \frac{\partial z}{\partial x} = \frac{\partial z}{\partial X}\cdot\frac{\partial X}{\partial x} + \frac{\partial z}{\partial Y}\cdot\frac{\partial Y}{\partial x} = \frac{\partial z}{\partial X} + y\frac{\partial z}{\partial Y}$

and $q = \frac{\partial z}{\partial y} = \frac{\partial z}{\partial X}\cdot\frac{\partial X}{\partial y} + \frac{\partial z}{\partial Y}\cdot\frac{\partial Y}{\partial y} = \frac{\partial z}{\partial X} + x\frac{\partial z}{\partial Y}$

in the given equation, we get

$$(y-x)(y-x)\frac{\partial z}{\partial X} = (y-x)^2\left(\frac{\partial z}{\partial Y}\right)^2$$

$\Rightarrow \frac{\partial z}{\partial X} = \left(\frac{\partial z}{\partial Y}\right)^2$, which is the standard form (1).

Then, the complete solution is $z = aX + bY + C$, where $a = b^2$.

$$\Rightarrow \qquad z = b^2(x+y) + bxy + C.$$

Example 4. *Find the complete integral of* $p^3 + q^3 = 27z$.

Solution. The given equation is $p^3 + q^3 = 27z$ which is in the standard form $f(p,q,z) = 0$.

Put $X = x + ay \quad \Rightarrow z = f(X) = f(x+ay)$

$$\Rightarrow p = \frac{\partial z}{\partial x} = \frac{dz}{dX} \text{ and } q = \frac{\partial z}{\partial y} = a\frac{dz}{dX}.$$

We may take $\frac{dz}{dX}$ in place of $\frac{\partial z}{\partial x}$ because z is a function of x only.

Hence, the given equation reduces to

$$(1+a^3)\left(\frac{dz}{dX}\right)^3 = 27z$$

$$\Rightarrow \quad (1+a^3)^{1/3}\frac{dz}{dX} = 3z^{1/3}$$

$$\Rightarrow \quad (1+a^3)^{1/3}.\frac{2}{3}z^{-1/3}dz = 2dX.$$

On integrating, we get

$$z^{2/3}(1+a^3)^{1/3} = 2X + c = 2(X+b)$$

$$\Rightarrow \quad (1+a^3)z^2 = 8(x+ay+b)^3 \quad \ldots(1)$$

which is the complete integral of the given equation.

To find the singular integral, differentiating (1) partially with respect to a and b, we get

$$3a^2z^2 = 24y(x+ay+b)^2 \quad \ldots(2)$$

and

$$0 = 24(x+ay+b)^2 \quad \ldots(3)$$

By eliminating a, b from (1), (2) and (3), we get $z = 0$, which is the required singular solution.

Example 5. *Find complete solution of* $z^2(p^2z^2 + q^2) = 1$.

(Bhopal–2008)

Solution. The given equation is $z^2(p^2z^2 + q^2) = 1$.

Putting $z = f(x+ay) = f(X)$,

where $X = x + ay$

So that $p = \frac{\partial z}{\partial x} = \frac{dz}{dX}$ and $q = \frac{\partial z}{\partial y} = a\frac{dz}{dX}$

In the given equation, we get

$$z^2\left[\left(\frac{dz}{dX}\right)^2.z^2 + a^2\left(\frac{dz}{dX}\right)^2\right] = 1$$

or

$$z^2(z^2+a^2)\left(\frac{dz}{dX}\right)^2 = 1.$$

On separating the variables, we get

$$z\sqrt{(z^2+a^2)}dz = dX.$$

On integrating, we get $\frac{1}{3}(z^2+a^2)^{3/2} = X + b$.

$$\Rightarrow \quad 9(x+ay+b)^2 = (z^2+a^2)^3 \quad \ldots(1)$$

Equation (1) gives the required complete solution.

To find the singular integral differentiating (1), partially with respect to a and b, we get

$$18(x+ay+b)y = 6(z^2+a^2)^2.a \quad \ldots(2)$$

and

$$18(x+ay+b) = 0 \quad \ldots(3)$$

From (2) and (3), we get $a = 0$. Hence, from (1), (3) and (4), we get $z = 0$.

Now since $z = 0 \Rightarrow p = 0, q = 0$ does not satisfy the given equation, therefore, the given equation does not have any singular solution.

EXERCISE 39.3

1. Solve $q = 3p^2$.
2. Solve $(x+y)(p+q)^2 + (x-y)(p-q)^2 = 1$ (IAS–1991, Bhopal–2006, Rajasthan–2006, VTU–2003)
3. Find the complete integral of $p^m \sec^{2m} x + z^l q^n \operatorname{cosec}^{2n} y = z^{lm/m-n}$.
4. Solve $(x^2+y^2)(p^2+q^2) = 1$.
5. Solve $p^2 + q^2 = npq$.
6. Solve $\sqrt{p} + \sqrt{q} = 1$.
7. Find the complete integral of $p^2 = zq$.
8. Solve $pz = (1+q^2)$.
9. Solve $9(p^2z + q^2) = 4$. (IAS–1988)
10. Solve $p(1+q^2) = q(z-a)$.
11. Solve $p^2 = z^2(1-pq)$.
12. Solve $pq = x^m y^n z^l$. (IAS–1989, 94, 2000)
13. Solve $z^2(p^2+q^2+1) = c^2$.
14. Find a complete integral of $p^3 + q^3 - 3pqz = 0$.

Hint to Selected Problems

1. Put $p = a, q = b$ in the given equation.
2. Put $x + y = X^2, x - y = Y^2$.

   Therefore $p = \frac{\partial z}{\partial x} = \frac{\partial z}{\partial X}.\frac{\partial X}{\partial x} + \frac{\partial z}{\partial Y}.\frac{\partial Y}{\partial x}$

   $$= \frac{1}{2X}\frac{\partial z}{\partial X} + \frac{1}{2Y}.\frac{\partial z}{\partial Y}.$$

   Similarly $q = \frac{1}{2X}\frac{\partial z}{\partial X} - \frac{1}{2Y}.\frac{\partial z}{\partial Y}$. Finally, putting all these values in the given equation.
3. The given equation can be written as

   $$\left(\frac{z^{-l/(m-n)}}{\cos^2 x}.\frac{\partial z}{\partial x}\right)^m + \left(\frac{z^{-l/(m-n)}}{\sin^2 y}.\frac{\partial z}{\partial y}\right)^n = 1.$$

Now putting

$$z^{-l/(m-n)}dz = dZ \Rightarrow Z = \frac{m-n}{(m-n-l)} z^{(m-n-l)/(m-n)}$$

$$\cos^2 x dx = dX \Rightarrow X = \frac{1}{2}\left(x + \frac{1}{2}\sin 2x\right)$$

and $\sin^2 y dy = dY \Rightarrow Y = \frac{1}{2}\left(y - \frac{1}{2}\sin 2y\right).$

7. Putting $z = f(x+ay) = f(x)$ in the given equation.

8. Do same as (7).

9. Putting $p = \frac{dz}{dX}$ and $q = a\frac{dz}{dX}$ in the given equation.

12. Putting $\frac{x^{m+1}}{m+1} = X, \frac{y^{n+1}}{n+1} = Y$,i.e., $p = x^m \frac{\partial z}{\partial X}$ and $q = y^n \frac{\partial z}{\partial Y}$ in the given equation.

14. Putting $\frac{dz}{dX} = p$ and $a\frac{dz}{dX} = q$ in the given equation.

ANSWERS

1. $z = ax + 3a^2 y + c.$

2. $z = a\sqrt{(x+y)} + \sqrt{(1-a^2)}\sqrt{(x-y)} + c.$

3. $\frac{(m-n)}{(m-n-l)} z^{(m-n-l)(m-n)} = \frac{a}{2}\left(x + \frac{1}{2}\sin 2x\right) + (1-a^m)^{1/n} \cdot \frac{1}{2}\left(y - \frac{\sin 2y}{2}\right) + c$

4. $z = \frac{a}{2}\log(x^2+y^2) + \sqrt{(1-a^2)}\tan^{-1}\frac{y}{x} + c$

5. $z = ax + \left(\frac{n \pm \sqrt{(n^2-4)}}{2}\right).ay + c$

6. $z = ax + (1-\sqrt{a})^2 y + c$

7. $z = be^{(ax+a^2 y)}$

8. $z^2 \mp [z\sqrt{(z^2-4a^2)} - 4a^2\log\{z + \sqrt{(z^2-4a^2)}\}] = 4x + 4ay + 2c$

9. $(z+a^2)^3 = (x+ay+b)^2$

10. $4a(z-a) = 4 + (x+ay+b)^2$

11. $\frac{1}{\sqrt{a}}\log[z\sqrt{a} + \sqrt{(1+az^2)}] + \sqrt{(1+az^2)} = X + C$

12. $\frac{z^{-l/2}+1}{-l/2+1} = \frac{1}{\sqrt{a}}\left[\frac{x^{m+1}}{m+1} + a\frac{y^{n+1}}{n+1}\right] + b$

13. $(1+a^2)(c^2-z^2) = (x+ay+b)^2$

14. $3a(x+ay) + c = (1+a^3)\log z.$

39.9.3 STANDARD FORM III : EQUATION OF THE FORM $f_1(x,p) = f_2(y,q)$

If the given equation is of the type $f_1(x,p) = f_2(y,q)$ ...(1)

then, first write $f_1(x,p) = f_2(y,q) = c_1$ ...(2)

Now, solving (2) for q and p, we get

$$p = \frac{\partial z}{\partial x} = g_1(x,c_1) \quad \text{and} \quad q = \frac{\partial z}{\partial y} = g_2(y,c_1).$$

Now $dz = pdx + qdy = g_1(x,c_1)dx + g_2(y,c_1)dy$

which gives $z = \int g_1(x,c_1)dx + g_2(y,c_1)dy + b.$

The general solution may be obtained from this complete integral. Also, there is no singular solution.

39.9.4 STANDARD FROM IV : EQUATION OF THE FORM $z = px + qy + f(p,q)$

The equation $z = px + qy + f(p,q)$ ...(1)

which is analogous to Clairaut's form, has its complete integral

$$z = ax + by + f(a,b) \quad ...(2)$$

For $\frac{\partial z}{\partial x} = p = a$ and $\frac{\partial z}{\partial y} = q = b$

In order to obtain the general solution, put $b = g(a)$

Therefore, $z = ax + y\,g(a) + f\{a, g(a)\}$ ...(3)

Differentiating (3) with respect to a, we get

$$0 = x + yg'(a) + f'(a) \quad ...(4)$$

Now, eliminate a from (3) and (4) and get the required general solution.

To obtain the singular solution, differentiating (2) with respect to a and b, which gives

$$0 = x + \frac{\partial f}{\partial a} \quad ...(5)$$

$$0 = y + \frac{\partial f}{\partial b} \quad ...(6)$$

Now, eliminate a and b between the equation (2), (5) and (6).

Solved Examples

Example 1. *Solve* $p^2 + q^2 = x + y$. (Bhopal–2006, Madras–2003)

Solution. The given equation can be written as

$$p^2 - x = y - q^2.$$

Let us write $p^2 - x = y - q^2 = a \Rightarrow p = \sqrt{(x+a)}$ and $q = \sqrt{(y-a)}$.

Now, putting the values of p and q in $dz = pdx + qdy$, we get

$$dz = \sqrt{(x+a)}\,dx + \sqrt{(y-a)}\,dy.$$

On integrating, we have

$$z = \frac{2}{3}(x+a)^{3/2} + \frac{2}{3}(y-a)^{3/2} + b.$$

Example 2. *Solve* $q = xyp^2$.

Solution. The given equation can be written as $p^2 x = q/y$.

Let us write $p^2 x = \frac{q}{y} = a$, which gives $p = \sqrt{(a/x)}$ and $q = ay$.

Now, putting the values of p and q in $dz = pdx + qdy$, we get

$$dz = \sqrt{(a/x)}\,dx + (ay)\,dy.$$

On integrating, we have $z = 2\sqrt{a}\sqrt{x} + \frac{a}{2}y^2 + b$

$$\Rightarrow \quad (2z - ay^2 - 2b)^2 = 16ax.$$

Example 3. *Solve* $z^2(p^2 + q^2) = x^2 + y^2$.

Solution. The given equation is

$$z^2(p^2 + q^2) = x^2 + y^2.$$

Replace $zdz = dZ \quad \Rightarrow \quad \frac{z^2}{2} = Z.$

Therefore, the given equation becomes

$P^2 + Q^2 = x^2 + y^2$, where $P = \frac{dZ}{dx}$ and $Q = \frac{dZ}{dy}$

$$\Rightarrow \quad P^2 - x^2 = y^2 - Q^2.$$

Let us write $P^2 - x^2 = y^2 - Q^2 = a$

$$\Rightarrow \quad P = \sqrt{(a + x^2)};\quad Q = \sqrt{(y^2 - a)}.$$

Now, putting the values of P and Q in $dZ = P\,dx + Q\,dy$, we get

$$dZ = \sqrt{(a + x^2)}dx + \sqrt{(y^2 - a)}\,dy.$$

On integrating, we have

$$Z = \frac{x}{2}\sqrt{(a+x^2)} + \frac{a}{2}\log\{x + \sqrt{(a+x^2)}\} + \frac{y}{2}\sqrt{(y^2-a)} - \frac{a}{2}\log\{y + \sqrt{y^2 - a)}\} + b$$

$$\Rightarrow \quad z^2 = x\sqrt{(a+x^2)} + a\log\{x + \sqrt{(a+x^2)}\} + y\sqrt{(y^2-a)} - a\log\{y + \sqrt{(y^2-a)}\} + c$$

Example 4. *Solve* $z = px + qy + c\sqrt{(1 + p^2 + q^2)}$. (Anna–2009)

Solution. The given equation is of the standard form IV. Therefore, the complete solution is

$$z = ax + by + c\sqrt{(1 + a^2 + b^2)} \qquad \ldots(1)$$

To find the singular solution, differentiating (1) partially with respect to a and b, we have

$$0 = x + \frac{ac}{\sqrt{(1+a^2+b^2)}}$$

$$\Rightarrow \quad a = \frac{-x}{\sqrt{(c^2 - x^2 - y^2)}} \qquad \ldots(2)$$

and

$$0 = y + \frac{bc}{\sqrt{(1+a^2+b^2)}}$$

$$\Rightarrow \quad b = \frac{-y}{\sqrt{(c^2 - x^2 - y^2)}} \qquad \ldots(3)$$

which gives $x^2 + y^2 = \frac{(a^2+b^2)\,c^2}{1 + a^2 + b^2}$

$$\Rightarrow \quad (c^2 - x^2 - y^2) = \frac{c^2}{1 + a^2 + b^2}$$

$$\Rightarrow \quad (1 + a^2 + b^2) = \frac{c^2}{(c^2 - x^2 - y^2)} \qquad \ldots(4)$$

Now using (2), (3) and (4), (1) becomes

$$z = \frac{-x^2}{\sqrt{(c^2 - x^2 - y^2)}} - \frac{y^2}{\sqrt{(c^2 - x^2 - y^2)}} + \frac{c^2}{\sqrt{(c^2 - x^2 - y^2)}}$$

$$= \frac{(c^2 - x^2 - y^2)}{\sqrt{(c^2 - x^2 - y^2)}} = \sqrt{(c^2 - x^2 - y^2)}$$

$$\Rightarrow \quad z^2 = c^2 - x^2 - y^2 \Rightarrow x^2 + y^2 + z^2 = c^2.$$

Example 5. *Find the singular solution of* $z = px + qy + \log pq$.

Solution. The given equation is of the standard form IV. Therefore, the complete integral is

$$z = ax + by + \log ab \qquad \ldots(1)$$

To find the singular solution differentiating (1) partially with respect to a and b, we get

$$0 = x + \frac{1}{a} \Rightarrow a = -\frac{1}{x} \text{ and } 0 = y + \frac{1}{b} \Rightarrow b = -\frac{1}{y}.$$

Now, putting the values of a and b in (1), the required singular integral is

$$z = -1 - 1 + \log\frac{1}{xy} \Rightarrow z = -2 - \log xy.$$

Example 6. *Solve* $z = px + qy - 2\sqrt{(pq)}$.

Solution. Here, the given equation is of the standard form IV.

Therefore, the complete integral

$$z = ax + by - 2\sqrt{(ab)} \quad \ldots(1)$$

To find the singular solution, differentiating (1) partially with respect to a and b, we get

$$0 = x - \frac{2}{2\sqrt{(ab)}} \Rightarrow x = \sqrt{(b/a)} \quad \ldots(2)$$

and $$0 = y - \frac{2}{2\sqrt{(ab)}} \Rightarrow y = \sqrt{(a/b)} \quad \ldots(3)$$

From (2) and (3), we get $xy = 1$, which is the required singular solution.

EXERCISE 39.4

Solve the following equations :

1. $\sqrt{p} + \sqrt{q} = 2x$
2. $pe^y = qe^x$
3. $pq = xy$
4. $py = 2yx + \log q$
5. $z(p^2 - q^2) = (x - y)$
6. Find the complete integral of $z = px + qy + p^2 + q^2$
7. $z = px + qy - 2p - 3q$
8. $z = px + qy - p^2q$
9. $z = px + qy + pq$ (GBTU(AG)–2012)
10. $x(1+y)p = y(1+x)q$
11. $z = px + qy + \sqrt{\alpha p^2 + \beta q^2 + \gamma}$
12. $x^2y^3z^{-3}p^2q = 1$
13. $p - 3x^2 = q^2 - y$
14. $\sin px \cos y - \cos px \sin y = p$

Hint to Selected Problems

1. The given equation can be written as $\sqrt{p} - 2x = -\sqrt{q}$.
   Now, let $\sqrt{p} - 2x = a = -\sqrt{q}$.
2. Putting $pe^{-x} = qe^{-y} = a$.
3. Putting $p = ax, q = y/c$.
4. Putting $p - 2x = \dfrac{1}{y}\log q = a$.
5. The given equation can be written as
   $$\left(\sqrt{z}\frac{dz}{dx}\right)^2 - \left(\sqrt{z}\frac{\partial z}{\partial x}\right)^2 = x - y.$$
   Then putting $\sqrt{z}dz = dZ$.
12. The given equation can be written as $x^2y^3p^2q = z^3$.
    Now putting $Z = \log z \Rightarrow \dfrac{\partial Z}{\partial X} = \dfrac{1}{z}\dfrac{\partial z}{\partial X}, \dfrac{\partial Z}{\partial Y} = \dfrac{1}{z}\dfrac{\partial z}{\partial Y}$
13. Let $p - 3x^2 = q^2 - y = a$
    $$\Rightarrow \quad p = a + 3x^3,\ q = \sqrt{a+y}.$$
    Putting the values of p and q in $dz = p\,dx + q\,dy$.

ANSWERS

1. $z = \frac{1}{6}(a+2x)^3 + a^2y + b$
2. $z = ae^x + ae^y + b$
3. $z = \frac{1}{2a}(a^2x^2 + y^2 + 2ab)$
4. $z = \frac{1}{a}(ax^2 + a^2x + e^{ay} + a.b)$
5. $z^{3/2} = (x+a)^{3/2} + (y+a)^{3/2} + c$
6. $z = ax + by + a^2 + b^2$
7. $z = ax + by - 2a - 3b$
8. $z = ax + by - a^2b$
9. $z = ax + by + ab$
10. $z + c = a(x + y + \log xy)$
11. $z = ax + by + \sqrt{(\alpha a^2 + \beta b^2 + \gamma)}$
12. $\log z - \sqrt{a}\log x = \dfrac{1}{2ay^2} + c$
13. $z = x^3 + ax \pm \frac{2}{3}(y+a)^{3/2} + b$
14. $y = x\cos\sin^{-1}c$

Objective Evaluations

Fill in the Blanks

1. A PDE must contain __________ derivative.
2. The order of the PDE is the order of its __________ derivative.
3. The degree of the PDE is the degree of its __________ derivative.
4. A PDE is said to be __________ if it is of the first degree in the dependent variable and its partial derivative.
5. Quasi-linear equation is also called __________ .
6. The solution of a PDE involves arbitrary __________ .
7. If each term of a PDE contains either the dependent variable or one of its derivative is known as __________ .
8. The complete integral of the equation of the type $f(p, q) = 0$ is $z = ax + by + c$, where a and b are connected by the relation __________ .
9. The equations $f_1(x, y, z, p, q) = 0$ and $f_2(x, y, z, p, q) = 0$ are said to be compatible if $(f_1, f_2) = 0$ __________.
10. The equation of the type $f_1(x, p) = f_2(y, p)$ does not have any __________ solution.

True/False

Write 'T' for True and 'F' for False statement.

1. A partial differential equation does not contain any partial derivative. **(T/F)**
2. The second order partial differential equation may also contain first order terms. **(T/F)**
3. A linear PDE does not contain the product of dependent variables and either of its partial derivatives. **(T/F)**
4. Quasi-linear equation is a linear equation. (T/F)
5. The equation of the type $f(p, q) = 0$ having so singular solution. **(T/F)**

Multiple Choice Questions

Choose the most appropriate one.

1. The equation of the envelope of the surfaces represented by the complete integral of the given PDE is called :
   (a) particular integral
   (b) singular integral
   (c) general solution
   (d) none of these
2. The complete integral of $z = px + qy + p^2 + q^2$ is :
   (a) $z = ax + by$
   (b) $z = a^2 + b^2$
   (c) $z = ax + by + a^2 + b^2$
   (d) none of these
3. The complete integral of $p = e^q$ is :
   (a) $a = e^b$
   (b) $b = e^a$
   (c) $z = e.a$
   (d) $z = ax + y \log a + c$
4. The solution of a PDE contains :
   (a) arbitrary functions
   (b) arbitrary constant
   (c) both (a) and (b)
   (d) none of these
5. If the number of arbitrary constants is more than the number of independent variables, then the PDE thus obtained will be of :
   (a) lower degree than the first
   (b) higher degree than the first
   (c) equal degree
   (d) none of these

Answers

Fill in the Blanks

1. partial 2. highest 3. highest 4. linear 5. uniform non-linear equation
6. functions 7. homogeneous 8. $f(a, b) = 0$ 9. 0 10. singular

True/False

1. F 2. T 3. T 4. F 5. T

Multiple Choice Questions

1. (b) 2. (c) 3. (d) 4. (a) 5. (b)

FFFFFF

CHAPTER

40 Linear and Non-Linear Partial Differential Equations

40.1 INTRODUCTION

Partial differential equation arise in connection with various type of physical and geometrical problems when the functions are implicit, that is, functions of two or more than two independent variables.

Partial differential equation is an equation having one or more partial derivatives of an unknown function of two or more than two independent variables, we say that a partial differential equation is linear if it is of the first degree in the dependent variable and its partial derivatives. If each term of the differential equation contains either the dependent variable or one of its derivatives, then the equation is called homogeneous, otherwise it is called non-homogeneous.

In this chapter, we shall discuss some methods : Lagrange's and Charpit's to solve partial differential equations.

40.2 LAGRANGE'S LINEAR DIFFERENTIAL EQUATION

Consider the partial differential equation of the type $Pp + Qq = R$, where P, Q, R are the functions of x, y and z and $p = \partial z / \partial x, q = \partial z / \partial y$. Then this partial differential equation of order one is called Lagrange's Linear Differential Equation.

40.2.1 LAGRANGE'S AUXILIARY EQUATION

Let u and v be two functions of x, y, z which are related by the relation

$$f(u, v) = 0 \qquad \ldots(1)$$

Differentiating (1) partially w.r.t. x and y, we get

$$\frac{\partial f}{\partial u}\left(\frac{\partial u}{\partial x}+\frac{\partial u}{\partial z}\cdot\frac{\partial z}{\partial x}\right)+\frac{\partial f}{\partial v}\left(\frac{\partial v}{\partial x}+\frac{\partial v}{\partial z}\cdot\frac{\partial z}{\partial x}\right)=0 \quad \text{or} \quad \frac{\partial f}{\partial u}\left(\frac{\partial u}{\partial x}+\frac{\partial u}{\partial z}p\right)+\frac{\partial f}{\partial v}\left(\frac{\partial v}{\partial x}+\frac{\partial v}{\partial z}.p\right)=0 \qquad \ldots(2)$$

and $$\frac{\partial f}{\partial u}\left(\frac{\partial u}{\partial y}+\frac{\partial u}{\partial z}\cdot\frac{\partial z}{\partial y}\right)+\frac{\partial f}{\partial v}\left(\frac{\partial v}{\partial y}+\frac{\partial v}{\partial z}\cdot\frac{\partial z}{\partial y}\right)=0 \quad \text{or} \quad \frac{\partial f}{\partial u}\left(\frac{\partial u}{\partial y}+\frac{\partial u}{\partial z}q\right)+\frac{\partial f}{\partial v}\left(\frac{\partial v}{\partial y}+\frac{\partial v}{\partial z}.q\right)=0 \qquad \ldots(3)$$

Eliminating $\frac{\partial f}{\partial u}$ and $\frac{\partial f}{\partial v}$ from (2) and (3), we get

From (2), $$\frac{\partial f / \partial u}{\partial f / \partial v} = -\frac{\left(\frac{\partial v}{\partial x}+\frac{\partial v}{\partial z}p\right)}{\left(\frac{\partial u}{\partial x}+\frac{\partial u}{\partial z}p\right)} \qquad \ldots(4)$$

From (3), $$\frac{\partial f / \partial u}{\partial f / \partial v} = -\frac{\left(\frac{\partial v}{\partial y}+\frac{\partial v}{\partial z}q\right)}{\left(\frac{\partial u}{\partial y}+\frac{\partial u}{\partial z}q\right)} \qquad \ldots(5)$$

From (4) and (5), we get

$$\left(\frac{\partial u}{\partial x}+\frac{\partial u}{\partial z}p\right)\left(\frac{\partial v}{\partial y}+\frac{\partial v}{\partial z}q\right)=\left(\frac{\partial u}{\partial y}+\frac{\partial u}{\partial z}q\right)\left(\frac{\partial v}{\partial x}+\frac{\partial v}{\partial z}p\right)$$

Solving this equation, we get

$$\left(\frac{\partial u}{\partial y}\cdot\frac{\partial v}{\partial z}-\frac{\partial v}{\partial y}\cdot\frac{\partial u}{\partial z}\right)p+\left(\frac{\partial v}{\partial x}\cdot\frac{\partial u}{\partial z}-\frac{\partial u}{\partial x}\cdot\frac{\partial v}{\partial z}\right)q=\left(\frac{\partial u}{\partial x}\cdot\frac{\partial v}{\partial y}-\frac{\partial u}{\partial y}\cdot\frac{\partial v}{\partial x}\right) \quad \text{or} \quad Pp+Qq=R \qquad ...(6)$$

where $P=\frac{\partial u}{\partial y}\cdot\frac{\partial v}{\partial z}-\frac{\partial v}{\partial y}\cdot\frac{\partial u}{\partial z}=\frac{\partial(u,v)}{\partial(y,z)}$, (Jacobian of u and v w.r.t. y and z)

$$Q=\frac{\partial v}{\partial x}\cdot\frac{\partial u}{\partial z}-\frac{\partial u}{\partial x}\cdot\frac{\partial v}{\partial z}=\frac{\partial(u,v)}{\partial(z,x)} \quad \text{and} \quad R=\frac{\partial u}{\partial x}\cdot\frac{\partial v}{\partial y}-\frac{\partial u}{\partial y}\cdot\frac{\partial v}{\partial x}=\frac{\partial(u,v)}{\partial(x,y)}$$

Thus $f(u, v) = 0$ is the general integral of the differential equation $Pp+Qq=R$. Now we shall determine the values of u and v. For this, let $u = a$ and $v = b$ be two equations, where a and b are arbitrary constants. That is $u(x,y,z) = a$ and $v(x, y, z) = b$.

This implies $du = 0$ and $dv = 0$

But $\qquad du=\frac{\partial u}{\partial x}dx+\frac{\partial u}{\partial y}dy+\frac{\partial u}{\partial z}dz \quad$ and $\quad dv=\frac{\partial v}{\partial x}dx+\frac{\partial v}{\partial y}dy+\frac{\partial v}{\partial z}dz$

Thus, we obtained

$$\frac{\partial u}{\partial x}dx+\frac{\partial u}{\partial y}dy+\frac{\partial u}{\partial z}dz=0 \qquad ...(7)$$

and

$$\frac{\partial v}{\partial x}dx+\frac{\partial v}{\partial y}dy+\frac{\partial v}{\partial z}dz=0 \qquad ...(8)$$

Solving, (7) and (8) by cross multiplication method for dx, dy and dz, we get

$$\frac{dx}{\frac{\partial u}{\partial y}\cdot\frac{\partial v}{\partial z}-\frac{\partial v}{\partial y}\cdot\frac{\partial u}{\partial z}}=\frac{dy}{\frac{\partial u}{\partial z}\cdot\frac{\partial v}{\partial x}-\frac{\partial u}{\partial x}\cdot\frac{\partial v}{\partial z}}=\frac{dz}{\frac{\partial u}{\partial x}\cdot\frac{\partial v}{\partial y}-\frac{\partial u}{\partial y}\cdot\frac{\partial v}{\partial x}} \text{ or } \frac{dx}{\frac{\partial(u,v)}{\partial(y,z)}}=\frac{dy}{\frac{\partial(u,v)}{\partial(z,x)}}=\frac{dz}{\frac{\partial(u,v)}{\partial(x,y)}} \text{ or } \frac{dx}{P}=\frac{dy}{Q}=\frac{dz}{R} \qquad ...(9)$$

Thus equation (9) is known as Lagrange's auxiliary equation or Lagrange's subsidiary equations.

40.2.2 GEOMETRICAL INTERPRETATION OF LAGRANGE'S LINEAR DIFFERENTIAL EQUATION

Lagrange's Linear differential equation is $Pp+Qq=R$ ...(1)

where $p=\frac{\partial z}{\partial x}, q=\frac{\partial z}{\partial y}$ and P,Q,R are the functions of x, y and z.

Equation (1) can be written as

$$Pp+Qq-R=0 \quad \text{or} \quad Pp+Qq+R(-1)=0 \qquad ...(2)$$

Lagrange's auxiliary equations are

$$\frac{dx}{P}=\frac{dy}{Q}=\frac{dz}{R}. \qquad ...(3)$$

These equation represent a family of curves and P,Q,R are the direction ratio of the tangent drawn at any point on the curves.

Since $f(u, v) = 0$ represents a surface through these curves, where $u = a$ (constant) and $v = b$ (constant) are the two particular integrals of the equation (3) and are the function of x, y and z. Further since, we know that the direction cosines of the normal to the surface $f(x,y,z)=0$ at any point on it are proportional to $\frac{\partial f}{\partial x}:\frac{\partial f}{\partial y}:\frac{\partial f}{\partial z}$

Dividing by $\frac{\partial f}{\partial z}$, we get $\frac{\partial f/\partial x}{\partial f/\partial z}:\frac{\partial f/\partial y}{\partial f/\partial z}:1$ ...(4)

Since $p=\frac{\partial z}{\partial x}=-\frac{\partial f/\partial x}{\partial f/\partial z}$ and $q=\frac{\partial z}{\partial y}=-\frac{\partial f/\partial y}{\partial f/\partial z}$, then (4) becomes $-p:-q:1$ or $p:q:-1$

Thus equation (2) represents that the normal at any point on the surface is perpendicular to the tangent to the curve obtained by equation (3) through which this surface passes. Hence, we say that the equations (1) and (3) give the same equivalent surfaces.

40.3 LAGRANGE'S LINEAR DIFFERENTIAL EQUATION WITH MORE THAN TWO INDEPENDENT VARIABLES

Let us consider a Lagrange's linear differential equation with u_1 independent variables :

$$P_1p_1+P_2p_2+...+P_np_n=R \qquad ...(1)$$

where $p_i=\frac{\partial z}{\partial x_i}$ for $i=1,2,3,...,n$ and $P_1,P_2,...,P_n$ and R are the functions of $x_1,x_2,...,x_n$ and z. Lagrange's auxiliary equations are

$$\frac{dx_1}{P_1}=\frac{dx_2}{P_2}=...=\frac{dx_n}{P_n}=\frac{dz}{R} \qquad ...(2)$$

Let $u_1 = a_1$ (constant), $u_2 = a_2$ (constant), ..., $u_n = a_n$ (constant) be n independent integrals of (2). Then the general integral of (1) is

$$f(u_1, u_2, ..., u_n) = 0.$$

WORKING PROCEDURE

To find the solution of the partial differential equation, use the following steps :

Step 1. *First change the linear partial differential equation into a standard form $Pp + Qq = R$.*

Step 2. *Find the Lagrange's auxiliary equations as follows :*

$$\frac{dx}{P} = \frac{dy}{Q} = \frac{dz}{R}.$$

Step 3. *Now find two independent integrals say $u = a$ (constant) and $v = b$ (constant) from auxiliary equations. Then we obtained the general integral of the given differential equation $f(u, v) = 0$.*

REMARKS

- Sometimes we use a set of multipliers to find $u = a$ and $v = b$.
- Set of multipliers may be functions of $x, y, z, \ldots$ etc. or may be constants.

Solved Examples

Example 1. *Solve the differential equation $yzp + zxq = xy$.*

Solution. Compare the given partial differential equation with $Pp + Qq = R$

we get $P = yz, Q = zx$ and $R = xy$.

Then the subsidiary equations are

$$\frac{dx}{P} = \frac{dy}{Q} = \frac{dz}{R} \quad \text{or} \quad \frac{dx}{yz} = \frac{dy}{zx} = \frac{dz}{xy} \qquad \ldots(1)$$

Taking the first two members of (1), we get

$$\frac{dx}{yz} = \frac{dy}{zx} \quad \text{or} \quad xdx - ydy = 0.$$

Integrating, we get $\quad x^2 - y^2 = c_1 \qquad \ldots(2)$

Now taking second and third members of (1), we get

$$\frac{dy}{zx} = \frac{dz}{xy} \quad \text{or} \quad ydy - zdz = 0.$$

Integrating, we get $\quad y^2 - z^2 = c_2 \qquad \ldots(3)$

Thus the general solution is

$$f(x^2 - y^2, y^2 - z^2) = 0.$$

Example 2. *Solve the partial differential equation*

$$pz - qz = z^2 + (x + y)^2.$$

Solution. Compare the given partial differential equation with the standard partial differential equation $Pp + Qq = R$, we get

$P = z, Q = -z$, and $R = z^2 + (x + y)^2$

The subsidiary equations are given by

$$\frac{dx}{P} = \frac{dy}{Q} = \frac{dz}{R} \Rightarrow \frac{dx}{z} = \frac{dy}{-z} = \frac{dz}{z^2 + (x + y)^2} \qquad \ldots(1)$$

Taking first and second ratio of (1), we get

$$\frac{dx}{z} = \frac{dy}{-z} \Rightarrow dx = -dy \Rightarrow dx + dy = 0$$

$$\Rightarrow x + y = c_1 \qquad \text{(On integrating)}$$

Now taking first and third ratio of (1), we get

$$\frac{dx}{z} = \frac{dz}{z^2 + (x + y)^2} \quad \text{or} \quad dx = \frac{zdz}{z^2 + (x + y)^2}$$

$$\text{or} \quad dx = \frac{zdz}{z^2 + c_1^2} \qquad [\because x + y = c_1]$$

On integrating, we get

$$2x = \log(z^2 + c_1^2) + \log c_2 \quad \text{or} \quad e^{2x} = c_2(z^2 + c_1^2)$$

$$e^{2x} = c_2[z^2 + (x + y)^2]$$

$$\Rightarrow \quad c_2 = \frac{e^{2x}}{x^2 + y^2 + 2xy + z^2}$$

Thus the general integral is given by

$$f\left(x + y, \frac{e^{2x}}{x^2 + y^2 + z^2 + 2xy}\right) = 0.$$

Example 3. *Solve $xzp + yzq = xy$.*

Solution. Compare the given differential equation with Lagrange's linear differential equation $Pp + Qq = R$, we get

$$P = xy, Q = yz, R = xy.$$

Then, the Lagrange's subsidiary equations are

$$\frac{dx}{P} = \frac{dy}{Q} = \frac{dz}{R} \Rightarrow \frac{dx}{xz} = \frac{dy}{yz} = \frac{dz}{xy} \qquad \ldots(1)$$

Taking first and second ratio of (1), we get

$$\frac{dx}{xz} = \frac{dy}{yz} \Rightarrow \frac{dx}{x} = \frac{dy}{y} \Rightarrow \frac{dx}{x} - \frac{dy}{y} = 0.$$

On integrating, we get $\log x - \log y = \log c_1$

or $\dfrac{x}{y} = c_1$.

Now, taking second and third ratio of (1), we get

$$\frac{dy}{yz} = \frac{dz}{xy} \quad \Rightarrow \quad \frac{dy}{z} = \frac{dz}{x}$$

$$\Rightarrow \quad xdy = zdz \quad \Rightarrow \quad c_1 ydy = zdz \quad [\because x = c_1 y]$$

On integrating, we get $c_1 y^2 - z^2 = c_2$

or $\left(\frac{x}{y}\right) y^2 - z^2 = c_2$

or $\quad xy - z^2 = c_2 \quad \left[\because c_1 = \frac{x}{y}\right]$

Thus the general integral is $f\left(\frac{x}{y}, xy - z^2\right) = 0$.

Example 4. *Solve* $p \tan x + q \tan y = \tan z$.

Solution. Compare the given differential equation with Lagrange's Linear differential equation $Pp + Qq = R$, we get

$$P = \tan x, Q = \tan y, R = \tan z.$$

Then, the Lagrange's subsidiary equations are

$$\frac{dx}{P} = \frac{dy}{Q} = \frac{dz}{R} \Rightarrow \frac{dx}{\tan x} = \frac{dy}{\tan y} = \frac{dz}{\tan z} \quad \ldots(1)$$

Taking first and second ratio of (1), we get

$$\frac{dx}{\tan x} = \frac{dy}{\tan y} \quad \Rightarrow \quad \frac{\cos x}{\sin x} dx = \frac{\cos y}{\sin y} dy.$$

On integrating, we get

$$\log \sin x - \log \sin y = \log c_1$$

or $\quad \log \frac{\sin x}{\sin y} = \log c_1$ or $\frac{\sin x}{\sin y} = c_1$.

Now taking second and third ratio of (1), we get

$$\frac{dy}{\tan y} = \frac{dz}{\tan z} \quad \Rightarrow \quad \frac{\cos y}{\sin y} dy = \frac{\cos z}{\sin z} dz.$$

On integrating, we get

$$\log \sin y - \log \sin z = \log c_2$$

or $\log \frac{\sin y}{\sin z} = \log c_2$ or $\frac{\sin y}{\sin z} = c_2$.

Thus, the general integral is $f\left(\frac{\sin x}{\sin y}, \frac{\sin y}{\sin z}\right) = 0$

or $\quad \frac{\sin x}{\sin y} = f\left(\frac{\sin y}{\sin z}\right)$.

Example 5. *Solve the following differential equation*

$$(z^2 - 2yz - y^2)p + (xy + zx)q = xy - zx.$$

(Kerala–2005)

Solution. Compare the given differential equation with Lagrange's differential equation $Pp + Qq = R$, we get $P = z^2 - 2yz - y^2, Q = xy + zx, R = xy - zx$.

Then, the Lagrange's subsidiary equations are

$$\frac{dx}{P} = \frac{dy}{Q} = \frac{dz}{R}$$

$$\Rightarrow \quad \frac{dx}{z^2 - 2yz - y^2} = \frac{dy}{xy + zx} = \frac{dz}{xy - zx} \quad \ldots(1)$$

Taking second and third ratio of (1), we get

$$\frac{dy}{xy + zx} = \frac{dz}{xy - zx} \quad \text{or} \quad \frac{dy}{y + z} = \frac{dz}{y - z}$$

or $\quad (y - z)dy = (y + z)dz$

or $\quad ydy - zdy = ydz + zdz$

or $\quad ydy - zdy - ydz - zdz = 0$

or $ydy - (zdy + ydz) - zdz = 0$

or $\quad ydy - d(yz) - zdz = 0$.

On integrating, we get $y^2 - 2yz - z^2 = c_1$.

Now taking the multipliers x, y, z, we get

$$\frac{dx}{z^2 - 2yz - y^2} = \frac{dy}{xy + zx} = \frac{dz}{xy - zx}$$

$$= \frac{xdx + ydy + zdz}{x(z^2 - 2yz - y^2) + y(xy + zx) + z(xy - zx)}$$

$$= \frac{xdx + ydy + zdz}{0}.$$

$$\therefore \; xdx + ydy + zdz = 0.$$

Integrating, we get $x^2 + y^2 + z^2 = c_2$.

Thus the general integral is

$$f(y^2 - 2yz - z^2, x^2 + y^2 + z^2) = 0.$$

Example 6. *Solve* $(y + z)p + (z + x)q = x + y$.

Solution. Compare the given differential equation with Lagrange's differential equation $Pp + Qq = R$, we get $P = y + z, Q = z + x, R = x + y$.

Then Lagrange's auxiliary equations are

$$\frac{dx}{P} = \frac{dy}{Q} = \frac{dz}{R} \Rightarrow \frac{dx}{y + z} = \frac{dy}{z + x} = \frac{dz}{x + y} \quad \ldots(1)$$

Equation (1) can also be taken as

$$\frac{dx - dy}{y - x} = \frac{dy - dz}{z - y} = \frac{dx + dy + dz}{2(x + y + z)} \quad \ldots(2)$$

Taking first and second ratio of (2), we get

$$\frac{dx - dy}{y - x} = \frac{dy - dz}{z - y} \quad \text{or} \quad \frac{dx - dy}{x - y} = \frac{dy - dz}{y - z}$$

Let $u = x - y$ and $v = y - z$

$\therefore du = dx - dy$ and $dv = dy - dz$

Then $\frac{dx - dy}{x - y} = \frac{dy - dz}{y - z}$ becomes $\frac{du}{u} = \frac{dv}{v}$

Integrating, we get

$$\log u = \log v + \log c_1$$

$$\log u - \log v = \log c_1$$

$$\log \frac{u}{v} = \log c_1 \quad \Rightarrow \quad \frac{u}{v} = c_1$$

$$\Rightarrow \quad \frac{x - y}{y - z} = c_1 \quad [\because u = x - y, v = y - z]$$

Now taking second and third ratio of (2), we get

$$\frac{dy - dz}{z - y} = \frac{dx + dy + dz}{2(x + y + z)} \quad \ldots(3)$$

Let $t = y - z$ and $s = x + y + z$

$\therefore \; dt = dy - dz$ and $ds = dx + dy + dz$

Hence, equation (3) becomes $\frac{dt}{-t} = \frac{ds}{2s}$

or $2\frac{dt}{t} + \frac{ds}{s} = 0.$

Integrating, we get

$\log t^2 + \log s = \log c_2$

or $\log(t^2 s) = \log c_2$ or $t^2 s = c_2$

$\Rightarrow (y-z)^2(x+y+z) = c_2$

$[\because t = y - z, s = x + y + z]$

Hence, the general integral is

$$f\left(\frac{x-y}{y-z}, (y-z)^2(x+y+z)\right) = 0.$$

Example 7. *Find the general integral of the following differential equation*

$$(y^2 + z^2 - x^2)p - 2xyq + 2zx = 0.$$

Solution. Compare this differential equation with Lagrange's differential equation $Pp + Qq = R$, we get $P = y^2 + z^2 - x^2, Q = -2xy, R = -2zx.$

Then Lagrange's auxiliary equations are

$$\frac{dx}{P} = \frac{dy}{Q} = \frac{dz}{R}$$

$$\Rightarrow \frac{dx}{y^2 + z^2 - x^2} = \frac{dy}{-2xy} = \frac{dz}{-2zx} \quad \ldots(1)$$

Taking second and third ratio of (1), we get

$$\frac{dy}{-2xy} = \frac{dz}{-2zx} \quad \text{or} \quad \frac{dy}{y} = \frac{dz}{z}.$$

Integrating, we get $\log y = \log z + \log c_1$

or $\log y - \log z = \log c_1$

or $\log\left(\frac{y}{z}\right) = \log c_1$ or $\frac{y}{z} = c_1.$

Taking the multipliers x, y, z then (1) becomes

$$\frac{dx}{y^2 + z^2 - x^2} = \frac{dy}{-2xy} = \frac{dz}{-2zx} = \frac{xdx + ydy + zdz}{-x(x^2 + y^2 + z^2)} \quad \ldots(2)$$

Now, taking third and fourth ratio of (2), we get

$$\frac{dz}{-2xz} = \frac{xdx + ydy + zdz}{-x(x^2 + y^2 + z^2)}$$

or $$\frac{dz}{2z} = \frac{xdx + ydy + zdz}{x^2 + y^2 + z^2}$$

or $$\frac{dz}{z} = \frac{2xdx + 2ydy + 2zdz}{x^2 + y^2 + z^2}$$

Let $u = x^2 + y^2 + z^2$

$\therefore \quad du = 2xdx + 2ydy + 2zdz$

$\therefore \quad \frac{dz}{z} = \frac{du}{u}$

Integrating, we get

$\log z + \log c_2 = \log u$

or $\frac{u}{z} = c_2$ or $\frac{x^2 + y^2 + z^2}{z} = c_2$

Hence, the general integral is

$$f\left(\frac{y}{z}, \frac{x^2 + y^2 + z^2}{z}\right) = 0.$$

Example 8. *Find the general solution of the following differential equation*

$$(mz - ny)p + (nx - lz)q = ly - mx.$$

(GBTU(AG)–2012, VTU–2010, SVTU–2009)

Solution. Compare the given differential equation with Lagrange's differential equation $Pp + Qq = R$, we get $P = mz - ny, Q = nx - lz, R = ly - mx.$

Then, Lagrange's auxiliary equations are

$$\frac{dx}{P} = \frac{dy}{Q} = \frac{dz}{R}$$

$$\Rightarrow \frac{dx}{mz - ny} = \frac{dy}{nx - lz} = \frac{dz}{ly - mx} \quad \ldots(1)$$

Taking the multipliers x, y, z then (1) becomes

$$\frac{dx}{mz - ny} = \frac{dy}{nx - lz} = \frac{dz}{ly - mx} = \frac{xdx + ydy + zdz}{0}$$

$\therefore \quad xdx + ydy + zdz = 0.$

Integrating, we get $x^2 + y^2 + z^2 = c_1.$

Again taking the multipliers l, m, n then (1) becomes

$$\frac{dx}{mz - ny} = \frac{dy}{nx - lz} = \frac{dz}{ly - mx} = \frac{ldx + mdy + ndz}{0}$$

$\therefore ldx + mdy + ndz = 0.$

Integrating, we get

$lx + my + nz = c_2.$

Hence, the general solution is

$$f(x^2 + y^2 + z^2, lx + my + nz) = 0.$$

Example 9. *Find the general solution of the following differential equation*

$$x(y^2 + z)p - y(x^2 + z)q = z(x^2 - y^2).$$

(UPTU–2008)

Solution. Comparing the given differential equation with Lagrange's differential equation $Pp + Qq = R$, we get

$P = x(y^2 + z), Q = -y(x^2 + z), R = z(x^2 - y^2).$

Then the Lagrange's auxiliary equations are

$$\frac{dx}{P} = \frac{dy}{Q} = \frac{dz}{R}$$

$$\Rightarrow \frac{dx}{x(y^2 + z)} = \frac{dy}{-y(x^2 + z)} = \frac{dz}{z(x^2 - y^2)} \quad \ldots(1)$$

Taking the multipliers $x, y -1$, then (1) becomes

$$\frac{dx}{x(y^2+z)} = \frac{dy}{-y(x^2+z)} = \frac{dz}{z(x^2-y^2)}$$
$$= \frac{xdx+ydy-dz}{0}$$

$\therefore \quad xdx + ydy - dz = 0.$

Integrating, we have $x^2 + y^2 - 2z = c_1$.

Again taking the multipliers $\frac{1}{x}, \frac{1}{y}, \frac{1}{z}$ then (1) becomes

$$\frac{dx}{x(y^2+z)} = \frac{dy}{-y(x^2+z)} = \frac{dz}{z(x^2-y^2)}$$
$$= \frac{\frac{1}{x}dx + \frac{1}{y}dy + \frac{1}{z}dz}{0}$$

$\therefore \quad \frac{1}{x}dx + \frac{1}{y}dy + \frac{1}{z}dz = 0.$

Integrating, we have

$$\log x + \log y + \log z = \log c_2$$

or $\quad xyz = c_2.$

Thus, the general solution is

$$f(x^2 + y^2 - 2z, xyz) = 0.$$

Example 10. *Find the general solution of the following differential equation*

$$xp + yq = z - a\sqrt{(x^2+y^2+z^2)}.$$

Solution. Compare the given equation with Lagrange's differential equation $Pp + Qq = R$, we have

$$P = x, Q = y, R = z - a\sqrt{(x^2+y^2+z^2)}.$$

Then, the Lagrange's auxiliary equations are

$$\frac{dx}{P} = \frac{dy}{Q} = \frac{dz}{R}$$

$$\Rightarrow \quad \frac{dx}{x} = \frac{dy}{y} = \frac{dz}{z - a\sqrt{(x^2+y^2+z^2)}} \quad \ldots(1)$$

Taking first and second ratio of (1), we have

$$\frac{dx}{x} = \frac{dy}{y}.$$

Integrating, we get $\log x = \log y + \log c_1$

or $\quad \log x - \log y = \log c_1$

or $\log\left(\frac{x}{y}\right) = \log c_1$ or $\frac{x}{y} = c_1$.

By taking the multipliers x, y, z, (1) becomes

$$\frac{dx}{x} = \frac{dy}{y} = \frac{dz}{z - a\sqrt{(x^2+y^2+z^2)}}$$
$$= \frac{xdx + ydy + zdz}{(x^2+y^2+z^2) - az\sqrt{(x^2+y^2+z^2)}} \quad \ldots(2)$$

Let $\quad u^2 = x^2 + y^2 + z^2$.

$\therefore \quad udu = xdx + ydy + zdz.$

Then (2) becomes

$$\frac{dx}{x} = \frac{dy}{y} = \frac{dz}{z - au} = \frac{udu}{u^2 - auz} \quad \ldots(3)$$

Now (3) can also be written as

$$\frac{dx}{x} = \frac{dy}{y} = \frac{dz}{z - au} = \frac{udu}{u^2 - auz} = \frac{dz + du}{(1-a)(z+u)}$$

Taking first and last ratio, we get

$$\frac{dx}{x} = \frac{dz + du}{(1-a)(z+u)} \quad \text{or} \quad (1-a)\frac{dx}{x} = \frac{dz+du}{z+u}$$

Let $\quad t = z + u$

$\therefore \quad dt = dz + du$

$\therefore \quad (1-a)\frac{dx}{x} = \frac{dt}{t}.$

Integrating, we get

$(1-a)\log x = \log t + \log c_2$ or $\log x^{1-a} = \log tc_2$

or $\quad x^{1-a} = c_2 t \quad$ or $\quad x^{1-a} = c_2(z+u)$

$[\because t = z + u]$

$$x^{1-a} = c_2[z + \sqrt{(x^2+y^2+z^2)}]$$

$[\because u^2 = x^2 + y^2 + z^2]$

Hence, the general solution is

$$x^{1-a} = [z + \sqrt{(x^2+y^2+z^2)}]\, f(x/y).$$

Example 11. *Find the general solution of the following differential equation* $p + 3q = 5z + \tan(y - 3x)$.

Solution. Compare the given differential equation with Lagrange's differential equation $Pp + Qq = R$, we get $P = 1, Q = 3, R = 5z + \tan(y - 3x)$

Then the Lagrange's auxiliary equations are

$$\frac{dx}{1} = \frac{dy}{3} = \frac{dz}{5z + \tan(y-3x)} \quad \ldots(1)$$

Taking first and second ratio of (1), we get

$$\frac{dx}{1} = \frac{dy}{3} \quad \text{or} \quad dy - 3dx = 0$$

Integrating, we get $y - 3x = c_1$.

Now taking first and third ratio of (1), we get

$$\frac{dx}{1} = \frac{dz}{5z + \tan(y-3x)} \quad \text{or} \quad \frac{dx}{1} = \frac{dz}{5z + \tan c_1}$$

$[\because y - 3x = c_1]$

or $\quad 5dx = \frac{5dz}{5z + \tan c_1}$

Integrating, we get

$$5x = \log(5z + \tan c_1) + \log c_2$$

$5x = \log c_2(5z + \tan c_1)$ or $e^{5x} = c_2(5z + \tan c_1)$

or $e^{5x} = c_2[5z + \tan(y - 3x)] \quad [\because c_1 = y - 3x]$

Hence, the general solution is

$$e^{5x} = [5z + \log(y - 3x)]\, f(y - 3x).$$

Example 12. *Find the general integral of the following differential equation*

$$x^2(y - z)p + (z - x)y^2 q = z^2(x - y).$$

(GBTU(AG)–2012, PTU–2009, Bhopal–2008, SVTU–2007)

Solution. Compare the given differential equation with Lagrange's differential equation $Pp + Qq = R$, we get

$P = x^2(y-z),\ Q = (z-x)y^2,\ R = z^2(x-y)$

Then the Lagrange's auxiliary equations are

$$\frac{dx}{P} = \frac{dy}{Q} = \frac{dz}{R}$$

$$\frac{dx}{x^2(y-z)} = \frac{dy}{(z-x)y^2} = \frac{dz}{z^2(x-y)} \quad \ldots(1)$$

Taking the multipliers $\frac{1}{x^2}, \frac{1}{y^2}, \frac{1}{z^2}$, then (1) becomes

$$\frac{dx}{x^2(y-z)} = \frac{dy}{(z-x)y^2} = \frac{dz}{z^2(x-y)}$$

$$= \frac{\frac{1}{x^2}dx + \frac{1}{y^2}dy + \frac{1}{z^2}dz}{0}$$

$$\therefore \quad \frac{dx}{x^2} + \frac{dy}{y^2} + \frac{dz}{z^2} = 0.$$

Integrating, we get $\frac{1}{x} + \frac{1}{y} + \frac{1}{z} = c_1$.

Now taking the multipliers $\frac{1}{x}, \frac{1}{y}, \frac{1}{z}$ then (1) becomes

$$\frac{dx}{x^2(y-z)} = \frac{dy}{(z-x)y^2} = \frac{dz}{z^2(x-y)}$$

$$= \frac{\frac{1}{x}dx + \frac{1}{y}dy + \frac{1}{z}dz}{0}$$

$$\therefore \quad \frac{1}{x}dx + \frac{1}{y}dy + \frac{1}{z}dz = 0.$$

Integrating, we get $\log x + \log y + \log z = \log c_2$

or $\log xyz = \log c_2$ or $xyz = c_2$.

Hence, the general integral is

$$\frac{1}{x} + \frac{1}{y} + \frac{1}{z} = f(xyz).$$

Example 13. *Solve* $y^2p - xyq = x(z-2y)$. (SVTU–2008)

Solution. Compare the given differential equation with Lagrange's differential equation $Pp + Qq = R$, we get $P = y^2, Q = -xy, R = x(z-2y)$.

Then the Lagrange's auxiliary equations are

$$\frac{dx}{P} = \frac{dy}{Q} = \frac{dz}{R}$$

$$\Rightarrow \quad \frac{dx}{y^2} = \frac{dy}{-xy} = \frac{dz}{x(z-2y)} \quad \ldots(1)$$

Taking first and second ratio of (1), we get

$$\frac{dx}{y^2} = \frac{dy}{-xy} \quad \text{or} \quad \frac{dx}{y} = \frac{dy}{-x} \quad \text{or} \quad xdx + ydy = 0.$$

Integrating, we get $x^2 + y^2 = c_1$.

Taking second and third ratio of (1), we get

$$\frac{dy}{-xy} = \frac{dz}{x(z-2y)} \quad \text{or} \quad (z-2y)dy = -ydz$$

or $zdy + ydz - 2ydy = 0$ or $d(yz) - 2y\,dy = 0$.

Integrating, we get $yz - y^2 = c_2$.

Hence, the general integral is

$$f(x^2 + y^2, yz - y^2) = 0.$$

Example 14. *Solve* $(y^2 + z^2)p - xyp = -zx$. (MTU–2011, PTU–2009, VTU–2009)

Solution. Compare the given differential equation with Lagrange's differential equation $Pp + Qq = R$, we get $P = y^2 + z^2, Q = -xy, R = -zx$.

Then the Lagrange's auxiliary equations are

$$\frac{dx}{P} = \frac{dy}{Q} = \frac{dz}{R}$$

$$\frac{dx}{y^2 + z^2} = \frac{dy}{-xy} = \frac{dz}{-zx} \quad \ldots(1)$$

Taking second and third ratio of (1), we get

$$\frac{dy}{-xy} = \frac{dz}{-zx} \quad \text{or} \quad \frac{dy}{y} = \frac{dz}{z}$$

Integrating, we get $\log y = \log z + \log c_1$

or $\frac{y}{z} = c_1$.

Now taking the multipliers x, y, z then (1) becomes

$$\frac{dx}{y^2 + z^2} = \frac{dy}{-xy} = \frac{dz}{-zx} = \frac{xdx + ydy + zdz}{0}$$

$$\therefore \quad xdx + ydy + zdz = 0.$$

Integrating, we get $x^2 + y^2 + z^2 = c_2$.

Hence, the general integral is

$$f\left(\frac{y}{z}, x^2 + y^2 + z^2\right) = 0.$$

Example 15. *Find the general solution of following differential equation :*

$$(x+2z)p + (4zx - y)q = 2x^2 + y.$$

Solution. Compare the given differential equation with Lagrange's differential equation $Pp + Qq = R$, we get $P = x + 2z, Q = 4zx - y, R = 2x^2 + y$.

Then, the Lagrange's auxiliary equations are

$$\frac{dx}{P} = \frac{dy}{Q} = \frac{dz}{R}$$

$$\Rightarrow \quad \frac{dx}{x+2z} = \frac{dy}{4zx - y} = \frac{dz}{2x^2 + y} \quad \ldots(1)$$

Taking the multipliers $y, x, -2z$ then (1) becomes

$$\frac{dx}{x+2z} = \frac{dy}{4zx - y} = \frac{dz}{2x^2 + y}$$

$$= \frac{ydx + xdy - 2zdz}{0}$$

$$\therefore \quad ydx + xdy - 2zdz = 0.$$

Integrating, we get $xy - z^2 = c_1$.

Now taking the multipliers $2x, -1, -1$ then (1) becomes

$$\frac{dx}{x+2z} = \frac{dy}{4zx - y} = \frac{dz}{2x^2 + y} = \frac{2xdx - dy - dz}{0}$$

$\therefore\ 2xdx - dy - dz = 0.$

Integrating, we get $x^2 - y - z = c_2$.

Hence, the general solution is

$$f(xy - z^2, x^2 - y - z) = 0.$$

Example 16. *Solve the differential equation*

$$x\frac{\partial u}{\partial x} + y\frac{\partial u}{\partial y} + z\frac{\partial u}{\partial z} = xyz.$$

Solution. The given differential can be written as

$$xp_1 + yp_2 + zp_3 = xyz.$$

Now compare this differential equation with Lagrange's differential equation

$P_1p_1 + P_2p_2 + P_3p_3 = R$, we get

$$P_1 = x, P_2 = y, P_3 = z, R = xyz$$

Then the Lagrange's auxiliary equations

$$\frac{dx}{P_1} = \frac{dy}{P_2} = \frac{dz}{P_3} = \frac{du}{R}$$

$$\Rightarrow \frac{dx}{x} = \frac{dy}{y} = \frac{dz}{z} = \frac{du}{xyz} \quad ...(1)$$

Taking first and second ratio of (1), we get $\frac{dx}{x} = \frac{dy}{y}$.

Integrating, we get $\log x = \log y + \log c_1$

or $\log x - \log y = \log c_1$

or $$\frac{x}{y} = c_1.$$

Taking second and third ratio of (1), we get $\frac{dy}{y} = \frac{dz}{z}$.

Integrating, we get $\log y = \log z + \log c_2$

or $$\log y - \log z = \log c_2$$

or $$\frac{y}{z} = c_2.$$

Finally, taking the multiplier $yz, zx, xy, -3$, then (1) becomes

$$\frac{dx}{x} = \frac{dy}{y} = \frac{dz}{z} = \frac{du}{xyz}$$

$$= \frac{yzdx + zxdy + xydz + -3du}{0}$$

$\therefore\ yzdx + zxdy + xydz - 3du = 0$

or $$d(xyz) - 3du = 0.$$

Integrating, we get $xyz - 3u = c_3$.

Hence, the general integral is

$$f\left(\frac{x}{y}, \frac{y}{z}, xyz - 3u\right) = 0.$$

Example 17. *Solve* $(x^2 - y^2 - z^2)p + 2xyp = 2xz$.

(VTU–2010, Anna–2009, SVTU – 2008)

Solution. The subsidiary equation of the given equation is

$$\frac{dx}{x^2 - y^2 - z^2} = \frac{dy}{2xy} = \frac{dz}{2xz}$$

From last two fractions, we get $\frac{dy}{y} = \frac{dz}{z}$

$\Rightarrow\ \log y = \log z + \log c_1$

$$\Rightarrow \frac{y}{z} = c_1$$

Further using multipliers x, y, z in (1), we get

each fractions $= \dfrac{xdx + ydy + zdz}{x(x^2 + y^2 + z^2)}$

$$\Rightarrow \frac{2xdx + 2ydy + 2zdz}{x^2 + y^2 + z^2} = \frac{dz}{z}$$

On integrating we get

$$\log(x^2 + y^2 + z^2) = \log z + \log c_2$$

$$\Rightarrow \frac{x^2 + y^2 + z^2}{2} = c_2$$

Hence, the general solution is given by

$$f\left(\frac{y}{z}, \frac{x^2 + y^2 + z^2}{2}\right) = 0.$$

Example 18. *Solve* $x(x^2 - z^2)p + y(z^2 - x^2)q = z(x^2 - y^2)$

(UPTU–2008)

Solution. Lagrange's subsidiary equation of the given equation is

$$\frac{dx}{x(x^2 - z^2)} = \frac{dy}{y(z^2 - x^2)} = \frac{dz}{z(x^2 - y^2)}$$

Using multiplers, x, y, z, we get each fraction

$$= \frac{xdx + ydy + zdz}{x^2(y^2 - z^2) + y^2(z^2 - x^2) + z^2(x^2 - y^2)}$$

$$= \frac{xdx + ydy + zdz}{0}$$

$$\Rightarrow xdx + ydy + zdz = 0$$

$$\Rightarrow x^2 + y^2 + z^2 = c_1$$

Again using multipliers $\frac{1}{x}, \frac{1}{y}, \frac{1}{z}$ we get

each fraction $= \dfrac{\frac{1}{x}dx + \frac{1}{y}dy + \frac{1}{z}dz}{y^2 - z^2 + z^2 - x^2 + x^2 - y^2}$

$$= \frac{\frac{1}{x}dx + \frac{1}{y}dy + \frac{1}{z}dz}{0}$$

$$\Rightarrow \frac{1}{x}dx + \frac{1}{y}dy + \frac{1}{z}dz = 0$$

On integration, we get

$$\log x + \log y + \log z = \log c_2$$

$$xyz = c_2$$

Hence, the general solution is given by

$$f(x^2 + y^2 + z^2, xyz) = 0$$

Example 19. *Solve* $(x^2 - yz)p + (y^2 - zx)q = z^2 - xy$

(GBTU–2010, Bhopal–2008, VTU–2006, Madras–2000)

Solution. The lagrange's subsidiary equation is given by

$$\frac{dx}{x^2 - yz} = \frac{dy}{y^2 - zx} = \frac{dz}{z^2 - xy}$$

$$\Rightarrow \quad \frac{dx-dy}{(x-y)(x+y+z)} = \frac{dy-dz}{(y-z)(x+y+z)} = \frac{dz-dx}{(z-x)(x+y+z)}$$

Consider

$$\frac{dx-dy}{(x-y)(x+y+z)} = \frac{dy-dz}{(y-z)(x+y+z)}$$

$$\Rightarrow \quad \frac{dx-dy}{(x-y)} = \frac{dy-dz}{(y-z)}$$

$$\Rightarrow \quad \log(x-y) = \log(y-z) + \log c_1$$

$$\Rightarrow \quad \log\left(\frac{x-y}{y-z}\right) = \log c_1$$

$$\Rightarrow \quad \frac{x-y}{y-z} = c_1$$

Similarly, taking the last two members, we get

$$\frac{y-z}{z-x} = c_2$$

Hence, general solution is given by

$$f\left(\frac{x-y}{y-z}, \frac{y-z}{z-x}\right) = 0.$$

EXERCISE 40.1

Find the general integrals of the linear partial differential equations :

1. $\left(\frac{y-z}{yz}\right)p + \left(\frac{z-x}{zx}\right)q = \left(\frac{x-y}{xy}\right).$ (Bhopal–2007)
2. $\frac{y^2 z}{x}p + zxq = y^2.$
3. $p+q = \frac{z}{a}.$
4. $px(z-2y^2) = (z-qy)(z-y^2-2x^3).$
5. $(y^3x-2x^4)p + (2y^4-x^3y)q = 9z(x^3-y^3).$
6. $px(x+y) - qy(x+y) = -(x-y)(2x+2y+z).$
7. $(y+zx)p - (x+yz)q = x^2-y^2.$
8. $x(x^2+3y^2)p - y(3x^2+y^2)q = 2z(y^2-x^2).$
9. $(x+y)(p-q) = z.$
10. $x^2p + y^2q = z^2.$
11. $p+q = 1.$
12. $p_2 + p_3 = 1 + p_1$ where $p_i \frac{\partial z}{\partial x_i}$, $i = 1,2,3.$
13. $x_2x_3zp_1 + x_3x_1zp_2 + x_1x_2zp_3 = x_1x_2x_3.$
14. $x\frac{\partial z}{\partial x} + y\frac{\partial z}{\partial y} + t\frac{\partial z}{\partial t} = az + \frac{xy}{t}.$
15. $(y+z+t)\frac{\partial t}{\partial x} + (z+x+t)\frac{\partial t}{\partial y} + (x+y+t)\frac{\partial t}{\partial z} = x+y+z.$
16. $\frac{1}{\sqrt{z}}(p_1 + x_1p_2 + x_1x_2p_3) = x_1x_2x_3.$
17. $(3x+y-z)p + (x+y-z)q = 2(z-y).$
18. Solve $x\frac{\partial u}{\partial x} + y\frac{\partial u}{\partial y} + z\frac{\partial u}{\partial z} = xyz.$
19. Solve $p_1 + p_2 + p_3 = 4z$, where $p_1 = \frac{\partial z}{\partial x_1}, p_2 = \frac{\partial z}{\partial x_2}, p_3 = \frac{\partial z}{\partial x_3}.$
20. Solve $(2x^2+y^2+z^2-2yz-2x-xy)p + (x^2+2y^2+z^2-yz-2zx-xy)q = x^2+y^2+2z^2-yz-zx-2xy.$

Hint to Selected Problems

1. The Lagrange's subsidiary equation is given by

$$\frac{dx}{\left(\frac{y-z}{yz}\right)} = \frac{dy}{\left(\frac{z-x}{zx}\right)} = \frac{dx}{\left(\frac{x-y}{xy}\right)}.$$

Now taking the following two sets of multipliers

(i) 1,1,1 (ii) yz, zx, xy

5. Lagrange's auxiliary equations are given by

$$\frac{dx}{y^3x-2x^4} = \frac{dy}{2y^4-x^3y} = \frac{dz}{9z(x^3-y^3)}.$$

Now firstly, taking first and second ratio. For second integral, taking $\frac{1}{x}, \frac{1}{y}, \frac{1}{3z}$ as a set of multipliers.

6. Lagrange's auxiliary equations are

$$\frac{dx}{x(x+y)} = \frac{dy}{-y(x+y)} = \frac{dz}{-(x-y)(2x+2y+z)}. \quad ...(1)$$

By taking first and second ratio, we get $xy = C_1$.

Also, from (1)

$$\frac{dx+dy}{(x-y)(x+y)} = \frac{dx+dy+dz}{(x-y)(x+y)-(x-y)(2x+2y+z)} = \frac{dx+dy+dz}{(x-y)\{(x+y)-2x-2y-z\}}$$

$$\Rightarrow \quad \frac{dx+dy}{x+y} = \frac{dx+dy+dz}{-(x+y+z)}.$$

Now substituting $x+y=u$ and $x+y+z=v$.

7. Here, we have $\frac{dx}{y+zx} = \frac{dy}{-(x+yz)} = \frac{dz}{x^2-y^2}$. Then taking following two set of multipliers

(i) $y, x, 1$ (ii) $x, y, -z.$

13. The Lagrange's auxiliary equations are

$$\frac{dx_1}{x_2x_3z} = \frac{dx_2}{x_3x_1z} = \frac{dx_3}{x_1x_2z} = \frac{dx_1}{x_1x_2x_3}.$$

ANSWERS

1. $f(x+y+z, xyz) = 0$
2. $f(x^3-y^3, x^2-z^2) = 0$
3. $z = e^{y/a}f(x-y)$
4. $z - y^2 + x^3 = x f\left(\frac{y}{z}\right)$
5. $f\left(\frac{x^3+y^3}{x^2y^2}, xyz^{1/3}\right) = 0$
6. $(x+y)(x+y+z) = f(xy)$
7. $f(x^2+y^2-z^2, xy+z) = 0$

8. $(x^2+y^2)z = f\left(\frac{xy}{z}\right)$ **9.** $(x+y)\log z = x + f(x+y)$ **10.** $\frac{1}{x}-\frac{1}{y} = f\left(\frac{1}{x}-\frac{1}{z}\right)$ **11.** $f(x-z, y-z)=0$

12. $f(x_1+z, x_1+x_2, x_1+x_3)=0$ **13.** $f(x_2^2-x_1^2, x_3^2-x_1^2, z^2-x_1^2)=0$ **14.** $f\left(\frac{y}{x},\frac{t}{x}, zx^{-a}-\frac{y}{t},\frac{x^{1-a}}{1-a}\right)=0$

15. $f\left[\frac{x-y}{y-z},\frac{z-t}{y-z},(z-t)(x+y+z+t)^{1/3}\right]=0$ **16.** $f(4\sqrt{z}-x_3^2, 2x_3-x_2^2, 2x_2-x_1^2)=0.$ **17.** $f\left(x-3y-z, \frac{x-y+z}{\sqrt{x+y-z}}\right)=0$

18. $f\left(\frac{x}{y},\frac{y}{z}, xyz-34\right)=0$ **19.** $f\left(\frac{z}{e^{4x_1}}\cdot\frac{z}{e^{4x_2}}\cdot\frac{z}{e^{4x_3}}\right)=0$ **20.** $f\left(\frac{x-y}{y-z},\frac{y-z}{z-x}\right)=0$

40.4 CHARPIT'S METHOD

The Charpit's method of solving the partial differential equation

$$f(x,y,z,p,q)=0 \quad \ldots(1)$$

is based on the introduction of a second partial differential equation of the first order. Let this second partial differential equation be

$$F(x,y,z,p,q,a)=0 \quad \ldots(2)$$

which contains an arbitrary constant 'a'. Now solving (1) and (2), to get $p=p(x,y,z,a)$ and $q=q(x,y,z,a)$ and substitute p and q into

$$dz = pdx + qdy \quad \ldots(3)$$

The equation (3) is integrable and the integral given by (3) will satisfy the given equation (1), for the values of p and q obtained from it will be same as the values of p and q given in (1). Thus z, p and q may be the functions of x and y and are satisfied by both (1) and (2), Therefore we have

$$\frac{df}{dx}=0,\quad \frac{df}{dy}=0 \text{ and } \frac{dF}{dx}=0,\quad \frac{dF}{dy}=0.$$

Thus, differentiating (1) and (2) with respect to x and y, we get

$$\frac{\partial f}{\partial x}+\frac{\partial f}{\partial z}\cdot\frac{\partial z}{\partial x}+\frac{\partial f}{\partial p}\cdot\frac{\partial p}{\partial x}+\frac{\partial f}{\partial q}\cdot\frac{\partial q}{\partial x}=0 \quad \text{or} \quad \frac{\partial f}{\partial x}+\frac{\partial f}{\partial z}.p+\frac{\partial f}{\partial p}\cdot\frac{\partial p}{\partial x}+\frac{\partial f}{\partial q}\cdot\frac{\partial q}{\partial x}=0 \quad \ldots(4)$$

and
$$\frac{\partial f}{\partial y}+\frac{\partial f}{\partial z}\cdot\frac{\partial z}{\partial y}+\frac{\partial f}{\partial p}\cdot\frac{\partial p}{\partial y}+\frac{\partial f}{\partial q}\cdot\frac{\partial q}{\partial y}=0 \quad \text{or} \quad \frac{\partial f}{\partial y}+\frac{\partial f}{\partial z}.q+\frac{\partial f}{\partial p}\cdot\frac{\partial p}{\partial y}+\frac{\partial f}{\partial q}\cdot\frac{\partial q}{\partial y}=0 \quad \ldots(5)$$

and
$$\frac{\partial F}{\partial x}+\frac{\partial F}{\partial z}\cdot\frac{\partial z}{\partial x}+\frac{\partial F}{\partial p}\cdot\frac{\partial p}{\partial x}+\frac{\partial F}{\partial q}\cdot\frac{\partial q}{\partial x}=0 \quad \text{or} \quad \frac{\partial F}{\partial x}+\frac{\partial F}{\partial z}.p+\frac{\partial F}{\partial p}\cdot\frac{\partial p}{\partial x}+\frac{\partial F}{\partial q}\cdot\frac{\partial q}{\partial x}=0 \quad \ldots(6)$$

and
$$\frac{\partial F}{\partial y}+\frac{\partial F}{\partial z}\cdot\frac{\partial z}{\partial y}+\frac{\partial F}{\partial p}\cdot\frac{\partial p}{\partial y}+\frac{\partial F}{\partial q}\cdot\frac{\partial q}{\partial y}=0 \quad \text{or} \quad \frac{\partial F}{\partial y}+\frac{\partial F}{\partial z}.q+\frac{\partial F}{\partial p}\cdot\frac{\partial p}{\partial y}+\frac{\partial F}{\partial q}\cdot\frac{\partial q}{\partial y}=0 \quad \ldots(7)$$

Now eliminating $\frac{\partial p}{\partial x}$ from (4) and (6). For this multiplying (4) by $\frac{\partial F}{\partial p}$ and (6) by $\frac{\partial f}{\partial p}$ and subtract, we get

$$\left(\frac{\partial f}{\partial x}+\frac{\partial f}{\partial z}p+\frac{\partial f}{\partial q}\cdot\frac{\partial q}{\partial x}\right)\frac{\partial F}{\partial p}-\left(\frac{\partial F}{\partial x}+\frac{\partial F}{\partial z}p+\frac{\partial F}{\partial q}\cdot\frac{\partial q}{\partial x}\right)\frac{\partial f}{\partial p}=0$$

and
$$\frac{\partial f}{\partial x}\cdot\frac{\partial F}{\partial p}-\frac{\partial F}{\partial x}\cdot\frac{\partial f}{\partial p}+\left(\frac{\partial f}{\partial z}\cdot\frac{\partial F}{\partial p}-\frac{\partial F}{\partial z}\cdot\frac{\partial f}{\partial p}\right)p+\left(\frac{\partial f}{\partial q}\cdot\frac{\partial F}{\partial p}-\frac{\partial F}{\partial q}\cdot\frac{\partial f}{\partial p}\right)\frac{\partial q}{\partial x}=0 \quad \ldots(8)$$

and eliminating $\partial q/\partial y$ from (5) and (7), so multiplying (5) by $\partial F/\partial q$ and multiplying (7) by $\frac{\partial f}{\partial q}$ and subtract, we get

$$\left(\frac{\partial f}{\partial y}+\frac{\partial f}{\partial z}.q+\frac{\partial f}{\partial p}\cdot\frac{\partial p}{\partial y}\right)\frac{\partial F}{\partial q}-\left(\frac{\partial F}{\partial y}+\frac{\partial F}{\partial z}q+\frac{\partial F}{\partial p}\cdot\frac{\partial p}{\partial y}\right)\frac{\partial f}{\partial q}=0$$

or
$$\frac{\partial f}{\partial y}\cdot\frac{\partial F}{\partial q}-\frac{\partial F}{\partial y}\cdot\frac{\partial f}{\partial q}+\left(\frac{\partial f}{\partial z}\cdot\frac{\partial F}{\partial q}-\frac{\partial F}{\partial z}\cdot\frac{\partial f}{\partial q}\right)q+\left(\frac{\partial f}{\partial p}\cdot\frac{\partial F}{\partial q}-\frac{\partial F}{\partial p}\cdot\frac{\partial f}{\partial q}\right)\frac{\partial f}{\partial y}=0 \quad \ldots(9)$$

Since, we know that $q=\frac{\partial z}{\partial y} \Rightarrow \frac{\partial q}{\partial x}=\frac{\partial^2 z}{\partial x\partial y}$ or $p=\frac{\partial z}{\partial x} \Rightarrow \frac{\partial p}{\partial y}=\frac{\partial^2 z}{\partial y\partial x}$. But $\frac{\partial^2 z}{\partial x\partial y}=\frac{\partial^2 z}{\partial y\partial x} \Rightarrow \frac{\partial q}{\partial x}=\frac{\partial p}{\partial y}$

Now adding (8) and (9), we get

$$\frac{\partial f}{\partial x}\cdot\frac{\partial F}{\partial p}-\frac{\partial F}{\partial x}\cdot\frac{\partial f}{\partial p}+\left(\frac{\partial f}{\partial z}\cdot\frac{\partial F}{\partial p}-\frac{\partial F}{\partial z}\cdot\frac{\partial f}{\partial p}\right)p+\frac{\partial f}{\partial y}\cdot\frac{\partial F}{\partial q}-\frac{\partial F}{\partial y}\cdot\frac{\partial f}{\partial q}+\left(\frac{\partial f}{\partial z}\cdot\frac{\partial F}{\partial q}-\frac{\partial F}{\partial z}\cdot\frac{\partial f}{\partial q}\right)q=0$$

or $$\left(\frac{\partial f}{\partial x}+p\frac{\partial f}{\partial z}\right)\frac{\partial F}{\partial p}+\left(\frac{\partial f}{\partial y}+q\frac{\partial f}{\partial z}\right)\frac{\partial F}{\partial q}+\left(-p\frac{\partial f}{\partial p}-q\frac{\partial f}{\partial q}\right)\frac{\partial F}{\partial z}+\left(-\frac{\partial f}{\partial p}\right)\frac{\partial F}{\partial x}+\left(-\frac{\partial f}{\partial q}\right)\frac{\partial F}{\partial y}=0 \quad ...(10)$$

Thus, the equation (10) is a linear differential equation for the dependent variable F, so to determine F use the subsidiary equations

$$\frac{dx}{-\partial f/\partial p}=\frac{dy}{-\partial f/\partial q}=\frac{dz}{-p\frac{\partial f}{\partial p}-q\frac{\partial f}{\partial q}}=\frac{dp}{\frac{\partial f}{\partial x}+p\frac{\partial f}{\partial z}}=\frac{dq}{\frac{\partial f}{\partial y}+q\frac{\partial f}{\partial z}} \quad ...(11)$$

The equations given in (11) are known as Charpit's auxiliary equations. It should be noted that there is no need to use all of Charpit's auxiliary equations, therefore, we find a solution of (11) in which either p or q must occur for $F \equiv 0$, then from $f \equiv 0$ and $F \equiv 0$, find the values of p and q and substitute these values of p and q into

$$dz = pdx + qdy$$

and then integrating, we obtain the solution of (1).

WORKING PROCEDURE

Step 1. *Putting all terms of the equation in L.H.S and denote the whole expression by f.*
Step 2. *Write Charpit's auxiliary equation.*
Step 3. *Find the values of different partial derivatives of f and put in Charpit's auxiliary equation.*
Step 4. *Select two proper fractions such that the resulting integral may come out in the simplest form involving at least one of p and q*
Step 5. *Solving this simplest relation (obtained in (4)) with the given equation to find p and q.*
Step 6. *Putting the values of p and q in $dz = pdx + qdy$.*

REMARKS

- This method is also applicable for the solution of the partial differential equations of order one but of any degree.
- This method is applied only when other methods are failed for finding the solution.
- From (11), we obtain $p = a$ (constant) and $q = b$ (constant), then after putting these values in (1), we get the complete integral.

Solved Examples

Example 1. *Find the complete integral of the following partial differential equation $z = px + qy + p^2 + q^2$ by Charpit's method.*

Solution. Let us take $f \equiv z - px - qy - p^2 - q^2 = 0$ ...(1)

and finding partial derivatives of w.r.t. x, y, z, p and q respectively,

$$\frac{\partial f}{\partial x}=-p, \frac{\partial f}{\partial y}=-q, \frac{\partial f}{\partial z}=1, \frac{\partial f}{\partial p}=-x-2p,$$

$$\frac{\partial f}{\partial q}=-y-2q.$$

Thus, the Charpit's auxiliary equations are

$$\frac{dx}{-\partial f/\partial p}=\frac{dy}{-\partial f/\partial q}=\frac{dz}{-p\frac{\partial f}{\partial p}-q\frac{\partial f}{\partial q}}$$

$$=\frac{dp}{\frac{\partial f}{\partial x}+p\frac{\partial f}{\partial z}}=\frac{dq}{\frac{\partial f}{\partial y}+q\frac{\partial f}{\partial z}}$$

$$\Rightarrow \frac{dx}{x+2p}=\frac{dy}{y+2q}=\frac{dz}{xp+2p^2+yq+2q^2}$$

$$=\frac{dp}{0}=\frac{dq}{0} \quad ...(2)$$

From (1), we get $dp = 0$.

$\Rightarrow$ $p = a$ (constant) and $dq = 0$

$\Rightarrow$ $q = b$ (constant).

Substituting these values of p and q into (1), we get $z = ax + by + a^2 + b^2$.

This is the required complete integral.

Example 2. *Find the complete integral of*

$$2zx - px^2 - 2qxy + pq = 0. \quad \text{(Rajasthan–2006)}$$

Solution. Assume $f \equiv 2zx - px^2 - 2qxy + pq = 0.$ ...(1)

Now finding partial derivatives of f with respect to x, y, z, p and q respectively, we get

$$\frac{\partial f}{\partial x}=2z-2px-2qy, \frac{\partial f}{\partial y}=-2qx, \frac{\partial f}{\partial z}=2x,$$

$$\frac{\partial f}{\partial p}=-x^2+q, \frac{\partial f}{\partial q}=-2xy+p.$$

Then the Charpit's auxiliary equations are

$$\frac{dx}{-\frac{\partial f}{\partial p}}=\frac{dy}{-\frac{\partial f}{\partial q}}=\frac{dz}{-p\frac{\partial f}{\partial p}-q\frac{\partial f}{\partial q}}$$

$$=\frac{dp}{\frac{\partial f}{\partial x}+p\frac{\partial f}{\partial z}}=\frac{dq}{\frac{\partial f}{\partial y}+q\frac{\partial f}{\partial z}}$$

$$\Rightarrow \frac{dx}{x^2-q}=\frac{dy}{2xy-p}=\frac{dz}{px^2-pq+2xyq-pq}$$

$$=\frac{dp}{2z-2qy}=\frac{dq}{0} \quad ...(2)$$

From (2), $dq = 0$. Integrating, $q = a$ (constant).
Putting the value of $q = a$ into (1), we get

$$2zx - px^2 - 2axy + pa = 0 \quad \text{or} \quad p = \frac{2x(z - ay)}{x^2 - a}.$$

Now substituting these values of p and q into $dz = pdx + qdy$, we get

$$dz = \frac{2x(z-ay)}{x^2 - a}dx + ady$$

or $$dz - ady = \frac{2x(z-ay)}{x^2-a}dx$$

or $$\frac{dz - ady}{z - ay} = \frac{2x\,dx}{x^2 - a}.$$

Integrating, we get

$$\log(z - ay) = \log(x^2 - a) + \log b$$

or $$z - ay = b(x^2 - a)$$

or $$z = ay + b(x^2 - a).$$

This is the required complete integral.

Example 3. *Solve* $p = (z + qy)^2$.

Solution. Assuming $f \equiv (z + qy)^2 - p = 0$.

Now finding the partial derivatives of f w.r.t. x, y, z, p and q.

$$\frac{\partial f}{\partial x} = 0, \frac{\partial f}{\partial y} = 2q(z+qy), \frac{\partial f}{\partial z} = 2(z+qy),$$

$$\frac{\partial f}{\partial p} = -1, \frac{\partial f}{\partial q} = 2y(z+qy).$$

Then the Charpit's auxiliary equations are

$$\frac{dx}{-\frac{\partial f}{\partial p}} = \frac{dy}{-\frac{\partial f}{\partial q}} = \frac{dz}{-p\frac{\partial f}{\partial p} - q\frac{\partial f}{\partial q}} = \frac{dp}{\frac{\partial f}{\partial x} + p\frac{\partial f}{\partial z}} = \frac{dq}{\frac{\partial f}{\partial y} + q\frac{\partial f}{\partial z}}$$

$$\Rightarrow \quad \frac{dx}{1} = \frac{dy}{-2y(z+qy)} = \frac{dz}{p - 2qy(z+qy)} = \frac{dp}{2p(z+yq)} = \frac{dq}{4q(z+qy)} \quad \ldots(2)$$

Taking second and fourth ratio of (2), we get

$$\frac{dy}{-2y(z+qy)} = \frac{dp}{2p(z+yq)}$$

$$\Rightarrow \quad \frac{dp}{p} + \frac{dy}{y} = 0.$$

Integrating, we get

$$\log p + \log y = \log a \quad \text{or} \quad py = a \quad \text{or} \quad p = \frac{a}{y}.$$

Substituting the value of p into (1), we get

$$(z + qy)^2 = \frac{a}{y} \quad \text{or} \quad (z+qy) = \sqrt{\frac{a}{y}}$$

or $$q = \frac{\sqrt{a}}{y^{3/2}} - \frac{z}{y}.$$

Now substituting the values of p and q into $dz = pdx + qdy$, we get

$$dz = \frac{a}{y}dx + \left(\frac{\sqrt{a}}{y^{3/2}} - \frac{z}{y}\right)dy$$

or $$ydz = adx + \sqrt{\frac{a}{y}}dy - zdy$$

or $$ydz + zdy = adx + \sqrt{a/y}\,dy$$

or $$d(yz) = adx + \sqrt{a/y}\,dy.$$

Integrating, we get $yz = ax + 2\sqrt{ay} + b$.
This is the required complete integral.

Example 4. *Apply Charpit's method to find complete integral of* $z^2(p^2z^2 + q^2) = 1$.

Solution. Assuming $f \equiv p^2z^4 + z^2q^2 - 1 = 0 \quad \ldots(1)$

Now finding the partial derivatives of f w.r.t. x, y, z, p and q, we get

$$\frac{\partial f}{\partial x} = 0, \frac{\partial f}{\partial y} = 0, \frac{\partial f}{\partial z} = 4p^2z^3 + 2zq^2,$$

$$\frac{\partial f}{\partial p} = 2pz^4, \frac{\partial f}{\partial q} = 2z^2q.$$

Then the Charpit's auxiliary equations are

$$\frac{dx}{-\frac{\partial f}{\partial p}} = \frac{dy}{-\frac{\partial f}{\partial q}} = \frac{dz}{-p\frac{\partial f}{\partial p} - q\frac{\partial f}{\partial q}} = \frac{dp}{\frac{\partial f}{\partial x} + p\frac{\partial f}{\partial z}} = \frac{dq}{\frac{\partial f}{\partial y} + q\frac{\partial f}{\partial z}}$$

$$\Rightarrow \frac{dx}{-2pz^4} = \frac{dy}{-2z^2q} = \frac{dz}{-2p^2z^4 - 2z^2q^2} = \frac{dp}{4p^3z^3 + 2zpq^2} = \frac{dq}{4p^2qz^3 + 2zq^3} \quad \ldots(2)$$

Taking fourth and fifth ratio of (2), we get

$$\frac{dp}{4p^3z^3 + 2zpq^2} = \frac{dq}{4p^2qz^3 + 2zq^3}$$

or $$\frac{dp}{p} = \frac{dq}{q}.$$

Integrating, we get $\log p = \log q + \log a$, where 'a' is a constant of integration.

or $\quad p = aq.$

Putting the value of p in (1), we get

$$a^2q^2z^4 + z^2q^2 - 1 = 0$$

or $$q^2 = \frac{1}{z^2(a^2z^2+1)} \quad \text{or} \quad q = \frac{1}{z\sqrt{(a^2z^2+1)}}$$

and $$p = aq = \frac{a}{z\sqrt{(a^2z^2+1)}}.$$

Now substitute the values of p and q into

$dz = pdx + qdy$, we get

$$\therefore\ dz = \frac{a}{z\sqrt{(a^2z^2+1)}}dx + \frac{1}{z\sqrt{(a^2z^2+1)}}dy$$

or $z\sqrt{(a^2z^2+1)}\,dz = adx + dy$.

Let $\quad u^2 = a^2z^2 + 1$

$\therefore \quad udu = a^2zdz$

Then, $\quad \dfrac{u^2du}{a^2} = a\,dx + dy.$

Integrating, we get

$$\frac{u^3}{3a^2} = ax + y + b$$

or $\quad \dfrac{(a^2z^2+1)^{3/2}}{3a^2} = ax + y + b$

or $\quad (a^2z^2+1)^{3/2} = 3a^2(ax+y+b)$

or $\quad (a^2z^2+1)^3 = 9a^4(ax+y+b)^2.$

This is the required complete integral.

Example 5. *Solve* $px + qy = pq$. (GBTU(AG)–2012)

Solution. Let us assume $f \equiv px + qy - pq = 0$ ...(1)

Differentiating (1) partially w.r.t. x, y, z, p and q, we get

$$\frac{\partial f}{\partial x} = p, \frac{\partial f}{\partial y} = q, \frac{\partial f}{\partial z} = 0, \frac{\partial f}{\partial p} = x - q, \frac{\partial f}{\partial q} = y - p.$$

Then the Charpit's auxiliary equations are

$$\frac{dx}{-\frac{\partial f}{\partial p}} = \frac{dy}{-\frac{\partial f}{\partial q}} = \frac{dz}{-p\frac{\partial f}{\partial p} - q\frac{\partial f}{\partial q}} = \frac{dp}{\frac{\partial f}{\partial x} + p\frac{\partial f}{\partial z}} = \frac{dq}{\frac{\partial f}{\partial y} + q\frac{\partial f}{\partial z}}$$

$$\Rightarrow \frac{dx}{q-x} = \frac{dy}{p-y} = \frac{dz}{2pq - xp - yq} = \frac{dp}{p} = \frac{dq}{q} \quad ...(2)$$

Taking last two ratio of (2), we get $\dfrac{dp}{p} = \dfrac{dq}{q}$.

Integrating, $\log p = \log q + \log a$ or $p = aq$.

Putting this value of p into (1), we get

$aqx + qy - aq^2 = 0$ or $ax + y = aq$

or $\quad q = \dfrac{ax+y}{a}$ and $p = aq = ax + y$.

Substitute the values of p and q into the following equation $dz = pdx + qdy$, we have

$$\therefore\ dz = (ax+y)dx + \frac{(ax+y)}{a}dy$$

or $adz = (ax+y)(adx+dy)$.

Let $u = ax + y$, then $du = adx + dy$.

Then above equation becomes $adz = udu$.

Now integrating, we get

$$2az = u^2 + b \quad \text{or} \quad 2az = (ax+y)^2 + b$$

$[\because u = ax + y]$

This is the required complete integral.

Example 6. *Solve* $(p^2+q^2)y = qz$.

(UPTU–2006, VTU–2007, Hissar–2005, 11)

Solution. $f(x, y, z, p, q) = 0$ is $(p^2+q^2)y - qz = 0$.

$$\frac{\partial f}{\partial x} = 0, \frac{\partial f}{\partial y} = p^2 + q^2, \frac{\partial f}{\partial z} = -q, \frac{\partial f}{\partial p} = 2py,$$

$$\frac{\partial f}{\partial q} = 2qy - z.$$

Now, Charpit's equations are

$$\frac{dx}{-\frac{\partial f}{\partial p}} = \frac{dy}{-\frac{\partial f}{\partial q}} = \frac{dz}{-p\frac{\partial f}{\partial p} - q\frac{\partial f}{\partial q}} = \frac{dp}{\frac{\partial f}{\partial x} + p\frac{\partial f}{\partial z}} = \frac{dq}{\frac{\partial f}{\partial y} + q\frac{\partial f}{\partial z}}$$

$$\Rightarrow \frac{dx}{-2py} = \frac{dy}{-2q+z} = \frac{dz}{-2p^2y - 2q^2y + qz} = \frac{dp}{-pq} = \frac{dq}{p^2+q^2-q^2} = \frac{dp}{0}$$

We have to choose the simplest integral involving p and q.

$$\frac{dp}{-pq} = \frac{dq}{p^2} \Rightarrow -\frac{dp}{q} = \frac{dq}{p} \Rightarrow pdp + qdq = 0.$$

Integrating $p^2 + q^2 = a^2$ (say).

Putting for $p^2 + q^2$ in the equation (1), we get

$$a^2y = qz \Rightarrow q = \frac{a^2y}{z},$$

so $\quad \sqrt{a^2 - q^2} = \sqrt{a^2 - \dfrac{a^4y^2}{z^2}}$

$$p = \frac{a}{z}\sqrt{z^2 - a^2y^2}$$

Now, $\quad dz = p\,dx + q\,dy$.

Putting for p and q in (2), we get

$$dz = \frac{a}{z}\sqrt{z^2 - a^2y^2}\,dx + \frac{a^2y}{z}dy$$

$$\Rightarrow \quad z\,dz = a\sqrt{z^2 - a^2y^2}\,dx + a^2y\,dy$$

$$\frac{zdz - a^2ydy}{\sqrt{z^2 - a^2y^2}} = a\,dx$$

Integrating, we get

$$\frac{1}{2}\cdot\frac{2}{1}\sqrt{z^2 - a^2y^2} = ax + b,$$

putting $z^2 - a^2y^2 = 0$.

On squaring, $z^2 - a^2y^2 = (ax+b)^2$.

Example 7. *Find the complete integral of the following differential equation* $pxy + pq + qy = yz$.

(JNTU–2006, Kurukshetra–2006)

Solution. Assuming $f \equiv pxy + pq + qy - yz = 0$ ...(1)

Now differentiating partially w.r.t. x, y, z, p and q, we get

$$\frac{\partial f}{\partial x} = py, \frac{\partial f}{\partial y} = px + q - z, \frac{\partial f}{\partial z} = -y,$$

$$\frac{\partial f}{\partial p} = xy + q, \frac{\partial f}{\partial q} = p + y.$$

Thus the Charpit's auxiliary equations are

$$\frac{dx}{-\frac{\partial f}{\partial p}} = \frac{dy}{-\frac{\partial f}{\partial q}} = \frac{dz}{-p\frac{\partial f}{\partial p} - q\frac{\partial f}{\partial q}} = \frac{dp}{\frac{\partial f}{\partial x} + p\frac{\partial f}{\partial z}} = \frac{dq}{\frac{\partial f}{\partial y} + q\frac{\partial f}{\partial z}}$$

$$\Rightarrow \frac{dx}{-xy - q} = \frac{dy}{-p - y} = \frac{dz}{-xyp - 2pq - yq} = \frac{dp}{py - py} = \frac{dq}{px + q - z - qy} \quad ...(2)$$

Taking fourth ratio with any one ratio, we get

$$dp = 0$$

Integrating we get $p = a$.

Putting the value of $p = a$ in (1), we get

$$axy + aq + qy = yz$$

or $$q = \frac{yz - axy}{(a + y)} = \frac{-y(ax - z)}{(a + y)}.$$

Now substitute the value of p and q into the following equation $dz = pdx + qdy$

we get $$dz = adx + \frac{-y(ax - z)dy}{a + y}$$

or $$dz - adx = \frac{y(z - ax)dy}{a + y}$$

or $$\frac{dz - adx}{z - ax} + \frac{-y}{a + y}dy = 0.$$

Let $u = z - ax$ then $du = dz - adx$, above equation is then becomes

$$\frac{du}{u} + \frac{-y\,dy}{a + y} = 0 \quad \text{or} \quad \frac{du}{u} - dy + \frac{ady}{a + y} = 0.$$

Integrating, we get

$$\log u - y + a\log(a + y) = \log b$$

or $$\log u(a + y)^a - \log b = y$$

or $$\log\frac{u(a + y)^a}{b} = y \quad \text{or} \quad u(a + y)^a = be^y$$

or $$(z - ax)(a + y)^a = be^y$$

or $$z - ax = be^y(a + y)^{-a}.$$

This is the required complete integral.

Example 8. *Solve* $z^2 = pqxy$. (Anna–2009, VTU–2004)

Solution. Assuming $f \equiv z^2 - pqxy = 0$ ...(1)

Differentiating (1) partially w.r.t. x, y, z, p and q, we get,

$$\frac{\partial f}{\partial x} = -pqy, \frac{\partial f}{\partial y} = -pqx, \frac{\partial f}{\partial z} = 2z,$$

$$\frac{\partial f}{\partial p} = -qxy, \frac{\partial f}{\partial q} = -pxy.$$

Then, the Charpit's auxiliary equations are

$$\frac{dx}{-\frac{\partial f}{\partial p}} = \frac{dy}{-\frac{\partial f}{\partial q}} = \frac{dz}{-p\frac{\partial f}{\partial p} - q\frac{\partial f}{\partial q}} = \frac{dp}{\frac{\partial f}{\partial x} + p\frac{\partial f}{\partial z}} = \frac{dq}{\frac{\partial f}{\partial y} + q\frac{\partial f}{\partial z}}$$

$$\Rightarrow \frac{dx}{qxy} = \frac{dy}{pxy} = \frac{dz}{2pqxy} = \frac{dp}{-pqy + 2pz} = \frac{dq}{-pqx + 2qz} \quad ...(2)$$

From (2), we obtain

$$\frac{pdx + xdp}{pqxy - pqxy + 2pxz} = \frac{qdy + ydq}{pqxy - pqxy + 2qzy}$$

or $$\frac{pdx + xdp}{2pxz} = \frac{qdy + ydq}{2qzy} \quad \text{or} \quad \frac{d(px)}{px} = \frac{d(qy)}{qy}.$$

Integrating, we get $\log px = \log qy + \log a$.

or $$px = aqy \quad \text{or} \quad p = \frac{ay}{x}q.$$

Putting this value of p in (1) we get

$$z^2 = aq^2y^2$$

or $$q^2 = \frac{z^2}{ay^2} \quad \text{or} \quad q = \frac{z}{\sqrt{a}\,y}$$

and $$p = \frac{ay}{x}q = \frac{ay}{x}\cdot\frac{z}{\sqrt{a}y} \Rightarrow p = \sqrt{a}\frac{z}{x}.$$

Substituting the values of p and q, into the following equation

$$dz = pdx + qdy$$

we get $$dz = \sqrt{a}\frac{z}{x}dx + \frac{z}{\sqrt{a}}\cdot\frac{dy}{y}$$

or $$\frac{dz}{z} = \sqrt{a}\frac{dx}{x} + \frac{1}{\sqrt{a}}\cdot\frac{dy}{y}$$

or $$\sqrt{a}\frac{dz}{z} = a.\frac{dx}{x} + \frac{dy}{y}.$$

Integrating, we get

$$\sqrt{a}\log z = a\log x + \log y + \log b$$

$$z^{\sqrt{a}} = bx^a y.$$

This is the required complete integral.

EXERCISE 40.2

Using Charpit's method, find the complete integral of the following differential equations:

1. $(p+q)(px+qy)-1=0.$
2. $px^5-4q^3x^2+6x^2z-2=0.$
3. $yzp^2=q.$
4. $2(pq+py+qx)+x^2+y^2=0.$
5. $2z+p^2+2y^2+qy=0.$ (JNTU–2005, Kurukshetra–2005)
6. $p^2-y^2q+x^2=y^2.$
7. $z=pq.$
8. $p^2+q^2-2px-2qy+2xy=0.$
9. $q-3y^2=0.$
10. $p^2+q^2-2px-2qy+1=0.$
11. $z(p^2-q^2)=x-y.$
12. $yp=2yx+\log q.$
13. $z^2(p^2+q^2+1)=c^2.$
14. $q^2y^2=z(z-px).$
15. $p^2=z^2(1-pq).$
16. $p^3+q^3=27z.$
17. $(x^2+y^2)(p^2+q^2)=1.$
18. $x^2p^2+y^2q^2=z^2.$
19. $16p^2z^2+9q^2z^2+4z^2=0.$
20. $p(1+q^2)+(b-z)q=0.$
21. $p^2x+q^2y=z.$
22. $2[z+px+qy]=yp^2.$
23. $px+qy=z[1+pq]^{1/2}.$
24. $(x^2-y^2)pq-xy(p^2-q^2)-1=0.$
25. $2(y+zq)=q(xp+yq).$
26. $xp+3yq=2(z-x^2q^2).$

Hint to Selected Problems

2. The Charpit is auxiliary equation becomes

$$\frac{dx}{-x^5}=\frac{dy}{12q^2x^2}=\frac{dz}{-px^5+12q^3x^2}$$

$$=\frac{dp}{5px^4-8q^3+12xz+6x^2p}=\frac{dq}{6qx^2}$$

Then take first and last ratio.

8. The Charpit's auxiliary equation becomes

$$\frac{dx}{2x-2p}=\frac{dy}{2y-2q}=\frac{dz}{2px-2p^2+2qy-2q^2}$$

$$=\frac{dp}{-2p+2y}=\frac{dq}{-2q+2x}$$

$$\Rightarrow \frac{dp+dq}{2(x+y-p-q)}=\frac{dx+dy}{2(x+y-p-q)}\Rightarrow (p-x)+(q-y)=a.$$

Then put this value in the given equation.

11. The Charpit's auxiliary equation becomes

$$\frac{dx}{-2zp}=\frac{dy}{2qz}=\frac{dz}{-2zp^2+2q^2z}=\frac{dp}{-1+p^3-pq^2}=\frac{dq}{1+p^2q-q^3}$$

12. The Charpit's auxiliary equation becomes

$$\frac{dx}{-y}=\frac{dy}{1/q}=\frac{dz}{-yp+1}=\frac{dp}{-2y}=\frac{dq}{p-2x}$$

Then taking first and fourth ratios.

17. The Charpit's auxiliary equation becomes

$$\frac{dx}{-2p(x^2+y^2)}=\frac{dy}{-2q(x^2+y^2)}=\frac{dz}{-(2p^2+2q^2)(x^2+y^2)}$$

$$=\frac{dp}{2x(p^2+q^2)}=\frac{dq}{2y(p^2+q^2)}.$$

Then, we have $pdx+qdy=-xdp-ydq \Rightarrow px+qy=0.$

ANSWERS

1. $z\sqrt{1+a}=2\sqrt{ax+y}+b$ **2.** $z=-\frac{1}{3}a^3e^{9/x^2}+\frac{1}{9}+\frac{1}{3x^2}+(ay+b)e^{3/x^2}$ **3.** $z^2=\frac{(x+b)^2}{(a-y^2)}$

4. $2z=ax-x^2+ay-y^2+\frac{1}{2}(x-y)\sqrt{2(x-y)^2+a^2}+\frac{a^2}{2\sqrt{2}}\log[\{\sqrt{2}(x-y)\}+\sqrt{2(x-y)^2+a^2}]+b$

5. $y^2\{(x-a)^2+y^2+2z\}=b$ **6.** $z=\frac{x}{2}\sqrt{a^2-x^2}+\frac{a^2}{2}\sin^{-1}\left(\frac{x}{a}\right)-\frac{a^2}{y}-y+b$ **7.** $2\sqrt{z}=\sqrt{a}.x+\frac{1}{\sqrt{a}}y+b$

8. $2z=x^2+y^2+ax+ay+\frac{1}{\sqrt{2}}\left[(x-y)\sqrt{(x-y)^2-\frac{a^2}{2}}-\frac{a^2}{2}\log\left\{(x-y)+\sqrt{(x-y)^2\frac{a^2}{2}}\right\}\right]+b$; **9.** $z=ax+y^3+b.$

10. $(a^2+1)z=\frac{u^2}{2}+\frac{u}{2}\sqrt{u^2-(a^2+1)}-\frac{a^2+1}{2}\log[u+\sqrt{u^2-(a^2+1)}]+b$, where $u=ax+y$

11. $z^{3/2}=(x+a)^{3/2}+(y+a)^{3/2}+b$ **12.** $az=ax^2+a^2x+e^{ay}+ab$ **13.** $(1+a^2)(c^2-z^2)=(x+ay+b)^2$ **14.** $z^{2a^2/(-1\pm\sqrt{(1+4a^2)})}=bxy^a$

15. $\log\left[\frac{\sqrt{1+az^2}-1}{z\sqrt{a}}\right]+\sqrt{1+az^2}=x+b+ay$ **16.** $(1+a^3)z^2=8(x+ay+b)^3$ **17.** $z=\frac{a}{2}\log(x^2+y^2)+\sqrt{1-a^2}\tan^{-1}\frac{y}{x}+b$

18. $\sqrt{1+a^2}.\log z=a\log x+\log y+b$ **19.** $\sqrt{(16a^2+9)}\sqrt{1-z}+2(ax+y)=b$ **20.** $2\sqrt{[c(z-b)-1]}=x+cy+c$

21. $\sqrt{(1+a)}.z=\sqrt{ax}+\sqrt{y}+b$ **22.** $z=\frac{ax}{y^2}-\frac{a^2}{4y^3}+\frac{b}{y}$ **23.** $v=\frac{ax+y}{\sqrt{a}.z}$ **24.** $z=\frac{a}{2}\log(x^2+y^2)+\frac{1}{a}\tan^{-1}\left(\frac{y}{x}+b\right)$

25. $(2z-ax^2)+\sqrt{(2z-ax^2)^2-8y^2}=by^2$ **26.** $z=a^2+\frac{ay}{x}+bx^2$

Objective Evaluations

Fill in the Blanks

1. The Lagrange's method can be used to solve _________ order PDE.
2. The general method to solve PDE is known as _________ method.
3. The complete integral of $px + qy = pq$ is _________.
4. The multipliers for $\frac{dx}{3x+y-z} = \frac{dy}{x+y-z} = \frac{dz}{z(z-y)}$ are _________.

True/False

Write 'T' for True and 'F' for False statement.

1. The Charpit's method can be used to solve non-linear equation. **(T/F)**
2. The complete method of $4z = pq$ is $az = (x + ay + b)^2$. **(T/F)**
3. The complete integral of $zpq = p + q$ is $z^2 = 2(a+1)(x + y/a) + b$. **(T/F)**

Multiple Choice Questions

Choose the most appropriate one.

1. The complete integral of $f(p, q) = 0$ is :
   (a) $z = ax + b$
   (b) $z = ax + by + c$
   (c) $z = ax + f(a).y + b$
   (d) none of these
2. The complete integral of $z = pq$ is :
   (a) $2\sqrt{z} = \sqrt{ax} + b$
   (b) $2\sqrt{z} = \sqrt{ax} + \frac{1}{\sqrt{a}} y$
   (c) $z = \sqrt{ax} + y$
   (d) $2\sqrt{z} = \sqrt{ax} + \frac{1}{\sqrt{a}} y + b$
3. The complete integral of $q = 3y^2$ is :
   (a) $z = ax + b$
   (b) $z = ax + y$
   (c) $z = ax + y^3 + b$
   (d) none of these
4. The complete integral of $pq = p + q$ is :
   (a) $z = a + by + c$
   (b) $z = ax + \frac{a}{a-1} y + c$
   (c) $z = ax + \frac{a}{a-1} y$
   (d) none of these

ANSWERS

Fill in the Blanks

1. First
2. Charpit's
3. $az = \frac{1}{2}(y + ax)^2 + b$
4. $-1, 3, 1$

True/False

1. T
2. F
3. T

Multiple Choice Questions

1. (c)
2. (d)
3. (c)
4. (b)

FFFFFF

CHAPTER

41 Partial Differential Equations of Second Order

41.1 INTRODUCTION

A partial differential equation is said to be of the second order when it includes at least one of the partial differential coefficients r, s, t but none of higher order, where

$$r = \frac{\partial^2 z}{\partial x^2} = \frac{\partial p}{\partial x},\quad s = \frac{\partial^2 z}{\partial x \partial y} = \frac{\partial p}{\partial y} = \frac{\partial q}{\partial x},\quad t = \frac{\partial^2 z}{\partial y^2} = \frac{\partial q}{\partial y}$$

Solved Examples

Example 1. *Solve* $xr + p = 9x^2y^3$.

Solution. The given equation can be written as

$$\frac{\partial p}{\partial x} + \frac{1}{x}.p = 9xy^3 \quad \ldots(1)$$

which is a linear equation in p.

$\therefore$ I.F. $= e^{\int (1/x)dx} = e^{\log x} = x$.

Therefore, solution of (1) is given by

$$px = \int 9x^2y^3\,dx + f(y)$$

$$\Rightarrow \quad px = 3x^3y^3 + f(y)$$

$$\Rightarrow \quad p = \frac{\partial z}{\partial x} = 3x^2y^3 + \frac{1}{x}f(y).$$

On integrating, we get

$$z = x^3y^3 + \log x f(y) + F(y).$$

Example 2. *Solve* $t - xq = x^2$.

Solution. The given equation can be written as

$$\frac{\partial q}{\partial y} - xq = x^2 \quad \ldots(1)$$

which is linear in q.

$\therefore$ I.F. $= e^{\int -x\,dy} = e^{-xy}$.

The solution of (1) is given by

$$qe^{-xy} = \int x^2 e^{-xy}\,dy + f(x) = -xe^{-xy} + f(x)$$

$$\Rightarrow \quad q = \frac{\partial z}{\partial y} = -x + f(x)e^{xy}.$$

On integrating, we get

$$z = -xy + f(x)\int e^{xy}\,dy + F(x)$$

$$= -xy + \frac{1}{x}f(x)e^{xy} + F(x).$$

Example 3. *Solve* $t = \sin xy$.

Solution. The given equation can be written as

$$\frac{\partial^2 z}{\partial y^2} = \sin xy.$$

On integrating, we have $\frac{\partial z}{\partial y} = -\frac{1}{x}\cos xy + f(x)$.

Again integrating, we get

$$z = -\frac{1}{x^2}\sin xy + y\,f(x) + F(x).$$

Example 4. *Solve* $yt - q = xy$.

Solution. The given equation can be written as

$$\frac{\partial q}{\partial y} - \frac{1}{y}q = x \quad \ldots(1)$$

which is linear in q.

$$\text{I.F.} = e^{-\int \left(\frac{1}{y}\right)dy} = e^{-\log y} = \frac{1}{y}$$

Therefore, solution of (1) is given by

$$q.\frac{1}{y} = \int x.\frac{1}{y}dy + f(x) = x\log y + f(x)$$

$$\Rightarrow \quad q = \frac{\partial z}{\partial y} = xy\log y + y\,f(x).$$

On integrating, w.r.t. y we get

$$z = x\int y\log y\,dy + f(x)\int y\,dy + F(x)$$

$$= x\left[\frac{y^2}{2}.\log y - \int \frac{1}{y}.\frac{y^2}{2}dy\right] + \frac{y^2}{2}f(x) + F(x)$$

$$= \frac{1}{2}xy^2 \log y - \frac{1}{4}xy^2 + \frac{y^2}{2}f(x) + F(x).$$

Example 5. *Solve* $s - t = \dfrac{x}{y^2}$.

Solution. The given equation can be written as

$$\frac{\partial p}{\partial y} - \frac{\partial q}{\partial y} = \frac{x}{y^2}. \qquad \ldots(1)$$

Integrating w.r.t. y, we get $p - q = -\dfrac{x}{y} + f(x)$.

Now by Lagrange's method, we have

$$\frac{dx}{1} = \frac{dy}{-1} = \frac{dz}{-\frac{x}{y} + f(x)}.$$

Taking first two members, we get $dx + dy = 0$

$\Rightarrow \quad x + y = c_1$

Now, taking first and last member, we get

$$dz = -\frac{x}{y}dx + f(x)dx$$

$$\Rightarrow \quad dz = -\frac{x}{c_1 - x}dx + f(x)dx$$

$$= \left(1 - \frac{c_1}{c_1 - x}\right)dx + f(x)dx.$$

On integrating, we get

$$z = x + c_1 \log(c_1 - x) + \phi(x) + c_2$$

$$= c_1 \log(c_1 - x) + \psi(x) + c_2$$

$$= (x + y)\log y + \psi(x) + F(x + y).$$

Example 6. *Find the surface passing through the parabolas* $z = 0, y^2 = 4ax$ *and* $z = 1, y^2 = -4ax$ *and satisfying the equation* $xr + 2p = 0$.

Solution. The given equation can be written as

$$x\frac{\partial p}{\partial x} + 2p = 0 \quad \Rightarrow \quad x^2\frac{\partial p}{\partial x} + 2xp = 0. \qquad \ldots(1)$$

On integrating w.r.t. x, we have $x^2 p = f(y)$

$$\Rightarrow \quad p = \frac{\partial z}{\partial x} = \frac{1}{x^2}f(y).$$

$$\Rightarrow \quad z = -\frac{1}{x}f(y) + F(y). \qquad \ldots(2)$$

Putting $z = 0, x = \dfrac{y^2}{4a}$, we get

$$0 = -\frac{4a}{y^2}f(y) + F(y). \qquad \ldots(3)$$

Also, putting $z = 1$ and $x = -y^2/4a$ in (2),

we get $1 = \dfrac{4a}{y^2}f(y) + F(y). \qquad \ldots(4)$

Adding (3) and (4), we have

$$2F(y) = 1 \Rightarrow F(y) = \frac{1}{2}.$$

Put in (3), we get $f(y) = y^2/8a$.

Putting, all these values in (2), we get the required surface $z = -\dfrac{y^2}{8ax} + \dfrac{1}{2}$.

$\Rightarrow \quad 8axz = 4ax - y^2.$

Example 7. *Find the satisfying* $t = 6x^3 y$, *containing the two lines* $y = 0 = z, y = 1 = z$.

Solution. The given equation can be written as $\dfrac{\partial q}{\partial y} = 6x^3 y$.

On integrating w.r.t. y, we get

$$q = 3x^3 y^2 + f(x)$$

$$\Rightarrow \quad \frac{\partial z}{\partial y} = 3x^3 y^2 + f(x) \qquad \ldots(1)$$

$$\Rightarrow \quad z = x^3 y^3 + yf(x) + F(x). \qquad \ldots(2)$$

If (1) contains the line $y = 0 = z$ and $y = 1 = z$.

Then $\quad 0 = F(x) \qquad \ldots(3)$

and $\quad 1 = x^3 + f(x) + F(x) \;\Rightarrow\; f(x) = 1 - x^3.$

Hence, the required surface is given by

$$z = x^3 y^3 + y(1 - x^3).$$

Example 8. *Show that a surface satisfying* $r = 6x + 2$ *and touching* $z = x^3 + y^3$ *along its section by the plane* $x + y + 1 = 0$ *is* $z = x^3 + y^3 + (x + y + 1)^2$.

Solution. The given equation can be written as

$$\frac{\partial p}{\partial x} = 6x + 2.$$

On integrating w.r.t. x, we get $p = 3x^2 + 2x + f(y)$.

$$\Rightarrow \quad \frac{\partial z}{\partial x} = 3x^2 + 2x + f(y)$$

$$\therefore \quad z = x^3 + x^2 + xf(y) + F(y). \qquad \ldots(1)$$

$$\Rightarrow \quad p = \frac{\partial z}{\partial x} = 3x^2 + 2x + f(y)$$

and $\quad q = \dfrac{\partial z}{\partial y} = xf'(y) + F'(y).$

Given that $z = x^3 + y^3$. $\qquad \ldots(2)$

Therefore, we get $p = 3x^2, q = 3y^2$

$[\because dz = pdx + qdy]$

Now, if (1) touches (2) along its section by the plane $x + y + 1 = 0$, then the values of p and q for any point on this plane should be equal *i.e.*,

$$3x^2 + 2x + f(y) = 3x^2 \qquad \ldots(3)$$

$$xf'(y) + F'(y) = 3y^2 \qquad \ldots(4)$$

and $\quad x + y + 1 = 0 \qquad \ldots(5)$

From (3) and (5), we get $f(y) = -2x = 2(y + 1)$.

$\therefore \quad f'(y) = 2$

Putting this value in (4), we get

$$F'(y) = 3y^2 - 2x = 3y^2 + 2(y + 1).$$

On integrating, we have

$$F(y) = y^3 + y^2 + 2y + C.$$

From equation (1)

$$z = x^3 + x^2 + x.\ 2(y + 1) + y^3 + y^2 + 2y + C. \qquad \ldots(6)$$

Equating the values of z from (2) and (6), when $y = -(x+1)$, we have

$$x^3 - (x+1)^3 = x^3 + x^2 + 2x(-x) - (x+1)^3 + (x+1)^2 - 2(x+1) + C.$$

$\Rightarrow \quad C = 1.$

Hence, the required surface is

$$z = x^3 + x^2 + 2x(y+1) + y^3 + y^2 + 2y + 1$$
$$= x^3 + y^3 + (x+y+1)^2.$$

EXERCISE 41.1

Solve the following differential equations:

1. $xys = 1$.
2. $s = 2x + 2y$.
3. $r = 2y^2$.
4. $\log s = x + y$.
5. $s = \dfrac{x}{y} + a$.
6. $rx = (n-1)p$.
7. $ys + p = \cos(x+y) - y\sin(x+y)$.
8. $p + r + s = 1$.
9. Find a surface satisfying $r + s = 0$ and touching the elliptic paraboloid $z = 4x^2 + y^2$ along its section by the plane $y = 2x + 1$.

Hint to Selected Problems

1. The given equation can be written as $\dfrac{\partial^2 z}{\partial x \partial y} = \dfrac{1}{xy}$. Now integrate w.r.t. y and x separately.
2. Do same as (1).
3. The given equation can be written as $\dfrac{\partial p}{\partial x} = 2y^2$. Then integrating w.r.t. x.
4. Do same as (1).
5. Do same as (1).
8. The given equation can be written as $\dfrac{\partial z}{\partial x} + \dfrac{\partial p}{\partial x} + \dfrac{\partial q}{\partial x} = 1$. On integrating w.r.t. x we get $p + q = x + f(y) - z$. Then use Lagrange's subsidiary equation.
9. Here, we have $\dfrac{\partial p}{\partial x} + \dfrac{\partial q}{\partial x} = 0$. On integrating w.r.t. x, we have $p + q = f(y)$. Now, using Lagrange's subsidiary equation.

ANSWERS

1. $z = \log x \log y + \phi(x) + F(y)$
2. $y = x^2 y + xy^2 + \phi(x) + F(y)$
3. $z = x^2 y^2 + xf(y) + F(y)$
4. $z = e^{x+y} + \phi(y) + F(x)$
5. $z = \dfrac{x^2}{2} \log y + axy + \phi(y) + F(x)$
6. $z = \dfrac{1}{n} x^n f(y) + F(y)$
7. $yz = y\sin(x+y) + \phi(y) + F(x)$
8. $z = x - y + e^{-y}\phi(y) + e^{-y}F(x-y)$
9. $z + 4x^2 + y^2 - 8xy + 8x - 4y + 2 = 0$.

41.2 LINEAR HOMOGENEOUS PARTIAL DIFFERENTIAL EQUATION WITH CONSTANT COEFFICIENTS

A partial differential equation, which is linear with respect to the dependent variable and its derivatives and in which the coefficient are not function of the dependent variable but merely constants is called a linear partial differential equation with constant coefficients, *i.e.*, an equation of the form

$$a_0 \frac{\partial^n z}{\partial x^n} + a_1 \frac{\partial^n z}{\partial x^{n-1} \partial y} + a_2 \frac{\partial^n z}{\partial x^{n-2} \partial y^2} + \ldots + a_n \frac{\partial^n z}{\partial y^n} = F(x, y) \qquad \ldots(1)$$

where $a_0, a_1, a_2, \ldots a_n$ are constants and $F(x, y)$ is a function of nth order with constant coefficients where all the partial derivative are of nth order is known as linear homogeneous partial differential equation with constant coefficients.

Putting $\dfrac{\partial}{\partial x} = D$ and $\dfrac{\partial}{\partial y} = D'$, (1) can be written as

$$(a_0 D^n + a_1 D^{n-1} D' + a_2 D^{n-2} D'^2 + \ldots + a_n D'^n) z = F(x, y) \quad \Rightarrow f(D, D')z = F(x, y). \qquad \ldots(2)$$

41.3 SOLUTION OF LINEAR PARTIAL DIFFERENTIAL EQUATION

The complete solution of (2) consists of two parts :

(a) The complementary function (C.F.) : which is the complete solution of the equation $f(D, D')z = 0$. It must contain n arbitrary functions, where n is the order of the differential equation.

(b) The particular integral (P.I.) : which is the particular solution of $f(D, D')z = F(x, y)$. Therefore complete solution of (2) is given by $z = \text{C.F.} + \text{P.I.}$

41.4 RULE FOR FINDING THE COMPLEMENTARY FUNCTION (C.F.)

Consider the equation $a_0 \dfrac{\partial^2 z}{\partial x^2} + a_1 \dfrac{\partial^2 z}{\partial x \partial y} + a_2 \dfrac{\partial^2 z}{\partial y^2} = 0$ or $(a_0 D^2 + a_1 DD' + a_2 D'^2)z = 0.$ ...(1)

WORKING PROCEDURE

Step 1. *Putting $D = m$ and $D' = 1$ in (1), we get*

$$a_0 m^2 + a_1 m + a_2 = 0. \quad ...(2)$$

This is the auxiliary equation.

Step 2. *Solve the auxiliary equation (2). Two cases will arise :*

Case (i). *If the auxiliary equation has the real and different roots, say m_1, m_2.*
Then, C.F. $= f_1(y + m_1 x) + f_2(y + m_2 x)$.

Case (ii). *If the auxiliary equation has equal (repeated) roots*
Then C.F. $= f_1(y + mx) + x\, f_2(y + mx)$.

Solved Examples

Example 1. *Solve* $2r + 5s + 2t = 0$.

Solution. We know that $r = \dfrac{\partial^2 z}{\partial x^2} = D^2 z, s = \dfrac{\partial^2 z}{\partial x \partial y} = DD'z$

and $\dfrac{\partial^2 z}{\partial y^2} = D'^2 z$.

Then the given equation can be written as

$$(2D^2 + 5DD' + 2D'^2)z = 0.$$

The auxiliary equation is $2m^2 + 5m + 2 = 0$

or $(2m+1)(m+2) = 0 \quad \therefore \quad m = -\dfrac{1}{2}, -2.$

Hence, the required solution is

$$z = f\left(y - \frac{1}{2}x\right) + \psi(y - 2x)$$

or $z = \phi(2y - x) + \psi(y - 2x)$.

Example 2. *Solve* $25r - 40s + 16t = 0$.

Solution. The given equation can be written as

$$(25D^2 - 40DD' + 16D'^2)z = 0.$$

It's A.E. is $25m^2 - 40m + 16 = 0$

or $(5m - 4)^2 = 0 \quad \therefore \quad m = \dfrac{4}{5}, \dfrac{4}{5}.$

Hence, the solution is

$$z = f_1\left(y + \frac{4}{5}x\right) + x f_2\left(y + \frac{4}{5}x\right)$$

or $z = \phi(5y + 4x) + x\psi(5y + 4x)$.

Example 3. *Solve* $(D^4 - D'^4)z = 0.$ (UPTU(Q. Bank)-2002)

Solution. The given equation is $(D^4 - D'^4)z = 0$.

It's A.E. is $m^4 - 1 = 0$ or $(m^2 - 1)(m^2 + 1) = 0$.

$\therefore \quad m = 1, -1, \pm i.$

Hence, the solution is

$$z = f_1(y + x) + f_2(y - x) + f_3(y + ix) + f_4(y - ix).$$

Example 4. *Solve* $(D^3 - 6D^2 D' + 11DD'^2 - 6D'^3)z = 0$.

Solution. The given equation is

$$(D^3 - 6D^2 D' + 11DD'^2 - 6D'^3)z = 0.$$

It's A. E. is $m^3 - 6m^2 + 11m - 6 = 0$.

or $(m - 1)(m - 2)(m - 3) = 0.$

$\therefore \quad m = 1, 2, 3.$

Hence, the solution is

$$z = f_1(y + x) + f_2(y + 2x) + f_3(y + 3x).$$

Example 5. *Solve* $\dfrac{\partial^3 z}{\partial x^3} - 4\dfrac{\partial^3 z}{\partial x^2 \partial y} + 4\dfrac{\partial^3 z}{\partial x \partial y^2} = 0.$

Solution. The equation can be written as

$$(D^3 - 4D^2 D' + 4DD'^2)z = 0.$$

It's A.E. is $m^3 - 4m^2 + 4m = 0$

or $m(m - 2)^2 = 0 \quad \Rightarrow \quad m = 0, 2, 2.$

Hence, the solution is

$$z = f_1(y) + f_2(y + 2x) + x f_3(y + 2x).$$

Example 6. *Solve* $r - 4s + 4t = 0$ (GBTU (AG)-2010)

Solution. Its A.E is $m^2 - 4m + 4 = 0 \Rightarrow m = 2, 2$

C.F. $= f_1(y + 2x) + x f_2(y + 2x)$

P.I. $= 0$

Hence the complete solution is

$$z = \text{C.F.} + \text{P.I} = f_1(y + 2x) + x f_2(y + 2x)$$

EXERCISE 41.2

Solve the following differential equation :

1. $r = a^2 t$.
2. $(4D^2 + 12DD' + 9D'^2)z = 0.$ (PTU-2010)
3. $\dfrac{\partial^3 z}{\partial x^3} - 3\dfrac{\partial^3 z}{\partial x^2 \partial y} + 2\dfrac{\partial^3 z}{\partial x \partial y^2} = 0.$
4. $(D^3 - 6D^2 D' + 12DD'^2 - 8D'^3)z = 0.$
5. $\dfrac{\partial^3 z}{\partial x^3} - 7\dfrac{\partial^3 z}{\partial x \partial y^2} + 6\dfrac{\partial^3 z}{\partial y^3} = 0.$
6. $D^4 - 2D^3 D' + 2DD'^3 - D'^4 = 0$. (UPTU(Q. Bank)-2002)

Hint to Selected Problems

1. The given equation can be written as $(D^2 - a^2D'^2)z = 0$
Auxiliary equation is given by $m^2 - a^2 = 0 \Rightarrow m = \pm a$
$\Rightarrow \quad z = f_1(y+ax) + f_2(y-ax)$.
2. Auxiliary equation is given by $4m^2 + 12m + 9 = 0$.
3. The given equation can be written as $(D^3 - 3D^2D' + 2DD'^2)z = 0$
Auxiliary equation is $m^3 - 3m^2 + 2m = 0 \Rightarrow m = 0, 1, 2$.
4. Auxiliary equation is given by
$m^3 - 6m^2 + 12m - 8 = 0 \quad \Rightarrow m = 2, 2, 2$.
5. The given equation can be written as $(D^3 - 7DD'^2 + 6D'^3)z = 0$
A.E. is given by $m^3 - 7m^2 + 6 = 0 \quad \Rightarrow \quad m = 1, 2, -3$.

ANSWERS

1. $z = f_1(y+ax) + f_2(y-ax)$
2. $z = f_1(2y-3x) + xf_2(2y-3x)$
3. $z = f_1(y) + f_2(y+x) + f_3(y+2x)$
4. $z = f_1(y+2x) + xf_2(y+2x) + x^2 f_3(y+2x)$
5. $z = f_1(y+x) + f_2(y+2x) + f_3(y-3x)$
6. $z = f_1(y-x) + f_2(y+x) + xf_3(y+x) + x^2 f_4(y+x)$

41.5 METHOD OF FINDING PARTICULAR INTEGRAL OF A LINEAR HOMOGENEOUS PARTIAL DIFFERENTIAL EQUATION

The given partial differential equation is $f(D, D')z = F(x, y)$.

Particular integral (P.I.) $= \dfrac{1}{f(D, D')} F(x, y)$.

Case 1. When $F(x, y)$ is the form $x^m y^n$ or a rational integer algebraic function of x and y.

$$\text{P.I.} = \frac{1}{f(D, D')} x^m y^n = [f(D, D')]^{-1} x^m y^n.$$

Expand $[f(D, D')]^{-1}$ in ascending power of D or D' and operate on $x^m y^n$ term by term.

REMARK

- Here : $\dfrac{1}{D}$ means integration w.r.t. x, $\dfrac{1}{D'}$ means integration w.r.t. y and so on.

Solved Examples

Example 1. *Solve* $(D^2 - 2DD' + D'^2) = 12xy$.

Solution. The auxiliary equation is $m^2 - 2m + 1 = 0$.
or $(m-1)^2 = 0 \quad \therefore \quad m = 1, 1$.
$\therefore$ C.F. $= f_1(y+x) + xf_2(y+x)$.

$$\text{Now P.I.} = \frac{1}{D^2 - 2DD' + D'^2} 12xy$$
$$= \frac{1}{(D-D')^2} .12xy$$
$$= \frac{1}{D^2}\left(1 - \frac{D'}{D}\right)^{-2} .12xy$$
$$= \frac{1}{D^2}\left(1 + \frac{2D'}{D} + \dots\right) .12xy$$
$$= \frac{1}{D^2}(12xy) + \frac{2}{D^3} D'(12xy)$$
$$= 2x^3y + \frac{2}{D^3} 12x = 2x^3y + x^4.$$

Hence, the required general solution is
$z = \text{C.F.} + \text{P.I.}$
or $\quad z = f_1(y+x) + xf_2(y+x) + 2x^3y + x^4$.

Example 2. *Solve* $(D^2 + 3DD' + 2D'^2)z = x + y$. (IAS-1986, 94)

Solution. The A.E. is $m^2 + 3m + 2 = 0$ or $(m+1)(m+2) = 0$
$\therefore \quad m = -1, -2$.
$\therefore$ C.F. $= f_1(y-x) + f_2(y-2x)$.

$$\text{Now P.I.} = \frac{1}{D^2 + 3DD' + 2D'^2}(x+y)$$
$$= \frac{1}{D^2}\left(1 + \frac{3D'}{D} + 2\frac{D'^2}{D^2}\right)^{-1}(x+y)$$
$$= \frac{1}{D^2}\left(1 - \frac{3D'}{D} + \dots\right)(x+y)$$
$$= \frac{1}{D^2}(x+y) - \frac{3}{D^3} D'(x+y)$$
$$= \frac{x^3}{6} + \frac{x^2}{2} y - \frac{3}{D^3}.1 = \frac{x^3}{6} + \frac{1}{2}x^2 y - \frac{3x^3}{6}$$
$$= -\frac{1}{3}x^3 + \frac{1}{2}x^2 y.$$

Hence, the complete solution is
$$z = f_1(y-x) + f_2(y-2x) - \frac{1}{3}x^3 + \frac{1}{2}x^2 y.$$

Example 3. *Solve* $r + (a+b)s + abt = xy$. (GBTU(AG)-2011)

Solution. The given equation can be written as
$$\{D^2 + (a+b)DD' + abD'^2\}z = xy.$$
It's A. E. is $m^2 + (a+b)m + ab = 0$
or $(m+a)(m+b) = 0 \Rightarrow \quad m = -a, -b$.
$\therefore$ C.F. $= f_1(y-ax) + f_2(y-bx)$.

Now $\text{P.I.} = \frac{1}{D^2+(a+b)DD'+abD'^2}.xy$

$$= \frac{1}{D^2}\left\{1+(a+b)\frac{D'}{D}+ab\frac{D'^2}{D^2}\right\}^{-1}.xy$$

$$= \frac{1}{D^2}\left\{1-(a+b)\frac{D'}{D}+...\right\}.xy$$

$$= \frac{1}{D^2}(xy)-(a+b)\frac{1}{D^3}D'.xy$$

$$= \frac{1}{6}x^3y-(a+b)\frac{1}{D^3}x$$

$$= \frac{1}{6}x^3y-\frac{(a+b)x^4}{24}.$$

Hence, the complete solution is

$$z = f_1(y-ax)+f_2(y-bx)$$
$$+\frac{1}{6}x^3y-\frac{1}{24}(a+b)x^4.$$

Example 4. *Find a real function V of x and y, reducing to zero when y = 0 and satisfying*

$$\frac{\partial^2 V}{\partial x^2}+\frac{\partial^2 V}{\partial y^2} = -4\pi(x^2+y^2).$$

Solution. The real function V will be given by the P.I. of the given equation.

Now $\text{P.I.} = \frac{1}{D^2+D'^2}\{-4\pi(x^2+y^2)\}.$

$$= \frac{1}{D^2}\left[1+\frac{D'^2}{D^2}\right]^{-1}\{-4\pi(x^2+y^2)\}$$

$$= \frac{1}{D^2}\left[1-\frac{D'^2}{D^2}+...\right]\{-4\pi(x^2+y^2)\}$$

$$= \frac{1}{D^2}\{-4\pi(x^2+y^2)\}-\frac{1}{D^4}\{-4\pi.2\}$$

$$= -4\pi\left\{\frac{x^4}{12}+\frac{x^2y^2}{2}\right\}-\left\{-8\pi\frac{x^4}{24}\right\}$$

$$= -\frac{\pi x^4}{3}-2\pi x^2y^2+\frac{\pi x^4}{3} = -2\pi x^2y^2.$$

Hence, the required general solution is

$$V = -2\pi x^2y^2.$$

EXERCISE 41.3

Solve the following differential equations:

1. $r-a^2t = x^2.$
2. $\frac{\partial^2 z}{\partial x^2}+3\frac{\partial^2 z}{\partial x\,\partial y}+2\frac{\partial^2 z}{\partial y^2} = 6(x+y).$
3. $(2D^2-5DD'+2D'^2)z = 24(y-x).$
4. $\frac{\partial^2 z}{\partial x^2}+3\frac{\partial^2 z}{\partial x\,\partial y}+2\frac{\partial^2 z}{\partial y^2} = 2x+3y.$
5. $\frac{\partial^3 z}{\partial x^3}-\frac{\partial^3 z}{\partial y^3} = x^3y^3.$
6. $\frac{\partial^2 z}{\partial x^2}+2\frac{\partial^2 z}{\partial x\partial y}+\frac{\partial^2 z}{\partial y^2} = x^2+xy+y^2.$
7. $\frac{\partial^2 z}{\partial x^2}-2\frac{\partial^2 z}{\partial x\,\partial y}+\frac{\partial^2 z}{\partial y^2} = x^2+y.$

Hint to Selected Problems

1. The given equation can be written as $(D^2-a^2D'^2)z = x^2$

A.E. $= m^2-a^2 = 0 \Rightarrow m = \pm a$

C.F. $= f_1(y+ax)+f_2(y-ax)$

$$\text{P.I.} = \frac{1}{(D^2-a^2D'^2)}.x^2 = \frac{1}{D^2\left(1-\frac{a^2D'^2}{D^2}\right)}.x^2$$

$$= \frac{1}{D^2}\left[1+\frac{a^2D'^2}{D^2}+\frac{a^4D'^4}{D^4}+...\right]x^2 = \frac{1}{D^2}(x^2) = \frac{1}{12}x^4.$$

2. A.E. $= m^2+3m+2 = 0 \Rightarrow m = -1,-2$

C.F. $= f_1(y-x)+f_2(y-2x)$

$$\text{P.I.} = \frac{1}{D^2+3DD'+2D'^2}6(x+y)$$

$$= \frac{1}{D^2}\left[\left(1+\frac{3D'}{D}+\frac{2D'^2}{D^2}\right)^{-1}\right]6(x+y).$$ Now expanding.

4. $m = -1,-2 \Rightarrow$ C.F. $= f_1(y-x)+f_2(y-2x)$

$$\text{P.I.} = \frac{1}{D^2+3DD'+2D'^2}(2x+3y)$$

$$= \frac{1}{D^2\left(1+\frac{3D'}{D}+\frac{2D'^2}{D^2}\right)}(2x+3y)$$

$$= \frac{1}{D^2}\left[1-\frac{3D'}{D}-\frac{2D'^2}{D^2},...\right](2x+3y).$$

ANSWERS

1. $z = f_1(y+ax)+f_2(y-ax)+\frac{1}{12}x^4$
2. $z = f_1(y-x)+f_2(y-2x)-2x^3+3x^2y$
3. $z = f_1(2y+x)+f_2(y+2x)+6x^2y+3x^3$
4. $z = f_1(y-x)+f_2(y-2x)-\frac{7}{6}x^3+\frac{3}{2}x^2y$
5. $z = f_1(y+x)+f_2(y+\omega x)+f_3(y+\omega^2x)+\frac{x^6y^3}{120}+\frac{x^9}{10080}$ where ω is one of the cube roots of unity.
6. $z = f_1(y-x)+f_2(y-x)+\frac{1}{4}(x^4-2x^3y+2x^2y^2)$
7. $z = f_1(y+x)+xf_2(y+x)+\frac{x^4}{12}+\frac{x^2y}{2}+\frac{x^3}{3}$

Case 2. When $F(x,y)$ is of the form e^{ax+by} [provided $f(a,b) \neq 0$]. Then, $\text{P.I.} = \frac{1}{f(D,D')} e^{ax+by} = \frac{e^{ax+by}}{f(a,b)}$.

(Put $D = a$ and $D' = b$)

Solved Examples

Example 1. *Solve* $\frac{\partial^3 z}{\partial x^3} - 3\frac{\partial^3 z}{\partial x^2 \partial y} + 4\frac{\partial^3 z}{\partial y^3} = e^{x+2y}$.

(UPTU-2007, GBTU(AG)-2011, Burdwan–2003)

Solution. The given equation can be written as

$$(D^3 - 3D^2D' + 4D'^3)z = e^{x+2y}.$$

It's A. E. is $m^3 - 3m^2 + 4 = 0$

or $(m+1)(m-2)^2 = 0$

or $m = -1, 2, 2.$

$\therefore$ C.F. $= f_1(y-x) + f_2(y+2x) + xf_3(y+2x)$.

Now, $\text{P.I.} = \frac{1}{D^3 - 3D^2D' + 4D'^3} e^{x+2y}$

$= \frac{1}{1^3 - 3.1^2.2 + 4.2^3} e^{x+2y}$

$= \frac{1}{1-6+32} e^{x+2y}$

$= \frac{1}{27}.e^{x+2y}$.

Hence, the complete solution is

$$z = f_1(y-x) + f_2(y+2x) + xf_3(y+2x) + \frac{1}{27}e^{x+2y}.$$

Example 2. *Solve* $(D^2 - 2DD')z = e^{2x} + x^3y$.

Solution. It's A. E. is $m^2 - 2m = 0$

or $m(m-2) = 0$ or $m = 0, 2$.

$\therefore$ C.F. $= f_1(y) + f_2(y+2x)$.

Now $\text{P.I.} = \frac{1}{D^2 - 2DD'}(e^{2x} + x^3y)$

$= \frac{1}{D^2 - 2DD'}.e^{2x} + \frac{1}{D^2 - 2DD'}.x^3y$

$= \frac{1}{2^2 - 2(2)(0)}.e^{2x} + \frac{1}{D^2}\left(1 - \frac{2D'}{D}\right)^{-1}.x^3y$

$= \frac{1}{4}e^{2x} + \frac{1}{D^2}\left(1 + \frac{2D'}{D} + \ldots\right)x^3y$

$= \frac{1}{4}e^{2x} + \frac{1}{D^2}(x^3y) + \frac{2}{D^3}x^3$

$= \frac{1}{4}e^{2x} + \frac{x^5y}{20} + \frac{x^6}{60}$

Hence, the complete solution is

$$z = f_1(y) + f_2(y+2x) + \frac{1}{4}e^{2x} + \frac{x^5y}{20} + \frac{x^6}{60}.$$

Case 3. (Short method). *When* $F(x,y)$ *is of the form* $\phi(ax+by)$.

If $f(D,D')$ is a homogeneous function of D or D' of degree n, then we have

$$\text{P.I.} = \frac{1}{f(D,D')}\phi(ax+by) = \frac{1}{f(a,b)}\iint\ldots\int \phi(v)\,dv^n \text{ where } v = ax+by.$$

After integrating $\phi(v)$ n times w.r.t. 'v', v must be replaced by $ax+by$.

Case 4. (Exceptional case) *When* $F(x,y)$ *is of the form* $\phi(ax+by)$ *and* $f(a,b) = 0$.

If $f(D,D')$ is a homogeneous function D or D' of degree n, then we have

$$\text{P.I.} = \frac{1}{(bD - aD')^n}\phi(ax+by) = \frac{x^n}{b^n\, n!}\phi(ax+by) \text{ [provided } f(a,b) = 0]$$

Example 3. *Solve* $(D^2 + 3DD' + 2D'^2)z = x + y$.

(UPTU(Q. Bank)-2002, UPTU(CO)-2008)

Solution. It's A. E. is $m^2 + 3m + 2 = 0$

or $(m+1)(m+2) = 0$ or $m = -1, -2$.

$\therefore$ C.F. $= f_1(y-x) + f_2(y-2x)$.

Now, $\text{P.I.} = \frac{1}{D^2 + 3DD' + 2D'^2}.(x+y)$

$[\because D = a = 1,\ D' = b = 1]$

$= \frac{1}{1^2 + 3.1.1 + 2.1^2}\iint V\,dV^2$

where $V = x + y$

$= \frac{1}{6}.\frac{1}{6}V^3 = \frac{1}{36}(x+y)^3$.

Hence, the complete solution is

$$z = f_1(y-x) + f_2(y-2x) + \frac{1}{36}(x+y)^3.$$

Example 4. *Solve* $\frac{\partial^2 z}{\partial x^2} + \frac{\partial^2 z}{\partial y^2} = \cos mx \cos ny$.

Solution. The given equation can be written as

$$(D^2 + D'^2)z = \frac{1}{2}[\cos(mx+ny) + \cos(mx-ny)].$$

It's A. E. is $m^2+1=0$ or $m=\pm i$,

$\therefore$ C.F. $= f_1(y+ix)+f_2(y-ix)$.

Now,

$$\text{P.I.} = \frac{1}{D^2+D'^2}\left[\frac{1}{2}\{\cos(mx+ny)+\cos(mx-ny)\}\right]$$

$$= \frac{1}{2}\frac{1}{D^2+D'^2}.\cos(mx+ny) + \frac{1}{2}.\frac{1}{D^2+D'^2}.\cos(mx-ny)$$

$$= \frac{1}{2}\frac{1}{m^2+n^2}\iint \cos u du.du + \frac{1}{2}\frac{1}{m^2+n^2}\iint \cos v.dv.dv$$

where $u = mx+ny$ and $v = mx-ny$

$$= \frac{1}{2}.\frac{1}{m^2+n^2}(-\cos u)+\frac{1}{2}\frac{1}{m^2+n^2}(-\cos v)$$

$$= \frac{1}{2}\frac{1}{m^2+n^2}\{-\cos(mx+ny)\} + \frac{1}{2(m^2+n^2)}\{-\cos(mx-ny)\}$$

$$= -\frac{1}{2(m^2+n^2)}[\cos(mx+ny)+\cos(mx-ny)]$$

$$= -\frac{1}{m^2+n^2}\cos mx \cos ny.$$

Hence, the complete solution is

$$z = f_1(y+ix)+f_2(y-ix) - \frac{1}{m^2+n^2}\cos mx.\cos ny.$$

Example 5. *Solve the linear partial differential equation.*

$$\frac{\partial^2 z}{\partial x^2}+2\frac{\partial^2 z}{\partial x\partial y}+\frac{\partial^2 z}{\partial y^2} = \sin(2x+3y)$$

(UPTU-2006, Q. Bank–2002)

Solution. We have $\frac{\partial^2 z}{\partial x^2}+2\frac{\partial^2 z}{\partial x\partial y}+\frac{\partial^2 z}{\partial y^2} = \sin(2x+3y)$

$\Rightarrow (D^2+2DD'+D'^2)z = \sin(2x+3y)$,

where $D = \frac{\partial}{\partial x}$ and $D' = \frac{\partial}{\partial y}$

Put $D=m,\ D'=1$

The auxiliary equation is $m^2+2m+1=0$

$(m+1)^2 = 0 \Rightarrow m=-1,-1$.

C.F. $= f_1(y-x)+xf_2(y-x)$.

$$\text{P.I.} = \frac{1}{D^2+2DD'+D'^2}\sin(2x+3y)$$

$$= \frac{1}{-4+2(-6)-9}\sin(2x+3y)$$

$$= \frac{1}{-25}\sin(2x+3y).$$

Hence, the complete solution is

z = C.F. + P.I.

$$z = f_1(y-x)+xf_2(y-x)-\frac{1}{25}\sin(2x+3y)$$

Example 6. *Solve* $(D^3-4D^2D'+4DD'^2)z = 4\sin(2x+y)$.

Solution. It's A. E. is $m^3-4m^2+4m=0$

or $m(m-2)^2=0$.

$\therefore$ $m=0,2,2$.

$\therefore$ C.F. $= f_1(y)+f_2(y+2x)+xf_3(y+2x)$.

$$\text{Now P.I.} = \frac{1}{D^3-4D^2D'+4DD'^2}.4\sin(2x+y) \quad [\because f(a,b)=0]$$

$$= \frac{1}{D(D-2D')^2}.4\sin(2x+y)$$

$$= \frac{1}{(D-2D')^2}.\frac{1}{D}.4\sin(2x+y)$$

$$= \frac{1}{(D-2D')^2}\{-2\cos(2x+y)\}$$

$$= \frac{x^2}{2!}\{-2\cos(2x+y)\} \text{ (Applying Case : 4)}$$

$$= -x^2\cos(2x+y).$$

Hence, the complete solution is

$$z = f_1(y)+f_2(y+2x) + xf_3(y+2x)-x^2\cos(2x+y).$$

Example 7. *Solve* $(D^2-6DD'+9D'^2)z = 6x+2y$.

Solution. It's A. E. is $m^2-6m+9=0$.

or $(m-3)^2=0$ or $m=3,3$.

$\therefore$ C.F. $= f_1(y+3x)+xf_2(y+3x)$

$$\text{Now P.I.} = \frac{1}{(D-3D')^2}.(6x+2y) \quad [\because f(a,b)=0]$$

$$= 2\frac{1}{(D-3D')^2}.(3x+y) = 2.\frac{x^2}{2}.(3x+y)$$

$$= x^2(3x+y).$$

Hence, the complete solution is

$$z = f_1(y+3x)+xf_2(y+3x)+x^2(3x+y).$$

Example 8. *Solve* $4r-4s+t = 16\log(x+2y)$.

Solution. The equation can be written as

$$(4D^2-4DD'+D'^2) = 16\log(x+2y).$$

It's A. E. is $4m^2-4m+1=0$

or $(2m-1)^2 = 0$ or $m = \frac{1}{2},\frac{1}{2}$.

$\therefore$ C.F. $= f_1(2y+x)+xf_2(2y+x)$.

$$\text{Now, P.I.} = \frac{1}{4D^2-4DD'+D'^2}.16\log(x+2y)$$

$$= \frac{1}{(2D-D')^2}.16\log(x+2y)$$

$$= \frac{x^2}{2^2.\,2!}.16\log(x+2y) \quad [\because f(a,b)=0]$$

$$= 2x^2\log(x+2y).$$

Hence, the complete solution is

$$z = f_1(2y+x) + xf_2(2y+x) + 2x^2\log(x+2y).$$

Example 9. *Solve* $2r - s - 3t = 5e^x / e^y$.

Solution. The given equation can be written as

$$(2D^2 - DD' - 3D'^2)z = 5e^{x-y}.$$

It's A. E. is $2m^2 - m - 3 = 0$

or $(2m-3)(m+1) = 0$ or $m = \frac{3}{2}, -1$.

$\therefore$ C.F. $= f_1(2y+3x) + f_2(y-x)$.

$$\text{Now, P.I.} = \frac{1}{2D^2 - DD' - 3D'^2}.5e^{x-y}$$

$$= \frac{1}{D+D'}.\frac{1}{D-3D'}.5e^{x-y}$$

$$= \frac{1}{D+D'}\frac{1}{2+3}.5e^{x-y}$$

$$= \frac{1}{D+D'}e^{x-y} = \frac{x}{1!}e^{x-y}$$

$[\because f(a,b)=0]$

$$= xe^{x-y}.$$

Hence, the complete solution is

$$z = f_1(2y+3x) + f_2(y-x) + xe^{x-y}.$$

Example 10. *Solve* $(D^3 - 4D^2D' + 4DD'^2)z = \cos(2x+y)$.

Solution. It's A. E. is $m^3 - 4m^2 + 4m = 0$

or $m(m-2)^2 = 0$ or $m = 0, 2, 2$.

$\therefore$ C.F. $= f_1(y) + f_2(2x+y) + xf_3(2x+y)$.

$$\text{Now, P.I.} = \frac{1}{D(D-2D')^2}.\cos(2x+y)$$

$$= \frac{1}{(D-2D')^2}.\frac{1}{D}\cos(2x+y)$$

$$= \frac{1}{(D-2D')^2}.\frac{\sin(2x+y)}{2}$$

$$= \frac{x^2}{2!}.\frac{\sin(2x+y)}{2} = \frac{1}{4}x^2\sin(2x+y).$$

$[\because f(a,b)=0]$

Hence, the complete solution is

$$z = f_1(y) + f_2(2x+y) + xf_3(2x+y) + \frac{1}{4}x^2\sin(2x+y).$$

Example 11. *Solve*

$$(D^3 - 7DD'^2 - 6D'^3)z = \sin(x+2y) + e^{2x+y}.$$

Solution. The given equation is

$$(D^3 - 7DD'^2 - 6D'^3)z = \sin(x+2y) + e^{2x+y}.$$

It's A. E. is $m^3 - 7m - 6 = 0$

or $(m+1)(m+2)(m-3) = 0$

or $m = -1, -2, 3$

$\therefore$ C.F. $= f_1(y-x) + f_2(y-2x) + f_3(y+3x)$

Now,

$$\text{P.I.} = \frac{1}{D^3 - 7DD'^2 - 6D'^3}[\sin(x+2y) + e^{2x+y}]$$

$$= \frac{1}{D^3 - 7DD'^2 - 6D'^3}\sin(x+2y) + \frac{1}{D^3 - 7DD'^2 - 6D'^3}.e^{2x+y}$$

$$= \frac{1}{1^3 - 7.1.2^2 - 6.2^3}\iiint \sin v\, dv^3 + \frac{1}{2^3 - 7.2.1^2 - 6.1^3}e^{2x+y},$$

where $v = x + 2y$

$$= -\frac{1}{75}\cos v - \frac{1}{12}e^{2x+y}$$

$$= -\frac{1}{75}\cos(x+2y) - \frac{1}{12}e^{2x+y}.$$

Hence, the complete solution is

$$z = f_1(y-x) + f_2(y-2x) + f_3(y+3x) - \frac{1}{75}\cos(x+2y) - \frac{1}{12}e^{2x+y}.$$

Case 5. (General Method).

Consider the equation $f(D,D')z = F(x,y)$ ...(1)

when $f(D,D')$ is homogeneous function of D and D'. We use the following result :

$$\frac{1}{(D-mD')}F(x,y) = \int F(x, a-mx)dx, \text{ where } a = y + mx.$$

After performing integration a must be replaced by $y + mx$ respectively.

To find the P.I., factorize $f(D,D')$ into linear factors. Thus from (1)

$$\text{P.I.} = \frac{1}{(D-m_1D')(D-m_2D')...(D-m_nD')}F(x,y). \quad ...(2)$$

The value of P.I. is obtained by applying the operations indicated by the factors.

Example 12. *Solve* $(D^2 - 2DD' - 15D'^2)z = 12xy$.

Solution. It's A. E. is $m^2 - 2m - 15 = 0$

or $(m+3)(m-5) = 0$ or $m = -3, 5$.

$\therefore$ C.F. $= f_1(y-3x) + f_2(y+5x)$.

$$\text{Now P.I.} = \frac{1}{D^2 - 2DD' - 15D'^2} 12xy$$

$$= \frac{1}{(D+3D')} \cdot \frac{1}{(D-5D')} 12xy$$

$$= \frac{12}{(D+3D')} \int x(a-5x)dx,$$

where $y + 5x = a$

$$= \frac{12}{(D+3D')} \left(\frac{1}{2}ax^2 - \frac{5}{3}x^3 \right)$$

$$= \frac{12}{(D+3D')} \left\{ \frac{1}{2}(y+5x)x^2 - \frac{5}{3}x^3 \right\}$$

$$= \frac{2}{(D+3D')} (3x^2y + 5x^3)$$

$$= 2\int \{3x^2(3x+b) + 5x^3\} dx,$$

where $y - 3x = b$

$$= 2\int (14x^3 + 3x^2b)dx = 2\left(\frac{14}{4}x^4 + x^3b \right)$$

$$= 7x^4 + 2x^3(y-3x) = x^4 + 2x^3y.$$

Hence, the complete solution is

$$z = f_1(y-3x) + f_2(y+5x) + x^4 + 2x^3y.$$

Example 13. *Solve* $r + s - 6t = y\cos x$.

(UPTU-2003, Anna-2005S, Bhopal-2008, SVTU-2008)

Solution. The equation can be written as

$(D^2 + DD' - 6D'^2)z = y\cos x$.

It's A. E. is $m^2 + m - 6 = 0$

or $(m-2)(m+3) = 0$ or $m = 2, -3$.

$\therefore$ C.F. $= f_1(y+2x) + f_2(y-3x)$.

$$\text{Now, P.I.} = \frac{1}{D^2 + DD' - 6D'^2} \cdot y\cos x$$

$$= \frac{1}{(D-2D')(D+3D')} y \cos x$$

$$= \frac{1}{(D-2D')} \cdot \int (3x+a) \cos x \, dx$$

where $y - 3x = a$

$$= \frac{1}{(D-2D')} [(3x+a)\sin x + 3\cos x]$$

$$= \frac{1}{D-2D'} [y\sin x + 3\cos x]$$

$$= \int [(b-2x)\sin x + 3\cos x]dx$$

where $y + 2x = b$

$$= -b\cos x - 2(-x\cos x + \sin x) + 3\sin x$$

$$= -(y+2x)\cos x + 2x\cos x + \sin x$$

$$= -y\cos x + \sin x.$$

Hence, the complete solution is

$$z = f_1(y+2x) + f_2(y-3x) - y\cos x + \sin x.$$

Example 14. *Solve* $(D^2 - DD' - 2D'^2)z = (y-1)e^x$.

(UPTU(Q Bank)-2002, Bhopal-2006)

Solution. It's A. E. is $m^2 - m - 2 = 0$

or $(m-2)(m+1) = 0$ or $m = 2, -1$.

$\therefore$ C.F. $= f_1(y+2x) + f_2(y-x)$.

$$\text{Now, P.I.} = \frac{1}{D^2 - DD' - 2D'^2} \cdot (y-1)e^x$$

$$= \frac{1}{(D-2D')} \cdot \frac{1}{(D+D')} (y-1)e^x$$

$$= \frac{1}{(D-2D')} \int (x+a-1)e^x dx,$$

where $y - x = a$

$$= \frac{1}{(D-2D')} [(a-1)\int e^x \, dx + \int xe^x \, dx]$$

$$= \frac{1}{(D-2D')} [(a-1)e^x + (x-1)e^x]$$

$$= \frac{1}{(D-2D')} [a+x-2]e^x$$

$$= \frac{1}{(D-2D')} (y-2)\, e^x = \int (b-2x-2)e^x dx,$$

where $y - 2x = b$

$$= (b-2)\int e^x \, dx - 2\int xe^x \, dx$$

$$= (b-2)e^x - 2(x-1)e^x = (b-2x)e^x = ye^x.$$

Hence, the complete solution is

$$z = f_1(y+2x) + f_2(y-x) + ye^x.$$

Example 15. *Solve* $\frac{\partial^2 z}{\partial x^2} - 4\frac{\partial^2 z}{\partial y^2} = \frac{4x}{y^2} - \frac{y}{x^2}$.

Solution. The given equation can be written as

$$(D^2 - 4D'^2)z = \frac{4x}{y^2} - \frac{y}{x^2}.$$

It's A. E. is $m^2 - 4 = 0$ or $m = 2, -2$.

$\therefore$ C.F. $= f_1(y+2x) + f_2(y-2x)$.

$$\text{Now, P.I.} = \frac{1}{(D+2D')} \cdot \frac{1}{(D-2D')} \left(\frac{4x}{y^2} - \frac{y}{x^2} \right)$$

$$= \frac{1}{(D+2D')} \int \left[\frac{4x}{(a-2x)^2} - \frac{(a-2x)}{x^2} \right] dx$$

where $y + 2x = a$

$$= \frac{1}{(D+2D')} \int \left[\frac{-2}{(a-2x)} + \frac{2a}{(a-2x)^2} - \frac{a}{x^2} + \frac{2}{x} \right] dx$$

$$= \frac{1}{(D+2D')}\left[\log(a-2x)+\frac{a}{(a-2x)}+\frac{a}{x}+2\log x\right]$$

$$= \frac{1}{(D+2D')}\left[\log y+2\log x+\frac{y+2x}{y}+\frac{y+2x}{x}\right]$$

$$= \frac{1}{(D+2D')}\left[\log y+2\log x+\frac{2x}{y}+\frac{y}{x}+3\right]$$

$$= \int\left[\log(b+2x)+2\log x+\frac{2x}{b+2x}+\frac{b+2x}{x}+3\right]dx$$

where $y-2x=b$

$$= \int\left[\log(b+2x)+2\log x+\frac{2x}{b+2x}+\frac{b}{x}+5\right]dx$$

$$= x.\log(b+2x)-\int\frac{2x}{(b+2x)}dx+2x\log x$$

$$-2\int\frac{1}{x}.\,x\,dx+\int\frac{2x}{b+2x}dx+b\log x+5x$$

$$= x\log(b+2x)+(2x+b)\log x+3x$$

$$= x\log y+y\log x+3x.$$

Hence, the complete solution is

$$z = f_1(y+2x)+f_2(y-2x)$$
$$+x\log y+y\log x+3x.$$

Example 16. *Solve the partial differential equation.*

$$\frac{\partial^2 z}{\partial x^2}-3\frac{\partial^2 z}{\partial x\partial y}+2\frac{\partial^2 z}{\partial y^2}=e^{2x-y}+e^{x+y}+\cos(x+2y)$$

(UPTU-2006, GBTU-2012)

Solution. Given equation is

$$\frac{\partial^2 z}{\partial x^2}-3\frac{\partial^2 z}{\partial x\partial y}+2\frac{\partial^2 z}{\partial y^2}$$
$$= e^{2x-y}+e^{x+y}+\cos(x+2y)$$

Given equation can be written as

$$(D^2-3DD'+2D'^2)z = e^{2x-y}+e^{x+y}+\cos(x+2y)$$

The auxiliary equation is

$$m^2-3m+2=0$$
$$m^2-2m-m+2=0$$
$$m(m-2)-1(m-2)=0$$
$$m=1,2.$$

Hence, C.F. $= f_1(y+x)+f_2(y+2x)$

Now, $\quad$ P.I. $= \dfrac{1}{(D-D')(D-2D')}\left\{e^{2x-y}+e^{x+y}+\cos(x+2y)\right\}$

$$= \frac{1}{(D-D')(D-2D')}e^{2x-y}$$
$$+\frac{1}{(D-D')(D-2D')}e^{x+y}$$
$$+\frac{1}{(D-D')(D-2D')}\cos(x+2y)$$
$$= I_1+I_2+I_3.$$

Let, $\quad I_1 = \dfrac{1}{(D-D')(D-2D')}e^{2x-y}$

(Replacing D by 2 and D' by -1)

$$I_1 = \frac{1}{(2+1)(2+2)}e^{2x-y} = \frac{1}{12}e^{2x-y}$$

Now, $I_2 = \dfrac{1}{(D-D')(D-2D')}e^{x+y}$

$$= \frac{1}{(D-D')(-1)}e^{x+y} = \frac{-1}{(D-D')}e^{x+y}$$

$$= -x.\frac{1}{1}e^{x+y} = -xe^{x+y}.$$

Now, $I_3 = \dfrac{1}{(D-D')(D-2D')}\cos(x+2y)$

$$= \frac{1}{D^2-3DD'+2D'^2}\cos(x+2y)$$

$$I_3 = \frac{1}{-1-3(-2)+2(-4)}\cos(x+2y)$$

(Replacing D^2 by -1; DD' by -2, D'^2 by -4)

$$= \frac{1}{-1+6-8}\cos(x+2y)$$

$$= -\frac{1}{3}\cos(x+2y)$$

P.I. $= I_1+I_2+I_3$.

Thus, required

$$\text{P.I.} = \frac{1}{12}e^{2x-y}-xe^{x+y}-\frac{1}{3}\cos(x+2y)$$

Hence, the complete solution is

$$z = \text{C.F.} + \text{P.I.}$$

$$= f_1(y+x)+f_2(y+2x)+\frac{1}{12}e^{2x-y}$$
$$-xe^{x+y}-\frac{1}{3}\cos(x+2y).$$

Example 17. *Solve*

$$r-s-2t = (2x^2+xy-y^2)\sin xy-\cos xy.$$

Solution. The given equation can be written as

$$(D^2-DD'-2D'^2)z$$
$$= (2x^2+xy-y^2)\sin xy-\cos xy.$$

It's A. E. is $m^2-m-2=0$

or $(m+1)(m-2)=0$

or $\quad m=-1,2.$

$\therefore$ $\quad$ C.F. $= f_1(y-x)+f_2(y+2x)$.

Now, P.I. $= \dfrac{1}{(D^2-DD'-2D'^2)}.\{(2x^2+xy-y^2)\sin xy-\cos xy\}$

$$= \frac{1}{(D+D')}.\frac{1}{(D-2D')}.\{(2x^2+xy-y^2)\sin xy-\cos xy\}$$

$$= \frac{1}{(D+D')}\int[\{(2x^2+x(a-2x)-(a-2x)^2\}\sin x(a-2x)-\cos x(a-2x)]dx,$$

(where $y+2x=a$)

$$= \frac{1}{(D+D')}\int [(a-4x)(x-a)$$
$$\sin(ax-2x^2) - \cos(ax-2x^2)]dx$$
$$= \frac{1}{(D+D')}[\int (x-a).\{(a-4x)\sin(ax-2x^2)\}dx$$
$$-\int \cos(ax-2x^2)dx]$$
$$= \frac{1}{(D+D')}[(x-a)\{-\cos(ax-2x^2)\}$$
$$+\int \cos(ax-2x^2)dx - \int \cos(ax-2x^2)dx]$$
$$= \frac{1}{(D+D')}(a-x)\cos(ax-2x^2)$$
$$= \frac{1}{(D+D')}(x+y)\cos xy$$
$$= \int (2x+b)\cos x(x+b)\, dx,$$

where $y - x = b$

$$= \int (2x+b)\, \cos(x^2+xb)\, dx$$
$$= \sin(x^2+xb) = \sin xy.$$

Hence, the complete solution is

$$z = f_1(y-x) + f_2(y+2x) + \sin xy.$$

Example 18. *Solve* $r - t = \tan^3 x \tan y - \tan x \tan^3 y.$ (AGRA(BE)-2001)

Solution. The given equation can be written as

$$(D^2 - D'^2)z = \tan^3 x \tan y - \tan x \tan^3 y.$$

It's A. E. is $m^2 - 1 = 0$

or $(m-1)(m+1) = 0$ or $m = 1, -1.$

$\therefore$ C.F. $= f_1(y+x) + f_2(y-x).$

Now,

$$\text{P.I.} = \frac{1}{(D^2 - D'^2)}(\tan^3 x \tan y - \tan x \tan^3 y)$$
$$= \frac{1}{(D+D')}\frac{1}{(D-D')}(\tan^3 x \tan y - \tan x \tan^3 y)$$
$$= \frac{1}{(D+D')}\int [\tan^3 x \tan(a-x)$$

where $y + x = a$

$$= \frac{1}{(D+D')}\int [(\sec^2 x - 1)\tan x \tan(a-x)$$
$$-\tan x \tan(a-x)\{\sec^2(a-x) - 1\}]dx$$
$$= \frac{1}{(D+D')}[\int \tan(a-x)\tan x \sec^2 x dx$$
$$-\int \tan x \tan(a-x)\sec^2(a-x)dx]$$
$$= \frac{1}{(D+D')}\left[\tan(a-x).\frac{\tan^2 x}{2} + \frac{1}{2}\int \sec^2(a-x)\tan^2 x dx + \tan x.\frac{\tan^2(a-x)}{2} - \frac{1}{2}\int \sec^2 .\tan^2(a-x)dx\right]$$
$$= \frac{1}{2}\frac{1}{(D+D')}[\tan^2 x \tan(a-x) + \tan x \tan^2(a-x)$$
$$-\int [\sec^2 x\{\sec^2(a-x) - 1\} - \sec^2(a-x)(\sec^2 x - 1)dx]$$
$$= \frac{1}{2}\frac{1}{(D+D')}[\tan^2 x \tan(a-x) + \tan x \tan^2(a-x)$$
$$+\int \{\sec^2 x - \sec^2(a-x)\}dx]$$
$$= \frac{1}{2}\frac{1}{(D+D')}[\tan^2 x \tan y + \tan x \tan^2 y + (\tan x + \tan y)]$$
$$= \frac{1}{2}\frac{1}{(D+D')}[\tan x \sec^2 y + \tan y \sec^2 x]$$
$$= \frac{1}{2}\int [\tan x \sec^2(b+x) + \tan(b+x)\sec^2 x]dx,$$

where $y - x = b$

$$= \frac{1}{2}[\tan x \tan(b+x) - \int \sec^2 x.\tan(b+x)dx$$
$$+\int \tan(b+x)\sec^2 x\, dx]$$
$$= \frac{1}{2}\tan x \tan y.$$

Hence, the complete solution is

$$z = f_1(y+x) + f_2(y-x) + \frac{1}{2}\tan x \tan y$$

EXERCISE 41.4

Solve the following differential equations:

1. $(D^2 - 2DD' + D'^2)z = e^{x+2y} + x^3$
2. $\frac{\partial^2 z}{\partial x^2} + \frac{\partial^2 z}{\partial y^2} = 12(x+y)$ (UPTU-2005)
3. $(D^2 - 5DD' + 4D'^2)z = \sin(4x+y)$ (GBTU(AG)-2011)
4. $(D - 2D')(D - D')^2 = e^{x+y}$
5. $\frac{\partial^2 z}{\partial x^2} - 5\frac{\partial^2 z}{\partial x \partial y} + 6\frac{\partial^2 z}{\partial y^2} = \exp.(3x-2y)$
6. $(D^2 - DD')z = \cos x \cos 2y$ (Bhopal-2008S)
7. $\frac{\partial^2 z}{\partial x^2} - 3\frac{\partial^2 z}{\partial x \partial y} + 2\frac{\partial^2 z}{\partial y^2} = e^{2x+3y} + \sin(x-2y)$
8. $\frac{\partial^2 z}{\partial x^2} + \frac{\partial^2 z}{\partial x \partial y} - 6\frac{\partial^2 z}{\partial y^2} = y \sin x$ (MTU(SUM)-2011, MTU-2012)
9. $(2D^2 - 5DD' + 2D'^2)z = 5\sin(2x+y)$
10. $(D^2 + 2DD' + D'^2)z = 2\cos y - x \sin y$ (PTU-2005)
11. $(D^2 + DD' - 6D'^2)z = x^2 \sin(x+y)$
12. $(D^2 + 5DD' + 6D'^2)z = \frac{1}{y-2x}$
13. $\frac{\partial^3 z}{\partial x^2 \partial y} - 2\frac{\partial^3 z}{\partial x \partial y^2} + \frac{\partial^3 z}{\partial y^3} = \frac{1}{x^2}$

14. $r+s-2t=(2x+y)^{1/2}$

15. $\frac{\partial^2 z}{\partial x^2}-\frac{\partial^2 z}{\partial x\partial y}=\sin x\cos 2y$ (UPTU-2003, 08)

16. $(D^2-2DD'+D'^2)z=\sin x$ (UPTU-2004, 09)

17. $(D^2-DD')z=\cos 2y(\sin x+\cos x)$ (UPTU-2003, GBTU-2010)

18. $(D+D'-1)(D+2D'-3)z=4+3x+6y$ (UPTU(Q. Bank)-2002)

19. $\frac{\partial^2 z}{\partial x^2}+\frac{\partial^2 z}{\partial y^2}=\cos mx\cos ny+30(x+y)$ (GBTU(AG)-2011, 12, UPTU(SUM)-2008)

20. $\frac{\partial^2 z}{\partial x^2}-\frac{\partial^2 z}{\partial x\partial y}=\sin x\cos y$ (GBTU(SUM)-2010)

21. $r+2s+t=2(y-x)+\sin(x-y)$ (MTU(SUM)-2011)

22. $(D^2+2DD'+D'^2)z=e^{2x+3y}$ (GBTU(AG)-2012)

23. $(D^3-4D^2D'+4DD'^2)z=6\sin(3x+2y)$ (UKTU-2012)

24. $(D^2-4DD'+4D'^2)z=e^{2x+y}$ (GBTU(CO)-2010)

25. $(D^2-8DD'+7D'^2)z=\sin(7x+y)$ (UPTU(CO)-2009)

26. $(D^2-DD')z=\sin(x+2y)$ (GBTU(AG)-2012)

27. $(D^2+4DD'-5D'^2)z=\sin(2x+3y)$ (MadraS-2006)

28. $\frac{\partial^3 z}{\partial x^3}-2\frac{\partial^3 z}{\partial x^2\partial y}=2e^{2x}+3x^2y$ (SVTU-2007)

29. $\frac{\partial^2 z}{\partial x^2}-\frac{\partial^2 z}{\partial x\partial y}-6\frac{\partial^2 z}{\partial y^2}=\cos(2x+y)$ (PTU-2010, SVTU-2009)

30. $(D^2-2DD'+D'^2)z=e^{x+y}$ (Bhopal-2007)

31. $\frac{\partial^3 z}{\partial x^3}-4\frac{\partial^3 z}{\partial x^2\partial y}+5\frac{\partial^3 z}{\partial x\partial y^2}-2\frac{\partial^3 z}{\partial y^3}=e^{2x+y}$ (Bhopal-2008)

32. $\frac{\partial^2 z}{\partial x^2}-2\frac{\partial^2 z}{\partial x\partial y}+\frac{\partial^2 z}{\partial y^2}=\sin x$ (PTU-2009S)

33. $\frac{\partial^3 z}{\partial x^3}-4\frac{\partial^3 z}{\partial x^2\partial y}+4\frac{\partial^3 z}{\partial x\partial y^2}=2\sin(3x+2y)$ (SVTU-2007)

34. $(D^3-7DD'^2-6D'^3)z=\cos(x+2y)+4$ (Anna–2008)

35. $(D^2-D'^2)z=e^{x-y}\sin(x+2y)$ (Anna–2009)

Hint to Selected Problems

1. A.E. is $m^2-2m+1=0 \Rightarrow m=1,1$

$$\text{P.I.}=\frac{1}{(D-D')^2}e^{x+2y}+\frac{1}{(D-D')^2}x^3$$
$$=\frac{e^{x+2y}}{(1-2)^2}+\frac{1}{D^2}\left[1-\frac{D'}{D}\right]^{-2}x^3$$

2. A.E. is $m^2+1=0 \Rightarrow m\pm i$

$$\text{P.I.}=\frac{1}{(D^2+D'^2)}12(x+y)=\frac{1}{1^2+1^2}.12\frac{(x+y)^3}{6}$$

4. $m=1,1,2$

$$\text{P.I.}=\frac{1}{(D-2D')(D-D')^2}e^{x+y}$$
$$=\frac{1}{(1-2)}\frac{1}{(D-D')^2}e^{x+y}=-\frac{x^2}{2}e^{x+y}$$

6. A.E. is $m^2-m=0 \Rightarrow m=0,1$

$$\text{P.I.}=\frac{1}{(D^2-DD')}.\cos x\cos 2y$$
$$=\frac{1}{(D^2-DD')}.\frac{1}{2}[\cos(x+2y)+\cos(x-2y)]$$

8. A.E. $(m+3)(m-2)=0 \Rightarrow m=2,-3$

$$\text{P.I.}=\frac{1}{(D^2+DD'-6D'^2)}y\sin x=\text{I.P. of}\left[\frac{1}{(D^2+DD'-6D'^2)}ye^{ix}\right]$$
$$=\text{I.P. of}\left\{e^{ix}\left[\frac{1}{(D+i)^2+(D+i)D'-6D'^2}\right]\right\}y$$

10. $$\text{P.I.}=\frac{1}{(D+D')^2}(2\cos y-x\sin y)$$
$$=\frac{1}{(D+D')}\int[(2\cos(x+b)-x\sin(x+b)]dx,$$
where $y-x=b$
$$=\frac{1}{D+D'}[2\sin(x+b)+x\cos(x+b)-\sin(x+b)]$$
$$=\frac{1}{D+D'}[\sin(x+b)+x\cos(x+b)]$$
$$=\int \sin(x+b)+x\cos(x+b)dx=x\sin(x+b)=x\sin y$$

12. $$\text{P.I.}=\frac{1}{(D^2+5DD'+6D'^2)}\frac{1}{(y-2x)}$$
$$=\frac{1}{(D+3D')(D+2D')}(y-2x)^{-1}$$
$$=\frac{1}{(-2+3)(D+2D')}\int V^{-1}dV,\quad (V=y-2x)$$
$$=\frac{1}{(D+2D')}\log(y-2x)=x\log(y-2x)$$

ANSWERS

1. $z=f_1(y+x)+xf_2(y+x)+e^{x+2y}+\frac{1}{20}x^5$

2. $z=f_1(y-ix)+f_2(y+ix)+(x+y)^3$.

3. $z=f_1(y+x)+f_2(y+4x)-\frac{1}{3}x\cos(4x+y)$

4. $z=f_1(y+2x)+f_2(y+x)+xf_3(y+x)-\frac{x^2}{2}e^{x+y}$.

5. $z=f_1(y+2x)+f_2(y+3x)+\frac{1}{63}e^{3x-2y}$

6. $z=f_1(y)+f_2(y+x)+\frac{1}{2}\cos(x+2y)-\frac{1}{6}\cos(x-2y)$.

7. $z=f_1(y+x)+f_2(y+2x)+\frac{1}{4}e^{2x+3y}-\frac{1}{15}\sin(x-2y)$

8. $z=f_1(y+2x)+f_2(y-3x)-(y\sin x+\cos x)$;

9. $z=f_1(2y+x)+f_2(y+2x)-\left(\frac{5x}{3}\right)\cos(2x+y)$

10. $z=f_1(y-x)+xf_2(y-x)+x\sin y$

11. $z = f_1(y+2x) + f_2(y-3x) + \frac{1}{4}\left(x^2 - \frac{13}{8}\right)\sin(x+y) - \frac{3}{8}x\cos(x+y)$

12. $z = f_1(y-2x) + f_2(y-3x) + x\log(y-2x)$

13. $z = f_1(x) + f_2(y+x) + xf_3(y+x) - y\log x.$

14. $z = f_1(y+x) + f_2(y-2x) + \frac{1}{15}(2x+y)^{5/2}.$

15. $z = f_1(x) + f_2(y+x) + \frac{1}{2}\sin(x+2y) - \frac{1}{6}\sin(x-2y)$

16. $z = f_1(y+x) + xf_2(y+x) - \sin x$

17. $z = f_1(y) + f_2(y+x) + \frac{1}{2}[\sin(x+2y) + \cos(x+2y)] - \frac{1}{6}[\sin(x-2y) + \cos(x-2y)]$

18. $z = e^x f_1(y-x) + e^{3x} f_2(y-2x) + 6 + x + 2y$

19. $z = f_1(y+ix) + f_2(y-ix) - \frac{1}{m^2+n^2}\cos mx \cos ny + (2x+y)^3$

20. $z = f_1(y) + f_2(y+x) - \frac{x}{2}\cos(x+y) - \frac{1}{4}\sin(x-y)$

21. $z = f_1\left(y + \frac{1}{2}x\right) + xf_2\left(y + \frac{1}{2}x\right) + 2x^2\log(x+2y)$

22. $z = f_1(y-x) + xf_2(y-x) + \frac{1}{25}e^{2x+3y}$

23. $z = f_1(y) + f_2(y+2x) + xf_3(y+2x) + 2\cos(3x+2y)$

24. $z = f_1(y+2x) + xf_2(y+2x) + \frac{x^2}{2}e^{2x+y}$

25. $z = f_1(y+x) + f_2(y+7x) - \frac{x}{6}\cos(7x+y)$

26. $z = f_1(y) + f_2(y+x) + \sin(x+2y)$

27. $z = f_1(y+x) + f_2(y-5x) + \frac{1}{17}\sin(2x+3y)$

28. $z = f_1(y) + xf_2(y) + f_3(y+2x) + \frac{1}{60}(15e^{2x} + 3x^5y + x^6)$

29. $z = f_1(y-3x) + f_2(y+2x) + \frac{x}{5}\sin(2x+y) + \frac{1}{25}\cos(2x+y)$

30. $z = f_1(x+y) + xf_2(x+y) + \frac{x^2}{2}e^{x+y}$

31. $z = f_1(y+x) + zf_2(y+x) + f_3(y+2x) - e^{2x+y}$

32. $z = f_1(y+x) + xf_2(y+x) - \sin x$

33. $z = f_1(y) + f_2(y+2x) + xf_3(y+2x) + 3x\cos(3x+2y)$

34. $f_1(yx) + f_2(y-2x) + f_3(y+3x) + \frac{1}{75}\sin(x+2y) + \frac{2}{3}x^3$

35. $z = f_1(y+x) + f_2(y-x) + \frac{3}{28}e^{x-y}[\sin(x+2y) - 2\cos(x+2y)]$

41.6 NON-HOMOGENEOUS LINEAR PARTIAL DIFFERENTIAL EQUATIONS WITH CONSTANT COEFFICIENTS

Linear differential equations which are not homogeneous are called non-homogeneous linear equations.

For example, $\frac{\partial^2 z}{\partial x^2} + 2\frac{\partial^2 z}{\partial x \partial y} + \frac{\partial^2 z}{\partial y^2} + 2\frac{\partial z}{\partial x} + 2\frac{\partial z}{\partial y} + z = 0$ is a non-homogeneous equation.

It can be written in the form $f(D, D') = F(x, y)$. Its complete solution $=$ C.F. + P.I.

41.7 METHOD FOR FINDING THE C.F.

Let us consider the equation $(D - mD' - a)z = 0$ or $p - mq = az$.

Applying Lagrange's method, the auxiliary equations are $\frac{dx}{1} = \frac{dy}{-m} = \frac{dz}{az}$.

From the first two members, we have

$$dy + m\,dx = 0 \Rightarrow y + mx = c_1 \text{ (constant)} \quad \ldots(1)$$

Again from the first and last members, we get

$$dx = \frac{dz}{az} \Rightarrow \log z = ax + \log c_2 \Rightarrow z = c_2 e^{ax}. \quad \ldots(2)$$

From (1) and (2), we have $z = e^{ax} f(y + mx)$.

Similarly, the solution of $(D - m_1D' - a_1)(D - m_2D' - a_2)\ldots(D - m_nD' - a_n)z = 0$ is

$$z = e^{a_1x} f_1(y + m_1x) + e^{a_2x} f_2(y + m_2x) + \ldots + e^{a_nx} f_n(y + m_nx).$$

Note. When $f(D, D')$ cannot be factorized into linear factors, then we proceed as follows :

Let the equation be $(D - D'^2)z = 0$ ...(1)

The trial solution of (1) be $z = Ae^{hx+ky}$ where A, h and k are constants. ...(2)

From (2), we have

$$Dz = \frac{\partial z}{\partial x} = Ahe^{hx+ky}; \qquad D'z = \frac{\partial z}{\partial y} = Ake^{hx+ky}$$

$$D^2z = \frac{\partial^2 z}{\partial x^2} = Ah^2e^{hx+ky}; \qquad D'^2z = \frac{\partial^2 z}{\partial y^2} = Ak^2e^{hx+ky}.$$

With these values, (1) gives

$$Ahe^{hx+ky} - Ak^2e^{hx+ky} = 0 \quad \text{or} \quad A(h-k^2)e^{hx+ky} = 0 \quad \text{or} \quad h = k^2.$$

Putting the value of h in (2), we get $z = Ae^{k^2x+ky}$.

A more general solution of (1) is given by $z = \sum Ae^{k^2x+ky}$, where A and k are arbitrary constants in each term and any number of terms may be taken in the above summation.

41.8 METHOD FOR FINDING THE P.I. OF NON-HOMOGENEOUS EQUATION WITH CONSTANT COEFFICIENTS

The methods of finding particular integrals of non-homogeneous partial differential equation are very similar to those of ordinary differential equation with constant coefficients.

We are considering few cases of finding P.I. of $f(D,D')z = F(x,y)$.

Case I. When $F(x,y) = e^{ax+by}$ and $f(a,b) \neq 0$, then P.I. $= \dfrac{1}{f(D,D')}e^{ax+by} = \dfrac{1}{f(a,b)}e^{ax+by}$ (Putting a for D and b for D')

Case II. When $F(x,y) = \sin(ax+by)$ or $\cos(ax+by)$, then P.I. $= \dfrac{1}{f(a,b)}\sin(ax+by)$ is obtained by putting $D^2 = -a^2, DD' = -ab, D'^2 = -b^2$, provided the denominator is non-zero.

Case III. When $F(x,y) = x^m y^n$, where m and n are positive integers, then we have

$$\text{P.I.} = \frac{1}{f(D,D')}x^m y^n = [f(D,D')]^{-1}x^m y^n \text{ with ascending powers of } \frac{D'}{D} \text{ or } \frac{D}{D'} \text{ or } D \text{ or } D'.$$

Case IV. When $F(x,y) = e^{ax+by}.V$, where V is a function of x and y, then we have

$$\text{P.I.} = \frac{1}{f(D,D')}e^{ax+by}.V = e^{ax+by}\frac{1}{f(D+a,D'+b)}.V.$$

Solved Examples

Example 1. *Solve* $(D^2 - D'^2 + D - D')z = 0$.

Solution. The given equation is $(D^2 - D'^2 + D - D')z = 0$.

or $(D-D').(D+D'+1)z = 0$.

Hence, the solution is

$$z = f_1(y+x) + e^{-x}f_2(y-x).$$

Example 2. *Solve* $r + 2s + t + 2p + 2q + z = 0$.

Solution. The given equation can be written as

$$(D^2 - 2DD' + D'^2 + 2D + 2D' + 1)z = 0$$

or $(D+D'+1)^2 z = 0$

Hence, the solution is

$$z = e^{-x}f_1(y-x) + xe^{-x}f_2(y-x).$$

Example 3. *Solve* $(D-2D'-1)(D-2D'^2-1)z = 0$.

Solution. The given equation is

$$(D-2D'-1)(D-2D'^2-1)z = 0.$$

C.F. corresponding to first factor is

$$z = e^x f_1(y+2x).$$

Now C.F. corresponding to second factor is $\sum Ae^{hx+ky}$,

where $h - 2k^2 - 1 = 0$ or $h = 2k^2+1$.

Hence, the solution is

$$z = e^x f_1(y+2x) + \sum Ae^{ky+(2k^2+1)x}.$$

Example 4. *Solve* $(D^3 - 3DD' + D' + 1)z = e^{2x+3y}$.

Solution. Here, $D^3 - 3DD' + D' + 1$ cannot be resolved into linear factor in D and D'.

Let $z = Ae^{hx+ky}$.

$$\therefore \quad (D^3 - 3DD' + D' + 1)z = A(h^3 - 3hk + k + 1)e^{hx+ky}.$$

Then $(D^3 - 3DD' + D' + 1)z = 0$

if $\quad h^3 - 3hk + k + 1 = 0$.

$\therefore$ C.F. $= \sum Ae^{hx+ky}$, where $h^3 - 3hk + k + 1 = 0$.

$$\text{Now, P.I.} = \frac{1}{D^3 - 3DD' + D' + 1}.e^{2x+3y}$$

$$= \frac{e^{2x+3y}}{2^3 - 3.2.3 + 3 + 1} = -\frac{1}{6}e^{2x+3y}.$$

Hence, the complete solution is

$$z = \sum Ae^{hx+ky} - \frac{1}{6}e^{2x+3y},$$

where $h^3 - 3hk + k + 1 = 0$.

Example 5. *Solve* $s + p - q = z + xy$.

Solution. The equation can be written as

$$(DD' + D - D' - 1)z = xy$$

or $\quad (D-1)(D'+1) = xy$.

$\therefore \quad$ C.F. $= e^x f_1(y) + e^{-y}f_2(x)$.

$$\text{Now, P.I.} = \frac{1}{(D-1)(D'+1)}xy$$

$$= -(1-D)^{-1}(1+D')^{-1}.xy$$

$$= -(1+D+\ldots)(1-D'+\ldots)xy$$

$$= -(1+D-D'-DD'\ldots)xy$$

$$= -(xy+y-x-1).$$

Hence, the complete solution is

$$z = e^x f_1(y) + e^{-y}f_2(x) - xy - y + x + 1.$$

Example 6. *Solve* $(D^2 + DD' + D' - 1)z = \sin(x+2y)$.

(GBTU-2010, SVTU-2009)

Solution. The given equation can be written as

$$(D+1)(D+D'-1)z = \sin(x+2y).$$

$$\therefore \text{ C.F.} = e^{-x} f_1(y) + e^x f_2(y-x).$$

$$\text{Now, P.I.} = \frac{1}{D^2+DD'+D'-1}.\sin(x+2y)$$

$$= \frac{1}{-1^2-1.2+D'-1}.\sin(x+2y)$$

(On putting -1^2 for D^2 and -1.2 for DD')

$$= \frac{1}{(D'-4)}\sin(x+2y) = \frac{D'+4}{D'^2-16}\sin(x+2y)$$

$$= (D'+4)\frac{1}{-2^2-16}\sin(x+2y)$$

(On putting -2^2 for D'^2)

$$= -\frac{1}{20}(D'+4)\sin(x+2y)$$

$$= -\frac{1}{20}[2\cos(x+2y)+4\sin(x+2y)]$$

$$= -\frac{1}{10}[\cos(x+2y)+2\sin(x+2y)].$$

Hence, the complete solution is

$$z = e^{-x} f_1(y) + e^x f_2(y-x) - \frac{1}{10}[\cos(x+2y)+2\sin(x+2y)].$$

Example 7. *Solve* $(D - D'^2 - 3D + 3D')z = xy + e^{x+2y}$.

(UPTU-2002, GBTU(CO)-2011)

Solution. The given equation can be written as

$$(D-D')(D+D'-3)z = xy + e^{x+2y}.$$

Its C.F. $= f_1(y+x) + e^{3x} f_2(y-x)$.

Now, P.I. corresponding to xy

$$= \frac{1}{(D-D')(D+D'-3)}xy$$

$$= -\frac{1}{3D}\left(1-\frac{D'}{D}\right)^{-1}\left(1-\frac{D+D'}{3}\right)^{-1}.xy$$

$$= -\frac{1}{3D}\left(1+\frac{D'}{D}+...\right)\left[1+\frac{D+D'}{3}+\left(\frac{D+D'}{3}\right)^2+...\right]xy$$

$$= -\frac{1}{3D}\left(1+\frac{D'}{D}+...\right)\left(1+\frac{D+D'}{3}+\frac{2DD'}{9}+...\right)xy$$

$$= -\frac{1}{3D}\left(1+\frac{D}{3}+\frac{D'}{3}+\frac{D'}{D}+\frac{D'}{3}+\frac{2DD'}{9}+...\right)xy$$

$$= -\frac{1}{3D}\left(xy+\frac{y}{3}+\frac{2x}{3}+\frac{1}{D}x+\frac{2}{9}\right)$$

$$= -\frac{1}{3}\left(\frac{x^2y}{2}+\frac{xy}{3}+\frac{x^2}{3}+\frac{x^3}{6}+\frac{2}{9}x\right).$$

Now, P.I. corresponding to e^{x+2y}

$$= \frac{1}{(D+D'-3)(D-D')}e^{x+2y}$$

$$= \frac{1}{(D+D'-3)}.\frac{1}{(1-2)}e^{x+2y}$$

$$= -\frac{1}{(D+D'-3)}e^{x+2y}.1$$

$$= -e^{x+2y}\frac{1}{D+1+D'+2-3}.1$$

$$= -e^{x+2y}\frac{1}{D+D'}.1 = -e^{x+2y}\frac{1}{D}\left(1+\frac{D'}{D}\right)^{-1}.1$$

$$= -e^{x+2y}\frac{1}{D}\left(1-\frac{D'}{D}...\right).1 = -xe^{x+2y}.$$

Hence, the complete solution is

$$z = f_1(y+x) + e^{3x} f_2(y-x) - \frac{1}{3}\left(\frac{x^2y}{2}+\frac{xy}{3}+\frac{x^2}{3}+\frac{x^3}{6}+\frac{2}{9}x\right) - xe^{x+2y}.$$

Example 8. *Solve* $(D-3D'-2)^2 z = 2e^{2x}\tan(y+3x)$.

Solution. Its C.F. $= e^{2x} f_1(y+3x) + xe^{2x} f_2(y+3x)$.

$$\text{Now P.I.} = \frac{1}{(D-3D'-2)^2}2e^{2x}\tan(y+3x)$$

$$= 2e^{2x}.\frac{1}{(D+2-3D'-2)^2}.\tan(y+3x)$$

$$= 2e^{2x}\frac{1}{(D-3D')^2}\tan(y+3x)$$

$$= 2e^{2x}\frac{x^2}{1^2.2!}.\tan(y+3x)$$

$$= x^2e^{2x}\tan(y+3x).$$

Hence, the complete solution is

$$z = e^{2x} f_1(y+3x) + xe^{2x} f_2(y+3x) + x^2e^{2x}\tan(y+3x).$$

Example 9. *Solve* $2\frac{\partial^2 z}{\partial x\partial y} + \frac{\partial^2 z}{\partial y^2} - 3\frac{\partial z}{\partial y} = 5\cos(3x-2y)$.

Solution. The given equation can be written as

$$(2DD' + D'^2 - 3D')z = 5\cos(3x-2y)$$

or $$D'(2D+D'-3)z = 5\cos(3x-2y)$$

$$\therefore \text{ C.F.} = f_1(x) + e^{3x/2} f_2(2y-x).$$

$$\text{Now, P.I.} = \frac{1}{2DD'+D'^2-3D'}5\cos(3x-2y)$$

$$= 5\frac{1}{2\times6-4-3D'}\cos(3x-2y)$$

(On Putting $-3(-2)$ for DD' and $-(-2)^2$ for D'^2)

$$= 5.\frac{1}{8-3D'}\cos(3x-2y)$$

$$= \frac{8+3D'}{64-9D'^2}\cos(3x-2y)$$

$$= 5(8+3D')\frac{1}{64-9\times(-4)}\cos(3x-2y)$$
$$= \frac{5}{100}(8+3D')\cos(3x-2y)$$
$$= \frac{1}{20}[8\cos(3x-2y)-3\sin(3x-2y)(-2)]$$
$$= \frac{1}{10}[4\cos(3x-2y)+3\sin(3x-2y)].$$

Hence, the complete solution is

$$z = f_1(x)+e^{3x/2}f_2(2y-x) + \frac{1}{10}[4\cos(3x-2y)+3\sin(3x-2y)].$$

Example 10. *Solve* $(D^2-DD'+D'-1)z = \cos(x+2y)+e^y$. (UPTU(Q. Bank)-2002)

Solution. The given equation can be written as

$$(D-1)(D-D'+1)z = \cos(x+2y)+e^y.$$

Now, P.I. corresponding to cos $(x+2y)$

$$= \frac{1}{D^2-DD'+D'-1}\cos(x+2y)$$
$$= \frac{1}{-1^2-(-1.\,2)+D'-1}.\cos(x+2y)$$
$$= \frac{1}{D'}\cos(x+2y) = \frac{D'}{D'^2}\cos(x+2y)$$
$$= \frac{D'}{-2^2}\cos(x+2y) = -\frac{1}{4}\{-2\sin(x+2y)\}$$
$$= \frac{1}{2}\sin(x+2y).$$

P.I. corresponding to e^y

$$= \frac{1}{(D-D'+1)(D-1)}e^y$$
$$= \frac{1}{(D-D'+1).(0-1)}e^y.1$$
$$= -e^y\frac{1}{D-(D'+1)+1}.1 = -e^y\frac{1}{D-D'}.1$$
$$= -e^{-y}.\frac{1}{D}\left(1-\frac{D'}{D}\right)^{-1}.1$$
$$= -e^y.\frac{1}{D}.1 = -xe^y.$$

Hence, the solution is

$$z = e^x f_1(y)+e^{-x}f_2(y+x)+\frac{1}{2}\sin(x+2y)-xe^y.$$

EXERCISE 41.5

Solve the following equations:

1. $t+s+q=0$
2. $DD'(D-2D'-3)z = 0.$
3. $D(D-2D'-3)z = e^{x+2y}.$
4. $(D+D'-1)(D+2D'-3)z = 4+3x+6y.$
5. $r-s+2q-z = x^2y^2$ (IAS-1993)
6. $[D^2-DD'+2D'+2D-2D']z = e^{2x+3y}+\sin(2x+y)+xy.$ (UPTU(B. Tech.)-2003)
7. $(D-D'^2)z = \cos(x-3y).$
8. $(D-D'-1)(D-D'-2)z = \sin(2x+3y).$ (UPTU(Q. Bank)-2002, UKTU-2011)
9. $(D-D'+2)(D+D'-1)z = e^{x-y}-x^2y.$
10. $(D^2-DD'-2D)z = \sin(3x+4y)-e^{2x+y}.$
11. $(D^2-DD'-2D'^2+2D+2D')z = \sin(2x+y)$ (GBTU-2011)
12. $(D+D'-1)^2z = xy$ (GBTU(AG)-2011)
13. $(D-3D'-2)^3z = 6e^{2x}\sin(3x+y)$ (GBTU(SUM)-2010)
14. $(D^2+2DD'+D'^2-2D-2D')z = \sin(x+2y)$ (UPTU-2004)
15. $\frac{\partial^2 z}{\partial x^2}-\frac{\partial^2 z}{\partial x\partial y}+\frac{\partial z}{\partial y} = x^2+y^2$ (Madras-2000S)

Hint to Selected Problems

1. The given equation can be written as

$$(DD'+D'+D'^2)z = 0 \Rightarrow D'(D+D'+1)z = 0$$

Solution is $z = f_1(x)+e^{-x}f_2(y-x)$.

3. C.F. is $z = f_1(y)+e^{3x}f_2(y+2x)$

$$\text{P.I.} = \frac{1}{D(D-2D'-3)}e^{x+2y} = \frac{e^{x+2y}}{1.(1-2.2-3)} = -\frac{1}{6}e^{x+2y}$$

5. $$\text{P.I.} = \frac{1}{(D^2-DD'+2D'-1)}x^2y^2$$

$$= -[1-CD^2-DD'+2D')]^{-1}x^2y^2$$

Now expanding by binomial theorem.

7. Since $D-D'^2$ cannot be resolved into factors, so

C.F. $= \Sigma Ae^{hx+ky} = \Sigma Ae^{k^2x+ky}$, where $h-k^2=0$

$$\text{P.I.} = \frac{1}{D-D'^2}.\cos(x-3y) = \frac{1}{D+9}\cos(x+3y)$$
$$= \frac{D-9}{D^2-81}\cos(x-3y) = \frac{-\sin(x-3y)-9\cos(x-3y)}{-1-81}$$

ANSWERS

1. $z = f_1(x)+e^{-x}f_2(y-x).$
2. $z = f_1(y)+f_2(x)+e^{3x}f_3(y+2x).$
3. $z = f_1(y)+e^{3x}f_2(y+2x)-\frac{1}{6}e^{x+2y}.$
4. $z = e^x f_1(y-x)+e^{3x}f_2(y-2x)+6+x+2y.$
5. $z = \Sigma Ae^{hx+ky}-x^2y^2-2y^2+4xy-4x^2y-52-8x^2-16y+16x$ where $h^2-hk+2k-1=0$
6. $z = f_1(y-x)+e^{-2x}f_2(y+2x)-\frac{1}{10}e^{2x+3y}-\frac{1}{6}\cos(2x+y)+\frac{1}{4}x^2y-\frac{1}{4}xy+\frac{3}{8}x^2-\frac{1}{2}x-\frac{1}{12}x^3$

7. $z = \sum A e^{k^2 + ky} + \frac{1}{82}[\sin(x-3y) + 9\cos(x-3y)]$ where $h - k^2 = 0$

8. $z = e^x f_1(y+x) + e^{2x}(y+x) + \frac{1}{10}[\sin(2x+3y) - 3\cos(2x+3y)]$.

9. $z = e^{-2x} f_1(y+x) + e^x f_2(y-x) - \frac{e^{x-y}}{4} + \frac{1}{2}\left[x^2 y + xy + \frac{3y}{2} + 3x + \frac{21}{4}\right] + \frac{3}{4}x^2$.

10. $z = f_1(y) + e^{2x} f_2(y+x) + \frac{1}{15}\sin(3x+4y) + \frac{2}{15}\cos(3x+4y) + \frac{1}{2}e^{2x+y}$.

11. $z = f_1(y-x) + e^{-2x} f_2(2x+y) - 1/6 \cos(2x+y)$

12. $z = e^x f_1(y-x) + x e^x f_2(y-x) + xy + 2y + 2x + 6$

13. $z = e^{2x} f_1(y+3x) + x e^{2x} f_2(y+3x) + x^2 e^{2x} f_3(y+3x) + x^3 e^{2x} \sin(3x+y)$

14. $z = \phi_1(y-x) + e^{2x}\phi_2(y-x) + \frac{1}{39}[2\cos(x+2y) - 3\sin(x+2y)]$

15. $z = f_1(y) + e^{-x} f_2(y+x) + \frac{1}{3}x^3 - x^2 + xy^2 + 6x$

41.9 EQUATIONS REDUCIBLE TO LINEAR PARTIAL DIFFERENTIAL EQUATIONS WITH CONSTANT COEFFICIENTS

A partial differential equation of the form

$$f\left(x\frac{\partial}{\partial x}, y\frac{\partial}{\partial y}\right) = F(x, y)$$

having variable coefficients are reduced to linear form by putting $x = e^X, y = e^Y$ so that $X = \log x$ and $Y = \log y$

$$\therefore \quad \frac{\partial z}{\partial x} = \frac{\partial z}{\partial X} \cdot \frac{\partial X}{\partial x} = \frac{1}{x}\frac{\partial z}{\partial X} \quad \text{or} \quad x\frac{\partial z}{\partial x} = \frac{\partial z}{\partial X}$$

$$\therefore \quad x\frac{\partial}{\partial x} \equiv \frac{\partial}{\partial X} \equiv D \text{ (say)}$$

Now, $x\frac{\partial}{\partial x}\left(x^{n-1}\frac{\partial^{n-1} z}{\partial x^{n-1}}\right) = x^n \frac{\partial^n z}{\partial x^n} + (n-1)x^{n-1}\frac{\partial^{n-1} z}{\partial x^{n-1}}$ or $x^n \cdot \frac{\partial^n z}{\partial x^n} = \left(x\frac{\partial}{\partial x} - n + 1\right)x^{n-1}\frac{\partial^{n-1} z}{\partial x^{n-1}}$.

Putting $n = 2, 3, \ldots$ we get

$$x^2\frac{\partial^2 z}{\partial x^2} = (D-1)x\frac{\partial z}{\partial x} = D(D-1)z, \; x^3\frac{\partial^3 z}{\partial x^3} = (D-2)x^2\frac{\partial^2 z}{\partial x^2} = D(D-1)(D-2)z \text{ etc.}$$

Similarly, $y\frac{\partial z}{\partial y} = \frac{\partial z}{\partial Y} = D'z; \; y^2\frac{\partial^2 z}{\partial y^2} = D'(D'-1)z; \; y^3\frac{\partial^3 z}{\partial y^3} = D'(D-1)(D'-2)z$ and $xy\frac{\partial^2 z}{\partial x \partial y} = DD'z$

Substituting in the given equation, it reduced to the form $f(DD')z = V$ which is an equation with constant coefficient.

Solved Examples

Example 1. Solve $x^2\frac{\partial^2 z}{\partial x^2} + 2xy\frac{\partial^2 z}{\partial x \partial y} + y^2\frac{\partial^2 z}{\partial y^2} = 0$.

Solution. Let $x = e^X, \; y = e^Y$

So that $X = \log x$ and $Y = \log y$.

Denoting $\frac{\partial}{\partial X}$ and $\frac{\partial}{\partial Y}$ by D and D' respectively.

$$\therefore \quad x^2\frac{\partial^2}{\partial x^2} = D(D-1), xy\frac{\partial^2}{\partial x \partial y} = DD'$$

and $y^2\frac{\partial^2}{\partial y^2} = D'(D'-1)$.

With these substitutions, the given equation reduce to

$$[D(D-1) + 2DD' + D'(D'-1)]z = 0$$

or

$$(D+D')(D+D'-1)z = 0.$$

Hence, solution is

$$z = f_1(Y-X) + e^X f_2(Y-X)$$

$$= f_1(\log y - \log x) + x f_2(\log y - \log x)$$

$$= f_1\left(\log\frac{y}{x}\right) + x f_2\left(\log\frac{y}{x}\right)$$

$$= g_1\left(\frac{y}{x}\right) + g_2\left(\frac{y}{x}\right).$$

Example 2. Solve

$$x^2\frac{\partial^2 z}{\partial x^2} - 4xy\frac{\partial^2 z}{\partial x \partial y} + 4y^2\frac{\partial^2 z}{\partial y^2} + 6y\frac{\partial z}{\partial y} = x^3 y^4.$$

(UPTU(Q. Bank)-2002)

Solution. Substituting $x = e^X, \; y = e^Y$

And denoting $\frac{\partial}{\partial X}$ and $\frac{\partial}{\partial Y}$ by D and D' respectively, then the equation reduces to

$$[D(D-1) - 4DD' + 4D'(D'-1) + 6D']z = e^{3X+4Y}$$

or $(D-2D')(D-2D'-1)z = e^{3X+4Y}$.

$$\therefore \text{C.F.} = f_1(Y+2X) + e^X f_2(Y+2X)$$

$$= f_1(\log y + 2\log x) + x f_2(\log y + 2\log x)$$

$$= f_1(\log yx^2) + x f_2(\log yx^2)$$

$$= g_1(yx^2) + x g_2(yx^2).$$

$$\text{Now, P.I.} = \frac{1}{(D-2D')(D-2D'-1)}e^{3X+4Y}$$

$$= \frac{e^{3X+4Y}}{(3-8)(3-8-1)}$$

$$= \frac{1}{30}e^{3X+4Y} = \frac{1}{30}x^3y^4.$$

Hence, the complete solution is

$$z = g_1(yx^2) + xg_2(yx^2) + \frac{1}{30}x^3y^4.$$

Example 3. *Solve* $x^2r - y^2t + xp - yq = \log x.$ (IAS-1993, 97)

Solution. The given equation can be written as

$$x^2\frac{\partial^2 z}{\partial x^2} - y^2\frac{\partial^2 z}{\partial y^2} + x\frac{\partial z}{\partial x} - y\frac{\partial z}{\partial y} = \log x.$$

Substituting $x = e^X$, $y = e^Y$ and denoting $\frac{\partial}{\partial X}$ and $\frac{\partial}{\partial Y}$ by D and D' respectively.

Then equation reduces to

$$[D(D-1) - D'(D'-1) + D - D']z = X$$

or $(D^2 - D'^2)z = X$ or $(D-D')(D+D')z = X.$

Its C.F. $= \phi_1(Y+X) + \phi_2(Y-X)$

$$= \phi_1(\log y + \log x) + \phi_2(\log y - \log x)$$

$$= \phi_1(\log xy) + \phi_2(\log y/x)$$

$$= f_1(xy) + f_2(y/x).$$

Now,

$$\text{P.I.} = \frac{1}{(D^2 - D'^2)}.X = \frac{1}{D^2}\left(1 - \frac{D'^2}{D^2}\right)^{-2}.X$$

$$= \frac{1}{D^2}.X = \frac{X^3}{6} = \frac{1}{6}(\log x)^3.$$

Hence, the complete solution is

$$z = f_1(xy) + f_2(y/x) + \frac{1}{6}(\log x)^3.$$

EXERCISE 41.6

1. $(x^2D^2 - y^2D'^2)z = x^2y.$
2. $x^2\left(\frac{\partial^2 z}{\partial x^2}\right) - y^2\left(\frac{\partial^2 z}{\partial y^2}\right) = xy.$ (IAS-1987)
3. $(x^2D^2 + 2xyDD' + y^2D'^2)z = x^m y^n.$
4. $(x^2D^2 + 2xyDD' + y^2D'^2 - nxD - xyD' + n)z = x^2y^2$
5. $x^2r - 3xys + 2y^2t + px + 2qy = x + 2y.$
6. $x^2D^2 + 2xyDD' + y^2D'^2 = (x^2+y^2)^{3/2}.$
7. $(x^2D^2 - xyDD' - 2y^2D'^2 + xD - 2yD')z = \log(y/x) - 1/2$

Hint to Selected Problems

1. Putting $x = e^X$, $y = e^Y \Rightarrow X = \log x$, $Y = \log y$

So, $x^2D^2 = D_1(D_1 - 1)$, $y^2D'^2 = D_1'(D_1' - 1)$

where $D_1 = \frac{\partial}{\partial X}, D_1' = \frac{\partial}{\partial Y}$

Then, equation (1) reduces to

$$[D_1(D_1-1) - D_1'(D_1'-1)]z = e^{2X+Y}$$

$$\Rightarrow \quad (D_1 - D_1')(D_1 + D_1' - 1) = e^{2X+Y}$$

3. Proceed same as (1), we get

$$(D_1 + D_1')(D_1 + D_1' - 1)z = e^{mX+nY}$$

4. The reduced equation is

$$\left[(D_1 + D_1' - 1)(D_1 + D_1' - n)\right]z = e^{2X} + e^{2Y}$$

5. The reduced equation is $\left[(D_1 + D_1' - 1)(D_1 - 2D_1')\right]z = e^X + 2e^Y$

C.F. $= \phi_1(Y+X) + \phi_2(Y+2X)$

$$= \phi_1(\log xy) + \phi_2(\log x^2y) = f_1(xy) + f_2(x^2y)$$

$$\text{P.I.} = \frac{1}{(D_1 - D_1')(D_1 - 2D_1')}(e^X + 2e^Y)$$

$$= \frac{e^X}{(1-0)(1-0)} + \frac{2e^Y}{(0-1)(0-2)} = e^X + e^Y = x + y$$

ANSWERS

1. $z = xf_1(y/x) + f_2(xy) + \frac{1}{2}x^2y.$
2. $z = f_1(xy) + xf_2(y/x) + xy\log x.$
3. $z = f_1(y/x) + xf_2(y/x) + \frac{x^m y^n}{(m+n)(m+n-1)}$
4. $z = xf_1(y/x) + x^n f_2(y/x) + \frac{x^2+y^2}{2-n}.$
5. $z = f_1(xy) + f_2(x^2y) + x + y.$
6. $z = f_1(y/x) + xf_2(y/x) + \left\{\frac{1}{(n^2-n)}(x^2+y^2)\right\}^{m/n}$
7. $z = f_1(yx^2) + f_2(y/x) + \frac{1}{2}(\log x)^2\log y - \frac{1}{4}(\log x)^2$

41.10 SOME EXAMPLES UNDER GIVEN GEOMETRICAL CONDITIONS

Example 1. *Find the surface passing through two lines* $z = x = 0, z - 1 = x - y = 0$ *satisfying* $r - 4s + 4t = 0.$ (IAS-1996)

Solution. The given equation can be written as

$$(D^2 - 4DD' + 4D'^2)z = 0.$$

It's A. E. is $m^2 - 4m + 4 = 0$

or $(m-2)^2 = 0$

or $\quad m = 2, 2.$

$$\therefore \quad z = f_1(y+2x) + xf_2(y+2x). \quad \ldots(1)$$

Since the surface passes through the lines

$z = x = 0$ and $z - 1 = x - y = 0$.

$\therefore \quad 0 = f_1(y)$. ...(2)

and $1 = f_1(y + 2x) + x f_2(y + 2x)$. ...(3)

From (2) and (3), we get

$$f_2(y+2x) = \frac{1}{x} = \frac{3}{3x} = \frac{3}{2x+x} = \frac{3}{2x+y}$$

$[\because x - y = 0]$

Hence, the surface is

$$z = x.\frac{3}{2x+y} \quad \text{or} \quad z(2x+y) = 3x.$$

Example 2. *Find the surface satisfying the equation $r + t - 2s = 0$ and conditions that $bz = y^2$, when $x = 0$ and $az = x^2$ when $y = 0$.*

Solution. The given equation can be written as

$(D^2 - 2DD' + D'^2)z = 0 \Rightarrow (D - D')^2 z = 0.$

$\therefore$ Its solution is $z = f_1(y + x) + x f_2(y + x)$. ...(1)

Since $z = y^2/b$ when $x = 0$ then from (1),

$$\frac{y^2}{b} = f_1(y)$$

giving $f_1(y+x) = \dfrac{(y+x)^2}{b}$. ...(2)

Again since $z = \dfrac{x^2}{a}$ when $y = 0$, then from (1)

$$\frac{x^2}{a} = f_1(x) + x f_2(x). \quad \text{...(3)}$$

But from (2), $f_1(x) = x^2/b$. Then (3) gives

$\dfrac{x^2}{a} = x f_2(x) + \dfrac{x^2}{b}$ or $f_2(x) = \dfrac{b-a}{ab} x$ which gives

$$f_2(y+x) = \frac{b-a}{ab}(y+x). \quad \text{...(4)}$$

Putting, value of $f_1(y+x)$ and $f_2(y+x)$ from (2) and (4) in (1) the required surface is

$$z = \frac{b-a}{ab} x(y+x) + \frac{(y+x)^2}{b}$$

or $\quad z = (x+y)\left(\dfrac{x}{a} + \dfrac{y}{b}\right)$.

Example 3. *A surface is drawn satisfying $r + t = 0$ and touching $x^2 + z^2 = 1$ along its section by $y = 0$. Find its equation in the form*

$$x^2(x^2 + z^2 - 1) = y^2(x^2 + z^2).$$

Solution. We have $(D^2 + D'^2)z = 0$.

$\Rightarrow \quad (D + iD)(D - iD')z = 0.$ ...(1)

Solution of (1) is given by

$$z = \phi_1(y + ix) + \phi_2(y - ix). \quad \text{...(2)}$$

Now, the given surface is

$$x^2 + z^2 = 1 \Rightarrow z = \sqrt{1 - x^2}. \quad \text{...(3)}$$

Since (2) and (3) touch along their common section by $y = 0$. ...(4)

The value of p and q from (2) and (3)must be the same.

$$p = i\phi_1'(y+ix) - i\phi_2'(y-ix) = -\frac{x}{\sqrt{1-x^2}}$$

and $\quad q = \phi_1'(y+ix) + \phi_2'(y-ix) = 0.$

Now, using (4), we have

$$\phi_1'(ix) - \phi_2'(-ix) = \frac{ix}{\sqrt{1+x^2 i^2}} \quad [\because i^2 = -1]$$

$\Rightarrow \quad \phi_1'(ix) + \phi_2'(-ix) = 0.$

On solving, we get

$$\phi_1'(ix) = \frac{ix}{2\sqrt{1+x^2 i^2}}, \phi_2'(-ix) = \frac{-xi}{2\sqrt{1+x^2 i^2}}$$

Let $ix = X$ and $-ix = Y$. Then

$$\phi_1'(X) = \frac{X}{2\sqrt{1+X^2}} \text{ and } \phi_2'(Y) = \frac{Y}{2\sqrt{1+Y^2}}$$

On integrating, we get $\phi_1(X) = \dfrac{1}{2}\sqrt{1+X^2} + C_1$;

$$\phi_2(Y) = \frac{1}{2}\sqrt{1+Y^2} + C_2$$

$$\Rightarrow \quad \phi_1(y+ix) = \frac{1}{2}\sqrt{1+(y+ix)^2} + C_1$$

and $\quad \phi_2(y - ix) = \dfrac{1}{2}\sqrt{1+(y-ix)^2} + C_2$.

Putting all these values in (2) and writing $C_1 + C_2 = C$.

$$z = \frac{1}{2}\sqrt{1+(y+ix)^2} + \frac{1}{2}\sqrt{1+(y-ix)^2} + C \quad \text{...(5)}$$

Equating two values of z from (3) and (5) at $y = 0$, we have

$$\frac{1}{2}[\sqrt{1-x^2}) + \sqrt{1-x^2})] + C = \sqrt{(1-x^2)} \Rightarrow C = 0.$$

Then (5) gives

$$2z = \sqrt{\{1+(y+ix)^2\}} + \sqrt{\{1+(y-ix)^2\}}$$

$$\Rightarrow 4z^2 = \{1+(y+ix)^2\} + \{1+(y-ix)^2\} + 2\sqrt{\{1+(y+ix)^2\}\{1+(y-ix)^2\}}$$

$$\Rightarrow 2z^2 = (1 + y^2 - x^2) + [\sqrt{\{1+(y+ix)^2\}\{1+(y-ix)^2\}}]. \quad \text{...(6)}$$

Squaring both sides of (6), we have

$$4z^4 = (1+y^2-x^2)^2 + \{1+(y+ix)^2\}\{1+(y-ix)^2\}$$
$$+2(1+y^2-x^2)\sqrt{\{1+(y+ix)^2\}\{1+(y-ix)^2\}}$$
$$\Rightarrow \quad 4z^4 = (1+y^2-x^2)^2 + \{1+y^2-x^2)+2ixy\}$$
$$\{(1+y^2-x^2)-2ixy\}$$
$$+2(1+y^2-x^2)\{2z^2-(1+y^2-x^2)\}$$
$$= (1+y^2-x^2)^2 + (1+y^2-x^2)^2 + 4x^2y^2$$
$$+4z^2(1+y^2-x^2) - 2(1+y^2-x^2)^2$$
$$= 4x^2y^2 + 4z^2(1+y^2-x^2).$$

EXERCISE 41.7

1. Find a surface satisfying $r-2s+t=6$ and touching the hyperbolic paraboloid $z=xy$ along its section by the plane $y=x$.
2. Find a surface satisfying equation $2x^2r - 5xys + 2y^2t + 2(px + qy = 0)$ and touching the hyperbolic paraboloid $z = x^2 - y^2$ along its section by plane $y = 1$.
3. Find a surface satisfying $r+s=0$ and touching the elliptic paraboloid $z=4x^2+y^2$ along its section by the plane $y=2x+1$.

Hint to Selected Problems

1. The given equation can be written as $(D^2-2DD'+D'^2)z=6$, *i.e.*, $(D-D')^2z=6$, C.F. $= \phi_1(y+x)+x\phi_2(y+x)$, P.I. $= 3x^2$.
2. The given equation can be written as
$$2x^2\frac{\partial^2 z}{\partial x^2} - 5xy\frac{\partial^2 z}{\partial x\,\partial y} + xy^2\frac{\partial^2 z}{\partial y^2} + 2\left(x\frac{\partial z}{\partial x} + y\frac{\partial z}{\partial y}\right) = 0.$$
Now putting $x=e^u, y=e^v$ *i.e.*, $u=\log x$ and $v=\log y$.

ANSWERS

1. $z = x^2 - xy + y^2$
2. $3z = 4yx^2 - y^4x^2 - 6\log y - 3$
3. $4x^2 - 8xy + y^2 + 8x - 4y + z + 2 = 0$

Objective Evaluations

Fill in the Blanks

1. $\dfrac{\partial^2 z}{\partial x^2}$ = __________.
2. $\dfrac{\partial^2 z}{\partial x \partial y}$ = __________.
3. $\dfrac{\partial^2 z}{\partial y^2}$ = __________.
4. $\dfrac{\partial p}{\partial x}$ = __________.
5. $\dfrac{\partial q}{\partial x} = \dfrac{\partial p}{\partial y}$ = __________.

Multiple Choice Questions

Choose the most appropriate one.

1. The auxiliary equation can be obtained by putting :
   (a) $D = m$ (b) $D' = 1$
   (c) $D = m, D' = 1$ (d) none of these
2. If the given partial differential equation is $f(D, D')z = F(x, y)$, then P.I. is given by :
   (a) $f(D, D')F(x, y)$ (b) $\dfrac{1}{f(D, D')}F(x, y)$
   (c) $\dfrac{1}{F(D, D')}f(x, y)$ (d) none of these
3. If the PDE $f(D, D')z = F(x, y)$, and $F(x, y)$ is of the form e^{ax+by}, $[f(a, b) \neq 0]$, then P.I. is given by :
   (a) e^{ax+by} (b) $f(a, b)\, e^{ax+by}$
   (c) $\dfrac{1}{f(a, b)} e^{ax+by}$ (d) none of these
4. If $f(D, D')$ is a homogeneous function of D and D', then $\dfrac{1}{(D - mD')}F(x, y) =$:
   (a) $F(x, a - mx)$ (b) $F(x, a - mx)\, dx$
   (c) $\int F(x, a + mx) dx$ (d) none of these
5. When $F(x, y) = e^{ax+by}.V$, where V is a function of x and y, then we have P.I. = :
   (a) e^{ax+by} (b) $e^{ax+by}.\dfrac{1}{f(D + a, D' + b)}.V$
   (c) $e^{ax+by}\dfrac{1}{f(D, D')}.V$ (d) none of these

ANSWERS

Fill in the Blanks

1. r	2. s	3. t	4. r	5. s

Multiple Choice Questions

1. (c)	2. (b)	3. (c)	4. (b)	5. (b)

FFFFFF

Applications of Partial Differential Equation to Engineering Problems

42.1 INTRODUCTION

Many problems in science and engineering, when formulated mathematically, lead to partial differential equations involving one or more unknown functions together with certain prescribed conditions on the functions which arise from the physical situations.

The process of obtaining all solutions of a partial differential equations under given conditions is known as boundary value problem. If time t is regarded as an independent variable and the conditions are stated as $t = 0$, the problem is called initial value problem.

42.2 CLASSIFICATION OF PARTIAL DIFFERENTIAL EQUATION

(UPTU-2005, 07)

Any equation $A\frac{\partial^2 u}{\partial x^2} + B\frac{\partial^2 u}{\partial x \partial y} + C\frac{\partial^2 u}{\partial y^2} + F(x, y, u, p, q) = 0$ ... (1)

where A, B, C may be constants or functions of x and y.

This equation (1) will be

(i) elliptic if $B^2 - 4AC < 0$

(ii) parabolic if $B^2 - 4AC = 0$

(iii) hyperbolic if $B^2 - 4AC > 0$

1. Parabolic equation : $\frac{\partial u}{\partial t} = \frac{\partial^2 u}{\partial x^2}$

2. Elliptic equation : $\frac{\partial^2 u}{\partial x^2} + \frac{\partial^2 u}{\partial y^2} = 0$

3. Hyperbolic equation : The wave equation $\frac{\partial^2 u}{\partial t^2} = c^2 \frac{\partial^2 u}{\partial x^2}$

42.3 PRODUCT METHOD : SOLUTION OF BOUNDARY VALUE PROBLEMS BY THE METHOD OF SEPARATION OF VARIABLES

Let
$$A\frac{\partial^2 u}{\partial x^2} + B\frac{\partial^2 u}{\partial x \partial y} + C\frac{\partial^2 u}{\partial y^2} + D\frac{\partial u}{\partial x} + E\frac{\partial u}{\partial y} + Fu = G \quad \text{... (1)}$$

be the general linear partial differential equation, where $A, B, \ldots\ldots, G$ are functions of x and y. Here, it must be noted that

(i) If any partial differential equation of second order cannot be put in above form, then it is said to be non-linear.

(ii) If $G = 0$, then equation (i) is called homogeneous. Thus, for non-homogeneous equation $G \neq 0$.

Let $u(x, y) = X(x)\, Y(y)$ ....(2)

be the solution of (1). Here $X(x)$ and $Y(y)$ are respectively the functions of x and y alone.

Using (2), we can find the following values

$\frac{\partial^2 u}{\partial x^2}, \frac{\partial^2 u}{\partial y^2}, \frac{\partial^2 u}{\partial x \partial y}, \frac{\partial u}{\partial x}$ and $\frac{\partial u}{\partial y}$ in terms of X and Y.

Putting all these values in (1), we get $\frac{F(D).X}{X} = \frac{g(D').Y}{Y}$, where, $F(D)$ and $g(D')$ are functions of $D = \frac{\partial}{\partial x}$ and $D' = \frac{\partial}{\partial y}$ respectively.

Thus, we can find a relation in which LHS is a function of x alone while RHS is a function of y alone.

Then putting

$$\frac{F(D).X}{X} = \frac{g(D').Y}{Y} = \lambda \text{ (say)} \quad \dots (3)$$

Thus, the solution of (1) reduces to be solution of a pair of ODE given by (3).

Solved Examples

Example 1. *Solve the boundary value problem* $\frac{\partial u}{\partial x} = 4\frac{\partial u}{\partial y}$ *with* $u(0, y) = 8e^{-3y}$ *by the method of separation of variables.* (UPTU-2008, JNTU-2006)

Solution. The given equation is

$$\frac{\partial u}{\partial x} = 4\frac{\partial u}{\partial y} \quad \dots (1)$$

with boundary conditions $u(0, y) = 8e^{-3y}$.

Let $u(x, y) = X(x)\,Y(y)$ (where X and Y are respectively the functions of x and y alone) be the solution of (1).

Putting the value in (1), we get

$$X'Y = 4XY'$$

$$\Rightarrow \frac{X'}{4X} = \frac{Y'}{Y} \quad \dots (2)$$

Here, dashes denote derivatives with respect to the relevant variable. Equation (2) is true only when each side is equal to same constant say λ.

Therefore (2) gives $X' - 4\lambda X = 0$ and $Y' - \lambda Y = 0$.

On solving, we get $X(x) = Ae^{4\lambda x}$ and $Y(y) = Be^{\lambda y}$.

Thus, $u(x, y) = ABe^{4\lambda x + \lambda y} = C\,e^{\lambda(4x+y)}$,

where $C\ (= AB)$ is any arbitrary constant.

Using given boundary condition

$$u(0, y) = 8e^{-3y} = C\,e^{\lambda y},$$

we get $C = 8, \lambda = -3$.

Hence, the solution of given boundary value problem is $u(x, y) = 8e^{-3(4x+y)}$

Example 2. *Solve by the method of separation of variables* $\frac{\partial^2 u}{\partial x^2} - 2\frac{\partial u}{\partial x} + \frac{\partial u}{\partial y} = 0.$

(UPTU-2005, Q. Bank–2002, PTU-2009 S, Bhopal-2008)

Solution. The given equation is

$$\frac{\partial^2 u}{\partial x^2} - 2\frac{\partial u}{\partial x} + \frac{\partial u}{\partial y} = 0 \quad \dots (1)$$

Let $u(x, y) = X(x)\,Y(y)$ be the solution of this equation.

Putting the value of u in the given equation, we get

$$X''Y - 2X'Y + XY' = 0 \quad \dots (2)$$

The dashes denote derivatives with respect to the relevant variable.

Now, relation (2) can be written as

$$\frac{X'' - 2X'}{X} = \frac{Y'}{Y} \quad \dots (3)$$

This is true if each side is equal to the same constant, say λ. Then, we have

$$X'' - 2X' - \lambda X = 0 \text{ and } Y' - \lambda Y = 0 \quad \dots (4)$$

$$\Rightarrow (D^2 - 2D - \lambda)X = 0 \text{ and } \frac{dY}{dy} = \lambda Y \quad \dots (5)$$

The auxiliary equation of (4) is given by

$$m^2 - 2m - \lambda = 0,$$

i.e., $m = 1 \pm \sqrt{(1+\lambda)}$.

Thus, $X(x) = Ae^{[1+\sqrt{(1+\lambda)}]x} + Be^{[1-\sqrt{(1+\lambda)}]x}$

and solution of (5) is given by

$$Y(y) = Ce^{\lambda y}$$

Finally, we get

$$u(x, y) = \left\{ACe^{[1+\sqrt{(1+\lambda)}]x} + BC\,e^{[1-\sqrt{(1+\lambda)}]x}\right\}e^{\lambda y}$$

$$= \left\{A_1 e^{[1+\sqrt{(1+\lambda)}]x} + A_2\,e^{[1-\sqrt{(1+\lambda)}]x}\right\}e^{\lambda y}$$

Example 3. *Use the method of separation of variables to solve the equation* $\frac{\partial^2 u}{\partial x^2} - 2\frac{\partial u}{\partial x} + \frac{\partial u}{\partial y} = 0$

(UPTU-2009, GBTU (AG)-2012)

Solution. Let $u = xy \quad \dots (1)$

Where X is a function x only and Y is function of y only.

$$\frac{\partial u}{\partial x} = \frac{\partial}{\partial x}(XY) = Y\frac{dX}{dx}$$

$$\frac{\partial u}{\partial y} = \frac{\partial}{\partial y}(XY) = X\frac{dY}{dy}$$

$$\frac{\partial^2 u}{\partial x^2} = \frac{\partial^2}{\partial x^2}(XY) = Y\frac{d^2X}{dx^2}$$

Putting these values in (1), we get

$$Y\frac{d^2X}{dx^2} - 2Y\frac{dX}{dx} + X\frac{dY}{dy} = 0$$

$$\Rightarrow YX'' - 2YX' + XY' = 0$$

$$\Rightarrow \quad \frac{X''-2X'}{X}+\frac{Y'}{Y}=0$$

$$\Rightarrow \quad \frac{X''-2X'}{X}=-\frac{Y'}{Y}=-P^2 \text{ (say)}$$

(i) $$\frac{X''-2X'}{X}=-P^2$$

$$\Rightarrow \quad X''-2X'+P^2X=0$$

A. E. is $\quad m^2-2m+p^2=0$

$$\Rightarrow \quad m=\frac{2\pm\sqrt{4-4P^2}}{2}=1\pm\sqrt{1-P^2}$$

$$\therefore \text{ C. F.} = C_1e^{(1+\sqrt{1-P^2})x}+C_2e^{(1-\sqrt{1-P^2})x}$$

P. I. = 0

Hence X = C.F. + P. I.

$$= C_1e^{(1+\sqrt{1-P^2})x}+C_2e^{(1-\sqrt{1-P^2})x} \quad \dots (2)$$

(ii) $$\frac{-Y'}{Y}=-P^2 \Rightarrow \frac{dY}{dy}=p^2Y \Rightarrow \frac{dy}{Y}=p^2dy$$

On integrating, we get $\log Y = p^2y+\log C_3$

$$Y=C_3e^{P^2y}$$

$$\therefore \quad u(x,y)=[C_1e^{(1+\sqrt{1-p^2})x}+C_2e^{(1-\sqrt{1-p^2})x}]C_3e^{P^2y} \quad \dots (3)$$

EXERCISE 42.1

1. Use the method of separation of variables to solve the equation

$$\frac{\partial^2 V}{\partial x^2}=\frac{\partial V}{\partial t}.$$

Given that $V=0$ when $t\to\infty$ as well as $v=0$ at $x=0$ and $x=l$. (AMIE-1997)

2. Use the method of separation of variables to solve $\frac{\partial u}{\partial x}=2\frac{\partial u}{\partial t}+u$, where $u(x,0)=6e^{-3x}$.

(Kerala–2005, UPTU-2006, Kurukshetra-2006, VTU-2009, GBTU(Sum)-2010, GBTU(AG)-2011,AMIETE 1997)

3. Use the method of separation of variables to solve the equation

$$\frac{\partial^2 u}{\partial x^2}=\frac{\partial u}{\partial y}+2u.$$

4. Solve by the method of separation of variables

$$4\frac{\partial u}{\partial t}+\frac{\partial u}{\partial x}=3u, \quad u=3e^{-x}-e^{-5x} \text{ when } t=0.$$ (SVTU-2008)

5. Use the method of separation of variables, find the solution of the following differential equation:

$$3\frac{\partial u}{\partial x}+2\frac{\partial u}{\partial y}=0, \quad u(x,0)=4e^{-x}.$$

(AMIETE-2000, VTU-2008S)

6. Solve by the method of separtion of variables.

$$py^3+qx^2=0$$ (VTU-2011, SVTU-2008)

7. Solve by the method of separation of variables

$$x^2\frac{\partial u}{\partial x}+y^2\frac{\partial u}{\partial y}=0$$ (VTU-2008)

8. Find a solution of the equation $\frac{\partial^2 u}{\partial x^2}=\frac{\partial u}{\partial y}+2u$ in the form $u=f(x)g(y)$. Solve the equation subject to the conditions $u=0$ and $\frac{\partial u}{\partial x}=1+e^{-3y}$, when x=0 for all values of y.

(Andhra-2000)

9. Solve by method of separation of variables (UKTU-2012)

$$4\frac{\partial u}{\partial x}+\frac{\partial u}{\partial y}=3u; u(0,y)=4e^{-y}-e^{-5y}$$ (UKTU-2012)

10. Solve by method of separation of variables

$$y^3\frac{\partial u}{\partial x}+x^2\frac{\partial u}{\partial y}=0$$ (GBTU-2011)

11. Solve by method of separation of variables

$$\frac{\partial^2 u}{\partial x^2}-\frac{\partial u}{\partial y}=0$$ (UPTU (SUM)-2007)

12. Solve by method of separation of variables

$$\frac{\partial u}{\partial x}=\frac{\partial u}{\partial y}$$ (GBTU-2012)

13. Solve by method of separation of variables

$$\frac{\partial z}{\partial x}+\frac{\partial^2 z}{\partial y^2}=0; z(x,0)=0,\ z(x,\pi)=0, z(0,y)=4\sin 3y$$ (MTU-2012)

14. Solve by method of separation of variables

$$\frac{\partial^2 u}{\partial x\partial t}=e^{-t}\cos x$$

given that $u=0$ when $t=0$ and $\frac{\partial u}{\partial t}=0$ when $x=0$.

[UPTU (SUM)2008, GBTU-2012, GBTU (AG)-2012]

15. Solve by method of separation of variables

$$u_{xx}=uy+2u, u(0,y)=0, \frac{\partial}{\partial x}u(0,y)=1+e^{-3y}$$

(GBTU-2010, UPTU-2009)

Hint to Selected Problems

1. Let us assume that $V=XT$, where X is a function of x only and T that of t only.

$$\frac{\partial V}{\partial t}=X\frac{dT}{dt} \text{ and } \frac{\partial^2 V}{\partial x^2}=T\frac{d^2X}{dx^2}$$

Substitute in $\frac{\partial^2 V}{\partial x^2}=\frac{\partial V}{\partial t}$, we get $X\frac{dT}{dt}=T\frac{d^2X}{dx^2}$

$$\Rightarrow \quad \frac{1}{T}\frac{dT}{dt}=\frac{1}{X}\frac{d^2X}{dx^2} \quad \dots (1)$$

Let each side of (1) be equal to constant (p^2)

$$\frac{1}{T}\frac{dT}{dt} = -p^2 \qquad \frac{1}{X}\frac{d^2X}{dx^2} = -p^2$$

$$\frac{dT}{dt} + p^2T = 0 \qquad \Rightarrow \qquad \frac{d^2X}{dx^2} + p^2X = 0$$

$$DT + p^2T = 0 \qquad D^2X + p^2X = 0$$

$$\Rightarrow \quad (D + p^2)T = 0 \qquad (D^2 + p^2)X = 0$$

$$\Rightarrow \quad D = -p^2 \qquad D^2 + p^2 = 0$$

$$\Rightarrow \quad T = C_1 e^{-p^2t} \qquad D^2 = -p^2 \Rightarrow D = \pm ip$$

$$X = C_2 \cos px + C_3 \sin px$$

$$V = C_1 e^{-p^2t}(C_2 \cos px + C_3 \sin px) \qquad \text{... (2)}$$

On putting $x = 0,\ V = 0$ in (2), we get

$$0 = C_1 e^{-p^2t} C_2 \Rightarrow C_2 = 0, \text{ since } C_1 \neq 0.$$

On putting the value of C_2 in (2), we get

$$V = C_1 e^{-p^2t} C_3 \sin px \qquad \text{... (3)}$$

Since C_3 cannot be zero

$$\sin pl = 0 = \sin n\pi \Rightarrow pl = n\pi \Rightarrow p = \frac{n\pi}{l},\ n \in \mathbf{Z}$$

On putting the value of p in (3), we get

$$V = C_1C_3 e^{-\frac{n^2\pi^2 t}{l^2}} \sin\frac{n\pi x}{l}$$

$$V = b_n e^{\frac{-n^2\pi^2 t}{l^2}} \sin\frac{n\pi x}{l}, \text{ where } b_n = C_1C_2.$$

The most general solution is

$$V = \sum_{m=1}^{\infty} b_n e^{\frac{-n^2\pi^2 t}{l^2}} \sin\frac{n\pi x}{l}.$$

ANSWERS

1. $v = b_n e^{-n^2\pi^2 t/l^2} \sin\frac{n\pi x}{l}$, where $b_n = C_1C_2$
2. $u = 6e^{-(3x+2t)}$
3. $u(x, y) = (C_1 \cos px + C_2 \sin px)\, C_3\, e^{-(b^2+2)y}$
4. $u(x, t) = 3e^{-x+t} - e^{-5x+2t}$
5. $u = 4e^{-x+3/2y}$
6. $Z = Ce^{4ax^3}.e^{-3ay^4}$
7. $u = Ce^{k(1/y-1/x)}$
8. $u = \frac{1}{\sqrt{2}}\sinh\sqrt{2}x + e^{-3y}\sin x$
9. $u = 4e^{x-y} - e^{2x-5y}$
10. $u = C_1C_2 e^{p^2\left(\frac{y^2}{4} - \frac{x^3}{3}\right)}$
11. $u(x, y) = C_1 e^{-p^2y}(C_2 \cos px + C_3 \sin px)$
12. $u(x, y) = C_1C_2 e^{k(x+y)}$
13. $z(x, y) = 4e^{ax} \sin 3y$
14. $u(x, t) = \sin x(1 - e^{-t})$
15. $u(x, t) = \frac{1}{\sqrt{2}}\sinh\sqrt{2}x + e^{-3y}\sin x$

42.4 HYPERBOLIC DIFFERENTIAL EQUATIONS

We know that the most important homogeneous hyperbolic differential equation is the wave equation, which can be written as follows :

$$\frac{\partial^2 u}{\partial t^2} = C^2\frac{\partial^2 u}{\partial x^2}, \text{ where } C \text{ is the wave speed.}$$

In this section, we shall discuss the derivation and solution of wave equation of one, two and three dimensional.

42.5 SOLUTION OF ONE DIMENSIONAL WAVE EQUATION

Consider one dimensional wave equation

$$\frac{\partial^2 u}{\partial x^2} = \frac{1}{C^2}\frac{\partial^2 u}{\partial t^2} \qquad \text{... (1)}$$

Suppose that solution of (1) is of the form

$$u(x, t) = X(x)\, T(t) \qquad \text{... (2)}$$

Putting the values from (2) in (1), we get

$$X''T = \frac{1}{C^2} XT''$$

$$\Rightarrow \qquad \frac{X''}{X} = \frac{T''}{C^2T} = \mu \ (say) \qquad \text{... (3)}$$

where μ is the separation constant.

From (3), we can deduce that $\quad X'' - \mu X = 0 \qquad$ ... (4)

and $\quad T'' - C^2\mu T = 0 \qquad$ ... (5)

Solving (4) and (5), we get

(i) When $\mu = 0$, $X = a_1x + a_2,\ T = a_3t + a_4$

(ii) When $\mu = -\lambda^2$, $(\lambda \neq 0)$ then

$$X = b_1e^{\lambda x} + b_2e^{-\lambda x} \text{ and } T = b_3e^{C\lambda t} + b_4e^{-C\lambda t}$$

(iii) When $\mu = \lambda^2$, $\lambda \neq 0$ then

$$X = C_1 \cos\lambda x + C_2 \sin\lambda x \text{ and } T = C_3 \cos Cpt + C_4 \sin Cpt$$

Therefore, the various possible solutions are given by

$$u(x, t) = (a_1 x + a_2)(a_3 t + a_4) \quad \text{... (6)}$$

$$u(x, t) = (b_1 e^{\lambda x} + b_2 e^{-\lambda x})(b_3 e^{C\lambda t} + b_4 e^{-C\lambda t}) \quad \text{... (7)}$$

and $$u(x, t) = (C_1 \cos\lambda x + C_2 \sin\lambda x)(C_3 \cos Cpt + C_4 \sin Cpt) \quad \text{... (8)}$$

Finally, since we are dealing with problems on vibration, $u(x, t)$, must be a periodic function of t. Thus, $u(x, t)$ must involve trigonometric terms. Hence, the solution given by (8) is the only suitable solution.

42.6 D'ALEMBERT'S SOLUTION OF WAVE EQUATION

Let $$\frac{\partial^2\phi}{\partial x^2} = \frac{1}{C^2}\frac{\partial^2\phi}{\partial t^2} \quad \text{... (1)}$$

be the given wave equation.

Define two new independent variables u and v such that

$$u = x + Ct \text{ and } v = x - Ct \quad \text{... (2)}$$

Now, $$\frac{\partial\phi}{\partial x} = \frac{\partial\phi}{\partial u}.\frac{\partial u}{\partial x} + \frac{\partial\phi}{\partial v}.\frac{\partial v}{\partial x} = \frac{\partial\phi}{\partial u} + \frac{\partial\phi}{\partial v}$$

$\Rightarrow$ $$\frac{\partial}{\partial x} \equiv \frac{\partial}{\partial u} + \frac{\partial}{\partial v} \quad \text{... (3)}$$

Then $$\frac{\partial^2\phi}{\partial x^2} = \frac{\partial}{\partial x}\left(\frac{\partial\phi}{\partial x}\right) = \left(\frac{\partial}{\partial u} + \frac{\partial}{\partial v}\right)\left(\frac{\partial\phi}{\partial u} + \frac{\partial\phi}{\partial v}\right) \quad \text{... (4)}$$

$\Rightarrow$ $$\frac{\partial^2\phi}{\partial x^2} = \frac{\partial^2\phi}{\partial u^2} + 2\frac{\partial^2\phi}{\partial u.\partial v} + \frac{\partial^2\phi}{\partial v^2}$$

Also, $$\frac{\partial\phi}{\partial t} = \frac{\partial\phi}{\partial u}.\frac{\partial u}{\partial t} + \frac{\partial\phi}{\partial v}.\frac{\partial v}{\partial t} = C\frac{\partial\phi}{\partial u} - C\frac{\partial\phi}{\partial v} = C\left(\frac{\partial\phi}{\partial u} - \frac{\partial\phi}{\partial v}\right)$$

$\Rightarrow$ $$\frac{\partial}{\partial t} \equiv C\left(\frac{\partial}{\partial u} - \frac{\partial}{\partial v}\right) \quad \text{... (5)}$$

Thus, $$\frac{\partial^2\phi}{\partial t^2} = \frac{\partial}{\partial t}\left(\frac{\partial\phi}{\partial t}\right) = C^2\left(\frac{\partial}{\partial u} - \frac{\partial}{\partial v}\right)\left(\frac{\partial\phi}{\partial u} - \frac{\partial\phi}{\partial v}\right)$$

$\Rightarrow$ $$\frac{1}{C^2}\frac{\partial^2\phi}{\partial t^2} = \frac{\partial^2\phi}{\partial u^2} - 2\frac{\partial^2\phi}{\partial u \partial v} + \frac{\partial^2\phi}{\partial v^2} \quad \text{... (6)}$$

Putting the values from (4) and (6) in (1), we get

$$\frac{\partial^2\phi}{\partial u^2} + 2\frac{\partial^2\phi}{\partial u\,\partial v} + \frac{\partial^2\phi}{\partial v^2} = \frac{\partial^2\phi}{\partial u^2} - 2\frac{\partial^2\phi}{\partial u \partial v} + \frac{\partial^2\phi}{\partial v^2}$$

$\Rightarrow$ $$\frac{\partial^2\phi}{\partial u.\partial v} = 0$$

On integrating, we get

$$\frac{\partial\phi}{\partial u} = F(u) \text{ , where } F(u) \text{ is any arbitrary function of } u.$$

Again integrating w.r.t. u, we get

$$\phi = \int F(u)\, du + g(v) = f(u) + g(v)$$

$\Rightarrow$ $$\phi = f(x + Ct) + g(x - Ct)$$

which is the D'Alembert's solution of wave equation.

42.7 SOLUTION OF TWO DIMENSIONAL WAVE EQUATION

To find the solution of $\dfrac{\partial^2 u}{\partial t^2} = C^2\left(\dfrac{\partial^2 u}{\partial x^2} + \dfrac{\partial^2 u}{\partial y^2}\right)$ *subject to the boundary conditions* $u(0, y, t) = u(a, y, t) = u(x, 0, t)$ $= u(x, b, t) = 0$ *and initial condition* $u(x, y, 0) = f(x, y)$ and $\left(\dfrac{\partial u}{\partial t}\right)_{t=0} = g(x, y)$.

The given equation is
$$\frac{\partial^2 u}{\partial x^2}+\frac{\partial^2 u}{\partial y^2}=\frac{1}{C^2}\frac{\partial^2 u}{\partial t^2} \quad \text{... (1)}$$
with boundary conditions

(a) $u(0, y, t) = 0$, (b) $u(a, y, t) = 0$, (c) $u(x, 0, t) = 0$, and (d) $u(x, b, t) = 0$.

Also, $u(x, y, 0) = f(x, y)$ ... (2)

and $$\left(\frac{\partial u}{\partial t}\right)_{t=0} = g(x, y) \quad \text{... (3)}$$

Now, suppose that solution of (1) is of the form
$$u(x, y, t) = X(x)\,Y(y)\,T(t) \quad \text{... (4)}$$
Putting the values from (4) in (1), we get
$$X''YT + XY''T = \frac{1}{C^2}XYT''$$
$$\Rightarrow \quad \frac{X''}{X}+\frac{Y''}{Y}=\frac{T''}{C^2T} \quad \text{... (5)}$$
Now, since x, y and t are independent variables, (5) can only be true if each term on each side is equal to a constant, say μ_1.

Let $$\frac{X''}{X}=\mu_1 \Rightarrow X''-\mu_1X=0 \quad \text{... (6)}$$
Now, using (a) and (b), equation (4) gives
$$X(0)\,Y(y)\,T(t)=0 \text{ and } X(a)\,Y(y)\,T(t)=0 \quad \text{... (7)}$$
By assuming $Y(y) \neq 0$, $T(t) \neq 0$, we get
$$X(0) = 0 \text{ and } X(a) = 0 \quad \text{... (8)}$$
Therefore, we want to solve equation (6) subject to the boundary conditions (8). Now, there are following cases :

Case (I). Let $\mu_1 = 0$. In this case, solution of (6) is given by
$$X(x) = Ax + B \quad \text{... (9)}$$
Using (8) in (9), we get $B = 0$ and $Aa + B = 0$

$\Rightarrow$ $A = B = 0$ $\Rightarrow$ $X(x) = 0$

$\Rightarrow$ $u = 0$

which does not satisfy (2). Thus, we reject this case.

Case (II). Let $\mu_1 = \lambda_1^2, \lambda_1 \neq 0$. In this case, solution of (6) is given by
$$X(x) = Ae^{x\lambda_1} + Be^{-x\lambda_1} \quad \text{... (10)}$$
Using (8) in (10), we get
$$0 = A + B \text{ and } 0 = Ae^{a\lambda_1} + Be^{-a\lambda_1} \quad \text{... (11)}$$
On solving, we get $A = B = 0 \Rightarrow u = 0$

$\Rightarrow$ This case is also rejected.

Case III. Let $\mu_1 = -\lambda_1^2, \lambda_1 \neq 0$. In this case, solution of (6) is given by
$$X(x) = A\cos\lambda_1 x + B\sin\lambda_1 x \quad \text{... (12)}$$
Using (8) in (12), we get $0 = A$ and $0 = A\cos\lambda_1 a + B\sin\lambda_1 a$

$\Rightarrow$ $A = 0$ and $\sin\lambda_1 a = 0$.

Now, $\sin\lambda_1 a = 0 \Rightarrow \lambda_1 a = m\pi$
$$\Rightarrow \quad \lambda_1 = \frac{m\pi}{a}, \quad m = 1, 2, \ldots \quad \text{... (13)}$$
Thus, the non-zero solution $X_m(x)$ are given by
$$X_m(x) = B_n \sin\frac{m\pi x}{a}, \quad m = 1, 2, 3, \ldots \quad \text{... (14)}$$
Further, let
$$\frac{Y''}{Y}=\mu_2 \Rightarrow Y''-\mu_2Y=0 \quad \text{... (15)}$$
Using (c) and (d), (4) gives
$$Y(0) = 0 \text{ and } Y(b) = 0 \quad \text{... (16)}$$

On solving (15) under boundary condition (16), we get

$$Y_n(y) = D_n \sin\left(\frac{n\pi y}{b}\right), \quad n = 1,2,3,... \qquad ...(17)$$

where, $\mu_2 = -\lambda_2^2$ *and* $\lambda_2 = \frac{n\pi}{b}$, $n = 1,2,3,...$ ... (18)

Therefore, (5) reduces to

$$\frac{T''}{C^2 T} = \mu_1 + \mu_2 = -\lambda_1^2 - \lambda_2^2 = -\pi^2\left(\frac{m^2}{a^2} + \frac{n^2}{b^2}\right)$$

$$\Rightarrow \quad T'' + \lambda_{mn}^2 T = 0 \qquad ...(19)$$

where, $\lambda_{mn}^2 = C^2\pi^2\left(\frac{m^2}{a^2} + \frac{n^2}{b^2}\right)$ ... (20)

The solution of (19) is given by

$$T_{mn}(t) = E_{mn}\cos\lambda_{mn}t + F_{mn}\sin\lambda_{mn}t \qquad ...(21)$$

Hence, $u_{mn}(x,y,t) = X_m(x)\,Y_n(y)\,T_{mn}(t)$

$$= (A_{mn}\cos\lambda_{mn}t + B_{mn}\sin\lambda_{mn}t)\left(\sin\frac{m\pi x}{a}\right)\sin\frac{n\pi y}{b} \qquad ...(22)$$

are solution of (1) satisfying (a) to (d).

Hence, the more general solution is given by

$$u(x, y, t) = \sum_{m=1}^{\infty}\sum_{n=1}^{\infty}(A_{mn}\cos\lambda_{mn}t + B_{mn}\sin\lambda_{mn}t)\left(\sin\frac{m\pi x}{a}\right)\sin\frac{n\pi y}{b}$$

where $A_{mn} = B_m D_n E_{mn}$ *and* $B_{mn} = B_m D_n F_{mn}$.

Solved Examples

Example 1. *Find the general solution of one dimensional wave equation* $\frac{\partial^2 u}{\partial x^2} = \frac{1}{C^2}\frac{\partial^2 u}{\partial t^2}$ *and find the particular solution for which* $u = f(x)$, $\frac{\partial u}{\partial t} = g(x)$ *at* $t = 0$.

(UPTU (CO)-2008)

Solution. The given equation is $\frac{\partial^2 u}{\partial x^2} = \frac{1}{C^2}\frac{\partial^2 u}{\partial t^2}$ ... (1)

Let string be stretched between fixed points (0, 0) and (a, 0). So we are to find $u(x, t)$ subject to the boundary condition

$$u(0,t) = 0,\ u(a,t) = 0;\ \forall t \qquad ...(2)$$

and initial conditions

$$\left.\begin{aligned} u(x,0) &= f(x) \\ \left(\frac{\partial u}{\partial t}\right)_{t=0} &= u_t(x,0) = g(x) \end{aligned}\right] \qquad ...(3)$$

Suppose that solution of (1) is of the form

$$u(x,t) = X(x)\,T(t) \qquad ...(4)$$

Putting the values from (4) in (3), we get

$$X''T = \frac{1}{C^2}XT''$$

$$\Rightarrow \quad \frac{X''}{X} = \frac{1}{C^2}\frac{T''}{T} = \mu \text{ (say)}, \qquad ...(5)$$

where μ is a separation constant.

Using (5), we can deduce that

$$X'' - \mu X = 0 \qquad ...(6)$$

and $T'' - \mu C^2 T = 0$ ... (7)

Using (2) and (4), we get

$$X(0)\,T(t) = 0 \text{ and } X(a)\,T(t) = 0 \qquad ...(8)$$

$$\Rightarrow X(0) = 0, X(a) = 0\ [T(t) \neq 0] \qquad ...(9)$$

Now, we have to solve (6) under (9). There are following cases :

Case (I). Let $\mu = 0$. In this case, solution of (6) is given by $X(x) = Ax + B$ ... (10)

Using (9) in (10), we get

$$B = 0 \text{ and } Aa + B = 0$$

$$\Rightarrow A = B = 0 \Rightarrow X(x) = 0$$

$$\Rightarrow u = 0$$

which does not satisfy (3). Thus, we reject this case.

Case (II). Let $\mu = \lambda^2, \lambda \neq 0$. In this case, solution of (6) is given by

$$X(x) = Ae^{\lambda x} + Be^{-\lambda x} \qquad ...(11)$$

Using (9) in (11), we get

$$0 = A + B \text{ and } 0 = Ae^{\lambda a} + Be^{-\lambda a} \qquad ...(12)$$

On solving, we get $A = B = 0$

$$\Rightarrow X(x) = 0 \quad \Rightarrow \quad u = 0$$

Hence, we reject this case also.

Case (III). Let $\mu = -\lambda^2, \lambda \neq 0$. In this case, solution of (6) is given by

$$X(x) = A\cos\lambda x + B\sin\lambda x \quad \text{... (13)}$$

Using (9) in (13), we get

$$0 = A \text{ and } 0 = A\cos\lambda a + B\sin\lambda a$$

On solving, we get $A = 0$ and $\sin\lambda a = 0$ $(B \neq 0)$

Now, $\sin\lambda a = 0 \Rightarrow \lambda a = n\pi$

$$\Rightarrow \lambda = \frac{n\pi}{a}, \quad n = 1, 2, \ldots \quad \text{... (14)}$$

Thus, the non-zero solution of (6) is given by

$$X_n(x) = B_n \sin\left(\frac{n\pi x}{a}\right) \quad \text{... (15)}$$

Now, using (14) in (7), we get

$$T'' + \left(\frac{n^2\pi^2C^2}{a^2}\right)T = 0$$

whose general solution is given by

$$T_n(t) = C_n \cos\frac{n\pi Ct}{a} + D_n \sin\frac{n\pi Ct}{a} \quad \text{... (16)}$$

So, $u_n(x, t) = X_n(x)\, T_n(x)$

$$= \left(E_n \cos\frac{n\pi Ct}{a} + F_n \sin\frac{n\pi Ct}{a}\right)\sin\frac{n\pi x}{a}$$

are solutions of (1) satisfying (2).

Thus, more general solution is given by

$$u(x, t) = \sum_{n=1}^{\infty} u_n(x, t)$$

$$= \sum_{n=1}^{\infty}\left(E_n \cos\frac{n\pi Ct}{a} + F_n \sin\frac{n\pi Ct}{a}\right)\sin\frac{n\pi x}{a} \quad \text{... (17)}$$

$$\Rightarrow \frac{\partial u}{\partial t} = \sum_{n=1}^{\infty}\left(-\frac{n\pi CE_n}{a}\sin\frac{n\pi Ct}{a} + \frac{n\pi C}{a}F_n\cos\frac{n\pi Ct}{a}\right)\sin\frac{n\pi x}{a} \quad \text{... (18)}$$

Putting $t = 0$ in (17) and (18) and using (3), we get

$$f(x) = \sum_{n=1}^{\infty} E_n \sin\frac{n\pi x}{a} \quad \text{... (19)}$$

$$\text{and } g(x) = \sum_{n=1}^{\infty}\frac{n\pi CF_n}{a}\sin\frac{n\pi x}{a} \quad \text{... (20)}$$

which are Fourier sine series expansions for $f(x)$ and $g(x)$ respectively. Thus, we get

$$E_n = \frac{2}{\pi}\int_0^a f(x)\sin\frac{n\pi x}{a}dx \quad \text{... (21)}$$

$$\text{and } \frac{n\pi CF_n}{a} = \frac{2}{a}\int_0^a g(x)\sin\frac{n\pi x}{a}dx$$

$$\Rightarrow F_n = \frac{2}{n\pi C}\int_0^a g(x)\sin\frac{n\pi x}{a}dx \quad \text{... (22)}$$

Example 2. *Solve the wave equation* $\frac{\partial^2 u}{\partial t^2} = C^2\frac{\partial^2 u}{\partial x^2}$, *where* $u = p_0\cos pt$ (p_0 *is a constant*) *when* $x = l$ *and* $u = 0$ *when* $x = 0$. [AMIE 1997]

Solution. We know that solution of the given equation is

$$u(x, t) = (C_1\cos nx + C_2\sin nx).(C_3\cos nCt + C_4\sin nCt) \quad \text{... (1)}$$

Putting $u = 0$ when $x = 0$, we get

$$C_1(C_3\cos nCt + C_4\sin nCt) = 0$$

$$\Rightarrow C_1 = 0$$

Therefore, we have

$$u(x, t) = C_2\sin nx\,(C_3\cos nCt + C_4\sin nCt)$$
$$= \sin nx\cos nCt.b_1 + \sin nx\sin nCt.b_2$$

where $b_1 = C_2C_3$ *and* $b_2 = C_2C_4$.

Now, using $u = p_0\cos pt$, when $x = l$, we get

$$p_0\cos pt = \sin nl\cos pt.b_1 + \sin nl\sin Ct.b_2$$

$\Rightarrow p_0\cos pt = \sin nl\cos nCt.b_1$ and

$$0 = \sin nl\sin nCt.b_2$$

$\Rightarrow p = nC$. Therefore,

$$p_0\cos pt = \sin nl\cos pt.b_1$$

$$\Rightarrow b_1 = \frac{p_0}{\sin nl} = \frac{p_0}{\sin(pl/C)}$$

$$\text{Hence, } u(x, t) = \sin nx\cos nCt.b_1 = \frac{p_0}{\sin(pl/C)}\cos pt\sin(px/C)$$

Example 3. *A string is stretched to two fixed points distance l apart. Motion is started by displacing the string in the form* $u = a\sin\left(\frac{\pi x}{l}\right)$ *from which it is released at time* $t = 0$. *Show that the displacement at any point at a distance x from one end at time t is given by*

$$u(x, t) = a\sin\left(\frac{\pi x}{l}\right)\cos\left(\frac{\pi Ct}{l}\right)$$

(AMIETE-2003, UPTU-2004, 09, UPTU (Sum)-2009, UKTU-2011, 12, VTU-2010, SVTU-2008, Kerala-2005)

Solution. It is known that the vibrations of the string are governed by the equation

$$\frac{\partial^2 u}{\partial x^2} = C^2\frac{\partial^2 u}{\partial x^2}$$

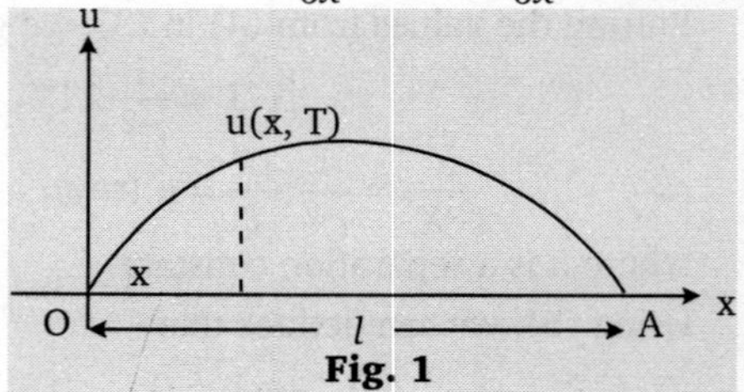

Fig. 1

As per given, the string is fastened at O and A respectively. Therefore,

$$u(0, t) = 0$$

and $u(l, t) = 0\ ;\ \forall t$ ... (2)

and $u(x, 0) = a \sin\dfrac{\pi x}{l}$ and $\left(\dfrac{\partial u}{\partial t}\right) = 0$ ... (3)

The solution of (1) is given by

$$u(x, t) = (C_1 \cos nx + C_2 \sin nx)(C_3 \cos Cnt + C_4 \sin Cnt)$$

Now, $u(0, t) = 0$ implies

$$0 = C_1(C_3 \cos Cnt + C_4 \sin Cnt)$$

$\Rightarrow$ $C_1 = 0$

$\Rightarrow$ $u(x, t) = \sin nx\,(b_1 \cos Cnt + b_2 \sin Cnt)$

Thus,

$$\frac{\partial u}{\partial t} = \sin nx\left[-Cn \sin Cnt\, b_1 + Cn.b_2 \cos Cnt\right]$$

$\Rightarrow$ $\left[\dfrac{\partial u}{\partial t}\right]_{t=0} = \sin nx\,[Cn.b_2] = 0\ ;\ \forall t$

$\Rightarrow$ $b_2 = 0$

Thus, (4) reduces to $u(x, t) = \sin x \cos Cnt\,.\,b_1$ ... (5)

Further, $u(l, t) = 0$ implies
$\sin nl \cos Cnt.b_1 = 0\ ;\ \forall t$

$\Rightarrow$ Either $\sin nl = 0$ or $b_1 = 0$.

Since $b_1 = 0$ makes the solution trivial, therefore, we take $\sin nl = 0$.

Now,

$$\sin nl = 0 \Rightarrow nl = m\pi \Rightarrow n = \frac{m\pi}{l}, m = 1, 2, ...$$

Then, from (5), we get

$$u(x, t) = \sin\frac{m\pi}{l}x \cos\left(\frac{Cm\pi}{l}.t\right).\, b_1$$

Since $u(x, 0) = a \sin\dfrac{\pi x}{l}$,

therefore $a \sin\dfrac{\pi x}{l} = \sin\dfrac{m\pi x}{l}.b_1$

which is true only when $m = 1$ and $b_1 = a$.

Hence, $u(x, t) = a \sin\left(\dfrac{\pi x}{l}\right)\cos\left(\dfrac{\pi Ct}{l}\right)$

Example 4. *A string is stretched and fastened to two points distance l apart. Motion is started displacing the string into the form $y = k(lx - x^2)$ from which it is released at time $t = 0$. Find the displacement of any point on the string at a distance of x from one end at time t.*

[UPTU Q. Bank 2002; IAS 1990,93,97]

Or

If the string is released from rest in the position $y = \dfrac{4t}{l^2}x(l - x)$, find the displacement y.

[AMIE-1991,97,2003]

Solution. It is known that the vibrations of the string are governed by the equation

$$\frac{\partial^2 y}{\partial t^2} = C^2\frac{\partial^2 y}{\partial x^2} \quad ....(1)$$

where $y(0, t) = 0$ and $y(l, t) = 0$ ...(2)

Also, the initial transverse velocity of any point of the string is zero. Thus, the initial condition is given by

$$\left.\begin{aligned}\left(\frac{\partial y}{\partial t}\right)_{t=0} &= 0\\ y(x, 0) &= k(lx - x^2)\end{aligned}\right] \quad ... (3)$$

We know that solution of (1) is given by

$$y(x, t) = (C_1 \cos px + C_2 \sin px)(C_3 \cos Cpt + C_4 \sin Cpt) \quad ... (4)$$

Now $y(0, t) = 0$

$\Rightarrow 0 = C_1(C_3 \cos Cpt + C_4 \sin Cpt) \Rightarrow C_1 = 0$

$\therefore$ $y(x, t) = \sin px\,(C_1' \cos Cpt + C_2' \sin Cpt)$ ... (5)

Therefore,

$$\left(\frac{\partial y}{\partial t}\right)_{t=0} = 0 \Rightarrow 0 = C_2' \sin px.(Cp) \quad [\because C_p \neq 0]$$

$$\Rightarrow C_2' = 0$$

Then, (5) gives $y(x, t) = C_1' \sin px\,(\cos Cpt)$

Now, $y(l, t) = 0 \Rightarrow C_1' \sin pl \cos Cpt = 0$

$\Rightarrow$ $C_1' = 0$ or $\sin pl = 0$

Since $C_1' = 0$ makes the solution trivial, thus, we have

$\sin pl = \sin n\pi,\ n = 1, 2, ... \Rightarrow p = \dfrac{n\pi}{l}$

$$\therefore \quad y_n(x, t) = C_1' \sin\frac{n\pi x}{l}\cos\frac{n\pi C}{l}.t = b_n \sin\frac{n\pi x}{l}\cos\frac{n\pi Ct}{l}$$

Therefore, the complete solution will be of the form

$$f(x, t) = \sum_{n=1}^{\infty} b_n \sin\frac{n\pi x}{l}\cos\frac{n\pi Ct}{l} \quad ... (6)$$

Further, $f(x, 0) = \sum\limits_{n=1}^{\infty} b_n \sin\dfrac{n\pi x}{l}$

$$\Rightarrow \quad lx - x^2 = \sum_{n=1}^{\infty} b_n \sin\frac{n\pi x}{l} \quad ... (7)$$

which is the expansion of $f(x)$ in the form of Fourier sine series. Thus we have

$$b_n = \frac{2}{l}\int_0^l f(x)\sin\frac{n\pi x}{l}dx = \frac{2}{l}\int_0^l (lx - x^2)\sin\frac{n\pi x}{l}dx$$

$$= \frac{2}{l}\left[(lx - x^2)\left(-\cos\frac{n\pi x}{l}\right)\frac{l}{n\pi} - (l - 2x)\right]$$

$$\left[\left(-\sin\frac{n\pi x}{l}\right)\frac{l^2}{n^2\pi^2}+(-2)\left(\cos\frac{n\pi x}{l}\right)\frac{l^3}{n^3\pi^3}\right]_0^l$$

$$=\frac{2}{l}\left[(-1)^{n+1}\frac{2l^3}{n^3\pi^3}+\frac{2l^3}{n^3\pi^3}\right]=\frac{8l^2}{n^3\pi^3},\ \text{if } n \text{ is odd}$$

$= 0$, if n is even.

Putting the value of b_n in (7), we get

$y(x, t) = \Sigma\frac{8l^2}{n^3\pi^3}\sin\frac{n\pi x}{l}\cos\frac{n\pi Ct}{l}$, when n is odd.

Example 5. *A tightly stretched string with fixed end points $x = 0$ and $x = 1$ is initially in a position given by $u = u_0\sin^3\left(\frac{\pi x}{l}\right)$. If it is released from rest from this position, find the displacement $u(x, t)$.*

[AMIE-1997, GBTU (CO)-2011, Rajasthan-2006, VTU-2003; JNTU-2002]

Solution. It is known that the equation of vibrating string is given by

$$\frac{\partial^2 u}{\partial t^2}=C^2\frac{\partial^2 u}{\partial x^2} \quad \text{... (1)}$$

with boundary conditions $u(0, t) = 0,\ u(l, t) = a$

$$\left(\frac{\partial u}{\partial t}\right)_{t=0} = 0;\quad u(x, 0) = u_0\sin^3\left(\frac{\pi x}{l}\right)$$

We know that solution of (1) is given by

$$u(x, t) = (C_1\cos nx + C_2\sin nx)(C_3\cos nCt + C_4\sin nCt) \quad \text{... (2)}$$

Now, $u(0, t) = 0$

$\Rightarrow\quad 0 = C_1(C_3\cos nCt + C_4\sin nCt)$

$\Rightarrow\quad C_1 = 0$

Then, (2) reduces to

$$u(x, t) = C_2\sin nx\,(C_3\cos nCt + C_4\sin nCt) \quad \text{... (3)}$$

$\Rightarrow\quad u(x, t) = \sin nx\,(b_1\cos nCt + b_2\sin nCt)$

where, $\ b_1 = C_2C_3,\ \ b_2 = C_2C_4$

Further,

$u(l, t) = 0 \Rightarrow \sin nl\,(b_1\cos nCt + b_2\sin nCt) = 0$

$\Rightarrow \sin nl = 0 \Rightarrow \sin nl = \sin n\pi \ \Rightarrow\ n = \frac{m\pi}{l}$

Therefore, $\ u(x,t) = \sin\frac{m\pi x}{l}$

$$\left(b_1\cos\frac{m\pi Ct}{l}+b_2\sin\frac{m\pi Ct}{l}\right) \quad \text{... (4)}$$

$$\Rightarrow\quad \frac{\partial u}{\partial t}=\sin\left(\frac{m\pi x}{l}\right)$$

$$\left(-\frac{m\pi C}{l}b_1\sin\frac{m\pi C}{l}.t+b_2\frac{mxC}{l}\cos\frac{m\pi C}{l}.t\right)$$

Further, $\frac{\partial u}{\partial t}=0$

$$\Rightarrow\quad 0 = \sin\frac{m\pi x}{l}.b_2\frac{m\pi C}{l} \Rightarrow b_2 = 0$$

Then, (4) gives $\ u(x, t) = \sin\frac{m\pi C}{l}\cos\frac{m\pi Ct}{l}.b_1$

$$\Rightarrow u_m(x, t) = b_m\sin\frac{m\pi x}{l}\cos\frac{m\pi Ct}{l}$$

Now, $\ u(x, t) = \sum_{m=1}^{\infty} u_m(x, t)$

$$= \sum_{m=1}^{\infty} b_m\sin\frac{m\pi x}{l}\cos\frac{m\pi Ct}{l} \quad \text{... (5)}$$

Also, $\ u(x,0) = u_0\sin^3\left(\frac{\pi x}{l}\right)$

$$=\frac{u_0}{4}\left(3\sin\frac{\pi x}{l}-\sin\frac{3\pi x}{l}\right)$$

Putting $t = 0$ in (5), we get

$$\sum_{m=1}^{\infty} b_m\sin\frac{m\pi x}{l}=\frac{u_0}{4}\left(3\sin\frac{\pi x}{l}-sin\frac{3\pi x}{l}\right)$$

where $\ b_1 = \frac{3u_0}{4}$ and $b_3 = \frac{-u_0}{4}$ remaining b_i's are all zero.

Hence, $\ u(x, t) = \frac{u_0}{4}$

$$\left(3\sin\frac{\pi x}{l}.\cos\frac{C\pi t}{l}-\sin\frac{3\pi x}{l}\cos\frac{3\pi Ct}{l}\right).$$

Example 6. *Find the solution of the wave equation $\frac{\partial^2 y}{\partial t^2}=C^2\frac{\partial^2 y}{\partial x^2}$ such that $y = p_0\cos pt$, when $x = l$ and $y = 0$ when $x = 0$ (b_0 is a constant).*

(IAS 1992)

Solution. The given equation is

$$\frac{\partial^2 y}{\partial t^2}=C^2\frac{\partial^2 y}{\partial x^2} \quad \text{... (1)}$$

The solution of (1) is given by

$$y(x, t) = (C_1\cos C\sqrt{k}\,t + C_2\sin C\sqrt{k}\,t)\text{–}(C_3\cos\sqrt{k}\,x + C_4\sin\sqrt{k}\,x) \quad \text{... (2)}$$

Now, $\ y(0, t) = 0$ implies

$0 = (C_1\cos C\sqrt{k}\,t\ +\ C_2\sin C\sqrt{k}\,t)\,C_3 \Rightarrow C_3 = 0$

Therefore, (2) becomes

$$y(x, t) = (C_1\cos C\sqrt{k}\,t + C_2\sin C\sqrt{k}\,t)(C_4\sin\sqrt{k}\,x)$$

$$= C_1'\cos C\sqrt{k}\,t\sin\sqrt{k}\,x + C_2'\sin C\sqrt{k}\,t\sin\sqrt{k}\,x \quad \text{... (3)}$$

where, $\quad C_1' = C_1C_4,\ C_2' = C_2C_4$

Further, $y(x, t) = p_0\cos pt$, when $x = l$.

Also, $\ p_0\cos pt = C_1'\cos C\sqrt{k}\,t\,\sin\sqrt{k}\,l + C_2'\sin C\sqrt{k}\,t\,\sin\sqrt{k}\,l$

Equating the coefficients of sin and cos on both the sides, we get

$$p_0 = C_1'\sin\sqrt{k}\,l \ \Rightarrow\ C_1' = \frac{p_0}{\sin\sqrt{k}\,l}$$

$$0 = C_2' \sin\sqrt{k}\, l \Rightarrow \quad C_2' = 0$$

Hence, $y(x,t) = \dfrac{p_0}{\sin\sqrt{k}\, l} \cos C\sqrt{k}\, t . \sin\sqrt{k}\, x$

$$= \frac{p_0}{\sin\sqrt{k}\, t} . \cos pt \sin\frac{p}{C} x$$

$$= \frac{p_0}{\sin(pl/C)} \cos pt \sin\left(\frac{p}{C} x\right)$$

Example 7. *A string is stretched and fastened to two points at a distance l apart. Motion is started by displacing the string in the form $u = a\sin\frac{\pi x}{l}$ from which it is released at a time t = 0. Show that the displacement of any point at a distance x from one end at time t is given by*

$$u(x,t) = a\sin\left(\frac{\pi x}{l}\right)\cos\left(\frac{\pi C t}{l}\right).$$

Solution. It is known that the vibration of the string are governed by the equation

$$\frac{\partial^2 u}{\partial t^2} = C^2 \frac{\partial^2 u}{\partial x^2} \quad \text{... (1)}$$

Since, end points of the string are fixed for all t, therefore, we have

$$u(0,t) = 0 \quad \text{and} \quad u(l,t) = 0$$

Also, the initial transverse velocity of any point of the string is zero, therefore

$$\left(\frac{\partial u}{\partial t}\right)_{t=0} = 0 \quad \text{... (2)}$$

$$u(x,0) = a\sin\frac{\pi x}{l}$$

Since the vibration of the string is periodic, therefore, solution of (1) is of the following periodic form

$$u(x,t) = (C_1 \cos px + C_2 \sin px)$$

$$(C_3 \cos Cpt + C_4 \sin Cpt) \quad \text{... (3)}$$

Therefore,

$$u(x,t) = C_2 \sin px\,(C_3 \cos Cpt + C_4 \sin Cpt)$$

$$= \sin px\,(C_3' \cos Cpt + C_4' \sin Cpt) \quad \text{... (4)}$$

$$\Rightarrow \frac{\partial u}{\partial t} = \sin px$$

$$\left[C_3'(-Cp\sin Cpt) + C_4'(Cp\cos Cpt)\right] \quad \text{... (5)}$$

$$\Rightarrow \left(\frac{\partial u}{\partial t}\right)_{t=0} = \sin px\,(C_4' C_p) = 0$$

$$\Rightarrow \quad C_4' = 0 \Rightarrow u(x,t) = \sin px\, C_3' \cos Cpt$$

Also, $u(l,t) = 0 \Rightarrow \sin pl = 0 \qquad (C_3' \neq 0)$

Further, $\sin pl = 0$

$$\Rightarrow \quad pl = n\pi \Rightarrow p = \frac{n\pi}{l}$$

$$\Rightarrow u(x,t) = \sin\left(\frac{n\pi x}{l}\right) C_3' \cos\left(\frac{Cn\pi t}{l}\right) \quad \text{... (6)}$$

Also, $u(x,0) = a\sin\dfrac{\pi x}{l}$

Thus, (6) gives $a\sin\dfrac{\pi x}{l} = \sin\dfrac{n\pi x}{l} . C_3'$

Let us take $C_3' = a$ and $n = 1$.

Hence, the required solution is given by

$$u(x,t) = a\sin\frac{\pi x}{l}\cos\left(\frac{C\pi t}{l}\right).$$

Example 8. *A string is stretched between the fixed points (0, 0) and (1, 0) and released at rest from the position $u = A\sin\pi x$. Obtain the formulae for its subsequent displacement u(x, t).*

Solution. It is known that the vibration of the string are governed by one-dimensional wave equation $\dfrac{\partial^2 u}{\partial t^2} = C^2 \dfrac{\partial^2 u}{\partial x^2}$ subject to the boundary conditions $u(0,t) = 0$, $u(1,t) = 0$ and initial conditions are

$$u(x,0) = A\sin\pi x, \left(\frac{\partial u}{\partial t}\right)_{t=0} = 0$$

Therefore, $u(x,t) = \sum_{n=1}^{\infty} C_n \cos nC\pi t \sin\pi x$

where, $\quad C_n = 2\int_0^1 A\sin\pi x \sin n\pi x\, dx$

$\Rightarrow \quad C_1 = 2\int_0^1 A\sin\pi x \sin\pi x\, dx$

$$= 2A\int_0^1 \sin^2\pi x\, dx = A\int_0^1 2\sin^2\pi x\, dx$$

$$= A\int_0^1 (1 - \cos 2\pi x)\, dx = A$$

All other coefficients $C_2 = C_3 = C_4 = ... = 0$

Hence,

$$u(x,t) = C_1 \cos C\pi t \sin\pi x = A\cos C\pi t \sin\pi x.$$

Example 9. *A string is stretched between the fixed points (0, 0) and (1, 0) and released at rest from the initial deflection given by*

$$f(x) = \frac{2l}{l} . x, \text{ when } 0 < x < l/2;$$

$$f(x) = \frac{2l}{l} . (l - x), \text{ when } l/2 < x < l$$

Find the deflection of the string at any time t.

[UPTU Q. Bank-2002]

Solution. We know that the deflection $u(x, t)$ of the string is given by

$$u(x,t) = \sum_{n=1}^{\infty} C_n \cos\frac{Cn\pi}{l} t \sin\frac{n\pi x}{l}$$

where, $\quad C_n = \dfrac{2}{l}\int_0^l f(x)\sin\dfrac{n\pi x}{l}\, dx$

$$= \frac{2}{l}\left[\int_0^{l/2} f(x)\sin\frac{n\pi x}{l}\, dx + \int_{l/2}^{l} f(x)\sin\frac{n\pi x}{l}\, dx\right]$$

$$= \frac{2}{l}\left[\int_0^{l/2} \frac{2k}{l} \sin\frac{n\pi x}{l} dx + \int_{l/2}^{l} \frac{2k}{l}(l-x)\sin\frac{n\pi x}{l} dx\right]$$

$$= \frac{4k}{l^2}\left[\int_0^{l/2} x\sin\frac{n\pi x}{l} dx\right] + \frac{4k}{l^2}\int_{l/2}^{l}(l-x)\sin\frac{n\pi x}{l} dx\right]$$

$$= \frac{4k}{l^2}\left[\left[-x\frac{l}{n\pi}\cos\frac{n\pi x}{l}\right]_0^{l/2} + \frac{l}{\pi x}\int_0^{l/2} 1.\cos\frac{n\pi x}{l} dx + \frac{4k}{l^2}\left[-(1-x)\frac{l}{n\pi}\cos\frac{n\pi x}{l}\right]_{l/2}^{l} - \frac{l}{n\pi}\int_{l/2}^{l}\cos\frac{n\pi x}{l} dx\right]$$

$$= \frac{4k}{l^2}\left[-\frac{l^2}{2n\pi}\cos\frac{n\pi}{2} + \frac{l}{n\pi}\left(\frac{l}{n\pi}\sin\frac{n\pi x}{l}\right)_0^{l}\right] + \frac{4k}{l}\left[-0 + \frac{l}{2}\frac{l}{n\pi}\cos\frac{n\pi}{2} - \frac{l}{n\pi}\left(\frac{l}{n\pi}\sin\frac{n\pi x}{l}\right)_{l/2}^{l}\right]$$

$$= \frac{4k}{l^2}\left[-\frac{l^2}{2n\pi}\cos\frac{n\pi}{2} + \frac{l^2}{n^2x^2}\sin\frac{n\pi}{2}\right] + \frac{4k}{l^2}\left[\frac{l^2}{2n\pi}\cos\frac{n\pi}{2} - \frac{l^2}{n^2\pi^2}\left(\sin n\pi - \sin\frac{n\pi}{2}\right)\right]$$

$$= \frac{4k}{l^2}\left[-\frac{l^2}{2n\pi}\cos\frac{n\pi}{2} + \frac{l^2}{n^2x^2}\sin\frac{n\pi}{2} + \frac{l^2}{2n\pi}\cos\frac{n\pi}{2} - \frac{l^2}{n^2\pi^2}\sin n\pi + \frac{l^2}{n^2\pi^2}\sin\frac{n\pi}{2}\right]$$

$$= \frac{8k}{n^2\pi^2}\sin\frac{n\pi}{2}$$

Thus,

$$u(x,t) = \sum_{n=1}^{\infty}\frac{8k}{n^2\pi^2}\sin\frac{n\pi}{2}\cos\frac{Cn\pi}{l}.t.\sin\frac{n\pi x}{l}$$

$$= \frac{8k}{\pi^2}\sin\frac{\pi}{2}\cos\frac{C\pi}{l}.t.\sin\frac{\pi x}{l} + \frac{8k}{2^2\pi^2}\sin\pi\cos\frac{2C\pi}{l}.t.\sin\frac{2\pi x}{l} + \frac{8k}{3^2\pi^2}\sin\frac{3\pi}{2}\cos\frac{3C\pi}{l}.t.\sin\frac{3\pi x}{l} + \ldots$$

$$= \frac{8k}{\pi^2}\left[\sin\frac{\pi x}{l}\cos\frac{\pi C}{l} - \frac{1}{3^2}\sin\frac{3\pi x}{l}\cos\frac{3\pi C}{l}.t + \ldots\ldots\right]$$

Example 10. *The point of trisection of a string are pulled aside through a distance h on opposite sides of the position of equilibrium, and the string is released from rest. Find an expression for the displacement of the string at any subsequent time and show that the mid point of the string always remains at rest.*

[Kerala-2005, UPTU Q. Bank-2002]

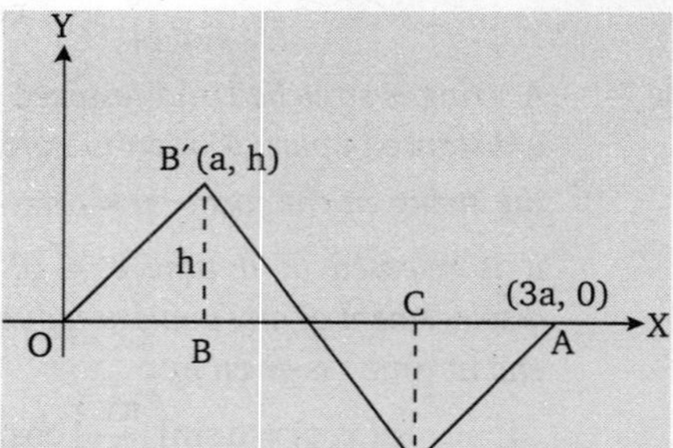

Fig. 2

Solution. Suppose $OBCA$ be the position of the equilibrium of the string of length $l = 3a$. Further, let B and C be the points of trisection of the string be pulled through a distance h on opposite sides and then released. It is known that this is the case in which vibrations are governed by one-dimensional wave equation

$$\frac{\partial^2 u}{\partial t^2} = C^3\frac{\partial^2 u}{\partial x^2} \qquad \ldots (1)$$

Now, equation of the line joining $O(0, 0)$ to $B'(a, h)$ is given by

$$y - 0 = \frac{h-0}{a-0} \Rightarrow y = \frac{h}{a}x$$

Also, the equation of joining $B'(a, h)$ to $C'(2a, -h)$ is given by

$$y - h = \frac{-h-h}{2a-a}(x-a) \Rightarrow y = \frac{h(3a-2x)}{a}$$

Further, equation of the line joining $C'(2a - h)$ to $A(3a- 0)$ is

$$y-(-h) = \frac{0-(-h)}{3a-2a}(x-2a)$$

$$\Rightarrow \quad y = \frac{h(x-3a)}{a}$$

Then, its initial deflection is given by

$$f(x) = \begin{cases} hx/a & ; \ 0 \le x \le a \\ \dfrac{h(3a-2x)}{a} & ; \ a \le x \le 2a \\ \dfrac{h(x-3a)}{a} & ; \ 2a \le x \le 3a \end{cases}$$

and initial velocity is $g(x) = 0$.

So, deflection $u(x, t)$ is given by

$$u(x,t) = \sum_{n=1}^{\infty} C_n \cos\frac{Cn\pi}{3a}.t\sin\frac{n\pi x}{3a}$$

where, $C_n = \frac{2}{3a}\int_0^{3a} f(x)\sin\frac{n\pi x}{3a}dx$

$$= \frac{2}{3a}\left[\int_0^a \frac{hx}{a}\sin\frac{n\pi x}{3a}dx + \int_a^{2a}\frac{h(3a-2x)}{a}\sin\frac{n\pi x}{3a}dx + \int_{2a}^{3a}\frac{h(x-3a)}{a}\sin\frac{n\pi x}{3a}.dx\right]$$

or $$C_n = \frac{2h}{3a^2}\left[\int_0^a x\sin\frac{n\pi x}{3a}dx + \int_a^{2a}(3a-2x)\sin\frac{n\pi x}{3a}dx + \int_{2a}^{3a}(x-3a)\sin\frac{n\pi x}{3a}dx\right]$$

$$= \frac{2h}{3a^2}\left[\left(-x.\frac{3a}{n\pi}\cos\frac{n\pi x}{3a}\right)_0^a + \frac{3a}{n\pi}\int_0^a\cos\frac{n\pi x}{3a}dx + \left[-(3a-2x)\frac{3a}{n\pi}\cos\frac{n\pi x}{3a}\right]_a^{2a} - \frac{6a}{n\pi}\int_a^{2a}\cos\frac{n\pi x}{3a}dx + \left[-(x-3a)\frac{3a}{n\pi}\cos\frac{n\pi x}{3a}\right]_{2a}^{3a} + \frac{3a}{n\pi}\int_a^{2a}\cos\frac{n\pi x}{3a}dx\right]$$

$$= \frac{2h}{3a^2}\left[\frac{27a^2}{n^2\pi^2}\sin\frac{n\pi}{3} - \frac{27a^2}{n^2\pi^2}\sin\frac{2n\pi}{3}\right]$$

[using $\sin n\pi = 0$]

$$= \frac{18h}{n^2\pi^2}\left[\sin\frac{n\pi}{3} - \sin\left(n\pi - \frac{n\pi}{3}\right)\right]$$

$$= \frac{18h}{n^2\pi^2}\left[1+(-1)^n\right]\sin\frac{n\pi}{3}$$

Thus, $$C_n \begin{cases} \frac{18h}{n^2\pi^2}.2\sin\frac{n\pi}{3} & ; \text{ if } n \text{ is even} \\ 0 & ; \text{ if } n \text{ is odd}\end{cases}$$

Further, let $n = 2m$. Then

$$u(x,t) = \frac{36h}{\pi^2}\sum_{m=1}^{\infty}\frac{1}{(2m)^2}\sin\frac{2m\pi}{3}\cos\frac{2m\pi Ct}{3}\sin\frac{2m\pi x}{3a}$$

$$= \frac{9h}{\pi^2}\sum_{m=1}^{\infty}\frac{1}{m^2}\sin\frac{2m\pi}{3}\sin\frac{2m\pi}{2a}x\cos\frac{2m\pi Ct}{3a}$$

Finally, to get the displacement of the mid point of the string, we put $x = \frac{3a}{2}$ in the above expression, then we get

$$u(x,t) = \frac{9h}{\pi^2}\sum_{m=1}^{\infty}\frac{1}{m^2}\sin\frac{2m\pi}{3}\sin m\pi\cos\frac{2m\pi Ct}{3} = 0$$

Hence, we conclude that the mid point of the string always remains at rest.

EXERCISE 42.2

1. Find the deflection $u(x,t)$ of the vibrating string whose length is π^2 and $C^2 = 1$ corresponding to zero initial velocity and initial deflection
$$f(x) = k(\sin x - \sin 2x).$$ (VTU-2011)

2. A string of length l has its ends $x = 0$ and $x = l$ fixed. It is released from rest in the position $y = \frac{4\lambda x\,[l-x]}{l^2}$. Find an expression for the displacement of the string at any subsequent time.

3. Solve completely the equation $\frac{\partial^2 y}{\partial t^2} = C^2\frac{\partial^2 y}{\partial x^2}$, representing the vibration of a string of length l, fixed at both ends, given that $y(0,t) = 0$, $y(l,t) = 0$; $y(x,0) = f(x)$ and $\frac{\partial}{\partial t}y(x,0) = 0$, $0 < x < l$. [UPTU-2005]

4. If a string of length l is initially at rest in equilibrium position and each of its points is given the velocity $\left(\frac{\partial y}{\partial t}\right)_{t=0} = b\sin^3\frac{\pi x}{l}$, find the displacement $y(x,t)$. [UPTU-2001,03,06]

5. The vibration of an elastic string is governed by the P.D.E. $\frac{\partial^2 u}{\partial t^2} = \frac{\partial^2 u}{\partial x^2}$. The length of the string is π and the ends are fixed. The initial velocity is zero and the initial deflection is $u(x,0) = 2(\sin x + \sin 3x)$. Find the deflection $u(x,t)$ of the vibrating string for $t > 0$.

6. Solve the wave equation
$$\frac{\partial^2 u}{\partial t^2} = a^2\frac{\partial^2 u}{\partial x^2}$$
under the conditions $u = 0$, when $x = 0$ and $x = \pi$
$\frac{\partial u}{\partial t} = 0$, where $t = 0$ and $u(x,0) = x$, $0 < x < \pi$.

7. A string is stretched and fastened to two points apart from a distance l. Motion is started by displacing the string into the form $y = k(lx - x^2)$ from which it is released at time $t = 0$. Find the displacement of any point on the string at a distance x from one end at time t. [UPTU-2002, IAS-1993]

8. Find the deflection $u(x,y,t)$ of the square membrane with $a = b = 1$ and $c = 1$, if the initial velocity is zero and the initial deflection is $f(x,y) = A\sin\pi x\sin 2\pi y$. [Kurukshetra -2004]

9. Solve the following boundary value problem
$$\frac{\partial^2 u}{\partial t^2} = C^2\frac{\partial^2 u}{\partial x^2},\quad 0 \le x \le 1,\ t \ge 0$$
subjected to the boundary conditions
$u(0,t) = 0$, $t > 0$, $u_x(1,t) = 0$, $t > 0$
and initial conditions
$$u(x,0) = \begin{cases} x, & 0 < x < 1/4 \\ \frac{1}{2} - x, & 1/4 < x < 1/2 \\ 0, & 1/2 < x < 1\end{cases}$$
and $u_t(x,0) = 0$, $0 < x < 1$.

10. Show how the wave equation $c^2 \frac{\partial^2 y}{\partial x^2} = \frac{\partial^2 y}{\partial t^2}$ can be solved by the method of separation of variables. If the initial displacement and velocity of a string stretched between $x = 0$ and $x = l$ are given by $y = f(x)$ and $\partial y / \partial t = g(x)$. Determine the constants in the series solution. [UPTU Q. Bank-2002]

11. A tightly stretched violin string of length l and fixed at both ends is plucked at $x = l/3$ and assumed initially the shape of a triangle of height a. Find the displacement y at any distance x and any time t after the string is released from rest. [UPTU Q. Bank-2002]

12. Transform the equation $\frac{\partial^2 y}{\partial t^2} = c^2 \frac{\partial^2 y}{\partial x^2}$ to its normal form using the transformation $u = x + Ct,\ v = x - Ct$ and hence solve it. Show that the solution may be put in the form

$$y = \frac{1}{2}[f(x+Ct) + f(x-Ct)]$$

Assume initial conditions $y = f(x)$ and $(\partial y / \partial t) = 0$ at $t = 0$. [UPTU-2003]

13. A tightly stretched string with fixed end points $x = 0$ and $x = l$ is initially at rest in its equilibrium position. If it is set vibrating by giving to each of its points an initial velocity $\lambda x(1 - x)$, find the displacement of the string at any distance x from one end at any time t. (MTU(SUM)-2011, (Bhopal-2008, Madras-2006, JNTU-2005, Anna-2009, UPTU-2002, PTU-2005)]

14. Find the deflection of the vibrating string which is fixed at the ends $x = 0$ and $x = 2$ and the motion is started by displacing the string into the form $\sin^3\left(\frac{\pi x}{2}\right)$ and releasing it with zero initial velocity at $t=0$. (MTU-2012)

15. Find the deflection of the vibrating string of unit length whose end points are fixed if the initial velocity is zero and the initial deflection is given by

$$u(x,0) = \begin{cases} 1 & 0 \le x \le 1/2 \\ -1 & \frac{1}{2} < x \le 1 \end{cases}$$ (GBTU-2012)

Hint to Selected Problems

1. By taking $a = \pi$ and $c^2 = 1$, the required deflection $u(x,t)$ is given by

$$u(x,t) = \sum_{n=1}^{\infty} \cos nt \sin nx \quad \text{... (1)}$$

where,

$$E_n = \frac{2}{\pi}\int_0^{\pi} f(x) \sin nx\, dx$$

$$= \frac{2k}{\pi}\int_0^{\pi}(\sin x - \sin 2x)\sin nx\, dx$$

$$= \frac{2k}{\pi}\left[\int_0^{\pi} \sin x \sin nx dx - \int_0^{\pi} (\sin x - \sin 2x)\sin nx\, dx\right] \quad \text{... (2)}$$

We know that

$$\int_0^{\pi} \sin px \sin qx\, dx = 0 \text{ if } p \ne q$$
$$= \pi/2 \text{ if } p = q.$$

Using (2) and (3), we get

$$E_1 = k,\ E_2 = -k, \ldots, E_n = 0 \quad \forall\, n \ge 3$$

Putting all these values in (1), we get

$$u(x,t) = k \cos t \sin x - \cos 2t \sin 2x$$

2. It is known that the displacement function $y(x, t)$ is the solution of wave equation given by

$$\frac{\partial^2 y}{\partial x^2} = \frac{1}{c^2}\frac{\partial^2 y}{\partial t^2} \quad \text{... (1)}$$

with boundary conditions

$$y(0,t) = y(l,t) = 0 \text{ for all } t \ge 0 \quad \text{... (2)}$$

and initial conditions $\left(\frac{\partial y}{\partial t}\right)_{t=0} = 0$ for $0 \le x \le l$

and $$y(x,0) = f(x) = \{4\lambda x(l-x)\}/l^2 \quad \text{... (3)}$$

Proceeding same as above, the solution of (1) satisfying the above boundary and initial conditions is

$$y(x,t) = \sum_{n=1}^{\infty} E_n \sin\frac{n\pi x}{l}\cos\frac{n\pi Ct}{l} \quad \text{... (4)}$$

where, $$E_n = \frac{2}{l}\int_0^l f(x)\sin\frac{n\pi x}{l}dx \quad \text{... (5)}$$

Putting the values of $f(x)$ given by (3) in (5), we get

$$E_n = \left(\frac{2}{l}\right)\times\left(\frac{4\lambda}{l^2}\right)\int_0^l (lx - x^2)\sin\frac{n\pi x}{l}dx$$

$$= \frac{8\lambda}{l^3}\left[(lx - x^2)\left(-\frac{l}{n\pi}\cos\frac{n\pi x}{l}\right) - (l - 2x)\left(-\frac{l^2}{n^2\pi^2}\sin\frac{n\pi x}{l}\right) + (-2)\left(\frac{l^3}{n^3\pi^3}\cos\frac{n\pi x}{l}\right)\right]_0^l$$

$$= \frac{8\lambda}{l^3}\left[-\frac{2l^3(-1)^n}{n^3\pi^3} + 2\frac{l^3}{n^3\pi^3}\right]$$

$$= \frac{16\lambda}{n^3\pi^3}\left[1 - (-1)^n\right] \qquad [\because \cos n\pi = (-1)^n]$$

$$= \begin{cases} 0 & ;\ \text{if } n = 2m,\ m = 1,2,3,\ldots \\ \frac{32\lambda}{(2m-1)^3\pi^3} & ;\ \text{if } n = 2m-1,\ m = 1,2,\ldots \end{cases}$$

Finally putting the value of E_n in (4), we get

$$y(x,t) = \frac{32\lambda}{\pi^3}\sum_{m=1}^{\infty}\frac{1}{(2m-1)^3}\sin\frac{(2m-1)\pi x}{l}\cos(2m-1)\frac{\pi Ct}{l}$$

3. Here, $$\frac{\partial^2 y}{\partial t^2} = c^2\frac{\partial^2 y}{\partial x^2} \quad \text{... (1)}$$

Let $$y = X(x)\,T(t) \quad \text{... (2)}$$

$$\frac{\partial^2 y}{\partial t^2} = X\frac{d^2 T}{dt^2}$$

$$\frac{\partial^2 y}{\partial x^2} = T\frac{d^2 X}{dx^2}$$

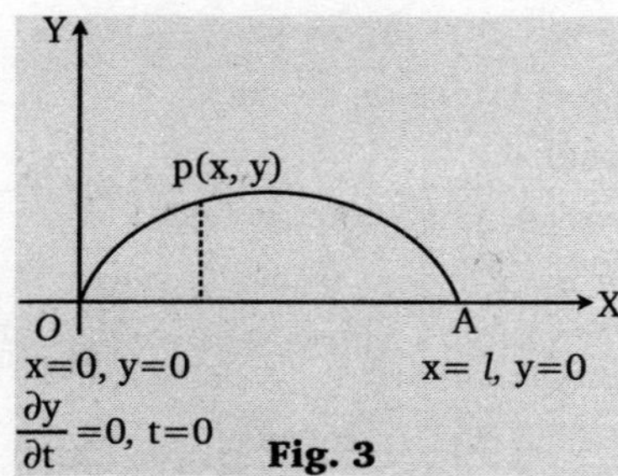

Fig. 3

Equation (1) becomes

$$X\frac{d^2T}{dt^2}=C^2T\frac{d^2X}{dx^2}$$

Separating the variables

$$\frac{1}{X}\frac{d^2X}{dx^2}=\frac{1}{C^2T}\frac{d^2T}{dt^2}=-p^2 \text{ (let)}$$

$$\frac{1}{X}\frac{d^2X}{dx^2}=-p^2 \qquad \text{If} \quad \frac{1}{C^2T}\frac{\partial^2T}{\partial t^2}=-p^2$$

$$\Rightarrow \quad \frac{d^2X}{dx^2}=-Xp^2 \qquad \Rightarrow (D^2+b^2C^2)T=0$$

$$\text{A.E. is } d^2+p^2C^2=0$$

$$\Rightarrow \quad \frac{d^2X}{dx^2}+Xp^2=0 \qquad \Rightarrow D^2=-p^2C^2 \Rightarrow D=\pm pC_i$$

$$\Rightarrow \quad (D^2+p^2)X=0 \qquad T=(C_2\cos pCt+C_4\sin pCt)$$

$$\Rightarrow \quad X=(C_1\cos px+C_2\sin px)$$

Putting the values of X and T in equation (2), we get

$$y=(C_1\cos px+C_2\sin px)(C_3\cos pCt+C_4\sin pCt) \quad \text{... (3)}$$

Now applying the boundary condition $x=0,\ y=0$

Putting these values in (3), we get

$$0=C_1(C_3\cos pCt+C_4\sin pCt) \Rightarrow C_1=0$$

Equation (3) becomes $y=C_2\sin px\,(C_3\cos pCt+C_4\sin pCt)$ ... (4)

Putting $x=l$ and $y=0$ in (4), we get

$$0=C_2\sin pl\,(C_3\cos pCt+C_4\sin pCt)$$

$$\Rightarrow \quad \sin pl=0=\sin n\pi$$

$$\Rightarrow \quad pl=n\pi$$

$$\Rightarrow \quad p=\frac{n\pi}{l}$$

On putting $p=\frac{n\pi}{l}$, (4) becomes

$$y=C_2\sin\frac{n\pi x}{l}\left(C_3\cos\frac{n\pi Ct}{l}+C_4\sin\frac{n\pi Ct}{l}\right) \quad \text{... (5)}$$

On differentiating (5) w.r.t. t, we get

$$\frac{\partial y}{\partial t}=C_2\sin\frac{n\pi x}{l}\left(-\frac{n\pi C}{l}C_3\sin\frac{n\pi Ct}{l}+\frac{n\pi C}{l}C_4\frac{n\pi C}{l}t\right) \quad \text{... (6)}$$

On putting $\frac{\partial y}{\partial t}=0$ and $t=0$ in (6), we get

$$0=C_2\sin\frac{n\pi x}{l}\cdot\frac{n\pi C}{l}C_4$$

$$\Rightarrow \quad C_4=0$$

On putting $C_4=0$, (5) becomes

$$y=C_2C_3\sin\frac{n\pi x}{l}\cos\frac{n\pi Ct}{l} \quad [\text{let } C_2C_3=b_n] \quad \text{... (7)}$$

Now applying $y=f(x)$ and $t=0$, (7) becomes

$$f(x)=b_n\sin\frac{n\pi x}{l}$$

C_2C_3 can be calculated using Fourier sine series as

$$b_n=\frac{2}{l}\int_0^1 f(x)\sin\frac{n\pi x}{l}\,dx$$

The required solution for the given equation is

$$y=b_n\sin\frac{n\pi x}{l}\cos\frac{n\pi Ct}{l}$$

4. The equation for the vibrations of the string is

$$\frac{\partial^2y}{\partial t^2}=C^2\frac{\partial^2y}{\partial x^2} \quad \text{... (1)}$$

The solution of equation (1) is

$$y(x,t)=(C_1\cos Cpt+C_2\sin Cpt)(C_3\cos px+C_4\sin px) \quad \text{... (2)}$$

Boundary conditions are

$$y(0,t)=0 \quad \text{... (3)}$$

$$y(l,t)=0 \quad \text{... (4)}$$

$$y(x,0)=0 \quad \text{... (5)}$$

$$\left(\frac{\partial y}{\partial t}\right)=b\sin^3\frac{\pi x}{l} \text{ at } t=0 \quad \text{... (6)}$$

Putting $x=0$ and $y=0$ in (2), we get

$$0=(C_1\cos Cpt+C_2\sin Cpt)C_3 \Rightarrow C_3=0$$

Putting the value of C_3 in (2), we get

$$y=(C_1\cos Cpt+C_2\sin Cpt)C_4\sin px \quad \text{... (7)}$$

Putting $x=l$ and $y=0$ in (7), we get

$$0=(C_1\cos Cpt+C_2\sin Cpt)\,C_4\sin pl$$

$$\Rightarrow \quad \sin pl=0=\sin n\pi\ (n\in I) \Rightarrow p=\frac{n\pi}{l}$$

Putting the value of p in (7), we get

$$y=\left(C_1\cos\frac{n\pi Ct}{l}+C_2\sin\frac{n\pi Ct}{l}\right)C_4\sin\frac{n\pi x}{l} \quad \text{... (8)}$$

Putting $t=0$ and $y=0$ in (8), we get

$$0=C_1C_4\sin\frac{n\pi x}{l} \Rightarrow C_1=0$$

Putting the value of C_1 in (8), we get

$$y=C_2C_4\sin\frac{n\pi Ct}{l}\sin\frac{n\pi x}{l}$$

$$y=b_n\sin\frac{n\pi Ct}{l}\sin\frac{n\pi x}{l} \quad (\text{where } C_2C_4=b_n)$$

The general solution is $y=\sum_1^\infty b_n\sin\frac{n\pi Ct}{l}\sin\frac{n\pi x}{l}$ ... (9)

On differentiating (9) w.r.t. x^t, we get

$$\frac{\partial y}{\partial t}=\sum_1^\infty b_n\frac{n\pi C}{l}\cos\frac{n\pi Ct}{l}\sin\frac{n\pi x}{l} \quad \text{... (10)}$$

On putting the values of $\frac{\partial y}{\partial t}=b\sin^3\frac{\pi x}{l}$ and $t=0$ in (10), we get

$$b\sin^3\frac{\pi x}{l}=\sum_1^\infty b_n\frac{n\pi C}{l}\sin\frac{n\pi x}{l}$$

$$\Rightarrow \quad \frac{b}{4}\left[3\sin\frac{\pi x}{l}\frac{3\pi x}{l}\right]=b_1\frac{\pi C}{l}\sin\frac{\pi x}{l}+\frac{2b_2\pi C}{l}\sin^2\frac{\pi x}{l}$$

$$+3b_3\frac{\pi C}{l}\sin\frac{3\pi C}{l}+\ldots$$

Equating the coefficients of $\sin\frac{\pi x}{l}$, $\sin\frac{2\pi x}{l}$, we get

$$\frac{3b}{4}=b_1\frac{\pi C}{l} \Rightarrow b_1=\frac{3bl}{4\pi C}$$

$b_2 = 0$ and $\frac{3b_3\pi C}{l} = \frac{b}{4} \Rightarrow b_3 = -\frac{bl}{12\pi C}$.

Also, $b_4 = 0 = b_5 = \ldots\ldots$ etc.

Putting the values of b's in equation (9), we get

$$y(x,t) = \frac{3bl}{4\pi C}\sin\frac{\pi Ct}{l}\sin\frac{\pi x}{l} - \frac{bl}{12\pi C}\sin\frac{3\pi Ct}{l}\sin\frac{3\pi x}{l}$$

$$= \frac{bl}{12\pi C}\left[9\sin\frac{\pi x}{l}\sin\frac{\pi Ct}{l} - \sin\frac{3\pi x}{l}\sin\frac{3\pi Ct}{l}\right].$$

5. The solution of given equation is

$$u(x,t) = (C_1\cos nx + C_2\sin nx)(C_3\cos nt + C_4\sin nt) \quad \ldots (1)$$

where, $u(0,t) = 0,\ u(\pi,t) = 0;\ \forall t$

$u(0,t) = 0 \Rightarrow (C_3\cos nt + C_4\sin nt)C_1 = 0$

$$\left(\frac{\partial u}{\partial t}\right)_{t=0} = 0;\quad u(x,0) = 2(\sin x + \sin 3x) \quad \ldots (2)$$

which gives $C_1 = 0$.

Thus $u(x,t) = (C_3\cos nt + C_4\sin nt)C_2\sin nx$
$= (b_1\cos nt + b_2\sin nt)\sin nx$

Again, $u(\pi,t) = 0 \Rightarrow (b_1\cos nt + b_2\sin nt)\sin n\pi = 0$

$\Rightarrow \sin n\pi = 0$

$\Rightarrow n$ must be an integer.

Thus, $u(x,t) = (b_1\cos nt + b_2\sin nt)\sin nx$

Then, $\left(\frac{\partial u}{\partial t}\right)_{t=0} = b_2 n\sin x \Rightarrow b_2 = 0$

$\Rightarrow u(x,t) = b_1\cos nt\sin nx$

Also, $u(x,0) = 2(\sin x + \sin 3x)$

$\Rightarrow 2(\sin x + \sin 3x) = \sin nx\, b_1$

$\Rightarrow 4\sin 2x\cos x = \sin x . b_1$

which is true when $b_1 = 4\cos x$ *and* $n = 2$.

Hence, we have $u(x,t) = 4\cos x\cos 2t\sin 2x$.

6. The solution is of the form

$$u(x,t) = (C_1\cos px + C_2\sin px)(C_3\cos apt + C_4\sin apt) \ldots (1)$$

On putting $u = 0$ and $x = 0$ in (1), we get

$0 = C_1(C_3\cos apt + C_4\sin apt) \Rightarrow C_1 = 0$.

On putting $C_1 = 0$ in (1), we get

$$u(x,t) = C_2\sin px(C_3\cos apt + C_4\sin apt) \quad \ldots (2)$$

On putting $x = \pi$ and $u = 0$ in (2), we get

$0 = C_2\sin p\pi(C_3\cos apt + C_4\sin apt)$

$\Rightarrow \sin px = 0 = \sin n\pi$

$\Rightarrow p = n$

On putting $p = n$ in equation (2), we get

$u(x,t) = C_2\sin nx(C_3\cos ant + C_4\sin ant)$

and $u(x,t) = \sin nx(b_1\cos ant + b_2\sin ant)$

$$[C_2C_3 = b_1 \text{ and } C_2C_4 = b_2] \quad \ldots (3)$$

On differentiating w.r.t. t, we get

$$\frac{\partial u}{\partial t} = \sin nx\left[-ab_1 n\sin ant + ab_2 n\cos ant\right] \quad \ldots (4)$$

On putting $\frac{\partial u}{\partial t} = 0$ and $t = 0$ in (4), we have

$0 = \sin n\pi(ab_2 n) \Rightarrow b_2 = 0$

On putting $b_2 = 0$ in (3), we get

$$u(x,t) = \sin nx\,(b_1\cos ant) \quad \ldots (5)$$

General solution is

$$u(x,t) = \sum_{n=1}^{\infty} b_n\sin nx\cos ant \quad \ldots (6)$$

On putting $u = x$ and $t = 0$ in (6), we get

$x = \sum_{n=1}^{\infty} b_n\sin nx$, where $b_n = \frac{2}{\pi}\int_0^{\pi} x\sin nx\,dx$.

$$= \frac{2}{\pi}\left[x\left(-\frac{\cos nx}{n}\right) - (-1)\left(-\frac{\sin nx}{n^2}\right)\right]_0^{\pi}$$

$$= \frac{2}{\pi}\left[-\frac{\pi}{n}\cos n\pi\right] = -\frac{2}{n}(-1)^n$$

On putting the value of b_n in (6), we get

$$u(x,t) = -2\sum_{n=1}^{\infty}\frac{(-1)^n}{n}\sin nx\cos nat$$

7. The vibrations of the string are governed by the equation

$$\frac{\partial^2 u}{\partial t^2} = C^2\frac{\partial^2 u}{\partial x^2} \quad \ldots (1)$$

The boundary conditions are

$$u(0,t) = 0 \text{ and } u(l,t) = 0 \quad \ldots (2)$$

Also, the initial transverse velocity of any point of the string is zero, therefore we get

$$\left(\frac{\partial u}{\partial t}\right)_{t=0} = 0\ ;\ \text{Also, } u(x,0) = k(lx - x^2) \quad \ldots (3)$$

Proceeding same as usual, the solution of (1) under (2) and (3) is given by

$u(x,t) = (C_1\cos px + C_2\sin px)(C_3\cos Cpt + C_4\sin Cpt)$

Now, $u(0,t) = 0 \Rightarrow (C_1\cos 0 + C_2\sin 0)(C_3\cos Cpt + C_4\sin Cpt) = 0$

$\Rightarrow C_1(C_3\cos Cpt + C_4\sin Cpt) = 0 \Rightarrow C_1 = 0$

Therefore, $u(x,t) = C_2\sin px(C_3\cos Cpt + C_4\sin Cpt)$
$= \sin px\,(C_3'\cos Cpt + C_4'\sin Cpt)$

$\Rightarrow \sin px = 0 = \sin n\pi$, n being integer.

$\Rightarrow p = \frac{n\pi}{l}$.

Therefore,

$$u(x,t) = \sin\frac{n\pi x}{l}\left[C_3'\cos\frac{Cn\pi}{l}.t + C_4'\sin\frac{Cn\pi}{l}.t\right]$$

$$\Rightarrow \frac{\partial u}{\partial t} = \sin\frac{n\pi x}{l}\left[C_3'\frac{Cn\pi}{l}\sin\frac{Cn\pi}{l}t + C_4'\frac{Cn\pi}{l}.\cos\frac{Cn\pi}{l}.t\right]$$

$$\Rightarrow \left(\frac{\partial u}{\partial t}\right)_{t=0} = C_4'\frac{Cn\pi}{l}\sin\frac{n\pi x}{l} = 0 \Rightarrow C_4' = 0$$

$$\Rightarrow u(x,t) = C_3'\sin\frac{n\pi x}{l}\cos\frac{Cn\pi t}{l}$$

Hence, the general solution of (1) is given by

$$u(x,t) = \sum_{n=1}^{\infty} u_n(x,t) = \sum_{n=1}^{\infty} b_n\sin\frac{n\pi x}{l}\cos\frac{Cn\pi t}{l}$$

Also, $u(x,0) = \sum_{n=1}^{\infty} b_n\sin\frac{n\pi x}{l} = k(lx - x^2)$, where $C_3' = b_n$

So, $b_n = \frac{2}{l}\int_0^l k(lx - x^2)\sin\frac{n\pi x}{l}dx$

8. The deflection of the square membrane is given by the two dimensional wave equation

$$\frac{\partial^2 u}{\partial t^2} = C^2\left(\frac{\partial^2 u}{\partial x^2} + \frac{\partial^2 u}{\partial y^2}\right)$$

The boundary conditions are
$u(x, 0, t) = 0 = u(x, 1, t)$;
$u(0, y, t) = 0 = u(1, y, t)$

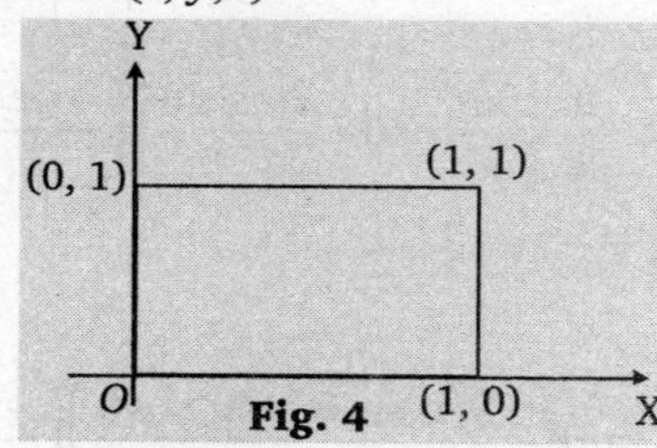

Fig. 4

Also, the initial conditions are

$$u(x, y, 0) = f(x, y) = A \sin \pi x \sin 2\pi y$$

$$\left(\frac{\partial u}{\partial t}\right)_{t=0} = 0$$

Thus, the deflection is given by

$$u(x, y, t) = \sum_{m=1}^{\infty} \sum_{n=1}^{\infty} A_{mn} \cos k_{mn} t \sin m\pi x \sin n\pi y$$

where, $A_{mn} = 4\int_0^1\int_0^1 f(x, y) \sin m\pi x \sin n\pi y \, dx\, dy$ ... (1)

$$= 4 \int_0^1 \int_0^1 \sin \pi x \sin 2\pi y \sin m\pi x \sin n\pi y \, dxdy$$

On integrating, we find that $A_{m1} = A_{m3} = A_{m4} = \ldots\ldots = 0$

But $A_{m2} = 4A \int_0^1\int_0^1 \sin \pi x \sin m\pi x \sin^2 2\pi y \, dx\, dy$

$$= 2A \int_0^1\int_0^1 \sin \pi x \sin m\pi x \,(1 - \cos 4\pi y)\, dx\, dy$$

$$= 2A \int_0^1 \sin \pi x \sin m\pi x \left(y - \frac{1}{4\pi} \sin 4\pi y \right)_0^1 dx$$

$$= 2A\int_0^1 \sin \pi x \sin m\pi x \, dx$$

Again, on integrating, we get $A_{22} = A_{32} = \ldots\ldots = 0$.

Further, we find that

$$A_{12} = 2A\int_0^1 \sin \pi x \sin \pi x \, dx = A \int_0^1 2\sin^2 \pi x \, dx$$

$$= A \int_0^1 (1 - \cos 2\pi x)\, dx = A \left[x - \frac{1}{2\pi} \sin 2\pi x \right]_0^1 = A$$

Hence, from (1) we get

$$u(x, y, t) = A_{12} \cos k_{12} t \sin \pi x \sin 2\pi y$$

$$= A \cos \sqrt{5}\, \pi t \sin \pi x \sin 2\pi y$$

$$\left[\because k_{12}^2 = \pi^2(m^2 + n^2) = \pi^2(1^2 + 2^2) = \sqrt{5}\, \pi\right]$$

ANSWERS

1. $u(x, t) = k \cos t \sin x - \cos 2t \sin 2x$

2. $y(x, t) = \frac{32\lambda}{\pi^3} \sum_{m=1}^{\infty} \frac{1}{(2m-1)^3} \sin \frac{(2m-1)\pi x}{l} \cos(2m-1)\frac{\pi C t}{l}$

3. $y = b_n \sin \frac{n\pi x}{l} \cos \frac{n\pi C t}{l}$, where $b_n = \frac{2}{l} \int_0^l f(x) \sin \frac{n\pi x}{l} dx$.

4. $y(x, t) = \frac{bl}{12\pi c}\left[9 \sin \frac{\pi x}{l} \sin \frac{\pi c t}{l} - \sin \frac{3\pi x}{l} \sin \frac{3\pi c t}{l} \right]$

5. $u(x, t) = 4 \cos x \cot 2t . \sin 2x$

6. $u(x, t) = -2 \sum_{n=1}^{\infty} \frac{(-1)^n}{n} \sin nx . \cos nat$

7. $u(x, t) = \sum_{n=1}^{\infty} u_n(x, t) = \sum_{n=1}^{\infty} b_n \frac{n\pi x}{l} \cos \frac{cn\pi t}{l}$; $u(x, 0) = \sum_{n=1}^{\infty} b_n \sin \frac{n\pi x}{l} = k(lx - x^2)$ where, $b_n = \frac{2}{l} \int_0^l k(lx - x^2) \sin \frac{n\pi x}{l} dx$.

8. $u(x, y, t) = A_{12} \cos k_{12} t \sin \pi x \sin 2\pi y = A \cos \sqrt{5}\, \pi t \sin \pi x \sin 2\pi y$

9. $u(x, t) = \frac{8}{\pi^2} \sum_{n=0}^{\infty} \frac{\sin \frac{1}{4}(2n-1)\pi}{(2n-1)^2} \cos \frac{(2n-1)\pi C t}{2} \sin \frac{(2n-1)\pi x}{2}$

10. $y(x, t) = \sum_{1}^{\infty} \left(a_n \cos \frac{n\pi C t}{l} + b_n \sin \frac{n\pi C t}{l} \right) \sin \frac{n\pi x}{l}$

$a_n = \frac{2}{l} \int_0^l f(x) \sin \frac{n\pi x}{l} dx$; $b_n = \frac{2}{n\pi C} \int_0^l g(x) \sin \frac{n\pi x}{l} dx$

11. $y(x, t) = \frac{9a}{\pi^2} \sum_{1}^{\infty} \frac{1}{n^2} \sin \frac{n\pi}{3} \cos \frac{n\pi C t}{l} \sin \frac{n\pi x}{l}$

13. $y(x, t) = \frac{8\lambda l^3}{c\pi^4} \sum_{m=1}^{\infty} \frac{1}{(2m-1)^4} \sin \frac{(2m-1)\pi c t}{l} \sin \frac{(2m-1)\pi x}{l}$

42.8 LAPLACE EQUATION

The heat equation is given by $\frac{\partial U}{\partial t} = c^2 \nabla^2 U$

If temperature are in steady state (*i.e.*, does not depend upon time *t*), then the heat equation reduces to

$$\nabla^2 u = 0$$

i.e., $\frac{\partial^2 U}{\partial x^2} + \frac{\partial^2 U}{\partial y^2} + \frac{\partial^2 U}{\partial z^2} = 0$. This is known as Laplace equation.

42.9 LAPLACE EQUATION IN TERMS OF POLAR COORDINATES

Let us suppose that the boundary of the region ∂R is a circle. Then, we use the polar coordinates as follows :

Let $x = r\cos\theta, \ y = r \sin \theta$

Then, $r^2 = x^2 + y^2$ and $\theta = \tan^{-1} y/x$

Therefore, we can find $\frac{\partial r}{\partial x} = \cos\theta, \ \frac{\partial r}{\partial y} = \sin\theta, \ \frac{\partial \theta}{\partial x} = \frac{-\sin\theta}{r}$ and $\frac{\partial \theta}{\partial y} = \frac{\cos\theta}{r}$ ... (1)

$u = u(r, \theta) \Rightarrow \quad \dfrac{\partial u}{\partial x} = \dfrac{\partial u}{\partial r}\dfrac{\partial r}{\partial x} + \dfrac{\partial u}{\partial \theta}\dfrac{\partial \theta}{\partial x} = \left(\dfrac{\partial u}{\partial r}\cos\theta - \dfrac{\partial u}{\partial \theta}\dfrac{\sin\theta}{r}\right)$

Also, $\dfrac{\partial u}{\partial y} = \dfrac{\partial u}{\partial r}\dfrac{\partial r}{\partial y} + \dfrac{\partial u}{\partial \theta}\dfrac{\partial \theta}{\partial y} = \left(\dfrac{\partial u}{\partial r}\sin\theta + \dfrac{\partial u}{\partial \theta}\dfrac{\cos\theta}{r}\right)$ ... (2)

and
$$\frac{\partial^2 u}{\partial x^2} = \frac{\partial}{\partial x}\left(\frac{\partial u}{\partial x}\right) = \frac{\partial}{\partial r}\left\{\frac{\partial u}{\partial r}\cos\theta - \frac{\partial u}{\partial \theta}\frac{\sin\theta}{r}\right\}\cos\theta + \frac{\partial}{\partial \theta}\left\{\frac{\partial u}{\partial r}\cos\theta - \frac{\partial u}{\partial \theta}\frac{\sin\theta}{r}\right\}\left(\frac{-\sin\theta}{r}\right)$$
$$= \left(\frac{\partial^2 u}{\partial r^2}\cos\theta - \frac{\partial^2 u}{\partial\theta\partial r}\frac{\sin\theta}{r} + \frac{\partial u}{\partial \theta}\frac{\sin\theta}{r^2}\right)\cos\theta$$
$$+\left(\frac{\partial^2 u}{\partial r\partial\theta}\cos\theta - \frac{\partial u}{\partial r}\sin\theta - \frac{\partial^2 u}{\partial\theta^2}\frac{\sin\theta}{r} - \frac{\partial u}{\partial\theta}\frac{\cos\theta}{r}\right)\left(\frac{-\sin\theta}{r}\right) \quad \text{... (3)}$$

Similarly,
$$\frac{\partial^2 u}{\partial y^2} = \left(\frac{\partial^2 u}{\partial r^2}\sin\theta + \frac{\partial^2 u}{\partial r\partial\theta}\frac{\cos\theta}{r} - \frac{\partial u}{\partial\theta}\frac{\cos\theta}{r^2}\right)\sin\theta$$
$$+\left(\frac{\partial^2 u}{\partial r\partial\theta}\sin\theta + \frac{\partial u}{\partial r}\cos\theta + \frac{\partial^2 u}{\partial\theta^2}\frac{\cos\theta}{r} - \frac{\partial u}{\partial\theta}\frac{\sin\theta}{r}\right)\left(\frac{\cos\theta}{r}\right) \quad \text{... (4)}$$

Adding (3) and (4), we get
$$\frac{\partial^2 u}{\partial x^2} + \frac{\partial^2 u}{\partial y^2} = \frac{\partial^2 u}{\partial r^2} + \frac{1}{r}\frac{\partial u}{\partial r} + \frac{1}{r^2}\frac{\partial^2 u}{\partial\theta^2} = 0$$
which is the required Laplace equation in polar coordinates.

REMARK

- The Laplace equation in Cartesian coordinates has constant coefficients only, whereas in polar coordinates, it has variable coefficients.

42.10 LAPLACE EQUATION IN CYLINDRICAL COORDINATES (r, θ, z)

Let (x, y, z) be the cartesian coordinates of the point P whose cylindrical coordinates are given by (r, θ, z) such that
$$x = r\cos\theta, \qquad y = r\sin\theta, \qquad z = z$$
which implies $r^2 = x^2 + y^2$ and $\theta = \tan^{-1} y/x$

Therefore,
$$\frac{\partial r}{\partial x} = \frac{x}{r} = \cos\theta, \quad \frac{\partial r}{\partial y} = \frac{y}{r} = \sin\theta, \quad \frac{\partial\theta}{\partial x} = \frac{1}{1 + y^2/x^2}\left(-\frac{y}{x^2}\right) = \frac{-\sin\theta}{r} \text{ and } \frac{\partial\theta}{\partial y} = \frac{\cos\theta}{r}$$

Now, we have
$$\frac{\partial u}{\partial x} = \frac{\partial u}{\partial r}\frac{\partial r}{\partial x} + \frac{\partial u}{\partial\theta}\cdot\frac{\partial\theta}{\partial x} + \frac{\partial u}{\partial z}\cdot\frac{\partial z}{\partial x} = \frac{\partial u}{\partial r}.\cos\theta + \frac{\partial u}{\partial\theta}.\left(-\frac{\sin\theta}{r}\right)$$
$$\Rightarrow \quad \frac{\partial}{\partial x} = \cos\theta\frac{\partial}{\partial r} - \frac{\sin\theta}{r}\frac{\partial}{\partial\theta}$$

Therefore,
$$\frac{\partial^2 u}{\partial x^2} = \frac{\partial}{\partial x}\left(\frac{\partial u}{\partial x}\right) = \left(\cos\theta\frac{\partial}{\partial r} - \frac{\sin\theta}{r}\frac{\partial}{\partial\theta}\right)\left(\cos\theta\frac{\partial u}{\partial r} - \sin\theta\frac{\partial u}{\partial\theta}\right)$$
$$= \cos^2\theta\frac{\partial^2 u}{\partial r^2} - 2\frac{\sin\theta\cos\theta}{r}\frac{\partial^2 u}{\partial r\,\partial\theta} + \frac{2\sin\theta\cos\theta}{r^2}\cdot\frac{\partial u}{\partial\theta} + \frac{\sin^2\theta}{r^2}\frac{\partial u}{\partial r} + \frac{\sin^2\theta}{r^2}\frac{\partial^2 u}{\partial\theta^2} \quad \text{... (1)}$$

In a similar manner, we can find
$$\frac{\partial^2 u}{\partial y^2} = \sin^2\theta\frac{\partial^2 u}{\partial r^2} + 2\frac{\sin\theta\cos\theta}{r}\frac{\partial^2 u}{\partial r\,\partial\theta} - \frac{2\sin\theta\cos\theta}{r^2}\cdot\frac{\partial u}{\partial\theta} + \frac{\cos^2\theta}{r}\frac{\partial u}{\partial r} + \frac{\cos^2\theta}{r^2}\frac{\partial^2 u}{\partial\theta^2} \quad \text{... (2)}$$

and
$$\frac{\partial^2 u}{\partial z^2} = \frac{\partial^2 u}{\partial z^2} \quad \text{... (3)}$$

On adding (1), (2) and (3), we get
$$\frac{\partial^2 u}{\partial x^2} + \frac{\partial^2 u}{\partial y^2} + \frac{\partial^2 u}{\partial z^2} = \frac{\partial^2 u}{\partial r^2} + \frac{1}{r^2}\frac{\partial^2 u}{\partial\theta^2} + \frac{1}{r}\frac{\partial u}{\partial r} + \frac{\partial^2 u}{\partial z^2} = 0$$
which is the required Laplace equation in cylindrical form.

REMARK

- The above equation can also be written as

$$\frac{1}{r}\frac{\partial}{\partial r}\left(r\frac{\partial u}{\partial r}\right)+\frac{1}{r^2}\frac{\partial^2 u}{\partial \theta^2}+\frac{\partial^2 u}{\partial z^2}=0$$

42.11 LAPLACE EQUATION IN SPHERICAL COORDINATES

Let (x, y, z) be the cartesian coordinates and (r, θ, ϕ) be the spherical coordinates at P such that

$$x = r\sin\theta\cos\phi, \qquad y = r\sin\theta\sin\phi, \qquad z = r\cos\theta$$

then, clearly we have

$$r^2 = x^2+y^2+z^2, \qquad \phi = \tan^{-1}(y/x) \text{ and } \theta = \tan^{-1}\left(\frac{\sqrt{x^2+y^2}}{z}\right)$$

Now,

$$\frac{\partial r}{\partial x}=\frac{x}{r}=\sin\theta\cos\phi,\ \frac{\partial r}{\partial y}=\frac{y}{r}=\sin\theta\sin\phi,\ \frac{\partial r}{\partial z}=\frac{z}{r}=\cos\theta,$$

$$\frac{\partial\theta}{\partial x}=\frac{\cos\theta\cos\phi}{r},\quad \frac{\partial\theta}{\partial y}=\frac{\cos\theta\sin\phi}{r},\quad \frac{\partial\theta}{\partial z}=\frac{-\sin\theta}{r},\ \frac{\partial\phi}{\partial x}=-\frac{\sin\phi}{r\sin\theta}$$

Therefore

$$\frac{\partial u}{\partial x}=\frac{\partial u}{\partial r}\frac{\partial r}{\partial x}+\frac{\partial u}{\partial \theta}.\frac{\partial \theta}{\partial x}+\frac{\partial u}{\partial \phi}.\frac{\partial \phi}{\partial x}$$

$$=\frac{\partial u}{\partial r}.\sin\theta\cos\phi+\frac{\partial u}{\partial\theta}.\frac{\cos\theta\cos\phi}{r}+\frac{\partial u}{\partial\phi}\left(-\frac{\sin\phi}{r\sin\theta}\right)$$

$$\Rightarrow \quad \frac{\partial}{\partial x}\equiv\sin\theta\sin\phi\frac{\partial}{\partial r}+\frac{\cos\theta\cos\phi}{r}\frac{\partial}{\partial\theta}-\frac{\sin\phi}{r\sin\theta}\frac{\partial}{\partial\phi}$$

Now,

$$\frac{\partial^2 u}{\partial x^2}=\frac{\partial}{\partial x}\left(\frac{\partial u}{\partial x}\right)=\left(\sin\theta\cos\phi\frac{\partial}{\partial r}+\frac{\cos\theta\cos\phi}{r}\frac{\partial}{\partial\theta}-\frac{\sin\phi}{r\sin\theta}\frac{\partial}{\partial\phi}\right)$$

$$\left(\sin\theta\cos\phi\frac{\partial u}{\partial r}+\frac{\cos\theta\cos\phi}{r}\frac{\partial u}{\partial\theta}-\frac{\sin\phi}{r\sin\theta}\frac{\partial u}{\partial\phi}\right)$$

$$=\sin^2\theta\cos^2\phi\frac{\partial^2 u}{\partial r^2}+2\frac{\sin\theta\cos\theta\cos^2\phi}{r}\frac{\partial^2 u}{\partial r\partial\theta}$$

$$-\frac{2\sin\theta\cos\theta\cos^2\phi}{r^2}.\frac{\partial u}{\partial\phi}-\frac{2\sin\phi\cos\phi}{r}.\frac{\partial^2 u}{\partial r\partial\phi}$$

$$+\frac{\sin\phi\cos\phi}{r^2}.\frac{\partial u}{\partial\phi}+\frac{\cos^2\theta\cos^2\phi}{r}.\frac{\partial u}{\partial r}+\frac{\cos^2\theta\cos^2\phi}{r^2}.\frac{\partial^2 u}{\partial\theta^2}$$

$$-2\frac{\cos\theta\sin\phi\cos\phi}{r^2\sin\theta}.\frac{\partial^2 u}{\partial\theta\partial\phi}+\frac{\cos^2\theta\sin\phi\cos\phi}{r^2\sin^2\theta}.\frac{\partial u}{\partial\phi}$$

$$+\frac{\sin^2\phi}{r}.\frac{\partial u}{\partial r}+\frac{\cos\theta\sin^2\phi}{r^2\sin\theta}.\frac{\partial u}{\partial\theta}+\frac{\sin^2\phi}{r^2\sin^2\theta}.\frac{\partial^2 u}{\partial\phi^2}+\frac{\sin\phi\cos\phi}{r^2\sin^2\theta}\frac{\partial u}{\partial\phi} \quad \ldots (1)$$

Also,

$$\frac{\partial u}{\partial y}=\frac{\partial u}{\partial r}\frac{\partial r}{\partial y}+\frac{\partial u}{\partial\theta}.\frac{\partial\theta}{\partial y}+\frac{\partial u}{\partial\phi}.\frac{\partial\phi}{\partial y}$$

$$=\frac{\partial u}{\partial r}.\sin\theta\sin\phi+\frac{\partial u}{\partial\theta}.\frac{\cos\theta\sin\phi}{r}+\frac{\partial u}{\partial\phi}\left(\frac{\cos\phi}{r\sin\theta}\right)$$

$$\Rightarrow \quad \frac{\partial}{\partial y}\equiv\sin\theta\sin\phi\frac{\partial}{\partial r}+\frac{\cos\theta\sin\phi}{r}\frac{\partial}{\partial\theta}+\frac{\cos\phi}{r\sin\theta}\frac{\partial}{\partial\phi}$$

which gives

$$\frac{\partial^2 u}{\partial y^2}=\frac{\partial}{\partial y}\left(\frac{\partial u}{\partial y}\right)=\left(\sin\theta\sin\phi\frac{\partial}{\partial r}+\frac{\cos\theta\sin\phi}{r}\frac{\partial}{\partial\theta}+\frac{\cos\phi}{r\sin\theta}\frac{\partial}{\partial\phi}\right)$$

$$\left(\sin\theta\sin\phi\frac{\partial u}{\partial r}+\frac{\cos\theta\sin\phi}{r}\frac{\partial u}{\partial\theta}+\frac{\cos\phi}{r\sin\theta}\frac{\partial u}{\partial\phi}\right)$$

$$= \sin^2\theta\sin^2\phi\frac{\partial^2 u}{\partial r^2} + 2\frac{\sin\theta\cos\theta\sin^2\phi}{r}\frac{\partial^2 u}{\partial r\,\partial\theta}$$

$$-\frac{2\sin\theta\cos\theta\sin^2\phi}{r^2}\cdot\frac{\partial u}{\partial\theta} + \frac{2\sin\phi\cos\phi}{r}\cdot\frac{\partial^2 u}{\partial r\partial\phi}$$

$$-\frac{\sin\phi\cos\phi}{r^2}\cdot\frac{\partial u}{\partial\phi} + \frac{\cos^2\theta\sin^2\phi}{r}\cdot\frac{\partial u}{\partial r} + \frac{\cos^2\theta\sin^2\phi}{r^2}\cdot\frac{\partial^2 u}{\partial\theta^2}$$

$$+2\frac{\cos\theta\sin\phi\cos\phi}{r^2\sin\theta}\cdot\frac{\partial^2 u}{\partial\theta\partial\phi} - \frac{\cos^2\theta\sin\phi\cos\phi}{r^2\sin^2\theta}\cdot\frac{\partial u}{\partial\phi}$$

$$+\frac{\cos^2\phi}{r}\cdot\frac{\partial u}{\partial r} + \frac{\cos\theta\cos^2\phi}{r^2\sin\theta}\cdot\frac{\partial u}{\partial\theta} + \frac{\cos^2\phi}{r^2\sin^2\theta}\cdot\frac{\partial^2 u}{\partial\phi^2} - \frac{\sin\phi\cos\phi}{r^2\sin^2\theta}\frac{\partial u}{\partial\phi} \quad \text{... (2)}$$

Now, $$\frac{\partial u}{\partial z} = \frac{\partial u}{\partial r}\frac{\partial r}{\partial z} + \frac{\partial u}{\partial\theta}\cdot\frac{\partial\theta}{\partial z} + \frac{\partial u}{\partial\phi}\cdot\frac{\partial\phi}{\partial z} = \frac{\partial u}{\partial r}.\cos\theta + \frac{\partial u}{\partial\theta}\left(-\frac{\sin\theta}{r}\right)$$

$\Rightarrow$ $$\frac{\partial}{\partial z} \equiv \cos\theta\frac{\partial}{\partial r} - \frac{\sin\theta}{r}\frac{\partial}{\partial\theta}$$

Therefore $$\frac{\partial^2 u}{\partial z^2} = \frac{\partial}{\partial z}\left(\frac{\partial u}{\partial z}\right) = \left(\cos\theta\frac{\partial}{\partial r} - \frac{\sin\theta}{r}\frac{\partial}{\partial\theta}\right)\left(\cos\theta\frac{\partial u}{\partial r} + \left(\frac{-\sin\theta}{r}\right)\frac{\partial u}{\partial\theta}\right)$$

$$= \cos^2\theta\frac{\partial^2 u}{\partial r^2} - 2\frac{\sin\theta\cos\theta}{r}\frac{\partial^2 u}{\partial r\,\partial\theta}$$

$$+\frac{2\sin\theta\cos\theta}{r^2}\cdot\frac{\partial u}{\partial\theta} + \frac{\sin^2\theta}{r^2}\cdot\frac{\partial u}{\partial r} + \frac{\sin^2\theta}{r^2}\cdot\frac{\partial^2 u}{\partial\theta^2} \quad \text{... (3)}$$

On adding (1), (2) and (3), we get

$$\frac{\partial^2 u}{\partial x^2} + \frac{\partial^2 u}{\partial y^2} + \frac{\partial^2 u}{\partial z^2} = \frac{\partial^2 u}{\partial r^2} + \frac{2}{r}\frac{\partial u}{\partial r} + \frac{1}{r^2}\frac{\partial^2 u}{\partial\theta^2} + \frac{\cot\theta}{r^2}\frac{\partial u}{\partial\theta} + \frac{1}{r^2\sin^2\theta}\frac{\partial^2 u}{\partial\phi^2} = 0$$

which is the required Laplace equation in spherical coordinates.

REMARK

- The above equation can also be written as

$$\frac{1}{r^2}\frac{\partial}{\partial r}\left(r^2\frac{\partial u}{\partial r}\right) + \frac{1}{r^2\sin\theta}\frac{\partial}{\partial\theta}\left(\sin\theta\frac{\partial u}{\partial\theta}\right) + \frac{1}{r^2\sin^2\theta}\frac{\partial^2 u}{\partial\phi^2} = 0$$

42.12 SOLUTION OF TWO DIMENSIONAL LAPLACE EQUATION : SEPARATION OF VARIABLES

Consider a two-dimensional Laplace equation in cartesian coordinates

$$\nabla^2 u = \frac{\partial^2 u}{\partial x^2} + \frac{\partial^2 u}{\partial y^2} = 0 \quad \text{... (1)}$$

Let $$u(x, y) = X(x)\ Y(y) \quad \text{... (2)}$$

be the solution of (1).

Using (2) in (1), we get

$$X''Y + Y''X = 0$$

$\Rightarrow$ $$\frac{X''}{X} = -\frac{Y''}{Y} = \lambda \text{ (say), where } \lambda \text{ is a separation constant.}$$

Now, we have the following cases :

Case (i) - Let $\lambda = \mu^2$, μ is real. Then $\frac{d^2X}{dx^2} - \mu^2 X = 0$ and $\frac{d^2Y}{dy^2} + \mu^2 Y = 0$

The solution of above equations are given by

$$X = C_1 e^{\mu x} + C_2 e^{-\mu x} \text{ and } Y = C_3\cos\mu y + C_4\sin\mu y$$

Thus, in this case, the required solution is given by

$$u(x, y) = X(x)\ Y(y) = (C_1 e^{\mu x} + C_2 e^{-\mu x})(C_3\cos\mu y + C_4\sin\mu y)$$

Case (ii) - If $\lambda = 0$

Then the equation reduces to $\dfrac{d^2X}{dx^2} = 0$ and $\dfrac{d^2Y}{dy^2} = 0$

Integrating twice, we get $X = d_1x + d_2$ and $Y = d_3y + d_4$

Thus, in this case, the required solution is given by

$$u(x, y) = (d_1x + d_2)(d_3y + d_4)$$

Case (iii) - Let $\lambda = -\mu^2$

Proceeding in the same way as in case (i), we get

$$X = e_1 \cos\mu x + e_2 \sin\mu x \text{ and } Y = e_3\, e^{\mu y} + e_4\, e^{-\mu y}$$

Hence, in this case, the required solution is given by

$$u(x, y) = (e_1 \cos\mu x + e_2 \sin\mu x)(e_3 e^{\mu y} + e_4\, e^{-\mu y}).$$

REMARK

- In all the above cases, C_i, d_i and e_i $(i = 1, \ldots, 3)$ are integration constants, which can be calculated by using the given boundary conditions.

42.12.1 SOME PARTICULAR PROBLEMS

(1) Interior Dirichlet's Problem for a Rectangle

The interior Dirichlet's problem for a rectangle is defined as follows :

PDE : $\nabla^2 u = 0, \quad 0 \le x \le a, \quad 0 \le y \le b$ **BCs :** $u(x, b) = u(a, y) = 0, u(0, y) = 0, u(x, 0) = f(x)$

(2) The Neumann Problem for a Rectangle

The Neumann problem for a rectangle is defined as follows :

PDE : $\nabla^2 u = 0, 0 \le x \le a, 0 \le y \le b$ **BCs :** $u_x(0, y) = u_x(a, y) = 0, u_y(x, 0) = 0, u_y(x, b) = f(x)$

(3) Interior Dirichlet's Problem for a Circle

The interior Dirichlet's problem for a circle is defined as follows :

PDE : $\nabla^2 u = 0, 0 \le r \le a, 0 \le \theta \le 2\pi$ **BC :** $u(a, \theta) = f(\theta), 0 \le \theta \le 2\pi$

(4) Exterior Dirichlet's Problem for a Circle

The exterior Dirichlet's problem for a circle is defined as follows :

PDE : $\nabla^2 u = 0$ **BC :** $u(a, \theta) = f(\theta)$

Here, u must be bounded as $r \to \infty$.

42.13 SOLUTION OF LAPLACE EQUATION OF THREE DIMENSIONAL

Consider the three dimensional Laplace equation

$$\frac{\partial^2 u}{\partial x^2} + \frac{\partial^2 u}{\partial y^2} + \frac{\partial^2 u}{\partial z^2} = 0 \quad \ldots (1)$$

Let

$$u(x, y, z) = X(x)\; Y(y)\; Z(z) \quad \ldots (2)$$

be the solution of (1), where X, Y and Z are functions of x, y and z, respectively.

Putting the value of u [From (2)] in (1), we get

$$\frac{X''}{X} + \frac{Y''}{Y} = -\frac{Z''}{Z} \quad \ldots (3)$$

Further, since x, y and z are independent variables, equation (3) can be true if each term on each side is equal to a constant.

Now, we have the following three cases :

Case (i) . If each term in (3) is zero.

In this case, we have

$$X'' = 0, Y'' = 0, Z'' = 0$$

On integrating each twice, we get

$$X = Ax + B, \quad Y = Cy + D, \quad Z = Ez + F$$

Hence, we get the solution of the form

$$u(x, y, z) = (Ax + B)\,(Cy + D)\,(Ez + F) \quad \ldots (4)$$

Case (ii). Suppose

$$\frac{X''}{X} = \lambda_1^2, \frac{Y''}{Y} = \lambda_2^2 \text{ such that } \lambda_1^2 + \lambda_2^2 = \lambda^2$$

Then, from equation (3), we can find

$$X'' - \lambda_1^2 X = 0,\ Y'' - \lambda_2^2 Y = 0,\ Z'' + \lambda^2 Z = 0$$

On solving, we get

$$X = Ae^{x\lambda_1} + Be^{-x\lambda_1}\ ;\ Y = Ce^{y\lambda_2} + De^{-y\lambda_2} \text{ and } Z = E\cos\lambda z + F\sin\lambda z$$

Hence, we get the solution of (1) is of the form

$$u(x, y, z) = (Ae^{x\lambda_1} + Be^{-x\lambda_1})(Ce^{y\lambda_2} + De^{-y\lambda_2})(E\cos\lambda z + F\sin\lambda z)$$

Case (iii). In this case, suppose that

$$\frac{X''}{X} = -\lambda_1^2,\ \frac{Y''}{Y} = -\lambda_2^2, \text{ and } -(\lambda_1^2 + \lambda_2^2) = \lambda^2$$

Then from (3), we can find

$$X'' + \lambda_1^2 X = 0,\ Y'' + \lambda_2^2 Y = 0,\ Z'' - \lambda^2 Z = 0$$

On solving, we get

$$X = A\cos\lambda_1 x + B\sin\lambda_1 x$$
$$Y = C\cos\lambda_2 y + D\sin\lambda_2 y$$

and

$$Z = Ee^{\lambda z} + Fe^{-\lambda z}$$

Hence, the general solution of (1) is given by

$$u(x, y, z) = \sum_{\lambda_1}\sum_{\lambda_2} (A\cos\lambda_1 x + B\sin\lambda_1 x)(C\cos\lambda_2 y + D\sin\lambda_2 y)(Ee^{\lambda z} + Fe^{-\lambda z})$$

Solved Examples

Example 1. *If u be a harmonic function in the interior of a rectangle $0 \le x \le a,\ 0 \le y \le b$ in the XY - plane satisfying Laplace equation*

$$\frac{\partial^2 u}{\partial x^2} + \frac{\partial^2 u}{\partial y^2} = 0 \quad \text{... (1)}$$

with boundary conditions

$$u(0, y) = 0,\ u(a, y) = 0 \quad \text{... (2)}$$
$$u(x, b) = 0,\ u(x, 0) = f(x) \quad \text{... (3)}$$

Obtain the solution of above problem.

[UPTU-2008]

Solution. By the method of separation of variables, we can find a function $u(x, y)$ such that

$$u(x, y) = X(x)\,Y(y) \quad \text{... (4)}$$

Putting this value of u in equation (1), we get

$$X''Y + XY'' = 0$$

$$\Rightarrow \quad \frac{X''}{X} = -\frac{Y''}{Y} \quad \text{... (5)}$$

For independent x, y, each side of equation (5) must be equal to the same constant say k.

Then (5) reduces to

$$X'' - kX = 0 \quad \text{... (6)}$$

and

$$Y'' + kY = 0 \quad \text{... (7)}$$

Now, using the given boundary conditions (2) in (4), we get

$$X(0)\,Y(y) = 0 \text{ and } X(a)\,Y(y) = 0$$

$$X(0) = 0 \text{ and } X(a) = 0 \quad \text{... (8)}$$

[$Y(y) \ne 0$ have been taken, because otherwise $u = 0$]

Now, we have the following cases :

Case (i). Let $k = 0$. Then solution of (6) is given by

$$X(x) = Ax + B$$

Using the boundary conditions given by (8), we get $B = 0$ and $Aa + B = 0$

i.e., $\Rightarrow A = B = 0 \Rightarrow X(x) = 0$

$\Rightarrow u = 0$

which does not satisfy the given boundary condition $u(x, 0) = f(x)$.

Hence, we reject this case (*i.e.*, $k = 0$).

Case (ii). Let $k = \lambda^2, \lambda \ne 0$. In this case, solution of equation (6) is given by

$$X(x) = Ae^{\lambda x} + Be^{-\lambda x}$$

Using the boundary conditions given by (6), we get

$$A + B = 0 \text{ and } Ae^{a\lambda} + Be^{-\lambda a} = 0$$

On solving, we get

$A = B = 0 \Rightarrow X(x) = 0 \Rightarrow u = 0$

Hence, again, we reject this case.

Case (iii). Let $k = -\lambda^2, \lambda \ne 0$. In this case, the solution of (6) is given by

$$X(x) = A\cos\lambda x + B\sin\lambda x$$

Using the boundary conditions given by (8), we get

$$A = 0 \text{ and } A\cos\lambda a + B\sin\lambda a = 0$$

On solving, we get $A = 0$ and $\sin\lambda a = 0$

Here, we have taken $B \ne 0$ because otherwise $X(x) = 0$.

Further, $\sin\lambda a = 0$

$\Rightarrow \lambda a = n\pi$, *i.e.*, $\lambda = \frac{n\pi}{a}$; $n = 1, 2, 3, ...$

Therefore, in this case, non-zero solution $X_n(x)$ of (6) is given by

$$X_n(x) = B_n \sin\left(\frac{n\pi x}{a}\right)$$

Further, using $\mu = -\lambda^2 = -\frac{n^2\pi^2}{a^2}$, equation (7) becomes

$$Y'' - \left(\frac{n^2\pi^2}{a^2}\right)Y = 0$$

whose solution is given by

$$Y_n(y) = C_n e^{n\pi y/a} + D_n e^{-n\pi y/a} \quad \text{... (9)}$$

Using the given boundary conditions, we get

$$X(x)\ Y(b) = 0 \Rightarrow Y(b) = 0$$

$$\Rightarrow Y_n(b) = 0 \quad \text{... (10)}$$

Putting $y = b$ in (9) and using (10), we get

$$0 = C_n e^{n\pi b/a} + D_n e^{-n\pi b/a}$$

$$C_n = -\left\{D_n e^{-n\pi b/a}\right\}\Big/ e^{n\pi b/a} \quad \text{... (11)}$$

Putting this value in (9), we get

$$Y_n(y) = \frac{D_n\left(e^{-n\pi y/a}\, e^{n\pi b/a} - e^{n\pi y/a}\, e^{-n\pi b/a}\right)}{e^{n\pi b/a}}$$

$$= \frac{D_n\left(e^{n\pi(b-y)/a} - e^{-n\pi(b-y)/a}\right)}{e^{n\pi b/a}}$$

which can also be written as

$$Y_n(y) = 2D_n e^{-n\pi b/a} \sinh\{n\pi(b-y)/a\}$$

$$\left[\because \sinh\theta = \frac{e^{\theta} - e^{-\theta}}{2}\right]$$

Therefore, $U_n(x, y) = X_n(x)\, Y_n(y)$

$$= F_n \sin(n\pi x/a) \sinh\{n\pi(b-y)/a\}$$

Also, the more general solution is given by

$$U(x, y) = \sum_{n=1}^{\infty} F_n \sin\left(\frac{n\pi x}{a}\right) \sinh\left[\frac{n\pi(b-y)}{a}\right]$$

Putting $y = 0$ and using given boundary conditions $u(x, 0) = f(x)$, we get

$$f(x) = \sum_{n=1}^{\infty} \left\{F_n \sin\left(\frac{n\pi b}{a}\right)\right\} \sin\left[\frac{n\pi x}{a}\right]$$

which is the half range Fourier sine series of $f(x)$ in $(0, a)$. Hence, we get

$$\Rightarrow F_n \sin\frac{n\pi b}{a} = \frac{2}{a}\int_0^a f(x)\frac{n\pi x}{a}dx$$

$$\Rightarrow F_n = \frac{2}{a \sinh\left(\frac{n\pi b}{a}\right)}\int_0^a f(x) \sin\frac{n\pi x}{a}\, dx$$

Example 2. *Find the steady temperature distribution in a thin plate bounded by the lines $x = 0$, $x = a$, $y = 0$, $y = \infty$. Assuming that heat can not escape from either surface; the sides $x = 0$, $x = a$ being kept at temperature zero. The lower edge $y = 0$ is kept at $f(x)$ and the edge $y = \infty$ at temperature zero.*

Solution. For steady state, we know that $\frac{\partial u}{\partial t} = 0$

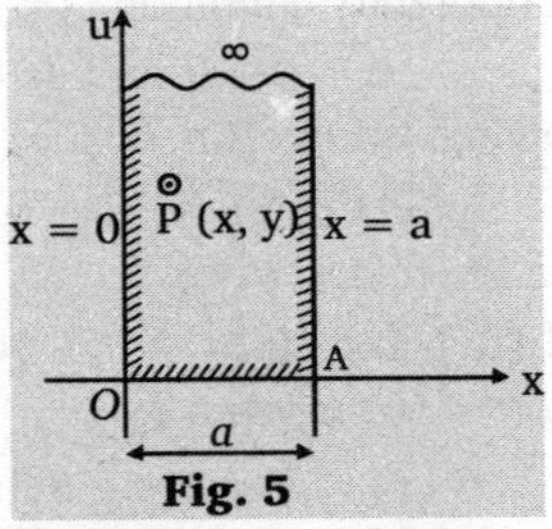

Fig. 5

Therefore the heat equation

$$\frac{\partial^2 u}{\partial x^2} + \frac{\partial^2 u}{\partial y^2} = \frac{1}{k}\frac{\partial u}{\partial t}$$

reduces to

$$\frac{\partial^2 u}{\partial x^2} + \frac{\partial^2 u}{\partial y^2} = 0 \quad \text{... (1)}$$

The given boundary conditions can be written as

$$u(0, y) = 0, \quad u(a, y) = 0 \quad \text{... (2)}$$

$u(x, y) \to 0$ as $y \to \infty$ and $u(x, 0) = f(x)$ ... (3)

Using the method of separation of variables, let us suppose solution of (1) is of the form

$$u(x, y) = X(x)\ Y(y) \quad \text{... (4)}$$

Putting this value in (1), we get

$$X''Y + XY'' = 0$$

$$\Rightarrow \frac{X''}{X} = -\frac{Y''}{Y} \quad \text{... (5)}$$

Now, since x and t are independent, therefore each side of (5) must be equal to the same constant say k. Then (5) reduces to

$$X'' - kX = 0 \quad \text{... (6)}$$

and $$Y'' + kY = 0 \quad \text{... (7)}$$

Using (2) in (4), we get

$$X(0)\, Y(y) = 0 \text{ and } X(a)\ Y(y) = 0$$

$$\Rightarrow X(0) = 0 \text{ and } X(a) = 0$$

(By taking $Y(y) \neq 0$) ... (8)

Now, we have to solve equation (6) using the boundary conditions (8) under the following three cases :

Case (i). Let $k = 0$. In this case, solution of (6) is given by

$$u = Ax + B \quad \text{... (9)}$$

Using boundary conditions given by (8) in (9), we get

$$B = 0 \text{ and } Aa + B = 0$$

On solving, we get

$$A = B = 0 \Rightarrow X(x) = 0 \quad u = 0$$

So, we reject this case.

Case (ii). Let $k = -\lambda^2, \lambda \neq 0$. In this case, solution of (6) is given by

$$X(x) = Ae^{x\lambda} + Be^{-x\lambda}$$

Using boundary conditions (8), we get

$$0 = A + B \text{ and } 0 = Ae^{a\lambda} + Be^{-a\lambda}$$

On solving, we get $A = B = 0$

$\Rightarrow \quad X(x) \equiv 0 \Rightarrow u = 0$

Hence, we reject $k = \lambda^2$.

Case (iii). Let $\mathbf{k = -\lambda^2, \lambda \neq 0}$. In this case, solution of (6) is given by

$$X(x) = A\cos\lambda x + B\sin\lambda x$$

Using boundary conditions (8), we get

$0 = A$ and $0 = A\cos\lambda a + B\sin\lambda a$

On solving, we get $A = 0$ and $\sin\lambda a = 0$

(By taking $B \neq 0$)

So, $\quad \sin\lambda a = 0$

$\Rightarrow \quad \lambda a = n\pi$

$\Rightarrow \quad \lambda = n\pi / a; \quad n = 1, 2, 3, ...$

Therefore, non-zero solutions $X_n(x)$ of (6) are given by

$$X_n(x) = B_n \sin\left(\frac{n\pi x}{a}\right) \quad ... (10)$$

Using $k = -\lambda^2 = -\dfrac{n^2\pi^2}{a^2}$, equation (7) becomes

$$Y'' - \left(\frac{n^2\pi^2}{a^2}\right) Y = 0 \quad ... (11)$$

On solving (11), we get

$$Y_n(y) = C_n e^{n\pi y/a} + D_n e^{-n\pi y/a} \quad ... (12)$$

with $Y_n(y) \to 0$ as $y \to \infty$ [Using (3)]

Therefore, we must take $C_n = 0$. Then, from (12), we have

$$Y_n(y) = D_n e^{-n\pi y/a}$$

$$\Rightarrow u_n(x, y) = X_n Y_n = E_n \sin\frac{n\pi x}{a} e^{-n\pi y/a}$$

$(E_n = B_n D_n)$

are solutions of (1) satisfying (2) and (3).

The more general solution is given by

$$u(x, y) = \sum_{n=1}^{\infty} E_n \sin\frac{n\pi x}{a} e^{-n\pi y/a}$$

Now, putting $y = 0$ and using $u(x, 0) = f(x)$, we get

$$f(x) = \sum_{n=1}^{\infty} E_n \sin\frac{n\pi x}{a}$$

which is a Fourier sine series and therefore

$$E_n = \frac{2}{a}\int_0^a f(x)\sin\frac{n\pi x}{a} dx.$$

Example 3. *A square plate is bounded by the lines $x = 0$, $y = 0$, $x = 10$ and $y = 10$. Its faces are insulated. The temperature along the upper horizontal edge is given by $u(x, 10) = x(10 - x)$ while the other three faces are kept at 0°C. Find the steady state temperature in the plate.*

Solution. The steady state temperature $u(x, y)$ is governed by the equation

$$\frac{\partial^2 u}{\partial x^2} + \frac{\partial^2 u}{\partial y^2} = 0 \quad ... (1)$$

subject to the boundary conditions

$$u(0, y) = u(10, y) = 0, \; 0 \le y \le a \quad ... (2)$$

$$u(x, 0) = 0, \quad 0 \le x \le a$$

$$u(x, 10) = 10x - x^2, \quad 0 \le x \le a \quad ... (3)$$

Now proceed same as example (2), and using $u(x, 10) = 10x - x^2$ in place of $u(x, b) = 100$ and $a = b = 10$,

we get $u(x,y) = \displaystyle\sum_{n=1}^{\infty} E_n \sin\frac{n\pi x}{10}\sinh\frac{n\pi y}{10}$ ... (4)

Putting $y = 10$ in (4) and using (3), we get

$$u(x, 10) = 10x - x^2 = \sum_{n=1}^{\infty} (E_n \sinh n\pi)\sin\frac{n\pi x}{10}$$

which is a half range Fourier series of $(10x - x^2)$ in (0, 10)

Therefore, we have

$$E_n \sinh n\pi = \frac{2}{10}\int_0^{10} (10x - x^2)\sin\frac{n\pi x}{10} dx$$

$$= \frac{1}{5}\left[(10x - x^2)\left(-\frac{10}{n\pi}\right)\cos\frac{n\pi x}{10} - (10 - 2x)\left(-\frac{100}{n^2\pi^2}\right)\sin\frac{n\pi x}{10} + (-2)\left(\frac{1000}{n^3\pi^3}\right)\cos\frac{n\pi x}{10}\right]_0^{10}$$

$$= \frac{1}{5}\left[-\frac{2000(-1)^n}{n^3\pi^3} + \frac{2000}{n^3\pi^3}\right]$$

$$= \frac{400}{n^3\pi^3}[1 - (-1)^n]$$

Thus, $\quad E_n = \dfrac{400\,\text{cosech}\, n\pi}{n^3\pi^3}[1 - (-1)^n]$

$$E_n = \begin{cases} 0, & \text{if } n = 2m \text{ and } m = 1, 2, 3, ... \\ \dfrac{800\,\text{cosech}\,(2m-1)\pi}{(2m-1)^3\pi^3}, & \text{if } n = 2m - 1, \; m = 1, 2, 3, ... \end{cases}$$

Hence, the required temperature is given by

$$u(x, y) = \frac{800}{\pi^3}\sum \frac{1}{(2m-1)^3}\sin\frac{(2m-1)\pi x}{10} \sinh\frac{(2m-1)\pi y}{10}\,\text{cosec}(2m-1)\pi.$$

Example 4. *Find the steady state temperature distribution in a rectangular plate of sides a and b insulated at the lateral surface and satisfying the boundary conditions*

$$u(0, y) = u(a, y) = 0, \quad 0 \le y \le b$$

and $u(x, 0) = 0$ *and* $u(x, b) = f(x), 0 \le y \le a$

Solution. We know that the heat flow in a body for a two dimensional case is governed by the equation

$$\frac{\partial^2 u}{\partial x^2} + \frac{\partial^2 u}{\partial y^2} = \frac{1}{C^2}\frac{\partial u}{\partial t}$$

For steady flow, we have $\frac{\partial u}{\partial t} = 0$.

Therefore, we have the following Laplace's equation

$$\frac{\partial^2 u}{\partial x^2} + \frac{\partial^2 u}{\partial y^2} = 0 \quad \text{... (1)}$$

subject to the boundary conditions

$$u(0, y) = u(a, y) = 0, \text{ for } 0 \le y \le b \quad \text{... (2)}$$

$$\left.\begin{array}{l} u(x, 0) = 0, \quad \text{for } 0 \le x \le a \\ \text{and} \quad u(x, b) = f(x), \text{ for } 0 \le x \le a \end{array}\right\} \quad \text{... (3)}$$

Now proceeding same as in example (2), and taking $u(x, b) = f(x)$ for $u(x, b) = 100$, we may get

$$u(x, y) = \sum_{n=1}^{\infty} E_n \sin\frac{n\pi x}{a} \sinh\frac{n\pi y}{a} \quad \text{... (4)}$$

Putting $y = b$ in (4) and using (3), we get

$$f(x) = \sum_{n=1}^{\infty}\left(E_n \sinh\frac{n\pi b}{a}\right)\sin\frac{n\pi x}{a}$$

which is a half range Fourier series of $f(x)$ in $(0, a)$.

Hence, $E_n = \dfrac{2}{a \sin\left(\dfrac{n\pi b}{a}\right)} \int_0^a f(x) \sin\frac{n\pi x}{a} dx$

Example 5. *Find the steady state temperature in a rectangular plate bounded by the lines x = 0, x = a, y = 0 and y = b if the edge y = 0 is insulated, the edge x = 0 and x = a are kept at* 0^oC *and the edge y = b is kept at temperature f(x).*

Solution. We know that the temperature $u(x,y)$ in steady state in two-dimensional plate is governed by

$$\frac{\partial^2 u}{\partial x^2} + \frac{\partial^2 u}{\partial y^2} = 0 \quad \text{... (1)}$$

As per given, the boundary conditions are

$$u(0, y) = u(a, y) = 0, \text{ for } 0 \le y \le b \quad \text{....(2)}$$

Also $\left(\dfrac{\partial u}{\partial y}\right)_{y=0} = 0$

(Because the edge $y = 0$ is insulated for $0 \le x \le a$) ... (3a)

and $u(x, b) = f(x)$, for $0 \le x \le a$ ... (3b)

Now proceeding same as in example (4), we get

$$u(x, y) = \sum_{n=1}^{\infty} u_n(x, y)$$

$$= E_n \sin\left(\frac{n\pi x}{a}\right)\cosh\left(\frac{n\pi y}{a}\right) \quad \text{... (4)}$$

Putting $y = b$ in (4) and using (3b), we get

$$f(x) = \sum_{n=1}^{\infty}\left(E_n \cosh\frac{n\pi b}{a}\right)\sin\frac{n\pi x}{a}$$

which is a half range Fourier sine series of $f(x)$ in $(0, a)$.

Hence, we get

$$E_n \cos\frac{n\pi b}{a} = \frac{2}{a}\int_0^a f(x) \sin\frac{n\pi x}{a} dx$$

$$\Rightarrow E_n = \frac{2}{a \cosh\left(\dfrac{n\pi b}{a}\right)}\int_0^a f(x) \sin\frac{n\pi x}{a} dx.$$

Example 6. *A rectangular metal plate is bounded by the lines x = 0, x = a, y = 0 and y = b. The three sides x = 0, x = a and y = b are insulated and the side y = 0 is kept at temperature* $u_0 \cos\left(\frac{\pi x}{a}\right)$. *Find the steady state temperature at any point of the plate.*

Solution. The governing equation is given by

$$\frac{\partial^2 u}{\partial x^2} + \frac{\partial^2 u}{\partial y^2} = 0 \quad \text{... (1)}$$

Since, $x = 0, x = a$ and $y = b$ are insulated.

Therefore, we have

$$\left(\frac{\partial u}{\partial x}\right)_{x=0} = 0, \quad \left(\frac{\partial u}{\partial x}\right)_{x=a} = 0 \quad \text{... (2)}$$

$$\left(\frac{\partial u}{\partial y}\right)_{y=0} = 0$$

and $u(x, a) \le u_0 \cos\left(\dfrac{\pi x}{a}\right)$ ... (3)

Using the method of separation of variables, suppose (1) has a solution of the form

$$u(x, y) = X(x)\ Y(y) \quad \text{... (4)}$$

Using (4) in (1), we get $X''Y + XY'' = 0$

$$\Rightarrow \frac{X''}{X} = -\frac{Y''}{Y} = \text{k (say)} \quad \text{... (5)}$$

$$\Rightarrow X'' - kX = 0 \quad \text{... (6)}$$

and $Y'' + kY = 0$ ... (7)

Also, from (4)

$$\frac{\partial u}{\partial x} = X'(x)\,Y(y) \quad \text{... (8)}$$

Using (2) in (8), we get

$$X'(0) = 0 \text{ and } X'(a) = 0 \quad \text{... (9)}$$

Now, we have the following three cases :

Case (i). Let $k = 0$. In this case, solution of equation (6) is given by

$$X(x) = Ax + B \quad \text{... (10)}$$

$$\Rightarrow \quad X'(x) = A$$

Using (9), we get $A = 0$.

Therefore $X(x) = B$

Case (ii). Let $k = \lambda^2, \lambda \neq 0$. In this case, the solution of (6) is given by

$$X(x) = Ae^{x\lambda} + Be^{-x\lambda}$$

$$\Rightarrow \quad X'(x) = \lambda(Ae^{\lambda x} - Be^{-\lambda x})$$

Using (9) in the above equation, we get

$$\lambda(A - B) = 0 \text{ and } \lambda(Ae^{a\lambda} - Be^{-a\lambda}) = 0$$

On solving, we get $A = B = 0$

$$\Rightarrow \quad X(x) \equiv 0$$

$$\Rightarrow \quad u \equiv 0$$

Case (iii) Let $k = -\lambda^2, \lambda \neq 0$. In this case, solution of (6) is given by

$$X(x) = A\cos\lambda x + B\sin\lambda x \quad \text{... (10)}$$

$$\Rightarrow \quad X'(x) = -A\lambda\sin\lambda x + B\lambda\cos\lambda x \quad \text{... (11)}$$

Using (9) in (10), we get

$$0 = B\lambda$$

and $\quad 0 = -A\lambda\sin\lambda a + B\lambda\cos\lambda a$ ... (12)

Now, letting $\lambda \neq 0$ and $A \neq 0$, equation (12) gives

$B = 0$ and $\sin\lambda a = 0$

$$B = 0 \text{ and } \lambda = \frac{n\pi}{a}, \; n = 1, 2, 3, \ldots \quad \text{... (13)}$$

Therefore, non-zero solutions of (10) is given by

$$X(x) = A\cos\left(\frac{n\pi x}{a}\right); \; n = 1, 2, 3, \ldots \quad \text{... (14)}$$

$\therefore$ All non-zero solution of (6) are given by

$$X_n(x) = A_n\cos\left(\frac{n\pi x}{a}\right), \; n = 0, 1, 2, 3, \ldots \quad \text{... (15)}$$

Now putting $k = -\lambda^2 = -\dfrac{n^2\pi^2}{a^2}$ in (7), we get

$$Y'' - \left(\frac{n^2\pi^2}{a^2}\right)Y = 0$$

whose general solution is given by

$$Y_n(y) = C_n e^{n\pi x/a} + D_n e^{-n\pi y/a} \quad \text{... (16)}$$

Also, from (4), we have

$$\frac{\partial u}{\partial y} = X(x)\,Y'(y) \quad \text{... (17)}$$

Putting $y = b$ in (17) and using (3), we get

$$0 = X(x)\,Y'(b)$$

$$\Rightarrow \quad Y'(b) = 0 \quad \text{[By letting } X(x) \neq 0]$$

$$\Rightarrow \quad Y_n'(b) = 0 \quad \text{... (18)}$$

From (16), we have

$$Y_n'(y) = C_n\left(\frac{n\pi}{a}\right)e^{n\pi y/a} - D_n\left(\frac{n\pi}{a}\right)e^{-n\pi y/a} \quad \text{... (19)}$$

Putting $y = b$ in (19) and using (18), we get

$$0 = \left(\frac{n\pi}{a}\right)(C_n e^{n\pi b/a} - D_n e^{-n\pi b/a})$$

$$\Rightarrow \quad D_n = \left(C_n e^{n\pi b/a}\right) \Big/ e^{-n\pi b/a}$$

Putting this value in (16), we get

$$Y_n(y) = C_n\left[\frac{(e^{n\pi y/a}e^{-n\pi b/a} + e^{-n\pi y/a}e^{n\pi b/a})}{e^{-n\pi b/a}}\right]$$

$$\Rightarrow Y_n(y) = C_n e^{n\pi b/a}\{e^{n\pi(b-y)/a} + e^{-n\pi(b-y)/a}\}$$

$$= 2C_n e^{n\pi b/a}\cosh\left\{\frac{n\pi(b-y)}{a}\right\}$$

Thus, $u_n(x, y) = X_n(x)\,Y_n(y)$

$$= E_n\cos\frac{n\pi x}{a}\cosh\left\{\frac{n\pi(b-y)}{a}\right\}$$

Further, consider the more general solution given by

$$u(x, y) = \sum_{n=0}^{\infty} u_n(x, y)$$

$$= \sum_{n=0}^{\infty} E_n\cos\left(\frac{n\pi x}{a}\right)\cosh\left\{\frac{n\pi(b-y)}{a}\right\} \quad \text{... (20)}$$

Putting $y = b$ and using (3), we get

$$u_0\cos\left(\frac{\pi x}{a}\right) = \sum_{n=0}^{\infty} E_n\cos\left(\frac{n\pi x}{a}\right)\cosh\left\{\frac{n\pi(b-a)}{a}\right\}$$

$$= E_0 + E_1\cos\left(\frac{\pi x}{a}\right)\cosh\left\{\frac{\pi(b-a)}{a}\right\}$$

$$+ E_2\cos\left(\frac{2\pi x}{a}\right)\cosh\left\{\frac{2\pi(b-a)}{a}\right\}$$

$$+ E_3\cos\left(\frac{3\pi x}{a}\right)\cosh\left\{\frac{3\pi(b-a)}{a}\right\} + \ldots$$

On equating the coefficients of like terms, we get $E_0 = 0$.

$$E_1\cosh\left\{\frac{\pi(b-a)}{a}\right\} = u_0$$

and $\quad E_n = 0 \quad$ for $\quad n \geq 1$.

Finally, putting this value in (20), we get

$$u(x, y) = u_0\,\text{sech}\left\{\frac{(b-a)\pi}{a}\right\}$$

$$\cos\left(\frac{\pi x}{a}\right)\cosh\left\{\frac{(b-y)\pi}{a}\right\}.$$

Example 7. *An infinitely long uniform plate is bounded by two parallel edges and an end at right angles to them. The breadth is π. .The end is maintained at $100^\circ C$ at all points and the other edges are at $0^\circ C$. Find the steady state temperature function $u(x, y)$.*

(UPTU–2008, PTU-2005, JNTU-2005, 08)

Solution. The governing equation for the steady state temperature is given by

$$\frac{\partial^2 u}{\partial x^2}+\frac{\partial^2 u}{\partial y^2}=0 \quad \text{... (1)}$$

with boundary conditions

$$u(0,y) = u(\pi, y)=0 \text{ for all } y\geq 0 \quad \text{... (2)}$$

$$\left.\begin{array}{l} u(x,y)\to 0 \text{ as } y\to\infty, \text{for } 0\leq x\leq\pi \\ \text{and } u(x,0)=f(x)=100, \quad \text{for } 0\leq x\leq\pi \end{array}\right\} \text{... (3)}$$

Now proceeding same as in example (2) by taking $a=\pi$, we get

$$u(x,y) = \sum_{n=1}^{\infty} E_n \sin nx\, e^{-ny} \quad \text{... (4)}$$

$$\text{where } E_n = \frac{2}{\pi}\int_0^{\pi} f(x) \sin nx\, dx \quad \text{... (5)}$$

$$= \frac{2}{\pi}\int_0^{\pi} 100 \sin nx\, dx$$

$$= \frac{200}{\pi}[-\cos nx]_0^{\pi}$$

$$= \frac{200}{\pi}[1-(-1)^n]$$

$$= \begin{cases} 0 \text{ if } n \text{ is even} \\ \dfrac{400}{(2m-1)\pi}, \text{ if } n=2m-1, \ m=1,2,3,\ldots \end{cases}$$

Finally, putting all these values in (4), we get

$$u(x,y)=\frac{400}{\pi}\sum_{n=1}^{\infty}\frac{\sin(2m-1)x}{2m-1}e^{-(2m-1)y}.$$

Example 8. *A thin rectangular homogeneous thermally conducting plate lies in the xy-plane defined by $0\leq x\leq a$, $0\leq y\leq b$. The edge $y = 0$ is held at its temperature $Tx(x - a)$, where T is a constant, while the remaining edges are held at 0^oC. The other faces are insulated and no internal sources and sinks are present. Find the steady state temperature inside the plate.*

Solution. As per given, we have no heat sources and sinks are present in the plate. Thus, the steady state temperature function is the solution of

$$\frac{\partial^2 u}{\partial x^2}+\frac{\partial^2 u}{\partial y^2}=0 \quad \text{... (1)}$$

with boundary conditions

$$u(0, y) = 0, \quad u(a, y) = 0, \quad u(x, b) = 0,$$

$$u(x, 0) = Tx(x-a) \quad \text{....(2)}$$

Now proceeding same as example (1), we get

$$u(x,y) = \sum_{n=1}^{\infty} A_m \sin\left(\frac{n\pi x}{a}\right)\sinh\left(\frac{n\pi(y-b)}{a}\right)$$

where,

$$A_n \sinh\left(\frac{-n\pi b}{a}\right) = \frac{2}{a}\int_0^a f(x)\sin\left(\frac{n\pi x}{a}\right)dx$$

Using the boundary conditions

$$U(x, 0) = Tx(x-a) = f(x)$$

we get

$$A_n \sinh\left(\frac{-n\pi b}{a}\right) = \frac{2}{a}\int_0^a Tx(x-a)\sin\left(\frac{n\pi}{a}x\right)dx$$

$$= \frac{2T}{a}\int_0^a x(x-a)\sin\left(\frac{n\pi x}{a}\right)dx$$

$$= \frac{-a}{n\pi}\cdot\frac{2T}{a}\left[\int_0^a x(x-a)\,.\,d\left\{\cos\left(\frac{n\pi x}{a}\right)\right\}\right]$$

$$= -\frac{2T}{n\pi}\left[(x-a)\cos\left(\frac{n\pi x}{a}\right)\right]_0^a$$

$$-\frac{a}{n\pi}\int_0^a (2x-a)\,d\left\{\sin\left(\frac{n\pi x}{a}\right)\right\}$$

$$= \frac{2aT}{n^2\pi^2}\left[(2x-a)\sin\left(\frac{n\pi x}{a}\right)\right]_0^a$$

$$-\int_0^a 2\sin\left(\frac{n\pi x}{a}\right)dx$$

$$= \frac{2aT}{n^2\pi^2}\left\{a\sin n\pi + \frac{2a}{n\pi}\left[\cos\left(\frac{n\pi x}{a}\right)\right]_0^a\right\}$$

$$= \frac{2aT}{n^2\pi^2}\cdot\frac{2a}{n\pi}\cos(n\pi-1) = \frac{4a^2T}{n^3\pi^3}[(-1)^n-1]$$

Hence, the required temperature function is given by

$$u(x,y) = \sum_{n=1}^{\infty}\frac{4Ta^2}{n^3\pi^3}\operatorname{cosech}\left(-\frac{n\pi b}{a}\right)$$

$$[(-1)^n-1]\sin\left(\frac{n\pi x}{a}\right)\sinh\left[\frac{n\pi(y-b)}{a}\right].$$

Example 9. *Solve* $\dfrac{\partial^2 u}{\partial x^2}+\dfrac{\partial^2 u}{\partial y^2}=0$, $0\leq x\leq a$, $0\leq y\leq b$ *subject to the boundary conditions* $u(0, y) = 0$, $u(x, 0) = 0$, $u(x, b) = 0$, $\dfrac{\partial u}{\partial x}(a,y) = T\sin^3\dfrac{\pi y}{a}$.

Solution. Proceeding same as example (3), we get

$$u(x,y) = (C_1e^{\lambda x}+C_2e^{-\lambda x})$$

$$(C_3\cos\lambda y + C_4\sin\lambda y)$$

Using $u(x, 0) = 0$, we get

$$u(x,0) = 0 = C_4\sin\lambda y\,(C_1e^{\lambda x}+C_2e^{-\lambda x})$$

Again, using $u(x, b) = 0$, we get

$$0 = C_4\sin\lambda b\,(C_1e^{\lambda x}+C_2e^{-\lambda x})$$

Now, $C_4\neq 0 \Rightarrow \sin\lambda b = 0$

$$\Rightarrow \lambda b = n\pi \Rightarrow \lambda = \frac{n\pi}{b}; \quad n=1,2,3,\ldots\ldots$$

$$\Rightarrow u(x,y) = C_4\sin\left(\frac{n\pi y}{b}\right)(C_1e^{\lambda x}+C_2e^{-\lambda x})$$

which can also be written as (By renaming the constants)

$$u(x,y) = \sin\left(\frac{n\pi y}{b}\right)$$

$$\left[A\exp\left(\frac{n\pi x}{b}\right)-B\exp\left(\frac{-n\pi x}{b}\right)\right].$$

$n = 1, 2, 3, \ldots$

Now, using(0, y) = 0, we get

$$0 = \sin\left(\frac{n\pi y}{b}\right)(A+B)$$

$$\Rightarrow A + B = 0, \quad i.e., \; A = -B$$

Therefore,

$$u(x, y) = A\sin\left(\frac{n\pi y}{b}\right)\left[\exp\left(\frac{n\pi x}{b}\right) - \exp\left(\frac{-n\pi x}{b}\right)\right]$$

$$= 2A\sin\left(\frac{n\pi y}{b}\right)\sinh\left(\frac{n\pi x}{b}\right), \quad n = 1, 2, 3, \ldots \quad \ldots (1)$$

$$\Rightarrow \frac{\partial u}{\partial x} = 2A\frac{n\pi}{b}\sin\left(\frac{n\pi y}{b}\right)\cosh\left(\frac{n\pi x}{b}\right)$$

Putting $x = a$ and using last boundary condition, we have

$$T\sin^3\frac{\pi y}{a} = 2A\frac{n\pi}{b}\sin\left(\frac{n\pi y}{b}\right)\cosh\left(\frac{n\pi a}{b}\right)$$

$$\Rightarrow \quad 2A = \frac{T\sin^3\frac{\pi y}{a}}{\frac{n\pi}{b}\sin\left(\frac{n\pi y}{b}\right)\cosh\left(\frac{n\pi a}{b}\right)}$$

Putting this value in (1), the required steady state temperature function is given by

$$u(x, y) = \frac{bT}{n\pi}\sin^3\frac{\pi y}{a}\,\mathrm{sech}\frac{n\pi}{b}a\sinh\left(\frac{n\pi x}{b}\right)$$

Hence, the general solution is given by

$$u(x, y) = \sum_{n=1}^{\infty}\frac{bT}{n\pi}\sin^3\frac{\pi y}{a}\,\mathrm{sech}\left(\frac{n\pi a}{b}\right)\sinh\left(\frac{n\pi x}{b}\right).$$

Example 10. *Find the potential function u(x, y, z) in a rectangular box defined by $0 \le x \le a$; $0 \le y \le b$; $0 \le z \le c$ if the potential is zero on all sides and the bottom, while $u = f(x, y)$ on the top of the box.*

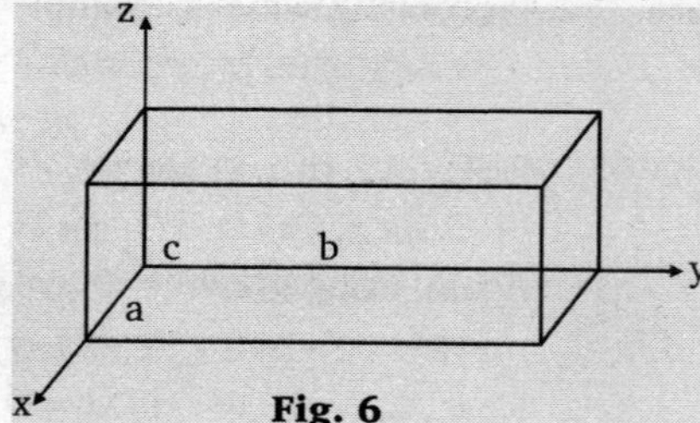

Fig. 6

Solution. Since, the potential distribution in the rectangular box satisfies the Laplace equation given by

$$\frac{\partial^2 u}{\partial x^2} + \frac{\partial^2 u}{\partial y^2} + \frac{\partial^2 u}{\partial z^2} = 0 \quad \ldots (1)$$

As per given, the boundary conditions are

$$\left.\begin{aligned} u(0, y, z) &= u(a, y, z) = 0 \\ u(x, 0, z) &= u(x, b, z) = 0 \\ u(x, y, 0) &= 0 \\ \text{and } u(x, y, c) &= f(x, y) \end{aligned}\right\} \quad \ldots (2)$$

Assume that

$$u(x, y, z) = X(x)\,Y(y)\,Z(z) \quad \ldots (3)$$

be the form of the solution of (1).

Putting the value from (3) into (1), we get

$$X''(x)\,Y(y)\,Z(z) + X(x)Y''(y)\,Z(z) + X(x)Y(y)Z''(z) = 0$$

$$\Rightarrow \frac{Y''(y)}{Y(y)} + \frac{Z''(z)}{Z(z)} = -\frac{X''(x)}{X(x)} = \lambda_1^2 \text{ (say)}$$

Now, we get

$$X''(x) + \lambda_1^2 X(x) = 0 \quad \ldots (4)$$

After the second separation, we have

$$\frac{Z''(z)}{Z(z)} = \lambda_3^2, \quad -\frac{Y''(y)}{Y(y)} = \lambda_2^2$$

$$\Rightarrow \quad Y''(y) + \lambda_2^2 Y(y) = 0 \quad \ldots (5)$$

$$\text{and} \quad Z''(z) - \lambda_3^2 Z(z) = 0 \quad \ldots (6)$$

$$\text{where} \quad \lambda_3^2 = \lambda_1^2 + \lambda_2^2$$

The general solution of (4), (5) and (6) are given by

$$X(x) = C_1\cos\lambda_1 x + C_2\sin\lambda_1 x$$

$$Y(y) = C_3\cos\lambda_2 y + C_4\sin\lambda_2 y$$

$$Z(z) = C_5\cosh\lambda_3 z + C_6\sinh\lambda_3 z$$

Using the boundary condition given by (2), we get

$$X(0) = X(a) = 0$$

$$Y(0) = Y(b) = 0$$

$$Z(0) = 0$$

Now, $X(0) = 0 \Rightarrow C_1 = 0$

$$X(a) = 0 \Rightarrow \lambda_1 a = m\pi$$

$$\Rightarrow \quad \lambda_1 = \frac{m\pi}{a}; \quad m = 1, 2, 3, \ldots$$

In a similar manner, we can say that

$Y(0) = 0$ gives $C_3 = 0$

$Y(b) = 0$ gives $\lambda_2 b = n\pi$

Therefore,

$$\lambda_2 = \frac{n\pi}{b}; \quad n = 1, 2, \ldots$$

and $Z(0) = 0 \Rightarrow C_5 = 0$

Also,

$$\lambda_3^2 = \lambda_1^2 + \lambda_2^2 = \pi^2\left(\frac{m^2}{a^2} + \frac{n^2}{b^2}\right) = \lambda_{mn}^2 \quad \text{(say)}$$

Then,

$$\lambda_3 = \pi\sqrt{\frac{m^2}{a^2} + \frac{n^2}{b^2}} = \lambda_{mn}$$

Therefore, the required general solution is given by

$$u(x, y, z) = X(x)\,Y(y)\,Z(z)$$

$$= \sum_{m=1}^{\infty}\sum_{n=1}^{\infty} C_{mn}\sin\frac{m\pi x}{a}\sin\frac{n\pi y}{b}\sinh\lambda_{mn}\cdot z$$

Finally, using the boundary condition $f(x, y) = u(x, y, c)$, we get

$$f(x,y) = \sum\sum C_{mn} \sinh \lambda_{mn} C \sin\frac{m\pi x}{a} \sin\frac{n\pi y}{b}$$

which is a double Fourier sine series.

Hence,

$$C_{mn} \sinh \lambda_{mn} C = \frac{4}{ab} \int_0^a \int_0^b f(x,y) \sin\frac{m\pi x}{a} \sin\frac{n\pi y}{b} dxdy$$

EXERCISE 42.3

1. Find the steady state temperature distribution in a thin rectangular plate bounded by the lines $x = 0, x = a, y = 0, y = b$. The edges $x = 0, x = a, y = 0$ are kept at temperature zero while the edge $y = b$ is kept at 100°C.
2. Evaluate the steady temperature in a rectangular plate of length a and width b, the sides of which are kept at temperature zero, the lower end is kept at temperature $f(x)$ and the upper edge is kept insulated.
3. Show that the velocity potential for an irrational flow of an incompressible fluid satisfies the Laplace equation.
4. Find the steady state temperature distribution in a rectangular plate of sides a and b insulated at the lateral surface and satisfying the boundary conditions
$u(0, y) = u(a, y) = 0$ for $0 \le y \le b$
$u(x, b) = 0$
and $u(x, 0) = x(a - x)$ for $0 \le y \le a$
5. A rectangular plate with insulated surface 8 cm wide and so long compared to its width that it can be considered infinite in the length without introducing an appreciable error. If the temperature along the short edge $y = 0$ is given by $u(x, 0) = 100 \sin\left(\frac{\pi x}{8}\right)$, in $0 < x < 8$, while the two long edges $x = 0$ and $x = 8$ as well as the other short edges are kept at 0°C. Find the steady state temperature function $u(x, y)$.
6. By separating the variables, show that $\frac{\partial^2 V}{\partial x^2} + \frac{\partial^2 V}{\partial y^2} = 0$ has solutions of the form $A \exp(\pm nx \pm ixy)$, where A and n are constants. Deduce that the function of the form
$$V(x, y) = \sum_r A_r \sin\left(\frac{r\pi y}{a}\right) e^{-(r\pi x)/a}, \quad x \ge 0,\ 0 \le y \le \infty.$$
where A_r are constants or plane harmonic functions satisfying the conditions $V(x, 0) = V(x, a) = 0$ and $V(x, y) \to 0$ as $x \to \infty$.
7. The diameter of a semi-circular plate of radius a is kept at 0°C and the temperature at the semi-circular boundary is T°C. Show that the steady state temperature in the plate is given by
$$u(r,\theta) = \frac{4T}{\pi} \sum_{n=1}^{\infty} \frac{1}{2n-1} \left(\frac{r}{a}\right)^{2n-1} \sin(2n-1)\theta$$ (GBTU (CO) 2011)
8. Solve: $\frac{\partial^2 V}{\partial r^2} + \frac{1}{r}\frac{\partial V}{\partial r} + \frac{1}{r^2}\frac{\partial^2 V}{\partial \theta^2} = 0$ with boundary conditions.
(i) V is finite when $r \to 0$
(ii) $V = \sum C_n \cos n\theta$ on $r = a$ (GBTU-2010)
9. Solve the Laplace equation $\frac{\partial^2 u}{\partial x^2} + \frac{\partial^2 u}{\partial y^2} = 0$ subject to the conditions
$u(0, y) = u(l, y) = u(x, 0) = 0$ and $u(x, a) = \frac{\sin n\pi x}{l}$
(VTU-2011, JNTU-2006, Kerala M. Tech.-2005, UPTU-2004)

Hint to Selected Problems

1. In this problem, we consider the steady state temperature, *i.e.*, it does not depend upon time. Thus the flow is governed by the Laplace equation given by
$$\frac{\partial^2 V}{\partial x^2} + \frac{\partial^2 V}{\partial y^2} = 0 \quad \text{... (1)}$$
As per given, the boundary conditions are
$$u(0, y) = 0,\ u(a, y) = 0 \quad \text{... (2)}$$
$$u(x, 0) = 0,\ u(x, b) = 100 \quad \text{... (3)}$$
Using the method of separation of variables, suppose (1) has a solution of the form
$$u(x, y) = X(x)\, Y(y) \quad \text{... (4)}$$
Putting this value in equation (1), we get
$$X''Y + XY'' = 0$$
$$\frac{X''}{X} = -\frac{Y''}{Y} \quad \text{... (5)}$$
Now, for independent x and t, each side of (5) must be equal to the constant, say k. Then we have
$$X'' - kX = 0 \quad \text{... (6)}$$
and
$$Y'' + kY = 0 \quad \text{... (7)}$$
Also, from (2) and (4), we have
$$X(0)Y(a) = 0 \text{ and } X(a)\ Y(y) = 0$$
$$X(0) = 0 \text{ and } X(a) = 0 \quad \text{... (8)}$$
Here, we must take $Y(y) \ne 0$ because otherwise $u = 0$, which does not satisfy the given conditions.

Now, we shall discuss the following three cases :

Case (I). Let $k = 0$. Then solution of equation (6) is given by
$$X(x) = Ax + B$$
Using boundary conditions (8), we get
$$B = 0 \text{ and } Aa + B = 0$$
$$A = B = 0 \Rightarrow X(x) = 0 \Rightarrow u = 0$$
Which does not satisfy given condition (3). Thus, we reject this case, *i.e.*,, $k = 0$.

Case (II). Let $k = \lambda^2, \lambda \ne 0$. In this case, solution of (6) is given by
$$X(x) = Ae^{x\lambda} + Be^{-x\lambda}$$
Using the boundary conditions (8), we get
$$A + B = 0 \text{ and } 0 = e^{a\lambda} + e^{-a\lambda}.$$

On solving, we get

$$A = B = 0 \Rightarrow X(x) = 0 \Rightarrow u = 0$$

Thus, we reject this case also.

Case (III). Let $k = -\lambda^2, \lambda \neq 0$. In this case, the solution of equation (6) is given by

$$X(x) = A\cos\lambda x + B\sin\lambda x$$

Using the boundary conditions (8), we get

$$A = 0 \text{ and } A\cos\lambda a + B\sin\lambda a = 0$$

On solving, we get

$$A = 0 \text{ and } \sin\lambda a = 0 \quad \text{[By taking } B \neq 0\text{]}$$

Further, $\sin\lambda a = 0$

$$\Rightarrow \quad \lambda = \frac{n\pi}{a}; \quad n = 1, 2, 3, \ldots$$

Thus, the non-zero solution $X_n(x)$ of (6) is given by

$$X_n(x) = B_n \sin\left(\frac{n\pi x}{a}\right)$$

Further, using $k = -\lambda^2 = -\frac{n^2\pi^2}{a^2}$, equation (7) reduces to

$$Y'' - \left(\frac{n^2\pi^2}{a^2}\right)Y = 0$$

The general solution of this equation is given by

$$Y_n(y) = C_n e^{n\pi y/a} + D_n e^{-n\pi y/a} \quad \ldots (9)$$

Using (3) and (4)

$$0 = X(x)\,Y(0)$$

$$\Rightarrow \quad Y(0) = 0 \quad \text{Now,} \quad Y(0) = 0$$

$$\Rightarrow \quad Y_n(0) = 0 \quad \ldots (10)$$

Putting $y = 0$ in (9) and using (10), we get

$$0 = C_n + D_n$$

$$D_n = -C_n$$

Then equation (7) becomes

$$Y_n(y) = C_n(e^{n\pi y/a} - e^{-n\pi y/a}) = 2\sinh\left(\frac{n\pi y}{a}\right)$$

Therefore,

$$u(x, y) = X_n(x)\,Y_n(y) = E_n \sin(n\pi x/a)\sinh(n\pi y/a)$$

where, $E_n = 2B_nC_n$ are arbitrary constants.

The more general solution is given by

$$u(x, y) = \sum_{n=1}^{\infty} E_n \sin\left(\frac{n\pi x}{a}\right)\sinh\left(\frac{n\pi y}{a}\right) \quad \ldots (11)$$

Putting $y = b$ and using given condition $u(x, b) = 100$, we get

$$100 = \sum_{n=1}^{\infty} E_n \sin\left(\frac{n\pi x}{a}\right)\sinh\left(\frac{n\pi b}{a}\right)$$

which is a Fourier sine series.

Thus,

$$E_n \sinh\frac{n\pi b}{a} = \frac{2}{a}\int_0^a 100.\sin\left(\frac{n\pi x}{a}\right)dx$$

$$= \frac{200}{a}\left[-\frac{\cos(n\pi x/a)}{(n\pi/a)}\right]_0^a$$

So,

$$E_n = \frac{200}{n\pi}\left[1 - (-1)^n\right]\operatorname{cosech}\left(\frac{n\pi b}{a}\right)$$

$$= \begin{cases} \dfrac{400\operatorname{cosech}\{(2m-1)\pi b/a}{(2m-1)\pi} & \text{if } n = 2m \\ & \text{if } n = 2m-1 \end{cases}$$

Putting this value in (11), we get

$$u(x, y) = \sum_{m=1}^{\infty} E_{2m-1}\sin\frac{(2m-1)\pi x}{a}\sinh\frac{(2m-1)\pi y}{a}$$

$$\Rightarrow \quad u(x, y) = \frac{400}{\pi}\sum_{m=1}^{\infty}\frac{1}{2m-1}\sin\frac{(2m-1)\pi x}{a}$$

$$\sinh\frac{(2m-1)\pi y}{a}\operatorname{cosech}\frac{(2m+1)\pi b}{a}$$

2. We know that temperature $u(x, y)$ in steady state in two-dimensional plate is governed by the Laplace equation.

$$\frac{\partial^2 u}{\partial x^2} + \frac{\partial^2 u}{\partial y^2} = 0 \quad \ldots (1)$$

According to the given conditions, we have

$$u(0, y) = 0, \; u(a, y) = 0 \quad \ldots (2)$$

$$\left(\frac{\partial u}{\partial y}\right)_{y=b} = 0 \text{ and } u(x, 0) = f(x) \quad \ldots (3)$$

Using the method of separation of variables, suppose

$$u(x, y) = X(x)\,Y(y) \quad \ldots (4)$$

is solution of (1).

Putting this value of u from (4) in (1), we get

$$X''Y + XY'' = 0$$

$$\Rightarrow \quad \frac{X''}{X} = -\frac{Y''}{Y} \quad \ldots (5)$$

Now, since x and t are independent, therefore each side of (5) must be equal to the same constant, say k. Then (5) reduces to

$$X'' - KX = 0 \quad \ldots (6)$$

and

$$Y'' + KY = 0 \quad \ldots (7)$$

with

$$X(0) = 0 \text{ and } X(a) = 0. \quad \ldots (8)$$

Now, there are following three cases :

Case (I). Let $k = 0$. In this case, solution of (6) is given by

$$X(x) = Ax + B. \quad \ldots (9)$$

Using given conditions (8), we get

$$A = B = 0.$$

Hence, we reject this case.

Case (II). **Let $k = \lambda^2, \lambda \neq 0$.** In this case, solution of (6) is given by

$$X(x) = Ae^{x\lambda} + Be^{-x\lambda} \quad \ldots (10)$$

Again, using boundary conditions (8) in (10), we get

$$A + B = 0 \text{ and } Ae^{a\lambda} + Be^{-a\lambda} = 0$$

$$\Rightarrow \quad A = B = 0 \Rightarrow X(x) = 0 \Rightarrow u = 0.$$

Hence, we reject this case also.

Case (III). Let $k = -\lambda^2, \lambda \neq 0$. Then solution of (6) is given by

$$X(x) = A\cos\lambda x + B\sin\lambda x$$

Using boundary conditions (8), we get

$$A = 0 \text{ and } A\cos\lambda a + B\sin\lambda a = 0$$

$$A = 0 \text{ and } \sin\lambda a = 0 \quad \text{[By taking } B \neq 0\text{]}$$

Now, $\sin\lambda a = 0$

$$\Rightarrow \quad \lambda a = n\pi \Rightarrow \lambda = \frac{n\pi}{a}; \quad n = 1, 2, 3, \ldots$$

Thus, non-zero solution $X_n(x)$ of (6) is given by

$$X_n(x) = B_n \sin\left(\frac{n\pi x}{a}\right)$$

Now, $k = -\lambda^2 = -\frac{n^2\pi^2}{a^2}$,

Then, equation (7) reduces to $Y'' - \left(\frac{n^2\pi^2}{a^2}\right)Y = 0$

whose general solution is given by

$$Y_n(y) = C_n e^{n\pi y/a} + D_n e^{-n\pi y/a} \quad \text{... (11)}$$

From (4), we can find

$$\frac{\partial u}{\partial y} = X(x)\, Y'(y) \Rightarrow \left(\frac{\partial u}{\partial y}\right)_{y=b} = X(x)\ Y'(b)$$

Using (3), we get

$X(x).Y'(b) = 0$

$$\Rightarrow \quad Y'(b) = 0 \quad \text{... (12)}$$

Differentiating (11) w.r.t. y, we get

$$Y_n'(y) = \left(\frac{n\pi}{a}\right) C_n\, e^{n\pi y/a} - \left(\frac{n\pi}{a}\right) D_n e^{-n\pi y/a} \quad \text{... (13)}$$

Putting $y = b$ and using (12), we get

$$0 = \left(\frac{n\pi}{a}\right)\left(C_n e^{n\pi b/a} - D_n e^{-n\pi b/a}\right)$$

$$D_n = C_n e^{n\pi b/a} \,/\, e^{-n\pi b/a}$$

Putting this value in (11), we get

$$Y_n(y) = \frac{C_n(e^{n\pi y/a} - e^{-n\pi y/a} + e^{-n\pi y/a} e^{n\pi b/a}}{e^{-n\pi b/a}}$$

$$Y_n(y) = C_n e^{n\pi b/a}\left\{e^{n\pi(b-y)/a} + e^{-n\pi(b-y)/a}\right\}$$

$$= 2C_n\, e^{n\pi b/a} \cosh\left\{\frac{n\pi(b-y)}{a}\right\}$$

Thus, $u_n(x, y) = X_n(x).Y_n(y)$

$$= E_n.\sin\frac{n\pi x}{a} \cosh\left\{\frac{n\pi(b-y)}{a}\right\} \quad \text{... (14)}$$

Now, the more general solutions are given by

$$u(x,y) = \sum_{n=1}^{\infty} u_n(x,y)$$

$$= \sum_{n=1}^{\infty} E_n \sin\left(\frac{n\pi x}{a}\right)\cosh\left(\frac{n\pi(b-y)}{a}\right) \quad \text{... (15)}$$

Now, putting $y = 0$ in (15) and using (3), we get

$$f(x) = \sum_{n=1}^{\infty}\left\{E_n \cosh\left(\frac{n\pi b}{a}\right)\right\}\sin\left(\frac{n\pi x}{a}\right)$$

which is the half range Fourier sine series of $f(x)$ in $(0, a)$. Therefore, we get

$$E_n = \frac{2}{a\cosh\left(\frac{n\pi b}{a}\right)}\int_0^a f(x).\sin\frac{n\pi x}{a}\,dx$$

3. Let S be a closed surface enclosed a volume V.

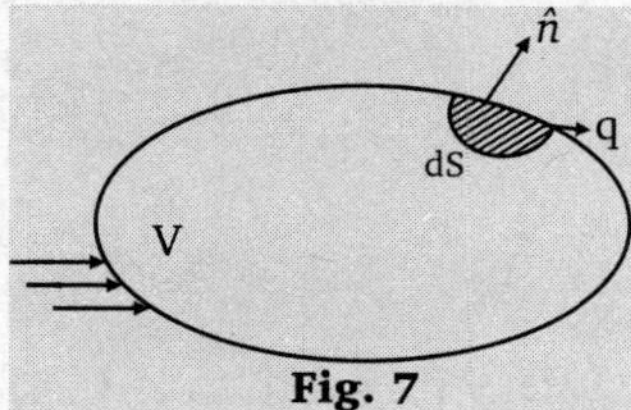

Fig. 7

Let ρ be the density of the fluid. If $\hat{n}$ be a unit normal vector drawn from outside to the surface element ds and $\bar{q}$ be the velocity of the fluid at that point. Then

Inward normal velocity $= -\bar{q}\,\hat{n}$

$\therefore$ Mass of the fluid entering per unit time through the element dS is $(-\bar{q}.\hat{n})\,dS$.

Thus the mass of the fluid entering in the surface S in the unit time is

$$= -\int\int_S \rho\,(\bar{q}.\hat{n})\,dS.$$

Further, the mass of the fluid within S is

$$= \int\int\int_S \rho dV$$

Rate of mass of fluid increasing in S is given by :

$$\frac{\partial}{\partial t}\iiint_V \rho\, dV = \iiint_V \frac{\partial\rho}{\partial t}\,dV.$$

Using the law of conservation of mass, we have

$$\iiint_V \frac{\partial\rho}{\partial t}\,dV = -\iint_S \rho(\bar{q}\ \hat{n})\,dS$$

$$= \iiint_V \nabla(\rho\bar{q})\,dV$$

[By Gauss divergence theorem]

$$\Rightarrow \iiint_V\left[\frac{\partial\rho}{\partial t} + \nabla(\rho\bar{q})\right]dV = 0$$

$\Rightarrow \dfrac{\partial\rho}{\partial t} + \nabla(\rho\bar{q}) = 0$, which is known as equation of continuity.

For an incompressible fluid :

$$\rho = \text{constant}$$

$$\nabla.\bar{q} = 0$$

Also, if flow is irrotational

$\bar{q} = -\nabla\phi$, where ϕ is a potential function.

$\nabla.q = \nabla.\nabla\phi = \nabla\ \phi = 0$, which is a Laplace equation.

4. Proceeding same as example (1) and using $f(x) = x(a-x)$, we get the required temperature.

$$u(x, y) = \sum_{n=1}^{\infty} E_n \sin\frac{n\pi x}{a}\sinh\frac{n\pi(b-y)}{a}$$

where $$E_n = \frac{2}{a\sinh\left(\frac{n\pi b}{a}\right)}\int_0^a f(x)\sin\frac{n\pi x}{a}dx$$

$$= \frac{2}{a\sinh\left(\frac{n\pi b}{a}\right)}\int_0^a (ax - x^2)\sin\frac{n\pi x}{a}dx$$

$$= \frac{2}{a\sinh\left(\frac{n\pi b}{a}\right)}\left[(ax-x^2)\left(-\frac{a}{n\pi}\right)\cos\frac{n\pi x}{a} - (a-2x)\left(-\frac{a^2}{n^2\pi^2}\right)\sin\frac{n\pi x}{a} + (-2)\left(\frac{a^3}{n^3\pi^3}\right)\cos\frac{n\pi x}{a}\right]_0^a$$

$$= \frac{2}{a\sinh\left(\frac{n\pi b}{a}\right)}\left[\frac{-2a^3(-1)^n}{n^3\pi^3} + \frac{2a^3}{n^3\pi^3}\right]$$

$$= \frac{4a^2}{n^3\pi^3}[1-(-1)^n]\text{cosec}\frac{n\pi b}{a}$$

$$= \begin{cases} 0, & \text{if } n = 2m \text{ (even) and } m = 1,2,3,... \\ \left\{\dfrac{8a^2}{\pi^3(2m-1)^3}\right\}\text{cosech}\left\{\dfrac{(2m-1)\pi b}{a}\right\}, & \text{if } n = 2m-1 \text{ and } m = 1,2,3,... \end{cases}$$

Putting this value in (1), we get

$$u(x, y) = \frac{8a^2}{\pi^3} \sum_{n=1}^{\infty} \frac{1}{(2m-1)^3} \sin\frac{(2m-1)\pi x}{a}$$

$$\sinh\frac{(2m-1)(b-y)\pi}{a} \operatorname{cosech}\frac{(2m-1)\pi b}{a}$$

5. Steady state temperature is governed by the following Laplace's equation

$$\frac{\partial^2 u}{\partial x^2} + \frac{\partial^2 u}{\partial y^2} = 0 \qquad \text{... (1)}$$

As per given, the boundary conditions are

$u(0, y) = u(8, y) = 0$ for $0 < y < \infty$

$$\left.\begin{aligned} &u(x,y) \to 0 \text{ as } y \to \infty \text{ for } 0 < x < 8 \\ \text{and} \quad &u(x,0) = 100 \sin\left(\frac{\pi x}{8}\right) \text{ for } 0 < x < 9 \end{aligned}\right\} \qquad \text{... (3)}$$

By taking $a = 8$, we get

$$u(x, y) = \sum_{n=1}^{\infty} E_n \sin\frac{n\pi x}{8} e^{-n\pi y/8} \qquad \text{... (4)}$$

Putting $y = 0$ in (4) and using (3), we get

$$100 \sin\left(\frac{\pi x}{8}\right) = \sum_{n=1}^{\infty} E_n \sin\left(\frac{n\pi x}{8}\right)$$

$$= E_1 \sin\left(\frac{\pi x}{8}\right) + E_2 \sin\left(\frac{2\pi x}{8}\right) + \ldots$$

Now comparing the coefficients of like terms of both the sides, we get

$$E_1 = 100 \text{ and } E_n = 0 \text{ for } n \geq 2.$$

Hence, the required temperature function is given by

$$u(x, y) = 100 \sin\left(\frac{\pi x}{8}\right) e^{-\pi y/8}$$

6. The given equation is

$$\frac{\partial^2 V}{\partial x^2} + \frac{\partial^2 V}{\partial y^2} = 0 \qquad \text{... (1)}$$

Suppose the solution of (1) is of the form

$$V(x, y) = X(x)\, Y(y) \qquad \text{... (2)}$$

Putting the value from (2) in (1), we get $X''Y + XY'' = 0$

$$\Rightarrow \quad \frac{X''}{X} = -\frac{Y''}{Y} = n^2 \text{ (say)} \qquad \text{... (3)}$$

From (3), we can find

$$X'' - n^2 X = 0 \qquad \text{... (4)}$$

and

$$Y'' + n^2 Y = 0 \qquad \text{... (5)}$$

whose solutions are given by

$$X = Ae^{nx} + Be^{-nx}$$

and

$$Y = C\cos ny + D\sin ny$$

Putting these values in (2), we get

$$V(x,y) = (Ae^{nx} + Be^{-nx})(C\cos ny + D\sin ny) \qquad \text{... (6)}$$

$$= A.\exp(\pm nx \pm iny) \qquad \text{... (7)}$$

Here, the given boundary conditions are

$$V(x, y) \to 0 \text{ as } x \to \infty \qquad \text{... (8)}$$

and $\quad V(x, 0) = V(x, a) = 0 \qquad \text{... (9)}$

Thus, we can write

$$V(x, y) = (E\cos ny + F\sin ny)\, e^{-nx} \qquad \text{... (10)}$$

Putting $y = 0$ in (10) and using (9), we get

$$0 = Ee^{-nx} \Rightarrow E = 0$$

Further, putting $y = a$ in (10) and using (9), we get

$$0 = F \sin na e^{-nx} \text{ for all }.$$

$$\Rightarrow \quad \sin na = 0 \Rightarrow n = \frac{r\pi}{a};\ r = 1,2,3,\ldots$$

Therefore, non-zero solution of (1) are given by (10) in the form

$$V(x, y) = F \sin\left(\frac{r\pi y}{a}\right) e^{-\left(\frac{r\pi x}{a}\right)}, \quad r = 1,2,3,\ldots$$

Hence, the more general solution of (1) is of the form

$$V(x, y) = \sum_{r=1}^{\infty} A_r \sin\left(\frac{r\pi y}{a}\right) e^{-\left(\frac{r\pi x}{a}\right)}$$

ANSWERS

1. $u(x, y) = \frac{400}{\pi} \sum_{m=1}^{\infty} \frac{1}{2m-1} \sin\frac{(2m-1)\pi x}{a} \sinh\frac{(2m-1)\pi y}{a} \operatorname{cosech}\frac{(2m+1)\pi b}{a}$

2. $u(x,y) = \sum_{n=1}^{\infty} u_n(x,y) = \sum_{n=1}^{\infty} E_n \sin\left(\frac{n\pi x}{a}\right) \cosh\left\{\frac{n\pi(b-y)}{a}\right\}$

$f(x) = \sum_{n=1}^{\infty} \left\{E_n \cosh\left(\frac{n\pi b}{a}\right)\right\} \sin\frac{n\pi x}{a}$, where, $E_n = \frac{2}{a\cosh\left(\frac{n\pi b}{a}\right)} \int_0^a f(x) \sin\frac{n\pi x}{a} dx$

4. $u(x, y) = \frac{8a^2}{\pi} \sum_{m=1}^{\infty} \frac{1}{(2m-1)^3} \sin\frac{(2m-1)nx}{a} \sinh\frac{(2m-1)(b-y)\pi}{a} \operatorname{cosech}\frac{(2m-1)\pi b}{a}$

5. $u(x, y) = 100 \sin\left(\frac{\pi x}{8}\right) e^{-\pi y/8}$

6. $V(x, y) = \sum_{r=1}^{\infty} A_r \sin\left(\frac{rxy}{a}\right) . e^{-\left(\frac{r\pi x}{a}\right)}$

7. $u(r,\theta) = \frac{4T}{\pi} \sum_{n=1}^{\infty} \frac{1}{2n-1} \left(\frac{r}{a}\right)^{2n-1} \sin(2n-1)\theta$

8. $V = \sum C_n \left(\frac{r}{a}\right)^n \cos n\theta$

9. $u(x,y) = \frac{\sinh(n\pi y / l)}{\sinh(n\pi a / l)} \sin\frac{n\pi x}{l}$

42.14 PARABOLIC DIFFERENTIAL EQUATIONS

Here, we shall consider a few problems dealing with the simplest of all parabolic equations, namely the one-dimensional heat equation. Under suitable physical assumptions and choice of units, this equation governs the distribution of temperature on a homogeneous thin rod occupying part of all x-axis, the variable t denoting the time.

42.15 ONE DIMENSIONAL HEAT EQUATION

(UPTU-2007, UPTU(CO)-2009, A.M.I.E.-1995,97)

Let us consider the flow of heat by conduction in a bar OA. Consider an element $PQQ'P'$ of the bar. The temperature $u(x, t)$ of the bar at any point P is function of x and time t.

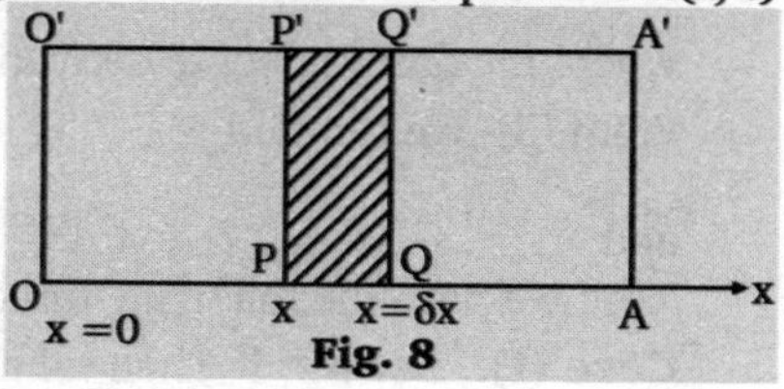

Fig. 8

Now, we have the following assumptions :

(i) The position of the bar coincides with the x-axis.

(ii) The bar is homogeneous.

(iii) It is sufficiently thin so that the heat is uniformly distributed over its cross-section at a given time t.

(iv) The surface of the bar is insulated to prevent any loss of heat through the boundary.

(v) $u(x, t)$ is the temperature at the point x at time t.

(vi) The amount of heat crossing any section of the bar is given by $kA\left(\frac{\partial u}{\partial x}\right)\delta t$

where, A = Area of the cross-section of the bar

$\frac{\partial u}{\partial x}$ = Temperature gradient at the section

δt = Time of the flow of heat

k = Thermal conductivity of the material of the bar.

The quantity of heat flowing into the element across the section PP' in time δt.

$$= -kA\left(\frac{\partial u}{\partial x}\right)_x \delta t$$

(The negative sign has been taken because heat flows in the direction of decreasing temperature.)

Also, the quantity of heat flowing out of the element across the section QQ' in time δt.

$$= -kA\left(\frac{\partial u}{\partial x}\right)_{x+\delta x} \delta t$$

Therefore the quantity of heat retained by the element is

$$= -kA\left(\frac{\partial u}{\partial x}\right)_x \delta t + kA\left(\frac{\partial u}{\partial x}\right)_{x+\delta x} \delta t = kA\,\delta t\left\{\left(\frac{\partial u}{\partial x}\right)_{x+\delta x} - \left(\frac{\partial u}{\partial x}\right)_x\right\} \quad \text{... (1)}$$

Now, suppose that this heat raises the temperature of the element by a small quantity δu. Therefore, the same quantity of heat is given by

$$= (\rho A\delta x)\sigma\,\delta u \quad \text{... (2)}$$

where σ is specific heat of the bar.

Since (1) and (2) are equal, therefore, we have

$$kA\delta t\,\{u(x+\delta x, t) - u(x, t)\} = (\rho A\delta x)\,\sigma\,\delta u$$

$$\Rightarrow \quad k\frac{u(x+\delta x, t) - u(x, t)}{\delta x} = \rho\sigma\frac{\delta u}{\delta t} \quad \text{... (3)}$$

As $\delta x \to 0$ and $\delta t \to 0$, we get

$$k\frac{\partial^2 u}{\partial x^2} = \rho\sigma\frac{\partial u}{\partial t}$$

$$\Rightarrow \quad \frac{\partial u}{\partial t} = k_1\frac{\partial^2 u}{\partial x^2} \quad \text{... (4)}$$

where, $k_1 = \frac{k}{\rho\sigma}$ is called the diffusivity of the material of the bar. Here, equation (4) is known as one-dimensional heat equation.

REMARK

- Heat equation is also known as Diffusion equation.

42.16 SOLUTION OF ONE DIMENSIONAL HEAT EQUATION

(GBTU (CO)-2011, UPTU-2007)

Consider the equation $$\frac{\partial u}{\partial t} = k\frac{\partial^2 u}{\partial x^2} \qquad \ldots (1)$$

For the method of separation of variables, let us assume that solution of (1) is of the form

$$u(x, t) = X(x)\,T(t) \qquad \ldots (2)$$

Putting the values from (2) in (1), we get

$$\frac{X''}{X} = \frac{T'}{kT} = \mu \text{ (say), a seperation constant.} \qquad \ldots (3)$$

where the dashes denote derivatives with respect to the relevant variable.

From (3), we can find $$X'' - \mu X = 0 \qquad \ldots (4)$$

and $$T' = \mu k T \qquad \ldots (5)$$

Now, we have the following three cases :

Case (I). Let $\mu = 0$ Then solutions of (4) and (5) are given by

$$X = a_1 x + a_2 \text{ and } T = a_3 \qquad \ldots (6)$$

Case (II). Let $\mu = \lambda^2, \lambda \neq 0$. Then (4) and (5) reduce to

$$X'' - \lambda^2 X = 0 \text{ and } T' = \lambda^2 kT$$

On solving these equations, we get

$$\left.\begin{aligned} X &= b_1 e^{\lambda x} + b_2 e^{-\lambda x} \\ T &= b_3 e^{\lambda^2 kt} \end{aligned}\right] \qquad \ldots (7)$$

Case (III). Let $\mu = -\lambda^2, \lambda \neq 0$. Then (4) and (5) reduce to

$$X'' + \lambda^2 X = 0 \text{ and } T' = -\lambda^2 kT$$

On solving, we get

$$\left.\begin{aligned} X &= C_1 \cos\lambda x + C_2 \sin\lambda x \\ T &= C_3 e^{-\lambda^2 kt} \end{aligned}\right] \qquad \ldots (8)$$

Hence, the various possible solutions are

$$u(x, t) = A_1 x + A_2$$

$$u(x, t) = (B_1 e^{\lambda x} + B_2 e^{-\lambda x})e^{\lambda^2 kt}$$

and $$u(x, t) = (C_1 \cos\lambda x + C_2 \sin\lambda x)e^{-\lambda^2 kt}$$

42.17 SOLUTION OF TWO DIMENSIONAL HEAT EQUATION

The two dimensional heat equation is given by

$$\frac{1}{k}\frac{\partial u}{\partial t} = \left(\frac{\partial^2 u}{\partial x^2} + \frac{\partial^2 u}{\partial y^2}\right) \qquad \ldots (1)$$

Let us suppose (1) has solution of the form

$$u(x, y, t) = X(x)\,Y(y)\,T(t) \qquad \ldots (2)$$

Putting the values from (2) in (1), we get

$$X''YT + XY''T = \frac{1}{k}XYT'$$

$$\Rightarrow \qquad \frac{X''}{X} + \frac{Y''}{Y} = \frac{1}{k}\frac{T'}{T} \qquad \ldots (3)$$

Now, since x,y and t are independent variables, thus (3) is true if each term on each side is equal to a constant such that

$$\frac{X''}{X} = -n^2, \quad \frac{Y''}{Y} = -m^2 \quad \text{and} \quad \frac{T'}{kT} = -p^2 \qquad \ldots (4)$$

with $$n^2 + m^2 = p^2$$

The constants may be chosen such that the solution u has the property that $u \to 0$ as $t \to \infty$.

Solving (4), we get

$$X_n(x) = A_n \cos nx + B_n \sin nx\,; \quad Y_m(y) = C_m \cos my + D_m \sin my$$

and $$T_p(t) = E_p e^{-p^2 kt} = F_{nm} e^{-(n^2+m^2)\,kt}$$

Hence, a suitable solution of (1) is given by

$$u_{nm}(x, y, t) = F_{nm}(A_n \cos nx + B_n \sin nx)(C_m \cos my + D_m \sin my)\, e^{-(n^2+m^2)kt}$$

42.18 SOLUTION OF THREE DIMENSIONAL HEAT EQUATION

Three dimensional heat equation is given by

$$\frac{\partial^2 u}{\partial x^2} + \frac{\partial^2 u}{\partial y^2} + \frac{\partial^2 u}{\partial z^2} = \frac{1}{k}\frac{\partial u}{\partial t} \quad \text{... (1)}$$

Let solution of (1) be of the form

$$u(x, y, z, t) = X(x)\, Y(y) Z(z)\, T(t) \quad \text{... (2)}$$

where X, Y, Z, T are respectively the function of x, y, z and t alone.

Putting the values from (2) in (1), we get

$$X''YZT + XY''ZT + XYZ''T = \frac{1}{k} XYZT'$$

$$\Rightarrow \quad \frac{X''}{X} + \frac{Y''}{Y} + \frac{Z''}{Z} = \frac{1}{k}\frac{T'}{T} \quad \text{... (3)}$$

Now, since x, y, z and t are independent variables, equation (3) is true only when each term on each side is a constant such that

$$\frac{X''}{X} = -n^2, \quad \frac{Y''}{Y} = -m^2, \quad \frac{Z''}{Z} = -l^2 \text{ and } \frac{T'}{kT} = -p^2 \quad \text{... (4)}$$

with

$$n^2 + m^2 + l^2 = p^2$$

Further, we have to choose the constants such that solution $u(x, y, z, t)$ has the property that $u \to 0$ as $t \to \infty$.

On solving (4), we get

$$X_n(x) = A_n \cos nx + B_n \sin nx$$

$$Y_m(y) = C_m \cos my + D_m \sin my$$

$$Z_l(z) = E_l \cos lz + F_l \sin lz$$

and

$$T_p(t) = G_p\, e^{-p^2 kt} = H_{mnl}\, e^{-(n^2+m^2+l^2)kt}$$

which gives $u_{nml}(x, y, z, t) = H_{nml}(A_n \cos nx + B_n \sin nx)(C_m \cos my + D_m \sin my).(E_l \cos lz + F_l \sin lz)\, e^{-(n^2+m^2+l^2)kt}$

which are the required suitable solution of (1).

The general solution of (1) can be obtained by putting $u(x, y, z, t) = \sum_{n=1}^{\infty}\sum_{m=1}^{\infty}\sum_{l=1}^{\infty} u_{nml}(x, y, z, t)$.

42.19 TRANSMISSION LINE EQUATIONS

(i) Telegraph Equation : $\frac{\partial^2 V}{\partial x^2} = RC\frac{\partial V}{\partial t}$ and $\frac{\partial^2 i}{\partial x^2} = RC\frac{\partial i}{\partial t}$ (UPTU (CO)-2009)

(ii) Radio Equation : $\frac{\partial^2 V}{\partial x^2} = LC\frac{\partial^2 V}{\partial t^2}$ and $\frac{\partial^2 i}{\partial x^2} = LC\frac{\partial^2 i}{\partial t^2}$ (UPTU (CO)-2009)

where V = potential, i = current, C = capacitance and L = inductance.

Solved Examples

Example 1. *Determine the solution of one dimensional heat equation*

$$\frac{\partial u}{\partial t} = C^2 \frac{\partial^2 u}{\partial x^2}$$

subject to the boundary conditions $u(0, t) = 0$, $u(l, t) = 0$ $(t > 0)$ and the initial condition $u(x, 0) = x$, l being the length of the bar.

[UPTU-2006]

Solution. We have $\frac{\partial u}{\partial t} = C^2 \frac{\partial^2 u}{\partial x^2}$ ... (1)

Boundary conditions are

$$u(0, t) = 0$$

$$u(l, t) = 0 \ (t > 0)$$

$$u(x, 0) = x$$

On solving (1), we get

$$u = C_1\, e^{-p^2C^2t}(C_2 \cos px + C_3 \sin px) \quad \text{... (2)}$$

Putting $x = 0$ and $u = 0$ in (2), we get

$$0 = C_1\, e^{-p^2C^2t}(C_2) \Rightarrow C_2 = 0$$

Putting $C_2 = 0$ in (2), we get

$$u = C_1 e^{-p^2C^2t} C_3 \sin px \quad \text{... (3)}$$

Again putting $x = l$, $u = 0$ in (3), we get

$$0 = C_1 e^{-p^2C^2t} C_3 \sin pl$$

$\Rightarrow \quad \sin pl = 0 = \sin n\pi$

$\Rightarrow \quad pl = n\pi$

$\Rightarrow \quad p = \frac{n\pi}{l}$, n is any integer.

Hence, (3) becomes

$$u = C_1C_3 e^{-\frac{n^2C^2\pi^2}{l^2}t} . \sin\frac{n\pi x}{l}$$

$$= b_n e^{-\frac{n^2C^2\pi^2}{l^2}t} \sin\frac{n\pi}{l}x \quad \text{... (4)}$$

On putting $t = 0$ and $u = x$ in (4), we get

$$x = b_n \sin\frac{n\pi}{l}x$$

General solution is $x = \sum_{n=1}^{\infty} b_n \sin\frac{n\pi}{l}x$.

Now, $b_n = \frac{2}{l} x \sin\frac{n\pi x}{l} dx$

$$= \frac{2}{l}\left[x.\frac{l}{n\pi}\left(-\cos\frac{n\pi x}{l}\right) - (1)\left(-\frac{l^2}{n^2\pi^2}\sin\frac{n\pi x}{l}\right)\right]_0^l$$

$$= \frac{2}{l}\left[\left(l.\frac{l}{n\pi}(-\cos n\pi) + \frac{l^2}{n^2\pi^2}\sin n\pi\right) - 0\right]$$

$$= \frac{2}{l}\left[-\frac{l^2}{n\pi}(-1)^n\right] = (-1)^{n+1}.\frac{2l}{n\pi}.$$

Putting the value of b_n in (4), we get

$$u = \frac{2l}{\pi}\sum_{n=1}^{\infty}\frac{(-1)^{n+1}}{n}\sin\frac{n\pi x}{l} e^{-\frac{n^2C^2\pi^2}{l^2}t}.$$

Example 2. *An insulated rod of length l has its ends A and B maintained at 0°C and 100° C respectively until steady state conditions prevail. If B is suddenly reduced to 0°C and maintained at 0°C, find the temperature at a distance x from A at time t.*

[UPTU-2004, 05; GBTU (AG)-2011; UKTU-2011]

Solution. The initial temperature of the rod can be written as

$$u(x, t) = 0 + \frac{100}{l}x = \frac{100}{l}x$$

While in steady state, the temperature distribution can be written as

$$u(x, t) = 0 + \frac{0}{l}x = 0$$

To find u in the intermediate period, calculating time from the instant when the end temperature were changed

$$u = u_1(x) + u_2(x)$$

where $u_2(x)$ is temperature after a sufficient long time and $u_1(x, t)$ is the transient temperature distribution tending to zero as $t \to \infty$. Hence, $u_2(x) = 0$.

Also, $u_1(x, t)$ satisfies one-dimensional heat flow

$$C^2\frac{\partial^2 u}{\partial x^2} = \frac{\partial u}{\partial t}$$

Thus, $u = (C_1 \cos px + C_2 \sin px)e^{-C^2p^2t}$ ... (1)

On putting $x = 0, u = 0$ in (1), we get

$$0 = C_1 e^{-p^2C^2t} \Rightarrow C_1 = 0$$

On putting $C_1 = 0$ in (1), we get

$$u = C_2 \sin px e^{-C^2p^2t} \quad \text{... (2)}$$

On putting $x = l, u = 0$ in (2), we get

$$0 = C_2 \sin pl\, e^{-p^2C^2t}$$

$\Rightarrow \quad \sin pl = 0 = \sin n\pi$

$\Rightarrow \quad pl = n\pi$

$\Rightarrow \quad p = \frac{n\pi}{l}$

On putting the value of P in (2), we get

$$u = C_2 \sin\frac{n\pi x}{l} e^{-\frac{n^2\pi^2C^2}{l^2}t} \quad \text{... (3)}$$

On putting $t = 0$, $u = \frac{100}{l}x$ in (3), we get

$$\frac{100x}{l} = C_2 \sin\frac{n\pi x}{l}$$

$$C_2 = \frac{2}{l}\int_0^l \frac{100}{l}.x.\sin\frac{n\pi x}{l}dx$$

$$C_2 = \frac{200}{l^2}\int_0^l x\frac{\sin n\pi x}{l}dx$$

$$C_2 = \frac{200}{l^2}\left[-\frac{xl}{n\pi}\cos\frac{n\pi x}{l} - (-1)\frac{l^2}{n^2\pi^2}\sin\frac{n\pi x}{l}\right]_0^l$$

$$C_2 = \frac{200}{l^2}\left[-\frac{l^2}{n\pi}\cos n\pi\right]$$

$$\Rightarrow \quad C_2 = -\frac{200}{n\pi}(-1)^n$$

On putting the value of in (3), we get

$$u = -\frac{200}{n\pi}(-1)^n \sin\frac{n\pi x}{l} e^{-\frac{n^2\pi^2C^2}{l^2}t}$$

$$u = (-1)^n \frac{200}{n\pi}.\sin\frac{n\pi x}{l} e^{-\frac{n^2\pi^2C^2}{l^2}t}.$$

Example 3. *Find by method of separation of variables, the solution u(x, t) of separation of variables,*

$$\frac{\partial u}{\partial t} = 3\frac{\partial^2 u}{\partial x^2},\ t > 0,\ 0 < x < 2,\ u(0, t) = 0,$$

$$u(2, t) = 0,\ t > 0,\ u(x, 0) = x, 0 < x < 2.$$

Solution. The given equation is

$$\frac{\partial u}{\partial t} = 3\frac{\partial^2 u}{\partial x^2} \quad \text{... (1)}$$

Suppose (1) has the solution of the form

$$u = X(x)\, T(t) \quad \text{... (2)}$$

where X and T respectively the function of x and t alone. Putting the value from (2) in (1), we get

$$XT' = 3X''T$$

$$\Rightarrow \quad \frac{X''}{X} = \frac{T'}{3T} = p^2 \text{ (say)} \qquad \dots (3)$$

where p is a separation constant.

From (3), we can deduce that

$$\frac{X''}{X} = \frac{T'}{3T} = p^2$$

$$\Rightarrow \quad X'' - p^2X = 0 \qquad \dots (4)$$

$$T' - 3p^2T = 0 \qquad \dots (5)$$

On solving (4) and (5), we get

$$T = C_3e^{3p^2t} \text{ and } X = C_1e^{px} + C_2e^{-px}$$

which gives $u(x, t) = e^{3p^2t}(C_1'e^{px} + C_2'e^{-px})$

where $C_1' = C_1C_3$ and $C_2' = C_2C_3$

Now, we shall use the given boundary conditions

$$u(0, t) = e^{3p^2t}(C_1' + C_2') = 0$$

$$\Rightarrow \quad C_2' = -C_1'$$

$$\therefore \quad u(x, t) = C_1'e^{3p^2t}(e^{-px} - e^{-px})$$

$$\dots (6)$$

Further, $u(x, t) = e^{3p^2t}(C_1'e^{2p} + C_2'e^{-2p}) = 0$

$$\Rightarrow \quad C_1'e^{2p} + C_2'e^{-2p} = 0$$

$$\Rightarrow \quad C_1'e^{2p} - C_1'e^{-2p} = 0$$

$$\Rightarrow \quad C_1' = 0, \quad C_2' = 0$$

$\Rightarrow \quad u(x, t) = 0$, which is absurd.

Further, let $p = 0$, then we have

$$\frac{X''}{X} = \frac{T'}{3T} = 0$$

$$\Rightarrow \quad X'' = 0 \text{ and } T' = 0$$

$$\Rightarrow \quad X = C_1x + C_2, \quad T = C_3$$

$$\Rightarrow \quad u(x, t) = C_3(C_1x + C_2) = C_1'x + C_2$$

$\Rightarrow \quad u(x, t)$ is independent of t, which is not admissible.

Now, let p be negative, *i.e.*, $p = -k^2$.

Then, we have

$$\frac{X''}{X} = \frac{T'}{3T} = -k^2$$

$$\Rightarrow \quad X'' + k^2X = 0$$

and $T' + 3k^2T = 0$

On solving these equations, we have

$$X = C_1\cos kx + C_2\sin kx \text{ and } T = C_3e^{-3k^2t}$$

$$\Rightarrow \quad u(x, t) = e^{-3k^2t}(C_1'\cos kx + C_2'\sin kx)$$

Now, $u(x, t) = 0$

$$\Rightarrow \quad 0 = e^{-3k^2t}C_1' \Rightarrow C_1' = 0$$

$$\Rightarrow \quad u(x, t) = C_1'\sin kx\, e^{-3k^2t}$$

$$\Rightarrow \quad u(2, t) = 0 = C_2'\sin 2ke^{-3k^2t}$$

$$\Rightarrow \quad \sin 2k = 0$$

$$\Rightarrow \quad 2k = n\pi; \quad n = 1, 2, \dots\dots$$

$$k = \frac{n\pi}{2}; \quad n = 1, 2, \dots\dots$$

Therefore, $u(x, t) = C_2'\sin\frac{n\pi x}{2}e^{-3n^2\pi^2t/4}$

$$= b_n\sin\frac{n\pi x}{2}e^{-3n^2\pi^2t/4}$$

$$= u_n(x, t) \text{ (say)}$$

Hence, the general solution of (1) is given by

$$u(x, t) = \sum_{n=1}^{\infty} b_n\sin\frac{n\pi x}{2}e^{-3n^2\pi^2t/4}.$$

Example 4. *A rod of length l with insulated sides is initially at a uniform temperature μ_0. Its ends are suddenly cooled to 0°C and are kept at that temperature. Show that the temperature function $u(x, t)$ is given by*

$$u(x, t) = \sum_{n=1}^{\infty} b_n\sin\frac{n\pi x}{l}e^{-C^2\pi^2n^2t/l^2},$$

where b_n is given by $u_0 = \sum_{n=1}^{\infty} b_n\sin\frac{n\pi x}{l}$

(GBTU-2010, 11)

Solution. The heat equation is given by

$$\frac{\partial u}{\partial t} = C^2\frac{\partial^2 u}{\partial x^2} \qquad \dots (1)$$

Suppose that solution of (1) is of the form

$$u(x, t) = X(x)\,T(t) \qquad \dots (2)$$

where X and T are respectively the functions of x and t alone. Putting the value from (2) in (1), we get

$$X\frac{dT}{dt} = C^2T\frac{d^2X}{dx^2}$$

$$\Rightarrow \quad \frac{1}{C^2T}\frac{dT}{dt} = \frac{1}{X}\frac{d^2X}{dx^2} = -p^2 \text{ (say)} \qquad \dots (3)$$

where p is a separation constant.

From (3), we can find

$$\frac{1}{C^2T}\frac{dT}{dt} = -p^2$$

$$\Rightarrow \quad \frac{dT}{dt} = -p^2C^2T \qquad \dots (4)$$

$$\text{and} \quad \frac{1}{X}\frac{d^2X}{dx^2} = -p^2 \qquad \dots (5)$$

Solving (4) and (5), we get

$$T = C_1e^{-p^2C^2t} \text{ and } X = (C_2\cos px + C_3\sin px)$$

which gives

$$u(x, t) = e^{-p^2C^2t}(C_1'\cos px + C_2'\sin px)$$

Now, $u(0, t) = 0$

$$\Rightarrow \quad 0 = C_1'e^{-p^2C^2t} \Rightarrow C_1' = 0$$

$$\Rightarrow \quad u(x, t) = e^{-p^2C^2t}\,.\,C_2'\sin px$$

Now,

$$u(l, t) = 0 \Rightarrow 0 = e^{-p^2C^2t}(C_1'\cos pl + C_2'\sin pl)$$

$$\Rightarrow \quad \sin pl = 0 \quad \Rightarrow \quad p = \frac{n\pi}{l}; \quad n = 1, 2, \dots\dots$$

Therefore,

$$u(x, t) = e^{-p^2C^2t} \sin \frac{n\pi x}{l} C_2'$$

$$= e^{-n^2\pi^2C^2t/l^2} \sin \frac{n\pi x}{l} C_2'$$

$$= b_n\, e^{-n^2\pi^2C^2t/l^2} \sin \frac{n\pi x}{l}$$

Hence, the general solution is given by

$$u(x, t) = \sum_{n=1}^{\infty} u_n(x, t)$$

$$= \sum_{n=1}^{\infty} b_n e^{-n^2\pi^2C^2t/l^2} \sin \frac{n\pi x}{l}$$

Finally, using $u = u_0$ when $t = 0$, we get

$$u_0 = \sum_{n=1}^{\infty} b_n \sin \frac{n\pi x}{l}.$$

Example 5. *Find the temperature in a bar of length 2 whose ends are kept at zero and lateral surface insulated if the initial temperature is* $\sin\frac{\pi x}{2} + 3\sin\frac{5\pi x}{2}$.

[MTU-2011, UPTU (CO)-2007, 09]

Solution. Let $u(x, t)$ be the temperature in the bar. The boundary conditions are

$$u(0, t) = 0 = u(2, t) \text{ for any } t \quad \ldots (1)$$

The initial condition is

$$u(x,0) = \sin\frac{\pi x}{2} + 3\sin\frac{5\pi x}{2} \quad \ldots (2)$$

One dimensional heat flow equation is

$$\frac{\partial u}{\partial t} = C^2 \frac{\partial^2 u}{\partial x^2} \quad \ldots (3)$$

Its solution is

$$u(x,t) = (C_1 \cos px + C_2 \sin px)C_3 e^{-c^2p^2t} \quad \ldots (4)$$

$$u(0,t) = 0 = C_1C_3e^{-C^2p^2t} \quad \text{(On using (1))}$$

$$\Rightarrow C_1 = 0$$

$\therefore$ From (4) $u(x,t) = C_2C_3 \sin px\, e^{-C^2p^2t}$ ... (5)

$$u_1(2,t) = 0 = C_2C_3 \sin 2p\, e^{-C^2p^2t}$$

(On using (1))

$$\Rightarrow \quad \sin 2p = 0 = \sin n\pi$$

$$\therefore \quad p = \frac{n\pi}{2}, n \in I$$

Hence from (5)

$$u(x,t) = b_n \sin\frac{n\pi x}{2} e^{\frac{-n^2\pi^2C^2t}{4}}$$

$$(\because C_2C_3 = b_n)$$

The most general solution is

$$u(x,t) = \sum_{n=1}^{\infty} b_n \sin\frac{n\pi x}{2} e^{\frac{-n^2\pi^2C^2t}{4}} \quad \ldots (6)$$

$$u(x,0) = \sin\left(\frac{\pi x}{2}\right) + 3\sin\left(\frac{5\pi x}{2}\right)$$

$$= \sum_{n=1}^{\infty} b_n \sin\frac{n\pi x}{2}$$

$$= b_1 \sin\left(\frac{\pi x}{2}\right)$$

$$+ b_2 \sin\left(\frac{2\pi x}{2}\right) + \ldots + b_5 \sin\left(\frac{5\pi x}{2}\right) + \ldots$$

Comparing, we get $b_1 = 1$ and $b_5 = 3$

Hence from (6),

$$u(x,t) = \sin\left(\frac{\pi x}{2}\right) e^{-\pi^2C^2t/4}$$

$$+ 3\sin\left(\frac{5\pi x}{2}\right) e^{-25\pi^2C^2t/4}$$

Example 6. *Solve* $\frac{\partial^2 u}{\partial x^2} + \frac{\partial^2 u}{\partial y^2} = 0$, *which satisfies the conditions* $u(0, y) = u(l, y) = u(x, 0) = 0$ *and* $u(x,a) = \sin\frac{n\pi x}{l}$.

(UPTU-2004, (CO)-2009, GBTU (AG)-2012)

Solution. Consider the heat flow in a metal plate of uniform thickness in the directions parallel to length and breadth of the plate. There is no heat flow along the normal to the plane of the rectangle.

Let $u(x,y)$ be the temperature at any point (x,y) of the plate at time t is given by

$$\frac{\partial u}{\partial t} = C^2\left(\frac{\partial^2 u}{\partial x^2} + \frac{\partial^2 u}{\partial y^2}\right)$$

In the steady state, u does not change with t.

$$\frac{\partial u}{\partial t} = 0$$

Let $\quad u = X(x) \,.\, Y(y) \quad \ldots (1)$

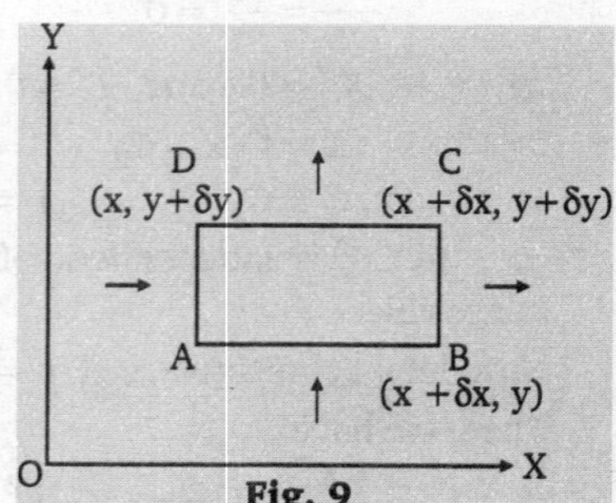

Fig. 9

Putting the values of $\frac{\partial^2 u}{\partial x^2}$ and $\frac{\partial^2 u}{\partial y^2}$ in (1),

we have $X''Y + XY'' = 0$ ... (2)

$$\Rightarrow \quad \frac{X''}{X} = \frac{-Y''}{Y} = -p^2 \quad \text{(say)}$$

$$D^2X = -p^2X \qquad \Rightarrow D^2Y = p^2Y$$

$$\Rightarrow \quad D^2X + p^2X = 0 \quad \Rightarrow \quad D^2Y - p^2Y = 0$$

$$\Rightarrow \quad (D^2 + p^2)X = 0 \quad \Rightarrow \quad (D^2 - p^2)Y = 0$$

A.E. is $D^2 + p^2 = 0$ A.E. is $D^2 - p^2 = 0$

$$\Rightarrow \quad D^2 = -p^2 \quad \Rightarrow \quad D^2 = p^2$$

$\Rightarrow \quad D = \pm ip \quad \Rightarrow \quad D = \pm p$

$X = C_1 \cos px + C_2 \sin px \quad Y = C_3 e^{py} + C_4 e^{-py}$

Putting the values of X and Y in (1), we have

$$u = (C_1 \cos px + C_2 \sin px)(C_3 e^{py} + C_4 e^{-py}) \quad \text{... (3)}$$

Putting $x = 0, u = 0$ in (3), we have

$$0 = C_1(C_3 e^{py} + C_4 e^{-py}) \Rightarrow C_1 = 0$$

(3) is reduced to $u = C_2 \sin px(C_3 e^{py} + C_4 e^{-py})$ ... (4)

On putting $x = l,\ u = 0$ in (4), we have

$$0 = C_2 \sin pl(C_3 e^{py} + C_4 e^{-py}) \Rightarrow C_2 \neq 0$$

$\therefore \quad \sin pl = 0 = \sin n\pi \Rightarrow Pl = n\pi$

or $\quad P = \dfrac{n\pi}{l}$

Now, (4) becomes

$$u = C_2 \sin\frac{n\pi x}{l}\left[C_3 e^{\frac{n\pi y}{l}} + C_4 e^{-\frac{n\pi y}{l}}\right] \quad \text{... (5)}$$

On putting $u = 0$ and $y = 0$ in (5), we have

$$0 = C_2 \sin\frac{n\pi x}{l}(C_3 + C_4)$$

$C_3 + C_4 = 0$

or $\quad C_3 = -C_4$

(5) becomes $u = C_2 C_3 \sin\dfrac{n\pi x}{l}\left(e^{\frac{n\pi y}{l}} - e^{-\frac{n\pi y}{l}}\right)$

On putting $y = a$ and $u = \sin\dfrac{n\pi x}{l}$ in (6), we have

$$\sin\frac{n\pi x}{l} = C_2 C_3 \sin\frac{n\pi x}{l}\left(e^{\frac{n\pi a}{l}} - e^{-\frac{n\pi a}{l}}\right)$$

i.e., $\quad C_2 C_3 = \dfrac{1}{e^{\frac{n\pi a}{l}} - e^{-\frac{n\pi a}{l}}}$

Putting this value in (6), we have

$$u = \sin\frac{n\pi x}{l} = \frac{e^{\frac{n\pi y}{l}} - e^{-\frac{n\pi y}{l}}}{e^{\frac{n\pi a}{l}} - e^{-\frac{n\pi a}{l}}}$$

or $$u = \sin\frac{n\pi x}{l}\,\frac{\sinh\frac{n\pi y}{l}}{\sin\frac{n\pi a}{l}}.$$

Example 7. *A thin rectangular plate whose surface is impervious to heat flow, has at t = 0 an arbitrary distribution of temperature f(x, y). If four edges x = 0, x = a, y = 0, y = b are kept at zero temperature, find the temperature at a point of the plate as t increases.* [UPTU-2002]

Solution. We know that the two dimensional heat equation is

$$\frac{\partial u}{\partial t} = C^2\left(\frac{\partial^2 u}{\partial x^2} + \frac{\partial^2 u}{\partial y^2}\right) \quad \text{... (1)}$$

As per given, the initial temperature of the plate is $f(x, y)$ and the temperature of the four edges of the plate are kept at 0°.

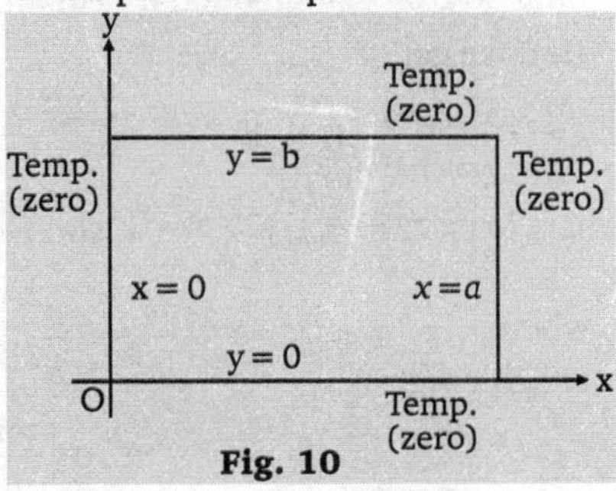

Fig. 10

Therefore, the required boundary conditions are

(i) $u(0, y, t) = 0$, (ii) $u(a, y, t) = 0$,

(iii) $u(x, 0, t) = 0$, (iv) $u(x, b, t) = 0$.

Also, the initial condition is given by

$$u(x, y, 0) = f(x, y) \quad \text{... (2)}$$

Suppose that solution of (1) is of the form

$$u(x, y, t) = X(x)\,Y(y)\,T(t) \quad \text{... (3)}$$

where X, Y and T are respectively the functions of x, y and t alone. Putting the values of (3) in (1), we get

$$\frac{1}{C^2 T} = \frac{1}{X}\frac{d^2X}{dx^2} + \frac{1}{Y}\frac{d^2Y}{dy^2} \quad \text{... (4)}$$

If (3) satisfies (1), we have the following three possibilities

(a) $\dfrac{1}{X}\dfrac{d^2X}{dx^2} = 0, \quad \dfrac{1}{Y}\dfrac{d^2Y}{dy^2} = 0, \quad \dfrac{1}{C^2T}\dfrac{dT}{dt} = 0$

(b) $\dfrac{1}{X}\dfrac{d^2X}{dx^2} = p_1^2, \dfrac{1}{Y}\dfrac{d^2Y}{dy^2} = p_2^2, \dfrac{1}{C^2T}\dfrac{dT}{dt} = p^2$

(c) $\dfrac{1}{X}\dfrac{d^2X}{dx^2} = -p_1^2, \dfrac{1}{Y}\dfrac{d^2Y}{dy^2} = -p_2^2, \dfrac{1}{C^2T}\dfrac{dT}{dt} = -p^2$

where $p^2 = p_1^2 + p_2^2$.

It can be easily verified that differential equation (c) only gives the solution. In this case, the general solution is given by

$X = A_1 \cos p_1 x + B_1 \sin p_1 x$;

$Y = A_2 \cos p_2 y + B_2 \sin p_2 y$; $\quad T = A_3\, e^{-C^2p^2t}$

Therefore

$$u(x, y, t) = (A_1 \cos p_1 x + B_1 \sin p_1 x)\,(A_2' \cos p_2 y + B_2' \sin p_2 y)\,e^{-C^2p^2t} \quad \text{... (5)}$$

Using boundary condition (i), we get

$$u(0, y, t) = A_1(A_2' \cos p_2 y + B_2' \sin p_2 y)\,e^{-C^2p^2t} = 0$$

$\Rightarrow \quad A_1 = 0$

Now, using boundary condition (ii), we get

$$u(a, y, t) = B_1 \sin p_1 a(A_2 \cos p_2 y + B_2' \sin p_2 y)\,e^{-C^2p^2t} = 0$$

$\Rightarrow \quad \sin p_1 a = 0 \Rightarrow p_1 a = m\pi$

$$p_1 = \frac{m\pi}{a};\ m = 1, 2, 3, \ldots$$

Similarly, by using boundary condition (iii) and (iv), we get

$$A_2' = 0 \text{ and } p_2 = \frac{n\pi}{b},\ n = 1,2,3,...$$

Thus, we have

$$u_{mn}(x,y,t) = A_{mn}\, e^{-C^2 p_{mn}^2 t} \cdot \sin\frac{m\pi}{a} \sin\frac{n\pi}{b} y$$

where, $p^2 = p_{mn}^2 = \pi^2\left(\frac{m^2}{a^2} + \frac{n^2}{b^2}\right)$

which gives

$$u(x, y, t) = \sum_{m=1}^{\infty}\sum_{n=1}^{\infty} A_{mn}\, e^{-C^2 p_{mn}^2 t} \cdot \sin\frac{m\pi}{a} x \sin\frac{n\pi}{b} y$$

Finally, to find the solution which satisfies the initial conditions also, we proceed as follows :

$$u(x, y, 0) = \sum_{m=1}^{\infty}\sum_{n=1}^{\infty} A_{mn}\, e^{-C^2 p_{mn}^2 t} \cdot \sin\frac{m\pi}{a} x \sin\frac{n\pi}{b} y = f(x, y)$$

The LHS is the double Fourier sine series of $f(x, y)$. Therefore,

$$A_{mn} = \frac{2}{a} \cdot \frac{2}{b} \int_{x=0}^{a} \int_{y=0}^{b} f(x, y) \sin\frac{m\pi}{a} x \sin\frac{n\pi}{b} y\, dx\, dy$$

Example 8. *A uniform rod, 20 cm in length is insulated over its sides. Its ends are kept at 0°C. Its initial temperature is* $\sin\left(\frac{\pi x}{20}\right)$ *at a distance x from an end, find temperature u(x, t) at time t. Given that* $\frac{\partial u}{\partial t} = a^2\left(\frac{\partial^2 u}{\partial x^2}\right)$.

Solution. The given equation is

$$\frac{\partial u}{\partial t} = a^2\left(\frac{\partial^2 u}{\partial x^2}\right) \quad ...(1)$$

As per given, the boundary conditions are

$$u(0, t) = u(20, t) = 0,\ \forall t \quad ...(2)$$

$$u(x, 0) = \sin\left(\frac{\pi x}{20}\right) \quad ...(3)$$

Suppose that the solution of (1) is of the form

$$u(x, t) = X(x)\, T(t) \quad ...(4)$$

Putting the values from (4) in (1), we get

$$\frac{X''}{X} = \frac{T'}{a^2 T} = \mu \ (say) \quad ...(5)$$

where μ is a separation constant.

From (5), we can deduce that

$$X'' - \mu X = 0 \quad ...(6)$$

and

$$T' = \mu a^2 T \quad ...(7)$$

The new boundary conditions are

$$X(0) = 0 \text{ and } X(20) = 0 \quad ...(8)$$

Now, we want to solve (6) using boundary conditions (8).

We shall discuss the following three cases :

Case (I). Let $\mu = 0$. In this case, solution of (6) is given by

$$X(x) = Ae^{\lambda x} + Be^{-\lambda x}$$

Using (8), we have $B = 0$

and $20A + B = 0 \Rightarrow A = 0$

$\Rightarrow X = 0$

$\Rightarrow u = 0$, which does not satisfy (3). Hence, we reject this case.

Case (II). Let $\mu = \lambda^2, \lambda \neq 0$. In this case, solution of (6) is given by

$$X(x) = Ae^{\lambda x} + Be^{-\lambda x}$$

Using boundary condition (8), we have

$$A + B = 0 \text{ and } Ae^{20\lambda} + De^{-20\lambda} = 0$$

On solving we get $A = B = 0 \Rightarrow X = 0 \Rightarrow u = 0$.

Therefore, we reject this case also.

Case (III). Let $\mu = -\lambda^2, \lambda \neq 0$. In this case, solution of (6) is given by

Using boundary condition (8), we get

$$A = 0 \text{ and } A\cos 20\lambda + B\sin 20\lambda = 0$$

$\Rightarrow \sin 20\lambda = 0$

$\Rightarrow 20\lambda = n\pi,\ n = 1, 2, ...$

$$\Rightarrow \lambda = \frac{n\pi}{20},\ n = 1, 2, ... \quad ...(9)$$

Thus, the non-zero solution of equation (6) is given by

$$X_n(x) = B_n \sin\left(\frac{n\pi x}{20}\right)$$

Using (9) in (7), we get

$$\frac{dT}{T} = -\frac{n^2\pi^2 a^2}{400} dt$$

whose general solution is given by

$$T_n(t) = D_n\, e^{-(n^2\pi^2 a^2/400)t}$$

Therefore, $u_n(x, t) = X_n(x)\, T_n(t)$

$$= E_n \sin\left(\frac{n\pi x}{20}\right) e^{-\frac{(n^2\pi^2 a^2)}{400}t}$$

are solutions of (1) satisfying (2).

The more general solution of (1) is given by

$$u(x,t) = \sum_{n=1}^{\infty} u_n(x,t) = \sum_{n=1}^{\infty} E_n \sin\left(\frac{n\pi x}{20}\right) e^{-\frac{(n^2\pi^2 a^2)}{400}t}$$

Putting $t = 0$ and using boundary condition (3), we get

$$\sin\frac{\pi x}{20} = \sum_{n=1}^{\infty} E_n \sin\left(\frac{n\pi x}{20}\right)$$

which is a Fourier sine series. Thus the constants E_n are given by

$$E_n = \frac{1}{20}\int_0^{20} \sin\frac{\pi x}{20} \sin\left(\frac{n\pi x}{20}\right) dx,\ n = 1,2,3,... \quad ...(10)$$

Here, we discuss the following two cases :

Case I. If $n \neq 1$, then (10) gives

$$E_n = \frac{1}{20}\int_0^{20}\left[\cos\frac{(n-1)\pi x}{20} - \cos\frac{(n+1)\pi x}{20}\right]dx$$

$$= \frac{1}{20}\left[\frac{20}{(n-1)\pi}\sin\frac{(n-1)\pi x}{20} - \frac{20}{(n+1)\pi}\sin\frac{(n+1)\pi x}{20}\right]_0^{20} = 0$$

$\Rightarrow E_n = 0$ for $n = 2, 3, 4, \ldots$

Case (II). If $n = 1$, then (10) gives

$$E_1 = \frac{1}{20}\int_0^{20}\left(2\sin^2\frac{\pi x}{20}\right)dx$$

$$= \frac{1}{20}\int_0^{20}\left(1-\cos\frac{2\pi x}{20}\right)dx$$

$$= \frac{1}{20}\left[x - \frac{10}{x}\sin\frac{\pi x}{10}\right]_0^{20} = 1$$

Finally, the general solution is given by

$$u(x,t) = E_1 \sin\frac{\pi x}{20}e^{-\frac{(\pi^2a^2)}{400}t} = \sin\frac{\pi x}{20}e^{-\frac{(\pi^2a^2)}{400}t}$$

Example 9. *Solve the one-dimensional heat equation* $\frac{\partial^2 u}{\partial x^2} = \frac{1}{k}\frac{\partial u}{\partial t}$ *in the region* $0 \le x \le \pi,\ t > 0$ *when*

(i) *u remains finite as* $t \to \infty$,

(ii) $u = 0$ *if* $x = 0$ *or* π, $\forall t$

(iii) *At* $t = 0$, $u = x$ *for* $0 \le x \le \pi/2$ *and*

$$u = \pi - x \text{ for } \frac{\pi}{2} < x \le \pi.$$

Solution. The given equation is

$$\frac{\partial u}{\partial t} = a^2\left(\frac{\partial^2 u}{\partial x^2}\right) \quad \ldots (1)$$

As per given, we have

$u(0,t) = u(20,t) = 0$, $\forall t$ finite quantity as $t \to \infty$ ... (2)

and boundary conditions are

$$u(0,t) = u(\pi,t) = 0;\ \forall t \quad \ldots (3)$$

Also, the initial condition is given by

$$u(x,0) = \begin{cases} x\ ;\ \text{when } 0 \le x \le \pi/2 \\ \pi - x\ ;\ \text{when } \frac{\pi}{2} \le x \le \pi \end{cases} \quad \ldots (4)$$

Suppose that solution of (1) is of the form

$$u(x,t) = X(x)\,T(t) \quad \ldots (5)$$

where X and T are respectively the functions of x and t alone. Putting the values from (5) in (1), we get

$$\frac{X''}{X} = \frac{T'}{kT} = \lambda \text{ (say)} \quad \ldots (6)$$

where λ is a separation constant.

In view of (2), we choose, $\lambda = -n^2$, $n \neq 0$. Then (6) reduces to

$$\frac{X''}{X} = -n^2 \Rightarrow (D^2 + n^2)X = 0$$

and $\frac{T'}{kT} = -n^2 \Rightarrow \frac{dT}{T} = -n^2kdT$

On solving, we get

$$X_n(x) = A_n\cos nx + B_n\sin nx$$

and $T_n(t) = C_n e^{-n^2kt}$

Therefore

$$u(x,t) = (D_n\cos nx + E_n\sin nx)\,e^{-n^2kt} \quad \ldots (7)$$

Putting $x = 0$ and $x = \pi$ in (7) and using (3), we get $0 = D_n$ and $0 = D_n\cos n\pi + E_n\sin n\pi$

$\Rightarrow D_n = 0$ and $\sin n\pi = 0$

Now, $\sin n\pi = 0 \Rightarrow n$ must be an integer.

Therefore, from (7), we have

$$u_n(x,t) = E_n\sin nx \,.\, e^{-n^2kt},\quad n = 1,2,3,\ldots \quad \ldots (8)$$

which are solutions of (1) satisfying the boundary condition (3).

Now, consider the series

$$u(x,t) = \sum_{n=1}^{\infty} u_n(x,t) = \sum_{n=1}^{\infty} E_n\sin nx\, e^{-n^2kt} \quad \ldots (9)$$

$$u(x,0) = \sum_{n=1}^{\infty} E_n\sin nx,\ n = 1,2,3,\ldots \quad \ldots (10)$$

which is a half range Fourier sine series in $(0, \pi)$, therefore E_n is given by

$$E_n = \frac{2}{\pi}\int_0^{\pi} u(x,0)\sin nx\,dx$$

$$= \frac{2}{\pi}\left[\int_0^{\pi/2} u(x,0)\sin nx\,dx + \int_{\pi/2}^{\pi} u(x,0)\sin nx\,dx\right]$$

$$= \int_0^{\pi/2}\frac{2\pi}{x}\sin nxdx + \int_{\pi/2}^{\pi}\left(\frac{2}{\pi}\right)(\pi - x)\sin nxdx$$

$$= \left[\left(\frac{2x}{\pi}\right)\left(-\frac{\cos nx}{n}\right) - \left(\frac{2}{\pi}\right)\left(-\frac{\sin nx}{x^2}\right)\right]_0^{\pi/2} + \left[\left(\frac{2}{\pi}(\pi - x\right)\left(-\frac{\cos nx}{n}\right) - \left(-\frac{2}{\pi}\right)\left(-\frac{\sin nx}{n^2}\right)\right]_{\pi/2}^{\pi}$$

$$= -\left(\frac{1}{n}\right)\cos\left(\frac{n\pi}{2}\right) + \left(\frac{2}{\pi n^2}\right)\sin\left(\frac{n\pi}{2}\right) + \frac{1}{n}\cos\left(\frac{n\pi}{2}\right) + \frac{2}{\pi n^2}\sin\left(\frac{n\pi}{2}\right)$$

Thus $E_n = \frac{4}{\pi n^2}\sin\frac{n\pi}{2}$

$$= \begin{cases} 0 & ;\ \text{if } n = 2m \text{ and } m = 1,2,3,\ldots \\ \frac{4(-1)^{m+1}}{\pi(2m-1)^2} & ;\ \text{if } n = 2m-1 \text{ and } m = 1,2,3,\ldots \end{cases}$$

Hence, the required solution is given by

$$u(x,t)=\frac{4}{\pi}\sum_{m=1}^{\infty}\frac{(-1)^{m+1}}{(2m-1)^2}\sin(2m-1)\,x\,e^{(2m-1)^2kt}\quad m=1,2,\ldots$$

Example 10. *Solve the boundary value problem* $\frac{\partial^2 u}{\partial x^2}=\frac{l}{k}\frac{\partial u}{\partial t}$ *satisfying the condition* $u(0,t)=u(l,t)=0$ *and* $u(x,0)=lx-x^2$.

Solution. Proceeding same as in example (4) by taking $a=l$ and $f(x)=u(x,0)=lx-x^2$, we get

$$u(x,t)=\sum_{n=1}^{\infty}E_n\sin\left(\frac{n\pi x}{l}\right)e^{-(n^2\pi^2kt)/l^2}\quad\ldots(1)$$

where $E_n=\frac{2}{l}\int_0^l f(x)\sin\frac{n\pi x}{l}dx$

$$=\frac{2}{l}\int_0^l(lx-x^2)\sin\frac{n\pi x}{l}dx$$

$$=\frac{2}{l}\left[(lx-x^2)\left\{\frac{-\cos(n\pi x/l)}{n\pi/l}\right\}-(l-2x)\left\{\frac{-\sin(n\pi x/l)}{(n\pi)^2/l^2}\right\}+(-2)\left\{\frac{\cos(n\pi x)/l}{(n\pi)^3/l^3}\right\}\right]_0^l$$

$$=\frac{2}{l}\left[-\left(\frac{2l^3}{n^3\pi^3}\right)\cos n\pi+\left(\frac{2l^3}{n^3\pi^3}\right)\right]$$

$$=\frac{4l^2}{n^3\pi^3}[1-(-1)^n]$$

Thus, $E_n=\begin{cases}(8l^2)/(2m-1)^3\pi^3; & \text{if } n=2m-1,\ m=1,2,3,\ldots\\ 0; & \text{if } n=2m(\text{even}),\ \text{where } m=1,2,3,\ldots\end{cases}$

Hence, $u(x,t)=\frac{8l^2}{\pi^3}\sum_{m=1}^{\infty}\frac{1}{(2m-1)^3}\sin\frac{(2m-1)\pi x}{l}e^{-\left[(2m-1)^2\pi^2kt\right]/l^2}$

Example 11. *Solve the boundary value problem* $\frac{\partial^2 u}{\partial x^2}=\frac{l}{k}\frac{\partial u}{\partial t}$ *satisfying the conditions* $u(0,t)=u(l,t)=0$ and $u(x,0)=x$ *when* $0\le x\le l/2$ $u(x,0)=(l-x)$ *when* $l/2\le x\le l$.

Solution. Proceeding same as in Example (4) by taking $a=l$ and we get

$$u(x,t)=\sum_{n=1}^{\infty}E_n\sin\left(\frac{n\pi x}{l}\right)e^{-(n^2\pi^2kt)/l^2}\quad\ldots(1)$$

It is also given that

$$u(x,0)=\begin{cases}x; & \text{when } 0\le x\le l/2\\ l-x; & \text{when } l/2\le x\le l\end{cases}\quad\ldots(2)$$

Putting $t=0$ in (1), we get

$$u(x,0)=\sum_{n=1}^{\infty}E_n\sin\frac{n\pi x}{l}$$

which is a half range Fourier sine series in $(0,l)$. Thus, E_n is given by

$$E_n=\frac{2}{l}\int_0^l u(x,0)\sin\frac{n\pi x}{l}dx$$

$$=\frac{2}{l}\left[\int_0^{l/2}u(x,0)\sin\frac{n\pi x}{l}dx+\int_{l/2}^{l}u(x,0)\sin\frac{n\pi x}{l}dx\right]$$

$$=\int_0^{l/2}\frac{2x}{l}\sin\frac{n\pi x}{l}dx+\int_{l/2}^{l}\frac{2}{l}(l-x)\sin\frac{n\pi x}{l}dx$$

$$=\left[\left(\frac{2x}{l}\right)\left(-\frac{\cos(n\pi x)/l}{(n\pi)/l}\right)-\left(\frac{2}{l}\right)\left(-\frac{\sin(n\pi x)/l}{(n\pi)^2/l^2}\right)\right]_0^{l/2}+\left[\left(\frac{2(l-x)}{l}\right)\left(-\frac{\cos(n\pi x)/l}{(n\pi)/l}\right)-\left(-\frac{2}{l}\right)\left(-\frac{\sin(n\pi x)/l}{(n\pi)^2/l^2}\right)\right]_{l/2}^{l}$$

$$=-\left(\frac{l}{n\pi}\right)\cos\left(\frac{n\pi}{2}\right)+\left(\frac{2l}{n^2\pi^2}\right)\sin\left(\frac{n\pi}{2}\right)+\left(\frac{l}{n\pi}\right)\cos\left(\frac{n\pi}{2}\right)+\left(\frac{2l}{n^2\pi^2}\right)\sin\frac{n\pi}{2}$$

Thus, $E_n=\frac{4l}{n^2\pi^2}\sin\frac{n\pi}{2}$

$$=\begin{cases}0; & \text{if } n=2m \text{ and } m=1,2,3,\ldots\\ \frac{4l}{(2m-1)^2\pi^2}; & \text{if } n=2m-1,\ m=1,2,\ldots\end{cases}$$

Hence, by (1), we have

$$u(x,t)=\frac{4l}{\pi^2}\sum_{m=1}^{\infty}\frac{1}{(2m-1)^2}\sin\frac{(2m-1)\pi x}{l}e^{-\left[(2m-1)^2\pi^2kt\right]/l^2}$$

Example 12. *Find the solution of one dimensional heat equation satisfying the following boundary conditions*

(i) *T is bounded as* $t\to\infty$

(ii) $\left.\frac{\partial T}{\partial x}\right|_{x=0}=0\ ;\ \forall t$

(iii) $\left.\frac{\partial T}{\partial x}\right|_{x=a}=0\ ;\ \forall t$

(iv) $T(x,0)=x(a-x);\ 0<x<a$

Solution. The general acceptable solution of the given partial differential equation is :

$$T(x,t)=\exp(-\alpha\lambda^2t)(A\cos\lambda x+B\sin\lambda x)$$

$$\Rightarrow\frac{\partial T}{\partial x}=\exp(-\alpha\lambda^2t)(-A\lambda\sin\lambda x+B\lambda\cos\lambda x)\quad\ldots(1)$$

Using boundary condition (ii) in (1), we get $B = 0$.
By using boundary condition (iii), we get

$\sin \lambda a = 0 \Rightarrow \lambda a = n\pi\,;\ n = 0, 1, 2,...$

Therefore, the more general solution is given by

$$T(x,t) = \sum A_n \exp(-\alpha\lambda^2 t)\cos \lambda x$$

$$= \sum_{n=0}^{\infty} A_n \exp\left[-\alpha\left(\frac{n\pi}{a}\right)^2 t\right]\cos\left(\frac{n\pi}{a}\right)x$$

Using boundary condition (iv), we have

$$T(x,0) = x(a-x)$$

$$= A_0 + \sum_{n=1}^{\infty} A_n \exp\left[-\alpha\left(\frac{n\pi}{a}\right)^2 t\right]\cos\left(\frac{n\pi}{a}\right)x$$

where, $A_0 = \frac{2}{\pi}\int_0^a (ax - x^2)\,dx = \frac{a^2}{6}$

$$A_n = \frac{2}{\pi}\int_0^a (ax - x^2)\cos\left(\frac{n\pi x}{a}\right)dx$$

$$= \frac{2a^2}{n^2\pi^2}(1+\cos n\pi) = \frac{2a^2}{n^2\pi^2}[1+(-1)^n]$$

$$\Rightarrow \quad A_n = \begin{cases} -\dfrac{4a^2}{n^2\pi^2}\,; & \text{if } n \text{ is even} \\ 0\,; & \text{if } n \text{ is odd} \end{cases}$$

Hence, the required solution is given by

$$T(x,t) = \frac{a^2}{6} - \frac{4a^2}{\pi^2}$$

$$\sum_{n=2,4,...}^{\infty} \frac{1}{n^2}\cos\left(\frac{n\pi}{a}\right)x.\exp\left[-\alpha\left(\frac{n\pi}{a}\right)^2 t\right]$$

Example 13. *Find the current i and voltage V in a transmission line of length l, t seconds after the ends are suddenly grounded given that* $i(x,0) = i_0$, $V(x,0) = V_0 \sin\left(\frac{\pi x}{l}\right)$ *and that R and G are negligible.* [UPTU-2008]

Solution. We have $\frac{\partial^2 V}{\partial x^2} = LC\frac{\partial^2 V}{\partial t^2}$

Let $V = XT$, where X and T are the functions of x and t respectively.

$$\frac{\partial^2 V}{\partial x^2} = T\frac{\partial^2 X}{\partial x^2} \text{ and } \frac{\partial^2 V}{\partial t^2} = X\frac{d^2 T}{dt^2}$$

or $\frac{\frac{d^2X}{dx^2}}{X} = LC\frac{\frac{d^2T}{dt^2}}{T} = -p^2$ (say)

Since the initial conditions suggest the values of V and i are periodic functions.

$$X = C_1 \cos px + C_2 \sin px$$

and $T = C_3 \cos\frac{pt}{\sqrt{LC}} + C_4 \sin\frac{pt}{\sqrt{LC}}$

so $V = (C_1 \cos px + C_2 \sin px)$

$$\left(C_3 \cos\frac{pt}{\sqrt{LC}} + C_4 \sin\frac{pt}{\sqrt{LC}}\right) \quad ...(1)$$

where, $t = 0$, $V = V_0 \sin\frac{\pi x}{l}$

$$V_0 \sin\frac{\pi x}{l} = (C_1 \cos px + C_2 \sin px)\,C_3 \quad ...(2)$$

On equating the coefficients, we get

$C_1C_3 = 0 \Rightarrow C_1 = 0$ and $C_2C_3 = V_0$, $p = \pi/l$

becomes (1)

$$V = \sin\frac{\pi x}{l}\left[V_0 \cos\frac{pt}{\sqrt{LC}} + C_2C_4 \sin\frac{pt}{\sqrt{LC}}\right] \quad ...(3)$$

Now, when $t = 0$, $i = i_0$ (constant).

Hence, $\frac{\partial i}{\partial x} = 0$

$$\frac{\partial i}{\partial x} = -C\frac{\partial V}{\partial t}$$

$\therefore \quad \frac{\partial V}{\partial t} = 0$ when $t = 0$

Now, $\frac{\partial V}{\partial t} = \sin\frac{\pi x}{l}\left(\frac{p}{\sqrt{LC}}\right)$

$$\left[-V_0 \sin\frac{pt}{\sqrt{LC}} + C_2C_4 \cos\frac{pt}{\sqrt{LC}}\right] \quad ...(4)$$

On putting $\frac{\partial V}{\partial t} = 0$ and $t = 0$ in (4), we get

$$C_2C_4 = 0 \Rightarrow C_4 = 0$$

Now (3) becomes, $V = V_0 \sin\frac{\pi x}{l}\cos\frac{\pi t}{\sqrt{LC}}$

$$\frac{\partial V}{\partial x} = \frac{\pi}{l}V_0 \cos\frac{\pi x}{l}\cos\frac{\pi t}{\sqrt{LC}} = -L\frac{\partial i}{\partial t} \quad ...(5)$$

$$\frac{\partial V}{\partial t} = -\frac{V_0\pi}{l\sqrt{LC}}\sin\frac{\pi x}{l}\sin\frac{\pi t}{l\sqrt{LC}} = -\frac{1}{C}\frac{\partial t}{\partial x} \quad ...(6)$$

Integrating (5) and (6), we get

$$i = -V_0\sqrt{\frac{C}{L}}\cos\frac{\pi x}{l}\sin\frac{\pi t}{l\sqrt{LC}} + f(x)$$

and $$i = -V_0\sqrt{\frac{C}{L}}\cos\frac{\pi x}{l}\sin\frac{\pi t}{l\sqrt{LC}} + F(t)$$

$\therefore$ $f(x)$ and $F(t)$ must be constant only, since $i = i_0$ when $t = 0$

$\therefore$ Constant $= i_0 = f(x)$

Hence, $$i = i_0 - V_0\sqrt{\frac{C}{L}}\cos\frac{\pi x}{l}\sin\frac{\pi t}{l\sqrt{LC}}.$$

EXERCISE 42.1

1. A rod of length l with insulated sides, is initially at a uniform temperature u_0. Its ends are suddenly cooled at 0°C and are kept at that temperature. Find the temperature distribution $u(x,t)$.

2. Find temperature distribution $y(x, t)$ in a uniform bar of unit length, whose one end is kept at 10°C and the other end is insulated. It is given that $y(x, 0) = 1 - x, 0 < x < 1$.

[UPTU Q. Bank-2002]

3. The faces $x = 0$ and $x = a$ of an infinite slab are maintained at zero temperature. Given that the temperature $u(x, t) = f(x)$ at $t = 0$. Find the temperature at time t.

4. Find the solution of one dimensional heat equation $\frac{\partial \theta}{\partial t} = a^2 \frac{\partial^2 \theta}{\partial x^2}$ under the boundary conditions $\theta(0, t) = 0, \theta(l, t) = 0$. When $t > 0$ and the initial condition $\theta(x, 0) = x$ when $0 < x < l$, l being the length of bar.

5. The four edges of a thin square plate of area π^2 are kept at temperature zero and the faces are perfectly insulated. The initial temperature is assumed to be $u(x, y, 0) = xy(n - x)(\pi - y)$. Find the temperature $u(x, y, t)$ in the plate.

6. Obtain the temperature $u(x, t)$ in a slab where ends $x = 0$ and $x = l$ are kept at temperature zero and whose initial temperature is given by

$$f(x) = \begin{cases} A \ ; \text{ when } 0 < x < l/2 \\ 0 \ ; \text{ when } l/2 < x < l \end{cases}$$

[GBTU-2010, GBTU (CO)-2010)

7. Solve $\frac{\partial^2 u}{\partial x^2} = h^2\left(\frac{\partial u}{\partial t}\right)$ when $u(0, t) = u(l, t) = 0$ and $u(x, 0) = \sin\frac{\pi x}{l}$.

8. Solve the onedimensional heat equation $\frac{\partial^2 u}{\partial x^2} = \frac{1}{k}\frac{\partial u}{\partial t}$ in the range $0 \le x \le 2\pi, t \ge 0$, subject to the boundary condition $u(x, 0) = \sin^3 x$, for $0 \le x \le 2\pi$ and $u(0, t) = u(2\pi, t) = 0$ for $t \ge 0$.

9. Solve $\frac{\partial^2 z}{\partial x^2} = \frac{1}{k}\frac{\partial z}{\partial t}$ with the condition $z = 0$ when $x = 0$ and $x = 1$ for all values of t.

10. Find a solution of $\frac{\partial V}{\partial t} = k\frac{\partial^2 V}{\partial x^2}$ such that $V \ne \infty$ if $t \to \infty$, $V = 100$ if $x = 0$ or π for all values of t; $V = 0$ if $t = 0$ for all values of x between 0 and π.

11. The heat flow equation in a homogeneous rod is $\frac{\partial^2 T}{\partial x^2} = \frac{1}{\alpha^2}\frac{\partial T}{\partial t}$, where T is the temperature and α^2 thermal diffusivity. The rod is of length L with insulated sides. Solve it with boundary conditions $T(x, 0) = f(x)$, $0 < x < L, T_x(0, t) = T_x(L, t) = 0$. What is the steady temperature of the rod.

12. A conducting bar of uniform cross-section lies among the x-axis with ends at $x = 0$ and $x = L$. It is kept initially at temperature 0° and its lateral surface is insulated. There are no heat sources in the bar. The end $x=0$ is kept at 0° and heat is suddenly applied at the end $x = L$ so that there is a constant flux q_0 at $x = L$. Find the temperature distribution in the bar for $t > 0$.

13. Show that the solution of $\frac{\partial u}{\partial t} = k\frac{\partial^2 u}{\partial x^2}$ subject to the conditions

(i) u is not infinite for $t \to \infty$.

(ii) $\frac{\partial u}{\partial x} = 0$ for $x = 0$ and $x = l$.

(iii) $u = lx - x^2$ for $t = 0$ between $x = 0$ and $x = l$ is

$$u = \frac{1}{6}l^2 - \frac{l^2}{\pi^2}\sum_{n=1}^{\infty}\frac{l}{n^2}\cos\frac{n\pi x}{l}e^{-(4\pi^2 n^2 kt)/l}$$

14. A rectangular plate bounded by the lines $x = 0$, $y = 0$, $x = a$, $y = b$ has an initial distribution given by $V = A\sin\left(\frac{\pi x}{a}\right)\sin\left(\frac{\pi y}{b}\right)$. The edges are kept at zero temperature and the plane faces are impervious to heat. Find the temperature at any point. [UPTU-2002]

15. A square plate with sides of unit length has its faces insulated and its sides kept at 0°C. If the initial temperature is specified, determine the subsequent temperature at any point of the plate.

16. The temperature distribution in a bar of length. π which is perfectly insulated at ends $x = 0$ and $x = \pi$ is governed by partial differential equation.

$$\frac{\partial u}{\partial t} = \frac{\partial^2 u}{\partial x^2}$$

Assuming the initial temperature distribution as $u(x, 0) = f(x) = \cos 2x$, find the temperature distribution at any instant of time. [MTU-2011]

17. Solve $\frac{\partial u}{\partial t} = a\frac{\partial^2 u}{\partial x^2}$; a constant, subject to the boundary conditions $u(0, t) = 0$, $u(\pi, t) = 0$ and the initial condition $u(x, 0) = \sin 2x$ [MTU-2012]

18. Find the temperature distribution in a rod of length 2m whose end points are fixed at temperature zero and the initial temperature distribution is $f(x) = 100x$. [GBTU-2012]

19. Find the temperature $u(x, t)$ in a homogeneous bar of heat conducting material of length l cm with its ends kept at zero temperature and initial temperature given by $\frac{x(L - x)d}{L^2}$.

[UPTU (SUM)-2009]

20. Find the temperature in a thin metal rod of length L with both ends insulated (so that there is no passage of heat through the ends) and with initial temperature $\sin\frac{\pi x}{L}$ in the rod.

[UPTU (SUM)-2009]

21. A homogeneous rod of conducting material of length '1' has its ends kept at zero temperature. The temperature at the centre is T and falls uniformly to zero at the two ends. Find the temperature distribution. [UKTU-2012]

22. An infinitely long plane uniform plate is bounded by two parallel edges and an end at right angles to them. The breadth is π. This end is maintained at temperature u_0 at all points and the other edges are at zero temperature. Determine the temperature at any point of the plate in the steady state.

[GBTU-2012]

23. Solve $\frac{\partial^2 u}{\partial x^2} + \frac{\partial^2 u}{\partial y^2} = 0, 0 < x < \pi, 0 < y < \pi$ which satisfies the conditions:

$u(0, y) = u(\pi, y) = u(x, \pi) = 0$ and $u(x, 0) = \sin^2 x$

[UKTU-2011]

24. Solve the Laplace equation $\frac{\partial^2 u}{\partial x^2}+\frac{\partial^2 u}{\partial y^2}=0$ in a rectange in the xy-plane with $u(x, 0) = 0$, $u(x, b)= 0$, $u(0, y)=0$ and $u(a, y) = f(y)$ plarallel to y-axis. [UPTU (SUM)-2008]

25. Find the steady state temperature distribution in a rectangular thin plate with its two surfaces insulated and with the conditions $u(0, y)=0$, $u(x, 0) = 0$, $u(a, y)=g(y)$, $u(x, b) = f(x)$. [UPTU (SUM)-2007]

26. Solve the following Laplace equation $\frac{\partial^2 u}{\partial x^2}+\frac{\partial^2 u}{\partial y^2}=0$ in a rectangle with $u(0, y) = 0$, $u(a, y)=0$, $u(x, b) = 0$ and $u(x, 0) = f(x)$ along x-axis. [UPTU-2008]

27. Solve the boundary value problem.

$$\frac{\partial^2 u}{\partial x^2}+\frac{\partial^2 u}{\partial y^2}=0,\ 0\le x\le a, 0\le y\le b$$

with the boundary conditions

$$u_x(0,y)=u_x(a,y)=u_y(x,0)=0 \text{ and } u_y(x,b)=f(x).$$

[MTU (SUM)-2011]

28. Neglecting R and G, find the emf $v(x, t)$ in a line of length l, t seconds after the ends were suddenly grounded, given that $i(x, 0) = i_0$ and $v(x,0)=e_1\sin\frac{\pi x}{l}+e_5\sin\frac{5\pi x}{l}$ [SVTU-2008, MTU-2012]

29. Solve $\frac{\partial^2 V}{\partial x^2}=LC\frac{\partial^2 V}{\partial t^2}$ assuming that the initial voltage is $V_0\sin\frac{\pi x}{l}$; $V_t(x_0)=0$ and $V=0$ at the ends $x=0$ and $x=l$ for all t. [UPTU (SUM)-2007]

30. A steady voltage distribution of 20 volts at the sending end and 12 volts at the receiving end is maintained in a telephone wire of length l. A time t=0, the receiving end is grounded. Find the voltage and current t sec later. Neglect leakance and inductance. [MTU-2011, UPTU (SUM)-2008]

31. A homogeneous rod of conducting malerial of length 100 cm has its ends kept at zero temperature and the temperature initially is

$$u(x,0)=x \qquad 0\le x\le 50$$
$$=100-x \qquad 50\le x\le 100$$

Find the temperature $u(x, t)$ at any time. [Bhopal-2007, SVTU-2007, Kurukshetra-2006]

Hint to Selected Problems

3. The temperature $u(x, t)$ in the given solid is governed by the one dimensional heat equation

$$C\frac{\partial^2 u}{\partial x^2}=\frac{\partial u}{\partial t} \qquad \text{... (1)}$$

As per given, the boundary conditions are

$$u(0,t)=0,\ u(a_1-1)=0,\ \forall t \qquad \text{... (2)}$$
$$u(x,0)=f(x) \qquad \text{... (3)}$$

Now, suppose that (1) has the solution of the form

$$u(x,t)=X(x)\,T(t) \qquad \text{... (4)}$$

where X and T are respectively the functions of x and t alone. Using the values of (4) in (1), we get

$$\frac{X''}{X}=\frac{T'}{CT}=\mu \text{ (say)} \qquad \text{... (5)}$$

where μ is a separation constant.

From equation (5), we can deduce that

$$X''-\mu X=0 \qquad \text{... (6)}$$

and

$$T'=\mu CT \qquad \text{... (7)}$$

Using (2) in (4), we get

$$X(0)=0 \text{ and } X(a)=0 \qquad \text{... (8)}$$

Now, we want to solve (6) subject to the boundary condition (8). Here, we have the following cases :

Case (I). Let $\mu = 0$. Then solution of (6) is given by

$$X(x) = Ax + B \qquad \text{... (9)}$$

Using (8), we get

$$A = B = 0$$
$$X(x) = 0$$

$u = 0$, which does not satisfy (3).

Case (II). Let $\mu=\lambda^2, \lambda\ne 0$: In this case, solution of (6) is given by

$$X(x)=Ae^{\lambda x}+Be^{-\lambda x}.$$

Using (8), we get $A+B=0$ and $0=Ae^{0\lambda}+De^{-a\lambda}$

$$\Rightarrow \quad A=B=0$$
$$\Rightarrow \quad X=0$$
$$\Rightarrow \quad u=0$$

Thus, we reject this case also.

Case (III). Let $\mu=-\lambda^2, \lambda\ne 0$. In this case, solution of (6) is given by

$$X(x)=A\cos\lambda x+B\sin\lambda x$$

Using (8), we get

$$A=0 \text{ and } A\cos\lambda a+B\sin\lambda a=0$$

Let $B\ne 0$, then $\sin\lambda a=0$

$$\lambda a=n\pi,\ n=1,2,.....$$
$$\lambda=\frac{n\pi}{a},\quad n=1,2,...$$

Therefore, non-zero solution of (6) is given by

$$X_n(x)=B_n\sin\left(\frac{n\pi x}{a}\right) \qquad \text{... (10)}$$

Putting $\lambda=\frac{n\pi}{a}$ in (7), we get

$$\frac{dT}{T}=-\frac{n^2\pi^2 C}{a^2}dt$$
$$\Rightarrow \quad \frac{dT}{T}=-C_n^2 dt$$

whose solution is given by

$$T_n(t)=D_n e^{-C_n^2 t} \qquad \text{... (11)}$$

Thus, we have

$$u_n(x,t)=X_n(x)\,T_n(t)=E_n\sin\left(\frac{n\pi x}{a}\right)e^{-C_n^2 t} \qquad \text{... (12)}$$

The more general solution of (1) is given by

$$u(x,t)=\sum_{n=1}^{\infty}u_n(x,t)=\sum_{n=1}^{2}E_n\sin\left(\frac{n\pi x}{a}\right)e^{-C_n^2t} \quad \dots (13)$$

Putting $t=0$ in (13) and using (3), we get

$$f(x)=\sum_{n=1}^{\infty}E_n\sin\left(\frac{n\pi x}{a}\right)$$

which is a Fourier sine series, thus the constants E_n are given by

$$E_n=\frac{2}{a}\int_0^a f(x)\sin\left(\frac{n\pi x}{a}\right)dx,\quad n=1,2,3,\dots$$

4. Proceeding same as example 3, we get

$$u\,u(x,y,t)=\sum_{m=1}^{\infty}\sum_{n=1}^{\infty}F_{mn}\sin mx\sin ny\; e^{-\lambda_{mn}^2t} \quad \dots (1)$$

where, $F_{mn}=\frac{4}{\pi^2}\int_{x=0}^{\pi}\int_{y=0}^{\pi}xy\,(\pi-x)(\pi-y)\sin mx\sin ny\,dx\,dy$

$$=\frac{4}{\pi^2}\left[\int_0^{\pi}(\pi x-x^2)\sin mx\,dx\right]\left[\int_0^{\pi}(\pi y-y^2)\sin ny\,dy\right] \quad \dots (2)$$

Using chain rule of integration by parts, we get

$$\int_0^{\pi}(\pi x-x^2)\sin mx\,dx=\left[(\pi x-x^2)\left(-\frac{\cos mx}{m}\right)-(\pi-2x)\left(-\frac{\sin mx}{m^2}\right)+(-2)\left(\frac{\cos mx}{m^2}\right)\right]_0^{\pi}$$

$$=\frac{2}{m^3}\left[1-(-1)^m\right]$$

In a similar manner, we get

$$\int_0^{\pi}(ny-y^2)\sin ny\,dy=\frac{2}{n^3}\left[1-(-1)^n\right]$$

Thus,
$$F_{mn}=\frac{16}{\pi^2m^3n^3}\left[1-(-1)^m\right]\left[1-(-1)^n\right]$$

$$=\begin{cases}0, & \text{when } m=2p \text{ or } n=2q\\ \dfrac{64}{\pi^2(2p-1)^3(2q-1)^3}, & \text{when } m=2p-1,\ n=2q-1\end{cases}$$

Hence, the required solution is given by

$$u(x,y,t)=\sum_{p=1}^{\infty}\sum_{q=1}^{\infty}F_{pq}\sin[(2p-1)x]\,\sin[(2q-1)y]\,e^{-\lambda_{pq}^2t}$$

where, $\lambda_{pq}^2=C^2\left[(2p-1)^2+(2q-1)^2\right]$

and $F_{pq}=\frac{64}{\pi^2}(2p-1)^3(2q-1)^3$.

8. Proceeding same as example (3) by taking $a=2\pi$ and $f(x)=5m^3x$, we get

$$u(x,t)=\sum_{n=1}^{\infty}E_n\sin\left(\frac{nx}{2}\right)e^{-(n^2kt/4)},\quad n=1,2,3,\dots \quad \dots (1)$$

Putting $t=0$ in (1) and using $u(x,0)=\sin^3x$, we get

$$\sum_{n=1}^{\infty}E_n\sin\left(\frac{nx}{2}\right)=\sin^3x=\frac{3}{4}\sin x-\frac{1}{4}\sin 3x.$$

Because $\sin 3x=3\sin x-4\sin^3x$

$$\Rightarrow\quad \sin^3x=\frac{1}{4}(3\sin x-\sin 3x)$$

Putting these values in (1), we get the required solution

$$u(x,t)=\frac{1}{4}(3\sin xe^{-kt}-\sin 3x\,e^{-9kt})$$

12. The given initial boundary value problem is given as follows :

$$\frac{\partial T}{\partial t}=\alpha\frac{\partial^2T}{\partial x^2}$$

Subject to the boundary conditions

$$T(0,t)=0,\ t>0,\quad \frac{\partial T}{\partial x}(L,t)=q_0,\ t>0$$

with initial conditions

$$T(x,0)=0,\quad 0\le x\le L.$$

Let $T(x,t)=T_s(x)+T_1(x,t)$

Where, T_s is a steady part and T is the transient part of the solution.

Thus, we have

$$\frac{\partial^2T_s}{\partial x^2}=0\ \Rightarrow\ T_s=Ax+B.$$

Using $x=0$, $T_s=0$, we get $B=0$, therefore, $T_s=Ax$.

Further using the boundary condition $\frac{\partial T_s}{\partial x}=q_0$, we get $A=q_0$.

Thus the steady solution is given by

$$T_s=q_0x$$

For the transient part, the boundary and initial condition can be redefined as follows :

(i) $T_1(0,t)=T(0,t)-T_s(0)-0-0=0$

(ii) $\frac{\partial T_1(x,t)}{\partial x}=\frac{\partial T(x,t)}{\partial x}-\frac{\partial T_s(L,t)}{\partial x}=q_0-q_0=0$

(iii) $T_1(x,0)=T(x,0)-T_s(x)=-q_0x,\ 0<x<L.$

Therefore, for the transient part, we have to solve the given PDE subject to these conditions. We know that the solution of given PDE is

$$T_1(x,t)=e^{-a\lambda^2t}(A\cos\lambda x+B\sin\lambda x)$$

Using (i), we get $A=0$

$$T_1(x,t)=Be^{-a\lambda^2t}\sin\lambda x$$

Again using boundary conditions (ii), we get

$$\left.\frac{\partial T}{\partial x}\right|_{x=L}=B\lambda\,e^{-a\lambda^2t}\cos\lambda L=0.$$

$$\Rightarrow\quad \lambda L=(2n-1)\frac{\pi}{2},\quad n=1,2,\dots$$

Thus, the general solution is given by

$$T_1(x,t)=\sum_{n=1}^{\infty}B_n\exp\left[-\alpha\left\{\frac{(2n-1)}{L}\right\}^2\pi^2t\right]\sin\left(\frac{2n-1}{2L}\pi x\right)$$

Applying initial condition (iii), we get

$$T_1(x,0)=-q_0x=\sum_{n=1}^{\infty}B_n\sin\left(\frac{2n-1}{2L}\pi x\right).$$

Now, multiplying both sides by $\sin\left(\frac{2n-1}{2L}\pi x\right)$, integrating between 0 to L and using

$$\int_0^L B_n\sin\left(\frac{2n-1}{2L}\right)\pi x\sin\left(\frac{2m-1}{2L}\pi x\right)dx=\begin{cases}0, & n\ne m\\ \dfrac{B_mL}{2}, & n=m\end{cases}$$

We get $-q_0\dfrac{4L^2}{(2m-1)^2\pi^2}\left[\sin\left(\dfrac{2m-1}{2}\pi\right)\right]=B_m.\dfrac{L}{2}.$

$$\Rightarrow -q_0\frac{4L^2}{(2m-1)^2\pi^2}(-1)^{m-1}=B_m.\frac{L}{2}$$

$$\Rightarrow\quad B_m=\frac{(-1)^m8Lq_0}{(2m-1)^2\pi^2}$$

Hence, the required temperature distribution is given by

$$T(x,t) = q_0 x + \frac{8Lq_0}{\pi^2}\sum_{m=1}^{\infty}\left[\frac{(-1)^m}{(2m-1)^2}\exp\left[-\alpha\{(2m-1)/2\}\right]^2\pi^2 t\right] - \sin\left(\frac{2m-1}{2L}\pi x\right)$$

20. $(u_x)_{x=0} = 0 = (u_x)_{x=L} = L; u(x,0) = \sin\frac{\pi x}{L}$

21. $u(x,0) = \begin{cases} 2Tx & 0 \le x \le \frac{1}{2} \\ 2T(1-x) & \frac{1}{2} \le x \le 1 \end{cases}$

ANSWERS

1. $u(x,t) = x + 2 + \sum_{n=1}^{\infty} E_n \sin n\pi x\, e^{-n^2\pi^2 C^2 t}$;

2. $y(x,t) = 10 + \sum_{n=1}^{\infty} E_n \sin\frac{(2n-1)}{2}\pi x\, e^{-C_n^2 t}$

3. $u(x,t) = \sum_{n=1}^{\infty} u_n(x,t) = \sum_{n=1}^{\infty} E_n \sin\left(\frac{n\pi x}{a}\right)e^{-C_n^2 t}$; $f(x) = \sum_{n=1}^{\infty} E_n \sin\frac{n\pi x}{a}$, where $E_n = \frac{2}{a}\int_0^a f(x)\sin\left(\frac{n\pi x}{a}\right)dx,\ n = 1,2,3,\ldots$

4. $E_n = \frac{2}{l}\int_0^l x \sin\frac{n\pi x}{l}dx = \begin{cases} \frac{2l}{n\pi} & ; n \text{ is odd} \\ -\frac{2l}{n\pi} & ; n \text{ is even} \end{cases}$

5. $u(x,y,t) = \sum_{p=1}^{\infty}\sum_{q=1}^{\infty} F_{pq}\sin[(2p-1)x]\sin[(2q-1)y]e^{-\lambda p^2 qt}$ where $\lambda_{pq}^2 = C^2\left[(2p-1)^2 + (2q-1)^2\right]$ and $F_{pq} = \frac{64}{\pi^2}(2p-1)^3(2q-1)^3$

6. $u(x,l) = \frac{4A}{\pi}\sum_{n=1}^{\infty}\frac{1}{n}\sin^2\frac{n\pi}{4}\sin\frac{n\pi x}{l}e^{-(n^2\pi^2 kt)/l^2}$

7. $u(x,t) = \sin\left(\frac{\pi x}{l}\right)e^{-\pi^2 t/h^2 l^2}$

8. $u(x,t) = \frac{1}{4}\left[3\sin xe^{-kt} - \sin 3xe^{-akt}\right]$

12. $T(x,t) = q_0 x + \frac{8Lq_0}{\pi^2}\sum_{m=1}^{\infty}\left[\frac{(-1)^m}{(2m-1)^2}\exp\left[-\alpha\{(2m-1)/2\}^2\pi^2 t\right]\sin\left(\frac{2m-1}{2L}\pi x\right)\right]$

14. $u(x,y,t) = A\sin\left(\frac{\pi x}{a}\right)\sin\left(\frac{\pi y}{a}\right)e^{-\pi^2 kt\left(\frac{1}{a^2}+\frac{1}{b^2}\right)}$

16. $u(x,t) = e^{-4t}\cos 2x$

17. $u(x,t) = \sin 2xe^{-4at}$

18. $u(x,t) = \frac{-400}{\pi}\sum_{n=1}^{\infty}\frac{\cos n\pi}{n}\frac{\sin n\pi x}{2}e^{-\left(\frac{c^2 n^2\pi^2 t}{4}\right)}$

19. $u(x,t) = \frac{8d}{\pi^3}\sum_{n=1}^{\infty}\frac{1}{(2n-1)^3}\frac{\sin(2n-1)\pi x}{L}e^{\frac{-(2n-1)^2\pi^2 C^2 t}{L^2}}$

20. $u(x,t) = \frac{2}{\pi} - \frac{4}{\pi}\sum_{m=1}^{\infty}\frac{1}{(4m^2-1)}\cos\left(\frac{2m\pi x}{L}\right)e^{\frac{-4m^2\pi^2 C^2 t}{L^2}}$

21. $u(x,t) = \frac{8T}{\pi^2}\sum_{m=1}^{\infty}\frac{(-1)^{m+1}}{(2m-1)^2}\sin(2m-1)\pi x e^{-[(2m-1)^2\pi^2 c^2 t]}$

22. $u(x,t) = \frac{4u_0}{\pi}\sum_{n=1}^{\infty}\frac{1}{(2n-1)}\sin(2n-1)xe^{-(2n-1)y}$

23. $u(x,t) = \frac{-8}{\pi}\sum_{m=1,2,3,\ldots}^{\infty}\frac{\sin(2m-1)x\sinh(2m-1)(\pi-y)}{(2m-1)\{(2m-1)^2-4\}\sinh(2m-1)\pi}$

24. $u(x,t) = \sum_{n=1}^{\infty} b_n \sin\frac{n\pi y}{b}\sinh\frac{n\pi x}{b}$ where, $b_n = \frac{2}{b\sinh\left(\frac{n\pi a}{b}\right)}\int_0^b f(y)\sin\frac{n\pi y}{b}dy$

25. $u(x,y) = \sum_{n=1}^{\infty}\left[b_n\sin\left(\frac{n\pi x}{a}\right)\sinh\left(\frac{n\pi y}{a}\right) + B_n\sin\left(\frac{n\pi y}{b}\right)\sinh\left(\frac{n\pi x}{b}\right)\right]$ where $b_n = \frac{2}{a\sinh\left(\frac{n\pi b}{a}\right)}\int_0^a f(x)\sin\frac{n\pi x}{a}dx$

and $B_n = \frac{2}{b\sinh\left(\frac{n\pi a}{b}\right)}\int_0^a g(y)\sin\frac{n\pi y}{b}dy$

26. $u(x,y) = \sum_{n=1}^{\infty} B_n\sin\left(\frac{n\pi x}{a}\right)\sinh\frac{n\pi}{a}(b-y)$, where, $B_n = \frac{2}{a\sinh\left(\frac{n\pi b}{a}\right)}\int_0^a f(x)\sin\frac{n\pi x}{a}dx$

27. $u(x,y) = \sum_{n=1}^{\infty} b_n\cos\frac{n\pi x}{a}\left(e^{\frac{n\pi y}{a}} - e^{\frac{-n\pi y}{a}}\right)$ where $b_n = \frac{1}{n\pi\cosh\frac{n\pi b}{a}}\int_0^a f(x)\cos\frac{n\pi x}{a}dx$

28. $V = e_1\sin\frac{\pi x}{l}\cos\frac{\pi t}{l\sqrt{2C}} + e_5\sin\frac{5\pi x}{l}\cos\frac{5\pi t}{l\sqrt{LC}}$

29. $V(x,t) = V_0\sin\frac{\pi x}{l}\cos\frac{\pi t}{l\sqrt{LC}}$

30. $V(x,t) = \frac{20(l-x)}{l} + \frac{24}{\pi}\sum_{n=1}^{\infty}\frac{(-1)^{n+1}}{n}\sin\frac{n\pi x}{l}e^{-n^2\pi^2 t/RCl^2}$ $i(x,t) = \frac{20}{lR} + \frac{24}{lR}\sum_{n=1}^{\infty}(-1)^n\cos\frac{n\pi x}{l}e^{-n^2\pi^2 t/RCl^2}$

31. $u(x,t) = \frac{400}{\pi^2}\sum_{n=1}^{\infty}\frac{(-1)^n}{(2n+1)^2} - e^{-[(2n+1)c\pi/100]^2 t}\sin\frac{(2n+1)\pi x}{100}$

Objective Evaluations

Multiple Choice Questions

Choose the most appropriate one.

1. The general second order linear differential equation in two independent variables x, y is given by $Au_{xx} + Bu_{xy} + Cu_{yy} + Du_x + Eu_y + Fu = G$, where A, B, C, D, E and F are given functions of x and y or constant. Then equation is hyperbolic if :

(a) $B^2 - 4AC > 0$ (b) $B^2 - 4AC = 0$
(c) $B^2 - 4AC < 0$ (d) none of these

2. The general second order linear differential equation in two independent variables x, y is given by $Au_{xx} + Bu_{xy} + Cu_{yy} + Du_x + Eu_y + Fu = G$, where A, B, C, D, E and F are given functions of x and y or constant. Then equation is parabolic if :

(a) $B^2 - 4AC > 0$ (b) $B^2 - 4AC = 0$
(c) $B^2 - 4AC < 0$ (d) none of these

3. The general second order linear differential equation in two independent variables x, y is given by $Au_{xx} + Bu_{xy} + Cu_{yy} + Du_x + Eu_y + Fu = G$, where A, B, C, D, E and F are given functions of x and y or constant. Then equation is elliptic if :

(a) $B^2 - 4AC > 0$ (b) $B^2 - 4AC = 0$
(c) $B^2 - 4AC < 0$ (d) none of these

4. The general second order linear differential equation in two independent variables x, y is given by $Au_{xx} + Bu_{xy} + Cu_{yy} + Du_x + Eu_y + Fu = G$, where A, B, C, D, E and F are given functions of x and y or constant. Then characteristic equation is given by :

(a) $\frac{dy}{dx} = (B \pm \sqrt{B^2 - 4AC})$ (b) $\sqrt{B^2 - 4AC}$
(c) $\frac{dy}{dx} = \frac{1}{2A}(B \pm \sqrt{B^2 - 4AC})$ (d) none of these

5. In Dirichlet conditions :

(a) u is prescribed by each point of a boundary ∂D of a domain D.
(b) where value of normal derivative $\frac{\partial u}{\partial n}$, in the boundary ∂D are specified.
(c) $\left(\frac{\partial u}{\partial n} + au\right)$ is specific on ∂D
(d) none of these

6. In Neumann conditions :

(a) u is prescribed by each point of a boundary ∂D of a domain D.
(b) where value of normal derivative $\frac{\partial u}{\partial n}$, in the boundary ∂D are specified.
(c) $\left(\frac{\partial u}{\partial n} + au\right)$ is specific on ∂D.
(d) none of these

7. Wave equation is represented by :

(a) $u_{tt} - c^2\nabla^2 u = 0$, where c is a constant.
(b) $u_t - k\nabla^2 u = 0$, where k is a constant.
(c) $\nabla^2 u = 0$
(d) $\nabla^2 u = f(x, y, z)$

8. Heat equation is represented by :

(a) $u_{tt} - c^2\nabla^2 u = 0$, where c is a constant.
(b) $u_t - k\nabla^2 u = 0$, where k is a constant.
(c) $\nabla^2 u = 0$
(d) $\nabla^2 u = f(x, y, z)$

9. Laplace equation is represented by :

(a) $u_{tt} - c^2\nabla^2 u = 0$, where c is a constant.
(b) $u_t - k\nabla^2 u = 0$, where k is a constant.
(c) $\nabla^2 u = 0$
(d) $\nabla^2 u = f(x, y, z)$

10. Poisson equation is represented by :

(a) $u_{tt} - c^2\nabla^2 u = 0$, where c is a constant.
(b) $u_t - k\nabla^2 u = 0$, where k is a constant.
(c) $\nabla^2 u = 0$
(d) $\nabla^2 u = f(x, y, z)$

ANSWERS

1. (a) **2.** (b) **3.** (b) **4.** (c) **5.** (a) **6.** (b) **7.** (a) **8.** (b) **9.** (a) **10.** (a)

Fourier Transforms

43.1 INTRODUCTION

If a function $f(x)$ defined on the interval $]-\infty,\infty[$, and piecewise continuous in each finite partial interval and absolutely integrable in $]-\infty,\infty[$, then

$$F(f(x)) = \tilde{f}(p) = \int_{-\infty}^{\infty} e^{ipx} f(x)dx$$

is defined as Fourier transform of $f(x)$.

The inverse formula for Fourier transform is given by

$$f^{-1}[\tilde{f}(p)] = f(x) = \frac{1}{2\pi}\int_{-\infty}^{\infty} \tilde{f}(p)e^{-ipx}dp\,.$$

REMARK

- We can also define $\tilde{f}(p) = F(f(x)) = \frac{1}{\sqrt{2\pi}}\int_{-\infty}^{\infty} e^{-ipx} f(x)dx$ and $F^{-1}[\tilde{f}(p)] = f(x) = \frac{1}{\sqrt{2\pi}}\int_{-\infty}^{\infty} e^{-ipx}\tilde{f}(p)dp$.

43.2 FOURIER SINE AND COSINE TRANSFORMS

Definition (1): *The infinite Fourier sine transform of the function $f(x)$, $0 < x < \infty$ is defined by $F_s[f(x)]$ or $\tilde{f}_s(p)$ and defined by*

$$\tilde{f}_s(p) = F_s[f(x)] = \sqrt{\frac{2}{\pi}}\int_0^{\infty} f(x)\sin pxdx$$

The inverse formula for infinite Fourier sine transform is given by

$$f(x) = F_s^{-1}[\tilde{f}_s(p)] = \sqrt{\frac{2}{\pi}}\int_0^{\infty} \tilde{f}_s(p)\sin pxdx\,.$$

Definition (2): *The infinite Fourier cosine transform of $f(x)$, $0 < x < \infty$ is denoted by $F_c[f(x)]$ or $\tilde{f}_c(p)$ and is defined by*

$$\tilde{f}_c[f(x)] = \tilde{f}_c(p) = \sqrt{\frac{2}{\pi}}\int_0^{\infty} f(x)\cos pxdx$$

The inversion formula for Fourier cosine transform is given by

$$f(x) = \sqrt{\frac{2}{\pi}}\int_0^{\infty} \tilde{f}_c(p)\cos pxdx\,.$$

43.2.1 LINEARITY PROPERTY OF FOURIER TRANSFORMS

Let $\tilde{f}(p)$ and $\tilde{g}(p)$ are Fourier transform of $f(x)$ and $g(x)$ respectively. Then

$F\{af(x)+bg(x)\} = a\tilde{f}(p)+b\tilde{g}(p)$ where a and b constants.

43.2.2 CHANGE OF SCALE PROPERTY

Theorem 1. **(For Complex Fourier Transform).** *If $\tilde{f}(p)$ is the complex Fourier transform of $f(x)$, the complex Fourier transform of $f(ax)$ is given by*

$$\frac{1}{a}\tilde{f}\left(\frac{p}{a}\right).$$

Proof. By definition, we have

$$\int_{-\infty}^{\infty} e^{ipx} f(x)dx = \tilde{f}(p). \qquad ...(1)$$

Consider $\tilde{f}(ap) = \int_{-\infty}^{\infty} e^{ipx} f(x)dx$.

Putting $ax = t \Rightarrow dx = dt$, we get

$$\tilde{f}(ap) = \frac{1}{a}\int_{-\infty}^{\infty} e^{ip(t/a)} f(t)dt = \frac{1}{a}\int_{-\infty}^{\infty} e^{i(p/a)x} f(x)dx = \frac{1}{a}\tilde{f}\left(\frac{p}{a}\right).$$

REMARK

- In a similar way, we can prove that :

(a) If $\tilde{f}_s(p)$ is the Fourier sine transform of $f(x)$, then Fourier sine transform of $f(ax)$ is given by $\frac{1}{a}\tilde{f}_s\left(\frac{p}{a}\right)$.

(b) If $\tilde{f}_c(p)$ is the Fourier cosine transform of $f(x)$, then Fourier cosine transform of $f(ax)$ is given by $\frac{1}{a}\tilde{f}_c\left(\frac{p}{a}\right)$.

Theorem 2. **(Shifting Property).** *If $\tilde{f}(p)$ is the complex Fourier transform of $f(x)$, then complex Fourier transform of $f(x-a)$ is $e^{ipa}\tilde{f}(p)$.*

Proof. By definition, we have

$$\int_{-\infty}^{\infty} e^{ipx} f(x)dx = \tilde{f}(p) \qquad ...(1)$$

Consider $\tilde{f}(x-a) = \int_{-\infty}^{\infty} e^{ipx} f(x-a)dx = \int_{-\infty}^{\infty} e^{ip(t+a)} f(t)dt$ [Putting $x-a=t$]

$$= e^{ipx}\int_{-\infty}^{\infty} e^{ipt} f(t)dt = e^{ipa}\tilde{f}(p)$$

43.2.3 SOME IMPORTANT INTEGRALS (TO BE USED DIRECTLY):

1. $\int e^{ax}\sin bx dx = \frac{e^{ax}}{a^2+b^2}(a\sin bx - b\cos bx)$

2. $\int e^{ax}\cos bx dx = \frac{e^{ax}}{a^2+b^2}(a\cos bx + b\sin bx)$

3. $\int_0^{\infty} e^{-ax}\sin bx dx = \frac{b}{a^2+b^2}$

4. $\int_0^{\infty} e^{-ax}\cos bx dx = \frac{a}{a^2+b^2}$

5. $\frac{d^n}{dx^n}\left(\frac{x}{a^2+x^2}\right) = \frac{(-1)^n.n!}{(a^2+x^2)^{(n+1)/2}}\cos\left[(n+1)\tan^{-1}\left(\frac{a}{x}\right)\right]$

6. $\frac{d^n}{dx^n}\left(\frac{a}{a^2+x^2}\right) = \frac{(-1)^n.n!}{(a^2+x^2)^{(n+1)/2}}\sin\left[(n+1)\tan^{-1}\left(\frac{a}{x}\right)\right]$

7. $\int_0^{\infty}\frac{\sin px}{x}dx = \begin{bmatrix}\pi/2 & ; & \text{if } p>0\\ -\pi/2 & ; & \text{if } p<0\end{bmatrix}$

8. $\int_0^{\infty} e^{-x^2}dx = \sqrt{\pi}, \int_0^{\infty} e^{-x^2}dx = \frac{\sqrt{\pi}}{2}$

9. $\int_0^{\infty}\frac{e^{ax}-e^{-ax}}{e^{\pi x}-e^{-\pi x}}dx = \frac{1}{2}\tan\frac{a}{2}, \int_0^{\infty}\frac{e^{ax}+e^{-ax}}{e^{\pi x}-e^{-\pi x}}dx = \frac{1}{2}\sec\frac{a}{2}$

10. $\int_0^{\infty}\frac{\cos px}{1+p^2}dp = \frac{\pi e^{-x}}{2}, \int_0^{\infty}\frac{p\sin px}{1+p^2}dp = \frac{\pi}{2}e^{-x}$

Theorem 3 **(Modulation Theorem).** *If $\tilde{f}(p)$ is the complex Fourier transform of $f(x)$ then, the Fourier transform of $f(x)\cos ax$ is $\frac{1}{2}[\tilde{f}(p-a)+\tilde{f}(p+a)]$.*

Proof: By definition, we have

$$\tilde{f}(p) = \int_{-\infty}^{\infty} e^{ipx} f(x)dx$$

Now, $F[f(x)\cos ax] = \int_{-\infty}^{\infty} e^{ipx} f(x)\cos ax dx$

$$= \int_{-\infty}^{\infty} e^{ipx} f(x)\frac{e^{iax}+e^{-iax}}{2}dx$$

$$= \frac{1}{2}\int_{-\infty}^{\infty} e^{i(p+a)x} f(x)dx + \frac{1}{2}\int_{-\infty}^{\infty} e^{i(p-a)x} f(x)dx$$

$$= \frac{1}{2}f(p+a) + \frac{1}{2}f(p-a)$$

REMARKS

- In a similar way, we can prove that
  If $\tilde{f}_s(p)$ is the complex Fourier transform of $f(x)$, then
  1. Fourier transform of $f(x)$ cos ax is $\frac{1}{2}[f_s(p+a)+f_s(p-a)]$
  2. Fourier transform of $f(x)$ sin ax is $\frac{1}{2}[f_s(p+a)-f_s(p-a)]$
  3. Fourier transform of $f(x)$ sin ax is $\frac{1}{2}[f_c(p-a)-f_c(p+a)]$

43.2.4 SOME MORE IMPORTANT RESULTS

1. Multiple Fourier Transforms

Let $f(x, y)$ be a function of two variables x and y. Temporarily, as a function of x, its Fourier transform is given by

$$\tilde{f}(p,y)=\frac{1}{\sqrt{2\pi}}\int_{-\infty}^{\infty} f(x,y)e^{ipx}\,dx$$

Also, regarding $\tilde{f}(p,y)$ as a function of y, its Fourier transform is given by

$$\tilde{f}(p,q)=\frac{1}{\sqrt{2\pi}}\int_{-\infty}^{\infty}\tilde{f}(p,y)e^{iqy}\,dy=\frac{1}{2\pi}\int_{-\infty}^{\infty}\int_{-\infty}^{\infty} f(x,y)e^{i(px+qy)}\,dxdy$$

2. Inversion Formula : If

$$f(x,y)=\frac{1}{\sqrt{2\pi}}\int_{-\infty}^{\infty}\tilde{f}(p,y)e^{-ipx}\,dp$$

and

$$\tilde{f}(p,y)=\frac{1}{\sqrt{2\pi}}\int_{-\infty}^{\infty}\tilde{f}(p,q)e^{-iqy}\,dq$$

Then

$$f(x,y)=\frac{1}{2\pi}\int_{-\infty}^{\infty}\tilde{f}(p,q)e^{-i(px+qy)}\,dpdq$$

3. Convolution

The function $H(x)=F*G=\frac{1}{\sqrt{2\pi}}\int_{-\infty}^{\infty}F(u)G(x-u)du$ is called the convolution or falting of two integrable functions F and G over the interval $]-\infty,\infty[$.

Sometime it can also be defined as $F*G=\int_{-\infty}^{\infty}F(u)G(x-u)du$.

4. The convolution of Falting Theorem for Fourier Transforms

If F[f(x)] and F[g(x)] are the Fourier transforms of the functions f(x) and g(x) respectively, then the Fourier transform of the convolution of f(x) and g(x) is the product of their Fourier transforms, i.e.,

$$F\{f(x)*g(x)\}=F\{f(x)\}.F\{g(x)\}$$

Proof. Consider LHS $= F\{f(x)*g(x)\}$

$$=F\int_{-\infty}^{\infty}f(u)g(x-u)du \quad \text{(By definition of convolution)}$$

$$=\int_{-\infty}^{\infty}e^{ipx}\left\{\int_{-\infty}^{\infty}f(u)g(x-u)du\right\}dx=\int_{-\infty}^{\infty}f(u)\left\{\int_{-\infty}^{\infty}e^{ipx}g(x-u)dx\right\}du$$

$$=\int_{-\infty}^{\infty}f(u)\left\{\int_{-\infty}^{\infty}e^{-ip(u+v)}g(v)dv\right\}du \quad [\text{Putting } x-u=v \Rightarrow dx=dv]$$

$$=\int_{-\infty}^{\infty}e^{ipu}f(u)\left\{\int_{-\infty}^{\infty}e^{ipv}g(v)dv\right\}du$$

$$=\int_{-\infty}^{\infty}e^{ipu}f(u)F\{g(x)\}du=F\{g(x)\}\int_{-\infty}^{\infty}e^{ipu}f(u)du=F\{g(x)\}.F\{f(x)\}.$$

5. Parseval's identity for Fourier Transform: Rayleigh Theorem or Plancharel's Theroem.

If f(p) and g(p) are the complex Fourier transform of F(x) and G(x) respectively, then

(i) $\frac{1}{2\pi}\int_{-\infty}^{\infty}f(p)\bar{g}(p)dp=\int_{-\infty}^{\infty}F(x)\overline{G(x)}dx$ (ii) $\frac{1}{2\pi}\int_{-\infty}^{\infty}|f(p)|^2\,dp=\int_{-\infty}^{\infty}|F(x)|^2\,dx$

where bar denotes the complex conjugate.

Proof. (i) By inversion formula, we have

$$G(x)=\frac{1}{2\pi}\int_{-\infty}^{\infty}g(p)e^{-ipx}\,dp \quad \ldots(1)$$

$$\Rightarrow \quad \overline{G(x)}=\frac{1}{2\pi}\int_{-\infty}^{\infty}\overline{g(p)}e^{ipx}\,dp \quad \ldots(2)$$

Thus, $$\int_{-\infty}^{\infty} F(x)\overline{G(x)}dx = \int_{-\infty}^{\infty} F(x)\left\{\frac{1}{2\pi}\int_{-\infty}^{\infty}\overline{g(p)e^{ipx}}dp\right\}dx$$

$$= \frac{1}{2\pi}\int_{-\infty}^{\infty}\overline{g(p)}\left\{\int_{-\infty}^{\infty} F(x)e^{ipx}dx\right\}dp = \frac{1}{2\pi}\int_{-\infty}^{\infty}\overline{g(p)}f(p)dp$$

(ii) Taking $G(x) = F(x)$ in part (i), we get

$$\frac{1}{2\pi}\int_{-\infty}^{\infty} f(p)\overline{f(p)}dp = \int_{-\infty}^{\infty} F(x)\overline{F(x)}dx$$

Hence, $$\frac{1}{2\pi}\int_{-\infty}^{\infty}|f(p)|^2 dp = \int_{-\infty}^{\infty}|F(x)|^2 dx$$

REMARKS

- In a similar way, we can prove that

(a) $\frac{2}{\pi}\int_0^{\infty} f_c(p)g_c(p)dp = \int_0^{\infty} F(x)G(x)dx$ (b) $\frac{2}{\pi}\int_0^{\infty} f_s(p)g_s(p)dp = \int_0^{\infty} F(x)G(x)dx$

(c) $\frac{2}{\pi}\int_0^{\infty}[f_c(p)]^2dp = \int_0^{\infty}[F(x)]^2dx$ (d) $\frac{2}{\pi}\int_0^{\infty}[f_s(p)]^2dp = \int_0^{\infty}[F(x)]^2dx$

43.3 RELATION BETWEEN FOURIER AND LAPLACE TRANSFORM

Consider the function $$f(t) = \begin{cases} e^{-xt}g(t) & ; \ t<0 \\ 0 & ; \ t>0 \end{cases} \quad \text{... (1)}$$

Then the Fourier transform of $f(t)$ is given by

$$F[f(t)] = \int_{-\infty}^{\infty} e^{ipt}f(t)dt$$

$$= \int_{-\infty}^{0} 0e^{ipt}dt + \int_0^{\infty} e^{-xt}g(t)e^{ipt}dt = \int_0^{\infty} e^{(ip-x)t}g(t)dt$$

$$= \int_0^{\infty} e^{-xst}g(t)dt = L\{g(t)\} \quad \text{[Putting } x-ip=s]$$

Hence, Fourier transformation of the function f(t) defined by (1) is the Laplace transform of function g(t).

43.4 FOURIER TRANSFORMS OF THE DERIVATIVE OF A FUNCTION.

The Fourier transform of the function $\frac{d^nF}{dx^n}$ *is* $(-ip)^n$ *times the Fourier transform of the function f(x) provided that the first* $(n-1)$ *derivatives of F(x) vanish as* $x \to \pm\infty$, *i.e,* $F\{f^n(x)\} = (-ip)^n F\{f(x)\}$

Proof. By definition, we have

$$F\{f^n(x)\} = \int_{-\infty}^{\infty} f^n(x)e^{ipx}dx = \left[f^{n-1}(x)e^{ipx}\right]_{-\infty}^{\infty} - \int_{-\infty}^{\infty} f^{n-1}(x)ipe^{ipx}dx$$

$$= -ip\int_{-\infty}^{\infty} f^{n-1}(x)e^{ipx}dx \quad \left[\because \lim_{x\to\pm\infty} f^{n-1}(x) = 0\right]$$

Repeating this process of integration by parts $(n-1)$ times more and assuming that

$$\lim_{x\to\pm\infty} f^r(x) = 0 \text{ for } r = 1,2,3,\ldots,n-1$$

Thus, finally we have

$$F\{f^n(x)\} = (-ip)^n\int_{-\infty}^{\infty} f(x)e^{ipx}dx$$

Hence, $$F[f^n(x)] = (-ip)^n F[f(x)].$$

REMARK

- In a similar way, we can prove the following results:

(a) $\tilde{f}_c^{2n}(p) = -\sum_{r=0}^{n-1}(-1)^r\alpha_{2n-2r-1}p^{2r} + (-1)^n p^{2n}\tilde{f}_c(p)$ (b) $\tilde{f}_c^{2n+1}(p) = -\sum_{r=0}^{n-1}(-1)^r\alpha_{2n-2r}p^{2r} + (-1)^n p^{2n+1}\tilde{f}_s(p)$

(c) $\tilde{f}_c^{2n}(p) = -\sum_{r=1}^{n}(-1)^r\alpha_{2n-2r}p^{2r-1} + (-1)^{n+1}p^{2n}\tilde{f}_s(p)$ (d) $\tilde{f}_s^{2n+1}(p) = -\sum_{r=1}^{n}(-1)^r\alpha_{2n-2r+1}p^{2r-1} + (-1)^{n+1}p^{2n+1}\tilde{f}_c(p)$

Provided that first $(n-1)$ derivatives of $f(x)$ vanish as $x\to\infty$ and $\frac{d^{n+1}f}{dx^{n-1}} \to \alpha_{n-1}$, etc. $x\to 0$.

Solved Examples

Example 1. *Find the complex Fourier transform of $f(x) = e^{-a|x|}$, where $a > 0$ and x belongs to $]-\infty, \infty[$.* (VTU–2010, SVTU–2008, Kottayam–2005)

Solution. We have $\tilde{f}(p) = \int_{-\infty}^{\infty} f(x)e^{ipx}dx = \int_{-\infty}^{\infty} e^{ipx}e^{-a|x|}dx$

$$= \int_{-\infty}^{0} e^{ipx}e^{ax}dx + \int_{0}^{\infty} e^{ipx}e^{-ax}dx$$

$$= \int_{-\infty}^{0} e^{(a+ip)x}dx + \int_{0}^{\infty} e^{-(a-ip)x}dx$$

$$= \left[\frac{e^{(a+ip)x}}{a+ip}\right]_{-\infty}^{0} + \left[\frac{e^{-(a-ip)x}}{-(a-ip)}\right]_{0}^{\infty}$$

$$= \frac{1}{a+ip} + \frac{1}{a-ip} = \frac{2a}{a^2+p^2}$$

Example 2. *Find the Fourier transform of $f(x)$ defined by*

$$f(x) = \begin{cases} 1 \ ; & |x| < a \\ 0 \ ; & |x| > a \end{cases}$$ *and hence evaluate*

(i) $\int_{-\infty}^{\infty} \frac{\sin pa \cos px}{p}dx$ *and* (ii) $\int_{0}^{\infty} \frac{\sin p}{p}dp$

(WBTU–2005, Madras–2003, PTU–2003, Kottayam–2005)

Solution. We have $|x| < a \Rightarrow -a < x < a$

and $|x| > a \Rightarrow x < -a$ or $x > a$

Thus, $$f(x) = \begin{cases} 1 \ ; & \text{if } -a < x < a \\ 0 \ ; & \text{if } x < -a \text{ or } x > a \end{cases} \quad \ldots(1)$$

So, $\tilde{f}(p) = F[f(x)]$

$$= \int_{-\infty}^{\infty} e^{ipx} f(x)dx$$

$$= \int_{-\infty}^{-a} e^{ipx} f(x)dx + \int_{-a}^{a} e^{ipx} f(x)dx + \int_{a}^{\infty} e^{ipx} f(x)dx$$

$$= 0 + \int_{-a}^{a} e^{ipx}dx + 0 \qquad \text{[Using (1)]}$$

$$= \left[\frac{e^{ipx}}{ip}\right]_{-a}^{a} = \frac{1}{ip}(e^{ipa} - e^{-ipa}) = \frac{1}{ip}(2i\sin pa)$$

$$\Rightarrow \tilde{f}(p) = [Ff(x)] = \frac{2\sin pa}{p} \quad \ldots(2)$$

Using corresponding inversion formula, we get

$$\frac{1}{2\pi}\int_{-\infty}^{\infty} \tilde{f}(p)e^{-ipx}dp = f(x)$$

$$\Rightarrow \int_{-\infty}^{\infty} \frac{2\sin pa}{p}e^{-ipx}dp = 2\pi f(x)$$

$$\Rightarrow 2\int_{-\infty}^{\infty} \frac{\sin pa}{p}(\cos px - i\sin px)dp = 2\pi f(x)$$

$$\Rightarrow \int_{-\infty}^{\infty} \frac{1}{p}(\sin pa \cos px - i\sin pa \sin px)dx = \pi f(x).$$

Equating real parts of both the sides, we get

$$\int_{-\infty}^{\infty} \frac{\sin pa \cos px}{p}dx = \pi f(x)$$

$$\Rightarrow \int_{-\infty}^{\infty} \frac{\sin pa \cos px}{p}dx = \begin{cases} \pi \ ; & |x| < a \\ 0 \ ; & |x| > a \end{cases}$$

Putting $x = 0$ and $a = 1$, we get

$$\int_{-\infty}^{\infty} \frac{\sin p}{p}dp = \pi \text{ or } 2\int_{-\infty}^{\infty} \frac{\sin p}{p}dp = \pi$$

Hence, $\int_{0}^{\infty} \frac{\sin p}{p}dp = \frac{\pi}{2}$.

Example 3. *Find Fourier sine and cosine transforms of $f(x) = x$*

Solution. By definition

$$\tilde{f}_c(p) = \int_{0}^{\infty} x\cos px dx \text{ and } \tilde{f}_s(p) = \int_{0}^{\infty} x\sin px dx$$

$$\Rightarrow \tilde{f}_c(p) - i\tilde{f}_s(p) = \int_{0}^{\infty} x(\cos px - i\sin px)dx$$

$$= \int_{0}^{\infty} xe^{-ipx}dx = \int_{0}^{\infty} \frac{y}{ip}e^{-y}\frac{dy}{ip}$$

[By putting $ipx = y$]

$$= -\frac{1}{p^2}\int_{0}^{\infty} y^{2-1}.e^{-y}dy = -\frac{\Gamma(2)}{p^2}$$

$$\Rightarrow \tilde{f}_c(p) - i\tilde{f}_s(p) = -\frac{1}{p^2} + i.0$$

Equation real and imaginary parts of both the sides, we get

$$\tilde{f}_c(p) = -\frac{1}{p^2} \text{ and } \tilde{f}_s(p) = 0$$

Example 4. *Find $f(x)$, if its Fourier cosine transform is $\frac{1}{(1+p^2)}$.*

Solution. Let $f(x) = F_c^{-1}\left\{\frac{1}{1+p^2}\right\}$. By inversion formula, we have

$$F(x) = \frac{2}{\pi}\int_{0}^{\infty} \frac{1}{1+p^2}\cos px dp \quad \ldots(1)$$

$$\Rightarrow \frac{dF}{dx} = \frac{2}{\pi}\int_{0}^{\infty} \frac{-p\sin px}{1+p^2}dp$$

$$= -\frac{2}{\pi}\int_{0}^{\infty} \frac{-p\sin px}{1+p^2}dp$$

$$= -\frac{2}{\pi}\int_{0}^{\infty} \frac{(1+p^2)-1}{1+p^2}.\frac{\sin px}{p}dp$$

$$= -\frac{2}{\pi}\int_{0}^{\infty} \left(1 - \frac{1}{1+p^2}\right).\frac{\sin px}{p}dp + \frac{2}{\pi}\int_{0}^{\infty} \frac{\sin px}{p(1+p^2)}dp$$

$$= -\frac{2}{\pi}.\frac{\pi}{2} + \frac{2}{\pi}\int_{0}^{\infty} \frac{\sin px}{p(1+p^2)}dx$$

$$\left[\because \int_{0}^{\infty} \frac{\sin px}{p}dp = \frac{\pi}{2}\right]$$

$$= -1 + \frac{2}{\pi}\int_{0}^{\infty} \frac{\sin px}{p(1+p^2)}dx \quad \ldots(2)$$

Also, $\dfrac{d^2F}{dx^2} = \dfrac{2}{\pi}\int_0^\infty \dfrac{p\cos px}{p(1+p^2)}dp$

$= \dfrac{2}{\pi}\int_0^\infty \dfrac{\cos px}{(1+p^2)}dx$

$\Rightarrow \dfrac{d^2F}{dx^2} = F(x)$ [Using (1)]

$\Rightarrow (D^2-1)F = 0$ ...(3)

$\Rightarrow F(x) = Ae^{-x} + Be^x$ ...(4)

Now, $F(0) = \dfrac{2}{\pi}\int_0^\infty \dfrac{dp}{1+p^2} = \dfrac{2}{\pi}\left[\tan^{-1}p\right]_0^\infty$

$= \dfrac{2}{\pi}.\dfrac{\pi}{2} = 1$

and $F(0) = A+B$ ...(5)

$\therefore A+B = 1$ ...(6)

From (4) $\dfrac{dF}{dx} = -Ae^{-x} + Be^x$.

Putting $x = 0$ in (2) and (6), we get

$\dfrac{dF}{dx} = -1$ and $\dfrac{dF}{dx} = -A+B$

$\Rightarrow -A+B = 1$ ...(7)

On solving (5) and (7), we get $A = 1$, $B = 0$. Hence from (4) $F(x) = e^{-x}$

Example 5. *If* $f(x) = \begin{cases} 1 \; ; & |x| < a \\ 0 \; ; & |x| > a \end{cases}$ *and* $\tilde{f}(p) = \dfrac{2\sin pa}{p}$, *where* $p \neq 0$ *then show that* $\int_0^\infty \dfrac{\sin^2 ax}{x}dx = \dfrac{\pi a}{2}$.

(VTU–2010, SVTU–2009, UPTU–2008)

Solution. Using Parseval's identity for Fourier transform, we get

$$\int_{-\infty}^{\infty}[f(x)]^2dx = \frac{1}{2\pi}\int_{-\infty}^{\infty}[\tilde{f}(p)]^2dp$$

$$\Rightarrow \int_{-\infty}^{-a}[f(x)]^2dx + \int_{-a}^{a}[f(x)]^2dx + \int_a^\infty[f(x)]^2dx = \frac{1}{2\pi}\int_{-\infty}^{\infty}[\tilde{f}(p)]^2dp$$

$$\Rightarrow 0 + \int_{-a}^{a}1^2dx + 0 = \frac{1}{2\pi}\int_{-\infty}^{\infty}\frac{4\sin^2 pa}{p^2}$$

$$\Rightarrow 2a = \frac{2}{\pi}\int_{-\infty}^{\infty}\frac{\sin^2 ax}{x^2}dx \text{ or } \int_{-\infty}^{\infty}\frac{\sin^2 ax}{x^2}dx = \pi a$$

or $2\int_0^\infty \dfrac{\sin^2 ax}{x^2}dx = \pi a$ or $\int_0^\infty \dfrac{\sin^2 ax}{x^2}dx = \dfrac{\pi a}{2}$.

Example 6. *Find the Fourier transform of f(x), if*

$$f(x) = \begin{cases} \dfrac{\sqrt{2\pi}}{2\varepsilon} & , |x| \le \varepsilon \\ 0 & , |x| > \varepsilon \end{cases}$$

Solution. We have,

$$F\{f(x)\} = \frac{1}{2\pi}\int_{-\infty}^{\infty}e^{ipx}f(x)dx$$

$$= \frac{1}{\sqrt{2\pi}}\int_{-\varepsilon}^{\varepsilon}e^{ipx}.\frac{\sqrt{2\pi}}{2\varepsilon}dx = \frac{1}{2\varepsilon}\left[\frac{e^{ipx}}{ip}\right]_{-\varepsilon}^{\varepsilon}$$

$$= \frac{e^{ipt} - e^{-ipt}}{2ip\varepsilon} = \frac{\sin p\varepsilon}{p\varepsilon}.$$

Example 7. *Find the Fourier transform of f(x), if*

$$f(x) = \begin{cases} x^2 & , |x| < a \\ 0 & , |x| > a \end{cases}$$

(SVTU–2008)

Solution. We have

$$\tilde{f}(p) = \frac{1}{\sqrt{2\pi}}\int_{-\infty}^{\infty}f(x)e^{ipx}dx = \frac{1}{2\pi}\int_{-a}^{a}x^2e^{ipx}dx$$

$$= \frac{1}{\sqrt{2\pi}}.\left[x^2.\frac{e^{ipx}}{ip} - 2x.\frac{e^{ipx}}{(ip)^2} + 2.\frac{e^{ipx}}{(ip)^3}\right]_{-a}^{a}$$

$$= \frac{1}{\sqrt{2\pi}}\left[\frac{a^2}{ip}(e^{ipa} - e^{-ipa}) + \frac{2a}{p^2}(e^{ipa} + e^{-ipa}) - \frac{2}{ip^3}(e^{ipa} - e^{-ipa})\right]$$

$$= \frac{1}{p^3\sqrt{2\pi}}\left[\frac{a^2p^2}{i}.2i\sin pa + 2ap.2\cos pa - \frac{2}{i}.2i\sin pa\right]$$

$$= \frac{1}{p^3}\sqrt{\frac{2}{\pi}}\left[(a^2p^2 - 2)\sin pa + 2ap.\cos pa\right]$$

Example 8. *Find the Fourier transform of*

$$F(x) = \begin{bmatrix} 1 - x^2 & , |x| \le 1 \\ 0 & , |x| > 1 \end{bmatrix}$$

(VTU–2011, Anna–2005, 12, Mumbai –2005)

and hence evaluate

$$\int_0^\infty\left(\frac{x\cos x - \sin x}{x^3}\right)\cos\frac{x}{2}dx$$

Solution. We have

$$\tilde{F}(p) = \frac{1}{\sqrt{2\pi}}\int_{-\infty}^{\infty}e^{ipx}F(x)dx$$

$$= \frac{1}{\sqrt{2\pi}}\int_{-1}^{1}(1-x^2)e^{ipx}dx$$

$$= \frac{1}{\sqrt{2\pi}}\left(\frac{1-x^2}{ip}e^{ipx}\right)_{-1}^{1} + \frac{2}{\sqrt{2\pi}}\int_{-1}^{1}x.\frac{e^{ipx}}{ip}dx$$

$$= \frac{\sqrt{2}}{ip\sqrt{\pi}}\left[\left(\frac{xe^{ipx}}{ip}\right)_{-1}^{1} - \int_{-1}^{1}1.\frac{e^{ipx}}{ip}dx\right]$$

$$= \frac{\sqrt{2}}{ip\sqrt{\pi}}\left\{\frac{e^{ip} + e^{-ip}}{ip} - \frac{(e^{ipx}) - 1}{(ip)^2}\right\}$$

$$= \frac{\sqrt{2}}{ip\sqrt{\pi}}\left[\frac{2\cos p}{ip} + \frac{e^{ip} - e^{-ip}}{p^2}\right]$$

$$= \frac{\sqrt{2}}{\sqrt{\pi}}\left[-\frac{2\cos p}{p^2} + \frac{2i\sin p}{ip^3}\right]$$

$$= -2\sqrt{\frac{2}{\pi}}\cdot\left[\frac{p\cos p - \sin p}{p^3}\right]$$

If $\tilde{F}(p) = \frac{1}{\sqrt{2\pi}}\int_{-\infty}^{\infty} F(x)e^{ipx}dx$

then $F(x) = \frac{1}{\sqrt{2\pi}}\int_{-\infty}^{\infty} \tilde{F}(p)e^{-ipx}dp$

$$\therefore \frac{1}{\sqrt{2\pi}}\int_{-\infty}^{\infty} \frac{2\sqrt{2/\pi}(p\cos p - \sin p)}{p^3}.e^{-ipx}dp$$

$$= \begin{cases} 1 - x^2, & |x| < 1 \\ 0, & |x| > 1 \end{cases}$$

or $-\int_{-\infty}^{\infty}\left(\frac{p\cos p - \sin p}{p^3}\right)\cos pxdp$

or $+i\int_{-\infty}^{\infty}\left(\frac{p\cos p - \sin p}{p^3}\right)\sin pxdp$

$$= \begin{cases} \frac{\pi}{2}(1 - x^2), & |x| < 1 \\ 0, & |x| > 1 \end{cases}$$

or $-\int_{-\infty}^{\infty}\frac{p\cos p - \sin p}{p^3}\cos pxdp$

$$= \begin{cases} \frac{\pi}{2}(1 - x^2), & |x| < 1 \\ 0, & |x| > 1 \end{cases}$$

Now taking $x = \frac{1}{2}$, we have

$$-\int_{-\infty}^{\infty}\frac{p\cos p - \sin p}{p^3}.\cos\frac{p}{2}dp = \frac{\pi}{2}\left(1 - \frac{1}{4}\right) = \frac{3\pi}{8}$$

or $2\int_0^{\infty}\frac{p\cos p - \sin p}{p^3}.\cos\frac{p}{2}dp = -\frac{3\pi}{8}$

or $\int_0^{\infty}\left(\frac{x\cos x - \sin x}{x^3}\right).\cos\frac{x}{2}dx = -\frac{3\pi}{16}$.

Example9. *Find the cosine transform of a function of x which is unity for $0 < x < a$ and zero for $x \geq a$. What is the function whose cosine transform is $\sqrt{\frac{2}{\pi}}\frac{\sin ap}{p}$?*

(VTU–2010, SVTU–2009 UPTU–2008)

Solution. It is given that $f(x) = \begin{cases} 1, & 0 < x < a \\ 0, & x \geq a \end{cases}$

We have $\tilde{f}_c(p) = \sqrt{\frac{2}{\pi}}\int_0^{\infty} f(x)\cos pxdx$

$$= \sqrt{\frac{2}{\pi}}\int_0^{a}\cos pxdx = \sqrt{\frac{2}{\pi}}\frac{\sin pa}{p}$$

Again, $f(x) = \sqrt{\frac{2}{\pi}}\int_0^{\infty}\tilde{f}_c(p)\cos pxdp$

$$= \sqrt{\frac{2}{\pi}}\int_0^{\infty}\sqrt{\frac{2}{\pi}}\frac{\sin pa}{p}\cos pxdp$$

$$= \frac{1}{\pi}\int_0^{\infty}\frac{\sin(a+x)p + \sin(a-x)p}{p}dp$$

$$= \frac{1}{\pi}\int_0^{\infty}\frac{\sin(a+x)p}{p}dp + \frac{1}{\pi}\int_0^{\infty}\frac{\sin(a-x)p}{p}dp$$

$$= \frac{1}{\pi}\left(\frac{\pi}{2} + \frac{\pi}{2}\right) = 1 \text{ if } x < a$$

and $= \frac{1}{\pi}\left(\frac{\pi}{2} - \frac{\pi}{2}\right) = 0 \text{ if } x > a$

$$\left[\because \int_0^{\infty}\frac{\sin ax}{x}dx = \frac{\pi}{2}, \text{ if } a > 0\right]$$

Example 10. *Find the Fourier sine and cosine transform of f(x), if*

$$f(x) = \begin{cases} x, & 0 < x < 1 \\ 2 - x, & 1 < x < 2 \\ 0, & x > 2 \end{cases}$$

[JNTU–2006]

Solution. We have

$$\tilde{f}_s(p) = \sqrt{\frac{2}{\pi}}\int_0^{\infty} f(x)\sin pxdx$$

$$= \sqrt{\frac{2}{\pi}}.\left[\int_0^1 x\sin pxdx + \int_1^2(2-x)\sin pxdx\right]$$

$$= \sqrt{\frac{2}{\pi}}.\left[\left(-\frac{x}{p}\cos px + \frac{1}{p^2}\sin px\right)_0^1 + \left\{-\left(\frac{2-x}{p}\right)\cos px - \frac{1}{p^2}\sin px\right\}_1^2\right]$$

$$= \sqrt{\frac{2}{\pi}}.\left[\frac{2}{p^2}\sin p - \frac{1}{p^2}\sin 2p\right]$$

$$= 2\sqrt{\frac{2}{\pi}}.\frac{\sin p}{p^2}(1 - \cos p)$$

$$\tilde{f}_c(p) = \sqrt{\frac{2}{\pi}}\int_0^{\infty} f(x)\cos pxdx$$

$$= \sqrt{\frac{2}{\pi}}\left[\int_0^1 x\cos pxdx + \int_1^2(2-x)\cos pxdx\right]$$

$$= \sqrt{\frac{2}{\pi}}\left[\left(\frac{x}{p}\sin px + \frac{1}{p^2}\cos px\right)_0^1 + \left\{\left(\frac{2-x}{p}\right)\sin px - \frac{\cos px}{p^2}\right\}_1^2\right]$$

$$= \sqrt{\frac{2}{\pi}}\frac{1}{p^2}[2\cos p - 1 - \cos 2p]$$

$$= 2\sqrt{\frac{2}{\pi}}.\frac{\cos p}{p^2}(1 - \cos p)$$

Example 11. *Find Fourier sine and cosine transforms of e^{-x} and using the inversion formulae recover the original functions, in both the cases.*

Solution. Let $f(x) = e^{-x}$

Then

$$\tilde{f}_s(p) = \sqrt{\frac{2}{\pi}}\int_0^\infty f(x)\sin px dx$$

$$= \sqrt{\frac{2}{\pi}}\int_0^\infty e^{-x}\sin px dx$$

$$= \sqrt{\frac{2}{\pi}}\left[\frac{e^{-x}}{1+p^2}(-\sin px - p\cos px)\right]_0^\infty$$

$$= \frac{p}{1+p^2}\sqrt{\frac{2}{\pi}}$$

and $\tilde{f}_c(p) = \sqrt{\frac{2}{\pi}}\int_0^\infty f(x)\cos px dx$

$$= \sqrt{\frac{2}{\pi}}\int_0^\infty e^{-x}\cos px dx$$

$$= \sqrt{\frac{2}{\pi}}\left[\frac{e^{-x}}{1+p^2}(-\cos px + p\sin px)\right]_0^\infty$$

$$= \frac{1}{1+p^2}\sqrt{\frac{2}{\pi}}.$$

Applying inversion to the sine transform, we have

$$f(x) = \sqrt{\frac{2}{\pi}}\int_0^\infty \tilde{f}_s(p).\sin px dp$$

$$= \frac{2}{\pi}\int_0^\infty \frac{p\sin px}{1+p^2}dp$$

and applying inversion to the cosine transform, we have

$$f(x) = \sqrt{\frac{2}{\pi}}\int_0^\infty \tilde{f}_c(p)\cos px dp$$

$$= \frac{2}{\pi}\int_0^\infty \frac{\cos px}{1+p^2}dp$$

Therefore, form Fourier integral theorem, we have

$$f(x) = \frac{1}{\pi}\int_0^\infty dp\int_{-\infty}^\infty f(v)\cos p(x-v)dv$$

or $$f(x) = \frac{1}{\pi}\int_0^\infty \cos px dp\int_{-\infty}^\infty f(v)\cos pv dv$$

$$+\frac{1}{\pi}\int_0^\infty \sin px dp\int_0^\infty f(v)\sin pv dv \quad \text{... (3)}$$

Case I: Define $f(x)$ in $[-\infty, 0]$ such that $f(x)$ in an even function of x, from (3), we have

$$f(x) = \frac{2}{\pi}\int_0^\infty \cos px dp\int_0^\infty f(v)\cos pv dv$$

Taking $f(x) = e^{-x}$, we have

$$e^{-x} = \frac{2}{\pi}\int_0^\infty \cos px dp\int_0^\infty e^{-v}\cos pv dv$$

$$= \frac{2}{\pi}\int_0^\infty \cos px\left[\frac{e^{-x}}{1+p^2}(-\cos pv + p\sin pv)\right]_0^\infty dp$$

$$= \frac{2}{\pi}\int_0^\infty \frac{\cos px}{1+p^2}dp$$

$$\therefore \quad \int_0^\infty \frac{\cos px}{1+p^2}dp = \frac{\pi}{2}e^{-x}$$

$\therefore$ From(2), we have $f(x) = \frac{2}{\pi}.\frac{\pi}{2}e^{-x} = e^{-x}$

Case II. Now, we define $f(x)$ in $(-\infty, 0)$ such that $f(x)$ in an odd function of x from (2), we have

$$f(x) = \frac{2}{\pi}\int_0^\infty \sin px dp\int_0^\infty f(v)\sin pv dv$$

Taking $f(x) = e^{-x}$ and simplifying, we have

$$\int_0^\infty \frac{p\sin px}{1+p^2}dp = \frac{\pi}{2}e^{-x}$$

$\therefore$ from (1), $f(x) = \frac{2}{\pi}.\frac{\pi}{2}e^{-x} = e^{-x}$.

Example 12. *Find Fourier cosine transform of* $f(x) = \frac{1}{1+x^2}$ *and hence find Fourier sine transform of* $F(x) = \frac{x}{1+x^2}$.

Solution. We have

$$\tilde{f}_c(p) = \sqrt{\frac{2}{\pi}}\int_0^\infty f(x)\cos px dx = \sqrt{\frac{2}{\pi}}\int_0^\infty \frac{\cos px}{1+x^2}dx$$

Differentiating both sides w.r.t p, we get

$$\frac{d}{dp}\tilde{f}_c(p) = -\sqrt{\frac{2}{\pi}}\int_0^\infty \frac{x\sin px}{1+x^2}dx$$

$$= -\sqrt{\frac{2}{\pi}}\int_0^\infty \frac{(x^2+1-1)\sin px}{x(1+x^2)}dx$$

$$= -\sqrt{\frac{2}{\pi}}\int_0^\infty \frac{\sin px}{x}dx + \sqrt{\frac{2}{\pi}}\int_0^\infty \frac{\sin px}{x(1+x^2)}dx$$

$$= -\sqrt{\frac{2}{\pi}}.\frac{\pi}{2} + \sqrt{\frac{2}{\pi}}\int_0^\infty \frac{\sin px}{x(1+x^2)}dx$$

Differentiating again w.r.t p, we get

$$\frac{d^2}{dp^2}\tilde{f}_c(p) = \sqrt{\frac{2}{\pi}}\int_0^\infty \frac{\cos px}{1+x^2}dx = \tilde{f}_c(p)$$

or $(D^2-1)\tilde{f}_c(p) = 0$

whose general solution is $\tilde{f}_c(p) = Ae^p + Be^{-p}$ ... (1)

Now when $p = 0$, $\tilde{f}_c(p) = \sqrt{\frac{2}{\pi}}\int_0^\infty \frac{dx}{1+x^2}$

$$= \sqrt{\frac{2}{\pi}}\left[\tan^{-1}x\right]_0^\infty = \frac{\pi}{2}\sqrt{\frac{2}{\pi}} = \sqrt{\frac{\pi}{2}}$$

and $\frac{d}{dp}\tilde{f}_c(p) = -\sqrt{\frac{\pi}{2}}$

$\therefore$ from (1), we have

$$\sqrt{\frac{\pi}{2}} = A + B \text{ and } -\sqrt{\frac{\pi}{2}} = A - B$$

Solving, $A = 0, B = \sqrt{\frac{\pi}{2}}$

$\therefore$ from (1), we have $\tilde{f}_c(p) = \sqrt{\frac{\pi}{2}}e^{-p}$

Now, we have

$$\tilde{f}_c(p) = \sqrt{\frac{2}{\pi}}\int_0^\infty \frac{\cos px}{1+x^2}dx = \sqrt{\frac{\pi}{2}}.e^{-p}$$

Now differentiating both sides w.r.t. p, we have

$$-\int_0^\infty \frac{x\sin px}{1+x^2}dx = -\frac{\pi}{2}.e^{-p}$$

or $\tilde{F}_s(p) = \sqrt{\frac{2}{\pi}}\int_0^\infty \frac{x}{1+x^2}\sin pxdx = \sqrt{\frac{\pi}{2}}.e^{-p}$

Example 13. *Find the sine and cosine transform of* $x^n e^{-ax}$

Solution. Let $f(x) = x^n e^{-ax}$

$$\therefore \quad \tilde{f}_s(p) = \sqrt{\frac{2}{\pi}}\int_0^\infty f(x)\sin pxdx$$

$$= \sqrt{\frac{2}{\pi}}\int_0^\infty x^n e^{-ax}\sin pxdx \qquad \ldots(1)$$

We have

$$\int_0^\infty e^{-ax}\sin pxdx$$

$$= \left[\frac{e^{-ax}}{a^2+p^2}(-a\sin px - p\cos px)\right]_0^\infty$$

$$= \frac{p}{a^2+p^2} = \frac{1}{2i}\left(\frac{1}{a-ip} - \frac{1}{a+ip}\right)$$

Differentiating both sides w.r.t. a, n times, we have

$$(-1)^n\int_0^\infty x^n e^{-ax}\sin pxdx$$

$$= \frac{1}{2i}\left[\frac{d^n}{da^n}(a-ip)^{-1} - \frac{d^n}{da^n}(a+ip)^{-1}\right]$$

$$= \frac{1}{2i}(-1)^n n!\left[(a-ip)^{-(n+1)} - (a+ip)^{-(n+1)}\right]$$

$$= \frac{1}{2i}(-1)^n n!\left[2ir^{-(n+1)}\sin(n+1)\theta\right]$$

Putting $a = r\cos\theta$, $p = r\sin\theta$

$$= (-1)^n n!(1/r)^{n+1}\sin(n+1)\theta$$

$$\therefore \int_0^\infty x^n e^{-ax}\sin pxdx = n!.[1/(a^2+p^2)^{(n+1)/2}]\sin\left\{(n+1)\tan^{-1}(p/a)\right\}$$

$$[\because r = (a^2+p^2)^{1/2} \text{ and } \theta = \tan^{-1}(p/a)]$$

Hence, from (1)

$$\tilde{f}_s(p) = \sqrt{\frac{2}{\pi}}\frac{n!\sin\{(n+1)\tan^{-1}(p/a)\}}{(a^2+p^2)^{(n+1)/2}}$$

Also, $\tilde{f}_c(p) = \sqrt{\frac{2}{\pi}}\int_0^\infty f(x)\cos pxdx$

$$\tilde{f}_c(p) = \sqrt{\frac{2}{\pi}}\int_0^\infty x^n e^{-ax}\cos pxdx \qquad \ldots(2)$$

We have

$$\int_0^\infty e^{-ax}\cos pxdx$$

$$= \left[\frac{e^{-ax}}{a^2+p^2}(-a\cos px + p\sin px)\right]_0^\infty$$

$$= \frac{a}{a^2+p^2} = \frac{1}{2}\left(\frac{1}{a-ip} + \frac{1}{a+ip}\right)$$

Differentiating both sides w.r.t. a, n times, we have

$$(-1)^n\int_0^\infty x^n e^{-ax}\cos pxdx$$

$$= \frac{1}{2}(-1)^n n!\left[(a-ip)^{-(n+1)} + (a+ip)^{-(n+1)}\right]$$

or $\int_0^\infty x^n e^{-ax}\cos pxdx$

$$= (-1)^n n!(1/r)^{n+1}\cos(n+1)\theta$$

Putting $a = r\cos\theta$, $p = r\sin\theta$ and on simplification

$$= (-1)^n n!\frac{\cos\{(n+1)\tan^{-1}(p/a)\}}{(a^2+p^2)^{(n+1)/2}}$$

Hence from (2), we get

$$\tilde{f}_c(p) = \sqrt{\frac{2}{\pi}}\frac{n!\cos\{(n+1)\tan^{-1}(p/a)\}}{(a^2+p^2)^{(n+1)/2}}.$$

Example 14. *Find Fourier sine transform of* $e^{-ax/x}$.

[PTU–2006, VTU–2010, Rohtak–2005]

Solution. If $f(x) = \frac{e^{-ax}}{x}$, then we have

$$\tilde{f}_s(p) = \sqrt{\frac{2}{\pi}}\int_0^\infty \frac{e^{-ax}}{x}\sin pxdx$$

Differentiating both sides w.r.t. p, we have

$$\frac{d}{dp}\tilde{f}_s(p) = \sqrt{\frac{2}{\pi}}\int_0^\infty e^{-ax}\cos pxdx$$

$$= \sqrt{\frac{2}{\pi}}\left[\frac{e^{-ax}}{a^2+p^2}(-a\cos px + p\sin px)\right]_0^\infty$$

$$= \frac{a}{a^2+p^2}.\sqrt{\frac{2}{\pi}}$$

$$\therefore \quad \tilde{f}_s(p) = a\sqrt{\frac{2}{\pi}}\int\frac{dp}{a^2+p^2} + c$$

$$= \sqrt{2/\pi}\tan^{-1}(p/a) + c$$

But when $p = 0$, $\tilde{f}_s(p) = 0$

$\therefore \quad c = 0$

Hence, $\tilde{f}_s(p) = \sqrt{2/\pi}.\tan^{-1}(p/a)$.

Example 15. *Find the Fourier sine transform of*

$$f(x) = \frac{1}{x(a^2+x^2)} \qquad \text{(UPTU–2008)}$$

Solution. We have

$$\tilde{f}_s(p) = \sqrt{\frac{2}{\pi}}\int_0^\infty \frac{1}{x(a^2+x^2)}\sin pxdx \quad ...(1)$$

Let $$I = \int_0^\infty \frac{1}{x(a^2+x^2)}\sin pxdx \quad ...(2)$$

Then $$\frac{dI}{dp} = \frac{d}{dp}\int_0^\infty \frac{\sin px}{x(a^2+x^2)}dx$$

$$= \int_0^\infty \left[\frac{\partial}{\partial p}\left\{\frac{\sin px}{x(a^2+x^2)}\right\}\right]dx$$

$$= \int_0^\infty \frac{\cos px}{a^2+x^2}dx \quad ...(3)$$

$$\therefore \quad \frac{d^2I}{dp^2} = -\int_0^\infty \frac{x\sin px}{a^2+x^2}dx = -\int_0^\infty \frac{x^2 \sin px}{x(a^2+x^2)}dx$$

$$= -\int_0^\infty \frac{(x^2+a^2)-a^2}{x(a^2+x^2)}\sin pxdx$$

$$= -\int_0^\infty \frac{\sin px}{x}dx + a^2\int_0^\infty \frac{\sin px}{x(a^2+x^2)}dx$$

$$= -\frac{\pi}{2} + a^2 I. \quad \left[\because \int_0^\infty \frac{\sin px}{x}dx = \frac{\pi}{2}\right]$$

$$\therefore \quad \frac{d^2I}{dp^2} - a^2 I = -\frac{\pi}{2}$$

or $(D^2 - a^2)I = -\dfrac{\pi}{2}$, where $D \equiv \dfrac{d}{dp}$

The solution of the above differential equation is

$$I = Ae^{-ap} + Be^{ap} + \frac{\pi}{2a} \quad ...(4)$$

$$\therefore \quad \frac{dI}{dp} = -Aae^{-ap} + Bae^{-ap} \quad ...(5)$$

Now from (1), when $p = 0$, we have $I = 0$ and from (2), when $p = 0$, we have

$$\frac{dI}{dp} = \int_0^\infty \frac{1}{a^2+x^2}dx = \frac{1}{a}\left[\tan^{-1}\frac{x}{a}\right]_0^\infty = \frac{\pi}{2a}$$

So putting $p = 0$ in (4) and (5), we get

$$A + B = -\frac{\pi}{2a^2} \quad ...(6)$$

and $a(-A+B) = \dfrac{\pi}{2a^2}$ i.e., $-A+B = \dfrac{\pi}{2a^2}$ ...(7)

Solving (6) and (7), we get $B = 0, A = -\dfrac{\pi}{2a^2}$ ...(8)

Putting the values of A and B in (4), we get

$$I = \int_0^\infty \frac{\sin px}{x(a^2+x^2)}dx = -\frac{\pi}{2a^2}e^{-ap} + \frac{\pi}{2a^2}$$

$$= \frac{\pi}{2a^2}(1-e^{-ap}).$$

Now putting the value of I in (1), we get

$$\tilde{f}_s(p) = \sqrt{\frac{2}{\pi}}\cdot\frac{\pi}{2a^2}(1-e^{-ap}) = \frac{1}{a^2}\sqrt{\frac{\pi}{2}}.(1-e^{-ap})$$

Example 16. *Find the Fourier cosine transform of e^{-x^2}.*

[VTU–2010, Rajasthan–2006, Anna–2009]

Solution. We have

$$\tilde{F}_c\{e^{-x^2}\} = \sqrt{\frac{2}{\pi}}\int_0^\infty e^{-x^2}\cos pxdx = I$$

Differentiating w.r.t. 'p', we have

$$\frac{dI}{dp} = -\sqrt{\frac{2}{\pi}}\int_0^\infty xe^{-x^2}\sin pxdx$$

$$= \frac{1}{2}\sqrt{\frac{2}{\pi}}\int_0^\infty (-2xe^{-x^2}).\sin pxdx$$

$$= \frac{1}{2}\sqrt{\frac{2}{\pi}}\left[(e^{-x^2}\sin px)_0^\infty - p\int_0^\infty e^{-x^2}\cos pxdx\right]$$

(Integrating by parts taking sin px as first function.)

$$= -\frac{p}{2}I$$

$$\therefore \quad \frac{dI}{I} = -\frac{p}{2}dp$$

Integrating, we have

$$\log I = -\frac{p^2}{4} + \log A \text{ or } I = Ae^{-p^2/4} \quad ...(2)$$

But when $p = 0$, from (1),

$$I = \sqrt{\frac{2}{\pi}}\int_0^\infty e^{-x^2}dx = \frac{1}{\sqrt{2}}$$

$\therefore$ From (2), $A = 1/\sqrt{2}$

Hence $I = F_c\{e^{-x^2}\} = (1/\sqrt{2})e^{-p^2/4}$

Example 17. *Find f(x) if its sine transform is $\dfrac{\pi}{2}$.*

Solution. We have $\tilde{f}_s(p) = \dfrac{\pi}{2}$

$\therefore$ Applying Fourier sine inversion formula, we have

$$f(x) = \sqrt{\frac{2}{\pi}}\int_0^\infty \tilde{f}_s(p)\sin pxdp$$

$$f(x) = \sqrt{\frac{2}{\pi}}\int_0^\infty \frac{\pi}{2}.\sin pxdp = \sqrt{\frac{\pi}{2}}\int_0^\infty \sin pxdp \quad ...(1)$$

Now, we have

$$\int_0^\infty e^{-ipx}dp = \left[\frac{e^{-ipx}}{(-ix)}\right]_{p=0}^\infty = 1/(ix) = -i/x$$

or $\int_0^\infty (\cos px - i\sin px)dp = -i/x$.

Equating imaginary parts on both sides, we have

$$\int_0^\infty \sin pxdp = 1/x$$

Hence form (1) $f(x) = \sqrt{(\pi/2)}.(1/x)$

Example 18. *Find the inverse Fourier transform of $\tilde{f}(p) = e^{-|p|y}$.*

Solution. We have $|p| = \begin{cases} -p & , \quad p \le 0 \\ p & , \quad p \ge 0 \end{cases}$

$$\therefore \; f(x) = \frac{1}{\sqrt{2\pi}} \int_{-\infty}^{\infty} \tilde{f}(p).e^{-ipx} dp$$

$$= \frac{1}{\sqrt{2\pi}} \int_{-\infty}^{\infty} e^{-|p|y}.e^{-ipx} dp$$

$$= \frac{1}{\sqrt{2\pi}} \int_{-\infty}^{0} e^{py}.e^{-ipx} dp + \frac{1}{\sqrt{2\pi}} \int_{0}^{\infty} e^{-py}.e^{-ipx} dp$$

$$= \frac{1}{\sqrt{2\pi}} \int_{-\infty}^{0} e^{(y-ix)p} dp + \frac{1}{\sqrt{2\pi}} \int_{0}^{\infty} e^{-p(y+ix)} dp$$

$$= \frac{1}{\sqrt{2\pi}} \left[\frac{e^{(y-ix)p}x}{(y-ix)} \right]_{-\infty}^{0} + \frac{1}{\sqrt{2\pi}} \left[\frac{e^{-p(y+ix)}}{-(y+ix)} \right]_{0}^{\infty}$$

$$= \frac{1}{\sqrt{2\pi}} \left[\frac{1}{(y-ix)} + \frac{1}{(y+ix)} \right] = \frac{y\sqrt{2}}{\sqrt{\pi}(y^2+x^2)}.$$

Example 19. *Find f(x) if $\tilde{f}_c(p) = p^n e^{-ap}$.*

Solution. Using Fourier cosine inversion formula, we have

$$f(x) = \sqrt{\frac{2}{\pi}} \int_0^{\infty} p^n e^{-ap} \cos pxdp \qquad \text{... (1)}$$

We have $\int_0^{\infty} e^{-ap} \cos pxdp = \dfrac{a}{a^2+x^2}$.

Differentiating both sides w.r.t. a, n times, we have

$$(-1)^n \int_0^{\infty} p^n e^{-ap} \cos pxdp = \frac{d^n}{da^n}\left(\frac{a}{a^2+x^2}\right)$$

$$= \frac{1}{2}\frac{d^n}{da^n}\left(\frac{1}{a-ix} + \frac{1}{a+ix}\right)$$

$$= \frac{1}{2}[(-1)^n n!(a-ix)^{-n-1} + (-1)^n n!(a+ix)^{-n-1}]$$

$$= \frac{(-1)^n n!}{x^{n+1}} \cos(n+1)\theta \sin^{n+1}\theta$$

Where $\theta = \tan^{-1}\left(\dfrac{x}{a}\right)$

$$\therefore \; \int_0^{\infty} p^n e^{-ap} \cos pxdp$$

$$= \frac{n!\cos(n+1)\theta}{x^{r+1}} . \frac{x^{n+1}}{(a^2+x^2)^{(n+1)/2}}$$

$$= \frac{n!\cos(n+1)\theta}{(a^2+x^2)^{(n+1)/2}}$$

$$\therefore \text{ From (1)} \quad f(x) = \sqrt{\frac{2}{\pi}} \frac{n!\cos(n+1)\theta}{(a^2+x^2)^{(n+1)/2}}$$

Example 20. *Use the sine inversion formula to obtain f(x) if*

$$\tilde{f}_s(p) = \frac{p}{1+p^2}.$$

Solution. Using Fourier sine inversion formula, we have

$$f(x) = \sqrt{\frac{2}{\pi}} \int_0^{\infty} \frac{p}{1+p^2}.\sin pxdp$$

$$= \sqrt{\frac{2}{\pi}} \int_0^{\infty} \frac{p^2}{p(1+p^2)} \sin pxdp$$

$$= \sqrt{\frac{2}{\pi}} \int_0^{\infty} \frac{(p^2+1)-1}{p(1+p^2)} \sin pxdp$$

$$= \sqrt{\frac{2}{\pi}} \int_0^{\infty} \frac{\sin px}{p} dp - \sqrt{\frac{2}{\pi}} \int_0^{\infty} \frac{\sin px}{p(1+p^2)} dp$$

$$\text{or} \quad f(x) = \sqrt{\frac{\pi}{2}} - \sqrt{\frac{2}{\pi}} \int_0^{\infty} \frac{\sin px}{p(1+p^2)} dp \qquad \text{... (1)}$$

$$\left[\because \int_0^{\infty} \frac{\sin px}{p} dp = \frac{\pi}{2} \right]$$

$$\therefore \quad \frac{df}{dx} = -\sqrt{\frac{2}{\pi}} \int_0^{\infty} \frac{\cos px}{1+p^2} dp \qquad \text{... (2)}$$

$$\text{and} \quad \frac{d^2 f}{dx^2} = \sqrt{\frac{2}{\pi}} \int_0^{\infty} \frac{p \sin px}{1+p^2} dp$$

$$\text{or} \quad \frac{d^2 f}{dx^2} - f = 0$$

Whose solution is $f = Ae^{x} + Be^{-x}$ ... (3)

$$\therefore \quad \frac{df}{dx} = Ae^{x} - Be^{-x} \qquad \text{... (4)}$$

Now when $x = 0, f = \sqrt{\dfrac{\pi}{2}}$, from (1)

and $\dfrac{df}{dx} = -\sqrt{\dfrac{2}{\pi}} \int_0^{\infty} \dfrac{dp}{1+p^2} = -\sqrt{\dfrac{\pi}{2}}$ [from (2)]

$\therefore$ From (3) and (4)

$$\sqrt{\frac{\pi}{2}} = A + B \text{ and } -\sqrt{\frac{\pi}{2}} = A - B$$

Solving, $A = 0, B = \sqrt{\dfrac{\pi}{2}}$

Hence $f(x) = \sqrt{\dfrac{\pi}{2}} e^{-x}$.

EXERCISE 43.1

1. Find the Fourier complex transform of $f(x)$ if
$$f(x) = \begin{cases} e^{i\omega x} & a < x < b \\ 0 & x < a, x > b \end{cases}.$$

2. Find the sine transform of a function of x which is equal to $\sin x$ for $0 < x < a$ and 0 for $x > a$.

3. Find the cosine transform of $f(x)$ if $f(x) = \begin{cases} \cos x & 0 < x < a \\ 0 & x > 0. \end{cases}$

4. Find the Fourier sine and cosine transform of $f(x)$ if
$$f(x) = \begin{cases} 1 & ; \ 0 \le x < 1 \\ 0 & ; \ x > 1 \end{cases}$$
[JNTU–2004, Kottayam–2005]

5. Find the Fourier sine transform of $\frac{x}{1+x^2}$

6. Find the sine and cosine transform of $\frac{e^{ax}+e^{-ax}}{e^{\pi x}-e^{-\pi x}}$

7. Find the Fourier sine transform of $f(x)$ if
$$f(x) = \begin{cases} 0 & ; \ 0 < x < a \\ x & ; \ a \le x \le b \\ a & ; \ x > b \end{cases}$$

8. Find the cosine transform of $\frac{1}{x^2+a^2}$.
[VTU–2011, Anna–2009]

9. Find the Fourier transform of $f(x) = e^{-x^2/2}$.

10. Show that $\int_0^\infty \frac{\cos xt}{1+t^2}dt = \frac{\pi}{2}e^{-x}, x \ge 0$. [VTU–2003]

11. Find $f(x)$ if its cosine transform is $\frac{1}{1+p^2}$.

12. Find $f(x)$ if (i) its sine transform is e^{-ap} (ii) its cosine transform is e^{-ap}. (SRM–2009)

13. Find $f(x)$ of its cosine transform is $\begin{cases} \sqrt{\frac{1}{2\pi}}\left(a-\frac{p}{2}\right) & , \ \text{if } p < 2a \\ 0 & , \ \text{if } p \ge 2a \end{cases}$

14. Find $f(x)$ if $\tilde{f}_s(p) = \frac{e^{-ap}}{p}$. Hence deduce that $F_s^{-1}\left\{\frac{1}{p}\right\}$.

15. Find the Fourier transform of $e^{-a^2x^2}$, $a < 0$. Hence deduce that $e^{-x^2/2}$ is self reciprocal in respect to Fourier transform.
(Kottayam–2005, Madras–2006)

16. Find the Fourier sine and cosine transform of $x^{n-1}, n > 0$.
(Madras–2006)

17. Solve the integral equation
$$\int_0^\infty f(\theta)\cos\alpha\theta d\theta = \begin{cases} 1-\alpha & ; \ 0 \le \alpha \le 1 \\ 0 & ; \ \alpha > 1 \end{cases}$$
(VTU–2011, Kurukshetra–2005)

18. If the Fourier sine transform of $f(x)$ is $\frac{1-\cos n\pi}{n^2\pi^2}, 0 \le x \le \pi$, find $f(x)$. (Delhi–2002)

19. Find the Fourier transform of $f(x)$, $f(x) = \begin{cases} a^2 - x^2 & ; \ |x| \le a \\ 0 & ; \ x > a \end{cases}$
(VTU–2007)

20. Find Fourier sine transform of $f(x) = \begin{cases} 4x & \text{for } 0 < x < 1 \\ 4-x & \text{for } 1 < x < 4 \\ 0 & \text{for } x > 4 \end{cases}$
(VTU–2006)

21. Find the finite Fourier sine and cosine transform of $f(x) = 2x$, $0 < x < 4$. (VTU–2011)

22. Find the Fourier cosine transform of $\left(1-\frac{x}{\pi}\right)^2$. (PTU–2006)

23. Solve the integral equation $\int_0^\infty f(x)\sin tx dx = \begin{cases} 1 & 0 \le t < 1 \\ 2 & 1 \le t < 2 \\ 0 & t \ge 2 \end{cases}$.
(Kottayam–2005)

24. Solve the integral equation $\int_0^\infty f(x)\cos\alpha x dx = e^{-a}$.
(SVTU–2009, Rohtak–2004)

25. Show that the inverse finite Fourier sine transform of $f(x) = \frac{1}{\pi}\left[1+\cos x\pi - 2\cos\frac{x\pi}{2}\right]$ is
$$f(x) = \begin{cases} 1 & ; \ 0 < x < \pi/2 \\ -1 & ; \ \pi/2 < x < \pi \end{cases}$$
(VTU–2008)

ANSWERS

1. $\frac{1}{\sqrt{2\pi}}\left[\frac{e^{i(p+\omega)a}-e^{i(p+\omega)b}}{p+\omega}\right]$
2. $\frac{1}{\sqrt{2\pi}}\left[\frac{\sin(p-1)a}{p-1}-\frac{\sin(p+1)a}{p+1}\right]$
3. $\frac{1}{\sqrt{2\pi}}\left[\frac{\sin(1+p)a}{1+p}+\frac{\sin(1-p)a}{1-p}\right]$
4. $\sqrt{\frac{2}{\pi}}\frac{(1-\cos p)}{p}, \sqrt{\frac{2}{\pi}}\left(\frac{\sin p}{p}\right)$
5. $\sqrt{\frac{\pi}{2}}e^{-p}$
6. $\frac{e^p - e^{-p}}{\sqrt{2\pi}(e^p+e^{-p}+2\cos a)}, \sqrt{\frac{1}{2\pi}}\frac{\cos a/2.(e^{p/2}+e^{-p/2})}{2\cos a + e^p + e^{-p}}$
7. $\sqrt{\frac{2}{\pi}}\left[-\frac{b\cos pb + a\cos pa}{p}+\frac{\sin pb - \sin pa}{p^2}\right]$
8. $\sqrt{\frac{2}{\pi}}.\frac{\pi}{2a}e^{-ap}$
9. $e^{-p^2/2}$
11. $\sqrt{\frac{\pi}{2}}.e^{-x}$
12. $\sqrt{\frac{2}{\pi}}.\frac{a}{a^2+x^2}$
13. $\frac{1-\cos 2ax}{2\pi x^2}$
14. $\sqrt{\frac{2}{\pi}}\tan^{-1}\left(\frac{x}{a}\right).\sqrt{\frac{\pi}{2}}$
15. $\frac{\sqrt{\pi}}{a}e^{-p^2/4a^2}$
16. $\frac{\Gamma(n)}{p^n}\sin\frac{n\pi}{2}$
17. $f(\theta) = \frac{2(1-\cos\theta)}{\pi\theta^2}$
18. $\frac{2}{\pi^3}\sum_{n=1}^{\infty}\left[\frac{1-\cos n\pi}{n^2}\right]n\pi$
19. $\frac{4}{p^3}[\sin p\alpha - p\alpha\cos p\alpha]$
20. $(2\cos p - \cos 4p - 1)/p^2 - \frac{2\sin p}{p}$
21. $-32(-1)^p/p\pi, \frac{32(-1)^p - 1}{p^2x^2}$
22. $\frac{2}{\pi p^2}$
23. $f(x) = (2+2\cos x - 4\cos 2x)/\pi x$
24. $\frac{2}{\pi(1+x^2)}$

43.5 FINITE FOURIER TRANSFORMS

43.5.1 FINITE FOURIER SINE TRANSFORMS

Let $f(x)$ be a function, which is sectionally continuous over some finite interval $]0, 1[$ for the variable x. Then finite Fourier sine transforms of $f(x)$ on this interval is given by $\tilde{f}(s) = \int_0^1 f(x)\sin\frac{p\pi x}{l}dx$, where p is any integer.

If the end points of the interval become $x = 0$ and $x = \pi$, then $\tilde{f}_s(p) = \int_0^{\pi} f(x)\sin px dx$.

REMARKS

- $\tilde{f}_s(p)$ is always zero when $p = 0$.
- The function $f(x)$ is called the inverse finite Fourier sine transform of $\tilde{f}_s(p)$, *i.e.*, $f(x) = F_s^{-1}\{\tilde{f}_s(p)\}$.

43.5.2 INVERSION FORMULA FOR SINE TRANSFORM

If $\tilde{f}_s(p)$ is the finite Fourier sine transform of $f(x)$ over the interval $]0, l[$, then the inversion formula for sine transform is given by

$$f(x) = \frac{2}{l}\sum_{p=1}^{\infty} \tilde{f}_s(p)\sin\frac{p\pi x}{l}$$

or in the interval $]0, \pi[$, we have

$$f(x) = \frac{2}{\pi}\sum_{p=1}^{\infty} \tilde{f}_s(p)\sin px.$$

43.5.3 FINITE FOURIER COSINE TRANSFORM

Let $f(x)$ be a function which is sectionally continuous over some finite interval $]0, l[$ of the variable x. Then the finite Fourier cosine transform of $f(x)$ on this interval is defined as

$$\tilde{f}_c(p) = \int_0^l f(x)\cos\frac{p\pi x}{l}dx, \text{ where } p \text{ is any integer.}$$

43.5.4 INVERSION FORMULA FOR COSINE TRANSFORM

If $\tilde{f}_c(p)$ is the finite Fourier cosine transform of $f(x)$ over the interval $]0, l[$ then the inversion formula for cosine transform is given by

$$f(x) = \frac{1}{l}\tilde{f}_c(0) + \frac{2}{l}\sum_{p=1}^{\infty}\tilde{f}_c(p)\cos\frac{p\pi x}{l}, \text{ where } \tilde{f}_c(0) = \int_0^l f(x)dx.$$

Also, if π is taken as the upper limit for the finite Fourier cosine transform, then inversion formula is given by

$$f(x) = \frac{1}{\pi}\tilde{f}_c(0) + \frac{2}{\pi}\sum_{p=1}^{\infty}\tilde{f}_c(p)\cos px, \text{ where } \tilde{f}_c(0) = \int_0^{\pi} f(x)dx.$$

43.5.5 MULTIPLE FINITE FOURIER TRANSFORMS

Let $f(x,y)$ be a function of two variables x and y defined in the square $0 \le x \le \pi$ and $0 \le y \le \pi$. Let us consider $f(x, y)$ temporarily as a function of x, the finite sine transform is given by

$$\tilde{f}_s(p, y) = \int_0^{\pi} f(x, y)\sin px dx$$

and now the finite sine transform of $\tilde{f}_s(p, y)$ which is a function of y is given by

$$\tilde{F}_s(p,q) = \int_0^{\pi}\tilde{f}_s(p, y)\sin qy dy$$

Thus,

$$\tilde{F}_s(p,q) = \int_0^{\pi}\int_0^{\pi} f(x, y)\sin px\sin qy dx dy$$

43.5.6 SOME OPERATIONAL PROPERTIES OF FINITE FOURIER TRANSFORM

(A) Operational Property of Finite Fourier Sine Transform:

The finite Fourier sine transforms resolves the differential form $f''(x)$ into a linear algebraic form in the transform $\tilde{f}_s(p)$ and the boundary value $f(0)$ and $f(\pi)$ such that

$$F_s[f''(x)] = -p^2\tilde{f}_s(p) + p\{f(0) - (-1)^p f(\pi)\}$$

whenever $f(x)$ and $f'(x)$ are continuous and $f''(x)$ is sectionally continuous on the interval $0 \le x \le \pi$

(B) Operational Property of Finite Fourier Cosine Transform :

If $f(x)$ and $f'(x)$ are continuous and if $f''(x)$ is sectionally continuous, the finite Fourier transformation resolved the differential form $f''(x)$ into an algebraic form is $\tilde{f}_c(p)$ and the boundary value $f'(0)$ and $f'(\pi)$ such that

$$F_c[f''(x)] = -p^2\tilde{f}_c(p) - f'(0) + (-1)^p f'(\pi)$$

REMARK

- If $\tilde{f}_c(p)$ is the cosine transform of a sectionally continuous function $f(x)$, $0 \le x \le \pi$ then

$$F_c^{-1}\left\{\frac{\tilde{f}_c(p)}{p^2}\right\} = \int_0^x\int_t^{\pi} f(r)dr dt = \frac{\tilde{f}_c(0)}{2\pi}(x - \pi^2) + A$$

where A is any arbitrary constant.

43.5.7 COMBINED PROPERTIES OF FINITE FOURIER SINE AND COSINE TRANSFORMS:

If $f(x)$ is continuous and $f'(x)$ is sectionally continuous, then

(i) $F_s[f'(x)] = -pF_c\{f(x)\}; p = 1,2,3...$ and (ii) $F_c[f'(x)] = pF_s\{f(x) - f(0) + (-1)^p f(\pi)\}; p = 0,1,2,3...$

REMARK

- If $H(x)$ is sectionally continuous function, then

$$F_s[H(x)] = -pF_c\left[\int_0^x H(r)dr\right], p = 1,2...$$

and $$F_c\left[H(x) - \frac{1}{\pi}\tilde{H}_c(0)\right] = pF_s\left[\int_0^x H(r)dr - \frac{x}{\pi}\tilde{H}_c(0)\right], p = 0,1,2...$$

43.5.8 CONVOLUTION

Let $F(x)$ and $G(x)$ be two functions defined on the interval $-2\pi < x < 2\pi$, then the function $F(x) * G(x) = \int_{-\pi}^{\pi} F(x-y)G(y)dy$ is called the convolution of $F(x)$ and $G(x)$ on the interval $-\pi < x < \pi$.

Solved Examples

Example 1. *Find the finite Fourier sine and cosine transforms of $f(x) = x$.*

Solution. We have

$$\tilde{f}_s(p) = \int_0^\pi f(x)\sin pxdx = \int_0^\pi x\sin pxdx$$

$$= \left(\frac{-x\cos px}{p}\right)_0^\pi + \frac{1}{p}\int_0^\pi \cos pxdx$$

$$= \frac{\pi(-1)^{p+1}}{p} + \frac{1}{p}\left[\frac{\sin px}{p}\right]_0^\pi = \frac{\pi(-1)^{p+1}}{p}$$

Similarly,

$$\tilde{f}_c(p) = \int_0^\pi f(x)\cos pxdx = \int_0^\pi x\cos pxdx$$

$$= \left(\frac{x\sin px}{p}\right)_0^\pi - \frac{1}{p}\int_0^\pi \sin pxdx$$

$$= \frac{1}{p}\left[\frac{\cos px}{p}\right]_0^\pi = \frac{(-1)^p - 1}{p^2}, p = 1,2,3...$$

Also, if $p = 0$, then $\tilde{f}_c(p) = \int_0^\pi x.1dx = \frac{\pi^2}{2}$.

Example 2. *Find the finite sine transform of f(x) if*

(i) $f(x) = \begin{cases} x & ; \ 0 \le x \le \pi/2 \\ \pi - x & ; \ \pi/2 \le x \le \pi \end{cases}$

(ii) $f(x) = \begin{cases} -x & ; \ x < c \\ \pi - c & ; \ x > c, \end{cases}$ where $0 \le c \le \pi$

Solution. (i) We have

$$\tilde{f}_s(p) = \int_0^\pi f(x)\sin pxdx$$

$$= \int_0^{\pi/2} x\sin pxdx + \int_{\pi/2}^\pi (\pi - x)\sin pxdx$$

$$= \left[x\frac{(-\cos px)}{p} + \frac{\sin px}{p^2}\right]_0^{\pi/2} + \left[(\pi - x)\left(\frac{-\cos px}{p}\right) - \frac{\sin px}{p^2}\right]_{\pi/2}^\pi$$

$$= \frac{2}{p^2}\sin\left(\frac{p\pi}{2}\right).$$

(ii) We have

$$\tilde{f}_s(p) = \int_0^\pi f(x)\sin pxdx$$

$$= \int_0^C -x\sin pxdx + \int_C^\pi (\pi - x)\sin pxdx$$

$$= \left[x\frac{\cos px}{p}\right]_0^C - \int_0^C 1.\frac{\cos px}{p}dx + \left[(\pi - x)\left(-\frac{\cos px}{p}\right)\right]_C^\pi - \frac{1}{p}\int_C^\pi \cos pxdx$$

$$= \frac{C}{p}\cos pC - \frac{1}{p^2}[\sin px]_0^C + \frac{\pi - C}{p}\cos pC - \frac{1}{p^2}[\sin px]_C^\pi$$

$$= \frac{C}{p}\cos pC - \frac{1}{p^2}\sin pC + \frac{\pi - C}{p}\cos pC + \frac{1}{p^2}\sin pC$$

Example 3. *Find the finite Fourier sine and cosine transforms of the function $f(x) = 2x, 0 < x < 4$.*

Solution. We have

$$\tilde{f}_s(p) = \int_0^l f(x)\sin(p\pi x/l)dx$$

$$= \int_0^4 2x\sin(p\pi x/4)dx, \ l = 4 \text{ (Given)}$$

$$= \left[\frac{-2x\cos(p\pi x/4)}{p\pi/4}\right]_0^4 + 2\int_0^4 \frac{\cos(px\pi/4)}{p\pi/4}dx$$

$$= -\frac{32}{p\pi}\cos p\pi + \frac{8}{p\pi}\left[\frac{\sin(p\pi x/4)}{p\pi/4}\right]_0^4$$

$$= -\frac{32}{p\pi}\cos p\pi$$

Also, $\tilde{f}_c(p) = \int_0^l f(x)\cos(p\pi x/l)dx$

$$= \int_0^4 2x\cos(p\pi x/4)dx, \quad \text{as } l = 4$$

$$= \left[\frac{2x\sin(p\pi x/4)}{p\pi/4}\right]_0^4 - 2\int_0^4 \frac{\sin(p\pi x/4)}{p\pi/4}dx$$

$$= \frac{8}{p\pi}\left[\frac{\cos(p\pi x/4)}{p\pi/4}\right]_0^4 = \frac{32}{p^2\pi^2}(\cos p\pi - 1),$$

if $p > 0$

and if $p = 0$, then $\tilde{f}_c(p) = \int_0^4 2x.1dx = 16$

Example 4. *Find the finite Fourier sine transforms of f(x) if (i) $x(\pi - x)$ and (ii) $x(\pi^2 - x^2)$*

Solution. (i) $\tilde{f}_s\{x(\pi - x)\} = \int_0^\pi x(\pi - x)\sin pxdx$

$$= \left[x(\pi - x)\left(-\frac{\cos px}{p}\right)\right]_0^\pi + \int_0^\pi (\pi - 2x)\left(\frac{\cos px}{p}\right)dx$$

$$= \left[(\pi - 2x)\left(\frac{\sin px}{p^2}\right) - 2.\left(\frac{\cos px}{p^3}\right)\right]_0^\pi$$

$$= \frac{2}{p^3}[1 - (-1)^p].$$

(ii) $\tilde{f}_s\{x(\pi^2 - x^2)\} = \int_0^\pi x(\pi^2 - x^2)\sin pxdx$

$$= \left[x(\pi^2 - x^2)\left(-\frac{\cos px}{p}\right)\right]_0^\pi + \int_0^\pi (\pi^2 - 3x^2)\left(\frac{\cos px}{p}\right)dx$$

$$= \left[(\pi^2 - 3x^2)\left(\frac{\sin px}{p^2}\right)\right]_0^\pi - \int_0^\pi (-6x)\left(\frac{\sin px}{p^2}\right)dx$$

$$= 6\left[x\left(\frac{\cos px}{p^3}\right) + \left(\frac{\sin px}{p^4}\right)\right]_0^\pi = \frac{6\pi}{p^3}(-1)^{p+1}.$$

Example 5. *Find the finite Fourier sine transform of f(x) if f(x) = sin nx.*

Solution. We have $\tilde{f}_s(p) = \int_0^\pi \sin nx \sin pxdx$

$$= \frac{1}{2}\int_0^\pi [\cos(p - n)x - \cos(p + n)x]dx$$

$$= \frac{1}{2}\left[\frac{\sin(p - n)x}{p - n} - \frac{\sin(p + n)x}{p + n}\right]_0^\pi = 0,$$

if $p \neq n$.

If $p = n$, then

$$\tilde{f}_s(p) = \int_0^\pi \sin nx \sin pxdx = \int_0^\pi \sin^2 nxdx$$

$$= \frac{1}{2}\int_0^\pi (1 - \cos 2nx)dx = \frac{1}{2}\left[x - \frac{\sin 2nx}{2n}\right]_0^\pi$$

$$= \frac{\pi}{2}.$$

Hence, $\tilde{f}_s(p) = 0$ if $p \neq n$, and

$\tilde{f}_s(p) = \pi/2$ if $p = n$

Example 6. *Find the finite cosine transform of f(x) if*

$$f(x) = -\frac{\cos k(\pi - x)}{k \sin k\pi}.$$

Solution. We have

$$\tilde{f}_c(p) = -\int_0^\pi \frac{\cos\{k(\pi - x)\}}{k\sin k\pi}\cos pxdx$$

$$= -\frac{1}{2k\sin k\pi}\int_0^\pi [\cos\{k(\pi - x) + px\} + \cos\{k(\pi - x) - px\}]dx$$

$$= -\frac{1}{2k\sin k\pi}\left[\frac{\sin(k\pi - kx + px)}{p - k} - \frac{\sin(k\pi - kx - px)}{p + k}\right]_0^\pi$$

$$= -\frac{1}{2k\sin k\pi}\left[\frac{\sin p\pi}{p - k} - \frac{\sin(-p\pi)}{p + k} - \frac{\sin k\pi}{p - k} + \frac{\sin k\pi}{p + k}\right]$$

$$= \frac{1}{2k}\left(\frac{1}{p - k} - \frac{1}{p + k}\right)$$

$$= \frac{1}{p^2 - k^2}, k \neq 0, 1, 2, 3...$$

Example 7. *Find finite Fourier sine transform of f(x) if*

$$f(x) = \frac{\sin k(\pi - x)}{\sin(k\pi)}.$$

Solution. We have

$$\tilde{f}_s(p) = \int_0^\pi f(x)\sin pxdx$$

$$= \int_0^\pi \frac{\sin\{k(\pi - x)\}}{\sin(k\pi)}.\sin pxdx$$

$$= \frac{1}{2\sin(k\pi)}\int_0^\pi [\cos\{px - k(\pi - x)\} - \cos\{px + k(\pi - x)\}]dx$$

$$= \frac{1}{2\sin(k\pi)}.\left[\frac{\sin\{px - k(\pi - x)\}}{p + k} - \frac{\sin\{px + k(\pi - x)\}}{p - k}\right]_0^\pi$$

$$= \frac{1}{2\sin(k\pi)}.\left[\frac{\sin p\pi + \sin k\pi}{p + k} - \frac{\sin p\pi - \sin k\pi}{p - k}\right]$$

$$= \frac{1}{2\sin(k\pi)}.\left(\frac{1}{p + k} + \frac{1}{p - k}\right)\sin(k\pi)$$

$$= \frac{p}{p^2 - k^2}, k \neq 0, 1, 2...$$

Example 8. *Find f(x) if its finite sine transform is given by*

$$\tilde{f}_s(p) = \frac{1 - \cos p\pi}{p^2\pi^2}, \text{ where } 0 < x < \pi.$$

Solution. We have

$$f(x) = \frac{2}{\pi}\sum_{p=1}^{\infty} \tilde{f}_s(p)\sin px$$

$$= \frac{2}{\pi}\sum_{p=1}^{\infty}\left(\frac{1-\cos p\pi}{p^2\pi^2}\right)\sin px.$$

$$= \frac{2}{\pi^3}\sum_{p=1}^{\infty}\left(\frac{1-\cos p\pi}{p^2}\right)\sin px$$

Example 9. *When f(x) = sin mx, where m is a positive integer, show that*

$\tilde{f}_s(p) = 0$ if $p \neq m$ *and* $\tilde{f}_s(p) = \pi/2$ if $p = m$

Solution. We have

$$\tilde{f}_s(p) = \int_0^{\pi} f(x)\sin px\,dx$$

$$= \int_0^{\pi} \sin mx \sin px\,dx$$

$$= \frac{1}{2}\int_0^{\pi}[\cos(m-p)x - \cos(m+p)x]dx$$

$$= \frac{1}{2}\left[\frac{\sin(m-p)x}{m-p} - \frac{\sin(m+p)x}{m+p}\right]_0^{\pi}$$

$$= 0, \text{ if } m \neq p$$

If $m = p$, then

$$\tilde{f}_s(p) = \int_0^{\pi}\sin^2 px\,dx = \frac{1}{2}\int_0^{\pi}(1-\cos 2px)dx$$

$$= \frac{1}{2}\left[x - \frac{\sin 2px}{2p}\right]_0^{\pi} = \frac{\pi}{2}.$$

Example 10. *Find the finite Fourier sine transform of f(x) if*

(i) $f(x) = \frac{2}{\pi}\tan^{-1}\frac{b\sin x}{1-b\cos x}$

(ii) $f(x) = \frac{2}{\pi}\tan^{-1}\frac{2b\sin x}{1-b^2}$

Solution. (i) If $f(x) = \frac{2}{\pi}\tan^{-1}\frac{b\sin x}{1-b\cos x}$, then

$$\tilde{f}_s(p) = \int_0^{\pi}\frac{2}{\pi}\left[\tan^{-1}\frac{b\sin x}{1-b\cos x}\right]\sin px\,dx$$

Now let $\tan\theta = \frac{b\sin x}{1-b\cos x}$

Then $\frac{i\sin\theta}{\cos\theta} = \frac{ib\sin x}{1-b\cos x}$

Applying componendo and dividend, we have

$$\frac{\cos\theta + i\sin\theta}{\cos\theta - i\sin\theta} = \frac{1-b\cos x + ib\sin x}{1-b\cos x - ib\sin x}$$

or $$\frac{e^{i\theta}}{e^{-i\theta}} = \frac{1-b(\cos x - i\sin x)}{1-b(\cos x + i\sin x)} = \frac{1-be^{-ix}}{1-be^{ix}}$$

$$\therefore\ e^{2i\theta} = \frac{1-be^{-ix}}{1-be^{ix}}$$

$$\therefore\ 2i\theta = \log(1-be^{-ix}) - \log(1-be^{ix})$$

$$= -\left\{be^{-ix} + \frac{b^2e^{-2ix}}{2} + \frac{b^3e^{-3ix}}{3} + \ldots\right\}$$

$$+\left\{be^{ix} + \frac{b^2e^{2ix}}{2} + \frac{b^3e^{3ix}}{3} + \ldots\right\}, \text{ if } |b| \le 1$$

$$= b(e^{ix} - e^{-ix}) + \frac{b^2}{2}(e^{2ix} - e^{-2ix})$$

$$+ \frac{b^3}{3}(e^{i3x} - e^{-3ix}) + \ldots$$

$$= 2ib\sin x + \frac{b^2}{2}.(2i\sin 2x)$$

$$+ \frac{b^3}{3}.(2i\sin 3x) + \ldots$$

$$\therefore\ \theta = \tan^{-1}\frac{b\sin x}{1-b\cos x} - b\sin x$$

$$+ \frac{b^2}{2}\sin 2x + \frac{b^3}{3}\sin 3x + \ldots$$

$\therefore$ From (1),

$$\tilde{f}_s(p) = \int_0^{\pi}\frac{2}{\pi}.\left[b\sin x + \frac{b^2}{2}\sin 2x + \frac{b^3}{3}\sin 3x + \ldots\right]\sin px\,dx$$

$$= \int_0^{\pi}\frac{2}{\pi}.\frac{b^p}{p}\sin^2 px\,dx$$

since all other integrals vanish as $\int_0^{\pi}\sin mx\sin nx\,dx = 0$ if m and n are integers and $m \neq n$

$$= \frac{b^p}{\pi p}\int_0^{\pi}(1-\cos 2px\,dx) = \frac{b^p}{\pi p}\left[x - \frac{\sin 2px}{2p}\right]_0^{\pi}$$

$$= \frac{b^p}{\pi p}.\pi = \frac{b^p}{p}, |b| \le 1$$

(ii) Let $\tan\theta = \frac{2b\sin x}{1-b^2}$

then $\frac{i\sin\theta}{\cos\theta} = \frac{2ib\sin x}{1-b^2}$.

Applying componendo and dividend, we have

$$\frac{\cos\theta + i\sin\theta}{\cos\theta - i\sin\theta} = \frac{1-b^2+2ib\sin x}{1-b^2-2ib\sin x}$$

$$= \frac{1-b^2+b(e^{ix}-e^{-ix})}{1-b^2-b(e^{ix}-e^{-ix})}$$

$$\therefore\ \frac{e^{i\theta}}{e^{-i\theta}} = \frac{(1+be^{ix}) - be^{-ix}(1+be^{ix})}{(1-be^{ix}) + be^{-ix}(1-be^{ix})}$$

or $$e^{2i\theta} = \frac{(1+be^{ix})(1-be^{-ix})}{(1-be^{ix})(1+be^{-ix})}$$

$$\therefore\ 2i\theta = \log(1+be^{ix}) + \log(1-be^{-ix})$$

$$- \log(1-be^{ix}) - \log(1+be^{-ix})$$

$$= \{\log(1+be^{ix}) - \log(1-be^{ix})\} - \{\log(1+be^{-ix}) - \log(1-be^{-ix})\}$$

$$= 2\left\{be^{ix} + \frac{b^3}{3}e^{3ix} + \frac{b^5}{5}e^{5ix}\ldots\right\} - 2\left\{be^{-ix} + \frac{b^3}{3}e^{-3ix} + \frac{b^5}{5}e^{-5ix} + \ldots\right\}$$

if $|b| \le 1$

$$= 2\left[b(e^{ix} - e^{-ix}) + \frac{b^3}{3}(e^{i3x} - e^{i3x}) + \frac{b^5}{5}(e^{i5x} - e^{-i5x}) + \ldots\right]$$

$$= 2\left[b(2i\sin x) + \frac{b^3}{3}(2i\sin 3x) + \frac{b^5}{5}(2i\sin 5x) + \ldots\right]$$

$$\therefore \quad \theta = \tan^{-1}\frac{2b\sin x}{1-b^2} = 2\left[b\sin x + \frac{b^3}{3}\sin 3x + \frac{b^5}{5}\sin 5x + \ldots\right] \quad \ldots (1)$$

Now if $f(x) = \dfrac{2}{\pi}\tan^{-1}\dfrac{2b\sin x}{1-b^2}$, then

$$\tilde{f}_s(p) = \int_0^\pi \frac{2}{\pi}\left[\tan^{-1}\frac{2b\sin x}{1-b^2}\right]\sin pxdx$$

$$= \int_0^\pi \frac{2}{\pi}.2\left[b\sin x + \frac{b^3}{3}\sin 3x + \frac{b^5}{5}\sin 5x\ldots\right]\sin pxdx$$

[from (1)]

Now if m and n are integers and $m \neq n$ then $\int_0^\pi \sin mx \sin nxdx = 0$

If p is even, then $\tilde{f}_s(p) = 0$

Again if p is odd, then

$$\tilde{f}_s(p) = \frac{2}{\pi}\int_0^\pi \frac{b^p}{p}.2\sin^2 pxdx$$

$$= \frac{2}{\pi}\frac{b^p}{p}\int_0^\pi (1-\cos 2px)dx$$

$$= \frac{2}{\pi}\frac{b^p}{p}\left[x - \frac{\sin 2px}{2p}\right]_0^\pi = \frac{2}{\pi}\frac{b^p}{p}.\pi = 2\frac{b^p}{p}.$$

Hence, $\tilde{f}_s(p) = \dfrac{1-(-1)^p}{p}b^p, |b| \le 1$.

EXERCISE 43.2

1. Find the finite Fourier sine and cosine transform of $f(x) = 1$
2. Show that the finite Fourier sine transform of $\dfrac{x}{\pi}$ is $\dfrac{1}{p}(-1)^{p+1}$.
3. Find the finite Fourier sine transform of $\left(1-\dfrac{x}{\pi}\right)$ and $\dfrac{x}{4\pi}$.
4. Find the finite sine and cosine transforms of $f(x) = x^2, 0 < x < \pi$.
5. Find the finite cosine transform of $\left(1-\dfrac{x}{\pi}\right)^2$.
6. Find the finite Fourier sine transforms of $f(x) = \dfrac{\pi}{3} - x + \dfrac{x^2}{2\pi}$.
7. Find the finite sine transforms of $f(x)$ if $f(x) = \cos kx$.
8. Find the finite Fourier cosine transforms of $f(x) = \sin nx$.
9. Find finite Fourier cosine transform of $f(x) = \dfrac{\cosh[c(\pi - x)]}{\sinh(\pi c)}$.
10. Find $f(x)$ if its finite sine transforms is given by $\tilde{f}_s(p) = \dfrac{2\pi(-1)^{p-1}}{p^3}, p = 1,2,3\ldots 0 < x < \pi$.

ANSWERS

1. $\dfrac{1}{p}[1-(-1)^p], 0$
3. $\dfrac{1}{p}, \dfrac{(-1)^{p+1}}{4p}$
4. $\pi^2(-1)^{p+1}/p + 2\{(-1)^p - 1\}/p^3, \dfrac{2\pi(-1)^p}{p^2}, p > 0$
5. $\dfrac{\pi}{3}$
6. $0, \dfrac{\pi}{2}$
7. $\dfrac{p}{p^2 - k^2}[1-(-1)^p \cos k\pi]$
8. 0 or $\dfrac{2n}{n^2 - p^2}$
9. $\dfrac{c}{c^2 + p^2}$
10. $4\sum_{p=1}^{\infty}\dfrac{(-1)^{p-1}}{p^3}\sin px$

43.6 APPLICATIONS OF FOURIER TRANSFORM

The finite sine and cosine transforms can be applied when the range of the variable selected for exclusion is 0 to ∞. The choice of sine and cosine transform is decided by the form of the boundary conditions at the lower limit of the variable selected for exclusion.

Hence, we have $F_s\left\{\dfrac{\partial^2 u}{\partial x^2}\right\} = \int_0^\infty \dfrac{\partial^2 u}{\partial x^2}\sin pxdx = \left[\dfrac{\partial u}{\partial x}\sin px\right]_0^\infty - p\int_0^\infty \dfrac{\partial u}{\partial x}\cos pxdx$

$$= -p\int_0^\infty \frac{\partial u}{\partial x}\cos pxdx \text{ if } \frac{\partial u}{\partial x} \to 0 \text{ as } x \to \infty$$

$$= -p\left\{[u\cos px]_0^\infty + p\int_0^\infty u\sin px\,dx\right\} = p(u)_{x=0} - p^2\bar{u}_s \qquad \text{[By assuming } u \to 0 \text{ as } x \to \infty]$$

Therefore, $$F_s\left\{\frac{\partial^2 u}{\partial x^2}\right\} = pu(0,t) - p^2\bar{u}_s(p,t) \qquad \ldots(1)$$

Where $u(x, t)$ is a function of two variable x and t and $\bar{u}_s(p,t)$ is the Fourier sine transform of $u(x, t)$ with respect to x.

Further, $$F_c\left\{\frac{\partial^2 u}{\partial x^2}\right\} = \int_0^\infty \frac{\partial^2 u}{\partial x^2}\cos px\,dx = \left[\frac{\partial u}{\partial x}\cos px\right]_0^\infty + p\int_0^\infty \frac{\partial u}{\partial x}\sin px\,dx$$

$$= -\left(\frac{\partial u}{\partial x}\right)_{x=0} + p\int_0^\infty \frac{\partial u}{\partial x}\sin px\,dx \qquad \left[\text{Assuming } \frac{\partial u}{\partial x} \to 0 \text{ as } x \to \infty\right]$$

$$= -\left(\frac{\partial u}{\partial x}\right)_{x=0} + p\left\{[u\sin px]_0^\infty - p\int_0^\infty u\cos px\,dx\right\} = -\left(\frac{\partial u}{\partial x}\right)_{x=0} - p^2\int_0^\infty u(x,t)\cos px\,dx.$$

Then, $$F_s\left\{\frac{\partial^2 u}{\partial x^2}\right\} = -\left(\frac{\partial u}{\partial x}\right)_{x=0} - p^2\bar{u}_c(p,t) \qquad \ldots(2)$$

Where, $\bar{u}_c(p,t)$ is the Fourier cosine transform of $u(x, t)$ with respect to x.

Remarks

- It must be noted that the successful use of a sine transform in removing a term $\frac{\partial^2 u}{\partial x^2}$ required $u(0, t)$ *i.e* ,u at $x = 0$ while the use of a cosine transform for the same purpose requires, $u_x(0,t)$, *i.e*, $\frac{\partial u}{\partial x}$ at $x = 0$.
- The terms $\frac{\partial u}{\partial x}$ or any partial derivative of odd order cannot be removed with the help of sine or cosine transforms.
- When one of the variables in a differential equation ranges form $-\infty$ to $+\infty$ then that variable can be excluded with the help of complex Fourier transforms.

Solved Examples

Example 1. *Solve* $\frac{\partial u}{\partial t} = 2\frac{\partial^2 u}{\partial x^2}$ *if* $u(0,t) = 0, u(x,0) = e^{-x}$, $x > 0$, *$u(x, t)$ is bounded where* $x > 0$, $t > 0$.

[Rohtak 2006]

Solution. As per given $$\frac{\partial u}{\partial t} = 2\frac{\partial^2 u}{\partial x^2} \qquad \ldots(1)$$

Subject to the boundary conditions

$$u(0,t) = 0, u(x,t) \text{ is bounded} \qquad \ldots(2)$$

and initial condition

$$u(x,0) = e^{-x}, x > 0 \qquad \ldots(3)$$

Since, $u(0, t)$ is given, taking the Fourier sine transform of both sides of (1), we get

$$\int_0^\infty \frac{\partial u}{\partial t}\sin px\,dx = 2\int_0^\infty \frac{\partial^2 u}{\partial x^2}\sin px\,dx$$

$$\Rightarrow \frac{d}{dt}\int_0^\infty u(x,t)\sin px\,dx$$

$$= 2\left\{\left(\frac{\partial u}{\partial x}\sin px\right)_0^\infty - \int_0^\infty \frac{\partial u}{\partial x}p\cos px\,dx\right\}$$

$$\Rightarrow \frac{d\bar{u}_s}{dt} = -2p\int_0^\infty \frac{\partial u}{\partial x}\cos px\,dx \text{ if } \frac{du}{dx} \to 0 \text{ as } x \to \infty$$

$$\left[\text{Assume } \bar{u}_s(p,t) = \int_0^\infty u(x,t)\sin px\,dx\right]$$

$$= -2p\left\{[u(x,t)\cos px]_0^\infty - \int_0^\infty u(x,t)(-p\sin px)dx\right\}$$

$$= -2p\left\{0 - u(0,t) + p\int_0^\infty u(x,t)\sin px\,dx\right\}$$

$$[\because u(x,t) \to 0, \cos x \to \infty]'$$

$$= 2pu(0,t) - 2p^2\bar{u}_s$$

$$\Rightarrow \frac{d\bar{u}_s}{dt} = -2p^2\bar{u}_s \qquad [\because u(0, t) = 0]$$

On separating the variables, we get

$$\frac{d\bar{u}_s}{du_s} = -2p^2dt \Rightarrow \log\bar{u}_s - \log C = -2p^2t$$

$$\Rightarrow \log\left(\frac{\bar{u}_s}{C}\right) = -2p^2t \Rightarrow \bar{u}_s(p,t) = Ce^{-2p^2t} \qquad \ldots(4)$$

Now, taking the Fourier sine transform of both sides of (3), we get

$$\int_0^\infty u(x,0)\sin px\,dx = \int_0^\infty e^{-x}\sin px\,dx$$

$$\Rightarrow \bar{u}_s(p,0) = \left[\frac{e^{-x}}{1+p^2}(-\sin px - p\cos px)\right]_0^\infty$$

$$= \frac{p}{1+p^2} \qquad \ldots(5)$$

Putting $t = 0$ in (4) and (5), we get

$$\frac{p}{1+p^2} = C$$

$$\therefore \quad \bar{u}_s(p,t) = \frac{p}{1+p^2}e^{-2p^2t}.$$

Taking the inverse Fourier sine transform, we get

$$u(x,t) = \frac{2}{\pi}\int_0^\infty \frac{p}{1+p^2}e^{-p^2t}\sin pxdx$$

Example 2. *Solve* $\frac{\partial u}{\partial t} = \frac{\partial^2 u}{\partial x^2}, x > 0, t > 0$ *subject to the conditions* $u(0, t) = 0, u(x,0) = \begin{cases} 1 & ; \quad 0 < x < 1 \\ 0 & ; \quad x > 1 \end{cases}$, *$u(x, t)$ is bounded.* (UPTU–2003)

Solution. Taking the Fourier sine transform of both side of given PDE, we get

$$\int_0^\infty \frac{\partial u}{\partial t}\sin pxdx = \int_0^\infty \frac{\partial^2 u}{\partial x^2}\sin pxdx$$

or $$\frac{d}{dt}\int_0^\infty u\sin pxdx$$

$$= \left[\frac{\partial u}{\partial x}\sin px\right]_0^\infty - p\int_0^\infty \frac{\partial u}{\partial x}\cos pxdx$$

$$\therefore \quad \frac{d\bar{u}_s}{dt} = -p\int_0^\infty \frac{\partial u}{\partial x}\cos pxdx$$

if $\frac{\partial u}{\partial x} \to 0$ as $x \to \infty$

$$= -p\left\{[u\cos px]_0^\infty + p\int_0^\infty u\sin pxdx\right\}$$

$$= -pu(0,t) - p^2\bar{u}_s;$$

if $u \to 0$ as $x \to \infty$

$$= -p^2\bar{u}_s.$$

On separating the variables, we get

$$\frac{d\bar{u}_s}{\bar{u}_s} = -p^2dt$$

whose solution is given by

$$\bar{u}_s(p,t) = Ce^{-p^2t} \quad \ldots(1)$$

Putting $t = 0$, we get $C = \bar{u}_s(p,0)$ ...(2)

Now, $\bar{u}_s(p,0) = \int_0^\infty u(x,0)\sin pxdx$

$$= \int_0^1 u(x,0)\sin pxdx + \int_1^\infty u(x,0)\sin pxdx$$

$$= \int_0^1 \sin pxdx$$

Now, from (2)

$$C = \int_0^1 \sin pxdx = \left[\frac{\cos px}{-p}\right]_0^1 = \frac{1-\cos p}{p}.$$

Thus, (1) gives $\bar{u}_s(p,t) = \left[\frac{(1-\cos p)}{p}\right]e^{-p^2t}$.

Finally, taking the inverse Fourier sine transform, we get

$$u(x,t) = \frac{2}{\pi}\int_0^\infty \frac{1-\cos p}{p}e^{-p^2t}\sin pxdp$$

which is the required solution.

Example 3. *Using the Fourier sine transform, solve the partial differential equation* $\frac{\partial V}{\partial t} = k\frac{\partial^2 V}{\partial x^2}$ *for* $x > 0$, $t > 0$ *under the boundary conditions* $V = V_0$ *when* $x = 0$, $t > 0$ *and the initial condition* $v = 0$ *when* $t = 0, x > 0$.

Solution. Taking the Fourier sine transform of both the sides of the given equation, we get

$$\sqrt{\frac{2}{\pi}}\int_0^\infty \frac{\partial V}{\partial t}\sin pxdx = k\sqrt{\frac{2}{\pi}}\frac{\partial^2 V}{\partial x^2}.\sin pxdx$$

$$\Rightarrow \frac{d}{dt}\sqrt{\frac{2}{\pi}}\int_0^\infty V\sin pxdx = -kp^2\tilde{V}_s + k_p\sqrt{\frac{2}{\pi}}V(0,t)$$

$$\Rightarrow \frac{d\tilde{V}_s}{dt} + kp^2\tilde{V}_s + k_p\sqrt{\frac{2}{\pi}}V_s \qquad [\because V(0,t) = V_0]$$

which is a linear differential equation of first order

$$\text{I.F.} = e^{\int kp^2dt} = e^{kp^2t}$$

Its solution is given by

$$\tilde{V}_se^{kp^2t} = C + \sqrt{\frac{2}{\pi}}\int kpV_0.e^{kp^2t}dt$$

$$= C + \sqrt{\frac{2}{\pi}}\frac{V_0}{p}e^{kp^2t} \quad \ldots(1)$$

But when $t = 0, \tilde{V}_s = 0$ ($\because$ $V = 0$ when $t = 0$)

Using (1), we have $0 = C + \frac{V_0}{p}\sqrt{\frac{2}{\pi}}$

$$\Rightarrow \quad C = -\frac{V_0}{p}\sqrt{\frac{2}{\pi}}.$$

Putting this value in (1), we get

$$\tilde{V}_se^{kp^2t} = \frac{V_0}{p}\sqrt{\frac{2}{\pi}}(e^{kp^2t} - 1)$$

$$\Rightarrow \quad \tilde{V}_s = \sqrt{\frac{2}{\pi}}\frac{V_0}{p}(1 - e^{-kp^2t}).$$

Applying the inverse Fourier transform, we have

$$V = \frac{2}{\pi}V_0\int_0^\infty \frac{(1-e^{-kp^2t})}{p}\sin pxdp$$

$$= \frac{2V_0}{\pi}\left[\int_0^\infty \frac{\sin pxdx}{p}dp - \int_0^\infty \frac{e^{-kp^2t}}{p}.\sin pxdp\right]$$

$$= \frac{2V_0}{\pi}\left[\frac{\pi}{2} - \int_0^\infty \frac{e^{-kp^2t}}{p}.\sin pxdp\right]$$

Hence, $$V(x,t) = V_0\left[1 - \frac{2}{\pi}\int_0^\infty \frac{e^{-kp^2t}}{p}.\sin pxdp\right].$$

Example 4. *Use the method of Fourier transform to determine the displacement y(x,t) of an infinite string, given that the string is initially at rest and that the initial displacement is f(x), $-\infty < x < \infty$ show that* $y(x,t) = \frac{1}{2}[f(x+Ct) + f(x-Ct)]$.

[Rohtak–2000]

Solution. It is known that the displacement of a string is governed by one dimensional wave equation

$$\frac{\partial^2 y}{\partial t^2} = C^2 \frac{\partial^2 y}{\partial x^2} \quad \text{... (1)}$$

where $y(x, t)$ is the displacement at any time, $t, -\infty < x < \infty, t > 0$, and $C^2 = \frac{T}{\rho}$

Taking the Fourier transform of both sides of (1), we get

$$\frac{1}{\sqrt{2\pi}}\int_{-\infty}^{\infty} \frac{\partial^2 y}{\partial t^2} e^{ipx} dx = C^2 \frac{1}{\sqrt{2\pi}}\int_{-\infty}^{\infty} \frac{\partial^2 y}{\partial x^2} e^{ipx} dx$$

$$\Rightarrow \frac{d^2}{dt^2}\frac{1}{\sqrt{2\pi}}\int_{-\infty}^{\infty} y e^{ipx} dx = C^2(-ip)^2 \tilde{y}(p,t)$$

$$\Rightarrow \frac{d^2\tilde{y}(p,t)}{dt^2} + C^2p^2\tilde{y}(p,t) = 0$$

whose solution is given by

$$\tilde{y}(p,t) = A\cos Cpt + B\sin Cpt \quad \text{...(2)}$$

As per given, the string is initially at rest, *i.e.*, $\frac{\partial y}{\partial t} = 0$ at $t = 0$.

Thus, $$\frac{1}{\sqrt{2\pi}}\int_{-\infty}^{\infty} \frac{\partial y}{\partial t} e^{ipx} dx = \frac{d}{dt}\frac{1}{\sqrt{2\pi}}\int_{-\infty}^{\infty} y e^{ipx} dx$$

$$= \frac{d\tilde{y}(p,t)}{dt} = 0 \text{ at } t = 0.$$

Now from (2), we have $0 = BC_p \Rightarrow B = 0$

Also, at $t = 0, y = f(x)$.

So, at $t = 0$,

$$\tilde{y}(p,0) = \frac{1}{\sqrt{2\pi}}\int_{-\infty}^{\infty} f(u)e^{ipu} du = \tilde{f}(p)$$

$$\therefore \tilde{y}(p,t) = \tilde{f}(p)\cos Cpt$$

Now, taking the inverse Fourier transform, we have

$$y(x,t) = \frac{1}{\sqrt{2\pi}}\int_{-\infty}^{\infty} \tilde{f}(p)\cos Cpte^{-ipx} dp$$

$$= \frac{1}{2\pi}\int_{-\infty}^{\infty}\left[\int_{-\infty}^{\infty} f(u)e^{ipu} du\right]\cos Cpte^{-ipx} dp$$

$$= \frac{1}{4\pi}\int_{-\infty}^{\infty}\left[\int_{-\infty}^{\infty} f(u)e^{ipu} du\right](e^{iCpt} + e^{-iCpt})e^{-ipx} dp$$

$$= \frac{1}{2}\left[\frac{1}{2\pi}\int_{-\infty}^{\infty}\left\{\int_{-\infty}^{\infty} f(u)e^{-i\alpha u} du\right\}(e^{-iC\alpha t} + e^{iC\alpha t})e^{i\alpha x} d\alpha\right]$$

Putting $p = -\alpha \Rightarrow dp = -d\alpha$

We get

$$y(x,t) = \frac{1}{2}\left[\frac{1}{2\pi}\int_{-\infty}^{\infty} f(u)e^{-i\alpha u}\left\{\int_{-\infty}^{\infty} e^{i\alpha(x+Ct)}dx\right\}du + \frac{1}{2\pi}\int_{-\infty}^{\infty} f(u)e^{-i\alpha u}\left\{\int_{-\infty}^{\infty} e^{i\alpha(x-Ct)}d\alpha\right\}du\right].$$

Finally, using Fourier integral formula, we get

$$y(x,t) = \frac{1}{2}[f(x+Ct) + f(x-Ct)].$$

Example 5. *A thin membrane of great extent is release from rest in the position z = f(x, y). Show that the displacement at any subsequent time is given by*

$$z(x,y) = \frac{1}{2\pi}\int_{-\infty}^{\infty}\int_{-\infty}^{\infty} F(p,q)\cos\{Ct\sqrt{p^2+q^2}\}e^{-i(px+qy)}dpdq$$

where F(p, t) is double Fourier's transform of f(x, y).

(UPTU–2005)

Solution. It is known that the displacement of the membrane is governed by two dimensional wave equation

$$\frac{\partial^2 z}{\partial t^2} = C^2\left(\frac{\partial^2 z}{\partial x^2} + \frac{\partial^2 z}{\partial y^2}\right), \text{ when } C^2 = \frac{T}{\rho}$$

Taking the Fourier transforms of both the sides, we get

$$\frac{1}{2\pi}\int_{-\infty}^{\infty}\int_{-\infty}^{\infty} \frac{\partial^2 z}{\partial t^2} e^{i(px+qy)}dxdy$$

$$= \frac{C^2}{2\pi}\int_{-\infty}^{\infty}\int_{-\infty}^{\infty}\left(\frac{\partial^2 z}{\partial x^2} + \frac{\partial^2 z}{\partial y^2}\right)e^{i(px+qy)}dxdy$$

$$\Rightarrow \frac{d^2 z}{dt^2} = C^2(-p^2 - q^2)\tilde{z} = 0, \text{ where}$$

$$\tilde{z} = \frac{1}{2\pi}\int_{-\infty}^{\infty}\int_{-\infty}^{\infty} z.e^{i(px+qy)}.dxdy$$

So, $\frac{d^2\tilde{z}}{dt^2} + C^2(p^2 + z^2)\tilde{z} = 0$ whose solution is given by

$$\tilde{z} = A\cos\{C\sqrt{p^2+q^2}.t\} + B\sin[C\sqrt{p^2+q^2}.t]$$

As per given, the initial conditions are $z = f(x,y), \frac{\partial z}{\partial t} = 0$ at $t = 0$.

Taking the Fourier transforms of these conditions, we get

$$\tilde{z} = \frac{1}{2\pi}\int_{-\infty}^{\infty}\int_{-\infty}^{\infty} f(x,y)e^{i(px+qy)}dxdy = F(p,q)$$

and $\frac{d\tilde{z}}{dt} = 0$ at $t = 0$.

$$\therefore \quad 0 = \left(\frac{d\tilde{z}}{dt}\right)_{t=0} = BC\sqrt{p^2+q^2}$$

$$\Rightarrow B = 0$$

Thus, the solution is given by $z = F(p,q)\cos\{C\sqrt{p^2+q^2}t\}$.

Applying the inversion Formula for double Fourier transform, we get

$$z(x,y,t) = \frac{1}{2\pi}\int_{-\infty}^{\infty}\int_{-\infty}^{\infty} F(p,q)$$

$$\cos\{Ct\sqrt{p^2+q^2}\}e^{-i(px+qy)}dpdq.$$

Example 6. *Use a cosine transform to show that the steady temperature in the semi-finite solid $y > 0$ when a temperature on the surface $y = 0$ is kept at unity over the strip $|x| < a$ and at zero outside the strip is*

$$\frac{1}{\pi}\left\{\tan^{-1}\left(\frac{a+x}{y}\right)+\tan^{-1}\left(\frac{a-x}{y}\right)\right\}$$

The result

$$\int_{-\infty}^{\infty} e^{-sx}x^{-1}\sin rxdu = \tan^{-1}\frac{r}{s}, r>0, s>0$$

may be assumed.

Solution. It is known that the steady temperature $u(x,y)$ in the semi-finite solid is governed by two dimensional Laplace equation

$$\frac{\partial^2 u}{\partial x^2}+\frac{\partial^2 u}{\partial y^2} = 0, 0<y<\infty, -\infty<x<\infty \quad \text{... (1)}$$

subject to the conditions

$u = 1, y = 0, -a < x < a$ and $u = 0, y = 0$ and $|x| < |a|$.

Taking Fourier cosine transform of both sides of (1), we get

$$\sqrt{\frac{2}{\pi}}\int_0^{\infty}\frac{\partial^2 u}{\partial x^2}\cos pxdx + \sqrt{\frac{2}{\pi}}\int_0^{\infty}\frac{\partial^2 u}{\partial y^2}\cos pxdx = 0$$

$$\Rightarrow \sqrt{\frac{2}{\pi}}\left[\frac{\partial u}{\partial x}.\cos px\right]_0^{\infty} + p\sqrt{\frac{2}{\pi}}\int_0^{\infty}\sin pxdx + \frac{d^2\tilde{u}_c}{dy^2} = 0$$

$$\left[\frac{\partial u}{\partial x}\to 0 \text{ as } x\to\infty \text{ and } \frac{\partial u}{\partial x}\to 0 \text{ as } x\to 0 \text{ by symmetry}\right]$$

$$\Rightarrow p\sqrt{\frac{2}{\pi}}[u\sin px]_0^{\infty} - p^2\sqrt{\frac{2}{\pi}}\int_0^{\infty}u\cos pxdx + \frac{d^2\tilde{u}}{dy^2} = 0$$

$$\Rightarrow \frac{d^2\tilde{u}_c}{dy^2} - p^2\tilde{u}_c = 0$$

whose solution is given by

$$\tilde{u} = Ae^{py} + Be^{-py} \quad \text{...(2)}$$

As $y\to\infty, \tilde{u}_c \to 0$ therefore $A = 0$

$$\Rightarrow \tilde{u}_c = Be^{-by} \quad \text{...(3)}$$

When $y = 0$,

$$\tilde{u}_c = \sqrt{\frac{2}{\pi}}\int_r^a 1.\cos pxdx + \sqrt{\frac{2}{\pi}}\int_0^{\infty} 0.\cos pxdx$$

$$= \sqrt{\frac{2}{\pi}}\left(\frac{\sin px}{p}\right)_0^{\infty} = \sqrt{\frac{2}{\pi}}\frac{\sin pa}{p}.$$

Also, from (3),

$$\sqrt{\frac{2}{\pi}}\frac{\sin pa}{p} = B \Rightarrow \tilde{u}_c = \sqrt{\frac{2}{\pi}}\frac{\sin pa}{p}e^{-py}$$

Taking the inverse Fourier cosine transform, we get

$$u(x,y) = \sqrt{\frac{2}{\pi}}\int_0^{\infty}\sqrt{\frac{2}{\pi}}\frac{\sin pa}{p}e^{-py}\cos pxdp$$

$$= \frac{1}{\pi}\int_0^{\infty}\frac{e^{-py}}{p}[\sin(a+x)p + \sin(a-x)p]dp$$

$$= \frac{1}{\pi}\int_0^{\infty}e^{-py}p^{-1}\sin(a+x)pdp + \frac{1}{\pi}\int_0^{\infty}e^{-py}p^{-1}\sin(a-x)pdp$$

$$= \frac{1}{\pi}\left[\tan^{-1}\left(\frac{a+x}{y}\right)+\tan^{-1}\left(\frac{a-x}{y}\right)\right]$$

Example 7. *Solve* $\dfrac{\partial^4 V}{\partial x^4}+\dfrac{\partial^2 V}{\partial y^2} = 0, \quad -\infty<x<\infty, y\ge 0$

subject to the conditions

(i) V and its partial derivative tends to zero as $x\to\pm\infty$.

(ii) $V = f(x), \dfrac{\partial V}{\partial y} = 0$ *on* $y = 0$.

Solution. Taking the Fourier transform of both the sides of the given equation, we get

$$\frac{1}{\sqrt{2\pi}}\int_{-\infty}^{\infty}\frac{\partial^4 V}{\partial x^4}e^{ipx}dx + \frac{1}{\sqrt{2\pi}}\int_{-\infty}^{\infty}\frac{\partial^2 V}{\partial y^2}e^{ipx}dx = 0$$

$$\Rightarrow \frac{1}{\sqrt{2\pi}}\left(\frac{\partial^3 V}{\partial x^3}e^{ipx}\right)_{-\infty}^{\infty} - \frac{ip}{\sqrt{2\pi}}\int_{-\infty}^{\infty}\frac{\partial^3 V}{\partial x^3}e^{ipx}dx + \frac{d^2}{dy^2}\frac{1}{\sqrt{2\pi}}\int_{-\infty}^{\infty}V.e^{ipx}dx = 0$$

$$\Rightarrow \frac{-ip}{\sqrt{2\pi}}\left[\left(\frac{\partial^2 V}{\partial x^2}e^{ipx}\right)_{-\infty}^{\infty} - ip\int_{-\infty}^{\infty}\frac{\partial^2 V}{\partial x^2}e^{ipx}dx\right] + \frac{d^2\tilde{V}}{dy^2} = 0$$

$$\left[\because \frac{\partial^3 V}{\partial x^3}\to 0 \text{ as } x\to\pm\infty\right]$$

$$\Rightarrow \frac{(ip)^2}{\sqrt{2\pi}}\left[\left(\frac{\partial V}{\partial x}e^{ipx}\right)_{-\infty}^{\infty} - ip\int_{-\infty}^{\infty}\frac{\partial V}{\partial x}e^{ipx}dx\right] + \frac{d^2\tilde{V}}{dy^2} = 0$$

$$\Rightarrow -\frac{(ip)^3}{\sqrt{2\pi}}\left[\left(Ve^{ipx}.dx\right)_{-\infty}^{\infty} - ip\int_{-\infty}^{\infty} Ve^{ipx}dx\right] + \frac{d^2\tilde{V}}{dy^2} = 0$$

$$\Rightarrow \frac{(ip)^4}{\sqrt{2\pi}}\int_{-\infty}^{\infty} Ve^{ipx}.dx + \frac{d^2\tilde{V}}{dy^2} = 0$$

$$\Rightarrow \frac{d^2\tilde{V}}{dy^2} + p^4\tilde{V} = 0$$

whose general solution is given by

$$\tilde{V} = C_1 \cos p^2 y + C_2 \sin p^2 y \quad \ldots (1)$$

Since, on $y = 0$, $V = f(x)$ and $\frac{\partial V}{\partial y} = 0$

$$\therefore \tilde{V} = \frac{1}{\sqrt{2\pi}}\int_{-\infty}^{\infty} f(x)e^{ipx}dx = \tilde{f}(p)$$

and $\frac{1}{\sqrt{2\pi}}\int_{-\infty}^{\infty} \frac{\partial V}{\partial y}e^{ipx}dx = 0 \Rightarrow \frac{d\tilde{V}}{dy} = 0$

Also, from (1), $\tilde{f}(p) = C_1$ and $d\tilde{V} = 0 = C_2$

$\Rightarrow C_2 = 0$

Then , from (1), we have $\tilde{V} = \tilde{f}(p)\cos p^2 y$.

Finally, applying inversion theorem for Fourier transform, we get

$$V = \frac{1}{\sqrt{2\pi}}\int_{-\infty}^{\infty} \tilde{f}(p)\cos(p^2 y)e^{-ipx}dp.$$

Example 8. *Use the finite cosine transform to solve* $\frac{\partial V}{\partial t} = k\frac{\partial^2 V}{\partial x^2}$ *with boundary conditions* $\frac{\partial V}{\partial x} = 0$, *when* $x = 0$ *and* $x = \pi, t > 0$ *the initial conditions* $V = f(x)$, *when* $t = 0, 0 < x < \pi$.

Solution. Taking the finite Fourier cosine transform (with $l = \pi$) of both the sides of the given equation, we get

$$\int_0^\pi \frac{\partial V}{\partial t}\cos pxdx = k\int_0^\pi \frac{\partial^2 V}{\partial x^2}\cos pxdx$$

$$\Rightarrow \frac{d\tilde{V}_c}{\partial t} = k[-p^2\tilde{V}_c - \{V_x(0,t) - V_x(\pi,t)\cos p\pi\}]$$

where $\tilde{V}_c$ is finite Fourier cosine transform of V.

$$\therefore \frac{d\tilde{V}_c}{dt} = -kp^2\tilde{V}_c \Rightarrow \tilde{V}_c = Ae^{-kp^2 t} \quad \ldots (1)$$

Taking finite Fourier cosine transform of initial conditions, we have at $t = 0$

$$\tilde{V}_c = \int_0^\pi f(y)\cos pydy$$

From (1), we have

$$A = \int_0^\pi f(x)\cos pxdx$$

$$\Rightarrow \tilde{V}_c = e^{-kp^2 t}\int_0^\pi f(y)\cos pydy$$

On taking the inverse finite Fourier cosine transform, we have

$$V(x,t) = \frac{1}{\pi}\tilde{V}_c(0) + \frac{2}{\pi}\sum_{p=1}^{\infty}\tilde{V}_c(p)\cos px$$

$$= \frac{1}{\pi}\int_0^\pi f(y)dy + \frac{2}{\pi}\sum_{p=1}^{\infty}\left[e^{-kp^2 t}\cos px.\int_0^\pi f(y)\cos pydy\right]$$

Example 9. *Show that the solution of Laplace's equation for r inside the semi-infinite strip* $x = 0, 0 < y < b$, *such that*

$$r = f(x) \;;\; \text{when } y = 0, 0 < x < \infty$$
$$r = 0 \;;\; \text{when } y = b, 0 < x < \infty$$
$$r = 0 \;;\; \text{when } x = 0, 0 < y < b$$

is given by

$$V = \frac{2}{\pi}\int_0^\infty f(u)du\int_0^\infty \frac{\sinh(b-y)p}{\sinh bp}\sin xp.up.dp.$$

Solution. Two dimensional Laplace's equation is given by

$$\frac{\partial^2 V}{\partial x^2} + \frac{\partial^2 V}{\partial y^2} = 0,\ 0 < x < \infty, 0 < y < b \quad \ldots (1)$$

Taking Fourier sine transform of (1) of both the sides, we get

$$\sqrt{\frac{2}{\pi}}\int_0^\infty \frac{\partial^2 V}{\partial x^2}\sin pxdx + \sqrt{\frac{2}{\pi}}\int_0^\infty \frac{\partial^2 V}{\partial y^2}\sin pxdx = 0$$

$$\sqrt{\frac{2}{\pi}}\left[\frac{\partial V}{\partial x}\sin px\right]_0^\infty - p\sqrt{\frac{2}{\pi}}\int_0^\infty \frac{\partial V}{\partial x}\cos pxdx + \frac{d^2\tilde{V}_s}{dy^2} = 0$$

$$\Rightarrow -p\sqrt{\frac{2}{\pi}}[V.\cos px]_0^\infty - p^2\sqrt{\frac{2}{\pi}}\int_0^\infty V\sin pxdx + \frac{d^2\tilde{V}_s}{dy^2} = 0$$

$$\left[\frac{\partial V}{\partial x} \to 0 \text{ as } x \to \infty\right]$$

$$\Rightarrow \frac{d^2\tilde{V}_s}{dy^2} - p^2\tilde{V}_s = 0 \qquad [V \to 0 \text{ as } x \to \infty]$$

whose solution is given by

$$\tilde{V}_s = A\cosh py + B\sinh py \quad \ldots (2)$$

Further, taking Fourier sine transform of given conditions, we have

When $y = 0$, $\tilde{V}_s = \sqrt{\frac{2}{\pi}}\int_0^\infty f(u)\sin pudu$

and when $y = b, \tilde{V}_s = 0$ So, from (2), we have

$$\sqrt{\frac{2}{\pi}}\int_0^\infty f(u)\sin pudu = A \text{ and}$$

$$0 = A\cosh pb + B\sinh pb$$

$$\Rightarrow B = -\frac{\cosh pb}{\sinh pb}.\sqrt{\frac{2}{\pi}}\int_0^\infty f(u)\sin pudu$$

Therefore,

$$\tilde{V}_s = \sqrt{\frac{2}{\pi}}\cosh py\int_0^\infty f(u)\sin pu du$$
$$-\sqrt{\frac{2}{\pi}}\frac{\cosh pb}{\sinh pb}\sinh py\int_0^\infty f(u)\sin pu du$$
$$= \sqrt{\frac{2}{\pi}}\int_0^\infty f(u)\sin pu \left[\cosh py - \frac{\cosh pb\sinh py}{\sinh pb}\right]du$$
$$= \sqrt{\frac{2}{\pi}}\int_0^\infty f(u)\sin pu \left[\frac{\sinh pb\cosh py - \cosh pb\sinh py}{\sinh pb}\right]du$$
$$= \sqrt{\frac{2}{\pi}}\int_0^\infty f(u)\sin pu.\frac{\sinh(b-y)p}{\sinh pb}du\,.$$

Finally, taking the inverse Fourier sine transform, we get

$$V = \sqrt{\frac{2}{\pi}}\int_0^\infty \tilde{V}_s \sin pxdp$$
$$= \frac{2}{\pi}\int_0^\infty f(u)du\int_0^\infty \frac{\sinh(b-y)p}{\sinh pb}.\sin pu.\sin pxdp\,.$$

Example 10. *Use finite Fourier transform to solve*

$$\frac{\partial u}{\partial t} = \frac{\partial^2 u}{\partial x^2},\; u(0,t) = 0, u(4,t) = 0, u(x,0) = 2x$$

Solution. Taking the finite Fourier sine transform of both sides of the given partial differential equation, we get

$$\int_0^4 \frac{\partial u}{\partial t}\sin\frac{p\pi x}{4}dx = \int_0^4 \frac{\partial^2 u}{\partial x^2}\sin\frac{p\pi x}{4}dx$$
$$\Rightarrow \frac{d\tilde{u}_s}{dt} = -\frac{p^2\pi^2}{4}\tilde{u}_s + \frac{p\pi}{4}[u(0,t) - u(4,t)\cos p\pi]$$

where $\tilde{u}_s$ is finite Fourier sine transform of u.

$$\therefore\; \frac{d\tilde{u}_s}{dt} = \frac{-p^2\pi^2}{15}\tilde{u}_s \Rightarrow \tilde{u}_s = Ae^{-p^2\pi^2 t/16}$$

Now, since $u(x, 0) = 2x$, where $0 < x < 4$, taking finite Fourier sine transform, we have

$$\text{at } t = 0,\; \tilde{u}_s = \int_0^4 2x.\sin\frac{px\pi}{4}dx$$
$$= \left[-2x\frac{4}{p\pi}\cos\frac{p\pi x}{4} + 2.\frac{4}{2\pi}.\frac{4}{p\pi}\sin\frac{p\pi x}{4}\right]_0^4$$
$$= \frac{32(-\cos p\pi)}{p\pi} = \frac{32[-(-1)]^p}{p\pi}$$
$$\therefore\; \frac{32(-1)^{p+1}}{p\pi} = A,$$

$$\text{Hence, } \tilde{u}_s = \frac{32(-1)^{p+1}}{p\pi}e^{-tp^2\pi^2/16}$$

Finally, taking the inverse finite Fourier sine transform, we get

$$u(x,t) = \frac{2}{4}\sum_{p=1}^\infty \frac{32(-1)^{p+1}}{p\pi}e^{-p^2\pi^2 t/16}.\sin\frac{p\pi x}{4}$$
$$= \frac{16}{\pi}\sum_{p=1}^\infty \frac{(-1)^{p+1}}{p}e^{-p^2\pi^2 t/16}.\sin\frac{p\pi x}{4}$$

Example 11. *Using finite Fourier transform. Find the solution of the wave equation* $\frac{\partial^2 u}{\partial t^2} = 4\frac{\partial^2 u}{\partial x^2}$ *subject to the conditions* $u(0,\; t) = 0.$ $u(\pi,t) = 0, u(x,0) = 0.1\sin x + 0.1\sin 4x$ and $u_t(x,0) = 0$ for $0 < x < \pi, t > 0$

Solution. Taking the finite Fourier transform on both the sides of the given equation, we get

$$\int_0^\pi \frac{\partial^2 u}{\partial t^2}\sin pxdx = 4\int_0^\pi \frac{\partial^2 u}{\partial x^2}\sin pxdx$$
$$\Rightarrow \frac{d^2\tilde{u}_s}{dt^2} = -4p^2\tilde{u}_s + 4p[u(0,t) - u(\pi,t)\cos p\pi]$$
$$\Rightarrow \frac{d^2\tilde{u}_s}{dt^2} + 4p^2\tilde{u}_s = 0$$
$$\Rightarrow \tilde{u}_s = A\cos 2pt + B\sin 2pt$$

Now, at $t = 0, u = (0.1)\sin x + (0.01)\sin 4x$ and $\frac{\partial u}{\partial t} = 0$

Taking finite sine transforms, we get

At $t = 0$,

$$\tilde{u}_s = \int_0^\pi \{(0.1)\sin x + (0.01)\sin 4x\}\sin pxdx$$
$$= (0.1)\int_0^\pi \sin x\, px\, dx + (0.01)\int_0^\pi \sin 4x \sin px\, dx$$

and $\frac{d\tilde{u}_s}{dt} = 0$.

Also, from (1), we have

$$(0.1)\int_0^\pi \sin x\sin px\, dx + (0.01)\int_0^\pi \sin 4x\sin px\, dx = A$$
$$\Rightarrow \frac{d\tilde{u}_s}{dt} = 0 = 2B_p \Rightarrow B = 0.$$

Now,

$$\tilde{u}_s = \left[\begin{array}{l}(0.1)\int_0^\pi \sin pxdx + (0.01)\\ \int_0^\pi \sin 4x\sin pxdx\end{array}\right]\cos 2pt$$

Finally, taking the inverse finite Fourier sine transform, we get

$$u(x,t) = \frac{2}{\pi}\sum_{p=1}^\infty \tilde{u}_s \sin px$$

$$= \frac{2}{\pi}(0.1)\sum_{p=1}^{\infty}\left\{\int_0^{\pi}\sin x \sin pxdx\right\}\cos 2pt \sin px$$

$$+\frac{2}{\pi}(0.01)\sum_{p=1}^{\infty}\left\{\int_0^{\pi}\sin 4xpxdx\right\}\cos 2pt \sin px$$

$$= \frac{2}{\pi}(0.1)\left[\int_0^{\pi}\sin x \sin xdx\right]\cos 2t \sin x$$

$$+\frac{2}{\pi}(0.01)\left[\int_0^{\pi}\sin 4x.\sin 4xdx\right]\cos 8t \sin 4x$$

$$\Rightarrow u(x,t) = (0.1)\cos 2t \sin x + (0.01)\cos 8t \sin 4t.$$

Example 12. *Find the steady temperature $V(x, y)$ in a long square bar of side π when one face is kept at constant and the other faces at zero temperature. Also $V(x, y)$ is bounded.*

Or

Find a function $V(x, y)$ which is harmonic in the open square $0 < x < \pi$, $0 < y < \pi$ takes a constant value V_0 on the edges $y = \pi$ and vanishes on the other edges of the square.

Solution. It is known that the steady temperature is governed by two dimensional Laplace equation

$$\frac{\partial^2 V}{\partial x^2}+\frac{\partial^2 V}{\partial y^2} = 0 \qquad \ldots(1)$$

subject to the boundary conditions a constant $V(0, y) = 0 = V(\pi, y), V(x, 0) = 0$ and $V(x, \pi) = V_0$ a constant.

Taking the Fourier sine transform of both sides of (1), we get

$$\int_0^{\pi}\frac{\partial^2 V}{\partial x^2}\sin pxdx + \int_0^{\pi}\frac{\partial^2 V}{\partial y^2}\sin pxdx = 0$$

$$\Rightarrow \left[\frac{\partial V}{\partial x}\sin px\right]_0^{\pi} - p\int_0^{\pi}\frac{\partial V}{\partial x}\cos pxdx$$

$$+\frac{\partial^2}{\partial y^2}\int_0^{\pi}V \sin pxdx = 0$$

$$\left[\because \frac{\partial V}{\partial x}\text{ is finite at } x = 0 \text{ and } x = \pi\right]$$

$$\Rightarrow -p\left[V\cos px\right]_0^{\pi} - p^2\int_0^{\pi}V.\sin pxdx + \frac{d^2\tilde{V}_s}{dy^2} = 0$$

$$\Rightarrow \frac{d^2\tilde{V}_s}{dy^2} - p^2\tilde{V}_s = 0$$

$$[\because V = 0 \text{ at } x = 0 \text{ and } x = \pi]$$

$$\Rightarrow \tilde{V}_s = A\cosh py + B\sinh py \qquad \ldots(2)$$

As per given when $y = 0$, $V = 0$ and when $y = \pi, V = V_0$

Taking finite Fourier sine transform of above boundary conditions, we get when $y = 0, \tilde{V}_s = 0$

and when $y = \pi, \tilde{V}_s = \int_0^{\pi}V_0 \sin pxdx$

$$= V_0\left(-\frac{\cos px}{p}\right)_0^{\pi} = V_0\left(\frac{1-\cos p\pi}{p}\right)$$

Thus, from (2), we have $0 = A$

and $V_0\left(\frac{1-\cos p\pi}{p}\right) = B\sinh p\pi$

$$\Rightarrow B = \frac{V_0}{\sinh p\pi}\left(\frac{1-\cos p\pi}{p}\right).$$

Putting these values in (2), we get

$$\tilde{V}_s = \frac{V_0}{\sinh p\pi}\left(\frac{1-\cos p\pi}{p}\right)\sinh py$$

Finally, taking the inverse finite Fourier sine transform, we get

$$V(x, y) = \frac{2}{\pi}\sum_{p=1}^{\infty}\frac{V_0}{\sinh p\pi}\left(\frac{1-\cos p\pi}{p}\right)\sinh py \sin px$$

$$= \frac{4V_0}{\pi}\sum_{n=0}^{\infty}\frac{\sinh(2n+1)y\sin(2n+1)x}{(2n+1)\sinh(2n+1)\pi}$$

Example 13. *The cross section of a long bar of diffusivity k is the square $0 < x < \pi, 0 < y < \pi$. If the four faces of the bar are maintained at zero temperature and the initial temperature is unity, use respective finite transforms to show that the temperature at time t is $\phi(x)\ \phi(y)$ where*

$$\phi(x) = \frac{4}{\pi}\sum_{n=0}^{\infty}(2n+1)^{-1}\sin(2n+1)xe^{[-k(2n+1)^2t]}$$

Solution. It is known that the temperature in the bar is governed by two dimensional heat equation

$$\frac{\partial^2 V}{\partial x^2}+\frac{\partial^2 V}{\partial y^2} = \frac{1}{k}\frac{\partial V}{\partial t} \qquad \ldots(1)$$

where k is the diffusivity of the bar and $V(x, y, t)$ is the temperature at any point and at any time t.

Here, boundary conditions are

$$V(0, y, t) = 0,\quad V(\pi, y, t) = 0, \qquad \ldots(2)$$

$$V(x, 0, t) = 0,\quad V(x, \pi, t) = 0, \qquad \ldots(3)$$

and initial condition is given by

$$V(x, y, 0) = 1 \qquad \ldots(4)$$

Taking the finite Fourier sine transform of both sides of (1), we get

$$\int_0^{\pi}\frac{\partial^2 V}{\partial x^2}\sin pxdx + \int_0^{\pi}\frac{\partial^2 V}{\partial y^2}\sin pxdx = \frac{1}{k}\int_0^{\pi}\frac{\partial V}{\partial t}\sin pxdx$$

$$\Rightarrow \quad -p^2\tilde{V}_s + \frac{\partial^2 V}{\partial y^2} = \frac{1}{k}\frac{\partial \tilde{V}_s}{\partial t} \quad \ldots (5)$$

Further, the finite Fourier sine transform of boundary and initial conditions are on $x = 0, x = \pi, \tilde{V}_s = 0$ on $y = 0, y = \pi, \tilde{V}_s = 0$ and

at $t = 0$ $\quad \tilde{V}_s = \int_0^\pi 1.\sin pxdx = \left(-\frac{\cos px}{p}\right)_0^\pi$

$$= \frac{1-\cos p\pi}{p} = 0 \text{ or } \frac{2}{p}$$

[As p is even or odd.]

Taking again the finite Fourier sine transform of (5), we get

$$-p^2\int_0^\pi \tilde{V}_s \sin qydy + \int_0^\pi \frac{\partial^2 \tilde{V}_s}{\partial y^2}\sin qydy = \frac{1}{k}\int_0^\pi \frac{\partial \tilde{V}_s}{\partial t}\sin qydy$$

$$\Rightarrow \quad -p^2\tilde{V}_s' - q^2\tilde{V}_s' = \frac{1}{k}\frac{d\tilde{V}_s'}{dt}$$

subject to the boundary condition (3).

Here $\tilde{V}_s'$ is finite Fourier sine transform of $\tilde{V}_s$

or $\frac{d\tilde{V}_s'}{dt} = -k(p^2+q^2)\tilde{V}_s'$

$$\Rightarrow \tilde{V}_s' = Ae^{-k(p^2+q^2)t} \quad \ldots (6)$$

Taking finite Fourier sine transform of $\tilde{V}_s$, we get

at $t = 0, \tilde{V}_s' = 0$, if p is even

and $\tilde{V}_s' = \int_0^\pi \frac{2}{p}\sin qydy$ if p is odd

$$= \frac{2}{pq}(1-\cos q\pi) = 0 \text{ or } \frac{4}{pq}$$

[According as q is even or odd]

Thus, when p and q are both even, we have at $t = 0, \tilde{V}_s' = 0$

So, from (6), $A = 0 \Rightarrow \tilde{V}_s' = 0 \Rightarrow V = 0$

Further, when p, q are both odd, we have at $t = 0$,

$\tilde{V}_s' = \frac{4}{pq}$ therefore, from (6), we have $\frac{4}{pq} = A$

$$\Rightarrow \tilde{V}_s' = \frac{4}{pq}e^{-k(p^2+q^2)t}$$

$$\Rightarrow \tilde{V}_s' = \frac{4}{(2n+1)(2m+1)}e^{-k\{(2n+1)^2+(2m+1)^2\}t}$$

Where $p = 2n+1, q = 2m+1$.

Taking again inverse finite sine transform, we get

$$V(x,y,t) = \frac{2}{\pi}\frac{8}{\pi}\sum_{n=0}^{\infty}\sum_{m=0}^{\infty}\frac{e^{-k\{(2n+1)^2+(2m+1)^2\}t}}{(2n+1)(2m+1)}\sin(2m+1)y\sin(2n+1)x$$

$$\Rightarrow \quad \phi(x) = \frac{4}{\pi}\sum_{n=0}^{\infty}(2n+1)^{-1}e^{-k(2n+1)^2t}\sin(2n+1)x$$

Example14. *Determine the displacement $u(x, t)$ in a horizontal string stretched from origin to the point $(\pi, 0)$ when the motion is due to the weight of the string alone. The string may be taken to be initially at rest in the position $u = 0$.*

Solution. Let the displacement $u(x, t)$ be taken positive along the vertical downward line.

Suppose that string is released from the position $u = 0$ along horizontal line. Then the displacement $u(x,t)$ of the string is governed by the differential equation.

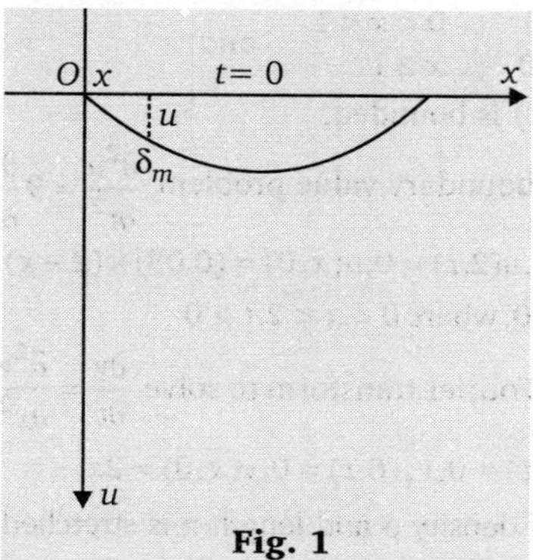

Fig. 1

$$\frac{\partial^2 u}{\partial t^2} = C^2\frac{\partial^2 u}{\partial x^2} + g, \; 0 < x < \pi, t > 0,$$

where $C^2 = T/\rho$. $\quad \ldots (1)$

The boundary conditions are

$u(0,t) = 0 = u(\pi,t), t > 0$ initial conditions are $u(x,0) = 0 = u_t(x,0), 0 < x < \pi$.

Taking the finite Fourier sine transform of both sides of (1), we have

$$\int_0^\pi \frac{\partial^2 u}{\partial t^2}\sin pxdx = C^2\int_0^\pi \frac{\partial^2 u}{\partial x^2}\sin pxdx + g\int_0^\pi \sin pxdx$$

$$\Rightarrow \quad \frac{d^2}{dt^2}\tilde{u}_s = -C^2p^2\tilde{u}_s + C^2p[u(0,t) - u(\pi,t)\cos p\pi] + \frac{q}{p}[1-(-1)^p]$$

$$\Rightarrow \quad \frac{d^2}{dt^2}\tilde{u}_s + C^2p^2\tilde{u}_s = \frac{q}{p}[1-(-1)^p]$$

Whose solution is given by

$$\tilde{u}_s(p,t) = A\cos pat + B\sin pat + \left(\frac{q}{C^2p^3}\right)[1-(-1)^p]. \quad \ldots (2)$$

Now, taking the finite Fourier sine transform of the given initial conditions, we have

When $t = 0, \tilde{u}_s = 0$ and $\frac{d^2}{dt^2}\tilde{u}_s = 0$.

Now, $\frac{d}{dt}\tilde{u}_s = -paA\sin pat + paB\cos pat$

When $t = 0, \tilde{u}_s = 0$

$\Rightarrow \quad A + \frac{q}{C^2 p^3}[1-(-1)^p][1-\cos pat].$

Taking the inverse Fourier sine transform, we get

$$u(x,t) = \frac{2}{\pi}\sum_{p=1}^{\infty}\tilde{u}_s(p,t)\sin px$$

$$= \frac{2q}{\pi C^2}\sum_{p=1}^{\infty}\frac{1}{p^3}[1-(-1)^p](1-\cos pat)\sin px$$

EXERCISE 43.3

1. Solve $\frac{\partial u}{\partial t} = \frac{\partial^2 u}{\partial x^2}, x > 0, t > 0$ subject to the conditions

(i) $u = 0$ when $x = 0, t > 0$

(ii) $u = \begin{cases} 1 & ; \ 0 < x < 1 \\ 0 & ; \ x \geq 1 \end{cases}$ and

(iii) $u(x, t)$ is bounded.

2. Solve the boundary value problem $\frac{\partial^2 u}{\partial t^2} = 9\frac{\partial^2 u}{\partial x^2}$ subject to all conditions
$u(0,t) = 0, u(2,t) = 0, u(x,0) = (0.05)\times(2-x)$
$u_t(x,0) = 0$, where $0 < x < 2, t > 0$

3. Use finite Fourier transform to solve $\frac{\partial v}{\partial t} = \frac{\partial^2 v}{\partial x^2}, 0 < x < 6, t > 0$ and $v_x(0,t) = 0, v_x(6,t) = 0, v(x,0) = 2x$.

4. A string of density ρ and length π is stretched to a tension ρC^2. At time $t = 0$, one end $(x = 0)$ is given a small oscillation $a\sin\omega t$. If the other end remains fixed, use a finite sine transform to show that the displacement of the point x at time t is

$$a\sin\omega t \sin\frac{\omega(\pi - C)}{C}\operatorname{cosec}\frac{\pi\omega}{C} + \left(\frac{2aC\omega}{\pi}\right)\sum_{p=0}^{\infty}(\omega^2 - p^2C^2)^{-1}\sin px \sin pct\,.$$

5. Solve $\frac{\partial u}{\partial t} = \frac{\partial^2 u}{\partial x^2}$ if $u_x(0,t) = 0, u(x,0) = \begin{cases} x & ; \ 0 \leq x \leq 1 \\ 0 & ; \ x > 1 \end{cases}$ and $u(x, t)$ is bounded where $x > 0, t > 0$.

6. Use the finite cosine transform to solve $\frac{\partial V}{\partial t} = k\frac{\partial^2 V}{\partial x^2}$, one dimensional heat equation with the boundary conditions $\frac{\partial V}{\partial x} = 0$ when $x = 0$ and $x = \pi, t > 0$ and the initial condition $V = f(x)$ when $t = 0, 0 < x < \pi$.

7. Solve $\frac{\partial u}{\partial t} = 2\frac{\partial^2 u}{\partial x^2}$ if $u(0,t) = 0, u(x,0) = e^{-x}, x > 0, u(x,t)$ is bounded where $x > 0$, $t > 0$.

8. A string is stretched between the two fixed points (0, 0) and (C, 0). If it is displaced into the curve $y = b\sin\left(\frac{n\pi x}{C}\right)$ and released from rest in that position at time $t = 0$. Solve the boundary value problem for the displacement $y(x, t)$.

9. A tightly stretched flexible string has its ends fixed at $x = 0$ and $x = l$. At time $t = 0$, the string is given a shape defined by $F(x) = \mu x(l - x)$ where μ is a constant and then released. Find the displacement of any point x of the string at any point $t = 0$ (VTU(ME) – 2006)

10. If the initial temperature of an infinite bar is given by

$$\theta(x) = \begin{cases} \theta_0 & \text{for } |x| < a \\ 0 & \text{for } |x| > a \end{cases}$$

Show that the temperature at any point x at any instant t is given by

$$\theta(x,t) = \frac{\theta_0}{\pi}\left[erf\frac{(a+x)}{2c\sqrt{t}} + erf\frac{(a-x)}{2c\sqrt{t}}\right]$$

(SVTU – 2008, Rohtak –2004)

FFFFFF

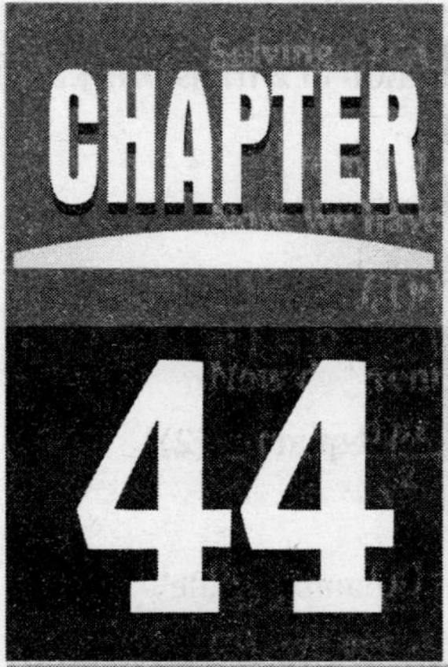

Z-Transforms

44.1 INTRODUCTION

Z-transforms plays an important role in discrete analysis as Laplace and Fourier transforms in continuous system. It has many properties similar to those of the Laplace transforms. The z-transforms operate on sequences of the discrete integer valued arguments, not on function of continuous arguments. For every operational rule of Laplace transforms, there is a corresponding operational rule of z-transforms and for every application of the Laplace transforms, there is a corresponding application of z-transforms.

44.2 Z-TRANSFORMS

If the function f_n is defined for discrete values ($n = 0, 1, 2, \ldots$) and $f_n = 0$, for $n < 0$ then its z-transform is defined by

$$Z(f^n) = F(z) = \sum_{n=-\infty}^{\infty} f(n)z^{-n} = \sum_{n=-\infty}^{\infty} \frac{f(n)}{z^n}$$

whenever the infinite series converges. Here z is a complex number. Z is an operator and $F(z)$ is the Z-transforms of $\{f(n)\}$.

44.2.1 SOME STANDARD Z-TRANSFORMS

(1) $Z(a^n) = \dfrac{z}{z-a}$ (Kotayam–2005) **(2)** $Z(n^p) = -z\dfrac{d}{dz}Z(n^{p-1})$, where p is a positive integer.

(3) $Z(1) = \dfrac{z}{z-1}$ **(4)** $Z(1) = \dfrac{z}{(z-1)^2}$

(5) $Z(n^2) = \dfrac{z^2+z}{(z-1)^3}$ **(6)** $Z(n^3) = \dfrac{z^3+4z^2+z}{(z-1)^4}$ **(7)** $Z(n^4) = \dfrac{z^2+11z^3+11z^2+z}{(z-1)^5}$

44.3 PROPERTIES OF Z-TRANSFORMS

44.3.1 LINEAR PROPERTY

If $<f(n)>$ and $<g(n)>$ are two sequences such that they can be added and a, b are any two scalars, then

$$Z[< a\,f(n) + b\,g(n) >] = aZ[< f(n) >] + bZ[< g(n) >]$$

Proof. Consider

$$Z[<af(n)+b\,g(n)>] = \sum_{n=-\infty}^{\infty} [a\,f(n)+b\,g(n)]z^{-n} \quad \text{[By definition of z-transform]}$$

$$= \sum_{n=-\infty}^{\infty} [af(n)z^{-n} + bg(n)z^{-n}] = a\sum_{n=-\infty}^{\infty} f(n)z^{-n} + b\sum_{n=-\infty}^{\infty} g(n)z^{-n}$$

$$= aZ[< f(n) >] + b\,Z[< g(n) >]$$

44.3.2 CHANGE OF SCALE PROPERTY

If $Z[<f(n)>] = F(z)$, then

(i) $z[< a^n f(n) >] = F\left(\dfrac{z}{a}\right)$ (ii) $Z[<a^{-n} f(n)>] = F(az)$

Proof. (i) We have

$$F\left(\frac{z}{a}\right)=\sum_{n=-\infty}^{\infty} f(n)\left(\frac{z}{a}\right)^{-n} \quad \text{[By definition of z-transform]...(1)}$$

Also we have

$$Z[<a^n f(n)>]=\sum_{n=-\infty}^{\infty} a^n f(n) z^{-n}=\sum_{n=-\infty}^{\infty} f(n)\left(\frac{z}{a}\right)^{-n}=F\left(\frac{Z}{a}\right)$$

(ii) We have

$$F(az)=\sum_{n=-\infty}^{\infty} f(n)(az)^{-n}$$

Also, we have

$$Z[\{a^{-n} f(n)\}]=\sum_{n=-\infty}^{\infty} a^{-n} f(n) z^{-n}=\sum_{n=-\infty}^{\infty} f(n)(az)^{-n}=F(az) \quad \text{(Using equation (2))}$$

REMARK

- Here, the geometric factor a^{-n} when $|a|<1$ damps the function u_n, hence, above rule is also known as 'damping rule'. (Madras–2006)

44.3.3 APPLICATION OF DAMPING RULE

The application of damping rule gives the following important results :

(i) $Z(na^n)=\dfrac{az}{(z-a)^2}$ (Madras–2000) (ii) $Z(n^2a^n)=\dfrac{az^2+a^2z}{(z-a)^3}$ (iii) $Z(\cos n\theta)=\dfrac{z(z-\cos\theta)}{z^2-2z\cos\theta+1}$

(iv) $Z(\sin n\theta)=\dfrac{z\sin\theta}{z^2-2z\cos\theta+1}$ (v) $Z(a^n\cos n\theta)=\dfrac{z(z-a\cos\theta)}{z^2-2az\cos\theta+a^2}$ (vi) $Z(a^n\sin n\theta)=\dfrac{az\sin\theta}{z^2-2az\cos\theta+a^2}$

44.3.4 SHIFTING PROPERTIES

If $Z[\{(f(n)\}]=F(z)$, *then*

(i) $Z[<f(n\pm k)>]=z^{\pm k}F(z)$ *and if* $n\geq 0$, *then*

(ii) $Z[<f(n+k)>]=z^nF(z)-\sum_{r=0}^{k-1} f(r)z^{k-r}$ (iii) $Z[<f(n-k)>]=z^{-n}F(z)-\sum_{m=1}^{k} f(-m)z^{-k+m}$

Proof. (i) Consider

$$Z[<f(n\pm k)>]=\sum_{n=-\infty}^{\infty} f(n\pm k)z^{-n}$$

$$=\sum_{n=-\infty}^{\infty} f(n\pm k)z^{-(n\pm k)}.z^{\pm n}=z^{\pm k}\sum_{n=-\infty}^{\infty} f(r)z^{-r} \quad \text{(where } n\pm k=r\text{)}$$

$$=z^{\pm k}F(z)$$

Hence,

$$Z[<f(n\pm k)>]=z^{\pm k}F(z)$$

(ii) Consider

$$Z[<f(n\pm k)]=\sum_{n=0}^{\infty} f(n+k)z^{-n}$$

$$=\sum_{n=0}^{\infty} f(n+k)z^{-(n+k)}.z^n=z^k\sum_{r=k}^{\infty} f(r)z^{-r} \quad \text{(where } n+k=r\text{)}$$

$$=z^k\sum_{r=0}^{\infty} f(r)z^{-r}-z^k\sum_{r=0}^{k-1} f(r)z^{-r}=z^kF(z)-\sum_{r=0}^{k-1} f(r)z^{-r}=z^kF(z)-\sum_{r=0}^{k-1} f(r)z^{k-r}$$

Hence,

$$Z[<f(n+k)>]=z^kF(z)-\sum_{r=0}^{k-1} f(r)z^{k-r}$$

(iii) Consider

$$Z[<f(n-k)>]=\sum_{n=0}^{\infty} f(n-k)z^{-n}=\sum_{n=0}^{\infty} f(n-k)z^{-(n+k)}.z^{-k}$$

$$=z^{-k}\sum_{r=0}^{\infty} f(r)z^{-r}+z^{-k}\sum_{r=-k}^{-1} f(1)z^{-r} \quad \text{(when } n+k=r\text{)}$$

$$=z^{-k}F(z)+\sum_{r=-k}^{-1} f(r)z^{-k-r}=z^{-k}F(z)+\sum_{m=1}^{k} f(-m)z^{-k+m}$$

Hence,

$$Z[<f(n-k)>]=z^{-k}F(z)+\sum_{m=1}^{k} f(-m)z^{-k+m}$$

44.3.5 SOME IMPORTANT DEDUCTIONS

(1) $Z[< f(n+1) >] = z F(z) - z f(0) = z [F(z) - f(0)]$

(2) $Z[< f(n+2) >] = z^2 F(z) - \sum_{r=0}^{1} f(r) z^{2-r}$

$\Rightarrow$ $Z[< f(n+2) >] = z^2 F(z) - z^2 f(0) - z f(1) = z^2 [F(z) - f(0) - z^{-1} f(1)]$

(3) $Z[< f(n+3) >] = z^3 F(z) - \sum_{r=0}^{2} f(r) z^{3-r}$

$\Rightarrow$ $Z[< f(n+3) >] = z^3 F(z) - z^3 f(0) - z^2 f(2) - z f(2)$

$\Rightarrow$ $Z[< f(n+3) >] = z^3 [F(z) - f(0) - z^{-1} f(1) - z^{-2} f(2)]$ and so on.

Shifting to the Right

If $Z(f_n) = F(z)$, *then* $Z(f(n-k)) = z^{-k} F(2)$ $(k > 0)$

Proof. By definition, we have

$$Z[f_{n-k}] = \sum_{n=0}^{\infty} f_{n-k} z^{-n} = f_0 z^{-k} + f_1 z^{-(k+1)} + \ldots$$

$$= z^{-k} \sum_{n=0}^{\infty} f_n z^{-n} = z^{-k} F(z)$$

Shifting to the left

If $Z(f_n) = F(z)$, *then* $Z(f(n+k)) = z^k [F(z) - f_0 - f_1 z^{-1} - f_2 z^{-2} - \ldots - f_{k-1} z^{-(k-1)}]$ (JNTU–2002)

Proof. We have

$$Z[f_{n+k}] = \sum_{n=0}^{\infty} f_{n+k} z^{-n} = z^k \left[\sum_{n=0}^{\infty} f_n z^{-n} - \sum_{n=0}^{k-1} u_n z^{-n} \right]$$

Hence, $$Z(f(n+k)) = z^k [F(z) - f_0 - f_1 z^{-1} - f_2 z^{-2} - \ldots - f_{k-1} z^{-(k-1)}]$$

44.4 Z-TRANSFORM OF nf(n)

If $z[< f(x) >] = F(z)$, *then* $Z[< nf(n) >] = -z \dfrac{dF(z)}{dz}$

Proof. We have

$$Z[< nf(n) >] = \sum_{n=-\infty}^{\infty} n f(n) z^{-n} = -z \sum_{n=-\infty}^{\infty} [-nf(n)] z^{-n-1}$$

$$= -z \sum_{n=-\infty}^{\infty} f(n)(-nz^{-k-1}) = -z \sum_{n=-\infty}^{\infty} f(n) \frac{d(z^{-n})}{dz}$$

$$= -z \frac{d}{dz} \sum_{n=-\infty}^{\infty} f(n) z^{-n} = -z \frac{d}{dz} F(z)$$

i.e., $$Z[< n f(n) >] = -z \frac{d}{dz} F(z)$$

REMARK

- The above result can be generalized as follows

$$Z[< n^k f(n) > 1] = \left(-z \frac{d}{dz} \right)^k F(z)$$

44.5 Z-TRANSFORM OF f(n)/n

If $z[< f(n) >] = F(z)$, *then* $z\left[\dfrac{f(n)}{n}\right] = \int_z \dfrac{1}{z} F(z) dz$

Proof. We have

$$z\left[\frac{f(n)}{n}\right] = \sum_{n=-\infty}^{\infty} \frac{f(n)}{n} z^{-n} = \sum_{n=-\infty}^{\infty} f(n) \left(\frac{1}{n} z^{-n} \right) = -\sum_{n=-\infty}^{\infty} f(n) \int_z z^{-n-1} dz$$

$$= -\int_z \sum_{n=-\infty}^{\infty} f(n) z^{-n-1} dz = -\int_z \frac{1}{z} \sum_{n=-\infty}^{\infty} f(n) z^{-n} dz = -\int_z z^{-1} F(z) dz$$

Hence, $$Z\left[\left\{\frac{f(n)}{n}\right\}\right] = -\int_z z^{-1} F(z) dz$$

44.6 INITIAL VALUE THEOREM

If $Z[<f(n)>] = F(z), n > 0,$ *then* $\quad f(0) = \lim_{z\to\infty} F(z)$

Proof. Here we have

$$Z[< f(n) >] = \sum_{n=0}^{\infty} f(n) z^{-n} = F(z)$$

or $\quad f(0) + f(1)z^{-1} + f(2)z^{-2} + \ldots = F(z)$

Taking limits of both the sides, as $z \to \infty$, we get

$$f(0) = \lim_{z\to\infty} F(z)$$

44.7 FINAL VALUE THEOREM

If $Z[<f(n)>] = F(z), n > 0,$ *then* $\quad \lim_{z\to\infty} F(n) = \lim_{z\to 1} (z-1) F(z)$

Proof. We have $\quad Z[< f(n+1) > - f(n)] = \sum_{n=0}^{\infty} [f(n+1) - f(n)] z^{-n}$

or $\quad Z[<f(n+1)>] - Z[<f(n)>] = \sum_{n=0}^{\infty} [f(n+1) - f(n)] z^{-n}$

$$zF(z) - zF(0) - F(z) = \lim_{k\to\infty} \sum_{n=0}^{k} [f(n+1) - f(n)] z^{-n}$$

Taking the limits of both the sides as $z \to 1$, we get

$$\lim_{z\to 1}(z-1)F(z) = f(0) + \lim_{z\to 1}\lim_{k\to\infty} \sum_{n=0}^{k} [f(n+1) - f(n)] z^{-n}$$

$$= f(0) + \lim_{k\to\infty} \sum_{n=0}^{k} \lim_{z\to 1} [f(n+1) - f(n)] z^{-n} \quad \text{(On changing the order of limits)}$$

$$= f(0) + \lim_{k\to\infty} \sum_{n=0}^{k} [f(n+1) - f(n)]$$

$$= \lim_{k\to\infty} [f(0) - f(0) + f(1) - f(1) + f(2) - f(2) + \ldots + f(n+1) - f(n)]$$

$$= \lim_{k\to\infty} (k+1) = \lim_{k\to\infty} f(k) = \lim_{n\to\infty} f(n)$$

Hence, $\quad \lim_{n\to\infty} f(n) = \lim_{z\to 1} (z-1) F(z)$

44.8 PARTIAL SUM THEOREM

If $Z[\{f(n)\}] = F(z),$ *then* $\quad Z\left[\left\langle \sum_{k=-\infty}^{n} f(k) \right\rangle\right] = \frac{F(z)}{1 - z^{-1}}$

Proof. Let $\{u(n)\}$ be a sequence such that $\quad u(n) = \sum_{k=-\infty}^{n} f(k) \quad \ldots(1)$

We want to find the value of $Z[\{u(n)\}]$.

From (1), we can write $\quad u(n) - u(n-1) = \sum_{k=-\infty}^{n} f(x) - \sum_{k=-\infty}^{n-1} f(x)$

or $\quad u(n) - u(n-1) = f(n)$

Taking Z-transforms of both the sides, we get

$$Z[\{u(n)\}] - Z[\{u(n-1)\}] = Z[\{f(n)\}]$$

or $\quad U(z) - z^{-1} U(z) = F(z)$, where $U(z) = Z[\{u(n)\}]$

$\Rightarrow \quad (1 - z^{-1}) U(z) = F(z)$

$\Rightarrow \quad U(z) = \frac{F(z)}{1 - z^{-1}}$ *i.e.,* $Z[\{u(n)\}] = \frac{F(z)}{1 - z^{-1}}$

Hence, $\quad Z\left[\left\langle \sum_{k=-\infty}^{n} f(k) \right\rangle\right] = \frac{F(z)}{1 - z^{-1}}$

44.9 CONVOLUTION THEOREM OF Z-TRANSFORMS

Definition : *Let $< f(n) >$ and $< g(n) >$ be two sequence and let the convolution of $< f(n) >$ and $< g(n) >$ be $< h(n) >$ where $< h(n) > = < f(n) >* < g(n) >$, then $< h(n) >$ is defined as*

$$< h(n) > = < f(n) >* < g(n) >$$

$$= \sum_{n=-\infty}^{\infty} f(n)g(n-k) = \sum_{n=-\infty}^{\infty} g(n)f(n-k) = < g(n) > * < f(n) >$$

Theorem. *If $Z[< f(n) >] = F(z)$ and $Z[< g(n) >] = G(z)$, then $Z[< h(n) >] = Z[< f(n) >* < g(n) >] = F(z)\,G(z)$ where the region of convergence of $Z[<h(n)>]$ is the common region of convergence of $F(z)$ and $G(z)$.*

Proof. We have, by defintion $Z[< h(n) >] = \sum_{n=-\infty}^{\infty}\left[\sum_{n=-\infty}^{\infty} f(k)g(n-k)\right]z^{-n} = \sum_{n=-\infty}^{\infty}\sum_{n=-\infty}^{\infty} f(k)g(n-k)z^{-n}$

$$= \sum_{k=-\infty}^{\infty} f(k)z^{-k}\sum_{r=-\infty}^{\infty} g(r)z^{-r} \qquad (n-k = r)$$

$$= \sum_{k=-\infty}^{\infty} f(k)z^{-k}G(z) = F(z).G(z)$$

44.10 INVERSE Z-TRANSFORMS

If $F(z)$ is the Z-transform of the sequence $< f(k) >$, then $<f(k)>$ is called the inverse Z-transform of $F(z)$. The operator for inverse Z-transform is denoted by Z^{-1}. Thus, if $F < f(k) > = F(z)$, then

$$Z^{-1}[F(z)] = < f(k) >$$

Some Useful Z-Transforms

S.No.	Sequence f(n), n ≥ 0	Z-transform : F(z)	S.No.	Sequence f(n), n ≥ 0	Z-transform : F(z)
1.	n	$z/(z-1)^2$	12.	$a^n \cos n\theta$	$\frac{z(z-a\cos\theta)}{z^2-2az\cos\theta+a^2}$
2.	n^2	$(z^2+z)/(z-1)^3$	13.	$\sinh n\theta$	$\frac{z\sinh\theta}{z^2-2z\cosh\theta+1}$
3.	n^p	$-z d/dz\,[Z(n^{p-1})]$, positive integer.	14.	$\cosh n\theta$	$\frac{z(z-\cosh\theta)}{z^2-2z\cosh\theta+1^2}$
4.	$\delta(n)=\begin{cases}1, & n=0\\ 0, & n\neq 0\end{cases}$	1	15.	$a^n \sinh n\theta$	$\frac{az\sinh\theta}{z^2-2az\cosh\theta+a^2}$
5.	$f(n)=\begin{cases}1, & n<0\\ 0, & n\geq 0\end{cases}$	$z/(z-1)$	16.	$a^n \cosh n\theta$	$\frac{z(z-1\cosh\theta)}{z^2-2az\cosh\theta+a^2}$
6.	a^n	$z/(z-a)$	17.	$a^n f(n)$	$(F(z/a))$
7.	na^n	$az/(z-a)^2$	18.	f_{n+1} f_{n+2} f_{n+3}	$z[F(z)-f_0]$ $z^2[F(z)-f_0-f_1z^{-1}]$ $z^3[F(z)-f_0-f_1z^{-1}-f_2z^{-2}]$
8.	n^2a^n	$(az^2+a^2z)/(z-a)^3$	19.	f_{n-k}	$z^{-k}F(z)$
9.	$\sin n\theta$	$\frac{z\sin\theta}{z^2-2z\cos\theta+1}$	20.	$n f_n$	$-z d/dz\,[F(z)]$
10.	$\cos n\theta$	$\frac{z(z-\cos\theta)}{z^2-2z\cos\theta+1}$	21.	$\frac{1}{n}f_n$	$-\int_0^z z^{-1}[F(z)]dz$
11.	$a^n \sin n\theta$	$\frac{az\sin\theta}{z^2-2az\cos\theta+a^2}$	22.	f_0	$\lim_{z\to\infty} F(z)$
			23.	$\lim_{n\to\infty} f_n$	$\lim_{z\to 1}[(z-1)F(z)]$

Some Useful Inverse Z-Transforms

S.No.	F(z)	Inverse Z-transform f(n)	S.No.	F(z)	Inverse Z-transform f(n)
1.	$\frac{z}{z-a}$	$a^n u(n)$	4.	$\frac{1}{z-a}$	$a^{n-1} u(n-1)$
2.	$\frac{z^2}{(z-a)^2}$	$(n+1)\, a^n u(n)$	5.	$\frac{1}{(z-a)^2}$	$(n-1)\, a^{n-2} u(n-2)$
3.	$\frac{z^3}{(z-a)^3}$	$\frac{1}{2!}(n+1)(n+2)a^n u(n)$	6.	$\frac{1}{(z-a)^3}$	$\frac{1}{2}(n-1)(n-2)a^{n-3}u(n-3)$

Solved Examples

Example 1. *Find the Z-transforms of the following sequences :*

(i) $< f(n) > = \{15, 10, 7, 4, 1, -1, 0, 3, 6\}$

(ii) $< f(n) > = \{15, 10, 7, 4, 1\}$

(iii) $< f(n) > = 1/3^n$

(iv) $< f(n) > = \frac{1}{2^n}, -3 \le n \le 2$

(v) $< f(n) > = \{a^n\}, n \ge 0$

Solution. (i) We have $Z[< f(n) >]$

$$= F(z) = \sum_{n=-3}^{5} f(n) z^{-n}$$

$$= 15z^3 + 10z^2 + 7z + 4 + \frac{1}{z} - \frac{1}{z^2} + 0 + \frac{3}{z^4} + \frac{6}{z^5}$$

Therefore, $Z[< f(n) >]$

$$= 15z^3 + 10z^2 + 7z + 4 + \frac{1}{z} - \frac{1}{z^2} + 0 + \frac{3}{z^4} + \frac{6}{z^5}$$

(ii) We have $Z[<f(n)>] = F(z)$

$$= \sum_{n=0}^{4} f(n) z^{-n} = 15 + \frac{10}{z} + \frac{7}{z^2} + \frac{4}{z^3} + \frac{1}{z^4}$$

(iii) $Z[<f(n)>] = F(z)$

$$= \sum_{n=-\infty}^{\infty} f(n) z^{-n} = \sum_{n=-\infty}^{\infty} \frac{1}{3^n} z^{-n}$$

$$= \ldots + 27z^3 + 9z^2 + 3z + 1 + \frac{1}{3z} + \frac{1}{9z^2} + \frac{1}{27z^3} + \ldots$$

(iv) $Z[<f(n)>] = F(z) = \sum_{n=-3}^{2} \frac{1}{2^n} z^{-n}$

$$= 8z^3 + 4z^2 + 2z + 1 + \frac{1}{2z} + \frac{1}{4z^2}$$

(v) $Z[\{f(n)\}] = F(z)$

$$= \sum_{n=0}^{\infty} a^n z^{-n} = 1 + \frac{a}{z} + \frac{a^2}{z^2} + \frac{a^3}{z^3} + \ldots$$

$$= \frac{1}{1 - \frac{a}{z}} = \frac{z}{z-a}$$

Example 2. *Find the Z-transforms of the following sequences :*

(i) $< f(n) > = < a^{|n|} >$

(ii) $< f(n) > = \left\{\frac{a^n}{n!}\right\}, n \ge 0$ (SVTU–2009)

(iii) $< f(n) > = \{C^n \cos(\alpha_n)\}, n \ge 0$ (UPTU–2004)

(iv) $< f(n) > = \{\sin(3n+5)\}, n \ge 0$ (VTU–2009, Kottayam–2005)

(v) $< f(n) > = \{C^n \cosh(\alpha_n)\}, n \ge 0$

Solution. (i) We have $Z[< a^{|n|} >] = \sum_{n=-\infty}^{\infty} a^{|n|} z^{-n}$

$$= \sum_{n=-\infty}^{-1} a^{-n} z^{-n} + \sum_{n=0}^{\infty} a^n z^{-n}$$

$$= [\ldots + a^3z^3 + a^2z^2 + az] + [1 + az^{-1} + a^2z^{-2} + a^3z^{-3} + \ldots]$$

$$= \frac{az}{1-az} + \frac{1}{1-az^{-1}} = \frac{az}{1-az} + \frac{z}{z-a}$$

$$= \frac{az(z-a) + z(1-az)}{(1-az)(z-a)} = \frac{z(1-a^2)}{(1-az)(z-a)}$$

(ii) $Z\left[\left\langle \frac{a^n}{n!} \right\rangle\right] = \sum_{n=0}^{\infty} \left(\frac{a^n}{n!}\right) z^{-n} = \sum_{n=0}^{\infty} \frac{(az^{-1})}{n!} . z^{-n}$

$$= \sum_{n=0}^{\infty} \frac{(az^{-1})^n}{n!}$$

$$= 1 + \frac{az^{-1}}{1!} + \frac{(az^{-1})^2}{2!} + \frac{(az^{-1})^3}{3!}$$

$$= e^{az^{-1}} = e^{a/z}$$

(iii) We have $Z\left[\left\langle C^n \cos(\alpha n)\right\rangle\right]$

$$= \sum_{n=0}^{\infty} [C^n \cos(\alpha n)] z^{-n}$$

$$= \sum_{n=0}^{\infty} C^n \left[\frac{e^{i\alpha n} + e^{-i\alpha n}}{2}\right] z^{-n}$$

$$= \frac{1}{2}\sum_{n=0}^{\infty} (C^n e^{i\alpha n}) z^{-n} + \frac{1}{2}\sum_{n=0}^{\infty} (C^n e^{-i\alpha n}) z^{-n}$$

$$= \frac{1}{2}\sum_{n=0}^{\infty} (C^n e^{i\alpha} z^{-1})^n + \frac{1}{2}\sum_{n=0}^{\infty} (C^n e^{-i\alpha} z^{-1})^n$$

$$= \frac{1}{2}\left[1 + Ce^{i\alpha} z^{-1} + (Ce^{i\alpha} z^{-1})^2 + (Ce^{i\alpha} z^{-1})^3 + \ldots\right]$$

$$+ \frac{1}{2}\left[1 + Ce^{-i\alpha} z^{-1} + (Ce^{-i\alpha} z^{-1})^2 + (Ce^{-i\alpha} z^{-1})^3 + \ldots\right]$$

$$= \frac{1}{2}\left[\frac{1}{1 - Ce^{i\alpha} z^{-1}}\right] + \frac{1}{2}\left[\frac{1}{1 - Ce^{-i\alpha} z^{-1}}\right] \quad [\because |z| > |C|]$$

$$= \frac{1}{2}\left[\frac{1 - Ce^{-i\alpha} z^{-1} + 1 - Ce^{i\alpha} z^{-1}}{(1 - Ce^{i\alpha} z^{-1})(1 - Ce^{-i\alpha} z^{-1})}\right]$$

$$= \frac{1}{2}\left[\frac{2 - Cz^{-1}(e^{i\alpha} + e^{-i\alpha})}{1 - Ce^{-i\alpha} z^{-1} - Ce^{i\alpha} z^{-1}) C^2 z^{-2}}\right]$$

$$= \frac{1}{2}\left[\frac{2 - Cz^{-1}(e^{i\alpha} + e^{-i\alpha})}{1 - Ce^{-i\alpha} z^{-1} - Ce^{i\alpha} z^{-1} + C^2 z^{-2}}\right]$$

$$= \left[\frac{1 - C\left(\dfrac{e^{i\alpha} + e^{-i\alpha}}{2}\right) z^{-1}}{1 - C(e^{i\alpha} + e^{-i\alpha}) z^{-1} + C^2 z^{-2}}\right]$$

$$= \frac{1 - (C\cos\alpha) z^{-1}}{1 - (2C\cos\alpha) z^{-1} + C^2 z^{-2}}$$

$$= \frac{z^2 - Cz\cos\alpha}{z^2 - 2Cz\cos\alpha + C^2}$$

(iv) We have

$$Z[< \sin(3n+5) >] = \sum_{n=0}^{\infty} \sin(3n+5) z^{-n}$$

$$= \sum_{n=0}^{\infty}\left[\frac{e^{i(3n+5)} - e^{-i(3n+5)}}{2!}\right] z^{-n}$$

$$= \frac{1}{2!}\sum_{n=0}^{\infty} e^{i(3n+5)} . z^{-n} - \frac{1}{2!}\sum_{n=0}^{\infty} e^{-i(3n+5)} z^{-n}$$

$$= \frac{e^{5!}}{2!}\sum_{n=0}^{\infty} (e^{3i} - z^{-1})^n - \frac{e^{-5i}}{2!}\sum_{n=0}^{\infty} (e^{-3i} z^{-1})^n$$

$$= \frac{e^{5i}}{2!}[1 + e^{3i} z^{-1} + (e^{3i} z^{-1})^2 + (e^{3i} z^{-1})^3 + \ldots]$$

$$- \frac{e^{-5i}}{2!}[1 + e^{-3i} z^{-1} + (e^{-3i} z^{-1})^2 + (e^{-3i} z^{-1})^3 + \ldots]$$

$$= \frac{e^{5i}}{2!}\left[\frac{1}{1 - e^{3i} z^{-1}}\right] - \frac{e^{-5i}}{2i}\left[\frac{1}{1 - e^{-3i} z^{-1}}\right] \quad (|Z| > 1)$$

$$= \frac{1}{2!}\left[\frac{e^{5i}(1 - e^{-3i} z^{-1}) - e^{-5i}(1 - e^{3i} z^{-1})}{(1 - e^{3i} z^{-1})(1 - e^{-3i} z^{-1})}\right]$$

$$= \frac{1}{2!}\left[\frac{(e^{5i} - e^{-5i}) - e^{2i} z^{-1} + e^{-2i} z^{-1}}{1 - e^{3i} z^{-1} - e^{-3i} z^{-1} + z^{-2}}\right]$$

$$= \frac{\left(\dfrac{e^{5i} - e^{-5i}}{2!}\right) - \left(\dfrac{e^{2i} - e^{-2i}}{2!}\right) z^{-1}}{1 - (2\cos 3) z^{-1} + z^{-2}}$$

$$= \frac{\sin 5 - (\sin 2) z^{-1}}{1 - (2\cos 3) z^{-1} + z^{-2}}$$

$$= \frac{z^2 \sin 5 - z\sin 2}{z^2 - 2z\cos 3 + 1}$$

(v) We have $Z[< C^n \cosh(\alpha n) >]$

$$= \sum_{n=0}^{\infty} [C^n \cosh(\alpha n)] z^{-n}$$

$$= \sum_{n=0}^{\infty} C^n \left[\frac{e^{\alpha n} + e^{-\alpha n}}{2}\right] z^{-n}$$

$$= \frac{1}{2}\sum_{n=0}^{\infty} (Ce^{\alpha} z^{-1})^n + \frac{1}{2}\sum_{n=0}^{\infty} (Ce^{-\alpha} z^{-1})^n$$

$$= \frac{1}{2}[1 + Ce^{\alpha} z^{-1} + (Ce^{\alpha} z^{-1})^2 + (Ce^{\alpha} z^{-1})^3 + \ldots]$$

$$+ \frac{1}{2}[1 + Ce^{-\alpha} z^{-1} + (Ce^{-\alpha} z^{-1})^2 + (Ce^{-\alpha} z^{-1})^3 + \ldots]$$

$$= \frac{1}{2}\left[\frac{1}{1 - Ce^{\alpha} z^{-1}}\right] + \frac{1}{2}\left[\frac{1}{1 - Ce^{-\alpha} z^{-1}}\right]$$

$$[\because |Z| > |Ce^{\alpha}| \text{ and } |Z| > |Ce^{-\alpha}|]$$

$$= \frac{1}{2}\left[\frac{1 - Ce^{-\alpha} z^{-1} + 1 - Ce^{\alpha} z^{-1}}{1 - Ce^{\alpha} z^{-1} - Ce^{-\alpha} z^{-1} + C^2 z^{-2}}\right]$$

$$= \frac{1}{2}\left[\frac{2 - C(e^{\alpha} + e^{-\alpha}) z^{-1}}{1 - C(e^{\alpha} + e^{-\alpha}) z^{-1} + C^2 z^{-2}}\right]$$

$$= \frac{1 - (C\cosh\alpha) z^{-1}}{1 - (2C\cosh\alpha) z^{-1} + C^2 z^{-2}}$$

$$= \frac{z^2 - Cz\cosh\alpha}{z^2 - 2Cz\cosh\alpha + C^2}$$

$$= \frac{z(z - C\cosh\alpha)}{z^2 - 2Cz\cosh\alpha + C^2}$$

Example 3. *Show that* $Z\left(\frac{1}{n!}\right) = e^{1/z}$.

Hence, evaluate

$$Z\left[\frac{1}{(n+1)!}\right] \text{ and } Z\left[\frac{1}{(n+2)!}\right].$$

Solution. We have

$$Z\left[\frac{1}{n!}\right] = \sum_{n=0}^{\infty} \frac{1}{n!} z^{-n}$$

$$= 1 + \frac{z^{-1}}{1!} + \frac{z^{-2}}{2!} + \frac{z^{-3}}{3!} + \ldots = e^{1/z}$$

Now, shifting $\left(\frac{1}{n!}\right)$ one unit to the left gives

$$Z\left[\frac{1}{(n+1)!}\right] = z\left[z\left(\frac{1}{n!}\right) - 1\right] = z(e^{1/z} - 1)$$

Again, shifting $\left(\frac{1}{n!}\right)$ two units to the left gives

$$Z\left[\frac{1}{(n+2)!}\right] = z^2(e^{1/z} - 1 - z^{-1})$$

Example 4. *Find the Z-transforms of discrete unit step function*

$$F(n) = \begin{cases} 1, & n < 0 \\ 0, & n \geq 0 \end{cases}$$

Solution. We have $Z = [\{F(n)\}] = \sum_{n=0}^{\infty} F(n) z^{-n}$

$$= \left[1 + z^{-1} + z^{-2} + z^{-3} + \ldots\right]$$

$$= \frac{1}{1 - z^{-1}} = \frac{1}{1 - \frac{1}{z}} = \frac{z}{z - 1}$$

Example 5. *Find the Z-transforms of {f(n)}, where*

$$F(n) = \begin{cases} 5^n, & n < 0 \\ 3^n, & n \geq 0 \end{cases}$$

Solution. We have

$$Z = [\{F(n)\}]$$

$$= \sum_{n=0}^{-1} 5^n z^{-n} + \sum_{n=0}^{\infty} 3^n z^{-n}$$

$$= [\ldots. + 5^{-3}z^3 + 5^{-2}z^2 + 5^{-1}z^1]$$

$$+ \left[1 + \frac{3}{z} + \frac{9}{z^2} + \frac{27}{z^3} + \ldots\right]$$

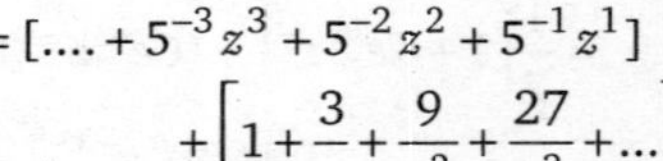

Fig. 1

$$= [5^{-1}z + 5^{-2}z^2 + 5^{-3}z^3 + \ldots] + \left[1 + \frac{3}{z} + \frac{9}{z^2} + \ldots\right]$$

$$= \frac{5^{-1}z}{1 - 5^{-1}z} + \frac{1}{1 - \frac{3}{z}} = \frac{z}{5 - z} + \frac{z}{3 - z}$$

where $\left|\frac{z}{5}\right| < 1$ and $\left|\frac{3}{z}\right| < 1$

$$= \frac{z^2 - 3z + 5z - z^2}{(5 - z)(z - 3)} = \frac{-2z}{z^2 - 2z + 15}$$

Remark

- The above two series are convergent in annulus, where $|z| < 3$ and $|z| < 5$.

Example 6. *Find the Z-transform of*

$$\left\{\cos\left(\frac{n\pi}{8} + \alpha\right)\right\}.$$

Solution. We have

$$Z\left\{\cos\left(\frac{n\pi}{8} + \alpha\right)\right\} = \sum \cos\left(\frac{n\pi}{8} + \alpha\right) z^{-n}$$

$$= \sum\left[\cos\frac{n\pi}{8}\cos\alpha - \sin\frac{n\pi}{8}\sin\alpha\right] z^{-n}$$

$$= \sum \cos\frac{n\pi}{8}\cos\alpha\, z^{-n} - \sum \sin\frac{n\pi}{8}\sin\alpha z^{-n}$$

$$= \cos\alpha \sum \cos\frac{n\pi}{8} z^{-n} - \sin\alpha \sum \frac{n\pi}{8} z^{-n}$$

$$= \cos\alpha \frac{z^2 - z\cos\frac{\pi}{8}}{z^2 - 2z\cos\frac{\pi}{8} + 1} - \sin\alpha \frac{z\sin\frac{\pi}{8}}{z^2 - 2z\cos\frac{\pi}{8} + 1}$$

$$= \frac{\left(z^2 - z\cos\frac{\pi}{8}\right)\cos\alpha - z\sin\frac{\pi}{8}\sin\alpha}{z^2 - 2z\cos\frac{\pi}{8} + 1}$$

$$= \frac{z^2\cos\alpha - z\cos\left(\frac{\pi}{8} - \alpha\right)}{z^2 - 2z\cos\frac{\pi}{8} + 1}$$

Example 7. *Find the Z-transform of* $\cosh\left(\frac{n\pi}{2} + \alpha\right)$.

(VTU–2011, UPTU–2008)

Solution. We have

$$F(z) = \sum_{n=0}^{\infty} \cosh\left(\frac{n\pi}{8} + \alpha\right) z^{-k}$$

$$= \sum_{n=0}^{\infty}\left[\frac{e^{(n\pi/2)+\alpha} - e^{-(n\pi/2)+\alpha)}}{2}\right] z^{-n}$$

$$= \frac{1}{2}e^{\alpha} \sum_{n=0}^{\infty} (e^{\pi/2}z^{-1})^n + \frac{1}{2}e^{-\alpha} \sum_{n=0}^{\infty} (e^{-\pi/2}z^{-1})^n$$

$$= \frac{1}{2}e^{\alpha}\left[1 + \left(e^{\pi/2}z^{-1}\right) + \left(e^{\pi/2}z^{-1}\right)^2 + \ldots\right]$$

$$+ \frac{1}{2}e^{-\alpha}\left[1 + \left(e^{-\pi/2}z^{-1}\right) + \left(e^{-\pi/2}z^{-1}\right)^2 + \ldots\right]$$

[Sum of two G.P's]

$$= \frac{1}{2}e^{\alpha}\left[\frac{1}{e^{-\pi/2}z^{-1}}\right] + \frac{1}{2}e^{-\alpha}\left[\frac{1}{1 - e^{-\pi/2}z^{-1}}\right]$$

$$= \frac{1}{2}\left[\frac{e^{\alpha}(1 - e^{-\pi/2}z^{-1}) + e^{-\alpha}(1 - e^{\pi/2}z^{-1})}{(1 - e^{-\pi/2}z^{-1})(1 - e^{-\pi/2}z^{-1})}\right]$$

$$= \frac{\frac{e^{\alpha}+e^{-\alpha}}{2} - \frac{e^{\alpha-\frac{\pi}{2}}+e^{-\alpha+\frac{\pi}{2}}}{2}}{1-e^{\pi/2}z^{-1}-e^{-\pi/2}z^{-1}+z^{-2}}$$

$$= \frac{\cosh\alpha - \cosh(\alpha-\pi/2).z^{-1}}{1-\left(2\cosh\frac{\pi}{2}\right)z^{-1}+z^{-2}}$$

$$= \frac{z^2\cosh\alpha - z\cosh\left(\frac{\pi}{2}-\alpha\right)}{z^2-2z\cosh\frac{\pi}{2}+1}$$

Example 8. *Find the Z-transform of*

(i) ^nC_k *(ii)* $^{k+n}C_n$ *(iii)* $^{k+n}C_n\,a^k$

Solution. (i) We have

$$Z\left[\{^nC_k\}\right] = \sum_{k=0}^{n} {}^nC_k z^{-k}$$

$$= 1 + {}^nC_1 z^{-1} + {}^nC_2 z^{-2} + \ldots + {}^nC_n z^{-k}$$

$$= (1+z^{-1})^n$$

(ii) We have

$$Z\left[\left\{^{k+n}C_n\right\}\right] = \sum_{k=0}^{\infty} {}^{k+n}C_n z^{-k}$$

$$= \sum_{k=0}^{\infty} {}^{k+n}C_k z^{-k} \quad [\because {}^nC_r = {}^nC_{n-r}]$$

$$= 1 + {}^{n+1}C_1 z^{-1} + {}^{n+2}C_2 z^{-2} + {}^{n+3}C_3 z^{-3} + \ldots$$

$$= 1 + (n+1)z^{-1} + \frac{(n+2)(n+1)}{2!}z^{-2}$$

(iii) We have

$$Z\left[\left\{^{k+n}C_n a^k z^{-k}\right\}\right] = \sum_{k=0}^{\infty} {}^{k+n}C_k a^k z^{-k}$$

$$(\because {}^nC_r = {}^nC_{n-r})$$

$$= \sum_{k=0}^{\infty} {}^{k+n}C_k (az^{-1})^k$$

$$|z| > a$$

$$= (1-az^{-1})^{-(n+1)}$$

Example 9. *Find Z-transforms of the following sequences by using the change of scale property*

(i) $<a^n>,\ n \ge 0$ *(ii)* $<C^n \sin(\alpha n)>,\ n \ge 0$

Solution. (i) We have

$$Z[<1>] = \sum_{n=0}^{\infty} 1.z^{-n}$$

$$= 1 + \frac{1}{z} + \frac{1}{z^2} + \frac{1}{z^3} + \ldots \text{ (Infinite G.P.)}$$

$$= \frac{1}{1-\frac{1}{z}} = \frac{z}{z-1}$$

Using change of scale property, *i.e.*, if $Z[<\{f(n)>] = F(z)$, then $Z[<a^n f(n)>] = F(z/a)$, we have

$$Z\left[<a^n>\right] = z\left[<a^n - 1>\right] = \frac{(z/a)}{(z/a)-1} = \frac{z}{z-a}$$

(ii) We have

$$Z[<\sin\alpha n>] = \sum_{n=0}^{\infty} (\sin\alpha n) z^{-n}$$

$$= \sum_{n=0}^{\infty}\left[\frac{e^{i\alpha n}-e^{-i\alpha n}}{2i}\right]z^{-n}$$

$$= \frac{1}{2i}\sum_{n=0}^{\infty}(e^{i\alpha}z^{-1})^n - \frac{1}{2!}\sum_{n=0}^{\infty}(e^{-i\alpha}z^{-1})^n$$

$$= \frac{1}{2i}\left[1+e^{i\alpha}z^{-1}+(e^{i\alpha}z^{-1})^2+\ldots\right]$$

$$-\frac{1}{2i}\left[1+e^{-i\alpha}z^{-1}+(e^{-i\alpha}z^{-1})^2+\ldots\right]$$

$$= \frac{1}{2i}\left[\frac{1}{1-e^{i\alpha}z^{-1}}\right] - \frac{1}{2i}\left[\frac{1}{1-e^{-i\alpha}z^{-1}}\right]$$

$$= \frac{1}{2i}\left[\frac{1-e^{-i\alpha}z^{-1}-(1-e^{i\alpha}z^{-1})}{1-e^{i\alpha}z^{-1}-e^{-i\alpha}z^{-1}+z^{-2}}\right]$$

$$= \frac{1}{2i}\left[\frac{(e^{i\alpha}-e^{-i\alpha})z^{-1}}{1-(e^{i\alpha}+e^{-i\alpha})z^{-1}+z^{-2}}\right]$$

$$= \frac{(\sin\alpha)z^{-1}}{1-(2\cos\alpha)z^{-1}+z^{-2}}$$

$$= \frac{z\sin\alpha}{z^2-2z\cos\alpha+1}$$

Now, using the change of scale property, we have

$$Z\left[<c^n\sin\alpha n>\right] = \frac{(z/c)\sin\alpha}{(z/c)^2-(2z/c)\cos\alpha+1}$$

$$= \frac{cz\sin\alpha}{z^2-2cz\cos\alpha+c^2}$$

Example 10. *Find the Z-transforms of the sequence*

$$<f(n)> = \sum_{n=0}^{\infty} z^n \sum_{n=0}^{\infty} 4^n$$

Solution. We know that

$$Z[<3^n>] = \frac{1}{1-3z^{-1}}$$

$$\text{and } Z[<4^n>] = \frac{1}{1-4z^{-1}}$$

Therefore

$$Z[<f(n)>] = Z[<3^n>]\,.\,Z[<4^n>]$$

(By convolution property)

$$= \left(\frac{1}{1-3z^{-1}}\right)\left(\frac{1}{1-4z^{-1}}\right)$$

$$= \frac{z^2}{(z-3)(z-4)} = \frac{z^2}{z^2-7z+12}$$

Example 11. *If* $F(z) = Z[< f(n) >] = \dfrac{2z^2+5z+14}{(z-1)^4}$, *then find the value of* $f(0)$.

Solution. By initial value theorem, we have

$$f(0) = \lim_{z\to\infty} F(z)$$

$$= \lim_{z\to\infty}\left[\frac{2z^2+5z+14}{(z-1)^4}\right]$$

$$= \lim_{z\to 0}\frac{1}{z^2}\left[\frac{2+5z^{-1}+14z^{-2}}{(1-z^{-1})^4}\right] = 0$$

44.11 EVALUATION OF INVERSE Z-TRANSFORMS

(1) Power Series Method : This is the simplest method of finding the inverse *Z*-transforms. If $F(z)$ is expressed as the ratio of two polynomials which cannot be factorized, we simply divide the numerator by the denominator and take the inverse *Z*-transforms of each term in the quotient.

(2) Partial Fractional Method : In this method, we decompose $F(z)/z$ into partial fractions and multiply the resulting expansion by z and then inverting the same. We use this method in a similar fashion as that of finding the inverse Laplace transforms, using partial fractions.

(3) Inversion Integral Method : The inverse Z-transforms of $F(z)$ is given by the following formula

$$f(n) = \frac{1}{2\pi i}\int_c F(z)\, z^{n-1} dz$$

= sum of residue of $F(z)\, z^{n-1}$ at the poles of $F(z)$ which are inside the contour c drawn.

Solved Examples

Example 1. *Using convolution theorem, to find*

$$Z^{-1}\left\{\frac{z^2}{(z-a)(z-b)}\right\}$$

Solution. Since, we know that

$$Z^{-1}\left\{\frac{z}{z-a}\right\} = a^n$$

and

$$Z^{-1}\left\{\frac{z}{z-a}\right\} = b^n$$

Therefore

$$Z^{-1}\left\{\frac{z^2}{(z-a)(z-b)}\right\} = Z^{-1}\left\{\frac{z}{z-a}\frac{z}{z-b}\right\}$$

$$= a^n + b^n$$

$$= \sum_{m=0}^{n} a^m . b^{n-m} = b^n \sum_{m=0}^{n}\left(\frac{a}{b}\right)^m, \quad \text{which in a G.P.}$$

$$= b^n . \frac{(a/b)^{n+1}-1}{\frac{a}{b}-1} = \frac{a^{n+1}-b^{n+1}}{a-b}$$

Example 2. *Find the inverse Z-transforms of* $\log\left(\dfrac{z}{z+1}\right)$ *by power series method.*

Solution. Putting $z = \dfrac{1}{t}$, $F(z) = \log\left(\dfrac{\frac{1}{t}}{\frac{1}{t}+1}\right)$

$$= -\log(1+t)$$

$$= -t + \frac{t^2}{2} - \frac{t^3}{3} + \ldots$$

$$= -z^{-1} + \frac{1}{2}z^{-2} - \frac{1}{3}z^{-3} + \ldots.$$

Hence, $f(n) = \begin{cases} 0, & \text{for } n = 0 \\ (-1)^n / n, & \text{otherwise} \end{cases}$

Example 3. *Find the inverse Z-transform of* $\dfrac{z}{(z+1)^2}$.

Solution. We have

$$F(z) = \frac{z}{z^2+2z+1} = z^{-1} - \frac{2+z^{-1}}{z^2+2z+1}$$

(By actual division)

$$= z^{-1} - 2z^{-2} + \frac{3z^{-1}+2z^{-2}}{z^2+2z+1}$$

$$= z^{-1} - 2z^{-2} + 3z^{-3} - \frac{4z^{-2}+3z^{-3}}{z^2+2z+1}$$

Continuing the process of division, we get an infinite series, *i.e.*,

$$F(z) = \sum_{n=0}^{\infty} (-1)^{n-1} . nz^{-n}$$

Hence, $f(n) = (-1)^{n-1} . n$

Example 4. *Find the inverse Z-transform of*

(i) $\dfrac{2z^2+3z}{(z+2)(z-4)}$ (VTU–2008, SVTU–2007)

(ii) $\dfrac{z^3-20z}{(z-2)^3(z-4)}$ (VTU–2011)

Solution. (i) We have

$$F(z)=\frac{2z^2+3z}{(z+2)(z-4)}$$

$$\Rightarrow \frac{F(z)}{z}=\frac{2z+1}{(z+2)(z-4)}=\frac{A}{z+2}+\frac{B}{z-4}$$

Using method of partical fraction, we get $A = 1/6$ and $B = 11/6$

Therefore

$$F(z)=\frac{1}{6}.\frac{z}{z+2}+\frac{11}{6}.\frac{z}{z-4}$$

On inversion, we have

$$f(n)=\frac{1}{6}(-2)^n+\frac{11}{6}(4)^n$$

(ii) We have

$$F(z)=\frac{z^3-20z}{(z-2)^3(z-4)}$$

$$\Rightarrow \frac{F(z)}{z}=\frac{z^2-20}{(z-2)^3(z-4)}=\frac{A+Bz+Cz^2}{(z-2)^3}+\frac{D}{z-4}$$

Then, we get $D=\frac{1}{2}$

Mutliplying throughout by $(z-2)^3$ $(z-4)$, we get

$$z^2-20=(A+Bz+Cz)^2(z-4)+D(z-2)^3$$

Putting $z = 0, 1, -1$ successively and solving the resulting simultaneous equations, we get

$$A=6, B=0, C=1/2$$

Hence, $$F(z)=\frac{1}{2}.\frac{12z+z^3}{(z-2)^3}-\frac{z}{z-4}$$

$$=\frac{1}{2}\frac{z(z-2)^2+4z^2+8z}{(z-2)^3}-\frac{z}{z-4}$$

$$=\frac{1}{2}\left\{\frac{z}{z-2}+2\frac{2z^2+4z}{(z-2)^3}\right\}-\frac{z}{z-4}$$

On inversion, we get

$$f(n)=\frac{1}{2}(2^n+2.n^2 2^n)-4^n$$

$$=2^{n-1}+n^2 2^n-4^n$$

Example 5. *Find the inverse Z-transform of* $\dfrac{2(z^2-5z+6.5)}{[(z-2)(z-3)^2]}$ *for* $2<|z|<3$.

Solution. We have

$$F(z)=\frac{2(z^2-5z+6.5)}{[(z-2)(z-3)^2]}$$

$$=\frac{A}{z-2}+\frac{B}{z-3}+\frac{C}{(z-3)^2}$$

Using the method of partial fraction, we get

$$A=B=C=1$$

Therefore

$$F(z)=\frac{1}{z-2}+\frac{1}{z-3}+\frac{1}{(z-3)^2}$$

$$=\frac{1}{2}\left(1-\frac{2}{z}\right)^{-1}-\frac{1}{3}\left(1-\frac{z}{3}\right)^{-1}+\frac{1}{9}\left(1-\frac{z}{3}\right)^{-2}$$

$$\left(\frac{2}{z}<1, \frac{z}{3}<1, \text{ i.e., } 2<|z|<3\right)$$

$$=\frac{1}{2z}\left[1+\frac{2}{z}+\frac{4}{z^2}+\frac{8}{z^3}+...\right]$$

$$-\frac{1}{3}\left[1+\frac{z}{3}+\frac{z^2}{9}+\frac{z^3}{27}+...\right]$$

$$+\frac{1}{9}\left[1+\frac{2z}{3}+\frac{3z^2}{9}+\frac{4z^3}{27}+...\right]$$

$$=\left(\frac{1}{2}+\frac{2}{z^2}+\frac{2^2}{z^3}+\frac{z^3}{z^4}+....\right)$$

$$-\left(\frac{1}{3}+\frac{z}{3^2}+\frac{z^2}{3^3}+\frac{z^3}{3^4}\right)$$

$$+\left(\frac{1}{3^2}+\frac{2z}{3^3}+\frac{3z^2}{3^4}+\frac{4z^3}{3^5}+....\right)$$

$$=\sum_{n=1}^{\infty}2^{n-1}z^{-n}-\sum_{n=0}^{\infty}\left(\frac{1}{3}\right)^{n-1}z^n$$

$$+\sum_{n=0}^{\infty}(n+1)\left(\frac{1}{3}\right)^{n+2}.z^n$$

Finally, on inversion, we get

$$f(u)=2^{n-1},\ n\geq 1$$

and $$f(n)=-(n+2)3^{n-2},\ n\leq 0$$

Example 6. *Using the inversion integral method, find the inverse Z-transfrom of* $\dfrac{3z}{(z-1)(z-2)}$

Solution. Let

$$F(z)=\frac{3z}{(z-1)(z-2)}$$

Its poles are at $z = 1$ and $z = 2$.

Using $F(z)$ in the inversion integral, we have

$$f(n)=\frac{1}{2\pi i}\int_c (z)z^{n-1}dz,$$

where c is a circle large enough to enclose both the poles of $F(z)$.

= sum of residues of $F(z)^{n-1}$ at $z = 1$ and $z = 2$

Now, we have

$$\text{res}[F(z).\, z^{n-1}]_{z=1}$$
$$= \lim_{z\to 1}\left\{(z-1).\frac{3z^n}{(z-1)(z-2)}\right\} = -3$$
$$\text{res}[F(z).\, z^{n-1}]_{z=2}$$
$$= \lim_{z\to 2}\left\{(z-2).\frac{3z^n}{(z-1)(z-2)}\right\} = 3.2^n$$

Hence, the requird inverse Z-transform

$f(n) = 3(2^n - 1),\ n = 0, 1, 2, 3, ...$

Example 7. *Find the inverse Z-transform of* $\dfrac{2z}{[(z-1)(z^2+1)]}$

Solution. Let

$$F(z) = \frac{2z}{(z-1)(z+i)(z-i)}$$

It has three poles at $z = 1, z = \pm i$.

Now, using $F(z)$ in the inversion integral, we have

$$f(n) = \frac{1}{2\pi i}\int_c f(z)\, z^{n-1} dz,$$

= sum of residues of $U(z).\, z^{n-1}$ at $z=1, z=\pm i$.

Now we have

$$\text{res}[F(z).\, z^{n-1}]_{z=1}$$
$$= \lim_{z\to 1}\left\{(z-1).\frac{2z^n}{(z-1)(z^2+1)}\right\} = 1$$
$$\text{res}[F(z).\, z^{n-1}]_{z=i}$$
$$= \lim_{z\to i}\left\{(z-i).\frac{2z^n}{(z-1)(z+i)(z-i)}\right\} = \frac{(-i)^n}{1+i}$$
$$\text{res}[F(z).\, z^{n-1}]_{z=-i}$$
$$= \lim_{z\to -i}\left\{(z+i).\frac{2z^n}{(z-1)(z+i)(z-i)}\right\} = \frac{(-i)^n}{i-1}$$

Hence, $$f(n) = 1 - \frac{(i)^n}{1+i} - \frac{(-i)^n}{1-i}$$

Example 8. *Find* $Z^{-1}\left\{\dfrac{3z^2-18z+26}{(z-2)(z-3)(z-4)}\right\}$. (Anna–2005)

Solution. Clearly, the poles are given by

$(z-2)\ (z-3)\ (z-4)=0 \Rightarrow z = 2, 3, 4$

Now, residue at $z = 2$

$$= \left[\frac{(z-2)z^{n-1}(3z^2-18z+26)}{(z-2)(z-3)(z-4)}\right]_{z=2}$$
$$= \left[\frac{3z^{n+1}-18z^n+26z^{n-1}}{(z-3)(z-4)}\right]_{z=2}$$
$$= \frac{3.2^{n+1}-18.2^n+26.2^{n-1}}{(-1)(-2)}$$
$$= 3.\ 2^n - 9.\ 2^n + 13.\ 2^{n-1} = 2^{n-1}$$

Residue at $z = 3$

$$= \left[\frac{(z-3)z^{n-1}.(3z^2-18z+26)}{(z-2)(z-3)(z-4)}\right]_{z=3}$$
$$= \left[\frac{3z^{n+1}-18z^n+26z^{n-1}}{(z-2)(z-4)}\right]_{z=3}$$
$$= \frac{3.3^{n+1}-18.3^n+26.3^{n-1}}{1(-1)}$$

and residue at $z = 4$

$$= \left[(z-4)\left[\frac{z^{n-1}(3z^2-18z+26)}{(z-2)(z-3)(z-4)}\right]\right]_{z=4}$$
$$= \left[\frac{3z^{n+1}-18z^n+26z^{n-1}}{(z-2)(z-3)}\right]_{z=4}$$
$$= \left[\frac{3.4^{n+1}-18.4^n+26.4^{n-1}}{2.1}\right] = 4^{n-1}$$
$$= z^{-1} F(z) = F(n) = \text{sum of residues}$$
$$= 2^{n-1} + 3^{n-1} + 4^{n-1},\ n > 0$$

44.12 SOLUTION OF DIFFERENCE EQUATION WITH CONSTANT COEFFICIENTS BY Z-TRANSFORM

WORKING PROCEDURE

For solving a linear difference equation with constant coefficient by using Z-transforms, we use the following steps :

Step 1. Take the Z-transforms of both sides of the given difference equation using the formulae of shifting property and given conditions.

Step 2. Transpose all terms without $F(z)$, to the right hand side.

Step 3. Divide the coefficient of $F(z)$, getting $F(z)$ as a function of z.

Step 4. Express this function in terms of the Z-transforms of known functions and take the inverse Z-transforms of both sides, which gives $f(k)$, *i.e.*, f_k as a function of k which is the required solution.

Solved Examples

Example 1. *Solve the difference equation $6f_{n+2}-f_{n+1}+f_n=0$, which $f(0)=0, f(1)=1$, by Z-transform.*

Solution. The given difference equation is

$$6f_{n+2}-f_{n+1}+f_n=0 \quad ...(1)$$

Taking Z-transform of both sides of (1), we get

$$Z[6f_{n+2}-f_{n+1}+f_n]=0$$

$$\Rightarrow \quad Z(6f_{n+2})-Z(f_{n+1})+Z(f_n)=0$$

$$=6[z^2F(z)-z^2f(0)-zf(1)]$$

$$-[zF(z)-zF(0)]+F(z)=0 \quad ...(2)$$

Using the given initial condition in (2), we get

$$6z^2F(z)-6z-zF(z)+F(z)=0$$

$$\Rightarrow \quad (6z^2-z+1)F(z)=6z$$

$$\Rightarrow \quad F(z)=\frac{6z}{6z^2-z+1}=\frac{6z}{(3z+1)(2z-1)}$$

$$=\frac{z^{-1}}{\left(1+\frac{z^{-1}}{3}\right)\left(1-\frac{z^{-1}}{2}\right)}$$

$$=\frac{6/5}{1-\frac{z^{-1}}{2}}-\frac{6/5}{1+\frac{z^{-1}}{3}}$$

$$\Rightarrow \quad f(n)=Z^{-1}\left[\frac{6/5}{1-z^{-1}/2}\right]-Z^{-1}\left[\frac{6/5}{1+z^{-1}/3}\right]$$

$$=\frac{6}{5}\left(\frac{1}{2}\right)^n-\frac{6}{5}\left(-\frac{1}{3}\right)^n$$

$$=\frac{6}{5}\left[\left(\frac{1}{2}\right)^n-\left(-\frac{1}{3}\right)^n\right]$$

Example 2. *Solve by Z-transforms*

$$f_{n+1}+\frac{1}{4}f_n=\left(\frac{1}{4}\right)^4, \; n\geq 0, \; f(0)=0$$

Solution. The given difference equation is

$$f_{n+1}+\frac{1}{4}f_n=\left(\frac{1}{4}\right)^4 \quad ...(1)$$

Taking the Z-transforms of both sides of (1), we get

$$Z\left[f_{n+1}+\frac{1}{4}f_n\right]=Z\left[\left(\frac{1}{4}\right)^n\right]$$

$$\Rightarrow \quad Z[f_{n+1}]+z\left(\frac{1}{4}f_n\right)=Z\left[\left(\frac{1}{4}\right)^n\right]$$

$$\Rightarrow zF(z)-zf(0)+\frac{1}{4}F(z)=\frac{1}{1-\frac{1}{4}z^{-1}},$$

where $|Z|>\frac{1}{4}$

$$\Rightarrow \quad zF(z)-0+\frac{1}{4}F(z)=\frac{1}{1-\frac{1}{4}z^{-1}}$$

$[\because f(0)=0]$

$$\Rightarrow \quad \left(z+\frac{1}{4}\right)F(z)=\frac{1}{1-\frac{1}{4}z^{-1}}$$

Therefore

$$F(z)=\frac{1}{z+\frac{1}{4}}\times\frac{1}{1-\frac{1}{4}z^{-1}}$$

$$=\frac{z^{-1}}{1+\frac{1}{4}z^{-1}}+\frac{2}{1-\frac{1}{4}z^{-1}}$$

$$\Rightarrow \quad f(n)=Z^{-1}\left[\frac{-2}{1+\frac{1}{4}z^{-1}}\right]+Z^{-1}\left[\frac{2}{1-\frac{1}{4}z^{-1}}\right]$$

$$=Z^{-1}\left[-2\left(1+\frac{1}{4}z^{-1}\right)^{-1}\right]$$

$$+Z^{-1}\left[2\left(1-\frac{1}{4}z^{-1}\right)^{-1}\right]$$

$$=-2\left(-\frac{1}{4}\right)^n+2\left(\frac{1}{4}\right)^n$$

which is the required solution.

Example 3. *Solve the difference equation*

$$f_{n+3}-3f_{n+2}+3f_{n+1}-f_n=F(n)$$

where $f(0)=f(1)=f(2)=0$, by Z-transforms.

Solution. The given difference equation is

$$f_{n+3}-3f_{n+2}+3f_{n+1}-f_n=F(n) \quad ...(1)$$

Taking *Z*-tranforms of both sides of (1), we get

$$Z(f_{n+3}-3f_{n+2}+3f_{n+1}-f_n)=Z[F(n)]$$

$$\Rightarrow Z[f_{n+3}]-3Z[f_{n+2}]+3Z[f_{n+1}]-Z[f_n]$$

$$=Z[F(n)]$$

$$\Rightarrow [z^3F(z)-z^3f(0)-z^2f(1)-zf(2)]$$

$$-3[z^2F(z)-z^2f(0)-zf(1)]$$

$$+3[zF(z)-zf(0)]-F(z)=Z[F(n)] \quad ...(2)$$

Using the given initial conditions in (2), we get

$$z^2F(z)-3z^2F(z)+3z-F(z)=\frac{1}{1-z^{-1}}$$

$$\Rightarrow \quad (z^2-3z^2+3z-1)\,F(z) = \frac{1}{1-z^{-1}}$$

$$\Rightarrow \quad (z-1)^3 F(z) = \frac{1}{1-z^{-1}}$$

i.e., $$F(z) = \frac{1}{(z-1)^3(1-z^{-1})}$$

$$= \frac{1}{z^3(1-z^{-1})^3(1-z^{-1})}$$

$$= z^{-3}\,(1-z^{-1})^{-4}$$

$$\Rightarrow \quad f(n) = \text{Ceofficients of } z^{-n} \text{ in } z^{-3}\,(1-z^{-1})^{-4}$$

$$= \text{Ceofficients of } z^{-n-3} \text{ in } (1-z^{-1})^{4}$$

$$= \frac{(n-2)(n-1)n}{6},\ n \geq 3$$

Example 4. *Solve* $f_n + \frac{1}{4}f_{n-1} = F(n) + \frac{1}{3}F(n-1)$

Solution. The given difference equation is

$$f_n + \frac{1}{4}f_{n-1} = F(n) + \frac{1}{3}F(n-1) \quad \ldots(1)$$

Taking Z-transforms of both sides of (1), we get

$$Z[\{f_n\}] + \frac{1}{4}Z[\{f_{n-1}\}]$$

$$= Z[\{F(n)\}] + \frac{1}{3}Z[\{F(n-1)]$$

$$\Rightarrow \quad F(z) + \frac{1}{4}z^{-1}F(z) = \left[1 + \frac{1}{3}z^{-1}\right]$$

$$\Rightarrow \quad F(z) = \frac{1+\frac{1}{3}z^{-1}}{1+\frac{1}{4}z^{-1}} = \frac{z+\frac{1}{3}}{z+\frac{1}{4}}$$

Clearly, there is only one simple pole at $z = -\frac{1}{4}$.

Consider the contour $|z| > \frac{1}{4}$

Residue at $z = -\frac{1}{4}$

$$= \left[\left(z+\frac{1}{4}\right).z^{n-1}.\frac{z+\frac{1}{3}}{z+\frac{1}{4}}\right]_{z=-\frac{1}{4}}$$

$$= \left[z^n + \frac{z^{n-1}}{3}\right]_{z=-\frac{1}{4}}$$

$$= \left(-\frac{1}{4}\right)^n + \frac{1}{3}\left(-\frac{1}{4}\right)^{n-1}$$

$$= -\frac{1}{4}\left(-\frac{1}{4}\right)^{n-1} + \frac{1}{3}\left(-\frac{1}{4}\right)^{n-1}$$

$$= \frac{1}{12}\left(-\frac{1}{4}\right)^{n-1}$$

Hence, $f(n) = \text{Residue} = \frac{1}{12}\left(-\frac{1}{4}\right)^{n-1}$

Example 5. *Using Z-transform, solve*

$f_{n+2} + 4f_{n+1} + 3f_n = 3^n$ *with* $f(0)=0, f(1) = 1.$

(UPTU–2003)

Solution. If $Z(f_n) = F(z)$, then $Z[f_{n+1}] = z[F(z) - f(0)]$

$$Z[f_{n+2}] = z^2[F(z) - f(0) - f(1)z^{-1}]$$

Also, $Z(2^n) = \frac{z}{z-2}$

Taking Z-transforms of both sides of the given equation, we get

$$z^2[F(z) - f(0) - f(1)z^{-1}] + 4z[F(z) - f(0)] + 3F(z) = \frac{z}{z-3} \quad \ldots(1)$$

Using the given condition, (1) reduces to

$$F(z) = (z^2 + 4z + 3) \ = z + \frac{z}{z+3}$$

Therefore,

$$\frac{F(z)}{z} = \frac{1}{(z+1)(z+3)} + \frac{1}{(z-3)(z+1)(z+3)}$$

$$= \frac{3}{8}.\frac{1}{z+1} + \frac{1}{24}\frac{1}{z-3} - \frac{5}{12}\frac{1}{z+3}$$

(On resolving into partial fractions)

$$\therefore \ F(z) = \frac{3}{8}\frac{z}{z+1} + \frac{1}{24}\frac{z}{z-3} - \frac{5}{12}\frac{z}{z+3}$$

On inversion, we get

$$f_n = \frac{3}{8}Z^{-1}\left(\frac{z}{z+1}\right) + \frac{1}{24}Z^{-1}\left(\frac{z}{z-3}\right) - \frac{5}{12}Z^{-1}\left(\frac{z}{z+3}\right)$$

$$= \frac{3}{8}(-1)^n + \frac{1}{24}3^n - \frac{5}{12}(-3)^n$$

Example 6. *Solve* $f_{n+2} + 6f_{n+1} + 9f_n = 2^n$ *with* $f(0) = f(1) = 0$ *using Z-tranforms.*

(VTU–2011, Anna–2009, SVTU–2009)

Solution. If $Z(f_n) = F(z)$, then

$$Z(f_{n+1}) = z[F(z) - f(0)]$$

and $Z(f_{n+2}) = Z[F(z) - f(0) - f(1)z^{-1}]$

Also, $Z(2^n) = \frac{z}{z-2}$

Taking Z-transform of both sides of the given equation, we get

$$z^2[F(z) - f(0) - f(1)Z^{-1}] + 6z[F(z) - f(0)] + 9F(z) = \frac{z}{z-2} \quad \ldots(1)$$

Using the given condition in (1), we get

$$F(z)(z^2 + 6z + 9) = \frac{z}{z-2}$$

$$\Rightarrow \quad \frac{F(z)}{z} = \frac{1}{(z-2)(z-3)^2}$$

$$= \frac{1}{25}\left[\frac{1}{z-2} - \frac{1}{z+3} - \frac{5}{[z+3]^2}\right]$$

Therefore

$$F(z) = \frac{1}{25}\left[\frac{z}{z-2} - \frac{z}{z+3} - 5.\frac{z}{(z+3)^2}\right] \quad ...(2)$$

On taking inverse Z-transform of both sides of (2), we get

$$f(n) = \frac{1}{25}\left[Z^{-1}\left(\frac{z}{z-2}\right) - Z^{-1}\left(\frac{z}{z+3}\right) + \frac{5}{3}Z^{-1}\left(-\frac{3z}{(z+3)^2}\right)\right]$$

$$= \frac{1}{25}\left[2^n - (-3)^n + \frac{5}{3}n(-3)^n\right]$$

$$\left[\because \; Z^{-1}\left\{\frac{az}{(z-a)^2}\right\} = na^n\right]$$

Example 7. *Using the Z-transform, solve*

$$f_{n+2} - 2f_{n+1} + f_n = 3n+5 \quad \text{(SVTU–2007)}$$

Solution. The given difference equation is

$$f_{n+2} - 2f_n+1 + f_n = 3n+5 \quad ...(1)$$

Taking Z-transform of both sides of (1), we get

$$z^2[F(z) - f(0) - f(1)z^{-1}] - 2z[F(z) - f(0)] + F(z)$$

$$= 3.\frac{z}{(z+1)^2} + 5.\frac{z}{z-1}$$

or

$$F(z) = \frac{5z^2 - 2z}{(z-1)^2} + f(0).\frac{z^2 - 2z}{(z-1)^2} + f(1).\frac{z}{(z-1)^2}$$

Taking inverse Z-transform, we get

$$f(n) = Z^{-1}\left[\frac{5z^2 - 2z}{(z-1)^4}\right] + f(0)Z^{-1}\left[\frac{z^2 - 2z}{(z-1)^2}\right] + f(1)Z^{-1}\left[\frac{z}{(z-1)^2}\right] \quad ...(2)$$

Notting that

$$Z(1) = \frac{z}{z-1}, \quad Z(n) = \frac{z}{(z-1)^2}$$

$$Z(n^2) = \frac{z^2+z}{(z-1)^3}, \quad Z(n^3) = \frac{z^3 + 4z^2 + z}{(z-1)^4}$$

We write

$$\frac{5z^2 - 2z}{(z-1)^4} + A.\frac{z^3 + 4z^2 + z}{(z-1)^4} + B.\frac{z^2 + z}{(z-1)^3} + C.\frac{z}{(z-1)^2} + D.\frac{z}{z-1}$$

Equating coefficient of like powers of z, we find

$$A = \frac{1}{2}, \; B = 1, \; C = -\frac{3}{2}, \; D = 0.$$

Therefore,

$$Z^{-1}\left\{\frac{5z^2 - 2z}{(z-1)^4}\right\} = \frac{1}{2}n^3 + n^2 - \frac{3}{2}n$$

$$= \frac{1}{2}n(n-1)(n+3)$$

also,

$$Z^{-1}\left\{\frac{z^2 - 2z}{(z-1)^2}\right\} = Z^{-1}\left\{\frac{z}{z-1}\right\} - Z^{-1}\left\{\frac{z}{(z-1)^2}\right\}$$

$$= 1 - n$$

and $Z^{-1}\left\{\frac{z}{(z-1)^2}\right\} = n$

Putting these values in (2), we get

$$f(n) = \frac{1}{2}n(n-1)(n+3) + f(0)(1-n) + f(1).n$$

$$= \frac{1}{2}n(n-1)(n+3) + C_0 + C_1 n$$

where $C_0 = f(0)$, $C_1 = f(1) - f(0)$

EXERCISE 44.1

1. Find the Z-transforms of the following functions for $n \geq 0$:

(i) 2^n (ii) $\sin \alpha n$

(iii) $\sin(3n+5)$ (iv) $\sinh \frac{n\pi}{2}$

(v) $\cosh n\theta$ (VTU–2011)

2. Find the Z-transforms of the following functions for $n \geq 0$:

(i) $\sin\left(\frac{n\pi}{2} + \alpha\right)$ (ii) $C^n \sinh \alpha n$

(iii) $\cos\left(\frac{n\pi}{2} + \frac{\pi}{4}\right)$ (iv) $\frac{a^n}{n!}$

3. Find the Z-transforms of the following functions for $n > 0$:

(i) $\sin 5n$ (ii) $e^{\alpha n}$

(iii) $3n \cosh 5n$

4. Find the inverse of Z-transforms of the following functions:

(i) $\frac{1}{(z-3)(z-2)}$ for $|z| < 2$

(ii) $\frac{1}{(z-5)^3}, |z| > 5$ (iii) $\frac{3z^2 + 4z}{z^2 - z + 1}, |z| > 1$

(iv) $\frac{ze^{-a}}{(z-e^{-a})^2}, |z| > |e^{-a}|$

(v) $\dfrac{z^3 - 20z}{(z-2)^3(z-4)}$ (VTU–2011)

(vi) $\dfrac{4z^2 - 2z}{z^3 - 5z^2 + 8z - 4}$ (VTU–2011)

(vii) $\dfrac{18z^2}{(2z-1)(4z+1)}$ (SVTU–2009)

5. Solve the following difference equations by Z-transform:

(i) $f_{n+1} - 2f_n + f_{n-1} = a_n,\ a \neq 1$

(ii) $f_n - \dfrac{5}{6} f_{n-1} + \dfrac{1}{6} f_{n-2} = F(n)$

(iii) $f_n + \dfrac{1}{a} f_{n-2} = \left(\dfrac{1}{3}\right)^n \cos\dfrac{n\pi}{2};\ (n \geq 0)$

(iv) $6f_{n+2} + 5f_{n+1} - f_n = 6F(n)$

(v) $6y_{k+2} - y_{k+1} - y_k = 0,\ y(0) = y(1) = 1$ (Kottayam–2005)

6. Show that $Z(\sinh n\theta) = \dfrac{z \sinh\theta}{z^2 - 2z\cosh\theta + 1}$ (VTU–2011)

7. Show that $Z(e^{-an} \sin n\theta) = \dfrac{ze^a \sin\theta}{z^2 e^{2a} - 2ze^a \cos\theta + 1}$ (SVTU–2007)

8. Show that $Z({}^{n+p}C_p) = \left(1 - \dfrac{1}{z}\right)^{-(p+1)}$. Using the damping rule deduce that $Z({}^{n+p}C_p a^n) = \left(1 + \dfrac{a}{z}\right)^{-(p+1)}$ (SVTU–2009)

9. Show that $Z\left(\dfrac{1}{n}\right) = z \log \dfrac{z}{z-1}$ (Madras–2003)

ANSWERS

1. (i) $\dfrac{z}{z-2}, |z| > 2$ (ii) $\dfrac{z \sin\alpha}{z^2 - 2z\cos\alpha + 1}$ (iii) $\dfrac{z^2 \sin 5 - z\sin 2}{z^2 - 2z\cos 3 + 1}$ (iv) $\dfrac{z \sinh\frac{\pi}{2}}{z^2 - 2z\cosh\frac{\pi}{2} + 1}$ (v) $\dfrac{z(z - a\cosh\theta)}{z^2 - 2az\cosh\theta + a^2}$

(vi) $2 + (2)^n + 3(n-1)2^n, n > 1$ (vii) $\dfrac{3}{4}\left(\dfrac{1}{2^{n-1}} + \dfrac{1}{(-4)^n}\right)$

2. (i) $\dfrac{z^2 \sin\alpha + z\cos\alpha}{z^2+1}, |z| > 1$ (ii) $\dfrac{Cz \sinh\alpha}{z^2 - 2Cz\cosh\alpha + 1}$ (iii) $\dfrac{z^2 - z}{\sqrt{2}(z^2+1)}$ (iv) $e^{a/z}$

3. (i) $\dfrac{z\sin 5}{z^2 - 2z\cos 5 + 1}$ (ii) $\dfrac{1}{1 - z^{-1}e^{an}}$ (iii) $\dfrac{z(z - 3\cosh 5)}{z^2 - 2z\cosh 5 + 9}$

4. (i) $\left(-\dfrac{1}{3} - \dfrac{z}{3^2} - \dfrac{z^2}{3^3} - \dfrac{z^3}{3^4} - \ldots\right) + \left(\dfrac{1}{2} + \dfrac{z}{2^2} + \dfrac{z^2}{2^3} + \dfrac{z^3}{2^4} + \ldots\right)$ (ii) $\begin{cases} \dfrac{(n-2)(n-1)5^{n-3}}{2}, & n \geq 3 \\ 0, & n < 3 \end{cases}$ (iii) $\left[3\left\{\cos\dfrac{n\pi}{3}\right\} + \dfrac{11}{\sqrt{3}} \sin\dfrac{n\pi}{3}\right] F(n)$

(iv) $[ne^{-an}]$ (v) $2^{n-1} + n^2 2^n - 4^n$

5. (i) $f_n = \dfrac{1}{a}(n+1)F(n) - \dfrac{a}{(a-1)^2}F(n) + \dfrac{a}{(a-1)^2}a^n F(n) + \dfrac{1}{1-a} nF(n-1)$ (ii) $f_n = \left[3 - \left(\dfrac{1}{2}\right)^n + \left(\dfrac{1}{3}\right)^n\right] F(n)$

(iii) $f_n = \left(\dfrac{n+2}{2}\right)\left(\dfrac{1}{3}\right)^n \cos\dfrac{n\pi}{2} F(n)$ (iv) $f_n = \left[\dfrac{6}{7}(n+1) - \dfrac{78}{49} + \dfrac{36}{49}\left(-\dfrac{1}{6}\right)^n\right] F(n)$ (v) $y_k = \dfrac{8}{5}\left(\dfrac{1}{2}\right)^k - \dfrac{3}{5}\left(-\dfrac{1}{3}\right)^k$

FFFFFF

GATE tutor

KEY Terms and Results

- **General Solution :** If the solution of nth order ODE contains n arbitrary constants, then it is called general solution.
- **Particular Solution :** A solution obtained from the general solution by giving particular values to the arbitrary constants is called particular solution.
- **Homogeneous Equation :** A differential equation of the form $\frac{dy}{dx} = \frac{f_1(x, y)}{f_2(x, y)}$ is said to be homogeneous if $f_1(x, y)$ and $f_2(x, y)$ are homogeneous function of same degree. To find the solution of homogeneous equation, put $y = vx$.
- **Linear Differential Equation :** A differential equation in which the dependent variable and all its derivatives are of first degree only and not multiplied together is called a linear differential equation.
- **Bernaulli's Equation :** A differential equation of the form $\frac{dy}{dx} + Py = Qy^n$ is called Bernoulli's equation.
- **Exact Equation :** A differential equation which can be derived from its primitive by direct differentiation without any further transformation is called an exact equation.
- **Integrating Factor :** If the given equation is not exact, then it can be made exact by multiplying some functions of x and y. This multiplier is known as integrating factor.
- **Auxiliary Equation :** The equation obtained by equating to zero the symbolic coefficient of y is called the auxiliary equation, briefly written as A.E.
- **Homogeneous Linear Differential Equation :** Any differential equation of the form
$$x^n \frac{d^n y}{dx^n} + A_1 \frac{d^{n-1}y}{dx^{n-1}} + \ldots + A_n y = X$$
is called a homogeneous linear differential equation of nth order, where $A_1, A_2, \ldots, A_n$ are constant and X is a function of x or a constant.
- **Simultaneous Linear Differential Equations :** The differential equations in which there is one independent variable and two or more than two independent variables. Such equations are called simultaneous linear differential equations.
- **Power Series :** An infinite series of the form $\sum_{m=0}^{\infty} a_m x^m$ is called a power series.
- The equation of the form
$$a_0 \frac{d^n y}{dx^n} + a_1 \frac{d^{n-1}y}{dx^{n-1}} + a_2 \frac{d^{n-2}y}{dx^{n-2}} + \ldots + a_n y = aB$$
where $a_0, a_1, a_2, \ldots, a_n$ are all constants and B is a function of x alone is called a linear differential equation of nth order with constant coefficients.
- The operator D stands for $\frac{d}{dx}$, D^2 for $\frac{d^2}{dx^2}$ and so on.
- The operator D^{-1} stands for integration.
- The solution of differential equation $f(n)y = B$, contains as many arbitrary constants as is the order of the given differential equation is called the complimentary function (C.F.).
- The solution of differential equation of $f(0)y = B$, with no arbitrary constant which is called the particular integral (P.I.).
- When the roots of the A.E are real and distinct, *i.e.*, $m_1, m_2, \ldots, m_n$, then
$$\text{C.F.} = C_1 e^{m_1 x} + C_2 e^{m_2 x} + \ldots + C_n e^{m_n x}.$$
- When the roots of auxiliary equation are equal, *i.e.*, $m_1 = m_2 = m_3$
Then C.F. $= (C_1 + C_2 x + C_3 x^2)\, e^{m_1 x} + C_4 e^{m_4 x} + \ldots + C_n e^{m_n x}$.
- When the roots of the auxiliary equation are imaginary, *i.e.*, $m_1 = \alpha + i\beta$ and $m_2 = \alpha - i\beta$
then, C.F. $= e^{\alpha x}(C_1 \cos\beta x + C_2 \sin\beta x) + C_3 e^{m_3 x} + \ldots + C_n e^{m_n x}$.
- When the roots of auxiliary equation are repeated imaginary
$$m_1 = m_2 = \alpha + i\beta, \quad m_3 = m_4 = \alpha - i\beta$$
then C.F. $= e^{\alpha x}\,[(C_1 + C_2 x)\cos\beta x + (C_3 + C_4 x)\sin\beta x] + C_5 e^{m_5 x} + \ldots + C_n\, e^{m_n x}$.
- When roots of auxiliary equation are irrational, *i.e.*, $m_1 = \alpha + \sqrt{\beta}$ and $m_2 = \alpha - \sqrt{\beta}$, then,
C.F. $= e^{\alpha x}(C_1 \cosh\sqrt{\beta}x + C_2 \sinh\sqrt{\beta}\,x) + C_3 e^{m_3 x} + \ldots + C_n e^{m_n x}$
- When roots of auxiliary equation are repeated irrational
i.e., $m_1 = m_2 = \alpha + \sqrt{\beta}$, $m_3 = m_4 = 2 - \sqrt{\beta}$, then,
C.F. $= e^{\alpha x}\left[(C_1 + C_2 x)\cosh\sqrt{\beta}x + (C_3 + C_4 x)\sinh\sqrt{\beta}\,x) + C_5 e^{m_5 x} + \ldots C_n e^{m_n x}\right]$
- When $B = e^{ax}$, then P.I. $= \frac{1}{f(0)} e^{ax} = \frac{1}{f(a)} e^{ax}$, provided $f(a) \neq 0$.
- When $f(a) = 0$, P.I. $= \frac{1}{f(0)} e^{ax} = \frac{1}{f(a)} e^{ax}$, provided $f'(a) \neq 0$.
- When $B = \sin(ax + b)$ or $\cos(ax + b)$,
then, P.I. $= \frac{1}{f(a^2)}\sin(ax+b) = \frac{1}{f(-a^2)}\sin(ax+b)$, provided $f(-a^2) \neq 0$
and P.I. $= \frac{1}{f(D^2)}\cos(ax+b) = \frac{1}{f(-a^2)}\cos(ax+b)$, provided $f(-a^2) \neq 0$.
- When $B = x^m$, m being positive integer P.I. $= \frac{1}{f(D)} x^m$.
- When $B = e^{ax}.V$, where V is a function of x, then,

$$\text{P.I.} = \frac{1}{f(D)}(e^{ax}.V) = e^{ax}\frac{1}{f(D+a)}.V.$$

- The homogeneous linear equations, are also known as Cauchy-Euler equations.
- To solve the homogeneous linear differential equation, we change the independent variable x to z by substitution $x = e^z$.
- If the given differential equation be written as $F(D_1)y = f(x)$, then we use the following results :

(a) $\frac{1}{D_1-\alpha}f(x) = x^{\alpha}\int x^{-\alpha-1} f(x)\,dx$

(b) $\frac{1}{D_1+\alpha}f(x) = x^{-\alpha}\int x^{\alpha-1} f(x)\,dx$

- Any differential equation of the form

$$(a+bx)^n\frac{d^ny}{dx^n} + A_1(a+bx)^{n-1}\frac{d^{n-1}y}{dx^{n-1}} + \ldots + A_{n-1}(a+bx)\frac{dy}{dx} + A_ny = X(x) \quad \ldots(1)$$

where the coefficients $A_1, A_2, \ldots, A_n$ are constants, and can be transformed into homogenous linear equation with constant coefficients by changing independent variable from x to z, by a suitable substitution $z = a + bx$.
- Sometimes, the given equation can be solved easily by making the substitution $e^z = a + bx$.
- Let x, y be the two dependent variables and t the independent variable. Consider the simultaneous equations :

$$f_1(D)x + f_2(D)y = T_1 \quad \ldots(1)$$

and $$g_1(D)x + g_2(D)y = T_2 \quad \ldots(2)$$

where $D \equiv \frac{d}{dt}$ and T_1, T_2 are functions of t.

To eliminate y, operating on both sides of (1) by $g_2(D)$ and on both sides (2) by $f_2(D)$ and subtracting

$$[f_1(D)g_2(D) - g_1(D)f_2(D)]x = g_2(D)T_1 - f_2(D)T_2$$

or $f(D)x = T$

which is a linear equation in x and t.
- The equations of the type

$$P_1dx + Q_1dy + R_1dz = 0$$
$$P_2dx + Q_2dy + R_2dz = 0,$$

where P_1, P_2, Q_1, Q_2, R_1 and R_2 are functions of x, y and z. The equations can be put in the form

$$\frac{dx}{R_1R_2 - Q_2R_1} = \frac{dy}{R_1P_2 - R_2P_1} = \frac{dz}{P_1Q_2 - P_2Q_1}$$

or $\frac{dx}{P} = \frac{dy}{Q} = \frac{dz}{R}$, where P, Q and R be function of x, y and z.
- This power series is said to be convergent at a point x if $\lim_{n\to\infty}\sum_{m=0}^{n} a_m x^m$ exists.
- The power series of the type $y = \sum_{m=0}^{\infty} a_m x^{m+n}$ is known as Quasi Power Series.
- Let $m = n$ be two equal roots of indical equation. Then putting $m = n$ in y and in $\frac{\partial y}{\partial m}$, we may get the two independent solutions, in which second solution always consists of a numerical multiple of the product of the first solution and log x added to another series.
- If the indical equation has two unequal roots $m = m_1$ and m_2 which do not differ by an integer, then by putting $m = m_1$ and m_2 in this series, we get two independent solutions.
- If $m = m_1$ and m_2 be two roots of indical equation which differ by an integer and some of the coefficients of power of K in the series for y infinity for $m = m_2$, put $C(m - m_2)$ for C_0. . Then, we get two independent solutions for $m = m_2$.
- If $m = m_1$ and m_2 $(m_1 > m_2)$ are two roots of the indical equation which differ by an integer and if one of the coefficient of the series for y becomes indeterminate when $m = m_2$, the complex solution is given by putting $m = m_2$ in y, which have two arbitrary constants.
- The series obtained by putting $m = m_1$ in y is merely a numerical multiple of one of the series contained in the first solution.
- The Legendre's differential equation is given by :

$$(1-x^2)\frac{d^2y}{dx^2} - 2x\frac{dy}{dx} + n(n+1)y = 0$$

- The first and second kind of Legendre's polynomial ($P_n(x)$ and $Q_n(x)$) are respectively given by

$$P_n(x) = \sum_{m=0}^{N}\frac{(-1)^m(2n-2m)!}{2^n.m!(n-m)!(n-2m)!}x^{n-2m},$$

where $N = \begin{cases} n/2, & \text{if } n \text{ is even} \\ \frac{n-1}{2}, & \text{if } n \text{ is odd}\end{cases}$

and $$Q_n(x) = \frac{n!}{1.3.5\ldots(2n+1)}\left[x^{-n-1} + \frac{(n+1)(n+2)}{2(2n+3)}x^{-n-3} + \frac{(n+1)(n+2)}{2.4.(2n+3)}\frac{(n+3)(n+4)}{(2n+5)+\ldots}x^{-n-5}\right] + \ldots$$

- The function of the type $\frac{1}{\sqrt{1-2xt+t^2}} = \sum_{n=0}^{\infty} P_n(x).t^n$ is known as generating function.
- The expression $P_n(x) = \frac{1}{2^n.n!}\frac{d^n}{dx^n}(x^2-1)^n$ is called Rodrigue's formula.
- Laplace's first integral for $P_n(x)$ is given by

$$P_n(x) = \frac{1}{\pi}\int_0^{\pi}\left[x \pm \sqrt{(x^2-1)}.\cos\theta\right]^n, \quad n \in Z^+$$

- Laplace's second integral for $P_n(x)$ is given by

$$P_n(x) = \frac{1}{\pi}\int_0^{\pi}\frac{d\theta}{[x \pm \sqrt{(x^2-1)}.\cos\theta]^{n+1}}, \quad n \in Z^+$$

- Orthogonal properties of Legendre's polynomial are given by $\int_{-1}^{1} P_n(x)\,P_m(x)\,dx = 0$, when $m \neq n$; $\int_{-1}^{1}[P_n(x)]^2\,dx = \frac{2}{2n+1}$, when $m = n$.
- List of Recurrence relations and formulae of Lagendre's function :

1. $(2n+1)xP_n = (n+1)P_{n+1} + nP_{n-1}$

2. $(2n-1)xP_{n-1} = nP_n + (n-1)P_{n-2}$

3. $nP_n = xP'_n - P'_{n-1}$
4. $(2n+1)P_n = P'_{n+1} - P'_{n-1}$
5. $(n+1)P_n = P'_{n+1} - xP'_n$
6. $(1-x^2)P'_n = n(P_{n-1} - xP_n)$
7. $(1-x^2)P'_n = (n+1)(xP_n - P_{n+1})$
8. $(2n+1)(x^2-1)P'_n = n(n+1)(P_{n+1} - P_{n-1})$

(Beltrami's relation)

9. $\sum_{k=0}^{n} (2k+1)P_k(x)P_k(y)$

$$= (n+1)\left[\frac{P_{n+1}(x)P_n(y) - P_{n+1}(y)P_n(x)}{(x-y)}\right]$$

[Christoffel's summation formula]

10. $Q'_{n+1} - Q'_{n-1} = (2n+1)Q_n$
11. $nQ'_{n+1} + (n+1)Q'_{n-1} = (2n+1)\,xQ'_n$
12. $(2n+1)xQ_n = (n+1)Q_{n+1} + nQ_{n-1}$
13. $(2n+1)(1-x^2)Q'_n = n(n+1)\,(Q_{n-1} - nQ_{n-1})$
14. $xQ'_n - Q'_{n+1} = nQ_n$
15. $Q'_n - xQ'_{n-1} = nQ_{n-1}$
16. $(x^2-1)\,Q'_n = nxQ_n - nQ_{n-1}$
17. $(x^2-1)Q'_n = (n+1).Q_{n+1} - (n+1)xQ_n$
18. $(x^2-1)(Q_nP'_n - P_nQ'_n) = C$
19. $\dfrac{Q_n(x)}{P_n(x)} = \int_x^\infty \dfrac{dx}{(x^2-1)\,P_n^2(x)}$
20. $Q_0(x) = \dfrac{1}{2}\log\dfrac{x-1}{x+1}$
21. $Q_1(x) = \dfrac{x}{2}\log\dfrac{x+1}{x-1} - 1$
22. $n\left[P_nQ_{n-1} - Q_nP_{n-1}\right] = 1$
23. $P_nQ_{n-2} - Q_nP_{n-2} = \dfrac{2n-1}{n(n-1)}\,x$

- The Bessel's differential equation is given by :

$$x^2\frac{d^2y}{dx^2} + x\frac{dy}{dx} + (x^2 - n^2)y = 0$$

- The solution of Bessel's function, denoted by $J_n(x)$ is given by

$$J_n(x) = \sum_{m=0}^{\infty} (-1)^m \frac{x^{n+2m}}{2^{n+2m}\, m!\,\Gamma(m+n+1)}$$

Generating function for $J_n(x)$ is given by

$$\left[\frac{1}{2}x\left(t - \frac{1}{t}\right)\right] = \sum_{n=0}^{\infty} t^n\, J_n(x)\,.$$

- We have $J_0(x) = 1 - \dfrac{x^2}{2^2} + \dfrac{x^4}{2^2.4^2} - \dfrac{x^6}{2^2.4^2.6^2} + \ldots$

and $\quad J_1(x) = \dfrac{x}{2} - \dfrac{x^3}{2^2.4} + \dfrac{x^5}{2^2.4^2.6} \ldots$

- List of Recurrence relations of Bessel's function :

1. $xJ'_n(x) = n\,J_n(x) - xJ_{n+1}(x)$
2. $(x^{-n}J_n)' = -x^{-n}J_{n+1}$
3. $xJ'_n(x) = -nJ_n(x) + xJ_{n-1}(x)$
4. $(x^nJ_n)' = x^nJ_{n-1}$
5. $2J'_n(x) = J_{n-1}(x) - J_{n+1}(x)$
6. $2nJ_n(x) = x\left[J_{n-1}(x) + J_{n+1}(x)\right]$
7. $\left[x\,J_n\,J_{n+1}\right]' = (xJ_n^2 - J_{n+1}^2)$
8. $x^2J''_n(x) = (n^2 - n - x^2)J_n(x) + xJ_{n+1}(x)$
9. $\left[J_{1/2}(x)\right]^2 + \left[J_{-1/2}(x)\right]^2 = \dfrac{2}{\pi x}$

- An integral of the type $\int_{-\infty}^{\infty} k(p,t)\,F(t)dt$ is defined as the integral transform of $F(t)$, provided it is convergent. It is denoted by $f(p)$ or $T[F(t)]$.
- If $F(t)$ is a function which is piecewise continuous on every finite interval in the range $t \ge 0$ and satisfies $|F(t)| \le M\,e^{at}$ for all $t \ge 0$ and for constant a and M, then the Laplace transform of $f(t)$ exists for all $p \ge a$.
- If the Laplace transform of a function $F(t)$ is $f(p)$, *i.e.*, $L\{F(t)\} = f(p)$, then $F(t)$ is known as inverse Laplace transform of $f(p)$.

  Symbolically, $F(t) = L^{-1}\{f(p)\}$, where L^{-1} is called the inverse Laplace transformation operator.
- A function $N(t)$ of t such that $\int_0^t N(t)\,dt = 0, \quad \forall t < 0$ is called Null function.
- If we restrict ourselves to functions $F(t)$ which are sectionally continuous in every finite interval $0 \le t \le N$ and of exponential order for $t > N$, then the inverse Laplace transform of $f(p)$, *i.e.*, $L^{-1}\{f(p)\} = F(t)$ is unique.
- The equation of the type $F\left(\dfrac{\partial z}{\partial x}, \ldots, \dfrac{\partial^2 z}{\partial x^2}, \ldots, \dfrac{\partial^2 z}{\partial x \partial y}, \ldots\right) = 0$ is called a partial differential equation.
- The order of the partial differential equation is the order of its highest derivative.
- The degree of partial differential equation is the power of the highest derivative in the equation.
- A partial differential equation is said to be linear if
  (i) it is of the first degree in the dependent variable and its partial derivatives.
  (ii) it does not contain the product of dependent variables and either of its partial derivatives, and
  (iii) it does not contain any transcendental function.
- A partial differential equation, which is not linear is called non-linear equation.
- If each term of a partial differential equation contains either the dependent variable (or unknown function) or one of its partial derivatives, it is said to be homogeneous.
- An equation which is not homogeneous, is called non-homogeneous linear equation.
- Let us consider the partial differential equation $f(x,y,z,p,q) = 0$ where x, y are independent variable and z is dependent, while $p = \dfrac{\partial z}{\partial x},\ q = \dfrac{\partial z}{\partial y}$, then a relation of type $F(x,y,z,a,b) = 0$ containing as many arbitrary constants as the number of independent variable in the above partial differential equation is called complete integral.
- In the complete integral $F(x,y,z,a,b) = 0$ giving the particular values to the constants a and b, we get the particular integral.
- The envelope of the surfaces given by the complete integral

$F(x, y, z, a, b) = 0$ is called singular integral. Therefore, the singular integral is obtained by eliminating a and b from

$F(x, y, z, a, b) = 0, \ \frac{\partial F}{\partial a} = 0$ and $\frac{\partial F}{\partial b} = 0$.

- Let $u = u(x, y, z)$ and $v = v(x, y, z)$ be two functions of x, y and z, then the solution of the differential equation $Pp + Qq = R$ of the types $f(u, v) = 0$ is called the general integral.
- The partial differential equation of the form $Pp + Qq = R$, where P, Q, R are functions of x, y, z in the standard form of the linear partial differential equation of the order one is called Lagrange's linear equation.
- The partial differential equation of the form $Pp + Qq = R$, where P, Q, R are functions of x, y and z and $p = \frac{\partial z}{\partial x}$, $q = \frac{\partial z}{\partial y}$. Then this partial differential equation of order one is called Lagrange's linear differential equation.
- The Charpit's method of solving the partial differential equation $f(x, y, z, p, q) = 0$ ...(1) is based on the introduction of a second partial differential equation of the first order.
- The procedure of solving the partial differential equation by Charpit's method is as follows :
  (i) Putting terms of the equation in L.H.S. and denote the whole expression by f.
  (ii) Write Charpit's auxiliary equation.
  (iii) Find the values of different partial derivatives of f and in Charpit's auxiliary equation.
  (iv) Select two proper fractions such that the resulting integral may come out in the simplest form involving at least one of p and q.
  (v) Solving this simplest relation [obtained in (iv)] with the given equation to find p and q.
  (vi) Putting the values of p and q in $dz = pdx + qdy$.
- Charpit's method is also applicable for the solution of the partial differential equation of order one but of any degree.
- Charpit's method is applied only when other methods are failed for finding the solution.
- $r = \frac{\partial^2 z}{\partial x^2} = \frac{\partial p}{\partial x}, \ s = \frac{\partial^2 z}{\partial x \partial y} = \frac{\partial p}{\partial y} = \frac{\partial q}{\partial x}, \ t = \frac{\partial^2 z}{\partial y^2} = \frac{\partial q}{\partial y}$.
- A partial differential equation is said to be of the second order when it includes at least one of the partial differential coefficients r, s, t but none of higher order.
- A partial differential equation which is linear with respect to dependent variable and its derivatives and in which the coefficients are not function of the dependent variable but merely constants is called a P.D.E. with constant coefficients.
- To find auxiliary equation put $D = m$ and $D' = 1$ in the given equation.
- If the auxiliary equation has the real and unequal roots, say m_1, m_2, then C.F. $= f_1(y + m_1 x) + f_2(y + m_2 x)$ and if the A.E. has repeated roots, then

  C.F. $= f_1(y + mx) + x f_2(y + mx)$.
- If $f(D, D')z = F(x, y)$ be the given equation and $F(x, y)$ is of the form $x^m y^n$, then

  P.I. $= \frac{1}{f(D, D')} x^m y^n = [f(D, D')]^{-1} x^m y^n$
- If $F(x, y)$ is of the form e^{ax+by}, then

  P.I. $= \frac{1}{f(D, D')} e^{ax+by} = \frac{e^{ax+by}}{f(a, b)}$, provided $f(a, b) \neq 0$.
- If $f(D, D')$ is a homogeneous function of D and D' of degree n, then

  P.I. $= \frac{1}{f(D, D')} \phi(ax + by) = \frac{1}{f(a, b)} \int \int ... \int \phi(V) dV^n$,

  $V = ax + by$, provided $f(a, b) \neq 0$.

  If $f(a, b) = 0$, then

  P.I. $= \frac{1}{(bD - aD')} \phi(ax + by) = \frac{x^n}{b^n . n!} \phi(ax + by)$.
- If $f(D, D')$ is a homogeneous function of D and D', then

  $\frac{1}{(D - mD')} F(x, y) = \int F(x, a - mx) dx$, where $a = y + mx$.
- If the given equation is non-homogeneous, then P.I. of $f(D, D')z = F(x, y)$ is given by
  (i) When $F(x, y) = e^{ax+by}$ and $f(a, b) \neq 0$, then

  P.I. $= \frac{1}{f(a, b)} e^{ax+by}$.

  (ii) When $F(x, y) = \sin(ax + by)$ or $\cos(ax + by)$, then

  P.I. $= \frac{1}{F(a, b)} \sin(ax + by)$ [or $\cos(ax + by)$]

  is obtained by putting $D^2 = -a^2, DD' = -ab, D'^2 = -b^2$

  (iii) When $F(x, y) = x^m y^n$, then

  P.I. $= \frac{1}{f(D, D')} x^m y^n = [f(D, D')]^{-1} x^m y^n$

  (iv) When $F(x, y) = e^{ax+by} . V$, where V is a function of x and y, then we have

  P.I. $= \frac{1}{f(D, D')} e^{ax+by} . V = e^{ax+by} \frac{1}{f(D + a, D' + b)} . V$

Objective Type Questions

PROBLEM SET – I

1. The order and degree of the differential equation $\frac{d^3 y}{dx^3} + 4\sqrt{\left(\frac{dy}{dx}\right)^3} + y^2 = 0$ are respectively : **[GATE(CE)–2010]**

(a) 3 and 3 (b) 3 and 1

(c) 3 and 2 (d) 2 and 3

2. The degree of the differential equation $\frac{d^2 y}{dt^2} + 2x^3 = 0$ is : **[GATE(CE)–2007]**

(a) 3 (b) 1

(c) 0 (d) 2

3. The order of the differential equation $\frac{d^2 y}{dt^2} + \left(\frac{dy}{dt}\right)^3 + y^4 = e^{-t}$ is : **[GATE(EC)–2009]**

(a) 1 (b) 3

(c) 2 (d) 4

4. The following differential equation has

$$3\left(\frac{d^2y}{dt^2}\right)+4\left(\frac{dy}{dt}\right)^3+y^2+2=x:$$ **[GATE(EC)–2005]**

(a) degree = 2, order = 1
(b) degree = 2, order = 3
(c) degree = 1, order = 2
(d) degree = 4, order = 3

5. The Blasius equation, $\frac{d^3y}{d\eta^3}+\frac{f}{2}\frac{d^2y}{d\eta^2}=0$, is a : **[GATE(ME)–2010]**

(a) third order non-linear ordinary differential equation
(b) second order non-linear ordinary differential equation
(c) third order linear ordinary differential equation
(d) none of the above

6. The partial differential equation $\frac{\partial u}{\partial t}+u\frac{\partial u}{\partial x}=\frac{\partial^2 u}{\partial x^2}$ is a : **[GATE(ME)–2013]**

(a) linear equation of order 2
(b) non-linear equation of order 2
(c) linear equation of order 1
(d) non-linear equation of order 1

7. Biotransformation of an organic compound having concentration (*X*) can be modeled using an ordinary differential equation where *K* is the reaction rate constant. If $x = a$ at $t = 0$, the solution of the equation is : **[GATE(CE)–2004]**

(a) $\frac{1}{x}=\frac{1}{a}+Kt$ (b) $x = ae^{-Kt}$
(c) $x = a + Kt$ (d) none of these

8. The solution for the differential equation $\frac{dy}{dx}=x^2y$ with the condition that $y = 1$ at $x = 0$ is : **[GATE(CE)–2007]**

(a) $y=e^{1/2x}$ (b) $y=e^{x^3/3}$
(c) $\log(y)=\frac{x^2}{2}$ (d) none of these

9. The solution to the differential equation $\frac{d^2u}{dx^2}-k\frac{du}{dx}=0$ is where *k* is constant, subjected to the boundary conditions $u(0) = 0$ and $u(L) = U$, is : **[GATE(ME)–2013]**

(a) $u=U\frac{x}{L}$ (b) $u=U\left(\frac{1+e^{kx}}{1+e^{kL}}\right)$
(c) $u=U\left(\frac{1-e^{kx}}{1-e^{kL}}\right)$ (d) none of these

10. A spherical naphthalene ball exposed to the atmosphere loses volume at the rate proportional to its instantaneous surface area due to evaporation. If the initial diameter of the ball is 2 cm and the diameter reduces to 1 cm after 3 months, the ball completely evaporates in : **[GATE(CE)–2006]**

(a) 9 months (b) 6 months
(c) 10 months (d) none of these

11. A body originally at 60°C cools down to 40°C in 15 minutes when kept in air at a temperature of 25°C. What will be the temperature of the body at the end of 30 minutes ? **[GATE(CE)–2007]**

(a) 31.5°C (b) 28.7°C
(c) 15°C (d) none of these

12. Solution of the differential equation $3y\frac{dy}{dx}+2x=0$ represents a family of : **[GATE(CE)–2009]**

(a) circles (b) ellipses
(c) parabolas (d) hyperbolas

13. Solution of $\frac{dy}{dx}=\frac{-x}{y}$ at $x = 1$ and $y=\sqrt{3}$ is : **[GATE(CE)–2008]**

(a) $x^2 + y^2 = 4$ (b) $x^2 + y^2 = 6$
(c) $x^2 + y^2 = 2$ (d) none of these

14. The solution of $\frac{dy}{dx}=y^2$ with initial value $y(0) = 1$ bounded in the interval : **[GATE(ME)–2007]**

(a) $-\infty \le x \le \infty$ (b) $x < 1, x > 1$
(c) $-2 \le x \le 2$ (d) $0 \le x \le 2$

15. The type of the partial differential equation $\frac{\partial f}{\partial t}=\frac{\partial^2 f}{\partial x^2}$ is : **[GATE(IN)–2013]**

(a) elliptic (b) parabolic
(c) non-linear (d) hyperbolic

16. The solution of the first order differential equation $x'(t) = -3x(t)$, $x(0) = x_0$ is : **[GATE(EE)–2005]**

(a) $x(t) = x_0e^{-3}$ (b) $x(t) = x_0e^{1/3}$
(c) $x(t) = x_0e^{-3t}$ (d) none of these

17. The solution of the differential equation $\frac{dy}{dx}+y^2=0$ is : **[GATE(ME)–2003]**

(a) ce^x (b) $y=\frac{1}{x+c}$
(c) unsolvable (d) none of these

18. Consider the differential equation $\frac{dy}{dx}=(1+y^2)x$. The general solution with constant *c* is : **[GATE(ME)–2011]**

(a) $y=\tan\left(\frac{x^2}{2}+c\right)$ (b) $y=\tan\frac{x^2}{2}+\tan c$
(c) $y=\tan^2\left(\frac{x}{2}+c\right)$ (d) none of these

19. Which of the following is a solution to the differential equation $\frac{dx(t)}{dt}+3x(t)=0$? **[GATE(EC)–2008]**

(a) $x(t) = 3e^{-t}$ (b) $x(t)=\frac{3}{2}t^2$
(c) $x(t) = 2e^{-3t}$ (d) none of these

20. With *k* as constant the possible for the first order differential equation $\frac{dy}{dx}=e^{-3x}$ is : **[GATE(EE)–2011]**

(a) $-\frac{1}{3}e^{3x}+k$ (b) $-\frac{1}{3}e^{-3x}+k$
(c) $-3e^{-x} + k$ (d) none of these

21. Match List–I with List–II and select the correct answer using the codes given below the lists : **[GATE(EC)–2009]**

List–I	List–II
A. $\frac{dy}{dx}=\frac{y}{x}$	1. Circles
B. $\frac{dy}{dx}=-\frac{y}{x}$	2. Straight lines

C. $\frac{dy}{dx} = \frac{x}{y}$ 3. Hyperbolas

D. $\frac{dy}{dx} = -\frac{x}{y}$

Codes :

	A	B	C	D
(a)	2	3	3	1
(b)	1	3	2	1
(c)	2	1	3	3
(d)	3	2	1	2

22. With initial condition $x(1) = 0.5$, the solution of the differential equation $t\frac{dx}{dt} + x = t$ is : **[GATE(EC, EE, IN)–2012]**

(a) $x = \frac{t}{2}$ (b) $x = \frac{t^2}{2}$

(c) $x = t^2 - \frac{1}{2}$ (d) none of these

23. The solution of the differential equation $\frac{dy}{dx} = Ky, y(0) = c$ is : **[GATE(EC)–2011]**

(a) $y = ce^{Kx}$ (b) $x = ce^{-Ky}$

(c) $y = ce^{-Kx}$ (d) none of these

24. Consider the differential equation $x^2\frac{d^2y}{dx^2} + x\frac{dy}{dx} - 4y = 0$ with the boundary conditions of $y(0) = 0$ and $y(1) = 1$. The complete solution of the differential equation is : **[GATE(ME)–2012]**

(a) $\sin\left(\frac{\pi x}{2}\right)$ (b) x^3

(c) $e^x \sin\left(\frac{\pi x}{2}\right)$ (d) none of these

25. The solution of the ordinary differential equation $\frac{dy}{dx} + 2y = 0$ for the boundary condition, $y = 5$ at $x = 1$ is : **[GATE(CE)–2012]**

(a) $y = e^{-2x}$ (b) $y = 2e^{-2x}$

(c) $y = 36.95e^{-2x}$ (d) none of these

26. Transformation to linear form by substituting $V = y^{1-n}$ of the equation $\frac{dy}{dt} + P(t)y = q(t)y^n; n > 0$ will be : **[GATE(CE)–2005]**

(a) $\frac{dV}{dt} + (1-n)PV = (1+n)q$

(b) $\frac{dV}{dt} + (1+n)PV = (1-n)q$

(c) $\frac{dV}{dt} + (1-n)PV = (1-n)q$

(d) none of the above

27. The maximum value of the solution $y(t)$ of the differential equation $y(t) + y''(t) = 0$ with initial conditions $\dot{y}(0) = 1$ and $y(0) = 1$, for $t \geq 0$ is : **[GATE(IN)–2013]**

(a) $\sqrt{2}$ (b) 2

(c) π (d) 0

28. The solution of the differential equation $\frac{dy}{dx} + \frac{y}{x} = x$, with the conditions that $y = 1$ at $x = 1$ is : **[GATE(CE)–2011]**

(a) $y = \frac{2}{3x^2} + \frac{x}{3}$ (b) $y = \frac{2}{3x} + \frac{x^2}{3}$

(c) $y = \frac{x}{2} + \frac{1}{2x}$ (d) $y = \frac{2}{3} + \frac{x}{3}$

29. The solution of $x\frac{dy}{dx} + y = x^4$ with the condition $y(1) = \frac{6}{5}$ is : **[GATE(ME)–2009]**

(a) $y = \frac{x^5}{5} + 1$ (b) $y = \frac{x^4}{5} + \frac{1}{x}$

(c) $y = \frac{x^4}{5} + 1$ (d) none of these

30. The solution of the differential equation $\frac{dy}{dx} + 2xy = e^{-x^2}$ with $y(0) = 1$ is : **[GATE(ME)–2006]**

(a) $(1+x)e^{-x^2}$ (b) $(1+x)e^{+x^2}$

(c) $(1-x)e^{+x^2}$ (d) $(1-x)e^{-x^2}$

31. The solution of $\frac{d^2y}{dx^2} + 2\frac{dy}{dx} + 17y = 0; y(0) = 1, \frac{dy}{dx}\left(\frac{\pi}{4}\right) = 0$ in the range $0 < x < \frac{\pi}{4}$ is given by : **[GATE(CE)–2005]**

(a) $e^x(\cos 4x - \frac{1}{4}\sin 4x)$ (b) $e^{-x}(\cos 4x + \frac{1}{4}\sin 4x)$

(c) $e^{-4x}(\cos x - \frac{1}{4}\sin x)$ (d) none of these

32. The solution to the ordinary differential equation $\frac{d^2y}{dx^2} + \frac{dy}{dx} - 6y = 0$: **[GATE(CE)–2010]**

(a) $y = c_1e^{3x} + c_2e^{-2x}$ (b) $y = c_1e^{-3x} + c_2e^{2x}$

(c) $y = c_1e^{3x} + c_2e^{2x}$ (d) none of these

33. The general solution of $\frac{d^2y}{dx^2} + y = 0$ is : **[GATE(CE)–2008]**

(a) $y = P\cos x$ (b) $y = P\sin x$

(c) $y = P\cos x + Q\sin x$ (d) none of these

34. Which of the following is solution of the differential equation $\frac{d^2y}{dx^2} + p\frac{dy}{dx} + (q+1) = 0$? **[GATE(ME)–2005]**

(a) xe^{-x} (b) xe^{-2x}

(c) e^{-3x} (d) none of these

35. The complete solution of the ordinary differential equation $\frac{d^2y}{dx^2} + p\frac{dy}{dx} + qy = 0$ is $y = c_1e^{-x} + c_2e^{-3x}$. Then p and q are : **[GATE(ME)–2005]**

(a) $p = 3, q = 3$ (b) $p = 4, q = 3$

(c) $p = 4, q = 4$ (d) none of these

36. It is given that $y'' + 2y' + y = 0, y(0) = 0, y(1) = 0$, what is $y(0.5)$? **[GATE(ME)–2008]**

(a) 0.37 (b) 0

(c) 0.62 (d) 1

37. Given that $x'' + 3x = 0$, and $x(0) = 1, \dot{x}(0) = 0$. What is $x(1)$? **[GATE(ME)–2008]**

(a) 0.99 (b) 0.16

(c) 0.32 (d) –0.99

38. A solution of the following differential equation is given by $\frac{d^2y}{dx^2} - 5\frac{dy}{dx} + 6y = 0$ **[GATE(EC)–2005]**

(a) $y = e^{2x} + e^{-3x}$ (b) $y = e^{-2x} + e^{3x}$
(c) $y = e^{-2x} + e^{-3x}$ (d) $y = e^{2x} + e^{3x}$

39. For the differential equation $\frac{d^2x}{dt^2} + 6\frac{dx}{dt} + 8x = 0$ with initial conditions $x(0) = 1$ and $\left.\frac{dx}{dt}\right|_{t=0} = 0$, the solution is : **[GATE(EE)–2010]**

(a) $x(t) = 2e^{-2t} - e^{-4t}$ (b) $x(t) = -e^{-6t} + 2e^{-4t}$
(c) $x(t) = e^{-2t} + 2e^{-4t}$ (d) none of these

40. A function $n(x)$ satisfies the differential equation $\frac{d^2n(x)}{dx^2} - \frac{n(x)}{l^2} = 0$ where L is a constant. The boundary conditions are : $n(0) = K$ and $n(\infty) = 0$. The solution to this equation is : **[GATE(EC)–2010]**

(a) $n(x) = K\exp(x/L)$ (b) $n(x) = K^2\exp(-x/L)$
(c) $n(x) = K\exp(-x/\sqrt{L})$ (d) $n(x) = K\exp(-1/Lx)$

41. A system described by a linear constant coefficient, ordinary, first order differential equation has an exact solution given by $y(t)$ for $t > 0$, when the forcing functions is $x(t)$ and the initial conditions is $y(0)$. If one wishes to modify the system that the solution becomes $-2y(t)$ for $t > 0$, we need to : **[GATE(EC)–2013]**

(a) change the initial condition to $-2y(0)$ and the forcing function to $-2x(t)$
(b) change the initial condition to $-y(0)$ and the forcing function to $2x(t)$
(c) change the initial condition to $2y(0)$ and the forcing function to $-x(t)$
(d) none of the above

42. For $\frac{d^2y}{dx^2} + 4\frac{dy}{dx} + 3y = 3e^{2x}$, the particular integral is : **[GATE(ME)–2006]**

(a) $\frac{1}{15}e^{2x}$ (b) $3e^{2x}$
(c) $\frac{1}{5}e^{2x}$ (d) none of these

PROBLEM SET – II

1. If L defines the Laplace Transform of a fuinction, $L[\sin(at)]$ will be equal to : **[GATE(CE)–2008]**

(a) $\frac{a}{s^2+a^2}$ (b) $\frac{a}{s^2-a^2}$
(c) $\frac{s}{s^2-a^2}$ (d) $\frac{s}{s^2+a^2}$

2. Laplace transform of the function sin wt is : **[GATE(ME)–2003]**

(a) $\frac{s}{s^2-w^2}$ (b) $\frac{w}{s^2-w^2}$
(c) $\frac{w}{s^2+w^2}$ (d) $\frac{s}{s^2+w^2}$

3. Lapalce transform for the function $f(x) = \cosh(ax)$ is : **[GATE(CE)–2009]**

(a) $\frac{s}{s^2-a^2}$ (b) $\frac{s}{s^2+a^2}$
(c) $\frac{a}{s^2-a^2}$ (d) $\frac{a}{s^2+a^2}$

4. The function $f(t)$ satisfies the differential equation $\frac{d^2f}{dt^2} + f = 0$ and the auxiliary conditions, $f(0) = 0$, $\frac{df}{dt}(0) = 4$. The Laplace transform of $f(t)$ is given by : **[GATE(ME)–2013]**

(a) $\frac{4}{s^2+1}$ (b) $\frac{2}{s^2+1}$
(c) $\frac{4}{s+1}$ (d) $\frac{2}{s^2-1}$

5. In what range should Re(s) remain so that the Laplace transform of the function $e^{(a+2)t+5}$ exists : **[GATE(EC)–2005]**

(a) Re$(s) < 2$ (b) Re$(s) > a + 2$
(c) Re$(s) > a + 7$ (d) Re$(s) < a + 5$

6. Evaluate $\int_0^\infty \frac{\sin t}{t}dt$: **[GATE(CE)–2007]**

(a) $\pi/2$ (b) π
(c) $\pi/4$ (d) $\pi/3$

7. If $F(s)$ is the Laplace transform of the function $f(t)$, then Laplace transform of $\int_0^1 f(\tau)d\tau$ is : **[GATE(MC)–2007]**

(a) $\frac{1}{S}F(S) - f(t)$ (b) $\frac{1}{S}F(S)$
(c) $\int F(S)dS$ (d) none of these

8. The Laplace transform of a function $f(t)$ is $\frac{1}{s^2(s+1)}$. The function $f(t)$ is : **[GATE(ME)–2010]**

(a) $1 + t + e^{-t}$ (b) $-1 + e^{-1}$
(c) $t - 1 + e^{-t}$ (d) $2t + e^{t}$

9. The inverse Laplace transform of $\frac{1}{(s^2+s)}$ is : **[GATE(ME)–2009]**

(a) $1 - e^{-t}$ (b) $1 + e^{-t}$
(c) $1 + e^{t}$ (d) $1 - e^{t}$

10. Given $L^{-1}\left[\frac{3s+1}{s^3+4s^2+(K-3)s}\right]$. If $\lim_{t\to\infty} f(t) = 1$, then value of K is : **[GATE(EC)–2010]**

(a) 1 (b) 4
(c) 2 (d) 3

11. The inverse Laplace transform of the function $F(s) = \frac{1}{s(s+1)}$ is given by : **[GATE(ME)–2012]**

(a) $f(t) = 1 - e^{-t}$ (b) $f(t) = e^{-t}$
(c) $f(t) = \sin t$ (d) none of these

12. A delayed unit step function is defined as $u(t-a)=\begin{cases}0, & \text{for } t<a\\ 1, & \text{for } t\ge a\end{cases}$. Its Laplace transform is : [GATE(ME)–2004]

(a) $\frac{e^{as}}{s}$ (b) $a.e^{-as}$

(c) $\frac{e^{-as}}{s}$ (d) $\frac{e^{as}}{a}$

13. The Laplace transform of $g(t)$ is : [GATE(EE)–2010]

(a) $\frac{e^{-3s}}{s}(1-e^{-2s})$ (b) $\frac{1}{s}(e^{-5s}-e^{-3s})$

(c) $\frac{1}{s}(e^{3s}-e^{5s})$ (d) none of these

14. $g(t)$ can be expressed as : [GATE(EE)–2010]

(a) $g(t)=f\left(2t-\frac{3}{2}\right)$ (b) $g(t)=f\left(\frac{t}{2}-\frac{3}{2}\right)$

(c) $g(t)=f(2t-3)$ (d) $g(t)=f\left(\frac{t}{2}-3\right)$

15. Consider the differential equation $\frac{d^2y(t)}{dt^2}+\frac{2dy(t)}{dt}+y(t)=\delta(t)$ with $y(t)|_{t=0}=-2$ and $\left.\frac{dy}{dt}\right|_{t=0}=0$. The numerical value of $\left.\frac{dy}{dt}\right|_{t=0}$ is : [GATE(EC, IN)–2012]

(a) 1 (b) –2

(c) –1 (d) 0

16. A solution for the differential equation $\dot{x}(t)+2x(t)=\delta(t)$ with initial condition $x(0) = 0$ is : [GATE(EC)–2006]

(a) $e^{2t}u(t)$ (b) $e^{-2t}u(t)$

(c) $e^{-t}u(t)$ (d) $e^{t}u(t)$

ANSWERS

PROBLEM SET – I

1. (c)	**2.** (b)	**3.** (c)	**4.** (c)	**5.** (a)	**6.** (b)	**7.** (a)	**8.** (b)	**9.** (c)	**10.** (b)
11. (a)	**12.** (b)	**13.** (a)	**14.** (b)	**15.** (b)	**16.** (c)	**17.** (b)	**18.** (a)	**19.** (c)	**20.** (b)
21. (a)	**22.** (a)	**23.** (a)	**24.** (b)	**25.** (c)	**26.** (c)	**27.** (a)	**28.** (b)	**29.** (b)	**30.** (a)
31. (b)	**32.** (b)	**33.** (c)	**34.** (b)	**35.** (b)	**36.** (b)	**37.** (a)	**38.** (d)	**39.** (a)	**40.** (b)
41. (a)	**42.** (c)								

PROBLEM SET – II

1. (a)	**2.** (c)	**3.** (a)	**4.** (a)	**5.** (b)	**6.** (a)	**7.** (b)	**8.** (c)	**9.** (a)	**10.** (b)
11. (a)	**12.** (c)	**13.** (a)	**14.** (b)	**15.** (a)	**16.** (b)				

Hint to Selected Problems

PROBLEM SET – I

1. $\frac{d^3y}{dx^3}+4\sqrt{\left(\frac{dy}{dx}\right)^3+y^2}=0$

Removing radicals we get $\left(\frac{d^3y}{dx^3}\right)^2=16\left[\left(\frac{dy}{dx}\right)^3+y^2\right]$

$\therefore$ The order is 3 since highest differential is $\frac{d^3y}{dx^3}$.

The degree is 2 since power of highest differential is 2.

2. Degree of a differential equation is the power of its highest order derivative after the differential equation is made free of radicals and the fractions if any, in the derivative power.

Hence, here the degree is 1, which is the power of $\frac{d^2x}{dt^2}$.

3. Highest derivative of the differential equation is 2.

4. Order is highest derivative term, so order = 2, degree is power of highest derivative term. So degree = 1

5. $\frac{d^3f}{dn^3}+\frac{f}{2}\frac{d^2f}{dn^2}=0$ is third order $\left(\frac{d^3f}{dn^3}\right)$ and it is non-linear, since the product $f\times\frac{d^2f}{dn^2}$ is not allowed in linear differential equation.

6. A differential equation in the form $\frac{dy}{dx}+Py=Q$ where, P and Q are function of x, *i.e.*, $f(x)$ is said to be linear equation.

$$\frac{\partial u}{\partial t}+u\frac{\partial u}{\partial x}=\frac{\partial^2 u}{\partial x^2}$$

The given equation is not complying with the definition of the linear equation, therefore it is a non-linear equation of order 2.

7. $\frac{dx}{dt}=-kx^2 \quad\Rightarrow\quad \frac{dx}{x^2}=-kdt$

Integrating both sides, we get

$$\int\frac{dx}{x^2}=-\int kdt$$

$$-\frac{1}{x}=-kt+c \quad\Rightarrow\quad \frac{1}{x}=kt+c'$$

At $t = 0, x = a$

$$\Rightarrow \frac{1}{a}=k\times 0+c' \quad\Rightarrow\quad c'=\frac{1}{a}$$

$$\therefore \frac{1}{x}=kt+\frac{1}{a}$$

8. $\frac{dy}{dx}=x^2y \quad\Rightarrow\quad \frac{dy}{y}=x^2dx$

$$\int\frac{dy}{y}=\int x^2dx \quad\Rightarrow\quad \log_e y=\frac{x^3}{3}+c_1$$

$$\Rightarrow y=e^{x^3/3+c_1}=e^{c_1}\times e^{x^3/3} \quad\Rightarrow\quad y=c\times e^{x^3/3}$$

Now at $x = 0, y = 1$

$$1 = c \times e^{0/3}$$

$$c_1 = 1$$

$\therefore \quad y = e^{x^3/3}$ is the solution.

10. $\dfrac{dV}{dt} = -kA$ ...(i)

where $V = \dfrac{4}{3}\pi r^3$, $A = 4\pi r^2$

$\Rightarrow \quad \dfrac{dV}{dt} = \dfrac{4}{3}\pi \times 3r^2 \dfrac{dr}{dt} = 4\pi r^2 \dfrac{dr}{dt}$

Putting these values in (i) we get,

$$4\pi r^2 \frac{dr}{dt} = -k(4\pi r^2)$$

$\dfrac{dr}{dt} = -k \quad \Rightarrow \quad dr = -kdt$

Integrating both sides, we get

$$r = -kt + c$$

At $t = 0$, $r = 1$

$\Rightarrow \quad 1 = -k \times 0 + c \Rightarrow \quad c = 1$

$\therefore \quad r = -kt + 1$

Now at $t = 3$ months $r = 0.5$ cm. ...(ii)

$\therefore \quad 0.5 = -k \times 3 + 1 \Rightarrow \quad k = \dfrac{0.5}{3}$

Now substituting this value of k in equation (ii), we get

$$r = -\frac{0.5}{3}t + 1$$

Putting $r = 0$ (ball completely evaporates)

in above and solving for t gives $0 = -\dfrac{0.5}{3}t + 1$

$\Rightarrow \quad t = 6$ months

11. $\dfrac{d\theta}{dt} = -k(\theta - \theta_0)$ (Newton's law of cooling)

This is in variable seperable form separating the variables, we get,

$$\frac{d\theta}{\theta - \theta_0} = -kdt \quad \Rightarrow \quad \int \frac{d\theta}{\theta - \theta_0} = \int -kdt$$

$\Rightarrow \quad \log(\theta - \theta_0) = -kt + c_1$

$\theta - \theta_0 = ce^{-kt}$ (where $c = e^{c_1}$)

$$\theta = \theta_0 + c.e^{-Kt}$$

given, $\theta_0 = 25°C$

now at $t = 0, \theta = 60°$

$60 = 25 + c.e^{o} \Rightarrow \quad c = 35$

$$\theta = 25 + 35e^{-Kt}$$

At $t = 15$ minutes $\theta = 40°C$

$$40 = 25 + 35\, e^{(-k \times 15)}$$

$\Rightarrow \quad e^{-15k} = \dfrac{3}{7}$ ...(i)

Now at $t = 30$ minutes

$\theta = 25 + 35\, e^{-30k} = 25 + 35(e^{-15k})^2$

Now substituting $e^{-15k} = \dfrac{3}{7}$ from (i), we get,

$$\theta = 25 + 35 \times \left(\frac{3}{7}\right)^2 = 31.428°C \approx 31.5°C$$

12. $3y\dfrac{dy}{dx} + 2x = 0 \quad \Rightarrow \quad \dfrac{dy}{dx} = -\dfrac{2x}{3y}$

$\Rightarrow \quad 3ydy = -2xdx \quad \Rightarrow \quad \int 3ydy = \int -2xdx$

$\Rightarrow \quad \dfrac{3}{2}y^2 = -2 \times \dfrac{x^2}{2} + c \quad \Rightarrow \quad 3y^2 + 2x^2 = c$

$\Rightarrow \quad \dfrac{x^2}{\left(\frac{1}{2}\right)} + \dfrac{y^2}{\left(\frac{1}{3}\right)} = c \quad \Rightarrow \quad \dfrac{x^2}{\left(\frac{1}{2}c\right)} + \dfrac{y^2}{\left(\frac{1}{3}c\right)} = 1$

which is the equation of a family of ellipses.

13. $\dfrac{dy}{dx} = -\dfrac{x}{y} \quad \Rightarrow \quad ydy = -xdx$

$\Rightarrow \quad \int ydy = \int -xdx \quad \Rightarrow \quad \dfrac{y^2}{2} = -\dfrac{x^2}{2} + c$

At $x = 1$, $\quad y = \sqrt{3}$

$\therefore \quad \dfrac{(\sqrt{3})^2}{2} = -\dfrac{1^2}{2} + c \quad \Rightarrow \quad c = 2$

$\therefore$ Solution is $\dfrac{y^2}{2} = -\dfrac{x^2}{2} + 2 \quad \Rightarrow \quad x^2 + y^2 = 4$

14. $\dfrac{dy}{dx} = y^2 \quad \Rightarrow \quad \int \dfrac{dy}{y^2} = \int dx$

$\Rightarrow \quad -\dfrac{1}{y} = x + c \quad \Rightarrow \quad \therefore \quad y = -\dfrac{1}{x+c}$

When $x = 0$, $\quad y = 1 \quad \therefore \quad c = -1$

$\therefore \quad y = -\dfrac{1}{x-1}$

y is bounded $x - 1 \neq 0$, *i.e.*, $x \neq 1$, *i.e.*, $x < 1$ or $x > 1$

16. Given, $x(t) = -3x(t)$

i.e., $\dfrac{dx}{dt} = -3x$

$$\frac{dx}{x} = -3dt$$

$$\int \frac{dx}{x} = \int -3dt$$

$\Rightarrow \quad \log x = -3t + c \quad \Rightarrow \quad x = e^{-3t+c} = e^c \times e^{-3t}$

Putting $e^c = c_1$

$$x = c_1 \times e^{-3t}$$

Now putting initial condition $x(0) = x_0$

$x_0 = c_1 e^0 = c_1 \; \therefore \quad c_1 = x_0$

$\therefore$ Solution is $x = x_0 e^{-3t}$ *i.e.*, $x(t) = x_0 e^{-3t}$

17. Given differential equation

$\dfrac{dy}{dx} + y^2 = 0 \quad \Rightarrow \quad -\dfrac{dy}{y^2} = dx$

On integrating, we get

$-\int \dfrac{dy}{y^2} = \int dx \quad \Rightarrow \quad \dfrac{1}{y} = x + c$

$\therefore \quad y = \dfrac{1}{x+c}$

18. $\dfrac{dy}{dx} = (1+y^2)x \quad \Rightarrow \quad \int \dfrac{dy}{(1+y^2)} = \int xdx$

$\tan^{-1} y = \dfrac{x^2}{2} + c \quad \Rightarrow \quad y = \tan\left(\dfrac{x^2}{2} + c\right)$

19. $\frac{dx}{dt} = -3x \Rightarrow \frac{dx}{x} = -3dt$

$\int \frac{dx}{x} = \int -3dt$

On integrating, we get

$\log x = -3t + c \Rightarrow x = e^{-3t} + c$

$\Rightarrow x = e^{c}.e^{-3t} = c_1 e^{-3t} \quad (c_1 = e^c)$

20. $\frac{dy}{dx} = e^{-3x} \Rightarrow \int dy = \int e^{-3x} dx$

$y = \frac{e^{-3x}}{-3} + K \qquad y = -\frac{1}{3}e^{-3x} + K$

21. A. $\frac{dy}{dx} = \frac{y}{x} \Rightarrow \frac{dy}{y} = \frac{dx}{x}$

$\Rightarrow \int \frac{dy}{y} = \int \frac{dx}{x}$

$\log y = \log x + \log c = \log cx$

$y = cx$ ...Equation of straight line

B. $\frac{dy}{dx} = -\frac{y}{x} \Rightarrow \frac{dy}{y} = -\frac{dx}{x}$

$\Rightarrow \int \frac{dy}{y} = -\int \frac{dx}{x}$

$\log y = -\log x + \log c$

$\log y + \log x = \log c \Rightarrow \log yx = \log c$

$yx = c$

$y = c/x$ ...Equation of hyperbola

C. $\frac{dy}{dx} = \frac{x}{y} \Rightarrow ydy = xdx$

$\Rightarrow \int ydy = \int xdx$

$\frac{y^2}{2} - \frac{x^2}{2} = \frac{c^2}{2}$ (constant)

$\frac{y^2}{c^2} - \frac{x^2}{c^2} = 1$ ...Equation of hyperbola

D. $\frac{dy}{dx} = -\frac{x}{y}$

$\Rightarrow \int ydy = -\int xdx$

$\Rightarrow \frac{y^2}{2} = -\frac{x^2}{2} + \frac{c^2}{2} = \frac{y^2}{2} + \frac{x^2}{2} = \frac{c^2}{2}$

$x^2 + y^2 = c^2$ ...Equation of circle

22. The given differential equation is $t\frac{dx}{dt} + x = t$ with initial condition which is same as $\frac{dx}{dt} + \frac{x}{t} = 1$

which is a linear differential equation.

$\frac{dx}{dt} + Px = Q$ where $P = \frac{1}{t}$ and $Q = 1$

Integrating factor $= e^{\int P dt} = e^{\int \frac{1}{t} dt} = e^{\log_e t} = t$

Solution is $x.\text{I.F.} = \int Q.(\text{I.F.})dt + c$

$x.\,t = \int 1.t.dt + c \Rightarrow xt = \frac{t^2}{2} + c$

$x = \frac{t}{2} + \frac{c}{t}$

Put $x(t) = \frac{1}{2} \Rightarrow \frac{1}{2} + \frac{c}{1} = \frac{1}{2} \Rightarrow c = 0$

so $x = \frac{1}{2}$ is the solution

24. $x^2\frac{d^2y}{dx^2} + x\frac{dy}{dx} - 4y = 0$

$y(0) = 0$ and $y(1) = 1$

Choice (b) satisfies the initial condition as well as equation as shown below.

If $y = x^2 \Rightarrow y(0) = 0^2 = 0$ and $y(1) = 1^2 = 1$

Also, $\frac{dy}{dx} = 2x, \frac{d^2y}{dx^2} = 2$

$\Rightarrow x^2\frac{d^2y}{dx^2} + x\frac{dy}{dx} - 4y = x^2 \times 2 + x \times 2x - 4 \times x^2$

$= 2x^2 + 2x^2 - 4x^2 = 0$

So $y = x^2$ is the solution to this equation with given boundary conditions.

25. Given $\frac{dy}{dx} + 2y = 0$ and $y(1) = 5$

$\frac{dy}{dx} = -2y$

$\int \frac{dy}{dx} = \int -2y$

$\Rightarrow \log y = -2x + c \Rightarrow y = e^{-2x}.e^{c}$

$\Rightarrow c_1 e^{-2x}$

$y(1) = c_1 e^{-2} = 5 \Rightarrow c_1 = \frac{5}{e^{-2}}$

So, $y = \frac{5}{e^{-2}}.e^{-2x} = 5e^2 e^{-2x} = 36.95e^{-2x}$

26. Given, $\frac{dy}{dt} + P(t)y = q(t)y^n; n > 0$

Putting $V = y^{1-n}$

$\frac{dV}{dt} = (1-n)y^{-n}\frac{dy}{dt} \Rightarrow \frac{dy}{dt} = \frac{1}{(1-n)y^{-n}}\frac{dV}{dt}$

substituting in the given differential equation, we get

$\frac{1}{(1-n)y^{-n}}\frac{dV}{dt} + P(t)y = q(t)y^n$

Multiplying by $(1-n)y^{-n}$, we get

$\frac{dV}{dt} + P(t)(1-n)y^{1-n} = q(t)(1-n)$

Now since $y^{1-n} = V$, we get

$\frac{dV}{dt} + (1-n)PV = (1-n)q$

(which is linear with V as dependent variable and t as independent variable.)

27. $y(t) + \ddot{y}(t) = 0$

$\therefore \quad 1 + D^2 = 0 \Rightarrow D = \pm i$

$\therefore \quad y = c_1 e^{ix} + c_2 e^{-ix} = A\cos x + B\sin x$

$y(0) = 1$

$1 = A \times 1 + B \times 0 \Rightarrow A = 1$

$\dot{y} = -A\sin x + B\cos x$

$\dot{y}(0) = 1$

$\therefore \quad 1 = -A \times 0 + B \times 1$

$\therefore \quad B = 1$

so $\quad y = \cos x + \sin x$

For maxima

$y' = -\sin x + \cos x = 0$

$\therefore \sin x = \cos x \quad \Rightarrow \quad x = 45°$

$y'' = -\cos x - \sin x$

$y'' < 0$ for $x = 45°$ $\therefore$ maxima

$y(\max) = \cos 45° + \sin 45° = \frac{1}{\sqrt{2}} + \frac{1}{\sqrt{2}} = \frac{2}{\sqrt{2}} = \sqrt{2}$.

28. $\frac{dy}{dx} + \frac{y}{x} = x, y(1) = 1$

This is a linear differential equation $\frac{dy}{dx} + Py = Q$ with $P = \frac{1}{x}$ and $Q = x$

$$\text{I.F.} = e^{\int P dx} = e^{\int \frac{1}{x} dx} = e^{\log x} = x$$

Solution is

$y.(\text{I.F.}) = \int Q.(\text{I.F.}) dx + C \quad \Rightarrow \quad y.x = \int (x.x) dx + C$

$\Rightarrow \quad yx = \int x^2 dx + C \quad \Rightarrow \quad yx = \frac{x^3}{3} + C$

$\Rightarrow \quad y = \frac{x^2}{3} + \frac{C}{x}$

Now $y(1) = 1 \quad \Rightarrow \quad \frac{1^2}{3} + \frac{C}{1} = 1 \Rightarrow C = \frac{2}{3}$

So the solution is $y = \frac{x^2}{3} + \frac{2}{3x}$

29. Given differential equation is

$x\frac{dy}{dx} + y = x^4 \quad \Rightarrow \quad \frac{dy}{dx} + \left(\frac{y}{x}\right) = x^3 \quad \ldots\text{(i)}$

Standard form of Leibnitz's linear equation is

$$\frac{dy}{dx} + Py = Q \quad \ldots\text{(ii)}$$

Where P and Q are function of x only and solution is given by $y.\,(\text{I.F.}) = \int Q(\text{I.F.}) dx + C$ where $\text{I.F.} = e^{\int P dx}$

by eqn (i) $\quad P = \frac{1}{x}$ and $Q = x^3$

$$\text{IF} = e^{\int \frac{1}{x} dx} = e^{\log x} = x$$

Solution is $y(x) = \int x^3.x dx + C$

$$yx = \frac{x^5}{5} + C$$

Given condition $y(1) = \frac{6}{5}$ means at $x = 1$; $y = \frac{6}{5}$

$\Rightarrow \quad \frac{6}{5} \times 1 = \frac{1}{5} + C \quad \Rightarrow \quad C = \frac{6}{5} - \frac{1}{5} = 1$

$\therefore \quad yx = \frac{x^5}{5} + 1 \quad \Rightarrow \quad y = \frac{x^4}{5} + \frac{1}{x}$

30. Given equation $\frac{dy}{dx} + 2xy = e^{-x^2}$

$\text{IF} = e^{\int 2x dx} = e^{x^2}$

Solution is $y(\text{I.F.}) = \int Q(\text{I.F.}) dx + C$

$$y.e^{x^2} = \int e^{-x^2}.e^{x^2} dx + C$$

$$y.e^{x^2} = x + C$$

At $x = 0, y = 1$ (Given)

$\therefore \quad 1.e^0 = 0 + c \quad \Rightarrow \quad c = 1$

So, the solution is $y.e^{x^2} = x + 1 \Rightarrow \quad y = e^{-x^2}(x+1)$

31. $\frac{d^2y}{dx^2} + 2\frac{dy}{dx} + 17y = 0$

$y(0) = 1 \quad \frac{dy}{dx}\left(\frac{\pi}{4}\right) = 0$

This is a linear differential equation.

$D^2 + 2D + 17 = 0$

$D = -1 \pm 4i$

$\therefore \; y = c_1 e^{(-1+4i)x} + c_2 e^{(-1-4i)x} = e^{-x} c_1 e^{4xi} + c_2 e^{-4xi}$

$= e^{-x}[c_1(\cos 4x + i \sin 4x)] + c_2[\cos(-4x) + i \sin(-4x)]$

$= e^{-x}[(c_1 + c_2)\cos 4x + (c_1 - c_2)i \sin 4x]$

Let $c_1 + c_2 = c_3$ and $(c_1 - c_2)i = c_4$

$y = e^{-x}[c_3 \cos 4x + c_4 \sin 4x]$

$\because y(0) = 1$

$\Rightarrow \quad 1 = e^{-0}(c_3 \cos 0 + c_4 \sin 0) \Rightarrow \quad c_3 = 1$

$\frac{dy}{dx} = e^{-x}(-4c_3 \sin 4x + 4c_4 \cos 4x) - e^{-x}(c_3 \cos 4x + c_4 \sin 4x)$

$= e^{-x}[(-4c_3 - c_4)\sin 4x + (4c_4 - c_3)\cos 4x]$

$\frac{dy}{dx}$ at $x = \frac{\pi}{4}$ is 0.

$\therefore \quad -(-4c_4 - c_3)e^{-\pi/4} = 0 \quad \Rightarrow \quad 4c_4 = c_3 \Rightarrow \quad c_4 = \frac{c_3}{4} = \frac{1}{4}$

$\therefore \quad c_3 = 1$ and $c_4 = \frac{1}{4}$

$y = e^{-x}(\cos 4x + \frac{1}{4}\sin 4x)$

32. $\frac{d^2y}{dx^2} + \frac{dy}{dx} - 6y = 0 \quad \Rightarrow \quad D^2 + D - 6 = 0$

$(D + 3)(D - 2) = 0 \quad \Rightarrow \quad D = -3$ or $D = 2$

$\therefore$ Solution is $y = c_1 e^{-3x} + c_2 e^{2x}$

33. $\frac{d^2y}{dx^2} + y = 0 \quad \Rightarrow \quad D^2 + 1 = 0$

$D = \pm i = 0 \pm 1.i$

$\therefore$ General solution is

$y = e^{0x}[c_1 \cos(1 \times x) + c_2 \sin(1 \times x)] = c_1 \cos x + c_2 \sin x$

$= P \cos x + Q \sin x$

34. Given equation is

$$\frac{d^2y}{dx^2} + p\frac{dy}{dx} + (q+1) = 0$$

$\Rightarrow \quad [D^2 + pD + (q + 1)]y = 0$

Put $p = 4, q = 3$

$\therefore \quad [D^2 + 4D + 4]y = 0$

$D^2 + 4D + 4 = 0$

$(D + 2)^2 = 0$

$\therefore \quad D = -2, -2$

$\Rightarrow \quad y = (c_1 x + c_2)e^{-2x}$

Out of choices given, $y = xe^{-2x}$ is the only answer in the required form (*i.e.*, $(c_1 x + c^2)e^{-2x}$ putting $c_1 = 1$ and $c_2 = 0$).

35. Given equation is

$$\frac{d^2y}{dx^2}+p\frac{dy}{dx}+qy=0$$

$$(D^2+pD+q)y=0$$

$$\therefore \quad D^2+pD+q=0$$

Solution is $y=c_1e^{-x}+c_2e^{-3x}$

So the roots of $D^2+pD+q=0$ are $\alpha=-1$ and $\beta=-3$

Sum of roots $=-p=-1-3 \quad \Rightarrow \quad P=4$

Product of roots $=q=(-1)(-3) \quad \Rightarrow \quad q=3$

36.

$$y''+2y'+y=0$$

$$(D^2+2D+1)y=0 \quad \Rightarrow \quad D^2+2D+1=0$$

$$(D+1)^2=0 \quad \Rightarrow \quad D=-1,-1$$

$\therefore \; y=(c_1+c_2)e^{-x}$

$y(0)=0 \quad \Rightarrow \quad 0=(c_1+c_2(0))e^{-x}$

$c_1=0$

$y(1)=0 \quad \Rightarrow \quad 0=(c_1+c_2)e^{-1}$

$\Rightarrow \quad c_1+c_2=0 \quad \Rightarrow \quad c_2=0$

$\therefore \; y=(0+0x)e^{-x}=0$ is the solution

$\therefore \; y(0.5)=0$

37.

$$x''+3x=0$$

A.E. is $D^2+3=0$ *i.e.,* $D=\pm\sqrt{3}i$

$\therefore \; x=A\cos\sqrt{3}t+B\sin\sqrt{3}t$

At $t=0, x=1$

$\Rightarrow \; A=1$

Now $\dot{x}=\sqrt{3}(B\cos\sqrt{3}t-A\sin\sqrt{3}t)$

At $t=0, \dot{x}=0 \quad \Rightarrow \quad B=0$

So $\; x=\cos\sqrt{3}t \quad \Rightarrow \quad x(1)=\cos\sqrt{3}=0.99$

38. A.E. $\Rightarrow D^2-5D+6=0$

$$(D-2)(D-3)=0 \quad \Rightarrow \quad D=2,3$$

$\therefore \; y=e^{2x}+e^{3x}$

39. Given $\dfrac{d^2x}{dt^2}+6\dfrac{dx}{dt}+8x=0$

$$x(0)=1 \text{ and } \left.\frac{dx}{dt}\right|_{t=0}=0$$

$$D^2+6D+8=0 \quad \Rightarrow \quad (D+4)(D+2)=0$$

$$D=-2 \text{ and } D=-4$$

$\therefore$ Solution is $x=c_1e^{-2t}+c_2e^{-4t}$

Since $x(0)=1$

We have $c_1+c_2=1$ ...(i)

$$\frac{dx}{dt}=-2c_1e^{-2t}-4c_2e^{-4t}$$

Since $\left[\dfrac{dx}{dt}\right]_{t=0}=0$, we have

$-2c_1-4c_2=0$ ...(ii)

Solving (i) and (ii) we have

$c_1=2$ and $c_2=-1$

Solution is $x(t)=2e^{-2t}-e^{-4t}$.

40. $\dfrac{d^2n(x)}{dx^2}-\dfrac{n(x)}{L^2}=0 \quad \Rightarrow \quad D^2-\dfrac{1}{L^2}=0$

$\Rightarrow \quad D^2=\dfrac{1}{L^2} \quad \Rightarrow \quad D=\pm\dfrac{1}{L}$

Solution is

$$n(x)=c_1e^{-1/L}+c_2e^{1/Lx}$$

$$n(0)=c_1+c_2=k$$

$$n(\infty)=c_1e^{-\infty}+c_2e^{\infty}=0$$

$\Rightarrow \quad c_2e^{\infty}=0 \quad \Rightarrow \quad c_2=0$

$\therefore \quad c_1=k$

$\therefore$ The solution is $n(x)=ke^{-1/Lx}$.

41. $\dfrac{dy(t)}{dt}+ky(t)=x(t)$

Taking Laplace transform of both sides, we have

$$y(s)-y(0)+ky(s)=x(s)$$

$$y(s)[s+k]=x(s)+y(0)$$

$$\Rightarrow \quad y(s)=\frac{x(s)}{s+k}+\frac{y(0)}{s+k}$$

Taking inverse Laplace transform, we have

$$y(t)=e^{-Kt}x\,(t)+y(0)e^{-Kt}$$

So if we want $-2y(t)$ as a solution both $x(t)$ and $y(0)$ has to be multipled by -2, hence change $x(t)$ by $-2x(t)$ and $y(0)$ by $-2y(0)$.

42. $\dfrac{d^2y}{dx^2}+4\dfrac{dy}{dx}+3y=3e^{2x}$

$\Rightarrow \quad (D^2+4D+3)y=3e^{2x}$

$$\text{P.I.}=\frac{1}{D^2+4D+3}3e^{2x}$$

Now, since $\dfrac{1}{f(D)}e^{ax}=\dfrac{1}{f(a)}e^{ax}$

$$\text{P.I.}=3\frac{e^{2x}}{(2)^2+4(2)+3}=\frac{3e^{2x}}{15}=\frac{e^{2x}}{5}$$

PROBLEM SET – II

1. $L[f(t)]=\int_0^{\infty}e^{-st}f(t)dt$

2. $L[\sin wt]=\dfrac{w}{s^2+w^2}$

5. $f(t)=e^{(a+2)t+5}=e^5.e^{(a+2)t}$

$$F(s)=\left[\frac{1}{5-(a+2)}\right]e^5$$

$\therefore$ For L.T. to exist, Re$(s)>a+2$

7. $L\left[\int_0^t\int_0^t\ldots\int_0^t f(t)dt^n\right] = \frac{1}{s^n}F(s)$

In this problem

So, $L\left[\int_0^t f(\tau)d\tau\right] = \frac{1}{s}F(s)$

8. $f(t) = L^{-1}\left[\frac{1}{s^2(s+1)}\right]$

$$\frac{1}{s^2(s+1)} = \frac{A}{s} + \frac{B}{s^2} + \frac{C}{s+1}$$

$$\frac{1}{s^2(s+1)} = \frac{As(s+1) + B(s+1) + C(s^2)}{s^2(s+1)}$$

Matching the coefficient of s^2, s and constant in numerator, we get

$$A + C = 0 \quad \ldots(i)$$

$$A + B = 0 \quad \ldots(ii)$$

$$B = 0 \quad \ldots(iii)$$

On solving, we get $A = -1, B = 1, C = 1$

So, $f(t) = L^{-1}\left[-\frac{1}{s} + \frac{1}{s^2} + \frac{1}{s+1}\right] = -1 + t + e^{-t} = t - 1 + e^{-t}$

$$\frac{1}{s^2+s} = \frac{1}{s} - \frac{1}{s+1}$$

$$L^{-1}\left(\frac{1}{s^2+s}\right) = L^{-1}\left(\frac{1}{s}\right) - L^{-1}\left(\frac{1}{s+1}\right) = 1 - e^{-t}$$

10. $\lim_{t\to\infty} f(t) = \lim_{S\to\infty} sF(s)$

Given that, $F(s) = \left[\frac{3s+1}{s^3 + 4s^2 + (K-3)s}\right]$

$\lim_{t\to\infty} f(t) = 1 \Rightarrow \lim_{s\to\infty}\left[\frac{3s+1}{s^3+4s^2+(K-3)s}\right] = 1 \Rightarrow K = 4$

12. $L[U(t-a)] = \int_0^\infty e^{-st}U(t-a)dt$

13. $L\{f(t)\} = \int_0^\infty e^{-st}f(t)dt$

$L\{f(t)\} = \int_0^3 e^{-st}f(t)dt + \int_3^5 e^{-st}f(t)dt\int_5^\infty e^{-st}f(t)dt$

14. We need $g(3) = f(0)$ and $g(5) = f(1)$

Only choice (b) satisfies both these conditions

$$g(t) = f\left(\frac{t}{2} - \frac{3}{2}\right), g(3) = f\left(\frac{3}{2} - \frac{3}{2}\right) = f(0)$$

and $g(5) = f\left(\frac{5}{2} - \frac{3}{2}\right) = f(1)$

15. $\frac{d^2y}{dt^2} + 2\frac{dy}{dt} + y(t) = \delta(t)$

Taking Laplace transform on both sides,

$$s^2y(s) + 2sy(s) + 4 + y(s) = 1$$

$$(s^2 + 2s + 1)y(s) = -(2s + 3)$$

$$y(s) = -\frac{(2s+3)}{(s+1)^2} \Rightarrow y(s) = -\left[\frac{2}{(s+1)} + \frac{1}{(s+1)^2}\right]$$

$$y(t) = -\left[2e^{-t} + e^{-t} - te^{-t}\right]u(t)$$

$$\left.\frac{dy}{dt}\right|_{\text{at } t=0+} = -[-2+1-0]$$

$$\left.\frac{dy}{dt}\right|_{\text{at } t=0+} = 1$$

16. $\dot{x}(t) + 2x(t) = \delta(t)$

Taking L.T. on both sides

$$sx(s) - x(0) + 2x(s) = 1$$

$$x(s)[s + 2] = 1$$

$$x(s) = \frac{1}{S+2}$$

$$x(t) = e^{-2t}u(t)$$

Self Assessment Test

Solve the following differential equations :

1. $y''+4y'+4y=3\sin x+4\cos x, y(0)=1$ and $y'(0)=0$ (JNTU–2003)

2. $\frac{d^2y}{dx^2}-4y=x\sin x$ (Madras–2009)

3. $\frac{d^2y}{dx^2}-6\frac{dy}{dx}+9y=6e^{3x}+7e^{-2x}-\log 2$ (VTU–2005)

4. $\frac{d^2y}{dx^2}+3\frac{dy}{dx}+2y=4\cos^2 x$ (Bhopal–2002S)

5. $(D^2-4D+3)y=\sin 3x\cos 2x$ (Madras–2000)

6. $\frac{d^3y}{dx^3}+2\frac{d^2y}{dx^2}+\frac{dy}{dx}=e^{-x}+\sin 2x$ (VTU–2004)

7. $\frac{d^2y}{dx^2}+2\frac{dy}{dx}+y=e^{2x}-\cos^2 x$ (Delhi–2002)

8. $(D^3-5D^2+7D-3)y=e^{2x}\cosh x$ (Nagarjuna–2008)

9. $\frac{d^2y}{dx^2}-y=e^x+x^2e^x$ (Nagpur–2009)

10. $(D^3-D)y=2x+1+4\cos x+2e^x$ (Mumbai–2006)

11. $\frac{d^2y}{dx^2}-6\frac{dy}{dx}+25y=e^{2x}+\sin x+x$ (VTU–2006)

12. $(D^2+1)^2y=x^4+2\sin x\cos 3x$ (Madras–2006)

13. $\frac{d^2y}{dx^2}+5\frac{dy}{dx}+6y=e^{-2x}\sin 2x$ (Bhopal–2008)

14. $\frac{d^2y}{dx^2}+2\frac{dy}{dx}+3y=e^x\cos x$ (VTU–2010)

15. $(D^2+4D+3)y=e^{-x}\sin x+xe^{3x}$ (Raipur–2005, Anna–2002S)

16. $(D^3+2D^2+D)y=x^2e^{2x}+\sin^2 x$ (PTU–2003)

17. $\frac{d^2y}{dx^2}+16y=x\sin 3x$ (VTU–2010S)

18. $\frac{d^2y}{dx^2}+3\frac{dy}{dx}+2y=e^{e^x}$ (SVTU–2009)

19. $(D^2+a^2)y=\tan ax$ (VTU–2005)

20. $x^2\frac{d^2y}{dx^2}-3x\frac{dy}{dx}+4y=(1+x^2)$ **(SVTU-2007)**

21. $x\frac{d^2y}{dx^2}-\frac{2y}{x}=x+\frac{1}{x^2}$ **(VTU-2005S)**

22. $x^2\frac{d^2y}{dx^2}+4x\frac{dy}{dx}+2y=\log x$ **(BHOPAL 2009)**

23. $x^2y''+xy'+y=2\cos^2(\log x)$ **(VTU-2011)**

24. $x^2\frac{d^2y}{dx^2}+5x\frac{dy}{dx}+4y=x\log x$ **(UPTU-2004)**

25. $x^2\frac{d^2y}{dx^2}+2x\frac{dy}{dx}-12y=x^3\log x$ **(BHOPAL-2007)**

26. $(2x+3)^2\frac{d^2y}{dx^2}-(2x+3)\frac{dy}{dx}-12y=6x$ **(VTU-2007, KERALA-2005, ANITA-2025)**

27. $(x-1)^3\frac{d^3y}{dx^3}+2(x-1)^2\frac{d^2y}{dx^2}-4(x-1)\frac{dy}{dx}+4y=4\log(x-1)$ **(SVTU 2009)**

28. $(1+x)^2\frac{d^2y}{dx^2}+(1+x)\frac{dy}{dx}+y=\sin[2\log(1+x)]$ **(PTU-2006, UTU 2007)**

ANSWERS

1. $y=(1+x)e^{-2x}+\sin x$

2. $y=C_1e^{2x}+C_2e^{-2x}-\frac{x}{3}\sinh x-\frac{2}{9}\cosh x$

3. $y=(C_1+C_2x)e^{3x}+3x^2e^{3x}+\frac{7}{25}e^{-2x}-\frac{1}{9}\log 2$

4. $y=C_1e^{-x}+C_2e^{-2x}+1+\frac{1}{10}(3\sin 2x-\cos 2x)$

5. $y=C_1e^x+C_2e^{3x}+\frac{1}{884}(10\cos 5x-11\sin 5x)+\frac{1}{20}(\sin x+2\cos x)$

6. $y=C_1+(C_2+C_3x)e^{-x}-\frac{x^2}{2}e^{-x}+\frac{3}{50}\cos 2x-\frac{2}{25}\sin 2x$

7. $y=(C_1+C_2x)e^{-x}+\frac{1}{2}+\frac{1}{5}(2\sin 2x+\cos 2x)$

8. $y=(C_1+C_2x)e^x+C_3e^{3x}+\frac{1}{8}(xe^{3x}-x^2e^x)$

9. $y=C_1e^x+C_2e^{-x}+\frac{e^x}{12}(2x^3-3x^2+9x)$

10. $y=C_1+C_2e^x+C_3e^{-x}+xe^x-(x^2+x)-2\sin x$

11. $y=e^{3x}(C_1\cos 4x+C_2\sin 4x)+\frac{1}{17}e^{2x}+\frac{1}{565}(23\sin x+6\cos x)+\frac{x}{25}+\frac{6}{625}$

12. $y=(C_1+C_2x)\cos x+(C_3+C_4x)\sin x+x^4-24x^2+72+\frac{1}{225}\sin 4x-\frac{1}{9}\sin 2x$

13. $y=C_1e^{-2x}+C_2e^{-3x}-\frac{e^{-2x}}{10}(\cos 2x+2\sin 2x)$

14. $y=e^{-x}(C_1\cos\sqrt{2x}+C_2\sin\sqrt{2x})+\frac{e^x}{41}(4\sin x+5\cos x)$

15. $y=C_1e^{-x}+C_2e^{-3x}-\frac{e^{-x}}{5}(\sin x+2\cos x)+\frac{e^{3x}}{22}\left(x-\frac{5}{11}\right)$

16. $y = C_1 + (C_2 + C_3x)e^{-x} + \frac{e^{2x}}{18}\left(x^2 - \frac{7x}{8} + \frac{11}{6}\right) + \frac{1}{100}(3\sin 2x + 4\cos 2x)$

17. $y = C_1\cos 4x + C_2\sin 4x + \frac{1}{7}\left(x\sin 3x - \frac{6}{7}\cos 3x\right)$

18. $y = C_1e^{-x} + C_2e^{-2x} + C_3e^{-3x} + e^{-2x}.e^{e^x}$

19. $y = C_1\cos ax + C_2\sin ax - \frac{1}{a^2}\cos ax\log(\sec ax + \tan ax)$

20. $y = (c_1 + c_2\log x)x^2 + \frac{1}{4} + 2x + \frac{1}{2}x^2(\log x)^2$

21. $y = C_1x^2 + C_2x^{-1} + \frac{1}{3}(x^2 - 1/x)\log x$

22. $C_1x^{-1} + C_2x^{-2} + \frac{1}{2}\log x - \frac{3}{4}$

23. $y = C_1x^{-2} + x[C_2\cos(\sqrt{3}\log x) + C_3\sin(\sqrt{3}\log x)] + 8\cos(\log x) - \sin(\log x)$

24. $y = x^{-2}(C_1 + C_2\log x) + \frac{x}{9}\left(\log x - \frac{2}{3}\right)$

25. $y = C_1x^3 + C_2x^{-4} + \frac{x^3}{98}\log x(7\log x - 2)$

26. $y = C_1(2x+3)^a + C_2(2x+3)^b - \frac{3}{14}(2x+3) + \frac{3}{4}$, where $a, b = \frac{3 \pm \sqrt{57}}{4}$

27. $y = C_1(x-1) + C_2(x-1)^2 + C_3(x-1)^{-2} + \log(x+1) + 1$

28. $y = C_1\cos\log(1+x) + C_2\sin\log(1+x) - \frac{1}{3}\sin[2\log(1+x)]$

CCCCCC

Index